ROBERT CAROLA

JOHN P. HARLEY
Professor of Physiology
Eastern Kentucky University

CHARLES R. NOBACK
Professor of Anatomy and Cell Biology
College of Physicians and Surgeons
Columbia University

McGRAW-HILL PUBLISHING COMPANY

New York • St. Louis • San Francisco
Auckland • Bogotá • Caracas
Hamburg • Lisbon • London
Madrid • Mexico • Milan
Montreal • New Delhi • Oklahoma City
Paris • San Juan • São Paulo
Singapore • Sydney • Tokyo • Toronto

Human
Anatomy
—AND—
Physiology

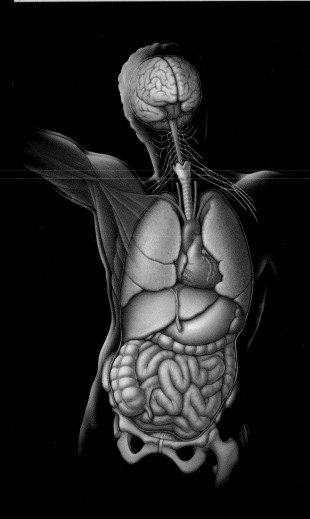

HUMAN ANATOMY AND PHYSIOLOGY

Copyright © 1990 by McGraw-Hill, Inc. All rights reserved.
Printed in the United States of America. Except as permitted under
the United States Copyright Act of 1976, no part of this publication
may be reproduced or distributed in any form or by any means, or
stored in a data base or retrieval system, without the prior written
permission of the publisher.

1 2 3 4 5 6 7 8 9 0   RMK   RMK   8 9 4 3 2 1 0 9

ISBN 0-07-557937-5

This book was set in Meridien by Waldman Graphics, Inc.
The editors were Denise T. Schanck and Holly Gordon;
the designer/art director was Gayle Jaeger;
the production supervisor was Sandy Moore.

The cover illustration was drawn by Robert Demarest.
The anatomy illustrations were drawn by Marsha J. Dohrmann,
Carol Donner, John V. Hagen, Neil O. Hardy,
Steven T. Harrison, Jane Hurd, Joel Ito, and George Schwenk.
Flowcharts were done by Network Graphics.
Pre-press was done by Lehigh Colortronics.

Lehigh Press was cover printer.
Rand McNally and Company was printer and binder.

Library of Congress Cataloging-in-Publication Data

Carola, Robert.
    Human anatomy and physiology / Robert Carola, John P. Harley,
Charles R. Noback.
        p.   cm.
    Includes index.
    ISBN 0-07-557937-5
    1. Human physiology.   2. Anatomy, Human.   I. Harley, John P.
II. Noback, Charles Robert, (date).   III. Title.
    [DNLM: 1. Anatomy.   2. Physiology.   QS 4 C292h]
QP36.C28   1990
612—dc20
DNLM/DLC
for Library of Congress   85-81849

This book is dedicated to
Leslie, Maria, and Matthew Carola;
Daniele D. Harley and Christopher M. Young;
Lindsay Barton and Eleanor Noback.

CONTENTS IN BRIEF

CONTENTS

PART **I**   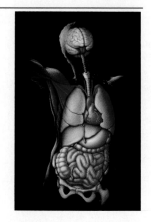   **HOW THE BODY IS ORGANIZED**   **1**

1   INTRODUCTION TO ANATOMY AND PHYSIOLOGY   3

4 EPITHELIAL AND CONNECTIVE TISSUES   89

5 THE INTEGUMENTARY SYSTEM   114

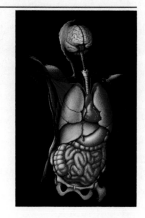

PART II   SUPPORT AND MOVEMENT   131

6   BONES AND OSSEOUS TISSUE   133

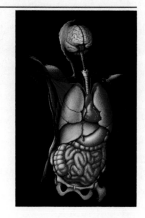

7 THE SKELETAL SYSTEM 152

8   ARTICULATIONS   207

9   MUSCLE TISSUE   235

10   THE MUSCULAR SYSTEM   260

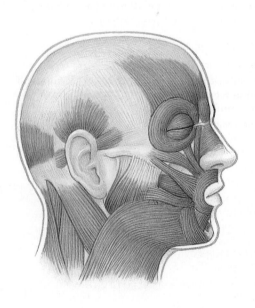

PART **III**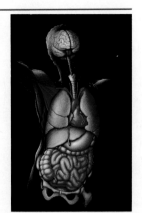

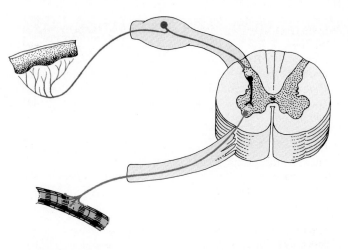

13 THE BRAIN AND CRANIAL NERVES 362

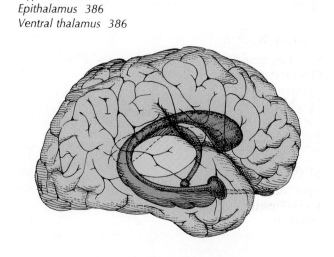

14 — THE AUTONOMIC NERVOUS SYSTEM — 411

15 — THE SENSES — 426

16 THE ENDOCRINE SYSTEM   474

PART **IV**

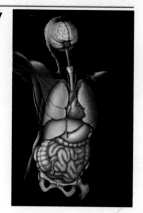

TRANSPORTATION AND MAINTENANCE 505

17

THE CARDIOVASCULAR SYSTEM: BLOOD 507

18   THE CARDIOVASCULAR SYSTEM: THE HEART   532

19 THE CARDIOVASCULAR SYSTEM: BLOOD VESSELS 569

20 THE LYMPHATIC SYSTEM AND IMMUNITY 614

21 THE RESPIRATORY SYSTEM 647

24   THE URINARY SYSTEM   766

25   REGULATION OF BODY FLUIDS, ELECTROLYTES, AND ACID-BASE BALANCE   801

PART V   REPRODUCTION AND DEVELOPMENT   821

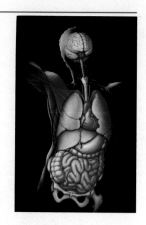

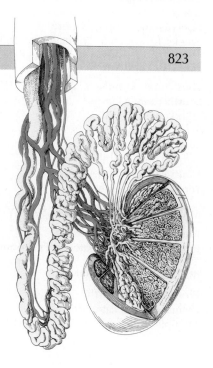

26   THE REPRODUCTIVE SYSTEMS   823

27   HUMAN GROWTH AND DEVELOPMENT   857

LIST OF TABLES

List of Tables

PREFACE

In 1977, Louise Brown was conceived in a Petrie dish and then implanted in her mother's uterus to be born nine months later, in 1978. The press was charged with excitement, but the real amazement should not have been about Baby Louise's conception and birth, but about the millions of normal conceptions and births that preceded and would follow hers. The wonder is not in a Petrie dish, but in the human body itself.

The human body is full of surprises. Have you ever wondered why a baby who hasn't breathed air for nine months in its mother's uterus can suddenly start breathing the moment it is born? Or why your finger-tips look like raisins after you've stayed in the bathtub too long? Or why you are allergic to strawberries? Or why you sometimes see floating spots in front of your eyes? Or why you dream? The study of human anatomy and physiology is filled with such fascinating questions—this book answers them.

Most anatomy and physiology textbooks cover roughly the same material, so the authors had to decide from the outset how they would make this book the best choice for teachers and students. Without hesitation, we knew we had to explain concepts clearly and present them in a manner understandable to the student and useful to the instructor—combining the text and art to encourage understanding, not merely memorization. Repetition is used in the narrative in the same way that a teacher uses repetition in the classroom, to reinforce an important point.

Another important difference lies in a book's comprehensiveness. Almost every reviewer asked that this book be more comprehensive than the others, imploring the authors at every opportunity to tell the *whole* story, clearly, carefully, and always with the reader in mind. What resulted is the book that you have asked for, being both comprehensive *and* readable. The material is extremely up to date, and includes many topics other texts do not even mention.

A carefully planned and organized text helps to make the student comfortable in a sea of new terms and concepts. We believe that students reading our book will always know where they are and where the book is taking them. The introduction to each chapter uses an overview technique, an excellent teaching and learning device.

We have used three unifying themes to enrich the students' understanding of how the body operates as a dynamic mechanism:

Homeostasis, which emphasizes the body's self-regulating ability.

The *interrelatedness* of body systems.

The *compatibility of anatomy and physiology* (form and function).

These themes not only help unify the overall text, but also help the student to see relationships between one body part and another, the shape of a body part to its function, and the dynamic self-regulating nature of the human body.

We carried our concern for clarity to the design and illustrations, carefully coordinating the art with the text, and breaking down complex physiological processes into simple, sequential steps. Anatomical drawings are large, color-coded throughout the book, and have a clarity that allows the authors to show more detail than usual.

We spent over nine years producing this textbook, always being certain to update each draft, and checking the accuracy over and over again. Our reviewers have told us that our textbook is the most accurate, up to date, comprehensive, and well written they have seen. As one reviewer said, "Many texts are so concerned with 'stand-alone' chapters that they fail to present the wonder and beauty of how body systems carry out integrated functions. You have captured this beauty."

The author team is unique in this field. *Robert Carola* is a science writer who has written six textbooks, as well as writing for the Smithsonian Institution, Fisher-Price, Exxon Corporation, Michigan Bell Telephone, IBM, and many other corporate clients. *John P. Harley* is a physiologist who has taught Human Anatomy and Physiology at Eastern Kentucky University for 20 years, and is the coauthor of *Microbiology* (Wm. C. Brown Publishers, 1990). *Charles R. Noback* is the coauthor of *The Human Nervous System*, Third Edition, and *The Nervous System: Introduction and Review*, Third Edition (both published by McGraw-Hill), and has been a Professor of Anatomy and Cell Biology at the College of Physicians and Surgeons, Columbia University for 40 years. He has contributed a section on the Human Nervous System to the *Encyclopaedia Britannica*.

During many years of teaching and textbook writing, the authors have never experienced such a conscientious and well-informed group of reviewers. As a result of their efforts, each draft of the manuscript produced refinements that improved the book immeasurably. Their names are listed on the facing page. A special note of thanks is due to Professor William W. Farrar of Eastern Kentucky University for his careful reading of the entire manuscript.

The authors would also like to thank the following medical illustrators who produced the remarkable illustrations in their specialized areas that contribute so much to the clarity and attractiveness of the book:

Robert J. Demarest
Cover/part-opening illustration
Marsha J. Dohrmann
Nervous system
Carol Donner
Reproductive system

John V. Hagen
Muscular system
Neil O. Hardy
Digestive system
Steven T. Harrison
Cardiovascular and urinary systems
Jane Hurd
Respiratory system
Joel Ito
Integumentary, endocrine, and lymphatic systems; cells and embryology
George Schwenk
Skeletal system

We are also grateful for the efforts of Professor Michael C. Kennedy of Hahnemann University School of Medicine, Philadelphia, for his review of the accuracy and conceptual focus of the illustrations.

Finally, the authors are grateful to the many publishing people who have contributed their talents and support, especially Seibert Adams, Edith Beard Brady, Ruth Gillies, Holly Gordon, Gayle Jaeger, Kent Porter, and Denise Schanck.

We would be happy to hear from any users of this book about how we can improve it for the next edition.

Robert Carola
John P. Harley
Charles R. Noback

MANUSCRIPT REVIEWERS

Robert A. Altbaum, M.D.
Westport, CT

Dean A. Beckwith
Illinois Central College

Jeffrey H. Black
East Central University

Robert J. Boettcher
Lane Community College

William Bonaudi
Truckee Meadows Community College

Clifton F. Bond
Highland Park Community College

James Bridger
Prince George's Community College

Jerry Button
Portland Community College

Kenneth H. Bynum
University of North Carolina, Chapel Hill

Lu Anne Clark
Lansing Community College

Glenna M. Cooper
South Plains College

Irene M. Cotton
Lorain County Community College

Darrell T. Davies
Kalamazoo Valley Community College

Clementine A. De'Angelis
Tarrant County Junior College

Edward A. DeSchuytner
North Essex Community College

William E. Dunscombe, Jr.
Union County College

William W. Farrar
Eastern Kentucky University

Julian Wade Farrior, Jr.
Gwynedd Mercy College

Douglas Fonner
Ferris State University

Rose B. Galiger
University of Bridgeport

Gregory Gillis
Bunker Hill Community College

Harold E. Heidtke
Andrews University

William J. Higgins
University of Maryland

H. Kendrick Holden, Jr.
Northern Essex County Community College

Gayle D. Insler
Adelphi University

David A. Kaufmann
University of Florida

Donald S. Kisiel
Suffolk County Community College

William G. Klopfenstein
Sinclair Community College

Thomas E. Kober
Cincinnati Technical College

Joseph R. Koke
Southwest Texas State University

Gordon Locklear
Chabot College

James A. Long
Boise State University

Dorothy R. Martin
Black Hawk College

Elden W. Martin
Bowling Green State University

Johnny L. Mattox
Northeast Mississippi Community College

R. J. McCloskey
Boise State University

Daniel McEuen
Mesa Community College

Robert C. McReynolds
San Jacinto College

Anne M. Miller
Middlesex Community College

William W. Miller, III
Grambling State University

Gordon L. Novinger
San Bernardino Valley College

Dennis E. Peterson
DeAnza College

Joseph S. Rechtschaffen, M.D.
New York, NY

Ralph E. Reiner
College of the Redwoods

Jackie S. Reynolds
Richland College

Donald D. Ritchie (deceased)
Barnard College

Robert L. Ross, Optometrist
Westport, CT

Louis C. Renaud
Prince George's Community College

David Saltzman
Santa Fe Community College

David S. Smith
San Antonio College

Stan L. Smith
Bowling Green State University

Tracy L. Smith
Housatonic Community College

Alexander A. Turko
Southern Connecticut State University

Donald H. Whitmore
University of Texas at Arlington

Barry James Wicklow
St. Anselm College

Leonard B. Zaslow, D.D.S.
Westport, CT

Stephen W. Ziser
Austin Community College

OVERVIEW OF THE SUPPLEMENTS

FOR THE INSTRUCTOR

INSTRUCTOR'S RESOURCE MANUAL

by John P. Harley

This manual has been prepared to facilitate the use of *Human Anatomy and Physiology*. It contains a variety of supplementary teaching aids, such as enrichment sections, listings of pertinent anatomy and physiology films, listings of software programs in anatomy and physiology, topics for discussion and library research, and answers to the text sections Understanding the Facts and Understanding the Concepts. The manual also contains alternate chapter sequences to accommodate the needs of students' varied backgrounds.

TEST BANK MANUAL

by John P. Harley

Available upon request to adopters, this manual contains 75 test items per chapter for each of the 29 text chapters. The test questions are also available on diskette for IBM PC, Apple II, and Macintosh computers.

OVERHEAD TRANSPARENCIES

Over 200 color transparencies of important illustrations, photographs, and electron micrographs from the text are also available, free to adopters.

SLIDE PACKAGE

Over 200 color 35-mm slides are available to adopters. These are the same images as the overhead transparencies.

LECTURE OUTLINES/TRANSPARENCY MASTERS MANUAL

by John P. Harley

Complete lecture outlines for each chapter are available and can be used as classroom handouts or transparency masters. Over 150 transparency masters containing graphs, flowcharts, and figures from the text are available, free to adopters.

INSTRUCTOR'S MANUAL TO ACCOMPANY THE LABORATORY MANUAL

by Ted Namm, Barbara Cocanour, Alease S. Bruce, and Joseph P. Farina

This manual provides the laboratory instructor with valuable suggestions for the most efficient utilization of exercises within the time frame of the laboratory. Each chapter lists supplies and equipment, includes comments on teaching the exercises, and gives answers to the Study Questions.

SOFTWARE

The *Human Anatomy and Physiology Image Library* is available on Macintosh CD-ROM and on a Macintosh 3.5-inch floppy disk. A database of over 600 images will allow instructors to create their own HyperCard tutorials, transparencies, tests, and classroom handouts.

VIDEO CASSETTES

With adoption of the text, the instructor will be provided with video cassettes that illustrate a variety of anatomical and physiological concepts.

FOR THE STUDENT

STUDENT'S STUDY MANUAL

by John P. Harley

This manual contains thought-provoking activities to help master each chapter's learning objectives and gain a thorough understanding of human anatomy and physiology.

LABORATORY MANUAL

by Joseph P. Farina, Ted Namm, Alease S. Bruce, and Barbara Cocanour

This lab manual presents student-tested laboratory exercises designed to accommodate laboratory sessions of various lengths and focuses. The 27 chapters are organized into five major categories: levels of organization; protection, movement, and support; control and integration; homeostatic systems; and continuity of life. Each chapter begins with a complete list of objectives followed by a general introduction. The material in each chapter, designed for laboratories of 3-hour length, is subdivided into exercises that may be adapted to shorter laboratory sessions. Each chapter ends with a selection of comprehensive Study Questions.

FLASH CARDS

by Barbara Cocanour

Two-hundred flash cards are packaged in a separate box to self-test the understanding of the muscles, skeletal parts, nerves, and blood vessels.

ACTIVITIES MANUAL

by Barbara Cocanour and William Farrar

This self-study tool offers a coloring book, crossword puzzles, quotation puzzles, and anatomical flash cards. These are designed to help achieve mastery of anatomical and physiological information. The coloring book includes modified illustrations from the text to be colored in and labeled. Brief definitions and descriptions as well as a self-test are included for each illustration. The crossword puzzles appear within each section and provide an entertaining way to recall important definitions. A quotation puzzle appears at the end of each unit and combines clues from all the crossword puzzles in that unit. The flash cards are designed to review the muscles, skeletal parts, nerves, and blood vessels.

RADIOGRAPHIC ANATOMY: A WORKING ATLAS

by Harry W. Fischer, M.D.

A complete atlas of human anatomy as it is seen through today's imaging technologies. Almost 200 beautifully reproduced radiographs, late-generation CAT scans, and high-resolution NMR images are accompanied by clearly labeled line drawings that highlight the structures that every physician, radiologist, and radiographic technician must know. This book was especially designed to help students relate their knowledge of gross anatomy to the clinical images they will see in the practice of medicine. The emphasis throughout is on frequently used images and imaging modalities, making *Radiographic Anatomy* an indispensable aid for all students of human gross anatomy.

A GUIDED TOUR TO *Human Anatomy and Physiology*

CHAPTER-OPENING MATERIAL

Each chapter opening begins with a Chapter Outline and Learning Objectives to give an overview of the chapter.

891

CHAPTER OUTLINE

PHYSIOLOGY OF HUMAN AGING 892
Hypotheses of Aging
Effects of Aging on the Body
A Final Word About Growing Old

CANCER: CELLS OUT OF CONTROL 901
Neoplasms
How Cancer Spreads
Causes of Cancer
Some Common Cancer Sites
Treatment of Cancer

MEDICAL TERMINOLOGY 909

28
The Body in Transition: Aging and Cancer

LEARNING OBJECTIVES

1 Describe some of the changes that occur in cells as they die.

2 Distinguish between senescence and senility.

3 Discuss the major hypotheses about aging.

4 Explain some effects of aging on each of the body systems, and list the major causes of death in older persons.

5 Distinguish between benign and malignant neoplasms.

6 Name the three major types of malignant neoplasms, and the kind of tissue where each originates.

7 Explain how cancer spreads from its original site in the body.

8 Name five or more carcinogens, and indicate the parts of the body most often affected by each.

9 Briefly explain the main hypotheses about how cancer develops.

10 Name the most common sites in the body where cancer develops, and discuss the incidence and survival rates for these cancers.

11 Describe the major treatments for cancer.

12 Name the four categories of approved cancer-fighting drugs.

A UNIQUE CHAPTER

There is a separate chapter on aging and cancer, two areas of practical concern to future health care professionals.

INTRODUCTORY OVERVIEW

Each chapter begins with an informative framework and introduces basic ideas that will be explored in the chapter.

REINFORCED PRESENTATION

Carefully organized textual material is reinforced in a variety of formats to enhance understanding.

In order to maintain homeostasis, the body is constantly reacting and adjusting to changes in the outside environment and within the body itself. Such *stimuli* (or environmental changes) are sensed and conveyed via nerves to the brain and spinal cord, where the messages (input) are analyzed, combined, compared, and coordinated by a process called *integration*. After being sorted out, messages are conveyed by nerves to the muscles and glands of the body. The nervous system expresses itself visibly through muscles and glands, causing muscles to contract or relax, and glands to secrete or not secrete their products.

Under normal conditions, activities of the muscles and glands are coordinated, so that our body parts work in harmony toward directed homeostatic goals. Examples of such efforts to maintain homeostasis are the maintenance of a relatively constant body temperature and the coordinated activities of muscles during movements. Homeostasis allows us to function normally despite constant changes in the environment.

The nervous system and the endocrine system are the two major regulatory systems of the body, and both are specialized to make the proper responses to stimuli. Both systems work together continuously to maintain homeostasis, but the nervous system is the faster of the two. Stimuli received by the nervous system are processed rapidly through a combination of electrical impulses and chemical substances called *neurotransmitters* for communication between two nerve cells, between a nerve cell and a muscle cell, or between a nerve cell and glandular cells. The endocrine system (the subject of Chapter 16) must depend on slower chemical transmissions, using chemical substances called hormones. The nervous system has been compared to the telephone system of a large city, while the endocrine system is more like its postal service. The endocrine system typically regulates such long-term processes as growth and reproductive ability, instead of the short, quick responses to stimuli controlled by the nervous system. However, stress can produce almost immediate reactions from the endocrine system. In fact, both systems are so closely interrelated that they can be considered a single regulatory agency.

ORGANIZATION OF THE NERVOUS SYSTEM

Peripheral Nervous System

The *peripheral nervous system* (PNS) is composed of the cranial nerves associated with the brain, and the spinal nerves associated with the spinal cord, as well as *groups* of nerve cell bodies called ganglia (see Figure 11.1). In general, we can say that the peripheral nervous system is made up of the nerve cells and their fibers that lie *outside* the brain and spinal cord. This system allows the brain and spinal cord to communicate with the rest of the body.

Two types of nerve cells are present in the peripheral nervous system: (1) *afferent* (L. *ad*, toward + *ferre*, to bring), or *sensory*, nerve cells carry nerve impulses from sensory receptors in the body *to* the central nervous system, where the information is processed; (2) *efferent* (L. *ex*, away from), or *motor*, nerve cells convey information *away from* the central nervous system to the effectors (muscles and glands).

The peripheral nervous system may be further divided, on a purely functional basis, into the somatic nervous system and the visceral nervous system (Table 11.1). Each of these systems is composed of an afferent (sensory) division and an efferent (motor) division.

Somatic nervous system The *somatic nervous system* is composed of afferent and efferent divisions. The *somatic afferent (sensory) division* consists of nerve cells that receive and process sensory input from the skin, voluntary muscles, tendons, joints, eyes, tongue, nose, and ears. This input is conveyed to the spinal cord and brain via the spinal and cranial nerves, and utilized by the nervous system at an unconscious level. On a conscious level, the sensory input is perceived as sensations such as touch, pain, heat, cold, balance, sight, taste, smell, and sound.

The *somatic efferent (motor) division* is composed of neuronal pathways that descend from the brain through the brainstem and spinal cord to influence the lower motor neurons of the cranial and spinal nerves. When these lower motor neurons are stimulated, they always excite (never inhibit) the skeletal muscles to contract. This system regulates the "voluntary" contraction of skeletal muscles. (As you saw in Chapter 9, not all such activity is actually voluntary or under our conscious control.)

Visceral nervous system The *visceral nervous system* is composed of afferent and efferent divisions. The *afferent (sensory) division* includes the neural structures involved in conveying sensory information from sensory receptors in the visceral organs of the cardiovascular, respiratory, digestive, urinary, and reproductive systems. Input from these sensory receptors is utilized on a conscious level, and is perceived as sensations such as pain, intestinal discomfort, urinary bladder fullness, taste, and smell. The *efferent (motor) division*, more commonly known as the *autonomic nervous system*, includes the neural structures involved in the motor activities that influence the smooth (involuntary) muscles, cardiac (heart) muscle, and glands of the skin and viscera.

Autonomic nervous system The *autonomic nervous system* (visceral efferent motor division) is made up of nerve fibers from the brain and spinal cord that may either inhibit

---

Chapter 11: The Action of Nerve Cells

TABLE 11.1 ORGANIZATION OF THE NERVOUS SYSTEM

Division	Components	Functions
Central nervous system (CNS)	Brain, spinal cord.	Body's central control system. Receives stimuli, relays "messages" for action to muscles and glands. Interpretive functions involved in thinking, learning, memory, etc.
Peripheral nervous system (PNS)	Cranial and spinal nerves, with afferent (sensory) and efferent (motor) nerve cells.	Enables brain and spinal cord to communicate with entire body. Afferent (sensory) cells: carry impulses from receptors to CNS. Efferent (motor) cells: carry impulses from CNS to effectors (muscles and glands).
Somatic nervous system	Axons (nerve fibers) of lower motor neurons that go directly from CNS to effector muscle without crossing junctions (synapses).	Afferent (sensory) division: receives and processes sensory input from skin, voluntary muscles, tendons, joints, eyes, ears. Efferent (motor) division: excites skeletal muscles.
Visceral nervous system	Nerve fibers that go from CNS to interact with other nerve cells within a ganglion located outside CNS; nerve fibers of second nerve cells that go to effectors.	Afferent (sensory) division: receives and processes input from internal organs of cardiovascular, respiratory, digestive, urinary, and reproductive systems. Efferent (motor) division (autonomic nervous system): may inhibit or excite smooth muscle, cardiac muscle, glands.
Autonomic nervous system (visceral motor division) Sympathetic nervous system		Relaxes intestinal wall muscles; increases sweating, heart rate, blood flow to voluntary muscles.
Parasympathetic nervous system		Contracts intestinal wall muscles; decreases sweating, heart rate, blood flow to voluntary muscles.

or excite smooth muscle, cardiac muscle, and glands. This system is the modulator, adjuster, and coordinator of "involuntary" visceral activities such as the heart rate and the secretions of glands. Many of these visceral activities can be carried out even if the organs are deprived of innervation by the autonomic nervous system. For example, the heart continues to contract without innervation. (This is why a transplanted heart continues to contract.) Although the heart can contract without innervation, the autonomic nervous system can either increase or decrease the strength of contraction and heart rate.

The autonomic nervous system is divided into two subsystems, the sympathetic and parasympathetic systems, which complement each other. The *sympathetic nervous system* stimulates activities that are mobilized during emergency and stress situations, the so-called fight, fright, and flight responses. These responses include an acceleration of the heart rate and strength of contraction, an increase in the concentration of blood sugar, and an increase in blood pressure. In contrast, the *parasympathetic nervous system* directs activities associated with the conservation and restoration of body re-

sources. These activities include a decrease in the heart rate and strength of contraction, and the rise in gastrointestinal activities associated with increased digestion and absorption of food.

The autonomic nervous system is dealt with in greater detail in a chapter of its own, Chapter 14.

Ask Yourself

1 *What are the main components of the central nervous system?*

2 *Based on the direction of the nerve impulse, what two types of nerve cells are present in the peripheral nervous system?*

3 *What are the major differences between the somatic and visceral nervous systems?*

4 *What are the two subdivisions of the autonomic nervous system?*

TABLES

Material is synthesized in tabular form for easy reference.

ASK YOURSELF QUESTIONS

Each section is followed by Ask Yourself questions that encourage students to review what they have just read. This serves as an additional study aid.

309

Neurons: Functional Units of the Nervous System

NEURONS: FUNCTIONAL UNITS OF THE NERVOUS SYSTEM

The nervous system contains about 100 billion nerve cells. These nerve cells, called *neurons* (Gr. nerve), are specialized to transmit impulses from short to relatively long distances, from one part of the body or central nervous system to another. Neurons have two important properties: (1) *excitability*, or the ability to respond to stimuli, and (2) *conductivity*, or the ability to conduct a signal.

Parts of a Neuron

Neurons are among the most specialized types of cells. Although they vary greatly in shape and size, all neurons contain three principal parts, each associated with a specific function: the cell body, dendrites, and an axon (Figure 11.2).

Cell body The *cell body* of a neuron may also be called a *soma* (Gr. body) or a *perikaryon* (per-ih-KAR-ee-on; Gr. *peri*,

near, around + *karyon*, kernel, nut). It may be star-shaped (stellate), roundish, oval, or even pyramid-shaped, but its distinguishing structural features are its complex, spreading processes (branches or fibers) that reach out to send or receive impulses to or from other cells. Besides varying in shape, cell bodies may vary in size from about half the size of a red blood cell to almost 17 times the size of a red blood cell. A cell body has a large central nucleus, which contains a prominent nucleolus, as well as several organelles that are responsible for metabolism, growth, and repair of the neuron. These organelles include chromatophilic substance (Nissl bodies), endoplasmic reticulum, mitochondria, neurofilaments (microfilar...

Chromatophili...
plasmic reticulum...
the rough ER is c...
with the Golgi ap...
substance contain...
in one to three...
neuron. Proteins...
rough ER, release...

FLOWCHARTS

Complex physiological processes are illustrated in easy-to-follow flowcharts.

DYNAMIC BODY ESSAYS

The Dynamic Body essays at the beginning of many chapters introduce dynamic aspects of the body in simple, nontechnical language.

262

THE DYNAMIC BODY

The Smallest Muscles

One of the most powerful muscles in the body is the one we sit on, the *gluteus maximus*. The longest one is the *sartorius*, which runs from the hip to the knee, and the largest in surface area is the *latissimus dorsi*, the broad muscle of the back. When we show our "muscle," we flex the *biceps brachii* in our upper arm, probably the best known muscle of all. The large muscles move our arms and legs, and, of course, they are more widely known and evident than some of our smaller muscles. But does the size of a muscle determine its importance?

Some of the tiniest muscles are those that help us communicate with one another. The smallest muscles in the body, like the smallest bones, are in the middle ear (drawing). Without them our delicate hearing apparatus could not move, and the sounds we hear would be muted. Without the tiny muscles within and around our eyeballs, we could not focus on objects, move our eyeballs without moving our head, or open and close our eyelids. Speaking requires the delicate coordination of the small muscles of the larynx, pharynx, palate, tongue, and mouth.

The tongue muscles also enable us to move food around within our mouths during chewing and swallowing. (Try chewing and swallowing *without* using your tongue.) One of those muscles, the *genioglossus*, performs an even more important role: it prevents suffocation by keeping the tongue from falling backward toward the throat. The entire act of swallowing is controlled by small muscles, and food is prevented from entering the breathing passages in several ways during swallowing. (If food did enter these passages, we could choke to death.) First muscles raise the soft palate to keep food out of the nasal cavity, and then muscles at the rear of the mouth form a thin slit, which keeps large food particles from passing into the breathing portion of the throat.

Circular muscles called *sphincters* are present throughout our bodies to open and close body openings such as the mouth and anus. The *pubococcygeus* muscle and the *external anal sphincter* prevent the involuntary release of feces from the body. A similar muscle, the *external sphincter of the urethra*, prevents urine from constantly flowing through the urethra to the outside. As we get older, these muscles may lose their tone and lead to *incontinence*, the inability to control urination or defecation.

Small muscles play important roles in all the body systems. For example, muscles all along the digestive tract help to move food from the throat to the anus, and sphincters at several important junctures allow food to pass systematically from one digestive compartment to the next and prevent food and digestive juices from backing up. Blood vessels that carry blood away from the heart (most arteries and arterioles) contain smooth muscle cells that can contract to assist the movement of blood through the vessels, as well as to control the amount of blood flow.

The life of the species also depends partly on certain small muscles. The *bulbocavernous* muscle in the male helps to propel semen through the penis during ejaculation. Two other small, relatively unheralded muscles also keep our reproductive function operating properly. Sperm cells inside the testicles remain alive and healthy only if they are kept at a well-regulated temperature. The *cremasteric* and *dartos* muscles help to maintain the temperature of sperm at a homeostatic level by lowering the testicles away from the heat of the body during hot conditions and raising them toward the body when it is cold.

All in all, small muscles may not make us walk or run, but they certainly help us *live*.

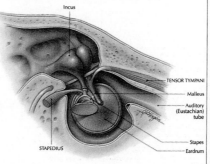

The tiny tensor tympani and stapedius muscles of the middle ear are essential elements of the hearing apparatus.

Artificial Pacemaker

The **artificial pacemaker** is a battery-operated electronic device that is implanted in the chest, with electrical leads to the heart of a person whose natural pacemaker (the SA node) has become erratic. In a relatively simple operation, electrode leads (catheters) from the pacemaker are passed beneath the skin, through the external jugular vein (or other neck vein), into the superior vena cava, into the right atrium, through the tricuspid valve, and into the myocardium of the right ventricle (see drawing). If the patient's veins are damaged or too narrow to receive the typical chest implant of the pacemaker with its connecting wires, the pacemaker is implanted in the left abdominal area, with a connecting lead inserted into the epicardium.

Three basic types of artificial pacemakers are available. The first type delivers impulses when the patient's heart rate is slower than that set for the pacemaker, and shuts off when the natural pacemaker is working adequately. The second is a fixed-rate model that delivers constant electrical impulses at a preset rate. The third is a transistorized model

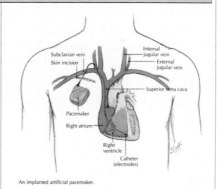

An implanted artificial pacemaker.

that picks up impulses from the patient's SA node and operates at 72 beats per minute when the natural pacemaker fails.

BOXED ESSAYS

Interesting essays provide practical examples of textual material.

forcing blood out through the pulmonary artery (to the lungs) and aorta (to the rest of the body).

3 During *atrial diastole* (0.7 sec), or relaxation of the atria, the ventricles remain contracted, and the atria begin refilling with blood from the large veins leading to the heart from the body.

4 *Ventricular diastole* (0.5 sec), or relaxation of the ventricles, begins before atrial systole, allowing the ventricles to fill with blood from the atria.

Path of blood through the heart Before we present the mechanical events of the cardiac cycle, we will describe the path of blood flow through the heart:

1 Blood enters the atria (Figure 18.12A). Oxygen-poor blood from the body flows into the right atrium at about the same time as newly oxygenated blood from the lungs flows into the left atrium: (a) the *superior vena cava* returns blood from all body structures above the diaphragm (except the heart and lungs). (b) The *inferior vena cava* returns almost all blood to the right atrium from all regions below the diaphragm. (c) The *coronary sinus* returns about 85 percent of the blood from the heart muscle to the right atrium. (d) The *pulmonary veins* carry oxygenated blood from the lungs into

the left atrium. The blood entering the right atrium (blue in Figure 18.12A) is low in oxygen and high in carbon dioxide because it has just returned from supplying oxygen to the body tissues. The blood entering the left atrium (red in Figure 18.12A) is rich in oxygen because it has just passed through the lungs, where it has picked up a new supply of oxygen and released its carbon dioxide. (This is the only time or place where *venous* blood is highly *oxygenated*, because it is coming to the heart directly from the lungs.)

2 Blood is forc
heart's natural pa
impulse that coor
systole). Blood is t
valves into the re

3 The ventricle
(Figure 18.12c).

4 The ventricle
lungs (Figure 18.
pressure that clos
atria and ventrick
leading out of the
low in oxygen ou
teries to the lung:

genated blood through the aortic semilunar valve into the aorta. The aorta branches into the ascending and descending arteries that carry oxygenated blood to all parts of the body (see Figure 18.12D). The left and right ventricles pump almost simultaneously, so that equal amounts of blood enter and leave the heart. By this time, the atria have already started to refill, preparing for another cardiac cycle.

Mechanical events of the cardiac cycle The heart beats in a more or less regular fashion about 2.5 billion times during an average lifetime. In order for such regularity to exist, the mechanical events of the cardiac cycle must be coordinated precisely.

The heart functions as a pump by contracting its chambers in order to generate the pressure that forces blood through

FIGURE 18.12

The cardiac cycle and the path of blood through the heart. (A) Blood enters the atria. (B) Blood is pumped into the ventricles. (C) The ventricles relax. (D) The ventricles contract, pumping blood through the pulmonary artery and aorta to the lungs and body.

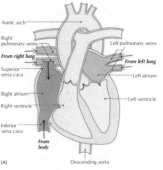

[A]

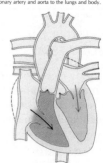

[B]

[C]

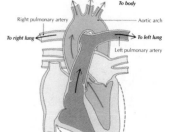

[D]

SEQUENTIAL PHYSIOLOGICAL EVENTS

Itemized text is reinforced with step-by-step drawings.

COLOR-CODED AND RIGHT-ORIENTED ILLUSTRATIONS

For clarity, each drawing is color-coded. For example, nerves are always yellow, lymph is always green, arteries are red, veins are blue, muscles are dark red. All anatomical drawings are drawn facing to the right for consistent orientation, as shown here.

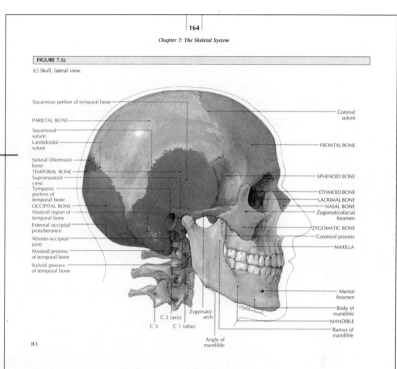

FIGURE 7.5c

(c) Skull, lateral view.

Squamous portion of temporal bone

PARIETAL BONE

Squamosal suture

Lambdoidal suture

Sutural (Wormian) bone

TEMPORAL BONE

Supramastoid crest

Tympanic portion of temporal bone

OCCIPITAL BONE

Mastoid region of temporal bone

External occipital protuberance

Atlanto-occipital joint

Mastoid process of temporal bone

Styloid process of temporal bone

Coronal suture

FRONTAL BONE

SPHENOID BONE

ETHMOID BONE

LACRIMAL BONE

NASAL BONE

Zygomaticofacial foramen

ZYGOMATIC BONE

Coronoid process

MAXILLA

Mental foramen

Body of mandible

MANDIBLE

Ramus of mandible

C 2 (axis) C 1 (atlas)

C 3

Zygomatic arch

Angle of mandible

[C]

through voltage-gated sodium and potassium *open ion channels* (see Figure 11.11). These selective channels are impermeable to large protein molecules, but are always open to Na^+, K^+, and Cl^-. When the concentration of Na^+ or K^+ ions becomes too high on either side, the channels open selectively to reestablish the distribution of ions that produces the resting membrane potential of -70 mV.

Mechanism of Nerve Action: Changing the Resting Membrane Potential into the Action Potential (Nerve Impulse)

The *change* in the electrical potential across a plasma membrane is the key factor in the creation and subsequent conduction of a nerve impulse. The process of conduction, although basically similar in unmyelinated and myelinated nerve fibers, differs somewhat. The following steps describe conduction in unmyelinated fibers:

1 A stimulus that is strong enough to initiate an impulse in a nerve cell is called a ***threshold stimulus.*** When such a stimulus is applied to a polarized resting plasma membrane of the axon of a neuron, the permeability of the membrane to sodium ions increases dramatically at the point of stimulation. For about half a millisecond, the sodium ion channels open, and sodium ions rush into the cell, reversing the relative electrical charges at the point of stimulus. This reversal of charges, giving the *inner side* of the plasma membrane a positive charge (of about $+30$ mV) relative to the *outer side,* is called ***depolarization.*** When a stimulus is strong enough to cause depolarization the neuron is said to *fire.*

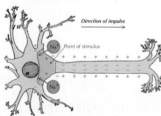

2 The depolarized patch on the plasma membrane produces a flow of current that stimulates the adjacent polarized patch. As a result, this new site becomes depolarized; sodium ions rush in through voltage-gated Na^+ channels, and the electrical charges become reversed as they did at the original point of stimulus. Once a patch on the axon is depolarized, an ***action potential (nerve impulse)*** is initiated (Figure 11.12). This continuous spread of the nerve impulse is characteristic of unmyelinated axons. The membrane voltage of the ongoing action potential operates to open the sodium and potassium channels just ahead of the action potential.

3 Shortly after the sodium ions move into the cell, the membrane at the original point of stimulus becomes more permeable to potassium ions. The voltage-gated K^+ channels open, and a number of potassium ions diffuse *out* from the cell. The outside of the membrane regains its original positive charge (the resting potential), and the original balance of sodium and potassium ions is restored. Also restored at the point where potassium ions rush out is the relative negative charge inside the cell and the relative positive charge outside the cell. Thus, the membrane is said to be ***repolarized.*** The transmission of a nerve impulse along the cell membrane may be visualized as a *wave* of depolarization and repolarization. At any given spot on the membrane, the sequence from resting potential to action potential and back to resting potential takes only a few milliseconds. Only a tiny fraction of the available stored sodium and potassium ions is used in the operation, and the original concentrations are restored by the sodium-potassium pump. As a result, many action potentials can occur before the concentrations of Na^+ and K^+ on either side of the membrane change significantly.

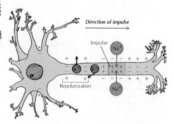

4 Successive acts of depolarization (reversal of the potentials) are repeated as the nerve impulse travels along the length of the axon. After each firing, there is an interval of from one-half to one millisecond before it is possible for an

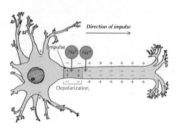

PROCESS DIAGRAMS

Complex processes are presented in segments, with each paragraph keyed to an accompanying drawing that directly follows the text.

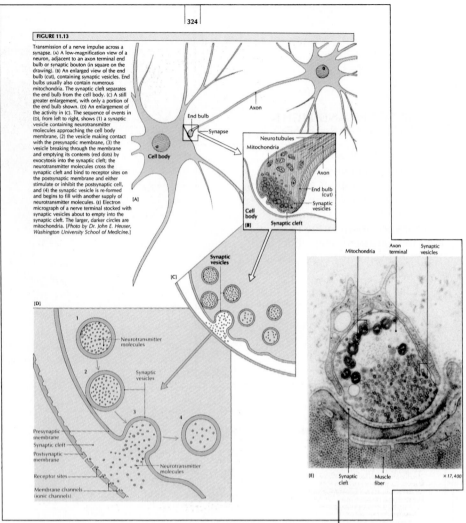

324

FIGURE 11.13

Transmission of a nerve impulse across a synapse. (A) A low-magnification view of a neuron, adjacent to an axon terminal end bulb or synaptic bouton (in square on the drawing). (B) An enlarged view of the end bulb (cut), containing synaptic vesicles. End bulbs usually also contain numerous mitochondria. The synaptic cleft separates the end bulb from the cell body. (C) A still greater enlargement, with only a portion of the end bulb shown. (D) An enlargement of the activity in (C). The sequence of events in (D), from left to right, shows (1) a synaptic vesicle containing neurotransmitter molecules approaching the cell body membrane, (2) the vesicle making contact with the presynaptic membrane, (3) the vesicle breaking through the membrane and emptying its contents (red dots) by exocytosis into the synaptic cleft; the neurotransmitter molecules cross the synaptic cleft and bind to receptor sites on the postsynaptic membrane and either stimulate or inhibit the postsynaptic cell, and (4) the synaptic vesicle is re-formed and begins to fill with another supply of neurotransmitter molecules. (E) Electron micrograph of a nerve terminal stocked with synaptic vesicles about to empty into the synaptic cleft. The larger, darker circles are mitochondria. [Photo by Dr. John E. Heuser, Washington University School of Medicine.]

Functions of the Au

Within the hypothalamus are many neural circuits, called *control centers*, which control such vital autonomic activities as body temperature, heart rate, blood pressure, blood osmolarity, and the desire for food and water. The hypothalamus is also involved with behavioral expressions associated with emotion, such as blushing.

Clearly, the hypothalamus is a critical participant in maintaining homeostasis. For example, the autonomic responses that control body temperature are initiated because the hypothalamus acts as a thermostat that monitors the temperature of blood flowing through a hypothalamic control center. Neurons in the hypothalamus respond to temperature changes, activating either heat-dissipating or heat-conserving control systems to maintain the desired body temperature. The heat-dissipating center in the anterior hypothalamus activates the responses of increased sweating and dilation of skin blood vessels, thus cooling the body. The heat-conserving center in the posterior hypothalamus causes shivering and the constriction of skin blood vessels, thus generating heat.

Cerebral Cortex and Limbic System

Structures of the limbic system, such as the limbic lobe, amygdala, and hippocampus, are connected to the hypothalamus, and use the hypothalamus to express their activities. These expressions include many visceral and behavioral responses associated with self-preservation (such as feeding and fighting) and preservation of the species (such as mating and care of the offspring). Electrical stimulation of the limbic lobe and hippocampus produces changes in the cardiovascular system, including alterations in the heart rate and tone of the blood vessels. Stimulation of the amygdala and limbic lobe may alter the secretory activity of digestive glands.

Even the cerebral cortex, which is usually considered the center of thought processes, utilizes the limbic system and hypothalamus, through its connections with the autonomic nervous system, to express some of our emotions. For example, when a person experiences anxiety, pleasure, or other emotional feelings, the cerebral cortex and limbic system become active, and relay the influences to the hypothalamus. The hypothalamus responds by relaying neural influences via the descending autonomic pathways to the cardiovascular centers in the brainstem. These influences are then projected to the pools of preganglionic neurons of the cranial nerves, and to the spinal cord. Depending upon which centers of the hypothalamus are stimulated, the resulting expressions can be sympathetic or parasympathetic.

Visceral Reflex Arc

A *visceral reflex* innervates cardiac muscle, smooth muscle, or glands. When stimulated, smooth muscles or cardiac muscles contract, and glands release their secretions. Such a reflex, like a somatic motor reflex, does not involve the cerebral cortex, and most visceral adjustments are made through regulatory centers, for example, in the medulla or spinal cord, without our conscious control or knowledge.

bladder, muscular contraction of the intestines, and constriction or dilation of blood vessels. Examples of reflex arcs in the medulla include the regulation of blood pressure, heart rate, respiration, and vomiting.

Ask Yourself

1 *What are some of the centers in the central nervous system that are involved in regulating the autonomic nervous system?*

2 *How does the central nervous system cooperate with the autonomic nervous system to regulate body temperature?*

3 *What are the components of an autonomic visceral reflex arc?*

FUNCTIONS OF THE AUTONOMIC NERVOUS SYSTEM

In this section, we provide an overall picture of the autonomic nervous system as a two-part regulatory system by looking at the way the sympathetic and parasympathetic divisions balance their influences to help us react to changes and maintain our internal homeostasis (Table 14.3). As an example, we show how the system operates during a downhill ski race.

Example of the Operation of the System: A Ski Race

An Olympic skier on a twisting downhill slope is concentrating every part of the body to negotiate the course faster than anyone else in the world. The skier's heart, beating as much as three times faster than yours is right now, is also pumping more blood, faster, to the skeletal muscles than yours is now. The skier's pupils are dilated. The blood vessels of the skin, body organs, and salivary glands—all but those of the skeletal muscles—are constricted. The sweat glands are stimulated. Epinephrine (adrenaline) and norepinephrine (noradrenaline) virtually pour out of the adrenal glands. Obviously ready for action, the skier shows the so-called "fight or flight" response, a state of heightened readiness.

MORE VIEWS TOWARD BETTER UNDERSTANDING

Several enlargements and photographs reinforce the main drawing.

HOW THE BODY OPERATES

An interesting, practical example of how the autonomic body system functions during a stressful ski race enhances the text.

Osseous (Bone) Tissue

FIGURE 6.4

A scanning electron micrograph of trabeculae in cancellous bone. Trabeculae in this area of the bone are the first to be resorbed during maturity, allowing for growth of the marrow cavity. [*Reprinted with permission from Weiss, Leon (ed.), Histology: Cell and Tissue Biology, 5th ed. New York: Elsevier, 1983.*]

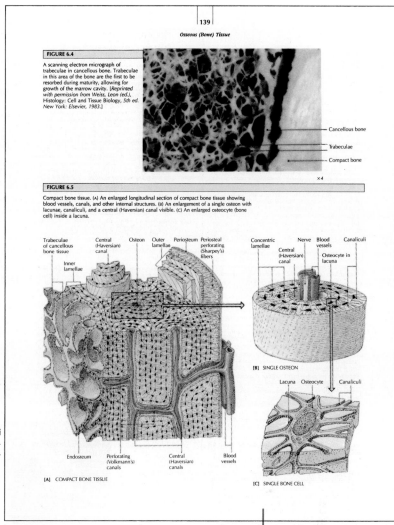

Cancellous bone

Trabeculae

Compact bone

×4

FIGURE 6.5

Compact bone tissue. (A) An enlarged longitudinal section of compact bone tissue showing blood vessels, canals, and other internal structures. (B) An enlargement of a single osteon with lacunae, canaliculi, and a central (Haversian) canal visible. (C) An enlarged osteocyte (bone cell) inside a lacuna.

Trabeculae of cancellous bone tissue Central (Haversian) canal Osteon Outer lamellae Periosteum Periosteal perforating (Sharpey's) fibers

Inner lamellae

Concentric lamellae Nerve Blood vessels Canaliculi

Central (Haversian) canal Osteocyte in lacuna

[B] SINGLE OSTEON

Lacuna Osteocyte Canaliculi

Endosteum Perforating (Volkmann's) canals Central (Haversian) canals Blood vessels

[A] COMPACT BONE TISSUE

[C] SINGLE BONE CELL

Vision

Muscles of the eye and eyelid A set of six muscles moves the eyeball in its socket. The action of these muscles is described in Table 15.5 (see also Figure 10.7). The muscles are the four *rectus muscles* and the *superior* and *inferior oblique muscles.* They are called extrinsic or *extraocular* muscles because they are outside the eyeball (*extra* = outside). One end of each muscle is attached to a skull bone, and the other end is attached to the sclera of the eyeball. The extraocular muscles are coordinated and synchronized so that both eyes move together in order to center on a single image. These movements are called the *conjugate movements* of the eyes.

Other muscles move the eyelid. The *orbicularis oculi* lowers the eyelid to close the eye, and the *levator palpebrae superioris* raises the eyelid to open the eye. The *superior tarsal* (Muller's) muscle is a smooth muscle innervated by the sympathetic nervous system. It helps to raise the upper eyelid, and when it is paralyzed (as in Horner's syndrome, see Chapter 14) it causes a slight drooping (ptosis) of the upper eyelid.

Inside the eyes are several smooth *intrinsic muscles.* The *ciliary muscle* eases tension on the suspensory ligaments of the lens and allows the lens to change its shape in order for the eye to focus (accommodate) properly. The *circular muscle* of the iris contracts the pupil, and the *radial muscle* dilates it.

Physiology of Vision

The visual process can be subdivided into five phases:

1 Refraction of light rays entering the eye.

2 Focusing of images on the retina by accommodation of the lens and convergence of the images.

3 Conversion of light waves by photochemical activity into neural impulses.

4 Processing of neural activity in the retina, and transmission of coded impulses through the optic nerve.

5 Processing in the brain, culminating in perception.

Let us follow the process through each phase in more detail.

Refraction Light waves travel parallel to each other, but they bend when they pass from one medium to another with a different density.* Such bending is called **refraction.** Light waves that enter the eye from the external air are refracted, so that they converge at the retina as a sharp, focused point called the *focal point* (Figure 15.21).

Before light reaches the retina, it passes through (1) the cornea, (2) the aqueous humor of the anterior chamber between the iris and lens, (3) the lens, and (4) the gelatinous vitreous humor in the vitreous chamber behind the lens. Refraction takes place as the light passes through both surfaces of the cornea (which is a convex, nonadjustable lens) and

*Fishermen have learned that when they try to grab a fish swimming below the surface, they must reach a little to the side of the image to compensate for the bending of light waves from air to water and vice versa.

TABLE 15.5	ACT
Muscle	
SKELETAL MUSCLES	
Medial rectus	
Lateral rectus	
Superior rectus	
Inferior rectus	
Superior oblique	
Inferior oblique	
Orbicularis oculi	Lowers eyelid (closes eye).
Levator palpebrae superioris	Raises eyelid (opens eye).
SMOOTH MUSCLES	
Ciliary muscle	Eases tension on suspensory ligament of lens, permits focusing (accommodation).
Circular muscle of iris	Contracts pupil.
Radial muscle of iris	Dilates pupil.
Superior tarsal muscle of upper eyelid	Raises eyelid.

again as it passes through the anterior and posterior surfaces of the lens (which is a convex, adjustable lens).

A normal eye can bring distant objects more than 6 m (20 ft) away to a sharp focus on the retina. When parallel light rays are focused exactly on the retina, and vision is perfect, the condition is called *emmetropia* (Gr. "in measure"). Nearsightedness, or *myopia* (Gr. "contracting the eyes"), occurs when light rays come to a focus *before* they reach the retina.* As a result, when the rays do reach the retina, they form an unfocused circle instead of a sharp point, and distant objects appear blurred (see Figure 15.21B). Farsightedness, or *hypermetropia* (Gr. "beyond measure"), occurs when light rays are focused *beyond* the retina, and as a result near objects appear blurred (see Figure 15.21c).†

Both myopia and hypermetropia can be corrected by wearing prescription eyeglasses or contact lenses, which are specially ground lenses placed in front of the eye to change the angle of refraction.

Myopia is so named because a nearsighted person often squints through narrowed eyelids in an effort to focus better. Although the resultant tiny opening requires little or no focusing, the amount of light entering the eye is decreased, and strain on the relevant eye muscles may cause headaches.

†Some textbooks show a corrective lens for myopia as) , and a corrective lens for hypermetropia as) . Such lenses are capable of refracting light rays, according to the fundamental laws of physics, but are actually never used to correct vision defects. The corrective lenses shown in Figures 15.21B and c are meant to approximate the actual shapes of corrective *eyeglass* lenses. In any given lens, the relationship of one curve to another changes depending on the specific prescription, but the basic shapes for correcting myopia and hypermetropia remain the same, as shown in Figures 15.21B and c, with the center of the lens being the thinnest point for myopia, and the thickest point for hypermetropia.

FINELY DETAILED ART

A greater amount of detail than usual is presented with enlarged portions of main illustrations.

FOOTNOTES

Footnotes contain many interesting, practical examples as well as illustrations which clear up age-old inaccuracies.

498

Chapter 16: The Endocrine System

affect the sleep cycle. The typical lack of light during long winters may contribute to *seasonal affective disorder* (SAD), a condition characterized by lack of energy and mood swings that border on depression. Some researchers speculate that the pineal gland is involved in SAD, but no firm evidence is available.

Thymus Gland

The *thymus gland* is a double-lobed lymphoid organ located behind the sternum in the anterior mediastinum (see Figure 16.1). It has an outer cortex containing many lymphocytes, and an inner medulla containing fewer lymphocytes as well as clusters of cells called thymic (Hassall's) corpuscles, whose function is unknown. The thymus gland is well supplied with blood vessels, but has only a few nerve fibers.

The thymus gland is large and active only during childhood, reaching its maximum effectiveness during early adolescence. After that time, the gland begins to atrophy because of the action of sex hormones, and is replaced by fatty tissue. Prolonged stress usually hastens atrophy. This happens because stress factors release adrenocortical hormones that have a destructive effect on thymus tissue. The thymus gland finally ceases activity altogether after 50 years or so, and it may therefore play an important role in the process of aging and the accompanying decrease in function of the immune system.

The main function of the thymus gland seems to be the processing of T cells (T lymphocytes). These cells are responsible for one type of immunity, called *cellular immunity* (see Chapter 20). Other lymphocytes, called B cells (B lymphocytes), are processed in the fetal liver before a child is born. B cells are responsible for a type of immunity called *humoral immunity* (see Chapter 20). There is some evidence that the thymus gland may be a true endocrine gland, since it produces thymic hormones, or "factors," that play a role in the development of T cells in the thymus, and their maintenance within other lymphoid tissue. Some of the hormones and factors include *thymosin alpha, thymosin B₁ to B₅, thymopoietin I and II, thymic humoral factor* (THF), *thymostimulin,* and *factor thymic serum* (FTS). The thymic hormones also play a role in the development of some B cells into plasma cells, which produce antibodies. There is also a possibility that the thymus gland may influence the secretion of reproductive hormones from the pituitary gland.

It had been thought until recently that thymic hormones were produced exclusively in the thymus gland, and that the main function of thymic hormones was to assist in the processing of bone marrow cells into T cells, infection-fighting cells of the immune system. Recent discoveries, however, indicate that thymosin B₄ and B₅ influence hormones of the reproductive system, and that thymosin B₄ is also synthesized by macrophages in the immune system.

Heart

Recent findings have revealed that in addition to being the complex pump that maintains circulation, the heart also acts as an endocrine organ. Cardiac muscle cells within both atria (the upper chambers of the heart) contain secretory granules that produce, store, and secrete a peptide hormone called *atriopeptin* (formerly called *atrial natriuretic factor,* or *ANF*).

Atriopeptin is secreted continuously in minute amounts, and is circulated throughout the body via the bloodstream. Secretion increases when excess salt accumulates in the body, when blood volume increases enough to stimulate stretch receptors in the atria, or when blood pressure rises significantly. Special target-cell receptors have been found in blood vessels, kidneys, and adrenal glands. Atriopeptin also affects neurons in the brain, especially the hypothalamus, where control and regulation occurs for blood pressure and the excretion of sodium, potassium, and water by the kidneys.

Current evidence suggests that atriopeptin helps to maintain a proper balance of fluid and electrolytes by increasing the output of sodium in urine; relaxes blood vessels directly, thus lowering blood pressure by reducing resistance to blood flow; lowers blood pressure by blocking the actions of hormones such as aldosterone, which tends to raise blood pressure; and reduces blood volume by stimulating the kidneys to filter more blood and produce more urine. Scientists speculate that atriopeptin complements the actions of other hormones, rather than acting on its own. It is hoped that carefully administered doses of atriopeptin will be useful in regulating blood pressure and electrolyte balance.

Digestive System

Among the major hormones of the digestive system are gastrin, secretin, and cholecystokinin. *Gastrin* is a polypeptide secreted by the mucosa (lining) of the stomach. Its function is to stimulate the production of hydrochloric acid and the digestive enzyme pepsin when food enters the stomach. Thus the stomach is both the producer and the target organ of gastrin.

Secretin is a polypeptide secreted by the mucosa of the duodenum. It stimulates a bicarbonate-rich secretion from the pancreas that neutralizes stomach acid as the acid passes to the small intestine. Secretin was the first hormone to be discovered (by the British scientists William M. Bayliss and Ernest H. Starling in 1902), and the first substance to actually be called a "hormone."

Cholecystokinin (CCK; koh-lee-sis-ᴛᴏᴇ-kine-in) is secreted from the wall of the duodenum. It stimulates the contraction of the (particularly fats secretion of enz

Research scie hormones," sub hormones and a of these substan tract. Two that stimulates the which stimulate mones are *bomi* stomach and in *tory polypeptide* ach.

EXTREMELY UP-TO-DATE MATERIAL

Recent research findings are included.

848

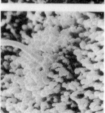

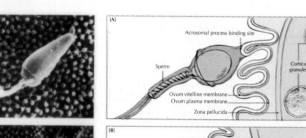
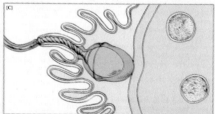
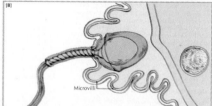
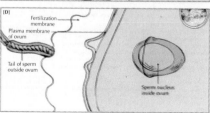

COMPLEMENTARY FIGURES

A sequence of vivid scanning electron micrographs is reinforced by comparable drawings with clear labels.

INNOVATIVE ILLUSTRATION TECHNIQUES

New, colorful techniques are used.

FIGURE 6.8
A gamma-ray scintigram of a normal child, with active growth zones at the epiphyses of long bones (shown as red clusters). Gamma-ray scintigrams are produced after radioactive isotopes are injected into the body. Then gamma-ray emissions are detected by a scintillating crystal detector and recorded as a scintigram, or "bone scan." [*Gruppo Editionale, Fabbri Milano.*]

and differentiate into osteoblasts. Most probably, they serve as an ion barrier around bone tissue. This barrier contributes to mineral homeostasis by regulating the movement of calcium and phosphate into and out of the bone matrix, which in turn helps control the deposition of hydroxyapatite in the bone tissue.

Ask Yourself

1 What is the diaphysis? The epiphysis? The metaphysis?

2 Does red bone marrow have a specific function?

3 What are the five kinds of bone cells?

DEVELOPMENT OF BONES

Bones develop through a process known as **ossification** (osteogenesis). Since the primitive "skeleton" of the human embryo is composed of either hyaline cartilage or fibrous membrane, bones can develop in the embryo in two ways: *intramembranous ossification* or *endochondral ossification* (also called *intracartilaginous ossification*). However, in both cases, bones are formed from a pre-existing "connective tissue skeleton." Bone is the same no matter how it develops. Only the bone-making sequence is different.

Figure 6.9 compares the fetal and adult skeletons. Some bones, such as the skull, develop by intramembranous ossification, and other bones, such as those in the arms and legs, develop by endochondral ossification. The ages when particular bones develop are shown on the skeletons.

Intramembranous Ossification

If bone tissue (spongy or compact) develops directly from mesenchymal (embryonic connective) tissue, the process is called **intramembranous ossification.*** The vault (arched part) of the skull, flat bones of the face (including those lining the oral and nasal cavities), and part of the clavicle (collarbone) are formed this way. The skull is formed relatively early in the embryo to protect the developing brain. Some flexible tissue still exists between the flat skull bones at birth, so when the baby is born its skull is flexible enough to pass unharmed through the mother's birth canal.

From the time of the initial bone development, intramembranous ossification spreads rapidly from its center until large areas of the skull are covered with protecting and supporting bone. The first rapid phase begins when an *ossification center* (see item 3 in the following list) first appears (from the eighth fetal week through the twelfth fetal week) and lasts through the end of the fifteenth fetal week, when the area is entirely covered (Figure 6.10).

*Intramembranous means "within the membrane," and *ossification* means "bone formation." The term *intramembranous ossification* was originally used because this layer of mesenchyme was thought to be a sheet of membrane.

Anatomy of a Runner

Muscle physiologist Larry Stewart of Ball State University's Human Performance Laboratory has found an important difference in the leg muscles of sprinters and long-distance runners. Most people have approximately equal numbers of fast-twitch and slow-twitch fibers in running muscles such as the gastrocnemius. Sprinters, however, have more fast-twitch fibers, and distance runners have more slow-twitch fibers.

Fast-twitch fibers are perfect for short, fast bursts of speed. They burn stored glycogen quickly, without using oxygen. In the process, however, lactic acid accumulates, and the muscles become fatigued when they run out of fuel.

The leg muscles of long-distance runners contain mostly slow-twitch fibers. (Bill Rodgers has about 75 percent slow-twitch fibers, and Alberto Salazar has 93 percent.) Slow-twitch muscles take a lit-

---

FIGURE 9.14
Duration of isometric contractions of three different muscles. The lateral rectus is a very fast muscle, the soleus is considered slow, and the gastrocnemius is in between.

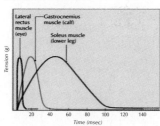

In contrast, *slow-twitch muscle fibers* are suited for prolonged and steady contractions. They have plenty of myoglobin and red blood cells, and they do not tire as easily as fast-twitch muscle fibers do. The large back muscles that help us to maintain an erect posture all through the day are an example of slow-twitch muscle fibers. Slow-twitch muscle fibers do not have, or need, as large a supply of calcium ions as fast-twitch muscle fibers do.

The color of fast-twitch and slow-twitch muscle fibers is different as well. Slow-twitch muscle fibers contain large amounts of myoglobin, which has a reddish color, and many capillaries with red blood cells. Thus slow-twitch muscle fibers are also called **red muscles**. Fast-twitch muscle fibers, with relatively little myoglobin and fewer capillaries with red blood cells, have a pale appearance and are referred to as **white muscles**.

By-products of Mechanical Work

When a muscle contracts, it converts chemical energy into mechanical energy. It also releases heat, uses up oxygen (sometimes faster than it can be replaced), and, if it works too hard for too long, it becomes fatigued.

Oxygen debt and muscle fatigue Ordinarily, when you play tennis or jog at a leisurely pace, your body uses extra oxygen. It is made available when your heart rate increases, carrying oxygen to your muscles, and when you begin to breathe harder than usual. But sometimes, when a runner is sprinting, for instance, the skeletal muscles work so hard that they use up oxygen faster than it can be supplied. When this happens, the muscles must find a way to get more oxygen.

The body receives most of the energy required to produce ATP from the glucose, amino acids, and fats in food. Ordinarily, glucose provides most of this energy, but in the process it also produces pyruvic acid, which must be catabolized. A resting muscle fiber receives enough oxygen to break down pyruvic acid:

$$\text{Pyruvic acid} + O_2 \longrightarrow CO_2 + H_2O + \text{energy}$$

Because oxygen is necessary for this process, it is called *aerobic* ("with oxygen") *respiration.*

Why is some meat dark and some light?	The reddish color of dark meat comes from the iron-rich protein myoglobin in the blood-rich dark meat. The oxygen that the muscle needs for long periods of movement is in the muscle cells next to the myoglobin. The dark meat of a turkey, for instance, is found mostly in the legs, where long-term energy is needed for running. White meat, in contrast, is found in the muscles that have short bursts of activity, such as the breast muscles of a turkey, which are adapted for fast, short flights.

THOUGHT-PROVOKING QUESTIONS

Everyday questions and answers enliven the text.

UNIFYING THEMES

Chapter concepts are tied together through the use of unifying themes; here, the interrelatedness of two body systems such as the neuroendocrine system and the immune system is described.

END-OF-CHAPTER MATERIAL

Concise Discussions of Disorders

INTERACTION BETWEEN THE NEUROENDOCRINE SYSTEM AND THE IMMUNE SYSTEM

It has been known since the 1950s that adrenal glucocorticoid hormones secreted in response to stress may cause the immune system to become suppressed and ineffective. Many other possible interactions between the neuroendocrine system and the immune system have been uncovered recently. Karen Bulloch of the State University of New York at Stony Brook has shown how nerve fibers penetrate into lymphoid tissue, especially in bone marrow and the thymus gland, where the production and maturation of lymphocytes take place. Lewis T. Williams of the University of California School of Medicine at San Francisco has found special receptor sites on lymphocytes for neurotransmitters released by nerve fibers. Eli K. Hazum of the Weizmann Institute of Science in Israel has discovered that neuropeptides such as *beta*-endorphin bind to specific receptor sites on lymphocytes. Manfred L. Karnovsky of the Harvard School of Medicine has provided evidence that macrophages are activated by serotonin. J. Edwin Blalock of the University of Alabama at Birmingham has demonstrated that lymphocytes produce hormones such as ACTH, which affect the adrenal glands. In addition, interferon has been shown to affect the neuroendocrine system.

Steven E. Keller and his coworkers at the Mount Sinai School of Medicine in New York City have found that lymphocytes are inhibited in response to stress. Janice K. Kiecolt-Glaser of the Ohio State University Medical Center has shown that DNA repair in the lymphocytes of emotionally disturbed patients is diminished. Ronald Glaser and his coworkers at the Ohio State University Medical Center have confirmed that while students are taking examinations, their immune systems do not function at peak level, with killer T cells being impaired, other cells producing subnormal amounts of interferon, and blood levels of the Epstein-Barr virus (which causes infectious mononucleosis) rising.

It is possible that some people enjoy good health because they are able to cope well with stress, while others react poorly under stress, causing ill health related to a breakdown of the immune system. Such interactions between the neuroendocrine system and the immune system continue to be studied intensively, and there is hope that the more we learn about them, the better we can prevent and treat disorders of the immune system.

PHYSIOLOGICAL AND ANATOMICAL ABNORMALITIES

Acquired immune deficiency syndrome (AIDS)

As its name suggests, *acquired immune deficiency syndrome*, or *AIDS*, is a disease that cripples the body's immune system. Victims of AIDS show a deficiency in two kinds of lymphocytes: natural killer T cells that kill virus-infected and tumor cells, and virus-specific killer T cells that kill cells containing a particular virus. Without an intact immune system, AIDS victims may die from microbial infections that are ordinarily minor problems. Also, relatively uncommon forms of cancers of the blood and lymphatic system seem to be unusually prevalent among AIDS victims. In fact, the first sign that a new disease had emerged was the sudden appearance of Kaposi's sarcoma (a cancer seen in blood vessels) in the 1970s among young, middle-class, white males. Until then, the disease appeared mainly among older Italian and Jewish men, and in Africa. The young male victims turned out to be predominantly homosexual, and the disease was also spreading among drug users and people who received frequent blood transfusions. (In the United States in 1986, 73 percent of AIDS victims were active male homosexuals or bisexuals, 17 percent were intravenous drug users, 5 percent were Haitian immigrants, and 1 percent were hemophiliacs who were dependent on blood transfusions.)

In 1984, it was discovered by Robert Gallo and his coworkers at the National Cancer Institute that AIDS is caused by *human immunodeficiency virus* (HIV). It is spread mainly by sexual contact and shared hypodermic needles, especially when the virus infects groups of people

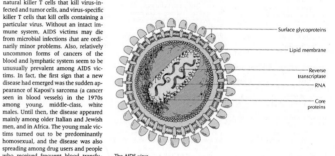

Surface glycoproteins

Lipid membrane

Reverse transcriptase

RNA

Core proteins

The AIDS virus.

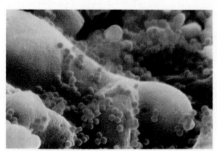

AIDS viruses (blue spheres) attack a helper T cell. ×80,000
[Courtesy of Lennart Nilsson. © Boehringer Ingelheim International, GmbH.]

living in close contact. Anal intercourse is a likely method of transmission because rectal capillaries are so close to the skin that infection may occur through microscopic breaks in the skin. However, researchers stress that AIDS is not exclusively a disease of homosexuals, bisexuals, or drug addicts. In Africa, it is a disease of heterosexuals.

HIV is a *retrovirus*, a virus that transmits genetic information in reverse fashion, from RNA to DNA instead of from DNA to RNA. As with other retroviruses, RNA is the genetic material (see drawing). The host cell is often a helper T cell (T4 lymphocyte), which plays a crucial role in regulating the immune system. The DNA assembled by the enzyme reverse transcriptase from the RNA template may remain latent among the chromosomes of the host lymphocyte until that cell is stimulated by a second infection. Then the viral DNA is reproduced rapidly, and new viruses leaving the host cell eventually kill it by destroying the plasma membrane.

Apparently, the AIDS virus is able to cross the blood-brain barrier to cause the proliferation of glial cells in the central nervous system. Victims of AIDS are also more susceptible to Kaposi's sarcoma, carcinomas (including skin cancers in the mouth or rectum of infected homosexuals), and B-cell lymphomas (tumors originating in lymphocytes).

AIDS was first described in 1981. It probably began in central Africa some time in the 1950s and spread first to the Caribbean and then to the United States and Europe. As many as 2 million Americans may be infected, and the situation is much worse in Africa and the Caribbean. Current treatment concentrates on interrupting the reverse transcriptase as it assembles the viral DNA. Initial use of azidothymidine (AZT), an unsuccessful anticancer drug, has been promising, and the search for an effective vaccine is underway. Prevention has been helped considerably by the development of blood tests that can detect the presence of antibodies to the AIDS virus before transfusions are given.

Hodgkin's disease

Hodgkin's disease is a form of cancer, typified by the presence of large, multinucleate cells (Reed-Sternberg cells) in the affected lymphoid tissue. The first sign of Hodgkin's disease is usually a painless swelling of the lymph nodes, most commonly in the cervical area. As lymphocytes, eosinophils, and other cells proliferate, there is a progressive enlargement of the lymph nodes, the spleen, and other lymphoid tissues. Late

Infectious mononucleosis

Infectious mononucleosis is a viral disease, caused by the Epstein-Barr virus, a member of the herpes group. It appears most often in 15- to 25-year-olds. It is usually accompanied by fever, increased lymphocyte production, and enlarged lymph nodes in the neck. Secondary symptoms include dysfunction of the liver, and increased numbers of monocytes. Infectious mononucleosis is not contagious in adults. In adolescents, however, the Epstein-Barr virus may be spread during close contact or via the exchange of body fluids, as when drinking glasses are shared or through kissing. The virus may also be spread from infant to infant in saliva, from the mother's breast, or food.

Systemic lupus erythematosus

Systemic lupus erythematosus (SLE) is considered a collagen disease because it affects the lining of joints and other connective tissue. Although the origin is uncertain, it is thought that SLE may be hereditary, and may be caused by a breakdown in the body's immune system. Women are infected about nine times more than men are. The first signs of the disease, which can be fatal in extreme cases, are a general malaise, fever, a so-called butterfly rash on the face, appetite loss, sensitivity to sunlight, and pains in the joints. SLE causes B cells to produce excess antibodies, called autoantibodies, which attack healthy cells. The basal layer of skin begins to deteriorate, and any organ can be affected. Because the body is unable to remove the autoantibodies, they may settle in such vital organs as the brain, heart, kidneys, and lungs, ultimately causing serious tissue damage.

644

Chapter 20: The Lymphatic System and Immunity

MEDICAL TERMINOLOGY

ANAPHYLAXIS (an-uh-fuh-LACK-sis; L. *ana*, intensification + Gr. *phulassein*, to guard) Hypersensitivity to a foreign substance.

ELEPHANTIASIS Obstruction (caused by a parasitic filarial worm) of the return of lymph to the lymphatic ducts that causes enlargement of a limb, usually a lower limb, or the genital area.

HYPERSPLENISM A condition in which the spleen is abnormally active and blood cell destruction is increased.

LYMPHADENECTOMY Removal of lymph nodes.

LYMPHADENOPATHY A disease of lymph nodes, resulting in enlarged glands.

LYMPHANGIOMA Benign tumor of lymph tissue.

LYMPHANGITIS Inflammation of lymphatic vessels.

LYMPHATIC METASTASIS A condition in which a disease travels around the body via the lymphatic system.

LYMPHOMA Tumor of the lymph nodes.

LYMPHOSARCOMA Malignant tumor of lymph tissue.

SPLENECTOMY Total removal of the spleen.

SPLENOMEGALY Enlargement of the spleen, following infectious diseases such as scarlet fever, typhus, and syphilis.

VACCINE General term for the immunization preparation used against specific diseases.

Practical Medical Terminology

SUMMARY

Comprehensive Summary

The *lymphatic system* returns to the blood excess fluid and proteins from the spaces around cells, plays a major role in the transport of fats from the tissue surrounding the small intestine to the blood, filters and destroys microorganisms and other foreign substances, and aids in providing long-term protection for the body.

The lymphatic system

1 The lymphatic system consists of vessels, lymph nodes, lymph, leukocytes, lymphatic organs (spleen and thymus gland), and specialized lymphoid tissues.

2 Excess interstitial fluid is drained from tissues by the lymphatic system. Once inside the lymphatic capillaries, the fluid is called *lymph*.

3 The composition of lymph is similar to that of blood, except that lymph lacks red blood cells and most of the blood proteins.

4 Leukocytes in lymph include monocytes, which develop into macrophages, and two types of lymphocytes, *B cells* and *T cells*. Specific defenses are accomplished by *lymphocytes*, a type of leukocyte. Lymphocytes are the basis of the immune response.

5 *Lymphatic capillaries* are one-way vessels with a closed terminal end. They join together to form *lymphatics*, which drain into the *right lymphatic duct* and the *thoracic duct*. The ducts return fluid to circulation through the right and left subclavian veins.

6 *Lymph nodes* are small masses of lymphoid tissue scattered along the lymphatic vessels. Lymphocytes in the nodes filter out harmful substances from the lymph, and are the initiating sites for the specific defenses of the immune system.

7 The three *tonsils* (pharyngeal, palatine, and lingual) form a band of lymphoid tissue that prevents foreign substances from entering the body through the throat.

8 The *spleen* is the largest lymphoid organ. Its major functions are filtering blood and manufacturing phagocytic lymphocytes.

9 The *thymus gland* forms antibodies in the newborn, and plays a major role in the early development of the body's immune system.

10 *Aggregated lymph nodules* (Peyer's patches, gut-associated lymphoid tissue, or GALT) are clusters of lymphoid tissue in the intestine and appendix. They are thought to respond to antigens from the intestine by generating plasma cells that secrete antibodies in large quantities.

Nonspecific defenses of the body

1 The *nonspecific defenses* of the body are those that do not involve the production of antibodies. They include the skin, mucous membranes and mucus, lacrimal glands, nasal hairs, cilia in the respiratory tract, and the acidity of the stomach.

2 When tissues are damaged, the normal result is the *inflammation response*,

which initiates healing and prevents further damage.

3 *Phagocytosis* is the destruction of foreign substances and dead tissue by leukocytes.

4 The body produces nonspecific chemical substances such as *complement*, *properdin*, and *interferon* that attack bacteria and viruses.

Specific defenses: the immune response

1 The *specific defenses* of the body, which constitute the *immune response*, discriminate among foreign substances (*antigens*) by forming specific proteins (*antibodies*) to react with the foreign substances and destroy them.

2 An *antigen* causes the production of specific antibodies, and reacts chemically with the antibody to form a stable complex called the *antigen-antibody complex*.

646

UNDERSTANDING THE FACTS	UNDERSTANDING THE CONCEPTS	SUGGESTED READING
1 List the four major functions of the lymphatic system.	1 Since the lymphatic system lacks a "pump," how do you explain the movement of lymph?	ADA, GORDON L., AND GUSTAV NOSSAL, "The Clonal-Selection Theory." *Scientific American*, August 1987.
2 Define lymph.	2 Although lymph contains all the coagulation factors found in blood and is capable of clotting to a small degree, why does lymph lack clotting ability?	BUISSERET, PAUL D., "Allergy." *Scientific American*, August 1982.
3 Describe the structure of a lymph node.	3 Why is the reticuloendothelial system important to the body?	BURNET, F. M., ed., *Immunology: Readings from Scientific American*. San Francisco: Freeman, 1976.
4 What is the main structural difference between lymphatic capillaries and blood capillaries?	4 Why are valves so important to the lymphatic system?	COHEN, IRUN R., "The Self, the World and Autoimmunity." *Scientific American*, April 1988.
5 Name and describe the vessels of the lymphatic system.	5 How do complement, properdin, and interferon help the body to combat infection?	COLLIER, R. JOHN, AND DONALD A. KAPLAN, "Immunotoxins." *Scientific American*, August 1987.
6 What is the major function of the lacteals?	6 How does the skin defend against microbes?	EDELSON, RICHARD L., AND JOSEPH M. FINK, "The Immunologic Function of Skin." *Scientific American*, June 1985.
7 Into which large veins does lymph drain?	7 What is the difference between active and passive immunity?	GALLO, ROBERT C., "The AIDS Virus." *Scientific American*, January 1987.
8 What portion of the body is drained by the thoracic duct?	8 What is the relationship of the lock-and-key model and the great specificity normally shown by antibodies?	LAURENCE, JEFFREY, "The Immune System in AIDS." *Scientific American*, December 1985.
9 Which lymphatic organ serves as a blood reservoir?	9 Why is it important for the IgG antibodies to pass freely through the placenta during pregnancy?	MARRCK, PHILIPPA, AND JOHN KAPPLER, "The T Cell and Its Receptor." *Scientific American*, February 1986.
10 Name the five classes of antibodies, and describe their functions.	10 How are monoclonal antibodies produced? What are some important uses?	MILSTEIN, CESAR, "Monoclonal Antibodies." *Scientific American*, October 1980.
11 How are lysosomes involved in phagocytosis?	11 Describe the desensitization procedure used to combat severe allergy.	OSMOND, D. G., AND P. K. LALA, "The Immune System," *American Journal of Anatomy*, 3 (1984).
12 Antigens are made up of what type of organic molecule?	12 Discuss the functions of the thymus gland.	RAFF, MARTIN C., "Cell-Surface Immunology." *Scientific American*, May 1976.
13 What is the basic shape of the antibody molecule?	13 Discuss the two mechanisms of autoimmunity.	ROITT, I. M., J. BROSTOFF, AND D. MALE, *Immunology*. St. Louis: Mosby, 1985.
14 Where do B cells and T cells originate?		SILBERNER, JOANNE, "Survival of the Fetus." *Science News*, October 11, 1986.
15 Which type of lymphocytes are active in cell-mediated immunity?		SINGER, A., AND R. HODGES, "Mechanism of B-cell and T-cell Interaction." *Annual Review of Immunology*, 1 (1983): 211.
16 Which cell type is the major cause of rejection in tissue or organ transplants?		SITES, D. P., J. D. STUBO, H. H. FUDENBERG, AND J. V. WELLS, *Basic and Clinical Immunology*, 5th ed. Los Altos, Calif.: Lange Medical Publications, 1984.
17 Why may anaphylactic shock be fatal?		TONEGAWA, SUSUMU, "The Molecules of the Immune System." *Scientific American*, October 1985.
18 Why are corneal transplants generally not rejected?		YOUNG, JOHN DING-E, AND ZANVIL A. COHN, "How Killer Cells Kill." *Scientific American*, January 1988.
19 Give the location of the following (be specific): a lymph nodes b aggregated lymph nodules c flap valves d lysozyme e lacteals f interferon (production site)		

Factual Questions

Thought-Provoking Conceptual Questions

Accessible Suggested Readings for the Interested Student

I
HOW THE BODY IS ORGANIZED

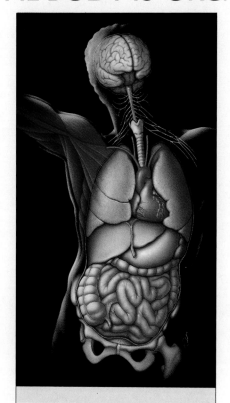

1
Introduction to Anatomy and Physiology

LEARNING OBJECTIVES

1 Define anatomy and physiology, and explain how structure and function are related in the body.

2 Describe some of the subdivisions of anatomy.

3 Describe the principles of homeostasis and negative feedback, and explain how they operate in the body.

4 Explain how stress affects homeostasis.

5 Identify the organizational levels of the body, and explain how they are related.

6 Briefly describe the four tissue groups that make up the body.

7 Identify the systems of the body, and explain a major function of each system.

8 Describe the anatomical position, and discuss its relationship to the directional terms of the body.

9 Distinguish among the relative directional terms, regions, planes, and cavities of the body.

10 Describe the major noninvasive techniques of exploring the body.

How does our body temperature remain the same when the external temperature changes? Why does skin wrinkle? Why does a healing wound itch? Why does a foot "fall asleep"? Why can't we taste food when we have a cold? If millions of sperm cells are released, why does only one fertilize an egg cell?

These are just a few of the many questions that are answered in this book, questions that many of us have wondered about. Although we have learned a great deal about the human body and what makes it function, there are still many unanswered questions. And so we continue studying, hoping that someday we will be able to understand what sleep is, what causes cancer, and what makes one cell develop into a pancreas while a seemingly identical cell becomes a liver. Although the questions are many, the search is tireless, because nothing fascinates the human mind more than the human body.

WHAT ARE ANATOMY AND PHYSIOLOGY?

Anatomy (from the Greek word meaning *to cut up*) is the study of the structure of a plant or animal, or any of its parts. Physiology (from the Latin word meaning the *study of nature*) is the study of all the vital processes (functions) of a plant or animal. In relation to the human body, **anatomy** is the study of the many *structures* that make up the body, and how those structures relate to each other. **Physiology** is the study of the *functions* of those structures. Anatomy and physiology are studied together to give the student a full appreciation and understanding of the body.

The field of human anatomy has become very large, and a number of subdivisions have been recognized and named. One subdivision is *regional anatomy,* the study of specific regions of the body, such as the head, neck, thorax (chest), or upper and lower limbs. Another is *systemic anatomy,* the study of different systems of the body, such as the reproductive system or the digestive system. Regional and systemic anatomy are both examples of *gross anatomy,* any branch of anatomy that can be studied without a microscope. In contrast, *microscopic anatomy* requires the use of a microscope. Other examples of anatomy are *developmental anatomy,* the study of human growth and development from fertilized egg to mature adult; *pathological anatomy,* or pathology, the study of changes in diseased cells and tissues; *histology,* the microscopic study of tissues; *cytology,* the study of cells; *radiographic anatomy,* the study of the structures of the body using x rays; and *splanchnology,* the study of internal organs.

As we begin to study anatomy and physiology, we can see that form and function go together. Parts of the body have specific shapes that make them suitable to perform their specific functions. Think of teeth. Three different kinds of teeth do three very different jobs. Flat molars grind food, sharp-pointed canines tear, and incisors cut. As you progress through this book, try to keep in mind the crucial connection between anatomy and physiology—form and function—and find as many examples of this connection as you can.

HOMEOSTASIS: COORDINATION CREATES STABILITY

When structure and function are coordinated, the body achieves a relative stability of its internal environment called **homeostasis** (ho-mee-oh-STAY-siss; Gr. "staying the same"). Although the outside environment changes constantly, the internal environment of the healthy body remains the same, within normal limits. Under normal conditions, homeostasis is maintained by adaptive mechanisms in the body, ranging from control centers in the brain to chemical substances called hormones that are secreted by various organs directly into the bloodstream. Some functions controlled by homeostatic mechanisms are blood pressure, body temperature, breathing, and heart rate.

For a body to remain healthy, a chemical balance inside and outside its cells must be carefully maintained. To achieve this balance, the composition of the **extracellular fluid**—the fluid outside the cells that surrounds and bathes them—must remain fairly constant. Extracellular fluid circulates throughout the body, and many materials are passed into and out of cells by way of the extracellular fluid. Thus, the fluid is instrumental in helping the body's systems maintain optimal temperature and pressure levels, as well as the proper balance of acids and bases, oxygen and carbon dioxide, and the concentrations of water, nutrients, and the many chemicals that circulate in the blood.

Practically everything that goes on in the body helps to maintain homeostasis. For instance, the kidneys filter blood and remove a carefully regulated amount of water and wastes. The lungs work together with the heart, blood vessels, and blood to distribute oxygen throughout the body and to remove carbon dioxide. From the digestive system, nutrients pass through the walls of the small intestine into the bloodstream, which carries them to parts of the body where they are needed. In general, all the systems of the body work together to contribute to the well-being that comes with inner stability.

Homeostasis and Negative Feedback

The whole regulation process of homeostasis is made possible by the coordinated feedback action of many organs and tissues under the direct or indirect control of sensitive networks in the nervous and hormonal systems. If all the systems are operating properly, we feel fine—homeostasis is taken for granted. If the coordination among the systems breaks down, we begin to feel uncomfortable. The less coordination there is, the worse we feel. If homeostasis is not restored by the body itself or with the help of outside intervention, we die.

Feedback occurs whenever an adjustment that a person makes in the present affects the future state of the body. For example, when we are too hot, our sweat glands are activated, and we perspire. The sweat evaporates and cools us. In contrast, when we are too cold, the muscles under our skin receive a message from the nervous system to contract and relax, contract and relax—we shiver—and the action gives off heat. Sweating and shivering are both involuntary responses initiated by part of the nervous system that is linked to a heat-sensitive regulating area in the brain. Although external temperatures may vary, our bodies must remain within a few degrees of the normal 37°C (98.6°F) to remain alive, let alone healthy (Figure 1.1). (See Chapter 23 for a discussion of the regulation of body heat.)

In this case, a feedback system produces a response of warmth that is *opposite* to the initiating stimulus of cold. This is called a **negative feedback system** (Figure 1.2). If blood sugar decreases, the response of a negative feedback system is to raise it; if blood sugar increases, the response is to lower it. Each of these responses is a *negative* action, or *not the same as* the initial stimulus of low or high blood sugar.

In contrast, a **positive feedback system** operates where the initial stimulus is *reinforced* rather than changed. For example, if blood sugar decreases, the response of a positive feedback system is to lower it still further. If continued without restraint, such a response would lower the blood sugar level until the person eventually dies. Positive feedback is relatively rare in our bodies, mainly because it disrupts homeostasis.

Stress and Homeostasis

To maintain homeostasis, the mind and body must adjust to the various imbalances that arise in the internal or external

FIGURE 1.1

Some typical body temperature ranges.

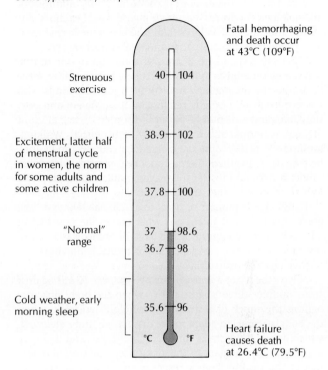

FIGURE 1.2

(A) A generalized negative feedback system. (B) A specific negative feedback system, set in motion to balance the effects of a mild hemorrhage.

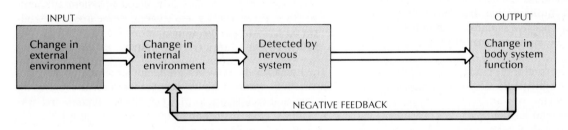

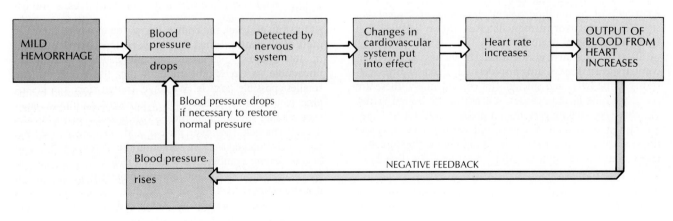

environment. The factors that put pressure on the body to make an adaptive change are collectively called *stress.* Stress may be physical (heat, noise), chemical (food, hormones), microbiological (viruses, bacteria), physiological (tumors, abnormal functions), developmental (old age, genetic changes), or psychological (emotional and mental disturbances).

The body responds to a stressful condition in one of two ways; it either adjusts to it constructively with negative feedback (thereby maintaining homeostasis), or it responds with positive feedback (thereby reinforcing the problem and possibly initiating a mental or physical disorder). Some amount of stress is normal and may actually be beneficial. Walking, for instance, places some stress on bones, muscles, and joints, helping to strengthen them. Too little exercise (or stress) causes bones and muscles to weaken, while too much physical stress breaks bones and tears muscles.

Hans Selye, a pioneer of stress research, has said that "the great capacity for adaptation is what makes life possible on all levels of complexity. It is the basis of homeostasis and of resistance to stress. . . . Adaptability is probably the most distinctive characteristic of life."*

Selye's theory of a *general adaptive syndrome* (GAS) against stress outlines three stages: *alarm,* when the body first recognizes the shock of a stressful disruption; *resistance,* when the body attempts to adapt to the disruption; and *exhaustion,* which occurs if the stress continues unchecked and the body loses its ability to adapt to the disruption. The overall objective of the healthy body's response to stress is always the same: to maintain the inner stability of the body within normal limits. Ordinarily, the more flexible a person is, the easier it is to adapt to stress and counteract it.

A person's mental attitude is very important in coping with stress. Loud rock music, for example, usually causes less hearing loss in rock musicians and their fans (who like the music and treat it as a positive stimulus) than in their *parents* (who may not like the music and may treat it as a negative stimulus).

People who cannot control the mood or pace of their lives increase their susceptibility to disease. Robert Karasek, a professor of industrial engineering at Columbia University, has reported that people who have little control over their jobs probably run the same risk of heart disease as do people who are heavy smokers or have high levels of blood cholesterol. Studies have shown that long-term stress decreases the effectiveness of the immune system, and some experts feel that stress is a key factor in most illnesses. (Selye believed that aging is the natural result of stress.)

Stress is implicated, either directly or indirectly, in cancer, coronary heart disease, lung disorders, accidental injuries, cirrhosis of the liver, and suicide. Not coincidentally, these are six of the nine leading causes of death in the United States. The three best-selling prescription drugs in the United States are also stress-related: Tagamet and Zantec, ulcer medications; and Dyazide, for the relief of high blood pressure. For many years the tranquilizer Valium was one of the three best-selling prescription drugs.

*H. Selye, *The Stress of Life* (New York: McGraw-Hill, 1976), p. 74.

The chemistry of stress The effects of stress on the body are complex and far-reaching. Initially, stress affects the brain, which then triggers additional responses in other parts of the nervous system, the immune system, and several other systems of the body. A typical sequence begins when the hypothalamus in the brain is stimulated by stress to release a chemical called CRF (corticotropin releasing factor). CRF stimulates the pituitary gland, which is actually attached to the brain, to release ACTH (adrenocorticotropic hormone), a powerful hormone that enters the bloodstream and reaches the adrenal glands. ACTH stimulates the adrenal glands to release hormones called glucocorticoids that increase the stress response of the immune system, which combats infections and other forms of stress. However, a continuing overload of stress can have the opposite effect, *reducing* the effectiveness of the immune system, and possibly allowing diseases such as AIDS (acquired immune deficiency syndrome; see Chapter 20) to progress and manifest their systems.

Ask Yourself

1 *What is homeostasis, and why is it important to the body?*

2 *What is the difference between negative and positive feedback?*

3 *What is stress, and how can it affect homeostasis?*

4 *How is the pituitary gland involved in the stress response?*

FROM ATOM TO ORGANISM: STRUCTURAL LEVELS OF THE BODY

The human body has several different structural levels of organization, starting with its constituent atoms, and increasing in size and complexity to the cells, tissues, organs, and systems that compose the complete organism (Figure 1.3).

Atoms

At its simplest level, the body is composed of **atoms,** the basic units of all matter. When the Greek philosopher Democritus (c. 460–370 B.C.) proposed the name *atomos* (which means uncuttable in Greek), he thought that the atom was the smallest possible particle. We have known since the beginning of this century, however, that atoms are made up of even smaller subatomic particles. The *nucleus* consists of electrically charged (positive) *protons* and uncharged *neutrons;* negatively charged *electrons* surround the nucleus. When two or more atoms combine, they form a *molecule* (such as when two oxygen atoms combine to form an oxygen molecule, O_2). If a molecule is made up of more than one element, it is a *compound,* such as water (H_2O), carbon dioxide (CO_2), or any of the carbohydrates, proteins, and fats that are so important to our bodies.

FIGURE 1.3

The structural levels of the body, as applied to the respiratory system. Specialized cells and tissues help the lungs and other parts of the system to perform their functions of breathing and gas exchange at an optimum level of efficiency.

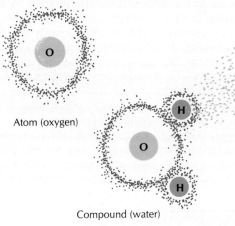

Atom (oxygen)

Compound (water)

CHEMICAL LEVEL

A hypothetical cell

CELLULAR LEVEL

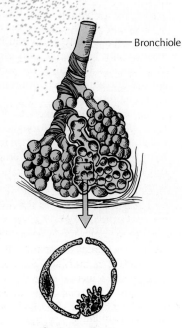

Bronchiole

Epithelial tissue of lung

TISSUE LEVEL

ORGANISM LEVEL

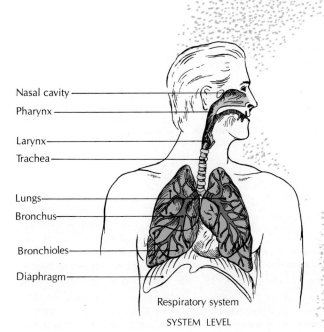

Nasal cavity

Pharynx

Larynx

Trachea

Lungs

Bronchus

Bronchioles

Diaphragm

Respiratory system

SYSTEM LEVEL

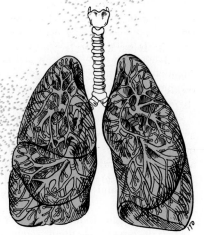

Lungs and bronchial trees

ORGAN LEVEL

Cells

Cells are the smallest independent units of life, and all life as we know it depends on the many chemical activities of cells. The cell theory, which describes the basic characteristics of cells, points out that (1) the cell is the basic living unit of structure and function of all organisms, (2) all cells come from the reproduction of pre-existing cells, and (3) all living things are made up of one or more cells.

Body cells can vary in size from the tiny head of a sperm cell at 5 micrometers, or 5 μm (five-millionths of a meter), long to a nerve cell with long, thin fibers that may be more than a meter long (although the fibers are too thin to be seen with the naked eye). These variations in size and shape are one example of how form and function work together. For instance, the long nerve cells carry messages over considerable distances throughout the body. They are constructed like a telephone wire, which also transmits electrical information over long distances.

Some of the basic functions of cells are growth, reproduction, and energy transformations. Within cells, structures called organelles are specialized to carry out specific activities. Mitochondria, for example, produce usable energy for the cell, and ribosomes are important for the manufacture of proteins. A typical body cell takes in and releases water, uses food and oxygen for certain body processes, and removes carbon dioxide and other waste products. Chapter 3 discusses human cell structure and function in detail.

Tissues

Tissues, which are made up of many similar cells that perform a specific function, are generally divided into four groups: epithelial, connective, muscle, and nervous. (Tissues are discussed in detail in Chapters 4, 6, 9, and 11.)

Epithelial tissue is found in the outer layer of the skin and in the linings of organs, vessels, and body cavities. It is well suited as a protective shield because its cells are closely packed and arranged in sheets (Figure 1.4A), and because it can add new cells when the old ones are worn or damaged. Epithelial tissue is also a secretory tissue, releasing specific chemical substances.

Connective tissue may also be called supportive tissue because, in addition to other functions, it connects and supports most parts of the body. It is found in skin, and also comprises the major portions of tendons and bones. Connective tissue frequently contains fibers that form a strong supportive mesh (Figure 1.4B). It is the most widely distributed of all body tissues.

Muscle tissue's most important feature is its ability to contract. For this reason, muscle tissue is often called contractile tissue. The main function of muscle tissue is to produce movement. It is found as *skeletal muscle* in the limbs, trunk, jaw, and face; as *smooth muscle* in the digestive tract, eyes, blood vessels, and ducts; and as *cardiac muscle* in the heart (Figure 1.4C).

Nervous tissue is found in the brain, spinal cord, and nerves. It responds to stimuli and transmits impulses or messages, and its long nerve fibers are well adapted to those functions. Nervous tissue is sometimes called conductile tissue because of its ability to conduct electrical impulses. Long nerve fibers can be seen clearly in Figure 1.4D.

Organs

The next higher level of complexity above the tissue is the organ. An *organ* is an integrated collection of two or more kinds of tissue that work together to perform a specific function. The stomach is one of the best examples of an organ that is made up of a combination of different types of tissue. Epithelial tissue lines the stomach and helps protect it; muscle tissue churns food and slowly moves it into the small intestine; nervous tissue transmits signals that are recognized as hunger pains; and connective tissue helps hold all the tissues together.

Systems

A *system* is a group of organs that work together to perform a major body function. The respiratory system, for example, contains several organs that provide a mechanism for breathing, and for the exchange of oxygen and carbon dioxide between the air outside the body and the blood inside. All the body systems are specialized within themselves and coordinated with each other to produce a dynamic and efficient *organism.*

Ask Yourself

1 *What are the four major types of body tissues?*

2 *What are the differences between organs and tissues?*

3 *How does a system depend on organs?*

FIGURE 1.4

Representative types of body tissues. (A) A highly magnified photograph of epithelial tissue. The arrangement of flat, overlapping sheets is shown clearly. Such a structure is well suited for the shingles of a roof or the skin of your body, both of which protect the inside from the outside environment. One hair is visible. (B) The meshed fibers of connective tissue are seen in this photograph of interwoven collagen fibers. Connective tissue is abundant in the body and can resist stress in several directions. (C) Cardiac muscle tissue is found only in the heart. It consists of separate cells, each with its own centrally located nucleus. (D) Nervous tissue consists of branched and slender nerve cells that span relatively long distances to go to and from muscles and other effectors such as glands. Motor nerves go to muscles; sensory nerves lead away from them. [(1.4A) Centre National des Recherches Iconographiques, Paris, France. (1.4B) D. M. Phillips/Taurus Photos. (1.4C) C. Abrahams, M. D./Custom Medical Stock. (1.4D) Lennart Nilsson, Behold Man. Boston, Little, Brown, 1974.]

From Atom to Organism: Structural Levels of the Body

[A] EPITHELIAL TISSUE   Hair   ×150

[B] CONNECTIVE TISSUE   Collagen fibers   ×3100

[C] CARDIAC MUSCLE TISSUE   ×200

Branched nerve cells

[D] NERVOUS TISSUE   ×200

BODY SYSTEMS

The systems of the body are constantly adjusting to changes inside and outside the body. Although each system has its own specialized function, none operates without help from the others. As you will see, the structure of each system is closely related to its particular function.

Integumentary System

The *integumentary system* (L. "to cover") consists of the skin and all the structures derived from it, including hair, nails, sweat glands, and oil glands (Figure 1.5). The main purpose of skin is to envelop the body with a protective barrier between the organs inside and the changing environment outside. This function is served admirably by the skin's layers of epithelial and connective tissues (see Figures 1.4A, B), which screen out dirt, microorganisms, and harmful chemicals and rays. Sweat glands in the skin help regulate body temperature and eliminate wastes, and nerve endings within the skin make it sensitive to touch, pain, pressure, heat, and cold. Among other functions, the skin is involved in the production of vitamin D.

Skeletal System

The *skeletal system* consists of bones, certain cartilages (connective tissue) and membranes (thin layers of tissue), and

FIGURE 1.5	FIGURE 1.6

The integumentary system of the body includes the skin, hair, nails, sweat glands, and oil glands.

The human skeletal system, showing some of the major bones.

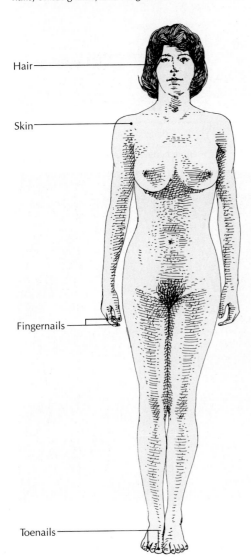

Hair
Skin
Fingernails
Toenails

INTEGUMENTARY SYSTEM

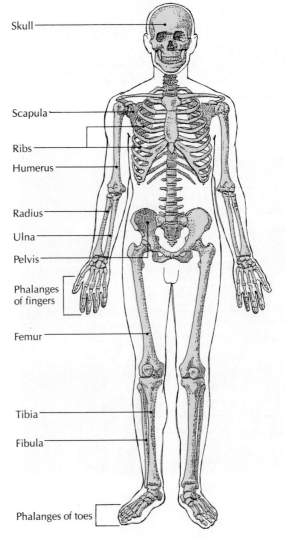

Skull
Scapula
Ribs
Humerus
Radius
Ulna
Pelvis
Phalanges of fingers
Femur
Tibia
Fibula
Phalanges of toes

SKELETAL SYSTEM

joints (Figure 1.6). It supports the body and protects the organs, provides a system of levers and a point of attachment for muscles that enable the body to move, and manufactures red blood cells and some white blood cells in the bone marrow. Bone tissue also stores the body's main supply of reserve calcium and phosphorus.

Muscular System

The *muscular system* consists of three different types of muscles: skeletal (Figure 1.7), smooth, and cardiac (heart). It also includes tendons, the fibrous cords of connective tissue that attach muscles to bones, and the motor nerves that stimulate muscle contractions. Muscles allow movement; help us to maintain a correct posture; and help move blood, food, urine, and other materials through various parts of the body. Finally, muscles produce much of our body heat.

Nervous System

The *nervous system* consists of the *central nervous system* (the brain and spinal cord) and the *peripheral nervous system* (nerves to and from the brain and spinal cord) (Figure 1.8). It also includes the sensory organs, which provide vision, hearing, and the other senses. The nervous system is the body's main control and regulating system. It coordinates sensations from the sensory organs and instructions for action to the muscles and glands, and regulates most body activities. Part of the peripheral nervous system is the *autonomic* (or *involuntary*) nervous system, which controls the smooth mus-

---

FIGURE 1.7

The human muscular system, showing some of the major skeletal muscles.

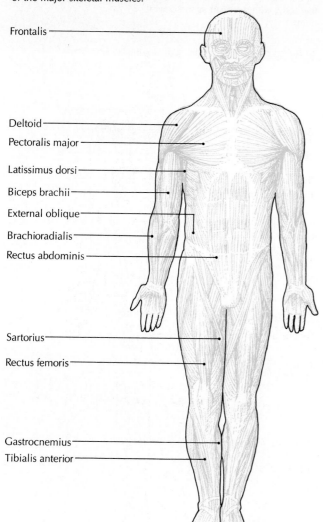

Frontalis

Deltoid
Pectoralis major
Latissimus dorsi
Biceps brachii
External oblique
Brachioradialis
Rectus abdominis

Sartorius
Rectus femoris

Gastrocnemius
Tibialis anterior

MUSCULAR SYSTEM

FIGURE 1.8

The nervous system, showing the brain and spinal cord (central nervous system) and some of the branching peripheral nerves (peripheral nervous system).

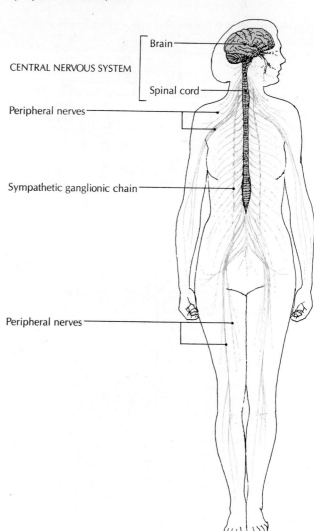

Brain

CENTRAL NERVOUS SYSTEM

Spinal cord

Peripheral nerves

Sympathetic ganglionic chain

Peripheral nerves

NERVOUS SYSTEM

cle in organs, cardiac muscle in the heart, and glands, so that you don't have to think about breathing, digesting, making your heart beat, secreting hormones from glands, and other involuntary activities.

Endocrine System

The *endocrine system,* the second major regulating system in the body, works in close conjunction with the nervous system. It is composed of ductless glands, each with a distinctive function (Figure 1.9). Ductless glands are so called because they release (secrete) their products directly into the bloodstream without the use of a separate system of ducts or vessels. The secretions of endocrine glands are called *hormones.* Hormones regulate chemical reactions within cells (metabo-

lism), growth and development, stress and injury responses, reproduction, and many other critical functions.

Cardiovascular System

The *cardiovascular system* consists of the heart, blood, and blood vessels (Figure 1.10). The heart pumps blood through a complex system of blood vessels (arteries, capillaries, and veins). Blood transports oxygen throughout the body and carbon dioxide to the lungs for removal, distributes dissolved nutrients and hormones to cells, and acts as a defense against infection. It also helps maintain an even body temperature, a stable metabolism, and water balance. Finally, it transports waste products to the kidneys and the lungs for excretion.

FIGURE 1.9

The human endocrine system, showing the location of the major ductless glands.

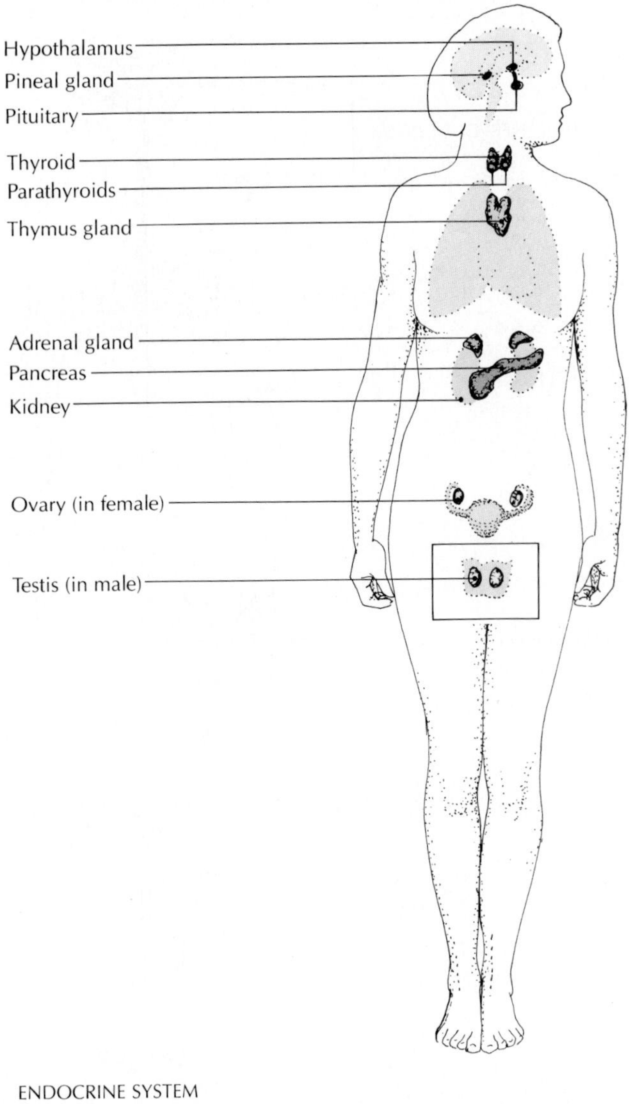

Hypothalamus
Pineal gland
Pituitary
Thyroid
Parathyroids
Thymus gland
Adrenal gland
Pancreas
Kidney
Ovary (in female)
Testis (in male)

FIGURE 1.10

The arteries (red) and veins (blue) of the cardiovascular system branch throughout the body, with the heart as the central pump.

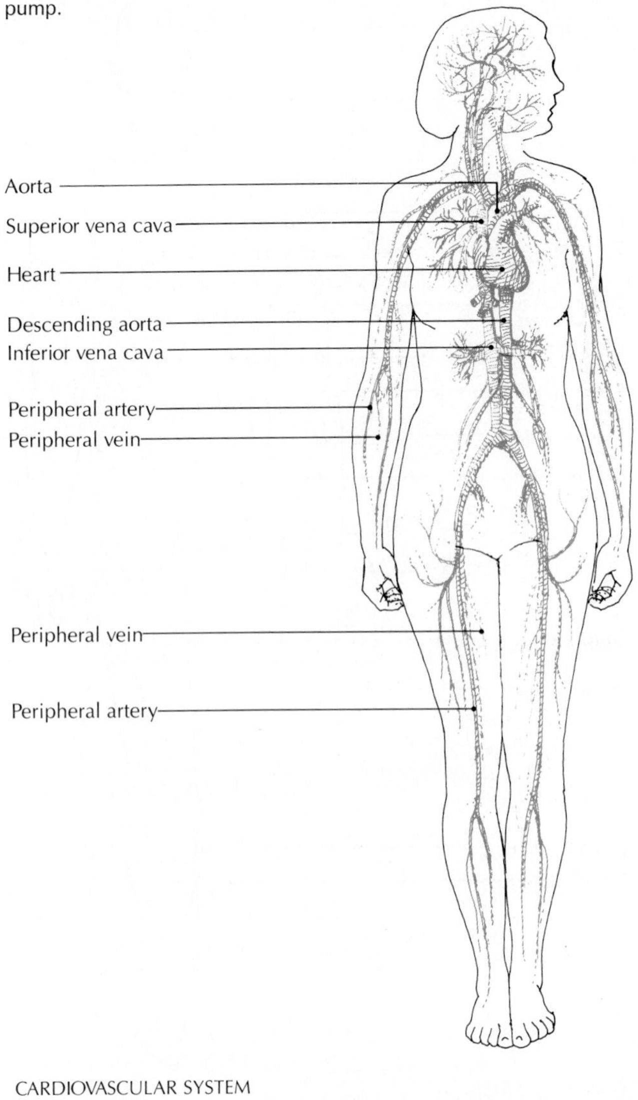

Aorta
Superior vena cava
Heart
Descending aorta
Inferior vena cava
Peripheral artery
Peripheral vein
Peripheral vein
Peripheral artery

Lymphatic System

The **lymphatic system** is made up of glands (including the spleen, tonsils, and thymus gland) and a network of thin-walled vessels that carry a clear fluid called *lymph* (Figure 1.11). The lymphatic system helps to defend the body against harmful microorganisms and tumor cells, produces certain white blood cells that act as disease fighters, returns excess fluid and proteins to the blood, and is an important part of the immune system, which combats infections.

Respiratory System

The **respiratory system** is composed of the nose; a system of

airways including the pharynx, larynx, and trachea; the lungs; and the muscles that help move air into and out of the body, the most important of which is the diaphragm (Figure 1.12). The respiratory system is concerned with the mechanics of breathing, and also provides a mechanism for the exchange of gases between blood and air. In the respiratory system, oxygen from the air moves into the blood and is carried to the tissues. In a reverse process, waste carbon dioxide from the tissues is carried by the blood to the lungs, where it is eventually exhaled from the body into the air outside.

Digestive System

The **digestive system** includes the teeth, tongue, salivary glands, esophagus (a tube leading from the mouth to the

FIGURE 1.11

The many glands and vessels of the lymphatic system help it to protect the body against harmful agents.

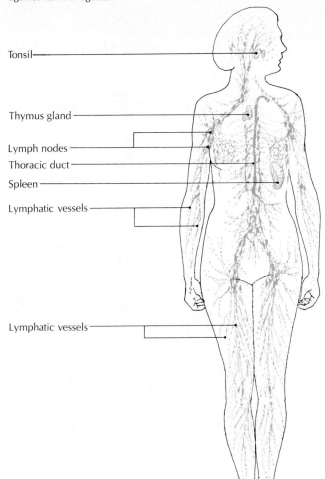

Tonsil

Thymus gland

Lymph nodes

Thoracic duct

Spleen

Lymphatic vessels

Lymphatic vessels

FIGURE 1.12

The respiratory system works in harmony with the cardiovascular system to circulate oxygen throughout the body and to expel waste carbon dioxide.

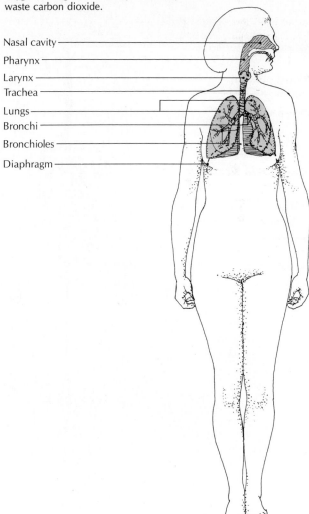

Nasal cavity

Pharynx

Larynx

Trachea

Lungs

Bronchi

Bronchioles

Diaphragm

stomach), stomach, small and large intestines, liver, gallbladder, and pancreas (Figure 1.13). The digestive system is compartmentalized, with each part adapted to a specific function. For example, the saclike shape of the stomach is well suited to store food before it moves into the small intestine. The overall function of the digestive system is to break down large molecules of food physically and chemically until they are small enough to be absorbed into the bloodstream from the small intestine. Solid, undigested wastes are removed from the body through the anus.

Urinary System

The kidneys, their drainage tubes (ureters), the urinary bladder, and the urethra are the main parts of the **urinary system**

(Figure 1.14). This system produces and eliminates urine. In so doing, it rids the body of wastes, helps regulate blood pressure and the composition and volume of blood, and helps to maintain the body's acid-base and water-salt balance. The urinary system, not the digestive system, is the body's main avenue for removing metabolic wastes produced by the cells of the body.

Reproductive Systems

Each of the sexes has reproductive organs (testes or ovaries) that secrete sex hormones and produce reproductive cells (sperm or eggs), and a set of ducts and accessory glands and organs such as the prostate gland, penis, uterus, and vagina. These and other structures make up the **reproductive systems**

FIGURE 1.13

The human digestive system is a series of separate compartments (organs), each with its own digestive function.

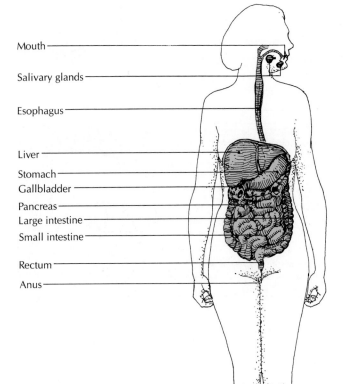

Mouth
Salivary glands
Esophagus
Liver
Stomach
Gallbladder
Pancreas
Large intestine
Small intestine
Rectum
Anus

DIGESTIVE SYSTEM

FIGURE 1.14

The urinary system consists of the kidneys that produce urine, ducts (ureters) that carry urine to the bladder, where it is stored, and the urethra, which conveys urine to the outside.

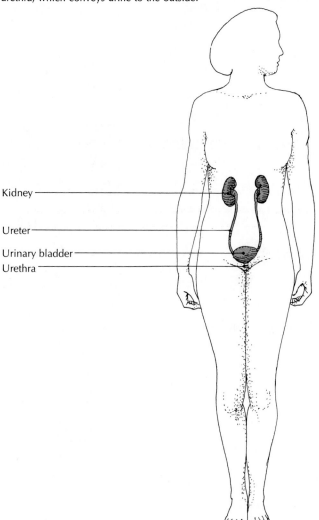

Kidney
Ureter
Urinary bladder
Urethra

URINARY SYSTEM

(Figure 1.15), which are responsible for maintaining the human species through reproduction and heredity.

Ask Yourself

1 *What does the skeletal system do besides support the body?*

2 *What are the two main divisions of the nervous system?*

3 *Why is the autonomic nervous system also called the involuntary nervous system?*

4 *What are hormones?*

5 *Which system is the body's main avenue of metabolic waste disposal?*

ANATOMICAL TERMINOLOGY

The language of anatomy will probably be unfamiliar to you at first. But once you have an understanding of the basic word roots, combining word forms, prefixes, and suffixes, you will find that anatomical terminology is not as difficult as first imagined. For instance, if you know that *cardio* refers to the heart and that *myo* means muscle, you can figure out that *myocardium* refers to the muscles of the heart. See Appendix C at the back of this book for a detailed list of prefixes, suffixes, and combining word forms. The Glossary at the back of this book can also help you learn the meanings of anatomical terms. It explains how these terms are derived from their Greek and Latin roots, and defines them.

FIGURE 1.15

(A) The female reproductive system.
(B) The male reproductive system.
The reproductive systems have been enlarged for emphasis.

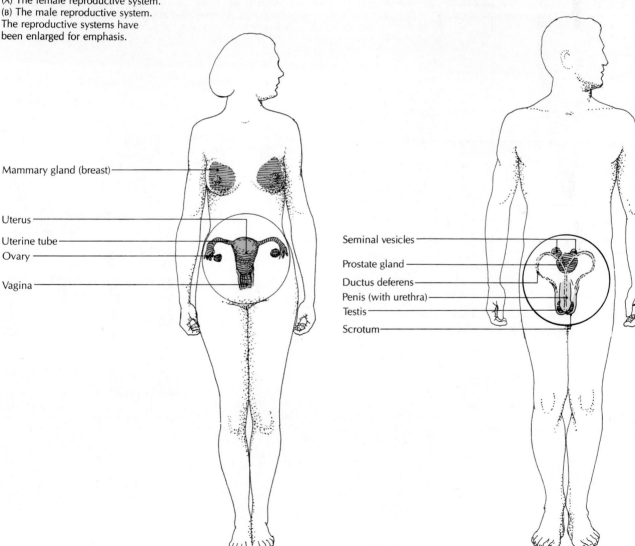

Mammary gland (breast)

Uterus
Uterine tube
Ovary

Vagina

Seminal vesicles

Prostate gland

Ductus deferens
Penis (with urethra)
Testis
Scrotum

[A] FEMALE REPRODUCTIVE SYSTEM

[B] MALE REPRODUCTIVE SYSTEM

FIGURE 1.16

Some of the directional terms of the body are identified in these drawings. Note the imaginary, but important, midline of the body. (A) The anatomical position of the body. Note that the body stands erect, and the palms face forward, with the thumbs away from the body. (B) Lateral view of the body showing directional terms.

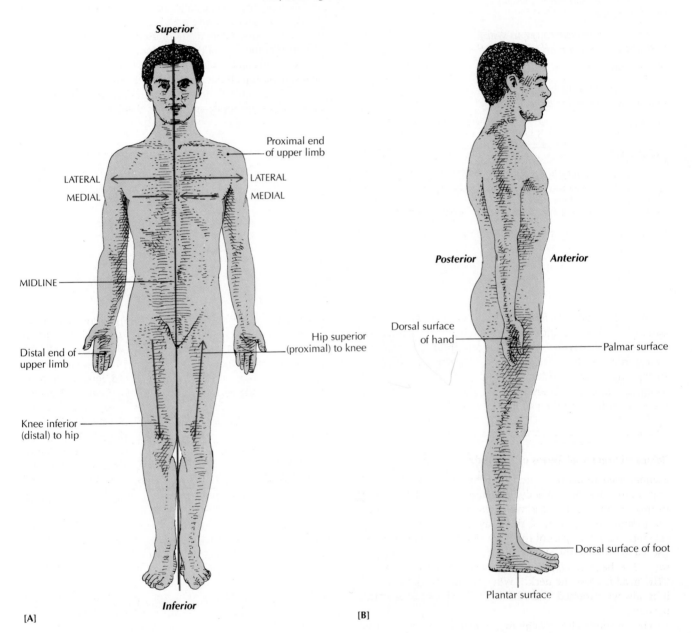

[A]

[B]

TABLE 1.1

Term	Definition	Example
Superior (cranial)	Toward the head.	The leg is superior to the foot.
Inferior (caudal)	Toward the feet or tail region.	The foot is inferior to the leg.
Anterior (ventral)	Toward the front of the body.	The toes are anterior to the heel.
Posterior (dorsal)	Toward the back of the body.	The heel is posterior to the toes.
Medial	Toward (nearer) the midline of the body.	The nose is medial to the eyes.
Lateral	Away (farther) from the midline of the body.	The eyes are lateral to the nose.
Proximal	Toward (nearer) the trunk of the body; toward the attached end of a limb.	The shoulder is proximal to the wrist.
Distal	Away (farther) from the trunk of the body; away from the attached end of a limb.	The wrist is distal to the forearm.
Superficial	Nearer the surface of the body.	The ribs are more superficial than the heart.
Deep	Farther from the surface of the body.	The heart is deeper than the ribs.
Peripheral	Away from the central axis of the body.	Peripheral nerves radiate away from the brain and spinal cord.

In order to be able to describe the location of particular parts of the body, anatomists have defined the **anatomical position** (Figure 1.16A), which has been universally accepted as a starting point for the positional references of the body. In the anatomical position, the body is standing erect and facing forward, the feet are together, and the arms are hanging at the sides with the palms facing forward.

Relative Directional Terms of the Body

Standardized terms of reference are used when anatomists explain the location of a body part. Notice that in the heading of this section we used the word *relative*, which means that the location of one part of the body is always described in relation to another part of the body. For instance, when you use standard anatomical terminology to locate the head, you say, ''The head is *superior* to the neck,'' instead of saying, ''The head is *above* the neck.'' When using directional terms, it is always assumed that the body is in the anatomical position.

Like so much else in anatomy, directional terms are used in pairs. If there is a term that means ''above,'' there is also a term that means ''below'' (Table 1.1). If the thigh is **superior** to the knee, the knee is **inferior** to the thigh. The term **anterior** (or *ventral*) means toward the front of the body, and **posterior** (or *dorsal*) means toward the back of the body (see Figure 1.16B). The toes are anterior to the heel, and the heel is posterior to the toes.

Medial means nearer to the imaginary midline of the body or a body part (see Figure 1.16A), and **lateral** means farther from the midline. The nose is medial to the eyes, and the eyes are lateral to the nose.

The terms *proximal* and *distal* are used mostly for body extremities, such as the arms, legs, and fingers. **Proximal** means nearer the trunk of the body (toward the attached end of a limb), and **distal** means farther from the trunk of the body (away from the attached end of a limb). The shoulder is proximal to the wrist, and the wrist is distal to the forearm.

Superficial means nearer the surface of the body, and **deep** means farther from the surface. **External** means outside, and **internal** means inside; they are not the same as superficial and deep.

Peripheral is used at times to describe structures other than internal organs that are located or directed away from the central axis of the body. Peripheral nerves and blood vessels, for instance, radiate away from the brain and spinal cord.

The sole of the foot is called the **plantar** surface, and the upper surface of the foot is called the **dorsal** surface. The palm of the hand is the **palmar** (or *volar*) surface, and the back of the hand is referred to as the dorsal surface (see Figure 1.16B).

The term **parietal** (puh-RYE-uh-tuhl) refers to the walls of a body cavity or the membrane lining the walls of a body cavity, and **visceral** (VIHSS-er-uhl) refers to an internal organ or a body cavity (such as the abdominal cavity), or describes a membrane that covers an internal organ.

Body Regions

With the body in the anatomical position, the regional approach can be used to describe general areas of the body. The main divisions of the body are the **axial** part, consisting of the head, neck, thorax (chest), abdomen, and pelvis; and the **appendicular** part, which includes the *upper extremities* (shoulders, upper arms, forearms, wrists, and hands) and the *lower extremities* (thighs, legs, ankles, and feet) (Figure 1.17A). (The extremities are also called *limbs*.) Figures 1.17B and C present additional technical terms for the body regions.

It is customary to subdivide the abdominal region with two vertical lines and two horizontal lines into nine regions, as shown in Figure 1.18A. (The two vertical lines are drawn downward from the centers of the collarbones; one horizontal line is drawn at the lower edge of the rib cage, and another

horizontal line is placed at the upper edges of the hipbones.) The abdominal region may also be divided into four quadrants, as shown in Figure 1.18B. All the technical terms used with the abdominal region are described in detail in Chapter 22, The Digestive System. To help you remember the divisions of the abdominal region, study the following list:

1 *Upper abdomen:* right hypochondriac region, epigastric region, left hypochondriac region; roughly the upper third of the abdomen.

2 *Middle abdomen:* right lumbar (lateral) region, umbilical region, left lumbar (lateral) region; roughly the middle third of the abdomen.

3 *Lower abdomen:* right iliac (inguinal) region, hypogastric (pubic) region, left iliac (inguinal) region; roughly the lower third of the abdomen.

FIGURE 1.17

(A) The basic regions of the body, as located within the axial and appendicular parts. (B) Ventral and (C) dorsal views of the body present a more detailed list of the many terms used to locate specific parts of the body.

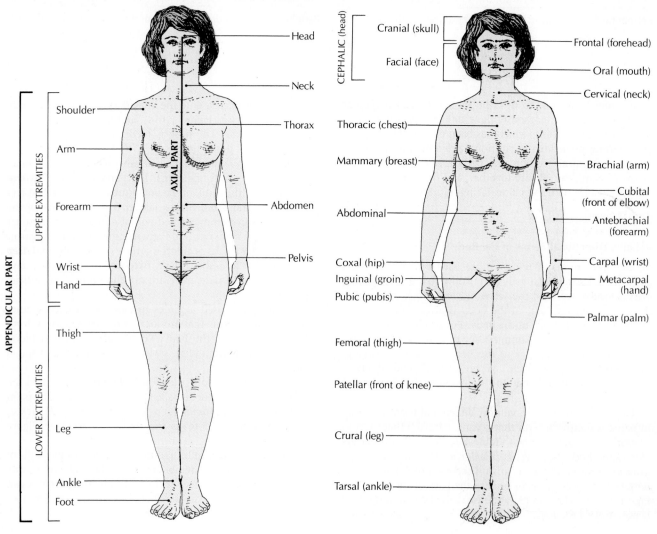

[A]

[B] VENTRAL (front)

FIGURE 1.18

(A) The nine subdivisions of the abdominal region. (B) The four quadrants of the abdominal region.

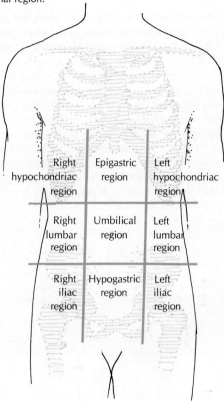

Right hypochondriac region | Epigastric region | Left hypochondriac region

Right lumbar region | Umbilical region | Left lumbar region

Right iliac region | Hypogastric region | Left iliac region

[A]

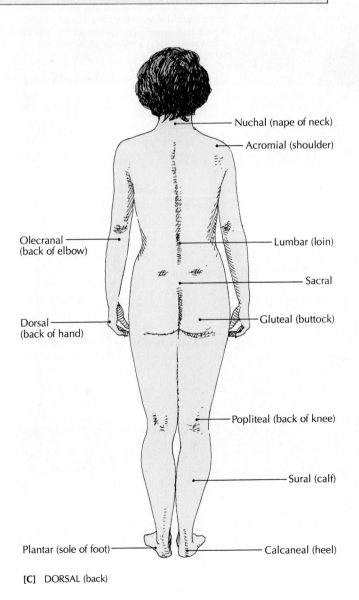

Nuchal (nape of neck)

Acromial (shoulder)

Olecranal (back of elbow)

Lumbar (loin)

Sacral

Dorsal (back of hand)

Gluteal (buttock)

Popliteal (back of knee)

Sural (calf)

Plantar (sole of foot)

Calcaneal (heel)

[C] DORSAL (back)

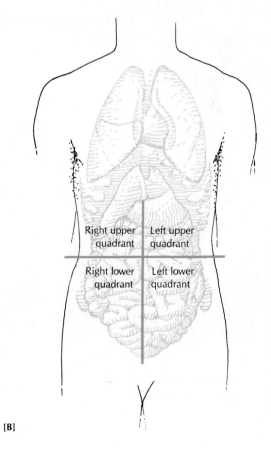

Right upper quadrant | Left upper quadrant

Right lower quadrant | Left lower quadrant

[B]

Body Planes

For further identification of specific areas, the body can be divided into imaginary flat surfaces, or *planes* (Figure 1.19). The *midsagittal plane* divides the left and right sides of the body lengthwise along the midline into externally symmetrical sections. If a longitudinal plane is placed off-center and separates the body into asymmetrical left and right sections, it is called a *sagittal plane.* If you were to face the side of the body and make a lengthwise cut at right angles to the midsagittal plane, you would make a *frontal* (or *coronal*) *plane,* dividing the body into asymmetrical anterior and posterior sections (see Figure 1.19B). If you divide the body horizontally into upper (superior) and lower (inferior) sections, you get a *transverse* (or *horizontal*) *plane* that is at right angles to

the midsagittal, sagittal, and frontal planes (see Figure 1.19A). Transverse planes do not produce symmetrical halves.

The system of planes is also used with *parts* of the body, including internal parts. Figure 1.20 shows how cross sections, oblique sections, and longitudinal sections of internal parts are made.

If your laboratory manual or any other book refers to a drawing of a sagittal *section,* a frontal *section,* or a transverse *section,* you should be aware of what is actually being shown and how it relates to its corresponding plane. Figure 1.21A shows a cut along the midsagittal plane of the head. Such a cut produces an exposed surface of the head called a *midsagittal section.* A cut along a frontal plane produces a *frontal section;* Figure 1.21B shows a frontal section of the brain. A cut along the transverse plane—in the case of Figure 1.21C, across the abdomen—produces a *transverse section.*

FIGURE 1.19	FIGURE 1.20

The imaginary body planes are an additional source of identification and location. (A) Representations of the midsagittal, sagittal, and transverse planes. (B) Representation of the frontal plane.

Body parts are often cut into sections for microscopic examination. In this figure the femur (the long bone of the thigh) has been sliced three different ways to produce a cross section (cut across), an oblique section (cut on a diagonal), and a longitudinal section (cut vertically).

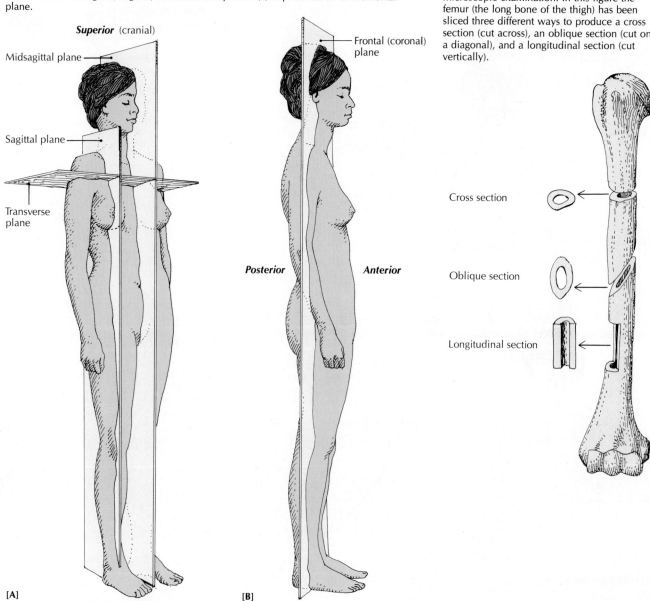

Superior (cranial)

Midsagittal plane

Sagittal plane

Transverse plane

Frontal (coronal) plane

Posterior   *Anterior*

Cross section

Oblique section

Longitudinal section

[A]   [B]

Inferior (caudal)

FIGURE 1.21

Sections that result from planar cuts are illustrated in these drawings. (A) A cut along the midsagittal plane of the head (smaller drawing) produces a midsagittal section of the head. (B) A cut along the frontal plane of the brain (smaller drawing) produces a frontal section. (C) A cut along the transverse plane of the abdomen (smaller drawing) results in a transverse section, looking down into the lower portion of the cut body.

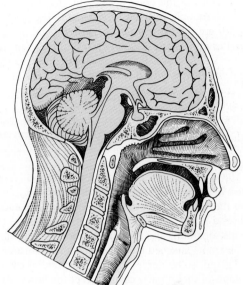

[A] MIDSAGITTAL

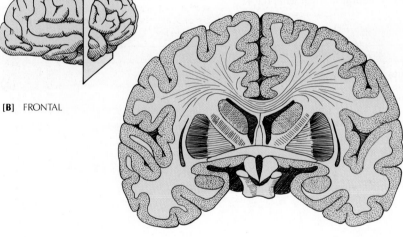

[B] FRONTAL

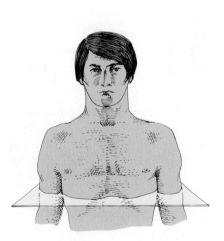

[C] TRANSVERSE

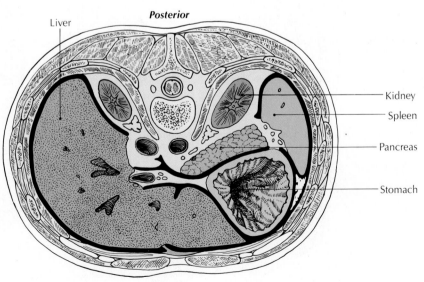

Posterior

Liver

Kidney

Spleen

Pancreas

Stomach

Anterior

Body Cavities

The cavities of the body house and protect the internal organs, commonly referred to as *viscera*. The two main body cavities are the **ventral (anterior) cavity** and the **dorsal (posterior) cavity** (Figure 1.22A). Each of these main cavities has further subdivisions. For example, the dorsal cavity contains the *cranial cavity,* which contains and protects the brain, and the *spinal* (or *vertebral) cavity,* which contains and protects the spinal cord.

The ventral body cavity is larger than the dorsal cavity, and is separated into the superior *thoracic cavity* and the inferior *abdominopelvic (peritoneal) cavity* by the diaphragm (Figure 1.22B).

The **thoracic cavity** is protected by the rib cage, and divided by membranes into the *pericardial cavity,* which contains the heart; the right and left *pleural cavities,* each of which houses a lung; and the *mediastinum,* the mass of tissues and organs separating the two lungs. The mediastinum contains the heart, its attached blood vessels, trachea, esophagus, thymus gland, some nerves, and the pericardial cavity—all the contents of the thoracic cavity except the lungs. Each pleural cavity is located within a sac lined by serous membranes called the *pleura.* Each sac encloses a lung.

Below the diaphragm lies the **abdominopelvic cavity,** which is divided by an imaginary line from the top of the lowest spinal bone (sacral promontory) to the top of the pubic bone (pubis symphysis) into the *abdominal cavity* and the *pelvic cavity* (see Figure 1.22A). The abdominal cavity contains the liver, gallbladder, stomach, pancreas, spleen, and intestines. The pelvic cavity contains the urinary bladder, rectum, anus, and the reproductive organs of the female and internal reproductive structures of the male. In the abdominopelvic cavity, a continuous membrane lines the cavity's walls and surrounds part or all of each organ. The potential space enclosed by this membrane is the *peritoneal cavity.*

Body Membranes

Membranes are layers of epithelial and connective tissue that line the body cavities and cover or separate certain regions, structures, and organs. The four main types of membranes are *mucous, serous, synovial,* and *cutaneous.* They are described in Chapter 4.

FIGURE 1.22

(A) The dorsal and ventral cavities, and the smaller cavities within them. (B) This frontal view of the body shows the thoracic, abdominopelvic, pericardial, pleural, and abdominal (peritoneal) cavities, and the mediastinum.

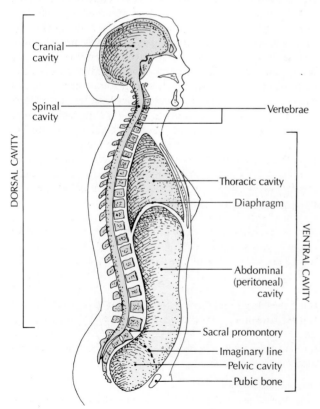

[A] MIDSAGITTAL SECTION

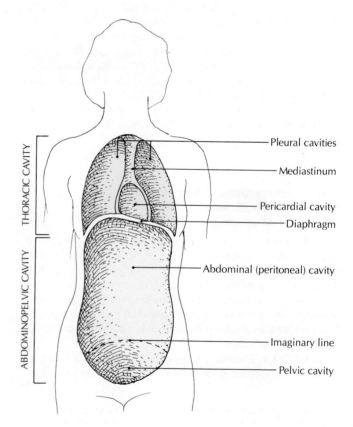

[B] FRONTAL SECTION

Ask Yourself

1 *What is the basic anatomical position of the body?*

2 *What is the difference between superior and anterior?*

3 *How does a sagittal plane differ from a midsagittal plane?*

4 *What are the various organs within the two main body cavities?*

5 *How is the thoracic cavity separated from the abdominopelvic cavity?*

NEW WAYS OF EXPLORING THE BODY

Until just a few years ago, a physician who wanted to look *inside* the body in order to make an accurate diagnosis was limited by relatively old-fashioned equipment. X rays can assess the damage to broken bones and help to diagnose diseases such as lung cancer that would be difficult or impossible to detect without surgery. But because x rays can only show a flat picture, without indicating depth or isolating different layers of an organ, they cannot always distinguish between healthy and diseased tissue.

Instruments for peering into the body, such as the fluoroscope and laparascope, are basically periscopelike tubes with viewing devices at both ends. The latest versions of such scopes project the image of the internal organ being viewed onto a television screen. Because such instruments can be inserted into body openings or tiny surgical incisions, the need for major exploratory surgery is reduced drastically. Now, even more thorough and dramatic noninvasive diagnostic techniques are available. These techniques have taken much of the guesswork out of the art of diagnosis, allowing physicians to actually see what they could only imagine before.

CAT (Computer-Assisted Tomography) and PET (Positron-Emission Tomography) Scanning

In 1973, the CAT (computer-assisted tomography*) scanner revolutionized the diagnosis of upper-body and brain disorders by being able to project a cross section of the internal body part being examined onto a television screen. Barely 10 years later, a second revolution came with the PET (positron-emission tomography) scanner, which not only detects physical disorders as a CAT scan does, but also diagnoses some metabolic imbalances.

A *CAT scan* combines x rays with computer technology to show cross-sectional views of internal body structures. Be-

FIGURE 1.23

A patient undergoing a CAT scan. [*Dan McCoy/Rainbow.*]

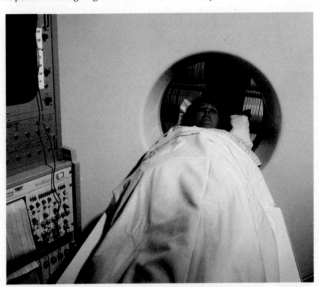

fore the actual scanning procedure begins, the patient is injected with a low-level radioactive tracer through an intravenous catheter in the arm. The patient lies on a table, and is passed slowly through a circular opening in the scanning device (Figure 1.23). A low-intensity x-ray beam then rotates 180 degrees across the width of the body. The degree to which the x ray is absorbed by various body tissues (which have different absorption rates) is recorded by detectors placed opposite the x-ray beam. The detectors convert the modified beams into electronic signals, which are fed into a computer. The computer analyzes the changes in the x-ray beams, and shows them in high-resolution black-and-white or color images on a television screen as a *tomogram* (Gr. a cut + *gramma*, something written), a picture of a predetermined plane section of an organ or part of the body (Figure 1.24). The video images are preserved on film, and are examined one "slice" at a time to obtain a three-dimensional picture in the physician's mind. Although computers that produce three-dimensional images are still in their developmental stages, some dramatic three-dimensional images can be obtained (Figure 1.25).

A CAT scan helps physicians to detect blood clots, tumors, and other physical damage, but it does not reveal how an organ is functioning. A *PET scan,* in contrast, reveals the metabolic state (level of chemical activity) of the organ by indicating the rate at which its tissues consume injected biochemicals such as glucose[†] (also called blood sugar). PET scans have been used to detect cancers (since some malignant tumors consume glucose at a faster rate than healthy tissue), and to study the metabolic patterns of the cardiovascular system. But their most dramatic application has been the examination of the brain.

---

*The word *tomography* means "a technique for making a picture of a section or slice of an object," from the Greek *tomos,* a cut or section + *graphein,* to write or draw.

[†]The brain's uptake of glucose is sometimes difficult to measure, so deoxyglucose is often used instead. However, some physicians feel that when deoxyglucose is used the results are distorted.

FIGURE 1.24

A CAT scan showing an optic nerve lesion. [*Dan McCoy/Rainbow.*]

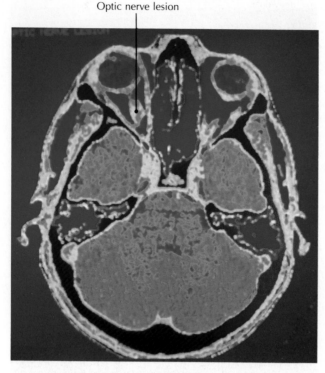

Optic nerve lesion

FIGURE 1.25

A three-dimensional CAT scan image of a twisted and compressed spinal bone (vertebra) (center, blue) following a motorcycle accident. Note the ruptured disks between the vertebrae. [*Dimensional Medicine, Inc.*]

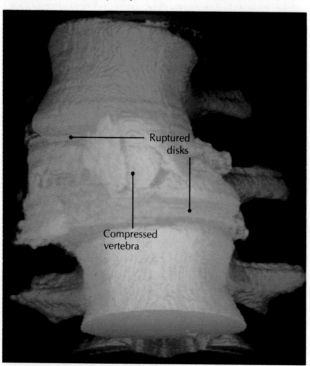

Ruptured disks

Compressed vertebra

FIGURE 1.26

(A) A patient undergoing a PET scan. The PET scanner is fringed with gamma-ray detectors. (B) PET detectors relay information about gamma rays to a computer. [(*A*) *Dan McCoy/Rainbow.*]

[A]

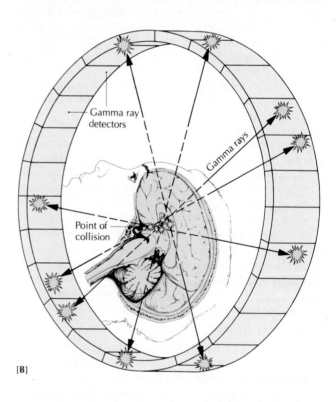

Gamma ray detectors

Gamma rays

Point of collision

[B]

The glucose that is administered to the patient contains a radioactive tracer. The "tagged" glucose travels to the brain, and is consumed by brain cells. Once inside the brain, the radioactive tracer begins to decay, and emits positively charged subatomic particles called *positrons*. When the positrons meet negatively charged subatomic particles called electrons in the brain cells, the positrons emit energy in the form of gamma rays that are detected by scanning devices that surround the patient's head (Figures 1.26A, B). The detectors

FIGURE 1.27

(A) A PET scan of a patient with senile dementia (Alzheimer's disease) shows a decreased rate of metabolism (blue and green on the color scale). (B) A PET scan of normal individual. [(*A and B*) *Dan McCoy/Rainbow*.]

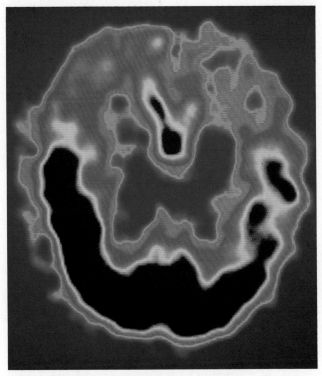

[A]

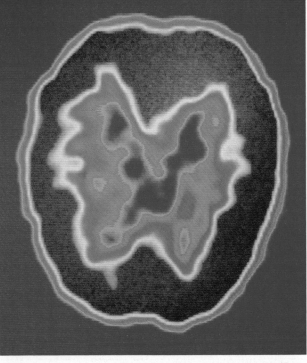

[B]

relay the gamma-ray information to a computer, which reconstructs color-coded, cross-sectional images that indicate the rate of metabolic activity (Figure 1.27). The processing of each image takes about 5 minutes.

Normal brain activity can be seen in reaction to light or movement; the extent of damage due to a stroke or epileptic seizure can be determined; and mental disorders such as schizophrenia, manic-depression, and senile dementia can be diagnosed (see Figure 1.27). Early studies of depression indicate a decreased brain metabolism, especially on the left side of the brain, where language and analytic functions are processed.

Dynamic Spatial Reconstructor (DSR)

The *dynamic spatial reconstructor,* or *DSR,* produces three-dimensional computer-generated images of the active brain. The DSR has 28 revolving x-ray machines, each firing 60 times a second. But what sets DSR apart from other new exploratory techniques is not its speed. A CAT scan produced in 5 minutes is hardly antiquated just because a DSR image takes less than a second. What *is* noteworthy about DSR is that it can look under, over, around, and *inside* an organ (Figure 1.28). It can view the flow of blood through an organ, and then replay it in slow motion or stop action. By watching the flow of blood through the brain, a physician may predict

FIGURE 1.28

A dynamic spatial reconstructor (DSR) produced these oblique views of branching arteries within the lung. [*E. L. Ritman,* Innovation et Technologie en Biologie et Medecine *8 (N Special 1): 37–50, 1987.*]

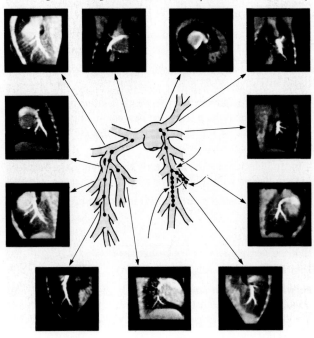

an impending stroke, providing a giant stride toward the *prevention* of disease instead of its treatment.

Magnetic Resonance Imaging (MRI)

Another tool for diagnosing disorders is ***magnetic resonance imaging,*** or **MRI.** (This technique is sometimes called *nuclear magnetic resonance*, or *NMR.*) Although the basic principles of MRI have been known for decades, it was not until fairly recently that it was used to produce diagnostic pictures of the human body. Like the CAT scanner, MRI produces basically anatomical images, but the images are so clear and discriminate so well between healthy and diseased tissue that they provide physicians with cross-sectional pictures better than any seen before.

To produce an image with MRI, the patient lies on a non-magnetic stretcher and is slipped inside the magnetic scanner, where the body is enveloped in a magnetic field thousands of times stronger than the earth's own field. The strong magnetic field causes positively charged subatomic particles called protons in the body's many hydrogen atoms (the body is mostly water, H_2O) to line up in rows parallel to the magnetic charge. Then radio waves of a specific frequency are introduced at right angles to the magnetic field, causing the protons to ''wobble'' out of line for a fraction of a second. When the radio signal is turned off, the protons return to their upright position. The lag between the two upright positions is called the *relaxation time*. It is different for each type of tissue, so it is possible to tell which part of the body the signals come from. (Cancerous tissue, for example, has a longer relaxation time than normal tissue.) Each relaxation time is read differently by a computer, which assigns to each part of the object a different shade of gray or color in the computer-generated image (Figure 1.29).

Magnetic resonance imaging is very expensive, but it has several advantages over CAT scans and other x-ray techniques:

1 It does not use potentially dangerous x rays, dyes, or radioactive tracers.

2 It does not register bone, which typically obscures the soft tissue in x rays.

3 It is able to differentiate between healthy and damaged myelin sheaths (which surround nerves), making possible an early diagnosis of multiple sclerosis, a disease that exhibits damaged myelin sheaths.

4 It can detect atherosclerosis early by revealing the build-up of fatty tissues within blood vessels.

5 It is able to detect some brain tumors that are invisible on CAT scans.

Digital Subtraction Angiography (DSA)

Digital subtraction angiography (Gr. *angeion*, vessel), or **DSA,** is a technique that produces three-dimensional pictures of blood vessels. It is especially effective in revealing blocked

FIGURE 1.29

A color-enhanced MRI scan of a 7-month-old child shows a malignant tumor between the right kidney and spinal column pressing on the spinal cord. [*UCLA School of Medicine.*]

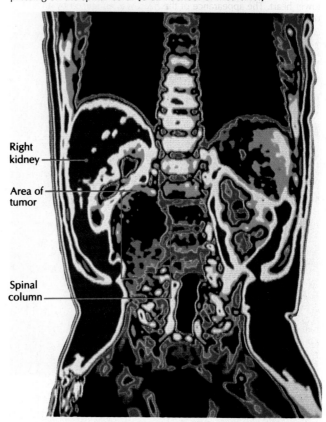

Right kidney

Area of tumor

Spinal column

vessels (Figure 1.30). After an image of the heart is produced by a digital x-ray scanner, blood vessels leading to the large arteries of the heart are injected with a substance that shows up in x rays. A second x-ray picture of the heart shows the opaque substance flowing through the arteries. A computer then subtracts the first image from the second, producing a final image that indicates any blocked vessels. Photographs are made from several angles, and the computer converts the images into a digital code and analyzes the extent of blockage in a vessel by comparing the various images. The computer also measures the rate of blood flow to the heart. With this information, physicians are often able to predict an imminent heart attack, and take appropriate action to prevent it.

Ultrasound (Sonography)

The noninvasive exploratory technique known as ***ultrasound*** (also called *sonography*) sends pulses of ultra-high-frequency sound waves into designated body cavities. From the echoes that result, pictures can be constructed of the object under investigation. The latest ultrasound systems include a computer that presents moving pictures on a television screen.

FIGURE 1.30

A picture of the heart and its left coronary artery made by digital subtraction angiography reveals a constriction (arrow). The constriction has blocked about 60 percent of the blood flow to the lower heart. The appearance of the heart was enhanced by an opaque dye. [*Fischer Imaging Corporation.*]

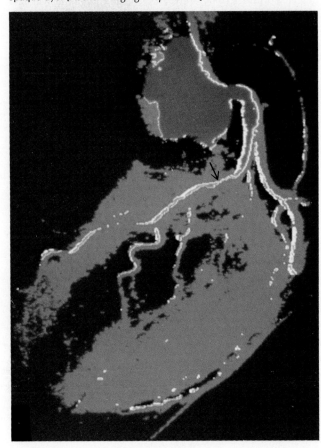

FIGURE 1.31

An ultrasound image of the head and neck of a 19-week-old fetus. [*Dan McCoy/Rainbow.*]

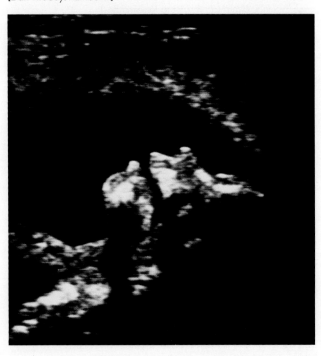

FIGURE 1.32

Ultrasound images of (A) a normal breast and (B) one with a benign cyst. The cyst shows up as a clear light area to the left of center. [(*A and B*) *Thomas Jefferson University Hospital, Division of Ultrasound and Radiology Imaging.*]

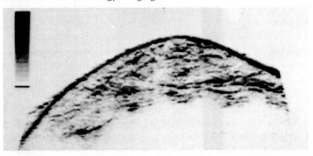

[A]

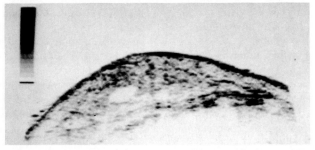

[B]

Recently improved ultrasound images, such as the one of a fetus shown in Figure 1.31, not only allow physicians to have a clear look at a developing fetus, they also permit a differentiation of tissues for the accurate diagnosis of heart disease and cancer (Figure 1.32).

Ultrasound offers an apparently harmless alternative to amniocentesis,* which presents some known risk to the fetus. Ultrasound can also be used in connection with amniocentesis, however, to help the physician guide the needle used in amniocentesis away from the placenta and fetus. Some practical applications of ultrasound include the detection of ectopic pregnancy (the implantation of a fertilized egg away from its normal location in the uterus) and multiple pregnancy, the assessment of fetal growth and development, the disclosure of some birth defects and probable miscarriages, and the assistance of physicians during fetal surgery.

*Diagnostic ultrasound may be capable of transforming normal cells into a precancerous state. Investigations on this subject are continuing. Amniocentesis is a technique that obtains representative cells of a developing fetus by sampling the fluid in which the fetus floats. It is used to test for genetic defects.

FIGURE 1.33

This thermogram reveals cancer in the left breast of a woman (red areas) by indicating a marked temperature increase in the cancerous area. [*Science Photo Library/Photo Researchers.*]

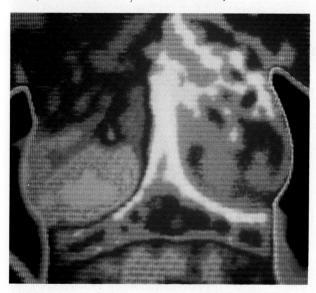

Thermography

Thermography ("heat writing") is a technique that reveals chemical reactions that are taking place within the body, based on heat changes on the skin. The thermographic camera can detect these normally invisible infrared radiations, convert them into electrical impulses, and record them as color-coded thermograms (Figure 1.33). Some malignant tumors give off more heat than normal tissues do, and can be detected with thermography. In fact, thermography was first used in 1965 to detect breast cancer. Arthritis, blood circulation problems, and other disorders can also be detected with thermography.

Ask Yourself

1 What is an important diagnostic limitation of x rays?

2 How does a PET scan provide different information than a CAT scan about the state of an organ?

3 How can the dynamic spatial reconstructor help to prevent disorders?

4 What are some of the advantages of magnetic resonance imaging over CAT scans?

5 What is the significance of the word subtraction *in the term digital subtraction angiography?*

6 How is ultrasound helpful in detecting cancer?

SUMMARY

What are anatomy and physiology?

1 The study of *anatomy* deals with the *structure* of the body; *physiology* explains the *functions* of the parts of the body.

2 Some subdivisions of anatomy are *regional* anatomy, *systemic* anatomy, *developmental* anatomy, *pathological* anatomy, *histology, cytology, radiographic* anatomy, and *splanchnology*. Any branch of anatomy that can be studied without a microscope is called *gross* anatomy; *microscopic* anatomy requires the use of a microscope.

3 In the human body, structure (anatomy) and function (physiology) work together to make the parts of the body operate at peak efficiency.

Homeostasis: coordination creates stability

1 *Homeostasis* is an inner stability of the body that exists even if the environment outside the body changes. Homeostasis is achieved when structure and function are properly coordinated and all the body systems work together.

2 In a healthy body, homeostasis exists on the cellular level, where a chemical balance inside and outside the cell is carefully regulated.

3 Practically everything that goes on in the body helps to maintain homeostasis, and the entire process is made possible by the coordinated action of many organs and tissues under the control of the nervous and endocrine systems.

4 When homeostasis breaks down, we become sick or even die. One way to unbalance homeostasis is to introduce *stress,* any internal or external factor that upsets the environment of the body. When the body is confronted by stress, it usually attempts to repair any damage and restore homeostasis as quickly as possible.

From atom to organism: structural levels of the body

1 At its simplest level, the body is composed of *atoms,* the basic units of all matter. Atoms are made up of a *nucleus,* which contains *protons* and *neutrons,* and *electrons* that surround the nucleus. When two or more atoms combine, they form a *molecule.* If a molecule is composed of more than one element, it is a *compound.*

2 *Cells* are the smallest independent units of life. Some of the basic functions of cells are growth, reproduction, and energy transfer.

3 *Tissues* are composed of many similar cells that perform a specific function. Tissues are classified into four types: *epithelial, connective, muscle,* and *nervous.*

4 An *organ* is an integrated collection of two or more kinds of tissues that combine to perform a specific function.

5 A *system* is a group of organs that work

together to perform a major body function. All the body systems are specialized within themselves and coordinated with each other to produce an *organism.*

Body systems

1 The *integumentary system* consists of the skin and all the structures derived from it. The main purpose of the skin is to protect the internal organs from the external environment.

2 The *skeletal system* consists of bones, certain cartilages and membranes, and joints. It supports the body, protects the organs, enables the body to move, manufactures blood cells in the marrow within the bone, and stores calcium and phosphorus.

3 The *muscular system* consists of muscles and tendons. It allows for movement and generates a large amount of body heat.

4 The *nervous system* consists of the *central nervous system* and the *peripheral nervous system;* it also includes special sensory organs. The nervous system is the body's main control and regulatory system.

5 The *endocrine system* comprises a group of ductless glands that secrete *hormones.* Hormones regulate chemical reactions within cells *(metabolism),* growth and development, stress and injury responses, reproduction, and many other critical functions.

6 The *cardiovascular system* consists of the heart, blood, and blood vessels. An important function of the cardiovascular system is to transport oxygen throughout the body and transport wastes such as carbon dioxide to the lungs for removal. Many other functions that help maintain homeostasis are influenced by this system.

7 The *lymphatic system* helps protect the body and produces antibodies, returns excess fluid and proteins to the blood, and helps the body build an immunity to disease.

8 The *respiratory system* accomplishes the process of breathing and also provides a mechanism for the exchange of gases between blood and air.

9 The *digestive system* breaks down food chemically and physically into molecules small enough to be absorbed from the small intestine into the bloodstream. Solid, undigested wastes are removed through the anus.

10 The *urinary system* consists of the kidneys, their ureters, the urinary bladder, and the urethra. The urinary system is an important contributor to homeostasis and is the body's main regulator of wastes produced by cells.

11 The *reproductive systems* have organs that produce specialized reproductive cells that make it possible to maintain the human species.

Anatomical terminology

1 When the body is in the *anatomical position,* it is in the erect position, facing forward, feet together, arms hanging at the sides, and palms facing forward.

2 Some basic pairs of anatomical terms used to describe the relative position of parts of the body include *superior/inferior, anterior/posterior, medial/lateral, proximal/distal, external/internal,* and *superficial/deep.*

3 The main *regions* of the body are the *axial* part, consisting of the head, neck, thorax, abdomen, and pelvis; and the *appendicular* part, which includes the upper and lower extremities. The abdominal region is divided into nine subregions.

4 The body, and parts of the body, may be divided into imaginary *planes,* including the *midsagittal, sagittal, frontal,* and *transverse.*

5 The *body cavities* house and protect the internal organs. The two main body cavities are the *ventral* and the *dorsal.* The dorsal cavity contains the *cranial* and *spinal* cavities, and the ventral body cavity is separated by the diaphragm into the superior *thoracic* and inferior *abdominopelvic* cavities.

6 *Membranes* are layers of epithelial and connective tissue that line the body cavities and cover or separate certain regions, structures, and organs. The four main types of membranes are *mucous, serous, synovial,* and *cutaneous.*

New ways of exploring the body

1 Several noninvasive techniques are replacing x rays and major exploratory surgery as effective diagnostic tools.

2 *Computer-assisted tomography,* or *CAT scanning,* combines x rays with computer technology to produce cross-sectional views of internal body structures. A CAT scan may reveal blood clots, tumors, and other disorders, but it cannot reveal how an organ is functioning metabolically.

3 *Positron-emission tomography,* or *PET scanning,* reveals the metabolic state of an organ by measuring the rate at which tissues consume chemical substances such as glucose. Some disorders that can be diagnosed using PET scans include cancer and schizophrenia.

4 The *dynamic spatial reconstructor (DSR)* produces three-dimensional computer-generated images to reveal the flow of blood through the brain. Such images may be used to prevent an impending stroke.

5 *Magnetic resonance imaging (MRI)* envelops the patient in a strong magnetic field to detect differences in healthy and nonhealthy tissue. It has several diagnostic advantages over CAT scans.

6 *Digital subtraction angiography (DSA)* uses a digital computer to produce three-dimensional pictures that effectively show blockages of blood vessels. The computer also measures the rate of blood flow to the heart. DSA has been useful in predicting heart attacks.

7 *Ultrasound* is an apparently harmless exploratory technique that sends pulses of ultra-high-frequency sound waves into designated body cavities. One use of this technique is to form images of developing fetuses. It can also be used to differentiate between healthy and diseased tissue, detect ectopic and multiple pregnancies, reveal certain birth defects and probable miscarriages, and assist in fetal surgery.

8 *Thermography* reveals chemical reactions that are taking place within the body, based on heat changes on the skin. It is useful in detecting cancer, arthritis, and circulatory problems.

UNDERSTANDING THE FACTS

1 What are anatomy and physiology, and how are they related?

2 Define homeostasis.

3 What is stress?

4 What is a compound?

5 What does a body system do?

6 Which body systems are involved in waste disposal?

7 What is the anatomical position?

8 What are the meanings of superior, inferior, anterior, and posterior?

9 Define parietal and visceral.

10 What are the two main regions of the body?

11 What is a transverse plane? What is a frontal plane?

12 Name the two main subdivisions of the ventral cavity.

13 What is one major advantage of PET scanning over CAT scanning?

14 How does dynamic spatial reconstruction differ from CAT scanning?

15 Why is ultrasound sometimes preferable to amniocentesis?

UNDERSTANDING THE CONCEPTS

1 What does the expression "function is determined by structure" mean in relation to the body?

2 What would be a reason for the size variations of different types of body cells?

3 How are the different structural levels of the body related?

4 How does stress affect homeostasis?

5 If a bullet entered the right side of the body just below the armpit and exited at the similar point on the left side, which cavities would be pierced (in order) and which organs probably damaged?

6 Explain why some noninvasive exploratory techniques can be used to *prevent* heart attacks and other disorders.

SUGGESTED READING

AREY, LESLIE BRAINERD, *Human Histology: A Textbook in Outline Form,* 4th ed. Philadelphia: Saunders, 1974.

BASMAJIAN, JOHN V., *Primary Anatomy,* 8th ed. Baltimore: Williams & Wilkins, 1982.

BEVELANDER, GERRIT, AND JUDITH A. RAMALEY, *Essentials of Histology,* 8th ed. St. Louis: Mosby, 1979.

FAWCETT, DON W., *Bloom and Fawcett's Textbook of Histology,* 11th ed. Philadelphia: Saunders, 1986.

GARDNER, ERNEST, DONALD J. GRAY, AND RONAN O'RAHILLY, *Anatomy: A Regional Study of Human Structure,* 4th ed. Philadelphia: Saunders, 1975.

GELLER, STEPHEN A., "Autopsy." *Scientific American,* March 1983.

HAM, ARTHUR W., AND DAVID H. CORMACK, *Histology,* 8th ed. Philadelphia: Lippincott, 1979.

KELLY, DOUGLAS, RICHARD L. WOOD, AND ALLEN C. ENDERS, *Bailey's Textbook of Microscopic Anatomy,* 18th ed. Baltimore: Williams & Wilkins, 1984.

KESSEL, RICHARD G, AND RANDY H. KARDON, *Tissues and Organs: A Text-Atlas of Scanning Electron Microscopy.* San Francisco: Freeman, 1979 (paperback).

LOCKHART, R. D., G. F. HAMILTON, AND F. W. FYFE, *Anatomy of the Human Body.* Winchester, Mass.: Faber & Faber, 1981 (paperback).

MONTGOMERY, ROYCE L., *Basic Anatomy for the Allied Health Professions.* Baltimore: Urban & Schwarzenberg, 1981.

MOORE, KEITH L., *Clinically Oriented Anatomy,* 2nd ed. Baltimore: Williams & Wilkins, 1985.

WOODBURNE, RUSSELL T., *Essentials of Human Anatomy,* 7th ed. New York: Oxford University Press, 1983.

2
Essentials of Body Chemistry

LEARNING OBJECTIVES

1 Define matter.

2 Explain the structure of an atom.

3 Define atomic number and atomic weight.

4 Describe an isotope.

5 Explain how an atom becomes stable through bonding.

6 Compare and contrast ionic, covalent, and hydrogen bonds.

7 Give examples of combination and decomposition reactions.

8 Describe a single- and a double-replacement reaction.

9 Give an example of an oxidation-reduction reaction.

10 Distinguish between hydrolysis and condensation.

11 List some of the important chemical properties of water.

12 Give examples of water as a solvent, transporter, temperature regulator, and lubricant.

13 Define an acid, a base, and a salt.

14 Define and describe pH.

15 Describe the role of a buffer system in maintaining pH and homeostasis.

16 Describe the different kinds of carbohydrates and some of their uses.

17 Describe the different kinds of lipids and some of their uses.

18 Describe the four different structural levels of proteins.

19 Describe the role of enzymes during chemical reactions.

20 Describe the structure and function of nucleic acids.

21 Explain the importance of ATP to life's processes.

Everything that goes on in the human body depends on some kind of chemical activity. Without chemical activity your muscles would be useless, your brain and nerves would be as ineffective as an unplugged television set, and the food you eat could not be chewed, swallowed, or digested. This chapter presents the chemistry you will need to understand the basic chemical processes that enable the human body to function.

MATTER, ELEMENTS, AND ATOMS

Everything in the universe is composed of *matter,* which is anything that occupies space and has mass. *Mass* refers to the amount of matter in an object. Mass is not the same as weight. *Weight* is the measurable gravitational attraction of the earth for an object. Mass refers only to the *amount* of matter, not its weight.

Matter is composed of *elements,* chemical substances that cannot be broken down into simpler substances by ordinary chemical means. Elements are represented by either one- or two-letter symbols. For instance, H is the symbol for the element hydrogen, O stands for oxygen, and Na stands for sodium. Currently 106 elements are known, 92 of which occur naturally. New elements are still being created in laboratories.

About 24 elements are found in living organisms, and four of these—oxygen, carbon, hydrogen, and nitrogen—account for 96 percent of the body's weight: oxygen makes up about 65 percent, carbon about 18 percent, hydrogen about 10 percent, and nitrogen about 3 percent. Several of the elements most important to human beings are listed in Table 2.1, along with the role they play in various vital chemical processes.

Each element is composed entirely of its own chemically distinct kind of atom. An *atom* is the smallest unit of an element that retains the chemical characteristics of that element.

How big is an atom?

Atoms are so small that the only way we can visualize their size is to relate it to objects in our own everyday lives. For instance, a child's balloon filled with hydrogen contains about 100 million million billion hydrogen atoms. A drop of water contains more than 100 billion billion hydrogen and oxygen atoms. If an atom were as large as the head of a pin, all the atoms in a grain of sand would fill a cube 1 mile (1.6 km) high.

TABLE 2.1   SOME CHEMICAL ELEMENTS ESSENTIAL TO PROPER BODY FUNCTION

Element	Symbol	Approximate percentage of human body	Significance to human body
Oxygen	O	65.0	Part of water and organic compounds that are essential to many life processes.
Carbon	C	18.5	Basic component of all organic compounds.
Hydrogen	H	9.5	Part of water and most organic compounds.
Nitrogen	N	3.2	Part of all protein molecules.
Calcium	Ca	1.5	Necessary for healthy bones and teeth, muscle contraction, blood clotting, hormone production.
Phosphorus	P	1.0	Component of nucleic acids and ATP; especially important in nervous tissue, bones, and teeth.
Potassium	K	0.4	Essential for body growth; important for nerve conduction, muscle contraction, water-ion balance in body fluids.
Sulfur	S	0.3	Component of many proteins.
Chlorine	Cl	0.2	Important for water movement between cells.
Sodium	Na	0.2	Component of extracellular fluid; critical to nerve and muscle response; maintains proper water balance in blood.
Magnesium	Mg	0.1	Aids in muscle contraction and nerve transmission.
Iodine*	I	<0.1	Necessary for production of thyroid hormone by thyroid gland.
Iron*	Fe	<0.1	Basic component of hemoglobin.

*Elements that make up less than (<) 0.1 percent of the body are known as *trace elements.* Some of the more common ones are cobalt, copper, fluorine, iodine, iron, manganese, silicon, and zinc (see Chapter 23).

Structure of Atoms

Atoms are not solid bits of matter. In fact, even atoms of the most dense substances consist mostly of space. Atoms have two main parts—a central *nucleus* and the surrounding *electron field* (Figure 2.1). The *nucleus* contains two kinds of particles, the positively charged *protons* (p^+) and the uncharged *neutrons* (n^0). (The hydrogen atom is the only one without any neutrons in the nucleus.) Moving around the nucleus are negatively charged particles called *electrons* (e^-). Modern atomic theory describes these electrons as existing in an electron field, or "cloud" (see Figure 2.1), which is a way of saying that any one electron cannot be found at any given point at any particular moment in time.

Atomic number and atomic weight If all atoms (except hydrogen) are composed of protons, neutrons, and electrons, how do we account for the apparent differences in the elements? The chemical and physical properties of an atom are determined by the number of protons and neutrons in its nucleus and by the number and arrangement of electrons surrounding the nucleus. The number of protons is critical. The *atomic number* of an element is the number of protons in the nucleus of one of its atoms. The atomic number alone can identify an element. For instance, if an atom has one proton, it is hydrogen; if an atom has eight protons, it is oxygen. Figure 2.2 shows how protons, neutrons, and electrons are distributed in hydrogen, carbon, and oxygen atoms (see also Table 2.2).

Note that in Figure 2.2 and Table 2.2 the number of protons in any atom of a given element is always the same as the number of electrons. In other words, the number of *positive* charges (protons) and the number of *negative* charges (electrons) are equal. As a result, the negative and positive charges cancel each other, and the atom is neutral—that is, it has no overall electrical charge.

FIGURE 2.2

Simplified diagrams of the structure of hydrogen, carbon, and oxygen atoms, showing the number of electrons (e^-) in orbit, and the number of protons (p^+) and neutrons (n^0) in the nucleus. Note that the number of protons equals the atomic number, the number of protons plus neutrons always equals the atomic weight, and the number of electrons equals the number of protons in any given atom.

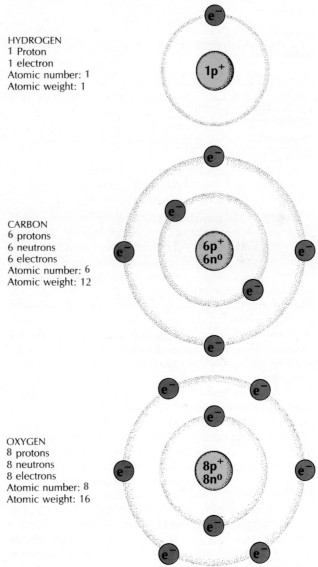

HYDROGEN
1 Proton
1 electron
Atomic number: 1
Atomic weight: 1

CARBON
6 protons
6 neutrons
6 electrons
Atomic number: 6
Atomic weight: 12

OXYGEN
8 protons
8 neutrons
8 electrons
Atomic number: 8
Atomic weight: 16

FIGURE 2.1

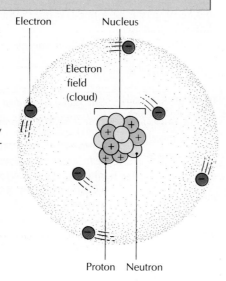

The structure of an atom. Negatively charged electrons moving around the nucleus form an electron field (or "cloud"), indicating that no one electron can be located at a particular place at any given time. The nucleus is composed of electrically neutral neutrons (gray) and positively charged protons (red).

Another measure of an atom is its *atomic weight,* which expresses the relative weight of an element compared with carbon. The atomic weight of carbon has been designated as 12. Thus, an atom of hydrogen (with an atomic weight of 1) is about one-twelfth as "heavy" as an atom of carbon, and an atom of calcium (with an atomic weight of 40) is about 3.3 times as heavy as a carbon atom. The approximate atomic weights of a few elements are given in the last column of Table 2.2; note that the atomic weight equals the number of protons plus the number of neutrons.

TABLE 2.2   ATOMIC STRUCTURE OF SOME COMMON ELEMENTS

Element	Atomic number	Number of Protons	Number of Neutrons	Electrons	Approximate atomic weight
Hydrogen (H)	1	1	0	1	1
Carbon (C)	6	6	6	6	12
Nitrogen (N)	7	7	7	7	14
Oxygen (O)	8	8	8	8	16
Sodium (Na)	11	11	12	11	23
Phosphorus (P)	15	15	16	15	31
Sulfur (S)	16	16	16	16	32
Chlorine (Cl)	17	17	18	17	35
Potassium (K)	19	19	20	19	39
Calcium (Ca)	20	20	20	20	40

Isotopes Most naturally occurring elements are actually a mixture of slightly different forms of that element. All atoms of an element have the same number of protons in the nucleus, but some have differing numbers of neutrons, and thus different atomic weights. But because all forms of an element have the same number of electrons, they *all have the same chemical properties.* The different atomic forms of an element are called **isotopes.** Some elements have as many as 20 naturally occurring isotopes, while others have as few as two. For example, about 99 percent of all carbon atoms have 6 protons and 6 neutrons, but other carbon atoms have 7 or 8 neutrons, giving them atomic weights of 13 and 14, respectively.* Because *most* of the carbon in nature has 6 neutrons, giving that isotope an atomic weight of 12, the overall atomic weight for carbon is set for convenience at 12 (see Table 2.2).

Some isotopes are naturally radioactive, and others are made radioactive in the laboratory by bombarding their nuclei with subatomic particles. An element is considered radioactive when its nucleus undergoes changes called *disintegration,* or "decay." In this process, nuclear particles and radiation are emitted. Because an isotope's nucleus is constantly changing, it is unstable. Many *radioisotopes* (short for radioactive isotopes) are used in biology and medicine. As their nuclei break down, or disintegrate, the isotopes emit high-energy x rays. Experiments have shown that radioisotopes emit one or more of three major types of radioactivity: *alpha particles* (α), *beta particles* (β), and *gamma rays* (γ). An alpha particle consists of two neutrons (2n) and two protons (2p). These particles do not travel very far and are not very penetrating. By contrast, beta particles are fast-moving electrons that are capable of producing dense ionization tracts. They have about 100 times more penetrating power than

alpha particles. Finally, gamma rays are not particles but a form of high-energy electromagnetic radiation. They travel at the speed of light and are so penetrating that they can be stopped only by lead or concrete, but may also be captured eventually by nuclear collision.

Radioactive isotopes can be introduced into the body and then traced with detectors such as the Geiger-Müller counter to help detect certain kinds of disorders. Radioactive rays can be used to detect and treat diseases. For example, radioactive iodine ($^{131}_{53}I$) is used to determine how the thyroid gland is functioning and to detect brain tumors. Radioactive cobalt ($^{60}_{27}Co$) is used to treat malignant cancers; radioactive phosphorus ($^{32}_{15}P$) is used in treating leukemia and bone cancer and in localizing breast cancers; radioactive sodium ($^{24}_{11}Na$) is used to study body fluids and circulation rates; radioactive carbon ($^{14}_{6}C$) traces the absorption and cellular use of food and drugs; and radioactive iron ($^{59}_{26}Fe$) and chromium ($^{54}_{24}Cr$) are helpful in studying red blood cells and hemoglobin. Technetium ($^{99}_{43}Tc$) is an isotope that is little known to the general public, but it is used extensively to help locate brain tumors before surgery is performed. It may soon replace radioactive iodine in that role.

*In a chemical atomic symbol, the atomic number is placed in an inferior position and the atomic weight (sometimes called the *mass number*) in a superior position. For example, the isotopes of an atom of carbon would be written $^{12}_{6}C$, $^{13}_{6}C$, and $^{14}_{6}C$. Carbon-14 ($^{14}_{6}C$) is radioactive.

How does a radioactive isotope such as cobalt help to destroy cancer cells?

The radiation emitted from radioactive cobalt can change the chemical nature of cancer cells enough to kill them. Cells that reproduce at a fast rate, such as cancer cells, are most susceptible to radiation, since radiation interferes with cellular division; this is why active cancer cells are supposed to die before healthy cells do. But radiation of any kind can be dangerous to all cells, especially those in bone marrow and those that line the intestines, which reproduce at a fairly fast rate.

Ask Yourself

1 *What are the basic components of an atom?*

2 *What are the electrical charges of neutrons, protons, and electrons?*

3 *How is the number of protons in the nucleus of an atom related to the atom's atomic number?*

4 *What is an isotope?*

Energy-Level Shells

The electrons of an atom are distributed around its nucleus in layers called *energy-level shells.* Seven energy-level shells are known to exist. Each shell can hold only a certain number of electrons. There are never more than two electrons in the

shell nearest the nucleus; as many as eight electrons can be in the second and third shells; larger numbers fill the more distant shells.

The number of electrons and protons in an atom determines its chemical character, but the way the electrons are arranged in the energy-level shells, especially the outer shell (the one farthest away from the nucleus), is most important. Only the electrons in the *outer* shell take part in a chemical reaction. When an atom has a complete outer shell—that is, the shell holds the maximum number of electrons possible— the shell is stable. An atom with an incomplete, or unstable, outer shell tends to gain, lose, or share electrons with another atom.

For instance, an oxygen atom has two electrons in its first shell but only six in its second shell. As a result, an oxygen atom tends to *gain* two more electrons in order to fill its outer shell with the stable number of eight electrons (Figure 2.3A). A sodium atom—which has two electrons in its first shell, eight in the second, and only one in the third—tends to *lose* the one electron in its outer (third) shell, leaving the second

FIGURE 2.3

(A) How atoms gain electrons. (1) An oxygen atom with six electrons in its second shell has two "available spaces" for electrons (represented by dashed circles) to fill the shell. (2) The shell becomes stable with the addition of two electrons. (B) How atoms lose electrons. (1) A sodium atom with only one electron in its third shell tends to lose it in order to have a stable, complete second shell. In this case, it is easier to lose one electron to achieve a stable second shell than to gain 17 electrons to complete the third shell. (2) The shell becomes stable after it loses the single electron. (C) How carbon shares electrons. (1) An atom of carbon has four electrons in its second shell and needs four more to complete the shell. (2) Carbon forms a stable compound by sharing four electrons with other atoms. The partial shells indicate that the electrons are being shared with other atoms.

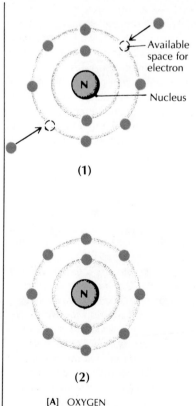
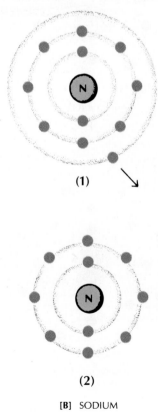
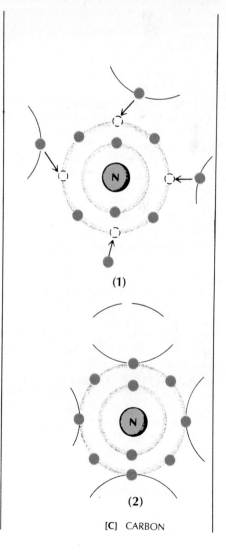

[A] OXYGEN

[B] SODIUM

[C] CARBON

shell with its stable number of eight electrons (Figure 2.3B). A final example is carbon, which has four electrons in its second shell. In order for the shell to be stable, it must have eight electrons, which occurs when the atom *shares* electrons with other atoms (Figure 2.3C). The ability to share electrons allows carbon to enter into many complex reactions.

Ask Yourself

1 *What are the layers of electrons surrounding an atom called?*

2 *Which energy-level shell is critical for chemical reactions?*

3 *How do shells of atoms become stable?*

HOW ATOMS COMBINE

Elements rarely exist in nature by themselves. Instead, atoms of elements combine chemically to form *molecules,* such as oxygen, O_2. (The subscript 2 indicates that the oxygen molecule contains two oxygen atoms.) *Compounds* are molecules that are made up of atoms of two or more elements (such as carbon dioxide, or CO_2, sodium chloride, or NaCl, and glucose, or $C_6H_{12}O_6$).

When atoms interact chemically to form molecules, the atoms are held together by electrical forces called *chemical bonds.* The kinds of bonds most important to the chemistry of the human organism are the ionic, covalent, and hydrogen bonds.

FIGURE 2.4

Ionic bonding. (A) A sodium atom has an available electron to give up to a chlorine atom with an incomplete outer shell. (B) After losing its "extra" electron, the sodium atom acquires a positive charge and is now a sodium *ion,* represented by Na^+; the chlorine atom, having received an electron to stabilize its outer shell, acquires a negative charge and becomes a chlor*ide ion* (Cl^-). (C) The oppositely charged sodium and chloride ions are attracted to each other and form an ionic bond. The electrical attraction is indicated by a fuzzy electron field.

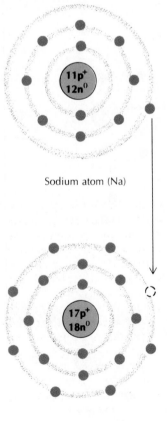

Sodium atom (Na)

Chlorine atom (Cl)

[A] FORMATION OF IONS

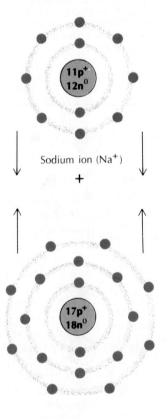

Sodium ion (Na^+)

+

Chloride ion (Cl^-)

−

[B] ATTRACTION OF OPPOSITE CHARGES

[C] IONIC BOND IS FORMED

Ionic Bonds

As we saw earlier, electrons have a negative electrical charge, and protons have a positive electrical charge. If an atom *loses* one or more electrons, it becomes *positively* charged, because there are now more positively charged protons in the nucleus than negatively charged electrons surrounding the nucleus. This positive charge is expressed by one or more "plus" signs. On the other hand, if an atom *gains* one or more electrons, it becomes *negatively* charged, because there are now more negative electrons than positive protons. This negative charge is expressed by one or more "minus" signs. Once an atom gains or loses electrons it acquires an electrical charge, and it is called an **ion.** A positive ion is known as a *cation*, and a negative ion is an *anion.**

The ions of elements are very different from the atoms from which they are formed. For example, sodium (Na) in its atomic form is a metal that reacts violently with water; chlorine (Cl) is a highly reactive gas. As ions, both are stable and do not react with water, although they dissolve in water. Together they form common table salt, NaCl. The human body is composed primarily of water, which has special properties (discussed later) that promote the breaking of ionic bonds and thus the formation of such important ions as potassium (K^+), required for the activation of many biological processes, and calcium (Ca^{2+}), necessary for the contraction of muscles.

Ionic bonds are formed when an atom or group of atoms develops an electrical charge and then becomes attracted to an atom or group of atoms with an opposite charge. Figure 2.4 shows how an ionic bond is formed between sodium and chlorine to produce sodium chloride.

Covalent Bonds

When atoms form ionic bonds, electrons are lost or gained. When atoms *share* electrons with other atoms, the chemical bond that is formed is called a **covalent bond** (Figure 2.5). In covalent bonding, electrons are always shared in pairs. When one pair of electrons is shared, a *single bond* is formed; when two pairs are shared, a *double bond* is formed.

Because covalent bonds are very strong, molecules and compounds with covalent bonds are very stable. There are two basic types of covalent bonds. In a *nonpolar* covalent bond, one or more electron pairs are distributed, or shared, equally between atoms. In a *polar* covalent bond, one or more electron pairs are distributed unequally between two atoms. In other words, the shared electrons are drawn more to one atom than to the other, creating a slight charge at each end of the molecule: positive for that portion in which the electrons spend less time, and negative where they spend more time.

---

*The most important cations in the body are sodium (Na^+), potassium (K^+), hydrogen (H^+), magnesium (Mg^{2+}), calcium (Ca^{2+}), and iron (Fe^{3+}). The most important anions are chloride (Cl^-), hydroxyl (OH^-), bicarbonate (HCO_3^-), sulfate (SO_4^{2-}), phosphate (PO_4^{3-}), and carboxyl (COO^-).

FIGURE 2.5

Covalent bonding. (A) An oxygen atom needs two electrons to complete its outer shell, and a hydrogen atom needs one electron to complete its outer shell. (B) A water molecule is formed when an oxygen atom and two hydrogen atoms share their electrons. Two single bonds, each with a pair of electrons, are formed.

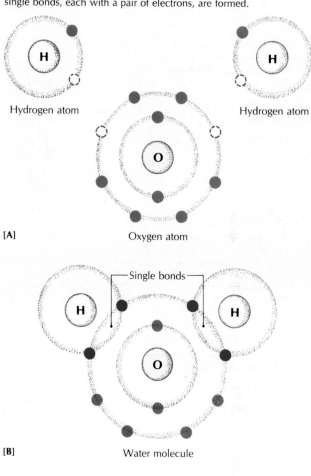

[A]

Hydrogen atom

Hydrogen atom

Oxygen atom

Single bonds

[B]   Water molecule

Hydrogen Bonds

In compounds containing covalently bonded hydrogen, the hydrogen electron is drawn to another atom, leaving a proton behind. As a result, the hydrogen atom gains a slight positive charge. The remaining proton is attracted to negatively charged atoms of oxygen or nitrogen in nearby molecules. When this happens, a weak **hydrogen bond** is formed. The hydrogen atom in one water molecule, for example, forms a hydrogen bond with the oxygen atom in another water molecule, and so forth, until many water molecules are bonded together (Figure 2.6). Hydrogen bonding is the intermolecular attractive force that gives water its special properties: a high melting point, a high evaporating point, an ability to retain heat better than most liquids, and an exceptional ability to dissolve ionic substances. Although hydrogen bonds are weaker than ionic or nonpolar covalent bonds, the many hydrogen bonds found in such substances as water, proteins, the genetic material called nucleic acid, enzymes, and hemoglobin encourage the formation of stable three-dimensional molecules.

FIGURE 2.6

The formation of hydrogen bonds (dashed lines) between polar covalently bonded water molecules. A single hydrogen bond is relatively weak, but several bonds work together to maintain the molecular shape of water.

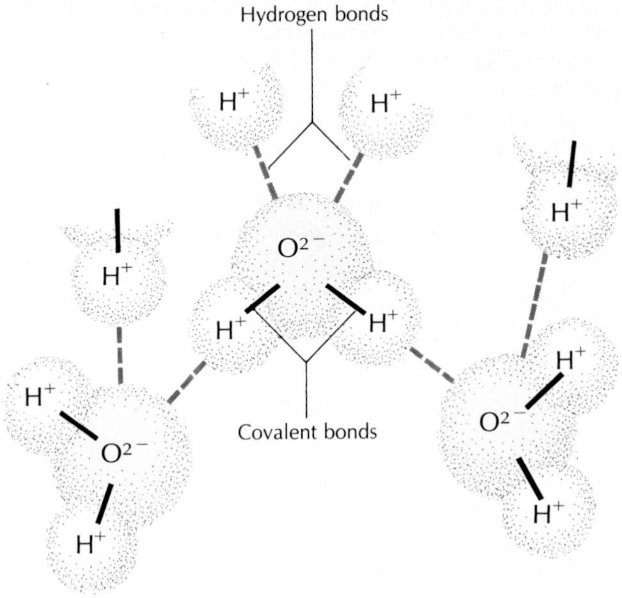

Ask Yourself

1 *What is an ion? A cation? An anion?*

2 *What is a covalent bond?*

3 *What is the difference between polar and nonpolar covalent bonds?*

4 *How is it possible for an atom that is already covalently bonded to form a hydrogen bond at the same time?*

CHEMICAL REACTIONS

As you just saw, atoms combine by forming bonds. The combining of two or more atoms causes a chemical change called a chemical reaction. Because chemical reactions also occur when bonds are broken, we can say that in a *chemical reaction,* bonds between atoms are broken or joined, and different combinations of atoms or molecules are formed. There are five basic types of chemical reactions: (1) combination reactions, (2) decomposition reactions, (3) replacement reactions, (4) oxidation-reduction reactions, and (5) hydrolysis and condensation reactions.

Another way of talking about chemical reactions is to consider them as part of the overall process of *metabolism,* which includes all the chemical activities that go on in the body (see Chapter 23). Metabolism either builds up substances (synthesis), or it breaks them down (decomposition). The synthesis process is called *anabolism,* and the decomposition process is called *catabolism.*

Combination Reactions

When two or more atoms, ions, or molecules combine to form a more complex substance, the process is called a *combination reaction:* $A + B \rightarrow AB$. (Another useful word for "combination" is *synthesis* [Gr. "to put together"].) The substances that are combined are the *reactants,* and the new molecule they form is the *product:*

$$C \quad + \quad O_2 \quad \longrightarrow \quad CO_2$$

Carbon   plus   oxygen   produces   carbon dioxide
(reactant)   (reactant)   (product)

The arrow means "produces," or "yields." It shows the direction in which the reaction is moving.

Some chemical reactions are *reversible.* That is, after the reactants combine, the product can be decomposed in a later reaction to produce the original reactants. Such a reversible reaction is shown by special arrows going in both directions:

$$A + B \rightleftharpoons AB$$

Other reactions are *not reversible.* That is, the products are not normally decomposed or rearranged to re-form the original reactants:

$$A + B \longrightarrow C + D\uparrow$$

The long single arrow means that the reaction proceeds in only one direction. In this equation, one of the products, shown as $D\uparrow$, is an escaping gas that is no longer available for a reversible reaction. Other factors also make a reaction irreversible. (We will show all chemical reactions with only one arrow, unless a special point is being made about their reversibility.)

All chemical reactions either release or absorb energy. Combination (anabolic) reactions require energy, and, as you will see below, chemical reactions that break the chemical bonds of large molecules release energy.

Decomposition Reactions

Decomposition reactions are the opposite of combination reactions. They result in the breakage of chemical bonds to form two or more products (atoms, ions, or molecules): $AB \rightarrow A + B$.

$$2HgO \quad \longrightarrow \quad 2Hg \quad + \quad O_2$$

Mercuric oxide   is decomposed   mercury   and   oxygen
            to produce

The digestion of food—which is simply the process of breaking down large, complex food molecules into smaller, simpler molecules—is an example of many decomposition reactions.

Replacement Reactions

Replacement reactions occur when atoms, ions, or molecules change places with other atoms, ions, or molecules to produce new arrangements of atoms, ions, or molecules. In other words, the bonds are broken between substances, the substances change places, and new bonds are formed between new substances. Replacement reactions are sometimes called *exchange* reactions.

In a *single-replacement reaction* (A + BC → AC + B), only one substance changes places with another:

$$Zn \quad + \quad H_2SO_4 \quad \longrightarrow \quad ZnSO_4 \quad + \quad H_2$$

| Zinc | changes places with | the hydrogen in sulfuric acid | to produce | zinc sulfate | and | hydrogen |

In a *double-replacement reaction* (AB + CD → AD + CB), two or more reactants change places as chemical bonds are broken and then reunited:

$$AgNO_3 \quad + \quad HCl \quad \longrightarrow \quad AgCl \quad + \quad HNO_3$$

| The silver (Ag) in silver nitrate | changes places with | the hydrogen in hydrochloric acid | to produce | silver chloride | and | nitric acid |

Oxidation-Reduction Reactions

Oxidation occurs when an atom or molecule *loses* electrons or hydrogen atoms:

$$Na \quad - \quad e^- \quad \longrightarrow \quad Na^+$$

| When a sodium atom | loses | a negatively charged electron | it is oxidized and produces | a positively charged sodium ion |

When an atom or molecule *gains* electrons or hydrogen atoms, the process is called **reduction:**

$$Cl_2 \quad + \quad 2e^- \quad \longrightarrow \quad 2Cl^-$$

| When a molecule of chlorine | gains | two negatively charged electrons | it is reduced and produces | two negatively charged chloride ions |

When an *oxidation-reduction reaction* takes place, the oxidizing agent is reduced (gains electrons), and the reducing agent is oxidized (loses electrons). So *oxidation and reduction always occur together:*

$$O_2 \quad + \quad 2H_2 \quad \longrightarrow \quad 2H_2O$$

| One oxygen molecule (the oxidizing agent) | reacts with | two hydrogen molecules (the reducing agent) | to produce | two water molecules, in which the oxygen has been reduced and the hydrogen has been oxidized |

Oxidation-reduction reactions are especially important to the body because they provide energy that is used in cellular activities such as muscle contractions.

Hydrolysis and Condensation

When **hydrolysis** (Gr. "water loosening") takes place, a molecule of water interacts with the reactant to break up the reactant's bonds. The original reactant is then rearranged, together with the water, into different molecules:

$$C_{12}H_{22}O_{11} \quad + \quad H_2O \quad \longrightarrow \quad C_6H_{12}O_6 \quad + \quad C_6H_{12}O_6$$

| Sucrose | plus | water | produces | glucose | and | fructose |

The bonds of sucrose (table sugar) are broken (hydrolyzed) by water to form two simple sugars, glucose and fructose. Glucose and fructose have the same numbers and kinds of atoms, but in different arrangements (isomers).

Hydrolysis is important to the body. By means of hydrolysis, large molecules of proteins, nucleic acids, and fats are broken down into simpler, smaller, more usable molecules. Generally speaking, almost all digestive and degradative processes in the body occur by hydrolysis.

Condensation reactions are essentially hydrolytic reactions in reverse: small molecules are united into larger molecules, and one or more molecules of water are eliminated. For example, when ammonia reacts with acetic acid, acetamide and water are produced via condensation:

$$NH_3 \quad + \quad C_2H_3O_2H \quad \longrightarrow \quad C_2H_5NO \quad + \quad H_2O$$

| Ammonia | plus | acetic acid | produces | acetamide | and | water |

Ask Yourself

1 *How do chemical reactions occur?*

2 *What is a combination reaction?*

3 *What is a decomposition reaction?*

4 *Why do oxidation and reduction always occur together?*

5 *What is hydrolysis? Condensation?*

WATER

Chemical compounds are generally divided into two categories, inorganic and organic. **Inorganic compounds** are composed of relatively small molecules that are usually bonded ionically. **Organic compounds** contain carbon and hydrogen and are usually bonded covalently. Certain salts—such as sodium chloride (NaCl) and potassium chloride (KCl)—are examples of inorganic compounds, but the major example is *water* (H_2O), without which the other inorganic compounds would be useless, since many of their reactions occur only in the presence of water.

Water is one of the most plentiful compounds on earth. It is the most common component of the body, making up about 62 percent of the body weight of an adult (Table 2.3). The percentage is usually higher in young children and lower in the elderly. Every day we take in about 2700 mL* (2¾ quarts) of water from various sources, and varying amounts of water are constantly leaving our cells and our bodies (Table 2.4). Water is one of the chief regulators of homeostasis, and the body cannot function without it.

*The abbreviation for milliliters is mL.

TABLE 2.3   APPROXIMATE WATER CONTENT OF THE ADULT HUMAN BODY

Tissues	Body weight (%)	Water (%)	Water in tissues (% body weight)	Liters of water
Muscle	41.7	75.6	31.53	22.10
Skin	18.0	72.0	12.96	9.07
Blood	8.0	83.0	6.64	4.65
Skeletal	15.9	22.0	3.50	2.45
Brain	2.0	74.8	1.50	1.05
Liver	2.3	68.3	1.57	1.10
Intestines	1.8	74.5	1.34	0.94
Fat tissue	8.5	10.0	0.01	0.70
Lungs	0.7	79.0	0.55	0.39
Heart	0.5	79.2	0.40	0.28
Kidneys	0.4	82.7	0.33	0.23
Spleen	0.2	75.8	0.15	0.11
Total body	100.0%	62.0%	60.48%	43.07 L

Source: Robert L. Vick, *Contemporary Medical Physiology* (Reading, Mass.: Addison-Wesley, 1984), p. 540. Used with permission.

TABLE 2.4   COMPONENTS OF WATER BALANCE IN THE HUMAN BODY

Intake (mL/24 hr)		Output (mL/24 hr)	
Source	Amount	Route	Amount
Preformed water in food	750	Skin and lungs	840
Water of metabolism	320	Urine	1760
Drinking	1630	Feces	100
Total intake	2700 mL	Total output	2700 mL

Source: Robert L. Vick, *Contemporary Medical Physiology* (Reading, Mass.: Addison-Wesley, 1984), p. 621. Used with permission.

Chemical Properties of Water

The special properties of water depend on its bonding structure: two hydrogen atoms covalently bonded to one oxygen atom (Figure 2.7A). The oxygen atom attracts electrons to itself and away from the hydrogen atoms. As a result, the oxygen atom picks up a slight negative charge, while the two hydrogen atoms bear slight positive charges. Because the two ends of a water molecule have two different charges, it is said to be a *polar molecule* (Figure 2.7B). Water can form hydrogen bonds with other water molecules and with a variety of other compounds.

Water is able to dissolve other polar substances because of its own polarity. A general rule for predicting solubility is: "Likes dissolve one another; unlikes do not"—where the terms *like* and *unlike* refer to the polarity of the substances.

For example, nearly all large molecules, such as carbohydrates and proteins, have areas that are slightly charged. Because water molecules are attracted to those charged areas, they help keep these other molecules dissolved. Also, many particles within cells are charged; thus, water molecules are attracted to the surfaces of those particles (Figure 2.7C). The neatly packed water molecules add stability to the surfaces of cell membranes and proteins.

Water as a Solvent

A *solvent* is a liquid or gas capable of dissolving another substance. A *solute* is the substance being dissolved. A *solution* is the homogeneous mixture of a solvent and the dissolved solute. All of our body fluids (including blood, urine, lymph, sweat, and digestive juices) are mostly water, and practically all the chemical reactions in the body take place in water. In fact, water is considered the "universal solvent" because it dissolves not only "like" polar substances, but many nonpolar substances as well. Water is used during digestion in a series of hydrolysis reactions that break down large compounds into smaller, more easily assimilated ones. Once the dissolved nutrients are in solution they can be absorbed through the walls of the small intestine into the bloodstream. Inhaled oxygen can be used only when it is in so-

FIGURE 2.7

The structure of a water molecule. (A) Two atoms of hydrogen are bonded to one atom of oxygen to form a molecule of water. (B) A simple model of a water molecule, showing the different electrical charges at either end. (C) The positive ends of water molecules are attracted to a small particle with a negatively charged surface, forming an orderly and stable condition.

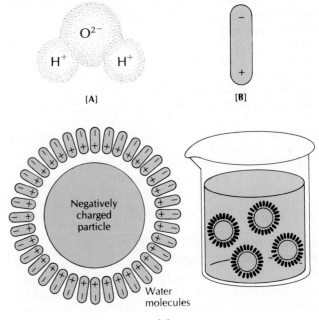

lution with the water in your blood, and waste carbon dioxide from cells is carried to the lungs in solution before it is exhaled.

Water as a Transporter

Water in the blood carries nutrients from the foods we eat to tissues throughout the body. It also transports waste products from the cells to the lungs, kidneys, and skin to be discarded as carbon dioxide, urine, or sweat. (Urine and sweat, which help eliminate excess nitrogen and salt, are both over 95 percent water.) Besides nutrients and wastes, the liquid portion of blood (plasma) carries hormones to their specific target sites throughout the body. And, as you just saw, water is essential for blood circulation in general. Water also transports materials within cells. About 60 percent of the water in the body is contained within tissue cells (Table 2.3).

Water as a Temperature Regulator

Because water retains heat better than most liquids, it takes a great deal of heat to turn liquid water into water vapor (gas). For this reason, a lot of heat is lost when we perspire, and the body is cooled by losing only a little water. (Perspiring is part of a negative feedback mechanism that helps maintain homeostasis.)

Even when temperatures are not very hot, water is constantly evaporating from the skin and lungs. This water loss, called *insensible perspiration,* because we don't feel or see it, balances the heat produced through metabolic processes in the cells in order that our body temperature remains constant (this is another example of homeostasis).

Water as a Lubricant

The lubricating fluid (called *synovial fluid*) that helps joints such as knees and elbows to move easily contains a great deal of water. The thin spaces around internal organs are also lubricated by fluids that contain water. In the chest cavity, for instance, lubrication is important, as the rib cage slides over several internal organs during breathing. Water moistens food as it passes along the digestive tract, starting with saliva in the mouth and throat that makes swallowing easy, and ending with the passage of water-softened feces through the anus. Water also keeps body cells moist inside and out and is a principal component of mucus and every other lubricating fluid in the body.

| *How long can a person live without water?* | *We can live without food for weeks, but without water we would be dead in about three days. An infant might not live that long. Death usually occurs when water loss is over 10 percent of a person's body weight.* |

Ask Yourself

1 *How do organic compounds differ from inorganic compounds?*

2 *What is a polar molecule?*

3 *What is a solvent? A solution?*

4 *Why is water called the "universal solvent"?*

5 *How does water act as a transporter, a regulator of body temperature, and a lubricant?*

ACIDS, BASES, SALTS, AND BUFFERS

As you have seen, most of the chemicals in your body are not "dry," but are dissolved in a water solution. This is not unusual when you realize that more than half of your body is water. The idea of substances being chemically active only when they are dissolved in water, or when they are "in solution," certainly applies to acids and bases, two biologically important substances.

Any substance, such as table salt (NaCl), whose solution conducts electricity is called an *electrolyte.* Many fluids present in our bodies contain strong electrolytes that break down into ions. Most water-soluble ionic compounds and most acids and bases are electrolytes.

An *acid* is a substance that releases hydrogen ions (H^+) when dissolved in water. For example, when hydrochloric acid (HCl) is placed in water, the following reaction takes place:

$$HCl \xrightarrow{\text{Water}} H^+ + Cl^-$$

| One molecule of hydrochloric acid | dissolved in water produces | one hydrogen ion | and | one chloride ion |

In contrast, a *base* is a substance that releases hydroxyl ions (OH^-) when dissolved in water. The following reaction is typical:

$$NaOH \xrightarrow{\text{Water}} Na^+ + OH^-$$

| One molecule of sodium hydroxide | dissolved in water produces | one sodium ion | and | one hydroxyl ion |

A *salt* is an ionic substance that contains an anion other than OH^- or O^{2-}. An acid reacts with a base to produce a salt. The reaction is called *neutralization:*

$$HCl + NaOH \longrightarrow NaCl + H_2O$$

| An acid | plus | a base | produce | a salt | and | water |

Acids have a sour taste and react with certain metals to release hydrogen gas. They can neutralize bases. Bases have a bitter taste and feel slimy. They can neutralize acids. Some common acids, bases, and salts found in the body are listed in Table 2.5.

TABLE 2.5 — SOME COMMON ACIDS, BASES, AND SALTS FOUND IN THE HUMAN BODY

ACID	FORMULA
Carbonic acid	H_2CO_3
Hydrochloric acid	HCl
Lactic acid	$CH_3CHOHCOOH$
Phosphoric acid	H_3PO_4
Sulfuric acid	H_2SO_4

BASE	
Ammonium hydroxide	NH_4OH
Magnesium hydroxide	$Mg(OH)_2$
Potassium hydroxide	KOH
Sodium bicarbonate	$NaHCO_3$
Sodium hydroxide	NaOH

SALT	
Magnesium sulfate	$MgSO_4$
Potassium chloride	KCl
Sodium carbonate	Na_2CO_3
Sodium chloride	NaCl

Dissociation

The tendency for some molecules to break up into charged ions in water is called **dissociation.** (When ions recombine to form a stable molecule, the process is called *association.*) Not all acids dissociate completely. Hydrochloric acid is considered a *strong acid* because it dissociates entirely in water. But acetic acid, the familiar ingredient of vinegar, is a *weak acid* because it does not completely dissociate in water. Combinations of weak acids are important in maintaining the concentration of hydrogen ions in cells.

Measuring Acidity and Alkalinity

The higher the concentration of hydrogen ions (H^+), the more acidic a solution is. Also, the higher the concentration of H^+ ions, the lower the concentration of hydroxyl ions (OH^-). Table 2.6 shows the relationship between hydrogen ion (H^+) concentration and hydroxyl (OH^-) concentration. A solution is considered neutral when the number of H^+ ions equals the number of OH^- ions.

The numerical scale that measures acidity and alkalinity is called the **pH scale.** It runs from 0 to 14, with neutrality at 7, and indicates the concentration of free hydrogen ions (H^+) in water (Figure 2.8). Each whole number on the pH scale represents a tenfold change (logarithmic) in acidity. So a cup of black coffee at pH 5 has ten times the concentration of H^+ ions that peas, at pH 6, have.

What do the letters pH stand for?	*The letter* p *in pH stands for* potentia, *the Latin word for "power," and the letter* H *stands for "hydrogen." So pH means "the power (or concentration) of hydrogen ions in solution."*

FIGURE 2.8

The pH scale. Note that pure distilled water is neutral, rather than being either acidic or basic; this condition occurs because distilled water has equal numbers of hydrogen ions and hydroxyl ions. Each number on the pH scale represents a tenfold change in acidity. Black coffee at pH 5 contains 10 times the acidity (hydrogen ions) of peas at pH 6.

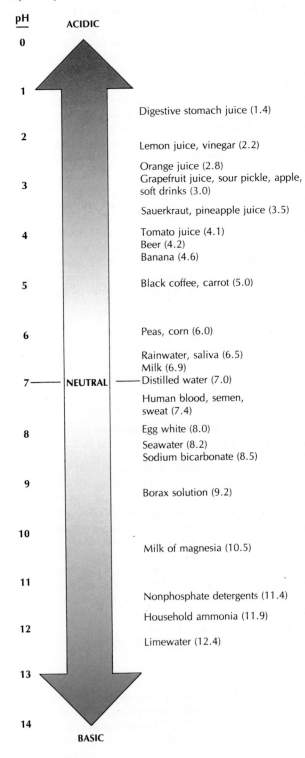

TABLE 2.6 RELATIONSHIP BETWEEN HYDROGEN ION (H^+) CONCENTRATION AND HYDROXYL ION (OH^-) CONCENTRATION

H^+ (hydrogen ion)	pH	OH^- (hydroxyl ion)
$10^0 = 1$	0	$10^{-14} = 0.00000000000001$
$10^1* = 0.1$	1	$10^{-13} = 0.0000000000001$
$10^2 = 0.01$	2	$10^{-12} = 0.000000000001$
$10^3 = 0.001$	3	$10^{-11} = 0.00000000001$
$10^4 = 0.0001$	4	$10^{-10} = 0.0000000001$
$10^5 = 0.00001$	5	$10^{-9} = 0.000000001$
$10^6 = 0.000001$	6	$10^{-8} = 0.00000001$
$10^7 = 0.0000001$ (Neutrality)	7 (Neutrality)	$10^{-7} = 0.0000001$
$10^8 = 0.00000001$	8	$10^{-6} = 0.000001$
$10^9 = 0.000000001$	9	$10^{-5} = 0.00001$
$10^{10} = 0.0000000001$	10	$10^{-4} = 0.0001$
$10^{11} = 0.00000000001$	11	$10^{-3} = 0.001$
$10^{12} = 0.000000000001$	12	$10^{-2} = 0.01$
$10^{13} = 0.0000000000001$	13	$10^{-1} = 0.1$
$10^{14} = 0.00000000000001$	14	$10^0 = 1$

*The superscripts are called *exponents;* they indicate a tenfold difference in concentration.

Controlling pH with Buffers

Homeostasis can be maintained only if there is a relatively constant pH of the blood and other body fluids. Too much of a strong acid or base can destroy the stability of cells. Also, too sudden a change of the pH balance may be destructive. Human blood, for example, cannot tolerate drastic changes from its normal pH of 7.4. In fact, the pH of blood rarely goes below 7.35 or above 7.45 because it contains chemical substances that regulate the acid-base balance. These substances, called *buffers,* are combinations of weak acids or weak bases and their respective salts in solution. Buffer solutions help blood and other body fluids to resist changes in pH when small amounts of strong acids or bases are added. So, even though black coffee has a pH of 5, the 7.4 pH of your blood does not change for more than a few seconds when you drink coffee. This quick return to normalcy is produced by the stabilizing action of buffers.

The buffer control systems are one of the body's major ways of maintaining pH homeostasis. Some of these systems are the fastest of the body's regulators of acid-base balance, but other mechanisms provide more effective long-term control. For example, buffer systems are also present in the respiratory system, where the acid-base balance is controlled by regulating the exhalation rate of carbon dioxide. In the urinary system, excess acid is neutralized by the removal of hydrogen ions from the blood. Buffering systems in the lungs and kidneys are explained in Chapter 25.

Ask Yourself

1 *What is an acid? A base? A salt?*

2 *What does the pH scale indicate?*

3 *How do buffers function?*

SOME IMPORTANT ORGANIC COMPOUNDS

When we speak of the chemistry of the living body, we usually think of the essential body nutrients: carbohydrates, lipids (fats), proteins, and nucleic acids (the hereditary material). These substances are the organic chemicals of the body. *Organic compounds* always contain carbon and hydrogen. They are typically composed of large and complex molecules that are held together by covalent bonds. Organic compounds, along with water, are the main materials that form the human body.

Carbohydrates

Carbohydrates are the major source of energy for most cells, since cells have the chemical machinery to break down these compounds easier than others. Carbohydrates are made of carbon, hydrogen, and oxygen, with twice as much hydrogen as oxygen. This ratio is shown in the general formula for a carbohydrate, CH_2O. A carbohydrate may contain three or more CH_2O units, depending upon its complexity.

Kinds of carbohydrates *Carbohydrates* are generally classified according to their molecular structure. The simplest types are called *monosaccharides* (Gr. *monos*, single + *sakkharon*, sugar). They are the building blocks of more complex sugars. Monosaccharides are so simple that they cannot be broken down by adding water (hydrolysis; see page 39). Three common monosaccharides are glucose, fructose, and ribose* (Figure 2.9).

*The suffix *-ose* usually stands for sugar. Prefixes such as *tri-* (three), *pent-* (five), or *hex-* (six) indicate the number of carbon atoms in the molecule. Pentoses, for example, are monosaccharides containing five carbons.

FIGURE 2.9

Molecular structures of glucose, fructose, and ribose. Glucose and fructose both have the same molecular formula, $C_6H_{12}O_6$, but different molecular *structures*. Each sugar can be represented in several different ways. Here, each is shown in open-chain and ring forms, and by its molecular formula.

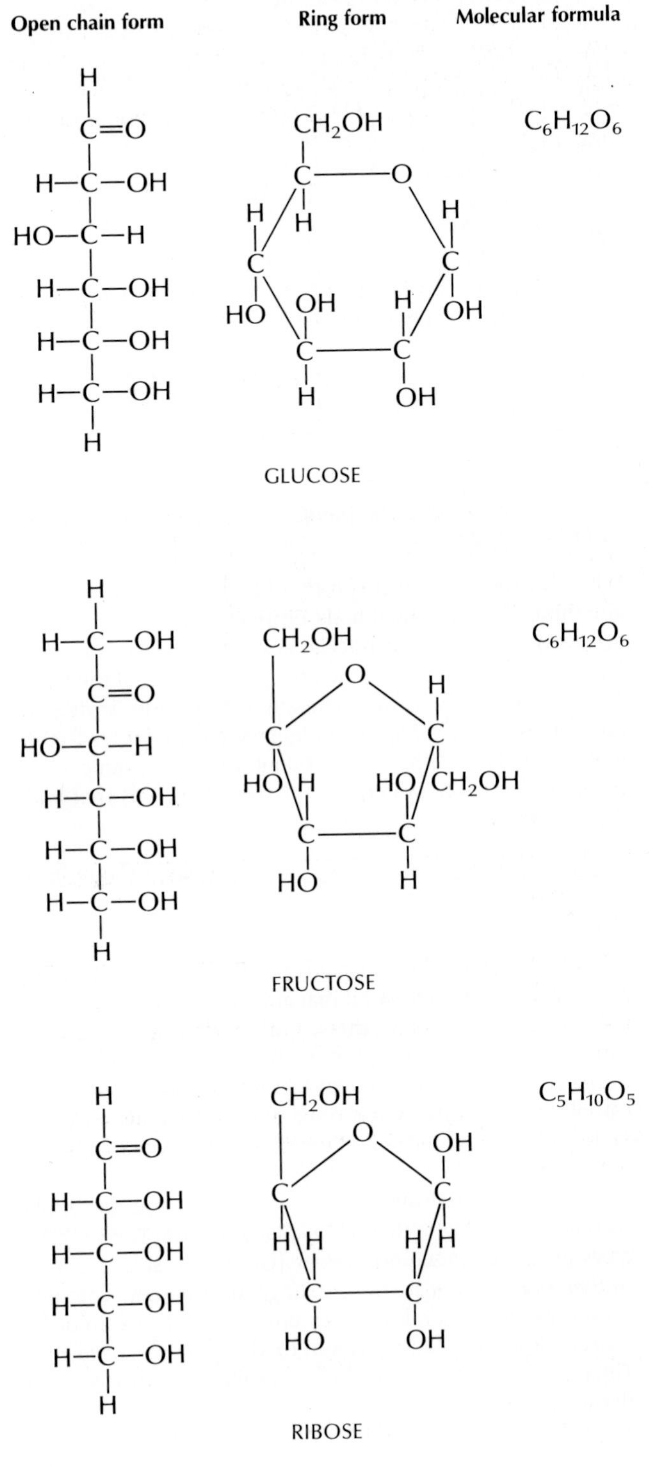

Open chain form   **Ring form**   **Molecular formula**

GLUCOSE $C_6H_{12}O_6$

FRUCTOSE $C_6H_{12}O_6$

RIBOSE $C_5H_{10}O_5$

Two monosaccharides can be combined to form a *disaccharide* (*di*, two). Disaccharides all have the molecular formula $C_{12}H_{22}O_{11}$. Many compounds that have the same molecular formulas have different molecular *structures*. Such compounds are called *isomers*. See Figure 2.9, which shows the structural differences between glucose and fructose, both of which have the formula $C_6H_{12}O_6$. Disaccharide isomers are sucrose, maltose, and lactose.

The kind of chemical reaction in which a molecule of water is lost when large organic molecules are synthesized from simpler organic molecules is called *dehydration synthesis*, or *condensation* (see page 39). In Figure 2.10 you can see how two monosaccharide glucose molecules unite to form the disaccharide maltose (malt sugar). Other typical reactions in which monosaccharides combine to form disaccharides are:

Glucose + Fructose ⟶ Sucrose + Water

Glucose + Galactose ⟶ Lactose + Water

Note that in each case a molecule of water is formed in the dehydration synthesis. The water is given up when the two monosaccharides combine, as shown in Figure 2.10. The reverse chemical reaction, hydrolysis, takes place when water is added. Disaccharides can thus be broken down into simpler monosaccharides by adding water:

Maltose + Water ⟶ Glucose + Glucose

The entire reversible processes of dehydration, the loss of a water molecule, and hydrolysis, the addition of a water molecule, may be written in a simple form as follows:

Mono-saccharide + Mono-saccharide ⇌ (Dehydration synthesis / Hydrolysis) Disaccharide + Water

Carbohydrate molecules containing more than two monosaccharides are called *polysaccharides* (*poly*, many). Polysaccharides are formed by dehydration synthesis and can be broken down into their many monosaccharide components by hydrolysis. Polysaccharide chains may contain hundreds or thousands of monosaccharide molecules. These long chains may be branched or unbranched. The general formula for a polysaccharide is $(C_6H_{10}O_5)_n$, where n equals the number of glucose units in the molecule. Such long chains of repeating units are called *polymers*.

Uses of carbohydrates Carbohydrates are important in the body mainly as energy sources for cellular activities. When the body needs energy, it breaks down and uses carbohydrates first, fats second, and proteins last (see Chapter 23). Some of the more important uses of carbohydrates are described in Table 2.7.

Lipids

Body fats belong to a diverse group of organic compounds called *lipids*. Lipids are insoluble in water but can be dissolved in organic solvents such as ether, alcohol, and chloroform. Like carbohydrates, they are composed mainly of carbon, hydrogen, and oxygen, but they may contain other elements, especially phosphorus and nitrogen. Also, lipids usually have more than twice as many hydrogen atoms as oxygen atoms.

Familiar lipids are fats and oils. Less well known but equally important are phospholipids, steroids, and prostaglandins.

Fats *Fats* are important mainly because they are energy-rich molecules (yielding about twice as much energy as carbohydrates) and are therefore a suitable source of reserve food, or long-term fuel. They are stored in the body in the form of *triacylglycerols*, also called neutral fats (because they are not electrically charged), or triglycerides. Fats also provide the body with insulation, protection, and cushioning.

Each fat molecule is composed of **glycerol**, also called *glycerin*, and **fatty acids** (Figure 2.11). The glycerol molecule contains three hydroxyl (OH) groups, and the fatty-acid molecules contain carboxyl (COOH) groups, shown in Figure 2.11 as:

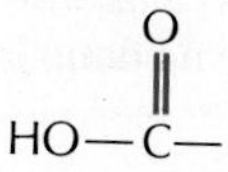

$$HO - \overset{\overset{\displaystyle O}{\|}}{C} -$$

The double covalent bond (in color) represents two pairs of electrons being shared by two atoms.

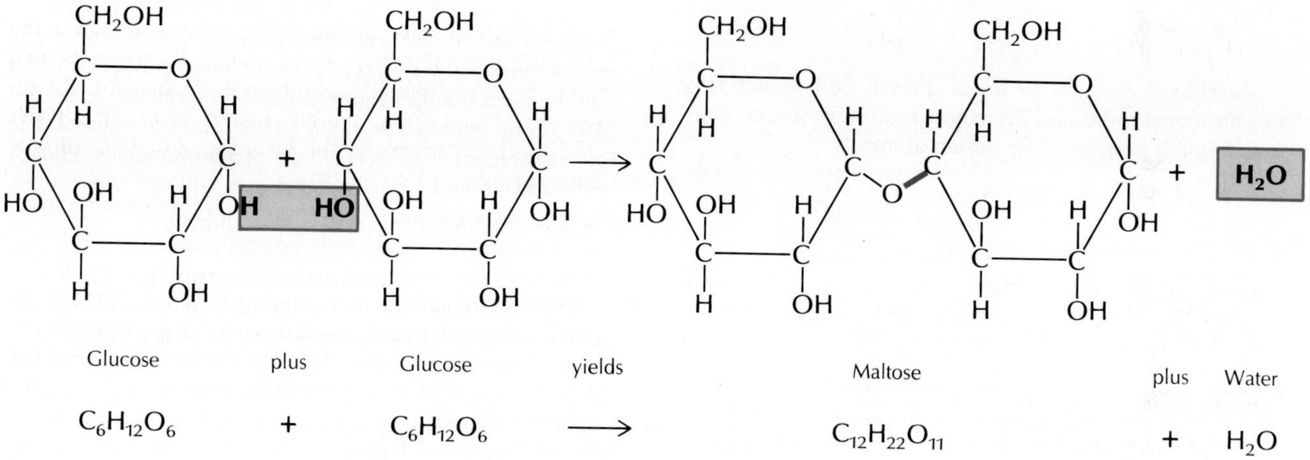

FIGURE 2.10

A dehydration synthesis. When a molecule of water (shown in blue on the left) is removed during the reaction, a bond (shown in red) is formed between the two monosaccharide molecules, producing a single molecule of the disaccharide maltose and a molecule of water (shown in blue on the right).

Glucose	plus	Glucose	yields	Maltose	plus	Water
$C_6H_{12}O_6$	$+$	$C_6H_{12}O_6$	$\longrightarrow$	$C_{12}H_{22}O_{11}$	$+$	H_2O

TABLE 2.7 SOME USES OF CARBOHYDRATES BY THE BODY

Name	Type	Description
Deoxyribose	Monosaccharide	Constituent of hereditary material, DNA.
Fructose	Monosaccharide	Important in cellular metabolism of carbohydrates. Found in fruits, it is the sweetest sugar.
Galactose	Monosaccharide	Present in brain and nerve tissue.
Glucose (dextrose)	Monosaccharide	Main energy source for the body. Breakdown of glucose produces ATP, which is used in almost all energy-requiring cellular activities. Used in intravenous feeding; does not have to be digested to be used by the body. The brain requires a constant supply.
Ribose	Monosaccharide	Constituent of RNA.
Lactose	Disaccharide	Milk sugar. Aids absorption of calcium.
Sucrose (saccharose)	Disaccharide	Common cane or beet sugar. Yields glucose and fructose upon hydrolysis.
Cellulose	Polysaccharide	Undigestible by body but serves as important fiber that provides bulk for the proper movement of food through the intestines.
Glycogen	Polysaccharide	Main form of carbohydrate storage; stored in the liver and muscles until needed as energy source and converted to glucose.
Heparin	Polysaccharide	Prevents excessive blood clotting.
Starch	Polysaccharide	Chief food carbohydrate in human nutrition.

Hydroxyl and carboxyl groups make the formation of fat possible. Each hydroxyl group may bind a fatty-acid carboxyl group. A molecule of fat is formed when the three hydroxyl (OH) groups in a molecule of glycerol combine with the carboxyl (COOH) portion of three fatty-acid molecules (see Figure 2.11). This reaction is a dehydration synthesis, similar to the dehydration reaction that produces disaccharides from the simpler monosaccharides. The reaction is reversible, and during digestion, fats are broken down into their glycerol/fatty-acid building blocks by hydrolysis, the addition of water.

Apparently, there is a connection between dietary fats and diseases of the heart and circulatory system. Dietary fats include both saturated and unsaturated fatty acids. *Saturated fatty acids* are solid at room temperature. They have no double bonds between their carbon atoms:

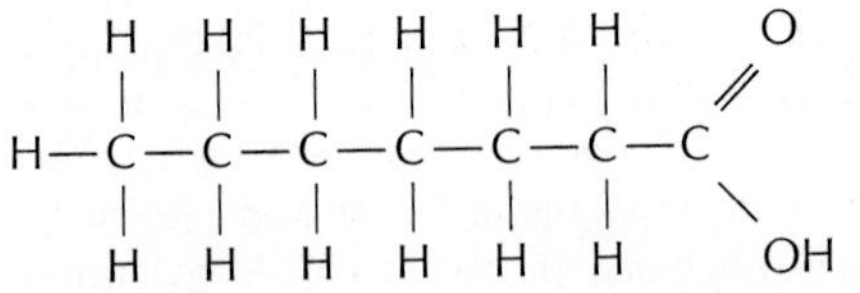

Since *unsaturated fats* are liquid at room temperature, they are considered to be *oils*. Unsaturated fats have one or more double bonds between their carbon atoms:

FIGURE 2.11

The dehydration synthesis of a fat. Three water molecules (shown in blue on the left) are removed from a glycerol molecule and three fatty-acid molecules. The $(CH_2)_{14}$ represents the 14 CH_2 units in each fatty-acid molecule. The bonding (shown in red) that results from this dehydration synthesis yields a fat (triacylglycerol) and three molecules of water (shown in blue on the right). The water molecules are produced by the removal of hydrogen atoms (H^+) from the glycerol molecule and hydroxyl (OH) groups from the fatty-acid molecules.

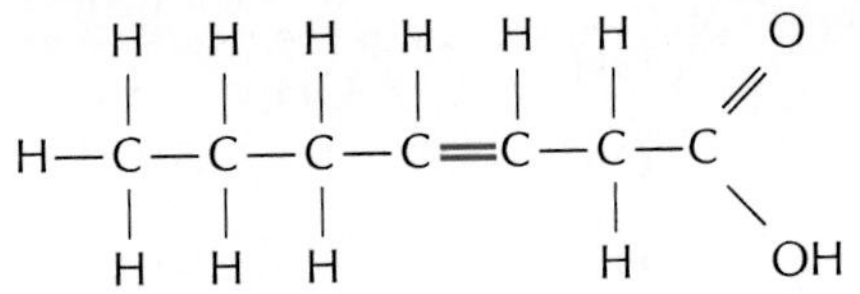

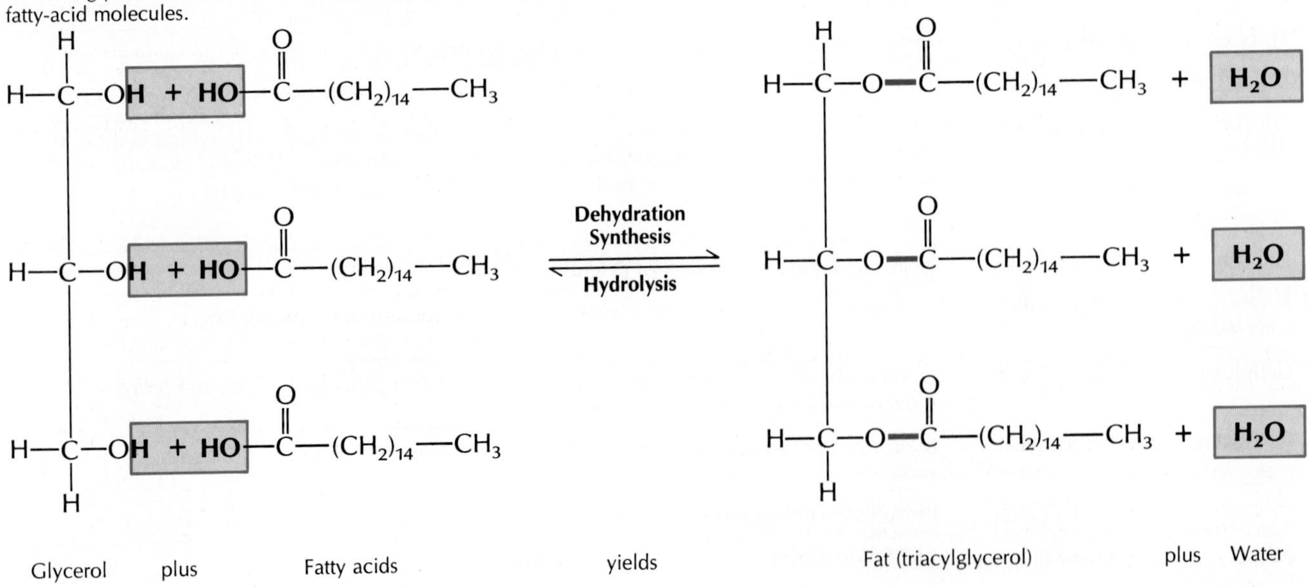

As the number of double bonds increases, the fats become more unsaturated, or oily. For this reason, unsaturated vegetable oils, which have many double bonds, are referred to as *poly*unsaturated fats. Animal fats generally contain more saturated fatty acids than vegetable fats do. Both saturated fats and cholesterol (an ingredient of fats) are thought to be involved in various cardiovascular diseases.

Phospholipids *Phospholipids* are fatty compounds that contain alcohol, fatty acids, and a phosphate group (PO_4^{3-}) that is often linked with a nitrogen-containing group. The phosphate group is bonded to the glycerol at the point where a third fatty acid would be in a neutral fat:

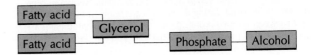

The polar end of a phospholipid molecule attracts water and thus is soluble in water, while the nonpolar end repels water and thus is insoluble (Figure 2.12). This tendency to be soluble at one end and insoluble at the other is important in cell membranes, which are built of phospholipids, cholesterol, and proteins. In Chapter 3 you will see further how phospholipids give membranes their distinct sheetlike structure and many of their functional properties.

Steroids *Steroids* are naturally occurring, fat-soluble compounds composed of four bonded carbon rings (Figure 2.13). Three of the rings are six-sided, and the fourth is five-sided. There are a total of 17 carbons in the four rings. Among the important steroids in the human body are cholesterol, bile salts, the male and female sex hormones, and some of the hormones secreted by the adrenal glands. Steroid hormones help regulate certain phases of metabolism in the body.

Glycerol   plus   Fatty acids   yields   Fat (triacylglycerol)   plus   Water

Some Important Organic Compounds

FIGURE 2.12

A simplified drawing of a phospholipid. The water-attracting polar end is shown by a sphere, and the water-repelling nonpolar end, in the form of hydrocarbon "tails," is shown by wavy lines.

Polar end (hydrophilic)

Hydrocarbon tails

Nonpolar end (hydrophobic)

FIGURE 2.13

Cholesterol, an example of a steroid, formed with four interlocking carbon rings. Note that one of the rings is five-sided, while the other three are six-sided.

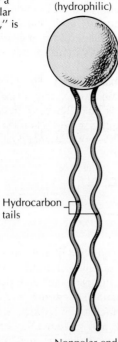

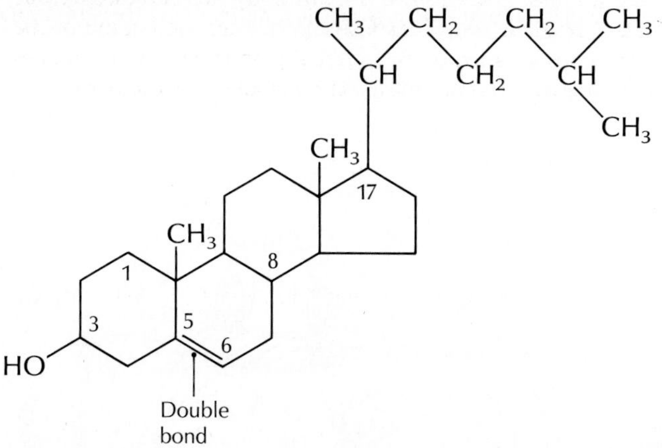

Double bond

Steroids that have a hydroxyl group are referred to as *sterols*. The sterol cholesterol is an important component of cell membranes (see Figure 2.13).

Prostaglandins *Prostaglandins* are fatty acids found in many kinds of tissues throughout the body (see page 499). Their varied uses are still being discovered, but among other things they can raise or lower blood pressure and cause the contraction of smooth muscle in the uterus.

Proteins

Proteins are extremely important to the structure and function of the body. They are the components of most of the tissues of the body. Like carbohydrates and lipids, proteins are constructed from relatively simple building blocks, but their structural possibilities are almost limitless. Proteins always contain carbon, hydrogen, oxygen, and nitrogen, and sometimes sulfur, phosphorus, and iron. Proteins form enzymes that control chemical activity, protect us against disease in the form of antibodies, carry oxygen throughout the body in hemoglobin molecules, make up several hormones, act as buffers, help blood to clot properly, and express genetic information.

Amino acids: building blocks of proteins Proteins are large, complex molecules composed of smaller structural subunits called **amino acids.** Amino acids always contain an amino group (NH_2) at one end of the molecule and a carboxyl group (COOH) at the other end (Figure 2.14). Each amino acid has one variable component, shown as R in Figure 2.14. This variable group contains an atom or group of atoms in different chemical combinations. The simplest amino acid, glycine, has a single hydrogen atom in the R position. A more complex amino acid, tyrosine, has the following structure in the R position:

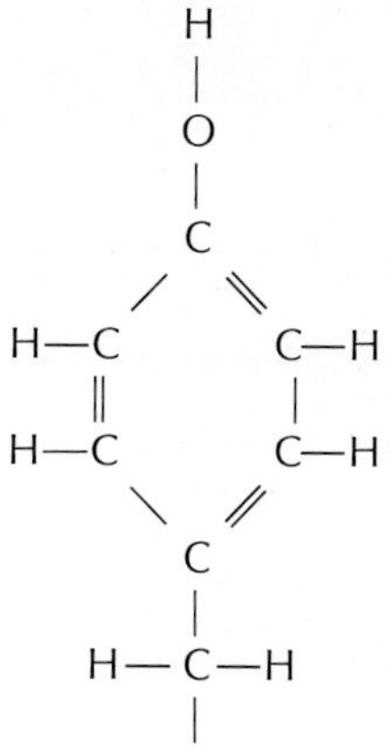

Only 22 amino acids are known in living systems, but they can combine in many different ways to produce all of the 50,000 or so proteins in the body. (In the same way, the 26 letters—amino acids—in our alphabet can be arranged to form all the words—proteins—in the dictionary.)

FIGURE 2.14

The structure of an amino acid. (A) The general structure of an amino acid, with R representing a variable group of atoms. (B) The same general amino acid structure, showing the amino group, variable group, and carboxyl group.

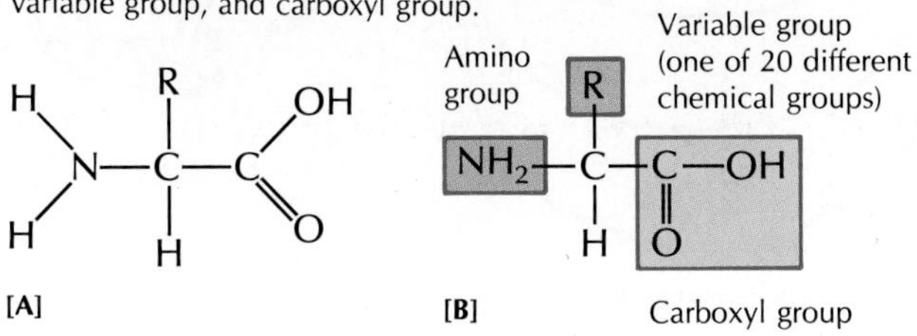

Amino group

Variable group (one of 20 different chemical groups)

Carboxyl group

[A]   [B]

FIGURE 2.15

A protein-formation reaction. A peptide bond (red) is formed at the point where the OH⁻ and H⁺ (blue, left) are removed, and a molecule of water (blue, right) is produced along with the dipeptide molecule. Note that there is still a carboxyl group on the right-hand end of the new dipeptide molecule and an amino group on the left end.

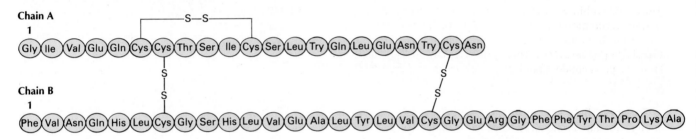

Amino acid   plus   Amino acid   produces   Dipeptide   plus   Water

FIGURE 2.16

The four organizational levels of proteins. (A) The *primary structure* of the two polypeptide chains—linked by disulfide bonds (—S—S—)—that make up the human insulin molecule. The abbreviations (for example, Gly for glycine) refer to individual amino acids. (B) The *secondary* (alpha-helical) *structure* of a portion of a protein, showing the hydrogen bonds between amino acids. (C) The coiled and folded *tertiary structure* of a globin molecule, made up of a chain of amino acids. (D) The *quaternary structure* of human hemoglobin, made up of four tertiary-structure molecules.

Chain A

1

Gly–Ile–Val–Glu–Gln–Cys–Cys–Thr–Ser–Ile–Cys–Ser–Leu–Try–Gln–Leu–Glu–Asn–Try–Cys–Asn

Chain B

1

Phe–Val–Asn–Gln–His–Leu–Cys–Gly–Ser–His–Leu–Val–Glu–Ala–Leu–Tyr–Leu–Val–Cys–Gly–Glu–Arg–Gly–Phe–Phe–Tyr–Thr–Pro–Lys–Ala

[A]   PRIMARY STRUCTURE

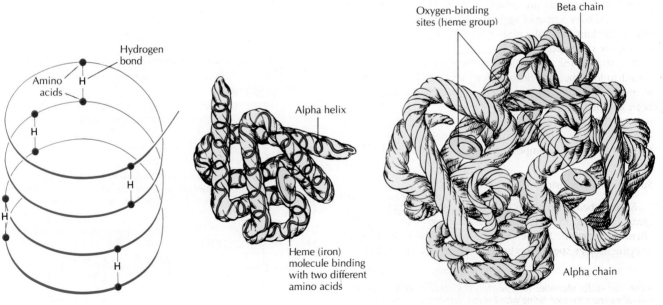

[B]   SECONDARY STRUCTURE   **[C]   TERTIARY STRUCTURE**   **[D]   QUATERNARY STRUCTURE**

Protein formation When two amino acids are bound together, they form a protein unit called a *dipeptide;* three amino acids bonded together make a *tripeptide.* When many amino acids bond together, the protein unit they form is a chain called a *polypeptide.* A protein usually consists of one or more polypeptide chains, folded into a complex, three-dimensional shape.

Figure 2.15 shows the condensation of two amino acids into a single dipeptide molecule. This reaction takes place when the C in the carboxyl (COOH) group of one amino acid bonds to the N in the amino (NH_2) group of another amino acid. In this now familiar dehydration synthesis, a water molecule splits out, and a dipeptide is formed. Such a union between amino acids results in the formation of a strong covalent bond called a ***peptide bond.***

Levels of protein structure Within a protein molecule, several different levels of structure can be distinguished. The *primary structure* is the linear sequence of amino acids in the polypeptide chains comprising the protein molecule. The sequence of amino acids determines the overall three-dimensional shape of the molecule, which in turn determines how the protein functions. Figure 2.16A illustrates the primary structure of the two polypeptide chains that make up the human insulin molecule. The way in which the chains are arranged in space is the result of rotation around the hydrogen bonds within each chain of amino acids. This *secondary structure* is commonly a helix (Figure 2.16B). The *tertiary structure* results from the folding of the helix into an overall globular shape (Figure 2.16C). Those proteins that consist of more than one polypeptide subunit form a *quaternary structure* (Figure 2.16D).

Enzymes The energy required to weaken chemical bonds enough to start a chemical reaction is called the *activation energy.* ***Enzymes*** are proteins that lower the amount of activation energy needed, increasing the rate of a chemical reaction without permanently entering into the reaction themselves. Such proteins are called *catalysts.* The same enzyme can be used over and over again to activate the same reaction. Some reactions may take place without enzymes, but only at a uselessly slow pace.

The shape—the molecular structure—of an enzyme permits it to join specific reactants (or *substrates*) together, atom to atom, and then withdraw after the chemical union has been completed (Figure 2.17). The action of some enzymes is reversible; they are thus able to drive a reaction in either direction. Under proper conditions, the same enzyme is capable of breaking a compound into its component parts:

$$\text{Substrate X + Substrate Y} \xrightleftharpoons{\text{Enzyme}} \text{Product Z}$$

Besides being capable of reversible action, enzymes are *specific.* That is, there is one specific enzyme for any given chemical reaction. In the small intestine, for example, the enzyme *lactase** speeds the breakdown of lactose (which has

*The *-ase* suffix identifies an enzyme. The prefix, *lact-* in this case, describes the substrate being acted upon. *Lact-* means milk, and *-ose* means sugar, so *lactose* (the substrate) is milk sugar.

FIGURE 2.17

The catalytic action of an enzyme on two substrate molecules, X and Y, to make a new compound, molecule Z. Enzymes may also be used to split compounds apart. In this *induced-fit model,* the shape of the enzyme changes slightly when it is in contact with the substrates, but it springs back to its original form after the action is complete. In a different model, known as the *lock-and-key model,* the enzyme is unchanged throughout the process.

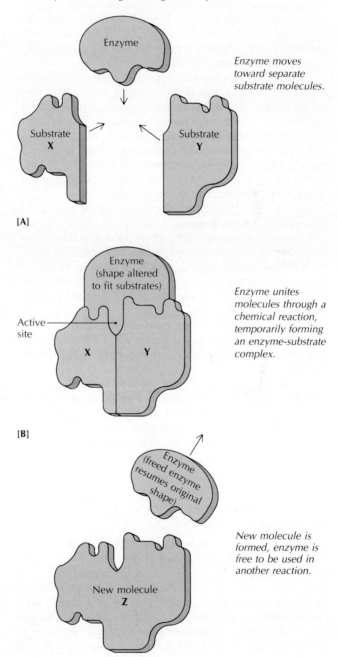

Enzyme moves toward separate substrate molecules.

[A]

Enzyme unites molecules through a chemical reaction, temporarily forming an enzyme-substrate complex.

[B]

New molecule is formed, enzyme is free to be used in another reaction.

[C]

12 carbon atoms) into two six-carbon molecules, glucose and galactose. This reaction takes place only in the presence of lactase.

An enzyme succeeds by physically binding at a specific site on the enzyme (the *active site*) the reactant molecules, actually altering their chemical bonds. Sometimes this results in breaking one molecule into two, as when lactose is converted into two monosaccharides. It may also result in bringing two compounds together, as when two amino acids are combined. *Enzyme inhibitors,* such as certain poisons, may prevent normal enzyme action by clogging the space usually occupied by the reactant or reactants.

Even though all enzymes are proteins, many of them are conjugated proteins, that is, they contain nonprotein components called *cofactors* that are bound tightly to enzymes and allow enzymes to function properly. Cofactors may be inorganic (metals), organic (*coenzymes*), or both. Many vitamins in the diet are converted into coenzymes in the body, which explains why vitamins are so important in the diet. Cofactors, like enzymes, are not changed or used up by a chemical reaction, and they can be used many times.

Several environmental requirements must be met if enzymes are to operate properly:

1 The temperature must remain within certain limits. Human enzymes generally function at optimum efficiency when the body temperature is 37°C (98.6°F). Cold slows down the reaction rate. When temperatures are too high, the hydrogen bonds of the enzyme are broken, and the enzyme becomes useless.

2 The pH of the cell environment affects an enzyme's rate of action. Each enzyme has an *optimal pH* at which it works best. Most enzymes operate effectively at pH levels between 5 and 9.

3 A sufficient amount of substrate must be present.

4 A certain amount of energy is required. Enzymes tend to lower the level of activation energy needed, and allow energy-rich molecules such as glucose to supply usable energy slowly. Thus, all of the available energy does not dissipate as heat all at once.

Nucleic Acids

Like proteins, nucleic acids are very large, complex molecules. The Swiss physician Friedrich Miescher first isolated them in 1869 from the nuclei of human pus cells. He called them nucleic acids because of their acidity and their location in the cell's nucleus. There are two types of nucleic acids: **deoxyribonucleic acid (DNA)** and **ribonucleic acid (RNA).** DNA makes up the chromosomes within the cell's nucleus and is the main repository for the genetic information of the cell. RNA may be present in either the nucleus or the cytoplasm, the part of the cell outside the nucleus.

The DNA in a cell determines which proteins will be synthesized. The proteins (all enzymes are proteins) determine the shape and structure of the cell as well as what functions the cell carries on. Thus, DNA controls the cell's activities by determining which proteins are synthesized (Figure 2.18).

FIGURE 2.18

General scheme by which DNA controls cellular functions.

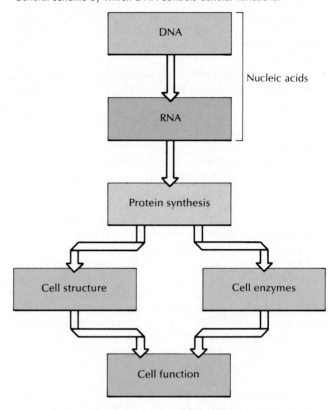

FIGURE 2.19

The structure of a nucleotide. Each nucleotide contains a phosphate group, a sugar, and a nitrogenous base; the nitrogenous base is covalently bonded to the sugar. Nucleotides bond together to form the nucleic acids RNA and DNA.

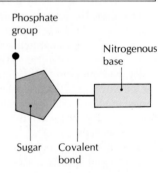

Structure of nucleic acids Nucleic acids are the largest organic molecules in the body. Their size results from the bonding together of small units called **nucleotides.** Each nucleotide has three components: a phosphate group, a five-carbon sugar, and a nitrogenous *base* (Figure 2.19). The nitrogenous base is bonded to the sugar by a covalent bond. DNA and RNA each have their own specific sugar. RNA contains ribose, a sugar that has five atoms of oxygen. Because DNA contains a sugar with one fewer oxygen atom, it is called *deoxy*ribose.

DNA contains four kinds of nitrogenous bases: the double-ring *purines* (adenine and guanine) and the single-ring *pyrimidines* (thymine and cytosine). Adenine, guanine, thymine, and cytosine are usually referred to as A, G, T, and C (Figure

Some Important Organic Compounds

FIGURE 2.20

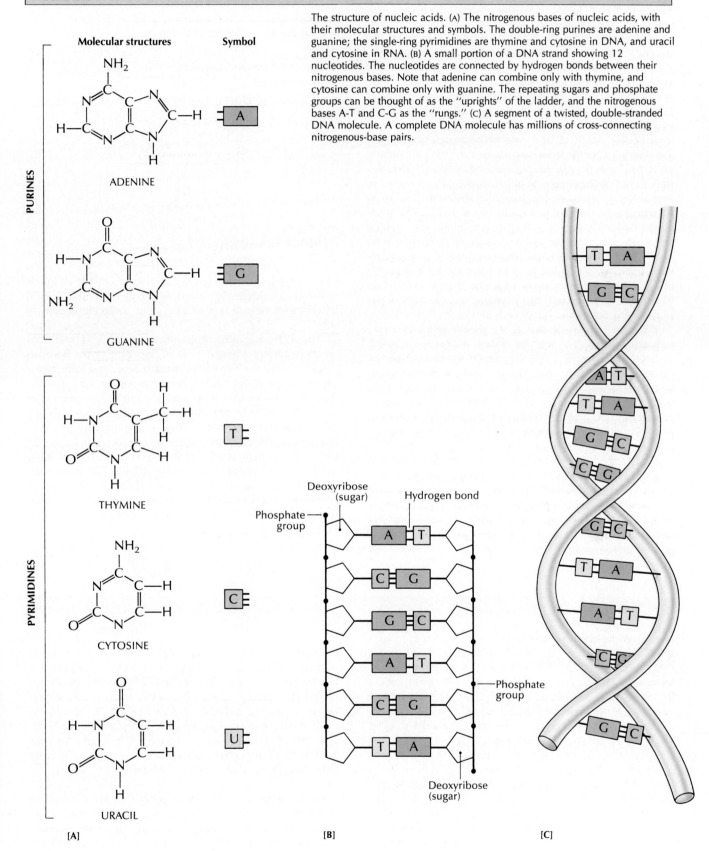

The structure of nucleic acids. (A) The nitrogenous bases of nucleic acids, with their molecular structures and symbols. The double-ring purines are adenine and guanine; the single-ring pyrimidines are thymine and cytosine in DNA, and uracil and cytosine in RNA. (B) A small portion of a DNA strand showing 12 nucleotides. The nucleotides are connected by hydrogen bonds between their nitrogenous bases. Note that adenine can combine only with thymine, and cytosine can combine only with guanine. The repeating sugars and phosphate groups can be thought of as the "uprights" of the ladder, and the nitrogenous bases A-T and C-G as the "rungs." (C) A segment of a twisted, double-stranded DNA molecule. A complete DNA molecule has millions of cross-connecting nitrogenous-base pairs.

Molecular structures   Symbol

PURINES

ADENINE

GUANINE

PYRIMIDINES

THYMINE

CYTOSINE

URACIL

[A]

Phosphate group — Deoxyribose (sugar) — Hydrogen bond

Phosphate group

Deoxyribose (sugar)

[B]

[C]

2.20A). An additional nitrogenous base, uracil (U), is also shown in Figure 2.20A; RNA contains uracil instead of thymine. In both double-stranded DNA and single-stranded RNA, the nucleotides are connected by hydrogen bonds between the nitrogenous bases (Figure 2.20B).

The components of DNA have been known for almost 100 years, but the *structure* of a DNA molecule was not discovered until 1953, when it was proposed by James D. Watson and Francis H. C. Crick. DNA has the shape of a twisted ladder, or *double helix* (Figure 2.20C).

A double strand of DNA may be compared to a ladder. The "uprights" of the ladder are made of chains of alternating phosphate groups and deoxyribose sugars. The ladder also has "rungs" made of pairs of nitrogenous bases connected to each other by hydrogen bonds and bonded to the sugars on the uprights by covalent bonds (see Figure 2.20B). The nitrogenous bases can join only in specific relationships: adenine and thymine can form bonds, while guanine can form a bond only with cytosine. In RNA, uracil, instead of thymine, always bonds with adenine. In both DNA and RNA, a purine always binds to a pyrimidine. Thus, the "rungs" are always of equal length. Overall, the sequence of base pairs determines the genetic make-up of each person.

Before we leave this section, we should define two very important words: *gene* and *chromosome.* A ***gene*** is a segment of DNA that controls a specific cellular function, either by determining which proteins will be synthesized or by regulating the action of other genes. Genes are the hereditary units that carry hereditary traits. A ***chromosome*** is a nucleoprotein (nucleic acid + protein) structure in the nucleus of a cell that contains the genes.

Functions of nucleic acids Nucleic acids determine and regulate the formation of proteins in cells. They also carry the hereditary information that makes it possible to transmit and maintain genetic information within the body so that as new cells are formed, they have exactly the same structure and function as their parent cells. Hereditary information can also be passed on from one generation to the next through nucleic acids.

Ask Yourself

1 *What are the components of a carbohydrate?*

2 *How does a monosaccharide differ from a polysaccharide?*

3 *What are some familiar kinds of lipids?*

4 *What are the structural subunits of proteins?*

5 *What are the two types of nucleic acids?*

ENERGY FOR LIVING: ATP

The energy you need to live comes from chemicals and their reactions. The immediate source of energy for most biological activities in the cell is a small organic molecule known as ATP.

The ***ATP*** (adenosine triphosphate) molecule is built from smaller units of adenine (one of the nitrogenous bases in nucleic acids), the five-carbon sugar ribose, and three phosphate groups (Figure 2.21). Notice in Figure 2.21 that the three phosphates that make up the triphosphate portion of the ATP molecule are connected by high-energy bonds (shown as wavy red lines). These bonds are capable of releasing much more stored energy than the bond that connects the first phosphate to ribose (shown as a straight red line). When energy is needed for cellular activity, ATP reacts with water (hydrolysis), and the last of the three phosphate bonds is broken. The last phosphate group is separated, yielding adenosine *di*phosphate (ADP), inorganic phosphate (P_i), and energy:

$$\text{ATP} \rightleftharpoons \text{ADP} + P_i + \text{Energy}$$

FIGURE 2.21

A structural diagram of an ATP (adenosine triphosphate) molecule. Note the high-energy bonds (shown as wavy red lines) between the phosphate groups.

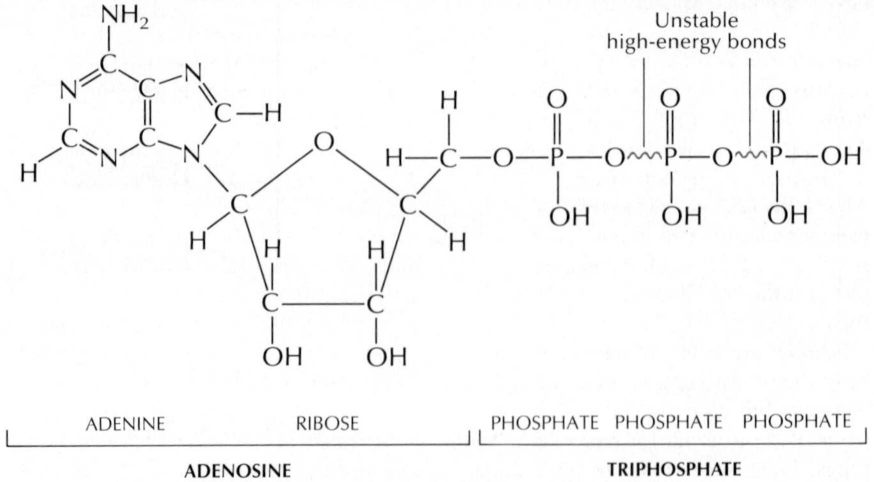

Most of the energy released from ATP can be used for the cell's immediate needs.

Note that the chemical reaction above requires an enzyme and is reversible. Converting ADP into ATP is a way of storing energy until it is needed. Although ATP stores energy, ATP itself is not produced and stored in ever-growing quantities. It is replaced as it is needed by the energy from decomposition reactions that take place within cells. This energy is combined with ADP and P$_i$ to synthesize additional ATP. The synthesis of ATP is an example of an *endergonic* ("energy in") reaction because energy is *required* for the reaction to take place. Activities that *liberate* energy, such as the breaking of ATP's high-energy bonds, are called *exergonic* ("energy out").

It should be understood that even though the bonds between the last two phosphate groups of an ATP molecule are usually referred to as being "energy-rich," the ATP molecule actually contains no more energy than many other organic molecules. ATP is a ready supplier of energy because it is a relatively unstable compound. What is most important about

ATP is the ready *availability* and *transferability* of the energy stored in its bonds. Indeed, a molecule of sucrose (ordinary table sugar) has much more potential energy than an ATP molecule does, but cells cannot use the energy from sucrose directly. They can, however, readily convert the energy in ATP bonds into such useful forms as movement, heat, or chemical reactions.

Body cells need a slow, steady energy source, especially one that can be controlled and regulated, since the chemical reactions in body cells can use energy only in small packages. The readily available energy in ATP provides cells with an effective way of regulating energy release.

Ask Yourself

1 *What is the main function of ATP?*

2 *How does ATP store and release energy?*

SUMMARY

Matter, elements, and atoms

1 *Matter* is anything that occupies space and has mass. *Mass* refers to the amount of matter in an object. Matter is composed of *elements,* basic substances that cannot be broken down into simpler substances by ordinary chemical means. An *atom* is the smallest unit of an element that retains the chemical characteristics of that element.

2 Some of the elements critical to proper body function are oxygen, carbon, hydrogen, nitrogen, calcium, and phosphorus.

3 The two basic parts of an atom are the nucleus and its surrounding electron field. The *nucleus* contains electrically charged (positive) *protons* and uncharged *neutrons.* Orbiting around the nucleus are negatively charged *electrons.*

4 An atom's *atomic number* indicates the number of protons in its nucleus. *Atomic weight* is an expression of an element's weight compared to carbon. It is calculated by adding the number of protons and the number of neutrons in the nucleus.

5 *Isotopes* are atoms of an element that have the typical number of protons and electrons but different numbers of neutrons. Radioisotopes (or radioactive isotopes) have nuclei that are undergoing

changes called "decay." Radioisotopes are used in biology and medicine.

6 The electrons of atoms are distributed around the nucleus in layers called *energy-level shells.*

7 The number of electrons in the outermost shell determines how one atom will react with another atom. Atoms that have a complete outer shell have a *stable* shell.

How atoms combine

1 Atoms of elements combine chemically to form *molecules; compounds* are composed of molecules with more than one kind of element. When atoms interact chemically to form molecules or compounds, the atoms are held together by electrically attractive forces called *chemical bonds.* Once an atom or molecule gains or loses electrons, it acquires an electrical charge and is called an *ion.*

2 *Ionic bonds* are formed when an atom or group of atoms develops an electrical charge and then becomes attracted to an atom or group of atoms with an opposite electrical charge.

3 When atoms gain stability by *sharing* one or more pairs of electrons with other atoms, their resultant chemical bond is called a *covalent bond.*

4 *Hydrogen bonds* are formed when the hydrogen atom already covalently

bonded in a compound acquires a slight positive charge and becomes attracted to negatively charged atoms nearby.

Chemical reactions

1 The combining or breaking apart of two or more atoms that brings about a chemical change is called a *chemical reaction.*

2 When two or more atoms, ions, or molecules combine to form a more complex substance, the process is called a *combination reaction.*

3 *Decomposition reactions* break chemical bonds in molecules to form two or more products.

4 *Replacement reactions* occur when atoms, ions, or molecules trade places with other atoms, ions, or molecules to produce new arrangements of atoms, ions, or molecules.

5 *Oxidation* occurs when an atom or molecule *loses* electrons or hydrogen atoms. When an atom or molecule *gains* electrons or hydrogen atoms, the process is called *reduction.*

6 When *hydrolysis* takes place, a molecule of water interacts with the reactant to break up the reactant's bonds. The original reactant is then rearranged, together with the water, into different molecules. *Condensation reactions* are hydrolytic reactions in reverse.

Water

1 *Inorganic compounds* are composed of relatively small molecules that are usually ionically bonded. The major example of an inorganic compound is water.

2 *Water* performs many important functions because it is a *polar molecule;* each end of the molecule bears a different electrical charge. Water is a solvent, transporter, temperature regulator, and lubricant.

Acids, bases, salts, and buffers

1 An *acid* is a substance that releases hydrogen ions when dissolved in water. A *base* is a substance that releases hydroxyl ions when dissolved in water.

2 A *salt* is the compound (other than water) formed during a neutralization reaction between acids and bases.

3 The tendency for molecules to break up into charged ions in water is called *dissociation.*

4 The degree of acidity or alkalinity of a solution is measured with the *pH scale.* This scale indicates the concentration of free hydrogen ions in water.

5 *Buffers* are chemical substances that regulate the body's acid-base balance and maintain pH homeostasis.

Some important organic compounds

1 *Organic* compounds always contain carbon and hydrogen. They are typically large and complex, and are held together by covalent bonds.

2 *Carbohydrates* are made of carbon, hydrogen, and oxygen, with hydrogen and oxygen in a 2:1 ratio; they are an important source of energy. Carbohydrates are classified as *monosaccharides, disaccharides,* or *polysaccharides,* depending upon how many monosaccharides they contain.

3 The type of chemical reaction in which a molecule of water is lost when large organic molecules are synthesized from simpler organic molecules is called condensation, or *dehydration synthesis.* The reverse reaction is hydrolysis.

4 Body fats belong to a diverse group of organic compounds called *lipids.* Fats and oils are familiar lipids. *Phospholipids, steroids,* and *prostaglandins* are less well known but equally important.

5 *Proteins* are large, complex molecules composed of smaller structural subunits called *amino acids.* The amino acids are linked together in chains called *polypeptides.* A protein consists of one or more polypeptide chains.

6 *Enzymes* are proteins that increase the rate of a chemical reaction without permanently entering into the reaction themselves.

7 *Nucleic acids* are very large molecules composed of bonded units called *nucleotides.* The two nucleic acids are *deoxyribonucleic acid (DNA)* and *ribonucleic acid (RNA).* Nucleic acids carry the body's hereditary messages and regulate the synthesis of proteins.

Energy for living: ATP

1 The energy for most biological activities comes from *ATP* (adenosine triphosphate), which is composed of smaller units of adenine, ribose, and three phosphate groups.

2 When the high-energy bonds in ATP are broken, they release stored energy that is usable for a cell's immediate needs.

UNDERSTANDING THE FACTS

1 Define matter.

2 Define atom, element, and isotope.

3 What are the differences among ionic, covalent, and hydrogen bonds?

4 How does water act as a solvent, transporter, temperature regulator, and lubricant?

5 What is an acid? A base?

6 What is meant by the term pH?

7 What is a carbohydrate? State its primary function.

8 What is the difference between a saturated and an unsaturated fat?

9 What is an amino acid?

10 How do enzymes affect chemical reactions?

11 What is DNA? RNA?

12 Define ATP and describe how it is used by cells.

UNDERSTANDING THE CONCEPTS

1 Distinguish between atomic number and atomic weight.

2 Why do electrons exist in different energy-level shells?

3 How do atoms combine?

4 Describe hydrolysis and condensation.

5 How is pH controlled with buffers?

6 Explain how enzymes function.

7 How do nucleic acids control cellular function?

SUGGESTED READING

BAKER, JEFFREY W., AND GARLAND E. ALLEN, *Matter, Energy, and Life: An Introduction to Chemical Concepts,* 4th ed. Reading, Mass.: Addison-Wesley, 1981 (paperback).

BLOOMFIELD, MOLLY M., *Chemistry and the Living Organism,* 3rd ed. New York: Wiley, 1984.

CARAFOLI, ERNESTO, AND JOHN T. PENNISTON, "The Calcium Signal." *Scientific American,* November 1985.

DOOLITTLE, RUSSELL F., "Proteins." *Scientific American,* October 1985.

HINKLE, PETER C., AND RICHARD E. MCCARTY, "How Cells Make ATP." *Scientific American,* March 1978.

LEHNINGER, ALBERT L., *Principles of Biochemistry.* New York: Worth, 1982.

MOROWITZ, HAROLD J., *Energy Flow in Biology.* Woodbridge, Conn.: Ox Bow, 1979.

PHILLIPS, DAVID C., "The Three-Dimensional Structure of an Enzyme Molecule." *Scientific American,* November 1966.

SHARON, NATHAN, "Carbohydrates." *Scientific American,* November 1980.

SHERMAN, ALAN, SHARON SHERMAN, AND LEONARD RUSSIKOFF, *Basic Concepts of Chemistry,* 3rd ed. Boston: Houghton Mifflin, 1984.

3
Cells: The Basic Units of Life

LEARNING OBJECTIVES

1 State the four parts of the cell theory.

2 Describe, locate, and list the general functions of the principal structures of a cell.

3 Diagram the fluid-mosaic model of membrane structure.

4 Describe the different types of passive-transport mechanisms that operate to move materials through cell membranes.

5 Explain some of the unique characteristics of the active-transport process.

6 Compare endocytosis and exocytosis.

7 Explain the structure and function of the endoplasmic reticulum.

8 Briefly explain the purpose of ribosomes.

9 Tell how the Golgi apparatus functions.

10 Explain how lysosomes and microbodies function.

11 Explain the role of mitochondria in supplying energy to the cell.

12 Describe the function of the centrioles.

13 Compare and contrast cilia and flagella.

14 List several examples of cytoplasmic inclusions.

15 Describe a chromosome.

16 Explain the function of the nucleolus.

17 Describe DNA replication.

18 Discuss the events associated with each stage of mitosis.

19 Explain the significance of cytokinesis.

20 Tell how proteins are synthesized.

We all begin life when a single sperm unites with a single egg—a fusion of two highly specialized cells. The resulting fertilized egg is the progenitor of all future cells in the body. The adult human body consists of more than 50 trillion (50 million million) cells. Most of these cells are specialized. They have different structures to perform different functions. No matter what its specific function, however, each cell is capable of carrying on such life-giving activities as digesting complex food molecules, replicating itself, building up chains of polypeptides, generating energy-rich ATP, engulfing foreign particles, creating new parts, or expelling old ones. And each cell works together with other cells to provide an environment that is compatible with all the processes of life.

WHAT ARE CELLS?

As you saw in Chapter 1, cells are the smallest independent units of life. About 150 years ago, the basic principles of the **cell theory** were first formulated by Theodor Schwann and Matthias Jakob Schleiden, and later formalized by Rudolf Virchow in 1855. Although these principles have been revised and updated since then, the four most important points remain the same:

1 All living things are composed of cells and cell products.

2 The cell is the basic unit of structure and function of all living things.

3 All cells come from the division of pre-existing cells.

4 An organism as a whole can be understood through the collective activities and interactions of its cells.

Cells vary enormously. They can range in size from microscopically tiny (a millionth of a meter for some bacteria) to very long (almost 5 m long for some giraffe nerve cells). The largest cells are ostrich eggs, which are about the size of a grapefruit. Your liver cells are more typical. Several thousand liver cells would fit into a cube smaller than the small letters in this sentence.

Your body cells have different sizes, shapes, and colors. These differences are related to the cells' functions. Early cell physiologists thought that the interior of cells consisted of a homogeneous fluid, which they called **protoplasm.** Today, the word protoplasm is used only in a very general way, and scientists divide cells into four basic parts:

1 The **nucleus,** the control center of the cell. It includes the chromosomes, which contain the genes that direct the reproduction and heredity of new cells. The nucleus is a clearly defined body that is separated from the surrounding cytoplasm by its own nuclear envelope, which is composed of an inner and outer membrane.

2 **Nucleoplasm,** the material within the nucleus.

3 **Cytoplasm,** the portion of the cell outside the nucleus. Metabolic reactions take place here. The fluid portion of the cytoplasm is called *cytosol.* Suspended within the cytosol and

nucleoplasm are various subcellular structures called *organelles* ("little organs"). Most organelles are bound by membranes, and have a specific structure and function. Some organelles are more prominent than others in certain cells.

4 The **plasma membrane,** the outer boundary of the cell.

Since cells vary so much in form and function, no "typical" cell exists. But to help you learn as much as possible about cells, Figure 3.1 shows an idealized version of a cell and all its parts. See also Table 3.1 for a summary of cell structures and their functions.

Ask Yourself

1 *What are the four main points of the cell theory?*

2 *What are the four basic parts of a cell?*

3 *How do nucleoplasm and cytoplasm differ?*

4 *What is cytosol?*

CELL MEMBRANES

The thin double membrane that forms the *outermost* layer of a cell is called the **plasma membrane.** Far from being a simple barrier between the inside and outside of a cell, the plasma membrane is a *selective* screen that allows only certain substances to enter and leave the cell. The plasma membrane maintains the boundary and integrity of the cell itself by keeping the cell and its contents separate and distinct from the surrounding environment. Due to the presence of specific molecules (receptors) on its surface, the plasma membrane is able to interact with other cells and the external environment.

Membranes inside the cell that enclose some organelles (and actually make up other organelles such as the endoplasmic reticulum) are similar in properties to the plasma membrane. They also help regulate cellular activities (see the sections on specific organelles).

Structure of Cell Membranes

Only in the past few decades have scientists begun to understand membrane structure and function. During that time, several descriptions (*models*) have been proposed to explain the microscopic structure of cell membranes, and it is now generally accepted that they are composed of a double phospholipid layer in which protein molecules are embedded. Some protein molecules are on the outside of the membrane, some are on the interior side, and others extend completely through both phospholipid layers. Based on this model, it is now believed that the transport of most substances through the cell membrane takes place via protein-lined channels.

TABLE 3.1

CELL STRUCTURES AND THEIR FUNCTIONS

Component	Description	Main functions
Centrioles	Located within centrosome; contain nine triple microtubules (see Figure 3.18).	Assist in cell reproduction; form basal body of moving cilia and flagella.
Cilia, flagella	Cilia are short and threadlike; flagella are much longer (see Figure 3.19).	Cilia move fluids or particles past fixed cells; help move free-swimming cells. Flagella provide means of movement for sperm cells.
Cytoplasm	Semifluid enclosed within plasma membrane; consists of fluid cytosol and intracellular structures such as organelles (see Figure 3.1).	Dissolves soluble proteins and substances necessary for cell's metabolic activities; houses organelles, vesicles, inclusions, and lipid droplets.
Cytoplasmic inclusions	Substances temporarily in cytoplasm.	Basic food material or stored products of cell's metabolic activities; include pigments (hemoglobin, melanin), stored fats and carbohydrates, and secretions (mucus). Foreign substances such as bacteria and dust.
Cytoplasmic vesicles (endosomes)	Membranous sacs (see Figure 3.1).	Store and transport cellular materials.
Cytoskeleton	Flexible cellular framework, interconnecting microfilaments, intermediate filaments, microtubules, other organelles (see Figure 3.17).	Provides support, assists in cell movement, site for binding of specific enzymes.
Cytosol	Fluid portion of cytoplasm; enclosed within plasma membrane; surrounds nucleus.	Houses organelles; serves as transporting medium for secretions and metabolic activities.
Deoxyribonucleic acid (DNA)	Nucleic acid that makes up the chromosomes.	Controls heredity and cellular activities.
Endoplasmic reticulum (ER)	Complex membrane system extending throughout cytoplasm (see Figure 3.12).	Internal transport and storage; rough ER serves as point of attachment for ribosomes; smooth ER produces steroids.
Golgi apparatus	Flattened stacks of disklike membranes (see Figure 3.14).	Packages proteins for secretion.
Intermediate filaments	Elongated, fibrillar structures composed of protein subunits (see Figure 3.17).	Key elements of cytoskeleton in most cells.
Lysosomes	Small, membrane-bound spheres (see Figure 3.1).	Digest materials; decompose harmful particles; play a role in cell death.
Microfilaments	Solid, rodlike structures containing the protein actin (see Figure 3.17).	Provide structural support and assist cell movement.
Microtubules	Hollow, slender, cylindrical structures (see Figures 3.16, 3.17).	Support; assist movement of cilia and flagella, and chromosomes. Possible transport system.
Mitochondria	Sacs with folded, double membranes (see Figure 3.15).	Produce most of the energy (ATP) required for cell; site of cellular respiration.
Nucleolus	Rounded mass within nucleus; contains RNA and protein (see Figure 3.20).	Preassembly point for ribosomes.
Nucleus	Large spherical structure surrounded by a double membrane; contains nucleolus and DNA (see Figure 3.20).	Contains DNA that controls cell's genetic program and activities.
Peroxisomes	Membranous sacs containing oxidative enzymes.	Carry out metabolic reactions and destroy hydrogen peroxide, which is toxic to cell.
Plasma membrane	Outer bilayered boundary of cell; composed of lipids and proteins (see Figures 3.2, 3.3).	Protection; regulates passage of substances into and out of cell; cell-to-cell recognition.
Ribonucleic acid (RNA)	Three types of nucleic acid involved in transcription and translation of genetic code.	Messenger RNA (mRNA) carries genetic information from DNA; transfer RNA (tRNA) is involved in amino-acid activation during protein synthesis; ribosomal RNA (rRNA) is involved in ribosome structure.
Ribosomes	Small structures containing RNA and protein; some attached to rough ER (see Figure 3.13).	Sites of protein synthesis.

FIGURE 3.1

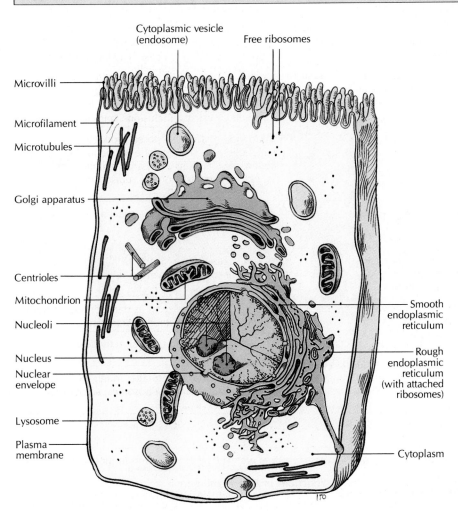

Cytoplasmic vesicle (endosome)

Free ribosomes

Microvilli

Microfilament

Microtubules

Golgi apparatus

Centrioles

Mitochondrion

Nucleoli

Nucleus

Nuclear envelope

Lysosome

Plasma membrane

Smooth endoplasmic reticulum

Rough endoplasmic reticulum (with attached ribosomes)

Cytoplasm

A highly simplified version of an "ideal" cell would look something like this. It is sliced open to show the various structures, or organelles (several of which are also sliced open), that function within the plasma membrane. The nucleus is sliced open to show the nucleoli inside. Details of the organelles are illustrated and described in the text, and the parts of a cell are also summarized in Table 3.1.

The Fluid-Mosaic Model

In 1972, two cell physiologists, S. Jonathan Singer and Garth Nicolson, formulated the *fluid-mosaic model* of membrane structure. According to this model, the membrane is a double layer (bilayer) composed mainly of phospholipids and proteins, and is fluid rather than solid. The lipid bilayer forms a fluid "sea," similar in consistency to vegetable oil, in which specific proteins float like icebergs (Figure 3.2). Being *fluid,* the membrane is in a constant state of flux, shifting and changing, yet at the same time retaining its basic structure and properties. The word *mosaic* refers to the many different proteins embedded on or within the phospholipid bilayer.

The following are the important points about the fluid-mosaic model (refer to Figures 2.12, 3.2, and 3.3):

1 *Phospholipids* are represented as looking like balloons on a string. They are arranged in a fatty double layer, consisting of sheets two molecules thick. This arrangement is called a *bimolecular layer,* or *bilayer.* The bimolecular layer resembles a sandwich, with the spherical heads as the "bread" of the sandwich and the inward-facing tails as the "meat" (Figure 3.3).

2 Phospholipids have one charged end and one uncharged end (see Figure 3.3). The charged ends can stick into the watery cytoplasm inside the cell and into the similarly watery external environment, while the uncharged ends face each other in the middle of the double layer.

3 The "tails" (the *lipid* portion) of the phospholipid molecules are attracted to each other, and are repelled by water (or are *hydrophobic*). As a result, the polar (charged) spherical "heads" (the *phosphate* portion) of the phospholipid molecules line up over the entire cell surface (the heads are *hydrophilic,* or "water-attracting"), and the fatty portions compose the center of the membrane between the two phosphate layers. The hydrophobic lipid area inhibits the movement of most water-soluble substances through cell membranes.

4 Figure 3.2 shows how the *protein* portions are embedded in the cell membranes like tiles in a mosaic or like icebergs floating in the fluid phospholipids. Some proteins protrude only into the phospholipid sandwich, some extend all the way into the cytoplasm, some extend entirely through the membrane, some are partially sunken in the outer layer, and some are partially sunken in the inner layer. All the proteins are capable of moving about in the flexible double layer of phospholipid molecules, however.

FIGURE 3.2

The fluid-mosaic model shows the double-layered membrane composed of phospholipids with polar (charged) "heads" and uncharged fatty-acid "tails." Protein and glycoprotein molecules float in the liquid membrane. (The "fluid" part of the fluid-mosaic model is the liquid membrane; the "mosaic" part is the proteins and glycoproteins.) Surface carbohydrates protrude from the glycopro-

teins and the phospholipid bilayer on the outside of the membrane. These surface carbohydrates play a key role in cell recognition, including immune reactions, and act as receptor sites for hormones such as insulin. Some glycoproteins extend completely through the membrane, while others are only partially embedded. Note the glycoprotein with a channel extending throughout its length.

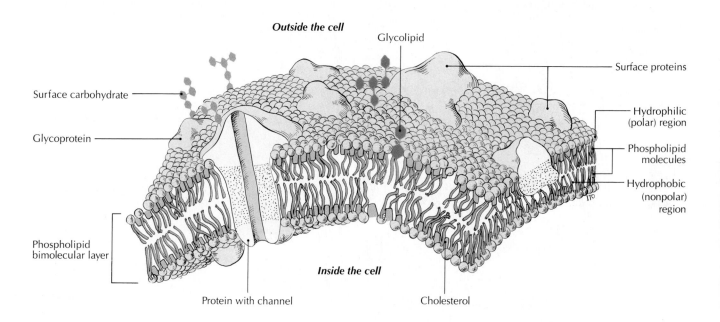

5 Protruding from some of the floating proteins are antennalike *surface carbohydrates,* most of which are small, complex, nonrepeating sequences of monosaccharides. These molecules are often linked covalently to surface proteins to form *glycoproteins,* and to surface lipids to form *glycolipids* (see Figure 3.2). Overall, these sugars and the portions of the proteins on the surface of the plasma membrane make up the *glycocalyx* (cell coat), the carbohydrate-rich peripheral zone of the cell surface. The glycoproteins and glycolipids are necessary for one cell to recognize or reject another cell. Apparently, glycoproteins of similar cells "fit together" chemically, whereas a foreign cell is not "recognized," and is repelled or even destroyed. This principle undoubtedly helps blood cells accept transfusions of the same blood type and reject blood that is not compatible. It also applies to skin grafts, organ transplants, and even mating on a cellular level. That is why sperm and eggs of different species generally cannot unite. Without compatible surface glycoproteins, cells would not form tissues, tissues would not form organs, organs would not form systems, and systems would not form your body.

6 Finally, the plasma membrane of human cells contains *cholesterol.* The flexibility (fluidity) of the membrane is determined by the ratio of cholesterol to phospholipids. The plasma membranes of human cells contain relatively large amounts of cholesterol (as much as one molecule for every phospholipid molecule), and as a result, it enhances the mechanical stability of the bilayer.

Functions of Cell Membranes

In general, cell membranes serve several important functions: (1) they separate the inside of a cell from the outside environment, (2) they separate cells from one another, (3) they separate the various parts within cells, (4) they provide an abundant surface on which chemical reactions occur, and (5) they regulate the passage of materials into and out of cells, and from one part of a cell to another.

Some of the proteins that are embedded in the cell membrane provide structural support. Other proteins are enzymes, and are involved in the many chemical reactions that take place at their particular site on the membrane. Other proteins regulate the movement of water-soluble substances through channels; some are special receptors for specific chemical signals and for specific molecules. Still other proteins play an important role in cell-to-cell recognition, and provide surface receptors for hormones and other regulatory molecules. The amount of a protein in a cell membrane may be a measure of that membrane's metabolic activity. The membranes of mitochondria, for example, contain more protein than the membranes of less active organelles.

Membranes are not like the solid walls of your house that protect the inside from the outside. Cellular membranes, including plasma membranes, are more like window screens that let the breeze in but keep the flies out. That quality of letting some things in and keeping others out is called **selec-**

FIGURE 3.3

A bimolecular layer, consisting of two phospholipid molecules. The yellow balls represent the charged, hydrophilic ("water-loving") portions of the molecules, and the orange strings the uncharged, hydrophobic ("water-hating") portions. Polar water molecules are attracted to the charged polar ends of the phospholipid molecules.

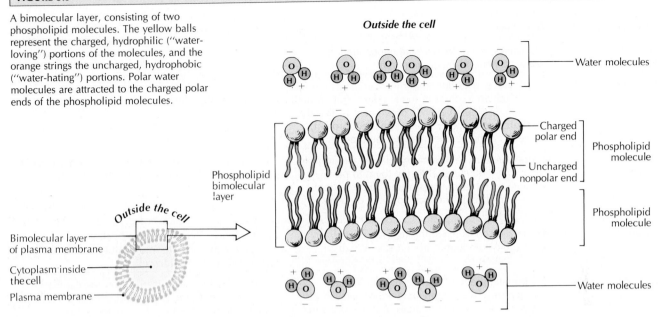

FIGURE 3.3

tive permeability (L. *permeare: per,* through + *meare,* pass). It is essential for maintaining cellular homeostasis. Some substances, such as water and fats, pass through cell membranes readily. Others, such as potassium or calcium ions, pass only with considerable difficulty.

In some cells, the plasma membrane has a greatly increased surface area. This specialization has resulted from the formation of many *microvilli* (see Figure 3.1), slender extensions of the plasma membrane that function in either the absorption or the secretion of molecules. The overall increase in surface area allows many more molecules to be involved in cellular transport. However, before we can fully understand how substances pass into and out of cells and organelles, we must understand how the molecules of those substances are able to move from one place to another. The basic types of such molecular movement are discussed in the following sections.

Ask Yourself

1 *To what do the terms fluid and mosaic refer in the fluid-mosaic model of cell membranes?*

2 *Why is the structure of a cell membrane frequently thought of as a "sandwich"?*

3 *What are some of the functions of membrane proteins?*

4 *What is a selectively permeable membrane?*

PASSIVE MOVEMENT ACROSS MEMBRANES

When molecules use their own energy to pass through a cell membrane, *without the use of cell energy,* the movement is called *passive transport.* Examples of passive transport across membranes are simple diffusion, facilitated diffusion, osmosis, filtration, and dialysis (Table 3.2).

Simple Diffusion

Unless a substance is at absolute zero, −273°C, its molecules are in constant motion, especially those in liquids and gases. Molecules tend to move randomly from areas where they are heavily concentrated to areas of lesser concentration, until they are evenly distributed in a state of *equilibrium.* This property of molecules spreading out randomly until they are evenly distributed is called **simple diffusion** (L. *diffundere,* to spread). For example, Figure 3.4A shows how evaporating perfume molecules diffuse through a room full of air molecules until the perfume molecules are evenly distributed.

The rate of diffusion depends on several factors:

1 *The state of matter of the diffusing molecules.* Gases diffuse rapidly and liquids diffuse more slowly, at a rate measurable in milliliters per week.

2 *The external temperature.* The higher the temperature, the faster the rate of diffusion.

3 *Molecular size.* Small molecules of glycerol, for example, diffuse faster than those of a large fatty acid.

TABLE 3.2		COMPARISON OF PASSIVE AND ACTIVE MOVEMENT ACROSS CELL MEMBRANES
Type of movement	**Description**	**Example in body**
PASSIVE MOVEMENT	Molecules move "down" the concentration gradient. No cell energy required.	
Simple diffusion	Molecules spread out randomly from areas of high concentration to areas of lower concentration until they are distributed evenly. Movement is directly through membrane or integral channel protein; however, a membrane need not be present.	Inhaled oxygen is transported into lungs and diffuses through lung cells into bloodstream.
Facilitated diffusion	Carrier proteins in plasma membrane temporarily bind with molecules, allowing them to pass through membrane, via protein channels, from areas of high concentration to areas of lower concentration.	Specific amino acids combine with carrier proteins to pass through plasma membrane into cell.
Osmosis	Water molecules move through selectively permeable membranes from areas of high concentration to areas of lower concentration.	Water moves into a red blood cell when concentration of water molecules outside cell is greater than it is inside cell.
Filtration	Hydrostatic pressure forces small molecules through selectively permeable membranes from areas of high pressure to areas of lower pressure.	When blood pressure is higher inside capillaries than in surrounding tissue fluid, pressure forces water and dissolved small particles out of capillaries through capillary walls. Blood pressure forces water and dissolved wastes into kidney tubules during formation of urine.
Dialysis	Small molecules move through a selectively permeable membrane, leaving larger molecules behind.	Does not occur naturally in body; process is used by artificial kidney machine (kidney dialysis).
ACTIVE MOVEMENT	Cell energy allows substances to move through selectively permeable membranes from areas of low concentration to areas of higher concentration. Requires a living cell and cell energy in the form of ATP.	
Active transport	Carrier proteins in plasma membrane bind with ions or molecules to assist them across membrane "against" the concentration gradient.	Movement of sodium ions from inside to outside cell during the conduction of nerve impulses (the sodium-potassium pump).
Endocytosis	Membrane-bound vesicles enclose large molecules, draw them into cytoplasm, and release them.	
Pinocytosis	Plasma membrane encloses small amounts of fluid droplets and takes them into cell.	Kidney cells take in tissue fluids in order to maintain fluid balance.
Receptor-mediated endocytosis	Extracellular molecules bind with specific receptor on plasma membrane, causing the membrane to invaginate and draw molecules into cell.	Intestinal epithelial cells take up large molecules.
Phagocytosis	Plasma membrane forms a pocket around a solid particle or cell and draws it into cell.	White blood cells engulf and digest harmful bacteria.
Exocytosis	Vesicle (with undigested particles) fuses with plasma membrane and expels particles from cell through plasma membrane.	Nerve cells release chemical messengers.

Certain small molecules (solutes) can move across the lipid bilayer of plasma membranes simply in response to a difference in the concentration of the molecule on either side of the membrane (that is, inside and outside the cell). This difference in solute concentration is called the *concentration gradient*. A molecule moving from an area of high concentration to one of lower concentration moves "down" the concentration gradient.

For small nonpolar molecules, diffusion occurs directly across the plasma membrane. In the human body, such diffusion occurs when air is inhaled and is rapidly transported into the small chambers of the lungs. Oxygen in the inhaled air passes from the cells of the lungs, by diffusion, into the bloodstream and is transported throughout the body. Oxygen from the bloodstream diffuses into cells, and waste carbon dioxide leaves cells and diffuses into the bloodstream, where it is transported to the lungs to be exhaled (Figure 3.4B).

For other small molecules—polar molecules not soluble in lipids—diffusion occurs not through the plasma membrane, but through an *integral channel protein* embedded in the lipid bilayer (Figure 3.4C). The mechanism for this form

of passive transport is not yet fully understood. However, it is generally accepted that the integral channel protein offers a continuous pathway for specific molecules to get across the plasma membrane so that they never come in contact with the hydrophobic layer of the membrane or its polar surface.

Facilitated Diffusion

Larger molecules, especially those not soluble in lipids, require assistance to pass through the protein channels of the plasma membrane. The process used by such substances is called *facilitated diffusion*. Amino acids, for example, are insoluble in lipids. But since they are instrumental in determining cell function, they must have a way of getting inside the cell. To pass through the membrane, a specific amino acid temporarily binds with a *carrier protein* in the plasma membrane to form a new compound that is soluble in lipids (Figure 3.5). The amino acid, in the form of the combination compound, moves through the plasma membrane and is released into the cytoplasm inside. As needed, the carrier pro-

FIGURE 3.4

Simple diffusion. (A) Evaporating molecules of perfume (blue) start from an area of high concentration (1), and begin to move randomly among the air molecules (red) in a room (2). After a short while, the perfume molecules are spread evenly throughout the room; a state of equilibrium has been reached (3). (B) Inhaled oxygen moves freely across the respiratory membrane of the lungs as carbon dioxide moves into the lungs to be exhaled. (C) Small polar molecules that cannot readily move across the hydrophobic interior of the plasma membrane diffuse through a channel protein.

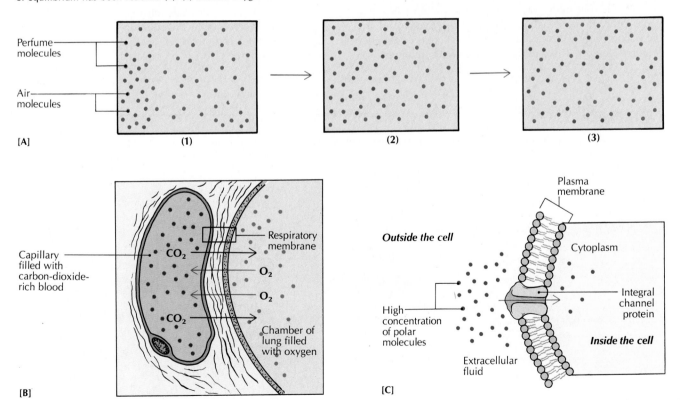

tein then repeats the process for another amino acid. No direct cell energy in the form of ATP is required. Instead, only the internal structure of the protein changes.

Although little is known about carrier proteins, several factors are known to affect the rate at which they move a substance through a membrane by facilitated diffusion. One factor is the number of proteins involved in each transport, and another is the difference in concentration between the substances inside and outside the cell. A third factor is how quickly the carrier protein combines with the substance to be transported. Facilitated diffusion is faster than simple diffusion.

Like simple diffusion, facilitated diffusion transports substances only *from areas where they are in high concentration to areas where they are in a lower concentration,* or "down" the concentration gradient.

Osmosis

The passage of water (or another solvent) through a selectively permeable membrane from an area of high concentration to an area of lower concentration is called **osmosis** (Gr. "pushing"). The relative levels of water concentration are determined by the amount of solute dissolved in the water (the solvent) on either side of the membrane. A higher concentration of solute (for example, sugar) on one side of the membrane means that less space is available there for the water molecules (Figure 3.6A). Water passes through the membrane from one side to the other; the solute cannot. The net effect is that more water moves into the area of lower water concentration than leaves, until the pressure of the increasing volume of solution counterbalances this tendency (Figures 3.6B, C). The **osmotic pressure** of the solution is the

force (hydrostatic pressure) that is required to stop the net flow of water across a selectively permeable membrane when the membrane separates solutions of different compositions.

The pressure on plasma membranes can be great enough to burst cells, so it is important for cells to have relatively constant internal and external pressures to maintain homeostasis. This principle can be seen in red blood cells (Figure 3.7A), which exist in an osmotically stable condition only in an *isotonic* (Gr. *isos,* equal + *tonos,* tension) solution, such as that in blood plasma. In an isotonic solution, the solute concentration is the *same* inside and outside the cell. The concentration of water molecules is also the same inside and outside the cell; thus water molecules move through the plasma membrane at the same rate in both directions, and there is no net movement of water in either direction.

In a *hypotonic* solution (Gr. *hypo,* under), the solute concentration is lower, or *less,* outside the cell than inside. The concentration of water molecules is greater outside the cell than inside. If the cell is a red blood cell, water moves *into* the cell until the cell bursts (or *lyses*) and loses its contents (Figure 3.7B).

In a *hypertonic* solution (Gr. *hyper,* above), the solute concentration is higher, or *more,* outside the cell than inside. Because there is a greater concentration of water molecules inside the cell than outside, the water moves *out* of the cell, causing the cell to shrivel like a raisin (Figure 3.7C). This condition is called *crenation* (L. *crena,* notch) in a red blood cell, and *plasmolysis* in other types of cells. Some body cells in a hypertonic solution can regain their shape and not die if the lost water is replaced quickly. When people drink salt water, for example, their cells release water in an effort to balance the excess salt in the solution outside the cells. If there is too much salt, however, death may result before the water lost from the cells can be replaced.

FIGURE 3.5

Facilitated diffusion. (A) Amino acids in the fluid outside the cell reach the plasma membrane. (B, C) A carrier protein inside the membrane binds temporarily with an amino acid and accelerates its movement across the membrane into the cytoplasm. (D) After the carrier protein deposits the first amino acid inside the cell, it returns to its original shape to pick up another amino acid. No energy is required from the cell.

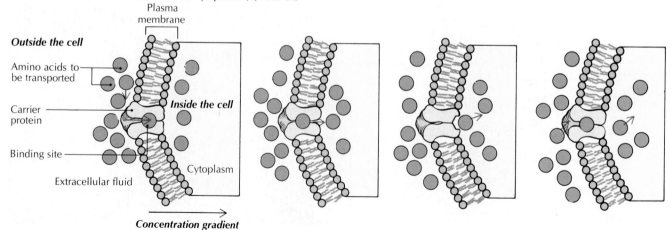

[A]　　　　　　　　　　[B]　　　　　　[C]　　　　　　[D]

FIGURE 3.6

Osmosis. (A) A container of water is divided in half by a selectively permeable membrane. Sugar molecules are dissolved in the left-hand side of the container, producing a condition where there is a higher concentration of water molecules on the right-hand side than on the left. (B) The sugar molecules are too large to pass through the membrane, but water molecules pass from the area of higher water concentration (right side) into the area of lower water concentration (left side)—"down" the concentration gradient. (C) The water level continues to increase on the left side until the pressure from the greater volume of solution on the left equals the tendency of the water on the right to move to an area of lower concentration. At this point an equal number of water molecules move from side to side.

Selectively permeable membrane

Sugar molecules

Water molecules

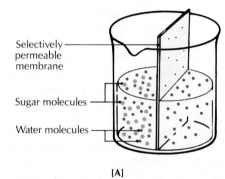

[A]

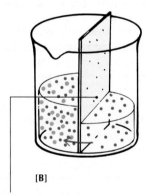

[B]

Net movement of water molecules across the selectively permeable membrane

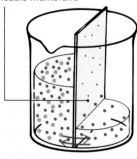

[C]

FIGURE 3.7

Human red blood cells in solutions of various solute concentrations. (A) In an isotonic solution, where the pressure is the same inside and outside the cells, the cells remain normal. (B) Cells in a hypotonic solution finally burst when the low concentration of solute outside forces too much liquid into the cells. (C) Cells in a hypertonic solution have a higher concentration of solute outside than inside, which causes them to lose their liquid and collapse. [(A to C) Dr. Peck-Sun Lin, Department of Therapeutic Radiology, New England Medical Center Hospitals.]

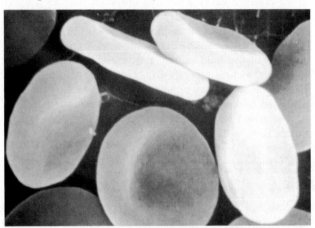

[A] Isotonic solution: normal water concentration   ×3000

[B] Hypotonic solution: water in   ×5000

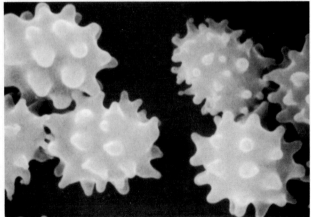

[C] Hypertonic solution: water out   ×3000

Filtration

Filtration is a process that *forces* small molecules through selectively permeable membranes with the aid of hydrostatic (water) pressure (or some other externally applied force). As usual with passive transport processes, movement is from areas of high pressure or concentration to areas of lower pressure or concentration. In the body, filtration is evident when blood pressure forces water and dissolved particles through the highly permeable walls of small blood vessels called *capillaries* (Figure 3.8). Filtration in this case works only when the blood pressure inside the blood vessel is greater than the pressure of the fluid outside the vessel. In filtration, large molecules such as proteins do not pass through the smaller membrane pores; they remain in the capillaries. Filtration is also used in the kidneys, where blood pressure forces water and dissolved wastes such as urea into the kidney tubules during the process of urine formation.

Dialysis

Dialysis is a process in which small particles are separated from large ones when the small particles diffuse through an artificial semipermeable membrane, leaving the larger particles behind. This process does not occur naturally in the body, but it is the basis of the artificial kidney, which mechanically removes impurities from the blood when the kidneys fail and are unable to filter out wastes.

The artificial kidney is a mechanical device that pumps blood containing impurities from the body through an artificial membrane. The blood is rinsed in a cleansing solution, and waste products (which are usually composed of small molecules) and excess water diffuse through the pores of the artificial membrane. If necessary, nutrients can be pumped into the rinsed blood at this time. Large-molecule proteins, which the body needs, do not diffuse through the membrane; they are retained in the body. The cleansed blood is then returned to the body. Artificial kidney machines will be discussed in greater detail in Chapter 24.

FIGURE 3.8

Filtration. Blood pressure in a capillary forces water and small dissolved molecules through the capillary wall into the outside tissue fluid. The process works only when the pressure inside the capillary is greater than the pressure in the tissue fluid.

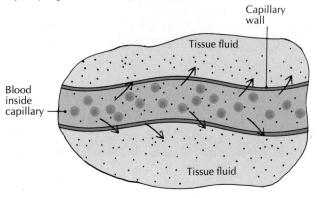

Ask Yourself

1 *How does passive transport work?*

2 *How do simple diffusion and facilitated diffusion differ?*

3 *Where do carrier proteins function?*

4 *What is osmosis?*

5 *Where in the body does filtration occur?*

6 *What is the principle of dialysis?*

ACTIVE MOVEMENT ACROSS MEMBRANES

The basic difference between passive and active processes of movement is that **active processes** move substances through a selectively permeable membrane *from areas of low concentration* on one side of the membrane *to areas of high concentration* on the other side, "against" the concentration gradient (see Table 3.2). Because of this, active movement across membranes requires energy. If passive movement can be compared to letting a car coast downhill, active movement can be compared to driving a car uphill, with ATP acting as the gasoline for the engine. Among the active processes are active transport, endocytosis, and exocytosis.

Active Transport

Passive transport is satisfactory when substances enter a cell, because the concentration of molecules outside the cell is greater than the concentration inside. Then the small molecules diffuse naturally through the plasma membrane into the cell. But when a state of equilibrium is reached, and still more molecules are needed within the cell, they must be transported (pumped) through the membrane *against diffusion*, or "against" the concentration gradient. This movement of small molecules or ions is accomplished by **active transport**, which requires energy from the cell in the form of ATP.

In the human body, four specific active-transport systems have been studied extensively, although exactly how they work is still not completely understood: (1) the *sodium-potassium pump* is common to many cells, especially nerve cells, where sodium and potassium concentration gradients are generated in order to produce an electrical charge; (2) the *calcium pump* is vital in muscle contraction, where it transports calcium ions needed for the contractile mechanism; (3) *sodium-linked co-transport* actively transports sugars and amino acids, while sodium ions tag along passively; (4) in *hydrogen-linked co-transport*, hydrogen ions tag along while sugars are being actively transported.

These four active-transport mechanisms all use a similar process (Figure 3.9):

1 A molecule outside the cell binds with a carrier protein at the membrane boundary.

FIGURE 3.9

Active transport. In this diagram, a molecule is transported through the plasma membrane by a carrier protein when (A) the "gates" of the carrier protein in the membrane open, (B, C) change their interior shape to conform to the shape of the passenger molecule, and (D) carry the molecule through. The carrier protein requires energy from the cell in the form of ATP.

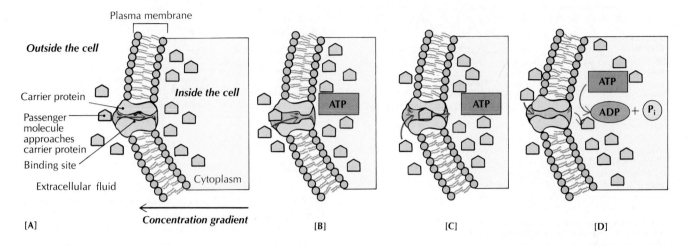

Outside the cell

Plasma membrane

Carrier protein

Passenger molecule approaches carrier protein

Binding site

Inside the cell

Extracellular fluid

Cytoplasm

Concentration gradient

[A]   [B]   [C]   [D]

FIGURE 3.10

Two types of endocytosis. Endocytosis may involve (A) pinocytosis, where extracellular molecules are taken into a cell through an invaginated vesicle, or (B) receptor-mediated endocytosis, where external molecules (called *ligands*) are bound to specific receptors on the plasma membrane before they are taken into the cell.

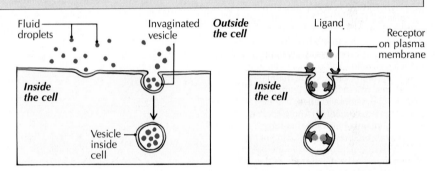

Fluid droplets

Invaginated vesicle

Outside the cell

Ligand

Receptor on plasma membrane

Inside the cell

Vesicle inside cell

Inside the cell

[A] PINOCYTOSIS

[B] RECEPTOR-MEDIATED ENDOCYTOSIS

2 The molecule-carrier protein complex moves through the membrane.

3 Assisted by at least one enzyme and energy available from ATP, the molecule and carrier protein separate, and the molecule is released.

4 The carrier protein returns to its original shape, and can repeat the process with another molecule.

Endocytosis

So far we have seen how small molecules can pass through plasma membranes. How do large molecules or particles move through? Large molecules such as lipids and proteins, as well as small amounts of fluids, can enter cells when the plasma membrane forms a pocket around the material, enclosing it and drawing it into the cytoplasm. This process of active movement is called **endocytosis** (Gr. *endon*, within). During endocytosis, only a small region of the plasma membrane folds inward, or *invaginates*, until it has formed a new intracellular vesicle that contains large molecules from outside the cell. Three types of endocytosis are known (Figure 3.10):

1 *Pinocytosis* ("cell drinking," from Gr. *pinein*, to drink + *cyto*, cell) is the nonspecific uptake of small droplets of extracellular fluid. Any material dissolved in the fluid is also taken into the cell, including low-molecular-weight nutrients, amino acids, glucose, vitamins, and other substances. Kidney cells are an example of the many types of cells that use pinocytosis to help regulate their fluid environment.

2 *Receptor-mediated endocytosis* involves a specific receptor on the plasma membrane that "recognizes" an extracellular macromolecule and binds with it. The substance bound to the receptor is called the *ligand* (L. *ligare*, to bind). The region of the plasma membrane that contains the receptor-ligand complex undergoes endocytosis.

FIGURE 3.11

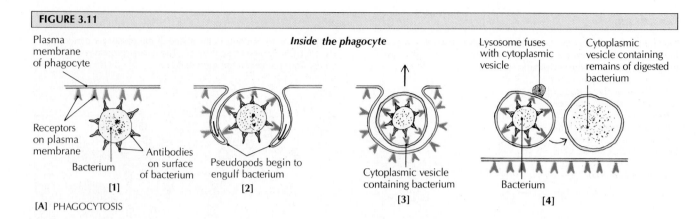

[A] PHAGOCYTOSIS

Phagocytosis and exocytosis.
(A) Phagocytosis. A possible "zipper interaction" mechanism for phagocytosis, showing (1) how antibodies on the surface of a bacterium bind to receptors on the plasma membrane of a phagocyte. (2) The bacterium is engulfed by pseudopods of the phagocyte. If only part of the surface of the bacterial cell is coated with antibodies, the pseudopods will not completely surround the cell, and the cell will not be engulfed. (3) The bacterium is enclosed within a cytoplasmic vesicle. (4) A lysosome fuses with the cytoplasmic vesicle and releases digestive enzymes that break down the bacterium. (B) Exocytosis. Unwanted particles engulfed and digested within a vesicle are transported to the plasma membrane, where the vesicle membrane and plasma membrane open, and the particles are expelled. (1) The cytoplasmic vesicle containing unwanted particles approaches the plasma membrane of the cell. (2) The vesicle attaches to the plasma membrane. (3) The membranes of the cell and vesicle fuse, and the vesicle membrane begins to "unzip." (4) The point of contact between the two membranes widens slightly, allowing the hydrophobic regions of both membranes to be in contact. (5) The fused membranes open. (6) The fusion of membranes is complete, and the unwanted particles are expelled from the cell.

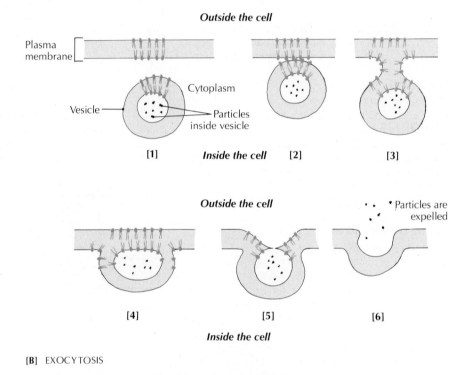

[B] EXOCYTOSIS

3 The active movement that allows large molecules and particles to pass through the plasma membrane into the cytoplasm is called **phagocytosis** ("cell eating," from Gr. *phagein*, to eat + *cyto*, cell). The particle brought into a cell by phagocytosis is not always a nutrient. Probably the most familiar example of phagocytosis is that of a white blood cell engulfing a harmful bacterium before digesting it.

Figure 3.11A shows what happens during phagocytosis. When a bacterium or other foreign particle makes contact with the plasma membrane of a *phagocyte* (Gr. "a cell that eats"), the membrane begins to form a pocket around the particle by sending out pseudopods (Gr. "false feet"). When the pocket is closed, it is called a *cytoplasmic vesicle*, or *endosome* (formerly called a *vacuole*). The particle and endosome enter the cytoplasm. At least one lysosome (see p. 70) attaches itself to the endosome and releases its digestive enzymes inside the endosome to digest the bacterium.

Exocytosis

Exocytosis (Gr. *exo*, outside) is basically the opposite process of endocytosis (see Figure 3.11B). The endosome, with its undigested particles, fuses with the plasma membrane and releases the unwanted particles from the cell. For example, nerve cells release their chemical messengers, and various gland cells secrete proteins by exocytosis.

Ask Yourself

1 *How do active and passive processes of movement across plasma membranes differ?*

2 *What is active transport?*

3 *What is the difference between endocytosis, phagocytosis, and exocytosis?*

FIGURE 3.12

The endoplasmic reticulum. (A) Note that the endoplasmic reticulum (blue) actually touches the outer nuclear envelope, where a transfer of materials takes place. Rough ER and smooth ER (without attached ribosomes) are shown. Ribosomes can be seen in both (B) the electron micrograph and (C) the drawing as tiny spheres that are liberally attached to the rough ER. [(B) Don Fawcett, M.D./Photo Researchers.)

[A]   Endoplasmic reticulum

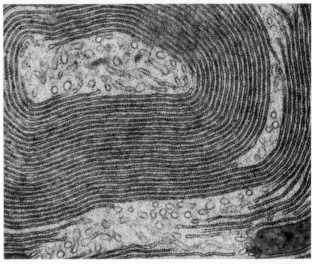

[B]   × 25,000

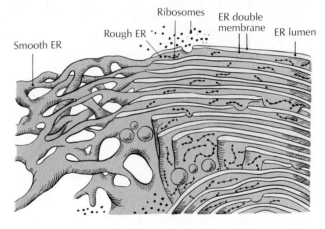

[C]

CYTOPLASM AND ORGANELLES

The cytoplasm of a cell is composed of two distinct parts, or phases. The ***particulate phase*** consists of well-defined structures, including membrane-bound vesicles (small fluid-filled structures) and organelles, lipid droplets, and *inclusions* (solid particles such as glycogen granules). The ***aqueous phase*** consists of the fluid *cytosol*, in which the organelles and vesicles are suspended, and in which are dissolved various soluble proteins and substances required for the cell's metabolic activities. Cytosol accounts for about 54 percent of a cell's total volume.

Before modern electron microscopes and other devices and techniques were available, cytoplasm was believed to be a thick gel containing a nucleus and organelles such as the Golgi apparatus, mitochondria, and centrioles. Now we know that cytoplasm also contains many other separate but integrated organelles, including the endoplasmic reticulum, ribosomes, and a *cytoskeleton*—a flexible lattice arrangement of microtubules, intermediate filaments, and microfilaments that supports the organelles and provides the machinery for the movement of cells and their organelles. We also know that each organelle has a specific function. Like most other body materials, the cytosol portion of cytoplasm is mostly water (70 to 85 percent), but its other typical components are proteins (10 to 20 percent), lipids (about 2 percent), complex carbohydrates, nucleic acids, amino acids, vitamins, and electrolytes (substrates such as potassium that dissolve in water and provide inorganic chemicals for cellular reactions).

Cells need a steady supply of oxygen and specific nutrients for fuel, for building new cell substances, and for synthesizing secretory products. The nutrients, which originate in the food we eat, pass from the digestive tract into the blood, into tissue fluids, and finally into cells. (Waste products leave cells through a reverse route.) Once inside the cytoplasm, food can be "burned," or oxidized, to supply the energy for the many kinds of work performed by the various organelles (see Chapter 2 for a discussion of oxidation reactions). Like the organelles, some stored substances, pigments, and various inclusions are also contained in the cytosol and supported by the cytoskeleton.

Endoplasmic Reticulum

Inward from the plasma membrane, and physically connected with it in places, is the ***endoplasmic reticulum***, frequently called simply the ***ER***. The ER is a complex labyrinth of flattened sheets, sacs, tubules, and double membranes that branch and spread throughout the cytoplasm and are connected with the nucleus of the cell in some places. Figure 3.12A shows the position of the endoplasmic reticulum in a simplified cell (see also Figure 3.1).

The ER functions as a series of channels that help various materials to circulate throughout the cytoplasm. It also serves as a storage unit for enzymes and other proteins, and as a point of attachment for structures that play an important part in forming proteins (Figures 3.12B, C).

The space between the double membranes of the endoplasmic reticulum is called the *ER lumen* or *ER cisternal space* (Figure 3.12C). Ribosomes are not attached to the inner sides of these membranes. When endoplasmic reticulum does not have attached ribosomes, it is called *smooth ER*. Smooth ER serves as a site for the production of steroids, detoxifies a wide variety of organic molecules, and stores calcium in skeletal muscle cells. *Rough ER* refers to endoplasmic reticulum that does have attached ribosomes on its outer face, as shown in Figure 3.1.

FIGURE 3.13

A ribosome. (A) A current model of a ribosome, with its two subunits. Apparently, the mRNA molecule is held in a channel between the large and small subunits. This representation is based on electron micrographs. (B) Simplified version of a ribosome, showing its A-site and P-site on the large subunit for the binding of tRNA molecules, and its binding site for mRNA on the small subunit. (C) Electron micrograph of a string of ribosomes (a polyribosome) connected by a strand of mRNA. [*Photo Researchers.*]

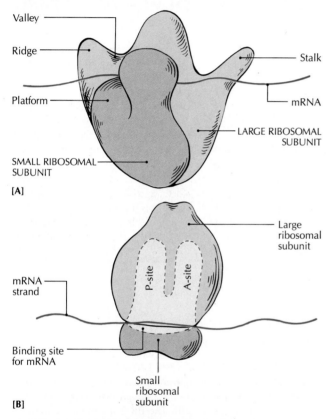

[A]

[B]

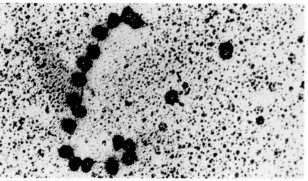

[C]   × 125,000

Ribosomes

Ribosomes are necessary for the synthesis of proteins from amino acids. They contain almost equal amounts of protein and a special kind of ribonucleic acid, ribosomal RNA (rRNA). Ribosomes are actually composed of two subunits with complex shapes (Figure 3.13A), but are usually pictured as slightly flattened spheres with "tucked-in" sections around the middle (Figure 3.13B). The role of ribosomes in protein synthesis is detailed later in this chapter.

Not all ribosomes are attached to the ER. Some float freely within the cytoplasm (see Figure 3.1). Whether ribosomes are free or attached, they are usually grouped in clusters that are connected by a strand of a single molecule of specialized nucleic acid called *messenger ribonucleic acid* (usually abbreviated as *messenger RNA*, or *mRNA*). Such clusters are called *polyribosomes*, or *polysomes* (Figure 3.13C).

Golgi Apparatus

The *Golgi apparatus,* or *Golgi complex* (named for Camillo Golgi, who discovered it in 1898), is a collection of membranes associated physically and functionally with the ER in the cytoplasm. Golgi apparatuses are usually located near the nucleus, and are found in almost all cells except red blood cells (Figure 3.14). They are composed of flattened stacks of membrane-bound *cisternae,* closed spaces serving as fluid reservoirs (Figure 3.14B). The Golgi apparatus is thought to specialize in the "packaging" and secretion of glycoproteins. Many of these glycoproteins are probably important components of the *glycocalyx,* or outer coating of the cell, especially as part of a system that allows one cell to recognize another cell of the same kind.

As proteins are synthesized by ribosomes attached to the rough ER, some of the proteins are sealed off in little packets called *transfer vesicles.* These vesicles pass along the ER to the Golgi apparatus and fuse with it (Figure 3.14C). Within the Golgi apparatus, the proteins and/or carbohydrates attached to them can be concentrated, modified, or compacted. Eventually, the proteins are packaged into relatively large *secretory vesicles* (also called *secretory granules*), which are released from the cisterna closest to the plasma membrane. When the vesicles reach the plasma membrane, they fuse with it and release their contents to the outside of the cell by exocytosis.

Golgi apparatuses are most apparent and abundant in cells that secrete chemical substances, such as pancreas cells secreting digestive enzymes, and nerve cells secreting transmitter substances.

Lysosomes and Microbodies

Lysosomes (Gr. "dissolving bodies") are small, membrane-bound organelles that may contain about 40 different digestive enzymes (called *acid hydrolases*) capable of digesting proteins, nucleic acids, lipids, and carbohydrates under acidic conditions. The building blocks of the destroyed substances are discharged into the cytoplasm, where they are used in

FIGURE 3.14

Golgi apparatus. (A) A simplified diagram of a cell, cut open to show the position of a Golgi apparatus (green). (B) An electron micrograph of a Golgi apparatus, showing secretory vesicles. (C) A drawing of a Golgi apparatus in three dimensions, showing how the transfer vesicles from the ER merge with the Golgi apparatus, and how secretory vesicles bud off from the opposite side of the Golgi apparatus on their way to delivering secretions. [*(B) L. S. Khawkinea/Biophoto Associates/ Photo Researchers.*]

Golgi apparatus

[A]

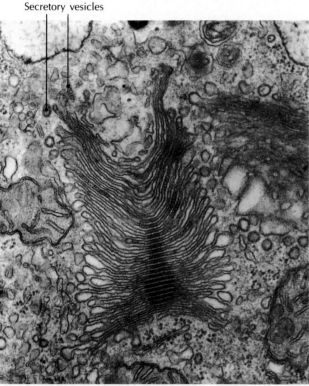

Secretory vesicles

[B]   × 50,000

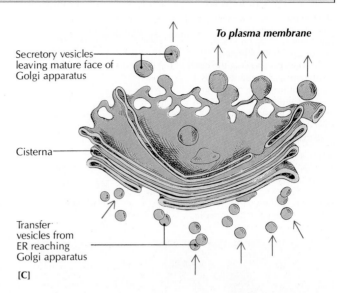

To plasma membrane

Secretory vesicles leaving mature face of Golgi apparatus

Cisterna

Transfer vesicles from ER reaching Golgi apparatus

[C]

the synthesis of new materials. The hydrolases are synthesized in the ER, transported to the Golgi apparatus for processing, and then transported to the lysosomes via secretory vesicles.

Lysosomes are found in almost all cells, especially those in tissues that experience rapid changes, such as the lungs, liver, and spleen, and in white blood cells. They are sometimes considered to be the digestive system of the cell, because their membranes can engulf—and their acid hydrolases can digest—complex molecules, microorganisms, and "worn-out" or damaged cell parts. (See the discussion of phagocytosis on p. 68.) Serious genetic diseases can occur if only a single hydrolase is absent, because then the foreign material inside a cell cannot be digested. For example, the human

genetic disorder *Tay-Sachs disease* (see Chapter 29) results when lipids accumulate in the brain and other tissues instead of being hydrolyzed.

The existence of lysosomes was first postulated in 1955 by the French cytologist Christian de Duve, who called them "suicide bags" because they are capable of destroying the cell containing them by bursting and releasing their digestive enzymes. Other people refer to them as "garbage-disposal units." It was not until the 1960s, however, that scientists verified the existence of lysosomes and showed that different lysosomes have different enzymes and functions.

Lysosomes are often confused with *microbodies*, since both appear as dense, granular structures. There are two types of microbodies, but only one type, **peroxisomes,** is found in animal cells. Peroxisomes are membrane-bound organelles that contain oxidative enzymes, which use oxygen to carry out their metabolic reactions. Peroxisomes are so named because they can both form hydrogen peroxide as they oxidize various substrates and then destroy the peroxide with the enzymes they contain. The destruction of peroxides is crucial to the cell because they are toxic products of many metabolic processes, especially fatty-acid oxidation. Peroxisomes are numerous in liver and kidney cells, where they are important in detoxifying certain compounds, such as alcohol.

Mitochondria

Mitochondria are double-membraned, saclike organelles found throughout the cytoplasm; in a typical liver cell they may occupy about 22 percent of the total volume of the cell (Figure 3.15A). Their main function is the conversion of energy stored in carbon-containing molecules into the high-energy bonds of ATP. ATP is an energy source for many cellular activities. (See the essay on p. 86.) In fact, mitochondria are sometimes called the "powerhouses" of the cell because they provide about 95 percent of the cell's energy supply.

FIGURE 3.15

Mitochondria. (A) A drawing of mitochondria (sliced open) in a simplified cell. (B) An electron micrograph of a sectioned mitochondrion showing the outer membrane and the inner folds (cristae) of the inner membrane. (C) A three-dimensional cutaway drawing of a mitochondrion. [(B) Don Fawcett, M.D./Keith Porter/Photo Researchers.]

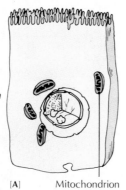

[A]   Mitochondrion

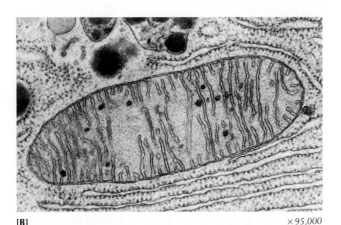

[B]   ×95,000

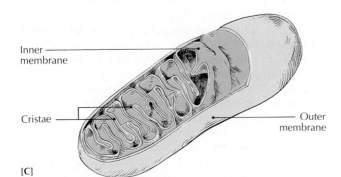

Inner membrane

Cristae

Outer membrane

[C]

The internal structure of a mitochondrion is compatible with its function. A complex inner membrane folds and doubles in upon itself to form incomplete partitions called *cristae,* or crests (Figures 3.15B, C). The space between the cristae, called the *matrix,* is filled with a fairly thick liquid. The cristae greatly increase the available surface area for chemical reactions that produce usable energy. The more sites for chemical reactions, the more energy that can be produced. (These important reactions will be described in detail in Chapter 23.) Enzymes for these chemical reactions are located on the cristae and in the matrix. It is estimated that a typical mitochondrion contains at least 100 different enzymes. (A complete cell may use 1000 different enzymes to carry on its usual metabolic activities.)

The number of mitochondria per cell varies from as few as 100 to as many as several thousand, depending upon the energy needs of the cell. Muscle cells, which convert chemical energy into mechanical energy, have a greater number of mitochondria than relatively inactive cells. A typical liver cell may contain about 1700 mitochondria.

Some cells have mitochondria in specialized locations. A mature sperm, for instance, has tight rows of numerous mitochondria wrapped around the upper end of its tail. These mitochondria provide energy for the vigorous movement of the tail as the sperm swims. Mitochondria are also involved in many other activities, including the movement of cilia, the metabolism of calcium, the active transport of materials across plasma membranes, and the respiration of cells.

Mitochondria, which contain their own DNA and ribosomes, are able to reproduce. They usually multiply when a cell needs increased amounts of ATP.

Microtubules, Intermediate Filaments, and Microfilaments

The most conspicuous elements of the cytoskeleton, *microtubules,* are hollow, slender, cylindrical structures found in most cells (Figures 3.16A, B). Each microtubule is made up of spiraling subunits of a protein called *tubulin* (Figure 3.16C). Microtubules have several functions within a cell:

1 They are usually associated with organelle movement, especially in cilia and flagella, and with chromosome movement while the cell nucleus is dividing.

2 They seem to be part of a transport system within the cell—in nerve cells, for example, they help move materials through the long fibers in the cytoplasm.

3 They form a supportive cytoskeleton within the cytoplasm.

4 They may be involved in the overall shape changes that cells go through during periods of cell specialization.

Intermediate filaments are a chemically heterogeneous group of structures whose protein subunits can be divided into five classes: *keratin-containing intermediate filaments,* found in epithelial cells; *glial filaments,* found in glial cells in the brain; *neurofilaments,* found in nerve cells; *desmin filaments,* found in muscle cells; and *vimentin filaments,* found in almost all cells. Intermediate filaments help to maintain the shape of cells and the spatial organization of organelles, and promote the mechanical activities in the cytoplasm.

In addition to the hollow microtubules and variable intermediate filaments, solid cytoplasmic *microfilaments* are also found in most cells. These filaments are composed of strings of protein molecules, including the protein actin. Actin microfilaments in nonmuscle cells provide mechanical support for various cellular structures, and help form contractile systems, which are thought to be responsible for many cellular movements.

In most cells, the microfilaments, intermediate filaments, and microtubules form the flexible cellular framework called the *cytoskeleton* (Figure 3.17). This latticed framework extends throughout the cytoplasm, interconnecting the micro-

FIGURE 3.16

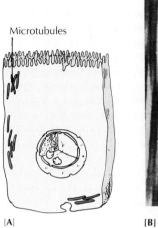

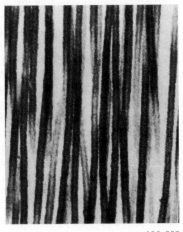

Microtubules. (A) Microtubules in a simplified cell. (B) An electron micrograph showing microtubules from the brain. (C) A drawing of a portion of a microtubule showing the arrangement of its spiraling protein units. [(B) *Dr. Joel Rosenbaum, Department of Biology, Yale University.*]

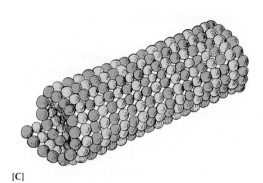

Microtubules

[A]

[B]   × 100,000

[C]

FIGURE 3.17

This diagram of the cytoskeleton in a cell shows how fine microtrabecular strands and other structural units form a three-dimensional network to support the principal structures of the cytoskeleton and other cellular structures.

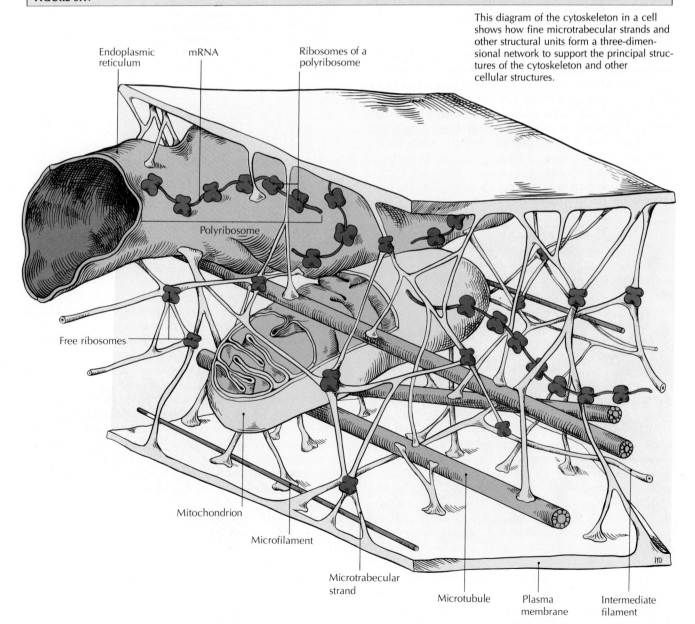

Endoplasmic reticulum

mRNA

Ribosomes of a polyribosome

Polyribosome

Free ribosomes

Mitochondrion

Microfilament

Microtrabecular strand

Microtubule

Plasma membrane

Intermediate filament

trabecular strands, microfilaments, intermediate filaments, microtubules, and other organelles in a structural and functional entity. In addition to holding organelles and particles more or less in place, the cytoskeleton plays a role in cell movement, and is a site for the binding of specific enzymes.

Centrioles

Near the nucleus of a cell lies a specialized region of cytoplasm called the *centrosome*, which contains two small organelles called ***centrioles*** (Figure 3.18A). Centrioles are bundles of cylinders made up of microtubules that resemble drinking straws. Each centriole is composed of nine triplet microtubules that radiate from the center like the spokes of a wheel (Figures 3.18B, C). The two centrioles lie at right angles to each other (Figures 3.18A, C), and are duplicated when the cell undergoes cell division. In fact, the centrioles are involved with the movement of chromosomes during cell division. (See the discussion of mitosis on p. 80.)

Cilia and Flagella

Cilia (L. originally "eyelids," later "eyelashes"; sing. *cilium*) and ***flagella*** (L. "small whips"; sing. *flagellum*) are threadlike appendages of some cells. The shafts of cilia and flagella are composed of nine pairs of outer microtubules and two single microtubules in the center (Figure 3.19A). Both types of shaft are anchored to the plasma membrane by a *basal body*, which is the specialized structure that acts as the template for the nine-by-two arrangement of microtubules and gives rise to the double microtubules in the center. The basal body is structurally identical to a centriole; it protrudes into the cytoplasm of the cell to which it is attached.

Groups of cilia beat in unison, creating a rhythmic, wavelike movement in only one direction (Figure 3.19B). Ciliated cells are found in the upper portions of the respiratory tract, where they sweep along mucus and foreign particles. They also appear in portions of the female reproductive system (Figure 3.19C), where they help move mucus, sperm, and eggs along the uterine tube (oviduct).

Flagella have the same internal structure as cilia, but are much longer. They usually number no more than one or two per cell and are used to propel the cell through a fluid environment. Their motion is more random than that of cilia, although it is probably based on a similar pattern of an active forward stroke followed by relaxation. In human beings, the only cells with flagella are sperm, which normally have only one flagellum each. The flagellated tail of a sperm enables the free-swimming cell to move relatively great distances to reach an egg deep within the female reproductive system.

FIGURE 3.18

Centrioles. (A) The position of centrioles near the cell nucleus. (B) An electron micrograph of a cross section of a centriole. (C) A drawing of a pair of centrioles. Each has nine triplets of microtubules and lies at a right angle to the other. [(B) *Don Fawcett, M.D./Photo Researchers.*]

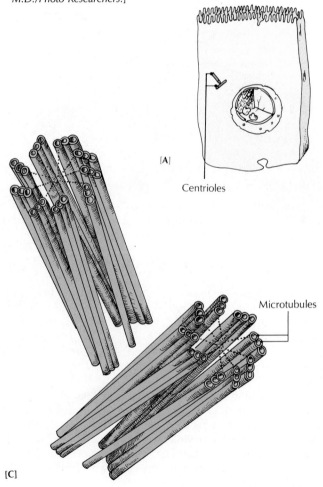

[A]

Centrioles

Microtubules

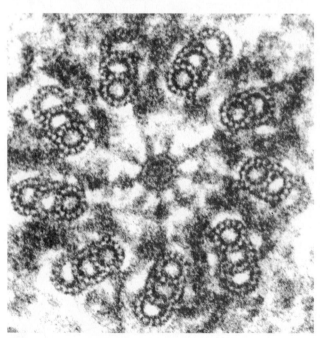

[B]   ×400,000

[C]

FIGURE 3.19

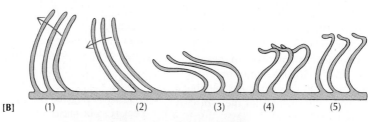

[B]   (1)   (2)   (3)   (4)   (5)

Two inner microtubules

Nine pairs of outer microtubules

[A]   Basal body

[C]

Cilia

Cilia and flagella. (A) A drawing of a portion of the shaft of a flagellum shows the rotating, propellerlike motion of the two central microtubules. Apparently, the nine pairs of outside microtubules slide back and forth, one at a time, creating a snapping motion that moves the cell forward. The outer microtubules also act as anchors. (B) The active stroke of the cilia (1,2) moves fluid at the cell surface forward. The cilia relax and return to their original position (3–5) to begin the process again. (C) A scanning electron micrograph of ciliated cells in a uterine tube. The cilia help sweep egg cells from the ovary to the uterus. [*Don Fawcett, M.D./Penelope Gaddum-Rosse/ Photo Researchers.*]

Cytoplasmic Inclusions

In addition to organelles, cytoplasm also contains **cytoplasmic inclusions,** chemical substances that are usually either food material or the stored products of the cell's metabolic activities. Unlike organelles, inclusions are not permanent components of a cell, and are constantly being destroyed and replaced.

Carbohydrates, for example, are stored mainly in liver and muscle cells in the form of *glycogen inclusions.* When the body needs energy, the liver or muscle cells convert the glycogen inclusions into usable, energy-rich glucose, the body's main source of energy. Lipids are also stored in cells as a source of energy, and may be used when carbohydrates are not available.

Also present in cytoplasm are several *pigments* (substances that produce color), probably the best known being *melanin,* the pigment that protects the skin by causing it to tan when exposed to excessive ultraviolet radiation from the sun. Melanin is also found in the hair and eyes. *Hemoglobin* is another well-known pigment. Its major function is to carry oxygen in red blood cells.

Mucus is an important lubricating fluid, secreted by some cells in the cavities of the body (including the nasal cavities). Besides being a lubricant, mucus helps trap and rid the body of a final example of inclusions, *foreign substances* such as microorganisms, dust, and cellular debris.

Ask Yourself

1 *What is the function of ribosomes?*

2 *How does a Golgi apparatus receive protein from the endoplasmic reticulum?*

3 *What do lysosomes do?*

4 *Why are mitochondria called the "powerhouses" of the cell?*

5 *How is the internal structure of a mitochondrion compatible with its function?*

6 *What is the function of cilia and flagella, and how do they accomplish it?*

7 *What is a cytoplasmic inclusion?*

THE NUCLEUS

The *nucleus* of a cell contains DNA, and is the control center of the cell. It has two important roles:

1 To direct the chemical reactions that occur in the cell by transcribing genetic information in the DNA into specific proteins (enzymes) that determine the cell's particular metabolic activities. The nucleus contains information for the synthesis of approximately 30,000 different proteins.

2 To store genetic information and to transfer it during cell division from one generation of cells to the next, and eventually from one generation of organisms to the next.

Obviously, the nucleus is an important structure, and it is present in almost every cell. Mature red blood cells, which do not have nuclei, are incapable of producing mRNA. Thus they cannot synthesize additional proteins, cannot duplicate themselves, and are unable to perform the typical wide range of cellular activities. They are destined to have a relatively short life.

Nuclei can vary in shape, location,. and number within a cell. They can have elongated shapes and be found at the base of tall cells, or they can be flattened against the cell membrane of fat cells. But nuclei are generally somewhat spherical and located near the center of the cell, firmly embedded in the cytoplasm (Figure 3.20A).

Nuclear Envelope

The boundary of the nucleus appears somewhat like the cell membrane, but with some important differences. It is not one bilayered membrane, but two, with a discernible "space" (compartment) between them, leading biologists to call it a *nuclear envelope.* Nor is the nuclear envelope continuous. Rather, it contains many openings called *nuclear pores* (Figures 3.20B to D), giving the nucleus something of the appearance of a whiffle ball (Figure 3.20C). The pores make it possible for materials to enter and leave the nucleus, and for the nucleus to be in direct contact with the endoplasmic reticulum (Figure 3.20E).

Chromosomes

The interior mass of the nucleus is the *nucleoplasm,* which contains, in a cell that is not dividing, genetic material called *chromatin* (see Figures 3.20C, D). Chromatin consists of a combination of DNA and protein. Two forms of chromatin exist. *Euchromatin* is the thin, active form of DNA in a non-dividing cell, and *heterochromatin* is condensed, inactive DNA. During a process of cell division called mitosis (described later in this chapter), strands of protein-rich chromatin become arranged in coiled threads called *chromosomes,* which store the hereditary material in segments of DNA called *genes.* Genes are located along the chromosomes in a specific sequence and position. In fact, the sequence and position of nucleotides in the genes create the genetic code that determines heredity and is responsible for protein structure.

FIGURE 3.20

Nucleus and nuclear envelope. (A) A simplified drawing of a cell showing the nucleus and nucleoli. (B) A scanning electron micrograph of a nucleus shows the nuclear pores in its envelope clearly. (C) A drawing and (D) an electron micrograph of the nucleus and its constituents. The drawing (C) shows two nucleoli floating in the nucleoplasm. The nuclear pores in the electron micrograph (D) are marked with arrows. (E) A simplified drawing of a two-dimensional thin section showing the nucleus, the synthesis-transport network (endoplasmic reticulum and Golgi apparatus), and the energy source (mitochondrion) that act together in the process of protein synthesis. [(B) Keiichi Tanaka, M.D. (D) Don Fawcett, M.D./Photo Researchers.]

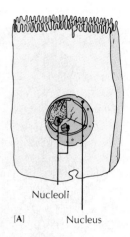

Nucleoli

[A]   Nucleus

Human beings have 23 pairs of chromosomes in their body cells (except the sperm and ova, which have a total of 23), for a total of 46. In Chapter 29 you will see what happens when a mistake occurs and a person has one extra chromosome, or when other genetic abnormalities occur. Chromosomes are also discussed in Chapters 26 and 27.

Nucleolus

Cell nuclei usually contain at least one dark, somewhat spherical mass called a *nucleolus* (new-KLEE-oh-luhss; "little nucleus"; pl. *nucleoli*) (see Figures 3.20A, D). Besides containing DNA and RNA, the nucleolus also contains protein, which is probably brought into the nucleus from the cytoplasm. The nucleolus can be thought of as a preassembly point for ribosomes. Ribosomes may be partially synthesized in the nucleolus and completed after they have entered the cytoplasm through the pores of the nuclear envelope. (The nucleolus does not have a membrane.) Some of the ribosomes become attached to the endoplasmic reticulum, where they assist in the synthesis of proteins (see How Proteins Are Synthesized, p. 82).

Ask Yourself

1 *What are the functions of the nucleus?*

2 *Why are the pores in the nuclear envelope important?*

3 *Distinguish between chromatin, nucleoplasm, and chromosomes.*

4 *What is the function of a nucleolus? What role does it play in the production of ribosomes?*

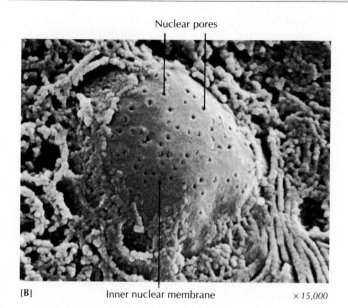

Nuclear pores

Inner nuclear membrane

[B]   ×15,000

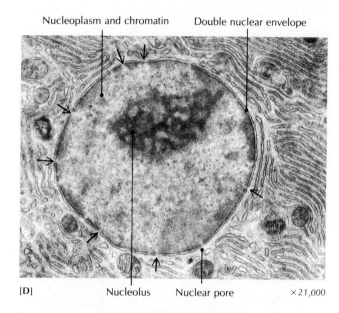

Nucleoplasm and chromatin   Double nuclear envelope

[D]   Nucleolus   Nuclear pore   ×21,000

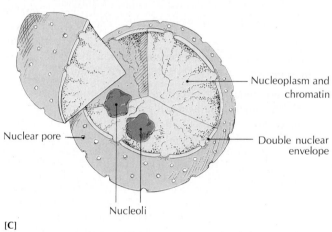

Nucleoplasm and chromatin

Nuclear pore

Double nuclear envelope

Nucleoli

[C]

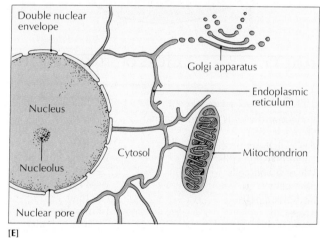

Double nuclear envelope

Golgi apparatus

Nucleus

Endoplasmic reticulum

Nucleolus

Cytosol

Mitochondrion

Nuclear pore

[E]

THE CELL CYCLE

In the next minute, about three billion cells in your body will die. If all is well, three billion new cells will be created during that same minute. In this way, homeostasis is maintained. (In Chapter 28, when we discuss cancer and aging, you will see what happens when too many or too few cells are produced.) The correct number and kind of new cells are produced when existing, healthy cells divide. Each original (parent) cell becomes two daughter* cells, both genetically identical to the parent.

This reproduction through duplication and division is the key to development and growth—it is the formula for life. The single fertilized egg that marks the beginning of your life

*The use of the word "daughter" has nothing to do with gender. A "daughter" cell is not necessarily a *female* cell, but merely an *offspring*, a new cell formed from a parent cell or cells.

doubles, then each of its daughter cells doubles, and so on until a predetermined number is reached. During the earlier stages of human growth the doubling process is accompanied by cell specialization. By the time we are adults we have about 100 different kinds of cells. The process of specializing in structure and function is called **differentiation.** Once cells specialize, their functions seldom change. For example, when a cell differentiates into an epithelial cell, it will never be any other kind of cell, and neither will its daughter cells.

The process of *cell division* marks the beginning and end of the **cell cycle,** or life span, of a single cell (Figure 3.21). The cell cycle is that period from the beginning of one cell division to the beginning of the next cell division (Table 3.3). Before cell division can begin, the parent cell must double its mass and contents. For cell division to be complete, both the nucleus and the cytoplasm must divide in such a way that each daughter cell gets one complete set of genetic material and all the necessary cytoplasmic constituents and organelles.

FIGURE 3.21

The cell cycle.

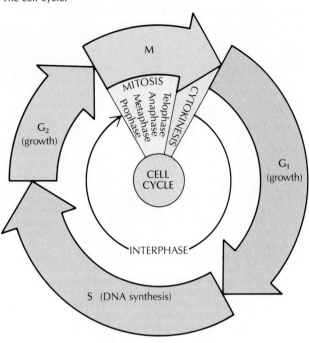

TABLE 3.3   CYCLES OF SPECIALIZED CELLS

Type of cell	Approximate life span
Bone marrow cells (blood-forming)	10 hours
Stomach cells	2 days
Egg cells, sperm cells	2–3 days
Large-intestine cells	3–4 days
White blood cells	13 days
Skin cells	19–34 days
Red blood cells	120 days
Liver cells	18 months
Nerve cells*	Lifetime of body

*Nerve cells, except for olfactory (smell) neurons in the nose, do not reproduce after early childhood.

Thus the cell cycle involves three major processes: (1) *interphase*—a period of cell growth during which the DNA in the nucleus replicates; (2) *mitosis*—the period during which the nucleus, with its genetic material, divides; and (3) *cytokinesis*—the division of the cytoplasm into two distinct but genetically identical cells. All three processes are described in the following sections.

DNA Replication

Before a cell divides, the DNA in its nucleus must produce a perfect copy of itself. This process is called **replication**, because the double strand of DNA makes a *replica*, or duplicate, of itself. The accurate replication of DNA before actual cell division is essential to ensure that each daughter cell receives the same genetic information as the parent cell. This feat is

accomplished, at the rate of about 50 nucleotides per second, with the assistance of several replication enzymes.

The replication of a twisted, double-stranded DNA molecule begins with the untwisting and separation of its two complementary strands (Figure 3.22). The "unzipping" of the double strands takes place when enzymes break the hydrogen bonds between the purine and pyrimidine nitrogenous bases. Each adenine-thymine base pair separates, and each cytosine-guanine base pair separates. Once the bases on each strand are exposed, the enzyme *DNA polymerase* attaches to each of the separated strands and moves along it one base at a time. The enzyme selects available bases (together with their sugar-phosphate backbones) from the nucleoplasm, and forms new hydrogen bonds with the exposed complementary bases on the unzipped strands. Each base thus couples with its complementary base partner: adenine bonds with thymine, and cytosine bonds with guanine. It is this principle of *complementary bonding* that ensures exact replication.

In this way, each separated strand of the original twisted DNA molecule acts as a mold, or *template*, to make an exact copy, not of itself, but of the *other* template strand of the unzipped DNA molecule. (The *sequence* of nucleotides is crucial in producing an exact copy.) The untwisted strand apparently begins to twist again as soon as the new bonds are formed. So the complicated activities of untwisting, unzipping, rezipping, and retwisting all take place at the same time.

Interphase

Often erroneously described as a resting period between cell divisions, **interphase** is actually a period of great metabolic activity, occupying about 90 percent of the total duration of the cell cycle. It is the period during which the normal activities of the cell take place: growth, cellular respiration, and RNA and protein synthesis. It also sets the stage for cell division, since *DNA replication is completed during interphase.*

INTERPHASE

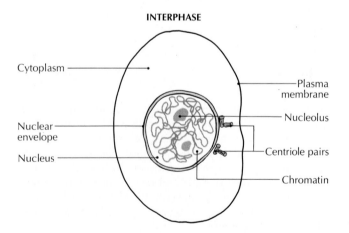

How often do mistakes occur during DNA replication?	*Less than one mistake is made for every billion nucleotides added. Genetic mistakes, or* mutations, *can have far-reaching effects, but most have little or no effect on the body.*

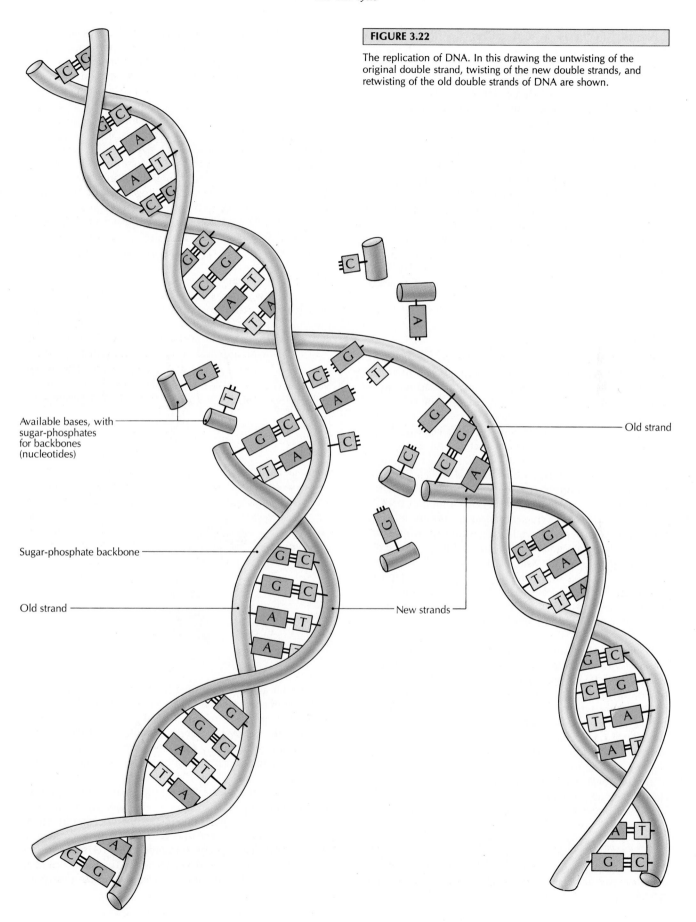

FIGURE 3.22

The replication of DNA. In this drawing the untwisting of the original double strand, twisting of the new double strands, and retwisting of the old double strands of DNA are shown.

Available bases, with sugar-phosphates for backbones (nucleotides)

Sugar-phosphate backbone

Old strand

New strands

Old strand

Visually, interphase is distinguished by the presence of thick chromatin threads and one or more nucleoli in the nucleus. There are three distinct stages of interphase:

1 *G1:* Immediately after a new daughter cell is produced, it enters a period of growth also known as the "first gap," because it represents the "gap" between cell division and DNA replication. Following cell division, one centriole pair separates and begins to replicate, as do other organelles.

2 *S:* After G1 the cell enters the S phase (for *synthesis,* because DNA is synthesized then), and the DNA of the chromosome doubles. Important here is the exact replication of DNA, which is not just a doubling of the quantity, but a doubling in which every chromosome duplicates itself. Two pairs of centrioles appear during this phase.

3 *G2:* With the S phase past and the chromosomes doubled, the cell goes into a "second gap" phase. The centriole pairs start to move apart as a prelude to the next cell division, and structures directly associated with mitosis are assembled.

Mitosis

Once the parent cell has duplicated its DNA, the actual division of the cell can begin. This division has two parts; the first is nuclear division, or *mitosis,** and the second, to be described below, is *cytokinesis,* the division of the cytoplasm. Mitosis and cytokinesis together are referred to as the *M phase* (see Figure 3.21). **Mitosis** accomplishes two things: (1) the arrangement of all cellular material for equal distribution between the daughter cells, and (2) the actual division of the nuclear material (DNA), and the *equal distribution of DNA to each new cell.* Mitosis is triggered by the replication of DNA, but it is not known what initiates the replication of DNA.

Although mitosis is a single, continuous process, it is usually described in four sequential stages known as *prophase, metaphase, anaphase,* and *telophase.* Each of these will be discussed in detail. For the sake of clarity, the drawings that accompany the text show only four chromosomes (and no organelles except the centrioles). Human cells contain 46 chromosomes, but the principles of mitosis are the same no matter how many chromosomes are involved.

Prophase After interphase, the cell enters the first stage of mitosis, **prophase.** In *early prophase,* the chromatin threads begin to coil, becoming shorter and thicker. (Before coiling begins, the total length of the DNA strands in a single human nucleus is greater than the length of the entire human body, but the strands are too thin to be seen with an ordinary light microscope.) The nucleoli and nuclear envelope begin to break up, and a burst of microtubules, called *asters* (L. stars), begins to radiate from the centrioles.

By *late prophase,* the chromatin threads coil again, forming clearly defined chromosomes. Each chromosome has two strands, or *chromatids,* each strand having a full complement

*The words *mitosis* and *mitochondrion* are both derived from the same Greek word, *mitos,* which means "thread." This root word refers to the threadlike appearance of chromosomes during mitosis, and to the threadlike shapes of mitochondria as viewed by early microscopists.

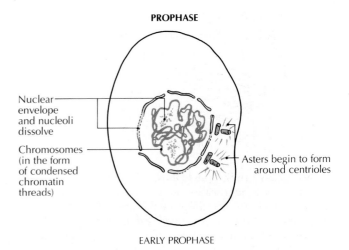

PROPHASE

Nuclear envelope and nucleoli dissolve

Chromosomes (in the form of condensed chromatin threads)

Asters begin to form around centrioles

EARLY PROPHASE

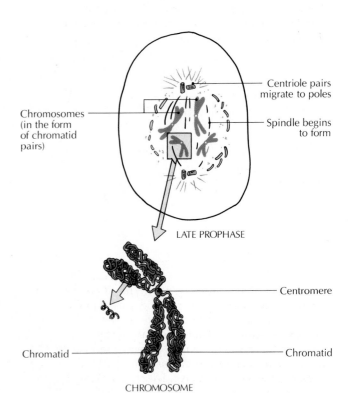

Centriole pairs migrate to poles

Chromosomes (in the form of chromatid pairs)

Spindle begins to form

LATE PROPHASE

Centromere

Chromatid

Chromatid

CHROMOSOME

of the replicated DNA formed in the S stage of interphase. Each pair of chromatids is joined somewhere along its length by a small spherical structure called a *centromere.* The fragments of the nuclear envelope and nucleoli disperse in the endoplasmic reticulum, and newly formed microtubules move in among the chromatid pairs. The two centriole pairs move apart.

By the *end of prophase,* the centriole pairs have moved to opposite ends, or *poles,* of the nucleoplasm. The position of the centrioles at the poles determines the direction in which the cell divides. Between the centrioles, the microtubules form a *spindle* that extends from pole to pole. The chromatid pairs move to the center of the spindle. (The asters, spindle, centrioles, and microtubules are called the *mitotic apparatus.*)

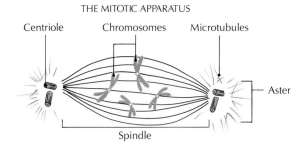

THE MITOTIC APPARATUS

Centriole　　Chromosomes　　Microtubules

Aster

Spindle

Telophase In *telophase,* the chromosomes arrive at the poles. In *early telophase,* the chromosomes lose their distinctive rodlike form and begin to uncoil. They appear as they did during interphase. The spindle and asters dissolve. By *late telophase,* fragments from the endoplasmic reticulum spread out around each set of chromosomes, forming a new nuclear envelope. The cell begins to pinch in at the middle. New nucleoli are formed from the nucleolar regions of the chromosomes. *Mitosis* is over, but *cell division* is not.

Metaphase During **metaphase,** the centromere of each chromatid pair attaches to one of the microtubules making up the spindle. The centrioles are pushed apart as the spindle lengthens, and the double-armed chromosomes are pulled to the center of the nucleoplasm, lining up across the spindle. Toward the end of metaphase, the centromeres double, so that each chromatid has its own. At this point, each chromatid may be considered a complete chromosome, each with its double-stranded DNA molecule.

TELOPHASE

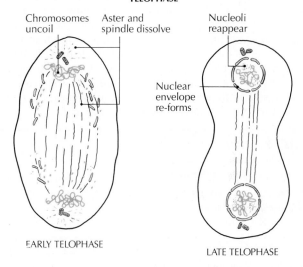

Chromosomes uncoil　　Aster and spindle dissolve

Nucleoli reappear

Nuclear envelope re-forms

EARLY TELOPHASE

LATE TELOPHASE

METAPHASE

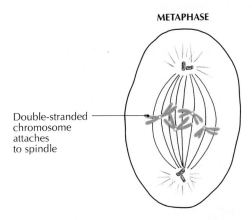

Double-stranded chromosome attaches to spindle

Cytokinesis

The third major phase of the cell cycle, and the final stage of cell division, is **cytokinesis** (Gr. *cyto,* cell + *kinesis,* movement), the separation of the cytoplasm into two parts. Before cytokinesis, the two newly formed nuclei still share the same cytoplasmic compartment. The separation is accomplished by *cleavage,* a pinching of the plasma membrane. The *cleavage furrow,* where the pinching occurs, looks as though someone tied a cord around the middle of the cell and pulled it tight. Two new, genetically identical daughter cells are formed, each about half the size of the original parent cell.

Anaphase The shortest stage of mitosis is **anaphase.** In *early anaphase,* the chromosome pairs separate, and the members of each pair begin to move toward opposite poles. Although the actual mechanism is not completely understood, the microtubules are generally acknowledged to be instrumental in moving the chromosomes toward the poles. By *late anaphase,* the poles themselves have moved farther apart. At this stage in human mitosis, 46 chromosomes would be near one pole, and 46 would be near the opposite pole.

CYTOKINESIS
(produces two cells)

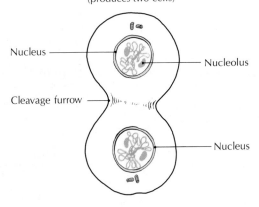

Nucleus

Nucleolus

Cleavage furrow

Nucleus

ANAPHASE

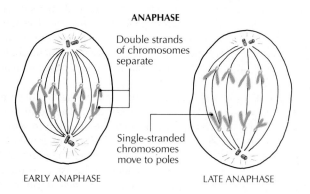

Double strands of chromosomes separate

Single-stranded chromosomes move to poles

EARLY ANAPHASE

LATE ANAPHASE

The Cell Cycle as a Continuum

Although we have described the cell cycle in neatly defined stages, the events of the cycle actually occur in a dynamic environment. Under laboratory conditions, some events of metaphase and anaphase may seem to occur almost simultaneously. Cytokinesis may begin during telophase or anaphase, and the doubling of the centrioles may occur before cytokinesis is complete.

Most cells, such as skin cells and cells in the lining of the digestive system, divide continually. The new cells replace old or damaged ones. The life span of a single cell might be 24 hours, with the cell division occurring in the first hour, and the rest of the cycle spent in interphase. Some cells, however, rarely divide. Muscle cells seldom do, and except for olfactory neurons (nerve cells involved with the sense of smell), after birth nerve cells never do. Apparently, the more specialized a cell is, the less frequently it divides.

Ask Yourself

1 *What is cellular differentiation?*

2 *Why is it necessary for DNA to replicate before mitosis begins?*

3 *How does complementary bonding occur in DNA?*

4 *What occurs during interphase?*

5 *What are the stages of mitosis?*

6 *What does mitosis accomplish?*

7 *What is cytokinesis?*

HOW PROTEINS ARE SYNTHESIZED

Protein synthesis—the making of proteins from amino acids—provides not only the structural proteins the body needs, but also the enzymes (which are also proteins) that control a cell's metabolism. By regulating protein synthesis, a cell regulates its own metabolism.

The many proteins the body needs to maintain homeostasis are produced from amino acids that must be assembled according to each individual's own genetic blueprint. This genetic program for the specific linking sequence of amino acids is contained in DNA, which is located in the cell's nucleus. The DNA itself does not assemble the amino acids in

FIGURE 3.23

Transcription: the formation of RNA from DNA. The double strand of DNA in the nucleus unwinds, and the strands separate to make a template for a complementary strand of mRNA, which migrates out of the nucleus into the cytoplasm.

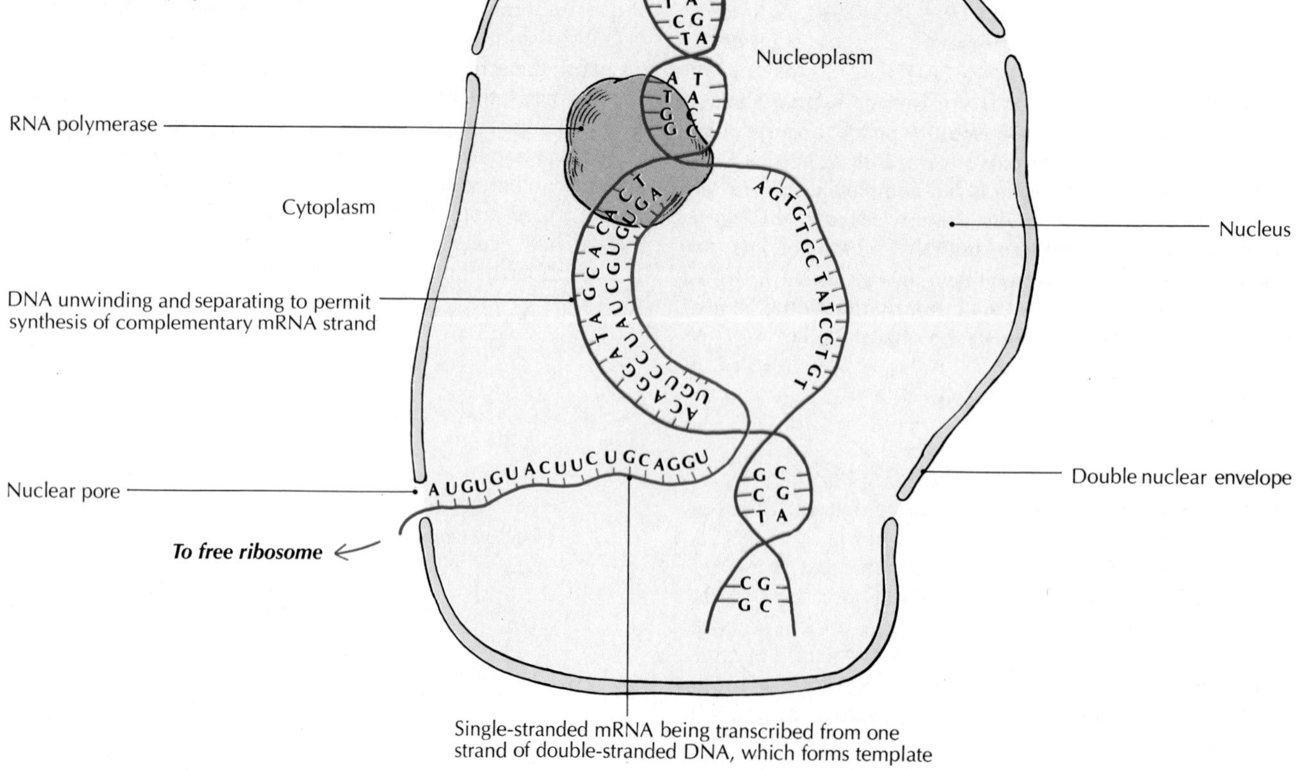

Single-stranded mRNA being transcribed from one strand of double-stranded DNA, which forms template

the correct sequence. Instead, DNA works together with RNA, which is present in the cytoplasm as well as in the nucleus—an important difference. *RNA is not restricted to the nucleus,* but is free to move about the cytoplasm, where protein synthesis actually occurs. The following drawings and descriptions depict the process of protein synthesis. (The illustrations are schematic and not drawn to scale; they represent only the principal steps of protein synthesis. Also, ribosomes are shown in a simplified form for clarity.)

Transcription

The first step in protein synthesis takes place in the nucleus and is called *transcription* (Figure 3.23). During transcription, single-stranded RNA is synthesized under the direction of DNA. With the aid of an enzyme, *RNA polymerase,* a portion of the double helix of DNA temporarily uncouples, and the sequence of nucleotide bases from one of the single strands of DNA is transcribed, in the form of complementary nucleotide bases, to a single strand of RNA. This RNA molecule is built up from nucleotides in the nucleoplasm, and has the base uracil (U) substituted for thymine (T). The RNA molecule is appropriately called *messenger RNA (mRNA),* because it is carrying the genetic message from the DNA in the nucleus to the ribosomes in the cytoplasm, where it directs the synthesis of protein.

Translation

The transcription of RNA from DNA is just the first step in the complex process of protein synthesis. After mRNA receives the genetic message from DNA, the DNA and mRNA uncouple, and the DNA again assumes a double helix form. The mRNA strand is now free to carry that message out of the nucleus and into the cytoplasm, where the genetic message must be read by ribosomes to produce the polypeptide chains that make up proteins. This part of the overall process of protein synthesis—the formation of a polypeptide under the control of mRNA in the cytoplasm—is called *translation.*

The transcription of mRNA from DNA is the equivalent of keeping the same language, but changing the medium—from an audiotape to a typewritten page, for example. In contrast, the translation of RNA to protein is the same as changing languages, or *translating* languages. The process of translation may be thought of as occurring in three stages: initiation, elongation, and termination.

Initiation The newly transcribed strand (molecule) of mRNA leaves the nucleus and enters the cytoplasm. The front end of the mRNA strand attaches to a ribosome and begins to move across it. The ribosome, in effect, decodes the message by reading the sequence of nucleotide bases as a series of triplets, or *codons,* each coding for a specific amino acid. Translation also involves another type of RNA found in cytoplasm, *transfer RNA (tRNA).* The function of tRNA is to pick up amino acids in the cytoplasm and put them in position opposite the appropriate codons of mRNA. Because the tRNA triplet is the complementary match of the mRNA codon, it is called an *anticodon.*

The molecules of tRNA are specialized; they provide the link between a specific codon on mRNA and a specific amino acid. Activated by an enzyme and energy from ATP, a specific tRNA molecule bonds to one of the 20 amino acids found in cytoplasm. For example, the tRNA specific for the amino acid proline bonds only to that amino acid; the tRNA specific for cysteine bonds only to cysteine, and so on. The mRNA codon for proline (CCU) pairs with the tRNA anticodon for proline

---

FIGURE 3.24

Initiation stage of protein synthesis. (A) After being transcribed from a DNA template in the nucleus, the mRNA strand moves out of the nucleus through a nuclear pore. An initiator tRNA molecule forms a covalent bond with its specific amino acid (methionine), and the tRNA anticodon (UAC) binds to the small ribosomal subunit. Note that the simplified version of a ribosome is being used in this and the following drawings. (B) The initiator tRNA-amino acid complex is hydrogen-bonded to its complementary mRNA initiation ("start") codon on the mRNA strand, and the small ribosomal subunit binds to mRNA. (C) The large ribosomal subunit binds to the small subunit, and the initiator tRNA, together with its amino acid, moves into the P-site on the large subunit.

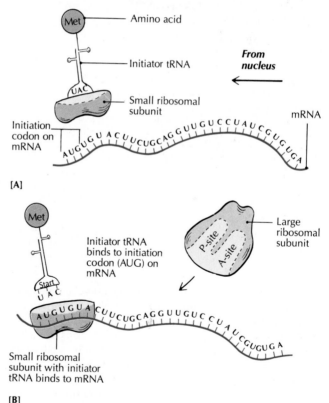

(GGA); the codon for cysteine (UGU) pairs with the anticodon for cysteine (ACA).

The first mRNA codon is always AUG. It codes for the amino acid methionine (Met), which functions as a "start" signal. ***Initiation*** of protein synthesis actually begins when a special *initiator tRNA* molecule carrying the amino acid methionine binds to the small ribosomal subunit (Figure 3.24A). The initiator tRNA anticodon (UAC), still attached to the small ribosomal subunit, recognizes the AUG initiation ("start") codon on the mRNA molecule and binds to it (Figure 3.24B). Next, the large ribosomal subunit binds to the small subunit, and the initiator tRNA moves into the P-site (Figure 3.24C; see also Figure 3.13B). Initiation is now complete, and the elongation of a polypeptide chain is about to begin:

INITIATION COMPLETE

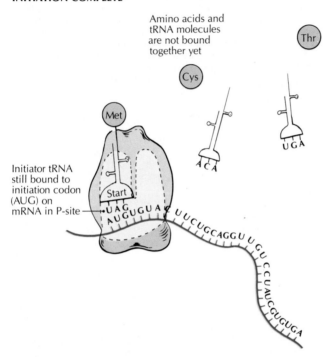

Elongation During the process of ***elongation***, amino acids are linked together to form a polypeptide chain. This process begins in the cytoplasm, away from the ribosome, when molecules of tRNA and their specific amino acids are activated by energy from ATP and become linked by a covalent bond. Each new amino acid-tRNA complex binds to the A-site on the large ribosomal subunit. The incoming amino acid forms a peptide bond with the first amino acid (methionine) as the polypeptide chain is begun (see Elongation 1).

The initiator tRNA, still bonded to its amino acid (methionine), moves out of the P-site and separates from its amino acid. At the same time, the second tRNA-amino acid complex moves into the P-site, exposing the A-site to the next incoming tRNA and its amino acid. As the mRNA strand continues to move across the ribosome one codon at a time, another tRNA, with its attached amino acid, binds to the A-site. The incoming amino acid links up with the previous one, continuing the polypeptide chain. These tRNA molecules can be used again after they have deposited their amino-acid "pas-

ELONGATION 1

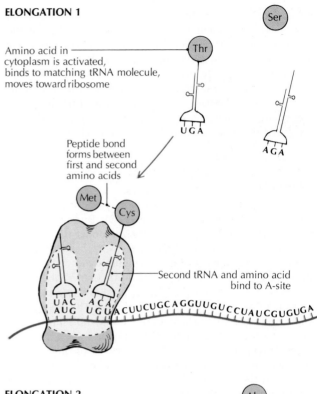

ELONGATION 2

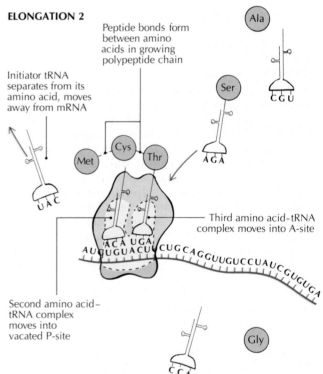

sengers." Additional tRNA molecules and their attached amino acids approach the mRNA strand (see Elongation 2).

The mRNA strand moves steadily onward, and more links are made between mRNA codons and tRNA anticodons. One after another, amino acids are bound together in an ever-lengthening chain until a complete protein molecule is built up according to the codon order on the mRNA (see Elongation 3).

ELONGATION 3

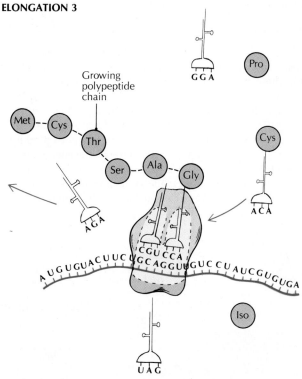

After the first ribosome has moved well past the AUG "start" codon on the mRNA, another ribosome may bond to the strand and initiate the synthesis of a second polypeptide chain. This process may occur repeatedly, so that a string of five or more ribosomes is involved, forming a polyribosome (see Figure 3.13C). Each ribosome in the group operates independently, each one synthesizing a polypeptide chain.

TERMINATION 1

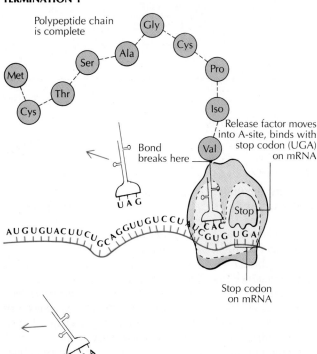

Termination The *termination* of the translation process occurs when one of the three mRNA "stop" codons (UAA, UAG, or UGA) reaches the A-site on the ribosome. A protein called a *release factor, not* a tRNA, binds directly to the stop codon, causing the hydrolysis of the bond linking the polypeptide chain to the peptide-tRNA at the P-site (see Termination 1).

The completed polypeptide chain is released from the ribosome (the chain shown here is greatly shortened for simplicity), the empty tRNA at the P-site is ejected, the release factor is released from the A-site, and the ribosomal subunits separate and move into the cytoplasmic pool, where they will be available for another cycle of protein synthesis (see Termination 2).

TERMINATION 2

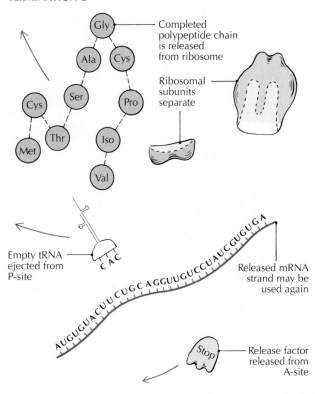

The polypeptide chain may combine with other polypeptide chains if a more complex protein molecule is to be formed. The completed protein then performs its job. Some proteins remain inside the cell, and others, such as hormones, leave the cell for their assigned destinations.

Ask Yourself

1 *Where is the genetic program located?*

2 *Why is it important for RNA to be able to leave the cell's nucleus?*

3 *What roles in protein synthesis are played by mRNA and tRNA?*

4 *What is a codon?*

Cellular Metabolism: How a Cell Obtains Usable Energy

The original energy source for a cell is the food we eat, which eventually is used to generate ATP in the cell after a series of chemical reactions. The chemical processes that transform food into living tissue and energy are called *metabolism.*

The simplified drawing below shows how metabolism works: (1) Food is eaten and moved into the digestive system. (2) In the digestive system, food is bro-ken down into smaller molecules of simple sugars such as glucose, and fatty acids and amino acids. (3) These small molecules are absorbed from the small intestine into the bloodstream, and are carried into individual cells. (4) The simple sugars, fatty acids, and amino acids enter the cytoplasm of a cell. (5) Here, the molecules are converted into smaller molecules of pyruvic acid, and a small amount of ATP is produced. (6) The py-ruvic acid, fatty acids, and amino acids then enter the mitochondria. (7) In the mitochondria, in the presence of oxygen, the molecules are broken down (reduced) further in a complicated series of chemical reactions (described in detail in Chapter 23). (8) The reactions produce enough ATP to provide the energy that a cell needs to function properly. Also produced at the same time are molecules of waste carbon dioxide and water.

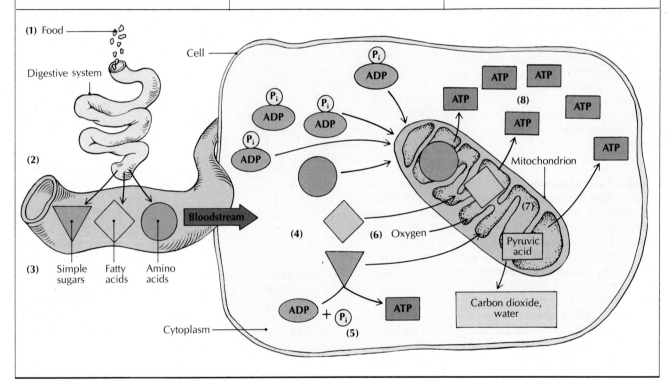

SUMMARY

What are cells?

1 The *cell theory* states that (a) all living things are composed of cells and cell products; (b) the cell is the basic unit of structure and function in all living things; (c) all cells come from the division of pre-existing cells; and (d) an organism as a whole can be understood through the collective activities and interactions of its cellular units.

2 Cells can range widely in size, color, shape, and function.

3 All cells have four basic parts: the *nucleus, nucleoplasm, cytoplasm,* and the *plasma membrane.*

4 The part of a cell that is not cytoplasm is the *nucleus,* the control center of the cell. The nucleus contains the chromosomes that direct the reproduction of, and contain the genetic blueprint for, new cells. The nucleus consists of *nucleoplasm.*

5 The portion of the cell surrounding the nucleus is *cytoplasm.* The fluid portion of the cytoplasm is called *cytosol;* it contains various subcellular structures called *organelles.*

6 The *plasma membrane* forms the outer boundary of the cell.

Cell membranes

1 Cell membranes are selective screens that allow certain substances to get into and out of cells. That quality is called *selective permeability.*

2 Cell membranes are composed mainly

of phospholipids and proteins, and form a dynamic *fluid-mosaic model.*

Passive movement across membranes

1 *Passive transport* occurs when molecules use their own energy to pass through a cell membrane from areas of high concentration to areas of lower concentration.
2 Examples of passive processes are *simple diffusion, facilitated diffusion, osmosis, filtration,* and *dialysis.*

Active movement across membranes

1 Active processes of movement across membranes require energy from the cell, and can move substances through a selectively permeable membrane from areas of low concentration to areas of higher concentration.
2 Active processes include *active transport, endocytosis, phagocytosis,* and *exocytosis.*

Cytoplasm and organelles

1 The cytoplasm of a cell is composed of two parts, or phases. The *particulate phase* consists of membrane-bound vesicles and organelles, lipid droplets, and inclusions. The *aqueous phase* consists of fluid cytosol. The typical components of cytosol are water, nucleic acids, proteins, complex carbohydrates, lipids, and inorganic substances.
2 The *endoplasmic reticulum (ER)* creates a series of channels for transport, stores enzymes and other proteins, and provides a point of attachment for ribosomes. Smooth ER does not contain ribosomes.
3 *Ribosomes* are the sites of protein synthesis.
4 The *Golgi apparatus* aids in the synthesis of glycoproteins. It also aids in the secretion of these products.
5 *Lysosomes* digest nutrients, and clean away dead or damaged cell parts.
6 *Mitochondria* produce most of the energy (in the form of ATP) required by a cell for metabolic activities.

7 *Microtubules, intermediate filaments,* and *microfilaments* (the *cytoskeleton*) provide a transport system, a supportive framework, and assist with organelle and chromosome movement.
8 *Centrioles* assist in cell reproduction and are involved with the movement of chromosomes during cell division.
9 *Cilia* are appendages that help move solids and liquids past fixed cells, and flagella help free-swimming cells move.
10 *Cytoplasmic inclusions* are chemical substances that are usually either basic food material or the stored products of the cell's metabolic activities.

The nucleus

1 The *nucleus* of a cell contains DNA, which controls the cell's genetic program, including heredity, protein structure, and other metabolic activities.
2 The *nuclear envelope* contains many *nuclear pores* that allow material to enter and leave the nucleus.
3 Within the nucleus, *chromosomes* store the segments of DNA called *genes,* which contain the hereditary information for protein synthesis.
4 The *nucleolus* is a preassembly point for ribosomes.

The cell cycle

1 The process of cells specializing in structure and function is called *differentiation.*
2 When DNA *replicates,* a copy of itself is made to ensure that each new cell receives the same genetic information as the parent cell.
3 *Mitosis* distributes all cellular material from the parent cell to the daughter cells equally; it includes the actual division of the DNA. It is divided into four sequential stages: *prophase, metaphase, anaphase,* and *telophase.*
4 *Cytokinesis* is the process of cytoplasmic division that produces two daughter cells after the nucleus divides during mitosis.

How proteins are synthesized

1 The many proteins that the body needs

are produced from amino acids, which are assembled according to each person's genetic blueprint. This genetic program is contained in the DNA in the nuclei of cells.
2 Several kinds of RNA carry out the program contained in DNA. *Messenger RNA (mRNA)* carries the genetic message out of the nucleus to the ribosomes in the cytoplasm. *Transfer RNA (tRNA)* links up with specific amino acids, and a growing polypeptide chain of amino acids eventually develops into a complete protein molecule.
3 Protein synthesis occurs in two major steps, transcription and translation, with translation consisting of initiation, elongation, and termination.
4 The chromosomes in the nucleus contain DNA, whose sequence of nucleotides codes for the synthesis of specific proteins. The message is transcribed from one of the strands of the DNA to a single-stranded mRNA molecule. This is *transcription.* The DNA molecule serves as a template, or mold, so that the mRNA strand is complementary to the DNA strand.
5 The mRNA carries the message out of the nucleus to the ribosomes in the cytoplasm. The formation of a polypeptide from mRNA in the cytoplasm is *translation.* The mRNA attaches itself to a ribosome. *Initiation* begins when the specific tRNA triplet of nucleotide bases (an *anticodon*) on an *initiator tRNA* molecule, bound to the amino acid methionine, binds to the mRNA triplet (*codon*) that codes for methionine, which signals "start." During *elongation,* the polypeptide chain lengthens as amino acids are carried to the ribosome by their matching tRNA, and each tRNA binds to its appropriate codon triplet on the mRNA strand. Amino acids are bonded together in a growing chain until a protein molecule is built up according to the codon order on the mRNA molecule. The appearance of a "stop" codon signals the end of the polypeptide chain, and releases the chain from the ribosome. This is *termination.*

UNDERSTANDING THE FACTS

1 What is the most abundant substance in cytosol?

2 What is the relationship between the quantity of protein in a plasma membrane and its metabolic activity?

3 In which method of movement of material through a membrane is your blood pressure a key factor?

4 Distinguish between smooth and rough ER.

5 Distinguish between the centrosome and centrioles.

6 What is an important difference between DNA and RNA as far as their locations within a cell are concerned?

7 Define *mitosis*.

8 Does DNA replication occur during mitosis? Explain.

9 During the process of cell division, when are the chromosomes first visible?

10 Which structures of a dividing cell are composed of microtubules?

UNDERSTANDING THE CONCEPTS

1 Is a cell composed entirely of living material? Explain.

2 What would happen if your cells lacked peroxisomes? Why?

3 What clues can you derive as to the function of a cell from the number of mitochondria it contains? What about the lysosomes?

4 Why is a "cell" that lacks a nucleus destined to have a relatively short life? Give one example of such a cell.

SUGGESTED READING

ALBERTS, BRUCE, DENNIS BRAY, JULIAN LEWIS, MARTIN RAFF, KEITH ROBERTS, AND JAMES D. WATSON, *Molecular Biology of the Cell,* 2nd ed. New York: Garland, 1988.

ALLEN, ROBERT DAY, "The Microtubule as an Intracellular Engine." *Scientific American,* February 1987.

DARNELL, JAMES, HARVEY LODISH, AND DAVID BALTIMORE, *Molecular Cell Biology.* New York: Scientific American Books, 1986.

DUSTIN, PIERRE, "Microtubules." *Scientific American,* August 1980.

HANAWALT, PHILIP C., ed. *Molecules to Living Cells: Readings from Scientific American.* San Francisco: Freeman, 1980 (paperback).

LAKE, JAMES A, "The Ribosome." *Scientific American,* August 1981.

"The Molecules of Life." *Scientific American,* October 1985. The entire issue is devoted to this subject.

PORTER, KEITH R., AND JONATHAN B. TUCKER, "The Ground Substance of the Living Cell." *Scientific American,* March 1981.

ROTHMAN, J. E., "The Compartmental Organization of the Golgi Apparatus." *Scientific American,* September 1985.

RUBENSTEIN, EDWARD, "Diseases Caused by Impaired Communication Among Cells." *Scientific American,* March 1980.

WEBER, KLAUS, AND MARY OSBORN, "The Molecules of the Cell Matrix." *Scientific American,* October 1985.

WOLFE, STEPHEN L., *Biology of the Cell,* 2nd ed. Belmont, Calif.: Wadsworth, 1981.

4
Epithelial and Connective Tissues

LEARNING OBJECTIVES

1 Define a tissue.

2 Give an overview of the major functions of epithelial tissues.

3 Present some general characteristics of epithelial tissues.

4 Describe the general classification of epithelial tissues.

5 List three exceptions to the general classification of epithelial tissues.

6 Describe the fibers, ground substance, and cells of connective tissues.

7 Discuss the main types of connective tissues, including cartilage, and give their functions.

8 Describe the four types of membranes found within the body.

9 Explain the role of collagen in several disorders of the connective tissues.

In a multicellular organism, such as the human body, individual cells differentiate during development to perform special functions. These specialized cells carry out their functions as multicellular aggregates called *tissues.* The study of tissues, *histology* (Gr. *histos*, web), provides many excellent examples of how structures in the body have cellular arrangements that are closely related to their functions. In this chapter we look at epithelial and connective tissues. Osseous (bone) tissue is described in Chapter 6, muscle tissue (including heart muscle) in Chapter 9, and nervous tissue in Chapter 11.*

EPITHELIAL TISSUES: FORM AND FUNCTION

Epithelial tissue exists in many structural forms, but in general it either *covers* or *lines* something. The skin covers the external surface of the body, and epithelial tissue lines the gastrointestinal tract, ducts of glands, the mouth, and other body cavities. Another common function of epithelial tissue is the secretion of substances that lubricate parts of the body, or take part in vital chemical reactions within the body. To suit these functions, epithelial tissues are typically made up of renewable flat sheets of cells that have surface specializations adapted for their specific roles.

Epithelial tissues (or simply, *epithelia*) can be arranged in two different ways, which correspond to their two different functions:

1 Most epithelial tissues are composed of cells arranged in sheets one or more layers thick. The function of these sheets is to cover surfaces or line body cavities or *ducts* (passages) that often connect with the surface of the body. Some tissues line the surfaces of internal organs, and are shaped to carry out the functions of absorption and protection.

2 Other epithelial tissues are organized into glands adapted for secretion. These modified epithelia are classified as *glandular epithelia.*

Functions of Epithelial Tissues

The typical functions of epithelial tissues are *absorption* (by the lining of the small intestine, for example), *secretion* (by glands), *transport* (by kidney tubules), *excretion* (by sweat glands), *protection* (by the skin), and *sensory reception* (by the taste buds). The size, shape, and arrangement of epithelial cells are directly related to these specific functions. For example, flattened sheets of epithelial cells in the outer tissues covering the body are well suited to protect it from injury and the loss of body fluids. These tissues also contain many sensitive, specialized receptors that allow us to feel the touch of a sharp nail before it punctures us, to taste, to smell, and in general to be aware of the world around us.

*Several other kinds of tissues are described with their relevant systems in separate chapters rather than being placed in one chapter. The authors have chosen this organization because they believe that it gives the reader a better understanding of how anatomy and physiology work together, and how the parts of the body make up a cohesive whole.

Although epithelial tissues prevent most of the physical objects of the outside world from entering the body, and keep body fluids from getting out, they also are the gateway for all the substances that *do* enter or leave the body. Internal body surfaces are lined with epithelial tissue that is specialized for absorption or secretion. Through the process of *absorption* (L. *absorbere: ab,* away from + *sorbere,* to suck), for example, the body takes in digested nutrients from the small intestine. Likewise, some epithelial cells are adapted for *secretion* (L. *secernere,* to separate), which is the process of forming various substances that are either eliminated or used by the body to carry on special functions. For example, the goblet cells that line parts of the digestive system secrete mucus that lubricates the tissue surface. Some secretory epithelial tissue is modified into glands that secrete their substances through ducts (such as salivary glands) or directly into the blood (such as the thyroid or adrenal glands).

The process of *excretion* (L. *excernere,* to sift out), unlike secretion, eliminates waste matter from the body via epithelial tissues in the kidneys and sweat glands. Part of the excretory function is the transport of wastes, water, and dissolved substances from inside the body to the outside. Throughout this chapter you will see how these varied functions are enhanced by the different structures of epithelial tissues.

General Characteristics of Epithelial Tissues

A covering or lining epithelial tissue may be single-layered or multilayered, and its cells may have different shapes. However, all kinds of epithelial tissues, except glandular tissues, share certain characteristics.

Cell shapes and junctions The cells that make up epithelial tissues are relatively regular in shape and are closely packed in continuous sheets. There is little or none of the extracellular material known as *matrix* between epithelial cells. The framework of the matrix that does exist for epithelial tissue is composed of *ground substance,* consisting of glycoproteins. Instead of depending on a matrix for support, the tight-fitting epithelial cells are held in place by strong adhesions formed between the plasma membranes of adjacent cells. The specialized parts that hold the cells together are known as *junctional complexes;* they enable groups of cells to function as a unit. There are three main kinds of junctional complexes: tight junctions, desmosomes, and gap junctions.

A *tight junction* is the site of close connection between two plasma membranes, with no extracellular space between the cells (Figure 4.1A). This "nonleaky" junction creates a permeability barrier that stretches across a continuous layer of epithelial cells, keeping material either in or out. Tight junctions are found in the epithelial tissues of the urinary bladder, where they hold urine within the bladder.

A *desmosome,* or *spot desmosome* (Gr. *desmos,* binding), is a junction with no direct contact between adjacent plasma membranes (Figure 4.1B). It is joined by an intercellular bonding material, in the form of threads or filaments, and many fibers, called tonofilaments, that extend into the plasma

FIGURE 4.1

A unique feature of epithelial cells is the presence of various functionally and structurally distinct intercellular junctions between their plasma membranes. The three main kinds of junctions are shown here in simplified form. (A) A tight junction. (B) A spot desmosome. (C) A gap junction.

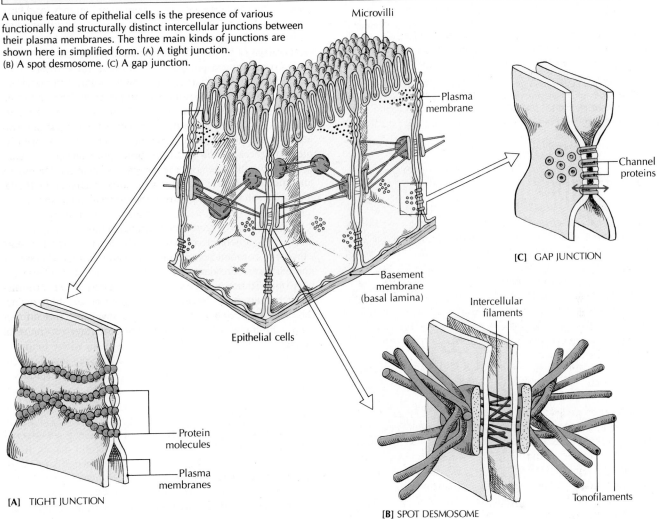

Microvilli

Plasma membrane

Channel proteins

[C] GAP JUNCTION

Basement membrane (basal lamina)

Epithelial cells

Intercellular filaments

Protein molecules

Plasma membranes

[A] TIGHT JUNCTION

Tonofilaments

[B] SPOT DESMOSOME

membrane to produce extremely strong "spot welds." Desmosomes are common in skin, where great stress is constantly being applied to the junctions.

A *gap junction* is formed from several links of channel protein connecting two plasma membranes (Figure 4.1c). Gap junctions are found in intestinal epithelial cells, where they allow the flow of ions and small molecules between adjacent cells.

Basement membranes Epithelial tissue is anchored to the underlying connective tissue by a basement membrane. The extracellular **basement membrane** (also called the *basal lamina*) is a thin layer composed of tiny fibers and nonliving polysaccharide material produced by the living epithelial cells (see Figures 4.1 and 4.2A). By firmly attaching the basal surface of the epithelial tissue to the underlying connective tissue, the basement membrane assures that the epithelium is held in position, reducing the possibility of tearing. Basement membranes vary in thickness throughout the body. They are quite thin in the skin and intestines and rather thick in the trachea. The principal roles of the basement membrane are

to provide elastic support and to act as a partial barrier for diffusion and filtration.

Lack of blood vessels Epithelial tissues do not contain blood vessels; they are *avascular*. Oxygen and other nutrients diffuse through their selectively permeable basement membranes from capillaries in the underlying connective tissue. Wastes from the epithelial tissues diffuse into the connective tissue capillaries. Although epithelia have no blood vessels, they may contain nerves. However, epithelial tissues in the stomach, intestines, and cervix of the uterus do *not* contain nerves.

Surface specializations Epithelial tissues usually have several types of surface specializations. Some epithelial cells, like the ones that line the ventral body cavities and blood vessels, have smooth surfaces, but most epithelial cells have irregular surfaces that result from the complex folding of their outer membranes. This extended folding produces **microvilli,** fingerlike projections that greatly increase the absorptive area of the cell (Figures 4.2A, B). Microvilli usually appear on cells

FIGURE 4.2

Microvilli. (A) Elaborate folding of the plasma membrane is evident not only in the microvilli (the outer surface) but also on the lateral and basal surfaces. The clusters of mitochondria at the base indicate an area of much cellular activity. Notice the tight junction between cells. (B) A scanning electron micrograph of the small intestine, where the extra surface area created by the microvilli aids the absorption of digested food into the bloodstream. The microvilli resemble tufts of a shag rug, an appropriate image, since *villi* is Latin for "shaggy hairs." [*Don Fawcett, M.D./Hirokawa/John Heuser, M.D./Photo Researchers.*]

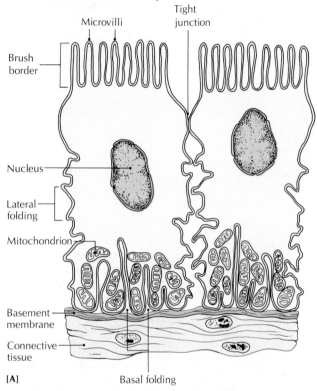

[A]

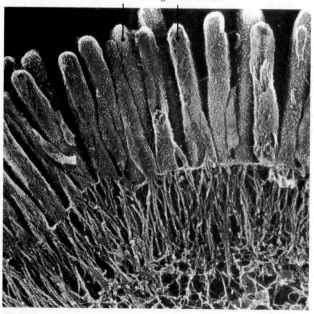

Microvilli forming brush border

[B]   × 30,000

that are involved in absorption or secretion. Some evidence suggests that the outer surface of microvilli on intestinal epithelium contains enzymes that break down sugar phosphates and reduce disaccharides to monosaccharides. The increased surface area of microvilli, coupled with their ability to break down complex compounds, aids the absorptive process so necessary for the efficient digestion of food. The shape of microvilli on the free surface of epithelial cells gives rise to the term **brush border**. As seen in Figure 4.2A, the elaborate folding of the plasma membrane may occur on lateral and basal surfaces of cells as well as on free surfaces. Other surface projections called **cilia** occur on some epithelial cells, such as in the respiratory tract, where their rapid lashing propels fluid and mucus over the surface of the epithelium. The beating motion of cilia creates a forceful paddling effect (see p. 74).

Regeneration Because epithelial tissue is located on surfaces, it is subject to constant injury. Fortunately, it is also capable of constant regeneration through cell division. When epithelial cells are damaged, they are shed and replaced by new cells. These cells are specialized to divide and move into the damaged area, where they can take on the function of the cells they replace. The rate of renewal depends on the type of epithelium. For example, the outer layer of the skin and the epithelial lining of the intestinal villi, where frequent friction is inevitable, are entirely replaced every few days. In contrast, the cells in glands and the lining of the respiratory tract are usually replaced every 5 to 6 weeks.

General Classification of Epithelial Tissues

Epithelial tissues are generally classified according to (1) the *arrangement of cells* and the *number of cell layers*, and subdivided further by (2) the *shape of the cells* in the superficial (top or outer) layer. In this section we discuss the two main groupings of epithelial tissues: *simple* and *stratified*.

If epithelial cells are one layer thick, they are **simple**; if they are arranged in two or more layers, they are described as **stratified**. The basic shapes of the superficial cells are **squamous** (flat), **cuboidal** (like a cube), and **columnar** (elongated). Using these descriptions, we may classify epithelial tissues according to their arrangement and shape as follows:

1 *Simple squamous epithelium:* a single layer of flat cells.

2 *Simple cuboidal epithelium:* a single layer of approximately cube-shaped cells.

3 *Simple columnar epithelium:* a single layer of tall, thin cells.

4 *Stratified squamous epithelium:* a multilayered arrangement of cells, with the superficial layer made up of flat cells.

5 *Stratified cuboidal epithelium:* a multilayered arrangement of cells, with the superficial layer made up of cube-shaped cells.

6 *Stratified columnar epithelium:* a multilayered arrangement of cells, with the superficial layer made up of tall, thin cells.

Type of tissue	Main locations	Description	Major functions
SIMPLE EPITHELIUM			
Simple squamous	Lining of lymph vessels, blood vessels, and heart (endothelium); glomerular (Bowman's) capsule in kidneys; lung air sacs (alveoli); small excretory ducts of many glands; inner ear membranes; serous membranes lining peritoneal, pleural, pericardial, and scrotal cavities (Figure 4.3A).	A single layer of flat cells, with scaly appearance. Nuclei are flattened, parallel to surface.	Permits diffusion or filtration through selectively permeable surfaces: for example, blood is filtered in kidneys to form urine; oxygen in lung alveoli diffuses into blood, and waste carbon dioxide from blood diffuses into alveoli.
Simple cuboidal	Lining of many glands and their ducts; surface of ovaries; inner surface of eye lens; pigmented epithelium of eye retina; some kidney tubules (Figure 4.3B).	A single layer of approximately cube-shaped cells. Nucleus of each cell is large and centrally located.	Secretion and absorption.
Simple columnar	Stomach, intestines (with microvilli), digestive glands, and gallbladder. Goblet cells are found in digestive tract, upper respiratory tract, uterine tube, and uterus (Figure 4.3C).	Single layer of cells taller than wide. Large, oval-shaped nuclei usually located at base of cell. May be ciliated or nonciliated; may secrete mucus (goblet cells); may have microvilli on free surfaces of cells.	Secretion, absorption, protection, lubrication; cilia and mucus combine to sweep away foreign substances; cilia may also help to move objects through a duct, as an egg cell in oviduct.
STRATIFIED EPITHELIUM			
Stratified squamous	Epidermis, vagina, mouth and esophagus, anal canal, distal end of urethra (Figure 4.3D).	Several layers of cells; only superficial layer composed of flat squamous cells. Underlying basal cells replace dying superficial squamous cells.	Protection.
Stratified cuboidal	Ducts of sweat glands, sebaceous (oil) glands, and developing epithelium in ovaries and testes (Figure 4.3E).	A multilayered arrangement of cells, with superficial layer composed of cuboidal cells.	Secretion.
Stratified columnar	Moist surfaces such as larynx, nasal surface of soft palate, parts of pharynx, urethra, and excretory ducts of salivary and mammary glands (Figure 4.3F).	A multilayered arrangement of cells, with superficial layer composed of tall, thin columnar cells. Sometimes ciliated.	Secretion and movement.
ATYPICAL EPITHELIUM			
Pseudostratified columnar	Large excretory ducts, most of male reproductive tract, nasal cavity and other respiratory passages, and part of ear cavity (Figure 4.3G).	Single layer of cells varying in height and shape; nuclei at different heights give false impression of cells being multi-layered. All cells in contact with basement membrane, but not all cells reach superficial layer. May be ciliated.	Protection, secretion, and movement of substances across surfaces.
Transitional	Urinary tract, where it lines bladder, ureters, urethra, and kidney calyxes (Figure 4.3H).	Stratified epithelium; surface cells cannot be classified by shape because it changes as tissue is distended. Usually no distinct basement membrane.	Allow for changes in shape.
Glandular	Various sites throughout body, including skin glands such as sweat and mammary glands, digestive glands such as salivary glands, and endocrine glands such as thyroid (Figure 4.4).	Epithelial cells modified to perform secretion. Main types are exocrine and endocrine; both are secretory, but only exocrine glands have ducts. Endocrine glands secrete hormones directly into bloodstream.	Synthesis, storage, and secretion of product.

These general types of epithelia, as well as some exceptions to this classification, are summarized in Table 4.1 and illustrated in Figures 4.3 and 4.4.

Simple squamous epithelium The main characteristic of *simple squamous epithelium* is its arrangement of flat cells in a single layer (Figure 4.3A). *Squamous* comes from the Latin word for scaly, as in the scaly skin of a reptile. The cells are very thin (that is, they are always wider than they are thick). This characteristic makes simple squamous epithelium suitable for areas where diffusion or filtration through a selectively permeable surface is important, as in the *endothelium,* the lining of the lymph vessels, blood vessels, and the heart.

Simple squamous epithelium is also found in (1) the glomerular (Bowman's) capsule in the kidneys, where blood is filtered to form urine; (2) lung air sacs (alveoli), where oxygen diffuses into the blood and waste carbon dioxide diffuses out; (3) smallest excretory ducts of many small glands; (4) membranes of the inner ear; and (5) serous membranes lining the peritoneal, pleural, pericardial, and scrotal body cavities—where it is called *mesothelium* and has microvilli. Clearly, simple squamous epithelium is not well suited as a protective covering because of the ease with which certain substances pass through its thin layer, but that same thinness makes it very appropriate for diffusion and filtration.

Simple cuboidal epithelium Looking down at a sheet of tightly packed cells of *simple cuboidal epithelium,* you may think you are seeing a neat arrangement of hexagonal floor tiles (Figure 4.3B). Actually, all epithelial tissues have a family resemblance when you observe them from above, and simple cuboidal epithelium is merely smaller and neater than

---

FIGURE 4.3A

The general types of epithelium. (A) Simple squamous epithelium. (1) A sheet of simple squamous epithelium usually shows a slight bulge in the center of the cell, where the large nucleus is located. (2) A stained preparation of simple squamous epithelium from the human abdominal cavity. Notice the centrally located nuclei and the irregular shape of the cells joined together at the interlocking edges to form a mosaic, like bits of a stained-glass window. [*Ed Reschke/Peter Arnold.*]

FIGURE 4.3B

(B) Simple cuboidal epithelium. (1) A simplified drawing of hexagonal cuboidal cells in neat rows. (2) This vertical section from the kidney reveals the large, round nuclei. [*Ed Reschke/Peter Arnold.*]

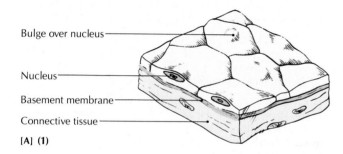

[A] (1)

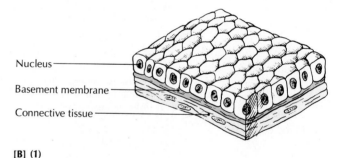

[B] (1)

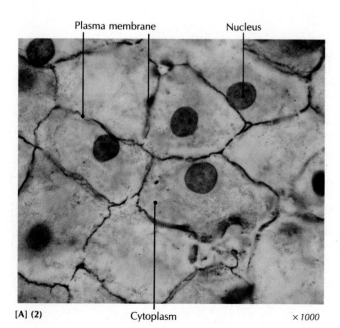

[A] (2)   Cytoplasm   ×1000

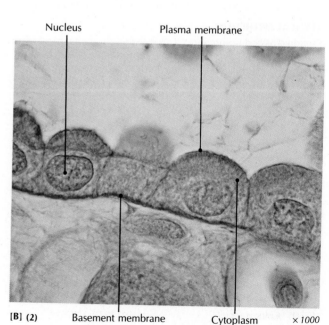

[B] (2)   Basement membrane   Cytoplasm   ×1000

simple squamous epithelium. In a vertical section it is possible to see that the nucleus of each cell is large, spherical, and centrally located. Depending on how cells are packed in different parts of the body, the individual cells may be shaped more like a trapezoid or a pyramid. Simple cuboidal epithelium is found in the lining of many glands and their ducts, the surface of the ovaries, the inner surface of the lens of the eye, the pigmented epithelium of the retina of the eye, and some kidney tubules, where it has microvilli on the free surface of the cell. Overall, simple cuboidal cells are active in absorption and secretion.

FIGURE 4.3C

(c) Simple columnar epithelium. (1) The simplified drawing indicates the usual position of the elongated nuclei near the base of the cells. (2) A vertical section of the tissue from the bile duct. (3) The goblet cells lining the digestive tract are modified simple columnar epithelial cells. Mucus accumulates at the top of the goblet cells, where it is secreted. (4) A scanning electron micrograph of a goblet cell, with microvilli, from the epithelial lining of the small intestine of a rat. [(2) Fred Hossler/Visuals Unlimited. (4) Courtesy of Mr. Bret Connors and Dr. Andrew P. Evan, Department of Anatomy, Indiana University.]

Simple columnar epithelium *Simple columnar epithelium* looks like simple cuboidal epithelium with the cells stretched into elongated columns (Figure 4.3C). The nuclei are large and oval-shaped, and are usually located near the base of the cell. Simple columnar epithelium may be ciliated or nonciliated, depending on its location and function. In the uterine tube (oviduct), for instance, ciliated cells help sweep the egg cell toward the uterus after it leaves the ovary.

Elsewhere, simple columnar epithelia may also have microvilli on their free surfaces. Cells in the small intestine are covered with microvilli that increase the surface area greatly and aid the absorption of food from the intestine into the bloodstream. Simple columnar epithelium in the stomach, intestines, digestive glands, and gallbladder protects the delicate linings and plays a role in absorption and secretion.

Some cells are further modified to secrete mucus, which protects and lubricates the walls of the digestive tract as food passes through. These modified cells, interspersed in a layer of columnar cells, are called *goblet cells*. Their secreted mucus accumulates near the top of the cell and causes the cell to bulge in the shape of a goblet (Figure 4.3C). Goblet cells are found in the digestive tract, upper respiratory tract, uterine tube, and uterus.

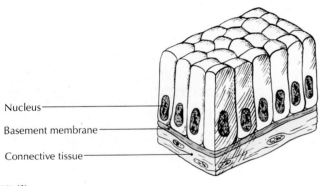

Nucleus ——
Basement membrane ——
Connective tissue ——

[C] (1)

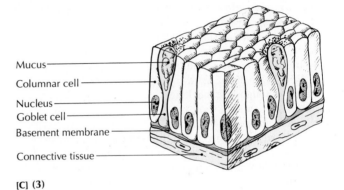

Mucus ——
Columnar cell ——
Nucleus ——
Goblet cell ——
Basement membrane ——
Connective tissue ——

[C] (3)

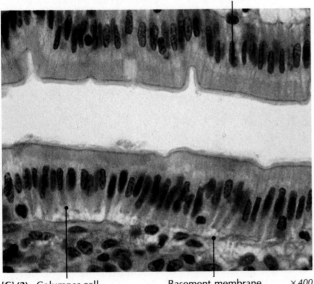

Nucleus

[C] (2)   Columnar cell   Basement membrane   ×400

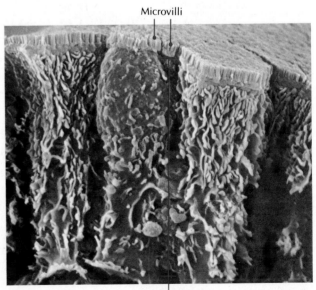

Microvilli

[C] (4)   Goblet cell   ×3000

Stratified squamous epithelium *Stratified squamous epithelium* usually has several layers of cells, but only the superficial layer contains squamous cells. The underlying germinative cells, known as **basal cells,** are modified cuboidal or columnar cells that lie next to the basement membrane (Figure 4.3D).* These basal cells have the ability to replace superficial squamous cells when the outer layer becomes worn or damaged. In a specific display of adaptability, growing basal cells undergo cell division and are pushed upward by newly formed basal cells beneath them. At the same time, the parent basal cells are changing their shapes to become flatter and larger. Eventually they become new squamous cells and replace the superficial squamous cells that are constantly dying and being sloughed off. This shedding process in the outer layer is called *desquamation.*

Because of its regenerative powers, stratified squamous epithelium appears in areas where a great deal of friction or the possibility of cellular injury or drying occurs. It is found in the epidermis (outermost layer of the skin), vagina, mouth, esophagus, anal canal, and distal end of the urethra.

We get a new epithelial layer of skin every few days, since developing basal cells in the epidermis are constantly dividing and being pushed to the surface. As these replacement cells get nearer to the surface, they produce **keratin,** a tough protein. The increasing transformation of the cells into keratin breaks down the cells' nuclei and organelles until they are no longer distinguishable. Once the nuclei have broken down, the cells cannot carry out their metabolic functions. So by the time the squamous epithelial cells reach the superficial layer of the skin, the cells are dead and composed mainly of tough, durable keratin. The cells have become *keratinized,* or *cornified,* and the tissue is known as *keratinized stratified squamous epithelium,* which has a protective function. These dead cells soon flake off unnoticed as the next batch of new cells is pushed to the surface. When the superficial squamous cells are in a moist environment, as they are in the mouth, esophagus, vagina, and cornea, they retain their nuclei and organelles, and the tissue is called *nonkeratinized stratified squamous epithelium.*

In the past, keratin was thought to be a single substance, but scientists now believe that there may be as many as 18 different varieties. Keratin has recently been discovered in epithelia that cover the intestinal tract and the nasal and bronchial passages. A better understanding of keratin and its origins may lead to a better means of distinguishing between two forms of cancer that are often confused. Keratin is present in *thymoma,* a form of cancer that arises in tissue of the thymus gland, but it is not present in *lymphoma,* a cancer that originates in lymphoid tissue. Keratin is also found in *carcinomas* (tumors that arise in epithelial tissue), but not in *sarcomas* (tumors of the connective tissue).

---

*The cells are said to be modified because they have the same relative thickness as cuboidal or columnar cells, but do not really look like "typical" cuboidal or columnar cells.

FIGURE 4.3D

(D) Stratified squamous epithelium. (1) The simplified drawing shows the top layer of simple squamous cells and underlying basal cells, which are modified cuboidal and columnar cells. (2) A vertical section of stratified squamous epithelium from the vagina. [*Ed Reschke/Peter Arnold.*]

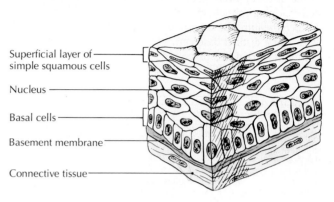

Superficial layer of simple squamous cells

Nucleus

Basal cells

Basement membrane

Connective tissue

[D] (1)

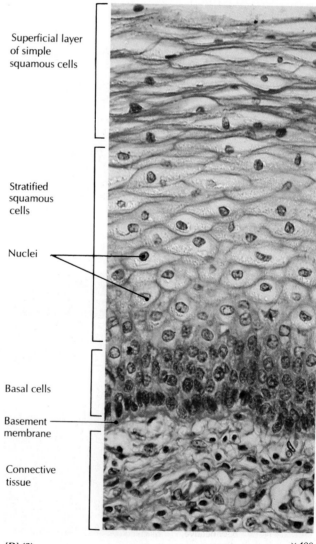

Superficial layer of simple squamous cells

Stratified squamous cells

Nuclei

Basal cells

Basement membrane

Connective tissue

[D] (2)   ×400

Stratified cuboidal epithelium *Stratified cuboidal epithelium,* which ideally should look like a neat pile of marshmallows stacked in even rows one atop the other, is found only in a few places in the body, and its cells are rarely shaped like true cubes (Figure 4.3E). This type of epithelium is found in the ducts of sweat glands, in sebaceous (oil) glands, and in developing epithelium in the ovaries and testes. As is usual with epithelium, the superficial cells are larger than the basal cells, and a distinct basement membrane is evident. The primary function of this tissue type is secretion.

Stratified columnar epithelium *Stratified columnar epithelium* is characterized by a regular arrangement of columnar cells at the superficial layer, with underlying smaller, cuboidal-type cells in contact with the basement membrane (Figure 4.3F). Stratified columnar epithelium is sometimes ciliated, as in the larynx and nasal surface of the soft palate. It is also found on the moist surfaces of the pharynx, urethra, and excretory ducts of the salivary and mammary glands. Its major functions are secretion and movement.

FIGURE 4.3E

(E) Stratified cuboidal epithelium. (1) The simplified drawing shows that the superficial cells are not true cubes. (2) A vertical section of stratified cuboidal epithelium from the thyroid gland. [*Ed Reschke/ Peter Arnold.*]

FIGURE 4.3F

(F) Stratified columnar epithelium. (1) Columnar cells can be seen at the superficial layer, with cuboidal-like cells in deeper layers. (2) A vertical section of stratified columnar epithelium from the male urethra. [*Michael H. Ross, M.D.*]

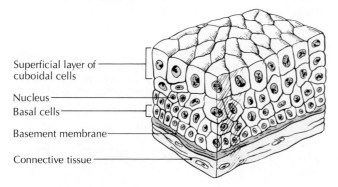

Superficial layer of cuboidal cells
Nucleus
Basal cells
Basement membrane
Connective tissue

[E] (1)

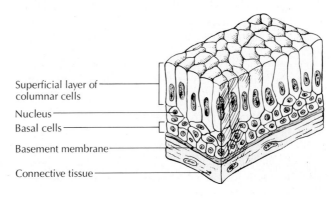

Superficial layer of columnar cells
Nucleus
Basal cells
Basement membrane
Connective tissue

[F] (1)

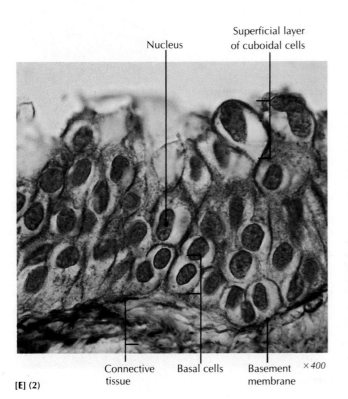

Nucleus
Superficial layer of cuboidal cells
Connective tissue
Basal cells
Basement membrane
×400

[E] (2)

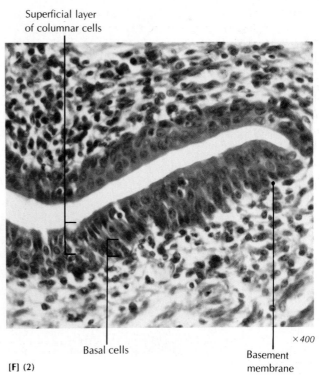

Superficial layer of columnar cells
Basal cells
Basement membrane
×400

[F] (2)

Exceptions to the General Classification of Epithelial Tissues

Three important kinds of epithelial tissues cannot be classified easily as typical epithelia. They are pseudostratified columnar epithelium, transitional epithelium, and glandular epithelium.

Pseudostratified columnar epithelium The distinctive feature of *pseudostratified columnar epithelium* is that all of its cells are in contact with the prominent basement membrane, but not all the cells reach the surface (Figure 4.3G). Pseudostratified columnar epithelium gives the appearance of being stratified, but it is actually a single layer of cells that vary in height and shape. Because of these irregularities, the nuclei appear at several different levels and give the false (*pseudo* means "false") impression of a multilayered stratification.

This type of epithelium is found in the trachea, bronchi and large bronchioles, and parts of the male reproductive tract. Cells in the respiratory system produce mucus, which helps to trap foreign particles. When cilia cover the free surface of this tissue, it is called *pseudostratified ciliated columnar epithelium.* Much of the upper respiratory tract contains pseudostratified ciliated columnar tissue. The beating cilia combine with secretions of mucus to trap foreign particles and carry them along to the throat, where they are either coughed out or swallowed and later eliminated harmlessly in the feces.

Transitional epithelium The cells of *transitional epithelium* surrounding an organ vary in shape, depending on how distended they are by the fluid the organ contains. This flexibility is a useful feature in the urinary bladder, where the bladder becomes larger or smaller depending on how much urine it is holding. Transitional epithelium is stratified and lines the urinary tract, including the ureters, urinary bladder, urethra, and calyxes of the kidneys. When the surface cells are not being stretched, they appear round. But when the bladder is full, the surface cells are stretched out and assume a flat, squamous appearance. Transitional epithelium is well adapted to changes of shape because its ample membranes are pleated like an accordion, allowing the organ it lines to stretch or contract without breaking. When transitional epithelium is in a relaxed condition, the basal cells have a cuboidal or slightly columnar shape (Figure 4.3H). The layer

---

FIGURE 4.3G

(G) Pseudostratified columnar epithelium. (1) Notice that all the cells touch the basement membrane, but some cells, such as the tapering cells, do not reach the surface. (2) A vertical section of ciliated tissue from the respiratory system (olfactory epithelium). [*Biophoto Associates/Photo Researchers.*]

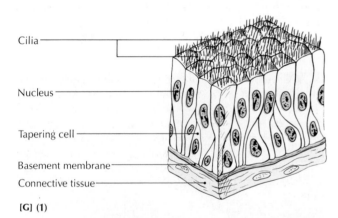

[G] (1)

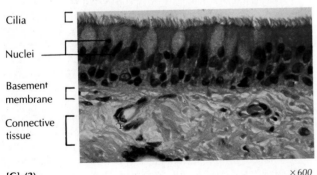

[G] (2)   ×600

FIGURE 4.3H

(H) Transitional epithelium. (1) When transitional epithelium stretches, as when the urinary bladder is full, the cells flatten out. (2) A vertical section of unstretched transitional epithelium tissue from the urinary bladder. [*Biophoto Associates/Photo Researchers.*]

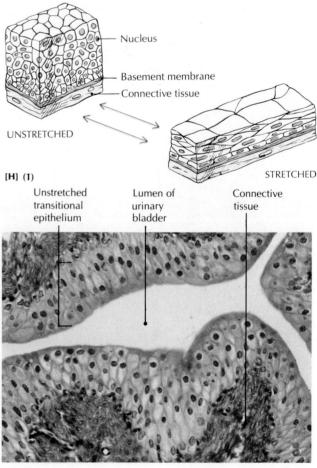

[H] (2)   ×400

above the basal cells contains irregularly shaped elongated cells; the superficial layer is made up of large, rounded cells.

Glandular epithelium As you have seen, epithelial tissue may be classified as either (1) covering or lining, or (2) glandular. Generally, cells of *glandular epithelium* are specialized for the synthesis, storage, and secretion of chemical substances.* The two main types of glands are *exocrine* and *endocrine*. *Endocrine glands* have specialized secretory cells, but the glands *do not have ducts*. Endocrine secretions (hormones) are released directly into adjacent *sinusoids* (large, extracellular spaces around the secretory cells), which pass the secretions into capillaries to be quickly transported by the bloodstream to target cells. Figure 4.4 shows how exocrine and endocrine glands develop from epithelium. Endocrine glands are discussed in detail in Chapter 16, and exocrine glands are described in the sections below.

*All glands are derived embryologically from epithelium.

FIGURE 4.4

The development of exocrine and endocrine glands from epithelium. (A, B) Cells from the surface epithelium grow down into the underlying connective tissue. (C) When an exocrine gland develops, the cells that connect the surface epithelium to the gland form a duct; the deepest cells become secretory. (D) When an endocrine gland develops, the connecting cells disappear; the deepest cells secrete directly into sinusoids, which pass the secretions into capillaries.

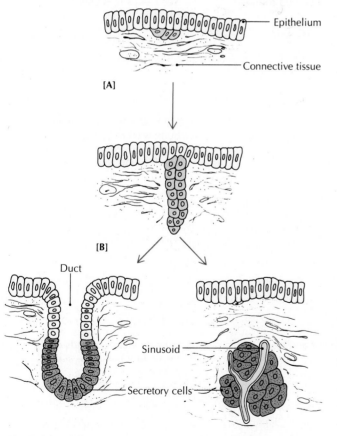

Epithelium

Connective tissue

[A]

[B]

Duct

Sinusoid

Secretory cells

[C] EXOCRINE GLAND   [D] ENDOCRINE GLAND

Exocrine Glands

Exocrine glands have specialized cells that produce the glands' secretions (for example, saliva or digestive juices), and ducts that carry the secretions to the body surfaces. Exocrine glands may be classified in several ways. The most typical ways are described below.

Unicellular and multicellular glands The only example of a *unicellular* (one-celled) *gland* in the human body is the *goblet*, or *mucous*, *cell*, found in the lining of the intestines, other parts of the digestive system, the respiratory tract, and the conjunctiva of the eye. A goblet cell produces a carbohydrate-rich glycoprotein called *mucin*, which later is secreted in the form of *mucus*, a thick lubricating fluid. A *multicellular gland* contains many cells. (See Table 4.2 for a classification of multicellular exocrine glands.)

Simple and compound glands A gland with only one *unbranched* duct is a *simple gland*. A gland with a *branched* duct system, resembling a tree trunk and its branches (although upside down), is a *compound gland*.

Tubular, alveolar, and tubuloalveolar glands All three of these categories are concerned with the shapes of the secretory portion of the glands. If the secretory portion of a gland is *tubular*, the gland is called a *tubular gland*. If the secretory portion is *rounded*, the gland is an *alveolar* (L. *alveolus*, hollow cavity), or *acinar* (L. *acinus*, grape, berry), *gland*. Glands whose secretory portions have both tubular and alveolar shapes are called *tubuloalveolar glands*. The secretory portion of a *simple coiled tubular gland* is coiled, and the secretory portion of a *simple branched tubular gland* is branched and tubular. Table 4.2 lists the types of glands developed from combinations of tubular, alveolar, simple, and compound forms, and shows their general shapes.

Mucous, serous, and mixed glands Glands may also be classified according to their secretions. *Mucous glands* secrete thick mucus; *serous glands* secrete a thinner, watery substance not unlike whey, the watery liquid left over after the curd and cream are removed from milk. (*Serous* comes from the Latin word for whey, *serum*.) The secretions of serous glands generally contain enzymes. *Mixed glands* contain both mucous and serous cells, and produce mucus and serous secretions.

Merocrine, holocrine, and apocrine glands Glands may also be classified according to how they release their secretions. *Merocrine glands* release their secretions without breaking the cell membrane, using the process of *exocytosis* (see Chapter 3). An example of a merocrine gland is the pancreas, which secretes digestive enzymes. When *holocrine glands* release their secretions, whole cells become detached, die, and actually become the secretion. Examples are rare, and the sebaceous (oil) glands in the skin may be the only holocrine glands in the body. *Apocrine glands* lose a small portion of their cytoplasm and plasma membrane along with their secretions but are able to restore the damaged parts so that the whole cell is not destroyed. Before the widespread

TABLE 4.2   CLASSIFICATION OF MULTICELLULAR EXOCRINE GLANDS

Type of gland	General shape	Shape of secretory portion	Location
MULTICELLULAR SIMPLE GLANDS			
Simple tubular		Straight and tubular	Intestinal glands (crypts of Lieberkühn)
Simple coiled tubular		Coiled	Sweat glands
Simple branched tubular		Branched and tubular	Mouth, tongue, and esophagus
Simple alveolar (acinar)		Rounded	Seminal vesicle glands (in male reproductive system)
Simple branched alveolar		Branched and rounded	Sebaceous (oil) glands
MULTICELLULAR COMPOUND GLANDS			
Compound tubular		Tubular	Mucous glands of mouth, bulbourethral glands (in male reproductive system), kidney tubules, and testes
Compound alveolar (acinar)		Rounded	Mammary glands
Compound tubuloalveolar		Tubular and rounded	Salivary glands, glands of respiratory passages, and pancreas

use of the electron microscope, the loss of cytoplasm and the plasma membrane was thought to be much greater than it actually is. Probably no loss of cytoplasm occurs in some glands that are still classified as apocrine, such as certain sweat glands. In the mammary glands, which are also considered to be apocrine glands, the destruction of cytoplasm and the membrane is very slight.

Ask Yourself

1 *What are some of the general characteristics of epithelial tissues?*

2 *How are epithelial tissues usually classified?*

3 *What are the three basic shapes of epithelial cells?*

4 *What are three exceptions to the basic system of classifying epithelial tissues?*

5 *How does an endocrine gland differ from an exocrine gland?*

6 *How does a simple exocrine gland differ from a compound one?*

CONNECTIVE TISSUES

Connective tissues are aptly named because they connect other tissues together. Unlike epithelial tissues, connective tissues have a great deal of extracellular, fibrous material that helps to support the cells of the other tissues. In fact, when cartilage and bone are included in the overall classification of connective tissues, the whole group is often referred to as *supporting* tissues. Besides their supportive function, connective tissues also form a protective sheath around hollow organs and are involved in storage, transport, and repair.

Connective tissues vary greatly in their structure and function. For this reason, we have grouped together the more generalized types—such as loose, dense, and elastic—under the section Connective Tissues Proper. The more specialized forms—cartilage, bone, and blood—are treated separately. In this chapter we discuss cartilage in some detail but give only a brief overview of osseous tissue and blood. Bones and osseous tissue are described in Chapter 6, the skeletal system in Chapter 7, and blood in Chapter 17.

All connective tissues consist of fibers, ground substance, cells, and some extracellular fluid (Figure 4.5 and Table 4.3). The extracellular fibers and ground substance are known as the *matrix*. The arrangement, function, and composition of elements in the matrix vary in different kinds of connective tissues, as you will see in this and the following sections.

Fibers of Connective Tissues

Connective tissues are usually classified according to the arrangement and density of their extracellular fibers, and the nature and consistency of the ground substance in which the fibers are embedded. Fibers of connective tissue may be collagenous, reticular, or elastic.

Collagenous fibers *Collagenous fibers* are also called *white fibers* because they appear whitish. They are composed mainly of the protein **collagen** (G. *kolla,* glue + *genes,* born, produced). Sturdy, flexible, and unstretchable collagenous fibers are the most common, and are found in all kinds of connective tissue (see Figure 4.9). They are arranged randomly, ranging from loose and pliable, as in the loosely woven connective tissue that supports most of the organs, to tightly packed and resistant, as in tendons. Generally, they look wavy when they are not under pressure. No matter how collagenous fibers are arranged, they are extremely well suited for support and protection. A collagenous fiber is actually a bundle of parallel *fibrils,* slender fibers composed of even smaller microfibrils. The number of fibrils in a bundle varies.*

Reticular fibers *Reticular fibers* form delicately branched networks (*rete* is Latin for net), as compared to the thicker bundles of collagenous fibers (see Figure 4.13). Reticular fibers are not elastic, but otherwise they are similar to collagenous fibers. They have the same molecular structure as collagenous fibers and practically the same chemical composition, and they also contain fibrils and microfibrils. Reticular fibers support fat cells, capillaries, nerves, muscle fibers, spleen cells, lymph nodes, bone marrow, and secretory cells in the liver.

Elastic fibers *Elastic fibers* appear singly, never in bundles, though they do branch and form networks. They contain microfibrils, but do not have the coarseness of collagenous fibers. Because they have a yellowish color, elastic fibers are sometimes called *yellow fibers.* Like rubber bands, elastic fibers stretch easily when they are pulled, and return to their original shape when the force is removed. The protein *elastin* in elastic fibers gives them the resilience that organs like the skin need to move, stretch, and contract.

Ground Substance of Connective Tissues

Ground substance of connective tissue is a homogeneous, extracellular material that ranges in consistency from a semifluid to a thick gel. Whereas fibers give connective tissue its strength and elasticity, ground substance provides a suitable medium for the passage of nutrients and wastes between the cells and bloodstream. The main ingredients of ground substance are glycoproteins and proteoglycans.† Examples in-

---

*A high number of collagenous fibers can make meat tough. But when collagen is boiled in water, it turns into gelatin, which is soft. This is the reason why tough meat simmered in a soup or stew becomes more tender. In contrast to this softening process, leather is toughened, or *tanned,* by the addition of tannic acid, which converts collagen into a firm, insoluble material.

†Proteoglycans have a protein core, with covalently bonded sulfate side chains.

TABLE 4.3 MAJOR COMPONENTS OF CONNECTIVE TISSUES (FIGURE 4.5)

Type of fiber or cell	Description	Major functions
EXTRACELLULAR COMPONENTS (MATRIX)		
Collagenous fibers	Thick bundles of whitish fibrils composed of collagen. Most common connective tissue fibers. Arranged randomly, ranging from loose and pliable to tightly packed and resistant.	Support and protect organs; connect muscles to bones, and bones to bones.
Reticular fibers	Delicately branched networks of inelastic fibrils. Have same chemical composition and molecular structure as collagenous fibers, but are thinner.	Support fat cells, capillaries, nerves, muscle fibers, and secretory liver cells. Form reticular framework of spleen, lymph nodes, and bone marrow.
Elastic fibers	Yellow fibers that appear only singly; branch to form networks. Less coarse than collagenous fibers. Contain protein elastin.	Allow hollow organs and other structures to stretch and recoil; support; suspension.
Ground substance	Homogeneous, extracellular material. Main ingredients are glycoproteins and proteoglycans.	Provides medium for passage of nutrients and wastes between cells and bloodstream; lubricant; shock absorber.
FIXED CELLS		
Fibroblasts	Large, long, flat, branching cells, with large, pale nuclei. Most common cells in connective tissue.	Synthesize matrix materials (fibers and ground substance); assist wound healing.
Adipose cells (fat cells)	May appear singly or in clusters. Single cells are bloated spheres containing fat. Nuclei and cytoplasm are flattened against side of cell.	Synthesize and store lipids.
Macrophage cells*	Irregularly shaped cells, with short processes. Cells are normally fixed along bundles of collagenous fibers but become free-moving amoeboid scavengers in cases of inflammation. Free macrophages are part of reticuloendothelial system.	Engulf and destroy foreign bodies in bloodstream and tissues.
Reticular cells	Flat, star-shaped cells resembling fibroblasts; associated with reticular fibers and such lymphoid organs as lymph nodes, bone marrow, and spleen.	Involved with immune response and presumably with formation of reticular fibers, macrophages, blood cells, and fat cells.
WANDERING CELLS		
Plasma cells	Oval-shaped cells, with large, dark nuclei off-center. Found where bacteria may enter through breaks in epithelium, and in serous membranes, lymphoid tissue, and areas of chronic infection.	Main producers of antibodies, which help to protect body against microbial infection.
Mast cells	Relatively large cells, with irregular shapes and small, pale nuclei. Secretory granules abundant in cytoplasm.	Produce heparin, which reduces blood clotting, and histamine, which increases vascular permeability and contracts smooth muscle.
Leukocytes (white blood cells)	Roundish cells, with various-shaped nuclei. Five kinds of leukocytes exist.	Protect against infection by engulfing and destroying harmful microorganisms. Also destroy tissue debris and play a role in inflammatory response.

*Macrophage cells can also be wandering cells. See page 104.

FIGURE 4.5

A diagrammatic rendering of connective tissue, showing a representative sample of cells and fibers bathed in tissue fluid.

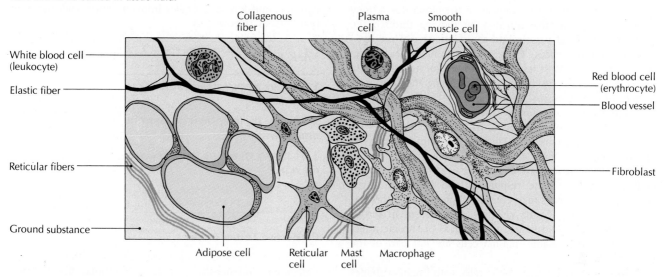

clude hyaluronic acid, chondroitin sulfate, and other specific sulfates. These compounds not only bind water and other tissue fluids that are needed for the exchange of nutrients and wastes, but also act as a lubricant and shock absorber. Hyaluronic acid forms a tough, protective mesh that helps prevent invasion by microorganisms and other foreign particles. Interestingly, some bacteria can produce an enzyme, called *hyaluronidase,* that breaks down the mesh of hyaluronic acid and allows the bacteria to enter. (See Chapter 26 for a discussion of this enzyme's important role in reproduction.)

Cells of Connective Tissues

The cells of connective tissues can usually be classified as either *fixed cells,* such as fibroblasts and adipose cells, or *wandering cells,* such as plasma cells, mast cells, and granular leukocytes (white blood cells). Macrophage cells may be either fixed or wandering.

Fixed cells *Fixed cells* have a permanent site and are usually concerned with long-term functions such as synthesis, maintenance, and storage:

1 Fibroblasts (L. *fibra,* fiber; Gr. *blastos,* growth) are large, long, flat, branching cells that appear spindle-shaped (see Figures 4.5, 4.8, 4.10, and 4.11). Their nuclei are light-colored and larger than in any other connective-tissue cell. Fibroblasts are the most common cells in connective tissue and the only cells found in tendons. They synthesize the *matrix* materials (fibers and ground substance) and are considered to be secretory cells. In this way, fibroblasts assist wound healing.

2 Adipose (fat) cells synthesize and store lipids. A mature cell accumulates so much fat that it forces the nucleus and cytoplasm to be flattened against the sides of the cell into a thin film around the rim. Adipose cells may appear singly or in clusters. Single cells are so bloated with one large droplet of stored fat that they look like spherical drops of oil (see Figures 4.5 and 4.12). Another type of fat cell contains fat in the form of small, multiple droplets. This fat is called *brown fat,* or multilocular fat (L. *locus,* place). Some brown fat is found along the aorta (the large blood vessel leaving the heart), in the skin between the shoulder blades (scapulas), and in the thorax.

3 Macrophage cells (macrophages) are active phagocytes ("cell eaters"). The name *macrophage* is descriptive: *macro* means large, and *phage,* as you have seen, comes from the Greek word *phagein,* to eat; so a macrophage is literally a "big eater." Macrophages have an irregular shape and short cytoplasmic extensions (processes). They are normally fixed along the bundles of collagenous fibers and are then referred to as *fixed macrophages.* But in cases of inflammation, they become detached from the fibers, change their shape to resemble amoebas (tiny, one-celled animals), and begin to move more actively as *free macrophages.* Free macrophages are scavengers that engulf and destroy foreign bodies in the bloodstream and tissues. They are very active in both phagocytosis and pinocytosis.

4 Reticular cells are flat, star-shaped cells with long processes. The cells come in contact with each other through these processes. The cells form the cellular framework in such netlike structures as bone marrow, lymph nodes, the spleen, and other lymphoid tissues. These cells are presumed to be involved in the formation of reticular fibers, phagocytic macrophages, blood cells, and fat cells, although these functions are being seriously challenged at the present time. There is agreement, however, that reticular cells play a role in the immune response.

Wandering cells *Wandering cells* in connective tissues are usually involved with short-term activities such as protection and repair:

1 *Leukocytes (white blood cells)* are roundish cells with nuclei that have various shapes (Figure 4.5). The cytoplasm may or may not contain secretory granules (see Chapter 17). Leukocytes drastically increase in number in times of infection. Although leukocytes circulate in the bloodstream, they perform their protective function by moving through the walls of tiny blood vessels into the connective tissue by a process called *diapedesis*. There are five kinds of leukocytes (see Chapter 17). Most leukocytes are phagocytes; they help protect the body against infection by engulfing and destroying harmful microorganisms.

2 *Lymphocytes,* or *plasma cells,* are one specific type of leukocyte. They are oval-shaped, with large, dark nuclei located off-center (Figure 4.5). They are the main producers of the antibodies that help defend the body against microbial infection. They are found in connective tissue under the moist epithelial linings of the respiratory and intestinal tracts, where microorganisms may enter through breaks in the epithelial membrane. Plasma cells are also present in serous membranes, lymphoid tissue, and in areas of chronic infection. Despite their name, plasma cells are not part of the blood plasma.

3 *Mast cells* are relatively large cells, with irregular shapes and small, pale nuclei (Figure 4.5). They are often found near blood vessels. Their cytoplasm is crowded with secretory granules, which are bound by membranes. These granules produce *heparin*, a polysaccharide that prevents the blood from clotting as it circulates throughout the body, and *histamine*, a protein that increases vascular permeability and contracts smooth muscle. In addition to heparin and histamine, mast-cell granules produce and store other sulfated compounds. Mast cells may also be effective in fighting long-term infections and inflammations, although their overall function is not completely understood.

4 *Macrophages* become mobile when stimulated by inflammation; they are listed here as a reminder of their frequent wanderings as phagocytes. Free macrophages are part of the extensive *reticuloendothelial system,* which is made up of an army of specialized cells concerned with producing antibodies and removing dead cells, tissue debris, microorganisms, and foreign particles from the fluids and matrix of body tissues (Figure 4.6).

Ask Yourself

1 *What are the major components of connective tissues?*

2 *What are the three kinds of connective tissue fibers?*

3 *What types of cells are found in connective tissue?*

Connective Tissues Proper

Connective tissues are usually classified according to the arrangement and density of their extracellular fibers (Table 4.4). *Loose (areolar) connective tissue* has irregularly arranged fibers and more tissue fluid and cells than fibers. In *dense (collagenous) connective tissue*, the number of fibers increases, and their arrangement becomes more regular. The following types of connective tissue are usually grouped under the heading connective tissues proper: embryonic, loose, dense, elastic, adipose, and reticular. Except for *embryonic* connective tissue, all the connective tissues we consider here are known as *adult connective tissues*.

FIGURE 4.7

A photomicrograph of mesenchymal tissue in an embryo. Notice the irregularly shaped cells lying in a homogeneous, jellylike matrix. [*Dr. N. Hoffman/Phototake.*]

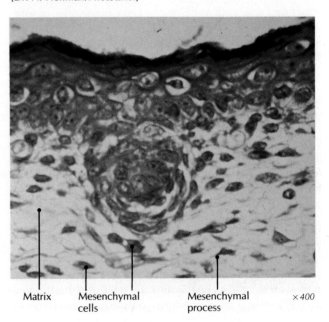

Matrix   Mesenchymal cells   Mesenchymal process   × 400

FIGURE 4.6

A scanning electron micrograph of a macrophage ingesting red blood cells. The macrophage is covered with microvilli. [*Manfred Kage/Peter Arnold.*]

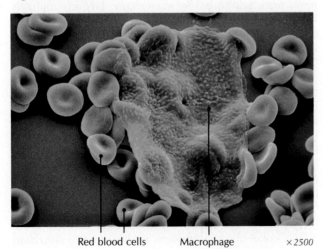

Red blood cells   Macrophage   × 2500

Embryonic connective tissue *Embryonic connective tissue* is found mostly in the unborn child, and only occasionally in adults. It is called *mesenchyme* when it is found in the embryo (during the first two months of prenatal development), and *mucous connective tissue* when it appears in the fetus (the third through final prenatal month). Mesenchymal cells vary in shape from stars to spindles (Figure 4.7). They gradually differentiate into cells of supporting tissues, vascular (blood vessel) tissues, smooth muscle, and blood. One of the most interesting aspects of mesenchymal tissue is that a small part of it is thought to be held in reserve for use in adult tissues. It reappears as fibroblasts, for example, when new adult connective tissue is needed for healing wounds. Mucous connective tissue is found only in the umbilical cord of the fetus, where it is sometimes called *Wharton's jelly.* It provides a flexible padding that protects the interior blood vessels and prevents the cord from snarling and being shut off when the fetus twists and turns inside the uterus.

Loose (areolar) connective tissue *Loose connective tissue* is also called *areolar tissue* because it contains tiny extracellular spaces (*areola* is Latin for "open place") that are usually filled with ground substance and tissue fluids (Figure 4.8). The spaces in areolar tissue are formed by its irregular, loosely woven fibers, which are not as prevalent as the tissue fluids and ground substance. The tissue fluids carry nutrients, gases, and wastes to the exchange point between the capillaries and tissues, where metabolism takes place. When too much tissue fluid accumulates, it usually causes swelling in the affected area. This condition is called *edema.* The ground substance in loose connective tissue acts as a barrier against harmful microorganisms.

Areolar tissue is the most common type of connective tissue, appearing nearly everywhere in the body. It provides support throughout the body and holds tissues and organs in place while allowing them some freedom of movement. It supports blood vessels and nerves, forms the subcutaneous layer that connects skin to muscles, and also acts as a protective packing material in the spaces between organs. Many different kinds of cells are found in areolar connective tissue, but the most common are fibroblasts and macrophages.

Dense (collagenous) connective tissue Compared with the relative openness of areolar tissue, *dense connective tissue* is tightly packed with coarse, collagenous fibers; tissue fluid and ground substance take up relatively little space. Dense connective tissue is classified into two types, based on the arrangement of fibers. Both types are distinctly white in appearance. They are dense *irregular* connective tissue and dense *regular* connective tissue:

1 *Dense irregular connective tissue* has thick fiber bundles that are arranged randomly in a tough, resilient meshwork (Figure 4.9). Collagenous fibers are very prevalent, with only scattered cells and little extracellular fluid and ground substance. Because the fibers run in many different directions, the tissue can be stretched in more than one way, as in the dermis of the skin, thus providing both support and protection. Dense irregular connective tissue is also found in the capsules around the spleen, testes, and other organs; in the deep fibrous coverings of muscles; in the covering sheaths of nerves, tendons, the brain, and the spinal cord; and in many other parts of the body. The most common cells in dense irregular tissue are fibroblasts, which can synthesize new fibers when necessary.

FIGURE 4.8

A scanning electron micrograph of loose (areolar) connective tissue. A fibroblast, a network of collagenous fibers, and red blood cells can be seen. [*Courtesy of Mr. Bret Connors and Dr. Andrew P. Evan, Indiana University.*]

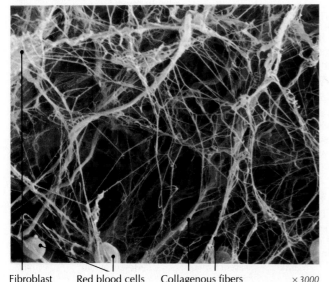

Fibroblast   Red blood cells   Collagenous fibers   ×3000

FIGURE 4.9

A scanning electron micrograph of dense irregular connective tissue. The dense bundles of collagenous fibers leave little room for cells, ground substance, or tissue fluid. [*D. M. Phillips/Taurus Photos.*]

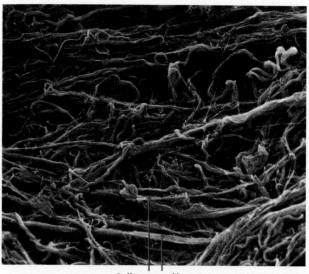

Collagenous fibers   ×3000

TABLE 4.4 SUMMARY OF CONNECTIVE TISSUES*

Type of tissue	Main locations	Description	Major functions
Embryonic	Mesenchyme: around developing bones and under skin of embryo. Mucous connective tissue: in umbilical cord of fetus (Figure 4.7).	Unspecialized packing material. Cells vary in shape from stars to spindles. Very little remains after cell differentiation.	Mesenchyme differentiates into supporting tissues, tissues of blood vessels, blood, and smooth muscle. Small amount held in reserve for use in adult tissues (reappears as fibroblasts during wound healing, for example). Mucous connective tissue forms padding for blood vessels in umbilical cord and prevents cord from snarling.
Loose (areolar)	Most parts of body, especially around and between organs. Nerves and blood vessels are wrapped in this tissue (Figure 4.8).	Irregular, loosely woven fibers in a semifluid base. Contains extracellular spaces filled with ground substance and tissue fluids. Most common connective tissue.	Supports tissues, organs, blood vessels, and nerves. Forms subcutaneous layer that connects muscles to skin, and allows for movement.
Dense (collagenous)	Irregular: dermis of skin, capsules of many organs; covering sheaths of nerves, tendons, brain, spinal cord; and deep fibrous coverings of muscles. Regular: tendons, ligaments, and aponeuroses (Figures 4.9 and 4.10).	Tightly packed with coarse, collagenous fibers; distinctly white. May be classified as either regular or irregular, depending on arrangement of fibers.	Provides support and protection; connects muscles to bones (tendons) and bones to bones (ligaments).
Elastic	Walls of hollow organs such as stomach; walls of largest arteries; some parts of heart; trachea, bronchi; ligaments between neural arches of spinal vertebrae; vocal cords; and suspensory ligament of penis (Figure 4.11).	Yellow elastic fibers branch freely; spaces between are filled by fibroblasts and some collagenous fibers.	Allows stretching and provides support and suspension.
Adipose (fat)	Beneath epidermis, around organs, and in many other sites (Figure 4.12).	Consists of clustered adipocytes (cells specialized for fat storage). Supported by collagenous and reticular fibers.	Provides reserve food supply, cushioning, protection, and insulation against loss of body heat.
Reticular	Interior of liver, spleen, lymph nodes, stomach, intestines, trachea, and bronchi (Figure 4.13).	Lattice of fine, interwoven threads that branch freely; spaces between fibers are filled by reticular cells.	Forms connecting and supporting frameworks of reticular fibers for adipose cells, lymph nodes, liver, spleen, capillaries, nerves, and muscle fibers.
Hyaline cartilage	Trachea, larynx, bronchi, nose, ventral ends of ribs, and ends of long bones (Figure 4.14).	Network of collagenous fibers filled in by ground substance. Translucent, pearly blue-white appearance. Enclosed within fibrous sheath.	Forms major part of embryonic skeleton; reinforces respiratory passages; aids free movement of joints; assists growth of long bones; and allows rib cage to move during breathing.
Fibrocartilage	Intervertebral disks, fleshy pad between pubic bones, and type of cartilage formed after injury (Figure 4.15).	Bundles of resilient collagenous fibers, which leave little room for ground substance. Usually merges with hyaline cartilage or fibrous connective tissue. No sheath.	Provides support and protection.
Elastic cartilage	External ear, epiglottis, auditory (Eustachian) tubes, and nasopharynx (Figure 4.16).	Appears dark because of dense elastic fibers, which are scattered in ground substance. Cells can produce elastic and collagenous fibers. Enclosed within sheath.	Provides flexibility and lightweight support.
Bone (osseous)	See Chapter 6.		

*Blood and lymph are sometimes classified as liquid connective tissues. In this book we treat them separately, in Chapters 17 and 20, respectively. Bone tissue is described in more detail in Chapter 6.

FIGURE 4.10

Dense regular connective tissue. Fibroblasts can be seen squeezed between the tightly packed collagenous fibers in this scanning electron micrograph. [*Don Fawcett, M.D./Fujiwara/Photo Researchers.*]

Collagenous fibers   Fibroblasts

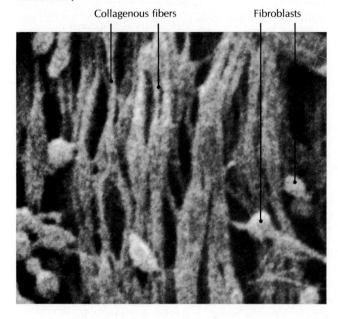

2 Dense regular connective tissue has fibers that are neatly arranged in parallel rows of thick bundles (Figure 4.10). This arrangement produces bundles of great strength that can withstand enormous tension in the direction the fibers are stretched. Such a strong structure is well suited for *tendons,* which connect muscles to bones; *ligaments,* which connect bones to other bones; and *aponeuroses,* which are broad bands of connective tissues. (A tendon as thick as a pencil can hold 1000 pounds [453 kg] without breaking, and tendons can usually stand stresses up to 9 tons [8165 kg] per square inch.

FIGURE 4.11

A scanning electron micrograph of elastic connective tissue. [*Centre National des Recherches Iconographiques.*]·

Elastic fibers

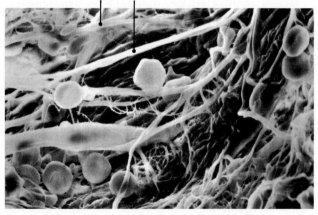

Bones almost always break before tendons do.) Cells in dense regular connective tissue are mostly fibroblasts, which are flattened out between the compressed layers of fibers.

Elastic connective tissue *Elastic connective tissue* is found only in certain specialized areas. Its yellow fibers dominate. It can be found in the walls of hollow organs like the stomach that need to stretch, in the walls of the largest arteries, in parts of the heart, in the trachea and bronchi, in ligaments between the neural arches of vertebrae in the spine, in the vocal cords, in the ligament suspending the penis, and in other parts of the body where stretching, support, and suspension are required. Elastic fibers branch freely and resemble a fishing net (Figure 4.11). The spaces between the elastic fibers are filled with fibroblasts and some collagenous fibers.

Adipose connective tissue Some adipose cells are normally found in loose connective tissue. But when any tissue is composed almost entirely of adipose cells arranged in clusters, it is considered to be *adipose tissue.* As shown in Figure 4.12, adipose tissue contains little or no extracellular substance; clusters of spherical, fat-bloated cells are supported by strands of collagenous fibers and some reticular fibers. By synthesizing and storing fat, adipose tissue provides a reserve food supply, cushions and protects organs, and helps to maintain homeostasis by preventing excessive loss of body heat. Adipose tissue usually makes up about 10 percent of the adult's body weight; it is distributed differently in males and females. Adipose tissue is found beneath the epidermis, surrounding organs, and in many other sites throughout the body.

FIGURE 4.12

A scanning electron micrograph of adipose connective tissue in the human heart. Notice the spherical adipose cells surrounded by collagenous fibers. [*C. Abrahams/Custom Medical Stock Photo.*]

Collagenous fibers

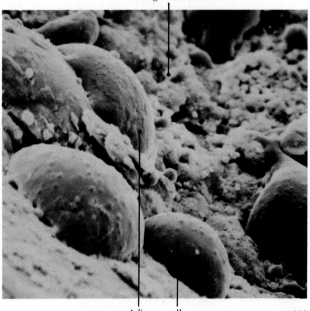

Adipose cells   × 1000

FIGURE 4.13

Reticular connective tissue. This section shows its characteristic supporting meshwork of fibers. [*Alexander Tsiaras/Science Source-Photo Researchers.*]

Reticular cells   Reticular fibers   Collagen
in fibers

Reticular connective tissue *Reticular connective tissue* consists of a lattice of interwoven threads that form connecting and supporting frameworks (Figure 4.13). Besides helping to support adipose cells (see Figure 4.12), reticular fibers also support capillaries, nerves, muscle fibers, and secretory cells of the liver. They are part of the framework of lymph nodes, the liver, and the spleen. Reticular fibers are also found in the linings of the stomach, intestines, trachea, and bronchi, and are present near basement membranes. The wide spaces between the branched fibers are filled with many reticular cells.

Ask Yourself

1 *What are the main kinds of connective tissues?*

2 *What is the most common kind of connective tissue?*

3 *How do regular and irregular dense connective tissues differ?*

4 *Why is elastic connective tissue well suited to the stomach?*

Cartilage

Cartilage is a specialized type of connective tissue that provides support and aids movement at the joints. Like other connective tissues, it consists of cells, fibers, and ground substance. The cartilage cells, which are called **chondrocytes** (Gr. *khondros,* cartilage), are embedded in small cavities within the matrix called **lacunae** (luh-KYOO-nee; L. *lacuna,* cavity, pool) (see Figure 4.14). Although cartilage is not as strong as bone, it is usually capable of supporting weight.

Unlike other types of connective tissue, cartilage does not contain blood vessels. Instead, oxygen, nutrients, and cellular wastes diffuse through the selectively permeable matrix. Because there is no blood supply, it is fairly easy to perform successful cartilage transplants. Without blood vessels, foreign proteins in the cells of transplanted cartilage have no way of entering the host body's circulation and causing an immune response. However, the lack of blood vessels interferes with the healing process of cartilage.

Cartilage appears to be the only type of human tissue that is completely immune to cancer. Recently, a substance called *anti-invasion factor (AIF)* has been extracted from cartilage. This substance is a collection of enzyme inhibitors. It is activated when cancerous cells approach the cartilage and release enzymes that begin to destroy healthy tissue. At the first sign of enzyme activity, the cartilage releases its enzyme inhibitors, and the cancer-causing enzymes are neutralized. Some biochemists believe that when AIF can be synthesized in the laboratory, it may prove useful in combating other diseases where harmful cells invade healthy tissue. Two examples are the blindness caused by diabetes, where blood vessels grow into the retina of the eye, and some forms of arthritis, where bone-destroying cells called osteoclasts go out of control and begin to attack the tissues in joints.

There are three types of cartilage, each with its distinctive matrix mixture of ground substance and fibers. The number and type of fibers in the matrix give each type of cartilage its characteristic strength and/or resiliency. The most common type is *hyaline cartilage,* which is strong and able to support weight. The other types are *fibrocartilage,* which is resilient and pliable because of its many collagenous fibers, and *elastic cartilage,* which is richly supplied with elastic fibers that make it flexible and elastic.

Hyaline cartilage *Hyaline cartilage* is the most prevalent type of cartilage in the body. Its name comes from the Greek word for "glassy," *hyalos,* because it has a translucent, pearly blue-white appearance. The main components of the cartilage matrix are collagen and proteoglycans. Hyaline cartilage is the most rigid type of cartilage, with its collagenous fibers scattered in a network that is completely filled in by ground substance (Figure 4.14). It is usually enclosed within a fibrous covering called a **perichondrium.** The perichondrium is composed of an outer layer of connective tissue and an inner layer of cells that can differentiate into chondrocytes.

Hyaline cartilage has many functions. It forms the major part of the skeleton of the embryo; reinforces respiratory passageways in the trachea, larynx, and bronchi; covers the portion of the bone facing the joint cavity, and thus aids the free

FIGURE 4.14

A cross section of cartilage tissue. The nuclei of the chondrocytes appear as dark areas, and are surrounded by cytoplasm. The cartilage matrix contains elastic fibers. [*Lennart Nilsson*, Behold Man. *Boston, Little, Brown, 1974.*]

Cartilage matrix   Chondrocyte

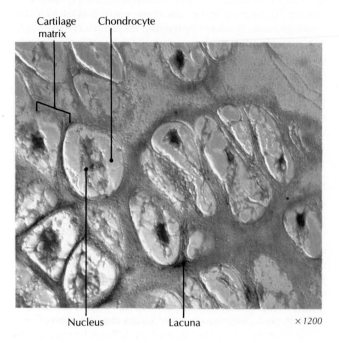

Nucleus   Lacuna   × 1200

FIGURE 4.15

A scanning electron micrograph of fibrocartilage, showing chondrocytes in lacunae and collagenous fibers. [*Ed Reschke/Peter Arnold.*]

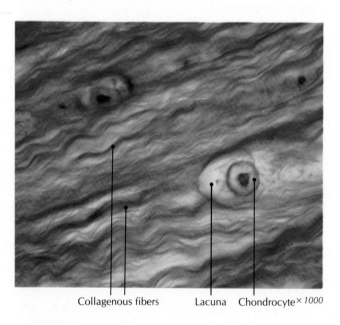

Collagenous fibers   Lacuna   Chondrocyte × 1000

FIGURE 4.16

A scanning electron micrograph of elastic cartilage, showing individual chondrocytes and elastic fibers. [*Don Fawcett, M.D.*]

Elastic fibers   Chondrocytes

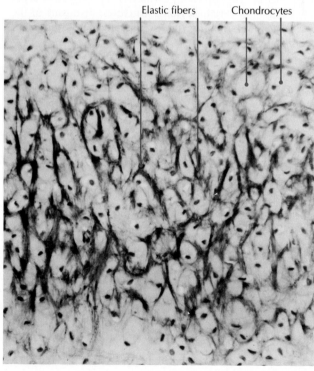

× 160

movement of joints at the ends of long bones; and is essential in the growth of long bones. This cartilage also appears in the nose, and in the ventral ends of ribs, where it allows the rib cage to expand and contract during breathing.

Fibrocartilage *Fibrocartilage* consists of bundles of thick collagenous fibers that give it durability and the strength to resist tension (Figure 4.15). The collagenous fibers leave little room for ground substance. The cells are like those in hyaline cartilage, but they are not as numerous. Fibrocartilage has no perichondrium and does not occur alone; it usually merges with hyaline cartilage or fibrous connective tissue. It resembles dense connective tissue and is found between the bodies of the vertebrae of the spinal column, in the disk of the pubic joint, in the lining of tendon grooves, in the rims of some sockets, and in some areas where tendons and ligaments are inserted into the ends of long bones.

Elastic cartilage *Elastic cartilage* is more flexible and elastic than hyaline cartilage. It is found in areas where lightweight support and flexibility are required, such as the external ear and the epiglottis in the throat (Figure 4.16). Elastic cartilage is also found in the auditory (Eustachian) tubes (which connect the middle ear to the upper throat) and in the larynx. Its chondrocytes are identical in appearance to those in hyaline cartilage, but they are capable of producing elastic fibers as well as collagenous fibers. Varying amounts of elastic fiber are scattered throughout the ground substance. As with hyaline cartilage, an enveloping perichondrial sheath is present. Because of the density of its branching elastic fibers, elastic cartilage appears black.

Bone as Connective Tissue

Bone is also a type of connective tissue. It is very similar to cartilage in that it consists mostly of matrix material containing lacunae and specific cell types. However, unlike cartilage, bone is a highly vascular tissue. Bone histology is treated in detail in Chapter 6.

Blood as Connective Tissue

Blood cells and blood-forming tissues are mentioned in this section on connective tissue not because they connect anything—they do not—but because they have the same embryonic origin (mesenchyme) as the more typical connective tissues, and because blood has the three components of any connective tissue (cells, fibers, and ground substance). Blood is treated extensively in Chapter 17.

Ask Yourself

1 *What is the function of cartilage?*

2 *What are the three types of cartilage?*

3 *What is a perichondrium?*

MEMBRANES

Membranes are thin, pliable layers of epithelial and/or connective tissue. They line body cavities, cover surfaces, or separate or connect certain regions, structures, or organs of the body. The four kinds of membranes are mucous, serous, synovial, and cutaneous.

Mucous Membranes

Mucous membranes line body passageways that open to the outside of the body, such as the nasal and oral cavities, and tubes of the respiratory, digestive, urinary, and reproductive systems. These membranes are made up of a layer of loose connective tissue covered by varying kinds of epithelial tissue. For example, the small intestine is lined with simple columnar epithelium, and the oral cavity is lined with stratified squamous epithelium. Glands in the mucous membranes secrete the protective, lubricating mucus, which consists of water, salts, and a sticky protein called mucin. Besides providing lubrication, mucus also helps trap and dispose of invading microorganisms.

Serous Membranes

Serous membranes are double membranes of loose connective tissue covered by a layer of simple squamous epithelium (known as mesothelium). They line some of the walls of the closed thoracic and abdominopelvic cavities, and cover the organs that lie within these cavities (Figure 4.17; see also Figure 1.22). The part of a serous membrane that lines a body wall (such as the thoracic wall) is the *parietal layer* (*parietal* means "wall"), and the part that covers organs within the cavity is the *visceral layer* (*viscera* means "body organ").

FIGURE 4.17

Serous membranes covering (A) the heart and (B) lining the thoracic and abdominopelvic cavities (sagittal section, female).

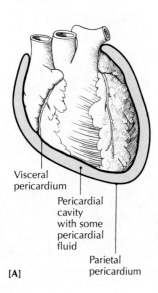

Visceral pericardium

Pericardial cavity with some pericardial fluid

Parietal pericardium

[A]

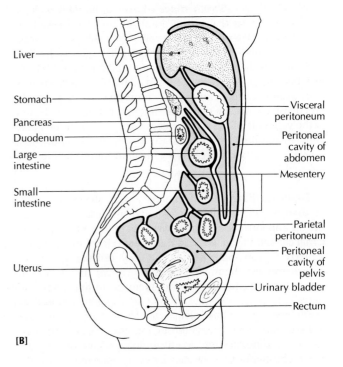

Liver

Stomach

Pancreas

Duodenum

Large intestine

Small intestine

Uterus

Visceral peritoneum

Peritoneal cavity of abdomen

Mesentery

Parietal peritoneum

Peritoneal cavity of pelvis

Urinary bladder

Rectum

[B]

Serous membranes include the *peritoneum, pericardium,* and *pleura:*

The **peritoneum** lines the peritoneal cavity and covers the organs inside the cavity. It also lines the abdominal and pelvic walls.

The **pericardium** lines the pericardial cavity with parietal pericardium, and covers the heart with a sac of visceral pericardium. (The heart is not inside the pericardial cavity.)

The **pleura** lines the pleural cavity and covers the lungs. It also lines the wall of the thorax. (*Pleura* means "side" or "rib," but is normally used in connection with the lungs.)

The peritoneum, pericardium, and pleura are all double-layered membranes with a thin space between. These spaces are the peritoneal, pericardial, and pleural cavities, respectively. This space between serous membranes receives secretions of *serous fluids* that act as protective lubricants around organs and help to remove harmful substances through the lymphatic system. *Pleurisy* is a disease that occurs when the pleural membranes become inflamed and stick together, making breathing painful.

Organs in the abdominopelvic cavity are suspended from the cavity wall by fused layers of visceral peritoneum. These fused tissues are called **mesenteries,** or visceral ligaments. Besides providing a point of attachment for organs, mesenteries also permit vessels and nerves to connect with their organs through the otherwise impenetrable lining of the abdominopelvic cavity. Many abdominal and pelvic organs such as the stomach, spleen, jejunum and ileum portions of the small intestine, uterus, and ovaries are connected to the abdominopelvic wall by a mesentery. (In the case of the stomach, this mesentery is called an omentum; in the uterus it is known as the broad ligament.) But the kidneys, pancreas, and some other structures are attached to the posterior wall of the abdominopelvic cavity by a mesentery *behind* the peritoneal cavity. For this reason, the kidneys and pancreas are considered to be *retroperitoneal* organs, or located behind the peritoneum (*retro* means "behind").

Synovial Membranes

Synovial membranes line the cavities of joints such as the knee or elbow, and other similar areas where friction needs to be reduced. They are composed of loose connective and adipose tissues covered by fibrous connective tissue. No epithelial tissue is present. The inner surface of synovial membranes is generally smooth and shiny, and the synovial fluid is a thick liquid with the consistency of egg white. It helps to reduce friction at the movable joints, and in fact, the synovial fluid together with the covering of the bones produces a mechanism that is practically frictionless. (Joints are described in Chapter 8.)

Cutaneous Membranes

Cutaneous membrane is another name for skin. The *epidermis* (outermost layer of the skin) consists of a layer of stratified squamous epithelium, and the underlying *dermis* is made up of a thicker layer of connective tissue, which contains collagenous, reticular, and elastic fibers. The skin is part of the integumentary system and is such an important membrane (as well as an organ of the body) that it is discussed separately in Chapter 5.

Ask Yourself

1 *Where are mucous membranes found?*

2 *What are the three kinds of serous membranes?*

3 *What are mesenteries?*

4 *How do synovial membranes aid movement?*

PHYSIOLOGICAL AND ANATOMICAL ABNORMALITIES

The prevention and treatment of certain diseases requires an understanding of the structure and function of **collagen,** the important protein that forms the fibrous network in almost all tissues of the body.

The *pathogenesis* (development of disease conditions) of certain inherited and acquired diseases of connective tissues is now understood with respect to collagen synthesis and metabolic dysfunction caused by a lack of vitamin C (ascorbic acid) in the diet. In *scurvy,* collagen fibers are not formed. As a result, bone growth is abnormal, capillaries rupture easily, and wounds and fractures to not heal. In **rheumatoid arthritis,** collagen in articular cartilage (cartilage within a joint) is destroyed, due to the release of specific enzymes from the inflamed synovial membranes. In **arteriosclerosis,** some of the damage to blood vessels is caused by the secretion of large amounts of collagen by the smooth muscle cells.

Many diseases that have an inflammatory component inevitably affect collagen and collagenous structures. As a result, abnormally large amounts of collagen and collagen fibers are deposited as part of the tissue repair process. The resulting collagenous scars limit tissue function, upset homeostasis, and contribute to the disease process.

SUMMARY

Epithelial tissues: form and function

1 The cells of *epithelial tissues* are arranged in two ways. (1) They may be sheets that cover surfaces or line body cavities or ducts that often connect with the surface of the body. Some tissues line the surfaces of internal organs. (2) The cells of other epithelial tissues may be modified to become *glandular epithelia* that are organized into glands.

2 The typical functions of epithelial tissues are absorption, secretion, transport, excretion, protection, and sensory reception.

3 All types of epithelial tissues except glandular tissues share certain characteristics: (1) the cells are *regular in shape* and *packed in continuous sheets,* (2) the tissue is usually anchored to the underlying connective tissue by a *basement membrane,* (3) the tissue is *avascular,* (4) there are several types of *surface specializations* (microvilli and cilia), and (5) the cells are capable of *regeneration.*

4 Epithelial tissues are classified according to the *arrangement of cells,* the *number of cell layers,* and the *shape of the cells* in the superficial layer.

5 Epithelial tissues one-cell-layer thick are *simple;* tissues with two or more cell layers are *stratified.* The basic shapes of the superficial cells are *squamous, cuboidal,* and *columnar.*

6 *Simple squamous epithelium* has an arrangement of flat cells in a single layer, suited for diffusion or filtration.

7 *Simple cuboidal epithelium* has a single layer of approximately cube-shaped cells. It functions in secretion and absorption.

8 *Simple columnar epithelium* is a single layer of elongated cells, with some cells modified to secrete mucus and others modified to accomplish absorption, protection, and movement.

9 *Stratified squamous epithelium* usually has at least three layers of cells; only the superficial layer contains flat squamous cells. Underlying *basal cells* can replace damaged superficial squamous cells. Primarily protective.

10 *Stratified cuboidal epithelium* is a multilayered arrangement of somewhat cube-shaped secretory cells.

11 *Stratified columnar epithelium* has a regular arrangement of columnar cells at the superficial layer, and smaller cuboidal cells underneath and in contact with the basement membrane. Its functions are secretion and movement.

12 Exceptions to the basic system of classification are *pseudostratified columnar epithelium, transitional epithelium,* and *glandular epithelium.*

13 *Pseudostratified columnar epithelium* has all of its cells in contact with the basement membrane, but not all reach the surface. Functions include protection, secretion, and movement.

14 *Transitional epithelium* is composed of cells capable of stretching as needed, thus allowing for changes in shape.

15 *Glandular epithelium* has cells specialized for the synthesis, storage, and secretion of their product. The two main types of glands are *exocrine* and *endocrine.*

16 *Endocrine glands* do not have ducts. They release their secretions into the bloodstream. *Exocrine glands* have specialized cells that produce and release the glands' secretions into ducts.

17 Exocrine glands are classified as *unicellular* or *multicellular; simple* or *compound; tubular, alveolar,* or *tubuloalveolar; mucous, serous,* or *mixed;* and *merocrine, holocrine,* or *apocrine.*

Connective tissues

1 Besides their supportive and protective functions, connective tissues also form a sheath around hollow organs; involved in storage, transport, repair.

2 All connective tissues consist of fibers, ground substance, and cells. Fibers and ground substance make up the *matrix.*

3 Types of connective tissue fibers are *collagenous, reticular,* and *elastic.*

4 *Ground substance* is a homogeneous extracellular material providing a medium for the passage of nutrients and wastes between cells and the blood.

5 The cells of connective tissues are usually classified as either *fixed cells (fibroblasts, adipose cells,* and *reticular cells)* or *wandering cells (plasma cells, mast cells,* and *leukocytes). Macrophage cells* may be either fixed or wandering cells. Fixed cells are usually concerned with *long-term functions.* Wandering cells are usually involved with *short-term activities.*

6 The *reticuloendothelial system* contains specialized cells that are active in defending the body against microorganisms, and are involved with other cells in producing antibodies. These cells remove undesirable matter from the fluids and matrix of body tissues.

7 Connective tissues are usually classified according to the arrangement and density of their fibers. The typical connective tissues are *loose* (areolar), *dense* (collagenous), *elastic, adipose,* and *reticular.*

8 *Embryonic connective tissue* is called *mesenchyme* in the embryo and *mucous connective tissue* in the fetus.

9 *Cartilage* is a specialized connective tissue that provides support and facilitates movement at the joints.

10 *Chondrocytes* are cartilage cells housed in *lacunae,* cavities within the matrix.

11 The three types of cartilage are *hyaline cartilage, fibrocartilage,* and *elastic cartilage.*

12 Hyaline cartilage and elastic cartilage are usually enclosed within a fibrous sheath called a *perichondrium.*

Membranes

1 *Membranes* are thin, pliable layers of epithelial or connective tissue that line body cavities, cover surfaces, or separate or connect certain regions, structures, or organs of the body.

2 *Mucous membranes* line body passageways that open to the outside.

3 *Serous membranes* line the pericardial, pleural, and peritoneal cavities within the closed thoracic and abdominopelvic cavities, and cover the organs that lie within these cavities. Serous membranes include the *peritoneum, pericardium,* and *pleura.*

4 *Synovial membranes* line the cavities of joints and other similar areas.

5 *Cutaneous membrane* is another name for skin.

Physiological and anatomical abnormalities

1 *Collagen,* a protein that forms the fibers in many body tissues, plays a role in scurvy, rheumatoid arthritis, and arteriosclerosis.

2 Diseases involving inflammation often result in collagenous scars, which limit tissue function.

UNDERSTANDING THE FACTS

1 Is pseudostratified epithelium correctly named? Why?

2 Describe the relationship of structure and function of transitional epithelium.

3 How is connective tissue classified?

4 What is the main ingredient of ground substance, and what are its functions?

5 What is the reticuloendothelial system? How important is it in the body?

6 What is "Wharton's jelly"?

7 Compare the three types of cartilage.

8 In what parts of the body are serous membranes found?

UNDERSTANDING THE CONCEPTS

1 Where in the body would you expect to find epithelial tissue?

2 We speak of cube-shaped cells, but as you observe these cells, what do you note? How many right angles can you discover in or on your body? Why do human beings use right angles so often in construction?

3 Is there a correlation between location and function of epithelial tissues and their ability to be renewed? Explain.

4 The anterior surface of the lens of the eye is covered with simple squamous epithelium. Is this a proper tissue type for this location?

5 What would happen if the body were deficient in the wandering cells of connective tissue? Could this situation be life-threatening?

6 Suppose that you are suffering from an injury and need corrective surgery that involves a cartilage transplant, among other things. What are the chances of recovery? Why?

7 How is collagen involved in scurvy, rheumatoid arthritis, and arteriosclerosis?

SUGGESTED READING

CAPLAN, ARNOLD I., "Cartilage." *Scientific American,* October 1981.

CUSHMAN, S. W., "Structure-Function Relationships in the Adipose Cell." *Journal of Cell Biology,* 46 (1970):326–341.

KAFELIDES, N. A., R. ALPER, AND C. C. CLARKE, "Biochemistry and Metabolism of Basement Membranes." *International Review of Cytology,* 61 (1979):167–228.

ROSS, RUSSELL, AND PAUL BORNSTEIN, "Elastic Fibers in the Body." *Scientific American,* June 1971.

STAEHELIN, L. ANDREW, "Structure and Function of Intercellular Junctions." *International Review of Cytology,* 39 (1974):191–283.

STAEHELIN, L. ANDREW, AND BARBARA E. HULL, "Junctions Between Living Cells." *Scientific American,* May 1978.

WEINSTOCK, M., AND C. P. LEBLOND, "Formation of Collagen." *Federation Proceedings,* 33 (1974):1205–18.

5
The Integumentary System

LEARNING OBJECTIVES

1 Describe the major functions of human skin.

2 Name the five layers of the epidermis.

3 Describe the structures that are found in the dermis.

4 Explain why skin has different colors.

5 Tell how a wound heals.

6 Describe the structure and function of a sweat gland.

7 Describe the structure and function of an oil gland.

8 List some functions of hair.

9 Describe the structure of hair.

10 Explain how hair develops and grows.

11 Describe the structure and function of nails.

12 Contrast first-, second-, and third-degree burns.

13 Describe several common skin disorders and one or more types of skin cancer.

The **integumentary system** (L. *integumentum,* cover) consists of the skin and its derivatives, which include several types of glands, hair, and nails. The system functions in protection, in the regulation of body temperature, in the excretion of waste materials, in the conversion of the sun's rays into vitamin D_3, and in the reception of stimuli such as pain, pressure, and temperature.

SKIN

The *skin* is the largest organ in the body, occupying almost 2 sq m (18 sq ft) of surface area. It varies in thickness on different parts of the body, from less than 0.5 mm on the eyelids to more than 5 mm on the middle of the upper back. A typical thickness is 1 to 2 mm.

Skin has two main parts: the *epidermis* is the outermost layer of epithelial tissue; the *dermis,* or *corium* (Gr. leather), is a thicker layer of connective tissue beneath the epidermis (Figure 5.1).

FIGURE 5.1A

The skin. (A) A "textbook" drawing of the skin's structure, showing its many components in an ideal arrangement. The stratum lucidum would not appear on hairy skin. An enlarged portion of the epidermis is shown in the small drawing (top). (Figure 5.1B is on the next page.)

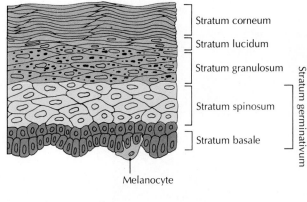

Stratum corneum
Stratum lucidum
Stratum granulosum
Stratum spinosum
Stratum basale
Stratum germinativum
Melanocyte

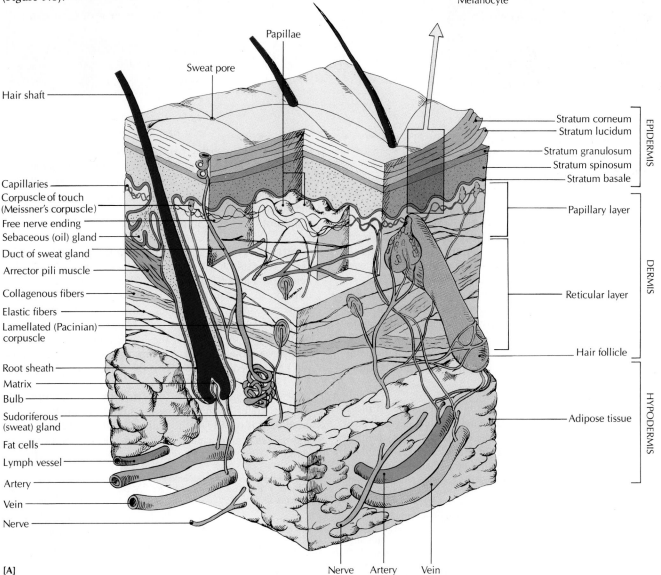

Papillae
Sweat pore
Hair shaft
Capillaries
Corpuscle of touch (Meissner's corpuscle)
Free nerve ending
Sebaceous (oil) gland
Duct of sweat gland
Arrector pili muscle
Collagenous fibers
Elastic fibers
Lamellated (Pacinian) corpuscle
Root sheath
Matrix
Bulb
Sudoriferous (sweat) gland
Fat cells
Lymph vessel
Artery
Vein
Nerve

Stratum corneum
Stratum lucidum
Stratum granulosum
Stratum spinosum
Stratum basale
EPIDERMIS
Papillary layer
Reticular layer
DERMIS
Hair follicle
Adipose tissue
HYPODERMIS

[A]

Nerve   Artery   Vein

FIGURE 5.1B

(B) A scanning electron micrograph of human skin. [*C. Ward Kischer, Ph.D., Department of Anatomy, University of Arizona College of Medicine.*]

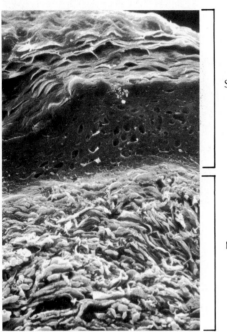

Stratum corneum

Nucleated layers

[B]

FIGURE 5.2

A vertical section of the thick epidermis of the fingertip. [*Biophoto Associates/Science Source-Photo Researchers.*]

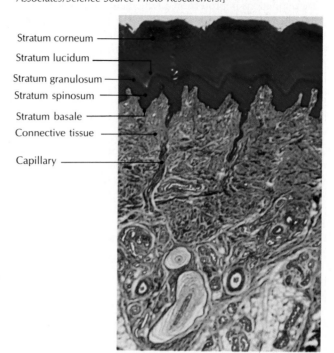

Stratum corneum
Stratum lucidum
Stratum granulosum
Stratum spinosum
Stratum basale
Connective tissue
Capillary

× 100

Epidermis

The outer layer of the skin, called the **epidermis** (Gr. *epi,* over + *derma,* skin), is stratified squamous epithelium. Because it contains no blood vessels, you can usually rub off dead skin without bleeding.* Most of the epidermis is so thin, however, that even minor cuts reach the dermis and draw blood. In the thick epidermis of the palms of the hands and the soles of the feet, there are five typical layers (strata). Starting with the outermost layer, they are *stratum corneum, stratum lucidum, stratum granulosum, stratum spinosum,* and *stratum basale* (Figure 5.2). (The stratum spinosum and stratum basale together are known as the *stratum germinativum* because they generate new cells.) In parts of the body other than the palms and soles, only the stratum corneum and stratum germinativum are regularly present.

Stratum corneum The **stratum corneum** (L. *corneus,* horny, cornified tissue) is a flat, relatively thick layer of dead cells arranged in parallel rows (see Figure 5.2). This cornified layer of the epidermis consists of soft keratin (as compared with the hard keratin in fingernails and toenails), which helps keep the skin elastic. The soft keratin also protects living cells

underneath from being exposed to the air and drying out. The cells in the stratum corneum are constantly being shed through normal abrasion. They are replaced by new cells that are formed by cell division and pushed up from the germinative layers below, as described in Chapter 4. Cells on the soles of the feet are completely replaced every month or so.

Stratum lucidum The **stratum lucidum** (L. *lucidus,* bright, clear) consists of flat, translucent layers of dead cells that contain a protein called *eleidin.* This protein is probably a transitional substance between the soft keratin of the stratum corneum and the precursor of soft keratin, *keratohyaline,* of the layer below (see Figure 5.2). The stratum lucidum appears only in the palms of the hands and soles of the feet, acting as a protective shield against the ultraviolet rays of the sun and preventing sunburn on the palms and soles.

Stratum granulosum The **stratum granulosum** (L. *granum,* grain) lies just below the stratum lucidum (see Figure 5.2). It is usually two to four cells thick. The cells contain granules of keratohyaline. This layer initiates the process of keratinization, associated with the dying process of cells.

---

*The red color of the lips is not due to a rich blood supply, as some people think. In fact the lips do *not* contain blood vessels of their own. They appear red because their stratified squamous epithelium is noncornified and relatively translucent. This translucency allows us to see the reddish color of the capillaries in the connective tissue papillae (fingerlike projections) that extend into the epithelium.

Why is a tattoo so difficult to remove?

Tattoo dyes are injected below the stratum germinativum, into the dermis, where the dyed cells do not move outward toward the surface as the skin is shed and replaced. In the same way, scars and stretch marks do not disappear easily because they involve the dermis.

THE DYNAMIC BODY

The Growth, Loss, and Replacement of Hair

Most of us think we have too little hair, or too much, or hair where we don't want it. We regularly wash it, cut it, curl it, or straighten it, without much further thought. But our hair is more than something to cut, curl, remove, or straighten. Hair is a modification of the skin, and like the skin, it has many different forms and functions. It also grows at different rates and to different lengths.

We are constantly shedding and (we hope) replacing our hair. A hair is shed when its growth is complete. After a short period of rest, the root produces the beginning of a new hair.

Hair is not static, and each type of hair has its own life cycle. Even the same type of hair (hair on the scalp, for example) grows at staggered rates, so no area of the body sheds all of its old hair at the same time. Hair grows faster at night than during the day, and faster in warm weather than in cold. The rate of hair growth also varies with age, usually slowing down as we get older. The fastest growth rate occurs in women between ages 16 and 24. Hair textures and locations also affect growth rates. Coarse, black hair grows faster than fine, blond hair. Scalp hairs last much longer (3 to 5 years) than eyebrow and eyelash hairs (about 10 weeks).

Human hair follicles alternate between growing and resting phases. Scalp hairs, which grow about 0.4 mm a day, usually last 3 to 5 years before they are shed and the follicles go into a resting phase of 3 to 4 months. Plenty of scalp hair is visible at any given time, however, because 80 to 90 percent of the follicles in the scalp are in the growing phase at the same time. (The average scalp contains about 125,000 hairs.) Because healthy, active follicles keep producing new hair, it is not unusual to lose 50 to 100 hairs a day without becoming bald. A person might lose and replace more than 1.5 million hairs in a lifetime. Some hairs have a longer resting phase than a growing phase. Eyebrow hairs, for instance, grow 1 to 2 months before the follicles rest for the next 3 to 4 months. Eyelashes have an even shorter growing phase.

Although it seems that men have more body hair than women do, they actually have about the same number of hair follicles, about two million. Male hair is coarser, and therefore more obvious. Baldness in men seems to occur when there is a genetic predisposition toward baldness. If the father is bald, the son will probably be bald too, with a type of baldness called *patterned baldness*.

Baldness also results from an abnormally large amount of testosterone (the male sex hormone) and an abnormally small amount of estrogen (a female sex hormone). Baldness can also be caused by disease, stress, malnutrition, and many other external factors.

× 1500

Scanning electron micrograph of human hair. [Manfred Kage/Peter Arnold.]

If scalp hair were left uncut, it would grow to a length of about 1 m (3.25 ft) in a lifetime. (Many exceptions of longer hair have been recorded.) If it did not grow in cycles that include inactivity and shedding, scalp hair would reach about 7.5 m (24.5 ft). Contrary to what some people believe, hair does not grow after a person dies. Instead, the skin shrinks, making the hair look longer.

Stratum spinosum The **stratum spinosum** (L. *spinosus,* spiny) is composed of several layers of polyhedral (many-sided) cells that have delicate "spines" protruding from their surface (see Figure 5.2). For this reason, these cells are sometimes called "prickle cells." The interlocking spinelike projections help to give support to this binding layer. Active protein synthesis takes place in the cells of the stratum spinosum, indicating that cell division and growth are occurring. Some new cells are formed here and are pushed to the surface to replace the cornified cells of the stratum corneum.

What is a blister? — *When the skin is burned or irritated severely, tiny blood vessels in the dermis widen and some clear plasma leaks out. The plasma accumulates between the dermis and epidermis in a fluid-filled pocket, or blister.*

Stratum basale The **stratum basale** (L. *basis,* base) rests on the basement membrane next to the dermis (see Figure 5.2). It usually consists of a single layer of columnar or cuboidal cells. Like the stratum spinosum, it is capable of undergoing cell division to produce new cells to replace those being shed in the exposed superficial layer.

Dermis

Most of the skin is composed of **dermis** ("true skin"), the strong, flexible connective-tissue meshwork of collagenous, reticular, and elastic fibers. *Collagenous* fibers, which are formed from the protein collagen, are very thick and give the skin much of its toughness. Although *reticular* fibers are thinner, they provide a supporting network. *Elastic* fibers give the

We All Get Wrinkles

Everybody's skin gets wrinkled. The length and depth of a wrinkle may differ from person to person, but no one is exempt. In fact, wrinkling is so predictable that we can expect to wrinkle in the same place, and at the same time, as the generation ahead of us. It is relatively easy to estimate a person's age merely by studying the wrinkles on the face and neck.

Wrinkling usually starts in the mid-twenties in the areas of the greatest facial expression: around the eyes, mouth, and brow. The familiar "crow's feet," lines radiating from the corner of each eye, usually appear by the age of 30. During the thirties and forties, new lines appear between the eyebrows and in front of the ears, followed in the fifties by wrinkles on the chin and bridge of the nose. New wrinkles appear on the upper lip in the sixties and seventies, and all of the established wrinkles become more pronounced. In the eighties, the ears become elongated, and we can see the "long ears" of old age.

Wrinkling happens when the protein *elastin* in the elastic tissue of the dermis loses its resiliency and degenerates into *elacin*. This causes the dermis to become more closely bound to the underlying tissue. Also, the layer of fat beneath our skin diminishes, causing the skin to sag into wrinkles. The wrinkling process is reinforced because an adult's aging epidermis does not produce new cells as readily as it did during childhood and adolescence. The epidermal cells are simply not as vital as they used to be.

Exposure to the ultraviolet rays in sunlight speeds up the wrinkling process, causing the collagen and elastin in the dermis to degenerate. As a result, the skin becomes slack and prematurely aged.

skin flexibility. The cells of the dermis are mostly *fibroblasts,* fat cells, and *macrophages* (which digest foreign particles). Blood vessels, lymphatic vessels, nerve endings, hair follicles, and glands are also present. The dermis is composed of two layers that are not clearly separated. The thin *papillary* layer is directly beneath the epidermis; the deeper, thicker layer is called the *reticular* layer.

Papillary layer The *papillary* (L. dim. *papula,* pimple) *layer* of the dermis consists of fairly loose connective tissue, with thin bundles of collagenous fibers. It is also known as the *subepithelial layer* because it lies just under the epithelial layer of the epidermis. The papillary layer is so named because of the papillae (tiny, fingerlike projections) that join it to the ridges of the epidermis (see Figure 5.1A). Most of the papillae contain loops of capillaries that nourish the epidermis; others have special nerve endings, called corpuscles of touch (Meissner's corpuscles), that serve as sensitive touch receptors (see Chapter 15). A double row of papillae in the palms and soles produce ridges that help keep the skin from tearing, improve the gripping surfaces, and produce distinctive fingerprint patterns in the finger pads. (Fingerprint patterns in the epidermis follow the corrugated contours of the dermis underneath. No two people, not even identical twins, have exactly the same fingerprints.)

Reticular layer The *reticular* ("netlike") *layer* of the dermis is made up of dense connective tissue, with coarse collagenous fibers and fiber bundles that crisscross to form a strong and elastic network (Figure 5.3). You can understand how tough the reticular layer is when you realize that it is the part of an animal's skin that is processed commercially to make leather.

Although the collagenous fibers of the dermis appear to be arranged randomly, there is actually a dominant pattern. In fact, different directional patterns are found in each area of the body (Figure 5.4). The resulting lines of tension over the skin are known as *cleavage (Langer's) lines.* They are very important during surgery because an incision made *across* the lines causes a gaping wound that heals more slowly and leaves a deeper scar than an incision made *parallel* to the directional lines. An incision made parallel to the cleavage lines heals faster because it disrupts the collagenous fibers only slightly, and the wound needs only a small amount of scar tissue to heal.

Embedded in the reticular layer are many blood and lymph vessels, nerves, free nerve endings, fat (adipose) cells, oil glands, and hair roots (see Figure 5.1A). Receptors for deep pressure (lamellated or Pacinian corpuscles) are distributed throughout the dermis and subcutaneous layer (see

FIGURE 5.3

Scanning electron micrograph of the meshed collagenous fibers of the skin. Collagen makes up about a third of the protein in the body. It is a major component of skin, tendons, ligaments, cartilage, and bone. [*Dr. Jerome Gross, The Developmental Biology Lab, Lovett Memorial Group, Massachusetts General Hospital.*]

×38,000

Chapter 15). The deepest region of the reticular layer contains smooth muscle fibers, especially in the genital and nipple areas, and at the base of hair follicles.

Although the dermis is highly flexible and resilient, it can be stretched beyond its limits. During pregnancy, for example, collagenous and elastic fibers may be torn. Characteristic abdominal "stretch marks" result from the repairing scar tissue.

Hypodermis Beneath the dermis lies the *hypodermis* (Gr. *hypo*, under), or *subcutaneous layer* (L. *sub*, under + *cutis*, skin). It is composed of loose, fibrous connective tissue such as adipose tissue. The hypodermis is generally much thicker than the dermis and is richly supplied with lymphatic and blood vessels, and nerves. Also within the hypodermis are the coiled ducts of sweat glands and the bases of hair follicles (see Figure 5.1A). The boundary between the epidermis and dermis is distinct; that between the dermis and the hypodermis is not. Fibrous elements within the dermis interface with the underlying loose connective tissue.

Where the skin is freely movable, hypodermis tissue fibers are scarce. Where it is attached to underlying bone or muscle, the hypodermis contains tightly woven fibers. In some areas of the hypodermis where extra padding is desirable, as in the breasts or heels of the feet, thick sheets of fat cells are present. The distribution of fat in this layer is largely responsible for the characteristic body contours of the female.

| *Where does the expression "getting cold feet," meaning being frightened, come from?* | *When you become frightened, the blood vessels in your skin constrict. As a result of the decreased blood flow, especially to your extremities, you get "cold feet."* |

FIGURE 5.4

Cleavage (Langer's) lines, in anterior (A) and posterior (B) views.

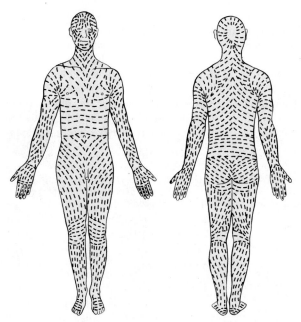

[A] ANTERIOR   [B] POSTERIOR

Functions of Skin

The obvious function of the skin is to cover and protect the inner organs. But this is only one of its many functions.

Protection The skin acts as a stretchable protective shield that prevents harmful microorganisms and foreign material from entering the body, and prevents the loss of body fluids. These functions are made possible by two features: (1) layered sheets of flat epithelial tissue that act like shingles on a roof, and (2) a nearly waterproof layer of keratin in the outer layer of the skin.

But skin is not simply a passive covering for our bones, muscles, and internal organs. Recent evidence shows that the skin plays an active role in defending the body against disease. Briefly (see Chapter 20 for a more detailed explanation), some disease-fighting cells, called lymphocytes, that are produced in the bone marrow are processed in the dermis before they are fully functional.

Temperature regulation Although the skin is almost waterproof and solid enough to keep out the water when you take a bath, it is also porous enough to allow some chemicals in* and to allow sweat to escape from sweat glands through ducts ending at the skin surface as tiny pores. (The word *pore* [Gr. *poros*, passage] is itself an indication that the skin is not completely solid.)

When sweat is excreted through the pores and then evaporates from the surface of the skin, a cooling effect occurs. On very humid days you are not cooled easily because the outside air is already almost full of water, and sweat merely builds up on your skin instead of evaporating. On cold days your skin acts as a sheet of insulation that helps retain body heat and keep your body warm. In addition, the underlying layer (dermis) contains a dense bed of blood vessels. On warm days the vessels dilate (enlarge), and heat radiation from the blood is increased; in this way, body heat is lost.

Through perspiration and the opening and closing of pores, the skin is an effective regulator of body temperature. In fact, heat loss through radiation, convection, conduction, and evaporation (mostly from the lungs but also from the skin) accounts for about 95 percent of the body's heat loss.

Excretion Through perspiration, small amounts of waste materials such as urea are excreted through the skin. Up to 1 g of waste nitrogen may also be eliminated through the skin every hour.

Synthesis The skin helps to screen out harmful excessive ultraviolet rays from the sun, but it also lets in some necessary ultraviolet rays, which convert a chemical in the skin called

---

*Unfortunately, several harmful substances can be absorbed through the skin and introduced into the bloodstream. Among the most common intruders are metallic nickel, mercury, many pesticides and herbicides, and a skin irritant called urushiol, found in poison ivy, poison oak, and poison sumac. Even the smoke from burning these plants may be enough to produce skin eruptions.

Diagnostic Information from Skin

Any physical examination includes a close look at the skin, because the skin provides many clues to the general health of the body. The skin color of a white person can reveal several abnormal conditions. For example, yellow-orange skin indicates jaundice (which means *yellow*), a condition arising when bile pigments enter the blood. A person who has been poisoned by silver salts has bluish-gray skin; cyanide poisoning turns the skin blue, indicating a lack of oxygen in the blood. The skin is a bronze tone when the cortex (outer layer) of the adrenal gland is underactive; and a person with unusually white skin is probably anemic, a condition often caused by an insufficient number of red blood cells.

Different types of rashes are identifiable in diseases such as measles, chicken pox, scarlet fever, ringworm, and syphilis. A deficiency of vitamin A is usually shown by patchy hair loss and a roughening of the skin. An allergic reaction is often identified by the eruption of a specific skin rash. Flushed skin means that the underlying blood vessels are dilated (enlarged); cold, clammy skin is a sign that the blood vessels have been constricted. An experienced physician can tell more from a handshake and a direct look than you could possibly imagine.

7-dehydrocholesterol into vitamin D_3 *(cholecalciferol).** Vitamin D in the diet is vital to the normal growth of bones and teeth. A lack of ultraviolet light and vitamin D impairs the absorption of calcium from the intestine into the bloodstream, no matter what the diet is. When children are deprived of sunshine, they generally become deficient in vitamin D. Unless they receive cholecalciferol from another source, they develop rickets, a disease that may deform the bones permanently. Laboratory animals on a high-sugar diet had five times as many cavities when they were deprived of ultraviolet light (and therefore vitamin D). Also, the body shows greater signs of mineral deprivation in the winter than in the summer, when sunlight is usually abundant.

Sensory reception The skin is an important sensory organ, containing sensory receptors for heat, cold, touch, pressure, and pain (see Chapter 15). Skin helps to protect us through its many nerve endings, which keep us responsive to things in the environment that might harm us: a hot stove, a sharp blade, a heavy weight. These nerve endings also help us to sense and enjoy the outside world, so that adjustments can be made to maintain homeostasis.

An itch or a tickle is simply a feeling of low-level pain, one that is not associated with the unpleasantness of pain. All pain sensations travel identical nerve pathways to the brain. We itch after being stung by a mosquito because the insect injects saliva that contains a chemical that dilates our blood vessels and keeps our blood from clotting before the mosquito has drained the blood it needs. The chemical in the saliva seeps into surrounding tissue, causing an allergic reaction that produces an itch.

Color of Skin

Skin gets its color from three factors: the presence of **melanin** (Gr. *melas*, black), a dark pigment produced by specialized cells called *melanocytes;* the yellow pigment *carotene;* and the color of the blood reflected through the epidermis. Mel-

*The general term "vitamin D" actually refers to a group of steroid vitamins, including vitamin D_3 (cholecalciferol) and vitamin D_2 (calciferol, or ergocalciferol).

anocytes are usually located in the deepest part of the stratum basale, where they extend their long processes under and around the neighboring cells (see Figure 5.1A). Melanin is found in all areas of the skin, but the skin is darker in the external genitals, the nipples and the dark area around them, the anal region, and the armpits. In contrast, there is hardly any pigment in the palms and soles. Melanin is present not only in skin but also in hair and in the iris and retina of the eyes.

The skin must absorb a certain amount of ultraviolet radiation from the sun to be able to manufacture vitamin D_3, but too much radiation damages the skin and can even cause skin cancer. The main function of melanin is to screen out excessive ultraviolet rays, especially protecting the nucleus and its genetic material. Extra protection is provided when melanin is darkened by the sun and transferred to the outer skin layers, producing a "suntan" that is less sensitive to the sun's rays than previously unexposed skin.* In a sense, "freckles" can be thought of as a permanent tan, produced by sun-darkened spots of melanin.

All races have some melanin in their skins. Although the darker races have slightly more melanocytes than light-skinned people do, the main reason for their darkness may be the wider distribution of melanin in the skin beyond the deepest portion of the stratum basale into higher levels of the epidermis. Also, the melanocytes of a dark-skinned person probably produce more melanin than the melanocytes of a light-skinned person. A person who is genetically unable to produce *any* melanin is an *albino* (L. *albus*, white) (see Chapter 29).

*Ironically, suntanned skin permits less vitamin D_3 synthesis than normal skin does, and it is possible that dark-skinned people produce less vitamin D_3 than light-skinned people do.

What causes the skin of Orientals to appear yellow?

In addition to melanin, skin contains a yellow pigment called carotene. It is usually overshadowed by melanin, but since Orientals have little melanin, their skin has a yellow tint.

How a Wound Heals

When the skin is cut, the healing process begins immediately. This process, which can take from a week to a month (depending on the severity of the cut), proceeds step by step:

1 Blood vessels are severed along with the dermis and epidermis. Red and white blood cells from the severed vessels leak into the wound. Blood cells called *platelets* (thrombocytes) and a blood-clotting protein called *fibrinogen* help to start a blood clot (see Chapter 17). A network of cells forms, and the edges of the wound begin to join together again. At the same time, cellular debris from the epidermis drifts among the blood cells. Tissue-forming cells called *fibroblasts* begin to approach the wound (Figure 5.5A).

2 Less than 24 hours later, the clotted area becomes dehydrated, and a scab forms. White blood cells called *neutrophils* move from blood vessels into the wound area and ingest microorganisms, cellular debris, and other foreign material. Epidermal cells at the edge of the wound begin to divide and start to build a bridge across the wound. Other white blood cells called *monocytes* migrate toward the wound from surrounding tissue (Figure 5.5B).

3 Two or three days after the wounding, monocytes enter the wound and ingest the remaining foreign material. Epidermal cells complete the bridge of new skin under the scab (Figure 5.5C). When a totally new epidermal surface has been formed, the protective scab is sloughed off. Fibroblasts build scar tissue with *collagen*.

4 About 10 days after the wounding, the epidermis has been restored, and the scab is gone (Figure 5.5D). Some monocytes and neutrophils remain in the area, but their work is usually completed by this time. Tough scar tissue continues to form, and bundles of collagen build up along the stress lines of the original cut.

As part of the overall healing process, an ***inflammatory response*** occurs, which includes redness, pain, swelling, scavenging by neutrophils and monocytes, and tissue repair by fibroblasts. The following events occur during the inflammatory response:

1 Blood vessels dilate and blood flow increases, probably because *histamine* is liberated. (Histamine, a powerful dilator of capillaries, may also be responsible for activating phagocytes and for increasing the flow of extracellular fluid to damaged tissues.)

2 Cell death occurs as lysosome enzymes are released by injured cells in the immediate area.

3 Blood vessels become more permeable to protein, white blood cells, and the blood-clotting agent fibrinogen.

4 Swelling *(edema)* is caused by the increased movement of fluids, which dilute the irritant.

5 White blood cells (neutrophils and monocytes) enter the area from nearby blood vessels and begin to remove cell debris and to phagocytize microorganisms.

6 Dead white blood cells and tissue debris form *pus*.

7 Tissue is repaired with the aid of regenerative fibroblasts.

Why does a healing wound itch? | When the skin is injured there is usually some damage to superficial nerves. After about a week or so, the nerves begin to fire electrical impulses as new nerve endings begin to grow. The firings feel like an itch on the skin, which usually subsides after a few days.

FIGURE 5.5

How a wound heals. (A) At the site of a wound, the epidermis, dermis, and blood vessels are severed. Fibroblasts move toward the wound area. (B) A scab forms. Epithelial cells at the edge of the wound begin to divide. Neutrophils begin to ingest cellular debris; monocytes move toward the wound area. (C) New skin is formed under the scab. Monocytes ingest the remaining foreign material in the wound area. Fibroblasts begin to build scar tissue. (D) The scab is gone. Scar tissue continues to form.

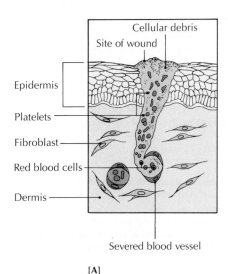
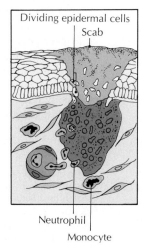
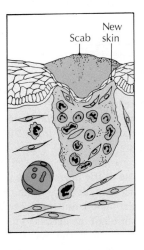
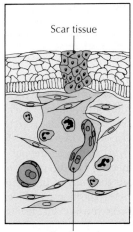

[A]   [B]   [C]   [D]

No Fun in the Sun

Why is sunbathing potentially dangerous? The process of tanning is activated by the ultraviolet rays in sunlight (called *ultraviolet B,* or *UVB*). These tanning rays can kill some skin cells, damage others so that normal secretions are stopped temporarily, increase the risk of skin cancer and mutations by affecting the genetic material in the nuclei of cells, and cause the immune system to falter. Ultraviolet rays can also damage enzymes and cell membranes, and interfere with cellular metabolism. If tissue destruction is extensive, toxic waste products and other cellular debris enter the bloodstream and produce the fever associated with sun poisoning.

Ultraviolet rays in sunlight cause tiny blood vessels below the epidermis to widen, allowing an increased flow of blood. The abundance of blood colors the skin red. Ultraviolet rays also stimulate melanocytes in the dermis to produce melanin. The cells begin to divide more rapidly than usual, creating new cells that travel toward the surface in an attempt to repair the damaged skin. Ordinarily, these new cells take about four weeks to reach the surface, where they die and shed unnoticed in the course of a normal day's activities. New cells produced in response to skin damage, however, are so numerous that they reach

the surface in four or five days. At this point, sheets of old, sunburned skin begin to peel off, and the pigment-rich melanocytes, which have traveled to just under the surface of the skin, produce the first signs of a tan. The tanned skin helps to prevent further ultraviolet damage by absorbing and scattering the harmful rays. Unfortunately, many people think that you have to burn before you can tan, but dermatologists (physicians who specialize in treating skin problems) disagree. Instead, they suggest short periods of sunbathing (about 10 min) until the skin has a protective tan.

Ask Yourself

1 *What are the components of the integumentary system?*

2 *What are the main functions of skin?*

3 *What are the two main layers of skin?*

4 *What are the layers of the epidermis called?*

5 *What are the two layers of dermis?*

6 *How does melanin protect the skin?*

7 *How does a wound heal?*

GLANDS OF THE SKIN

The glands of the skin are the sudoriferous (sweat) glands and the sebaceous (oil) glands.

Sudoriferous (Sweat) Glands

Sudoriferous glands (L. *sudor,* sweat) are also known as *sweat glands.* Two types of sweat glands exist, eccrine (EKK-rihn) and apocrine (APP-uh-krihn). *Eccrine glands* (Gr. *ek-krinein,* to exude, secrete) are small sweat glands. They are distributed over nearly the entire body surface. (There are no sweat glands on the nail beds, margins of the lips, eardrums, inner lips of the vulva, or the tip of the penis.) These glands are generally of the simple, coiled tubular type, with the secretory portion embedded in the hypodermis, and a hollow, corkscrewlike duct leading up through the dermis to the surface of the epidermis (see Figure 5.1A). The duct generally straightens out somewhat as it reaches the surface. All of these ducts combined would be about 9.5 km (6 mi) long.

Most eccrine glands secrete sweat by a physiological process called *perspiration** when the temperature rises. Eccrine glands that respond to psychological stress are most numerous on the palms, fingers, and soles. Sweat from eccrine glands consists of a colorless aqueous fluid, holding in solution neutral fats, albumin and urea, lactic acid, sodium and potassium chloride, and traces of sugar and ascorbic acid. Its excretion helps in body temperature regulation and homeostasis, largely by the cooling effect of evaporation (see Chapter 23). Sweat glands are efficient. They secrete sweat even when the skin appears to be dry, and the constant combination of perspiration and evaporation keeps the body from overheating.†

Apocrine, or *odiferous,* *glands* are found in the armpits, the dark region around the nipples, the outer lips of the vulva, and the anal and genital regions. These are larger and more deeply situated than the eccrine glands. The female breasts are apocrine glands that have become adapted to secrete and release milk instead of sweat. Apocrine glands become active at puberty and enlarge just before menstruation. They respond to stress (including sexual activity), not heat, by secreting sweat of a characteristic odor. Secretions of apocrine glands have specific human pheromones. (A *pheromone* is any substance secreted by an animal that provides communication with other members of the same species to elicit certain behavioral responses.) But a smell that we commonly call "body odor" results when bacteria on the skin begin to decompose the secretions as they feed on them.

---

*The word *perspiration* literally means "to breathe through" (L. *per,* through + *spirare,* to breathe), in reference to the ancient belief that the skin breathed.

†An inactive person produces about 70 mL (2.1 fl oz) of sweat a day, and an athlete or manual laborer can produce a liter or more during an hour of vigorous physical activity. The inactive person loses sweat by simple diffusion, through a process called *insensible perspiration* (see page 806); the active person perspires in a more noticeable way.

The *ceruminous glands* in the outer ear canals are also apocrine skin glands. They secrete the watery component of *cerumen* (ear wax). Cerumen helps to trap foreign particles before they can enter the deeper portions of the ear.

Sebaceous (Oil) Glands

Sebaceous, or *oil, glands* (L. *sebum,* tallow, fat) are simple, branched alveolar glands found in the dermis. Their main functions are lubrication and protection. They are connected to hair follicles (see Figure 5.1A). The secretory portion is made up of a cluster of cells, polyhedral in the center of the cluster and cuboidal on the edges. Secretions are produced by the breaking down of the interior cells, which become the oily secretion called *sebum,* found at the base of the hair follicle. Sebum is a semifluid substance composed almost entirely of lipid. About 60 percent of sebum consists of glycerides and free fatty acids, but some waxes are also present. Destroyed cells are replaced from the outer layer of cuboidal cells.

The exact role sebum plays in the physiology of the skin is not known. Possibly it serves as a permeability barrier, an emollient (skin-softening agent), or a protective agent against bacteria and fungi. Sebum can also act as a pheromone.

When sebaceous glands become inflamed and accumulate sebum, the gland opening becomes plugged, and a **blackhead** results. If the plugging is not relieved, a pimple or even a sebaceous cyst may develop. The "blackness" of blackheads is caused by air-exposed sebum, not dirt. Blackheads and pimples often accompany hormonal changes during puberty, when an oversecretion of sebum may enlarge the gland and plug the pore. The resulting skin disease is called *acne vulgaris.* (See Physiological and Anatomical Abnormalities, p. 127.) According to the National Center for Health Statistics, over 85 percent of all 17-year-olds have acne to some degree. Only 28 percent of adolescents from ages 12 to 17 are free from it.

Sebaceous glands are found all over the surface of the skin except in the palms and soles, where they would be a nuisance.

Why do your fingertips and toes pucker when you take a bath?

The thick skin on the underside of your fingers and toes is capable of absorbing a great deal of water, and because your palms and soles are not coated with sebum, water soaks in easily through the epidermis. The dermis does not expand, however, so the epidermis becomes corrugated temporarily. Also, when you get out of the tub the excess water begins to evaporate. When it does, some of the water that was there before your bath is also removed, and your skin becomes dried out and wrinkled until the body fluid is replaced. (The same thing happens to a grape when it is dried out to make a raisin.)

Ask Yourself

1 *What are the two types of skin glands?*

2 *What are the functions of sweat glands?*

3 *What are the functions of sebaceous glands?*

HAIR

Hair is composed of cornified threads of cells, a specialization that develops from the epidermis. Because hair arises from the skin, it is considered to be an appendage of the skin. It covers the entire body, except for the palms, soles, lips, tip of the penis, inner lips of the vulva, and nipples.

Functions of Hair

Obviously, humans do not use hair as an insulative covering the way many other mammals do, but hair does provide some protective functions. Scalp hair provides some insulation against cold air and the heat of the sun. Like most hair, scalp hair also protects us from bumps. Eyebrows act as cushions in protecting the eyes and also help reduce glare and prevent sweat from running into the eyes. Eyelashes act as screens against foreign particles. Tiny hairs in the nostrils help keep out dust particles. Other openings in the body, such as the ears, anus, and vagina, are also protected by hair.

The functions of axillary (armpit) and pubic hair are disputed. Is the function of this hair to collect odor-producing sweat as a sexual attractant? Our odor-conscious society may not find such an idea appealing, but in other human societies, and in animal societies, body odors are used as fundamental sexual signals. Another possible function of axillary hair may be to provide lubrication when the arms are moving at the sides, especially during walking or running; and pubic hair may provide the same function during sexual intercourse. Pubic hair also serves as a protective cushion for the pubic area.

The newly sprouted facial hair of the male teenager serves to advertise his sexual maturity in the same way that the mane of a lion or the colorful plumage of a male bird does. Both males and females in our society use scalp hair as a sexual attractant, with hair styles varying almost as often as clothing styles.

Structure of Hair

Hair consists of epithelial cells arranged in three layers. From the inside out of a hair *shaft* (the portion that protrudes from the skin), these are the *medulla, cortex,* and *cuticle* (Figure 5.6). The medulla is composed of soft keratin, and the cuticle and cortex are composed of hard keratin. A strand of hair is stronger than an equally thick strand of nylon or copper.

The *medulla* forms the central core of the hair, and it usually contains loosely arranged cells separated by air "cells" or liquid in its extracellular spaces. The *cortex* is the thickest layer of hair, consisting of several layers of cells. Pigment in the cortex gives hair its color. When hair pigment fades from the cortex and the medulla becomes completely filled with air, the hair appears to be gray to white, with whiteness representing a total loss of pigment. The *cuticle* is made up of thin squamous cells that overlap to create a scale-like appearance (see Figure 5.6). The cuticle can be softened or even dissolved by chlorine in pool water. The greatest damage occurs over sustained periods of time, as the effects of the chlorine accumulate.

The portion of hair that protrudes from the skin is the *shaft,* and the portion embedded beneath the skin is the *root.* The lower portion of the root, located in the hypodermis, enlarges to form the *bulb.* The bulb is composed of a *matrix* of epithelial cells. The bulb pushes inward along its bottom surface to form a papilla of blood-rich connective tissue (Figure 5.7). The entire bulb is enclosed within a tubular *follicle* (L. *folliculus,* little bag). The hair follicle consists of three sheaths: (1) an inner epithelial root sheath, (2) an outer epithelial root sheath, and (3) a connective-tissue sheath.

Just before a hair is to be shed, the matrix cells gradually become inactive and eventually die. The root of the hair becomes completely keratinized and detaches from the matrix. This cornified root bulb begins to move along the follicle until it stops near the level of the sebaceous gland. The papilla atrophies, and the outer root sheath collapses.

A new hair begins to grow when new cells from the outer root sheath start to develop near the old papilla. A new ma-

FIGURE 5.7

Drawing of a longitudinal section of human hair. The hair root includes the follicle and the bulb.

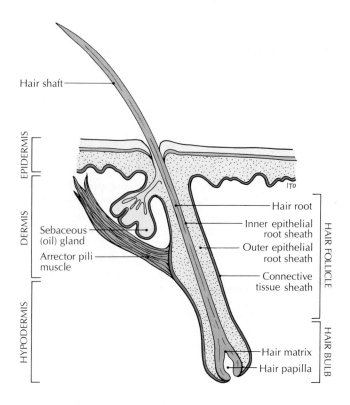

FIGURE 5.6

Hair. (A) The microscopic structure of hair. Notice how the scales of the cuticle all face upward. (B) A scanning electron micrograph of a hair shaft on the scalp. [*Biophoto Associates/Photo Researchers.*]

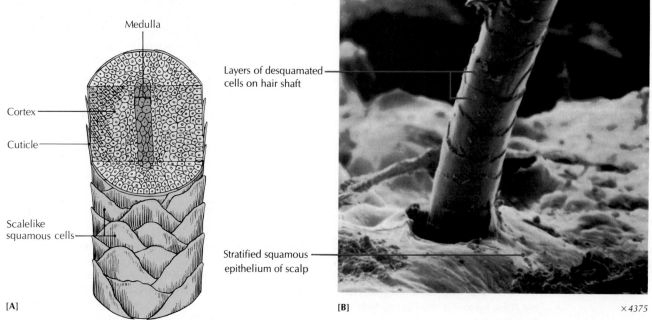

[A]   [B]   ×4375

trix develops, and a new hair starts to grow up the follicle, pushing the old hair out of the way if it hasn't been shed already.

Since the hair follicle arises from the epidermis, projecting at an angle, the hair shaft usually points away from the bulb. For this reason, hair covers the scalp more efficiently than if it grew at right angles to the scalp. The shape of the follicle openings determines whether your hair is straight, wavy, or curly. A round opening (in cross section) produces straight hair; an oval opening produces wavy hair; and a spiral-shaped opening produces curly hair.

An interesting part of the hair follicle is the bundle of smooth muscle attached about halfway down the follicle. As you saw in Figures 5.1A and 5.7, a sebaceous gland nestles between the hair follicle and the muscle like a child sitting in the crook of a tree. This muscle is the **arrector pili** muscle. When it contracts, it pulls the follicle and its hair to an erect position, elevates the skin above, and produces a "goose bump" on the skin. The contracting muscle also forces sebum from the sebaceous gland.

In furry animals, this action helps to warm the animal by producing an insulating layer of warm body air between the erect hair and the skin. If the "goose bump" reaction is the result of an animal being frightened instead of being cold, the erect hairs also make the animal look larger and less vulnerable to attack. In people, the benefits of "goose bumps" are minimal. The muscle contraction, or shivering, that accompanies the "goose bumps" does generate some heat, however, and the coating of sebum on the skin may help keep the body warm by limiting perspiration.

Development of Hair

The first hairs begin to develop during the third fetal month, when the epidermis begins to project downward into the dermis. Eventually these downgrowths form hair follicles that produce hairs. By the fifth month, the fetus is covered with fine, downy hair called **lanugo** (L. *lana,* fine wool). Later in the fifth month, eyebrows and head hairs become apparent. Lanugo hair is shed before birth, except on the eyebrows and scalp, where it becomes thicker. After the baby is five or six months old, this hair is replaced by coarser hair, and the rest of the body becomes covered with a film of delicate hair called **vellus** (L. "fleece" or "coarse wool").

Coarse hair appears at puberty in the genital area and armpits, and young men also develop chest and facial hair. Body hair usually thickens during puberty. The early coarse hair of the eyebrows and scalp, as well as the hair that appears at puberty, is called **terminal hair.** Although the vellus and terminal hairs seem to be "new," no new hair follicles are developed after birth.

Ask Yourself

1 *What are some of the functions of hair?*

2 *What are the three layers of a hair called?*

3 *What is a hair follicle?*

4 *How are "goose bumps" produced?*

NAILS

Nails, like hair, are modifications of the epidermis. Also, like the cuticle and cortex of a hair, they are made of hard *keratin.* Nails are composed of flat, horny plates on the dorsal surface of the distal segment of the fingers and toes (Figure 5.8). They appear pink because the nail is translucent, allowing the red color of the vascular tissue underneath to show through. The proximal part of the nail, the **lunula** (L. "little moon"; commonly called "half-moon"), is white because the "red" capillaries in the underlying dermis do not show through.* The lunula usually does not appear on the little finger.

*Some anatomists contend that the lunula is white because it is the portion of the nail formed first. Relatively young nail substance like the lunula is generally whiter, thicker, and more opaque than the more mature nail that extends over the nail bed.

FIGURE 5.8

(A) The basic components of a fingernail. (B) A drawing of a sagittal section of a fingernail.

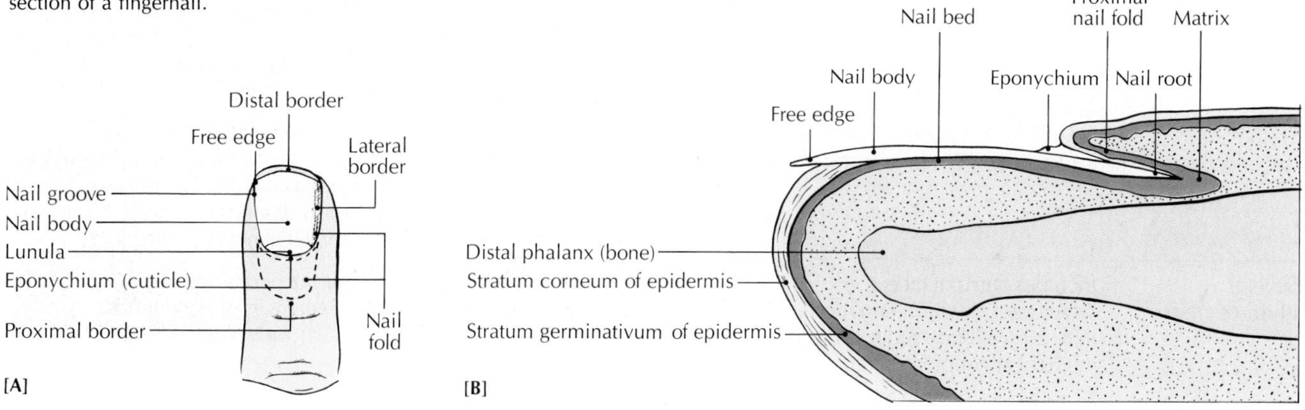

The body of the nail consists of cornified dead cells. It is the part that shows; the root is the part hidden under the skin folds of the nail groove (Figure 5.8A). The nail ends with a *free edge* that overhangs the tip of the finger or toe. The nail rests on an epithelial layer of skin called the **nail bed,** and the thicker layer of skin beneath the nail root is the **matrix,** the area where new cells are generated for nail growth and repair (Figure 5.8B). If the nail is injured, a new one will grow as long as the living matrix is intact. After birth, fingernails grow faster than toenails, with an average growth rate of 0.5 mm a week. Both fingernails and toenails grow faster in warm weather than in cold.

The developing nail is originally covered by thin layers of epidermis called the **eponychium** (epp-oh-NICK-ee-uhm; Gr. *epi,* upon + *onyx,* nail). The eponychium remains at the base of the mature nail and is commonly called the *cuticle.*

The functions of human nails are much more limited than the functions of claws, horns, and hooves of other animals. Our nails protect our fingers and toes, and allow us to pick up and grasp objects. And we use them to scratch.

PHYSIOLOGICAL AND ANATOMICAL ABNORMALITIES

Burns

Burns occur when skin tissues are damaged by heat, electricity, radioactivity, or chemicals. The seriousness of burns can be classified according to (1) *extent* (how big an area of the body is involved), and (2) *depth* (how many layers of tissue are injured) (Table 5.1).

A *first-degree burn* (such as a sunburn) may be red and painful, but it is not serious. Generally, it damages only the epidermis but does not destroy it. Such a burn responds to simple first-aid treatment, including cold water and sterile bandages. Butter and other greasy substances should never be used on *any* burn, because they may actually help to "cook" the damaged skin even further. Instead, the burn should be flushed or immersed in cold water (not ice water), or cold compresses (not ice) should be applied. Cold water helps to reduce pain, swelling, fluid loss, and infection, and also limits the extent of the damage.

A *second-degree burn* destroys the epidermis, and also causes some cell destruction in the dermis. Oozing blisters and scarring usually result. After a second-degree burn, the body may be able to regenerate new skin. Second-degree burns require prompt medical attention. If left untreated, a second-degree burn can progress to a third-degree burn. First- and second-degree burns are also called *partial-thickness burns* because only the epidermis is damaged seriously.

A *third-degree burn* involves the epidermis, dermis, and underlying tissue. Because the skin cannot be regenerated, this kind of burn must be treated with surgery and skin grafting. Ordinarily, the victim is in shock but feels no pain because nerve endings in the burned area have been destroyed. The damaged area is charred or pearly white, and fluid loss is severe. A third-degree burn is called a *full-thickness burn* because all skin layers are destroyed.

A burn is the most traumatic injury the body can receive. Besides causing obvious tissue damage, serious burns expose the body to microorganisms, hamper blood circulation and urine production, and create a severe loss of body water, plasma, and plasma proteins that can produce shock. In fact, a major burn causes homeostatic imbalances in every system of the body. A severe burn leaves the skin more vulnerable to microbial infection than other types of wounds do. This happens because neutrophils, the skin's specialized infection-fighting cells (see Chapters 17 and 20), are practically immobilized by the burn. Instead of rushing to the infection site and releasing disease-fighting enzymes, the traumatized neutrophils release their enzymes prematurely. Because the enzymes interfere with the chemical signal from the infection site, the neutrophils do not know where to go, and only a few make it to the infected area.

TABLE 5.1 A CLASSIFICATION OF BURNS

Type of burn	Surface area affected	Depth of tissue damage	Major effects
Minor (first-degree)	Less than 10% of body surface.	Epidermis damaged but not destroyed.	Mild swelling, reddening, pain; injured cells peel off and skin heals without scarring, usually within 2 weeks.
Serious (second-degree)	More than 15% of body surface for an adult, 10% for a child.	Epidermis and part of dermis destroyed. New skin may regenerate.	Red or mottled appearance, blisters, swelling, wet surface due to plasma loss. Greater pain than third-degree burn (which destroys sensitive nerve endings).
Severe (third-degree)	Includes burns of face, eyes, hands, feet, genitals, and more than 20% of body surface. Prompt medical attention required.	All skin layers destroyed; deep tissue destruction. Nerve endings in skin destroyed. Skin cannot be regenerated. Surgery and skin grafts necessary.	White or charred appearance, severe loss of body fluids.

TABLE 5.2   TYPES OF SKIN LESIONS (TISSUE ALTERATIONS)

Name	Example	Description
Bleb, bulla	Second-degree burn	A fluid-filled elevation of the skin.
Cyst	Epidermal cyst	A mass of fluid-filled tissue, extending to dermis or hypodermis.
Macule	Freckle; flat, pigmented mole	A discolored spot, not elevated or depressed.
Papule	Acne, measles	Raised, red area resembling small pimples.
Pustule	Acne vulgaris, smallpox	Raised, pus-filled pimple.
Tumor	Dermatofibroma*	Raised mass extending into dermis.
Vesicle	Blister, chicken pox, herpes simplex	A small sac filled with serous fluid.
Wheal	Mosquito bite, hives	Local swelling, itching.

*A dermatofibroma is a fibrous tumor of the dermis.

Some common skin disorders

Acne *Acne vulgaris* ("common acne") is most common among adolescents, when increased hormonal activity causes the sebaceous glands to overproduce sebum. When the flow of sebum is increased, dead keratin cells may become clogged in a follicle. These plugs at the skin opening are called either *blackheads* (open *comedones*—sing. *comedo*—which protrude from the follicle and are not covered by the epidermis) or *whiteheads* (closed comedones, which do not protrude from the follicle and are covered by the epidermis). The blocked follicle may become infected with bacteria, which secrete enzymes that convert the clogged sebum into free fatty acids. These acids irritate the lining of the follicle and eventually cause the follicle to burst. When the acid and sebum seep into the dermis, they cause an inflammation that soon appears on the surface of the skin as a pus-filled papule called a "pimple" (Table 5.2). Picking and scratching merely spread the infection and may produce scarring.

Acne appears mostly on the face, chest, upper back, and shoulders. The problem generally affects young men more severely than young women, probably because the causative hormones are *androgens*, male hormones found in much greater abundance in males than in females.

The most advanced form of acne is **cystic acne**, which produces deep skin lesions called *cysts*. It is produced when sebaceous glands secrete excessive amounts of oil that nourish the infectious bacteria that cause acne in the first place.

Bedsores *Bedsores* (*decubitus ulcers*) are produced when bony, unprotected areas of the skin undergo constant pressure, usually from the weight of the body itself. The pressure causes blood vessels to be compressed, depriving the affected tissue of oxygen and nutrition, and often leading to cell death. The most typical problem areas are the hips, elbows, tailbone, knees, heels, ankles, and shoulder blades. The first signs of bedsores are warm, reddened spots on the skin. Later, the spot may become purplish, indicating that blood vessels are being blocked and circulation is impaired. Actual breaks in the skin may follow, and bacterial infection is common if the lesions are left untreated. Cleanliness and dryness are important in preventing bedsores, as is changing the position of the patient frequently.

Birthmarks and moles *Birthmarks* and *moles* are common skin lesions. The technical name for a birthmark is a **vascular nevus** (L. "birthmark"; plural *nevi*). A **nevus flammeus**, or *port-wine stain*, is a pink to bluish-red lesion that usually appears on the back of the neck. The mucous membrane, as well as the skin, may be discolored. The cause of nevi is not known. A **hemangioma**, or *strawberry mark*, affects only the superficial blood vessels. Strawberry marks are usually present at birth, but they may also appear any time after birth. The most common sites are the face, shoulders, scalp, and neck. The mark may grow slowly, remain the same size, or become smaller or even disappear altogether through the years.

The **common mole**, or **nevus**, is a benign lesion (in most cases) that usually appears before the age of 5 or 6, but it may appear any time up to about 30 years of age. Moles that darken, enlarge, bleed, or appear after a person is 30 should be checked by a physician, since an occasional mole may be transformed into a cancerous growth. Moles start out as flat brown or black spots, and typically enlarge and become raised later, especially during adolescence and pregnancy. Most people have some moles, and many people have 50 or even more small moles, but by the age of 60 there may be only five or six moles left. The tendency to have moles is thought to be an inherited characteristic. It is not unusual for a mole to contain hair.

Psoriasis The cause of **psoriasis** is unknown, but there is general agreement that heredity plays a role. Attacks of psoriasis can be brought on by trauma, cold weather, pregnancy, hormonal changes, and emotional stress. The disease occurs when skin cells move from the basal layer to the stratum corneum in only four days instead of the usual 28. As a result, the cells do not mature, and the stratum corneum becomes flaky. Lesions are red, dry, and elevated, and are covered with silvery, scaly patches. The most usual sites are the elbows and knees, scalp, face, and lower back. Psoriasis is most common in adults, but may occur at any age.

Allergic responses *Poison ivy, poison oak,* and *poison sumac* all cause skin irritations when contact is made with those plants, which contain *urushiol,* a powerful skin irritant. (It is interesting that these plants have no effect the first time a person is exposed to them.) Exposed parts of the body usually begin to redden several hours (or even several days) after exposure. Redness, itching, and swelling generally progress to vesicles (raised, red sacs), blisters, and finally a dry crust after serous fluid oozes from the blisters. Some people are so allergic to urushiol that they become affected by the smoke of the burning plants or by touching tools or pets that have touched the plants.

Warts *Warts (verrucae)* are benign epithelial tumors caused by various papilloma viruses. Although they may appear anywhere on the body and on people of all ages, they are most common on the hands of children. This is probably so because the skin of the hands is likely to be irritated often, and a child's immune system is not yet effective against the virus. A wart is a raised area of the skin that has a pitted surface. It is usually no darker than the skin color, except on the soles of the feet (plantar warts), where it is often yellowish. (Warts may appear darker than the skin because dirt becomes lodged in the tiny crevices between the fibers that make up the wart.) Warts are transmitted by direct contact, and they may be spread to other parts of the body by scratching and picking. They usually disappear after a year or so, but may be removed by a surgeon if no complications are expected.

Skin cancer

The two most common forms of skin cancer are *basal-cell epithelioma (-oma* means tumor), also called basal-cell carcinoma, and *squamous-cell carcinoma.* The most serious type of skin cancer is *malignant melanoma.* All three forms can be prevented to a great degree by avoiding overexposure to the ultraviolet rays in sunlight. Other causes include arsenic

poisoning, radiation, and burns. It is believed that people who have moles may have an increased risk of developing melanomas.

Basal-cell epithelioma *Basal-cell epithelioma* generally appears on the face, where sweat glands, oil glands, and hair follicles are abundant. It occurs most frequently in fair-skinned males over 40. Three types of lesions are typical: (1) *nodulo-ulcerative lesions* are small and pinkish during the early stage; eventually they enlarge and become ulcerated and scaly. These lesions usually do not *metastasize* (spread to other tissues) and can be treated locally with good results. If neglected, however, they may extend to surrounding normal skin tissue and produce infection and hemorrhage. (2) *Superficial basal-cell epitheliomas* frequently erupt on the back and chest. These lightly pigmented areas are sharply defined and slightly elevated. They are associated with exposure to substances that contain arsenic. (3) *Sclerosing basal-cell epitheliomas* are waxy, yellowish-white patches that appear on the head and neck.

Squamous-cell carcinoma *Squamous-cell carcinoma* usually appears as premalignant lesions, typically in the keratinizing epidermal cells of the lips, mouth, face, and ears. Unlike basal-cell epithelioma, squamous-cell carcinoma may metastasize actively, especially when the lesions occur on the ears and lower lip. If the lymph nodes are affected, the symptoms of pain, malaise, fatigue, weakness, and anorexia (absence of appetite) generally occur. Squamous-cell carcinoma is most common in fair-skinned males over 60.

Malignant melanoma *Malignant melanoma* involves the pigment-pro-

ducing melanocytes. It usually starts as small, dark growths resembling moles that gradually become larger, change color, become ulcerated, and bleed easily. As with basal-cell epithelioma and squamous-cell carcinoma, the incidence of malignant melanoma is highest among fair-skinned persons, and it is slightly more common among women than among men. Besides the usual causes, malignant melanomas seem to be stimulated by hormonal changes during pregnancy. Surgery is always necessary to remove the tumors. Untreated deep lesions may metastasize to nearby lymph nodes, the liver, lungs, and the brain and spinal cord. Most melanomas can be cured if they are treated early.

Bruises

Bruises (Middle Eng. *brusen,* to crush) appear "black-and-blue" because a hard blow to the surface of the skin breaks capillaries and releases blood into the dermis. Although the blood is red, it creates a black-and-blue mark on the surface because the skin filters out all but the blue light that reflects off the bruise and makes it appear dark blue or purplish. Ordinarily, the darker the bruise, the deeper the blood has seeped. Bruises sometimes turn yellow or green after several days. This is usually an indication that the spilled red blood cells have begun to break down into their components. Iron in the blood often gives the bruise a greenish color, as the decaying red pigment hemoglobin is transformed into a yellowish substance called *hemosiderin.* These color changes indicate that the bruise is in its final stages. Scavenger white blood cells move into the affected area and ingest the hemosiderin and other debris, and the tissue returns to its normal color.

MEDICAL TERMINOLOGY

ALOPECIA (L. "mange of a fox," "baldness") Loss of hair resulting in baldness.

ANTIPRURITIC DRUGS Drugs used to counteract itchiness.

ATHLETE'S FOOT (tinea pedis) A fungal infection that produces lesions of the foot, often accompanied by itching and pain.

CAFÉ AU LAIT SPOTS Light brown, flat spots on the skin; also known as von Recklinghausen's disease.

CALLUS (L. "hard") An area of the skin that has become hardened by repeated external pressure or friction.

CELLULITIS Inflammation of the skin and subcutaneous tissue, with or without the formation of pus.

CORN (L. *cornu,* horn) An area of the skin, usually on the feet, where the horny cells of the epidermis have become hardened by external pressure, from ill-fitting shoes, for instance.

DANDRUFF A scaly collection of dried sebum on the scalp.

DERMABRASION A surgical technique to scrape away areas of the epidermis, usually to remove acne scars. Facial skin is anesthetized and frozen at −30°C (−86°F) for 15 to 30 min, and the skin is smoothed with a high-speed diamond wheel. A typical smoothing procedure takes about 20 min.

DERMATOLOGIST A physician who specializes in the treatment of skin disorders.

DERMATOLOGY The medical study of the physiology and pathology of the skin.

DERMATOME (Gr. *derma,* skin + *tomos,* a cutting) An instrument used in cutting thin slices of the skin, as in skin grafting.

ECZEMA (Gr. eruption) An inflammatory condition of the skin, producing red, papular, and vesicular lesions, crusts, and scales.

ELECTROLYSIS Destruction of hair follicles by inserting a needle conducting an electric current into the follicle.

HIVES A skin eruption of wheals, usually associated with intense itching.

IMPETIGO *(impetigo contagiosa)* (ihm-puh-TIE-go; L. "an attack") A contagious infection of the skin, caused by staphylococci or streptococci bacteria and characterized by small red macules that become pus-filled.

KELOID (Gr. clawlike) A growth of scar tissue.

KERATOSIS (Gr. *keras,* horn) A thickening and overactivity of the horny cells of the epidermis.

LIVER SPOTS Dark, flat patches on the skin, surrounded by lighter areas. The patches signal the skin's attempt to protect itself from ultraviolet damage. The condition occurs most often in people over 30. (Also called *age spots.*)

MELASMA (Gr. *melas,* black) A condition characterized by dark patches on the skin, usually brought on by stress, sensitivity to chemicals or sunlight, physical injury, and hormonal changes (such as in the "mask of pregnancy").

POLYP (Gr. "many-footed") A growth protruding from the mucous lining of an organ, such as the nose.

RINGWORM (tinea) A fungal infection, also known as *dermatophytosis,* characterized by red papules that spread in a circular pattern.

SCABIES A contagious skin disease caused by the mite *Sarcoptes scabiei,* which lays eggs under the skin, causing itching and the eruption of vesicles.

VESICATION The formation of blisters.

SUMMARY

Skin

1 Skin is the major part of the *integumentary system,* which also includes the hair, nails, and glands of the skin.

2 The *epidermis* of the skin is the outermost layer of epithelial tissue, and the *dermis* is a thicker layer of connective tissue beneath the epidermis.

3 A typical arrangement of layers (strata) in the epidermis, from the outermost stratum to the deepest one, is: *stratum corneum, stratum lucidum, stratum granulosum, stratum spinosum,* and *stratum basale.* The stratum spinosum and stratum basale are collectively known as the *stratum germinativum.*

4 The dermis is composed of the thin *papillary* layer and the deeper, thicker *reticular* layer.

5 The *subcutaneous layer,* or *hypodermis,* lies beneath the dermis.

6 Skin serves as a stretchable *protective shield,* a *regulator of body temperature,* a source of *excretion* of waste materials, a *screen* against harmful ultraviolet rays, a *converter* of the sun's rays into vitamin D_3, and a *sensory receptor.*

7 Skin color results from the presence of *melanin,* the pigment *carotene,* and the color of the underlying blood.

8 The wound-healing process involves the formation of blood clots by platelets and fibrinogen, phagocytosis by neutrophils and monocytes, and the growth of new epidermal and collagenous tissue. The *inflammatory response* is also part of the overall healing process.

Glands of the skin

1 The glands of the skin are the *sudoriferous (sweat) glands* and the *sebaceous (oil) glands.* Sudoriferous glands are typically classified as *eccrine* glands, small sweat glands that secrete sweat, or *apocrine* glands, which become active at puberty and produce thicker secretions.

2 The major role of the sudoriferous glands is temperature regulation, while the function of sebaceous glands is lubrication.

Hair

1 *Hair* is composed of cornified threads of cells that develop from the epidermis. Because of its origin, hair is considered to be an appendage of the skin.

2 Hair consists of epithelial cells arranged in three layers. From the inside out of a hair shaft, these are the *medulla, cortex,* and *cuticle.*

3 The hair *shaft* protrudes from the skin, the *root* is embedded beneath the skin, and the entire structure is enclosed within a *follicle.*

4 The first hairs begin to develop during the third fetal month. Terminal hairs appear at puberty in the genital area and armpits, and young men also develop chest and facial hair.

5 Human hair is constantly being shed and replaced; hair follicles alternate between growing and resting phases.

Nails

1 *Nails,* like hair, are modifications of the epidermis.

2 Composed of hard keratin, nails are flat, horny plates on the dorsal surface of the distal segment of the fingers and toes.

Physiological and anatomical abnormalities

1 *Burns* occur when skin tissues are damaged by heat, electricity, radioactivity, or chemicals. They are classified according to how big an area of the body is involved and how many layers of tissue are injured. Third-degree burns are the most severe.

2 *Acne vulgaris* occurs when sebum clogs follicles, which become infected with bacteria, burst, and cause inflammations on the surface of the skin.

3 *Bedsores* are produced when unprotected areas of the skin undergo constant pressure.

4 *Birthmarks (vascular nevi)* and *moles (nevi)* are common skin lesions that are usually benign.

5 *Psoriasis* is produced when immature skin cells reach the surface and begin to flake off. Its cause is unknown.

6 *Allergic responses* usually occur when the skin is in contact with urushiol, a powerful skin irritant in poison ivy, oak, and sumac.

7 *Warts* are benign epithelial tumors caused by a viral infection.

8 The most common forms of skin cancer are *basal-cell epithelioma* and *squamous-cell carcinoma.* The most serious type is *malignant melanoma.*

9 *Bruises* occur when broken capillaries release blood into the dermis.

UNDERSTANDING THE FACTS

1 What structures are found in the dermis?

2 What three factors determine skin color?

3 What is the inflammatory response?

4 What are some of the functions of hair?

5 What determines whether your hair is straight, wavy, or curly?

6 What are some functions of nails?

7 What is the relationship of blackheads and acne vulgaris to sebaceous glands? Why is acne vulgaris most common during adolescence?

8 What is the most serious form of skin cancer?

UNDERSTANDING THE CONCEPTS

1 How does the structure of the skin relate to its functions of protection, temperature control, waste removal, radiation protection, vitamin production, and environmental responsiveness?

2 When you receive a hypodermic injection, the needle passes through several layers of skin. List the basic layers and all subdivisions, in sequence from the outside in, that would be penetrated.

3 What is the purpose of the stratum germinativum?

4 Compare the structures of sudoriferous and sebaceous glands.

5 Explain how fibroblasts, monocytes, fibrinogen, collagen, neutrophils, and histamine work together in the healing of a wound.

6 Explain how hair develops and grows.

7 Assume that you are a student nurse, assigned to help with a severe-burn victim. Because of the knowledge you have gained from your anatomy and physiology class, you know that this patient faces several life-threatening problems. Explain these problems.

SUGGESTED READING

BRAVERMAN, IRWIN M., *Skin Signs of Systemic Disease,* 2nd ed. Philadelphia: Saunders, 1981.

EPSTEIN, ERVIN, AND ERVIN EPSTEIN, JR., eds. *Skin Surgery,* 5th ed. 2 vols. Springfield, Ill.: Charles C Thomas, 1982.

FENSKE, NEIL A., "How Age Affects Skin Structure, Disease and Treatment." *Modern Medicine,* 50:11 (November 1982):198–210.

LEVINE, N., "The Skin and Sports, Part 1: Managing the Damage of Physical Play." *Modern Medicine* 48:7 (April 1980):15–30, 37–43.

MARPLES, MARY J., "Life on the Human Skin." *Scientific American,* January 1969.

MAUGH, T. H., "Hair: A Diagnostic Tool to Complement Blood, Serum and Urine." *Science* 20 (1978):1271.

MONTAGNA, WILLIAM, "The Skin." *Scientific American,* February 1965.

MOOLTEN, SYLVAN E., M.D., "Bedsores: An Update." *Hospital Medicine* 18:8 (August 1982):64A–64GG.

PILLSBURY, DONALD M., AND CHARLES L. HEATON, *A Manual of Dermatology,* 2nd ed. Philadelphia: Saunders, 1980.

ROSS, RUSSELL, "Wound Healing." *Scientific American,* June 1969.

SCHEUPLEIN, R. J., "Permeability of the Skin." *Physiological Review* 51 (1971):702.

TREGEAR, R., *Physical Functions of the Skin.* New York: Academic Press, 1966.

WAGNER, M. W., "Emergency Care of the Burned Patient." *American Journal of Nursing* 77 (1977):1788.

WILHELM, D. L., "Inflammation and Healing." *Pathology* Vol. 1. Edited by W. A. D. Anderson and John M. Kissane. St. Louis: Mosby, 1977.

II

SUPPORT AND MOVEMENT

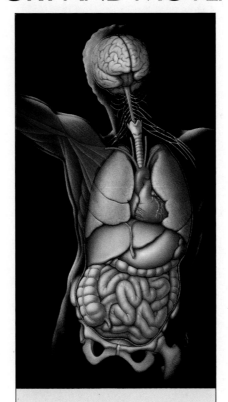

6
Bones and Osseous Tissue

LEARNING OBJECTIVES

1 Explain how bones are classified by shape, and relate their forms to their functions.

2 Describe the composition of bone matrix.

3 Describe the composition of spongy (cancellous) bone tissue.

4 Describe the composition of compact bone tissue.

5 Relate the structure of the osteon to its function.

6 Describe the gross anatomy of a typical bone.

7 State the functions of osteogenic cells, osteoblasts, osteocytes, osteoclasts, and bone-lining cells.

8 Describe intramembranous and endochondral ossification.

9 Describe how bones grow in diameter.

10 Explain how calcium is stored and released.

11 Explain the role of bone tissue in the production of blood cells.

12 Describe how specific hormones and bones work together to maintain homeostasis.

13 Explain several effects of nutrition on bones.

14 Describe the following bone disorders: osteogenesis imperfecta, osteomalacia, rickets, osteomyelitis, osteosarcomas, Paget's disease, and osteoporosis.

From a structural point of view, the human skeletal system consists of two main types of supportive connective tissue: cartilage and bone. The histology of cartilage was described in Chapter 4. This chapter deals specifically with the histology, gross anatomy, and physiological and mechanical functions of bone.

Osseous (bone) tissue is specialized connective tissue that has the strength of cast iron and the lightness of pine wood. We usually think of bone tissue as comprising the skeleton that supports and protects our internal organs, but its functions go far beyond that. Bone tissue is the storehouse and main supply of reserve calcium and phosphate, and bone marrow serves as the site for the manufacture of red blood cells and some white blood cells. Bones themselves aid movement by providing a point of attachment for muscles and transmitting the force of muscular contraction from one part of the body to another during movement. Bone is not dry, brittle, or dead. It is a living, changing, productive tissue that is continually resorbed (dissolved and assimilated), re-formed, and remodeled (replaced or renewed).

TYPES OF BONES AND THEIR MECHANICAL FUNCTIONS

The varied functions of bones may be classified as either mechanical or physiological. Bones make up the skeleton and provide the rigid framework that *supports* the body. They *protect* vulnerable internal organs such as the brain, heart, lungs, and organs of the pelvis by forming the sturdy walls of body cavities. They also make body *movement* possible by providing anchoring points for muscles and by acting as levers at the joints. These are the *mechanical aspects* of bone function.

Bones are usually classified by shape as flat, irregular, long, sesamoid, or short (Figure 6.1). Accessory bones are included here as a separate group. The shapes of bones generally tell something about their mechanical functions. For example, a long bone acts as a lever, a short bone is usually a connecting bridge between other bones, and a flat bone is a protective shell.

Long Bones

A bone is classified as a *long bone* when its length is greater than its width. The most obvious long bones are in the arms and legs (the longest is the femur, or thighbone), but some are relatively short, as in the fingers and toes. Long bones act as levers that are pulled by contracting muscles. It is this lever action that makes it possible for the body to move.

Short Bones

Short bones have about the same dimensions in length, width, and thickness, but they are shaped irregularly. They occur only in the wrists (carpal bones) and ankles (tarsal bones), where only limited movement is required. Short bones are almost completely covered with articular surfaces, where one bone moves against another in a joint. However, there are some nonarticular areas on short bones where nutrient blood vessels enter, where tendons attach bones to muscles, and where ligaments connect bones.

Flat Bones

These bones are actually thin or curved more often than they are flat. *Flat bones* include the ribs, scapulae (shoulder blades), sternum (breastbone), and bones of the cranium (skull). (See Figure 7.1 for illustrations of the ribs, scapulae, and sternum.) The shape of flat bones usually facilitates muscle attachment, and the gently curved bones of the skull form a protective enclosure for the brain.

Irregular Bones

Irregular bones do not fit neatly into any other category. Examples are the vertebrae, many facial bones, and the hipbones. The vertebrae have extensions that protrude from their main bony elements and serve as sites for muscle attachment. These bones support the spinal cord and protect it against compression forces. *Pneumatic bones,* such as the maxillary bone of the skull, have air-filled cavities that help warm and humidify the air.

Sesamoid Bones

Sesamoid bones are small bones embedded within certain tendons, the fibrous cords that connect muscles to bones. Sesamoid bones usually occur where tendons pass over the joint of a long bone, as in the wrist or knee. The patella, or kneecap, within the tendon of one of the thigh muscles (quadriceps femoris) and the pisiform carpal bone within the tendon of a wrist muscle (flexor carpi ulnaris) are typical sesamoid bones. Besides helping to protect the tendon, sesamoid bones help the tendon overcome compression forces, thus increasing the mechanical efficiency of joints. Sesamoid bones are so called because they sometimes resemble sesame seeds. Their number varies from person to person.

Accessory Bones

Accessory bones (also called *supernumerary bones*) are most commonly found in the feet. They usually occur when developing bones do not fuse completely. Unfused accessory bones may look like extra bones or broken bones in x rays, and it is important not to confuse them with actual fractures. *Sutural (Wormian*) bones* are small bone clusters that occur between the joints of the flat bones of the skull (see Figure 7.5B). The number of sutural bones in an individual varies. In general, accessory bones add some slight support and protection to the area of the skeleton where they are found.

*Wormian bones are named after Olaus Worm, the seventeenth-century Danish anatomist.

Types of Bones and Their Mechanical Functions

FIGURE 6.1

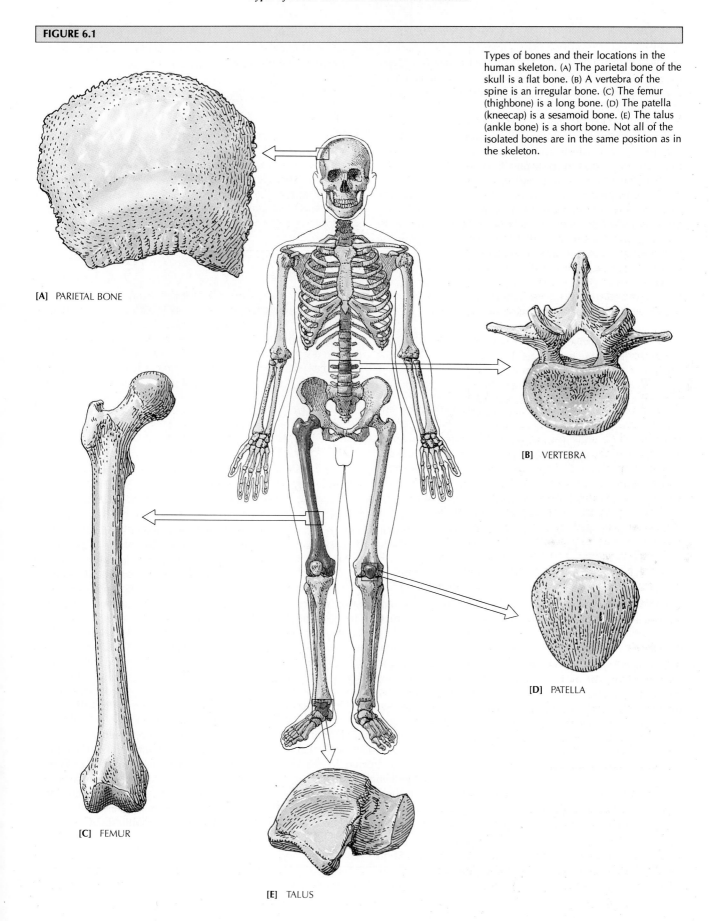

Types of bones and their locations in the human skeleton. (A) The parietal bone of the skull is a flat bone. (B) A vertebra of the spine is an irregular bone. (C) The femur (thighbone) is a long bone. (D) The patella (kneecap) is a sesamoid bone. (E) The talus (ankle bone) is a short bone. Not all of the isolated bones are in the same position as in the skeleton.

[A] PARIETAL BONE

[B] VERTEBRA

[C] FEMUR

[D] PATELLA

[E] TALUS

THE DYNAMIC BODY

Bone Modeling and Remodeling

Most people think of bones as dry, rigid, and unchanging. Even the word *skeleton* comes from a Greek word meaning "dried up." The truth about living bones, however, is that they are like dynamic factories where bone cells (the workers on the production line) are continuously engaged in the modeling and remodeling of bones. These cells enable bone tissue to adapt its size and shape to the demands of an ever-changing environment.

Bone *modeling* is the alteration of bone size and shape during the bone's developmental growth. It alters the amount and distribution of bone tissue in the skeleton, and determines the form of bones. Modeling occurs on different bone surfaces and at different rates during growth. Although bone is dense and hard, it is able to replace or renew itself through a process called *remodeling*. Remodeling occurs throughout adult life as a response to a variety of stresses.

Modeling

A dramatic example of bone modeling is the shaping of the skull bones. Bone yields to such soft structures as blood vessels and nerves by forming grooves, holes, and notches, not only in the fetus, when bone forms around blood vessels and nerves, but even later on, especially when vessels or nerves are rerouted because of injuries or other traumatic changes.

As we grow to adulthood, the brain and cranium continue to grow. As the cranium enlarges, the curvature of its four major bones must decrease. The changing of the curvature is accomplished by the growth of successive layers of new bone on the outer surface of the skull bones and, at the same time, the *resorption* (dissolution and assimilation) of old bone at the inner surface of the skull bones. When the cranium begins to shrink in old age (the brain mass also diminishes), bone is re-

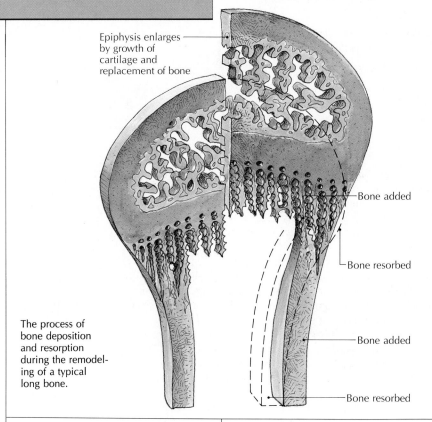

Epiphysis enlarges by growth of cartilage and replacement of bone

Bone added

Bone resorbed

Bone added

Bone resorbed

The process of bone deposition and resorption during the remodeling of a typical long bone.

sorbed without being replaced. Most growth patterns are under hormonal control.

Remodeling

Just as for many other tissues of the body, the structural units of bone tissue must either be replaced or renewed in order to maintain homeostasis. This remodeling occurs through the selective resorption of old bone and the simultaneous production of new bone. For example, new bone may grow and its internal patterns may change to accommodate added body weight or other stresses. In contrast, bone that is not used tends to lighten by losing bone cells. Remodeling occurs at specific locations on the bone called *foci*. Within each focus are specialized cells that make up a bone-remodeling unit. Together,

the whole unit has a specific job: the erosion and refilling of the bone focus with new bone. Remodeling units travel through bone tissue continually doing their job of remodeling. Approximately 5 to 10 percent of the skeleton is remodeled each year.

An example of remodeling can be seen when braces are placed on teeth or when a tooth is extracted. When the teeth are repositioned during orthodontic treatment, force is placed on the bony tooth sockets. The slow shift of the teeth during their realignment takes place when some bone is resorbed on one side of the socket and added onto the other side. When a tooth is extracted, pressure on the socket is reduced, and some bone in the socket is resorbed. This reduction of pressure and bone mass allows the neighboring teeth to shift slightly.

Ask Yourself

1 *What are some of the mechanical functions of bones?*

2 *What are the basic shapes of bones? How do their shapes relate to their functions?*

3 *What are accessory bones?*

4 *Where are sutural bones found?*

OSSEOUS (BONE) TISSUE

The human body is about 62 percent water, but osseous (bone) tissue contains only about 20 percent water. As a result, your bones are stronger and more durable than your skin or your eyeballs, for instance. But even though most bones are as strong as cast iron, they are lighter and more flexible.

Like other types of connective tissue, osseous tissue is composed of cells embedded in a *matrix* of ground substance and fibers. However, bone tissue is more rigid than other tissues because its homogeneous organic ground substance also contains inorganic salts, mainly calcium phosphate and calcium carbonate. In the bones, these compounds plus others form hydroxyapatite crystals. (When the body needs the calcium or phosphate that is stored within the bones, the hydroxyapatite crystals ionize and release the required amounts.)

A network of collagenous fibers in the matrix gives bone tissue its strength and flexibility. Although the hardness of bone comes from the inorganic salts, its structure depends equally on the fibrous framework. When water and organic substances are removed from bone, it can crumble into a powdery chalk. On the other hand, if the inorganic salts are removed by a process called *decalcification*, the bone becomes so flexible that it can be tied into a knot.

The older we get, the less organic matter and the more inorganic salts we have in our bones. Because of this shifting proportion of matrix to salts, the bones of older people are less flexible and more breakable than the bones of children (see the essay on page 137).

Most bones have an outer shell of compact bone tissue enclosing an interior of spongy bone tissue, except where the spongy tissue is replaced by a marrow cavity or by air spaces, called *sinuses*. For example, the leg bones contain marrow cavities, and some of the irregular skull bones contain sinuses that make the bones light. Irregular bones vary in the amount of spongy and compact tissue present. Short bones consist of spongy bone and marrow cavities.

The flat bones of the cranium (the part of the skull enclosing the brain) consist of two thin plates, called *tables*, of compact bone tissue with a layer of spongy bone tissue sandwiched between them (Figure 6.2). The layer of spongy tissue contains marrow and veins (the diploic veins) and is called the **diploë** (DIHP-low-ee; Gr. "double"). Because of this protective arrangement, the outer table can be fractured without harming the inner table and brain.

FIGURE 6.2

(A) In the flat bones of the cranium, two tables (plates) of compact bone tissue surround a center of spongy bone tissue and marrow, the diploë. (B) An enlarged view showing more detail.

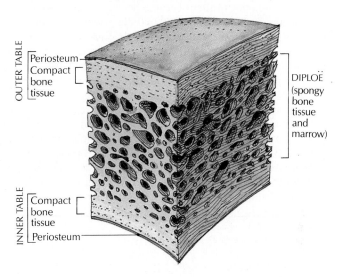

[A]

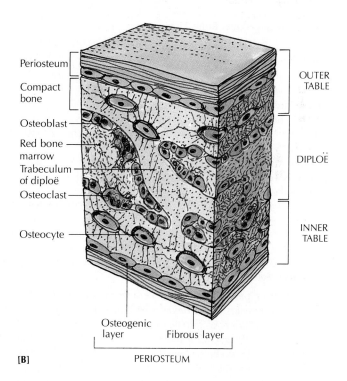

[B]

Grandma's Brittle Bones

Every year about 200,000 elderly Americans break a hip, and as many as 20 percent (40,000) of them become permanently disabled. Another 40,000 die of complications, such as pneumonia or blood clots. These problems, as well as muscular atrophy, are due to the patients' prolonged inactivity. The usual cause of such brittle bones is *osteoporosis* (Gr. *osteon*, bone + *poros*, passage), a loss of bone mass that can make the bones so porous that they crumble under the ordinary stress of moving about. People with this condition don't "fall and break a hip." They usually break their hip (or, more often, the head of the femur) while walking, and then fall when the hip support is gone.

Most victims of osteoporosis are women, usually over age 60. Women are more susceptible to the deterioration process, which seems to accelerate after menopause. After menopause, the ovaries produce very little, if any, estrogen, one of the female hormones. Without estrogen, old bone is destroyed faster than new bone tissue can be remodeled, so it becomes porous and brittle.

Besides a lack of estrogen, osteoporosis can be hastened by smoking, poor diet (calcium and vitamin D deficiency, for example), and lack of exercise.

Proper diet and exercise throughout one's life are very important in keeping the skeleton healthy and in preventing many of the structural problems associated with old age.

Osteoporosis is certainly not limited to aged hipbones. Vertebrae can crumble and produce what is called a "dowager's hump" in the upper back. In fact, most men and women over 70 are markedly shorter than they used to be, because osteoporosis causes the vertebrae and other small bones to collapse and press closer together when the intervertebral disks become thinner.

Spongy (Cancellous) Bone Tissue

Spongy, or *cancellous* (latticelike), *bone tissue* consists of an open, interlaced pattern of tissue designed to withstand maximum stress and to support shifts in weight distribution. An interior view of bone shows an outer shell of compact bone tissue around a center of spongy tissue (Figure 6.3). The spongy meshwork is constructed along the lines of greatest pressure, or stress. This structure provides the greatest strength with the least weight.

Prominent in the interior structure of spongy bone tissue are *trabeculae* (L. dim. *trabs*, beam), tiny spikes of bone tissue surrounded by bone matrix that has *calcified*, or become hard by the deposition of calcium salts (Figure 6.4). Spongy bone tissue is found inside most bones.

Compact Bone Tissue

The compact part of a bone is very hard and dense (like ivory), and appears to the naked eye to be solid, but it is not. It contains cylinders of calcified bone known as *osteons* (Gr. *osteon*, bone), or *Haversian systems*. These cylinders are made up of concentric layers, or *lamellae*, of bone (Figure 6.5). The term *lamellae* is derived from the Latin word for "thin plates." These lamellae are arranged like wider and wider drinking straws, each one nestled inside the next wider one.

In the center of the osteons are *central canals* (*Haversian canals**) longitudinal channels that contain nerves, lymphatic vessels, and blood vessels. Central canals usually have

*Note that a *Haversian system* and an osteon are the same structure; the *central (Haversian) canal* is just the longitudinal channel.

FIGURE 6.3

A section through a human hip joint. The porous, spongy (cancellous) portion of the bone has a streaked appearance, which indicates how the bone is built up in the direction of the greatest stress. [*Andreas Feininger, Life Magazine-Time, Inc.*]

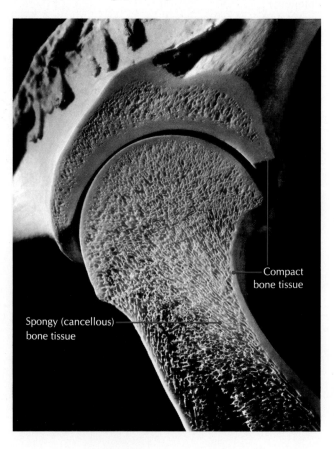

Compact bone tissue

Spongy (cancellous) bone tissue

FIGURE 6.4

A scanning electron micrograph of trabeculae in cancellous bone. Trabeculae in this area of the bone are the first to be resorbed during maturity, allowing for growth of the marrow cavity. [*Reprinted with permission from Weiss, Leon (ed.), Histology: Cell and Tissue Biology, 5th ed. New York: Elsevier, 1983.*]

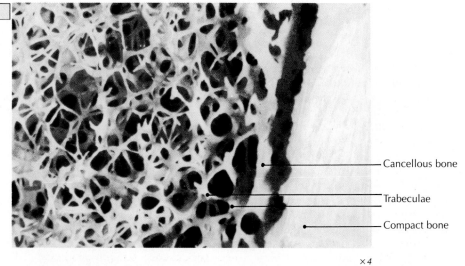

— Cancellous bone

— Trabeculae

— Compact bone

×4

FIGURE 6.5

Compact bone tissue. (A) An enlarged longitudinal section of compact bone tissue showing blood vessels, canals, and other internal structures. (B) An enlargement of a single osteon with lacunae, canaliculi, and a central (Haversian) canal visible. (C) An enlarged osteocyte (bone cell) inside a lacuna.

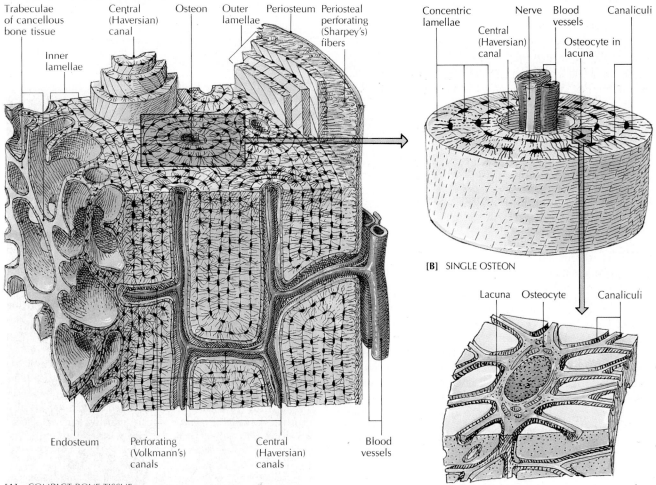

Trabeculae of cancellous bone tissue
Inner lamellae
Central (Haversian) canal
Osteon
Outer lamellae
Periosteum
Periosteal perforating (Sharpey's) fibers
Endosteum
Perforating (Volkmann's) canals
Central (Haversian) canals
Blood vessels

[A] COMPACT BONE TISSUE

Concentric lamellae
Central (Haversian) canal
Nerve
Blood vessels
Canaliculi
Osteocyte in lacuna

[B] SINGLE OSTEON

Lacuna
Osteocyte
Canaliculi

[C] SINGLE BONE CELL

FIGURE 6.6

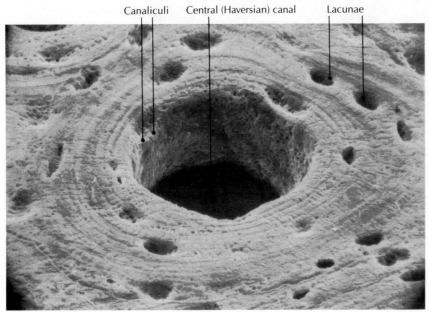

Canaliculi   Central (Haversian) canal   Lacunae

[A]   × 1000

Lamellae. (A) A scanning electron micrograph of compact bone tissue showing an osteon with its concentric lamellae housing lacunae and canaliculi. (B) In spongy bone tissue, lamellae are not arranged around osteons in a concentric pattern, but in slightly curved or flat sheets. This photomicrograph shows concentric lamellae forming an osteonlike structure (arrow) within the trabeculae of spongy tissue. The osteonlike structure permits nutrients to reach interior osteocytes within the trabeculae. [(A) *Richard Kessel and Randy Kardon,* Tissues and Organs: A Text-Atlas of Scanning Electron Microscopy. *San Francisco, W. H. Freeman, 1979.* (B) *P. Meunier. Reproduced with permission from Weiss, Leon (ed.).* Histology, Cell and Tissue Biology, *5th ed. New York: Elsevier Science Publishing Co., 1983.*]

[B]   × 1300

branches called ***perforating canals*** *(Volkmann's canals, nutrient canals)* that run at right angles to the central canals and extend the system of nerves and vessels outward to the *periosteum* (outer covering) and inward to the *endosteum* (inner lining) of the bony marrow cavity. Unlike central canals, perforating canals are not enclosed by concentric lamellae.

Lamellae contain *lacunae,* or little spaces, which house the *osteocytes,* or bone cells (Figure 6.6). Radiating like spokes from each lacuna are tiny *canaliculi* (L. dim. *canalis,* channel) that channel nutrients and wastes by diffusion into and out of blood vessels in the central canals.

The structure of osteons provides the great strength needed to resist typical, everyday compressive forces on long bones. Also, the branching structure of osteons establishes a continuous system that allows blood vessels and nerves to pass freely throughout the compact bone. Because osteons are usually less than about 0.4 mm in diameter, they allow compact bone and its osteocytes to be very close to their central blood vessels.

Ask Yourself

1 *What does the bone matrix contain?*

2 *What is the structure of spongy bone tissue? Of compact bone tissue?*

3 *Is there a difference between an osteon and a central canal?*

4 *What is the periosteum? The endosteum?*

GROSS ANATOMY OF A TYPICAL BONE

The long bones of the body (for example, the humerus, tibia, and radius) provide an excellent descriptive model for the gross anatomy of a typical bone. Most adult long bones have a tubular shaft, called the ***diaphysis*** (dye-AHF-uh-siss; Gr. "to grow between"). The diaphysis consists of a hollow cavity of compact bone tissue filled with marrow *(medullary cavity)* and a roughly spherical ***epiphysis*** (ih-PIHF-uh-siss; Gr. "to grow upon") of spongy bone tissue at each end of the bone (Figure 6.7). The epiphysis is the end of a long bone; it is usually wider than the shaft.*

Separating these two main sections at either end of the bone is the ***metaphysis.*** It is made up of the ***epiphyseal (growth) plate*** and the adjacent bony trabeculae of cancellous bone tissue on the diaphyseal side of the long bone. The growth plate is a thick plate of hyaline cartilage that provides the framework for construction of the cancellous bone tissue within the metaphysis. The epiphyseal plates and metaphyses are the only places where long bones continue to grow in length after birth (Figure 6.8).

The medullary cavity running through the length of the diaphysis contains *yellow* marrow, which is mostly fat (adipose tissue). The porous latticework of the spongy epiphyses is filled with *red* bone marrow. The red ***bone marrow,*** also known as *myeloid tissue* (Gr. *myelos,* marrow), manufactures red blood cells, which give the marrow its color. Lining the

---

*The flat bones and irregular bones of the trunk and limbs have many epiphyses. The long bones of the fingers and toes have only one epiphysis.

FIGURE 6.7

A typical long bone showing key anatomical features; the interior is partially exposed.

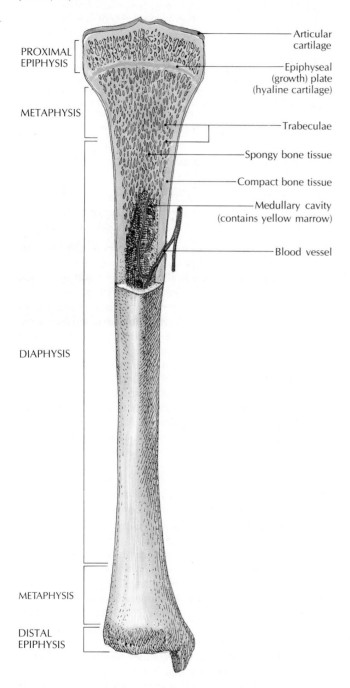

PROXIMAL EPIPHYSIS

METAPHYSIS

DIAPHYSIS

METAPHYSIS

DISTAL EPIPHYSIS

Articular cartilage

Epiphyseal (growth) plate (hyaline cartilage)

Trabeculae

Spongy bone tissue

Compact bone tissue

Medullary cavity (contains yellow marrow)

Blood vessel

medullary cavity of compact bone tissue, and covering the trabeculae of spongy bone tissue, is the **endosteum,** the membrane that lines the internal cavities of bones.

Covering the outer surface of the bone (except in the joint) is the **periosteum,*** a fibrous membrane that has the potential to form bone during growth periods and in fracture healing. The periosteum contains nerves, lymphatic vessels, and many capillaries that provide nutrients to the bone and give the distinctive pink color to living bone. Nutrients reach the marrow and spongy bone tissue by means of an artery that penetrates the compact bone tissue through a small opening called the **nutrient foramen.**

One of the few places where the periosteum is lacking is at the epiphyses. There it is replaced by articular cartilage, which provides a slick surface that reduces friction and allows the joint to work smoothly.

BONE CELLS

Bones contain five types of cells that are capable of changing their roles as the needs of the body change in the growing and adult skeletons.

1 Osteogenic cells are found mostly in the deep layers of the periosteum and in bone marrow. These cells are capable of being transformed into bone-forming cells *(osteoblasts)* or bone-destroying cells *(osteoclasts)* during times of stress and healing.

2 Osteoblasts (Gr. *osteon,* bone + *blastos,* bud or growth) synthesize and secrete unmineralized ground substance, called *osteoid* (Gr. *osteon,* bone + *eidos,* form). When calcium salts are deposited in the fibrous osteoid, it hardens, or calcifies, into bone matrix. Osteoblasts act as pump cells to move calcium and phosphate into and out of bone tissue, thereby calcifying or decalcifying it. Osteoblasts are usually found in the growing portions of bones, including the periosteum.

3 Osteocytes are the main cells of fully developed bones. They take on the shapes of their individual lacunae within the matrix. Osteocytes are derived from the osteoblasts that have secreted bone tissue around themselves. They play an active role in homeostasis by helping to release calcium from bone tissue into the blood, thereby regulating the concentration of calcium in the body fluids. Osteocytes also seem to keep the matrix in a stable and healthy state by secreting enzymes and maintaining its mineral content.

4 Osteoclasts (Gr. *klastes,* breaker) are multinuclear giant cells that are usually found where bone is resorbed (dissolved and assimilated) during its normal growth.[†]

5 Bone-lining cells are found on the surface of most bones in the adult skeleton. These cells are believed to be derived from osteoblasts that cease their physiological activity and flatten out on the bone surface. These cells may have several functions. They may serve as osteogenic cells that can divide

*As you can see in Figure 6.5A, the periosteum is often attached to the underlying bone by collagenous fibers called *periosteal perforating (Sharpey's) fibers.* These collagenous fibers penetrate the inner layer of the periosteum and become embedded in the matrix of the bone.

[†]To remember the difference between osteoblast and osteoclast, remember the *b* in osteoblast as standing for "building."

FIGURE 6.8

A gamma-ray scintigram of a normal child, with active growth zones at the epiphyses of long bones (shown as red clusters). Gamma-ray scintigrams are produced after radioactive isotopes are injected into the body. Then gamma-ray emissions are detected by a scintillating crystal detector and recorded as a scintigram, or "bone scan." [*Gruppo Editionale, Fabbri Milano.*]

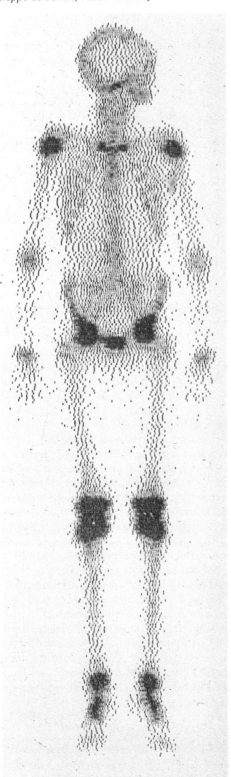

and differentiate into osteoblasts. Most probably, they serve as an ion barrier around bone tissue. This barrier contributes to mineral homeostasis by regulating the movement of calcium and phosphate into and out of the bone matrix, which in turn helps control the deposition of hydroxyapatite in the bone tissue.

Ask Yourself

1 *What is the diaphysis? The epiphysis? The metaphysis?*

2 *Does red bone marrow have a specific function?*

3 *What are the five kinds of bone cells?*

DEVELOPMENT OF BONES

Bones develop through a process known as *ossification* (osteogenesis). Since the primitive "skeleton" of the human embryo is composed of either hyaline cartilage or fibrous membrane, bones can develop in the embryo in two ways: *intramembranous ossification* or *endochondral ossification* (also called *intracartilaginous ossification*). However, in both cases, bones are formed from a pre-existing "connective tissue skeleton." Bone is the same no matter how it develops. Only the bone-making sequence is different.

Figure 6.9 compares the fetal and adult skeletons. Some bones, such as the skull, develop by intramembranous ossification, and other bones, such as those in the arms and legs, develop by endochondral ossification. The ages when particular bones develop are shown on the skeletons.

Intramembranous Ossification

If bone tissue (spongy or compact) develops directly from mesenchymal (embryonic connective) tissue, the process is called **intramembranous ossification.*** The vault (arched part) of the skull, flat bones of the face (including those lining the oral and nasal cavities), and part of the clavicle (collarbone) are formed this way. The skull is formed relatively early in the embryo to protect the developing brain. Some flexible tissue still exists between the flat skull bones at birth, so when the baby is born its skull is flexible enough to pass unharmed through the mother's birth canal.

From the time of the initial bone development, intramembranous ossification spreads rapidly from its center until large areas of the skull are covered with protecting and supporting bone. The first rapid phase begins when an *ossification center* (see item 3 in the following list) first appears (from the eighth fetal week through the twelfth fetal week) and lasts through the end of the fifteenth fetal week, when the area is entirely covered (Figure 6.10).

*Intramembranous means "within the membrane," and ossification means "bone formation." The term intramembranous ossification was originally used because this layer of mesenchyme was thought to be a sheet of membrane.

FIGURE 6.9

Some differences between the fetal and the adult skeletons. (A) Epiphyses begin to appear in the fetal skeleton about the fifth fetal week and continue to form until the sixteenth fetal week. (B) In the adult skeleton, zones of growth disappear and epiphyses fuse with diaphyses between the ages of about 13 and 25. (C) Skeletons of a newborn, an 8-year-old child, and an adult.

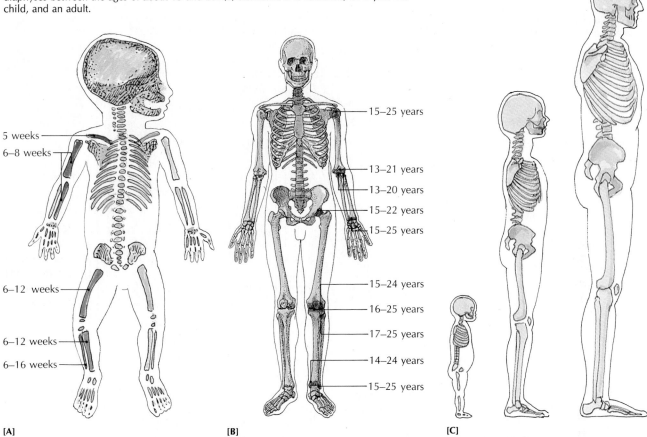

5 weeks
6–8 weeks

6–12 weeks

6–12 weeks
6–16 weeks

15–25 years

13–21 years
13–20 years
15–22 years
15–25 years

15–24 years
16–25 years
17–25 years
14–24 years
15–25 years

[A]   [B]   [C]

FIGURE 6.10

Photograph of the skull of a 3½-month-old fetus, showing the spongy bone radiating from the centers of intramembranous ossification. [*L. Belanger. Reproduced with permission from Weiss, Leon (ed.). Histology, Cell and Tissue Biology, 5th ed. New York: Elsevier Science Publishing Co., 1983.*]

During the subsequent growth of the skull, the bones develop at a slower rate. The second phase continues into adolescence. Because the brain reaches full size by about the tenth year, the development of the protective skull vault is completed early in adolescence. In contrast, the bones of the face do not reach their adult size until the end of the adolescent growth spurt, anywhere from the age of 14 to 17.

Because intramembranous ossification is studied most easily in the skull, we describe the process there:

1 A layer of embryonic connective tissue, called *mesenchyme,* forms between the developing scalp and brain.

2 A plentiful supply of blood vessels arises in the mesenchyme, where some mesenchymal cells are already connected to neighboring cells by long, thin fibers, or processes (Figure 6.11A).

3 Bone tissue development begins when thin strands that will eventually become branching trabeculae appear in the matrix. At about the same time, the mesenchymal cells become larger and more numerous, and their processes thicken and connect with other embryonic connective tissue cells, forming a ring of cells around a blood vessel (Figure 6.11B). The site of this ring formation is a ***center of osteogenesis*** (also known as an ***ossification center***). It begins about the second month of prenatal life.

FIGURE 6.11

Intramembranous ossification.
(A) The mesenchyme forms. (B) The mesenchymal cells enlarge and form a ring around a blood vessel. (C) The mesenchymal cells differentiate into osteoblasts that secrete osteoid. (D) As the osteoid calcifies into bone matrix, it entraps osteocytes in lacunae. (E) The osteoclasts aid bone resorption by removing small areas of bone (calcium) from the walls of the lacunae.

Mesenchymal cells   Center of osteogenesis   Osteoblast   Osteoid   Osteocyte   Calcified bone matrix   Osteoblast   Osteoclast

Processes of mesenchymal cells

Blood vessel

[A]   [B]   [C]   [D]   [E]

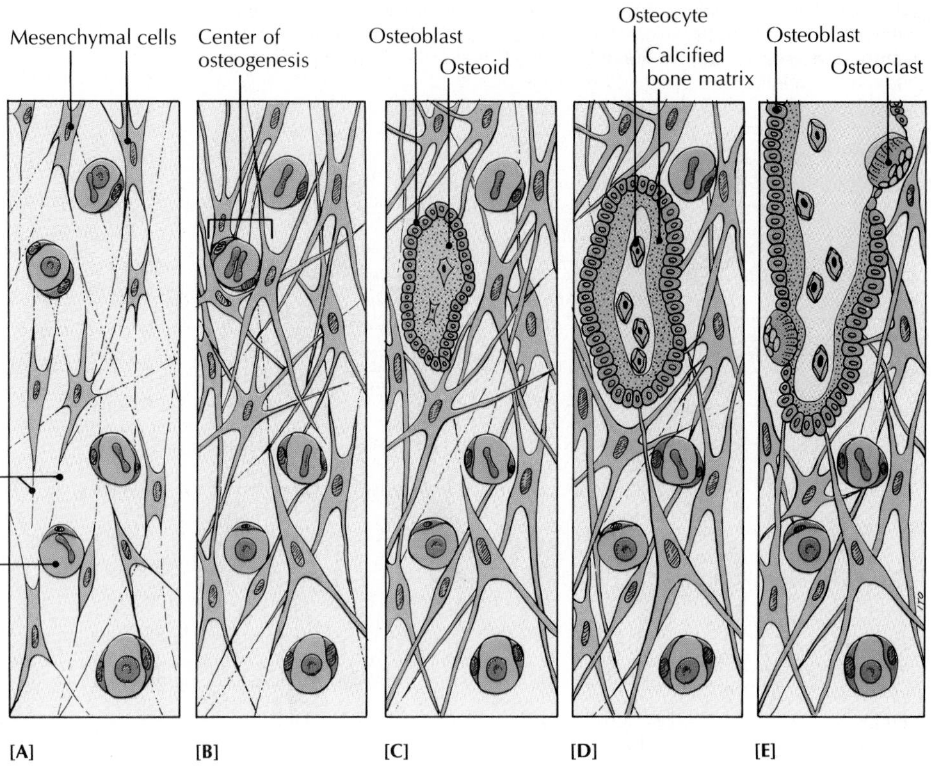

4 The mesenchymal cells differentiate into osteoid-secreting osteoblasts (Figure 6.11C). Then the osteoblasts begin to cause calcium salt deposits that form the spongy, latticelike bone matrix. (From the spongy matrix, the trabeculae of cancellous bone tissue will develop later.)

5 In this process of calcification, some osteoblasts become trapped within lacunae in the developing matrix. The entrapped osteoblasts lose their ability to form bone tissue; they are now called osteocytes (Figure 6.11D). These osteocytes remain in contact with the osteoblasts through the *canaliculi*, the tiny channels that are formed when the bone matrix is deposited around the long cell processes of the osteocytes. As bone-forming osteoblasts change into osteocytes, they are replaced by new osteoblasts that develop from the osteogenic cells in the surrounding connective tissue. In this way, the bone continues to grow.

6 While the osteoblasts are synthesizing and mineralizing the matrix, osteoclasts are playing a role in bone resorption by removing small areas of bone (calcium) from the walls of the lacunae (Figure 6.11E).

7 The trabeculae continue to thicken into the dense network that is typical of cancellous bone tissue (see Figure 6.4). The collagenous fibrils deposited on the trabeculae crowd nearby blood vessels, which eventually condense into blood-forming marrow.

8 The osteoblasts on the surface of the spongy bone tissue form the *periosteum*—the membrane that covers the outer surface of the bone. It is made up of an inner osteogenic layer (with osteoblasts) and a thick, fibrous outer layer. The inner layer eventually creates a protective layer of compact bone tissue over the cancellous-tissue interior. Once intramembranous ossification has stopped, the osteogenic layer becomes inactive, at least temporarily. It becomes active again when necessary—to repair a bone fracture, for example.

Endochondral Ossification

When bone tissue develops by replacing hyaline cartilage, the process is known as ***endochondral ossification.*** The term *endochondral* (Gr. *endo,* within + *khondros,* cartilage) means "inside the cartilage" and is used to describe this type of bone formation because it takes place where the cartilage model is eroded. Endochondral ossification produces long bones and all other bones not formed by intramembranous ossification. Remember, however, that the *cartilage itself is not converted into bone.* The cartilage model of the skeleton is completely destroyed and replaced by newly formed bone. The composition of cartilage and bone is compared in Table 6.1.

Figure 6.12 shows the stages of endochondral bone formation at the cellular level:

1 The ossification center is a cartilaginous matrix that includes *chondrocytes* (cartilage cells) in lacunae (Figure 6.12A).

2 As the ossification process begins, the chondrocytes and lacunae enlarge, crowding out the cartilaginous matrix. As the cells mature, they secrete alkaline phosphatase, which triggers a chemical reaction in the matrix that causes mineralization, or calcification (Figure 6.12B).

TABLE 6.1 COMPARISON OF BONE AND CARTILAGE

Feature	Bone	Cartilage
Components	Bone cells (osteocytes), collagenous fibers, ground substance, and mineral components.	Cartilage cells (chondrocytes), collagenous fibers, ground substance. (Elastic fibers are found in elastic cartilage.)
Location of cells	Housed in lacunae within matrix.	Housed in lacunae within matrix.
Outer covering of tissue	Periosteum (except at joints), a fibrous membrane containing nerves, lymphatic vessels, and capillaries.	Perichondrium (except at joints); composed of outer layer of dense connective tissue and inner layer of cells that differentiate into chondrocytes. (Fibrocartilage has no perichondrium.)
Derivation	Embryonic mesenchyme.	Embryonic mesenchyme.
Blood vessels	Contains blood vessels.	Has no blood vessels.
Strength	Stronger than cartilage because of minerals and abundant fibers in matrix.	Not as strong as bone but capable of supporting weight; matrix contains fewer fibers than bone matrix.
Nutrients	Bone cells obtain nutrients via canaliculi that channel nutrients by diffusion into and out of blood vessels.	Cartilage cells receive nutrients by diffusion through selectively permeable intercellular matrix (and perichondrium).

FIGURE 6.12

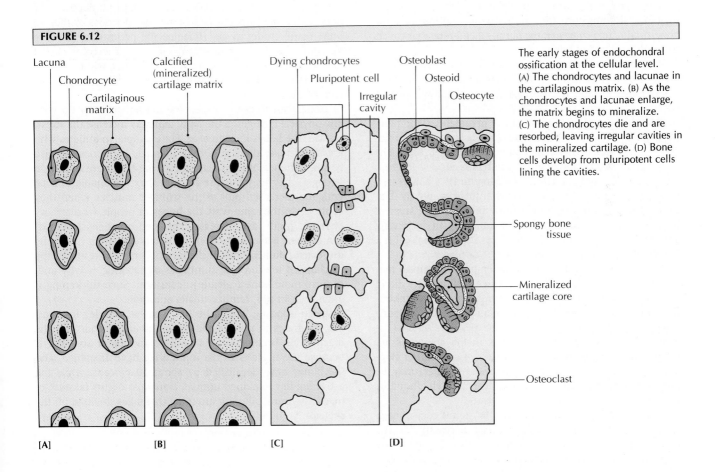

The early stages of endochondral ossification at the cellular level. (A) The chondrocytes and lacunae in the cartilaginous matrix. (B) As the chondrocytes and lacunae enlarge, the matrix begins to mineralize. (C) The chondrocytes die and are resorbed, leaving irregular cavities in the mineralized cartilage. (D) Bone cells develop from pluripotent cells lining the cavities.

[A]   [B]   [C]   [D]

3 The calcified matrix blocks the diffusion of nutrients to the chondrocytes, which begin to die. Eventually, they are resorbed, leaving irregular cavities in the hardened matrix (Figure 6.12C). Once the bone is formed, these cavities contain the bone marrow.

4 *Pluripotent cells* (which have the potential to divide into several distinct cell types) lining the cavities begin to differentiate into osteoblasts and osteocytes. The osteoblasts deposit osteoid on the mineralized cartilage cores, forming a thin layer of spongy bone tissue (Figure 6.12D). The osteoclasts begin the process of bone resorption during this stage.

Endochondral ossification is slower than intramembranous ossification. Figure 6.13 depicts endochondral ossification at the tissue level. The most typical sequence takes place in long bones and begins in the diaphyseal area:

1 Bone tissue begins to develop in a limb bud in the embryo. About six to eight weeks after conception, mesenchymal cells multiply rapidly and bunch together in a dense central core of precartilage tissue, which eventually forms the cartilage model (Figure 6.13A).

2 Soon after, an outline of a primitive *perichondrium* (the membrane that covers the surface of cartilage) appears (Figure 6.13B).

3 As it grows, the cartilage model is invaded by capillaries, triggering the transformation of the perichondrium into the bone-producing *periosteum,* the fibrous membrane that covers the outer surfaces of bones.

4 As intramembranous ossification occurs in the periosteum, a hollow cylinder of trabecular bone, called the *bone collar,* forms around the cartilage of the diaphysis (Figure 6.13C).

5 By the second or third month of prenatal development, the **primary center of ossification** is established near the middle of what will become the diaphysis (Figure 6.13D). The cartilaginous matrix mineralizes, and bone cells develop rapidly in the diaphyseal area. At the same time, blood vessels develop in the periosteum and branch into the diaphysis (Figures 6.13D, E). The diaphysis is now bony, with a layer of bone tissue just under the periosteum and an increasing amount of marrow in the center of the shaft (Figure 6.13F). The epiphyses, however, still consist of cartilage.

6 After birth, chondrocytes in the epiphyses begin to mature and enlarge in the same way earlier chondrocytes did in the diaphysis. Here, too, the matrix mineralizes and the chondrocytes die. Blood vessels and osteogenic cells from the periosteum enter the epiphyses, where the cells develop into osteoblasts. The centers of this activity in the epiphyses are called **secondary** (or **epiphyseal**) **centers of ossification** (Figures 6.13G, H). Although a few secondary centers may be present before birth, most do not appear until childhood or even adolescence.

7 Some cartilage remains on the outer (joint) edge of each epiphysis to form the *articular cartilage* necessary for the smooth operation of the joints (Figure 6.13I). The *epiphyseal (growth) plate,* the thicker plate of cartilage between the epiphysis and diaphysis on either end of the bone, also remains throughout the growth period. It provides the framework for development of the cancellous bone tissue inside the *metaphysis,* the region of the epiphyseal plate and bony trabeculae on the diaphyseal side of the long bone.

The longitudinal growth of bone after birth takes place in small spaces in the cartilaginous epiphyseal plates. During growth, cartilage cells here continue to grow and move into the metaphysis, where they are resorbed and replaced by bone tissue. The epiphyseal plates always remain the same thickness, even though they generate new cartilage cells until the age of about 17. Depending on the bone, the growing period stops altogether any time from adolescence until the early twenties.

8 As growth in length of the long bone slows, the distal epiphyseal plate disappears. At this point—the *fusion,* or *closure,* of the epiphysis—all that remains of the epiphyseal plate is a thin epiphyseal line (Figure 6.13J). The proximal epiphyseal plate disappears somewhat later (Figure 6.13K). By the time longitudinal growth ends, the epiphyseal cartilage is completely replaced by osseous tissue.

How Bones Grow in Diameter

Although no new longitudinal bone growth takes place after the age of 25, the *diameter* of bones may continue to increase throughout most of our lives. Through intramembranous ossification, the osteogenic cells in the periosteum deposit new bone tissue beneath the periosteum, while most of the old bone tissue erodes away. (Imagine the periosteum growing outward like the bark of a growing tree.) The combination of cell deposition and erosion widens the bone without thickening its walls. The width of the marrow cavity may be increased also.

Bones can thicken, or become denser, to keep up with any physical changes in the body that may increase the stress, or load, the bones have to bear. For example, in athletes who greatly increase the size of their muscles, the bones can also be strengthened at the same time. If the bones are not made stronger, they may fracture because they are unable to cope with the increased pull of the stronger muscles when they contract. In the same way, the bones in people who are overweight (and have increased the load on their skeletons) can thicken enough to offset the additional stress caused by the extra pounds. This thickening occurs as a result of the mechanical tension that stimulates osteoblastic activity and, in turn, more bone (calcium) deposition. Such thickening is often seen in ribs, femurs, tibiae, and radii.

In an opposite way, a person who has one leg in a cast but continues to put weight on the other leg finds that the inactive leg becomes thin in only a month or so. That is because the muscles in the injured leg have atrophied, and the bones have decalcified by about 30 percent, while the active leg has remained normal. Bone loss occurs because of inactivity, immobility, or anything that takes the load off the skeleton.*

---

*When astronauts are subjected to prolonged weightlessness in space, their bones begin to degenerate and lose calcium and other minerals. Dietary supplements of calcium, and even attempts at strenuous exercise, do not offset the degenerative effects of a lack of normal, everyday loads (stress) on bones.

FIGURE 6.13

Endochondral ossification in a typical long bone. Cartilage is shown in light blue, mineralized cartilage in darker blue. (A) Cartilage model. (B) The perichondrium forms. (C) The perichondrium is replaced by the periosteum. The bone collar forms. (D) Blood vessels enter the matrix. The primary center of ossification is established. (E) Blood vessels spread through the matrix. (F) The marrow cavity forms. (G) Blood vessels enter the proximal epiphyseal cartilage, creating a secondary center of ossification. (H) Another secondary center of ossification forms in the distal epiphyseal cartilage. (I) Some cartilage remains on the joint edge of each epiphysis to form articular cartilage. (J) As growth in length of the bone slows, the distal epiphyseal plate disappears, and the epiphyseal line forms on the interior of the epiphysis. (K) The proximal plate disappears somewhat later.

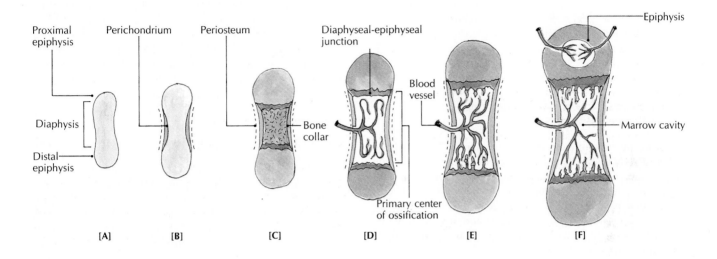

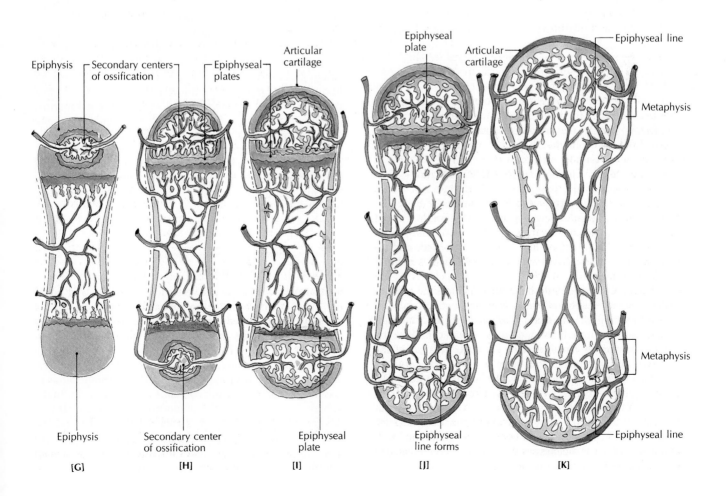

Ask Yourself

1 *What is ossification?*

2 *What is the basic difference between intramembranous and endochondral ossification?*

3 *Can cartilage be converted into bone? Explain.*

4 *What are ossification centers?*

5 *How do bones grow in diameter?*

HOMEOSTASIS AND THE PHYSIOLOGICAL FUNCTIONS OF BONES

Less obvious than the mechanical functions of bones are the *physiological* ones. The red bone marrow in the ends of some long bones, ribs, and vertebrae, and in the diploë of the skull bones manufactures red blood cells, some white blood cells, and blood platelets. Bone is also the body's main supplier and storehouse of calcium, phosphate, and magnesium salts. In fact, about 99 percent of all the calcium in the body is found in bones, as well as 86 percent of the phosphate and 54 percent of the magnesium.

Calcium Storage and Release

Without calcium, some enzymes could not function, cells would come apart, the permeability of cell membranes would be affected, muscles (including the heart) could not contract, nerve functions would be hampered, and blood would not clot. Of course, without calcium there would be no bones in the first place, since bones consist mainly of calcium.

If the diet does not provide enough calcium, the bones release it, and if there is too much calcium in the body, the bones store it. Apparently, the actual storage and release of calcium takes place in relatively young osteons. As osteons become more mature, they lose their ability to store and release calcium, and become more involved with structural functions instead.

To maintain homeostasis, bones help regulate the amount and consistency of extracellular fluid by either adding calcium to it or by taking calcium out of it. Small decreases of calcium in plasma and extracellular fluid *(hypocalcemia)* can cause the nervous system to become more excitable because of increased neuronal membrane permeability, with resultant muscular spasms *(tetany)*. On the other hand, too much calcium *(hypercalcemia)* in body fluids depresses the nervous system and causes muscles to become sluggish and weak because of the effects of calcium on the muscles' cell membranes. Also, increased calcium causes constipation and lack of appetite because it depresses the muscular contractility of the gastrointestinal tract.

Phosphate Storage and Release

Bones also help regulate the amount of phosphate in the body. Under hormonal control, bones can release phosphate salts when needed by the body. Changing the level of phosphate in either the extracellular fluid or blood does not cause significant immediate effects on the body. However, the proper amount of phosphate is vital to the body's acid-base balance. Too much acid can depress the nervous system to the point of coma or death, and too much alkali can cause death by overstimulating the nervous and muscular systems. Obviously, the physiological functions of bones are a vital factor in maintaining the body's overall homeostasis.

Production of Blood Cells

Besides helping to regulate the levels of minerals in the blood, the marrow within bones contributes to homeostasis by actually manufacturing some blood cells. Specifically, adult red bone marrow is the body's main blood-making, or *hemopoietic tissue* (Gr. *haima*, blood + *poiein*, to make). It produces *all* of the red blood cells (erythrocytes), platelets (cells involved in blood clotting), and certain white blood cells (granular leukocytes and immature lymphocytes and monocytes). At birth, all the bone marrow in the body is red marrow, but by adolescence most of it is replaced by yellow fat cells, which form the *yellow marrow*. Although yellow marrow is not active hemopoietic tissue, it has the potential to become active under stress and produce red blood cells.

In the adult, red marrow is found only in the proximal epiphyses of certain bones such as the femur and humerus, in some short bones, and in the vertebrae, sternum, ribs, and cranium. Besides producing red blood cells, the red marrow contains macrophages and manufactures white blood cells, some of which help to protect the body from disease. The continuing production of red blood cells is an important function because these cells live for approximately 120 days, and need to be replaced as they die. About 2.5 million red blood cells are produced per second, and about 200 billion daily (see Chapter 17).

Effects of Hormones on Bones

Several hormones have a direct effect on bones, and bones also have an effect on hormone secretion. For instance, the secretion of *parathyroid hormone (PTH)* and a group of recently discovered hormones that are metabolites of vitamin D appear to be stimulated by low calcium levels in the blood. The mechanism is as follows: in response to hypocalcemia, PTH is secreted by the parathyroid glands. This hormone binds to specific membrane receptors on bone and kidney cells. In the kidney, PTH stimulates the production of a vitamin D metabolite called $1,25\text{-}(OH)_2D_3$. This metabolite acts on the small intestine to stimulate the absorption of dietary calcium and, along with PTH, promotes the mobilization of calcium from bone. At the same time, $1,25\text{-}(OH)_2D_3$ and PTH cause the kidneys to resorb more calcium ions. As a result, the plasma and extracellular levels of calcium rise to the normal level (normocalcemia), which inhibits the secretion of PTH through a negative feedback loop, as described in Chapter 1.

Parathyroid hormone also aids in the production of bone-destroying osteoclasts and speeds up bone remodeling. As part of this resorption process, calcium is made available to the body in a usable form. Balancing the action of parathyroid hormone is *calcitonin* (a hormone released from the thyroid gland), which lowers the calcium level in the blood and slows down bone resorption. Parathyroid hormone and calcitonin thus appear to work together to maintain calcium homeostasis in the blood.

Besides the hormones just mentioned, growth hormone, thyroxin, adrenal cortical hormones, and vitamins D, A, and C are important in bone maturation. Growth hormones and thyroxin stimulate the endochondral ossification process. Both male and female sex hormones (from the gonads) regulate growth rates by controlling the appearance of centers of ossification and the rate of bone maturation. (Hormones are discussed more fully in Chapters 16 and 26.)

Effects of Nutrition on Bones

Proper nutrition is essential for normal bone development and maintenance. The most obvious nutritional needs are for calcium and phosphorus, since they constitute almost half of the content of bone. If the body is deficient in either mineral, the bones become brittle and break easily.

If a diet is too low in vitamin D, the normal ossification process at the epiphyseal growth plates is upset, and the bones are easily deformed. A deficiency in vitamin A may cause an imbalance in the ratio of osteoblasts and osteoclasts, thereby slowing down the growth rate. Abnormally low amounts of vitamin C also inhibit growth by causing an insufficient production of collagen and bone matrix, a condition that also delays the healing of broken bones.

Ask Yourself

1 *What are some of the physiological functions of bones?*

2 *How is calcium stored and released in bone?*

3 *How does calcium in the bones contribute to homeostasis?*

4 *In the adult, where are most red blood cells produced?*

5 *How do hormones and nutrition affect bones?*

PHYSIOLOGICAL AND ANATOMICAL ABNORMALITIES

Osteogenesis imperfecta

Osteogenesis imperfecta is an inherited condition in which the bones are abnormally brittle and subject to fractures. The basic cause of this disorder is a decrease in the activity of the osteoblasts during bone formation (osteogenesis). In some cases, the fractures occur during prenatal life, and so the child is born with the deformities. In other cases, the fractures occur as the child begins to walk. However, the tendency to fracture bones is reduced after puberty.

Osteomalacia and rickets

Osteomalacia (ahss-teh-oh-muh-LAY-shee-uh) (Gr. *osteon*, bone + *malakia*, soft) and *rickets* (variant of Gr. *rhakhitis*, disease of the spine) are skeletal defects caused by a deficiency of vitamin D, which leads to a widening of the epiphyseal growth plates, an increased number of cartilage cells, wide osteoid seams, and a decrease in linear growth. A deficiency of vitamin D may result from an inadequate diet, an inability to absorb vitamin D, or from too little exposure to sunlight.

Rickets is a childhood disease. It occurs less frequently than it used to, primarily because of improved dietary habits. It is most common in black children, not necessarily because of inadequate diets but because highly pigmented skin absorbs fewer of the ultraviolet rays in sunlight, which are needed to convert 7-dehydrocholesterol in the skin to vitamin D (cholecalciferol). Skeletal deformities such as bowed legs, knock-knees, and a bulging forehead are typical in a young child with rickets. Leg deformities occur especially if the child is old enough to attempt to walk, because walking puts excessive pressure on the soft leg bones.

Osteomalacia is the adult form of rickets and is sometimes referred to as "adult rickets." It leads to *demineralization*, an excessive loss of calcium and phosphorus. Although the skeletal deformities of rickets may be permanent, the similar skeletal abnormalities of osteomalacia may disappear with proper administration of large doses of vitamin D.

Osteomyelitis

Osteomyelitis (Gr. *osteon*, bone + *myelos*, marrow) is an inflammation of bone, and/or bone-marrow infection that can be either chronic or acute. It is frequently caused by *Staphylococcus aureus* and other bacteria, which can invade the bones or elsewhere in the body. Bacteria may reach the bone through the bloodstream or through a break in the skin from an injury. Although the disease often remains localized, it can spread to the marrow, cancellous tissue, and periosteum. *Acute osteomyelitis* is usually a blood-carried disease that most often affects rapidly growing children. *Chronic osteomyelitis*, more prevalent in adults, is characterized by draining sinuses and spreading lesions. Prompt use of antibiotics, such as vancomycin, is effective in treating the disease.

Osteosarcomas

Osteosarcomas (Gr. *sark*, flesh + *oma*, tumor), or *osteogenic sarcomas,* are forms of bone cancer. Such malignant bone tumors are rare. Because the incidence of osteosarcomas is higher in growing adolescents than in children or adults, and because the adolescents affected are often taller than average, there is some speculation that areas of rapid growth are most vulnerable. No definite cause is known. Localized pain and tumors are common signs of malignancy.

The most common form of bone cancer is a *myeloma,* in which malignant tumors in the bone marrow interfere with the normal production of red blood cells. Anemia, osteoporosis, and fractures may occur. Myeloma occurs more frequently in women than in men.

Paget's disease

Paget's disease (*osteitis deformans*) is a progressive bone disease in which a pattern of excessive bone destruction followed by bone formation contributes to thickening of bones. This deformity usually involves the skull, pelvis, and lower extremities. The cause is unknown. Paget's disease occurs after the age of 40 and most typically in the sixties.

SUMMARY

Types of bones and their mechanical functions

1 Bones may be classified according to their shape as *long, short, flat, irregular,* or *sesamoid. Accessory bones* are a minor category.

2 The shapes of bones are related to their functions. The mechanical functions of bones are to support the body, protect it, and make movement possible.

Osseous (bone) tissue

1 *Osseous (bone) tissue* is composed of cells embedded in a matrix of ground substance, inorganic salts, and collagenous fibers. The inorganic salts give bone its hardness, and the organic fibers and ground substance give it strength and flexibility.

2 Most bones have an outer shell of compact bone tissue surrounding spongy bone tissue.

3 *Spongy (or cancellous) bone tissue* has a lacy pattern designed to withstand stress and support shifts in weight. Tiny spikes of bone tissue called *trabeculae,* surrounded by calcified matrix, give spongy bone its latticelike appearance.

4 *Compact bone tissue* includes *osteons,* concentric cylinders of calcified bone. Within the osteons are *central (Haversian) canals* that carry nerves, lymphatic vessels, and blood vessels. These canals are connected with the outer surfaces of bones through *perforating (Volkmann's) canals.*

5 The *periosteum* is a fibrous membrane that covers the outer surfaces of bones, except in joints. It contains bone-forming cells, nerves, and vessels.

Gross anatomy of a typical bone

1 Most long bones consist of a tubular shaft called a *diaphysis,* with an *epiphysis* at either end of the bone.

2 Separating the diaphysis and epiphysis at each end of the bone is the *metaphysis.* It is made up of the *epiphyseal (growth) plate* and adjacent bony trabeculae of spongy bone tissue.

3 The epiphyseal plates and metaphyses are the only places where long bones continue to grow in length after birth.

Bone cells

1 Bones contain five types of cells: *osteogenic cells,* which can be transformed into osteoblasts or osteoclasts; *osteoblasts,* which synthesize and secrete new bone matrix as needed; *osteocytes,* which help maintain homeostasis; *osteoclasts,* giant cells responsible for the resorption of bone; and *bone-lining cells,* which may have several diverse functions.

2 Bone cells change their roles as the needs of the body change in the growing and adult skeletons.

Development of bones

1 Bones develop through *ossification* (osteogenesis). If bone is formed from mesenchymal (embryonic connective) tissue, the process is called *intramembranous ossification.*

2 If bone develops by replacing a cartilage model, the process is called *endochondral ossification.*

3 Bones grow in diameter through intramembranous ossification, as osteogenic cells deposit new bone tissue beneath the periosteum and old bone tissue erodes.

Homeostasis and the physiological functions of bones

1 Bones help maintain homeostasis by storing calcium and other minerals and releasing them as needed to maintain proper levels of those minerals in the blood and other tissues.

2 Red *bone marrow* produces red blood cells, contains macrophages, and manufactures some white blood cells that help the body fight disease.

3 Several hormones have a direct effect on bones, and bones have an effect on hormone secretion.

4 Calcium and phosphorus make up about half of the content of bone, and they must be supplied in a well-balanced diet. Adequate levels of vitamins A, C, and D are essential for the proper growth, mending, and strength of bones.

Physiological and anatomical abnormalities

1 Several common bone disorders include *osteogenesis imperfecta, osteomalacia* and *rickets, osteomyelitis, osteosarcomas,* and *Paget's disease.*

2 *Osteoporosis* occurs most often in the elderly, as bones grow porous and crumble under ordinary stress.

UNDERSTANDING THE FACTS

1 What are some shapes of bones and their related functions?

2 What is the function of inorganic salts in bone tissue?

3 Compare the structure and function of central (Haversian) canals and perforating (Volkmann's) canals.

4 Distinguish between a diaphysis and an epiphysis.

5 What is the epiphyseal (growth) plate, and what are its functions?

6 What main roles do osteocytes play?

7 How do the functions of osteoblasts and osteoclasts differ?

8 Describe the bone disorders of osteomalacia, osteomyelitis, and osteoporosis.

UNDERSTANDING THE CONCEPTS

1 Compare the composition of spongy and compact bone tissue.

2 Relate the structure of an osteon to its function.

3 Why does the skull vault reach its final size and development before the face does?

4 What is the relationship between physical activity and bone development? Between body weight and bone development?

5 Explain some ways in which hormones and vitamins can affect your bones.

SUGGESTED READING

BOURNE, GEOFFREY H., Vol. I–IV. *The Biochemistry and Physiology of Bone,* 2nd ed. New York: Academic Press, 1972.

DELUCA, HECTOR, *Osteoporosis: Recent Advances in Pathogenesis and Treatment Series.* Baltimore: University Park Press, 1981.

EVANS, FRANCES, ed. *Studies on the Anatomy and Function of Bone and Joints.* New York: Springer-Verlag, 1966.

HALL, BRIAN K., *Developmental and Cellular Skeletal Biology.* New York: Academic Press, 1978.

HANCOX, N. M., *Biology of Bone.* Cambridge: Cambridge University Press, 1972.

HARRIS, W. H., AND R. P. HEANEY, *Skeletal Renewal and Metabolic Bone Disease.* Boston: Little, Brown, 1970.

LOOMIS, W. F., "Rickets." *Scientific American,* December 1970.

MCLEAN, F. C., AND M. URIST, *Bone: An Introduction to the Physiology of Skeletal Tissue,* 3rd ed. Chicago: University of Chicago Press, 1968.

PARFIIT, A. M., "The Physiological and Clinical Significance of Bone Histomorphometric Data." *Bone Histomorphometry Techniques and Interpretation.* Edited by Robert Recher. Boca Raton, Fla.: CRC Press, 1983.

SHIPMAN, PAT, ALAN WALKER, AND DAVID BISHELL, *The Human Skeleton.* Cambridge, Mass.: Harvard University Press, 1985.

TRUETA, J., *Studies in the Development and Decay of the Human Frame.* Philadelphia: Saunders, 1968.

VAUGHN, JANET M., *The Physiology of Bone,* 3rd ed. New York: Oxford University Press, 1981.

7
The Skeletal System

LEARNING OBJECTIVES

1 Describe the physiological and structural/mechanical functions of the skeleton.

2 Identify the general features and surface markings of bones, and relate them to function.

3 Distinguish between the axial and appendicular divisions of the skeleton, and identify their components.

4 Define the terms *suture, fontanel,* and *foramen,* and give examples of each.

5 Identify the bones of the skull, and describe their locations, main features, and functions.

6 Describe the main parts of a typical vertebra.

7 Identify the five sections of the vertebral column, and describe their locations, main features, and functions.

8 Identify the bones of the thorax, and describe their locations, main features, and functions.

9 Describe the parts of a typical rib.

10 Identify the bones of the upper extremities, and describe their locations, features, and functions.

11 Identify the bones of the lower extremities, and describe their locations, features, and functions.

12 Explain the major differences between the male and female pelvis.

13 Describe some types of fractures of the human skeleton.

14 Distinguish between compression and extension fractures of the vertebral column.

15 Explain the meaning of a herniated disk, hydrocephalus, and spina bifida.

16 Identify the four normal curves of the spinal column and three types of abnormal spinal curvatures.

We are born with as many as 300 bones, but several of them fuse during childhood to form the adult skeleton of 206 bones. There are exceptions. For example, 1 person in 20 has an extra rib, and the number of small sutural (Wormian) bones in the skull varies from person to person. But for the most part, the bone count of 206 is the accepted one.

Other than the bones that come in pairs, no two are alike. They may differ in size, shape, weight, and even composition. This diversity of form is directly related to many structural or mechanical functions of the skeleton.

The most obvious function of the skeleton is to hold up the body. Without a bony skeleton to support our bag of skin and inner organs, we would collapse into a formless heap.

Besides supporting the inner organs, the skeleton protects many of them within bony cavities. For instance, the heart and lungs are safely enclosed within a roomy rib cage that offers protection and freedom of movement at the same time, and the brain is cushioned within the cranium, a shock-absorbing bone case that is designed to protect the brain from the many bumps of everyday life. The skeleton also protects many passageways. For instance, the bony scaffolding of the nasal region supports and protects the airway for the passage of air during breathing.

The bones of the skeleton also act as a system of levers for the pulley action of muscles. This lever-pulley arrangement provides attachment sites on bones for muscles, tendons, and ligaments, and allows us to move our entire bodies, or just one finger or toe.

The previous chapter showed that bone is a living tissue. This chapter presents an introduction to the skeletal system itself, concentrating on its two main parts: the central "anchor" of the axial skeleton, and the peripheral limbs of the appendicular skeleton. Throughout the chapter, each specific bone bears the same color in each illustration. For example, the temporal bone is always blue.

GENERAL FEATURES AND SURFACE MARKINGS OF BONES

Just as a bone's shape may indicate its purpose (see Chapter 6), the surface features of a bone often give clues about the bone's function. Table 7.1 describes some of the more important features of bones and their functions. For each of these features there is an example, illustrated in this chapter.

Some outgrowths or *processes* (such as tuberosities) may be attachment sites for the tendons of muscles or the ligaments that connect bones. A long, narrow ridge (linea aspera of femur) or a more prominent one (iliac crest of pelvis) may be the site where broad sheets of muscles are attached. Large, rounded ends (condyle of femur) or large depressions (glenoid fossa of scapula) indicate adaptations of joints between bones. Grooves, foramina (holes), or notches are usually

formed on bones to accommodate such structures as blood vessels, nerves, and tendons.

For example, a muscle is attached to a trochanter, tuberosity, or tubercle on a ridge. If the muscle is removed, the ridge will eventually disappear. Processes are a response to the continuous force exerted by muscles on a bone. All the foramina in the skull, through which nerves or blood vessels pass, were present before the bone was in the vicinity. The bone responded by not invading the territory of the nerve or blood vessel (for example, the foramen ovale for the mandibular nerve). The medial groove on the humerus is present because the humerus yields as the radial nerve presses upon it slightly.

Ask Yourself

1 What is a process on a bone? Give examples of processes, openings, and depressions.

2 Give an example of how a particular feature or marking on a bone is related to its function.

DIVISIONS OF THE SKELETON

The skeleton is divided into two major parts: the axial skeleton and the appendicular skeleton (Figure 7.1, Table 7.2). The *axial skeleton* is so named because it forms the longitudinal *axis* of the body. It is made up of the skull, vertebral column, sternum, and ribs. The *appendicular skeleton* is composed of the upper and lower extremities, which include the pectoral (shoulder) and pelvic girdles that attach the upper and lower appendages to the axial skeleton.

Ask Yourself

1 What are the two main divisions of the skeleton, and where are they joined together?

2 What bones make up the axial skeleton?

3 How is the appendicular skeleton put together?

4 What are the basic parts of each upper extremity? Each lower extremity?

5 How many bones make up the axial skeleton? The appendicular skeleton?

TABLE 7.1   SOME GENERAL FEATURES OR MARKINGS OF BONES

Feature or marking	Description	Example
PROCESSES (OUTGROWTHS) WHERE A BONE FORMS A JOINT WITH AN ADJACENT BONE		
Condyle (L. knuckle)	Rounded, knuckle-shaped projection; concave or convex.	Condyles of femur (Figure 7.30).
Facet (Fr. little face)	Small, flat surface.	Head and tubercle of ribs (Figure 7.31).
Head (caput) (L. head)	Expanded, rounded surface at proximal end of a bone; often joined to shaft by a narrowed neck, and bearing the ball of a ball-and-socket joint.	Head of femur, humerus, radius (Figures 7.24, 7.25, 7.30).
Trochlea (L. pulleys)	Grooved surface serving as a pulley.	Trochlea of humerus (Figure 7.24).
PROCESSES OF CONSIDERABLE SIZE ATTACHED TO CONNECTIVE TISSUE		
Cornu (L. horn)	Curved, hornlike protuberance.	Cornu of hyoid bone (Figure 7.11).
Crest or lip ridge	Wide, prominent ridge, often on the long border of a bone.	Iliac crest (Figure 7.29).
Line or linea	Narrow, low ridge.	Linea aspera of femur (Figure 7.30).
Eminence (L. to stand out)	Projecting part of bone, especially one upon the surface of a bone.	Intercondylar eminence of tibia (Figure 7.32).
Epicondyle	Eminence upon a bone above its condyle.	Epicondyles of humerus (Figure 7.24).
Malleolus (L. hammer)	Hammer-shaped, rounded process.	Malleoli of tibia and fibula (Figure 7.32).
Spine (spinous process)	Sharp, elongated process.	Spine of scapula and of a vertebra (Figures 7.13, 7.23).
Sustentaculum	Process that supports.	Sustentaculum of foot (Figure 7.33).
Trochanter	Either of the two large, roughly rounded processes found below the neck of the femur.	Trochanters of femur (Figure 7.30).
Tubercle (L. small lump)	Small, roughly rounded process.	Tubercles of humerus (Figure 7.24).
Tuberosity (L. lump)	Medium, roughly rounded, elevated process.	Ischial tuberosity (Figure 7.29).
OPENINGS (HOLES) TO BONES		
Canal or meatus (L. channel)	Relatively narrow tubular channel, opening to a passageway.	Hypoglossal canal, carotid canal (Figure 7.5E); external auditory meatus (Figure 7.6).
Fissure	Groove or cleft.	Superior orbital fissure (Figure 7.5A).
Foramen (L. opening)	Natural opening into or through a bone.	Foramen magnum in occipital bone (Figure 7.5E).
DEPRESSIONS ON BONES		
Fossa (L. trench)	Shallow depressed area.	Fossa of scapula, mandibular fossa (Figures 7.2, 7.5E, 7.10).
Groove or sulcus (L. groove)	Deep furrow on the surface of a bone or other structure.	Intertubercular and radial groove of humerus (Figure 7.24).
Notch	Deep indentation, especially on the border of a bone.	Trochlear notch of ulna (Figure 7.25).
Paranasal sinus (L. hollow)	Air cavity within a bone in direct communication with nasal cavity.	Maxillary sinus (Figure 7.9).

FIGURE 7.1

(A) Anterior view of the skeleton. The axial skeleton is shown in orange.
(B) Posterior view of the skeleton. The appendicular skeleton is shown in green.

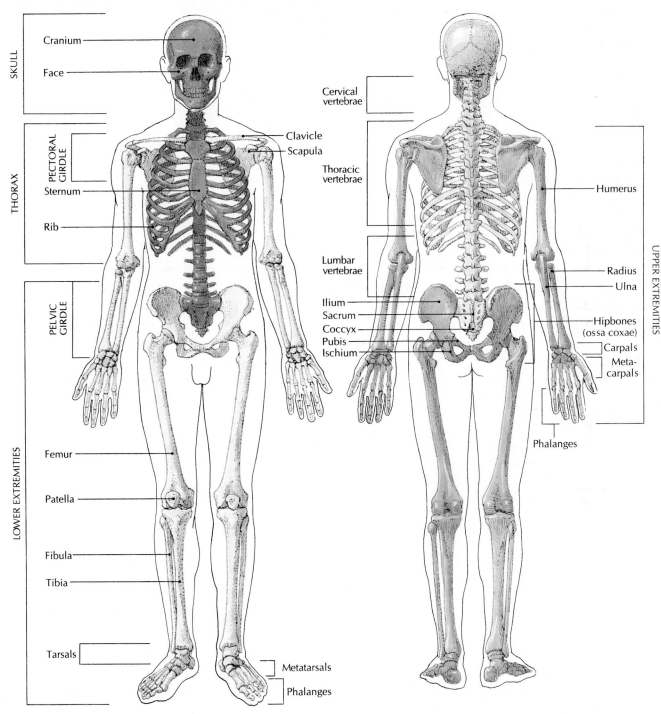

SKULL
Cranium
Face

THORAX
PECTORAL GIRDLE
Clavicle
Scapula
Sternum
Rib

PELVIC GIRDLE

LOWER EXTREMITIES
Femur
Patella
Fibula
Tibia
Tarsals
Metatarsals
Phalanges

Cervical vertebrae
Thoracic vertebrae
Lumbar vertebrae
Ilium
Sacrum
Coccyx
Pubis
Ischium

Humerus
Radius
Ulna
Hipbones (ossa coxae)
Carpals
Meta-carpals
Phalanges

UPPER EXTREMITIES

[A] ANTERIOR

[B] POSTERIOR

TABLE 7.2 DIVISIONS OF THE ADULT SKELETON (206 BONES)

AXIAL SKELETON (80 BONES)		APPENDICULAR SKELETON (126 BONES)	
Skull (29 bones)*		**Upper extremities** (64 bones)	
Cranium	8	Pectoral (shoulder) girdle	4
Parietal (2)		Clavicle (2)	
Temporal (2)		Scapula (2)	
Frontal (1)		Arm†	2
Ethmoid (1)		Humerus (2)	
Sphenoid (1)		Forearm	4
Occipital (1)		Ulna (2)	
Face	14	Radius (2)	
Maxillary (2)		Wrist	16
Zygomatic (malar) (2)		Carpals (16)	
Lacrimal (2)		Hand and fingers	38
Nasal (2)		Metacarpals (10)	
Inferior nasal concha (2)		Phalanges (28)	
Palatine (2)			
Mandible (1)		**Lower extremities** (62 bones)	
Vomer (1)		Pelvic girdle	2
Ossicles of ear	6	Fused ilium, ischium, pubis	
Malleus (hammer) (2)		Thigh	4
Incus (anvil) (2)		Femur (2)	
Stapes (stirrup) (2)		Patella (2)	
Hyoid	1	Leg†	4
		Tibia (2)	
Vertebral column (26 bones)		Fibula (2)	
Cervical vertebrae	7	Ankle	14
Thoracic vertebrae	12	Tarsals (14)	
Lumbar vertebrae	5	Foot and toes	38
Sacrum (5 fused bones)	1	Metatarsals (10)	
Coccyx (3 to 5 fused bones)	1	Phalanges (28)	
Thorax (25 bones)†			**126**
Ribs	24		
Sternum	1		
	80	**Total (Axial and Appendicular)**	**206**

*The number of skull bones is sometimes listed as 22, when the ossicles of the ears (6 bones) and the single hyoid bone are counted separately.
†The thoracic vertebrae are sometimes included in this category.

†Technically, the term *arm* refers to the upper extremity between the shoulder and elbow; the *forearm* is between the elbow and wrist. The upper part of the lower extremity, between the pelvis and knee, is the *thigh*; the *leg* is between the knee and ankle.

THE AXIAL SKELETON

THE SKULL

At the top of the axial skeleton is the **skull,** usually defined as the skeleton of the head, with or without the mandible (lower jaw) (Figures 7.2 and 7.3; see also Table 7.4 on p. 160). Other bones closely associated with the skull are typically pharyngeal (throat) bones; besides the mandible, they include the auditory bones (ear ossicles) and the hyoid bone. (In this book we will treat the mandible, ear ossicles, and hyoid as associated bones of the skull.)

The skull can be divided into (1) the *cranial skull,* which supports and protects the brain, its surrounding membranes (meninges), and the cerebrospinal fluid (see Figure 7.2), and

(2) the *facial skull,* which forms the framework for the nasal cavities and the oral (mouth) cavity.

The bones of the skull form a supporting framework that combines a minimum of bony substance and weight with a maximum of strength and support. Bony capsules surround and protect the brain, the eyes, inner ear, and nasal passages. The skull is lightened by small cavities called *paranasal sinuses.*

Bony buttresses (like the delicate but strong buttresses that support the walls of many Gothic cathedrals) within the skull can sustain enormous pressures exerted by the teeth during biting and chewing. (The grinding molars can exert as much pressure as 500 kg/sq cm, or 40,000 lb/sq in., about 1000 times more than the pressure under the tires of a large automobile.) To prevent such pressures from crushing the facial

FIGURE 7.2

An exploded view of the 8 bones of the cranial skull, the 14 bones of the facial skull, and the separate hyoid bone and the ossicles of the ear. (The ossicles are shown much larger than they are, and cannot be shown in their actual location.) In the views of the skull (in Figures 7.2 to 7.4), each bone is shown in the same color.

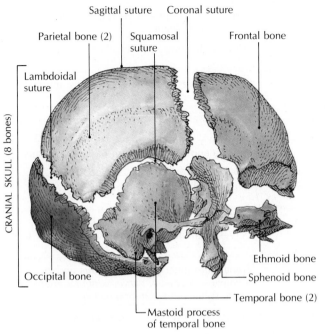

CRANIAL SKULL (8 bones)

Sagittal suture
Coronal suture
Parietal bone (2)
Squamosal suture
Frontal bone
Lambdoidal suture
Occipital bone
Mastoid process of temporal bone
Temporal bone (2)
Sphenoid bone
Ethmoid bone

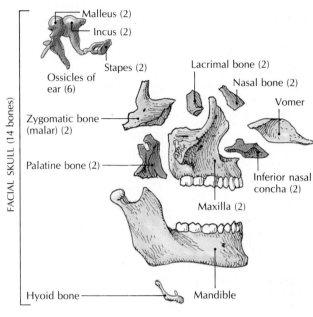

FACIAL SKULL (14 bones)

Malleus (2)
Incus (2)
Stapes (2)
Ossicles of ear (6)
Lacrimal bone (2)
Nasal bone (2)
Vomer
Zygomatic bone (malar) (2)
Palatine bone (2)
Inferior nasal concha (2)
Maxilla (2)
Hyoid bone
Mandible

FIGURE 7.3

(A) Coronal (transverse) section through the skull, illustrating the bony capsules surrounding the brain (cranial cavity), eye (orbit), maxillary sinus, and nose (nasal skeleton surrounding nasal cavity). The sinuses serve to lighten the amount of bone in the skull.
(B) Right lateral view of skull, indicating buttresses (arrows).

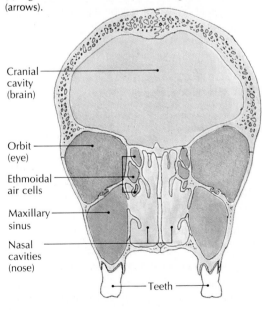

Cranial cavity (brain)
Orbit (eye)
Ethmoidal air cells
Maxillary sinus
Nasal cavities (nose)
Teeth

[A]

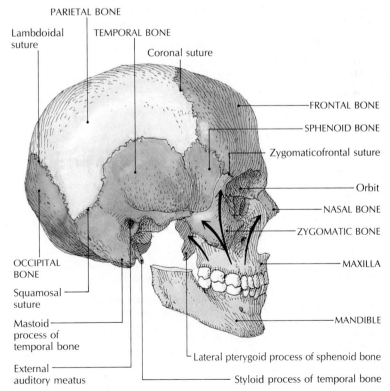

PARIETAL BONE
TEMPORAL BONE
Lambdoidal suture
Coronal suture
FRONTAL BONE
SPHENOID BONE
Zygomaticofrontal suture
Orbit
NASAL BONE
ZYGOMATIC BONE
MAXILLA
MANDIBLE
Lateral pterygoid process of sphenoid bone
Styloid process of temporal bone
OCCIPITAL BONE
Squamosal suture
Mastoid process of temporal bone
External auditory meatus

[B]

TABLE 7.3 MAJOR FORAMINA OF THE SKULL

Foramen	Location	Structures passing through foramen
Carotid canal (Figure 7.5E)	Petrous portion of temporal bone.	Internal carotid artery, sympathetic plexus.
Condylar canal (condyloid) (Figure 7.5E)	Posterior to each occipital condyle.	Vein from the sigmoid venous sinus to the suboccipital plexus.
Cribriform plate, foramen of (Figures 7.5F, 7.8)	Ethmoid bone.	8 to 10 filaments of each olfactory nerve.
External acoustic meatus (canal) (Figure 7.6)	Superior to jugular foramen and within petrous bone.	Facial and vestibulocochlear nerves; internal auditory artery.
Hypoglossal canal (Figure 7.5F)	Above anterior part of occipital condyle.	Hypoglossal nerve, meningeal artery.
Incisive (Figure 7.5E)	Bony palate behind incisor teeth.	Nasopalatine nerve, branch of palatine artery.
Infraorbital (Figure 7.5A)	Maxillary bone just below inferior margin of orbit.	Infraorbital nerve and artery.
Jugular (Figure 7.5E)	Posterior to carotid canal between occipital and petrous portion of tympanic bone.	Glossopharyngeal, vagus, spinal accessory nerves; sigmoid sinus joins internal jugular vein.
Magnum (Figure 7.5E)	Base of occipital bone between occipital condyles.	Spinal cord junction with medulla and its meninges, accessory nerve, vertebral arteries to brain, spinal, and vertebral veins.
Mandibular (Figure 7.10)	Medial part of ramus of mandible.	Inferior alveolar nerve and blood vessels.
Mental (Figures 7.5A, C)	Outer surface of mandible below second premolar tooth.	Mental nerve, blood vessels.
Nasolacrimal	Maxillary bone (from lacrimal bone to inferior meatus of nasal cavity).	Nasolacrimal duct.
Optic (Figure 7.5F)	Base of lesser wing of sphenoid bone.	Optic nerve, ophthalmic artery.
Ovale (Figures 7.5E, F, 7.7)	Greater wing of sphenoid bone.	Mandibular branch of trigeminal nerve.

Palatine
  Greater (Figure 7.5E)
  Lesser (Figure 7.5E)
	Through the posterior hard palate.
	Posterior to greater palatine foramen.
		Greater palatine nerve and artery.
		Lesser palatine nerve and artery.

Parietal (Figure 7.5B)	Near sagittal suture within parietal bone.	An artery and vein.
Rotundum (Figures 7.5F, 7.7)	Base of greater wing of sphenoid just behind superior orbital fissure.	Maxillary branch of trigeminal nerve.
Spinosum (Figure 7.5C)	Posterior angle of sphenoid bone.	Middle meningeal vessels.
Stylomastoid (Figure 7.5E)	Temporal bone between styloid and mastoid processes.	Facial nerve, stylomastoid artery.
Superior orbital fissure (Figure 7.5F)	Between greater and lesser wings and sphenoid bone near apex of orbit.	Oculomotor, trochlear, abducent, and ophthalmic nerves; ophthalmic veins.
Supraorbital (Figure 7.5A)	Superior margin of orbit.	Supraorbital nerve and blood vessels.
Zygomaticofacial (Figure 7.5A)	Zygomatic bone.	Zygomaticofacial nerve, zygomatic blood vessels.

skeleton, three buttresses extend from the teeth upward through the facial skull (Figure 7.3B).

Besides containing buttresses, the skull has numerous *foramina* (L. openings), openings of various sizes through which nerves and blood vessels pass. (The singular form is *foramen,* fuh-RAY-muhn.) The main foramina of the skull are described in Table 7.3 and shown in Figures 7.5, 7.6, 7.7, 7.8, and 7.10.

Sutures and Fontanels

The skull contains 29 bones, 11 of which are paired (see Table 7.2). Except for the mandible, ear ossicles, and hyoid, they are joined by *sutures* (L. *sutura,* seam), wriggly, seamlike joints that make the skull bones of an adult immovable (see Figure 7.4). Four major cranial skull sutures are:

FIGURE 7.4

(A) Skull of a newborn infant, illustrating the location of the fontanels. Superior and lateral views. (B) Sutures in an adult skull, superior and lateral views. Note the absence of fontanels, which have closed.

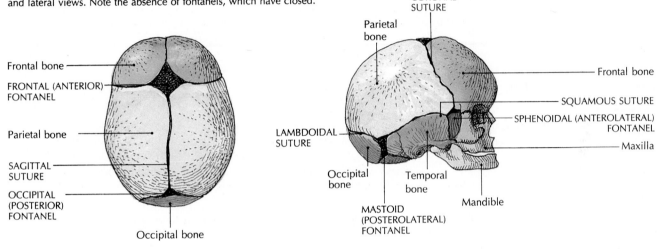

[A]

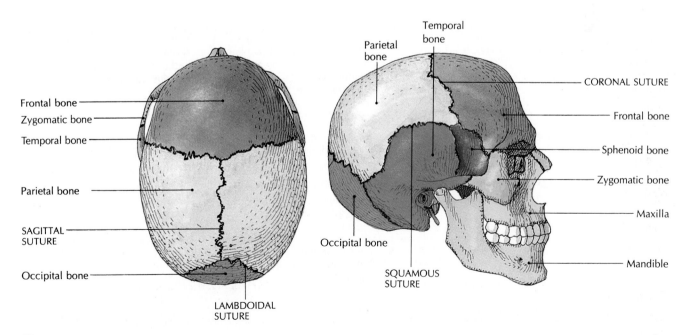

[B]

1 The *coronal,* between the frontal and parietal bones.

2 The *lambdoidal,* between the parietal and occipital bones.

3 The *sagittal,* between the right and left parietal bones.

4 The *squamous,* between the temporal and parietal bones.

Because the skull bones are joined by pliable membranes rather than tight-fitting sutures during fetal life and early childhood, it is relatively easy for the skull bones to move and overlap as the infant passes through the narrow birth canal. Some of the larger membranous areas between such incompletely ossified bones are called **fontanels** (Figure 7.4).

They allow the skull to expand as the child's brain completes its growth and development during the first few years of postnatal life. The membrane-filled fontanels (often called "soft spots") are:

1 The *frontal (anterior) fontanel,* located between the angles of the two parietal bones and the two sections of the frontal bones. This diamond-shaped fontanel is the largest, and usually does not close until 18 to 24 months after the child is born. The pulsing flow in the arteries of the brain can be felt at the frontal fontanel. If childbirth is difficult, a pulse monitor may be placed at this site to monitor the baby's heartbeat.

TABLE 7.4			BONES OF THE SKULL (29 BONES)
Bone	**Location**	**Description**	**Function**
CRANIAL SKULL (8 BONES)			
Ethmoid (1)	Base of cranium, anterior to body of sphenoid (Figure 7.8).	Made up of horizontal cribriform plate, median perpendicular plate, paired lateral masses. Contains ethmoidal sinuses, crista galli, superior and middle conchae.	Forms roof of nasal cavity and septum, part of cranium floor. Site of attachment for membranes covering brain.
Frontal (1)	Anterior and superior parts of cranium, forehead, brow areas (Figures 7.5A–F).	Shaped like large scoop; smooth exterior and internal depressions. Frontal squama forms forehead; orbital plate forms roof of orbit; supraorbital ridge forms brow ridge. Contains frontal sinuses, supraorbital foramen.	Forms forehead, upper part of orbits, brow ridge. Protects front of brain. Contains passageway for nerves and blood vessels.
Occipital (1)	Posterior part of cranium, including base of cranium (Figures 7.5A–F).	Slightly curved plate, with turned-up edges. Made up of squamous, base, and two lateral parts. Contains foramen magnum, occipital condyles, hypoglossal canals, atlanto-occipital joint, external occipital crest and protuberance.	Protects posterior part of brain. Forms passageways (foramina) for spinal cord and nerves for tongue and muscles. Site of attachment for muscles, ligaments.
Parietal (2)	Superior sides and roof of cranium, between frontal and occipital bones (Figures 7.5A–F).	Broad, slightly convex plates; smooth exteriors and internal depressions.	Protect top, sides of brain. Passageway for blood vessels, meninges.
Sphenoid (1)	Base of cranium, anterior to occipital and temporal bones (Figure 7.7).	Wedge-shaped. Made up of body, greater and lesser lateral wings, pterygoid processes. Contains sphenoidal sinuses, sella turcica, optic foramen, superior orbital fissure, foramen ovale, foramen rotundum, foramen spinosum.	Forms anterior part of base of cranium. Houses pituitary gland. Contains passageway for optic and other cranial nerves, meningeal artery to brain.
Temporal (2)	Sides and base of cranium at temples (Figure 7.5).	Made up of squamous, petrous, tympanic, mastoid areas. Contain zygomatic process, mandibular fossa, ear ossicles, mastoid sinuses.	Form temples, part of cheekbones. Articulate with lower jaw. Protect ear ossicles. Site of attachment for neck muscles.

2 The *occipital (posterior) fontanel,* located between the occipital bone and the two parietal bones. This smaller, diamond-shaped fontanel closes about two months after birth.

3 The two *sphenoidal (anterolateral) fontanels,* situated at the junction of the frontal, parietal, temporal, and sphenoid bones. They usually close about three months after birth.

4 The two *mastoid (posterolateral) fontanels,* situated at the junction of the parietal, occipital, and temporal bones. They begin to close about two months after birth, but do not close completely until about 12 months after birth.

Bone	Location	Description	Function
FACIAL SKULL (14 BONES)			
Inferior nasal conchae (2)	Lateral walls of nasal cavities, below superior and middle conchae of ethmoid bone (Figure 7.5D).	Thin, cancellous, shaped like curved leaves.	Permit air to be filtered before entering lungs.
Lacrimal (2)	Medial wall of orbit, behind frontal process of maxilla (Figure 7.5).	Small, thin, rectangular. Contains depression for lacrimal sacs, nasolacrimal tear duct.	Lacrimal sacs collect excess tears from eye surface. Nasolacrimal ducts drain tears from sacs to opening into inferior meatus of nasal cavity.
Mandible (1)	Lower jaw, extending from chin to mandibular fossa of temporal bone (Figure 7.10).	Largest, strongest facial bone. Horseshoe-shaped horizontal body, with two perpendicular rami. Contains tooth sockets, coronoid, condylar, alveolar processes, mental foramen.	Forms lower jaw, part of temporomandibular joint. Contains sockets for lower teeth. Site of attachment for muscles. Passageway for nerves, blood vessels.
Maxillae (2)	Upper jaw and anterior part of hard palate (Figure 7.5).	Made up of zygomatic, frontal, palatine, alveolar processes. Contain infraorbital foramen, maxillary sinuses, tooth sockets.	Form upper jaw, front of hard palate, part of eye sockets. Contain sockets for upper teeth. Passageway for sensory nerve, blood vessels.
Nasal (2)	Upper bridge of nose between frontal processes of maxillae (Figure 7.5).	Small, oblong; attached to lower nasal cartilage.	Form supports for bridge of upper nose.
Palatine (2)	Posterior part of hard palate, floor of nasal cavity and orbit; posterior to maxillae (Figure 7.5).	L-shaped, with horizontal and vertical plates. Contain greater and lesser palatine foramina.	Horizontal plate forms posterior part of hard palate; vertical plate forms part of wall of nasal cavity, floor of orbit. Passageway for nerves.
Vomer (1)	Posterior and inferior part of nasal septum (Figure 7.5).	Thin, shaped like plowshare.	Forms posterior and inferior nasal septum dividing nasal cavities.
Zygomatic or malar (2)	Cheekbones below and lateral to orbit (Figure 7.5).	Curved lateral part of cheekbones. Made up of temporal process, zygomatic arch. Contain zygomaticofacial and zygomaticotemporal foramina.	Form cheekbones, outer part of eye sockets. Passageway for nerves.
OTHER SKULL BONES (7)			
Hyoid (1)	Below root of tongue, above larynx (Figures 7.2, 7.11).	U-shaped, suspended from styloid process of temporal bone, attached to tongue and its muscles.	Supports tongue. Site of attachment for muscles used in speaking, swallowing.
Ossicles of ear Incus (anvil) (2) Malleus (hammer) (2) Stapes (stirrup) (2)	Inside cavity of petrous portion of temporal bone (middle ear), between eardrum and oval window (Figure 7.6).	Tiny bones, shaped like anvil, hammer, stirrup, articulating with one another and attached to tympanic membrane.	Bones act as a series of levers conveying sound vibrations from eardrum to oval window (see Chapter 15).

Bones of the Cranium

The eight bones of the cranium are the frontal bone, the two parietal bones, the occipital bone, the two temporal bones, the sphenoid bone, and the ethmoid bone. The sutural bones are also considered as part of the cranium.

The cranium consists of a roof and a base. The roof, called the ***calvaria*** (not calvarium), is made up of the squamous (L. *squama,* scale, which indicates that the bone is thin and relatively flat) part of the frontal bone (frontal squama), the parietal bones, and the occipital bone above the occipital protuberance (Figures 7.5A to D). The ***base*** of the cranium, as viewed internally from above (Figure 7.5F), has three depressions, or *fossae.*

Frontal bone As shown in Figures 7.5A and C, the large, scoop-shaped ***frontal bone*** forms the forehead and the upper part of the orbits (eye sockets). It forms the shell of the forehead as the *frontal squama,* the roof of the orbits as the *orbital*

plate, and protrudes over the eyes as the *supraorbital ridges* (or *margins*) to form the eyebrow ridges where the squamous and orbital portions meet.

The supraorbital nerve and blood vessels pass through the small supraorbital foramen (often only a notch) in the supraorbital ridge. Lining the inner surface of the frontal bone are many depressions, which follow the convolutions of the brain. At birth, the frontal bone has two parts, separated by a *metopic* or frontal *suture.* This suture generally disappears by the age of six, when the frontal bones become united (see Figure 7.4).

Parietal bones Figure 7.5 shows the two ***parietal bones*** (L. *paries,* wall), which form the superior, sides, roof, and part of the back of the skull. These broad, slightly convex bones have smooth exteriors, but like the frontal bone, they have internal depressions that accommodate the convolutions of the brain, and blood vessels that supply the thin membranes (meninges) of the brain. It is seldom mentioned that the parietal bones also form part of the back of the cal-

---

FIGURE 7.5A

(A) Skull, anterior view.

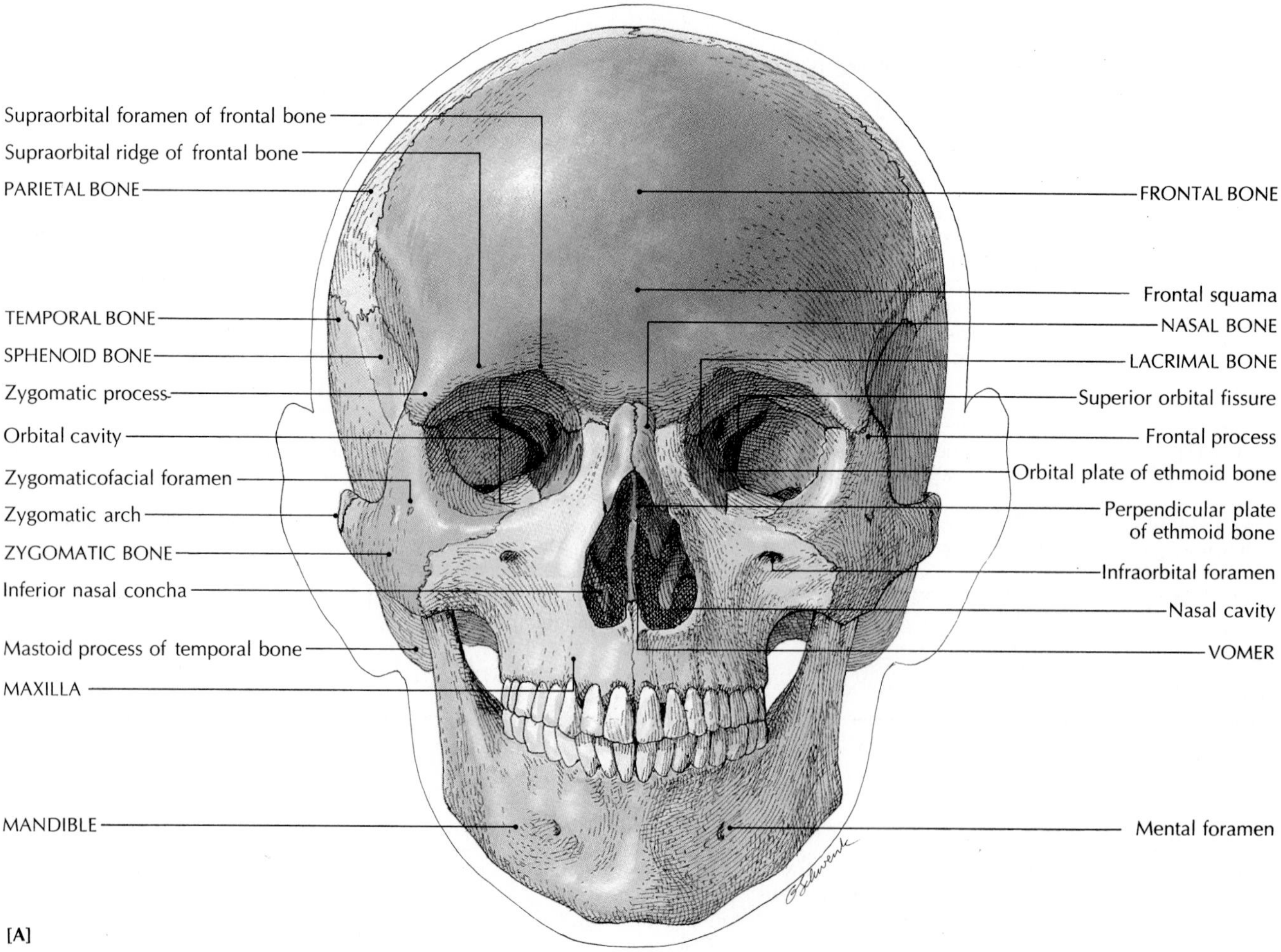

Supraorbital foramen of frontal bone
Supraorbital ridge of frontal bone
PARIETAL BONE
TEMPORAL BONE
SPHENOID BONE
Zygomatic process
Orbital cavity
Zygomaticofacial foramen
Zygomatic arch
ZYGOMATIC BONE
Inferior nasal concha
Mastoid process of temporal bone
MAXILLA
MANDIBLE

FRONTAL BONE
Frontal squama
NASAL BONE
LACRIMAL BONE
Superior orbital fissure
Frontal process
Orbital plate of ethmoid bone
Perpendicular plate of ethmoid bone
Infraorbital foramen
Nasal cavity
VOMER
Mental foramen

[A]

varia. In each parietal bone is a parietal foramen. Passing through the foramen is an emissary vein that connects the superior sagittal dural sinus inside the skull with scalp veins outside the skull. (Emissary veins act as safety valves.)

Occipital bone Figure 7.5 shows the location of the *occipital bone* (L. *occiput,* in back of the head, *ob,* in back of + *caput,* head). This bone forms the posterior part of the cranial skull.

The occipital bone is a slightly curved plate with edges that turn in on themselves at right angles. It consists of a squamous part, a base part, and two lateral parts. At its base is a large oval opening called the *foramen magnum,* where the spinal cord passes through to form a junction with the medulla oblongata of the brain. On either side of the foramen magnum are the *occipital condyles* and the paired *hypoglossal canals,* through which the hypoglossal nerves pass. These nerves stimulate the muscles of the tongue. The *atlanto-oc-cipital joint* between the occipital condyles and the first cervical vertebra (the atlas) permits the head to nod up and

down in the "yes" movement.

The *external occipital crest* and the *external occipital pro-tuberance* can be felt easily at the base of the occipital bone. Both landmarks are sites of attachment for muscles and ligaments.

Temporal bones Figure 7.5 shows the paired *temporal bones* (pertaining to the temples at the sides of the head). Together with portions of the sphenoid bone, they form part of the sides and base of the cranium.

Each temporal bone has four parts: the squamous, petrous, and tympanic portions, and the mastoid region. The largest, and the superior part, is the *squamous portion.* The *zygomatic process* of the temporal bone joins the temporal process of the zygomatic bone to form the slender *zygomatic arch* in the lateral part of the cheekbone (Figures 7.5A, C, E). The *mandibular fossa* is located posterior to the zygomatic arch. The fossa forms the socket that together with the condyle of the mandible, forms the temporomandibular (jaw) joint.

FIGURE 7.5B

(B) Skull, posterior view.

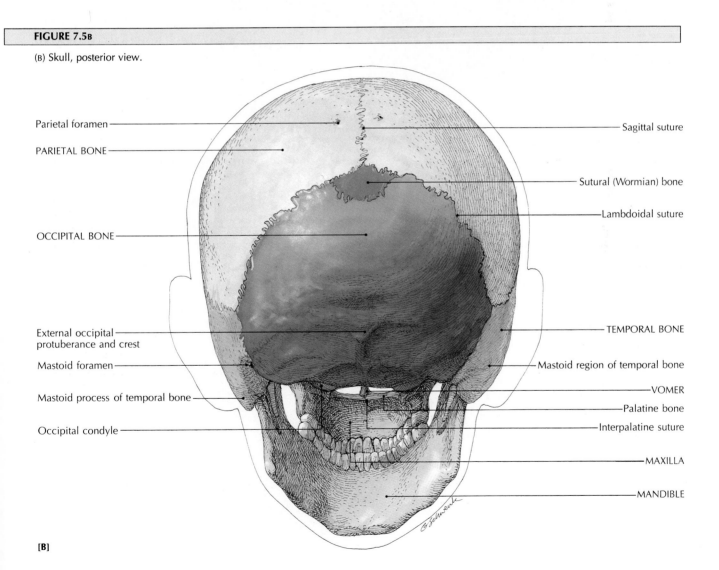

[B]

FIGURE 7.5c

(c) Skull, lateral view.

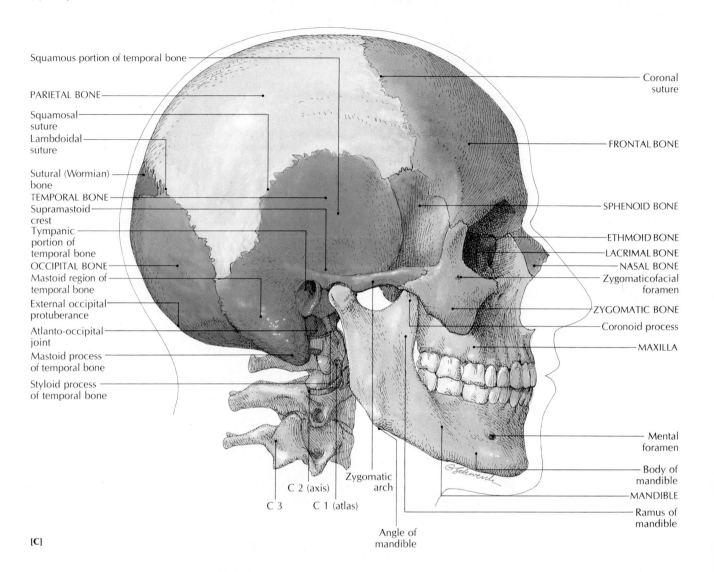

Squamous portion of temporal bone

PARIETAL BONE

Squamosal suture

Lambdoidal suture

Sutural (Wormian) bone

TEMPORAL BONE

Supramastoid crest

Tympanic portion of temporal bone

OCCIPITAL BONE

Mastoid region of temporal bone

External occipital protuberance

Atlanto-occipital joint

Mastoid process of temporal bone

Styloid process of temporal bone

C 2 (axis)

C 3

C 1 (atlas)

Zygomatic arch

Angle of mandible

Coronal suture

FRONTAL BONE

SPHENOID BONE

ETHMOID BONE

LACRIMAL BONE

NASAL BONE

Zygomaticofacial foramen

ZYGOMATIC BONE

Coronoid process

MAXILLA

Mental foramen

Body of mandible

MANDIBLE

Ramus of mandible

[C]

FIGURE 7.5D

(D) Skull, right median (sagittal) view.

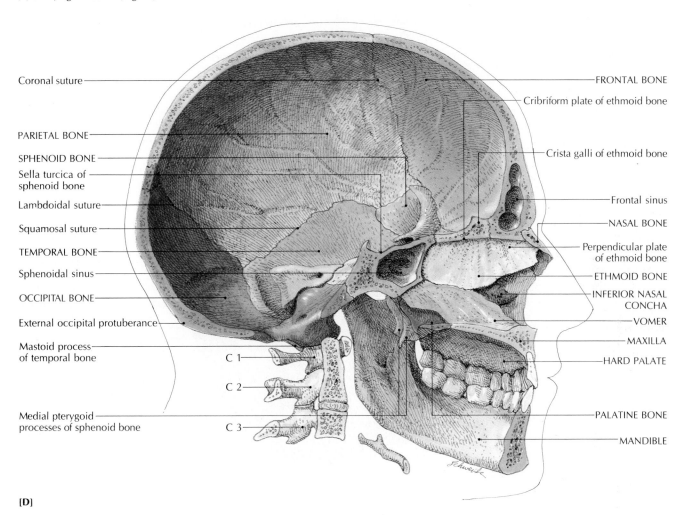

Coronal suture

PARIETAL BONE

SPHENOID BONE

Sella turcica of sphenoid bone

Lambdoidal suture

Squamosal suture

TEMPORAL BONE

Sphenoidal sinus

OCCIPITAL BONE

External occipital protuberance

Mastoid process of temporal bone

Medial pterygoid processes of sphenoid bone

C 1

C 2

C 3

FRONTAL BONE

Cribriform plate of ethmoid bone

Crista galli of ethmoid bone

Frontal sinus

NASAL BONE

Perpendicular plate of ethmoid bone

ETHMOID BONE

INFERIOR NASAL CONCHA

VOMER

MAXILLA

HARD PALATE

PALATINE BONE

MANDIBLE

[D]

FIGURE 7.5E

(E) Skull, inferior view.

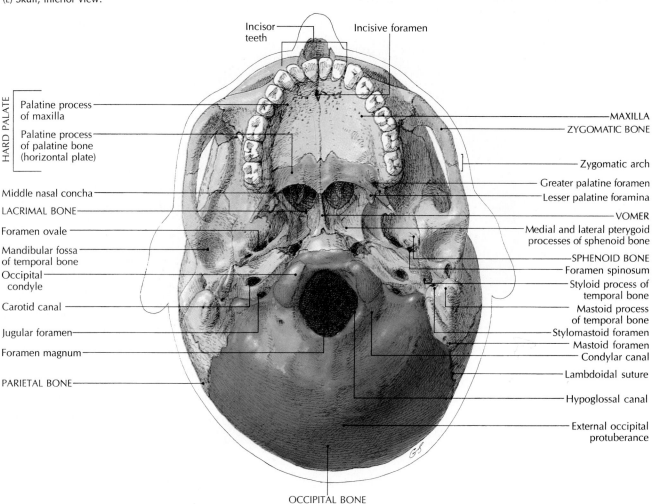

Incisor teeth

Incisive foramen

HARD PALATE

Palatine process of maxilla

Palatine process of palatine bone (horizontal plate)

Middle nasal concha

LACRIMAL BONE

Foramen ovale

Mandibular fossa of temporal bone

Occipital condyle

Carotid canal

Jugular foramen

Foramen magnum

PARIETAL BONE

MAXILLA

ZYGOMATIC BONE

Zygomatic arch

Greater palatine foramen

Lesser palatine foramina

VOMER

Medial and lateral pterygoid processes of sphenoid bone

SPHENOID BONE

Foramen spinosum

Styloid process of temporal bone

Mastoid process of temporal bone

Stylomastoid foramen

Mastoid foramen

Condylar canal

Lambdoidal suture

Hypoglossal canal

External occipital protuberance

OCCIPITAL BONE

[E]

The **petrous portion** of the temporal bone is in the floor of the *middle cranial fossa*, wedged between the occipital and sphenoid bones (see Figures 7.5F, 7.7A). Projecting downward from the inferior surface of the petrous portion is the **styloid process.** The elongated styloid process is the site for the attachment of muscles and ligaments involved with some movements of the lower jaw, tongue, and pharynx (throat) during speaking, swallowing, and chewing. Encased within the temporal bone are the three tiny auditory (hearing) bones

of the middle ear, the *malleus, incus,* and *stapes* (Figure 7.6), and the sensory receptors for hearing (cochlea) and balance (see Chapter 15).

The **tympanic portion** of the temporal bone forms part of the wall of the **external acoustic meatus** (the external opening of the ear) and part of the wall of the tympanic cavity, both of which are involved with the transmission and resonance of sound waves. The tympanic portion is the site for the attachment of the tympanic membrane (eardrum).

FIGURE 7.5ꜰ

(ꜰ) Skull, superior view. The floor of the cranium and the anterior, middle, and posterior fossae are viewed from above.

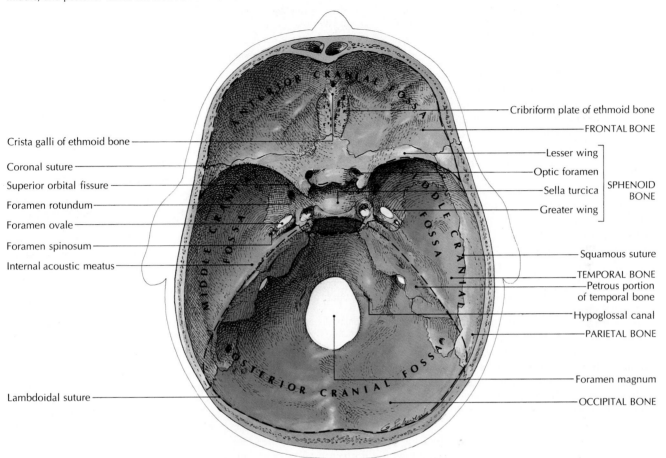

Crista galli of ethmoid bone

Coronal suture

Superior orbital fissure

Foramen rotundum

Foramen ovale

Foramen spinosum

Internal acoustic meatus

Lambdoidal suture

Cribriform plate of ethmoid bone

FRONTAL BONE

Lesser wing

Optic foramen

Sella turcica   SPHENOID BONE

Greater wing

Squamous suture

TEMPORAL BONE

Petrous portion of temporal bone

Hypoglossal canal

PARIETAL BONE

Foramen magnum

OCCIPITAL BONE

ANTERIOR CRANIAL FOSSA

MIDDLE CRANIAL FOSSA

MIDDLE CRANIAL FOSSA

POSTERIOR CRANIAL FOSSA

[F]

FIGURE 7.6

Middle-ear cavity within the petrous portion of the temporal bone, illustrating the ear bones (ossicles): malleus, incus, and stapes. (See Figure 7.2 and Chapter 15 for other illustrations of the ear bones.)

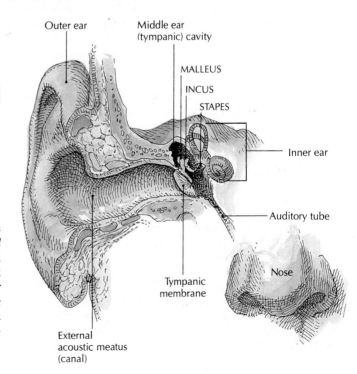

Outer ear

Middle ear (tympanic) cavity

MALLEUS

INCUS

STAPES

Inner ear

Auditory tube

Tympanic membrane

Nose

External acoustic meatus (canal)

The **mastoid region** is actually the posterior squamous portion of the temporal bone. Each prominent **mastoid process** in the posterior part of the temporal bone is located behind the ear; it provides the point of attachment for a neck muscle. These processes contain air spaces called *mastoid air cells* or *sinuses,* which connect with the middle ear. Because of this connection, infections within the middle ear can spread to the sinuses and cause an inflammation called *mastoiditis.*

The Changing Shape of the Skull

In the normal newborn infant, with its relatively massive brain, the cranial skull is very large compared to the facial skull. In fact, the facial skull is only one-eighth of the entire skull at birth, whereas it is one-half of the skull in the adult. The postnatal growth of the skull consists primarily of increases in the dimensions of the calvaria (the roof of the cranium, making up most of the cranial vault), the base of the cranial skull, and the facial skull. These three parts grow at different rates over different periods of time.

The size and contour of the vault are determined largely by the response of the calvaria to the fast growth of the brain during prenatal and early postnatal life and, in part, by genetic factors (which account for differences in head shapes among population groups). The vault expands rapidly during the first postnatal year and then more slowly, until it reaches almost adult size by the seventh year.

The growth of the skull base results primarily in the lengthening of the skull, which continues until some time during adolescence. This lengthening is largely independent of the growth of the brain.

The small size of the skull of the neonate (newborn) is related to the underdeveloped mandible, maxillae, and primary teeth, as well as the small nasal cavity and maxillary sinuses. The growth of the facial skull takes place over a longer span of time than the growth of the cranial skull. The "baby teeth" erupt between 6 and 30 months of age. During childhood the jaws enlarge as the permanent teeth develop,

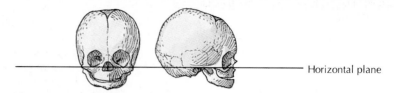

SKULL OF NEWBORN

Horizontal plane

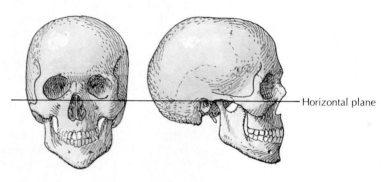

SKULL OF ADULT

Horizontal plane

The horizontal plane passes through the lower part of the orbit and the external auditory meatus. The cranial skull and orbit are above the plane, and the facial skull is below the plane.

and the alveolar processes (tooth sockets) form before and during the eruption of the permanent teeth from the sixth through the thirteenth year (except for the later-erupting wisdom teeth). By the seventh year, the orbits are almost as large as in the adult. Thus during childhood and puberty there is a rapid growth of the facial skeleton, largely associated with the eruption of the permanent teeth and the enlargement of the paranasal sinuses.

The disappearance of many sutures of the cranial skull takes place later. This process may begin between the ages of 30 and 40, and continues for many years. In old age, the skull generally becomes thinner and lighter. Most noticeable is the decrease in the size of the mandible and maxillae associated with the loss of teeth and the resorption of the tooth sockets. In these toothless skulls, there is a marked reduction in the vertical dimension of the face.

FIGURE 7.7

Sphenoid bone. (A) In the floor of the cranium, superior view. (B) Posterior view.

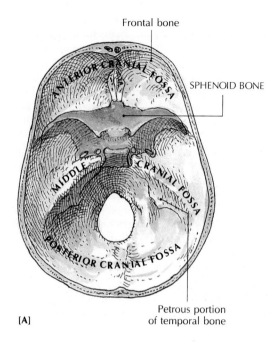

Frontal bone

SPHENOID BONE

ANTERIOR CRANIAL FOSSA

MIDDLE CRANIAL FOSSA

POSTERIOR CRANIAL FOSSA

[A]

Petrous portion
of temporal bone

Sphenoid bone As shown in Figure 7.7 the **sphenoid bone** (SFEE-noid; Gr. *sphen,* wedge) forms the anterior part of the base of the cranium. It is generally described as being "wedge-shaped" or "wing-shaped," because it looks somewhat like a bat or a butterfly with two pairs of outstretched wings. It has a cube-shaped central *body,* two **lesser lateral wings** and two **greater lateral wings,** and a pair of **pterygoid processes** (Gr. *pteron,* wing), which project downward.

The pterygoid processes form part of the walls of the nasal cavity, and the undersurface of the body forms part of the roof of the nasal cavity. A key feature of the sphenoid is a deep depression within its body that houses and protects the important and delicate pituitary gland (see Chapter 16). This depression is the **sella turcica** (SEH-luh TUR-sihk-uh), so called because it is said to resemble a Turkish saddle (L. *sella,* saddle).

The lesser wings form a part of the floor of the *anterior cranial fossa.* At the base of each lesser wing is the **optic foramen,** through which pass the optic nerve and ophthalmic artery into the orbit.

The greater wings form a major part of the **middle cranial fossa** (see Figure 7.7A). Between the greater and lesser wings is the **superior orbital fissure** (see Figure 7.5A), located between the middle cranial fossa and the orbit. Through the fissure pass three cranial nerves for the eye muscles and the sensory nerve to the orbital and forehead regions. Three important foramina in the greater wing are the **foramen rotundum** and the **foramen ovale** for different divisions of the trigeminal nerve, and the **foramen spinosum** for the middle meningeal artery to the side and roof of the skull (see Figure 6.7B).

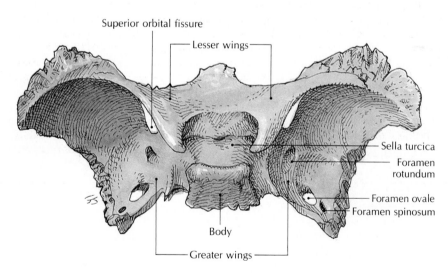

Superior orbital fissure

Lesser wings

Sella turcica

Foramen rotundum

Foramen ovale
Foramen spinosum

Body

Greater wings

[B]

Ethmoid bone The ***ethmoid bone*** (Gr. *ethmos*, strainer) forms the roof of the nasal cavity and the medial part of the floor of the anterior cranial fossa between the orbits (Figure 7.8). The light and delicate ethmoid bone consists of a horizontal cribriform plate (L. *cribrum*, strainer), the median perpendicular plate, and paired lateral masses, or labyrinths.

The inferior surface of the ***cribriform plate*** forms the roof of the nasal cavity, and the superior surface is part of the floor of the cranial cavity. The cribriform plate may have as many as 20 foramina. Through these tiny openings the filaments of the olfactory (smell) nerve pass from the mucous membrane of the nose to the brain.

The ***median perpendicular plate*** forms a large part of the vertical nasal septum between the two nasal cavities. The ***crista galli,*** or cockscomb, is a vertical protuberance in the center of the cribriform plate. It provides a site of attachment for a membrane (dura mater) that covers the brain.

Each lateral mass has a plate that forms the lateral wall of a nasal cavity and the orbital plate of an orbit. It also contains the superior and middle ***conchae*** (KONG-kee; L. "conch shell"), curved scrolls of bone that extend into the nasal cavity and are part of the lateral wall. In each lateral mass are ***ethmoidal air cells*** (paranasal sinuses) that drain into the nasal cavity.

Sutural bones ***Sutural bones*** (also called *Wormian* bones) are separate small bones found in the sutures of the calvaria (see Figure 7.5B). They are found most often in the lambdoidal suture. Their number may vary.

Paranasal Sinuses

In the interior of the ethmoid, maxillary, sphenoid, and frontal bones are four pairs of air cavities called ***paranasal sinuses*** (Figure 7.9). They are named after the bones where they are located. The sinuses have two main functions. They act as resonating chambers for the voice and, being air-filled cavities, they lighten the weight of the skull bones.

The paranasal sinuses are formed after birth as outgrowths from the nasal cavity, when the spongy part of the bone is resorbed and replaced by an air-filled cavity. The ***maxillary*** and ***sphenoidal sinuses*** are not fully formed until adolescence.

The walls of the paranasal sinuses are lined with mucous membranes, which communicate with the mucous membranes of the nasal cavity through small openings. An inflammation of the mucosa of the nasal cavity is called rhinitis, or a "head cold." When the inflammation occurs in the mucous membranes of the sinuses, or when the sinuses react adversely to foreign substances (allergy), the result is *sinusitis.* The ethmoidal air cells (paranasal sinuses) are made up of many small cavities in the ethmoid bone, between the orbit and the nasal cavity.

The ***maxillary sinuses*** are the largest of the paranasal sinuses. Mucus within the sinus drains into the nasal cavity through a single opening, which is high enough so that drainage is difficult when the nasal mucosa is swollen. Thus, fluid may accumulate and cause maxillary sinusitis.

FIGURE 7.8

Ethmoid bone. (A) In the cranium, right median view. (B) Right lateral view. (C) Anterior view.

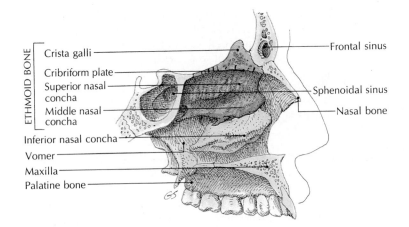

[A]

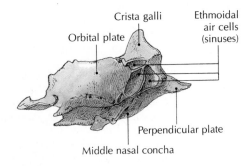

[B]

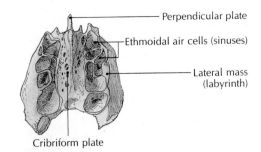

[C]

FIGURE 7.9

Paranasal sinuses: frontal, ethmoidal, sphenoidal, and maxillary. The ethmoidal air cells, referred to as sinuses, look like honeycombs. (A) Anterior view. (B) Right median view.

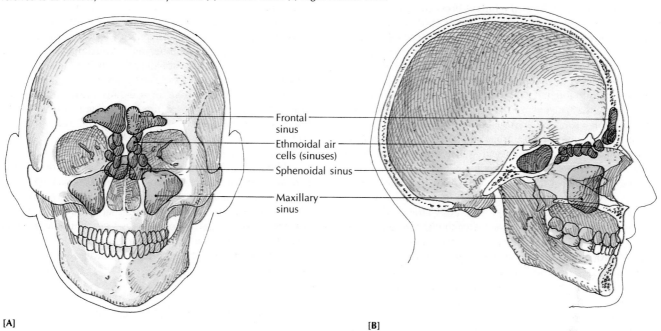

Frontal sinus

Ethmoidal air cells (sinuses)

Sphenoidal sinus

Maxillary sinus

[A]   [B]

The *sphenoidal sinuses* are contained within the body of the sphenoid bone. The size of the sphenoidal sinuses varies greatly, and they may extend into the occipital bone.

The *frontal sinuses* are separated by a bony septum that is often bent to one side. Each sinus extends above the medial end of the eyebrow and anteriorly into the medial portion of the roof of the orbit (see Figure 7.9).

Bones of the Face

The facial skull is anterior and inferior to most of the cranial skull, to which it is attached. Its main functions are to support and protect the structures associated with the nasal, oral, orbital, and pharyngeal (throat) cavities. It also provides a pair of joints between the condyles of the mandible and temporal bones (the temporomandibular, or TM, joints), which permit the lower jaw to open and close. Except for the mandible, all facial bones are united by sutures, and are immovable.

The facial skull is composed of 14 irregularly shaped bones (see Figures 7.2, 7.5C, and 7.8). They include the two inferior nasal conchae, vomer, two palatine bones, two maxillae, two zygomatic (malar) bones, two nasal bones, two lacrimal bones, and mandible. The single hyoid bone and the six ossicles of the ear are considered separately from cranial or facial bones.

Inferior nasal conchae Figures 7.5D and 7.8 show the location of the *inferior nasal conchae* in the lateral wall of each nasal cavity, below the superior and middle conchae of the ethmoid bone. These conchae are thin, bony plates, shaped somewhat like the curved leaves of a scroll. They

increase the surface area of the nasal mucosa, a ciliated membrane with many blood vessels. The fluid portion of the blood leaks out of these blood vessels, continuously moistening (humidifying) the nasal cavity. Cilia in the nasal cavities filter the air, and the blood vessels warm it.

Vomer The *vomer* (L. plowshare), shown in Figures 7.5A, D, and E, forms part of the nasal septum. The single vomer, as its name suggests, is shaped like a plowshare. It is very thin, and is often bent to one side. If the septum is pushed too far to one side and one nasal cavity is much smaller than the other, the condition is called a *deviated septum.* With such a condition, an allergy attack or a severe head cold may swell the nasal mucosa enough to close the smaller cavity or even both nasal cavities.

Palatine bones Figures 7.5D and E show the paired *palatine bones* (referring to the *palate,* or roof of the mouth), which lie behind the maxillae (upper jaws), and form the posterior part of the hard palate and parts of the floor and walls of the nasal cavity and floor of the orbit. The hard palate is composed of the palatine processes of the palatine and maxillary bones.

Each palatine bone is L-shaped, and has a horizontal plate and a vertical plate. The horizontal plate (palatine process) forms the posterior part of the hard palate. The vertical plate forms the posterior lateral wall of the nasal cavity and part of the floor of the orbit. Nerves that serve the palate pass through the *greater* and *lesser palatine foramina* in the posterior lateral corner of the horizontal plate. In some dental procedures, dentists infiltrate these foramina with anesthetics.

Maxillae Figures 7.5A, C, and E show the two *maxillae* (L. upper jaw), which join to form the upper jaw. Each maxilla has a hollow body containing a large maxillary sinus and four processes. (1) The *zygomatic process* extends along the lateral orbital border, and articulates with the zygomatic bone to form the *zygomatic arch.* (2) The *frontal process* extends along the medial wall of the orbit, where it articulates with the frontal bone. (3) The *palatine process* extends horizontally to meet the palatine process of the other maxilla and form the anterior part of the hard palate. (4) The *alveolar process* joins the alveolar process of the other maxilla to form the *alveolar arch,* which contains the bony sockets for the upper teeth. Beneath the orbital margin is the *infraorbital foramen,* through which pass the sensory infraorbital nerve and blood vessels. The lengthening of the face just before adolescence is caused by the growth of the maxillae.

Zygomatic (malar) bones The *zygomatic bones,* also known as *malar* bones (L. *mala,* cheekbone), are the cheekbones (Figures 7.5A, C). They lie below and lateral to the orbit. Each of the two zygomatic bones acts as a tie by con-necting the maxilla with the frontal bone above and the temporal bone behind. Through the *zygomaticofacial* and *zygomaticotemporal foramina* in this bone pass the nerves bearing the same names.

Nasal bones As shown in Figures 7.5A, C, and D, the two *nasal bones* lie side by side between the frontal processes of the maxillae. These two small, oblong bones unite to form the supportive bridge of the upper nose. In addition, they articulate with the frontal, ethmoid (perpendicular plate), and maxillary bones (frontal process), and are attached to the lower nasal cartilage.

Lacrimal bones Each *lacrimal bone* (L. *lacrima,* tear) is a thin bone located in the medial wall of the orbit behind the frontal process of the maxilla (Figures 7.5A, C, E). The rectangular lacrimal bones are the smallest facial bones. In a depression of each bone is a lacrimal sac, which collects excess tears from the surface of the eye. Tears from the lacrimal sacs drain through the nasolacrimal ducts and foramen into the nasal cavity, sometimes causing a runny nose.

FIGURE 7.10

(A) Mandible, right lateral view. (B) The mandible in the infant at birth (top), in the adult (center), and in the aged (bottom). Note how the alveolar process disappears in old age after the teeth have been lost.

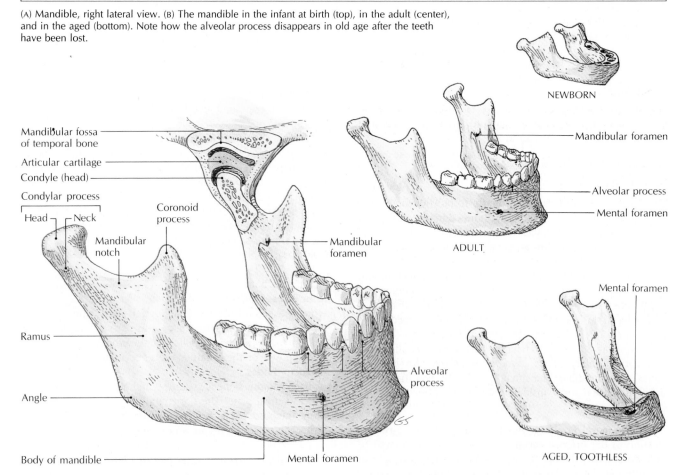

[A]

[B]

Mandible As shown in Figure 7.10, the **mandible** (L. *mandere*, to chew) is the skeleton of the lower jaw. It is a single bone that extends from the mandibular fossa of one temporal bone to the mandibular fossa of the other, forming the chin. It is the largest and strongest facial bone. The right and left halves of the mandible fuse together in the center at the *symphysis menti* during the first or second year of life.

Except for the ossicles of the ear and the hyoid bone, the mandible is the only movable bone in the skull. It can be raised, lowered, drawn back, pushed forward, and moved from side to side. Try it the next time you are chewing food.*

The mandible consists of a horseshoe-shaped horizontal body joined to two perpendicular upright portions called **rami** (RAY-mye; L. branches; singular, *ramus*). The site where the body joins each ramus is known as the **angle** of the mandible. At the superior end of each ramus are two processes separated by a deep depression called the **mandibular notch:** (1) the **coronoid process** is the attachment site for the temporalis muscle, and (2) the head of the **condylar process** articulates with the mandibular fossa of the temporal bone to form the temporomandibular (TM) joint. The inferior alveolar nerve enters the bone through the **mandibular foramen,** passes through the mandible, and a branch emerges as the mental nerve through the **mental foramen,** below the first molar tooth. Branches of the inferior alveolar nerve supply the teeth of the lower jaw. Blood vessels also pass through the mental foramen. The superior edge of the body of the mandible is the **alveolar process,** which contains the sockets for the lower teeth.

Hyoid Bone

Figure 7.11 shows the location of the U-shaped **hyoid bone** located below the root of the tongue and above the larynx.

*Because the joint between the mandible and the temporal bone is slightly loose, it can be dislocated forward by a severe blow to the jaw. The head of the mandible may even slip forward during a yawn, locking the jaw in an open position. The condylar process can be realigned with the mandibular fossa of the temporal bone by depressing the jaw.

When the chin is held up, the bone can be felt above the "Adam's apple," or thyroid cartilage. The hyoid bone does not articulate directly with any other bone. Instead, it is held in position by muscles and the stylohyoid ligaments, which extend from the styloid process of each temporal bone to the hyoid.

In the center of the hyoid bone is a **body,** and projecting backward and upward, a pair of **lesser** and **greater horns** or **cornua** (sing. *cornu*). The hyoid supports the tongue and provides attachment sites for muscles used in speaking and swallowing. In fact, the bone moves in a rotary motion (forward, up, back, and down) during the swallowing sequence. Because it is somewhat freely suspended, the hyoid bone can be held between the index finger and thumb and moved gently from side to side.

FIGURE 7.11

Hyoid bone. (A) Anterior view, in position in the neck, superior to the larynx. Note how the hyoid is suspended from the stylohyoid ligaments. (B) Anterior view. (C) Right lateral view.

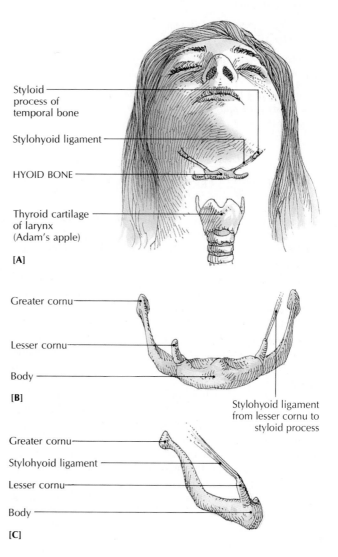

Styloid process of temporal bone

Stylohyoid ligament

HYOID BONE

Thyroid cartilage of larynx (Adam's apple)

[A]

Greater cornu

Lesser cornu

Body

[B]

Stylohyoid ligament from lesser cornu to styloid process

Greater cornu

Stylohyoid ligament

Lesser cornu

Body

[C]

What is a "cleft palate"?

"Cleft palate" is the common name for a defect that occurs when the structures that form the palate do not fuse before birth. As a result, there is a gap in the midline of the roof of the mouth. Such a gap creates a continuity between the oral and nasal cavities. When a gap is forward, on the upper lip, the lip is separated, and a "hare lip" ("cleft lip") results. A hare lip never appears at the midline, and may be paired. An infant with a cleft palate may have difficulty suckling. In most cases, the defect can be repaired, at least partially, by surgery.

Ossicles of the Ear

Within the petrous portion of the temporal bone is the middle ear cavity, containing three pairs of tiny auditory bones, or *ossicles* (see Figure 7.6). The ossicles are connected, and transmit sound waves from the tympanic membrane (eardrum) to the inner ear. These bones are named according to their shapes. The outermost and largest bone is the *malleus* (L. hammer), or hammer. It is attached to the tympanic membrane. The middle bone is the *incus* (L. anvil), or anvil. The innermost bone, the *stapes* (L. stirrup), is shaped like a stirrup. The stapes is the smallest bone in the body. It fits into a tiny membranous oval window, which separates the middle and inner ears. In Chapter 15 you will learn more about the anatomy and physiology of hearing.

Ask Yourself

1 *How is the skull usually defined? What are the two typical divisions of the skull?*

2 *What are sutures? Name the four major skull sutures.*

3 *What are fontanels? Give some examples.*

4 *What are the eight bones of the cranium?*

5 *Not counting the bones of the ear, what are the six pairs and three unpaired bones of the face?*

THE VERTEBRAL COLUMN

The skeleton of the trunk of the body consists of the vertebral column (commonly called the spine or the spinal column), the ribs, and the sternum (breastbone). These bones, along with the skull, make up the axial skeleton.

The *vertebral column* is actually more like an S-shaped spring than a column (Figure 7.12). It extends from the base of the skull through the entire length of the trunk. The spine is composed of 26 separate bones called *vertebrae* (VER-tuh-bree; L. something to turn on, from *vertere*, to turn), which are united by a sequence of fibrocartilaginous *intervertebral disks* to form a strong but flexible support for the neck and trunk. The vertebral column is stabilized by ligaments and muscles that permit twisting and bending movements, but limit some other movements that might be harmful to either the spinal column or the spinal cord.

In addition to protecting the spinal cord and spinal nerve roots and providing a support for the weight of the body, the spinal column helps the body keep an erect posture. The resilient intervertebral disks act as shock absorbers when the load upon the spinal column is increased, and they also allow the vertebrae to move without damaging each other (see Figure 7.19). The vertebral column is also the point of attachment for the muscles of the back.

FIGURE 7.12

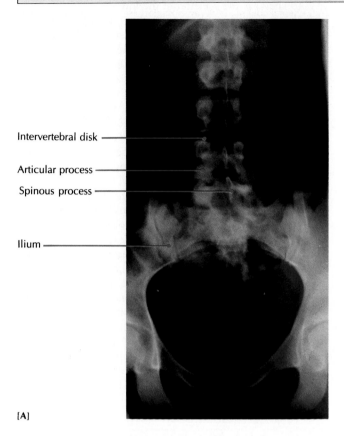

Intervertebral disk

Articular process

Spinous process

Ilium

[A]

The vertebral column varies in length depending upon the height of the individual. The average length is 70 cm (28 in.) in adult males and 60 cm (24 in.) in adult females. The disks account for about one-quarter of the length of the vertebral column, and are thickest in the cervical and lumbar areas, where movement is greatest.

The adult vertebral column has 24 movable vertebrae, plus the fused vertebrae of the sacrum and coccyx, arranged as follows:

Cervical (neck) vertebrae	7
Thoracic (chest) vertebrae	12
Lumbar (back) vertebrae	5
Sacral (5 fused vertebrae)	1
Coccygeal (3 to 5 fused vertebrae)	1
Total	**26**

In a child there are 33 separate vertebrae, the 9 in the sacrum and coccyx not yet being fused.

Curvatures of the Vertebral Column

When viewed from the side, the adult vertebral column exhibits four curves: (1) a forward cervical curve, (2) a back-

The vertebral column. (A) X ray, posterior view. (B) Anterior view. Note that the vertebrae are numbered in sequence within each region, starting from the head. (C) Right lateral view. Note the four spinal curves. [(A) *Larry Mulvehill/Photo Researchers.*]

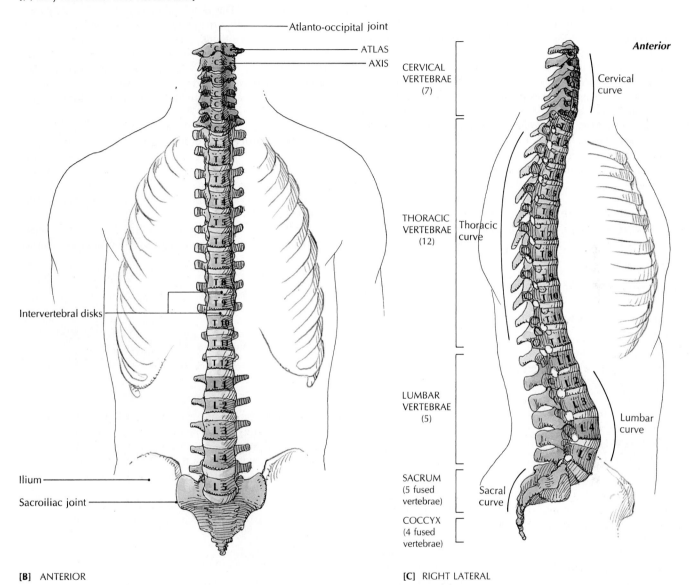

[B] ANTERIOR

[C] RIGHT LATERAL

ward thoracic curve, (3) a forward lumbar curve, and (4) a backward sacral curve (see Figure 7.12c). The thoracic and sacral curves are known as ***primary curves*** because they are present in the fetus. The cervical and lumbar curves are ***secondary,*** because they do not appear until after birth. The cervical curve appears about three months after birth when the infant begins to hold up its head, and the lumbar curve appears during the latter half of the first year when the infant begins to stand and sit upright.

The curves of the vertebral column provide the spring and resiliency necessary to cushion such ordinary actions as walking, and they are critical for maintaining a balanced center of gravity in the body.

During the later stages of pregnancy some women tend to increase the degree of the lumbar curve in an effort to maintain a balanced center of gravity. Backaches sometimes occur because of this exaggerated curve and the increased pressure on the posterior lumbar region. (See Physiological and Anatomical Abnormalities at the end of this chapter for a further discussion of abnormal curvatures of the spine.)

A Typical Vertebra

Vertebrae are irregularly shaped bones, but they are shaped similarly within their regions. Except for the first and second cervical vertebrae, the sacrum, and the coccyx, they all have a similar structure. A "typical" vertebra consists of a body, a vertebral (neural) arch, and several processes or projections (Figure 7.13). The body provides weight and support, the arch supplies protection for the spinal cord, and the processes allow movement.

Vertebral body The disk-shaped *vertebral body* is located anteriorly in each vertebra. It functions mainly to give strength and to support weight. The bodies of vertebrae from the third cervical to the first sacral become progressively larger as they need to bear heavier weights (see Figure 7.12). The upper and lower ends of each body are slightly larger than the middle. These roughened ends articulate with the fibrocartilaginous intervertebral disk located between the bodies of the vertebrae above the sacrum. The anterior edge of the body contains tiny holes through which blood vessels enter.

Vertebral (neural) arch The *vertebral arch* is posterior to the vertebral body. Each arch has two thick *pedicles* (L. little feet), which form the lateral walls, and two *laminae* (L. thin plates), which form the posterior walls of the arch (see Figure 7.13). The arch and the body together create the *vertebral foramen.* The sequence of all the vertebral foramina forms the *vertebral* (spinal) *canal,* which encloses the spinal cord and its surrounding meninges, nerve roots, and blood vessels (see Figure 7.13). The vertebral arch protects the spinal cord in the same way that the cranium protects the brain.

Each pedicle contains one vertebral notch on its inferior border and one on its superior border (see Figure 7.13B). These notches or "gutters" are arranged so that together they form an *intervertebral foramen,* through which the spinal nerves and their accompanying blood vessels pass.

Vertebral processes Seven *vertebral processes* (bony projections) extend from the lamina of a vertebra (see Figure 7.13). Three of these processes, the *spinous process* and the paired *transverse processes,* are the attachment sites for vertebral muscles. These processes act like levers, helping the attached muscles and ligaments move the vertebrae.

The other four processes are two superior and two inferior *articular processes.* All four processes join directly to other

FIGURE 7.13

(A) Drawing of a typical vertebra, illustrating its parts. The vertebral arch (dotted lines) is made up of two pedicles and two laminae. Superior view. (B) Right lateral view. The sequence of vertebral foramina forms the vertebral canal, through which passes the spinal cord. The cauda equina is located caudal to the spinal cord. Both drawings are shown approximately actual size.

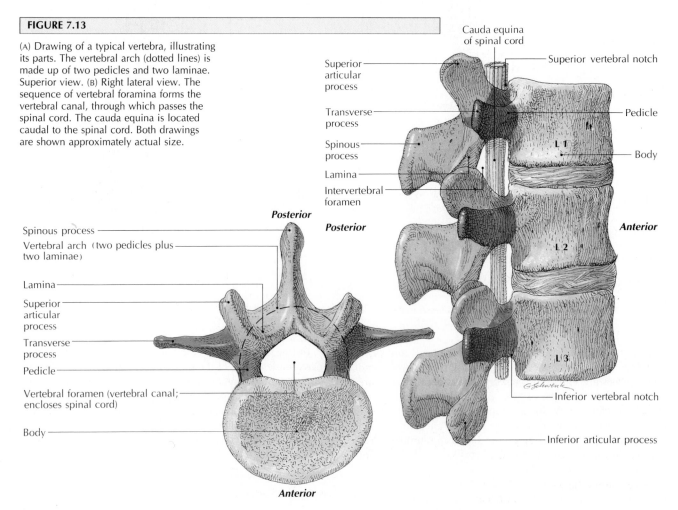

[A] SUPERIOR

[B] RIGHT LATERAL

bones. The superior articular processes of one vertebra articulate with the inferior articular process of the vertebra above (see Figure 7.13B). The articular processes prevent the vertebrae from slipping forward, and they restrict movement between two vertebrae. Movement between two vertebrae occurs at the intervertebral disk and at the paired joints between the articular processes.

Cervical Vertebrae

The *cervical vertebrae* (L. *cervix,* neck) are the seven (C1 to C7) small neck vertebrae between the skull and the thoracic vertebrae (Figure 7.14; see also Figure 7.12). They support the head, and enable it to move up, down, and sideways.

FIGURE 7.14

Cervical vertebrae. (A) The atlas, or first cervical vertebra, superior view. (B) The axis, or second cervical vertebra, anterior view. (C) The articulation of the dens of the atlas with the atlas, and their relationship to the other cervical vertebrae; posterior view. (D) A typical cervical vertebra, superior view. Note that the vertebral artery passes through the transverse foramen, and the spinal cord through the vertebral foramen.

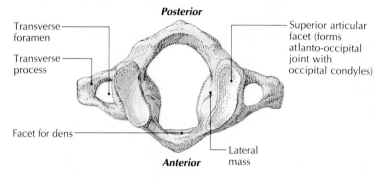

Transverse foramen · Transverse process · Facet for dens · *Posterior* · *Anterior* · Superior articular facet (forms atlanto-occipital joint with occipital condyles) · Lateral mass

[A] ATLAS (FIRST CERVICAL VERTEBRA)

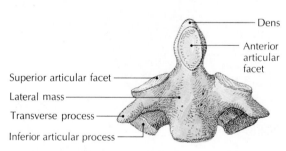

Superior articular facet · Lateral mass · Transverse process · Inferior articular process · Dens · Anterior articular facet

[B] AXIS (SECOND CERVICAL VERTEBRA)

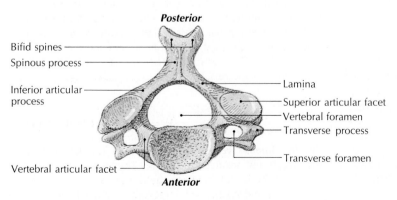

Posterior · Bifid spines · Spinous process · Inferior articular process · Vertebral articular facet · *Anterior* · Lamina · Superior articular facet · Vertebral foramen · Transverse process · Transverse foramen

[D] TYPICAL CERVICAL VERTEBRA

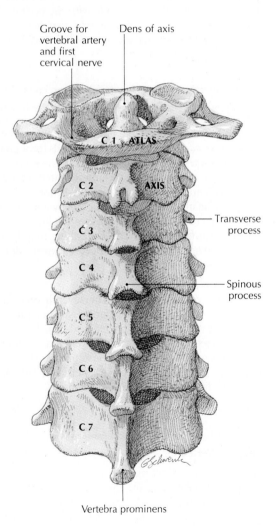

Groove for vertebral artery and first cervical nerve · Dens of axis · C 1 ATLAS · C 2 AXIS · C 3 · C 4 · C 5 · C 6 · C 7 · Transverse process · Spinous process · Vertebra prominens

[C] CERVICAL VERTEBRA

Each of these vertebrae has a pair of triangular openings called the **transverse foramina,** which are found only in these vertebrae. In all but the seventh, each foramen is large enough for the passage of a vertebral artery. The paired vertebral arteries carry a critical amount of blood to the brain. The third to sixth cervical vertebrae are similar in shape, with a small, broad body, and short, forked (bifid) spines. Their articular facets are always positioned in the same way. However, the first, second, and seventh cervical vertebrae are not typical.

First cervical vertebra (atlas) The first cervical vertebra is called the **atlas** because it supports the head, as the Greek god Atlas supported the heavens on his shoulders. It is ringlike, with a short anterior and a long posterior arch, with large lateral masses on each side (see Figure 7.14A). Although the atlas is the widest of the cervical vertebrae, it has no spine and no body. It articulates above with the occipital condyles and below with the second cervical vertebra (axis). These *atlanto-occipital joints* make it possible to make the nodding "yes" motion with the head.

Second cervical vertebra (axis) The second cervical vertebra is known as the **axis** because it forms a pivot point for the atlas to move the skull in a twisting "no" motion. The pivot is formed by a peglike protrusion from the body called the **dens,** or *odontoid process,* which extends through the opening in the atlas (see Figures 7.14B, C). The dens (Gr. tooth) is held in position against the anterior arch and away from the spinal cord by a strong transverse ligament of the atlas.

Quick death from hanging occurs when the transverse ligament snaps and the dens crushes the lower medulla and the adjacent spinal cord. The same thing may happen with *whiplash,* when the head is snapped backward during a violent automobile accident.

Seventh cervical vertebra The seventh cervical vertebra has an exceptionally long, unforked spinous process with a tubercle at its tip that can be felt, and usually seen, through the skin (see Figure 7.14C). This vertebra is known as the **vertebra prominens.** The transverse process is large, and the transverse foramina in the cervical vertebrae are the passageways for the vertebral arteries to the brain.

Thoracic Vertebrae

All 12 **thoracic vertebrae** (Gr. *thorax,* breastplate) articulate with ribs (see Figure 7.20). The thoracic vertebrae (T1 to T12) are larger than the cervical ones, and have distinctive articular facets, called *costal facets* (L. *costa,* rib), since they join the ribs (Figure 7.15).

Because there are 12 thoracic vertebrae and 12 corresponding flexible intervertebral disks, the total mobility of this region is considerable, even with the presence of the ribs and sternum. The ribs articulate with the bodies and transverse processes of the thoracic vertebrae (Table 7.5).

FIGURE 7.15

Thoracic vertebrae. (A) Lateral view. (B) Superior view. The intervertebral notches between adjacent vertebrae form the intervertebral foramen, through which passes a spinal nerve.

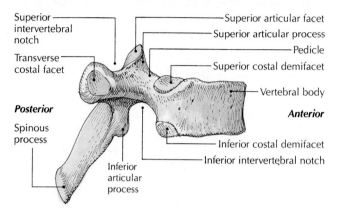

[A]　RIGHT LATERAL

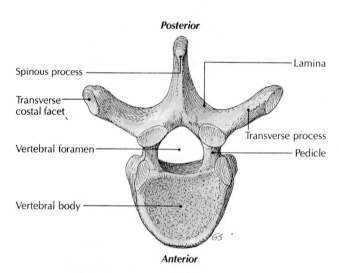

[B]　SUPERIOR

The thoracic vertebrae increase in size as they progress downward. The first four vertebrae are something like the cervical vertebrae, and the last four have certain features of lumbar vertebrae. So only the middle four thoracic vertebrae are considered typical. Their bodies, viewed from above, are heart-shaped, and they have circular vertebral foramina (see Figure 7.15). Three articular facets on each side provide attachments for the ribs.

Lumbar Vertebrae

The five **lumbar vertebrae** (L1 to L5) (L. *lumbus,* loin) are the largest and strongest vertebrae. They are situated in the "small of the back," between the thorax and the pelvis (see Figures 7.1 and 7.12). Their large kidney-shaped bodies have

FIGURE 7.16

Lumbar vertebrae. (A) Right lateral view. (B) Superior view.

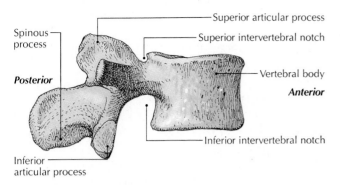

Spinous process

Posterior

Inferior articular process

Superior articular process

Superior intervertebral notch

Vertebral body

Anterior

Inferior intervertebral notch

[A] RIGHT LATERAL

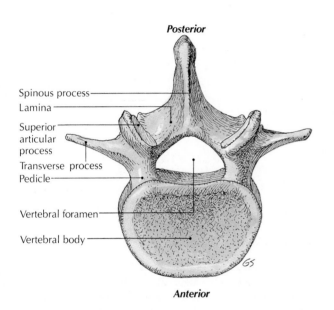

Posterior

Spinous process
Lamina
Superior articular process
Transverse process
Pedicle

Vertebral foramen

Vertebral body

Anterior

[B] SUPERIOR

short, blunt, four-sided spinous processes, which are adapted for the attachment of the lower back muscles (Figure 7.16). The arrangement of the facets on the articular processes of each vertebra maximizes forward and backward bending, but lateral bending is limited, and rotation is practically eliminated. The lumbar vertebrae have no facets for articulations with ribs, and no transverse foramina. The transverse processes are long and slender, and the vertebral foramina are oval or triangular.

In most people, the spinal cord ends between the first and second lumbar vertebrae. Therefore, a **lumbar puncture** (commonly called a *spinal tap*) can usually be made safely just below this point in order to obtain *cerebrospinal fluid*. This clear fluid, which bathes the brain and spinal cord and acts as a shock absorber for the central nervous system, is useful in diagnosing certain diseases.

Sacrum and Coccyx

The wedge-shaped *sacrum* not only gives support to the vertebral column, it also provides strength and stability to the pelvis. It is composed of five vertebral bodies fused in an adult into one bone by four ossified (bony) intervertebral disks (Figure 7.17). The sacrum is curved, forming a concave surface anteriorly and a convex surface posteriorly. The four *transverse lines* on the otherwise smooth concave anterior surface indicate the fusion sites of the originally separate vertebrae. At the lateral ends of these lines are two parallel rows of four pelvic (ventral) *sacral foramina* (see Figure 7.17). On the posterior convex surface of the sacrum are four pairs of *posterior* (dorsal) *foramina*. The bones lateral to these foramina are known as the *lateral masses.*

The sacrum articulates above with the last lumbar vertebra (L5), below with the coccyx, and laterally from the *auricular surfaces* with the two iliac bones of the hip to form the *sacroiliac joints.*

The fused laminae and short spines form the roof of the *sacral canal.* Because the laminae of the lower sacral vertebrae (S4 and S5) are often absent or do not meet in the midline, a gap called the *sacral hiatus* (see Figure 7.17) is present at these levels. Caudal anesthesia, which spreads upward and acts directly on the spinal nerves, can be injected into the sacral hiatus.

The projecting anterior edge of the first sacral vertebra is called the *sacral promontory* (see Figure 7.17). It is used as a landmark for making pelvic measurements. Just above the promontory is the *lumbosacral joint* between the fifth lumbar vertebra and the sacrum. The sacrum is tilted at this point, so that the articulation forms the *lumbosacral angle.*

The *coccyx,* or tailbone, consists of three to five (usually four) fused vertebrae, the vestiges of an embryonic tail that usually disappears about the eighth fetal week.* Triangular in shape, it has an *apex, base, pelvic* and *dorsal surfaces,* and two *lateral borders* (see Figure 7.17). Its base articulates with the lower end of the sacrum by means of a fibrocartilaginous intervertebral disk.

Coccyx is the Greek name for a cuckoo. The coccyx is so named because it was thought to resemble a beak.

Ask Yourself

1 *How many vertebrae are there in the vertebral column?*

2 *What are the major functions of the vertebral column? What are the purposes of the intervertebral disks?*

3 *What is the purpose of the four natural curves of the vertebral column?*

4 *What are the parts of a typical vertebra?*

5 *From the neck down, what are the five vertebral regions called?*

FIGURE 7.17

Sacrum and coccyx. (A) Ventral surface. (B) Dorsal surface.

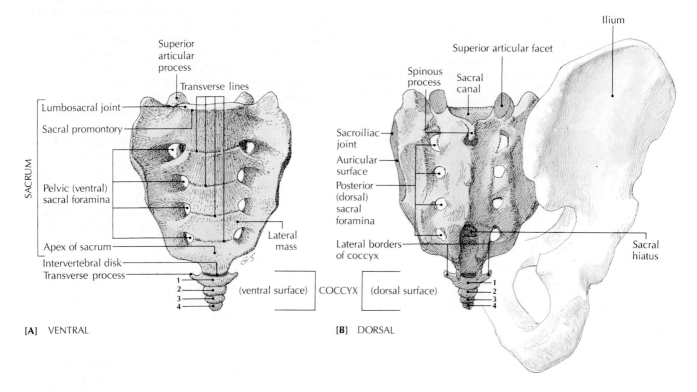

[A] VENTRAL

[B] DORSAL

TABLE 7.5 BONES OF THE VERTEBRAL COLUMN (26 BONES)*

Bone or group of bones	Location	Description	Functions
Cervical vertebrae (7 bones; C1–C7)	Between skull and thorax, in neck (Figure 7.14).	First (atlas), second (axis), and seventh vertebrae are modified; third through sixth are typical. All contain transverse foramina.	Atlas supports head, permits "yes" motion of head at the joint between skull and atlas. Axis permits "no" motion at the joint between axis and atlas.
Thoracic vertebrae (12 bones; T1–T12)	Between cervical and lumbar vertebrae, in thorax (Figure 7.15).	Larger than cervical, but smaller than lumbar, vertebrae.	Articulate with ribs. Allow some movement of spine in thoracic area.
Lumbar vertebrae (5 bones; L1–L5)	Between thorax and pelvis, in "small of back" (Figure 7.16).	Largest, strongest vertebrae; adapted for attachment of back muscles.	Support back muscles. Allow forward and backward bending of spine.
Sacrum (5 fused bones)	Between lumbar vertebrae and coccyx, in lower spine (Figure 7.17).	Wedge-shaped, made up of five fused bodies united by four intervertebral disks.	Support vertebral column. Give strength and stability to pelvis.
Coccyx (3 to 5 fused bones)	End of vertebral column, below sacrum (Figure 7.17).	Triangular tailbone, united with sacrum by intervertebral disk.	Vestige of an embryonic tail.

*The total number of bones in the vertebral column is given here as 26 because the coccyx and sacrum are counted as one each. It is not uncommon to count the 24 movable vertebrae of the cervical, thoracic, and lumbar regions together with 5 fused sacral vertebrae and 4 fused coccygeal vertebrae, for a grand total of 33.

THE THORAX

The *thorax* (Gr. breastplate) is the chest, which is part of the axial skeleton. The thoracic skeleton is formed posteriorly by the bodies and intervertebral disks of 12 thoracic vertebrae, and anteriorly by 12 pairs of ribs, 12 costal cartilages, and the sternum. The thorax is fairly narrow at the top and broad below, and it is wider than it is deep (Figure 7.18).

The cagelike thoracic skeleton is a good example of a functional structure. It protects the heart, lungs, and some abdominal organs. It supports the bones of the shoulder girdle and arm. The lever arms formed by the ribs and costal cartilages provide a flexible mechanism for breathing. Also, the sternum provides a point of attachment for the ribs.

Sternum

The *sternum* (Gr. *sternon,* breast), or breastbone, is the midline bony structure of the anterior chest wall (see Figure 7.18). It resembles a dagger about 17.5 cm (7 in.) long in the adult, and consists of a manubrium, body, and xiphoid process.

The *manubrium* (L. handle) has a pair of *clavicular notches,* which articulate with the clavicle (collarbone). This *sternoclavicular joint* is the site of direct attachment of the pectoral appendage to the axial skeleton. It permits some bending movements, though in the elderly the movements become restricted because of ossification. At the lateral *costal notches*

the sternum articulates with the costal cartilages of the first ribs and part of the second ribs. On its upper border is the *jugular* (suprasternal) *notch.* The manubrium is united with the body of the sternum at the movable *manubriosternal joint,* which acts as a hinge to allow the sternum to move forward during inhalation. This joint forms a slight angle, called the *sternal angle,* which can be felt through the skin. Because this angle is opposite the second rib, it is a fairly reliable starting point for counting the ribs.

The *body* of the sternum is about twice as long as the manubrium. It articulates with the second through tenth pairs of ribs.

The *xiphoid process* (ZIFF-oid; Gr. sword-shaped) is the smallest, thinnest, and most variable part of the sternum. It is cartilaginous during infancy, but usually is almost completely ossified and joined to the body of the sternum by the fortieth year. The xiphoid process does not articulate with any ribs or costal cartilages, but several ligaments and muscles (including abdominal muscles) are attached to it.

Because the sternum is relatively accessible, a physician may insert a bore needle into its marrow cavity to obtain a specimen of red blood cells that are developing there. Such a procedure is called a *sternal puncture.*

Ribs

The *ribs* are curved, slightly twisted strips of bone that form the widest and major part of the thoracic cage (Figure 7.19 and Table 7.6). There are usually 12 pairs of ribs, all of which articulate posteriorly with the vertebral column. The ribs in-

---

FIGURE 7.18

Skeleton of the thorax. (A) Anterior view. (B) Posterior view. Note the sternum. Ribs and vertebrae are numbered.

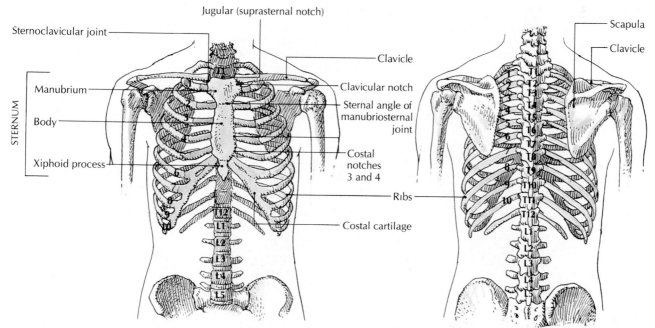

[A] ANTERIOR

[B] POSTERIOR

TABLE 7.6   BONES OF THE THORAX (25 BONES)

Bone	Location	Description	Functions
Ribs (12 pairs)	Encircle thorax, articulating posteriorly with thoracic vertebrae and anteriorly with sternum (Figure 7.19).	Long, curved, varying in length and width. Ribs 1–7 (true vertebrosternal ribs) attach directly to sternum. Ribs 8–12 (false ribs) do not attach directly to sternum. Ribs 8–10 (vertebrochondral ribs) attach to rib 7 cartilage. Ribs 11–12 attach only to vertebrae.	With sternum and thoracic vertebrae, form strong but lightweight cage to protect heart, lungs, other organs. With costal cartilages, provide flexible mechanism for breathing. Support bones of upper extremities.
Sternum	At midline of anterior chest wall (Figure 7.19).	Dagger-shaped, about 15 cm long. Made up of manubrium, body, xiphoid process. Articulates with ribs 1–7 directly, ribs 8–10 indirectly.	Provides anterior attachment site for ribs. With ribs, forms protective cage for heart, lungs, other organs. Supports bones of upper extremities.

FIGURE 7.19

The ribs as part of the thorax, anterior view. Note that ribs consist of bony ribs and costal cartilage.

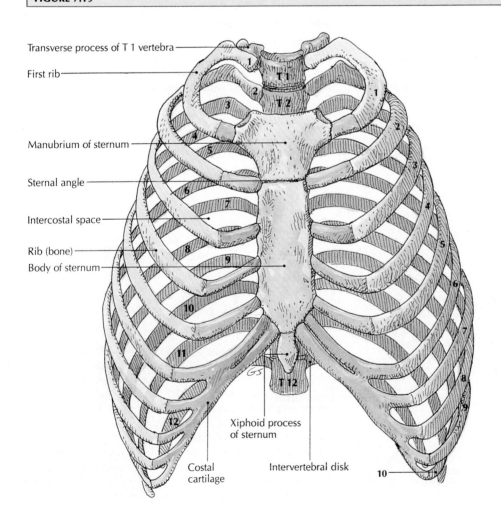

Transverse process of T 1 vertebra

First rib

Manubrium of sternum

Sternal angle

Intercostal space

Rib (bone)

Body of sternum

Xiphoid process of sternum

Costal cartilage

Intervertebral disk

FIGURE 7.20

Typical rib, costal cartilage, thoracic vertebra, and sternum, illustrating the articulations of those structures.

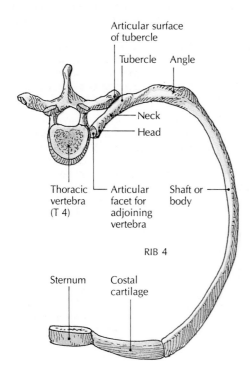

Articular surface of tubercle
Tubercle   Angle
Neck
Head
Thoracic vertebra (T 4)   Articular facet for adjoining vertebra   Shaft or body
RIB 4
Sternum   Costal cartilage

or *vertebrosternal ribs*, because they attach to both the vertebrae and the sternum. The lower five pairs of ribs (8 to 12) are known as **false ribs** because they are attached to the sternum only indirectly. Of the false ribs, 8 to 10 have costal cartilages that connect with each other and also with the cartilage of rib 7. They are called *vertebrochondral ribs*. Ribs 11 and 12 are called **floating ribs** because they attach only to the vertebral column (see Figure 7.19).

Ribs 3 to 9 are known as **typical ribs** (Figure 7.20). Each typical rib has a wedge-shaped *head* on the end next to the spine. The head articulates with an intervertebral disk and body of an adjacent vertebra. The short, flattened *neck* is located between the head and the tubercle. The *tubercle* is located between the neck and the body of a typical rib. It forms an articular surface with a transverse process on a vertebra. The *shaft*, or *body*, is the main part of the rib. It curves sharply forward after its junction with the tubercle to form a distinct *angle*, and then arches downward until it joins the costal cartilage. On the lower border of the rib is a *costal groove*, forming a protective passageway for intercostal blood vessels and nerves (see Figure 7.20).

Ask Yourself

1 *What are the components of the thorax?*

2 *What are the functions of the thoracic cage?*

3 *What are the main parts of the sternum? Where are the axial skeleton and appendicular skeleton connected to each other?*

4 *How many pairs of ribs are there? What is a true rib? A false rib? A floating rib?*

5 *What are the main parts of a typical rib?*

crease in length from rib 1 to rib 7 and decrease from 8 to 12. The space between the ribs is the *intercostal space*. It contains the intercostal muscles.

The upper seven pairs (ribs 1 to 7) connect directly with the sternum by their attached strips of *costal cartilage*, which are made of hyaline cartilage. These ribs are called **true ribs**

THE APPENDICULAR SKELETON

The **appendicular skeleton,** although its name is derived from the Latin word *appendere*, to hang from, should not be thought of as consisting of only the hanging parts—the arms and legs.* In fact, the appendicular skeleton is composed of the *upper extremities*, which include the scapula (shoulder blade) and clavicle (collarbone) of the upper limb or shoulder girdle, in addition to the arms, forearms, and hands, and the *lower extremities*, which include the hipbone of the lower limb or pelvic girdle, as well as the thighs, legs, and feet.

*In terms of the appendicular skeleton, an *extremity* and a *limb* mean the same thing; a limb is not an arm or leg. Remember that the upper limb, or extremity, includes the scapula and clavicle *as well as* the arm, forearm, wrist, and hand.

THE UPPER EXTREMITIES (LIMBS)

The skeleton of the **upper extremity** or **limb** consists of 64 bones (Figure 7.21 and Table 7.7). These include the scapula and clavicle of the upper limb (shoulder) girdle, the humerus of the arm, the radius and ulna of the forearm, the carpals of the wrist, the metacarpals of the palm, and the phalanges of the fingers. The upper extremity is connected to and supported by the axial skeleton by only one joint and many muscles. The joint is the sternoclavicular joint between the manubrium of the sternum and the clavicle. The muscles form a complex of suspension bands from the vertebral column, ribs, and sternum to the shoulder girdle.

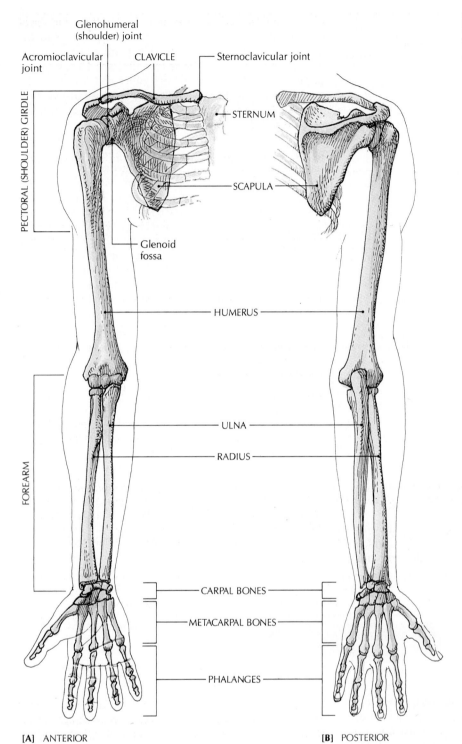

Acromioclavicular joint

Glenohumeral (shoulder) joint

CLAVICLE

Sternoclavicular joint

PECTORAL (SHOULDER) GIRDLE

STERNUM

SCAPULA

Glenoid fossa

HUMERUS

FOREARM

ULNA

RADIUS

CARPAL BONES

METACARPAL BONES

PHALANGES

[A] ANTERIOR

[B] POSTERIOR

FIGURE 7.21

The upper extremity, which includes the pectoral (shoulder) girdle. (A) Anterior view. (B) Posterior view.

Pectoral Girdle

The upper limb girdle is known as the ***shoulder girdle*** or ***pectoral girdle*** (see Figure 7.21). It consists of the clavicle and scapula. The clavicle is a long bone that extends from the sternum to the scapula in front of the thorax. The triangular scapula is located behind the thoracic cage.

The pectoral girdle is designed more for mobility than for stability. The pelvic girdle, in contrast, is designed for stability, sacrificing mobility. The pectoral girdle is held in place, and surrounded by, muscles and ligaments. The stability of the shoulder is provided by these muscles and ligaments, rather than by the shape of the bones and joints. Because of this arrangement, the shoulder is often the site of dislocation injuries such as shoulder separations.

TABLE 7.7   BONES OF THE UPPER EXTREMITIES (64 BONES)

Bone	Location	Description	Functions
SHOULDER (PECTORAL) GIRDLE			
Clavicle (2)	At root of neck, extending horizontally from manubrium of sternum to acromion of scapula (Figure 7.22).	Collarbone. Double-curved, long bone, with rounded medial end and flattened lateral end. Held in place by ligaments.	Holds shoulder joint and arm away from thorax, so upper limb can swing freely.
Scapula (2)	On posterior thoracic wall, between rib 2 and rib 7 (Figure 7.23).	Shoulder blade. Flat, triangular bone, with horizontal spine separating fossae.	Site of attachment for muscles of arm and chest.
ARM			
Humerus (2)	Upper arm, between shoulder and elbow (Figure 7.24).	Longest, largest bone of upper limb. Forms ball of the ball-and-socket joint with the glenoid fossa of the scapula.	Site of attachment for muscles of shoulder and arm, permitting arm to flex and extend at elbow.
FOREARM			
Radius (2)	Between elbow and wrist, lateral to ulna (on thumb side) (Figure 7.25).	Smaller of two bones in forearm.	Allows forearm to rotate in radial motion.
Ulna (2)	Between elbow and wrist, medial to radius (on little finger side) (Figure 7.25).	Larger of two bones in forearm. Large proximal end consists of olecranon process (prominence of elbow).	Forms hinge joint at elbow.
WRIST			
Carpals (16)	Between forearm and palm of hand (Figure 7.26).	Small short bones. In each wrist, 8 carpals in 2 transverse rows of 4.	With attached ligaments, allow slight gliding movement between wrist bones.
HANDS AND FINGERS			
Metacarpals (10)	Palm of hand, between wrist, thumb, fingers (Figure 7.26).	5 miniature long bones in each hand in fanlike arrangement. Articulate with fingers at metacarpophalangeal joint (the knuckle).	Aid opposition movement of thumb by high mobility of its metacarpal. Enable cupping of hand by slight mobility of metacarpals of little and ring fingers.
Phalanges (28)	Fingers, thumbs (Figure 7.26).	Miniature long bones, 2 in each thumb, 3 in each finger. Articulate with each other at interphalangeal joint.	Allow fingers to participate in stable grips because the short phalanges and their interphalangeal joints can only bend and straighten (flex and extend).

Clavicle The paired *clavicles* (L. key), or collarbones, are located at the root of the neck (see Figures 7.18, 7.19, and 7.21). The entire length of the clavicle can be felt through the skin. The medial or *sternal* end of the clavicle articulates with the manubrium just above the first rib at the **sternoclavicular joint,** connecting the axial and appendicular skeletons. The lateral or *acromial* end articulates with the **acromion process.** This highly mobile **acromioclavicular joint** involves the clavicle in all movements of the scapula.

The clavicle is a horizontal double-curved, long bone, with a rounded medial end and a flattened lateral end (Figure 7.22). The medial part of the clavicle is curved anteriorly, and the lateral part is curved posteriorly. The clavicle is held in place by strong ligaments at both ends. At the medial end is the **costal tuberosity,** where the costoclavicular ligament is attached (Figure 7.22B). At the lateral end is the **conoid tubercle,** where the coracoclavicular ligament is attached. The clavicle is also a point of attachment for muscles of the shoulder girdle and neck. Large blood vessels and some nerves for the upper limb pass below the anterior curvature.

The main function of the clavicle is to act as a strut to hold the shoulder joint and arm away from the thoracic cage, allowing the upper limb much freedom of movement. Because of its vulnerable position and relative thinness, the clavicle is broken more often than any other bone in the body, and when it is, the whole shoulder is likely to collapse.

FIGURE 7.22

Clavicle. (A) Superior view. (B) Inferior view.

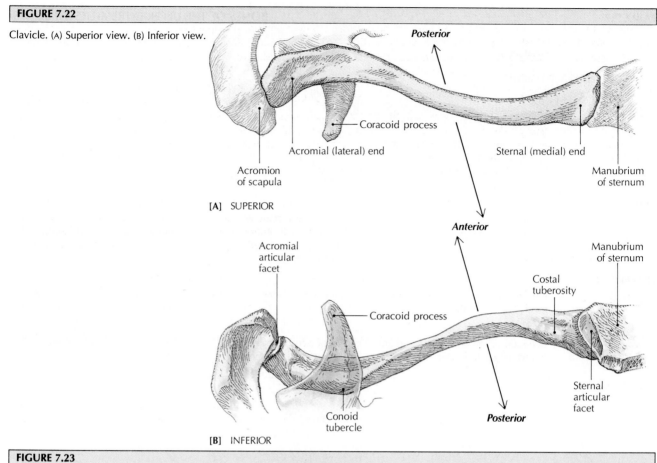

Posterior

Coracoid process

Acromial (lateral) end

Sternal (medial) end

Manubrium of sternum

Acromion of scapula

[A] SUPERIOR

Anterior

Acromial articular facet

Manubrium of sternum

Coracoid process

Costal tuberosity

Conoid tubercle

Sternal articular facet

Posterior

[B] INFERIOR

FIGURE 7.23

Scapula. (A) Anterior view, costal surface. (B) Posterior view, dorsal surface. (C) Lateral view.

Coracoid process

Articular facet

Acromion

Glenoid fossa (cavity)

Lateral (axillary) border

Superior border

Superior angle

Medial (vertebral) border

[A] ANTERIOR

Acromion

Supraspinous fossa

Superior angle

Lateral (axillary) border

Scapular spine

Infraspinous fossa

Inferior angle

[B] POSTERIOR

Acromion Coracoid process

Glenoid fossa (cavity)

Scapular spine

Lateral (axillary) border

Inferior angle

[C] LATERAL

Scapula The *scapula* (L. shoulder), or shoulder blade (because of its resemblance to the blade of a shovel), is located on the posterior thoracic wall between the second and seventh ribs (see Figure 7.18).

The flat, triangular *body* of the scapula has a horizontally oriented *spine* that can be felt on the posterior surface (Figure 7.23). The prominent ridge of the spine separates the supraspinous fossa from the infraspinous fossa. The spine and fossae provide attachment sites for muscles that move the arm. The spine ends in the large, flat *acromion,* which forms the point of the shoulder. It articulates with the clavicle and is an attachment site for chest and arm muscles. Below the acromion is the shallow *glenoid fossa* or *cavity,* which acts as a socket for the head of the humerus. This articulation forms the *glenohumeral* (shoulder) *joint.* Projecting over the glenoid fossa is the anteriorly directed *coracoid process,* which serves as an attachment for the coracoclavicular ligament and several muscles of the arm and chest.

Figure 7.23 shows the location of the three borders and two angles of the scapula. The *medial* (or vertebral) *border* runs parallel to the vertebral column. The *superior* and *inferior angles* are located at the ends of the medial border. Extending upward from the inferior angle to the glenoid fossa is the *lateral* (or axillary) *border.* Running horizontally from the glenoid fossa to the superior angle is the *superior border.*

Bones of the Arm, Forearm, and Hand

The bones of the arm, forearm, and hand are the humerus, ulna, radius, carpals (wrist), metacarpals, and the phalanges of the fingers.

Humerus The *humerus* (L. upper arm) is the arm bone located between the shoulder and elbow. It is the longest and largest bone of the upper limb (Figure 7.24).

The *shaft* (or body) of the humerus is cylindrical in its upper half and flattened from front to back in its lower half. The *head* of the humerus is the ball of the ball-and-socket glenohumeral joint with the glenoid fossa of the scapula. Close to the head are the *greater* and *lesser tubercles,* which are attachment sites for muscles originating on the scapula. Passing along the *intertubercular* (bicipital) *groove* between the two tubercles is the tendon of the long head of the biceps muscle for the forearm. The *anatomical neck* is located between the head and the tubercles. The *surgical neck,* so named because it is the site of frequent fractures of the upper end of the humerus, is located just below the tubercles. The *deltoid tuberosity,* halfway down the lateral side of the shaft, is the attachment site for the deltoid (shoulder) muscle. On the posterior surface is the *radial groove* spiraling from medial to lateral. Along this groove passes the radial nerve.

At the distal end of the humerus are the *trochlea* (L. pulley), which is connected like a pulley with the olecranon process of the ulna, and the *capitulum* (L. little head), which articulates with the head of the radius (see Figure 7.25). Some muscles of the forearm and fingers are attached to the

FIGURE 7.24

Right humerus. (A) Anterior view. (B) Posterior view.

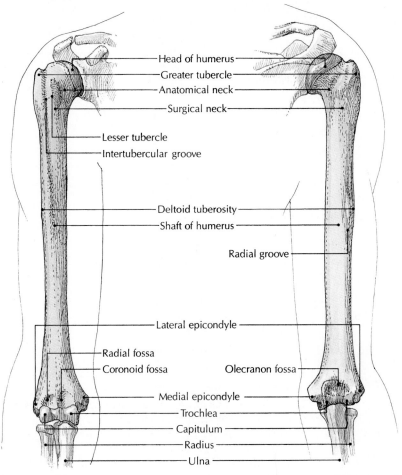

Head of humerus
Greater tubercle
Anatomical neck
Surgical neck
Lesser tubercle
Intertubercular groove
Deltoid tuberosity
Shaft of humerus
Radial groove
Lateral epicondyle
Radial fossa
Coronoid fossa
Olecranon fossa
Medial epicondyle
Trochlea
Capitulum
Radius
Ulna

[A] ANTERIOR

[B] POSTERIOR

lateral and *medial epicondyles.* On the anterior surface are the *radial fossa* and *coronoid fossa,* which accommodate the head of the radius and the coronoid process of the ulna, respectively, when the arm is bent at the elbow. On the posterior surface is the *olecranon fossa,* which accommodates the olecranon process of the ulna when the arm is straightened out.

Ulna The *ulna* (L. elbow) is the longer of the two long bones of the forearm, between the elbow and the wrist. It is medially located (on the side of the little finger).

The large proximal end of the ulna consists of the *olecranon process,* which curves upward and forward to form the *semilunar* or *trochlear notch* (Figure 7.25A). This half-moon-shaped depression articulates with the trochlea of the humerus to form the hinged elbow joint. The *coronoid process* on the anterior surface of the ulna forms the lower border of the trochlear notch. The *radial notch* on the lateral surface of the coronoid process is the articulation site for the head of the radius. The ulna is narrow at its distal end, with a small

head and a short, blunt peg called the *styloid process,* which is attached to a fibrocartilaginous disk that separates the ulna from the carpus.

Radius The *radius* (L. spoke of a wheel) is the long bone located lateral to the ulna (on the thumb side).

At the proximal end of the radius, its disk-shaped *head* articulates with the capitulum of the humerus (see Figure 7.24A) and the radial notch of the ulna (see Figure 7.25A). The *radial tuberosity* on the medial side is the attachment site for the biceps muscle. An interosseous membrane connects the shafts of the ulna and radius. The *shaft* becomes broader toward its large distal end, which articulates with two of the lunate carpal bones in the hand (see Figure 7.26). The wrist joint is called the *radiocarpal* joint because it joins the forearm (radius) and wrist (carpus). At the lower end of the radius are the prominent *styloid process* (which may be felt on the outside of the wrist where it joins the hand) and the U-shaped *ulnar notch* into which fits the head of the ulna.

FIGURE 7.25

Right radius and ulna. (A) Anterior view. (B) Posterior view. (C) Right lateral view of articulations with the humerus at the elbow.

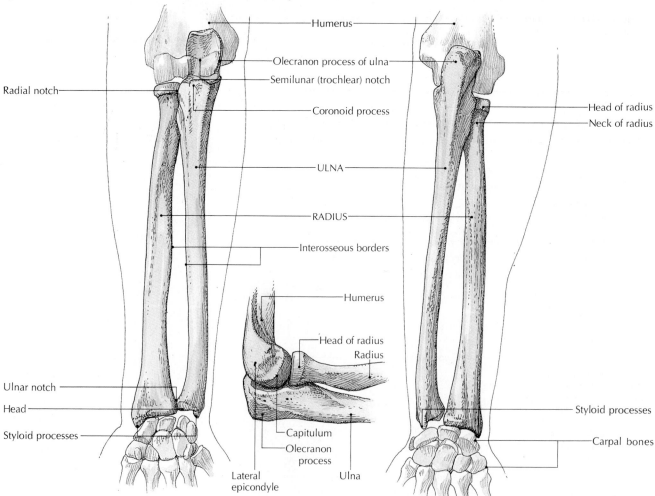

[A]   ANTERIOR   [C]   RIGHT LATERAL   [B]   POSTERIOR

Carpus (wrist) The wrist or *carpus* (Gr. *karpos*, wrist) is composed of eight short bones, connected to each other by ligaments that restrict their mobility primarily to a gliding movement (see Chapter 8).* As shown in Figure 7.26, the carpals are arranged in two transverse rows of four bones each. In the proximal row, from lateral to medial position, are the *scaphoid, lunate, triquetrum,* and *pisiform.* In the distal row are the *trapezium, trapezoid, capitate,* and *hamate.*

The carpal bones as a unit are shaped so that the back of the carpus is convex, and the palmar side is concave. In this concave area, the nine long tendons and a nerve pass from the forearm to the fingers of the hand.

The easily felt small pisiform is a clinically useful landmark. It is actually a sesamoid bone within the tendon of the flexor carpi ulnaris muscle.

Metacarpus and phalanges The 5 *metacarpal* bones make up the skeleton of the palm of the hand, or *metacarpus,* and the 14 *phalanges* are the finger bones (see Figure 7.26).

The *metacarpal* (L. behind the wrist) *bones* are miniature long bones. They are numbered from the lateral (thumb) side as metacarpals I to V. They are arranged as a fan from their proximal ends (bases), which articulate with the distal row of carpal bones, to their distal ends (heads). Each head artic-

*The wrist is the region between the forearm and the hand (i.e., distal to the forearm). A "wrist" watch is not usually worn around the wrist; instead it encircles the distal end of the forearm, just proximal to the head of the ulna.

ulates with the proximal phalanx of a digit. The *meta-carpophalangeal joint* (referred to as the MP joint) forms a "knuckle."

The bones of the digits (fingers) are the 14 *phalanges* (fuh-LAN-jeez; Gr. line of soldiers; singular, *phalanx,* FAY-langks). Each of these bones has a base, shaft, and head. As shown in Figure 7.26, the thumb (digit I) has two phalanges (proximal and distal), and each finger (digits II to V) has three phalanges (proximal, middle, and distal). Except for the thumb, which has only one *interphalangeal joint* (referred to as the IP joint), the digits have a proximal interphalangeal (PIP) joint and a distal interphalangeal (DIP) joint.

Ask Yourself

1 *What bones comprise the upper extremities?*

2 *What bones make up the shoulder girdle?*

3 *What is the main function of the shoulder girdle?*

4 *What are the bones of the arm and hand, and how many of each type of bone are there?*

5 *What are the functions of the ulna and radius?*

FIGURE 7.26

Bones of the right hand. (A) Palmar (ventral, anterior) aspect. (B) Dorsal (posterior) aspect.

Key: **S** = Scaphoid   **L** = Lunate   **TRI** = Triquetral   **P** = Pisiform   **TRU** = Trapezium   **TRO** = Trapezoid   **C** = Capitate   **H** = Hamate

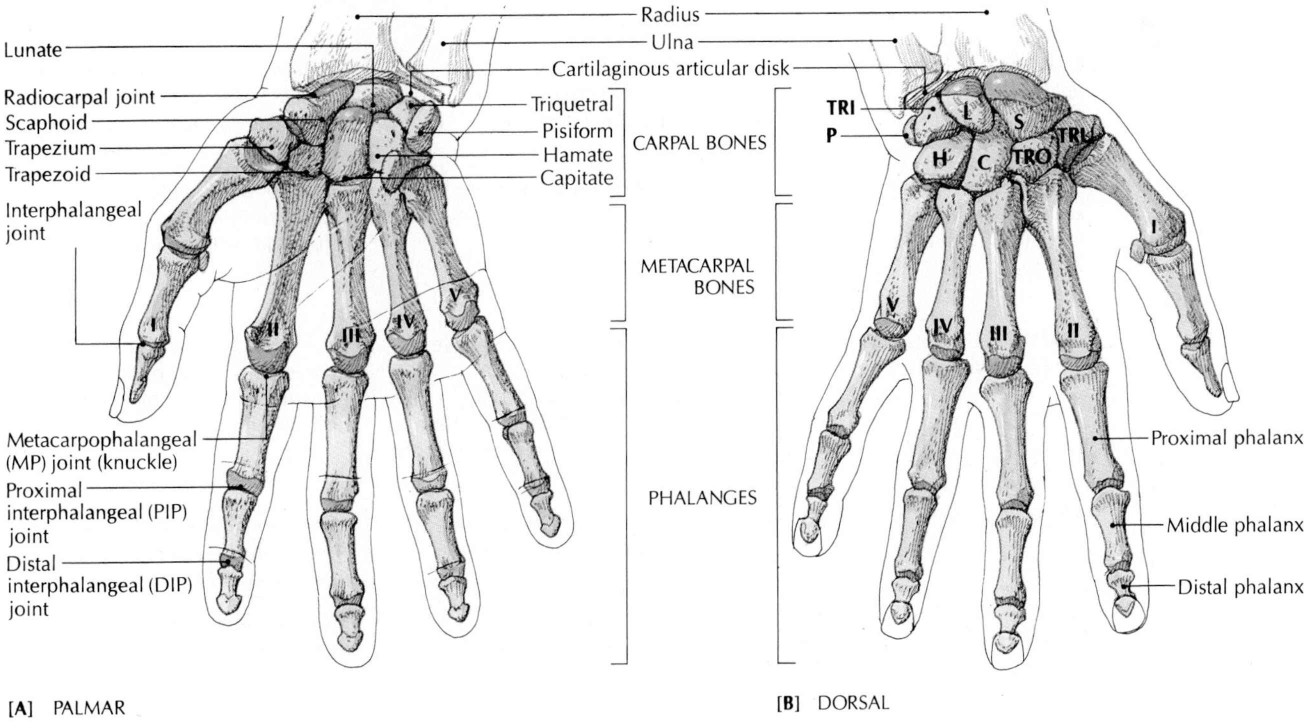

[A] PALMAR

[B] DORSAL

THE LOWER EXTREMITIES (LIMBS)

As part of the appendicular system, the lower extremities or limbs consist of 62 bones (Table 7.8, Figure 7.27; see also Figure 7.1B). These include the hipbones of the lower limb (pelvic) girdle, the femur of the thigh, the tibia and fibula of the leg, the tarsal bones of the ankle, the metatarsals of the foot, and the phalanges of the toes.

Pelvic Girdle and Pelvis

The lower limb girdle, called the **pelvic girdle,** is formed by the hipbones, which are also known as the ***ossa coxae*** (L. hipbones; sing. *os coxa*) or *innominate bones* (see Figure 7.27). The paired hipbones are the broadest bones in the body. They are formed in the adult by the fusion of the ilium, ischium, and pubis. These three bones are separate in infants, children, and young adolescents, but generally fuse between the ages

TABLE 7.8   BONES OF THE LOWER EXTREMITIES (62 BONES)

Bone	Location	Description	Functions
PELVIC GIRDLE			
Hipbone (os coxa, innominate bone) (2)	Hip, between sacrum and femur (Figure 7.27).	Flat bone, the broadest in the body, formed by fusion of ilium, ischium, pubis. With sacrum and coccyx forms pelvis. Forms socket portion of ball-and-socket joint with femur.	Holds body upright. Site of attachment for trunk and lower limb muscles. Transmits body weight to femur. Supports, protects organs.
THIGH			
Femur (2)	Thigh, between hip and knee (Figures 7.27, 7.30).	Typical long bone. Longest, strongest, and heaviest bone in body. Forms ball portion of ball-and-socket joint with pelvic bones. Provides articular surface for knee.	Supports body, especially during running or walking.
Patella (2)	Kneecap, within tendon of extensor of knee joint (Figures 7.30, 7.31).	Sesamoid bone within quadriceps femoris tendon. Shaped like a small plate.	Increases leverage for action of quadriceps muscle by keeping tendon away from axis of rotation of knee. Protects knee.
LEG			
Fibula (2)	Lower leg, between knee and ankle, parallel and lateral to tibia (Figures 7.27, 7.32).	Smaller long bone of lower leg. Articulates proximally with tibia and distally with talus.	Bears little, if any, body weight, but gives strength to ankle joint and limits its movement somewhat.
Tibia (2)	Lower leg (shinbone), between knee and ankle; parallel and medial to fibula (Figures 7.27, 7.32).	Larger long bone of lower leg. Articulates with femur, fibula, talus.	Supports body weight, transmitting it from femur to talus. Aid in locomotion.
ANKLE			
Tarsals (14)	Ankle, heel bones between tibia, fibula, metatarsals (Figure 7.33).	Short bones. 7 in each ankle including talus (ankle), calcaneous (heel), cuboid, navicular, 3 cuneiforms. With metatarsals, form arches of foot.	Bear body weight. Work together as a lever to raise body and transmit thrust during running and walking.
FOOT AND TOES			
Metatarsals (10)	Between ankle, heel, toes (Figure 7.33).	Miniature long bones. 5 in each foot. Form sole; with tarsals, form arches of feet.	Improve stability while standing. Absorb shocks. Bear weight. Aid in locomotion.
Phalanges (28)	Toes, at end of lower limb, next to metatarsals (Figure 7.33).	Miniature long bones. 2 in each big toe, 3 in each other toe. Arranged as in hand.	Provide stability during locomotion.

FIGURE 7.27

The lower extremity, which includes the pelvic (hip) girdle. (A) Anterior view. (B) Posterior view.

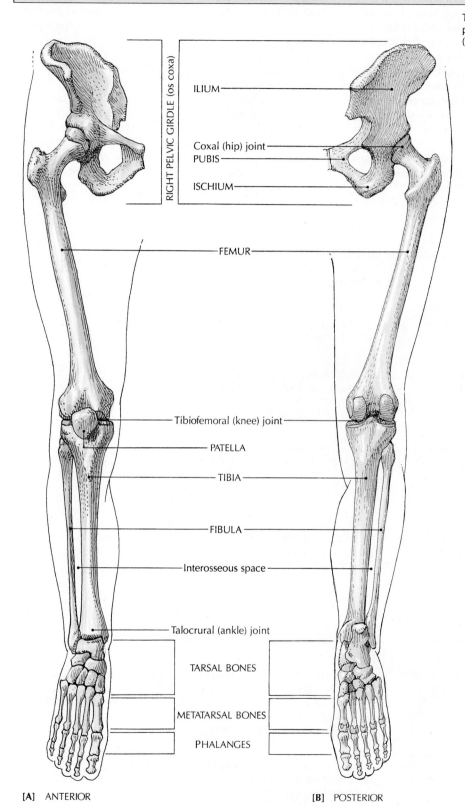

RIGHT PELVIC GIRDLE (os coxa)

ILIUM

Coxal (hip) joint
PUBIS

ISCHIUM

FEMUR

Tibiofemoral (knee) joint

PATELLA

TIBIA

FIBULA

Interosseous space

Talocrural (ankle) joint

TARSAL BONES

METATARSAL BONES

PHALANGES

[A] ANTERIOR

[B] POSTERIOR

of 15 and 17. On the lateral surface of each hipbone is a deep cup, called the *acetabulum* (ass-eh-TAB-yoo-luhm; L. vinegar cup), which is the socket of the ball-and-socket joint with the head (ball) of the femur (see Figure 7.28). The acetabulum is formed by parts of the ilium, ischium, and pubis.

Although the bones of the pelvic and shoulder girdles bear some resemblance, their functions are rather different. Because they have to hold the body in an upright position, the bones of the pelvic girdle are built more for support than for the exceptional degree of movement of the upper extremities.

The bowl-shaped *pelvis* (L. basin) is formed by the sacrum and coccyx posteriorly, and the two hipbones anteriorly and laterally (Figure 7.28). The pelvis is bound into a structural unit by ligaments at the lateral pairs of *sacroiliac joints* between the sacrum and the two ilia, at the *symphysis pubis* (pubic symphysis) between the bodies of the pubic bones, and at the *sacrococcygeal joints* between the sacrum and coccyx. The pubic symphysis is especially important for the structural security of the pelvis.

The basic functions of the pelvis are (1) to provide attachment sites for muscles of the trunk and lower limbs, (2) to transmit and transfer the weight of the body from the vertebral column to the femur of the lower limb, and (3) to support and protect the organs within the pelvis. The size and proportions of the female pelvis should be large enough to

FIGURE 7.28

Pelves, anterior views. (A) Male pelvis. (B) Female pelvis. Note the wider pelvic aperture in the female.

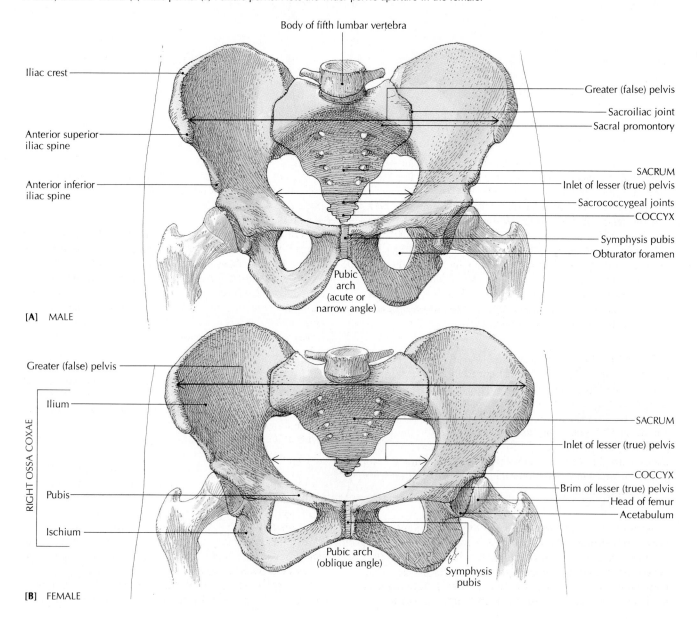

[A] MALE

[B] FEMALE

FIGURE 7.29

Right hipbone (os coxa). (A) Medial view. The borders of the different colored areas illustrate the lines of fusion of sutures at the junctions of the pubis, ischium, and ilium. (B) Lateral view.

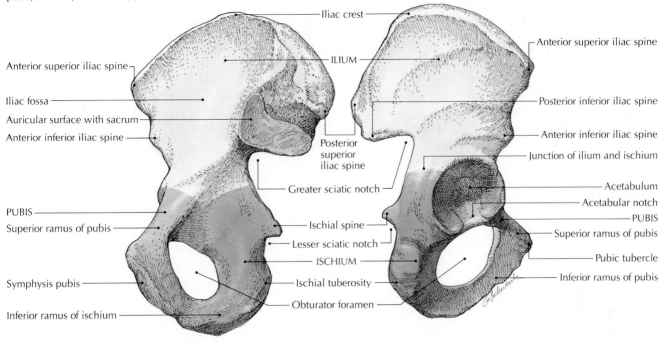

[A] MEDIAL

[B] LATERAL

allow for the passage of a full-term fetus from the uterus through the vaginal canal.

The pelvis is usually divided into the *greater* or *false pelvis* and the *lesser* or *true pelvis,* as shown in Figure 7.28. Obstetricians often refer to the lesser pelvis in the female as the "obstetric pelvis" because it is the critical region during childbirth, providing the opening through which the baby must pass. The junction between the two pelves is the *pelvic brim,* which surrounds the superior pelvic aperture (pelvic inlet). This brim is the bony ring extending from the *sacral promontory* to the top of the symphysis pubis.

Ilium The *ilium* (L. flank) is the largest of the three fused hipbones (Figure 7.29). It forms the easily felt lateral prominence of the hip. (Think of the *l* in ilium; it will help you to remember that it occupies the *l*ateral position in the ossa coxae.) On its superior border is the *iliac crest,* which ends anteriorly as the *anterior superior iliac spine.* On the medial posterior part of the ilium is the *auricular surface* that articulates with the sacrum. Below this auricular surface is the *greater sciatic notch.* The internal surface of the ilium is the concave *iliac fossa.*

Ischium The *ischium* (IHSS-kee-uhm; L. hip joint) is the lowest and strongest bone of the ossa coxae. It is formed by the lower lateral portion of the acetabulum and the *ischial tuberosity,* the bony prominence that bears the weight of the body when we are seated (see Figure 7.29). Extending from the body of the ischium are the slender superior and inferior

rami. Above the body is the *ischial spine,* which is located between the greater sciatic notch of the ilium and the *lesser sciatic notch* of the ischium.

Pubis The bilateral body of the *pubis* (L. *pubes,* adult) is joined together in front to form the *symphysis pubis* (see Figures 7.28 and 7.29). Extending from the body of the pubis are the *superior* and *inferior rami,* which join with the ilium and ischium. The *pubic tubercle* is located on the body of the pubis. A large opening called the *obturator foramen* is bounded by the rami and bodies of the pubis and ischium. Nerves and blood vessels pass through this foramen into the thigh.

Pelves of the male and female Because of the structure of the female pelvis, a woman is able to carry and deliver a child (see Figure 7.28). The female pelvis usually shows the following differences from the male pelvis:

1 The bones are lighter and thinner, and the bony markings are less prominent because the muscles are smaller.

2 The sacrum is less curved and is set more horizontally, which increases the distance between the coccyx and the symphysis, and makes the sacrum broader. The pubic rami are longer, and the ischial tuberosities are set further apart and turned outward. As a result, the *pubic arch* has a wider angle. (This angle is the easiest criterion for distinguishing male from female skeletons.) The combination of all these features creates wider and shallower hips.

3 The pelvis has larger openings. The true pelvis and these openings surround and define the size of the birth canal.

FIGURE 7.30

Right femur. (A) Anterior view. (B) Posterior view.

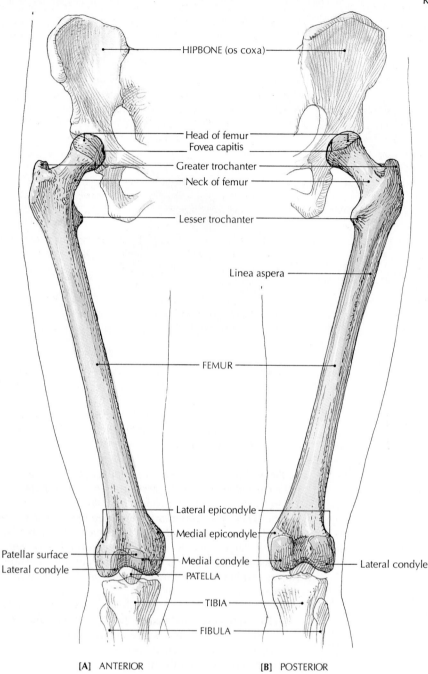

HIPBONE (os coxa)

Head of femur
Fovea capitis
Greater trochanter
Neck of femur
Lesser trochanter

Linea aspera

FEMUR

Lateral epicondyle
Medial epicondyle
Patellar surface
Lateral condyle
Medial condyle
PATELLA
Lateral condyle
TIBIA
FIBULA

[A] ANTERIOR   [B] POSTERIOR

Bones of the Thigh, Leg, and Foot

Completing our study of the lower extremities are the bones of the thigh, leg, and foot, including the femur, patella, tibia, fibula, tarsal bones, metatarsal bones, and phalanges.

Femur The *femur* (FEE-mur; L. thigh), or thighbone, is located between the hip and the knee. It is the strongest,

heaviest, and longest bone in the body. (A person's height is usually about four times the length of the femur.) This strong bone plays an important part in supporting the body, and provides mobility via the hip and knee joints. It can withstand a pressure of 3500 kg/sq cm (1200 lb/sq in.), more than enough to cope with the pressures involved in normal walking, running, or jumping.

The proximal end of the femur consists of a head, neck, and greater and lesser trochanters (Figure 7.30). The *head,* which forms slightly more than half a sphere, articulates with the acetabulum (socket) of the hipbone. In the center of the head is a small depression called the *fovea capitis,* where the ligament and a blood vessel of the head of the femur are attached. If the blood vessel is ruptured due to trauma, the head of the femur may deteriorate. The thin *neck,* connecting the head to the shaft, is a common site for fractures of the femur, especially in elderly people.* At the junction between the head and neck, laterally the *greater trochanter* (Gr. "to run") and medially the *lesser trochanter* are the sites of attachment for some large thigh and buttock muscles.

The *shaft* or body of the femur is slightly bowed anteriorly, and is fairly smooth except for a longitudinal posterior ridge called the *linea aspera* (L. rough line), which provides attachment sites for muscles. At the end of the femur toward the knee are *medial* and *lateral condyles.* The condyles articulate with the tibia. Above the condyles are medial and lateral *epicondyles.* Epicondyles are for the attachment muscles, and condyles function in the movement of joints. The medial condyle is larger than the lateral condyle, so that when the knee is planted during walking, the femur rotates medially to "lock" the knee. Between the condyles on the anterior surface is a slight groove that separates the articular surface from the *patellar surface.* When the leg is bent or extended, the patella (kneecap) slides along this groove (patellar surface).

Patella The *patella* (L. little plate), or kneecap, is located within the quadriceps femoris tendon. This is the tendon of the muscle that extends the leg from the knee. Facets on the deep surface of the patella fit into the groove between the condyles of the femur.

*Nowadays a patient is encouraged to walk with a walker the day after an artificial hip is implanted. In the past, elderly patients often died after a hip fracture because they remained bedridden and developed pneumonia or other serious diseases.

FIGURE 7.31

Patella. (A) Anterior view. (B) Posterior view.

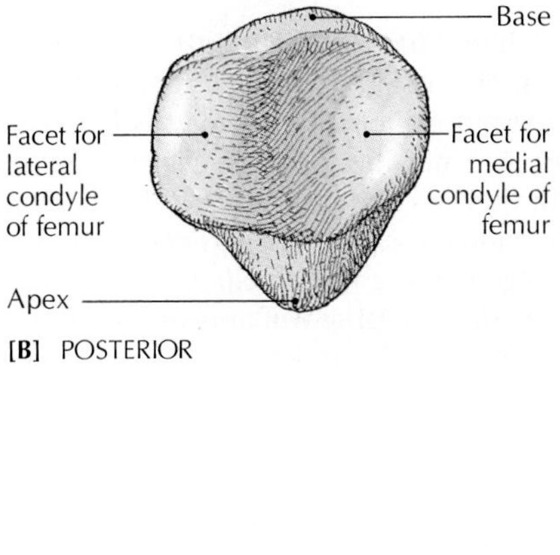

Base

Apex

[A] ANTERIOR

Base

Facet for lateral condyle of femur

Facet for medial condyle of femur

Apex

[B] POSTERIOR

The patella is the largest sesamoid bone in the body. It protects the knee, but more importantly, it increases the leverage for the action of the quadriceps muscle by keeping its tendon further away from the axis of rotation of the knee. The slightly pointed *apex* of the patella lies at the inferior end, and the rounded *base* is at the superior border (Figure 7.31).

Tibia The *tibia* (L. pipe), or shinbone, is located on the anterior and medial side of the leg, between the knee and the ankle. It is the second longest and heaviest bone in the body (the femur is first). This bone supports the body weight, transmitting it from the femur to the talus bone at the ankle joint.

At the proximal end of the tibia the *medial* and *lateral condyles* articulate with the condyles of the femur at the knee joint (Figure 7.32). On the proximal anterior surface is the prominent *tibial tuberosity,* where the patellar ligament is attached.

At the distal end of the tibia is the *medial malleolus,* which articulates medially with the head of the talus. The junction of the talus and medial malleolus forms the easily felt prominence on the medial side of the ankle.

The tibia and fibula are attached throughout their lengths by an *interosseous membrane.* They articulate at both the proximal and the distal tibiofibular joints. The site of the distal articulation is the *fibular notch* on the tibia.

Fibula The *fibula* (L. pin or brooch) is a long, slender bone parallel and lateral to the tibia (see Figure 7.32). It probably is so named because together with the tibia it somewhat resembles the clasp of a pin. The head at its proximal end articulates with the lateral condyle of the tibia, but not with the femur. It articulates distally with the talus.

The slender shaft of the fibula bears little, if any, body weight, and it is not involved in the knee joint. However, the security of the ankle joint depends largely upon the seemingly delicate fibula.

The medial malleolus of the tibia and the prominent *lateral malleolus* of the fibula both articulate tightly with the head of the talus. However, the fibula and talus are even more firmly bound together by ligaments to form a *mortise* (socket), which strengthens the ankle joint *(talocrural joint)* but limits the movement there to bending the foot up or down. Because the head of the talus is slightly wider anteriorly, and the mortise is at the widest part of the talus, the

What are "shin splints"?

Some joggers and long-distance runners suffer from shin splints, a painful condition of the anterior part of the tibia. The pain is caused when the sheaths of swollen leg muscles block the normal blood circulation of the muscles, which in turn causes more swelling and pain. This condition typically afflicts poorly conditioned people who overuse their muscles, and trained athletes who do not warm up properly or go too far beyond their usual limits.

FIGURE 7.32

Right tibia and fibula. (A) Anterior view. (B) Posterior view.

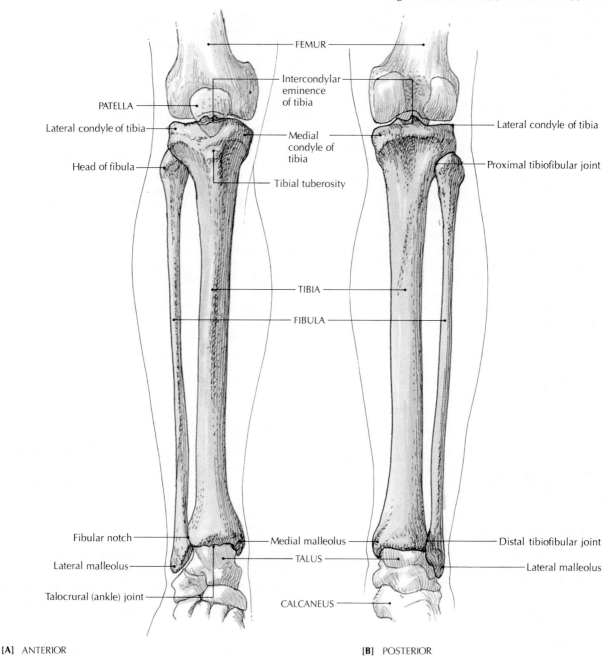

[A] ANTERIOR

[B] POSTERIOR

ankle is most stable when the foot is flexed (as when a skier is crouching down). When the foot is extended, as when standing on tiptoe, the ankle joint is less stable because the mortise is at a narrower part of the talus.

Tarsus The *tarsus* is composed of the seven proximally located tarsal bones of the foot. They are classified as short bones.

The foot, unlike the hand, has relatively little free movement between its bones. The fingers, with their manipulative and gripping roles, are the functionally dominant structures in the hand. However, it is not the toes, but the tarsal and metatarsal bones, with their weight-bearing and locomotive roles, that are the functionally significant structures of the foot. The bones work together as a lever, helping to raise the body and to transmit the thrust during walking and running.

The tarsal bones are the talus, calcaneus, cuboid, navicular, and three cuneiforms (Figure 7.33).

1 The *talus* (TAY-luhss; L. ankle), or anklebone, is the central and highest foot bone. It articulates with the tibia and fibula to form the ankle joint. Together with the calcaneus (heelbone) it receives the weight of the body.

2 The *calcaneus* (kal-KAY-nee-us; L. heel) is commonly called the heel bone. It is the largest tarsal bone and is suited to supporting weight and adjusting to irregularities of the ground. It acts as a lever, providing a site of attachment for the large calf muscles through the *calcaneals* (Achilles) *tendon.*

3 The *cuboid* bone is usually, but not necessarily, shaped like a cube. It is the most laterally placed tarsal bone.

4 The *navicular* (L. little ship) is a flattened oval bone that some people think looks like a boat because of the depression on its posterior surface that houses the head of the talus. The three cuneiform bones line up on its anterior surface.

5 The three *cuneiform* bones (KYOO-nee-uh-form; L. wedge) are referred to as *medial* (first), *intermediate* (second), and *lateral* (third). They line up along the anterior surface of the navicular bone. Their wedgelike shapes contribute to the structure of the transverse arch of the foot.

Metatarsus and phalanges The 5 metatarsal (L. behind the ankle) bones form the skeleton of the sole of the foot, and the 14 phalanges are the toe bones.

The *metatarsus* consists of five metatarsal bones, numbered I to V from the medial side (see Figure 7.33). They are miniature long bones, consisting of a proximal base, a body or shaft, and a distal head. The tarsus and metatarsus are arranged as arches, primarily to improve stability during standing, while the toes provide a stable support during locomotion.

The *phalanges* (fuh-LAN-jeez; Gr. line of soldiers), or toes, like the metatarsals, are miniature long bones, with a base, shaft, and head. There are 14 phalanges in each foot, arranged similarly to the phalanges of the hand. The phalanges of the foot, however, are much shorter, and are quite different functionally from those in the hand, contributing to stability rather than to precise movements. The big toe has two phalanges (proximal and distal) and the other four toes each have three phalanges (proximal, middle, distal).

Arches of the Foot

The sole of your foot is arched for the same reason that your spine is curved: the elastic spring created by the arched bones of the foot absorbs enormous everyday shocks just the way the spring-curved spine does. A running step may flatten the arch by as much as half an inch. When the pressure is released as the foot is raised, the arch springs back into shape, returning about 17 percent of the energy of one step to the next. The arches combine with the calcaneus tendons to reduce by about half the amount of work the muscles need to expend when we run. Without arches, our feet would be unable to move properly or to cushion the normal pressure of several thousand pounds per square inch every time we

take a step. In addition, the arches prevent nerves and blood vessels in the sole of the foot from being crushed.

Longitudinal and transverse arches There are two longitudinal arches of the foot: (1) the medial longitudinal arch, and (2) the lateral longitudinal arch. The so-called *transverse arch* is located roughly along the distal tarsal bone and the tarsometatarsal joints. It runs across the foot between the heel and the ball of the foot. Technically, it is not an arch because only one side contacts the ground. The high *medial longitudinal arch* consists, in order, of the calcaneus, talus, navicular, three cuneiforms, and metatarsals I, II, and III (Figure 7.33c). It runs from the heel to the ball of the foot on the inside. The talus is the keystone, and the calcaneus and the distal ends of the metatarsals are the pillars in contact with the ground. The low *lateral longitudinal arch* consists of the calcaneus, cuboid, and metatarsals IV and V. It runs parallel to the medial longitudinal arch from the heel to the ball of the foot on the outside. The talus is the keystone and the calcaneus and the distal ends of the metatarsal bones are the pillars that contact the ground.

The arches are held in place primarily by strong ligaments, and in part by attached muscles and the bones themselves. The ligaments are largely responsible for the resiliency of the arches.

The *plantar* (underside of the foot) *calcaneonavicular ligament,* extending from the front of the calcaneus to the back of the navicular, is known as the "spring" ligament. It is important because it keeps the calcaneus and navicular bones together, and holds the talus in its keystone position. The security of the high arch and, in a way, of the whole foot, depends on this ligament. If it weakens, and the calcaneus and navicular spread apart, the talus will sag and occupy the space between them; the result is a "flat foot" ("fallen arch"). The tendon of the powerful tibialis posterior muscle comes from the medial side of the foot and passes immediately below the "spring" ligament on its way to its lateral insertion. This tendon is presumed to act as an additional spring, helping to reinforce the ligament during excessive strain.

Weight bearing When you are standing up, half of your body weight is directed to each talus. In turn, each talus transmits half of this weight backward through the calcaneus to the heel. The remaining weight is directed forward to the arches, to be distributed equally at the weight-bearing heads of the metatarsals.

When you walk, your body weight is applied first to your heel as it strikes the ground. As your body leans forward, your weight shifts quickly to the lateral border of your foot and across the metatarsal heads (weight-bearing points) to the first metatarsal bone, associated with the big toe. At this point, your heel leaves the ground, and your big toe imparts a forward thrust. At the same time, your lower limb is extended at the hip and knee joints, and your ankle and big toe are bent downward.

When you run, your weight is borne initially only by the distal ends of the longitudinal arches (balls of the feet). Then the arches spring back, helped by the thrust from the medial toes.

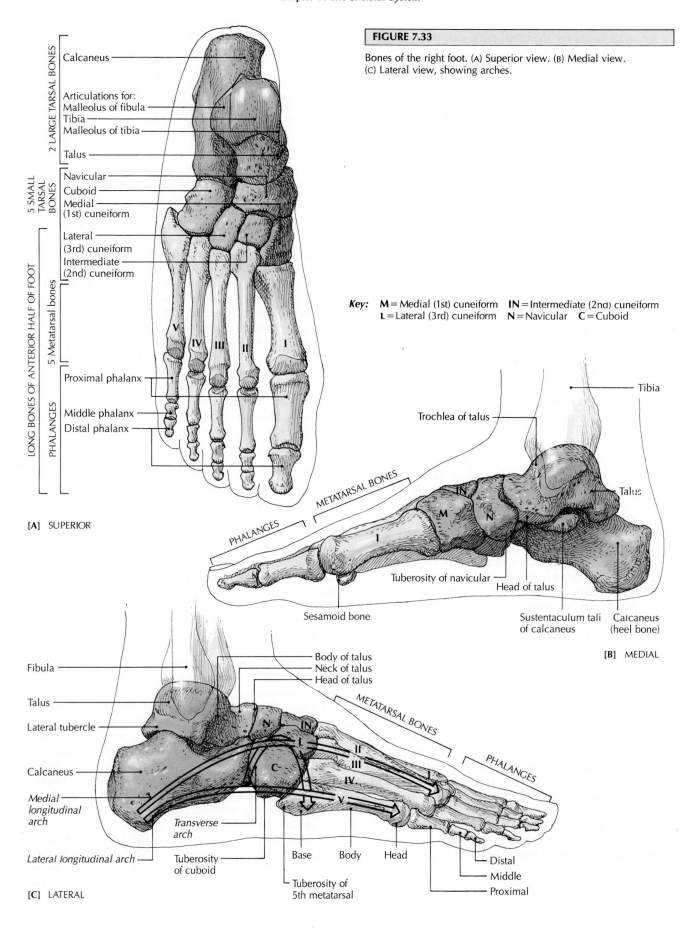

FIGURE 7.33

Bones of the right foot. (A) Superior view. (B) Medial view. (C) Lateral view, showing arches.

2 LARGE TARSAL BONES

Calcaneus

Articulations for:
Malleolus of fibula
Tibia
Malleolus of tibia

Talus

5 SMALL TARSAL BONES

Navicular
Cuboid
Medial (1st) cuneiform
Lateral (3rd) cuneiform
Intermediate (2nd) cuneiform

LONG BONES OF ANTERIOR HALF OF FOOT

5 Metatarsal bones

V IV III II I

PHALANGES

Proximal phalanx
Middle phalanx
Distal phalanx

[A] SUPERIOR

Key: **M** = Medial (1st) cuneiform **IN** = Intermediate (2nd) cuneiform
L = Lateral (3rd) cuneiform **N** = Navicular **C** = Cuboid

Tibia

Trochlea of talus

Talus

METATARSAL BONES

PHALANGES

IN
M
N

I

Sesamoid bone

Tuberosity of navicular

Head of talus

Sustentaculum tali of calcaneus

Calcaneus (heel bone)

[B] MEDIAL

Fibula

Talus

Lateral tubercle

Calcaneus

Medial longitudinal arch

Lateral longitudinal arch

Transverse arch

Tuberosity of cuboid

Tuberosity of 5th metatarsal

Body of talus
Neck of talus
Head of talus

METATARSAL BONES

PHALANGES

N IN
L
II
III
IV
V
C

I

Base Body Head

Distal
Middle
Proximal

[C] LATERAL

PHYSIOLOGICAL AND ANATOMICAL ABNORMALITIES

Fractures

A *fracture* is a broken bone. Children have fractures more often than adults because children have slender bones and are more active. Fortunately, the supple, healthy bones of children mend faster and better than the more brittle bones of older people. (A femur broken at birth is fully united within three weeks, but a similar break in a person over 20 may take four or five months to heal completely.) Usually, broken bones that are reset soon after injury have an excellent chance of healing perfectly because the living tissue and adequate blood supply at the fracture actually stimulate a natural repositioning.

In elderly people, bones contain relatively more calcified bone and less organic material. Consequently, old bones lose their elasticity and they break more easily. A fall that a child hardly notices can be serious in an elderly person. "To fall and break a hip," a common disaster among the elderly, could frequently be better stated, "to break a hip and fall," because the fragile old bones may crack merely under the strain of walking, making the legs give way. The hip (actually the neck of the femur, the most fragile part in elderly people) may be broken before the body hits the ground.

A fractured bone goes through several stages of healing. But even before healing can begin properly, the fragments of the broken bone must be manipulated, or *reduced*, back into their original positions by a physician. Usually the bone is immobilized by a cast, splint, or traction, and in severe cases, surgery and a continuing program of physical therapy may be necessary. Typically, the following steps take place:

1 Adjacent tissue and blood vessels in the periosteum, osteon system, and marrow cavity that cross the fracture are damaged, and hemorrhaging occurs. Some osteocytes of the osteon system, periosteum, and marrow tissue in the immediate vicinity of the fracture die, and their lacunae appear empty. A blood clot, or **hematoma** (L. *hemato*, blood + *oma*, tumor), forms a few hours after the fracture occurs (see part A of the drawing on page 202).

2 Fibroblasts enter the area, and after a few days new fibrous tissue, called a **callus** (L. hard skin), develops at the fracture site. The callus eventually closes the gap between the fragmented bones altogether. The callus that forms around the break is known as the *external callus;* the callus between the broken ends and in the marrow cavity is the *internal callus* (Part B).

3 After a week or so, rapidly developing osteogenic cells from the surrounding periosteum differentiate into osteoblasts and form new bony trabeculae that begin to knit the fragments together. (When surgeons operate on a broken bone they are careful to replace the periosteum, since it is crucial to proper healing.) Osteogenic cells farther from the fracture differentiate into cartilage-making **chondroblasts,** depositing new cartilage in the outer collar around the break (Part C).

4 The cartilage is gradually replaced with new bone. When the cartilage is completely replaced, the callus begins to be **remodeled.** Spongy bone appears first in the remodeling process, but under stress it is eventually converted into compact bone that firmly repairs the fracture (Part D). Fractures heal faster when some pressure is placed on the mending bone.

5 When the break is cemented, *osteoclasts* resorb the bony external callus and any part of the internal callus that interferes with the normal functioning of the marrow cavity. Usually, the only remaining sign of a healed simple fracture is a slight bump on the bone directly over the break point, where extra osteoblasts have been deposited. (The next time you're riding on a highway, notice how the breaks in the road are sealed with a little too much asphalt, usually leaving a slight bump.)

Fractures of the vertebral column Many fractures of the vertebral column may be serious in themselves, but the real danger lies in injury to the spinal cord, which can result in paralysis or death. (Spinal cord injuries will be discussed in Chapter 12.)

The most common type of fracture is a *compression fracture,* which crushes the body of one or more vertebrae. Compressions often occur where there is the greatest spinal mobility: the middle or lower regions of the vertebral column, and near the point where the lumbar and thoracic regions meet. Although a compression fracture crushes the body of a vertebra, the vertebral arches and the ligaments of the spine remain intact. As a result, the spinal cord is not injured.

In contrast to compression fractures, *extension fractures* and dislocations involve a pulling force, usually affecting the posterior portions of the vertebral column. When the neck is severely hyperextended (bent backward), as in a "whiplash" injury, the atlas may break at several points, and further extension may break off the arch of the axis at the isthmus. An even greater force may rupture one of the ligaments and *annulus fibrosus* (outer ring) of the C2/C3 intervertebral disk. Such a great force separates the skull, atlas, and axis from the rest of the vertebral column, and the

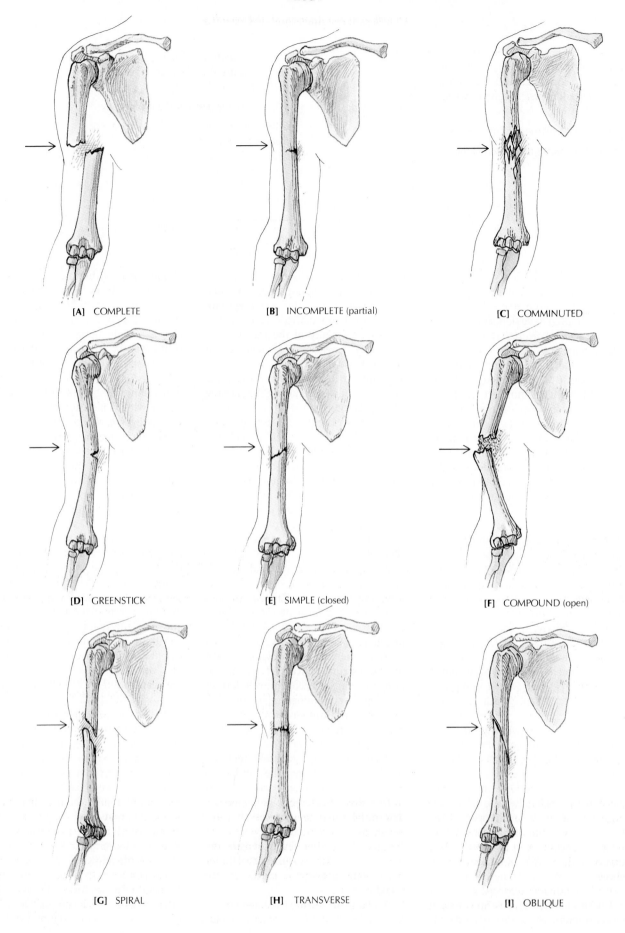

[A]   COMPLETE

[B]   INCOMPLETE (partial)

[C]   COMMINUTED

[D]   GREENSTICK

[E]   SIMPLE (closed)

[F]   COMPOUND (open)

[G]   SPIRAL

[H]   TRANSVERSE

[I]   OBLIQUE

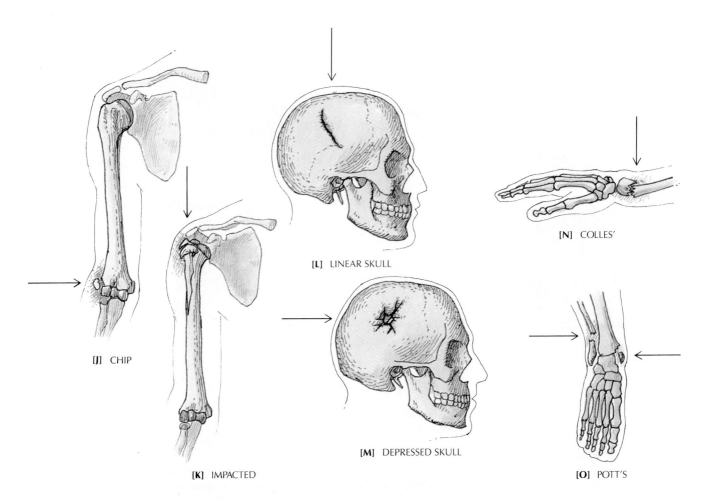

[J] CHIP

[K] IMPACTED

[L] LINEAR SKULL

[M] DEPRESSED SKULL

[N] COLLES'

[O] POTT'S

Fractures can be classified according to the type and complexity of the break, the location of the break, and certain other special features. The following commonly used types and classifications are shown in the respective illustrations.

(A) *Complete.* The bone breaks completely into two pieces.
(B) *Incomplete (partial).* The bone does not break completely into two or more pieces.
(C) *Comminuted.* The bone is splintered or crushed into small pieces.
(D) *Greenstick.* The bone is broken on one side and bent on the other. Common in children.
(E) *Simple (closed).* The bone is broken, but it does not break through the skin.
(F) *Compound (open).* The bone is broken and cuts through the skin.

(G) *Spiral.* The bone is broken by twisting.
(H) *Transverse.* The bone is broken directly across, at a right angle to the long-bone axis.
(I) *Oblique.* The bone is broken at a slant, at approximately a 45-degree angle to the long-bone axis.
(J) *Chip.* The bone is chipped where a protrusion is exposed.
(K) *Impacted.* The bone is broken when one part is forcefully driven into another, as at a shoulder or hip.
(L) *Linear skull.* The skull is broken in a line, lengthwise on the bone.
(M) *Depressed skull.* The skull is broken by a puncture, causing a depression below the surface.
(N) *Colles'.* The distal end of the radius is displaced posteriorly.
(O) *Pott's.* The distal part of the fibula and medial malleolus are broken.

spinal cord is usually severed in the process. Hyperextension injuries usually do not occur in the thoracic region because of the support of the ribs and the relative immobility of the thoracic vertebrae.

Fracture usually accompanies dislocations of vertebrae because the thoracic and lumbar articular processes interlock. The spinal cord is not necessarily severed when cervical vertebrae are dislocated because the vertebral canal in the cervical region is wide enough to allow some displacement without damaging the cord.

The primary treatment for injuries of the vertebral column is usually immobilization, which allows the bone to heal and prevents damage to the spinal cord. Surgery may be necessary to relieve pressure or repair severely damaged vertebrae or tissues. Exercises to strengthen back muscles are ordinarily prescribed after the fracture is healed.

HOW A FRACTURE HEALS

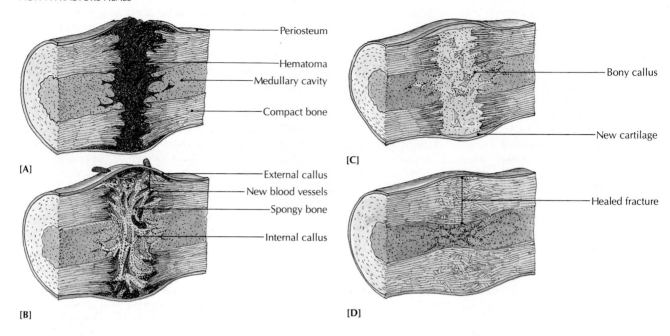

[A]

Periosteum
Hematoma
Medullary cavity
Compact bone

[B]

External callus
New blood vessels
Spongy bone
Internal callus

[C]

Bony callus
New cartilage

[D]

Healed fracture

Herniated disk

Herniated disk (also called ruptured or slipped disk) occurs when the soft, pulpy center *(nucleus pulposus)* of an intervertebral disk protrudes through a weakened or torn surrounding outer ring *(annulus fibrosus)* on the posteriolateral side of the disk. The nucleus pulposus pushes against a spinal nerve, or occasionally, on the spinal cord itself. This produces a continuous pressure on the spinal cord, which may cause permanent injury. Actually, nothing "slips"; the nucleus pulposus pushes out. Herniated disks occur most often in adult males. They may be caused by a straining injury or by degeneration of the intervertebral joint. The sacral or lumbar regions are usually affected, but herniation may occur anywhere along the spine. Sharp pain usually accompanies a herniated disk, and because roots of spinal nerves can be involved, the pain may radiate beyond the primary low back area to the buttocks, legs, and feet.

Treatment usually consists of bed rest, sometimes including traction. Heat applications, a regulated exercise program, and muscle-relaxing or pain-killing drugs may be prescribed. If such traditional treatment is ineffective, surgery is done to remove the protruding nucleus pulposus. To gain access to the hernia, a portion of the vertebral arch (lamina) is removed. Such an operation is called a *laminectomy*.

If a spinal fusion is required, bone chips from the ilium are placed over the laminae. The bone that develops from the chips fuses to form a splint.

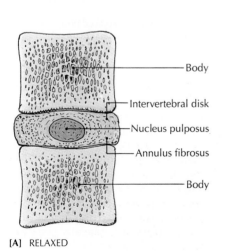

[A] RELAXED

Body
Intervertebral disk
Nucleus pulposus
Annulus fibrosus
Body

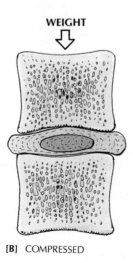

WEIGHT

[B] COMPRESSED

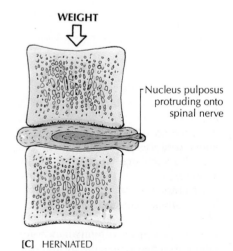

WEIGHT

[C] HERNIATED

Nucleus pulposus protruding onto spinal nerve

Hydrocephalus

Sometimes the calvarial bones, along with the fontanels, close earlier or later than expected. If they fuse too early, brain growth may be retarded by the excessive pressure. This condition is called *microcephalus* (Gr. *mikros*, small + *kephale*, head). This condition can be alleviated by removing the bone and widening the sutures. A contrasting problem is **hydrocephalus** (Gr. *hudor*, water + head), commonly called "water on the brain." It is usually a congenital condition in which an abnormal amount of cerebrospinal fluid accumulates around the brain and brain cavities, causing an enlargement of the skull and pressure on the brain. Fusion of the skull sutures is delayed by the increased volume of the cranial cavity.

Physicians attend to the problem by inserting replaceable plastic tubes with pressure valves in the brain, which drain away excess fluid into a vein or body cavity. The valve prevents excessive rises in pressure and allows sufficient fluid to be retained. Even with drainage procedures, mental retardation and vision loss may occur. Without drainage, the condition is usually fatal.

Spina bifida

Spina bifida (SPY-nuh BIFF-ih-duh; L. *bifidus*, split into two parts), or cleft spine, affects one of every 1000 children. In its severe form, it is the most common crippler of newborns. Spina bifida is a condition in which the neural arches of one or more vertebrae do not close completely during fetal development. In seri-

ous cases, the spinal cord and nerves in the area of the defective vertebrae form a fluid-filled sac, called a myelomeningocele, which protrudes through the skin. Because the myelomeningocele is covered by only a thin membrane, the protruding spinal nerves are easily damaged or infected.

Until the mid-1970s, many newborns with spina bifida died of a cerebral infection shortly after birth, or suffered paralysis, mental retardation, and a loss of function in the nerves controlling the urinary bladder, intestines, and legs. Now, however, many spina bifida babies can be operated on a day or two after they are born, with a high rate of success.

Infants born with spina bifida generally suffer from hydrocephalus also.

Spinal curvatures

Three abnormal spinal curvatures are kyphosis, lordosis, and scoliosis. **Kyphosis** (Gr. hunchbacked) is a condition where the spine curves backward abnormally, usually at the thoracic level. A characteristic "hunchbacked" or "roundbacked" appearance results. Adolescent kyphosis is the most common form. It generally results from infection or other disturbances of the vertebral epiphysis during the active growth period. Adult kyphosis (hunchback) is generally caused by a degeneration of the intervertebral disks, resulting in collapse of the vertebrae, but many other causes, such as poor posture and tuberculosis of the spine, may be responsible.

Lordosis (Gr. bent backward), also known as "swayback," is an exagger-

ated forward curvature of the spine in the lumbar area. Among the causes of lordosis are the great muscular strain of advanced pregnancy, an extreme "potbelly" or general obesity that places abnormal strain on the vertebral column, tuberculosis of the spine, rickets, and poor posture.

The most common spinal curvature is **scoliosis** (Gr. crookedness), an abnormal lateral curvature of the spine in the thoracic, lumbar, or thoracolumbar portion of the vertebral column. It is interesting to note that curves in the thoracic area are usually convex to the right, and lumbar deformities are usually convex to the left.

Functional, or postural, *scoliosis* is caused by uneven leg length or simply by poor posture, but *structural scoliosis* has more complicated and serious causes. Among the causes are birth defects (*hemivertebra* is a congenital condition characterized by the absence of the lateral half of a vertebra), polio (which causes a paralysis of muscles on one side of the body, so that the spine curves toward the other side), and cerebral palsy and muscular dystrophy (which destroy or distort vertebrae). The most common form of structural scoliosis is called *idiopathic* (Gr. "a disease having its own origin") *scoliosis* because its cause is unknown. Beginning in early childhood or adolescence, it becomes progressively worse until the skeleton matures. If scoliosis is untreated, it may lead to breathing difficulties as the lungs are crowded. Back pain and degenerative arthritis of the spine may also result.

MEDICAL TERMINOLOGY

ABLATION (L. *ablatus*, removed) The surgical removal of part of a structure, such as part of a bone.

BONE BIOPSY (Gr. *bios*, life + *opsia*, examination) The surgical removal of a small piece of bone for microscopic examination.

BONE-MARROW TEST The withdrawal of bone marrow from the medullary cavity of a bone for microscopic examination.

BUNION An inflamed protrusion of the inner side of the joint of the big toe.

CERVICAL RIB An overdevelopment of the costal projection of the seventh cervical vertebra. It resembles a rib, and can be a separate bone.

COLLES' FRACTURE A specialized type of fracture that occurs at the distal end of the radius. It often occurs when a person extends the hands to break a fall.

CRANIOTOMY The surgical cutting or removal of part of the cranium.

CREPITATION (L. *crepitare*, to crackle) The grating sound caused by the movement

of fractured bones or by other bones rubbing together.

GENU VALGUM (L. *genu*, knee + *valgus*, bowlegged) Bowleggedness; a deformity typical of rickets.

GENU VARUM (L. *varus*, bent inward) Knock-kneed.

KINESIOLOGY (Gr. *kinema*, motion) The study of movement and the active and passive structures involved.

OSTECTOMY The surgical excision of a bone.

OSTEOCLASIS The surgical refracture of an improperly healed broken bone.

OSTEOPLASTY The surgical reconstruction or repair of a bone.

PELVIMETRY The measurement of the pelvic cavity and birth canal by a physician prior to the birth of a child. The procedure determines if the opening of the mother's lesser pelvis is large enough to allow the passage of the child's head and shoulders.

POTT'S DISEASE Tuberculosis of the spine, which may result in a partial destruction of vertebrae and a spinal curvature.

POTT'S FRACTURE A fracture and dislocation of the lower fibula. It is usually caused by the forceful turning outward of the foot, which destabilizes the ankle joint.

REDUCTION The nonsurgical manipulation of fractured bones to return (reduce) them to their normal positions.

REPLANTATION The reattachment of a severed limb.

SHOULDER SEPARATION A separation of the acromioclavicular joint, not the shoulder. It is usually a serious injury only when the accompanying ligaments are torn.

SPINAL FUSION The fusion of two or more vertebrae.

SPONDYLITIS (Gr. *spondulos*, vertebra) Inflammation of one or more vertebrae.

SYNDACTYLISM (Gr. *syn*, together + *daktulos*, finger) The whole or partial fusion of two or more fingers or toes.

SUMMARY

The most obvious function of the skeleton is to hold the body up. It also protects the inner organs and passageways, and it acts as a system of levers that allows us to move. The bones supply reserve calcium and phosphate, and red blood cells are produced within the bone marrow.

General features and surface markings of bones

The markings on the surface of any bone give clues about the bone's function. Some important features of bones are *processes,* or outgrowths on bones, such as a condyle, crest, or trochanter; *openings* to bones, such as a foramen or meatus; and *depressions* on bones, such as a fossa, groove, or notch (see Table 7.1).

Divisions of the skeleton

1 The skeleton (206 bones) is divided into two major portions: the axial skeleton (80 bones) and the appendicular skeleton (126 bones). They are joined together at the shoulder girdle and pelvic girdle to form the overall skeleton.

2 The *axial skeleton* forms the longitudinal axis of the body. It is made up of the skull, vertebral column, sternum, and ribs.

3 The *appendicular skeleton* is composed of the upper and lower extremities, which include the shoulder and pelvic girdles.

4 Each *upper extremity* consists of the pectoral (shoulder) girdle, upper arm bone, two forearm bones, and the wrist and hand bones. Each *lower extremity* consists of the pelvic (hip) girdle, upper leg (thigh) bone, two lower leg bones, and the ankle and foot bones.

THE AXIAL SKELETON

The skull

1 The *skull* is usually defined as the skeleton of the head, with or without the mandible (lower jaw). The skull can be divided into the *cranial skull* and the *facial skull.* The skull *protects* many structures, including the brain and eyes, *provides points of attachment* for muscles involved in eye movements, chewing, swallowing, and other movements, and *supports* various structures such as the mouth, pharynx, and larynx.

2 Except for the mandible (and the ear ossicles and hyoid), the skull bones are joined together by *sutures,* seamlike joints that make the bones of an adult skull immovable. The four major skull sutures are the *coronal, lambdoidal, sagittal,* and *squamous* sutures.

3 The membrane-covered spaces between incompletely ossified sutures are the four *fontanels: anterior, posterior, anterolateral, posteriolateral.*

4 The 8 *cranial bones* are the *frontal,* two *parietal, occipital,* two *temporal, sphenoid,* and *ethmoid.* The sutural bones are also considered part of the cranium.

5 In the interior of the ethmoid, maxillary, sphenoid, and frontal bones are four pairs of air cavities called *paranasal sinuses.*

6 The 14 *facial bones* are two *inferior nasal conchae, vomer,* two *palatines,* two *maxillae,* two *zygomatic,* two *nasal,* two *lacrimal,* and *mandible.* The *hyoid* and the 6 *ear ossicles* are considered with the facial bones.

The vertebral column

1 The *vertebral column,* or spine, is the skeleton of the back. The spine is composed of 26 separate bones called *vertebrae.*

2 The main functions of the vertebral column are to protect the spinal cord and nerves, support the weight of the body, and keep the body erect. Resilient *intervertebral disks* act as shock absorbers, and protect the vertebrae.

3 The adult vertebral column has four *curves,* which provide spring and resiliency.

4 A "typical" vertebra consists of a body, a vertebral (neural) arch, and several processes. The arch and the body meet to form an opening called the *vertebral foramen.* The sequence of foramina forms the *vertebral canal,* which encloses the spinal cord. The processes are attachment sites for muscles and ligaments.

5 The *cervical vertebrae* are the 7 between the skull and the thorax. The *atlas* supports the head and permits the "yes" motion; the *axis* permits the "no" motion. The 12 *thoracic vertebrae* articulate with the ribs. The 5 *lumbar vertebrae* are the largest and strongest vertebrae, and provide attachments for lower back muscles. The adult *sacrum,* composed of 5 fused vertebral bodies, supports both the spinal column and the pelvis. The *coccyx* consists of 3 to 5 fused vertebrae.

The thorax

1 The *thorax,* or chest, is formed by the bodies and intervertebral disks of 12 thoracic vertebrae posteriorly, 12 pairs of ribs, 12 costal cartilages, and the sternum anteriorly.

2 The thoracic cage protects inner organs, provides a point of attachment for some bones and muscles of the upper extremities, and provides a flexible breathing mechanism.

3 The *sternum,* or breastbone, consists of a manubrium, body, and xiphoid process. The articulation of the manubrium with the clavicle is the upper attachment of the axial skeleton to the appendicular skeleton.

4 The *ribs* are usually composed of 12 pairs, all of which articulate posteriorly with the vertebral column. The *true ribs* (1 to 7) attach to the vertebrae and sternum, but the *false ribs* (8 to 12) attach directly only to the vertebral column. Ribs 11 and 12 are further called *floating ribs,* because they are not even indirectly attached to the sternum or ribs above.

5 A *typical rib* is composed of a head, neck, and shaft.

THE APPENDICULAR SKELETON

The upper extremities (limbs)

1 Each *upper extremity* of the appendicular skeleton includes the scapula and clavicle of the upper limb (shoulder) girdle, humerus of the arm, radius and ulna of the forearm, carpal bones of the wrist, metacarpals of the palm, and phalanges of the fingers. Some functions of the upper extremity include *balancing* while the body is moving, *grasping* of objects, and the *manipulation* of objects.

2 The upper limb girdle is also known as the *shoulder girdle* or *pectoral girdle.* It consists of the clavicle and scapula.

3 The (upper) arm bone is the humerus, and the forearm bones are the ulna and radius. In each wrist are 8 carpals, in each palm 5 metacarpals, in each thumb 2 phalanges, and in each finger 3 phalanges.

The lower extremities (limbs)

1 The skeleton of the *lower extremity* or limb consists of the hipbones of the lower limb (pelvic) girdle, the femur of the thigh, the tibia and fibula of the leg, the tarsal bones of the ankle, the metatarsals of the foot, and the phalanges of the toes. Among the functions of the lower extremity are *movement,* such as in walking, and *balancing,* such as in standing.

2 The *pelvic girdle* is formed by the hipbone (os coxa or innominate bone), which helps hold the body in an upright position. The *pelvis* is formed by the sacrum, coccyx, and hipbones.

3 The *hipbones* (ossa coxae) are formed in the adult by the fusion of the ilium, ischium, and pubis.

4 The female pelvis is lighter and wider than the male pelvis to enable a woman to carry and deliver a child.

5 The bones of the legs and feet are the femur (2), patella (2), tibia (2), fibula (2), tarsus (14), metatarsus (10), and phalanges (28).

6 The *arches* of the foot provide strength and resiliency. The two true arches are the *medial longitudinal arch* and the *lateral longitudinal arch.* The so-called transverse arch contacts the ground on only one side.

Physiological and anatomical abnormalities

1 A *fracture* is a broken bone. It goes through several stages of healing, including the formation of a fibrous callus, which eventually closes the gap between the broken bones. New bone cells replace the callus during the final stages of healing.

2 Commonly used classification of fractures include *complete* and *incomplete, simple* and *compound, transverse* and *oblique.* The broken bone of a compound fracture breaks through the skin.

3 The most common type of vertebral fracture is a *compression fracture,* which crushes the body of one or more vertebrae. An *extension fracture* involves a pulling force, usually affecting the posterior portions of the vertebral column.

4 A *herniated disk* is usually called a ruptured or slipped disk. It occurs when the soft center of an intervertebral disk protrudes through a weakened portion of the disk.

5 *Hydrocephalus,* commonly called "water on the brain," is a condition in which an abnormal amount of cerebrospinal fluid accumulating in the skull causes the skull to enlarge.

6 *Spina bifida* is a condition in which the neural arches of one or more vertebrae do not close completely during fetal development.

7 Three abnormal spinal curvatures are *kyphosis, lordosis,* and *scoliosis.*

UNDERSTANDING THE FACTS

1 What are the unpaired bones of the skull?

2 What is included in an extremity?

3 What are the two main divisions of the skull?

4 Which fontanel persists longest after birth?

5 Which are the movable bones of the skull?

6 What are the main functions of the spinal column?

7 What is the difference between the pelvic girdle and the pelvis?

8 What is spina bifida?

9 Match the following surface features of bones with their descriptions:

a	crest	**1**	A large tubular opening, not necessarily through a bone.
b	condyle	**2**	A groove or cleft between adjacent bones.
c	meatus	**3**	A wide prominent ridge, often on the border of a bone, attached to connective tissue.
d	fossa	**4**	A rounded, knuckle-shaped projection that forms a joint.
e	fissure	**5**	A shallow depression on a bone.

Choose the correct answers for the following:

10 The skull bone that protects the posterior part of the brain and forms a passageway for the spinal cord is the
a temporal **c** occipital **e** ethmoid
b parietal **d** sphenoid

11 The strongest and largest vertebrae are the
a cervical **c** lumbar **e** coccygeal
b thoracic **d** sacral

12 The bone that holds the shoulder and arm away from the thorax, enabling the arm to swing freely, is the
a carpal **c** humerus **e** scapula
b metacarpal **d** clavicle

13 The bone that enables the lower arm to rotate is the
a humerus **c** ulna **e** scapula
b radius **d** clavicle

UNDERSTANDING THE CONCEPTS

1 What are some of the differences in the skeleton of the child and adult? The adult male and female?

2 How is timing important in the development of the skull?

3 In what ways do cervical, thoracic, and lumbar vertebrae differ?

4 What problems would you have if your spinal column lacked its normal curvature?

SUGGESTED READING

BASMAJIAN, JOHN V., *Primary Anatomy*, 7th ed. Baltimore: Williams & Wilkins, 1976.

BROWN, B. J., *Complete Guide to Prevention and Treatment of Athletic Injuries.* Englewood Cliffs, N.J.: Parker, 1972.

CLEMENTE, C. D., ed., *Gray's Anatomy*, 30th American ed. Philadelphia: Lea & Febiger, 1985.

EVANS, F. G., ed., *Studies in the Anatomy and Function of Bones and Joints.* New York: Springer-Verlag, 1966.

GARDNER, ERNEST, DONALD J. GRAY, AND RONAN O'RAHILLY, *Anatomy: A Regional Study of Human Structure,* 4th ed. Philadelphia: Saunders, 1975.

HOGAN, L., AND I. BELAND., "Cervical Spine Syndrome." *American Journal of Nursing* 76 (1976):1104.

LANGEBARTEL, D. A., *The Anatomical Primer: An Embryological Explanation of Human Gross Morphology.* Baltimore: University Park Press, 1977.

MOORE, KEITH L., *Clinically Oriented Anatomy.* Baltimore: Williams & Wilkins, 1980.

ROMANES, G. J., ed., *Cunningham's Textbook of Anatomy,* 12th ed. Oxford: Oxford University Press, 1981.

SNELL, RICHARD S., *Clinical Anatomy for Medical Students.* Boston: Little, Brown, 1973.

THOMPSON, C. W., *Manual of Structural Kinesiology,* 9th ed. St. Louis: Mosby, 1981.

TRUETA, J., *Studies in the Development and Decay of the Human Frame.* Philadelphia: Saunders, 1968.

WILSON, F. C., *The Musculoskeletal System: Basic Processes and Disorders,* 2nd ed. Philadelphia: Lippincott, 1983.

ZORAB, P. A., *Scoliosis.* New York: Academic Press, 1973.

8
Articulations

LEARNING OBJECTIVES

1 Define articulation, and explain the basis for the two principal methods of classifying articulations.

2 Describe the three types of fibrous joints, and give an example of each.

3 Describe the two types of cartilaginous joints, and give an example of each.

4 Describe the four basic structural features of a synovial joint and their functions.

5 Give the functions of bursae and tendon sheaths.

6 Explain those factors that inhibit movement at a synovial joint.

7 Define the basic terms used in describing movements that take place at synovial joints.

8 Differentiate among uniaxial, biaxial, and multiaxial joints.

9 Describe the structure and movements of the six major types of synovial joints.

10 Describe the basic structure and possible movements of the jaw, hip, shoulder, and knee joints.

11 Compare the stability and flexibility of the hip and shoulder joints.

12 Explain why the knee joint is injured so often in active sports.

13 Explain how nerves and blood vessels reach the joints and how they function.

14 Describe the major forms of arthritis.

Bones give the body its structural framework, and muscles give it its power, but movable joints or *articulations* (L. *articulus,* small joint) provide the mechanism that allows the body to move. (Not all joints are movable, however.) An articulation is the place where two adjacent bones meet, or where adjacent cartilages or adjacent bones and cartilages are joined, *even if the joint does not allow movement.* The effectiveness of the articular system involves the exquisite coordination of the nervous, muscular, and skeletal systems. To fully appreciate the usefulness of movable joints, try to walk or sit without bending your knees, or eat dinner without bending your elbow or wrist.

CLASSIFICATION OF JOINTS

Joints are classified by two methods. One way to classify joints is by the extent of their *function,* that is, their *degree of movement.* According to this system, a joint may be immovable, slightly movable, or freely movable. An immovable joint, called a **synarthrosis** (Gr. *syn.,* together + *arthrosis,* articulation), is an articulation in which the bones are rigidly joined together. A slightly movable joint, called an **amphiarthrosis** (Gr. *amphi,* on both sides), allows limited motion. A freely movable joint is known as a **diarthrosis** (Gr. *dia,* between).

Another way to classify joints is by their *structure.* This classification is based on the presence or absence of a joint cavity, and the kind of supporting tissue that binds the bones together. Based on structure, three types of joints are recognized: fibrous, cartilaginous, and synovial.

Table 8.1 sums up the classification of joints based on structure. As you can see, fibrous joints are generally synar-

throses, cartilaginous joints are generally amphiarthroses, and synovial joints are generally diarthroses. However, the structural and functional categories are *not always* equivalent. Where can you see inconsistencies between the two ways of classifying joints?

FIBROUS JOINTS

Fibrous joints lack a joint cavity, and fibrous connective tissue unites the bones. Because they are joined together tightly, fibrous joints are mainly immovable in the adult, although some of them do allow slight movement. Three types of fibrous joints are generally recognized: sutures, syndesmoses, and gomphoses.

Sutures

A *suture* is usually such a tight union in an adult that movement rarely occurs between the two bones. Because sutures are found only in the skull, they are sometimes called "skull type" joints (see Figure 7.4B). Movement at sutures can occur in fetuses and young children, since the joints have not yet grown together. In fact, the flexibility of the skull is necessary for a newborn baby to be able to pass through its mother's narrow birth canal. The flexibility of cranial sutures in fetuses and children is also important to allow for the growth of the brain. In adults, the fibers of connective tissue between bones are replaced by bone, and the bones become permanently fused. This fusion provides complete protection for the brain from external factors. Such a sealed joint is called a **synostosis** (Gr. *syn,* together + *osteon,* bone).

TABLE 8.1   CLASSIFICATION OF JOINTS

Structural classification	Extent of movement (functional classification)*	Type of structure	Type of movement†	Examples
FIBROUS Fibrous connective tissue unites articulating bones. No joint cavity.	Mostly immovable, some slightly movable; usually synarthroses.	**Suture:** found only in skull; fibrous tissue between articulating bones in children, but permanently fused in adults.	Some movement in fetuses and young children; none in adults.	Cranial sutures, such as coronal suture between frontal and parietal bones.
		Syndesmosis: articulating bones held together (but not touching) by fibrous or interosseous ligaments.	Slight movement: twisting of forearm (pronation, supination).	Inferior tibiofibular joint; interosseous ligament between shafts of radius and ulna.
		Gomphosis: a peg fitting into a socket.	Mostly immovable; very slight movement of teeth in their sockets.	Roots of teeth in alveolar processes of mandible and maxillae.

Structural classification	Extent of movement (functional classification)*	Type of structure	Type of movement†	Examples
CARTILAGINOUS Articulating bones united by plate of hyaline cartilage or fibrocartilaginous disk.	Mostly slightly movable, some immovable; usually amphiarthroses.	*Synchondrosis:* temporary joint composed of hyaline cartilage joining diaphysis and epiphysis of growing long bones.	Immovable; permits growth, not movement, of long bones.	Epiphyses of femur.
		Symphysis: bony surfaces bridged by flattened plates or disks of fibrocartilage.	Slight movement.	Symphysis pubis, manubriosternal joint, intervertebral joints between bodies of vertebrae.
SYNOVIAL Articulating bones moving freely along smooth, lubricated articular cartilage; enclosed within flexible articular capsule.	Freely movable; usually diarthroses.	*Uniaxial:* movement of bone about one axis of rotation (includes hinge, pivot, biaxial, condyloid, multiaxial, gliding, saddle, and ball-and-socket).		
		Hinge: convex surface of one bone fitted into concave surface of other.	Flexion, extension.	Elbow, interphalangeal joints, knee, ankle.
		Pivot: central bony pivot surrounded by collar of bone and ligament.	Supination, pronation, rotation.	Proximal radioulnar joint, atlantoaxial joint.
		Biaxial: movement of bone about two axes of rotation.		
		Condyloid (ellipsoidal): modified ball-and-socket.	Flexion, extension, abduction, adduction, circumduction.	Metacarpophalangeal (knuckle) joints, except thumb.
		Multiaxial: movement of bone about three axes.		
		Gliding: essentially flat articular surfaces.	Simple gliding movement within narrow limits.	Between articular processes of vertebrae, acromioclavicular joint, some carpal and tarsal bones.
		Saddle: opposing articular surfaces with both concave and convex surfaces that fit into one another.	Abduction, adduction, opposition, reposition.	Carpometacarpal joint of thumb.
		Ball-and-socket: globelike head of one bone fitted into cuplike concavity of another bone.	Most movable type of joint. Flexion, extension, internal rotation, lateral rotation, abduction, adduction, circumduction.	Shoulder joint, hip joint.

*Note that synarthrosis and fibrous, amphiarthrosis and cartilaginous, and diarthrosis and synovial are not *always* synonymous.

†See Table 8.2 for explanation of the terms for different types of movement.

FIGURE 8.1

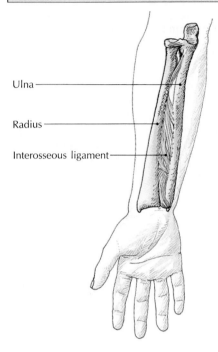

A syndesmosis. The interosseous ligament joins the ulna and radius of the forearm.

Ulna

Radius

Interosseous ligament

FIGURE 8.2

Cartilaginous joints throughout the body (red). Only some are labeled.

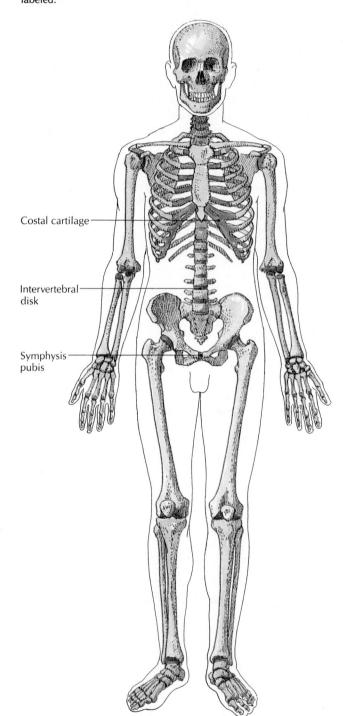

Costal cartilage

Intervertebral disk

Symphysis pubis

Syndesmoses

When bones are close together but not touching, and are held together by collagenous fibers or interosseous ligaments, the joint is called a *syndesmosis* (Gr. to bond together: *syn* + *desmos,* bond). The amount of movement, if any, in a syndesmosis depends on the distance between the bones and the amount of flexibility of the fibrous connecting tissue. Ligaments (composed of collagenous connective tissue) make a firm articulation at the inferior tibiofibular joint, so that very little movement occurs there. This limited movement adds strength to the joint. In contrast, the interosseous ligaments between the shafts of the radius and the ulna allow much more movement, including the twisting of the forearm (Figure 8.1).

Gomphoses

A *gomphosis* (Gr. *gomphos,* bolt) is a fibrous joint made up of a peg and a socket. The root of each tooth (the peg) is anchored into its socket by the fibrous periodontal ("around the tooth") ligament, which extends from the root of the tooth to the alveolar processes of the maxillae and mandible.

CARTILAGINOUS JOINTS

In *cartilaginous joints,* the bones are united by a plate of hyaline cartilage or a fibrocartilaginous disk (Figure 8.2). They lack a joint cavity, and permit little or no movement.

The two types of cartilaginous joints are the synchondrosis and the symphysis.

Synchondroses

A *synchondrosis* (Gr. *syn* + *chondros*, cartilage) is also called a *primary cartilaginous joint*. The chief function of a synchondrosis is to permit growth, not movement. It is a temporary joint composed of an epiphyseal plate of hyaline cartilage that joins the diaphysis and epiphysis of a growing long bone. A synchondrosis is eventually replaced by bone when the long bone stops growing, and it then becomes a synostosis. However, a few synchondroses are not replaced by synostoses and are still present in the adult. One such articulation is the sternoclavicular joint, where the clavicle, the first costal cartilage, and the manubrium of the sternum are joined (see Figure 7.18).

Symphyses

A *symphysis* (Gr. growing together) is sometimes called a *secondary synchondrosis*. In such an articulation the two bony surfaces are covered by thin layers of hyaline cartilage. Between them are disks of fibrocartilage that serve as shock absorbers. Fibrocartilage is a dense mass of collagenous fibers filled with cartilage cells and a scant cartilage matrix, so it has the firmness of cartilage and the toughness of a tendon. One of these slightly movable joints is the symphysis pubis, the midline joint between the bodies of the pubic portions of the paired hipbones. During pregnancy, the symphysis pubis is relaxed somewhat to allow for the necessary displacement of the mother's hipbones as the fetus grows. Two other examples are the slightly movable manubriosternal joint at the sternal angle, and the intervertebral disks between the bodies of the vertebrae.

SYNOVIAL JOINTS

Most of the permanent joints in the body are synovial. Of all the types of joints, *synovial joints* allow the greatest range of movement. Such free movement is possible because the ends of the bones are covered with a smooth hyaline *articular cartilage*, the joint is lubricated by thick fluid called *synovial fluid* (Gr. *syn* + *ovum*, egg)* or *synovium*, and the joint is enclosed by a flexible *articular capsule*. A synovial joint has a joint cavity.

Typical Structure of Synovial Joints

Because synovial joints allow more free movement than any other type of joint, they are also more complicated in structure. In general, *the more flexible a joint is, the less stable it is.*

*Synovial fluid gets it name because its thick consistency resembles the white of an egg.

FIGURE 8.3

Synovial joints (diarthroses). A typical synovial joint (A) without an articular disk and (B) with an articular disk.

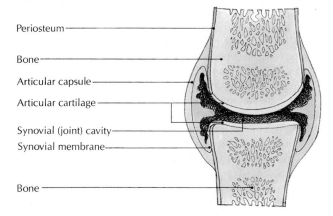

Periosteum
Bone
Articular capsule
Articular cartilage
Synovial (joint) cavity
Synovial membrane
Bone

[A]

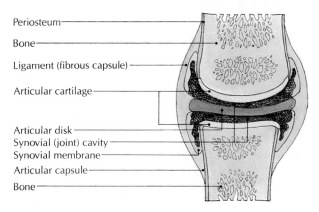

Periosteum
Bone
Ligament (fibrous capsule)
Articular cartilage
Articular disk
Synovial (joint) cavity
Synovial membrane
Articular capsule
Bone

[B]

You will see how this rule applies when we look at some examples of major joints later in this chapter. The four essential features of a synovial joint (Figure 8.3) are the synovial cavity, the articular cartilage, the articular capsule, and ligaments.

Synovial cavity A *synovial cavity*, or joint cavity, is the space between the two articulating bones, but does not include the articular cartilage. It contains folds of *synovial membrane* that sometimes contain pads of fat. These fatty pads help to fill spaces between articulating bones, and also reduce friction. The synovial membrane secretes the thick synovial fluid that lubricates the synovial cavity. Combined with the articular cartilage, synovial fluid provides an almost friction-free surface for the easy movement of joints.

Articular cartilage An *articular (hyaline) cartilage* caps the surface of the bones facing the synovial cavity. In a living body, the articular cartilage has a silvery-blue luster, and appears as polished as a pearl. Because of its thickness and elasticity, the articular cartilage acts like a shock absorber. If it is worn away (which occurs in some joint diseases), move-

Knee Injuries—The "Achilles Heel" of Athletes

One of the most frequent sports injuries is commonly called torn cartilage (usually the medial meniscus) of the knee. When the cartilage is torn, it may become wedged between the articular surfaces of the femur and tibia, causing the joint to "lock." Until recently, such an injury might sideline an athlete for months, or even end his or her career altogether. Fifteen or twenty years ago, a football player recovering from surgery to repair knee cartilage that was torn on the opening day of the season would remain on crutches for about two months, and would be unable to play for the remainder of the season. Now, a diagnostic and surgical technique called **arthroscopy** ("looking into a joint") makes it possible for an athlete with torn cartilage to return to the football field in about half the usual time.

Designed in Japan in 1970, arthroscopy permits a surgeon to place a lighted scope about the size of a pencil into the joint capsule to view the structural damage. If the damage is local, and there do not seem to be complications, the surgeon makes another quarter-inch incision and inserts microsurgical instruments to clear away damaged cartilage. Most incisions are closed with a single stitch, or even just a Band-Aid. The patient is usually encouraged to walk lightly on the day of the operation.

Such rapid recovery does not give leg muscles time to atrophy. So besides having minor surgery instead of major tissue damage during conventional open surgery, the patient does not have to worry about rehabiliting weakened muscles.

Another common knee injury (probably the most common) is a tear of the anterior and posterior cruciate ligaments. When the cruciate ligaments are torn, the knee joint becomes nonfunctional. To repair the cruciate ligaments, holes are drilled through the femur. Sutures are stitched to the damaged ligaments, and the sutures are passed out through the holes (see drawings). The ligaments are secured to the bone by knotting the sutures outside the femur. If the collateral ligament is separated from the bone, it is reattached by being sutured to the bone. Six to eight weeks of healing is usually followed by about a year of rehabilitative therapy.

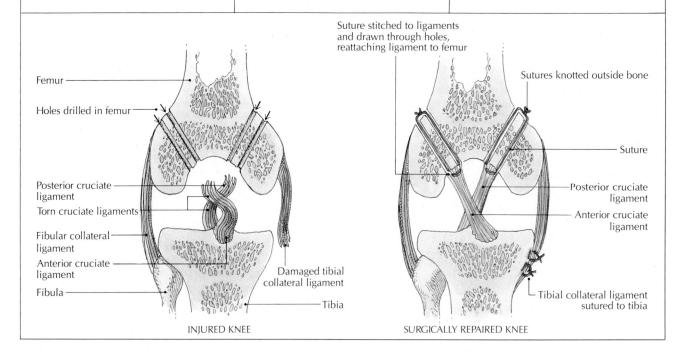

Suture stitched to ligaments and drawn through holes, reattaching ligament to femur

Femur

Holes drilled in femur

Posterior cruciate ligament

Torn cruciate ligaments

Fibular collateral ligament

Anterior cruciate ligament

Fibula

Damaged tibial collateral ligament

Tibia

INJURED KNEE

Sutures knotted outside bone

Suture

Posterior cruciate ligament

Anterior cruciate ligament

Tibial collateral ligament sutured to tibia

SURGICALLY REPAIRED KNEE

ment becomes restricted and painful. The cartilage itself is insensitive to feeling, since it has no nerve supply, but the other portions of the joint are supplied with pain-receptor nerves.

Within several synovial joints there are **articular disks,** or fibrocartilaginous disks (see Figure 8.3B). Their roles vary depending upon the joint. These disks may act as shock absorbers to reduce the effect of shearing (twisting) upon a joint and to prevent jarring between bones. They also adjust the unequal articulating surfaces of the bones so that the surfaces fit together more evenly. Finally, they allow two kinds of movements to take place at once, as when the jaw moves backward or forward and side to side at the same time.

In some joints, the fibrocartilaginous disk forms a complete partition, dividing the joint into two cavities. In the knee joint, the fibrocartilages, called the medial and lateral **menisci** (muh-NISS-eye; singular, **meniscus;** Gr. *meniskos*, crescent), are crescent-shaped wedges that form incomplete partitions (see Figure 8.17C). The menisci serve to cushion as well as guide the articulating bones. Many athletes, especially football players, tear these menisci, commonly referred to as torn cartilages. (See the essay on this page.)

Articular capsule An **articular capsule** lines the synovial cavity in the noncartilaginous parts of the joint. This fibrous capsule is lax, and pliable, permitting considerable movement. The inner lining of the capsule is the synovial membrane, which extends from the margins of the articular cartilages. The outer layer of the capsule is a fibrous membrane, which extends from bone to bone across a joint, and reinforces the capsule. "Double-jointed" people have loose articular capsules.

Ligaments The *ligaments* are fibrous thickenings of the articular capsule that join one bone to its articulating mate. They vary in shape, and even in strength, depending upon their specific roles. Most ligaments are considered inelastic, and yet they are pliable enough to permit movement at the joints. However, they will tear rather than stretch under excessive stress, as when an ankle is sprained. A sprained ankle results from excessive inversion of the foot, which causes a partial tearing of the anterior talofibular ligament and the calcaneofibular ligament. Torn ligaments are extremely painful, and are accompanied by immediate local swelling. In general, however, ligaments are strong enough to prevent any excessive movement and strain, and a rich supply of sensory nerves prevents a person from stretching the ligaments excessively when the joints are being overworked.

Bursae and Tendon Sheaths

Two other structures associated with joints, but not part of them, are bursae and tendon sheaths. *Bursae* (BURR-see; singular, bursa; Gr. purse) resemble flattened sacs. They are filled with synovial fluid. Bursae are found wherever it is necessary to eliminate the friction that occurs when a muscle or tendon rubs against another muscle, tendon, or bone (Figures 8.4A, B). They also function to cushion certain muscles and to facilitate the movement of muscles over bony surfaces. Bursitis results when bursae become inflamed.

A modification of a bursa is the *tendon* (synovial) *sheath* surrounding long tendons that are subjected to constant friction. Such sheaths surround the tendons of the wrist, palm, and finger muscles (Figure 8.4C). Tendon sheaths are long, cylindrical sacs filled with synovial fluid. Like bursae, tendon sheaths reduce friction and permit tendons to slide easily.

Ask Yourself

1 *How do synarthroses, amphiarthroses, and diarthroses compare?*

2 *What are the three types of fibrous joints?*

3 *Where are the two kinds of cartilaginous joints found in the body?*

4 *What are the four main components of a synovial joint?*

5 *What are the functions of bursae and tendon sheaths?*

FIGURE 8.4

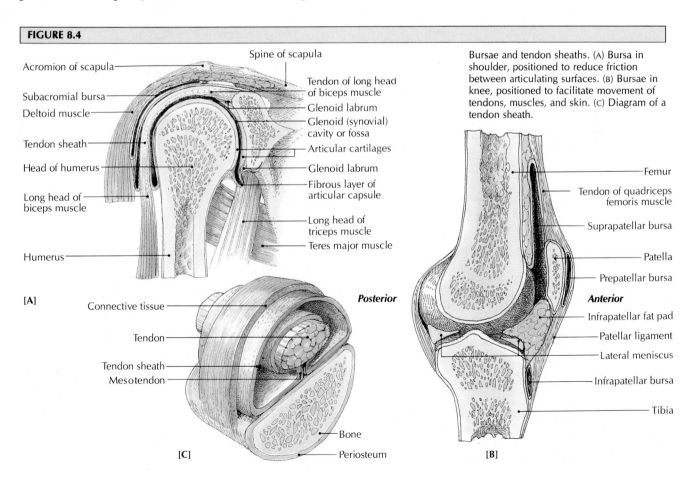

Bursae and tendon sheaths. (A) Bursa in shoulder, positioned to reduce friction between articulating surfaces. (B) Bursae in knee, positioned to facilitate movement of tendons, muscles, and skin. (C) Diagram of a tendon sheath.

MOVEMENT AT SYNOVIAL JOINTS

Any movement produced at a synovial joint is limited in some way. Among the limiting factors are the following:

1 *Interference by other structures.* For example, lowering the shoulder is limited by the presence of the thoracic cage.

2 *Tension exerted by ligaments of the articular capsule.* For example, the thigh cannot be hyperextended at the hip joint because the iliofemoral ligament becomes taut as it passes in front of the hip joint.

3 *Muscle tension.* For example, when the knee is straight, it is more difficult to raise the thigh than when the knee is bent, because the stretched hamstring muscles exert tension on the back of the thigh.

Terms of Movement

Muscles and bones work together to allow different types of movement at different joints (Figures 8.5 to 8.12). Terms used to describe particular movements are defined in Table 8.2.

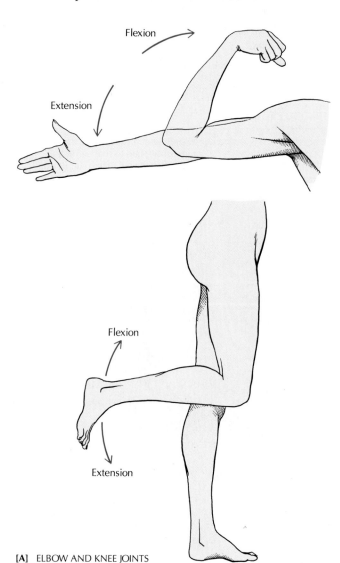

[A]   ELBOW AND KNEE JOINTS

FIGURE 8.5

Movements of joints. (A) Flexion and extension at elbow and knee joints. (B) Flexion, extension, and hyperextension at neck. (C) Hyperextension, extension, and palmar flexion of hand at wrist. (D) Dorsiflexion and plantar flexion of ankle.

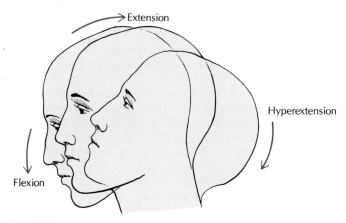

[B]   NECK

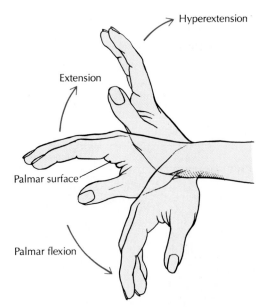

[C]   WRIST JOINT

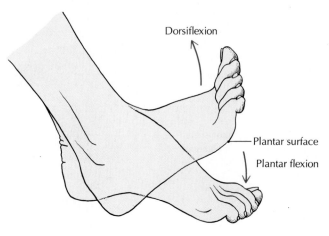

[D]   ANKLE JOINT

FIGURE 8.6

Movements of joints. (A) Abduction of limbs from body. (B) Adduction of limbs toward body. (C) Abduction of fingers. (D) Adduction of fingers.

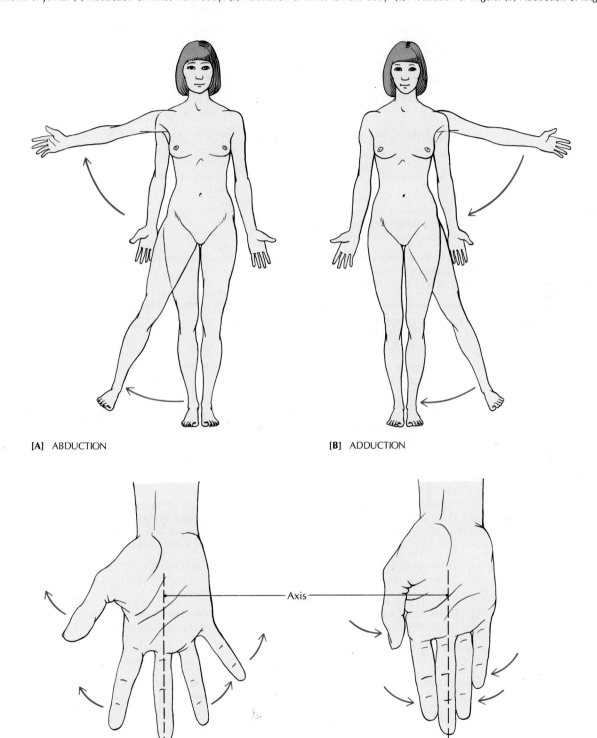

[A] ABDUCTION

[B] ADDUCTION

[C] ABDUCTION

[D] ADDUCTION

TABLE 8.2   TYPES OF MOVEMENTS AT SYNOVIAL JOINTS

Action	Definition	Examples
Flexion	Bending motion in which angle between two bones is decreased (Figure 8.5).	Bending forearm at elbow or leg at knee.
Extension	Straightening motion in which angle between two bones is increased; opposite of flexion (Figure 8.5).	Straightening forearm at elbow or leg at knee.
Hyperextension	Extension beyond straight (anatomical) position (Figures 8.5B, C).	Bending back (extending) hand at wrist beyond anatomical position.
Dorsiflexion*	Flexion of foot at ankle joint (Figure 8.5D).	Bending foot up at ankle.
Palmar flexion	Flexion of hand at wrist (Figure 8.5C).	Bending hand in direction of palm.
Plantar flexion*	Extension of foot at ankle (Figure 8.5D).	Bending foot in direction of sole, as in standing on tiptoes.
Abduction	Movement of limb away from midline of body; movement of fingers or toes away from longitudinal axis of hand or foot (Figures 8.6A, C).	Raising upper extremity to side; spreading fingers.
Adduction	Movement of limb toward or beyond midline of body; movement of fingers or toes toward longitudinal axis of hand or foot (Figures 8.6B, D).	Moving upper extremity down toward side.
Opposition	Angular movement in which thumb pad is brought to touch and to oppose a finger pad of the extended fingers; occurs only at carpometacarpal joint of thumb (Figure 8.7A).	One of the movements used in gripping pen.
Reposition	Movement that returns thumb to anatomical position; opposite of opposition (Figure 8.7B).	One of the movements used in releasing grip on object.
Circumduction	Movement in which distal end of bone moves in circular motion while proximal end remains stable; accomplished by successive flexion, abduction, extension, and adduction (Figure 8.8).	Movement at the shoulder (glenohumeral) joint in baseball pitcher's windup.
Rotation	Movement of body part (usually entire extremity) around its own axis without any displacement of its axis (Figure 8.9).	Shaking head "no."
Medial (internal)	Movement in which ventral surface of extremity directed toward midline of body.	Movement of humerus at shoulder joint, with hand directed toward body.
Lateral (external)	Movement in which ventral surface directed outward away from midline.	Movement of humerus at shoulder joint, with hand directed away from body.
Supination	Pivoting movement of forearm in which radius is "rotated" to become parallel to ulna (Figure 8.10A).	Movement using in tightening a screw with a screwdriver, so that palm faces forward at end of movement.
Pronation	Pivoting movement of forearm in which radius is "rotated" diagonally across ulna (Figure 8.10B).	Pivot of forearm that turns palm backward, 180 degrees away from anatomical position.
Inversion	Movement of sole of foot inward (medially) (Figure 8.11A).	Movement of foot in which big toe is turned upward and away from midline of body.
Eversion	Movement of sole of foot outward (laterally) (Figure 8.11B).	Movement of foot in which big toe is turned downward and toward midline of body.
Protraction	Forward movement (Figure 8.12A).	Pushing jaw forward (protrusion); forward movement of clavicle in rounding shoulders.
Retraction	Backward movement (Figure 8.12B).	Pulling jaw backward; backward movement of clavicle in assuming military brace posture.
Elevation	Raising a body part (Figure 8.12C).	Shrugging shoulders; closing mouth.
Depression	Lowering a body part (Figure 8.12D).	Drooping shoulders; lowering jaw to open mouth.

*At the ankle, where the normal position is angular, dorsiflexion and plantar flexion are equivalent to flexion and extension.

FIGURE 8.7

Movements of joints. (A) Opposition of thumb. (B) Reposition of thumb.

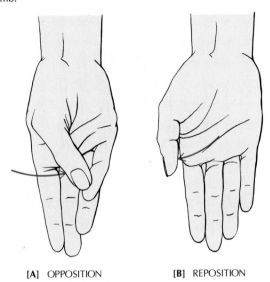

[A]   OPPOSITION   [B]   REPOSITION

FIGURE 8.9

Movements of joints. Rotation of head in "no" movement.

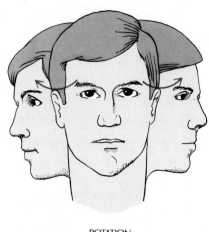

ROTATION

FIGURE 8.8

Movements of joints. Circumduction of limbs at shoulder and hip joints. Note how the limbs describe imaginary "cones of movement."

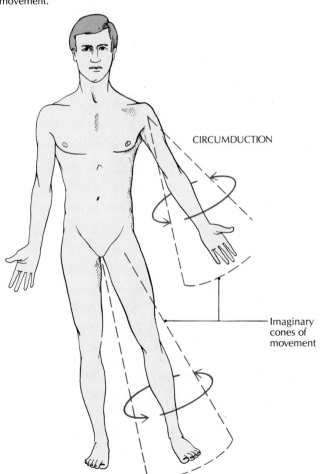

CIRCUMDUCTION

Imaginary cones of movement

FIGURE 8.10

Movements of joints. (A) Supination of forearm. (B) Pronation of forearm.

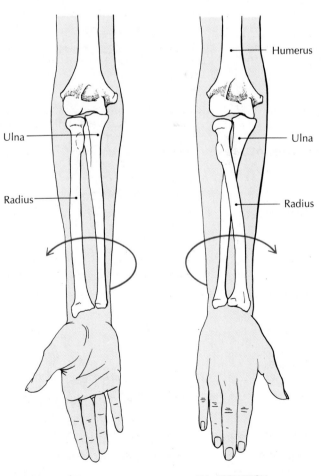

Humerus

Ulna

Radius

Ulna

Radius

[A]   SUPINATION   [B]   PRONATION

FIGURE 8.11

Movements of joints. (A) Inversion of foot. (B) Eversion of foot.
Inversion is like walking on the outsides of your soles, and eversion
is like walking on the insides of your soles.

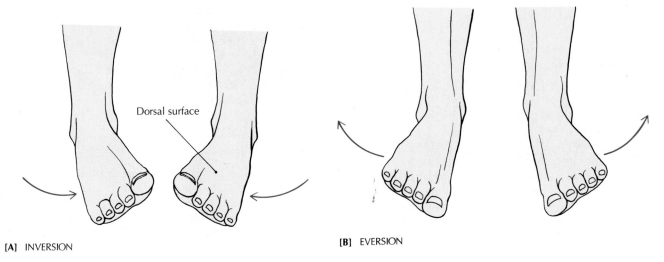

[A] INVERSION

[B] EVERSION

FIGURE 8.12

Movements of joints. (A) Protraction of jaw. (B) Retraction of jaw. (C) Depression of jaw. (D) Elevation of jaw.

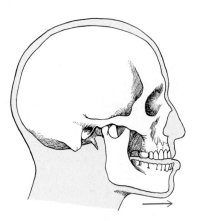 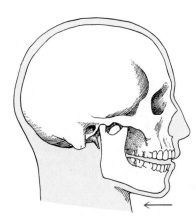

[A] PROTRACTION

[B] RETRACTION

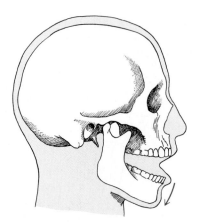 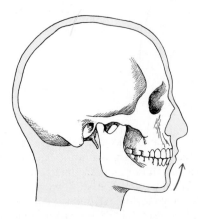

[C] DEPRESSION

[D] ELEVATION

Ask Yourself

1 *What three factors restrict movement at synovial joints?*

2 *How does circumduction form a "cone of movement"?*

3 *How does rotation differ from supination and pronation?*

4 *What movement is necessary to open the mouth?*

Uniaxial, Biaxial, and Multiaxial Joints

In any movement about a joint, one member of a pair of articulating bones moves in relation to the other, with one bone maintaining a fixed position, and the other bone moving about an axis (or axes). For example, at the shoulder joint (the glenohumeral joint between the scapula and the humerus), the scapula is fixed in relation to the movement of the humerus. The motions of all diarthrodial joints can be reduced to rotations in one, two, or three planes (or combinations of several planes). These planes are perpendicular to each other if there are more than one plane of rotation in the movements. In essence, the movements of the bones can be described in most joints in terms of the following pairs of movements in one of three planes: (1) flexion and extension, (2) abduction and adduction, and (3) lateral (external) and medial (internal) rotation.

When the movement of a bone at a joint is limited to rotation about one axis, as is the elbow joint, the joint is said to be **uniaxial**, and to have *one degree of freedom* (Figure 8.13A). In other words, the forearm may be flexed or extended from the elbow, moving in only one plane. When two movements can take place about two axes of rotation, as in the radiocarpal (wrist) joint, the joint is **biaxial**, and has *two degrees of freedom* (Figure 8.13B). The hand can be flexed or extended in one plane and moved from side to side (abduction and adduction) at the wrist in a second plane. When three independent rotations occur about three axes, as in the glenohumeral (shoulder) joint, the joint is **multiaxial** or **triaxial**, and has *three degrees of freedom* (Figure 8.13C). The arm movements from the shoulder occur in three planes: flexion and extension, abduction and adduction, and medial and lateral rotation.

TYPES OF SYNOVIAL JOINTS

Synovial joints are freely movable joints that are classified according to the shape of their articulating surfaces and the types of joint movements those shapes permit. (Once again, structure influences function.) Six types of synovial joints are recognized: hinge, pivot, condyloid (ellipsoidal), gliding, saddle, and ball-and-socket. (See Table 8.1 for a complete classification of joints.)

Hinge Joints

Hinge joints roughly resemble the hinges on the lid of a box. The convex surface of one bone fits into the concave surface of another bone, so that only a uniaxial, back-and-forth movement occurs around a single (transverse) axis (see Figure 8.13A). These uniaxial joints have strong collateral ligaments in the capsule around the joint. The rest of the capsule is thin and lax, permitting flexion and extension, the only movements possible in these joints. Hinge joints are found in the elbow, finger, knee, and ankle.

Pivot Joints

Another type of uniaxial joint is the **pivot joint** (also called *trochoid*, from a Greek word meaning wheel), which is only able to rotate around a central axis. Pivot joints are composed of a central bony pivot surrounded by a collar made partly of a bone and partly of a ligament. Pivot joints have a rotational movement around a long axis through the center of the pivot (like the hinges of a gate).

The atlantoaxial joint between the atlas and the axis is a pivot joint in which the collar (composed of the anterior bony arch of the atlas and the transverse ligament) rotates around the pivot, which is the dens (odontoid process) of the axis. The "no" movement of the head occurs at the atlanto-axial pivot joint.

Condyloid Joints

Condyloid, *condylar* (Gr. *condylus*, knuckle), or *ellipsoidal joints* are modifications of the multiaxial ball-and-socket joint (see p. 221). However, because the ligaments and muscles around the joint limit the rotation to two axes of movement, the joint is classified as *biaxial*. Examples of condyloid joints are the metacarpophalangeal joints (knuckles) of the fingers (except the thumb). In condyloid joints the axes are at right angles to each other, permitting the usual flexion and extension of a hinge joint, as well as abduction, adduction, and circumduction. However, rotational movement is not permitted.

| *What causes the "popping" sound when you crack your knuckles?* | *There are at least three likely possibilities: (1) When a joint is contracted, small ligaments or muscles may pull tight and snap across the bony protuberances of the joint. (2) When the joint is pulled apart, air can pop out from between the bones, creating a vacuum that produces a popping sound. (3) When the fluid pressure of the synovial fluid is reduced by the slow articulation of a joint, tiny gas bubbles in the fluid may burst, producing the sound. Because bones are not completely hardened until the age of 18 or so, a child or teenager who cracks knuckles may deform and enlarge the knuckle bones.* |

FIGURE 8.13

Planes of movement in synovial joints. (A) Uniaxial movement in elbow joint, illustrating axis for flexion and extension. (B) Biaxial movement in radiocarpal (wrist) joint, illustrating axis for flexion and extension and axis for abduction and adduction. (C) Multiaxial movement in glenohumeral (shoulder) joint, illustrating axis for flexion and extension, axis for abduction and adduction, and axis for medial rotation and lateral rotation.

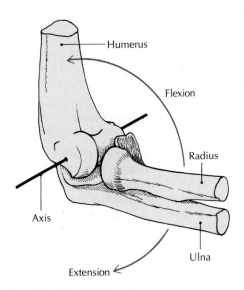

[A] UNIAXIAL (elbow)

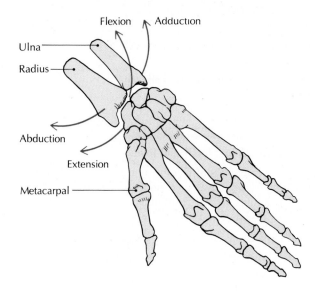

[B] BIAXIAL (wrist)

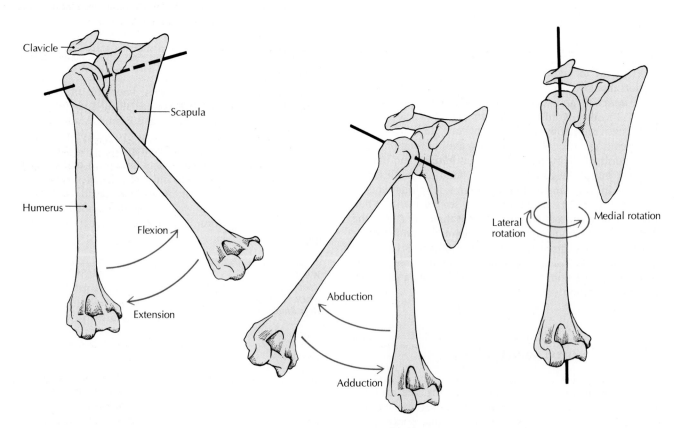

[C] MULTIAXIAL (shoulder)

Gliding Joints

Although some *gliding* (plane) *joints* have only one axis of rotation, side-to-side and back-and-forth movements are possible. Because two planes are involved, gliding joints are considered *biaxial*. The numerous gliding joints are almost always small, and are formed by essentially flat articular surfaces. Examples include the joints between the articular processes of adjacent vertebrae, the acromioclavicular joint at the lateral end of the clavicle, and the joints between some carpal and tarsal bones.

Saddle Joints

The *saddle joint* is so named because the opposing articular surfaces of both bones are shaped like a saddle, that is, they have both concave and convex areas. Movement is permitted in several directions, and the joint is considered *multiaxial*. Movements are abduction, adduction, opposition, and reposition. The carpometacarpal joint of the thumb is the best example of a saddle joint in the body.

Ball-and-Socket Joints

Ball-and-socket joints are composed of a globelike head of one bone that fits into a cuplike concavity of another bone.

This is the most freely movable of all joints, permitting movement along three axes of rotation. Actually, an almost infinite number of axes are available, and the joint is classified as *multiaxial*. The movements of a ball-and-socket joint are flexion, extension, medial (internal) rotation, lateral (external) rotation, abduction, adduction, and circumduction. The shoulder and hip joints are ball-and-socket joints.

Ask Yourself

1 How can you describe uniaxial, biaxial, and multiaxial joints?

2 What is the difference in the direction of hinge motion in hinge joints and pivot joints?

3 What is an example of a condyloid joint?

4 Which type of joint is the most movable?

DESCRIPTION OF SOME MAJOR JOINTS

The jaw, shoulder, hip, and knee are major joints of the body, each with its own distinctive anatomical characteristics. The following sections describe and illustrate these joints in detail. In addition, Table 8.3 outlines some major articulations of the body, including the ones mentioned above.

TABLE 8.3   SOME MAJOR ARTICULATIONS OF THE HUMAN BODY

Joint	Articulation	Classification of joint	Type of structure	Type of movement
JOINTS ASSOCIATED WITH THE SKULL				
Temporomandibular (jaw, TM) joint	Between head of mandible and mandibular fossa of temporal bones; divided into two compartments by fibrocartilaginous articular disk.	Diarthrosis (synovial).	Hinge (lower compartment). Gliding (upper compartment).	Opening and closing jaw results from simultaneous hinge and gliding movements. Protraction, retraction, grinding movements.
Atlanto-occipital joint	Between atlas and occipital bone of skull.	Diarthrosis (synovial).	Hinge.	"Yes" movement of head.
JOINTS OF THE VERTEBRAL COLUMN				
Atlantoaxial joints (paired)				
Medial	Between dens of axis and anterior arch of atlas and its transverse ligament.	Diarthrosis (synovial).	Pivot.	"No" movement of head.
Lateral	Between articular processes of atlas and axis.	Diarthrosis (synovial).	Gliding.	"No" movement of head.
Intervertebral joints	Paired joints between articular processes of adjacent vertebrae.	Diarthrosis (synovial).	Gliding.	*Neck:* considerable variety and range of movement.
	Unpaired joints (intervertebral disks) between adjacent bodies of vertebrae.	Amphiarthrosis (fibrocartilaginous).	Synchondrosis.	*Thorax:* considerable variety, limited range. *Lumbar:* essentially flexion, extension, lateral bending.

(Table 8.3 continues on the following pages.)

TABLE 8.3   SOME MAJOR ARTICULATIONS OF THE HUMAN BODY (Continued)

Joint	Articulation	Classification of joint	Type of structure	Type of movement
JOINTS ASSOCIATED WITH THE RIBS AND STERNUM				
Costovertebral joints	Between articular facets on heads and tubercles of ribs and costal facets on transverse processes of thoracic vertebrae.	Diarthrosis (synovial).	Gliding.	Rotation.
Sternocostal joints	Between costal cartilage of rib 1 and sternum.	Synarthrosis (fibrous).	Syndesmosis (rib 1).	None at rib 1.
	Between ends of costal cartilages of ribs 2–7 and concavities on sides of sternum.	Diarthrosis (synovial).	Gliding (true ribs).	Rotation of true ribs (2–7).
Interchondral joints	Between successive costal cartilages 5–9.	Diarthrosis (synovial).	Gliding.	Adjustment during respiration.
Manubriosternal joint	Between manubrium and body of sternum.	Amphiarthrosis (cartilaginous).	Symphysis.	Slight movement because of fibrocartilaginous disk between manubrium and sternal angle.
Xiphisternal joint	Between xiphoid process and body of sternum.	Amphiarthrosis (cartilaginous).	Symphysis.	Slight movement.
JOINTS OF THE CLAVICLE AND PECTORAL (SHOULDER) GIRDLE				
Sternoclavicular	Between medial end of clavicle, manubrium, and first costal cartilage; divided into two compartments by fibrocartilaginous articular disk.	Diarthrosis (synovial).	Double gliding.	Elevation, depression, protraction, retraction.
Acromioclavicular	Between lateral end of clavicle and medial surface of acromion.	Diarthrosis (synovial).	Gliding.	Essentially an action of accommodation between movement of clavicle and scapula.
Coracoclavicular	Between clavicle and coracoid process of scapula; connected by corococlavicular ligament.	Synarthrosis (fibrous).	Syndesmosis.	Prevents separation of clavicle from scapula.
JOINTS OF THE ARM AND FOREARM				
Glenohumeral (humeroscapular, shoulder) joint	Between head (ball) of humerus and glenoid cavity (socket) of scapula.	Diarthrosis (synovial).	Ball-and-socket.	Flexion, extension, abduction, adduction, medial and lateral rotation, circumduction.
Elbow	Between trochlea and capitulum of humerus, trochlear notch of ulna, and head of radius.	Diarthrosis (synovial).	Hinge.	Flexion, extension.
Radioulnar articulation				
Proximal	Between head of radius and radial notch of ulna.	Diarthrosis (synovial).	Pivot.	Pronation, supination.
Distal	Between head of ulna and ulnar notch of radius.	Diarthrosis (synovial).	Pivot.	Pronation, supination.
Radiocarpal (wrist) joint	Between distal end of radius (and articular disk) and proximal row of carpal bones (scaphoid, lunate, triquetrum).	Diarthrosis (synovial).	Ellipsoidal.	Radial abduction, ulnar adduction, flexion, extension, hyperextension, circumduction.

Description of Some Major Joints

Joint	Articulation	Classification of joint	Type of structure	Type of movement
JOINTS OF THE HAND				
Midcarpal joints	Between proximal row and distal row.	Diarthrosis (synovial).	Hinge.	Slight flexion, extension.
Carpometacarpal joint of thumb	Between trapezium and proximal end of first metacarpal bone.	Diarthrosis (synovial).	Saddle.	Abduction, adduction, opposition, reposition.
Metacarpophalangeal (knuckle) joints	Between heads of metacarpal bones and bases of proximal phalanges.	Diarthrosis (synovial).	Condyloid.	Flexion, extension, abduction, adduction, circumduction.
Interphalangeal joints	Between heads of phalanges and concave base of adjacent phalanges.	Diarthrosis (synovial).	Hinge.	Flexion, extension.
JOINTS OF THE PELVIS				
Sacroiliac joint				
Anterior	Between sacrum and ilium on anterior side.	Diarthrosis (synovial).	Gliding.	Slight gliding and rotary movement; gives resilience to joint.
Posterior	Between sacrum and ilium on posterior side.	Synarthrosis (fibrous).	Symphysis.	Slight movement; gives security to joint.
Symphysis pubis	Between bodies of pubic bones.	Amphiarthrosis (fibrocartilaginous).	Symphysis.	Practically immovable, but accommodates during childbirth.
Hip (coxal) joint	Between head (ball) of femur and acetabulum (socket) of os coxa (hipbone).	Diarthrosis (synovial).	Ball-and-socket.	Flexion, extension, abduction, adduction, medial and lateral rotation, circumduction.
JOINTS OF THE LEG AND ANKLE				
Tibiofemoral (knee) joint	Between medial and lateral condyles of distal femur and medial and lateral condyles of proximal femur.	Diarthrosis (synovial).	Modified hinge.	Flexion, extension, some rotation ("screw-home" action at end of extension).
Talocrural (ankle) joint	Between socket for talus (distal tibia flanked by medial and lateral malleolus of tibia) and upper surface of talus.	Diarthrosis (synovial).	Hinge.	Dorsiflexion, plantar flexion (hyperextension).
JOINTS OF THE FOOT				
Subtalar	Posterior joint between talus and calcaneus.	Diarthrosis (synovial).	Three joints articulate as a unit; axis of rotation forms a line called the subtalar axis (axis of Henke).	Eversion (combination of abduction and pronation), inversion (combination of adduction and supination).
Talocalcaneonavicular	Combined anterior joint between talus and calcaneus and joint between talus and navicular.	Diarthrosis (synovial).		
Transverse tarsal	Combined joint between calcaneus and cuboid and joint between talus and navicular.	Diarthrosis (synovial).		
Tarsometatarsal	Between four anterior tarsal bones and bases of metatarsal bones.	Diarthrosis (synovial).	Gliding.	Slight movement.
Metatarsophalangeal	Between heads of metatarsal bones and bases of proximal phalanges.	Diarthrosis (synovial).	Condyloid.	Flexion, extension, abduction, adduction, circumduction.
Interphalangeal	Between heads of phalanges and concave bases of adjacent phalanges.	Diarthrosis (synovial).	Hinge.	Flexion, extension.

Temporomandibular (Jaw) Joint

The *temporomandibular* or *TM joint* (jaw) is the only movable joint of the head. It is a synovial joint with a combination of hinge and gliding structures (Figure 8.14, see Table 8.3).

The structure of the jaw is well fitted to the movements involved in chewing, biting, speaking, and so on. The articular surface of the temporal bone has a concave back part, called the *mandibular fossa,* and a convex part, called the *articular tubercle.* The joint cavity between the mandible and the temporal bone is divided into two compartments by a fibrocartilaginous *articular disk,* which is fused to the *articular capsule.* The capsule is attached above to the temporal articular surface and below to the neck of the mandible.

The TM joint is movable in all three planes: up and down (elevation and depression), backward and forward (retraction and protraction), and from side to side (an alteration of protraction and retraction from one side to the other). The lateral motion is possible due to the articular disk.

Two movements occur at the TM joint while the mandible is being depressed and the mouth is opening. The first movement is the forward gliding that takes place in the upper compartment of the joint cavity as the disk and the head of the mandible move onto the articular tubercle. When the mouth is closed, the convex head of the mandible fits into the mandibular fossa; as the mouth is opened, the head rides forward onto the articular tubercle. The second movement is a hingelike one that takes place in the lower compartment as the head of the mandible rotates on the articular head (condyle).

In the acts of protraction (protrusion) and retraction of the mandible, the head of the mandible and the articular disk slide forward and then backward on the articular surface of the temporal bone. The grinding action of the teeth is produced when protrusion and retraction of the mandible alternate from one side to the other.

Shoulder Joint

The *shoulder joint* is a multiaxial, ball-and-socket, synovial joint (diarthrosis). It is also called the *glenohumeral* or *humeroscapular joint* because the head of the humerus (the "ball") articulates with the shallow, concave *glenoid fossa* (the "socket") of the scapula (Figure 8.15; see Table 8.3, Figure 8.4A). As is typical of a ball-and-socket joint, the movements of flexion, extension, abduction, adduction, medial rotation, lateral rotation, and circumduction occur at the shoulder joint.

FIGURE 8.14

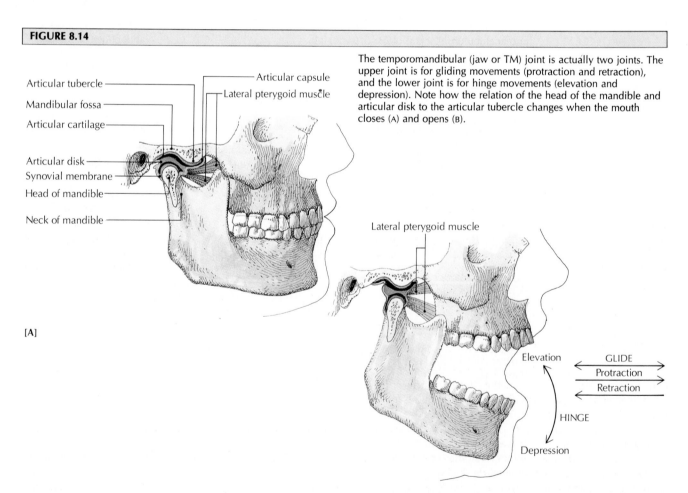

The temporomandibular (jaw or TM) joint is actually two joints. The upper joint is for gliding movements (protraction and retraction), and the lower joint is for hinge movements (elevation and depression). Note how the relation of the head of the mandible and articular disk to the articular tubercle changes when the mouth closes (A) and opens (B).

FIGURE 8.15

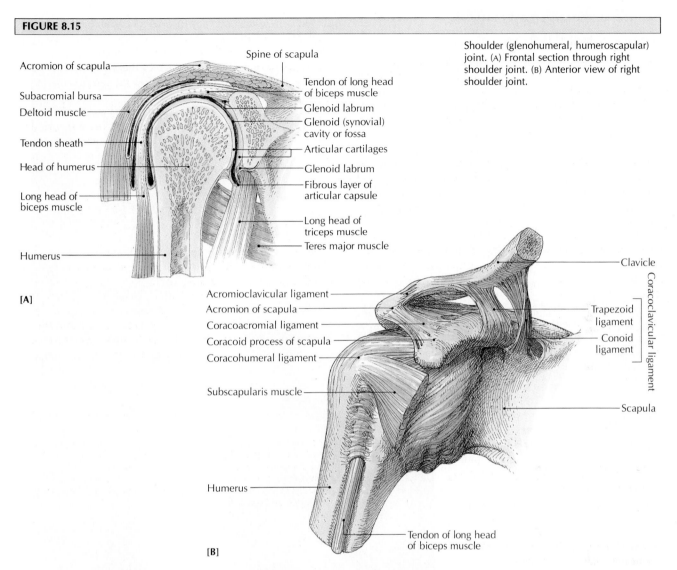

Spine of scapula

Acromion of scapula

Subacromial bursa

Deltoid muscle

Tendon sheath

Head of humerus

Long head of biceps muscle

Humerus

Tendon of long head of biceps muscle

Glenoid labrum

Glenoid (synovial) cavity or fossa

Articular cartilages

Glenoid labrum

Fibrous layer of articular capsule

Long head of triceps muscle

Teres major muscle

[A]

Shoulder (glenohumeral, humeroscapular) joint. (A) Frontal section through right shoulder joint. (B) Anterior view of right shoulder joint.

Acromioclavicular ligament

Acromion of scapula

Coracoacromial ligament

Coracoid process of scapula

Coracohumeral ligament

Subscapularis muscle

Humerus

Tendon of long head of biceps muscle

[B]

Clavicle

Coracoclavicular ligament

Trapezoid ligament

Conoid ligament

Scapula

The shoulder joint has more freedom of movement than any other joint in the body, but what it gains in mobility it loses in stability and security. As a result, the shoulder joint is dislocated more than any other joint. (See Physiological and Anatomical Abnormalities at the end of this chapter.) The extensive freedom of movement results mainly from the laxity of the articular capsule and the shallowness of the glenoid fossa.

The shallow glenoid fossa is deepened slightly by a fibrocartilaginous rim called the **glenoid labrum.** The **articular capsule,** looser than in any other important joint, completely envelops the articulation. It is attached to the glenoid labrum of the scapula, to the articular margin of the head, and to part of the anatomical neck of the humerus. The articular capsule is under tension during adduction and lateral rotation, but the **coracohumeral ligament** between the scapula and humerus becomes taut during those movements, strengthening the joint considerably.

The strength and stability of the shoulder joint depend also on four muscles, which nearly encircle the joint and hold the head of the humerus in the glenoid fossa. These joint-reinforcing muscles—the *supraspinatus, infraspinatus, teres minor,*

and *subscapularis* muscles (sometimes referred to as SITS muscles, for the first letter of each)—are collectively called the **rotator cuff** or musculotendinous cuff muscles. Rotator cuff injuries are common in baseball pitchers because pitching motions strain the rotator cuff muscles and their tendons.

Of the several bursa in the shoulder region, the large subacromial bursa is important because it is associated with subacromial bursitis. (See Physiological and Anatomical Abnormalities at the end of this chapter.)

Hip Joint

The **hip (coxal) joint** is a multiaxial, ball-and-socket, synovial joint (diarthrosis). The head of the femur articulates with the acetabulum of the hipbone (innominate bone, os coxa) (Figure 8.16).

The bones of the hip are arranged in such a way that the joint is one of the most secure, strong, and stable articulations in the body. The large, globular **head** (ball) of the femur fits snugly into the deep hemispherical **acetabulum** (socket). The

Two Million Replaceable Joints

Physicians have been trying to create successful replaceable joints since the nineteenth century, when German surgeons used hip ball joints made of ivory. Infection and rejection by the body were the typical results. It wasn't until 1968 that the first successful total hip replacement was performed in the United States, thanks mainly to the work of British orthopedic surgeon Sir John Charnley, who perfected the technique. He used a metal ball in a socket of high-density polyethylene and an acrylic cement to hold the ball and socket in place. By 1980, replaceable parts had been used successfully for practically every joint in the body.

Arthritis (inflammation of the joints) has been the major cause of damaged joints. About 30 million Americans suffer from arthritis, and between 1970 and 1980 about 2 million of them received artificial joints. The most successful joint replacement has been the hip; about 80,000 such operations are performed in the United States each year.

Physicians and engineers have combined their skills to design artificial joints that not only provide a normal range of movement, but also resist rejection by the surrounding body tissue. Eventually it is hoped that the artificial joints will actually be held in place by new bone growth. An artificial knee that induces bone material to grow into it has already been developed at Johns Hopkins University. The new knee joint is coated with beads of cobalt that stimulate bone growth, and such growth should form a more permanent bond than cement.

Following a two- or three-week hospitalization for the surgical procedure and several weeks of physical therapy, most hip replacement patients become reasonably independent in about three months. General physical fitness and normal daily activities are usually encouraged, but vigorous sports such as jogging and tennis are not.

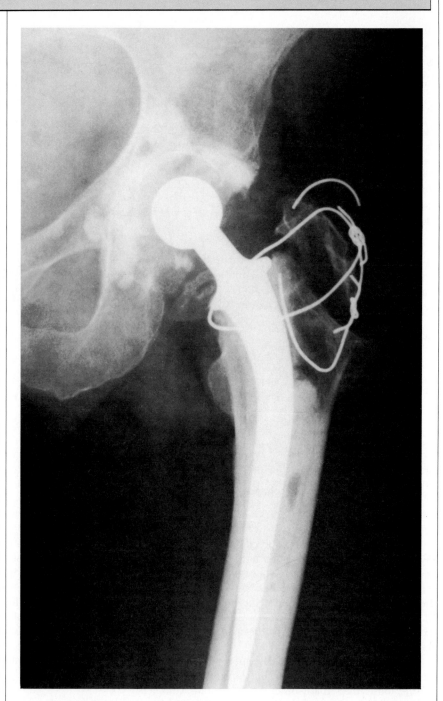

An x ray of an artificial hip joint that is replacing a joint severely damaged by arthritis. The implantation of hip joints is the most successful and widely performed joint-replacement operation. [*Biophoto Associates/Photo Researchers.*]

FIGURE 8.16

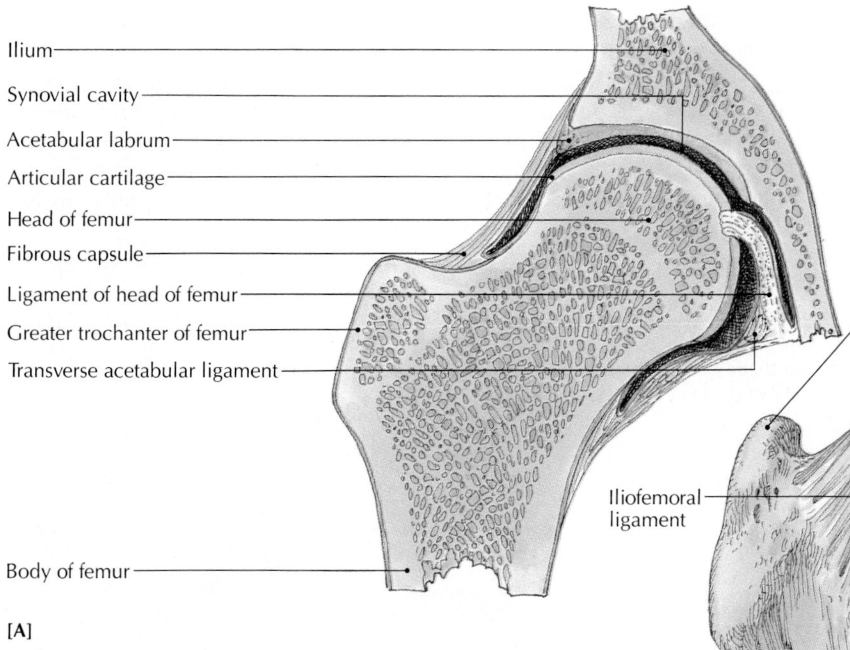

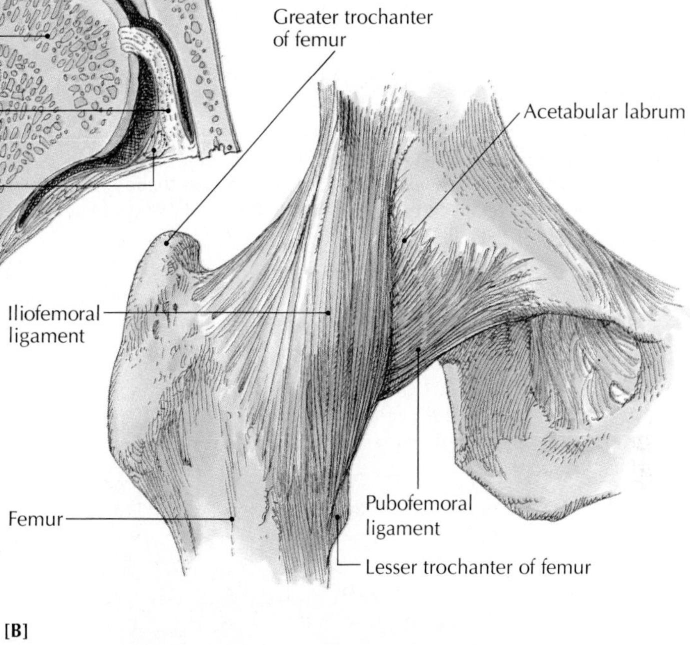

Hip (coxal) joint. (A) Diagrammatic coronal section. Note the trabecular organization of the spongy bone of the femur and ilium. The arrangement of the trabeculae indicates the lines of pressure exerted on the bones. (B) Dissection of hip joint, anterior view, showing iliofemoral (Y) ligament.

Ilium
Synovial cavity
Acetabular labrum
Articular cartilage
Head of femur
Fibrous capsule
Ligament of head of femur
Greater trochanter of femur
Transverse acetabular ligament
Body of femur

[A]

Greater trochanter of femur
Acetabular labrum
Iliofemoral ligament
Femur
Pubofemoral ligament
Lesser trochanter of femur

[B]

acetabulum is made deeper by the fibrocartilaginous **acetabular labrum,** which forms a complete circle around the socket. The deep socket holds the head of the femur securely, but because of such stability the movement at the hip joint is not as free as it is at the ball-and-socket shoulder joint. A notch on the inferior aspect of the acetabulum is bridged by the **transverse acetabular ligament.** The acetabulum faces downward, laterally, and forward. The head of the femur is mounted on the neck of the femur so that it forms a 125-degree angle with the long shaft of the femur (see Figure 8.16). The neck helps to transform the rotary movements of the head of the femur at the hip joint into the angular movements of the shaft.

The fibrous capsule in the hip is thick and tense compared with the relatively thin and lax capsule of the shoulder. The outer fibrous layer of the articular capsule is strengthened by three ligaments: the **iliofemoral, pubofemoral,** and **ischiofemoral.** The most important, the iliofemoral ligament, is shaped like an inverted Y, and covers the anterior surface of the joint. The Y ligament is one of the strongest in the body. It prevents hyperextension (backward bending) of the hip joint, becoming taut and resisting the tensile stresses placed upon it when the hip is fully extended. As a result, when the body is in a standing position, with the body weight centered slightly behind the hip joint, erect stance can be maintained without any muscular effort in the hip region.

The movements at the hip joints are flexion, extension, abduction, adduction, medial rotation, lateral rotation, and circumduction.

Knee Joint

The **knee (tibiofemoral) joint** is the largest and most complex joint in the body, as well as one of the weakest and most vulnerable to injury (Figure 8.17). It is a synovial joint (diarthrosis) with modified hinge structure. It is capable of some rotational movements.

The knee joint is actually a composite of three synovial joints: (1) the articulation between the medial femoral and the medial tibial condyles, (2) the articulation between the lateral femoral and the lateral tibial condyles, and (3) the articulation between the patella and the femur (medial and lateral tibiofemoral joints and patellofemoral joint).

Flexion and extension are the primary movements at the knee joint, with a slight amount of medial and lateral rotation. The main flexors are the **hamstring muscles,** and the key extensor is the **quadriceps femoris.** With the knee flexed, the leg can be rotated laterally and medially. A rotational "screw-home" movement occurs just as the knee assumes full extension. If this "screw-home" action takes place when the foot is not free, as in the act of standing up, the femur is the bone that rotates medially in relation to the tibia until the knee is locked. If the "screw-home" action takes place when the foot is free, as when a punter kicks a football, the tibia rotates laterally in relation to the femur. The unlocking of the extended knee (the reverse of the "screw-home" action) is initiated by the *popliteus muscle*.

The "screw-home" phase that locks the knee occurs because the articular surface of the medial condyle is longer than that of the lateral condyle. As a result, the lateral condyle uses up its articular surface just before full extension is realized. The completion of extension occurs as the medial

condyle continues to rotate on its longer articular surface, accompanied by the "screw-home" action and the locking of the knee. During this final phase, the lateral condyle acts as a pivot.

Several anatomical features are basic to knee movements. The curved condyles of the femur articulate with the flattened condyles of the tibia, allowing some rotation along with flexion and extension. Because the pelvis is wide, the shaft of the femur is medially directed to assume an oblique set at the knee joint. Except in bowlegged people, the tibias of both legs are usually parallel.

FIGURE 8.17

Knee (tibiofemoral) joint. (A) Anterior view of right knee joint. The patella has been removed, and the patellar ligament has been cut. (B) Sagittal section showing bursae of knee. (C) Articular surface of tibia in knee joint. Note the form and shape of the semilunar cartilages (menisci). (D) Cruciate ligaments of the knee joint in flexed knee, lateral view. The anterior cruciate ligament is slack during flexion.

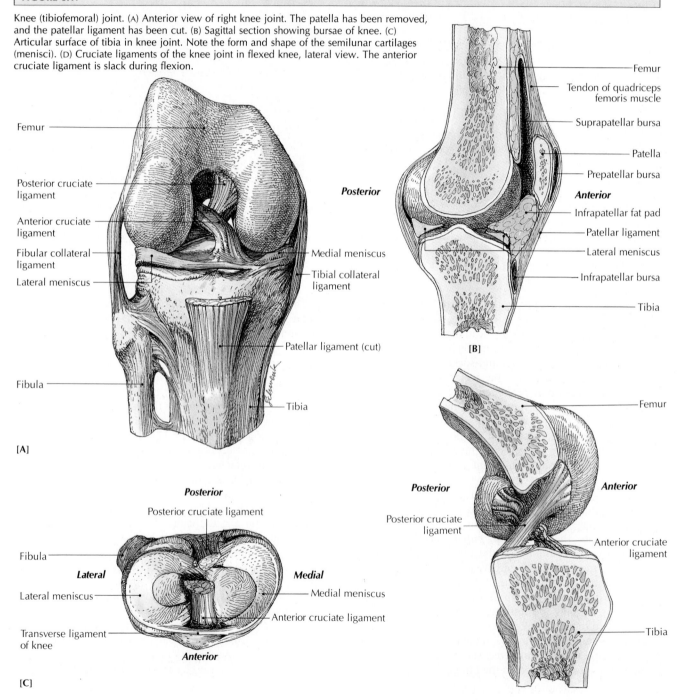

In front, the synovial membrane is lax when the knee is extended. The membrane is extensive enough so that no strain is exerted on it when the knee is flexed. Thickenings within the articular capsule include the *tibial* (medial) *collateral ligament* and the *fibular* (lateral) *collateral ligament* extending from the sides of the femur. Both are strong ligaments that are slack during flexion and taut during extension.

A common knee injury to football players is caused by a blow on the lateral side of the leg, which produces excessive stress on the medial side of the knee that may cause a strained (stretched) or torn tibial collateral ligament. (See the essay on p. 212.) Such an injury may be accompanied by damage to the medial meniscus. Injuries to the knee usually involve the three Cs: collateral ligaments, cruciate ligaments, and cartilages (menisci).

In the center of the knee joint, between the condyles of the femur and tibia, are two ligaments: the *anterior* and *posterior cruciate ligaments.* They are strong cords that cross each other like an X. The anterior cruciate ligament is taut when the knee is fully extended, and slack when the knee is flexed. It prevents the backward dislocation of the femur, forward dislocation of the tibia, and excessive extensor or rotational movements, especially hyperextension. The posterior cruciate ligament becomes tighter as flexion proceeds. It prevents the forward dislocation of the femur, backward dislocation of the tibia, and hyperflexion of the knee.

Of the two ligaments, the anterior cruciate ligament is torn much more often. The damage may come from a blow driving the tibia forward on the femur, from a blow driving the femur back on the tibia, or from excessive hyperextension of the knee. The resulting instability of the knee is so serious that use of the limb may be impaired severely.

Within the knee joint are two crescent-shaped fibrocartilaginous plates called the *lateral* and *medial menisci* (muh-NISS-eye; singular, meniscus; Gr. *meniskos,* crescent, moon). They act as a cushion between the ends of bones that meet in a joint. The menisci at the knee joint deepen the articular surfaces between the femoral and tibial condyles, and increase the security of the joint by adjusting the nonmatching surfaces of the tibial and femoral condyles.

Injury to the medial meniscus occurs about 20 times more often than to the lateral meniscus because the medial meniscus is attached to the medial collateral ligament. A sudden twist of the flexed knee that is bearing weight can tear the medial meniscus. This injury is most common in athletes, especially football players (see the essay on p. 212).

In addition to the ligaments, the muscles and tendons surrounding the knee joint also play a role in stabilizing it. Of the muscles, the most important is the quadriceps femoris.

Several bursae cushion the knee joint. They are (1) the *suprapatellar bursa* above the patella, between the quadriceps femoris muscle and the femur, (2) the *prepatellar bursa* between the skin and the patella, and (3) the *infrapatellar bursa* between the skin and the tibial tuberosity. The prepatellar bursa, which allows for the free movement of the skin over the patella, is subject to a friction bursitis, known as "housemaid's knee" or "water on the knee." (Repeated blows or falls may also cause "water on the knee.") This condition occurs in miners and others who frequently work on all fours. Bursitis of the infrapatellar bursa is often found in roofers and others who kneel with their trunks upright.

Ask Yourself

1 *What types of movement are possible in the temporomandibular joint?*

2 *What major structures make up the ball-and-socket joint of the shoulder?*

3 *What are the SITS muscles?*

4 *How does the arrangement of the articulating bones of the hip joint contribute to its stability?*

5 *What three joints make up the composite knee joint?*

6 *What are the primary movements at the knee joint?*

7 *What purposes do menisci serve?*

NERVE SUPPLY AND NUTRITION OF SYNOVIAL JOINTS

The nerve and blood supplies of synovial joints are discussed together because of the general rule "Where there is a nerve, there is also an artery." In other words, nerves cannot function without a readily available nourishing blood supply, and to go one step further, joint movement cannot be regulated properly, and without injury, without the effective stimulation of the joints by sensory nerves.

Nerve Supply of Joints

The same nerves supplying the muscles that move a joint also send out branches that supply the skin over the muscle attachments and the joint itself. One joint may be supplied by the branches of several nerves. Many *sensory* nerve fibers of these nerves terminate as nerve endings in the fibrous capsules, ligaments, and synovial membranes of the joints. Sensory nerve fibers relay information about the joint activity to the spinal cord and brain. After this information is processed in the spinal cord and brain it contributes to the information conveyed by *motor* fibers of these nerves to the muscles controlling the movements of the joints.

Sensory nerve fibers are usually considered to be either algesic or proprioceptive. *Algesic* fibers are pain receptors. They are particularly responsive when the capsule surrounding the joint is worn, or the connecting ligament is overstretched or torn. Injuries to joint ligaments and nonarticulating parts of the cartilage and disks usually produce intense pain because of the many algesic nerve endings in those areas. (The part of the cartilage that is directly involved with articulations has no nerve supply.)

The *proprioceptive* nerve fibers transmit information about the reflex control of body posture and the exact degree of movement at a joint. Impulses from nerve endings in the articular capsule pass to the spinal cord and brain, where they activate the mechanisms that control the muscles that move joints.

Blood and Lymph Supply of Joints

All the tissues of synovial joints receive nutrients either directly or indirectly from blood vessels, except the articulating portions of the articular cartilages, disks, and menisci. (No articulating cartilage has a direct blood supply, for example.) The articulating areas are nourished indirectly by the synovial fluid that is distributed over the surface of the articular cartilage. The flow of synovial fluid is stimulated by movement at the joints, and this is why physical activity is essential for the maintenance of healthy joints. Under normal conditions, only enough synovial fluid is secreted to produce a thin film over the joint surfaces. If a joint becomes inflamed, however, the secretion of synovial fluid may become overstimulated, causing the fluid to accumulate and producing swelling and discomfort at the joint.

Arteries near a synovial joint send out branches that join together freely on the outer surface of the joint. These branches penetrate the fibrous capsule and ligaments, and form another branching network of capillaries that spreads throughout the synovial membrane. These capillaries are so numerous, and so close to the surface of the synovial membrane, that it is relatively easy for a hemorrhage to occur into the articular cavity, even as the result of only a minor injury. The arteries also extend into the fatty pads and the non-articulating portions of the articular cartilage, disks, and menisci.

When ligaments and tendons are severed or seriously damaged, they heal slowly but in due course very well, and with the proper surgical treatment they can become as strong as they were before. The repair is associated with the generation of new fibroblasts and a good capillary blood supply, both of which grow into the damaged region. The fibroblasts become oriented parallel to the long axis of the ligament or tendon, and form new collagenous fibers.

In addition to the blood vessels around a joint, there are many lymphatic vessels that form a network near the flexor part of the joint. These lymphatic vessels join adjacent lymphatic vessels of the body wall or relevant limb. They drain excessive tissue fluids from the region of the joint and return them to the bloodstream.

Ask Yourself

1 *What are algesic nerve fibers? Proprioceptive nerve fibers?*

2 *How is the nonarticulating portion of a joint nourished?*

3 *Why is a hemorrhage at a joint likely after an injury?*

PHYSIOLOGICAL AND ANATOMICAL ABNORMALITIES

Arthritis

The word *arthritis* means "inflammation (*-itis*) of a joint (*arthro-*)." Arthritis may affect any joint in the body, but the most common sites are the shoulders, knees, neck, hands, low back, and hips. About 30 million Americans have arthritis seriously enough to require medical treatment. Arthritis is most common in men over 40, postmenopausal women, blacks, and people in low-income groups. Elderly people in general are most seriously affected. There are as many as 25 different specific forms of arthritis. The most common types are described in the following paragraphs.

Gouty arthritis *Gouty arthritis*, or simply *gout*,* is a metabolic disease resulting from chemical processes in the body. One product of the metabolism of

*The word *gout* comes from the Latin word *gutta*, meaning "drop," because it was once thought that gout was caused by drops of unhealthy humors, or poisons, dripping into the joints.

Gouty arthritis, one of the most painful forms of arthritis, often affects the big toe. [*Carroll Weiss/Camera M.D.*]

purine is uric acid, which is normally excreted in urine. Sometimes, however, the body produces too much uric acid or the kidneys do not remove enough of it. The excess uric acid combines with sodium to form needle-sharp crystals of sodium urate salt that settle in soft body tissues and cause inflammation and pain. Any joint can be affected by gout, but the most common sites are the big toe (see photo) and, oddly, since it is not a joint at all, the cartilage of the rim of the ear, just above the earlobe. The pain is sudden and intense, usually starting in the metatarsophalangeal (first) joint of the big toe. The pain may progress to the instep, ankle, heel, knee, or wrist joints. Crystal deposits in the kidneys may lead to kidney stones or even kidney failure.

Osteoarthritis *Osteoarthritis,* or *degenerative arthritis,* is the most common form of arthritis. It is a chronic, progressive degenerative disease, usually caused by a breakdown of chondrocytes during the normal "wear and tear" of everyday living. Osteoarthritis often affects the weight-bearing joints such as the hip, knee, and lumbar region of the vertebral column. It occurs equally in men and women. Osteoarthritis usually begins as a normal part of aging, but it may also develop as a result of damage to joints, including infection or metabolic disorders. Typical symptoms are pain in the joints, morning stiffness, grating of the joints, and restricted movement. Osteoarthritis of interphalangeal joints produces irreversible bony growths in the distal joints (Heberden's nodes; see photo) and in the proximal joints (Bouchard's nodes).

Rheumatoid arthritis *Rheumatoid arthritis* is the most debilitating form of chronic arthritis. It is an inflammatory disease, usually involving matching joints on opposite sides of the body and their surrounding bursae, tendons, and tendon sheaths. Most movable joints can be affected, but the fingers, wrists, and knees are most susceptible. Rheumatoid arthritis affects three times as many women as men, and is most prevalent in women between 35 and 45. The disease usually starts with general symptoms such as fatigue, low-grade fever, and anemia before it begins to affect the joints. Eventually the tissue in the joint capsule becomes thickened (the tissue is then called a *pannus*) as a result of inflammation and the accumulation of

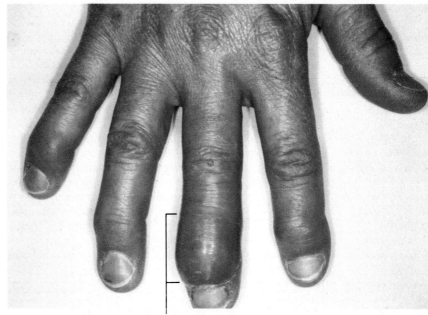

Heberden's node

An irreversible effect of osteoarthritis is Heberden's nodes in the distal joints. [*Arthritis and Rheumatism Association.*]

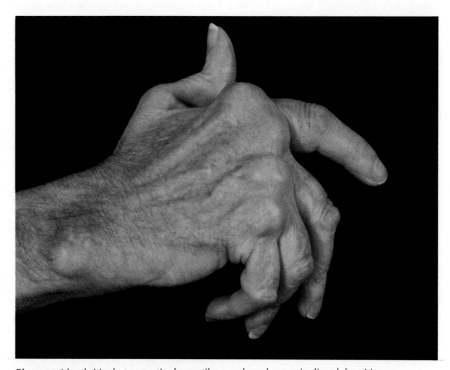

Rheumatoid arthritis destroys articular cartilage and produces crippling deformities. [*SIU—Peter Arnold.*]

synovial fluid. Soon the disease invades the interior of the joint and interferes with the nourishment of the articular cartilage. Slowly, the cartilage is destroyed, and crippling deformities are produced (see photo). Fibrous tissue then develops on the articulating surfaces of the joint and makes movement difficult (a condition called *fibrous ankylosis;* Gr. *ankulos,* bent). In the final stage, the fibrous tissue becomes calcified and forms a solid fusion of bone, so that the joint is completely nonfunctional *(bony ankylosis).*

Bursitis

Bursitis is an inflammation of one or more *bursae*. Such an inflammation produces pain and swelling, and restricts movement. It occurs most frequently in the subacromial bursa near the shoulder joint, the prepatellar bursa near the knee joint, and the olecranon bursa of the elbow.

Bursitis may be caused by an infection or by physical stress from repeated friction related to a person's activities (miner's knee, tennis elbow) or from repeated blows or falls.

Dislocation

A *dislocation,* or *luxation* (L. *luxare,* to put out of joint), is a displacement movement of bones in a joint that causes two articulating surfaces to become separated. A partial dislocation is called a *subluxation*. A dislocation usually is the result of physical injury, but it can also be congenital (as in hip dysplasia) or the side effect of a disease such as Paget's disease (see Chapter 6).

The *shoulder joint* is dislocated more than any other joint, partly because of the shallowness of the glenoid cavity, which holds the head of the humerus, and the loose capsule. The *hip joint,* in contrast to the shoulder, is seldom dislocated because of the stability produced by the secure fit of the head of the femur inside the deep acetabulum, the strong articular capsule, and the strong ligaments and muscles that surround the joint.

The *jaw joint* may also become dislocated. Following a vigorous laugh or yawn, the head of the mandible may move so far forward that it glides in front of the articular tubercle. As a result of such a dislocation, the mouth cannot be closed, except by a knowledgeable second party who understands how to relocate the joint.

Immediate reduction (relocation) of the displaced joint prevents edema, muscle spasm, and further damage to tissues, nerves, and blood vessels. After reduction, the injured joint may need to be immobilized for two to eight weeks. (However, see New Therapy for Damaged Joints on this page.)

Bunion

A *bunion* (Old Fr., *buigne,* bump on the head) is a lateral deviation of the big toe toward the second toe, accompanied by the formation of a bursa and callus at the bony prominence of the first metatarsal. It may be caused by poorly fitted shoes that compress the toes. Bunions are most common among women.

Sprain

A *sprain* is a tearing of ligaments that follows the sudden wrench of a joint. It is usually followed by pain, loss of mobility (which may not occur until several hours after the injury), and a "black-and-blue" discoloration of the skin, caused by hemorrhaging into the tissue surrounding the joint. In a slight sprain, the ligaments heal within a few days, but severe sprains (usually called "torn ligaments") may take weeks or even months to heal. In some cases, the ligament must be repaired surgically. A sprained ankle is the most common joint injury.

New therapy for damaged joints

Ordinarily, damaged joints are immobilized in a cast for several weeks after surgery. According to recent reports, however, joints heal faster after surgery when they are kept in constant motion, rather than being immobilized. When joints move, they stimulate the flow of synovial fluid, which prevents the synovial membrane from adhering to cartilage in the joint. Constant motion at the joint can be accomplished by a new technique called *continuous passive motion* (CPM), in which a motorized apparatus moves the joint backward and forward gently. Usually, the apparatus does not disturb sleep after a short period of adjustment, and the muscles are not fatigued because the machine does all the work.

Generally, CPM patients are ready for physical therapy about 7 to 10 days after surgery. Immobilized patients usually begin therapy about six weeks after surgery, when their plaster cast is removed.

MEDICAL TERMINOLOGY

ANKYLOSIS (Gr. *ankulos,* bent + *osis,* condition of) Stiffness or crookedness in joints.

ARTHRODESIS (Gr. *arthro,* joint + *desis,* binding) The surgical fusion of the bones of a joint.

ARTHROGRAM (joint + Gr. *grammos,* picture) An x-ray picture of a joint taken after the injection into the joint of a dye opaque to x rays.

ARTHROPLASTY (joint + Gr. *plastos,* molded) The surgical repair of a joint or the replacement of a deteriorated part of a joint with an artificial joint.

BURSECTOMY (bursa + *-ectomy,* removal of) The surgical removal of a bursa.

CAPSULORRHAPHY (capsule + Gr. *raphe,* suture) The surgical repair of a joint capsule to prevent recurrent dislocations.

CHRONDRITIS (Gr. *khondros,* cartilage + *-itis,* inflammation) Inflammation of a cartilage.

MENISCECTOMY (meniscus + *-ectomy,* removal of) The surgical removal of the menisci of the knee joint.

OSTEOPLASTY (Gr. *osteon,* bone + *plastos,* molded) The scraping away of deteriorated bone.

OSTEOTOMY (bone + *-tomy,* cutting of) The surgical cutting of bone, for example, the realignment of bone to relieve stress.

RHEUMATISM (Gr. *rheumatismos,* to suffer from a flux or stream) Any of several diseases of the muscles, tendons, joints, bones, or nerves. Generally replaced by the term *arthritis*.

RHEUMATOLOGY The study of joint diseases, especially arthritis.

SUBLUXATION (L. *sub,* less than + *luxus,* dislocated) An incomplete dislocation.

SYNOVECTOMY (L. *synovia,* lubricating liquid + *-ectomy,* removal of) The surgical removal of the synovial membrane of a joint.

SYNOVITIS (*synovia* + *-itis,* inflammation) An inflammation of the synovial membrane.

SUMMARY

Bones give the body its structural framework, and muscles give it its power, but joints (articulations) provide the mechanism that allows the body to move.

Classification of joints

1 Joints may be classified by function and degree of movement: an immovable joint is a **synarthrosis;** a slightly movable joint is an **amphiarthrosis;** a freely movable joint is a **diarthrosis.**

2 Based on structure, or the type of tissue that connects the bones, joints may be classified as *fibrous, cartilaginous,* or *synovial.*

Fibrous joints

1 Fibrous connective tissue unites the bones in *fibrous joints.*

2 Three types of fibrous joints are **sutures, syndesmoses,** and **gomphoses.**

Cartilaginous joints

1 In **cartilaginous joints,** bones are united by a plate of hyaline cartilage or a softer fibrocartilaginous disk.

2 The two types of cartilaginous joints are **synchondroses** and **symphyses.**

Synovial joints

1 **Synovial joints** are the articulations where the bones move easily upon each other. Such free movement takes place because the ends of the bones are plated with a smooth **articular cartilage,** lubricated by **synovial fluid,** and bound together by an **articular capsule.**

2 The synovial joint is composed of the **synovial cavity, articular cartilage, articular capsule,** and **ligaments.**

3 **Bursae** are sacs filled with synovial fluid that help eliminate friction when a muscle or tendon rubs against another muscle, tendon, or bone. A **tendon sheath** is a modification of a bursa that helps reduce friction around tendons.

Movement at synovial joints

1 Movement at synovial joints may be restricted by interference by other structures, tension exerted by ligaments of the articular capsule, and muscle tension.

2 Some basic terms of movement are described in Table 8.2.

3 In any movement about a joint, one member of a pair of articulating bones maintains a fixed position, and the other bone moves in relation to it about an axis (or axes).

4 When movement is restricted to rotation about one axis, the joint is said to be **uniaxial.** When two movements can take place about two axes of rotation, the joint is **biaxial.** A **multiaxial** (triaxial) joint has three independent rotations about three axes.

Types of synovial joints

1 *Synovial joints* are freely movable joints that are classified on the basis of the shape of their articulating surfaces and the types of joint movements those shapes permit.

2 *Hinge joints* are uniaxial. The convex surface of one bone fits into the concave surface of another.

3 *Pivot joints* are uniaxial joints that have a rotational movement around a long axis through the center of the pivot.

4 *Condyloid* (condylar) or *ellipsoidal joints* are modifications of ball-and-socket joints, with ligaments and muscles limiting movement to only two axes of rotation.

5 *Gliding* (plane) *joints* are small biaxial joints formed by essentially flat surfaces.

6 *Saddle joints* are multiaxial joints in which both articulating bones have saddle-shaped concave and convex areas.

7 *Ball-and-socket joints* are the most movable type. The globelike head (ball) of one bone fits into the cuplike concavity (socket) of another bone.

Description of some major joints

1 The *jaw (temporomandibular) joint* is the only movable joint of the head. It is a synovial joint composed of the articular head of the mandible and the mandibular fossa and articular tubercle of the temporal bone. It is movable in all three planes.

2 The *shoulder (glenohumeral* or *humeroscapular) joint* is a multiaxial ball-and-socket joint. The head of the humerus articulates with the shallow glenoid cavity of the scapula. The shoulder joint has more freedom of movement than any other joint in the body.

3 The *hip (coxal) joint* is a multiaxial, ball-and-socket, synovial joint. The head of the femur articulates with the acetabulum of the hipbone (os coxa). The hip joint is one of the most stable articulations in the body.

4 The *knee (tibiofemoral) joint* is the largest, most complex, and one of the weakest joints in the body. It is a synovial, modified hinge joint. The knee joint is a composite of the articulation between (a) the medial femoral and medial tibial condyles, (b) the lateral femoral and lateral tibial condyles, and (c) the patella and the femur.

Nerve supply and nutrition of synovial joints

1 Joints cannot function without the effective stimulation of nerves, which in turn need a nourishing blood supply.

2 Some nerves of a joint are *sensory* nerves that convey information to the brain and spinal cord. Others are *motor* nerves that convey processed information from the brain and spinal cord to the muscles controlling joint movements.

3 All the tissues of synovial joints receive nutrients directly from blood vessels, except the articulating cartilages, disks, and menisci, which are nourished by the synovial fluid.

Physiological and anatomical abnormalities

1 *Arthritis* is an inflammation of a joint. It most commonly affects the shoulders, knees, neck, hands, low back, and hips. The major forms of arthritis are **gouty arthritis, osteoarthritis,** and **rheumatoid arthritis.**

2 *Bursitis* is an inflammation of one or more bursae.

3 A *dislocation,* or *luxation,* is a displacement movement of bones in a joint that causes two articulating surfaces to become separated.

4 A *bunion* is a lateral deviation of the big toe toward the second toe, accompanied by the formation of a bursa and callus at the bony prominence of the first metatarsal.

5 A *sprain* is a tearing of ligaments that follows the sudden wrench of a joint.

6 A new technique used in treating damaged joints is **continuous passive motion** (CPM) in which a motorized apparatus keeps the joint in constant motion until it heals sufficiently for physical therapy.

UNDERSTANDING THE FACTS

1 What is an articulation?

2 Which of the three major joint types by function is the most movable? The least movable?

3 What are the three major joint types by structure?

4 What is a syndesmosis? A gomphosis? A synchondrosis?

5 What type, by structure, are most permanent joints?

6 Give three functions of articular disks.

7 What are the functions of synovial fluid?

8 Perform the following movements, and name the type of movement that is occurring:
  a move thumb pad to touch a finger pad
  b move arm so that it describes the surface of a cone
  c bring wrist close to shoulder
  d rotate hand and forearm, turning palm forward
  e move head indicating "no"
  f spread fingers
  g open mouth

9 Give an example of a uniaxial, biaxial, and multiaxial joint.

10 Which joint in the body is the most movable?

11 Which is the only movable joint in the head?

12 Why is the shoulder joint also called the glenohumeral or humeroscapular joint?

13 What movements are possible at the shoulder joint?

14 How is the iliofemoral ligament important to the hip joint?

15 A torn cartilage of the knee, a common athletic injury, usually involves which specific structure?

16 Which parts of synovial joints have no nerve supply?

17 Why is physical activity so essential to proper nourishment of joints?

18 Why do injured ligaments heal relatively slowly?

19 List the six most common sites of arthritis.

UNDERSTANDING THE CONCEPTS

1 What is the difference between a suture and a synostosis?

2 The articulating surfaces of joints are often almost friction-free. What is responsible for this smoothness?

3 What relationship does the depth of the socket of a joint have to its range of movement?

4 How does the structure of a pivot joint relate to its movement?

5 Why is it important for the hip joint to be very secure, strong, and stable?

6 Why do injuries to joint ligaments and nonarticulating parts of a joint often produce intense pain while injuries to the articulating portions do not?

7 Why is arthritis most common in the elderly?

8 The statement "Structures that encourage movement do not promote stability" appears in one form or another throughout this chapter. Now that you have studied this material on articulations, what does this statement mean to you?

SUGGESTED READING

AREHART-TREICHEL, J., "The Joint Destroyers." *Science News,* September 1982.

AUFRANC, O. E., AND R. H. TURNER, "Total Replacement of the Arthritic Hip." *Hospital Practice,* October 1971.

CAPLAN, ARNOLD I., "Cartilage." *Scientific American,* October 1984.

ENIS, J. E., "The Painful Knee." *Hospital Medicine,* December 1980.

HAMERMAN, D., AND ROSENBERG, L. C., "Diarthrodial Joints Revisited." *Journal of Bone and Joint Surgery,* 52A (1970): 725.

NAPIER, J., "The Antiquity of Human Walking." *Scientific American,* April 1967.

NATIONAL INSTITUTES OF HEALTH, "Total Hip Joint Replacement." *Conference Summary,* vol. 4, no. 4. U.S. Department of Health and Human Services, 1982.

ROSSE, C., AND D. K. CLAWSON, *The Musculoskeletal System in Health and Disease.* New York: Harper & Row, 1980.

SIMON, W. H., *The Human Joint in Health and Disease.* Philadelphia: University of Pennsylvania Press, 1978.

SONSTEGARD, D. A., LARRY S. MATTHEWS, AND HERBERT KAUFER, "The Surgical Replacement of the Human Knee Joint." *Scientific American,* January 1978.

THOMPSON, C. W., *Manual of Structural Kinesiology,* 9th ed. St. Louis: Mosby, 1981.

"Total Hip-Joint Replacement in the United States." *Journal of the American Medical Association,* October 15, 1982.

WILSON, F. C., ed., *The Musculoskeletal System: Basic Processes and Disorders,* 2nd ed. Philadelphia: Lippincott, 1983.

9
Muscle Tissue

LEARNING OBJECTIVES

1 Identify the four important physiological properties and three general functions of muscle tissue.

2 List the three types of muscle tissue, and relate each to its special functions and characteristics.

3 Describe the cell structure and organization of skeletal muscle.

4 Identify three forms of fascia and five other connective tissue elements in skeletal muscle.

5 Describe the blood and nerve supply to skeletal muscles.

6 Define motor unit, neuromuscular junction, and synaptic cleft, and explain their functions in the nervous control of muscles.

7 Explain the all-or-none principle as it relates to muscular contraction.

8 Describe how the strength of a muscle contraction can be controlled.

9 Outline the major steps in the process of muscular contraction, according to the sliding-filament theory.

10 Explain how muscle relaxation is accomplished.

11 Describe the sources of energy for muscular contraction.

12 Distinguish between isotonic and isometric contractions, and define twitch, treppe, and tetanic contractions.

13 Explain the differences between fast- and slow-twitch muscle fibers.

14 Define oxygen debt and muscle fatigue, and tell how they are related.

15 Explain the function of heat production by a muscle in the body.

16 Describe the basic physiological properties of smooth muscle.

17 Distinguish between single-unit and multi-unit smooth muscle.

18 Describe the basic structure and function of cardiac muscle tissue.

19 Describe the effects of muscular dystrophy, myasthenia gravis, and tetanus on the muscular system.

J oints make a skeleton potentially movable, and bones provide a basic system of levers, but bones and joints cannot move by themselves. The driving force, the power behind movement, is muscle tissue.

The basic physiological property of muscle tissue is ***contractility,*** the ability to *contract,* or shorten. In addition, muscle tissue has three other important physiological properties. ***Excitability*** (or irritability) is the capacity to receive and respond to a *stimulus,* ***extensibility*** is the ability to stretch, and ***elasticity*** is the ability to return to its original shape after being stretched or contracted. These four properties of muscle tissue are related, and all involve movement.

In the process of contracting, important work is done. For example, food is passed along the digestive tract by a series of rhythmic waves of *smooth muscle* contractions. The contractions of *cardiac muscle* pump blood with remarkable force and consistency from the heart to all parts of the body. As certain *skeletal muscles* in your body contract, your lower limbs move at the ankle, knee, and hip, and you walk or run. Muscular contractions also help to maintain body posture in a standing or sitting position, even when there is no obvious motion. Finally, the contractions of skeletal muscles produce much of the heat used to keep body temperature at a normal level. To sum up, the general functions of muscle tissue are *movement, posture,* and *heat production.*

The muscular system presents no exception to the general rule that in the human body structure and function work together to accomplish the most advantageous results. Different types of muscles do different types of jobs. Each of the actions described above is carried out by a specialized type of muscle: *skeletal* (walking), *smooth* (digestion), or *cardiac* (heart beating). In the following sections you will see how the anatomy of each muscle type allows it to perform its own special physiological function.

SKELETAL MUSCLE

Skeletal muscle tissue acquired its name because most of the muscles involved are attached to the skeleton, and make it move.* It is also called *striated muscle* because its fibers or cells are composed of alternating light and dark stripes, or striations. Skeletal muscle has yet another descriptive name, *voluntary muscle,* because we can contract it when we want to. For example, you can clench your fist tightly by voluntarily contracting the muscles in your forearm. Although skeletal muscles can be contracted voluntarily, they are also capable of contraction *without* conscious control (involuntarily). Muscles are usually in a partially contracted state, which gives them *tonus,* or what we commonly refer to as "muscle tone." Tonus is necessary to keep a muscle ready to react to the stimulus preceding a complete contraction, to hold parts of the body such as the head erect, and to aid in the return of blood to the heart.

*The word *muscle* is based on the Latin *musculus,* which means "little mouse." Muscles were so named because the movement of muscles under the skin was thought to resemble a running mouse.

FIGURE 9.1

The anatomy of a skeletal muscle, from gross structure to molecular structure. (A) Muscle in arm. (B) Muscle fascicle, or bundle of muscle fibers. (C) Muscle fiber. (D) Myofibril. Each fiber is composed of myofibrils, which reveal the banding that gives the designation *striated* to the muscle. (E) Sarcomere, the functional unit of muscle contraction, extending from Z line to Z line. (F) Electron micrograph showing the bands and lines illustrated in (D) and (E). (G) The two main muscle proteins, myosin (thick myofilament) and actin (thin myofilament). Their overlapping or failure to overlap gives the myofibril its banded appearance. (H) Cross section of a myofibril through an A band. The thin protein myofilaments are actin, and the thick ones are myosin. [(F) *W. Bloom and D. W. Fawcett,* A Textbook of Histology, *11th ed. Philadelphia: W. B. Saunders, 1986, p. 283.*]

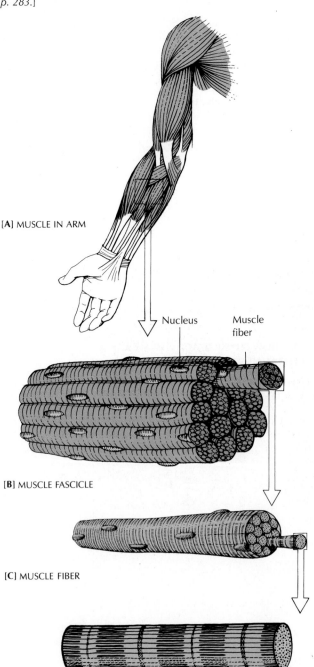

[A] MUSCLE IN ARM

Nucleus   Muscle fiber

[B] MUSCLE FASCICLE

[C] MUSCLE FIBER

[D] MYOFIBRIL

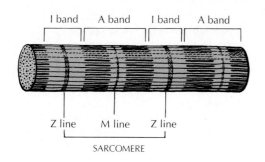

I band   A band   I band   A band

Z line   M line   Z line

SARCOMERE

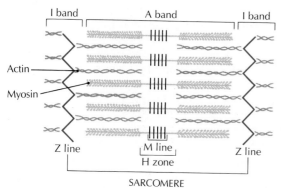

I band   A band   I band

Actin

Myosin

Z line   M line   Z line

H zone

SARCOMERE

[E]

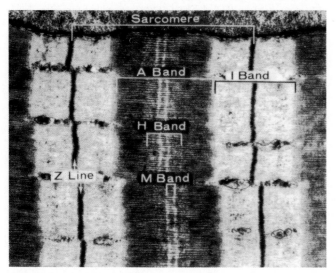

Sarcomere

A Band   I Band

H Band

Z Line   M Band

[F]   ×20,000

MYOSIN (thick myofilament)

ACTIN (thin myofilament)

[G]   [H]

Myosin   Actin

In the following sections, we will examine the structure of skeletal muscles, starting with the muscle cell and proceeding to the tissue level. After describing the nervous control of muscular contraction and the energy supply for it, we explain how a muscle contracts.

Cell Structure and Organization

Skeletal muscle is composed of individual, specialized cells called *muscle fibers* (Figure 9.1). These multinucleated cells are known as fibers because they have a long, cylindrical shape and several nuclei. Thus, they look more like fibers than "typical cells." Muscle fibers average 3.0 cm (1.2 in.) in length, but some may be more than 30 cm (12 in.) or as short as 0.1 cm (0.04 in.).* Diameters usually range from 0.01 cm (0.004 in.) to 0.001 cm (0.0004 in.).

Each skeletal muscle fiber is enclosed within a thin cell membrane called the *sarcolemma* (Gr. *sarkos,* flesh + *lemma,* husk). The fiber contains several nuclei, and a specialized type of cytoplasm called *sarcoplasm.* Within the sarcoplasm are many mitochondria and a large number of individual threadlike fibers known as *myofibrils* (*myo,* muscle), which run lengthwise and parallel to one another (Figure 9.2). Around each myofibril, and running parallel to it, is the *sarcoplasmic reticulum.* This network of tubes and sacs contains calcium ions, and is something like the smooth endoplasmic reticulum found in other types of cells. Crossing the sarcoplasmic reticulum at right angles are the *transverse tubules,* or *T tubules,* a series of tubes that run across the fiber to the outside. The sarcolemma continues within the muscle fiber as the lining of the T tubule. Together, a T tubule and a *terminal cisterna* of the sarcoplasmic reticulum on either side of the tubule make up a *triad.*

The myofibrils are made up of many thick and thin threads called *myofilaments.* The thick myofilaments are composed of a fairly large protein called *myosin.* The thin myofilaments are composed mainly of a smaller protein, *actin,* and they also contain the proteins troponin and tropomyosin. An overlapping of the thick myosin strands and the thin actin strands produces the dark *A bands;* the thin actin strands alone appear as the light *I bands* (see Figures 9.1D, E, F). The alternating arrangement of the light and dark bands gives skeletal muscle its striations, or stripes. Cutting across each I band, like a dime in a stack of pennies, is a dark *Z line.* Within the A band is a somewhat lighter *H zone,* consisting of thick myosin strands only. Extending across the H zone is a delicate *M line,* which connects adjacent thick filaments. The fundamental unit of muscle contraction is the *sarcomere* (Gr. *meros,* part), which is made up of a section of the muscle fiber that extends from one Z line to the next one.

*The sartorius muscle of the anterior thigh is the longest muscle in the body. It may contain muscle fibers over 30 cm (12 in.) long, reaching from the hip to the knee. The shortest fibers are those of the stapedius muscle in the inner ear, which is shorter than 1 mm (0.04 in.); these fibers attach to the tiniest of bones, the stapes of the ear. The most powerful muscle in the body is probably the gluteus maximus of the buttocks.

FIGURE 9.2

Three-dimensional drawing of a small portion of one muscle fiber (cell). The myofibril to the right has the sarcoplasmic reticulum peeled away to show the A and I bands, H zone, and M and Z lines. The transverse (T) tubules conduct depolarization (the reduction of the electrical charge across a cellular membrane at the beginning of a muscular contraction) from the fiber surface into the muscle fiber. The terminal cisternae of the sarcoplasmic reticulum are connected to the T tubules in such a way as to allow a change in the permeability of the sarcoplasmic reticulum to follow the depolarization conveyed by the T tubules.

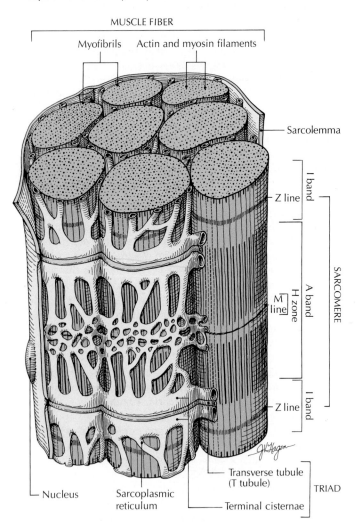

fasciae), and appears in three forms: superficial, deep, and subserous (or visceral).

The **superficial fascia,** also called the *subcutaneous layer,* lies just beneath the skin, but is not part of it (see Chapter 5). Varying in thickness, the superficial fascia contains fat deposits and is essentially a protective, insulating layer. Within this fascia are blood vessels, nerves, glands, lymphatic vessels, hair follicles, sebaceous glands, sweat glands, and the facial muscles.

Lying below the superficial fascia is the **deep fascia,** made up of several layers of dense connective tissue. An outer layer covers most of the body, and an inner layer covers the body walls. Other extensions of the deep fascia cover and separate muscles, permitting each one to contract independently. The deep fascia contains nerves and blood vessels, but no fat.

The **subserous** (visceral) **fascia** lies between the deep fascia and the serous membranes that line body cavities. Composed of loose areolar tissue, it covers and supports the organs of the thoracic, abdominal, and pelvic cavities.

Other Connective Tissue Associated with Muscles

Below the deep fascia that surrounds a muscle is another connective tissue sheath, called the **epimysium** (Figure 9.3). This sheath is sometimes continuous with the deep fascia. Extending inward from the epimysium is a layer of connective tissue, the **perimysium,** that encloses bundles of muscle fibers. These bundles of muscle fibers are called **fascicles.** Further extensions of the connective tissue, called **endomysium,** wrap around each muscle fiber. The three sheaths of connective tissue contain many blood vessels, lymphatic vessels, and nerves.

Besides serving as packing material around muscles, protecting and separating them, the connective tissue sheaths provide a point of attachment to bones and other muscles. Extending from the sheaths that cover muscles or muscle fibers are tendons and aponeuroses. A **tendon** is a strong cord of fibrous connective tissue that attaches muscle to the periosteum of bone (see Figure 8.3A). An **aponeurosis** is a broad, flat sheet of dense connective tissue that attaches to two or more muscles that work together, or to the coverings of a bone.

Fasciae

The fibrous connective tissue that covers the skeletal muscles and holds them together is part of a network extending throughout the body. It is called **fascia** (FASH-ee-uh; plural,

How much of a person's body weight is taken up by skeletal muscles?

Skeletal muscles comprise about 40 to 50 percent of an adult male's total body weight, and about 30 to 40 percent of an adult female's total body weight.

Why is one muscle larger than another?

Ordinarily, one muscle is larger than another because it contains more bundles of fibers. The largest muscles are found where large, forceful movements are common, such as in the back and legs. Delicate muscles, such as those in the eye, generally produce delicate movements. When a muscle is enlarged by exercise, the fibers increase in size, but the number of fibers remains the same.

FIGURE 9.3

(A) Cross section of a skeletal muscle showing how each muscle fiber is surrounded by the endomysium, each fascicle by the perimysium, and each muscle (group of fascicles) by the epimysium. (B) Enlarged version of the cross section in (A).

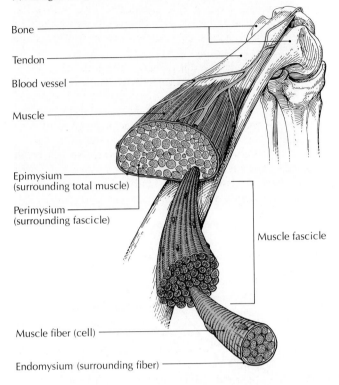

Bone
Tendon
Blood vessel
Muscle
Epimysium (surrounding total muscle)
Perimysium (surrounding fascicle)
Muscle fascicle
Muscle fiber (cell)
Endomysium (surrounding fiber)

[A]

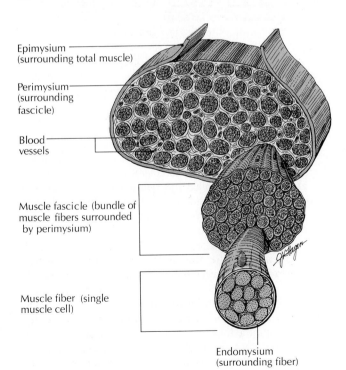

Epimysium (surrounding total muscle)
Perimysium (surrounding fascicle)
Blood vessels
Muscle fascicle (bundle of muscle fibers surrounded by perimysium)
Muscle fiber (single muscle cell)
Endomysium (surrounding fiber)

[B]

Blood and Nerve Supply

Muscles are supplied with blood by arteries that penetrate the connective tissue coverings. These arteries branch out into tiny, thin-walled blood vessels called *capillaries,* which carry an abundant supply of oxygen-rich blood to the muscles. In fact, each muscle fiber is supplied with oxygen and glucose by several capillaries that surround individual muscle cells. As you will see later, without a steady and adequate oxygen supply, the muscles would be unable to contract properly, and if oxygen is cut off for too long, the muscles weaken and die. Accumulated wastes are removed from the muscles via capillaries and carried in the blood toward the heart by veins. After reaching the heart, the blood is sent to the lungs, where the exchange of carbon dioxide and oxygen takes place. As a result of this exchange, fresh, oxygen-laden blood is sent back to the heart and recirculated through the arteries to the muscles.

Each skeletal-muscle fiber is contacted by at least one nerve ending. One motor nerve fiber can stimulate several muscle fibers at the same time (see Figure 9.4A). This important neuromuscular junction is described next.

What causes a hand or foot to "fall asleep"?

Pressure on a limb (as when you cross your legs, or fall asleep on your arm) may temporarily cut off its blood supply or the steady stream of nerve impulses that help to maintain muscle tone. Sensation may be lost as long as pressure remains on the nerve or blood vessel, but when the blockage is removed, the nerve impulses are effective once more, and the blood supply is returned. The combination of these renewed processes may cause an uncomfortable "pins and needles" feeling until normal functions resume completely.

Ask Yourself

1 *What are the four important properties of muscle tissue?*

2 *What are the three types of muscle tissue?*

3 *Why is skeletal muscle also known as striated muscle and voluntary muscle?*

4 *Where is a sarcomere located within a muscle?*

5 *What is the function of fasciae?*

6 *What is the difference between a fascicle and a fascia?*

7 *What are the main connective tissue sheaths associated with muscles?*

8 *How and why are muscles supplied with nerves and blood vessels?*

FIGURE 9.4

A neuromuscular junction. (A) Scanning electron micrograph of a motor neuron fiber terminating on several muscle fibers. The small branches of the neuron, called axon terminals, form flattened motor end plates on the surface of the muscle fiber.
(B) Schematic drawing of a neuromuscular junction. (C) Enlarged drawing of the portion of (B) enclosed in a rectangle, showing the motor end plate in detail. (D) Enlarged drawing of the portion of (C) enclosed in a rectangle, showing the synaptic cleft between the sarcolemma of the muscle fiber and the axon terminal. (E) Electron micrograph of a neuromuscular junction; compare with the drawing (D). [(A) *Lennart Nilsson*, Behold Man. *Boston, Little, Brown, 1974.* (E) *Photo by Dr. John E. Heuser, Washington University School of Medicine.*]

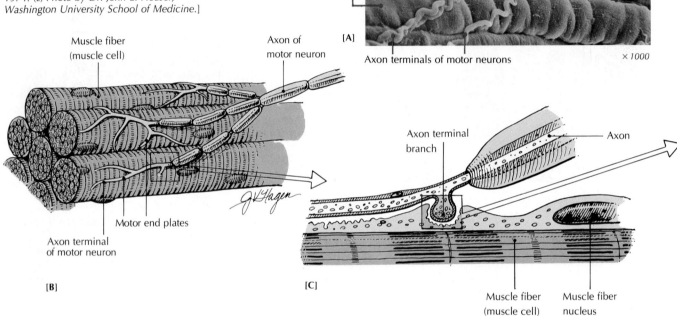

Axons of motor neurons

Muscle fibers

Axon terminals of motor neurons

[A]

×1000

Muscle fiber (muscle cell)

Axon of motor neuron

[B]

Motor end plates

Axon terminal of motor neuron

Axon terminal branch

Axon

[C]

Muscle fiber (muscle cell)

Muscle fiber nucleus

Nervous Control of Muscle Contractions

Although some small muscle fibers (such as those in the eye) may contract individually, muscle fibers usually contract in groups. Skeletal-muscle fibers are packed together into fascicles averaging about 150 fibers, and the fibers within each fascicle are controlled by a single *motor neuron* (a nerve cell that stimulates a muscle). In the powerful leg muscles, one motor neuron may stimulate up to 1600 fibers. A motor neuron, together with the muscle fibers it innervates, is called a *motor unit* (Figure 9.4).

Motor end plates The site where a motor nerve ending contacts a muscle fiber is called a ***neuromuscular*** (nerve + muscle) ***junction*** or a ***myoneural*** (muscle + nerve) ***junction.*** (Both junctions are the same; only the terminology is different.) The end branches of the motor neuron, known as *axon terminals,* gain access to the muscle fiber through the endomysium. At the point of contact between the muscle fiber

and the motor neuron, the muscle fiber membrane forms a ***motor end plate.*** The motor end plate is the specialized portion of the sarcolemma of a muscle fiber. It surrounds the terminal end of the axon. As shown in Figure 9.4D, mitochondria are particularly abundant near a motor end plate. The invaginated area of the sarcolemma under and around the axon terminal is called the ***synaptic gutter*** or ***synaptic trough,*** and the clefts inside the folds along the sarcolemma are called ***subneural clefts*** (see Figure 9.4D). These clefts greatly increase the surface area of the synaptic gutter.

At the motor end plate, nerve endings are separated from the sarcolemma of the muscle fiber by a tiny gap called a ***synaptic cleft*** (see Figures 9.4D, E). For many years neuroscientists have thought that when the chemical transmitter *acetylcholine* (uh-SEET-uhl-KOH-leen) is released from the synaptic vesicles of the nerve endings, it bridges the synaptic cleft and flows into the folds of the sarcolemma. Some acetylcholine then becomes attached to the receptor sites in the sarcolemma, initiating an electrochemical impulse across the sarcolemma of the muscle cell, so that sodium ions move *into* the sarcoplasm and potassium ions move *out.* This explana-

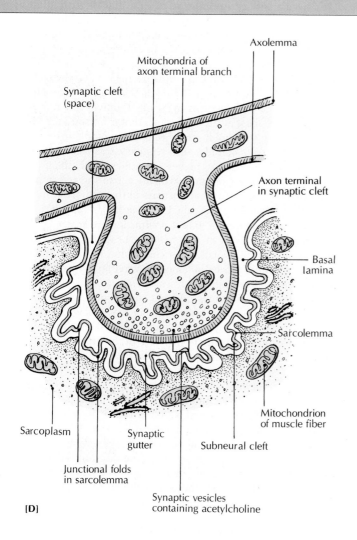

[D]

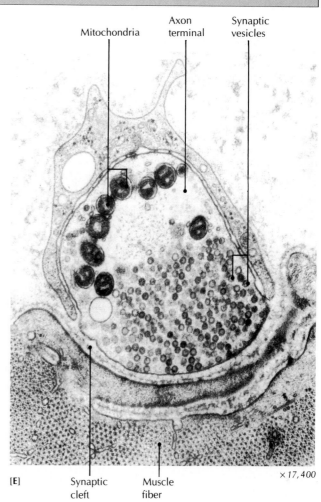

[E]

tion is known as the *classical vesicular hypothesis* (Figure 9.5A).

Recently, a new hypothesis has been proposed to account for the origin of the acetylcholine released by the stimulated nerve cell. This hypothesis agrees with the vesicular hypothesis that the vesicles store acetylcholine and play a role in its regulation within the nerve cell. However, it suggests that the acetylcholine released by the nerve does *not* originate in the vesicles, but is derived directly from the cytoplasm inside the neuron. This is known as the *cytoplasmic hypothesis* (Figure 9.5B).

Regardless of which hypothesis is correct, the final result is a temporary disturbance in the permeability of the sarcolemma. Sodium ions move into the cell while potassium ions move out. This upsets the balance of electrical charges on the inner and outer surface of the sarcolemma. During this process of *depolarization* (the reduction of the electrical charge across a membrane at the beginning of a muscular contraction), an electrical impulse is conveyed successively through the cell membrane and the transverse tubules to the sarcoplasmic reticulum, where it triggers the release of calcium ions, which stimulate the muscle to contract.

Acetylcholinesterase, an enzyme found on the muscle-fiber membranes, then breaks down acetylcholine into acetate and choline. As a result, depolarization ceases, and the muscle fiber relaxes.

All-or-none principle The minimal nervous stimulation needed to cause a muscle fiber to contract is called the *threshold (liminal) stimulus.* If the intensity of the stimulation is increased beyond the threshold level, the intensity of the contraction will *not* increase. If the stimulus is below the minimum required for contraction, it is considered to be a *subthreshold (subliminal) stimulus,* and the muscle fibers will not contract at all.

When a sufficiently strong stimulus is received by a motor neuron, the impulse is transmitted to the muscle fibers in a motor unit. Once stimulated by a sufficient nervous impulse, a muscle fiber contracts to its maximum capacity. This tendency to contract fully or not at all is called the **all-or-none principle.**

Although individual muscle fibers follow the all-or-none principle, whole muscles usually have graded contractions.

FIGURE 9.5

The origin and release of acetylcholine at a neuromuscular junction. (A) Vesicular hypothesis. (B) Cytoplasmic hypothesis.

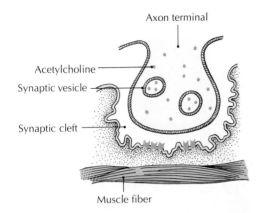

Axon terminal

Acetylcholine

Synaptic vesicle

Synaptic cleft

Muscle fiber

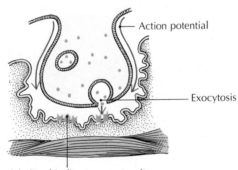

Action potential

Exocytosis

Acetylcholine binding to receptor site

[A] VESICULAR HYPOTHESIS

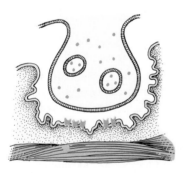

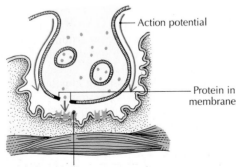

Action potential

Protein in membrane

Acetylcholine binding to receptor site

[B] CYTOPLASMIC HYPOTHESIS

(You need a stronger contraction to lift a suitcase than to lift a soupspoon, for example.) How is this possible if the all-or-none principle really works? The strength of a muscle contraction depends on *how many* fibers are stimulated, the frequency of stimulation, and how many motor units are activated. All of these processes involve *summation,* the contraction of varying numbers of muscle fibers all at once. For example, if the soupspoon is to be lifted, only a few motor units are contracted at the same time. Motor units do not all have the same threshold, and if only the motor units with low thresholds are stimulated, a relatively small number of motor units contract. At higher intensities of stimulation, other motor neurons respond, which leads to the activation of more motor units. Such an increase in the number of motor units being activated is called *recruitment of motor units.*

Different gradations of muscle contraction are achieved by two different methods of summation: (1) In *multiple motor unit summation,* all gradations of contraction between minimal and maximal can be obtained by varying the *number* of motor units contracting; (2) In *wave summation,* each motor unit contracts many times in rapid succession. Eventually, the contractions occur so close together that a new contraction occurs before the previous one is finished. In this way, each new contraction *adds* to the force of the preceding one, increasing the overall strength of contraction for the muscle.

Refractory period After a contraction, a skeletal-muscle fiber loses its irritability and cannot contract again for about 0.005 second. This recovery period is called the **refractory period.** (The refractory period for cardiac-muscle fibers is about 0.3 second.) The refractory period is generally divided into the *absolute refractory period* (from the threshold stimulus until repolarization is about one-third complete) and the *relative refractory period* (from the end of the absolute period to the start of a new depolarization). During the absolute refractory period, no stimulus, no matter how long or strong, will cause a muscle fiber to be stimulated. During the relative refractory period, stronger-than-normal stimuli are needed to stimulate a muscle fiber.

Sliding-Filament Theory of Muscle Contraction

The arrangement of the myosin and actin molecules in the myofilaments is crucial to the mechanism of muscular contraction. A complete actin myofilament is made up of not only actin but tropomyosin and troponin as well. Troponin is actually composed of three subunits (troponin I, T, and C), and is sometimes referred to as a *troponin complex.* The molecules of actin, tropomyosin, and troponin are arranged in thin, twisted strands (Figure 9.6A). The thicker myosin myofilament is composed of myosin molecules that have oval-shaped heads and long tails (Figure 9.6B). In the process leading to muscular contraction, the heads of the myosin molecules move toward the actin myofilaments, forming what are usually called *cross bridges* (Figure 9.7). The myosin cross bridges are active only when they are attached to the actin myofilaments.

FIGURE 9.6

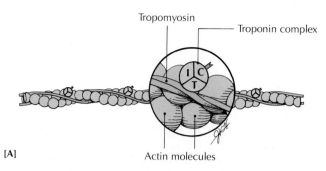

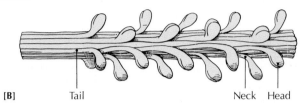

Actin and myosin myofilaments. (A) Thin actin myofilament in the resting state. The myofilament contains tropomyosin and troponin as well as actin. The troponin complex is represented by the divided blue sphere. (B) A thick myosin myofilament, showing the movable heads.

FIGURE 9.7

The initiation of muscle contraction, according to the sliding-filament theory. Drawings (A) through (H) show how myosin forms cross bridges with the actin myofilament. Drawing (I) shows how the actin binding site (shaded area) is covered by the tropomyosin filament. Drawing (J) shows the introduction of calcium ions from the sarcoplasmic reticulum. The troponin complex (blue sphere) moves and pulls the tropomyosin, leaving the actin binding site exposed. Now myosin is free to form a cross bridge with the actin myofilament, pull it, and cause a muscle contraction.

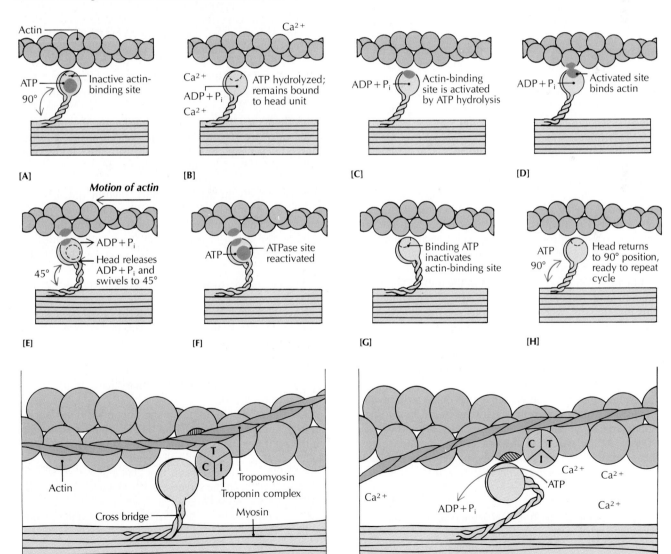

FIGURE 9.8

Summary of events in muscular contraction and relaxation, according to the sliding-filament theory.

CONTRACTION

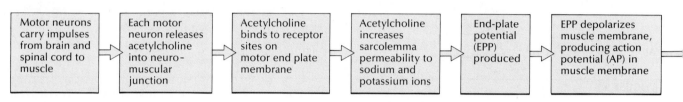

According to the **sliding-filament theory,*** the myosin cross bridges act as hooks to pull the actin myofilaments along, so that the actin and myosin myofilaments slide past each other (Figure 9.7). As the actin myofilaments from opposite ends of the sarcomere are pulled toward each other, the sarcomere becomes shorter. The same thing happens in other sarcomeres and other fibers, causing the muscle to contract.†

Apparently, the process of muscle *contraction* works in the following sequence (see Figures 9.7 and 9.8):

1 Nerve impulses from the brain and spinal cord are carried to the muscle fibers by *motor neurons.*

2 Each motor neuron releases *acetylcholine,* which diffuses across the neuromuscular junction.

3 Acetylcholine initiates a new wave of depolarization that spreads over the sarcolemma and the T tubules within the muscle fiber. The T tubule comes into contact with the sarcoplasmic reticulum (see Figure 9.2). This triggers the release of *calcium ions* (Ca^{2+}) from the cistern of the sarcoplasmic reticulum into the myofilament (Figure 9.7B).

4 Before calcium ions are present in the muscle fibers, the troponin and tropomyosin prevent actin from combining with myosin. But when calcium ions are introduced, they bind to the troponin on the actin myofilaments. This causes the troponin complex to shift, so that *active sites* are exposed on the actin strands (see Figure 9.7D). With the active sites exposed, the myosin cross bridges are free to attach to the actin myofilaments to form actomyosin.

5 At the same time that the active sites of the actin myofilaments are exposed, myosin is activated by calcium to perform its role as an enzyme (sometimes referred to as myosin ATPase). Acting as ATPase, it splits ATP (adenosine triphosphate) into ADP (adenosine diphosphate) and inorganic phosphorus (P_i), releasing *energy* (Figure 9.7B).

*This theory, proposed by H. E. Huxley in 1954, is the one most generally accepted today.

†If you block your ears with the tips of your index fingers, raise your elbows, and make a fist, you can actually hear the rumble of your hand muscles contracting. The rumble becomes louder as you clench your fist tighter.

6 The energy released when ATP is broken down is "stored" in the structural conformation of the myosin head, and is used to move the heads of the myosin molecules toward the actin myofilaments (Figure 9.7C). These heads form *cross bridges,* which attach to the exposed active sites on the actin myofilaments to form actomyosin (Figure 9.7D). The myosin heads then tilt and change shape, pulling the actin myofilaments along, so that the myosin and actin myofilaments slide past each other (Figure 9.7E). As the myofilaments slide, the cross bridges detach from one site and attach to the next site (Figures 9.7G, H). Skeletal-muscle contractions occur so rapidly that each myosin cross bridge may attach and release the active sites on the actin myofilaments as many as 100 times a second.

7 When the thin actin and thick myosin myofilaments slide by each other, the actin myofilaments from opposite ends of a sarcomere move toward each other, and the *muscle contracts.* The width of the A bands is unchanged, but the I bands shorten and the Z lines move closer together (Figure 9.9).

When a muscle *relaxes,* the following sequence occurs (see Figure 9.8):

1 Acetylcholine is broken down by *acetylcholinesterase,* which is released from the muscle. This breakdown prevents further stimulation of the muscle by the nerve ending.

2 Without the stimulation of a nerve impulse, the calcium ions move away from the myofilaments and are returned to the sarcoplasmic reticulum, by a primary active transport Ca-ATPase, to be stored.

3 Without calcium, troponin and tropomyosin prevent actin from combining with myosin by blocking the active sites on the actin myofilament. Thus the myosin cross bridges can no longer attach to actin myofilaments.

4 As the myosin and actin myofilaments return to their original positions in the sarcomere, the I bands become larger and the Z lines move farther apart (see Figure 9.9). The sarcomeres return to their original (resting) length, and the muscle fiber *relaxes.*

A new twist to the sliding-filament theory According to a recent hypothesis, the myosin cross bridges and the actin sites

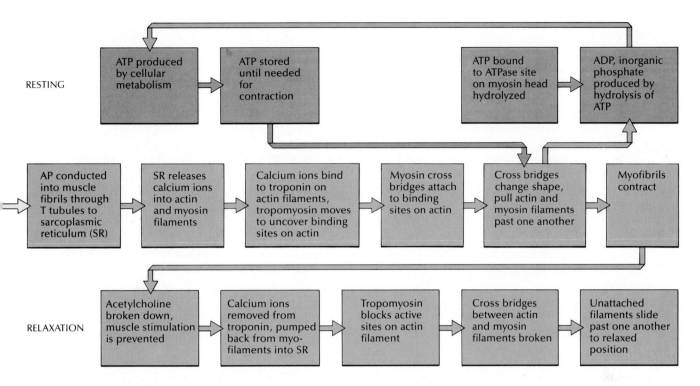

are both arranged in twisted spirals.* This new information has made it difficult for some to accept the ratchetlike parallel mechanism of the sliding-filament theory. Instead, during contraction, myosin myofilaments are believed to rotate on their long axis in a screwlike motion. Such a movement coincides perfectly with the positions of the active sites on the actin filament.

There are nine myosin cross bridges for every six available actin myofilaments. Apparently, the angular arrangement of the spiraling myosin cross bridges is such that every *other* actin myofilament is contacted (Figure 9.10). It seems that it is this successive attachment and detachment of the cross bridges that actually causes the myosin to rotate.

Energy for Contraction: ATP and Creatine Phosphate

The energy needed for a muscle fiber to contract comes directly from the hydrolysis of ATP into ADP and P_i. The indirect sources of energy for contraction—in the form of creatine phosphate, glucose, and glycogen—all depend on the hydrolysis of ATP.

ATP and muscle contraction Myosin is not only a major structural component of muscle tissue. The globular portions of the myosin myofilaments act as enzymes (in the form of ATPase). In the presence of free magnesium ions (Mg^{2+}), myosin catalyzes the hydrolysis of ATP into ADP and inorganic phosphate, forming a complex with actin, and releasing some free energy:

*This modification of Huxley's sliding-filament theory was proposed in 1982 by P. Obendorf of Victoria, Australia.

$$ATP + H_2O \xrightarrow[Mg^{2+}]{ATPase} ADP + P_i + \text{free energy}$$

The free energy derived from removing the terminal phosphate from ATP is used by a cell to accomplish functions that require energy; in the case of muscle cells, energy is used for the contraction of muscle fibers. Several forms of ATPase exist within cells. For example, muscle fibers contain a calcium ATPase in their sarcoplasmic reticulum membranes. This ATPase hydrolyzes ATP and uses the free energy released to move calcium ions from the cytosol back into the storage area inside the sarcoplasmic reticulum. Also, as you saw above, myosin ATPase functions during the contraction sequence.

Creatine phosphate and muscle contraction Enough ATP is present in a resting muscle fiber to allow it to contract for a few seconds. Since muscle fibers usually contract for more than a few seconds, where does the needed ATP come

---

What causes ''rigor mortis''?

Three or four hours after death, ATP in muscles breaks down, and is not replaced. Thus, there is no ATP to release the cross bridges between the actin and myosin myofilaments. The myofilaments become locked in place, and the muscles become rigid. This condition is called rigor mortis (''death stiffening''). It is complete in about 12 hours. About 15 to 25 hours later, the muscle proteins are destroyed by enzymes in the cells, and rigor mortis disappears.

FIGURE 9.9

The mechanism of muscle activity. Thin actin myofilaments change position as a muscle contracts (A), rests (B), and stretches (C).

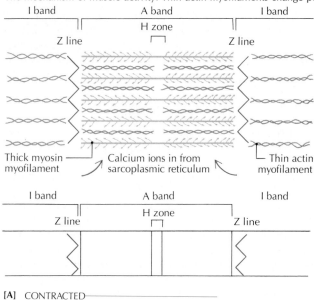

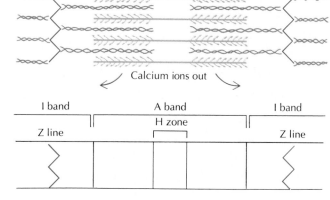

[A] CONTRACTED

[B] RESTING

from? The primary source of extra ATP is from the transfer of phosphate from **creatine phosphate** (also called *phosphocreatine*) to ADP during a muscle contraction:

$$\text{ADP} + \text{creatine phosphate} \xrightarrow[\text{phosphokinase}]{\text{creatine}} \text{ATP} + \text{creatine}$$

Although the concentration of creatine phosphate in muscle fibers is about five times that of ATP, it still supplies only enough ATP for a few additional seconds of contractions. The

ATP needed by muscle fibers for sustained contractions comes from the breakdown of stored muscle glycogen into glucose, and the subsequent breakdown of blood glucose into carbon dioxide and water by way of cellular metabolism (see Chapter 23). Each molecule of glucose provides up to 38 molecules of ATP. Also, the additional breakdown of fats and amino acids provides extra ATP for sustained contractions. When the muscle fibers return to a resting state, they continue the breakdown of glucose to form ATP until the resting level of creatine phosphate is restored. Glucose is then used to replenish the glycogen depleted during contractions.

FIGURE 9.10

New ideas about muscular contraction suggest that the myosin myofilaments rotate as they interact with every *other* actin myofilament.

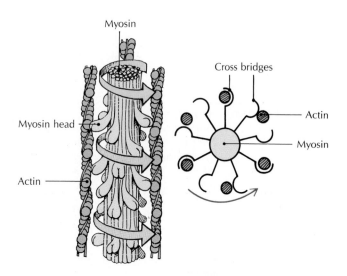

Ask Yourself

1 *Where is a neuromuscular junction located?*

2 *What structures make up a motor unit?*

3 *How does the all-or-none principle operate?*

4 *What happens during a refractory period?*

5 *What is the function of myosin cross bridges?*

6 *What roles do acetylcholine and acetylcholinesterase play in the sliding-filament theory?*

7 *How is muscle relaxation accomplished?*

8 *What roles do ATP and creatine phosphate play in muscle contraction?*

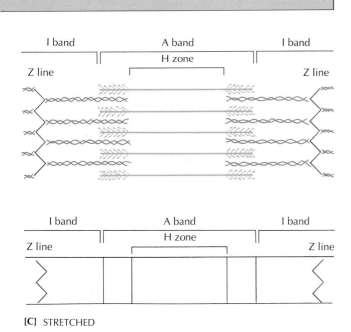

[C] STRETCHED

Types of Muscle Contraction

Several types of muscle contraction have been identified, including twitch, isotonic, isometric, treppe, and tetanic.

Twitch A momentary, spasmodic contraction of a muscle fiber in response to a single stimulus (such as an electric current or a direct stimulation of a motor neuron) is a **twitch**. It is the simplest type of recordable muscle contraction. The short, jerky action of a twitch is usually produced artificially in order to record the response on a graph called a *myogram* (Figure 9.11).

Isotonic and isometric contractions When a muscle contracts by becoming shorter and thicker, the contraction is called **isotonic** (Gr. *isos*, equal + *tonos*, tension) because the amount of force, or tension, remains constant as movement takes place (Figure 9.12A). For example, when you pull open a door, your arm muscle contracts and moves your arm, which moves the door. In contrast, if the load on a muscle is greater than the tension developed by the muscle, the muscle retains its original length, and the contraction is **isometric** (Gr. *metron*, length) (Figure 9.12B). Energy is used during an isometric contraction, and heat energy is produced, but a typical contraction, or shortening, does not occur. For example, holding a door open involves an isometric contraction, since your arm muscle doesn't shorten, but does become more tense without producing any movement. To use another example, running produces noticeable isotonic contractions, and standing produces isometric contractions. An isotonic contraction is the type we generally think of when a muscle is working actively. However, most body movements involve both isotonic and isometric contractions.

Treppe When a rested muscle receives repeated stimuli over a prolonged period, the first few contractions increase in strength, so that the myogram starts out looking like an upward staircase (Figure 9.13). This type of contraction is known as **treppe** (Ger. staircase). After several contractions, a steady tension for each contraction is reached, and the contractions level off. If stimuli and contractions continue, fatigue occurs. If controlled properly, however, as when athletes "warm up," treppe contractions increase blood flow to a muscle and prepare it for a maximum output of strength when it is needed. Treppe is thought to be caused by an increase in calcium ions that bind to troponin C.

Myograms of several types of contractions. (A) Twitch. The muscle is stimulated once at point A. There is a short latent period (A–B), only a few milliseconds, before the contraction actually begins at point B and peaks at point C. The contraction lasts about 0.04 second. The following relaxation period (C–D) lasts about 0.05 second. (B) Summation. Summation occurs when more than one stimulus is given before the muscle has relaxed completely after the first stimulus. (C) Incomplete tetanus. Incomplete tetanic contraction occurs when a muscle is just beginning to relax when another stimulus is received. (D) Complete tetanus. Complete tetanic contraction occurs when the rest periods between stimulations become so short that the muscle cannot relax at all.

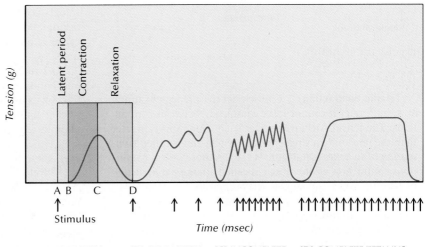

FIGURE 9.12

Comparison of an isotonic twitch (A) and an isometric twitch (B), as shown on graphs.

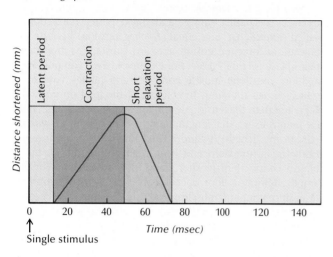

[A] ISOTONIC TWITCH

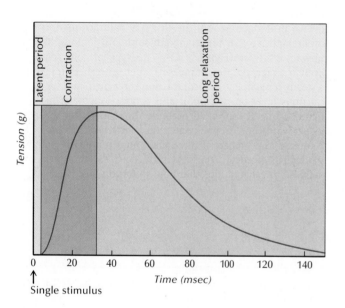

[B] ISOMETRIC TWITCH

FIGURE 9.13

Myogram of a treppe contraction, or "staircase" phenomenon. The stimulations occur repeatedly at about 0.5-second intervals. Note the upward staircase effect during the first few seconds and then the leveling off of the strength of the contractions.

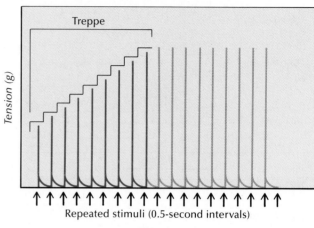

Tetanic contraction If a muscle receives repeated stimuli at a rapid rate, it cannot relax completely between contractions. The tension achieved under such conditions is greater than the tension of a single muscle twitch, and is called *summation of twitches* (see Figure 9.11B). A more or less continuous contraction of the muscle is called a *tetanic contraction* or *tetanus.* When incomplete relaxations are still evident between contractions, the muscle is said to be in a state of *incomplete tetanus* (see Figure 9.11C). *Complete tetanus* occurs when the muscle is in a steady state of contraction, with no relaxation at all between stimuli (see Figure 9.11D).

Fast-Twitch and Slow-Twitch Muscle Fibers

Although there is a fairly continuous range in the speed of contraction of the body's muscles, muscle fibers generally are classified as fast-twitch (type II fibers) or slow-twitch (type I fibers). The differences between these types of fibers are due to different myosin ATPase isoenzymes that are also designated as "fast" or "slow." The enzymes give the fibers their "fast" or "slow" characteristics.

The response of a muscle fiber or whole muscle can be expressed by the duration of a muscle twitch (contraction-relaxation cycle). For example, the muscles that move the eyes—all fast muscles—have contraction-relaxation cycles lasting about 30 milliseconds. A large leg muscle such as the soleus—a slow muscle—has a contraction-relaxation cycle lasting about 3000 milliseconds. The contraction-relaxation cycles of other muscles lie between these two extremes (Figure 9.14).

Fast-twitch muscle fibers, such as those in the delicate eye and hand muscles, are well adapted to produce rapid contractions. Fast-twitch muscle fibers contain many energy-producing mitochondria. They also have an extensive sarcoplasmic reticulum, providing a plentiful supply of calcium ions, which trigger a quick response to nerve impulses. However, these muscle fibers tire easily, since they contain relatively little *myoglobin,* an oxygen-binding protein that increases the rate of oxygen diffusion into a muscle fiber and provides a small store of oxygen within the fiber. Because fast-twitch muscle fibers also have fewer capillaries than slow-twitch muscles do, oxygen and other nutrients are supplied to them more slowly.

Anatomy of a Runner

Muscle physiologist Larry Stewart of Ball State University's Human Performance Laboratory has found an important difference in the leg muscles of sprinters and long-distance runners. Most people have approximately equal numbers of fast-twitch and slow-twitch fibers in running muscles such as the gastrocnemius. Sprinters, however, have more fast-twitch fibers, and distance runners have more slow-twitch fibers.

Fast-twitch fibers are perfect for short, fast bursts of speed. They burn stored glycogen quickly, without using oxygen. In the process, however, lactic acid accumulates, and the muscles become fatigued when they run out of fuel.

The leg muscles of long-distance runners contain mostly slow-twitch fibers. (Bill Rodgers has about 75 percent slow-twitch fibers, and Alberto Salazar has 93 percent.) Slow-twitch muscles take a lit-

tle time to reach a maximal contraction, but when they do, they can sustain it just about as long as oxygen is available.

Apparently, some runners with a preponderance of fast-twitch muscles can train them to become slow-twitch fibers. The slow-twitch fibers of long-distance runners, however, cannot be converted into muscle fibers suitable for sprinting.

FIGURE 9.14

Duration of isometric contractions of three different muscles. The lateral rectus is a very fast muscle, the soleus is considered slow, and the gastrocnemius is in between.

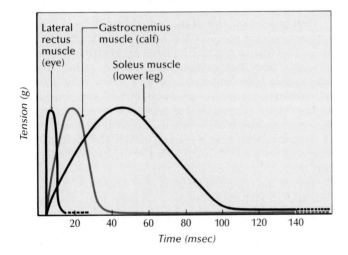

In contrast, **slow-twitch muscle fibers** are suited for prolonged and steady contractions. They have plenty of myoglobin and red blood cells, and they do not tire as easily as fast-twitch muscle fibers do. The large back muscles that help us to maintain an erect posture all through the day are an example of slow-twitch muscle fibers. Slow-twitch muscle fibers do not have, or need, as large a supply of calcium ions as fast-twitch muscle fibers do.

The color of fast-twitch and slow-twitch muscle fibers is different as well. Slow-twitch muscle fibers contain large amounts of myoglobin, which has a reddish color, and many capillaries with red blood cells. Thus slow-twitch muscle fibers are also called **red muscles**. Fast-twitch muscle fibers, with relatively little myoglobin and fewer capillaries with red blood cells, have a pale appearance and are referred to as **white muscles**.

By-products of Mechanical Work

When a muscle contracts, it converts chemical energy into mechanical energy. It also releases heat, uses up oxygen (sometimes faster than it can be replaced), and, if it works too hard for too long, it becomes fatigued.

Oxygen debt and muscle fatigue Ordinarily, when you play tennis or jog at a leisurely pace, your body uses extra oxygen. It is made available when your heart rate increases, carrying oxygen to your muscles, and when you begin to breathe harder than usual. But sometimes, when a runner is sprinting, for instance, the skeletal muscles work so hard that they use up oxygen faster than it can be supplied. When this happens, the muscles must find a way to get more oxygen.

The body receives most of the energy required to produce ATP from the glucose, amino acids, and fats in food. Ordinarily, glucose provides most of this energy, but in the process it also produces pyruvic acid, which must be catabolized. A resting muscle fiber receives enough oxygen to break down pyruvic acid:

$$\text{Pyruvic acid} + O_2 \longrightarrow CO_2 + H_2O + \text{energy}$$

Because oxygen is necessary for this process, it is called *aerobic* ("with oxygen") *respiration*.

Why is some meat dark and some light?

The reddish color of dark meat comes from the iron-rich protein myoglobin in the blood-rich dark meat. The oxygen that the muscle needs for long periods of movement is in the muscle cells next to the myoglobin. The dark meat of a turkey, for instance, is found mostly in the legs, where long-term energy is needed for running. White meat, in contrast, is found in the muscles that have short bursts of activity, such as the breast muscles of a turkey, which are adapted for fast, short flights.

How Physical Activity Affects Skeletal Muscle

The number of skeletal muscle cells in the body does not increase as a result of strenuous physical activity, but the *diameter* of the individual muscle fibers increases, as do the number of myofibrils and mitochondria, the amount of sarcoplasmic reticulum, and the blood supply. This is called muscular **hypertrophy** (Gr. *hyper*, over + *trophe*, nourishment). As a muscle fiber becomes larger and stronger through exercise, it is made even more powerful as the connective tissue around it begins to toughen. (You sometimes get a tough steak for the same reason.) In contrast, if a muscle is not used over a period of months, it will shrink and become weak. This is muscular **atrophy** (Gr. *a*, without + *trophe*, nourishment).

As an exercised muscle becomes larger, it also becomes stronger and uses oxygen and the available energy sources more efficiently. The efficiency of a trained athlete is usually about 20 to 25 percent, and some athletes have achieved 40 percent. This compares favorably with the efficiency of machines, which usually are about 20 percent efficient in converting fuel into usable energy.

The athlete increases his or her efficiency by determining the correct pace for various activities; by practicing a degree of neuromuscular control that eliminates wasteful contractions; by trimming excess fat from his or her body; by learning how to make the most use of the available oxygen, thereby making the oxygen debt smaller; and by warming up the muscles properly before exercising.

Exercise that involves lifting or moving a weight is called *isotonic*, and physical activity that involves pushing against a stationary object (such as a desk or a wall) or pushing one hand against the other is called *isometric*. Isotonic exercises produce larger muscles than isometric ones do. However, isometric exercises are much more effective in conditioning the cardiovascular system.

If enough oxygen is not immediately available, however, as when an athlete is running a sprint, the pyruvic acid cannot be catabolized completely into carbon dioxide and water, and most of it is reduced to *lactic acid* instead. This method is called *anaerobic* ("without oxygen") *respiration* because it does not require oxygen. Although anaerobic respiration allows muscles to contract, it is effective for only a short time. Most of the lactic acid is removed to the liver for conversion back to glucose. But excess lactic acid begins to accumulate in the skeletal muscles, and makes contraction more and more difficult. Eventually, lactic acid builds up to the point where it causes the tired feeling we call **muscle fatigue**.

To overcome fatigue, lactic acid must be removed from the muscle fibers by converting it into carbon dioxide and water, a process that requires oxygen. But because the body has just completed a spurt of vigorous physical activity, oxygen is in short supply. In a sense, the oxygen has been "borrowed" temporarily. Now the athlete must "pay it back" by taking deep, rapid breaths that rush oxygen to the fatigued muscle fibers. This heavy breathing is called paying back the **oxygen debt**.

During a state of fatigue, there is no change in nervous impulses, neuromuscular junctions, or stimulation of muscle fibers. Contractions merely become weaker and weaker, and the muscle finally stops contracting altogether, because (1) ATP is not present in sufficient amounts, (2) toxic products (carbon dioxide and lactic acid) accumulate, and (3) circulatory disturbances to the muscle do not allow the delivery of needed substances or the removal of waste products.

When a muscle is completely fatigued, it does not contract, but it does not relax either. For this reason, complete fatigue may be confused with a muscle cramp. It is important not to try to use a fatigued muscle, because if a muscle's glycogen supply is depleted, it will begin to use the protein contained in its own fibers.

Heat production One of the useful by-products of muscle contraction is the production of heat. Heat is usually considered a loss when a machine is producing mechanical energy. But in the body, heat production is necessary to help maintain a stable body temperature. Even a resting muscle gives off some heat, known as *resting heat. Initial heat* is generated during muscle contraction and relaxation. It results largely from the breakdown of phosphates, not from the breakdown of glycogen. Initial heat is released quickly (it is over in less than a second), and is usually effective for up to 30 minutes after the contraction ceases. *Recovery heat* is released only after the muscle has completely relaxed after a contraction. It may take up to 5 minutes to produce. Recovery heat comes as a result of the resynthesis of ATP and creatine phosphate, and includes the aerobic breakdown of pyruvic acid into water and carbon dioxide, and also the aerobic conversion of lactic acid into water and carbon dioxide.

Ask Yourself

1 *What are the three stages of the twitch contraction?*

2 *How do isotonic and isometric contractions differ?*

3 *Why is treppe also called a "staircase" contraction?*

4 *What are some of the basic differences between fast-twitch and slow-twitch muscle fibers?*

5 *When does an oxygen debt occur, and how does it relate to muscle fatigue?*

6 *What purpose does heat production by a muscle serve?*

SMOOTH MUSCLE

Smooth muscle tissue is so called because it is not striated, and therefore appears "smooth" under the microscope. It is sometimes called *involuntary muscle* because it is controlled by the autonomic nervous system, the involuntary division of the nervous system.* Smooth muscle tissue is found most commonly in the circulatory, digestive, respiratory, and urogenital systems.

Unlike skeletal muscle, smooth muscle is not connected to bones. Instead, slender smooth muscle fibers generally form sheets in the walls of large, hollow organs such as the stomach and urinary bladder. Although smooth muscle fibers are often arranged in parallel layers, the exact arrangement of the fibers varies from one location to another. In the walls of the intestines, for example, smooth muscle fibers are arranged at right angles to each other, one layer running longitudinally and the next wrapped around the circumference of the tubular intestine. These layers work together in a coordinated action to supply the constrictions that move the intestinal contents toward the anus prior to defecation. In the bladder and uterus (Figure 9.15A), the layers are poorly defined, and are oriented in several different directions. Connective tissue outside the layers extends into the spaces between muscle fibers and binds them into bundles. In contrast to the walls of hollow organs, smooth muscle fibers are wrapped around some blood vessels like tape around a rubber hose (Figure 9.15B). This arrangement is appropriate, since the smooth muscle fibers in blood vessels function to change the diameter of the vessels.

Properties of Smooth Muscle

The two main characteristics of smooth muscle are that (1) its contraction and relaxation periods are slower than those of any other type of muscle, and (2) its action is rhythmical. Its contractions may last for 30 seconds or more, but it does not tire easily. Such sustained contractions, plus the ability to stretch far beyond its resting state, make smooth muscle well suited to the muscular control of the stomach and intestines, the urinary bladder, and the uterus, especially during pregnancy.

More smooth muscle is found in the digestive system than in any other place in the body. Smooth muscle cells that line the walls of the stomach and intestines contract and relax rhythmically to help move food along the digestive tract. After we swallow food, all muscular contractions in the digestive system are involuntary until we consciously initiate the process of defecation.

---

*Although smooth muscle is usually controlled involuntarily, techniques such as biofeedback may actually allow a person to regulate some functions of smooth muscle. (See the essay on p. 413.)

(See the essay on p. 413.)

FIGURE 9.15

Smooth muscle. (A) Scanning electron micrograph of a cross section of a uterus. A thick layer of smooth muscle, called myometrium, surrounds the interior opening. The myometrium contains smooth muscle bundles, connective tissue, and blood vessels and is covered by a thin serosal layer that forms the visceral peritoneum. (B) Scanning electron micrograph of smooth muscle wrapped around a vein. Nerve axons can be seen as thin white lines. [(A) *Gene Shih and Richard Kessel,* Living Images: Biological Microstructures Revealed by SEM. *Boston, Science Books International, 1982.* (B) *Fawcett/Vehara/Photo Researchers.*]

Endometrium Myometrium Serosal layer

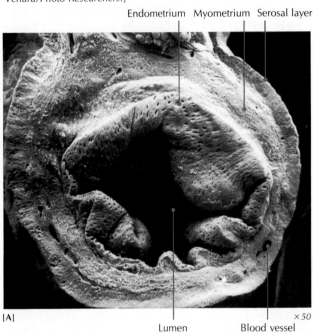

[A]   ×50

Lumen   Blood vessel

Smooth muscle surrounding vein   Neurolemmocyte

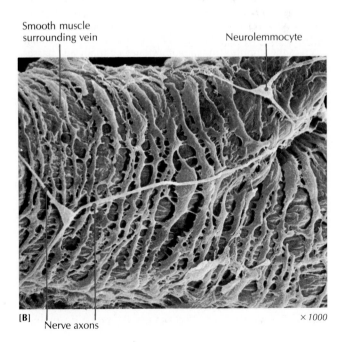

[B]   ×1000

Nerve axons

Structure of Smooth Muscle Fibers

Like the cells of skeletal muscle tissue, the cells of smooth muscle tissue are called *fibers*. Each fiber is long (but not nearly as long as skeletal muscle fibers), spindle-shaped, and slender. It contains only one nucleus, which is usually located near the center of the fiber, at its widest point. The shortest fibers (about 0.01 mm) are in the walls of blood vessels, and the longest (about 0.5 mm) are found in the uterus during the late stages of pregnancy. The fibers of the intestines have a more typical length of about 0.2 mm.

Although smooth muscle fibers are arranged differently from those in skeletal muscle, the basic contractile mechanism appears to be much the same. Actin, myosin, and tropomyosin are present, but troponin is absent. The actin and myosin myofilaments within the myofibrils are very thin, and arranged more randomly than in skeletal muscle cells. As a result, smooth muscle cells lack striations. Also, smooth muscle myosin is chemically distinct from skeletal muscle myosin. The sarcoplasmic reticulum is poorly developed, and T tubules and Z lines are not present. The slowness of contractions is thought to be due to a limited amount of ATPase activity at the cross bridges.

Based on their arrangements as separate bodies or in bundles, smooth muscle fibers are usually classified as either the single-unit type or the multi-unit type.

Single-unit smooth muscle Most smooth muscle is the *single-unit* type, which is generally arranged in large sheets of fibers (Figure 9.16A). It is sometimes called *visceral* (L. *viscera,* body organs) *smooth muscle* because it surrounds the hollow organs of the body, such as the uterus, stomach, intestines, and urinary bladder, as well as some blood vessels. The fibers are in contact with each other along many gap junctions (see Figure 4.1). When a muscle cell is stimulated, it contracts and spreads the stimulation to the adjacent cell. This method produces a steady wave of contractions, such as those that push food through the intestines. The smooth muscle fiber that receives the stimulus from a motor neuron initially and passes it on to adjacent fibers is known as the *pacemaker cell.*

Two types of contractions take place in single-unit smooth muscle: tonic and rhythmic. *Tonic contractions* cause the muscle to remain in a constant state of partial contraction, or *tonus.* Tonus is necessary in the stomach and intestine, for example, where food is moved along the digestive tract. It is also found in ring-shaped muscles called *sphincters,* which regulate the openings from one part of the digestive tract to the next. Tonus also prevents stretchable organs such as the stomach and urinary bladder from becoming stretched out of shape permanently.

Smooth muscle helps to retain the tension in the walls of expandable organs and tubes, such as the bladder and blood vessels. When a smooth muscle fiber expands or rests, its tension is decreased or increased, respectively, for only a few minutes. Then the normal tension is restored.

Rhythmic contractions (*rhythmicity*) are a pattern of repeated contractions produced by the presence of self-exciting muscle fibers from which spontaneous impulses travel. The digestive system provides excellent examples of the two types of rhythmic contractions: (1) *mixing movements,* which resemble the kneading of dough, blend the swallowed food with digestive juices, and (2) *propulsive movements,* or *peristalsis,* propel the swallowed food from the throat to the anus. The peristaltic contractions of smooth muscle in the intestines and other tubular parts of the digestive tract form a tightening ring of muscle that moves along the tract, pushing the contents of the tract forward. (You will see the importance of peristalsis when we discuss the digestive process in Chapter 22.) Figure 9.17 shows how smooth muscle contracts like the folds of an accordion.

Multi-unit smooth muscle *Multi-unit smooth muscle* is so named because each of its individual fibers can be stimulated by separate motor nerve endings. There are no connections between the fibers, and each multi-unit fiber can function independently (see Figure 9.16B). Multi-unit smooth muscle is found, for example, in the iris and ciliary muscles of the eye, where rapid muscular adjustments must be made in order for the eye to focus properly. The erector muscles in the skin that cause the hair to stand on end and that produce "goose bumps" are also of the multi-unit type, as are the muscles in the wall of the ductus deferens, the tube that carries sperm from the testes during ejaculation.

Biochemical Characteristics of Smooth Muscle Contraction

In smooth muscle cells, just as in skeletal muscle cells, changes in the intracellular calcium concentration indirectly control muscle contraction. However, smooth muscle cells do not have troponin, the calcium-binding protein that regulates contractile responses with respect to a change in calcium concentration within the cell.

In smooth muscle, calcium can come from three sources: (1) intracellular stores within vesicular structures, (2) intracellular stores within mitochondria, and (3) extracellular stores. Regardless of the calcium source, in smooth muscle, calcium controls cross-bridge activity by regulating an enzyme called *myosin light-chain kinase,* which adds a phosphate

---

FIGURE 9.16

Smooth muscle of the (A) single-unit, or visceral, type and (B) the multi-unit type.

Nerve fibers

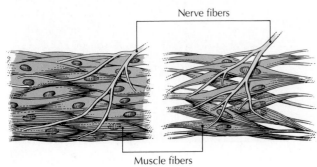

Muscle fibers

[A]   [B]

FIGURE 9.17

Smooth-muscle contraction. (A) Electron micrograph of partly contracted smooth muscle fibers. (B) Enlarged drawing of partly contracted fiber. (C) Electron micrograph of fully contracted smooth muscle fibers. (D) Enlarged drawing of fully contracted fiber, showing how bundles of myofilaments in the fiber contract. The dense bodies anchor the myofilaments and have characteristics similar to the Z lines in skeletal muscle. [(A and C) A. W. Ham and D. H. Cormack, Histology, 9th ed., J. B. Lippincott, Philadelphia, 1987.]

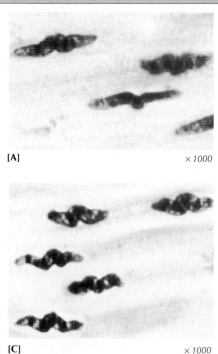
[A] × 1000

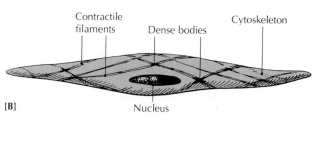
[B]

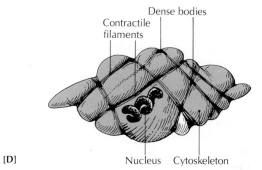
[D]

[C] × 1000

group to the myosin molecule. The effect of calcium ions (Ca^{2+}) is mediated by a calcium-binding protein called *calmodulin*. The complex of calmodulin with calcium activates myosin light-chain kinase so that it adds a phosphate group to the myosin molecule. Smooth muscle is thus activated to contract by an influx of calcium, but the triggering mechanism is not impulses from voluntary nerves (as in skeletal muscle), but rather inputs that either act on the sarcoplasmic reticulum or on specific calcium channels in the sarcolemma to increase calcium movement into the cell. These inputs include: (1) stretching of the smooth muscle myofibrils, (2) spontaneous electrical activity (*pacemaker potential*) within the sarcolemma, (3) specific neurotransmitters released by autonomic neurons, (4) hormones and hormone modulators such as prostaglandins, and (5) locally induced changes in the extracellular fluid surrounding the smooth muscle cell (such as pH, oxygen, and carbon dioxide levels). The cascading mechanism of smooth muscle contraction, when activated by calcium, is shown in Figure 9.18.

Like skeletal muscle, smooth muscle uses ATP as an immediate source of energy for contraction. However, smooth muscle does not have the energy reserves, such as creatine phosphate, found in skeletal muscle. This means that in smooth muscle, the short-term energy supply is very limited, and the muscle must rely on the synthesis of ATP from various substrates to meet its energy requirements for contraction. Thus, one major characteristic of smooth muscle is that it can use a wide variety of substrates for ATP production. Two examples are carbohydrates and fats. A second major characteristic of smooth muscle is that the speed of shortening is much slower than in skeletal muscle. Thus, the rate at which ATP must be made for the contractile process is also lower.

FIGURE 9.18

The cascade of specific reactions by which calcium ions (Ca^{2+}) activate the contraction of smooth muscle.

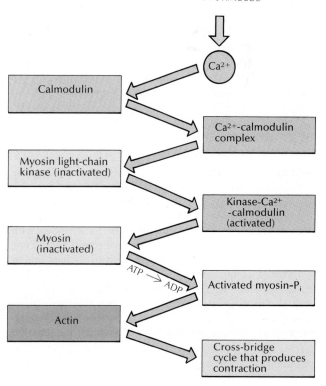

Finally, smooth muscle is a more efficient contractile unit than skeletal muscle is. As previously noted, smooth muscle is well suited for long-term maintenance of tension. The reason for this efficiency may involve a "latch" mechanism, whereby the cross bridge remains in the attached position for a long time, thus reducing the cycling rate and the rate of ATP consumption.

Ask Yourself

1 *Why is smooth muscle also called involuntary muscle?*

2 *How does the sheetlike arrangement of smooth muscle cells help the muscle to perform its jobs?*

3 *Why is the slow and rhythmic action of smooth muscle important?*

4 *What is the difference between mixing movements and propulsive movements?*

5 *What is single-unit smooth muscle? Multi-unit smooth muscle?*

6 *How does the chemical mechanism of contraction of smooth muscle differ from that of skeletal muscle?*

CARDIAC MUSCLE

Cardiac muscle tissue (Gr. *kardia,* heart) is found only in the heart. It contains the same type of myofibrils and protein components as skeletal muscle. Although the number of myofibrils varies in different parts of the heart, the contractile process is basically the same as that described for skeletal muscle. As expected in such hard-working tissue, cardiac muscle cells contain huge numbers of mitochondria.* The sarcoplasmic reticulum and T tubules are also evident.

Structure of Cardiac Muscle

Although cardiac muscle cells are closely packed, they are *separate,* each with its own nucleus. The term *cardiac muscle fiber* refers to a chain of cells joined end to end by cell junctions, *not* a single fiber as in skeletal and smooth muscle. The chain of cells forms a network within the muscle fiber, with the short cells branching freely. With an electron microscope you can see that cardiac muscle fibers have striations similar to those of skeletal muscle (Table 9.1). Cardiac fibers are crossed by dark bands that are wider than the Z lines in skeletal muscle (Figure 9.19). These bands are called **intercalated disks** (L. *intercalatus,* to insert between). They separate the cells within a muscle fiber, strengthen the junction

*Mitochondria in cardiac muscle cells occupy about 35 percent of the cell, whereas the mitochondria in skeletal muscle cells occupy only about 2 percent of the cell. This reflects cardiac muscle's extreme dependence on aerobic metabolism.

between cells, and help an impulse to pass quickly from one cell to the next. Intercalated disks always occur in place of Z lines.

Cardiac muscle. (A) Electron micrograph of cardiac muscle in longitudinal section, showing the typical appearance of intercalated disks between cardiac fibers. Notice the numerous mitochondria. (B) Detailed drawing of intercalated disks, showing the two types of junctions involved. [(A) *W. Bloom/D. W. Fawcett/Photo Researchers.*]

[A]   Intercalated disks   × 12,000

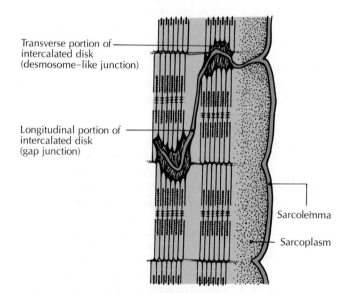

[B]

TABLE 9.1   SUMMARY OF MUSCLE TISSUE

Type of tissue	Locations	Type of contraction	Type of fibers	Type of striation	Functions
Skeletal	Skeletal muscles.	Voluntary.	Long, cylindrical.	Heavily striated.	Causes movement of skeleton, helps to maintain posture, produces heat when contracted.
Smooth	Walls of hollow internal organs, blood vessels, vessels of digestive, respiratory, reproductive, and urinary systems.	Involuntary.	Spindle-shaped.	Nonstriated.	Causes movement within internal organs and vessels.
Cardiac	Heart.	Involuntary.	Short, branching.	Finely striated.	Causes heart to pump blood.

Functioning of Cardiac Muscle

Cardiac muscle depends on nervous impulses to some extent but it is also able to contract rhythmically on its own. (Without a nerve supply, it contracts at only about half its normal rate.) As you will see in Chapter 18, when the heart is discussed in detail, cardiac muscle is designed to conduct impulses that help the upper and lower chambers of the heart beat in a carefully synchronized way. Instead of beating independently, the cells of the heart are all coordinated so that their rate and rhythm are appropriate for the job of pumping blood 24 hours a day. Because all the muscle cells of the heart are "connected" with each other by intercalated disks into a continuous network, the entire heart contracts when a single cell receives an impulse.

When cardiac muscle cells are at rest between contractions (beats), the cell membrane builds up an electrical charge in the process of *polarization*. The outside of the cell has a positive charge, and the inside has a negative charge. Just before a contraction, the electrical stimulus reaches a threshold level, and the cell membrane "leaks" calcium ions. The voltage difference is momentarily equalized in the process of *depolarization*. Contraction follows depolarization in one cell after another. After contraction, the original electrical charges of the resting state are restored. This is *repolarization*. Then the cycle begins again with another depolarization.

Cardiac muscle and skeletal muscle differ in certain ways. The sarcoplasmic reticulum is less extensive in cardiac muscle. Also, the calcium-ion sensitivity of intact cardiac muscle is much greater than that of skeletal muscle. Because a significant amount of calcium enters the cardiac muscle cell during contraction, the cell can actually contract for longer periods than a skeletal muscle cell can. However, since the

excitation-contraction coupling is so calcium-dependent in cardiac tissue, cardiac muscle is affected by calcium imbalances sooner than any other excitable tissue.

One major difference between the electrical potential of cardiac muscle and the other types of muscle tissue is that cardiac muscle has a built-in safety feature against developing a tetanic contraction. This protective device is crucial, because a tetanic contraction in cardiac muscle would be fatal. Tetanization is avoided because of the extended period of depolarization (the *refractory period*) in cardiac muscle. The refractory period in skeletal muscles is about 1 to 2 milliseconds, and the contraction itself takes 20 to 100 milliseconds. In cardiac muscle, however, the refractory period lasts about 200 milliseconds, almost as long as the contraction. As a result, a second contraction cannot be produced until the muscle relaxes, not fast enough to cause a tetanic contraction.

Probably the most important characteristic of cardiac muscle is its ability to resist prolonged wear as it strenuously and continuously pumps blood from the heart to all parts of the body. Consider that the heart pumps without stopping for an entire lifetime, about 100,000 beats a day. If you live to be 70 years old, your heart will beat about 2½ billion times.

Contractions of cardiac muscle last between 200 and 250 milliseconds, longer than those of smooth muscle. This pace allows the necessary time for blood to leave the chambers of the heart. Because the heart pumps all the time, cardiac muscle needs a constant supply of oxygen. If oxygen deprivation lasts longer than about 30 seconds, cardiac muscle cells may stop contracting, and the result is heart failure.

Can cardiac muscle be regenerated?	*When cardiac muscle becomes injured, as in a heart attack, there is little chance that the muscle fibers will be able to regenerate. Any healing that does take place produces scar tissue, not muscle tissue.*

Ask Yourself

1 *What is the purpose of intercalated disks?*

2 *Do cardiac muscle cells always need nervous impulses in order to contract? Explain.*

3 *Why is it important for cardiac muscle tissue to be wear-resistant?*

Embryonic Development of Muscles

The differentiation (specialization) of cells begins very early during embryonic development, forming three distinct layers of germ cells: the inner *endoderm*, the middle *mesoderm*, and the outer *ectoderm* (see Chapter 27). Endodermal cells develop into the epithelium of the digestive tract (including the liver and pancreas) and respiratory tract. The ectoderm develops into nervous tissue and outer body regions such as the epidermis of the skin and its derivatives, including hair, sweat glands, nails, and sebaceous glands. Muscle (except in the iris of the eye) and connective tissue (including cartilage and bone) are derived from **mesodermal cells.** Primitive muscle cells are called *myoblasts.* Muscle development begins about the fifth week of embryonic development, and the muscles differentiate into their final shapes and relations throughout the body during the seventh and eighth weeks.

Most *skeletal muscles* develop from mesodermal tissue arranged in paired, segmented cell masses called **somites** (drawing). The somites are located on both sides of the central neural tube of the primitive nervous system. The inner and outer walls of the somites differentiate into distinct layers that form different parts of the embryo: (1) The outer layer of a somite is the **dermatome** (Gr. skin slice). It develops into connective

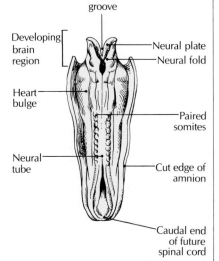

Neural groove

Developing brain region

Heart bulge

Neural tube

Neural plate
Neural fold

Paired somites

Cut edge of amnion

Caudal end of future spinal cord

tissue, including the dermis of the skin. (2) The inner layer, called the **sclerotome** (Gr. hard slice), breaks up into a mass of mesenchymal cells that migrate to the spaces surrounding the primitive spinal cord, where they develop into connective tissue. (3) The middle layer of a somite is the **myotome** (Gr. muscle slice). The cells of the myotome arrange themselves longitudinally, eventually differentiating into most of the striated skeletal muscle fibers of the body. The other skeletal muscles are derived from branchial (gill) arch mesodermal cells rather than myotomes.

Smooth muscle develops from primitive mesodermal cells that migrate to the developing linings of hollow organs and vessels, including those in the digestive, vascular, respiratory, urinary, and reproductive systems. It also includes the arrector pili muscle associated with hair. *Cardiac muscle* develops from mesodermal cells that specialize very early to eventually form the heart, migrating to the developing heart while it is still a simple tube (see p. 542).

PHYSIOLOGICAL AND ANATOMICAL ABNORMALITIES

Muscular dystrophy

Muscular dystrophy is a general name for a group of inherited diseases that result in progressive weakness due to the degeneration of muscles. It is usually limited to the skeletal muscles, but cardiac muscle may also be affected. The characteristic symptoms include the degeneration and reduction in size of muscle fibers, and an increase in connective tissue and fat deposits. Muscular dystrophy usually begins in childhood or early adolescence, but adults can also be afflicted.

Although the exact cause of muscular dystrophy is unknown, the defective gene that causes the *Duchenne* form of the disease has been identified.* Progressive deterioration cannot be stopped, at least not permanently. Treatment usu-

*The Duchenne form of muscular dystrophy affects only boys, generally of preschool age.

ally includes exercise, physiotherapy, braces, and, occasionally, surgery. It is known that patients show increased levels of creatine in the urine, and increased blood levels of the enzymes transaminase and creatine phosphokinase (CPK) are also detectable early.

Myasthenia gravis

Myasthenia gravis ("grave muscle weakness") is a disease in which even the slightest muscular exertion causes extreme fatigue. Although myasthenia may become progressively worse, it is usually not fatal unless the respiratory muscles fail and breathing becomes impossible. It is related to an improper transmission of nerve impulses at the neuromuscular junction, usually caused by too little acetylcholine, too much acetylcholinesterase (an enzyme that inhibits the activity of acetylcholine), or an

inadequate muscular response to the release of acetylcholine. The actual cause of the disease is not known. There is some evidence that myasthenia gravis may be caused by the production of antibodies that damage or destroy the neuromuscular junction sites on the sarcolemma. Although it can occur at any age, myasthenia gravis most commonly affects women between the ages of 20 and 40, and men over 40. It is about three times as common in women as in men.

Tetanus (lockjaw)

Technically, **tetanus** (Gr. *tetanos*, stretched) is a disease that affects the nervous system, but it is discussed here because it produces spasms and painful convulsions of the skeletal muscles. Tetanus, commonly called *lockjaw* because of the characteristic tightening of the jaw muscles, is caused by the extracellular toxin produced by the bacillus *Clostridium tetani*. This bacillus produces an exotoxin 50 times stronger than poisonous cobra venom and 150 times stronger than strychnine. (An *exotoxin* is a toxin produced in the cell and released into the environment.) Most often, *C. tetani* is introduced into a puncture wound, cut, or burn by infected soil, especially soil that contains horse or cattle manure. These bacilli multiply only in anaerobic conditions, so deep puncture wounds are ideal for their growth. As the bacilli multiply, they produce the exotoxin that destroys surrounding tissue, entering the central nervous system by way of branching spinal nerves.

The early indication of tetanus is local pain and stiffening, but once the exotoxin begins to spread, painful spasms are felt in the muscles of the face and neck, chest, back, abdomen, arms, and legs. Prolonged convulsions may cause sudden death by asphyxiation.

Most children today receive a DPT vaccine, which permanently immunizes them against *d*iphtheria, *p*ertussis (whooping cough), and *t*etanus. If a child who has been immunized suffers an injury that might be conducive to a tetanus infection, a booster shot of tetanus toxoid is usually given.

Trichinosis

Trichinosis is a muscle infection caused by larvae of the parasitic roundworm *Trichinella spiralis*. The disease occurs when infected and improperly cooked pork is eaten, and the ingested larvae mature and reproduce in the victim's intestine. New larvae are transported through the lymphatic system and bloodstream to skeletal muscles, especially in the diaphragm, chest, arms, and legs. Initial nausea, fever, and diarrhea may be followed by muscular stiffness, swelling, and pain. In severe cases the respiratory, nervous, and cardiovascular systems may be affected.

MEDICAL TERMINOLOGY

ASTHENIA Loss or lack of bodily strength; weakness.
CHARLEY HORSE A cramp or stiffness of various muscles in the body, especially in the arm or leg, caused by injury or excessive exertion.
CONVULSION An intense involuntary tetanic contraction (or series of contractions) of a whole group of muscles.
CRAMP A sudden, involuntary, complete tetanic muscular contraction causing severe pain and temporary paralysis, often occurring in the leg or shoulder as the result of strain or chill.
FIBRILLATION Uncoordinated twitching of individual muscle fibers with little or no movement of the muscle as a whole.
MYOCARDITIS Inflammation of cardiac muscle tissue.
MYOMA Benign tumor of muscle tissue.
MYOPATHY A general term for any disease of a muscle.
MYOSARCOMA Malignant tumor of muscle tissue.

MYOSITIS Inflammation of a muscle, usually a skeletal muscle.
PHYSIOTHERAPY Treatment of muscle weakness by physical methods such as heat, massage, and exercise.
SPASM A sudden, involuntary contraction of a muscle or group of muscles.
TIC A habitual, spasmodic muscular contraction, usually in the face, hands, or feet, and often of neurotic origin.
TREMOR A trembling or quivering of muscles.

SUMMARY

1 The basic properties of muscle tissue are *contractility, excitability, extensibility,* and *elasticity.*
2 The general functions of muscle tissue are movement, posture, and heat production. Muscles are specialized in their form to perform different functions. The three types of muscle tissue are *skeletal, smooth,* and *cardiac.*

Skeletal muscle

1 Most *skeletal muscle tissue* is attached to the skeleton. It is called *striated* muscle because it appears to be striped, and *voluntary* muscle because it can be contracted voluntarily.
2 Skeletal muscle tissue is composed of individual cells called *muscle fibers. Fascicles* are groups of fibers, and *muscles* are groups of fascicles.
3 Skeletal muscle fibers contain several nuclei and a specialized type of cytoplasm called *sarcoplasm,* which con-

tains many mitochondria and individual fibers called *myofibrils.* Each fiber is enclosed within a membrane, the *sarcolemma.*
4 Each myofibril is composed of myofilaments containing the proteins *myosin* and *actin,* and the proteins troponin and tropomyosin.
5 Myofibrils have alternating light and dark bands. The dark *A bands* contain myosin; the light *I bands* contain actin. Cutting across each I band is a *Z line.* Within the A band is a pale *H zone,* which contains the *M line.*
6 A section of myofibril from one Z line to the next makes up a *sarcomere,* the fundamental unit of muscle contraction.
7 *Fascia* is a sheet of fibrous tissue enclosing muscles or groups of muscles. The three major types of fasciae are *superficial, deep,* and *subserous.*
8 A sheath of connective tissue called *endomysium* surrounds each muscle fi-

ber, *perimysium* surrounds each fascicle, and *epimysium* encases muscles.
9 Muscles are supplied with blood by arteries. Each skeletal muscle fiber is contacted by at least one nerve ending.
10 A motor neuron, together with the muscle fiber it stimulates, is a *motor unit.* The motor neuron endings make contact with muscle fibers at the *motor end plate,* and the actual electrical impulse is transmitted to the fibers across a small space called a *synaptic cleft.* A nerve ending contacts a muscle fiber at a *neuromuscular junction.*
11 The tendency of a muscle fiber to contract fully or not at all is called the *all-or-none principle.* The minimal nervous stimulation required to produce a muscle contraction is the *threshold.*
12 The recovery period after a contraction is called the *refractory period.*
13 The *sliding-filament theory* describes the mechanism of muscle contraction.

Cross bridges between myosin and actin filaments help to slide the filaments past each other, producing contraction.

14 Impulses from the brain and spinal cord are carried to the muscle by *motor neurons*, which release *acetylcholine* at the neuromuscular junction. This produces an impulse in the muscle cell membrane (sarcolemma). This impulse is conveyed to the *transverse tubules* and then to the *sarcoplasmic reticulum*, where it triggers the release of calcium. The release of calcium sets the sliding-filament mechanism in motion.

15 The main energy source for muscle contraction is *ATP*. Expended ATP is replenished through the breakdown of *creatine phosphate (phosphocreatine)*.

16 Several types of muscle contraction include *twitch, isotonic, isometric, treppe,* and *tetanus*.

17 *Slow-twitch muscles* and *fast-twitch muscles* respond to stimuli at different speeds. Slow-twitch muscles contain large amounts of *myoglobin,* and are also called *red muscles;* fast-twitch muscles are also called *white muscles.*

18 *"Oxygen debt"* results during strenuous activity, when lactic acid is generated faster than oxygen can be brought to the muscle fibers. This oxygen debt is paid when the excess lactic acid is removed and sufficient oxygen is restored. Heavy exercise can also produce *muscle fatigue,* when contractions grow weaker because not enough ATP is available.

19 Heat given off during muscle contraction helps to maintain a stable body temperature.

Smooth muscle

1 *Smooth muscle tissue* is not striated. It is also known as *involuntary muscle* because it is controlled by the involuntary division of the nervous system.

2 The slow and rhythmic contractions of smooth muscle make it suitable for the contractile control of internal organs (but not the heart), especially the stomach, intestines, urinary bladder, and uterus. It also lines hollow vessels, including blood vessels.

3 *Single-unit smooth muscle* fibers contract as a single unit in response to nervous stimulation transmitted across the junctions by calcium and potassium ions. *Multi-unit smooth muscle* consists of individual fibers that are stimulated by separate motor nerve endings.

4 The contractile process of smooth muscle is essentially the same as that of skeletal muscle. However, biochemically, smooth muscle does not have troponin. Instead, the calcium-binding protein calmodulin mediates the phosphate group in myosin. Overall, smooth muscle is a more efficient contractile unit than is skeletal muscle.

Cardiac muscle

1 *Cardiac muscle tissue* is found only in the heart. It is striated and involuntary.

2 Cardiac muscle cells are closely packed end to end, but remain separate, each with its own nucleus. *Intercalated disks* strengthen the junction between cells, and facilitate the passing of an impulse from one cell to the next.

3 Cardiac muscle depends on nervous impulses to some extent, but it is also able to contract rhythmically on its own.

4 Cardiac muscle can resist wear as it pumps blood continuously.

Physiological and anatomical abnormalities

1 *Muscular dystrophy* is a general name for a group of inherited diseases that degenerate muscles and reduce their size. It is usually limited to skeletal muscles.

2 *Myasthenia gravis* is related to the improper transmission of nerve impulses at the neuromuscular junction.

3 *Tetanus* is a disease of the nervous system that produces spasms and painful convulsions of skeletal muscles. It is caused by a bacterial exotoxin.

4 *Trichinosis* is an infection caused by larvae of a parasitic roundworm. It affects skeletal muscles and may also affect the respiratory, nervous, and cardiovascular systems in severe cases.

UNDERSTANDING THE FACTS

1 What is the difference between extensibility and elasticity?

2 What are the general functions of muscle tissue?

3 What produces the alternating dark and light bands in striated skeletal muscle?

4 What is the fundamental unit of skeletal muscle contraction?

5 What is the difference between the epimysium and the perimysium?

6 What purposes do blood vessels and nerves serve in skeletal muscles?

7 What is a threshold stimulus?

8 Most cellular energy, with respect to muscles, comes from the breakdown of what substance?

9 How do isometric and isotonic contractions occur?

10 How can treppe contractions prepare a muscle for maximum output of strength?

11 In which parts of the body are fast-twitch muscles found? Slow-twitch muscles?

12 What are the basic properties of smooth muscle?

13 What is the most important characteristic of cardiac muscle?

14 Give a location of the following:
   a voluntary muscle tissue
   b fascia
   c endomysium
   d transverse tubules
   e sarcolemma
   f pacemaker cell
   g intercalated disks

UNDERSTANDING THE CONCEPTS

1 What roles do the sarcoplasmic reticulum and transverse tubules play in the process of muscular contraction?

2 What is the function of the synaptic cleft in muscular contraction?

3 How does the motor unit function in the nervous control of skeletal muscles?

4 What are the main steps in the contraction of a skeletal muscle?

5 Why is some muscle called red muscle and other muscle called white?

6 How does the body pay its "oxygen debt"?

7 How does the high mitochondria content of cardiac muscle relate to the function of the muscle?

8 In what important ways do the structure and function of skeletal, smooth, and cardiac muscle differ?

9 How do muscular dystrophy, myasthenia gravis, and tetanus affect the muscular system?

SUGGESTED READING

BULBRING, E., AND M. F. SHUBA, eds., *Physiology of Smooth Muscle.* New York: Raven Press, 1976.

CARAFOLI, ERNESTO, AND JOHN T. PENNISTON, "The Calcium Signal." *Scientific American,* November 1985.

COHEN, CAROLYN, "The Protein Switch of Muscle Contraction." *Scientific American,* November 1975.

DUNANT, YVES, AND MAURICE ISRAËL, "The Release of Acetylcholine." *Scientific American,* April 1985.

HOYLE, GRAHAM, "How Is Muscle Turned On and Off?" *Scientific American,* April 1970.

HUXLEY, H. E., "The Mechanism of Muscular Contraction." *Scientific American,* December 1965.

KELLY, D. E., R. L. WOOD, AND A. C. ENDERS, "Muscle." In Bailey's *Textbook of Microscopic Anatomy,* 18th ed. Baltimore: Williams & Wilkins, 1984.

LESTER, HENRY A., "The Response to Acetylcholine." *Scientific American,* February 1977.

MARGARIA, RODOLFO, "The Sources of Muscular Energy." *Scientific American,* March 1972.

MERTON, P. A., "How We Control the Contraction of Our Muscles." *Scientific American,* May 1972.

ONTEL, M., "The Growth and Metabolism of Developing Muscle." In C. T. Jones, ed., *Biochemical Development of the Fetus and Neonate.* Amsterdam: Elsevier, 1982.

OSTER, GERALD, "Muscle Sounds." *Scientific American,* March 1984.

SHEPHARD, R. J., *The Physiology and Biochemistry of Exercise.* New York: Praeger, 1981.

SMALL, J. V., AND A. SOBIESZEK, "The Contractile Apparatus of Smooth Muscle." *International Review of Cytology* 64 (1980):241–306.

SQUIRE, J., *The Structural Basis of Muscle Contraction.* New York: Plenum, 1981.

10
The Muscular System

LEARNING OBJECTIVES

1 Explain how muscles are attached to bones, and how the attachment sites are related to the skeleton.

2 Describe three patterns of muscle fascicles, and explain how each pattern is related to the power and range of movement of a muscle with that pattern.

3 Describe the action of the following types of muscles: flexor, extensor, abductor, adductor, pronator, supinator, rotator, levator, depressor, protractor, retractor, tensor, sphincter, evertor, and invertor.

4 Define the roles of agonist, antagonist, synergist, and fixator, and explain how they are coordinated in a group of muscles.

5 Explain the principles of first-class, second-class, and third-class levers, and give an example of each type.

6 Tell how leverage offers a mechanical advantage to the body.

7 Give examples of muscles named according to size, shape, attachment site, heads of origin, and action.

8 Identify the major skeletal muscles in each region of the body, and describe the origin, insertion, and action of each.

9 Identify the muscles used in performing a variety of everyday activities.

10 Describe the types of hernias and their causes.

11 Describe what goes wrong in tendinitis, tennis elbow, and tension headaches.

Bones operate as a system of levers, but muscles provide the power to make them move. A skeletal muscle (which is the only type of muscle described in this chapter) is attached to one bone at one end and, across a joint, to another bone at its other end. When the muscle contracts, the two bones are pulled *toward* each other, as when your elbow bends, *around* on each other, as when your neck twists, or *away* from each other, as when your elbow straightens.

Every movement of skeletal muscle, from lifting a heavy television set to scratching your nose, from running a mile to tapping your foot, involves at least two sets of muscles. When one muscle contracts, an opposite one stretches. During contraction, a muscle grows thicker, and shorter by about 15 percent of its resting length. A contracted muscle is brought back to its original condition by the contraction of its opposing muscle, and it can be stretched to 120 percent of its resting length.

There are about 600 *skeletal muscles* in your body, which make up the ***muscular system.**** In this chapter we examine the most important of them, and find out how they operate on a gross level.

*We say "about" 600 muscles rather than giving a definite number because anatomists disagree about whether to count some muscles as separate or as pairs. It is generally agreed, however, that we have more than 600 muscles.

ATTACHMENT OF MUSCLES

Although the form and actions of skeletal muscles vary greatly, all have certain basic features in common. All contain a fleshy center, often the widest part, called the **belly** of the muscle, and two ends that attach to other tissues. In Chapter 4, we described the connective tissues that form this attachment. It is important to note the continuity of the tissues that hold the muscle fibers together, bind the fibers into bundles, wrap an entire muscle, and attach muscle to bone.

Tendons and Aponeuroses

A muscle is usually attached to a bone (or a cartilage) by a **tendon,** a tough cord of connective tissue composed of closely packed collagen fibers.* A tendon is an extension of the deep fascia and/or the epimysium surrounding the muscle. It also extends into the periosteum that covers the bone. The thickest tendon in the body is the calcaneal (kal-KAY-nee-uhl) tendon, commonly called the Achilles tendon (Figure 10.1). It attaches the calf muscles (gastrocnemius; gas-trahk-NEE-mee-uss; Gr. "belly of the leg"; the soleus and plantaris are also calf muscles) to the heel bone (calcaneus).

In some parts of the body, such as the abdominal wall, the tendon spreads out in a broad, flat sheet called an **aponeurosis.** This sheetlike attachment is directly or indirectly connected with the various muscle sheaths and in some cases with the periosteum covering a bone. Another example of an aponeurosis is the fibrous sheath beneath the skin of the palm, called the *palmar aponeurosis.*

*Although most muscles are attached to bones, not all are. Some, such as the muscles of facial expression, are attached to skin, and the lumbrical muscles of the hand are attached to tendons of other muscles.

FIGURE 10.1

The muscle-tendon-bone relationship. (A) The calcaneal (Achilles) tendon connecting the gastrocnemius (calf) muscle and the calcaneus (heel bone). (B) Detail, showing collagenous periosteal perforating (Sharpey's) fibers passing into the connective tissue, periosteum, and bone.

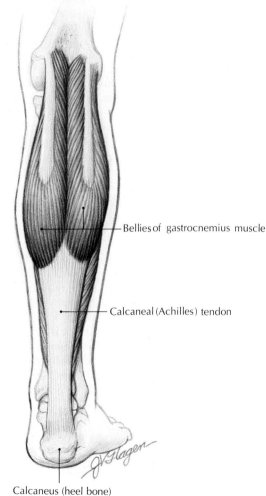

Bellies of gastrocnemius muscle

Calcaneal (Achilles) tendon

[A] Calcaneus (heel bone)

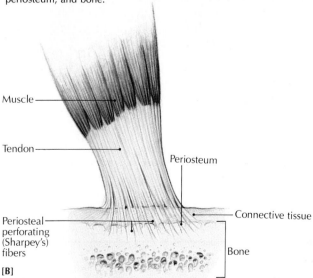

Muscle

Tendon

Periosteum

Connective tissue

Periosteal perforating (Sharpey's) fibers

Bone

[B]

THE DYNAMIC BODY

The Smallest Muscles

One of the most powerful muscles in the body is the one we sit on, the *gluteus maximus*. The longest one is the *sartorius*, which runs from the hip to the knee, and the largest in surface area is the *latissimus dorsi*, the broad muscle of the back. When we show our "muscle," we flex the *biceps brachii* in our upper arm, probably the best known muscle of all. The large muscles move our arms and legs, and, of course, they are more widely known and evident than some of our smaller muscles. But does the size of a muscle determine its importance?

Some of the tiniest muscles are those that help us communicate with one another. The smallest muscles in the body, like the smallest bones, are in the middle ear (drawing). Without them our delicate hearing apparatus could not move, and the sounds we hear would be muted. Without the tiny muscles within and around our eyeballs, we could not focus on objects, move our eyeballs without moving our head, or open and close our eyelids. Speaking requires the delicate coordination of the small muscles of the larynx, pharynx, palate, tongue, and mouth.

The tongue muscles also enable us to move food around within our mouths during chewing and swallowing. (Try chewing and swallowing *without* using your tongue.) One of those muscles, the *genioglossus*, performs an even more important role: it prevents suffocation by keeping the tongue from falling backward toward the throat. The entire act of swallowing is controlled by small muscles, and food is prevented from entering the breathing passages in several ways during swallowing. (If food did enter these passages, we could choke to death.) First muscles raise the soft palate to keep food out of the nasal cavity, and then muscles at the rear of the mouth form a thin slit, which keeps large food particles from passing into the breathing portion of the throat.

Circular muscles called *sphincters* are present throughout our bodies to open and close body openings such as the mouth and anus. The *pubococcygeus* muscle and the *external anal sphincter* prevent the involuntary release of feces from the body. A similar muscle, the *external sphincter of the urethra*, prevents urine from constantly flowing through the urethra to the outside. As we get older, these muscles may lose their tone and lead to *incontinence*, the inability to control urination or defecation.

Small muscles play important roles in all the body systems. For example, muscles all along the digestive tract help to move food from the throat to the anus, and sphincters at several important junctures allow food to pass systematically from one digestive compartment to the next and prevent food and digestive juices from backing up. Blood vessels that carry blood away from the heart (most arteries and arterioles) contain smooth muscle cells that can contract to assist the movement of blood through the vessels, as well as to control the amount of blood flow.

The life of the species also depends partly on certain small muscles. The *bulbocavernous* muscle in the male helps to propel semen through the penis during ejaculation. Two other small, relatively unheralded muscles also keep our reproductive function operating properly. Sperm cells inside the testicles remain alive and healthy only if they are kept at a well-regulated temperature. The *cremasteric* and *dartos* muscles help to maintain the temperature of sperm at a homeostatic level by lowering the testicles away from the heat of the body during hot conditions and raising them toward the body when it is cold.

All in all, small muscles may not make us walk or run, but they certainly help us *live*.

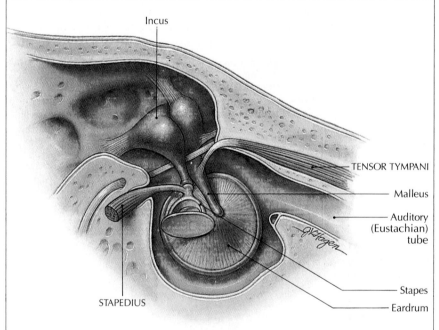

The tiny tensor tympani and stapedius muscles of the middle ear are essential elements of the hearing apparatus.

Besides merely being attachments, tendons add length and thickness to muscles, are especially important in reducing strain on muscles, and add strength to muscle action.

Origin and Insertion of Muscles

When a muscle contracts, one of the bones attached to it remains stationary while the other bone moves along with the contraction. In reality, the mechanics are hardly ever that simple, but for convenience we say that the end of a muscle attached to the bone that does not move is the *origin,* and the point of attachment of the muscle on the bone that moves is the *insertion* (Figure 10.2). Generally, the origin is the more proximal attachment (closer to the axial skeleton) and the insertion is the more distal attachment.

However, *origin* and *insertion* are only relative terms, and they can be reversed with the same muscle, depending upon the action. For example, muscles from the chest to the upper extremity generally move the extremity with the origin on the thoracic skeleton and the insertion on the extremity. But when a person climbs a rope, the same muscles pull the body up, so the extremity becomes the origin and the insertion is on the thoracic skeleton.

Ask Yourself

1 *What is the fleshy center of a muscle called?*

2 *What is the difference between a tendon and an aponeurosis?*

3 *What is the function of a tendon?*

4 *What is the difference between the origin and insertion of a muscle?*

ARCHITECTURE OF MUSCLES

The fibers of skeletal muscles are grouped in small bundles (fascicles), with the fibers within each bundle running parallel to one another. The bundles, however, are organized in various architectural patterns in different muscles. The specific pattern determines the range of movement and the power of the muscle. The greater the length of the belly, the greater the range of movement. Also, the greater the number of fibers, the greater the total force generated by the muscle. The arrangement of fascicles and their tendons of attachment creates patterns such as strap muscles, fusiform muscles, pennate muscles, and circular muscles.

Strap Muscles and Fusiform Muscles

A *strap muscle* has all its fascicles running parallel to the long axis of the muscle. This type of muscle generally has a wide range of motion, but is not very powerful. The sternohyoid of the neck and the rectus abdominis of the abdominal wall are strap muscles (Figure 10.3A). A *fusiform muscle* (L. *fusus,* spindle) is spindle-shaped, with a thick, fleshy belly that tapers at both ends into tendons (Figure 10.3B). The biceps brachii of the arm is a fusiform muscle.

Pennate Muscles

A *pennate muscle* (L. *penna,* feather) has many short fascicles set obliquely (at an angle) to a long tendon, so it resembles a feather. The tendon in a pennate muscle extends the length of the muscle.

A *unipennate* (*uni,* one) *muscle* has oblique fascicles arranged on one side of the tendon, making the muscle look like half a feather (Figure 10.3C). The pulling direction of a unipennate muscle is usually toward the side of the tendon that contains the fascicles. The flexor pollicis longus, which flexes the thumb, is a unipennate muscle.

A *bipennate* (*bi,* two) *muscle* has oblique fascicles on both sides of the tendon (Figure 10.3D). Such a muscle has equal pull from both sides of the tendon. The rectus femoris of the thigh is a bipennate muscle.

A *multipennate* (*multi,* many) *muscle* has many oblique fascicles arranged along several tendons in the central axis of the muscle (Figure 10.3E). The deltoid in the shoulder is a multipennate muscle.

In general, pennate muscles have more fascicles directly attached to tendons than nonpennate muscles do. As a result, pennate muscles have greater power than muscles of other

FIGURE 10.2

Origin and insertion of arm muscles. The two origins of the biceps muscle are on the scapula, and the insertion is on the radius. Contraction of the biceps flexes the elbow joint (arrow up).

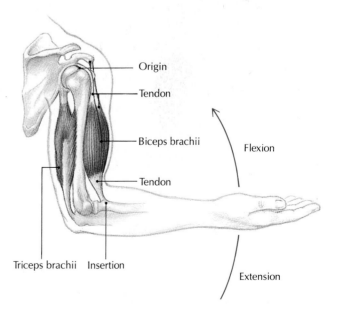

Origin

Tendon

Biceps brachii

Tendon

Flexion

Triceps brachii   Insertion

Extension

FIGURE 10.3

Architecture of muscle. (A) A parallel strap muscle, sternohyoid. (B) A fusiform muscle, biceps brachii. (C) A unipennate muscle, flexor pollicis longus. (D) A bipennate muscle, rectus femoris. (E) A multipennate muscle, deltoid. (F) A circular or sphincter muscle, orbicularis oris.

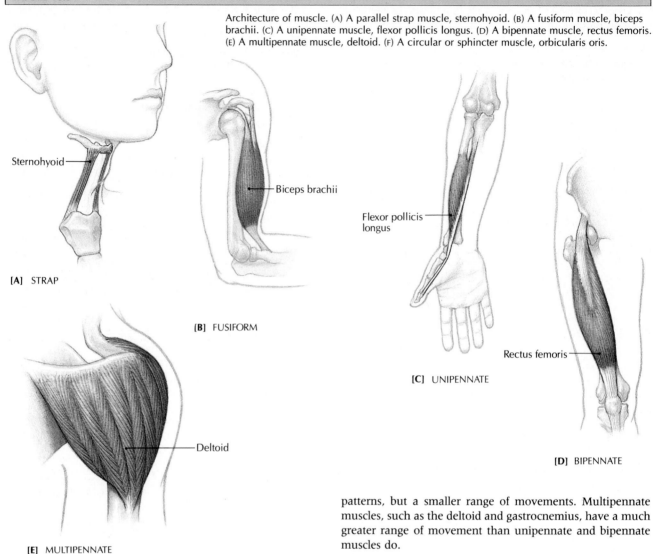

Sternohyoid

[A] STRAP

Biceps brachii

[B] FUSIFORM

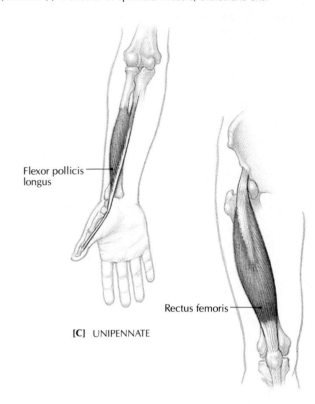

Flexor pollicis longus

[C] UNIPENNATE

Rectus femoris

[D] BIPENNATE

Deltoid

[E] MULTIPENNATE

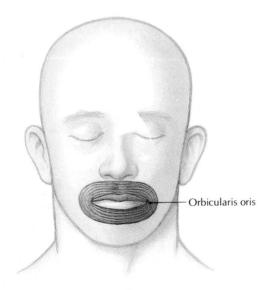

Orbicularis oris

[F] CIRCULAR (SPHINCTER)

patterns, but a smaller range of movements. Multipennate muscles, such as the deltoid and gastrocnemius, have a much greater range of movement than unipennate and bipennate muscles do.

Circular Muscles

A circular muscle has fascicles arranged in a circular pattern around an opening or a structure (Figure 10.3F). For example, the orbicularis oris encircles the mouth, and the orbicularis oculi surrounds the eye.

Ask Yourself

1 *What are the four main patterns of fascicle arrangement?*

2 *What are the three types of pennate muscles?*

3 *Where are circular muscles located?*

4 *Which arrangement of fascicles generally contributes to a greater range of movement?*

INDIVIDUAL AND GROUP ACTIONS OF MUSCLES

Skeletal muscles can be classified according to the types of movement that they perform. Table 10.1 defines these actions and gives examples, and Figure 10.4 shows some of the major muscles.

Most movements, even those that seem simple, actually involve the complex interaction of several muscles or, sometimes, groups of muscles. Muscles produce or restrict movement by acting as agonists (also called prime movers), antagonists, synergists, and fixators.

1 An *agonist* (Gr. *agonia*, contest, struggle) or **prime mover** is the muscle that is primarily responsible for producing a movement.

2 An *antagonist* (Gr. "against the agonist") opposes the movement of a prime mover, but only in a subtle way. It does not oppose the agonist while it is contracting, but only at the end of a strong contraction to protect the joint. The antagonist helps produce a smooth movement by slowly re-laxing as the agonist contracts, so it actually cooperates rather than "opposes."

3 A *synergist* (SIHN-uhr-jist; Gr. *syn*, together + *ergon*, work) works together with a prime mover by preventing movements at an "in-between" joint when a prime mover passes over more than one joint. In general, a synergist complements the action of a prime mover.

4 A *fixator* or *postural* muscle provides a stable base for the action of a prime mover. It usually steadies the proximal end (such as the arm), while the actual movement is taking place at the distal end (the hand).

In much the same way that the origin and insertion of a muscle can be reversed depending upon the specific activity involved, in different situations the same muscle can act in any of the roles described.

The coordination of an action by a group of muscles can be demonstrated when a heavy object is held firmly in a clenched fist. The *prime movers* in this case are the long flexors (see Figure 10.21) of the fingers (flexor digitorum superficialis, flexor digitorum profundus, and flexor pollicis longus).

TABLE 10.1   CLASSIFICATION OF MUSCLES BASED ON ACTION

Muscle type	Definition of action	Example	Figure showing muscle
Flexor	Bending so angle between two bones decreases.	Flexor pollicis longus.	Figure 10.21.
Extensor	Bending so angle between two bones increases.	Extensor carpi ulnaris.	Figure 10.4B.
Dorsiflexor	Bending hand or foot dorsally (toward back of hand or foot).	Tibialis anterior.	Figure 10.4A.
Palmar flexor	Bending (flexing) wrist ventrally (toward palm).	Flexor carpi ulnaris.	Figure 10.22.
Plantar flexor	Bending (extending) foot at ankle toward sole of foot.	Gastrocnemius.	Figure 10.4A, B.
Abductor	Movement away from midline of body or structure.	Abductor pollicis brevis.	Figure 10.25.
Adductor	Movement toward midline of body or structure.	Adductor pollicis.	Figure 10.25.
Pronator	Turning of forearm so palm faces downward.	Pronator teres.	Figure 10.21.
Supinator	Turning of forearm so palm faces upward.	Supinator.	Figure 10.23.
Rotator	Turning movement around a longitudinal axis.	Sternocleidomastoid.	Figures 10.9, 10.11.
Medial (internal) rotator	Turning movement so anterior surface faces median plane.	Subscapularis.	Figure 10.20A.
Lateral (external) rotator	Turning movement so anterior surface faces away from median plane.	Infraspinatus.	Figure 10.18.
Levator (elevator)	Movement in an upward direction.	Levator scapulae.	Figure 10.17.
Depressor	Movement in a downward direction.	Depressor labii inferioris.	Figure 10.6A, B.
Protractor	Movement in a forward direction.	Lateral pterygoid.	Figure 10.8B.
Retractor	Movement in a backward direction.	Temporalis (horizontal fibers).	Figure 10.8A.
Tensor	Makes a body part more tense.	Tensor fasciae latae.	Figure 10.26A.
Sphincter	Reduces size of an opening (orifice).	Orbicularis oris.	Figure 10.6.
Evertor	Turning movement of foot so sole faces outward.	Peroneus longus.	Figure 10.4A, B.
Invertor	Turning movement of foot so sole faces inward.	Tibialis anterior.	Figure 10.4A.

FIGURE 10.4

The muscular system. (A) Anterior view. (B) Posterior view.

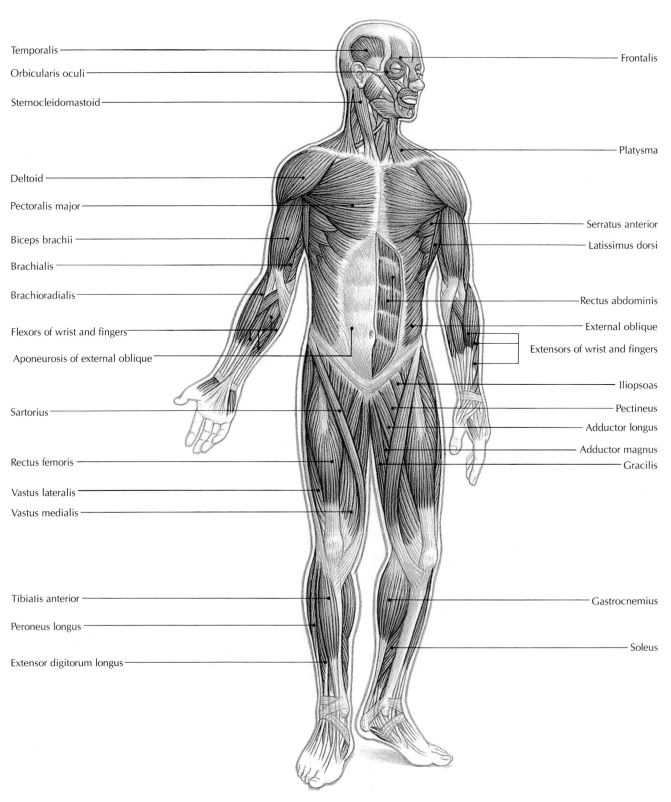

Temporalis

Orbicularis oculi

Sternocleidomastoid

Deltoid

Pectoralis major

Biceps brachii

Brachialis

Brachioradialis

Flexors of wrist and fingers

Aponeurosis of external oblique

Sartorius

Rectus femoris

Vastus lateralis

Vastus medialis

Tibialis anterior

Peroneus longus

Extensor digitorum longus

Frontalis

Platysma

Serratus anterior

Latissimus dorsi

Rectus abdominis

External oblique

Extensors of wrist and fingers

Iliopsoas

Pectineus

Adductor longus

Adductor magnus

Gracilis

Gastrocnemius

Soleus

[A] ANTERIOR

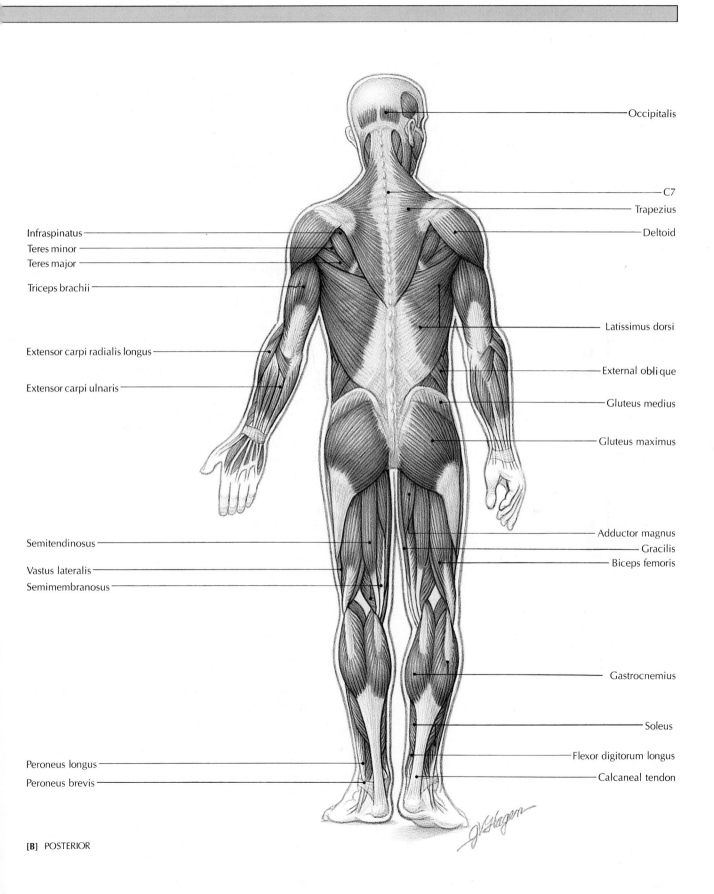

Occipitalis

C7

Trapezius

Deltoid

Infraspinatus

Teres minor

Teres major

Triceps brachii

Latissimus dorsi

Extensor carpi radialis longus

External oblique

Extensor carpi ulnaris

Gluteus medius

Gluteus maximus

Adductor magnus

Semitendinosus

Gracilis

Biceps femoris

Vastus lateralis

Semimembranosus

Gastrocnemius

Soleus

Peroneus longus

Flexor digitorum longus

Peroneus brevis

Calcaneal tendon

[B] POSTERIOR

When these muscles contract unopposed, the wrist also flexes. This undesired wrist flexion is eliminated, and the wrist is kept in a neutral or even hyperextended position by the contraction of the extensors (see Figure 10.22) of the wrist (extensor carpi radialis longus and brevis, and extensor carpi ulnaris muscles). These extensors of the wrist are the *synergists* in this action.

If the clenched fist is slowly flexed, the extensors of the wrist can act as *antagonists* in the control of the activity by relaxing at the same time the prime movers are contracting.

As the hand clenches to hold the heavy object, the shoulder and elbow joints are stabilized to a greater or lesser degree, depending upon the weight of the object. The shoulder girdle, arm, and forearm are stabilized by the integrated actions of such muscles as the pectoralis major, deltoid, supraspinatus, subscapularis, biceps brachii, brachialis, and triceps brachii. These muscles are the *fixators*.

Ask Yourself

1 *What is a pronator? A levator? A retractor? A sphincter?*

2 *What muscles are involved in pronation and supination?*

3 *What is the main job of an agonist muscle? Of a synergist?*

4 *Why do we say that an antagonist does not actually oppose an agonist?*

LEVER SYSTEMS AND MUSCLE ACTIONS

The movements of most skeletal muscles are accomplished through a system of levers. In a lever system, there is a *rigid lever arm* that pivots around a fixed point called a *fulcrum* (L. *fulcire,* to support). Also, within every lever are two different *forces:* (1) the weight to be moved (resistance to be overcome) and (2) the pull or effort applied (applied force). In the body, the bone acts as the lever arm, and the joint as the fulcrum. The weight of the body part to be moved is the resistance to be overcome. The applied force generated by the contraction of the muscle (or group of muscles) at the insertion is usually enough to produce a movement.

There are three recognized types of levers, referred to as first-, second-, and third-class levers.

First-Class Levers

In a *first-class lever,* the force is applied at one end of the lever arm, the weight to be moved (resistance) is at the other end, and the fulcrum is at a point between the two. A person using a crowbar is an example (Figure 10.5A). In the body, an example of the use of a first-class lever is raising the facial portion of the head. In lifting the head, the atlanto-occipital joint between the atlas and the occipital bone acts as a *fulcrum.* The vertebral muscles inserting at the back of the head generate the *applied force.* The facial portion of the head is the *weight* to be moved.

Second-Class Levers

In a *second-class lever,* the applied force is at one end of the lever arm, the fulcrum is at the other end, and the weight to be moved is at a point between the two. A wheelbarrow is a classic example (Figure 10.5B). In the body, there are only a few examples of second-class levers. One example is raising the body by standing on "tiptoe" (the balls and toes of the feet). In this action, the ball of the foot is the *fulcrum,* the *applied force* is generated by the calf muscles (gastrocnemius and soleus) on the back of the leg, which insert on the calcaneus, and the *weight* to be moved is at the ankle joint.*

Third-Class Levers

In a *third-class lever,* the weight to be moved is at one end of the lever arm, the fulcrum is at the other end, and the applied force to move the weight is close to the fulcrum. Lifting a full shovel is an example (Figure 10.5C). A third-class lever is the most common lever in the body. It is effective because it permits the active muscle to be inserted close to the fulcrum at the joint, and it is well suited for lifting an object rapidly with minimal contraction of the muscle. Flexing the forearm at the elbow to lift a weight involves a third-class lever. The *applied force* is generated by the contraction of the biceps brachii muscle at the proximal ends of the radius and ulna. The *fulcrum* is the elbow joint, and the *weight* to be moved is the weight of the forearm, hand, and any object held in the hand (Figure 10.5D).

Leverage: Gaining a Mechanical Advantage

If you have ever used a screwdriver to pry open a can of paint, you know that the screwdriver (lever) made the work easier than if you just used your hands. When muscles use a lever system, they also gain such a mechanical advantage, which is called *leverage.*

The leverage of a muscle is improved as the distance from the insertion to the joint (fulcrum) is increased. In other words, a muscle with an insertion relatively far from a joint has *more power* than a muscle with an insertion closer to the joint, because the longer lever arm between the joint and insertion produces more power. However, the longer lever arm must move a greater distance to produce this power, and the movement is slower. Thus when one advantage is gained, another is lost. For example, the pectineus muscle, which is attached close to the axis of rotation at the hip joint, has less power than the adductor longus muscle, which is attached farther away from the axis (see Figure 10.26). But the pec-

*It is interesting to note that the powerful gastrocnemius (calf) muscle has no equally large antagonistic muscles to lower the body from the tiptoe position—gravity brings you down with no trouble. If you touch the front part of your leg, you can feel the tibia practically uncovered by muscle all the way down.

FIGURE 10.5

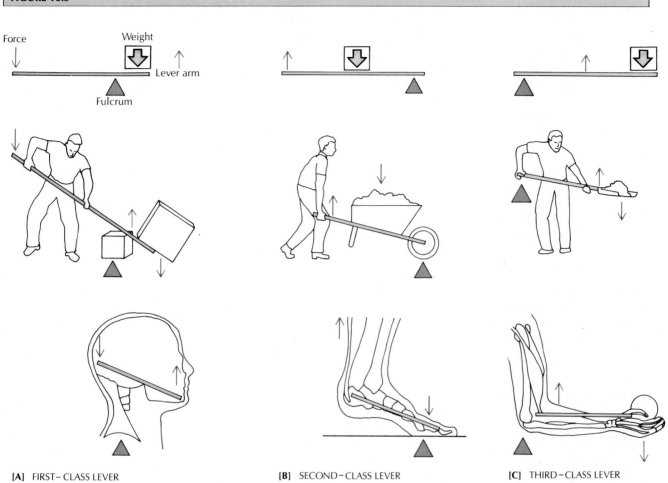

Force   Weight

Lever arm

Fulcrum

[A] FIRST–CLASS LEVER   [B] SECOND–CLASS LEVER   [C] THIRD–CLASS LEVER

tineus muscle can produce a quicker movement than the adductor longus because it moves through a shorter distance.

By using a lever system, muscles can produce amazing amounts of force. When you rise from a squatting position to a standing position, the thigh muscles that straighten your knee joint must produce about 10 pounds of force for every pound of body weight. If you weigh 120 pounds, it takes a force of 1200 pounds to lift you up. A great deal of heat energy is also produced, but most of it is dissipated because it is produced in such a quick burst.

Ask Yourself

1 *What two forces are involved in a lever system?*

2 *What are the main differences in the three classes of levers? Give an example of each of the three types of levers in the body.*

3 *How does leverage operate in the human body?*

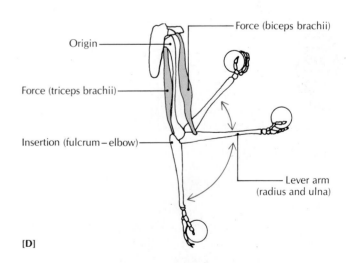

Force (biceps brachii)

Origin

Force (triceps brachii)

Insertion (fulcrum–elbow)

Lever arm (radius and ulna)

[D]

Classes of levers. (A) In a first-class lever, the fulcrum is between the force and the weight. (B) In a second-class lever, the weight is between the force and the fulcrum. (C) In a third-class lever, the force is between the weight and the fulcrum. The arrows indicate the applied force and the weight of the object to be moved. (D) A third-class lever in the body. When the forearm flexes or extends, the *lever arm* is formed by the radius and ulna, and the elbow joint forms the *fulcrum*.

HOW MUSCLES ARE NAMED

The name of a muscle generally tells something about its structure, location, or function. **Shape** is described in such muscles as the trapezius (shaped like a trapezoid), deltoid (triangular or delta-shaped), or gracilis (slender). **Size** is clearly indicated in the gluteus *maximus* (largest) and gluteus *minimus* (smallest) muscles. We know the **location** of the supraspinatus and infraspinatus muscles (above and below, *supra* and *infra,* the spine of the scapula) and the tibialis anterior (in front of the tibia).

Some muscles are named to indicate their **attachment sites.** For example, the sternohyoid muscle is attached to the sternum and hyoid bones. The number of **heads of origin** can be determined in muscles such as the biceps (*bi,* two) and triceps (*tri,* three). The **action** or function of a muscle is plain in the extensor digitorum, which extends the fingers (digits), and the levator scapulae, which raises (elevates) the scapula. The names of some muscles indicate the **direction of their fibers** with respect to the structures to which they are attached, for example, *transversus* (across) and *obliquus* (slanted or oblique).

Most muscles are named by a combination of the above methods. An example is the flexor digitorum profundus, which means the deep (profundus) flexor of the fingers (digitorum). The name of this muscle thus tells its depth, location, and action.

SPECIFIC ACTIONS OF PRINCIPAL MUSCLES

The following descriptive sections, tables, and illustrations analyze the actions of the major muscles of the body.

Muscles of Facial Expression (Table 10.2, Figure 10.6)

The **muscles of facial expression** are located not only in the face, but in the scalp and neck. Two features characterize these muscles—they are all innervated by the facial (seventh) cranial nerve, and they are used in the display of human emotions. They can move parts of the face because their principal insertions extend into the deep layers of the skin. Their origins may be located in a tendon, aponeurosis, bone, or the skin itself. Some of the facial muscles are arranged around the eyes, mouth, nose, and ears, and act as either **sphincters** (ringlike muscles that normally constrict a bodily opening until relaxation is called for) or **dilators** (expanders).

Facial muscles are used to express a wide variety of emotions. For example, when you smile, the muscles radiating from the angles of your mouth raise the corners of your upper lip. When you frown, certain muscles lower your eyebrows and depress the corners of your lower lip. (The cliché that it takes more effort to frown than to smile is true; you use 43 muscles to frown and only 17 to smile.)

FIGURE 10.6

Muscles of facial expression. (A) Anterior view. (B) Lateral view.

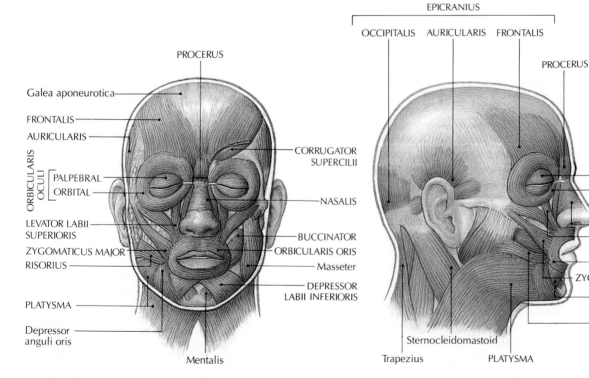

[A] ANTERIOR [B] LATERAL

TABLE 10.2 MUSCLES OF FACIAL EXPRESSION

Muscle	Origin	Insertion	Innervation	Action
MUSCLES OF SCALP				
Epicranius				
Frontalis	Galea aponeurotica.	Fibers of orbicularis oculi.	Temporal branches of facial nerve (VII).	Elevates eyebrows (surprised look); produces horizontal frown.
Occipitalis	Occipital bone.	Galea aponeurotica.	Posterior auricular branch of facial nerve (VII).	Draws scalp backward.
Auricularis	Galea aponeurotica.	Vicinity of outer ear.	Temporal branches of facial nerve (VII).	Wiggles ears in a few individuals.
MUSCLES OF EYELID AND ORBIT				
Orbicularis oculi				
Palpebral (eyelid)	Medially from medial palpebral ligament.	Sphincter fibers oriented in concentric circles around orbital margin and in eyelid.	Temporal and zygomatic branches of facial nerve (VII).	Closes eyelid (as in winking).
Orbital	Medial orbital margin (maxilla) medial palpebral ligament.	Sphincter muscle.	Temporal and zygomatic branches of facial nerve (VII).	Closes eyelid tightly (against bright light).
Corrugator supercilii	Brow ridge of frontal bone.	Skin of eyebrow.	Temporal branch of facial nerve (VII).	Pulls eyebrows together.
Levator palpebrae superioris*	Central tendinous ring around optic foramen.	Skin of upper eyelid.	Temporal and zygomatic branches of oculomotor nerve (III).	Raises upper eyelid.
MUSCLES OF NOSE				
Nasalis	Maxilla medial to orbit.	Lower region of cartilage of nose.	Facial nerve (VII).	Widens nasal aperture (as in deep breathing).
Procerus	Lower part of nasal bone, lateral nasal cartilage.	Skin between eyebrows.	Buccal branches of facial nerve (VII).	Pulls eyebrow downward and inward.
MUSCLES OF MOUTH				
Orbicularis oris	Sphincter muscle within lips, encircles mouth and merges with other muscles of mouth.		Buccal and mandibular branches of facial nerve (VII).	Closes mouth; purses lips; significant role in speech.
Levator labii superioris	Above infraorbital foramen of maxilla.	Skin of upper lip.	Buccal branch of facial nerve (VII).	Raises lateral aspect of upper lip.
Zygomaticus major	Zygomatic bone.	Skin at angle of mouth.	Buccal branch of facial nerve (VII).	Draws angle of mouth upward (as in smiling).
Risorius	Fascia over masseter muscle.	Skin at angle of mouth.	Buccal branch of facial nerve (VII).	Draws angle of mouth laterally (as in grinning).
Depressor labii inferioris	Mandible.	Skin of lower lip.	Mandibular branch of facial nerve (VII).	Lowers lateral aspect of lower lip.
MUSCLE OF CHEEK				
Buccinator	Alveolar processes of mandible and maxilla, from pterygomandibular raphe (connective tissue) between sphenoid and mandible.	Fibers of orbicularis oris.	Buccal branch of facial nerve (VII).	Pulls cheek against teeth to move food during chewing and to make blowing or sucking motion.
MUSCLE OF NECK				
Platysma	Fascia of upper thorax over pectoralis major, deltoid muscles.	Mandible, skin of lower face, fibers of orbicularis oris at angle of mouth.	Cervical branch of facial nerve (VII).	Draws outer part of lower lip down and back (as in frowning); tenses skin of neck.

*Usually associated with extrinsic muscles (those outside the eyeball); not a true muscle of facial expression. (Shown in Figure 10.7.)

Besides expressing emotion, the facial muscles have other uses. For example, the nasalis musculature dilates your nostrils as you inhale. The **buccinator muscle** (L. *bucca*, cheek) lining the cheek is essential for nudging food from between the cheek and teeth during chewing, and for blowing or sucking.* The scalp (epicranius) muscles originating from a broad,

*When the buccinator muscles lose their tone and elasticity, the cheeks bulge uncontrollably when a person makes a blowing motion. This condition is known as "Gillespie's pouches," named for the jazz trumpeter Dizzy Gillespie, whose ineffective buccinator muscles cause his cheeks to bulge every time he blows into his trumpet. (However, the buccinator muscle was known as the trumpeter's muscle long before Dizzy Gillespie came along.)

flat aponeurosis (galea aponeurotica) can wrinkle the forehead, producing a look of surprise or fright.

Muscles That Move the Eyeball (Table 10.3, Figure 10.7)

The movements of the eyeball are controlled by six rapidly responsive *extrinsic* (outside the eyeball) eye muscles. These muscles can rotate the eyeball around the horizontal, vertical, and sagittal axes passing through the center of the eyeball. The four *recti* ("upright") *muscles* insert in front of the horizontal axis, and the two *oblique muscles* insert behind the horizontal axis. The recti muscles pass slightly obliquely forward from their origin around the optic foramen. They form a cone as they spread out toward their insertions in the eyeball. The actions of the recti muscles can be understood by realizing that they pull on the eyeball in a plane slightly oblique to the visual (sagittal) axis. Squint (strabismus or cross-eye) is a disorder in which both eyes cannot be directed

FIGURE 10.7

Muscles that move the eyeball (extrinsic muscles). (A) View from above, with horizontal and sagittal axes of the eyeball indicated. (B) Lateral view, with vertical axis of the eyeball indicated.

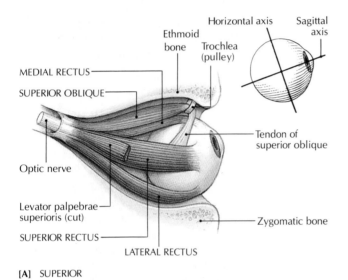

[A] SUPERIOR

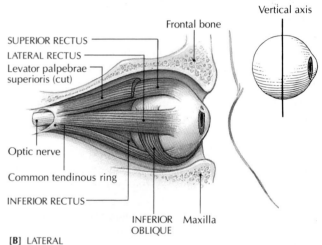

[B] LATERAL

TABLE 10.3 — MUSCLES THAT MOVE THE EYEBALL: EXTRINSIC OCULAR MUSCLES

Muscle	Origin	Insertion	Innervation	Action
Superior rectus	Tendinous ring anchored to bony orbit around optic foramen.	Superior and central part of eyeball.	Oculomotor nerve (III).	Rolls eye upward; also adducts and rotates medially.
Inferior rectus	Tendinous ring anchored to bony orbit around optic foramen.	Inferior and central part of eyeball.	Oculomotor nerve (III).	Rolls eye downward; also adducts and rotates laterally.
Lateral rectus	Tendinous ring anchored to bony orbit around optic foramen.	Lateral side of eyeball.	Abducens nerve (VI).	Rolls (abducts) eye laterally (out).
Medial rectus	Tendinous ring anchored to bony orbit around optic foramen.	Medial side of eyeball.	Oculomotor nerve (III).	Rolls (abducts) eye medially (in).
Superior oblique	Tendinous ring anchored to bony orbit around optic foramen.	Posterior and lateral to equator of eyeball under superior rectus.*	Trochlear nerve (IV).	Depresses, abducts, and rotates eye medially.
Inferior oblique	Tendinous ring anchored to bony orbit around optic foramen.	Posterior and lateral to equator of eyeball under lateral rectus.	Oculomotor nerve (III).	Elevates, abducts, and rotates eye laterally.

*The tendon of the superior oblique changes its direction abruptly when it passes through the trochlea (a fibrocartilaginous pulley) located in the upper front of the bony orbit.

FIGURE 10.8

Muscles of mastication. (A) Temporal and masseter muscles, lateral superficial view.
(B) Medial and lateral pterygoid muscles, lateral view deep to the mandible.

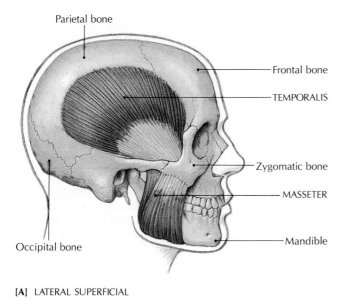

Parietal bone

Frontal bone

TEMPORALIS

Zygomatic bone

MASSETER

Occipital bone

Mandible

[A] LATERAL SUPERFICIAL

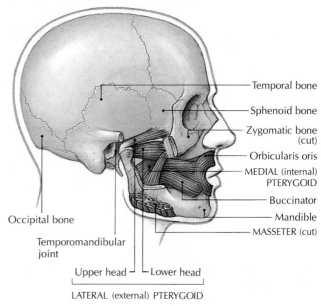

Temporal bone

Sphenoid bone

Zygomatic bone
(cut)

Orbicularis oris

MEDIAL (internal)
PTERYGOID

Buccinator

Mandible

MASSETER (cut)

Occipital bone

Temporomandibular
joint

Upper head ┐ ┌ Lower head

LATERAL (external) PTERYGOID

[B] LATERAL DEEP

at the same point or object at the same time. It can result from an imbalance in the power and length of the recti muscles.

Muscles of Mastication
(Table 10.4, Figure 10.8)

Four pairs of muscles produce biting and chewing movements: the ***masseter, temporalis, lateral pterygoid*** (Gr. "wing"; ter-uh-GOID), and ***medial pterygoid.*** The masseter (Gr. *maseter,* chewer) and the temporalis muscles lie close to the skin, and can be felt easily when the teeth are clenched.

In chewing, the mandible may move from side to side in a grinding motion. This movement is controlled by the masseter and temporalis muscles on the same side as the direction of the movement, and by the lateral and medial pterygoid muscles on the opposite side. The actions of the temporomandibular (jaw) joint, the joint involved in biting and chewing, are discussed in Chapter 8.

Muscles That Move the Hyoid Bone
(Table 10.5, Figure 10.9)

Many muscles associated with the mouth, throat (pharynx), and neck are attached to the hyoid bone. The ***suprahyoid***

muscles extend above the hyoid and attach it to the skull. The straplike ***infrahyoid muscles*** extend from the hyoid to skeletal structures below. The precise movements of the hyoid are controlled by coordinated muscular activity, and are especially evident during swallowing, when the hyoid bone moves upward and forward and then downward and backward to its original position. The hyoid is the main muscle attachment for the tongue and larynx. When the mandible does not move, the hyoid is raised by the ***mylohyoid, geniohyoid, digastric*** (suprahyoid), and ***stylohyoid*** muscles. The infrahyoid muscles pull the hyoid downward. The diaphragmlike paired mylohyoid muscles form the floor of the mouth below the tongue. Both suprahyoid and infrahyoid muscles can aid in depressing the mandible.

Muscles That Move the Tongue
(Table 10.6, Figure 10.10)

The muscles of the tongue are essential for normal speech and for the manipulation of food within the mouth. The *intrinsic muscles,* which are located within the tongue, are oriented in the horizontal, vertical, and longitudinal planes. They are able to fold, curve, and squeeze the tongue during speech and chewing. The action of the three *extrinsic muscles* of the tongue allow it to protrude (***genioglossus muscle***), retract (***styloglossus***), and depress (***hyoglossus***). All tongue muscles are innervated by the hypoglossal cranial nerve (XII).

TABLE 10.4   MUSCLES OF MASTICATION: MUSCLES THAT MOVE THE MANDIBLE

Muscle	Origin	Insertion	Innervation	Action
Masseter	Zygomatic arch (cheekbone).	Outer surface of angle and ramus of mandible.	Mandibular division of trigeminal nerve (V).	Elevates mandible (closes mouth); slightly protrudes mandible.
Temporalis	Temporal fossa of temporal bone.	Coronoid process of mandible.	Mandibular division of trigeminal nerve (V).	Elevates mandible; retracts mandible (pulls jaw back).
Medial pterygoid	Medial surface of lateral pterygoid plate of sphenoid bone, maxilla.	Inner surface of angle and ramus of mandible.	Mandibular division of trigeminal nerve (V).	Elevates mandible; slightly protrudes mandible; draws jaw toward opposite side in grinding movements.
Lateral pterygoid	Lateral surface of lateral pterygoid plate of sphenoid bone.	Condyle of mandible and fibrocartilage articular disk.	Mandibular division of trigeminal nerve (V).	Protrudes mandible; depresses mandible (opens mouth); produces side-to-side movements during chewing and grinding.

TABLE 10.5   MUSCLES THAT MOVE THE HYOID BONE

Muscle	Origin	Insertion	Innervation	Action
SUPRAHYOID MUSCLES (MUSCLES OF FLOOR OF MOUTH)				
Digastric				
Posterior belly	Mastoid process of temporal bone.	Common tendon attached to body of hyoid.	Facial nerve (VII).	Raises hyoid.
Anterior belly	Digastric fossa of mandible near symphysis.		Mandibular division of trigeminal nerve (V).	Depresses mandible.
Stylohyoid	Styloid process of temporal bone.	Body of hyoid.	Facial nerve (VII).	Raises hyoid; pulls hyoid backward.
Mylohyoid	Mylohyoid line on internal surface of mandible.	Hyoid bone, central raphe in floor of mouth.	Mandibular division of trigeminal nerve (V).	Raises hyoid; forms and elevates floor of mouth.
Geniohyoid	Adjacent to symphysis of mandible.	Body of hyoid.	First cervical nerve.	Pulls hyoid upward and forward; helps to open jaw. Depresses mandible when hyoid is fixed.
INFRAHYOID MUSCLES (STRAP MUSCLES)				
Sternohyoid	Manubrium of sternum.	Body of hyoid.	Cervical plexus.*	Depresses hyoid.
Sternothyroid	Manubrium of sternum.	Thyroid cartilage of larynx.	Cervical plexus.	Depresses thyroid and hyoid.
Thyrohyoid	Thyroid cartilage of larynx.	Greater horn of hyoid.	Cervical plexus.	Depresses hyoid; raises thyroid cartilage.
Omohyoid				
Inferior belly	Superior border of scapula (near coracoid process).	Intermediate tendon.	Cervical plexus.	Depresses hyoid.
Superior belly	Intermediate tendon attached to medial end of clavicle.	Body of hyoid.	Cervical plexus.	Depresses hyoid.

*The cervical plexus innervating these muscles is formed from nerves derived from the first three cervical nerves.

FIGURE 10.9

Muscles that move the hyoid bone. The sternocleidomastoid muscle of the neck and the digastric muscle attached to the hyoid bone are shown on the right but not on the left of this anterior view.

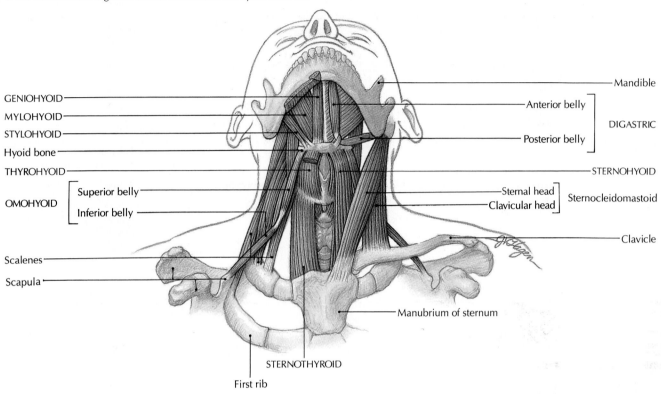

FIGURE 10.10

Muscles that move the tongue, lateral view.

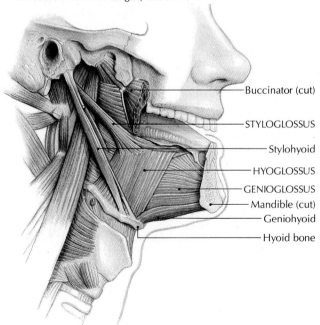

Normally, the genioglossus muscle prevents the tongue from falling backward toward the throat, where it can cause suffocation. But when the muscle is paralyzed or totally relaxed by an anesthetic, it can block the respiratory passage. Anesthetists keep the tongue in a forward position by pushing the mandible forward, and sometimes by inserting a tongue depressor as well.

Muscles That Move the Head
(Table 10.7, Figure 10.11)

In a sense, the head is balanced upon the vertebral column, with the atlas articulating with the occipital condyles. This *atlanto-occipital* articulation is involved in flexing (bending forward) and extending (holding erect) the head, and is located about midway between the back of the head and the tips of the nose and chin. The head is normally bent slightly forward, with the neck muscles inserted in the occipital region (such as the splenius capitis and semispinalis capitis) partially contracted to prevent the head from falling forward. When the head flexes, the posterior neck muscles relax while the **sternocleidomastoid** and other anterior neck muscles contract. The *bilateral* (both sides) contraction of the **splenius capitis, longissimus capitis,** and **semispinalis capitis** extends the head. The *unilateral* (one side only) contraction of these muscles rotates the head, tilts the chin up, and turns the face toward the contracted side.

TABLE 10.6 MUSCLES THAT MOVE THE TONGUE

Muscle	Origin	Insertion	Innervation	Action
Genioglossus	Tubercle of anterior part of mandible beside midline.	Fibers pass through tongue to insert in bottom of tongue.	Hypoglossal nerve (XII).	Protrudes and depresses tongue.
Styloglossus	Tip of styloid process of temporal bone.	Side of tongue.		Retracts and elevates tongue.
Hyoglossus	Greater horn of hyoid bone.	Side of tongue.		Depresses tongue.
Intrinsic muscles	Arranged in three planes within tongue in horizontal, vertical, and longitudinal planes.			Alter shape of tongue.

TABLE 10.7 MUSCLES THAT MOVE THE HEAD

Muscle	Origin	Insertion	Innervation	Action
Sternocleido-mastoid	Sternum, clavicle.	Mastoid process of temporal bone.	Spinal accessory nerve (XI).	*Bilateral:* flex vertebral column, bringing head down. *Unilateral:* bends vertebral column to same side, drawing head toward shoulder and rotating head so chin points to opposite side.
Semispinalis capitis	Vertebral arches of C7 and T1 to T6 vertebrae.	Occipital bone.	Dorsal rami of cervical and upper thoracic nerves.	*Bilateral:* extend head. *Unilateral:* extends head; turns face to opposite side.
Splenius capitis	Spines of C7 and T1 to T4 vertebrae, ligamentum nuchae (nape of neck).	Occipital bone, mastoid process of temporal bone.	Dorsal rami of middle cervical nerves.	*Bilateral:* extend head. *Unilateral:* bends head to same side; rotates face to same side.
Longissimus capitis	Transverse processes of T1 to T4 or T5 vertebrae.	Mastoid process of temporal bone.	Dorsal rami of middle and lower cervical nerves.	*Bilateral:* extend head. *Unilateral:* bends head to same side; rotates face to same side.

FIGURE 10.11

Muscles that move the head, posterior view.

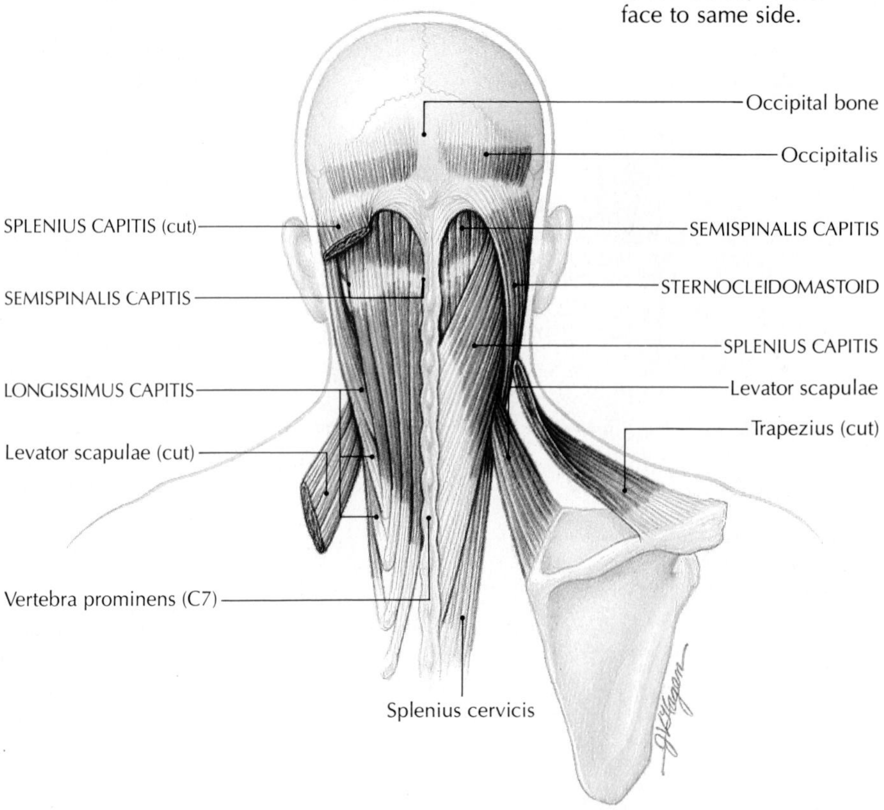

SPLENIUS CAPITIS (cut)
SEMISPINALIS CAPITIS
LONGISSIMUS CAPITIS
Levator scapulae (cut)
Vertebra prominens (C7)
Occipital bone
Occipitalis
SEMISPINALIS CAPITIS
STERNOCLEIDOMASTOID
SPLENIUS CAPITIS
Levator scapulae
Trapezius (cut)
Splenius cervicis

Intrinsic Muscles That Move the Vertebral Column (Figure 10.12)

The vast numbers of small muscles and muscle bundles that make up the intrinsic back muscles are located in a pair of broad, longitudinally oriented gutters on either side of the spines of the vertebrae. They are usually organized into two major groups: the *superficial* group, called the **erector spinae** or **sacrospinalis muscles**, and the *deep* group, called the **transversospinalis muscles**.

The superficial group consists of overlapping muscle fascicles. They mount "up" the vertebral column, and each bundle spans from origin to insertion over five or so vertebrae. This muscle group is subdivided into three columns: (1) the lateral column, the **iliocostalis**, originates on the iliac crest and after spanning five levels inserts on a rib or transverse process of a cervical vertebra; (2) the middle column, the **longissimus**, originates and inserts on the transverse processes of the thoracic and cervical vertebrae and mastoid process of the skull; and (3) the medial column, the **spinalis**, extends from spinous process to spinous process. Muscles of the superficial group are named **lumborum, thoracis,** and **cervicis,** depending on whether they are associated with the lumbar, thoracic, or cervical region.

The deep group consists of small bundles of shorter muscle fascicles called the **transversospinalis muscles.** Each bundle originates on a transverse process (located relatively laterally), extends upward and obliquely, and inserts on a spinous process. Some bundles (**rotatores**) run up one or two levels, others (**multifidus**) span two or three vertebrae, and still others (**semispinalis**) span four to five vertebrae. The semispinalis muscles are called **thoracis** or **cervicis,** depending on their location.

FIGURE 10.12

Intrinsic muscles of the vertebral column, posterior view.

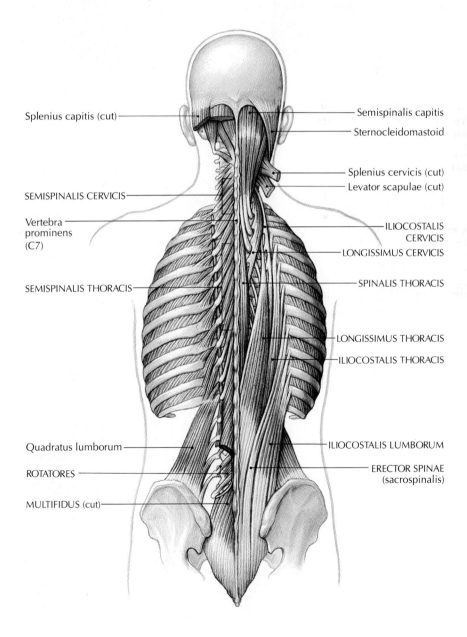

Splenius capitis (cut)

Semispinalis capitis

Sternocleidomastoid

Splenius cervicis (cut)

Levator scapulae (cut)

SEMISPINALIS CERVICIS

Vertebra prominens (C7)

ILIOCOSTALIS CERVICIS

LONGISSIMUS CERVICIS

SPINALIS THORACIS

SEMISPINALIS THORACIS

LONGISSIMUS THORACIS

ILIOCOSTALIS THORACIS

Quadratus lumborum

ILIOCOSTALIS LUMBORUM

ROTATORES

ERECTOR SPINAE (sacrospinalis)

MULTIFIDUS (cut)

FIGURE 10.13

Muscles used in quiet respiration. (A) External and internal intercostal muscles. Note that the direction of the fibers of the internal and external intercostal muscles are at approximate right angles to each other. (B) Scalene muscles, which extend from the cervical vertebrae to the first two ribs.

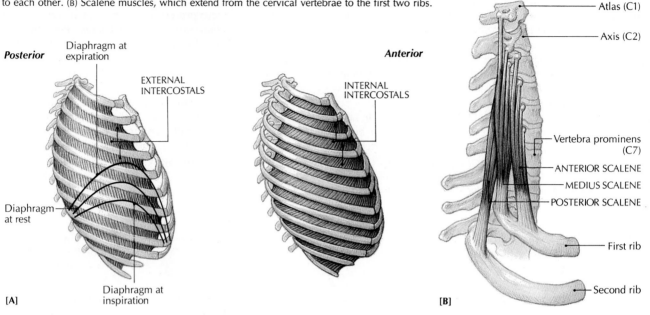

Posterior

Diaphragm at expiration

EXTERNAL INTERCOSTALS

Diaphragm at rest

Diaphragm at inspiration

[A]

Anterior

INTERNAL INTERCOSTALS

[B]

Atlas (C1)

Axis (C2)

Vertebra prominens (C7)

ANTERIOR SCALENE

MEDIUS SCALENE

POSTERIOR SCALENE

First rib

Second rib

FIGURE 10.14

How the ribs move during breathing.
(A) When the ribs are at rest, they are like handles hanging at the side of a bucket.
(B) During inspiration, the ribs twist outward and upward in the "bucket-handle" movement, which increases the width of the thoracic cage.
(C) The front-to-back dimension is also increased by the "pump-handle" movement, which elevates the sternal end of the upper ribs.

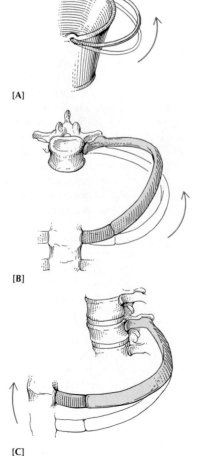

[A]

[B]

[C]

The intrinsic back muscles act to extend, bend laterally, and rotate the vertebral column and head. They also counteract the force of gravity, which tends to flex the vertebral column. The posture of the vertebral column is regulated by the vertebral muscles. Contrary to a common belief, these muscles are not active constantly. They are fairly relaxed during normal standing, and they are almost completely relaxed when the vertebral column is bent back (extension), because the ligaments of the vertebral column assume the load. The axiom to remember is: When ligaments suffice, the muscles will yield. In contrast, the vertebral muscles are most active when the vertebral column is bent forward. Extension and hyperextension of the column are used in forced inspiration (as in gasping for air). Gravity and the contraction of the rectus abdominus muscles flex the vertebral column.

Muscles Used in Quiet Respiration (Breathing)
(Table 10.8, Figures 10.13 and 10.14)

Quiet respiration, or normal breathing, involves the coordinated activity of several muscles or groups of muscles. Inspiration (breathing in) requires the enlargement of the thoracic cage, so that air pressure inside the thorax is less than the atmospheric pressure. Because of this imbalance, air is sucked into the lungs as the dome-shaped **diaphragm** contracts and moves down. This movement is accompanied by the relaxation of the muscles of the abdominal wall. The contraction of the **external intercostal muscles** raises the ribs, in what is called the "bucket-handle movement," which increases the transverse diameter of the thoracic cage.

TABLE 10.8   MUSCLES USED IN QUIET RESPIRATION

Muscle	Origin	Insertion	Innervation	Action
Diaphragm	Back to lumbar vertebrae, sides and front to lower six costal cartilages, front to xiphoid cartilage.	Central tendon of diaphragm.	Phrenic nerve (from C3 to C5 nerves), iliohypogastric and ilioinguinal nerves	Contraction pulls central tendon down to flatten dome and increase vertical length and volume of thorax, resulting in inspiration.
External intercostals	Each from lower border of each bony rib.	Upper border of next rib below.	Intercostal nerves.	Elevates ribs to increase all diameters of thorax, resulting in inspiration.
Internal intercostals	Each from upper border of each bony rib and costal cartilage.	Lower border of each bony rib.	Intercostal nerves.	Role not yet determined with certainty.
Scalenes Anterior	Fronts of transverse processes in cervical vertebrae.	Rib 1.	Branches of cervical nerves.	Steady ribs 1 and 2 during respiration; may assist in elevating them.
Medius and posterior	Backs of transverse processes in cervical vertebrae.	Medius to rib 1, posterior to rib 2.	Branches of cervical nerves.	Steady ribs 1 and 2 during respiration; may assist in elevating them.

In expiration (breathing out), the diaphragm relaxes and is raised to its higher resting position, arching up into the thorax, by the pressure generated in the abdominal cavity by the contractile tone of the muscles in the abdominal wall. The thoracic cage becomes smaller when the ribs are lowered, largely by elastic recoil, and expiration may also be aided by the contraction of the *internal intercostal muscles.* The *scalene muscles* also assist in respiration by steadying the first two ribs and possibly elevating them.

How can the rib cage change its shape and size to accommodate the shifting volume of the thorax? The answer lies in the "bucket-handle" and "pump-handle" movements of the ribs that result from the twisted shape of a typical rib, and its cartilaginous connections to the sternum and vertebral column (Figure 10.14). Although the increase in size of the rib cage during inspiration requires muscular effort, the decrease during expiration is merely an elastic recoil, produced by the lungs and the costal cartilages.

Muscles That Support the Abdominal Wall
(Table 10.9, Figure 10.15)

The muscles of the abdominal wall act to support and protect the internal organs. Several layers of muscles running in different directions add strength to the abdominal wall:

1 As the outside layer of the anterior and lateral walls, the *external abdominal oblique* muscle extends forward and down from the ribs until it becomes an aponeurosis, which attaches to the linea alba at the midline. The *linea alba* (L. white line) is a tendon running from the xiphoid process of the sternum to the pubic symphysis.

2 The next layer inward is the *internal abdominal oblique* muscle, which runs forward and upward from the iliac crest and inguinal ligament, until it too becomes an aponeurosis that connects with the linea alba.

3 The innermost layer is the *transversus abdominis* muscle, which extends horizontally from the dorsolumbar fascia to the inguinal ligament and ribs. After becoming an aponeurosis (passing deep to the rectus abdominis), it continues to meet the linea alba.

4 Running lengthwise on either side of the linea alba is the *rectus abdominis* muscle, which extends from the rib cage and sternum to the pubic crest. It is enveloped by the rectus sheath, which is actually made up of the aponeuroses of the other three layers of muscles. The transverse bands crossing the rectus abdominis muscle are called *tendinous inscriptions.*

Together these four abdominal muscles act as a dynamic corset around the abdomen, supplying the necessary pressure for respiration, urination, defecation, and parturition (childbirth). In addition, they also play roles in flexion and lateral bending of the vertebral column.

On the posterior wall of the abdomen, the *quadratus lumborum* muscle extends from the twelfth rib to the posterior iliac crest. Besides allowing the trunk to bend toward one side, this muscle gives the vertebral column stability and plays a role in respiration.

TABLE 10.9   MUSCLES SUPPORTING THE ABDOMINAL WALL

Muscle	Origin	Insertion	Innervation	Action
MUSCLES OF ANTERIOR AND LATERAL WALLS				
Rectus abdominis	Pubic crest, symphysis pubis.	Costal cartilages of ribs 5 to 7, xiphoid process.	Branches of intercostal nerves (T7 to T12).	Powerful flexor of lumbar vertebral column; depresses rib cage, plays role in stabilizing pelvis against leverage of thigh muscles during walking and running.
External abdominal oblique	Ribs 5 to 12.	Iliac crest, linea alba.	Branches of intercostal (T8 to T12), iliohypogastric, and ilioinguinal nerves.	Along with rectus abdominis, these muscles hold in and compress the abdominal contents, aiding in defecation, urination, and childbirth. Contract to protect abdominal contents against external blows; flex trunk, rotate vertebral column.
Internal abdominal oblique	Iliac crest, inguinal ligament, lumbodorsal fascia.	Costal cartilages and ribs 8 to 12, linea alba, xiphoid process.	Branches of intercostal (T8 to T12), iliohypogastric, and ilioinguinal nerves.	
Transversus abdominis	Iliac crest, inguinal ligament, lumbar fascia, cartilages of ribs 7 to 12.	Xiphoid process, linea alba, pubis.	Branches of intercostal (T8 to T12), iliohypogastric, and ilioinguinal nerves.	
MUSCLE OF POSTERIOR WALL				
Quadratus lumborum (see Figure 10.12)	Iliac crest, iliolumbar ligament, transverse processes of lower lumbar vertebrae.	Rib 12, transverse processes of upper lumbar vertebrae.	Ventral rami of T12 to L4 nerves.	*Unilateral:* flexes (bends) trunk toward same side. *Bilateral:* extends and stabilizes lumbar vertebral column; aids inspiration by increasing length of thorax.

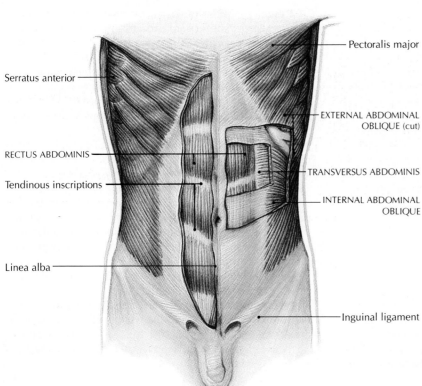

Serratus anterior

Pectoralis major

RECTUS ABDOMINIS

EXTERNAL ABDOMINAL OBLIQUE (cut)

Tendinous inscriptions

TRANSVERSUS ABDOMINIS

INTERNAL ABDOMINAL OBLIQUE

Linea alba

Inguinal ligament

FIGURE 10.15

Muscles supporting the abdominal wall. In this anterior view, some of the muscles have been cut to show the different layers of muscles, from superficial to deep.

Muscles That Form the Pelvic Outlet (Table 10.10, Figure 10.16)

The muscles of the pelvic outlet are often divided into two groups: those of the pelvic diaphragm and those of the perineum. The **pelvic diaphragm** (pelvic floor) is the funnel-shaped muscular floor of the pelvic cavity. Its role is to support the pelvic organs and thus prevent organs such as the urinary bladder, uterus, and rectum from falling down (undergoing prolapse) and moving through the diaphragm (as in hernia). The **levator ani muscle** (consisting of the **pubococcygeus** and **iliococcygeus muscles**), the **coccygeus muscle,** and fascia comprise the pelvic diaphragm. The *urogenital (UG) diaphragm* is composed of fascia and the **deep transverse perinei muscles.** The UG diaphragm also functions as part of the **external sphincter of the urethra,** and provides added reinforcement.

The urethra of both sexes and the vagina of the female pass through both the pelvic and UG diaphragms. The rectum passes through the pelvic diaphragm to become the anus. The tonus of the medial portion of the **pubococcygeus muscle,** called the *puborectalis,* acts as a major deterrent to fecal incontinence (the involuntary passage of feces). Certain muscles of these diaphragms participate in maintaining urinary continence (the ability to retain the contents of the bladder until conditions are proper for urination).

FIGURE 10.16

Muscles forming the pelvic outlet. (A) Male. (B) Female.

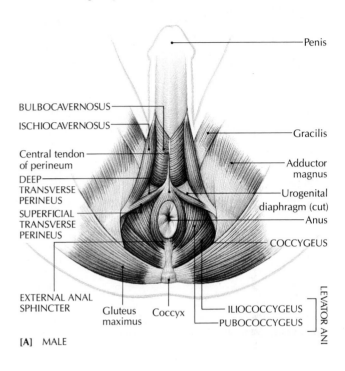

[A] MALE

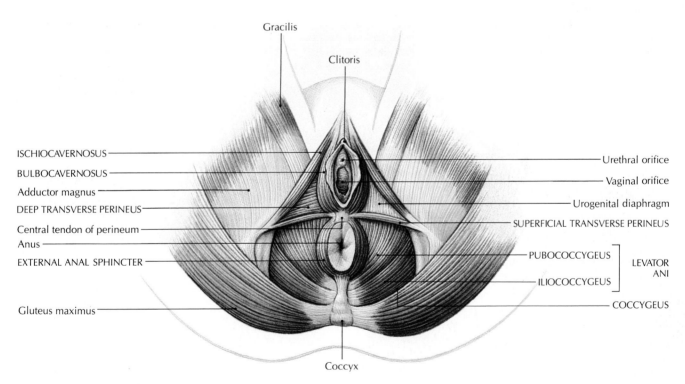

[B] FEMALE

TABLE 10.10 MUSCLES FORMING THE PELVIC OUTLET

Muscle	Origin	Insertion	Innervation	Action
MUSCLES OF PELVIC DIAPHRAGM (PELVIC FLOOR)				
Levator ani				
Pubococcygeus	Pubis.	Coccyx, anal canal, central tendon of perineum and urethra.	Branches of third and fourth sacral nerves and pudendal nerve (second, third, fourth sacral).	Supports pelvic viscera; helps maintain intra-abdominal pressure; acts as sphincter to constrict anus.
Iliococcygeus	Ischial spine.	Coccyx.	Branches of third and fourth sacral nerves and pudendal nerve (second, third, fourth sacral).	Same as pubococcygeus.
Puborectalis (pubic sling; most medial portion of levator ani; not illustrated)	Pubic arch.	Pubic arch (loops around rectum to form sling).	Branches of fourth sacral nerve and pudendal nerve.	Draws the rectum anteriorly (forward); acts as principal component for fecal continence.
Coccygeus	Ischial spine.	Fifth sacral vertebra, coccyx.	Branches of fourth and fifth sacral nerves and pudendal nerve (second, third, fourth sacral).	Same as pubococcygeus.
MUSCLES OF PERINEUM (UROGENITAL DIAPHRAGM)				
Superficial transverse perineus	Ischial tuberosity.	Central tendon of perineum.	Pudendal nerve (second, third, fourth sacral).	Pulls coccyx slightly forward after being pressed backward during defecation or childbirth.
External sphincter of urethra	Central tendon of perineum.	Midline in male; vaginal wall in female (not illustrated).	Pudendal nerve (second, third, fourth sacral).	Under voluntary control, prevents urination when bladder wall contracts; relaxes during urination; can contract to cut off stream of urine.
Deep transverse perineus	Ischial rami.	Central tendon.	Pudendal nerve (second, third, fourth sacral).	Helps support perineum by steadying central perineal tendon.
Ischiocavernosus	Ischial tuberosity and rami of pubis and ischium.	Corpus cavernosum of penis in males; clitoris in females.	Pudendal nerve (second, third, fourth sacral).	Contributes to maintaining erection of penis and clitoris.
Bulbocavernosus (bulbospongiossus)	Central tendon of perineum.	Fascia of urogenital triangle and penis in males; pubic arch, clitoris in females.	Pudendal nerve (second, third, fourth sacral).	In male, contracts to expel last drops of urine, contracts rhythmically during ejaculation to propel semen in urethra; in female, acts as sphincter at vaginal orifice.
MUSCLES OF PERINEUM (ANAL REGION)				
External anal sphincter	Anococcygeal raphe.	Central tendon of perineum.	Pudendal nerve (second, third, fourth sacral).	Constant state of tonic contraction to keep anal orifice closed; closes anus during efforts not associated with defecation.

The *perineum* (per-uh-NEE-uhm) is the diamond-shaped region extending from the pubic symphysis to the coccyx. The anterior half of the perineum, shaped like a triangle, is the urogenital diaphragm (or triangle). The posterior half, also triangular, is the anal region. The muscles of the urogenital triangle provide voluntary control of the urethra and the release of urine.

The male perineum is of special importance to proctologists (surgeons who treat disorders of the rectum and anus) and urologists (surgeons who treat disorders of the urogenital organs). The female perineum is of special importance to obstetricians and gynecologists, especially for the management of childbirth and many female disorders involving the bladder, urethra, vagina, rectum, and anus.

Occasionally, labor and delivery during childbirth weaken the pelvic floor, and the pelvic diaphragm may be stretched or torn. Such a condition may be accompanied by a hernia or prolapse of the uterus or rectum. The damage can usually be repaired by surgery.

Muscles That Move the Shoulder Girdle
(Table 10.11, Figures 10.17 and 10.18)

Three joints contribute to the movements of the shoulder girdle. They are (1) the sternoclavicular joint between the clavicle and the manubrium of the sternum, (2) the acromioclavicular joint between the acromion and clavicle, and (3) the scapulothoracic joint between some muscles attached to the scapula and the thoracic wall. The acromioclavicular joint is a sliding joint of accommodation. The movements of the two axes of the sternoclavicular joint are (1) elevation and depression and (2) protraction and retraction.

The muscles of the shoulder girdle contract together to stabilize the girdle for certain movements, such as lifting a heavy object held in the hand. Muscles active during *elevation* include the upper fibers of the *trapezius, levator scapulae,* and *rhomboids.* Those active during *depression* include the

---

TABLE 10.11   MUSCLES THAT MOVE THE SHOULDER GIRDLE

Muscle	Origin	Insertion	Innervation	Action
Trapezius	Occipital bone, ligamentum nuchae, spines of C7 to T12 vertebrae.	Lateral third of clavicle, acromion and spine of scapula.	Spinal accessory motor nerve (X1), C3 to C4 sensory nerves.	Steadies, elevates, retracts, and rotates* scapula.
Levator scapulae	C1 to C4 of vertebrae.	Upper vertebral border of scapula.	Dorsal scapula nerve.	Elevates and rotates* scapula downward.
Rhomboids: major and minor	Ligamentum nuchae, spines of C7 to T5 vertebrae.	Medial border of scapula.	Dorsal scapula nerve.	Retract and elevate scapula.
Serratus anterior	Ribs 1 to 8, midway between angles and costal cartilages.	Entire medial border of scapula.	Long thoracic nerve.	Steadies, rotates* upward, holds against chest wall, and protracts scapula.
Pectoralis minor	Ribs 3, 4, and 5.	Coracoid processes of scapula.	Medial pectoral nerve.	Steadies, depresses, rotates* downward, and protracts scapula.
Pectoralis major	Medial half of clavicle (clavicular head) and sternum, ribs 1 to 6 (sternal head).	Lateral crest of intertubercular groove of humerus.	Medial and lateral pectoral nerves.	Adducts and rotates* arm medially. Clavicular head flexes arm; sternal head extends arm.
Latissimus dorsi	Spines of L7 to sacral vertebrae, iliac crest, ribs 7 to 12.	Intertubercular groove of humerus.	Thoracodorsal nerve (middle subscapular nerve).	Depresses, rotates* downward, and retracts scapula; also extends, adducts, and medially rotates humerus at shoulder (glenohumeral joint).
Subclavius	Rib 1 (median).	Clavicle.	Nerve to subclavius.	Draws clavicle toward sternoclavicular joints; also steadies clavicle.

*The axis of rotation of the scapula is located in the center of the scapula. When the glenoid fossa (at shoulder joint) is directed downward, the scapula is said to rotate downward. When the glenoid fossa is directed upward, the scapula is said to rotate upward.

FIGURE 10.17

Muscles that move the shoulder girdle. Superficial muscles are shown on the left and deep muscles on the right of the illustration in this anterior view.

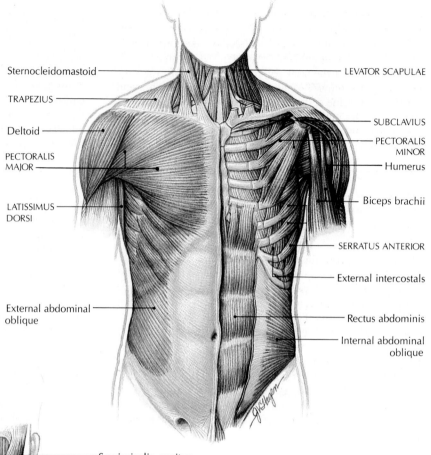

Sternocleidomastoid

TRAPEZIUS

Deltoid

PECTORALIS MAJOR

LATISSIMUS DORSI

External abdominal oblique

LEVATOR SCAPULAE

SUBCLAVIUS

PECTORALIS MINOR

Humerus

Biceps brachii

SERRATUS ANTERIOR

External intercostals

Rectus abdominis

Internal abdominal oblique

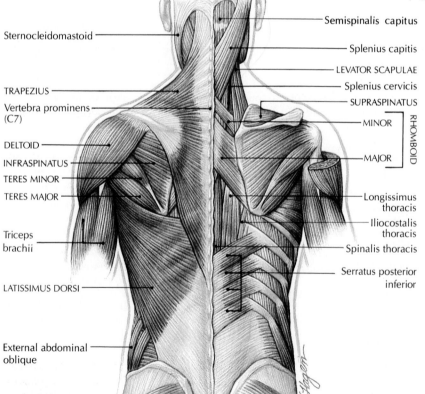

Sternocleidomastoid

TRAPEZIUS

Vertebra prominens (C7)

DELTOID

INFRASPINATUS

TERES MINOR

TERES MAJOR

Triceps brachii

LATISSIMUS DORSI

External abdominal oblique

Gluteus maximus

Left

Semispinalis capitus

Splenius capitis

LEVATOR SCAPULAE

Splenius cervicis

SUPRASPINATUS

MINOR

MAJOR

RHOMBOID

Longissimus thoracis

Iliocostalis thoracis

Spinalis thoracis

Serratus posterior inferior

FIGURE 10.18

Muscles of the back. The muscles acting on the shoulder girdle are emphasized. Superficial muscles are shown on the left and deep muscles on the right in this posterior view.

pectoralis major, pectoralis minor, and *latissimus dorsi.* Upward rotation of the girdle (as in reaching upward) uses both the upper and lower fibers of the *trapezius* and lower fibers of the *serratus anterior. Downward rotation* uses the *levator scapulae, rhomboids major* and *minor, pectoralis major* and *minor,* and the *latissimus dorsi. Upward* and *downward rotary* movements of the scapula rotate around an axis located roughly in the center of the scapula just below the middle of the spine. The function of the *subclavius* is to steady the clavicle.

Muscles That Move the Humerus at the Glenohumeral (Shoulder) Joint (Table 10.12, Figures 10.18, 10.19, and 10.20)

Many muscles contribute to the movements of the shoulder joint. The four SITS muscles—*supraspinatus, infraspinatus, teres minor,* and *subscapularis*—are the *rotator cuff muscles.* They are located adjacent to the articular capsule of the joint and act as "dynamic ligaments," preventing instability and dislocations in an otherwise potentially unstable joint. The deltoid muscle overrides the SIT muscles (supraspinatus, intraspinatus, teres minor), and helps form the roundness of the shoulder. Other contributing muscles are noted below.

The shoulder joint has three axes of rotation, resulting in (1) flexion and extension, (2) abduction and adduction, and (3) medial and lateral rotation. A combination of all these movements in succession is *circumduction.*

The *flexors* include the *pectoralis major* (clavicular head), *deltoid* (clavicular part), *coracobrachialis,* and *biceps brachii.* The *extensors* include the *latissimus dorsi, teres major, pectoralis major* (sternal head), *deltoid* (spinous part), and *triceps* (long head). The *abductors* are the *supraspinatus* and *deltoid* (acromial part), and the *adductors* are the *pectoralis major* (both heads), *latissimus dorsi, teres major, coracobrachialis,* and *triceps* (long head). The *lateral rotators* include the *infraspinatus, teres minor,* and *deltoid* (spinous part). The *medial rotators* include the *pectoralis major* (both clavicular and sternal heads), *teres major, latissimus dorsi, subscapularis,* and *deltoid* (clavicular part).

Muscles That Move the Forearm (Table 10.13, Figures 10.19, 10.21, and 10.22)

At the one-axis elbow joint, flexion and extension are the only movements. The *flexors* include the *biceps brachii, brachialis,* and *brachioradialis.* The *extensors* are the *triceps brachii* and the tiny *anconeus.*

At the proximal and distal radioulnar joints, pronation and supination are the only movements. The *pronators* include the *pronator teres* and the *pronator quadratus.* The *supinators* are the *supinator* and the *biceps brachii.* Supination is more powerful than pronation. That is why screws, caps on jars, and door handles are designed the way they are. (For left-handed people, the motion involved in moving these objects is *pronation.*)

FIGURE 10.19

Superficial muscles of the right upper limb. Muscles that move the humerus at the glenohumeral joint and muscles that move the forearm at the elbow are emphasized. (A) Anterior view. (B) Posterior view.

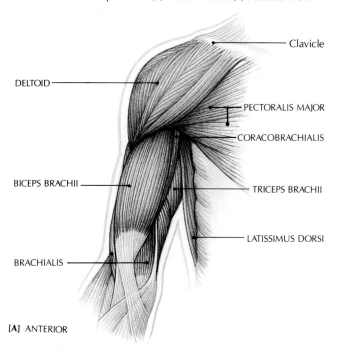

[A] ANTERIOR

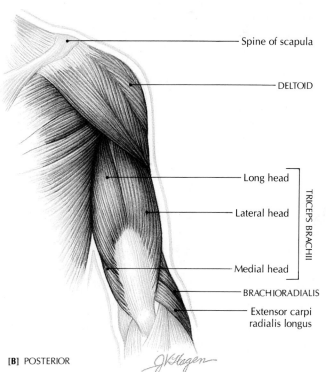

[B] POSTERIOR

TABLE 10.12   MUSCLES THAT MOVE THE HUMERUS AT THE GLENOHUMERAL JOINT

Muscle	Origin	Insertion	Innervation	Action
ROTATOR CUFF MUSCLES				
Supraspinatus	Supraspinatus fossa of scapula.	Greater tubercle of humerus.	Suprascapular nerve.	Stabilizes joint; abducts arm.
Infraspinatus	Infraspinatus fossa of scapula.	Greater tubercle of humerus.	Suprascapular nerve.	Stabilizes joint; laterally rotates arm.
Teres minor	Lower border of lateral scapula.	Greater tubercle of humerus.	Axillary nerve.	Stabilizes joint; laterally rotates arm.
Subscapularis	Subscapular fossa of scapula.	Lesser tubercle of humerus.	Upper and lower subscapular nerves.	Stabilizes joint; medially rotates arm.
OTHER CONTRIBUTING MUSCLES				
Teres major	Inferior angle of scapula.	Medial crest of intertubercular groove of humerus.	Lower subscapular nerve.	Stabilizes upper arm during adduction; adducts, extends, and medially rotates humerus.
Deltoid	Clavicle (lateral third), acromion process and spine of scapula.	Deltoid tuberosity of humerus.	Axillary nerve.	Clavicular part: flexes and medially rotates arm; acromial part: abducts arm; spinous part: extends and laterally rotates arm.
Pectoralis major	Medial half of clavicle (clavicular head) and sternum, and ribs 1 to 6 (sternal head).	Lateral aspect of intertubercular groove of humerus.	Medial and lateral pectoral nerves.	Clavicular head: flexes, adducts, and medially rotates arm; sternal head: extends, adducts, and medially rotates arm. (See Table 10.11 for action at shoulder girdle.)
Latissimus dorsi	Spines of lower thoracic, lumbar, and sacral vertebrae, iliac crest, and ribs 7 to 12.	Intertubercular groove of humerus.	Thoracodorsal (middle subscapular) nerve.	Extends, adducts, and medially rotates arm; also depresses, retracts, and rotates scapula downward.
Coracobrachialis	Coracoid process of scapula.	Middle third of humerus.	Musculocutaneous nerve.	Flexes and adducts arm.

TABLE 10.13   MUSCLES THAT MOVE THE FOREARM

Muscle	Origin	Insertion	Innervation	Action
Biceps brachii	*Long head:* supraglenoid tubercle of scapula. *Short head:* coracoid process of scapula.	Tuberosity of radius, bicipital aponeurosis.	Musculocutaneous nerve.	Flexes elbow; supinates forearm.
Brachialis	Anterior surface of humerus.	Tuberosity and coronoid process of ulna.	Musculocutaneous nerve.	Flexes elbow.
Triceps brachii	*Long head:* infraglenoid tubercle of scapula. *Lateral head:* posterior and lateral humerus above radial groove. *Medial head:* posterior humerus below radial groove.	Olecranon process of ulna.	Radial nerve.	Extends elbow.
Supinator	Lateral epicondyle of humerus, supinator crest of ulna.	Lateral surface of radius (distal to head).	Radial nerve.	Supinates forearm (rotates palm of hand anteriorly).
Pronator teres	Medial epicondyle of humerus, coronoid process of ulna.	Midlateral surface of radius.	Median nerve.	Pronates forearm; flexes elbow.
Pronator quadratus	Anterior distal end of ulna.	Anterior distal end of radius.	Median nerve.	Pronates forearm.
Brachioradialis	Lateral supracondylar ridge of humerus.	Lateral surface of radius near base of styloid process.	Radial nerve.	Flexes elbow.
Anconeus	Posterior lateral epicondyle of humerus.	Lateral surface of olecranon process of ulna.	Radial nerve.	Extends elbow.

FIGURE 10.20

Deep muscles of the right upper limb. Muscles that move the humerus at the glenohumeral joint and muscles that move the forearm at the elbow are emphasized. (A) Anterior view. (B) Posterior view.

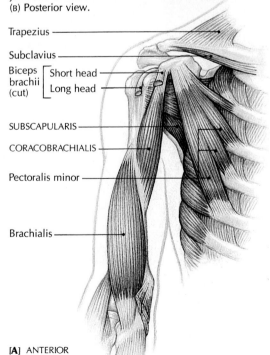

Trapezius
Subclavius
Biceps brachii (cut) — Short head — Long head
SUBSCAPULARIS
CORACOBRACHIALIS
Pectoralis minor
Brachialis

[A] ANTERIOR

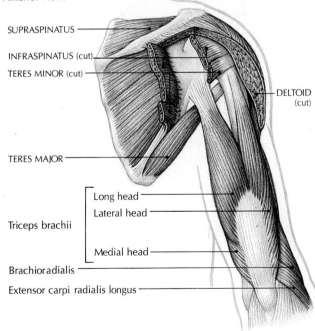

SUPRASPINATUS
INFRASPINATUS (cut)
TERES MINOR (cut)
DELTOID (cut)
TERES MAJOR
Triceps brachii — Long head — Lateral head — Medial head
Brachioradialis
Extensor carpi radialis longus

[B] POSTERIOR

Muscles That Move the Wrist and Hand at the Radiocarpal and Midcarpal Joints (Table 10.14, Figures 10.21, 10.22, and 10.23)

The wrist joints include the (1) radiocarpal joint (between the radius and proximal row of carpal bones), where most of the movements occur, and (2) the midcarpal joint (between the proximal and distal rows of carpal bones), where a slight amount of flexion and extension occurs. There is no movement between the distal carpal bones and the metacarpals. The radiocarpal joint is a two-axis condyloid joint where (1) flexion and extension (and hyperextension) and (2) abduction (radial deviation, toward the radius) and adduction (ulnar deviation, toward the ulna) occur.

FIGURE 10.21

Superficial muscles of the anterior forearm and hand. Muscles acting on the forearm and wrist are emphasized.

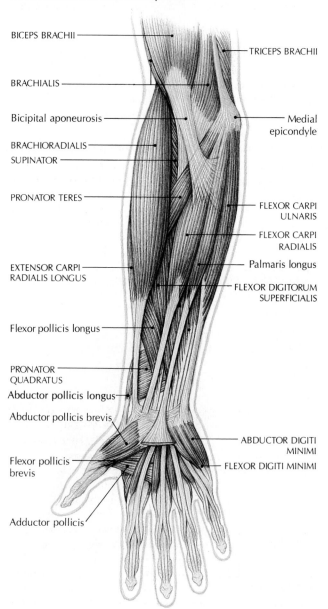

FIGURE 10.22

Superficial muscles of the posterior forearm and hand. Muscles acting on the forearm and wrist are emphasized.

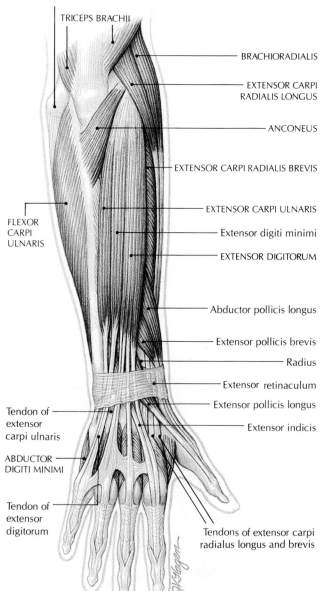

TABLE 10.14   MUSCLES THAT MOVE THE WRIST AND HAND AT THE RADIOCARPAL AND MIDCARPAL JOINTS

Muscle	Origin	Insertion	Innervation	Action
Flexor carpi radialis	Medial epicondyle of humerus.	Base of second and third metacarpals.	Median nerve.	Flexes and abducts wrist.
Flexor carpi ulnaris	Medial epicondyle of humerus, olecranon process of ulna.	Pisiform, hamate, base of fifth metacarpal.	Ulnar nerve.	Flexes and adducts wrist.
Extensor carpi radialis longus	Lateral epicondyle of humerus.	Base of second metacarpal.	Radial nerve.	Extends and abducts wrist.
Extensor carpi radialis brevis	Lateral epicondyle of humerus.	Base of third metacarpal.	Radial nerve.	Extends and abducts wrist.
Extensor carpi ulnaris	Lateral epicondyle of humerus.	Base of fifth metacarpal.	Radial nerve.	Extends and adducts wrist.

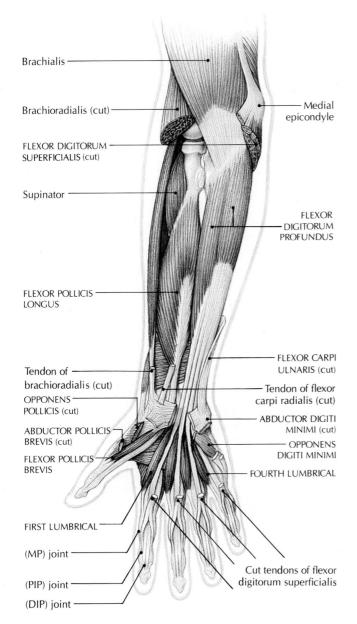

Brachialis

Brachioradialis (cut)

FLEXOR DIGITORUM SUPERFICIALIS (cut)

Supinator

FLEXOR POLLICIS LONGUS

Tendon of brachioradialis (cut)

OPPONENS POLLICIS (cut)

ABDUCTOR POLLICIS BREVIS (cut)

FLEXOR POLLICIS BREVIS

FIRST LUMBRICAL

(MP) joint

(PIP) joint

(DIP) joint

Medial epicondyle

FLEXOR DIGITORUM PROFUNDUS

FLEXOR CARPI ULNARIS (cut)

Tendon of flexor carpi radialis (cut)

ABDUCTOR DIGITI MINIMI (cut)

OPPONENS DIGITI MINIMI

FOURTH LUMBRICAL

Cut tendons of flexor digitorum superficialis

FIGURE 10.23

Deep muscles of the anterior forearm and hand. Muscles acting on the fingers and thumb are emphasized.

The *flexor* muscles include the **flexor carpi ulnaris, flexor carpi radialis,** and the **palmaris longus.** The *extensor* muscles are the **extensor carpi radialis longus, extensor carpi radialis brevis,** and the **extensor carpi ulnaris.** The *abductor* muscles include the **flexor carpi radialis, extensor carpus radialis longus** and **brevis,** and **abductor pollicis longus.** The *adductor* muscles are the **extensor carpi ulnaris** and the **flexor carpi ulnaris.** The contraction of any of these muscles in a sequential order can produce circumduction. Contracting in concert, these muscles act to stabilize the wrist for the effective use of the fingers.

Muscles That Move the Fingers (Except the Thumb) (Table 10.15, Figures 10.21, 10.22, 10.23, and 10.24)

The movements of the fingers (except the thumb) are based on a remarkable interplay of several muscle groups, and especially on the role of the lumbrical muscles. To fully appreciate this interplay we must understand the movements at each joint, and the anatomical relationships and actions of the tendons and muscles acting on those joints.

Each finger has three joints: (1) the *metacarpophalangeal joint* (MP or knuckle joint) between the metacarpal and proximal phalanges, (2) the *proximal interphalangeal joint* (PIP joint) between the proximal and middle phalanges, and (3) the *distal interphalangeal joint* (DIP joint) between the middle and distal phalanges. The metacarpophalangeal joint is a condyloid joint with two axes of rotation, permitting flexion and extension, abduction and adduction, and circumduction. The proximal and distal interphalangeal joints are hinge joints with one axis of rotation, permitting only flexion and extension.

The *flexors* of the metacarpophalangeal joints are the seven **interosseous muscles** and four **lumbrical muscles,** and the *extensor* is the **extensor digitorum** (including the extensor digitorum indicus and digitorum minimus). The *abductors* and *adductors* are the interosseous muscles and the abductor digiti minimi. The four **dorsal interossei** abduct (spread) the fingers (except the little finger*), and the three **palmar interossei** adduct the fingers (bring them back together). The movements of abduction and adduction of the fingers are defined in relation to the third (middle) finger, which acts as the axis of the hand. Thus, spreading of the fingers away from the third finger is abduction, and movement toward the middle finger is adduction. The **abductor digiti minimi** abducts (spreads) the little finger away from the fourth finger. Circumduction combines all of these actions in sequence. Because the **flexor digitorum profundus** and **flexor digitorum**

*Each finger has its own anatomical name. The thumb is the *pollex* (L. thumb), next to it is the *index* finger (L. pointer), then the *medius* or middle finger, *annularis* (L. ring) or fourth finger, and *minimus* (L. least) or little finger.

superficialis pass on the palmar side of the MP joint, they contribute to *flexion* at the joint. The *extension* of the MP joint is carried out by the **extensor digitorum muscles,** whose tendons attach to the base of each proximal phalanx on the dorsal side of the axis. The *extensors* for the index finger and little finger are the **extensor digitorum indicus** and the **extensor digitorum minimus,** respectively.

The *flexor* of the proximal interphalangeal joint is the **flexor digitorum superficialis,** with assistance from the **flexor digitorum profundus.** The *flexor* of the distal interphalangeal joint is the **flexor digitorum profundus.** The *extensors* of the PIP and DIP joints are the interossei and lumbricals, with minor assistance from the extensor digitorum.

The "paradox" of the lumbrical and interosseous muscles acting to flex the MP joint and to extend the PIP and DIP joints is explained by the relationship of the tendons of these muscles to the joints. The tendons of these muscles (1) pass on the palmar side of the MP joint, (flexing the joint), and (2) continue to join the extensor tendon on the dorsal side of each of the proximal phalanges II to IV.

TABLE 10.15   MUSCLES THAT MOVE THE FINGERS (EXCEPT THE THUMB)

Muscle	Origin	Insertion	Innervation	Action
Flexor digitorum superficialis	Medial epicondyle of humerus, coranoid process of ulna, anterior border of radius.	Both sides of middle phalanges of fingers 2 to 5.	Median nerve.	Flexes PIP and MP joints.*
Flexor digitorum profundus	Anterior and medial surface of ulna.	Palmar surfaces of distal phalanges of fingers 2 to 5.	Median nerve, ulnar nerve.	Flexes PIP, DIP, and MP joints.*
Interossei (4 dorsal interossei and 3 palmar interossei)	Metacarpal bones.	Bases of proximal phalanges and extensor expansions of fingers.	Ulnar nerve.	Dorsal interossei: abduct fingers. Palmar interossei: adduct fingers; extend PIP and DIP joints.†
Lumbricals (4 muscles)	Tendons of flexor digitorum profundus with palm of hand.	Extensor tendon (expansions) of digits distal to knuckles.	Lateral two by median nerve; medial two by ulnar nerve.	Flex MP joint (knuckles); extend PIP and DIP joints.†
Extensor digitorum	Lateral epicondyle of humerus.	Dorsal surface of phalanges in fingers 2 to 5.	Radial nerve.	Extends MP joint (and PIP slightly).
Flexor digiti minimi	Hook of hamate bone and adjacent region.	Medial side of proximal phalanx of little finger.	Ulnar nerve.	Flexes proximal phalanx of little finger.
Opponens digiti minimi	Hook of hamate bone and adjacent region.	Medial half of palmar surface of little finger.	Ulnar nerve.	Rotates little finger toward midline of hand (opposition).
Abductor digiti minimi	Pisiform bone and adjacent region.	Medial side of proximal phalanx of little finger.	Ulnar nerve.	Abducts little finger away from fourth finger.

*Because of the pull they can generate, the long flexor muscles can help flex joints more proximal than that of their primary action.
†The extensor digitorum is not the primary extensor of the PIP and DIP joints, because it expends most of its force at the MP joint. The lumbricals and interossei extend the PIP and DIP joints through their attachments to the extensor tendons.

FIGURE 10.24

Deep intrinsic muscles of the hand. Muscles that move the fingers are emphasized. (A) Three palmar interossei, palmar view. (B) Four dorsal interossei, dorsal view. (C) Lumbrical muscles, palmar view.

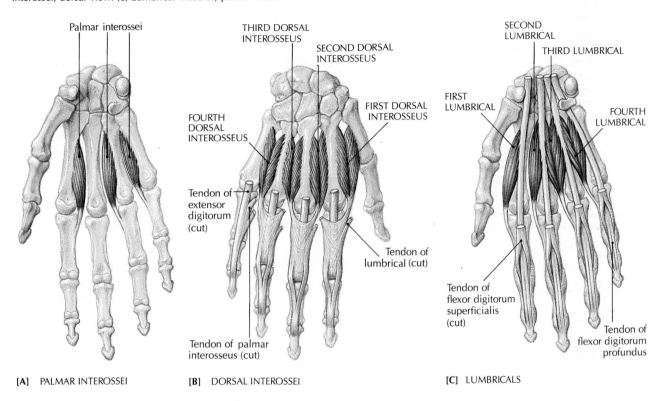

[A] PALMAR INTEROSSEI   [B] DORSAL INTEROSSEI   [C] LUMBRICALS

The roles of the lumbrical and interosseous muscles are critical in the extension of the PIP and DIP joints, and in the coordination of the finger movements. Although the tendons of the extensor digitorum muscles continue beyond the MP joint and attach to the base of the middle and distal phalanges on the dorsal sides, they contribute minimally to the extension of the PIP and DIP joints. This occurs because the force generated by the contraction of the attached muscles is almost completely expended at the MP joint. This complex arrangement between the flexors and extensors of the interphalangeal joints is all-important in producing delicate finger movements. The extensor activities of the lumbricals (largely) and interossei allow those muscles to act as effective antagonists in adjusting the activity of the flexor muscles during writing and other fine finger movements.

Muscles That Move the Thumb
(Table 10.16, Figures 10.23 and 10.25)

The movements of the thumb take place at (1) the carpometacarpal joint, which is a saddle joint with three axes of rotation, and (2) two hinge joints involving the phalanx. The movements include (1) flexion and extension, (2) abduction and adduction, and (3) opposition and reposition. Crucial to the opposability of the thumb are opposition and reposition, and abduction and adduction. In opposition the metacarpal bone rolls toward the midline of the hand so that the ball of the thumb's distal phalanx can touch the balls of the distal phalanges of the other fingers. The muscular mound at the base of the thumb, the *thenar eminence,* is formed by the three thenar muscles: ***abductor pollicis brevis, flexor pollicis brevis,*** and ***opponens pollicis.*** At the carpometacarpal joint, the flexor is the flexor pollicis brevis. The *extensors* are the ***extensor pollicis longus, extensor pollicis brevis,*** and ***abductor pollicis longus.*** The *adductor* is the ***adductor pollicis.*** The *opposition* movement is by the ***opponens pollicis*** and the *reposition* movements by the ***abductor pollicis longus*** and ***extensor pollicis longus*** and ***brevis.*** At the phalangeal joints, the *flexors* are the ***flexor pollicis longus*** and ***flexor pollicis brevis,*** and the *extensors* are the ***extensor pollicis longus*** and ***brevis.***

Several features of the thumb contribute immensely to the versatility of the hand. The length of the thumb and the presence of the flexor pollicis longus and the opponens muscles permit the opposition movement and the great flexibility in using the thumb along with the other fingers.

The little finger is associated with the *hypothenar eminence,* formed by the flexor digiti minimi, abductor digiti minimi, and opponens muscles. They have roles similar to their counterparts of the thenar eminence.

TABLE 10.16 MUSCLES THAT MOVE THE THUMB

Muscle	Origin	Insertion	Innervation	Action
Flexor pollicis longus	Ventral surface of radius; interosseous membrane.	Base of distal phalanx of thumb.	Median nerve.	Flexes thumb.
Flexor pollicis brevis	Trapezium and adjacent region.	Base of proximal phalanx of thumb.	Median nerve.	Flexes thumb; helps in opposition and reposition.
Opponens pollicis	Trapezium and adjacent region.	Radial half of palmar surface of thumb.	Median nerve.	Rolls thumb toward midline of palm (opposition).
Abductor pollicis brevis	Scaphoid and trapezium bones and adjacent region.	Base of proximal phalanx of thumb.	Median nerve.	Abducts thumb; helps in opposition.
Adductor pollicis	Second and third metacarpals and capitate bones.	Base of proximal phalanx of thumb.	Ulnar nerve.	Adducts thumb.
Extensor pollicis longus	Dorsal surface of ulna; interosseous membrane.	Base of distal phalanx of thumb.	Radial nerve.	Extends thumb; helps in reposition.
Extensor pollicis brevis	Dorsal surface of radius; interosseous membrane.	Base of proximal phalanx of thumb.	Radial nerve.	Extends thumb; helps in reposition.
Abductor pollicis longus	Dorsal surfaces of ulna and radius.	Base of first metacarpal bone.	Radial nerve.	Abducts and extends thumb; helps in reposition.

FIGURE 10.25

Muscles of the back of the thumb.

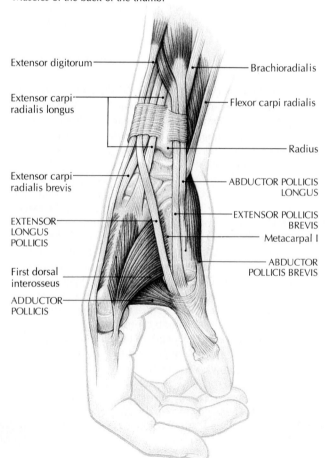

Extensor digitorum
Extensor carpi radialis longus
Extensor carpi radialis brevis
EXTENSOR LONGUS POLLICIS
First dorsal interosseus
ADDUCTOR POLLICIS

Brachioradialis
Flexor carpi radialis
Radius
ABDUCTOR POLLICIS LONGUS
EXTENSOR POLLICIS BREVIS
Metacarpal I
ABDUCTOR POLLICIS BREVIS

Muscles That Move the Femur at the Hip Joint
(Table 10.17, Figures 10.26 and 10.27)

The ball-and-socket hip joint, with its three axes of rotation, is capable of (1) flexion and extension, (2) abduction and adduction, (3) medial and lateral rotation, and (4) circumduction. The powerful muscles surrounding the hip joint are active during walking, running, and climbing. They also act to stabilize the hip joint.

The **hamstring muscles** (**semitendinosus, semimembranosus,** and **biceps femoris**) act as extensors of the hip and flexors of the knee during walking on the level; these muscles relax as the hip flexors contract. The **gluteus maximus** acts as an extensor of the hip only when power is required, as in moving against gravity. For example, it is used in rising from a seated position, in climbing upstairs or up a steep hill, and in running. The **psoas major** contributes to the stability of the hip and is a powerful flexor. (The psoas major and the **iliacus** are sometimes considered together as the **iliopsoas** muscle.) The **tensor fasciae latae** and gluteus maximus pull on the iliotibial tract (a modified fascial sheath located on the lateral thigh) extending beyond the knee joint. This muscle functions to brace the knee so the joint doesn't buckle while the other foot is off the ground during walking. The **adductor magnus** and **adductor longus** are powerful muscles used in kicking a soccer ball, for instance.

During locomotion the hip remains parallel to the ground through the action of the **gluteus medius** muscle. If this muscle is paralyzed on one side, a lurching gait results. If you

Why an Intramuscular Injection?

Because deep muscle tissue contains relatively few nerve endings for pain, irritating drugs are injected into a muscle instead of being taken orally or being injected under the skin. Also, when medication is injected intramuscularly, it is absorbed into the body rapidly because of the rich capillary beds in muscle tissue.

Intramuscular injections are usually given in the buttock (gluteus medius and gluteus maximus), the lateral part of the thigh, or the deltoid muscle of the arm. The most usual site for intramuscular injections is the gluteus maximus or gluteus medius muscle, about two or three inches below the iliac crest, in the upper outer quadrant of the buttock. When the gluteus maximus is used it is important to avoid injuring the nearby sciatic nerve. Also, drugs should not be injected directly into large blood vessels, no matter where they are located. A standing position tenses this muscle, and an injection into a tense muscle causes considerable pain. Therefore, the patient is usually placed in a prone position to relax the muscle before the injection.

TABLE 10.17   MUSCLES THAT MOVE THE FEMUR AT THE HIP JOINT

Muscle	Origin*	Insertion*	Innervation	Action
Iliopsoas[†]	Iliacus: iliac fossa of false pelvis. Psoas: lumbar vertebrae.	Lesser trochanter of femur.	Femoral nerve. Iliacus innervated by femoral nerve; psoas innervated directly by L2 and L3.	Flexes hip joint.
Pectineus	Superior ramus of pubis.	Pectineal line below lesser trochanter of femur.	Femoral and obturator nerves.	Flexes and adducts hip joint.
Adductor longus	Body of pubis.	Middle third of linea aspera of femur.	Obturator nerve.	Adducts and flexes hip joint.
Adductor brevis	Body and inferior ramus of pubis.	Proximal part of linea aspera of femur.	Obturator nerve.	Adducts and helps flex hip joint.
Adductor magnus	Inferior ramus of pubis to ischial tuberosity.	Middle of linea aspera, adductor tubercle of femur.	Obturator nerve and tibial portion of sciatic nerve.	Adducts and helps flex and extend hip joint.
Gluteus maximus	Iliac crest, sacrum, coccyx.	Iliotibial tract of tensor fasciae latae and gluteal tuberosity of femur.	Inferior gluteal nerve.	Extends and laterally rotates hip joint.
Gluteus medius	Ilium.	Greater trochanter of femur.	Superior gluteal nerve.	Abducts and medially rotates hip joint; steadies pelvis.
Gluteus minimus (not illustrated)	Ilium.	Greater trochanter of femur.	Superior gluteal nerve.	Abducts and medially rotates hip joint; steadies pelvis.
Tensor fasciae latae	Iliac crest, anterior superior iliac spine.	Lateral condyle of tibia through iliotibial tract.	Superior gluteal nerve.	Abducts, flexes, and medially rotates hip joint; steadies trunk; extends knee joint.
Small lateral rotators (piriformus, obturator internus, quadratus femoris)	Sacrum, obturator foramen, ischial tuberosity.	Greater trochanter of femur.	Branch from sacral flexus for each muscle.	Laterally rotates and adducts hip joint.

*The origin and insertion of a muscle acting on the hip and knee joints are reversible, depending on the movement. The insertion is on the femur when the thigh is moved and the pelvis is fixed in position, whereas the insertion is on the pelvis when the pelvis is moved and the thigh is fixed.
[†]The iliopsoas muscle is composed of an iliacus muscle and a psoas muscle.

FIGURE 10.26

Superficial muscles of the right lower limb. Muscles acting on the hip and knee joints are emphasized. (A) Anterior view. (B) Posterior view.

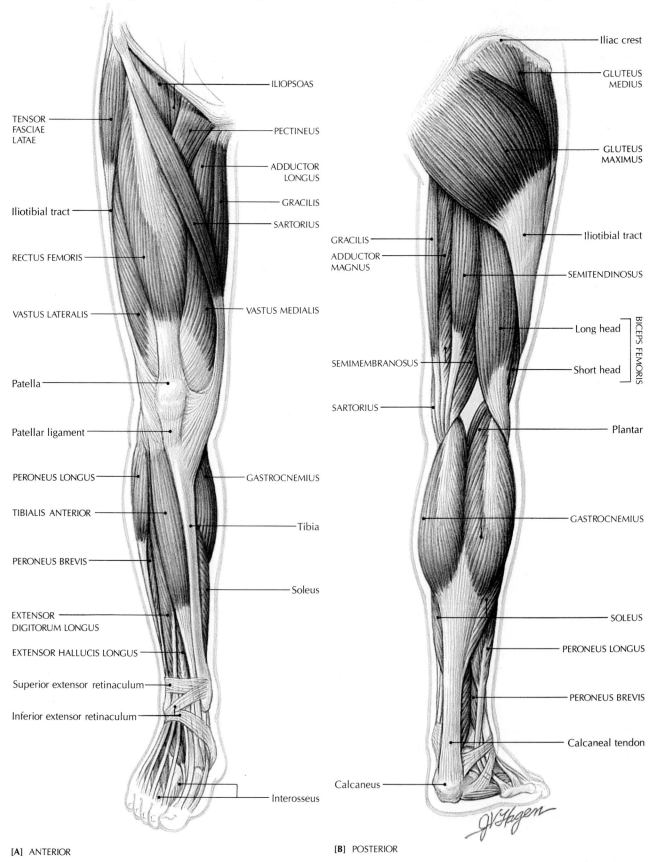

ILIOPSOAS

TENSOR FASCIAE LATAE

PECTINEUS

ADDUCTOR LONGUS

GRACILIS

SARTORIUS

Iliotibial tract

RECTUS FEMORIS

VASTUS LATERALIS

VASTUS MEDIALIS

Patella

Patellar ligament

PERONEUS LONGUS

GASTROCNEMIUS

TIBIALIS ANTERIOR

Tibia

PERONEUS BREVIS

Soleus

EXTENSOR DIGITORUM LONGUS

EXTENSOR HALLUCIS LONGUS

Superior extensor retinaculum

Inferior extensor retinaculum

Interosseus

Iliac crest

GLUTEUS MEDIUS

GLUTEUS MAXIMUS

Iliotibial tract

GRACILIS

ADDUCTOR MAGNUS

SEMITENDINOSUS

Long head

Short head

BICEPS FEMORIS

SEMIMEMBRANOSUS

SARTORIUS

Plantar

GASTROCNEMIUS

SOLEUS

PERONEUS LONGUS

PERONEUS BREVIS

Calcaneal tendon

Calcaneus

[A] ANTERIOR

[B] POSTERIOR

FIGURE 10.27

Deep muscles of the right lower limb. Muscles acting on the hip and knee joints are emphasized. (A) Anterior view. (B) Posterior view.

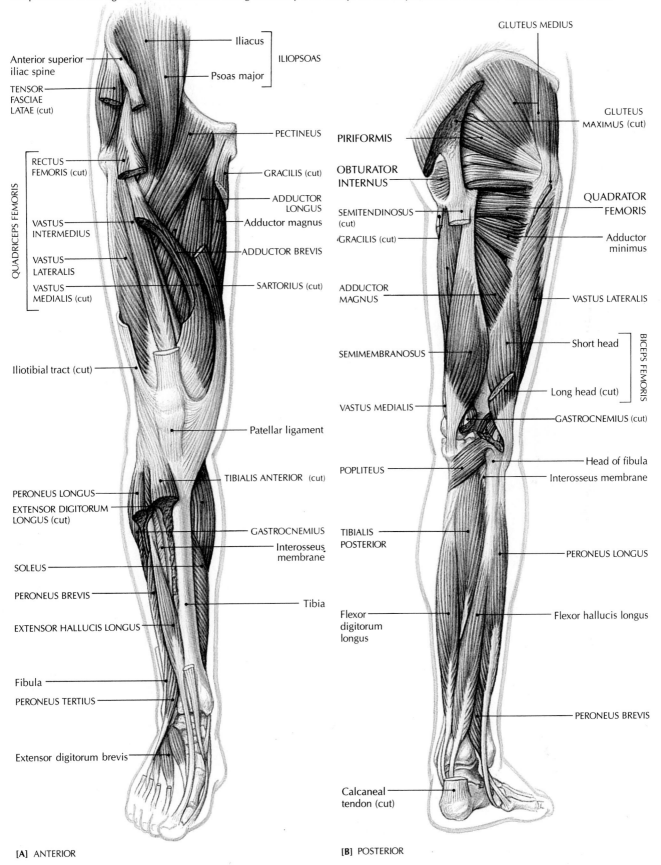

[A] ANTERIOR

[B] POSTERIOR

place your hand on the gluteus muscle while walking, you will note that it contracts while the limb is grounded and relaxes when the limb is free and in motion. The contraction pulls down on the pelvis and prevents the opposite side from dropping, thereby preventing a lurch.

Muscles That Act at the Knee Joint
(Table 10.18, Figures 10.26, 10.27, and 10.28)

The knee is a hinge joint capable of flexion and extension, accompanied by a slight amount of medial and lateral rotation, when the knee is not fully extended. During the final stage of full extension during walking, the femur rotates medially in relation to the tibia when the foot is planted on the ground. Try it. This is known as ''locking'' the knee to make

FIGURE 10.28

Muscles of the right leg and foot. Muscles acting on the ankle joint are emphasized. (A) Lateral view. (B) Medial view.

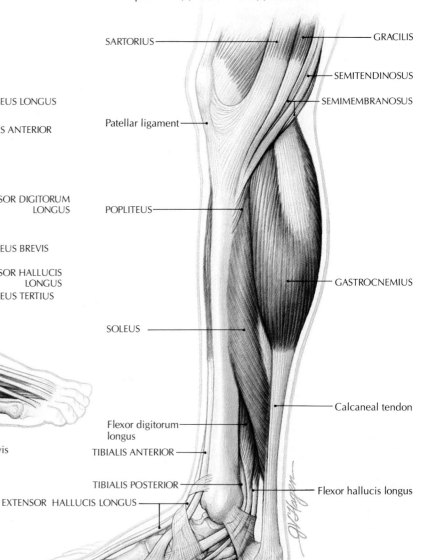

SOLEUS

PERONEUS LONGUS

TIBIALIS ANTERIOR

GASTROCNEMIUS

EXTENSOR DIGITORUM LONGUS

PERONEUS BREVIS

EXTENSOR HALLUCIS LONGUS

PERONEUS TERTIUS

Calcaneal tendon

Calcaneus

[A] LATERAL

Extensor digitorum brevis

SARTORIUS

GRACILIS

SEMITENDINOSUS

SEMIMEMBRANOSUS

Patellar ligament

POPLITEUS

GASTROCNEMIUS

SOLEUS

Calcaneal tendon

Flexor digitorum longus

TIBIALIS ANTERIOR

TIBIALIS POSTERIOR

Flexor hallucis longus

EXTENSOR HALLUCIS LONGUS

Flexor hallucis brevis

[B] MEDIAL

TABLE 10.18   MUSCLES THAT ACT AT THE KNEE JOINT

Muscle	Origin*	Insertion*	Innervation	Action
QUADRICEPS FEMORIS				
Rectus femoris	Anterior interior iliac spine.	Patella, through patellar ligament to tibial tuberosity.	Femoral nerve.	Extends leg at knee joint; flexes thigh at hip joint.
Vastus lateralis	Greater trochanter and linea aspera of femur.	Patella, through patellar ligament to tibial tuberosity.	Femoral nerve.	Extends leg at knee joint.
Vastus intermedius	Linea aspera of femur.	Patella, through patellar ligament to tibial tuberosity.	Femoral nerve.	Extends leg at knee joint.
Vastus medialis	Upper two-thirds of body of femur.	Patella, through patellar ligament to tibial tuberosity.	Femoral nerve.	Extends leg at knee joint.
HAMSTRING MUSCLES				
Biceps femoris	Long head: ischial tuberosity. Short head: linea aspera of femur.	Head of fibula and lateral condyle of tibia.	Long head: tibial division of sciatic nerve. Short head: Common peroneal nerve. Division of sciatic nerve.	Flexes and laterally rotates knee joint; extends thigh at hip joint.
Semitendinosus	Ischial tuberosity.	Medial surface of superior tibia.	Tibial division of sciatic nerve.	Flexes and medially rotates knee joint; extends thigh at hip joint.
Semimembranosus	Ischial tuberosity.	Medial condyle of tibia.	Tibial division of sciatic nerve.	Flexes and medially rotates knee joint; extends thigh at hip joint.
OTHER MUSCLES				
Gracilis	Body and inferior ramus of pubis.	Medial surface of proximal tibia.	Obturator nerve.	Flexes and medially rotates knee joint; adducts thigh at hip joint.
Sartorius	Anterior superior iliac spine.	Upper medial surface of tibia.	Femoral nerve.	Flexes leg at knee joint and thigh at hip joint.
Popliteus	Lateral condyle of femur.	Posterior surface of tibia.	Tibial nerve.	Flexes and rotates leg at knee joint; unlocks knee joint.

*In the knee joint, the origin and insertion can be reversed depending upon whether the leg and foot are moved in relation to the thigh, as when one is seated and the leg freely moves, or whether the foot is planted and the thigh does much of the moving, as in squatting exercises or in rising from the seated position.

it rigid. If the knee is extended while the limb is free, as when punting a football, the tibia rotates laterally in relation to the femur to lock the knee. Try extending the knee if the foot is off the ground. Unlocking of the knee by the *popliteus muscle* is the first stage of flexion of the knee joint. The *hamstring muscles* are flexors *(gracilis, sartorius,* and *popliteus),* lateral rotators *(biceps femoris),* and medial rotators *(semitendinosus* and *semimembranosus).* The *sartorius, gracilis,* and *gastrocnemius* contribute to flexion.

People with longer-than-average hamstrings can readily touch the ground with their fingers or palms without flexing their knees. Individuals with shorter-than-average ham-strings cannot touch the ground with their fingertips without flexing their knees. In athletes, long hamstrings are prone to muscle pulls. In addition to being the prime movers of the knee, the thigh muscles are important in stabilizing the knee joint when it is not locked.

The knee-jerk reflex used by physicians to test reflexes results from a tap on the tendon of the *quadriceps femoris muscle.* This muscle is made up of four parts, as shown in Table 10.18. The patella, the bone to which the quadriceps femoris is attached, acts to place the attachment of the patella and quadriceps femoris further away from the flexion-extension axis of rotation. Such an adjustment increases the length

FIGURE 10.29

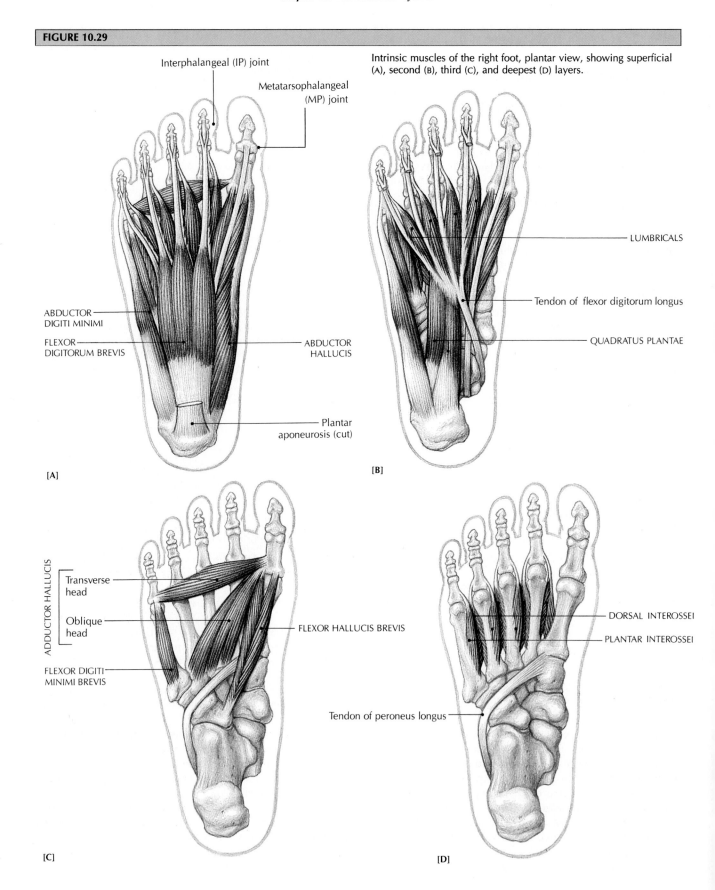

Intrinsic muscles of the right foot, plantar view, showing superficial (A), second (B), third (C), and deepest (D) layers.

Interphalangeal (IP) joint

Metatarsophalangeal (MP) joint

ABDUCTOR DIGITI MINIMI

FLEXOR DIGITORUM BREVIS

ABDUCTOR HALLUCIS

Plantar aponeurosis (cut)

[A]

LUMBRICALS

Tendon of flexor digitorum longus

QUADRATUS PLANTAE

[B]

ADDUCTOR HALLUCIS

Transverse head

Oblique head

FLEXOR DIGITI MINIMI BREVIS

FLEXOR HALLUCIS BREVIS

[C]

DORSAL INTEROSSEI

PLANTAR INTEROSSEI

Tendon of peroneus longus

[D]

of the "lever arm" of the joint by increasing the leverage and power of the strong quadriceps femoris.

Four bursae are located in the vicinity of the patella, and are important to knee movement. The *suprapatellar bursa* is located just above the patella, between the femur and tendon of the quadriceps femoris muscle. It permits free movement of the quadriceps tendon over the distal end of the femur, and facilitates the full range of extension and flexion of the knee joint. The *infrapatellar bursa* is located just below the patella, between the tibia and the patellar ligament. The *subcutaneous prepatellar bursa* lies between the skin and the anterior surface of the patella. It allows free movement of the skin over the underlying patella during flexion of the knee joint, for example. The subcutaneous prepatellar bursa can become the victim of *friction bursitis,* as a result of rubbing between the skin and patella. When inflammation is chronic, the bursa can become distended with fluid in such individuals as miners and housemaids, who frequently work on all fours. Such a bursitis is commonly known as *housemaid's knee.* The *subcutaneous infrapatellar bursa* lies between the skin and the fascia anterior to the tibial tuberosity. It allows the skin to glide over the tibial tuberosity and to withstand pressure when kneeling with the trunk upright (when praying, for example).

Muscles That Move the Foot at the Talocrural and Subtalar Joints
(Table 10.19, Figures 10.26, 10.27, and 10.28)

The ankle (talocrural) joint is a hinge joint capable of plantar flexion (moves sole of foot down) and dorsiflexion. The *soleus* and *gastrocnemius* muscles, both powerful plantar flexors, are active during the take-off phase of the foot in walking and running. Because of the shortness of its muscle belly, the gastrocnemius cannot flex the knee and plantar flex the ankle joint at the same time, but it can do each independently. With the knee flexed, the soleus is the great plantar flexor of the ankle. In many individuals, these muscles are not active during ordinary standing.

Muscles That Move the Foot at the Subtalar Axis (Axis of Henke)
(Table 10.19, Figures 10.27 and 10.28)

The movements of eversion of the foot (turning so sole faces outward) and inversion (turning so sole faces inward) use the subtalar axis, or axis of Henke (the joint between the talus and calcaneus), when we walk over rough terrain. When the foot is planted on the ground, it assumes an awkward position somewhere between eversion and inversion. The muscles responsible for these movements are the invertors and evertors. Should these muscles be "caught off guard," a sprained ligament might result. The *tibialis ante-*

rior and *posterior* muscles, whose tendons pass on the great-toe side of the subtalar axis, are invertors. The *peroneus longus, brevis,* and *tertius,* whose tendons pass on the little-toe side of the axis, are evertors.

Muscles of the Toes
(Figure 10.29)

The names and attachments of the small intrinsic muscles of the foot to the toes are somewhat similar to those in the hand. However, many of these muscles are ineffective in carrying out the presumed movements suggested by their names. Their primary functional role is to help maintain the stability of the resilient foot as it adjusts to the forces placed upon it.

The role of the short and long extensors of the toes is *dorsiflexion* (extension) at the metatarsophalangeal (MP) joints. The *lumbricals* are potential *extensors* of the interphalangeal (IP) joints. *Flexion* of the IP joints is accomplished by the *flexor digitorum longus* and the *quadratus plantae. Flexion* of the MP joints is the role of the *interossei, lumbricals,* and short flexors and abductors of the great and little toes. *Abduction* and *adduction* are accomplished by the *interossei, abductor* and *adductor hallucis,* and *abductor digiti minimi.* Apparently, the actions and functions of the lumbricals and interossei are not as important as their counterparts in the hand.

In walking, we plant the heel on the ground and then pass the weight forward on the lateral small-toe side of the foot, and then across the heads of the metatarsals to the metatarsal of the great toe for the take-off. Try it. During the take-off, the interossei muscles, by simultaneously flexing the MP joints and extending the IP joints, permit transfer of some weight from the metatarsal heads to the toes.

Specifically, there are four basic muscle layers of the foot. From the outside (superficial) to the inside (deep), the first layer (Figure 10.29A) contains the *abductor hallucis,* which abducts the great toe (hallux); the *flexor digitorum brevis,* which flexes the second phalanges of the four small toes; and the *abductor digiti minimi,* which abducts the small toe. In the second layer (Figure 10.29B) are the *quadratus plantae,* which flex the terminal phalanges of the four small toes, and the *lumbricals,* which flex the proximal phalanges and also extend the distal phalanges of the four little toes. The third layer (Figure 10.29C) consists of the *flexor hallucis brevis,* which flexes the proximal phalanx of the great toe; the *adductor hallucis,* which abducts the great toe; and the *flexor digiti minimi brevis,* which flexes the proximal phalanx of the little toe. The fourth or innermost muscle layer (Figure 10.29D) contains (1) the four *dorsal interossei,* which abduct the second, third, and fourth toes, and flex the proximal and extend the distal phalanges, and (2) the three *plantar interossei,* which adduct the third, fourth, and fifth toes. Remember that the axis of abduction and adduction of the foot is along the second toe. (The second toe is abducted by two dorsal interossei.)

TABLE 10.19   MUSCLES THAT MOVE THE FOOT AT THE TALOCRURAL AND SUBTALAR JOINTS

Muscle	Origin	Insertion	Innervation	Action
Soleus	Head of fibula, medial border of tibia.	Posterior surface of calcaneus.	Tibial nerve.	Plantar flexes ankle joint; steadies leg during standing.
Gastrocnemius	Posterior aspects of condyles of femur.	Posterior surface of calcaneus.	Tibial nerve.	Plantar flexes ankle joint; flexes knee joint.
Peroneus longus	Head and lateral surface of fibula.	Lateral side and plantar surface of first metatarsal and medial cuneiform.	Superficial peroneal nerve.	Plantar flexes ankle joint; everts foot.
Peroneus brevis	Lateral surface of fibula.	Dorsal surface of fifth metatarsal.	Superficial peroneal nerve.	Plantar flexes ankle joint; everts foot.
Tibialis posterior	Interosseous membrane, fibula, and tibia.	Bases of second, third, and fourth metatarsals, navicular, cuneiform, cuboid, calcaneus.	Tibial nerve.	Plantar flexes ankle joint; inverts foot.
Tibialis anterior	Lateral condyle and surface of tibia.	First metatarsal, medial cuneiform.	Deep peroneal nerve.	Dorsiflexes ankle joint; inverts foot.
Extensor digitorum longus	Lateral condyle of tibia, anterior surface of fibula.	Middle and distal phalanges of four outer toes.	Deep peroneal nerve.	Dorsiflexes ankle joint; extends toes at metatarsophalangeal and interphalangeal joints.
Extensor hallucis longus	Middle of fibula, interosseous membrane.	Dorsal surface of distal phalanx of great toe.	Deep peroneal nerve.	Dorsiflexes ankle joint; extends great toe.
Peroneus tertius	Distal third of fibula, interosseous membrane.	Dorsal surface of fifth metatarsal.	Deep peroneal nerve.	Dorsiflexes ankle joint; everts foot.

PHYSIOLOGICAL AND ANATOMICAL ABNORMALITIES

Hernias

A *hernia* (L. protruded organ), or rupture, is the protrusion of any organ or body part through the muscular wall that usually contains it. *Inguinal* (IHNG-gwuh-null; L. groin) hernias are the most common type, occurring most often in males. Most inguinal hernias are caused by a weakness in the fascial wall that allows the intestines to protrude in the area of the groin. *Hiatal* (hye-A-tuhl; L. *hiatus,* gap) hernias develop when a defective diaphragm allows a portion of the stomach to pass through the opening for the esophagus in the diaphragm into the thoracic cavity. *Femoral* hernias are most common in women. They occur just below the groin, where the femoral artery passes into the femoral canal. Usually a fatty deposit within the femoral canal creates a hole large enough for part of the bladder and peritoneum to bulge through. *Umbilical* hernias pro-trude at the navel. They are most common in newborns, obese women, and women who have had several pregnancies. *Incisional* hernias result from the weakening around a surgical wound that does not heal properly.

Tendinitis

Tendinitis (which you might think should be spelled tendonitis) is a painful inflammation of the tendon and tendon sheath, typically resulting from a sports injury or similar strain on the tendon. It may also be associated with musculoskeletal diseases such as rheumatism or with abnormal posture. Tendinitis occurs most often in the shoulder area, calcaneal tendon, or hamstring.

Tennis elbow

Tennis elbow is often thought of as a form of tendinitis, but it is not. Its medical name is *epicondylitis,* or inflammation of (1) the forearm extensor and supinator tendon fibers where they attach to the lateral humeral epicondyle or (2) the lateral collateral ligament of the elbow joint. Tennis elbow afflicts people who rotate their forearms frequently or have chronically weak joints.

Tension headache

Tension headaches, the most common type of headache, are often caused by emotional stress, fatigue, or other factors that produce painful muscular contractions in the scalp and back of the neck. Muscle spasms may constrict blood vessels, increasing general discomfort. Pain usually spreads from the back of the head to the area above the eyes, and may cause pain in the eyes themselves. The removal of toxic muscle wastes may be hampered by constricted blood vessels in the scalp, causing even further tenderness.

SUMMARY

Attachment of muscles

1 A muscle is usually attached to a bone or cartilage by a **tendon.** Some tendons are expanded into a broad, flat sheet called an *aponeurosis.*

2 In addition to acting as attachments, tendons add useful length and thickness to muscles, reduce muscle strain, and add strength to muscle action.

3 The *origin* is the place on the bone that does not move when the muscle contracts, and the *insertion* is the place on the bone that does move when the muscle contracts. Generally, the origin is the attachment closer to the axial skeleton. Some muscles have more than one origin or insertion.

4 The origin and insertion of some muscles can be reversed, depending on which bone moves.

Architecture of muscles

1 The fibers of skeletal muscles are grouped into small bundles called fascicles, with the fibers within each bundle running parallel to each other.

2 The specific architectural pattern of fascicles within a muscle determines the range of movement and power of the muscle.

3 Patterns of arrangement of fascicles and tendons of attachment include *strap, fusiform, pennate,* and *circular* muscles.

Individual and group actions of muscles

1 Most movements involve the complex interactions of several muscles or even groups of muscles.

2 An *agonist* or *prime mover* is the muscle that is primarily responsible for producing a movement. An *antagonist* muscle helps produce a smooth movement by slowly relaxing as the agonist contracts. A *synergist* works with a prime mover by preventing movements at an "in-between" joint when the prime mover passes over more than one joint. A *fixator* provides a stable base for the action of the prime mover.

3 In different situations, the same muscle can act as a prime mover, antagonist, synergist, or fixator.

Lever systems and muscle actions

1 Most skeletal muscle movements are accomplished through a system of levers. A lever system includes a *rigid lever arm* that pivots around a fixed point called a *fulcrum,* with the bone acting as lever arm and the joint as fulcrum. Within every lever system there are two different *forces:* the resistance to be overcome and the effort applied.

2 There are three types of levers, referred to as first-, second-, and third-class levers.

3 The mechanical advantage that muscles gain by using a lever system is called *leverage.*

How muscles are named

The name of a muscle generally tells something about its location, action (or function), or structure, such as its size, shape, attachment sites, number of heads of origin, or direction of fibers.

Specific actions of principal muscles

See Tables 10.2 through 10.19.

Physiological and anatomical abnormalities

1 A *hernia* is the protrusion of an organ or body part through the muscular wall that usually contains it. The most common types of hernias are *inguinal, hiatal, femoral, umbilical,* and *incisional.*

2 *Tendinitis* is a painful inflammation of a tendon and tendon sheath.

3 *Tennis elbow,* or epicondylitis, is inflammation of the forearm extensor and supinator tendon fibers where they attach to the lateral humeral epicondyle, or of the lateral collateral ligament of the elbow joint.

4 The most common type of headache is the *tension headache.* It is often caused by emotional stress, fatigue, or other factors that produce painful muscular contractions in the scalp and back of the neck.

UNDERSTANDING THE FACTS

1 Are all muscles attached to bones? Explain.

2 May a muscle have more than one origin and insertion?

3 How are the fascicles arranged in a pennate muscle?

4 What role do fixator muscles play in group action?

5 Which class of lever is used most in the human body?

6 Distinguish between extrinsic and intrinsic muscles.

7 List the muscles that are involved in the movement of the eyeball.

8 List the muscles of mastication.

9 Name the muscles involved in closing the mouth. Raising the upper eyelid. Rolling the eye downward.

10 Which muscles are used in swallowing? Nodding "yes"?

11 Define linea alba.

12 What is the perineum?

13 Name the muscles that act on the shoulder girdle.

14 Name the four rotator cuff muscles. What do they do?

15 Finger movement is achieved mainly by the contraction of muscles in what area?

16 What muscles form the "hamstring" group?

17 What are the three most common sites used for intramuscular injections?

18 What is a hiatal hernia? An inguinal hernia?

19 Distinguish between tendinitis and tennis elbow.

UNDERSTANDING THE CONCEPTS

1 Give one or more examples of muscles named for their shape, size, action, heads of origin, and attachment site. Explain the meaning in each case.

2 Why may the tongue pose a problem for the anesthetist? Name the muscle involved and explain how the problem may be overcome.

3 Explain the structure of the muscles of the abdominal wall and how their structure is related to their functions.

4 When an incision is made in the abdominal wall for, say, an appendectomy, three laminated muscles would be cut. List them in order.

5 Describe the muscles used and the actions performed in throwing a ball and writing with a pencil.

6 In what ways are the muscles of the hands and feet alike? In what ways are they different?

7 What anatomical fact is responsible for a pianist's forearms becoming tired before the fingers?

8 Which characteristic of the human thumb makes it so important? Give examples of its importance in several daily activities.

SUGGESTED READING

ASTRAND, P. O., AND K. RODAHL, *Textbook of Work Physiology: Physiological Basis of Exercise,* 2nd ed. New York: McGraw-Hill, 1977.

BASMAJIAN, J. V., *Muscles Alive,* 4th ed. Baltimore: Williams & Wilkins, 1978.

BASMAJIAN, J. V., AND M. A. MAC-CONAILL, *Muscles and Movements: A Basis for Human Kinesiology.* Baltimore: Williams & Wilkins, 1977.

GALLISTEL, C. R., "From Muscles to Motivation." *American Scientist,* 63 (1980):398.

GOSS, C. M., "On the Anatomy of Muscles for Beginners by Galen of Pergamon." *The Anatomical Record,* 145 (1963):477–502.

HINSON, M. M., *Kinesiology,* 2nd ed. Dubuque, Iowa: Brown, 1981.

NAPIER, J., "The Antiquity of Human Walking," *Scientific American,* April 1967.

PAULY, J. E., " An Electromyographic Analysis of Certain Movements and Exercises. I. Some Deep Muscles of the Back." *The Anatomical Record,* 115 (1966):223–234.

PLATZER, W., *Locomotor System.* Vol. 1, of W. Kahle, H. Leonhardt, and W. Platzer, *Color Atlas and Textbook of Human Anatomy.* Chicago: Year Book Medical Publishers, 1978.

ROSSEM, C., AND D. KAY CLAWSON, *The Musculoskeletal System in Health and Disease.* New York: Harper & Row, 1980.

WILSON, F. C., *The Musculoskeletal System: Basic Processes and Disorders,* 2nd ed. Philadelphia: Lippincott, 1983.

III

CONTROL, COMMUNICATION, AND COORDINATION

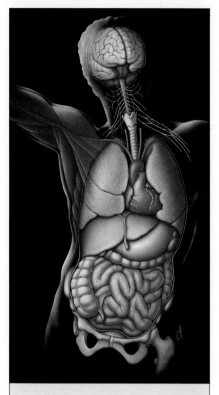

11
The Action of Nerve Cells

LEARNING OBJECTIVES

1 Explain the gross anatomical divisions of the nervous system, and list the main components of each division.

2 Describe the functional divisions and subdivisions of the peripheral nervous system.

3 Identify the two basic properties of the neuron.

4 Describe the parts of a neuron and their functions.

5 Identify the three factors that affect the speed of conduction of a nerve impulse.

6 Compare the three types of neurons based on function, and classify the functional components of neurons.

7 Identify the three types of neurons based on structure, and describe their four major segments.

8 Identify the associated cells of the central and peripheral nervous systems, and describe their major functions.

9 Explain the necessary conditions for regeneration of peripheral nerve fibers, and how the process works.

10 Explain the functions of the sodium-potassium pump and ion channels.

11 Describe the steps in the process of initiating and conducting an impulse in a neuron.

12 Explain how the all-or-none law applies to the firing of a neuron.

13 Describe the process of saltatory conduction.

14 Distinguish between electrical and chemical synapses.

15 Explain how a nerve impulse is conducted across a synapse.

16 Describe three types of graded potentials.

17 Describe the four major groups of neurotransmitters.

18 Distinguish between diverging and converging neuronal circuits, and identify five types of neuronal circuits.

19 Describe one or more disorders involving neurons or the transmission of nerve impulses.

In order to maintain homeostasis, the body is constantly reacting and adjusting to changes in the outside environment and within the body itself. Such *stimuli* (or environmental changes) are sensed and conveyed via nerves to the brain and spinal cord, where the messages (input) are analyzed, combined, compared, and coordinated by a process called *integration.* After being sorted out, messages are conveyed by nerves to the muscles and glands of the body. The nervous system expresses itself visibly through muscles and glands, causing muscles to contract or relax, and glands to secrete or not secrete their products.

Under normal conditions, activities of the muscles and glands are coordinated, so that our body parts work in harmony toward directed homeostatic goals. Examples of such efforts to maintain homeostasis are the maintenance of a relatively constant body temperature and the coordinated activities of muscles during movements. Homeostasis allows us to function normally despite constant changes in the environment.

The nervous system and the endocrine system are the two major regulatory systems of the body, and both are specialized to make the proper responses to stimuli. Both systems work together continuously to maintain homeostasis, but the nervous system is the faster of the two. Stimuli received by the nervous system are processed rapidly through a combination of electrical impulses and chemical substances called *neurotransmitters* for communication between two nerve cells, between a nerve cell and a muscle cell, or between a nerve cell and glandular cells. The endocrine system (the subject of Chapter 16) must depend on slower chemical transmissions, using chemical substances called hormones. The nervous system has been compared to the telephone system of a large city, while the endocrine system is more like its postal service. The endocrine system typically regulates such long-term processes as growth and reproductive ability, instead of the short, quick responses to stimuli controlled by the nervous system. However, stress can produce almost immediate reactions from the endocrine system. In fact, both systems are so closely interrelated that they can be considered a single regulatory agency.

ORGANIZATION OF THE NERVOUS SYSTEM

Although the nervous system is a single, unified communications network, it is usually divided on a gross anatomical basis into the central nervous system and the peripheral nervous system.

Central Nervous System

The *central nervous system* (CNS) consists of the brain and spinal cord (Figure 11.1), which are surrounded and protected by the bony skull and vertebral column, respectively. It may be thought of as the body's central control system, receiving and interpreting or integrating all stimuli, and relaying nerve impulses to muscles and glands, where the designated actions actually take place.

Peripheral Nervous System

The *peripheral nervous system* (PNS) is composed of the cranial nerves associated with the brain, and the spinal nerves associated with the spinal cord, as well as *groups* of nerve cell bodies called ganglia (see Figure 11.1). In general, we can say that the peripheral nervous system is made up of the nerve cells and their fibers that lie *outside* the brain and spinal cord. This system allows the brain and spinal cord to communicate with the rest of the body.

Two types of nerve cells are present in the peripheral nervous system: (1) *afferent* (L. *ad*, toward + *ferre*, to bring), or *sensory*, nerve cells carry nerve impulses from sensory receptors in the body *to* the central nervous system, where the information is processed; (2) *efferent* (L. *ex*, away from), or *motor*, nerve cells convey information *away from* the central nervous system to the effectors (muscles and glands).

The peripheral nervous system may be further divided, on a purely functional basis, into the somatic nervous system and the visceral nervous system (Table 11.1). Each of these systems is composed of an afferent (sensory) division and an efferent (motor) division.

Somatic nervous system The *somatic nervous system* is composed of afferent and efferent divisions. The *somatic afferent (sensory) division* consists of nerve cells that receive and process sensory input from the skin, voluntary muscles, tendons, joints, eyes, tongue, nose, and ears. This input is conveyed to the spinal cord and brain via the spinal and cranial nerves, and utilized by the nervous system at an unconscious level. On a conscious level, the sensory input is perceived as sensations such as touch, pain, heat, cold, balance, sight, taste, smell, and sound.

The *somatic efferent (motor) division* is composed of neuronal pathways that descend from the brain through the brainstem and spinal cord to influence the lower motor neurons of the cranial and spinal nerves. When these lower motor neurons are stimulated, they always excite (never inhibit) the skeletal muscles to contract. This system regulates the "voluntary" contraction of skeletal muscles. (As you saw in Chapter 9, not all such activity is actually voluntary or under our conscious control.)

Visceral nervous system The *visceral nervous system* is composed of afferent and efferent divisions. The *afferent (sensory) division* includes the neural structures involved in conveying sensory information from sensory receptors in the visceral organs of the cardiovascular, respiratory, digestive, urinary, and reproductive systems. Input from these sensory receptors is utilized on a conscious level, and is perceived as sensations such as pain, intestinal discomfort, urinary bladder fullness, taste, and smell. The *efferent (motor) division,* more commonly known as the *autonomic nervous system,* includes the neural structures involved in the motor activities that influence the smooth (involuntary) muscles, cardiac (heart) muscle, and glands of the skin and viscera.

Autonomic nervous system The *autonomic nervous system* (visceral efferent motor division) is made up of nerve fibers from the brain and spinal cord that may either inhibit

FIGURE 11.1

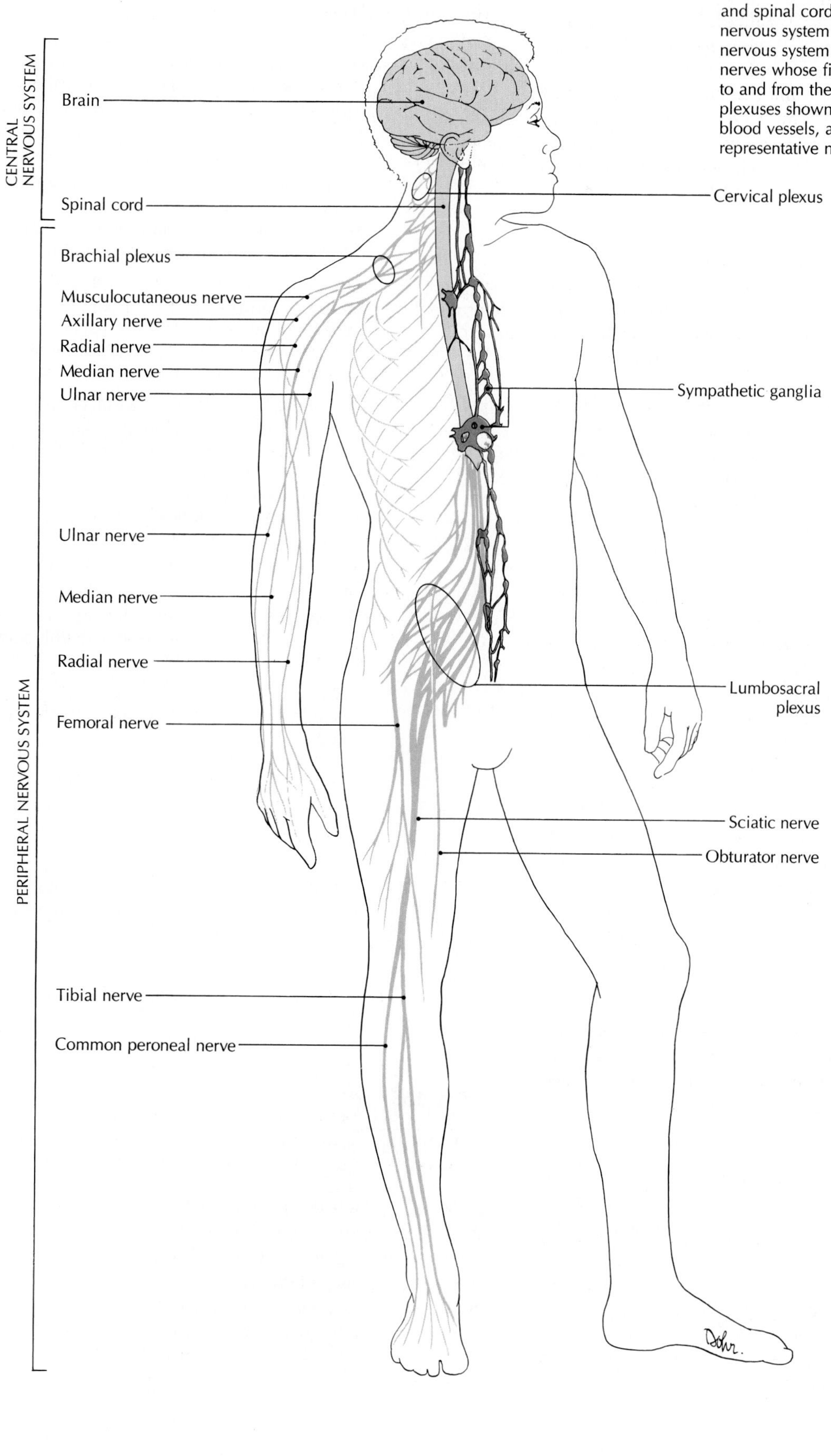

The human nervous system. The brain and spinal cord make up the central nervous system (beige). The peripheral nervous system (yellow) is composed of nerves whose fibers convey nerve impulses to and from the brain and spinal cord. The plexuses shown are networks of nerves, blood vessels, and lymphatic vessels. Only representative nerves are shown.

CENTRAL NERVOUS SYSTEM

Brain

Spinal cord

Cervical plexus

PERIPHERAL NERVOUS SYSTEM

Brachial plexus

Musculocutaneous nerve
Axillary nerve
Radial nerve
Median nerve
Ulnar nerve

Sympathetic ganglia

Ulnar nerve

Median nerve

Radial nerve

Femoral nerve

Lumbosacral plexus

Sciatic nerve
Obturator nerve

Tibial nerve

Common peroneal nerve

TABLE 11.1   ORGANIZATION OF THE NERVOUS SYSTEM

Division	Components	Functions
Central nervous system (CNS)	Brain, spinal cord.	Body's central control system. Receives stimuli, relays ''messages'' for action to muscles and glands. Interpretive functions involved in thinking, learning, memory, etc.
Peripheral nervous system (PNS)	Cranial and spinal nerves, with afferent (sensory) and efferent (motor) nerve cells.	Enables brain and spinal cord to communicate with entire body. Afferent (sensory) cells: carry impulses from receptors to CNS. Efferent (motor) cells: carry impulses from CNS to effectors (muscles and glands).
Somatic nervous system	Axons (nerve fibers) of lower motor neurons that go directly from CNS to effector muscle without crossing junctions (synapses).	Afferent (sensory) division: receives and processes sensory input from skin, voluntary muscles, tendons, joints, eyes, ears. Efferent (motor) division: excites skeletal muscles.
Visceral nervous system	Nerve fibers that go from CNS to interact with other nerve cells within a ganglion located outside CNS; nerve fibers of second nerve cells that go to effectors.	Afferent (sensory) division: receives and processes input from internal organs of cardiovascular, respiratory, digestive, urinary, and reproductive systems. Efferent (motor) division (autonomic nervous system): may inhibit or excite smooth muscle, cardiac muscle, glands.
Autonomic nervous system (visceral motor division) Sympathetic nervous system		Relaxes intestinal wall muscles; increases sweating, heart rate, blood flow to voluntary muscles.
Parasympathetic nervous system		Contracts intestinal wall muscles; decreases sweating, heart rate, blood flow to voluntary muscles.

or excite smooth muscle, cardiac muscle, and glands. This system is the modulator, adjuster, and coordinator of ''involuntary'' visceral activities such as the heart rate and the secretions of glands. Many of these visceral activities can be carried out even if the organs are deprived of innervation by the autonomic nervous system. For example, the heart continues to contract without innervation. (This is why a transplanted heart continues to contract.) Although the heart can contract without innervation, the autonomic nervous system can either increase or decrease the strength of contraction and heart rate.

The autonomic nervous system is divided into two subsystems, the sympathetic and parasympathetic systems, which complement each other. The *sympathetic nervous system* stimulates activities that are mobilized during emergency and stress situations, the so-called fight, fright, and flight responses. These responses include an acceleration of the heart rate and strength of contraction, an increase in the concentration of blood sugar, and an increase in blood pressure. In contrast, the *parasympathetic nervous system* directs activities associated with the conservation and restoration of body re-

sources. These activities include a decrease in the heart rate and strength of contraction, and the rise in gastrointestinal activities associated with increased digestion and absorption of food.

The autonomic nervous system is dealt with in greater detail in a chapter of its own, Chapter 14.

Ask Yourself

1 *What are the main components of the central nervous system?*

2 *Based on the direction of the nerve impulse, what two types of nerve cells are present in the peripheral nervous system?*

3 *What are the major differences between the somatic and visceral nervous systems?*

4 *What are the two subdivisions of the autonomic nervous system?*

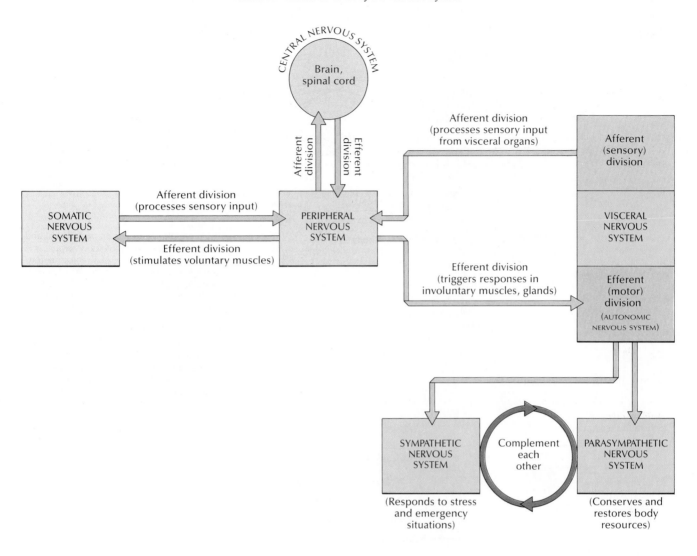

NEURONS: FUNCTIONAL UNITS OF THE NERVOUS SYSTEM

The nervous system contains about 100 billion nerve cells. These nerve cells, called **neurons** (Gr. nerve), are specialized to transmit impulses from short to relatively long distances, from one part of the body or central nervous system to another. Neurons have two important properties: (1) *excitability,* or the ability to respond to stimuli, and (2) *conductivity,* or the ability to conduct a signal.

Parts of a Neuron

Neurons are among the most specialized types of cells. Although they vary greatly in shape and size, all neurons contain three principal parts, each associated with a specific function: the cell body, dendrites, and an axon (Figure 11.2).

Cell body The *cell body* of a neuron may also be called a *soma* (Gr. body) or a *perikaryon* (per-ih-KAR-ee-on; Gr. *peri,*

near, around + *karyon,* kernel, nut). It may be star-shaped (stellate), roundish, oval, or even pyramid-shaped, but its distinguishing structural features are its complex, spreading processes (branches or fibers) that reach out to send or receive impulses to or from other cells. Besides varying in shape, cell bodies may vary in size from about half the size of a red blood cell to almost 17 times the size of a red blood cell. A cell body has a large central nucleus, which contains a prominent nucleolus, as well as several organelles that are responsible for metabolism, growth, and repair of the neuron. These organelles include chromatophilic substance (Nissl bodies), endoplasmic reticulum, mitochondria, neurofilaments (microfilaments), neurotubules, and Golgi apparatus.

Chromatophilic substance is made up of rough endoplasmic reticulum (ER) and free ribosomes. As in other cells, the rough ER is continuous with the smooth ER and, in turn, with the Golgi apparatus (see Chapter 3). The chromatophilic substance contains RNA, and manufactures as much protein in one to three days as the whole protein content of the neuron. Proteins are synthesized by ribosomes located on the rough ER, released into the lumen of the ER, and conveyed

FIGURE 11.2

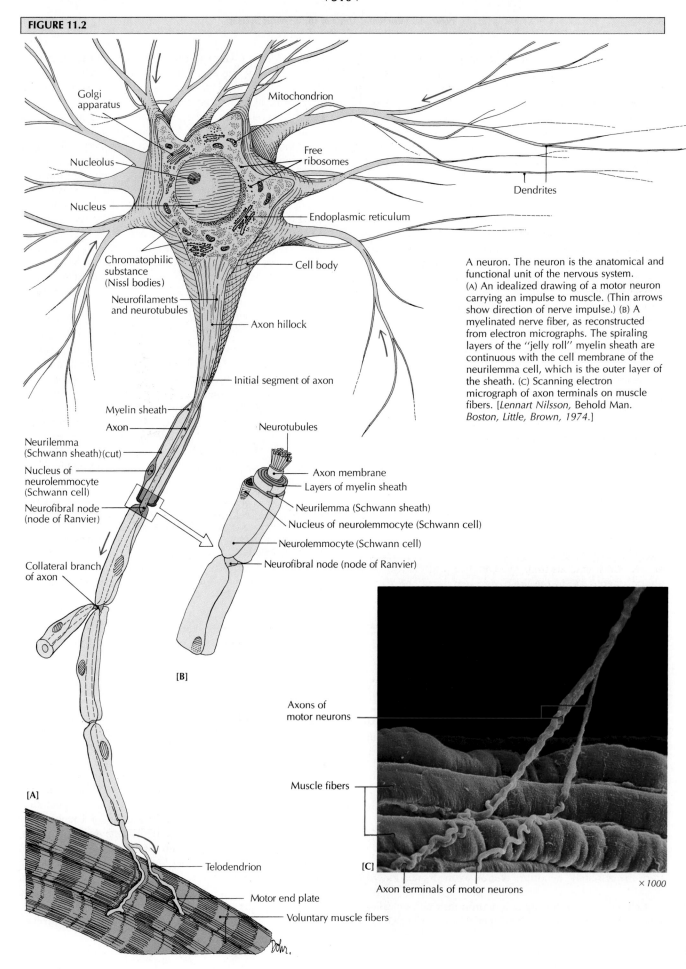

Golgi apparatus

Mitochondrion

Nucleolus

Free ribosomes

Nucleus

Dendrites

Endoplasmic reticulum

Chromatophilic substance (Nissl bodies)

Cell body

Neurofilaments and neurotubules

Axon hillock

Initial segment of axon

Myelin sheath

Axon

Neurotubules

Neurilemma (Schwann sheath)(cut)

Axon membrane

Nucleus of neurolemmocyte (Schwann cell)

Layers of myelin sheath

Neurofibral node (node of Ranvier)

Neurilemma (Schwann sheath)

Nucleus of neurolemmocyte (Schwann cell)

Neurolemmocyte (Schwann cell)

Collateral branch of axon

Neurofibral node (node of Ranvier)

[B]

Axons of motor neurons

Muscle fibers

[A]

[C]

Telodendrion

Motor end plate

Voluntary muscle fibers

Axon terminals of motor neurons

×1000

A neuron. The neuron is the anatomical and functional unit of the nervous system. (A) An idealized drawing of a motor neuron carrying an impulse to muscle. (Thin arrows show direction of nerve impulse.) (B) A myelinated nerve fiber, as reconstructed from electron micrographs. The spiraling layers of the "jelly roll" myelin sheath are continuous with the cell membrane of the neurilemma cell, which is the outer layer of the sheath. (C) Scanning electron micrograph of axon terminals on muscle fibers. [*Lennart Nilsson, Behold Man. Boston, Little, Brown, 1974.*]

to the Golgi apparatus. Within the Golgi apparatus, the proteins are sorted out and packaged into vesicles that contain the precursors of neurotransmitters, other vesicles that contain replacement proteins for the plasma membrane, and lysosome proteins.

The neurotubules and neurofilaments of a cell body are threadlike protein structures that actually extend into and throughout the cell body and processes of the neuron, and run parallel to the long axis of each process. *Neurotubules* play a role in the intracellular transport of proteins and other substances from the cell body to the ends of the processes and also in the opposite direction. Vesicles may even be transported in both directions at the same time (Figure 11.3). The proteins that are transported throughout the cell are needed to replace used-up protein in the neuron, and for the regeneration of nerve fibers in the peripheral nervous system. *Neurofilaments* (also called microfilaments) are semirigid tubular structures that provide a skeletal framework for the axon. In Alzheimer's disease (see p. 404), many neuron cell bodies in the cerebral cortex of the brain contain "tangles" of neurotubules and neurofilaments.

Dendrites and axons Neurons have two types of processes, dendrites and axons. Either process is sometimes referred to as a *nerve fiber.* As shown in Figure 11.2, the typical neuron has many short, threadlike branches called *dendrites* (Gr. *dendron,* tree), which are actually extensions of the cell body. As a result, the organelles in the cell body are continuous with those in the dendrites. Dendrites, which may be excited by excitatory synapses or inhibited by inhibitory synapses, conduct nerve impulses *toward* the cell body. A neuron may have as many as 200 dendrites.

A nerve cell generally has just one **axon,** or *axis cylinder,* a specialized process that is typically long. The axon may extend one or more meters from the cell body, as in the sciatic nerve, which extends from the spinal cord to the foot. (The shortest axons are in the brain, and may be only a few micrometers long.) An axon carries nerve impulses *away* from the cell body, to the next nerve cell, muscle cell, or gland.

The cytoplasm of an axon is called *axoplasm,* and the plasma membrane of an axon is known as the *axolemma.* Unlike the cell body, the axon does not contain chromatophilic substance, ribosomes, or Golgi apparatus.

In most neurons, the axon originates from a cone-shaped elevation of the cell body called the *axon hillock* (see Figure 11.2A). The thin part of the axon immediately after the axon hillock, called the *initial segment,* is important in generating the nerve impulse.* The axon may have some side branches, called *collateral branches,* that leave the main axon at right angles. The axon ends distally in a spray of tiny branches called *telodendria.* The branches of the telodendria usually have tiny swellings called **end bulbs** or **synaptic boutons** (Fr. button). A **synapse** (Gr. *synapsis,* a connection) is formed where an end bulb associates with another receptor membrane.

An axon may be covered with a laminated lipid sheath called **myelin** (MY-ih-linn; Gr. *myelos,* marrow), which forms a thick pad of insulation called a **myelin sheath** or *medullary sheath* (see Figures 11.2A, B). A nerve fiber that has a myelin sheath is called a *myelinated fiber.* In the peripheral nervous system, the myelin sheath is made up of rolled layers of the plasma membrane of a type of peripheral nerve cell called a **neurolemmocyte** (Schwann cell). The outer layer, or sheath, of these cells is known as the **neurilemma** (Gr. nerve + rind or husk) or *Schwann sheath* (Figure 11.4). The myelin sheaths of axons in the central nervous system are formed by oligodendroglia.

*Since the membrane of the initial segment has the lowest excitability threshold, it is the site of nerve-impulse initiation.

What are the relative sizes of the parts of a neuron?	If the cell body of a motor neuron were enlarged to the size of a baseball, the axon would extend about a mile, and the dendrites would fill a large field house.

FIGURE 11.3

Intracellular transport. Neurotransmitters and other substances to be transported along an axon between the cell body and synaptic terminal of a neuron are contained within vesicles that are moved the length of the axon along neurotubules (microtubules) that act like conveyor belts. Precursors of neurotransmitters are manufactured by the Golgi apparatus in the cell body. Surplus substances are packaged into multivesicular bodies and transported back to the cell body, where they are degraded by lysosomes. Mitochondria are transported in either direction (arrows), in response to the energy needs of the cell.

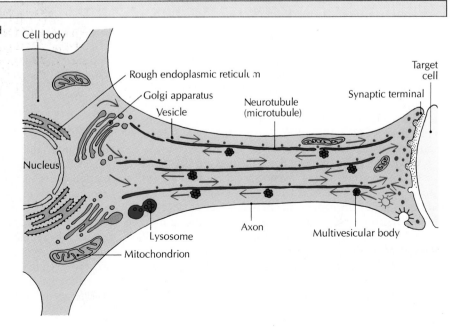

FIGURE 11.4

The myelin sheath. (A) Three stages in the development of a myelin sheath, beginning with a single axon surrounded by a neurilemma sheath. Eventually the axon is completely insulated by overlapping layers of myelin; arrows indicate the direction of growth. (B) Electron micrograph of a rat sciatic nerve in cross section, which shows the layers of the myelin sheath around the axon; the boxed area is enlarged in the inset (top). [*Dr. Geoffrey McAuliffe, Department of Anatomy, Robert Wood Johnson Medical School.*]

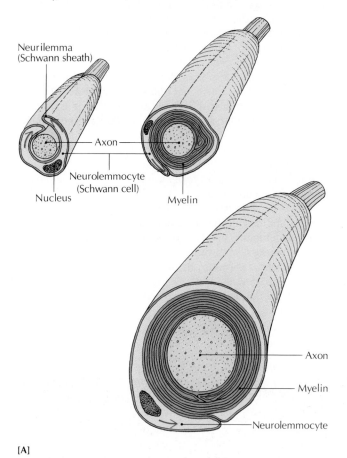

[A]

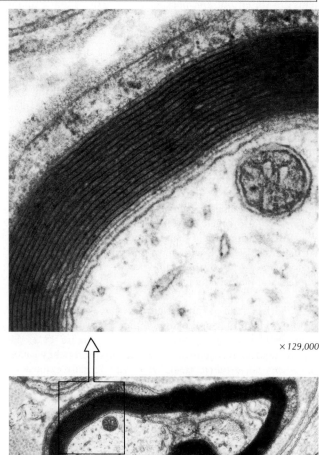

×129,000

×29,000

[B] Axon Myelin sheath

The myelin sheath is segmented, interrupted at regular intervals by gaps called **neurofibral nodes** (nodes of Ranvier). The distance from one node to the next is an *internode*. The longer the internode, the thicker is the diameter of the sheath. The myelin sheath of each internode is formed by one neurolemmocyte. The myelin is absent at each neurofibral node. The maximum speed of the human nerve impulse in a myelinated fiber is about 120 m/sec (360 km/hr), with a minimum speed of about 3 m/sec.

A nerve fiber in a peripheral nerve that has a neurilemma sheath, but does not have a myelin sheath is called an *unmyelinated fiber*. Unmyelinated fibers are usually protected and nourished by the organ tissue in which they are located. These fibers conduct at relatively slow speeds, ranging from 0.7 to 2.3 m/sec. As stated previously, a collection of cell bodies of neurons of peripheral nerves is called a *ganglion*. These cell bodies are encapsulated by satellite cells, which are the equivalents of the neurolemmocytes of the nerves.

In common usage, a **nerve** is simply a bundle of fibers enclosed in a sheath, like many telephone wires in a cable (Figure 11.5). The correct technical term for bundles of fibers and their sheaths in the central nervous system is *tracts*, and for those in the peripheral nervous system is *nerves*.

Types of Neurons: Based on Function

The neurons and nerve fibers of the peripheral nervous system may be classified according to the direction in which they transmit nerve impulses. Neurons and nerve fibers conveying information from sensory receptors in the body *to* the central nervous system are called **afferent** or **sensory neurons** or **fibers**. Their cell bodies are in ganglia located close to the central nervous system. The distal endings of these afferent fibers (dendrites) are the *sensory receptors* (or are connected to the sensory receptors), which are responsive to stimulation from either internal or external environments.

Efferent or **motor neurons** and **fibers** convey nerve impulses *away from* the central nervous system to the ***effectors*** (muscles and glands), where the response actually takes place. Your hand pulls away from a hot stove after arm muscles receive a nerve impulse from efferent neurons. Remember that most cranial and spinal nerves are called *mixed nerves* because they are composed of hundreds, or even thousands, of afferent *and* efferent nerve fibers.

Interneurons, also called *association neurons, connector neurons,* or *internuncial neurons* (L. *nuntius,* messenger), lie mostly within the central nervous system, but a few are distributed between the central and peripheral nervous systems. Their functions are to carry impulses from sensory neurons to motor neurons, and to process incoming neural information. The interneurons with long axons, called *relay neurons,* convey signals over long distances. Interneurons with short, usually branching, axons are called *local circuit neurons.* They convey signals locally, over short distances. Local circuit neurons play an important role in learning, emotions, and language. Complex functions such as learning and memory may depend on thousands of local circuit neurons. Less complicated activities, such as the simpler reflexes, may involve only a few or no interneurons, since sensory and motor neurons may be directly connected.

Classification of Functional Components of Nerves

The functional components of the neurons and their fibers of the peripheral nervous system may be classified as follows:

1 ***General somatic afferent*** fibers carry sensory information from the skin, voluntary muscles, joints, and connective tissues to the central nervous system.

2 ***General visceral afferent*** fibers carry information from the visceral organs to the central nervous system.

3 ***General somatic efferent*** fibers carry nerve impulses from the central nervous system to most of the voluntary (skeletal) muscles of the body. These impulses result in the contraction of voluntary muscles. These muscles are derived embryologically from masses of cells called *myotomes* (see p. 256).

4 ***General visceral efferent*** fibers carry impulses from the central nervous system that modify the activities of the heart, smooth muscles, and glands. These are the fibers of the autonomic nervous system, discussed in Chapter 14.

FIGURE 11.5

A nerve in cross section. Scanning electron micrograph showing bundles of nerve fibers, single nerve fibers within a bundle, and blood vessel that supplies the nerve fibers with nutrients. Nerve bundles are held together in large groups by a connective tissue sheath called epineurium, which divides into another sheath called perineurium when it encloses individual bundles. An even smaller sheath is the endoneurium, which surrounds each single nerve fiber. Note how a bundle of nerve fibers resembles a tightly bound telephone cable. [*Gene Shih and Richard Kessel,* Living Images: Biological Microstructures Revealed by SEM. *Boston, Science Books International, 1982.*]

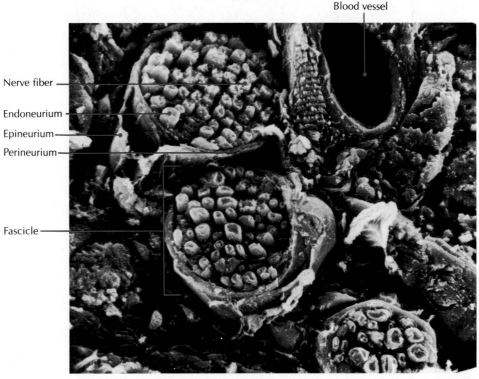

Blood vessel

Nerve fiber

Endoneurium

Epineurium

Perineurium

Fascicle

×*900*

5 *Special visceral efferent* fibers carry impulses from the brain to the voluntary muscles derived from embryonic branchiomeric (gill) arches, not myotomes (see p. 256). This "visceral" musculature is found in the jaw, muscles of facial expression, pharynx (throat), and larynx.

6 *Special afferent* fibers carry neural information from the receptors of the olfactory (smell), optic (sight), auditory (hearing), vestibular (balance), and gustatory (taste) systems to the central nervous system. These special senses are discussed in Chapter 15.

Types of Neurons: Based on Structure

Neurons may also be classified according to their structure, or more specifically, according to the number of their processes. *Multipolar neurons* have many dendrites radiating from the cell body, but only one axon (Figure 11.6A). Most of the neurons of the brain and spinal cord are multipolar. *Bipolar neurons* have only two processes (Figure 11.6B). Essentially, all neurons in the adult develop from bipolar cells, which are the neurons of the embryonic nervous system. In the adult, bipolar neurons are located in only a few structures, including the retina of the eye, cochlear and vestibular nerves of the ear, and the olfactory nerve in the upper nasal cavity. *Unipolar neurons* are the most common sensory neurons in the peripheral nervous system (Figure 11.6C). The single process of a unipolar neuron divides into two branches. One branch extends into the brain or spinal cord, conducting nerve impulses away from the cell body, and the other branch extends to a peripheral sensory receptor in some distal part of the body.

For all of their structural differences, most neurons are composed of four distinct functional segments, as shown in Figure 11.7:

FIGURE 11.6

Classification of neurons based on their structure. (A) Multipolar neuron. (B) Bipolar neuron. (C) Unipolar neuron.

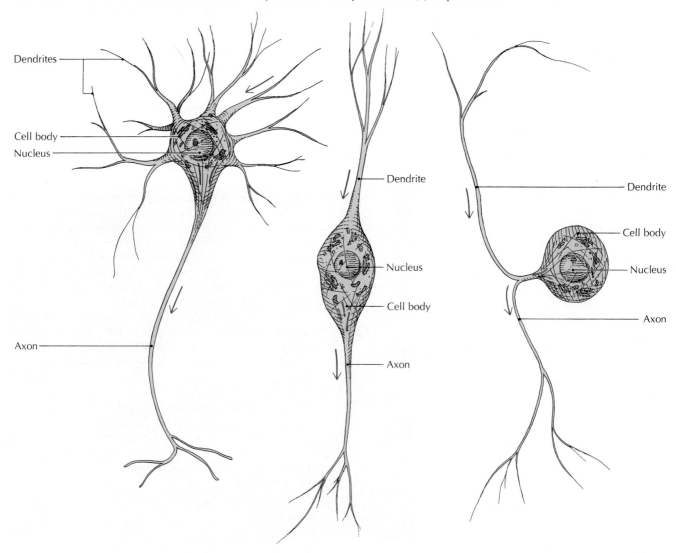

[A] MULTIPOLAR   [B] BIPOLAR   [C] UNIPOLAR

FIGURE 11.7

Functional organization of a typical neuron. (A) Most neurons contain a receptive segment, an initial segment, a conductive segment, and a transmissive segment. The cell body may be located within the receptive segment, as in the lower motor neuron shown in (B), or within the conductive segment, as in the sensory neuron shown in (C).

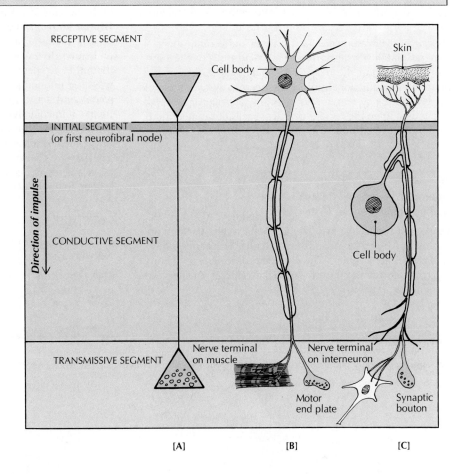

RECEPTIVE SEGMENT

Skin

Cell body

INITIAL SEGMENT
(or first neurofibral node)

Direction of impulse

CONDUCTIVE SEGMENT

Cell body

TRANSMISSIVE SEGMENT

Nerve terminal on muscle

Nerve terminal on interneuron

Motor end plate

Synaptic bouton

[A]   [B]   [C]

1 The *receptive segment* in a multipolar neuron is composed of the cell body and its dendrites. It is the segment that receives a continuous bombardment of synaptic inputs from numerous other neurons on its plasma membrane. These complex inputs are processed in the receptive segment, and if sufficiently stimulated, the resolution of this processing is conveyed to the initial segment at the junction of the cell body and axon (axon hillock).

2 The *initial segment* is the trigger zone of the neuron, where the processed neural information of the receptive segment is converted into a nerve impulse. As a result, this segment, with its low threshold, is critical for the generation of a nerve impulse.

3 The *conductive* (or conductile) *segment* (the axon) is specialized for the conduction of neural information as nerve impulses over relatively long distances in the body. It acts to convey results of the neural processing of the receptive segment via nerve impulses to the terminal transmissive segment.

4 The *transmissive* (effector) *segment* contains the axon terminals that convert the stimulation of the action potential to release chemical neurotransmitters at its synapses. These chemicals exert influences upon an effector cell (another neuron, a muscle fiber, or a glandular cell).

In unipolar and bipolar neurons, both of which are afferent neurons, the receptive segment is located within a peripheral sensory receptor. The initial segment is the first neurofibral node, where the conversion to a nerve impulse occurs, and the conductive segment conveys all-or-none (see p. 323) nerve impulses to the transmissive segment in the synaptic nerve terminal. The cell bodies of these neurons are located within the conductive segments.

The cell body of a neuron is sometimes classified as the *trophic segment* (Gr. *trophe*, food or nourishment), so named because it sustains the neuron metabolically.

Ask Yourself

1 *What are the basic properties of neurons?*

2 *What are the main parts of a neuron?*

3 *What is myelin?*

4 *What is the difference between afferent and efferent neurons?*

5 *What is the main function of an interneuron?*

6 *What is the function of the transmissive segment of a neuron? The receptive segment?*

ASSOCIATED CELLS OF THE NERVOUS SYSTEM

Associated cells are specialized cells in the central and peripheral nervous systems other than neurons. For the most part, they support the neurons in some way. Although the neuron is the functional unit of the nervous system, most of the cells in the system are not neurons, but neuroglial cells. Depending upon the region, neuroglial cells outnumber neurons by 10 to 50 times.

Neuroglia: Associated Cells of the Central Nervous System

In place of connective tissue, which is relatively sparse in the central nervous system, are the neuroglia. The **neuroglia**, or *glial cells* ("nerve glue"; Gr. *glia*, glue), are nonconducting cells that protect and nurture, as well as support, the nervous system.

Certain glial cells (astrocytes and ependymal cells), along with the capillary bed of the central nervous system, form what is known as the *blood-brain barrier*. This barrier permits certain chemical substances to gain access to the neurons, and slows down or even prevents other substances, such as the antibiotic penicillin, from reaching the neurons. In this way, the nonneural cells and blood vessels of the brain and spinal cord act to maintain the homeostasis of the environment immediately surrounding each neuron and its processes. (See the essay on p. 364 for a further discussion of the blood-brain barrier.)

There are four types of neuroglia: astrocytes, oligodendrocytes, microglia, and ependyma (Figure 11.8). They are described in Table 11.2.

Associated Cells of the Peripheral Nervous System

The peripheral nervous system consists mainly of nerves, nerve endings, organs of special sense, and *ganglia* (singular,

FIGURE 11.8

(A) Astrocyte with foot plates on a capillary. (B) Oligodendrocyte. (C) Microglial cell. (D) Ependyma.

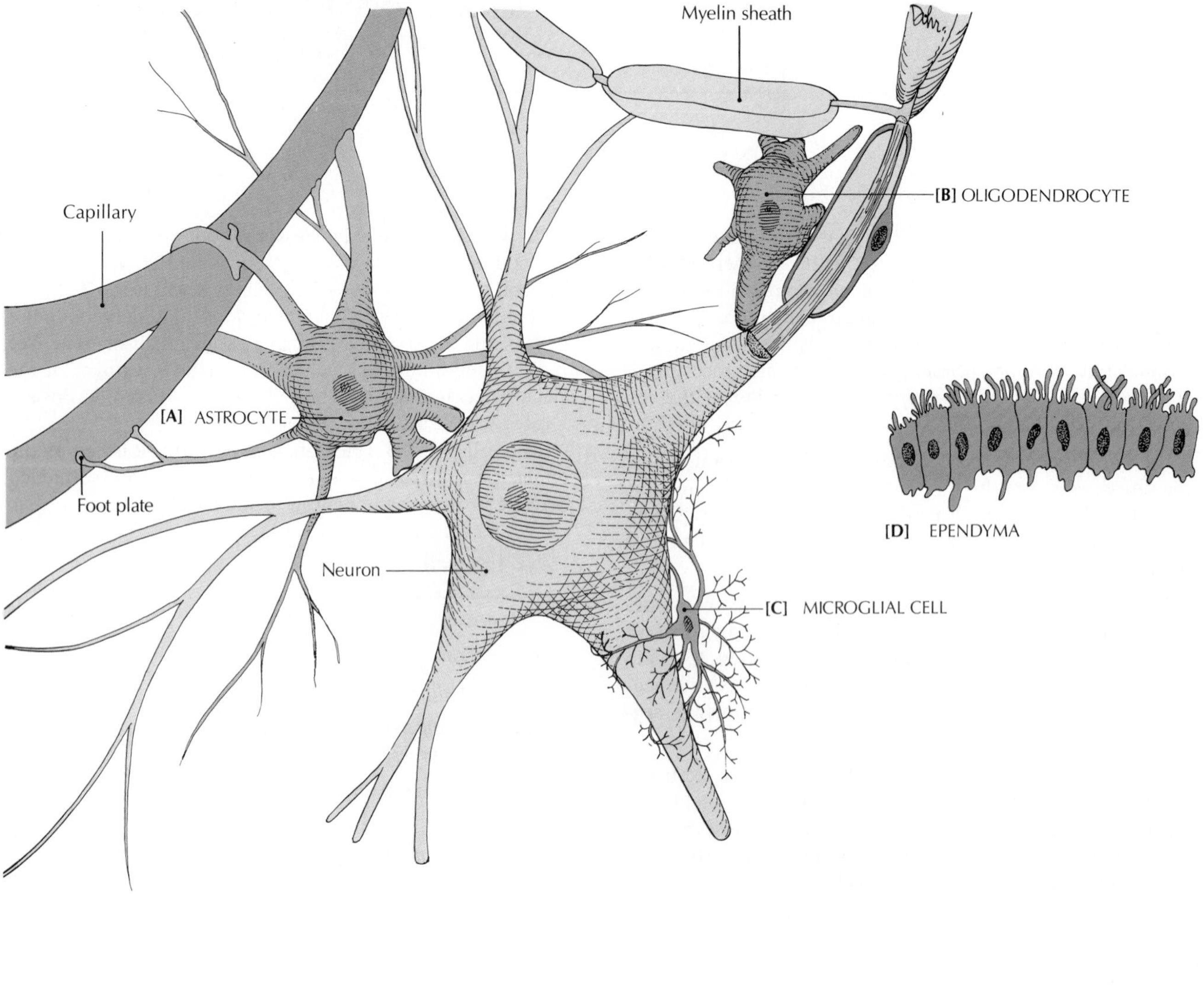

TABLE 11.2   CELLS OF THE NERVOUS SYSTEM

Type of cell	Description	Functions
Neuron	Functional unit of nervous system, with properties of excitability and conductivity. Composed of cell body, dendrites, axon.	Specialized to transmit nerve impulses over long distances in the body.
Afferent neuron of peripheral nervous system	Sensory neuron.	Receives impulses from receptor cells, carries impulses toward CNS.
Efferent neuron of peripheral nervous system	Motor neuron.	Carries neural information away from CNS to effectors.
Interneuron	Connector neuron that lies almost entirely within CNS.	Carries impulses from sensory neurons to motor neurons. Involved in processing of incoming neural information and complex activities such as learning, emotions, and language.
Neuroglia	Associated cells of CNS. Nonconducting.	Protect, nourish, support cells of CNS.
Astrocyte	Largest, most numerous glial cell, with long, starlike processes. May be protoplasmic (in gray matter of CNS) or fibrous (in white matter of CNS).	Sustains neurons nutritionally, helps maintain homeostasis of concentration of chemicals for transmission of impulses, provides packing material and structural support, helps regulate transfer of substances from capillaries to nervous tissue, keeps impulses in proper channels.
Oligodendrocyte	Relatively small, with several treelike processes. Found in gray and white matter of CNS.	Similar to neurolemmocyte. Produces and nurtures myelin sheath segments (internodes) of many nerve fibers (each process forming one internodal segment). Provides supportive framework, supplies nutrition for neurons, may increase speed of electrical conduction in a saltatory fashion.
Microglial cell	Smallest glial cell. Usually found between neurons and along blood vessels. Macrophage, not a true glial cell.	Under stressful conditions, such as injury, removes disintegrating products of neurons.
Ependymal cell	Elongated cell. Arranged in single layer to line central canal of spine and ventricles of brain.	Helps to form part of inner membrane of neural tube during embryonic growth, secretes cerebrospinal fluid.
Peripheral glial cells	Associated cells of PNS.	Provide various types of sheaths. Help to protect, nourish, maintain cells of PNS.
Satellite cell (capsule cell)	Essentially the same structure as neurolemmocyte.	Forms capsule around cell bodies of neurons of peripheral ganglia; separates cell bodies from connective tissue framework of ganglion.
Neurolemmocyte	Flattened cell. Arranged in single layer along nerve fibers.	Forms myelin sheaths of peripheral nerve fibers. Associated with both myelinated and unmyelinated fibers.

The Guiding "Glue" of the Nervous System

In 1982, a group of Rockefeller University researchers, led by Nobel laureate Gerald M. Edelman, discovered a cellular "glue" that may be involved in holding cells together. The glue in this "glue control model" is a glycoprotein of the plasma membrane of a neuron called a *cell adhesion molecule (CAM)*. The CAM found in neurons is called *N-CAM*, and is thought to be present in the plasma membranes of embryonic and adult neurons and their processes. N-CAMs are basically recognition molecules with the ability to recognize each other in ad-

jacent neurons. During early development, neurons aggregate in prescribed groupings when N-CAM is turned on, and may migrate and have their processes grow when it is turned off. This enables the neurons to assemble into their final organization, and the processes make synaptic contact with other neurons and muscles when the N-CAM is turned on again. Developmental errors usually result when the mechanism that turns the N-CAM on and off is faulty.

Subsequent research has indicated

that N-CAM may be the substance that guides the specific embryonic development and growth of neurons by providing binding sites between neurons and other structures such as muscles. Experiments with animals suggest that a pathway containing N-CAM guides the growth of axons from the eye to the brain. In contrast, a region of the brain where axons never grow (between the areas receiving input related to sight and smell) does not contain any N-CAM. N-CAM also seems to guide nerve growth at neuromuscular junctions.

---

ganglion; Gr. cystlike tumor, hence, nerve bundle). *Ganglia* are groups of cell bodies located *outside* the brain and spinal cord. They include (1) the *cranial* and *spinal (dorsal root) sensory ganglia,* which are located near the central nervous system, and (2) the *motor ganglia* of the autonomic nervous system, which are located both near the central nervous system and near or within the visceral organs.

The peripheral ganglia and nerves are composed of connective tissue, neurons, and associated cells. The *associated cells* are the **satellite cells** (or **capsule cells**) of the ganglia, and the **neurilemma cells** (Schwann sheath) of the nerves (see p. 311). A sequence of these associated cells forms a single continuous layer that ensheaths each neuron. Satellite cells surround each cell body, and neurilemma cells ensheath the nerve processes (axon and dendrites). Satellite cells and neurilemma cells both nurture the neurons, and are similar in structure and function. Neurilemma cells form the myelin sheath of myelinated nerve fibers (see Figure 11.2). They also ensheath the nerve processes of the unmyelinated fibers, but do not form myelin. The cell bodies are unmyelinated, with the exception of the myelinated cell bodies of the bipolar cells of the vestibulocochlear (ear) cranial nerve (see Chapter 13). The nerve fibers and ganglia are bound together in groups by connective tissue, which contains blood vessels that nourish the fibers and ganglia.

Degeneration and Regeneration of Nerve Fibers

The production of neurons begins before birth, and is not completed until infancy. Before this completion, some neurons die, and they continue to die throughout our lives, never again equaling the number at infancy. (Such a loss of brain cells, for instance, is usually no problem in a healthy person, because we start out with many more neurons than we actually need.) But even though nerve cells are not replaced

when they die, severed peripheral nerve *fibers* are sometimes able to regenerate if the cell body is undamaged, and if the neurilemma is intact.

When a peripheral nerve fiber is cut, the motor neuron has the capacity to regenerate its axon, and a sensory neuron has the capacity to regenerate its dendrites. First, the part of the axon severed from its cell body begins to degenerate in the stump distal to the cut, and in a few weeks disappears altogether (Figure 11.9A).

As part of the process of *regeneration,* the cell body enlarges and chromatophilic substance increases its activity and produces extra protein, which is needed to support the growth of new branches, called *terminal sprouts,* from the proximal portion of the axon still connected to the cell body. At the same time, neurolemmocytes in the distal stump divide and arrange themselves into continuous neurilemma cords of cells. These cords extend from the cut end of the stump to the sensory and motor nerve endings at the distal end of the nerve (Figure 11.9B).

Each fiber can form 50 or more terminal sprouts. The sprouts grow along the neurilemma cords, which act as guides the way a tree or a trestle guides the path of a growing vine. The actual guide is the *basal lamina* of each neurilemma cell. The fiber proceeds to grow into the newly formed neurilemma cord (also called tube) (Figure 11.9B). Most sprouts grow along cords, and do not form viable functional contacts. These sprouts will degenerate eventually, as, for example, sprouts of motor fibers in a cord leading to sensory endings, or sprouts of sensory fibers in a cord leading to motor endings. A regenerating sprout that finally reaches a nerve ending compatible with its own functional role can make a physiological connection. A sensory fiber may connect with a similar sensory ending, or a fiber of a motor neuron may connect with a motor end plate or fibers of a voluntary muscle (see Figure 11.9C). Such "successful" sprouts become myelinated by surrounding neurolemmocytes.

Severed fibers may also receive support from adjacent un-

FIGURE 11.9

Degeneration and regeneration of severed peripheral somatic motor nerve fibers. The regenerated portions of the fibers conduct nerve impulses more slowly than before, because the distance between nodes (internodes) is shorter in regenerated nerves.

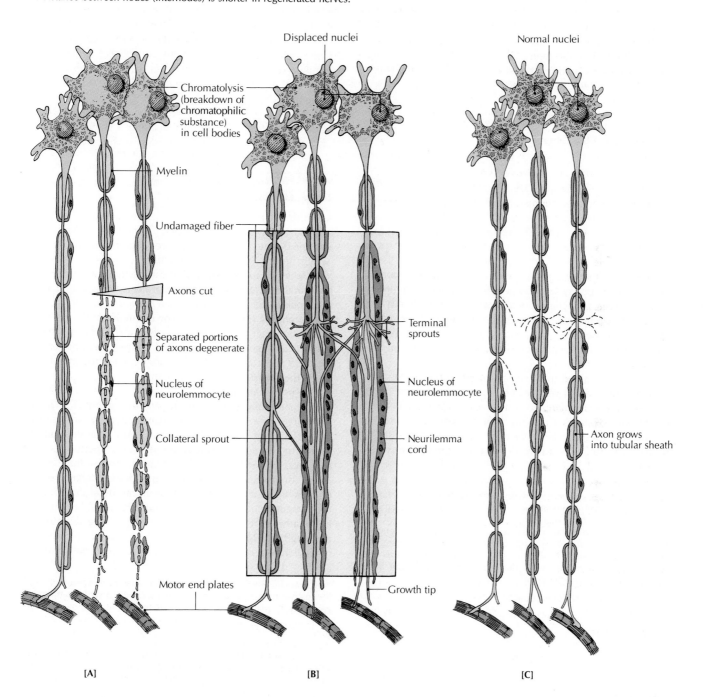

damaged fibers that send out *collateral sprouts* from neurofibral nodes (see Figure 11.9ʙ). These collateral sprouts are attracted to the axonless neurilemma cords. The degenerating nerve fibers somehow stimulate the healthy fibers to form collateral sprouts, which are helpful in reinnervating degenerated muscle fibers.

Depending upon the severity of the damage to the nerve, and the distance of the injury from the nerve endings at the motor end plate, functional recovery of nerve fibers may take from about a month to more than a year.

Regeneration in the Central Nervous System

When axons of the central nervous system of mature human beings and other mammals are severed completely, they do not regenerate to a functional degree, except in a few exceptional cases.* Damaged neurons make an attempt to regenerate axonal sprouts at the severed end, but they are unsuccessful in reaching their original terminals, probably for three reasons: (1) the sprouts cannot penetrate the glial scar tissue formed at the site of injury; (2) there is no basal lamina in the central nervous system to guide the regeneration of fibers; and (3) oligodendroglia are not arranged longitudinally (as are neurolemmocytes), and so cannot guide the fibers in an organized regeneration. Unfortunately, this lack of functional regeneration can have serious effects, such as those that follow some strokes.

It should be noted that since mature neurons are not capable of cell division, the cell body must be viable in order for regeneration to take place both in the central and peripheral nervous systems. Currently, many research projects are attempting to resolve the problem of the structural and functional regeneration of neurons of the mammalian nervous system.

---

*The severed axons of neurons with cell bodies in the hypothalamus will regenerate into the neurohypophysis of the pituitary gland (see Chapter 16), and functional recovery will occur. Also, severed unmyelinated axons of certain central nervous system neurons that contain neurotransmitters such as dopamine and norepinephrine can regenerate successfully. Finally, central nervous system axons can regenerate into nerve segments and nonnervous tissues that have been transplanted into the brain.

Ask Yourself

1 *What are the main functions of neuroglia?*

2 *What are the four types of neuroglia?*

3 *What are the associated cells of the peripheral nervous system?*

4 *Compare the functions of terminal and collateral sprouts.*

5 *Why are severed neurons in the central nervous system unable to regenerate?*

PHYSIOLOGY OF NEURONS

Neurons are specialized to respond to stimulation by generating several types of electrical signals. These are expressed as changes in the electrical potentials that are conducted along the plasma membranes of the dendrites, cell body, and axon of each neuron. The difference in potential across the semipermeable plasma membrane of the neuron results from differences in the concentration of certain ions on either side of the membrane.*

In the following sections we will explain how electrical signals are generated, conducted along a nerve fiber, and passed on to an adjacent neuron, muscle fiber, or gland cell.

Resting Membrane Potential

A "resting" neuron is one that is not conducting a nerve impulse, but *is* electrically charged. (It may be compared with a charged battery that does not conduct an impulse until it is switched on.) The plasma membrane of such a "resting" neuron is said to be **polarized,** meaning that the intracellular fluid on the inner side of the membrane is negatively charged *in relation to* the positively charged extracellular fluid outside the membrane (Figure 11.10). The difference in the electrical charge between the inside and outside of the plasma membrane at any given point is called the *potential difference*. It creates a potential for electrical activity along the membrane called the **resting membrane potential.**

The potential difference is measured in volts or millivolts (mV).† Normally, the resting membrane potential of about -70 mV is due to the unequal distribution of various electrically charged ions (sodium, potassium, chloride, and organic protein molecules) and to the selective permeability of the plasma membrane to the different ions. Sodium (Na^+) and chloride (Cl^-) ions in the vicinity of the charged neuron are in a higher concentration in the extracellular fluid outside the plasma membrane, and potassium (K^+) and protein ions are in a higher concentration in the intracellular fluid inside. Although the Na^+, K^+, and Cl^- ions are all able to pass readily through the cell's membrane, it is the Na^+ and K^+

---

*Comparisons between a nerve impulse and the flow of electricity are interesting, but tend to divert the discussion away from the main issue of the relative concentration of ions. Although the nervous system is most conveniently excited by electrical stimulation, a nerve impulse is not the same as a flow of electricity. Instead, it is an electrochemical impulse, and it travels at a much slower rate than electricity: about 100 m/sec or less for an electrochemical impulse, as compared with about 300,000 km/sec (the speed of light) for an electrical current. Another difference between a nerve impulse and an electrical current is that the nerve impulse does not diminish in power as it is propagated along an axon, the way an electrical current does.

†The resting membrane potential of -70 mV (less than one-tenth of a volt) means that the electrical charge on the inside of the plasma membrane measures $\frac{70}{1000}$ of a volt *less* than that on the outside. By convention, a minus sign denotes a negative charge inside the membrane.

FIGURE 11.10

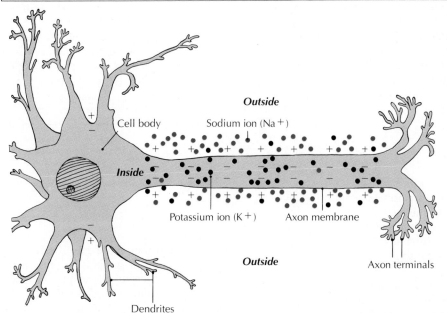

Outside

Cell body   Sodium ion (Na⁺)

Inside

Potassium ion (K⁺)   Axon membrane

Outside

Axon terminals

Dendrites

Resting membrane potential of an unmyelinated nerve fiber. Sodium ions (red) and chloride ions are highly concentrated on the outside of the plasma membrane; potassium ions (purple), negatively charged protein molecules (not shown), and some chloride ions (not shown) are highly concentrated on the inside. When the neuron is in this resting condition, it is said to be polarized. The movement of ions through selective channels across the plasma membrane is strictly regulated to maintain the proper concentration and the resultant electrical charge at any given point along the axon.

Key

● = Sodium ion (Na⁺)

● = Potassium ion (K⁺)

FIGURE 11.11

Sodium-potassium pump and voltage-gated ion channels within the semipermeable axon membrane. These mechanisms maintain a balance beween sodium ions (red dots) and potassium ions (purple dots) on both sides of the axon membrane. The gated ion channels, shown here in their closed state, open and close their gates in sequence when a nerve impulse is being transmitted along the axon. Note that the ion pump transports two potassium ions into the axon for every three sodium ions it pumps out, and in this way it ensures a slightly negative charge inside a resting neuron.

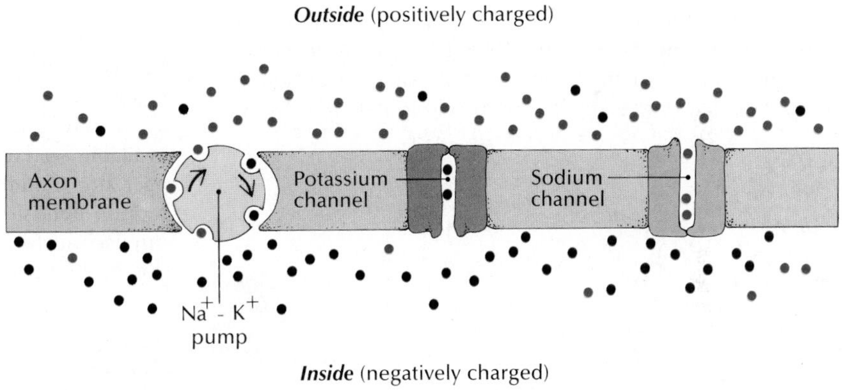

Outside (positively charged)

Axon membrane   Potassium channel   Sodium channel

Na⁺ - K⁺ pump

Inside (negatively charged)

ions that have the most important effect on the resting membrane potential.

The Na⁺ and K⁺ ions are constantly diffusing and leaking through the semipermeable plasma membrane, moving from regions of high concentration to regions of lower concentration. (The large negative protein ions cannot move easily from the inside of the neuron to the outside.) Yet the concentration of Na⁺ and K⁺ ions in the inner and outer sides of the membrane is remarkably constant. The extracellular fluid usually contains about 10 Na⁺ ions for every K⁺ ion. The ratio inside is reversed, with K⁺ ions outnumbering Na⁺ ions by at least 10 to 1.

Such ionic consistency is necessary if a neuron is to remain excitable and be able to respond to changes in its surroundings. How is this homeostasis maintained? Part of the answer lies in a self-regulating transport system, located within the plasma membrane, known as the ***sodium-potassium pump*** (Figure 11.11), which is powered by ATP and other energy-rich compounds. The more sodium that leaks into the neuron through the plasma membrane, the more active the pump becomes, pumping actively to restore the ionic concentrations that maintain the resting membrane potential. The sodium-potassium pump transports three sodium ions out of the cell for every two potassium ions it brings into the cell. So even though both ions are positive, more positive ions are moving out through the membrane than are moving in. Partly for this reason, the inner side of the plasma membrane is more negative than the outer side.

In addition to the pump, the plasma membrane of a neuron is selectively permeable to Na⁺, K⁺, and Cl⁻ ions through voltage-gated sodium and potassium *open ion channels* (see Figure 11.11). These selective channels are impermeable to large protein molecules, but are always open to Na$^+$, K$^+$, and Cl$^-$. When the concentration of Na$^+$ or K$^+$ ions becomes too high on either side, the channels open selectively to reestablish the distribution of ions that produces the resting membrane potential of -70 mV.

Mechanism of Nerve Action: Changing the Resting Membrane Potential into the Action Potential (Nerve Impulse)

The *change* in the electrical potential across a plasma membrane is the key factor in the creation and subsequent conduction of a nerve impulse. The process of conduction, although basically similar in unmyelinated and myelinated nerve fibers, differs somewhat. The following steps describe conduction in unmyelinated fibers:

1 A stimulus that is strong enough to initiate an impulse in a nerve cell is called a **threshold stimulus.** When such a stimulus is applied to a polarized resting plasma membrane of the axon of a neuron, the permeability of the membrane to sodium ions increases dramatically at the point of stimulation. For about half a millisecond, the sodium ion channels open, and sodium ions rush into the cell, reversing the relative electrical charges at the point of stimulus. This reversal of charges, giving the *inner side* of the plasma membrane a positive charge (of about $+30$ mV) relative to the *outer side,* is called **depolarization.** When a stimulus is strong enough to cause depolarization the neuron is said to *fire.*

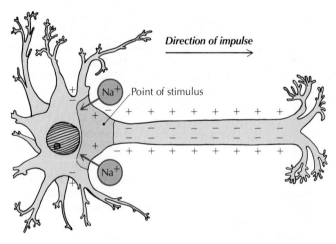

2 The depolarized patch on the plasma membrane produces a flow of current that stimulates the adjacent polarized patch. As a result, this new site becomes depolarized; sodium ions rush in through voltage-gated Na$^+$ channels, and the electrical charges become reversed as they did at the original point of stimulus. Once a patch on the axon is depolarized, an **action potential (nerve impulse)** is initiated (Figure 11.12). This continuous spread of the nerve impulse is characteristic of unmyelinated axons. The membrane voltage of the ongoing action potential operates to open the sodium and potassium channels just ahead of the action potential.

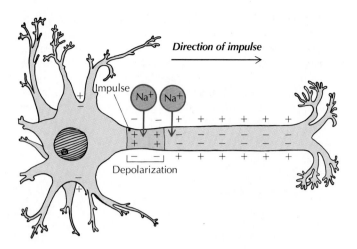

3 Shortly after the sodium ions move into the cell, the membrane at the original point of stimulus becomes more permeable to potassium ions. The voltage-gated K$^+$ channels open, and a number of potassium ions diffuse *out* from the cell. The outside of the membrane regains its original positive charge (the resting potential), and the original balance of sodium and potassium ions is restored. Also restored at the point where potassium ions rush out is the relative negative charge inside the cell and the relative positive charge outside the cell. Thus, the membrane is said to be **repolarized.** The transmission of a nerve impulse along the cell membrane may be visualized as a *wave* of depolarization and repolarization. At any given spot on the membrane, the sequence from resting potential to action potential and back to resting potential takes only a few milliseconds. Only a tiny fraction of the available stored sodium and potassium ions is used in the operation, and the original concentrations are restored by the sodium-potassium pump. As a result, many action potentials can occur before the concentrations of Na$^+$ and K$^+$ on either side of the membrane change significantly.

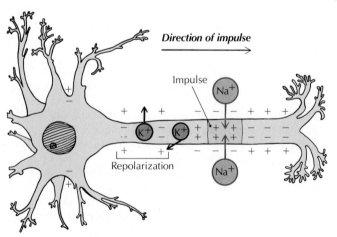

4 Successive acts of depolarization (reversal of the potentials) are repeated as the nerve impulse travels along the length of the axon. After each firing, there is an interval of from one-half to one millisecond before it is possible for an adequate stimulus to generate another action potential. This brief period is called the **refractory period.** During this refractory period the resting potential of the fiber membrane is being restored at the part of the membrane where the impulse has just passed. Afterward, the fiber is "recharged" and is ready to transmit another impulse. Although a few nerve fibers are theoretically capable of generating and conducting 2000 impulses per second, the typical number for most fibers is about 300 per second.

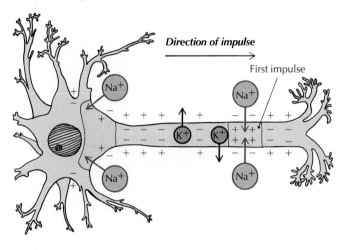

FIGURE 11.12

Voltage and ionic changes during an *action potential* as it would appear on an oscilloscope. During the *depolarization* phase (pink) of the action potential, there is a rapid increase in the flow of sodium ions (Na^+) into the axon. The *repolarization* phase (yellow) is characterized by a rapid increase in the flow of potassium ions (K^+) out of the axon. The action potential is sometimes called a *spike* because of its shape on an oscilloscope.

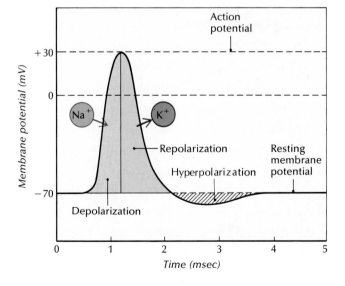

How fast do people react to stimuli?   *Olympic male sprinters react to the starter's gun in about 12/100 of a second.*

All-or-None Law

A minimum stimulus is necessary to instigate a nerve impulse, but an increase in the intensity of the stimulus does not increase the strength of the impulse. The principle is the same as when a gun is fired: if the trigger is not pulled hard enough, nothing happens, but the minimum required pull on the trigger will fire the gun. Pulling the trigger even harder, however, will not make the gun fire harder. The principle that states that a neuron will fire at full power or not at all is known as the **all-or-none law.** It applies primarily to action potentials in axons, but also applies to muscle contraction as discussed in Chapter 9.

But what about the difference we can feel between a light touch and a strong one, or between a soft sound and a loud one? Don't those differences show that the all-or-none law is wrong? No, they do not. Such differences can be perceived when the *frequency* of the impulse, not its strength, is changed. Some nerve fibers can conduct at different frequencies per second. The more frequent the impulses, the higher the level of excitation. Other fibers respond to *different thresholds*, but all impulses carried on any given fiber are of the same strength. Also, the *number of neurons* involved can make a difference in how strong the stimulus is perceived to be. You feel the difference between a light push and a strong one, for example, because the strong push affects more of your neurons. But when it comes to a single axon in an experimental situation, the all-or-none law still applies.

Saltatory Conduction

In a myelinated nerve, the action potential appears to "jump" from one neurofibral node to the next. For this reason, conduction along myelinated fibers is known as **saltatory conduction** (L. *saltare*, to jump). It takes less time for an impulse to jump from node to node along a myelinated fiber than if it traveled smoothly along an unmyelinated fiber. The nerve impulse actually travels almost instantaneously from an active depolarized node of the axon to the adjacent node, which has a resting potential. Voltage-gated sodium channels are highly concentrated in the tiny portion of the axon exposed to the extracellular fluid at the neurofibral node. The inactive node is hypersensitive to voltage, and is most responsive to becoming depolarized by the active node; it is stimulated to fire, producing an electrical current that fires the next node, and so on, all along the fiber.

Synapses

After an action potential travels along an axon, it reaches the branching axon terminals in the transmissive segment of the neuron. If the nerve impulse is to be effective, its influences must be conveyed to another nerve, muscle, or gland cell. In this discussion we will concentrate on the junction between the axon terminal of one neuron and the dendrite, cell body, or axon of the next neuron. This junction between neurons is called a **synapse** (SIN-aps; Gr. connection) (Figure 11.13).

FIGURE 11.13

Transmission of a nerve impulse across a synapse. (A) A low-magnification view of a neuron, adjacent to an axon terminal end bulb or synaptic bouton (in square on the drawing). (B) An enlarged view of the end bulb (cut), containing synaptic vesicles. End bulbs usually also contain numerous mitochondria. The synaptic cleft separates the end bulb from the cell body. (C) A still greater enlargement, with only a portion of the end bulb shown. (D) An enlargement of the activity in (C). The sequence of events in (D), from left to right, shows (1) a synaptic vesicle containing neurotransmitter molecules approaching the cell body membrane, (2) the vesicle making contact with the presynaptic membrane, (3) the vesicle breaking through the membrane and emptying its contents (red dots) by exocytosis into the synaptic cleft; the neurotransmitter molecules cross the synaptic cleft and bind to receptor sites on the postsynaptic membrane and either stimulate or inhibit the postsynaptic cell, and (4) the synaptic vesicle is re-formed and begins to fill with another supply of neurotransmitter molecules. (E) Electron micrograph of a nerve terminal stocked with synaptic vesicles about to empty into the synaptic cleft. The larger, darker circles are mitochondria. [*Photo by Dr. John E. Heuser, Washington University School of Medicine.*]

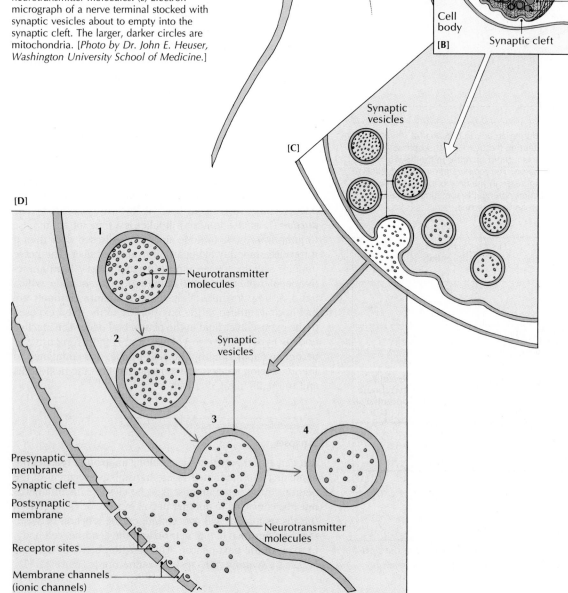

The nerve cell carrying the impulse *toward* a synapse is called the **presynaptic** ("before the synapse") neuron. It initiates a response in the receptive segment of a **postsynaptic** ("after the synapse") neuron leading *away from* the synapse (Figure 11.14). The presynaptic cell is always a neuron, but the postsynaptic cell can be a neuron, muscle cell, or gland cell.*

The transmission of a nerve impulse at a synapse can be either chemical or electrical, but chemical synapses are far more common than electrical ones. In the following sections we will discuss electrical synapses briefly, and then describe chemical synapses in greater detail.

Electrical synapses An *electrical synapse* is a *gap junction* in which the two communicating cells are electrically coupled by tiny intercellular channels called *connexons*. The free movement of ions occurs through the connexons between the presynaptic and postsynaptic cells. Because of the extremely low resistance associated with connexons, the nerve impulse virtually travels directly from the presynaptic to the postsynaptic cell. Also, since there are no chemical neurotransmitters that

*A synapse from an axon to a dendrite is an *axodendritic* synapse, and one from an axon to a cell body is an *axosomatic* synapse. A synapse from one axon to another is an *axoaxonic* synapse. Such a synapse can be formed only where the postsynaptic axon is not insulated with myelin, such as at the axon hillock and the end bulb regions.

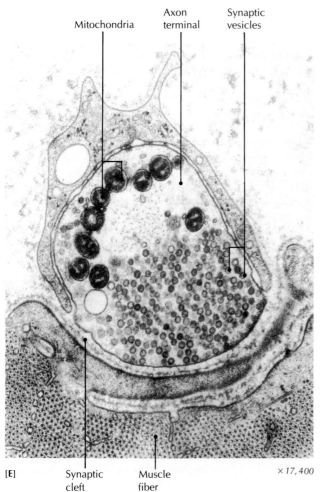

Mitochondria   Axon terminal   Synaptic vesicles

[E]   Synaptic cleft   Muscle fiber   ×17,400

FIGURE 11.14

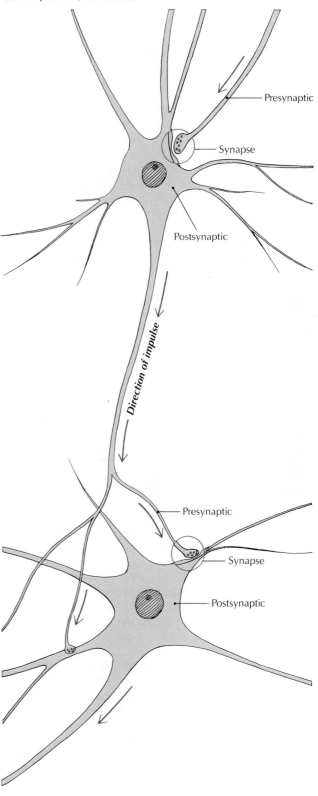

Highly simplified representation of the relationship of presynaptic and postsynaptic neurons, and the direction of the nerve impulse (action potential) transmission.

Presynaptic

Synapse

Postsynaptic

Direction of impulse

Presynaptic

Synapse

Postsynaptic

require time to react, the transmission of an electrical synapse is almost instantaneous.

In human beings (and other mammals), electrical synapses are found between cardiac cells, between smooth muscle cells of the intestinal tract, and between a few neurons—in the retina of the eye, for example.

Chemical synapses In a *chemical synapse,* two cells communicate by way of a chemical agent called a *neurotransmitter,* which is released by the transmissive segment of the presynaptic neuron. A neurotransmitter is capable of changing the resting potential in the plasma membrane of the receptive segment of the postsynaptic cell.

When the nerve impulse of a presynaptic neuron reaches the terminal synaptic endings (boutons) of the axon, it depolarizes the presynaptic plasma membrane, causing voltage-sensitive calcium channels to open. As a result, calcium ions (Ca^{2+}) diffuse into the presynaptic terminal. As calcium ions rush through Ca^{2+} channels of the plasma membrane of the terminal ending, they cause the storage vesicles to fuse with the plasma membrane, releasing the neurotransmitter from the vesicles by exocytosis into the narrow *synaptic cleft** (see Figure 11.13D). One such neurotransmitter is the excitatory chemical *acetylcholine;* another is *norepinephrine.* (More than 50 other possible neurotransmitters are known; see p. 328.)

The synapse is like a one-way valve, with the neurotransmitter stored in the axons of the presynaptic neuron, and all of the receptor sites for that transmitter in the membrane of the postsynaptic neuron. The transmitters can pass only from axon terminals across the synaptic cleft to the receptor cell, but not the other way. This directional control prevents neural effects from traveling in all directions at once. If they did, neural messages would be garbled and scrambled, probably producing static.

The short trip across the synaptic cleft takes about a millisecond. That interval results in a *synaptic delay.* Some of the released neurotransmitter binds with *receptor protein sites* in the postsynaptic membrane, causing changes in the membrane's permeability to certain ions. The remaining neurotransmitter in the synaptic cleft may diffuse away and be deactivated by enzymes, and its products may be recycled by being taken up by the presynaptic ending and resynthesized. As a result of the change in permeability in the postsynaptic membrane, sodium, potassium, and/or chloride ions rush through their open ion channels and permit the same kind of depolarization that the presynaptic cell experienced. Waves of excitation pass along to the postsynaptic neuron, and the resulting nerve impulse is able to continue its path to an eventual effector.

Once the neurotransmitter has bridged the synaptic cleft, the acetylcholine is quickly broken down into acetate and choline by the enzyme *acetylcholinesterase* (also called *cholinesterase*). Some chemical substances, such as certain insecticides and ''nerve gases,'' work by inhibiting the chemical action of this enzyme. Since acetycholine is not destroyed, nerve impulses take place continuously. Muscle cells remain in a state of contraction, and the breathing muscles are paralyzed. Death by asphyxiation results.

*This narrow gap is usually less than 1/500 the width of a human hair (about 25 nanometers).

Postsynaptic Potentials (Graded Local Potentials)

Neurotransmitters interact with the receptor sites of chemically gated channels on the postsynaptic (receptor) membranes of the receptive segments of neurons, muscle fibers, and gland cells, producing a *postsynaptic potential (PSP)* (Figure 11.15). Postsynaptic potentials are not typical all-or-none potentials, but are *graded local potentials,* which spread their effects passively, and fade out a short distance from the site of stimulation. Each postsynaptic potential propagates along the cell membrane for short distances, and lasts for 1 to a few milliseconds. The two basic types of responses that can be produced at different synaptic sites on the postsnyaptic membrane are excitatory postsynaptic potentials and inhibitory postsynaptic potentials, both of which can vary in intensity.

Excitatory postsynaptic potentials The excitatory response is associated with sodium ions diffusing into the neuron, and potassium ions diffusing out through the postsynaptic membrane. This combined action leads to depolarization. An *excitatory postsynaptic potential (EPSP)* is a partial depolarizing effect that lowers the membrane potential, but not all the way to generate an action potential. An EPSP lasts only a few milliseconds and can spread only a short distance. Because an EPSP makes the postsynaptic membrane more excitable, it does increase the chance of generating an action potential. Such an effect is called *facilitation.* If enough EPSPs bombard the receptive segment of the neuron over a short period, the resulting EPSPs can add up, causing *summation.* If there is sufficient summation, the partial depolarizing effect can reach the initial segment of the neuron, where an action potential can be triggered.

Inhibitory postsynaptic potentials Whereas *excitation* (EPSP) brings a neuron to a state where it is more likely to generate an action potential, *inhibition* makes the neuron *less* likely to trigger an action potential. Inhibition occurs when the neurotransmitter interacts with a postsynaptic receptor site, increasing the negative potential inside the membrane above the resting level to as much as -90 mV (see Figure 11.15).* This *inhibitory postsynaptic potential (IPSP)* occurs when potassium and chloride channels are opened, and chloride ions rush into the cell, and potassium ions rush out through the postsynaptic membrane. Such an inhibitory graded potential tends to prevent the postsynaptic potential from reaching the initial segment, where a threshold potential would trigger an action potential.

*Because the polarity of the membrane has been increased (and thus becomes more negative), the cell is said to be *hyperpolarized.*

How many synapses may a neuron have?

Some neurons have only a few synapses, but a motor neuron may have as many as 2000, and a Purkinje cell (conducting myofiber) of the cerebellum may have as many as 200,000 on its dendrites alone.

FIGURE 11.15

Electrical potential changes recorded across the plasma membrane at various sites of a motor neuron. A cathode-ray oscilloscope records the electrical potential (as voltage differences) across the plasma membrane, with one electrode outside the neuron and the other inside the neuron. The resting membrane potential ranges between -60 and -70 mV, with the inside negative relative to the outside. When the membrane potential moves toward zero, the membrane becomes *depolarized,* and when the membrane potential increases above the resting value the membrane becomes *hyperpolarized.* On the surface of the dendrites and cell body are excitatory and inhibitory synapses, which, when stimulated, produce local, graded, nonpropagating potentials. These are exhibited as an excitatory or depolarizing postsynaptic potential (EPSP) and an inhibitory or hyperpolarizing postsynaptic potential (IPSP). These local potentials are summated at the axon hillock and, if adequate, may trigger an integrated potential at the initial segment and an all-or-none action potential, which is conducted along the axon to the motor end plate.

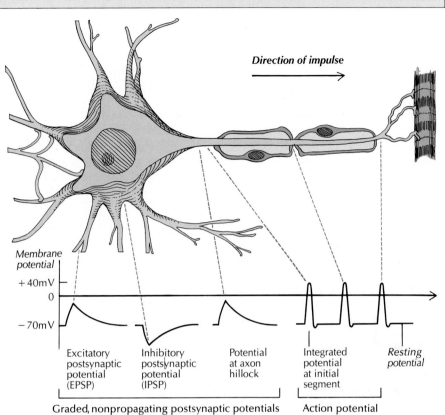

Direction of impulse

Membrane potential

$+40$mV
0
-70mV

Excitatory postsynaptic potential (EPSP)

Inhibitory postsynaptic potential (IPSP)

Potential at axon hillock

Integrated potential at initial segment

Resting potential

Graded, nonpropagating postsynaptic potentials

Action potential

In a sense, the plasma membrane of the receptive segment of each neuron is a miniature integration center, where the EPSPs and IPSPs summate either *spatially* (occurring in space) or *temporally* (through time). The summing of the synaptic inputs from different neurons upon the dendrites and cell body of one neuron is called *spatial summation,* indicating that each of these many synaptic inputs occupies a slightly different site on the receptive plasma membrane of the postsynaptic membrane. *Temporal summation* occurs when one presynaptic neuron can increase its effect on one postsynaptic neuron by firing repeatedly. These two types of summation are important because synaptic input from presynaptic neurons activates the postsynaptic neuron.

If the resulting activity from spatial and temporal summations is sufficient to reach and excite the initial segment, then an action potential can be triggered in the conductile segment of the axon. In effect, the action potential is an expression of part of a code message that conveys information about the resolution of the graded potentials in the receptive segment of one neuron to another neuron.

Receptor Potentials

The sensory receptors of the body (such as those involved with sensing touch pressure and monitoring blood pressure) are associated with the terminal of a sensory nerve fiber. When the terminal is stimulated, it can exhibit a graded local potential called a *receptor potential.* An example of a receptor

potential is the touch-pressure receptor of a lamellated (Pacinian) corpuscle in the skin (see Chapter 15). This ending has sodium and potassium channels that are *modality gated.* (The word *modality* refers to a sensory component such as touch-pressure.) A stimulus on the nerve ending causes an inflow of sodium ions through Na$^+$ channels into the unmyelinated nerve terminal, and an outflow of potassium ions through K$^+$ channels, producing a graded potential similar to an EPSP. This potential spreads along the plasma membrane to reach the first neurofibral node (low-threshold site of the myelinated portion of the nerve fiber), where an action potential is generated. The first neurofibral node is the initial segment.

Neurotransmitters

Neurotransmitters are chemical agents synthesized by nerve cells, stored in secretory vesicles, and released when calcium ions from the synaptic cleft flood the vesicles. The effect of a neurotransmitter on the postsynaptic membrane may be to produce either an excitatory or inhibitory response. Either response is a result of the properties of the receptor, triggered by the transmitter. For example, the neurotransmitter acetylcholine evokes an excitatory response at the motor-end-plate synapse in a voluntary muscle cell, but it evokes an inhibitory response at its synapse with a cardiac muscle cell.

Many possible neurotransmitters are being investigated. At least 50 are present in the nervous system, and there may

TABLE 11.3   SOME MAJOR NEUROTRANSMITTERS

Transmitter group	Transmitter compound	Probable functions within the nervous system
Acetylcholine	Acetylcholine (ACh)	Excitatory and inhibitory. Neuromuscular, neuroglandular transmission. Also involved in memory.
Amino acids	Gamma-aminobutyric acid (GABA)	Inhibitory. Evokes IPSPs in brain neurons.
	Glutamate	Excitatory. Evokes EPSPs in brain neurons.
	Aspartate	Excitatory. Evokes EPSPs in brain neurons.
	Glycine	Inhibitory. Evokes IPSPs in brain neurons. Important in spinal cord.
Monoamines	Dopamine	Modulates activity of adrenergic neurons. Involved in arousal, motor activity. Evokes EPSPs in brain neurons.
	Histamine	Evokes EPSPs in brain neurons.
	Norepinephrine (NE)	Excitatory and inhibitory. Involved in arousal, motor activity, visceral functions (such as heat regulation, reproduction). Acts as hormone when secreted into the bloodstream.
	Serotonin (5-hydroxy-tryptamine, or 5-HT)	Involved in sleep, mood, appetite, pain. Evokes EPSPs in brain neurons.
Neuropeptides	Somatostatin	Inhibits secretion of growth hormone.
	Endorphins, enkephalins	Analgesic properties.
	Substance P	Involved in mediating pain.

be over 100. Those substances definitely established as transmitters are listed in Table 11.3. Neurotransmitters are classified in four groups: (1) acetylcholine; (2) amino acids, such as gamma-aminobutyric acid (GABA), glycine, and glutamate; (3) monoamines (each with an amine radical, —NH$_2$), such as norepinephrine (noradrenaline), dopamine, and serotonin; and (4) neuropeptides (basically chains of amino acids), such as somatostatin, endorphins, and enkephalins.

Acetylcholine One of the best-known neurotransmitters is *acetylcholine* (ACh). It is released by many neurons in both the central and peripheral nervous systems, although its role in the peripheral nervous system is more obvious. Such acetylcholine-releasing neurons are known as *cholinergic neurons,* and their fibers are called *cholinergic fibers.* From cell bodies deep within the nucleus basalis of the cerebrum, cholinergic fibers terminate diffusely in the cerebral cortex.

Acetylcholine is most important as the neurotransmitter between motor neurons and voluntary muscle fibers. Through the lower motor neurons it exerts excitatory effects on voluntary muscle contraction. Through the parasympathetic nervous system it has inhibitory effects on heart muscle, and excitatory effects on smooth muscles involved with the dilation of the pupils of the eyes.

Acetylcholine binds with excitatory receptor sites on the postsynaptic plasma membrane of the muscle of the motor end plate, and depolarizes the muscle cell membrane. The resulting action potential in the plasma membrane of the muscle fiber sets off the chain of activities that produces a muscle contraction. The contraction of the muscle is maintained by the continued restimulation of the axon, which generates additional action potentials. (The mechanism is essentially the same for neuron-neuron or neuron-gland cell junctions.)

The body has a built-in mechanism for resupplying acetylcholine. After its initial activity at the synapse, any acetylcholine within the synaptic cleft is inactivated by being broken down by the enzyme *acetylcholinesterase* into the free compounds acetate and choline. Most of the choline then diffuses back across the synaptic cleft to the pool of choline in the cytosol of the releasing neuron, where the choline is reunited with acetate by a second enzyme, choline acetylase, to synthesize fresh acetylcholine. This recycled acetylcholine is now ready to transmit another nerve impulse.

Amino acids Although *gamma-aminobutyric acid* (GABA) is not an amino acid, it is classified with the amino acids because it is synthesized from glutamic acid. It is the major inhibitory transmitter of the small local circuit neurons in such structures as the cerebral cortex, cerebellum, and upper brainstem. Neurons that secrete GABA are referred to as GABAergic. *Glycine,* the simplest amino acid, is the major inhibitory transmitter of local circuit neurons in the lower brainstem and spinal cord. By selectively inhibiting certain neurons, these transmitters act chemically like the chisels of a sculptor, refining and focusing brain activity into meaningful patterns by eliminating and suppressing nonessential activity. *Glutamate* and *aspartate* are the most potent excitatory transmitters in the central nervous system.

Monoamines The two important groups of *monoamines* are catecholamines and indoleamines. The *catecholamines* (which contain a catechol nucleus) include norepinephrine and dopamine. The prominent *indoleamine* (which contains an indole nucleus) in the brain is serotonin.

Norepinephrine (noradrenaline) is located in neurons with cell bodies in the brainstem and in peripheral nerves of the sympathetic nervous system. Neurons and fibers containing norepinephrine are called *noradrenergic neurons* and *noradrenergic fibers*. The fibers are distributed widely in such structures as the cerebral cortex, cerebellum, and spinal cord. Fibers that terminate in the cerebral cortex are involved with various levels of consciousness. The locus ceruleus in the brainstem is a nucleus of origin. It responds to novel stimuli in the environment (such as a familiar sound among many other sounds) and then, through its projections, acts to suppress irrelevant stimuli and to enhance the relevant stimuli, making them stand out enough to alert and maintain our state of arousal.

Norepinephrine is released at the neuromuscular junctions with smooth and cardiac muscles, and at the neuroglandular junctions with the cells of exocrine glands (glands that have ducts to carry their secretions). Moving across the synaptic cleft, it depolarizes the postsynaptic cell membrane. After combining with the receptor sites, most of the norepinephrine immediately detaches and re-enters the synaptic cleft. Some is recycled and stored in synaptic vesicles. Excess norepinephrine is inactivated by the enzyme monoamine oxidase (MAO) or by the enzyme catechol-O-methyl transferase (COMT). Finally, some norepinephrine diffuses away from the cleft.

''Pep pills'' such as those containing the drug amphetamine increase the level of norepinephrine in the brain by blocking its reuptake in the neuron, and by inhibiting the action of MAO. The euphoria and hallucinations produced by amphetamines are thought to be the result of the inhibition of MAO. An ''amphetamine psychosis'' quite similar to schizophrenia can result from the habitual use of amphetamines. Chlorpromazine is a chemical used in the treatment of schizophrenia and similar disturbances. It acts by blocking the reuptake of norepinephrine.

Dopamine is a neurotransmitter found in several locations in the brain. Dopaminergic neurons have critical roles in certain neural circuits involved with voluntary motor integration. Deficits in dopamine in certain of these circuits are associated with Parkinson's disease. Dopamine is the prolactin-inhibiting hormone of the hypothalamus, acting to inhibit the release of the hormone prolactin from the pituitary gland.

Serotonin (5-hydroxytryptamine, or 5-HT) is a neurotransmitter associated with various swings in moods, including depression, elation, insomnia, hallucinations, and mania. The relatively few serotoninergic neurons have their cell bodies in the brainstem. However, their fibers are widely distributed throughout the brain, including the cerebral cortex.

Neuropeptides Among the *neuropeptides* (neuroactive peptides) found in the central nervous system are somatostatin, the endorphins, and the enkephalins. *Somatostatin* is also known as the growth-hormone-inhibiting hormone of the hypothalamus. It inhibits the release of growth hormone from the pituitary gland. Apparently, somatostatin acts by altering the excitability of the postsynaptic membrane.

Endorphins (for ''endogenous morphinelike substances'') and *enkephalins* (for ''in-the-head substances'') are naturally occurring peptides found in several regions of the brain and spinal cord. They function at postsynaptic receptor sites in the pain pathways to suppress synaptic activity leading to pain sensation. These opiate drugs are not used therapeuti-

cally because of their relatively brief effectiveness, and because they are addictive.

Many neuropeptides are classified as *neuromodulators*, chemical agents that are capable of altering (modulating) the responsiveness of neurons to a neurotransmitter. Since the responses cannot be described as EPSPs or IPSPs, the term *modulator* is used to distinguish these agents. In general, neuromodulators bring about chemical changes in a neuron, and do not function as specific neurotransmitters. Some substances thought to function as neuromodulators include neuropeptides, histamine, prostaglandins, and the hormones cortisol and estrogen (see Chapter 16). Both neuromodulators and hormones act as specific receptors on target cells. Neuromodulators are released by neurons near target cells, whereas hormones are circulated through the bloodstream.

Ask Yourself

1 What is polarization?

2 What is a membrane potential?

3 What is a threshold stimulus?

4 What is an action potential?

5 What is the all-or-none law?

6 How do sodium-potassium pumps function?

7 What is a synapse?

8 How do presynaptic and postsynaptic neurons differ?

9 How does a neurotransmitter work?

10 What is a neuromodulator?

NEURONAL CIRCUITS

The nervous system is an exquisitely structured network of neurons arranged in synaptically connected sequences called **neuronal circuits,** or neuronal chains (Figure 11.16). The circuits are formed by input, intrinsic, and relay neurons:

1 An *input neuron* is a nerve cell that conveys information from one group of neurons to another. Such a group is called a *nucleus* (an anatomical term), or a *neuron pool* (a physiological term). A **neuron pool** is a group of cell bodies and their dendrites characterized by their physiological activity. When the neurons of a pool are stimulated, they tend to act together to express a functionally defined goal. For example, a stimulated pool produces flexion of a joint by exciting agonist muscles, and by inhibiting the antagonistic muscles.

2 An *intrinsic neuron* is an interneuron that is often located entirely within the nucleus or pool. It is sometimes called a local circuit neuron.

3 A *relay neuron* is a nerve cell projecting from one nucleus or pool to another. Within each nucleus or pool there is synaptic organization, where processing occurs.

Several patterns and types of neuronal circuits exist. The cirucits described in the following sections are the ones most commonly identified.

Divergence and Convergence

Some neurons in the central nervous system may synapse with as many as 25,000 other neurons. When the transmissive segment of a presynaptic neuron branches out to have many synaptic connections with the receptive segments of many other neurons, it is an example of *divergence* (L. *divergere*, to bend) (Figure 11.16A). In diverging synapses, one neuron may excite or inhibit many others. The principle of *convergence* (L. *convergere*, to come together, to merge) is illustrated when the receptive segment of a postsynaptic neuron is excited or inhibited by the axon terminals of many presynaptic neurons (Figure 11.16B).

FIGURE 11.16

Neuronal circuits. Arrows indicate direction in which nerve impulses are propagated. (A) Principle of divergence. One presynaptic neuron branches and synapses with several postsynaptic neurons. (B) Principle of convergence. Several presynaptic neurons synapse with one postsynaptic neuron. (C) Simple feedback circuit, in which the axon collateral branch of neuron A synapses with interneuron B, which, in turn, synapses with neuron A. (D) Parallel circuits, in which neuron A synapses in neuron pools B-C and D-E, with each, in turn, projecting as parallel circuits. (E) Two-neuron sequence of afferent neuron synapsing with efferent neuron, found in a two-neuron reflex arc. (F) Three-neuron sequence of afferent neuron, interneuron, and efferent neuron, found in a three-neuron reflex arc.

As many as several thousand presynaptic neurons may converge (spatial summation) on one postsynaptic neuron. If the excitatory influences on the receptive segment are sufficient, an action potential can be generated.

Feedback Circuit

A *feedback circuit* is a mechanism for returning some of the output of a neuron (or neurons) for the purpose of modifying the output of a prior neuron (or neurons). Feedback in the nervous system is called *negative feedback*, because the feedback modulates an effect by *inhibiting* the prior output. An example of negative feedback occurs in the regulation of body temperature. The negative feedback illustrated in Figure 11.16C occurs when the lower motor neuron, which innervates a voluntary muscle, is excited. Its collateral axonal branch stimulates an interneuron through excitatory synapses. This interneuron then inhibits the lower motor neuron, readying it for restimulation by its presynaptic neurons.

Parallel Circuits

As a result of the neural processing of convergence and divergence, information is often conveyed by relay neurons to other neural levels in *parallel circuits* (Figure 11.16D). Through these circuits, different forms of neural information can be relayed, and ultimately recombined at the same time, at other levels. For example, the sensory pathways for pain ascend to the brain from the spinal cord via two parallel systems: the spinothalamic pathway and the spinoreticulothalamic pathway. Much of the information conveyed via

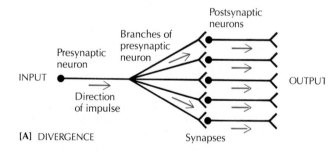

[A] DIVERGENCE

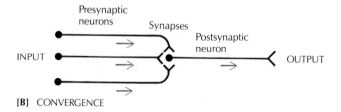

[B] CONVERGENCE

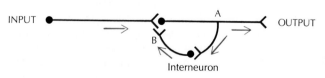

[C] FEEDBACK CIRCUIT

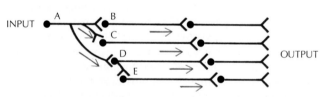

[D] PARALLEL CIRCUITS

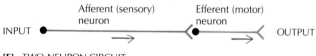

[E] TWO-NEURON CIRCUIT

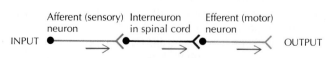

[F] THREE-NEURON CIRCUIT

these pathways is recombined for processing at higher cerebral levels. In another example, the several motor tracts from the brain are parallel projections that convey influences to the local spinal circuits and the lower motor neurons.*

Two-Neuron Circuit

The simplest neuronal circuit is the **two-neuron** (monosynaptic) **circuit**. It consists of an afferent (sensory) neuron, one *(mono-)* synapse, and an efferent (motor) neuron (Figure 11.16E). A familiar two-neuron sequence is the "knee-jerk" (patellar) extension reflex, where a tap on the patellar tendon of the flexed knee produces an extension of the knee.

Three-Neuron Circuit

A **three-neuron circuit** is a sequence of an afferent (sensory) neuron, an interneuron, and an efferent (motor) neuron (Fig-

*Lower motor neurons have their cell bodies in the central nervous system, with their axons extending into the peripheral nervous system in order to synapse through motor end plates on voluntary muscle cells.

ure 11.16F). Such a chain is used in flexor reflexes, such as the flexion of the elbow when you touch a hot stove and pull your hand away. The circuit involved in the elbow flexion consists of (1) sensory neurons from the pain receptors in the hand to the spinal cord, (2) interneurons, located entirely within the spinal cord, which connect the sensory neurons with (3) the lower motor neurons to the biceps muscle, which is stimulated to contract. Synapses are located between sensory neurons and the interneurons, and between the interneurons and the lower motor neurons. This type of circuit, where each neuron in the sequence is connected to a different neuron, is known as an *open circuit*.

Ask Yourself

1 *How do diverging and converging circuits differ?*

2 *How do parallel circuits operate?*

3 *Besides the number of neurons involved, what is the difference between a two-neuron circuit and a three-neuron circuit?*

PHYSIOLOGICAL AND ANATOMICAL ABNORMALITIES

Multiple sclerosis

Multiple sclerosis (MS) is a progressive demyelination of neurons that interferes with the conduction of nerve impulses and results in impaired sensory perceptions and motor coordination. Because almost any myelinated site in the brain and spinal cord may be involved, the symptoms of the disease may be diverse. With repeated attacks of inflammation at myelinated sites, scarring (sclerosis) takes place and some permanent loss of function occurs. The disease usually affects young adults between 18 and 40, and is five times more prevalent in whites than in blacks.

Although MS is very disabling, it progresses slowly, and most patients lead productive lives, especially during the recurring periods of remission. Among the typical symptoms are problems with vision, muscle weakness and spasms, urinary infections and bladder incontinence, and drastic mood changes. The specific cause of MS is not known, but several possibilities are being investigated, including a slow-acting viral infection, an allergic response of the nervous system to its own tissue (an au-

toimmune response), and an allergic reaction to an infectious agent. The onset of the disease is often preceded by emotional stress, fatigue, and acute respiratory infections.

Amyotrophic lateral sclerosis

Amyotrophic lateral sclerosis (ALS; commonly known as Lou Gehrig's disease) is a common motor neuron disease. Upper motor neurons in the brain are affected, as are lower motor neurons in the spinal cord. The disease causes muscles to atrophy.

In about 10 percent of the cases, ALS is inherited, affecting men and women almost equally. In the noninherited forms of the disease, it is most prevalent in individuals whose occupations require strenuous physical activity. (Lou Gehrig was a professional baseball player with the New York Yankees.) Amyotrophic lateral sclerosis generally affects men four times more often than women, and is more common among whites than blacks. Possible causes include an enzyme deficiency that produces malnutrition of motor neurons, a deficiency of vitamin E that causes dam-

age to plasma membranes, a metabolic interference with the production of nucleic acids by nerve fibers, and autoimmune disorders. The onset of the disease may be triggered by viral infection, trauma, and physical exhaustion.

Victims of ALS generally have weakened and atrophied muscles, especially in the hands and forearms. They may also exhibit impaired speech and difficulty in breathing, chewing, and swallowing, resulting in choking or excessive drooling. No effective treatment or cure exists, and the disease is invariably fatal.

Peripheral neuritis

Peripheral neuritis is a progressive degeneration of the axons and myelin sheaths of peripheral nerves, especially those that supply the distal ends of muscles of the limbs. It results in muscle atrophy and weakness, the loss of tendon reflexes, and some sensory loss. The disease is associated with infectious diseases such as syphilis and pneumonia, metabolic and inflammatory disorders such as gout and diabetes mellitus, nutritional deficiencies, and chronic intoxication, including lead and arsenic poisoning and chronic exposure to benzene, sulfonamides, and other chemicals. The prognosis is usually good if the underlying cause can be identified and removed.

Myasthenia gravis

Myasthenia gravis (*myo,* muscle + Gr. *asthenia,* weak) is an autoimmune disease caused by antibodies (molecules that defend against a foreign substance) directed against acetylcholine receptors. The antibodies reduce the number of functional receptors or impede the interaction between acetylcholine and its receptors. As a result, voluntary muscles become chronically weak.

Parkinson's disease

Parkinson's disease (Parkinsonism or "shaky palsy") is a motor disability characterized by symptoms such as stiff posture, tremors, and reduced spontaneity of facial expressions. It results from a deficiency of the neurotransmitter dopamine in certain brain neurons involved with motor activity. Parkinson's disease is discussed in greater detail in Chapter 13.

Huntington's chorea

Huntington's chorea is a fatal hereditary brain disease, which has been associated with insufficient amounts of GABA. In most cases, the onset of the disease occurs in the fourth and fifth decades of life, usually after the victim has married and had children, and hence too late to avoid passing the affliction on to children. Huntington's chorea is also discussed in Chapter 29.

Therapeutic use of drugs for nervous system disorders

The therapeutic use of drugs to treat various diseases of the nervous system is based on the effects of the drugs at synapses. Some drugs are effective because their actions are similar to those of the natural neurotransmitters. Drugs that mimic acetylcholine are called *cholinomimetic* or *parasympathomimetic* ("have parasympathetic effects") drugs. Those drugs that mimic norepinephrine and amphetamines are *sympathomimetic.* Epinephrine and amphetamines are sympathomimetic drugs. Because they act to constrict blood vessels and inhibit mucus secretion, they are effective as nasal decongestants.

Other drugs enhance the effects of natural neurotransmitters by inhibiting the enzyme that deactivates any excess neurotransmitter substance. For example, the drug *physostigmine* inhibits the enzyme acetylcholinesterase, which normally deactivates acetylcholine in the synaptic cleft. This inhibition results in a build-up of acetylcholine. Physostigmine and related drugs have been used therapeutically along with immunosuppressive drugs to treat myasthenia gravis. Physostigmine helps to overcome the problems of myasthenia gravis by inhibiting acetylcholinesterase, thereby increasing the usable amount of acetylcholine.

Some drugs, known as **blocking agents,** act by occupying some receptor sites on the postsynaptic membrane, and thus preventing (blocking) the normal neurotransmitter from exerting its full influence. *Curare,* a poison used on arrow tips by certain South American Indians to kill animals, is such a drug. A derivative called d-tubocurarine is used in small doses therapeutically as a muscle relaxant. Curare acts by blocking the acetylcholine receptors at the motor end plate. The subsequent paralysis of the respiratory muscles causes death.

Ophthalmologists use *homatropine* to dilate the pupils of the eyes during examinations. It acts by blocking the effects of acetylcholine at the synapses with the constrictor muscle of the pupil. As a result, the pupillary dilator acts without interference, and the pupil is dilated.

Two types of adrenergic receptors are called *alpha receptors* and *beta receptors.* *Chlorpromazine,* an alpha receptor blocker, is used as a tranquilizer. *Propranolol,* a beta receptor blocker, is used in the treatment of certain cardiovascular disorders.

SUMMARY

The body uses a combination of electrical impulses and chemical messengers to react and adjust to stimuli in order to maintain homeostasis.

Organization of the nervous system

1 The nervous system is usually divided on a gross anatomical basis into the **central nervous system** (the brain and spinal cord) and the **peripheral nervous system** (nerve structures located outside the central nervous system).

2 The peripheral nervous system contains **afferent** (sensory) nerve cells and **efferent** (motor) nerve cells.

3 The peripheral nervous system may be divided on a functional basis into the **somatic nervous system** and the **visceral nervous system,** each composed of afferent and efferent neurons and their fibers. The *somatic afferent division* conveys sensory information from the skin, muscles, joints, ears, eyes, and associated structures; the *visceral afferent division* conveys sensory information from the viscera, nose (smell), and taste buds to the central nervous system. The *somatic efferent division* conveys influences from the central nervous system for the regulation of the voluntary muscles. The *visceral efferent division,* also known as the *autonomic nervous system,* conveys influences from the central nervous system involved in the regulation of cardiac muscle, smooth (involuntary) muscle, and glands. The *special visceral efferent division* conveys influences from the brain to voluntary muscles of the jaw, facial expression, pharynx, and larynx. The autonomic nervous system is subdivided into the *sympathetic* and *parasympathetic* nervous systems, which complement each other.

Neurons: functional units of the nervous system

1 *Neurons* are nerve cells specialized to transmit impulses throughout the body. Their basic properties are **excitability** and **conductivity.**

2 A neuron is composed of a **cell body,** branching **dendrites,** and an **axon.** Dendrites conduct nerve potentials toward the cell body, and the axon usually carries impulses away from the cell body.

Axons may be coated with a sheath of *myelin,* which enhances the speed of conduction.

3 *Afferent* (sensory) *neurons* and *fibers* of the peripheral nervous system carry information from sensory receptor cells to the central nervous system. *Efferent* (motor) *neurons* and *fibers* of the peripheral nervous system carry neural information away from the central nervous system to muscles and glands. *Interneurons* are connecting neurons that carry impulses from sensory neurons to motor neurons.

4 Neurons and their peripheral fibers may be classified into *functional components* as *somatic, visceral, general, special, afferent,* and *efferent.*

5 Neurons may be classified according to their *structure* as *multipolar, bipolar,* and *unipolar.*

6 Most neurons are composed of specific functional segments known as *receptive, initial, conductive,* and *transmissive.*

Associated cells of the nervous system

1 *Associated cells* are specialized cells in the central and peripheral nervous systems other than neurons. They are involved in nourishing, protecting, and supporting neurons.

2 *Neuroglia* are nonconducting cells that protect, nourish, support, and help to maintain homeostasis within the central nervous system. The types of neuroglia are astrocytes, oligodendrocytes, microglia, and ependymal cells.

3 Associated cells of the peripheral nervous system provide various types of sheaths. They are *satellite cells* (capsule cells) and *neurilemma cells* (Schwann sheaths).

4 *Regeneration* of severed peripheral nerves is usually possible if the cell body is undamaged and the neurilemma is intact.

Physiology of neurons

1 *Nerve impulses* are conducted along a nerve fiber when specific changes occur in the electrical charges of the fiber membrane.

2 A resting cell that is not conducting an impulse is *polarized,* with the outside of its plasma membrane positively charged with respect to its negatively charged interior. The imbalance of electrical charges creates a potential for electrical activity *(resting membrane potential).*

3 The resting potential of a neuron is maintained by a *sodium-potassium pump* that regulates the concentration of sodium and potassium ions inside and outside the plasma membrane. The pumps are complemented by selective *ion channels.*

4 A *threshold stimulus* is the minimal stimulus required to instigate a nerve impulse.

5 *Depolarization* is the decrease of the cell membrane potential that can make the outside of the cell negative relative to the inside. When a stimulus is strong enough to cause sufficient depolarization, the nerve cell *fires.* Depolarization can produce a *nerve impulse,* also called an *action potential.*

6 When the original electrical charges are restored, the membrane is *repolarized.* The *refractory period* is the time during which a neuron will not respond to a second stimulus.

7 The phenomenon of a neuron firing at full power or not at all is the *all-or-none law.*

8 *Saltatory conduction* occurs when a nerve impulse jumps from node to node on a myelinated fiber.

9 A *synapse* is the junction between the axon terminal of one neuron and the dendrite, cell body, or specific parts of the axon of the next neuron. The cell carrying the impulse toward a synapse is a *presynaptic* cell; it initiates a response in the receptive segment of a *postsynaptic* neuron leading away from the synapse.

10 Synapses can be electrical or chemical. In a chemical synapse, two cells communicate by way of a chemical agent called a *neurotransmitter,* such as acetylcholine or norepinephrine. In a chemical synapse, the electrical activity of one cell spreads readily to the next cell.

11 Neurotransmitters interact with the receptor sites of the postsynaptic membrane, producing a *postsynaptic potential (PSP),* or graded potential. Postsynaptic potentials may be *excitatory (EPSP)* or *inhibitory (IPSP).* A *receptor potential* is the graded potential exhibited by the terminal segment of a nerve fiber located within a sensory receptor.

12 Neurotransmitters are classified in four groups: acetylcholine, amino acids, monoamines, and neuropeptides.

Neuronal circuits

1 Neurons are organized in networks arranged in synaptically connected sequences called *neuronal circuits* or neuronal chains.

2 The principle of *divergence* is shown when the transmissive segment of a neuron branches to have many synaptic connections with the receptive segments of many other neurons. When the axon terminals from many presynaptic neurons synapse with the receptive segment of only one postsynaptic neuron, it is an example of *convergence.*

3 Neuronal circuits may be *divergent, convergent, feedback circuits, parallel circuits, two-neuron circuits,* or *three-neuron circuits.*

Physiological and anatomical abnormalities

1 *Multiple sclerosis* is a progressive degeneration of some myelin sheaths in neurons of the brain and spinal cord. Other degenerative diseases include *amyotrophic lateral sclerosis* and *peripheral neuritis.*

2 Some of the diseases related to the improper release or utilization of neurotransmitters are *myasthenia gravis, Parkinson's disease,* and *Huntington's chorea.*

3 Drugs used for treatment of nervous system disorders are selected because they mimic the effects of, inhibit the deactivation of, or block the action of natural neurotransmitters.

UNDERSTANDING THE FACTS

1 Is it true or false that the somatic nervous system always excites skeletal muscles and never inhibits them?

2 Name the two types of effectors in the body.

3 What is another name for the sympathetic division of the autonomic system?

4 What is the function of chromatophilic substance?

5 What is the function of neurofibrils?

6 In what direction is a nerve impulse transmitted in the dendrite? The axon?

7 List the three factors that affect the speed of transmission of a nerve impulse along the nerve fiber.

8 Distinguish between a nerve and a tract.

9 Give the location of the following (be specific):
   a peripheral nervous system
   b chromatophilic substance
   c axolemma
   d neurofibral nodes
   e satellite cells
   f neuroglia

10 Distinguish between the three types of neurons by structure.

11 What are the main functions of acetylcholine and norepinephrine?

12 What are the basic functions of astrocytes?

13 What is saltatory conduction?

14 What are the two kinds of synapses between neurons?

15 Define IPSP, EPSP, and receptor potential.

UNDERSTANDING THE CONCEPTS

1 Why is it essential that the afferent neurons pass impulses *only* toward the CNS, and the efferent neurons pass impulses *only* away from the CNS?

2 How do the sympathetic and parasympathetic nervous systems complement each other?

3 In which direction would impulses be transmitted in a mixed nerve? Explain your answer.

4 What factors are essential to the regeneration of a severed peripheral nerve fiber? Explain how the process works.

5 Contrast the nerve impulse with an electric current.

6 Briefly explain the important series of events involved in the transmission of a nerve impulse.

7 Explain how we can perceive the difference in the intensity of stimuli.

8 How does the number of synapses in a neural circuit affect the total transmission time of a nerve impulse?

9 Explain why the synapse is so important in controlling the direction of a nerve impulse.

10 In what ways is the nervous system involved in multiple sclerosis?

SUGGESTED READING

BLOOM, FLOYD E., "Neuropeptides." *Scientific American,* October 1981.

CARPENTER, M. B., AND J. SUTIN, *Human Neuroanatomy,* 8th ed. Baltimore: Williams & Wilkins, 1983.

CAVE, L. J., "Brain's Unsung Cells." *Bioscience,* 33 (1983):614.

DUNANT, YVES, AND MAURICE ISRAËL, "The Release of Acetylcholine." *Scientific American,* April 1985.

FINE, ALAN, "Transplantation in the Central Nervous System." *Scientific American,* August 1986.

GOTTLEIB, DAVID I., "GABAergic Neurons." *Scientific American,* February 1988.

KEYNES, RICHARD D., "Ion Channels in the Nerve-Cell Membrane." *Scientific American,* March 1979.

LLINÁS, RODOLFO R., "Calcium in Synaptic Transmission," *Scientific American,* October 1982.

MORELL, PIERRE, AND WILLIAM T. NORTON, "Myelin." *Scientific American,* May 1980.

NOBACK, CHARLES R., AND ROBERT J. DEMAREST, *The Human Nervous System: Basic Principles of Neurobiology,* 3rd ed. New York: McGraw-Hill, 1981.

NOBACK, CHARLES R., AND ROBERT J. DEMAREST, *The Nervous System: Introduction and Review,* 3rd ed. New York: McGraw-Hill, 1986. (Paperback)

PATTERSON, PAUL H., DAVID D. POTTER, AND EDWIN J. FURSHPAN, "The Chemical Differentiation of Nerve Cells." *Scientific American,* July 1978.

SCHWARTZ, JAMES H., "The Transport of Substances in Nerve Cells." *Scientific American,* April 1980.

SHEPHERD, GORDON M., "Microcircuits in the Nervous System." *Scientific American,* February 1978.

STEINBERG, S., "Endorphins." *Science News,* 124 (1983):136.

STEVENS, CHARLES F., "The Neuron." *Scientific American,* September 1979.

THOMPSON, RICHARD F., *The Brain: An Introduction to Neuroscience.* New York: Freeman, 1985.

12
The Spinal Cord and Spinal Nerves

LEARNING OBJECTIVES

1 Describe the basic anatomy of the spinal cord and its protective coverings.

2 Explain the internal structure of the spinal cord, including the gray and white matter and nerve roots.

3 Describe the origin, termination, and functional roles of the major ascending and descending spinal tracts.

4 Describe how a processing center in the spinal cord functions.

5 Compare the pyramidal and extrapyramidal systems of the central nervous system.

6 Define reflex, and list the basic components of a simple reflex arc.

7 Explain how the stretch reflex works.

8 Describe the function of a gamma motor neuron reflex arc.

9 Describe the components and function of withdrawal reflexes.

10 Explain how reflexes may be used in diagnosing disorders of the nervous system.

11 Explain how spinal nerves are named.

12 Describe the structure of the spinal nerves, their roots, sheaths, branches, and plexuses.

13 Identify the location and function of the major spinal nerves in the cervical, brachial, lumbar, and sacral plexuses.

14 Describe the role of intercostal nerves.

15 Describe the arrangement and function of dermatomes.

16 Compare the effects of paraplegia, quadriplegia, and hemiplegia on bodily functions.

17 Explain how poliomyelitis, sciatica, shingles, and spinal meningitis affect the nervous system.

As you learned in Chapter 11, the nervous system is divided on a gross anatomical basis into the central and peripheral nervous systems. However, to give a clearer understanding of the relationships between the structure and function of the nervous system, we will focus here on the spinal cord and the spinal nerves, and in Chapter 13 on the brain and the cranial nerves.

The spinal cord, with its 31 pairs of spinal nerves, serves two important functions: (1) It is the connecting link between the brain and most of the body, and (2) it is involved in spinal reflex actions, both somatic and visceral. Somatic spinal reflexes involve a series of responses to receptors in the skin and muscles. These reflexes help us to move and maintain a correct posture. Visceral spinal reflexes occur in certain organs, for example, when the urinary bladder becomes distended and evokes the urge to urinate. Spinal reflexes also help to regulate blood pressure by affecting the smooth muscle in blood vessels, and also influence the action of glands. Thus, both voluntary and involuntary movement of the limbs, as well as certain visceral processes, depend on the spinal cord, a vital link between the brain and body.

BASIC ANATOMY OF THE SPINAL CORD

The *spinal cord* is the part of the central nervous system that extends from the foramen magnum of the skull downward (caudally) for about 45 cm (18 in.) to the level of the first lumbar vertebra (L1) in adults. Its upper end is continuous with the lowermost part of the brain (the medulla). Its lower end tapers off as the cone-shaped *conus terminalis,* also called the *conus medullaris,* located in the vicinity of the first lumbar vertebra (Figure 12.1). Extending caudally from the conus is a nonneural fiber called the *filum terminale,* which attaches to the coccyx. The filum consists mainly of fibrous connective tissue.

Surrounding and protecting the spinal cord is the bony vertebral column. The central opening, or foramen, in the vertebral column has a diameter about the size of your index finger (about 1 cm). Inside the column, the cylindrical cord is about as thick as a pencil. It is slightly flattened dorsally and ventrally, with two prominent enlargements known as the *cervical* and *lumbosacral* (or lumbar) *enlargements* (see Figure 12.1). Emerging from these enlargements are the spinal nerves that innervate the upper and lower limbs.

Of the 31 pairs of spinal nerves and roots, there are 8 cervical (C) nerve pairs, 12 thoracic (T), 5 lumbar (L), 5 sacral (S), and 1 coccygeal (Co). Each pair of spinal nerves typically passes through a pair of intervertebral foramina, located between two successive vertebrae, and then is distributed to a specific pair of segments of the body (Figure 12.2). Note that there is a pair of spinal nerves for each vertebra.

The roots of all nerves passing caudally below the conus terminalis (below L1 vertebral level) resemble flowing, coarse strands of hair. For this reason, the lumbar and sacral roots are collectively called the *cauda equina* (KAW-duh ee-KWY-nuh), which means "horse's tail" in Latin.

The spinal cord and the roots of its nerves are protected not only by the flexible bony vertebral column and its ligaments, but also by the spinal meninges and cerebrospinal fluid.

Spinal Meninges

The spinal cord is surrounded by three layers of protective membranes called *meninges* (muh-NIHN-jeez; Gr. plural of *meninx,* membrane) (Figure 12.3). They are continuous with the same layers that cover the brain. The outer layer, called the *dura mater* (L. hard mother), is a tough, fibrous membrane that merges with the filum terminale. The middle layer, the *arachnoid* (Gr. cobweblike), runs caudally to the S2 vertebral level, where it joins the filum terminale. The arachnoid is so named because it is delicate and transparent, and has some connective strands running across the space that separates it from the innermost layer, the *pia mater* (L. tender mother). The thin, highly vascular pia mater is tightly attached to the spinal cord and its roots. It contains blood vessels that nourish the spinal cord. A pair of fibrous bands of pia mater that extend from each side of the entire length of the spinal cord is the *denticulate ligament* (see Figure 12.3). This ligament is attached to the dura mater at regular intervals.

Between the dura mater and the periosteum of the vertebrae is the *epidural space* (see Figure 12.3), containing many blood vessels and some fat. Anesthetics can be injected into the epidural space below the S2 vertebral level to dull pain. This procedure, known as a saddle block, permits patients to be conscious during operations on the pelvic region, and also during painless childbirth.

Between the dura mater and the arachnoid is the *subdural space,* which is merely a slit, and contains no cerebrospinal fluid. The *subarachnoid space,* which separates the arachnoid and the pia mater, contains cerebrospinal fluid, blood vessels, and spinal roots.

Cerebrospinal Fluid

Cerebrospinal fluid (CSF) is essentially a clear, watery ultrafiltrate solution formed from blood in the capillaries of the brain.* The basic mechanism involves both an active transport system and passive diffusion. The fluid passes through small openings from the fourth ventricle into the subarachnoid spaces around the brain and spinal cord. Under normal conditions, it returns to the blood at about the same rate that it is formed (0.3 to 0.4 mL/min, or about 300 mL/day).

The cerebrospinal fluid provides a cushion that, together with the bony vertebral column, protects the delicate tissues of the spinal cord. It is also involved in the exchange of nutrients and wastes between the blood and the neurons of the brain and spinal cord, which helps in the maintenance of homeostasis.

The fluid is generally similar to blood plasma fluid. It consists of water; a small amount of protein; oxygen and carbon dioxide in solution; sodium, potassium, calcium, magnesium, and chloride ions; glucose; a few white blood cells; and many other organic materials. The subarachnoid space contains about 75 mL (60 percent) of the total amount (about 125 mL) of the cerebrospinal fluid in the central nervous system.

---

*Because cerebrospinal fluid is formed in the brain, and is a crucial part of the brain's physiology, it will also be discussed in the next chapter, which discusses the brain in detail.

FIGURE 12.1

The spinal cord and spinal nerves and their relation to the vertebral column, posterior view.

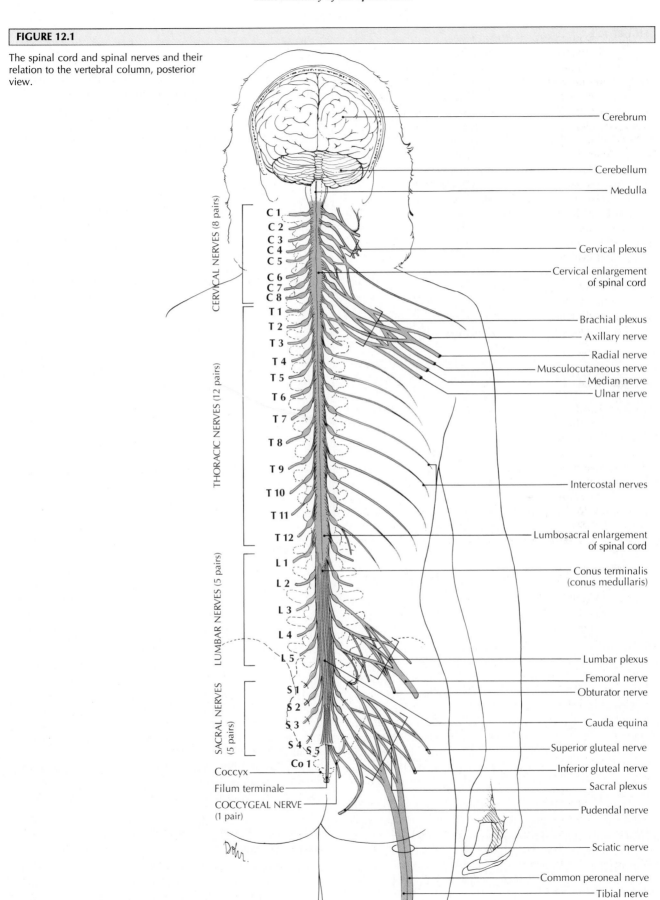

CERVICAL NERVES (8 pairs)

THORACIC NERVES (12 pairs)

LUMBAR NERVES (5 pairs)

SACRAL NERVES (5 pairs)

C 1
C 2
C 3
C 4
C 5
C 6
C 7
C 8
T 1
T 2
T 3
T 4
T 5
T 6
T 7
T 8
T 9
T 10
T 11
T 12
L 1
L 2
L 3
L 4
L 5
S 1
S 2
S 3
S 4 S 5
Co 1

Coccyx
Filum terminale
COCCYGEAL NERVE
(1 pair)

Cerebrum
Cerebellum
Medulla
Cervical plexus
Cervical enlargement
of spinal cord
Brachial plexus
Axillary nerve
Radial nerve
Musculocutaneous nerve
Median nerve
Ulnar nerve
Intercostal nerves
Lumbosacral enlargement
of spinal cord
Conus terminalis
(conus medullaris)
Lumbar plexus
Femoral nerve
Obturator nerve
Cauda equina
Superior gluteal nerve
Inferior gluteal nerve
Sacral plexus
Pudendal nerve
Sciatic nerve
Common peroneal nerve
Tibial nerve

FIGURE 12.2

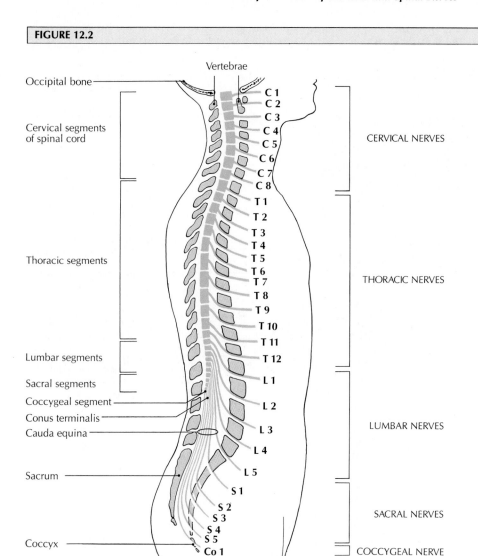

Right lateral view of the spinal cord and vertebral column, showing the various spinal segments and the emergence of spinal nerves from the vertebral column.

Cerebrospinal fluid has many important clinical uses. In certain neurological disorders, alterations can occur in its cellular and chemical content, as well as its pressure. The fluid can be removed for analysis by inserting a needle into the subarachnoid space between L3 and L4, with the patient usually lying curled up on one side (Figure 12.4). Such a *lumbar puncture*, or "spinal tap," is performed on this lower lumbar region to avoid injury to the spinal cord, which ends between L1 and L2. A *cisternal puncture* can be performed by inserting a needle between the occipital bone and atlas (C1), entering the cisterna cerebellomedullaris, and withdrawing spinal fluid (see Figure 12.4).

A marked increase in white blood cells in the cerebrospinal fluid occurs in acute bacterial meningitis, and a moderate increase may indicate the presence of a viral infection or cerebral tissue damage. The protein (gamma globulin) content is increased in multiple sclerosis, and glucose levels are reduced during active bacterial infections. The presence of red

blood cells indicates that blood has entered the subarachnoid space. These changes can be determined from an analysis of cerebrospinal fluid obtained from a lumbar puncture. An increase in intracranial pressure may be the result of a brain tumor that obstructs the normal circulation of cerebrospinal fluid, causing it to back up. A lumbar puncture is never attempted when a brain tumor is suspected because if the tumor blocks the subarachnoid space, the removal of cerebrospinal fluid may cause a drop in fluid pressure below the tumor. The resulting higher pressure above the block can force the medulla of the brain into the foramen magnum, compressing the medulla and killing the patient.

Internal Structure

If you were to cut the spinal cord in cross section, you would see a tiny central canal, which contains cerebrospinal fluid,

FIGURE 12.3

The spinal meninges. (A) A drawing of the spinal meninges. (B) The top layers are peeled away to show the triple-layered meninges and the spaces between them.

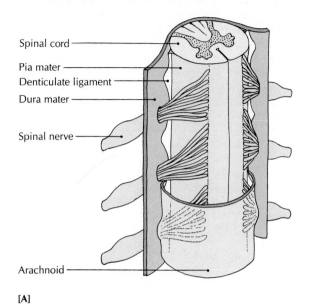

[A]

Spinal cord
Pia mater
Denticulate ligament
Dura mater
Spinal nerve
Arachnoid

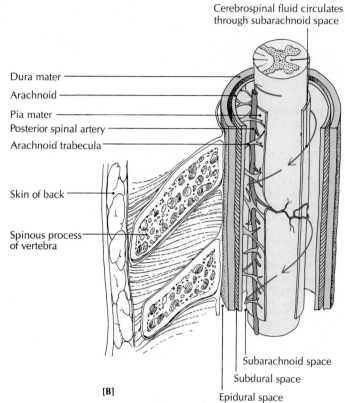

Cerebrospinal fluid circulates through subarachnoid space

Dura mater
Arachnoid
Pia mater
Posterior spinal artery
Arachnoid trabecula
Skin of back
Spinous process of vertebra

Subarachnoid space
Subdural space
Epidural space

[B]

FIGURE 12.4

Lumbar puncture, or spinal tap. The spine is in a flexed position, which separates the vertebral spinous processes, and allows the (A) lumbar puncture needle to enter the subarachnoid space well below the termination of the spinal cord (B).

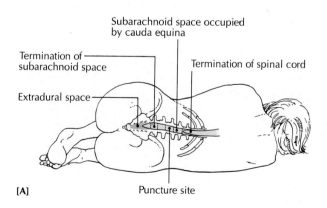

Subarachnoid space occupied by cauda equina
Termination of subarachnoid space
Termination of spinal cord
Extradural space
Puncture site

[A]

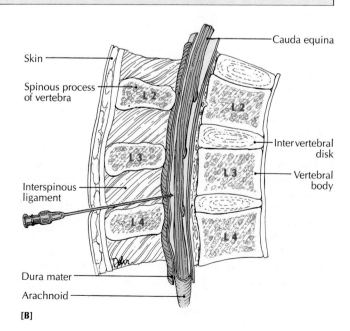

Skin
Spinous process of vertebra
Interspinous ligament
Dura mater
Arachnoid
Cauda equina
Intervertebral disk
Vertebral body
L2
L3
L4

[B]

and a dark portion of H-shaped or butterfly-shaped "gray matter" surrounded by a larger area of "white matter" (Figure 12.5). The spinal cord is divided into more-or-less symmetrical left and right halves by a deep groove, called the *anterior median fissure*, and a median septum, called the *posterior median septum*. Extending out from the spinal cord are the ventral and dorsal roots of the spinal nerves. (The spinal cord and nerves cut vertically are shown in Figure 12.6; the roots are shown in Figure 12.7.)

Gray matter The *gray matter* of the spinal cord consists of nerve cell bodies and dendrites of association and efferent neurons, unmyelinated axons of spinal neurons, sensory and motor neurons, and axon terminals of neurons. The gray matter forms an H shape, and is composed of three pairs of columns of neurons running up and down the length of the spinal cord from the upper cervical level to the sacral level (see Figure 12.5A). The pairs of columns that form the two vertical bars of the H are called *horns*. The two that run dorsally are the **posterior horns,** which function in afferent input, and the two that run ventrally are the **anterior horns,** which function in efferent somatic output. The pair that forms the cross bar of the H is known as the **gray commissure** (L. "joining together"), which functions in cross reflexes.

FIGURE 12.5

Internal structure of the spinal cord.
(A) Cross section of the spinal cord, with the bones that protect it, and enlarged view of cord and meninges. Spinal nerves connect with the spinal cord in *pairs* at the ventral and dorsal roots. (B) Cross section of the cord, showing some prominent internal features, including columns, or funiculi, of myelinated nerve fibers. The insets show the composition of white matter (right) and gray matter (left).

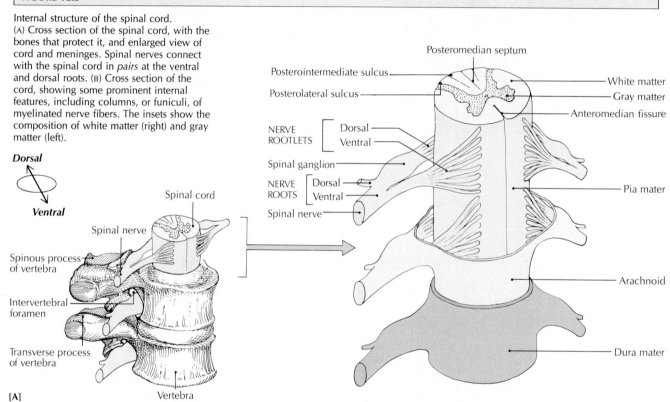

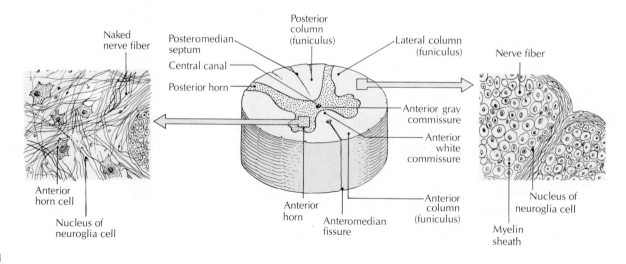

Photographs of (A) the spinal cord and (B) spinal nerves sectioned vertically, posterior view. [(A, B) *N. Gluhbegovic and T. H. Williams,* The Human Brain: A Photographic Guide. *New York, Harper & Row, 1980.*]

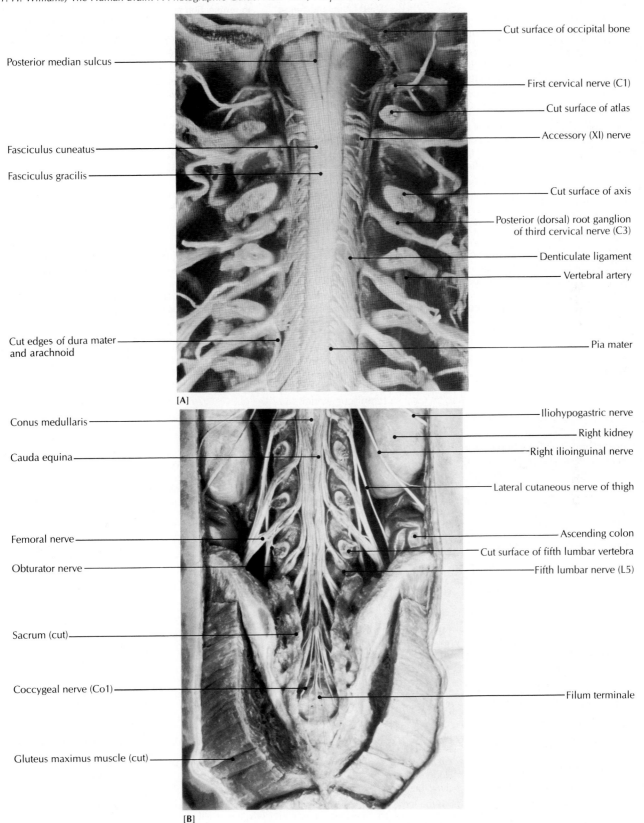

Posterior median sulcus

Fasciculus cuneatus

Fasciculus gracilis

Cut edges of dura mater and arachnoid

[A]

Cut surface of occipital bone

First cervical nerve (C1)

Cut surface of atlas

Accessory (XI) nerve

Cut surface of axis

Posterior (dorsal) root ganglion of third cervical nerve (C3)

Denticulate ligament

Vertebral artery

Pia mater

Conus medullaris

Cauda equina

Femoral nerve

Obturator nerve

Sacrum (cut)

Coccygeal nerve (Co1)

Gluteus maximus muscle (cut)

[B]

Iliohypogastric nerve

Right kidney

Right ilioinguinal nerve

Lateral cutaneous nerve of thigh

Ascending colon

Cut surface of fifth lumbar vertebra

Fifth lumbar nerve (L5)

Filum terminale

TABLE 12.1   MAJOR ASCENDING (SENSORY) TRACTS FROM SPINAL CORD TO BRAIN

Tract and location in spinal cord	Origin	Description and course to thalamus of brain	Course from thalamus to termination in cortex of brain	Sensations conveyed
Lateral spinothalamic, in anterior half of lateral column.*	Posterior horn of spinal cord gray matter. (Location of cell bodies of second-order neurons.)	Fibers of second-order neurons cross to the opposite side at each spinal cord level, ascend and terminate in thalamus.	Axonal fibers of third-order neurons with cell bodies in thalamus terminate in postcentral gyrus of cerebral cortex.	Pain and temperature.
Spinoreticulothalamic, in lateral and anterior columns.*	Posterior horn of spinal cord gray matter. (Location of cell bodies of second-order neurons.)	Crosses to opposite side in spinal cord, ascends to and terminates in reticular formation of brainstem. After a sequence of several neurons, fibers terminate in thalamus.	Fibers of neurons of cell bodies in thalamus continue to postcentral gyrus of cerebral cortex.	Pain.
Fasciculus gracilis and fasciculus cuneatus, in posterior column.	First-order neurons have cell bodies in dorsal root ganglia, and ascend on same side of cord.	Axons of first-order neurons terminate in nuclei gracilis and cuneatus in medulla of same side. Axonal fibers of second-order neurons with cell bodies in nuclei gracilis and cuneatus cross to opposite side to form medial lemniscus, which ascends to, and terminates in, thalamus.	Axonal fibers of third-order neurons with cell bodies in thalamus terminate in postcentral gyrus of cerebral cortex.	Touch pressure, two-point discrimination, vibratory sense, position sense (proprioception).†
Anterior spinothalamic, in anterior column.*	Posterior horn of spinal cord gray matter. (Location of cell bodies of second-order neurons.)	Crosses to opposite side at each spinal cord level. Joins medial lemniscus of brainstem, which terminates in thalamus.	Axonal fibers of third-order neurons with cell bodies in thalamus terminate in postcentral gyrus of cerebral cortex.	Light touch.‡

Tract and location in cerebellum of brain	Origin	Description and course to cerebellum	Termination	Modality conveyed
Posterior (dorsal) spinocerebellar, in posterior half of lateral column.	Posterior horn of spinal cord gray matter.	Uncrossed tract. Enters cerebellum via inferior cerebellar peduncle.	Cerebellum.	Unconscious proprioception.§
Anterior (ventral) spinocerebellar, in anterior half of lateral column.	Posterior horn of spinal cord gray matter.	Some fibers cross to opposite side in spinal cord. Enters cerebellum via superior cerebellar peduncle.	Cerebellum.	Unconscious proprioception.

*These three tracts are collectively called the *anterolateral system.*

†*Touch-pressure* (deep touch) sensation is obtained by deforming the skin with pressure. The sensation of discriminating between one and two points applied to the skin is called *two-point discrimination;* it is acute on the ball of a finger, but poor on the back of the body. *Vibratory sense* is the sensation of feeling vibrations when the stem of a vibrating tuning fork is placed on the bones of a joint. *Position sense* enables us to know where the various body parts are with our eyes closed.

‡*Light touch* is the sensation obtained by touching, but not deforming, the skin or by moving a hair.

§*Unconscious proprioception* (*proprio* = self) is the sensory information from the body, which does not result in any sensation. Such information about movement and limb position is continuously streaming in from the body, and is utilized by the nervous system in the coordination of voluntary muscles during movements.

TABLE 12.2 MAJOR DESCENDING (MOTOR) TRACTS FROM BRAIN TO SPINAL CORD

Tract and location in spinal cord	Origin	Description	Termination	Motor impulse conveyed
PYRAMIDAL SYSTEM				
Lateral (pyramidal) corticospinal, in lateral column.	Cerebral cortex.	Crosses to opposite side between medulla and spinal cord.	Gray matter, but primarily anterior horn.	Manipulative movements of extremities, especially delicate finger movements. Excites flexor muscles, inhibits extensor muscles.
Anterior (pyramidal) corticospinal.	Cerebral cortex.	Mainly uncrossed tracts; crosses near termination.	Spinal cord, where fibers enter anterior gray columns.	Skilled voluntary movements of extremities.
EXTRAPYRAMIDAL SYSTEM				
Rubrospinal, in lateral column.	Nucleus ruber of midbrain.	Crosses in midbrain to opposite side in midbrain.	Gray matter, but primarily anterior horn.	Manipulative movements of extremities. Excites flexor muscles, inhibits extensor muscles. Involved in posture.
Medullary and pontine reticulospinal, in anterior column and anterior half of lateral column.	Reticular formation of medulla and pons.	Mainly uncrossed tracts.	Anterior horn.	Maintenance of erect posture. Integrate movements of body and limbs. Pontine reticulospinals excite extensor muscles, inhibit flexor muscles. Medullary reticulospinals excite flexor muscles, inhibit extensor muscles.
Vestibulospinal, in anterior column.	Vestibular nuclei of medulla.	Uncrossed tract.	Anterior horn.	Maintenance of erect posture. Integrates movements of body and limbs. Excites extensor muscles, inhibits flexor muscles.

DESCENDING (MOTOR) TRACTS OF THE SPINAL CORD   ASCENDING (SENSORY) TRACTS OF THE SPINAL CORD

White matter The *white matter* of the spinal cord gets its name because it is composed mainly of myelinated nerve fibers, and myelin has a whitish color. The white matter is divided into three pairs of columns, or *funiculi* (fyoo-NICK-yoo-lie; L. little ropes), of myelinated fibers that run the entire length of the cord. The funiculi consist of the anterior (ventral) column, the posterior (dorsal) column, the lateral column, and a commissure area (see Figure 12.5B).

The bundles of fibers within each funiculus are subdivided into *tracts* called *fasciculi* (fah-SICK-yoo-lie; L. little bundles).

Ascending tracts are made up of sensory fibers that carry impulses *up* the spinal cord to the brain (Table 12.1); *descending tracts* of motor fibers transmit impulses from the brain *down* the spinal cord to the efferent neurons. The longer tracts carry nerve impulses up to the brain, or down through the cord to neurons that innervate the muscles or glands. The shorter tracts convey impulses from one level of the cord to another. The major ascending and descending tracts are described in Tables 12.1 and 12.2. Their functional roles are discussed on page 344.

FIGURE 12.7

Dorsal and ventral roots of a typical spinal nerve.

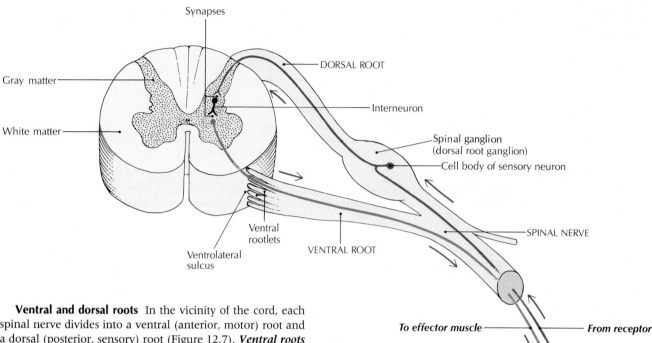

Ventral and dorsal roots In the vicinity of the cord, each spinal nerve divides into a ventral (anterior, motor) root and a dorsal (posterior, sensory) root (Figure 12.7). *Ventral roots* contain efferent nerve fibers* and convey motor information. *Dorsal roots* contain afferent nerve fibers, which enter the cord with sensory information. The ventral roots emerge from the spinal cord in a groove called the *ventrolateral sulcus*, and the dorsal roots enter the cord in the *dorsolateral sulcus* (see Figure 12.5B). The axons of anterior horn cells make up the ventral roots and lie *within* the gray matter of the cord. In contrast, groups of cell bodies whose axons make up the dorsal roots lie *outside* the cord, and are called **dorsal root ganglia,** or *spinal ganglia.*

FUNCTIONAL ROLES OF PATHWAYS OF THE CENTRAL NERVOUS SYSTEM

Each pathway of the central nervous system is basically composed of organized sequences of neurons. Some neurons have long axons (fibers) that terminate in processing centers, where the processing of neural information takes place. These **processing centers** may be considered as the ''computers'' of the pathways. They are called by one of several names: nucleus, ganglion, gray matter of spinal cord, or cortex.

Processing Centers

Each processing center consists of (1) the terminal branches of axons entering the center, (2) cell bodies and dendrites of neurons whose axons form the tract, and (3) intrinsic neurons whose dendrites, cell bodies, and axons are located wholly within the center. The interactions among neurons within the processing centers cause the input information to

*Some fibers of ventral roots are now known to be afferent.

be altered *(processed)* in some way before it is conveyed to the next center. Many centers receive input from nerve cells originating in more than one location.

Some pathways contribute to the complex processing within the brain itself. These are neuronal circuits of interconnected sequences of a number of processing centers (some are called feedback circuits). They integrate and process information at the higher levels of brain functions, including behavioral activities, complex voluntary movements, and thinking.

Sensory Pathways

Some sensory pathways have sequences that are made up of three neurons. In some pathways, the neurons in the sequence are called first-, second-, and third-order neurons. A **first-order neuron** extends from the sensory receptor to the central nervous system; a **second-order neuron** extends from the spinal cord or brainstem to a nucleus in the thalamus; and a **third-order neuron** extends from the thalamus to a sensory area of the cerebral cortex.

A critical feature of many pathways is that the axons of a tract (for example, the second-order neuron of many sensory pathways) cross over *(decussate)* from one side of the spinal cord or brainstem to the other side (see Tables 12.1 and 12.2). Because of this crossing over, one side of the body communicates with the opposite side of the brain. By knowing where a pathway crosses over, a physician can use this information to help locate the site of an injury (lesion) in the central nervous system. For example, the touch-pressure pathway

Embryonic Development of the Spinal Cord

The embryonic development of the nervous system begins with the formation of the *notochord,* a cellular rod that defines the vertical axis of the embryo. As the notochord continues to develop, the embryonic ectoderm thickens to form the *neural plate.* On about the nineteenth day, the neural plate folds in on itself to form a *neural groove* with *neural folds* on each side (drawing A). By day 21, the neural folds have fused, forming the **neural tube** (drawing B). The cephalic end of the tube develops into the brain, and the caudal portion becomes the spinal cord. At the site of the closure forming the neural tube, there develops a paired segmented series of outgrowths called the **neural crest** (drawing B). It contains the cells that form the future neurons of the spinal ganglia of the autonomic nervous system (drawing D).

The spinal-cord portion of the neural tube differentiates into three layers (drawing C): (1) The inner *matrix (ependymal) layer,* which lines the central canal of the cord, contains the cells that develop into all neurons and macroglial cells of the spinal cord. All mitosis of these cell occurs within the matrix layer. Once a potential neuron or macroglial cell leaves the matrix layer it migrates to its final destination and does not divide. (2) The middle *mantle layer* develops into the butterfly-shaped gray matter, which contains the cell bodies of neurons, and synapses between neurons and macroglia. (3) The outer *marginal layer* develops into the white matter of the spinal cord, which is composed of macroglial cells and myelinated and unmyelinated axons. The axons of the white matter ascend to other spinal levels and to the brain, and descend from the brain and to other levels of the spinal cord.

As shown in cross-sectional drawing C, the thickened inner side walls of the neural tube form the *lateral plates,* and the thin dorsal and ventral walls form the *roof plate* and *floor plate,* respectively. These plates surround a longitudinal opening, or sulcus, called the *sulcus limitans.* The sulcus divides the lateral plates into two portions: (1) a dorsal afferent (sensory) *alar plate,* whose neurons are involved with reflexes that convey signals to the brain, and (2) a ventral efferent (motor) *basal plate,* whose neurons are involved with reflexes that convey signals from the brain.

At the outset, the development of the spinal cord and vertebrae proceed at an even pace, and the spinal cord extends throughout the entire length of the bony vertebral column. But about the third fetal month the vertebral development moves ahead, and the spinal cord becomes shorter than the vertebral column. The spinal cord of a newborn child ends at about the third lumbar vertebra, and extends to between the first and second lumbar vertebrae during adolescence and adulthood. As a result of the differing developmental rates of the spinal cord and vertebral column, the spinal nerves do not align with the intervertebral spaces they pass through, with several of the spinal nerves projecting downward.

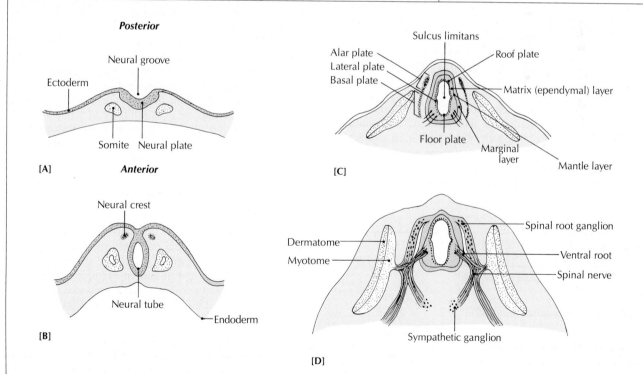

Development of the spinal cord, transverse sections. (A) 19 days after fertilization. (B) 21 days. (C) 26 days. (D) 30 days.

decussates in the medulla (lower brainstem). A lesion in this pathway on one side of the spinal cord results in the loss of sensation on the *same* side of the body as the lesion, whereas a lesion above the decussation in the brain results in the loss of sensation on the *opposite* side of the body.

Many tracts are named after their centers of origin and termination. For example, the lateral spinothalamic tract (a tract in a pain pathway) originates in the gray matter of the spinal cord *(spino-)*, terminates in the thalamus of the cerebrum *(thalamic)*, and is located laterally in the spinal cord *(lateral)*.

Pyramidal and Extrapyramidal Systems

The motor pathways and associated neural circuits are classified as pyramidal or extrapyramidal systems (see Table 12.2). Both systems convey influences from the motor area of the brain's cerebral cortex to the local circuits that regulate voluntary movements. These local circuits influence *lower motor neurons,* or motor neurons whose axons directly stimulate all the voluntary muscles.

The *pyramidal system* consists of corticobulbar fibers that influence the lower motor neurons of the cranial nerves from the brain. The corticospinal pyramidal tract influences the lower motor neurons of the spinal nerves of the spinal cord. It is called the pyramidal system because the fibers of the corticospinal tract run from their origin in the cortical motor area in the brain through an area called the pyramids of the medulla to the local circuits (see Figure 13.6).

The *extrapyramidal system* consists of many centers, complex circuits within the brain, and several descending pathways. This system includes all the pathways that influence and regulate the motor control of the lower motor neurons, except those of the pyramidal system. The motor pathways include the rubrospinal tract (from the nucleus ruber of the midbrain, also called the red nucleus) and the reticulospinal tracts (from the reticular nucleus of the lower brainstem). These tracts are influenced by the motor cortical areas, and hence the corticorubrospinal pathway, which is an indirect system with many centers. Malfunctioning of the extrapyramidal system leads to disorders characterized by abnormal movement, including Parkinson's disease and Huntington's chorea.

Ask Yourself

1 Where is the cauda equina located, and what is it?

2 What are the three spinal meninges called?

3 What are some functions of cerebrospinal fluid?

4 What does the gray matter of the spinal cord consist of?

5 How are the anterior and posterior horns of the spinal cord formed?

6 What are upper motor neurons?

7 What is the difference between funiculi and fasciculi?

8 What is the significance of the crossing over of spinal cord tracts?

SPINAL REFLEXES

In addition to linking the brain and most of the body, the spinal cord coordinates reflex action. A *reflex* (L. to bend back) is a predictable involuntary response to a stimulus, such as quickly pulling your hand away from a hot stove. A reflex involving the skeletal muscles is called a *somatic reflex.* A reflex involving responses of smooth muscle, cardiac muscle, or a gland is a *visceral* (autonomic) *reflex.* Visceral reflexes control the heart rate, respiratory rate, digestion, and many other body functions, as described in Chapter 14 (The Autonomic Nervous System). Both types of reflex action allow the body to respond quickly to internal and external changes in the environment in order to maintain homeostasis.

All *spinal reflexes,* that is, reflexes carried out by neurons in the spinal cord alone and not immediately involving the brain, are based on the sequence shown in Figure 12.8. Such a system is called a *reflex arc.* In a *monosynaptic* (one-synapse, two-neuron) *reflex arc,* the sensory and motor neurons synapse directly. More often, however, one or more *interneurons* (connecting neurons) synapse with the sensory and motor neurons in a *polysynaptic reflex arc.* An example of a simple polysynaptic reflex is pulling your foot away when you step on a tack (the stimulus).*

Most reflexes in the human body are more complex than the one shown in Figure 12.8. The more usual reflexes are actually *chains* of reflexes, with the possibility of several muscles being activated almost simultaneously.

Reflex actions save time because the "message" being transmitted by the impulse does not have to travel from the stimulus at a receptor all the way to the brain. Instead, most reflex actions never travel any further than the spinal cord, though some extend to the brainstem.

Stretch (Myotatic) Reflex

The *stretch (myotatic) reflex* is a two-neuron (monosynaptic) reflex arc. It acts to maintain our erect upright posture and stance by exciting the extensor muscles of the lower limbs, back, neck, and head. A well-known example of this reflex is the *knee jerk,* or *patellar reflex,* which is produced by tapping the patellar tendon of the relaxed quadriceps femoris muscle

---

*Actually, how "simple" is a so-called simple reflex? The puncture by the tack stimulates sensory cells, which stimulate other cells that carry the impulse that causes the foot to move away. Other impulses enable you to hop around on one foot without losing your balance and falling down. Additional impulses are involved when you yell "ouch." Still other impulses go beyond the spinal reflex arc and reach your brain, causing you to feel pain. The reflex response is "simple" only in comparison with other, more complicated responses.

of the thigh (Figure 12.9). Such a reflex is described as *ipsilateral* (*ipsi*, same + *lateral*, side) because the response occurs on the same side of the body and spinal cord where the stimulus is received.

In the knee jerk, a tap on the patellar tendon suddenly stretches the quadriceps muscle and some of the receptors within it (Figure 12.10). These receptors, called *neuromuscular spindles*, respond to changes in the length of a muscle. The stretched spindles excite sensitive nerve endings (stretch-gated receptors) of afferent fibers within the spindles, generating nerve impulses. The afferent fibers convey these nerve impulses to the L2 or L3 vertebral level of the spinal cord. Here the afferent neurons synapse with lower motor neurons called *alpha motor neurons*. (Alpha motor neurons are among the largest of spinal neurons.) The axons of the alpha neurons carry the impulse rapidly to the motor end plates of the quad-

riceps femoris muscle, stimulating it to contract, thus causing the lower leg to swing forward (extend) (see Figure 12.9).

Following a tendon tap, each voluntary muscle can passively increase the firing rate of the *alpha motor neuron arc*, as it does in the stretch reflex. The resulting muscle contraction is coupled with a passive shortening and a reduced firing rate of the spindle, followed by a decrease in the firing rate of the alpha motor neuron arc. Thus, when this arc functions on its own, the contractions are jerky and coarse. Also, it is impossible to sustain a steady contraction during an alpha motor neuron arc. For example, it is not possible to hold a heavy object with a steady hand when the elbow is flexed, because the shortened spindles in the contracted muscles reduce their firing rate quickly, and the muscle relaxes suddenly. The maintenance of a smooth and steady muscular contraction is discussed in the following section.

FIGURE 12.8

One possible pathway of a simple three-neuron reflex action, forming a reflex arc. A reflex always starts with a sensory neuron and ends with a motor neuron. The arc pictured here begins with a tack

pricking the skin surface of a big toe. The impulse (arrows) travels from the toe to the spinal cord and back to a muscle in the foot, which jerks away from the tack—a flexor or withdrawal reflex.

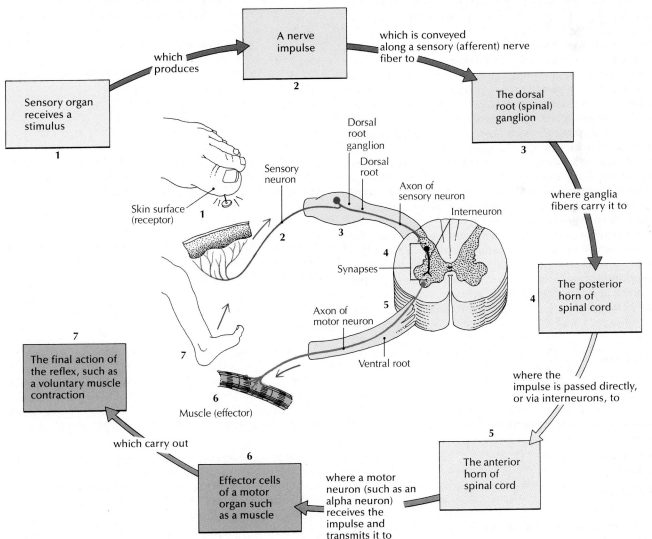

FIGURE 12.9

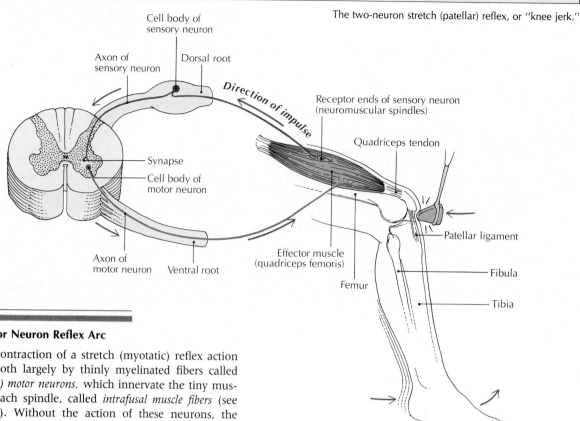

The two-neuron stretch (patellar) reflex, or "knee jerk."

Gamma Motor Neuron Reflex Arc

The muscle contraction of a stretch (myotatic) reflex action is made smooth largely by thinly myelinated fibers called *gamma (lower) motor neurons*, which innervate the tiny muscles within each spindle, called *intrafusal muscle fibers* (see Figure 12.10). Without the action of these neurons, the stretch (myotatic) reflex action would be jerky, as described above, because muscle fibers are unable to coordinate their activity when they are fully stretched.

Gamma motor neurons have their cell bodies in the anterior horn of the spinal cord. *Extrafusal muscle fibers* are the voluntary muscle fibers that do the work of contraction. They are called extrafusal because they are located *outside* the spindle (*extra* = outside).

The *gamma motor neuron reflex arc* acts to smooth out the movement of muscle contractions by controlling the intrafusal fibers in muscle spindles. It also acts to maintain muscular contraction when an object is being held or lifted. Depending upon the desired result, the gamma motor neurons can be stimulated to increase, maintain, or decrease the number of nerve impulses stimulating the intrafusal fibers. The firing rate of the gamma motor neurons is regulated exquisitely by the brain, primarily under our voluntary control.

If the desired result is to *increase* or *maintain* the contractile state of the voluntary muscle, there is an increase in the number of nerve impulses conveyed by the gamma motor neurons. Such an increase in impulses excites the intrafusal muscles to contract. The contracted intrafusal fibers stimulate the spindle to increase the number of excitatory nerve impulses conveyed by the afferent fibers of the spindle to the alpha motor neurons and the voluntary muscle (extrafusal muscle fibers). The result is an increase or a maintenance of the steady contraction of the voluntary muscle.

The opposite action occurs if the desired result is to *relax* the voluntary muscle. First comes a decrease in gamma motor nerve impulses, then the relaxation of the intrafusal muscle fibers, next a decrease in the number of impulses to the alpha motor neurons, and finally, the relaxation of the voluntary muscle.

Withdrawal Reflex Arc

The *withdrawal reflexes* are also known as protective or escape reflexes (see Figure 12.8). For example, when you touch a hot stove or step on a tack, you immediately pull away the injured hand or foot. These reflexes are polysynaptic, and the circuit comprises, in order, (1) sensory receptors, (2) afferent neurons, (3) spinal interneurons, (4) alpha motor neurons, and (5) voluntary muscles.

Withdrawal reflexes, which are basically flexor responses, are initiated by a wide variety of receptors, all of which are associated with afferent fibers. Such receptors, located in the skin, muscles, joints, and body organs, are called *flexor reflex afferents* (FRAs).

The flexor reflex afferents can activate circuits within the spinal cord that have synaptic connections with both sides and different levels of the spinal cord. When the interneurons within the circuit are excited by an intense stimulus, the flexor reflex activity can spread to (1) the opposite side of the spinal cord to influence circuits that evoke responses in the

FIGURE 12.10

A neuromuscular spindle, showing the encapsulated intrafusal muscle fibers. The extrafusal muscle fibers are outside the spindle.

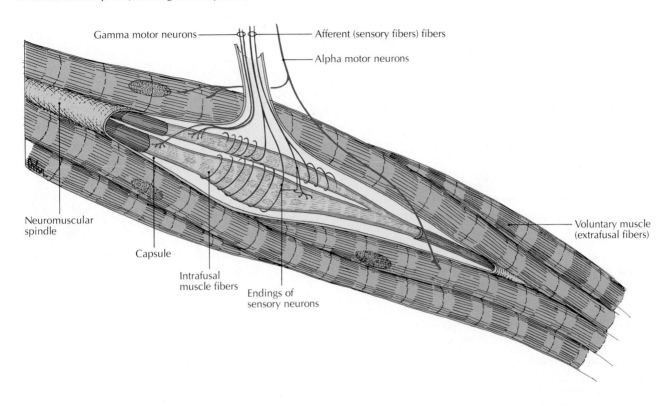

opposite extremity, and (2) other spinal segments that produce movements of the *other extremity on the same side.* The latter movements involve both flexor and extensor reflexes.

For example, if you pick up a hot frying pan with your right hand, you will most likely drop the pan, pull your injured right hand away, and make compensating movements with both your legs and your left arm as you move away from the stove. Such a complex of movements illustrates both the principle of reciprocal innervation and the crossed extensor reflex (Figure 12.11).

According to the principle of *reciprocal innervation,* the contraction of the prime mover (agonistic) muscles is synchronized with the relaxation of the antagonistic muscles. In our example, your right arm flexes when the contraction of the flexor muscles is accompanied by the relaxation of the extensor muscles. In this case, spinal interneurons integrate the excitatory and inhibitory stimuli to ensure the coordinated reciprocal innervation of the agonist with the antagonist.

The levels of the spinal cord are in communication through spinal interneurons. Some interneurons, known as *commissural interneurons,* convey influences from one side of the spinal cord to the other side at each spinal level. Within the lumbosacral enlargement levels, these commissural interneurons are involved in the **crossed extensor reflex.** The interneuron circuitry of this reflex is basic in coordinating the

activity of the two lower limbs during walking and running. Other spinal interneurons with long axons convey neural influences between the lumbosacral enlargement and the cervical enlargement (innervated upper limb). These *intersegmental interneurons* are involved with circuits coordinating the movements of the upper and lower limbs, as during walking and running.

The crossed extensor reflex is important because it helps the lower limbs support the body. When the right leg moves up from the ground by flexion, the left leg is activated to extend so that it supports the body. Such a reciprocal arrangement occurs when we walk or run. The spinal reflex circuits are also involved in the maintenance of body position and posture by stimulating motor nerves to regulate and sustain tonus of the body musculature.

Functional and Clinical Aspects of Reflex Responses

In a resting muscle, some of the fibers are always partially contracted because of the continual stimulation by receptors of certain reflex arcs, especially the stretch reflex. This continuous state of contraction is known as *muscle tone* (tonus). Muscle tone can also be described as the minimal degree of contraction exhibited by a normal muscle at "rest." When an

FIGURE 12.11

An ipsilateral reflex (A), and a contralateral, or crossed extensor reflex (B). In both reflexes there is a reciprocal innervation of the agonist and antagonist muscles.

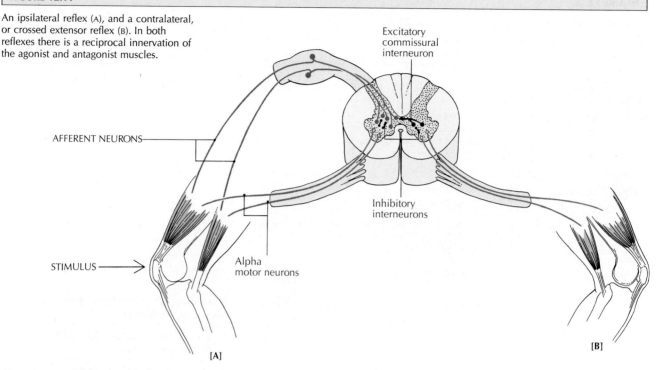

examiner passively manipulates a limb in a relaxed patient (flexion and extension), the muscle tone is expressed as the *amount of resistance* perceived by the examiner.

When some of the lower motor neurons or afferent fibers innervating neuromuscular spindles are injured so they cannot convey impulses, the muscle loses some of its tonus, a condition called *hypotonia*. During hypotonia a muscle has a decreased resistance to passive movement. Poliomyelitis (polio) is a viral infection of motor neurons. When this condition is followed by an absence of stimuli to the muscles, they lose much or all of their tonus. If a muscle is completely deprived of its motor innervation, it will lose all of its tonus, a condition called *atony*. *Hypertonia* is a condition in which a muscle has increased tonus. During hypertonia a muscle has an increased resistance to passive movement, as in the later stages following some strokes.

In addition to changes in tonus, lesions in the nervous system may result in changes in reflex activity (reflexia). *Hyporeflexia* is a condition in which a reflex is less responsive than normal, and *hyperreflexia* is one in which a reflex is more responsive than normal. Both conditions are associated with changes in the excitability of the stretch reflex. Hyporeflexia, like hypotonia, is caused by lesions of afferent fibers from the spindles, or of the lower motor neurons of a muscle. Hyporeflexia accompanies poliomyelitis, and may also indicate other disorders such as diabetes or syphilis.

Hyperreflexia and hypertonia are symptoms of lesions in certain descending motor pathways from the brain. It is believed that these lesions remove more inhibitory influences than excitatory ones. Many victims of a stroke exhibit both hypertonia and hyperreflexia, as well as Babinski's reflex, a reflex that is absent in the normal adult.

Any reflex response other than the normal one may indicate a disorder of the nervous system. Some of the major diagnostic reflexes are described in Table 12.3.

Ask Yourself

1 *What is a reflex action?*

2 *What are the components of a stretch (myotatic) reflex?*

3 *How does a gamma motor neuron arc coordinate muscular contraction?*

4 *What is a withdrawal reflex arc?*

5 *How are reflex responses used for diagnostic purposes?*

STRUCTURE AND DISTRIBUTION OF SPINAL NERVES

At each segment of the spinal cord, a pair of nerves branches and exits the H-shaped gray matter. One nerve of the pair exits to the left, entering the left side of the body. The other nerve of the pair exits to the right, entering the right side of the body. Each nerve has a **ventral** (anterior) **root** and a **dorsal** (posterior) **root** (see Figure 12.7), which meet shortly after leaving the spinal cord to form a single **mixed nerve**. *All spinal nerves are mixed*, containing both sensory and motor fibers that, together with cranial nerves, form the peripheral nervous system.

TABLE 12.3   SOME DIAGNOSTIC REFLEXES OF THE CENTRAL NERVOUS SYSTEM

Reflex	Description	Indication
Abdominal reflex	Anterior stroking of the sides of lower torso causes contraction of abdominal muscles.	Absence of reflex indicates lesions of peripheral nerves or in reflex centers in lower thoracic segments of spinal cord; may also indicate multiple sclerosis.
Plantar (Achilles) reflex	Tapping of calcaneal (Achilles) tendon of soleus and gastrocnemius muscles causes both muscles to contract, producing plantar flexion of foot.	Absence of reflex may indicate damage to nerves innervating posterior leg muscles or to lumbosacral neurons; may also indicate chronic diabetes, alcoholism, syphilis, subarachnoid hemorrhage.
Babinski's reflex	Stroking of the lateral part of sole causes toes to curl (plantar reflex).	Reflex indicates damage to upper motor neuron of pyramidal motor system.
Biceps reflex	Tapping of biceps tendon in elbow produces contraction of brachialis and biceps muscles, producing flexion at elbow.	Absence of reflex may indicate damage at the C5 or C6 vertebral level.
Brudzinski's reflex	Forceful flexion of neck produces flexion of legs, thighs.	Reflex indicates irritation of meninges.
Hoffmann's reflex	Flicking of index finger produces flexion in all fingers and thumb.	Reflex indicates damage to upper motor neuron of spinal cord.
Kernig's reflex	Flexion of hip, with knee straight and patient lying on back, produces flexion of knee.	Reflex indicates irritation of meninges or herniated intervertebral disk.
Patellar reflex (knee jerk)	Tapping of patellar tendon causes contraction of quadriceps femoris muscle, producing upward jerk of leg.	Absence of reflex may indicate damage at the L2, L3, or L4 vertebral level; may also indicate chronic diabetes, syphilis.
Romberg's reflex	Inability to maintain balance when standing with eyes closed.	Reflex indicates dorsal column injury.
Triceps reflex	Tapping of triceps tendon at elbow causes contraction of triceps muscle, producing extension at elbow.	Absence of reflex may indicate damage at C6, C7, or C8 vertebral level.

How Spinal Nerves Are Named

The spinal nerves are named for their associated vertebra (cervical, thoracic, lumbar, sacral, or coccygeal) and numbered (see Figure 12.2). Most spinal nerves pass through an intervertebral foramen, and then are distributed to a specific segment of the body. The first cervical nerve, however, passes between the occipital bone and the first cervical vertebra. The numbering of each cervical nerve corresponds to the vertebra *below* its exit. For example, the third cervical (C3) nerve emerges above the third cervical vertebra. But because there are only seven cervical vertebrae, the eighth cervical (C8) nerve passes through the intervertebral foramen between the seventh cervical vertebra and the first thoracic vertebra (see Figure 12.2). The numbering of each of the spinal nerves other than the cervical nerves corresponds to the vertebra *above* its exit from the vertebral column.

Structure of Spinal Nerves

Nerves of the peripheral nervous system are rougher and more cordlike than the tissue of the central nervous system because of three sheaths of connective tissue around the nerve fibers (Figure 12.12).

Each nerve is made up of nerve fibers enclosed in distinct bundles called *fascicles* (Figure 12.12A). Surrounding the entire peripheral nerve to bind together the large number of fascicles is a tube of connective tissue called the *epineurium* ("upon the nerve"). Also within the epineurium are blood vessels and lymph vessels (Figure 12.12B). Some small nerves do not have an epineurium. A thicker sheath of connective tissue, the *perineurium* ("around the nerve"), encases each fascicle of nerve fibers. Each of these fibers is also covered by the *endoneurium* ("within the nerve"), the interstitial connective tissue that separates individual nerve fibers.

FIGURE 12.12

A peripheral nerve. (A) The various parts of a spinal nerve, which contains both afferent (sensory) and efferent (motor) neurons. (B) Scanning electron micrograph, showing epineurium, perineurium, endoneurium, single nerve fibers, fascicles, and blood vessels. Nerve fibers within the fascicles are both myelinated and unmyelinated. [*Gene Shih and Richard Kessel,* Living Images: Biological Microstructures Revealed by SEM. *Boston, Science Books International, 1982.*]

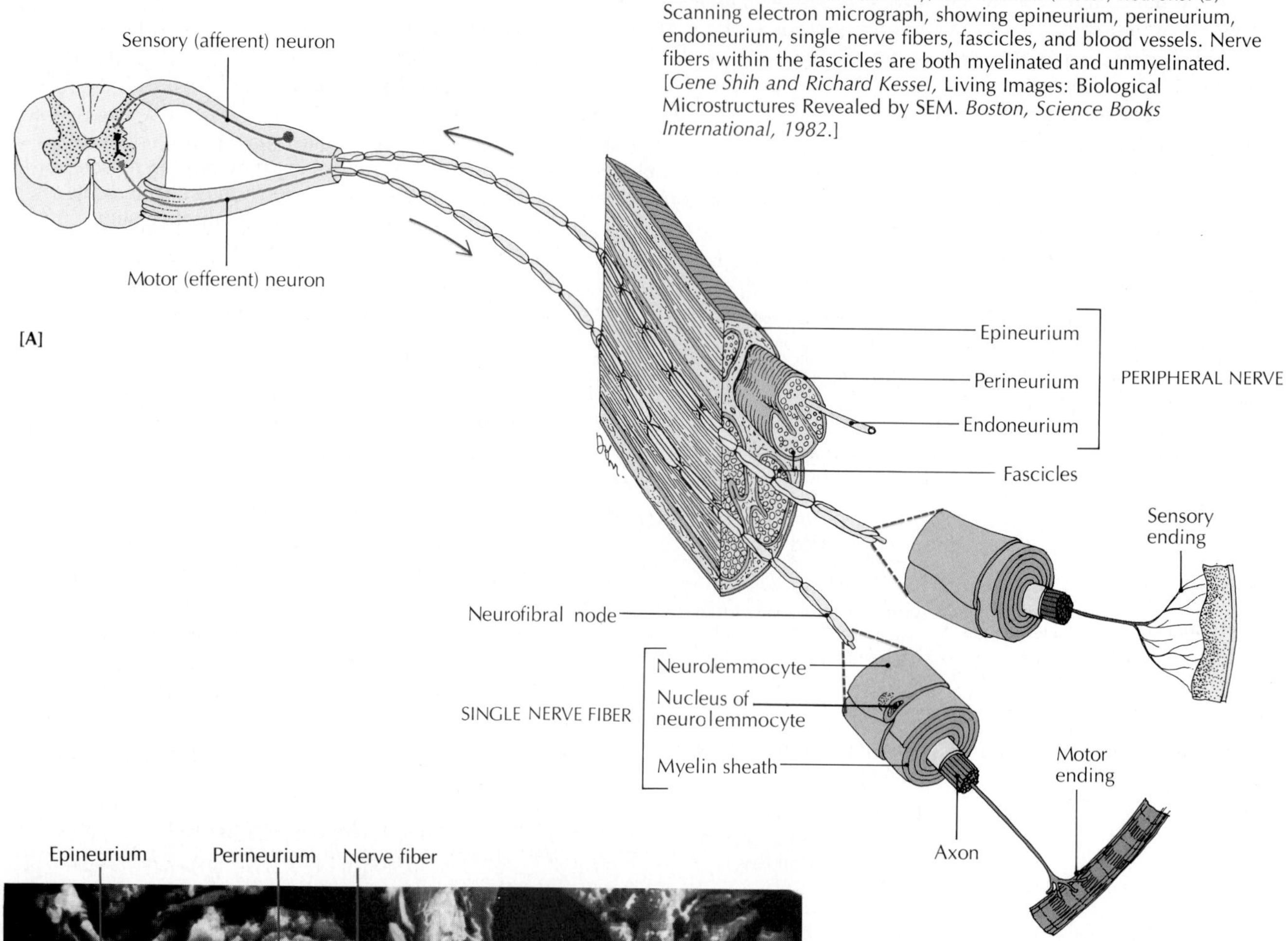

Sensory (afferent) neuron

Motor (efferent) neuron

[A]

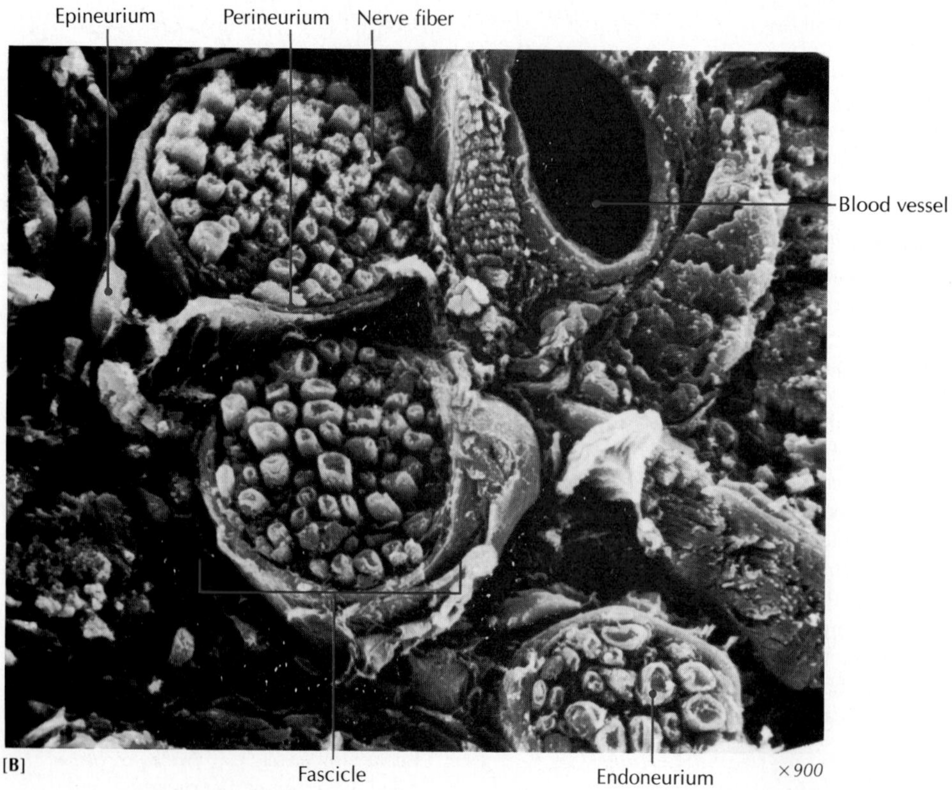

Epineurium
Perineurium   PERIPHERAL NERVE
Endoneurium

Fascicles

Sensory ending

Neurofibral node

Neurolemmocyte
Nucleus of neurolemmocyte
Myelin sheath

SINGLE NERVE FIBER

Axon

Motor ending

Epineurium   Perineurium   Nerve fiber

Blood vessel

[B]   Fascicle   Endoneurium   ×900

Branches of Spinal Nerves

A short distance after the dorsal and ventral roots join together to form a spinal nerve proper, the nerve divides into several branches called *rami* (RAY-mye; singular, *ramus*, RAY-muhss). These branches are the dorsal (posterior) ramus, ventral (anterior) ramus, meningeal ramus, and the rami communicantes (Figure 12.13).

The branches of the *dorsal ramus* (a mixed nerve) innervate the skin of the back, the skin on the back of the head, and the tissues and intrinsic (deep) muscles of the back. Branches of the *ventral ramus* (a mixed nerve) innervate the skin, tissues, and muscles of the neck, chest, abdominal wall, both pairs of limbs, and the pelvic area. The *meningeal ramus* innervates the vertebrae, spinal meninges, and spinal blood vessels. The *rami communicantes* (singular, *ramus communicans*) are composed of sensory (general visceral afferent) and motor nerve fibers associated with the autonomic nervous system innervating the visceral structure (see Chapter 14).

FIGURE 12.13

Rami and other parts of a spinal nerve, in this case a thoracic spinal nerve. Dorsal and ventral roots join to form a spinal nerve. The nerve divides into a dorsal (posterior) ramus, a ventral (anterior) ramus, a meningeal ramus, and white and gray rami communicantes. Each sympathetic ganglion of the autonomic nervous system near the spinal cord is a paravertebral ganglion (adjacent to the spinal column), and each one near the abdominal viscera is a prevertebral ganglion.

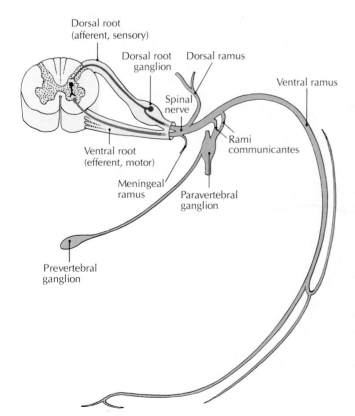

Plexuses

The ventral rami of the spinal nerves (except T2 through T12) are arranged to form several complex networks of nerves called *plexuses* (L. braided) (Figure 12.14). In a plexus, the nerve fibers of the different spinal nerves are sorted and recombined, so that fibers associated with a particular peripheral nerve are composed of the fibers from several different rami. Plexuses include: (1) the *cervical plexus,* (2) the *brachial plexus,* (3) the *lumbar plexus,* (4) the *sacral plexus* (sometimes the lumbar and sacral plexuses are referred to collectively as the *lumbosacral plexus*), and (5) the *coccygeal plexus.* Note that the ventral rami of T2 through T12 do not form plexuses. Instead, each ramus innervates a segment of the thoracic and abdominal walls.

Cervical plexus The *cervical plexus* is composed of the ventral rami of spinal nerves C1 through C4 (Table 12.4). Its branches can be placed into four groups:

1 The *cutaneous* (skin) sensory nerves innervate the skin and scalp near the external ear, the skin of the sides and front of the neck, and the skin of the upper portions of the thorax and shoulder.

2 The branches of the *ansa* (looplike) *cervicalis* and its superior and inferior roots provide the motor innervation to the strap muscles attached to the hyoid bone in the neck, and also the shoulder muscles.

3 The *phrenic* nerves from C3, C4, and C5 descend into the thorax to innervate the diaphragm.

4 Some nerves supply motor innervation to the prevertebral muscles of the neck. These muscles assist in flexion.

Lesions of both phrenic nerves cause paralysis of the diaphragm. The major symptoms are shortness of breath and difficulty in coughing or sneezing. A severe case of hiccups (spasmodic, sharp contractions of the diaphragm) can be relieved by surgically crushing the phrenic nerve in the lower neck, which temporarily paralyzes the diaphragm. (The nerve fibers regenerate in time.) A less drastic method is the injection of an anesthetic solution around the phrenic nerve. The anesthesia lies on the anterior surface of the middle third of the scalene muscle, producing a temporary paralysis of half of the diaphragm.

What is a "broken neck"?

A "broken neck" occurs when the spinal cord is severed above the fourth cervical nerve. The lesion severs the motor tracts (upper motor neurons) from the brain. As a result, all muscles of respiration including the diaphragm (innervated by cervical nerves 4 and 5), and all thoracic and abdominal muscles (innervated by intercostal nerves) are paralyzed. Death results because the victim cannot breathe.

TABLE 12.4   PLEXUSES OF THE SPINAL NERVES

Plexus	Components	Location	Major nerve branches	Regions innervated	Result of damage to specific nerves
Cervical	Ventral rami of C1 to C4 nerves.	Neck region; origin covered by sternocleidomastoid muscle.	Cutaneous, muscular, communicating, phrenic, ansa cervicalis.	Skin and some muscles of back of head, neck; diaphragm.	*Phrenic:* paralysis of respiratory muscles.
Brachial	Ventral rami of C5 to C8 and T1 nerves.	Lower neck, axilla.	Axillary, ulnar, median, radial, musculocutaneous.	Muscles and skin of neck, shoulder, arm, forearm, wrist, hand.	*Axillary:* impaired ability to abduct and rotate arm. *Ulnar:* "clawhand" (fingers partially flexed). *Median:* "ape hand" (thumb adducted against index finger); impaired ability to oppose thumb. *Radial:* "wristdrop" (inability to extend hand at wrist).
Lumbar*	Ventral rami of L1 to L4 nerves.	Interior of posterior abdominal wall.	Femoral, obturator.	Muscles, skin of abdominal wall.	*Femoral:* inability to extend leg; marked weakness in flexing hip; inflammation leads to lumbago (back pain).
Sacral*	Ventral rami of L4, L5, and S1 to S3 nerves.	Posterior pelvic wall.	Superior gluteal, inferior gluteal; sciatic nerve branches: tibial; common peroneal; pudendal.	Buttocks, medial thigh, muscles and skin of posterior thigh, posterior leg, plantar (sole) of foot. Muscles and skin of lateral posterior thigh, anterior leg, dorsal foot. Voluntary sphincters of urethra, anus.	*Superior gluteal:* walk with lurching gait; incontinence in urination, defecation. *Inferior gluteal:* difficulty walking up stairs. *Sciatic:* severely impaired ability to extend hip, flex knee; impaired ability to plantar flex foot. *Common peroneal:* "footdrop" (inability to dorsiflex foot, toes). *Pudendal:* incontinence.
Coccygeal	Co1 nerve, plus communications from S4 and S5 nerves.	Coccyx region.	A few fine filaments.	Skin in coccyx region.	None.

*Both are part of the lumbosacral plexus.

Figure 12.14

Spinal plexuses, with some representative
spinal nerves, anterior view.

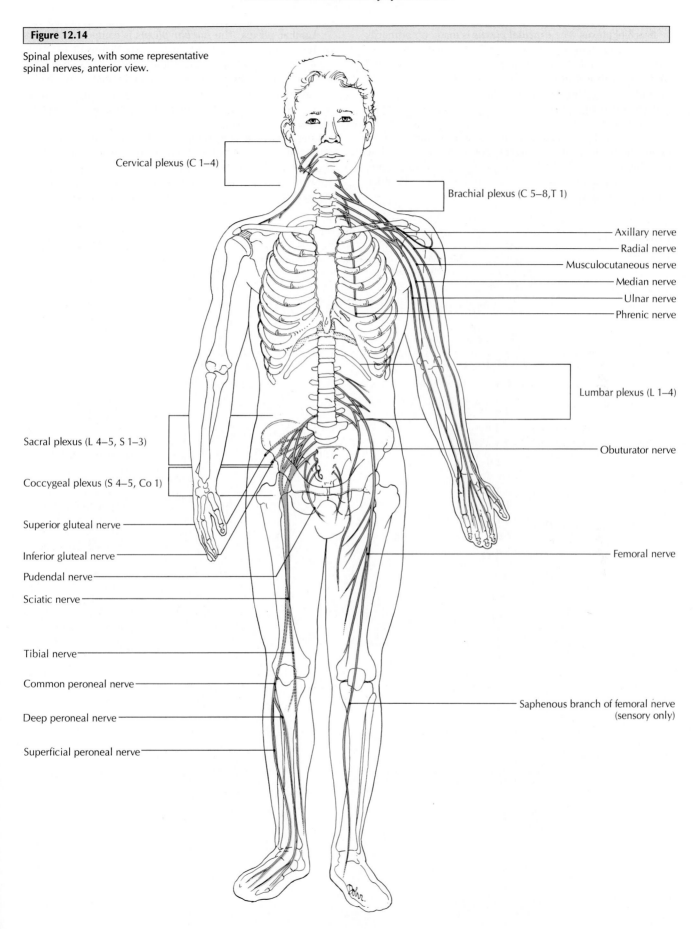

Cervical plexus (C 1–4)

Brachial plexus (C 5–8, T 1)

Axillary nerve
Radial nerve
Musculocutaneous nerve
Median nerve
Ulnar nerve
Phrenic nerve

Lumbar plexus (L 1–4)

Sacral plexus (L 4–5, S 1–3)

Obuturator nerve

Coccygeal plexus (S 4–5, Co 1)

Superior gluteal nerve

Inferior gluteal nerve

Femoral nerve

Pudendal nerve

Sciatic nerve

Tibial nerve

Common peroneal nerve

Saphenous branch of femoral nerve
(sensory only)

Deep peroneal nerve

Superficial peroneal nerve

Brachial plexus The ***brachial plexus*** is made up primarily of the ventral rami of C5 to C8 and T1 spinal nerves (see Figure 12.14, Table 12.4). The plexus extends downward and laterally, passing behind the clavicle and into the armpit. In its course downward, the brachial plexus consists of branches and recombinations that are called, in order, roots, trunks, divisions, and cords.

The brachial plexus gives rise to a number of nerves to the upper limb. Five major nerves constitute the terminal branches of the cords. These nerves—the axillary, ulnar, median, radial, and musculocutaneous—innervate the shoulder, arm, forearm, and hand.

Lumbar plexus The ***lumbar plexus*** is composed of fibers of the ventral rami of L1 to L4 nerves (see Figure 12.14, Table 12.4). It supplies the anterior and lateral abdominal wall, external genitals, and the thigh. Two major nerves, the femoral and obturator, and some lesser nerves (genitofemoral, ilioinguinal, iliohypogastric) are derived from the recombining branches.

The *femoral nerve* is composed of fibers from L2, L3, and L4. It innervates the muscles that flex the hip joint and extend the knee joint (knee jerk). Inflammation of the femoral nerve leads to lumbago (pain in the lumbar region). The *obturator nerve,* also composed of fibers from L2, L3, and L4, innervates

FIGURE 12.15

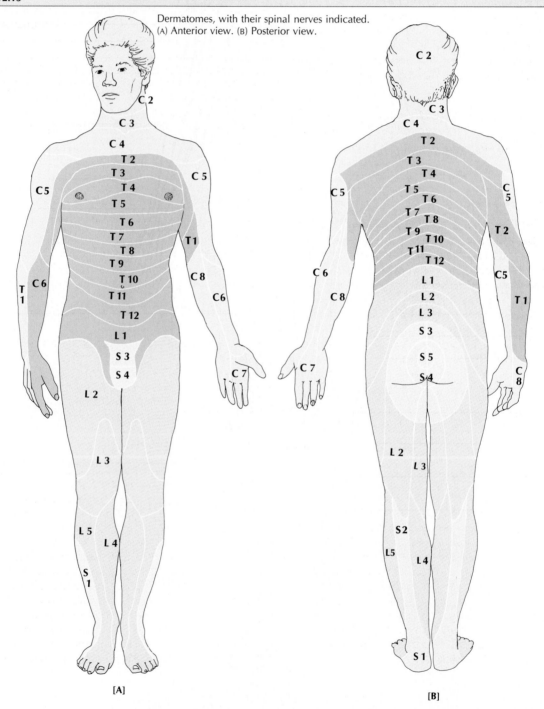

Dermatomes, with their spinal nerves indicated. (A) Anterior view. (B) Posterior view.

[A]

[B]

the adductor muscles of the thigh, as well as the skin.

Sacral plexus The *sacral plexus* is composed of fibers of the ventral rami of L4, L5, and S1 through S3 spinal nerves (see Figure 12.14, Table 12.4). It passes in front of the sacrum and into the regions of the buttocks. Several important nerves are derived from the plexus. The *sciatic nerve*, the thickest and longest nerve in the body, extends from the pelvic area to the foot, dividing above the knee into the *common peroneal* and *tibial nerves*. Together, those nerves innervate the thigh, leg, and foot muscles. Inflammation of the sciatic nerve, or *sciatica*, is described in Physiological and Anatomical Abnormalities at the end of the chapter.

The *superior gluteal nerve* innervates the gluteus medius, gluteal minimus, and tensor fascia lata muscles, and the *inferior gluteal nerve* innervates the gluteus maximus muscle. The *pudendal nerve* innervates the voluntary muscles of the perineum, especially the sphincters of the urethra and anus. It is the pudendal nerve, among others, that is blocked by local anesthesia in the ''saddle block'' procedure during childbirth.

Coccygeal plexus The *coccygeal plexus* is formed by the coccygeal nerve (Co1) and sacral nerves S4 and S5. A few fine nerve filaments supply the skin in the coccyx region.

Intercostal Nerves

The **intercostal nerves** are the second through twelfth thoracic nerves (T2 to T12). The ventral rami of the intercostal nerves innervate muscles and skin in the thoracic and abdominal walls. After these nerves leave the intervertebral foramina, they take a course parallel to the bony ribs, but continue past them. Intercostal nerves T2 through T6 innervate intercostal muscles, plus the skin on the lateral and anterior thoracic walls. Nerves T7 through T12 innervate the intercostal muscles, plus the abdominal wall and its overlying skin.

Dermatomes

The dorsal root of each spinal nerve is distributed to a specific region or segment of the body, and supplies the sensory innervation to a segment of the skin known as a **dermatome** (Gr. *derma*, skin + *tomos*, cutting). There are 30 dermatomes, one for each spinal nerve, except C1 (Figure 12.15). The face and scalp in front of the ears are innervated by a cranial nerve called the trigeminal, whose three divisions (ophthalmic, maxillary, and mandibular nerves) each innervate a separate region.

When the function of even a single dorsal nerve root is interrupted, there is a faint but definite decrease of sensitivity in the dermatome. The area of diminished sensitivity is detected by using a light pin scratch to stimulate a pain sensation. Using this method, a map of the dermatomes can be used in locating the sites of injury to dorsal roots of the spinal cord.

The sensory innervations of adjacent dermatomes overlap in such a way that if a spinal nerve is cut, the loss of sensation in its dermatome is minimal. However, if one dorsal root is irritated, as in shingles (described in Physiological and Anatomical Abnormalities), the resultant pain does spread to adjacent, overlapping dermatomes.

Ask Yourself

1 *How are spinal nerves named?*

2 *What are the epineurium, perineurium, and endoneurium?*

3 *What is a ramus? A plexus?*

4 *How do the thoracic nerves differ from the other spinal nerves?*

5 *What nerves make up the cervical plexus?*

6 *What nerves make up the brachial plexus?*

7 *What nerves make up the lumbar and sacral plexuses?*

8 *What is a dermatome?*

What is the ''funny bone''?	*The ''funny bone'' is actually the exposed part of the ulnar nerve, which extends the entire length of the upper limb into the hand. The nerve is well protected by tissue everywhere but at the elbow, just behind the medial epicondyle of the humerus. When you hit your ''funny bone,'' you are actually stimulating your ulnar nerve, and so you feel a strange twinge. (Because the bone of the upper arm is the humerus, could the term ''funny bone'' have come from a play on words—humorous for humerus?)*
Why does your foot sometimes ''fall asleep''?	*When you cross your legs or otherwise put pressure on one of your legs, you may press down on a nerve and temporarily cut off the feeling in your foot. When the local pressure is removed, you may feel a numbness we often call ''pins and needles'' as the nerve endings become reactivated. The same thing may happen when a blood vessel that supplies a peripheral nerve is temporarily forced shut. In that case, the ''pins and needles'' feeling occurs when the many tiny capillaries in the foot restore the nerve's blood supply.*

PHYSIOLOGICAL AND ANATOMICAL ABNORMALITIES

Injury and disease can severely impair the functioning of the spinal cord and spinal nerves. Only a few examples are given here.

Spinal cord injury

Spinal cord injury is any lesion of the spinal cord that bruises, cuts, or otherwise damages the neurons of the cord. Each year in the United States there are over 10,000 such injuries, most of them the result of motor vehicle accidents.

Transection, or severing, of the spinal cord may be complete or incomplete. Complete transections produce total flaccid paralysis of all skeletal muscles, bowel or bladder incontinence, loss of reflex activity *(areflexia)*, and a total loss of sensation below the level of the injury. Injuries or diseases causing incomplete transections produce partial loss of voluntary movements and sensations below the level of the injury.

Immediately following a spinal cord injury, *spinal shock* occurs, lasting for several hours to several weeks. Spinal shock involves paralysis, areflexia, and loss of sensation above the level of the injury. In the case of partial transection, spinal shock is followed by a period of spasticity, exaggerated spinal reflexes, and a decreased sensitivity to pain and temperature. Babinski's reflex may also appear.

People with incomplete lesions may recover partially, but until recently there was no hope of any restoration of motor and sensory functions with complete transections. Researchers are presently experimenting with nerve grafts, nerve regeneration, enzymes, hormones, steroids, growth-associated proteins, and various drugs that offer some hope to an estimated 500,000 American victims of spinal cord injury.

Paraplegia is the motor or sensory loss of function in both lower extremities. It results from transection of the spinal cord in the thoracic and upper lumbar regions. When the spinal cord is severed completely at a spinal level below the cervical enlargement and above the upper lumbosacral enlargement, paralysis below the lesion generally occurs immediately afterward, impairing excretory and sexual functions. If damage

to the spinal cord is incomplete, however, some sensory and motor capability will remain below the lesion.

Quadriplegia is a paralysis of all four extremities, as well as any part of the body below the level of injury to the spinal cord. It usually results from injury at the C8 to T1 level. Quadriplegia is more complicated than paraplegia because it affects other body systems. For example, the cardiovascular and respiratory systems may be unable to function properly because of insufficient muscle action.

Hemiplegia is paralysis of upper and lower limbs on one side of the body. It is usually the result of damage to only one side of the spinal cord above C5, or serious brain damage on the opposite side.

Poliomyelitis

Poliomyelitis is a contagious viral infection that affects both the brain and spinal cord, and sometimes causes the destruction of neurons. The poliomyelitis virus shows a preference for infiltrating the lower motor neurons of the spinal cord and brainstem. The initial symptoms may be sore throat and fever, diarrhea, or painful back and limbs. In cases of *nonparalytic polio*, these symptoms disappear in less than a week. When motor neurons in the spinal cord are damaged, there is obvious paralysis of muscles within a few days. Paralysis may be limited to the limbs (especially the lower limbs), or it may also affect the muscles used for breathing and swallowing. The mortality rate for all types of poliomyelitis is 5 to 10 percent.

Treatment is supportive rather than curative. Medication is ineffective against the polio virus except as a preventive measure. The Salk vaccine, which became available in 1955, virtually eliminated the disease among those immunized, and the Sabin oral vaccine has been shown to be even more effective, and easier to use.

Sciatica

Sciatica (sye-AT-ih-kuh) is a form of *neuritis* (nerve inflammation) characterized by sharp pains along the sciatic nerve and its branches. Pain usually ex-

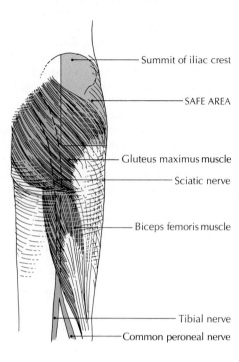

The safe area for intramuscular injection is indicated in green, and the area where the sciatic nerve is vulnerable is shown in red.

tends from the buttocks to the hip, back, and posterior thigh, leg, ankle, and foot. One of the most common causes is pressure from a herniated posterior intervertebral disk on a dorsal or ventral root of the sciatic nerve or one of its branching nerves, the tibial or common peroneal. Sciatic pain may also be caused by a tumor, by inflammation of the nerve or its sheath, or by disease in an adjacent area such as the sacroiliac joint.

The sciatic nerve may be injured by an intramuscular injection into the buttocks. The safe area for injection is the upper, outside quadrant (see drawing). An injection into the apparently safe area may pierce a nerve if the patient is standing up instead of lying face down.

Shingles

Shingles, or *herpes zoster*, is an acute inflammation of the dorsal root ganglia. It is caused by the same virus that causes chickenpox. Unlike chickenpox, shingles usually affects adults over 50, but there is a connection between the two diseases. Shingles occurs when the virus that caused childhood chickenpox lies dormant in the ganglia of cranial nerves or the ganglia of posterior nerve roots, and then becomes reactivated and attacks the root ganglia.

Shingles begins with fever and a general feeling of illness, followed in two to four days by severe deep pain in the trunk, and occasionally in the arms and legs. A rash typically appears in a band (dermatome) around one side of the chest, trunk, or abdomen, and the pain and rash always progress along the course of one or more spinal nerves (usually intercostal nerves) beneath the skin. Sometimes the ganglion of the trigeminal cranial nerve is affected, producing pain in the eyeball and a rash from the eyelid to the hairline. The rash

usually disappears within two or three weeks, but some pain may persist for months. Shingles seldom recurs.

Spinal meningitis

Spinal meningitis is an inflammation of the spinal meninges, especially the arachnoid and pia mater, which increases the amount of cerebrospinal fluid and alters its composition. The cause may be either viral or bacterial. Meningitis may also follow a penetrating wound, or an infection in another area. Meninges infected by bacteria produce large amounts of pus, which infiltrates the cerebrospinal fluid in the subarachnoid space between the arachnoid and pia mater. Pus is not usually formed when the infection is viral.

The first sign of meningitis is usually a headache, accompanied by fever, chills, and vomiting. Also evident may be rigidity in the neck region, positive Brudzinski's and Kernig's reflexes (see Table 12.3), exaggerated deep tendon reflexes, and back spasms that cause the body to arch upward. Coma may develop. Diagnosis is usually made by analyzing cerebrospinal fluid withdrawn by a lumbar puncture. The fluid is examined for its content of protein, sugar, and bacteria. The chances of recovery are excellent if the disease is diagnosed early, and if the infecting microorganisms respond to antibiotics. The recovery rate is less encouraging with infants and the elderly, and most untreated victims of any age will die.

MEDICAL TERMINOLOGY

ATAXIA (Gr. "not ordered") Loss of motor coordination due to disease of the nervous system or certain genetic disorders.

CORDOSTOMY (Gr. *tomos,* a cut) Cutting into a nerve fiber tract (usually spinothalamic) of the spinal cord, usually to relieve pain.

CRYONEUROSURGERY (Gr. *kryos,* icy cold) The surgical destruction of nerve tissue by exposure to extreme cold.

EPIDURAL ANESTHESIA (caudal block) An injection of anesthesia into the epidural space outside the dura mater, which anesthetizes the nerves of the cauda equina.

FRIEDREICH'S ATAXIA Degenerative disorder of the cerebellum and dorsal columns of the spinal cord.

HEMATOMYELIA Hemorrhaging within or upon the spinal cord.

MYELITIS (Gr. *myelo,* marrow, spinal cord) A general term for inflammation of the spinal cord. Various forms of myelitis range from poliomyelitis and leukomyelitis, which produce some motor and sensory dysfunction, respectively, to acute transverse myelitis, which affects the entire thickness of the cord and usually produces rapid degeneration.

MYELOGRAM X ray of the spinal cord, which is obtained by injecting contrast fluid into the subarachnoid space surrounding the cord, a technique called *myelography.*

NERVE BLOCK An injection of anesthesia near a nerve that supplies the area to be treated.

NEURALGIA (*algia* = pain) Pain along the length of a nerve.

NEURECTOMY The removal of part of a nerve.

NEURITIS Inflammation of a nerve.

NEUROLYSIS (Gr. *lys,* loosening) Relief of tension upon nerve caused by adhesions.

NEUROTRIPSY (Gr. *trips,* friction) The surgical crushing of a nerve.

PALSY (L. paralysis, disabled) A condition marked by some paralysis, weakness, and loss of muscle coordination.

POSTLATERAL SCLEROSIS Degeneration of white matter in dorsal and lateral funiculi, caused by insufficient vitamin B_{12}.

REFLEX TESTING Stimulation of a nerve to determine if the appropriate reflex is operative.

SPINA BIFIDA (L. *bi,* two + *fidus,* split) Certain congenital defects producing the absence of a neural arch, with varying degrees of protrusion of a portion of the spinal cord or meninges.

SPINAL ANESTHESIA The injection of anesthesia into the epidural space to block the nerves below and numb the lower part of the body.

SUMMARY

The spinal cord is the connecting link between the brain and most of the body. It also controls many reflex actions. Thus, many voluntary and involuntary actions depend on it.

Basic anatomy of the spinal cord
1 The *spinal cord* extends caudally from the brain for about 45 cm. Its upper end is continuous with the brain.
2 There are 31 pairs of spinal nerves and

roots: 8 cervical, 12 thoracic, 5 lumbar, 5 sacral, and 1 coccygeal. The lumbar and sacral roots are called the *cauda equina.*
3 The spinal nerves emerging from the

cervical enlargement of the cord innervate the upper limbs, and the nerves emerging from the *lumbosacral enlargement* innervate the lower limbs.

4 The spinal cord and spinal nerve roots are protected by the bony vertebral column and its ligaments, the triple-layered *spinal meninges* (inner pia mater, arachnoid, and outer dura mater), and the cerebrospinal fluid.

5 The *gray matter* of the spinal cord consists primarily of nerve cells. Three pairs of nerve cell columns in the gray matter are the *posterior horns, anterior horns,* and a median connecting column that forms the *gray commissure.*

6 Each spinal nerve emerges from the cord as ventral rootlets that form a *ventral root,* and as dorsal rootlets that form a *dorsal root.* The cell bodies of neurons whose axons comprise the dorsal roots are located in *dorsal root ganglia.*

7 The *white matter* of the cord is composed mainly of bundles of myelinated nerve fibers. These bundles are *ascending* (sensory) *tracts* and *descending* (motor) *tracts.* The white matter is divided into longitudinal columns called *funiculi.*

Functional roles of pathways of the central nervous system

1 Each pathway of the central nervous system can be viewed as consisting of sequences of *processing centers,* each center having terminal branches of axons entering the center, cell bodies and dendrites of neurons whose axons form the tract, and intrinsic neurons whose dendrites, cell bodies, and axons are located within the center.

2 The motor pathways and associated neural circuits are classified as either *pyramidal* or *extrapyramidal systems.*

Spinal reflexes

1 The spinal cord is involved with spinal reflex actions. A *reflex* is a predictable, involuntary response to a stimulus which enables the body to adapt quickly to environmental changes. The system of a sensory cell, an effector cell, and usually one or more connecting nerve cells is a *reflex arc.*

2 The *stretch (myotatic) reflex* is a two-neuron, ipsilateral reflex. An example is the knee-jerk reflex. It consists of a neuromuscular spindle, an afferent neuron, an alpha motor neuron, and a voluntary muscle. It plays a role in maintaining body position, and is integrated with other spinal reflexes in normal voluntary motor activities.

3 A *gamma motor neuron reflex arc* consists of gamma motor neurons, a neuromuscular spindle, afferent neurons, an alpha motor neuron, and a voluntary muscle. The gamma motor neuron arc acts to smooth out the movement of muscle contractions or to maintain the contraction when an object is being held or lifted.

4 The *withdrawal flexor reflexes* are protective, escape reflexes. The flexor reflex circuit is composed of sensory receptors, afferent neurons, spinal interneurons, alpha motor neurons, and voluntary muscles.

5 Reflexes used for diagnostic purposes include the patellar reflex (knee jerk), plantar (Achilles) reflex, and Babinski's reflex.

Structure and distribution of spinal nerves

1 At each segment of the spinal cord a pair of nerves is distributed to each side of the body. Each nerve has a *ventral* (anterior) *root* and a *dorsal* (posterior) *root.* Motor fibers emerge from the ventral root, and sensory fibers emerge as the dorsal root. Both roots meet to form a single *mixed nerve.*

2 Each nerve is made up of nerve fibers enclosed in bundles of connective tissue called *fascicles.* Several fascicles are held together by a sheath called the *epineurium,* each fascicle is encased by the *perineurium,* and each nerve fiber is covered by the *endoneurium.*

3 After a spinal nerve leaves the vertebral column, it divides into initial branches called *rami.* The ventral rami (except T2 to T12 nerves) are arranged to form networks of nerves called the cervical, brachial, lumbar and sacral (lumbosacral), and coccygeal *plexuses.*

4 The *intercostal nerves* are the T2 through T12 spinal nerves.

5 The dorsal root of each spinal nerve is distributed to a specific region or segment of the body and supplies sensory innervation to a segment of the skin called a *dermatome.*

Physiological and anatomical abnormalities

1 *Spinal cord injury* is a lesion of the spinal cord that damages the neurons of the spinal cord. *Transection* of the spinal cord may be complete or incomplete. *Spinal shock* occurs immediately following a spinal cord injury.

2 *Paraplegia* is the motor or sensory loss of function in both lower extremities. *Quadriplegia* is a paralysis of all four extremities, as well as any part of the body below the level of injury to the cord. *Hemiplegia* is a paralysis of upper and lower limbs on one side of the body.

3 *Poliomyelitis* is a contagious viral infection that affects both the brain and spinal cord, and sometimes causes the destruction of neurons.

4 *Sciatica* is a form of nerve inflammation characterized by sharp pains along the sciatic nerve and its branches.

5 *Shingles* is an acute inflammation of the dorsal root ganglia. It occurs when the virus that caused childhood chickenpox lies dormant in the ganglia of cranial nerves or the ganglia of posterior nerve roots, and then becomes reactivated during adulthood and attacks the root ganglia.

6 *Spinal meningitis* is an inflammation of the spinal meninges, which increases the amount of cerebrospinal fluid and alters its composition. The cause may be either viral or bacterial.

UNDERSTANDING THE FACTS

1 What are the major functions of the spinal cord?

2 What is the basic anatomy of the spinal cord?

3 What is the filum terminale?

4 At what level does the adult spinal cord terminate?

5 What are the three layers of spinal meninges?

6 Each spinal nerve is composed of two roots. Name them and give their functions.

7 Name the major ascending and descending spinal tracts. Where are they located?

8 List in sequence the components of a simple reflex arc.

9 Define a mixed nerve.

10 Distinguish between epineurium, perineurium, and endoneurium.

11 Identify the four rami of a spinal nerve, and give their functions.

12 What two nerves are divisions of the sciatic nerve?

13 When a spinal nerve is cut, why may the sensation in its dermatome still be felt?

14 Give the location of the following (be specific):
   a conus medullaris
   b fasciculi
   c arachnoid
   d longest nerve in the body
   e "funny bone"
   f pyramidal system
   g extrapyramidal system.

UNDERSTANDING THE CONCEPTS

1 What is the significance of the subarachnoid space?

2 How do the stretch (myotatic) reflex and the flexor reflexes help the body to maintain homeostasis?

3 Describe the type of sensation conveyed by each of the major ascending spinal tracts, and the type of motor impulse conveyed by each of the major descending tracts.

4 Describe how a processing center in the spinal cord functions.

5 What general regions are innervated by the cervical, brachial, lumbar, and sacral plexuses?

6 How do the following disorders affect the nervous system?
   a poliomyelitis   **c** shingles
   b sciatica   **d** spinal meningitis.

SUGGESTED READING

BANNISTER, L. H., "Sensory Terminals of Peripheral Nerves." In D. H. Landon, ed., *The Peripheral Nerve.* London: Chapman and Hall, 1976.

BARR, M., *The Human Nervous System.* New York: Harper & Row, 1979.

BOWSHER, D., *Introduction to the Anatomy and Physiology of the Nervous System.* Philadelphia: Davis, 1974.

BRODAL, A., *Neurological Anatomy in Relation to Clinical Medicine.* New York: Oxford University Press, 1981.

BROWN, DONALD R., *Neurosciences for Allied Health Therapies.* St. Louis: Mosby, 1980.

CARPENTER, M. B., AND J. SUTIN, *Human Neuroanatomy,* 8th ed. Baltimore: Williams & Wilkins, 1983.

HOYLE, G., *Muscles and Their Nervous Control.* New York: Wiley, 1982.

HUBBARD, J. I., *The Peripheral Nervous System.* New York: Plenum, 1974.

HUBEL, D. H., "Special Issues on Neurobiology." *Scientific American,* September 1979.

NOBACK, CHARLES R., AND ROBERT J. DEMAREST, *The Human Nervous System: Basic Principles of Neurobiology,* 3rd ed. New York: McGraw-Hill, 1981.

NOBACK, CHARLES R., AND ROBERT J. DEMAREST, *The Nervous System: Introduction and Review,* 3rd ed. New York: McGraw-Hill, 1986. (Paperback)

PEARSON, K., "The Control of Walking." *Scientific American,* December 1976.

PHILLIPS, C. G., AND R. PORTER, *Corticospinal Neurons: Their Role in Movement.* New York: Academic Press, 1977.

SNELL, R. S., *Clinical Neuroanatomy for Medical Students.* Boston: Little, Brown, 1980.

VALLO, A. B., K. E. HAGBARTH, H. TOREBJORK, AND B. G. WALLIN, "Somatosensory, Proprioceptive and Sympathetic Activity in Human Peripheral Nerves." *Physiology Review,* 59 (1979):919–957.

WILLIAMS, P. L., AND R. WARWICK, *Functional Neuroanatomy of Man.* Philadelphia: Saunders, 1975.

362

13
The Brain and Cranial Nerves

CHAPTER OUTLINE

GENERAL STRUCTURE OF THE BRAIN 363

MENINGES, VENTRICLES, AND CEREBROSPINAL FLUID 365
Cranial Meninges
Ventricles of the Brain
Cerebrospinal Fluid in the Brain

NUTRITION OF THE BRAIN 367
Effects of Deprivation

BRAINSTEM 368
Neural Pathways of Nuclei and Long Tracts
Reticular Formation
Reticular Activating System
Medulla Oblongata
Pons
Midbrain

CEREBELLUM 375
Anatomy of the Cerebellum
Functions of the Cerebellum

CEREBRUM 378
Anatomy of the Cerebrum
Functions of the Cerebrum
Cerebral Lobes
Diencephalon

LEARNING AND MEMORY 386

SLEEP AND DREAMS 390

CRANIAL NERVES 392
Cranial Nerve I: Olfactory
Cranial Nerve II: Optic
Cranial Nerve III: Oculomotor
Cranial Nerve IV: Trochlear
Cranial Nerve V: Trigeminal
Cranial Nerve VI: Abducens
Cranial Nerve VII: Facial
Cranial Nerve VIII: Vestibulocochlear
Cranial Nerve IX: Glossopharyngeal
Cranial Nerve X: Vagus
Cranial Nerve XI: Accessory
Cranial Nerve XII: Hypoglossal

CHEMICALS AND THE NERVOUS SYSTEM 401

PHYSIOLOGICAL AND ANATOMICAL ABNORMALITIES 404

MEDICAL TERMINOLOGY 407

THE DYNAMIC BODY: **THE BLOOD-BRAIN BARRIER 364**

ESSAY: **LEFT BRAIN, RIGHT BRAIN 384**

ESSAY: **EMBRYONIC DEVELOPMENT OF THE BRAIN 388**

ESSAY: **OPEN WIDE 397**

LEARNING OBJECTIVES

1 Describe the major divisions and subdivisions of the brain.

2 Compare the meninges of the brain with those of the spinal cord.

3 Explain the ventricular system and the circulation of cerebrospinal fluid in the brain.

4 Describe the nutritional needs of the brain and the blood-brain barrier.

5 Describe the three main anatomical divisions of the brainstem and their functions.

6 Explain the roles of the reticular formation and its neural pathways.

7 Describe the structure and functions of the cerebellum.

8 Describe the structure and functions of the cerebrum.

9 Describe the location and roles of the six cerebral lobes.

10 Explain the function of the limbic system.

11 Describe the major divisions of the diencephalon and their functions.

12 Outline one major theory of how learning takes place in the brain.

13 Distinguish between short-term and long-term memory, and describe some areas of research on memory.

14 Describe the four distinct wave patterns of the brain and their relationship to levels of consciousness.

15 Explain the four stages of sleep, REM sleep, and their relationship to sleep cycles.

16 Name the 12 pairs of cranial nerves, and describe their locations and functions.

17 Describe the effects of stimulants, depressants, antidepressants, psychedelic and hallucinogenic drugs, analgesics, and antianxiety drugs on the nervous system.

18 Define senility, Alzheimer's disease, cerebral palsy, cerebrosvascular accident, epilepsy, Parkinson's disease, dyslexia, encephalitis, and Bell's palsy and other disorders of cranial nerves.

The brain weighs only about 1400 g (3 lb), and yet it contains approximately 100 billion neurons, about the same number as stars in the Milky Way.* In addition, each neuron may have from 1000 to 10,000 synaptic connections with other nerve cells. There may be as many as 100 *trillion* synapses in the brain. It is tempting to compare the human brain with a computer. But there is really no comparison. Nothing we know can match the exquisite complexity of the brain.

Your brain does much more than help you to think and make decisions. It is your body's main key to homeostasis, regulating body processes from cell metabolism to the overall functioning of organs and systems. Your nerves and their specialized receptors may *receive* the sensations of sound, touch, vision, smell, and taste, but you actually *experience* the sensation in your brain. Your brain is working even when you are sleeping to activate, coordinate, and regulate the body's many functions and their relationships to the outside world.

*To get some idea of the enormity of 100 billion, think of this: you would have to spend almost *$5.5 million a day* to spend $100 billion in 50 years. If you wanted to spend $100 billion in *one* year, you would have to spend $274 million *a day.* One hundred billion is a lot of neurons.

FIGURE 13.1

Several views of the brain, showing some major subdivisions. (A) Right lateral view of the external surface of the brain, showing the cerebellum and four of the six cerebral lobes. (B) Right sagittal section. (C) Right sagittal section, showing the major subdivisions of the brain and its connection to the spinal cord. [(A *and* B) *Martin M. Rotker, M.D./Taurus.*]

GENERAL STRUCTURE OF THE BRAIN

The brain is technically called the *encephalon* (en-SEFF-uh-lon; Gr. *en*, in + *kephale*, head). It has four major divisions: brainstem, cerebellum, cerebrum, and diencephalon (Figure 13.1). The major divisions and structures of the brain are summarized in Tables 13.1 and 13.2 (pp. 365, 376).

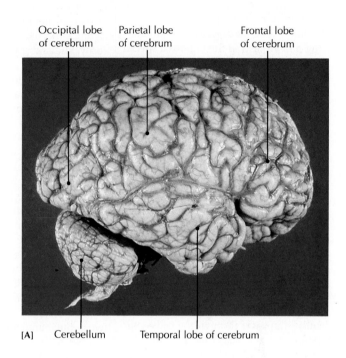

[A]

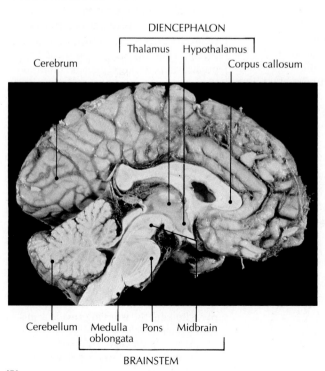

[B]

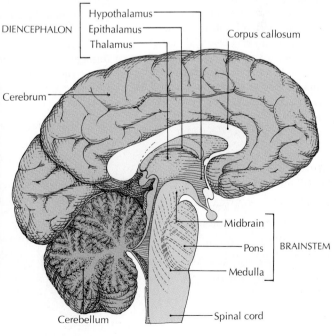

[C]

THE DYNAMIC BODY

The Blood-Brain Barrier

Nowhere in the body is the need for homeostasis greater than in the brain, and nowhere else in the body is an organ so well protected against chemical imbalances. The mechanism for maintaining homeostasis in the brain is a special one, focusing on capillaries that supply blood to the brain. The structure of brain capillaries is different from capillaries in the rest of the body. The walls of other capillaries contain penetrable gaps that allow most substances to pass through. The walls of brain capillaries, however, are formed by endothelial cells that are joined by tight junctions (drawing). The junctions actually merge the outer membrane layers of adjoining cells. The relatively solid walls of brain capillaries make up the **blood-brain barrier** *(hematoencephalic barrier).*

Because of the blood-brain barrier, all substances entering the brain from the blood must either diffuse through the endothelial cells or be conveyed via active transport through the cell membranes rather than passing through spaces between cells. The blood-brain barrier helps to maintain the delicate homeostatic balance of neurons in the brain by restricting the entrance of potentially harmful substances from the blood, by allowing essential nutrients to enter, by removing excess substances, and by helping to regulate the concentration of ions in the extracellular fluid.

Lipid-Soluble Molecules Pass

Because the membranes of the endothelial cells are composed primarily of lipid molecules, lipid-soluble substances pass through the blood-brain barrier rather easily. Such substances include nicotine, caffeine, ethanol, and heroin. (Heroin is more often abused than morphine, for example, which has a relatively low lipid solubility.) In contrast, water-soluble molecules such as sodium, potassium, and chloride ions are unable to cross the barrier without assistance. Molecules of essential water-soluble substances such as glucose (the brain's main energy source) and those amino acids that the brain cannot synthesize are "recognized" by carrier proteins and transported across the barrier. Some other substances, such as large protein molecules and most antibiotics cannot enter at all. For example, tetracycline crosses the barrier easily, but penicillin is admitted only in trace amounts, to mention two popular antibiotics.

Essential nutrients pass through the blood-brain barrier easily, assisted by carrier proteins that "recognize" the molecules of specific substances and bind to them before transporting them through the barrier. Transport systems not only carry essential substances through the barrier into the brain, they also remove surplus material. In this way, the transport systems contribute directly to the homeostasic regulation of the brain.

Some drugs pass through the blood-brain barrier easily. They include barbiturates; anesthetics such as sodium pentothal, ether, and nitrous oxide (laughing gas); carbon monoxide; cyanide; strychnine; hallucinogenic drugs such as LSD and mescaline; and alcohol in large quantities.

Brain capillaries are almost completely surrounded by the foot processes of astrocytes (see drawing), which synthesize and degrade compounds that are important for neurons. Astrocytes also help to control the ionic composition of the extracellular fluid surrounding neurons.

Another Barrier Has Been Discovered

Recently, a "metabolic" blood-brain barrier was discovered. It complements the cellular blood-brain barrier by enzymatically altering substances that enter the endothelial cells of brain capillaries, converting them into a chemical form that cannot actually enter the brain. One such substance is L-dopa, a precursor of the important chemicals dopamine and norepinephrine.

How can scientists modify the barriers to allow antibiotics and other useful drugs to reach the brain in order to treat disorders such as Parkinson's disease and cancer? Most anticancer drugs fail to destroy brain tumors because insufficient amounts of the drugs are able to reach the tumors. One experimental approach aims to target specific brain tumors by linking anticancer drugs with lipid-soluble antibodies, immunologic molecules that can "recognize" and bond to cancer cells. When an antibody-drug complex attacks the tumor, the anticancer drug is also delivered at the precise tumor site. Another promising idea is to bond water-soluble drugs to lipid-soluble carrier molecules that can penetrate the blood-brain barrier. Once inside the brain tissue, the entire complex is modified enzymatically to become water-soluble, so that it is unable to escape from the brain tissue. At this point, the drug would be activated by enzymes in the brain by separating it from its carrier, allowing the drug to provide a sustained release.

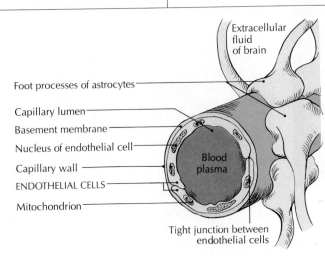

Foot processes of astrocytes

Capillary lumen

Basement membrane

Nucleus of endothelial cell

Capillary wall

ENDOTHELIAL CELLS

Mitochondrion

Extracellular fluid of brain

Blood plasma

Tight junction between endothelial cells

TABLE 13.1   MAJOR DIVISIONS OF THE BRAIN

Brainstem	**Cerebrum**
Medulla oblongata	
Pons	**Diencephalon**
Midbrain	Thalamus
	Hypothalamus
Cerebellum	Epithalamus
	Ventral thalamus (subthalamus)

The *brainstem* is composed of the midbrain, pons, and medulla. The *cerebellum* is the coordinating center for muscular movement. It is located posterior to the brainstem. The *diencephalon* is composed of the thalamus, hypothalamus, epithalamus, and ventral thalamus (subthalamus).

Probably the most obvious feature of the brain is the large pair of hemispheres that make up about 85 percent of the brain tissue. These are the two hemispheres of the *cerebrum.* Another obvious feature of the brain is the outer portion of the cerebrum, called the *cerebral cortex,* with its many folds or convolutions. Lying below the gray matter of the cerebral cortex are both white matter and deep, large masses of gray matter called the *basal ganglia.* Connecting the two cerebral hemispheres is a bundle of nerve fibers called the *corpus callosum,* which relays nerve impulses between the hemispheres. Emerging from the cerebrum are 2 of the 12 cranial nerves. The other 10 arise from the brainstem.

The brain is covered by the same three meninges that protect the spinal cord. Cerebrospinal fluid also flows through the brain from a series of cavities called *ventricles* into the subarachnoid space.

MENINGES, VENTRICLES, AND CEREBROSPINAL FLUID

The human brain is mostly water (about 75 percent in the adult). It has the consistency of gelatin, and if it were not supported somehow, the brain would slump and sag. Fortunately, it has ample support. The brain is protected by the scalp, with its hair, skin, fat, and other tissues, and by the reinforced bony cranium, one of the strongest structures in the body. It also floats shockproof in cerebrospinal fluid, and is encased by three layers of cranial meninges.

Cranial Meninges

The dura mater, arachnoid, and pia mater are basically the same in the brain and spinal cord. However, there are some minor anatomical differences.

Dura mater   The outermost cranial meninx is the *dura mater.* It consists of two fused layers: an inner dura mater that is continuous with the spinal dura mater, and an outer dura mater, which is actually the periosteal layer of the skull bones. The outer cranial dura mater is a tough, fibrous layer containing veins and arteries that nourish the bones. The inner dura mater extends into the fissure that divides the left and right hemispheres of the cerebrum (the *falx cerebri*), and reaches into the fissure between the cerebrum and cerebellum (*tentorium cerebelli*) (Figure 13.2). By dividing the cranial cavity into three distinct compartments, the dura mater adds considerable support to the brain.

The fused inner and outer cranial dura mater is intimately attached to the bone of the cranial cavity. As a result, there is no epidural space between the membrane and the bone, as there is in the spine. Between the inner dura mater and the arachnoid is a potential space called the **subdural space.** It does not contain cerebrospinal fluid.

In certain locations, the two layers of dura mater separate to form channels (see Figure 13.2A), which are lined with endothelium and contain venous blood. These spaces are the *dural sinuses of the dura mater,* which drain venous blood from the brain.

Arachnoid   The middle layer of the meninges, between the dura mater and pia mater, is the **arachnoid,** a delicate connective tissue. Between the arachnoid and the pia mater is a network of trabeculae (see Figure 13.2) and the *subarachnoid space,* which contains cerebrospinal fluid. The arachnoid contains no blood vessels of its own, but blood vessels are present in the subarachnoid space.

Pia mater   The *pia mater* is the delicate innermost meningeal layer. It directly covers, and is attached to, the surface of the brain, and dips down into the fissures between the raised ridges of the brain. Most of the blood to the brain is supplied by the large number of small blood vessels in the pia mater.

In head injuries, blood may flow from a severed blood vessel into the potential space between the skull and cranial dura mater (an extradural or epidural hemorrhage), into the potential subdural space (a subdural hemorrhage), into the subarachnoid space (a subarachnoid hemorrhage), or into the brain itself.

Ventricles of the Brain

Within the brain is a series of connected cavities called *ventricles* (L. little bellies). Each cranial ventricle is filled with cerebrospinal fluid, and is lined by cuboidal or epithelial cells, known as *ependyma* or *ependymal epithelium.* A network of blood vessels called a *choroid plexus* is formed in several places where the ependyma contacts the pia mater. The four true ventricles are numbered from the top of the brain downward. They are the *left* and *right lateral ventricles* of the cerebral hemispheres, the *third ventricle* of the diencephalon, and the *fourth ventricle* of the pons and medulla (Figure 13.3).

Each lateral ventricle is connected to the third ventricle of the diencephalon through the small *interventricular foramen* (of Monro). The third ventricle is continuous with the fourth ventricle through a narrow channel called the *cerebral aqueduct* (of Sylvius) of the midbrain.

FIGURE 13.2

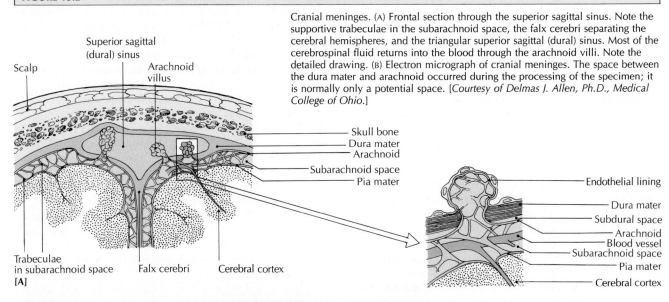

Cranial meninges. (A) Frontal section through the superior sagittal sinus. Note the supportive trabeculae in the subarachnoid space, the falx cerebri separating the cerebral hemispheres, and the triangular superior sagittal (dural) sinus. Most of the cerebrospinal fluid returns into the blood through the arachnoid villi. Note the detailed drawing. (B) Electron micrograph of cranial meninges. The space between the dura mater and arachnoid occurred during the processing of the specimen; it is normally only a potential space. [*Courtesy of Delmas J. Allen, Ph.D., Medical College of Ohio.*]

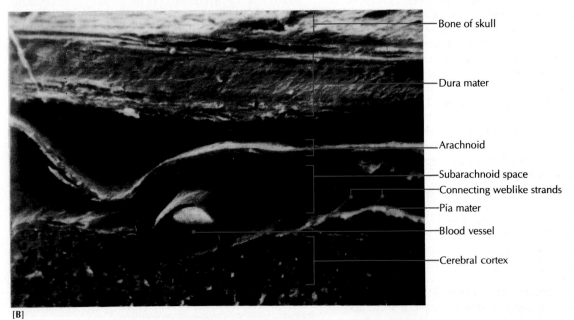

Cerebrospinal Fluid in the Brain

Cerebrospinal fluid (CSF) is a clear, colorless liquid, which is essentially an ultrafiltrate of blood. The fluid in the subarachnoid space provides a special environment in which the brain floats, cushioning it against hard blows and sudden movements. The weight of the brain floating in CSF is only about 14 percent of its actual weight.

Besides providing a protective buoyancy for the brain, the cerebrospinal fluid maintains homeostasis by helping to control the chemical environment of the central nervous system. It conveys excess components and unwanted substances away from the extracellular fluid and into the venous portion of the blood circulatory system.

An adult has about 125 mL of cerebrospinal fluid, with about 50 mL in the ventricles and 75 mL in the subarachnoid space around the brain and spinal cord.

Does the weight of the brain increase after childhood?

Although the number of brain cells does not increase after infancy, they do grow in size and degree of myelination as the body grows. In addition, the number of glial cells increases after birth. These changes account, in large measure, for the fact that the adult brain is about three times as heavy as it was at birth. After the age of about 20, the brain begins to lose about one gram a year as neurons die and are not replaced.

Formation of cerebrospinal fluid The ependymal cells that line the ventricles and the pia mater, with its rich blood supply, form the *choroid plexuses,* with their intricate networks of capillaries. The choroid plexuses are considered to be components of the blood-brain barrier (see p. 364 and Figure 13.3). Most of the cerebrospinal fluid is formed continuously at the choroid plexuses of the lateral, third, and fourth ventricles by a combination of diffusion and active transport.

The choroid plexuses are able to produce cerebrospinal fluid because of the selective permeability of their blood vessels. The choroid plexuses do not allow blood cells or the largest protein molecules to pass into the ventricles. However, they do permit the passage of traces of protein; oxygen and carbon dioxide in solution; sodium, potassium, calcium, magnesium, and chloride ions; glucose; and a few white blood cells.

Circulation of cerebrospinal fluid Ordinarily, cerebrospinal fluid moves from the ventricles inside the brain to the subarachnoid space outside the brain. The fluid flows slowly from the two lateral ventricles of the brain, where much of the fluid is formed, through the paired interventricular foramina to the third ventricle (see Figure 13.3). From there, it passes through the cerebral aqueduct into the fourth ventricle. The fluid leaves the ventricular system through three openings, called apertures, in the roof of the fourth ventricle. From there, the fluid oozes into the subarachnoid space (the *cisterna magna*) behind the medulla.

Next, some cerebrospinal fluid slowly makes its way down the spinal cord to the lumbar cistern. Most of the fluid, however, circulates slowly toward the top of the brain through the subarachnoid space. Here the spongelike *arachnoid villi*, which contain pressure-sensitive valves, permit a one-way bulk flow of cerebrospinal fluid into the superior sagittal sinus, which drains venous blood from the brain (see Figures 13.2A and 13.3).

The total volume of cerebrospinal fluid is formed and renewed about three times a day. When the pressure of the cerebrospinal fluid in the subarachnoid space exceeds the venous pressure, the small channels open, and the fluid flows into the superior sagittal sinus, where it joins the venous blood. When the venous pressure exceeds the pressure in the subarachnoid space, the channels close.

What is the difference between a concussion and a contusion?

A concussion (L. shake violently) is an abrupt and momentary loss of consciousness after a violent blow to the head. It results from the sudden movement of the brain within the skull. The loss of consciousness may be due to the sudden pressure upon neurons essential to the conscious state, or to sudden changes in the polarization of certain neurons. A contusion (L. a bruising) is a cerebral injury that produces a bruising of the brain due to blood leakage. Unconsciousness follows a contusion, and may last minutes or hours. A concussion is rarely serious, but a contusion can be.

Ask Yourself

1 *How does the dura mater nourish the cranial bones?*

2 *What is contained in the subarachnoid space?*

3 *Which meningeal layer envelops the brain directly?*

4 *What are cranial ventricles?*

5 *What are the functions of cerebrospinal fluid?*

6 *Where is cerebrospinal fluid formed?*

7 *Describe the circulation of cerebrospinal fluid in the brain.*

NUTRITION OF THE BRAIN

The nutrients needed by the brain can reach it only through the blood, and about one pint of blood is circulated to the brain every minute. Blood reaching the brain contains glucose as well as oxygen. A steady supply of glucose is necessary, not only because it is the body's chief source of usable energy, but also because the brain cannot store it. The brain also requires about 20 percent of all the oxygen used by the body, and the need for oxygen is high even when the brain is at rest.

Effects of Deprivation

A lack of either oxygen or glucose will damage brain tissue faster than any other tissue. Deprivation for even a few minutes may produce permanent brain damage. Lack of oxygen in body tissues *(hypoxia)* can occur even before birth, especially during the last four months of prenatal development. It is also common during the first year after birth.

Proper and continuous nourishment is so important to the brain that it has built-in regulating devices that make it almost impossible to constrict blood vessels that would reduce the incoming blood supply. It is likely that the brain's blood vessels are prevented from constricting by the products of cell metabolism, which are either formed in the brain itself or carried to the brain by the blood. An increase in carbon dioxide, or a decrease in oxygen, dilates the blood vessels leading to the brain. The opposite condition, *hyperventilation* (too much oxygen and too little carbon dioxide), is one of the few

---

How much energy does the brain require during a two-hour exam?

During two hours of intense mental strain, the brain uses up about as much glucose as there is in a single peanut. This does not mean that you can get by with only a peanut for lunch or breakfast before an exam, because some glucose is used for other metabolic needs.

FIGURE 13.3

Ventricles of the brain, and circulatory path of cerebrospinal fluid through the cranial pathways. The arrows indicate the direction of the flow.

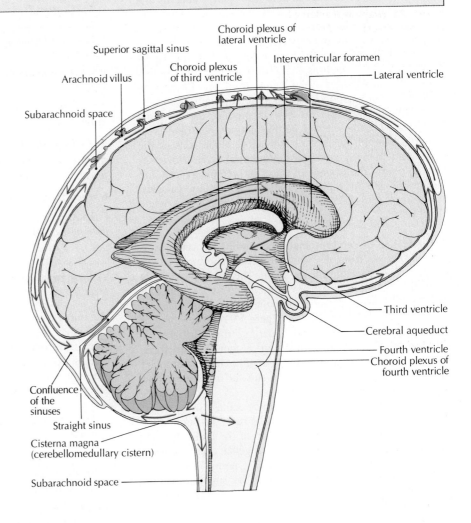

Superior sagittal sinus

Choroid plexus of lateral ventricle

Arachnoid villus

Choroid plexus of third ventricle

Interventricular foramen

Subarachnoid space

Lateral ventricle

Third ventricle

Cerebral aqueduct

Fourth ventricle
Choroid plexus of fourth ventricle

Confluence of the sinuses

Straight sinus

Cisterna magna (cerebellomedullary cistern)

Subarachnoid space

conditions that allows *constriction* of cerebral blood vessels. A hyperventilated person may suffer dizziness, mental confusion, convulsions, or even unconsciousness until the brain's regulatory system balances the supply of oxygen and carbon dioxide.

How long can the brain be deprived of oxygen before it becomes damaged?

If the brain is deprived of oxygen for more than 5 seconds we lose consciousness. After 15 to 20 seconds, the muscles begin to twitch in convulsions, and after 9 minutes, brain cells are damaged permanently. (There is a big difference between holding your breath for 15 seconds, and depriving the brain of oxygen for the same amount of time.)

Ask Yourself

1 *What two substances does the brain need continuously?*

2 *Why is it important that blood vessels to the brain not be constricted easily?*

BRAINSTEM

The **brainstem** is the stalk of the brain, and it relays messages between the spinal cord and the brain. Its three segments are the *medulla oblongata, pons,* and *midbrain* (Figure 13.4; see also Figure 13.1). The brainstem is continuous with the diencephalon above and the spinal cord below. It narrows slightly as it leaves the skull, passing through the foramen magnum to merge with the spinal cord.

Other structures of the brainstem that are very important functionally include the *long tracts of ascending and descending pathways* between the spinal cord and parts of the brain, which all pass through the brainstem. A network of nerve cell bodies and fibers called the *reticular formation* is also located throughout the core of the entire brainstem. It plays a vital role in maintaining life. Finally, all *cranial nerves* except I (olfactory) and II (optic) emerge from the brainstem (see Figure 13.17). The sensory nuclei in which terminate the sensory fibers of these cranial nerves, and in which originate the motor nuclei from the motor fibers of these cranial nerves, are located in the brainstem.

Neural Pathways of Nuclei and Long Tracts

In the nervous system, the word **nucleus** means a collection of nerve cells *inside* the central nervous system. A similar collection of nerve cells *outside* the central nervous system is a **ganglion.**

As you learned in Chapter 11, adequate stimulation of the sensory receptors results in a nerve impulse, or action potential. The action potential is conveyed to the spinal cord and brain by afferent fibers. This afferent input is processed within the central nervous system in *nuclei* and *centers* (collections of cell bodies and dendrites). (The basal ganglia of the cerebrum are actually nuclei.) It is then conveyed through *tracts* (bundles of axons) to other nuclei and centers for further processing. These sequences of nuclei and tracts are called **pathways.**

Sensory (ascending) pathways extend upward from the spinal cord through the brainstem to the cerebrum and cerebellum. Such ascending pathways include pain, touch, and visual pathways. *Motor (descending) pathways* extend from the cerebrum and cerebellum to the brainstem and spinal cord. Table 12.1 on page 342 summarizes some of the major pathways. We will consider them in more detail in Chapters 14 and 15.

Reticular Formation

Deep within the brainstem is a slender but complex network of nerve cells and fibers called the **reticular formation** (L. *reticulum,* netlike). It is organized into (1) *ascending (sensory) pathways* from ascending spinal cord tracts and from the cerebellum; (2) *descending (motor) pathways* from the cerebral cortex and hypothalamus; and (3) *cranial nerves* (Figure 13.5). The reticular formation runs through the entire length of the brainstem (including the midbrain, pons, and medulla), with axons extending into the spinal cord and diencephalon. Many neurons in the reticular formation have as many as 30,000 synaptic connections with other fibers in the central nervous system. A lesion in the reticular formation of the upper brainstem may result in loss of consciousness *(coma),* which may last for months or even years.

Neurons within the reticular formation are organized into several groups, each having a specific and life-sustaining function. The reticular formation contains the respiratory and cardiovascular centers, which help to regulate such functions as breathing, heart rate, and the changing diameter of blood vessels. It also helps to regulate our level of awareness. When the effects of sensory stimuli pass through the brainstem on their way to the highest centers of the brain, they stimulate the reticular formation, which in turn results in the increased activity of the cerebral cortex.

| *Why do we eventually stop hearing a clock ticking?* | *The ticking of a clock, or any stimulus that is repeated 10 to 15 times at approximately 1-second intervals, usually ceases to arouse neurons in the reticular formation. If the same clock stops, however, we usually notice the silence. People who live near a highway rarely notice the repetitive traffic noise. The phenomenon is called habituation.* |

FIGURE 13.4

Brainstem, dorsal view. All cranial nerves except the olfactory and optic nerves emerge from the brainstem. Only the cranial nerves on the right side of the body are shown here. The oculomotor and abducens nerves emerge on the ventral side.

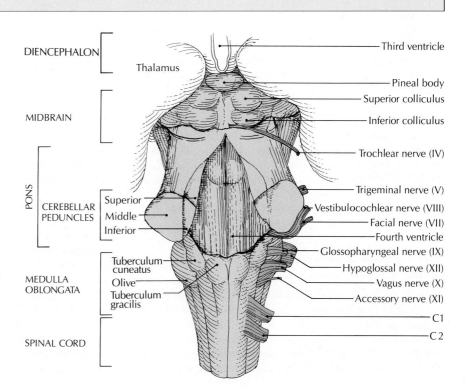

FIGURE 13.5

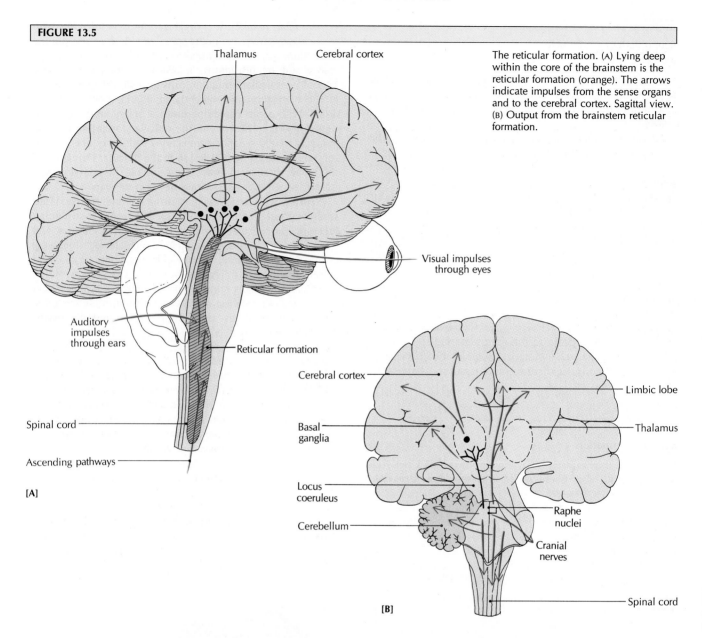

The reticular formation. (A) Lying deep within the core of the brainstem is the reticular formation (orange). The arrows indicate impulses from the sense organs and to the cerebral cortex. Sagittal view. (B) Output from the brainstem reticular formation.

[A]

[B]

The ascending pathways to the cerebrum are integrated in the ascending *reticular activating system,* which influences the brain's state of arousal or wakefulness. The reticular formation also contains the spinoreticulothalamic pain pathway, which is associated with the dull and diffuse qualities of pain. Descending motor pathways include the reticulospinal pathways, which convey motor impulses to the spinal cord.

Reticular Activating System

The reticular formation of the brainstem receives a great variety of sensory information as input from our internal and external environments. Projections from the reticular formation are widespread. They influence most of the processing centers and cortical areas of the central nervous system. Two specialized nuclear groups in the brainstem are especially in-

teresting: the *raphe nuclei,* located in the midsagittal region, and the *locus coeruleus,* located in the reticular formation of the pons (see Figure 13.5B). The neurons of the raphe nuclei contain the neurotransmitter serotonin, and those of the locus coeruleus contain norepinephrine (noradrenaline). The neurons of the raphe nuclei and locus coeruleus are unusual because their axons project throughout the central nervous system and directly to the cerebral cortex and cerebellum.

The influences from the reticular formation are expressed functionally by the **reticular activating system** (RAS) (sometimes referred to as the *arousal system*), which has roles associated with many behavioral activities. These roles include adjusting certain aspects of the sleep-wake cycle, awareness, alertness, levels of sensory perception, emotions, and motivation. The reticular activating system also helps the cerebellum to coordinate selected motor units to produce smooth, coordinated contractions of skeletal muscles, as well as maintaining muscle tonus.

Medulla Oblongata

The lowermost portion of the brainstem is the ***medulla oblongata***, or simply, the ***medulla***. About the same length (3 cm) as the midbrain and pons, it is situated in the inferior part of the cranial cavity.

The medulla is continuous with the spinal cord, and extends from the foramen magnum to the pons (see Figures 13.1 and 13.4). Within the reticular formation of the medulla are the bundles of nuclei that make up the vital cardiac, vasomotor (constriction and dilation of blood vessels), and respiratory centers. The medulla monitors carbon dioxide levels (as well as hydrogen-ion concentration) in the body so closely that it will cause respiration rates to double if carbon dioxide concentrations in the blood rise by as little as .03 percent. The medulla also regulates vomiting, sneezing, coughing, and swallowing.

The medulla is connected to the pons by longitudinal bundles of nerve fibers. It is joined to the cerebellum by the paired bundles of fibers called the *inferior cerebellar peduncles* (peh-DUNG-kuhlz; L. little feet).

The ventral surface of the medulla contains bilateral elevated ridges called the ***pyramids*** (see Figures 13.1 and 13.4). The pyramids are composed of the fibers of motor tracts from the motor cerebral cortex to the spinal cord. These ***pyramidal*** (corticospinal) ***tracts*** cross over, or *decussate*, in the lower part of the medulla to the opposite side of the spinal cord, forming an X.* This crossing over of the motor nerve fibers in the medulla is called the ***pyramidal decussation*** (Figure 13.6).

Because almost all motor and sensory pathways cross over, *each side of the brain controls the opposite side of the body.* The left cerebral hemisphere affects the muscles on the right side of the body, and the right cerebral hemisphere is similarly linked to the left side of the body. (See the essay on p. 384.)

On the lateral anterior surface of each pyramidal tract is an olive-shaped swelling, appropriately called the *olive* (see Figure 13.4). It is formed by the inferior olivary nucleus, whose fibers project and convey excitatory input signals through the inferior cerebellar peduncle to the cerebellum.

On the dorsal surface of the medulla are two pairs of bumps: the *clava* or *tuberculum gracilis* (L. lump + slender) and the *cuneate tubercle* or *tuberculum cuneatus* (L. wedge) (see Figure 13.4). These tubercula are formed by the relay nuclei (nuclei gracilis and cuneatus) in the posterior column-medial lemniscus pathway that convey touch and related sensations.

Several cranial nerves emerge from the medulla (Figure 13.7). The rootlets of cranial nerve XII emerge from the groove located anterior to the olive. The rootlets of nerves IX, X, and XI emerge from the groove located posterior to the olive. Cranial nerves VI, VII, and VIII emerge from the junction between the medulla and pons.

Pons

Just above the medulla is the ***pons*** (L. bridge), so named because it forms a connecting bridge between the medulla and the midbrain, the uppermost portion of the brainstem (see Figures 13.1 to 13.4). The posterior portion of the pons is called the *dorsal pons*, and the anterior portion is the *ventral pons*.

Dorsal pons The ***dorsal pons*** consists of the reticular formation, some nuclei associated with cranial nerves, ascending (sensory) pathways, and some fibers of descending (motor) pathways. Within the reticular formation are the *pneumotoxic* and *apneustic centers*, which help to regulate breathing. These respiratory centers are integrated with the respiratory centers of the medulla. The locations of the cranial nerve nuclei are shown in Figure 13.7.

The ascending (sensory) pathways include the neurons of the ascending reticular system, and such sensory tracts as the medial lemniscus (touch pressure) and spinothalamic and trigeminothalamic tracts (pain and temperature). The descending (motor) pathways are made up of the corticobulbar fibers, corticoreticular fibers, and rubrospinal tract. The corticobulbar fibers are pathways influencing the motor nuclei of the cranial nerves. The corticoreticular fibers regulate the activity of the reticulospinal tracts that project to the spinal cord. The rubrospinal tract conveys motor impulses from one side of the midbrain to the other side of the body. Nerve V emerges from the ventrolateral aspect of the pons.

Ventral pons The ***ventral pons*** contains the pontine nuclei, which are the relay nuclei of the corticopontocerebellar pathway. The pontocerebellar fibers of this pathway convey excitatory influences to the cerebellum via the middle cerebellar peduncle (see Figure 13.6c). The corticopontocerebellar pathway is the means by which the cerebral cortex communicates with the cerebellum.

Midbrain

The ***midbrain,*** or mesencephalon, is the segment of the brainstem located between the diencephalon and the pons (see Figures 13.1 and 13.4). It connects the pons and cerebellum with the cerebrum (forebrain). On the ventral surface of the midbrain is a pair of ***cerebral peduncles,*** made up of the pyramidal tract (fibers to the motor nuclei of the spinal nerves within the spinal cord), corticobulbar fibers (motor fibers to the cranial-nerve motor nuclei), and corticopontine fibers to the pons (see Figure 13.10). Emerging from the fossa between the peduncles on its ventral side is the pair of oculomotor nerves (cranial nerve III).

Passing through the midbrain is the cerebral aqueduct. The dorsal portion of the midbrain, situated above the aqueduct, is called the roof or *tectum* (L. roof). The tectum contains four elevations called *colliculi* (L. little hills); the colliculi are known collectively as the ***corpora quadrigemina*** (L. bodies of four twins). The *superior* pair of colliculi are reflex centers that help to coordinate the movements of the eyeballs and

---

Decussate comes from the Latin word *decussare*, which in turn is derived from *dec,* meaning ''10.'' The Latin symbol for 10 is X, which represents the crossing over of the pyramidal tracts.

FIGURE 13.6

Decussation (crossing over) of motor and sensory nerve fibers in the medulla. (A) Schematic drawing (anterior view) of ascending touch pathway and descending motor (corticospinal) tract. (B) Detail of decussation in the pyramidal (corticospinal) tracts. The ascending touch pathway is composed of a sequence of three neurons. The second neuron in the sequence decussates in the medulla. (C) The following pathways decussate: (1) corticopontine pathway from cerebral cortex to pontine nuclei to cerebellum; (2) cerebellothalamic pathway from cerebellum to thalamus to motor cerebral cortex; (3) corticospinal pyramidal tract from motor cortex to spinal cord.

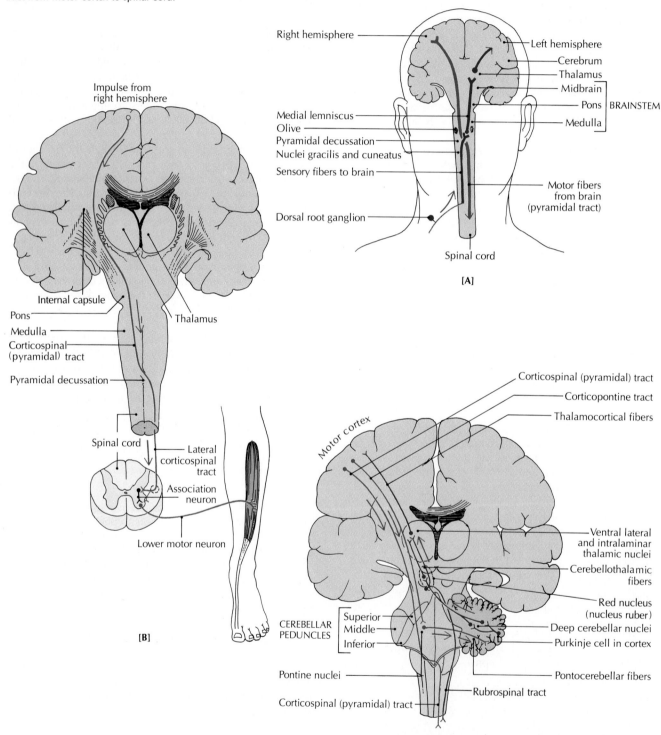

FIGURE 13.7

Location of cranial nerve nuclei within the brainstem. (A) Afferent (sensory) cranial nerve nuclei. (B) Efferent (motor) cranial nerve nuclei.

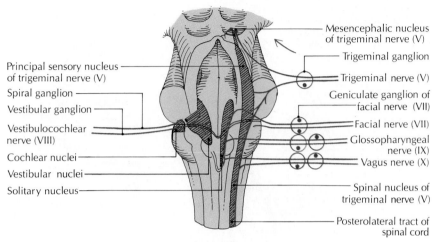

[A]

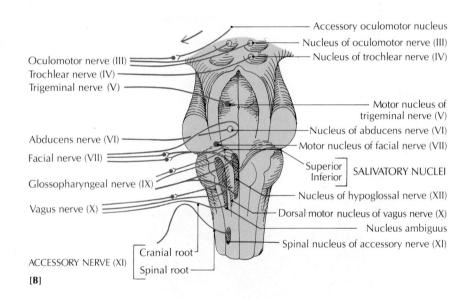

[B]

head, regulate the focusing mechanism in the eyes, and adjust the size of the pupils in response to certain visual stimuli. The trochlear nerves (cranial nerve IV) emerge from the roof of the midbrain. Just posterior are the *inferior colliculi,* which are relay nuclei of the auditory pathways conveying influences to the thalamus and eventually to the auditory cortex.

The large **red nucleus** *(nucleus ruber)* is a major motor nucleus of the reticular formation (see Figure 13.6B). (It is so named because it has a reddish appearance in the fresh state.) This nucleus is the termination point for nerve fibers from the cerebral cortex and cerebellum. It also gives rise to the rubrospinal tract to the spinal cord. The heavily black-pigmented nucleus called the **substantia nigra** (L. *nigra,* black) is integrated into neural circuits with the basal ganglia of the cerebrum (see Figure 13.11). The corticorubrospinal

pathway, along with the corticospinal tract, is involved in somatic motor activities. The substantia nigra is a basal ganglion. It has a role in Parkinson's disease (see p. 405).

Ask Yourself

1 *What are the three major components of the brainstem?*

2 *What is the reticular formation?*

3 *What are cerebral peduncles?*

4 *How do ganglia and nuclei differ?*

5 *What are pyramidal tracts?*

6 *What is pyramidal decussation?*

FIGURE 13.8

Cerebellum. (A) Sagittal view. (B) Superior surface, showing the lobes and vermis. (C) Inferior surface. (D) Sagittal section, showing the arbor vitae. [(B–D) *Murray L. Barr and John A. Kiernan,* The Human Nervous System: An Anatomical Viewpoint, *5th ed., Philadelphia, J. B. Lippincott, 1988.*]

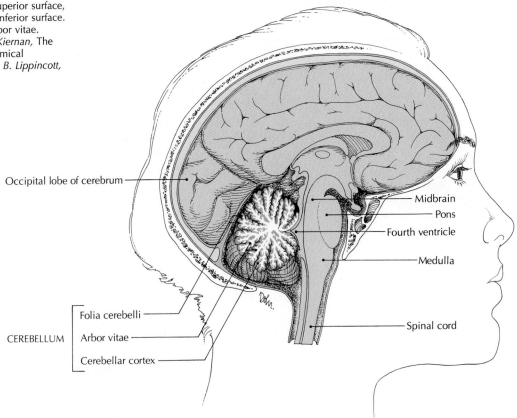

Occipital lobe of cerebrum

Midbrain

Pons

Fourth ventricle

Medulla

Spinal cord

CEREBELLUM

Folia cerebelli

Arbor vitae

Cerebellar cortex

[A]

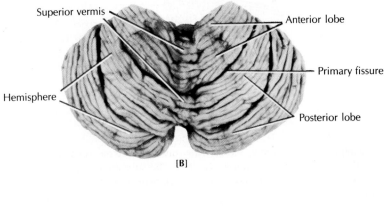

Superior vermis

Anterior lobe

Primary fissure

Hemisphere

Posterior lobe

[B]

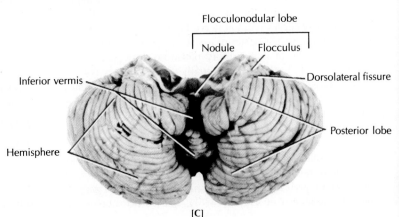

Flocculonodular lobe

Nodule   Flocculus

Inferior vermis

Dorsolateral fissure

Posterior lobe

Hemisphere

[C]

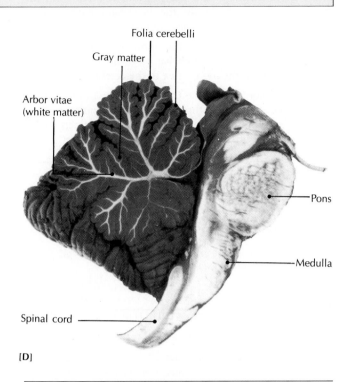

Folia cerebelli
Gray matter
Arbor vitae (white matter)
Pons
Medulla
Spinal cord

[D]

CEREBELLUM

The main role of the *cerebellum* (L. little brain) is to regulate balance, timing and precision of body movements, and body positions. It processes input from sensory receptors in the head, body, and limbs. Through connections with the cerebral cortex, vestibular system, and reticular formation, the cerebellum refines and coordinates muscular movements. It does not initiate any movements, and is not involved in the conscious perception of sensations.

Anatomy of the Cerebellum

The cerebellum is located behind the pons in the posterior cranial fossa (Figure 13.8). It is the second largest part of the brain, the cerebrum being the largest. The cerebellum is separated from the occipital lobes of the cerebrum by a fold of dura mater called the *tentorium cerebelli,* and by the transverse cerebral fissure.

The cerebellum may be divided into three parts: (1) a midline portion, called the *vermis* (L. wormlike), (2) two small flocculonodular lobes (vestibular cerebellum), and (3) two large lateral lobes. The *flocculonodular lobes* (L. *flocculus,* little tuft of wool), together with the centrally placed wormlike **vermis,** play a role in maintaining muscle tone, equilibrium, and posture through their influence on the motor pathways that regulate the activity of the muscles of the trunk.

The much larger **lateral lobes** or *hemispheres* of the cerebellum help to smooth out muscle movement, and they synchronize the delicate and precise timing of the many muscles involved with any complex activity. Such synchronization is

especially apparent in the movements of the upper and lower extremities.

The lobes of the cerebellum are covered by a surface layer of gray matter called the **cerebellar cortex** (L. bark or shell), which is composed of a network of tens of billions of neurons. The cerebellar cortex is corrugated, with long, parallel ridges called *folia cerebelli,* which are more regular than the gyri of the cerebral cortex (see Figure 13.8D). The folia are separated by *fissures* (deep folds). Under the cortex is a mass of white matter composed of nerve fibers. Lying deep within the white matter are the deep cerebellar nuclei, from which axons project out of the cerebellum to the cerebral cortex. A median section of the vermis reveals a branched arrangement of white matter, called the **arbor vitae** (VYE-tee; L. tree of life) (see Figures 13.8A, D).

The cerebellum is attached to the brainstem by three **cerebellar peduncles.** The *inferior cerebellar peduncle,* which connects the medulla to the cerebellum, is a bundle of fibers originating in the spinal cord, medulla, and vestibular system, and terminating in the cerebellum. In addition, some fibers from the cerebellum pass through this peduncle to the vestibular nuclei. A major source of input to the cerebellum is via fibers from the inferior olivary nucleus (which forms the olive of the medulla). The fibers cross over and pass through the inferior cerebellar peduncle. The *middle cerebellar peduncle* is composed of pontocerebellar fibers passing from the pontine nuclei of the opposite side to the cerebellum (see Figure 13.6C). The *superior cerebellar peduncle,* which connects the cerebellum to the midbrain, is composed of fibers from the deep nuclei of the cerebellum. These fibers cross over in the midbrain and terminate in the brainstem reticular formation, red nucleus, and thalamus.

Functions of the Cerebellum

The cerebellum integrates the contractions of muscles in relation to each other as they participate in a movement or series of movements. It is especially involved with coordinating agonists and antagonists in a cooperative way. The cerebellum smoothes out the action of each muscle group by regulating and grading muscle tension and tone in a precise and delicate way. Although the cerebellum *does not initiate any movements,* it participates in each movement through connections to and from the cerebral cortex.

The cerebellum continuously monitors sensory input from muscles, tendons, joints, and vestibular (balance) organs. These sensory *proprioceptive inputs* (the sense of the relative position of one body part to another) are derived on an unconscious level from neuromuscular spindles, tendon (Golgi) organs, the vestibular system, and other sensory endings. They are conveyed primarily through the inferior cerebellar peduncle. Some supplementary information is derived from the auditory and visual systems.

One way to understand the function of the cerebellum is to observe the results of a cerebellar lesion, such as might be caused by a stroke or a tumor on one side of the cerebellum. In such a lesion, the reflexes are diminished, and the absence of perfect coordination is shown through tremors and jerky, puppetlike movements. The condition is called *ataxia* (Gr.

lack of order). The patient has the symptoms on the same side of the body as the lesion because the fibers cross at two sites, making a "double cross" instead of the usual single crossing over (see Figures 13.6A, C). The double-cross expression of cerebellar activity is first conveyed (1) from the cerebellum to the cerebral cortex via a crossing in the midbrain; then (2) the control of the movement is conveyed by way of the pyramidal tract via a second crossing in the lower medulla to the opposite side of the body (the same side as the cerebellar lesion).

In addition to the cerebellum's involvement with ongoing motor activities, it is also involved with the neural events *prior to* voluntary movement. This is accomplished in the following way:

1 The idea to initiate and make a movement is generated in the sensory and association areas of the cerebral cortex.

2 One of the several neuronal circuits through which this decision is routed is the corticopontocerebellar pathway to the cerebellum.

3 This input interacts with the continuous input from the proprioceptive sensors from the head and body coming through the inferior cerebellar peduncle.

4 The cerebellum influences the motor systems by sending impulses via the superior cerebellar peduncle, first to the thalamus and then to the motor area of the cerebral cortex.

5 The motor cortex uses the input just prior to and during the ongoing movements through the corticospinal, corticobulbar, corticorubrospinal, and corticoreticular tracts to the brainstem and spinal cord.

Ask Yourself

1 *What are the parts of the cerebellum?*

2 *What is the cerebellar cortex?*

3 *What are the cerebellar peduncles?*

4 *What are the main functions of the cerebellum?*

TABLE 13.2   MAJOR STRUCTURES OF THE BRAIN

Structure	Description	Major functions
Basal ganglia	Large masses of gray matter contained deep within each cerebral hemisphere.	Help to coordinate muscle movements by relaying information via thalamus to motor area of cerebral cortex to influence descending motor tracts.
Brainstem*	Stemlike portion of brain, continuous with diencephalon above and spinal cord below. Composed of midbrain, pons, medulla oblongata.	Relays messages between spinal cord and brain, from brainstem cranial nerves to cerebrum. Helps control heart rate, breathing rate, blood pressure. Involved with hearing, taste, other senses.
Cerebellum	Second largest part of brain. Located behind pons, in posterior section of cranial cavity. Composed of cerebellar cortex, two lateral lobes, central flocculonodular lobes, medial vermis, some deep nuclei.	Processing center involved with coordination of muscular movements, balance, precision, timing, body positions. Processes sensory information used by motor systems.
Cerebral cortex	Outer layer of cerebrum. Composed of gray matter and arranged in raised ridges (gyri), grooves (sulci), depressions (fissures).	Involved with most conscious activities of living. (See major functions of cerebral lobes.)
Cerebral lobes	Major divisions of cerebrum, consisting of frontal, parietal, temporal, occipital lobes (named for bones under which they lie), insula. Also include limbic lobe.	Frontal lobe involved with motor control of voluntary movements, control of emotional expressions and moral behavior. Parietal lobe involved with general senses, taste. Temporal lobe involved with hearing, equilibrium, emotion, memory. Occipital lobe organized for vision and associated forms of expression. Insula may be involved with gastrointestinal and other visceral activities. Limbic lobe (along with the limbic system) is involved with emotions, behavioral expressions, recent memory, smell.
Cerebrospinal fluid	Fluid that circulates in ventricles and subarachnoid space.	Supports and cushions brain. Helps control chemical environment of central nervous system.
Cerebrum*	Largest part of brain. Divided into left and right hemispheres by longitudinal fissure and divided into cerebral lobes. Also contains cerebral cortex (gray matter), white matter, basal ganglia, diencephalon.	Controls voluntary movements, coordinates mental activity. Center for all conscious living.

*In this book, the following terminology is used: The *brainstem* is made up of the midbrain, pons, and medulla oblongata; the *cerebrum* includes the cerebral hemispheres and diencephalon. Some textbooks use variations of these definitions.

Structure	Description	Major functions
Corpus callosum	Bridge of nerve fibers that connects one cerebral hemisphere with the other.	Connects cerebral hemispheres, relaying sensory information between them. Allows left and right hemispheres to share information, helps to unify attention.
Cranial nerves	Twelve pairs of peripheral cranial nerves. Olfactory (I) and optic (II) arise from cerebrum; others (III through XII) arise from brainstem. May be sensory, motor, or mixed.	Concerned with senses of smell, taste, vision, hearing, balance. Also involved with specialized motor activities, including eye movement, chewing, swallowing, breathing, speaking, facial expression.
Diencephalon	Deep portion of brain. Composed of thalamus, hypothalamus, epithalamus, ventral thalamus.	Connects midbrain with cerebral hemispheres. (See major functions of thalamus and hypothalamus.)
Hypothalamus	Small mass below the thalamus; forms floor and part of lateral walls of third ventricle.	Highest integrating center for autonomic nervous system. Controls most of endocrine system through its relationship with the pituitary gland. Regulates body temperature, water balance, sleep-wake patterns, food intake, behavioral responses associated with emotion.
Medulla oblongata	Lowermost portion of brainstem. Connects pons and spinal cord. Site of decussation of descending direct corticospinal (motor) tract and an ascending sensory (touch, etc.) pathway from spinal cord to thalamus; emergence of cranial nerves VI through XII; movement of cerebrospinal fluid from ventricle to subarachnoid space.	Contains vital centers that regulate heart rate, breathing rate, constriction and dilation of blood vessels, blood pressure, swallowing, vomiting, sneezing, coughing.
Meningeal spaces	Spaces associated with meninges. Potential epidural space between skull and dura mater; potential subdural space between dura mater and arachnoid; subarachnoid space between arachnoid and pia mater (contains cerebrospinal fluid).	Provide subarachnoid circulatory paths for cerebrospinal fluid, protective cushion.
Meninges	Three layers of membranes covering brain. Outer tough dura mater; arachnoid; inner delicate pia mater, which adheres closely to brain.	Dura mater adds support and protection. Arachnoid provides space between the arachnoid and pia mater for circulation of cerebrospinal fluid. Choroid plexuses, at places where pia mater and ependymal cells meet, are site of formation of much cerebrospinal fluid. Pia mater provides blood vessels that supply blood to brain.
Midbrain	Located at upper end of brainstem. Connects pons and cerebellum with cerebrum. Site of emergence of cranial nerves III, IV.	Involved with visual reflexes, movement of eyes, focusing of lens, dilation of pupils.
Pons	Short, bridgelike structure composed mainly of fibers that connect midbrain and medulla, cerebellar hemispheres, and cerebellum and cerebrum. Lies anterior to cerebellum and between midbrain and medulla. Site of emergence of cranial nerve V.	Controls certain respiratory functions. Serves as relay station from medulla to higher structures in brain.
Reticular formation	Complex network of nerve cells organized into ascending (sensory) and descending (motor) pathways. Located throughout core of entire brainstem.	Specific functions for different neurons, including involvement with respiratory and cardiovascular centers, regulation of brain's level of awareness.
Thalamus	Composed of two separate bilateral masses of gray matter. Located in center of cerebrum.	Intermediate relay structure and processing center for all sensory information (except smell) going to cerebrum.
Ventricles	Cavities within brain that are filled with cerebrospinal fluid. Left and right lateral ventricles in cerebral hemispheres, third ventricle in diencephalon, fourth ventricle in pons and medulla.	Provide circulatory paths for cerebrospinal fluid. Choroid plexuses associated with ventricles are site of formation of most cerebrospinal fluid.

CEREBRUM

The largest and most complex structure of the nervous system is the ***cerebrum*** (suh-REE-bruhm; L. brain). It consists of two cerebral hemispheres (see Figure 13.1). Both hemispheres are composed of a cortex (gray matter), white matter, and basal ganglia. The cortex is further divided into six lobes: the frontal, parietal, temporal, occipital, limbic, and insula (central lobe) (Figure 13.9). Each of the first four lobes contains special functional areas, including speech, hearing, vision, movement, and the appreciation of general sensations. The olfactory nerve and bulb are located beneath the frontal lobe (see Figure 13.13).

All of our conscious living depends on the cerebrum. Popularly, it is considered the region where thinking is done, but no specific parts of the cerebrum have been identified as the exact areas of consciousness or intellectual learning.

Anatomy of the Cerebrum

The cerebrum has a surface mantle of gray matter called the ***cerebral cortex.*** The cortex is a thin (about 4.5 mm), convoluted covering containing about 50 billion neurons and 250 billion glial cells. The raised ridges of the cortex are called convolutions or ***gyri*** (JYE-rye; singular, *gyrus*), which are separated by slitlike grooves called ***sulci*** (SUHL-kye; singular, *sulcus*). The gyri and sulci increase the surface area of the cerebral cortex, resulting in a 3:1 proportion of cortical gray matter to the underlying white matter (see Figure 13.9c). Although each person has a specific pattern of gyri and sulci, the overall developmental design of the cerebrum produces a fairly consistent pattern of gyri and sulci.

Extremely deep grooves or depressions are called ***fissures*** (L. *fissus*, crack). The cerebral hemispheres are separated by the *longitudinal fissure*, and the cerebrum is separated from the cerebellum by the *transverse cerebral fissure.*

Beneath the cortex lies a thick layer of white matter. The white matter consists of interconnecting groups of axons projecting in two basic directions. One group extends *from the cortex* to other cortical areas of the same and opposite hemisphere, to the thalamus, to the basal ganglia, to the brainstem, or to the spinal cord. The other group extends *from the thalamus* to the cortex. Actually, the thalamus is functionally integrated with the cerebral cortex in the highest sensory and motor functions of the nervous system.

The white matter consists of three types of fibers (association fibers, commissural fibers, and projection fibers), all of which originate in cell bodies in the cerebral cortex (Figure 13.10):

1 *Association fibers* link one area of the cortex to another area of the cortex of the *same hemisphere.*

2 *Commissural fibers* are the axons that project from a cortical area of one hemisphere to a corresponding cortical area of the *opposite hemisphere.* The two major cerebral commissures are the *anterior commissure* and the massive bundle of axons called the ***corpus callosum*** (L. hard body). Both connect the two cerebral hemispheres, and relay nerve impulses between them (see Figure 13.1). The corpus callosum contains about 200 million nerve fibers.

FIGURE 13.9

Gyri, sulci, and fissures of the cerebral hemispheres. (A) Superior view. (B) Right lateral view.

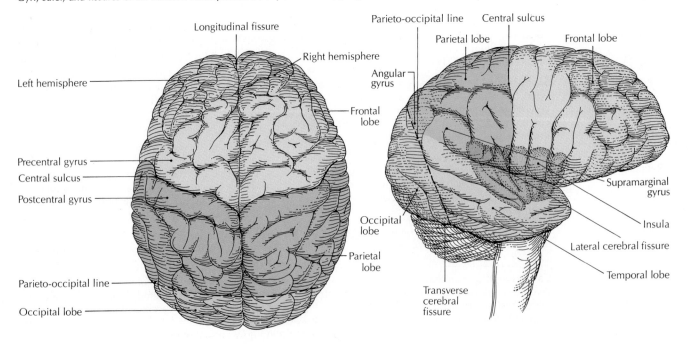

[A]   [B]

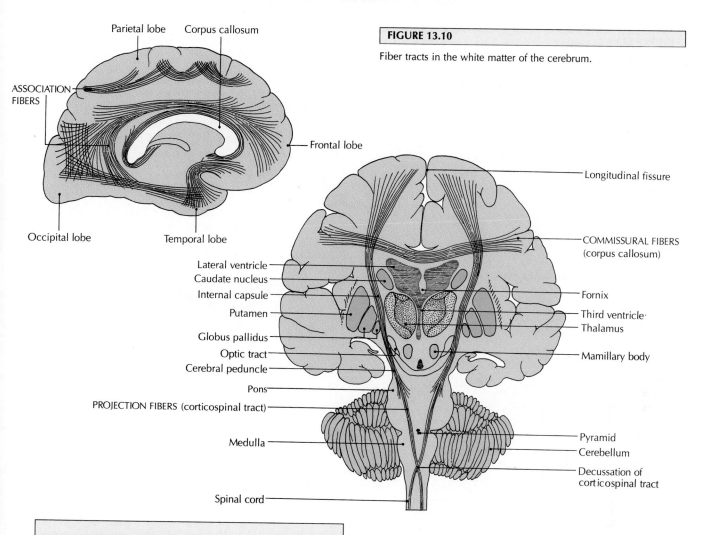

Parietal lobe · Corpus callosum

ASSOCIATION FIBERS

Frontal lobe

Occipital lobe · Temporal lobe

FIGURE 13.10

Fiber tracts in the white matter of the cerebrum.

Longitudinal fissure

COMMISSURAL FIBERS (corpus callosum)

Lateral ventricle
Caudate nucleus
Internal capsule
Putamen
Globus pallidus
Optic tract
Cerebral peduncle
Pons
PROJECTION FIBERS (corticospinal tract)
Medulla
Spinal cord

Fornix
Third ventricle·
Thalamus
Mamillary body
Pyramid
Cerebellum
Decussation of corticospinal tract

(c) In this photo, the cranial meninges have been removed to show the convolutions of the cerebral cortex more clearly. [*Martin M. Rotker, M.D./Taurus.*]

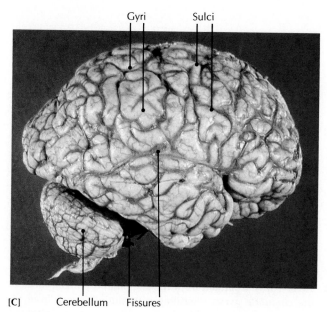

Gyri   Sulci

[C]   Cerebellum   Fissures

3 *Projection fibers* are the axons that project from the cerebral cortex *to other structures* of the brain, such as the basal ganglia, thalamus, and brainstem, and to the spinal cord. The corticospinal (pyramidal) tract from the motor cortex to the spinal cord is composed of projection fibers.

Beneath the white matter of each cerebral hemisphere is a large core of gray matter called the **basal ganglia** (Figure 13.11). The paired basal ganglia (which are actually nuclei) are the cell bodies of neuron clusters that help to coordinate muscle movements by relaying information from the cerebral cortex back to the motor cortex of the cerebrum. The inner core of the basal ganglia is called the **lentiform** ("lens-shaped") **nucleus.** It is made up of the *globus pallidus* ("pale ball") and the *putamen* (pyoo-TAY-muhn; L. prunings). Following the contour of each lateral ventricle is a **caudate** ("tail-shaped") **nucleus.** The caudate nucleus and lentiform nucleus together form the **corpus striatum** (L. furrowed body), the largest mass of gray matter in the basal ganglia. The caudate nucleus and putamen are called the *striatum.* The *subthalamus* (ventral thalamus) of the diencephalon and the *substantia nigra* (a small bundle of gray matter) of the midbrain are now considered to be basal ganglia.

FIGURE 13.11

Basal ganglia. (A) Section through the brain, showing the parts of the basal ganglia (orange). (B) Structures that make up the basal ganglia.

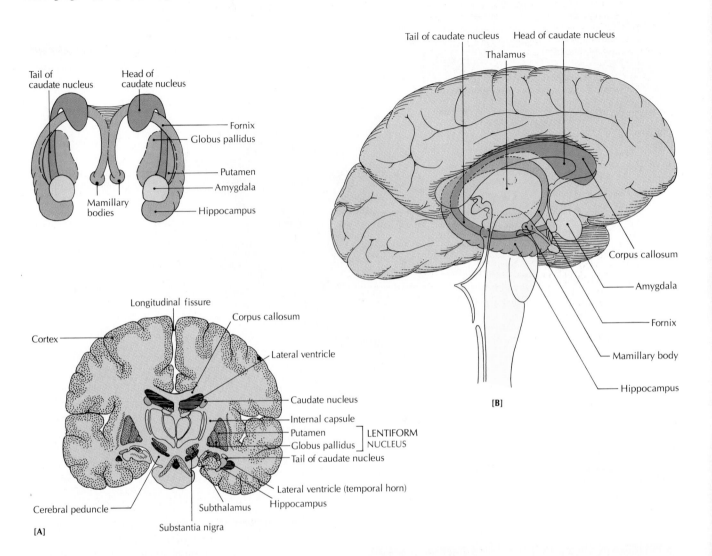

[A]

[B]

Functions of the Cerebrum

The two cerebral hemispheres have the same general appearance, but each has slightly different functions. One of the hemispheres, usually the left, is active in speech, writing, calculation, language comprehension, and analytic thought processes (Figure 13.12). The other hemisphere, usually the right, is more specialized for the appreciation of spatial relationships, conceptual nonverbal ideas, simple language comprehension, and general thought processes. (See the essay on p. 384.) The left hemisphere sorts out the *parts* of things, while the right hemisphere concentrates on the *whole*. In a manner of speaking, the left hemisphere sees the trees, but not much of the forest, while the right hemisphere sees the forest, but not too many of the trees.

The specific functions of the cerebrum, and their localized areas, are discussed further in the following sections on the cerebral lobes.

Cerebral Lobes

Each cerebral hemisphere is subdivided into six lobes: the frontal, parietal, temporal, occipital, insula (central lobe), and limbic lobes (see Figures 13.9 and 13.12). The first four lobes are named for the skull bones covering them. The frontal lobe is separated from the parietal lobe by the **central sulcus,** or *fissure of Rolando.* The **lateral cerebral sulcus** *(fissure of Sylvius)* divides the frontal and parietal lobes from the temporal lobe. The parietal and temporal lobes are separated from the occipital lobe by the arbitrary *parieto-occipital line* (see Figure 13.9). Buried deep in the central sulcus is the small central lobe, or *insula.*

Another subdivision of each hemisphere, the **limbic lobe,** is on the medial surface of the cerebral hemisphere (Figure 13.13). It is separated from the frontal and parietal lobes by the **cingulate gyrus** ("girdling convolution") and from the temporal lobe by the **rhinal sulcus.** The **parieto-occipital sul-**

FIGURE 13.12

Cerebrum. (A) A partial Brodmann's numbered map of the right lateral cerebral hemisphere. Brodmann's maps are useful mainly to show regions of different neural structure, but some areas also have a functional designation. (B) A functional map of the right cerebral hemisphere, showing some of the major areas of specialization. The four major cerebral lobes are also shown.

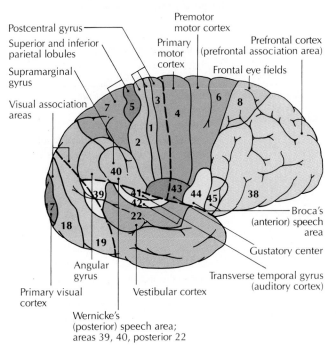

[A]

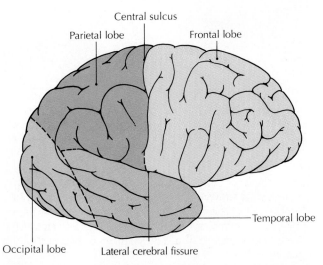

[B]

FIGURE 13.13

Limbic system. (A) Simplified right lateral view. (B) Major components of the limbic system.

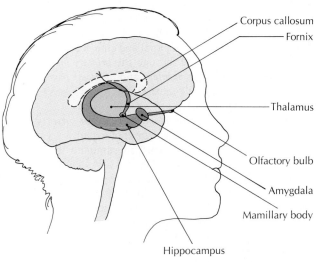

[A]

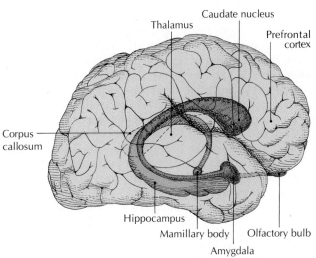

[B]

cus on the medial surfaces separates the occipital lobe from the parietal lobe.

Frontal lobe The *frontal lobe* (also called the "motor lobe") is involved with two basic cerebral functions. One is the motor control of voluntary movements, including those associated with speech. The other is the control of a variety of emotional expressions and moral and ethical behavior. The motor activity is expressed through the *primary motor cortex* (Brodmann area 4), *supplementary* and *premotor motor cortex* (area 6), the *frontal eye field* (area 8), and *Broca's speech area** (area 44) (see Figure 13.12).

From the **primary** and **supplementary motor cortex,** nerve impulses are conveyed through the motor pathways to the brain, brainstem, and spinal cord to stimulate the motor nerves of the voluntary muscles. Figure 13.14 shows the body drawn in proportion to the cortical space allotted to sensations and movement patterns of different parts of the body. Relatively large cortical areas are devoted to the face, larynx, tongue, lips, and fingers, especially the thumb. The thumb is so important for our dexterity that more of the brain's gray matter is devoted to manipulating the thumb than to controlling the thorax and abdomen. The disproportionate allotment of the cortical areas to these body parts reflects the delicacy with which facial expressions, vocalizations, and manual manipulations can be controlled.

The *frontal eye field* is a cortical area that regulates the scanning movements of the eyes, such as searching the sky to locate an airplane. **Broca's speech area** *(anterior speech area)* is critically involved with the formulation of words.

The cortex in front of the motor areas is called the *prefrontal cortex.* It has a role in the various forms of emotional expression. The functional role of the prefrontal cortex has been deduced largely from patients who have had the white matter of various areas of the prefrontal lobe surgically removed or cut, a procedure called a prefrontal lobotomy. Following the operation, the patients are usually less excitable and less creative than before, but vent their feelings frankly, without the typical restraint. Physical drive is lowered, but intelligence is not. An awareness of pain remains, but normal feelings associated with the intensity of pain are somehow lost.

Parietal lobe The *parietal lobe* is concerned with the evaluation of the general senses, and of taste. It integrates and processes the general information that is necessary to create an awareness of the body and its relation to its external environment. The parietal lobe is composed of the *postcentral gyrus* (areas 3, 1, 2), which integrates general sensations, the *superior* and *inferior parietal lobules* (areas 5, 7), and the *supramarginal gyrus* (area 40) (see Figure 13.12A). The postcentral gyrus is called the **primary somesthetic association area** (general sensory area) because it receives information about the general senses from receptors in the skin, joints, muscles, and body organs (see Figure 13.12B).

A sensation is perceived and localized on a general, vague level of awareness by the thalamus, but in the parietal lobe the sensory modalities are perceived on a more complex and subtle level. As a result, the source of touch or pain can be localized more accurately. At a still higher level, areas 5, 7, and 40 are involved in the appreciation of the texture, weight, shape, and form of an object held in the hand. Area 40 has a critical role in the processing of general sensory information, which produces our perception of our "body image."

A gustatory (taste) area (area 43) is located within, and near the vicinity of, the base of the postcentral gyrus.

| *Do the most intelligent people have the largest brains?* | *Not necessarily. Some of the most intelligent people in history had smaller-than-average brains, and one of the largest brains ever measured belonged to an idiot. Brain surface area may be a better indicator of intelligence, but it is by no means infallible.* |

FIGURE 13.14

The human body is distorted in this model, so that the body parts are proportional to the space allotted in the brain to sensations from different parts of the body. If the eyes were drawn to scale, they would be larger than the entire body. [*British Museum of Natural History.*]

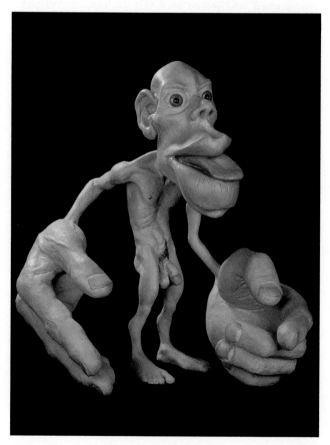

---

*This area was named for Pierre Paul Broca, who first described it in the late nineteenth century.

Temporal lobe The *temporal lobe* is the lobe located closest to the ears. It has critical functional roles in hearing, equilibrium, and to a certain degree, emotion and memory. The *auditory cortex* is located in the *transverse temporal gyrus* (of Heschl) (areas 41, 42), just anterior to the vestibular (equilibrium) area (see Figure 13.12). Electrical stimulation of the auditory cortex elicits such elementary sounds as clicks and roaring, while the same stimulation of the vestibular area may produce feelings of dizziness and vertigo (a sense of rotation). The understanding of the spoken word involves the participation of the *angular gyrus* (area 39), located in the zone between the temporal and occipital lobes.

The temporal lobe contains another visual cortex known as the *inferotemporal cortex.* This area is thought to be involved in gaining attention and detecting movement within the visual fields.

The *anterior temporal lobe,* together with certain structures in the limbic lobe, is involved in some way with emotion, hallucinations, and short-term memory (recall of recent events, from a few minutes to weeks or months ago). The bilateral removal of the anterior temporal lobe results in the loss of recent memory, but long-term memory and intellect are retained.

Occipital lobe Although the *occipital lobe* is relatively small, it is important because it contains the visual cortex. It is made up of several areas organized for vision and its associated forms of expression. The general region where the parietal, temporal, and occipital lobes mesh on the lateral surface is essential for the comprehension of the written and spoken word, and for an appreciation of the scheme and image of the body.

Visual images from the retina of the eye are transmitted and projected to the *primary visual cortex* (area 17) (see Figure 13.12). Each specific site in the retina is represented by a specific site in area 17. Visual information from this area is conveyed for further processing and elaboration to the visual association area (areas 18 and 19), and then to other areas such as the *angular gyrus* (area 39). Areas 18 and 19 are involved with assembling the features of a visual image and making it meaningful.

Wernicke's area (posterior speech area) (areas 39, 40, 22) is concerned with the understanding of the written word (area 39) and the spoken word (area 22), and with the ability to conceive the symbols of language. Lesions on the dominant side of Wernicke's area 39 (usually the left side) can produce *alexia,* a failure to recognize written words. (*Dyslexia* is associated with reading and writing disabilities; see p. 406.) At the same time, the primary visual perception remains intact, and the patient may accurately describe the shape, color, and size of an object.

Insula Little is known about the role of the *insula* (L. island), or *central lobe,* which appears to be an island in the midst of the rest of the brain. It seems to be associated with gastrointestinal and other visceral activities.

Limbic lobe and limbic system The *limbic lobe* (L. *limbus,* border) is the ring of the cortex, located on the medial surface of each cerebral hemisphere, and surrounding the central core of the cerebrum (see Figure 13.13B). This lobe is composed of the cingulate gyrus, isthmus, parahippocampal gyrus, and uncus.

The *olfactory cortex,* which is involved with the perception of odors, is located in the region of the uncus. Irritation of this region may produce hallucinations of an odor (usually foul), along with fear and feelings of unreality. The cingulate sulcus separates the cingulate gyrus from the frontal and parietal lobes. The rhinal sulcus separates the temporal lobe from the parahippocampal gyrus and uncus.

The *limbic system* is defined in functional terms as an assemblage of cerebral, diencephalic, and midbrain structures that are actively involved in memory and emotions and the visceral and behavioral responses associated with them. This physiologically defined system includes such structures as the limbic lobe, amygdala, hippocampus, and parts of other structures such as the thalamus and midbrain. Two of these structures are closely associated with the limbic lobe. One is the *hippocampus* (Gr. sea horse; so named because it is S-shaped), which is located medial to the parahippocampal gyrus, in the floor of the lateral ventricle. The other is *amygdala,* a complex of nuclei located deep within the uncus (see Figure 13.13).

The amygdala and hippocampus are integrated with the functional expressions of the limbic system. When the amygdala of a cat is stimulated electrically, the cat may begin to sniff, lick, bite, or show other behavioral patterns associated with getting food or drink. It may also show attack patterns associated with "fight or flight," including dilated pupils, extended claws, or erect back hair. Such expressions stop as soon as the electrical stimulation is discontinued.

The hippocampus is involved with memory traces for recent events. Patients with Korsakoff's syndrome, in which lesions are present in the hippocampus, have a loss of short-term memory, and lose their sense of time. They tend to become confused easily, forget questions that were just asked, and may reply to questions with irrelevant answers. All such expressions of the amygdala and hippocampus are associated with an interaction with other centers. For example, the impact of inputs of the sensory systems from the sensory cortical areas upon these structures can trigger a variety of emotional and behavioral responses.

The limbic system responds to sensory stimuli by way of the hypothalamus, and in fact, the main outlet for the overt expression of the activities of the limbic system is the hypothalamus. Pathways from the limbic system project to the hypothalamus. In turn, the hypothalamus exerts widespread effects upon the entire body via (1) the autonomic nervous system, and (2) the endocrine system through the anterior pituitary gland. Autonomic responses include an increase in salivation, changes in the digestive process, and alterations in the blood pressure. As you saw above, somatic motor activities include responses associated with eating and with self-protection.

In an experimental situation, limbic stimulation can alter blood hormone levels and cause other dramatic changes. The stimulation of certain areas of limbic structures produces feelings of intense pleasure, while the stimulation of other sites leads to feelings of extreme pain. Still other sites, known as punishing centers, evoke fear and even terror.

Left Brain, Right Brain

Because the corticospinal pathway crosses in the lower medulla at the base of the brain, each side of the brain controls actions on the *opposite* side of the body. This much has been known for at least 100 years. But recent experiments have begun to refine our knowledge of the inner workings of the brain.

Connecting the left and right cerebral hemispheres, and relaying messages between them, is the corpus callosum. In normal people, the corpus callosum makes it possible for both sides of the brain to work together.

Split-Brain Experiments

In the early 1950s, Roger W. Sperry and his coworkers at the California Institute of Technology began experiments that strengthened the case for the "split-brain" theory. In 1981, Sperry received the Nobel Prize in Medicine for his work with split-brain patients, individuals whose corpus callosum had been severed in an effort (usually successful) to relieve epileptic seizures. (This procedure did not alter the patients' normal personality or behavior.)

In one experiment, when Sperry told split-brain patients to hold a spoon in their right hand (without looking at the spoon or their hand), the patients usually could describe the spoon without difficulty. If the spoon was held in the *left* hand, however, the patients could *not* describe it. Why? Because the speech center is located in the *left* hemisphere, which also controls the actions of the *right* hand. Without a corpus callosum, objects held in the left hand could not be described because there was no way to transmit the information to the main speech center in the left hemisphere.

In a similar experiment, subjects faced a split picture screen, focusing on the midline. A picture of a spoon was projected onto the left side of the screen, and the image of a knife was shown on the right side. When asked to describe what they saw, the subjects reported seeing a knife. (They actually saw both pictures, but the verbal side of the split brain—the left—could report verbally on only the picture on the right.) Then the subjects were asked to pick up the same object they saw on the screen—the knife. Instead of a knife, however, they picked up a spoon. When asked to identify the object, the split-brain subjects said they had picked up a knife. Although they *knew* it was a spoon, they were unable to *say* it. Without an intact corpus callosum to unify their attention and allow the two hemispheres to *share* information, the subjects were hopelessly confused. Each half of the brain was competing for control, asserting its own sense of the outside world. (Most split-brain individuals learn to adjust, reducing their conflict enough so that they can begin to live a fairly normal life.)

Many experiments by Sperry and others have helped to prove that the left and right cerebral hemispheres are involved with different skills, including:

Left cerebral hemisphere

Right-hand control
Spoken language
Written language
Scientific skills
Numerical skills
Reasoning
Sorts out parts

Right cerebral hemisphere

Left-hand control
Music awareness
Recognition of faces and other three-dimensional shapes

Diencephalon

The *diencephalon* (L. "between brain") is the deep part of the cerebrum, connecting the midbrain with the cerebral hemispheres. It houses the third ventricle, and is composed of the thalamus, hypothalamus, epithalamus, and ventral thalamus (subthalamus). The pituitary gland (hypophysis) is connected to the hypothalamus (see Chapter 16).

Flanking the diencephalon laterally is the *internal capsule*, a massive structure of white matter made up of ascending fibers that convey information to the cerebral cortex, and descending fibers *from* the cerebral cortex. Many fibers of the internal capsule continue caudally into the cerebral peduncles of the midbrain (see Figure 13.10).

Thalamus The *thalamus* (Gr. inner chamber, so named because early anatomists thought it was hollow, resembling a room) is composed of two egg-shaped masses of gray matter covered by a thin layer of white matter. It is located in the center of the cranial cavity, directly beneath the cerebrum and above the hypothalamus. It forms the lateral walls of the third ventricle (see Figure 13.3).

The thalamus is the intermediate relay point and processing center for all sensory impulses (except the sense of smell) ascending to the cerebral cortex from the spinal cord, brainstem, cerebellum, basal ganglia, and other sources. After processing the input, the thalamus relays its output to the cerebral cortex. The thalamus is involved with four major areas of activity:

1 *Sensory systems.* Fibers from the thalamic nuclei project into the sensory areas of the cerebral cortex, where the sensory input is "decoded" and translated into the appropriate sensory reaction. For example, light is "seen," and sound is "heard." Crude sensations and some aspects of the general senses may also be brought to consciousness in the thalamus.

2 *Motor systems.* The thalamus has a critical role in influencing the motor cortex. Some thalamic nuclei receive neural input from the cerebellum and basal ganglia, and then project into the motor cortex. The motor pathways that regulate the voluntary muscles innervated by the cranial and spinal nerves originate in the motor cortex.

3 *General neural background activity.* Background neurophysiological activities of the brain, such as the sleep-wake cycles and the electrical brain waves, are expressed in the cerebral cortex. These cortical rhythms are generated and

Art awareness
Insight and imagination
Grasps the whole

Although the latest research supports the idea that each hemisphere controls the actions of the opposite side of the body, it disputes the notion that the brain is divided into neat compartments that totally control separate body functions. Apparently, such functions as speaking, for example, are not as localized as they were once thought to be. Radioactive tracers in the brains of experimental volunteers who were asked to speak out loud indicated that the flow of blood to the left hemisphere increased, as expected, but blood flow also increased in the right side. This is an important point. We must understand that the brain probably functions as a whole. It would be simplistic to believe that any given function exists in only one isolated part of the brain, without any relation to any other part.

Some Interesting Split-Brain Facts

Young children have speech potential in both hemispheres, and if the left hemisphere is badly damaged, they develop a speech center in the right hemisphere. If there are no unusual circumstances, speech dominance is fixed in the left hemisphere by the time a child is about 8 years old.

People usually turn their eyes and head to the right when they are thinking deeply.

When a person is playing the piano and humming the music at the same time, the right hand is more accurate than the left. When one melody is being played and another is being hummed, however, the *left* hand makes fewer mistakes than the right. This happens because if an activity calls for two parts of the brain to perform closely related tasks, a person functions best when both control centers are located in the same hemisphere. If the task demands less-related activities at the same time, a person performs best when the tasks are controlled by different hemispheres.

About 10 percent of the population is left-handed. A disproportionate number of great artists have been left-handed, showing a possible advantage in using the left hand for a creative action that is controlled by the opposite, right, hemisphere. Left-handed people also show less difference between their left and right hemispheres than right-handers do. Also, left-handers vary greatly among themselves in the way their brains process information.

Which face is happier? Most people choose the top one, even though the faces are mirror images. According to one theory, more people choose the top face because they are making the decision with their *right* brains, the side that controls emotions.

---

monitored by thalamic nuclei, which receive much of their input from the ascending reticular systems in the brainstem.

4 *Expression of the cerebral cortex.* The thalamus, through its connections with the limbic system, helps to regulate many expressions of emotions and uniquely human behaviors. In fact, it is linked with the highest expressions of the nervous system, such as thought, creativity, interpretation and understanding of the written and spoken word, and the identification of objects sensed by touch. Such accomplishments are possible because of the two-way communication between the thalamus and the association area of the cortex.

Hypothalamus The *hypothalamus* ("under the thalamus") lies directly under the thalamus (see Figure 13.1). It is a small region (about the size of a lump of sugar, and only 1/300 of the brain's total volume) located in the floor of the diencephalon, forming the floor and part of the lateral walls of the third ventricle. Extending from the hypothalamus is the *hypophysis (pituitary gland),* which is neatly housed within the sella turcica in the body of the sphenoid bone.

The hypothalamus regulates body functions to maintain homeostasis, including the regulation of body fluids, metabolism, blood sugar levels, body temperature, and repro-

ductive cycles. When the hypothalamus detects chemical changes in the blood, it instructs the pituitary gland to release the appropriate hormones to correct the imbalance. (The relationship of the hypothalamus to the pituitary gland is discussed in depth in Chapter 16, The Endocrine System.) It is the chief regulator of the autonomic nervous system and many visceral functions. It controls, modulates, and influences many physiological and endocrine activities, and it contains centers that regulate basic behavioral patterns, such as those involved in feeding, fighting, reproduction, and escape.

The most important functions of the hypothalamus are the following:

1 *Integration with the autonomic nervous system.* The hypothalamus is considered to be the highest integrative center associated with the autonomic nervous system. Through the autonomic nervous system, it adjusts the activities of other regulatory centers, such as the cardiovascular centers in the brainstem. The hypothalamus modifies blood pressure, peristalsis (muscular movements that push food along the digestive tract) and glandular secretions in the digestive system, secretion of sweat glands and salivary glands, control of the urinary bladder, and rate and force of the heartbeat.

2 *Temperature regulation.* The hypothalamus plays a vital role in the regulation and maintenance of body temperature. Specialized nuclei within the hypothalamus monitor the temperature of the blood to within 1/100 of a degree Celsius. A decrease in temperature activates a neuronal center in the posterior hypothalamus. This triggers mechanisms for heat production and heat conservation, such as increased metabolic activity, shivering, constriction of blood vessels in the skin, and decreased sweating. A rise in blood temperature activates a neuronal center in the anterior hypothalamus that triggers activities to prevent a rise in the body temperature, such as the dilation of blood vessels of the skin, increased sweating (which produces cooling), and panting.

3 *Water and electrolyte balance.* The hypothalamus has a "thirst center" and a "thirst-satiety (satisfaction) center" that help to produce a balance of fluids and electrolytes in the body. These centers regulate the intake of water (through drinking) and its output (through kidneys and sweat glands). Certain neurons in the hypothalamus, known as *osmoreceptors*, maintain homeostasis by sensing and monitoring the osmotic concentration of the blood. As the blood passes through the nuclei, the osmoreceptor neurons trigger activities that help maintain a normal extracellular fluid volume. For example, when the water level in the blood is reduced, causing the osmotic concentration to rise, a hormone called *antidiuretic hormone* (ADH) is produced by the hypothalamus and released into the bloodstream by the posterior lobe of the pituitary gland. ADH stimulates the kidneys to absorb water from newly formed urine, and return the water to the bloodstream (see Chapter 16).

4 *Sleep-wake patterns.* The hypothalamus is integrated with the neural circuitry that regulates sleep-wake patterns and the state of awareness (see p. 390 for a discussion of sleep).

5 *Food intake.* The hypothalamus has a crucial role in food consumption. When neural centers collectively called the "appestat," or "hunger or feeding centers" are stimulated by low levels of glucose, fatty acids, and amino acids in the blood and possibly other factors, they activate appropriate body responses to satisfy the deficiency: you experience hunger. After food has been eaten, the "satiety (satisfaction) center" is stimulated to inhibit the feeding center. New research speculates that the hypothalamus contains numerous neurons that release the neurotransmitters serotonin, dopamine, and norepinephrine, all of which are suspected of being involved with appetite. The *exact hypothalamic location* of the release sites is crucial. For example, norepinephrine in the mid-hypothalamus *stimulates* appetite, but norepinephrine in the lateral hypothalamus *inhibits* appetite. This portion of the hypothalamus also helps to regulate peristaltic movements of the gastrointestinal tract, as well as some glandular secretions related to digestion.

6 *Behavioral responses associated with emotion.* The subjective feelings of emotion (pleasure, pain, anger, fear, love) are expressed as visible physiological and physical changes when the cerebral cortex activates the autonomic nervous system by way of the hypothalamus. The autonomic nervous system, in turn, is responsible for changes in the heart rate and blood pressure, blushing, dryness of the mouth, clammy hands, crying, gastrointestinal discomfort, fidgeting and many other emotional expressions.

7 *Endocrine control.* The hypothalamus produces the hormones oxytocin and ADH, released by the posterior pituitary lobe, as well as certain other hormones (neurosecretory chemicals) that control the secretions of the anterior pituitary lobe hormones (see Chapter 16).

8 *Sexual responses.* The dorsal region of the hypothalamus contains specialized nuclei that respond to sexual stimulation of the genital organs (see Chapter 26). The sensation of an orgasm involves nerve activity within this center.

Epithalamus The *epithalamus* is the dorsal portion of the diencephalon, located near the third ventricle. It contains three parts: the trigonum habenulae, pineal body, and posterior commissure. The *trigonum habenulae* contains nerve fibers that extend to the midbrain. It acts as a relay center. The *pineal body* (epiphysis) is a glandlike structure that extends outward from the posterior end of the epithalamus by way of a stalk. (The neuroendocrine nature of this body is discussed in Chapter 16.) The *posterior commissure* consists of commissural fibers that connect with the superior colliculi of the midbrain.

Ventral thalamus The *ventral thalamus* or *subthalamus,* located ventrally in the diencephalon, contains the subthalamic nucleus, which is a basal ganglion.

Ask Yourself

1 *What are the basal ganglia?*

2 *What is the cerebral cortex?*

3 *What is the benefit of cerebral gyri and sulci?*

4 *What is the function of the corpus callosum?*

5 *What are the six lobes of the cerebrum? Where is each located?*

6 *What are the roles of the limbic system and the limbic lobe?*

7 *What are the main parts of the diencephalon?*

8 *What are the chief functions of the thalamus? The hypothalamus? The epithalamus?*

LEARNING AND MEMORY

Although we know a great deal about learning patterns in animals, our knowledge about the neuronal processes of learning is still rudimentary. It now seems certain, contrary to what was once thought, that *no single area or structure of the brain controls learning.* In fact, long-term studies of children with learning disabilities due to brain damage have shown that alternate areas of the brain can be trained to "relearn" what had once been learned by the damaged portion. Apparently, the brain can make such learning adjustments as long as there is a memory trace of what had been learned in the first place.

Young children recover from an injury to the cerebral cortex more completely than teenagers or adults do because the cerebral hemispheres of a child have not yet become specialized. As a result, an undamaged area of the cortex is able to learn to some degree what had been lost in the damaged portion. Apparently, the dominance of one cerebral hemisphere over the other develops gradually during childhood, and does not become relatively well fixed until about age 10. Before that age, it appears that language and speech capabilities are found relatively equally in both hemispheres, and either hemisphere can assume the functional role in language or speech. The natural "twin-brain" of a child explains why a right-handed child can be taught to write with the left hand, and why some athletes who have been trained since childhood to use both hands remain ambidextrous as adults. In contrast, adult patients with cerebral damage usually find it difficult or impossible to switch from using one hand to another, or to relearn lost language and speech abilities.

One theory of learning suggests that when one nerve circuit is used over and over again, synaptic connections are improved, and "preferred pathways" are created. Learning may also be a matter of changing pre-existing pathways, thus producing new pathways of behavior. (Learning is often described as a change in behavior based on experience.) Other theories of learning suggest that frequently used neurons grow in size, become more efficient, and exhibit an increase in the number of glial cells surrounding them. (Glial cells continue to reproduce throughout a person's lifetime.) While none of these theories has been proved, none can be dismissed. More and more scientists believe that learning involves a variety of mechanisms, not just one.

Memory is related to learning because in order to remember something, we have to "learn" it first. Apparently, **memory traces** are formed during learning and are imprinted on the brain when neurons record and store information. **Short-term memory** is the process by which we remember recent events that have no permanent importance (the score of yesterday's football game or a telephone number that can be forgotten as soon as the call is made). **Long-term memory** is the process by which we remember information that for some reason is interpreted as being important enough to store for a lifetime.

Scientists are not certain whether short-term and long-term memory use the same basic mechanism, but it has become clear that all memory is related more to electrochemical activity in the brain than electrical activity alone, as was once thought. Recent experiments with rats have shown that a flow of norepinephrine accompanies the process of laying down a memory trace. If norepinephrine is blocked during the process, there is no memory of recently learned events. Also, it appears that memory is achieved only when chemical changes in the brain are given more than a few seconds to occur. Some researchers feel that such chemical changes represent the formation of "neuronal loops," closed circuits of neurons and synapses that register short-term memory. Synapses may be related to memory, but no definite evidence is available.

The ability to convert short-term memory into long-term memory is lost with the removal of the medial portions of both temporal lobes (which includes the hippocampus and amygdala) and/or the nearby diencephalon (Figure 13.15). Patients who lose these portions of the brain retain their long-term memory, but are unable to form and store *new* long-term memories. Neither short-term nor long-term memory is impaired when only one temporal lobe is removed.

Considerable experimental evidence supports the theory that the neural connections of the hippocampus and amygdala with the diencephalon, medial temporal lobe, and prefrontal cortex have a crucial role in the storage of recently acquired memory. After the perception and awareness of short-term memory sensations are formed in the highest levels of the neocortex, including the temporal lobe, they are conveyed from the temporal lobe neocortex via neural circuits to the hippocampus and amygdala. These two structures are considered to be the neural processing centers through which the sensation of short-term memories are processed and conveyed into the long-term memory bank.

What do our earliest memories recall?	Earliest memories are usually of visual events that occurred between our third and fourth years.

FIGURE 13.15

Sites of brain damage that can lead to memory loss, indicating that these sites are important for memory formation.

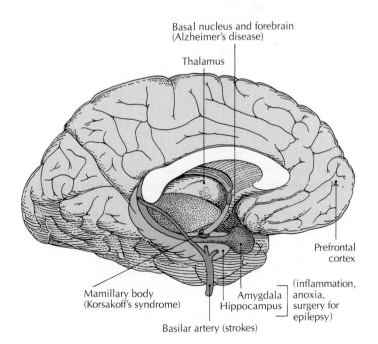

Basal nucleus and forebrain (Alzheimer's disease)

Thalamus

Prefrontal cortex

(inflammation, anoxia, surgery for epilepsy)

Amygdala
Hippocampus

Basilar artery (strokes)

Mamillary body (Korsakoff's syndrome)

Embryonic Development of the Brain

The central nervous system develops from cells that first form a flat plate. This *neural plate* is soon transformed by the rolling of its edges into a tubelike structure called the *neural tube,* which will develop into the brain and spinal cord. The hollow part of the tube will become the ventricular system of the brain.

By about the fourth fetal week, the neural tube enlarges to form three cavities, or vesicles, called the *prosencephalon* (forebrain), *mesencephalon* (midbrain), and the *rhombencephalon* (hindbrain), which grow and differentiate according to their own patterns of development. By the fifth week, cranial nerves III through XII of the now 3.5-mm embryo have already begun to develop as offshoots from the brainstem. In the seven-week embryo (about 11 mm in length), the prosencephalon has divided into two secondary vesicles, known as the *telen-* cephalon (cerebrum) and the *diencephalon*; the rhombencephalon has also divided into two vesicles, called the *metencephalon* (cerebellum and pons) and the *myelencephalon* (medulla).

By the end of the third fetal month, the overall form of the brain is recognizable, but the surface is still smooth. Sulci begin to develop in the cerebrum during the fourth month, and the folds of the cerebellum are clearly formed in a six-

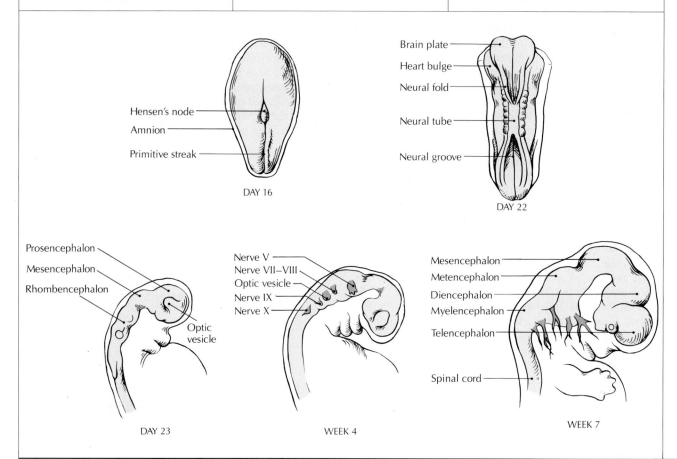

Apparently, long-term memories are recalled when the anterior temporal cortex is stimulated, activating a specific neural combination within the long-term memory bank. This phenomenon has been demonstrated by applying mild electrical stimulation to the anterior temporal cortex, which elicits a specific experience from the past.

One theory to explain the mechanism of long-term memory suggests that the reverberating circuitry of short-term memory may persist as a dynamic activity. Another suggestion is that long-term memory is associated with some permanent structural and functional alteration in the brain. Following deep anesthesia, anoxia (deficiency of oxygen), or

cooling of the brain, recent short-term memory is abolished, but long-term memory is not. These circumstances have led neuroscientists and psychologists to speculate that physical and biochemical changes lead to alterations in the connections between neurons in the brain that account for the persistence of long-term memory. Whatever the nature of the memory traces, it is certain that they are not localized in a single brain structure.

The temporal lobes probably function as a directory that permits the retrieval of stored memories, but it is not known exactly where memory is stored in the brain. In fact, rather than being stored in any one place, memory information is

month fetus. By the eighth month, the major sulci are present, and the occipital lobe dominates the underlying cerebellum.

Although the cranial portion of the newborn skull is only about three-quarters of the adult skull, the infant's head is disproportionately large. At birth the brain is about 10 percent of body weight, while the adult brain is about 2 percent of body weight. Although the brain weighs about 350 g at birth, early growth is so rapid that it weighs about 1000 g after one year. Growth of the brain slows down considerably after the first year, but continues until full growth (usually between 1250 and 1375 g) is attained during puberty.

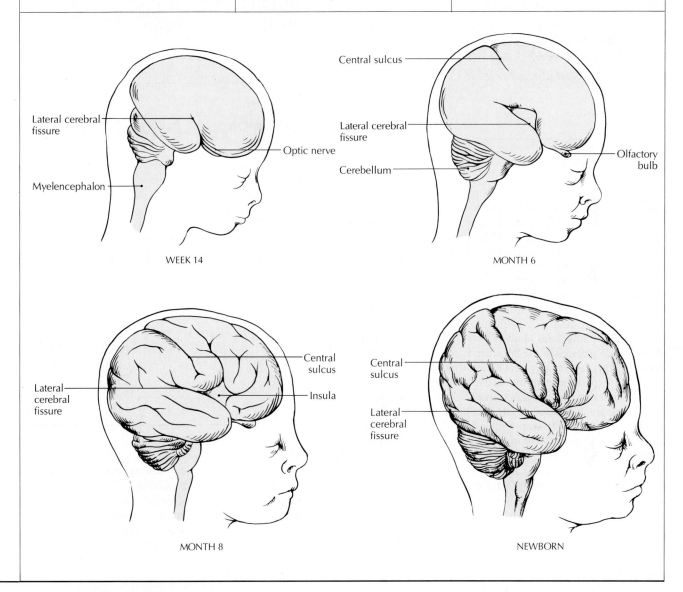

probably contained in large groups of brain cells that work together somehow to store and retrieve memories.

It has been shown that memory traces can be erased by chemicals that are known to inhibit protein synthesis, and some researchers have speculated that RNA, which codes for the synthesis of specific proteins, is also involved in coding for memory. The RNA theory arose when it was observed that RNA increases in proportion to increased brain activity, and decreases when brain activity slows down. Unfortunately, upsurges of RNA also occur during most normal activity of neurons, and other organs and glands besides the brain (such as the kidneys, thyroid, and pancreas) also show increases in RNA content whenever they are stimulated by their appropriate hormones.

On the basis of research with snails and slugs, it seems likely that the learning of even simple tasks results in changes in receptor cell membranes and nerve transmission, rather than from new patterns in the wiring of nerve fibers. Although there is a huge distance between snails and human beings, researchers hope that their growing knowledge of the lower forms of life will prove to be useful in understanding our own learning mechanism. Obviously, much additional research is needed before we can fully explain the mechanism of memory.

FIGURE 13.16

Alpha, beta, delta, and theta brain waves, as recorded by an electroencephalograph.

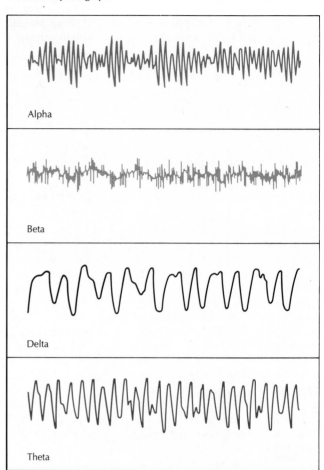

Ask Yourself

1 *Does one area of the brain control learning?*

2 *What is a ''preferred pathway''?*

3 *How is norepinephrine involved with memory?*

4 *What is the difference between short-term and long-term memory?*

5 *How is RNA involved with memory?*

SLEEP AND DREAMS

The brain is always active; in fact, some cortical areas of the human brain are more active and require a greater blood supply when the body is asleep than when it is awake. Whether a person is asleep or awake, brain activity in the form of waves may be recorded with an instrument called an *electroencephalograph*. The tracing this instrument produces, the *electroencephalogram* (''electric writing in the head''), is referred to by the initials *EEG*. The EEG shows the changing levels of electrical activity in the brain during various levels of consciousness, and it is frequently used to detect brain damage by locating areas of altered wave patterns. The four distinct wave patterns in the brain are known as alpha, beta, delta, and theta waves (Figure 13.16). *Alpha waves* (8 to 12 cycles per second), evident in relaxed adults whose eyes are closed, usually indicate a state of well-being. *Beta waves* (5 to 10 cps) are typical of an alert, stimulated brain. *Delta waves* (0.5 to 2 cps) are seen during deep sleep, in damaged brains, and in infants. *Theta waves* (3 to 7 cps) usually occur during sleep.

It is generally agreed that sleep is one of the more important activities of human beings (lack of sleep will cause death faster than lack of food), but the exact benefits of sleep are not known, and the causes of sleep are equally unknown. A group of scientists at Harvard University recently isolated an extremely potent sleep-inducing hormone, appropriately named factor S, for sleep. Factor S is a small peptide that appears to build up gradually while we are awake, finally reaching the point where the amount is great enough to cause sleep. The hypothesis has yet to be confirmed.

It is particularly ironic that so little is known about sleep, because if you are 20 years old, you have already spent about 8 years of your life asleep. By the time you are 60, you will have slept about 20 years.

The EEG has made possible the detection of at least four separate stages of sleep:

1 During *stage 1 sleep* the rhythm of alpha waves slows down, and slower theta waves appear (Figure 13.17). Heart and breathing rates decrease slightly, the eyes roll slowly from side to side, and the individual experiences a floating sensation. This stage is usually not classified as true sleep, and individuals awakened from stage 1 are quick to agree, in-sisting that they were merely ''resting their eyes.'' Stage 1 sleep usually lasts less than 5 minutes.

2 *Stage 2 sleep* is indicated by the appearance on the EEG of short bursts of waves known as ''sleep spindles'' (12 to 14 cps), along with ''K complexes,'' which are high-voltage bursts that occur before and after a sleep spindle. Eyes are generally still, and heart and breathing rates decrease only slightly. Sleep is not deep.

3 Slow delta waves appear during *stage 3 sleep*. This stage of intermediate sleep is characterized by steady, slow breathing, slow pulse rate (about 60), and decline in temperature and blood pressure.

4 *Stage 4 sleep,* also known as oblivious sleep, is the deepest stage. It usually begins about an hour after falling asleep. Delta waves become even slower, and heart and breathing rates drop to 20 or 30 percent below those in the waking state. Although the EEG indicates that the brain acknowledges outside noises and other external stimuli, the sleeper is not awakened by such disturbances.

Sleep proceeds in cycles 80 to 120 minutes long. An important condition known as *REM* (rapid eye movement) *sleep* takes place during entry into stage 1 of a new cycle.

FIGURE 13.17

Sleep patterns. (A) Sleep cycles vary during the night, ending with lengthening REM periods and less deep sleep than in the first hours of sleep. (B) Brain-wave patterns during the various stages of sleep, as shown in EEGs.

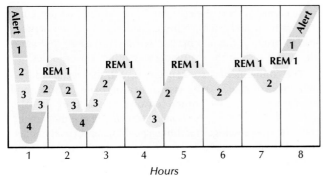

[A]

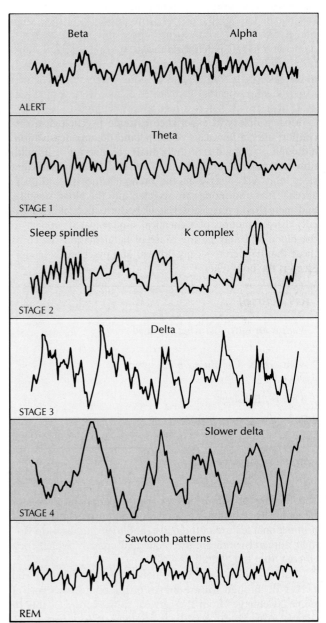

[B]

Toward morning there is usually less stage 3 and stage 4 sleep, and more REM sleep. For children as well as adults, REM sleep accounts for about 20 to 25 percent of total sleep time, although the percentage may be as high as 40 or 50 percent in infants. It is during the REM period that dreams occur, and the brain operates almost independently from the outside environment. Breathing and heart rate may increase or decrease, testosterone secretion increases and penile erections occur, and twitching body movements and perspiration are common. (Such physiological changes during REM sleep do not affect males only. In females, the clitoris also becomes erect during the REM period, and estrogen secretion probably increases as well.)

The REM period is thought to be essential for a general relaxation of normal build-ups of stress and tension. Dreams may serve to release tensions or fears that build up during the day. This theory is given credence by studies that show that seven out of eight adult dreams are somewhat unpleasant, and about 40 percent of the dreams of children are actually nightmares, although night terror in children does not usually occur during REM sleep. Another theory suggests that dreams are an attempt by the brain to bring coherence to the accelerated firings of neurons in the brainstem during REM sleep. According to this theory, dreams occur when the arbitrary firings of the neurons are translated into images as close to our actual experiences as possible. The brain is attempting to take random impulses and direct them into themes with a "plot." The more bizarre dreams may be caused when the brain is unable to connect the neuron activity with any logical experiences or memories.

Babies, whose central nervous system is not fully developed, spend more than half their sleeping hours in the REM state, but adults spend only about 15 percent of their total sleep in the REM condition. The pituitary gland releases growth hormone during deep sleep, especially during childhood and adolescence, when the proper regulation of growth is vital. Once we reach 45 or so, we spend less and less time in the deepest stage of sleep, and the secretion of growth hormone slows down more and more until it practically stops. Perhaps this reduction of growth hormone is an important connecting link to the aging process, but researchers can only speculate at this point.

Sleep is a highly individual process. No two people have precisely the same pattern of the sleep-wake cycle, nor do they require the same amount of sleep. (Generally, we need less sleep as we get older.) The average person has three or four dreams each night, with each dream lasting 10 minutes or more. Dreams occur during the REM stage. When awakened during this stage, four out of five people will describe having had a vivid, active dream, embellished with imagery and some fantasies. When awakened during the other stages of sleep (non-REM sleep), the aroused person may sometimes describe a dream, but unlike the REM dream, it is concerned with experiences related to thought processes.

It appears that learning and memory are reinforced during periods of REM sleep. Apparently, new stored information is consolidated and differentiated more during REM sleep than non-REM sleep. (On a practical level, this means that students who have studied for a morning exam would do well to get a full night's sleep, allowing for the final dream portion of the sleep cycle, rather than trying to wake up earlier than usual, thereby cheating themselves of the retentive qualities of REM sleep just before a natural awakening.)

Most people will experience extreme psychological discomfort after a few days of sleep (and dream) deprivation, although occasional rare individuals can live for years with only two or three hours of sleep per day. Without sufficient sleep, the ATP reserve in the brain declines precariously, adrenal stress hormones are secreted into the blood steadily, and mental and physical levels of performance falter markedly. Large amounts of a chemical similar to LSD appear in the blood and may be the cause of hallucinations (such as those that happen to sleepy drivers) and pyschotic behavior. Sleep is the only relief.

Ask Yourself

1 What is an EEG, and what is it used for?

2 What are the four distinct brain-wave patterns?

3 What is REM sleep?

4 Do we dream every night?

CRANIAL NERVES

The 12 pairs of **cranial nerves** are the peripheral nerves of the brain. Their names are an indication of some anatomical or functional feature of the nerve, and their numbers (in Roman numerals) indicate the sequential order in which they emerge from the brain.

Cranial nerves I and II are nerves of the cerebrum, and nerves III through XII are nerves of the brainstem. (Part of nerve XI emerges from the cervical spinal cord.) Of the 10 brainstem nerves, one (VIII) is a purely sensory nerve, five are primarily motor nerves with some proprioceptive fibers (III, IV, VI, XI, and XII), and four are mixed nerves containing both sensory and motor fibers (V, VII, IX, and X) (Figure 13.18, Table 13.3).

The *motor (efferent) fibers* of the cranial nerves emerge from the brainstem. They arise from bundles of neurons called **motor nuclei,** which are stimulated by nerve impulses from many outside sources, including the cortex of the cerebrum and the sense organs. Axons of motor neurons in cranial motor nuclei are in the form of cranial nerves. These axons have two roles. They either (1) stimulate voluntary muscles, or (2) synapse with ganglia of the autonomic nervous system. These ganglia, in turn, relay nerve impulses to cardiac (heart) muscle, smooth (visceral) muscle, and glands.

The *sensory (afferent) fibers* of cranial nerves emerge from neurons with cell bodies outside the brain. These neurons are either **sensory ganglia,** which are groups of neurons situated on the trunks of cranial nerves, or are located in peripheral sense organs such as the eye. Axons of sensory neurons enter the brain, synapse with bundles of neurons called **sensory nuclei,** produce the appropriate sensation (vision, for example, in the case of the eye), and play other sensory roles.

The cranial nerves are concerned with the *specialized (special) senses* of smell, taste, vision, hearing, and balance, and the *general senses,* and other sensory inputs. They are also involved with the *specialized motor activities* of eye movement, chewing, swallowing, breathing, speaking, and facial expression. The vagus nerve is an exception, projecting fibers to organs in the abdomen and thorax. In the following sections you will see how individual nerves attend to these diverse functions.

Cranial Nerve I: Olfactory

The **olfactory nerve** (L. *olfacere,* to smell) is strictly a sensory nerve. The 10 to 20 million bipolar neurons *(olfactory cells)* of the olfactory nerve are located high in the nasal cavities within nasal epithelium. They act both as chemoreceptors that sense odors, and as conductors of impulses that ultimately result in the perception called smell. The unmyelinated axons of olfactory cells pass through the foramina of the cribriform plate of the ethmoid bone and terminate in the **olfactory bulb,** which is actually an ''appendage'' of the brain. The nerves in the bulb synapse with neurons that form the *olfactory tract,* and convey impulses to the primary olfactory cortex of the uncus of the cerebrum.

Cranial Nerve II: Optic

The **optic nerve** is a special sensory nerve. It conveys impulses that result in vision and in reflexes associated with vision. Each optic nerve is actually a tract, composed of about a million axons that arise from the ganglion cells of the retina. In the retina, rods and cones (the photoreceptive cells, which will be described in Chapter 15), interneurons, and bipolar neurons form a circuitry that interacts with ganglion neurons.

Each optic nerve passes out of the orbit through the optic foramen into the cranial cavity (see Figure 13.18). While passing through the X-shaped **optic chiasma** (kye-AZ-muh; so named because its X shape reminded early anatomists of the Greek letter *chi* [KYE], which also has an X shape), the axons of the medial half of the retinas of both eyes cross the midline. The axons from the lateral half of the retinas of both eyes do not cross the midline (see Figure 15.25). One-half of the fibers of each optic nerve of each eye then cross over to the other side to form the optic chiasma. After passing through the optic chiasma, the axons continue, forming the two *optic tracts.*

The axons of each optic tract terminate mainly in the lateral geniculate body of the thalamus. They synapse with neurons that project as the optic radiations to the primary visual cortex in the occipital lobe of the cerebrum. Other fibers terminate in the superior colliculus of the midbrain. The mid-

FIGURE 13.18

Origins of the cranial nerves, basal view of the brain.

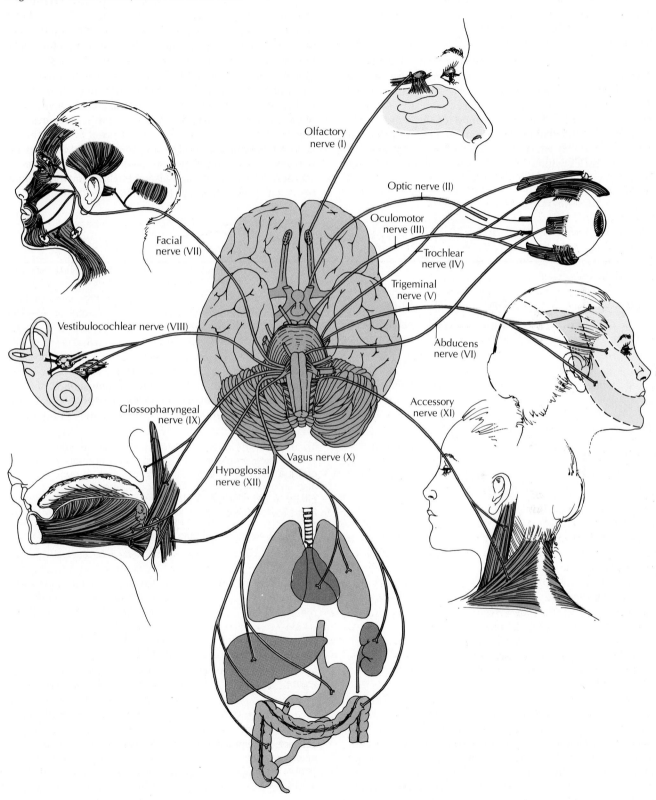

Olfactory
nerve (I)

Optic nerve (II)

Oculomotor
nerve (III)

Facial
nerve (VII)

Trochlear
nerve (IV)

Trigeminal
nerve (V)

Vestibulocochlear nerve (VIII)

Abducens
nerve (VI)

Accessory
nerve (XI)

Glossopharyngeal
nerve (IX)

Vagus nerve (X)

Hypoglossal
nerve (XII)

brain, through its connections with cranial nerves III, IV, and VI, is involved with subconscious visual reflexes (eye movements, pupillary responses, and focusing).

Cranial Nerve III: Oculomotor

Cranial nerves III (oculomotor), IV (trochlear), and VI (abducens) are classified as the *extraoculomotor nerves* because they innervate the extraocular (extrinsic) muscles that move the eyeball. The *oculomotor nerve* (L. *oculus*, eye + motor, or "eye movement") innervates most of the muscles that move the eye. In addition, it innervates the levator palpebrae superioris muscle, which raises the upper eyelid, and the smooth muscles within the eyeball that contract the pupil (usually in response to bright light) and regulates the focusing of the lens.

The brainstem motor nucleus, including the parasympathetic fiber of the autonomic nervous system, of the oculomotor nerve is located in the midbrain (Figure 13.19). The oculomotor nerve emerges from the ventral surface of the midbrain, passes through the superior orbital fissure, and enters the orbit. It innervates the superior rectus, inferior rectus, medial rectus, levator palpebrae superioris, and inferior oblique muscles. The sensory fibers from the proprioceptors (neuromuscular spindles) in the muscles innervated by cranial nerves III, IV, and VI join the trigeminal nerve (V) to terminate in a brainstem trigeminal nucleus (see Figures 13.7 and 13.19).

The parasympathetic fibers of the oculomotor nerves innervate the constrictor muscles of the pupil (usually a response to bright light) and also the ciliary muscles, which regulate the tension on the lens for focusing.

TABLE 13.3   CRANIAL NERVES

Nerve	Type	Origin	Distribution	Function
I Olfactory	Sensory (special sense).	Nasal mucous membrane high in nasal cavities.	Terminates in olfactory bulb of cerebrum.	Smell (olfaction).
II Optic	Sensory (special sense).	Retina of eye.	Terminates in lateral geniculate body of thalamus and superior colliculus of midbrain.	Vision. Afferent limb of reflex of focusing and constricting pupil.
III Oculomotor*	Motor.	Midbrain.	To all extrinsic muscles of eyeball except superior oblique and lateral rectus; also autonomic fibers to ciliary muscles of lens and constrictor muscles of iris.	Movements of eyeball, elevation of upper eyelid, constriction of pupil, focusing of lens.
IV Trochlear*	Motor.	Caudal midbrain.	Innervates superior oblique muscle of eye.	Eye movements (down and out).
V Trigeminal Ophthalmic nerve (V1)	Sensory.	Pons.	General area of forehead, eyes.	Conveys general senses from cornea of eyeball, upper nasal cavity, front of scalp, forehead, upper eyelid, conjunctiva, lacrimal (tear) gland.
Maxillary nerve (V2)	Sensory.	Pons.	General area of maxillary region.	Conveys general senses from cheek, upper lip, upper teeth, mucosa of nasal cavity, palate, parts of pharynx.
Mandibular nerve (V3)	Mixed (sensory and motor).	Pons.	Sensory: general area of mandibular region. Motor: innervates muscles of mastication.	Sensory: conveys general senses from tongue (not taste), lower teeth, skin of lower jaw. Motor: chewing.
VI Abducens*	Motor.	Caudal pons.	Innervates lateral rectus muscle of eye.	Abduction of eye (lateral movement).

Cranial Nerve IV: Trochlear

The *trochlear nerve* (L. pulley) is so named because it innervates the superior oblique muscle, whose tendon passes through a cartilaginous pulleylike sling on its way to its insertion into the sclera (outer covering) of the eyeball. The nerve arises from its nucleus in the caudal midbrain, emerges from the dorsal surface of the midbrain, and passes forward through the superior orbital fissure into the orbit, where it innervates the superior oblique muscle. The trochlear nerve is long and slender, allowing it to follow a winding course (see Figure 13.19).

When the trochlear nerve is injured, double vision may result, especially when looking downward. As a result, a person with an injured trochlear nerve might have trouble walking downstairs or stepping off a curb into the street. (See the table in Physiological and Anatomical Abnormalities on p. 406 for an outline of some major disorders of the cranial nerves.)

Cranial Nerve V: Trigeminal

The *trigeminal nerve* (L. *tri,* three + *geminus,* twin, referring to its three major branches and two roots) is the largest, but not the longest, of the cranial nerves (Figure 13.20). It is a mixed nerve, being the chief facial sensory nerve and the motor nerve of the chewing muscles. The sensory nerves convey impulses of the general senses, similar to those found in the spinal nerves. All the sensory fibers terminate in trigeminal nuclei in the brainstem. The motor fibers originate from the trigeminal motor nucleus in the pons.

Nerve		Type	Origin	Distribution	Function
VII	Facial	Mixed (sensory, special sense, and motor).	Pons.	Sensory: innervates taste buds of tongue. Motor: innervates muscles of facial expression; autonomic fibers to salivary glands, lacrimal glands.	Sensory: taste. Motor: salivation, lacrimation, movement of facial muscles.
VIII	Vestibulocochlear				
	Cochlear (auditory)	Sensory (special sense).	Medulla, pons.	Cochlea of inner ear.	Hearing.
	Vestibular	Sensory (special sense).	Medulla, pons.	Semicircular ducts, utricle and saccule of inner ear.	Equilibrium.
IX	Glossopharyngeal	Mixed (sensory, special sense, and motor).	Medulla.	Sensory: conveys taste from posterior third of tongue, general senses from upper pharynx. Motor: innervates stylopharyngeus muscle; autonomic fibers stimulate parotid gland.	Taste, other sensations of tongue. Secretion of saliva; swallowing.
X	Vagus	Mixed.	Medulla.	Voluntary muscles of soft palate, cardiac muscle, smooth muscle in respiratory, cardiovascular, digestive systems.	Swallowing. Monitors oxygen and carbon dioxide levels in blood, senses blood pressure, other visceral activities of affected systems.
XI	Accessory (spinal accessory)[†]	Motor.	Medulla, cervical spinal cord.	Muscles of larynx, sternocleidomastoid, trapezius.	Voice production (larynx); muscle sense; movement of head, shoulders.
XII	Hypoglossal*	Motor.	Medulla.	Tongue muscles.	Movements of tongue during speech, swallowing; muscle sense.

*Contains some sensory proprioceptive fibers from extraocular muscles of eyes and tongue muscles, which leave nerve and join trigeminal nerve (see Figure 13.18).

†Fibers of spinal accessory nerve innervating muscles of the larynx join the vagus nerve; hence, these muscles are usually said to be innervated by the vagus nerve.

FIGURE 13.19

Distribution of the oculomotor, trochlear, and abducens nerves (III, IV, VI).

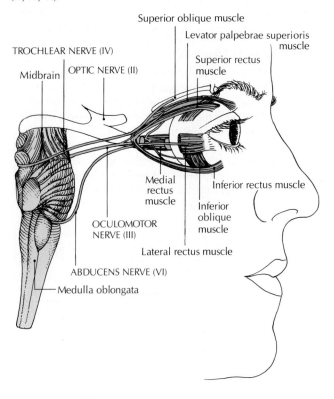

TROCHLEAR NERVE (IV)

Midbrain

OPTIC NERVE (II)

Superior oblique muscle

Levator palpebrae superioris muscle

Superior rectus muscle

Medial rectus muscle

Inferior rectus muscle

Inferior oblique muscle

OCULOMOTOR NERVE (III)

Lateral rectus muscle

ABDUCENS NERVE (VI)

Medulla oblongata

Cranial Nerve VI: Abducens

The *abducens nerve* (L. *abducere,* to lead away) is an extra-oculomotor nerve, together with cranial nerves III and IV. The brainstem nucleus is located in the caudal pons. The nerve emerges at the junction of the pons and medulla, and passes forward through the superior orbital fissure into the orbit, where it innervates the lateral rectus muscle to *abduct* (turn outward) the eye, hence the name *abducens* (see Figure 13.19).

Cranial Nerve VII: Facial

The *facial nerve* is a mixed nerve. It emerges from the junction of the pons and medulla, passes through the internal auditory meatus and the facial canal of the temporal bone, and leaves the skull from the stylomastoid foramen and other foramina. The *geniculate ganglion,* in the petrous portion of the temporal bone, is the sensory ganglion of the facial nerve.

The sensory fibers of the nerve innervate the taste buds of the anterior two-thirds of the tongue. The fibers terminate in

The sensory root has a sensory ganglion, located close to where the nerve emerges from the pons within the middle cranial fossa. The cell bodies of the sensory nerves are located in this *trigeminal ganglion* (the equivalent of the dorsal root ganglion of a spinal nerve).

After the trigeminal nerve emerges from the pons, it divides into three branches, the ophthalmic, maxillary, and mandibular nerves (see Table 13.3).

1 The *ophthalmic nerve* (V^1), a sensory nerve, passes through the superior orbital fissure, through the orbit, and into the upper head. Its branches convey the general senses from the front of the scalp, forehead, upper eyelid, conjunctiva (inner membrane of the eyelid), cornea of the eyeball, and the upper nasal cavity.

2 The *maxillary nerve* (V^2), a sensory nerve, passes through the foramen rotundum. Its branches convey general sensations from the skin of the cheek, upper lip, upper teeth, mucosa of the nasal cavity, palate, and parts of the pharynx. It covers the general area of the maxillary bone.

3 The *mandibular nerve* (V^3), a mixed nerve, passes from the skull through the foramen ovale. Its branches convey general senses from the mucosa of the mouth (including the tongue, but not sensations of taste), lower teeth, and skin around the lower jaw. The motor fibers innervate the chewing muscles. The mandibular nerve is distributed to the general region of the mandible. (See the essay on p. 397.)

FIGURE 13.20

Distribution of the trigeminal nerve (V).

Motor branches to muscles of mastication (temporalis, masseter, external pterygoid, internal pterygoid, mylohyoid and anterior belly of digastric)

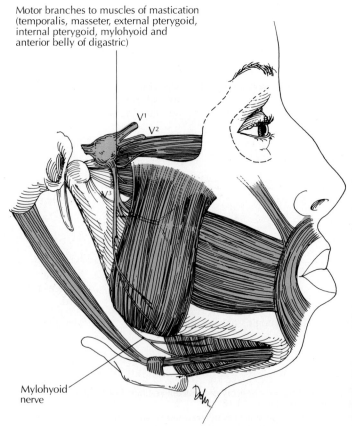

V^1

V^2

V^3

Mylohyoid nerve

Open Wide

A toothache on the lower left-hand side of your mouth brings you to your dentist. After being x-rayed, you learn that the filling in your tooth is cracked and that you may have an abscess (a pus-filled localized area of infection).

Before repairing the tooth, your dentist gives you some anesthetic. The dentist first applies a topical anesthetic where a needle will be inserted to administer a major anesthetic. The anesthetic desensitizes the *inferior alveolar branch of the trigeminal nerve* (see Figure 13.20). The dentist takes care not to actually touch the nerve. Touching the nerve would give you an "electric shock," because the nerve would be depolarized instantly. The dentist also avoids the blood vessels adjacent to the

nerve trunk, pulling back once or twice on the syringe to make sure no blood is drawn out.

Now the anesthetic diffuses through the soft tissue along the forward path of the inferior alveolar nerve. Branches from the nerve radiate to each tooth on the lower left-hand side, the chin, and the lower lip (see Figure 13.20). The anesthetic will also reach the nearby *lingual nerve*, anesthetizing the left lateral edge of your tongue. You can feel the numbness move from the back of your mouth toward the front, following the course of the nerve to the midline of your lower lip.

In a few minutes, your tooth is completely insensitive to pain, and your den-

tist takes care of the problem without causing you any discomfort. Because your dentist suspected an abscess (which did not materialize) that would have required delicate root canal surgery, you were given an anesthetic that would be effective for two to four hours so that you would not be uncomfortable later in the evening when the anesthetic wore off and the sensitivity of the surgery remained. Your only problem now is being able to feel the normal sensations again. As you get up to leave, your dentist reminds you not to test the numbness by biting your tongue or grinding your teeth before the feeling returns, since either action could cause a serious injury in the absence of pain.

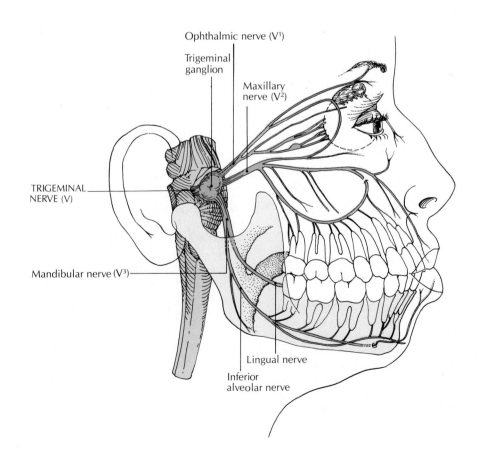

Ophthalmic nerve (V¹)

Trigeminal ganglion

Maxillary nerve (V²)

TRIGEMINAL NERVE (V)

Mandibular nerve (V³)

Lingual nerve

Inferior alveolar nerve

the solitary nucleus in the brainstem. Fibers project from this nucleus into the thalamus, and then to the gustatory (taste) area of the cerebral cortex.

The motor fibers originate from the facial nucleus in the pons and innervate all the muscles of facial expression. Parasympathetic efferent fibers, originating from a salivatory nucleus, belong to the autonomic nervous system and innervate the lacrimal gland and the submandibular and sublingual salivary glands.

Cranial Nerve VIII: Vestibulocochlear

The *vestibulocochlear nerve* (L. entranceway + snail shell) is a sensory nerve composed of two nerves, the cochlear (auditory) nerve and the vestibular nerve. The cochlear nerve conveys impulses concerned with hearing, and the vestibular nerve conveys information about equilibrium or balance and the position and movements of the head.

The fibers of the *cochlear nerve* arise at their synapses with the hair cells of the spiral organ in the snail-shaped cochlea of the inner ear (Figure 13.21). The cell bodies of the cochlear nerve are located in the spiral ganglion near the cochlea. The axons of the cochlear neurons terminate in the cochlear nuclei of the upper medulla. Auditory pathways from the cochlear nuclei terminate in the medial geniculate body of the thalamus. From the thalamus, impulses are relayed via fibers of the auditory radiations to the auditory cortex in the temporal lobe of the cerebrum.

The *vestibular nerve* arises at its synapse with the hair cells in the semicircular canals, utricle, and saccule of the inner ear. The cell bodies of the vestibular nerve are located in the vestibular ganglion within the petrous part of the temporal bone. Its axons terminate in the vestibular nuclei in the upper medulla. It is thought that a vestibular pathway projects to the thalamus, where fibers project to the vestibular area of the temporal lobe of the cerebrum.

Cranial Nerve IX: Glossopharyngeal

The *glossopharyngeal nerve* (Gr. tongue + throat) is a small mixed nerve. After the nerve emerges from the upper medulla, it leaves the posterior cranial fossa through the jugular foramen (Figure 13.22). Its branches are distributed to the region of the posterior third of the tongue, and the upper pharynx. The sensory fibers terminate in several sensory nuclei in the medulla, and its motor fibers originate from motor nuclei in the medulla.

Sensory fibers convey taste and general senses from the posterior third of the tongue, and general senses from adjacent structures of the upper pharynx. When the sensory nerve in the back of the mouth is stimulated, the act of swallowing and the gag reflex are triggered. A small branch of the nerve terminating in the carotid sinus (in the neck) monitors blood pressure. Motor fibers innervate the stylopharyngeus muscle, which helps to elevate the pharynx during swallowing. Autonomic (parasympathetic) fibers in the glossopharyngeal nerve stimulate the parotid gland (anterior to the ear) to secrete.

Cranial Nerve X: Vagus

The *vagus nerve* (L. wandering) is a mixed nerve. It is so named because it has the most extensive distribution of any cranial nerve (Figure 13.23). The vagus nerve is the longest cranial nerve, innervating structures in the head, neck, thorax, and abdomen. Two different types of motor fibers are apparent:

1 The fibers that innervate the voluntary (striated) muscles of the soft palate, pharynx, and larynx originate in a motor nucleus in the medulla. These muscles are involved in swallowing and speaking.

2 Originating from another nucleus in the medulla are fibers of the parasympathetic system (see Chapter 14), which convey impulses involved with the activity of cardiac muscle, smooth muscle, and exocrine glands (glands with ducts) of the cardiovascular, respiratory, and digestive systems.

The sensory fibers of the vagus nerve convey impulses from all the structures of the systems innervated by the motor fibers. The sensory fibers terminate in the solitary nucleus in

FIGURE 13.21

Distribution of the vestibulocochlear nerve (VIII).

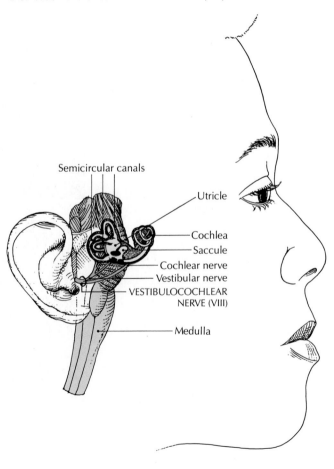

Semicircular canals

Utricle

Cochlea

Saccule

Cochlear nerve

Vestibular nerve

VESTIBULOCOCHLEAR NERVE (VIII)

Medulla

the medulla. Some of the afferent fibers monitor the blood levels of oxygen and carbon dioxide, and sense blood pressure changes in the aorta, carotid sinus, and carotid body.

Cranial Nerve XI: Accessory

The **accessory** (or *spinal accessory*) **nerve** is a mixed nerve, originating from both the medulla and the cervical spinal cord (Figure 13.24). Both the bulbar (medullary) and spinal roots are composed primarily of motor fibers.

The *bulbar root* arises from a motor nucleus in the medulla, joins the spinal root, and passes through the jugular foramen. The fibers of the bulbar root leave the fibers of the spinal root, and join the vagus nerve. Thus the larynx is actually innervated by the accessory nerve, which joins the vagus nerve. The fibers eventually form the recurrent laryngeal nerve, which innervates the muscles of the larynx that regulate the vocal cords. The laryngeal nerve also supplies the sensory innervation of the upper trachea and the region around the vocal cord.

The *spinal root* originates from neurons in the anterior gray horn of the first five cervical spinal levels. The fibers of the spinal root emerge from the spinal cord, ascend within the vertebral canal, pass through the foramen magnum, join the bulbar root, and pass through the jugular foramen. The spinal fibers leave the bulbar root fibers and pass into the neck to innervate the sternocleidomastoid and trapezius muscles. Both muscles are involved with movements of the head. The proprioceptive fibers from these muscles pass through the cervical nerves to the spinal cord. Other afferent fibers, for example those from the region of the vocal cords, terminate in trigeminal nuclei of the medulla.

Cranial Nerve XII: Hypoglossal

The **hypoglossal nerve** ("under the tongue") is a motor nerve. It innervates the muscles of the tongue. It also contains some proprioceptive fibers of muscle sense. The motor fibers originate from a motor nucleus in the medulla, emerge from the medulla, and pass through the hypoglossal canal into the region of the floor of the mouth (Figure 13.25).

FIGURE 13.22

Distribution of the glossopharyngeal nerve (IX).

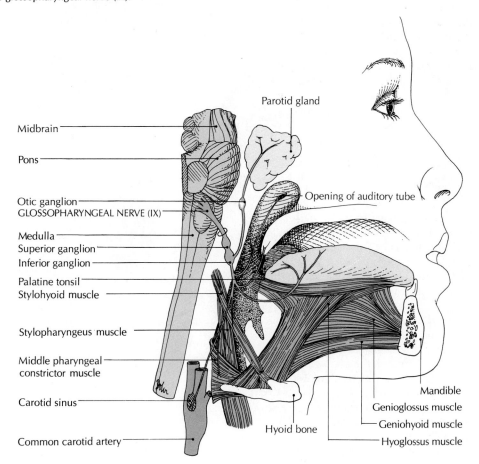

FIGURE 13.23

Distribution of the vagus nerve (X).

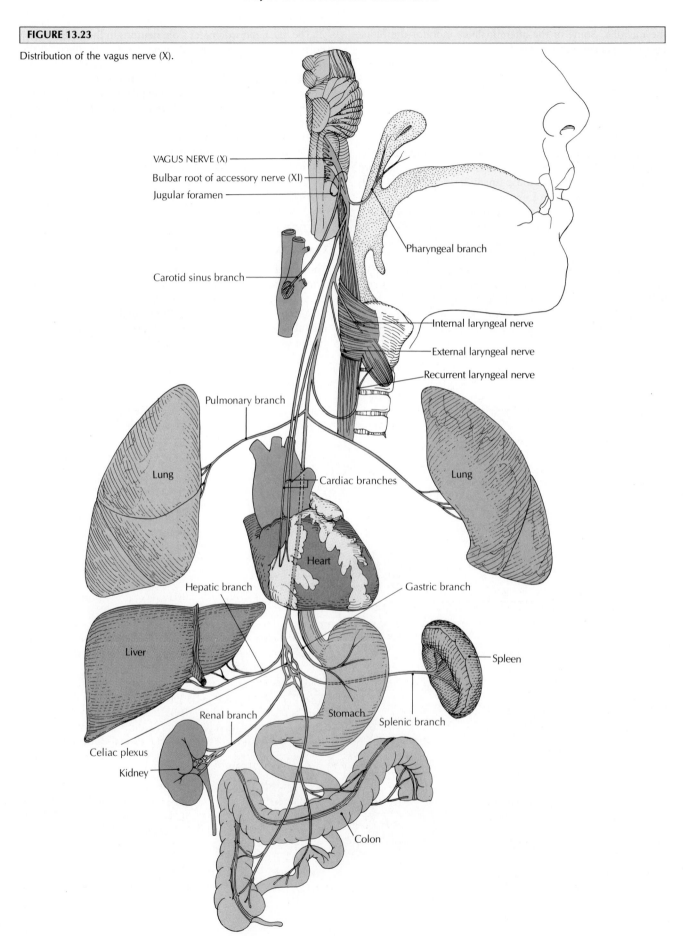

VAGUS NERVE (X)

Bulbar root of accessory nerve (XI)

Jugular foramen

Pharyngeal branch

Carotid sinus branch

Internal laryngeal nerve

External laryngeal nerve

Recurrent laryngeal nerve

Pulmonary branch

Lung

Lung

Cardiac branches

Heart

Hepatic branch

Gastric branch

Liver

Spleen

Renal branch

Stomach

Splenic branch

Celiac plexus

Kidney

Colon

FIGURE 13.24

Distribution of the accessory nerve (XI).

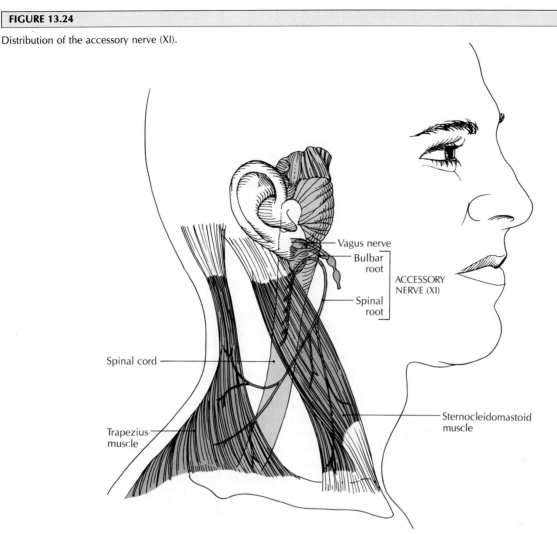

The branches of the hypoglossal nerve innervate the intrinsic muscles within the tongue, and three extrinsic muscles: hyoglossal, genioglossal, and styloglossal. These muscles are important in manipulating food within the mouth, in speaking, and in swallowing.

Ask Yourself

1 *How many cranial nerves are there?*

2 *Which cranial nerve is actually a tract?*

3 *Which are the extraoculomotor nerves?*

4 *What are the three main branches of the trigeminal nerve?*

5 *What are the two main functions of the vestibulocochlear nerve?*

6 *Which cranial nerve has the widest distribution?*

7 *Which cranial nerves are mixed nerves? Motor nerves? Sensory nerves?*

CHEMICALS AND THE NERVOUS SYSTEM

Certain chemicals are essential for the proper functioning of the nervous system. Among these are the **neurotransmitters** and **neuromodulators,** which were discussed in Chapter 11. At least 50 neurotransmitters or transmitterlike substances have been identified in the brain, and many other transmitter substances are under investigation (see Table 11.3).

Chemicals that are not natural constituents of the nervous system may be used to alter the normal functions of the nervous system or to correct abnormal functions. These chemicals or drugs fall into several major categories (Table 13.4):

1 Stimulants increase the activity of the central nervous system, especially the brain. Among these drugs are amphetamines, caffeine, cocaine, and nicotine. Depending on the type and strength of the particular drug taken, their effects may range from mild (for caffeine) to powerful (for methedrine or "speed").

2 Depressants are drugs that inhibit the activity of the central nervous system, especially the brain. Among the depres-

FIGURE 13.25

Distribution of the hypoglossal nerve (XII).

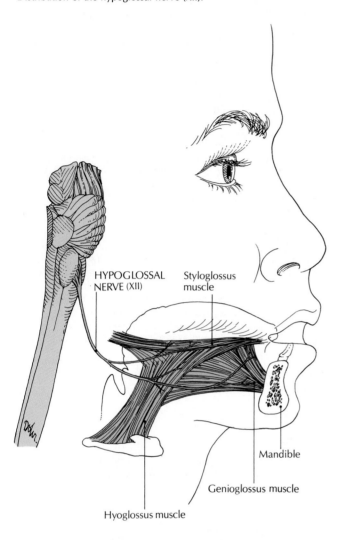

HYPOGLOSSAL
NERVE (XII)

Styloglossus
muscle

Mandible

Genioglossus muscle

Hyoglossus muscle

sants are anesthetics, barbiturates, alcohol, opiates, and tranquilizers. In small amounts, barbiturates induce sleep, alcohol reduces tension, opiates relieve pain, and tranquilizers reduce anxiety. But in larger doses, all depressants may cause damage and abnormal reactions. Certain stimulants and depressants may be addictive.

3 *Antidepressants* are drugs that relieve the symptoms of psychological depression. The most commonly used types are the tricyclic antidepressants, such as Tofranil and Elavil, which block the uptake of norepinephrine and serotonin at nerve endings.

4 *Psychedelic* and *hallucinogenic drugs* alter perception and mood. Extreme effects include *hallucinations,* or distortions of sense perceptions, and even *psychoses,* or severe mental illnesses.

5 *Analgesics* are pain-relieving drugs, which may be classified as *nonopioid* analgesics, or *opiates* and *opioid* analgesics. Opiates are derived from natural substances, while opioids are synthetic drugs. They both have similar pharmacological properties, including the ability to relieve pain, and the likelihood of dependence (addiction) with prolonged use. The *nonopioids* include *aspirin,* which is used to relieve mild pain and to reduce fever and inflammation; *acetaminophen* (Tylenol and Datril), which is used to relieve mild pain and to reduce fever; and *ibuprofen* (Nuprin and Advil), which is used to relieve pain. *Opiates* and *opioids* include *codeine, meperidine hydrochloride* (Demerol), and *methadone hydrochloride* (Dolophine). *Morphine* is an opiate used to relieve severe pain, and also to allay the fear and anxiety of pain. Opiates and opioids tend to produce *tolerance,* the necessity of increasing the dosage in order to maintain the effect of the drug.

6 *Antianxiety drugs* are used to suppress anxiety (tranquilizers) or to induce sleep (sleeping pills). They are among the most commonly used drugs. If antianxiety drugs are taken alone, they are relatively safe, although they have the potential to be addictive when abused. When taken in combination with central nervous system depressants or alcohol, however, they can be dangerous. Antianxiety drugs include the groups called benzodiazopines, three of which are *diazepam* (Valium), *alprazolam* (Xanax), and *chlordiazepoxide hydrochloride* (Librium). Barbiturates are also used as antianxiety drugs, but their primary action produces drowsiness. Barbiturates are addictive, and an overdose may be fatal.

TABLE 13.4   EFFECTS OF DRUGS ON THE NERVOUS SYSTEM

Type of drug	Major effects	Mechanism of action
STIMULANTS		
Amphetamines ("uppers"): Benzedrine, Dexedrine, Methedrine ("speed"), phenmetrazine hydrochloride, Preludin	Elevate mood, produce sense of increased energy, decrease appetite, increase anxiety and irritability. *Powerful reaction:* accelerate heart rate, dilate pupils, increase blood sugar.	Stimulate sympathetic nervous system.
Caffeine: coffee, tea, cocoa, cola drinks	Relieves drowsiness, muscle fatigue; prolongs physical and intellectual activity. *Mild reaction:* accelerates heart rate, dilates pupils, increases blood sugar.	Stimulates sympathetic nervous system. Facilitates synaptic transmission.

Type of drug	Major effects	Mechanism of action
Cocaine	More powerful than amphetamines, but similar reactions. Temporary sense of well-being and alertness.	Stimulates central nervous system. Inhibits uptake of norepinephrine.
Nicotine: tobacco	Similar to caffeine. *Medium reaction:* accelerates heart rate, dilates pupils, increases blood sugar.	Stimulates sympathetic nervous system. Facilitates synaptic transmission.
DEPRESSANTS Anesthetics: ether, chloroform, benzene, toluene, carbon tetrachloride	Induce unconsciousness. Similar to alcohol intoxication.	Depress central nervous system. Block transmission of electrical impulse that triggers contraction of ventricles.
Barbiturates: Nembutal, Seconal, Amytal, Tuinal, phenobarbital	Sedative action, induce sleep. Addictive. High doses may induce respiratory failure. Combination of barbiturates and alcohol may produce extreme depression of central nervous system, and may cause death.	Depress reticular formation and, in large doses, medulla oblongata. Interfere with synthesis and secretion of norepinephrine and serotonin.
Ethyl alcohol: whiskey, gin, beer, wine	Small amounts reduce tension, produce feeling of well-being, stimulation. Additional amounts depress the lower brain centers, impair motor coordination, produce insensitivity to touch, distorted vision, difficulty with speech. Prolonged use may lead to cirrhosis of liver, unconsciousness, coma, brain damage.	Exact mechanism unknown. May inhibit activity of thalamus and interfere with action of acetylcholine, norepinephrine, serotonin. Reduces neuron function in brain. Inhibits motor and sensory regions of cortex. Depresses visual, auditory, speech centers of cortex. Inhibits cerebellum. Depresses reticular formation.
Opiates and opioids: heroin, morphine, codeine, Demerol, methadone, opium	Reduce pain, tension, anxiety; induce muscle relaxation, lethargy; decrease physical drive; produce drowsiness. May produce loss of appetite, constipation, fatal depression of respiratory center. Highly addictive.	Depress thalamus.
Tranquilizers: meprobamate (Miltown, Equanil), chlorpromazine (Thorazine), chlordiazepoxide (Librium), diazepam (Valium)	Reduce anxiety, tensions. Antipsychotic.	Block receptors of adrenalin and acetylcholine. Depress activity of reticular formation.
ANTIDEPRESSANTS Dibenzapines: Tofranil, Elavil	Reverse effects of psychological depression.	Increase levels of norepinephrine in brain.
MAO (monoamine oxidase) inhibitors: Nardil, Parnate	Reverse effects of psychological depression.	Increase levels of norepinephrine in brain.
Ritalin	Reverses effects of psychological depression.	Increases levels of norepinephrine in brain.
PSYCHEDELICS, HALLUCINOGENS Cannabis (marijuana), hashish	Alters perception of sensory phenomena. Usually produces a sense of well-being. May produce anxiety, lassitude.	Unknown.
LSD (lysergic acid diethylamide)	Distorts visual and auditory imagery. Produces sense of increased sensory awareness.	Inhibits brain serotonin, perhaps by decreasing activity of serotonin-producing cells in reticular system.
Mescaline, psilocybin, DMT (dimethyltryptamine), DET (diethyltryptamine), DOM or STP (diemthroxymethylamphetamine), Ditran, phenylcyclidine (Sernyl)	Produce hallucinations, increased sensory awareness, psychoses.	Mimic molecular structure of serotonin. Duplicate effects of nervous system activity. May enhance effects of biogenic amines (naturally occurring amines formed in body), which they resemble.

TABLE 13.5 SOME REPRESENTATIVE ANALGESIC, ANTI-INFLAMMATORY, AND ANTIANXIETY DRUGS

Type of drug	Representative drugs	Type of drug	Representative drugs
ANALGESICS		**SLEEPING PILLS, ANTIANXIETY DRUGS**	
Narcotic analgesic (opiates and opioids)	Codeine, morphine, Demerol.	Minor tranquilizers	Alprazolam (Xanax), Atarax, chlordiazepoxide (Librium), meprobamate (Miltown, Equanil), phenobarbital, diazepam (Valium), Serax, Tranxene, Vistaril.
Nonnarcotic pain reliever	Aspirin, acetaminophen (Tylenol), Darvon, Ibuprofen (Nuprin, Advil).	Major tranquilizers	Haldol, Mellaril, Navane, Prolixin, Sparine, Stelazine, Thorazine, Trilafon.
Anti-inflammatory (antiarthritic)	Aspirin, Indocin, Butazolidin, Motrin, Tolectin, Naprosyn, Nalfon.	Sleeping pills	Butisol, chloral hydrate, Dalmane, Doriden, Nembutal, Quaalude, Placidyl, Seconal.
		Antidepressants and lithium	Tricyclic antidepressants: Elavil, Sinequan, Tofranil. Lithium: Eskalith, Lithane.

PHYSIOLOGICAL AND ANATOMICAL ABNORMALITIES

The nervous system expresses itself through the sensory system and associated sensations or through the motor system and the activity of muscles and glands. Malfunctioning of the system due to injury or disease can be noted through clinical signs and symptoms. The signs may be an increase, decrease, or distortion of a sensation, as, for example, greater pain, less or absent pain, or a distortion of pain, as in a phantom limb. (Following the amputation of a foot, for example, the patient can often feel excruciating pain where the foot used to be.) A malfunction in the motor system may be indicated by greater, less, or distorted activity by the muscular and glandular systems. The enhancement of sensations and motor activities following a lesion can be explained in the following way: the injured structure may contain centers that regulate other centers by inhibiting them. Following the injury, the inhibition is decreased, and the ordinarily inhibited centers express themselves without the usual controls. The result may be a more intense feeling of sensations, and greater motor activity.

Because disturbances of the brain are so numerous, we will give only a brief description of some of the major ones here. We will also summarize some of the major disorders of cranial nerves.

Senility

Brain cells, which are irreplaceable,

begin to deteriorate at a relatively early age. Because of a surplus of brain cells, however, a shrinkage (atrophy) of the brain and a slowing down of mental processes usually does not occur until after the age of 60; sometimes it does not occur at all. Severe atrophy of the brain is commonly called "senility." The more technical name is **senile dementia** (L. *senex*, old man + *demens*, out of mind). It is characterized by progressive mental deterioration, including anxiety, irritability, difficulty with speech, and irrationality. In most cases, senility does not occur until after the age of 70. However, it should be emphasized that senility is a disease, and definitely *not* a normal condition of aging.

Alzheimer's disease

In a few cases, brain atrophy may occur much earlier than usual during the normal course of aging, even as early as 30 or 40 years of age, with the symptoms being identical to senile dementia. Such a condition is known as **Alzheimer's disease,** a neurodegenerative disease characterized by a progressive loss of memory and intellectual function. Like senility, Alzheimer's disease is not a part of the normal aging process. Alzheimer's disease afflicts more than 3 million Americans and kills about 120,000 of them a year, making it the fourth leading cause of death among the elderly (after cardiovascular disease, cancer, and

stroke). Death usually occurs less than 10 years after the first symptoms appear.

The death of certain brain cells appears to be the basis for many of the symptoms associated with Alzheimer's disease. Autopsies of victims invariably reveal significant changes in the brain, including a reduced number of neurons, especially in areas related to thought processes, but also in neurons with cell bodies in the base of the cerebrum, and axons that release acetylcholine at their terminals in the cerebral cortex; an increased amount of a protein called *amyloid* ("starchlike") in and around blood vessels; and an accumulation of tangled neurofilaments and *neuritic plaques* (cellular debris and amyloid). Victims of Alzheimer's disease also have a reduced amount of the enzyme *protein kinase C,* which aids in the proper uptake of phosphate and the cascade of cellular events that follow. It is possible that increased amounts of calcium in the elderly brain help to inhibit protein kinase C. Finally, there is a marked reduction in the release of acetylcholine. This reduction occurs when certain cholinergic neurons with their cell bodies in the basal nucleus (of Meynert, located at the base of the cerebrum) and terminating fibers in wide areas of the cerebral cortex are functionally impaired. Presumably, the loss of acetylcholine inhibits memory formation.

The drug clonidine may help victims

of Alzheimer's disease retain their memory. Clonidine acts on neuron receptors, possibly improving the transmission of nerve impulses to the prefrontal cortex, an area generally affected by lesions in the elderly. But as yet, no effective treatment, cure, or satisfactory diagnostic test exists.

Alzheimer's disease has been discovered to have a genetic basis, with a gene defect located on chromosome 21, the same chromosome involved in Down's syndrome (see Chapter 29). The disorder is not necessarily inherited, however.

Cerebral palsy

Cerebral palsy* actually comprises a group of neuromuscular disorders that usually result from damage to a child either before it is born, during childbirth, or shortly after birth. The major types of cerebral palsy are *spastic, anthetoid,* and *ataxic.* All three forms involve an impairment of voluntary motor activity to some degree, ranging from muscular weakness to complete paralysis. Related disorders such as mental retardation and speech difficulties may accompany the disease. Causes are varied, from infection and malnutrition of the mother to prolonged labor, brain infection, or circulatory problems. There is no known cure.

Cerebrovascular accident (CVA)

A *cerebrovascular accident* (CVA), commonly called a **stroke,** is a sudden withdrawal of sufficient blood supply to the brain, caused by the impairment of blood vessels to the brain. The resulting oxygen deficiency causes brain tissue to be damaged or even destroyed *(infarction).* Unconsciousness results if the blood and oxygen supply to the brain is insufficient for as little as 10 seconds. Irreversible brain damage can occur if the brain is deprived of oxygen for 5 minutes or more. CVA kills about half of the people it strikes, and about half of those who survive are permanently disabled. Cerebrovascular accident is the most common brain disorder, and the third most common cause of death in the United States (after cardiovascular disease and cancer).

**"Palsy" is a distorted form of *paralysis,* which comes from the Greek word meaning "to loosen," because the muscles seem to fall loose.*

The major causes of CVA are *thrombosis* (Gr. a clotting), the clotting of blood in a blood vessel; *embolism* (Gr. "to throw in"), a blockage in a blood vessel; and *hemorrhage,* a rupture of a cerebral artery as the result of an *aneurysm,* local dilation or ballooning of an artery. As a result, the blood supply to the brain is reduced, and blood leaks through the dura mater and puts harmful pressure on brain tissue, sometimes destroying it. The risk of thrombosis, embolism, or hemorrhage is increased by atherosclerosis (artery disease), hypertension (high blood pressure), diabetes mellitus, a high-fat diet, cigarette smoking, lack of exercise, and oral contraceptives.

Epilepsy

Epilepsy (Gr. *epilambanein,* seize upon) is a nervous disorder characterized by recurring attacks of motor, sensory, or psychological malfunction, with or without unconsciousness or convulsive movements. In *symptomatic* epilepsy, seizures can be traced to one of several known causes, including a brain tumor or abscess, diseases that affect central blood vessels, and poisons. Epilepsy is often caused by brain damage before, during, or shortly after birth. The more common occurrence is *idiopathic* (Gr. a disease of unknown cause) epilepsy, in which brain cells act abnormally for no apparent reason. The disease is known as *grand mal* (Fr. great illness) when the motor areas of the brain are affected and severe spasms and loss of consciousness are involved. It is called *petit mal* (small illness) when the sensory areas are affected, without convulsions and prolonged unconsciousness. During an epileptic seizure, neurons in the brain fire at unpredictable times, even without a stimulus.

Headache

Headaches related to muscle tension were discussed in Chapter 10. Here we consider headaches that are connected with cranial nerves, blood vessels, and meninges.

The brain itself is not sensitive to pain, but the veins on the surface of the brain, the cerebral arteries, cranial nerves, and parts of the dura mater are. If any of these areas is disturbed, a **headache** may result. Pressure on cranial veins, arteries, or meninges is sometimes produced by tumors, hemorrhage, men-

ingitis, or an inflamed trigeminal nerve root. However, some of the more typical causes are emotional stress, increased blood pressure, and food allergies that make blood vessels dilate or constrict, stimulating the pain-sensitive nerve endings in the vessels. The resulting headache may be accompanied by dizziness or vertigo.

Migraine headaches are severe, recurring headaches that usually affect only one side of the head. They are often preceded by fatigue, nausea, vomiting, and the vision of zigzag lines or brightness; they may be accompanied by intense pain, nausea, vomiting, and sensitivity to light and noise. Among the causes are emotional stress, hypertension, menstruation, and certain foods such as chocolate, animal fats, and alcohol. Migraine headaches frequently occur within families (suggesting a genetic basis in some cases), in compulsive, tense people, and, interestingly, on weekends and holidays, when the normal rhythm is disrupted.

Trigeminal neuralgia (tic douloureux) is a stabbing pain in one side of the face, along the path of one or more branches of the trigeminal nerve. It usually occurs sporadically. The cause is unknown.

Cranial arteritis is marked by intense pain and tenderness in the temples when the arteries in that region become inflamed. Untreated cranial arteritis may block the artery leading to the retina (central retinal artery) and cause blindness, especially in the elderly.

Parkinson's disease

Parkinson's disease (also called Parkinsonism, paralysis agitans, and shaking palsy) is a progressive neurological disease characterized by stiff posture, an expressionless face, slowness in voluntary movements, tremor at "rest," and a shuffling gait. The involuntary tremor, especially of the hands, often disappears when the upper limb moves. The tremor is accompanied by "pill-rolling" actions of the thumb and fingers. Parkinson's disease results from a deficiency of the neurotransmitter *dopamine,* released by neurons with their cell bodies in the substantia nigra, and their axons terminating in the striatum (caudate nucleus and putamen). Such a shortage of dopamine usually occurs when the neurons projecting from the substantia nigra to the striatum of the basal ganglion degenerate. The deficiency prevents brain cells from performing their usual inhibitory functions within the central nervous system.

There is some evidence that the degeneration of the substantia nigra and striatum may result from a failure of repair mechanisms related to aging rather than from a primary degenerative process. It is also postulated that environmental toxins may cause the most common form of Parkinson's disease.

Parkinson's disease affects more men than women, typically those over 60. It usually progresses for about 10 years, at which time death generally occurs from other causes, such as pneumonia or another infection.

At present, there is no known cure. Treatment consists of drugs and physical therapy that keep the patient functional as long as possible. Short-term drug therapy usually includes *levodopa* (L-dopa), a dopamine substitute, although unpleasant side effects, including nausea and vomiting, are common. Dopamine itself cannot be administered as a drug because it does not penetrate the blood-brain barrier. Impassable drugs may be combined with fat-soluble drugs that allow them to pass through the barrier and then separate after the drug has entered the brain. New drugs that help to counter the side effects of levodopa are currently being used in conjunction with that drug.

Dyslexia

Dyslexia (L. *dys*, faulty + Gr. *lexis*, speech) is an extreme difficulty in learning to identify printed words. It is most commonly seen as a reading and writing disability, and it is usually identified in children when they are learning to read and write. Typical problems include letters that appear to be backwards, words that seem to move on the page, and transposed letters. The disorder is not related to intelligence.

Reading and writing are usually dominated by the left cerebral hemisphere, but dyslexics appear to have an overdeveloped right hemisphere that competes with the left hemisphere for the control of language skills. Appar-

---

TABLE 13.6 SOME DISORDERS OF CRANIAL NERVES

Nerve		Disorder
I	Olfactory	Fracture of cribriform plate of ethmoid bone, or lesions along olfactory pathway may produce total inability to smell *(anosmia);* as a result, food tastes flat.
II	Optic	Trauma to orbit or eyeball, or fracture involving optic foramen produces inability of pupil to constrict. Certain poisons (such as wood alcohol), or infections (such as syphilis) may damage nerve fibers. Increase in eye fluid (aqueous humor) pressure produces glaucoma, causing partial or complete loss of eyesight. Pressure on optic pathway or injury or clot in temporal, parietal, or occipital lobes of cerebrum may cause blindness in certain regions of visual fields.
III	Oculomotor	Lesion or pressure on nerve produces dilated pupil, drooping upper eyelid *(ptosis),* absence of direct pupil reflex, squinting *(strabismus),* double vision *(diplopia).*
IV	Trochlear	Damage to nerve causes some inability to look down with one eye, producing double vision. Paralysis may produce strabismus and diplopia.
V	Trigeminal	Injury to terminal branches or trigeminal ganglion causes loss of pain and touch sensation, abnormal tingling, itching, numbness *(paresthesias),* inability to contract chewing muscles, deviation of mandible to side of lesion when mouth is opened. Severe spasms in nerve branches cause nerve pain *(trigeminal neuralgia,* or *tic douloureux).*
VI	Abducens	Nerve lesion causes inability of eye to move laterally, with eye cocked in; diplopia; strabismus.
VII	Facial	Damage causes Bell's palsy, loss of taste on anterior two-thirds of tongue on side of lesion. Cerebral stroke may produce paralysis of lower muscles of facial expression (below eye) on opposite side of lesion. Sounds are louder on side of injury because stapes muscle is paralyzed.
VIII	Vestibulocochlear	Tumor or other injury to nerve produces progressive hearing loss, noises in ear *(tinnitus),* involuntary rapid eye movement *(nystagmus),* whirling dizziness *(vertigo).*
IX	Glossopharyngeal	Injury to nerve produces loss of gag reflex, difficulty in swallowing *(dysphagia),* loss of taste on posterior third of tongue (not noticed by patient unless tested), loss of sensation on affected side of soft palate, decrease in salivation.
X	Vagus	Unilateral lesion of nerve produces sagging of soft palate, hoarseness due to paralysis of vocal fold, dysphagia.
XI	Accessory	Laceration of neck produces inability to contract sternocleidomastoid and upper fibers of trapezius muscles, drooping shoulders, inability to rotate head *(wry neck).*
XII	Hypoglossal	Damage to nerve produces protruded tongue deviated toward affected side, moderate difficulty in speaking *(dysarthia),* chewing, and dysphagia.

ently, the abnormal distribution of nerve cells takes place during the middle third of pregnancy, when the outer cortex of the fetal brain is formed. Prenatal disturbances such as a small stroke, maternal stress, or a viral infection may underly the abnormal fetal development. Since more boys than girls are dyslexic, another possible cause may be abnormal levels of testosterone, the male hormone. An early discovery of reading and writing problems and improved remedial teaching techniques appear to be the main factors in treating dyslexics.

Encephalitis

Encephalitis is an inflammation of the brain, usually involving a virus transmitted by a mosquito or tick. Several other viral causes are known. Brain tissue is infiltrated by an increased number of infection-fighting white blood cells called lymphocytes, cerebral edema occurs, ganglion cells in the brain degen-

erate, and neuron destruction may be widespread. Symptoms generally include the sudden onset of fever, progressing to headache and neck pain (indicating meningeal inflammation), drowsiness, coma, and paralysis. Extreme cases may be fatal, especially in herpes encephalitis.

Bell's palsy and other disorders of cranial nerves

Bell's palsy results from the dysfunction the facial (VII) nerve. The disorder may also be initiated by an inflammation, generally in the area of the internal auditory meatus. Such an inflammation is commonly caused by an infection associated with hemorrhage, tumor, meningitis, or local trauma. Symptoms occur on the same side as the lesion. They include weakness or paralysis on one side of the face, facial distortion, the inability

to close the eye (the eye rolls upward when closing is attempted) or wrinkle the forehead, a loss of taste on the anterior two-thirds of the tongue, and a drooping mouth (which leads to excessive drooling). A cerebral stroke may produce paralysis of the muscles of facial expression below the eye on the side opposite the lesion. Bell's palsy occurs most often in people over 60, although it can strike at any age. Recovery is usually spontaneous and complete in less than two months, but partial recovery occurs in about 10 percent of the cases. Treatment includes the use of prednisone, a steroid drug that reduces tissue swelling near the nerve and improves blood flow and nerve conduction.

Disorders involving cranial nerves range from infections that may damage nerve fibers to serious injuries, cerebral strokes, or tumors that may produce paralysis and the loss of some sensations (Table 13.6).

MEDICAL TERMINOLOGY

AGNOSIA (Gr. *a*, without + *gnosis*, knowledge) Inability to recognize objects.

ANALGESIA (Gr. *an*, without + *algesia*, sense of pain) Inability to feel pain.

ANESTHESIA (Gr. *an*, without + *aisthesis*, feeling) Partial or complete loss of sensation, especially tactile sensibility, induced by disease or an anesthetic.

APHASIA (Gr. *a*, without + *phasis*, speech) Partial or total loss of the ability to speak and understand words, resulting from brain damage.

APOPLEXY (Gr. *apoplessein*, to cripple by a stroke) Sudden loss of muscular control, with partial or total loss of sensation and consciousness, resulting from rupture or blocking of a blood vessel in the brain.

ATAXIA (Gr. *a*, without + *taktos*, order) Lack of coordination of voluntary muscles.

BRADYKINESIA (Gr. *bradus*, slow + *kinesis*, movement) A condition characterized by abnormally slow movements.

CAT SCAN (computerized axial tomography scan) A two-dimensional picture taken at various levels of tissue (*tomography* = a picture of a section) by an x ray apparatus that circles around the body on an axis. The x ray is interpreted as a picture by a computer.

CEREBRAL ANEURYSM (Gr. *aneurusma*, dilation) A bubblelike sac formed by the enlargement of a cerebral blood vessel.

COMA (Gr. *koma*, deep sleep) A state of deep, prolonged, unconsciousness, usually the result of injury, disease, or poison.

CONVULSION (L. *convellere*, to pull violently) Violent involuntary muscular contractions and relaxations.

CRANIECTOMY The removal of a portion of the skull bone.

CRANIOSTOMY (Gr. *tomos*, cut) An incision into the cranium to open the skull in order to have access to the brain.

DYSARTHIA (L. *dys*, faulty + Gr. *arthron*, joint) Lack of coordination of the muscles that control speech.

DYSKINESIA (L. *dys*, faulty + Gr. *kinesis*, movement) Difficulty in performing voluntary muscular movements due to brain lesion.

ENCEPHALOPATHY A general term for any brain disease.

LETHARGY (Gr. *lethargos*, forgetful) A state of sluggish indifference or unconsciousness resembling deep sleep into which a person relapses after being roused.

MENINGIOMA A meningeal tumor.

MENINGOCELE A protrusion of the me-

ninges through an opening in the skull or spinal column.

NARCOLEPSY (Gr. *narke*, numbness + *lepsia*, to seize) A condition of uncontrollable sleepiness and sleep.

PARALYSIS (Gr. to loosen) Loss of muscle function caused by damage to the brain, spinal cord, or nerves.

PARALYSIS AGITANS (Gr. to loosen + L. *agitare*, to agitate) A technical (and contradictory) term for Parkinson's disease.

PARESIS (Gr. *parienai*, to let go) Partial or complete paralysis, with the muscles appearing to be relaxed.

SCISSORS GAIT A manner of walking in which the legs are crossed over; occurs in cerebral palsy.

SPASTIC (Gr. *spastikos*, to pull) Pertaining to hyperactive muscular contractions, spasms, or convulsions.

STUPOR (L. *stupere*, to be stunned) A state of reduced sensibility or mental confusion; lethargy.

SUBDURAL HEMATOMA (Gr. *haimato*, blood + *oma*, tumor) A localized pool of venous blood in the subdural space.

TORPOR (L. *torpere*, to be stiff) A condition of mental or physical inactivity or insensibility; lethargy.

TREMOR (L. *tremere*, to tremble) An involuntary shaking or trembling.

SUMMARY

General structure of the brain

1 The major parts of the brain are the **cerebrum, diencephalon, cerebellum,** and **brainstem.** The cerebrum consists of the two *cerebral hemispheres.* The diencephalon is composed primarily of the thalamus, hypothalamus, epithalamus, and ventral thalamus (subthalamus). The brainstem is composed of the midbrain, pons, and medulla oblongata.

2 Each hemisphere of the cerebrum consists of outer gray matter, deep white matter, and deep gray matter *(basal ganglia).* The outer gray matter, called the *cerebral cortex,* has many folds and convolutions. The *corpus callosum,* a bundle of nerve fibers, connects the hemispheres and relays messages between them.

Meninges, ventricles, and cerebrospinal fluid

1 The brain is covered by the same three *meninges* that protect the spinal cord. The outermost *dura mater* consists of an inner dura mater that is continuous with the spinal dura mater, and an outer dura mater, which is actually the periosteum of the skull bones. The cranial dura mater contains blood vessels. The potential space between the dura mater and the arachnoid is the subdural space.

2 The middle *arachnoid* is a thin layer between the dura mater and pia mater. Between the arachnoid and pia mater is the *subarachnoid space,* which contains cerebrospinal fluid.

3 The innermost *pia mater* directly covers the surface of the brain. In certain locations, the pia mater joins together with modified ependymal cells to form the *choroid plexuses,* networks of small blood vessels.

4 *Cerebrospinal fluid* supports and cushions the brain and helps control the chemical environment of the central nervous system.

5 Cerebrospinal fluid circulates through the *ventricular system* of the brain, which consists of a series of connected cavities called **ventricles.**

6 Cerebrospinal fluid is essentially an ultrafiltrate of blood. Most of it is formed continuously at the choroid plexuses of the lateral, third, and fourth ventricles.

7 Cerebrospinal fluid circulates from the choroid plexuses through the ventricles and then the subarachnoid space in the vicinity of the medulla, until it reaches the subarachnoid villi, through which it passes back into the venous blood of the superior sagittal sinus.

Nutrition of the brain

1 A lack of oxygen or glucose will damage brain tissue faster than any other tissue. Glucose, the body's chief source of usable energy, cannot be stored in the brain.

2 The brain has built-in regulating devices that make it almost impossible to constrict the blood vessels that carry the incoming blood supply.

3 The *blood-brain barrier* is an anatomical and physiological network of selectively permeable membranes that prevent some substances from entering the brain, while allowing relatively free passage to others.

Brainstem

1 The *brainstem* is the stalk of the brain, relaying messages between the spinal cord and the cerebrum.

2 Within the brainstem are ascending and descending pathways between the spinal cord and parts of the brain.

3 Contained within the core of the brainstem is the *reticular formation,* with its pathways and integrative functions.

4 Of the 12 cranial nerves all but the olfactory and optic nerves emerge from the brainstem.

5 The lowermost portion of the brainstem is the *medulla oblongata.* It contains vital cardiac, vasomotor, and respiratory centers and also regulates vomiting, sneezing, coughing, and swallowing.

6 The *pons* forms a connecting bridge between the medulla oblongata and the midbrain. It also contains fibers that project to the hemispheres of the cerebellum and link the cerebellum with the cerebrum. Within the reticular formation of the pons are respiratory centers that are integrated with those in the medulla.

7 The *midbrain* connects the pons and cerebellum with the cerebrum. It contains all the ascending fibers projecting to the cerebrum. It also contains oculomotor nerves, which are involved with movements of the eyes, focusing, and constriction of the pupils of the eyes.

Cerebellum

1 The *cerebellum* is located behind the pons in the posterior cranial fossa. It is mainly a coordinating center for muscular movement, involved with balance, precision, timing, and body positions. It has no role in perception, conscious sensation, or intelligence.

2 The cerebellum is divided into the unpaired midline **vermis,** the small bilateral *flocculonodular lobes,* and the two large *lateral lobes* or hemispheres.

3 The cerebellum is covered by a surface layer of gray matter called the *cerebellar cortex,* which is composed of a neuronal network of circuits.

4 The cerebellum is connected to the brainstem by bundles of nerve fibers called *cerebellar peduncles.*

Cerebrum

1 The *cerebrum* is the largest and most complex structure of the nervous system. It is composed of the cerebral *hemispheres, white matter,* and *basal ganglia.* Each hemisphere is further divided into six lobes—the frontal, parietal, occipital, temporal, limbic, and insula.

2 The surface of the cerebrum is a mantle of gray matter called the *cerebral cortex.* The cortex contains raised ridges *(gyri),* slitlike grooves *(sulci),* and deep *fissures,* all of which increase the surface area of the cerebrum.

3 A deep groove called the *longitudinal fissure* runs between the cerebral hemispheres. The *corpus callosum,* a bridge of nerve fibers between the hemispheres, relays nerve impulses between them.

4 Beneath the cortex is a thick layer of white matter consisting of association, commisural, and projection fibers.

5 Each cerebral hemisphere contains a large core of *gray matter* called the *basal ganglia,* which help to coordinate muscle movements by relaying neural inputs from the cerebral cortex to the thalamus and finally back to the motor cortex of the cerebrum.

6 The *frontal lobe* is involved with the motor control of voluntary movements and the control of a variety of emotional expressions and moral behavior. The *parietal lobe* is concerned with the evaluation of the general senses, and of taste. The *temporal lobe* has critical roles in hearing, equilibrium, and to a certain degree, emotion and memory. The *occipital lobe* is involved with vision and its associated forms of expression. The *insula* (central lobe) appears to be associated with gastrointestinal and other visceral activities.

7 The limbic lobe is an integral part of the *limbic system,* which is involved with emotions, behavioral expressions, recent memory, and smell.

8 The *diencephalon* is located deep to the cerebrum; it connects the midbrain with the cerebral hemispheres. It is composed of the thalamus, hypothalamus, epithalamus, and ventral thalamus (subthalamus).

9 The *thalamus* is the intermediate relay point and processing center for all sensory impulses (except smell) going to the cerebral cortex.

10 The *hypothalamus* is the highest integrating center for the autonomic nervous system and regulates many physiological and endocrine activities through its relationship with the *pituitary gland* (hypophysis). It is involved with the regulation of body temperature, water balance, sleep-wake patterns, food intake, behavioral responses associated with emotion, endocrine control, and sexual responses.

11 The *epithalamus* acts as a relay station to the midbrain. It contains the *pineal body* (epiphysis).

12 The *ventral thalamus* or *subthalamus* is functionally integrated with the basal ganglia.

Learning and memory

1 Little is known about the neuronal processes of human learning, but it has been shown that specific areas of the brain are involved with different types of learning. No single area of the brain controls learning, and scientists believe that learning involves a variety of mechanisms, not just one.

2 It appears that "memory traces" are formed during learning and are imprinted on the brain when neurons record and store information.

3 *Short-term memory* refers to the remembrance of recent events, and *long-term memory* refers to the remembrance of information that may last a lifetime. It is not known if both types of memory use the same basic mechanism, but all memory is related to electrical and chemical activity in neuronal circuits.

4 The temporal lobes probably function as a directory that permits the retrieval of stored memories, but it is not known exactly where memory is stored in the brain.

5 Learning probably results from changes in receptor cell membranes and nerve transmission, and possibly from new patterns in the wiring of nerve fibers.

Sleep and dreams

1 Brain activity may be recorded on an *electroencephalograph* or *EEG.* The four distinct brain-wave patterns are *alpha, beta, delta,* and *theta* waves.

2 The EEG has made it possible to detect four separate stages of sleep, and an important condition known as *REM* (rapid eye movement) *sleep,* which occurs during reentry into a new stage 1 phase. Dreams occur during REM sleep. It is thought that REM sleep is essential for relaxation of normal stress, and for brain maturation. Learning and memory are probably reinforced during REM sleep.

3 The factor that produces sleep is unknown.

Cranial nerves

1 The 12 pairs of *cranial nerves* are the peripheral nerves of the brain. Their numbers indicate the sequential order in which they emerge from the brain and its meninges.

2 The *motor (efferent) fibers* of the cranial nerves emerge from the brainstem. They arise from groups of neurons called *motor nuclei.* The *sensory (afferent) fibers* of cranial nerves arise from cell bodies located in *sensory ganglia* or in peripheral sense organs such as the eye. Axons of sensory neurons synapse with *sensory nuclei.* Fibers transmit input to the cerebrum, where sensations are brought to a conscious level.

3 The cranial nerves, in ascending numerical order, are the *olfactory* (I), *optic* (II), *oculomotor* (III), *trochlear* (IV), *trigeminal* (V), *abducens* (VI), *facial* (VII), *vestibulocochlear* (VIII), *glossopharyngeal* (IX), *vagus* (X), *accessory* (XI), and *hypoglossal* (XII).

Chemicals and the nervous system

1 Neurotransmitters and other chemicals that are natural constituents of the body are essential for proper functioning of the nervous system. Drugs may alter the normal workings of the nervous system. Most drugs operate at the synaptic level.

2 Some major categories of drugs are *stimulants, depressants, antidepressants, psychedelics and hallucinogens, analgesics,* and *antianxiety drugs.*

Physiological and anatomical abnormalities

1 *Senility,* the atrophy of the brain, is a disease, not a normal condition of aging.

2 *Alzheimer's disease* is a neurodegenerative disease characterized by progressive loss of memory and intellectual function.

3 *Cerebral palsy* comprises a group of neuromuscular disorders that involve an impairment of voluntary motor activity.

4 A *cerebrovascular accident* (CVA), or *stroke,* is a sudden withdrawal of sufficient blood supply to the brain. The major causes are *thrombosis, embolism,* and *hemorrhage.*

5 *Epilepsy* is a nervous disorder characterized by recurring attacks of motor, sensory, or psychological malfunction. During an epileptic seizure, neurons in the brain fire abnormally at unpredictable times.

6 *Headaches* may occur when cerebral blood vessels, cranial nerves, or the meninges are disturbed. The brain itself is not sensitive to pain. Intense pain is associated with *migraine headaches, trigeminal neuralgia,* and *cranial arteritis.*

7 *Parkinson's disease* is thought to be caused by a diminution of dopamine, a neurotransmitter substance in the brain. Such a shortage leads to a degeneration of certain neurons of a basal ganglion.

8 *Dyslexia* is an extreme difficulty in learning to read and write. Dyslexics appear to have an overdeveloped right cerebral hemisphere that competes with the left hemisphere, which normally controls reading and writing.

9 *Encephalitis* is a severe brain inflammation, usually caused by a virus.

10 Disorders involving *cranial nerves* range from infections that may damage nerve fibers to serious injuries, cerebral strokes, or tumors that may produce paralysis and the loss of some sensations.

UNDERSTANDING THE FACTS

1 Name the four major parts of the brain, and their main subdivisions.

2 What is the function of the corpus callosum?

3 What features of the brain provide support and protection?

4 In which specific part of the brainstem are the respiratory and cardiovascular centers located?

5 What is the blood-brain barrier?

6 Name the six cerebral lobes.

7 Give the location of the following (be specific):
   a corpus callosum
   b cerebral aqueduct
   c nuclei (of nervous system)
   d corpora quadrigemina
   e insula
   f thalamus
   g reticular formation

8 Which of the cerebral lobes is largely involved with vision?

9 What two major functions are associated with the limbic system?

10 Which part of the brain could be described as the intermediate relay point for almost all sensory impulses to the cerebral cortex?

11 Which area of the cerebrum controls learning?

12 What is a memory trace? Short-term memory? Long-term memory?

13 What are the four stages of sleep?

14 Which cranial nerve(s) originate from the cerebrum?

15 Name the cranial nerves that are purely sensory in function.

16 What is the optic chiasma?

17 Which cranial nerves are involved with vision, including the movement of the eyeballs?

18 Name and define three or more classes of drugs.

19 Is senility a normal aspect of aging?

20 Briefly define Alzheimer's disease, cerebral palsy, cerebrovascular accident, epilepsy, Parkinson's disease, dyslexia, encephalitis, and Bell's palsy.

UNDERSTANDING THE CONCEPTS

1 Describe the circulation of cerebrospinal fluid through the brain.

2 What is the importance of a proper oxygen and glucose supply to the brain?

3 If an abnormality occurred within the medulla, what body functions might be affected?

4 What are the physiological implications of the pyramidal decussation?

5 If a cerebellar lesion occurs in the right side of the brain, on which side of the body will symptoms occur? How do you explain this fact?

6 What is the physiological impact of the gyri and sulci of the cerebral cortex?

7 How would you explain the fact that the two cerebral hemispheres are similar in appearance but dissimilar in function?

8 Compare the proportion of the brain's gray matter devoted to manipulating the thumb to that allotted to controlling the thorax and abdomen. Explain.

9 How are the four distinct brain-wave patterns related to levels of consciousness?

10 Why is your hypothalamus so important to your well-being, and, in fact, to your survival?

11 Why is REM sleep important?

SUGGESTED READING

AOKI, CHIYE, AND PHILIP SIEKEVITZ, "Plasticity in Brain Development." *Scientific American,* December 1988.

COWAN, W. MAXWELL, "The Development of the Brain." *Scientific American,* September 1979.

GAZZANIGA, MICHAEL S., "The Split Brain in Man." *Scientific American,* December 1979.

GILMAN, SID, AND SARAH S. WINANS, *Manter and Gatz's Essentials of Clinical Neuroanatomy and Neurophysiology,* 7th ed. Philadelphia: Davis, 1987.

GOLDSTEIN, GARY W., AND A. LORRIS BETZ, "The Blood-Brain Barrier." *Scientific American,* September 1986.

HUBEL, DAVID H., "The Brain." *Scientific American,* September 1979.

KETY, SEYMOUR S., "Disorders of the Human Brain." *Scientific American,* September 1979.

KUFFLER, S. W., W. J. G. NICHOLLS, AND A. ROBERT MARTIN, *From Neuron to Brain,* 2nd ed. Sunderland, Mass.: Sinauer, 1984.

MISHKIN, MORTIMER, AND TIM APPENZELLER, "The Anatomy of Memory." *Scientific American,* June 1987.

NATHANSON, JAMES A., AND PAUL GREENGARD, "Second Messengers in the Brain." *Scientific American,* August 1977.

NAUTA, WALLE J. H., AND MICHAEL FEIRTAG, "The Organization of the Brain." *Scientific American,* September 1979.

NOBACK, CHARLES R., AND ROBERT J. DEMAREST, *The Nervous System: Introduction and Review,* 3rd ed. New York: McGraw-Hill, 1986. (Paperback)

NOLTE, JOHN, *The Human Brain.* St. Louis: Mosby, 1981.

Scientific American, September 1979. (Available in book form.) The entire issue is devoted to the brain.

SHEPHERD, G. M., *Neurobiology.* New York: Oxford University Press, 1983.

VELLUTINO, FRANK R., "Dyslexia." *Scientific American,* March 1987.

WURTMAN, RICHARD J., "Alzheimer's Disease." *Scientific American,* January 1985.

14
The Autonomic Nervous System

LEARNING OBJECTIVES

1 State the primary function of the autonomic nervous system.

2 Compare structurally and functionally the somatic motor system with the autonomic nervous system.

3 Identify the two divisions of the peripheral autonomic nervous system, explain their basic functions, and describe their structural features.

4 Identify three major groups of autonomic ganglia and five important autonomic plexuses.

5 Describe the main pathways followed by sympathetic neurons between the central nervous system and the effectors.

6 Describe adrenergic and cholinergic receptors.

7 Compare the neurotransmitters released by the sympathetic and parasympathetic divisions, and the effects of stimulation on each division.

8 Describe the pathways followed by parasympathetic neurons between the central nervous system and the effectors.

9 Explain how certain parts of the central nervous system exercise control over the autonomic nervous system.

10 Identify the five components of an autonomic visceral reflex arc.

11 Describe the specific actions of each division of the autonomic nervous system on the major visceral effectors.

12 Explain how the two divisions of the autonomic nervous system work together to maintain homeostasis.

13 Describe the following disorders related to the autonomic nervous system: Horner's syndrome, autonomic dysreflexia, and achalasia.

When you see your favorite food, your digestive juices begin to flow. You don't have to do anything except see the food to start the secretion. This response is considered "involuntary" because you don't control it on a conscious level. Such responses occur in *cardiac muscle,* in *smooth muscle* of the internal organs, and in *glands,* such as salivary and sweat glands.

The **autonomic nervous system** (ANS) innervates these three types of effectors. It is divided into the peripheral autonomic nervous system and the central autonomic control centers. The autonomic nervous system is also known as the **visceral efferent motor system,** because it is concerned with the internal organs, or *viscera.* Its primary function is to regulate visceral activities to maintain homeostasis.

Two important points about the autonomic nervous system must be understood at the outset:

1 The autonomic nervous system is exclusively the *motor* system, involved with influencing the activity of cardiac muscle, smooth muscle, and glands of the body.

2 Sensory input to the autonomic nervous system is derived from sensory receptors, and is conveyed by *both* the somatic afferent system and the visceral afferent system.

Although the autonomic nervous system does not operate independently from the rest of the nervous system, its structure and functions are often contrasted with those of the somatic nervous system. The basic function of the somatic motor (efferent) system is to regulate the coordinated skeletal muscle activities involved with movement and the maintenance of posture. These activities are associated with adjustments to the *external* environment, and are under our conscious control. In contrast, the major role of the autonomic nervous system is to regulate circulation, respiration,* digestion, and other functions *not* usually subject to our conscious control, in order to maintain a relatively stable *internal* environment. The two systems often work together. For example, when you are exposed to the cold, you shiver, and the blood vessels in your skin constrict. Shivering is a muscular reaction of the somatic motor system that produces heat. Constriction of blood vessels is an action of the autonomic nervous system that conserves heat.

The two systems differ significantly in the way their neurons connect to their effectors. The neurons of the somatic nervous system *directly* innervate the effectors (voluntary muscles). Only *one* set of neurons links the central nervous system with these effectors, and *no* synapses are involved outside the central nervous system. The neurons of the autonomic nervous system, in contrast, link the central nervous system with the effectors via *two* sets of neurons. The first neuron originates in the central nervous system, and synapses with a second neuron outside the central nervous system, which finally reaches the effectors. The synapses occur *outside* the central nervous system in clusters of cell bodies and their dendrites called **autonomic ganglia.** Autonomic ganglia may be thought of as relay stations between the central nervous system and the effectors.

*Inhalation and exhalation are controlled by the somatic motor system, while the activities of the trachea, bronchi, and bronchioles are controlled by the autonomic nervous system.

STRUCTURE OF THE PERIPHERAL AUTONOMIC NERVOUS SYSTEM

The **peripheral autonomic nervous system** is a motor system consisting of two divisions, the *sympathetic* and *parasympathetic.* Each division sends efferent nerve fibers to the muscle, organ, or gland it innervates. In general, but not always, the two divisions have opposite effects. In the broadest terms, we can say that the sympathetic division helps the body adjust to stressful situations, and the parasympathetic division is active when the body is operating under normal conditions. In reality, the functions of the nervous system are much too complex to set up such a clear distinction between the two divisions. They are more easily distinguished anatomically, as you will see in the following sections.

Anatomical Divisions of the Peripheral Autonomic Nervous System

The efferent nerve fibers of the autonomic nervous system that innervate the visceral organs emerge from the central nervous system at several different levels. The visceral efferent fibers that emerge through the thoracic and lumbar spinal nerves constitute the **thoracolumbar** (thuh-RASS-oh-LUM-bar) **division.** At the *cranial* level, visceral efferent fibers leave the central nervous system by way of cranial nerves III, VII, IX, and X from the brainstem. At the *sacral* level, other visceral efferent fibers leave by way of sacral spinal nerves 3 and 4. These two groups make up the **craniosacral division.** *The craniosacral division is the parasympathetic division of the autonomic nervous system, and the thoracolumbar division is the sympathetic division.*

Preganglionic and Postganglionic Neurons

The linkage between the central nervous system and the effectors within the body differs in the somatic nervous system and the autonomic nervous system. As we have said, the somatic system has a *one-neuron linkage,* and the autonomic nervous system has a *two-neuron linkage.* Each somatic (lower) motor neuron has an axon that courses from its cell body in the brainstem or spinal cord through a cranial or spinal nerve, directly innervating a voluntary muscle. Hence, there is a one-neuron linkage between the central nervous system and an effector.

In contrast, the first neuron in the autonomic nervous system, called a **preganglionic neuron,** has its cell body in the brainstem or spinal cord, and a *myelinated* axon that courses through a cranial or spinal nerve and terminates by synapsing with the dendrites and cell body of one or more neurons (postganglionic neurons and interneurons) in an autonomic ganglion located outside the central nervous system. (Located wholly within each ganglion are small interneurons.) The second neuron in a two-neuron linkage is called a **postganglionic neuron.** It has its cell body in an auto-

THE DYNAMIC BODY

Biofeedback and the Autonomic Nervous System

The activity of the autonomic nervous system in response to an ever-changing internal environment goes on without our conscious awareness or control. Some of these automatic and involuntary functions involve heart rate, blood pressure, respiration, blood-glucose levels, and body temperature. We can have some *conscious* control over these activities through **biofeedback,** a technique that allows a person to monitor and control his or her own bodily functions.

In biofeedback, a mechanical device is used to register and display signs of physiological responses to make the person aware of them and to enable him or her to monitor changes in them. In general, any physiological function that can be recorded, amplified by electronic instruments, and fed back to the person through any of the five senses can be regulated to some extent by that person.

Biofeedback can work because every change in a physiological state is accompanied by a responsive change in a mental (or emotional) state. Conversely, every change in a mental state is accompanied by a change in the physiological state. The autonomic nervous system thus acts as the connecting link between the mental and physiological states.

Biofeedback is not new. Yoga and Zen masters of Eastern cultures have long demonstrated their ability to control bodily functions that were thought to be strictly involuntary. Now, Western culture has devised a technique through which, under the proper conditions, anyone can learn to duplicate these once "impossible" physiological feats.

Biofeedback control holds considerable promise as a means of treating some psychosomatic problems (illnesses caused mainly by stress or psychological factors) such as ulcers, anxiety, and phobias, as well as many other disorders. Biofeedback can also relieve muscle-tension headaches, reduce the pain of migraine headaches, achieve relaxation during childbirth, lower blood pressure, alleviate irregular heart rhythms, and control epileptic seizures. The greatest value of these biofeedback techniques is that they demonstrate that the autonomic nervous system is not entirely autonomous and automatic; some visceral responses can be controlled.

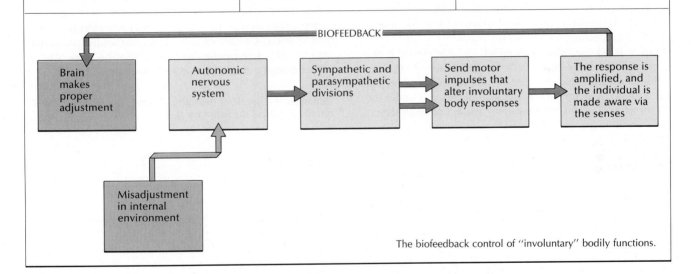

The biofeedback control of "involuntary" bodily functions.

nomic ganglion, and an *unmyelinated* axon courses through nerves and plexuses before terminating in a motor ending associated with cardiac muscle, smooth muscle, or a gland.

Postganglionic neurons do not always synapse directly with each effector. Many effectors, such as cardiac muscle fibers, are linked together by tight junctions. As a result, a stimulated effector can convey excitatory information to an adjacent effector.

Autonomic Ganglia

Autonomic ganglia are arranged in three groups: (1) paravertebral or lateral ganglia, (2) prevertebral or collateral ganglia, and (3) terminal or peripheral ganglia. The first two are part of the sympathetic division, and the third is part of the parasympathetic division:

1 The **paravertebral** (*para,* beside) **ganglia** (lateral ganglia) form beadlike rows of 21 or 22 swellings (3 cervical, 10 or 11 thoracic, 4 lumbar, 4 sacral) that run down both sides of the bony vertebral column, *outside* the spinal cord. These beadlike ganglia are connected by intervening nerve fibers to form two vertical ganglionic chains known as the **sympathetic trunks** (see Figure 14.2). These trunks and ganglia lie on the surface of the vertebral column and extend from the upper cervical vertebrae to the coccyx.

2 Sometimes a preganglionic fiber passes through the sympathetic trunk without forming a synapse. Instead, it joins with *prevertebral ganglia* (collateral ganglia), which lie in front of the vertebrae near the large thoracic, abdominal, and pelvic arteries that give them their names. The largest prevertebral ganglia are the *celiac* (solar) ganglion just below the diaphragm, the *superior mesenteric* ganglion in the upper abdomen, and the *inferior mesenteric* ganglion near the lower abdomen (see Figure 14.2).

3 Terminal ganglia (peripheral ganglia) are composed of small collections of ganglion cells, which are very close to or within the organs they innervate (see Figure 14.2). Terminal ganglia are especially common in the gastrointestinal tract and urinary bladder. They are considered to be part of the parasympathetic division.

Plexuses

Some postganglionic nerve fibers are distributed like cables to branching, interlaced networks following along the blood vessels in the thoracic, abdominal, and pelvic cavities, called *autonomic plexuses.* The autonomic plexuses include the following:

1 The *cardiac plexus* lies among the large blood vessels emerging from the base of the heart. It has a regulatory effect on the heart.

2 Most of the *pulmonary* (L. lung) *plexus* is located posterior to each lung. Sympathetic nerve fibers dilate the bronchi (delicate air passages leading from the windpipe to the lungs), and parasympathetic fibers constrict them.

3 The *celiac* (Gr. abdomen) *plexus,* or *solar plexus,* is the largest mass of nerve cells outside the central nervous system. It lies on the aorta in the vicinity of the celiac artery, behind the stomach. A sharp blow to the solar plexus, just under the diaphragm below the sternum, may cause unconsciousness by slowing the heartbeat and reducing the blood supply to the brain.

4 The *hypogastric* ("under the stomach") *plexus* connects the celiac plexus with the pelvic plexuses below. It innervates the organs and blood vessels of the pelvic region.

5 The *enteric* (Gr. intestine) *plexuses* receive both sympathetic and parasympathetic fibers. Located between the longitudinal and circular muscles of the digestive system, they help to regulate *peristalsis* (rhythmic contraction of the smooth muscles in the digestive system).

Sympathetic Division

The **sympathetic** (thoracolumbar, adrenergic) **division** of the autonomic nervous system arises from cell bodies in the lateral gray horn of the spinal cord. The myelinated nerve fibers emerge from the spinal cord in the ventral nerve roots of the 12 thoracic and first 2 or 3 lumbar spinal nerves. This emergence of fibers is known as the **thoracolumbar outflow.** These preganglionic fibers form small nerve bundles called **white rami communicantes** (singular, *ramus communicans*). They are white because the nerve fibers are myelinated. The fibers then pass to the paravertebral ganglia of the sympathetic trunk (Figure 14.1).

When a preganglionic neuron reaches the sympathetic trunk, it may take one of several pathways:*

1 It may synapse with a postganglionic neuron, which then terminates on an effector. A preganglionic neuron may synapse with a postganglionic neuron on the same level in the sympathetic trunk. Then the postganglionic neuron joins

*Although we speak of several pathways, the impulse actually spreads over all of the pathways at once.

FIGURE 14.1

Cross section of the spinal cord, comparing the pathways of (A) somatic nerves, and (B) visceral nerves. Afferent nerve fibers are purple, and efferent nerve fibers are in blue. Dotted lines represent efferent postganglionic fibers.

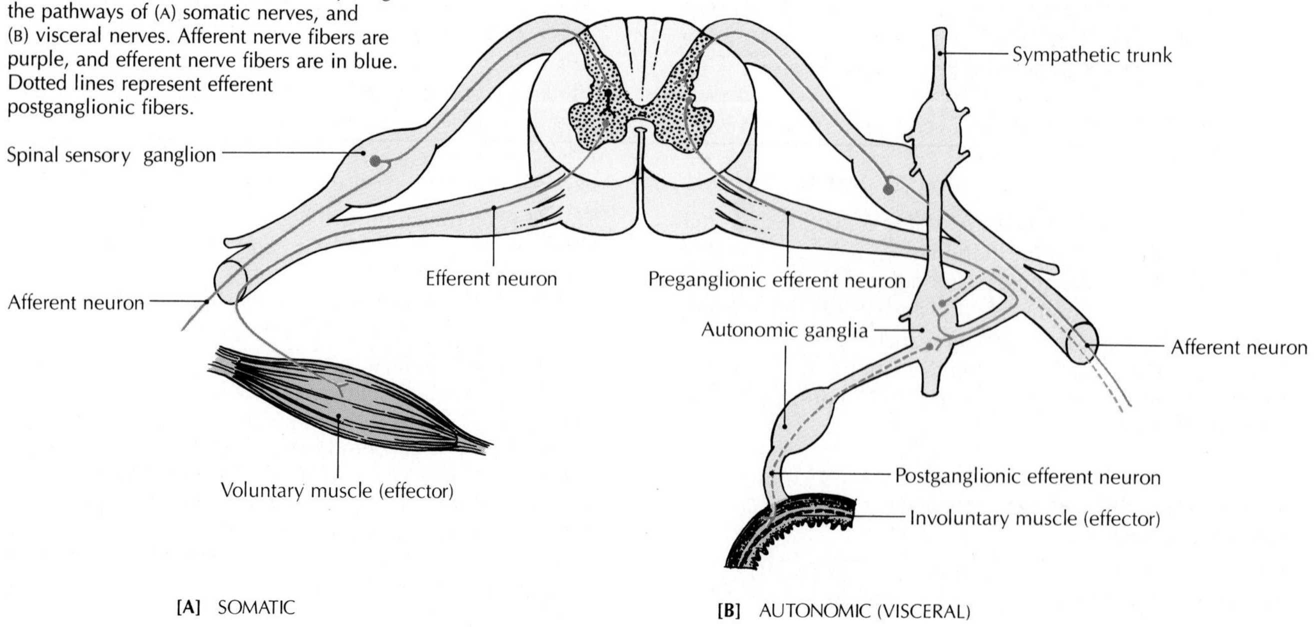

Spinal sensory ganglion

Sympathetic trunk

Afferent neuron

Efferent neuron

Preganglionic efferent neuron

Autonomic ganglia

Afferent neuron

Postganglionic efferent neuron

Involuntary muscle (effector)

Voluntary muscle (effector)

[A] SOMATIC

[B] AUTONOMIC (VISCERAL)

the spinal nerve by way of the *gray rami communicantes* before it terminates in an effector. These rami communicantes are gray because their fibers are unmyelinated.

2 In another pathway, the preganglionic neuron may course up or down to a different level of the sympathetic trunk before synapsing with a postganglionic neuron.

3 In still another pathway, it may pass directly through a ganglion (without synapsing) in the sympathetic trunk to synapse with a postganglionic neuron within a prevertebral (collateral) ganglion.

Each preganglionic fiber synapses with 20 or more postganglionic neurons in a ganglion. Some axons of the postganglionic neurons may pass directly to an effector (such as the heart or lungs), but most postganglionic axons pass through the gray rami communicantes and then travel back to the spinal nerve to be distributed for the innervation of sweat glands as well as for the smooth muscles of blood vessels, arrector pili muscles of the skin (which make the hair stand erect), the body wall, and the limbs. Other axons from the ganglia of the chain in the neck form nerve plexuses around blood vessels (*perivascular* plexuses) before terminating in the visceral structures of the head (such as the dilator muscles of the pupil and glands in the head). The postganglionic fibers from cells in the prevertebral ganglia form perivascular plexuses that innervate the viscera in the abdominal and pelvic regions. The sympathetic outflow from the spinal cord is distributed as shown in Figure 14.2.

Sympathetic neurotransmitters The neurotransmitter released by the preganglionic nerve terminals is *acetylcholine* (ACh). It is inactivated to choline and acetate by the enzyme *acetylcholinesterase*. The neurotransmitter released by the postganglionic nerve terminal is *norepinephrine* (NE). However, there are a few exceptions. Sympathetic postganglionic fibers to most sweat glands and to some blood vessels release acetylcholine instead of norepinephrine.

Only a small amount of the released norepinephrine stimulates the postganglionic neuron or effector. The excess neurotransmitter is rapidly deactivated by the enzyme *catechol-O-methyl transferase* (COMT) or taken up by the nerve terminal for recycling. The norepinephrine that is taken up is deactivated within the terminal by the enzyme *monoamine oxidase* (MAO).

The sympathetic division, its neurons, and its fibers comprise the *adrenergic* system, because most postganglionic neurons and fibers of this division release norepinephrine (also called nor*adrenaline*). The medulla of the adrenal gland secretes both norepinephrine and epinephrine into the bloodstream. Actually, the cells of the adrenal medulla should be considered as postganglionic neurons. The *preganglionic* neurons of the sympathetic division release acetylcholine. In contrast, both the preganglionic and postganglionic neurons of the parasympathetic system release acetylcholine. As a result, the parasympathetic division is called the *cholinergic* system.

Adrenergic receptors Located in or on the plasma membrane of the cells of effector organs are specific proteins that respond to epinephrine and norepinephrine. These two neurotransmitters are called *catecholamines* because they both have a six-carbon catechol ring and an amine group (Figure 14.3). There are two main types of receptors: alpha (α) and beta (β) (Figure 14.4). The type of receptor is determined by the specific physiological response it produces, and by its high specificity for certain drugs that either excite or inhibit it. Norepinephrine reacts mainly with alpha receptors, and epinephrine reacts mainly with beta receptors. In some receptors, however, overlap occurs.

Recently, beta receptors have been divided into β_1 and β_2 receptors. The β_1 receptors mediate an increased heart rate and strength of contraction and cause the release of the hormone renin from the kidneys. The β_2 receptors mediate almost all other beta-receptor effects. In like manner, the alpha receptors have been subdivided into α_1 and α_2 receptors. The effects for alpha and beta receptors are shown in Table 14.3.

Effects of sympathetic stimulation The sympathetic division of the autonomic nervous system is anatomically and physiologically organized to affect widespread regions of the body, or even the entire body, for sustained periods of time. For example, each preganglionic neuron synapses with numerous postganglionic neurons, which in turn have long axons that terminate in neuroeffector synapses over a large area. The widespread and long-lasting sympathetic effects are directly related to the slow deactivation of norepinephrine, and to the extensive distribution of norepinephrine and epinephrine released by the adrenal medulla into the bloodstream.

Parasympathetic Division

The axons of the *parasympathetic division* (Gr. *para*, beside, beyond; that is, located beside the sympathetic division) of the autonomic nervous system may also be called the craniosacral or cholinergic division. Its preganglionic neurons originate in the brainstem and sacral levels of the spinal cord. The cranial portion supplies parasympathetic innervation to the viscera of the head, neck, thorax, and most of the abdominal viscera (Figure 14.5). The sacral portion supplies parasympathetic innervation to the viscera of the lower abdomen and pelvis. The body walls and limbs do not have parasympathetic innervation (Tables 14.1 and 14.2).

The preganglionic fibers from the cranial portion of the parasympathetic division are known as the *cranial parasympathetic outflow*. These fibers emerge from the brainstem through cranial nerves III, VII, IX, and X. The preganglionic fibers from the sacral portion are the *sacral parasympathetic outflow*. They leave the spinal cord by way of the ventral roots of spinal nerves S3 and S4.

Preganglionic fibers from cell bodies located in the midbrain are conveyed by the oculomotor (III) cranial nerve to a synapse in the ciliary ganglion. The postganglionic axon terminals innervate the constrictor muscles of the pupil, as well as the ciliary muscles that change the shape of the lens to focus the eyes.

Preganglionic fibers from cell bodies located in the lower pons leave by way of the facial (VII) cranial nerve to either the pterygopalatine (sphenopalatine) or submandibular ganglion, where they synapse. The postganglionic axon terminals innervate the lacrimal glands, which secrete tears, and the glands of the nasal, oral, and pharyngeal cavities.

FIGURE 14.2

Sympathetic (thoracolumbar) division of the autonomic nervous system. Nerve fibers actually emerge from both sides of the cord.

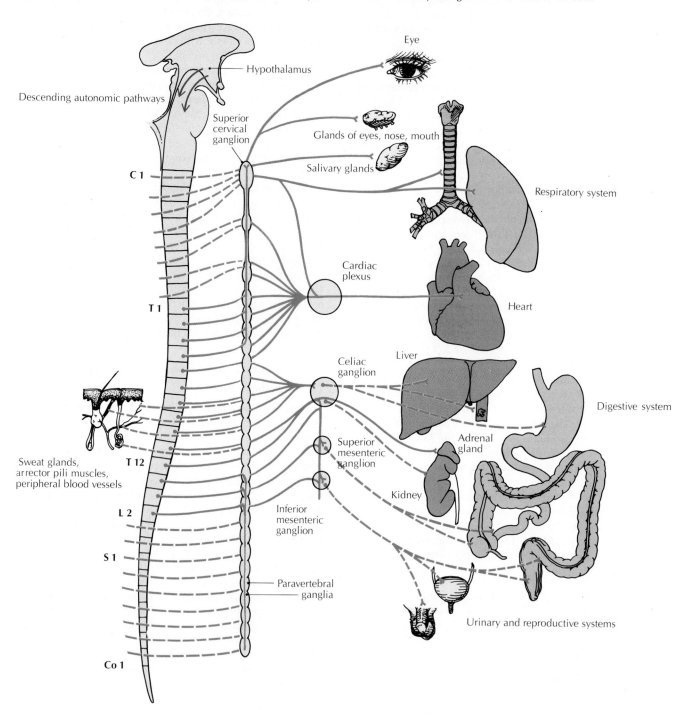

TABLE 14.1 COMPARISON OF SOME FEATURES OF THE SYMPATHETIC AND PARASYMPATHETIC DIVISIONS

Feature	Sympathetic division	Parasympathetic division
Distribution	Throughout the body.	Limited primarily to viscera of head, thorax, abdomen, pelvis.
Location of ganglia	Paravertebral and prevertebral ganglia close to CNS.	Terminal ganglia near effectors.
Outflow from CNS	Thoracolumbar levels.	Craniosacral levels.
Ratio of preganglionic to postganglionic neurons	Each preganglionic neuron synapses with many postganglionic neurons.	Each preganglionic neuron synapses with a few postganglionic neurons.
Neurotransmitter at preganglionic nerve terminals	Acetylcholine.	Acetylcholine.
Neurotransmitter at neuroeffector junction	Usually norepinephrine.	Acetylcholine.
Deactivation of neurotransmitter	Slow, by way of monoamine oxidase or catechol-O-methyl transferase.	Rapid, by way of acetylcholinesterase.

Adapted from Charles R. Noback and Robert J. Demarest, *The Human Nervous System*, 3rd ed. (New York: McGraw-Hill, 1981), pp. 224–225. Used with permission.

FIGURE 14.3

Two catecholamines. Note that the only difference between norepinephrine and epinephrine is a methyl group ($-CH_3$) in epinephrine (red), which replaces hydrogen.

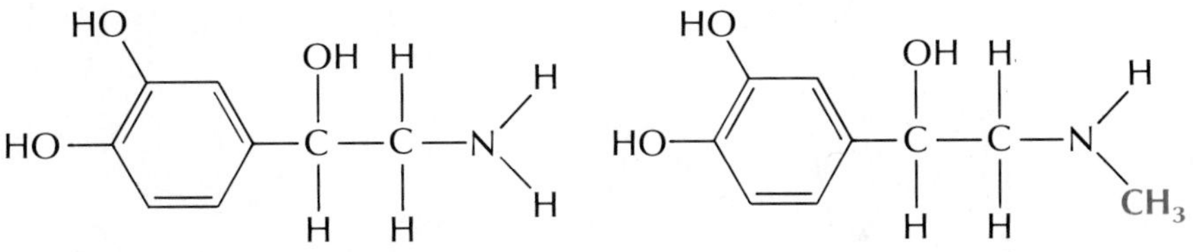

Catechol ring   Amine group   Methyl group

NOREPINEPHRINE   EPINEPHRINE

FIGURE 14.4

Plasma membrane of a cell of an effector organ showing alpha-receptor protein sites for norepinephrine and beta-receptor protein sites for epinephrine.

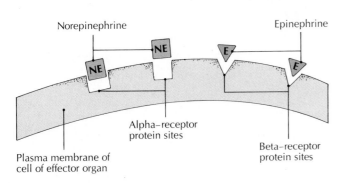

Outside cell

Norepinephrine   Epinephrine

Alpha–receptor protein sites

Beta–receptor protein sites

Plasma membrane of cell of effector organ

Inside cell

From cell bodies in the upper medulla, preganglionic fibers pass through the glossopharyngeal (IX) cranial nerve to the otic ganglia. After a synapse, the postganglionic neurons innervate the parotid glands, which secrete saliva.

Preganglionic fibers emerge from cell bodies located in the dorsal vagal nucleus of the medulla, and are conveyed by the vagus (X) nerve to a synapse with postganglionic neurons in the terminal ganglia. The axon terminals of the postganglionic fibers innervate the heart, lungs, blood vessels, kidneys, esophagus, stomach, small intestine, and the first half of the large intestine.

The parasympathetic fibers of the sacral outflow emerge from cell bodies in the gray matter of the spinal cord at the S3 and S4 levels. The preganglionic fibers pass through spinal nerves and their branches, called *pelvic splanchnics* (SPLANK-nicks), and synapse with postganglionic fibers in terminal ganglia close to the effectors. The axon terminals of the postganglionic fibers innervate the lower large intestine, uterus, genitals, and sphincter muscles of the urinary bladder and urethra.

FIGURE 14.5

Parasympathetic (craniosacral) division of the autonomic nervous system. Preganglionic and postganglionic neurons are cholinergic.

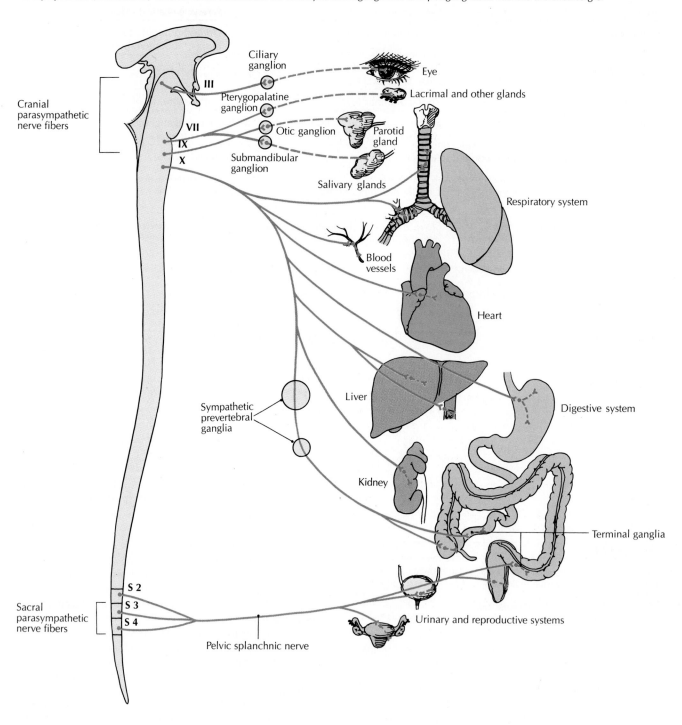

TABLE 14.2   SOME COMPLEMENTARY FUNCTIONS OF THE SYMPATHETIC AND PARASYMPATHETIC DIVISIONS

Part of body	Sympathetic effect	Parasympathetic effect
Eye (iris)	Dilates pupil.	Constricts pupil.
Salivary glands	Decreases secretion; thick, viscous saliva.	Increases secretion; thin, watery saliva.
Lungs (bronchial tubes)	Causes bronchodilation.	Causes bronchoconstriction.
Heart	Increases heart rate and blood pressure, dilates coronary arteries, increases blood flow to voluntary muscles.	Decreases heart rate, constricts coronary arteries, decreases blood flow to voluntary muscles.
Intestinal walls	Causes relaxation, decreases peristalsis, contracts sphincters.	Causes contraction, increases peristalsis, relaxes sphincters.
Urinary bladder	Inhibits constriction of bladder wall, contracts sphincters.	Stimulates contraction of bladder wall, relaxes sphincters.
Sweat glands	Increases secretion.	No innervation.
Peripheral blood vessels	Constricts.	No innervation for many.

Parasympathetic neurotransmitter   The neurotransmitter of the parasympathetic division is acetylcholine. Both the preganglionic and postganglionic neurons release it, and therefore the parasympathetic division and its fibers are classified as *cholinergic.*

Cholinergic receptors   All preganglionic nerve fibers of the autonomic nervous system release acetylcholine, which, in turn, excites the postganglionic nerve fibers. The postganglionic fibers of the parasympathetic division transmit impulses to its terminals to release acetylcholine, which interacts with receptor sites on the effector cells. The response by the effector may be either excitatory or inhibitory (see Table 14.3). The response by the effector is determined by the action of the receptor sites on the effector. The same principle applies to the response to drugs by neurons and effectors. As an example, let us look at the effects of the drugs muscarine and nicotine.

Muscarine and nicotine produce different actions because they interact differently with the receptor sites on the postsynaptic membranes. Muscarine is an alkaloid drug derived from a poisonous mushroom. Its activity, known as the *muscarinic effect,* partly resembles that of acetylcholine. Muscarine exerts a "cholinergic" effect on such effectors as smooth muscle, cardiac muscle, and glands, but has essentially no effect on cholinergic synapses within the ganglia and at the motor end plates of voluntary muscles. The activity of nicotine, known as the *nicotinic effect,* also partly resembles that of acetylcholine. Nicotine acts like acetylcholine at the cholinergic synapses in the autonomic ganglia and at the motor end plates of voluntary muscles, but has essentially no effect on smooth muscle, cardiac muscle, or glands.

Two examples of useful anticholinergic drugs and their derivatives are atropine and homatropine. Atropine is an antimuscarinic alkaloid agent derived from the belladonna plant (also called "deadly nightshade"). Its inhibitory action on the activity of acetylcholine is utilized to depress bronchial secre-

tions and sweating, or to increase the heart rate. Homatropine, a semisynthetic drug used by ophthalmologists when examining the eyes, blocks the sphincter muscle of the iris that regulates the size of the pupil, and the ciliary muscles that adjust the lens. As a result, the pupil is dilated, and the ability to accommodate (focus) is lost, producing an extreme sensitivity to light, and a lens that is fixed for distant vision only. (This is why your vision becomes blurred after your ophthalmologist puts drops in your eyes.)

Effects of parasympathetic stimulation   In general, the parasympathetic division is geared to respond to a specific stimulus in a discrete region for a short time. This effect occurs because the postganglionic neurons have relatively short axons that are distributed for short distances to specific areas (Figure 14.6). Also, the rapid deactivation of acetylcholine by acetylcholinesterase results in only a short-term effect by the neurotransmitter.

Ask Yourself

1 Why is the sympathetic division called the thoracolumbar division and the parasympathetic division called the craniosacral division?

2 What is the main function of autonomic ganglia?

3 What is the difference between a preganglionic and a postganglionic neuron?

4 What is the sympathetic trunk?

5 Where in the body is the parasympathetic division distributed?

6 What are the two main types of adrenergic receptors? Cholinergic receptors?

FIGURE 14.6

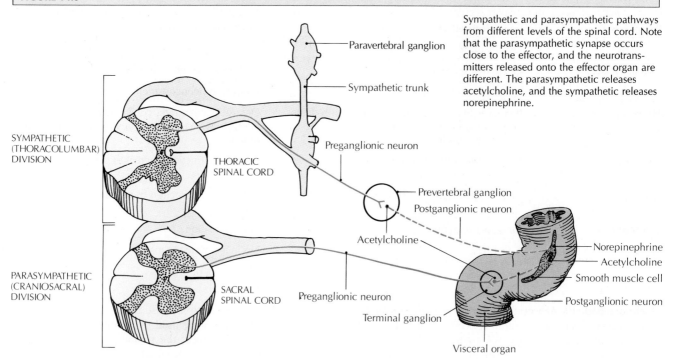

Sympathetic and parasympathetic pathways from different levels of the spinal cord. Note that the parasympathetic synapse occurs close to the effector, and the neurotransmitters released onto the effector organ are different. The parasympathetic releases acetylcholine, and the sympathetic releases norepinephrine.

CENTRAL AUTONOMIC CONTROL CENTERS

Neural centers in many regions of the central nervous system are involved with producing specific sympathetic and parasympathetic responses in the visceral organs. These higher centers are located in the brainstem, reticular formation, spinal cord, hypothalamus, cerebral cortex, and structures of the limbic system. The hypothalamus is considered to be the highest and main subcortical regulatory center of the autonomic nervous system.

The neural influences from the higher centers are exerted upon the activities of the autonomic nervous system through the visceral reflex arcs (see p. 421). Sensory inputs from the receptors in the viscera are conveyed via visceral afferent fibers to the central nervous system. There, after being processed and modulated by the higher centers, appropriate efferent influences are conveyed via the motor neurons of the sympathetic and parasympathetic outflow to the viscera.

Brainstem and Spinal Cord

Regulatory centers in the medulla oblongata of the brainstem are involved with visceral reflex arcs. For example, the cardiovascular center monitors heart rate, blood pressure, carbon dioxide level and the pH of blood, and blood vessel tone. After receiving input from sensory receptors, the cardiovascular center processes the neural information through the autonomic preganglionic neurons in the brainstem (ex-

pressed through the cranial nerve outflow) and the spinal cord (expressed through the thoracolumbar nerve outflow).

Respiratory centers in the brainstem are involved with reflex circuits that regulate the tone of the delicate vessels (bronchial trees) of the lungs. The neurons involved with respiration are integrated into somatic, not autonomic, reflex arcs. These arcs control somatic voluntary muscles such as the diaphragm and intercostal muscles, which are involved in breathing.

Sensory inputs from receptors in the urinary bladder to reflex centers in the sacral spinal cord regulate the tone of the muscles of the bladder, as well as the involuntary muscle sphincters. Influences from the brain are integrated into the activity that contracts the bladder when it is full.

Hypothalamus

The hypothalamus is the highest controlling neural *regulator,* and is regarded as the *coordinating center* of the autonomic nervous system. In fact, any autonomic response can be evoked by stimulating a site within the hypothalamus. (See the responses noted in Table 14.3.) Allowing for a considerable overlap, the stimulation of sites in the anterior and medial hypothalamus tends to be associated with parasympathetic responses, and the stimulation of sites in the posterior and lateral hypothalamus is usually associated with sympathetic responses. Certain somatic responses involving voluntary muscles, such as shivering, are also associated with the hypothalamus.

Within the hypothalamus are many neural circuits, called *control centers,* which control such vital autonomic activities as body temperature, heart rate, blood pressure, blood osmolarity, and the desire for food and water. The hypothalamus is also involved with behavioral expressions associated with emotion, such as blushing.

Clearly, the hypothalamus is a critical participant in maintaining homeostasis. For example, the autonomic responses that control body temperature are initiated because the hypothalamus acts as a thermostat that monitors the temperature of blood flowing through a hypothalamic control center. Neurons in the hypothalamus respond to temperature changes, activating either heat-dissipating or heat-conserving control systems to maintain the desired body temperature. The heat-dissipating center in the anterior hypothalamus activates the responses of increased sweating and dilation of skin blood vessels, thus cooling the body. The heat-conserving center in the posterior hypothalamus causes shivering and the constriction of skin blood vessels, thus generating heat.

Cerebral Cortex and Limbic System

Structures of the limbic system, such as the limbic lobe, amygdala, and hippocampus, are connected to the hypothalamus, and use the hypothalamus to express their activities. These expressions include many visceral and behavioral responses associated with self-preservation (such as feeding and fighting) and preservation of the species (such as mating and care of the offspring). Electrical stimulation of the limbic lobe and hippocampus produces changes in the cardiovascular system, including alterations in the heart rate and tone of the blood vessels. Stimulation of the amygdala and limbic lobe may alter the secretory activity of digestive glands.

Even the cerebral cortex, which is usually considered the center of thought processes, utilizes the limbic system and hypothalamus, through its connections with the autonomic nervous system, to express some of our emotions. For example, when a person experiences anxiety, pleasure, or other emotional feelings, the cerebral cortex and limbic system become active, and relay the influences to the hypothalamus. The hypothalamus responds by relaying neural influences via the descending autonomic pathways to the cardiovascular centers in the brainstem. These influences are then projected to the pools of preganglionic neurons of the cranial nerves, and to the spinal cord. Depending upon which centers of the hypothalamus are stimulated, the resulting expressions can be sympathetic or parasympathetic.

Visceral Reflex Arc

A *visceral reflex* innervates cardiac muscle, smooth muscle, or glands. When stimulated, smooth muscles or cardiac muscles contract, and glands release their secretions. Such a reflex, like a somatic motor reflex, does not involve the cerebral cortex, and most visceral adjustments are made through regulatory centers, for example, in the medulla or spinal cord, without our conscious control or knowledge.

Unlike the somatic reflex, which uses only one efferent neuron, a sequence of two efferent neurons is involved in an autonomic visceral reflex arc. A *visceral reflex arc* is made up of (1) an *afferent receptor,* (2) an *afferent neuron* that conveys sensory influences to the central nervous system, (3) *interneurons* within the gray matter of the central nervous system that connect with preganglionic neurons in the sympathetic division, (4) *two efferent neurons* that are part of the sequence composed of preganglionic neurons, a sympathetic ganglion, and a postganglionic neuron, and (5) a *visceral effector.*

Some examples of autonomic visceral reflex arcs that occur in the spinal cord are the contraction of a full urinary bladder, muscular contraction of the intestines, and constriction or dilation of blood vessels. Examples of reflex arcs in the medulla include the regulation of blood pressure, heart rate, respiration, and vomiting.

Ask Yourself

1 *What are some of the centers in the central nervous system that are involved in regulating the autonomic nervous system?*

2 *How does the central nervous system cooperate with the autonomic nervous system to regulate body temperature?*

3 *What are the components of an autonomic visceral reflex arc?*

FUNCTIONS OF THE AUTONOMIC NERVOUS SYSTEM

In this section, we provide an overall picture of the autonomic nervous system as a two-part regulatory system by looking at the way the sympathetic and parasympathetic divisions balance their influences to help us react to changes and maintain our internal homeostasis (Table 14.3). As an example, we show how the system operates during a downhill ski race.

Example of the Operation of the System: A Ski Race

An Olympic skier on a twisting downhill slope is concentrating every part of the body to negotiate the course faster than anyone else in the world. The skier's heart, beating as much as three times faster than yours is right now, is also pumping more blood, faster, to the skeletal muscles than yours is now. The skier's pupils are dilated. The blood vessels of the skin, body organs, and salivary glands—all but those of the skeletal muscles—are constricted. The sweat glands are stimulated. Epinephrine (adrenaline) and norepinephrine (noradrenaline) virtually pour out of the adrenal glands. Obviously ready for action, the skier shows the so-called "fight or flight" response, a state of heightened readiness.

TABLE 14.3   SOME EFFECTS OF THE AUTONOMIC NERVOUS SYSTEM

Visceral effector	Receptor	Sympathetic division	Parasympathetic division
Blood vessels		Generally constricted.	Generally dilated slightly.
External genitalia	α	Constricted.	Dilated.
Salivary glands	α	Constricted.	Dilated.
Skeletal muscles	α_1, β_2	Dilated (as a result of cholinergic neurons).	No innervation.
Skin	α_1, β_2	Constricted.	Dilated slightly.
Viscera	α	Constricted.	Dilated.
Eye			
Muscles of iris	α	Radial muscle contracted (pupil dilated).	Sphincter muscle contracted (pupil contracted).
Ciliary muscle	α	Relaxed for far vision.	Contracted for near vision.
Heart			
Rate	β_1	Increased.	Decreased.
Strength of contraction	β_1	Increased.	Decreased.
Gastrointestinal tract			
Motility and tone	α_2	Inhibited.	Stimulated.
Sphincters	α	Stimulated (contracted).	Inhibited (relaxed).
Secretion	—	Probably inhibited.	Stimulated.
Gallbladder	—	Inhibited.	Stimulated.
Liver	β_2	Increased glycogenolysis (resulting in increased blood sugar).	No innervation.
Lungs			
Bronchial tubes	α	Lumen dilated.	Lumen constricted.
Bronchial glands	—	No innervation.	Secretion stimulated.
Bronchial muscle	β_2	Relaxed.	Contracted.
Glands of head			
Lacrimal glands	—	No innervation.	Secretion stimulated.
Salivary glands	α	Scanty, viscous secretion.	Profuse, watery secretion.
Urinary bladder	β_2	Relaxed.	Contracted.
Skin			
Sweat glands	α	Stimulated to secrete.	No innervation.
Blood vessels	α_1, β_2	Constricted.	No innervation.
Piloerector muscles	α	Contracted (hair stands erect).	No innervation.
Sex organs	α	Vasoconstricted (orgasm).	Vasodilated (erection).
Adrenal medulla	—	Epinephrine, norepinephrine secreted (innervated by preganglionic cholinergic sympathetic neurons).	No innervation.

Adapted from Charles R. Noback and Robert J. Demarest, *The Human Nervous System*, 3rd ed. (New York: McGraw-Hill, 1981), pp. 224–225. Used with permission.

At the same time, systems not needed in the race are practically shut down. Digestion, urination, and defecation can wait until the race is over. Blood is reserved for the skeletal muscles. The *sympathetic division* is in almost total command.

When the race is over, and it is time to relax and enjoy a leisurely meal, the emphasis is on the "normal" or "maintenance" body functions and a restoration of the body's energy resources. The *parasympathetic division* is in command now. The heart rate and the force of its contractions are reduced, saliva and other digestive juices flow freely, the pupils are constricted to protect the eyes from excessive brightness, muscles are relaxed, and the body is free once again to devote time to ridding itself of wastes. Some of the glucose that was pouring out of the liver into the skeletal muscles during the race is now being diverted to other organs. So is blood. Blood vessels in the skin that were constricted to lessen the chance of serious bleeding in case of a wound are now back to normal. The extra perspiration that helped keep the skier cool during the race is not needed now, and the sweat glands relax. Blood vessels in the intestines dilate, while those in the skeletal muscles constrict to their normal diameter. The autonomic nervous system, together with the central nervous system, has kept the body in balance with its surroundings.

Coordination of the Two Divisions

The hypothalamus can increase or decrease the activity of the sympathetic and parasympathetic divisions. Small changes are constantly being made in one division or the other, with every change geared to promote homeostasis.

Many bodily activities involve either one autonomic division or the other. However, sexual activities require the coordinated, sequential involvement of both the sympathetic and parasympathetic divisions. Also, although the two divisions may be said to have opposite, or "antagonistic," effects on viscera and glands, not every structure innervated by the autonomic nervous system receives innervation from both divisions (Table 14.3). For example, piloerector muscles in the skin may be innervated by only one (sympathetic) of the two divisions.

Many organs receive a *dual innervation,* with apparent opposite responses to stimulation from the sympathetic and parasympathetic divisions. However, such responses do not mean that the two divisions are antagonistic in the sense that they are working against each other. Rather, like antagonistic muscles, they are coordinated to achieve a single functional goal. The eye, for example, shows an interesting dual response to the degree of light intensity affecting it. The pupil dilates when the radial (dilator) smooth muscle cells of the iris are stimulated to contract by sympathetic fibers. When the sphincter (constrictor) muscle cells of the iris are stimulated to contract by parasympathetic fibers, the pupil constricts.

Responses of Specific Organs

The autonomic nervous system does not control the basic activity of the organs it innervates, but it does alter that activity. The organs innervated by the autonomic nervous system are not fully dependent upon autonomic innervation. For example, if the heart is deprived of its autonomic innervation, it will still contract, but it will not respond to the changing demands of the body, and will not increase its rate when physical activity is increased. (This ability of the heart to contract without innervation is one of the reasons it can be transplanted to another person.) In contrast, a voluntary muscle deprived of its lower motor neuron innervation will not contract.

A response is due partly to the specific neurotransmitter and partly to the nature of the receptor site upon which the transmitter acts. The same neurotransmitter may not have the same effect on different effectors. For example, norepinephrine stimulates the smooth muscle of an arteriole (a small, terminal branch of an artery) to contract, but it stimulates the smooth muscle of the bronchial tubes to dilate, causing the lungs to relax. Acetylcholine stimulates voluntary muscle to contract, but it stimulates cardiac muscle to relax.

Ask Yourself

1 *Which division of the autonomic nervous system prepares a person for intense muscular activity?*

2 *Do both divisions operate simultaneously? Explain.*

3 *What is dual innervation?*

4 *Does a neurotransmitter always have the same effect? Explain.*

PHYSIOLOGICAL AND ANATOMICAL ABNORMALITIES

Horner's syndrome

In *Horner's syndrome,* sympathetic postganglionics are interrupted. Lesions typically occur from the superior cervical ganglion to the head, or the preganglionics from T1 to T4 to the cervical ganglion. The lesion blocks the flow of sympathetic activity to the head, resulting in a drooping eyelid, constricted pupil, sunken eyeball, and flushed, dry skin. All of these symptoms are unilateral to the lesion.

Autonomic dysreflexia

In *autonomic dysreflexia,* lesions occur in the spinal cord above T4 to T6 (above the sympathetic splanchnic outflow). Acute hypertension develops from the stimulation of sympathetic preganglionics by visceral or somatic afferents as they ascend the spinal cord. Blood pressure rises as high as 300 mm Hg.

Achalasia

Achalasia (Gr. failure to relax) can occur at any point along the gastrointestinal tract where the parasympathetics fail to relax the smooth muscle. In *achalasia of the esophagus*, the individual has difficulty swallowing, due to a persistent contraction of the esophagus where it enters the stomach.

In *Hirschsprung's disease,* the large intestine becomes enlarged and distended because a short segment of it is continuously constricted, thereby obstructing the passage of feces. The constricted segment contains the flaw. The neurons normally present in the intestinal wall are lacking in this segment. These neurons, which are essential for normal peristalsis, are necessary for the expression of the autonomic nervous system's influences.

SUMMARY

The *autonomic nervous system* may also be called the *visceral efferent motor system.* Most of the functions of our internal organs (viscera) are not under our conscious control.

Structure of the peripheral autonomic nervous system

1 The autonomic nervous system is divided into the *sympathetic* and *parasympathetic divisions.* The sympathetic division is also called the *thoracolumbar division* because its nerve fibers emerge from the thoracic and upper lumbar spinal nerves. The parasympathetic division is called the *craniosacral division* because its visceral efferent fibers leave the CNS via cranial nerves in the brainstem and spinal nerves in the sacral region of the spinal cord.

2 In the one-neuron linkage of the somatic motor system, each lower motor neuron connects directly with its effector, with no synapse outside the central nervous system. In the two-neuron linkage of the autonomic system, one neuron *(preganglionic)* synapses with a second neuron *(postganglionic)* in an *autonomic ganglion* before it stimulates its effector.

3 *Paravertebral* (lateral) and *prevertebral* (collateral) autonomic *ganglia* occur in the sympathetic division, and *terminal* (peripheral) *ganglia* in the parasympathetic division. Rows of paravertebral ganglia form the *sympathetic trunks.*

4 Some postganglionic nerve fibers are distributed to branching networks in the thoracic, abdominal, and pelvic cavities as *autonomic plexuses.*

5 The sympathetic division arises from cell bodies in the lateral gray horn of the spinal cord. The preganglionic nerve fibers emerge from the spinal cord in the ventral roots of the 12 thoracic and first 2 or 3 lumbar spinal nerves. This emer-gence of fibers is the *thoracolumbar outflow;* they form the *white rami communicantes.*

6 The neurotransmitter released by sympathetic preganglionic nerve terminals is *acetylcholine.* With a few exceptions, the neurotransmitter released by the sympathetic postganglionic nerve terminals is *norepinephrine* (nonadrenaline) and therefore the division is called the *adrenergic division.* The effects of norepinephrine are usually widespread and relatively long lasting.

7 *Adrenergic receptors* are specific protein-receptor sites in the plasma membrane of cells of effector organs. They respond to norepinephrine and epinephrine.

8 The preganglionic fibers from the cranial portion of the parasympathetic division are called the *cranial parasympathetic outflow.* The preganglionic fibers from the sacral portion comprise the *sacral parasympathetic outflow.*

9 The neurotransmitter of the parasympathetic division is *acetylcholine,* and the division is classified as *cholinergic.* The effects of acetylcholine are usually short range and relatively short term.

10 *Cholinergic receptors* respond to acetycholine at synapses. Preganglionic receptors are *nicotinic,* and postganglionic receptors are *muscarinic.*

Central autonomic control centers

1 *Neural centers* in the central nervous system, including the brainstem, reticular formation, spinal cord, hypothalamus, cerebral cortex, and structures of the limbic system, are involved with producing sympathetic and parasympathetic responses in the visceral organs.

2 Regulatory centers in the medulla of the *brainstem* are involved with influencing visceral reflex arcs.

3 The *hypothalamus* is the highest and main subcortical regulatory center of the autonomic nervous system. It modulates autonomic centers in the brainstem and spinal cord.

4 Structures of the *limbic system* utilize the hypothalamus to express their activities, which include many visceral and behavioral responses associated with self-preservation and preservation of the species.

5 The *cerebral cortex* utilizes the limbic system and hypothalamus to express some emotions.

6 A *visceral reflex arc* innervates cardiac muscle, smooth muscle, and glands. Its components are an afferent receptor, afferent neuron, interneurons, two efferent neurons, and visceral effector.

Functions of the autonomic nervous system

1 The main function of the autonomic nervous system is to promote homeostasis.

2 The sympathetic and parasympathetic divisions are coordinated into a balanced, complementary system that helps the body adjust to constantly changing environmental conditions in order to maintain homeostasis.

3 Generally, the sympathetic division prepares the body for stressful situations, and the parasympathetic division is active when the body is at rest.

4 Many organs receive a *dual innervation,* with opposite responses to stimulation by the sympathetic and parasympathetic divisions.

5 A response is due partly to the particular neurotransmitter and partly to the nature of the receptor site.

Physiological and anatomical abnormalities

1 The few disorders of the ANS include *Horner's syndrome, autonomic dysreflexia,* and *achalasia.*

UNDERSTANDING THE FACTS

1 What is the primary function of the autonomic nervous system?

2 What basic structural difference exists in the connection of somatic neurons to their respective effectors?

3 Name three groups of autonomic ganglia and five plexuses.

4 The neurotransmitters generally released by the sympathetic division's preganglionic fibers is _____, while the transmitter released by its postganglionic fibers is _____ .

5 Through which nerves do parasympathetic fibers emerge from the central nervous system?

6 What are catecholamines and how do they react with alpha and beta receptors?

7 Describe nicotinic and muscarinic modes of action.

8 Do autonomic visceral reflex arcs involve the cerebral cortex?

9 Does the autonomic nervous system need the cooperation of the brain for normal function? Explain.

10 Does the autonomic nervous system control the basic activity of the organs that it innervates? Explain.

11 How does biofeedback work? Explain.

UNDERSTANDING THE CONCEPTS

1 How do the general functions of the sympathetic and parasympathetic divisions differ?

2 Describe the pathways of sympathetic neurons from the spinal cord to the effectors.

3 What is the difference between the white rami communicantes and the gray rami communicantes?

4 Make an anatomical comparison between a somatic reflex arc and an autonomic reflex arc.

5 Compare the neurotransmitter substances of the sympathetic and parasympathetic divisions.

6 Why are the effects of sympathetic stimulation generally widespread and long lasting, while those of parasympathetic stimulation are generally local and short term?

7 When we say the sympathetic and parasympathetic systems are antagonistic, do we mean that they work against each other? Explain.

SUGGESTED READING

BURNSTOCK, G., AND M. COSTA, *Adrenergic Neurons: Their Organization, Function, and Development in the Peripheral Nervous System.* New York: Halsted Press, 1975.

CRYER, P. E., "Physiology and Pathophysiology of the Human Sympathoadrenal Neuroendocrine System." *New England Journal of Medicine,* 303 (1980):436.

DAVENPORT, H. W., "Epinephrin(e)." *Physiologist,* 25 (1983):76–82.

DECARA, L. V., "Learning in the Autonomic Nervous System." *Scientific American,* January 1970.

HOFFMAN, B. B., AND R. J. LEFKOWITZ, "Alpha-Adrenergic Receptor Subtypes." *New England Journal of Medicine,* 302 (1980):1390.

KOIZUMI, K., AND C. M. BROOKS, "The Autonomic System and Its Role in Controlling Body Functions." In V. B. Mountcastle, *Medical Physiology,* 14th ed. St. Louis: Mosby, 1980.

MAYER, S. E., "Neurohumoral Transmission and the Autonomic Nervous System." In *The Pharmacological Basis of Therapeutics,* 6th ed. New York: Macmillan, 1980.

OTTOSON, D., *Physiology of the Nervous System.* New York: Oxford University Press, 1983.

STARKE, P., "Presynaptic Receptors," *Annual Review of Physiology,* 21 (1981):7.

VAN TOLLER, C., *The Nervous Body: An Introduction to the Autonomic Nervous System and Behavior.* New York: Wiley, 1979.

15
The Senses

LEARNING OBJECTIVES

1 Define a sensory receptor, and describe its four basic features.

2 Name and define the types of sensory receptors based on location, type of stimulus, type of sensation, and type of nerve ending.

3 Describe the specific receptors and neural pathways for light touch, touch-pressure, heat, and cold.

4 Explain the function of pain, and describe the receptors and neural pathways for pain.

5 Define proprioception, itch, tickle, and stereognosis, and describe the receptors for each sensation.

6 Relate the structure of gustatory receptors to function, identify the basic taste sensations, and describe the neural pathways for taste.

7 Describe the structure and function of olfactory receptors, and neural pathways for olfaction.

8 Identify the major structures of the ear, and give their functions.

9 Describe the pathway of sound waves inside the ear, and explain how sound waves are converted into generator potentials.

10 Identify the major structures of the vestibular system, and explain their roles in dynamic and static equilibria.

11 Discuss the neural pathways for hearing and equilibrium.

12 Describe the structure and function of the layers, cavities, and receptor cells of the eyeball.

13 Give the functions of the major accessory structures of the eye.

14 Explain the principles of refraction, accommodation, and convergence.

15 Describe the roles played by rods and cones.

16 Describe how the eye adapts to light and dark.

17 Trace the neural pathways for vision from the retina to the brain.

18 Describe several disorders of the eye and ear, including symptoms and causes.

About 2000 years ago, Aristotle identified five senses—touch, taste, smell, hearing, and sight—and it is still common to refer to the "five senses" of the body. In fact, the skin alone is involved with the sensations of fine touch, touch-pressure (deep pressure), heat, cold, and pain. Also included in a more complete list of the senses are a sense of balance or equilibrium, and a sense of body movement. In addition, receptors in the circulatory system register changes in blood pressure and blood levels of carbon dioxide and hydrogen ions, and receptors in the digestive system are involved in the feelings of hunger and thirst.

Our impressions of the world are limited and defined by our senses. In fact, all knowledge and awareness depend on the reception and decoding of stimuli from the outside world and from within our bodies. In this chapter you will see how our senses work, and how we depend on them much more than we think we do.

SENSORY RECEPTION

All animals have some means of sensing and responding to changes in the environment. If they didn't, they couldn't survive. We are able to cope with change because part of our nervous system is specialized to make sure we have a suitable reaction to any stimulus. Structures that are capable of perceiving and changing such stimuli are called sense organs, or *receptors*. Receptors are the body's links to the outside world and the world within us.

In terms of the nervous system, a receptor is (or is associated with) the peripheral end of the dendrites of afferent neurons. Sensory receptors are stimulated by specific stimuli. The eye (the receptor) is stimulated by visible light waves (the stimulus), and specialized receptors inside the ear are stimulated by audible sound waves. Sound waves do not stimulate the sensory receptors that are specialized to receive *light* waves, and light waves have no effect on our ears.

A favorite question is: Do sounds occur when there is no living thing to hear them? The answer is no. A "sound" is something that is received *(sensed)* by the ear and "heard" *(perceived)* by the brain. If there is no ear to receive sound waves, and no brain to translate those sound waves into what we consciously recognize as the "sound" of thunder, for example, the thunder will send out sound waves, but there will be no perceived "sound." The same is true of the other senses.

All sensory receptors are structures that are capable of converting environmental information (stimuli) into nerve impulses. Thus, all receptors are *transducers*, that is, they convert one form of energy into another. Since all nerve impulses are the same, different types of receptors convert different kinds of stimuli (such as light or heat) into the same kind of impulse. The intensity of stimuli varies with the frequency of the stimulus, and the number of receptors stimulated at one time. A receptor may be (1) the terminal or receptor segment of a sensory neuron (such as a "pain" neuron with its cell body in a dorsal root ganglion) or (2) the terminal segment of a sensory neuron associated with specialized neuroepithelial cells (such as hair cells in the cochlea of the ear, and rods and cones in the retina of the eye). When a stimulus is strong enough, the stimulated neuroepithelial receptor cells or the terminal receptive segments of the sensory neurons (the unmyelinated portion) generate *receptor* or *generator potentials.* These receptor potentials are graded potentials. They can travel only short distances along the cell membrane, because unlike action potentials, they are not self-regenerating. However, the neuroepithelial cells can stimulate the receptive segment of sensory neurons through a transmitter. When the receptor potentials of the receptor segment of the neuron reach the first neurofibral node (the initial segment), they trigger an *action potential* (nerve impulse), which can travel long distances to a synapse. The different sensations are brought about when nerve fibers connect with specialized portions of the brain. We "see" a tree, for example, not when its image enters our eyes, but when that coded image stimulates the vision centers of our brain.

Basic Characteristics of Sensory Receptors

There are several types of sensory receptors, and many ways to classify them, but certain features are basic to all sensory receptors:

1 They *contain sensitive receptor cells* that respond to certain minimum (threshold) levels of intensity. That is, the stimulus must be strong enough to generate a receptor potential and then an action potential.

2 Their *structure is designed to receive a specific stimulus.* For example, the eye contains light-absorbing pigments, an elastic adjustable lens, a nonadjustable lens (cornea), and other structures suitable for capturing light waves in the visible spectrum.

3 Their primary receptor cells *synapse with afferent nerve fibers* that travel to the central nervous system along peripheral or cranial nerves. Some receptors in the skin are connected to neurons that have nerve fibers that extend for about a meter before they reach the spinal cord and form a synapse. In contrast, the primary receptor cells in the eye have short axons that synapse with other cells in the retina before projecting to the brain via the optic nerve.

4 After receptor cells synapse with afferent neurons, the nerve impulses are *conveyed along neural pathways* through the brainstem and diencephalon to the cerebral cortex of the brain. In the diencephalon and cerebral cortex, the original stimulus and nerve impulse are *translated* into a recognizable sensation such as sight or sound.

Classification of Sensory Receptors

Sensory receptors may be classified according to their location, type of sensation, type of stimulus, or structure (the presence or absence of a sheath).

Location of receptor Four kinds of sensory receptors are recognized on the basis of their location:

1 *Exteroceptors* (L. "received from the outside") respond to external environmental stimuli that affect the skin directly.

These stimuli result in the sensations of touch pressure, pain, and temperature.

2 *Teleceptors* (Gr. "received from a distance") are the exteroceptors located in the eyes, ears, and nose. They detect environmental changes (stimuli) that occur some distance away from the body. These stimuli are ultimately perceived as sight, sound, and smell.

3 *Interoceptors* (L. "received from the inside"), also called *visceroceptors* (L. "received from the viscera"), respond to stimuli from within the body, such as blood pressure, blood carbon dioxide, oxygen, and hydrogen ion levels, and the stretching action of smooth muscle in organs and blood vessels. Interoceptors are located within organs that have motor innervation from the autonomic nervous system. They help to maintain homeostasis.

4 *Proprioceptors* (L. "received from one's own self") respond to stimuli in such deep body structures as joints, tendons, muscles, and the vestibular apparatus of the ear. They are involved with sensing where parts of the body are in relation to each other, and the position of the body in space.

Sight, hearing, equilibrium, smell, and taste are known as the **special senses** because their receptors are found in restricted regions of the body. Their sensory receptors are also more specialized and complex than those of the **general senses** (also called the *somatic senses*), which include touch pressure, heat, cold, pain, and body position.

Type of sensation Sensory receptors can detect several types of sensations that are associated with the general senses. *Thermal* sensations include cold and warmth. *Pain* sensations are the feelings initiated by harmful stimuli. Like pain, both **light touch** and **touch-pressure** sensations are also produced by mechanical stimuli that come in contact with the body. Light touch involves a finer discrimination than touch pressure does. *Position sense* is elicited by the movement of joints and muscles. It includes both the sense of position when the body is not moving, and the sense of body movement, called *kinesthesia* (Gr. *kinesis*, motion + *esthesis*, perception).

Type of stimulus Another way to classify sensory receptors is by the stimuli to which they respond. *Thermoreceptors* (Gr. "heat receivers") respond to temperature changes. *Nociceptors* (NO-see; L. "injury receivers") respond to potentially harmful stimuli that produce pain. *Chemoreceptors* (L. "chemical receivers") respond to chemical stimuli that result in taste and smell, and to changes in the levels of carbon dioxide, oxygen, and hydrogen ions in the blood, as well as other chemical changes. *Photoreceptors* (Gr. "light receivers") in the retina of the eye respond to the visual stimuli of visible light waves. *Mechanoreceptors* (Gr. "mechanical receivers") respond to and monitor such physical stimuli as touch pressure, muscle tension, joint position changes, air vibrations in the cochlear system of the ear that produce hearing, and head movements detected by the vestibular system of the ear that result in sensing body equilibrium. Mechanoreceptors are the most widespread of all the sensory receptors, and are also the most varied in sensitivity and structure. *Baroreceptors* (Gr. "pressure receivers") are mechanoreceptors that respond to changes in blood pressure.

Sensory endings of receptors The neuronal terminals of spinal nerves end in sensory receptors, or *nerve endings*. Two kinds of nerve endings are usually distinguished:

1 Terminals that lack neurolemmocytes (Schwann cells), myelin, and other cellular coverings are called **free nerve endings** or *naked nerve endings*. Free nerve endings are the naked telodendria in the surface epithelium of the skin, connective tissues, blood vessels, and other tissues. They are the sensors for such perceived sensations as pain, light touch, and temperature.

2 Receptors that are covered with various types of capsules are known as **encapsulated endings**. Encapsulated endings are located in the skin, muscles, tendons, joints, and body organs. Two such endings are *lamellated (Pacinian) corpuscles*, which are involved with vibratory sense and touch pressure (deep pressure) on the skin, and *tactile (Meissner's) corpuscles*, which are skin sensors that detect light pressure (Figure 15.1).

Corpuscles of Ruffini and *bulbous corpuscles (of Krause)*, previously thought to be the skin receptors for temperature, are now believed to be sensors for touch pressure, position sense of a body part, and movement. They are probably variants of other encapsulated endings, such as lamellated corpuscles. Other important encapsulated receptors are the *neuromuscular spindles* and *tendon (Golgi) organs*, which monitor the tension and stretch in muscles and tendons, and are involved with voluntary muscle reflexes.

Sensory Receptors and the Brain

The main purpose of the senses is to inform us about environmental conditions and changes that may be beneficial or harmful. The sensory system is an effective survival mechanism because it converts information from the environment into appropriate reactions.

A stimulus is converted into a generator potential, and then into an action potential with the purpose of stimulating an effector: a muscle contracts, a gland secretes a hormone, a blood vessel constricts, and then dilates when the muscle relaxes and the blood pressure dilates the vessel. The role of the brain in this process is to receive the nerve impulse, discard what is irrelevant, compare it with information already stored, and coordinate the final impulse to the effector. There are some exceptions, such as during a reflex arc, when the impulse travels only as far as the spinal cord for spinal nerves, or to the brainstem for cranial nerves. But generally the brain mediates sensory impulses and translates them into perceptible sensations. The brain's involvement between the stimulus and the response enables us to have complex behavioral patterns.

It is the complexity of the brain, not the quality of the sensory receptor, that increases the power of perception. An octopus eye has many similarities to a human eye, and yet it is unlikely that an octopus is able to see what you see. Although both eyes are similar, the octopus lacks the necessary higher brain centers to process incoming stimuli as well as you do.

Individual nerve impulses are essentially identical, but an impulse for sight is distinguished from an impulse for taste

FIGURE 15.1

Nerve endings, sensory receptors, and root hair plexuses around hair follicles in the skin.

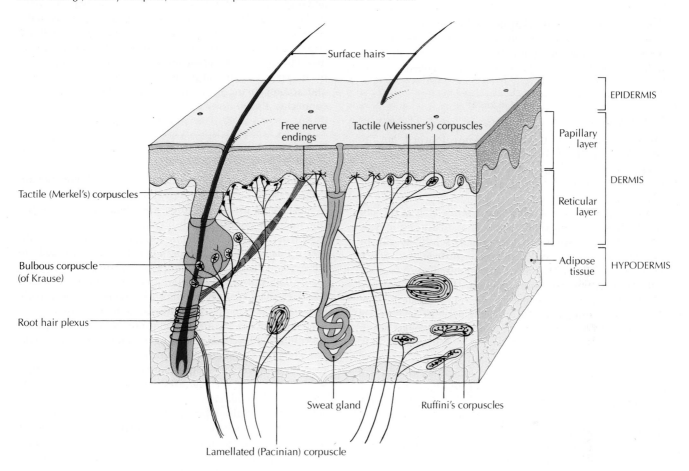

by *the relationship between where the impulse is coming from and the places in the brain where it is going.* For example, taste receptors in the tongue receive stimuli and convert them into nerve impulses. Afferent fibers carry the impulses to a location in the brain that is specialized to receive only taste impulses. The brain is impartial. The type of sensation elicited by the brain depends upon *where* the brain is stimulated, not *how* it is stimulated.

Only rarely is a stimulus interpreted incorrectly. If a sufficiently intense stimulus is applied to the skin, for instance, more than one receptor may be stimulated, and the brain may be confused. For example, a person stimulated by intense heat may interpret it wrongly as cold. The brief sensation of cold followed by heat that is perceived when you plunge your hand into very hot water results because cold receptors are stimulated initially, followed by the stimulation of warm receptors.

How much outside stimuli does the nervous system screen out as irrelevant or distracting?

The brain discards about 99 percent of the input it receives from sensory receptors.

Ask Yourself

1 *What is a sensory receptor?*

2 *What is a transducer?*

3 *How does a generator potential differ from an action potential?*

4 *What are the basic characteristics of a sensory receptor?*

5 *What are the minimal requirements for the perception of a sensation?*

6 *What are the types of sensory receptors based on location?*

7 *What are the types of sensory receptors based on the types of stimuli to which they respond?*

8 *What is the role of the brain in the sensory system?*

GENERAL SENSES

Sensory receptors in the skin detect stimuli that the brain interprets as light touch, touch-pressure (deep pressure), vibration, heat, cold, and pain. Several other general sensations, such as itch and tickle, will also be described.

Light Touch

Light touch is perceived when the skin is touched, but not deformed. It is also known as fine touch and cotton touch. Receptors for light touch are most numerous in the dermis, especially in the tips of the fingers and toes, the tip of the tongue, and lips.

In contrast to such sensitive areas of the skin as the fingertips, the torso (especially the back) and back of the neck are relatively insensitive to light touch. Sensitivity can be measured with a test called *two-point discrimination*, which measures the minimal distance that two stimuli must be separated to be felt as two distinct stimuli. Usually, one or two points of a compass are applied to the skin without the subject seeing how many points are being used. In areas where sensory receptors are abundant, two distinct compass points may be felt when they are separated by only 2 or 3 mm. Where there are few receptors far apart, the points may have to be separated by as much as 60 or 70 mm before they can be felt as two points. What this means is that there is virtually no spot on the fingertips that is insensitive to tactile stimuli, but it is possible to touch a pin to the skin of the back without initiating an impulse.

Receptors of light touch include free nerve endings and tactile (Merkel's) corpuscles in the epidermis, and tactile (Meissner's) corpuscles just below, in the uppermost (papillary) layers of the dermis (see Figure 15.1). *Free nerve endings* are the most widely distributed receptors in the body, and are involved with pain and thermal stimuli as well as light touch (Table 15.1). The next most numerous cutaneous receptors are those for touch, cold, and heat.

There are many sensory receptors called *root hair plexuses* around hair follicles. When hairs are bent, they act as levers, and the slight movement stimulates the free nerve endings surrounding the follicles, which act as detectors of touch and movement (see Figure 15.1). For this reason, a tiny insect crawling along a hairy arm will be felt even if its feet never touch the skin.

Tactile (Merkel's) corpuscles (also called Merkel's disks, Merkel cells, or Merkel endings) are modified epidermal cells with free nerve endings attached (see Figure 15.1). They are found in the deep epidermal layers of the palms of the hands and the soles of the feet. As shown in Figure 15.1, the "disk" portion of a tactile (Merkel's) corpuscle is formed when the unmyelinated terminal branches of myelinated afferent nerve fibers penetrate the basal layer of the epidermis. Once inside the epidermis, the fibers lose their covering of neurolemmocytes, and expand into a terminal disk attached to the base of a tactile (Merkel's) corpuscle.

Unlike tactile (Merkel's) corpuscles, *tactile (Meissner's) corpuscles* are egg-shaped encapsulated nerve endings (see Figure 15.1). They are found in abundance on the palms of the hands, soles of the feet, lips, eyelids, external genitals, and nipples. These receptors are situated in the papillary layer of the dermis, just below the epidermis. Tactile (Meissner's) corpuscles contain flattened cells that are probably modified connective tissue, nerve endings that intertwine among them, and spiraling terminal branches that lose their myelin sheaths before they enter the corpuscle at its base. The flattened cells and accompanying fibers are enclosed within a capsule of connective tissue.

Touch-Pressure (Deep Pressure)

The difference between light touch and touch-pressure (or deep pressure) on your skin can be shown by gently touching a pencil (light touch) and then squeezing it as hard as you can (touch-pressure). Touch-pressure results from a deformation of the skin, no matter how slight. Sensations of touch-pressure last longer than sensations of light touch, and are felt over a larger area. Receptors for touch-pressure are primarily **lamellated,** or **Pacinian** (pah-SIN-ee-an) **corpuscles,** (see Figure 15.1). They are mechanoreceptors that actually measure *changes* in pressure, rather than pressure itself. Lamellated corpuscles are distributed throughout the dermis and subcutaneous layer, especially in the fingers, external genitals, and breasts, but are also found in muscles, joint capsules, the wall of the urinary bladder, and other areas that are regularly subjected to pressure. They are also abundant in the mesenteries (membranes that attach the intestines to the dorsal abdominal wall), where their function is not known.

A long, myelinated nerve fiber enters at the base of the lamellated corpuscle, loses its myelin sheath inside the corpuscle, and proceeds as an unmyelinated fiber within the capsule. The naked fiber is covered by as many as 100 alternating layers of flattened, concentric connective tissue cells and extracellular fluid, so that a longitudinal section looks like a sliced onion (see Figure 15.1).

Vibration

Most tactile receptors are involved to some degree in the detection of vibration. The term **vibration** refers to the continuing periodic change in a displacement with respect to a fixed reference. This change, per unit time, is termed the *frequency*. Different receptors detect different frequencies. For example, lamellated corpuscles can detect vibrations (frequencies) as high as 700 cycles per second (cps). Tactile (Meissner's) corpuscles and corpuscles of Ruffini, on the other hand, respond to low-frequency vibrations up to 100 cps.

Heat and Cold

Until recently it was believed that the cutaneous receptors for heat were the corpuscles of Ruffini, and the receptors for cold were the bulbous corpuscles (of Krause), but further investigation has disproved those beliefs. Currently, the cutaneous receptors for heat and cold are considered to be naked nerve

TABLE 15.1   SOME GENERAL SENSORY RECEPTORS AND THEIR NEURAL PATHWAYS

Cutaneous sensation	Receptors	Neural pathways
Light touch (cotton touch; skin not deformed)	Free nerve endings, tactile (Merkel's) corpuscles, tactile (Meissner's) corpuscles.	Posterior column-medial lemniscal pathway, anterior spinothalamic tracts of anterolateral system (Figure 15.1).
Touch-pressure (deep pressure), two-point discrimination, vibratory sense (tuning fork)	Hair plexuses, lamellated (Pacinian) corpuscles, corpuscles of Ruffini, bulbous corpuscles (of Krause).*	Posterior column-medial lemniscal pathway, spinocervicothalamic pathway (for touch) (Figure 15.1).
Heat	Free nerve endings.	Anterolateral system, including lateral spinothalamic tract (Figure 15.1).
Cold	Free nerve endings.	Anterolateral system, including lateral spinothalamic tract (Figure 15.1).
Pain	Specialized free nerve endings.	Lateral spinothalamic tract, indirect spinoreticulothalamic pathway.
Proprioception	Specialized "spray" endings, lamellated corpuscles in synovia and ligaments.	Posterior column-medial lemniscal system (conscious), spinocerebellar system (unconscious), spinocervicothalamic pathway.

*Bulbous corpuscles (of Krause) (once thought to be cold detectors) and corpuscles of Ruffini (once thought to be heat receptors) are actually mechanoreceptors.

endings. Cold receptors respond to temperatures below skin temperature, and heat receptors respond to temperatures above skin temperature.

So-called *cold spots* and *warm spots* are found over the surface of the entire body, with cold spots being more numerous. A *spot* refers to a small area that, when stimulated, yields a temperature sensation of warmth or cold. A spot is associated with several nerve endings. The lips have both cold and warm spots, but the tongue is only slightly sensitive to warmth. Nerve endings that innervate teeth are usually sensitive to cold, but much less sensitive to heat. The face is less sensitive to cold than other parts of the body that are usually covered by clothing.

Pain

The subjective sensation we call pain is a warning signal that alerts the body of a harmful or unpleasant stimulus. The sensation may be initiated by receptors that are sensitive to mechanical, thermal, electrical, and chemical stimuli. Pain receptors are **specialized free nerve endings** (see Figure 15.1) that are present in most parts of the body (the intestines and brain tissue have no pain receptors). There are presumed to be about 3 million such *nociceptors* distributed over the surface of the body.

Types of pain include (1) fast-conducted, sharp, prickling pain, (2) slow-conducted, burning pain, and (3) deep, aching pain in joints, tendons, and viscera. Other distinctions are

sometimes made: *superficial somatic pain* originates from stimulation of skin receptors; *deep somatic pain* arises from stimulation of receptors in joints, tendons, and muscles; *visceral pain* originates from stimulation of receptors in body organs.

Some tissues are more sensitive to pain than others. A needle inserted into the skin produces great pain, but the same needle probed into a muscle produces little pain. An arterial puncture is painful, but a venous puncture is almost painless. A kidney stone that distends a ureter (the tube leading from the kidney to the urinary bladder) produces excruciating pain. In contrast, the intestines are not sensitive to pain if they are cut or burned, but *are* sensitive if they are distended or markedly contracted (cramps).

To some extent, the perception of pain is a matter of attention. Soldiers who are wounded in battle may experience little pain on the battlefield, only to complain of severe pain when the fighting has stopped. During the stress of battle (or other stress), the nervous system suppresses pain by increasing the release of opioid transmitters (also known as endorphins), especially along the pain pathways. Thus, pain perception often has two phases: the initial sensation and the reaction following that sensation.

We adapt to most of our senses so that they don't become a bother. If we didn't, we would be continuously aware of the touch pressure of clothing on our skin, or the hot water in our tub. (This phenomenon of *adaptation* refers to the decline in the response of receptors to a continuous, even stimulation.) However, it is to our benefit that we do not get completely used to pain. If pain is to be useful as a warning signal that prevents serious tissue damage, it must be felt each time it occurs, even to some extent when we are asleep. For example, who has not turned over in the middle of the night to relieve the pain caused by sleeping on a twisted arm?

Why does metal feel colder on the skin than wood does?   *Because metal absorbs heat from the skin more rapidly.*

The Body's Own Tranquilizers

Scientists have long searched for alternatives to such pain-killing drugs as morphine and opium, which are highly addictive. The answer may come from the body itself.

In the 1970s, American researchers discovered how opiates work. Opiate molecules lock onto special receptor sites of certain neurons in the central nervous system, slowing down the firing rate of those neurons. Apparently, the decreased firing rate decreases the sensation of pain. Many pain receptors are located in the spinal cord, where the pain impulse is first introduced into the central nervous system. It was also found that opiates had an especially strong effect on the thalamus, where pain is eventually processed into an actual sensation.

Within a short time after the discovery of the mechanism of opiates, teams of scientists around the world isolated natural short-chain neuropeptides in the brain and pituitary gland, which they called the **endorphins** (*end*ogenous + *morphine*). Also discovered were the breakdown products of endorphins, smaller peptides that were named the **enkephalins** (en-KEFF-uh-lihnz; "in the head").

Endorphins and enkephalins are morphinelike substances that occur naturally in the nervous system. They are the brain's own opiates, having the pain-killing effects of opiates such as morphine. It is likely that endorphins and enkephalins work by binding to the same neuronal receptors that bind opiate drugs. These receptor sites in the central nervous system are associated with the intrinsic pain pathways. In addition, the endorphins and enkephalins apparently act as neurotransmitters in the pain-inhibiting pathways. In addition to moderating pain, enkephalins in the limbic system seem to counteract psychological depression by producing a state of euphoria similar to the feelings produced by opiate drugs.

Some researchers believe that the "natural high" experienced by serious joggers is caused by the release of endorphins. Most scientists, while not disagreeing with the existence and function of endorphins as pain-moderating substances, feel that only a few athletes are consistently vigorous enough to secrete sufficient endorphins to produce a natural euphoria.

It is thought that other neurotransmitters are associated with the pain pathways. Some other neurotransmitters that play important roles in pain are substance P, serotonin, dopamine, norepinephrine, and dynorphin.

Referred pain Pain that originates in a body organ or structure is usually perceived to be on the body surface, often at a site away from the visceral source. A visceral pain felt subjectively in a somatic area is known as **referred pain.** For example, the pain of a coronary heart disease *(angina pectoris)* may be felt in the left shoulder, arm, and armpit; an irritation of the gallbladder may be felt under the shoulder blades.

One possible explanation for the brain's "misinterpretation" of most visceral pain is that certain neurons use a common dorsal root to innervate both the visceral and somatic locations involved in referred pain. It is thought that the visceral sensory fibers and somatic sensory fibers both discharge into a common pool of neurons in the central nervous system. In a sense, the brain interprets the source of visceral pain as a region of the skin because pain impulses originate in the skin more frequently than in the viscera.

Another explanation of referred pain is that the area of the body to which the pain is referred usually is part of the body that develops from the same embryological area *(dermatome)* as the real source of the pain. These dermatomes are supplied by branches of the same peripheral nerves. Using the example above, the heart and left arm are derived from the same dermatome. Thus, angina pectoris can be felt in a part of the left arm.

Phantom pain Another unusual phenomenon in the sensing of pain is the **phantom pain** that is felt in an amputated limb *(phantom limb).* Such pain may be intense, and it is actually felt. The sensations of pain, "pins and needles," and temperature change are often felt by amputees in their amputated limbs for several months. Ordinarily, the pain is felt more in the joints than in other regions of the phantom limb, and more in the distal portion of the amputated segment than in the proximal portion. Phantom pain usually persists longest in those regions that have the largest representation in the cerebral cortex: the thumb, hand, and foot.

The neural mechanism for phantom pain is not known completely. It appears that pools of neurons associated with the sensations of the missing limb are somehow activated, and result in the perception. Impulses in the pools of neurons may be triggered by the irritation of peripheral nerves in the proximal stump.

Proprioception

Receptors in muscles, tendons, and joints transmit impulses about our position sense up the dorsal columns of the spinal cord. These impulses help us to be aware of the position of our body and its parts without actually seeing them. This sense of position is called **proprioception** (L. *proprius*, one's self + receptor) or the *kinesthetic sense*. The receptors in or near joints that are responsible for proprioception are specialized "spray" endings. Lamellated corpuscles in the synovial membranes and ligaments may be involved. As you will see later, these proprioceptors are assisted by the semicircular canals of the ear.

Do analgesic pain-killers work by desensitizing the receptors?

No. Analgesics have no effect on sensory receptors. They modify the perception of pain or the emotional reaction to pain.

Itch and Tickle

Itch is probably produced by the repetitive, low-key stimulation of slow-conducting nerve fibers in the skin. *Tickle* is caused by a mild stimulation of the same type of fibers, especially when the stimulus moves across the skin. Receptors for both sensations are found almost exclusively in the superficial layers of the skin. It is thought that the sensations result from the activation of several sensory endings and that the information is conveyed via a combination of pathways. Like the areas most sensitive to pain, itch usually occurs where naked endings of unmyelinated fibers are abundant. Itch occurs on the skin, in the eyes, and in certain mucous membranes (such as in the nose and rectum), but not in deep tissues or viscera.

Although itching is usually produced by a repetitive, mechanical stimulation of the skin, it is often produced by chemical stimuli such as polypeptides known as *kinins,* and by the histamine that the body releases during an allergy attack or inflammatory response.

Stereognosis

The ability to identify unseen objects by handling them is *stereognosis* (STEHR-ee-og-NO-sis; Gr. *stereos,* solid, three-dimensional; *gnosis,* knowledge). This ability depends on the sensations of touch and pressure, as well as on sensory areas in the parietal lobe of the cerebral cortex. Damage to certain areas of the cortex of the parietal lobe usually impairs stereognosis, even if the cutaneous sensations are intact.

Corpuscles of Ruffini and Bulbous Corpuscles (of Krause)

Corpuscles of Ruffini are now considered to be variants of touch-pressure receptors. They are located deep within the dermis and subcutaneous tissue, especially in the soles of the feet (see Figure 15.1). They are thought to be mechanoreceptors that respond to the displacement of the surrounding connective tissue within the corpuscle, and appear to be sensors for touch-pressure, position sense, and movement. A large, myelinated afferent nerve fiber enters the corpuscle, loses its myelin sheath, and forms many treelike terminal branches that intertwine with the collagen fibers within the core of the corpuscle.

Bulbous corpuscles (of Krause) are found in the dermis of the conjunctiva (the covering of the whites of the eyes and the lining of the eyelids), tongue, and external genitals. They are thought to be mechanoreceptors.

Why does scratching relieve the itching of insect bites? Scratching an insect bite temporarily soothes the naked endings of unmyelinated fibers near the surface of the skin. However, scratching usually makes the inflammation worse by irritating the skin and releasing the chemical stimulant that caused the itch in the first place.

Neural Pathways for General Senses

The neural pathways involved in relaying data from specific general sensory receptors to the cerebral cortex include the *dorsal column-medial lemniscus tract,* the *spinothalamic tracts,* and the *trigeminothalamic tract.* The other ascending tracts and pathways for taste, smell, hearing, and vision are discussed in later sections of this chapter, and a summary of the basic ascending neural pathways is presented in Table 15.2.

In general, those afferent nerves that convey highly localized and discriminative sensations are larger, have more myelin, and conduct faster than those that convey less-defined sensations. They travel in the dorsal column-medial lemniscus and trigeminothalamic pathways. The afferent nerves that convey the less-defined sensations travel in the spinothalamic tracts.

Neural pathways for light touch The sensory area of the brain specialized for touch is located in the general sensory region of the parietal lobe of the cerebral cortex. Light touch is mediated by at least three neural pathways from the spinal cord to the cerebral cortex, including (1) the *dorsal column-medial lemniscal pathway,* (2) the *spinocervicothalamic pathway* of the dorsal column-medial lemniscal system, and (3) the *anterior spinothalamic tracts* of the anterolateral system.

Neural pathways for touch-pressure Touch-pressure is mediated by the *dorsal column-medial lemniscal pathway,* and probably by the *spinocervicothalamic pathway* of the posterior column-medial lemniscal system.

Neural pathways for temperature The sensory area of the brain for temperature is the same as for touch, the parietal lobe of the cerebral cortex. Crude sensations of temperature may be experienced in the thalamus. The sensation of temperature change is mediated by the *lateral spinothalamic tract* of the anterolateral system.

Neural pathways for pain It is thought that pain impulses are conveyed by two or more pathway systems, including (1) the *lateral spinothalamic tract,* which consists of a sequence of at least three neurons with long axonal processes that relay pain impulses from the spinal cord to the thalamus, and (2) the *indirect spinoreticulothalamic pathway,* which consists of a sequence of many neurons that relay pain impulses to the reticular formation and thalamus. The pathways are in the anterolateral system. Fibers from the spinotectal tract to the midbrain may also be involved. The perception of pain occurs in the thalamus, but the discrimination (judgment) of the type of pain and its intensity occur in the parietal lobe of the cerebral cortex.

Neural pathways for proprioception Many of the proprioceptive impulses are relayed to the cerebellum, but some are conveyed to the cerebral cortex through the *medial lemniscal pathway* and from *thalamic projections.* The sensory area for conscious position sense is located in the parietal lobe of the cerebral cortex. The neural pathways are the spinocerebellar tracts (unconscious perception) and dorsal column-medial

TABLE 15.2   BASIC ASCENDING NEURAL PATHWAYS

System and pathway or tract	Final destination	Sensations involved
Anterolateral system*	Cerebral cortex.	Pain, light touch, temperature.
Lateral spinothalamic tract	Cerebral cortex.	Pain, temperature.
Anterior spinothalamic tract	Cerebral cortex.	Light touch.
Indirect spinoreticulothalamic pathway	Cerebral cortex.	Pain.
Posterior column-medial lemniscal system*	Cerebral cortex.	Touch-pressure, proprioception, vibration.
Dorsal column-medial lemniscal pathway	Cerebral cortex.	Touch-pressure, light touch, stereognosis, associated tactile discrimination senses, vibration, two-point discrimination, conscious proprioception.
Spinocervicothalamic (spinocervicolemniscal) pathway	Cerebral cortex.	Touch-pressure, proprioception.
Spinocerebellar system	Cerebellar cortex.	Unconscious proprioception.
Anterior spinocerebellar tract	Cerebellar cortex.	Unconscious proprioception (from lower half of body).
Posterior spinocerebellar tract	Cerebellar cortex.	Unconscious proprioception (from lower half of body).
Cuneocerebellar tract	Cerebellar cortex.	Unconscious proprioception (from upper half of body).
Rostral spinocerebellar tract	Cerebellar cortex.	Unconscious proprioception (from upper half of body).

*The trigeminothalamic tracts convey all the sensations in this system from the head to the thalamus.

lemniscal pathway (conscious perception). Degenerative diseases of the dorsal column of the spinal cord produce *ataxia* (lack of muscular coordination associated with inadequate sensory input) because proprioceptive impulses are not conveyed to the cerebellum and to the thalamus and cerebral cortex.

Neural pathways for somesthetic sensations from head
Those afferent nerves that convey highly localized discriminative sensations from the face, mouth, nasal cavities, and associated structures, as well as those conveying cruder sensations, form the *trigeminothalamic tracts.* These tracts originate from the spinal trigeminal nucleus (pain and temperature) and principal sensory nuclei (touch-pressure), respectively. The tracts parallel the spinothalamic and the dorsal column-medial lemniscus tracts. The sensations are primarily felt on the skin and the mucosal membranes of the nasal and oral cavities. They involve the sensory region of the parietal lobe of the cerebral cortex.

Ask Yourself

1 *What are the receptors for light touch? Touch-pressure? Pain?*

2 *How do light touch and touch-pressure differ?*

3 *What types of stimuli can initiate pain?*

4 *What is the difference between somatic and visceral pain?*

5 *What is proprioception? Stereognosis?*

6 *What are the three ascending neural pathways involved in the general senses?*

TASTE (GUSTATION)

The receptors for taste, or **gustation** (L. *gustus,* taste), and smell are both chemoreceptors, and the two sensations are clearly interrelated. (A person whose nasal passages are "blocked" by a cold cannot "taste" food as effectively.*) But despite some similarities, taste and smell are separate and distinct senses, and will be treated separately here.

Structure of Taste Receptors

The surface of the tongue is covered with many small protuberances called **papillae** (puh-PILL-ee; singular, *papilla,* puh-PILL-uh; L. diminutive of *papula,* nipple, pimple). Papillae give the tongue its bumpy appearance (Figure 15.2). They are most numerous on the dorsal surface of the tongue, and are also found on the palate (roof of the mouth), throat, and posterior surface of the epiglottis.

The three main types of papillae are shown in Figure 15.2. *Fungiform* (L. mushroomlike) *papillae* are scattered singly, especially near the tip of the tongue. From 10 to 12 *circumvallate* (L. "wall around") *papillae* form two rows parallel to the V-shaped sulcus terminalis near the posterior third of the tongue. *Filiform* (L. threadlike) *papillae* are pointed structures near the anterior two-thirds of the tongue.

---

*It is interesting that some astronauts have had trouble tasting their food while traveling in space. Apparently, the loss of taste is caused by the lack of gravity. The heart, accustomed to pumping against gravity, forces more blood into the head than is necessary, and other body fluids accumulate there as well. The result is congestion, the same feeling we feel when we have a head cold.

FIGURE 15.2

Taste-sensitive areas and papillae of the tongue, dorsal surface. (A) The specific taste-sensitive areas are shown in color. Sour sensations (green) are perceived most acutely on the sides of the tongue, about halfway back, by the stimulation of the hydrogen ions in acids. Saltiness (red) is tasted mainly at the sides and tip of the tongue. Bitter sensations (yellow) are perceived mainly at the back of the tongue. Sweetness (purple) is tasted optimally at the tip of the tongue, where sugars react with the fatty substances in the nerve ending. The center of the tongue, with only a few taste buds, is relatively insensitive to taste. The photomicrographs show cross sections cut from representative regions of the mucous membrane of the tongue. Each fungiform papilla (B) contains 1 to 8 taste buds, and each circumvallate papilla (C) may contain 90 to 250 taste buds. Filiform papillae (D) do not necessarily contain taste buds. [(A) *John D. Cunningham/Visuals Unlimited.* (B) *Biophoto Associates/Photo Researchers.* (C) *M. I. Walker/Photo Researchers.*]

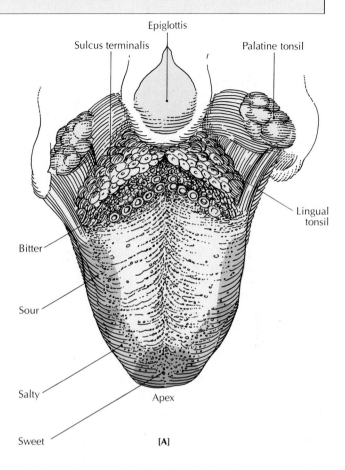

[A]

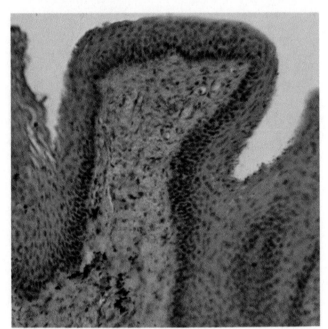

[B]   × 58

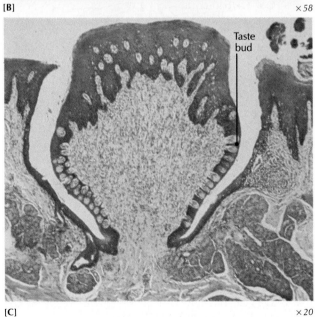

[C]   × 20

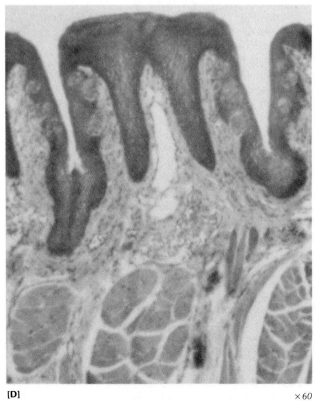

[D]   × 60

Located within the crevices of the papillae are approximately 10,000 receptor organs for the sense of taste, popularly called **taste buds**. They are barrel-shaped clusters of *chemoreceptor* (taste or gustatory) *cells* and *supporting* (sustentacular) *cells,* arranged like alternating segments of an orange (Figure 15.3). Each taste bud contains about 25 receptor cells. The more numerous supporting cells act as reserve cells, replenishing the receptor cells when they die. Mature receptor cells have a life of only about 10 days, and they usually can be replaced from reserve cells in about 10 hours. Taste receptor cells are replaced with decreasing frequency as we get older. This explains, in part, why our sense of taste may diminish with age, and may also explain why babies dislike spicy foods and tend to favor relatively bland baby foods.

Extending from the free end of each receptor cell are short *taste hairs* (microvilli) that project through the tiny outer opening of the taste bud, called the *taste pore,* into the surface epithelium of the oral cavity. It is thought that gustatory sensations are initiated on the taste hairs, but before a substance can be tasted it must be in solution. Saliva containing ions or dissolved molecules of the substance to be tasted enters the taste pore and interacts with receptor sites on the taste hairs.

Basic Taste Sensations

Although all taste cells are structurally identical, each cell has many different types of receptor sites. Because the proportion of different types varies from cell to cell, each taste cell can respond to a variety of stimuli. The four generally recognized basic taste sensations are sweet, sour, bitter, and salty. We can taste many subtle flavors because of combinations of the four basic sensations, complemented by an overlay of odors. Taste perception is also helped by information about the texture, temperature, spiciness, and odor of food. The areas of response to the four basic tastes are located on specific parts of the tongue (see Figure 15.2). Salt and sweet are perceived most acutely on the tongue, but bitter and sour are perceived more acutely on the palate.

Many substances other than sugar evoke a sweet taste, including compounds such as glycols, alcohols, amino acids, and certain salts of lead and beryllium. (Children may be poisoned seriously if they eat sweet-tasting, but toxic, peelings of lead-based paint.) Sour tastes are produced by the hydrogen ions in acids, and salty tastes are produced by the anions of ionized salts. Bitterness is due primarily to two major classes of organic compounds, alkaloids and long-chain organic substances, but also to some inorganic substances such as quinine, caffeine, strychnine, and nicotine. Bitterness is sensed most delicately, and may be related to a kind of primitive protective mechanism that warns us to avoid bitter-tasting (and presumably toxic) substances. Bitterness can be detected at a dilution of 1 part in 2 million, while sourness requires a dilution of 1 part in 135,000 to be detected, saltiness 1 part in 500, and sweetness 1 part in 250. The bitterest substance known to date is denatonium saccharide, which can be tasted even when diluted to 1 part in 20 million.

FIGURE 15.3

Taste buds on the tongue. Compare the scanning electron micrograph of a taste bud (A) with the section drawing through a taste bud (B). [(A) *Bruce Coleman.*]

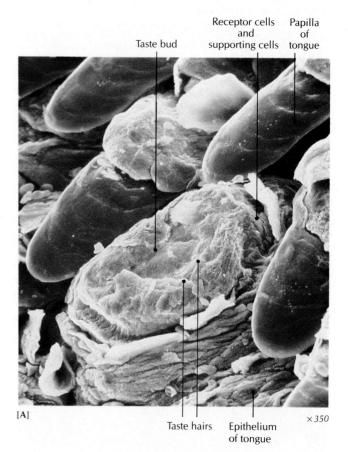

[A]   ×350

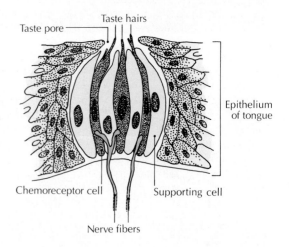

[B]

Is sugar the sweetest substance?	No. Saccharine (a synthetic organic compound), for example, is about 600 times more detectable as a sweetener than sugar.

The Taste Mechanism

The exact mechanisms that stimulate a taste cell are not known, although many theories have been proposed. Different types of potential gustatory stimuli may cause proteins on the surface of the receptor cell membrane to change the permeability of the membrane, in effect "opening and closing gates" to chemical stimuli. Variations in the intensity of tastes are produced by differences in the firing frequency of nerve impulses.

Neural Pathways for Gustation

Taste impulses are conveyed from the anterior two-thirds of the tongue to the brain by a branch of the facial nerve (cranial nerve VII). Impulses from the posterior third of the tongue are carried to the brain by the glossopharyngeal (IX) nerve, and from the palate and pharynx by the vagus (X) nerve. The taste fibers of all three cranial nerves terminate in the *nucleus solitarius* (also called the *nucleus gustatorius*) in the medulla. From there, axons project to the thalamus, and then to the "taste center" in the parietal lobe of the cerebral cortex.

Ask Yourself

1 *What are papillae?*

2 *Can you describe the structure of a taste bud?*

3 *What is the sensory role of taste hairs?*

4 *Which areas of the tongue respond to each of the four basic taste sensations?*

SMELL (OLFACTION)

Our sense of smell, or **olfaction** (L. *olere,* to smell + *facere,* to make), is perhaps as much as 20,000 times more sensitive than our sense of taste. For example, we can taste quinine in a concentration of 1 part in 2 million, but we can smell mercaptans (the type of chemical released by skunks) in a concentration of 1 part in 30 *billion.* Adults can usually sense up to 10,000 different odors, and children can do even better. (Unfortunately, our sense of smell is not perfect. Several poisonous gases, including carbon monoxide, are not detectable by our olfactory receptors.)

Structure of Olfactory Receptors

The **olfactory receptor cells** are located high in the roof of the nasal cavity, in specialized areas of the nasal mucosa called the *olfactory epithelium* (Figure 15.4). Each nostril contains a small patch of pseudostratified, columnar olfactory epithelium about 2.5 sq cm (about the size of a thumbnail). The epithelium consists of three types of cells: (1) *receptor cells,* which actually are the olfactory neurons, (2) *sustentacular*

(supporting) *cells,* and (3) a thin layer of small *basal cells.* These basal cells are capable of undergoing mitosis and replacing degenerating receptor and sustentacular cells. Each receptor cell has a lifetime of about 30 days. They are replaced by basal cells, which are continually differentiating into new olfactory neurons and forming new synaptic connections in the olfactory bulb.

Because olfactory cells are the only neurons exposed to the external environment, they can be damaged or destroyed rather easily by disease or other trauma. As a result, about 1 percent of our olfactory receptor cells (neurons) are lost every year, without replacement.

We have more than 25 million bipolar receptor cells (a hunting dog has about 220 million), each of which is surrounded by sustentacular cells.* Each thin receptor cell has a short dendrite extending from its superficial end to the surface epithelium. The receptor cell ends in a bulbous *olfactory vesicle* (see Figure 15.4C). From this swelling, 6 to 20 long *cilia* project through the mucuslike fluid that covers the surface epithelium. The fluid is secreted by the sustentacular cells and *olfactory (Bowman's) glands.* It is important because odoriferous substances need to be dissolved before they can stimulate receptor sites.

The receptive sites of the cilia are exposed to the molecules responsible for the odors. The axons of the bipolar receptive cells (neurons) pass through the *basal lamina* and join other axons to form fascicles of the olfactory nerve (cranial nerve I). These unmyelinated olfactory nerve fibers are among the smallest and slowest-conducting fibers of the nervous system. They pass through the foramina in the cribriform plate of the ethmoid bone on their way to the olfactory bulbs. The **olfactory bulbs** are specialized structures of gray matter, stemlike extensions of the olfactory region of the brain.

Once inside the olfactory bulbs, the terminal axons of the receptor cells synapse with dendrites of *tufted cells, granule cells,* and *mitral cells.* These complex, ball-like synapses are called **olfactory glomeruli** (gluh-MARE-you-lie; L. *glomus,* ball). Each glomerulus receives impulses from about 26,000 receptor cell axons. These impulses are conveyed along the axons of mitral and tufted cells, which form the *olfactory tract* running posteriorly to the olfactory cortex in the temporal lobe of the cerebrum. Olfaction is the only sense that does not project fibers into the thalamus before reaching the cerebral cortex.

How Odors Are Perceived

Olfactory impulses are sorted and integrated in the glomeruli before being relayed to the cortex. It is believed that the glomeruli are the critical sites where odors are first processed. The olfactory cortex is initially involved in distinguishing and determining the intensities of odors. Actually, little is known about the neural mechanism of olfaction. We do know that in order for substances to be sensed, they must be volatile,

*These olfactory neurons (receptor cells) are the most primitive neurons in the central nervous system. They are chemoreceptor cells. Each olfactory receptor cell is a chemical detector, a transducer of a stimulus, and the transmitter of the nerve impulse to the olfactory bulb.

FIGURE 15.4

Olfactory receptors. (A) Olfactory receptive area in the roof of the nasal cavity, medial view. (B) Enlarged drawing of the olfactory epithelium, showing receptor cells, sustentacular cells, basal cells, and the formation of the olfactory tract inside the olfactory bulb. (C) Scanning electron micrograph of olfactory epithelial surface, showing olfactory vesicle, cilia extending into the nasal mucosa, and microvilli of sustentacular cells. [*Gene Shih and Richard Kessel, Living Images: Biological Microstructures Revealed by SEM. Boston, 1982, Science Books International.*]

water-soluble, and lipid-soluble. Without these qualities, the odoriferous particles could not be carried into the nostrils by air currents, dissolve in the mucuslike coating on the olfactory epithelium, and penetrate the lipid barrier surrounding the olfactory receptor cell.

Many theories have been proposed to explain how we perceive odors. A likely suggestion is that odor molecules have some kind of a physical interaction with protein receptor sites on a cell membrane. Such an interaction somehow alters membrane permeability and leads to a generator potential in the receptor cell, which produces action potentials in the nerve fibers that synapse with neurons in the olfactory bulb. According to this theory, the discrimination of different odors results from the simultaneous but varying stimulation of receptor cells.

Although we can detect thousands of different odors, they are generally divided into seven primary categories: musky, camphoraceous, floral, pungent, pepperminty, ethereal, and putrid. Any particular odor can be described as one of these primary odors, or some combination of two or more of them.

Why do we "sniff" when we want to detect an odor?

Ordinarily, odors are carried to the nose by air currents. Sniffing creates eddy currents that greatly increase the amount of air that reaches the olfactory receptors high in the nasal cavities. Sniffing is a semireflexive response, usually occurring when an odor attracts our attention.

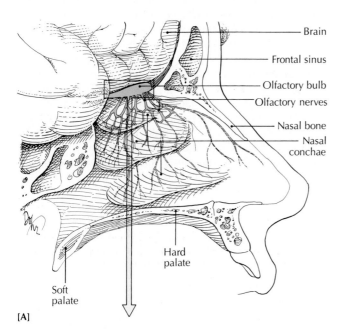

[A]

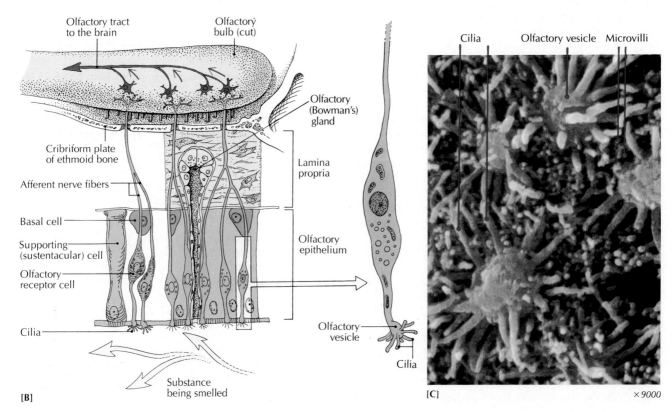

[B]

[C] ×9000

The quality of our sense of smell varies as conditions change. For example, when we have a cold, our sense of smell is inhibited because the mucous membranes of the upper nasal cavities are swollen, and the receptors are buried under a thickened layer of mucus. When we are hungry, our sense of smell is enhanced.

Neural Pathways for Olfaction

Mitral cells in the olfactory bulbs project axonal branches through the olfactory tract to the primary olfactory cortex. The *primary olfactory cortex* is composed of the cortex of the uncus and adjacent areas, located in the temporal lobe of the cerebral cortex.

Ask Yourself

1 *What are the three types of cells in the olfactory epithelium?*

2 *What are olfactory bulbs?*

3 *What happens in the olfactory glomeruli?*

4 *What is one current theory to explain the mechanism of olfaction?*

HEARING AND EQUILIBRIUM

Hearing (audition) and equilibrium are considered in the same section because both sensations are received in the same organ: the inner ear. The ear actually has two functional units: (1) the *auditory apparatus* (also called the *acoustic apparatus*), concerned with hearing, and (2) the *vestibular apparatus*, concerned with posture and balance. The auditory apparatus is innervated by the *cochlear nerve*, and the vestibular apparatus is innervated by the *vestibular nerve*. The two nerves are collectively known as the *vestibulocochlear nerve* (cranial nerve VIII).

Anatomy of Hearing

The auditory system is organized to detect several aspects of sound, including pitch, loudness, and direction. The anatomical components of this system are the external ear, the middle ear, and the inner ear (Table 15.3).

External ear The external ear is the part you can see. It is also called the **auricle** (L. *auris*, ear) or *pinna* (PIHN-uh; L. wing). It is composed of a thin plate of fibrocartilage covered by a close-fitting layer of skin. The funnel-like curves of the auricle are well designed to collect sound waves and direct them to the middle ear (Figure 15.5A). The deepest depression, the *concha* (KONG-kuh; L. conch shell), leads directly to the external auditory canal (meatus) (Figure 15.5B). The area of the concha is partly covered by two small pro-

TABLE 15.3   MAJOR ANATOMICAL COMPONENTS OF THE EAR

Structure	Description
EXTERNAL EAR	
Auricle (pinna)	Cartilaginous, exterior "flap" of ear, designed to convey sound waves to the middle ear.
External auditory canal (meatus)	Canal leading from floor of concha in outer ear to tympanic membrane.
MIDDLE EAR	
Tympanic membrane (eardrum)	Fibrous tissue extending across deep inner end of external auditory canal, forming partition between external and middle ears.
Tympanic cavity	Narrow, air-filled space in temporal bone; separated from external auditory canal by tympanic membrane, and from inner ear by bony wall containing round and oval windows.
Auditory tube (Eustachian tube)	Tube leading downward and inward from tympanic cavity to nasopharynx.
Auditory ossicles (ear bones)	Malleus (hammer), incus (anvil), and stapes (stirrup) form a lever chain from tympanic membrane to oval window of inner ear.
INNER EAR (LABYRINTH)	
Vestibule	Central chamber of inner ear; includes utricle and saccule filled with endolymph and surrounded by perilymph.
Semicircular ducts*	Located posterior to the vestibule are three small ducts lying at right angles to each other; suspended in perilymph. Each duct has an expanded end, the ampulla, which contains a receptor structure, the crista ampullaris.
Cochlea	Located anterior to the vestibule is a spiral structure containing perilymph-filled scala vestibuli and scala tympani, and endolymph-filled scala media (cochlear duct).
Spiral organ (of Corti)	Organ of hearing resting on the basilar membrane of the cochlea; a complex of supporting cells and hair cells.

*There are three semicircular canals surrounded by bone. Within each canal is a semicircular duct (about one-quarter the diameter of the canal), which contains endolymph. Surrounding each duct, and contained within each canal, is perilymph.

FIGURE 15.5

The ear. (A) Frontal cutaway diagram of the right ear, showing the external, middle, and inner ears. A section of the cochlear duct has been removed to show the spiral organ. (B) Structural components of the external ear. [(B) *Martin Dohrn/Science Photo Library/Photo Researchers.*]

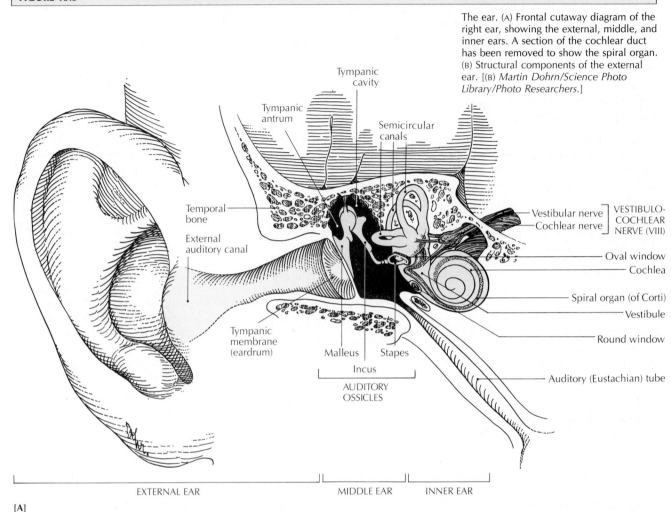

[A]

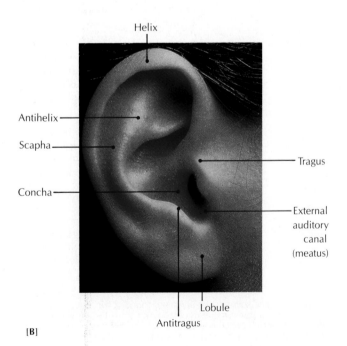

[B]

jections, the *tragus* in front (TRAY-guhss; Gr. *tragos,* goat, because hairs said to resemble a goat's beard may grow at the entrance of the external auditory canal) and the *antitragus* behind.

The *helix* is the prominent ridge that forms the rim of the uppermost portion of the auricle (see Figure 15.5B). The *antihelix* is a curved ridge, more or less concentric to the helix, that surrounds the concha. It is separated from the helix by a furrow called the *scapha.* The *lobule,* or earlobe, is the fatty, lowermost portion of the auricle. It is the only part of the external ear without any cartilage. Because the lobule contains few nerve endings and many capillaries, it is a frequently used as a source of blood samples for blood count "work-ups."

Why does your voice sound different on a tape recording?

When you hear yourself speak, you are hearing some extra resonance produced by the conduction of sound waves through the bones of your skull. Your voice as played by a tape recorder is the way it sounds to a listener, who receives the sound waves only through air conduction.

The ***external auditory canal*** (meatus) is a slightly curved canal, extending about 2.5 cm (1 in.) from the floor of the concha to the tympanic membrane, which separates the external ear from the middle ear. The outer third of the wall of the external auditory canal is composed of cartilage, and the inner two-thirds is carved out of the temporal bone (see Figure 15.5B). The canal and tympanic membrane are covered with skin. Fine hairs in the external ear are directed outward, and sebaceous glands and modified sweat glands (ceruminous glands) secrete *cerumen* (suh-ROO-muhn), or earwax. The hairs and wax make it difficult for tiny insects and other foreign matter to enter the canal. Cerumen also prevents the skin of the external ear from drying out. The canal also acts as a buffer against humidity and temperature changes that can alter the elasticity of the eardrum.

Middle ear The middle ear is a small chamber between the tympanic membrane and the inner ear. It consists of the tympanic cavity (Gr. *tumpanon,* drum) and contains the auditory ossicles (ear bones).

The ***tympanic membrane,*** popularly called the ***eardrum,*** forms a partition between the external ear and middle ear. It is a thin layer of fibrous tissue continuous externally with skin, and internally with the mucous membrane that lines the middle ear. Between its concave external surface and convex internal surface is a layer of circular and radial fibers that give the membrane its firm elastic tension. The tympanic membrane is attached to a ring of bone (the tympanic annulus) and vibrates in response to sound waves entering the external auditory canal. The tympanic membrane is well endowed with blood vessels and nerve endings, so a "punctured eardrum" usually produces considerable bleeding and pain.

The ***tympanic cavity*** (middle-ear cavity) is a narrow, irregular, air-filled space in the temporal bone. It is separated laterally from the external auditory canal by the tympanic membrane, and medially from the inner ear by the bony wall, which has two openings, the *oval window* and the *round window* (see Figure 15.5A). An opening in the posterior wall of the cavity leads into the *tympanic antrum,* a chamber that is continuous with the small air cells of the mastoid process. When an infection of the middle ear progresses through the tympanic antrum into the mastoid cells, it can cause *mastoiditis.*

In the anterior wall of the tympanic cavity is the ***auditory tube,*** commonly called the ***Eustachian tube*** (yoo-STAY-shun). It leads downward and inward from the tympanic cavity to the nasopharynx, the space above the soft palate that is continuous with the nasal passages (see Figure 15.5A). The mucous-membrane lining of the nasopharynx is also continuous with the membrane of the tympanic cavity. As a result, an infection may spread from the nose or throat into the middle ear, producing a middle ear infection, or *otitis media.** (See Physiological and Anatomical Abnormalities at the end of this chapter.)

The main purpose of the auditory tube is to maintain equal air pressure on both sides of the tympanic membrane by permitting air to pass from the nasal cavity into the middle ear. The pharyngeal opening of the tube remains closed when the external pressure is greater, but opens during swallowing, yawning, and nose blowing so that minor differences in pressure are adjusted without conscious effort. The tube may remain closed when the pressure change is sudden, as when an airplane takes off or lands. However, the pressure can usually be equalized, and the discomfort relieved, by swallowing or yawning. This maneuver stimulates the tensor veli palatini muscle to contract, pulling on a portion of the cartilage of the auditory tube, causing the tube to open.

The three ***auditory ossicles*** (ear bones) of the middle ear form a chain of levers extending from the tympanic membrane to the inner ear (see Figure 15.5A). This lever system transmits sound waves from the external ear to the inner ear. From the outside in, the tiny, movable bones are the ***malleus*** (hammer), ***incus*** (anvil), and ***stapes*** (STAY-peez) (stirrup) (Figure 15.6). The ear bones are the smallest bones in the body, with the stapes being the smallest of all.

The auditory ossicles are held in place and attached to each other by ligaments. Two tiny muscles, the tensor tympani and the stapedius, are attached to the ear bones. The *tensor tympani* is attached to the handle of the malleus. When this muscle contracts, it pulls the malleus inward, increasing the tension on the tympanic membrane and reducing the amplitude of vibrations transmitted through the chain of auditory ossicles. The *stapedius* (see figure on page 263) attaches to the neck of the stapes. Its contraction pulls the footplate of the stapes, decreasing the amplitude of vibrations at the oval window.

In response to a sequence of loud sounds, the stapedius (acoustic) reflex comes into play to reduce the loudness. In this reflex, the stapedius muscle, which is innervated by the facial nerve, contracts, and thereby dampens the amplitude of the vibrations of the stapes. The faint words of a whisper can be enhanced by concentrating mentally, and thereby relaxing the stapedius muscle.

Inner ear The inner ear is also called the ***labyrinth*** (Gr. maze) because of its intricate structure of interconnecting chambers and passages (Figure 15.7). It consists of two main structural parts, one inside the other: (1) the *bony labyrinth* is a series of channels hollowed out of the petrous portion of the temporal bone. It is filled with a fluid called *perilymph.* (2) The bony labyrinth surrounds the inner *membranous labyrinth,* which contains a fluid called *endolymph* and all the sensory receptors for hearing and equilibrium.

The membranous labyrinth consists of three semicircular ducts, as well as the utricle, saccule, and cochlear duct, all of which are filled with endolymph and contain various sensory receptors (cristae ampullaris maculae and spiral organ). The semicircular ducts are located within the semicircular canals of the bony labyrinth. They are about one-quarter the diameter of the canals. Perilymph is located in the space between the ducts and the bony walls of the canals. Because the membranous labyrinth fits inside the bony labyrinth, these two bony channels have the same basic shape.

The bony labyrinth consists of the vestibule, three semicircular canals, and the spirally coiled cochlea. The ***vestibule***

*It is a good idea to keep your mouth slightly open when you blow your nose. Blowing the nose hard with the mouth closed creates a pressure that may force infectious microorganisms into the middle ear.

FIGURE 15.6

The auditory ossicles: malleus, incus, and stapes.

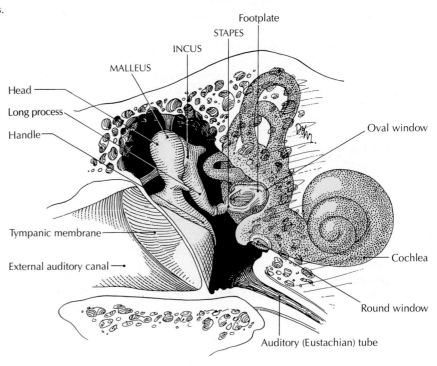

(L. entrance) is the central chamber of the labyrinth. Within the vestibule are the two endolymph-filled sacs of the membranous labyrinth, the *utricle* (YOO-trih-kuhl; L. little bottle) and the smaller *saccule* (SACK-yool; L. little sack). Each sac contains a sensory patch called a *macule* (see Figure 15.7).

The three *semicircular ducts* and *canals* are perpendicular to each other, allowing each one to be oriented in one of the three planes of space (see Figure 15.7). The ducts are lined by the membranous labyrinth, while the canals are surrounded and lined by bone. On the basis of their locations, the semicircular ducts are called *superior, lateral,* and *posterior.* Each duct has an expanded end called an *ampulla* (am-POOL-uh), which contains a receptor structure, the *crista ampullaris.* The utricle, saccule, and semicircular ducts are concerned with equilibrium, not hearing, and will be discussed in further detail later in the chapter.

Beyond the semicircular ducts is the spiral *cochlea* (KAHK-lee-uh; Gr. *kokhlos,* snail), so named because it resembles a snail's shell (see Figure 15.7). It may be thought of as a bony tube wound $2\frac{3}{4}$ times in the form of a spiral. The cochlea is divided longitudinally into three spiral ducts: (1) the *scala* (L. staircase) *vestibuli,* which communicates with the vestibule; (2) the *scala tympani,* which ends at the round window; and (3) the *scala media,* or *cochlear duct,* which lies beneath the other ducts (see Figure 15.7). The cochlear duct contains endolymph and the spiral organ, whereas the scala vestibuli and scala tympani contain perilymph. The three ducts, arranged in parallel, ascend in a spiral around the bony core, or *modiolus* (L. hub).

The cochlear duct is separated from the scala vestibuli by the *vestibular membrane,* and from the scala tympani by the *basilar membrane.* Resting on the basilar membrane is the **spiral organ (of Corti),** the organ of hearing (see Figure 15.7). The spiral organ is an organized complex of supporting cells and hair cells. The hair cells are arranged in rows along the length of the coil. The outer hair cells are arranged in three rows, and the inner hair cells are in a single row along the inner edge of the basilar membrane (Figure 15.8A). There are about 3500 inner hair cells and 20,000 outer hair cells.

Both the inner and outer hair cells have bristlelike *sensory hairs,* or *stereocilia,* which are specialized microvilli, and on one side, a basal body (Figure 15.8B). Each outer hair cell has 80 to 100 sensory hairs, and each inner hair cell has 40 to 60 sensory hairs. In each hair cell, the hairs are arranged in rows that form the letter W or U, with the base of the letter directed laterally. The tips of many hairs are embedded within, and firmly bound to, the *tectorial membrane* above the spiral organ.

Why do you cough when your ear is probed?

Stimulation of the external auditory canal can initiate a coughing reflex involving the auricular branch of the vagus nerve, which supplies the skin of the external auditory canal.

Physiology of Hearing

Physically, **sound** is the alternating compression and decompression of the medium (usually air) through which the sound is passing. In air that is disturbed by sound, waves of compression, in which air molecules are pushed together, are

FIGURE 15.7

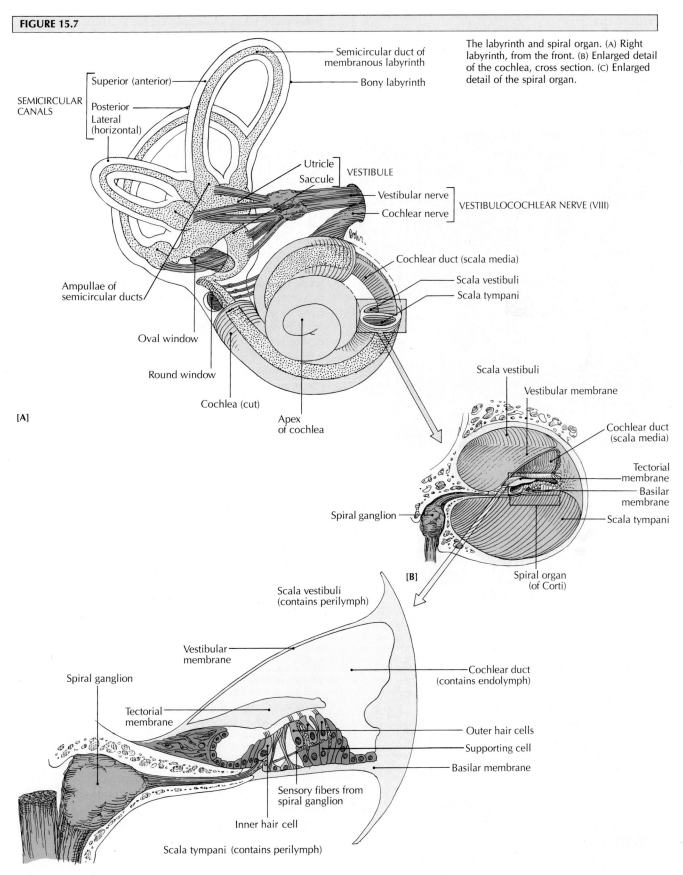

The labyrinth and spiral organ. (A) Right labyrinth, from the front. (B) Enlarged detail of the cochlea, cross section. (C) Enlarged detail of the spiral organ.

SEMICIRCULAR CANALS
- Superior (anterior)
- Posterior
- Lateral (horizontal)

Semicircular duct of membranous labyrinth

Bony labyrinth

Utricle

Saccule

VESTIBULE

Vestibular nerve

Cochlear nerve

VESTIBULOCOCHLEAR NERVE (VIII)

Ampullae of semicircular ducts

Cochlear duct (scala media)

Scala vestibuli

Scala tympani

Oval window

Round window

Cochlea (cut)

Apex of cochlea

[A]

Scala vestibuli

Vestibular membrane

Cochlear duct (scala media)

Tectorial membrane

Basilar membrane

Scala tympani

Spiral ganglion

Spiral organ (of Corti)

[B]

Scala vestibuli (contains perilymph)

Vestibular membrane

Spiral ganglion

Tectorial membrane

Cochlear duct (contains endolymph)

Outer hair cells

Supporting cell

Basilar membrane

Sensory fibers from spiral ganglion

Inner hair cell

Scala tympani (contains perilymph)

[C]

FIGURE 15.8

Sensory hair cells in the spiral organ. (A) Scanning electron micrograph of a single row of inner hair cells (top) and three rows of outer hair cells. (B) Drawing of inner and outer hair cells, showing the rows of stereocilia. [(A) *CNRI/Science Photo Library/ Photo Researchers*.]

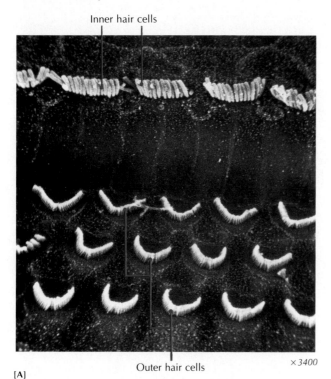

Inner hair cells

Outer hair cells

× 3400

[A]

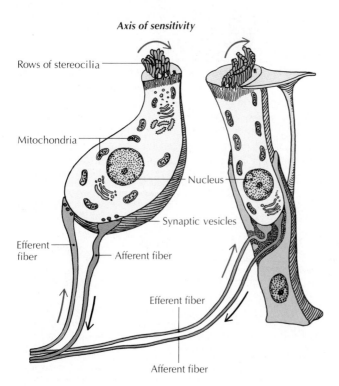

Axis of sensitivity

Rows of stereocilia

Mitochondria

Nucleus

Synaptic vesicles

Efferent fiber

Afferent fiber

Efferent fiber

Afferent fiber

[B]   INNER HAIR CELL   OUTER HAIR CELL

followed by waves of decompression, in which the air molecules are farther apart. **Frequency** is the number of sound waves per second. Very few naturally vibrating bodies produce simple vibrations or a "pure" tone, but instead produce combinations of frequencies. When the combinations are subjectively interpreted by the nervous system, they give various sounds their *timbre*, or "quality."

It is not known exactly how the brain interprets distinctive sounds, but we do know that *loudness* or intensity is determined by the size of the sound waves, *timbre* by their shape, and *pitch* by their frequency. We also understand that the difference between noise and music is in the synchronous regularity of musical sound waves.

Sound waves are conducted through air in the external ear, through solids in the middle ear, and through liquid in the inner ear. The transition is important, because sound waves in the air do not pass readily into a liquid medium, as you may know if you swim underwater. Although our three-phase conduction of sound waves works admirably, there is a potential loss of energy when sound waves have to pass from air to fluid, since fluid is more difficult to move than air. This energy loss is just about balanced by the lever action of the ossicles in the middle ear, however, and in fact there is no appreciable loss of energy. The efficiency of the energy transfer from the tympanic membrane (which responds to air vibrations) to the oval window (which transfers the vibration to the endolymph in the cochlea) is enhanced, because the surface area of the tympanic membrane is about 20 times greater than that of the foot of the stapes in the oval window. This contributes to overcoming much of the energy lost during the transfer from air to endolymph. The three-phase conduction system is estimated to work at 99.9 percent transmission efficiency. The ear is considered a more efficient energy converter than the eye.

The hair cells of the spiral organ are the mechanoreceptors in which the mechanical energy of sound is transduced into the generator potentials of the cochlear nerve endings. Each hair cell is innervated by several neurons. There are about 23,500 hair cells in the spiral organ, and about 30,000 neurons and fibers in the cochlear nerve. Sound waves are converted into generator potentials in the following way (Figure 15.9):

1 Sound waves enter the external ear. The waves reverberate against the sides of the external auditory canal and create waves of pressure. The waves reach the tympanic membrane.

2 Air molecules under pressure cause the tympanic membrane to vibrate.* Low-frequency sound waves produce slow vibrations, and high-frequency waves produce rapid vibrations. The vibrations move the malleus, on the other side of the membrane.

3 The handle of the malleus† strikes the incus, causing it to vibrate.

4 The vibrating incus moves the stapes into and out of the oval window.

5 The sound waves that reach the inner ear through the

*We can hear sounds when the tympanic membrane vibrates the width of a hydrogen atom.

†The tympanic cavity contains the malleus, incus, and stapes.

FIGURE 15.9

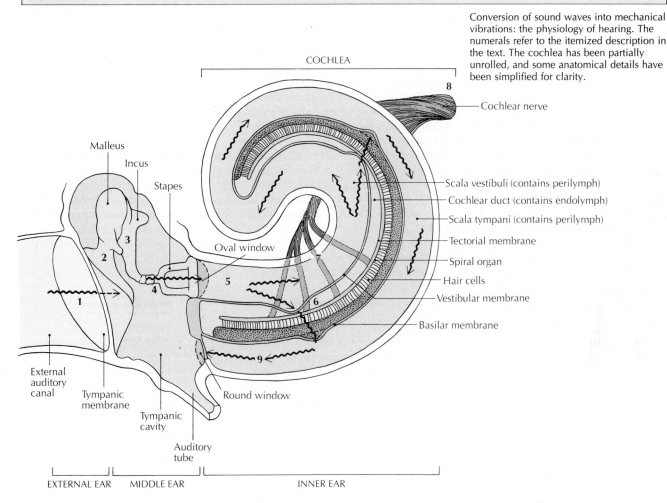

COCHLEA

Conversion of sound waves into mechanical vibrations: the physiology of hearing. The numerals refer to the itemized description in the text. The cochlea has been partially unrolled, and some anatomical details have been simplified for clarity.

8 — Cochlear nerve

— Scala vestibuli (contains perilymph)
— Cochlear duct (contains endolymph)
— Scala tympani (contains perilymph)
— Tectorial membrane
— Spiral organ
— Hair cells
— Vestibular membrane
— Basilar membrane

Malleus
Incus
Stapes
Oval window
Round window

External auditory canal
Tympanic membrane
Tympanic cavity
Auditory tube

EXTERNAL EAR   MIDDLE EAR   INNER EAR

oval window set up pressure changes that vibrate the perilymph in the scala vestibuli.

6 Vibrations in the perilymph are transmitted across the vestibular membrane to the endolymph of the cochlear duct, and also up the scala vestibuli and down the scala tympani. The vibrations are transmitted to the basilar membrane, causing the membrane to ripple.* The fundamental vibratory ripples result in the perception of pure tones. Overtones such as musical sounds, chords, and harmonics result from secondary vibrations superimposed on the fundamental vibrations of the spiral organ.

7 Receptor hair cells of the spiral organ that are in contact with the overlying tectorial membrane are bent, causing them to generate graded generator potentials that excite the cochlear nerve to generate action potentials, or nerve impulses. When the hairs are displaced toward the basal body (axis of sensitivity), the hair cells are excited; when the hairs are displaced away from the basal body, the hair cells are inhibited.

*The ripples in the long axis of the basilar membrane are concerned with the *frequency* and *intensity* of sound. The spiral organ is organized so that the high tones are near the base and the low tones near the apex. Loudness is associated with the amplitude of the vibrations (the amount of displacement of the basilar membrane).

8 The nerve impulses are conveyed along the cochlear branch of the vestibulocochlear nerve. These fibers activate the auditory pathways in the central nervous system, which terminate in the auditory area of the temporal lobe of the cerebral cortex, where the appropriate sound is perceived.

9 Vibrations in the scala tympani are dissipated out of the cochlea through the round window into the middle ear.

The nuances of timbre, pitch, and intensity of sounds are somehow maintained with each amplifying step from tympanic membrane through auditory ossicles, cochlea, and spiral organ to the cochlear nerve. As a result, when the sound impulses reach the auditory area of the cerebral cortex, we can distinguish a saxophone's note from a piano's, a child laughing from a dog barking, and the fine shadings of each other's speech.

How can you tell the direction of a sound?	*Depending on the position of the head, sound reaches the closer ear about 1/1500 of a second sooner than the other ear. Also, the sound is a little louder in the closer ear. These differences are recognized and analyzed by the brain to tell you from what direction a sound is coming.*

The Sounds We Hear

Human ears are generally responsive to frequencies from about 50 to 20,000 cycles per second (cps), but some people can hear from about 16 to 30,000, especially during their early years. (Some animals, notably bats and dogs, can hear much higher frequencies.) Hearing ability declines steadily from early childhood, probably because the basilar membrane loses some of its elasticity. In addition, harmful calcium deposits may form, and hair cells begin to degenerate. The human ear is most used to hearing frequencies ranging from 300 to 3000 cps, the approximate range of the human voice. (*Cycles per second* may also be referred to as *hertz*, abbreviated *Hz*.)

Current estimates are that some environmental noises are twice as intense now as they were in the 1960s, and their intensity is expected to continue to double every 10 years. Studies have shown that factory workers have twice as much hearing loss as white-collar office workers. Approximately 10 million Americans use hearing aids, and many of these individuals may have suffered hearing impairment through prolonged exposure to sounds that were not thought of as excessively loud. The accompanying table shows the decibel scale and the range of some common noise sources. A *decibel* represents the relative intensity of a sound, zero decibels being the faintest

sound the average person can hear. Each increment of 10 increases the intensity 10 times (exponentially). The ratio between 140 decibels and 0 decibels is about 100 trillion to 1. Ear damage depends on the length of exposure, as well as the decibel level.

It is interesting that you are more susceptible to hearing damage when you find the sound unpleasant. This is probably why rock musicians suffer less hearing impairment than would be expected. But generally, noise is a stressor, constricting veins and arteries, damaging ear tissue, reducing oxygen and nutrients to the ear, causing high blood pressure, and increasing the heartbeat.

Exposure level	Decibels	Common noise sources	Permissible exposure
Harmful to hearing	140	Jet engine 25 m away Shotgun blast	None (any exposure dangerous)
	130	Threshold of pain Jet takeoff 100 m away Air raid siren	1 min 3 min 5 min
	120	Propeller aircraft Discotheque	 10 min
Possible hearing loss	110	Live rock band Jet takeoff 600 m away	26 min 1 hr
	100	Power mower Jackhammer Subway station with train coming Farm tractor	2 hr 3½ hr
	90	Convertible ride on freeway Motorcycle 8 m away Heavy-duty truck Average street traffic Food blender Heavy traffic 5 m away	8 hr 16 hr
Very noisy	80	Alarm clock Garbage disposal	
Moderately noisy (urban area)	70	Vacuum cleaner Private car Noisy business office	No limit
	60	Conversation Singing birds	
Quiet (surburban and small town)	50	Light traffic 30 m away Quiet business office	
	40	Soft radio music Faucet dripping	
Very quiet	30	Library Soft whisper at 5 m	
	20	Quiet suburban dwelling Broadcasting studio	
	10	Leaves rustling	
	0	Threshold of hearing	

Neural Pathways for Hearing

Axons of neurons with cell bodies in the spiral ganglion of the cochlear nerve extend from the spiral organ and terminate centrally in the dorsal and ventral cochlear nuclei. The cochlear nerve enters the brainstem at the junction of the pons and medulla, and terminates in the cochlear nuclear complex. After entering, each fiber divides into two main branches.

Axons of neurons in the dorsal and ventral cochlear nuclei ascend as crossed and uncrossed fibers in the lateral lemnisci to the inferior colliculus. Thus, the auditory pathways ascend bilaterally. From cell bodies in the inferior colliculus, the pathway continues as the brachium of the inferior colliculus to the medial geniculate body of the thalamus. From cell bodies in the medial geniculate body, axons project via the auditory radiations to the primary auditory cortex (transverse temporal gyrus of Heschl, areas 41 and 42).

A lesion of the auditory pathway on one side results in a decrease of hearing acuity in both ears. This decrease is related to the fact that the ascending auditory pathways are composed of both crossed and uncrossed projections.

Vestibular Apparatus and Equilibrium

Specific parts of the inner ear help the body to cope with changes in position and acceleration. The purpose of this *ves-* *tibular apparatus* (Figure 15.10) is to signal changes in the *motion* of the head (**dynamic equilibrium,** also called *kinetic equilibrium*) and in the *position* of the head with respect to gravity (**static equilibrium,** or *posture*). The main components of the vestibular apparatus are the utricle, saccule, and the three fluid-filled semicircular ducts of the membranous labyrinth. The receptors of the utricle and saccule regulate static equilibrium, and the receptors in the ampullae of the semicircular ducts respond to movements of the head. The equilibrium system also receives input from the eyes and from some proprioceptors in the body, especially the joints. (Try standing on your toes with your eyes closed. Without your eyes to guide your body, you invariably begin to fall forward.)

Specialized proprioceptors of the vestibular apparatus, known as **hair cells,** are arranged in clusters of *hair bundles* (Figure 15.11). Hair cells are extremely sensitive receptors that convert a mechanical force applied to a hair cell into an electrical signal that is relayed to the brain. An extremely slight movement of a hair bundle can cause the hair cells to respond. Two types of sensory hairs are present in hair bundles: (1) *stereocilia,* which are actually modified microvilli, and (2) *kinocilia,* which are modified cilia. Each hair cell has about 100 stereocilia in a tuft, and 1 kinocilium at the edge of the tuft (see Figure 15.11B). Because of its asymmetry, a receptor hair cell is said to be polarized.

The electrical signal triggered by the movement of a hair bundle depends on the direction of the movement. When the hairs of the hair cell are bent in the direction of the kinocilium (known as the *axis of sensitivity,* see Figure 15.8), the hair cell can generate a receptor potential in the nerve ending, synapsing with the hair cell. When the hairs of the hair cell are bent in the direction opposite to the axis of sensitivity, the hair cell is not excited.

Static equilibrium: utricles and saccules The receptor region of the utricles and saccules, called the **macula,** contains receptor hair cells embedded in a jellylike **otolithic** (Gr. "ear stones") **membrane** (Figure 15.12). Loosely attached to the membrane, and piled on top of it, are hundreds of thousands of calcium carbonate crystals called **otoconia** or *statoliths* ("standing stones"). The utricles and saccules are both filled with endolymph. Hair cells in the utricle respond to the motion changes that occur during the straight-line acceleration and deceleration of the head (back-and-forth and up-and-down movements). The hair cells also monitor the position of the head in space, controlling posture. For example, the next time you dive into a pool, notice how you turn and swim upward without having to decide consciously which way is up. This is evidence of your utricles at work, telling you the position of your head in relation to gravity. The utricles are also responsible for initiating the "righting reflex," which we see when a cat is dropped upside down and lands on its feet.

The mechanism of the static equilibrium response depends upon the difference in density between the otoconia and the endolymph inside each utricle and saccule. Otoconia have a greater density than endolymph. As a result, when the head moves in a change of posture, the otoconia resist the external force and lag behind the motion of the endolymph. The otoconia and the otolithic membrane remain relatively still and

FIGURE 15.10

The vestibular system.

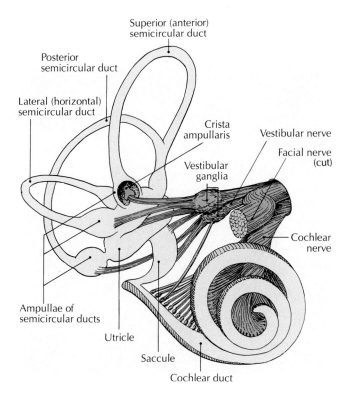

Superior (anterior) semicircular duct

Posterior semicircular duct

Lateral (horizontal) semicircular duct

Crista ampullaris

Vestibular nerve

Facial nerve (cut)

Vestibular ganglia

Cochlear nerve

Ampullae of semicircular ducts

Utricle

Saccule

Cochlear duct

FIGURE 15.11

Hair cells. (A) Scanning electron micrograph of a hair bundle from the inner ear of a bullfrog. Note the single kinocilium. (B) Drawing of a hair cell in cross section. [(A) *Courtesy of Robert S. Kimura, Ph.D., Massachusetts Eye and Ear Infirmary, Boston.*]

Kinocilium

Stereocilia

Microvilli

[A]   ×9000

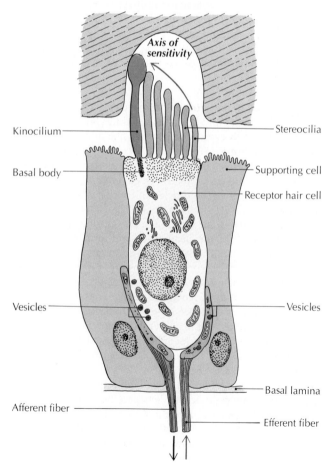

Axis of sensitivity

Kinocilium — — Stereocilia

Basal body — — Supporting cell

— Receptor hair cell

Vesicles — — Vesicles

— Basal lamina

Afferent fiber — — Efferent fiber

[B]

FIGURE 15.12

Macula, receptor region of utricle and saccules. (A) Receptor cells have bundles of hair that project into the gelatinous otolithic membrane; otoconia crystals rest on top of the membrane. (B) Scanning electron micrograph of otoconia crystals. [(B) *Lennart Nilsson*, Behold Man. *Boston, Little, Brown, 1974.*]

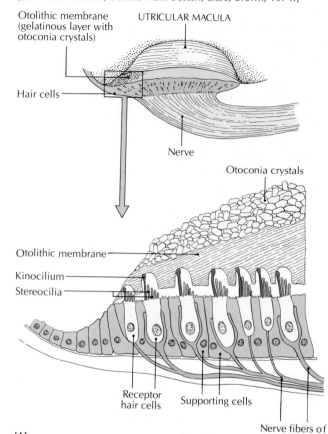

Otolithic membrane (gelatinous layer with otoconia crystals)   UTRICULAR MACULA

Hair cells

Nerve

Otoconia crystals

Otolithic membrane —

Kinocilium —
Stereocilia —

Receptor hair cells   Supporting cells

Nerve fibers of vestibular ganglion

[A]

Otoconia crystals

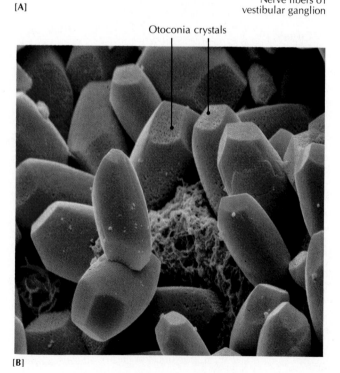

[B]

bend the hairs of the hair cell. Each macular hair cell responds to the gravitational force exerted upon the hairs by the dense otoconia (Figure 15.13). When the head is held horizontally, the gravitational force is directed downward upon the hair bundle. When the head is tilted to the side, the hair bundle of each cell is displaced along the axis of sensitivity, and can excite the afferent nerve fiber. A tilt of the head that bends the hair against the axis of sensitivity has the opposite effect (the hair cell inhibits the vestibular afferent neuron).

The bending of the hair cells alters the permeability of the cells to sodium and potassium ions. The resulting graded generator potential stimulates nerve endings of the vestibular nerve fibers to generate another graded generator potential, and subsequently an action potential (in a vestibular nerve fiber), which is transmitted to the vestibular nuclei in the brainstem and cerebellum (see Figure 15.14).

Evidence suggests that saccules may function like utricles, as well as serving as auditory receptors for low frequencies.

As efficient as human utricles and saccules are, they are not nearly as impressive as the static sensors of some other animals. This lack of efficiency is demonstrated when an airplane pilot does not realize that he or she is flying through clouds upside down. Also, the slowness of the utricle in registering deceleration can be seen in the delayed stumbling response of a standee in a bus when the bus stops suddenly.

Dynamic equilibrium: semicircular ducts The utricles and saccules are *organs of gravitation*, responding to movements of the head in a straight line: forward, backward, up, or down. In contrast, the crista ampullaris of the semicircular ducts responds to changes in acceleration in the *direction of head movements*, specifically turning and rotating. These movements are called angular movements, in contrast to straight-line movements.

Because each of the three ducts is situated in a different plane, at right angles to each other, at least one duct is affected by every head movement. Each duct has a bulge, the

---

FIGURE 15.13

Static equilibrium. (A) Displacement of the hair bundle stimulates the hair cells in the utricle and saccule. (B) As a hair cell responds to the displacement of the hair bundle, nerve impulses are transmitted to the brain along an afferent nerve fiber at the base of the cell. (1) When the hair bundle is displaced, (2) depolarization occurs and spreads through the cell. (3) Depolarization causes selectively permeable channels in the base of the cell to open. Vesicles release their neurotransmitter substance, which diffuses across the synaptic gap between the hair cell and the adjacent neuron, transmitting the impulse from one cell to the next. (4) The message is sent to the brain along an afferent fiber of the vestibulocochlear (VIII) nerve.

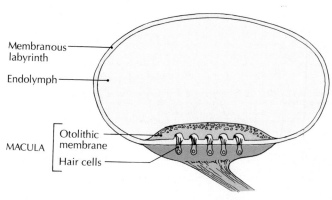

HEAD STATIONARY

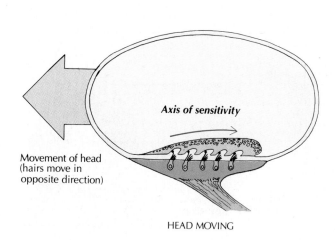

HEAD MOVING

[A]

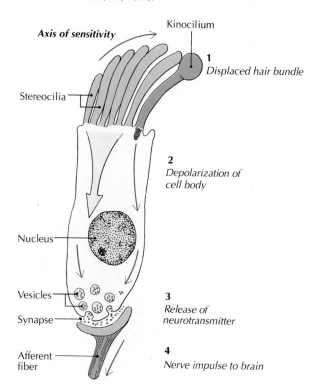

[B]

FIGURE 15.14

Dynamic equilibrium. (A) Displacement of the hair bundle stimulates the hair cells in the crista ampullaris of the semicircular ducts. (B) The ampulla in transparent full view. (C) The ampulla in cross section. Hairs anchored in the crista ampullaris project into the gelatinous cupula.

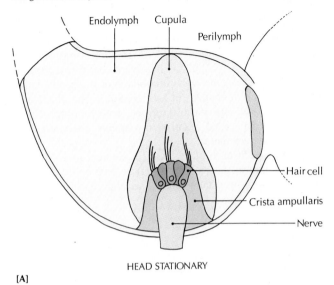

Endolymph   Cupula

Perilymph

Hair cell

Crista ampullaris

Nerve

HEAD STATIONARY

[A]

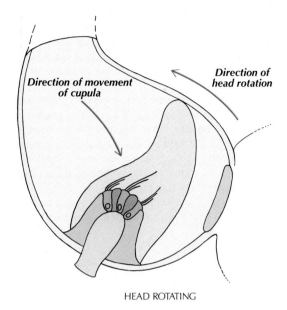

Direction of movement of cupula

Direction of head rotation

HEAD ROTATING

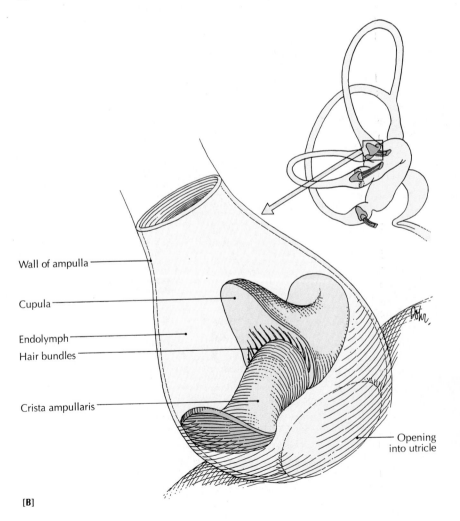

Wall of ampulla

Cupula

Endolymph

Hair bundles

Crista ampullaris

Opening into utricle

[B]

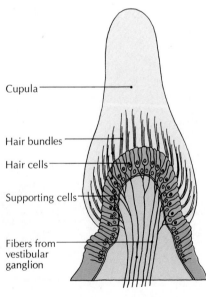

Cupula

Hair bundles

Hair cells

Supporting cells

Fibers from vestibular ganglion

[C]

ampulla, that contains a patch of hair cells and supporting cells embedded in the *crista ampullaris* (Figure 15.14). The hairs of the hair cells project into a gelatinous flap called the *cupula* (KYOO-pyuh-luh; L. little cask or tub). The cupula acts like a swinging door, with the crista as the hinge. The free edge of the cupula brushes against the curved wall of the ampulla (see Figure 15.14). When the head rotates, the endolymph in the semicircular ducts lags behind due to inertia, displacing the cupula and the hairs projecting into it in the opposite direction. (The semicircular ducts do not sense movement at slow, steady speeds because the head and the endolymph move at the same rate.) As a result of the slight displacement of the cupula, the hairs bend in the direction of the axis of sensitivity. This stimulates the nerve endings to generate a graded generator potential, and at the trigger zone, an action potential in each nerve fiber. The brain receives the impulse and signals the appropriate muscles to contract in order to maintain the body's equilibrium.

A feeling of dizziness occurs when you spin about or move violently and then stop suddenly. Because of inertia, the endolymph keeps moving (the way you keep moving forward when your car stops quickly), and it continues to stimulate the hair cells. Although you know you have stopped moving, the signals being sent from your inner ear to your brain make you feel that motion is still occurring, but in the reverse direction. In other words, while in motion the inertia of the endolymph bends the hairs in the direction opposite to the direction of the movement. Immediately after stopping, the endolymph continues to move relative to the ducts. As a result, the hairs are now bent in the direction of the prior movement. Because the inner ears of deaf mutes are not functional, they are immune to dizziness and motion sickness, and rely on visual cues for the maintenance of normal locomotion and posture. (Without visual cues, a swimmer who has lost the use of the labyrinth may navigate down instead of up to reach the surface.)

Neural Pathways for Equilibrium

The vestibular tracts consist of pathways to the brainstem, spinal cord, cerebellum, and cerebral cortex. The 19,000 nerve fibers of each vestibular nerve have their cell bodies in the vestibular ganglion near the membranous labyrinth. The primary vestibular fibers from the vestibular nerve pass into the upper medulla and terminate (1) in each of the four vestibular nuclei in the upper lateral medulla, and (2) in specific regions of the cerebellum.

The sensory signals from the vestibular sensors of the labyrinth are indicators of the position and movements of the head. These inputs from the vestibular receptors are critical in (1) generating compensatory movements to maintain balance and an erect posture in response to gravity, (2) produc-

ing the conjugate (coupled) movements of the eyes that compensate for changes in the position of the constantly moving head, and (3) supplying information for the conscious awareness of position, acceleration, deceleration, and rotation. The vestibular functions are supplemented by proprioceptive inputs from the muscles and joints, as well as the visual system.

Ask Yourself

1 *What are the main parts of the middle ear?*

2 *What is the spiral organ?*

3 *How are sound waves converted into nerve impulses?*

4 *What is the function of otoconia?*

5 *What is the importance of hair cells?*

6 *What is the difference between static equilibrium and dynamic equilibrium?*

VISION

We live primarily in a visual world, and sight is our most dominant sense. The specialized exteroceptors in our eyes comprise about 70 percent of the receptors of the entire body, and the optic nerves contain about one-third of all the afferent nerve fibers carrying information to the central nervous system.

Although we can rely on our eyes to bring us many of the sights of the external world, they are not able to reveal everything. We "see" only those objects that emit or are illuminated by light waves in our receptive range, representing only 1/70 of the entire electromagnetic spectrum. Some organisms, such as insects, are sensitive to shorter wavelengths in the range of ultraviolet, and other organisms can "see" longer wavelengths, in the infrared range of the spectrum.

Light reaches our light-sensitive "film," or *retina*, through a transparent window, the *cornea.* In addition to the basic and accessory structures of the eyeball, vision involves the brain and the optic nerve. We consider first the basic anatomy of the eye, and then the physiology of vision.

Structure of the Eyeball

The human eyeball can be compared to a simple, old-fashioned box camera. Instead of being a box, the eyeball is a sphere about 2.5 cm (1 in.) in diameter. In both cases, light passes through a lens. The external image is brought to a focus on the sensitive retina, which is roughly equivalent to the film in a camera. Over a hundred million specialized neurons convert light waves into electrochemical impulses, which are decoded by the brain. The retina is composed of layers of slender photoreceptors (rods and cones) and a complex of interacting, processing neurons. Some of these sensory neurons send axons to the brain via the optic nerve (Figure 15.15).

Do the vestibular organs work during space travel?	*The utricle and saccule do not work under zero gravity conditions, but the semicircular canals do. About 40 percent of astronauts become motion sick at some time during space travel.*

FIGURE 15.15

The human eye. (A) Horizontal section through the eye. (B) Horizontal section through the anterior portion of the eye (enlarged drawing).

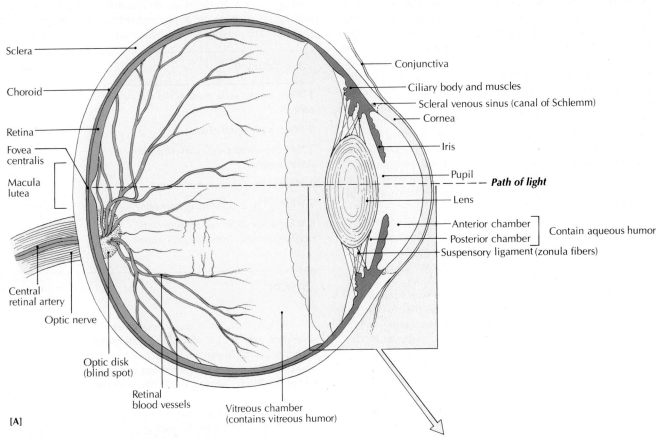

[A]

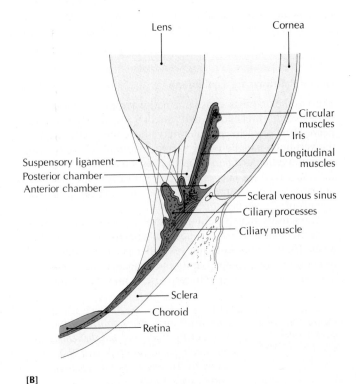

[B]

The wall of the eyeball consists of three coats or layers of tissue: the outer supporting layer, the vascular middle layer, and the inner retinal layer. The eyeball is divided into three cavities: the anterior chamber, the posterior chamber, and the vitreous chamber (Figures 15.15 and 15.16, Table 15.4).

Supporting layer The outer *supporting layer* of the eyeball consists mainly of a thick membrane of tough, fibrous connective tissue. The posterior segment, which comprises five-sixths of the tough outer layer, is the opaque white *sclera* (Gr. *skleros,* hard). The sclera forms the "white" of the eye, giving the eyeball its shape and protecting the delicate inner layers (see Figure 15.16A). The anterior segment of the supporting layer is the transparent *cornea* (L. *corneus,* horny tissue), which comprises the modified anterior one-sixth of the outer layer. The cornea bulges slightly. If you close your eyes, place your finger lightly on your eyelid, and move your eye, you will feel the bulge. Light enters the eye through the cornea. Although the tissues of the sclera and cornea are not identical, the two structures are continuous. The cornea of this layer contains no blood vessels.* The supporting layer

*Cornea transplants from one individual to another have been very successful. (About 30,000 cornea transplants were performed in the United States in 1988, with a success rate of approximately 95 percent.) The typical problem of tissue rejection is usually avoided because the cornea has no blood vessels or lymphatic vessels. As a result, antibodies that cause rejection cannot reach the cornea easily.

FIGURE 15.16

Three layers of the wall of the eyeball. (A) Supporting layer.
(B) Vascular layer. (C) Retinal layer.

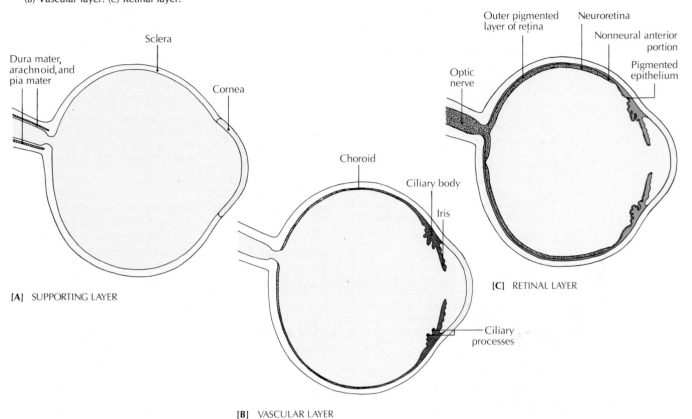

[A] SUPPORTING LAYER

[B] VASCULAR LAYER

[C] RETINAL LAYER

completely encloses the eyeball, except for the posterior portion, where small perforations in the sclera allow the fibers
of the optic nerve to leave the eyeball on their way to the
brain.

Vascular layer Because the middle layer of the eyeball
contains many blood vessels, it is called the **vascular layer**
(see Figure 15.16B). The dark color of the middle layer is
produced by pigments that help to lightproof the wall of the
eye by absorbing stray light and reducing reflection. The posterior two-thirds of the vascular layer consists of a thin membrane called the **choroid,** which is essentially a layer of blood
vessels and connective tissue sandwiched between the sclera
and the retina (see Figure 15.16B).

The vascular layer becomes thickened toward the anterior
portion to form the **ciliary body** (see Figure 15.16B). Extending inward from the ciliary body are the fine *ciliary
processes.* The smooth muscles in the ciliary body *(ciliary muscles)* contract to ease the tension on the suspensory ligament
of the lens, which consists of fibrils that extend from the
ciliary processes to the lens. The **lens** of the eye is a flexible,
transparent, colorless, avascular body of epithelial cells behind the iris, the colored part of the eye. The lens is held in
place by the *suspensory ligament of the lens,* and by the ciliary
processes. The shape of the lens can be adjusted so that objects at different distances can be brought into focus on the

retina. This mechanism is called *accommodation* (see p. 462).
The lens loses much of its elasticity with aging, making it
difficult to focus efficiently without corrective eyeglasses.

The anterior extension of the choroid is a thin muscular
layer called the **iris** (Gr. rainbow) because it is the colored
part of the eyeball that can be seen through the cornea. In
the center of the iris is an adjustable circular aperture, the
pupil (L. doll), so called because when you look into someone else's eyes, you can see a reflected image of yourself that
looks like a little doll. The pupil appears black because most
of the light that enters the eye is not reflected outward. The
iris, acting as a diaphragm, is able to regulate the amount of
light entering the eye because it contains smooth muscles that
contract or dilate in an involuntary reflex in response to the
amount of light available, causing the pupil to become larger
or smaller. The smaller the pupil, the less light entering the
eye. This mechanism is called *adaptation* (see p. 466). The
pupil may shrink with age, causing the iris to appear lighter.
This is why the eyes of some elderly people appear to be
lighter than they used to be.

The next time you are standing in bright sunshine with a
friend, notice how his or her pupils constrict to tiny black
dots, protecting the delicate retina from excessive light. In
contrast, when you enter a darkened movie theater your pupils dilate greatly in an effort to increase the amount of light
reaching the retina.

Structure	Description
SUPPORTING LAYER OF EYEBALL	
Sclera	Opaque layer of connective tissue over posterior five-sixths of outer layer of eyeball; "white" of the eye. Gives eyeball its shape, protects inner layers. Perforated to allow optic nerve fibers to exit.
Cornea	Transparent anterior portion of outer layer of eyeball. Light enters eye through cornea, a nonadjustable lens.
VASCULAR LAYER OF EYEBALL	
Choroid	Thin membrane of blood vessels and connective tissue between sclera and retina. Posterior two-thirds of vascular layer of eyeball.
Ciliary body	Thickened vascular layer in anterior portion of eyeball. Ciliary muscles help lens to focus by either increasing or decreasing tension on suspensory ligament of lens. Produces aqueous humor and some elements of vitreous humor.
Ciliary processes	Inward extensions of ciliary body. Help hold lens in place.
Lens	Elastic, colorless, transparent body of epithelial cells behind iris. Shape modified to focus on subjects at different distances (accommodation) through action of ciliary muscles. The lens is adjustable.
Iris	Colored part of eye. Thin, muscular layer; anterior extension of choroid. Size of pupil, and thus amount of light entering eye, regulated by the degree of constriction and relaxation of iris by reflexive activity of smooth muscle.
Pupil	Adjustable circular opening in iris. Opens and closes reflexively relative to amount of light available (adaptation).
RETINAL LAYER OF EYEBALL	
Retina	Multilayered, light-sensitive membrane; innermost layer of eyeball. Connected to brain by circuit of neurons in optic nerve. Consists of neural layer and pigmented layer, which prevents reflection from back of retina. Receives focused light waves, transduces them into nerve impulses that the brain converts into visual perceptions.

Structure	Description
Rods	Specialized photoreceptor cells, not color-sensitive. Respond in dim light and very light-sensitive. Function in peripheral vision.
Cones	Specialized color-sensitive photoreceptor cells, concentrated in fovea. Associated with visual acuity. Not as light-sensitive as rods.
Fovea (fovea centralis)	Depressed area in center of retina containing only cones. Area of most acute image formation and color vision.
CAVITIES OF EYEBALL	
Aqueous chambers Anterior chamber	Anterior to iris, posterior to cornea (between cornea and iris). Contains aqueous humor.
Posterior chamber	Lies between iris and lens. Contains aqueous humor.
Vitreous chamber	Largest cavity, fills entire space behind lens. Contains vitreous humor.
ACCESSORY STRUCTURES OF EYE	
Bony orbit	Bony socket surrounding and protecting eyeball. Composed of portions of several skull bones.
Eyelids	Folds of skin, forming almond-shape around open eye. Protect eye; glands in eyelid keep cornea moist.
Eyelashes	Short, thick hairs on eyelids. Prevent foreign materials from entering eye. Secretions from modified sebaceous glands keep eyelids from sticking together.
Eyebrows	Thickened ridges of skin over protruding frontal bone; covered with short, flattened hairs. Protect eye from perspiration, excessive light, foreign materials.
Conjunctiva	Transparent mucous membrane lining inner eyelid, extending over surface of eyeball and terminating at cornea.
Lacrimal apparatus	Consists of lacrimal gland, lacrimal ducts, lacrimal sac, nasolacrimal duct. Secretions from lacrimal gland keep eye moist, combat bacteria with a bactericidal enzyme, and distribute nutrients. Gland located laterally under upper eyelid. Tears flow from gland across cornea to sac (located medial to eye) to nasolacrimal duct leading to nasal cavity.

FIGURE 15.17

[A]

[B]

The retina. (A) Diagram of a section through the retina. Light first passes through the vitreous humor, the jellylike substance that fills the vitreous chamber of the eyeball. The light passes through several layers of cells before reaching the light-sensitive rods and cones. Beyond the rods and cones is the pigmented epithelial layer, which absorbs stray light, and prevents reflection from the back of the retina. When light energy stimulates a rod or cone, that energy is converted into the electrical energy of a receptor potential. The stimulus is sent from the receptor cells through an intermediate set of bipolar cells and finally to ganglion cells, whose axons form the optic nerve. (Other processing neurons are the horizontal and amacrine neurons.) (B) Scanning electron micrograph of a retina, showing the pigmented epithelial layer, outer and inner segments of rods, outer nuclear layer, and bipolar cells. [*Gene Shih and Richard Kessel*, Living Images: Biological Microstructures Revealed by SEM. *Boston, 1982, Science Books International.*]

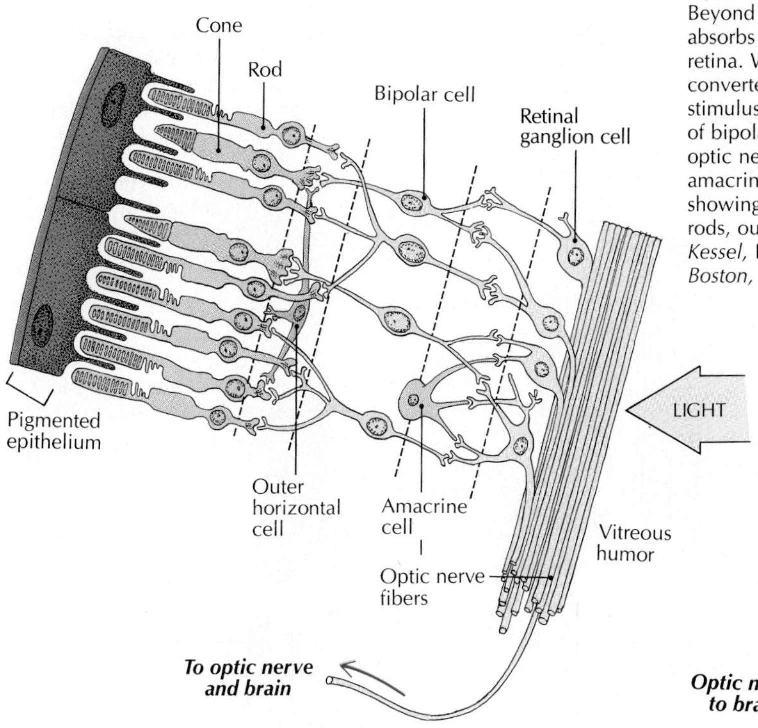

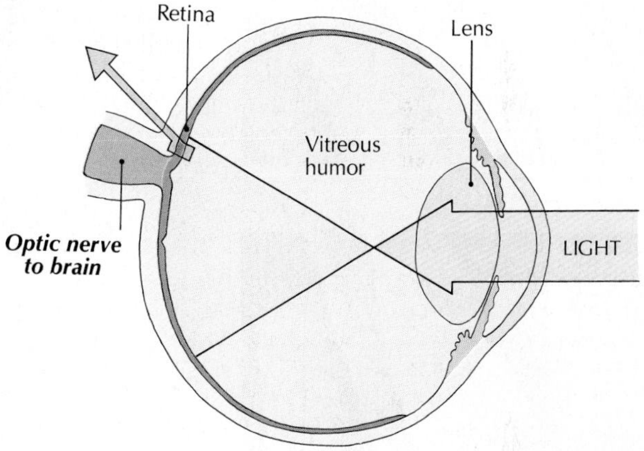

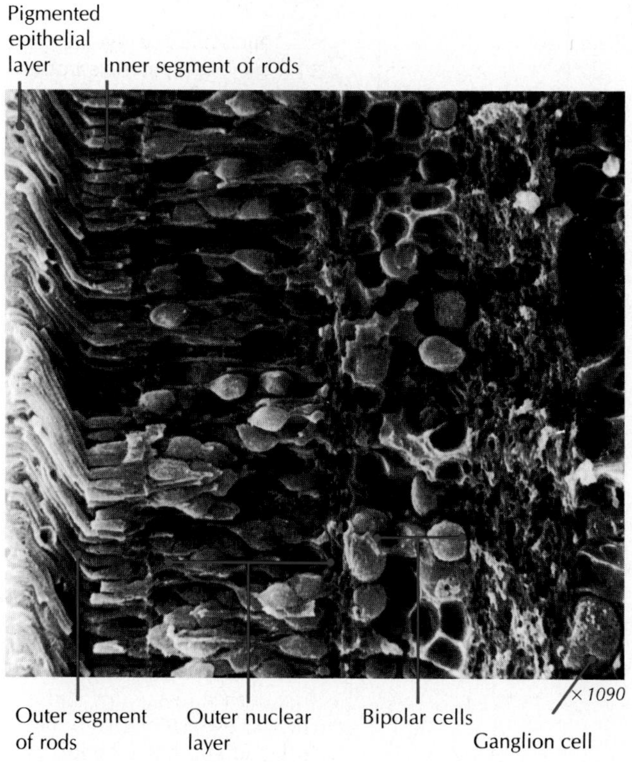

Retinal layer The innermost layer of the eyeball is the *retina* (REH-tin-uh; L. *rete,* net), an egg-shaped, multilayered, light-sensitive membrane containing a network of specialized nerve cells (Figure 15.17; see also Figure 15.25). It is connected to the brain by a circuit of over a million neurons in the optic nerve. The retina has a thick and a thin layer. The thick layer is nervous tissue, called the *neuroretina,* that connects with the optic nerve. Behind it is a thin layer of pigmented epithelium that prevents reflection from the back of the retina. The pigmented layer, along with the choroid, actually absorbs stray light (light that is not used by the photoreceptor cells) and prevents reflection back to the neuroretina. Stray light in the eye can restimulate the photoreceptors. Albinos, who have no eye pigment, are abnormally sensitive to light because the stray light is not absorbed by pigment.

The function of the neuroretina is to receive focused light waves and convert them into nerve impulses that can be projected to the brain and converted into visual perceptions. The neuroretina does not extend into the anterior portion of the eyeball, where light could not be focused on it (see Figure 15.16c).

FIGURE 15.18

Rods and cones. (A) Scanning electron micrograph of rods. (B) Scanning electron micrograph of cones. (C) Detailed drawing of rod cell. The outer segment of each rod is composed mainly of approximately 2000 disks stacked in an orderly pile. The disks contain most of the light-absorbing protein molecules that initiate the generation of a nerve impulse. (D) Detailed drawing of a cone cell. The shapes of the rods and cones give them their names.
[(A *and* B) *E. R. Lewis, Y. Y. Zeevi, and F. S. Werblin,* Brain Research *15 (1969): 559–562.*]

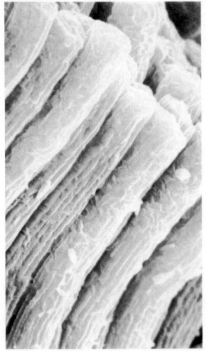

[A]

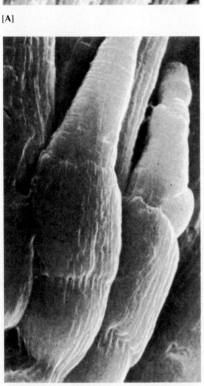

[B]

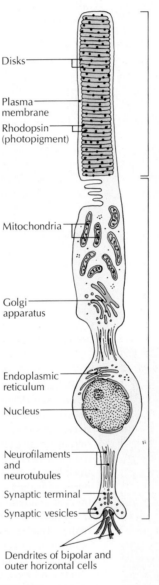

Disks

Plasma membrane

Rhodopsin (photopigment)

OUTER SEGMENT

Mitochondria

Golgi apparatus

INNER SEGMENT

Endoplasmic reticulum

Nucleus

Neurofilaments and neurotubules

Synaptic terminal

Synaptic vesicles

Dendrites of bipolar and outer horizontal cells

[C] ROD

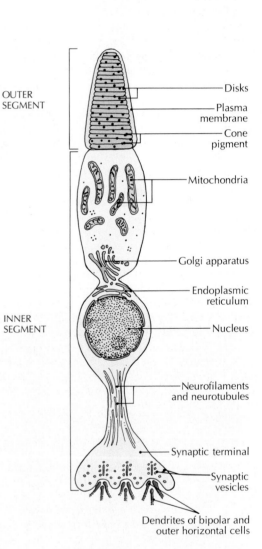

Disks

Plasma membrane

Cone pigment

Mitochondria

Golgi apparatus

Endoplasmic reticulum

Nucleus

Neurofilaments and neurotubules

Synaptic terminal

Synaptic vesicles

Dendrites of bipolar and outer horizontal cells

[D] CONE

The neuroretina consists of highly specialized photoreceptor nerve cells, the *rods* and *cones* (Figure 15.18). The outer segment of a rod or cone contains most of the elements necessary (including light-sensitive photopigments) to absorb light and produce a generator potential. The inner segment contains mitochondria, Golgi apparatuses, endoplasmic reticulum, the nucleus, and other organelles necessary for generating energy and renewing molecules in the outer segment. The inner segment also contains a synaptic terminal, which allows the photoreceptor cells to communicate electrochemically with other retinal cells.

In addition to rods and cones, the neuroretina contains several other cells, which are actually neurons. These include bipolar cells, outer horizontal cells, amacrine cells (inner horizontal cells), and ganglion neurons. These neurons form the complex neuronal circuitry for the processing of light waves within the retina (see Figure 15.17). The ganglion neurons contain axons that leave the eye and comprise the nerve fibers of the optic nerves, optic chiasma, and optic tracts. (See p. 468 for a more detailed description of the neural cells of the retina.)

Each eye has about 125 million rods and 7 million cones. Most of the cones are concentrated in the center of the retina directly behind the lens in an area called the *macula lutea* (L. "yellow spot"), especially in a small depressed rod-free area called the *fovea*, or *fovea centralis* (FOE-vee-uh; L. small pit) (see Figure 15.15). The rods, and some cones, are located in the remainder of the retina, called the *peripheral retina*. The rods are the sensors for the perception of black-to-white shades, and the cones are the sensors for the perception of color. Night vision is almost totally rod vision, since the color-sensitive cones require 50 to 100 times more stimulation than rods do. (In order to distinguish the functions of the rods and cones, remember that the *c* in *cone* can stand for *color*.)

The cones and their neural connections constitute the *cone system,* and the rods and their neural connections make up the *rod system.* The rod system is highly sensitive to light but has a low visual acuity, which makes it difficult to recognize small objects. The cone system is not as sensitive to light; it is coded for red, green, and blue color perception and requires good illumination. In the fovea, it has high visual acuity. There are three types of color cones: *red* cones are responsive in the red region of the visual spectrum, *green* cones are re-

sponsive in the green region, and *blue* cones are responsive in the blue region.

Vision is sharpest, and color perception is optimal, on the fovea of the macula lutea. The nonmacular retina (for peripheral vision) is sensitive to weak light intensities, and is associated with black and white vision. In normal light we can see best by looking directly at an object, so that the image falls on the cones in the fovea. However, in poor light we can see best by *not* looking directly at an object (as one does when star gazing), so that the image falls on the light-sensitive rods, located on the nonmacular portion of the retina. The portion of the retina where the optic nerve exits from the eyeball contains neither rods nor cones, and is called the *blind spot* or optic disk.

Cavities of the eyeball The eyeball is divided into three cavities. The region between the cornea and iris is the **anterior chamber** (see Figure 15.15). The **posterior chamber** lies between the iris and lens. Both chambers are filled with *aqueous humor,* a thin, watery fluid that is essentially an ultrafiltrate of blood similar to cerebrospinal fluid. Aqueous humor is largely responsible for maintaining a constant pressure within the eyeball. It also provides such essential nourishment as oxygen, glucose, and amino acids for the lens and cornea, which do not have blood vessels to nourish them. Aqueous humor is produced by capillaries in the ciliary body. It passes through the posterior and then the anterior chamber and diffuses into a drainage vein called the *scleral venous sinus* (canal of Schlemm) at the base of the cornea.

The third and largest cavity of the eyeball is the **vitreous chamber,** which occupies about 80 percent of the eyeball. It fills the entire space behind the lens. This chamber contains *vitreous humor,* a gelatinous substance with the consistency of raw egg white. The humor is actually a modified connective tissue. Its function is to keep the eyeball from collapsing as a result of external pressure. Except for the addition of collagen and hyaluronic acid, the chemical composition is similar to that of aqueous humor. The vitreous humor also provides another source of nourishment for the lens and possibly the retina. The vitreous humor is formed by the ciliary body.

Accessory Structures of the Eye

Most of the accessory structures of the eye are either protective devices or muscles. They include the bony orbits, eyelids, eyelashes, eyebrows, conjunctiva, lacrimal (tear) apparatus, and muscles that move the eyeball and eyelid (Figure 15.19).

Why do most newborn babies have blue eyes?

Eye color depends upon the number and placement of pigment cells (melanocytes) in the eye. Darker eye color is a result of a greater concentration of melanocytes. At birth, melanocytes are still being distributed in the eyes; they first appear at the back of the iris. However, the eyes appear brown only when melanocytes are deposited in the anterior part of the iris, in front of the muscles of the iris. This deposition does not occur until a few months after birth. Babies of dark-skinned, dark-eyed parents are usually born with dark eyes.

Why do we sometimes see ''spots'' in front of our eyes?

Such ''spots'' are called muscae volitantes, *which is Latin for ''flying flies.'' They are actually the shadows of either red blood cells that have escaped from the capillaries in the retina, or of collagenous particles. These particles and cells move slowly through the vitreous humor of the eye. Their presence is normal and harmless.*

Embryonic Development of the Eye

The eyes develop from three embryonic sources: neuroectoderm, surface ectoderm, and mesoderm. The first sign of eye development appears in the 22-day embryo when a pair of shallow optic sulci (grooves) form from the neuroectoderm in the forebrain at the cephalic end of the embryo (drawing A). The sulci develop into a pair of lateral outpockets called **optic vesicles** (drawing B). The distal portion of each optic vesicle dilates, while the proximal portion remains constricted to form the **optic stalk.** Eventually, axons of ganglion cells elongate and extend into the optic stalk to become the *optic nerve.* The optic vesicle soon reaches the surface ectoderm and induces* it to thicken into a **lens placode** (drawing C). By the beginning of the fifth week, the lens placode, in turn, induces each optic

*The principle of *induction* occurs when one tissue is caused, or *induced,* to change its developmental pattern as a result of the influence of a different tissue. These changes are probably produced by chemical stimuli.

vesicle to turn in on itself (invaginate), forming a double-layered **optic cup** (drawing D). The internal layer of the optic cup differentiates into the complex *neural layer* (neuroretina) of the retina (which includes rods, cones, bipolar cells, horizontal cells, amacrine cells, and ganglion cells), while the external layer develops into the *pigmented epithelium* of the retina. The retina is well developed at birth except for the central foveal region, which completes its development by the fourth month after birth.

The optic cup induces the lens placode to invaginate, and to form a **lens vesicle** that is partially surrounded by the optic cup (drawing D). The vesicle sinks below the outer surface and pinches off like a submerged bubble. This "bubble," now almost completely surrounded by the optic cup, forms the *lens,* while the surface ectoderm becomes the thin surface epithelial layer of the *cornea* (drawings E and F). The *iris* eventually forms from the rim of the optic cup, which

partially covers the lens.

Linear grooves called **optic fissures** develop on the inferior surface of the optic cups and optic stalks. Blood vessels develop within these fissures. The hyaloid artery supplies the optic cup and lens vesicle, and the hyaloid vein returns blood. The hyaloid vessels are eventually enclosed within the optic nerve when the edges of the optic fissure fuse. The open end of the optic cup now forms a circular opening, the future *pupil.* The distal portions of the hyaloid vessels eventually degenerate, but the proximal portions become the central artery and vein of the retina.

The optic cup is surrounded by mesenchymal (mesoderm) cells that differentiate into an inner vascular layer, comprising the *choroid, ciliary body, iris,* and an outer fibrous layer comprising the *sclera* and *cornea.* The choroid is continuous with the arachnoid layer of the brain, and the sclera is continuous with the dura mater.

Bony orbit The eye is enclosed in a socket, or **bony orbit,** which protects it from external buffeting (see Chapter 7). The floor of the orbit is composed of parts of the maxilla, zygomatic, and palatine bones. The roof is composed of the orbital plate of the frontal bone and the lesser wing of the sphenoid. Several openings in the bones of the orbit allow the passage of nerves and blood vessels. Between the bony orbit and the eyeball is a layer of fatty tissue that cushions the eyeball and permits its smooth rotation.

Eyelids, eyelashes, eyebrows The *eyelids* (palpebrae) are folds of skin that create an almond-shaped opening around the eyeball when the eye is open. The points of the almond, where the upper and lower eyelids meet, are called *canthi* (KAN-thigh). The medial (or inner) canthus is the one closest to the nose, and the lateral (or outer) canthus is the point closest to the ear. The eyelid may be divided into four layers: (1) a skin layer contains the eyelashes, (2) a muscular layer contains the orbicularis oculi muscle, which lowers the eyelid to close the eye, (3) a fibrous connective tissue layer contains many modified sebaceous glands, whose secretions keep the eyelids from sticking together, and (4) the innermost layer is composed of a portion of the lining of the eyelid, the conjunctiva.

Eyelids protect the eyeball from dust and other harmful external objects. In addition, the periodic blinking of the eyelids sweeps glandular secretions (tears) over the eyeball, keeping the cornea moist. During sleep, the closed eyelids prevent evaporation of the secretions. They also protect the eye by closing reflexively when an external object threatens the eye, as when a piece of paper is suddenly blown toward your face.

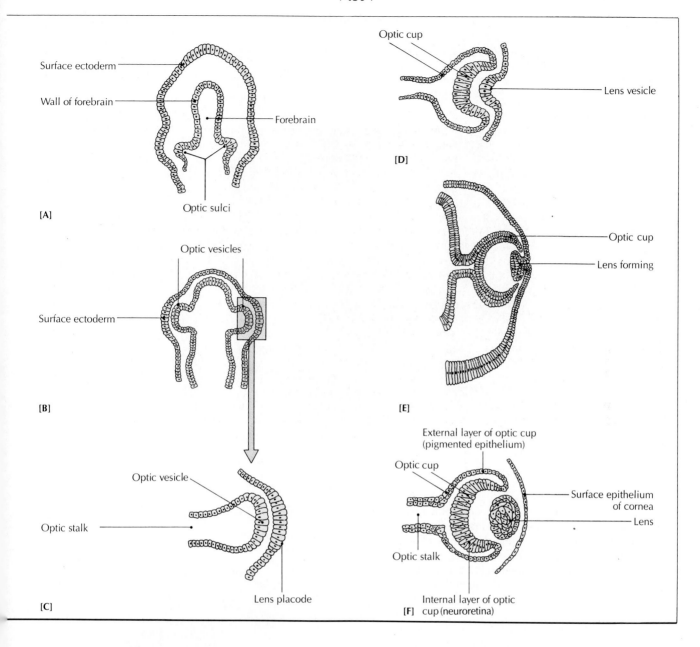

The edges of the eyelids are lined with short, thick hairs, the *eyelashes*. The eyelids of each eye contain about 200 eyelashes. Each eyelash lasts three to five months before it is shed and replaced. Eyelashes are the only hairs that do not whiten with age. Eyelashes act as strainers to prevent foreign materials from entering the eye. *Eyebrows* are thickened ridges of skin over the protruding frontal bone, covered with short, flattened hairs. They protect the eye from perspiration, excessive sunlight, and foreign materials, and also help to absorb the force of blows to the eye and forehead.

Conjunctiva The *conjunctiva* (L. connective) is a thin, transparent mucous membrane that lines the eyelids and bends back over the surface of the eyeball, terminating at the transparent cornea, which is uncovered. The portion that lines the eyelid is the *palpebral conjunctiva,* and the portion that covers the white of the eye is the *bulbar conjunctiva.* Between both portions of the conjunctiva are two recesses called the *conjunctival sacs.* The looseness of the sacs makes movement of the eyeball and eyelid possible. Your ophthalmologist usually pulls back your lower eyelid to place eyedrops in the inferior conjunctival sac.

Lacrimal apparatus The *lacrimal apparatus* (LACK-ruh-mull; L. "tear") is made up of the lacrimal gland, lacrimal sac, lacrimal ducts, and nasolacrimal duct (Figure 15.20). The eyeball is kept moist by the secretions of the *lacrimal gland,* or tear gland, located under the upper lateral eyelid and extending inward from the outer canthus of each eye.*

*The lacrimal glands of infants take about four months to develop fully. As a result, the eyes of a newborn baby should be protected from dust, bright light, and other irritants.

FIGURE 15.19

The external right eye, as seen from the front, showing some accessory structures.
[*Martin Dohrn/Science Photo Library/Photo Researchers.*]

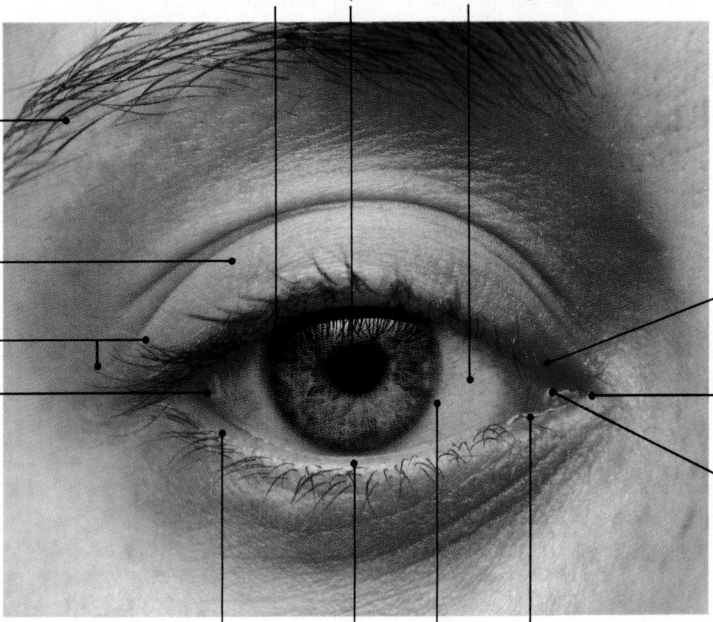

Iris Pupil

Bulbar conjunctiva covering sclera

Eyebrow

Upper eyelid (palpebra)

Eyelashes

Lateral canthus

Lacrimal punctum on upper lacrimal papilla

Medial canthus

Plica semilunaris

Lower eyelid (palpebra) Palpebral conjunctiva Sclerocorneal junction Lacrimal punctum

The eye blinks every 2 to 10 seconds, with each blink lasting only 0.3 to 0.4 seconds. Blinking stimulates the lacrimal gland to secrete a sterile fluid that serves at least four purposes. (1) It washes foreign particles off the eye. (2) It kills invading bacteria with a mild antibacterial enzyme, lysozyme. (3) It distributes water and nutrients to the cornea and lens. (4) It gives the eyeball a clear, moist, and smooth surface. Tears are composed of salts, water, and mucin (organic compounds produced by mucous membranes). In addition to the steady secretion of tears, *reflex* tears are produced in emergencies, as when the fumes from a sliced onion irritate the eyes.

Approximately 3 to 12 *lacrimal ducts* lead from each gland onto the superior conjunctival sac in the upper eyelid. From there tears flow down across the eye into small openings near the inner canthus called *lacrimal puncta*. The puncta open into the *lacrimal ducts,* which drain excess tears from the area of the inner canthus to the *lacrimal sac,* the dilated upper end of the *nasolacrimal duct* (see Figure 15.20). The nasolacrimal duct is a longitudinal tube that delivers excess tears into the nasal cavity.

Ordinarily, tears are carried away by the nasolacrimal duct to the nasal cavity, but when a person is crying or has conjunctivitis or hay fever, the tears form faster than they can be removed. In such cases, tears run down the cheeks, and the nasal cavity becomes overloaded. This is why you have to blow your nose when you cry. Also, a watery fluid sometimes flows out of the nose after blowing it. This fluid is tears flowing out of the unplugged nasolacrimal duct.

FIGURE 15.20

The lacrimal apparatus.

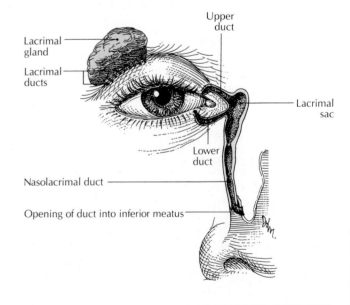

Upper duct

Lacrimal gland

Lacrimal ducts

Lacrimal sac

Lower duct

Nasolacrimal duct

Opening of duct into inferior meatus

What causes "bloodshot eyes"?

Although the bulbar conjunctiva is normally colorless, its blood vessels can become dilated and congested by infection or external irritants such as smoke. "Bloodshot eyes" is the result.

Muscles of the eye and eyelid A set of six muscles moves the eyeball in its socket. The action of these muscles is described in Table 15.5 (see also Figure 10.7). The muscles are the four *rectus muscles* and the *superior* and *inferior oblique muscles*. They are called extrinsic or *extraocular* muscles because they are outside the eyeball (*extra* = outside). One end of each muscle is attached to a skull bone, and the other end is attached to the sclera of the eyeball. The extraocular muscles are coordinated and synchronized so that both eyes move together in order to center on a single image. These movements are called the *conjugate movements* of the eyes.

Other muscles move the eyelid. The *orbicularis oculi* lowers the eyelid to close the eye, and the *levator palpebrae superioris* raises the eyelid to open the eye. The *superior tarsal* (Muller's) muscle is a smooth muscle innervated by the sympathetic nervous system. It helps to raise the upper eyelid, and when it is paralyzed (as in Horner's syndrome, see Chapter 14) it causes a slight drooping (ptosis) of the upper eyelid.

Inside the eyes are several smooth *intrinsic muscles*. The *ciliary muscle* eases tension on the suspensory ligaments of the lens and allows the lens to change its shape in order for the eye to focus (accommodate) properly. The *circular muscle* of the iris contracts the pupil, and the *radial muscle* dilates it.

Physiology of Vision

The visual process can be subdivided into five phases:

1 Refraction of light rays entering the eye.

2 Focusing of images on the retina by accommodation of the lens and convergence of the images.

3 Conversion of light waves by photochemical activity into neural impulses.

4 Processing of neural activity in the retina, and transmission of coded impulses through the optic nerve.

5 Processing in the brain, culminating in perception.

Let us follow the process through each phase in more detail.

Refraction Light waves travel parallel to each other, but they bend when they pass from one medium to another with a different density.* Such bending is called **refraction**. Light waves that enter the eye from the external air are refracted, so that they converge at the retina as a sharp, focused point called the *focal point* (Figure 15.21).

Before light reaches the retina, it passes through (1) the cornea, (2) the aqueous humor of the anterior chamber between the iris and lens, (3) the lens, and (4) the gelatinous vitreous humor in the vitreous chamber behind the lens. Refraction takes place as the light passes through both surfaces of the cornea (which is a convex, nonadjustable lens) and

*Fishermen have learned that when they try to grab a fish swimming below the surface, they must reach a little to the side of the image to compensate for the bending of light waves from air to water and vice versa.

TABLE 15.5   ACTIONS OF THE PAIRED EYE MUSCLES

Muscle	Action
SKELETAL MUSCLES	
Medial rectus	Rotates eyeball inward.
Lateral rectus	Rotates eyeball outward.
Superior rectus	Rotates eyeball upward, inward.
Inferior rectus	Rotates eyeball downward, inward.
Superior oblique	Rotates eyeball downward, outward.
Inferior oblique	Rotates eyeball upward, outward.
Orbicularis oculi	Lowers eyelid (closes eye).
Levator palpebrae superioris	Raises eyelid (opens eye).
SMOOTH MUSCLES	
Ciliary muscle	Eases tension on suspensory ligament of lens, permits focusing (accommodation).
Circular muscle of iris	Contracts pupil.
Radial muscle of iris	Dilates pupil.
Superior tarsal muscle of upper eyelid	Raises eyelid.

again as it passes through the anterior and posterior surfaces of the lens (which is a convex, adjustable lens).

A normal eye can bring distant objects more than 6 m (20 ft) away to a sharp focus on the retina. When parallel light rays are focused exactly on the retina, and vision is perfect, the condition is called *emmetropia* (Gr. "in measure"). Nearsightedness, or *myopia* (Gr. "contracting the eyes"), occurs when light rays come to a focus *before* they reach the retina.* As a result, when the rays do reach the retina, they form an unfocused circle instead of a sharp point, and distant objects appear blurred (see Figure 15.21B). Farsightedness, or *hypermetropia* (Gr. "beyond measure"), occurs when light rays are focused *beyond* the retina, and as a result near objects appear blurred (see Figure 15.21C).†

Both myopia and hypermetropia can be corrected by wearing prescription eyeglasses or contact lenses, which are specially ground lenses placed in front of the eye to change the angle of refraction.

Myopia is so named because a nearsighted person often squints through narrowed eyelids in an effort to focus better. Although the resultant tiny opening requires little or no focusing, the amount of light entering the eye is decreased, and strain on the relevant eye muscles may cause headaches.

†Some textbooks show a corrective lens for myopia as)) , and a corrective lens for hypermetropia as)) . Such lenses are capable of refracting light rays, according to the fundamental laws of physics, but are actually never used to correct vision defects. The corrective lenses shown in Figures 15.21B and C are meant to approximate the actual shapes of corrective *eyeglass* lenses. In any given lens, the relationship of one curve to another changes depending on the specific prescription, but the basic shapes for correcting myopia and hypermetropia remain the same, as shown in Figures 15.21B and C, with the center of the lens being the thinnest point for myopia, and the thickest point for hypermetropia.

FIGURE 15.21

Refraction. (A) In a normal (emmetropic) eye, light rays come to a focal point exactly on the retina. (B) In a myopic (nearsighted) eye, the focal point falls *in front* of the retina. This may be caused by an elongated eyeball or a thickened lens. Corrective lenses placed in front of the eye make the focal point fall on the retina. (C) In a hypermetropic (farsighted) eye, the focal point falls *behind* the retina. This may be caused by a shortened eyeball or a thinned lens. Corrective lenses redirect the light rays to produce a focal point on the retina.

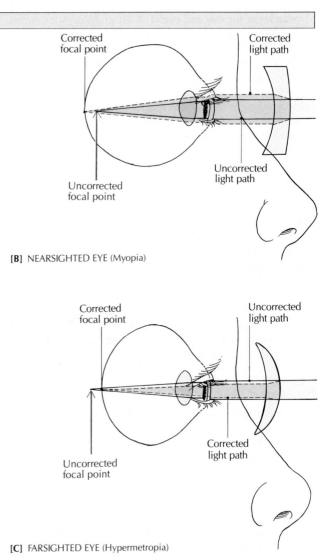

[B] NEARSIGHTED EYE (Myopia)

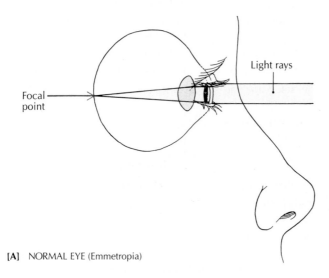

[A] NORMAL EYE (Emmetropia)

[C] FARSIGHTED EYE (Hypermetropia)

Astigmatism (Gr. "without focus") occurs when the curvature of the cornea or lens is not uniform. As a result, part of the image formed on the retina is unfocused. This condition can usually be corrected with lenses that have greater bending power in one axis than in others.

Accommodation Images from an object less than 6 m (20 ft) away would normally be focused behind the retina instead of on it. To bring images into perfect focus on the retina is the role of the adjustable lens. This is accomplished by a reflex called *accommodation.* When you want to focus your eyes on an object close to you (near-sight vision), you involuntarily contract the ciliary muscles in your eye, which pull the ciliary body slightly forward and inward, reducing the tension on the suspensory ligaments attached to the lens capsule. When the tension is reduced, the elastic lens becomes thicker (rounds up). The rounder lens is able to focus on a close object (Figure 15.22).

When distant objects are viewed, the ciliary muscles relax. As they relax, the tension on the suspensory ligaments becomes greater, so that the lens becomes thinner (flattened) (see Figure 15.22B). Looking at a distant building or tree is restful to your eyes when you are doing long-term close work because your ciliary muscles are relaxed.

As the lens becomes harder and less elastic with age, it becomes more and more difficult to focus on near objects, a condition known as *presbyopia* (Gr. *presbus,* old man + eye). In old age, the lens also gradually acquires a cloudy, yellow tint. When you were a young child you probably held a book

about 7 cm (3 in.) from your eyes.* By the time you are 40 years old, you will probably be holding your book about 16 cm (6 in.) or more away, and by 60 the distance may increase to about 1 m (39 in.). At that age, if not sooner, corrective convex lenses will probably be necessary to focus on close objects.

Convergence Human beings have *binocular vision,* meaning that although we have two eyes, we perceive only one image. Each eye receives an image from a slightly different angle, which creates the impression of distance, depth, and three-dimensionality. For this reason, binocular vision is sometimes called *stereoscopic vision* (Gr. *stereos,* solid + *skopein,* to see).

*Parents, who forget what it was like to read a book at ten years old, are constantly complaining that their children are ruining their eyes by reading in poor light and by holding the book too close. In fact, because of the acute sensitivity of their rods and cones and the elasticity of their lenses, healthy children are not hurting their eyes at all.

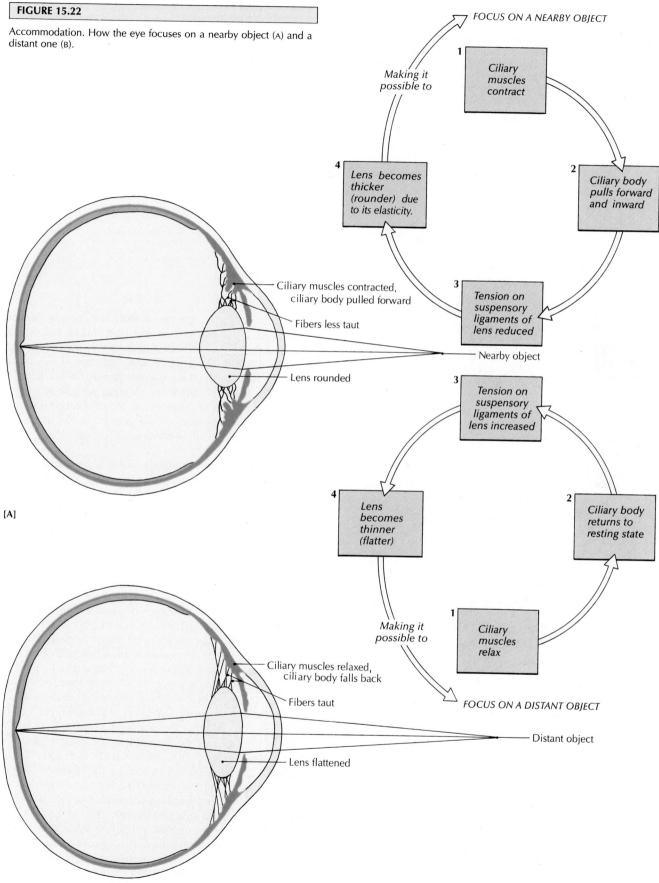

FIGURE 15.22

Accommodation. How the eye focuses on a nearby object (A) and a distant one (B).

FIGURE 15.23

Isomers of retinal. (A) The molecular structure of 11-*cis*-retinal is altered by light to convert it into all-*trans*-retinal. After the cycle of light transmission and perception is completed, all-*trans* is converted back to 11-*cis* retinal by enzymatic action. (B) A three-dimensional representation of how 11-*cis* retinal changes its shape when it absorbs light. In the top drawing, the hydrogen atoms (blue) attached to carbons-11 and -12 are on the same side of the carbon backbone, causing the backbone to bend. When light is absorbed, it causes the backbone to rotate between carbons-11 and -12, straightening out the chain to form all-*trans* retinal (bottom).

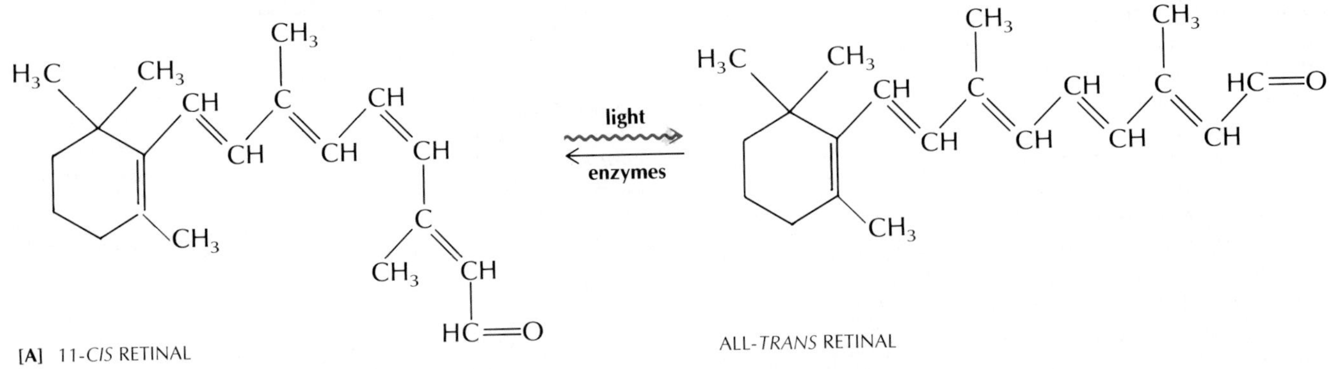

[A] 11-*CIS* RETINAL

ALL-*TRANS* RETINAL

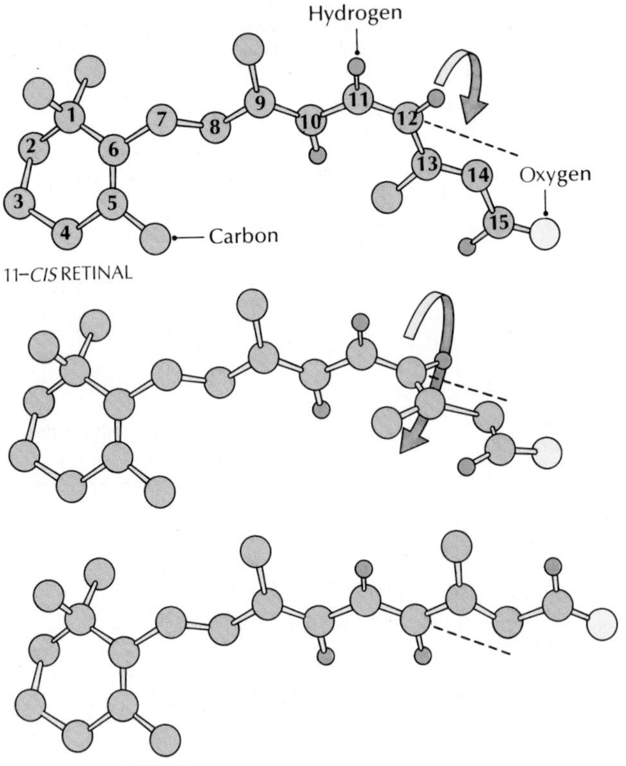

11–*CIS* RETINAL

ALL-*TRANS* RETINAL

[B]

In binocular vision, the two eyeballs turn slightly inward to focus on a close object, so that both images fall on the corresponding points of both retinas at the same time. This action is called **convergence** (L. *com*, together + *vergere*, to bend). In order to produce a single image, the six pairs of extraocular muscles must move together with perfect coordination. As a result of convergence, the simultaneous stimulation of both retinas produces the perception of a single image in the occipital lobe of the cerebral cortex.

Photochemical activity in rods Vision begins when ''packets'' of electromagnetic energy called *photons* are converted into neural signals that the brain can decode and analyze. (A photon is the smallest possible quantity of light.) The translation of photons into neural signals is accomplished by rods and cones, the photoreceptor cells of the eye. Each eye contains about 125 million rods, which are located in most of the neuroretina, but are absent in the fovea and blind spot. How is the absorption of light waves by rods translated into vision?

The extracellular fluid of rods contains a high concentration of positively charged sodium ions and a low concentration of positively charged potassium ions. The distribution of ions inside the cell is the opposite: abundant potassium ions and few sodium ions. These concentrations are maintained by a sodium-potassium pump. In a resting state, potassium ions tend to diffuse to the outside, creating a relatively negative charge inside the cell. Ordinarily, a photoreceptor cell in a dark environment has a permeability to sodium ions, which move from the extracellular fluid into the outer segment of the cell (see Figure 15.18). This inward flow of sodium ions into the outer segment is balanced by the outward flow of potassium ions from the rest of the cell.

When light is absorbed by a photoreceptor cell, however, sodium ions cease to flow into the cell. This causes the negative charge inside the cell to increase, producing a hyperpolarization at the outer segment that spreads to the synaptic ending. The hyperpolarization controls the flow of neural information across synapses to other retinal cells.

What does 20/20 vision mean?	*If you have ''20/20 vision,'' your eyes can see at 20 ft what the normal eye can see at that distance. The larger the second number, the worse the visual acuity (sharpness); 20/60 means that you can see at 20 ft what a normal eye sees at 60 ft. (Legal blindness is 20/200 or worse in both eyes, with or without corrective eyeglasses.)*

In rods, the photosensitive membrane is made up of an orderly pile of disks inside the surface membrane (see Figure 15.18). When the disks absorb light, they release a neurotransmitter that relays neural information to the surface membrane, where a nerve impulse is generated. The neurotransmitter is a nucleotide, *cyclic guanosine monophosphate* (cGMP). In darkness, rods contain large amounts of cGMP, which binds to pores in the surface membrane and opens them. Sodium ions rush in through the open pores. In light, the concentration of cGMP decreases, the pores close, and a hyperpolarization is produced as sodium ions are blocked from entering the cell.

The hyperpolarization of photoreceptor cells occurs as follows. Each rod contains about 100 million molecules of a reddish, light-sensitive pigment called **rhodopsin.** It contains a light-absorbing organic molecule called 11-*cis* retinal, which is an isomer of **retinal,** a derivative of vitamin A (Figure 15.23).* Retinal is coupled with a specific protein for each type of photopigment in receptor cells. The protein coupled with retinal is called **scotopsin.** When 11-*cis* retinal absorbs a photon of light, it changes its molecular configuration to form all-*trans* retinal, another isomer of retinal, and activates scotopsin's ability to act as an enzyme. The reaction produces large amounts of the activated protein called *transducin.* Transducin activates a specific enzyme called a *phosphodies-*

terase, which hydrolyzes cGMP molecules. Sodium ions are prevented from entering the rods, hyperpolarization occurs, neural signals are processed by bipolar, amacrine, horizontal, and ganglion cells in the retina, action potentials of the ganglion cells are conveyed to the brain via the optic nerve, and light is finally perceived. Afterward, retinal is enzymatically combined with scotopsin to synthesize rhodopsin again. Figure 15.24 illustrates the overall reaction in detail.

Photochemical activity in cones The events of visual excitation are similar in rods and cones. The major difference is that there are three different types of cones, each one containing separate photopigments that respond to red, green, and blue light. These photopigments all contain 11-*cis* retinal, and have the same basic molecular structure as rhodopsin, but they contain *photopsins,* which are slightly different from the scotopsin of rods.

The perception of color depends on which cones are stimulated. The final perceived color, which can combine all three types of cones, and is almost unlimited in its color possibilities, is determined by the combinations of different levels of excitation of each type of cone. The proper mix of all three basic colors* produces the perception of white, and the absence of all three colors produces the perception of black.

---

*A condition known as *night blindness*—the inability to see in the dark—can result from a deficiency of vitamin A, which is essential for the synthesis of retinal.

*Note that cone pigments respond to red, green, and blue light. These are not the same as the primary colors of paint pigments, red, yellow, and blue.

FIGURE 15.24

The physiology of light absorption and perception. The process begins when a photon of light is absorbed by 11-*cis* retinal in a rhodopsin molecule, and ends when impulses are sent to the brain and perceived as visual images. After the process is completed, the split molecule of rhodopsin is resynthesized.

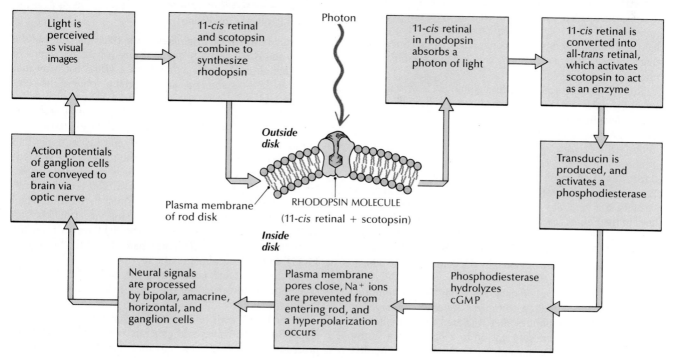

FIGURE 15.25

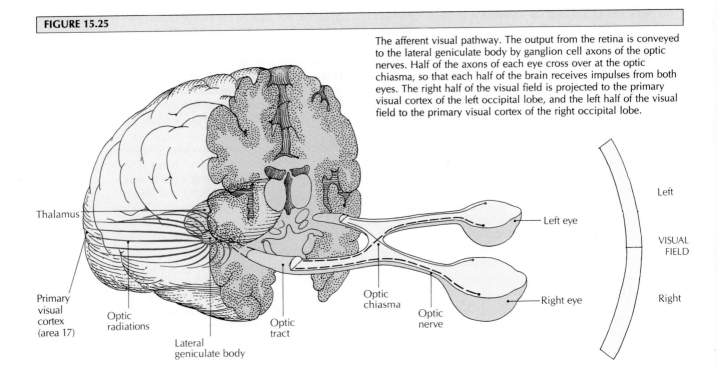

The afferent visual pathway. The output from the retina is conveyed to the lateral geniculate body by ganglion cell axons of the optic nerves. Half of the axons of each eye cross over at the optic chiasma, so that each half of the brain receives impulses from both eyes. The right half of the visual field is projected to the primary visual cortex of the left occipital lobe, and the left half of the visual field to the primary visual cortex of the right occipital lobe.

In "red," "green," and "blue" cones, the same retinal absorbs a different frequency of light. In the red cones, it absorbs long waves of the visible spectrum; in the green cones, middle waves; and in the blue cones, short waves.

Adaptation to light and dark Rods function best in dim light. One reason is that bright light "bleaches" (breaks down) rhodopsin, which decreases the amount of photopigment available to rods. As a result, the light-sensitive rods become overloaded, even in ordinary daylight, and they stop transmitting neural signals. The capability of rods in bright light is also decreased because cones have neural connections that inhibit rods when the light conditions are more appropriate for cone function.

Rods begin to regain their functional levels of sensitivity when the light source is diminished. Some sensitivity returns after a few minutes, as when we enter a darkened movie theater from a sunlit street. However, a complete sensitivity in dim light, or ***dark adaptation,*** usually takes from 20 to 30 minutes while the "bleached" rhodopsin is resynthesized. Dark adaptation consists of a rapid phase and a slow phase. The *rapid phase* (neural adaptation) is generally over in a few seconds. It is generally thought to take place at the neuronal level in the retina. The *slow phase* (photochemical adaptation) is the one that takes 20 to 30 minutes to complete as the "bleached" rhodopsin in the rods is resynthesized. Adaptation to the dark is complemented by the dilation of the pupils, a separate process.

The adjustment to bright light, known as ***light adaptation,*** may take only 5 minutes to complete. It occurs dramatically during the first few minutes after we enter a brightly lighted area, such as when we step out on to a beach on a sunny day. The glare may be uncomfortable initially, but the dis-comfort decreases gradually as the eyes decrease in sensitivity as they are adapting. One explanation for the phenomenon of light adaptation is associated with the bleaching of photopigments. Other authorities contend that light adaptation is not related to the photopigments, but takes place instead within the retina at the neuronal level.

Neural Pathways for Vision

All visual information originates in the rods and cones of the retina, and is conveyed to the brain by way of the axons of the ganglion cells. These axons form a visual pathway that begins in the eyes and ends in the occipital lobes of the cerebral cortex.

The field of vision (the environment viewed by the eyes) for each eye is divided into (1) an outer, lateral *(temporal)* half and an inner, medial *(nasal)* half, and (2) an upper half and a lower half. Rays of light entering the eye move diagonally across the eyeball. Because the lens in the eye acts like the lens in a camera, the field of vision is reversed in both the vertical and horizontal planes. Thus, in the horizontal plane, the light waves from the temporal visual field fall on the nasal half of each retina, and the light waves from the nasal field fall on the temporal half of the retina (Figure 15.25). In the vertical plane, the rays of light from the upper half of the visual field fall on the lower half of the retina, and rays from the lower half of the visual field fall on the upper half of the retina.

In the retina, rods or cones are the photoreceptors, which respond to light waves and generate neural signals that are processed by other retinal cells and stimulate the ganglion

FIGURE 15.26

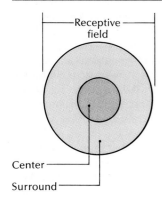

Receptive field

Center

Surround

[A]

Zones of receptive fields of neurons of the visual pathway. (A) A receptive field has an interior central portion called the *center,* and a peripheral concentric portion called the *surround.* (B) An on-center, off-surround receptive field. Photoreceptor cells, bipolar cells, and a ganglion cell form a direct pathway to the brain. Note that although light entering the retina passes through the ganglion cell and other neural cells, it stimulates only the photoreceptor cells, which then stimulate neural cells of the retina to send a message to the brain via the optic nerve. (C) The opposite type of receptive field, with an inhibitory center (off-center) and an excitatory surround (on-surround). Amacrine cells (not shown) and horizontal cells form lateral pathways to the brain. Horizontal cells may have an effect (small arrows) on the bipolar cells that is opposite to the direct effect of the rods and cones.

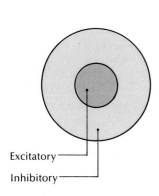

Excitatory

Inhibitory

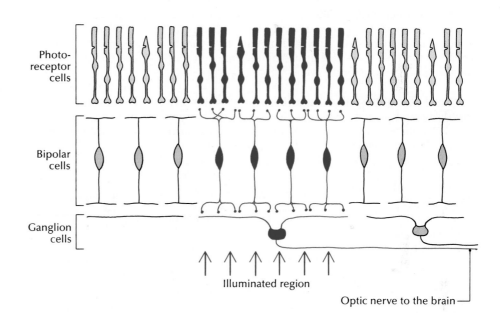

Photo-receptor cells

Bipolar cells

Ganglion cells

Illuminated region

Optic nerve to the brain

[B] ON-CENTER, OFF-SURROUND

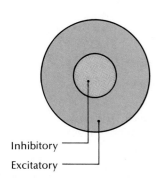

Inhibitory

Excitatory

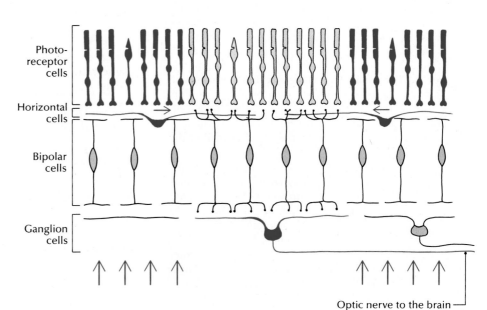

Photo-receptor cells

Horizontal cells

Bipolar cells

Ganglion cells

Optic nerve to the brain

[C] OFF CENTER, ON-SURROUND

cells to produce action potentials. Then, nerve fibers of the ganglion cells from both eyes carry the impulses along two optic nerves. These two nerves meet at the *optic chiasma,* where the fibers from the nasal half of each retina cross over; the fibers from the temporal half of each retina do not cross over. Because half the nerve fibers from each eye cross over at the optic chiasma, each side of the brain receives visual messages from both eyes. After passing through the optic chiasma, the nerve fibers are called the *optic tracts.* Each optic tract contains nasal fibers from the opposite side, and temporal fibers from the same side.

Each optic tract continues posteriorly until it reaches a nucleus in the thalamus called the *lateral geniculate body.* From there, the axons of neurons in the lateral geniculate body pass to the primary visual cortex in the occipital lobe of the cerebral cortex.

A reflex pathway proceeds from the retinas directly to the superior colliculus in the midbrain. This reflex system is involved in unconscious movements of the eye, including contraction and dilation of the pupil, and movements involving both eyes together. These coordinated movements are called the *conjugate movements* of the eye.

Neural cells of the retina The retina is composed of five main classes of neurons, which are interconnected by synapses. The three types of retinal neurons that form a *direct pathway* from the retina to the brain are *photoreceptor cells* (rods and cones), *bipolar cells,* and *ganglion cells* (Figure 15.26B). The two remaining classes of retinal neurons, *horizontal cells* and *amacrine cells,* synapse with the neural elements *laterally* (Figure 15.26C). These five classes of neurons have many subtypes of cells, bringing the total number of functional neural elements in the retina to about 60.

The neural elements in the retina interact with each other and begin processing the stimuli received by the photoreceptors. In the direct pathway, this evaluation is accomplished by the convergence of signals from a number of photoreceptors onto a single ganglion cell whose axon projects via the optic nerve to the brain. The horizontal and amacrine cells, acting laterally, connect adjacent neurons and allow them to modify the signals as they are conveyed along the direct pathway to the brain.

Direct pathway and lateral influences Photoreceptors are stimulated by the light from the visual field. The resulting activity is conveyed to the ganglion cells via the bipolar cells. The photoreceptors that converge on a single ganglion cell and influence its activity describe the *receptive field* of that neuron.

The receptive field of a ganglion cell is organized into two zones: a small circular zone called the *center,* and a surrounding, concentric zone called the *surround* (Figure 15.26A). The center may be likened to the hole of a doughnut, while the surround is the doughnut itself. Receptive fields may be classified as either (1) on-center, off-surround (excitatory center, inhibitory surround) (Figure 15.26B) or (2) off-center, on-surround (inhibitory center, excitatory surround) (Figure 15.26C). Stimulating the receptors in the center of an excitatory center, inhibitory-surround receptive field increases

the firing rate of the ganglion cell. Stimulating the surround of the same receptive field decreases the firing rate of the ganglion cell. The reverse is true for the inhibitory-center, excitatory-surround cell. When the entire field is stimulated, the response of the ganglion cell is the result of the interactions among inhibitory and excitatory photoreceptor signals.

Horizontal and amacrine cells modify communication among the elements of the direct pathway. During *lateral inhibition,* these lateral elements sharpen the contrast between different colors and different intensities of light projected from the visual field onto the retina. By allowing inhibitory signals to pass from a strongly stimulated area to an adjacent weakly stimulated area, the elements further depress the response from the weakly stimulated area. As a result, there is a greater distinction between the strongly and weakly stimulated areas. This ensures that contrast from the visual field passes, via these modified signals, along the direct pathway to the optic nerve and the brain.

In one sense, the retina may be thought of as a mosaic composed of one million transmitters, with each transmitter being one ganglion cell of the retina. Each transmitter receives its information from the interaction of a certain finite number of retinal cells as a center-surround. In turn, the ganglion cells as a mosaic transmit their influences from one million ganglion cells via their one million axons from each optic nerve to the brain.

Ask Yourself

1 *How much of the entire electromagnetic spectrum do we see?*

2 *What are the three layers of the eyeball?*

3 *What is the importance of the fovea?*

4 *What are the differences between aqueous humor and vitreous humor?*

5 *Why do we blink our eyes periodically?*

6 *What is refraction? Accommodation? Convergence?*

7 *What is the function of 11-cis-retinal?*

8 *How is the sodium-potassium pump involved in the physiology of vision?*

9 *What substance is the neurotransmitter in rods and cones?*

10 *What is the significance of the differences in cone pigments?*

11 *How do eyes adapt to sudden darkness?*

12 *What is the function of the optic chiasma?*

13 *What are the five main classes of retinal neurons?*

PHYSIOLOGICAL AND ANATOMICAL ABNORMALITIES

Otosclerosis

The stapes may become fused to the bone in the region of the oval window and become immobile, a condition called *otosclerosis* (Gr. *otos*, ear + *skleros*, hard). When this happens, the bones cannot vibrate properly, and the transmission of sound waves to the inner ear may be almost impossible. A hearing aid, which transmits sound waves to the inner ear by conduction through the bones of the skull rather than by air conduction along the auditory ossicles, can make hearing almost normal in people who have been deafened by otosclerosis. An operation that allows the stapes to move again has also had some success.

Labyrinthitis

Labyrinthitis, or inflammation of the inner ear, can cause discharge from the ear, vomiting, hearing loss, and vertigo (the feeling of spinning) and other forms of dizziness. Vertigo upsets the balance, and the patient tends to fall in the direction of the affected ear. Within three to six weeks the symptoms usually subside. The causes of labyrinthitis include bacterial infections, allergies, toxic drugs, severe fatigue, and overindulgence of alcohol.

Ménière's disease

Head noises, dizziness, and hearing loss are all characteristic of *Ménière's disease,* an inner ear disorder. It is thought to be caused by an excess of endolymph, and dilation of the labyrinth. The basilar membrane is distorted, the semicircular ducts are affected by pressure, and some cochlear hair cells degenerate. A patient may have residual tinnitus (ringing or whistling in the ears) and hearing loss after repeated attacks over many years.

Motion sickness

Many people know the familiar queasy feeling of *motion sickness,* which results from the sensation of motion or from repeated rhythmic movements. Symptoms of nausea, vomiting, pallor, and cold sweats are common when people travel on boats, planes, cars, and trains. However, suffering from one type of motion sickness, such as seasickness, does not necessarily mean that a person is susceptible to other forms. Motion sickness arises from the excessive stimulation of the labyrinthine receptors of the inner ear. Tension, fear, offensive odors, and visual stimuli are also important factors.

Another theory of motion sickness suggests that it results from conflicting perceptions. For example, while seated in the lounge of a ship, a person's vestibular system may detect the rocking motion of the ship, but the visual system may not. Alternatively, in a car, the vestibular system may not detect motion (since the forward motion of the car is constant), while the visual system detects motion by observing the passing scene through the side windows. According to this theory, matching the two perceptions should help to relieve the motion sickness. In other words, watch the horizon list with the ship, or look out of the front windshield of the car to perceive less motion.

Otitis media

Otitis media, inflammation of the middle ear, causes the tympanic membrane to redden and bulge out. If untreated, the membrane may rupture. Otitis media is commonly seen in children and is most prevalent between the ages of 6 and 24 months. If the inflammation is severe enough or prolonged, scarring, structural damage, and hearing loss may result. This disorder seems to arise from a malfunction of the auditory (Eustachian) tube, in which bacteria from the nasopharynx enters the middle ear. Otitis media may be acute or chronic. Both forms are bacterial infections, caused by organisms such as *Streptococcus pneumoniae* or *Hemophilus influenzae.*

Detached retina

Normally, no space exists between the two retinal layers, but sometimes the layers become separated. Such a condition is commonly called a *detached retina.* The portion of the neuroretina that is detached usually stops functioning because its blood supply is impaired.

New surgical procedures, including photocoagulation by laser beam and cryosurgery, can successfully reattach the retinal layers in about 90 percent of the cases, and thus arrest further detachment.

Cataract

One of the most common causes of blindness is *cataract formation,* in which the lens becomes opaque. As the opaque areas increase, there is a progressive loss of vision, and if the cataract is severe enough, total blindness may result. The molecular weight of the protein within the lens increases with age, and consequently the protein may become cloudy and then opaque. The result is a cataract. Cataracts are generally associated with people over 70, but they can occur at any age. There are various types of cataracts, including senile, congenital, traumatic, and toxic. Cataracts are also frequent in people with diabetes, since the abnormal sugar metabolism of diabetes may affect the vitality of the lens. Treatment varies from frequent changes in eyeglasses to compensate for gradual vision loss, to surgery, in which the lens is removed and replaced with an artificial lens. Accommodation and visual acuity are lost when the natural lens is removed. A combination of the artificial lens and bifocal glasses or contact lenses is used to restore focusing ability.

Conjunctivitis

Conjunctivitis, or inflammation of the conjunctiva, is a common eye disorder. It may have several causes. Bacteria, viruses, pollen, smoke, pollutants, and excessive glare all affect the conjunctiva, causing discharge, tearing, and pain. Vision, however, generally is not affected. "Pinkeye," a form of conjunctivitis caused by pneumococci or staphylococci bacteria, is contagious. In such cases, affected people should avoid spreading the infection by not rubbing the infected eye or sharing towels and pillows. Conjunctivitis can be acute or chronic, and treatment varies with the cause.

Glaucoma

Glaucoma (glaw-KOH-muh) is a

leading cause of blindness in the United States, affecting over 1 million people. The disease strikes people of all ages, but mainly those over 40; women are more susceptible than men. Glaucoma occurs when the aqueous humor does not drain properly. Since more is formed than is drained, the fluid builds up in the eyeball and increases the intraocular pressure. If the pressure continues, it destroys the neurons of the retina, and blindness usually results. Glaucoma may be chronic (90 percent of cases) or acute. Chronic forms result in progressively reduced vision. A common symptom is a vision defect in which lights appear to be surrounded by halos. Acute forms can occur suddenly at any age, causing pain, pressure, and blurring. Chronic glaucoma may be genetically linked. Close relatives of patients with glaucoma are five times more susceptible to developing the disease than those with no glaucoma in their family history.

MEDICAL TERMINOLOGY

The ear

EUSTACHITIS Inflammation of the auditory (Eustachian) tube.

EXTERNAL OTITIS (Gr. *ous*, ear + *itis*, inflammation) Inflammation of the outer ear.

IMPACTED CERUMEN An accumulation of cerumen, or earwax, that blocks the ear canal and prevents sound waves from reaching the tympanic membrane.

MYRINGITIS Inflammation of the tympanic membrane or eardrum. Also known as tympanitis.

OTOGLIA (Gr. *ous*, ear + *algia*, pain) Earache.

OTOPLASTY (Gr. *ous*, ear + *plastos*, molded) Surgery of the outer ear.

OTORRHEA (Gr. *ous*, ear + *rrhea*, discharge) Fluid discharge from the ear.

OTOSCOPE (Gr. *ous*, ear + *skopein*, to see) Instrument used to look into the ear.

The eye

ACHROMATOPSIA (Gr. *a*, not + *kroma*, color + *ope*, vision) Complete color blindness.

AMBLYOPIA (Gr. *amblus*, dim + *ops*, eye) Dullness of vision from not using the eye. Also called "lazy eye."

AMETROPIA (Gr. *ametros*, without measure + *ops*, eye) Inability of the eye to focus images correctly on the retina, resulting from a refractive disorder.

ANOPIA (Gr. *a*, no + *ops*, eye) No vision, especially in one eye.

BLEPHARITIS (Gr. *blepharon*, eyelid + *itis*, inflammation) Inflammation of the eyelid.

CHALAZION (Gr. *khalaza*, hard lump) Swelling in the tarsal or Meibomian glands of the eyelid.

DIPLOPIA (Gr. *di*, double + *ope*, vision) Double vision.

ESOTROPIA Medial deviation (turning inward) of the eyeball, resulting in diplopia, caused by a muscular defect or weakness in coordination. Also called "crosseye" and convergent strabismus.

EXOPHTHALMIA (Gr. *ex*, out + *ophalmos*, eye) Abnormal protrusion of the eyeball.

EXOTROPIA (Gr. *ex*, out + *tropos*, turn) Lateral deviation (turning) of the eyeball, resulting in diplopia. Also called "walleye" and divergent strabismus.

KERATITIS Inflammation of the cornea.

KERATOPLASTY Plastic surgery of the cornea, such as corneal transplant.

MIOTIC Drug that causes the pupil to contract.

MYDRIASIS Dilation of the pupil.

MYDRIATIC Drug that causes the pupil to dilate.

NYSTAGMUS (Gr. *nustagmos*, drowsiness) Usually, a side-to-side involuntary movement of the eye, with slow movement in one direction and fast movement in the opposite direction; probably the result of a central nervous system disease.

OPHTHALMOSCOPE Instrument used to see the interior of the eyeball.

OPTIC NEURITIS Inflammation of the optic nerve.

PTOSIS Drooping (prolapse) of an organ or part. Specifically, drooping of the eyelid.

RETINITIS Inflammation of the retina.

SCLERITIS Inflammation of the sclera.

SCOTOMA (Gr. *skotos*, darkness) A blind spot or dark spot seen as a result of vision loss in part of the visual field.

STRABISMUS (Gr. *strabos*, squinting) "Crossed eyes." An eye muscle defect in which the eyes are not aimed in the same direction, resulting from dysfunction of the extrinsic eye muscles.

TRACHOMA Disease caused by *Chlamydia trachomatis*, a bacterium spread by insects, body contact, poor hygiene, and contaminated water. It is a leading cause of blindness in many countries.

SUMMARY

Sensory reception

1 Sense organs, or **receptors**, are the body's link to the outside world and to changes within the body.

2 All receptors are *transducers*, that is, they convert one form of energy into another.

3 The receptor cells respond initially with *receptor (graded) potentials*, which are converted into action potentials. All stimulated receptors generate the same type of nerve impulse, the *action potential*. Different sensations occur when nerve fibers connect with specialized portions of the central nervous system.

4 Features basic to all types of sensory receptors are: (1) they contain *sensitive receptor cells* that respond to threshold levels of intensity, (2) they have a *structure* that is conducive to receiving a specific stimulus, (3) their primary receptor cells *synapse with afferent nerve fibers* that travel to the central nervous system along neural pathways, (4) after receptor cells synapse with afferent neurons, *neural pathways* are formed to carry neural influences through the brainstem to the cerebral cortex.

5 Sensory receptors may be classified according to their *location* (**exteroceptor,**

teleceptor, interoceptor, proprioceptor); type of sensation (thermal, pain, light touch, touch pressure, position sense); type of stimulus (thermoreceptor, nociceptor, chemoreceptor, photoreceptor, mechanoreceptor); or sensory ending (free nerve ending, encapsulated ending).

6 The *general senses* include light touch, touch-pressure (deep pressure), two-point discrimination, vibratory sense, heat, cold, pain, and body position. Taste, smell, hearing and equilibrium, and vision are called the **special senses** because they originate from sensors in restricted (special) regions of the head.

7 The *complexity of the brain,* not the quality of the sensory receptor, increases the range and quality of perception.

8 Individual nerve impulses are essentially identical, but an impulse for sight is distinguished from an impulse for taste by the *relationship between where the impulse is coming from and the place in the brain where it is going.*

General senses

1 Sensory receptors in the skin detect stimuli that the brain interprets as light touch, touch pressure, heat, cold, pain, proprioception, and stereognosis. Other miscellaneous cutaneous sensations include itch and tickle.

2 Two receptors for light touch are *tactile corpuscles (of Merkel)* and *tactile corpuscles (of Meissner).* Receptors for deep pressure are *lamellated corpuscles (of Pacini). Corpuscles (of Ruffini)* and *bulbous corpuscles (of Krause)* are mechanoreceptors for touch.

3 The cutaneous receptors for heat and cold are probably free nerve endings.

4 Pain receptors are **specialized free nerve endings** present in most parts of the body.

5 *Endorphins* and *enkephalins* are peptides in the brain that have analgesic effects similar to opiates such as morphine.

6 *Referred pain* is a visceral pain that is felt in a somatic area. *Phantom pain* is felt in an amputated limb.

7 The sense of position of body parts in relation to each other is called *proprioception. Itch* and *tickle* are probably produced by the activation of several sensory types of nerve endings. The information resulting in the sensation is probably conveyed by a combination of pathways. *Stereognosis* is the ability to identify unseen objects merely by handling them.

Taste (gustation)

1 The receptors for taste and smell are both chemoreceptors, and both sensations are interrelated, but taste *(gustation)* and smell *(olfaction)* are separate.

2 The surface of the tongue is covered with many small protuberances called *papillae,* including three types, filiform, fungiform, and circumvallate. Located within the crevices of papillae are **taste buds,** the receptor organs for the sense of taste.

3 The four basic taste sensations are sweet, sour, bitter, and salty.

Smell (olfaction)

1 The *olfactory receptor cells* are located high in the roof of the nasal cavity, in specialized areas of the nasal mucosa called the *olfactory epithelium.* The epithelium consists of *receptor cells, sustentacular (supporting) cells,* and *basal cells.*

2 The olfactory receptor cell (neuron) ends in a bulbous *olfactory vesicle,* from which extend *cilia* that project through the mucuslike fluid that covers the surface epithelium.

3 Olfactory nerve fibers extend through foramina in the cribriform plate of the ethmoid bone and terminate in the *olfactory bulbs,* from which axons of neurons project to the olfactory cortex of the cerebrum.

Hearing and equilibrium

1 The ear consists of two functional units, the *auditory* (acoustic) *apparatus,* concerned with hearing, and the *vestibular apparatus,* concerned with posture and balance.

2 The anatomical components of the auditory apparatus are the external ear, the middle ear, and the inner ear. The *external ear* is composed of the **auricle** and **external auditory canal;** the *middle ear* is made up of the **tympanic membrane** (eardrum), **tympanic cavity, auditory** (Eustachian) **tube,** and the three **auditory ossicles** (ear bones); the *inner ear,* or **membranous labyrinth,** is composed of the **vestibule** (which contains the **utricle** and **saccule), semicircular** canals and **ducts,** and **cochlea** (which contains the **spiral organ of Corti).**

3 *Sound* is the alternating compression and decompression of the medium through which the sound is passing.

4 Sound waves reach the spiral organ through a sequence of vibrations that start in the external ear and tympanic membrane, and progress into the inner ear. The displacement of the hairs of the hair cells in the spiral organ generates generator potentials and, subsequently, nerve impulses in the cochlear nerve. Their influences are conveyed to the auditory area of the temporal lobe via the auditory pathways.

5 The main receptors for *equilibrium* are the utricle, saccule, and semicircular ducts in the inner ear. The equilibrium system also receives input from the eyes and from some proprioceptors in the skin and joints.

6 The purpose of the **vestibular system** is to signal changes in the motion of the head *(dynamic equilibrium),* and in the position of the head with respect to gravity *(static equilibrium,* or posture).

7 Specialized receptor cells of the vestibular sense organs are **hair cells,** which are arranged in clusters called hair bundles. *Stereocilia* and a *kinocilium* are present in each hair bundle. When hairs are bent in the direction of the stereocilia, the hair cells convert a mechanical force into an electrical signal that is conveyed to the brain via the vestibular nerve.

8 When the head moves in a change of posture, calcium carbonate crystals *(otoconia)* in the inner ear respond to gravity, resulting in the bending of the hairs of hair cells. The bending stimulates nerve fibers to generate a generator potential and then an action potential, which is transmitted to the brain. The brain signals appropriate muscles to contract, and body posture is adjusted to follow the new head position.

9 The utricles and saccules are *organs of gravitation,* responding to movements of the head in a straight line: forward, backward, up, or down. In contrast, the **crista ampullaris** of the semicircular ducts responds to changes in the *direction of head movements,* including turning, rotating, and bending. Hair cells in the crista project into a gelatinous flap called the **cupula.** When the head rotates, the endolymph in the semicircular canals lags behind, displacing the cupula and the hairs projecting into it. The resulting action potentials are sent to the neural centers in the brain, which signals certain muscles to respond appropriately to maintain the body's equilibrium.

Vision

1 The human eye can "see" only those objects that emit or are illuminated by light waves representing about 1/70 of the electromagnetic spectrum.

2 The wall of the eyeball consists of three layers of tissue: the outer **supporting layer**, the **vascular layer**, and the inner **retinal layer**.

3 The posterior five-sixths of the supporting layer is the **sclera** ("white" of the eye). It gives the eyeball its shape, and protects the delicate inner layers. The anterior segment of the supporting layer is the transparent **cornea**, through which light enters the eye.

4 The vascular layer contains many blood vessels. The posterior two-thirds is a thin membrane, the **choroid**. The **ciliary body** is the thickened anterior portion of the choroid layer. The **lens** is an elastic body that changes shape to focus on objects at different distances.

5 The anterior extension of the choroid is a muscular layer, the **iris**, which is the colored part of the eye. An adjustable opening in the center of the iris is the **pupil**. It opens and closes in response to the amount of light available.

6 The innermost portion of the eyeball is the **retina**. It contains (1) a *layer of nervous tissue (neuroretina)* that receives focused light waves, transduces them, and processes their effects into neural impulses, which are converted into visual perceptions in the brain, and (2) a *pigmented layer* behind the neural layer that absorbs light not utilized by rods and cones, and thus prevents reflection of light that could restimulate the rods and cones.

7 The retina contains highly specialized photoreceptor nerve cells, the **rods** and **cones**, as well as other types of nerve cells and fibers. Rod cells respond to the entire visual spectrum and produce black-to-white vision. They function in dim light and peripheral vision. The color-sensitive cone cells are mainly concentrated in the center of the retina in the **fovea centralis**.

8 The hollow eyeball is divided into three cavities. The **anterior chamber** lies between the iris and cornea. The **posterior chamber** lies between the iris, lens, and vitreous chamber. Both chambers are filled with *aqueous humor*. The largest cavity of the eyeball is the **vitreous chamber** (the space behind the lens), which contains *vitreous humor.*

9 The accessory structures of the eye are mainly protective and supportive devices. They include the **bony orbits, eyelids, eyelashes, eyebrows, conjunctiva, lacrimal apparatus,** and **muscles** that act on the eyeball and eyelid.

10 *Refraction* is the bending of light waves as they pass from one medium to another with a different density. *Accommodation* is the mechanical process that involves the ciliary muscle and the lens to bring images to exact focus on the retina. *Convergence* is the movement of both eyes inward to focus on nearby objects in the visual field, so that the objects fall on corresponding points of the retinas of both eyes simultaneously.

11 Vision begins when a *photon* is absorbed by a photoreceptor cell (rod or cone). Sodium ions cease to flow into the cell, producing a hyperpolarization that spreads to the synaptic ending. When photoreceptor cells absorb light, they release the neurotransmitter *cyclic guanosine monophosphate* (cGMP).

12 The hyperpolarization in rods occurs as the light-absorbing 11-*cis* retinal (an isomer of **retinal**) in the photopigment *rhodopsin* absorbs light. 11-*cis* retinal is converted into all-*trans* retinal, which activates *scotopsin* (the protein in rhodopsin) to act as an enzyme. The activated protein *transducin* activates a phosphodiesterase enzyme, which hydrolyzes cGMP. Sodium ions are blocked from entering the rods, hyperpolarization occurs, neural signals are processed by bipolar, amacrine, horizontal, and ganglion cells in the retina, action potentials of the ganglion cells are conveyed to the brain, and light is perceived.

13 After light is absorbed by photoreceptor cells, and is finally perceived, retinal is enzymatically combined with scotopsin to resynthesize rhodopsin.

14 Three types of cones contain separate photopigments that respond to red, green, or blue light. Cone pigments contain 11-*cis*-retinal and *photopsins*. The final color is determined in the cerebrum by the combinations of different levels of excitation of different types of cones.

15 *Dark adaptation* occurs as eyes adapt slowly to darkness, and **light adaptation** occurs when eyes adjust to bright light.

16 The field of vision for each eye is divided into (1) an outer, lateral *(temporal)* half and an inner, medial *(nasal)* half, and (2) an upper half and a lower half. Nerve fibers of ganglion cells in the retina carry impulses from both eyes along two optic nerves. The nerves meet at the **optic chiasma,** where the fibers from the nasal half of each retina cross over; the temporal fibers do not cross over. As a result of the crossover, each side of the brain receives visual messages from both eyes.

17 After passing through the optic chiasma, the nerve fibers are called the **optic tracts.** Each optic tract continues until it reaches the **lateral geniculate body** in the thalamus. From there, axons pass to the primary visual cortex in the cerebral cortex.

18 The retina is composed of five main classes of synaptically interconnected neurons. **Photoreceptors, bipolar cells,** and **ganglion cells** form a direct pathway from the retina to the brain; **horizontal cells** and **amacrine cells** form lateral pathways.

Physiological and anatomical abnormalities

1 *Otosclerosis* is a condition in which the stapes become immobile, preventing the normal transmission of sound waves to the inner ear.

2 *Labyrinthitis* is an inflammation of the inner ear.

3 *Ménière's disease* is an inner ear disorder. It is thought to be caused by excess endolymph and dilation of the labyrinth.

4 *Motion sickness* results from excessive stimulation of the labyrinthine receptors of the inner ear, and perhaps conflicting perceptions of motion.

5 *Otitis media* is an inflammation of the middle ear, caused by a bacterium.

6 *Detached retina* is a separation of the two retinal layers.

7 A *cataract* is a clouding of the lens of the eye, sometimes causing blindness.

8 *Conjunctivitis* is an inflammation of the conjunctiva.

9 *Glaucoma* is a leading cause of blindness, occurring when the aqueous humor does not drain properly, increasing the pressure inside the eyeball.

UNDERSTANDING THE FACTS

1 Distinguish between the special senses and the general senses.

2 Define *proprioception.*

3 What are *free,* or *naked, nerve endings?*

4 With which stimuli are free nerve endings involved?

5 With what sensation are lamellated corpuscles (of Pacini) associated?

6 Which areas of the body lack pain receptors?

7 What is meant by *referred pain? Phantom pain?* Give examples.

8 Distinguish between an interoceptor and an exteroceptor, and give examples of each.

9 List the three qualities that a substance must have in order to be smelled.

10 Which nerve innervates the auditory apparatus? The vestibular apparatus?

11 What is the main purpose of the auditory tube?

12 The bony labyrinth is filled with _____ , whereas the membranous labyrinth is filled with _____ .

13 Which structure is involved in the "righting reflex"?

14 In what areas of the brain do the neural pathways for hearing and equilibrium terminate?

15 What is the function of the pupil? The retina?

16 What structures make up the lacrimal apparatus?

17 List four functions of tears.

18 What is the role of vitamin A in the visual process?

19 Distinguish among the visual conditions of myopia, emmetropia, and hypermetropia.

20 What parts of the ear are affected in motion sickness? In otitis media?

UNDERSTANDING THE CONCEPTS

1 What important features do all sensory receptors possess in common?

2 What is meant by the statement, "All receptors are transducers"?

3 What are some implications of two-point discrimination?

4 Describe the neural pathways for gustatory and olfactory sensations.

5 Why do we not adapt to pain to the extent that we adapt to most of our other senses?

6 Why does phantom pain persist longest in such regions as the thumb, hand, and foot?

7 Describe some of the protective features of the ear.

8 Trace the pathway of sound vibrations from the external environment to the receptor cells in the spiral organ.

9 List the structures in the body that play a role in equilibrium.

10 Why can you see best at night when you are not looking directly at an object?

11 Describe the origin, circulatory route, and final disposition of aqueous humor.

12 Trace the pathway of tears from their production to final disposition.

13 Name the intrinsic and extrinsic muscles of the eye and give their functions.

14 Describe the visual process from the reception of light rays in the eye to the perception of an image in the brain.

15 What are the causes of glaucoma? Of cataracts?

SUGGESTED READING

ANGEVINE, JAY B., JR., AND CARL W. COTMAN, *Principles of Neuroanatomy.* New York: Oxford University Press, 1981.

BARLOW, H. B., AND J. D. MOLLON, eds., *The Senses.* Cambridge: Cambridge University Press, 1982.

GILMAN, SID, AND SARAH S. WINANS, *Manter and Gatz's Essentials of Clinical Neuroanatomy and Neurophysiology,* 7th ed. Philadelphia: Davis, 1987.

GLICKSTEIN, MITCHELL, "The Discovery of the Visual Cortex." *Scientific American,* September 1988.

HUBEL, DAVID H., *Eye, Brain, and Vision.* New York: Scientific American Books (Freeman), 1988.

HUDSPETH, A. J., "The Hair Cells of the Inner Ear." *Scientific American,* January 1983.

KORETZ, JANE F., AND GEORGE H. HANDELMAN, "How the Human Eye Focuses." *Scientific American,* July 1988.

LOEB, GERALD E., "The Functional Replacement of the Ear." *Scientific American,* February 1985.

MASLAND, RICHARD H., "The Functional Architecture of the Retina." *Scientific American,* December 1986.

NASSAU, KURT. "The Causes of Color." *Scientific American,* October 1980.

NOBACK, CHARLES R., AND ROBERT J. DEMAREST, *The Human Nervous System,* 3rd ed. New York: McGraw-Hill, 1981.

PARKER, DONALD E., "The Vestibular Apparatus." *Scientific American,* November 1980.

PETTIGREW, JOHN D., "The Neurophysiology of Binocular Vision." *Scientific American,* August 1972.

POGGIO, TOMASO, AND CHRISTOF KOCH, "Synapses That Compute Motion." *Scientific American,* May 1987.

SCHNAPF, JULIE L., AND DENIS A. BAYLOR, "How Photoreceptor Cells Respond to Light." *Scientific American,* April 1987.

STRYER, LUBERT, "The Molecules of Visual Excitation." *Scientific American,* July 1987.

WOLFE, JEREMY M., ed., *The Mind's Eye: Readings from Scientific American.* New York: Freeman, 1986.

16
The Endocrine System

LEARNING OBJECTIVES

1 Describe the locations of the major endocrine glands.

2 Define a hormone, and identify its three chemical categories.

3 Explain how a negative feedback system helps to maintain a relatively constant basal metabolic rate.

4 Discuss the mechanisms of hormone operation in the body.

5 Explain the relationship between the pituitary gland and the hypothalamus.

6 Give the functions of the major hormones of each lobe of the pituitary gland.

7 Describe the anatomy of the thyroid gland and the functions of the thyroid hormones.

8 Explain the functions of parathyroid hormone.

9 Describe the anatomy of the adrenal cortex, and give the functions of its three types of hormones.

10 Explain how aldosterone is regulated by negative feedback.

11 Describe the anatomy of the adrenal medulla and the functions of its hormones.

12 Compare the effects of stress on the adrenal cortex and medulla.

13 Describe the anatomy of the pancreas, and explain how its two major hormones function.

14 Identify the male and female gonads and their major hormones and functions.

15 Give the functions of two hormonelike substances secreted by the kidneys.

16 Describe the pineal gland and one of its possible functions.

17 Explain the relationship of the thymus gland to the immune system.

18 Give the functions of three digestive hormones.

19 Discuss the functions of prostaglandins in the body.

20 Describe some disorders of the endocrine glands.

The endocrine system is concerned mainly with four major metabolic functions. First, it helps to maintain the homeostasis of the body by regulating such activities as the concentration of minerals and electrolytes in body fluids, the balance of enzymes and substrates, and the metabolism of proteins, lipids, carbohydrates, and other organic compounds. Second, the secretions of the endocrine system act in concert with the nervous system to help the body react properly to stress. Third, the endocrine system is a major regulator of body growth and development. Finally, it controls sexual development and reproduction.

The endocrine system is made up of tissues or organs called **endocrine glands** (Figure 16.1). These glands secrete chemicals into extracellular spaces, from which they enter the bloodstream and circulate throughout the body to their target areas. In contrast, *exocrine glands,* such as sweat glands and salivary glands, secrete into *ducts* through which the secretions are transported to specific locations.

HORMONES AND THEIR FEEDBACK SYSTEMS

A **hormone** is a specialized chemical "messenger," produced and secreted by an endocrine cell or tissue. It circulates through body fluids, and affects the metabolic activity of a target cell or tissue in a specific way. Only rarely does a hormone operate independently. More typically, one hormone influences, depends upon, and balances another in a controlling feedback network.

Biochemistry of Hormones

Hormones are usually steroids, derivatives of amino acids (amines), or proteins (peptides); a few hormones are glycoproteins, such as those secreted by the pancreas. Hormones secreted by the thyroid gland are amines, and steroids are secreted by the ovaries, testes, and adrenal glands. The type of secretion depends upon the embryonic origin of the gland. Glands derived from endoderm secrete proteins; glands derived from ectoderm secrete amines; and glands derived from mesoderm secrete steroids.

Hormones are synthesized from raw materials in the cell and secreted into the extracellular space, usually entering the bloodstream by way of capillaries that flow between the cells of the gland. Steroid hormones, being *lipid-soluble,* are poorly soluble in the water portion of plasma, and must be transported by blood in combination with plasma globular protein. Because protein hormones are *water-soluble,* they are transported freely in the circulating blood. Some amine hormones require *carrier proteins* to transport them to their target cells.

When liberated into the blood, hormones travel to all parts of the body, but they are effective only in specific **target cells**—cells with compatible receptors either located on the surface of the cell's plasma membrane or within the cell's cytoplasm. Water-soluble hormones react with specific receptors on the plasma membrane, while lipid-soluble hormones pass easily through the phospholipid plasma membrane and bind to cytoplasmic receptors. Hormones are effective in extremely small amounts. Only a few molecules

FIGURE 16.1

Location of the major endocrine glands.

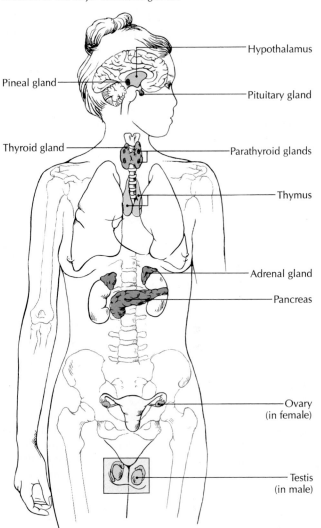

of a hormone may be enough to produce a dramatic response in a target cell.

Hormones themselves do not initiate biochemical reactions within cells, but they do help control the rate of those reactions. Like enzymes, hormones are not changed by the reaction they regulate. The secretion rate of hormones varies from one gland to another, and also for each gland, depending on environmental conditions. For example, the hormone epinephrine (commonly called adrenaline) is secreted by the adrenal glands in response to an emergency that calls for the immediate increase of blood flow to the muscles. In contrast, the hormones of the ovaries are secreted over extended periods of time to regulate the ongoing menstrual cycle.

Feedback Control System of Hormone Secretion

Although hormones are always present in some amount in endocrine glands, they are not secreted continuously. Instead,

FIGURE 16.2

The negative feedback system that helps to control the metabolic rate.

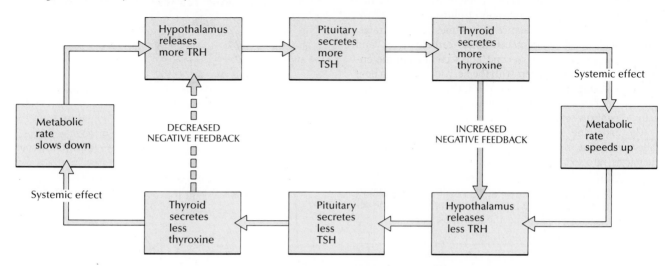

endocrine glands tend to secrete the amount of hormones that the body needs to maintain homeostasis. The regulation of homeostasis occurs through a *feedback control system,* where changes in the body or in the environment are *fed back* through a circular system into a central control unit (such as the brain), where the initial adjustments are made. A feedback system that produces a response that changes or reduces the initiating stimulus is called a *negative feedback system* (see Figure 16.2). The system is called negative because its response to the initial stimulus is a *negative* action (or *not the same as* the initial stimulus). In contrast, a *positive feedback system* is one in which the initial stimulus is reinforced. Positive feedback systems are relatively rare in our bodies, mainly because they usually lead to instability or pathologic states.

The growth of cancerous cells is a dramatic example of positive feedback. Ordinarily, dividing cells have built-in regulating devices that turn the dividing mechanism "on" or "off" at the proper intervals. In cancer cells, however, the regulating mechanism has gone awry, and once cells start dividing, they cannot stop on their own. Instead of being signaled to stop dividing when normal cell division is completed, cancer cells keep receiving signals to divide and grow, eventually causing the death of adjacent healthy cells.

With some endocrine glands, negative feedback systems monitor the amount of hormone secreted and alter the level of cellular activity as needed in order to maintain homeostasis. For example, suppose that the rate of chemical activity in the cells (the *metabolic rate*) drops (Figure 16.2). The hypothalamus responds to this lowered rate by releasing *thyrotropin-releasing hormone* (TRH), which causes the anterior pituitary gland to secrete *thyroid-stimulating hormone* (TSH). This hormone, in turn, causes the thyroid gland to secrete another hormone, *thyroxine.* Thyroxine increases the metabolic rate, restoring homeostatic equilibrium. Conversely, if the metabolic rate of the body were to increase, the hypothalamus would secrete less TRH, thus slowing the anterior pituitary's secretion of TSH. As a result, the thyroid gland would secrete

less thyroxine, and the metabolic rate would decrease. Normally, such negative feedback systems operate continuously to maintain the proper balance of endocrine secretions.

Ask Yourself

1 *What is the difference between an endocrine gland and an exocrine gland?*

2 *What is a hormone?*

3 *What is a target cell?*

4 *What do we call the self-regulating system that helps to correct hormone imbalances?*

5 *What is the main functional difference between negative and positive feedback?*

MECHANISMS OF HORMONE CONTROL

The function of a hormone is to modify the physiological activity of a target cell or tissue. Because hormones are secreted from endocrine glands into the blood and have their effects elsewhere, all hormones are considered to be chemical messengers. Much remains to be learned about how hormones work, but two schemes seem to operate for most hormones, and their basic mechanisms are generally accepted. The first scheme applies to hormones that are proteins or amines. Since they are water-soluble and cannot diffuse through the plasma membrane easily, they initiate their physiological response on specialized receptors on the plasma membrane, *outside* the target cell. The second scheme applies mainly to steroid hormones. Since they are lipid-soluble and diffuse freely into the cytoplasm, they initiate their physiological response by binding to cytoplasmic receptors *inside* the target cell.

Fixed-Membrane-Receptor Mechanism

It was not until the late 1950s that a suitable mechanism was proposed to show how water-soluble or amine hormones regulate cellular responses. These hormones are lipid-insoluble and do not enter cells readily. According to the hypothesis, which is called the *fixed-membrane-receptor mechanism* (formerly called the *second-messenger concept*), such a hormone is secreted by an endocrine gland and circulates through the bloodstream (step 1 in Figure 16.3). At the cells of the target organ, the hormone acts as a "first (or extracellular) messenger," binding to receptor sites for that hormone, located on the plasma membranes (step 2).

The hormone-receptor complex activates the enzyme *adenylate cyclase* in the membrane (step 3). The activated enzyme converts ATP into a nucleotide, **cyclic AMP** (cyclic adenosine 3',5'-monophosphate; cAMP), which becomes the "second (or intracellular) messenger" (step 4). A single molecule of a hormone may cause the production of thousands of molecules of cyclic AMP. (Cyclic AMP is called "cyclic" because of the ring structure that is formed from phosphate groups when ATP is converted into cAMP.)

Cyclic AMP diffuses throughout the cell and activates a cellular enzyme called *protein kinase,* which causes the cell to respond with its distinctive physiological function (step 5). Cyclic AMP may cause many biochemical changes. For example, it indirectly activates enzymes in the liver, decreases

the tension of smooth muscle, increases the contractility of cardiac muscle, increases the secretion of thyroid hormone from the thyroid gland, and causes many other biochemical changes. After inducing the target cell to perform its specific function, cyclic AMP is inactivated by one or more enzymes called phosphodiesterases (step 6). The receptor sites on the plasma membrane of the target cell now become available for new reactions.

In this example the second messenger is cyclic AMP, which was produced from ATP. Another cyclic nucleotide, called *cyclic GMP* (guanosine 3',5'-monophosphate), is also a second messenger. It is used by epinephrine and ACTH; ACTH also uses cyclic AMP. Cyclic GMP is produced from guanosine triphosphate (GTP), which is similar to ATP, and activates different enzymes than those activated by the cyclic AMP system. Having two different second messengers allows the same cell to respond differently to different hormones.

Mobile-Receptor Mechanism

Because steroid hormones are lipid-soluble and pass easily through the plasma membrane, their receptors are inside the target cells. The hormonal mechanism for steroid hormones, called the **mobile-receptor mechanism** (formerly called the *gene-activated mechanism*), involves the stimulation of protein synthesis (Figure 16.4).

Steroid hormones, which are synthesized from cholesterol, operate in the following way: After being released from a carrier protein in the blood, the steroid hormone enters the target cell by diffusion (step 1) and binds to a specific steroid-

---

FIGURE 16.3

Fixed-membrane-receptor mechanism of hormonal control, formerly called the second-messenger concept.

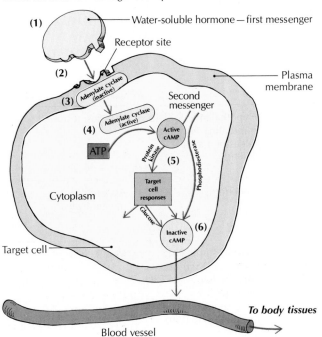

FIGURE 16.4

Mobile-receptor mechanism for the synthesis of protein. In this case, steroid hormones are involved, rather than the typical proteins and other amino acid derivatives that make up most hormones.

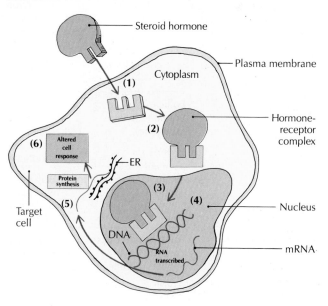

TABLE 16.1   MAJOR ENDOCRINE GLANDS: HORMONES AND FUNCTIONS

Gland	Hormone	Function of hormone	Means of control	Target area
Posterior pituitary (neurohypophysis)	Antidiuretic hormone (ADH; vasopressin).	Inhibits urine formation by increasing water absorption from kidneys. Constricts arterioles, raises blood pressure.	Synthesized in hypothalamus and released from neurohypophysis. Action potentials located in hypothalamic secretory neurons.	Kidney tubules, arterioles.
	Oxytocin.	Stimulates contractions of pregnant uterus. Stimulates milk ejection from breasts after childbirth.	Synthesized in hypothalamus and released from neurohypophysis. Action potentials located in hypothalamic secretory neurons.	Smooth muscle, especially uterus of pregnant woman; mammary glands (breasts).
Anterior pituitary (adenohypophysis)	Growth hormone (GH; somatotropin; somatotropic hormone, STH).	Stimulates growth. Promotes protein synthesis, fat mobilization. Slows carbohydrate metabolism.	Hypothalamic growth-hormone releasing factor (GHRF); growth-hormone inhibiting factor (GHIF).	Bone, muscle.
	Prolactin (lactogenic hormone; luteotropic hormone, LTH).	Promotes breast development during pregnancy. Stimulates milk production after childbirth.	Hypothalamic prolactin-inhibiting factor (PIF); prolactin-releasing factor (PRF).	Mammary glands.
	Thyroid-stimulating hormone (TSH; thyrotropin, thyrotropic hormone).	Stimulates production and secretion of thyroid hormones.	Hypothalamic thyrotropin-releasing hormone (TRH).	Thyroid gland.
	Adrenocorticotropic hormone (ACTH; corticotropin, adrenocorticotropin).	Stimulates production and secretion of adrenal cortex steroids.	Hypothalamic corticotropin-releasing hormone (CRH).	Adrenal cortex, skin, liver, mammary glands.
	Luteinizing hormone (LH; lutropin).	Female: stimulates development of corpus luteum, release of mature ovum, production of progesterone and estrogen. Male: stimulates secretion of testosterone, development of interstitial tissue of testes.	Hypothalamic gonadotropin-releasing hormone (GnRH).	Ovaries, testes.
	Follicle-stimulating hormone (FSH; follitropin).	Female: stimulates growth of ovarian follicle, ovulation. Male: stimulates sperm production.	Hypothalamic follicle-stimulating hormone-releasing factor (FSHRF).	Ovaries, testes.
	Melanocyte-stimulating hormone (MSH; melanotropin).	Apparently involved with skin color, in combination with ACTH.	Uncertain.	Apparently melanocytes.
Thyroid (follicular cells)	Thyroid hormones: thyroxine (T_4), triiodothyronine (T_3).	Increase body metabolism. Increase sensitivity of cardiovascular system to sympathetic nervous activity. Affect maturation and homeostasis of skeletal muscle.	Thyroid-stimulating hormone (TSH) from adenohypophysis. TSH regulated by thyrotropin-releasing hormone (TRH) from brain.	Most cell types.

Gland	Hormone	Function of hormone	Means of control	Target area
Thyroid (parafollicular cells)	Calcitonin (thyrocalcitonin).	Lowers calcium and phosphate levels in blood.	Blood calcium concentration.	Bone, kidneys, other cells.
Parathyroid	Parathormone (PTH; parathyroid hormone).	Increases calcium level, decreases phosphate level in blood.	Blood calcium concentration.	Bone, intestine, kidneys, other cells.
Adrenal cortex	Glucocorticoids, mainly cortisol (hydrocortisone), cortisone, corticosterone, 11-deoxycorticosterone.	Affect metabolism of all foods. Stimulate gluconeogenesis and regulate level of blood sugar. Act as anti-inflammatory drugs. Affect growth. Decrease effects of stress. Decrease ACTH secretion.	Corticotropin-releasing hormone (CRH) from hypothalamus, ACTH from adenohypophysis.	General.
	Mineralocorticoids, mainly aldosterone.	Control sodium retention and potassium loss in urine.	Angiotensin II, blood potassium concentration.	Kidney tubules.
	Gonadocorticoids (adrenal sex hormones).	Slight effect on gonads.	ACTH.	Ovaries, testes.
Adrenal medulla	Epinephrine (adrenaline).	Increases pulse rate, blood pressure, heart rate. Regulates diameter of arterioles. Stimulates contraction of smooth muscle. Increases blood sugar by stimulating breakdown of glycogen.	Sympathetic nervous system.	Heart, smooth muscle, arterioles, liver, skeletal muscle.
	Norepinephrine (noradrenaline).	Constricts arterioles. Increases metabolic rate.	Sympathetic nervous system.	Arterioles.
Pancreas (beta cells in pancreatic islets)	Insulin.	Facilitates glucose transport across plasma membranes. Lowers blood sugar by increasing liver and muscle glycogen.	Blood glucose concentration.	Muscle, liver, adipose tissue.
Pancreas (alpha cells in pancreatic islets)	Glucagon.	Increases blood sugar by decreasing liver glycogen.	Blood glucose concentration.	Liver.
Ovaries (follicle)	Estrogens.	Affect development of sex organs and female characteristics.	Follicle-stimulating hormone (FSH).	Reproductive system, other parts of the body.
Ovaries (corpus luteum)	Progesterone, estrogens.	Influence menstrual cycle. Stimulates growth of uterine wall, maintains pregnancy.	Luteinizing hormone (LH).	Uterus.
Placenta	Estrogens, progesterone, human chorionic gonadotropin (hCG).	Maintains pregnancy.	Uncertain.	Ovaries, mammary glands, uterus.
Testes	Androgens, mainly testosterone.	Affect development of sex organs and male characteristics.	Luteinizing hormone (LH).	Reproductive system, other parts of the body.

(Table 16.1 continues on the following page.)

TABLE 16.1 (Continued)

Gland	Hormone	Function of hormone	Means of control	Target area
Thymus	Thymosin alpha, thymosin B_1 to B_5, thymopoietin I and II, thymic humoral factor (THF), thymostimulin, factor thymic serum (FTS).	Help develop T cells in thymus, maintain T cells in other lymphoid tissue; involved in development of some B cells into antibody-producing plasma cells. May influence secretion of reproductive hormones from pituitary.	Uncertain.	T cells and B cells in lymphoid tissue.
Digestive system	Secretin.	Stimulates release of pancreatic juice to neutralize stomach acid.	Acid in small intestine.	Cells of pancreas.
	Gastrin.	Produces digestive enzymes and hydrochloric acid in stomach.	Food entering stomach.	Stomach mucosa.
	Cholecystokinin (CCK).	Stimulates release of pancreatic enzymes and gallbladder contraction.	Food in duodenum.	Pancreas, gallbladder.
Heart	Atriopeptin (atrial natritic factor, ANF).	Helps maintain homeostatic balance of fluids, electrolytes. Lowers blood pressure, blood volume.	Salt concentration, blood pressure, blood volume.	Blood vessels, kidneys, adrenal glands.

receptor protein in the cytoplasm (step 2). The hormone-receptor complex acquires an affinity for DNA that causes it to enter the nucleus of the cell, where it binds to DNA (step 3) and regulates the transcription of specific genes to form messenger RNA (step 4). The newly transcribed mRNA leaves the nucleus and moves to the rough endoplasmic reticulum in the cytoplasm, where it initiates the process of protein synthesis (step 5) (see Chapter 3). Some of these newly synthesized proteins may be enzymes whose effects on cellular metabolism constitute the cellular responses attributable to the specific steroid hormone in question (step 6).

Cortisol, progesterone, estrogen, testosterone, and thyroxine are mobile-receptor hormones, and have their receptors localized in the nucleus. All the other hormones usually use the fixed-membrane-receptor mechanism, but some nonsteroid hormones, such as prolactin, can also use endocytosis to enter the cytoplasm of a target cell. How the nonsteroid hormone combines with intracellular receptors is not yet known.

Ask Yourself

1 *If a hormone is the "first messenger," what is the "second messenger"?*

2 *What is the function of a receptor site on a plasma membrane?*

3 *What kind of hormones stimulate protein synthesis within a cell?*

PITUITARY GLAND (HYPOPHYSIS)

The *pituitary gland,** also known as the *hypophysis* (Gr. "undergrowth"), is located directly below the hypothalamus, and rests in the *sella turcica,* a depression in the sphenoid bone. This important gland is protected on three sides by bone, and on top by a tough membrane. It is about 1.0 cm long, 1.0 to 1.5 cm wide, and 0.5 cm thick—about the size and shape of a plump lima bean. Because of the closeness of the pituitary to the optic chiasma, an enlarged pituitary generally affects vision by impinging upon the optic pathways.

The pituitary gland has two distinct lobes: the anterior lobe (*adenohypophysis*) and the posterior lobe (*neurohypophysis*) (Figure 16.5). The adenohypophysis is the larger lobe, accounting for about 75 percent of the total weight of the gland. A more significant difference between the two lobes is the abundance of functional secretory cells in the adenohypophysis (Gr. *aden,* gland) and the presence of only supporting pituicytes in the neurohypophysis. The neurohypophysis, as its name suggests, has a greater supply of large nerve endings. Secretory cells produce and secrete hormones directly from the adenohypophysis, while the neurohypophysis obtains its hormones from neurosecretory cells in the

*The pituitary gland received its name, which means "mucus" in Latin, because it was thought to transfer mucus from the brain into the nose through the cribriform plate of the ethmoid bone.

FIGURE 16.5

The blood vessels that make up the hypothalamic-hypophyseal portal system provide the link between the hypothalamus and the adenohypophysis of the pituitary.

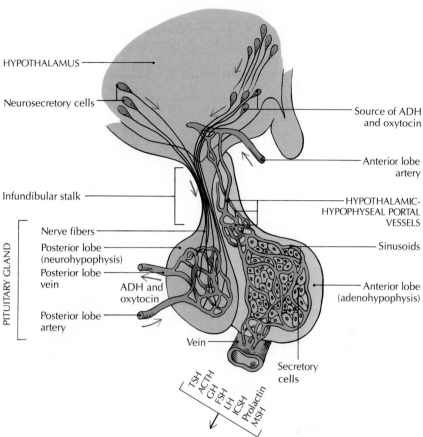

hypothalamus. These modified nerve cells project their axons down a stalk of nerve cells and blood vessels called the *infundibular* (L. funnel) *stalk,* or *infundibulum,* into the pituitary gland. In this way, a direct link exists between the nervous system and the endocrine system.

Other hypothalamic neurosecretory cells secrete releasing hormones into the portal vessels of the infundibular stalk. The portal vessels carry releasing hormones to the secretory cells in the adenohypophysis, and the secretory cells respond by secreting hormones.

Relationship between the Pituitary and Hypothalamus

Although the daily secretions of the pituitary gland are less than one-millionth of a gram, it was once called the "master gland" because of its control over most of the other endocrine glands and body organs (Figure 16.6). In truth, the *hypothalamus* might better deserve the title "master gland," since substances released from the hypothalamus control the secretions of the adenohypophysis, and hormones secreted from the neurohypophysis are synthesized and regulated by nerve centers in the hypothalamus.

The hypothalamus secretes at least nine releasing or inhibiting substances. Some of these substances are either *releasing factors* or *inhibiting factors,* while others are actually

hormones.* The hypothalamus and the adenohypophysis are connected by an extensive system of blood vessels called the *hypothalamic-hypophyseal portal system* (see Figure 16.5). Hormones produced in the hypothalamus are transported through the portal vessels to the adenohypophysis, where they either stimulate or inhibit the release of the appropriate pituitary hormones.

The link between the hypothalamus and the neurohypophysis relies on nerve impulses; hence, the name *neurohypophysis* for the posterior pituitary. The neurohypophysis is composed of unmyelinated axons of nerves whose cell bodies are in the hypothalamus, and pituicytes, which have a supporting rather than a secretory function.

Hormones of the Neurohypophysis

The neurohypophysis does not actually manufacture any hormones. Instead, hormones synthesized in the cell bodies of the hypothalamus move down the nerve fibers as secretory granules, and pass through the infundibular stalk to axon terminals lying on capillaries in the neurohypophysis. The

---

*A releasing or inhibiting *factor* is a releasing or inhibiting substance whose specific molecular structure is not known. In contrast, a *hormone* is a releasing or inhibiting substance whose molecular structure *is* known.

FIGURE 16.6

Pituitary gland, together with the hypothalamus. Target areas for each hormone are shown under the relevant box. Note the detailed drawing of the hypothalamus.

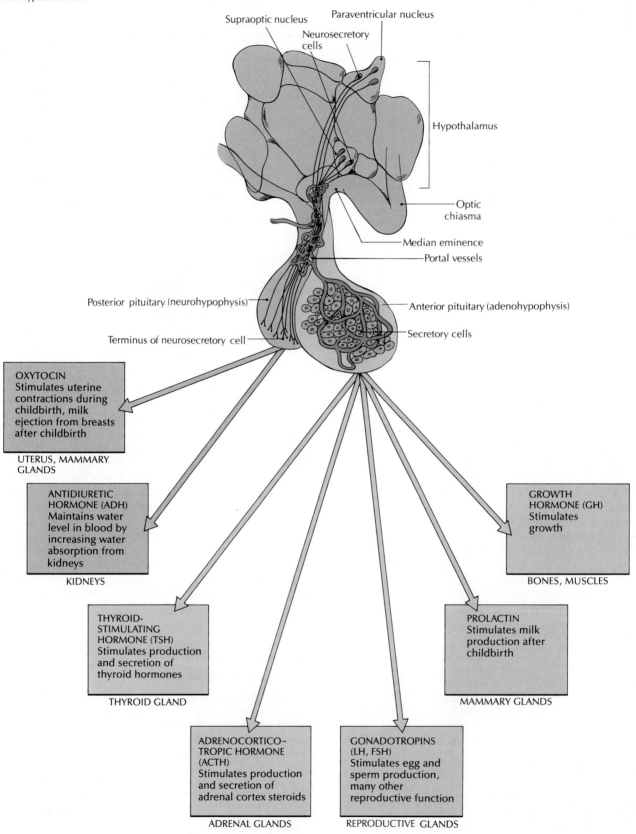

secretory granules are stored in the terminals until nerve impulses from the hypothalamus stimulate the secretion of hormones. Hypothalamic nerve cells are called *neurosecretory cells* because they are neurons with a secretory ability.

The nerve endings from the hypothalamus secrete two hormones, *antidiuretic hormone* (ADH) and *oxytocin,* into the capillaries. ADH is produced by the *supraoptic nucleus,* and oxytocin is produced by the *paraventricular nucleus* (see Figure 16.6). Thus, the hormones of the neurohypophysis are actually neurosecretions from hypothalamic nerve cells.

Antidiuretic hormone A *diuretic* is a substance that stimulates the excretion of urine, whereas an *antidiuretic* decreases urine excretion. **Antidiuretic hormone (ADH),** or *vasopressin,* is a peptide composed of nine amino acids (Figure 16.7). Its major role is to increase water absorption in the collecting ducts of the kidneys (kidney tubules) so that less urine is excreted. Thus, it plays a critical role in the overall regulation of the body's fluid balance. It also can stimulate the contraction of smooth muscles in the arterioles, small blood vessels that connect the arteries to capillaries. As a result of the action of ADH, blood vessels constrict, and blood pressure rises.

The main target organs of ADH are the kidneys. It works by increasing the permeability of the kidney tubules, thereby allowing water to be reabsorbed rather than excreted. Such a regulating mechanism is important if the body becomes dehydrated for any reason. For example, hemorrhaging causes an increase in ADH secretion so that the body's balance of fluids is maintained. Emotional or physical stress, increased plasma osmotic pressure,* decreased extracellular fluid volume, very strenuous exercise, and drugs such as nicotine or barbiturates all *increase* the secretion of ADH, which in turn produces a rise in blood pressure and an inhibition of urine formation.

The secretion of ADH is *decreased* by a drop in plasma osmotic pressure, by increased extracellular fluid volume, or by increased alcohol levels. If you consume large quantities of beer, for example, you will stimulate urine excretion, not only because you increase the liquid content of your body, but also because the alcohol in the beer inhibits the secretion of ADH, which results in an increase in urine excretion. In addition, another component of beer, lupulin from hops, is a diuretic.

A deficiency of ADH results in a disease known as *diabetes insipidus.*† Without sufficient ADH, large volumes of dilute urine are excreted, usually about 5 L a day, typically leading to dehydration and constant thirst.

Oxytocin *Oxytocin* (AHK-suh-TOE-sihn; Gr. ''sharp childbirth'') stimulates uterine contractions during childbirth, and milk ejection (for feeding the infant) after childbirth. Oxytocin works during childbirth by stimulating smooth muscle contractions in the uterus. The uterus becomes very sensitive

---

*Specific cells in the hypothalamus act as osmoreceptors that respond to changes in the osmotic pressure (caused by the presence of solutes such as electrolytes) of the interstitial fluid.

†This form of diabetes is not to be confused with *diabetes mellitus;* see Physiological and Anatomical Abnormalities at the end of this chapter.

FIGURE 16.7

Chemical structures of (A) ADH and (B) oxytocin. The hormones are similar, differing only in two amino acids (boxes).

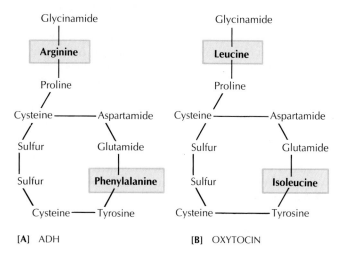

[A] ADH   [B] OXYTOCIN

to oxytocin during the late stages of pregnancy, and secretions increase during the actual process of childbirth. The attending physician may even inject oxytocin (actually synthetic oxytocin, called Pitocin) directly into the uterus through the vagina to stimulate labor.

The ejection of milk from the mammary glands occurs after childbirth, when oxytocin stimulates *myoepithelial cells* that surround the ducts of the breasts. The myoepithelial (basket) cells contract, forcing milk out of the network of milk-containing sinuses (alveoli) in the breasts, and through ducts in the nipples. Suckling by the newborn infant transmits impulses to the brain that stimulate nerve cells in the hypothalamus to cause the release of oxytocin. The infant does not receive any milk for about a minute after it starts suckling, but once oxytocin is secreted, the contraction of the myoepithelial cells begins, and milk flows from the nipple. Although oxytocin stimulates milk *ejection,* the actual *production* of milk is stimulated by *prolactin,* a pituitary hormone secreted by the adenohypophysis.

Like ADH, oxytocin is a peptide composed of nine amino acids. The structure of the two hormones is similar except for a difference in two of the amino acids (see Figure 16.7).

Hormones of the Adenohypophysis

The true endocrine portion of the pituitary is the adenohypophysis, which synthesizes seven separate hormones. All hormones secreted by the adenohypophysis are polypeptides, and all but two are considered to be *tropic* (TROE-pihk) *hormones,* hormones whose primary target is another endocrine gland. (''Tropic'' is an adjective derived from the Greek word *trophikos,* meaning ''turning toward'' or ''to change.'' It has nothing to do with the tropics.) The two nontropic hormones are growth hormone and prolactin, which are involved with growth and milk production, respectively.

Growth hormone *Growth hormone (GH)* is also called *somatotropin* or *somatotropic hormone* (STH). Rather than influencing a specific target organ, it affects all parts of the body that are associated with growth. Growth hormone has its most dramatic effect on the growth rate of children and adolescents, increasing tissue mass and stimulating cell division.

The most direct effect of growth hormone is to maintain the epiphyseal disks of the long bones (see Chapter 6). If a young person is deficient in growth hormone, there is a marked closure of the epiphyseal disks, the body stops growing, and *dwarfism* results. In contrast, if the secretion of growth hormone does not decrease toward the end of adolescence, as it usually does, *giantism* occurs, and the person continues to grow to 7 or even 8 ft tall. When growth hormone is overproduced after normal growth has stopped, a condition called *acromegaly* occurs, with bones in the head, hands, and feet thickening rather than lengthening.

Growth hormone stimulates the growth rate of body cells by increasing the formation of RNA, which speeds up the rate of protein synthesis. At the same time, growth hormone decreases the breakdown of proteins. It also causes a shift from the use of carbohydrates for energy to the use of fats. By increasing the use of the body's adipose tissue to produce fatty acids, and decreasing the body's use of glucose, growth hormone conserves carbohydrates. The overall effect is a change in the body's composition, so that muscle mass is increased, and deposits of fat are decreased. When blood glucose levels are increased by growth hormone, the result is *hyperglycemia.* This so-called *diabetogenic* ("producing diabetes") *effect* is misleadingly similar to diabetes mellitus.

The secretion of growth hormone is controlled by two factors produced by the hypothalamus and transported to the adenohypophysis. These are *growth-hormone releasing factor* (GHRF) and *growth-hormone inhibiting factor* (GHIF). GHRF was finally isolated and artificially reproduced in 1982 by researchers at the Salk Institute in La Jolla, California.

Prolactin The other hormone considered to be nontropic is **prolactin.** Prolactin has two functions in females. (1) Together with the female sex hormone estrogen, it stimulates the development of the duct system in the mammary glands during pregnancy. (2) It stimulates milk production (lactogenesis) from the mammary tissue after childbirth. (Recall that oxytocin stimulates milk *ejection* from the mammary glands during breastfeeding, but not milk *production.*)

Because a woman is not pregnant most of the time, the secretion of prolactin is usually inhibited. Inhibition is accomplished by the secretion of *prolactin-inhibiting factor* (PIF) by the hypothalamus. The inhibition of prolactin secretion begins to ease during pregnancy, and by the time the baby is born and the placenta is expelled from the uterus, most of the inhibitory effects of PIF are removed. In addition, the process of nursing a baby apparently causes the hypothalamus to secrete a *prolactin-releasing factor* (PRF), which *stimulates* the secretion of prolactin.

Thyroid-stimulating hormone *Thyroid-stimulating hormone* (TSH) is also known as *thyrotropin.* It stimulates the synthesis and secretion of thyroid hormones in several ways. The main effect of TSH is to stimulate the secretion of *thy-roxine,* the main thyroid hormone. An excessive amount of TSH increases the blood flow into the thyroid gland. As a result, the cells grow excessively, producing an enlarged thyroid gland, called a *goiter.*

The secretion of TSH is controlled by *thyrotropin-releasing hormone* (TRH) produced by the hypothalamus. The secretion of TRH depends on the amount of thyroxine circulating in the blood. Low blood levels stimulate an increase in TRH. When a normal level of thyroxine is reached, the production of TRH slows to a rate that merely maintains a stable condition. It has been found that cold temperatures also stimulate the production of TRH and the subsequent secretion of TSH.

Adrenocorticotropic hormone *Adrenocorticotropic hormone* (ACTH) is also called *corticotropin* or *adrenocorticotropin.* It stimulates the adrenal cortex to produce and secrete steroid hormones called *glucocorticoids.* Secretions of ACTH are regulated by the liberation of corticotropin-releasing hormones (CRHs) from the hypothalamus, which are in turn regulated by a feedback system influenced by such factors as stress, insulin, and ADH and other hormones.

Luteinizing hormone *Luteinizing hormone* (LH, lutropin) receives its name from the *corpus luteum,* a temporary endocrine tissue in the ovaries that secretes the female sex hormones progesterone and estrogen. LH, a gonadotropic hormone, stimulates ovulation, the monthly release of a mature egg from an ovary. In the male, the same hormone used to be called **interstitial cell-stimulation hormone** (ICSH), but now is called luteinizing hormone in both sexes. Its target cells are interstitial ("between spaces") cells in the testes that secrete the male hormone testosterone. The mechanism for the control of LH depends on a specific gonadotropin-releasing hormone (GnRH) from the hypothalamus, which is regulated by a typical negative feedback system involving levels of progesterone, estrogens, and testosterone.

Follicle-stimulating hormone *Follicle-stimulating hormone* (FSH), or *follitropin,* is also a gonadotropic hormone of the adenohypophysis. In females, FSH stimulates the growth of follicle cells in the ovaries that eventually develop into mature egg cells (oogenesis) during each menstrual cycle; it also stimulates the follicle cells to secrete estrogen. In the male, FSH stimulates the cells of the testes that produce sperm (spermatogenesis). The regulatory mechanism for FSH is similar to the negative feedback systems of other hormones of the adenohypophysis. The specific regulating factor from the hypothalamus is called *follicle-stimulating hormone-releasing factor* (FSHRF), which is released according to the blood levels of male and female sex hormones.

Does milk production cease once the menstrual cycle resumes after pregnancy?

The drive to produce food for offspring is so strong that prolactin levels remain high, and milk continues to be produced, not only after the menstrual cycle starts again, but even if a woman becomes pregnant again, just as long as she is still nursing her child.

Melanocyte-stimulating hormone The exact hormonal function of *melanocyte-stimulating hormone (MSH)* is uncertain. In fact, MSH may be classified more correctly as a precursor to an active hormone. The fetus produces MSH in the intermediate lobe (pars intermedia) of the pituitary gland, located between the anterior and posterior lobes. The intermediate lobe degenerates shortly before birth, and remains in the fully developed body only as a nonfunctional remnant. MSH is also present in small amounts in the anterior lobe.

The presence of MSH indirectly increases the activity of *melanocytes* (pigmented cells that affect skin, eye, and hair color). One form of MSH, called α-MSH, is produced by the stimulation of ACTH. When ACTH is released from the anterior pituitary, α-MSH is also released. Normally, the amount of MSH released is not enough to stimulate melanocytes. When ACTH is also produced, however, the combination of α-MSH and ACTH causes a change in pigmentation. The reason why α-MSH is not effective by itself is that it is secreted in small amounts, while the quantity of ACTH is large. Thus it is likely that ACTH is considerably more important than α-MSH in determining the amount of melanin in the skin, and in turn, skin color.

MSH is secreted by basophilic cells. Its secretion is stimulated by a hypothalamic regulating factor called melanocyte-stimulating hormone releasing factor (MRF), and inhibited by melanocyte-stimulating hormone inhibiting factor (MIF). A deficiency of MSH causes pale skin, and an excess causes the skin to darken.

Ask Yourself

1 What are the two lobes of the pituitary gland called?

2 What is the purpose of the hypothalamic-hypophyseal portal system?

3 What hormones are secreted by the adenohypophysis?

4 How do hypothalamic-releasing factors function?

How can the tissues of the adenohypophysis be specialized to synthesize so many different hormones?

There are three types of cells in the adenohypophysis: acidophils, basophils, and chromophobes. Acidophils, which stain red, secrete growth hormone and prolactin. Basophils, which stain blue, secrete TSH, FSH, LH, and MSH, and may be involved in the secretion of ACTH. The function of chromophobes is not firmly established, but they may be acidophils and basophils that have lost their granules after secreting, or reserve cells capable of differentiating into acidophils or basophils; they may also take part in the secretion of ACTH. Chromophobes stain little, if at all, hence their name, which means ''fear of color.''

THYROID GLAND

The *thyroid gland* is located in the neck, anterior to the trachea (Figure 16.8A). It consists of two lobes, one on each side of the junction between the larynx (voice box) and trachea (windpipe). The lobes are connected across the second and third tracheal rings by a bridge of thyroid tissue called an *isthmus*. In about half of the cases, there is a pyramidal process extending upward from the isthmus.

The thyroid gland has a complex circulatory system through which amino acids, iodine, gland secretions, and other substances are transported (see Figure 16.8B). The gland is composed of hundreds of thousands of spherical sacs, or *follicles*, which are hollow balls filled with a gelatinous colloid in which the thyroid hormones are stored (Figure 16.8C). The follicles are made up of a single layer of cuboidal epithelial cells.

Two types of cells make up the thyroid gland. The *follicular cells* are the principal cells. The *parafollicular cells*, though usually larger than follicular cells, are not as plentiful. Clusters of parafollicular cells are found between follicles. Single cells may appear within the wall of a follicle, but are not in contact with the colloid inside the lumen.

The thyroid is the only endocrine gland that is able to store its secretions outside its principal cells, and the stored form is different from the actual hormone that is secreted into the bloodstream. The stored chemical is broken down by several enzymes before it is released into the blood. The rich supply of blood vessels extending through the thyroid gland makes it relatively easy for the hormones to move through the follicle cells into the capillaries.

Thyroxine and Triiodothyronine

Follicular cells secrete the most abundant thyroid hormone, *thyroxine.* The thyroxine molecule, which contains four atoms of iodine, is often referred to as tetraiodothyronine, or T_4. Follicular cells also synthesize **triiodothyronine,** often called T_3 because its molecules contain only three atoms of iodine. Thyroxine and triiodothyronine are collectively called the *thyroid hormones*. The parafollicular cells (also called C cells) produce the hormone *calcitonin*, which mainly reduces the level of calcium in blood. Most of the triiodothyronine is

How can LH and FSH affect both male and female organs?

The same hormone is able to stimulate similar kinds of actions in two very different glands (testes and ovaries) because male and female sex organs have the same embryonic origin. Thus, FSH is associated with sex cell production in males and females, and LH is associated with sex-hormone secretion in males and females.

FIGURE 16.8

The thyroid gland. (A) Anterior view. (B) Scanning electron micrograph of thyroid tissue, showing the distinct capillary plexus covering each follicle. Amino acids, iodine, and other substances enter through the capillary plexus that covers each follicle. Secretions from the follicle cells enter the circulation through the capillary plexus. Veins are evident. (C) Cross section of a thyroid follicle. The cuboidal epithelial cells form a hollow lumen, in which thyroid hormones are stored. Some parafollicular cells lie outside the follicle. [(B) Fujita, Tanaka, and Tokunaga, *SEM Atlas of Cells and Tissues.* Tokyo, Igaku-Shoin, 1981.]

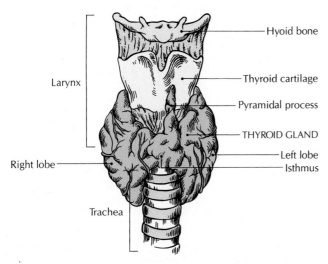

[A]

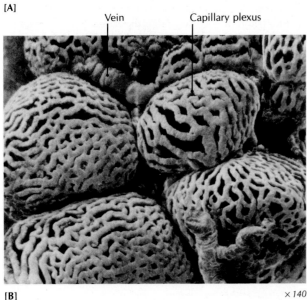

[B]   × 140

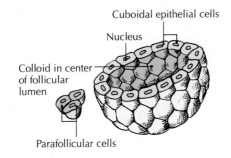

[C]

made from thyroxine, not in the thyroid itself, but in the surrounding tissues.

Although thyroxine accounts for about 90 percent of the thyroid's secretions, triiodothyronine is highly concentrated, and is just as effective. Both hormones consist mainly of iodine and the amino acid tyrosine. Iodine is obtained directly from dietary sources. People who live in inland areas where the iodine content of the water is low may become deficient in iodine if they do not have a supplementary source, such as iodized salt. Tyrosine is synthesized by the body from a wide variety of dietary sources, so there is no problem in obtaining it.

Both hormones appear to have the same endocrine functions. They accelerate cellular reactions in most body cells. They increase body metabolism (the rate at which cells use oxygen and food to produce energy and heat); cause the cardiovascular system to be more sensitive to sympathetic nervous activity, thus increasing cardiac output and heart rate; affect the maturation and homeostasis of the skeletal and central nervous systems; stimulate cellular differentiation and protein synthesis; and affect the growth rate, water balance, and several other physiological processes.

In general, thyroid hormones increase the oxygen consumption of most body cells, but not in the brain, spleen, or testes. In so doing, energy is produced and body heat is given off in a temperature-raising process called a *calorigenic effect.* Thyroid hormones are essential for proper skeletal growth in children, and they operate together with growth hormone from the pituitary to regulate overall growth and maturation, including the development of the nervous system.

Because the thyroid hormones affect most cells of the body rather than a specific target area, the possibilities of malfunctions are far-reaching and can be serious. Overactivity of the thyroid gland results in *hyperthyroidism,* which usually produces a goiter. Underactivity of the thyroid gland results in *hypothyroidism.* An underactive thyroid gland during prenatal development or infancy causes *cretinism,* which results in mental retardation and irregular development of bones and muscles. (Both conditions are discussed in Physiological and Anatomical Abnormalities at the end of this chapter.)

The ionized form of iodine, *iodide* (I⁻), is found in food. This trace element comes to the thyroid as free ions via the blood. The follicular cells of the thyroid remove iodide from the blood, using an ATP-driven iodine pump. The iodine is bound and collected by a large glycoprotein called *thyroglobulin* (also called "colloid protein"); it then is used to iodinate tyrosine residues of thyroglobulin to produce the thyroid hormone. When thyroid hormones are released, the iodinated tyrosine molecules are removed from the thyroglobulin and secreted into the blood.

The entire process of thyroid production and secretion is regulated by thyroid-stimulating hormone (TSH), secreted by the adenohypophysis (see Figure 16.2). The secretion of TSH is stimulated when thyroxine concentration in the blood is low and by such factors as cold, stress, or pregnancy that increase the need for energy. The flow of TSH is inhibited when the blood level of thyroxine is high. In such a case, thyroxine in the blood slows down the secretion of thyrotropin-releasing hormone (TRH) from the neurohypophysis until a normal maintenance level of throxine is reached.

Calcitonin

The parafollicular cells of the thyroid produce a polypeptide hormone called **calcitonin** (CT), or *thyrocalcitonin*. It lowers the calcium level in the blood, in direct contrast to the action of the hormone (parathormone) secreted by the parathyroid glands, which promotes a rise in blood calcium. As calcium concentration increases, so does the secretion of calcitonin, which then reduces the calcium blood level toward normal. Calcitonin acts directly on osteoclasts to reduce their effectiveness in remodeling bone and in the reabsorption of calcium. Apparently, it also increases the movement of calcium from the blood into the bones. Calcitonin acts for short periods only; the long-term regulation of calcium in the blood is a function of the parathyroid hormone. It is interesting that the secretion of calcitonin is directly regulated by the amount of calcium in the blood, rather than depending upon the higher control of a pituitary-hypothalamus feedback system. Calcitonin also lowers phosphate levels in the blood.

Ask Yourself

1 *Where is the thyroid gland located?*

2 *What is the most abundant thyroid hormone?*

3 *How does the regulation of calcitonin secretion differ from that of thyroxine?*

PARATHYROID GLANDS

The **parathyroid glands** are tiny, pea-sized glands embedded in the posterior of the thyroid lobes, usually two glands in each lobe (Figure 16.9A). The parathyroid glands are so small that they were not discovered until 1850. Before then, the parathyroids were frequently removed unknowingly during goiter surgery, and the patients died "mysteriously." The adult parathyroids are composed mainly of small *principal* (or *chief*) *cells,* which secrete most of the parathyroid hormone, and larger *oxyphilic cells,* which secrete a reserve capacity of hormone (Figure 16.9B).

The main function of **parathormone** (PTH), or *parathyroid hormone,* is to regulate the levels of calcium (Ca^{2+}) and phosphate (HPO_2^{4-}) in the blood. An improper balance of calcium and phosphate ions in the blood can cause faulty transmission of nerve impulses, destruction of bone tissue, hampered bone growth, and muscle tetany.

When the level of calcium in the blood is low, parathormone *increases* the level in several ways. (1) It stimulates the activity of osteoclasts, which break down bone tissue, releasing calcium ions from bones into the blood. (2) It enhances the absorption of calcium and phosphate from the small intestine into the blood. Adequate dietary vitamin D and the hormone *1,25-dihydroxyvitamin D_3* (produced in the kidneys) are important for this process. (3) It promotes the reabsorption of calcium by the kidney tubules, so that the amount of calcium excreted in urine is decreased.

FIGURE 16.9

Parathyroid glands. (A) Posterior view. (B) Section of a human parathyroid gland, showing the small principal cells and the larger oxyphilic cells. [*W. Bloom and D. W. Fawcett,* A Textbook of Histology, *11th ed. Philadelphia, W. B. Saunders, 1986.*]

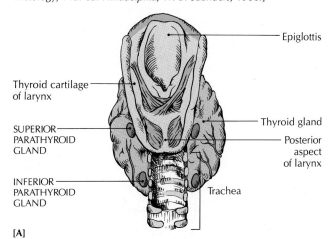

[A]

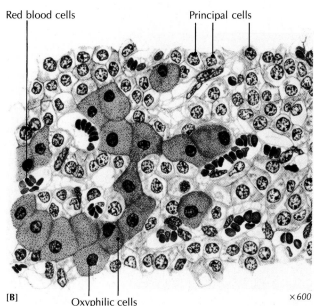

[B]   Oxyphilic cells   ×600

Parathormone *decreases* the concentration of phosphate in the blood by inhibiting its reabsorption by the kidney tubules, and thereby increasing its excretion in the urine. The decrease in blood phosphate tends to raise the level of calcium in the blood. The interaction between parathormone and calcitonin is illustrated in Figure 16.10.

In summary, three hormones play major roles in regulating the concentration of calcium in blood plasma: (1) Calcitonin from the thyroid gland decreases plasma calcium by inhibiting its reabsorption from bone tissue. (2) Parathormone from the parathyroid glands increases plasma concentration by releasing calcium ions from bone tissue by stimulating osteoclasts, and increasing its reabsorption in the kidney tubules. (3) 1,25-Dihydroxyvitamin D_3 from the kidneys increases the intestinal absorption of calcium ions into the blood plasma, and also mobilizes calcium ions from bone tissue.

FIGURE 16.10

Negative feedback mechanism of the thyroid gland (calcitonin) and the parathyroid glands (parathormone).

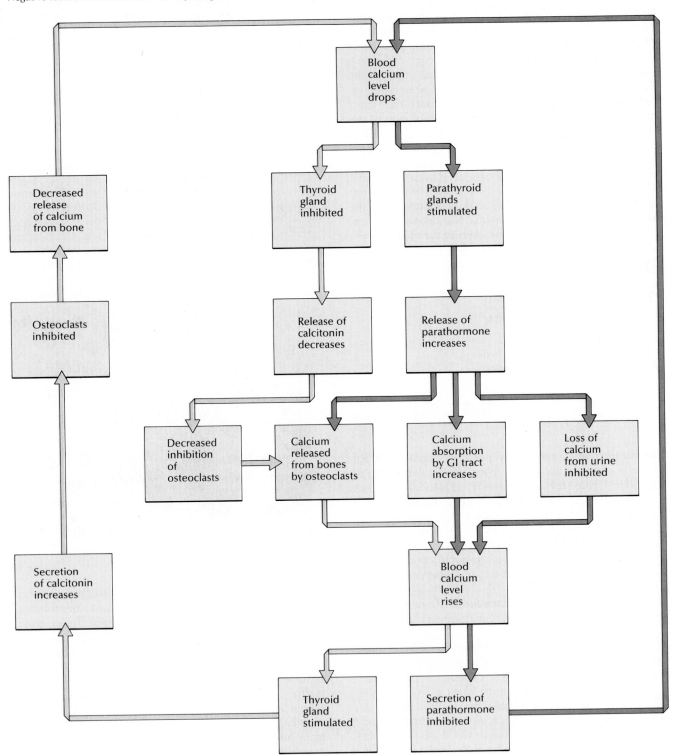

Ask Yourself

1 *What effect does parathormone have on the levels of calcium and phosphate in the blood?*

2 *How do calcitonin, parathormone, and 1,25-dihydroxyvitamin D_3 work together in the body to regulate calcium in the blood plasma?*

ADRENAL GLANDS

The two ***adrenal*** (uh-DREEN-uhl; L. "upon the kidneys") ***glands*** rest like tilted berets on the superior tip of each kidney (Figure 16.11A). Each adrenal gland is actually made up of two separate endocrine glands. The inner portion of each gland is the *medulla* (L. "marrow," meaning inside); the outer portion, which surrounds the medulla, is the *cortex* (L. "bark," as in the outer bark of a tree) (Figure 16.11B). The medulla and cortex not only produce different hormones, but also have separate target organs.

Adrenal Cortex

The ***adrenal cortex*** accounts for about 90 percent of the weight of the adrenal gland, which weighs from 5 to 7 g. Like other glands with a mesodermal embryonic origin, its hormones are steroids. The adrenal cortex secretes three classes of general steroid hormones: *glucocorticoids,* a single *mineralocorticoid* (aldosterone), and small quantities of *gonadocorticoids,* or sex hormones.

The tissue of the cortex has three distinct zones lying beneath the outermost capsule (Figure 16.11C). Directly beneath the capsule is the thin *zona glomerulosa,* which supplies cells for all three zones if regeneration is necessary. It also contains the enzymes necessary for the production of the mineralocorticoid hormone aldosterone. The next level down from the outer capsule is the thick *zona fasciculata,* which makes up the bulk of the adrenal cortex. The glucocorticoids corticosterone and cortisol, as well as a small amount of gonadocorticoids, are secreted by the zona fasciculata. Cholesterol, the precursor of steroid hormones, is more concentrated in the cells of the zona fasciculata than in any other part of the body. These cells also contain considerable amounts of vitamin C. The *zona reticularis* is the deepest layer of the adre-

FIGURE 16.11

Adrenal glands. (A) The adrenals rest on top of the kidneys. (B) Cross section. (C) Enlarged section of the adrenal cortex of an adult man, showing the layered regions from the outer capsule to the zona reticularis, which merges with the inner medulla.

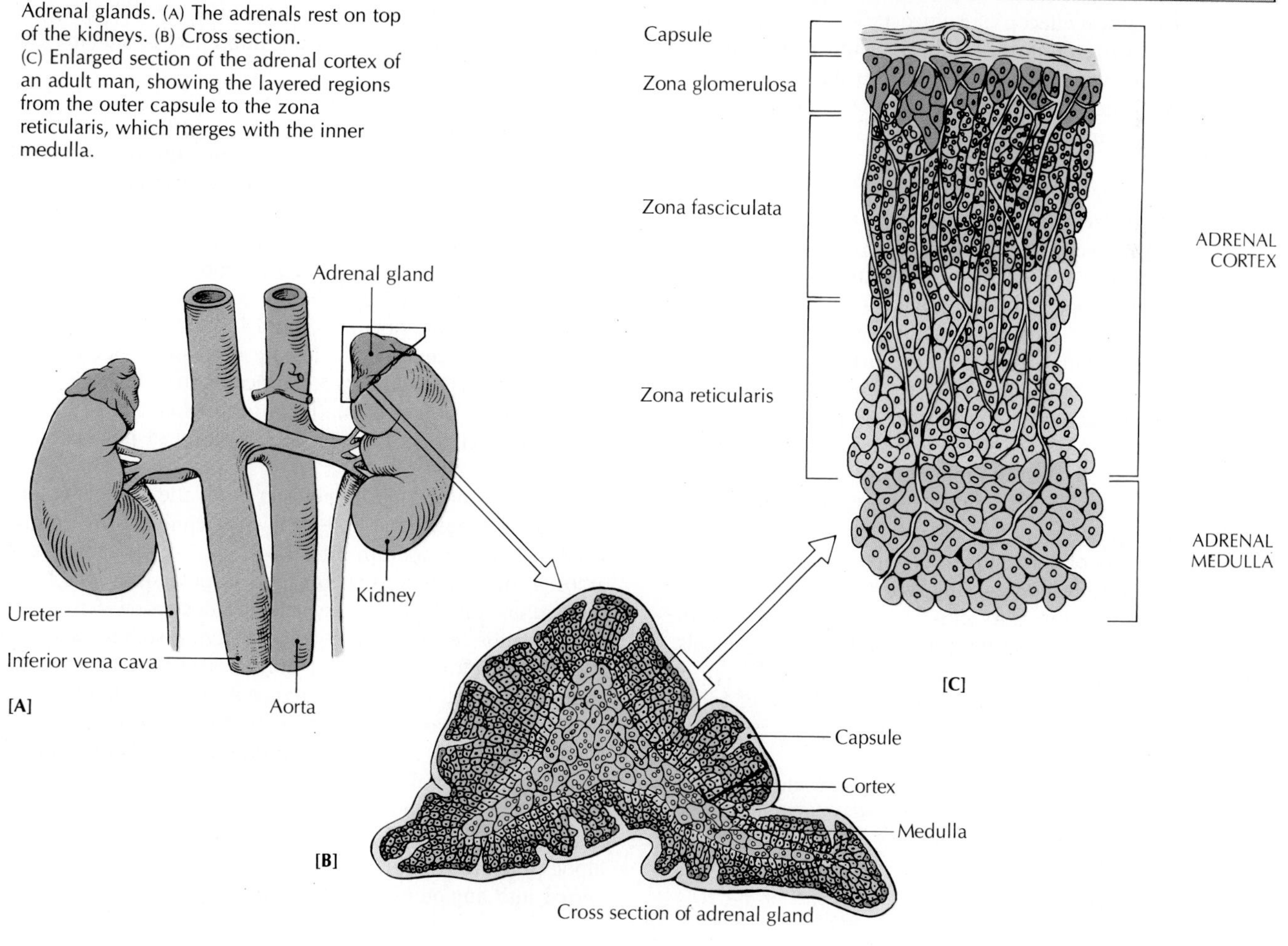

FIGURE 16.12

Negative feedback mechanism regulating the secretion of aldosterone.

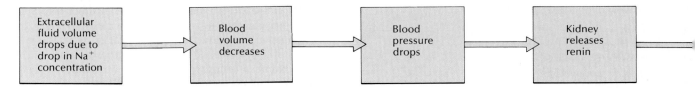

nal cortex. Although the cells here are not as orderly as those in the zona fasciculata, they are similar. The cells of the zona reticularis secrete dehydroepiandrosterone (DHEA), a substrate for the production of sex hormones.

Glucocorticoids The word *glucocorticoid* (gloo-koh-KORE-tih-koid) itself explains that the secretions come from the cortex (-*corticoid*) and that they help control the concentration of blood sugar (*gluco-* refers to glucose, or blood sugar). There are four types of glucocorticoids: *cortisol* (also called hydrocortisone), *cortisone, corticosterone,* and *11-deoxycorticosterone.* Cortisol accounts for about 95 percent of the activity of glucocorticoids. In addition to regulating the level of blood sugar, the glucocorticoids affect the metabolism of all types of foods, act as anti-inflammatory agents, affect growth, and decrease the effects of physical or emotional stress.

The most critical effect of the glucocorticoids is the stimulation of gluconeogenesis, the synthesis by the liver of glucose from noncarbohydrate substances such as amino acids and fats.* (*Gluco,* glucose + *neo,* new + *genesis,* production or generation; so gluconeogenesis is simply "the production of new glucose.")

Because cortisol stimulates the synthesis of glucose from amino acids or fats instead of from carbohydrates, cells do not need to remove glucose from the blood. As a result, the level of blood sugar rises, in what is known as a hyperglycemic or diabetogenic effect, called adrenal diabetes. Fortunately, this effect is usually balanced by the antidiabetogenic effect of insulin secreted by the pancreas. Enzymes that help convert amino acids into glucose become plentiful in the liver, and RNA increases also, probably taking part in the synthesis of the required enzymes. Besides facilitating the metabolism of proteins, cortisol promotes the metabolism of fats by removing fatty acids from adipose tissue and making them available as an energy source.

Another effect of cortisol is to suppress allergic reactions and inflammatory responses. One way it may operate is by decreasing the inflammatory effect of enzymes released by lysosomes. It also decreases the permeability of blood vessels, thereby reducing the flow of plasma that normally causes swelling and irritation. By decreasing the ability of white blood cells to engulf foreign substances, cortisol can slow down the inflammatory process. Also, by decreasing the efficiency of the thymus, spleen, and other lymphatic tissues related to the immune system, cortisol can reduce the number of antibodies that may cause tissue irritation. Finally, cortisol lowers fever, which in turn helps to reduce dilated blood vessels to their normal size.

The secretion of cortisol is controlled by a negative feedback system between the adrenal cortex and the adenohypophysis. When the blood level of cortisol drops, the hypothalamus is stimulated, and secretes the corticotropin-releasing hormone (CRH). Then the secretion of ACTH (corticotropin) from the adenohypophysis is increased, which stimulates the adrenal cortex to secrete additional cortisol. As the level of cortisol in the blood goes up, the secretion of ACTH goes down until the level of cortisol is normal. Pain, anxiety, and other types of stress induce the release of CRH in the same way that a lowered amount of cortisol does, and the same regulatory system is put into effect to stabilize the level of cortisol in the blood.

Mineralocorticoids Another type of adrenocortical hormone is a *mineralocorticoid* (mihn-uh-ruhl-oh-KORE-tih-koid). Mineralocorticoids are a group of steroid hormones produced in the cells of the zona glomerulosa. They control the concentration of minerals. The main mineralocorticoid is *aldosterone,* which controls the retention of sodium and the loss of potassium in urine. Mineralocorticoid activity also increases the reabsorption of sodium from sweat, saliva, and gastric juice. Mineralocorticoids are not as dependent upon CRH and ACTH as the glucocorticoids are. Instead, their means of control is a combination of angiotensin (a blood-vessel-constricting substance) and the level of potassium in the blood.

The main target area of aldosterone is the kidney, where the kidney tubules and collecting duct epithelia are stimulated to reabsorb sodium into the blood. At the same time, aldosterone stimulates the transport of potassium ions from the kidney tubules into the urine. Aldosterone acts upon sweat glands in the same way, conserving sodium and secreting potassium to create a normal acid-base balance, as well as a normal electrolyte balance in the body fluids.

The secretion of aldosterone is stimulated or inhibited by shifting levels of sodium and potassium, but the primary mechanism of regulation is the concentration of the octapeptide angiotensin II in the bloodstream. The entire mechanism starts when decreased sodium levels in the blood lower the blood pressure. The plasma globulin angiotensinogen is converted into angiotensin I when the enzyme renin, which is

*Another process promoted by cortisol is *glycogenesis,* the production of glycogen from other carbohydrates. Glycogen stored in the liver becomes effective as an energy source only when enzymes in the liver break down the large organic molecules of glycogen into smaller molecules of glucose that can circulate in the blood and become available to cells when they need energy.

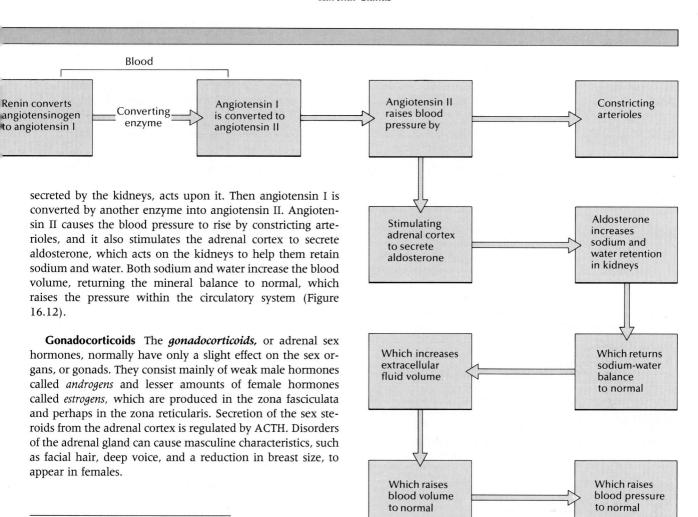

secreted by the kidneys, acts upon it. Then angiotensin I is converted by another enzyme into angiotensin II. Angiotensin II causes the blood pressure to rise by constricting arterioles, and it also stimulates the adrenal cortex to secrete aldosterone, which acts on the kidneys to help them retain sodium and water. Both sodium and water increase the blood volume, returning the mineral balance to normal, which raises the pressure within the circulatory system (Figure 16.12).

Gonadocorticoids The *gonadocorticoids,* or adrenal sex hormones, normally have only a slight effect on the sex organs, or gonads. They consist mainly of weak male hormones called *androgens* and lesser amounts of female hormones called *estrogens,* which are produced in the zona fasciculata and perhaps in the zona reticularis. Secretion of the sex steroids from the adrenal cortex is regulated by ACTH. Disorders of the adrenal gland can cause masculine characteristics, such as facial hair, deep voice, and a reduction in breast size, to appear in females.

Adrenal Medulla

The *adrenal medulla,* the inner portion of the adrenal gland, should be thought of as an entirely separate endocrine gland from the adrenal cortex. It secretes different hormones, has a different tissue structure, is derived from different embryonic tissue, and unlike the adrenal cortex, is not essential to life.

The cells of the adrenal medulla may extend somewhat into the innermost portion of the zona reticularis (see Figure 16.11). They are usually grouped in clumps, and are surrounded by blood vessels, especially capillaries. Granules in the cells store high concentrations of two hormones, epinephrine and norepinephrine, both of which are small catecholamines (Figure 16.13). Dopamine is also produced in the adrenal medulla.

The secretory cells of the adrenal medulla (sometimes called *chromaffin cells* because of their tendency to stain a dark color) are derived from the same embryonic ectodermal tissue as the ganglia of the sympathetic nervous system. Because of this common origin, it is not surprising that both types of cells have similar functions and effects. Adrenal chromaffin cells synthesize, store, and secrete a complex mixture of epinephrine and other hormones and a variety of proteins and peptides. These secretions have effects similar to those produced in the sympathetic nervous system.

Epinephrine and norepinephrine The main secretion of the adrenal medulla is *epinephrine* (ep-ih-NEFF-rihn), commonly known as adrenaline. The other hormone, *norepinephrine* (NE; noradrenaline), though closely related to epinephrine, is produced in much smaller amounts and has somewhat lesser effects. The hormones produce effects similar to stimulation by the sympathetic nervous system, but the hormonal effects last longer because the hormones remain in the blood for some time after the initial secretion. Also, epinephrine increases the rate of metabolism and the output of the heart even more than the sympathetic nervous system does. The major effects of epinephrine and norepinephrine are listed in Table 16.2.

The effectors on which epinephrine and norepinephrine act can be separated into two groups, based on their response to each hormone. Effector organs may contain either alpha or beta receptors on their plasma membranes, or both. Even though alpha and beta receptors both bind to epinephrine and norepinephrine, they do not initiate the same response. Alpha-receptor activation increases intracellular cyclic GMP, and beta-receptor stimulation activates intracellular cyclic AMP (see p. 477). Thus, the response of a tissue having alpha receptors for epinephrine and norepinephrine may be com-

TABLE 16.2   MAJOR EFFECTS OF EPINEPHRINE AND NOREPINEPHRINE

Organ or function	Effect of epinephrine	Effect of norepinephrine
Heart and blood vessels	Dilates coronary vessels; dilates arterioles in skeletal muscles and viscera; increases heart rate and cardiac output.	Dilates coronary vessels; causes vasoconstriction in other organs; increases heart rate and cardiac output.
Blood pressure	Raises blood pressure because of increased heart output and peripheral vasoconstriction.	Raises blood pressure because of peripheral vasoconstriction.
Muscles	Inhibits contraction of smooth muscles of digestive system, producing relaxation; dilates respiratory passageways; decreases rate of fatigue in skeletal muscle; increases respiratory rate and volume.	Relaxes smooth muscle of gastrointestinal tract.
Metabolism	Stimulates glycogenolysis (the breakdown of glycogen to glucose) in liver and muscles, which elevates levels of blood glucose and muscle lactic acid; increases oxygen consumption; enhances lipid metabolism; inhibits insulin release from pancreas, providing a ready supply of fatty acids for fuel by skeletal muscles and making glucose available, especially to the skeletal muscles and central nervous system during emergencies.	Enhances lipid metabolism and release of free fatty acids from adipose tissue.

FIGURE 16.13

Chemical structures of (A) epinephrine and (B) norepinephrine. Both hormones are catecholamines secreted by the adrenal medulla.

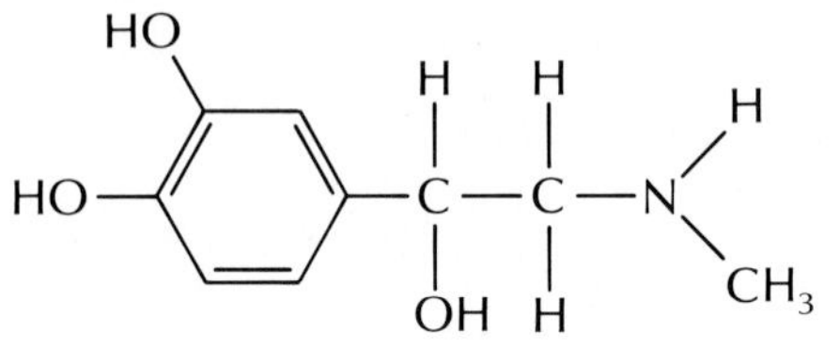

[A] EPINEPHRINE

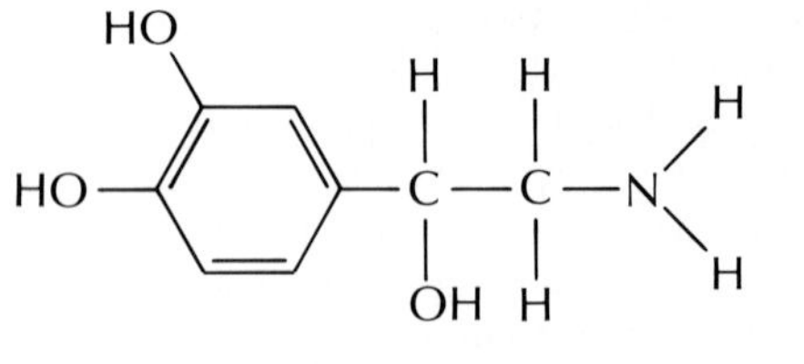

[B] NOREPINEPHRINE

pletely different from the response of a tissue having beta receptors for the same hormones.

Stress and the Adrenal Cortex

The initial perception of a stressful condition is made by the brain and other sensory receptors. The hypothalamus receives the message, and secretes ACTH-releasing factor, which in turn stimulates the adenohypophysis of the pituitary to secrete ACTH. The pituitary secretion quickly reaches its target organ, the cortex of the adrenal gland, which secretes cortisol (Figure 16.14).

Increased amounts of cortisol help the body deal with stress in the following ways:

1 The catabolism (breakdown) of protein is stimulated. This releases amino acids as an energy source, and also is a means of tissue repair in case of injury.

2 Amino acids are converted into glucose by the liver (gluconeogenesis). This provides a source of energy during stress, when eating and digestion would not normally take place. Without the new source of glucose, the body could suffer severe hypoglycemia. Other hormones such as glucagon are induced to stimulate gluconeogenesis.

Stress and the Adrenal Medulla

The adrenal medulla is stimulated to react to stress by the same pathway that stimulates the cortex. The secretion of epinephrine is stimulated by impulses from the sympathetic nervous system that originate in the hypothalamus. In times of stress, epinephrine produces a condition sometimes called the "fight-or-flight" effect, which permits the body to react quickly and strongly to emergencies. The secretion of epinephrine causes extreme effects that last a very short time. In fact, enzymes in the liver inactivate epinephrine in about three minutes.

After epinephrine has been secreted, some blood vessels constrict and others dilate, redistributing blood to such organs as the brain and muscles, where it is most needed. Digestion, an energy-consuming function, is halted during the emergency by the diversion of blood from the stomach and intestines to the muscles, where extra energy may be needed. Blood is moved away from the skin, reducing the danger from a possible surface wound. Blood pressure rises; the time needed for blood clotting is reduced; and circulation increases. Respiration rate increases and bronchioles dilate. Enzymes in the liver are activated to release glucose from glycogen for increased energy. In each case, the secretion of epinephrine favors the body's survival in times of stress.

Stressful situations bring about increased secretions of ACTH by the adenohypophysis, as well as the release of epinephrine and norepinephrine from the adrenal medulla. The secretion of ACTH subsequently increases the efficiency of the inflammatory reaction and adjusts the body's metabolism to react properly to its immediate problems.

FIGURE 16.14

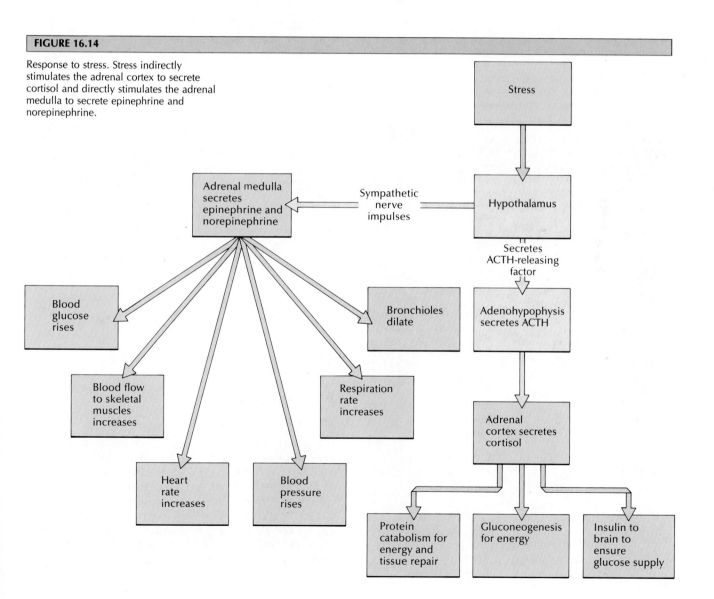

Response to stress. Stress indirectly stimulates the adrenal cortex to secrete cortisol and directly stimulates the adrenal medulla to secrete epinephrine and norepinephrine.

FIGURE 16.15

Pancreas. (A) Location and anatomy.
(B) Section of pancreatic tissue, showing a
pancreatic islet. (C) A highly magnified
human pancreatic islet, with alpha, beta,
and delta cells clearly delineated.
[(B) *Biophoto/Photo Researchers.* (C)
Courtesy of K. Kovacs and E. Horvath.]

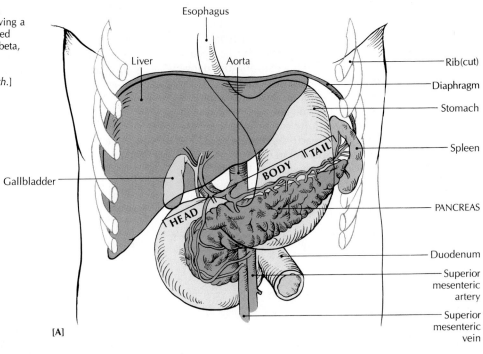

[A]

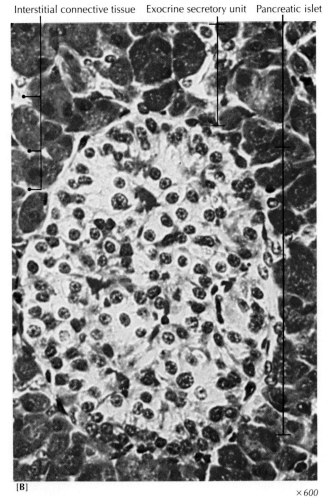

Interstitial connective tissue   Exocrine secretory unit   Pancreatic islet

[B]   × 600

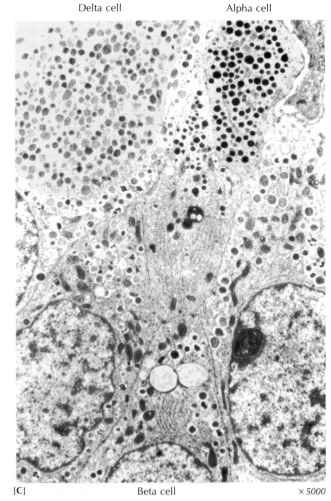

Delta cell   Alpha cell

[C]   Beta cell   × 5000

PANCREAS

The *pancreas* (Gr. "all flesh") is an elongated (12 to 15 cm), fleshy organ consisting of a head, body, and tail (Figure 16.15A). The area where the head and body join is the neck. The pancreas is located posterior to the stomach, with the head tucked into the curve of the duodenum (where stomach meets small intestine). The body and tail extend laterally to the left, with the tail making contact with the spleen.

The pancreas is considered a mixed gland because it functions both with ducts, as an exocrine gland, and without ducts, as an endocrine gland. As an exocrine gland, it acts as a digestive organ, secreting digestive enzymes and alkaline materials into a duct that empties into the small intestine. (We will discuss the pancreas as a digestive organ in Chapter 22.) As an endocrine gland, it secretes its hormones into the bloodstream. The endocrine portion of the pancreas makes up only about 1 percent of the total weight of the gland. This portion synthesizes, stores, and secretes hormones from clusters of cells called the *pancreatic islets* (islets of Langerhans) (Figure 16.15B).

The adult pancreas contains between 200,000 and 2,000,000 pancreatic islets scattered throughout the gland. The islets contain four special groups of cells, called alpha, beta, delta, and F cells (Figure 16.15C). *Alpha cells* produce glucagon, and *beta cells* produce insulin. Both hormones help to regulate metabolism, and are usually secreted simultaneously. Beta cells are the most common type. They are generally located near the center of the islet, and are surrounded by the other cell types. *Delta cells* secrete somatostatin, the hypothalamic growth-hormone inhibiting factor that also inhibits the secretion of both glucagon and insulin. *F cells* secrete pancreatic polypeptide, which is released into the bloodstream after a meal. The endocrine function of pancreatic polypeptide is not known.

Glucagon

Alpha cells in the pancreatic islets synthesize, store, and secrete the hormone *glucagon* (GLOO-kuh-gon). When the level of blood glucose falls, glucagon stimulates the liver to convert glycogen to glucose *(glycogenolysis)*, which causes the blood glucose level to rise. Glucagon also stimulates *gluconeogenesis*, the formation of glucose from noncarbohydrate sources such as amino acids and lactic acid. Of these two processes, glycogenolysis is the more important source of glucose. Glucagon increases the concentration of cyclic AMP from ATP in liver cells, causing the enzyme *phosphorylase a* to be produced. This enzyme separates glucose units from the large glycogen molecule, and the freed glucose units enter the bloodstream rapidly. As the result of a negative feedback system, the level of blood sugar is raised during periods of fasting or any other time that the level of blood sugar drops below normal. For this reason, glucagon is considered to be a *hyperglycemic factor*. Glucagon also stimulates the release of fatty acids and glycerol from adipose tissue.

Insulin

Like glucagon, *insulin* is a peptide hormone (Figure 16.16).* Beta cells, influenced by the concentration of blood sugar, secrete insulin when the blood sugar level rises, as it does immediately after a meal. The most important effect of insulin is to facilitate glucose transport across plasma membranes. It also enhances the conversion of glucose to glycogen (glycogenesis), which is then stored in the liver as a ready source of blood sugar.

*Insulin was the first protein to have its full structure determined. This was done in 1955 by Frederick Sanger, who received the Nobel Prize in 1958 in recognition of his work.

FIGURE 16.16

Steps in the production of insulin by the human body. (A) Long chains of amino acids are assembled by ribosomes to form *preproinsulin*. (B) As preproinsulin leaves the ribosome some of its amino acids are enzymatically cleaved, forming *proinsulin*. Proinsulin contains the material that will ultimately form two chains of amino acids connected by disulfide bridges. Before insulin is formed, the connecting peptide portion of the chain (green) must be enzymatically cleaved. (C) Once the connecting peptide section is deleted, all that remains is *insulin*, an A-chain linked to a B-chain by disulfide bridges. Pig insulin can be transformed into human insulin by replacing the final amino acid (arrow), alanine in pigs, by threonine. Purified pork insulin is a plentiful future source of insulin for diabetics, as well as insulin produced by genetically engineered bacteria.

[A] PREPROINSULIN

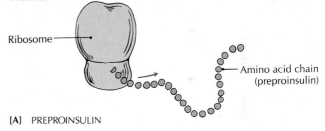

[B] PROINSULIN

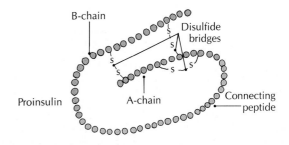

[C] INSULIN

The functions of insulin are opposite to those of glucagon, and the two hormones usually work in concert to maintain a normal blood glucose level. Figure 16.17 illustrates the negative feedback mechanisms that regulate the secretion of glucagon and insulin. When blood sugar is high, glucose moves into the liver cells easily. Insulin then stimulates liver enzymes to add phosphate groups to glucose molecules. The glucose 1-phosphate can now be added to the glycogen molecule for storage *(glycogenesis)*. Thus, insulin indirectly inhibits the conversion of glycogen to glucose and its release into the bloodstream. As a result, the concentration of blood sugar drops to a normal level. When beta cells do not produce enough insulin, diabetes mellitus results. Excessive amounts of insulin usually produce *hypoglycemia,* or ''low blood sugar.'' Insulin is also the primary hormone that stimulates the synthesis and storage of fats in adipose tissue, though their actual release from adipose tissue is brought about by glucagon. Insulin also promotes the uptake of glucose by fat cells. This glucose is then used to synthesize fatty acids and glycerol, which are stored in adipose tissue as neutral fats.

In muscle cells, insulin is thought to activate the membrane transport of glucose. If glucose is not needed immediately after it enters muscle cells, it is stored as glycogen by the same mechanism that operates in liver cells.

Nerve cells in the brain are freely permeable to glucose, which is fortunate, since glucose is the brain's only energy source.

Insulin produces an increase in stored fuel and body protein, and is considered an anabolic hormone. (Anabolism is the chemical process that builds up simple substances into more complex ones.) Some of the major effects of insulin secretion are summarized below:

1 Insulin increases glucose transport through some plasma membranes, which increases the metabolism of carbohydrates and decreases blood glucose (hypoglycemic effect).

2 Insulin increases amino acid transport through plasma membranes, which increases the synthesis of proteins and, in turn, growth.

3 Insulin increases the conversion of glucose into fatty acids, and enhances their transport through some plasma membranes, which increases fat build-up and storage.

Why do diabetics have to inject insulin instead of taking it orally?

Insulin is a peptide, which is a small protein molecule. It cannot be taken by mouth because it would be rapidly inactivated by protein-digesting enzymes in the digestive tract. Synthetic insulin, which is not a protein, can be taken orally, but it is not as effective as natural insulin injected subcutaneously.

Ask Yourself

1 *What is the function of the pancreatic islets?*

2 *Why is glucagon called a hyperglycemic factor?*

FIGURE 16.17

Negative feedback mechanism regulating the secretion of glucagon and insulin. Note that both the pancreas and liver are involved.

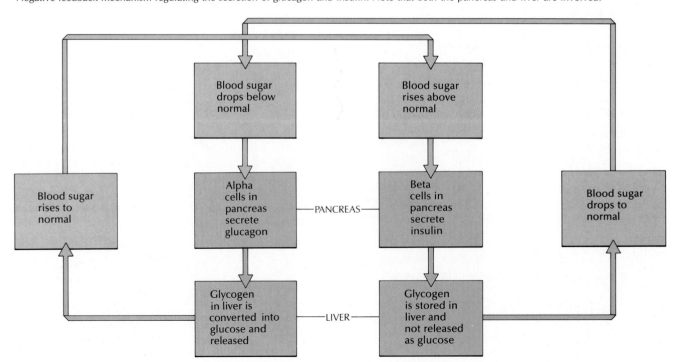

3 *How does glucagon function opposite to insulin?*

4 *How does insulin help to regulate carbohydrate metabolism?*

Ask Yourself

1 *What are the male and female gonads called?*

2 *What are the functions of the major male and female hormones?*

GONADS

The *gonads* (Gr. *gonos*, offspring), which are the *ovaries* in the female and the *testes* in the male, secrete hormones that help to regulate reproductive functions. These sex hormones include the male *androgens* and the female *estrogens, progestins,* and *relaxin.*

Male Sex Hormones

Three hormones help to regulate male reproductive functioning. The most important one is *testosterone,* which is produced by the interstitial cells of the testes. The others are *luteinizing hormone* (LH) and *follicle-stimulating hormone* (FSH), both produced by the adenohypophysis.

Testosterone acts with LH and FSH to stimulate the production of sperm (spermatogenesis). It is necessary for the growth and maintenance of the male sex organs, and promotes the development and maintenance of sexual behavior. It also stimulates the growth of facial and pubic hair, as well as the enlargement of the larynx, which causes the voice to deepen. The testes also produce the hormone *inhibin,* which inhibits the secretion of FSH (see Figure 26.10).

Female Sex Hormones

Three major classes of ovarian hormones help to regulate female reproductive functioning. *Estrogens* (estrin, estrone, and estradiol) help regulate the menstrual cycle and the development of the mammary glands and female secondary sex characteristics. The *progestins* (progesterone) also regulate the menstrual cycle and the development of the mammary glands, and aid in the formation of the placenta during pregnancy. *Relaxin,* which is produced in small quantities, is involved in childbirth, softening the cervix at the time of delivery, and causing the relaxation of ligaments of the symphysis pubis.

Overall, the secretion of hormones from the ovaries is regulated by two anterior pituitary hormones: follicle-stimulating hormone and luteinizing hormone. *Follicle-stimulating hormone* (FSH) initiates the monthly development of a follicle within the ovary. As a result, the level of estrogen rises. Estradiol, which is secreted by the corpus luteum of the ovary, inhibits the production of FSH. *Luteinizing hormone* (LH) initiates the production of progestin by the ovarian follicle. Progestin also inhibits FSH production.

The intricate interrelationships of hormones that regulate reproductive mechanisms are discussed fully in Chapters 26 and 27.

OTHER SOURCES OF HORMONES

In addition to the major endocrine glands, other glands and organs carry on hormonal activity. In the following sections we briefly discuss the kidneys, pineal gland, thymus gland, heart, digestive system, placenta, and the hormones known as prostaglandins.

Kidneys

The paired *kidneys* are primarily organs for the excretion of wastes, but they also produce several hormones, including erythropoietin, 1,25-dihydroxyvitamin D_3, prekallikreins, and prostaglandins. In addition, the kidneys produce *renin* (*renes* is Latin for "kidneys"), an enzyme whose natural substrate is the plasma protein angiotensinogen (see Figure 16.12).

The function of *erythropoietin* (Gr. *eruthros*, red + *poiesis*, making) is to stimulate the production of erythrocytes (red blood cells) by facilitating the synthesis of hemoglobin and the release of erythrocytes from bone marrow. However, it is still not known whether erythropoietin is produced in the kidneys, or if the kidneys elaborate a factor (renal erythropoietic factor) that causes it to be made from a precursor blood protein.

Pineal Gland

The *pineal gland* (PIHN-ee-uhl; L. *pinea*, pine cone) is also known as the *pineal body* or the *epiphysis cerebri.* It is a pea-sized body located in the midbrain, deep within the cerebral hemispheres of the brain, at the posterior end of the third ventricle (see Figure 16.1). The pineal gland has been called a "neuroendocrine transducer"—a system that converts a signal received through the nervous system (dark and light, for instance) into an endocrine signal (shifting levels of hormone secretion). Information about light and dark reaches the pineal gland through an indirect route. It is detected by the eyes, and then travels from the optic nerve to the brain, where sympathetic nerves carry the signal to the pineal.

Several substances that have hormonal activity have been isolated from the pineal, but the functions of these substances have not been fully established. One of these is *melatonin,* which is derived from serotonin. The pineal gland produces steady secretions of melatonin throughout the night; light inhibits the production of melatonin. It has been observed that melatonin causes drowsiness, and the pineal gland may affect the sleep cycle. The typical lack of light during long winters may contribute to *seasonal affective disorder* (SAD), a condition characterized by lack of energy and mood swings that border on depression. Some researchers speculate that the pineal gland is involved in SAD, but no firm evidence is available.

Thymus Gland

The *thymus gland* is a double-lobed lymphoid organ located behind the sternum in the anterior mediastinum (see Figure 16.1). It has an outer cortex containing many lymphocytes, and an inner medulla containing fewer lymphocytes as well as clusters of cells called thymic (Hassall's) corpuscles, whose function is unknown. The thymus gland is well supplied with blood vessels, but has only a few nerve fibers.

The thymus gland is large and active only during childhood, reaching its maximum effectiveness during early adolescence. After that time, the gland begins to atrophy because of the action of sex hormones, and is replaced by fatty tissue. Prolonged stress usually hastens atrophy. This happens because stress factors release adrenocortical hormones that have a destructive effect on thymus tissue. The thymus gland finally ceases activity altogether after 50 years or so, and it may therefore play an important role in the process of aging and the accompanying decrease in function of the immune system.

The main function of the thymus gland seems to be the processing of T cells (T lymphocytes). These cells are responsible for one type of immunity, called *cellular immunity* (see Chapter 20). Other lymphocytes, called B cells (B lymphocytes), are processed in the fetal liver before a child is born. B cells are responsible for a type of immunity called *humoral immunity* (see Chapter 20). There is some evidence that the thymus gland may be a true endocrine gland, since it produces thymic hormones, or "factors," that play a role in the development of T cells in the thymus, and their maintenance within other lymphoid tissue. Some of the hormones and factors include *thymosin alpha, thymosin B₁ to B₅, thymopoietin I and II, thymic humoral factor* (THF), *thymostimulin,* and *factor thymic serum* (FTS). The thymic hormones also play a role in the development of some B cells into plasma cells, which produce antibodies. There is also a possibility that the thymus gland may influence the secretion of reproductive hormones from the pituitary gland.

It had been thought until recently that thymic hormones were produced exclusively in the thymus gland, and that the main function of thymic hormones was to assist in the processing of bone marrow cells into T cells, infection-fighting cells of the immune system. Recent discoveries, however, indicate that thymosin B₄ and B₅ influence hormones of the reproductive system, and that thymosin B₄ is also synthesized by macrophages in the immune system.

Heart

Recent findings have revealed that in addition to being the complex pump that maintains circulation, the heart also acts as an endocrine organ. Cardiac muscle cells within both atria (the upper chambers of the heart) contain secretory granules that produce, store, and secrete a peptide hormone called *atriopeptin* (formerly called *atrial natriuretic factor,* or *ANF*).

Atriopeptin is secreted continuously in minute amounts, and is circulated throughout the body via the bloodstream. Secretion increases when excess salt accumulates in the body, when blood volume increases enough to stimulate stretch receptors in the atria, or when blood pressure rises significantly. Special target-cell receptors have been found in blood vessels, kidneys, and adrenal glands. Atriopeptin also affects neurons in the brain, especially the hypothalamus, where control and regulation occurs for blood pressure and the excretion of sodium, potassium, and water by the kidneys.

Current evidence suggests that atriopeptin helps to maintain a proper balance of fluid and electrolytes by increasing the output of sodium in urine; relaxes blood vessels directly, thus lowering blood pressure by reducing resistance to blood flow; lowers blood pressure by blocking the actions of hormones such as aldosterone, which tends to raise blood pressure; and reduces blood volume by stimulating the kidneys to filter more blood and produce more urine. Scientists speculate that atriopeptin complements the actions of other hormones, rather than acting on its own. It is hoped that carefully administered doses of atriopeptin will be useful in regulating blood pressure and electrolyte balance.

Digestive System

Among the major hormones of the digestive system are gastrin, secretin, and cholecystokinin. *Gastrin* is a polypeptide secreted by the mucosa (lining) of the stomach. Its function is to stimulate the production of hydrochloric acid and the digestive enzyme pepsin when food enters the stomach. Thus the stomach is both the producer and the target organ of gastrin.

Secretin is a polypeptide secreted by the mucosa of the duodenum. It stimulates a bicarbonate-rich secretion from the pancreas that neutralizes stomach acid as the acid passes to the small intestine. Secretin was the first hormone to be discovered (by the British scientists William M. Bayliss and Ernest H. Starling in 1902), and the first substance to actually be called a "hormone."

Cholecystokinin (CCK; koh-lee-sis-TOE-kine-in) is secreted from the wall of the duodenum. It stimulates the contraction of the gallbladder, which releases bile when food (particularly fats) enters the duodenum. It also stimulates the secretion of enzyme-rich digestive juices from the pancreas.

Research scientists are continually discovering "candidate hormones," substances that have many of the properties of hormones and are suspected of being "true" hormones. Some of these substances are peptides found in the gastrointestinal tract. Two that act on the intestines are *villikinin,* which stimulates the contraction of intestinal villi, and *motilin,* which stimulates intestinal motility. Two other possible hormones are *bombesin,* which stimulates acid secretion in the stomach and inhibits stomach motility, and *gastric-inhibitory polypeptide,* which inhibits acid secretion in the stomach.

Placenta

The *placenta* is a specialized organ that develops in a pregnant female as a source of nourishment for the developing fetus (see Chapter 27). It secretes the estrogens, progesterone, and human chorionic gonadotropin (hCG), which help to maintain pregnancy. Target areas are the ovaries, mammary glands, and uterus.

Prostaglandins

Prostaglandins are hormonelike substances that belong to a family of 20-carbon unsaturated fatty-acid derivatives. They are manufactured from the fatty acids that make up the structural parts of plasma membranes. (The specific type of prostaglandin produced depends on which fatty acids are present in the membrane.) Prostaglandins were originally thought to come from the prostate gland, hence their name. They are actually produced by all nucleated cells, and are found in practically all tissues of the body.

There are more than 16 different prostaglandins, belonging to 9 classes designated PGA to PGI. Further subdivisions within each class depend on the number of double bonds in the fatty-acid side chains. For example, the designation PGE_2 indicates two carbon-carbon double bonds outside the five-ring structure in a class E prostaglandin:

Prostaglandins are considered to be chemical modulators or messengers rather than true hormones, since they are not produced by specific endocrine glands and secreted into the blood for transport to target cells. Instead, they are produced locally, close to their site of action. Apparently, disturbances near a cell cause the release of fatty acids from the cellular membrane, and enzymes called cyclooxygenases convert the fatty acids into prostaglandins. Prostaglandins, unlike true hormones, are not stored. Once they are synthesized in response to a stimulus, they are released immediately, act quickly and locally, and are then inactivated and degraded.

The actions of prostaglandins are as varied as their origins. At the cellular level, prostaglandins appear to work along with specific hormones to increase or decrease the effect of cyclic AMP on target cells. At the tissue level, prostaglandins can raise or lower blood pressure, regulate digestive secretions, inhibit progesterone secretion by the corpus luteum, reduce infection by stimulating microorganism-destroying blood cells, and regulate blood clotting. In addition, prostaglandins cause the contraction of smooth muscle in the uterus, and can dilate air passages to the lungs.

Little is known about how prostaglandins work. These hormonal modulators may operate inside cells to turn chemical reactions on and off, or they may function by changing the composition of cellular membranes. Worldwide research on prostaglandins is proceeding extensively.

Prostaglandins have many clinical uses. Prostaglandin drugs may be useful in inducing labor and in treating asthma, arthritis, ulcers, and hypertension (high blood pressure). One of the first clinical uses of prostaglandins was to induce labor in pregnant women whose bodies were not promoting labor on their own, and prostaglandin drugs have also been used to induce abortions. There is also some evidence that prostaglandin drugs may inhibit the growth of viruses, and may be useful in treating multiple sclerosis and other diseases. Clinical studies indicate that separate forms of prostaglandins may be useful in treating diseases of the blood vessels by increasing blood flow, in promoting the healing of stomach and duodenal ulcers, and in protecting the stomach and small intestine from irritating drugs such as aspirin. A specialized use of prostaglandin drugs is to delay the natural closure of a blood vessel in newborn babies who require corrective surgery for a congenital heart disease. The prostaglandin called prostacyclin relaxes and dilates blood vessels, and also suppresses the action of blood-clotting structures called platelets. Prostacyclin may be useful in preventing heart attacks and strokes, and in preventing blood clots from forming during heart bypass operations, when the blood is circulated outside the body.

Drugs that inhibit the effects of prostaglandins are being tested, and some are already in use. Such a drug has been used to relieve menstrual cramps, excessive muscular contractions produced by high levels of uterine prostaglandins.

Ask Yourself

1 *What hormones do the kidneys secrete?*

2 *What is one possible function of the pineal gland?*

3 *Why do diseases become more prevalent as the thymus atrophies?*

4 *What appear to be the major functions of atriopeptin?*

5 *What are the main endocrine secretions of the digestive system?*

6 *Why are prostaglandins considered to be hormonal modulators rather than true hormones?*

7 *What are some of the clinical uses of prostaglandin drugs?*

What is the connection between prostaglandins and aspirin?	*Aspirin reduces pain, inflammation, and fever by inhibiting cyclooxygenase, the enzyme needed for the production of prostaglandins.*
Are prostaglandins involved in contraception?	*There is some evidence that contraceptive intrauterine devices (IUDs) work by irritating uterine tissue, which then produces a prostaglandin that causes uterine contractions, preventing a fertilized egg from becoming implanted in the uterus.*

PHYSIOLOGICAL AND ANATOMICAL ABNORMALITIES

Endocrine disorders are usually caused by underfunctioning (*hypo-*) or overfunctioning (*hyper-*), inflammation, or tumors of the endocrine glands. Certain disorders arise in the hypothalamus or pituitary; others develop in the endocrine glands that come under the influence of these higher control centers. Some of the major disorders of the endocrine system are reviewed here briefly.

Pituitary gland: adenohypophysis

Among the disorders of the adenohypophysis is *giantism,* caused by the oversecretion of growth hormone during the period of skeletal development. The body of a person with this disorder grows much larger than normal, sometimes to more than 2½ meters (8 ft) and 180 kilograms (400 lb). In most cases, death occurs before the age of 30.*

The effects of acromegaly, such as thickening of the jaw, nose, and hands, are shown here at three different stages of an afflicted woman's life. [*Reprinted with permission from "Clinico-pathologic Conference,"* The American Journal of Medicine, *Vol. 20, No. 1, January 1956.*]

A normal-sized person (left) walks next to a "giant" (center), whose pituitary gland was overactive, and a pituitary dwarf or midget, whose pituitary gland was underactive. [*Syndication International-Photo Trends.*]

Acromegaly (ak-roh-MEG-uh-lee; Gr. *akros,* extremity + *megas,* big) is a form of giantism that affects adults after

*The tallest person for whom there are authenticated records was Robert Wadlow of Illinois. He was 8 ft, 11 in. tall and still growing when he died at the age of 22.

the skeletal system is fully developed. It is often caused by a pituitary tumor. A tumor of a gland is called an *adenoma* (Gr. *aden,* gland + *oma,* tumor), and is usually benign. After maturity, oversecretion of growth hormone does not lengthen the skeleton, but does cause some cartilage and bone to thicken. The jaw, hands, feet, eyebrow ridges, and soft tissues may widen noticeably, and enlargement of the heart, liver, and other internal organs is also possible. Blood pressure may increase, and subsequent congestive heart failure is likely. Muscles grow weak, and osteoporosis and painful enlargements at the joints may occur. Acromegaly may be treated with some success by the surgical removal of the adenohypophysis, or by using radiation treatments.

Persons with undersecretion of growth hormone are known as *pituitary dwarfs,* or *midgets.* Although their intelligence and body proportions are normal, they do not grow any taller than a normal 6-year-old child. Some pituitary dwarfs become prematurely senile, and most die before the age of 50. Usually, their sexual organs and reproductive ability are not fully developed, but some pituitary dwarfs are capable of producing normal-sized children. Some dwarfism is caused by a deficiency of the thyroid hormones rather than by a deficiency of pituitary growth hormone. A treatment of hormonal replacement may increase growth somewhat and help retain fertility in sexually active patients.

Pituitary gland: neurohypophysis

When ADH (vasopressin) is undersecreted, excessively large amounts of water are excreted in the urine (*polyuria*). This condition, known as *diabetes insipidus,* is accompanied by dehydration and unrelenting thirst (*polydipsia*). Diabetes insipidus can often be treated successfully by administering controlled doses of ADH.

Thyroid gland

Overactivity of the thyroid gland, known as *hyperthyroidism* (Graves' disease), may be caused by long-acting thyroid stimulator (LATS). LATS is an antibody that acts on the thyroid gland in the same way that TSH does. Hyperthyroidism may also be caused by immunoglobulins such as human thyroid stimulator and LATS protector, or by a thyroid tumor. Among the symptoms are nervousness, irritability, increased heart rate and blood pressure, weakness, weight loss, and high oxygen use, even at rest. The bulging eyes (*exophthalmos*) typical of this condition are due partly to increased fluid behind the eyes caused by an exophthalmos-producing substance (EPS). Drug therapy that inhibits thyroxine production, and administration of radioactive iodine have successfully replaced surgery as treatments for hyperthyroidism.

Underactivity of the thyroid, or *hypothyroidism,* is often associated with *goiter* (L. "throat"), an enlarged thyroid. Such a swelling in the neck is caused

An example of a massive goiter. [*John Paul Kay/Peter Arnold.*]

when insufficient iodine in the diet forces the thyroid to expand in an attempt to produce more thyroxine. The adaptive responses are triggered, in large part, by an increased secretion of TSH, which attempts to stimulate the iodine-trapping mechanism and the subsequent steps in the metabolism of iodine. Most cases of goiter used to be found in areas away from the ocean, where iodine content in the soil and water supply is low. With the addition of minute amounts of iodine to ordinary table salt and drinking water in recent years, goiter has practically disappeared as a common ailment. Symptoms of hypothyroidism include decreased heart rate, blood pressure, and body temperature; lowered basal metabolic rate; and underactivity of the nervous system.

Underactivity of the thyroid during the development of the fetus after the twelfth week causes *cretinism* (CREH-tin-ism), characterized by mental retardation and irregular development of bones and muscles. The skin is dry, eyelids are puffy, hair is brittle, and the shoulders sag. Ordinarily, cretinism cannot be cured, but early diagnosis and treatment with L-thyroxine (a drug form of thyroxine) may arrest the disease before the nervous system is damaged.

If the thyroid becomes underactive during adulthood, *myxedema* results, producing swollen facial features, dry skin, low basal metabolic rate, tiredness, possible mental retardation, and intolerance to cold in spite of increased body weight. Like cretinism, myxedema may be corrected if it is treated early.

The familiar "circus dwarf" is the irreversible result of an underactive thyroid gland that interferes with the normal growth of cartilage. Such dwarfs are usually intelligent and active, and they have stubby arms and legs, a relatively large chest and head, and flattened facial features.

Parathyroid glands

The most common causes of underactivity of the parathyroid glands *(hypoparathyroidism)* are damage to or removal of the parathyroids during surgery and parathyroid adenoma. Hypoparathyroidism produces low levels of calcium in blood plasma and an overabundance of phosphorus. Such an imbalance may cause faulty transmission of nerve impulses, osteoporosis and hampered bone growth, and paralysis of muscles (tetany). Controlled doses of vitamin D and calcium salts may restore normal calcium levels. Low-phosphate diets and drugs that increase the excretion of phosphorus in urine may also be successful.

Overactivity of the parathyroid glands *(hyperparathyroidism)* is usually caused by an adenoma. It results in excessive amounts of calcium and lowered amounts of phosphorus. As in hypoparathyroidism, osteoporosis is evident, and many other general symptoms may occur, including loss of appetite, nausea, weight loss, personality changes, stupor, kidney stones, duodenal ulcers, kidney failure, increased blood pressure, and congestive heart failure. Treatment may include the surgical removal of excess parathyroid tissue, drugs that lower the calcium level, and in severe cases of kidney failure, an artificial kidney or kidney transplant.

Adrenal cortex

Two diseases caused by overactivity of the adrenal cortex *(hyperadrenalism)* are Cushing's disease and adrenogenital syndrome. *Cushing's disease* is usually caused by a cortical tumor that overproduces glucocorticoids. The tumor, in turn, is caused by an increased secretion of ACTH by the anterior pituitary. Symptoms include fattening of the face, chest, and abdomen (the limbs remain normal), accompanied by abdominal striations and a tendency toward diabetes caused by increased blood sugar. Protein is lost, and the muscles become weak. Surgical removal of the causative tumor usually brings about a remission and reduces the secretion of ACTH.

Adrenogenital syndrome is also caused by an overactive adrenocortical tumor, which stimulates excessive production of the cortical male sex hormones known as *androgens*. These androgens cause male characteristics to appear in a female, and accelerate sexual development in a male. Hormonal disturbances during the fetal development of a female child may cause a distortion of the genitals, so that the clitoris and labia become enlarged and resemble a penis and scrotum. In a mature woman, an extreme case of adrenogenital syndrome may produce a beard. (Such pronounced male characteristics in a female may be caused by other hormonal malfunctions besides defects in the adrenal cortex.) Adrenogenital syndrome is often treated with daily doses of cortisol, which usually reinstates the normal steroid balance.

Underactivity of the adrenal cortex *(hypoadrenalism)* produces **Addison's disease,** whose symptoms include anemia (deficiency of red blood cells), weakness and fatigue, increased blood potassium, and decreased blood sodium. Skin color becomes bronzed because excess ACTH, produced by the pituitary in an effort to restore a normal cortical hormone level, also induces alpha-MSH secretion, and the two hormones together induce abnormal deposition of skin pigment. Until recently, Addison's disease was usually fatal, but now it can be controlled with regular doses of cortisol and aldosterone.

Pancreas

About 4 percent of the United States population will develop *diabetes mellitus* at some time in their lives. It can occur either as *Type I, juvenile diabetes,* which usually begins early in life, or *Type II, maturity-onset diabetes,* which occurs later in life, mainly in overweight people. Heredity plays a major role in the development of both types. Diabetes results when beta cells do not produce enough insulin. When this happens, glucose accumulates in the blood and spills into the urine, but does not enter the cells. Excess glucose in the urine is a diuretic and causes dehydration. Because the cells are unable to use the accumulated glucose (the most readily available energy source in the body), the body actually begins to starve. Appetite may increase, but eventually the body consumes its own tissues, literally eating itself up.

Because the removal of glucose from the kidneys requires large amounts of water, the diabetic person produces excessive sugary urine and may excrete as much as 20 L of sugary urine per day.* In response to the increased urine production, with the possibility of serious

*The word *diabetes* comes from the Greek word for "siphon" or "to pass through," referring to the seemingly instant elimination of liquids. The word was actually used by the Greeks as early as the first century. In the seventeenth century, the sweetness of diabetic urine was discovered, and the name of the disease was lengthened to diabetes mellitus. *Mellitus* comes from Greek *meli*, honey.

body dehydration, diabetics become extremely thirsty and drink huge amounts of liquids.

The use of fats (to replace glucose) for energy production in the diabetic causes the accumulation of acetoacetic acid, β-hydroxybutyric acid, and keto acids in the blood and body fluids. This leads to acidosis, which can lead to coma and death.

Diabetes is incurable in any form, but mild diabetes can usually be controlled by strict dietary regulation and exercise. More serious cases may require treatment with regular injections of insulin. If the disease is untreated, almost every part of the diabetic's body will be affected, and gangrene, hardening of the

arteries (arteriosclerosis), other circulatory problems, and further complications may occur.

Low blood sugar, *hypoglycemia,* may be caused by the excessive secretion of insulin. Sugar is not released from the liver, and the brain is deprived of its necessary glucose. Hypoglycemia can be controlled by regulating the diet, especially carbohydrate intake. On a short-term basis, the sugar in a glass of orange juice may restore the normal glucose balance in the blood, but a long-term treatment usually consists of a reduction of carbohydrates in the diet. Carbohydrates tend to stimulate large secretions of insulin, thereby removing sugar from the bloodstream too quickly.

MEDICAL TERMINOLOGY

HYPERPLASIA (Gr. *hyper,* over + *plasis,* change, growth) A nontumorous growth in a tissue or organ.

HYPOPLASIA Incomplete or arrested development of a tissue or organ.

POSTPRANDIAL (L. *post,* after + *prandium,* late breakfast, meal) Usually pertaining to an examination of blood sugar content after a meal.

RADIOIMMUNOASSAY (RIA) (RAY-dee-oh-im-myu-noh-ASS-ay) A technique that measures minute quantities of a substance, such as a hormone, in the blood.

REPLACEMENT THERAPY A method of treatment where insufficient secretions of hormones are replaced with natural or synthetic chemicals.

SIMMONDS' DISEASE A disorder characterized by a total absence of pituitary hormones, usually caused by a tumor or blockage of the blood vessels supplying the pituitary gland.

STEROID THERAPY Use of steroids to treat certain endocrine disorders.

VIRILISM (L. *vir,* man) Masculinization in women.

SUMMARY

The endocrine system and the nervous system together constitute the two great regulatory systems of the body. The endocrine system is made up of tissues or organs called *endocrine glands,* which secrete chemicals directly into the bloodstream. These chemical secretions are called *hormones.*

Hormones and their feedback systems
1 Most hormones are steroids, derivatives of amino acids, or proteins. A few are glycoproteins.

2 Hormones travel through the bloodstream to all parts of the body, but affect only those *target cells* that have compatible receptors.

3 Hormonal secretions are usually kept at a normal level by *negative feedback systems* involving other glands and hormones.

Mechanisms of hormone control
1 Water-soluble hormones (amine and protein hormones) regulate cellular responses through a mechanism called the *fixed-membrane-receptor mechanism.* The end result of the chemical activation is the diffusion of *cyclic AMP* throughout the cell, which causes the cell to perform its distinctive function.

2 Lipid-soluble hormones (steroids) regulate cellular activity through the *mobile-receptor mechanism,* involving the synthesis of proteins that affect the cell's function.

Pituitary gland (hypophysis)
1 The *pituitary gland,* or *hypophysis,* consists of the anterior lobe *(adenohypophysis)* and the posterior lobe *(neurohypophysis).* The adenohypophysis has an abundance of secretory cells,

while the neurohypophysis contains many nerve endings.

2 The pituitary is connected to the hypothalamus by a stalk of nerve cells and blood vessels called the *infundibular stalk,* which provides a direct link between the nervous system and the endocrine system.

3 The *hypothalamus* releases regulating substances called *releasing* and *inhibiting factors* that control the secretions of the adenohypophysis. Nerve centers in the hypothalamus regulate secretions from the neurohypophysis. The connection between the hypothalamus and the adenohypophysis is facilitated by a system of blood vessels called the *hypothalamic-hypophyseal portal system,* whereas the hypothalamic link with the neurohypophysis relies on nerve impulses.

4 The hypothalamus synthesizes *antidiuretic hormone* (ADH) and *oxytocin,* which are stored in and secreted from the neurohypophysis.

5 The true endocrine portion of the pituitary is the adenohypophysis, which synthesizes and secretes seven separate hormones: *growth hormone* (GH), *prolactin, thyroid-stimulating hormone* (TSH), *adrenocorticotropic hormone* (ACTH), *luteinizing hormone* (LH), *follicle-stimulating hormone* (FSH), and *melanocyte-stimulating hormone* (MSH).

Thyroid gland

1 The *thyroid gland* secretes the thyroid hormones *thyroxine* and *triiodothyronine* from its follicular cells and *calcitonin* from its parafollicular cells. Thyroid hormones increase basal metabolism, accelerate growth, and stimulate cellular differentiation and protein synthesis.

2 Calcitonin lowers calcium and phosphate levels in the blood.

Parathyroid glands

Parathormone, the hormone of the *parathyroid glands,* increases the level of calcium in the blood and decreases the concentration of phosphate.

Adrenal glands

1 The *adrenal glands* are composed of an inner *medulla* and an outer *cortex.* The adrenal cortex secretes the three types of steroid hormones: glucocorticoids, mineralocorticoids, and gonadocorticoids.

2 *Glucocorticoids,* mainly *cortisol,* are essential to the proper metabolism of carbohydrates, proteins, and fats, most critically through the stimulation of *gluconeogenesis,* the synthesis by the liver of glucose from noncarbohydrate sources. They also act to suppress allergic reactions and inflammatory responses.

3 *Mineralocorticoids,* the most important of which is *aldosterone,* control mineral balance through the retention and loss of sodium and potassium.

4 *Gonadocorticoids,* or adrenal sex hormones, affect the sex organs, but only slightly.

5 The adrenal medulla secretes *epinephrine* and *norepinephrine,* which stimulate the sympathetic nervous system.

6 Stressful conditions stimulate increased production of ACTH by the pituitary gland, which stimulates the adrenal cortex and adrenal medulla to prepare the body's muscular, digestive, circulatory, and respiratory systems to cope with the stress.

Pancreas

1 The *pancreas* functions as an exocrine gland, secreting digestive enzymes into ducts, and as an endocrine organ, secreting hormones into the bloodstream. The endocrine portion synthesizes, stores, and secretes hormones from the *pancreatic islets.*

2 The pancreatic islets contain alpha cells that secrete *glucagon,* beta cells that secrete *insulin,* delta cells that secrete somatostatin, and F cells that secrete pancreatic polypeptide.

3 Glucagon raises the level of blood sugar and stimulates the release of fatty acids and glycerol from adipose tissue. The most important effect of insulin is to facilitate glucose transport across cellular membranes.

Gonads

1 The *gonads, ovaries* in a female and *testes* in a male, secrete hormones that control reproductive functions.

2 The major hormones are *testosterone* in males and *estrogens, progestins,* and *relaxin* in females.

Other sources of hormones

1 The *kidneys* secrete several hormones, including erythropoietin, 1,25-dihydroxyvitamin D_3, prekallikreins, and prostaglandins. In addition, the kidneys produce *renin,* an enzyme that causes blood pressure to increase.

2 The *pineal gland* contains several chemicals, whose functions have not been established. The secretion of *melatonin,* which may affect skin pigmentation, seems to be affected by light signals and may be involved with the sleep cycle, seasonal affective disorder, and puberty.

3 The main function of the *thymus gland* is the processing of T cells. The gland begins to atrophy with age, which may play a role in the aging process and the accompanying decline of the immune system.

4 The *heart* produces, stores, and secretes a peptide hormone called *atriopeptin.* It is believed that atriopeptin helps to maintain the proper balance of fluids and electrolytes and lowers excessively high blood pressure and volume.

5 The *digestive system* secretes several digestive hormones, especially *gastrin,* which aids digestion in the stomach; *secretin,* which helps to neutralize stomach acid; and *cholecystokinin,* which stimulates the release of digestive juices from the gallbladder and pancreas.

6 The *placenta* produces hormones that help maintain pregnancy.

7 *Prostaglandins* are fatty-acid hormonal mediators or messengers that are found in many parts of the body. They appear to be involved in regulating blood flow and pressure, blood clotting, digestive secretion, contraction of the uterus, and microorganism-destroying blood cells.

Physiological and anatomical abnormalities

1 Disorders of the adenohypophysis include *giantism, acromegaly,* and *pituitary dwarfism.* A disorder of the neurohypophysis is *diabetes insipidus.*

2 *Hyperthyroidism* and *hypothyroidism* result in metabolic dysfunction and conditions such as *goiter, cretinism,* and *myxedema. Hypoparathyroidism* and *hyperparathyroidism* cause abnormal calcium and phosphorus blood levels, resulting in defects in nerve impulse transmission, bone growth, and kidney function.

3 Overactivity of the adrenal cortex causes *Cushing's disease* and *adrenogenital syndrome.* Underactivity of the adrenal cortex produces *Addison's disease.*

4 *Diabetes mellitus* occurs when the pancreas does not produce sufficient insulin. Excessive secretion of insulin may lead to *hypoglycemia.*

UNDERSTANDING THE FACTS

1 Name the three chemical categories into which most hormones fit.

2 What are the functions of adenylate cyclase, cyclic AMP, and cyclic AMP phosphodiesterase?

3 Which glands secrete steroid hormones?

4 Which portion of the pituitary actually synthesizes and secretes hormones?

5 What is the natural function of oxytocin? What are its medical uses?

6 Which specific hormones affect the growth of children and adolescents?

7 Name the hormones that govern the production and ejection of milk, and give their source.

8 What are the gonadotropic hormones produced by the adenohypophysis?

9 Which is the only endocrine gland able to store its secretions outside its principal cells?

10 What are the basic functions of the thyroid hormones?

11 In general, what is the effect of thyroid hormones on the oxygen consumption of most body cells?

12 Glucocorticoids are essential to the proper metabolism of which food group(s)?

13 Why is the concentration of cholesterol so great in the cells of the zona fasciculata of the adrenal cortex?

14 What are neurosecretory cells?

15 What is the main function of aldosterone?

16 Locate the following structures:
  a infundibular stalk
  b pancreatic islets
  c chromaffin cells
  d follicular cells
  e pars intermedia

17 Name the specific secretion of the alpha, beta, delta, and F cells of the pancreatic islets.

UNDERSTANDING THE CONCEPTS

1 Compare the basic approach to body regulation of the endocrine system with that of the nervous system. Do they complement each other, and if so, in what way?

2 Explain what is meant by a negative feedback system.

3 In acromegaly, why do the long bones thicken but not lengthen? (You may have to think back to your study of the skeletal system.)

4 Exposure to cold temperatures will increase the secretion of TSH. What is the result of this response?

5 What significance is there in the fact that thyroid hormones are not specific for one target area?

6 Describe the relatively simple, yet very effective, method of maintaining a proper thyroxine level in the blood.

7 How is adequate vitamin D tied in with proper parathormone function?

8 Even though cortisol is used to combat severe allergic reactions and inflammation, it cannot be used over extended periods of time. Which of its functions are responsible for this fact?

9 How do the adrenal glands react to stress, and how do these reactions help the body meet stressful situations?

10 What are some of the natural effects of epinephrine that make it a useful medical drug?

11 Why do some feel that the thymus may play an important role in the aging process?

12 What is the significance of the fact that the entrance of fatty foods into the duodenum serves as a stimulant for the secretion of cholecystokinin?

13 Why is so much effort being made in prostaglandin research?

SUGGESTED READING

BINKLEY, SUE, "A Timekeeping Enzyme in the Pineal Gland." *Scientific American,* April 1979.

CARMICHAEL, STEPHEN W., AND HANS WINKLER, "The Adrenal Chromaffin Cell." *Scientific American,* August 1985.

EDWARDS, C. R. W., ed., *Endocrinology.* Chicago: Year Book Medical Publishers, 1986.

GREENSPAN, FRANCIS S., AND PETER H. FORSHAM, *Basic and Clinical Endocrinology,* 2nd ed. East Norwalk, CT: Appleton & Lange, 1986.

GUILLEMIN, ROGER, AND ROGER BURGUS, "The Hormones of the Hypothalamus." *Scientific American,* November 1972.

MARTIN, C. R., *Endocrine Physiology.* New York: Oxford University Press, 1985.

MCEWEN, BRUCE S., "Interactions Between Hormones and Nerve Tissue." *Scientific American,* July 1976.

NATHANSON, JAMES A., AND PAUL GREENGARD, "Second Messengers in the Brain." *Scientific American,* August 1977.

NOTKINS, ABNER LOUIS, "The Causes of Diabetes." *Scientific American,* November 1979.

O'MALLEY, BERT W., AND WILLIAM T. SCHRADER, "The Receptors of Steroid Hormones." *Scientific American,* February 1976.

ORCI, LELIO, JEAN-DOMINIQUE VASSALLI, AND ALAIN PERRELET, "The Insulin Factory." *Scientific American,* September 1988.

PASTAN, IRA, "Cyclic AMP." *Scientific American,* August 1972.

PIKE, J. E., "Prostaglandins." *Scientific American,* November 1971.

SCHALLY, ANDREW V., ABBA J. KASTIN, AND AKIRA ARIMURA, "Hypothalamic Hormones: The Link Between Brain and Body." *American Scientist,* November/December 1977.

SUTHERLAND, E. W., "Studies on the Mechanism of Hormone Action." *Science,* 177 (1972): 401.

WILLIAMS, R. H., ed., *Textbook of Endocrinology,* 6th ed. Philadelphia: Saunders, 1981.

IV

TRANSPORTATION AND MAINTENANCE

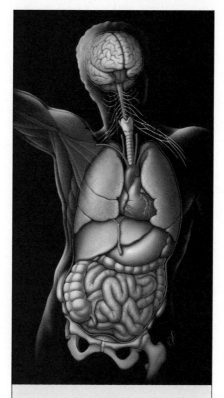

17
The Cardiovascular System: Blood

LEARNING OBJECTIVES

1 Discuss the major functions of blood, and tell how blood helps to maintain homeostasis in the body.

2 Describe some important properties of blood.

3 List the main components of blood plasma, and give their functions.

4 Explain how the structure of a red blood cell (erythrocyte) relates to its major functions.

5 Discuss the transport of oxygen and carbon dioxide by the blood.

6 Define erythropoiesis, and tell where it takes place.

7 Describe the life span, destruction, and removal of erythrocytes.

8 Name the two basic categories of leukocytes (white blood cells), and the types of leukocytes included in each.

9 Compare the five types of leukocytes as to number, structure, origin, and function.

10 Describe the structure, origin, and function of platelets.

11 Define hemostasis, and briefly describe three hemostatic processes.

12 Compare the extrinsic and intrinsic mechanisms of blood clotting.

13 Describe the factors and mechanisms that prevent blood from clotting under normal conditions.

14 Discuss ABO blood grouping, and define agglutinogens and agglutinins.

15 Tell how the Rh factor may be involved in hemolytic disease of the newborn.

16 Compare the conditions and causes of several types of anemia.

17 Describe the conditions and causes of hemophilia and leukemia.

For practical reasons, it is to our advantage to discuss human anatomy and physiology in terms of separate, functional body systems. But we can see that such a classification is artificial when we attempt to classify blood. In the strictest sense, the *cardiovascular system* is made up of the heart (as a pump) and blood vessels (as a means of transport). But to exclude blood from the cardiovascular system would be like considering the respiratory system without including the air we breathe. So we will include blood, at least nominally, as part of the cardiovascular system.

Blood can also be classified as part of the *circulatory system*, which is a more general term that includes not only the blood, blood vessels, and heart, but also the lymph and lymph vessels. Blood is also part of the *hematologic (blood) system*, which includes active bone marrow, the lymph nodes, spleen, and specialized cells known as *macrophages*.

Functionally, blood is classified as a specialized type of connective tissue, as we saw in Chapter 4. Actually, it is a liquid tissue. Just like other connective tissues, blood contains cellular elements, called *formed elements,* and a fluid matrix, called *plasma.* Both the formed elements and plasma play important roles in homeostasis. Throughout the chapter, emphasis is placed on showing the many ways that blood is specialized to help maintain homeostasis.

FUNCTIONS OF BLOOD

Although animals have evolved various ways of transporting food, gaseous waste products, and regulatory substances throughout their bodies, the most common method uses a circulating fluid called **blood.** As blood moves throughout the body, tissues are continuously adding to it their waste products, secretions, and metabolites, and taking from it vital nutrients, oxygen, hormones, and other substances. Overall, blood *transports* oxygen and carbon dioxide; *defends* the body against harmful microorganisms; *destroys* foreign particles; is involved in *inflammation, coagulation* (blood clotting), and the *immune response;* and helps *regulate* the pH of body fluids. The major functions of blood are described in greater detail in the following sections and throughout the chapter.

Blood Transports Gases of Respiration

As blood circulates throughout the body, it carries oxygen from the lungs to body cells, where oxygen diffuses out of the blood as it is needed. Blood also transports carbon dioxide from body cells to the lungs. Carbon dioxide is a waste product of cellular metabolism.

Blood Helps Regulate Acid-Base Balance

Some carbon dioxide is needed in the blood to maintain the pH balance, but too much carbon dioxide may lower the pH dangerously. A stable blood pH is maintained by efficient buffers, such as the plasma proteins, amino acids, and the bicarbonate ion (HCO_3^-).

Blood Aids in Nutrient, Hormone, and Enzyme Transport

Blood carries many types of nutrients, including simple sugars, amino acids, fats, and vitamins. The concentration of nutrient substances in the blood is regulated carefully. Hormones secreted by endocrine glands are also circulated to their target organs by the blood, and enzymes are also transported by the blood.

Blood Helps Transport Wastes

Blood carries the waste products of cellular metabolism to the kidneys, liver, lungs, and sweat glands for eventual removal from the body. The kidneys eliminate urea, uric acid, and water produced during cellular metabolism. Some excess water is eliminated by the lungs and skin. The liver removes bilirubin, the product of catabolized hemoglobin (the protein in red blood cells that binds to oxygen and carries it in the blood).

Blood Helps Regulate Body Temperature

Blood helps to cool and heat parts of the body, and keeps the body temperature relatively stable and uniform. The temperature-regulating center in the hypothalamus provides the necessary homeostatic feedback control. Blood flow to the skin can be increased or decreased. An increased blood flow to the skin, via the vasodilation of blood vessels, increases heat loss and lowers body temperature; a decreased blood flow to the skin, via the vasoconstriction of blood vessels, decreases heat loss and raises body temperature. In addition, water in the blood absorbs the heat produced by cellular metabolism, and excess heat is carried to the lungs and skin, where it is released into the outside environment.

Blood Aids the Stoppage of Bleeding

Blood can control its own flow through clotting. It contains "clotting factors" and other substances that enable it to clot when a blood vessel is damaged. A blood protein called *fibrin* helps form a clot by producing a stringy gel that traps blood cells. (See p. 519 for a detailed discussion of blood clotting.)

Blood Helps Regulate Body Fluids

Blood regulates the amount of water and electrolytes in body tissues. It bathes tissues in fluid, and maintains the osmotic concentration throughout the body. This, in turn, helps to regulate plasma osmotic pressure and blood volume.

Blood Defends Against Harmful Microorganisms and Toxins

Blood contains specialized cells and antibodies that aid the immune system by resisting or destroying foreign microorganisms (see Chapter 20). White blood cells called *leukocytes* engulf and destroy microorganisms, break down cellular wastes, and are predominant in areas of infection or in dead tissue. Some leukocytes, such as *lymphocytes*, produce antibodies that neutralize foreign substances. Lymphocytes are also associated with the acquired resistance to infections. The blood carries some toxic substances to the liver, where they are destroyed.

With all of this activity, the composition of the tissue fluids tends to change, and can only be maintained constant by a continuous exchange with the blood. In turn, the composition of the blood may be altered, and many complex systems are needed to keep it within acceptable narrow homeostatic limits. The composition of blood not only has a great influence on the functions of blood itself, it also affects the functions of many other tissues and organs. The rest of this chapter will show some of the ways blood helps to maintain a constant cellular environment that produces homeostasis.

Ask Yourself

1 What are the major functions of blood?

2 How does blood help to transport nutrients throughout the body and remove wastes?

3 How does blood help to stabilize the body's acid-base balance and the composition of body fluids?

4 How does blood help to regulate body temperature?

PROPERTIES OF BLOOD

The blood volume of a healthy person fluctuates very little. Even when blood is lost, it is replaced rapidly. The blood of an average adult is about 7 to 9 percent of total body weight, or 79 mL/kg of body weight. An average man has 5 to 6 L of blood, and an average woman has 4 to 5 L.

Because blood contains red blood cells, it is thicker, denser, and more adhesive than water, and flows four to five times more slowly. This comparative resistance to flow is called *viscosity*. The more red blood cells and blood proteins, the higher the viscosity. Blood viscosity also depends on flow. Slow-moving blood is more viscous than fast-moving blood. The viscosity of blood ranges between 3.5 and 5.5, compared to 1.000 for water. The *specific gravity*, or density, of blood is between 1.045 and 1.065, as compared with 1.000 for water.

The red color of arterial blood is due to oxygenated *hemoglobin*, a pigment carried by the red blood cells. When oxygen is removed, the blood appears bluish or darker. White blood cells and platelets in the blood are clear, and the plasma is yellowish.

Blood is slightly alkaline, with a pH that usually ranges between 7.35 and 7.45. Arterial blood is more alkaline than venous blood because it has less carbon dioxide. As the level of carbon dioxide increases, it reacts with water to form carbonic acid, which then lowers the blood pH (see p. 512).

The temperature of blood averages about 38°C (100.4°F).

Ask Yourself

1 What is the normal viscosity of blood? The pH? The temperature?

2 Why is blood usually red?

COMPONENTS OF BLOOD

Blood consists of a liquid part, known as **plasma** (about 55 percent), and a solid part, or the **formed elements** (about 45 percent), which are mostly blood cells or particles suspended in the plasma. The formed elements include red blood cells (erythrocytes), white blood cells (leukocytes), and fragmented cells called platelets (thrombocytes) (Figure 17.1).

FIGURE 17.1

Scanning electron micrograph of human blood cells. [*Bruce Wetzel and Harry Schaefer, National Cancer Institute.*]

Thrombocytes (platelets)   Erythrocytes (red blood cells)

Leukocytes (white blood cells)   ×4000

Plasma

Plasma is the liquid part of blood. It is about 90 percent water, and provides the solvent for dissolving and transporting nutrients. A group of proteins comprise another 7 percent of the plasma. The remaining 3 percent is composed of electrolytes, amino acids, glucose and other nutrients, various enzymes, hormones, metabolic wastes, and traces of many other organic and inorganic materials (Table 17.1).

Water in plasma The water circulated in blood plasma is readily available to cells, tissues, and extracellular fluids of the body as needed to maintain the normal state of hydration. Water is also the solvent for both extracellular and intracellular chemical reactions. Thus, many of the properties of blood plasma can be correlated with the basic properties of water (see Chapter 2 for properties of water as a solvent, transporter, temperature regulator, and lubricant). The water in blood plasma also contains many solutes whose concentrations are changing constantly to meet the needs of the body.

Because of the relatively high concentration of proteins in plasma, an osmotic effect is created that helps to move water into the blood from the fluid outside the blood vessels. Water is actually the medium through which all materials are exchanged between cells and circulating blood. The similar composition of blood plasma and extracellular fluid reflects the ease of water exchange between the two fluids in maintaining homeostasis.

Plasma proteins The proteins in blood plasma are referred to as *plasma proteins.* The total protein component of blood plasma can be divided into the *albumins, fibrinogen,* and the *globulins.*

The most abundant plasma proteins (about 60% of all plasma proteins) are the ***albumins,*** which are synthesized in the liver. The main function of albumins is to promote water retention in the blood by increasing the osmotic pressure, which, in turn, maintains the blood volume and pressure. If the amount of albumin in the plasma decreases, fluid leaves the bloodstream and accumulates in the surrounding tissue, causing a swelling known as *edema* (ih-DEE-muh). Albumins also act as carrier molecules by binding to some substances, such as hormones, that are transported in plasma.

Also produced by the liver is ***fibrinogen*** (about 4% of all plasma proteins), a plasma protein essential for blood clotting. When fibrinogen and several other proteins involved in clotting are removed from plasma, the remaining liquid is called **serum.** The role of fibrinogen and related proteins in blood clotting will be described later in this chapter.

The *globulins* (about 36% of all plasma proteins) are divided into three classes, based on their structure and function: alpha (α), beta (β), and gamma (γ). The alpha and beta globulins are produced by the liver. Their function is to transport lipids and fat-soluble vitamins in the blood. One form of these transport molecules, *low-density lipoproteins* (LDLP), transports cholesterol from its site of synthesis in the liver to various body cells. Another transporter, *high-density lipoproteins* (HDLP), removes cholesterol from arteries, preventing its deposition there.

Gamma globulins are the immunoglobulins, antibodies that

TABLE 17.1 REPRESENTATIVE PLASMA VALUES

Measurement	Normal range
Blood volume	80–85 mL/kg body weight
Blood osmolality	285–295 mOsm
Blood pH	7.35–7.45
Enzymes	
Creatine phosphokinase (CPK)	*Female:* 5–35 mU/mL *Male:* 5–55 mU/mL
Lactic acid dehydrogenase (LDH)	60–120 U/mL
Phosphatase (acid)	*Female:* 0.01–0.56 Sigma U/mL *Male:* 0.13–0.63 Sigma U/mL
Hematology values	
Hematocrit	*Female:* 37–48% *Male:* 45–52%
Hemoglobin	*Female:* 12–16 g/100 mL *Male:* 13–18 g/100 mL
Red blood cell count	4.2–5.9 million/mm^3
White blood cell count	4300–10,880/mm^3
Hormones	
Testosterone	*Male:* 300–1100 ng*/100 mL *Female:* 25–90 ng/100 mL
Adrenocorticotropic hormone (ACTH)	15–70 pg[†]/mL
Growth hormone	*Children:* over 10 ng/mL *Adult male:* below 5 ng/mL
Insulin	6–26 μU/mL (fasting)
Ions	
Bicarbonate	24–30 mmol/L
Calcium	2.1–2.6 mmol/L
Chloride	100–106 mmol/L
Potassium	3.5–5.0 mmol/L
Sodium	135–145 mmol/L
Organic molecules (other)	
Cholesterol	120–220 mg/100 mL
Glucose	70–110 mg/100 mL (fasting)
Lactic acid	0.6–1.8 mmol/L
Protein (total)	6.0–8.4 g/100 mL
Triglycerides	40–150 mg/100 mL
Urea nitrogen	8–25 mg/100 mL
Uric acid	3–7 mg/100 mL

*Nanograms.
[†]Picograms.
Source: New England Journal of Medicine, 1980. Used with permission.

help to prevent diseases such as measles, tetanus, and poliomyelitis. (An *antibody* is any one of millions of distinct proteins produced by the body that is capable of inactivating a specific bacterium, virus, protein, or cancer cell that it "recognizes" to be foreign. The antibody combines with the foreign body, known as an *antigen,* forming an *antigen-antibody complex.*) The five classes of antibodies that make up the immunoglobulins are designated as IgG, IgA, IgM, IgD, and IgE (see Chapter 20).

In addition to the albumins, fibrinogen, and globulins,

plasma also contains other regulatory and protective substances: hormones, enzymes, and vitamins. Hormones are secreted from endocrine glands into the plasma and transported throughout the body to their target organs. Enzymes and vitamins are found in the plasma in various concentrations in transit to and from various cells. Some enzymes, such as those essential for blood clotting, are found in plasma, and have specific functions in maintaining homeostasis. These functions are described later in this chapter.

Plasma electrolytes *Electrolytes* are inorganic compounds that separate into ions when they are dissolved in water. The ions are either positively charged (*cations*) or negatively charged (*anions*). The major cation of plasma is sodium (Na^+), which has an important effect on osmotic pressure and fluid movements, and helps determine the total volume of extracellular fluid. The principal anion is chloride (Cl^-). The other major ions in plasma are potassium (K^+), calcium (Ca^{2+}), phosphate (PO_4^{2-}), iodide (I^-), and magnesium (Mg^{2+}). As you saw in Chapter 2, these chemicals have many different uses in the body.

Nutrients and waste products Although *glucose* ("blood sugar") appears in plasma in low concentrations, it is the body's most readily available source of usable energy, and it is the *only* source of energy for red blood cells. Glucose enters the body in carbohydrate foods.

Like glucose, *amino acids* have a high turnover rate, that is, they are used rapidly by body cells, and usually appear in the plasma in low concentrations. Amino acids are also important because they provide the building blocks for protein synthesis.

Lipids are found in plasma in the form of phospholipids, triglycerides, free fatty acids, and cholesterol. They are important components of nerve cells and steroid hormones, and some serve the body as an excellent source of fuel. Lipids are introduced into the body through the diet.

Several *metabolic wastes* are transported by the plasma, especially lactic acid, and some nitrogenous waste products of protein metabolism.

Gases and buffers Oxygen, nitrogen, and especially carbon dioxide are the principal gases dissolved in plasma. Carbon dioxide is transported both in the dissolved state and in the form of the bicarbonate ion (HCO_3^-). Nitrogen is transported in the dissolved state by plasma.

Red Blood Cells (Erythrocytes)

Red blood cells, or *erythrocytes* (ih-RITH-roh-sites; Gr. *erythros*, red + cells), make up about half the volume of human blood. There are about 25 trillion erythrocytes in the body, and each cubic millimeter of blood contains 4 to 6 million erythrocytes. (The average man has about 5.5 million erythrocytes per cubic millimeter of blood, and the average woman has about 4.8 million.) Erythrocytes measure about 7 micrometers in diameter, and about 2 micrometers thick.* If 5

*A micrometer (μm) equals 1/1000 of a millimeter; it replaces the older term micron (μ).

or 6 red blood cells were placed in a row, they would reach across the period at the end of this sentence.

The erythrocyte is shaped like a disk and is slightly concave on top and bottom, like a doughnut without the hole poked completely through (Figure 17.2). This shape provides a larger surface area for gas diffusion than a flat disk or a sphere. When an erythrocyte is mature, it no longer has a nucleus or many organelles, such as mitochondria. Thus, it must rely on its store of already-produced proteins, enzymes, and RNA.

Hemoglobin Almost the entire weight of an erythrocyte consists of **hemoglobin**, Hb (HEE-moh-gloh-bihn; Gr. *haima*, blood + L. *globulus*, little globe), an oxygen-carrying globular protein.

Each adult hemoglobin molecule consists of 5 percent *heme*, an iron-containing pigment, and 95 percent *globin*, a polypeptide protein.* Males usually have more hemoglobin (14–18 g/100 mL peripheral blood) than females do (12–16 g/100 mL). Attached to each of hemoglobin's four polypeptide chains (two α and two β) is a heme group, which gives blood its color. Attached to each heme group is an iron atom (Figure 17.3), which is the binding site for oxygen. The iron atom plays a key role in hemoglobin's function as an oxygen-carrying substance by binding oxygen and releasing it to tissues at the appropriate times.

The function of hemoglobin depends on its ability to pick up oxygen that has been inhaled into the lungs, to transport it via the blood vessels to body tissues, and to release it as the tissues need it. Hemoglobin also carries waste carbon dioxide from the tissues to the lungs, where it is exhaled. By depositing oxygen and removing carbon dioxide, hemoglobin also helps to maintain a stable acid-base balance in the blood.

*Hemoglobin got its name in a roundabout way. Early microscopes showed the shape of a red blood cell as a sphere, not the biconcave disk we can identify today. As a result, the cells were called "globules," the proteins they contained were called "globulins" or "globins," and the substance was named "hemoglobin" or "blood protein."

FIGURE 17.2

Drawing of an erythrocyte, cut open to show its biconcave shape. A micrometer equals 1/1000 of a millimeter.

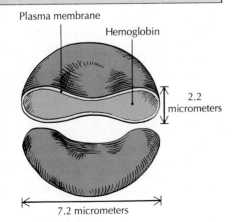

FIGURE 17.3

Hemoglobin. (A) Quaternary structure of a hemoglobin molecule. The folded chain represents globulin (protein), and the ovals represent the iron-containing heme groups. Note that the entire molecule consists of two alpha polypeptide chains and two beta chains. (B) Primary chemical structure of the heme group from a single hemoglobin monomer. Within each heme group is an iron atom (Fe^{2+}), where a single oxygen molecule (O═O) may bind.

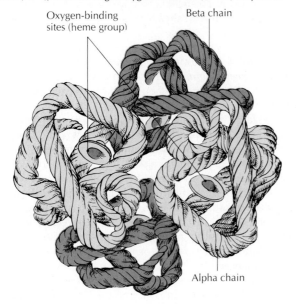

[A]

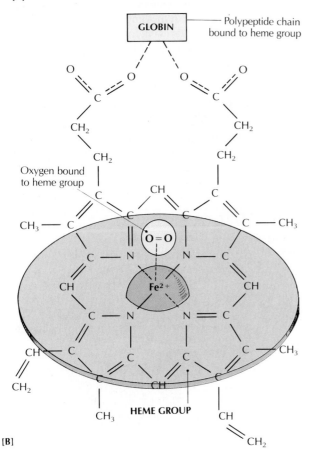

[B]

Hemoglobin has a bluish-purple color when it is deoxygenated, but it becomes red when it is loaded with oxygen. (Oxygenated hemoglobin is called *oxyhemoglobin,* HbO_2, and deoxygenated hemoglobin is called *reduced hemoglobin.*) As oxygenated blood circulates from the lungs to the rest of the body, it loses more and more oxygen to the cells, until it appears blue in the veins. The oxyhemoglobin in the erythrocytes of subcutaneous capillaries gives cheeks and lips their pink color.

Transport of oxygen in blood The major role of erythrocytes is to transport oxygen from the lungs to the body cells. The abundant oxygen in the alveoli (air sacs) of the lungs combines with hemoglobin to form oxyhemoglobin by bonding with the iron atom. The oxyhemoglobin is then transported in the red blood cells to the tissues. Because oxygen is constantly being used up by cellular oxidation, there is more oxygen in the blood coming from the lungs than there is in the tissues. Thus, oxygen in oxyhemoglobin is free to diffuse into the cells of the tissues.

Unfortunately, some toxic agents bind to hemoglobin even more readily than oxygen does. Air pollutants such as insecticides and sulfur dioxide bind to hemoglobin and prevent it from carrying oxygen effectively. Such poisons cause the numbness or dizziness that is characteristic of a lack of oxygen. Probably the best-known hemoglobin poison is carbon monoxide, found in automobile exhaust fumes and cigarette smoke.* Carbon monoxide binds to hemoglobin about 210 times faster than oxygen and forms a stable compound. Even at concentrations of 0.1 or 0.2 percent in the air, carbon monoxide is dangerous, and increased amounts may cause death by blocking the uptake of oxygen by hemoglobin. When this happens, tissues die of lack of oxygen.[†]

Transport of carbon dioxide in blood The other major function of erythrocytes is to help transport carbon dioxide from the tissues to the lungs, where it will be exhaled from the body. Carbon dioxide is more soluble in water than oxygen is, and it diffuses easily through capillary walls from the body tissues. Once in the blood, carbon dioxide is transported in three ways:

1 About 60 percent of the carbon dioxide reacts with water to form carbonic acid:

$$CO_2 + H_2O \rightleftharpoons H_2CO_3$$
$$\text{carbonic acid}$$

Carbonic acid is converted quickly into bicarbonate and hydrogen ions:

$$H_2CO_3 \rightleftharpoons HCO_3^- + H^+$$
$$\text{bicarbonate ion} \quad \text{hydrogen ion}$$

*Twenty percent of a cigarette smoker's hemoglobin is nonfunctional because it is bound to carbon monoxide.

[†]Hemoglobin that binds to carbon monoxide instead of oxygen is even a brighter red than oxyhemoglobin. It is ironic that victims of carbon monoxide poisoning, whose tissues are fatally starved for oxygen, have bright red lips.

Thus, the actual amount of carbonic acid in the blood at any time is very small. The two reactions above occur primarily in red blood cells, which contain large amounts of carbonic anhydrase (an enzyme that facilitates the reactions). Once the bicarbonate ion is formed, it moves out of the red blood cells, and is carried in the plasma. The removal of bicarbonate ions tends to produce a net positive charge inside the red blood cells. In order to restore an equilibrium, and homeostasis, negative chloride ions move into the red blood cells from the plasma. This exchange of chloride ions for bicarbonate ions is called the *chloride shift*. The process is reversed when blood reaches the lungs.

2 About 30 percent of the carbon dioxide reacts directly with hemoglobin to form the carbaminohemoglobin compound ($HbCO_2$), which is carried from the tissues to the lungs. When carbaminohemoglobin arrives in the lungs, the carbon dioxide is exchanged for oxygen.

3 The remaining 10 percent of the carbon dioxide is dissolved directly in the plasma and red blood cells as molecular carbon dioxide (CO_2).

Erythrocyte membranes and solute concentration Erythrocytes have a thin plasma membrane that is strong and flexible, allowing the erythrocytes to move easily through small blood vessels. It is permeable to water, oxygen, carbon dioxide, glucose, urea, and several other substances, but it is impermeable to hemoglobin and other large proteins.

Under normal conditions, the water and solute concentrations on both sides of the erythrocyte membrane are the same. Thus, the concentration in the plasma is *isotonic* to the fluid inside the erythrocyte. This relationship maintains a constant osmotic pressure on both sides of the plasma membrane, as well as a normal cell shape (see Figure 3.7A). If the plasma concentration of solutes increases above normal, the amount of water outside the erythrocyte *decreases* relative to the water inside, and the plasma becomes *hypertonic*. As a result, water leaves the cell faster than it enters, and the cell shrinks, or **crenates** (L. notch or cleft). In contrast, if the plasma concentration of solutes decreases below normal, the amount of water outside the erythrocyte *increases* relative to the water inside, and the plasma becomes *hypotonic*. As a result, water enters the cell faster than it leaves, and the cell swells and eventually bursts. The rupturing of erythrocytes is called *hemolysis* (Gr. blood + loosening) (see Figure 3.7B). Thus, the homeostasis of erythrocytes depends, to a large extent, on the concentration of solutes in the blood plasma.

Erythrocyte production (erythropoiesis) Before birth, the fetus produces blood cells progressively in the yolk sac, liver, and spleen. During the fifth fetal month, blood-cell production decreases in these sites, and increases in the bone marrow cavities. After birth, erythrocytes are manufactured primarily, and continuously, in the red marrow of certain bones, especially in the vertebrae, ribs, sternum, pelvis, and the upper ends of the femur and humerus. The process is called **erythropoiesis** (ih-RITH-roh-poy-EE-sis; Gr. red + to make).

Erythrocytes are derived from large embryonic cells in the bone marrow called **stem cells** or *hemocytoblasts*, which are destined or committed to become different kinds of blood cells (Figure 17.4). Some of the hemocytoblasts differentiate within the bone marrow into *myeloid progenitor cells*, and others become *lymphoid progenitor cells*. Among the different cells derived from the myeloid progenitor is the *rubriblast*. The rubriblast becomes the *prorubricyte*, which begins hemoglobin synthesis. Then the prorubricyte differentiates into a *rubricyte* as more hemoglobin appears. After several cell divisions, the nucleus grows smaller, and nucleoli are no longer visible. At this stage, the cell is called a *metarubricyte*. It loses its nucleus and becomes a *basophilic erythrocyte*. After a few days, the cell enters the bloodstream through the vascular channels of the bone marrow. The cell is now known as a **reticulocyte** (Gr. "network cell") because it contains an intricate netlike pattern of endoplasmic reticulum. Within a few hours, the endoplasmic reticulum disappears, and the reticulocyte becomes an **erythrocyte**, or red blood cell.

The production and destruction (in the spleen) of erythrocytes are maintained at an equal level. If red blood cells are lost from the circulatory system, the rate of erythropoiesis is increased until the normal erythrocyte level is regained.

Normal adult erythropoiesis produces about 10 billion cells an hour, enough in about a week for a pint of blood. (Blood donors are advised, however, to wait about eight weeks before donating again.) Certain nutrients are necessary to maintain this pace, especially amino acids for the production of hemoglobin and other proteins, and iron for the production of heme. Required in trace amounts are riboflavin, vitamin B_{12} and folic acid, which are necessary for the cell to mature, and vitamin B_6, which is required for the synthesis of hemoglobin.

The main controller of the rate of erythrocyte production is **erythropoietin**, a glycoprotein hormone produced mostly in the kidneys. It operates in a negative feedback system, as follows: the rate of erythropoiesis is sensitive to the pressure of oxygen in arterial blood, which is detected by the kidneys. Any process that decreases the delivery of oxygen to the tissues (such as hemorrhage) leads to *hypoxia* (low oxygen concentration), and a decreased oxygen pressure. This decrease in oxygen stimulates the production of erythropoietin, which then increases the number of hemocytoblasts committed to the production of erythrocytes. The increased number of erythrocytes increases oxygen pressure and the amount of oxygen that reaches the tissues. This relieves the hypoxia, which, in turn, inhibits the production of erythropoietin, and homeostasis is restored.

Destruction and removal of erythrocytes The life span of an erythrocyte is only about 120 days, primarily because it has no nucleus, and is unable to replace the enzymes and other proteins that it needs to function properly. Although erythrocytes are able to use some glucose as a source of energy, they cannot synthesize much protein. As the cells age, their protein goes through a normal process of degradation,

---

If the blood in our veins is bluish, why does it appear red when we cut a vein and begin to bleed?

The deoxygenated, venous blood turns bright red as soon as it is exposed to oxygen in the air.

FIGURE 17.4

Origin and development of blood cells. Dashed lines indicate the omission of some intermediate stages. The mature cells are shown enlarged.

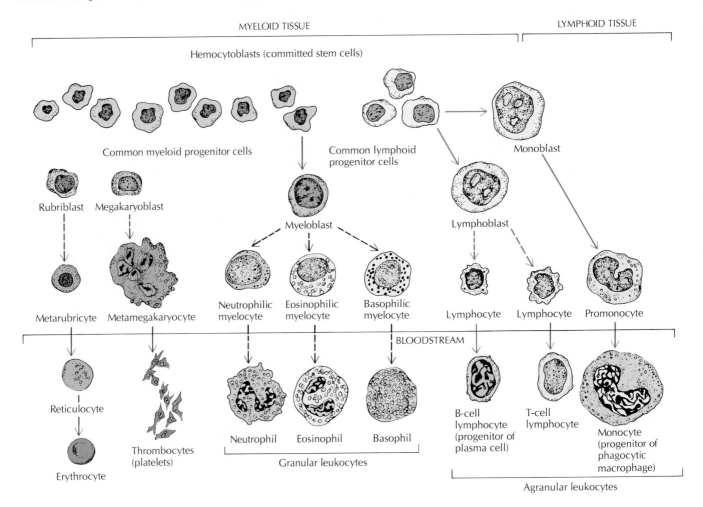

which cannot be repaired by the cells. As a result, the plasma membrane begins to leak, and leaks more and more as the integrity of the protein continues to be lost.

As these aged and fragile erythrocytes pass through the narrow capillaries (sinusoids) of the spleen, their leaky membranes rupture, and the cellular remnants are engulfed by phagocytic cells called *macrophages* (see Figure 4.6). The macrophages digest the hemoglobin into smaller amino acids, which are returned to the body's amino acid pool for the future synthesis of new proteins. The heme portion of the hemoglobin molecule is converted first into *biliverdin*, and then into the pigment *bilirubin*, which binds to plasma albumin and is transported to the liver. Within the liver, bilirubin is conjugated to glucoronic acid, and is eventually secreted in bile. (If the liver is faulty, as in alcoholism or malaria, bilirubin may accumulate in abnormally high amounts and cause the skin to turn yellow—a condition known as *jaundice*.) The iron portion of the heme is conjugated with protein and stored in the bone marrow as *ferritin* (L. *ferrum*, iron).

White Blood Cells (Leukocytes)

White blood cells, or **leukocytes** (LYOO-koh-sites; Gr. *leukos*, white + cells), serve as scavengers that destroy microorganisms at infection sites, help remove foreign chemicals, and remove debris that results from dead or injured tissue cells. Leukocytes range from slightly larger to much larger than erythrocytes (Table 17.2). Unlike erythrocytes, leukocytes do have nuclei, and the cells are able to move about independently and pass through blood vessel walls into the tissues.

Leukocytes are able to obtain a continuous supply of energy, and their anabolic and catabolic chemical processes are much more complex than those of erythrocytes. For example, they can synthesize protein, and are able to produce RNA in their nuclei.

In adults, there are about 1000 erythrocytes for every leukocyte. The normal adult leukocyte count is between 4000 and 12,000 per cubic millimeter, which may increase as a result of infection to 25,000 per cubic millimeter, about the same number that newborn infants have.

The production of white blood cells is called **leukopoiesis**. Some leukocytes originate from the same undifferentiated hemocytoblasts of blood-forming bone marrow that erythrocytes do. However, leukopoiesis also occurs throughout the body in lymphoid tissues such as lymph nodes, spleen, and tonsils.

The term *leukocyte* is used to cover a number of different cell types that circulate in the blood. The two basic classifications of leukocytes are granulocytes (polymorphonuclear leukocytes, or polymorphs) and agranulocytes (mononuclear leukocytes).

Granulocytes The most numerous of the white blood cells are **granulocytes**, so named because they contain large numbers of granules in the cytoplasm outside their multilobed nuclei (see Figure 17.4). The three types of granulocytes are *neutrophils, eosinophils,* and *basophils* (see Figure 17.4 and Table 17.2). When stained with Wright's stain, the granules of neutrophils appear light pink to blue-black, eosinophil granules appear red to red-orange, and granules of basophils are blue-black to red-purple.* Unstained leukocytes are ac-

*The names of granulocytes are derived from the type of stain with which they are most easily stained for laboratory preparations. The suffix *phil* comes from the Greek *philos,* which means "loving" or "having a preference for." Neutrophils "prefer" a *neutral* dye, eosinophils an *eosin* (acid) dye, and basophils a *basic* dye.

TABLE 17.2   TYPES OF BLOOD CELLS

Type of cell	Approximate number	Description	Function
Erythrocyte (red blood cell)	About 50 percent of total blood volume (4–6 million per mm³).	Biconcave disk. No nucleus, principal component hemoglobin. Diameter about 7 μm.	Transports oxygen to body cells, helps to remove waste carbon dioxide.
Leukocyte (white blood cell)	Less than 1 percent of total blood volume.	Several forms of cells. Capable of independent amoeboid movement.	Defends body against harmful microorganisms.
Granulocytes (polymorphonuclear)	About 52–75 percent of total leukocytes.	About two times larger than erythrocytes. Contain many granules in cytoplasm.	
Neutrophil	About 50–70 percent of total leukocytes.	Multilobed nucleus. Granules appear pink to blue-black in a neutral stain. Diameter about 10 μm.	Destroys microorganisms and other foreign particles by phagocytosis.
Eosinophil	About 1–4 percent of total leukocytes.	B-shaped nucleus. Granules appear red to red-orange in an acid stain. Diameter about 10 μm.	Helps modulate allergic inflammatory reactions, destroys antibody-antigen complexes.
Basophil	Less than 1 percent of total leukocytes.	Lobed nucleus. Granules appear blue-black to red-purple in basic stain. Diameter about 10 μm.	Involved in allergic reactions and inflammation; contains histamine, heparin, and SRS-A.
Agranulocyte (mononuclear)		Contains only occasional granules in cytoplasm.	
Monocyte	About 2–8 percent of total leukocytes.	Largest blood cells, approximately two to three times larger than erythrocytes, diameter about 20 μm.	Under stress, becomes macrophage, a large, mobile phagocyte that ingests and destroys harmful particles.
Lymphocyte	About 20–40 percent of total leukocytes.	Large, round nucleus that nearly fills the cell. Diameter about 10 μm.	Not a phagocyte like other leukocytes. Involved in immune response and synthesis of antibodies. Abundant in lymphoid tissue.
Thrombocyte (platelet)	Less than 1 percent of total blood volume (130,000 to 360,000 per mm³).	Fragments of megakaryocytes. No nucleus. Diameter about 2–4 μm.	Important in coagulation and hemostasis, releases serotonin.

tually clear (not even white), containing none of the pigment found in erythrocytes.

About 60 percent of the granulocytes are **neutrophils,** scavenger cells that move like amoebas and send out long projections called pseudopods ("false feet") to engulf and destroy microorganisms and other foreign particles. The granules inside the cytoplasm are packets of enzymes (lysozymes) that digest the intruders, break them down, and eventually destroy them. In the process, called *phagocytosis* ("cell eating"), the neutrophils may be destroyed also as their granules are depleted. Dead microorganisms and neutrophils make up the thick, whitish fluid we call *pus.*

Specific chemicals released at the site of infection or tissue injury attract neutrophils, as well as some other types of leukocytes such as monocytes and macrophages, and cause them to migrate quickly to the problem areas. This process of chemical attraction is called **chemotaxis.** Neutrophils (and leukocytes in general) are able to deform, elongate, and squeeze through the pores of capillary walls by a process called *diapedesis* (dye-uh-puh-DEE-sis; Gr. "a leaping through").

Eosinophils have B-shaped nuclei. Like neutrophils, eosinophils are phagocytes that have an amoeboid movement. Their granules are lysosomes (see Chapter 3) that release specific enzymes that destroy phagocytized material. Eosinophils are most likely to phagocytize antibody-antigen complexes. For unknown reasons, the eosinophil count increases during allergy attacks, with certain parasitic infections, some autoimmune diseases (the production of antibodies that attack one's own tissues), and in certain types of cancer. In addition, eosinophils contain the protein plasminogen, which helps dissolve blood clots.

Basophils are granulocytes with elongated, indistinctly lobed nuclei. They are the least numerous of all the granulocytes. Their granules contain *heparin* (an anticoagulant), *histamine* (which dilates general body blood vessels and constricts blood vessels in the lungs), and *slow-reacting substance* (SRS-A) of allergies. SRS-A produces some of the allergic symptoms, such as bronchial constriction. The exact function of basophils is not known, but they may play a role in phagocytosis, and can cause typical anaphylactic shock reactions or circulatory shock (see Physiological and Anatomical Abnormalities in Chapter 18, p. 564).

Granulocytes develop in the bone marrow from committed but as yet undifferentiated cells called *myeloblasts* (see Figure 17.4). When myeloblasts develop distinct granules, they become *promyelocytes.* The promyelocytes become either *basophilic, neutrophilic,* or *eosinophilic myelocytes* when the granules differentiate to such a degree that they react with the appropriate dye and can be identified by color. The respective myelocytes develop slightly indented nuclei and become *metamyelocytes.* As the metamyelocytes mature, nuclear indentation becomes more marked, and the cells are now called *band cells.* Once the nucleus separates into definite lobes, the cells are called basophils, neutrophils, or eosinophils.

The bone marrow contains millions of mature granulocytes that can be released into the blood when necessary. The bone marrow of an average person produces about 100 billion neutrophils a day. Most of the neutrophils in the bloodstream are not actually circulating, but instead are clinging to the inner walls of blood vessels, ready to move toward an infection or injury site. The number of neutrophils in the bloodstream may actually quintuple in the first few hours of a serious infection.

Most granulocytes have a life span of about 5 to 10 days, and survive only a few hours after they enter the bloodstream. The level of granulocytes is probably regulated hormonally, but the precise mechanism is not known. Recently, a plasma protein that stimulates *granulopoiesis* (formation of granulocytes) has been discovered. It may participate in the homeostatic regulation of granulocyte levels.

Agranulocytes Despite their name, **agranulocytes** (or *nongranular leukocytes*) usually have a few nonspecific lysosome granules in their cytoplasm. The two types of agranulocytes are monocytes and lymphocytes, which comprise about 5 and 30 percent of the leukocytes, respectively.

The largest blood cells are **monocytes,** which are mobile phagocytes. They have large, folded nuclei and often have fine granules in the cytoplasm (see Figure 17.4).

Monocytes develop in bone marrow from *monoblasts,* enter the bloodstream for about 30 to 70 hours, and then leave by diapedesis. Once in the tissue spaces, monocytes enlarge 5 to 10 times their normal size and become phagocytic macrophages. These macrophages form a key portion of the reticuloendothelial system, which lines the vascular portions of the liver, lungs, lymph nodes, thymus gland, and bone marrow. (The reticuloendothelial system is the system of macrophages throughout the body.) In the connective tissue of these regions, macrophages phagocytize microorganisms and cellular debris. Macrophages also play a role in the immune system by processing specific antigens. It is also thought that monocytes and macrophages produce a group of substances collectively called *colony-stimulating factor* (CSF), which stimulates the bone marrow to produce more monocytes and neutrophils.

Lymphocytes are small, mononuclear, agranular leukocytes with a large, round nucleus that occupies most of the cell (see Figure 17.4). They get their name from lymph, the fluid that transports them. Lymphocytes move sluggishly, and do not travel the same routes through the bloodstream as other leukocytes. They originate from the hemocytoblasts of bone marrow, and then invade lymphoid tissues, where they establish colonies. These colonies then produce additional lymphocytes *without* involving the bone marrow. Most lymphocytes are found in the body's tissues, especially in lymph nodes, the spleen, thymus, tonsils, adenoids, and the lymphoid tissue of the gastrointestinal tract. They also differ from other leukocytes in being able to leave as well as re-enter the circulatory system, and can leave the blood more easily than other cells to enter lymphoid tissue. Some lymphocytes live for years, recirculating between blood and lymphoid organs. The biggest difference between lymphocytes and other white blood cells is that lymphocytes are not phagocytes.

Two distinct types of lymphocytes are recognized: B cells and T cells (see Figure 17.4). **B cells** originate in the bone marrow, and colonize lymphoid tissue. In contrast, **T cells** are associated with, and influenced by, the thymus gland before they colonize lymphoid tissue. Both B cells and T cells regulate the cellular immune response that protects the body from its own defense system, and secrete chemicals that destroy harmful microscopic intruders such as bacteria, poisons, viruses, and tissue and chemical debris. When B cells are activated, they enlarge and become *plasma cells* (see Figure 17.4). Plasma cells have much more cytoplasm than B cells do, which enables them to accommodate the biochemical machinery necessary for the production of antibodies.

Platelets (Thrombocytes)

Platelets (so named because of their platelike flatness), or *thrombocytes* (Gr. *thrombus,* clot + cells), are disk-shaped, and about one-quarter the size of erythrocytes. Their main function is to start the intricate process of blood clotting. Platelets are much more numerous than leukocytes, averaging about 350,000 per cubic millimeter, but because they are so small, they occupy less area than leukocytes do (see Table 17.2). (An average adult has about a trillion platelets.) Platelets lack nuclei, and are incapable of cell division, but they have a complex metabolism and internal structure (Figure 17.5). Once in the bloodstream, platelets have a life span of 7 to 8 days.

About 200 billion platelets are produced every day. Like other blood cells, they originate from committed hemocytoblasts in the bone marrow (see Figure 17.4). The hemocytoblasts involved in platelet formation develop into myeloid progenitor cells from which large cells called **megakaryoblasts** arise. The megakaryoblasts differentiate into *promegakaryocytes, megakaryocytes,* and *metamegakaryocytes.* Platelets break off from the pseudopods of these cells in the bone marrow, and then enter the bloodstream. Thus, the platelets that appear in the circulating blood are not actually blood cells, but are cellular fragments of megakaryocytes. After entering the bloodstream, platelets begin to pick up and store chemical substances that can be released later to help seal vessel breaks.

Platelets adhere to each other, and to the collagen in connective tissue, but not to red or white blood cells, a property that is directly related to blood clotting and the overall process of *hemostasis,* the prevention and control of bleeding (Figure 17.5B). When a blood vessel is injured (capillaries rupture many times a day), platelets immediately move to the site and begin to clump together, attaching themselves to the damaged area. The platelets release granules that contain *serotonin,* which constricts broken or injured vessels and retards bleeding, and *adenosine diphosphate* (ADP), which attracts more platelets to the damaged area. If the break is small enough, it is repaired by the platelet plug. However, if there are not enough platelets to make the repair, they begin the process of blood clotting, or **coagulation** (L. *coagulare,* to curdle), described in a following section, Hemostasis (p. 519).

Blood Studies

Several blood studies are usually performed as part of a routine physical examination. Three commonly performed tests measure the amount of hemoglobin, number of red blood cells, and white blood cell differential count:

FIGURE 17.5

Platelets. (A) Flash-contact x-ray micrograph of human platelets. This new technique shows *live* cells; earlier techniques show cells that have been killed by the preparation. Note the forming pseudopod, which will become part of the meshwork essential for blood clotting. (B) Electron micrograph of cross section of an injured blood vessel (capillary). Note the platelet plugging a tiny break in the capillary wall. Larger breaks in blood vessels attract many platelets to the injured site. Sections of four erythrocytes are shown above the platelet. [(A) *Ralph Feder, IBM, Thomas J. Watson Research Center.* (B) *Hans R. Baumgartner, Hoffman LaRoche & Company, Basel.*]

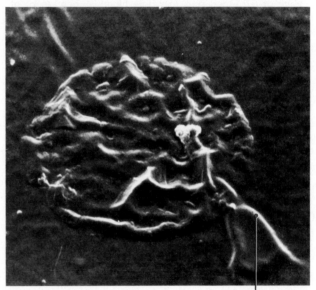

[A]   Pseudopod × 30,000

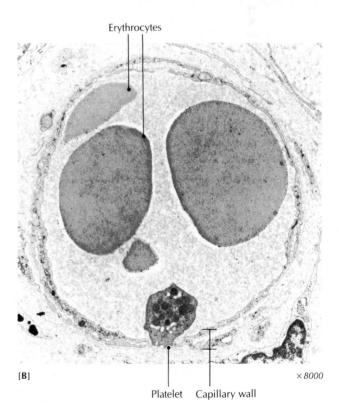

Erythrocytes

[B]   × 8000

Platelet   Capillary wall

1 The *amount of hemoglobin* in the blood is measured by a *hemoglobinometer* (hemometer), a specially adapted photometer. A sample of whole blood is treated chemically to form a stable pigment. The density (concentration) of hemoglobin is then measured and compared to a standard color scale. The normal range is 14 to 18 g Hb/100 mL of peripheral blood for men, and 12 to 16 g Hb for women.

2 The number of red blood cells is determined either by doing an actual red blood cell count (discussed below), or is estimated by centrifuging blood to obtain a hematocrit (HCT). The *hematocrit* (Gr. "to separate blood") is the volume percentage of red blood cells in whole blood. A hematocrit of 46 percent (Figure 17.6) means that in every 100 mL of whole blood there are 46 mL of red blood cells, 2 mL of white blood cells and platelets (known as the *buffy coat*), and 52 mL of fluid (plasma). The normal hematocrit for an adult male is 45 to 52 percent, and for an adult female 37 to 48 percent.

From the hematocrit and an estimate of the total blood volume in the body, the total volumes of plasma and erythrocytes can be determined as follows: total blood volume is approximately 8 percent of body weight. Thus, in a 70-kg person,

Total blood weight = 0.08 × 70 kg = 5.6 kg

Since 1 kg of blood occupies about 1 L,

Total blood volume = 5.6 L

Using our example of a 46-percent hematocrit, then

Total red blood cell volume = 0.46 × 5.6 L
= 2.58 L

and

Plasma volume = 5.6 L − 2.58 L
= 3.02 L

3 A *blood cell count* (CBC, complete blood count) can be determined by a *hemocytometer,* a device that includes a ruled glass slide for counting the number of cells per square as they are viewed under a microscope. Separate samples are used for erythrocytes and leukocytes. The number and shape of

FIGURE 17.6

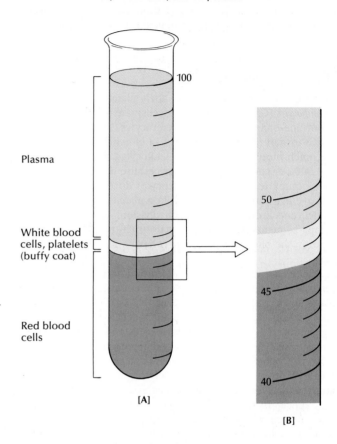

Hematocrit. (A) The test tube is filled with whole blood to the 100 mark, and then centrifuged. The red blood cells become packed at the bottom. (B) The percentage of red blood cells (hematocrit) can then be determined; in this case, it is 46 percent.

Plasma

White blood cells, platelets (buffy coat)

Red blood cells

[A]

[B]

the different types of leukocytes may also be noted, a procedure sometimes called the *white blood cell differential count* (Table 17.3). The cells are stained and counted under a microscope by a technician. A minimum of 100 cells is counted, and the percentage of each type of leukocyte is given. In most laboratories, automatic counting devices are now used to record the numbers in seconds.

TABLE 17.3	VALUES FOR A WHITE BLOOD CELL (LEUKOCYTE) DIFFERENTIAL COUNT

| Type of leukocyte | Differential count | |
	Normal range (%)	Actual count (%)*
Granulocytes		
Neutrophils	50–70	65
Eosinophils	1–4	3
Basophils	0.5–1	1
Agranulocytes		
Lymphocytes	20–40	25
Monocytes	2–8	6
	Total =	100

*In a differential white blood cell count the sum of the different leukocytes must equal 100 percent. The values listed are for adults.

Ask Yourself

1 *What are the major components of plasma?*

2 *What are the formed elements of blood?*

3 *What is the main function of hemoglobin?*

4 *Can you describe the process of erythropoiesis?*

5 *What are the three types of granulocytes? The two types of agranulocytes?*

6 *What are the functions of T cells and B cells?*

7 *Can you describe three major types of blood studies?*

HEMOSTASIS: THE PREVENTION OF BLOOD LOSS

One drawback of a circulatory system such as ours, where the liquid blood is under high pressure, is that serious bleeding can take place even after a slight injury. To prevent the possibility of uncontrolled bleeding, we have a three-part hemostatic mechanism consisting of (1) the constriction of blood vessels, (2) the clumping together (aggregation) of platelets, and (3) blood clotting. Overall, *hemostasis* is a specific type of homeostasis that prevents blood loss.

Vasoconstrictive Phase

Normally, when a tissue is damaged and blood escapes from a blood vessel, the vessel wall constricts in order to narrow the opening of the vessel and slow the flow of blood. This vasoconstriction is due to contraction of the smooth muscle of the vessel wall as a direct result of the injury, and the release of vasoconstrictor chemicals from platelets. Proper vasoconstriction is also enhanced by pain reflexes, producing constriction in proportion to the extent of the injury. Constriction of capillaries, which do not have muscular layers, is due to the vascular compression caused by the pressure of lost blood that accumulates in surrounding tissues. Injured blood vessels may continue to constrict for 20 minutes or more.

Platelet Phase

The next event in hemostasis is the escape from blood vessels of platelets, which swell and adhere to the collagen in adjacent connective tissues. The attachment somehow stimulates the platelet granules to release (degranulate) several chemicals, including serotonin, adenosine diphosphate (ADP), prostaglandins, and phospholipids. The serotonin and a prostaglandin (thromboxane A_2) stimulate vasoconstriction, and the phospholipids activate specific clotting factors. By now, the released ADP has caused the platelets to become very sticky, so that as more and more of them move into the injured area they stick together. In about a minute they can clog a small opening in the vessel with a platelet plug (Figure 17.7). The process is called *platelet aggregation.* It is important partly because it successfully stops hundreds of small hemorrhages every day, and partly because it triggers the blood-clotting mechanism.

Basic Mechanism of Blood Clotting (Coagulation Phase)

If the blood vessel damage is so extensive that the platelet plug cannot stop the bleeding, the complicated process of blood clotting—the *coagulation phase*—begins. The basic clotting mechanism involves the following events:

1 Aided by a plasma globulin called *antihemophilic factor*

(AHF), blood platelets disintegrate and release the enzyme *thromboplastinogenase* and platelet factor 3 (see Table 17.4).

2 Thromboplastinogenase combines with AHF to convert the plasma globulin *thromboplastinogen* into the enzyme *thromboplastin.*

3 Thromboplastin combines with *calcium ions* to convert the inactive plasma protein *prothrombin* into the active enzyme *thrombin.*

4 Thrombin acts as a catalyst to convert the soluble plasma protein *fibrinogen* ("giving birth to fibrin") into the insoluble, stringy plasma protein *fibrin.*

5 The fibrin threads entangle the blood cells and create a clot (Figure 17.8).

FIGURE 17.7

Scanning electron micrographs showing stages of platelet aggregation. (A) When exposed to ADP at the site of an injured blood vessel, platelets stick together. (B) Platelets begin to swell, become spiny, and adhere to even more platelets to form a plug to close the opening in the vessel. [(A and B) *James G. White, M.D.*]

[A]   ×30,000

[B]   ×30,000

FIGURE 17.8

Scanning electron micrograph of a portion of a blood clot. The tangled threads are fibrin, which bind the clot into an insoluble mass, and the circles are platelets and erythrocytes, enmeshed in the fibrin filaments. [*Courtesy of Dr. A. Hattori, First Department of Internal Medicine, Niigata University School of Medicine.*]

Erythrocytes

×2000

Platelet   Fibrin filaments

This basic process may be summarized as follows:

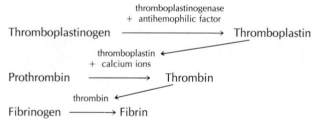

The conversions of prothrombin into thrombin and of fibrinogen into fibrin are well understood. What is not completely clear is how prothrombin is activated in the first place.

Extrinsic and intrinsic pathways Intense research on blood coagulation took place in the 1960s. At that time, two partially independent pathways were identified for the triggering of a blood clot. (1) The *extrinsic pathway* is a rapid clotting system activated when blood vessels are ruptured and tissues are damaged. (2) The *intrinsic pathway* is activated when the inner walls of blood vessels become damaged or irregular. (Remember that *extrinsic* means "outside" and *intrinsic* means "inside.")

Damaged tissue triggers the **extrinsic pathway,** which initiates blood clotting by the release of *thromboplastin,* known in this form as *tissue factor.* (A somewhat different form of thromboplastin is at work at the site of ruptured vessels, triggered by the disintegration of platelets.) Tissue factor combines with a mixture of enzymes and the phospholipids from damaged cell membranes released by the injured tissue to

produce another substance called **prothrombin activator** (Figure 17.9). At this point, the extrinsic system merges with the intrinsic system to activate yet another mechanism (sometimes called the *common pathway;* see Figure 17.9) that actually produces the clot. The common pathway includes steps 3, 4, and 5 described above.

The **intrinsic pathway** for initiating blood clotting uses only substances found in the blood. These substances are called *clotting factors,* and are described in Table 17.4. Injury to the inner wall of a blood vessel activates clotting factor XII, which triggers a series of rapid chemical reactions usually called the "cascade effect." Each step activates the next step in the sequence, until prothrombin activator is produced (see Figure 17.9). After prothrombin activator is formed in the extrinsic and intrinsic pathways, the basic blood-clotting process proceeds through the common pathway, as described in Figure 17.9 and steps 3, 4, and 5 on page 519. The extrinsic pathway usually produces a clot in as little as 15 seconds, while the intrinsic pathway requires 2 to 6 minutes.

At this point, the fibrin threads form only a weak mesh, and the clot must be strengthened if it is to hold. Platelets and plasma globulins release a *fibrin-stabilizing factor* (see blood coagulation factor XIII in Table 17.4) that responds to thrombin to create an interlacing pattern of fibrin threads (see Figure 17.8). Within a few minutes after the clot is formed, it begins to contract, squeezing out serum and helping the clot to solidify. The power to contract comes from platelets, which contain actin and myosin, the same proteins that make muscle contraction possible. (Platelets contain more actin and myosin than any tissue in the body except muscle.) Most of the serum is drained within an hour, and the solid clot is finally complete. A "scab" forms, dries up, and in a few days, falls off as the underlying tissue heals (see Figure 5.5).

A well-known dietary substance involved with blood clotting is vitamin K.* Found in leafy green vegetables, tomatoes, vegetable oils, and also produced by intestinal bacteria, vitamin K is necessary for the production of prothrombin and other clotting factors by the liver.

Hemostasis and the Nervous System

The sympathetic nervous system helps to control massive blood loss when a blood clot is inadequate to stop the flow. When the body loses more than 10 percent of its blood, there is a sudden drop in blood pressure, and the body goes into shock. The decreased blood pressure triggers reflexes in the sympathetic nervous system that constrict veins and the small terminal branches of arteries (arterioles) in an attempt to limit the decrease in blood pressure. Also, the heart rate may rise from a normal 72 to 84 beats a minute to as many as 200, increasing the blood flow to counteract the reduced blood pressure. Blood flow to the brain and heart is especially increased. Without the perfect functioning of the sympathetic reflexes, a person would probably die after losing 15 to 20 percent of the total blood volume. When the feedback system is operating properly, a person may still be alive after losing

*The K stands for *Koagulation,* the German word for "clotting." The discoverer of vitamin K originally named it the Koagulation-Vitamin.

FIGURE 17.9

The extrinsic pathway of blood clotting, and the ''cascade effect'' of the intrinsic pathway. Each step in the sequence is initiated by the previous reaction. Note how both pathways combine to form the basic clotting mechanism of the common pathway.

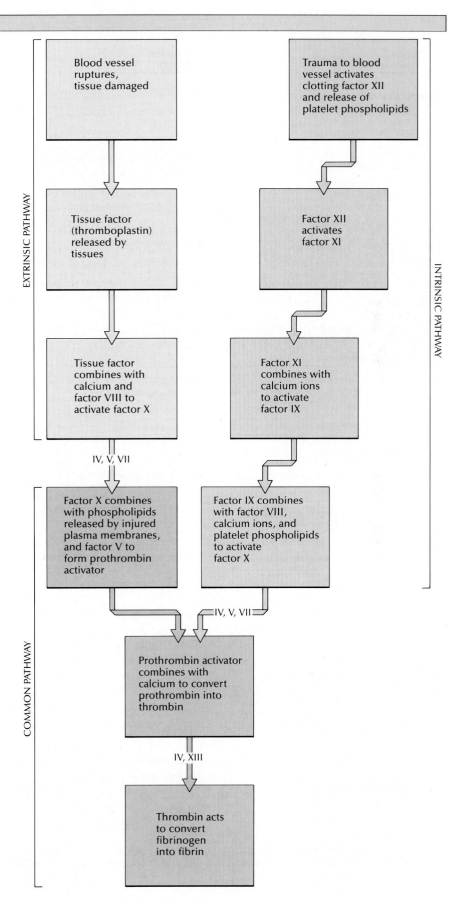

TABLE 17.4   BLOOD-CLOTTING FACTORS

Coagulation factor number and name*	Description and origin	Function
PLASMA COAGULATION FACTORS		
I Fibrinogen	Plasma protein synthesized in liver.	Precursor of fibrin; converted to fibrin in final stage of clotting. Serum is plasma minus fibrinogen.
II Prothrombin	Plasma protein synthesized in liver. Synthesis requires vitamin K.	Precursor of thrombin, the enzyme that converts fibrinogen into fibrin.
III Thromboplastin	Complex lipoprotein formed from disintegrating platelets or tissues.	Combines with calcium to convert prothrombin into active thrombin.
IV Calcium ions	Inorganic ion in plasma, acquired from bones and diet.	Necessary for formation of thrombin and for all stages of clotting.
V Proaccelerin, labile factor, or accelerator globulin	Plasma protein synthesized in liver.	Necessary for extrinsic and intrinsic pathways.
VI	No longer thought to be a separate entity; possibly the same as factor V.	
VII Serum prothrombin conversion accelerator (SPCA), stable factor, or proconvertin	Plasma protein synthesized in liver. Synthesis requires vitamin K.	Necessary for first phase of extrinsic pathway.
VIII Antihemophilic factor (AHF), antihemophilic factor A, or antihemolytic globulin (AHG)	Plasma protein (globulin) synthesized in liver and other tissues.	Necessary for first phase of intrinsic pathway. Deficiency causes hemophilia A, a genetic disorder.
IX Plasma thromboplastin component (PTC), Christmas factor, or antihemophilic factor B	Plasma protein synthesized in liver. Synthesis requires vitamin K.	Necessary for first phase of intrinsic pathway. Deficiency causes hemophilia B.
X Stuart-Prower factor or Stuart factor	Plasma protein synthesized in liver. Synthesis requires vitamin K.	Necessary for early phases of extrinsic and intrinsic pathways.
XI Plasma thromboplastin antecedent (PTA) or antihemophilic factor C	Plasma protein synthesized in liver.	Necessary for first phase of intrinsic pathway. Deficiency causes hemophilia C.
XII Hageman factor or glass factor	Plasma protein; source unknown.	Necessary for first phase of intrinsic pathway; activates plasmin; activated by contact with glass, probably involved with clotting outside the body.
XIII Fibrin-stabilizing factor (FSF) or Laki-Lorand factor	Protein present in plasma and platelets; source unknown.	Necessary for final phase of clotting.
PLATELET COAGULATION FACTORS		
Pf_1 Platelet factor 1 or platelet accelerator	Platelets.	Same as factor V; accelerates action of platelets.
Pf_2 Platelet factor 2 or thrombin accelerator	Platelets, phospholipid.	Accelerates thrombin formation at start of intrinsic pathway; accelerates conversion of fibrinogen into fibrin.
Pf_3 Platelet factor 3 or platelet thromboplastic factor	Platelets, phospholipid.	Necessary for first phase of intrinsic pathway.
Pf_4 Platelet factor 4	Platelets.	Binds the anticoagulant heparin during clotting.

*Coagulation (clotting) factors are substances in the plasma that are essential for the maintenance of normal hemostasis. Thirteen factors are recognized and identified by Roman numerals I to XIII. The platelet coagulation factors are identified as Pf_1 to Pf_4.

as much as 40 percent of the total blood volume. If the lost blood is not replaced quickly, however, death will follow.

Inhibition of Clotting: Anticoagulation

As many as 35 compounds may be required for blood coagulation. Such a complex system of checks and balances is necessary to prevent clotting when there is no bleeding. An unwanted clot in a blood vessel that cuts off the blood supply to a vital organ is one of the body's worst enemies. How are blood clots prevented and broken down if they do form?

Most of the body's anticoagulant substances circulate within the blood, and the blood vessels themselves help prevent clotting. The blood vessels contribute in two ways. First, the smoothness of the inner walls normally prevents activation of the intrinsic clotting mechanism. Second, a thin layer of negatively charged protein molecules attached to the inner walls repels the clotting factors, preventing the initiation of clotting. Injury to a blood vessel removes both of these safeguards, activating factor XII and the rest of the intrinsic pathway.

Heparin and antithrombin One of the most powerful anticoagulants in the blood is **heparin,** a polysaccharide produced by mast cells and basophils.* Heparin is concentrated mostly in the liver and lungs. Minute quantities of heparin in normal circulating blood also prevent clotting by combining with the *antithrombin-heparin cofactor* (also called *antithrombin* or *antithrombin III*) to induce the cofactor to combine with thrombin 1000 times more rapidly than usual. Such a rapid binding to thrombin removes it almost instantly from the bloodstream, and makes clotting almost impossible. Without heparin, antithrombin-heparin cofactor binds to thrombin molecule-for-molecule, removing it from the blood in about 15 minutes.

The combination of heparin and antithrombin-heparin cofactor also reacts with several clotting factors in the extrinsic and intrinsic pathways, further inhibiting blood clotting. Thrombin itself acts as an anticoagulant. When its concentration becomes too high, it destroys factor VIII to prevent clotting.

Fibrinolysis Clot prevention is important, but so is clot *destruction,* or **fibrinolysis** ("fibrin breaking"). Small blood clots form continually in blood vessels throughout the body. If they are not removed promptly, the blood vessels will become clogged. In the process of fibrinolysis, a blood protein called *plasminogen* is activated into an enzyme called *plasmin.* The plasmin digests the threads of fibrin by first making them soluble and then breaking them into small fragments. The fragments are removed from the bloodstream by phagocytic white blood cells and macrophages. Excessive amounts of coagulants are routinely removed by the liver.

*Mast cells arise from hemocytoblasts. They are not normally found in blood, but instead lodge in connective tissues throughout the body.

Anticoagulant drugs When used under medical supervision, anticoagulant drugs can sometimes remove blood clots in the body. The best-known anticoagulant drug is *aspirin* (acetylsalicylic acid), which works by preventing platelets from sticking together to form a plug. It also inhibits the release of clot-promoting substances from platelets.

Two drugs that digest the fibrin threads of a clot are urokinase and streptokinase. *Urokinase,* an enzyme produced and secreted by the kidneys and found abundantly in urine, activates plasminogen and enhances fibrinolysis. Urokinase can be obtained inexpensively from cultured kidney cells. *Streptokinase* is released by certain streptococcal bacteria. Like urokinase, it activates plasminogen to speed up fibrinolysis. It is used to dissolve blood clots (thrombi) in veins and arteries. Streptokinase also helps to dissolve the fibrin threads in a blood clot by converting plasminogen into plasmin, the fibrin-destroying enzyme.

When vitamin K is in short supply, the liver produces enough prothrombin and other clotting substances for normal clotting. *Dicumarol* is a compound that resembles vitamin K to such an extent that the liver enzymes that form prothrombin will pick up dicumarol instead of vitamin K. The anticoagulatory effect of dicumarol is often used to prevent clotting after surgery.

In addition to being used to remove blood clots and to keep blood from coagulating during surgery, anticoagulant drugs may be necessary to prevent clotting in blood that will be used later for blood transfusions. (Once blood clots, it cannot be used for transfusions.) To avoid such clotting, a dilute, sterile solution of a *citrate* or an *oxalate salt* is added to collected blood. Clotting does not occur because citrate ions or oxalate ions combine with the available calcium ions, making calcium unavailable for its usual blood-clotting functions.

Blood Coagulation Tests

Several tests are used to determine blood-clotting time. The most popular ones are platelet count, bleeding time, clotting time, and prothrombin time.

The **blood platelet count** must be greater than 150,000 per cubic millimeter in order for normal coagulation to take place. Also, if platelet function is not normal, normal coagulation may not occur.

A pierced fingertip or earlobe usually bleeds from 3 to 6 minutes. A longer **bleeding time** for this wound generally (but not always) indicates a platelet deficiency.

Clotting time is determined by placing blood in a test tube and tipping it back and forth every 30 seconds or until it clots. This usually occurs in 5 to 8 minutes. Because the condition and size of test tubes vary, standardization is necessary to obtain accurate results.

Why doesn't blood clot in the blood vessels?

Because the enzyme thrombin does not exist in the normal circulation, but is generated from a precursor, prothrombin, only in the vicinity of clumped platelets.

The test for **prothrombin time** (PT) indicates the amount of prothrombin in the blood. Immediately after blood is removed, oxalate is added to prevent the prothrombin from being converted into thrombin. Then calcium ions and tissue extract containing thromboplastin are added to the blood. The calcium offsets the effect of the oxalate, and the tissue extract activates the conversion of prothrombin. The time usually required for blood to clot, referred to as the *prothrombin time*, is about 12 seconds. (A longer prothrombin time may also mean a decreased quantity of some factor other than prothrombin.) Similar tests are used to determine the relative quantities of other clotting factors.

Ask Yourself

1 *Starting with prothrombin and ending with fibrin, what are the basic steps in blood clotting?*

2 *How do platelets assist in blood clotting?*

3 *What is the ''cascade effect''?*

4 *What are blood-clotting factors?*

5 *How is heparin involved with coagulation?*

6 *What is fibrinolysis?*

BLOOD TYPES

In 1900, a Viennese pathologist named Karl Landsteiner proved that there were individual differences in blood. He isolated two distinct glycoproteins (antigens) on the surface of red blood cells that could, in combination with certain samples of incompatible blood, cause the red blood cells to clump together. These two proteins were called A and B antigens and are now referred to as **agglutinogens** (uh-GLOOT-n-oh-jehnz). The clumping together of cells in general is called **agglutination,** and the clumping of red blood cells is called *hemagglutination.* Based on the possible combinations of the two agglutinogens, four types of blood were identified: a person with agglutinogen A is *blood type A,* someone with agglutinogen B is *type B,* someone with both agglutinogens is *type AB,* and someone with neither A nor B agglutinogen is *type O.* These blood types are inheritable characteristics, passed on from parents to their children.

Are blood types distributed evenly? | *No. In the United States about 44 percent of the white population has type O, 39.5 percent have type A, 11.5 percent have type B, and 5 percent have type AB. The distribution among American blacks is approximately 47 percent type O, 28 percent type A, 20 percent B, and 5 percent AB.*

Blood Grouping and Transfusions

The hemagglutination Landsteiner witnessed, which was the cause of sometimes fatal blood transfusions, was the result of an antibody-antigen reaction produced when two incompatible blood types are combined. This incompatibility is due to the presence of one of two antibodies in the blood plasma. These antibodies, or *isoantibodies,* are referred to as **isohemagglutinins** (eye-so-hem-uh-GLOO-tih-nihnz) because they agglutinate the red blood cells from other individuals whose blood is incompatible. Thus, the blood plasma of a person with blood type A has the isohemagglutinin against type B red blood cells. If blood type B or AB is introduced, the anti-B agglutinin reacts violently with the B isohemagglutinins present on the donated red blood cells, causing these "in-

FIGURE 17.10

ABO blood grouping. (A) Schematic representation of blood groups. (B) Schematic representation of antibody-antigen complex formed when type B blood is transfused into a recipient with type A blood. The agglutinins of the donated blood have little or no effect upon the red blood cells of the recipient.

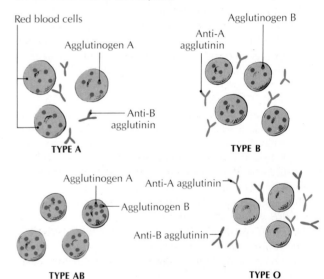

[A]

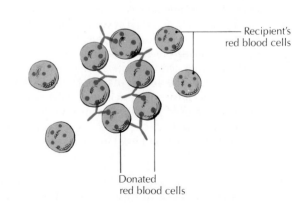

[B]

Blood Tests and Paternity Suits

Blood types are sometimes used in paternity suits, even though most tests of blood types cannot show who *is* the child's father, but can prove only who is *not*. A famous paternity case involved Charlie Chaplin. His blood type was O, the baby's was B, and the mother who made the claim had type A. Look at the table below to see if Mr. Chaplin could have been the father.

More recent tests have been used in paternity suits to determine who the father actually is. These tests are more costly than the earlier ones, but they can prove paternity by showing that the child and alleged father both have five rare blood factors that are not found in the mother. An even newer test called HLA (for human leukocyte antigen) matches substances in the leukocytes of the child and alleged father to determine paternity.

Another paternity test employs new genetic techniques with a high degree of accuracy. Called "DNA fingerprinting," the technique involves short stretches of DNA called intervening sequences, which provide distinct markers of human differences. DNA is extracted from white blood cells, and enzymes are used to cleave specific locations near the intervening sequences. The resulting fragments of DNA vary in length from one person to another. The DNA of at least one parent must match those of their child. If the known mother and her child have different sequences, and the alleged father does also, he cannot be the true father. Because DNA is more stable than blood, hair or skin samples from deceased people can also be tested for paternity.

INHERITANCE OF ABO BLOOD TYPE

Blood types of parents	Children's blood type possible	Children's blood type not possible
A + A	A, O	AB, B
A + B	A, B, AB, O	—
A + AB	A, B, AB	O
A + O	A, O	AB, B
B + B	B, O	A, AB
B + AB	A, B, AB	O
B + O	B, O	A, AB
AB + AB	A, B, AB	O
AB + O	A, B	AB, O
O + O	O	A, B, AB

vaders" to hemagglutinate (Figure 17.10). Table 17.5 shows the complementary distribution of agglutinins and isohemagglutinins for the four blood types.

The blood-matching system for blood types A, B, AB, and O is known as the **ABO blood-grouping system** (see p. 913 in Chapter 29). It operates as follows during blood transfusions from one person to another (see Figure 17.10): type A blood is incompatible with B, and these two blood types will form clumps if combined. Theoretically, a person with AB blood, containing no isohemagglutinins, can *receive* any type of blood (formerly called a universal recipient); and someone with type O blood, containing no agglutinogens, can safely *give* blood to any of the other types (formerly called a universal donor).

In a transfusion, the donor's blood (usually 0.5 L per transfusion) is diluted substantially when transfused into the recipient's circulatory system. Because of this dilution, the isohemagglutinins in the donor's blood cause little hemagglutination of the recipient's red blood cells. Thus, only antibodies in the recipient's blood and the type of red blood cells of the donor are important in transfusions. In practice, the blood types of the donor and recipient are tested, or *cross matched,* to be sure that they are compatible for a transfusion (see Table 17.5).

TABLE 17.5   ABO BLOOD GROUPING

Blood type	Agglutinogens (isoantigens) on erythrocytes	Agglutinins (isoantibodies) in blood plasma	Can donate blood to*	Can receive blood from*
A	B	Anti-B	A, AB	A, O
B	B	Anti-A	B, AB	B, O
AB	A, B	None	AB	A, B, AB, O
O	None	Anti-A, Anti-B	A, B, AB, O	O

*In practice, only matched blood types are used for transfusions.

Rh Factor

About 40 years after formulating the ABO blood-grouping system, Landsteiner and A. S. Wiener identified another factor in blood: the ***Rh factor.*** (The Rh factor was so named because it was first discovered in *Rhesus* monkeys.) Of the several genes responsible for different blood agglutinogens, this one is of clinical interest because it sometimes causes Rh disease, or ***hemolytic disease of the newborn*** (formerly called *erythroblastosis fetalis*) (see Figure 17.11).

About 85 percent of white Americans have the Rh factor (in addition to the ABO grouping), and are called ''Rh-positive.'' The remaining 15 percent, without the Rh factor, are ''Rh-negative.'' We now know that the Rh system contains six erythrocyte antigens (D, C, E, c, d, and e). Of these, the D antigen is most important in causing the production of antibodies. If blood from an Rh-positive person is transfused into an Rh-negative person, the Rh-negative blood will form antibodies against the Rh-positive red blood cells. During the next several months, antibodies build up in the Rh-negative blood plasma. This build-up is harmless. However, if the Rh-negative person receives additional Rh-positive blood, the newly formed antibodies will agglutinate the Rh-positive red blood cells.

The child of an Rh-positive father and Rh-negative mother has an equal chance of inheriting either factor. A potential problem arises if an Rh-negative mother conceives an Rh-positive child (Figure 17.11). Normally, the circulation of a fetus is separate from the mother's circulation, but sometimes some fetal blood cells leak into the mother's blood during late pregnancy or at birth. If some of the antigens (Rh, C, D, E) in the fetus's blood leak through the placenta into the mother's bloodstream, the mother will produce antibodies in response to the fetal Rh-positive antigen. These antibodies are not yet developed enough to harm the first baby, but if a later fetus is Rh-positive, the mother's antibodies can cross the placenta, enter the fetus's blood, and endanger the fetus (see Figure 17.11).

In hemolytic disease of the newborn, the fetus's agglutinated erythrocytes gradually undergo hemolysis and release hemoglobin into the blood of the fetus. Reticuloendothelial cells convert the hemoglobin into bilirubin, which causes jaundice (yellowing) of the fetus's skin. If born alive, the child must be given immediate blood transfusions of Rh-negative blood. The disease can now be prevented. An Rh-negative

---

FIGURE 17.11

Development of hemolytic disease of the newborn. (A) In an Rh-negative woman pregnant for the first time with an Rh-positive fetus, Rh-positive agglutinogens (antigens) (red) from the fetus's blood may diffuse through the placenta. (The fetus may be Rh-positive if its father is Rh-positive.) (B) Over time, the Rh-negative mother develops anti-Rh agglutinins, or antibodies (green). The first child will have been born before it could be affected by the antibodies. (C) A second Rh-positive child may receive some of its mother's anti-Rh agglutinins through the placenta, which may destroy that second child's red blood cells unless appropriate countermeasures are taken.

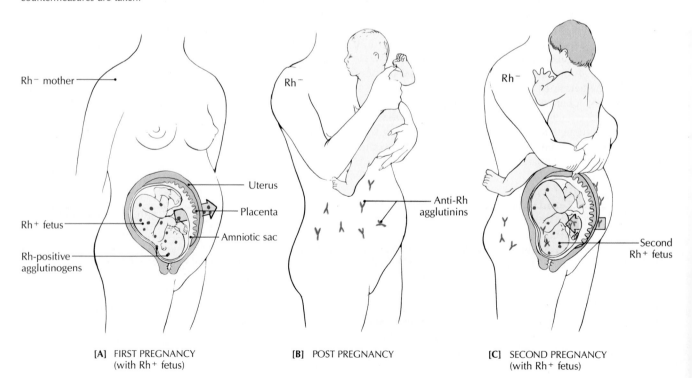

[A] FIRST PREGNANCY
(with Rh+ fetus)

[B] POST PREGNANCY

[C] SECOND PREGNANCY
(with Rh+ fetus)

woman who has an Rh-positive pregnancy can be given injections of a drug called Rho-GAM after delivery. The drug prevents her from making the anti-Rh antibodies, and thus protects a future Rh-positive fetus.

Other Blood Antigen Systems

Although only the ABO and the Rh systems have been included here, there are about 100 described blood antigen systems. When one considers the number of combinations of blood antigens possible, it appears likely that there are enough different arrangements to provide every human being with an individual blood type. Since only a few types seem to be medically important, those few are well known, but the rest are present and functioning, and are important in organ transplants.

Ask Yourself

1 *What is an agglutinogen? An agglutinin?*

2 *What is the Rh factor?*

3 *What is hemolytic disease of the newborn?*

PHYSIOLOGICAL AND ANATOMICAL ABNORMALITIES

Anemias

Anemia is a condition in which the number of red blood cells, the normal concentration of hemoglobin, or the hematocrit is below normal. In general, anemia decreases the blood's oxygen-carrying capacity. When anemia is due to the excessive destruction of erythrocytes, there is usually an abnormally high amount of bilirubin in the plasma, producing a characteristic jaundice of the skin and a darkening of the feces. Anemia is usually a symptom of some underlying disease. Descriptions of several types of anemia follow.

Hemorrhagic anemia results from heavy blood loss, and is sometimes seen in cases of heavy menstrual bleeding, severe wounds, or the sort of internal bleeding that accompanies a serious stomach ulcer or hookworm infection.

Iron-deficiency anemia is the most common type of anemia. It can be caused by a long-term blood loss, low intake of iron, or faulty iron absorption. If iron is not available to make erythrocytes, a *hypochromic* (low concentration in erythrocytes) *microcytic* (small erythrocytes) anemia develops.

Aplastic anemia is characterized by failure of the bone marrow to function normally. It is usually caused by poisons such as lead, benzene, or arsenic that hamper the production of red blood cells, or by radiation such as x rays or atomic radiation that damages the bone marrow. With the marrow not functioning properly, the total count of erythrocytes and leukocytes decreases drastically.

Hemolytic anemia is produced when an infecting organism such as the malar-

ial parasite enters red blood cells and reproduces until the cell actually bursts. Such a rupture is called *hemolysis* ("blood destruction"), and causes the anemia. (Other causes may be hereditary, or due to sickle-cell anemia or an adverse reaction to a drug.) *Thalassemia* is the name of a group of hereditary hemolytic anemias characterized by impaired hemoglobin production. As a result, the synthesis of erythrocytes is also impaired. It is most common among people of Mediterranean descent. Sickle-cell anemia (discussed below) is another hereditary hemolytic anemic condition.

Pernicious anemia is usually caused by an improper absorption of dietary vitamin B_{12}, which is required for the complete maturation of red blood cells. The majority of cases are associated with antibodies that destroy *intrinsic factor* in the stomach mucosa. Intrinsic factor is a protein that combines with vitamin B_{12}, and facilitates its absorption by the small intestine. Immature red blood cells accumulate in the bone marrow. The number of mature red blood cells circulating in the blood may drop as low as 20 percent of normal, and immature cells may enter the bloodstream. If diagnosed early, pernicious anemia is usually treated successfully with injections of vitamin B_{12} and an improved diet. If left untreated, the disease may affect the nervous system, causing decreased mobility, general weakness, and damage to the brain and spinal cord.

Sickle-cell anemia is a genetic hemolytic anemia that occurs most often in blacks; it affects 1 in 600 American blacks. It results from the inheritance of

the hemoglobin S-producing gene, which causes the substitution of the amino acid valine for glutamic acid in two of the four chains that compose the hemoglobin molecule. Sickle-cell anemia is so named because the abnormal red blood cells become rigid, rough, and crescent-shaped like a sickle (see photo). Sickled cells do not carry or release oxygen as well as normal erythrocytes. Such cells make the blood more viscous by clogging up capillaries and other small blood vessels, reducing blood supply to some tissues and producing swelling, pain, and tissue destruction. No effective permanent treatment has yet been discovered. Chronic fatigue and pain, increased susceptibility to infection, decreased bone-marrow activity, and even death are common results of sickle-cell anemia.

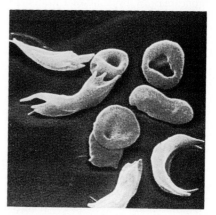

×2000

Scanning electron micrograph of human erythrocytes from a patient suffering from sickle-cell anemia. Instead of being normal biconcave disks, these cells are variously distorted. [*N. Calder/Science Source/Photo Researchers.*]

Hemophilia

Hemophilia is a hereditary disease in which blood fails to clot or clots very slowly. The gene that causes the condition is carried by females, who are not themselves affected, and is expressed mostly in males. Females can be born with the disease only when the mother is a carrier and the father already has hemophilia. *Hemophilia A* results from a deficient or inactive clotting factor VIII (see Table 17.4); *hemophilia B* is caused by a deficient or inactive clotting factor IX; and *hemophilia C* results from a deficient or inactive clotting factor XI.

Because these factors are needed to produce prothrombin and form blood clots, the untreated hemophiliac bleeds very easily, especially into joints. Until recently, hemophilia created a severe disability, but excellent transfusion therapy to reduce fatal bleeding is now available. However, there is no known permanent cure. Hemophilia as a sex-linked genetic disorder is discussed in Chapter 29.

Leukemia

Leukemia is a malignant (cancerous) disease characterized by uncontrolled leukocyte production. Leukocytes increase in number as much as 50- or 60-fold, and millions of abnormal, immature leukocytes are released into the bloodstream and lymphatic system. The white blood cells tend to use the oxygen and nutrients that normally go to other body cells, causing the death of otherwise healthy cells. Infiltration of the bone marrow by abnormal leukocytes prevents the normal production of white and red blood cells, causing anemia. Megakaryocytes are crowded out also, reducing the platelet count and sometimes causing internal hemorrhaging. Because almost all the new white blood cells are immature and incapable of normal function, victims of leukemia have little resistance to infection. In most cases, the combination of starving cells and a deficiency in the immune system is enough to cause death. The cause of leukemia is unknown, although a viral infection is considered likely.

The two major forms of leukemia are lymphogenous and myelogenous. *Acute lymphogenous leukemia* appears suddenly and progresses rapidly. It occurs most often in children, and is much more serious than *chronic lymphogenous leukemia,* which occurs later in life. Both of these leukemias are caused by the cancerous production of lymphoid cells in lymph nodes, which spread to other parts of the body. In *myelogenous leukemia,* the cancer begins in myelogenous cells such as early neutrophils and monocytes in the bone marrow, and then spreads throughout the body, where white blood cells can be formed in extraneous organs. Acute myelogenous leukemia can occur at any age, whereas chronic myelogenous leukemia occurs during middle age.

In 1983, scientists led by Dr. Takaji Miyake discovered a natural hormone that regulates the production of granulocytes. The hormone, called *granulopoietin,* was found in the urine of patients suffering from aplastic anemia. It is hoped that the discovery of granulopoietin will help to combat leukemia.

Bone-marrow transplantation

Another possible method of combating disorders of blood-forming tissues such as aplastic anemia, leukemia, and lymphoma, is *bone-marrow transplantation.* The purpose of bone-marrow transplantation is to introduce normal blood cells into the bloodstream of the patient in an attempt to counteract the effects of the disease. In order to ensure a high degree of compatibility between donor and recipient so that the recipient will not reject the donated marrow, physicians try to match six tissue factors called human leukocyte antigens (HLA). In bone-marrow transplantation, about 500 to 700 mL of marrow from a compatible donor (preferably a twin or sibling) are aspirated (removed by suction) from the donor's pelvic bones (usually the iliac crest) with a large hypodermic needle attached to a syringe. Only a small incision is necessary. The removed marrow is mixed with an anticoagulant (heparin) and then filtered to separate the donor's T cells, specialized cells of the immune system that would otherwise encourage rejection of the marrow by the recipient. The recipient receives the donor's marrow through an intravenous infusion. Two units of the donor's blood, which were removed during two sessions about a week before the transplantation, are transfused back into the donor to replace the blood aspirated along with the marrow.

A National Bone Marrow Registry has been established to link needy recipients with volunteer donors. About 60 centers are operative currently, with the number growing rapidly.

MEDICAL TERMINOLOGY

ANOXIA (*an-*, without + oxygen) An oxygen deficiency in the blood.

APHERESIS (uh-FUR-ee-sis; Gr. *aphairein,* to take away from) A medical technique for cleansing the blood, in which a portion (such as plasma) suspected of containing harmful substances is removed and replaced with fresh ingredients.

BLOOD PLASMA SUBSTITUTE A chemical substance that imitates plasma characteristics. It is used to keep up the blood volume temporarily during emergencies until the blood is matched, and to help replace fluids and electrolytes.

CITRATED WHOLE BLOOD Whole blood placed in a solution of acid citrate or a similar compound to prevent coagulation.

CYANOSIS (Gr. *kyanos,* blue) A bluish discoloration, especially of the skin and mucosa, as a result of an excessive amount of deoxyhemoglobin.

DIRECT (IMMEDIATE) TRANSFUSION Transfer of blood directly from one person to another, without exposure of the blood to air.

DRIED PLASMA Normal plasma that has been vacuum dried to prevent microorganisms from growing in it.

ERYTHROPENIA (Gr. red + poverty, lack) Decreased red blood cell count due to disease or hemorrhage.

EXCHANGE TRANSFUSION Direct transfer of blood from donor to replace blood as it is removed from recipient. This technique is used in poisonings and other conditions.

FRACTIONED BLOOD Blood separated into its components.

HEMORRHAGE (Gr. blood + bursting forth) A discharge of blood from a broken vessel.

HEPARINIZED WHOLE BLOOD Whole blood placed in a heparin solution to prevent coagulation.

INDIRECT (MEDIATE) TRANSFUSION Transfer of blood in which whole or fractioned donor blood is stored for later delivery to a recipient. The blood can be separated into components, and patients receive only the needed portions.

LEUKOPENIA (Gr. white + poverty, lack) Decreased white blood cell count due to disease or hemorrhage.

NORMAL PLASMA Plasma that has had blood cells removed, but retains the normal concentrations of solutes; employed to bring blood volume back to normal levels.

PACKED RED CELLS The concentrated solution of erythrocytes that remains when plasma is removed from whole blood.

PLATELET CONCENTRATE Platelets separated from fresh whole blood; used for platelet-deficiency blood disorders.

POLYCYTHEMIA (pahl-ee-sigh-THEE-mee-uh; "many cells in the blood") A condition in which the number of red blood cells is above normal.

PURPURA (Gr. *porphura*, shellfish yielding a purple dye) Purple spots on the skin resulting from escaped erythrocytes from capillaries or larger hemorrhagic areas.

SEPTICEMIA (sehp-tih-SEE-mee-uh; Gr. *septos*, rotten + blood) The presence of harmful substances, such as bacteria or toxins, in the blood. Also called *blood poisoning*.

THROMBOCYTOPENIA A decreased number of platelets, resulting from impaired production or increased destruction.

THROMBOCYTOSIS An increased number of platelets, usually due to increased production accompanying diseases such as leukemia and polycythemia, or following the removal of the spleen. Also called *thrombocythemia*.

VENESECTION Opening a vein to withdraw blood. Also called *phlebotomy*.

VON WILLEBRAND'S DISEASE An inherited antihemophilic blood factor disorder and capillary defect, characterized by abnormal bleeding of the nose, gums, and skin.

WHOLE BLOOD Blood that has all its components (formed elements, plasma, and plasma solutes) in natural concentration.

SUMMARY

Functions of blood

1 *Blood* is the circulating fluid that transports nutrients, oxygen, carbon dioxide, gaseous waste products, and regulatory substances throughout the body.

2 Blood also defends against harmful microorganisms, is involved in inflammation, coagulation, and the immune response, and helps regulate the pH of body fluids. In general, blood helps maintain homeostasis by providing a constant cellular environment.

Properties of blood

1 The average person has about 5 L of blood.

2 *Blood viscosity* and *specific gravity* are greater than those of water.

3 Blood *pH* ranges from 7.35 to 7.45.

4 Blood *temperature* averages about 38°C (100.4°F).

Components of blood

1 Blood consists basically of two parts, liquid *plasma* and solid *formed elements* (red blood cells, white blood cells, platelets) suspended in the plasma.

2 *Plasma* provides the solvent for dissolved nutrients. It is about 90 percent water, 7 percent dissolved plasma proteins (albumin, globulins, fibrinogen), and 3 percent electrolytes, amino acids, glucose and other nutrients, enzymes, antibodies, hormones, metabolic waste, and traces of other materials.

3 Red blood cells, or *erythrocytes*, make up about half the volume of blood. Their biconcave shape provides a large surface for gas diffusion. They contain the globular protein **hemoglobin**, which transports oxygen from the lungs to all body cells, and helps remove waste carbon dioxide. Hemoglobin contains a small amount of *heme*, an iron-containing pigment that binds to oxygen and releases it at appropriate times.

4 The production of erythrocytes in bone marrow is called **erythropoiesis**. The rate of erythropoiesis is controlled mainly in a negative feedback system involving the glycoprotein hormone *erythropoietin*.

5 White blood cells, or *leukocytes*, destroy microorganisms at infection sites and remove foreign substances and body debris. Leukocytes use an amoeboid movement to crawl along the inner walls of blood vessels and can pass through blood vessel walls into tissue spaces. They produce their own continuous supply of energy, and synthesize RNA and protein.

6 The two basic classifications of leukocytes are **granulocytes** and **agranulocytes**. The three types of granulocytes are neutrophils, eosinophils, and basophils. **Neutrophils** destroy harmful microorganisms and other foreign particles, **eosinophils** help destroy parasites and antibody-antigen complexes, and **baso-**phils are involved in allergic reactions and inflammation, although their specific function is unknown.

7 The two types of agranulocytes are monocytes and lymphocytes. **Monocytes** are capable of becoming macrophages that ingest and destroy harmful substances. **Lymphocytes** are involved in the immune response and the synthesis of antibodies.

8 Platelets, or **thrombocytes**, initiate the blood-clotting process when a blood vessel is injured, and are important in the overall process of **hemostasis**, the prevention and control of bleeding. They are fragmented from **megakaryocytes**.

9 Blood studies commonly performed for routine physical examinations include red blood cell count, hematocrit, and white blood cell differential count.

Hemostasis: the prevention of blood loss

1 The body has a three-phase hemostatic mechanism consisting of constriction of blood vessels, clumping (**aggregation**) of platelets, and blood clotting (**coagulation**).

2 The final product of coagulation is the conversion of the soluble plasma protein **fibrinogen** into the insoluble, stringy plasma protein **fibrin**. This reaction requires the enzyme **thrombin**. Fibrin threads entangle the escaping blood cells and form a **clot**.

3 The *extrinsic pathway* of blood clotting is a rapid clotting system activated when blood vessels are ruptured and tissues are damaged. The *intrinsic pathway* is activated when the inner walls of blood vessels become damaged or irregular. Either pathway may precede the basic clotting mechanism of the conversion of fibrinogen to fibrin.

4 *Clotting factors* are substances found in the blood that are specialized to enhance blood clotting.

5 Substances in the blood, such as heparin and the antithrombin-heparin cofactor, prevent it from clotting when there is no bleeding.

6 *Fibrinolysis,* a process in which fibrin is broken down, promptly removes naturally forming small clots from blood vessels.

7 Blood coagulation tests include platelet count, bleeding time, clotting time, and prothrombin time.

Blood types

1 Blood can be classified into groups based on the presence of an A or B antigen (*agglutinogen*) on the surface of erythrocytes. Protective antibodies in blood plasma are called **agglutinins.** Agglutinins of one type of blood react violently with agglutinogens of other types of blood, causing incompatible cells to clump together, a process called **agglutination.** The four blood types in the *ABO blood grouping* are A, B, AB, and O.

2 Another factor in blood is the *Rh factor.* It can cause **hemolytic disease of the newborn,** a disease that endangers a fetus if the fetus is Rh-positive, the mother is Rh-negative, and the mother has had a previous pregnancy in which the fetus was Rh-positive.

3 There are about 100 blood antigen systems in addition to the ABO and Rh systems.

Physiological and anatomical abnormalities

1 *Anemia* is a condition in which the number of erythrocytes, the normal concentration of hemoglobin, or the hematocrit is below normal. In general, anemia decreases the blood's oxygen-carrying capacity. The most common anemia is *iron-deficiency* anemia. Other anemias include *hemorrhagic, aplastic, hemolytic, pernicious,* and *sickle-cell.*

2 *Hemophilia* is a hereditary disease in which blood fails to clot, or clots very slowly. It is carried by females and transmitted to male offspring.

3 *Leukemia* is characterized by uncontrolled leukocyte production. The two major forms of leukemia are lymphogenous and myelogenous.

4 *Bone-marrow transplantation* involves introducing normal blood cells into the bloodstreams of patients with disorders of blood-forming tissues.

UNDERSTANDING THE FACTS

1 About how much blood does the average adult have?

2 Which of the formed elements in blood is the largest? The most abundant? The most important?

3 What is the most abundant plasma protein?

4 A plasma protein that is essential for blood clotting is _____ .

5 The major plasma cation is _____, while the major anion is _____ .

6 Name the primitive cells of the red bone marrow from which erythrocytes are derived.

7 Where and how are old red blood cells destroyed?

8 Define chemotaxis.

9 Which is the most abundant leukocyte?

10 Name the two basic categories of leukocytes and the types of leukocytes in each category.

11 What is the origin of platelets?

12 Define hemostasis.

13 List the three mechanisms involved in hemostasis.

14 What is antihemophilic factor (AHF), and what is its function?

15 Thrombin is an enzyme that functions in _____ .

16 An important function of mast cells is the production of _____ .

17 What is the function of plasmin?

18 How do citrates and oxalates prevent blood from clotting?

19 Do Rh-negative individuals have any anti-Rh factors? Can they produce them?

20 Name the vitamin that is important in preventing pernicious anemia, and explain its role.

UNDERSTANDING THE CONCEPTS

1 Discuss the major functions of blood and how they contribute to homeostasis.

2 Why is carbon monoxide such an effective hemoglobin poison?

3 Why is good nutrition, including dietary vitamins and minerals, so important to normal hemoglobin production?

4 How is oxygen transported from the lungs to body tissues? How is carbon dioxide transported from the tissues to the lungs?

5 What is the significance of the fact that most of the iron from old, destroyed erythrocytes is salvaged for reuse?

6 What is the significance of diapedesis?

7 Distinguish between the extrinsic and intrinsic mechanisms of blood coagulation, and discuss the steps in each process.

8 How does the structure of capillary walls help to prevent blood coagulation under normal circumstances?

9 List the five types of leukocytes, and compare their functions.

10 How do platelets contribute to hemostasis?

11 Why are the agglutinins of the recipient so crucial in blood transfusions?

12 Why can an individual with type AB blood receive blood from any of the other blood types?

13 List the conditions that must be present in order for hemolytic disease of the newborn to develop.

14 Describe the types of anemia studied, and give the basic causes of each.

15 How do you explain the fact that leukemia victims have difficulty fighting infections when they have a greatly increased number of white blood cells?

16 Why is vitamin K important to the body?

SUGGESTED READING

AREHART-TREICHEL, JOAN, "Artificial Blood." *Science News,* April 12, 1980.

CASTELLINO, F. J., "Plasminogen Activators." *Bioscience,* November 1983.

DOOLITTLE, RUSSELL F., "Fibrinogen and Fibrin." *Scientific American,* December 1981.

FIGUEROA, W. G., *Hematology.* New York: Wiley, 1981.

FREDA, V. J., et al., "Prevention of Rh-Hemolytic Disease with Rh-Immune Globin." *American Journal of Obstetrical Gynecology,* 128(1977):456–460.

GOLDE, DAVID W., AND JUDITH C. GASSON, "Hormones That Stimulate the Growth of Blood Cells." *Scientific American,* July 1988.

GRABER, S. E., AND S. B. KRANTZ, "Erythropoietin and the Control of Red Cell Production." *Annual Review of Medicine,* 29(1978):51–66.

LAWN, R. M., AND G. A. VEHAR, "The Molecular Genetics of Hemophilia." *Scientific American,* March 1986.

PERUTZ, M. F., "The Hemoglobin Molecule." *Scientific American,* November 1964.

PERUTZ, M. F., "Hemoglobin Structure and Respiratory Transport." *Scientific American,* December 1978.

ROSENBERG, R. D., AND K. A. BAUER, "New Insights into Hypercoaguable States." *Hospital Practice,* 21(1986):131.

TILL, J. E., "Cellular Diversity in the Blood-Forming System." *American Scientist,* 69(1981):522–527.

TURITTO, V. T., AND H. J. WEISS, "Red Blood Cells: Their Dual Role in Thrombosis Formation." *Science,* 207(1980):541.

WEISS, RICK, "Sanguine Substitutes." *Science News,* September 26, 1987.

WINTHROBE, M. M., et al., *Clinical Hematology,* 8th ed. Philadelphia: Lea & Febiger, 1981.

WOOD, W. B., "White Blood Cells Versus Bacteria." *Scientific American,* February 1971.

ZUCKER, MARJORIE B., "The Functioning of Blood Platelets." *Scientific American,* June 1980.

18

The Cardiovascular System: The Heart

LEARNING OBJECTIVES

1 Explain the basic functions of the heart, and give its size, shape, position, and location.

2 Describe the covering of the heart, the heart wall, and the cardiac skeleton.

3 Relate the structure of the heart chambers to their functions.

4 Describe the structure and functions of the different heart valves.

5 Name the veins and arteries that supply the heart muscle and the vessels that emerge from the heart.

6 Describe the two processes by which the heart regulates its own blood flow.

7 Describe how cardiac muscle differs from skeletal muscles.

8 Describe a typical action potential of a cardiac muscle cell.

9 Describe the impulse-conducting system of the heart.

10 Tell what activities occur during each part of an electrocardiogram.

11 Discuss the uses of the electrocardiogram.

12 Define cardiac cycle, and describe the path of the blood through the heart during the cycle.

13 Describe the mechanical events that occur during the cardiac cycle.

14 Discuss the cause and significance of the different heart sounds.

15 Define cardiac output, describe how it is calculated, and discuss factors that influence it.

16 Explain the role of the nervous and endocrine systems in the control of heart rate.

17 Describe the control of stroke volume.

18 Tell how the Frank-Starling law of the heart operates.

19 Describe the causes and symptoms of two degenerative heart disorders.

20 Describe some forms of valvular heart disease.

As you saw in the last chapter, blood is the connecting link between the outside environment and the cells of your body. In order to maintain homeostasis throughout the body, blood must flow continuously. Your heart is the pump that provides the energy to keep blood circulating, and your blood vessels are the pipelines through which blood is channeled to all parts of your body. This chapter describes the structure and function of the heart, and the next one discusses the blood vessels.

Human blood flows in a *closed system* of vessels, remaining essentially within the vessels that carry it. Blood travels in a circle, pumped from the muscular heart out through elastic arteries, through tiny capillaries, and then through veins back again to the heart. Each side of the heart contains an elastic upper chamber called an *atrium,* where blood enters the heart, and a lower pumping chamber called a *ventricle,* where blood leaves the heart. Thus, the heart is actually a *double pump.* The oxygen-poor blood returning to the right atrium from the body tissues is pumped by the right ventricle (the first pump) into the lungs, where carbon dioxide is exchanged for oxygen. The oxygenated blood moves from the lungs into the left atrium and then into the left ventricle (the second pump), which moves the blood throughout the body. The circulation to and from the lungs is called the **pulmonary circulation** (L. *pulmo,* lung), and the circulation to and from the body is called the **systemic circulation** (Figure 18.1).

The heart, together with its vessels, takes shape and begins to function long before any other major organ (see page 542). It begins beating in a human embryo during the fourth week of pregnancy, and continues throughout the life of a person at a rate of about 70 times a minute, 100,000 times a day, or about 2.5 billion times during a 70-year lifetime. The specialized cardiac muscle tissue that performs this extraordinary feat is found only in the wall of the heart.

The human body contains 4 to 6 L (about 1 to 1.5 gal) of blood, but the heart takes only a little more than a minute to pump a complete cycle of blood throughout the body. In times of strenuous exercise, the heart can quintuple this output. Yet, the heart is a relatively small organ. On the average, the adult heart pumps about 7500 L (2000 gal) of blood throughout the body every day.

STRUCTURE OF THE HEART

The **heart** (Gr. *kardia,* as in *cardiac,* L. *cor*)* is shaped like a blunt cone. It is about the size of the clenched fist of its owner. It averages about 12 cm long and about 9 cm wide (5.0 in. by 3.5 in.). The heart of an adult male weighs about 250 to 390 g (8.8 to 13.8 oz), and the heart of an adult female usually weighs between 200 and 275 g (7.0 to 9.7 oz).

---

*The word *coronary* is often used in reference to structures and events involving the heart. (A "coronary thrombosis" is a blockage of a coronary artery by a blood clot. The resultant heart attack is informally called a "coronary.") However, the term actually comes from the Latin word *corona,* which means "crown." If you look at Figure 18.8, you will see that the two coronary arteries sit, slightly askew, on top of the heart like a crown.

Location of the Heart

The heart is located in the center of the chest. It is slanted diagonally, with about two-thirds of its bulk to the left of the body's midline (Figure 18.2A). The heart is turned on its longitudinal axis so that the right ventricle is partially in front of the left, directly behind the sternum. The left ventricle faces the left side and the back of the thorax. The heart lies closer to the front of the thorax than the back.

The surfaces of the heart are called the *anterior* (sternocostal, next to the sternum and ribs), *inferior* (diaphragmatic, against the diaphragm), *left side* (pulmonary, next to the left lung), and *posterior* (base) surfaces.

The pointed end of the blunt cone is called the **apex** (see Figure 18.2A). It extends forward, downward, and to the left. Normally the apex is located between the fifth and sixth ribs (the fifth intercostal space) on the midclavicular line (a perpendicular line from the middle of the clavicle to the diaphragm). The uppermost part of the heart, the **base,** extends upward, backward, and to the right. Anteriorly, it lies just below the second rib.

The base is in a relatively fixed position because of its attachments to the great vessels, but the apex is able to move. When the ventricles contract, they change shape just enough so that the apex moves forward and strikes the left chest wall near the fifth intercostal space. This thrust of the apex is what we normally feel from the outside as a heartbeat.

Covering of the Heart: Pericardium

The heart does not hang freely in the chest. It hangs by the great blood vessels inside a protective sac called the **pericardium** ("around the heart") or *pericardial sac* (Figure 18.3). This sac is composed of an outer, fibrous layer of connective tissue, the *fibrous pericardium,* and an inner layer of serous tissue, the *serous pericardium.* The serous pericardium is divided into an outer, or *parietal,* layer, which lines the inner surface of the fibrous pericardium, and an inner, or *visceral,* layer, which covers the outer surface of the heart and the adjoining portions of the large blood vessels. The serous pericardium surrounds the *pericardial cavity,* which contains a small amount of serous *pericardial fluid.* Because the visceral pericardium forms the outer layer of the heart wall, it is usually called the *epicardium* ("upon the heart").

The heart is held securely in place by ligaments that bind the pericardium to the sternum, spinal column, and other parts of the chest cavity. The pericardium is tough and inelastic, yet loose-fitting enough to allow the heart to move in a limited way. The serous pericardial fluid moistens the sac, and minimizes friction between the membranes as the heart moves during its contraction-relaxation phases.

During a physical examination, why does a physician tap the chest wall while listening with a stethoscope?

A physician can estimate the size of the heart by tapping the chest wall progressively and listening for sound changes.

FIGURE 18.1

Systemic and pulmonary circulation. The systemic circulation starts in the heart, flows through the muscles, organs, and tissues of the body, and then returns to the heart. The pulmonary circulation flows only from the heart to the lungs and back to the heart. Oxygenated blood is shown in red, deoxygenated blood in blue.

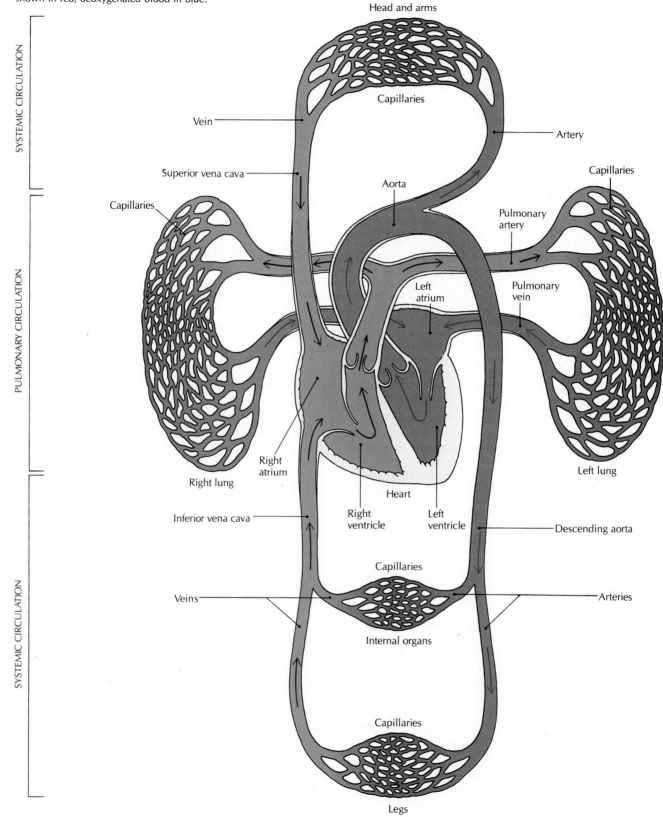

FIGURE 18.2

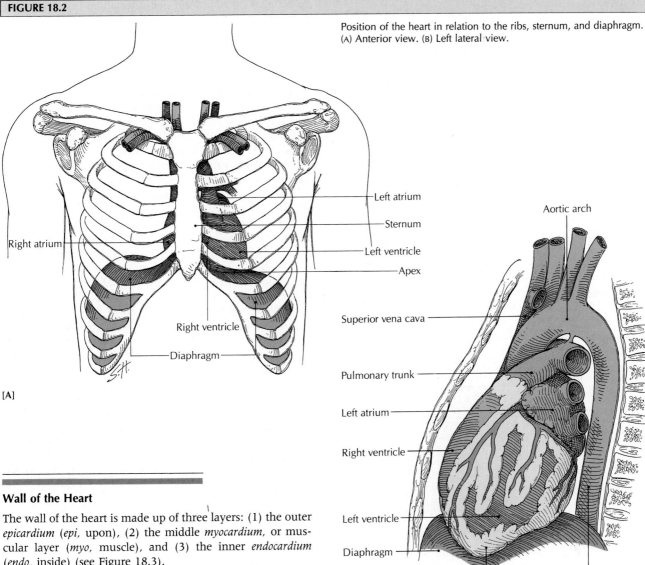

Position of the heart in relation to the ribs, sternum, and diaphragm. (A) Anterior view. (B) Left lateral view.

[A]

[B]

Wall of the Heart

The wall of the heart is made up of three layers: (1) the outer *epicardium* (*epi*, upon), (2) the middle *myocardium*, or muscular layer (*myo*, muscle), and (3) the inner *endocardium* (*endo*, inside) (see Figure 18.3).

If you were to cut away the parietal pericardium, you would see that the surface of the heart itself is reddish and shiny. This shiny membrane is the **epicardium.** The epicardium is continuous with the parietal pericardium. Inside the epicardium, and often surrounded with fat, are the main coronary blood vessels that supply and drain blood from the heart.

Directly beneath the epicardium is the middle layer, the **myocardium** ("heart muscle"), which is a thick layer of cardiac muscle that gives the heart its special pumping ability (see Chapter 9). The myocardium has three spiral layers of cardiac muscle, which are attached to a fibrous ring (fibrous trigone) that forms the cardiac skeleton. The spiral is the most effective arrangement for squeezing blood out of the heart's chambers.

The inside cavities of the heart, and all of the associated valves and muscles, are covered with the **endocardium** ("inside the heart"). It is a thin, fibrous layer lined with simple squamous epithelial tissue (endothelium), which is continuous with the endothelium of the blood vessels, and some connective tissue.

Cardiac Skeleton

When blood is pumped to the lungs and the rest of the body, it is wrung out of the ventricles like water from a wet cloth. To accomplish this wringing motion, the heart has a fibrous **cardiac skeleton** of tough connective tissue that provides attachment sites for the valves and muscular fibers.

The cardiac skeleton is made up of a *fibrous trigone* (Gr. *trigonos*, triangular) and four rings or cuffs, one surrounding each of the heart's four openings (see Figure 18.6). These heart openings are regulated by valves. Four rings of the cardiac skeleton support the valves and prevent them from stretching. This anatomical arrangement prevents blood from flowing backward from the ventricles into the atria, and from the arteries back into the ventricles.

536

Chapter 18: The Cardiovascular System: The Heart

FIGURE 18.3

Covering and wall of the heart. (A) Layers of the pericardial sac and the heart wall. (B) Enlarged view of the structure of the pericardium and the ventricular heart wall, internal view.

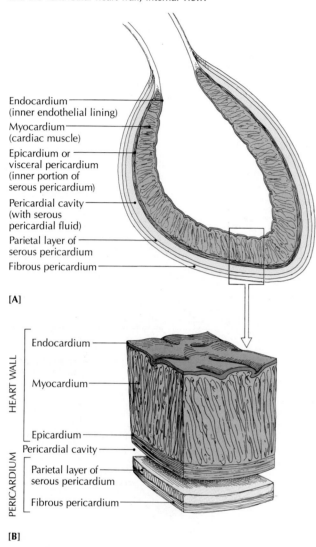

be before it closed at birth (see Adaptations in Fetal Circulation in Chapter 19). The lining of both atria is smooth except for some ridges called *musculi pectinati* (pectinate muscles), which are formed by parallel muscle bundles that look much like a comb. The musculi pectinati are located on both right and left auricles, and on the anterior wall of the right atrium.

The right and left ventricles have walls that contain ridges called *trabeculae carneae* (L. "little beams of flesh"), which are formed by coarse bundles of cardiac muscle fibers. Both right and left ventricles contain papillary muscles and supportive cords called *chordae tendineae,* which are attached to the atrioventricular valves. The upper part of the right ventricular wall, called the *conus* or *infundibulum,* is smooth and funnel-shaped. This area leads to the pulmonary artery. The left ventricle does not have this conus.

Because the heart is a closed system of chambers, when the left and right ventricles pump almost simultaneously, equal amounts of blood must enter and leave the heart. The wall of the left ventricle is thicker, however, because it must be strong enough to supply blood to all parts of the body, whereas the right ventricle supplies only the lungs. As a result, the left ventricular blood pressure is higher (120 mm Hg) during contraction than in the right ventricle (20 mm Hg). The left ventricle is thicker to accommodate the higher pressure required to pump blood a greater distance against a high resistance. The walls of the atria are thinner than the ventricular walls because less pressure is required to pump blood only a short distance (and against less resistance) into the ventricles.

On the surface of the heart are some *sulci* (depressions) that are helpful in locating certain features of the ventricles and atria, as well as coronary vessels (Figure 18.5). The **coronary sulcus,** encircling the heart, indicates the border between the atria and ventricles. Embedded in fat within the coronary sulcus are the *right* and *left coronary arteries.*

FIGURE 18.4

Schematic drawing of anterior view of interior of the heart, with blood vessels removed, to show the four chambers.

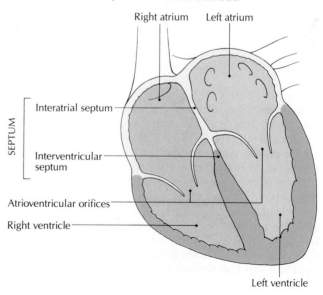

Chambers of the Heart

The heart is a hollow organ, containing four cavities, or chambers (Figure 18.4). Dividing the heart vertically down the middle into a **right heart** and a **left heart** is a wall of muscle called the **septum** (L. *sepire,* to separate with a hedge). At the top of each half of the heart is a chamber called an **atrium** (AY-tree-uhm; L. porch, antechamber). Below it is another chamber, the **ventricle** ("little belly"). Each atrium has a flaplike appendage of unknown function called an auricle ("little ear") (see Figure 18.5). The atria lead to the ventricles by way of openings called *atrioventricular orifices* (L. *or,* mouth + *facere,* to make). In a normal heart, blood flows from the atria to the ventricles, never the other way.

On the septal wall of the right atrium is an oval depression called the *fossa ovalis,* where the fetal foramen ovale used to

FIGURE 18.5

Anterior external view of the heart, showing surface features and great vessels.

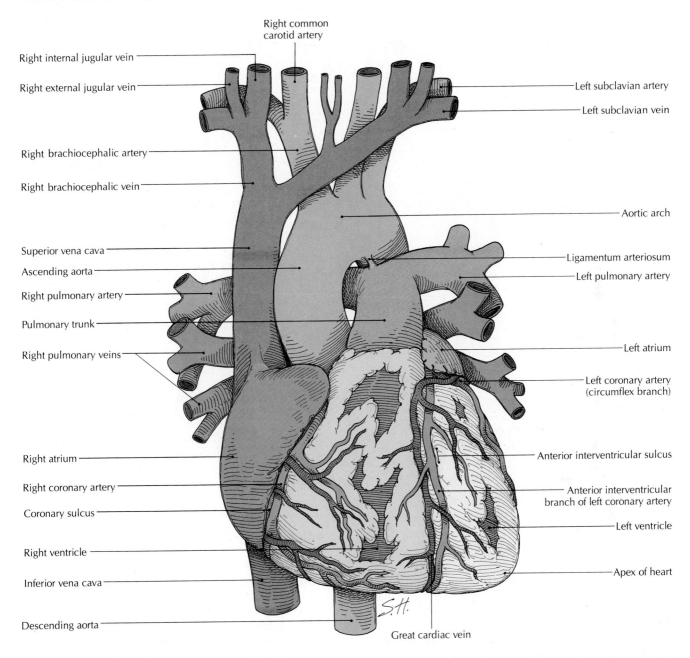

Right internal jugular vein

Right external jugular vein

Right common carotid artery

Left subclavian artery

Left subclavian vein

Right brachiocephalic artery

Right brachiocephalic vein

Aortic arch

Superior vena cava

Ascending aorta

Right pulmonary artery

Pulmonary trunk

Right pulmonary veins

Ligamentum arteriosum

Left pulmonary artery

Left atrium

Left coronary artery (circumflex branch)

Right atrium

Right coronary artery

Coronary sulcus

Right ventricle

Inferior vena cava

Descending aorta

Anterior interventricular sulcus

Anterior interventricular branch of left coronary artery

Left ventricle

Apex of heart

Great cardiac vein

On the anterior surface of the heart is the **anterior interventricular** ("between the ventricles") **sulcus,** which marks the location of the *interventricular septum,* the anterior part of the septum separating the ventricles. Embedded within the anterior interventricular sulcus are the *anterior interventricular artery* and the *great cardiac vein.* On the posterior surface of the heart is the **posterior interventricular sulcus,** which marks the location of the posterior interventricular septum. Embedded within the posterior interventricular sulcus are the *posterior interventricular descending artery* and the *middle cardiac vein.*

Valves of the Heart

The four heart valves direct the flow of blood through the heart in the proper direction. By maintaining a pressure gradient, they also prevent the backflow of blood. The two *atrioventricular valves* direct the flow of blood from the atria to the ventricles, and the two *semilunar valves* direct the flow of blood from the ventricles to the pulmonary artery (on the way to the lungs) and to the aorta (on the way to the rest of the body) (Figure 18.6).

FIGURE 18.6

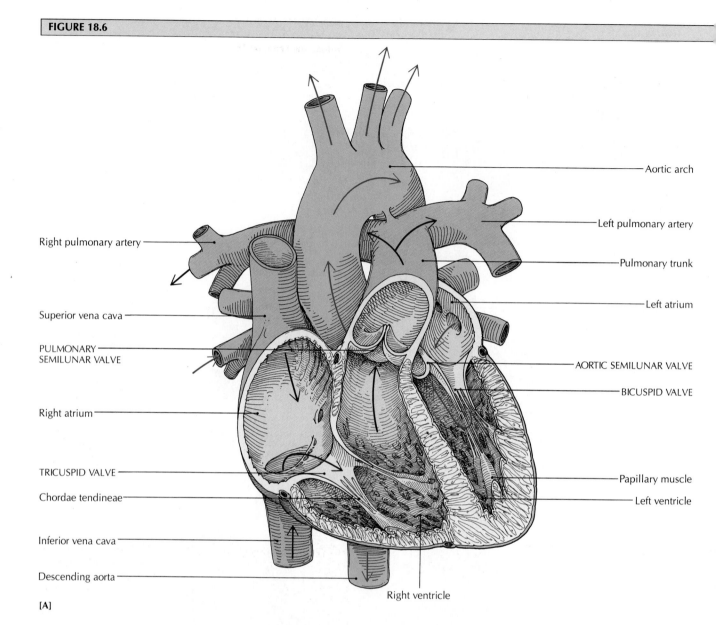

Aortic arch

Left pulmonary artery

Pulmonary trunk

Left atrium

AORTIC SEMILUNAR VALVE

BICUSPID VALVE

Papillary muscle

Left ventricle

Right pulmonary artery

Superior vena cava

PULMONARY SEMILUNAR VALVE

Right atrium

TRICUSPID VALVE

Chordae tendineae

Inferior vena cava

Descending aorta

Right ventricle

[A]

Atrioventricular valves The two *atrioventricular (AV) valves* differ in structure, but they operate in the same way. The right atrioventricular is known as the *tricuspid valve* of the right heart because it consists of three cusps or flaps (L. *tri*, three + *cuspis,* point) (Figure 18.6B). The left atrioventricular, or *bicuspid valve* of the left heart, has two cusps (*bi* = two). Because the cusps resemble a bishop's miter (a tall, pointed hat), the valve is also called the *mitral* (MY-truhl) *valve.*

Each cusp is a thin, strong, fibrous flap covered by endocardium. Its broad base is anchored into a ring of the cardiac skeleton. Attached to its free end are strong, yet delicate, tendinous cords called *chordae tendineae* (Figure 18.6D). The chordae tendineae, which resemble the cords of a parachute, are continuous with the nipplelike papillary muscles (L. *papilla,* nipple) in the wall of the ventricle (Figures 18.6D, E).

When blood is flowing from the atrium to the ventricle, the valve cusps lie open against the ventricular wall. When the ventricle contracts, the cusps are brought together by the increasing ventricular blood pressure, and the atrioventricular opening is closed (Figure 18.6C). At the same time, the papillary muscles contract, putting tension on the chordae tendineae. The chordae tendineae pull on the cusps, preventing them from being forced upward into the atria. Otherwise, blood would flow backward from the ventricle into the atrium.

Semilunar valves The *semilunar* ("half-moon") *valves* prevent blood in the pulmonary artery and aorta from flowing back into the ventricles. The left semilunar valve is larger and stronger than the right.

The right semilunar valve is in the opening between the right ventricle and the pulmonary artery, and is called the *pulmonary semilunar valve.* It allows oxygen-poor blood to enter the pulmonary artery on its way to the lungs from the right ventricle. The left semilunar valve is the *aortic semilu-*

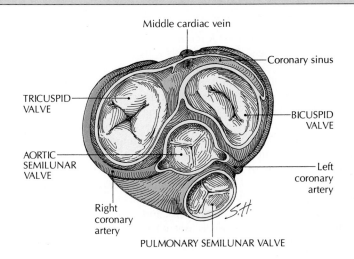

Middle cardiac vein

Coronary sinus

TRICUSPID VALVE

BICUSPID VALVE

AORTIC SEMILUNAR VALVE

Left coronary artery

Right coronary artery

PULMONARY SEMILUNAR VALVE

Valves of the heart. (A) The two atrioventricular valves (tricuspid and bicuspid) separate the atria and ventricles. The two semilunar valves (pulmonary and aortic) permit the flow of blood out of the heart to the lungs and body. (B) The closed valves of the heart, viewed from above; the atria have been removed. (C) The seemingly fragile pulmonary valve opens (left) and closes (right) about once a second. (D) Cusps of the tricuspid (right atrioventricular) valve remain open, allowing blood to flow from the right atrium into the right ventricle, when the chordae tendineae and papillary muscles are relaxed. The valve closes, blocking blood flow, when the chordae tendineae are taut and the papillary muscles are contracted. (E) A photo inside the right ventricle showing the branching chordae tendineae rising from a tough papillary muscle. [(C) *Dr. Wallace McAlpine.* (E) *Lennart Nilsson, Behold Man. Boston, Little, Brown, 1974.*]

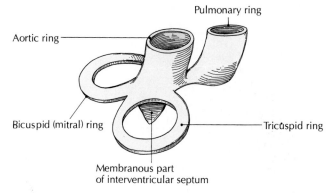

Pulmonary ring

Aortic ring

Bicuspid (mitral) ring

Tricuspid ring

Membranous part of interventricular septum

[B]

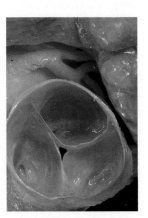

[C]   Valve open   Valve closed

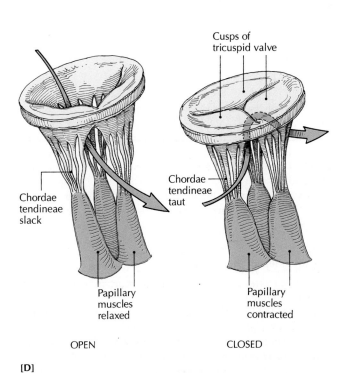

Cusps of tricuspid valve

Chordae tendineae taut

Chordae tendineae slack

Papillary muscles relaxed

Papillary muscles contracted

OPEN   CLOSED

[D]

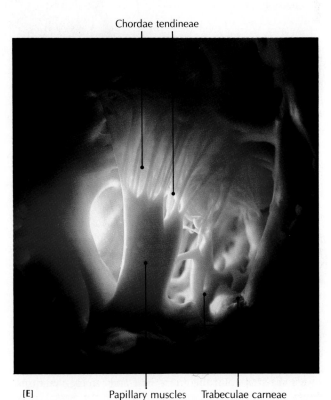

Chordae tendineae

[E]   Papillary muscles   Trabeculae carneae

Cardiac Catheterization

A precise way of looking *inside* the heart is via **cardiac catheterization** (KATH-uh-tuh-rihz-AY-shun). A *catheter* (Gr. something inserted) is a hollow, flexible tube of plastic. It is inserted into a vein, usually in the arm, thigh, or neck, and then moved along slowly with the aid of a fluoroscope into the chambers of the heart (see drawings), and, if necessary, into the lung.

With cardiac catheterization, it is possible to measure the blood flow through the heart and lungs, the volume and pressure of blood in the heart, the pressure of blood passing through a heart valve, and the oxygen content of blood in the heart and its major blood vessels. It is also possible to inject harmless radiopaque dyes through the catheter that can be viewed on an x ray to show septal openings and other congenital disorders in the aorta and pulmonary artery. This technique is called *angiocardiography.* Currently a motion-picture x-ray machine is commonly used with dye injected through cardiac catheterization. This technique is called *cineangiocardiography,* and can indicate a heart abnormality by tracing the path of dye through the heart and its blood vessels.

A recent application of cardiac catheterization is *balloon angioplasty.* In this procedure, a catheter with a small, un-inflated balloon on is tip is inserted into an obstructed artery and inflated. The material causing the obstruction is pushed aside, the catheter is withdrawn, and blood flow is improved. Also, clot-dissolving enzymes (such as streptokinase and urikinase) can be injected directly into coronary arteries after a heart attack. Cardiac catheterization does not affect the functioning of the heart, and is usually not uncomfortable for the patient.

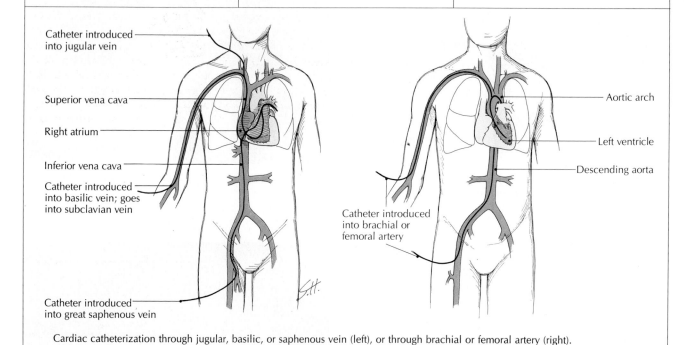

Catheter introduced into jugular vein

Superior vena cava

Right atrium

Inferior vena cava

Catheter introduced into basilic vein; goes into subclavian vein

Catheter introduced into great saphenous vein

Aortic arch

Left ventricle

Descending aorta

Catheter introduced into brachial or femoral artery

Cardiac catheterization through jugular, basilic, or saphenous vein (left), or through brachial or femoral artery (right).

nar valve, which allows freshly oxygenated blood to enter the aorta from the left ventricle.

Each valve contains three cusps. The base of each cusp is anchored to the fibrous ring of the cardiac skeleton. The free borders of the cusps curve outward, and extend into the artery. These cusps do not have chordae tendineae attached to their free margins, as the atrioventricular valves do. During a ventricular contraction, the blood rushes out of the ventricles and pushes the cusps up and against the wall of the artery, allowing blood to flow through the artery. When the ventricles relax, some blood flows back, filling the space between the cusps and the artery wall and forcing the cusps down and together. This motion closes off the artery com-

pletely, and prevents blood from flowing back into the relaxed ventricles due to gravity.

Great Vessels of the Heart

The largest arteries and veins of the heart function in pulmonary and systemic circulation. The vessels carrying oxygen-poor blood *from the heart to the lungs* are the **pulmonary arteries** (see Figure 18.1). Those returning oxygen-rich blood *from the lungs to the heart* are the **pulmonary veins.** Draining venous blood from the upper and lower parts of the body to the heart are the **superior vena cava** and the **inferior vena** *cava,* respectively.* The artery carrying highly oxygenated blood away from the heart is the *aorta.*

Blood Supply to the Heart

The heart muscle needs more oxygen than any organ except the brain. To obtain this oxygen, the heart must have a generous supply of blood. Like other organs, the heart receives its blood supply from arterial branches that arise from the aorta. The flow of blood that supplies the heart tissue itself is the *coronary circulation.* The heart needs its own separate blood supply because blood in pulmonary and systemic circulation cannot seep through the lining of the heart (endocardium) from the cardiac cavities to nourish cardiac tissue. The heart pumps about 380 L (100 gal) to its own muscle tissue every day, or about 5 percent of all the blood pumped by the heart.

Coronary arteries Blood is supplied to the heart by the right and left *coronary arteries,* which are the first branches off the aorta, close to the point where the aorta emerges from the heart. The aortic semilunar valve partially covers the openings of these arteries while blood is pumped into the

---

*The superior and inferior veins are called *cava* (L. hollow, cavern) because of their great size.

aorta from the heart. Blood can pass through the openings of the coronary arteries only when the left ventricle has relaxed, and the cusps do not cover the openings. Blood enters these arteries as a result of the elastic recoil force of the aorta.

The branching of the coronary arteries varies from person to person, but the following arrangement is the common one. The *left coronary artery* begins behind the left cusp of the aortic semilunar valve and divides into (1) an *anterior interventricular branch (artery),* which descends in the anterior interventricular sulcus, and (2) a *circumflex branch (artery),* which continues in the coronary sulcus.

The *right coronary artery* arises behind the right cusp of the aortic semilunar valve, and descends in the coronary sulcus. On the posterior surface a branch of it descends as the *posterior interventricular branch (artery)* in the posterior interventricular sulcus (Figure 18.7).

On the anterior surface of the heart, the right coronary artery gives off a *marginal branch (artery),* which feeds both the ventral and dorsal surfaces of the right ventricle. From the arterial crown, small branches distribute themselves over the surface of the right atrium.

Most of the heart's blood supply flows into the myocardium when the heart muscle is relaxed. The reason for this is that contraction of the heart muscle compresses the coronary vessels where they penetrate the muscular heart wall. This partially or totally occludes these vessels for a short time during contraction. Although some variation exists in coro-

---

FIGURE 18.7

Coronary veins and arteries, anterior view.

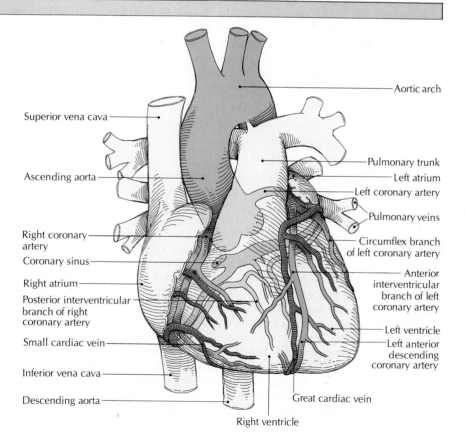

nary circulation, the largest amount of blood generally goes to the musculature of the left ventricle. The right coronary artery has branches that mainly supply the right side of the heart, but they also send some blood to the left ventricle. Branches of the left coronary artery send the main blood supply to the left side of the heart, but a small amount is also sent to the musculature of the right ventricle. The branches of the right and left coronary arteries *anastomose* (connect or join) on the surface of the heart and within the heart muscle, providing some collateral (additional) circulation.

Coronary sinus and veins Most of the cardiac veins drain into the ***coronary sinus,*** a large vein located in the coronary sulcus on the posterior surface of the heart (see Figure 18.7). The coronary sinus empties into the right atrium. The coronary sinus receives blood from the *great cardiac vein,* which drains the anterior portion of the heart, the *middle cardiac vein* and *oblique vein,* which drain the posterior aspect of the heart, and the *small cardiac vein,* which drains the right side of the heart. Among the other veins is the *anterior cardiac vein,* which drains directly into the right atrium. Tiny veins called *thebesian veins* open directly into each of the four heart chambers.

Regulation of blood flow Blood flow to the heart tissue is self-regulating, and two processes are thought to be involved. In the ***metabolic process of autoregulation,*** cellular metabolites (waste products) tend to build up in active heart tissue. When they reach a certain concentration, local blood vessels dilate, blood flow increases, the excess metabolites are removed, and homeostasis is restored. In the ***myogenic process of autoregulation,*** the stimulus arises in the smooth muscle within arterioles in the myocardium, which have a resting "tone," or tension. As blood pressure rises, blood flow increases, and the vessels become distended. This causes the smooth muscle fibers that surround the vessels to contract, decreasing blood flow, lowering blood pressure, and restoring a homeostatic blood flow.

Embryonic Development of the Heart

The embryonic circulatory system forms and becomes functional before any other system, and by the end of the third week after fertilization the system is fulfilling the nutritional and respiratory needs of the embryo. The heart begins as a network of cells, from which two longitudinal ***heart cords*** (also called *cardiogenic cords*) are formed about 18 or 19 days after fertilization. On about day 20, the heart cords develop canals, forming ***heart tubes*** (also called *endocardial heart tubes*). On days 21 and 22 the heart tubes fuse, forming a single median ***endocardial heart tube*** (Figures 18.8A, B). Immediately following, the endocardial heart tube elongates; dilations and constrictions develop, forming first the ***bulbus cordis, ventricle,*** and ***atrium,*** and then the ***truncus arteriosus*** and ***sinus venosus*** (Figure 18.8C). At this time (day 22), the sinus venosus develops *left and right horns,* the venous end of the heart is established as the ***pacemaker,*** heart contractions resembling peristaltic waves begin in the sinus venosus,

and on about day 28 coordinated contractions of the heart begin to pump blood in one direction only.

Each horn of the sinus venosus receives an *umbilical vein* from the chorion (primitive placenta), a *vitelline vein* from the yolk sac, and a *common cardinal vein* from the embryo itself (Figure 18.8D).* Eventually, the left horn forms the coronary sinus, and the right horn becomes incorporated into the wall of the right atrium.

Growth is rapid, and because the heart tube is anchored at both ends within the pericardium, by day 25 the tube bends back upon itself, forming first a U-shape and then an S-shape. The tube also twists around, beginning to resemble the fully developed heart (see Figures 18.9C, D). During the first weeks of embryonic growth, the heart is about nine times as large in proportion to the whole body as it is in the adult. Also, its position is higher in the thorax than the permanent position it will assume later.

Partitioning of the heart On about day 25, partitioning of the atrioventricular orifice, atrium, and ventricle begins, and is completed about 10 to 20 days later (Figure 18.8E). Most of the wall of the left atrium develops from the pulmonary vein, and as mentioned above, the right horn of the sinus venosus becomes incorporated into the wall of the right atrium. By about day 28, the heart wall has formed its three layers (endocardium, pericardium, epicardium). At about day 40, the ***interatrial septum*** forms, dividing the single atrium into left and right atria. An opening in the septum, the ***foramen ovale,*** closes at birth (see p. 563). The partitioning of the single ventricle into left and right ventricles begins with the formation of an upward fold in the floor of the ventricle, called the ***interventricular septum*** (see Figure 18.9E), and is completed about day 48. The development of the atrioventricular valves, chordae tendineae, and papillary muscles proceeds from about the fifth week until the fifth month. The external form of the heart continues to develop from about day 28 to day 60.

---

*The vitelline veins become broken up into sinusoids by rapidly developing cords of liver cells, and eventually become the hepatic (liver) sinusoids. The right umbilical vein and part of the left umbilical vein degenerate, and the remaining portion of the left umbilical vein transports all the blood from the placenta to the fetus. The right common cardinal vein becomes the superior vena cava, while most of the left common cardinal vein degenerates.

FIGURE 18.8

Early embryonic development of the heart. (A, B) The primitive heart tubes fuse together and form a single endocardial heart tube during days 21 and 22 after fertilization; ventral views. (C) On about day 22 or 23, the first major structures form, and the heart tube begins to bend and twist; ventral view. (D) On about day 25, the tube has formed an S-shape. (E) At about 32 days, the interior partitions can be seen clearly in this frontal section. (F) A frontal view at about 5 weeks shows the remaining three pairs of aortic arches. (G) A frontal section at 6 weeks shows the aorta and pulmonary trunk after the bulbis cordis has been incorporated into the ventricles to become the infundibulum. Black arrows indicate the movement of the heart, and red arrows indicate the flow of blood.

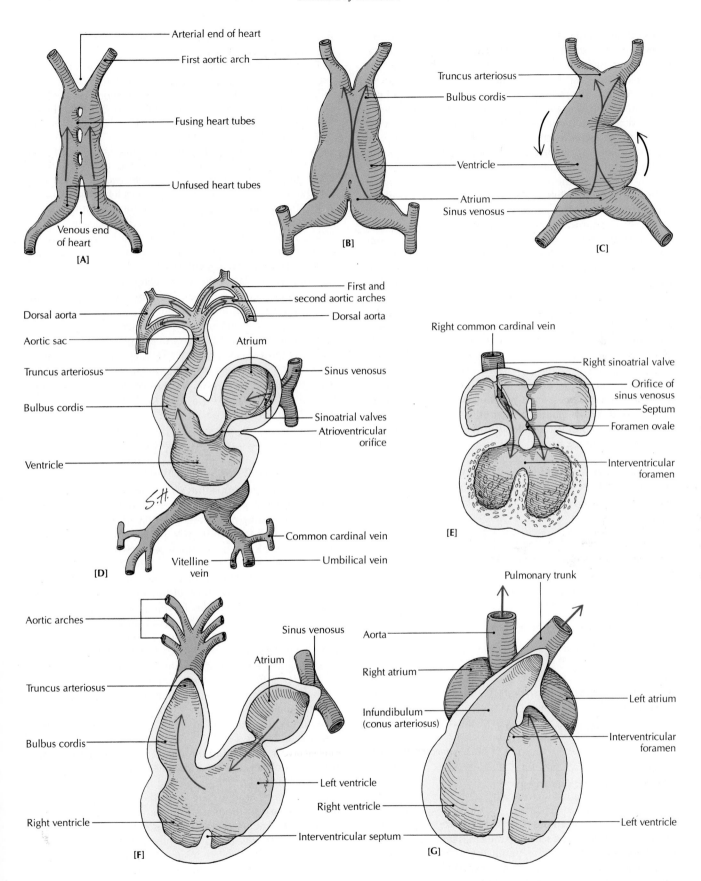

[A]
- Arterial end of heart
- First aortic arch
- Fusing heart tubes
- Unfused heart tubes
- Venous end of heart

[B]

[C]
- Truncus arteriosus
- Bulbus cordis
- Ventricle
- Atrium
- Sinus venosus

[D]
- Dorsal aorta
- Aortic sac
- Truncus arteriosus
- Bulbus cordis
- Ventricle
- First and second aortic arches
- Dorsal aorta
- Atrium
- Sinus venosus
- Sinoatrial valves
- Atrioventricular orifice
- Common cardinal vein
- Umbilical vein
- Vitelline vein
- *S.H.*

[E]
- Right common cardinal vein
- Right sinoatrial valve
- Orifice of sinus venosus
- Septum
- Foramen ovale
- Interventricular foramen

[F]
- Aortic arches
- Truncus arteriosus
- Bulbus cordis
- Right ventricle
- Sinus venosus
- Atrium
- Left ventricle
- Interventricular septum

[G]
- Pulmonary trunk
- Aorta
- Right atrium
- Infundibulum (conus arteriosus)
- Right ventricle
- Left atrium
- Interventricular foramen
- Left ventricle

Development of major blood vessels Blood leaves the embryonic heart through the arterial end (truncus arteriosus) and returns through the horns of the sinus venosus at the venous end, passing into the single atrium through the sinoatrial orifice. The orifice is equipped with two flaps of endocardium that function as primitive valves to prevent the backflow of blood. Blood from the atrium flows into the ventricle through the atrioventricular orifice. When the ventricle contracts, blood is pumped into the bulbis cordis. The bulbis cordis narrows to become the truncus arteriosus, which pierces the roof of the pericardium to open into the **aortic sac.** From the somewhat dilated aortic sac emerge the **aortic arches** (Figure 18.8F).

Six pairs of aortic arches develop, but only the left arch of the fourth pair forms part of the arch of the mature aorta. The first, second, and fifth pairs disappear during embryonic life. The third pair forms the common carotid arteries and the internal carotid arteries; the right arch of the fourth pair becomes the right subclavian artery; the proximal part of the left sixth aortic arch develops into the proximal part of the left pulmonary artery, and the distal part forms the ductus arteriosus; and the proximal part of the right sixth aortic arch develops into the proximal part of the right pulmonary artery, while the distal part of the arch degenerates.

At about day 35 to 42, the truncus arteriosus is divided into the "great arteries," the **pulmonary trunk** and **aorta** (Figure 18.8G). The pulmonary trunk carries blood to the lungs, and the aorta carries blood to the blood vessels that supply the rest of the body.

Ask Yourself

1 *In what way is the heart a double pump?*

2 *What are the layers of the heart wall, and how do they differ from the pericardium?*

3 *What is the purpose of the cardiac skeleton?*

4 *What are the separate functions of the four heart chambers?*

5 *What are the functions of the atrioventricular and semilunar valves?*

6 *Why does the heart need a separate blood supply?*

7 *How does the heart regulate its own blood flow?*

PHYSIOLOGY OF THE HEART

The heart has two purposes. One is to receive oxygen-poor blood from the body and send it to the lungs for a fresh supply of oxygen. The second is to pump the newly oxygenated blood to all parts of the body, where body cells can use it for their day-to-day metabolic work. The following sections explain how the heart accomplishes these tasks.

Structural and Metabolic Properties of Cardiac Muscle

Cardiac muscle cells function as a *single unit* in response to physiological stimulation, rather than as a group of separate units as skeletal muscle does. Cardiac muscle cells act in this way because they are connected end-to-end by *intercalated disks*, which contain gap junctions and desmosomes (see p. 254). *Gap junctions* allow action potentials to be transmitted from one cardiac cell to another. *Desmosomes* hold the cells together and serve as the attachment sites for myofibrils. This connection maintains cell-to-cell cohesion so that the "pull" of one contractile unit is transmitted to the next one. Overall, this series of interconnected cells forms a latticework called a *syncytium* (sihn-SIE-shum). The importance of a syncytial muscle mass is that when either the atrial or ventricular muscle mass is stimulated, the action potential spreads over the entire syncytium, and causes the muscle cells in the entire muscle mass to contract in unison.

Cardiac muscle contains a large number of mitochondria, which provide a constant source of energy for the hardworking heart muscle (see Figure 9.19). It also has an abundant blood supply and a high concentration of *myoglobin*, a muscle pigment that stores oxygen. A ready supply of oxygen is important because the heart cannot continue to beat after its oxygen is used up.

Electrical Properties of Cardiac Muscle

When an action potential travels through the heart, each cardiac muscle cell produces and conducts its own action potential. The resting membrane potential of individual cardiac muscle cells is about -90 mV (interior to exterior). As a result, the cardiac muscle **action potential** is similar to that in nerve and skeletal muscle tissue (see Chapters 9 and 11), but it lasts longer. (Recall that an action potential is a self-propagating wave of electrical disturbance on a plasma membrane.) As with skeletal muscle, stimulation produces a *propagated action potential* (one that travels in all directions) that initiates a contraction. The action potential can be divided into five phases: depolarization, early repolarization, plateau, repolarization, and resting potential (Figure 18.9).

Depolarization in the heart results from a large increase in the inward movement of sodium ions, which causes the membrane potential to reverse from its high resting potential of -90 mV to a potential of about $+30$ mV. Depolarization lasts about 2 msec.

The **early repolarization phase** is associated with the movement of negative chloride ions to the inside of the cell after depolarization. These chloride ions make the inside of the cell more negative, and drop the membrane potential from $+30$ mV to the plateau phase potential of $+20$ mV. Early repolarization lasts about 1 msec.

The **plateau phase** in cardiac muscle is much longer than in skeletal muscle or nerve tissue. Also, the heart muscle cell remains in an absolute *refractory period* (while it does not respond to a stimulus) during almost the entire action potential, and can respond to a large stimulus only after the action

Physiology of the Heart

FIGURE 18.9

A cardiac action potential from a ventricular muscle cell divided into five phases based on ion movement.

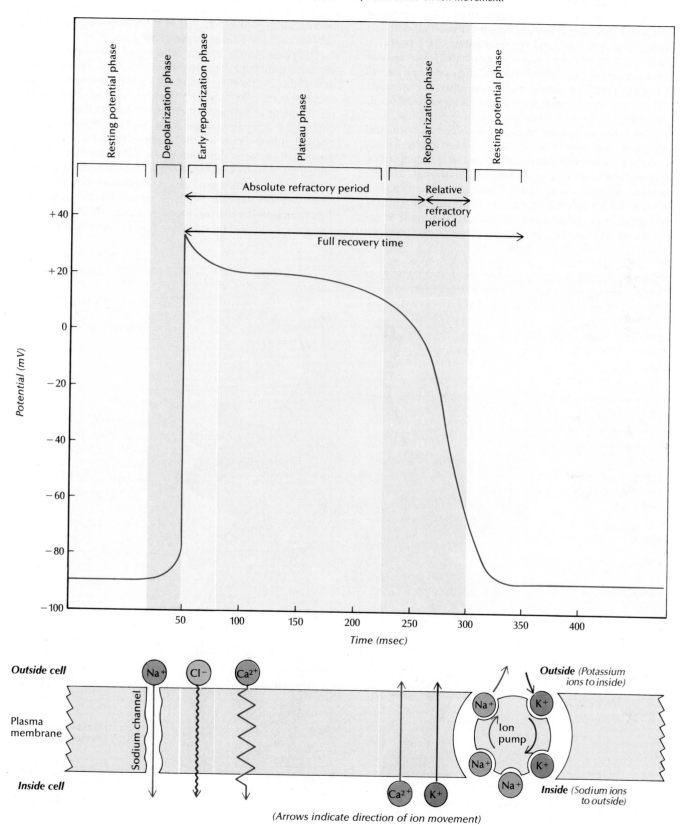

(Arrows indicate direction of ion movement)

potential is finished. (In order for this to occur, the inward movement of chloride ions must be countered. Positively charged calcium ions move into the cytoplasm from the sarcoplasmic reticulum and extracellular fluid. This movement prevents the membrane from returning to its normal electrical potential of -90 mV, and the repolarization levels off, or *plateaus.*) The absolute refractory period (0.25 sec) is nine times longer than in skeletal muscle. The extra time allows the heart to refill with blood, and ensures that no extra beats will occur as the electrical impulse travels through the heart. The plateau phase lasts about 200 msec.

During **repolarization,** potassium ion channels open, and calcium ion channels close, so that potassium ions move out of the heart cell, causing the inside of the cell to become more negative as the positive potassium ions move out. The increasing negativity inside the cell returns the membrane to its normal -90 mV. It is important to note that at this point the electrical potential across the plasma membrane is returning to normal, but the ion distributions are reversed:

more potassium ions are on the outside, while sodium ions are highly concentrated inside the cell. Active transport mechanisms now predominate to pump sodium out and potassium in. The active transport of these ions continues to maintain a -90 mV potential in the resting state, since ion diffusion occurs to some degree at all times.

The phases of the cardiac action potential just described represent only the excitation portion of cardiac muscle function. If this electrical excitation is to cause a contraction, the two events must be "coupled." Calcium ions are involved in the coupling of excitation and contraction. The role of calcium is the same as in skeletal muscle. The sarcoplasmic reticulum supplies most of the calcium. However, a second important source is diffusion across the plasma membrane during the action potential. From both of these sources, the concentration of calcium in the cytoplasm increases. Calcium then combines with the regulator muscle protein, troponin. This combination removes tropomyosin's inhibition of cross-bridge formation between the two muscle proteins, actin and

FIGURE 18.10

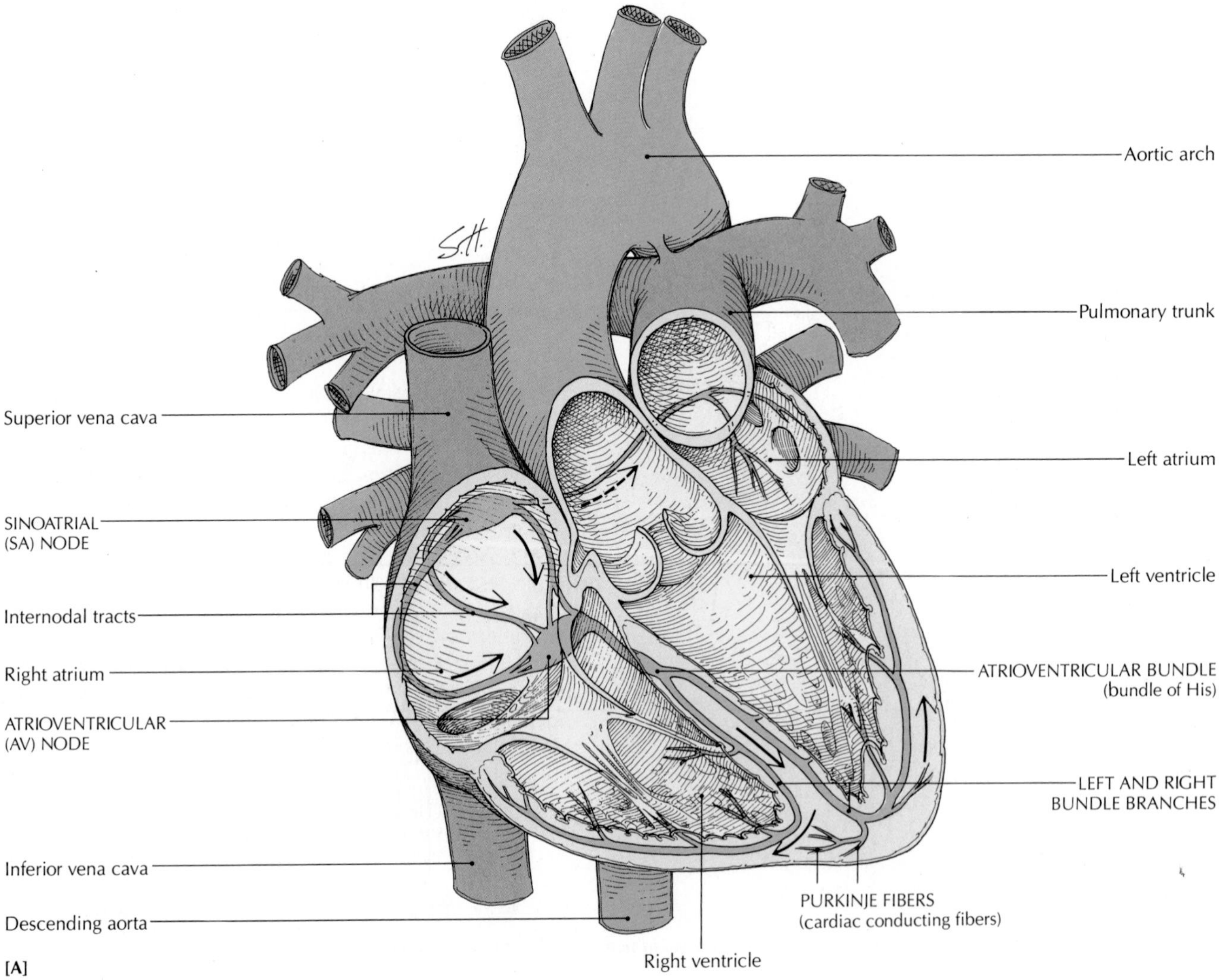

[A]

myosin. Cross-bridge formation occurs, and the cardiac muscle cells contract. During repolarization, the cytoplasmic calcium is restored to its low concentration by an active transport mechanism that transports calcium ions back into the sarcoplasmic reticulum and to the immediate vicinity of the plasma membrane.

As you have seen, under normal circumstances atrial and ventricular muscle cells have such a high stable negative *resting potential* that they show no spontaneous electrical activity. (However, if there is sufficient time or abnormal conditions, any myocardial cell will depolarize.) These cells are thus excited only by electrical impulses from adjacent cells. But the cells of a specialized conducting system in the heart—consisting of the sinoatrial node, the atrioventricular node, and the Purkinje fibers (cardiac conducting myofibers); see Figure 18.10—do not have a steady resting potential. Instead, they show what is called *pacemaker activity* or a *pacemaker potential,* which is discussed further in the next section. This activity allows heart muscle and smooth muscle to contract without nervous innervation.

The heart's natural pacemaker, anterior view. (A) Each electrical impulse from the sinoatrial (SA) node travels through the internodal tracts and causes the atria to contract. The impulse slows as it passes through the atrioventricular (AV) node and the atrioventricular bundle. It descends through the left and right bundle branches to the Purkinje fibers, where it causes a ventricular contraction. (B) The numbers on the drawing give the time, in seconds, required for the impulse from the sinoatrial node to reach various parts of the heart.

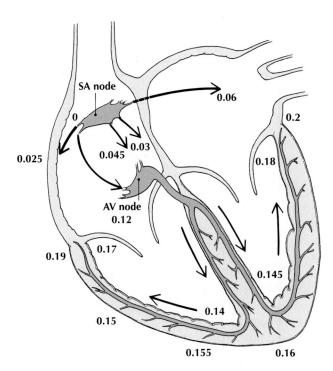

[B]

Impulse-Conducting System of the Heart

You saw in Chapter 9 that skeletal muscles of the body cannot contract unless they receive an electrochemical stimulation from the nervous system. Although the central nervous system does exert some control over the heart, cardiac muscle has its own built-in electrochemical activator, called a *pacemaker,* and can initiate a beat independently of the central nervous system (Figure 18.10).

The electrical stimulation that starts the heartbeat and controls its rhythm originates in the superior wall of the right atrium, near the entry point of the superior vena cava, in a mass of specialized heart muscle tissue called the *sinoatrial node,* or *SA node.** Although the muscles of the atria are not continuous with the muscles of the ventricles, the atria and ventricles must be coordinated at every beat of the heart. This coordination is made possible by the SA node.

The pacemaker activity causes the SA node to depolarize spontaneously at regular intervals, 70 to 80 times a minute. The SA node makes contact with adjacent atrial muscle cells, and causes them to be depolarized by conduction through the gap junctions of the intercalated disks. These atrial cells, in turn, cause their neighboring cells to start action potentials. In this way, a wave of electrical activity spreads throughout the right atrium and then the left atrium, much like the spreading ripples on a pond. Many experiments have shown that no conductile tissue exists in the atria. Electrical activity is simply spread from one cell to the next. The electrical stimulation causes the atria to contract, and blood is forced down into the ventricles.

A few hundredths of a second after leaving the SA node, the wave of electrical activity reaches the *atrioventricular node,* or *AV node,* which lies at the base of the right atrium, between the atrium and ventricle (see Figure 18.10). The AV node delays the electrical activity another few hundredths of a second before allowing it to pass into the ventricles. This delay allows time for the atria to force blood into the ventricles.

From the AV node, a group of conducting fibers in the interventricular septum called the *atrioventricular bundle (bundle of His)* divides into two branches that spread along the septum, one branch for each ventricle. Because a sheet of connective tissue separates the atria from the ventricles, the atrioventricular bundle is the only electrical link between the atria and ventricles. When the branches reach the apex of the ventricles, they divide into hundreds of tiny specialized cardiac muscle fibers called *Purkinje fibers* (cardiac conducting myofibers) that follow along the muscular walls of the ventricles.[†] Such an arrangement concentrates the electrical impulse in a definite pathway so that it can make con-

---

*The sinoatrial node is so named because it is located in the wall of the *sinus venosus* during early embryonic development, and is absorbed into the right atrium along with the sinus venosus (see p. 542).

[†]Purkinje fibers are actually modified cardiac muscle cells that conduct action potentials at higher velocities than normal cardiac muscle cells do. They are continuous with the myocardial muscle cells, but have only feeble contractile properties.

tact with all areas of the ventricular muscle. Therefore, an impulse traveling along the Purkinje fibers is conducted rapidly and directly into the cardiac muscle, and each syncytium contracts in unison with the others, producing a coordinated pumping effort.

Electrocardiogram

The rhythm of the heart, and the passage of an electrical current generated by an action potential from the SA node through the atria, down into the AV node, and through the atrioventricular bundle and Purkinje fibers of the ventricles, can be measured quite easily and accurately with an instrument called the *electrocardiograph,* which produces a recording of the electrical waves of the heart (Figure 18.11A). The recording is called an **electrocardiogram,** and the printed record is abbreviated as either **EKG** or **ECG** (K for the Greek *kardia,* or C for cardio).

The electrocardiograph has electrodes, which when placed at certain points on the body, can detect the electrical activity in the heart. (The body fluids are an excellent conductor of electricity.) To enhance electrical contact, a jelly containing an electrolyte is put on the skin where the electrodes (or leads) are to be attached. The three standard leads for an EKG are connected in the following way: lead I to the right wrist and left wrist, lead II to the right wrist and left ankle, and lead III to the left wrist and left ankle. In addition, six electrodes are placed on the chest in standard positions numbered V1 to V6 (Figure 18.11D).

Electrical activities recorded Different electrical impulses during the cardiac cycle are recorded in an EKG as distinct *deflection waves.* The first activity in the electrocardiogram is the **P wave** (see Figures 18.11A to C). It is caused by the electrical voltage generated by the passage of the impulse from the SA node, through the muscle fibers of the atria, and reaching the AV node. The P wave represents the depolarization (excitation) and contraction of both atria.

The passage of the wave between the atria and ventricles is marked by a short horizontal segment immediately following the P wave. This is the *P-R segment.* Next, the depolarization of the ventricles produces a short dip (Q), a tall spiked peak (R), and a sharp dip (S). This triple-wave activity is the **QRS complex** (wave), recorded as the ventricles are depolarized. The repolarization of the atria also takes place at this time, but is masked on the EKG by the stronger ventricular depolarization. After a short lull marked by a horizontal wave called the *S-T segment,* a recovery wave in the opposite direction (from ventricles to atria) is shown by a rounded peak called the **T wave.** It represents the repolarization of the ventricles.

A small upward deflection, the *U wave,* is sometimes recorded after the T wave in a normal EKG. It is speculated that the U wave is due to the slow repolarization of papillary muscles, but its origin is not known for certain. It has been noted that the U wave becomes higher when the heart is enlarged or the blood potassium level drops below normal.

FIGURE 18.11

The electrocardiogram. (A) Normal tracing of an electrocardiogram. In an actual EKG, the vertical lines are 0.04 second apart, and the horizontal lines are 1 mm apart. A 1-mV signal produces a deflection of 10 mm. (B) Events of the cardiac cycle as they are recorded on an EKG. (C) Summary of EKG events. (D) The three standard leads and the six chest leads for an EKG.

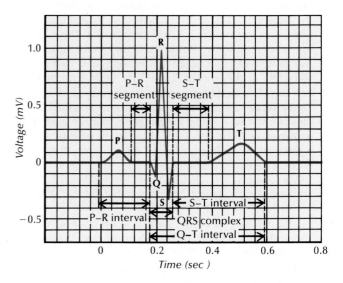

[A]

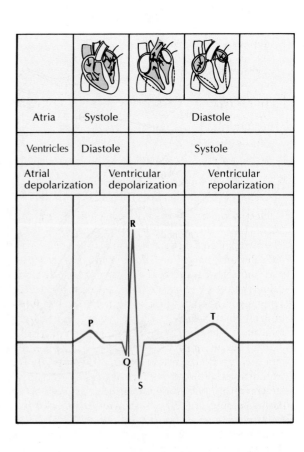

Atria	Systole	Diastole	
Ventricles	Diastole	Systole	
Atrial depolarization	Ventricular depolarization	Ventricular repolarization	

[B]

EKG event	Range of duration (seconds)	Corresponding physiological events in heart
P wave	0.06–0.11	Depolarization (excitation) of atria prior to their contraction. Impulse begins in SA node, spreads through muscles of atria to AV node.
P–R segment (wave)	0.06–0.10	Atrial depolarization and conduction through AV node.
P–R interval (onset of P wave to onset of QRS complex)	0.12–0.21	Time between onset of atrial depolarization and contraction, and onset of ventricular depolarization and contraction.
QRS complex (wave and interval)	0.03–0.10	Depolarization of ventricles; repolarization of atria is masked on EKG by ventricular depolarization.
S–T segment (wave) (end of QRS complex to onset of T wave)	0.10–0.15	End of ventricular depolarization to beginning of repolarization of ventricles.
T wave	Varies.	Repolarization of ventricles.
S–T interval (end of QRS complex to end of T wave)	0.23–0.39	Interval between completion of depolarization and end of repolarization.
Q–T interval (onset of QRS complex to end of T wave)	0.26–0.49	Ventricular depolarization plus ventricular repolarization.

[C]

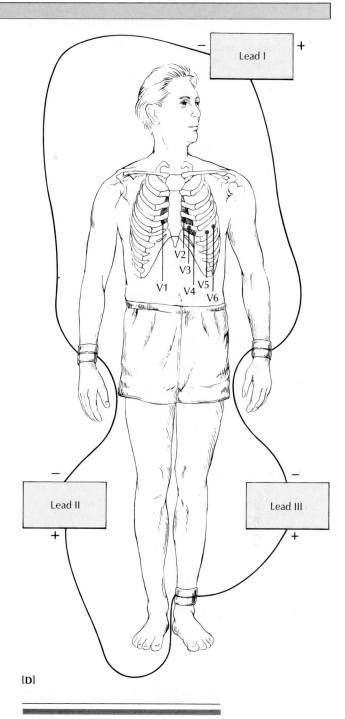

[D]

Electrocardiogram as a diagnostic tool Examining the frequency and duration of EKG deflection waves is helpful in evaluating heart function. For example, a heightened P wave indicates an enlarged atrium, a higher-than-normal Q wave may indicate a heart attack (myocardial infarction), and a heightened R wave usually indicates enlarged ventricles. An acute myocardial infarction raises the S-T segment above the horizontal, and an S-T segment below the horizontal may indicate an abnormally high blood potassium level.

Any deviation from the normal rate or sequence of excitation is called *cardiac arrhythmia*. It is caused by structural or functional disorders such as abnormal heart rhythm, enlargement of atria or ventricles, or damage to the myocardium. Because the electrical activity of the heart is sensitive to changes in ion concentration, the EKG can usually show an abnormal electrolyte regulation. Some of the more easily detectable abnormalities are described in Physiological and Anatomical Abnormalities at the end of the chapter.

Cardiac Cycle

The *cardiac cycle* is a carefully regulated sequence of steps that we think of as the beating of a heart. The cycle includes the *contraction*, or *systole* (SISS-toe-lee), of the atria and ventricles, and the *relaxation*, or *diastole* (die-ASS-toe-lee), of the atria and ventricles. The cardiac cycle proceeds in four stages:

1 During *atrial systole* (which lasts 0.1 sec), both atria contract, forcing blood into the ventricles.

2 During *ventricular systole* (0.3 sec), both ventricles contract,

Artificial Pacemaker

The **artificial pacemaker** is a battery-operated electronic device that is implanted in the chest, with electrical leads to the heart of a person whose natural pacemaker (the SA node) has become erratic. In a relatively simple operation, electrode leads (catheters) from the pacemaker are passed beneath the skin, through the external jugular vein (or other neck vein), into the superior vena cava, into the right atrium, through the tricuspid valve, and into the myocardium of the right ventricle (see drawing). If the patient's veins are damaged or too narrow to receive the typical chest implant of the pacemaker with its connecting wires, the pacemaker is implanted in the left abdominal area, with a connecting lead inserted into the epicardium.

Three basic types of artificial pacemakers are available. The first type delivers impulses when the patient's heart rate is slower than that set for the pacemaker, and shuts off when the natural pacemaker is working adequately. The second is a fixed-rate model that delivers constant electrical impulses at a preset rate. The third is a transistorized model

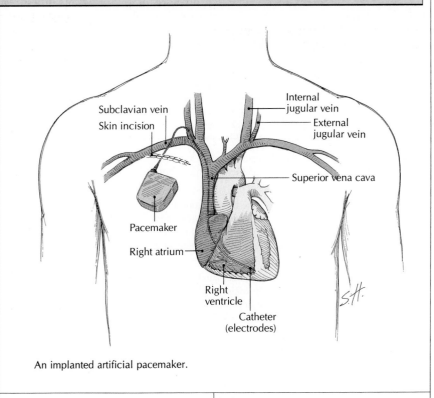

An implanted artificial pacemaker.

that picks up impulses from the patient's SA node and operates at 72 beats per minute when the natural pacemaker fails.

---

forcing blood out through the pulmonary artery (to the lungs) and aorta (to the rest of the body).

3 During *atrial diastole* (0.7 sec), or relaxation of the atria, the ventricles remain contracted, and the atria begin refilling with blood from the large veins leading to the heart from the body.

4 *Ventricular diastole* (0.5 sec), or relaxation of the ventricles, begins before atrial systole, allowing the ventricles to fill with blood from the atria.

Path of blood through the heart Before we present the mechanical events of the cardiac cycle, we will describe the path of blood flow through the heart:

1 *Blood enters the atria* (Figure 18.12A). Oxygen-poor blood from the body flows into the right atrium at about the same time as newly oxygenated blood from the lungs flows into the left atrium: (a) the *superior vena cava* returns blood from all body structures above the diaphragm (except the heart and lungs). (b) The *inferior vena cava* returns almost all blood to the right atrium from all regions below the diaphragm. (c) The *coronary sinus* returns about 85 percent of the blood from the heart muscle to the right atrium. (d) The *pulmonary veins* carry oxygenated blood from the lungs into

the left atrium. The blood entering the right atrium (blue in Figure 18.12A) is low in oxygen and high in carbon dioxide because it has just returned from supplying oxygen to the body tissues. The blood entering the left atrium (red in Figure 18.12A) is rich in oxygen because it has just passed through the lungs, where it has picked up a new supply of oxygen and released its carbon dioxide. (This is the only time or place where *venous* blood is highly *oxygenated*, because it is coming to the heart directly from the lungs.)

2 *Blood is forced into the ventricles* (Figure 18.12B). The heart's natural pacemaker (the SA node) fires an electrical impulse that coordinates the contractions of both atria (atrial systole). Blood is forced through the one-way atrioventricular valves into the relaxed ventricles.

3 *The ventricles, filled with blood, hesitate for an instant* (Figure 18.12C).

4 *The ventricles contract, sending blood to the body and lungs* (Figure 18.12D). The ventricular contraction creates a pressure that closes the atrioventricular valves between the atria and ventricles, while opening the two semilunar valves leading out of the ventricles. The right ventricle forces blood low in oxygen out through the right and left pulmonary arteries to the lungs. The left ventricle pumps the newly oxygenated blood through the aortic semilunar valve into the aorta. The aorta branches into the ascending and descending arteries that carry oxygenated blood to all parts of the body (see Figure 18.12D). The left and right ventricles pump almost simultaneously, so that equal amounts of blood enter and leave the heart. By this time, the atria have already started to refill, preparing for another cardiac cycle.

Mechanical events of the cardiac cycle The heart beats in a more or less regular fashion about 2.5 billion times during an average lifetime. In order for such regularity to exist, the mechanical events of the cardiac cycle must be coordinated precisely.

The heart functions as a pump by contracting its chambers in order to generate the pressure that forces blood through

---

FIGURE 18.12

The cardiac cycle and the path of blood through the heart. (A) Blood enters the atria. (B) Blood is pumped into the ventricles. (C) The ventricles relax. (D) The ventricles contract, pumping blood through the pulmonary artery and aorta to the lungs and body.

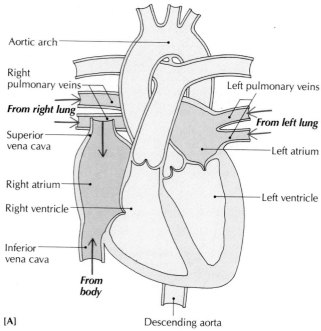

[A]

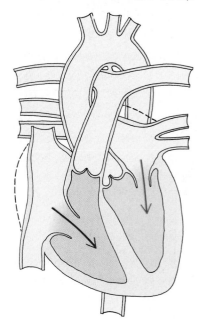

[B]

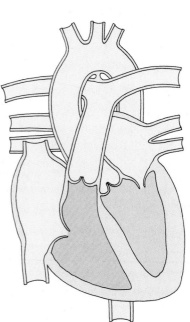

[C]

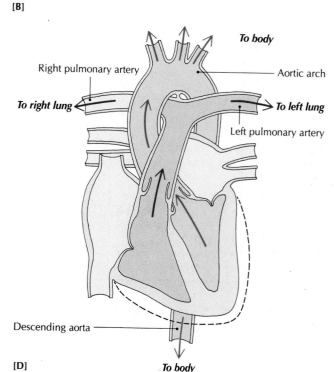

[D]

FIGURE 18.13

Diagram showing the changes that take place in the left heart and aorta during a single cardiac cycle, at a heart rate of 75 contractions per minute. The Roman numerals in (E) designate the first to fourth heart sounds.

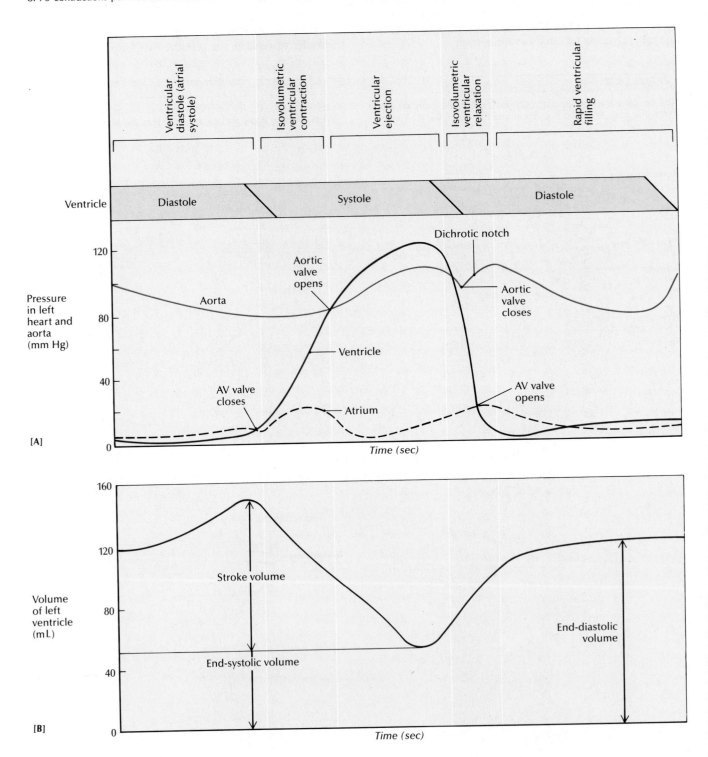

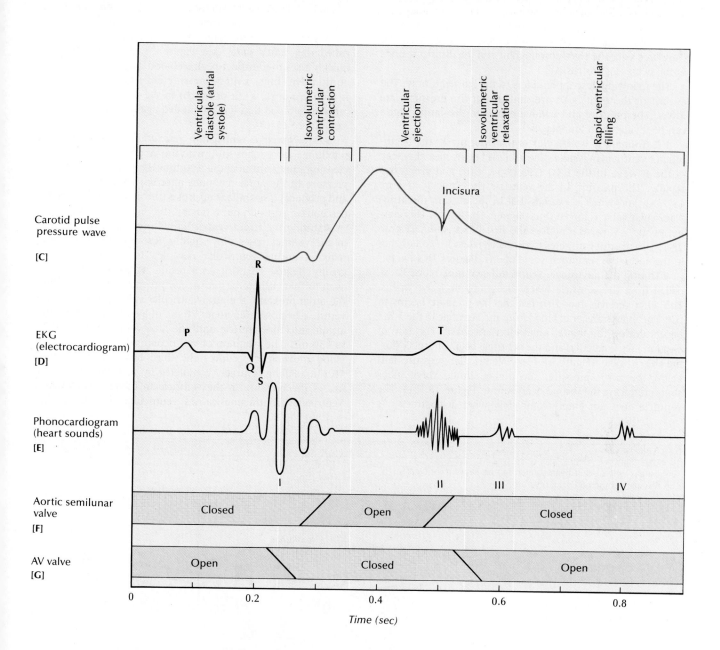

Ventricular diastole (atrial systole)

Isovolumetric ventricular contraction

Ventricular ejection

Isovolumetric ventricular relaxation

Rapid ventricular filling

Carotid pulse pressure wave

[C]

Incisura

EKG (electrocardiogram)

[D]

P   R   Q   S   T

Phonocardiogram (heart sounds)

[E]

I   II   III   IV

Aortic semilunar valve

[F]

Closed   Open   Closed

AV valve

[G]

Open   Closed   Open

0   0.2   0.4   0.6   0.8

Time (sec)

the heart, into the blood vessels throughout the body, and back to the heart. The events that occur during a single cardiac cycle can be shown by measuring the pressures and pressure differences in the chambers of the heart, and by measuring blood volume. These measurements can then be correlated with the electrocardiogram and the distinct sounds of the heart (Figure 18.13).

The following description refers to the left heart only. The events for the right heart are similar, but the pressures are lower. The pressure and volume events of the cardiac cycle can be divided into four stages:

1 *Isovolumetric ventricular contraction* marks the beginning of ventricular systole. Contraction begins near the peak of the R wave of the EKG (Figure 18.13D), and progresses rapidly. The pressure in the ventricle rises quickly (Figure 18.13A), and when it exceeds that in the atrium, the atrioventricular valve is forced shut (Figures 18.13A, G). This closing of the AV valve produces the first heart sound (Figure 18.13E). Pressure continues to rise, and when it exceeds that in the aorta, the semilunar valve opens (Figures 18.13A, F).

2 During the next stage, *ventricular ejection,* blood is expelled from the ventricle when the semilunar valve opens. The aorta receives blood so fast that its pressure begins to rise. The main transfer of blood from the ventricle to the aorta occurs during this initial rapid ejection phase. The rate of ejection falls gradually, and the ventricle begins to relax during the T wave (Figure 18.13D). Ventricular pressure falls below the aortic pressure, and the semilunar valve snaps shut, producing the second heart sound (Figure 18.13E). The ventricle does not empty completely when it contracts. As

will be seen later, the *stroke volume* is the difference between the end-diastolic and end-systolic volumes (Figure 18.13B). When blood is ejected from the left ventricle, a pressure wave spreads throughout the arterial system. When this wave is measured or felt in the large carotid artery in the neck, it is called the *carotid pulse* (see Figure 18.13C). The pressure quickly rises to a distinct peak as blood is ejected. When the aortic valve shuts, a sharp pressure deflection is noted. This is the *incisura* (L. a cut or notch) in the pressure curve, and represents blood that has rebounded against the aortic valve (see Figure 18.13A).

3 *Isovolumetric ventricular relaxation* occurs after the semilunar valve snaps shut, with the ventricular pressure decreasing rapidly from about 100 mm Hg to almost zero. The pressure in the aorta rebounds after the aortic valve closes, and produces a secondary upstroke (the *dichrotic notch*) in the descending part of a pulse tracing (see Figure 18.13A).

4 During the **rapid ventricular filling** stage, the pressure in the ventricle rises only slightly. Because the ventricle is relaxing during ventricular diastole, its volume increases greatly (Figure 18.13B). As a result, its pressure decreases even further. When the ventricular pressure decreases below the atrial pressure, the atrioventricular valve opens, and the ventricle begins to fill; about 70 mL of blood moves from the atrium into the ventricle until the final volume is about 120 to 130 mL. The vibration of ventricular walls caused by this blood movement produces the third heart sound (Figure 18.13E). Filling is largely completed by mid-diastole, and it causes the pressure in the ventricle to increase somewhat. Ventricular contraction causes ventricular pressure to in-

FIGURE 18.14

Chest areas where adult heart sounds can be heard most clearly. The ribs are numbered for clarity.

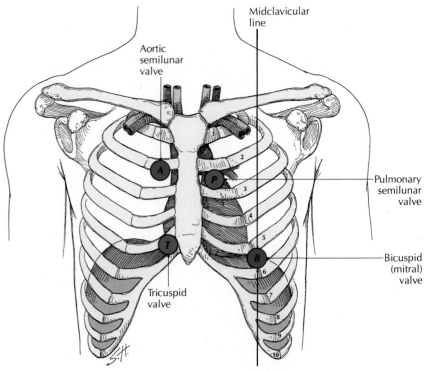

Noninvasive Cardiac Diagnosis

Several noninvasive techniques are available for exploring the heart without even the simplest form of surgery. Two of these are the ultrasound echocardiogram and the nuclear scanner. The *echocardiogram* passes high-frequency sound waves through the chest, and as the waves bounce off the heart, they are converted into electric signals and displayed on a television monitor. Echoes obtained from the walls of the ventricles and the flaps of the mitral valve can be used to estimate end-diastolic and end-systolic ventricular volumes. With this information, stroke volume and cardiac output can be calculated. With two-dimensional echocardiography, physicians can see cross sections of all four chambers in action, and can detect blood clots, tumors, defective valves, and other problems.

Nuclear scanning uses a specialized camera to pick up radioactive traces from isotopes that have been injected into the bloodstream. The scanner then creates a computer-assisted color-coded scan of the beating heart. When different isotopes are used, such a picture can re- veal the amount of blood pumped by the left ventricle, the motion of blood through the heart, and dead or damaged tissue following a heart attack. It can even locate tissue that is not receiving sufficient blood and oxygen.

Another innovation in noninvasive diagnosis is the *dynamic spatial reconstructor* (DSR). The DSR sends x rays through the body and produces images of the internal organs on a television screen. It produces a full-sized, three-dimensional image of the heart (or any other organ) in motion.

---

crease above atrial pressure, the atrioventricular valve closes, and the cardiac cycle starts over again.

Heart sounds Detectable *heart (valve) sounds* are produced with each heartbeat. Physicians first studied the sounds by putting their ears to the patient's chest, later by listening with a solid wooden stick shaped like a trumpet, and then by listening with a stethoscope. These sounds represent the *auscultatory events* (L. *auscultare*, to listen to) of the cardiac cycle, and can be heard best in the areas indicated in Figure 18.14. Heart sounds can be amplified and recorded by placing an electronically amplified microphone on the chest. The recording is called a *phonocardiogram*. It shows heart sounds as waves, as illustrated in Figure 18.13E.

There are four heart sounds associated with the cardiac cycle, although only the first and second sounds (traditionally referred to as *lubb* and *dupp*, respectively) can be heard easily with a stethoscope. The *first heart sound* is more complex, lower in pitch, and lasts longer than the second sound (see Figure 18.13E). The first sound occurs when the ventricles have been filled, and the atrioventricular valves of both atria close, the aortic and pulmonic valves open, and blood begins to be ejected into the aorta and pulmonary arteries. The first heart sound can be heard most clearly with a stethoscope in the area of the apex.

The *second heart sound* is high in pitch and lasts for only a short time. It is produced by the slamming of the semilunar valves after the ventricles have pumped their blood to the lungs and body, and have begun to contract. Ordinarily, the aortic valve closes a split second before the pulmonic valve, but in a healthy heart the two sounds are usually perceived as one. The second heart sound is heard best over the second intercostal space, where the aorta is closest to the surface. Immediately after the second sound there is a moment of silence.

A low-pitched *third heart sound* is heard occasionally. It is caused by the vibration of the ventricular walls after the atrioventricular valves open and the blood gushes into the ventricles. The sound is heard best in the tricuspid area.

A *fourth heart sound* is usually not heard with an unamplified stethoscope in normal hearts because of its low frequency. It is caused by blood rushing into the ventricles. It is best heard in the mitral area.

Heart sounds are an important tool in diagnosing valvular abnormalities. Any unusual sound is called a *murmur*, but not all murmurs indicate a valve problem, and many have no clinical significance. By listening carefully, a physician can detect *resting heart murmurs* that may be symptoms of valvular malfunctions, congenital heart disease, high blood pressure, and many other serious problems. If the atrioventricular valves are faulty, for instance, a gentle blowing or hissing sound can be heard between the first and second heart sounds: lubb-hiss-dupp—lubb-hiss-dupp—. Some whooshing sounds are not heart murmurs that are the result of heart disease. Instead, they are the sounds of blood swirling around sharp corners as it moves through the heart and lungs. Many adolescent and young adults have murmurs called *functional murmurs* (because the valves still function), but soon outgrow them.

Cardiac Output

The heart at rest pumps about 75 mL of blood with every beat. At an average rate of 70 beats per minute, the heart pumps more than 5.25 L a minute, 315 L an hour, 7560 L a day, and 2,759,000 L a year.

Cardiac output (CO) is the quantity of blood pumped by either ventricle (not both) in one minute. It is expressed in liters per minute. (The left ventricle is the one usually measured.)

The amount of blood expelled with each ventricular contraction (volume per beat) is called the *stroke volume* (SV), or *stroke output*. The stroke volume (see Figure 18.14B) is the difference between the volume of the left ventricle at the end of diastole (filling) and its volume at the end of systole (emptying):

SV = end-diastolic volume − end-systolic volume

Cardiac output is determined by multiplying the heart rate by the stroke volume. If the normal stroke volume is about 75 mL (0.075 L), and the normal heart rate is between 70 and 80 beats per minute, then

$$
\begin{aligned}
\text{Cardiac output} &= \text{heart rate} \times \text{stroke volume} \\
\text{(L/min)} &\quad\quad \text{(beats/min)} \quad\quad \text{(L/beat)} \\
&= 70 \text{ to } 80 \times 0.075 \text{ L} \\
&= 5.25 \text{ to } 6.0 \text{ L/min}
\end{aligned}
$$

The *cardiac reserve* is the difference between the actual volume of blood pumped and the volume the heart is capable of pumping under stressful conditions. Cardiac reserve measures the potential blood-pumping ability of the heart, while cardiac output measures the actual work done. For example, in the normal young adult, the cardiac reserve is between 300 to 400 percent. In a well-trained athlete, it is occasionally as high as 500 to 600 percent, whereas in a physically inactive person, it may be as low as 200 percent.

The *cardiac index* (CI) is a measurement of the cardiac output of a resting subject, as related to body surface area (BSA):

$$
CI = \frac{CO}{BSA}
$$

It is expressed in liters per minute per square meter. Since the rate of metabolism is related to body surface area, and the cardiac output is related to the rate of metabolism, the cardiac index provides an index for the comparison of different individuals to normal values. A normal cardiac index is between 2.5 and 4.0 L/min/sq m.

Nervous Control of the Heart

The major function of the heart is to pump blood through a closed system of vessels. This function is regulated at different levels by the cerebrum, hypothalamus, medulla oblongata, and the autonomic nerves (Figure 18.15). The effects of the autonomic nervous system on the heart are strictly *regulatory,* speeding up or slowing down the heart rate, and are not essential for the heart to beat. (If you were to sever all nerve connections from the brain and spinal cord to the heart, the heart would still beat.)

The main control center is located in the medulla oblongata, which receives sensory information about body temperature, emotions, feelings, and stress from the cerebrum and hypothalamus. It also receives special sensory information about the chemical composition of the blood from chemically sensitive chemoreceptors, and information about how arteries are stretched by changes in blood pressure from pressure-sensitive baroreceptors.

The upper part of the medulla contains an area called the *cardioacceleratory center (CAC)*, or *pressor center;* the lower part contains the *cardioinhibitory center (CIC)*, or *depressor center.*

Because the neurons of the pressor and depressor centers interact to maintain homeostasis, they are collectively called the *cardioregulatory center.*

Sympathetic nerve fibers arise from the cardioacceleratory center, travel down the spinal cord through specific tracts, emerge from the cord by way of the cardiac nerves, and innervate the heart, where neuronal branches release norepinephrine (Figure 18.15). Norepinephrine accelerates the heart rate and strength of contraction. Vagal nerve fibers arising from the cardioinhibitory center go directly to the SA and AV nodes of the heart, where neuronal branches release acetylcholine, which decreases the heart rate.

Baroreceptors Clusters of cells called *baroreceptors* or *pressoreceptors* are located in the walls of the aortic arch and the carotid artery sinuses. (The *carotid sinuses* are the enlarged areas of the common carotid arteries, where the arteries branch into the external and internal carotids.) These receptors help maintain a homeostatic flow of blood to the brain by responding to changes in blood pressure that stretch the blood vessels.

When blood pressure rises above normal, the aorta and carotid sinuses dilate. The baroreceptors are stimulated by the stretch of the sinuses, and send impulses to the medulla. The medulla responds by sending parasympathetic and sympathetic impulses to the heart and blood vessels. As a result, the heart rate decreases and the blood vessels dilate, creating a negative feedback loop that causes blood pressure to drop. Stimulation of the baroreceptors is reduced, and blood pressure returns to normal.

After the impulses have entered the medulla, secondary signals inhibit the *cardioacceleratory center* and excite the *cardioinhibitory center.* The net effects are vasodilation throughout the peripheral circulatory system, and decreased heart rate and strength of heart contraction. These physiological changes, together with a decrease in peripheral resistance and a decrease in cardiac output, cause the arterial pressure to decrease. In contrast, low blood pressure produces the opposite effects, causing the blood pressure to rise back toward normal, which restores homeostasis.

The carotid baroreceptors are especially important in stabilizing the blood pressure and blood flow to the brain; this feedback pathway is called the *carotid sinus reflex.* The baroreceptors located in the wall of the aortic arch are involved with the *aortic reflex.* It operates exactly as the carotid sinus reflex does, but helps regulate the overall systemic blood flow and pressure.

Chemoreceptors The carotid sinuses and aortic arch also contain chemically sensitive receptors, or *chemoreceptors,* that are part of the *chemoreceptor reflex.* These receptors respond to changes in the blood levels of oxygen, carbon dioxide, and hydrogen ions. An increase in the concentration of carbon dioxide or hydrogen ions, or a decrease in oxygen concentration results in an increased heart rate by way of the same pathways used by the baroreceptor reflexes. If the chemical stimulus is reversed, the response is a decrease in the heart rate.

Endocrine Control of the Heart

As you just saw, chemical transmitters are used by the nervous system to regulate the activity of the heart. Any generalized sympathetic activity within the autonomic nervous system affects the medullary region of the adrenal glands (see Chapter 16). When cells of the adrenal medulla are stimulated, they produce norepinephrine, some of which enters the blood. The major portion of norepinephrine, however, is converted in the medulla into epinephrine with the aid of the enzyme phenylethanolamine-N-methyl transferase. Epinephrine then enters the bloodstream, and along with norepinephrine, increases the heart rate and strength of contraction.

Regulation of Cardiac Output

We have shown that variations in cardiac output can be produced by changes in the heart rate, stroke volume, or both, depending upon the physiological situation. The major controlling factor is heart rate. Heart rate can change by more than three times, but stroke volume can vary by only about half that amount. Figure 18.16 summarizes the major factors that influence heart rate and stroke volume, and thus cardiac output.

Control of heart rate The normal heart rate is determined primarily by the rhythmic pacemaker potentials of the SA node. It is also influenced by the autonomic nervous system

FIGURE 18.15

Neuroregulation of the heart by cardioregulatory centers in the medulla. The vagus nerve usually sends steady (tonic) inhibitory impulses (green) to the heart, keeping the heart rate regular. Under stressful conditions, impulses from sympathetic nerves (black) from the cardioregulatory control center in the medulla accelerate the heart rate.

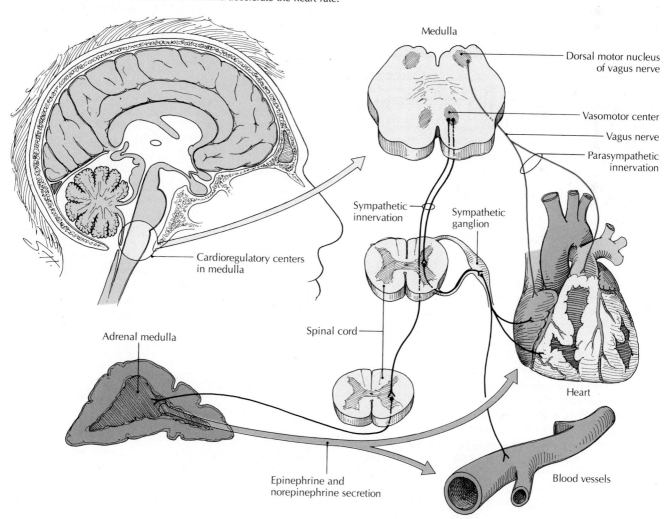

and certain hormones. Both sympathetic and parasympathetic fibers innervate the SA node, as well as other parts of the heart's conductive system.

Sympathetic stimulation increases the rate at which the pacemaker potential develops, causing the SA cells to reach their action potential threshold faster. As a result, the heart rate increases. Parasympathetic stimulation has the opposite effect, decreasing the pacemaker potential, causing the pacemaker cells to take longer to reach their action potential threshold, and decreasing the heart rate. In the resting homeostatic state, the heart rate *(inherent rate)* is set by the tone of the parasympathetic system, in the range of 60 to 100 beats per minute. Because the autonomic neurons innervate other parts of the heart's conductive system, sympathetic stimulation also accelerates the spread of the electrical impulse through the AV node, atrioventricular bundle, and Purkinje fibers. Parasympathetic stimulation slows the electrical impulse along this same conducting pathway.

The regulatory effects of the autonomic nervous system are produced by the release of acetylcholine from the parasympathetic neurons and norepinephrine from the sympathetic neurons. Acetylcholine affects the pacemaker potential by increasing the permeability of the SA node cells to potassium, which slows the rate of depolarization. Norepinephrine enhances the depolarization rate of SA node cells by increasing the flow of calcium ions into the cells.

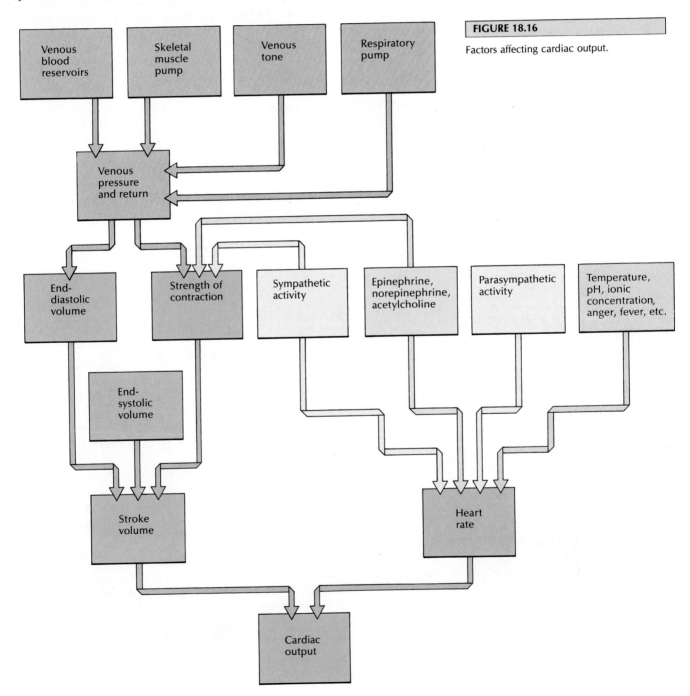

FIGURE 18.16

Factors affecting cardiac output.

The heart rate can be altered to a lesser degree by factors outside of autonomic control. For example, blood temperature, pH, ionic concentrations, hormones, anger, pain, exercise, fever, and grief all have an effect on the heart rate.

Control of stroke volume As you saw in the discussion of cardiac output, the stroke volume is the difference between the blood volumes in the left ventricle at the beginning and end of each contraction. By varying either of these volumes, the stroke volume and cardiac output are altered. The stroke volume is regulated by either increasing or decreasing the force of contraction. Changes in the contractile force can be produced by two major physiological factors: changes in the end-diastolic volume, and changes in sympathetic stimulation of the ventricles.

The relationship between the strength of a heart contraction and the resting length of its muscle fibers was first measured in 1895 by O. Z. Frank, and later by E. H. Starling. They found that the amount of blood returned to the heart by the veins is the *major factor* regulating end-diastolic volume. The **Frank-Starling law of the heart** states that within physiological limits, the more the ventricles are filled during diastole (relaxation), the more blood they will eject upon systole (contraction). In this way, the heart takes in and pumps out the same amount of blood during a cardiac cycle.

The mechanism behind the Frank-Starling law is based on the fact that cardiac muscle fibers increase their strength of contraction when they are stretched. (In the same way, the more you stretch a rubber band, the harder it snaps when you let it go.) An increase in venous return to the heart increases end-diastolic volume, which distends the ventricles, stretches cardiac muscle fibers, increases stroke volume, and finally, increases cardiac output.

The factors that influence venous return and pressure also have a direct effect on end-diastolic volume and stroke volume (see Figure 18.16). Since veins are more distensible than arteries, they can serve as *reservoirs* that store blood, which, in turn, can either increase or decrease the volume of blood returning to the heart. The contraction of skeletal muscles compresses veins during movement and squeezes blood toward the heart. This action is known as the *skeletal muscle pump*. Venous blood reserves can be actively mobilized by contracting the smooth muscle in the walls of the veins. Sympathetic stimulation causes this constriction *(venous tone)*, which increases venous pressure and, in turn, end-diastolic volume. Venous return is also influenced during the breathing process by a mechanism called the *respiratory pump*. During inspiration (taking air in), there is a decrease in pressure in the thoracic cavity and an increase in abdominal pressure. This forces venous blood from the abdominal area back to the heart by increasing venous pressure.

Ask Yourself

1 *What are some of the properties of cardiac muscle?*

2 *How does an electrical impulse flow through the heart?*

3 *What is the purpose of an electrocardiogram?*

4 *What events occur during the cardiac cycle?*

5 *What is the cardiac output, and how is it calculated?*

6 *Why is the SA node considered to be the heart's natural pacemaker?*

7 *What role does the autonomic nervous system play in regulating cardiac output?*

8 *How is stroke volume regulated?*

9 *What is the Frank-Starling law of the heart?*

10 *How does venous return affect stroke volume?*

11 *What are two mechanisms that affect venous pressure?*

PHYSIOLOGICAL AND ANATOMICAL ABNORMALITIES

Heart disease is a commonly used term for any disease that affects the heart. A more appropriate term is *cardiovascular disease,* which includes both heart and blood vessel disorders. About 40 million Americans have some form of cardiovascular disease, and it is responsible for more deaths than all other causes of death *combined.** Although the American death rate from cardiovascular disease is decreasing, it is still one of the highest in the world. This year more than 1.5 million Americans will suffer heart attacks, and of the 550,000 who survive, 100,000 will have another, fatal, attack within a year.

This section will concentrate on cardiovascular diseases that involve the heart primarily. The following chapter will discuss diseases that originate in the blood vessels.

Degenerative heart disorders

The first group of heart disorders results from the deterioration of the tissues or organs of the cardiovascular system.

Myocardial infarction (heart attack) When the blood flow through a coronary artery is reduced for any reason (usually because of a clot or plaque* build-up), the myocardium is deprived of oxygen, and begins to die. The result is a **myocardial infarction** (L. *infercire*, to stuff), or *heart attack*. An *infarct* is an area of tissue that has died because of an inadequate blood supply. Although a

*The cardiovascular diseases that cause the most deaths each year are heart attack (56.3 percent), stroke (17.4 percent), hypertensive disease (3.1 percent), rheumatic fever and rheumatic heart disease (0.8 percent), all other cardiovascular diseases (22.4 percent).

*Plaque in an artery is a build-up of cholesterol and other lipids.

Heart Surgery

Surgery on blood vessels and the heart has been done since the early 1930s. However, it was not until the heart-lung machine was introduced in 1953 that true open-heart surgery was possible. In December of 1982, the first operation in which an artificial heart was placed into a person took place.

Heart-Lung Machine

Open-heart surgery is possible only if the heart is quiet and empty of blood. The **heart-lung machine** takes over the job of pumping *and* oxygenating blood while open-heart surgery is in progress (see drawing A). Tubes are inserted into

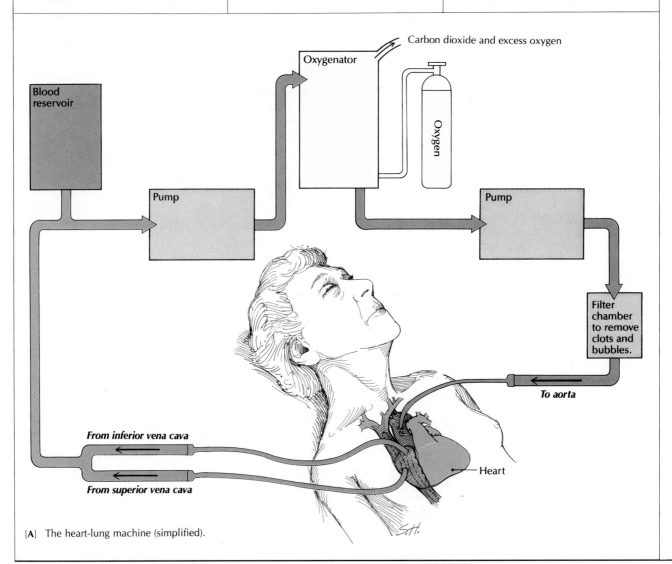

[A] The heart-lung machine (simplified).

heart attack is always serious, it is not always diagnosed easily. The usual symptoms include pain in the midchest (*angina pectoris*, described on p. 566), which travels up the neck or out through the shoulders and arms, especially the left. Sometimes there may be only a shortness of breath, or no symptoms at all.

Congestive heart failure *Congestive heart failure* (CHF) occurs when either ventricle fails to pump blood out of the heart as quickly as it enters the atria, or when the ventricles do not pump equal amounts of blood. For example, left heart failure occurs if the left ventricle is weakened and pumps less blood than normal; the right ventricle will then pump more blood *into* the lungs than can be pumped by the left ventricle. As a result, the lungs become engorged with blood in a condition called *pulmonary* *edema*, which also results in kidney failure. The old-fashioned term *dropsy* refers to the accumulation of fluid in the abdomen and legs that also accompanies heart failure.

Valvular heart diseases

As you saw earlier, the two major heart sounds provide information about the heart's valves. Abnormal heart sounds, or *murmurs*, may be indicative

the inferior and superior vena cavae to lead blood through a pump and oxygenator, where carbon dioxide is removed and oxygen is added, just as occurs in the lungs. A pump then returns the oxygenated blood into the arterial circulation by way of the aorta or one of its branches.

While surgery is taking place and blood is being circulated through the heart-lung machine, blood clotting is discouraged by introducing the anticoagulant heparin into the circulation. When the natural circulation through the heart is restored, the effect of heparin is reversed by the introduction of protamine sulfate.

An important technique used with the heart-lung machine is *hypothermia*, in which the patient's body is cooled enough to induce ventricular fibrillation, thus providing a quiet state of the heart. Recent techniques allow cooling of only the heart. The newer techniques eliminate all heart activity during the operation, and reduce metabolism, so that absolutely no strain is placed on the myocardium during surgery.

Coronary Bypass Surgery

The most common serious heart disease is obstruction or narrowing of the coronary arteries by atherosclerosis. When coronary arteries become blocked, the flow of blood to the heart is reduced or cut off completely, and angina pectoris or myocardial infarction may result. In **coronary bypass surgery**, the surgeon removes the diseased portion of the coronary artery and replaces it with a segment of the saphenous vein from the patient's own leg (see drawing B). (Arteries may also be used, but veins are more accessible to the surgeon.) (Because the replacement vessel is taken from the patient's body, there is virtually no danger of tissue rejection.) One end of the vein is stitched to the aorta, and the other end is stitched to the coronary artery beyond the point of obstruction. Thus, the diseased artery is "bypassed," and normal blood flow is re-established.

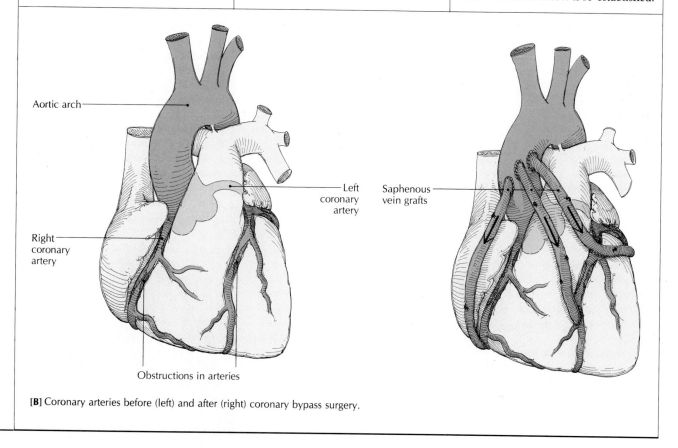

Aortic arch

Left coronary artery

Saphenous vein grafts

Right coronary artery

Obstructions in arteries

[B] Coronary arteries before (left) and after (right) coronary bypass surgery.

of a **valvular heart disease**, in which one or more cardiac valves operate improperly. Certain types of valvular malfunction may produce *regurgitation* (backflow of blood through an incompletely closed valve) or *stenosis* (an incompletely opened valve). Either type of malfunction can lead to congestive heart failure. Some forms of valvular heart disease are (1) *mitral insufficiency* (blood from the left ventricle flows back into the left atrium during systole), (2) *mitral stenosis* (narrowing of the valve reduces flow from the left atrium to the left ventricle), (3) *aortic insufficiency* (blood flows back into the left ventricle during diastole), (4) *aortic stenosis* (pressure in the left ventricle increases in response to a narrowed valve opening, and ischemia—localized tissue anemia—results because of an insufficient oxygen supply), (5) *pulmonary insufficiency* (blood flows back from the pulmonary artery into the right ventricle during diastole), (6) *pulmonary stenosis* (blood is prevented from leaving the right ventricle, causing ventricular hypertrophy and eventual failure), (7) *tricuspid insufficiency* (blood flows back into the right atrium during systole), and (8) *tricuspid stenosis* (blood is obstructed from flowing from the right atrium to the right ventricle). Valvular heart diseases may sometimes be corrected with

Heart Transplants and the Artificial Heart

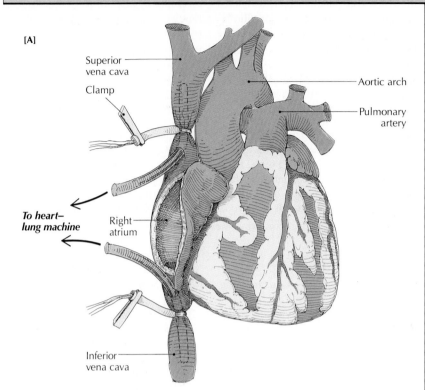

[A]

Superior vena cava
Clamp
Aortic arch
Pulmonary artery

To heart–lung machine

Right atrium

Inferior vena cava

(1) Front wall of right atrium of recipient's heart removed.

On December 3, 1967, Dr. Christiaan Barnard removed the diseased heart of Louis Washkansky and replaced it with a healthy human heart from a young woman who had been killed in an automobile accident. It was the first human **heart-transplant** operation. Since then, many such operations have been performed. The longest surviving heart-transplant patient died 18 years after the operation.

The main reason why few patients live as long as five years after a heart transplant is that the tissue of the donor heart is rejected by the host's body. Several drugs are currently used to suppress this immune reaction; some suppress the body's tendency to reject foreign tissue without inactivating the immune system altogether. Without a functional immune system, the body cannot overcome even a minor infection, and the patient may die of a secondary infectious disease such as pneumonia.

The usual procedure for a heart transplant is to remove most of the front of the recipient's heart, leaving the back walls of both atria, the pulmonary ar-

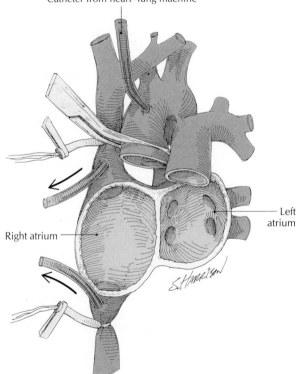

Catheter from heart–lung machine

Right atrium

Left atrium

(2) Front wall of recipient's heart removed, tubes for heart-lung machine inserted.

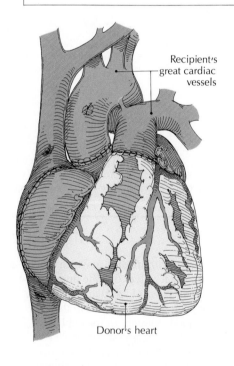

Recipient's great cardiac vessels

Donor's heart

(3) Trimmed donor heart is sutured in place. A fourth suture, around the left atrium, is not visible.

tery, and aorta (see drawing A). The donor heart is trimmed to fit, and sewn onto the recipient's heart, starting at the septum. Next, the junctions between the atria and the pulmonary artery of the donor heart and the recipient's heart are sewn together. The patient is connected to a heart-lung machine during the four-to five-hour operation.

The first permanent *artificial heart* was implanted in Barney Clark on December 2, 1982, almost 15 years to the day after the first heart transplant. After 112 days, Clark died of "circulatory collapse due to multiorgan failure." Although the artificial heart was in good condition at the time of his death, throughout the 112 days he was plagued with kidney failure, chronic respiratory problems, inflammation of the colon, and low blood pressure.

The artificial heart used in the early transplants was called the Jarvik-7, after its designer, Dr. Robert K. Jarvik (see drawing B). It consists of four artificial valves, two flexible polyurethane ventricles on aluminum bases, and two small tubes leading from the bottom of the ventricles through the chest wall and outside the body. During the operation, the patient is put on a heart-lung machine, the natural ventricles are removed, and polyurethane cuffs are sutured to the atria, pulmonary artery, and aorta. The separate right and left ventricles of the artificial heart are then snapped onto the cuffs and connected to an external driving system by means of the two tubes.

In the artificial heart, blood passes through the natural atria into the artificial ventricles, where it is circulated to the pulmonary artery from the right side and to the aorta from the left side, just as in the natural heart. Plastic rings support the tilting-disk inflow and outflow valves, and diaphragms inside the ventricles are powered by pulses of compressed air. Following the operation, the patient is tethered by 2-meter-long tubes to an external mobile driver system, which pumps and regulates air flow. Reserve air tanks are kept on hand in case the main compressor fails.

As of this writing, the use of the artificial heart was suspended indefinitely.

the surgical substitution of an artificial valve.

Congenital heart diseases

A congenital disease is one that is present at birth, but is not necessarily hereditary.

Ventricular septal defect *Ventricular septal defect* is the most common congenital heart disease. It is an opening that is present at birth in the ventricular septum that allows blood to move back and forth between the ventricles. Small openings usually either close naturally, before permanent damage occurs, or are repaired surgically. Large openings are not always reparable, and patients often die in their first year from biventricular congestive heart failure or secondary complications.

Interatrial septal defect In an *interatrial septal defect*, the foramen ovale between the two atria fails to close at birth, and the child is born with an opening between the right and left atria that allows blood to pass back and forth. After birth, blood usually moves from the left to the right, because atrial pressure is higher on the left. Eventually, the entire right side of the heart becomes enlarged, and heart failure occurs.

Tetralogy of Fallot *Tetralogy of Fallot* is a combination of four (Gr. *tetra*, four) congenital defects: (1) ventricular septal defect, (2) pulmonary stenosis, (3) enlargement of the right ventricle, and (4) emergence of the aorta from both ventricles. Blood usually moves from the right to the left ventricle, although it may move from the left to right when pulmonary stenosis is mild. The most common symptom of tetralogy of Fallot is *cyanosis* (Gr. dark blue), a bluish discoloration of the skin, resulting from inadequate oxygenation of the blood. Ventricular septal defect allows oxygenated blood to mix with deoxygenated blood in the ventricles. Pulmonary stenosis obstructs blood from leaving the right ventricle, producing the third defect, an enlarged right ventricle. The emergence of the aorta from within the ventricles, coupled with stenosis of the pulmonary artery, allows insufficient blood to reach the lungs.

Infectious heart diseases

Severe damage to the heart valves or heart walls can result from certain infectious diseases.

To aortic arch

To pulmonary artery

Tilting-disk inflow valve

From right atrium (natural)

From left atrium (natural)

Tilting-disk outflow valve

Rubber diaphragm

Right ventricle

Left ventricle

Rubber diaphragm

To driving system

[B] The Jarvik-7 artificial heart.

Rheumatic fever and rheumatic heart disease *Rheumatic fever* is a severe infectious disease occurring mostly in children. It is characterized by fever and painful inflammation of the joints, and frequently results in permanent damage to the heart valves. Rheumatic fever is a hypersensitive reaction to a specific streptococcal bacterial infection, in which antibodies cause inflammatory reactions in the joints and heart. *Rheumatic heart disease* refers to the secondary complications that affect the heart. They include *pancarditis* (*pan* = all, because it encompasses pericarditis, myocarditis, and endocarditis) in the early acute phase, and valvular heart disease later.

Pericarditis, myocarditis, endocarditis Inflammation of the three layers of the heart wall are called, from the outside in, pericarditis, myocarditis, and endocarditis.

Pericarditis is an inflammation of the parietal and/or visceral pericardium. When it is caused by a bacterial, fungal, or viral infection, it is known as *infectious pericarditis*. It may also be caused by uremia, high-dose radiation, rheumatic fever, cancer, drugs, and trauma such as myocardial infarction. Pain is typical, especially when the heart is in a position to press against its covering membranes.

Myocarditis is an inflammation of the myocardium. It may result from viral, bacterial, parasitic, or other infections; hypersensitive immune reactions; large doses of radiation; and poisons such as alcohol. Myocarditis usually proceeds without complications, but occasionally it leads to heart failure, arrhythmias, and other complications.

Endocarditis is caused by a bacterial or fungal infection of the endocardium, heart valves, and endothelium of adjacent blood vessels. Untreated, endocarditis is usually fatal.

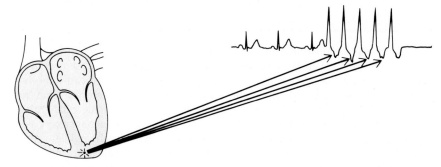

[A] PAROXYSMAL VENTRICULAR TACHYCARDIA

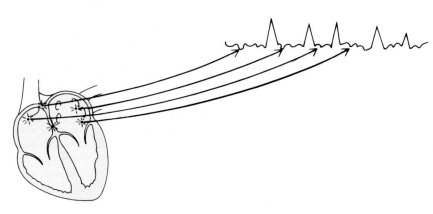

[B] ATRIAL FIBRILLATION

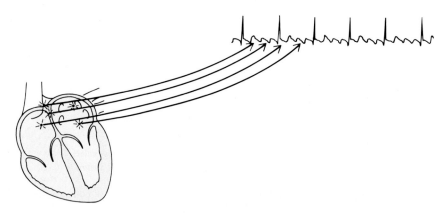

[C] ATRIAL FLUTTER

Cardiac complications

Among the complications that may affect the heart are circulatory shock, rapid rise in pressure, and abnormal rhythms.

Circulatory shock *Circulatory shock* refers to a generalized inadequacy of blood flow throughout the body. It may be caused by *reduced blood volume*, as a result of hemorrhage, severe diarrhea or vomiting, lack of water, or severe burns

that involve serious fluid loss. Another possible cause is *increased capacity of veins*, as a result of bacterial toxins. A third cause is *damage to the myocardium*, usually as result of a heart attack.

When blood volume drops below about four-fifths of normal, one of the results is *hypovolemic shock*, which impairs circulation and the flow of liquids to the tissues. Such a condition is usually caused by hemorrhage, severe burns, or

any other occurrence that reduces the fluid content of the body. Hypovolemic shock is usually accompanied by low blood pressure, rapid heart rate, rapid and shallow breathing, and cold, clammy skin.

In contrast to hypovolemic shock, *cardiogenic shock* occurs when the cardiac output is decreased, resulting in too little fluid reaching the tissues. It is caused by severe failure of the left ven-

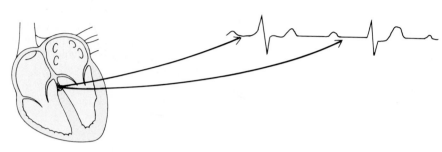

[D] ATRIOVENTRICULAR BLOCK

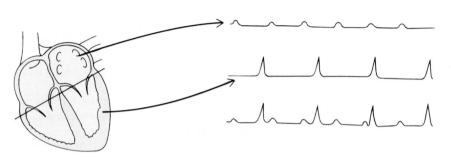

[E] COMPLETE HEART BLOCK

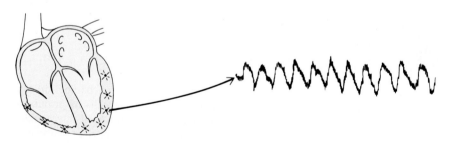

[F] VENTRICULAR FIBRILLATION

tricle, usually as a complication of myocardial infarction, and most patients die within 24 hours.

Cardiac tamponade In *cardiac tamponade* (Fr. *tampon*, plug), a rapid rise in pressure occurs inside the pericardial sac, usually because of the accumulation of blood or other fluid as a result of an infection or severe hemorrhage in the chest area.

Cardiac arrhythmias Abnormal elec-trical conduction in the conductive tissues or changes in heart rate and rhythm are called **cardiac arrhythmias** (see drawings). In *paroxysmal ventricular tachycardia* (Gr. *paroxunein*, to stimulate; *takhus*, swift), the heart suddenly starts to race at a steady rhythm of 200 to 300 beats per minute. In medical terms, *paroxysm* refers to something that starts and stops suddenly, and *tachycardia* is a heart rate faster than 100 beats per minute. The AV node may continue to fire abnormally for minutes, hours, or days.

Atrial fibrillation (L. *fibrilla*, fibril) occurs when the atria suddenly begin to beat with a fast, but feeble, twitching movement. The ventricles contract normally for a short time, but their beat soon becomes irregular also. Because of the irregular beat, the ventricles may contract when they are not full, reducing the cardiac output. Blood may also stagnate and clot in the atria. If these clots get into the circulation, they can clog arteries and cause heart attacks, stroke, or other serious problems. The leading cause of atrial fibrillation is mitral stenosis.

Atrial flutter occurs when the atria beat regularly, but at a very rapid, "flapping" pace of about 200 to 300 beats per minute. Ordinarily, every third or fourth atrial beat will stimulate a ventricular beat, but the relationship can vary from two to one to as much as six to one. The causes of atrial flutter are usually similar to those of atrial fibrillation.

Atrioventricular block (or *heart block*) occurs when cells of the atrioventricular node become diseased and cannot transmit adequate electrical impulses. The condition may progress from a pattern where the rhythm of the impulse is almost imperceptibly slowed to an irregular, slow atrial beat and a very slow ventricular beat that may be accompanied by dizziness. If the cells of the AV node totally lose their ability to conduct, the beat of the heart is taken over by one of the emergency pacemakers in the atria or ventricles. At this stage, called *complete heart block*, the atria and ventricles beat independently. If the ventricles stop beating for a few seconds, fainting and convulsions may occur.

Ventricular fibrillation is caused by a continuous recycling of electrical waves through the ventricular myocardium. As a result, abnormal contraction patterns with varied rates are set up. During ventricular fibrillation, the strong, steady contractions of the ventricles are replaced by a feeble twitching that pumps little or no blood. These effects can be produced by damage to the myocardium, which usually results from an inadequate blood supply. Without treatment, the victim usually dies in a matter of minutes.

Angina pectoris

Angina pectoris (L. strangling + chest) is an example of referred pain. It occurs when not enough blood gets to the heart muscle, due to a damaged or blocked artery. Sometimes exercising or stress will cause angina (pain). Stopping the exercising or the stress may relieve the pain, drugs may be prescribed, or surgery may be necessary to replace the damaged artery.

Angina does not always lead to a heart attack. Sometimes collateral circulation develops, more blood reaches the heart muscle, and the pain decreases. Angina may even disappear altogether if the heart muscle is receiving enough blood.

MEDICAL TERMINOLOGY

ANEURYSM A bulging of a portion of the heart, aorta, or other artery.

ASYSTOLE Failure of the myocardium to contract.

AUSCULTATION Listening for heart sounds.

BRADYCARDIA A lower-than-normal heartbeat.

CARDIAC ARREST Cessation of normal, effective heart action, usually caused by asystole or ventricular fibrillation.

CARDIAC MASSAGE Manual stimulation of the heart when asystole occurs.

CARDIOMEGALY Enlargement of the heart.

CARDIOTONIC DRUG Drug that strengthens the heart.

CHRONOTROPIC DRUG Drug that changes the timing of the heart rhythm.

A positive chronotropic drug increases the heart rate, and a negative drug decreases the heart rate.

COMMISSUROTOMY An operation to widen a heart valve that has been thickened by scar tissue.

CONSTRICTIVE PERICARDITIS A condition in which the heart muscle cannot expand and contract properly because the pericardium has shrunk and thickened.

CORONARY OCCLUSION Blockage in the circulation of the heart.

COR PULMONALE Heart disease resulting from disease of the lungs in which the right ventricle hypertrophies and there is pulmonary hypertension.

DEFIBRILLATOR Instrument that corrects abnormal cardiac rhythms by applying electric shock to the heart.

EMBOLISM (Gr. *emballaein*, to throw in) Obstruction or occlusion of a blood vessel by an air bubble, blood clot, mass of bacteria, or other foreign material.

PALPITATION Skipping, pounding, or racing heartbeats.

SINUS RHYTHM Normal cardiac rhythm regulated by the SA node.

STETHOSCOPE Instrument used to listen to sounds in the chest.

STOKES-ADAMS SYNDROME Sudden seizures of unconsciousness which may accompany heart attacks.

TACHYCARDIA A faster-than-normal heartbeat.

THROMBUS (Gr. a clotting) A blood clot obstructing a blood vessel or heart cavity; the condition is *thrombosis*.

SUMMARY

Human blood flows in a *closed system*, remaining essentially within the vessels that carry it. The heart is a *double pump*, pumping oxygen-poor blood to the lungs, and pumping oxygen-rich blood to the rest of the body. The circulation of the blood to and from the lungs is the *pulmonary circulation*, and the circulation to and from the body is the *systemic circulation*.

Structure of the heart

1 The cone-shaped heart is about the size of a fist. It is located in the center of the chest. It is oriented obliquely, with about two-thirds of its bulk to the left of the body's midline.

2 The heart lies within a protective sac called the *pericardium (pericardial sac)*. It is composed of an outer *fibrous pericardium* and an inner *serous pericardium*.

Pericardial fluid between the sac and the heart helps to minimize friction when the heart moves. The serous pericardium is divided into an outer *parietal layer* and an inner *visceral layer*, separated by the *pericardial cavity*. The visceral layer forms part of the heart wall.

3 The wall of the heart is composed of the *epicardium* or outer layer (the visceral pericardium), the *myocardium* or middle muscular layer, and the *endocardium* or inner layer.

4 The *cardiac skeleton* is a structure of tough connective tissue inside the heart. It provides attachment sites and support for the valves and muscular fibers that allow the heart to wring blood out of the ventricles.

5 The heart is made up of two separate, parallel pumps, often called the *right heart* and *left heart*. Each of the two

pumps has a receiving chamber on top called an *atrium*, and a discharge pumping chamber below called a *ventricle*. Separating the left and right hearts is a thick wall of muscle called the *septum*.

6 Visible features on the surface of the heart include some sulci (depressions) and coronary veins and arteries that carry blood to and from the heart.

7 The two *atrioventricular (AV) valves* permit blood to flow from the atria to the ventricles, and the two *semilunar valves* permit the flow from the ventricles to the pulmonary artery and aorta. The atrioventricular valves are the *tricuspid* valve of the right heart, and the *bicuspid*, or *mitral*, valve of the left heart. The right semilunar valve is the *pulmonary semilunar valve*, and the left semilunar valve is the *aortic semilunar valve*.

8 The great vessels of the heart are the *superior vena cava, inferior vena cava, pulmonary artery, pulmonary veins,* and *aorta.*

9 Circulation to and from the tissues of the heart is the *coronary circulation.* Blood is supplied to the heart by the right and left *coronary arteries.* Most of the cardiac veins drain into the *coronary sinus.*

10 The embryonic circulatory system forms and becomes functional before any other system, fulfilling the nutritional and respiratory needs of the embryo by the end of the third week after fertilization.

Physiology of the heart

1 Cardiac muscle cells function as a single unit in response to physiological stimulation because they are connected by *intercalated disks.* The interconnected cells form a latticework called a *syncytium.* Action potentials spread over a syncytium, causing all the cardiac muscle cells to contract in unison.

2 The *cardiac action potential* can be divided into five phases: *depolarization, early repolarization, plateau, repolarization,* and *resting potential.*

3 The impulse-conducting system of the heart consists of the *sinoatrial (SA) node (pacemaker)* in the right atrium, the *atrioventricular (AV) node* between the atrium and ventricle, a tract of conducting fibers called the *atrioventricular bundle* that divides into a branch for each ventricle, and modified nerve fibrils called *Purkinje fibers* in the walls of the ventricles.

4 An *electrocardiogram,* or *EKG,* is a recording of the electrical activity of the heart. A normal electrocardiogram shows a *P wave,* a *QRS complex,* and a *T wave.* It is a useful diagnostic tool.

5 The *cardiac cycle* is the carefully regulated sequence of steps that comprises a heartbeat. A complete cardiac cycle consists of an atrial contraction, or *systole,* a ventricular systole, an atrial relaxation, or *diastole,* and a ventricular diastole.

6 The path of blood through the heart proceeds as follows. (1) Oxygen-poor blood from the body enters the right atrium, and oxygen-rich blood from the lungs enters the left atrium. (2) Blood from the atria is forced into the ventricles. (3) After hesitating for an instant, the ventricles contract. The right ventricle pumps oxygen-poor blood to the lungs, and the left ventricle pumps oxygen-rich blood (which just entered the heart from the lungs) through the aorta to the body. By this time, the atria have started to refill, and another cardiac cycle is about to begin.

7 The mechanical events (pressure and volume changes) of the cardiac cycle can be divided into four stages: *isovolumetric ventricular contraction, ventricular ejection, isovolumetric ventricular relaxation,* and *rapid ventricular filling.*

8 *Heart sounds* are caused by the closing of the heart valves and vibrations in the heart wall. These sounds can be used to diagnose cardiovascular abnormalities. An unusual sound is called a *murmur.*

9 *Cardiac output* is the quantity of blood pumped by either ventricle in one minute. The amount of blood expelled with each ventricular contraction is the *stroke volume.* Cardiac output is determined by multiplying the heart rate by the stroke volume. *Cardiac reserve* is the difference between the actual volume of blood pumped and the volume the heart is capable of pumping under stressful conditions. *Cardiac index* is a measurement of cardiac output in relation to body surface area.

10 The central mechanism regulating the heartbeat, rate, and volume is the *cardioregulatory center* in the medulla. The autonomic system speeds or slows the heartbeat, but does not initiate it. Endocrine control is involved also.

11 Nervous control operates through a negative feedback system involving *baroreceptors* and *chemoreceptors* in the carotid sinuses and aorta.

12 *Cardiac output* is regulated primarily by controlling the heart rate and stroke volume.

13 The *Frank-Starling law* of the heart states that the heart will pump all the blood it receives.

Physiological and anatomical abnormalities

1 Two of the major types of degenerative heart disorders are *myocardial infarction,* or heart attack, and *congestive heart failure.*

2 Malfunctions produced by *valvular heart diseases* are regurgitation and stenosis, which may lead to heart failure.

3 Congenital heart diseases are those present at birth. Examples are *ventricular septal defect, interatrial septal defect,* and *tetralogy of Fallot.*

4 Infectious diseases, such as *rheumatic fever,* may have complications affecting the heart, such as *rheumatic heart disease,* and may damage heart valves. Inflammation of the heart walls is called *pericarditis, myocarditis, endocarditis,* or *pancarditis,* depending on the layer affected.

5 Among the complications that affect the heart are *circulatory shock, cardiac tamponade,* and various *cardiac arrhythmias.*

6 *Angina pectoris,* or pain in the chest area, results from insufficient blood reaching the heart muscle.

UNDERSTANDING THE FACTS

1 What is meant by a closed system of blood vessels?

2 Give the location of the following (be specific):
 a myocardium
 b chordae tendineae
 c fossa ovalis
 d atrioventricular bundle
 e mitral valve

3 Describe the shape, size, position, and location of the heart.

4 Why are the ventricular walls of the heart thicker than the atrial walls?

5 What are the functions of pericardial fluid?

6 Is there an advantage to having many collateral branches of the coronary arteries? Explain.

7 Distinguish between systole and diastole.

8 Specifically when does your heart rest?

9 Define cardiac reserve and cardiac index.

10 What is the actual cause of the heart sounds?

11 Name the great vessels of the heart.

12 In the electrocardiogram, what is indicated by the P wave? The QRS complex? The T wave?

13 Describe the mechanical events of the cardiac cycle.

14 Define cardiac output, and tell how it is calculated.

15 What factors regulate cardiac output?

16 The baroreceptors are sensitive to changes in ———.

17 How does the endocrine system help regulate cardiac output?

18 Define myocardial infarction.

19 Define stenosis.

20 What is the relationship of rheumatic fever to heart disease?

21 Define shock, and differentiate between two types of shock.

UNDERSTANDING CONCEPTS

1 Describe the basic functions of the heart.

2 Trace the pathway of blood through the heart, starting with the right atrium. Be sure to include all the valves in the pathway.

3 Describe the cardiac skeleton and its functions.

4 Give the location and describe the function of the papillary muscles.

5 Explain the impulse-conducting system of the heart.

6 Does the human heart need nervous stimulation in order to beat? Explain.

7 What is the refractory period in cardiac muscle, and what is its physiological significance?

8 How do the properties of cardiac muscle affect heart function?

9 Describe the Frank-Starling law of the heart.

SUGGESTED READING

ADOLPH, E. F., "The Heart's Pacemaker." *Scientific American*, March 1967.

DEBAKEY, M., AND A. GOTTON, *The Living Heart*. New York: McKay, 1977.

DHALLA, N. S., "Calcium Movements in Relation to Heart Function." *Basic Res. Cardiology*, 77 (1982):117–139.

HELLER, L. J., AND D. E. MOHRMANN, *Cardiovascular Physiology*. New York: McGraw-Hill, 1981.

HONIG, C. R., *Modern Cardiovascular Physiology*. Boston: Little, Brown, 1981.

HURLEY, R. E., ed., *The American Heart Association Heartbook*. New York: Dutton, 1980.

JARVIK, ROBERT K., "The Total Artificial Heart." *Scientific American*, January 1981.

KATZ, A. M., *Physiology of the Heart*. New York: Harper & Row, 1980.

LEVY, M. N., AND P. J. MARTIN, "Neural Regulation of the Heartbeat." *Annual Review of Physiology*, 43 (1981):443–453.

LITTLE, ROBERT C., *Physiology of the Heart and Circulation*, 2nd ed. Chicago: Year Book Medical Publishers, 1981. (Paperback)

OPIE, LIONEL, *The Heart: Physiology, Metabolism, Pharmacology, and Therapy*. Orlando, Fla.: Academic Press, 1984.

PHIBBS, BRENDAN, *The Human Heart: A Guide to Heart Disease*, 4th ed. St. Louis: Mosby, 1979. (Paperback)

ROBINSON, THOMAS F., STEPHEN M. FACTOR, AND EDMUND H. SONNENBLICK, "The Heart as a Suction Pump." *Scientific American*, June 1986.

SHEPARD, J. T., AND P. M. VANHOUTTE, *The Human Cardiovascular System*. New York: Raven Press, 1979.

STALLONES, RENEL A., "The Rise and Fall of Ischemic Heart Disease." *Scientific American*, November 1980.

WIGGERS, CARL J., "The Heart." *Scientific American*, May 1967.

WINFREE, ARTHUR T., "Sudden Cardiac Death: A Problem in Topology." *Scientific American*, May 1983.

19
The Cardio-vascular System: Blood Vessels

LEARNING OBJECTIVES

1 Describe the structure and function of arteries, arterioles, capillaries, venules, and veins.

2 Describe three different types of capillaries and relate their structure to their functions.

3 Compare the structure of the walls in the different types of blood vessels.

4 Explain microcirculation.

5 Describe the function of pulmonary circulation, and name the major vessels in this circuit.

6 Discuss the function and major components of each division of systemic circulation.

7 Describe the function of the portal systems, and name their major components.

8 Explain the main differences between fetal and adult circulation.

9 Identify the major arteries and veins, and describe the region supplied or drained by each.

10 Summarize the principles governing the flow of blood through the cardiovascular system.

11 Discuss the systems or mechanisms involved in the control of blood flow.

12 Describe the factors that help in maintaining normal blood pressure.

13 Explain how the body maintains a homeostatic arterial blood pressure.

14 Describe the movement of fluids and other materials across capillary walls.

15 Explain venous pressure and return.

16 Describe the different arterial blood pressure measurements and the procedure for measuring blood pressure.

17 Discuss the causes or effects of stroke, aneurysms, atherosclerosis, coronary artery disease, hypertension, thrombophlebitis, and varicose veins.

18 Explain the role of exercise in promoting a healthy cardiovascular system.

Although the center of the cardiovascular system is the heart, it is the blood vessels that carry the blood throughout the body. As the blood flows from the heart through these vessels, it supplies cells with substances needed for metabolism and homeostatic regulation. When cellular waste products accumulate, the blood carries them off through other vessels for excretion by the kidneys, skin, and lungs. (As you will see in Chapter 20, an alternate pathway for the return of tissue fluids to the bloodstream is the *lymphatic system,* which is sometimes regarded as a "second circulatory system.")

TYPES OF BLOOD VESSELS

Arteries are blood vessels that carry blood *away from* the heart to the organs and tissues of the body. Because different arteries contain varying amounts of elastic and muscle tissue, some are called *elastic,* and others are called *muscular.* The walls of elastic arteries expand slightly with each pulse of pressure from a heartbeat. The muscular arteries branch into smaller *arterioles.* These vessels play important roles in determining the amount of blood going to any organ or tissue, and in regulating blood pressure. The muscular arteries and arterioles can be either dilated or constricted by nervous control. Arterioles branch into smaller vessels called *metarterioles,* which carry blood either into venules (tiny veins), or into the smallest vessels in the body, the capillaries.

Capillaries are microscopically fine vessels, with walls mostly one cell thick. The thin capillary wall is porous, and allows the passage of water and small particles of dissolved materials. Capillaries are distributed throughout the body, except in the dead outer layers of skin, and in such special places as the lenses of the eyes.

Capillaries converge into larger vessels called *venules,* which merge to form larger vessels called veins. *Veins* carry blood *toward* the heart. They are generally softer and more flexible than arteries, and they collapse if blood pressure is not maintained.

Arteries

Most **arteries*** are efferent vessels that carry blood *away from* the heart to the capillary beds throughout the body (see Figure 19.7). In the adult, all arteries except the pulmonary artery carry oxygenated blood. The left and right pulmonary arteries carry *deoxygenated* blood from the heart to the lungs. Thus, the most reliable way to classify blood vessels is by the *direction* in which they carry blood, either toward the heart or away from it.

The great arteries that emerge from the heart are often called *trunks.* The major arterial trunks are the **aorta** from the left ventricle, and the **pulmonary trunk** from the right ventricle (see Figure 18.6).

*The word *artery* comes from the Greek word for "windpipe." When the ancient Greeks dissected corpses they found blood in the veins, but none in the arteries. As a result, they thought the arteries carried air.

The central canal of all blood vessels, including arteries, is called the *lumen.* Surrounding the lumen of an artery is a thick wall, composed of three layers, or *tunicae* (pl. of L. *tunica,* covering) (Figures 19.1 and 19.2).

The **tunica intima** ("innermost covering") has a lining of endothelial cells (simple squamous epithelium), a thin subendothelial layer of fine areolar tissue, and an internal elastic layer containing collagen fibers.

The **tunica media** ("middle covering") is the thickest layer of arterial wall in the large arteries. It is composed mainly of connective tissue, smooth muscle cells, and elastic tissue. The walls of the largest arteries (elastic arteries) have elastic tissue rather than smooth muscle. In the smaller arteries (muscular arteries), the elastic tissue in the tunica media is replaced by smooth muscle.

The **tunica adventitia** ("outermost covering") is composed mainly of collagen fibers and elastic tissue. Occasional smooth muscle fibers run longitudinally next to the outer border of the tunica media. Nerves and lymphatic vessels are also found within this layer. The walls of the large arteries (greater than 20 mm) are nourished by small blood vessels called the **vasa vasorum** ("vessels of the vessels"), which form capillary networks within the tunica adventitia and outer part of the tunica media.

The strong elastic walls of the largest arteries allow these vessels to adjust to the great pressure created by the contraction of the ventricles during systole. Blood is ejected into the aorta at a speed of about 30 to 40 cm per second (almost a mile an hour).

FIGURE 19.1

A typical elastic artery is shown in this cutaway drawing. Small arteries do not contain the vasa vasorum, but do have the three layers shown here.

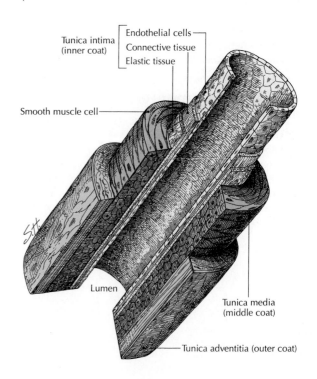

Tunica intima (inner coat)
Endothelial cells
Connective tissue
Elastic tissue
Smooth muscle cell
Lumen
Tunica media (middle coat)
Tunica adventitia (outer coat)

FIGURE 19.2

Scanning electron micrograph of a transverse section through an arteriole, showing the lumen, lining of endothelial cells, tunica intima, tunica media, and tunica adventitia. [*Fujita, Tanaka, and Tokunaga,* SEM Atlas of Cells and Tissues. *Tokyo, Igaku-Shoin, 1981.*]

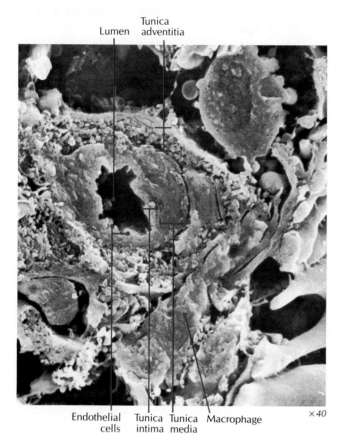

Lumen Tunica adventitia

Endothelial cells Tunica intima Tunica media Macrophage ×40

Arterioles

The arteries nearest the heart are the largest. As their distance from the heart increases, the arteries branch into smaller and smaller arteries, then into **arterioles** shortly before reaching the capillary networks. Arterioles are covered by the three tunicae (Figure 19.2). Because their walls contain smooth muscles, arterioles can dilate or constrict, thus controlling the flow of blood from arteries into capillaries and later into the organs as they require blood. If necessary, an arteriole can dilate to increase blood flow to capillaries by as much as 400 percent.

Since blood is already within large arteries and veins, why do these vessels need their own blood supply?

The walls of large arteries and veins are so thick that nutrients cannot diffuse far enough to reach all of their cells.

Terminal arterioles (those closest to a capillary) are muscular and well supplied with nerves, but they do not have an internal elastic layer or their own blood vessels (vasa vasorum), as larger arteries do. The tunica media is particularly well supplied with sympathetic nerves, which cause the smooth muscles to contract and the lumen to constrict.

Capillaries

Terminal arterioles branch out to form **capillaries** (L. *capillus,* hair), which connect the arterial and venous systems. Capillaries are generally composed of only a single layer (tunica intima) of endothelial cells on a thin basement membrane of glycoprotein (Figure 19.3). Capillaries are the smallest and most numerous blood vessels. If all the capillaries in an adult body were connected, they would stretch about 96,000 km

FIGURE 19.3

Electron micrograph of a cross section of a capillary. In this micrograph, the capillary wall is composed of two endothelial cells, and the large nucleus of one of the cells is prominent. Note the clefts at the junctions between the endothelial cells. Pinocytic vesicles in the cytoplasm may assist liquid materials to pass through the capillary wall. [*D. W. Fawcett/Photo Researchers.*]

Capillary wall Capillary lumen Nucleus of endothelial cell

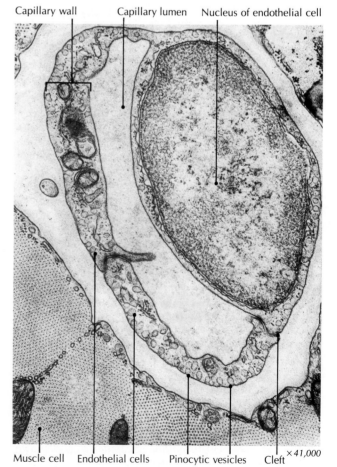

Muscle cell Endothelial cells Pinocytic vesicles Cleft ×41,000

Tumors and the Growth of Blood Vessels

Malignant tumors require a generous blood supply. They manage to acquire that supply by encouraging the sprouting and extension of new capillary networks, a process called *angiogenesis*. The new capillaries connect the tumor to nearby veins and arteries, keeping it well nourished as it grows. The cellular mechanism for angiogenesis was unknown until 1985, when Bert L. Vallee and his coworkers at Harvard Medical School announced the isolation of a protein that stimulates angiogenesis. The researchers, together with scientists at the University of Washington in Seattle, also succeeded in cloning the protein.

The growth-stimulating protein has been named *angiogenin*. It is about a thousand times more potent than normal growth factors. If future research is successful, scientists will learn what turns angiogenin on and off, and the protein may be used to stimulate the growth of new blood vessels after a heart attack, burns, or other serious tissue damage. Also, tumors may be controlled by inhibiting the action of angiogenin. The detection of angiogenin in the blood or urine may also provide an early warning of cancer.

TABLE 19.1 APPROXIMATE AVERAGE PHYSICAL CHARACTERISTICS OF BLOOD VESSELS

Type of vessel	Diameter (mm)	Wall thickness (mm)	Length (cm)	Internal pressure (mm Hg)	Cross-sectional area (sq cm)	Percentage of total body blood volume
Aorta	25.000	2.000	40.000	100	2.5	6
Medium-sized arteries	4.000	0.800	15.000	90	20.0	13
Arterioles	0.300	0.020	0.200	60	40.0	2
Capillaries	0.008	0.001	0.075	30	2500.0	5
Venules	0.020	0.002	0.200	20	250.0	5
Medium-sized veins	5.000	0.500	15.000	15	80.0	20
Large veins	15.000	0.800	20.000	10	20.0	39
Venae cavae	30.000	1.500	40.000	10	8.0	10

(60,000 mi). This abundance of capillaries makes an enormous surface area available for the exchange of gases, fluids, nutrients, and wastes between the blood and nearby cells.

Capillaries are about 600 times narrower than a medium-sized vein, and about 500 times narrower than a medium-sized artery (see Table 19.1). The diameter of capillaries varies with the function of the tissue.

Types of capillaries At least three different types of capillaries are recognized, and each one performs a specific function (Figure 19.4).

Continuous capillaries are found in muscle tissue. Their walls are made up of one continuous endothelial cell, with the ends overlapping in a tight endothelial cell junction (Figure 19.4A). Apparently, intracellular (pinocytic) vesicles help to move large soluble particles across the membrane of the cell by exocytosis and endocytosis. The walls of **fenestrated capillaries** (L. *fenestra,* window) consist of two or more adjacent endothelial cells connected by thin endothelial membranes called fenestrations or pores (Figure 19.4B). Fenestrated capillaries are usually found in the kidneys, endocrine glands, and intestines. **Discontinuous capillaries,** also known as **sinusoids** or **vascular sinuses,** have fenestrations and a much wider lumen than the other types (Figure 19.4C). Such an open, irregular structure is highly permeable. This type of capillary system is found in the liver and spleen. Liver sinusoids also contain active phagocytes called *stellate reticuloendothelial cells (Kupffer cells;* KOOP-fur), which are part of the reticuloendothelial system (see Chapter 4).

Capillary blood flow Blood leaves the heart traveling about 30 to 40 cm/sec (1 ft/sec), but it is slowed to only 2.5 cm/sec (1 in./sec) by the time it reaches the arterioles, and less than 1 mm/sec (0.04 in./sec) in the capillaries. Blood remains in the capillaries for only a second or two, but given the short length of capillaries (about 1 mm), that is long enough for the crucial exchanges of nutrients and wastes.

Because capillary walls are usually only one cell thick, certain materials pass through rather easily. The walls are *selectively permeable,* allowing some substances to filter through and holding back others. Small molecules, including gases such as oxygen and carbon dioxide, certain waste products, salts, sugars, and amino acids, pass through freely, but the large molecules of plasma proteins pass through only with difficulty, if at all. Red blood cells cannot pass through the capillary walls.

Microcirculation of the blood The capillaries and their associated structures (including terminal ateriole, metarterioles, and venules) constitute the ***microcirculation*** of the blood (Figure 19.5). This name reflects the extremely narrow diameters of the vessels (see Table 19.1).

Types of capillaries. (A) Continuous capillary in muscle, showing the overlapping ends of the single endothelial cell. (B) Fenestrated capillary of an endocrine gland, showing two endothelial cells connected by thin stretches of membrane called fenestrations. (C) Discontinuous capillaries (sinusoids), showing open, irregular structure and a phagocytic stellate reticuloendothelial (Kupffer) cell.

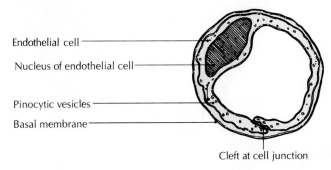

Endothelial cell

Nucleus of endothelial cell

Pinocytic vesicles

Basal membrane

Cleft at cell junction

[A]

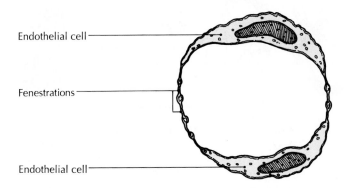

Endothelial cell

Fenestrations

Endothelial cell

[B]

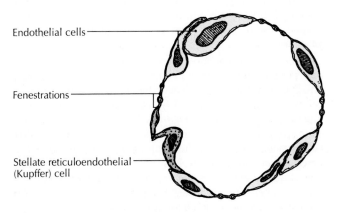

Endothelial cells

Fenestrations

Stellate reticuloendothelial (Kupffer) cell

[C]

A *metarteriole* is a vessel that emerges from an arteriole, traverses the capillary network, and empties into a venule. At the junction of the metarteriole and capillary is the *precapillary sphincter*, a ringlike smooth muscle cell that regulates the flow of blood into the capillaries. The smooth muscle cells of the precapillary sphincters are not innervated, but constrict or dilate in response to local changes in oxygen and carbon dioxide levels, pH, temperature, and circulating chemical agents. Such a response, which does not depend on hormonal or nervous stimulation, is called *autoregulation.*

Some metarterioles connect directly to venules by way of *thoroughfare* or *preferential channels.* Similar to thoroughfare channels are *collateral channels* (*capillary shunts* or *arteriovenous anastomoses*), which bypass the capillary beds and act as direct links between arterioles and venules. Collateral channels are numerous in the fingers, palms, and earlobes, where they control heat loss. Because these collateral channels have thick walls, they are not exchange vessels in the way that capillaries are. They are heavily innervated, and are muscular at the arteriole end and elastic at the venule end.

Venules

Blood drains from the capillaries into ***venules*** (VEHN-yoolz), tiny veins that unite to form larger venules and veins. The transition from capillaries to venules occurs gradually. The immediate postcapillary venules are called ***pericytic venules*** because contractile cells called *pericytes* are wrapped around them. Pericytic venules consist mainly of endothelium and a thin tunica adventitia. Postcapillary venules play an important role in the exchange between blood and interstitial fluid. Unlike capillaries, these venules are easily affected by inflammation and allergic reactions. They respond to these conditions by opening their pores, allowing water, solutes, and white blood cells to move out into the extracellular space.

Muscular venules generally accompany muscular arteries and arterioles. They have a very thin wall, a continuous epithelium, and no pericytes.

Veins

Venules join together to form ***veins.*** Superficial veins are found in areas where blood is collected near the surface of the body, and are especially abundant in the limbs. Veins become larger and less branched as they move away from the capillaries and toward the heart.

Why do we sometimes have dark circles under our eyes?

In the skin below your eyes are hundreds of tiny blood vessels, which help drain blood from the head. When you are ill or tired, blood circulation may slow down, concentrating blood in the vessels and causing them to swell. Because the skin under the eyes is very thin, the engorged vessels become visible. The dark circles are actually pools of blood.

FIGURE 19.5

Microcirculation of the blood. Metarterioles provide the path of least resistance between the arterioles and venules. Note the absence of smooth muscle in the true capillaries.

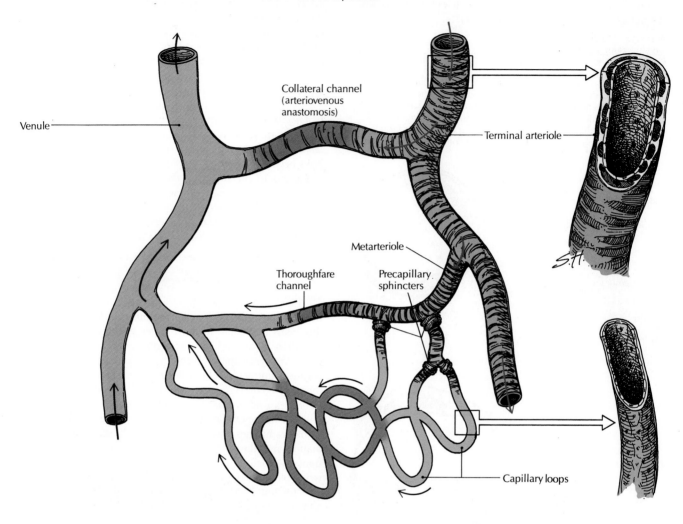

Most veins are relatively large vessels that carry *oxygen-deficient* blood from the body tissues to the heart. However, there are three exceptions:

1 The four *pulmonary veins* carry *oxygenated* blood from the lungs to the left atrium of the heart.

2 The *hepatic portal system* of veins carries blood from the capillaries of the intestines to the capillaries of the liver (see p. 576).

3 In the *hypophyseal portal system,* capillaries of the hypothalamus unite to form veins that divide into a second set of capillaries in the anterior pituitary gland.

The walls of veins contain the same three layers (tunicae) as arterial walls, but the tunica media is much thinner (Figure 19.6). Also, venous walls contain less elastic tissue, collage-nous tissue, and smooth muscle. As a result, veins are very distensible and compressible. The smooth muscle fibers that *are* found in veins are arranged in either a circular or longitudinal pattern. Like arteries, veins are nourished by small vasa vasorum.

Veins usually contain paired semilunar bicuspid valves that permit blood to flow in only one direction, restricting any backflow (see Figure 19.6A). Blood pressure in the veins is low, and the venous blood is helped along by the skeletal muscle pump (the compression of the venous walls by the contraction of surrounding skeletal muscles). The venous valves, which are derived from folds of the tunica intima, are especially abundant in the legs, where gravity cannot assist the return of blood to the heart. There are no valves in veins narrower than 1 mm or in regions of great muscular pressure, such as the thoracic and abdominal cavities.

FIGURE 19.6

Veins. (A) Cutaway drawing of a medium-sized vein, showing a one-way valve and the triple-layered wall. Compare the thickness of the layers with that of the arterial layers shown in Figure 19.1 Some veins contain one-way valves that permit the blood to flow only toward the heart. (B) Scanning electron micrograph of a medium-sized vein and artery, showing the larger lumen and thinner wall of the vein. Surrounding the vessels is connective tissue. [*Richard G. Kessel and Randy H. Kardon,* Tissues and Organs: A Text-Atlas of Scanning Electron Microscopy. *San Francisco: W. H. Freeman, 1979.*]

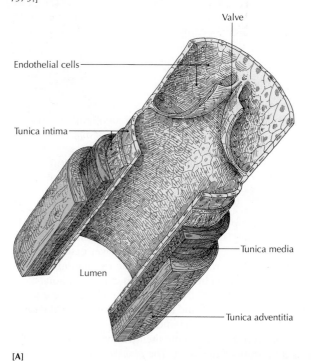

Valve
Endothelial cells
Tunica intima
Tunica media
Lumen
Tunica adventitia

[A]

Ask Yourself

1 *What are the basic functions of arteries? Veins?*

2 *What are the three layers of arterial walls?*

3 *What are arterioles? Venules?*

4 *Why are capillaries sometimes called exchange vessels?*

5 *What are the three types of capillaries?*

6 *What is meant by the microcirculation of the blood?*

7 *How do venous valves function?*

CIRCULATION OF THE BLOOD

Blood circulates throughout the body in two main circuits: the pulmonary and systemic circuits (Figure 19.7; see also Figure 18.1). In the following sections we will describe these two circuits, as well as some subdivisions or special areas of circulation. (Coronary circulation was described in Chapter 18.)

Pulmonary Circulation

The *pulmonary circulation* supplies blood only to the lungs. It carries deoxygenated blood from the heart to the lungs, where carbon dioxide is removed and oxygen is added. It

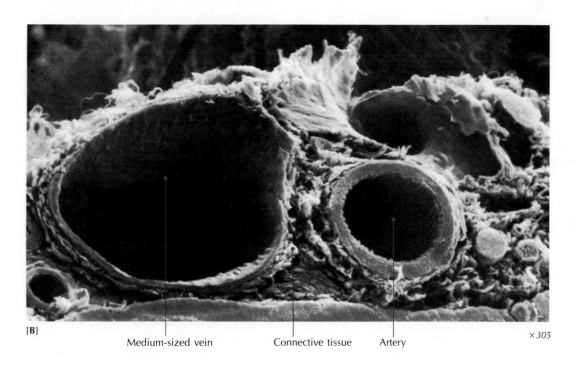

[B]
Medium-sized vein   Connective tissue   Artery   ×305

FIGURE 19.7

Diagrammatic representation of the circulatory system, showing the systemic circulation and the pulmonary circulation. Oxygenated blood is indicated by red arrows and deoxygenated blood by blue arrows.

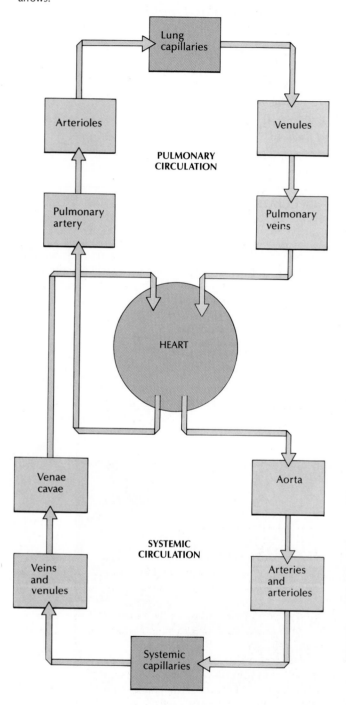

then returns the oxygenated blood to the heart for distribution to the rest of the body (Figure 19.8). Pulmonary circulation takes 4 to 8 seconds.

The major blood vessels of the pulmonary circulation are the *pulmonary trunk* and two *pulmonary arteries,* which carry deoxygenated blood from the right ventricle of the heart to the lungs; *pulmonary capillaries,* which are the site of the exchange of oxygen and carbon dioxide in the lungs; and the four *pulmonary veins* (two from each lung), which carry oxygenated blood from the lungs to the left atrium of the heart. Each of the pulmonary veins usually enters the left atrium through a separate opening.

Systemic Circulation

The **systemic circulation** supplies all the cells, tissues, and organs of the body with oxygen-rich blood, and also returns oxygen-poor blood (Figure 19.9). The systemic circuit from the heart and back again takes about 25 to 30 seconds. The systemic circulation is divided into the arterial and venous divisions. The main vessel of the **arterial division** is the *aorta,* which emerges from the left ventricle as the *ascending aorta,* curves backward over the top of the heart as the *aortic arch,* and continues down through the thorax and abdomen as the *descending aorta.* The descending aorta terminates in the two *common iliac arteries,* which supply the lower extremities.

The **venous division** of the systemic circulation is linked to the arterial system by capillary beds. All the venous blood from the upper part of the body eventually drains into the large *superior vena cava,* and the venous blood from the lower extremities, pelvis, and abdomen drains into the *inferior vena cava.* Both venae cavae empty their deoxygenated blood into the right atrium of the heart. The *coronary sinus* drains blood from the walls of the heart, and is part of the venous division of the systemic circulation.

Portal Systems

Most veins transport blood directly back to the heart from a capillary network, but in the case of a **portal system,** the blood passes through *two* sets of capillaries on its way to the venous system. The human body has two portal systems. The **hypophyseal portal system** moves blood from the capillary bed of the hypothalamus directly, by way of veins, to the capillary bed of the pituitary gland (hypophysis; see Figure 16.5). The **hepatic portal system** (Gr. *hepatikos,* liver) moves blood from the capillary beds of the intestines to the sinusoidal beds of the liver (Figure 19.10).

The hepatic portal system varies considerably from one person to the next, but it follows a basic pattern. The **hepatic portal vein** is formed behind the neck of the pancreas by the union of the **splenic** and **superior mesenteric veins.** The splenic mesenteric vein returns blood from the spleen, and the superior mesenteric vein returns blood from the small intestine. The hepatic portal vein also receives coronary, cystic, and pyloric branches.

The splenic vein receives the left *gastroepiploic vein* (or *short*

FIGURE 19.8

Pulmonary circulation. The freshest, most highly oxygenated blood in the body enters the left heart from the lungs through pulmonary *veins*, not arteries.

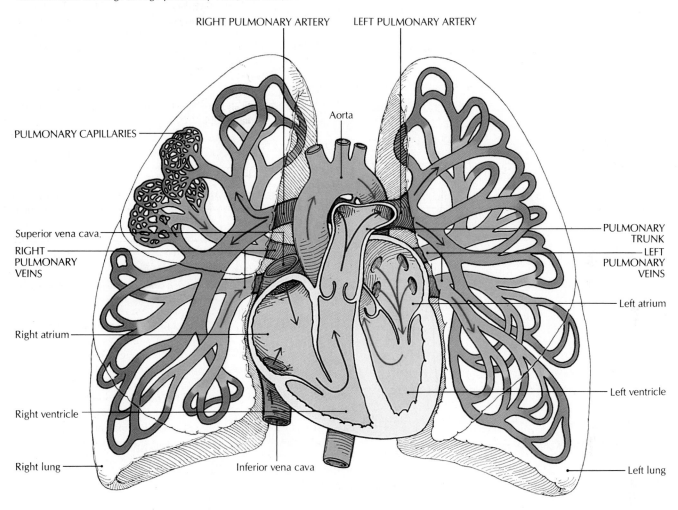

gastric vein) from the greater curvature of the stomach, and the *inferior mesenteric vein*, which receives blood from the veins of the large intestine.

The *cystic vein* brings blood from the gallbladder to the hepatic portal vein. The *right gastric (pyloric) vein* arises from the stomach and then enters the hepatic portal vein. The *left gastric (coronary) vein* also empties into the hepatic portal vein.

Because the veins of the hepatic portal system lack valves, the veins are vulnerable to the excessive pressures of venous blood. This situation can produce *hemorrhoids* or *piles* (dilated veins in the anal region) in the internal hemorrhoidal plexus (see Figure 19.10).

Much of the venous blood returning to the heart from the capillaries of the spleen, stomach, pancreas, gallbladder, and intestines contains products of digestion. This nutrient-rich blood is carried by the *hepatic portal vein* to the *sinusoids* of the liver. After leaving the sinusoids of the liver, the blood is collected in the **hepatic veins** and drained into the **inferior vena cava,** which returns the blood to the right atrium of the heart.

Cerebral Circulation

The brain is supplied with blood by four major arteries: two **vertebral arteries** and two **internal carotid arteries** (see Figure 19.14A). Blood from the vertebral arteries supplies the cerebellum, brainstem, and posterior part of the cerebrum. The internal carotid arteries supply the rest of the cerebrum and both eyes.

The vertebral arteries merge on the ventral surface of the brain to form the *basilar artery*, which terminates by forming the left and right *posterior cerebral arteries*. The internal carotid arteries enter the cranial cavity and then branch into the *ophthalmic arteries*, which go to the eyes, and the *anterior* and *middle cerebral arteries*, which go to the medial and lateral parts of the cerebral hemispheres, respectively.

All the blood entering the cerebrum must first pass through the **cerebral arterial circle** *(circle of Willis)* (see Figure 19.14B). This circular *anastomosis*, or shunt, at the base of the

FIGURE 19.9

Systemic circulation, showing the major arteries and veins. The circulation between the heart and lungs is not shown here (see Figure 19.8).

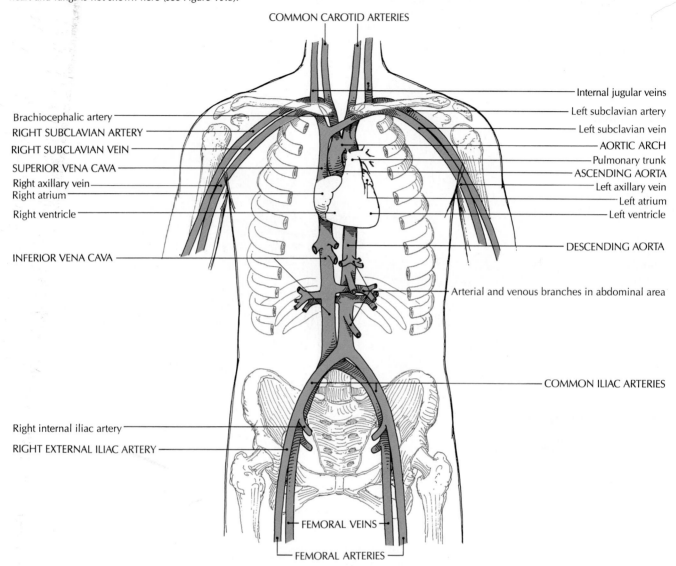

brain consists of the proximal portion of the posterior cerebral arteries, the posterior communicating arteries, the internal carotid arteries, the anterior cerebral arteries, and the anterior communicating artery. A blockage of blood within a vessel of the cerebral arterial circle may be partially overcome by a reversal of the blood flow.

Blood flows from the brain capillaries into large *venous sinuses* located in the folds of the dura mater. The sinuses then empty into the *internal jugular veins* on each side of the

neck. Blood returns to the heart by way of the *brachiocephalic veins* (see Figure 19.17). The junction of the left and right brachiocephalic veins forms the *superior vena cava*, which conveys blood to the right atrium of the heart.

Cutaneous Circulation

The arrangement of blood vessels in the skin allows for the increase or decrease of heat radiation from the integumentary system. When the body temperature increases, more blood flows to the superficial layers, from which heat radiates from the body. In contrast, when the body needs to conserve heat, blood is shunted away from the surface of the skin through deep arteriovenous anastomoses. In addition to these shunts, the skin has an extensive system of venous plexuses that can

Why do your nose and cheeks turn red on a cold day?

When the skin is very cold, oxygen in the capillaries is not needed for metabolism by the relatively inactive skin. Thus, high amounts of hemoglobin accumulate in the capillaries, showing through the skin.

FIGURE 19.10

Hepatic portal circulation. The hepatic portal system transports venous blood to the liver that has absorbed nutrients from the gastrointestinal tract. The liver also receives oxygenated blood from the hepatic artery. Arrows indicate the direction of flow.

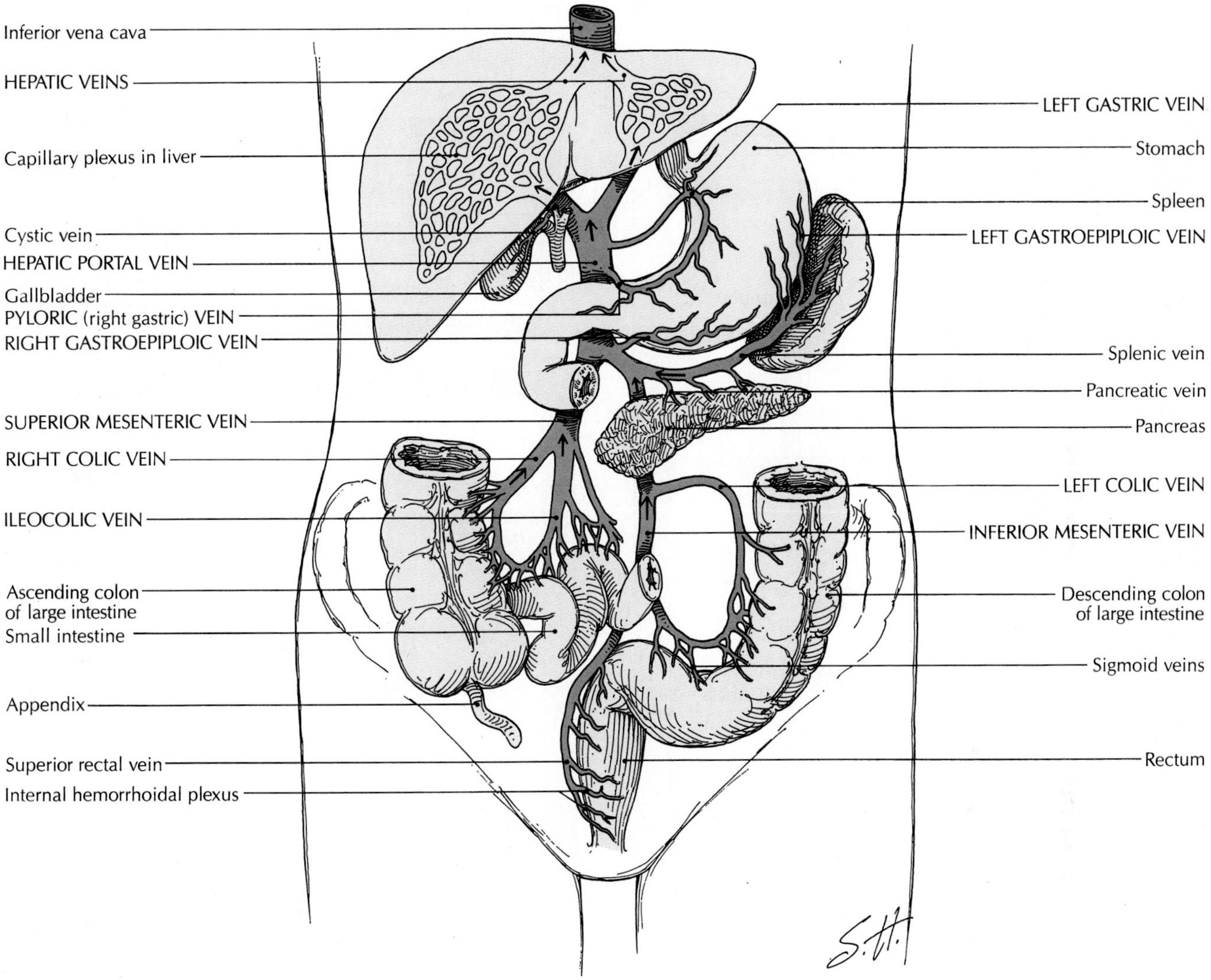
Inferior vena cava
HEPATIC VEINS
Capillary plexus in liver
Cystic vein
HEPATIC PORTAL VEIN
Gallbladder
PYLORIC (right gastric) VEIN
RIGHT GASTROEPIPLOIC VEIN
SUPERIOR MESENTERIC VEIN
RIGHT COLIC VEIN
ILEOCOLIC VEIN
Ascending colon of large intestine
Small intestine
Appendix
Superior rectal vein
Internal hemorrhoidal plexus
LEFT GASTRIC VEIN
Stomach
Spleen
LEFT GASTROEPIPLOIC VEIN
Splenic vein
Pancreatic vein
Pancreas
LEFT COLIC VEIN
INFERIOR MESENTERIC VEIN
Descending colon of large intestine
Sigmoid veins
Rectum

hold a great deal of blood. From these plexuses, heat is radiated to the surface of the skin. The amount of blood flowing through the plexuses can be controlled by either the constriction or dilation of the appropriate vessels. The diameter of the vessels is regulated by the hypothalamic temperature control center by way of sympathetic nerves to the vessels.

Skeletal Muscle Circulation

Blood nourishes skeletal muscles and also removes wastes during both physical activity and rest. Because the total body mass of skeletal muscle is so large, the blood vessels play an important role in homeostasis, especially during physical activity. The main controllers of blood flow are the sympathetic vasodilator (cholinergic) fibers that cause the blood vessels to dilate, increasing blood flow upon demand. The autoregulatory response of precapillary sphincters also dilates blood vessels in response to a decrease in the oxygen level in active muscles.

Adaptations in Fetal Circulation

The circulatory system of the fetus differs from that of a child and adult for two main reasons: (1) the fetus gets oxygen and nutrients and eliminates carbon dioxide and waste products from the mother's blood. (2) The fetal lungs, kidneys, and digestive system (except for the liver) are not functional. At birth, however, the baby must make several rapid physiological and anatomical adjustments as it shifts to an essentially adult circulation.

FIGURE 19.11

Fetal circulation. (A) Before birth, the foramen ovale permits the passage of deoxygenated blood from the right atrium to the left atrium in the fetal heart; the lungs are essentially bypassed. Also, the ductus arteriosus assists in bypassing the lungs by carrying blood from the pulmonary artery (from the right ventricle) directly into the aorta. Note that much of the blood bypasses the liver through the ductus venosus. (B) After birth, the ductus arteriosus is converted into the ligamentum arteriosum, and an adult-type circulation is in effect. The foramen ovale closes, the ductus venosus becomes the ligamentum venosum, the left umbilical vein becomes the round ligament, and the umbilical arteries become the lateral umbilical ligaments.

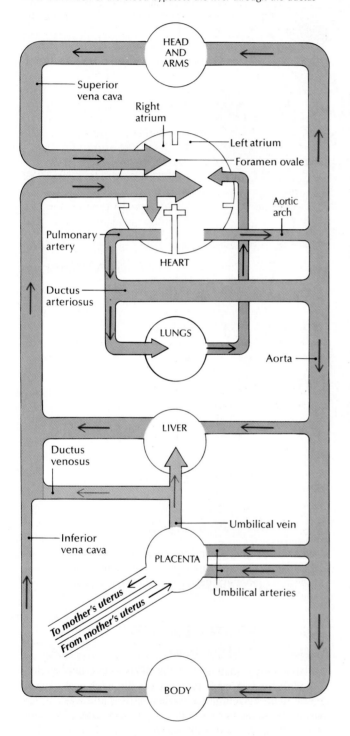

[A]

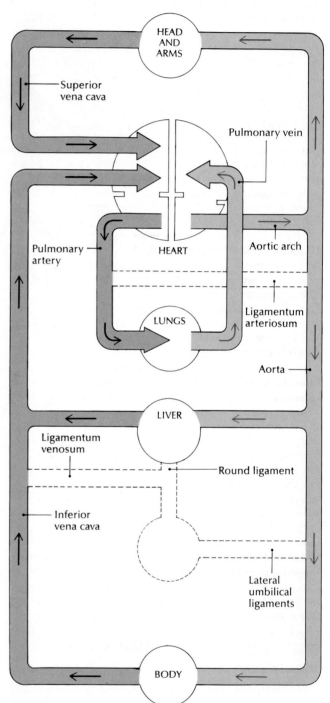

[B]

Fetal circulation The *placenta,* a thick bed of tissues and blood vessels embedded in the wall of the mother's uterus, provides an indirect connection between the mother and the fetus. It contains arteries and veins of both the mother and the fetus. These vessels intermingle, but do not join. As a result of this close intermingling, the fetus "breathes," obtains nutrients, and eliminates wastes through its mother's blood instead of its own organs. The **umbilical cord** connects the placenta to the fetus. It contains a single *umbilical vein,* which carries oxygenated, nutrient-rich blood from the placenta to the fetus (Figure 19.11A). Coiled around the umbilical cord are two *umbilical arteries,* which carry deoxygenated blood and waste material from the fetus to the placenta (see Figure 27.7).

The umbilical vein, carrying fully oxygenated blood, enters the abdomen of the fetus, where it branches. One branch joins the fetal portal vein, which goes to the liver, while the other branch, called the **ductus venosus,** joins the inferior vena cava and bypasses the liver (see Figure 19.11A). Before the oxygenated blood that enters the vena cava actually enters the heart, it becomes mixed with deoxygenated blood being returned from the lower fetal extremities.

The fetal inferior vena cava empties into the right atrium. A large opening called the **foramen ovale** ("oval window") connects the two atria. Most of the blood passes through the foramen ovale into the left atrium, and down into the left ventricle. From there it is pumped into the aorta for distribution throughout the body of the fetus.

Blood from the fetal head and upper extremities enters the right atrium via the superior vena cava, where it is deflected mainly into the right ventricle. The blood then leaves the right ventricle through the pulmonary artery. However, since the fetal lungs are still collapsed and nonfunctional, most of the blood passes into a shunt in the aortic arch called the **ductus arteriosus,** where it mixes with blood coming from the left ventricle. By connecting the pulmonary artery to the aorta, the ductus arteriosus allows fetal blood to bypass the lungs, and to flow from the aorta to the umbilical arteries to the placenta. Some blood, however, does go from the pulmonary arteries to the lungs, not only to supply lung cells with oxygen and other nutrients, but to ensure a circulation that will accept the blood flow when the newborn takes its first breaths. The blood is drained by the pulmonary veins.

Because of the mixing of oxygenated and deoxygenated blood, the fetal blood has a lower oxygen content than adult blood does.* The blood in the pulmonary artery is low in oxygen and nutrients, and high in waste products. It flows through the ductus arteriosus into the aorta, down toward the lower half of the body, and eventually to the two umbilical arteries. The umbilical arteries terminate in the blood vessels of the placenta, where carbon dioxide is exchanged for oxygen, and waste products are exchanged for nutrients.

---

*To help compensate for this deficiency, *fetal hemoglobin* can combine with oxygen more easily than adult hemoglobin can. This special property of fetal hemoglobin is lost within a few days after birth.

Circulatory system of the newborn At birth, several important changes must take place in the cardiovascular system of the newborn infant (Figure 19.11B). When an infant takes its first breath, the collapsed lungs inflate, and the pulmonary vessels become functional. As a result, the pressure on the right side of the heart is reduced, the foramen ovale closes, and the heart begins to function like a double pump. It takes several months for the foramen ovale to close completely. Eventually, only a depression called the *fossa ovalis* remains at the former site of the foramen ovale (see Figure 18.8). At the same time that the foramen ovale is closing, the ductus arteriosus is gradually constricting. In about six weeks it becomes the *ligamentum arteriosum.*

During the first few days after birth, the stump of the umbilical cord dries up and drops off. Inside the body, the umbilical vessels atrophy. The umbilical vein becomes the *round ligament* of the liver, the ductus venosus forms the *ligamentum venosum* (a fibrous cord embedded in the wall of the liver), and the umbilical arteries become the *lateral umbilical ligaments.*

If either the foramen ovale or ductus arteriosus does not close off, the oxygen content of the blood will be low, and the baby's skin will appear slightly blue. A newborn baby with this condition is therefore called a "blue baby." Most often, a defective ductus arteriosus must be corrected surgically, usually a few days after birth (see p. 611).

Ask Yourself

1 *How does the systemic circulation differ from the pulmonary circulation?*

2 *What is the hepatic portal system? The hypophyseal portal system?*

3 *What are some special areas of circulation?*

4 *What are the two main differences between fetal circulation and a young child's circulation?*

5 *What is the ductus arteriosus?*

6 *What causes the condition that produces a "blue baby"?*

MAJOR ARTERIES AND VEINS

Two arterial trunks emerge from the heart (see Figure 19.8). The **pulmonary trunk,** arising from the right ventricle, supplies blood to the respiratory portions of the lungs. The **aorta,** emerging from the left ventricle, supplies blood to the rest of the body. Figure 19.12A gives an overall view of the major arteries of the body. The main segments of the aorta are summarized in Table 19.2 and illustrated in Figures 19.12 and 19.13. The major arteries of specific parts of the body are presented in Tables 19.3 through 19.5 and in Figures 19.13 through 19.16.

FIGURE 19.12

Major arteries and veins, anterior view. (A) The aorta and its principal branches. (B) Principal veins.

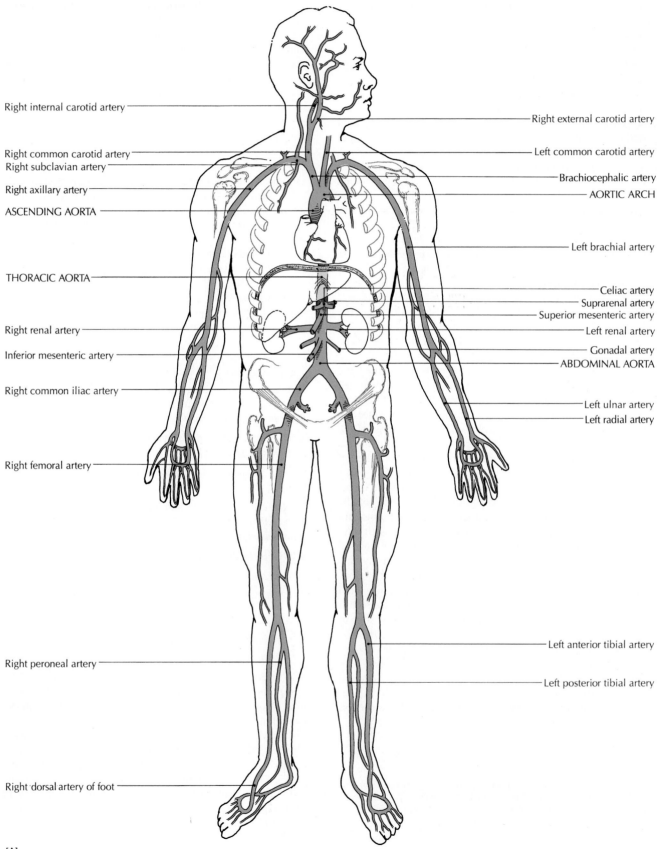

Right internal carotid artery

Right common carotid artery
Right subclavian artery

Right axillary artery

ASCENDING AORTA

THORACIC AORTA

Right renal artery

Inferior mesenteric artery

Right common iliac artery

Right femoral artery

Right peroneal artery

Right dorsal artery of foot

Right external carotid artery

Left common carotid artery

Brachiocephalic artery
AORTIC ARCH

Left brachial artery

Celiac artery
Suprarenal artery
Superior mesenteric artery
Left renal artery
Gonadal artery
ABDOMINAL AORTA

Left ulnar artery
Left radial artery

Left anterior tibial artery

Left posterior tibial artery

[A]

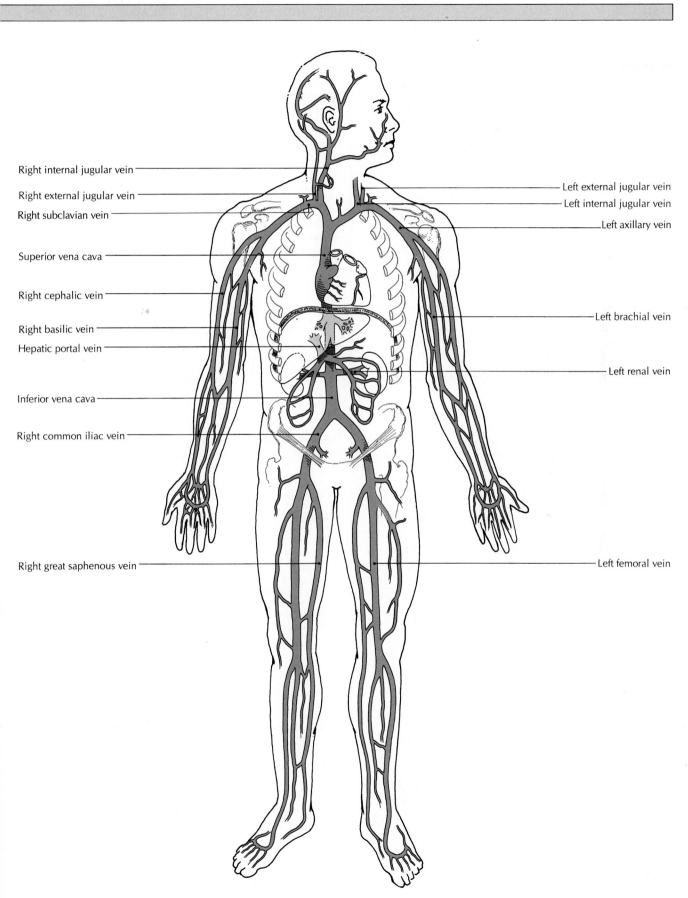

Right internal jugular vein

Right external jugular vein

Right subclavian vein

Superior vena cava

Right cephalic vein

Right basilic vein

Hepatic portal vein

Inferior vena cava

Right common iliac vein

Right great saphenous vein

Left external jugular vein

Left internal jugular vein

Left axillary vein

Left brachial vein

Left renal vein

Left femoral vein

[B]

TABLE 19.2   MAJOR ARTERIAL BRANCHES OF THE AORTA (FIGURE 19.13)

Artery	Major branches	Region supplied
ASCENDING AORTA		
Right coronary artery	Marginal and posterior interventricular branches.	Nutrient arteries to heart wall (myocardium, endocardium, epicardium).
Left coronary artery	Circumflex and anterior interventricular branches.	Nutrient arteries to heart wall (myocardium, endocardium, epicardium).
AORTIC ARCH		
Brachiocephalic trunk (innominate artery)	Right common carotid artery.*	Head, neck.
	Right subclavian artery.†	Right upper limb.
Left common carotid artery*		Head, neck.
Left subclavian artery†		Left upper limb.
THORACIC AORTA (thoracic portion of descending aorta)		
Bronchial arteries (unpaired)		Bronchi, bronchioles (not structures of lungs that oxygenate blood).
Esophageal arteries (unpaired)		Esophagus.
Intercostal arteries (9 pairs)	Dorsal branch.	Spinal cord, muscles and skin of back.
Posterior intercostal arteries (1 pair below rib 12)	Muscular branch.	Chest muscles.
	Cutaneous branch.	Skin of thorax.
	Mammary branch.	Breasts.
Superior phrenic arteries		Diaphragm.

Artery	Major branches	Region supplied
ABDOMINAL AORTA (abdominal portion of descending aorta)		
Inferior phrenic arteries		Diaphragm, lower esophagus.
Celiac artery (trunk) (unpaired)	Left gastric artery.	Stomach, esophagus.
	Common hepatic artery.	Liver, stomach, pancreas, duodenum, gall-bladder, bile duct.
	Splenic artery.	Stomach, pancreas, spleen, omentum.
Superior mesenteric artery (unpaired)	Inferior pancreaticoduo-denal artery.	Head of pancreas, duodenum.
	Jejuneal and ileal arteries.	Small intestine.
	Ileocolic artery.	Lower ileum, appendix, ascending and part of transverse colon.
	Right colic and middle colic arteries.	Transverse and ascending colon.
Inferior mesenteric artery (unpaired)	Left colic artery.	Transverse and descending colon.
	Sigmoid artery.	Descending and sigmoid colon.
	Superior rectal artery.	Rectum, anal region.
Renal arteries	Inferior suprarenal abdominal artery.	Kidneys, ureters, suprarenal (adrenal) gland.
Gonadal (testicular or ovarian) arteries		Testes or ovaries.
Lumbar arteries (4 or 5 pairs)		Skin, muscle, and vertebrae of lumbar region of back, spinal cord and meninges of lower back, caudal equina.
Median sacral (middle) artery		Sacrum, coccyx, rectum.
Common iliac arteries	External iliac arteries.	Lower limb.
	Internal iliac arteries (hypogastric artery).	Viscera, walls of pelvis, perineum, gluteal region.

*Right and left common carotid arteries ascend into the neck and then bifurcate at the upper part of the larynx into right and left external and internal carotid arteries. See Table 19.3 for branches.

†Branches of the subclavian arteries are listed in Table 19.4.

FIGURE 19.13

Major arterial branches of the aorta.

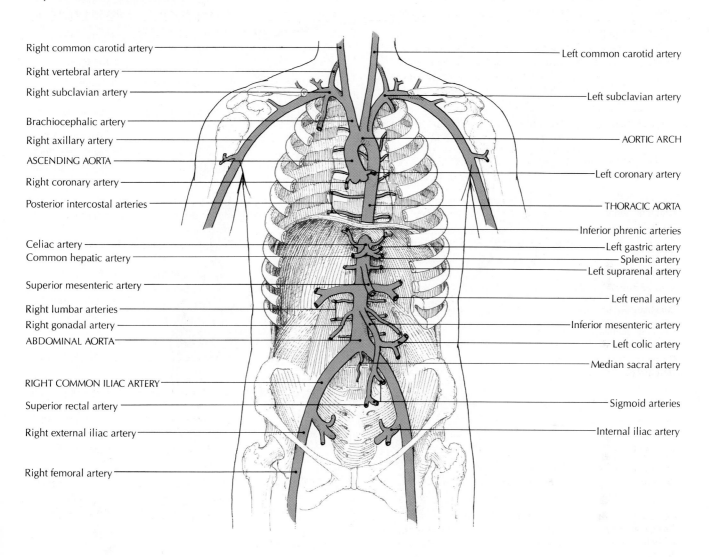

Right common carotid artery — Left common carotid artery

Right vertebral artery —

Right subclavian artery — Left subclavian artery

Brachiocephalic artery —

Right axillary artery — AORTIC ARCH

ASCENDING AORTA —

Right coronary artery — Left coronary artery

Posterior intercostal arteries — THORACIC AORTA

Inferior phrenic arteries

Celiac artery — Left gastric artery

Common hepatic artery — Splenic artery

Left suprarenal artery

Superior mesenteric artery — Left renal artery

Right lumbar arteries —

Right gonadal artery — Inferior mesenteric artery

ABDOMINAL AORTA — Left colic artery

Median sacral artery

RIGHT COMMON ILIAC ARTERY —

Superior rectal artery — Sigmoid arteries

Right external iliac artery — Internal iliac artery

Right femoral artery —

FIGURE 19.14

Major arteries of the head and neck. (A) Arteries of the head and neck, lateral view. (B) Ventral surface of the brain, showing the cerebral arterial circle (circle of Willis).

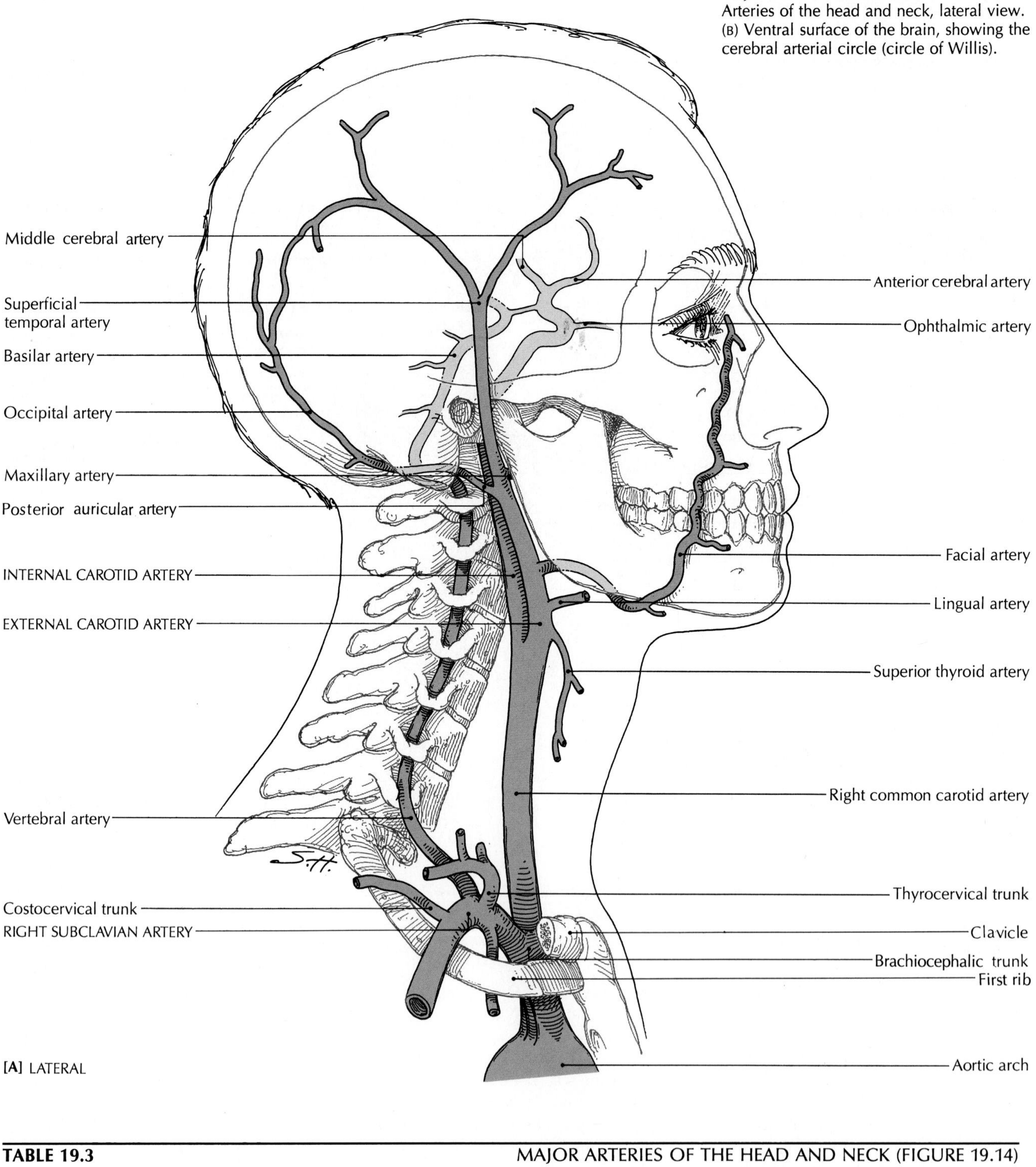

Middle cerebral artery

Superficial temporal artery

Basilar artery

Occipital artery

Maxillary artery

Posterior auricular artery

INTERNAL CAROTID ARTERY

EXTERNAL CAROTID ARTERY

Vertebral artery

Costocervical trunk

RIGHT SUBCLAVIAN ARTERY

Anterior cerebral artery

Ophthalmic artery

Facial artery

Lingual artery

Superior thyroid artery

Right common carotid artery

Thyrocervical trunk

Clavicle

Brachiocephalic trunk

First rib

Aortic arch

[A] LATERAL

TABLE 19.3   MAJOR ARTERIES OF THE HEAD AND NECK (FIGURE 19.14)

Artery	Major branches	Region supplied	Comments
Internal carotid artery	Anterior cerebral artery.	Brain.	Internal carotid artery has no branches in neck. It passes through carotid canal into cranial cavity. After giving off ophthalmic artery, it terminates by dividing into anterior and middle cerebral arteries.
	Middle cerebral artery.	Brain.	

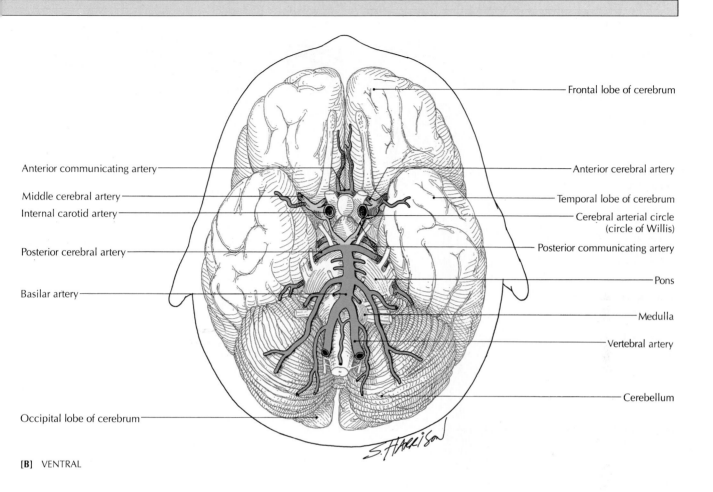

Frontal lobe of cerebrum

Anterior communicating artery

Middle cerebral artery

Internal carotid artery

Posterior cerebral artery

Basilar artery

Occipital lobe of cerebrum

Anterior cerebral artery

Temporal lobe of cerebrum

Cerebral arterial circle (circle of Willis)

Posterior communicating artery

Pons

Medulla

Vertebral artery

Cerebellum

[B] VENTRAL

Artery	Major branches	Region supplied	Comments
Subclavian artery	Vertebral artery.	Brain, spinal cord (cervical region), vertebrae, other deep neck structures.	Vertebral artery terminates as posterior cerebral artery.
	Thyrocervical trunk.	Trachea, esophagus, thyroid gland, neck muscles.	
	Costocervical trunk.	Muscles of back of neck, vertebrae of neck.	
External carotid artery	Superior thyroid artery.	Thyroid gland, neighboring muscles, larynx.	External carotid artery terminates in region of parotid gland by bifurcating into superficial temporal artery and maxillary artery.
	Lingual artery.	Floor of mouth, tongue, neighboring muscles (including oropharynx), salivary glands.	
	Facial artery.	Muscles and tissues of face below eyes, including soft palate, tonsils, and submandibular gland.	
	Occipital artery.	Skin, muscles, and associated tissues of back of head (scalp), brain meninges.	
	Posterior auricular artery.	Skin, muscles, and associated tissues of and near ear and posterior scalp.	
	Superficial temporal artery.	Outer ear, parotid gland, forehead, scalp and muscles in temporal region.	
	Maxillary artery.	Face, including upper and lower jaws, teeth, palate, nose, muscles of mastication, and brain dura mater.	

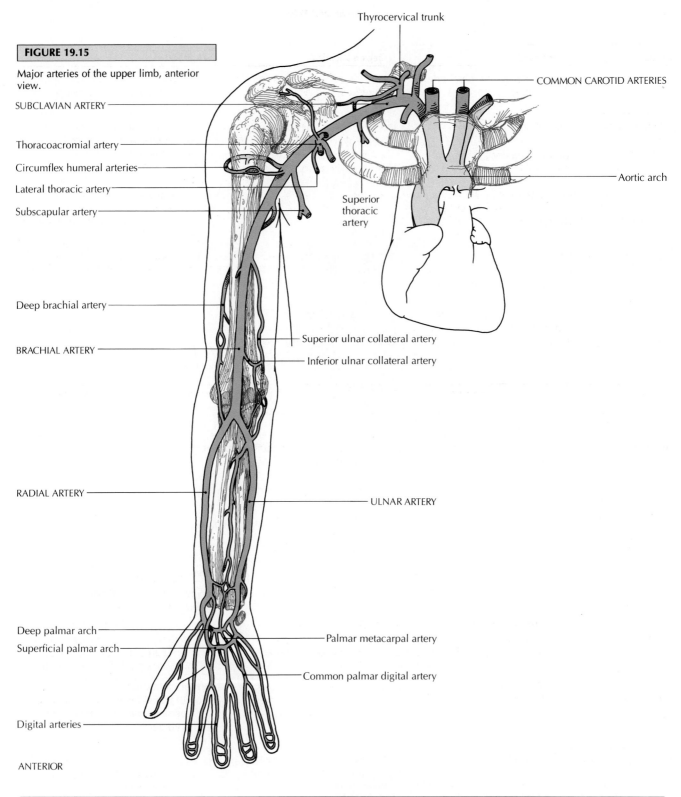

FIGURE 19.15

Major arteries of the upper limb, anterior view.

Thyrocervical trunk

SUBCLAVIAN ARTERY

COMMON CAROTID ARTERIES

Thoracoacromial artery

Circumflex humeral arteries

Lateral thoracic artery

Subscapular artery

Aortic arch

Superior thoracic artery

Deep brachial artery

Superior ulnar collateral artery

BRACHIAL ARTERY

Inferior ulnar collateral artery

RADIAL ARTERY

ULNAR ARTERY

Deep palmar arch

Superficial palmar arch

Palmar metacarpal artery

Common palmar digital artery

Digital arteries

ANTERIOR

TABLE 19.4   MAJOR ARTERIES OF THE THORAX AND UPPER LIMB (FIGURE 19.15)

Artery	Major branches	Region supplied	Comments
Axillary artery	Superior thoracic artery. Thoracoacromial artery. Lateral thoracic artery. Subscapular artery. Circumflex humeral artery.	Chest wall, muscles of chest wall, shoulder girdle, shoulder joint.	

Artery	Major branches	Region supplied	Comments
Brachial artery	Deep (profunda) brachial artery. Ulnar collateral arteries.	Muscles of arm, upper forearm, humerus, and elbow joint.	Brachial artery in anterior elbow is site used for measurement of blood pressure.
Radial artery	Branches in forearm. Continues as deep palmar arch and palmar branches. Arch (along with superior palmar arch) provides palmar membranes, which continue as digital arteries.	Elbow joint, radius, muscles of radial (thumb) side of forearm. Carpal and metacarpal bones and muscles in palm. Fingers.	Pulse of radial artery can be felt on radial side of wrist at base of thumb and on back of wrist between tendons of thumb muscles. Anastomoses among arteries of lower forearm and those of branches of arches of palm are substantial.
Ulnar artery	Branches in forearm. Continues as superficial palmar arch and palmar branches. Arch (along with deep palmar arch) provides palmar and digital branches.	Elbow joint, ulna, muscles of ulnar (small-finger) side of forearm. Carpal and metacarpal bones and muscles in palm. Structures of palm and fingers.	Pulse of ulnar artery can be felt on the ulnar side of wrist.

TABLE 19.5 MAJOR ARTERIES OF THE PELVIS AND LOWER LIMB (FIGURE 19.16)

Artery	Major branches	Region supplied	Comments
Common iliac artery	Internal iliac artery (hypogastric artery).	Pelvic viscera, including urinary bladder, prostate gland, ductus deferens, uterus, vagina, rectum. Peroneal region, including anal canal, external genitalia. Walls of pelvic cavity, including muscles. Gluteal region (buttocks), including muscles, hip joint.	
	External iliac artery.	Lower limb.	External iliac artery enters thigh behind inguinal ligament and becomes femoral artery.
Femoral artery	Muscular branches, including deep (profunda) femoral artery.	Skin and muscles of thigh, including anterior flexors, medial adductors, extensors (hamstrings).	Femoral artery, located on anterior thigh, passes backward through adductor magnus to become popliteal artery on back of knee (popliteal fossa).
Popliteal artery	Muscular branches.	Knee joint, gastrocnemius and soleus muscles.	Popliteal artery terminates just beyond knee joint by dividing into anterior tibial artery and posterior tibial artery.
Anterior tibial artery	Muscular branches.	Knee joint, muscles, and skin in front of leg, ankle joint.	Anterior tibial artery continues over ankle to become pedal artery of foot.
Dorsal pedal artery		Muscles, skin, and joints (including ankle) on dorsal side of foot.	
Posterior tibial artery	Muscular branches, including peroneal artery on lateral side of leg. Peroneal, medial malleolar, calcaneal, and plantar arteries.	Knee joint, muscles and skin on back of leg, ankle joint. Fibula, muscles of fibula, ankle joint, heel, toes.	

FIGURE 19.16

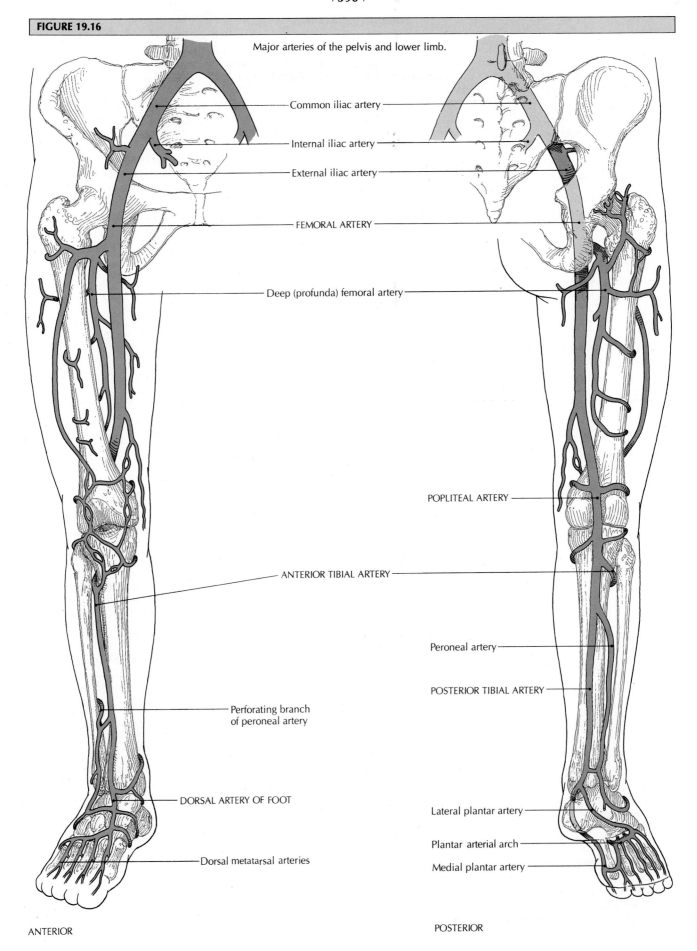

Major arteries of the pelvis and lower limb.

Common iliac artery

Internal iliac artery

External iliac artery

FEMORAL ARTERY

Deep (profunda) femoral artery

POPLITEAL ARTERY

ANTERIOR TIBIAL ARTERY

Peroneal artery

POSTERIOR TIBIAL ARTERY

Perforating branch
of peroneal artery

DORSAL ARTERY OF FOOT

Lateral plantar artery

Plantar arterial arch

Dorsal metatarsal arteries

Medial plantar artery

ANTERIOR

POSTERIOR

Four large veins drain blood from the body into the heart. The **pulmonary veins** drain blood from the lungs into the left atrium. The other three veins empty into the right atrium. The **superior vena cava** drains the head, neck, upper limbs, and thorax; the **inferior vena cava** drains the abdomen, pel- vis, and lower limbs; and the **coronary sinus** receives blood from the veins that drain the heart. Figure 19.12B is an over- all view of the major veins of the body. More detail about specific veins is given in Tables 19.6 through 19.9 and illus- trated in Figures 19.17 through 19.20.

FIGURE 19.17

Major veins of the head and neck.

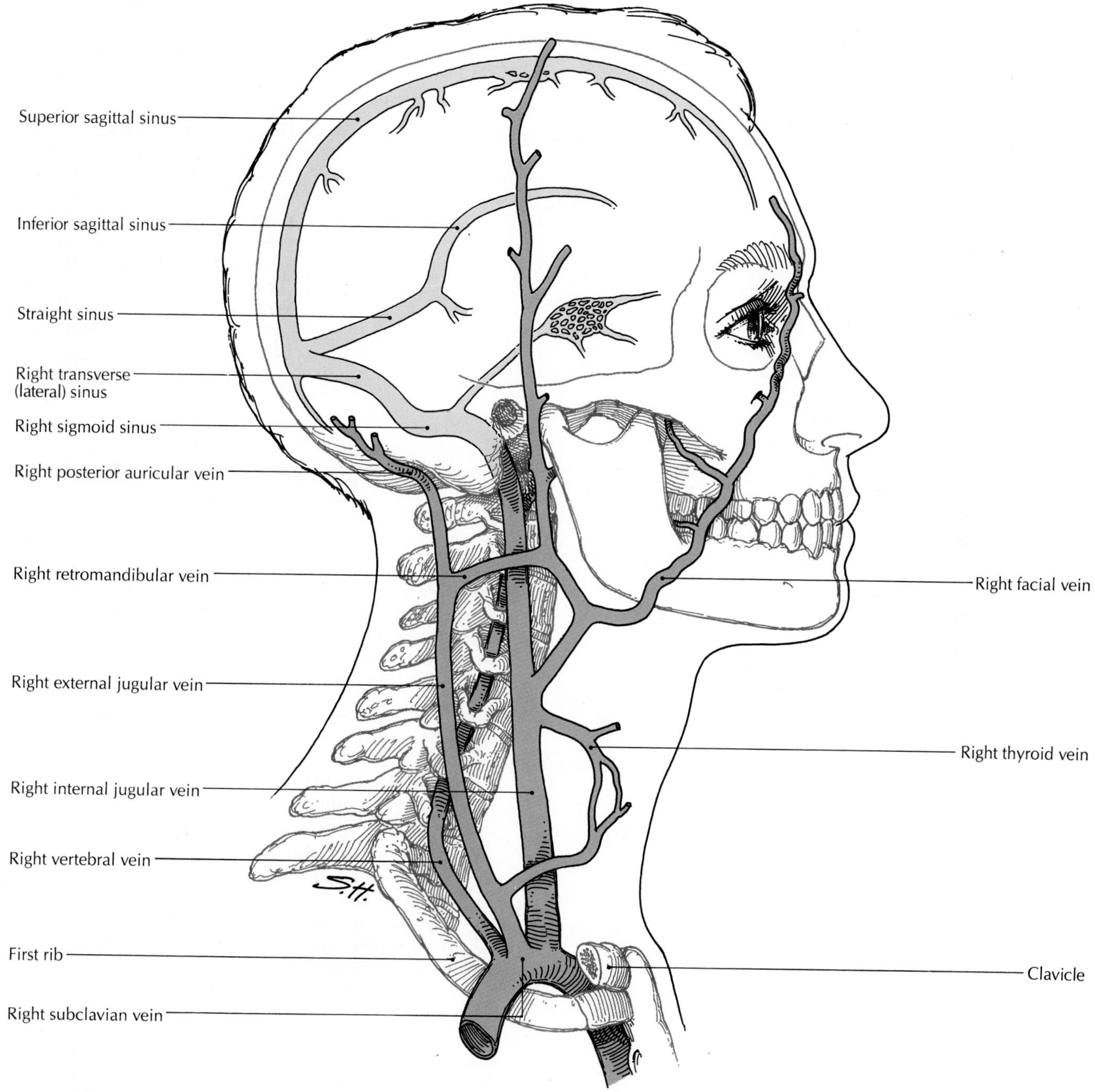

TABLE 19.6   MAJOR VEINS OF THE HEAD AND NECK (FIGURE 19.17)

Major vein	Tributary veins draining into major veins	Region drained	Comments
Internal jugular vein	Dural sinuses, including superior sagittal sinus, inferior sagittal sinus, straight sinus, transverse sinuses, sigmoid sinuses.	Brain, essentially all venous blood within cranial cavity.	Sigmoid sinuses at base of skull continue through jugular foramina and become paired internal jugular veins. Dural sinuses are venous blood vessels located between layers of dura mater.
	Facial and thyroid veins.	Face, neck.	Internal jugular veins join subclavian veins to form brachiocephalic veins, which drain into superior vena cava.
External jugular vein	Retromandibular vein. Posterior auricular vein. Other veins.	Posterior face. Scalp behind ear. Parotid glands, scalp, muscles of neck.	External jugular formed by union of retromandibular and posterior auricular veins and drains into subclavian vein. Catheter can be passed into external jugular (internal jugular or cephalic) veins and then successively through superior vena cava, right atrium, right ventricle, pulmonary artery into lung.

TABLE 19.7   MAJOR VEINS OF THE UPPER LIMB (FIGURE 19.18)

Type of vein	Tributary veins draining into major veins	Region drained	Comments
Deep veins	Brachial vein. Axillary vein. Subclavian vein.	Upper limb.	Deep veins may accompany an artery and may have transverse anastomoses *(venae comitantes)* with arteries. Axillary and subclavian veins accompany corresponding artery. Right and left subclavian veins join with internal jugular veins to form brachiocephalic veins.
Superficial veins	Dorsal venous arch.	Back of hand.	Superficial veins are cutaneous and have anastomoses with deep veins.
	Basilic vein.	Medial side of forearm and arm.	Basilic vein originates from dorsal venous arch, extends along medial side of limb, and terminates by joining brachial vein to form axillary vein.
	Cephalic vein.	Lateral side of forearm and arm.	Cephalic vein originates from dorsal venous arch, extends along lateral side of limb, and terminates by emptying into axillary vein near lateral end of clavicle.
	Medial cubital vein.	Anterior elbow.	Medial cubital vein communicates between cephalic and basilic veins. These veins form variable patterns. Cubital vein is often used when vein must be punctured for injection, transfusion, or withdrawal of blood.

FIGURE 19.18

Major veins of the upper limb.

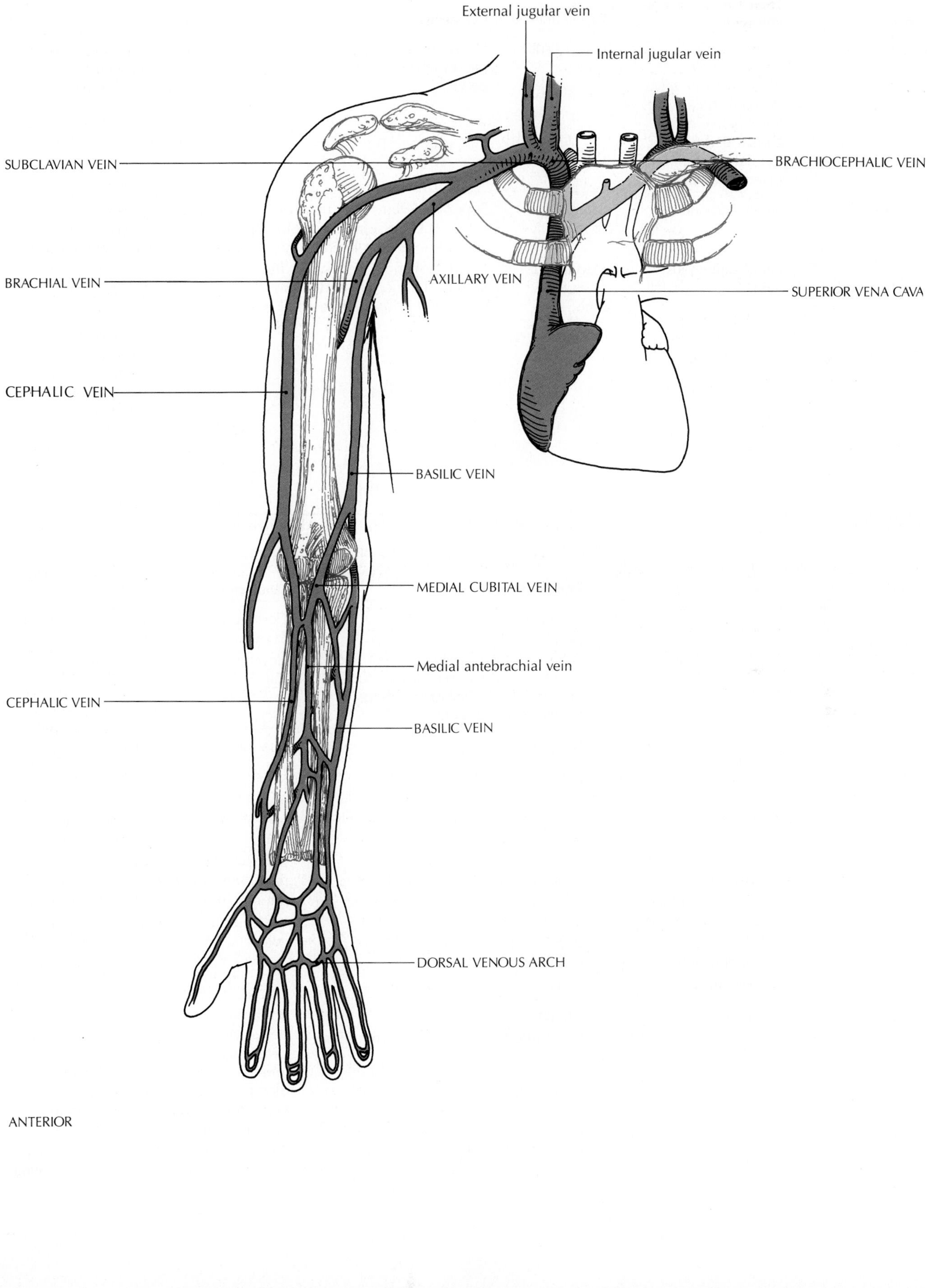

External jugular vein

Internal jugular vein

SUBCLAVIAN VEIN

BRACHIOCEPHALIC VEIN

AXILLARY VEIN

BRACHIAL VEIN

SUPERIOR VENA CAVA

CEPHALIC VEIN

BASILIC VEIN

MEDIAL CUBITAL VEIN

Medial antebrachial vein

CEPHALIC VEIN

BASILIC VEIN

DORSAL VENOUS ARCH

ANTERIOR

TABLE 19.8   MAJOR VEINS OF THE LOWER LIMB (FIGURE 19.19)

Type of vein	Tributary veins draining into major veins	Region drained	Comments
Deep veins	Femoral vein. Popliteal vein. Anterior tibial vein. Posterior tibial vein. Peroneal vein. Dorsal pedal vein. Medial plantar vein. Lateral plantar vein.	Lower limb.	Deep veins are organized as *venae comitantes*. Femoral vein passes under inguinal ligament to become external iliac vein of pelvis.
Superficial veins	Small saphenous vein.	Lateral side of leg.	Superficial veins are cutaneous and have anastomoses with deep veins. They arise at lateral side of dorsal venous arch of foot, ascend on posterolateral side of calf, and join deep popliteal vein.
	Great saphenous vein.	Medial side of lower limb.	Great saphenous vein arises from medial side of dorsal venous arch of the foot, ascends upward on medial aspect of leg and thigh, and joins deep femoral vein in groin.

TABLE 19.9   MAJOR VEINS OF THE THORAX, ABDOMEN, AND PELVIS (FIGURE 19.20)

Major vein	Tributary veins draining into major veins	Region drained	Comments
Common iliac vein	External iliac vein.	Lower limb.	Common iliac vein is continuation of femoral vein.
	Internal iliac vein.	Pelvis, buttocks, pelvic viscera, including gluteal muscles, rectum, urinary bladder, prostate, uterus, vagina, external genitalia.	Common iliac veins unite to form inferior vena cava.
Inferior vena cava (abdominal area)	Ascending lumbar veins.	Body wall.	Inferior vena cava is largest blood vessel in body (approximately 3.5 cm in diameter). It receives tributaries from lumbar, visceral, and hepatic areas, but not directly from digestive tract, pancreas, or spleen.
	Renal veins.	Kidneys.	
	Gonadal veins (ovarian and testicular veins).		Left gonadal vein drains directly into renal vein; right gonadal vein drains directly into inferior vena cava.
	Suprarenal veins.		Left suprarenal vein drains into renal vein.
	Inferior phrenic veins.	Diaphragm.	
	Hepatic vein.	Liver.	(See portal system below.)
Portal system (hepatic portal vein)	Splenic vein.	Spleen.	Portal system drains abdominal and pelvic viscera, which are supplied by celiac, superior mesenteric, and inferior mesenteric arteries. Venous blood of portal system drains into liver and its sinusoids and then into hepatic vein, which is tributary of inferior vena cava.
	Superior mesenteric vein.	Small intestine, part of large intestine.	
	Inferior mesenteric vein.	Large intestine.	

(Table 19.9 continues on page 596.)

FIGURE 19.19

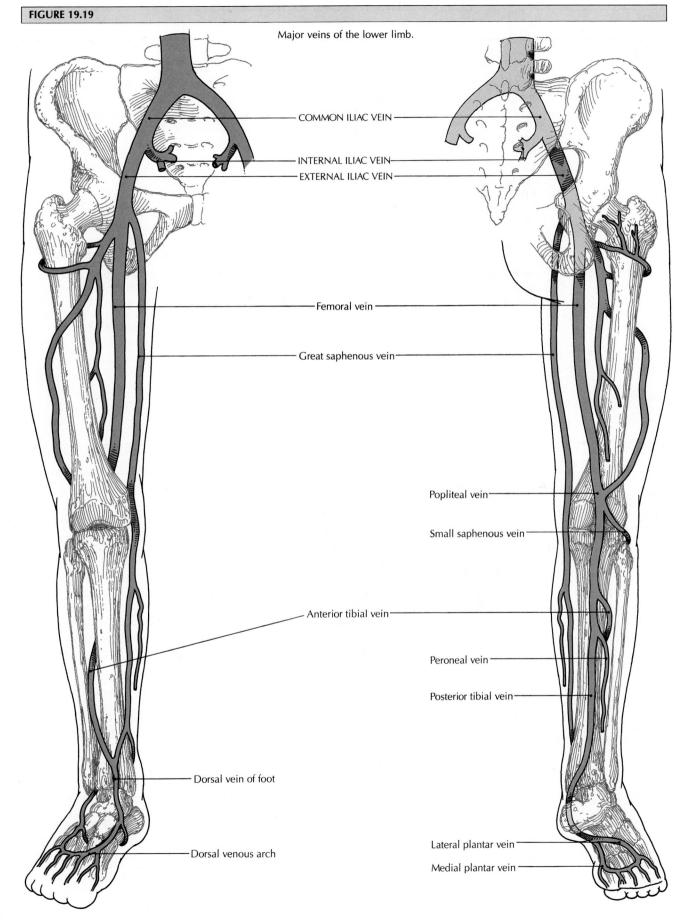

Major veins of the lower limb.

COMMON ILIAC VEIN

INTERNAL ILIAC VEIN
EXTERNAL ILIAC VEIN

Femoral vein

Great saphenous vein

Popliteal vein

Small saphenous vein

Anterior tibial vein

Peroneal vein

Posterior tibial vein

Dorsal vein of foot

Lateral plantar vein
Medial plantar vein

Dorsal venous arch

ANTERIOR

POSTERIOR

TABLE 19.9 (Continued)

Major vein	Tributary veins draining into major veins	Region drained	Comments
Azygos vein	Azygos vein.		Azygos vein drains into superior vena cava.
	Intercostal veins.	Chest wall.	
	Right ascending lumbar vein.	Abdominal body wall.	
	Right bronchial vein.	Bronchi.	
	Hemiazygos vein.		Hemiazygos vein drains into azygos vein at level of ninth thoracic vein.
	Intercostal veins (lower 4 or 5).	Chest wall.	
	Left ascending lumbar vein.	Abdominal body wall.	
	Left bronchial vein.	Bronchi.	
	Accessory hemiazygos vein.		Accessory hemiazygos vein joins azygos vein at level of eighth thoracic vein.
	Intercostal veins (upper 3 or 4).	Chest wall.	Azygos system drains blood from thoracic wall and posterior wall of abdomen. It has venous connections with veins draining inferior vena cava, including those of esophagus, mediastinum, pericardium, and bronchi. When inferior vena cava of hepatic portal vein is obstructed, venous blood can be diverted to azygos system and superior vena cava.
	Right bronchial vein.	Bronchi.	

FIGURE 19.20

Major veins of the thorax, abdomen, and pelvis.

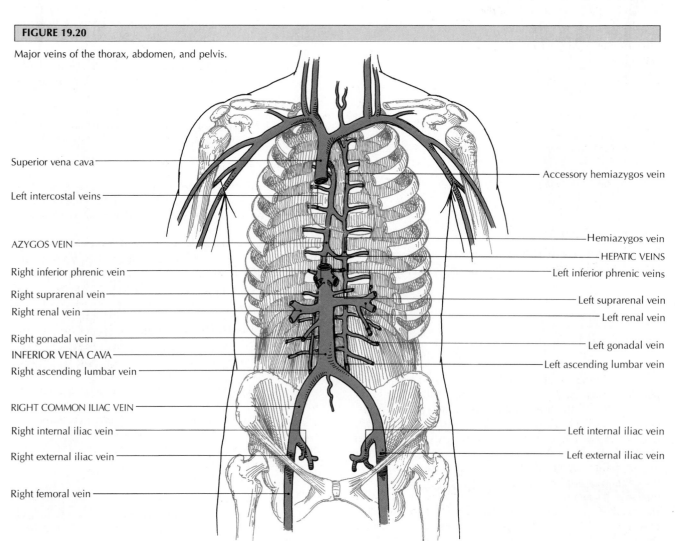

How Blood Vessels Are Named

The majority of blood vessels are named according to the major organ or anatomical site supplied (by arteries) or drained (by veins), or according to their location in the body. For example, the *renal artery* supplies the kidneys, and the *renal vein* drains the kidneys. The depth of the vessel is also used. For example, the *internal jugular vein* lies deeper in the neck than the *external jugular vein* does. In some instances, the name of a blood vessel changes as it passes into a different part of the body. For example, the *subclavian artery* runs underneath the clavicle in the neck region. When it enters the region of the armpit, its name changes to the *axillary artery,* and it becomes the *brachial artery* when it enters the arm. In general, veins run parallel to most arteries, and are given the same names, although there are exceptions.

Ask Yourself

1 *What are the major segments of the aorta?*

2 *Which arteries supply blood to the head and neck?*

3 *What region is supplied by the axillary artery?*

4 *What are the major arteries of the lower limb?*

5 *What are the major veins of the head and neck?*

6 *Which veins of the upper limb are deep veins? Superficial veins?*

7 *What region is drained by the azygos vein and its tributaries?*

8 *Can you describe some of the ways that blood vessels are named?*

PHYSIOLOGY OF CIRCULATION

The overall function of the cardiovascular system is to transport life-sustaining blood from one part of the body to another. Although the heart and all the blood vessels play important roles in circulating blood, the capillaries are the focal point of the system for nutrient and waste exchange. In fact, all of the physiological mechanisms that help regulate cardiac output and the diameter of blood vessels function to supply the capillaries with the precise amounts of blood that will maintain homeostasis in the body's cells. These cardiovascular activities are integrated to maintain an optimum blood pressure and blood flow through the capillaries. We examine these functions and properties in the rest of this chapter.

Principles of Blood Flow

Hemodynamics ("blood power") is the study of the physical principles that govern *blood flow* through the blood vessels and heart. Blood is forced out of the heart and through the blood vessels under great *pressure*. Blood flows through the narrow arterioles, capillaries, and venules with much difficulty. In other words, these vessels offer *resistance* to the flow of blood. The physiological regulation of blood flow, pressure, and resistance in maintaining homeostasis is based on some simple laws of hemodynamics.

Blood flow The term **blood flow** refers to the volume (quantity) of blood flowing through a vessel or group of vessels during a specific period of time. It is usually measured in milliliters per minute. Because arteries near the heart are elastic, blood flow is *pulsatile* (it does not flow at a constant rate) and *turbulent.*[*] Blood flow in the capillaries, venules, and veins is less pulsatile and turbulent than in the arteries.

Blood pressure *Blood pressure* is the force (energy) with which blood is pushed against the walls of blood vessels and circulated throughout the body when the heart contracts (Figure 19.21A). It is measured in millimeters of mercury. The rhythmic expansion and contraction of the vessels creates a pulsating pressure wave. Blood flows along a *pressure gradient,* from one end of a vessel, where a certain pressure (P_1) exists, to the other end, where the pressure (P_2) is somewhat less. It is not the blood pressure at any point in the blood vessel that determines blood flow, but the *difference* between any two points ($P_1 - P_2$). Blood pressure equals the blood flow multiplied by the resistance of the vessel:

Blood pressure (BP) = blood flow (F) × resistance (R)

From this equation it can be seen that as blood pressure ($P_1 - P_2$) increases, blood flow may also increase; conversely, as resistance increases, blood flow may decrease.

Blood velocity The distance blood flows along a vessel during a specific time period is its *velocity,* measured in centimeters per second (Figure 19.21B). The velocity of blood in the aorta is approximately 30 cm/sec, in arterioles 1.5 cm/sec, in capillaries 0.04 cm/sec, in venules 0.5 cm/sec, and in the venae cavae 8 cm/sec.

Generally, as the diameter of a vessel gets smaller and the total cross-sectional area increases as the vessels branch out like the limbs of a tree, the blood velocity decreases. In contrast, when blood returns to the heart and the diameter of a vessel increases and the total cross-sectional area decreases, the blood velocity increases. Furthermore, as the major arteries branch from the aorta, they are feeding the same volume of blood into more and more vessels. The cross-sectional area of all these vessels is much higher than that of the aorta. As a result, velocity decreases.

---

[*]The sounds produced by the whirlpool-like motion of blood flow are used to determine arterial blood pressure, and to detect heart murmurs (see Chapter 18).

FIGURE 19.21

Blood pressure and blood velocity. (A) Blood pressure and relative cross-sectional area of vessels ranging from the aorta (largest artery) to the tiny capillaries to the venae cavae (largest veins). Note that although the capillaries are the smallest vessels, as a group they have the greatest total cross-sectional area. (B) Blood velocity through various vessels.

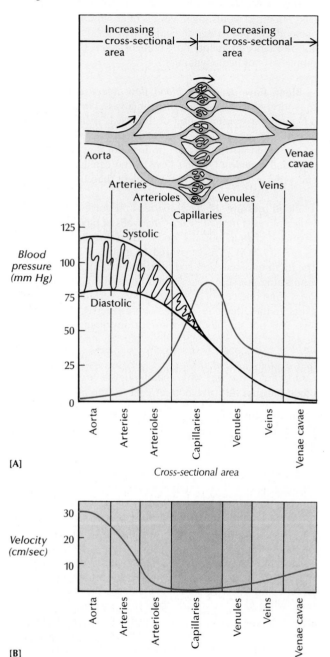

[A]

Blood pressure (mm Hg)

Cross-sectional area

Velocity (cm/sec)

[B]

The total cross-sectional area of the different segments of the circulation varies greatly (see Figure 19.21). The mean blood velocity of blood flow is inversely proportional to the cross-sectional area of any particular segment, provided that the total volume of blood flowing through each segment is constant.

Because capillaries are the most numerous type of blood vessels, they have the greatest total cross-sectional area. As a result, blood flow is slowest in capillaries. However, the velocity of blood flow through the entire capillary bed is very high, since the cross-sectional area represents the sum of that enormous number of individual capillaries. Blood remains in each capillary about 1 to 2 seconds. This slow blood flow allows just enough time for the exchange of gases, nutrients, and waste products between the blood and extracellular tissue fluid.

Peripheral resistance   As circulating blood cells touch vessel walls, and other blood cells, their flow is impeded by friction. The friction offered by the entire system of blood vessels is called *total peripheral resistance.* More specifically, the resistance offered by the pulmonary system is called the *total pulmonary resistance.* Resistance is expressed in CGS units (centimeters, grams, seconds). The amount of resistance depends upon the *length, radius,* and *total cross-sectional area* of the blood vessels, and the *viscosity* ("thickness" or "stickiness") of the blood. Longer vessels, smaller radii, larger cross-sectional area, and higher viscosity all produce greater resistance. The most important factors governing blood viscosity are the concentration of red blood cells (hematocrit) and the concentration of blood proteins. Blood viscosity increases with an increase in either hematocrit or protein concentration.

Overall, as total peripheral resistance, and thus friction, increases, the energy needed to propel blood forward is dissipated as heat. As this energy decreases, so does the blood pressure (see Figure 19.21). There is also a progressive decrease in blood velocity (Figure 19.21B) until blood reaches the capillaries, and then it increases slightly as blood returns to the heart.

Interrelationships   We have shown that the blood flow (F) is caused by blood pressure difference (BP, or $P_1 - P_2$) and is opposed by resistance to flow (R). This relationship can be expressed as:

$$\text{Blood flow (F)} = \frac{\text{pressure difference or driving force (BP)}}{\text{resistance to flow (R)}}$$

Recall that blood pressure within a vessel is actually the difference between causing (P_1) and opposing (P_2) pressures ($P_1 - P_2$), which can be abbreviated as ΔP, with the Greek delta (Δ) meaning "change in." With this in mind, we can write the equation above as:

$$F = \frac{\Delta P}{R}$$

Resistance to blood flow can be expressed as a ratio of the factors that govern blood flow:

$$R = \frac{\text{vessel length} \times \text{viscosity}}{\text{cross-sectional area}}$$

Blood flow in the arteries is pulsatile and turbulent. Each volume of blood near the heart is accelerated during systole, and slows down during diastole. In addition, within each volume, the blood cells rub against each other, creating friction. The force applied to overcome these two resistive components produces a sawtoothed pressure wave (see Figure 19.21A). This pulsatile flow continues into the large arteries until it smooths out and becomes nonpulsatile in the small arteries and veins.

The interrelationships among the pressure difference, vessel radius and length, and viscosity can be applied (with the exceptions noted above) to all blood vessels and the cardiac output of the heart. Because blood flow is equated to cardiac output, cardiac output can be expressed as

$$\text{Cardiac output} = \frac{P_1 - P_2}{\text{total peripheral resistance}}$$

In this equation, P_1 equals the pressure generated during the ejection from the left ventricle (systole) in the aorta, and P_2 equals the pressure in the right atrium during diastole. Thus, the total drop in pressure through the entire systemic circulation is equal to $P_1 - P_2$. Because the pressure in the right atrium during diastole is approximately 0 mm Hg, P_1, which represents the mean (the middle point between systolic and diastolic) aortic pressure, becomes the significant driving force at the beginning of the systemic circulation. The equation can finally be written as:

$$\text{Cardiac output} = \frac{\text{mean aortic pressure}}{\text{total peripheral resistance}}$$

We have already shown that the resistance in the major arteries is very low because of their large radii and large cross-sectional areas. Thus, because there is very little resistance, there is also very little pressure change in the arteries (see Figure 19.21A). As a result, the mean aortic pressure will approximately equal the mean arterial pressure. Thus, the cardiac output shows the following relationships:

$$\text{Cardiac output} = \frac{\text{mean arterial pressure}}{\text{total peripheral resistance}}$$

or

Mean arterial pressure = cardiac output

$$\times \text{ total peripheral resistance}$$

As you will see in the following sections, the body has various mechanisms for regulating vessel diameter (and total peripheral resistance) and cardiac output, which in turn help to maintain a homeostatic mean arterial pressure and a steady blood flow to the tissues.

How much thicker than water is blood? *Normal blood is about three times more difficult to move than water is.*

Regulation of Blood Flow

The most important factor in determining resistance to blood flow is the size of the lumen (radius or diameter) of a blood vessel. This opening is regulated primarily by nervous and hormonal mechanisms that affect the smooth muscle in the walls of arterioles and precapillary sphincters. Of lesser influence is the blood pressure within the vessel. By either constricting or dilating blood vessels, and thereby controlling the resistance to the flow of blood, blood flow is adjusted and homeostasis is maintained within the capillaries, and thus, within the body's cells.

A small change in the radius of a vessel produces a large change in the blood flow. For example, if the radius of a vessel is increased just from 1 mm to 2 mm, and all other values remain constant, there is approximately a 16-fold increase in blood flow.

The processes of constricting *(vasoconstriction)* and dilating *(vasodilation)* are collectively called **vasomotion**. Vasomotion affects *resistance* in arteries, and either the *capacitance* (unit volume a vessel will accommodate per unit change in pressure) or *passive stretching* of veins. Intense vasoconstriction of arterioles can completely stop the flow of blood to a capillary bed, whereas complete dilation can cause more than a 20-fold increase in the rate of blood flow through the same bed.

Nervous control Sympathetic stimuli cause the blood vessels (specifically, arterioles, metarterioles, and venules) of the skin and abdominal organs to constrict, producing a decrease in blood flow to those areas. In contrast, sympathetic stimuli cause blood vessels in the heart, brain, and skeletal muscles to dilate, increasing blood flow. Parasympathetic stimuli cause the blood vessels of the reproductive organs and digestive tract to dilate by releasing acetylcholine, which inhibits smooth muscle contraction.

As presented in Chapter 14, most postganglionic sympathetic nerve fibers release norepinephrine, which reacts with alpha-receptors in the blood vessel walls, causing constriction. Since the blood vessels of the heart and brain do not have any alpha-receptors, norepinephrine has no effect there. However, these vessels do have beta-receptors, which respond to epinephrine released from the adrenal medulla. The adrenal medulla is often stimulated by the sympathetic system. Epinephrine causes blood vessels in the heart and brain to dilate, thus ensuring that they receive an adequate blood flow during stressful situations.

Effect of Exercise on the Cardiovascular System

The human body works best when it is physically active. Exercise is especially beneficial to the cardiovascular system. To obtain a maximum and permanent benefit from exercise, however, it must be done regularly.

During exercise (especially aerobic exercise), changes in the microcirculation of the body increase the delivery of oxygen and other life-sustaining substances to the active skeletal muscles. Blood flow during rest averages 3 to 4 mL/min per 100 g of muscle tissue, but during exercise blood flow may increase to 80 mL or more (see graph A). The *extra oxygen supply to active muscles* is produced mostly by an increased extraction of oxygen from the blood, but also by an increase in cardiac output.

Cardiac output rises during exercise roughly in proportion to the increased oxygen demand by the active skeletal muscles (graph B). During rest, cardiac output is approximately 5 L/min. It can rise to about 35 L/min in well-conditioned athletes. A rise in cardiac output of 120 percent is quite common, even during moderate exercise. The amount of oxygen released during vigorous exercise can increase up to 15 mL/100 mL of blood (graph B).

The *stroke volume* increases during moderate exercise from approximately 70 mL/min to about double that volume (graph C). This increase is due primarily to an increased contractility of cardiac muscle caused by impulses from sympathetic nerves. These impulses increase the strength of the ventricular contraction and the expulsion of the reserve blood.

Heart rate increases along with the amount of exercise, and is the major factor in increasing the cardiac output. During vigorous exercise, the heart rate can reach approximately 200 beats per minute. As the physical conditioning of an individual improves, however, the amount of increase in heart rate becomes progressively less. This happens

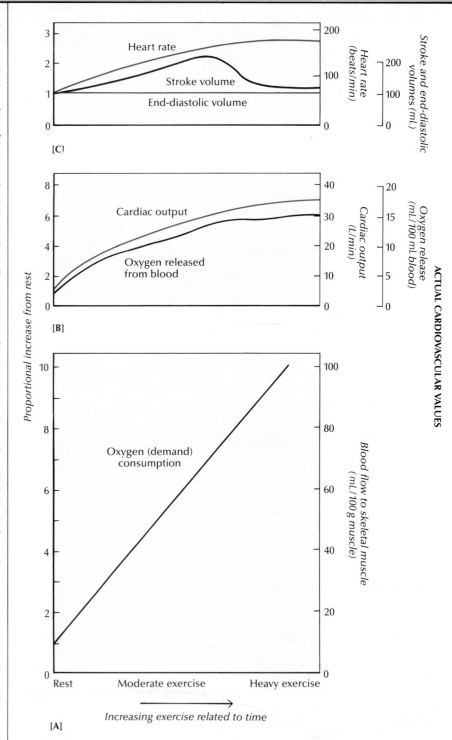

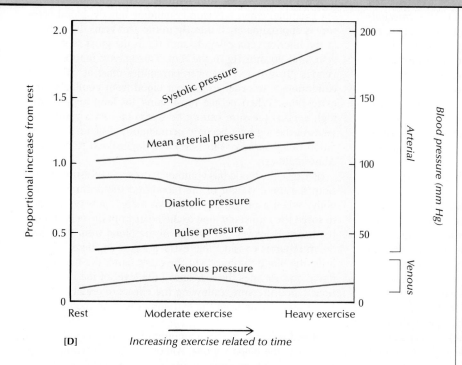

[D]   *Increasing exercise related to time*

Effect of exercise on various cardiovascular functions. (A) An increase in exercise causes a proportional increase in oxygen consumption by skeletal muscles. (B) The increased demand for oxygen causes an increased cardiac output, which increases blood flow to skeletal muscles and leads to a greater amount of oxygen extraction from the blood. (C) The increase in cardiac output is due primarily to an increase in heart rate and only secondarily to a small increase in stroke volume. Note that the stroke volume actually falls after prolonged exercise with an increased cardiac output due to less time during diastole for complete filling. (D) Arterial pressure changes more than venous blood pressure during exercise. The increase in pulse pressure (systolic − diastolic) is related primarily to the increase in systolic pressure. This is due to the quicker ejection of blood by the heart due to sympathetic nervous stimulation.

pressure increases more than the diastolic, the pulse pressure also increases proportionately. The increase in arterial pressure is, by comparison, somewhat less. There is only a slight increase in venous pressure at the beginning of exercise, but it drops eventually to resting values (averaging 15 mm Hg).

There is a marked *dilation of blood vessels in the skeletal muscles* during exercise. At the same time, *vasoconstriction occurs in the skin and other organs.* The increased vasoconstriction increases peripheral resistance, venous return, and the overall volume of blood returning to the heart. Also, there is an overall decrease in total peripheral resistance (by as much as 50 percent) during exercise because of increased cardiac output and the dilation of blood vessels in muscles.

The *mass of the heart* increases permanently after prolonged periods of exercise. This enlargement is caused by either the increased thickness of the ventricular walls (cardiac hypertrophy) or a lengthening of cardiac muscle fibers, depending on the exercise regimen (isometric or isotonic exercise, respectively). This increase in mass benefits the heart by strengthening it.

The *density of capillaries in cardiac muscle* increases after a period of regular exercise, and the *number of capillaries in skeletal muscle tissue* also increases, allowing more blood to flow to the muscle fibers. Both factors improve the transport of oxygen to active muscles.

Cardiovascular responses to exercise depend on the frequency, duration, and intensity of the exercise being done. Strenuous exercise is the most effective way to increase cardiac output. For example, aerobic exercise increases cardiac output by increasing the length of the cardiac muscle fibers and, in turn, the stroke volume. Other exercises increase the heart rate and venous return, which contribute to an increased cardiac output and to a more efficient cardiovascular system.

because the stroke volume is greater at the beginning of exercise, keeping the cardiac output the same. Because the cardiac output increases sevenfold, while the heart rate increases threefold, the stroke volume must be more than double (graph C). Note also that there is no change in the end-diastolic ventricular volume.

Blood pressure in the systemic circulatory system increases temporarily as exercise increases, while peripheral resistance decreases temporarily four- to fivefold (graph D). Systolic blood pressure increases more than diastolic pressure, but it seldom rises above 180 mm Hg during heavy exercise. In contrast, diastolic pressure does not change much. In fact, it actually *decreases* during moderate exercise. Because the systolic

Hormonal and chemical control Blood flow is also regulated by hormonal and other chemical substances that can cause constriction and dilation. For example, the *catecholamines* (epinephrine, norepinephrine, dopamine) that are released by specific nerve axons are generally vasoconstrictors, depending upon their concentrations and combinations. The hypothalamic peptides *vasopressin* and *oxytocin,* and the bloodborne *angiotensin* are powerful vascoconstrictors. Specific tissues also produce chemicals that have localized effects. For example, eosinophils and mast cells in most tissues can release *histamine,* which causes vessel dilation. The heart produces a hormone, atriopeptin (see Chapter 16), that is a powerful vasoconstrictor. Finally, some tissues, such as blood, contain *bradykinin,* which is a strong vasodilator.

Autoregulation Blood flow at the capillary level is often regulated by local changes in *cellular metabolism.* If a tissue becomes very active, not enough oxygen will be available to the smooth muscle in the walls of the arterioles and precapillary sphincters to allow them to continue contracting. As a result, these blood vessels dilate and increase the flow of blood, which brings more oxygen to the tissue. Many normal metabolites, such as hydrogen and potassium ions, inorganic phosphate, carbon dioxide, and some of the intermediates of purine metabolism (adenosine, for example) can accumulate enough to cause dilation in most vascular beds, and blood flow is increased.

The arterioles use autoregulation, for example, to help regulate body temperature. When the body is too warm, the muscular arteriole walls relax, allowing blood to flow to the skin, where heat dissipates. If the body is cold, the appropriate arteriole walls contract, directing the flow of blood *away* from the skin and toward the center of the body.

Arterioles also respond to emotional stimuli, increasing the flow of blood to the face and neck, producing what we commonly call a "blush." On the other hand, the flow of blood to the brain rarely varies. About 600 mL (1.3 pt) are delivered to the brain each minute whether we are awake or asleep.

Ask Yourself

1 How is the nervous system involved in the regulation of blood flow?

2 What are some of the hormonal and chemical controls of blood flow?

3 How are the arterioles involved in autoregulation?

Factors Affecting Blood Pressure

Several factors affect blood pressure in both the arteries and the veins. We consider these factors in the following sections before turning to a discussion of the regulation of blood pressure.

Hydrostatic pressure and gravity The pressure that results from gravity acting on the fluids of the body influences *hydrostatic pressure.* It determines the weight of the blood in the various vessels, and therefore affects venous return. For example, when a person is standing quietly, blood pressure is approximately 8 mm Hg in the arm veins, 22 mm Hg in the inferior vena cava, 40 mm Hg in the great saphenous vein, and 90 mm Hg in the foot. This gravity-induced high venous pressure in the lower extremities must be overcome continuously in order to prevent blood from pooling in the extremities. When people must stand for long periods, the high venous pressure cannot be overcome. As a result, the veins in the leg may become permanently distended, a condition called *varicose veins* (see Physiological and Anatomical Abnormalities).

Because the blood has volume, hydrostatic pressure results from the mere weight of the blood and the position of the body. When a person is standing or seated, gravity operates to speed the return of blood to the heart from the veins above the heart. However, the weight of the blood from the heart down requires considerable extra pressure in these veins to drive the blood "uphill." On the other hand, if a person lies down and elevates the legs above the level of the heart, venous blood pressure, following the effects of gravity, moves blood more easily from the extremities back to the heart.

Blood volume The total amount of blood in circulation is known as the *blood volume.* Almost 80 to 90 percent of the blood is in the systemic circulation at any given time, and the rest is in the pulmonary circulation. Of the blood in the systemic circulation, about 75 percent is in the veins, 20 percent in the arteries, and 5 percent in the capillaries (see Table 19.1). A severe loss of blood due to hemorrhage, shock, burns, or other dehydration factors will lower the blood pressure because there is less blood in the arteries. In contrast, an increase in blood volume causes the blood pressure to rise.

Vessel elasticity It is primarily the *vessel elasticity* (flexibility) of the large arteries that determines the pressure within a vessel when the ventricles contract. The elasticity results from the resilient nature of the arterial wall. If a vessel becomes less elastic, as often happens with aging, the blood pressure rises, and the recoil that usually pushes blood forward is lost.

Cardiac output The volume of blood ejected by the left ventricle per unit of time is the *cardiac output.* After leaving the left ventricle, the blood enters a closed system of tubes, which has a given capacity at a given time. This action creates a blood pressure wave that causes blood to flow. If the cardiac output increases and the volume of the vessels remains the same, blood pressure and blood flow will increase. If the cardiac output remains constant, but the blood volume increases, the blood pressure will decrease in order to maintain homeostasis.

Blood viscosity and resistance The viscosity of blood determines its resistance to flow. A thin fluid flows more easily than a thicker one, and less force (blood pressure) is required to move it along. In contrast, a thick fluid has more resistance than a thin one, and requires a greater blood pressure to make it flow.

Regulation of Arterial Blood Pressure

The blood pressure within the large systemic arteries must be maintained precisely to ensure an adequate blood flow to the tissues, and to keep the pressure from reaching dangerous highs and lows that can upset homeostasis. The following factors affect arterial blood pressure by way of mechanisms that regulate vessel resistance and cardiac output.

Vasomotor center The *vasomotor center* is located in the lower part of the pons and the medulla oblongata. One of its major functions is to regulate the diameter of blood vessels, especially arterioles. Sympathetic nerve impulses continually leave the vasomotor center and travel to blood vessels, where they cause varying degrees of constriction called *vasomotor tone*. This steady state of tonicity is central in maintaining a stable peripheral resistance that produces an acceptable mean arterial pressure.

The vasomotor center is stimulated by hormones and carbon dioxide. Also, its monitoring activity is affected by nerve impulses from specific receptors associated with the cardiovascular system, higher brain centers, and other areas.

Cardioregulatory center As you learned in Chapter 18, the medulla contains excitatory and inhibitory sympathetic nerve centers, collectively called the *cardioregulatory center,* that regulate the heart rate. This center also influences arterial pressure by affecting cardiac output. It is influenced by many of the same factors (for example, hormones and carbon dioxide) that influence the vasomotor center.

Baroreceptors Normal arterial pressure is maintained also because the medulla is continuously informed of the state of the circulation by specialized nerves that innervate pressure receptors called *baroreceptors.* These baroreceptors are located in the walls of the aortic arch and the carotid artery sinus, as well as in the walls of large arteries in the neck and thorax (Figure 19.22). When arterial pressure increases, baroreceptors are stretched, causing impulses to be sent to the medulla, where the cardioinhibitory center is stimulated and the cardioacceleratory center is inhibited. As a result, cardiac output is decreased, arterioles are dilated, and arterial pressure decreases. A decrease in arterial pressure produces the opposite response. These regulatory responses are reinforced through impulses sent to the vasomotor center by baroreceptors.

Higher brain centers and emotions The *higher brain centers* and *emotions* play important roles in affecting arterial pressure. For example, any excitatory emotion such as fear or rage that stimulates the sympathetic nervous system also stimulates the vasomotor center, which causes constriction of arterioles and a subsequent rise in pressure. Opposite emotions such as grief, loneliness, and depression cause a decrease in sympathetic stimulation, dilation of arterioles, and a decrease in arterial pressure. Many other physiological, psychological, and pathological factors can affect arterial pressure.

Hormones and chemicals Several hormones affect arterial pressure. For example, the *renin-angiotensin system* takes several hours to alter arterial pressure. When too little blood reaches the kidneys, they produce an excess of an enzyme called *renin* (L. *renes*, kidneys). Renin stimulates the production of the protein *angiotensin II*, a powerful but short-acting vasoconstrictor that raises arterial pressure. Angiotensin II also increases blood pressure by stimulating the release of aldosterone from the adrenal cortex, which increases water reabsorption and blood volume, and by stimulating neural centers in the hypothalamus that increase the intake of water to increase blood volume, blood flow, and blood pressure.

A constriction of vessels that leads to increased arterial pressure may be also caused by increased levels of other hormones, including vasopressin, atriopeptin, epinephrine, and norepinephrine.

Chemoreceptors *Chemoreceptors* respond to levels of arterial blood oxygen, carbon dioxide, and hydrogen ions (see Chapter 18). If the oxygen content decreases or the level of carbon dioxide or hydrogen ions increases, arterial pressure is lowered, and chemoreceptors are stimulated. Impulses are sent to the vasomotor center, which causes sympathetic impulses to travel to blood vessels. The vessels are constricted, and arterial pressure increases. The increased blood pressure, in turn, increases blood flow, and homeostasis is restored.

Regulation by the kidneys In addition to exerting a hormonal control, the kidneys help control arterial pressure by regulating blood volume. If the arterial pressure increases, the kidneys filter a greater amount of blood, which increases the quantity of fluids and solutes excreted in urine. As a result, blood volume is lowered and, in turn, arterial pressure drops also. In contrast, if the pressure decreases, fluid is retained, blood volume increases, and arterial pressure rises. Overall, the kidneys play a major role in the long-term regulation of arterial pressure.

Capillary fluid shift A mechanism called the *capillary fluid shift* regulates arterial pressure by altering blood volume (Figure 19.23). A rise in arterial pressure forces fluid out of the permeable capillaries into the extracellular fluid, lowering blood volume, and consequently, arterial pressure. A fall in arterial pressure causes more fluid to enter the capillaries, which raises blood volume and arterial pressure. This same mechanism plays an important role in the second-by-second exchange of materials between blood, extracellular tissue fluids, and the tissue cells themselves.

The mechanism responsible for the capillary fluid shift is called the *Starling-Landis concept*, so named after the individuals who first described it in 1896. Basically, there are five mechanisms that can move materials through capillary walls: *diffusion, filtration, osmosis, reabsorption,* and *pinocytosis.* Oxygen, for example, *diffuses* out of systemic capillaries from an area of high concentration to one of low concentration. Many capillaries have fenestrations, and small molecules can also diffuse through the endothelial cell membranes. Diffusion, along with pinocytosis to a limited degree, controls most of the metabolic exchange of nutrients, gases, and wastes that occurs in capillaries.

FIGURE 19.22

Nervous control of blood pressure, based on rises and drops in arterial blood pressure. The nervous system regulates blood pressure on a minute-to-minute basis. Arterial baroreceptors are located in the walls of the large arteries leading from the heart.

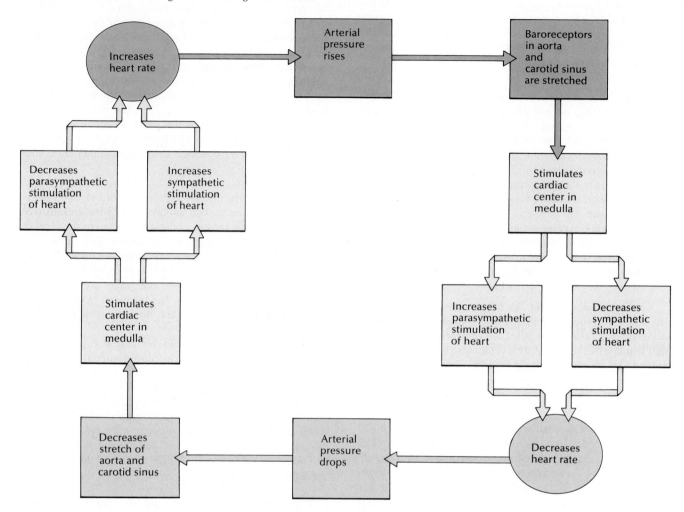

The force behind *filtration,* which pushes fluid out of the capillary, is plasma pressure, or **hydrostatic pressure (HP).** An opposing force called *osmotic pressure* helps prevent fluid from accumulating by returning it to the blood. Osmotic pressure is produced by the plasma proteins in the blood serum. Recall from Chapter 3 that water moves across a selectively permeable membrane from a region of lower osmotic pressure (higher water pressure) to a region of higher osmotic pressure (lower water pressure). This specific osmotic pressure, called **oncotic pressure (OP),** promotes *reabsorption,* which draws fluid back into the capillaries. Extracellular fluid contains a slight hydrostatic pressure that opposes the movement of fluid out of the capillary. It also has a slight oncotic pressure, due to the proteins in the fluid.

The **net force** for fluid movement, either out of or into the capillary, is the difference between those forces moving fluid *out* of the capillary (capillary HP + tissue OP) and those moving fluid *into* the capillary (capillary OP + tissue HP). If homeostasis is to be maintained, the net movement of fluid *outward* at the arterial end of a capillary must equal the net movement of fluid *inward* at the venous end.

Using typical values for the arterial end of a capillary (where t = tissue values and c = capillary values), we can calculate the net filtration *(fluid out)* at the arterial end of a capillary as follows:

$$
\begin{aligned}
\text{Net filtration out} &= (HP_c + OP_t) - (OP_c + HP_t) \\
&= (40 \text{ mm Hg} + 3 \text{ mm Hg}) \\
&\quad - (25 \text{ mm Hg} + 5 \text{ mm Hg}) \\
&= +13 \text{ mm Hg forcing fluid out}
\end{aligned}
$$

FIGURE 19.23

Forces involved in filtration and reabsorption in a capillary. The values are general, and do not apply to all capillaries. Values are in mm Hg, where HP_c = capillary hydrostatic pressure, OP_t = tissue oncotic pressure, OP_c = capillary oncotic pressure, and HP_t = tissue hydrostatic pressure. The difference in fluid movement (13 − 5 = 8 mm Hg) is accounted for by the lymphatic system. The thickness of the arrows indicates the relative amounts of fluid flowing in a particular direction.

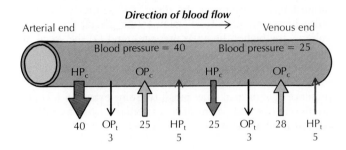

The inward movement of fluid results from the marked drop in blood pressure (from 40 to 25 mm Hg) as blood moves through the capillary, and a slight rise in blood oncotic pressure (from 25 to 28 mm Hg). This rise in blood OP occurs when water filters out of the capillary, leaving protein in. Thus, the net reabsorption (*fluid in*) at the venous end of a capillary can be calculated as follows:

Net
reabsorption in $\begin{aligned} &= (OP_c + HP_t) - (HP_c + OP_t) \\ &= (28 \text{ mm Hg} + 5 \text{ mm Hg}) \\ &\quad\quad - (25 \text{ mm Hg} + 3 \text{ mm Hg}) \\ &= -5 \text{ mm Hg (or 5 mm Hg) forcing fluid in} \end{aligned}$

Notice that the net pressure forcing fluid out (+13 mm Hg) is higher than the pressure forcing fluid in (−5 mm Hg). The fluid that is not reabsorbed, along with any protein that escapes from the plasma and cells, is returned to the venous circulation by the lymphatic vessels. This return action maintains fluid homeostasis.

Regulation of Venous Pressure

Because the heart pumps all the blood it receives, the pressure of venous blood and its return to the heart are important in maintaining a normal cardiac output and overall homeostasis. The return of venous blood is determined by five major factors: arterial pressure, right atrial pressure, resistance to blood flow through vessels, venous pumps, and hydrostatic pressure.

Pressure gradient for venous return The *venous pressure* is the pressure that forces blood back toward the heart. The *right atrial pressure* is the pressure against which the venous blood must flow. Therefore, venous return is determined to a large degree by the difference between the causing pressure (arterial pressure) and the opposing pressure (right atrial pressure). This difference is the *pressure gradient for venous return.*

Resistance to blood flow Since venous blood flow is determined by the same formula as arterial blood flow,

$$\text{Venous blood flow} = \frac{P_1 - P_2}{R}$$

any factor that increases *resistance to blood flow* in the veins also reduces the amount of blood returning to the right heart (*venous return*) and cardiac output. For example, because the walls of veins contain smooth muscle, sympathetic stimulation causes the vessel walls to constrict. This makes the vessel less distensible and raises the pressure within, which increases venous return.

Venous pumps Every time a muscle contracts, the flexible veins are compressed, moving the blood. Since most veins have valves that prevent backflow, blood is allowed to flow only *toward* the heart (see Figure 19.5). This one-way mechanism is called the *skeletal muscle venous pump.* During inspiration (breathing in), the diaphragm pushes on the abdominal viscera, increasing abdominal and thoracic pressure. As a result, the venous pressure gradient between the veins in the abdomen and thorax is increased. The final effect is an auxiliary *respiratory pump*, which increases venous return during inspiration.

If a person stands still for a long time, the leg muscles cannot push the venous blood up rapidly enough, and the venous blood tends to accumulate in the leg veins. Reduced return of blood to the heart may result in poor blood flow to the rest of the body. Decreased blood flow to the brain can cause fainting. (Fainting forces a person to lie down, helping blood to flow to the brain without the resistance of gravity.) Another potential problem is *varicose veins* (see Physiological and Anatomical Abnormalities, p. 610).

Measuring Arterial Blood Pressure

Arterial blood pressure depends on the volume of blood in the arteries and the elasticity of the arterial walls, as well as on the rate and force of ventricular contractions. If the arteries are elastic, they can be stretched by large volumes of blood without an appreciable rise in blood pressure.

When blood is ejected into the arteries by the ventricles during systole, an equal amount of blood is *not* simultaneously released out of the arteries. In fact, only about one-third of the blood leaves the arteries during systole, and the excess volume raises the arterial pressure. For example, when blood is ejected into the large arteries during ventricular systole, most of the stroke volume is used to stretch the walls of the arteries—almost like blowing up a balloon (Figure 19.24). During diastole, the elasticity of the arteries is used to keep the blood moving forward, even though the heart is not contracting, just as the elasticity of a balloon can be used to expel air out of the balloon. After systole, when the ventricular contraction is over, the arterial walls return to their unstretched condition as blood continues to leave the arteries. Pressure slowly decreases, but before all the blood has left the artery, the next ventricular contraction occurs, and the pressure begins to build up again. Because of this consistent rhythm, the arterial pressure never reaches zero, and there is always enough pressure to keep the blood flowing (see Figure 19.22).

Systolic and diastolic pressure Blood pressure levels are measured by two numbers, both expressed as millimeters of mercury (abbreviated as mm Hg). The first number, called the *systolic pressure,* represents the highest pressure reached during ventricular ejection, and the second number, called the *diastolic pressure,* represents the pressure during the interval between heartbeats. A normal young adult's blood pressure is 120/80 mm Hg or *less.* Blood pressure is considered high, or *hypertensive,* in an adult when the systolic reading exceeds 140 and the diastolic reading is higher than 95.

Blood pressure varies with age. The systolic pressure of a newborn baby may be only 40, rising to 80 after a month. During adolescence it may progress from 100 to 120, and continues to rise slightly throughout adulthood. The normal pressure of a 60-year-old man is about 140/90, depending on many factors. Most physicians agree that blood pressure need not rise above an acceptable middle-age level, even in old age, if good health is maintained. Blood pressure usually rises temporarily during exercise or stressful conditions, and a systolic reading of 200 would not be considered abnormal under those circumstances.

Pulse pressure The difference between systolic and diastolic pressure is called the *pulse pressure.* In the example above of a blood pressure of 120/80, the pulse pressure equals 40 mm Hg, as calculated below:

$$\text{Pulse pressure} = \text{systolic pressure} - \text{diastolic pressure}$$
$$= 120 - 80$$
$$= 40$$

Pulse pressure is determined by three factors: (1) arterial distensibility (the greater the distensibility, the higher the pulse pressure), (2) stroke volume (the greater the stroke volume, the higher the pulse pressure), and (3) the speed of the blood ejected from the left ventricle into the systemic circulation (the greater the speed, the higher the pulse pressure). Many cardiovascular diseases can be detected by the presence of abnormal pulse pressure. Some examples include arteriosclerosis and patent ductus arteriosus (see Physiological and Anatomical Abnormalities).

Mean arterial blood pressure Arterial blood pressure changes throughout the cardiac cycle. One of the most useful measurements indicating these pressure changes is the *mean arterial pressure (MAP).* Mean arterial pressure is the average pressure that drives the blood through the systemic circulatory system. It is determined by two factors: the cardiac

---

FIGURE 19.24

Pulse-wave propulsion of blood during systole and diastole. The elastic arteries help push the blood forward by contracting behind the systolic rush of blood that stretches the arterial walls.

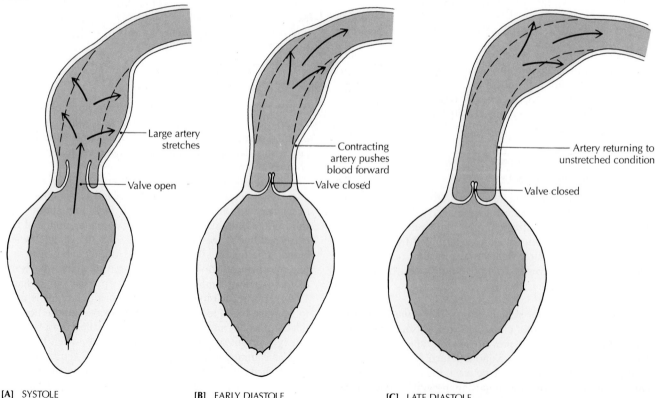

[A] SYSTOLE [B] EARLY DIASTOLE [C] LATE DIASTOLE

FIGURE 19.25

Some of the major factors involved in the homeostatic regulation of blood pressure.

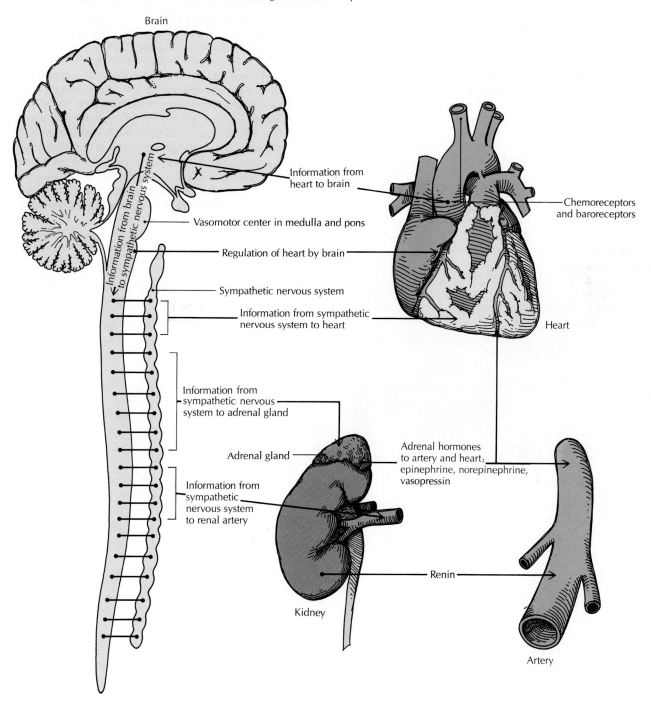

output and the rate at which blood drains from the arteries. Arterial drainage is controlled by the vascular resistance and blood flow. Therefore,

MAP = cardiac output × total peripheral resistance

Since diastole usually lasts longer than systole, MAP is approximately equal to the diastolic pressure plus one-third of the pulse pressure, or

$$MAP = \frac{\text{systolic pressure} + 2 \text{ (diastolic pressure)}}{3}$$

Using our previous example of 120/80, we have

$$MAP = \frac{120 + 2(80)}{3} = 93.3 \text{ mm Hg}$$

Under normal conditions, the mean arterial pressure is close to 90 mm Hg. This pressure is crucial for the maintenance of a steady blood flow to capillaries. Several regulatory mechanisms are available for achieving this stable blood pressure, including cardiac output, blood volume, and peripheral resistance. In Chapter 18 we examined specific neural, chemical, and hormonal factors that help regulate cardiac output by acting directly on the heart. These same factors also help regulate mean arterial blood pressure by acting directly on blood vessels. Figure 19.25 presents an overall view of this blood pressure regulation.

FIGURE 19.26

(A) Measurement of arterial blood pressure with a sphygmomanometer and stethoscope. (B) The materials needed for measuring blood pressure. Arterial events are shown. [(A) *Yoav/Phototake.*]

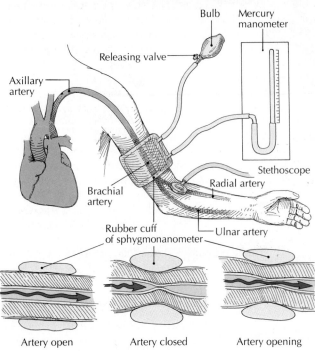

[A]

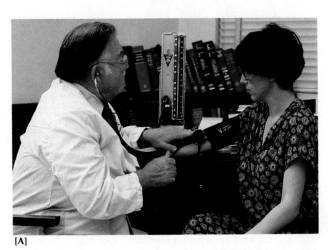

[B]

The sphygmomanometer The instrument used for measuring arterial blood pressure is the *sphygmomanometer* (sfig-moh-muh-NOM-ih-ter). It consists of an inflatable cuff, a rubber bulb, and either a column of mercury, an air gauge, or an electronic display (Figure 19.26). The hollow cuff is wrapped around the upper arm, and inflated with the rubber bulb to a pressure above the systolic pressure. (This high cuff pressure collapses the arteries under the cuff, stopping the flow of blood.) The pressure within the cuff is shown on the scale of the sphygmomanometer. A stethoscope is placed over the brachial artery just below the cuff, and the pressure in the cuff is slowly released until the blood begins to flow normally. At that point, a faint tapping sound in the stethoscope signals the high-velocity and turbulent release of blood. When the sound is first noted, the figure on the mercury column represents the *systolic* blood pressure. As the cuff pressure is lowered further, the sounds become progressively louder, and then softer. When the sound stops, the *diastolic* blood pressure is noted. The absence of sound indicates that blood is flowing freely through the artery.

Pulse When an artery lies close to the surface of the skin, a *pulse* can be felt that corresponds to the beating of the heart and the alternating expansion and elastic recoil of the arterial wall. The pulse is produced when the left ventricle forces blood against the wall of the aorta, and the impact creates a continuing pressure wave along the branches of the aorta and

FIGURE 19.27

The arterial pulse can be felt most easily at the points shown in the drawing.

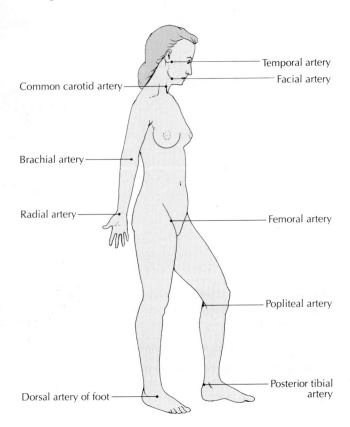

the rest of the elastic arterial walls (see Figure 19.21A). A venous pulse occurs only in the largest veins. It is produced by the changes in pressure that accompany atrial contractions. All arteries have a pulse, but it can be felt most easily at the points shown in Figure 19.27.

The most common site for checking the pulse rate is the radial artery on the underside of the wrist. The three middle fingers are usually used (but not the thumb, which may be close enough to the radial artery to reflect its own pulse beat). The pulse is checked for several reasons. For example, a physician can detect the number of heartbeats per minute (heart rate), the strength of the beat, the tension of the artery, the rhythm of the beat, and several other diagnostic factors. The average resting pulse rate can range between 70 to 90 beats per minute in adults and from 80 to 140 in children. When the pulse rate exceeds 100 beats per minute, the condition is called *tachycardia* (Gr. *takhus,* swift); when the rate is below 60 beats per minute, the condition is called *bradycardia* (Gr. *bradus,* slow).

The pulse rate normally decreases during sleep, and increases after eating or exercising. During a fever it may increase about five beats per minute for every degree Fahrenheit above the normal body temperature. Pulse rates tend to increase significantly after severe blood loss, and are usually high in cases of serious anemia.

Ask Yourself

1 How is arterial blood pressure calculated?

2 What are the major factors that help maintain normal blood pressure?

3 What is the function of the vasomotor center?

4 How is venous pressure regulated?

5 What is a venous pump?

6 How does gravity affect venous blood pressure?

7 What is meant by capillary fluid shift?

8 What is the physiological difference between systolic and diastolic blood pressure?

9 What is the difference between pulse pressure and pulse?

10 How does a sphygmomanometer work?

PHYSIOLOGICAL AND ANATOMICAL ABNORMALITIES

Aneurysms

An *aneurysm* (Gr. "to dilate") is a balloonlike, blood-filled dilation of a blood vessel. It occurs most often in the larger arteries, especially in the brachial artery, abdominal aorta, femoral artery, and popliteal artery. Cerebral aneurysms are also common. Aneurysms happen when the muscle fibers of the tunica media become elongated, and the wall weakens. The wall may also be weakened by degenerative changes related to such diseases as *arteriosclerosis* (hardening of the arteries). If an aneurysm is not treated, the vessel will eventually burst, and if a cerebral vessel is affected, a stroke may result.

Atherosclerosis

Atherosclerosis (Gr. "porridgelike hardening") is the leading cause of coronary artery disease. It is characterized by deposits of fat, fibrin, cellular debris, and calcium on the inside of arterial walls (see photo). These built-up materials, called *atheromatous plaques,* adhere to the tunica intima, narrowing the lumen and reducing the elasticity of the vessel.

Although no one really knows the basic cause of atherosclerosis, the earli-

est stage in its development is believed to be damage to the endothelial cells and tunica intima of the vessel wall. Once the damage occurs, the endothelial cells proliferate and attract lipid substances. Several factors increase its progress. Among them are cigarette smoking and the amount of animal fat and cholesterol in the diet. Other factors that may have an effect are hypertension (high blood pressure), diabetes, age, stress, heredity, and the male sex hormones.

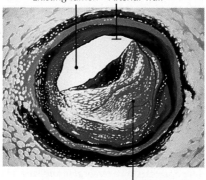

Existing lumen   Arterial wall

Plaque

Atherosclerotic plaque can narrow the lumen of an artery substantially. [*Courtesy of NIH.*]

There are three ways in which atherosclerosis can cause a heart attack. (1) It can completely clog a coronary artery. (2) It can provide a rough surface where a blood clot *(thrombus)* can form and grow so it closes off the artery and causes a *coronary thrombosis.* (3) It can partially block blood to the myocardium and cause the heart to stop beating rhythmically.

Coronary artery disease

The typical effect of *coronary artery disease* (CAD) is a reduced supply of oxygen and other nutrients to the myocardium. The condition is usually brought on by atherosclerosis. As a result, less blood reaches the heart, and *myocardial ischemia* (ih-SKEE-mee-uh; Gr. *iskhein,* to keep back + *haima,* blood) or local anemia in cardiac muscle, occurs.

Lipoproteins play an important role in atherosclerosis and CAD. They are formed when lipids combine with certain blood proteins. In fact, nearly all the lipid in the blood is carried as lipoproteins. They are classified according to their densities and functions. (1) *Chylomicrons* are aggregates of triacylglycerols and cholesterol, and are derived from the digestive system. Because they contain less than 2 percent protein, they are of low density. (2) *Very-low-density lipoproteins (VLDL)* are synthesized in the liver and transported by the blood to adipose tissue, where they are stored. (3) *Low-density lipoproteins (LDL)* constitute the primary supply of cholesterol to all tissues. (4) *High-density lipoproteins (HDL)* are rich in phospholipids and cholesterol. They are produced in many tissues of the body, and transported to the liver for storage.

The incidence of CAD is closely related to the concentration of these specific lipoproteins in the blood. On the one hand, there is a strong positive correlation between CAD and high blood levels of LDL as related to total cholesterol. On the other hand, there is a strong negative correlation between CAD and the ratio of HDL to the total cholesterol. That is, the higher the level of HDL, the less chance of developing CAD. Several ways to improve the HDL-LDL cholesterol ratio are exercising regularly, reducing animal fat in the diet, giving up cigarettes, and increasing the consumption of unsaturated omega-3 oils from cold-water fish.

In 1987, Joseph Goldstein and Michael Brown won the Nobel prize in physiology or medicine for discovering an LDL-receptor on the surface of cells. These receptors control the amount of cholesterol in the blood.

Several new drugs have been used successfully to lower cholesterol levels and dissolve arterial blood clots. *Lovastatin* (Mevacor) inhibits an enzyme in the liver that helps manufacture cholesterol. As a result, the liver produces more LDL-receptors that draw cholesterol from the blood, reducing blood cholesterol levels. Lovastatin combined with other drugs can reduce the blood cholesterol level by about 50 percent. *Gemifibrozil* reduces the risk of coronary heart disease by slightly lowering the total cholesterol level, while moderately increasing HDL and moderately reducing LDL. *Tissue-plasminogen activator* (t-PA) activates plasminogen and converts it into the active enzyme plasmin, the enzyme that destroys fibrin, which forms the threads of arterial blood clots that can lead to heart attacks. Currently, however, t-PA is extremely expensive.

Hypertension (high blood pressure)

Hypertension, commonly called *high blood pressure,* is systolic or diastolic pressure that is above normal all the time (not just as a result of specific activities or conditions). About 35 million Americans have hypertension, and more than a million die annually of related diseases. The two types of hypertension are the more common *primary* or *essential (idiopathic) hypertension,* an above-normal blood pressure that cannot be attributed to any particular cause, and *secondary hypertension,* which results from a disorder such as kidney disease or arteriosclerosis.

Hypertensive arterial walls are hard and thick, and their elasticity is reduced, forcing the heart to work harder to pump enough blood. If hypertension persists, the heart may become enlarged, a condition called *hypertensive heart disease.* High blood pressure can also cause a stroke if the extra force of the blood breaks an artery in the brain and a cerebral hemorrhage occurs.

Although hypertension may be accompanied by headache, dizziness, or other symptoms, one of the problems with treating hypertension is that often there are no external signs that blood pressure is abnormally high. Prognosis is optimistic if the disease is diagnosed and treated early, before complications develop. If untreated, hypertension is accompanied by a high mortality rate.

Hypertension may be an inherited problem, but it is also related to stress, obesity, a diet that is high in sodium and saturated fats, aging, lack of physical activity, race (it is most common in blacks), and the use of tobacco and oral contraceptives (especially if used together). Hypertension may also occur when regulatory mechanisms in the central nervous system break down.

Stroke

A *stroke* occurs when the brain's blood supply is cut off. There are several ways a stroke may happen. (1) Atherosclerosis in the arteries of the brain or neck may block the flow of blood. (2) A blood clot *(thrombus)* may form in the atherosclerotic vessel, closing off the artery and causing a *cerebral thrombosis.* (3) A traveling blood clot *(embolus)* can become wedged in a small artery of the brain or neck; this kind of stroke is an *embolism.* (4) A weak spot in a blood vessel may break; this is a *cerebral hemorrhage.* (When the weak spot bulges, it is called a cerebral *aneurysm.*) (5) In rare cases, a brain tumor may press on a blood vessel and shut off the blood supply.

Because the brain controls the body's movements, any part of the body can be affected by a stroke. The damage may be temporary or permanent. If a brain artery is blocked in the area that controls speaking, speech will be affected.

Thrombophlebitis

Thrombophlebitis (Gr. *thrombos,* clot + *phleps,* blood vessel + *itis,* inflammation) is an acute condition characterized by clot formation and inflammation of deep or superficial veins. Thrombophlebitis is usually the result of an alteration in the endothelial lining of the vein. Platelets begin to gather on the roughened surface, and the consequent formation of fibrin traps red blood cells, white blood cells, and more platelets. The result is a blood clot. If untreated, the clot *(thrombus)* may become detached from the vein and begin to move through the circulatory system. Such a mobile blood clot is called an *embolus.* The result may be pulmonary embolism, a blockage of the pulmonary veins that is a potentially fatal condition. Some causes of thrombophlebitis include surgery, trauma, childbirth, oral contraceptives, prolonged bed rest, infection, intravenous drug abuse, and chemical irritation.

Varicose veins

Varicose veins (L. *varix,* swollen veins) are abnormally dilated and twisted veins. The saphenous veins and their branches in the legs are often affected. They become permanently dilated and stretched when the one-way valves in the legs weaken. As a result, some blood flows backward and pools in the veins. Varicose veins can result from a hereditary weakness of the valves, but can also be caused or aggravated by vein inflammation (phlebitis), blood-clot formation (thrombophlebitis), pregnancy, lack of exercise, loss of elasticity in old age, smoking, low-fiber diet, and occupations that require long periods of standing. Untreated varicose veins may lead to edema in the ankles and lower legs, leg ulcers, dizziness, pain, fatigue, and nocturnal cramps.

Portal cirrhosis of the liver

Portal cirrhosis of the liver results from chronic alcoholism, the ingestion of poison, viral diseases (such as infectious hepatitis), or other infections in the bile ducts. It is a condition in which large amounts of fibrous tissue develop within the liver. This fibrous tissue destroys many of the functioning liver cells, and eventually grows around the blood vessels, constricting them and greatly impeding the flow of portal blood through the liver. This impediment blocks the return of blood from the intestines and spleen. As a result, blood pressure increases so much that fluid moves out through the capillary fluid shift into the peritoneal cavity, leading to *ascites,* the accumulation of free fluid in the peritoneal cavity. Because this fluid is mostly plasma, which contains large quantities of protein, a high colloid osmotic pressure is created in the abdominal fluid, pulling more fluid by osmosis from the blood, liver, and gastrointestinal tract, in a positive feedback cycle. The patient often dies because of excessive fluid loss.

Patent ductus arteriosus

Patent ductus arteriosus occurs when the ductus arteriosus of the fetus, which carries blood from the pulmonary artery to the descending aorta, does not shut down at birth (see p. 581). The fetal ductus arteriosus bypasses the nonfunctioning lungs. But when the lungs begin to function at birth (and the baby is no longer relying on the placenta for oxygen and the release of carbon dioxide), the blood normally passes through the pulmonary artery to the lungs. Patent ductus arteriosus creates a left-to-right movement of some arterial blood from the aorta to the pulmonary artery, which recirculates arterial blood through the lungs.

Transposition of the great vessels

In *transposition of the great vessels,* the arteries are reversed. The aorta emerges from the right ventricle, and the pulmonary artery emerges from the left ventricle. As a result, oxygenated blood returning to the left side of the heart is carried back to the lungs by the pulmonary artery, and unoxygenated blood traveling to the right side of the heart is carried into the systemic circulation by the transposed aorta.

MEDICAL TERMINOLOGY

ANGIOGRAM An x ray of blood vessels, taken after a radiopaque dye is injected into the vessels.

ARTERIOGRAPHY Technique of x-raying arteries, after a radiopaque dye is injected.

AVASCULAR NECROSIS Condition in which tissue dies from the lack of blood (from the blood vessels).

CLAUDICATION Improper circulation of blood in vessels of a limb, which causes pain and lameness.

CYANOSIS A bluish discoloration of tissue due to an oxygen deficiency in the systemic blood.

EDEMA Swelling due to abnormal accumulation of fluid in intercellular tissue spaces.

ENDARTERECTOMY Removal of an obstructing region of the inner wall of an artery.

HEMATOMA Tumor or swelling in tissue due to an accumulation of blood from a break in the wall of a blood vessel.

HYPOTENSION Low blood pressure.

OCCLUSION Clot or closure in the lumen of a blood vessel or other structure.

PHLEBOSCLEROSIS Thickening or hardening of walls of veins.

SHUNT Connection between two blood vessels or between two sides of the heart.

THROMBECTOMY Removal of a blood clot from a blood vessel.

VALVOTOMY Cutting into a valve.

VENIPUNCTURE Inserting a catheter or needle into a vein.

SUMMARY

Types of blood vessels

1 *Arteries* carry blood away from the heart to capillary beds throughout the body. Arterial blood is oxygenated, with the exception of the blood in the *pulmonary arteries,* which carry deoxygenated blood from the heart to the lungs.

2 The major arterial trunks are the *aorta* from the left ventricle and the *pulmonary trunk* from the right ventricle.

3 Arterial walls are composed of three layers: the inner *tunica intima,* the middle *tunica media,* and the outer *tunica adventitia.*

4 Arteries branch into smaller arteries and then into smaller *arterioles* shortly before reaching the capillary networks.

Terminal arterioles control the flow of blood from arteries into capillaries.

5 Arterioles enter the body tissues and branch out further to form *capillaries,* the bridge between the arterial and venous systems.

6 The *microcirculation* of the blood consists of the capillaries, terminal arterioles, metarterioles, and venules.

7 The three types of capillaries are *continuous capillaries, fenestrated capillaries,* and *sinusoids,* each differentiated to perform a specific function.

8 Blood drains from capillaries into *venules,* tiny veins that unite to form larger venules and veins. *Veins* carry oxygen-poor blood from the body tissues to the heart, with the exception of the *pulmonary veins,* the *hepatic portal system* that carries blood from the capillaries of the intestines to the capillaries of the liver, and the *hypophyseal portal system* in which veins formed from the capillaries of the hypothalamus divide into the capillaries of the anterior pituitary gland.

9 Veins usually contain paired semilunar bicuspid valves that permit blood to flow only toward the heart. Their walls contain the same layers as arterial walls.

10 Venous blood pressure is low, and blood is assisted toward the heart by the skeletal muscle pump and the respiratory pump.

Circulation of the blood

1 The *pulmonary circulation* carries oxygen-deficient blood from the heart to the lungs, where carbon dioxide is removed and oxygen is added. It then carries oxygenated blood back to the heart.

2 The *systemic circulation* supplies the tissues of the body with blood high in oxygen concentration, and also removes blood high in carbon dioxide. The main vessels of the *arterial division* are the aorta, ascending aorta, aortic arch, descending aorta, and common iliac arteries. The main vessels of the *venous division* are the superior vena cava, inferior vena cava, and coronary sinus.

3 Veins ordinarily transport blood directly back to the heart from a capillary network, but the two *portal systems* of the body (hepatic and hypophyseal) transport the blood to a second set of capillaries on its way to the venous system.

4 The brain is supplied with blood by two *vertebral arteries* and two *internal carotid arteries.* All the blood entering the cerebrum must first pass through the *cerebral arterial circle* (circle of Willis).

5 The arrangement of blood vessels in the skin allows for the increase or decrease of heat radiation from the integumentary system.

6 Blood nourishes skeletal muscles and also removes wastes. The main controllers of blood flow are the sympathetic vasodilator fibers that cause the blood vessels to dilate.

7 The *circulatory system of a fetus* differs from that of a child or adult in that the fetus gets nutrients and removes its wastes through the placenta, and its lungs, kidneys, and digestive system (except for the liver) do not function.

8 The fetus has an opening in the septum between the atria called the *foramen ovale,* and a vessel called the *ductus arteriosus* that bypasses the lungs by carrying blood from the pulmonary artery to the aorta. In a normal child, both close at birth.

Major arteries and veins

Tables 19.2 through 19.9 and accompanying Figures 19.12 through 19.20 summarize the major arteries and veins of the body.

Physiology of circulation

1 *Hemodynamics* is the study of the principles that govern blood flow.

2 *Blood flow* refers to the volume (quantity) of blood flowing through a vessel during a specific period of time. *Blood pressure* is the force with which blood is pushed against the walls of blood vessels. *Blood velocity* is the distance blood moves along a vessel during a specific time period. *Peripheral resistance* is the impediment to blood flow by friction. It depends on the length, radius, and total cross-sectional area of the vessel and the viscosity of the blood.

3 Blood-flow homeostasis is maintained by a combination of *nervous* control, *hormonal* and *chemical* control, and *metabolic* control.

4 Factors affecting blood pressure are gravity (hydrostatic pressure), blood volume, vessel elasticity, cardiac output, blood viscosity, and resistance.

5 Specific neural, chemical, and hormonal factors help regulate arterial blood pressure by acting directly on blood vessels. They include the *vasomotor center,* the *cardioregulatory center, baroreceptors, chemoreceptors,* various hormones, the *kidneys, capillary fluid shift,* and *higher brain centers.*

6 Because the heart pumps all of the blood it receives, the pressure of the venous blood and its return to the heart are important in maintaining homeostasis. The regulation of venous blood return is determined by *mean arterial pressure, right atrial pressure, resistance to blood flow* through vessels, *venous pumps,* and *hydrostatic pressure.*

7 Blood-pressure levels are expressed by two numbers. The first number is the *systolic pressure* and the second is the *diastolic pressure.*

8 The difference between systolic and diastolic pressure is called the *pulse pressure.* It is determined by *arterial distensibility, stroke volume,* and the *speed of blood ejected* from the left ventricle into the systemic circulation.

9 Arterial blood pressure changes throughout the cardiac cycle. *Mean arterial pressure* (MAP) is the average pressure that drives the blood through the systemic circulatory system.

10 An instrument used for measuring blood pressure is the *sphygmomanometer.*

11 A *pulse* is a beat felt on the surface of the skin over a nearby artery. It corresponds to the beat of the heart and the alternating expansion and recoil of the arterial wall.

Physiological and anatomical abnormalities

1 An *aneurysm* is a balloonlike dilation of a blood vessel that occurs most often in the large arteries.

2 *Atherosclerosis,* the leading cause of coronary artery disease, is characterized by deposits of fat and other substances on the inside of arterial walls.

3 *Coronary artery disease* is a condition in which there is a reduced supply of oxygen and other nutrients to the myocardium.

4 *Hypertension,* commonly called high blood pressure, is a continuously elevated blood pressure. There are two types: *primary (idiopathic)* and *secondary.*

5 A *stroke* occurs when the brain's blood supply is cut off.

6 *Thrombophlebitis* is an acute condition in which a blood clot forms and inflammation occurs in veins.

7 *Varicose veins* are abnormally dilated and twisted veins, usually located in the legs.

8 *Portal cirrhosis of the liver* involves the growth of fibrous tissue, which blocks blood flow to the liver and results in fluid accumulation in the peritoneal cavity.

9 *Patent ductus arteriosis* is a condition in which the ductus arteriosis does not shut down at birth.

10 *Transposition of the great vessels* is a reversal of the placement of the arterial trunks so that the aorta emerges from the right ventricle and the pulmonary artery emerges from the left ventricle.

UNDERSTANDING THE FACTS

1 Arteries branch into smaller vessels called _____ , which then branch into smaller _____ .

2 Define vasa vasorum.

3 Describe the sinusoids of the liver.

4 What is the function of stellate reticuloendothelial cells (Kupffer cells)?

5 What functions do precapillary sphincters perform?

6 The _____ drains blood from the walls of the heart.

7 List the major blood vessels of the pulmonary circulation.

8 Much of the blood in the fetus bypasses the liver by flowing through the _____ .

9 List the major arteries that branch from the aorta.

10 How do the walls of veins differ from the walls of arteries?

11 What is microcirculation?

12 Define mean arterial blood pressure.

13 What is the pulse?

14 How does exercise affect venous circulation?

15 Define aneurysm.

16 List some factors that can increase the progress of atherosclerosis.

17 What factors are often related to hypertension?

18 Give the location of the following (be specific):

   a external carotid artery
   b vasa vasorum,
   c capillary channels
   d common iliac vein
   e foramen ovale
   f chemoreceptors
   g great saphenous vein

UNDERSTANDING THE CONCEPTS

1 Why is it better to classify blood vessels by direction of blood flow rather than by the type of blood they carry?

2 Describe the structure of the arterial wall and how it relates to the function of arteries.

3 How do the arterioles help the body to cope with environmental heat?

4 Why is it important that the muscular system be nourished by the blood on a priority basis?

5 Your text states that the arterial system holds about 20 percent of the blood, while the venous system holds about 75 percent. How do you account for this difference?

6 In fetal circulation, the brain receives blood that is rich in oxygen. How is this accomplished?

7 Discuss the modifications of the circulatory system that occur in the fetus.

8 What are some of the arterial conditions that can bring about an increase in blood pressure?

9 What are the physiological implications of the arteries having walls that are elastic?

10 How will dehydration of the body affect circulation?

11 What is the role of the baroreceptors in maintaining normal blood pressure?

12 What is the role of renin in blood pressure?

13 Describe the process of taking blood pressure, and explain what is actually occurring within the arm.

14 Describe the major processes that permit fluid to move across capillary walls.

15 What factors specifically regulate mean arterial pressure?

16 Discuss the role of exercise in the promotion of a healthy cardiovascular system.

SUGGESTED READING

BENDITT, EARL P., "The Origin of Atherosclerosis." *Scientific American,* February 1977.

BROWN, MICHAEL S., AND JOSEPH L. GOLDSTEIN, "How LDL Receptors Influence Cholesterol and Atherosclerosis." *Scientific American,* November 1984.

DONALD, D. E., AND J. T. SHEPHERD, "Autonomic Regulation of the Peripheral Circulation." *Annual Review of Physiology,* 42 (1980): 429–439.

GUYTON, A. C., T. C. COLEMAN, AND J. MARKS, "Blood Pressure Regulation. Basic Concepts." *Federal Proceedings,* 40 (1981): 2252–2256.

JOHANSEN, KAJ, "Aneurysms." *Scientific American,* July 1982.

MURPHY, R. A., "Control of Tone in Vascular Smooth Muscle." *Archives of Internal Medicine,* 143 (1983): 1001–1006.

PICKERING, T. G., A. W. NELSON, AND H. L. ADAMS, "Blood Pressure During Normal Daily Activities, Sleep, and Exercise." *Journal of the American Medical Association,* 247 (1982): 992–996.

RODKIEWICZ, C. M., ed., *Arteries and Arterial Blood Flow: Biological and Physiological Aspects.* New York: Springer-Verlag, 1983.

SMITH, JAMES J., AND JOHN P. KAMPINE, *Circulatory Physiology: The Essentials,* 2nd ed. Baltimore: Williams & Wilkins, 1984.

WOOD, J. E., "The Venous System." *Scientific American,* January 1968.

ZWEIFACH, BENJAMIN W., "The Microcirculation of the Blood." *Scientific American,* January 1959.

ZWEIFACH, B. W., A. P. SHEPHERD, AND P. M. VANHOUETTE, "Selected Topics on Microcirculation." *Physiologist,* 25 (1982): 353–396.

20
The Lymphatic System and Immunity

LEARNING OBJECTIVES

1 Give the four major functions of the lymphatic system.

2 Compare the composition of lymph and blood.

3 Describe the structure and function of the lymphatic capillaries and other vessels.

4 Trace the circulation of lymph.

5 Describe the structure and the functions of the lymph nodes, tonsils, spleen, and thymus gland.

6 Explain the main differences between the nonspecific and specific defenses of the body.

7 Describe the sequence of events in the inflammation response and its role in the body.

8 Explain how phagocytosis helps the body combat infection.

9 Tell how complement, properdin, and interferon contribute to the body defenses.

10 Define antigen, antibody, and antigenic determinant, and describe their structures.

11 Name the classes of antibodies, and compare their functions.

12 Describe the formation of B cells and T cells.

13 Differentiate between humoral and cell-mediated immunity.

14 Describe the function of memory cells.

15 Explain how the clonal selection process works.

16 Describe the functions of helper T cells, suppressor T cells, and killer T cells.

17 Define hypersensitivity, and compare immediate and delayed hypersensitivity.

18 Compare autoimmune diseases with immune deficiency diseases.

19 Define innate resistance and acquired immunity, natural and artificial immunity, and active and passive immunity.

20 Explain some uses of monoclonal antibodies.

Because the cardiovascular system and the lymphatic system both circulate fluid throughout the body, the lymphatic system is sometimes called the "second circulatory" system. But there are distinct differences. The lymphatic system is not a closed, circular system, and it does not have a central pump. It is made up of a network of thin-walled vessels that carry a clear fluid that helps to maintain the proper fluid balance in the tissues and blood, conserves protein, and removes unwanted substances from the tissues. Basically, the lymphatic system performs four major functions:

1 It collects excess water and proteins from the fluids that bathe cells throughout the body, and returns them to the blood.

2 It transports fats from the tissue surrounding the small intestine to the blood.

3 It filters and destroys microorganisms and other foreign substances.

4 It provides long-term protection against microorganisms and other foreign substances.

THE LYMPHATIC SYSTEM

The **lymphatic system** begins with very small vessels called lymphatic capillaries, which are in direct contact with interstitial fluid and surrounding tissues (Figure 20.1). The system collects and drains most of the fluid that seeps from the bloodstream and accumulates in the spaces between cells. (A small amount of fluid also moves back into the blood capillaries.) The small lymphatic capillaries merge to form larger lymphatic vessels called lymphatics, which pass through specialized structures called lymph nodes. These larger vessels converge in two main drainage ducts that return the excess fluid to the blood circulation through the subclavian veins at the base of the neck. All the tissues of the body except those of the central nervous system and the cornea are drained by the lymphatic system.

In addition to lymphatic capillaries, lymphatics, and lymph nodes, the lymphatic system consists of lymphoid organs: the spleen and thymus gland (see Figure 20.1). The three pairs of tonsils are actually lymphatic tissue of the throat region. Lymphatic tissue is a special type of tissue, set apart from the fundamental tissue types already described. It is a variety of reticular connective tissue containing varying amounts of specialized white blood cells called lymphocytes (see Chapter 17 for a complete discussion of lymphocytes). The tissue is called either *loose lymphatic tissue* or *dense lymphatic tissue* depending on whether the lymphocytes are loosely or densely packed. All lymphatic organs contain relatively large numbers of lymphocytes within a framework of reticular cells and fibers. Finally, aggregated lymph nodules, which are scattered throughout the small intestine, help instigate the secretion of antibodies. (The components of the lymphatic system are summarized in Table 20.1.)

TABLE 20.1	MAJOR STRUCTURAL COMPONENTS OF THE LYMPHATIC SYSTEM
Structure	**Major functions**
Lymphatic capillaries	Collect excess interstitial fluid in tissues.
Lymphatics (collecting vessels)	Carry lymph from lymphatic capillaries to veins in the neck, where it is returned to the bloodstream.
Lymph nodes	Situated along collecting lymphatic vessels to filter foreign material from the lymph on its way to the bloodstream.
Tonsils	Destroy foreign substances at the upper entrances of the respiratory and digestive systems.
Spleen	Filters foreign substances from blood, manufactures phagocytic lymphocytes, stores red blood cells, releases blood to the body in case of extreme blood loss.
Thymus gland	Forms antibodies in newborn, is involved in initial development of immune system, site of differentiation of lymphocytes into T cells, produces thymosin.
Aggregated lymph nodules (Peyer's patches, gut-associated lymphoid tissue, or GALT)	Respond to antigens in the intestine by generating plasma cells that secrete antibodies.

Lymph

The blood hydrostatic pressure inside blood vessels, which is generated by ventricular contraction, causes water, small proteins (albumin), and other materials to be forced out of capillaries into the spaces between cells (see p. 602). This **interstitial fluid** bathes and nourishes surrounding body tissues. However, if there were not some means of draining it from the tissue, excess interstitial fluid would cause the tissues to swell, producing *edema*. Normally, however, excess fluid moves from the body tissues into the far-reaching small lymphatic capillaries and through the lymphatic system until it returns to the blood. Once inside the lymphatic capillaries, the fluid is called **lymph** (L. "clear water"). The flow of lymph is affected by the pressure of tissue fluids, the lymphatic capillary pump, the valves of the lymphatic vessels, the contraction of smooth muscle in the walls of lymphatic vessels, the pressure on lymphatic vessels by surrounding skeletal muscles, the pulsations of adjacent arteries, and the auxiliary respiratory pump. The flow in the lymphatic system most clearly resembles venous return.

FIGURE 20.1

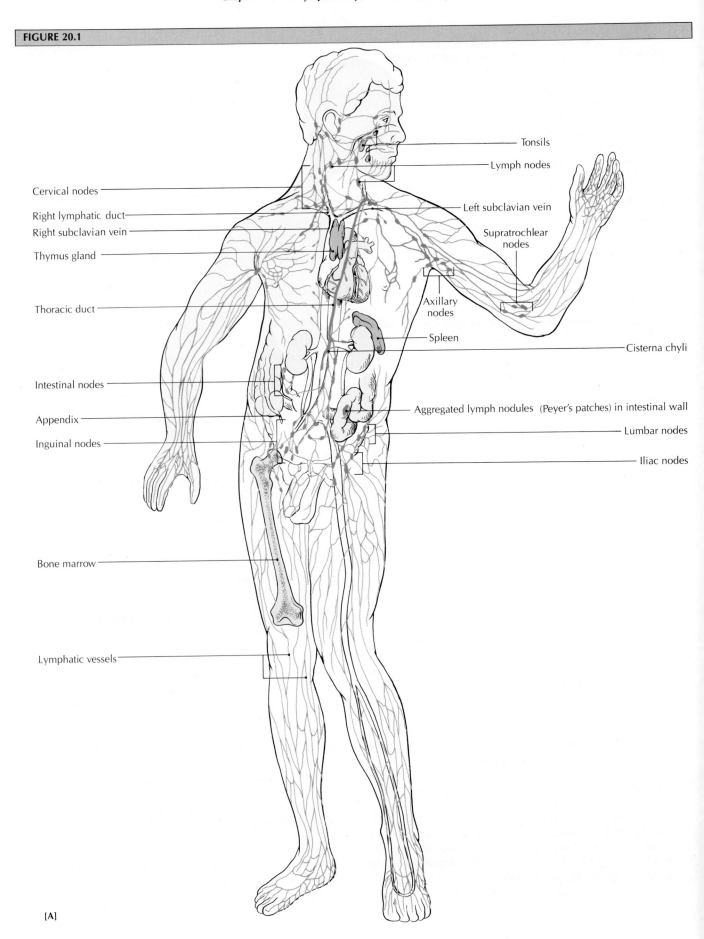

Tonsils
Lymph nodes
Cervical nodes
Right lymphatic duct
Right subclavian vein
Thymus gland
Left subclavian vein
Supratrochlear nodes
Thoracic duct
Axillary nodes
Spleen
Cisterna chyli
Intestinal nodes
Appendix
Aggregated lymph nodules (Peyer's patches) in intestinal wall
Inguinal nodes
Lumbar nodes
Iliac nodes
Bone marrow
Lymphatic vessels

[A]

The lymphatic system. (A) Major components of the lymphatic system, including central lymphoid tissues (bone marrow and thymus), and peripheral lymphoid tissues. The right lymphatic duct drains the upper right quadrant of the body (yellow), and the thoracic duct drains the rest of the body. (B) Relationship of the blood vessels and lymphatic vessels (green). Oxygenated blood in arteries is shown in red, and deoxygenated blood in veins is shown in blue. Excess fluid and proteins from the blood are collected in tissue spaces by lymphatic capillaries and returned to the venous system by lymphatic vessels. (C) Enlargement showing the relationship of lymphatic and blood capillaries. Arrows indicate the direction of fluid movement. Because of the higher blood hydrostatic pressure at the arterial end, the filtration of fluid occurs out of blood capillaries into the interstitial fluid. (D) Basic structure of lymphatic tissue.

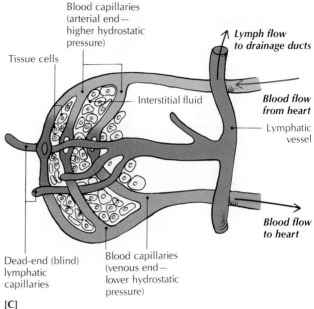

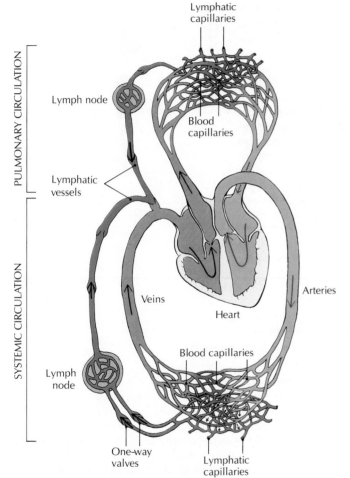

[B]

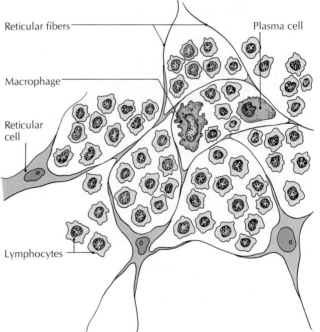

[D]

Composition and cells of lymph The body contains about 1 to 2 L of lymph, or about 1 to 3 percent of body weight. Like blood, lymph contains water, some plasma proteins, electrolytes, lipids, leukocytes, coagulation factors, antibodies, enzymes, sugars, nitrogen, urea, and amino acids. (It does not contain erythrocytes and most of the proteins found in blood.) The cells of the lymph fluid are *leukocytes,* and they perform much the same protective functions as leukocytes do in the blood (see Chapter 17). Most of the leukocytes essential to the function of lymph are produced in the bone mar-

row, and are found in the blood as well as in the organs and lymph nodes of the lymphatic system.

Monocytes are a class of leukocytes capable of developing into phagocytic **macrophage cells.** Macrophages are collectively referred to as the *reticuloendothelial system* ("network of endothelial cells"). These cells are found in lymph, in lymph nodes, and adhering to the walls of blood vessels and lymph vessels. Macrophages also migrate into and become attached to many other tissues, and are then referred to as *tissue macrophages* or **histocytes** ("tissue cells"). Histocytes are common

618

THE DYNAMIC BODY

The Skin as an Infection Fighter

Only recently has the skin, with all of its other protective functions, been recognized as an integral part of the immune system that protects the body from infection. This discovery was made after it was determined that the epithelial cells of the thymus gland and epidermis have genetic and structural similarities. (The thymus gland functions not only as part of the endocrine system, but also as an important component of the immune system.)

One of the most startling findings in the search for *functional* similarities between cells of the skin and the immune system was that keratinocytes in the skin secrete interleukin-1, a chemical also secreted by macrophages throughout the immune system. Prior to this discovery, keratinocytes were known to generate the horny outer layer of keratin that protects hair and skin. Now it could be shown that these cells also secreted hormonelike chemicals that affected the maturation of skin lymphocytes, the functional cells of the immune system.

Defensive Cells Are Mobilized

The secretion of interleukin-1 by keratinocytes is critical for the eventual production of lymphocytes in the skin. When molecules of a foreign substance (antigen) penetrate the skin's outer layer of keratin, they bind to the surface of nonpigmented granular dendrocytes (Langerhans' cells) in the skin, which then present the antigen molecules to skin lymphocytes that have surface receptors that bind to the antigen (see drawing). This binding exposes a second receptor on the lymphocytes, which receives molecules of interleukin-1 from nearby keratinocytes. The lymphocytes are then stimulated to release interleukin-2, which binds to antigen receptors on other lymphocytes. These lymphocytes are stimulated to start the proliferation of many identical lymphocytes, which attack the invading molecules. Defensive lymphocytes can enter vessels of the lymphatic system, traveling throughout the body to attack antigens that may have spread beyond the skin.

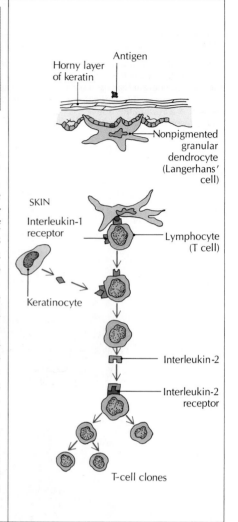

in the walls of the lung, the sinusoids of the liver (where they are called stellate reticuloendothelial cells or Kupffer cells), the spleen (the littoral cells), and the bone marrow. These cells group into large clusters that surround and isolate foreign particles that are too large to phagocytize. This "walling-off" process occurs in certain chronic infections, such as tuberculosis, and limits the spread of the microorganism.

Macrophages in the lymph nodes are active phagocytes, and play a major role in resistance against *all* invading particles. They also are important in resistance against a specific foreign microorganism or its toxin. However, specific defenses are accomplished by another class of leukocyte, the *lymphocyte.*

The two fundamental types of lymphocytes are **B cells** and **T cells** (also called **B lymphocytes** and **T lymphocytes**).* The B cells produce specific antibodies, and the T cells attack specific foreign cells. The human body contains about 2 trillion lymphocytes. They are the backbone of the immune system, and are the basis of the immune response. The complex functions of the B and T cells are discussed later in this chapter (see pp. 631 and 633).

*It was not until 1972 that a clear distinction could be made in human lymphocytes between B cells and T cells. (B and T cells were identified in mice in 1970.)

Lymphatic Capillaries and Other Vessels

Lymphatic capillaries and blood capillaries are somewhat similar in structure. Both consist of a single layer of endothelial tissue that permits fluid absorption and *diapedesis,* the passage of blood cells through vessel walls. The major structural difference is an important one: lymphatic capillaries are one-way vessels with a dead-end (blind) terminal end, not part of a circuit of vessels as the blood capillaries are.

Lymphatic capillaries are slightly wider than blood capillaries. They are formed by a single layer of endothelial cells. These cells regulate the passage of materials into and out of the lymph. The ends of adjacent endothelial cells overlap to form *flap valves* that open to permit fluid to enter the capillary, and close when the capillary contracts (Figure 20.2).

Lymphatic capillaries are most abundant near the innermost and outermost surfaces of the body, for example, the dermis of the skin, and the mucosal and submucosal layers of the respiratory and digestive systems. Lymphatic capillaries are also numerous beneath the mucous membrane that lines the body cavities and covers the surface of organs. Very few lymphatic capillaries are found in muscles, bones, or con-

FIGURE 20.2

Lymphatic capillaries. (A) Flap valves between adjacent endothelial cells in lymphatic capillaries open to permit tissue fluid to enter, and close to prevent leakage. (B) Electron micrograph of "loose junction" between the overlapping ends of two endothelial cells in the wall of a lymphatic capillary. [*Johannes A. G. Rhodin, M.D., Ph.D., Department of Anatomy, College of Medicine, Tampa, Florida.*]

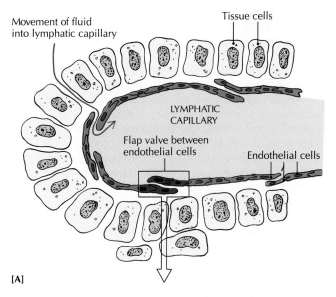

[A]

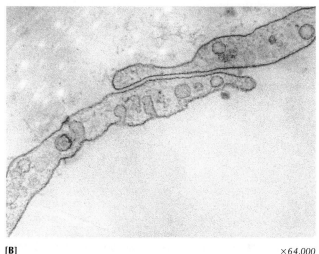

[B]   ×64,000

nective tissue. There are none in the central nervous system and cornea of the eyeball.

Specialized lymphatic capillaries called *lacteals* (L. *lacteus,* of milk) extend into the intestinal villi. Lacteals absorb fat from the small intestine and transport it into the blood for distribution throughout the body. The lymph in the lacteals takes on a milky appearance (hence the name *lacteal*) because of the presence of many small droplets of fat. At that point, the mixture of lymph and finely emulsified fat is known as *chyle* (KILE; L. *chylus,* juice).

The lymphatic capillaries join with other capillaries to become larger collecting vessels called *lymphatics.* Lymphatics resemble veins, but their walls are thinner than venous walls,

they contain more valves, and they pass through specialized masses of tissue (the lymph nodes). Lymphatic vessels are usually found in loose connective tissue, running parallel to blood vessels. Lymphatics are arranged into a superficial set and a deep set within the body (see Figure 20.1A), and pass through various lymph nodes. The superficial lymphatic vessels in the skin and subcutaneous tissue tend to follow the course of superficial veins, and the deeper vessels follow the deep veins and arteries.

Lymphatics join with one another to form two large ducts, the *right lymphatic duct* and the *thoracic duct,* that empty their contents into the subclavian veins above the heart.

Lymph travels in only one direction because of valves within the lymphatics that do not allow fluid to flow back (see Figure 20.1B). The lymphatic valves operate in the same way as the one-way valves in veins.

Circulation of Lymph

Since there is no central pump to circulate the lymph, the actual movement is accomplished primarily by three other forces. (1) The action of circular and some longitudinal smooth muscles in the lymphatic vessels other than capillaries. (2) The squeezing action of voluntary muscles during normal body movement helps to move lymph through the vessels. Lymph flow may increase 5- to 15-fold during vigorous exercise. (3) The lymphatic system runs parallel with the venous system in the thorax, where a subatmospheric pressure exists. This pressure gradient creates a "pull factor, called the auxiliary respiratory pump," that aids lymph flow.

All lymph vessels are directed toward the thoracic cavity. The upper right quadrant of the body contains lymphatics that drain their contents into the right subclavian vein through the *right lymphatic duct* (see Figure 20.1A). This drainage includes the right side of the head, right upper extremity, right thorax and lung, right side of the heart, and the upper portion of the liver. The remainder of the lymphatic system returns its fluid to the left subclavian vein through the *thoracic* (left lymphatic) *duct,* the largest of the lymphatics. The circulation of lymph through the lymphatic system may be summarized as follows:

1 The upper right quadrant of the body drains into the right lymphatic duct.

2 The right lymphatic duct empties into the right subclavian vein.

3 The three-quarters of the body not drained by the right subclavian vein is drained by the main thoracic (left lymphatic) duct.

4 The thoracic duct begins as a dilated portion called the *cisterna chyli* within the abdomen, below the diaphragm. It extends upward through the diaphragm, along the posterior wall of the thorax, into the left side of the base of the neck.

5 At the base of the neck, the thoracic duct receives the left jugular lymphatic vessels from the head and neck, the left subclavian vessels from the left upper extremity, and other lymphatic vessels from the thorax and related parts.

6 The thoracic duct then opens into the left subclavian vein, returning the lymph to the blood.

Lymph Nodes

Scattered along the lymphatic vessels, like beads on a string, are small (1 to 25 mm in length), bean-shaped masses of tissue called **lymph nodes** (Figure 20.3). The nodes are found in the largest concentrations at the neck, armpit, thorax, abdomen, and groin (see Figure 20.1B). Lesser concentrations are found behind the elbow and knee. The *superficial* lymph nodes are located near the body surface in the neck, armpit, and groin. The *deep* nodes are located deep within the groin area, near the lumbar vertebrae, at the base of the lungs, attached to the tissue surrounding the small intestines, and in the liver. Most lymph passes through at least one lymph node on its way back to the bloodstream.

Lymph nodes, which are sometimes incorrectly called glands,* filter out harmful microorganisms and other foreign substances from the lymph, trapping them in a mesh of reticular fibers. Lymph nodes are also the initiating sites for the specific defenses of the immune response. These functions are intimately related to the structure of the lymph node and the cells it contains.

Lymph nodes are covered by a *capsule* of fibrous connective tissue. Projections of connective tissue called *trabeculae* extend inward from the capsule toward the center of the lymph node, dividing it into compartments. The outer portion of each compartment is the *cortex* of the node. It contains lymphocytes in dense clusters called *lymph nodules.* In the middle of each nodule is the *germinal center,* where lymphocytes are produced by cell division. The lymph nodes produce about 10 billion lymphocytes every day. The inner part of a lymph node is the *medulla.* It contains strands of lymphocytes extending from the nodule. These strands are appropriately called the *medullary cords.*

Running from the cortex to the medulla, and surrounding the nodules and medullary cords, are *medullary sinuses,* through which the lymph flows before it leaves the node. The lymph node effectively funnels foreign materials in the lymph through the sinuses so that the lymph comes in contact with lymphocytes and macrophages. Afferent (*to* the node) lymphatic vessels with lymph enter the node at various points along the outer capsule. Efferent (*from* the node) lymphatic vessels leave the node from a small depressed area called the *hilus.* Blood vessels enter, as well as leave, through the hilus.

*Lymph nodes were originally called *lymph glands* (L. *glans,* acorn) because they seemed to resemble acorns. Soon, other small bits of tissue came to be called glands, even if they did not look anything like acorns. When it was discovered that some of these bits of tissue secreted various fluids, it was determined that a gland was a structure that formed secretions. Ironically, the original "glands," the lymph glands, do not secrete fluids, and were misnamed in the first place. The word *node,* derived from the Latin word for "knob," seems to describe the structure more appropriately.

Why are lymph nodes sometimes removed during cancer operations?	*Lymph nodes near a cancer site may contain viable cancer cells. On entering the lymph nodes, cancer cells can multiply and establish secondary cancers by dispersing cells throughout the body via the lymphatic system.*

Tonsils

The **tonsils** are aggregates of lymphatic nodules enclosed in a capsule of connective tissue. There are three pairs of tonsils: (1) the *pharyngeal* tonsils in the upper posterior wall of the pharynx behind the nose, (2) the *palatine* tonsils on each side of the soft palate, and (3) the *lingual* tonsils at the base of the tongue (see Figure 20.1A).

The tonsils lack afferent lymphatic vessels, a feature that distinguishes them from lymph nodes. The efferent lymphatics of the tonsils contribute many lymphocytes to the lymph. These cells are capable of leaving the tonsils and destroying invading microorganisms in other parts of the body. Together the tonsils form a band of lymphoid tissue that is strategically placed at the upper entrances to the digestive and respiratory systems, where foreign substances may enter easily. Most infectious microorganisms are either killed by lymphocytes at the surface of the pharynx, or are killed later, after the initial defenses are set in motion by the tonsils. The presence of plasma cells within the tonsils indicates the formation of antibodies.

Although tonsils usually function to prevent infection, they may become infected repeatedly themselves. In such cases, some physicians recommend their removal. The palatine tonsils are the ones most frequently removed (the familiar operation is called a *tonsillectomy*). The lingual tonsils are rarely removed. Pharyngeal tonsils, popularly called *adenoids,* may also become enlarged. If so, obstruction of the nasal pharynx caused by swelling can interfere with breathing, necessitating the removal of adenoids *(adenoidectomy).*

Spleen

The **spleen** is the largest lymphoid organ in the body. It is about the size and shape of a clenched fist, measuring about 12 cm in length, and is purplish in color. Located below the diaphragm on the left side of the body, it rests on portions of the stomach, kidney, and large intestine (see Figure 20.1A).

The main functions of the spleen are filtering blood and manufacturing phagocytic lymphocytes and monocytes. It also contributes to the functioning of the cardiovascular and lymphatic systems as follows:

1 Macrophages, abundant in the spleen, help remove damaged or dead erythrocytes and platelets, microorganisms, and other debris from the blood as it circulates through the spleen. Macrophages also remove iron from the hemoglobin of worn-out red blood cells, returning it to the circulation for use by the bone marrow in producing new red blood cells. The breakdown of hemoglobin also enables the production of bilirubin that can be circulated to the liver, where it becomes a constituent of bile.

2 Antigens in the blood of the spleen activate lymphocytes to develop into cells that either produce antibodies or are otherwise involved in the immune reaction.

3 The spleen produces red blood cells during fetal life. It can also store newly formed red blood cells and platelets in later life and release them into the blood as they are needed.

4 Because the spleen contains a large volume of blood, it

FIGURE 20.3

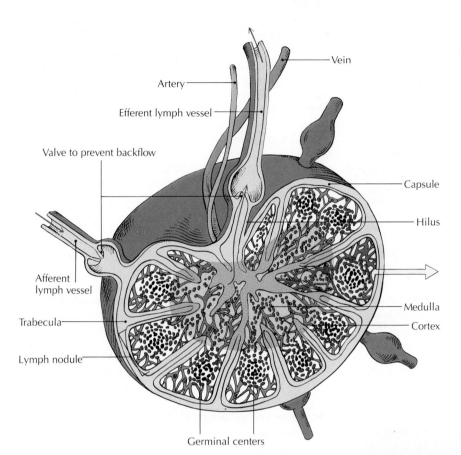

Vein

Artery

Efferent lymph vessel

Valve to prevent backflow

Afferent lymph vessel

Trabecula

Lymph nodule

Capsule

Hilus

Medulla

Cortex

Germinal centers

[A]

Lymph node. (A) Drawing of a node, showing how lymph flows through the sinuses surrounding the nodules. This arrangement brings potential antigens in contact with lymphocytes, monocytes, and macrophages, enhancing defensive reactions. The drawing on the right shows an enlarged portion of the lymph node. (B) Scanning electron micrograph of the interior of a lymph node. [*Fujita, Tanaka, and Togunaga,* SEM Atlas of Cells and Tissues. *Tokyo, Igaku-Shoin, 1981.*]

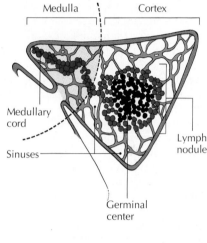

Medulla | Cortex

Medullary cord

Sinuses

Lymph nodule

Germinal center

serves as a blood reservoir. If the body loses blood suddenly, the spleen contracts and adds blood to the general circulation. It also relieves the venous pressure on the heart by releasing stored blood into the circulation during bursts of physical activity. (The spleen is capable of releasing approximately 200 mL of blood into the general circulation in one minute.)

The tissue structure of the spleen is similar to that of lymph nodes (Figure 20.4). Surrounded by a *capsule* of connective tissue, the spleen is divided by *trabeculae* into compartments called *lobules.* The functional part of the medulla consists of *splenic pulp,* which contains small islands of white pulp scattered throughout red pulp. The *white pulp* is made up of compact masses of lymphocytes surrounding small branches of the splenic artery. These masses, which occur in intervals, are called *splenic nodules (Malpighian corpuscles).* Within the *red pulp* are *venous sinusoids* filled with blood and lined with monocytes and macrophages. (The pulp is red because of the many erythrocytes in the blood.) This arrangement brings the blood and any foreign materials it may contain to the lymphocytes, monocytes, and macrophages for cleansing. Since the spleen does not receive lymphatics and lymph, it cannot be considered a filter in the same way that a lymph node is a filter. Efferent lymphatics, which leave the spleen, contribute lymphocytes, monocytes, and macrophages to the lymph.

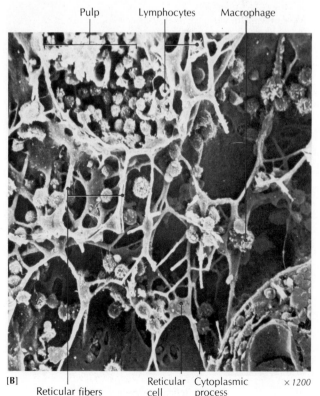

Pulp | Lymphocytes | Macrophage

[B]

Reticular fibers | Reticular cell | Cytoplasmic process | × 1200

FIGURE 20.4

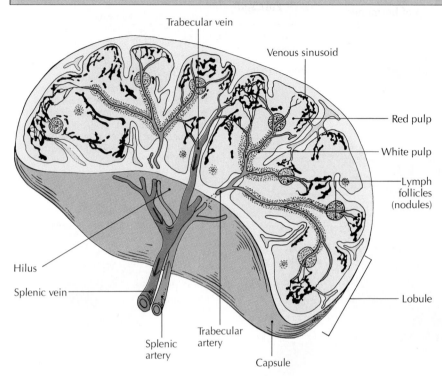

Trabecular vein

Venous sinusoid

Red pulp

White pulp

Lymph follicles (nodules)

Lobule

Hilus

Splenic vein

Splenic artery

Trabecular artery

Capsule

[A]

Spleen. (A) Drawing of the spleen, showing blood flow through the white pulp to the venous sinusoids in the red pulp. Note that the white pulp consists of nodules and lymphocytes, and that the red pulp is an open mesh with venous sinuses running through it. (B) Scanning electron micrograph of a section through the spleen. [*Fujita, Tanaka, and Togunaga.* SEM Atlas of Cells and Tissues. *Tokyo, Igaku-Shoin, 1981.*]

Venous sinuses   Splenic cords

×1100

[B]   Reticular fiber of sinus wall

Thymus Gland

The **thymus gland** is a ductless, pinkish-gray mass of flattened lymphoid tissue situated just behind the top of the sternum (see Figure 20.1A). It consists of two *lobes* joined by connective tissue, and surrounded by a *capsule* of connective tissue. Each lobe is divided into *lobules* by coarse *trabeculae*, with each lobule having an outer *cortex* and an inner *medulla*. The cortex contains many T cells, most of which degenerate before they ever leave the thymus gland. The only blood vessels within the thymus gland are capillaries. The medulla contains fewer lymphocytes than the cortex does, making the epithelial reticular cells in the medulla more obvious.

The fetal thymus gland is already involved with the production of lymphocytes, and assists in the formation of antibodies elsewhere. The thymus gland is relatively large at birth (about 12 to 15 g), forms antibodies in the newborn, and plays a major role in the early development of the body's immune system. It increases in size from birth to puberty, and is most active in childhood and early adolescence. The thymus gland shrinks progressively during adulthood, and degenerates into adipose and connective tissue in old age. Like the tonsils and spleen, the thymus gland has efferent lymph vessels, but no afferent ones. As a result, no lymph drains into it.

Undifferentiated lymphocytes migrate to the thymus gland to become specialized, or immune-competent T cells (T for "thymus-dependent") with defined roles in specific defenses against foreign cells. Other possible functions of the thymus gland are not understood completely, but it is clearly involved in the formation of a permanent immunity system of antibodies. The thymus gland also secretes a group of hormones collectively called *thymosin,* which may be necessary for the differentiation of T cells from stem cells. Other peptide hormones secreted by the thymus gland also appear to stimulate the activity of T cells.

Aggregated Lymph Nodules (Peyer's Patches)

Aggregated lymph nodules (Peyer's patches) are clusters of unencapsulated lymphoid tissue found in the intestine and appendix. Because of their location, they are also called *gut-associated lymphoid tissue* (GALT). Similar clusters of tissue are found along the respiratory tract.

Aggregated lymph nodules are thought to generate plasma cells that secrete antibodies in large quantities in response to antigens from the intestine (Figure 20.5). Such plasma cells do not remain clustered in aggregated lymph nodules, but are distributed along the length of the intestine in the following way. Inactive B cells migrate from bone marrow to aggregated lymph nodules. B cells activated by exposure to antigens leave the aggregated lymph nodules through efferent lymphatics, migrate to lymph nodes, and then later enter the blood through the thoracic duct. The B cells are carried by the bloodstream throughout the body, and eventually come to reside as plasma cells beneath the mucosal surface of the intestine. Here they recognize an enormous variety of specific antigens (foreign substances), and produce corresponding antibodies, primarily against bacteria and some viruses.

Ask Yourself

1 *What are some of the major components of lymph?*

2 *What are the functions of macrophages in lymph?*

3 *What are lymphatics?*

4 *What are the two main draining ducts of the lymphatic system?*

5 *What are the functions of lymph nodes?*

6 *How do the tonsils help prevent infection?*

7 *What are the major functions of the spleen?*

8 *Why is the thymus gland larger during childhood than in later life?*

9 *What are aggregated lymph nodules?*

NONSPECIFIC DEFENSES OF THE BODY

In this section and the next, we discuss the body's two main lines of defense: the nonspecific and specific defenses. The *specific defenses,* usually referred to as the immune response, involve the formation of antibodies, which help to destroy foreign substances. The *nonspecific defenses* do *not* involve

---

FIGURE 20.5

Aggregated lymph nodules (Peyer's patches). (A) Aggregated lymph nodules in the intestine expose lymphocytes to antigens in the intestinal lumen. (B) Lymphocytes leave the aggregated lymph nodules and move through the lymph and blood systems to locate as mature plasma cells beneath the intestinal epithelium (C). The plasma cells are believed to secrete immunoglobulin A (IgA).

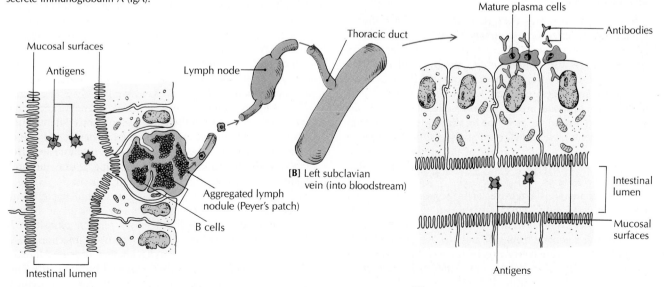

[A]   [C]

the production of antibodies. We begin with some familiar nonspecific defenses that are relatively uncomplicated in their action.

First Line of Defense: Restricting Entrance

The skin, hair follicles, and sweat glands contain many microorganisms that are barred from entering the body by the tightly packed cells of the dermis and epidermis. The acidic surface and high fat content of the skin also restrict the growth of microbes. Finally, microorganisms on the skin are removed as the outer layer of squamous cell epithelium is constantly worn away and replaced.

Some areas of the body that have openings to the outside, such as the nasal passages, mouth, lungs, and digestive system, are lined with *mucous membranes.* The stickiness of the secreted mucus traps many microorganisms, and prevents them from attaching to the epithelium or entering the tissues. Other nonspecific defensive factors are *sweat,* which washes microorganisms from the pores and skin surface, and tears from the *lacrimal glands,* which wash away foreign particles from the eyes. Both sweat and tears contain the enzyme *lysozyme,* which attacks bacterial cell walls to kill the invading organisms.

Many other general mechanical and chemical defenses exist. For example, *nasal hairs* filter the external air before it enters the upper respiratory tract, and *cilia* of the upper respiratory tract sweep bacteria and other particles trapped in mucus toward the digestive tract for eventual elimination. When microorganisms do enter the stomach, the extreme *acidity* of the stomach kills most of them. Finally, the normal microorganisms that live on the skin and mucous membranes provide a major barrier to potentially dangerous nonresident microorganisms, and also suppress their growth.

Inflammation Response

When tissues are damaged by punctures, abrasions, burns, foreign objects, infections, or toxins, the normal result is *inflammation* of the tissue. The sequence of events that follows such injuries is the **inflammation response.** Inflammation progresses through several phases. Localized effects of tissue damage and infection are *redness, heat, swelling,* and *pain.* The redness and heat are produced by vasodilation, which increases blood flow to the area. The localized edema (swelling) is caused by the increased amount and pressure of interstitial fluid. The pain is usually a result of the pressure of edema on receptors and nerves, and of chemical by-products that irritate nerve endings.

If tissue damage is extensive, or if the infection is widespread, inflammation will have *systemic effects,* such as changes in heart and breathing rates, increased body temperature, and a general feeling of malaise. When large numbers of neutrophils are fighting such an infection, they release *pyrogens,* chemical substances that raise the body temperature by acting on the temperature-regulation center in the hypothalamus. The resultant fever may help to fight infection by increasing the blood flow and leukocyte activity in the area, while inhibiting the growth of some infectious organisms. (See the box on p. 759.)

Sequence of events When tissues are damaged, *histamine* and other substances such as *kinins* are released from the injured cells, mast cells, and basophils. These substances cause vasodilation, which increases blood flow and the permeability of capillaries in the area (Figure 20.6). Any damage to the blood capillaries themselves also allows plasma and serum proteins such as globulins and fibrinogen to enter the interstitial space. The increased blood flow produces the characteristic redness and heat, and the accumulation of interstitial fluid in tissue spaces produces the characteristic edema and discomfort.

Healthy tissue is somewhat isolated from damaged tissue by the formation of fibrinogen clots at the site of the injury. As a result, tissue damage and toxic spread of infectious organisms are normally restricted to the immediate area of the wound.

Another early phase of inflammation begins with the migration of leukocytes to the affected area, where they accumulate. Soon after the initial damage, neutrophils begin to adhere to the inner walls of blood capillaries near the damage site. This action is called *margination.* It precedes the passage of the neutrophils through the vessel walls into the interstitial spaces by diapedesis. Neutrophils are attracted to the damage site by chemical substances that are released by the infecting microorganisms. This attraction is called *chemotaxis.* Monocytes, macrophages, and histocytes, which are also attracted to the site chemically, begin to arrive shortly after the neutrophils do.

A few hours after the damage, the tissues are greatly infiltrated by neutrophils. The neutrophils are mature cells, fully capable of phagocytizing bacteria and small particles of damaged tissue. Each neutrophil may consume up to 20 bacteria before dying.

When monocytes enter the tissues from the blood, they are still immature cells. After arriving in the tissues, they differentiate into macrophages, which migrate toward the damage site. Being larger and more active than neutrophils, macrophages can ingest as many as 100 bacteria, larger cells such as protozoa, and cell debris from dead or damaged tissues. As macrophages become more numerous and active, neutrophils become less and less active.

Drainage by lymphatics of tissue fluid, cell debris, proteins, and infectious cells from the damage site brings potential antigens into intimate contact with lymphocytes in the lymph nodes. This contact initiates the specific immune defenses involving the recognition of specific antigens and the production of antibodies.

Phagocytosis Nonspecific defenses rely on **phagocytosis** ("the process of eating cells") to combat infection. This process normally involves leukocytes that consume foreign substances or dead tissue, rather than healthy body cells. This distinction between "nonself" and "self" may be aided by the coating of foreign matter by a special group of globulin

FIGURE 20.6

The major events of inflammation.

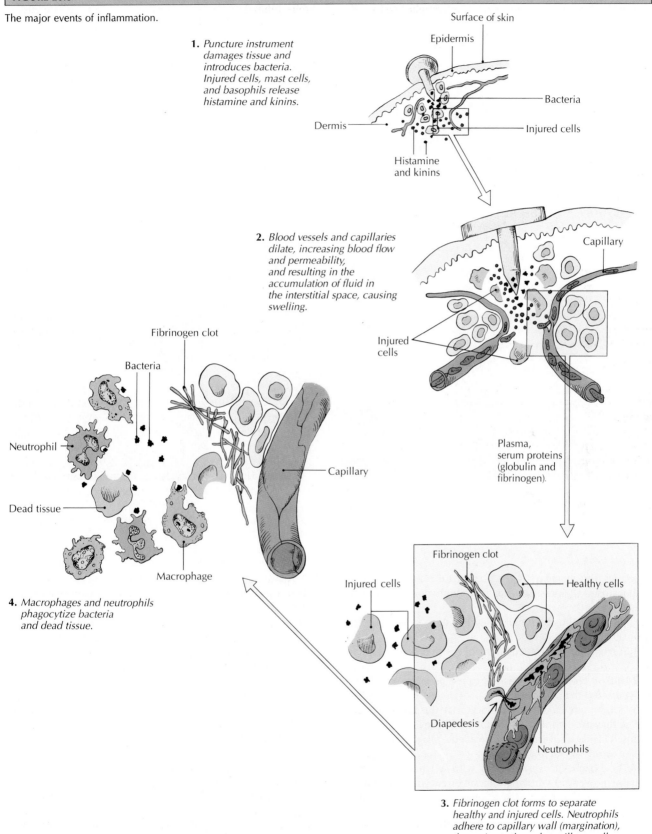

1. *Puncture instrument damages tissue and introduces bacteria. Injured cells, mast cells, and basophils release histamine and kinins.*

Surface of skin
Epidermis
Bacteria
Injured cells
Dermis
Histamine and kinins

2. *Blood vessels and capillaries dilate, increasing blood flow and permeability, and resulting in the accumulation of fluid in the interstitial space, causing swelling.*

Capillary
Injured cells
Plasma, serum proteins (globulin and fibrinogen).

Fibrinogen clot
Bacteria
Neutrophil
Dead tissue
Macrophage
Capillary

4. *Macrophages and neutrophils phagocytize bacteria and dead tissue.*

Fibrinogen clot
Injured cells
Healthy cells
Diapedesis
Neutrophils

3. *Fibrinogen clot forms to separate healthy and injured cells. Neutrophils adhere to capillary wall (margination), then migrate through capillary wall by diapedesis.*

FIGURE 20.7

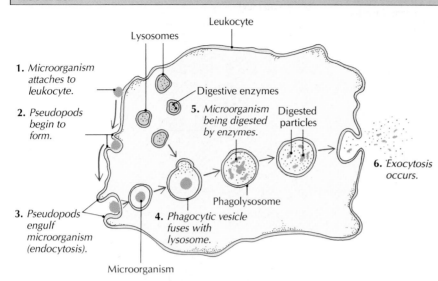

Leukocyte

Lysosomes

1. *Microorganism attaches to leukocyte.*

Digestive enzymes

5. *Microorganism being digested by enzymes.*

Digested particles

2. *Pseudopods begin to form.*

6. *Exocytosis occurs.*

3. *Pseudopods engulf microorganism (endocytosis).*

Phagolysosome

4. *Phagocytic vesicle fuses with lysosome.*

Microorganism

[A]

Phagocytosis. (A) Diagram of the events of phagocytosis in neutrophils and macrophages. (B) Scanning electron micrographs of a macrophage devouring a colony of bacteria. The phagocytic process usually takes less than a second. [*Lennart Nilsson.*]

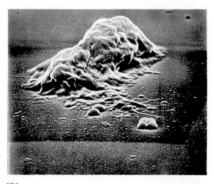

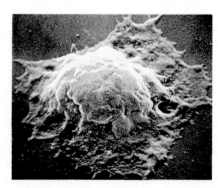

[B]   × 3000   × 3000   × 3000

molecules, the *opsonins* (Gr. "to prepare for eating"), prior to phagocytosis. Opsonins do not react with normal, healthy body cells.

Figure 20.7 illustrates the following steps in the process of phagocytosis:

1 Foreign particles or microorganisms attach to a leukocyte.

2 Leukocytes form protrusions called pseudopods ("false feet") around the particles.

3 Pseudopods engulf (by endocytosis) the foreign particles, so they are completely surrounded by the leukocyte, and enclosed inside a *phagocytic vesicle* (called a *phagosome*) within the leukocyte's cytoplasm.

4 The phagocytic vesicle fuses with a lysosome, which contains powerful hydrolytic (digestive) enzymes capable of breaking down proteins, lipids, nucleic acids, and complex polysaccharides.

5 After the phagocytic vesicle and lysosome fuse, the foreign particles and enzymes are contained within a single membrane contributed by both the phagocytic vesicle and

the lysosome membrane. The combined vesicles are now called a *phagolysosome*. The foreign particles are digested in the fused compartment.

6 Some of the digested contents may contribute nutrients to the leukocyte, while the rest are emptied outside the cell, where the phagolysosome fuses with the leukocyte's surface membrane. The expulsion process is known as *exocytosis*.

Nonspecific Antiviral and Antibacterial Substances

To protect itself from infection, the body is also able to produce several important chemical substances that attack bacteria and viruses: complement, properdin, and interferon. The actions of these substances are less specific than the antibody defenses that respond to an antigen, although complement does require a prior reaction of antibodies. This is an example of how nonspecific and specific defenses sometimes interact to maintain the body's homeostasis.

Complement *Complement* is a group of 11 proteins and 8 additional factors found in blood serum. It is named for its ability to enhance (or *complement*) the body's many defensive actions. Each protein in the complement group has the capacity to interact with the preceding complement protein, forming a "cascade reaction" like the one that operates during blood coagulation (see Figure 17.8). When the cascade reaction begins, the system is activated, and different proteins of the complement group perform either defensive or supportive roles.

The complement group may be activated by clusters of antibodies bound to antigens, pieces of DNA, fragments of cell membranes, polysaccharides, digestive enzymes similar to trypsin, and some bacterial proteins. All of these activating substances are present after damage to body cells, after the phagocytosis of microorganisms and the exocytosis of partially digested material, and during infection.

Some activated complement proteins are involved in the direct destruction of bacteria and viruses, and the prevention of viral attachment to cells. Other components increase the phagocytic activity of leukocytes, enhance the dilation of blood vessels, or play a role in chemotaxis. Certain complement proteins function only after antibodies have reacted with foreign material (see p. 628), and may promote the clumping of antigen-antibody complexes, making phagocytosis easier and more efficient. Phagocytosis is enhanced by the production of a specific complement fragment, C3b, which is an opsonin. Opsonins stimulate phagocytosis by adhering to bacteria, viruses, and antigen-antibody complexes, making them more susceptible to ingestion by phagocytes.

In the *classic pathway*, complement is activated by the interaction between antigen and antibody. However, in the *alternate pathway*, foreign substances can be attacked by a complement-mediated defense even before specific antibodies are formed.

Properdin *Properdin* (pro-PER-dihn; L. *pro*, acting as + L. *perdere*, to give away, hence "to destroy") is a protein found in blood serum. It exists in two forms, and the differences between them are not understood completely. Apparently, properdin functions to enhance the activation of complement by stabilizing and prolonging the existence of essential complement proteins. This antibody-independent activation may be an important initial defense against various types of infection.

Interferon Several different types of *interferons* are known. They are small protein molecules produced by virally infected cells. Interferons have the ability to protect other body cells by preventing viruses from multiplying within them. In viral infections, interferon is the first of the body defenses to appear. However, the interferon molecule itself does not engage in antiviral activity. Instead, when it binds to a body cell, it causes the cell to produce specific enzymes, which block the synthesis of specific proteins that the virus needs in order to multiply. This nonspecific defense of the body allows the rapidly acting interferon system to contain a viral infection until the slower-acting immune system can begin functioning. Interferon also stimulates lymphocytes that attack and kill virus-infected cells and cancer cells.

Ask Yourself

1 *How are the nonspecific defenses of the body characterized?*

2 *What are some ways that the body keeps out foreign substances?*

3 *How is the inflammation response initiated?*

4 *What are the events in the inflammation response?*

5 *How is phagocytosis accomplished?*

6 *What is the function of complement?*

7 *What are interferons? How do they function?*

SPECIFIC DEFENSES: THE IMMUNE RESPONSE

The nonspecific defenses of the body do not discriminate among types of foreign materials. The **specific defenses,** which constitute the **immune response,** do discriminate among foreign substances *(antigens)* by forming specific proteins, called *antibodies,* and/or specific cells that react with the foreign substances and help destroy them. The overall protective mechanism is known as **immunity** (L. *immunis,* exempt).

The so-called **immune system** is made up of lymphatic tissues and organs located all over the body. These separate tissues and organs communicate with each other via cells circulated by lymphatic and blood vessels. The function of the immune system is to protect the body against harm from the outside world, and to ensure that the defensive cells of the system do not turn against the body. The functional units of the immune system are lymphocytes. About 2 trillion lymphocytes in the body produce about 100 million trillion antibodies. What is even more impressive than this number is that *not all antibodies are alike.* For example, antibodies against the measles virus will not react with the polio virus, and those against the *Streptococcus* bacterium will not operate against *Staphylococcus.* Millions of different molecular configurations exist to cope with the huge number of possible antigens and combinations of antigens. It is estimated that the human body can recognize approximately 1 million antigenic determinants (small, identifiable portions of antigens). In 1987 Susumu Tonegawa won the Nobel Prize in physiology or medicine for discovering how the body can change its genes to fashion a seemingly unlimited number of antibodies, each specifically targeted at an invading microbe or foreign substance.

Antibodies enormously increase the body's ability to distinguish self (body) from nonself (substances that are not part of the body), and help maintain the integrity and homeostasis of the individual. Skin transplants from one area of a person's body to another will be accepted as self, but transplants from one person to another (except an identical twin) will be rejected as nonself.

After the body is exposed to a foreign substance for the

first time, the immune machinery has the capacity to "remember" this exposure and mount a rapid and specific antibody defense the next time the same substance is encountered. Both the killing and memory capacities of antibodies are derived from lymphocytes.

Antigens

An *antigen* is a substance against which an antibody is produced. It is a large protein or polysaccharide molecule that is ordinarily foreign to body tissues. When antigens enter the body, they cause the production of specific antibodies. This characteristic is called the *antigen immunogenicity*. They also react chemically with the specific antibody produced to form a stable complex called the **antigen-antibody complex.** This characteristic is known as the *antigen reactivity*. A substance that has both characteristics is called a *complete antigen*.

An antibody does not form against, or react with, the entire antigen. Instead, it combines with specific sites on the surface of the antigen. These combining sites, called **antigenic determinants,** or *epitopes* (see Figure 20.8A), are composed of distinctive proteins or polysaccharides. B and T cells not only can recognize antigens from their surface markings, but can also distinguish one specific antigen from among the over 100,000 known ones, and react accordingly.

When antibodies or specific T cells are said to form "against the *Streptococcus* bacterium," it means that the bacterium has specific sites that fulfill the requirements of an antigen; they are both *immunogenic* (they cause the production of antibodies) and *reactive* (they form a stable antigen-antibody complex). In the case of *Streptococcus,* these antigenic determinants on its surface are polysaccharides. A single cell, such as a bacterium, may carry several different antigenic determinants, with each one eliciting a response from a different specific antibody. In contrast, each type of B cell has only one type of antibody on its cell surface.

Almost any protein is capable of acting as an antigen. Egg albumin (egg white), incompatible blood cells, proteins on the surfaces of transplanted tissues, pollen, dust, animal danders (flaking skin), animal hair, and some components of foods are common examples of potential antigens. Also, because only antigenic determinants, not the entire organism, are necessary to produce an immune reaction, bacterial components such as flagella, capsules, cell walls, and toxins can be antigenic.

Certain small substances with very low molecular weights can be antigenic. Such substances are called *haptens,* and are immunogenic only when they are combined with a much larger *carrier molecule,* such as a protein. The antibiotic *penicillin* is an example of a hapten. In some patients who receive penicillin for the first time, it combines with a protein carrier to cause the formation of antibodies against penicillin. Subsequent doses of penicillin will cause the antibodies to react with the small penicillin molecule, producing a severe immune response (an allergy, in this case).

Antibodies

Antibodies are proteins produced by lymphocytes in response to an antigen. They are specialized to react with antigens, triggering a complex process, called *immunity,* that protects the body by destroying the invader. In contrast to phagocytes, which provide an immediate defense against infections, antibodies contribute an *active immunity* that provides relatively long-term protection against chronic infections. Antibodies cannot kill foreign organisms on their own, but they initiate the killing of such organisms by activating complement, phagocytes, and killer cells. Antibodies can also combine with viruses or bacterial toxins to prevent them from binding to receptors on their target cells.

Structure and specificity of antibodies The molecular structure of antibodies is a crucial part of their ability to react with specific antigens. An antibody molecule is Y-shaped and symmetrical, at least most of the time (Figure 20.8). (Because the molecule is flexible, it can also assume a T shape.) The molecule is composed of four polypeptide chains: two identical heavy chains and two identical light chains. Each chain is made up of amino acids. The light chains contain about 220 amino acids, and the heavy chains contain about 440 amino acids.

FIGURE 20.8

Fundamental unit of antibody molecular structure. (A) The variable regions have a different amino-acid structure for each specific antibody molecule. The function of the carbohydrate group is not known. (B) Schematic drawing of the structure of an IgG molecule. Each sphere represents an amino acid.

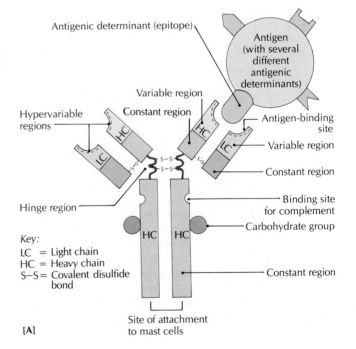

Antigenic determinant (epitope)

Antigen (with several different antigenic determinants)

Hypervariable regions

Variable region

Constant region

Antigen-binding site

Variable region

Constant region

Binding site for complement

Carbohydrate group

Hinge region

Key:
LC = Light chain
HC = Heavy chain
S—S = Covalent disulfide bond

Constant region

Site of attachment to mast cells

[A]

Why don't we become immune to colds?	*The common cold is caused by many strains of viruses, and an immunity to one strain does not convey immunity to another.*

The single stem of the Y is formed by two parallel heavy chains joined together by covalent disulfide bonds (—S—S—). The two "arms" of the Y are formed by a bending out of the heavy chains. Parallel to each bent portion of the heavy chain is a short light chain (see Figure 20.8A). The light chains are joined to the bent heavy chains by disulfide bonds.

The molecular structure of antibodies varies only in the regions where reactions with antigens can take place. This region, which is the tip of each arm of the Y, is therefore known as the *variable region*. It has the structural potential to bind to specific antigens like a lock (antibody) accepting a key (antigen) (see Figure 20.8A). The remainder of the arms of the Y, and the entire stem, are always identical from one antibody class to another. This portion of the antibody is the *constant region*. Each heavy chain also contains a flexible *hinge region*, so that antigen-binding sites need not remain a fixed distance apart. Because of the hinge region, the antibody molecule can change its shape to react with the maximum number of antigens.

An antibody is very specific, and will attempt to react with only the specific antigen that stimulated its synthesis in the first place. This *specificity* is an important property of antibodies. The exact fit of the "lock-and-key" model will not work if the antigen has a different molecular shape.

Classes of antibodies Because antibodies belong to the *globulin* group of proteins, and are involved with the *immune* response, they are called **immunoglobulins** (abbreviated as

Ig). Immunoglobulins are abundant in the body, making up about 20 percent of the total weight of plasma proteins. Five classes of immunoglobulins are known: IgA, IgD, IgE, IgG, and IgM. The constant region on the heavy chain of an antibody identifies its class. Immunoglobulins may have two or more identical reactive sites for an antigen. Those with two sites are called *bivalent;* those with more than two are called *multivalent.* Immunoglobulins D, E, and G are bivalent, and A and M are multivalent. All classes of immunoglobulins except IgM are composed of a single molecule with the basic Y structure. IgM immunoglobulins are composed of five identical molecules bonded together into a *pentamer* (a structure with five similar parts).

Each of the five classes of immunoglobulins has a separate defensive role:

1 IgA molecules are found mainly in the mucous membranes of the nose and throat, where they help fight respiratory allergens. In general, IgA molecules act as localized protective barriers against microorganisms at potential points of entry. IgA is the major class of antibody in saliva, tears, secretions from the intestinal and respiratory tracts, and a mother's milk.

2 IgD molecules are present only in small numbers. Although they are receptors for antigens, their specific function is still unknown. Current evidence suggests that IgD molecules play a role in antigen-triggered lymphocyte differentiation.

3 IgE molecules are responsible for immediate allergic reactions (such as asthma and hay fever). After the first ex-

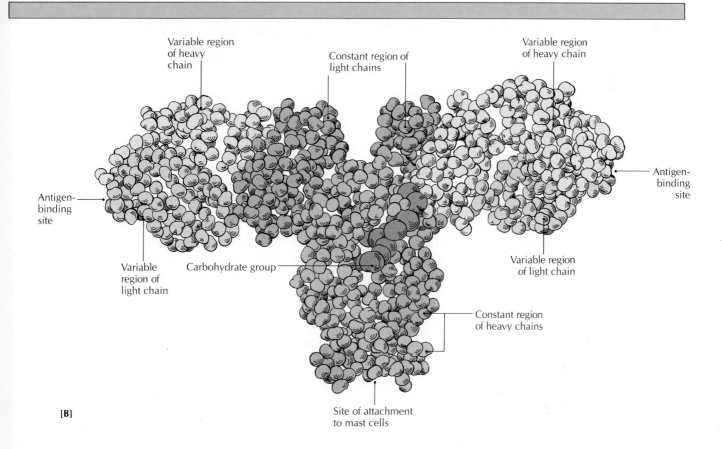

[B]

posure to an allergen, IgE molecules are synthesized and become tightly bound to the surface of mast cells. When these IgE molecules combine with their specific allergens, they trigger the release of histamine and other inflammatory substances from the mast cells, producing increases in capillary permeability and mucus flow. The release of histamine dilates small blood vessels, and drains plasma from the vessels into the surrounding tissue, sometimes producing a dangerously low blood pressure. IgE may also help leukocytes, antibodies, and complement components to reach inflammation sites.

4 *IgG molecules* are the most common type of immunoglobulin. They are produced in abundance the second and subsequent times the body is exposed to a specific antigen. IgG molecules are the only immunoglobulins that pass through the placenta during pregnancy from mother to fetus, providing the newborn with natural passive immunity. Such immunity is accomplished because placental cells have receptors that bind only to IgG molecules. Phagocytes that become coated with IgG molecules have an increased efficiency to ingest and destroy infectious microorganisms (Figure 20.9). IgG also activates the first component of the complement system. IgG molecules move through the walls of blood vessels into interstitial spaces most efficiently, and it is no coincidence that half of the body's IgG is found in the interstitial fluid, and the other half is found in the blood. The four other classes of circulating antibodies are restricted mainly to the bloodstream and local sites of the immune response.

5 *IgM molecules* are the largest antibodies, and are the major antigen fighters. They are the first antibodies to arrive at an infection site, and the predominant antibody secreted into the blood during the early stages of a first-time exposure. Because IgM is a pentamer, composed of 5 four-chain units with a total of 10 antigen-binding sites, it is even more efficient than IgG in activating the first component of the complement system. IgM molecules also stimulate the activity of macrophages.

Activation of complement by antibodies Complement is involved in several nonspecific defenses, but it also plays a crucial role, together with antibodies, in the defense against specific antigens. The sites for complement binding are exposed only after the antibody has reacted with the antigen. When two IgG molecules are bound side-by-side on an antigen cell, they can activate the first protein component of complement, which then binds to both antibodies. A single IgM molecule can also activate complement. The first active protein in the complement group releases enzymes that activate the next protein in the group. The complement proteins continue to bind to one another in sequence. Each protein that binds to the antigen-antibody complex changes it slightly, so that the next protein can bind to it.

| *Do men and women have the same immune mechanisms?* | *Apparently not. There is some evidence to support the theory that X chromosomes carry some of the genes that control immunity, and women have twice as many X chromosomes as men do. As a result, women seem to be less susceptible to viral and bacterial diseases.* |

FIGURE 20.9

IgG molecules and phagocytosis. A bacterium coated with IgG molecules is phagocytosed by a macrophage that has cell-surface receptors that bind to the IgG molecules. Phagocytosis is activated when the molecules on the bacterium bind to the receptors on the macrophage.

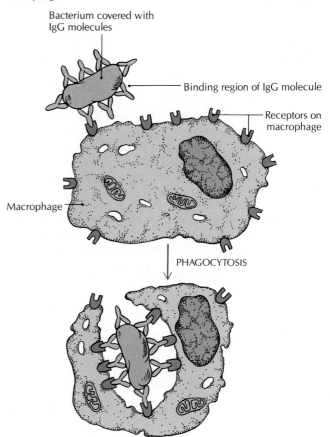

When all the proteins are attached to the antigen cell, they make a hole in its plasma membrane, which allows water and ions to flow freely into the cell, causing the cell to burst. The complement protein complex is appropriately called the *membrane attack complex* (MAC). In a system that is meant to minimize the possibility of destroying the wrong cell, all but the final complement protein must be set in place on the plasma membrane before the activation of the final protein finally blows a hole in the membrane. (Part of the overall complement system is involved in allergic reactions, causing the exploded cell to release histamine.)

Formation and Maturation of Lymphocytes

The specific defenses that recognize and respond to antigenic determinants are found in the closely related (but distinct) B cells and T cells. These two types of lymphocytes are fairly similar in origin. Both are formed from precursor hemopoietic **stem cells** in the fetal bone marrow. Some stem cells leave the marrow and migrate to the thymus gland, where they mature into inactive T cells.

The tissues of the thymus gland and bone marrow, where stem cells develop into lymphocytes, are called *central lymphoid tissues* (Figure 20.10; see also Figure 20.1A). The newly formed lymphocytes, still inactive, can eventually migrate via the blood to the *peripheral lymphoid tissues:* lymph nodes, spleen, the gut-associated lymphoid tissue (GALT), which include aggregated lymph nodules in the small intestine, appendix, tonsils, and adenoids (see Figure 20.1A). There they become activated and finally react with foreign antigens. Most of the migration of lymphocytes to peripheral lymphoid tissue occurs early in development.

T and B cells differentiate and proliferate fully only after they have been activated by a specific antigen. Macrophages are also essential for the proliferation and activation of lymphocytes. When antigens enter the body, they are phagocytized by macrophages, displayed on the surface of the macrophages, and then presented to T cells and B cells, which "recognize" the antigen. While this process is proceeding, the macrophages are also secreting *interleukin-1,* a protein that stimulates the production of more T cells and B cells. After the T and B cells are activated by their recognition of the specific antigen, they cause macrophages in the vicinity of the infection to become more efficient at phagocytizing and digesting invading antigens.

Activated T cells do not secrete antibody, but activated B cells develop into antibody-secreting cells, mostly *plasma cells.* (The *B* in B cells stands for *bursa of Fabricius,* a small patch of lymphoid tissue in the intestine of birds, where B cells of birds are processed.)

Humoral and Cell-Mediated Immunity

The two types of lymphocytes display different reactions to antigens, as explained in the following sections. (Table 20.2 outlines some of the major differences and similarities between B cells and T cells.)

B cells and humoral immunity When B cells come into contact with a specific antigen for the first time, they respond by first dividing, and then developing into *plasma cells,* which produce specific soluble antibodies and secrete them into the blood and lymph (Figure 20.11). The reaction is called the *primary immune response.*

Antibodies are not capable of independent movement. Instead they are carried to the site of an infection or injury by the blood and lymph, where they bind to the specific antigen that caused their production. Because such body fluids were once referred to as *humors,* this type of immunity involving B cells in the production of antibodies is called *humoral immunity.* Humoral immunity is most active against bacteria, viruses, toxins, and other soluble foreign proteins.

Not all activated B cells immediately develop into antibody-secreting plasma cells. Some retain their previous appearance as smaller, inactivated cells, and instead of circulating for a short time in the blood or lymph, they remain in lymphoid tissue for a long time. Such activated, but apparently inactive, B cells are called *B memory cells,* because they have the ability to "remember" the sensitizing antigen and

---

FIGURE 20.10

Development of T and B cells from hemopoietic stem cells in the fetal liver and postnatal bone marrow. Some stem cells mature in the bone marrow into bone-marrow lymphocytes (B cells). Thymus lymphocytes (T cells) mature only after the precursor stem cells migrate to the thymus gland via the bloodstream. T and B cells are activated when they come in contact with a specific antigen in peripheral lymphoid tissue.

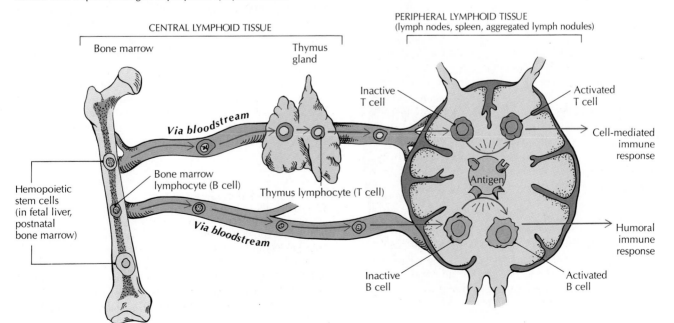

FIGURE 20.11

Activated B cell. Upon contact with a specific antigen, a B cell (A) differentiates into a larger plasma cell (B), which is capable of producing specific antibodies. Note that the plasma cell has a smaller, less dense nucleus and more endoplasmic reticulum and ribosomes. [(A) *Joseph Feldman.* (B) *Dorothea Zucker-Franklin.*]

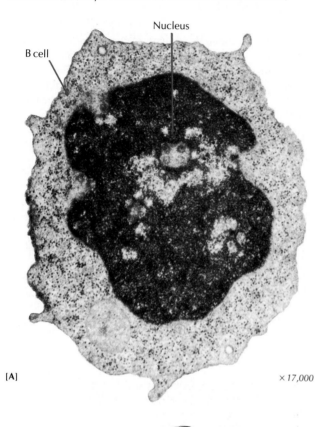

[A] × 17,000

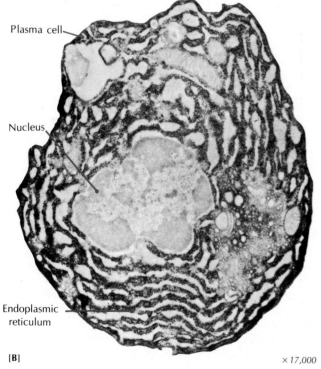

[B] × 17,000

TABLE 20.2 COMPARISON OF B CELLS AND T CELLS

Property	B cells	T cells
Site of production of undifferentiated cell	Bone marrow.	Bone marrow.
Site of differentiation	Specific site unknown; possibly bone marrow.	Thymus gland.
Response after binding to antigen	Become enlarged, multiply repeatedly to produce plasma cells; plasma cells release specific antibodies.	Become enlarged, multiply repeatedly, liberate lymphokines.
Antibody production	Synthesize and release specific antibodies.	Stimulate B cells to produce specific antibodies.
Type of immunity produced	Humoral (antibody-mediated) immunity.	Cell-mediated immunity.
Cytotoxic activity	None.	Activated T cells kill specific antigen-bearing cells on contact.
Factor causing response to antigens	Macrophages.	Macrophages.
Effect on macrophages	None.	Stimulate phagocytic activity.
Basic functions	Release specific antibodies.	Secrete specific toxins, stimulate production of specific antibodies by B cells, stimulate phagocytic activity of macrophages, produce cell-mediated immunity.

react to it the next time it appears. Such a *secondary immune response* or *anamnestic response* occurs at the second, and subsequent, exposures to the same antigen that produced the primary immune response.

A secondary response starts faster and releases many more antibodies than a primary response (Figure 20.12). For this reason, a second tissue or organ transplant from a donor is rejected much faster than the first transplant. Memory cells are capable of multiplying during a secondary response,

FIGURE 20.12

Primary and secondary (anamnestic) responses of contact with an antigen. The existence of memory cells is responsible for the faster initiation of synthesis of antibodies and a higher rate of production following the second contact. In the secondary response the antibody reaches a greater concentration in the blood. Exposure to a new antigen initiates a new primary response.

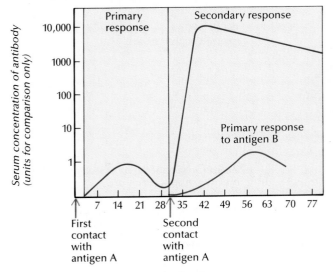

which makes them even more effective against a recurring antigen. If a totally new antigen is introduced, the primary immune response is elicited all over again, indicating that lymphocytes have an antigen-specific memory, reacting appropriately to each exposure to an antigen.

T cells and cell-mediated immunity T cells also respond to specific antigens, but they do not produce antibodies as B cells do. T cells give protection by (1) producing chemicals that destroy antigens, if the antigen is an infecting virus; (2) inducing macrophages or other host cells to destroy the antigen; (3) stimulating B cells to produce antibodies; or (4) regulating the immune response to make certain that the system does not overreact to the point where it damages the body. The protective method of T cells is called **cell-mediated immunity** because it involves direct contact between T cells and antigens.

Cell-mediated immunity operates against multicellular parasites, fungi, cancer cells, and tissue transplants. It is active against any cells that contain an infecting bacterium or virus. Such intracellular parasites are not attacked by the soluble antibodies prominent in humoral immunity because the antibodies are unable to cross the host-cell membranes to reach the parasites.

Clonal Selection Theory of Antibody Formation

The unifying theory of antibody formation was presented in the 1950s by Niels Jerne, Macfarlane Burnet, David Talmadge, and Joshua Lederberg. It is called the **clonal selection**

theory of antibody formation. Basically, it states that an antigen influences the *amount* of antibody produced, but it does not affect its three-dimensional structure (its combining sites). The theory also proposes that each B cell is genetically programmed to react *only to a particular antigen*. The clonal selection process works in the following ways:

1 Each B cell has a unique DNA pattern that makes it genetically programmed to respond to a specific antigen, even *before it meets that antigen* (Figure 20.13). Each lymphocyte produces only one type of antibody in response to its specific corresponding antigen.

2 Small amounts of specific antibody structures become bound to the plasma membrane surface of each lymphocyte. These structures act as receptors for specific antigens.

3 Any lymphocytes that have antibodies corresponding to body cells are killed during fetal life, so that all normal lymphocytes become tolerant of self.

4 When mature lymphocytes encounter their specific antigen, they become activated and begin to enlarge and divide.

5 Repeated cell divisions form a **clone,** a group of genetically identical cells descended from a single activated lymphocyte.

As cell division continues, some lymphocytes become plasma cells, and others become memory cells. The plasma cells produce the large amounts of antibody necessary to inactivate the antigen. All the newly formed plasma cells make the same specific antibody as their original lymphocyte ancestor. It is estimated that each plasma cell can secrete as many as 2000 identical antibody molecules each *second*. A plasma cell can maintain this pace for four or five days before it dies.

Memory cells persist in the body after the antigen disappears, and in fact, they may remain potentially active for a person's entire lifetime. Upon a second enounter with the antigen, memory cells proliferate, differentiate into plasma cells, and secrete antibody so rapidly that the symptoms of the disease may not even be observed. Because of this capability, memory cells are said to have an *immunological memory.*

The surface proteins that identify each cell as self or nonself are encoded by genes in a region of DNA called the *major histocompatibility* ("tissue compatibility") *complex* (MHC). Because there are millions of variants of the MHC genes that encode each protein, it is highly unlikely that two people (other than identical twins) will have the same MHC-encoded proteins. It has been suggested that antigenic determinants and MHC-encoded proteins must form a complex before a T cell can bind to them.

Differentiation of T Cells

The clonal selection theory also applies in part to T cells, which provide cell-mediated immunity (Figure 20.14). The major difference is that although some T cells that remain potentially active after an antigen is inactivated persist as *memory cells,* others differentiate further in lymphoid tissue into three types of cells with specialized functions: the helper T cell, the suppressor T cell, and the killer T cell.

Helper T cells are essential to the differentiation of B cells into plasma cells and their subsequent secretion of antibodies.

FIGURE 20.13

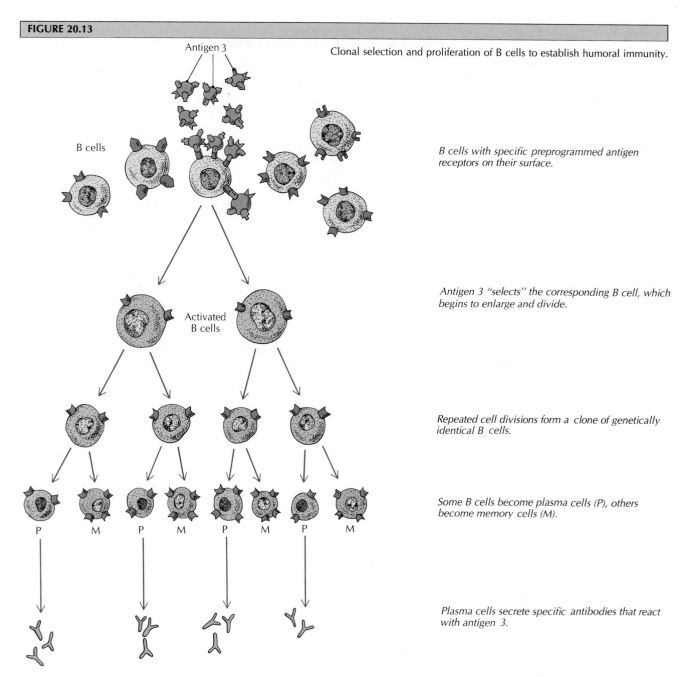

Clonal selection and proliferation of B cells to establish humoral immunity.

B cells with specific preprogrammed antigen receptors on their surface.

Antigen 3 "selects" the corresponding B cell, which begins to enlarge and divide.

Repeated cell divisions form a clone of genetically identical B cells.

Some B cells become plasma cells (P), others become memory cells (M).

Plasma cells secrete specific antibodies that react with antigen 3.

Each helper T cell is capable of activating hundreds of specific B cells. In order for a programmed B cell to become activated, two successive events must take place. First, the B cell must encounter its specific antigen. Second, it must be assisted by helper T cells, which release a *B-cell growth factor*, which somehow completes the activation of the B cell. Helper T cells are required for the appropriate response of killer T cells and suppressor T cells to an antigen, and they also activate macrophages, by activating chemical factors called lymphokines (see below).

Suppressor T cells, after being stimulated by helper T cells, suppress the response of B cells and other T cells to antigens. They inhibit the development of B cells into plasma cells, regulate the activity of killer T cells, and suppress the production of antibodies when they become excessive. Suppres-

sor T cells also suppress autoimmune responses, in which the body forms antibodies against its own antigens. Helper T cells that activate suppressor T cells are themselves inhibited in a feedback circuit that regulates the activity of both types of cells. Many such self-regulatory circuits are active within the complex interaction of lymphocytes.

Killer T cells have specific receptors for antigenic determinants. Killer T cells migrate from lymphoid tissue to the site of foreign-cell invasion. There they secrete a group of small proteins, called *lymphokines*, which attract phagocytes; prevent the reproduction of invading microorganisms, infected host cells, or viruses inside host cells; bind to and kill the antigen cells that activated them; keep macrophages near the site of the immune response; and increase the phagocytic activity of macrophages.

In order for a killer T cell to destroy an antigen cell, it must come into direct contact with it. A killer T cell destroys an antigen cell by secreting a pore-forming protein called *perforin* onto its plasma membrane. The pores cause the antigen cell to leak in the places directly contacted by the plasma membrane of the killer T cell, and the antigen cell soon dies (Figure 20.15). This reaction is accomplished without the use of complement.

Apparently, a killer T cell can dissociate itself from a decomposing antigen cell to attach to another. This reaction is very specific, and only cells that have the specific surface antigen are attacked. Killer T cells are the main cause of rejections of tissue or organ transplants from one person to another, and are also responsible for the activity against tumors and cancer cells.

Ask Yourself

1 What is the basic difference between specific and nonspecific defenses?

2 What is an antigen? An antibody? An antigen-antibody complex?

3 What are the five classes of antibodies?

4 What is the difference between humoral immunity and cell-mediated immunity?

5 How do antibodies activate complement?

6 What is the clonal theory of antibody formation?

7 What are some basic differences between B and T cells?

HYPERSENSITIVITY (ALLERGY)

About one out of every five people is **hypersensitive** or **allergic** to antigens such as dust, pollen, or certain foods and chemicals in amounts that do not affect most people. In the case of allergies, the antigen is known as an *allergen*. Allergens elicit the production of unnecessarily high levels of IgE, which coats mast cells and basophils. The allergen becomes attached to the IgE, and the coated cells release amines such as *histamines*. It may be that allergies develop when suppressor T cells, which normally suppress the production of IgE by plasma cells, are either absent or inefficient.

FIGURE 20.14

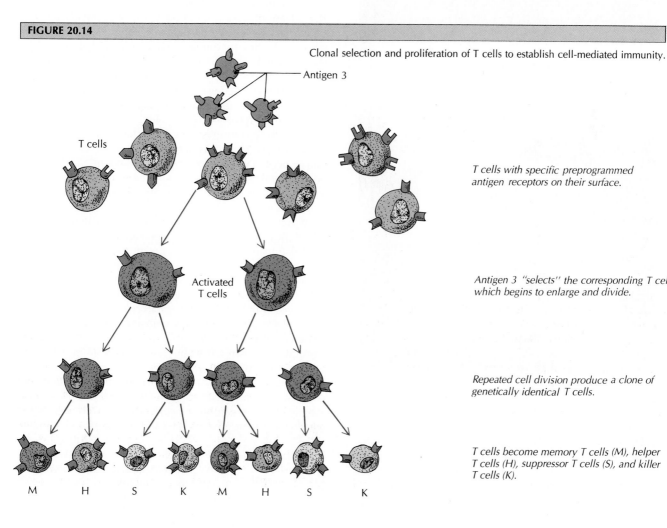

Clonal selection and proliferation of T cells to establish cell-mediated immunity.

Antigen 3

T cells

T cells with specific preprogrammed antigen receptors on their surface.

Activated T cells

Antigen 3 "selects" the corresponding T cell, which begins to enlarge and divide.

Repeated cell division produce a clone of genetically identical T cells.

T cells become memory T cells (M), helper T cells (H), suppressor T cells (S), and killer T cells (K).

M H S K M H S K

FIGURE 20.15

How killer T cells destroy antigen cells. (A) Scanning electron micrograph of killer T cells (bottom) contacting a larger antigen (cancer) cell. (B) The killer T cells have secreted perforin onto the plasma membrane of the antigen cell, producing holes that cause the cell to leak its contents. The antigen cell is larger than before because water has rushed in. The rough surface of the cancer cell indicates that the cell is dying. [(A *and* B) *Andrejs Liepins, Ph.D.*]

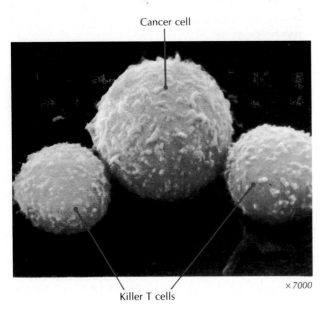

Cancer cell

×7000

Killer T cells

[A]

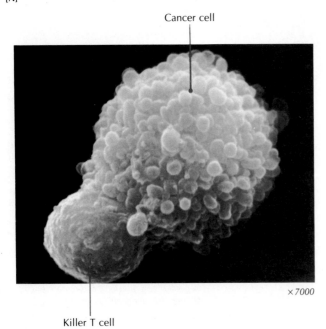

Cancer cell

×7000

Killer T cell

[B]

Allergens may enter the body through the respiratory or digestive tracts, where they trigger antigen-antibody reactions. The location of the meeting between allergen and antibody determines the severity of the allergic reaction. If an antibody encounters an allergen in the bloodstream, the allergen is neutralized without any side effects because there are no reactive cells there. In contrast, if the antibody meets the allergen in or on a reactive cell, neutralization occurs, but it disturbs the cell. It is this disturbance and release of toxic chemicals that produces the specific allergic reaction. The hypersensitivity response can be either immediate or delayed.

Immediate Hypersensitivity

When a person who has already been exposed to an allergen is exposed again, the allergic reaction sometimes occurs rapidly, within minutes or hours. Such a reaction to an allergen is called *immediate hypersensitivity.* The earlier primary response results in the production of IgE, which binds to the surface of mast cells (or circulating basophils). The IgE molecules have no effect on the mast cells until the IgE encounters the specific allergens that combine with adjacent IgE molecules (Figure 20.16A). When this combination occurs, the mast cells immediately release inflammatory substances such as histamine and leukotriene into the extracellular spaces. The release of these substances is called *degranulation. Histamine* increases the permeability of blood vessels and causes smooth muscles to contract. *Leukotriene* is synthesized by mast cells and basophils after their reaction with the allergen. Its effects are similar to those of histamine, but are longer lasting. Also released is *eosinophil chemotactic factor,* which attracts eosinophils to the area to help control the local responses to allergens.

The inflammation resulting from the release of chemicals from activated mast cells may be mild or severe. Inflammation associated with the contraction of smooth muscles in the bronchial tubes (bronchioles) may be serious enough to interfere with breathing. This is the cause of the condition known as *asthma.*

In most hypersensitivity reactions, the inflammation is *localized.* In the allergies to pollen (popularly called *hay fever*), the inflammation is localized in the mucous membrane of the nose, where the allergens first make contact with the responsive tissue (see Figure 20.16B). Some of the symptoms of hay fever, such as watery eyes, sneezing, or sniffling, may also be associated with the eyes or respiratory tract.

If an allergen enters the circulation, the effects may be distributed throughout the body, affecting several organ systems. Such a *systemic* reaction may result from insect stings, some therapeutic drugs (such as penicillin), and components of certain foods. In systemic reactions, inflammation and edema are extensive, lowering blood volume and blood pressure and possibly causing *anaphylactic shock.* Occasionally the reaction is severe enough to constrict air passages in the

If you have an allergy, will your children inherit it?

Possibly, but it is more likely that they will inherit the susceptibility to allergy rather than an allergy to a specific substance.

FIGURE 20.16

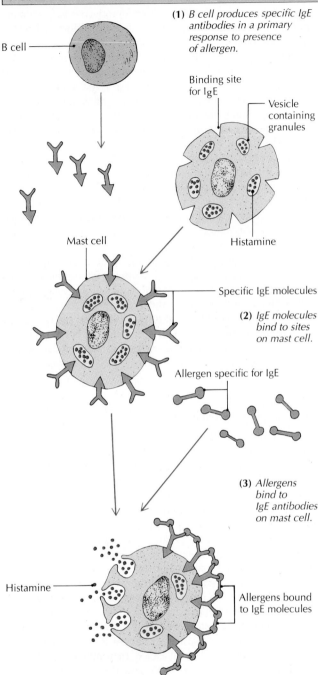

(1) *B cell produces specific IgE antibodies in a primary response to presence of allergen.*

Binding site for IgE

Vesicle containing granules

B cell

Mast cell

Histamine

Specific IgE molecules

(2) *IgE molecules bind to sites on mast cell.*

Allergen specific for IgE

(3) *Allergens bind to IgE antibodies on mast cell.*

Histamine

Allergens bound to IgE molecules

(4) *Binding of allergen to IgE antibodies on cell surface triggers the release of histamine, leukotriene, and other inflammatory substances. This process is called degranulation.*

(5) *Plasma escapes from capillaries when capillary permeability is increased in response to the release of inflammatory substances.*

[A]

(A) Role of IgE and mast cells in the allergic response (immediate hypersensitivity). (B) Scanning electron micrograph of a mast cell releasing histamine-containing granules that cause sneezing and watery eyes. [*Courtesy of Lennart Nilsson. © Boehringer Ingelheim International GmbH.*]

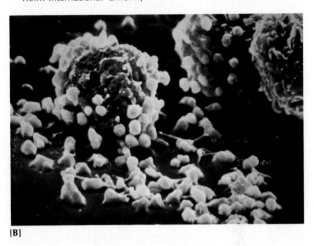

[B]

lungs. Suffocation may result unless it is treated with an injection of epinephrine or an antihistamine.

Allergies may be treated with *antihistamine drugs,* which reduce the effects of histamine on the skin and blood vessels; *immunosuppressive drugs,* such as steroids or cromoglycate, which suppress the formation of antibodies; or *desensitization,* a procedure in which first small doses and then progressively larger doses of allergens are injected into the body until the allergic response is minimized.

Delayed Hypersensitivity

When the effects of an allergic reaction take several days or longer to develop and be expressed, the reaction is called **delayed hypersensitivity.** T cells are responsible for delayed hypersensitivity, and the mechanism of sensitization and response is the same as described earlier for T cells. In fact, delayed hypersensitivity may be no different from the normal activity of T cells directed at intracellular infections, cancer cells, or tumors.

Delayed hypersensitivity is responsible in those cases when the body rejects tissues and organs transplanted from

Why doesn't a pregnant woman's uterine tissue reject the implanted fetus?	*Apparently, two factors are at work in a successful pregnancy. Immediately after conception, and during pregnancy, the uterus is less responsive to its own chemicals that normally attack foreign tissue. Also, although the fetus and most of its surrounding layer of cells contain proteins that would be recognized as nonself by cells in the mother's uterus, most of these proteins do not occur at the point where the placenta connects the fetus with its mother's uterus.*

another person. Rejection reactions are mediated mainly by T cells. The rejection of donor tissue is an immune response to foreign antigens on the surface of the cells in the transplanted tissue that are different from those of the host (except in the case of identical twins). The cornea is an exception. Because corneas do not contain blood vessels or lymphatic vessels (which carry T cells), they can usually be transplanted to another person without rejection.

After receiving nonself tissue from a donor, the host begins manufacturing T cells that are active against the foreign cells bearing surface antigens. The T cells migrate to the transplant site and attack the tissue. Such a reaction not only deprives the host of the normal, dependable use of the transplanted tissue, but also produces a continued inflammation that causes tissue damage and increased systemic stress.

Immunosuppressive Drugs

Drugs that help to suppress a typical immune reaction are called, logically enough, **immunosuppressive drugs.** Such drugs can be very useful in helping the body to accept a tissue transplant. The longer foreign tissue can remain in the host body without causing delayed hypersensitivity, the greater is the possibility that the tissue will be accepted as self, a phenomenon known as *tolerance.*

Several drugs have been used to suppress the immune response following transplant surgery. These drugs usually must be given when the foreign tissue (and its surface antigens) is introduced. Improved immunosuppressive drugs have made it possible to transplant kidneys, livers, hearts, and lungs with some degree of success.

Some immunosuppressive drugs, such as *Imuran* and *6-mercaptopurine,* work by inhibiting the synthesis of DNA, which consequently limits the proliferation of lymphocytes and their differentiation into killer T cells. Other drugs, such as *cortisone,* act to suppress inflammation. One of the newest immunosuppressive drugs is *cyclosporine,* which blocks either the production or the action of interleukin-2, the lymphokine produced by helper T cells. Interleukin-2 is necessary for the proliferation of antigen-activated T cells. Cyclosporine does not affect the B-cell system.

Ask Yourself

1. What is an allergen?

2. How do the two types of hypersensitivity differ?

3. What is the purpose of immunosuppressive drugs?

4. What is an autograft? A homograft?

AUTOIMMUNITY AND IMMUNODEFICIENCY

Early in the embryonic development of lymphocytes, the ones with receptors for antigenic determinants on self molecules are either eliminated or suppressed. This allows the immune system to respond only to foreign (nonself) antigens, and to show tolerance for its own (self) antigens, because fetal lymphocytes are exposed to the antigens of the fetus. When tolerance to self antigens breaks down, however, the body forms antibodies to its own antigens, resulting in **autoimmunity** ("protection against one's self") and **autoimmune diseases.** One such disease is *myasthenia gravis,* in which a person makes antibodies abnormally against the acetylcholine receptors on skeletal muscle cells. The receptors cease to function properly, causing, among other things, faulty breathing that may lead to death.

It is probable that the immune system actively inhibits its own suppressive responses, turning itself back on after it has been turned off after successfully fighting an infection. The process is called *contrasuppression,* and some researchers speculate that autoimmune diseases are caused by overaggressive contrasuppression.

Deficiency diseases involving the immune system can be subdivided into two categories. **Primary immunodeficiencies** are of genetic or developmental origin. They involve lymphoid tissue, and may affect the T-cell system, the B-cell system, or both. **Secondary immunodeficiencies** are caused by nonspecific factors such as drugs, malnutrition, x rays, and cancer. These deficiencies usually occur relatively late in life, and are not as severe as the primary immunodeficiencies.

Mechanism of Autoimmune Diseases

Autoimmune diseases may result from two different mechanisms. First, a body cell may synthesize or reveal a new antigen for which the body had not formed a tolerance because the antigen was not exposed to the fetal lymphocytes during embryonic development. An example of an autoimmune disease resulting from this mechanism is *sympathetic ophthalmia,* where one eye is badly injured and releases antigens that the immune system has not been exposed to previously. The new antigens cause antibodies to attack the uninjured eye, which may be destroyed or damaged severely.

A second mechanism for autoimmunity results when a mutation (alteration of the genetic program) in a B or T cell produces a "forbidden clone" that produces a mutant immunoglobulin that reacts with normal body antigens. *Rheumatoid arthritis* is a disease where B cells mutate, producing a mutant IgM (also called rheumatoid factor). The mutant IgM reacts against IgG, and the resultant complex causes inflammation in joints and swelling of lymph nodes. Other examples of autoimmune disease are listed in Table 20.3.

Immunodeficiencies

Conditions that decrease the effectiveness of the immune system, or that destroy its ability to respond to antigens altogether, are called **immunodeficiencies.** Such conditions may result from a complete absence of B or T cells, or their decreased number or activity. If B cells are absent or low in number, the person will be vulnerable to bacterial infections (but not viral infections if T cells are normal). If T cells are absent or low in number, the person is particularly vulnerable to viral infections, parasites, and tumor cells. *Acquired immune deficiency syndrome (AIDS)* is a disease in which the number of helper T cells is decreased, making the affected person vulnerable to infections such as *Pneumocystis pneumonia* and cancers such as *Kaposi's sarcoma* (see Physiological and Anatomical Abnormalities, p. 642).

Some people are born without either B or T cells, and have practically no protection against infection. This condition, known as *Swiss-type agammaglobulinemia*, is usually fatal in the first few years of life (Figure 20.17). In some cases, this disease has been treated with some success with bone marrow transplants.

TABLE 20.3   SOME AUTOIMMUNE DISEASES

Disease	Specificity of autoantibodies against	Result
Addison's disease	Adrenal gland.	Weakness, skin pigmentation changes, weight loss, electrolyte imbalance.
Autoimmune hemolytic anemia	Erythrocyte antigens.	Hemolysis, anemia.
Goodpasture's syndrome	Basement membrane (kidney, lung).	Pulmonary hemorrhage, kidney failure.
Hashimoto's thyroiditis	Thyroid antigens.	Goiter, abnormal changes in thyroid gland.
Idiopathic thrombocytopenic purpura	Platelets.	Hemorrhages in skin and mucous membranes.
Diabetes (type I)	T cells.	Beta cells in pancreas destroyed.
Multiple sclerosis	Macrophages, T cells.	Myelin progressively destroyed.
Myasthenia gravis	Acetylcholine receptors on skeletal muscle cells.	Progressive neuromuscular weakness, breathing difficulty.
Pernicious anemia	Intrinsic factor.	B_{12} absorption from intestine prevented, severe anemia.
Rheumatoid arthritis	Immunoglobulin.	Inflammation and deterioration of joints and connective tissue.
Systemic lupus erythematosus	DNA, nuclear antigens.	Facial rash, lesions of blood vessels, heart, and kidneys.

FIGURE 20.17

Immunodeficiencies. This 10-year-old boy was born with a deficient immune system that left him without B or T cells. He lived his entire life in a germ-free bubble to reduce the possibility of infection. He died at the age of 12 when he was removed from the bubble to undergo a bone-marrow transplant in the hope that it would allow him to live freely. Successful bone-marrow transplants may eventually permit such children to live a more normal life without the constant threat of infection. [*Jim deLeon.*]

Ask Yourself

1. *What causes autoimmunity?*

2. *What are immunodeficiencies?*

3. *How do primary and secondary immunodeficiencies differ?*

TYPES OF ACQUIRED IMMUNITY

Resistance to disease may be innate, or it may be acquired in several different ways. Some individuals and species are not susceptible to diseases that infect other individuals or species. This resistance to disease, which is present at birth, is called *innate resistance* (innate immunity or native resistance). For example, human beings are not susceptible to canine distemper, and dogs are not susceptible to tetanus. Some individuals are not susceptible to some diseases that other members of their species are susceptible to. In a sense, innate resistance is not an immunity, since it does not involve the specific defenses of the immune response.

Acquired immunity is the resistance to infection, either by humoral antibody production or cell-mediated immunity. Such immunity can be *active* or *passive*. It may be acquired *naturally*, through an encounter with the infectious organism, or *artificially*, through the injection of a noninfectious antigenic determinant such as a vaccine (Table 20.4).

Active immunity results when the body manufactures its own antibodies or T cells in response to a foreign antigen. *Passive immunity* results when antibodies, not antigens, are transferred from one person to another. This happens when maternal antibodies (IgM) cross the placenta and provide short-term (weeks to months) protection for the fetus. Maternal antibodies are also passed to the infant from the mother's milk during breastfeeding.

Artificial passive immunity occurs when specific antibodies are injected into the body. Such antibodies are taken from the blood of someone who has been infected or immunized, or from animals that have been deliberately immunized to provide antibodies. Artificial passive immunity is particularly useful when a person has been exposed to a dangerous disease, and must be immunized as quickly as possible.

Passive immunity is used against botulism, diphtheria, whooping cough, rabies, tetanus, and rattlesnake venom. The serum containing the antibodies is called the *immune serum*. Although the protection of artificial passive immunity is immediate upon injection, it lasts only a few weeks.

If an antigen is artificially prepared and injected as a vaccine into a person, the person is said to have an **artificial acquired active immunity**. Such prepared antigens are infectious organisms that are either dead or severely weakened by laboratory methods. Today, vaccination is available against many diseases, including polio, tetanus, diphtheria, whooping cough, German measles (rubella), mumps, and some forms of influenza.

Active immunity, whether it is acquired naturally or artificially, may last a lifetime for diseases such as measles and chicken pox. For other diseases, such as diphtheria or tetanus, immunity may last only a few years. In such cases, an additional ("booster") vaccination may be given to retain immunity. Certain viruses may occur in so many different forms that natural or artificial immunity to one form will not be effective against another form, in which case the disease may recur.

TABLE 20.4   TYPES OF IMMUNITY

Type of Immunity	Description
Innate (native) resistance (immunity)	Certain individuals or species are born with immunity to specific substances that affect other individuals or species. Permanent.
Active immunity	Antibodies are formed in the body in response to exposure to an antigen.
Natural active immunity	Antibodies are formed during the course of a disease, sometimes providing permanent immunity (measles, chicken pox, smallpox, or yellow fever). Long-lasting or permanent.
Artificial active immunity	Prepared dead or weakened antigens in vaccines are injected into individual to stimulate the production of specific antibodies. May require booster injections. Long-lasting or permanent.
Passive immunity	Antibodies are acquired from a source outside the body.
Natural passive immunity	Antibodies from an immune pregnant woman pass to the fetus through the placenta, or to the baby in milk during breastfeeding. The newborn baby receives temporary (several weeks or months) immunity against diseases to which the mother has active immunity. May be effective for several months.
Artificial passive immunity	Immune serum from immunized animals or human beings is injected into exposed individual, who receives specific antibodies (diphtheria, tetanus). Effective weeks to months.

Ask Yourself

1 *What is acquired immunity?*

2 *What is innate immunity?*

3 *How do active and passive immunity differ?*

4 *What is artificial immunity? Natural immunity?*

MONOCLONAL ANTIBODIES

In 1975, British scientists Georges Kohler and Cesar Milstein fused an activated lymphocyte from the spleen of a mouse with a cancer (myeloma) cell from the bone marrow. The result was a hybrid cell called a **hybridoma** that divided over and over again, producing identical copies (clones) of itself (Figure 20.18). The clones of the hybridoma produced a limitless supply of antibodies called **monoclonal antibodies.**

The technique of producing monoclonal antibodies is important because it provides the possibility of developing monoclonal antibodies from human cells that could be used to manufacture vaccines against cancer and other diseases. Also, when radioactive tracers are attached to circulating monoclonal antibodies, it is possible to locate cancer cells anywhere in the body, because the antibodies react with cancer cells as they are encountered. This is especially important to determine if a primary cancer has spread to other parts of the body.

Hybridoma technology also offers the possibility of attaching cancer-killing chemicals to monoclonal antibodies. This "targeted drug therapy" would permit chemotherapy to be directed *only* at cancer cells. Currently, cancer chemotherapy destroys some body cells as well as cancer cells, because it cannot be focused precisely on the cancer cells alone.

The availability of monoclonal antibodies makes it relatively easy to obtain sufficient antigen samples from body

FIGURE 20.18

Monoclonal antibodies. (A) The mechanism for the production of monoclonal antibodies.
(B) Scanning electron micrograph of a hybridoma in the process of cloning. [*David Scharf/Peter Arnold.*]

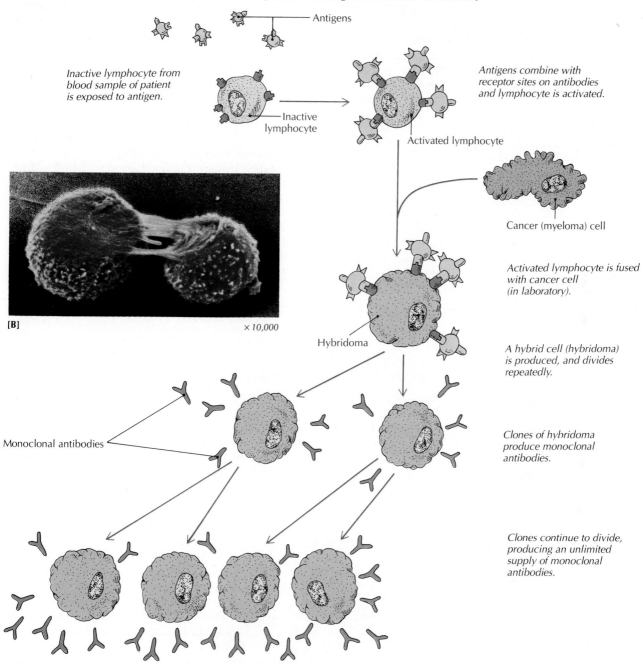

Antigens

Inactive lymphocyte from blood sample of patient is exposed to antigen.

Inactive lymphocyte

Activated lymphocyte

Antigens combine with receptor sites on antibodies and lymphocyte is activated.

Cancer (myeloma) cell

Activated lymphocyte is fused with cancer cell (in laboratory).

[B]   × 10,000

Hybridoma

A hybrid cell (hybridoma) is produced, and divides repeatedly.

Monoclonal antibodies

Clones of hybridoma produce monoclonal antibodies.

Clones continue to divide, producing an unlimited supply of monoclonal antibodies.

[A]

tissues. Such samples could be used to match the tissue types of host and donor before a tissue transplantation. The closer the types can be matched, the better the chance of reducing tissue rejection.

About 7000 kidney transplants are performed each year in the United States, and about 60 percent result in acute rejection. (The *hyperactive* rejection stage takes place immediately after a transplant. *Acute rejection* occurs days or weeks following the transplant. The *chronic* stage of rejection follows, lasting as long as the transplanted organ does.) In 1986, the Food and Drug Administration (FDA) approved the first monoclonal antibody for use in acute rejection following human kidney transplants. The newly approved monoclonal antibody, which is called OKT3, attacks T cells responsible for tissue rejection. In a recent trial, OKT3 reversed kidney-transplant rejections in 94 percent of transplant recipients. In comparable tests, treatment with cyclosporine produced a 75 percent success rate.

INTERACTION BETWEEN THE NEUROENDOCRINE SYSTEM AND THE IMMUNE SYSTEM

It has been known since the 1950s that adrenal glucocorticoid hormones secreted in response to stress may cause the immune system to become suppressed and ineffective. Many other possible interactions between the neuroendocrine system and the immune system have been uncovered recently. Karen Bulloch of the State University of New York at Stony Brook has shown how nerve fibers penetrate into lymphoid tissue, especially in bone marrow and the thymus gland, where the production and maturation of lymphocytes take place. Lewis T. Williams of the University of California School of Medicine at San Francisco has found special receptor sites on lymphocytes for neurotransmitters released by nerve fibers. Eli K. Hazum of the Weizmann Institute of Science in Israel has discovered that neuropeptides such as *beta*-endorphin bind to specific receptor sites on lymphocytes. Manfred L. Karnovsky of the Harvard School of Medicine has provided evidence that macrophages are activated by serotonin. J. Edwin Blalock of the University of Alabama at Birmingham has demonstrated that lymphocytes produce hormones such as ACTH, which affect the adrenal glands. In addition, interferon has been shown to affect the neuroendocrine system.

Steven E. Keller and his coworkers at the Mount Sinai School of Medicine in New York City have found that lymphocytes are inhibited in response to stress. Janice K. Kiecolt-Glaser of the Ohio State University Medical Center has shown that DNA repair in the lymphocytes of emotionally disturbed patients is diminished. Ronald Glaser and his coworkers at the Ohio State University Medical Center have confirmed that while students are taking examinations, their immune systems do not function at peak level, with killer T cells being impaired, other cells producing subnormal amounts of interferon, and blood levels of the Epstein-Barr virus (which causes infectious mononucleosis) rising.

It is possible that some people enjoy good health because they are able to cope well with stress, while others react poorly under stress, causing ill health related to a breakdown of the immune system. Such interactions between the neuroendocrine system and the immune system continue to be studied intensively, and there is hope that the more we learn about them, the better we can prevent and treat disorders of the immune system.

PHYSIOLOGICAL AND ANATOMICAL ABNORMALITIES

Acquired immune deficiency syndrome (AIDS)

As its name suggests, *acquired immune deficiency syndrome,* or *AIDS,* is a disease that cripples the body's immune system. Victims of AIDS show a deficiency in two kinds of lymphocytes: natural killer T cells that kill virus-infected and tumor cells, and virus-specific killer T cells that kill cells containing a particular virus. Without an intact immune system, AIDS victims may die from microbial infections that are ordinarily minor problems. Also, relatively uncommon forms of cancers of the blood and lymphatic system seem to be unusually prevalent among AIDS victims. In fact, the first sign that a new disease had emerged was the sudden appearance of Kaposi's sarcoma (a cancer seen in blood vessels) in the 1970s among young, middle-class, white males. Until then, the disease appeared mainly among older Italian and Jewish men, and in Africa. The young male victims turned out to be predominantly homosexual, and the disease was also spreading among drug users and people who received frequent blood transfusions. (In the United States in 1986, 73 percent of AIDS victims were active male homosexuals or bisexuals, 17 percent were intravenous drug users, 5 percent were Haitian immigrants, and 1 percent were hemophiliacs who were dependent on blood transfusions.)

In 1984, it was discovered by Robert Gallo and his coworkers at the National Cancer Institute that AIDS is caused by *human immunodeficiency virus* (HIV). It is spread mainly by sexual contact and shared hypodermic needles, especially when the virus infects groups of people

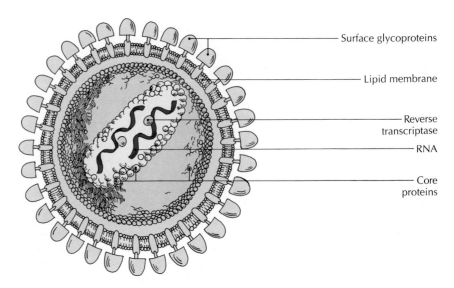

The AIDS virus.

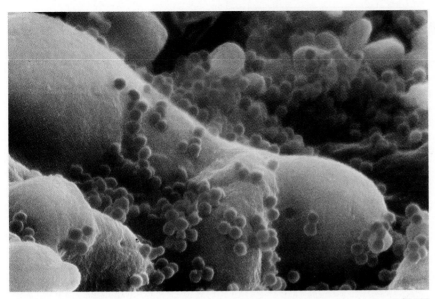

AIDS viruses (blue spheres) attack a helper T cell.
[*Courtesy of Lennart Nilsson. © Boehringer Ingelheim International, GmbH.*]

×*80,000*

living in close contact. Anal intercourse is a likely method of transmission because rectal capillaries are so close to the skin that infection may occur through microscopic breaks in the skin. However, researchers stress that AIDS is not exclusively a disease of homosexuals, bisexuals, or drug addicts. In Africa, it is a disease of heterosexuals.

HIV is a *retrovirus,* a virus that transmits genetic information in reverse fashion, from RNA to DNA instead of from DNA to RNA. As with other retroviruses, RNA is the genetic material (see drawing). The host cell is often a helper T cell (T4 lymphocyte), which plays a crucial role in regulating the immune system. The DNA assembled by the enzyme reverse transcriptase from the RNA template may remain latent among the chromosomes of the host lymphocyte until that cell is stimulated by a second infection. Then the viral DNA is reproduced rapidly, and new viruses leaving the host cell eventually kill it by destroying the plasma membrane.

Apparently, the AIDS virus is able to cross the blood-brain barrier to cause the proliferation of glial cells in the central nervous system. Victims of AIDS are also more susceptible to Kaposi's sarcoma, carcinomas (including skin cancers in the mouth or rectum of infected homo-

sexuals), and B-cell lymphomas (tumors originating in lymphocytes).

AIDS was first described in 1981. It probably began in central Africa some time in the 1950s and spread first to the Caribbean and then to the United States and Europe. As many as 2 million Americans may be infected, and the situation is much worse in Africa and the Caribbean. Current treatment concentrates on interrupting the reverse transcriptase as it assembles the viral DNA. Initial use of azidothymidine (AZT), an unsuccessful anticancer drug, has been promising, and the search for an effective vaccine is underway. Prevention has been helped considerably by the development of blood tests that can detect the presence of antibodies to the AIDS virus before transfusions are given.

Hodgkin's disease

Hodgkin's disease is a form of cancer, typified by the presence of large, multinucleate cells (Reed-Sternberg cells) in the affected lymphoid tissue. The first sign of Hodgkin's disease is usually a painless swelling of the lymph nodes, most commonly in the cervical area. As lymphocytes, eosinophils, and other cells proliferate, there is a progressive enlargement of the lymph nodes, the spleen, and other lymphoid tissues. Late

symptoms of the disease include edema of the face and neck, anemia, jaundice, and increased susceptibility to infection.

Hodgkin's disease occurs most often in young adults between the ages of 15 to 38, and in people over 50. Interestingly, it occurs in Japan only in people over 50. Its cause is unknown but it is potentially curable. It is usually treated with considerable success with a combination of radiation therapy and chemotherapy. Untreated, Hodgkin's disease is fatal.

Infectious mononucleosis

Infectious mononucleosis is a viral disease, caused by the Epstein-Barr virus, a member of the herpes group. It appears most often in 15- to 25-year-olds. It is usually accompanied by fever, increased lymphocyte production, and enlarged lymph nodes in the neck. Secondary symptoms include dysfunction of the liver, and increased numbers of monocytes. Infectious mononucleosis is not contagious in adults. In adolescents, however, the Epstein-Barr virus may be spread during close contact or via the exchange of body fluids, as when drinking glasses are shared or through kissing. The virus may also be spread from infant to infant in saliva, from the mother's breast, or food.

Systemic lupus erythematosus

Systemic lupus erythematosus (SLE) is considered a collagen disease because it affects the lining of joints and other connective tissue. Although the origin is uncertain, it is thought that SLE may be hereditary, and may be caused by a breakdown in the body's immune system. Women are infected about nine times more than men are. The first signs of the disease, which can be fatal in extreme cases, are a general malaise, fever, a so-called butterfly rash on the face, appetite loss, sensitivity to sunlight, and pains in the joints. SLE causes B cells to produce excess antibodies, called autoantibodies, which attack healthy cells. The basal layer of skin begins to deteriorate, and any organ can be affected. Because the body is unable to remove the autoantibodies, they may settle in such vital organs as the brain, heart, kidneys, and lungs, ultimately causing serious tissue damage.

MEDICAL TERMINOLOGY

ANAPHYLAXIS (an-uh-fuh-LACK-sis; L. *ana,* intensification + Gr. *phulassein,* to guard) Hypersensitivity to a foreign substance.

ELEPHANTIASIS Obstruction (caused by a parasitic filarial worm) of the return of lymph to the lymphatic ducts that causes enlargement of a limb, usually a lower limb, or the genital area.

HYPERSPLENISM A condition in which the spleen is abnormally active and blood cell destruction is increased.

LYMPHADENECTOMY Removal of lymph nodes.

LYMPHADENOPATHY A disease of lymph nodes, resulting in enlarged glands.

LYMPHANGIOMA Benign tumor of lymph tissue.

LYMPHANGITIS Inflammation of lymphatic vessels.

LYMPHATIC METASTASIS A condition in which a disease travels around the body via the lymphatic system.

LYMPHOMA Tumor of the lymph nodes.

LYMPHOSARCOMA Malignant tumor of lymph tissue.

SPLENECTOMY Total removal of the spleen.

SPLENOMEGALY Enlargement of the spleen, following infectious diseases such as scarlet fever, typhus, and syphilis.

VACCINE General term for the immunization preparation used against specific diseases.

SUMMARY

The *lymphatic system* returns to the blood excess fluid and proteins from the spaces around cells, plays a major role in the transport of fats from the tissue surrounding the small intestine to the blood, filters and destroys microorganisms and other foreign substances, and aids in providing long-term protection for the body.

The lymphatic system

1 The lymphatic system consists of vessels, lymph nodes, lymph, leukocytes, lymphatic organs (spleen and thymus gland), and specialized lymphoid tissues.

2 Excess interstitial fluid is drained from tissues by the lymphatic system. Once inside the lymphatic capillaries, the fluid is called *lymph.*

3 The composition of lymph is similar to that of blood, except that lymph lacks red blood cells and most of the blood proteins.

4 Leukocytes in lymph include monocytes, which develop into macrophages, and two types of lymphocytes, *B cells* and *T cells.* Specific defenses are accomplished by *lymphocytes,* a type of leukocyte. Lymphocytes are the basis of the immune response.

5 *Lymphatic capillaries* are one-way vessels with a closed terminal end. They join together to form *lymphatics,* which drain into the *right lymphatic duct* and the *thoracic duct.* The ducts return fluid to circulation through the right and left subclavian veins.

6 *Lymph nodes* are small masses of lymphoid tissue scattered along the lymphatic vessels. Lymphocytes in the nodes filter out harmful substances from the lymph, and are the initiating sites for the specific defenses of the immune system.

7 The three *tonsils* (pharyngeal, palatine, and lingual) form a band of lymphoid tissue that prevents foreign substances from entering the body through the throat.

8 The *spleen* is the largest lymphoid organ. Its major functions are filtering blood and manufacturing phagocytic lymphocytes.

9 The *thymus gland* forms antibodies in the newborn, and plays a major role in the early development of the body's immune system.

10 *Aggregated lymph nodules* (Peyer's patches, gut-associated lymphoid tissue, or GALT) are clusters of lymphoid tissue in the intestine and appendix. They are thought to respond to antigens from the intestine by generating plasma cells that secrete antibodies in large quantities.

Nonspecific defenses of the body

1 The *nonspecific defenses* of the body are those that do not involve the production of antibodies. They include the skin, mucous membranes and mucus, lacrimal glands, nasal hairs, cilia in the respiratory tract, and the acidity of the stomach.

2 When tissues are damaged, the normal result is the *inflammation response,* which initiates healing and prevents further damage.

3 *Phagocytosis* is the destruction of foreign substances and dead tissue by leukocytes.

4 The body produces nonspecific chemical substances such as *complement, properdin,* and *interferon* that attack bacteria and viruses.

Specific defenses: the immune response

1 The *specific defenses* of the body, which constitute the *immune response,* discriminate among foreign substances *(antigens)* by forming specific proteins *(antibodies)* to react with the foreign substances and destroy them.

2 An *antigen* causes the production of specific antibodies, and reacts chemically with the antibody to form a stable complex called the *antigen-antibody complex.*

3 An antibody reacts with specific sites on the surface of an antigen called *antigenic determinants,* which are formed by distinctive proteins or polysaccharides.

4 *Antibodies* are proteins produced by lymphocytes in response to an antigen. They contribute an *active immunity* that provides a relatively long-term protection against chronic infections.

5 An *antibody molecule* is composed of four polypeptide chains. The molecular structure of an antibody is crucial to its ability to react with specific antigens. The *variable region* of an antibody has the structural potential to bind to specific antigens like a lock (antibody) accepting a key (antigen).

6 Antibodies are *immunoglobulins.* Five known classes of immunoglobulins are IgA, IgD, IgE, IgG, and IgM.

7 The proteins that make up *complement* must bind in sequence to antibodies attached to antigens before the antigen cell membrane is ruptured and the antigen cell is destroyed.

8 B and T cells are formed from precursor *stem cells* in the fetal bone marrow. Some stem cells migrate to the thymus gland, where they mature into T cells. It is not certain where the remaining stem cells mature into B cells, but they may remain in the bone marrow.

9 *Humoral immunity* involves B cells, which differentiate into *plasma cells* that secrete antibodies into the blood or lymph. Some B cells become *memory cells* that react to an antigen the second time it appears. The direct contact between T cells and antigens is *cell-mediated immunity.*

10 According to the *clonal selection theory* of antibody formation, B cells are genetically programmed, before meeting an antigen, to react only to a particular antigen.

11 T cells differentiate into *helper T cells, suppressor T cells,* and *killer T cells.*

Hypersensitivity (allergy)

1 The two types of *hypersensitivity (allergy)* are *immediate* and *delayed.* Severe systemic reactions can produce *anaphylactic shock.* Delayed hypersensitivity is responsible for the body's

rejection of tissues or organs transplanted from another person.

2 *Immunosuppressive drugs* help to suppress a typical immune reaction. Such drugs usually cause the immune system to become less effective, lowering the body's resistance against infection.

Autoimmunity and immunodeficiency

1 *Autoimmunity* is the formation of antibodies against the body's own antigens.

2 Conditions that decrease the effectiveness of the immune system, or that destroy its ability to respond to antigens altogether, are called *immunodeficiencies.*

Types of immunity

1 Some individuals are not susceptible to diseases that infect other individuals. This inborn resistance to disease is called *innate resistance,* or native resistance.

2 *Acquired immunity* is the resistance to infection, either by humoral antibody production or cell-mediated immunity. Such immunity can be active or passive, acquired naturally or artificially.

3 *Active immunity* results when the body manufactures its own antibodies or T cells in response to a foreign antigen. *Passive immunity* results when antibodies, not antigens, are transferred from one person to another.

Monoclonal antibodies

1 The laboratory fusion of an activated lymphocyte with a cancer cell produces a hybrid cell called a *hybridoma.* Clones of the hybridoma produce a limitless

supply of antibodies called *monoclonal antibodies.*

2 Monoclonal antibodies may be useful in producing a vaccine against cancer, tracing the location of cancer cells in the body, and directing or transporting specific antibodies or cancer-killing chemicals to cancer cells.

Interaction between the neuroendocrine system and the immune system

1 Many interactions between the neuroendocrine and immune systems are known. For example, adrenal glucocorticoid hormones secreted in response to stress may suppress the immune system.

2 Apparently, the major interactions between the two systems involve the adverse effects of stress.

Physiological and anatomical abnormalities

1 *Acquired immune deficiency syndrome (AIDS)* arises from a weakening of natural cell-mediated immunity as a result of infection by the HIV virus.

2 *Hodgkin's disease* is a form of cancer that involves a painless, progressive enlargement of the lymph nodes, spleen, and other lymphoid tissues.

3 *Infectious mononucleosis* is a viral disease accompanied by enlarged lymph nodes and an increased lymphocyte production.

4 *Systemic lupus erythematosus* is a breakdown of the immune system in which B cells produce excess antibodies that attack healthy connective tissue cells throughout the body.

UNDERSTANDING THE FACTS

1 List the four major functions of the lymphatic system.

2 Define lymph.

3 Describe the structure of a lymph node.

4 What is the main structural difference between lymphatic capillaries and blood capillaries?

5 Name and describe the vessels of the lymphatic system.

6 What is the major function of the lacteals?

7 Into which large veins does lymph drain?

8 What portion of the body is drained by the thoracic duct?

9 Which lymphatic organ serves as a blood reservoir?

10 Name the five classes of antibodies, and describe their functions.

11 How are lysosomes involved in phagocytosis?

12 Antigens are made up of what type of organic molecule?

13 What is the basic shape of the antibody molecule?

14 Where do B cells and T cells originate?

15 Which type of lymphocytes are active in cell-mediated immunity?

16 Which cell type is the major cause of rejection in tissue or organ transplants?

17 Why may anaphylactic shock be fatal?

18 Why are corneal transplants generally not rejected?

19 Give the location of the following (be specific):
- **a** lymph nodes
- **b** aggregated lymph nodules
- **c** flap valves
- **d** lysozyme
- **e** lacteals
- **f** interferon (production site)

UNDERSTANDING THE CONCEPTS

1 Since the lymphatic system lacks a "pump," how do you explain the movement of lymph?

2 Although lymph contains all the coagulation factors found in blood and is capable of clotting to a small degree, why does lymph lack clotting ability?

3 Why is the reticuloendothelial system important to the body?

4 Why are valves so important to the lymphatic system?

5 How do complement, properdin, and interferon help the body to combat infection?

6 How does the skin defend against microbes?

7 What is the difference between active and passive immunity?

8 What is the relationship of the lock-and-key model and the great specificity normally shown by antibodies?

9 Why is it important for the IgG antibodies to pass freely through the placenta during pregnancy?

10 How are monoclonal antibodies produced? What are some important uses?

11 Describe the desensitization procedure used to combat severe allergy.

12 Discuss the functions of the thymus gland.

13 Discuss the two mechanisms of autoimmunity.

SUGGESTED READING

ADA, GORDON L., AND GUSTAV NOSSAL, "The Clonal-Selection Theory." *Scientific American,* August 1987.

BUISSERET, PAUL D., "Allergy." *Scientific American,* August 1982.

BURNET, F. M., ed., *Immunology: Readings from Scientific American.* San Francisco: Freeman, 1976.

COHEN, IRUN R., "The Self, the World and Autoimmunity." *Scientific American,* April 1988.

COLLIER, R. JOHN, AND DONALD A. KAPLAN, "Immunotoxins." *Scientific American,* August 1987.

EDELSON, RICHARD L., AND JOSEPH M. FINK, "The Immunologic Function of Skin." *Scientific American,* June 1985.

GALLO, ROBERT C., "The AIDS Virus." *Scientific American,* January 1987.

LAURENCE, JEFFREY, "The Immune System in AIDS." *Scientific American,* December 1985.

MARRCK, PHILIPPA, AND JOHN KAPPLER, "The T Cell and Its Receptor." *Scientific American,* February 1986.

MILSTEIN, CESAR, "Monoclonal Antibodies." *Scientific American,* October 1980.

OSMOND, D. G., AND P. K. LALA, "The Immune System," *American Journal of Anatomy,* 3 (1984).

RAFF, MARTIN C., "Cell-Surface Immunology." *Scientific American,* May 1976.

ROITT, I. M., J. BROSTOFF, AND D. MALE, *Immunology.* St. Louis: Mosby, 1985.

SILBERNER, JOANNE, "Survival of the Fetus." *Science News,* October 11, 1986.

SINGER, A., AND R. HODGES, "Mechanism of B-cell and T-cell Interaction." *Annual Review of Immunology,* 1 (1983): 211.

SITES, D. P., J. D. STUBO, H. H. FUDENBERG, AND J. V. WELLS, *Basic and Clinical Immunology,* 5th ed. Los Altos, Calif.: Lange Medical Publications, 1984.

TONEGAWA, SUSUMU, "The Molecules of the Immune System." *Scientific American,* October 1985.

YOUNG, JOHN DING-E, AND ZANVIL A. COHN, "How Killer Cells Kill." *Scientific American,* January 1988.

21
The Respiratory System

LEARNING OBJECTIVES

1 List the respiratory structures through which air passes before it reaches the bloodstream.

2 Describe the structure and the function of the nose within the respiratory system.

3 Name the parts of the pharynx, and describe their locations and functions.

4 Explain the function of the epiglottis, and the major function of the larynx.

5 Describe the trachea and its relationship to the respiratory tract.

6 Identify the parts of the "respiratory tree" from the largest to the smallest branches.

7 Relate the structure of the alveoli and surrounding capillary network to their functions.

8 Describe the structure of the lungs and their pleurae.

9 Summarize the processes of expiration and inspiration, naming the muscles and other structures involved.

10 Define total lung capacity, residual volume, tidal volume, vital capacity, minute respiratory volume, and alveolar ventilation.

11 Explain the significance of Boyle's law, Dalton's law, Henry's law, and specific solubilities in the transport of gases within the body.

12 Compare the processes of external and internal respiration.

13 Describe the role of oxyhemoglobin in oxygen transport and the main factors that affect it.

14 Explain how carbon dioxide is transported in the blood and the role of carbaminohemoglobin.

15 Discuss the roles of the medullary rhythmicity area, apneustic area, and pneumotaxic area in the control of breathing.

16 Describe how the concentrations of carbon dioxide and oxygen in the blood regulate breathing.

17 Explain how sensors in joints or muscles and proprioceptors in the lungs affect breathing.

A human body can survive without food for as much as several weeks, and without water for several days, but if breathing stops for three to six minutes, death is likely. Everyone requires a constant supply of oxygen to body tissues, especially to the heart and brain. The *respiratory system* delivers air containing oxygen to the blood and removes gaseous waste products of metabolism. It includes the lungs, the several passageways leading from outside to the lungs, and the muscles that move air into and out of the lungs. In this chapter we treat the anatomy, histology, and neurological control of the respiratory system, the chemical action of respiratory gases (oxygen and carbon dioxide) in blood circulation, and some disorders of the system.

The term *respiration* has several meanings and related functions in physiology. First, *cellular respiration* is the sum of biochemical events by which the chemical energy of foods is released to provide energy for life's processes. In cellular respiration, oxygen is utilized as a final electron acceptor in the cytochrome chain (see Chapter 23). Second, *external respiration* is a form of gas exchange in which oxygen from the lungs moves into the blood, and carbon dioxide and water move from the blood to the lungs. Third, *internal respiration* is the process in which body cells exchange carbon dioxide for oxygen in the blood. In contrast to these physiological processes, *ventilation,* or *breathing,* is the mechanical process that moves air into and out of the lungs. It includes two phases, inspiration and expiration.

It is helpful to think of the action of the respiratory system as a series of steps, even though in a living body they all take place at the same time and continuously. First, air, rich in oxygen and poor in carbon dioxide, travels through the respiratory tract deep into the terminal portions of the lungs; this is *inspiration.* There, oxygen diffuses from the blood into the lungs. From the lungs, oxygen-rich blood is carried by the systemic circulatory system to all parts of the body. Where cellular respiration is being carried on, oxygen moves from the blood into the cells, and carbon dioxide and other wastes are released from the cells into the blood. Finally, oxygen-poor venous blood, carrying its load of wastes, is forced back to the lungs, and some of the air is exhaled during *expiration* (see Figure 21.16).

Although the primary function of the respiratory system is the exchange of oxygen and carbon dioxide, it has many secondary functions. Although the lungs do not play an important role in the transfer of matter, some products are eliminated through the breath. The respiratory tract is the site of sound production. We also use our lungs and breathing muscles as a convenient high-volume, low-pressure pump for dozens of daily jobs such as inflating balloons, cooling hot coffee, warming cold hands, and playing wind instruments. Laughing and crying, as well as sneezing and coughing, are other examples of activities of the respiratory system. Also, as you will see in later chapters, the respiratory system is one of a number of body systems that help maintain homeostasis by regulating the pH of blood and other body fluids within a narrow functional range. Finally, the muscles of the respiratory system assist in abdominal compression during urination, defecation, and childbirth.

FIGURE 21.1

The respiratory system, with the interior of the right lung (not in scale) revealed.

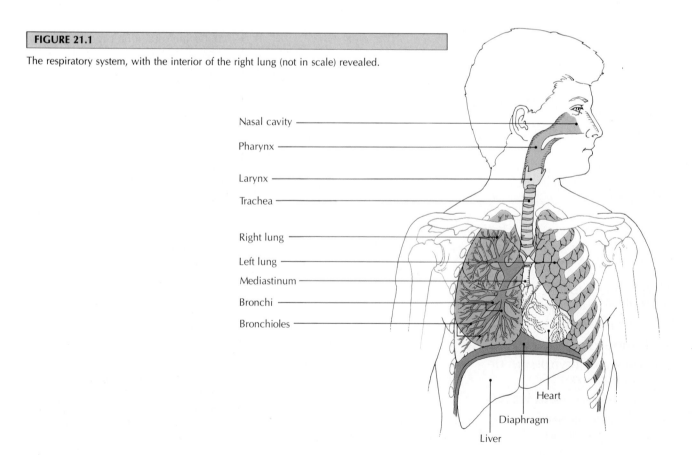

Nasal cavity

Pharynx

Larynx

Trachea

Right lung

Left lung

Mediastinum

Bronchi

Bronchioles

Heart

Diaphragm

Liver

RESPIRATORY TRACT

Air flows through respiratory passages because of differences in pressure produced by chest and trunk muscles during respiration. Except for the beating of cilia in the respiratory lining, the passageways are simply a series of openings through which air is forced. These passageways and the lungs make up the respiratory tract (Figure 21.1 and Table 21.1 on p. 662).

Nose

Air normally enters the respiratory tract through the nose. The external nose is supported at the bridge by **nasal bones.** The *septal cartilage* separates the two external openings of the *nostrils*, or **external nares** (NAIR-eez; L. nostrils). Along with the vomer and perpendicular plate of the ethmoid bone, the septal cartilage forms the **nasal septum,** which divides the nasal cavity into two bilateral halves. The surface of the nasal cavity is lined with two types of mucosa: (1) *respiratory mucosa* warms and moistens inhaled air, and (2) *olfactory mucosa* contains the receptor neurons of smell.

The **nasal cavity** fills the space between the base of the skull and the roof of the mouth. It is supported above by the

ethmoid bones, and on the sides by the *ethmoid, maxillary*, and *inferior conchae* bones (Figure 21.2). The superior, middle, and inferior conchae, also known as the turbinate bones, bear longitudinal ridges that are covered with highly vascular respiratory mucosa. The blood vessels in this area, called venous sinuses, bring a large quantity of blood to the mucous membrane that adheres to the underlying periosteum. The blood is a constant source of heat that warms the air inhaled through the nose. Since warm air holds more water than cold air, the air is also moistened as it moves through the nasal cavity.

Between the ridges of the conchae are folds called the **superior, middle,** and **inferior meatuses** (singular, *meatus*). These meatuses serve as air passageways. The *hard palate,* strengthened by the palatine and parts of the maxillary bones, forms the floor of the nasal cavity. If these bones do not grow together normally by the third month of prenatal development, the nasal cavity and mouth are not separated adequately. The condition is known as a *cleft palate*. The immediate problem facing a newborn infant with a cleft palate is that not enough suction can be created to nurse properly. A cleft palate can usually be corrected surgically.

The *soft palate* (see Figure 21.2) is a flexible muscular sheet that separates the oropharynx from the nasal cavity. It is continuous with the posterior border of the hard palate, and consists of several skeletal muscles covered by mucous mem-

FIGURE 21.2

Right median sagittal section of the head, showing the upper portion of the respiratory system: the nasal cavity, pharynx, larynx, and part of the trachea.

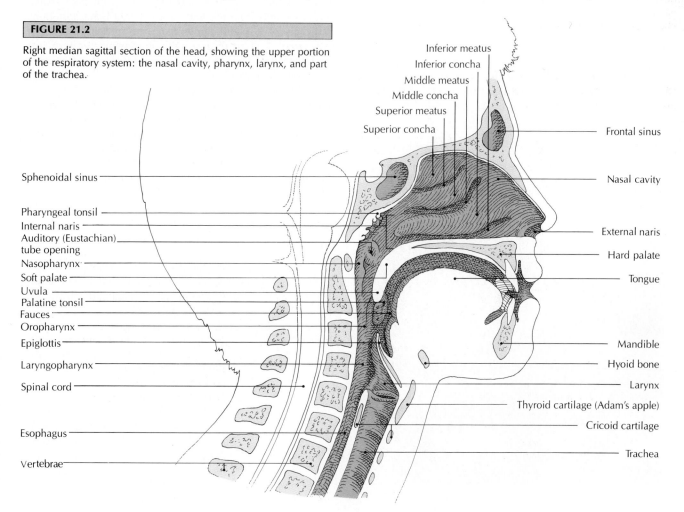

brane. Depression of this area during breathing enlarges the air passage into the pharynx. When the soft palate is elevated during swallowing, it prevents food from entering the nasal pharynx.

The *frontal, sphenoidal, maxillary,* and *ethmoid sinuses* are blind sacs that open into the nasal cavity. Together they are called the *paranasal sinuses.* They give the voice a full, rich tone. A **nasolacrimal** (L. "nose" and "tear") **duct** leads from each eye to the nasal cavity. Excessive secretion of tears in the eyes is drained into the nose; that is why weeping may be accompanied by a watery flow from the nose (runny nose). At the back of the nasal cavity, the two *internal nares* open into the pharynx (throat).

The lining of the nasal cavity is specialized. Most of the covering membrane is supplied with mucus-secreting *goblet cells,* which keep the surfaces wet (see Figure 4.3C). The mucus moistens air before it goes into the lungs, and catches many of the small dust particles carried in air that get past the coarse hairs in the nostrils. The nasal epithelium is a pseudostratified columnar tissue bearing millions of cilia, which also help capture minute particles (Figure 21.3).

The *olfactory epithelium* in the upper part of the nasal cavity contains sensory nerve endings that are connected to the olfactory nerve. Aromatic substances, caught on the surface and dissolved in the moisture, stimulate the nerve endings and cause the sensation of smell (see Chapter 15).

Pharynx

The **pharynx** (throat) is connected to the nasal cavity through the internal nares, and also to the mouth, or *buccal (oral) cavity* (see Figure 21.2). The pharynx leads to the rest of the respiratory passage and to the esophagus. The part of the pharynx above the soft palate is the **nasopharynx.** The **oropharynx,** at the back of the mouth, extends from the soft palate to the epiglottis, where the respiratory and alimentary tracts separate into the *trachea* (windpipe) and the *esophagus.* The lower part of the pharynx is the **laryngopharynx.** All parts of the pharynx are continuous with each other.

Like the nasal cavity, the *nasopharynx* is lined with ciliated epithelium that aids in cleaning inspired air (see Figure 21.3). The auditory (Eustachian) tubes open into the nasopharynx just behind and lateral to the internal nares. These tubes are air passages that equalize the air pressure on both sides of the eardrums; they must be kept open or the ears will ache. If these tubes are not open, changes in atmospheric pressure, such as those felt when one goes quickly up or down in an airplane, can cause the ears to "pop." The temporary imbalance can usually be alleviated by swallowing.

The *oropharynx* has muscular walls, and along with the laryngopharynx initiates the act of swallowing. It is separated from the mouth by a pair of membranous *fauces* (narrow passageways; singular, *faux*) that spread open during swallowing, or during panting, when most of the air entering the respiratory tract is drawn in through the mouth rather than the nostrils. Like most tubes that are likely to suffer abrasion, the oropharynx is lined with stratified squamous epithelium, which is constantly being renewed.

The lowermost portion of the pharynx is the *laryngopharynx.* It extends from the level of the hyoid bone posteriorly to the larynx. It is at that point that the respiratory and digestive systems become distinct, with air moving anteriorly into the larynx, and food moving posteriorly into the esophagus past the flexible glottis.

Tonsils are lymphoid structures that are part of both the pharynx and the immune system (see Chapter 20). *Lingual tonsils* are at the base of the tongue; *palatine tonsils* are at the sides of the oropharynx; and *pharyngeal tonsils* (adenoids) hang from the roof of the nasopharynx. If the adenoids become inflamed and swollen, they can interfere with breathing through the nose, may block the opening of the auditory tube, and even make speech difficult.

Larynx

The **larynx** (voice box) not only is an air passage from the pharynx to the rest of the respiratory tract, but also produces most of the sound used in speaking and singing (Figure 21.4).

FIGURE 21.3

Scanning electron micrograph of the ciliated epithelium that lines the nasal cavity. The cilia wave continuously, helping to clear the nasal passage of foreign particles. [*Lennart Nilsson,* Behold Man. *Boston, Little, Brown, 1974.*]

Cilia

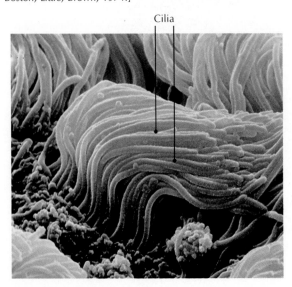

Why are nosebleeds so common?

Bleeding from the nose is common because the nose sticks out far enough to be struck easily, the blood supply is abundant in the nasal membranes, and the mucosal layer is delicate. Bleeding can usually be controlled by packing the external or internal nares, or both. Nosebleeds are more common in cold weather because the lower absolute humidity of the inspired air promotes drying, cracking, and bleeding of the nasal epithelium.

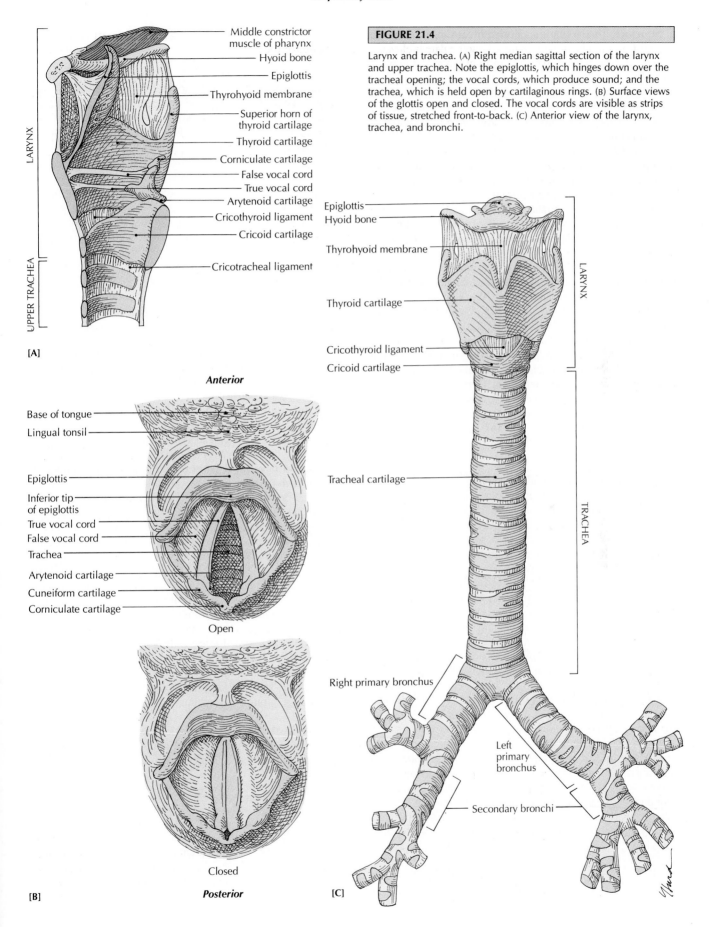

[A]

LARYNX

UPPER TRACHEA

Middle constrictor muscle of pharynx
Hyoid bone
Epiglottis
Thyrohyoid membrane
Superior horn of thyroid cartilage
Thyroid cartilage
Corniculate cartilage
False vocal cord
True vocal cord
Arytenoid cartilage
Cricothyroid ligament
Cricoid cartilage
Cricotracheal ligament

FIGURE 21.4

Larynx and trachea. (A) Right median sagittal section of the larynx and upper trachea. Note the epiglottis, which hinges down over the tracheal opening; the vocal cords, which produce sound; and the trachea, which is held open by cartilaginous rings. (B) Surface views of the glottis open and closed. The vocal cords are visible as strips of tissue, stretched front-to-back. (C) Anterior view of the larynx, trachea, and bronchi.

Epiglottis
Hyoid bone
Thyrohyoid membrane
Thyroid cartilage
Cricothyroid ligament
Cricoid cartilage
LARYNX
Tracheal cartilage
TRACHEA
Right primary bronchus
Left primary bronchus
Secondary bronchi

[B]

Anterior

Base of tongue
Lingual tonsil
Epiglottis
Inferior tip of epiglottis
True vocal cord
False vocal cord
Trachea
Arytenoid cartilage
Cuneiform cartilage
Corniculate cartilage

Open

Closed

Posterior

[C]

Structure The larynx is supported by cartilages, of which the most prominent is the roughly triangular *thyroid cartilage* (Adam's apple). It is visible in the front of the throat in men, but is less conspicuous in women. The larynx is supported above the thyroid cartilage by ligaments connected to the hyoid bone. Below the thyroid cartilage is the *cricoid cartilage,* a ring of cartilage that connects the thyroid cartilage with the trachea below.

The *epiglottis* is a flap of cartilage that folds down over the opening into the larynx during swallowing, and swings back up when the act of swallowing ceases. Since air must pass from the pharynx (which is *dorsal* to the mouth) to the larynx (which is *ventral* to the esophagus), there is a crossover between the respiratory and digestive tracts. The epiglottis works well most of the time, mainly because people involuntarily inhale before swallowing, getting air into the lungs, and then exhale after swallowing, thus clearing the air passage. If food does accidentally get into the larynx, it is usually forced out by a strong cough reflex.

Mammals produce sound mainly by vibrations of the *vocal cords* (Figure 21.4B). These paired strips of stratified squamous epithelium at the base of the larynx have a front-to-back slit between them, the *glottis.* Above and beside the vocal cords is a pair of *vestibular folds,* usually called *false vocal cords,* that protrude into the vestibule of the larynx, hence *vestibular folds.* The true vocal cords are held in place and regulated by a pair of *arytenoid cartilages,* which in turn are supported by a pair of *cuneiform cartilages.* Still another pair of cartilages, the *corniculate cartilages,* lies between the arytenoids and the epiglottis.

Sound production Sound production is the result of a complex coordination of muscles. The immediate source of most human sound is the vibration of the vocal cords, which oscillate rapidly with a frequency of about 50 hertz in a deep bass to about 1700 hertz in a high soprano.* The pitch is regulated mainly by the tension put on the cords by the arytenoid cartilages, with greater tension producing higher pitch. But the actual size of the cords is also effective. The longer, thicker cords of most men give lower pitches than the shorter, thinner cords of most women. The fundamental tone from the cords is only a part of the final quality of the sound, as overtones are added by changes in the positions of lips, tongue, and soft palate. Vowels are produced through the open throat and mouth, and can be altered by movements of the lips and tongue. Consonants are added, giving subtlety and variety to speech, by further changes in the lips and tongue. Two consonants, *m* and *n,* can be articulated only if the nasal passage is open.

The variable shapes of the nasal cavity and sinuses give voices their individual qualities. In fact, each person has such a distinct set of vocal overtones that every human voice is as unique as a set of fingerprints. The intensity, volume, or "loudness" of vocal sounds is regulated by the amount of air passing over the vocal cords, and that in turn is regulated by the pressure applied to the lungs, mainly by the abdominal muscles.

Trachea

The *trachea* (TRAY-kee-uh), or windpipe, is an open tube about 2.5 cm (1 in.) in diameter and 10 to 12 cm (4 to 5 in.) long. It extends from the base of the larynx to the top of the lungs, where it forks into two branches, the right and left *bronchi* (singular, *bronchus*) (Figure 21.4C). The trachea is kept open by 16 to 20 cartilaginous rings that are open on the dorsal side next to the esophagus. The rings are connected by fibroelastic connective tissue and longitudinal smooth muscle, making the trachea both flexible and extensible.

The inside surface of the trachea is lined with pseudostratified ciliated columnar epithelium, which produces moist mucus (Figure 21.5). It contains upward-beating cilia. Those dust particles that are not caught in the nose and pharynx may be trapped in the trachea and carried up to the pharynx by the cilia to be swallowed or spat out *(expectorated).*

Lungs: The Respiratory Tree

The trachea branches into the right and left bronchi, the *primary bronchi* that enter the right and left lungs. Each primary bronchus divides into smaller *secondary bronchi,* which in turn divide into *tertiary (segmental) bronchi.* These bronchi continue to branch into smaller and smaller tubes called **bronchioles** and then **terminal bronchioles** (Figures 21.6A, B). The whole system looks so much like an upside-down tree that it is commonly called the "respiratory tree."

Bronchi have cartilage in their walls, but bronchioles do not. Bronchioles contain more smooth muscle tissue than the bronchi, however. Both bronchioles and bronchi are lined with ciliated columnar epithelium.

The bronchioles continue to branch, finally ending in *respiratory bronchioles* (Figure 21.6C). They, in turn, branch into many *alveolar ducts,* which lead into microscopic air sacs called *alveoli,* where gas exchange takes place. Alveolar ducts and alveoli are lined with simple squamous epithelium.

In summary, it can be said that the *conducting portion* of the lung is composed of the bronchi and bronchioles. Beyond the terminal bronchioles (the most distal branches of the bronchial tree), the thinner-walled respiratory bronchioles begin the *respiratory portion* of the lung, where the exchange of respiratory gases occurs. The respiratory portion consists of the alveolar ducts, alveolar sacs, and alveoli, which lead from the respiratory bronchioles.

Alveoli The functional units of the lungs are the **alveoli** (singular, *alveolus*). Each lung contains over 350 million alveoli, each surrounded by many capillaries. Alveoli are clus-

---

Why does yelling make us hoarse? | *Yelling vibrates the vocal cords and slams them together with such strong force that the cords may swell or even bleed. Swollen cords do not close properly, and hoarseness results as air leaks between them.*

---

*A hertz (Hz) is a unit of frequency measured in cycles per second.

FIGURE 21.5

Trachea. (A) Scanning electron micrograph of a section through the wall of the trachea. The mucosa is made up of the epithelial layer and a thin connective-tissue layer, the lamina propria. A small portion of the basal lamina (basement membrane) is exposed. The submucosa contains many blood vessels and mucus-secreting glands. Also shown are the adventitia (the outermost layer of the connective tissue) and the hyaline cartilage between the submucosa and adventitia. (B) Cross section of the trachea. [(A) *Richard Kessel and Randy Kardon,* Tissues and Organs: A Text-Atlas of Scanning Electron Microscopy. *San Francisco, W. H. Freeman, 1979.*]

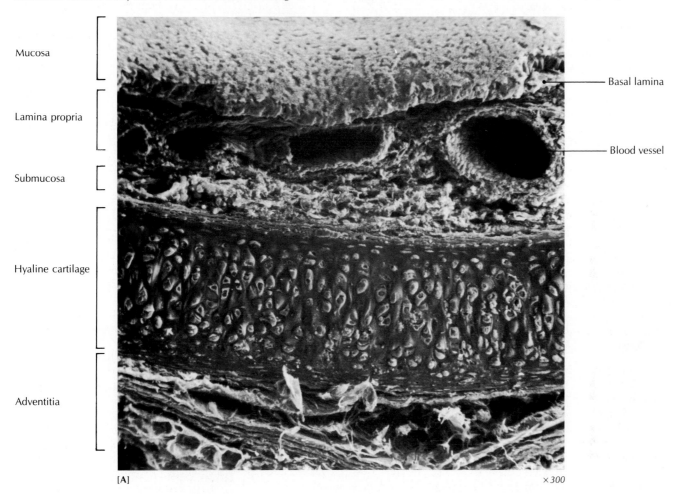

Mucosa

Lamina propria

Submucosa

Hyaline cartilage

Adventitia

Basal lamina

Blood vessel

[A]

×300

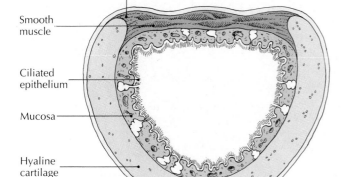

Posterior

Membranous portion of trachea

Smooth muscle

Ciliated epithelium

Mucosa

Hyaline cartilage

Adventitia

Anterior

[B]

tered in bunches like grapes (Figures 21.6C, D), and provide enough surface area to allow ample gas transfer. The total interior area of adult lungs provides about 60 to 70 sq m, 20 times greater than the surface area of the skin (Figure 21.7). Every time you inhale, you expose to fresh air an area of lung roughly equal to that of a tennis court.

A single alveolus looks like a bubble, which is supported by a *basement membrane (basal lamina)*. A group of several alveoli with a common opening into an alveolar duct is called an *alveolar sac.* The lining epithelium of alveoli consists mainly of a single layer of squamous cells (squamous pulmonary epithelial cells), also called *type I cells* (Figure 21.8). It also has *septal cells,* also called *type II cells,* which are smaller, scattered, cuboidal secretory cells. Type II cells secrete a detergentlike phospholipid called *surfactant* (dipalmitoyl lecithin), which helps keep alveoli inflated by reducing surface tension. Alveoli also contain phagocytic *alveolar macrophages* that adhere to the alveolar wall or circulate freely in the lumen of alveoli. These macrophages ingest and destroy microorganisms and other foreign particles that have penetrated

FIGURE 21.6

Lungs. (A) Lower portion of the respiratory tract: the respiratory tree. Note detail of the lobes of the lungs. The left lung, with no horizontal fissure, has only two lobes, the superior and inferior; the right lung also has a middle lobe. (B) Lung lobule, showing alveolar sacs. (C) Detail of alveolar sac. (D) Clusters of alveoli are surrounded by a capillary network, where the exchange of oxygen and carbon dioxide takes place.

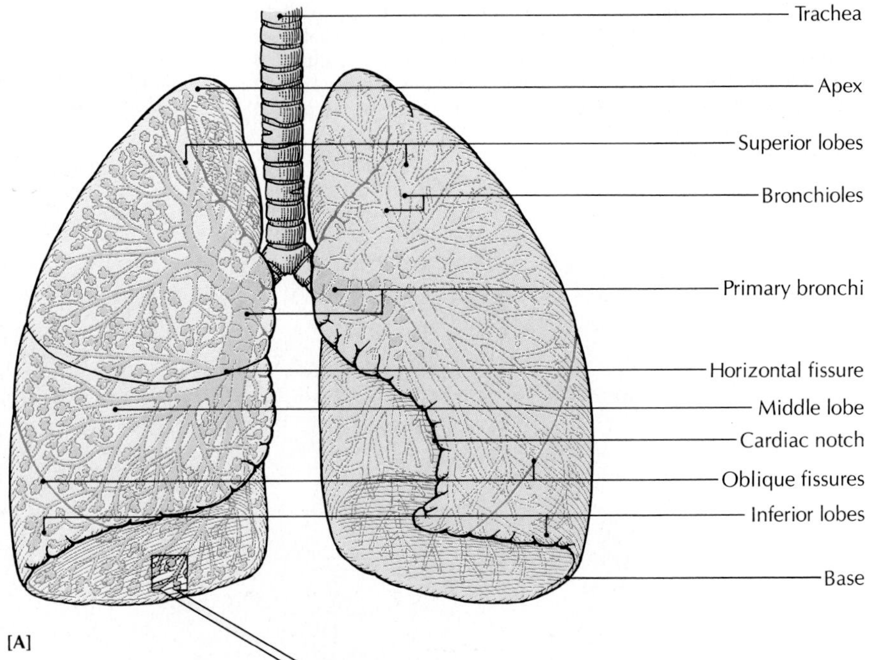

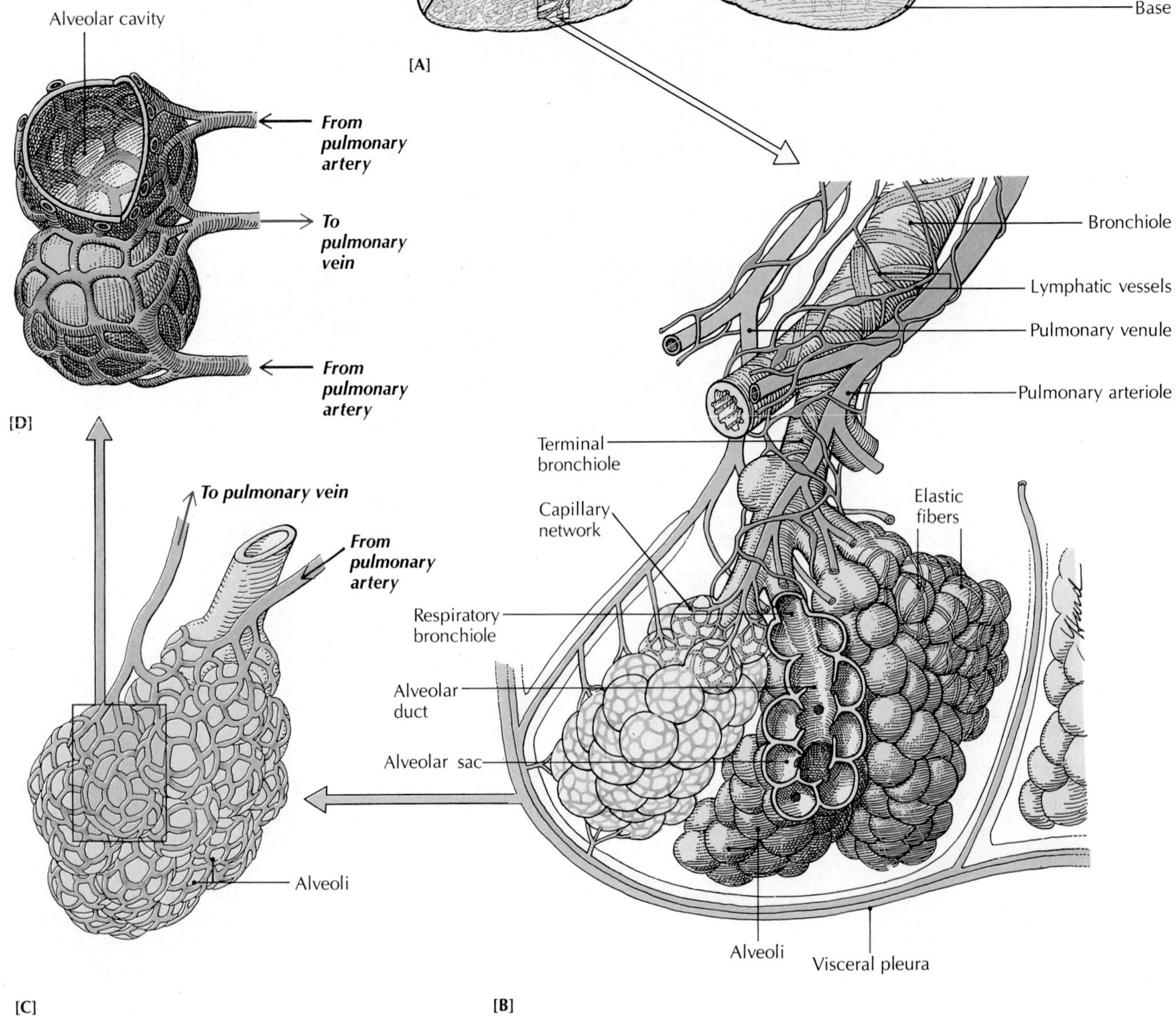

FIGURE 21.7

Scanning electron micrograph of the inside of a lung. The darker holes are alveoli and alveolar ducts, which lead into deeper alveoli. The micrograph gives an idea of the convolutions that add to the amount of surface available for gas exchange. [*Manfred Kage/Peter Arnold.*]

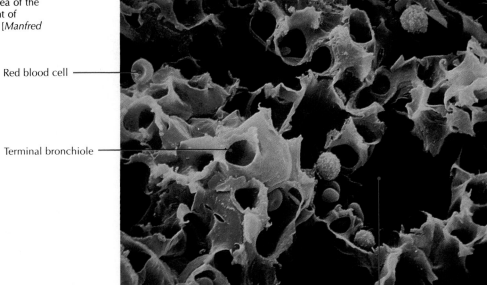

Red blood cell

Terminal bronchiole

Alveolus

×180

FIGURE 21.8

Scanning electron micrograph of the surface of an alveolus, showing a type II alveolar cell and a macrophage. [*Fujita, Tanaka, and Tokunaga,* SEM Atlas of Cells and Tissues. *Tokyo, Igaku-Shoin, 1981.*]

Type II cell

Macrophage

×2000

FIGURE 21.9

Schematic drawing of a section through a
portion of lung tissue, showing an alveolus
and surrounding capillaries.

Alveolus

Type II alveolar (septal) cell

Elastic fibers

Alveolus

Nucleus of type I
alveolar cell

Connective tissue fibers

Alveolus

Type II alveolar
(septal) cell

Connective tissue
(interstitial) cell

White blood cell
(monocyte)

Red blood cells

Nucleus of capillary
endothelium

Alveolus

Capillary
endothelium

Type I alveolar
(squamous epithelial) cell

Macrophages

the alveolar lining. The foreign material either is moved upward by ciliary action to be expelled by coughing, or enters
lymphatics to be carried to the lymph nodes at the hilum of
the lung.

Since the alveoli are the sites of gas transfer to the blood
by diffusion, the membranes of both alveolar walls and the
capillary walls that line them must be thin enough to give
maximum permeability, yet strong enough to hold open the
air cavities (Figure 21.9). Alveoli are about 25 μm in diameter, and their walls are about 4 μm thick, which is much
thinner than a sheet of paper. The capillaries surrounding
each alveolus are also thin-walled for gas exchange. The exchange is facilitated by the small size and the large number
of capillaries (Figure 21.10). Capillaries are so small that red
blood cells flow through them in single file, giving each cell
maximum exposure to the alveolar walls. There are so many
capillaries that at any one time almost a liter of blood is being
processed in the lungs.

Alveolar surface tension At the surface of a liquid, especially where air and water meet, the molecules of the liquid
are more attracted to one another than they are to the air.

FIGURE 21.10

Scanning electron micrograph of the extensive capillary network
surrounding alveoli. [*Courtesy of Professor T. Murakami,
Department of Anatomy, Okayama University School of Medicine.*]

Alveolar sacs   Capillaries

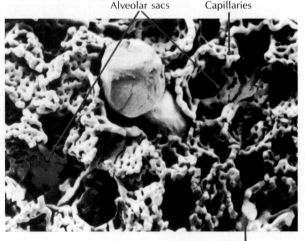

Connection between capillaries and arteriole $\times 100$

"Smoker's Cough"

The human body has several defenses against dirty air. Hairs in the nose filter out large particles. The mucus that is secreted continuously by the goblet cells in the upper respiratory tract washes out or dissolves smaller irritants, and traps particles that enter the respiratory tract. Also, the many cilia in the upper tract sweep away foreign particles and excessive mucus. But air pollutants (especially cigarette smoke, ozone, sulfur dioxide, and nitrogen dioxide) can stiffen, slow, or even destroy the cilia, causing the mucus to clog, a reaction frequently resulting in "smoker's cough." When cilia in the nose are impaired by pollutants, or are partially paralyzed by cold air, there is an overproduction of mucus, which is not swept back into the throat as usual. When this happens, the mucus drips forward and causes a runny nose.

As a result of inefficient cilia, microorganisms and other foreign particles can penetrate the lung tissue and cause respiratory infections, and possibly even lung cancer. As lung tissue becomes more irritated by air pollutants, mucus flows more freely to help remove the inhaled pollutants. A normal coughing mechanism expels the contaminated air and mucus. Smoking or air pollution can trigger excessive mucus flow that eventually blocks the air passages, which in turn causes more coughing—a type of positive feedback. Over time, the muscles surrounding the bronchial tree can weaken from prolonged, strong coughing, and breathing becomes progressively more difficult. If this positive feedback cycle continues, *chronic bronchitis* (a chronic inflammation of the mucous membranes of the respiratory tree) can occur.

The result is a thin film of greater density on the surface than beneath the surface of the liquid. This film of strongly attracted molecules produces a **surface tension** that tends to make the surface contract to its smallest possible area. (That is why small raindrops are spherical.) An alveolus is so wet that its liquid lining would tend to make the alveolus collapse if there were not some way to reduce the surface tension. That tendency is counteracted by surfactant, which reduces the surface tension of the water on the inner surface of each alveolus. As a result, the alveolar walls can be thin without collapsing.

Because babies have no need to breathe before they are born, they do not produce surfactant, but they do need to start making it just before or just after birth. Without it they may suffer from insufficient opening of the alveoli. If the alveoli fail to open, or if they do open and collapse, the condition is known as *atelectasis* (imperfect expansion). Because the lungs develop toward the end of gestation, premature babies are at a high risk to develop hyaline membrane disease (see Physiological and Anatomical Abnormalities). Atelectasis can also result from a number of pathological conditions, such as pneumonia, tuberculosis, or cancer.

Lungs: Lobes and Pleurae

The **lungs** fill the thoracic cavity except for a midventral region, called the *mediastinum,* where the heart and major blood vessels lie. Each lung is somewhat pointed at the *apex* (the top) and concave at the the *base,* where it lies against the diaphragm (see Figure 21.6A). Because the heart takes more space on the left side than on the right, and because the liver bulges up somewhat on the right side, the two lungs are not symmetrical. The left one is thinner and longer than the right.

The right lung has three main lobes, the *superior, middle,* and *inferior* (see Figure 21.6B). The left lung has no middle lobe, but has a concavity, the *cardiac notch,* into which the heart fits. Each lobe is further subdivided into 10 *bronchopulmonary segments,* which are served by individual bronchi and bronchioles, and which function somewhat independently. This partial independence makes it possible to remove one diseased segment without incapacitating the rest of the lung. Bronchopulmonary segments contain smaller subdivisions, called *lobules,* which are surrounded by elastic connective tissue (see Figure 21.6C). A lobule has hundreds of alveoli. It is served by a bronchiole carrying gases, a lymphatic vessel, a small vein (a *pulmonary venule*), and an artery (a *pulmonary arteriole*).

The lungs are covered by two pleural membranes that are continuous with each other: the inner *visceral pleura* directly on the surface of a lung, and the *parietal pleura* lining the chest cavity (Figure 21.11). The small moisture-filled potential space between the visceral and parietal pleura is the **pleural cavity.** With each breath, the visceral pleura slides on the parietal layer, with the necessary lubrication being provided by the thin film of fluid between the two layers.

Three functions are associated with the pleurae. (1) The thin film of moisture from the membranes within the pleural cavity acts as a lubricant for the lungs, which are in constant motion. (The pleural cavity does not contain a significant amount of fluid, as is sometimes thought.) (2) The air pressure in the pleural cavity is less than the atmospheric pressure, and thus aids in the mechanics of breathing. (3) The pleurae effectively separate the lungs from the medially located mediastinum, which contains the other thoracic organs. These organs include the heart, the esophagus, the thoracic duct, nerves, and major blood vessels (see Figure 1.22).

Nerve and Blood Supply

The smooth muscle of the tracheobronchial tree is innervated by the autonomic nervous system. Branches of the vagus

Embryonic Development of the Respiratory Tree

About four weeks after fertilization, at about the same time that the laryngotracheal tube and esophagus are separating into distinct structures, an endodermal *bronchial bud* forms at the caudal end of the developing laryngotracheal tube (drawing A). The single bronchial bud soon divides into two buds, the right one being larger (drawing B). These bronchial buds develop into the left and right primary bronchi of the future lungs during the fifth week (drawing C). The right bud divides to form three secondary bronchi, and the left bud forms only two. The right bronchus continues to be larger and more branched than the left during embryonic development, and by

the eighth week the bronchi show their basic adult form, with three secondary bronchi and three lobes in the right lung, but only two secondary bronchi and two lobes in the left lung (drawing D). The bronchi subsequently branch repeatedly to form more bronchi and bronchioles.

By week 24, the epithelium of the bronchioles thins markedly, and highly vascularized primitive *alveoli* (terminal air sacs) have formed. A baby born prematurely during week 26 has a fairly good chance to survive without the help of artificial respiratory devices because its lungs have developed an alveolar mechanism for gas exchange. The development of pulmonary blood vessels is as

important to the survival of premature infants as is the thinness of the alveolar epithelium, which allows an adequate gas exchange. A newborn infant has only about 15 percent of the adult number of alveoli. The number and size of alveoli continue to increase until a child is about eight years old. Also at about 24 weeks, type II alveolar epithelial cells begin to secrete *surfactant*, a liquid that prevents alveoli from collapsing by reducing their surface tension (see p. 657). After two or three weeks of secretion, the alveoli are strong enough to retain air and remain intact when breathing finally begins at birth.

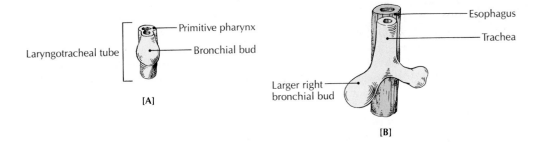

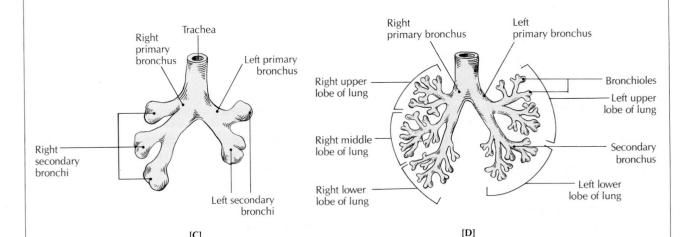

Embryonic development of the bronchi, ventral views. [A] 4 weeks (early), [B] 4 weeks (late), [C] 5 weeks, and [D] 8 weeks.

FIGURE 21.11

Pleurae. (A) Transverse section of the thorax as seen from above, showing the relationship of the lungs and pleurae. (B) Coronal section of thorax, showing the pleurae and the separation of right and left pleural membranes by the mediastinum. Portions of the parietal pleura are named according to the structures to which the portions fuse: the mediastinal pleura next to the mediastinum, the costal pleura next to the ribs, and the diaphragmatic pleura next to the diaphragm. Each lung receives a primary bronchus at the hilus of the lung.

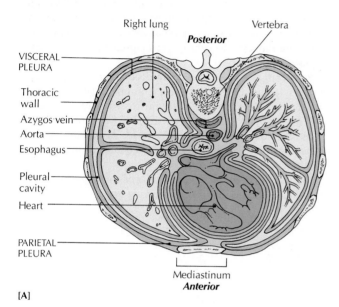

[A]

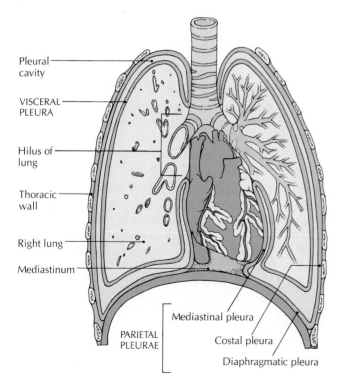

[B]

nerves run alongside the pulmonary blood vessels, carrying parasympathetic fibers that constrict the bronchi. Branches from the thoracic sympathetic ganglia carry fibers that dilate the bronchi. In addition, the bronchial walls contain parasympathetic ganglia. Bronchoconstriction may also be caused by acetylcholine, histamine, and some bacterial toxins that act directly on bronchial smooth muscle. In contrast, epinephrine and norepinephrine cause bronchodilation by relaxing the smooth muscle.

The respiratory tree is supplied with blood by the pulmonary and bronchial circulations. The pulmonary circuit begins in the right ventricle, and ends in the left atrium. Overall, this circuit includes the pulmonary arteries, the capillaries of the lungs, and the pulmonary veins. More specifically, the *bronchial circulation* consists of the blood supply to tissues of the lung's air-conducting passageways (terminal bronchioles, supporting tissues, and the outer walls of the pulmonary arteries and veins). The bronchial blood supply comes from the bronchial arteries, which branch off the thoracic aorta. Since the blood delivered to the alveoli comes from the right ventricle of the heart, it is low in oxygen.

Ask Yourself

1 *What are the basic components of the respiratory system?*

2 *Why is the lining of the nasal cavity important in respiration?*

3 *What is the major function of the larynx?*

4 *What is the function of the epiglottis?*

5 *What are bronchi and bronchioles?*

6 *Why are alveoli called the functional units of the lungs?*

7 *What is the function of surfactant?*

MECHANICS OF BREATHING

The layer of air covering the surface of the earth exerts a pressure of about 760 mm Hg, or approximately 1 atmosphere (atm). Most of the air that we breathe is inert (chemically unreactive) nitrogen, and has no importance in normal respiration. There is a small amount of equally inert argon, and rarer gases such as neon and helium are present in tiny amounts. The atmospheric gases of interest in respiration are oxygen and carbon dioxide. Air also contains varying amounts of water vapor, volatile hydrocarbons, compounds of sulfur and nitrogen, solid particles of dust, and aerosols (minute droplets of liquids other than water).

Changes in the size of the chest cavity, and thus of the lungs, allow us to inhale and exhale air, in a process called *pulmonary ventilation*, or more commonly, *breathing*. When the chest cavity expands, and the air pressure inside is lowered, the greater pressure outside causes a flow of air into the lungs. When the chest cavity shrinks, the increased pressure inside causes some contained air to flow out.

Boyle's Law

Robert Boyle, a seventeenth-century Irish chemist, discovered important relationships between the pressure and volume of a gas. *Boyle's law* predicts that when a gas is compressed to half its volume, its pressure doubles. The concept that decreasing the volume of a gas increases its pressure, and vice versa, is important for understanding respiratory phenomena. For example, when muscles in the rib cage and abdomen cause lung volume to expand during inspiration, pressure in the lungs falls below that of the atmosphere, and air flows from the area of higher pressure to enter the respiratory tree. In a similar way, air leaves the lungs when air pressure in the lungs becomes greater than that of the atmosphere by the compression of the chest wall and abdomen.

Muscular Control of Breathing

Pulmonary ventilation (breathing) requires continual work by the muscles that increase the volume of the thoracic cavity and expand the lungs. During *inspiration* (inhalation), air is brought into the lungs to equalize a reduction of air pressure caused by the enlarged thoracic cavity. The mechanism operates in the following way (Figure 21.12):

1 Several sets of muscles contract, the main ones being the *diaphragm* and the *external intercostal muscles.* The intercostal muscles stretch from rib to rib, and when they contract, they pull the ribs closer together. As the ribs are only loosely fastened along the anterior ends, near the sternum, they rise upward and outward, enlarging the thoracic cavity.

2 The thoracic cavity is enlarged further when the muscles of the diaphragm contract, lowering the diaphragm. Although the intercostal and abdominal muscles are important in the mechanics of breathing, the diaphragm is the principal organ of breathing.

3 To compensate for the compression of adominal organs by the lowered diaphragm, the abdominal muscles relax.

4 The increased size of the thoracic cavity causes the pressure in the cavity to drop 4 mm Hg below the atmospheric pressure to about 756 mm Hg, and air rushes through the respiratory passages into the lungs, equalizing the pressure (Figure 21.13).

Ordinary *expiration* (exhalation), or the expulsion of air from the lungs, occurs in the following way:

FIGURE 21.12

The chest at full inspiration (A) and at the end of expiration (B). Note the differences in the size of the thoracic cavity, the position of the sternum and ribs, and the shape of the diaphragm and abdominal wall.

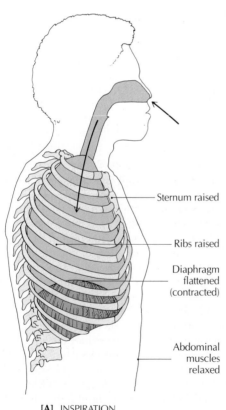

Sternum raised

Ribs raised

Diaphragm flattened (contracted)

Abdominal muscles relaxed

[A] INSPIRATION

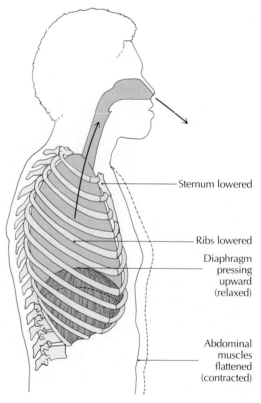

Sternum lowered

Ribs lowered

Diaphragm pressing upward (relaxed)

Abdominal muscles flattened (contracted)

[B] EXPIRATION

The Heimlich Maneuver

Choking on food is such a common accident, and so often causes death from asphyxiation, that many restaurants post notices on how to perform the **Heimlich maneuver**. If a person gets a particle stuck in the glottis or larynx and cannot either exhale or inhale, a second person can help dislodge the particle by using the Heimlich maneuver. The aim is to compress the remaining air in the lungs violently enough to blow the particle free.

If the victim can stand, the rescuer stands behind the victim, places a fist against the abdomen between the ribs and navel, covers the fist with the other hand, and gives a strong, sudden squeeze inward and upward (see photograph). If the victim is unable to stand, the rescuer kneels astride the victim and applies pressure with the heel of one hand between the ribs and navel, pressing down on that lower hand with the other.

Some danger accompanies the Heimlich maneuver because too forceful or badly placed pressure can injure ribs, the diaphragm, or the liver.

[Seth Resnick/Stock, Boston.]

1 External intercostal muscles and the diaphragm relax, allowing the thoracic cavity to return to its original, smaller size by *extrinsic elastic recoil* provided by the costal cartilages, increasing the pressure within the thoracic cavity. The decrease in volume of the cavity is due partly to the *intrinsic elastic recoil* (pulmonary compliance) of the lung tissues themselves, which were stretched during inspiration, and partly to the upward push of the diaphragm, which bulges passively under pressure of the viscera.

2 Abdominal muscles contract, pushing the abdominal organs against the diaphragm, and further increasing the pressure within the thoracic cavity.

3 The elastic lungs contract as air is expelled.

During *forced expiration,* which occurs during strenuous physical activity, other muscles contract. **Internal intercostal muscles,** which run at right angles to the external ones, reduce the lung space when they contract. Also, the abdominal muscles can contract forcibly, pressing the viscera against the passive diaphragm and further reducing lung space. Although the expiratory muscles work during forced expiration, the main ones are the abdominal muscles, which force additional air out of the lungs.

The expansibility of the thorax, and thus the lungs, is called **compliance.** Compliance is expressed as the volume increase of the lungs for each unit increase in intra-alveolar pressure. The greater the increase in volume for a given increase in pressure, the greater is the compliance. Compliance defines the relationship between changes in pressure and volume, and is an important concept in respiration. For a sealed container, the larger a change in volume for a given change in pressure, the greater is the compliance, and vice versa. A toy balloon, for example, has a large compliance because only a small increase in its internal pressure causes the balloon to expand greatly. A bicycle inner tube has less compliance than a balloon, and a stiff-walled box has much less compliance. Any factor that decreases lung compliance (such as edema, a condition in which the lungs stiffen because of the formation of fibrotic tissue) decreases the ease with which air is moved in and out of the lungs, and thus makes breathing more difficult.

The visceral pleura covering the lungs, and the parietal pleura lining the rib cage are separated only by the moisture from these membranes. Nevertheless, the pleurae are free to slide slightly past one another, but as no air is present between them, they do not separate when the rib cage expands. If the chest wall is punctured, and air enters the pleural cavity, the wet connection between the pleurae is broken, and a lung collapses. Such an air pocket is called a *pneumothorax.* A pleural space filled with blood is a *hemothorax* and one filled with pus is an *empyema.*

Amounts of Air in the Lungs

The events of pulmonary ventilation can be described by subdividing the amount of air in the lungs into four volumes and four capacities (Figure 21.14, Table 21.2). The apparatus used to measure these amounts is called a *respirometer*. The *residual volume* is the volume of air remaining in the lungs after the most forceful expiration. It is about 1.1 L in women and 1.2 L in men. During ordinary (quiet) breathing at rest, both men and women inhale and exhale about 0.5 L with each breath. This is the *tidal volume.* When the deepest possible breath is taken in, the excess beyond the usual tidal volume is the *inspiratory reserve volume.* It is about 1.9 L in women and 3.3 L in men. At the end of an ordinary exhalation, if all possible air is expelled, the quantity beyond the tidal volume is the *expiratory reserve volume,* usually about 0.7 L in women and 1.0 L in men.

When describing events of air movement, it is desirable at times to consider two or more of the above volumes together.

FIGURE 21.13

Pressure relations of the atmosphere, lungs, and pleural cavity at the end of an inspiration. After inspiration, the pressures in the lungs and in the atmosphere are equal. During inspiration, the pressure in the lungs (intrapulmonary pressure) drops to about 756 mm Hg. During expiration, it rises to about 760 mm Hg. The elasticity of the lungs maintains a negative pressure of about 4 mm Hg in the pleural cavity.

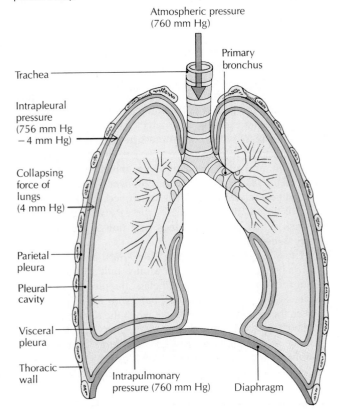

TABLE 21.1 — MAJOR RESPIRATORY STRUCTURES

Structure	Major functions
Nasal cavity	Filters, warms, moistens incoming air. Provides passageway for air to pharynx.
Paranasal sinuses	Produce mucus. Act as resonators for sound. Lighten skull.
Pharynx	Provides passageway for air between nose and larynx, and for food from mouth to esophagus.
Larynx (voice box)	Provides passageway for air between pharynx and rest of respiratory tract. Produces sound. Protects trachea from foreign objects.
Trachea (windpipe)	Provides passageway for air to and from thoracic cavity. Traps and expels foreign matter by motion of cilia.
Diaphragm	Enlarges thoracic cavity to allow for inspiration, returns to original position for expiration.
Bronchi	Provide passageways for air to and from lungs. Filter air.
Bronchioles	Provide passageways for air to and from alveoli.
Alveoli	Site of gas exchange; functional units of lungs.
Lungs	Major respiratory organs.
Pleurae	Protect, compartmentalize, and lubricate outer surfaces of lungs.

TABLE 21.2 — NORMAL LUNG VOLUMES AND CAPACITIES*

Function	Women	Men
Inspiratory reserve volume	1.9	3.3
Tidal volume	0.5	0.5
Inspiratory capacity (inspiratory reserve volume + tidal volume)	2.4	3.8
Expiratory reserve volume	0.7	1.0
Residual volume	1.1	1.2
Functional residual capacity (expiratory reserve volume + residual volume)	1.8	2.2
Vital capacity (inspiratory reserve volume + tidal volume + expiratory reserve volume)	3.1	4.8
Total lung capacity (vital capacity + residual volume)	4.2	6.0

*The maximum amount of air that the respiratory tree can take in is the *maximum respiratory minute rate*. It equals the inspiratory capacity multiplied by the maximal respiratory rate (2.4 × 35 = 84.0 L/min in women; 3.8 × 35 = 133 L/min in men).

FIGURE 21.14

Respirometer tracings showing amounts of air moved (by the average man) during different kinds of breathing. (A) Quiet breathing. (B) Maximal inhalation, normal exhalation. (C) Normal inhalation, maximal exhalation. (D) Maximal inhalation, maximal exhalation. (E) Comparison of average air volumes and capacities for men and women.

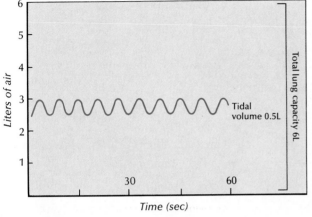

[A] QUIET BREATHING

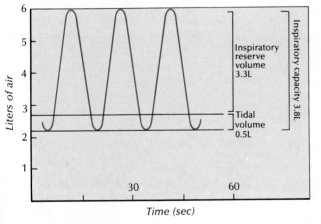

[B] MAXIMAL INHALATION, NORMAL EXHALATION

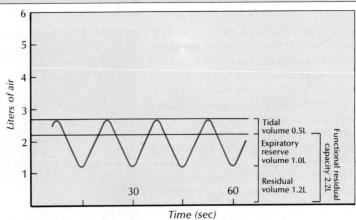

[C] NORMAL INHALATION, MAXIMAL EXHALATION

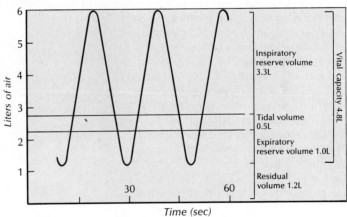

[D] MAXIMAL INHALATION, MAXIMAL EXHALATION

Total lung capacity Men 6L Women 4.2L	Vital capacity Men 4.8L Women 3.1L	Inspiratory reserve volume Men 3.3L Women 1.9L	Inspiratory capacity
		Tidal volume Men and Women 0.5L	
		Expiratory reserve volume Men 1.0L Women 0.7L	Functional residual capacity
	Residual volume Men 1.2L Women 1.1L	Residual volume	

[E] AVERAGE AIR VOLUMES AND CAPACITIES

Cardiopulmonary Resuscitation (CPR)

When breathing stops, speedy action by a properly trained person is essential. *Cardiopulmonary resuscitation (CPR)* is the re-establishment of respiration and circulation. The rescuer makes sure that (1) there is a free air passage to the lungs, (2) air is pumped into the lungs, and (3) blood circulation is maintained.

Free air passage is aided by removing any apparent obstruction in the mouth and throat, and by draining fluids, if any, from the mouth. The head must be allowed to fall back so as not to pinch shut the air passage, and the tongue should not obstruct it.

Air can be pumped into the lungs by artificial respirators, but these are rarely available in an emergency, and instead, mouth-to-mouth resuscitation, or exhaled air ventilation, is usually performed (see photograph, left). In this method, the rescuer should remember these critical points: (1) Keep the victim's head well back. (2) Make sure the mouth-to-mouth contact is airtight. (3) Hold the victim's nose closed. (4) Inflate the victim's lungs without blowing so hard as to damage the alveoli. (5) Allow the victim's lungs to deflate after each breath. (6) Keep up the procedure patiently even when it may seem futile.

If the pulse is weak or absent, or no heart sounds can be heard, *external cardiac compression* can be applied (see photograph, right). In this procedure, the operator places the base of the palm on the sternum of the supine victim, presses down firmly for half a second, and then releases the pressure. This action is repeated about 10 to 12 times a minute. When successful, such efforts can force enough blood from the heart to provide oxygen and stimulate the heart to beat under its own power. The action must be smooth and gentle, and the pressing palm of the hand must be directly over the heart, or there is danger of causing internal injury. Rough or misplaced pressure can break ribs, damage the liver or diaphragm, and disrupt blood flow. If possible, it is preferable to have two trained operators, one applying artificial respiration, and the other using cardiac compression.

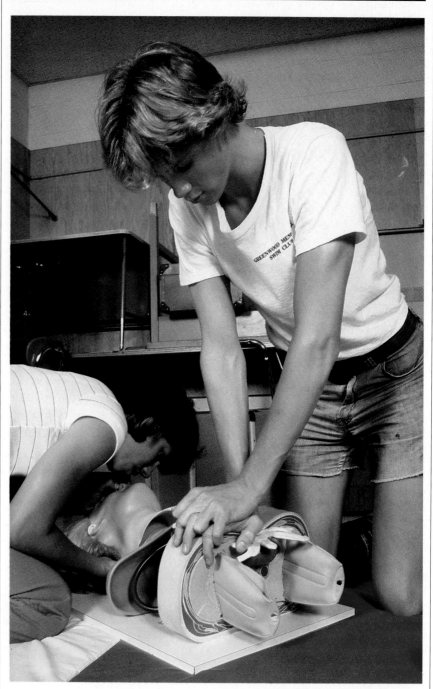

[Frank Siteman/Stock, Boston.]

Such combinations are called **pulmonary capacities.** The **inspiratory capacity** equals the inspiratory reserve volume plus the tidal volume. This is the amount of air (about 2.4 L in women, 3.8 L in men) that a person can breathe, beginning at the normal expiratory level and then distending the lungs the maximal amount. The **functional residual capacity** equals the expiratory reserve volume plus the residual volume. This is the amount of air that remains in the lungs at the end of normal expiration (about 1.8 L in women, 2.2 L in men).

The **total lung capacity** equals vital capacity plus residual volume. It is the maximum volume to which the lungs can be expanded with the greatest force (about 4.2 L in women, 6.0 L in men). The **vital capacity** equals the inspiratory reserve volume plus the tidal volume plus the expiratory reserve volume. It is the maximum amount of air that a person can expel from the lungs after the lungs have been filled to their maximum extent (about 3.1 L in women, 4.8 L in men).

About 0.15 L of space is taken up by the nasal passages, trachea, bronchi, and bronchioles. It is known as *anatomical dead space* because no gas exchange occurs there.

During breathing, the lungs are rarely expanded to their maximum size, and never completely emptied. As a result, they are never filled with fresh air. Instead, some of the carbon dioxide from the blood is retained in the lungs after every breath, and the inhaled oxygen is mixed with the contained air. This residual volume of contained air acts as a reservoir that provides for the exchange of gases, especially between breaths.

The amount of air in the lungs is also affected by how deeply and how fast we breathe. During heavy breathing, we might use the entire vital capacity, including the inspiratory and expiratory reserves as well as the tidal volume, for a total of 3.1 to 4.8 L. If such maximum breaths are taken once a second, a person might for a short time have a **minute respiratory volume** of 186 to 288 L/min (60 breaths × 3.1 or 4.8 L). During quiet breathing, which normally occurs 12 to 16 times a minute and uses only the tidal volume of 0.5 L, the minute respiratory volume is about 6 to 8 L/min (12 or 16 breaths × 0.5 L).

Although the minute respiratory volume is determined by both the respiratory rate (breaths per minute) and the depth of breathing, the two factors are not of equal importance. Because the air in the anatomical dead space is merely pushed back and forth during breathing, the last air expelled from the alveoli does not immediately reach the outside, and the fresh air inhaled does not immediately reach the alveoli. If a breath of 0.5 L is taken in, and 0.15 L is taken up by anatomical dead space, only 0.35 L of fresh air reaches the alveoli. If a person takes 15 breaths a minute and 0.5 L of air at each breath, the minute respiratory volume is 15 × 0.5, or 7.5 L/min. But since the alveoli receive only 0.35 L, the actual rate of **alveolar ventilation** is 15 × 0.35, or 5.25 L. Increasing the depth of breathing can be more effective, in terms of lung volume, than increasing the respiratory rate. The same minute respiratory volume may be achieved by increasing either factor, but deeper breathing is obviously more efficient than merely breathing faster. When unusually large volumes of air are needed, during strenuous exercise, for example, both respiratory rate and depth of breathing are increased.

Ask Yourself

1 *What is Boyle's law?*

2 *What mechanical events happen during inspiration? Expiration?*

3 *What is meant by the total lung capacity?*

4 *What is anatomical dead space?*

5 *What is vital capacity?*

6 *How are minute respiratory volume and alveolar ventilation calculated?*

FACTORS AFFECTING GAS MOVEMENT AND SOLUBILITY

Oxygen and carbon dioxide are gases at ordinary temperatures and pressures. In a study of respiration, therefore, we must be aware of the general principles governing the action of gases under different conditions.

Dalton's Law: Partial Pressure

Dalton's law, named for the English chemist John Dalton, states that each gas in a mixture is capable of exerting its own pressure in proportion to its concentration in the mixture, independently of the presence of other gases. Each gas exerts its own *partial pressure,* which is denoted as P_{O_2} for oxygen and P_{CO_2} for carbon dioxide. Dalton's law applies to the breathing process because oxygen and carbon dioxide can move in opposite directions in the same place at the same time, each following its own pressure gradient (moving from the area where the partial pressure is greater to the area where the partial pressure is less). Thus, oxygen at a high concentration in an alveolus can pass into the blood, where the oxygen concentration is lower. At the same time, carbon dioxide at a high concentration in the blood can pass into an alveolus, where the carbon-dioxide concentration is lower.

Solubility of Gases

Respiratory gases must be able to dissolve, first in the fluid lining the alveoli, and then in the water component of the blood. The **solubility of gases** in liquid is defined as the volume in milliliters of a gas that dissolves in 1 mL of water at 37°C (normal body temperature). Solubility is affected by three factors—type of molecules, temperature, and pressure: (1) *Type of molecules.* Different gas molecules have different chemical properties that affect their solubility in a liquid. As a general rule, larger molecules, which have a relatively high molecular weight, diffuse more slowly than smaller molecules with lower molecular weights. Carbon dioxide is about 24 times more soluble than oxygen. (At 37°C, only about 0.03 mL of oxygen dissolves in 1 mL of water, compared with 0.6 mL of carbon dioxide.) (2) *Temperature.* The solubility of gases in liquids is, in general, greater at lower temperatures. However, since respiratory gases are normally exchanged within the narrow range of physiological temperature (36°C to 39°C), the effects of temperature on gas solubility are relatively unimportant. (3) *Pressure.* Pressure can affect the solubility of gases greatly, and these effects are described in detail below.

Pressure and solubility: Henry's law According to *Henry's law* (named for the nineteenth-century chemist William Henry), a liquid can usually dissolve more gas at high pressure than it can at lower pressure. That is why champagne in a tightly corked bottle shows no bubbles, but fizzes when opened. It is under high pressure when corked, but when uncorked it is subject only to ordinary atmospheric pressure, and the carbon dioxide comes out of solution. Under normal physiological conditions, the lungs and blood are not subjected to great variation in pressure. However, when unusual pressure changes do occur, they can affect gas exchange.

The most noticeable effect occurs at high altitudes, where there is not enough oxygen to maintain normal respiration. At 3800 m (12,500 ft), where the partial pressure of oxygen (P_{O_2}) is the equivalent of 100 mm Hg instead of the 160 mm Hg at sea level, newcomers to the high altitude become sluggish and feel faint, nauseated, or uncomfortable, but people who live in high mountains compensate for the reduced oxygen pressure by producing more red blood cells. Altitudinal adaptations are also affected by changes in the level of 2,3-diphosphoglycerate (DPG), a compound that influences the binding and release of oxygen from hemoglobin (see p. 670).

Higher-than-normal pressures are less common, but they do occur. Deep-sea divers, for example, may have too much nitrogen dissolved in their blood as a result of the higher atmospheric pressures under water (where the air may contain up to 80 percent nitrogen) so that when they come up to the surface too quickly, the nitrogen comes out of solution in the form of gas bubbles. The bubbles may lodge in joints, creating the painful and incurable malady known as *bends* or decompression sickness.* The bends can be avoided by decompressing the body so slowly that the nitrogen diffuses out without bubbling, or by breathing a mixture of oxygen and helium instead of oxygen and nitrogen. Helium has the peculiar property of being *less* soluble in water at increased pressure.

Henry's law can be used to advantage when greater-than-usual quantities of oxygen are needed, such as when a heart is not supplying sufficient blood, or following carbon monoxide poisoning. A person suffering from such trouble can be put in a closed container called a *hyperbaric chamber* (Gr. overpressure), in which the P_{O_2} can be artificially raised from the normal 160 mm Hg to 480 to 640 mm Hg. Such pressures cause extra oxygen to dissolve in the blood, so the oxygen can be delivered to the tissues.

*The condition is also called *caisson disease* because it may affect people who work in caissons (pressurized, watertight chambers) during the construction of deep-sea tunnels and other underwater projects.

Diffusion of Gases

The transfer of oxygen and carbon dioxide across the alveolar-capillary membrane interface occurs by diffusion. The rate of gas diffusion (V_{gas}) through the blood-gas barrier found in the lungs depends upon the characteristics of the gas and the barrier. *Fick's law of gas diffusion* can be expressed as

$$V_{gas} = (P_1 - P_2) \times \frac{A}{T} \times D$$

where ($P_1 - P_2$) is the difference between partial pressures on the two sides of the alveolar membrane, A is the surface area available for diffusion, T is the thickness of the membrane, and D is a diffusion coefficient. Since

$$D = \frac{solubility}{\sqrt{molecular\ weight}}$$

and the molecular weight is related to the size of the molecule, Fick's law is usually written as:

$$V_{gas} = (P_1 - P_2) \times \frac{area}{thickness} \times \frac{solubility}{size}$$

These factors determine the exchange of oxygen and carbon dioxide across the alveolar-capillary membrane interface by diffusion.

Ask Yourself

1 *What are Dalton's law and Henry's law? How do they affect respiration?*

2 *Why is the solubility of oxygen and carbon dioxide important in respiration?*

GAS TRANSPORT

The physiologically important part of respiration is the exchange of gases between alveoli and capillaries in the lungs, and between cells and capillaries in the body tissues. The first exchange is called *external respiration*, and the second is called *internal respiration* (Figures 21.15 and 21.16).

External Respiration

The air in the lungs is richer in carbon dioxide and poorer in oxygen than the outside air (Table 21.3). As a result, the P_{CO_2} in alveoli is about equal to 40 mm Hg (instead of the 0.3 mm Hg outside), and the P_{O_2} is only about 104 mm Hg (instead of the 160 mm Hg outside) (Figure 21.17). Blood in the alveolar and bronchiolar capillaries coming from tissues where cellular metabolism is occurring has a high P_{CO_2} of 45 mm Hg. The important point is that the carbon dioxide in the blood is at a higher partial pressure, and thus is more

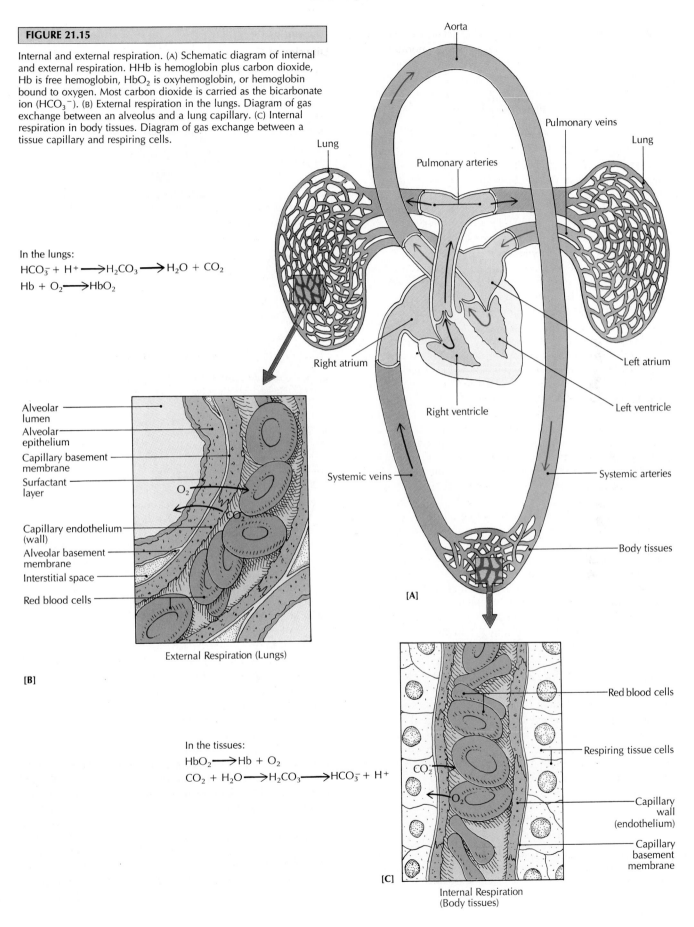

FIGURE 21.15

Internal and external respiration. (A) Schematic diagram of internal and external respiration. HHb is hemoglobin plus carbon dioxide, Hb is free hemoglobin, HbO_2 is oxyhemoglobin, or hemoglobin bound to oxygen. Most carbon dioxide is carried as the bicarbonate ion (HCO_3^-). (B) External respiration in the lungs. Diagram of gas exchange between an alveolus and a lung capillary. (C) Internal respiration in body tissues. Diagram of gas exchange between a tissue capillary and respiring cells.

In the lungs:

$$HCO_3^- + H^+ \longrightarrow H_2CO_3 \longrightarrow H_2O + CO_2$$

$$Hb + O_2 \longrightarrow HbO_2$$

External Respiration (Lungs)

[B]

Alveolar lumen
Alveolar epithelium
Capillary basement membrane
Surfactant layer
Capillary endothelium (wall)
Alveolar basement membrane
Interstitial space
Red blood cells

O_2
CO_2

Aorta
Lung
Pulmonary arteries
Pulmonary veins
Lung
Right atrium
Left atrium
Right ventricle
Left ventricle
Systemic veins
Systemic arteries
Body tissues

[A]

In the tissues:

$$HbO_2 \longrightarrow Hb + O_2$$

$$CO_2 + H_2O \longrightarrow H_2CO_3 \longrightarrow HCO_3^- + H^+$$

Red blood cells
Respiring tissue cells
Capillary wall (endothelium)
Capillary basement membrane

CO_2
O_2

[C]

Internal Respiration (Body tissues)

FIGURE 21.16

Gas exchange between body tissues and the outside environment.

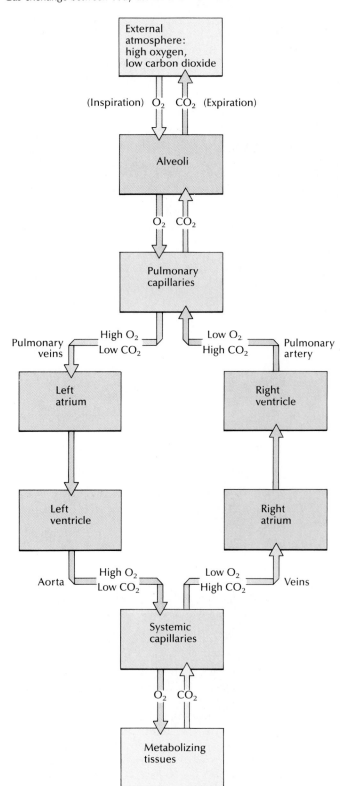

FIGURE 21.17

Gas exchange in lungs and tissues. Approximate partial pressures of oxygen and carbon dioxide are given. Red arrows represent oxygen-rich blood; purple arrows represent oxygen-poor blood.

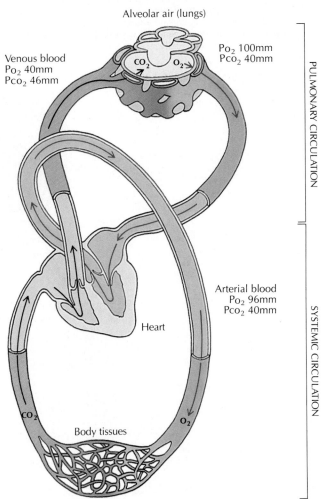

concentrated than that in the alveoli, while the partial pressure and concentration of oxygen in the blood are lower than those in the alveoli. As each gas flows down its own concentration gradient from a region of higher concentration to a region of lower concentration, effective exchange occurs.

As blood leaves the alveolar capillaries, it has lost carbon dioxide and gained oxygen; the concentrations of these gases are about equal to the concentrations in alveolar air, that is, the blood has a P_{CO_2} of 40 mm Hg and a P_{O_2} of 104 mm Hg. This exchange between the blood and alveoli is ***external respiration*** (see Figures 21.15A, B).

Internal Respiration

When oxygenated blood reaches active cells where *cellular* or *metabolic respiration* is taking place, ***internal respiration*** oc-

TABLE 21.3	APPROXIMATE COMPOSITION OF RESPIRATORY GASES ENTERING AND LEAVING LUNGS (STANDARD ATMOSPHERIC PRESSURE, YOUNG ADULT MALE AT REST)		
	Oxygen volume (%)/ partial pressure (mm Hg)	Carbon dioxide volume (%)/ partial pressure (mm Hg)	Nitrogen volume (%)/ partial pressure (mm Hg)
Inspired air	21/160	0.04/0.3	78.0/597
Expired air	16/120	4.0/27	79.2/566
Alveolar air	14/104	5.5/40	79.1/569

*Percentages do not add up to 100 because water is also a component of air.

FIGURE 21.18

Binding of hemoglobin with oxygen at three temperatures as a function of P_{O_2}. Within limits, a higher P_{O_2} allows a higher percent saturation (binding). This is termed an oxyhemoglobin dissociation curve. Notice the effects of temperature on percent saturation.

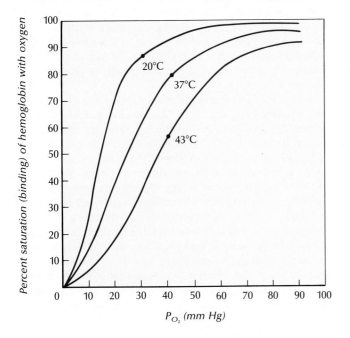

FIGURE 21.19

Binding of hemoglobin with oxygen at three levels of partial pressure of carbon dioxide. High pressure inhibits the formation of oxyhemoglobin by increasing the hydrogen ion concentration, that is, lowering pH.

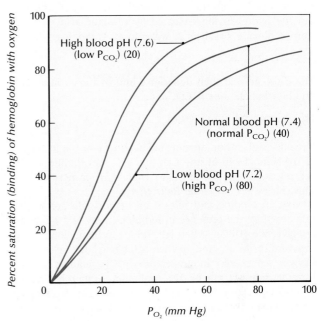

curs (see Figure 21.15C). It is the opposite, as far as gas exchange is concerned, of external respiration. In an active cell, the P_{O_2} is reduced to about 40 mm Hg, and the P_{CO_2} is raised to 45 mm Hg as a result of cellular metabolism (see Chapter 23). After blood in the tissue capillaries has moved away from the respiring cells, it has the high P_{CO_2} of 45 mm Hg, and the low P_{O_2} of 40 mm Hg.

Transport of Oxygen in the Blood

Blood carries oxygen as dissolved oxygen and as oxygen combined with hemoglobin. On its own, the water in blood can carry only about 1 percent of an adult's oxygen requirement. But *hemoglobin* (Hb), the respiratory pigment in red blood cells, has such an affinity for oxygen that it binds about 98 percent of the oxygen available in the lungs (see Chapter 17). In this way, hemoglobin acts as a carrier, and makes

possible the effective transport of oxygen throughout the body. The conditions under which hemoglobin binds oxygen effectively are so complex, and so various, that they must be considered in some detail.

When oxygen is bound to hemoglobin, a new compound is formed: *oxyhemoglobin* (HbO_2). The formation or breakdown of oxyhemoglobin is largely dependent on the P_{O_2} in the plasma. A high P_{O_2} favors binding, and a low P_{O_2} favors release (Figure 21.18).

The hemoglobin-oxyhemoglobin shift is also influenced by other factors. Increased temperature decreases the affinity of hemoglobin for oxygen. As the temperature rises, the blood holds less oxygen (see Figure 21.18). An increase in cellular metabolism raises the temperature of the cells. As a result, the most active tissues receive the most oxygen.

An increased P_{CO_2} and/or increased concentration of hydrogen ions (H^+) decreases the affinity of hemoglobin for oxygen (Figure 21.19). This condition is beneficial to the body because tissues that metabolize rapidly produce more carbon dioxide and hydrogen ions, which increases the P_{CO_2} and hydrogen-ion concentration in the blood, which in turn promotes the release of oxygen from oxyhemoglobin. For example, during strenuous exercise, the temperature, P_{CO_2}, and hydrogen-ion concentration all increase, while the affinity of hemoglobin for oxygen at lower P_{O_2} decreases. This means more oxygen is released to the tissues that need oxygen for their increased metabolism. It is important to remember that through all of these shifts, the body is attempting to maintain a stable pH. (The more hydrogen ions available, the lower will be the pH.)

The partial pressure of carbon dioxide is important in influencing the degree of saturation of hemoglobin, because in circulating blood an increased P_{CO_2} shifts the curve to the right, and allows the blood to carry more carbon dioxide (see Figure 21.19). The influence of carbon dioxide is an indirect one because it is not the carbon dioxide itself that is effective, but the hydrogen ions released by carbonic acid (H_2CO_3). Carbon dioxide dissolved in water forms carbonic acid:

$$CO_2 + H_2O \rightleftharpoons H_2CO_3$$

The carbonic acid then dissociates into hydrogen ions (H^+) and bicarbonate ions (HCO_3^-):

$$H_2CO_3 \rightleftharpoons H^+ + HCO_3^-$$

The effects of carbon dioxide and acidity favor the formation of oxyhemoglobin at low CO_2 and H^+ concentrations, and the release of oxygen at high CO_2 and H^+ concentrations. This shift is known as the **Bohr effect.**

Oxygen binding and release by hemoglobin are further influenced by the presence in red blood cells of 2,3-diphosphoglycerate (DPG). When DPG binds to hemoglobin, oxygen cannot do so, and oxyhemoglobin readily releases its oxygen. An increase in temperature, pH, or cellular metabolism increases the rate of DPG formation. Certain hormones, such as testosterone, thyroid hormone, and catecholamines, increase the formation of DPG within red blood cells. On the other hand, the presence of DPG in alveolar capillaries would seem to compete with oxygen for a place on the hemoglobin molecule. However, there is presumably a high enough P_{O_2} in the plasma of such capillaries for the DPG to be replaced by oxygen.

A change in the P_{O_2} of the atmosphere, as when one moves to high altitudes, affects the DPG level of blood. Within a few hours after ascending to a level of about 4500 m (15,000 ft), a person's DPG concentration increases and the affinity for oxygen decreases. Natives of high altitudes have the reverse response when they descend to low altitudes.

The entire system of oxygen uptake in the lungs, and the subsequent release in the tissue capillaries is thus regulated by a combination of P_{CO_2}, H^+ concentration, P_{O_2}, and DPG. Together they make the blood carry out effectively one of its main functions: the delivery of oxygen to the cells where it is needed for cellular respiration.

Transport of Carbon Dioxide in the Blood

Carbon dioxide is readily soluble in water. Therefore, when it is generated in respiring cells, it diffuses first into the cytoplasm, then through the plasma membrane into the extracellular fluid, and finally through capillary walls into the blood plasma. There it can be carried in three forms: as dissolved gas in the water of the plasma, as part of the bicarbonate ion, or bound to the amino groups of proteins (see Chapter 2).

Only about 10 percent of carbon dioxide is carried simply as a dissolved gas in the blood. The amount of carbon dioxide carried this way is related to the P_{CO_2} and the solubility of carbon dioxide. In the lung capillaries, the greater concentration of carbon dioxide in blood than in the alveolar spaces causes a net outward diffusion. However, carbon dioxide elimination by this method is only slightly effective.

The greatest proportion of carbon dioxide (68 percent of the total) is carried as a part of the bicarbonate ion. When carbon dioxide, activated by carbonic anhydrase, reacts with water, it forms carbonic acid (H_2CO_3), which undergoes partial dissociation to yield hydrogen ions (H^+) and bicarbonate ions (HCO_3^-). This reaction occurs in venous blood, mainly inside the erythrocytes, where the enzyme carbonic anhydrase is most abundant. (Plasma contains no carbonic anhydrase.) The resulting bicarbonate ions diffuse from the erythrocytes into the plasma.

The removal of bicarbonate ions from the erythrocytes develops a positive charge inside the cell, and chloride ions (Cl^-) from sodium chloride (NaCl) in the plasma diffuse into the cell in an exchange called the **chloride shift** (Figure 21.20).* The chloride shift greatly increases the ability of the plasma to carry carbon dioxide. Because of the movement of chloride ions, their concentration in erythrocytes of venous blood is higher than that in arterial erythrocytes. The total osmotic activity of erythrocytes in venous blood is also greater, because although the proteins inside the cell bind many of the hydrogen ions, the hydration of carbon dioxide also produces bicarbonate ions, which are exchanged for osmotically active chloride ions. Therefore, in order to maintain osmotic equilibrium within erythrocytes in venous blood, the shift of chloride ions is accomplished by a movement of water that slightly increases the size of the erythrocytes. As a result, the venous hematocrit (see p. 518) is slightly greater than

---

What substance do some athletes breathe on the sidelines?	*Fatigued football players and other athletes frequently breathe through a plastic mask from a tank of pure oxygen. Current research indicates that the athletes gasp for air because of increased carbon dioxide in their bloodstream, not because of a need for extra oxygen, and pure oxygen is no better than air in helping athletes to recover from fatigue. However, oxygen may help remove the excess lactic acid that builds up during intense physical activity.*

---

*The chloride shift is also known as the Hamburger shift, after the German physiologist who first described it.

FIGURE 21.20

The chloride shift. (A) In tissue capillaries, chloride ions move from plasma into red blood cells. (B) In alveolar capillaries, chloride ions pass from red blood cells back into plasma. The enzyme carbonic anhydrase is shown here, and not in the equations on page 670, because carbonic anhydrase is most abundant inside red blood cells; the equations represent a more general situation.

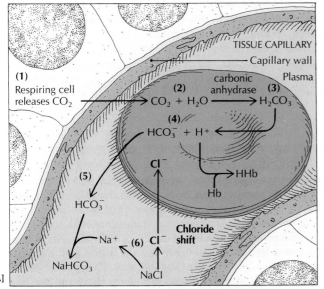

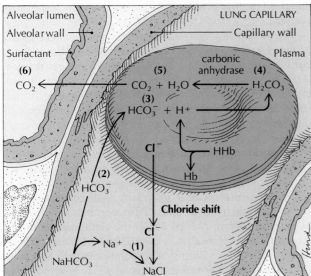

the arterial hematocrit. The reverse movement of chloride ions and water occurs in erythrocytes when they enter arterial blood.

Carbon dioxide combines readily with the amino groups of the amino acids (represented by R below) present in blood proteins, forming *carbamino compounds:*

$$RNH_2 + CO_2 \rightleftharpoons RNHCOO^- + H^+$$

Carbamino compounds are produced by many proteins, but mainly by hemoglobin. The result is *carbaminohemoglobin:*

$$HbNH_2 + CO_2 \rightleftharpoons HbNHCOO^- + H^+$$

While oxygen is being bonded to the heme groups in hemoglobin, carbon dioxide is being bonded to the amino groups of the same molecule. Although hemoglobin can carry oxygen and carbon dioxide at the same time, the presence of one reduces the bonding power of the other. Oxyhemoglobin is less likely to carry carbon dioxide than deoxygenated hemoglobin is, and deoxygenated hemoglobin carrying carbon dioxide is less likely to bind to oxygen. Recent studies indicate that DPG competes with carbon dioxide sites on the hemoglobin molecule. As a result, the amount of carbon dioxide transported as carbamino compounds may be less than half as much as was previously thought. This is a fortunate result for respiratory exchange; it makes it easier for blood in the tissue capillaries to release oxygen where it is needed, and to pick up carbon dioxide. In lung capillaries, it aids the release of carbon dioxide and the binding of oxygen as needed.

The amount of carbon dioxide transported by the blood is also influenced by the P_{O_2} in the blood. When the P_{O_2} decreases, the amount of dissolved carbon dioxide increases. Thus, at any P_{CO_2}, the greater the concentration of deoxygenated hemoglobin (deoxyhemoglobin), the higher the concentration of carbon dioxide in the blood. This phenomenon is called the *Haldane effect*. Also, because deoxyhemoglobin is a weaker acid than oxyhemoglobin, it can combine with more hydrogen ions, and in turn, more carbon dioxide can combine with deoxyhemoglobin.

Carbon Monoxide and Hemoglobin

Two of the chemical characteristics of carbon monoxide (CO) combine to make it a deadly poison. First, its affinity for hemoglobin is 210 times greater than oxygen's affinity for hemoglobin. As a result, very small amounts of carbon monoxide can compete successfully with oxygen for hemoglobin. Once carbon monoxide binds to hemoglobin as carboxyhemoglobin, it remains bound indefinitely, preempting binding sites that oxygen might have used. Second, when carbon monoxide combines with one or two of the four heme groups in the hemoglobin molecule, the affinity of the remaining heme groups for oxygen is greatly increased. This effect results in even less oxygen being released to the tissues.

Because of these two characteristics, a concentration of carbon monoxide as low as 0.1 percent in inspired air inactivates 50 percent of the body's hemoglobin in a few hours. Such an inactivation is usually fatal.

Ask Yourself

1 *What is the difference between internal and external respiration?*

2 *What is oxyhemoglobin?*

3 *What is the Bohr effect?*

4 *What is the chloride shift?*

5 *How are carbamino compounds formed?*

6 *What is the Haldane effect?*

7 *How does carbon monoxide compete with oxygen for hemoglobin?*

NEUROCHEMICAL CONTROL OF BREATHING

The rate and depth of breathing can be controlled consciously temporarily, but they are generally regulated directly by involuntary nerve impulses. These nerve impulses are in turn regulated by a number of secondary factors, including tension receptors in the breathing muscles, concentrations of gases in the blood, and certain self-regulating parts of the brain.

Deliberate attempts to breathe too little or not at all are soon frustrated by a build-up of carbon dioxide, which results in an irresistible need to exhale or inhale. Deliberate over-breathing, as when trying to start a fire by blowing on the coals, results in *hyperventilation*, lowering the carbon-dioxide concentration and raising the pH of the blood, which causes dizziness or even momentary unconsciousness. At this point, involuntary breathing controls take over, and normal breathing is resumed automatically.

During normal, quiet breathing, called *eupnea* (YOOP-nee-uh), the main nerve controls are in the respiratory centers of the brain in the medulla and pons (Figure 21.21A). When problems occur in these control centers, several disorders may result. *Apnea* (AP-nee-uh) is a temporary failure in breathing such as occurs in electrical shock or near-drowning. *Dyspnea* (DISP-nee-uh) is difficult, labored breathing such as occurs in women during natural childbirth. *Hyperpnea* (hi-PER-pnee-uh) is an increase in the rate and depth of breathing. *Tachypnea* (tahk-IP-nee-uh) is very rapid, shallow breathing. *Orthopnea* (or-THOP-nee-uh) is the inability to breathe in a horizontal position.

Medullary Rhythmicity Area

According to current understanding, which is still incomplete, the **medullary rhythmicity area** contains two circuits of neurons, one stimulating inhalation and the other stimulating exhalation. They work alternately. When an **inspiratory neuron** fires, it causes other similar neurons to fire until sufficient impulses are built up. During a period of about two seconds, the firing of the inspiratory neurons (via the phrenic nerves) causes contraction of the diaphragm and contraction of the external intercostal muscles (via the intercostal nerves). When these muscles contract, the rib cage expands, and air rushes into the lungs.

While the inspiratory neuron circuit is firing, it is also sending inhibitory impulses to a second circuit consisting of **expiratory neurons.** When the inspiratory neurons stop firing, they no longer send inhibitory impulses to the expiratory circuit. As the expiratory neurons start firing, the diaphragm and external intercostal muscles relax, and the rib cage becomes smaller. The increased pressure in the lungs forces air

What is a yawn? *Either excitement or boredom can result in insufficient breathing, so that carbon dioxide builds up and causes a sudden, all but irresistible need to take in an unusually deep breath through a gaping mouth, that is, to yawn.*

to be expelled. During the three seconds or so when the expiratory circuit is activated, its neurons send inhibitory impulses to the inspiratory circuit, preventing inspiration. When the expiratory circuit is "fatigued," the inspiratory circuit is once more ready to act. With inspiratory and expiratory circuits alternating, rhythmic breathing continues, even during periods of unconsciousness.

Such a regular breathing mechanism may not be completely developed in infants. A respiratory disorder affecting small children and some adults is *Cheyne-Stokes breathing*, in which short, shallow breaths are followed first by faster, deeper ones, then by temporary stoppage. This kind of breathing is sometimes called the death rattle, because it may occur just before the failure of some vital function.

Apneustic and Pneumotaxic Areas of the Pons

Two other breath-control areas are in the pons. One is the **apneustic area,** which, when experimentally stimulated, causes strong inhalations and weak exhalations. The other is the **pneumotaxic area,** which, when stimulated, stops such breathing actions (see Figure 21.21). The rhythmicity area of the medulla can function without the action of these areas in the pons, but may be less efficient. The whole system, including the inspiratory and expiratory circuits in the medulla and the two areas in the pons, is called the **respiratory center.**

The immediate cause of rhythmic breathing is alternating contraction and relaxation of the breathing muscles, and that is influenced largely by automatic control centers of the brain. But what controls the control center? Among the factors acting on the control center are chemical components in the blood and sensory receptors in the joints, lung tissues, and muscles.

Chemical controls Carbon-dioxide concentration in the blood has a powerful but indirect effect on the respiratory rate (Figure 21.22). When carbon dioxide is dissolved in water, it forms carbonic acid, which dissociates into bicarbonate ions and hydrogen ions. If the concentration of carbon dioxide rises above normal levels, the concentration of hydrogen ions increases, that is, the pH level is lowered. A lowered blood pH can directly affect the respiratory centers in the brain, increasing the respiratory rate. A lowered blood pH can also be sensed by chemoreceptors in the **carotid** and **aortic bodies** (see Chapter 19), which are located in the sinuses of the carotid arteries and the aorta. These bodies send impulses via sensory nerves to the medullary rhythmicity area, and thus increase the respiratory rate. The carotid and aortic bodies are also sensitive to blood-pressure changes.

Increased blood pressure and increased carbon-dioxide (or H^+) concentration act together to increase the rate and depth of breathing. When a stimulus is needed to make a person breathe more, carbon dioxide can be artificially added to the gases breathed, up to 4 percent. If the carbon-dioxide concentration in the blood is lower than normal, as during heavy breathing, the blood pH rises into what is called *respiratory alkalosis*, and the breathing rate falls.

FIGURE 21.21

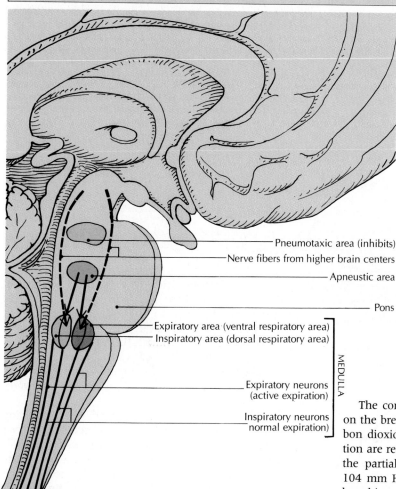

Respiratory center in the brain. (A) Mid-sagittal section of brainstem, containing the respiratory center. (B) Medullary rhythmicity area. (*Left*) When the inspiratory circuit is active, inspiration occurs, and the expiratory circuit is inhibited. (*Right*) When the expiratory circuit is active, expiration occurs and the inspiratory circuit is inhibited. The two circuits act alternately to regulate rhythmic breathing.

Pneumotaxic area (inhibits)
Nerve fibers from higher brain centers
Apneustic area
Pons
Expiratory area (ventral respiratory area)
Inspiratory area (dorsal respiratory area)
MEDULLA
Expiratory neurons (active expiration)
Inspiratory neurons normal expiration)

[A]

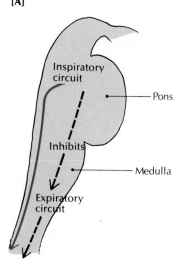

Inspiratory circuit
Pons
Inhibits
Medulla
Expiratory circuit

Inspiratory circuit
Inhibits
Expiratory circuit

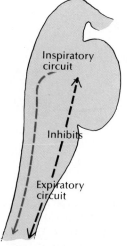

Contraction impulse to muscles of inspiration: lungs expand and air enters lungs.

Relaxation impulse to muscles of inspiration: lungs contract and air is pushed out of lungs.

[B]

The concentration of oxygen in the blood has less effect on the breathing centers than does the concentration of carbon dioxide. Relatively large changes in oxygen concentration are required to make a perceptible difference. A drop in the partial pressure of oxygen from the normal of about 104 mm Hg to about 70 mm Hg can cause an increase in breathing; if the P_{O_2} falls below 70 mm Hg, there is too little oxygen to support cellular respiration, and the whole system stops working. Death from lack of oxygen is called *asphyxia* (Gr. "no throbbing").

The lungs themselves have a subtle *self-regulating system* that keeps the flow of blood and the amount of air in the alveoli in balance. If the blood in a bronchiolar arteriole has a low concentration of oxygen and a high concentration of carbon dioxide, the muscles of the arteriole walls constrict, reducing the blood flow. At the same time, the alveolar membranes relax enough to allow the alveolar lumen to increase. Conversely, if the blood oxygen is high and carbon dioxide low, the arterioles relax and the alveoli constrict. The result of this system is that alveolar (that is, air) space is matched effectively with gas content of the blood, and maximal gas exchange results.

Sensors and reflexes Experience shows that exercise results in increased breathing. This is partly due not only to the increased output of carbon dioxide, but also to sensors in joints and muscle spindles. These sensors send impulses directly to the cerebral cortex, which can relay the impulses to the medullary breathing area. Thus, it has been shown that unconscious or otherwise immobilized people can be made to breathe more deeply if their muscles and joints are moved by someone else.

FIGURE 21.22

A summary of the control of oxygen and carbon dioxide content in the blood. (1) Oxygen is taken into the lungs during inspiration, and carbon dioxide is expelled during expiration. (2) Oxygen diffuses through the alveolar membrane into the blood, and carbon dioxide diffuses from the blood into the lumen of the alveolus. (3) Oxygen-rich blood passes from the lungs to the heart and from the heart to the body cells. (4) Oxygen diffuses from the blood into the body cells, and carbon dioxide diffuses from the body cells into the blood. (5) Blood poor in oxygen and rich in carbon dioxide passes from body cells to the heart, and then to the lungs. Steps 6 through 9 illustrate what happens when the concentration of carbon dioxide in the blood is increased. (6) Increased carbon dioxide in the blood stimulates chemoreceptors in the heart (aortic bodies) and in blood vessels (carotid bodies). (7) Carotid and aortic bodies send nerve impulses to the respiratory center in the brainstem. (8) The respiratory center in the brainstem sends impulses to the diaphragm, internal intercostal muscles, and heart. Impulses to diaphragm and intercostal muscles increase respiratory rate to excrete excess carbon dioxide. (9) Impulses to heart increase heart rate, which pumps more blood to the lungs in order to eliminate excess carbon dioxide.

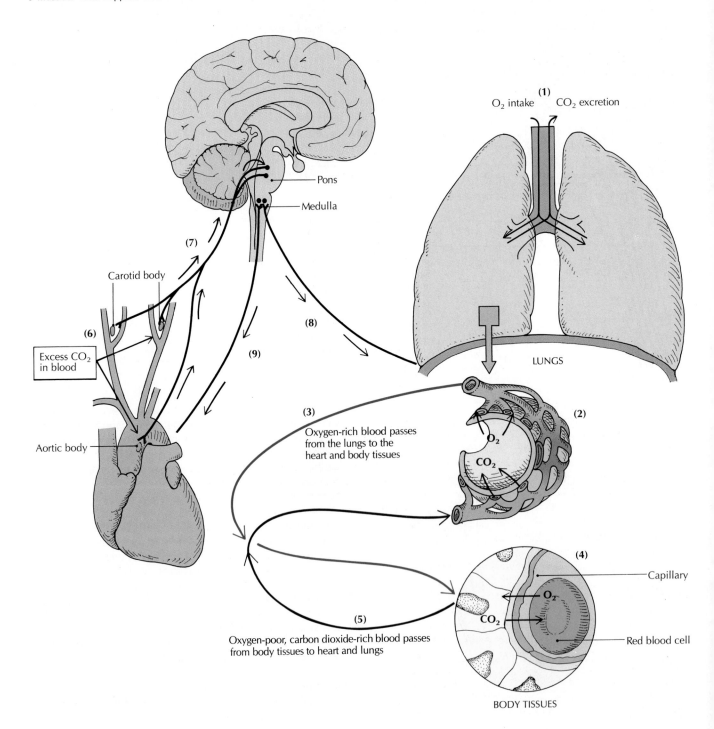

Inhalation and exhalation are affected also by proprioceptors in the visceral pleurae, bronchioles, and alveoli. When the lungs are full of air, these tissues are stretched, and the stretching causes **pulmonary stretch receptors** in the smooth muscle layer of the lung airways to send impulses by way of the vagus nerve (cranial nerve X) to the respiratory area in the medulla. The respiratory area then causes inspiration to stop. When the tissues are not stretched, the nerve impulses cease, and the respiratory center "calls for" an inspiration. The inhalation and exhalation reflexes are called the *Hering-Breuer reflex.* The stretch-receptor center prevents excessive stretching of the thin, delicate membranes in the lungs.

Ask Yourself

1 *What is the function of the medullary rhythmicity area?*

2 *What are the components of the respiratory center?*

3 *What is the function of stretch receptors in the lungs?*

OTHER ACTIVITIES OF THE RESPIRATORY SYSTEM

Several other activities involve the respiratory system. A **coughing reflex,** involving short, forceful exhalations through the mouth, occurs when sensitive parts of the air passages to the lungs are irritated, either by solids or liquids. The reflex persists even during deep coma.

A **sneeze** is an involuntary explosive expiration, mainly through nasal passages, and is usually caused by irritation of the mucous membranes in the nose. Particles ejected by a robust sneeze travel about 165 km/hr (103 mph).

A **hiccup** is a sudden, involuntary contraction of the diaphragm that usually results from a disruption in the normal pattern of breathing. Hiccups may be caused by the stimulation of nerves in the digestive system (see Chapter 22), but the respiratory system is involved. When the diaphragm is suddenly contracted, air is "sucked in" so abruptly that the epiglottis snaps shut, producing the sound of a hiccup. Hiccupping usually lasts 2 to 5 minutes, although some highly unusual cases persist for months or even years. It serves no known function.

Snoring can occur when muscles in the throat relax during sleep, and the loose tissue of the soft palate and uvula (the flap of tissue hanging down from the soft palate) partially obstruct the upper airway.

Sighing, sobbing, crying, yawning, and *laughing* are variants of ordinary breathing, and are so closely associated with subjective emotional conditions as to defy exact definition.

Why can a sneeze be stopped by pressing between the upper lip and the base of the nose?

The nerve receptor for the stimulation of sneezing is located there. Firm pressure deadens the nerve and suppresses the sneeze.

PHYSIOLOGICAL AND ANATOMICAL ABNORMALITIES

A constant supply of oxygen and the ready removal of waste carbon dioxide are so necessary to continued life that some sort of respiratory failure is the immediate cause of most deaths. Any interference with respiration is serious, and requires quick attention.

Rhinitis

Any inflammation of the nose can be called **rhinitis** (Gr. *rhin-*, nose). The most common nasal trouble is what is known as the "common cold." This virus-induced malady (or several maladies) causes excessive mucus secretion and swelling of the membranes of the nose and pharynx, and can spread into the nasolacrimal ducts, sinuses, auditory tubes, and larynx. It can cause fever, general malaise, and breathing difficulty, but is seldom grave except in people weakened by some other illness.

Laryngitis

A local infection of the larynx is **laryngitis**. It can totally incapacitate the vocal cords, making normal speech impossible, but is usually not painful. It may be *idiopathic,* that is, not associated with any other disease. Irritants such as tobacco smoke can cause swelling of the vocal cords to such an extent that chronic hoarseness results.

Asthma

Asthma (Gr. no wind) is a general term for difficulty in breathing. *Bronchial asthma* is a result of the constriction of smooth muscles in the bronchial and bronchiolar walls, accompanied by excess mucus secretion and insufficient contraction of the alveoli. Air is held in the lungs, and the victim cannot exhale normally. The condition is usually caused by an allergic reaction (see Chap-

ter 20), but may also be brought on by an emotional upset. Asthma is not always dangerous, though over a long period of time it may result in permanently damaged alveoli. An acute attack, if not treated immediately, can cause death by asphyxiation.

Hay fever

Hay fever is characterized by abundant tears and a runny nose. Like asthma, it is caused by an allergic reaction. So many people are affected that pollen counts of the atmosphere are regularly reported in the public media when pollen is abundant. The worst plant offenders are wind-pollinated trees, grasses, and ragweed.

Emphysema

Emphysema (Gr. blown up) is a condition in which the alveoli fail to con-

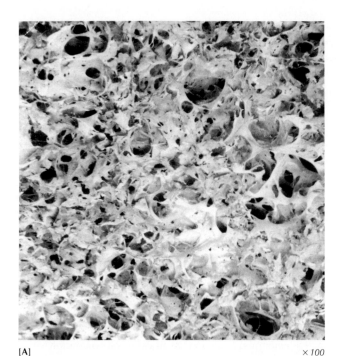

[A]   ×100

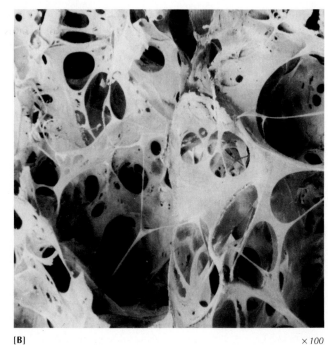

[B]   ×100

(A) Normal lung. (B) Emphysematous lung. [(A *and* B) *Kenneth A. Siegesmund, Ph.D., Department of Anatomy and Cellular Biology, Medical College of Wisconsin.*]

tract properly, and thus do not expel enough air. A person suffering from emphysema cannot exhale normal amounts of air even when trying forcibly to do so. As a result, undesirable levels of carbon dioxide may build up. In long-standing cases, the alveoli deteriorate, their walls harden, and alveolar capillaries are damaged (see photos). Fortunately, because a normal adult has about eight times as many alveoli as needed for routine activities, acute cases of emphysema are not as prevalent as they might be otherwise. As the disease progresses, the respiratory center in the brain loses its effectiveness, the right ventricle becomes overworked in its efforts to force blood through the constricted lung capillaries, and the victim runs short of breath even after slight exercise.

The most common cause of emphysema is air pollution, including tobacco smoke, industrial solvents, and agricultural dust. It can develop in old people as a result of hardening of alveolar walls, without any apparent environmental stimulus.

Pneumonia

Pneumonia is a general term for any condition that results in the filling of the aveoli with fluid. It can be caused by a number of factors: chemicals, bacteria, viruses, rickettsias, or fungi, but the usual infective agent is a *Pneumococcus* bacterium. Pneumonia most readily attacks people who are already weakened by illness, or whose lungs are damaged. Blood infections, chronic alcoholism, inhalation of fluids into the lungs, or even prolonged immobility in bed can predispose a person to bacterial infection of the lungs.

Tuberculosis

Tuberculosis, a disease caused by *Mycobacterium tuberculosis,* can occur in almost any part of the body, but because it most commonly affects the lungs, it is usually regarded as a respiratory ailment. The infectious bacterium is usually inhaled, and if it becomes established, the lungs can be damaged in a variety of ways.

Lung cancer

Cancer of the lungs, or *pulmonary carcinoma,* seems to be caused usually by environmental factors, cigarette smoking being the one most commonly mentioned. Any inhaled irritant apparently can stimulate some cells to begin abnormal growth (see Chapter 28). As has been pointed out frequently, early diagnosis is critical. Practically all sufferers from lung cancer cough, most breathe with difficulty, and many have chest pains and spit blood. Any such combination of symptoms calls for chest x rays, bronchoscopy, and biopsy. Damaged lung tissue shows up in x rays, sometimes as a discrete region and sometimes in such scattered spots as to suggest a "snowstorm." Lung cancer can spread to other organs, or other organs can send malignant cells to the lungs. In either case, the outlook is grave. Since it is known that the average survival time for a victim of lung cancer after diagnosis is about nine months, and that cigarette smokers run about 20 times the risk of nonsmokers, cigarette smoking is clearly a risky practice.

Pleurisy

Pleurisy is an infection of the pleurae. It is incapacitating, but is not necessarily dangerous in itself. There are several stages and types of the disease. *Fibrinous* or *dry pleurisy* causes intense pain in the parietal pleurae (the visceral pleurae are insensitive), and results in audible crackling or grating when the patient breathes. *Serofibrinous pleurisy* is characterized by deposits of fibrin and by an accumulation of watery fluid (up to 5 L in extreme cases) in the pleural cavity. It can be detected by percussion, but x rays are more useful. In *purulent pleuritis* or *empyema*, a pus-laden secretion accumulates in the pleural cavity. All types of pleurisy are caused by microorganisms.

Drowning

Probably the most common accidental cause of respiratory stoppage is *drowning*. A drowning person suffers a laryngeal spasm trying to inhale under water. Little water enters the lungs, but much may be swallowed. Death occurs within minutes unless oxygen is delivered to the tissues promptly.

Cyanosis

Whenever breathing is stopped and the pulse is weak or absent, *cyanosis* (Gr. dark blue) occurs. Cyanosis is the development of a bluish color of the skin, especially the lips, resulting from the build-up of deoxygenated hemoglobin, which is less crimson than oxygenated hemoglobin and appears bluish through the skin.

Hyaline membrane disease

Hyaline membrane disease, or "glassy-lung" disease, is a failure of newborn infants to produce enough surfactant to allow alveoli to fill with air properly (see p. 657). At birth, lungs contain no air, but they must quickly become inflated and stay inflated. In cases of too little pulmonary surfactant, the lungs may be lined with a hyaline (transparent) coating, hence the name of the disease.

Sudden infant death syndrome (SIDS)

Another malady of infants is "crib death," or *sudden infant death syndrome* (SIDS). It claims about 10,000 victims under 1 year of age annually in the United States, but its cause remains unknown. It is included here among respiratory ailments because one likely cause is respiratory failure, either from spasmodic closure of the air passages, or some malfunction of the respiratory center in the brain. The fact that crib death is most common in the autumn seems to indicate that an infectious organism, perhaps a virus, is responsible.

MEDICAL TERMINOLOGY

ANOXIA Absence of oxygen.

ANTITUSSIVE A drug that controls coughing.

BRONCHITIS Inflammation of the bronchi.

BRONCHODILATOR A drug that dilates the bronchi.

BRONCHOSCOPY Examination of the interior of the bronchi.

EXPECTORANT A drug that encourages the expulsion of mucus from the chest.

HYPERVENTILATION Excessive movement of air into and out of the lungs.

HYPOXIA Insufficient oxygen in organs and tissues.

LOBECTOMY Removal of one of the lobes of the lungs.

NASAL POLYPS Tumors caused by abnormal growth of the nasal mucous membranes.

PNEUMONECTOMY Removal of one of the lungs.

PULMONARY EDEMA A condition in which there is excessive fluid in the lungs.

SPUTUM (SPYOO-tuhm) Liquid substance containing mucus and saliva.

TRACHEOSCOPY Examination of the interior of the trachea.

TRACHEOTOMY A surgical procedure to make a hole and insert a breathing tube in the trachea.

SUMMARY

The *respiratory system* delivers oxygen to the body tissues, and removes gaseous wastes, mainly carbon dioxide.

Respiratory tract

1 The *respiratory tract* includes the *nose, nasal cavities, pharynx, larynx, trachea,* and *bronchi,* which lead by way of *bronchioles* to the *lungs.*

2 Air usually enters the body through the nose. The respiratory mucosa of the *nasal cavity* is specialized to moisten and warm air, and to capture particles like dust. The olfactory mucosa is specialized to sense odors.

3 The *pharynx* connects the nasal cavity and mouth with the rest of the respiratory tract and the esophagus. It is divided into the *nasopharynx, oropharynx,* and *laryngopharynx.*

4 The *larynx* contains the *vocal cords,* which are largely responsible for producing sound. At the opening of the larynx, the *glottis* closes during swallowing and opens to allow air to pass.

5 The *trachea,* or windpipe, carries air from the larynx to two *bronchi.* Its mucosal and ciliated lining traps and removes dust particles before they can enter the lungs.

6 These two bronchi divide into smaller and smaller bronchi, and then into even smaller *bronchioles,* forming the "respiratory tree." Around the tiniest branches, called respiratory bronchioles, are minute air sacs known as *alveoli.*

7 *Alveoli* are the functional units of the lungs. Exchange of oxygen and carbon dioxide takes place through the walls of the alveoli and the walls of the pulmonary capillaries. The walls of the alveoli are thin enough to permit gases to pass through, but thick enough to remain as open sacs. The tendency of alveolar walls to contract is offset by the action of a *surfactant* on their inner linings.

8 In the thoracic cavity, the two *lungs* are separated by the *mediastinum.* The right lung has three *lobes,* and the left lung has two. Each lobe is further divided into *bronchopulmonary segments* and then into *lobules.*

9 The lungs are covered by the *visceral pleura,* and the thoracic cavity by the

parietal pleura. Between these membranes is the *pleural cavity.*

10 The smooth muscle of the tracheobronchial tree is innervated by the autonomic nervous system. The respiratory tree is supplied with blood by the pulmonary and bronchial circulations.

Mechanics of breathing

1 *Ventilation* is the mechanical process that moves air into and out of the lungs.

2 During *inspiration,* air is pushed into the lungs by the pressure of the outside air when the size of the chest cavity is increased by contraction of the *external intercostal muscles* and the *diaphragm.*

3 During ordinary *expiration,* air is expelled from the lungs when the *respiratory muscles* are relaxed. During *forced expiration,* the *abdominal* and *internal intercostal muscles* are contracted. The subsequent reduction in the size of the chest raises the pressure of the air in the lungs and forces air out.

4 Four volumes of air movement are measured in assessing respiratory function. The *residual volume* is the amount of air retained in the lungs at the end of a maximal exhalation. The *tidal volume* is the amount of air inhaled and exhaled during one normal, quiet breath. The *inspiratory reserve volume* is the amount of air greater than the tidal volume that can be taken in during a maximal inhalation. The *expiratory reserve volume* is the amount of air greater than the tidal volume that can be expired during a maximal exhalation.

5 Four *pulmonary capacities* are measured. The *inspiratory capacity* is inspiratory reserve volume plus tidal volume. The *functional residual capacity* is expiratory reserve volume plus residual volume. The *total lung capacity* is vital capacity plus residual volume. The *vital capacity* is inspiratory reserve volume plus tidal volume plus expiratory volume.

6 The *anatomical dead space* is that part of the respiratory tract in which there is no exchange of respiratory gases: the nasal passages, trachea, bronchi, and bronchioles.

7 The *minute respiratory volume* is the amount of air inspired per minute, and is equal to the *respiratory rate* (the number of breaths per minute) times the volume of air inhaled per breath. *Alveolar ventilation* is the amount of fresh air delivered to the alveoli per breath.

Factors affecting gas movement and solubility

1 According to *Dalton's law,* each gas in a mixture exerts its own pressure, called its partial pressure, in proportion to its concentration in the mixture of gases.

2 The solubility of a gas, measured as the volume of the gas that will dissolve in 1 mL of water at 37°C, is affected by characteristics of the molecule of the gas, temperature, and pressure.

3 At equal pressures and temperatures, carbon dioxide and oxygen differ in solubility. Carbon dioxide is about 24 times more soluble in water than oxygen.

4 According to *Henry's law,* the quantity of a gas that will dissolve in a liquid is proportional to the partial pressure of the gas. The higher the pressure, the more gas will stay in solution.

5 According to *Fick's law of gas diffusion,* the factors that determine the rate of exchange of oxygen and carbon dioxide across the aveolar-capillary membranes are the partial pressure on either side of the membrane, the surface area, the thickness of the membrane, and the solubility and size of the molecules.

Gas transport

1 *External respiration* is the exchange of gases between alveoli and lung capillaries. Because of its higher partial pressure in the alveoli as compared to that in the capillaries, oxygen diffuses from the alveoli into the blood. Carbon dioxide moves in the opposite direction.

2 *Internal respiration* is the exchange of gases in body tissues. There the partial pressures enable the tissues to take up oxygen and release carbon dioxide.

3 About 98 percent of the oxygen carried in blood is bound to the *hemoglobin* of erythrocytes in the form of *oxyhemoglobin.* In tissue capillaries, oxyhemoglobin gives up its oxygen.

4 The binding and release of oxygen are determined by the partial pressure of the gas. A high partial pressure favors binding and low pressure favors release.

5 Oxygen uptake or release is affected by temperature, partial pressure of carbon dioxide, and presence in erythrocytes of 2,3-*diphosphoglycerate* (DPG).

6 Carbon dioxide is carried in blood in three forms: as dissolved carbon dioxide (10 percent), as bicarbonate ions (68 percent), and as *carbamino compounds,* chiefly *carbaminohemoglobin* (22 percent).

7 When carbonic acid in erythrocytes in tissue capillaries dissociates, the hydrogen ions in the erythrocytes are exchanged for chloride ions from the plasma. In alveolar capillaries, chloride ions are returned to the plasma. This exchange is the *chloride shift.*

Neurochemical control of breathing

1 The rate and depth of breathing are controlled by nerve impulses to the breathing muscles from the *respiratory center* in the brain, a system of circuits and controls in the medulla and pons.

2 The *medullary rhythmicity area* contains two circuits that operate alternately, one stimulating inspiration and the other stimulating expiration.

3 Other breathing controls are in the pons: an *apneustic area* and a *pneumotaxic area,* both of which influence the rhythmicity area.

4 If the concentration of carbon dioxide rises above normal levels, the concentration of hydrogen ions increases as a result of dissociation of carbonic acid. This lowering of blood pH directly affects respiratory centers in the brain and is also sensed by carotid and aortic chemoreceptors that send impulses to the medullary rhythmicity center.

5 A self-regulating system of contractile muscles in lung capillaries aids in distributing blood to alveoli where gas exchange is most effective.

6 Respiratory centers are affected by *pulmonary stretch receptors* that send impulses by way of the vagus nerve to the brain to call for a cessation of inhalation and thus prevent overinflation.

Other activities of the respiratory system

Other activities of the respiratory system include coughing, sneezing, crying, laughing, yawning, hiccuping, and snoring.

Physiological and anatomical abnormalities

1 Among the most common conditions and accidents affecting the respiratory system are *rhinitis, laryngitis, asthma, hay fever, emphysema, pneumonia, tuberculosis, lung cancer, pleurisy, drowning,* and *choking. Cyanosis* occurs when breathing stops.

2 Respiratory diseases peculiar to infancy are *hyaline membrane disease* and *sudden infant death syndrome.*

UNDERSTANDING THE FACTS

1 What do the conchae contribute functionally to the respiratory system?

2 What is meant by a cleft palate?

3 What factors often contribute to "smoker's cough"?

4 Name the three parts of the pharynx, and give their locations and functions.

5 What is the function of the trachea?

6 What are the two tubes into which the trachea divides?

7 What roles do the macrophages in the alveoli play?

8 Distinguish between visceral and parietal pleurae.

9 Approximately how many alveoli do your lungs contain? About what is their surface area?

10 Define atelectasis.

11 What muscles are used during inspiration? During expiration?

12 What is tidal volume?

13 What are the components of the respiratory center in the brain?

14 To which gas normally present in the body are the breathing centers the most sensitive?

15 What is the Hering-Breuer reflex?

16 What causes sneezing? Coughing?

17 How does the body compensate for the deficiency of oxygen at high altitudes?

18 What factor has the greatest effect on the formation and breakdown of oxyhemoglobin?

19 In what way is the greatest percentage of carbon dioxide carried?

20 Define pneumonia.

21 What are some of the causes of emphysema?

22 What is the usual cause of asthma?

23 Describe the Heimlich maneuver.

24 Give the location of the following (be specific):
   a buccal cavity
   b thyroid cartilage
   c mediastinum
   d intercostal muscles

UNDERSTANDING THE CONCEPTS

1 What are the functions of the respiratory system?

2 How do the structures of the larynx contribute to sound production?

3 What roles do the pleurae of the lungs play in respiration?

4 Why is the small size of the capillaries surrounding the alveoli important physiologically?

5 What is surfactant, and why is it important?

6 List in order the structures through which a molecule of oxygen in the atmosphere would pass in order to reach the bloodstream.

7 Compare the processes of internal and external respiration.

8 Why is the residual volume physiologically important?

9 What effect would an increase of carbon dioxide, decrease of oxygen, or decrease of carbon dioxide levels have on the respiratory rate?

10 Explain the significance of Boyle's law, Dalton's law, and Henry's law in gas transport within the body.

11 Describe the methods and mechanisms of carbon-dioxide transport.

12 How can hemoglobin transport both oxygen and carbon dioxide at the same time?

13 From what you have learned in this chapter, how do you explain the poor prognosis for an individual suffering from lung cancer?

14 Briefly discuss SIDS.

15 Discuss the correct procedures to follow (a) when breathing stops, and (b) when the pulse is weak or absent.

SUGGESTED READING

AVERY, M. E., N. WANG, AND H. W. TAEUSCH, "The Lung of the Newborn Infant." *Scientific American,* April 1973.

BRULEY, D., ed., *Oxygen Transport to Tissue—VI.* New York: Plenum, 1985.

CLEMENTS, JOHN A., "Surface Tension in the Lungs." *Scientific American,* December 1962.

COMROE, JULIUS H., JR., "The Lung." *Scientific American,* February 1966.

COMROE, JULIUS H., JR., *Physiology of Respiration,* 2nd ed. Chicago: Year Book Medical Publishers, 1974.

EGAN, D. F., *Fundamentals of Respiratory Therapy,* 3rd ed. St. Louis: Mosby, 1977.

FEN, W. O., "The Mechanism of Breathing." *Scientific American,* June 1960.

GUZ, A., "Regulation of Respiration in Man." *Annual Reviews of Physiology,* 37 (1975):303.

KING, R. J., "Pulmonary Surfactant." *Journal of Applied Physiology,* 53 (1982): 1–8.

LEVITZKY, MICHAEL G., *Pulmonary Physiology,* 2nd ed. New York: McGraw-Hill, 1986.

MINES, A. H., *Respiratory Physiology.* New York: Raven Press, 1981.

MOUNTCASTLE, V. B., ed., *Medical Physiology,* Vols. 1, 2, 14th ed. St. Louis: Mosby, 1980.

NAEYE, R. L., "Sudden Infant Death." *Scientific American,* April 1980.

SCHLESINGER, R. B., "Defense Mechanisms of the Respiratory System." *Biosciences,* 32 (1982):45–50.

SLONIM, N. B., AND L. H. HAMILTON, *Respiratory Physiology,* 4th ed. St. Louis: Mosby, 1981.

VON EULER, C., AND H. LAGERCRANTZ, eds., *Central Nervous Control Mechanisms in Breathing.* New York: Pergamon Press, 1980.

WEIBEL, E. R., "How Does Lung Structure Affect Gas Exchange?" *Chest,* 83 (1983):657–665.

WEST, J. B., *Respiratory Physiology,* 2nd ed. Baltimore: Williams & Wilkins, 1979.

22
The Digestive System

LEARNING OBJECTIVES

1 Differentiate between digestion and absorption.

2 List the major organs of the alimentary canal and the associated structures of digestion, and give the basic functions of each.

3 Describe the four layers of tissue in the wall of the digestive tract.

4 Discuss the digestive activities that occur in the mouth.

5 Describe the three main parts of a tooth, the composition of teeth, and the types of permanent teeth.

6 Name the three pairs of salivary glands, and explain the functions of saliva.

7 Describe the four stages in swallowing, along with the structures involved in the process.

8 Discuss the structure of the abdominopelvic cavity and the peritoneum.

9 Identify the four major regions of the stomach and its two openings, and describe its tissues as they relate to its functions.

10 Describe the types of digestive movements of the stomach and how they are regulated.

11 Discuss the composition and functions of gastric juice.

12 Name the three parts of the small intestine, and explain how the special features of its mucosa increase its absorptive surface.

13 Describe the digestive movements of the small intestine and the action of its major enzymes and hormones.

14 Describe the absorption of carbohydrates, proteins, and fats from the small intestine into the bloodstream.

15 Discuss the structure and functions of the pancreas.

16 Describe the anatomy and basic functions of the liver as they relate to digestion.

17 Explain the functions of the gallbladder in the digestive process.

18 Name the major functions and segments of the large intestine.

The digestive tract, which includes the mouth, throat, esophagus, stomach, and intestines, is like a tube within the body (Figure 22.1). Nutrients that are within this tube are still not inside the body proper until they are absorbed from the small intestine. After food is ingested, it undergoes *digestion,* the mechanical and chemical processes that break down large food molecules into smaller ones. But the small molecules are useless unless they can get into the individual cells of the body. This is accomplished when the small food molecules pass through the cells of the small intestine into the bloodstream and lymphatic system. This second process is called *absorption.*

After digestion and absorption have taken place, the small molecules are ready to be used by the body. Some of them are used as a source of energy. Others, such as amino acids, are used by cells to rebuild, repair, and reproduce themselves. Materials that are not digested and absorbed are finally eliminated from the body. The processes of digestion and absorption include:

1 *Ingestion,* or eating.

2 *Peristalsis,* or the involuntary, sequential muscular contractions that move ingested nutrients along the digestive tract.

FIGURE 22.1

The digestive system, showing structures and functions of the digestive tract and associated structures such as the liver and pancreas.

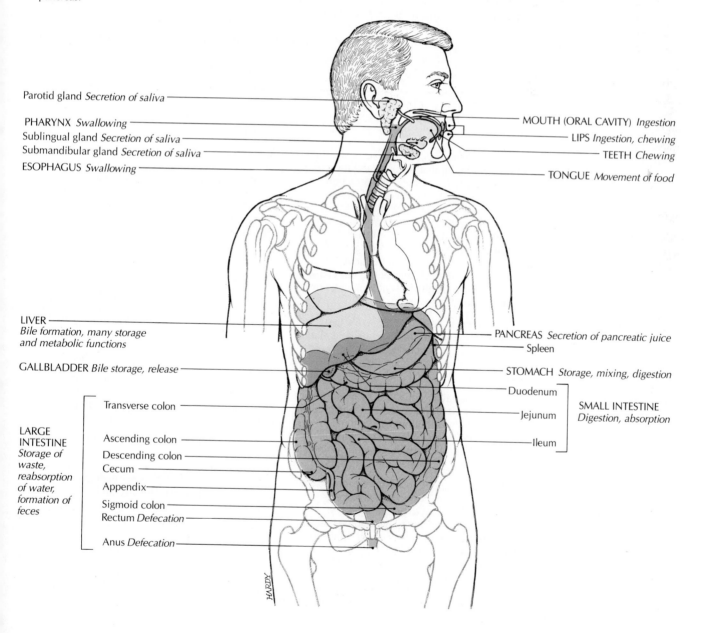

Parotid gland *Secretion of saliva*

PHARYNX *Swallowing*
Sublingual gland *Secretion of saliva*
Submandibular gland *Secretion of saliva*
ESOPHAGUS *Swallowing*

MOUTH (ORAL CAVITY) *Ingestion*
LIPS *Ingestion, chewing*
TEETH *Chewing*
TONGUE *Movement of food*

LIVER
Bile formation, many storage and metabolic functions

GALLBLADDER *Bile storage, release*

PANCREAS *Secretion of pancreatic juice*
Spleen
STOMACH *Storage, mixing, digestion*

Transverse colon

Duodenum
Jejunum

SMALL INTESTINE
Digestion, absorption

LARGE
INTESTINE
Storage of waste, reabsorption of water, formation of feces

Ascending colon
Descending colon
Cecum
Appendix
Sigmoid colon
Rectum *Defecation*
Anus *Defecation*

Ileum

3 *Digestion,* or the conversion of large nutrient particles into small molecules.

4 *Absorption,* or the passage of usable nutrient molecules from the small intestine into the bloodstream and lymphatic system, for the final passage into body cells.

5 *Defecation,* or the elimination from the body of undigested and unabsorbed material as solid waste.

In this chapter you will learn how the body obtains and assimilates nutrients, and how some waste products are removed.

physical means such as chewing, peristalsis, and the churning movements of the stomach and small intestine to mix the food with enzymes and digestive juices. The digestive process, assisted by enzymes, takes place in the **digestive tract,** or **alimentary canal.** The part of the digestive tract below the diaphragm is called the *gastrointestinal (GI) tract.*

The alimentary canal consists of the mouth, pharynx, esophagus, stomach, small intestine, large intestine, rectum, anal canal, and anus. From mouth to anus this canal is about 9 m (30 ft) long. The *associated structures* of the digestive system include the teeth, lips, tongue, cheeks, salivary glands, pancreas, liver, gallbladder, and bile duct.

INTRODUCTION TO THE DIGESTIVE SYSTEM

In simple terms, digestion is the process of breaking down large molecules of food that cannot be used by the body into small, soluble molecules that can be absorbed and used by cells. **Chemical digestion** breaks down food particles through a series of catabolic reactions. **Mechanical digestion** involves

Basic Functions

The digestive system is divided into compartments, each one adapted to perform a specific function (see Figure 22.1). In the *mouth,* the breakdown of food begins with chewing and enzymatic action. The *pharynx* (throat) performs the act of swallowing, and food passes through the *esophagus* to the

TABLE 22.1   STRUCTURE OF THE WALL OF THE DIGESTIVE TRACT (FIGURE 22.2)

Layer	Description	Functions
Mucosa (tunica mucosa, mucous membrane) Epithelium Lamina propria Muscularis mucosa	Innermost layer of digestive tract, consisting of epithelium, lamina propria, and muscularis mucosa. *Epithelium* is site of interaction between ingested food and body. *Lamina propria* is layer of connective tissue supporting epithelium. *Muscularis mucosa* consists of two thin muscular layers and contains a nerve plexus. Surface area of epithelial layer is increased by large folds (rugae in the stomach and plicae circulares and villi in the small intestine), indentations (crypts), and glands.	Acts as lubricating, secreting, absorbing layer. Lubricates solid contents and facilitates their passage through esophagus and digestive tract. In remainder of digestive tract, contains secretory cells that produce digestive juices (mucus, enzymes, various ions) and absorptive cells.
Submucosa (tunica submucosa)	Highly vascular layer of connective tissue between mucosa and muscle layers; contains many nerves. Submucosal glands are present in duodenum of small intestine and in esophagus.	Contains blood vessels, nerves, and in certain areas lymphatic nodules.
Muscularis externa (tunica muscularis) Circular muscle Longitudinal muscle Oblique muscle	Main muscle layer, consisting in most regions of inner *circular* layer and outer *longitudinal* layer of mostly smooth muscle. Stomach contains additional *oblique* layer internal to other layers. Upper esophagus and sphincters of anus consist of striated muscle fibers.	Moves food along lumen of digestive tract by antagonistic actions of smooth muscles. Wave of muscular contraction is called *peristalsis.*
Serosa (tunica serosa) Visceral peritoneum Parietal peritoneum Adventitia	Outermost lamina, consisting of thin connective tissue, and in many places epithelium covering digestive tube and digestive organs. Where epithelium is lacking (as in esophagus), serosa is called *adventitia.* Portion covering viscera is *visceral peritoneum;* portion lining abdominal wall is *parietal peritoneum.* Double-layered *mesentery* is portion of serous membrane that connects intestines to dorsal abdominal wall.	Attaches viscera to abdominal wall and contains blood vessels, nerves, and lymphatics.

FIGURE 22.2

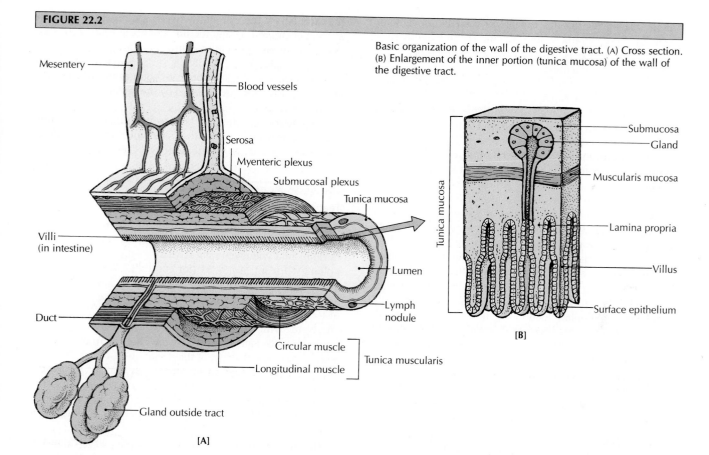

Basic organization of the wall of the digestive tract. (A) Cross section. (B) Enlargement of the inner portion (tunica mucosa) of the wall of the digestive tract.

stomach. The *stomach* stores food and breaks it down with acid and some enzymes. The *small intestine* is where most of the large molecules are chemically broken down into smaller ones, and where most of the water and nutrients are absorbed into the bloodstream and lymphatic system. The *large intestine* carries undigested food, removes additional water from it, and releases solid waste products through the *anus.* The *associated structures* assist in the breakdown and conversion of food particles in a variety of ways, both mechanical and chemical.

Tissue Structure

Despite the compartmentalization of the digestive system, the walls of the various portions of the tract have the same basic organization. The wall of the tube (from the inside out) consists of four main layers of tissue: *mucosa, submucosa, muscalaris externa,* and *serosa.* They are shown in Figure 22.2 and described in Table 22.1.

MOUTH

By following a mouthful of food through the digestive tract, we can observe the digestive process in detail. We begin by taking food into the **mouth** (also called the **oral cavity** or **buccal cavity**). The mouth has two parts: (1) The small, outer

vestibule is bounded by the lips and cheeks on the outside and the teeth and gums on the inside. (2) The **oral cavity proper** extends from behind the teeth and gums to the *fauces,* the opening that leads to the *pharynx* (Figure 22.3). The fauces has sensory nerve endings that trigger the involuntary phase of swallowing. If the sensory region of the fauces is stimulated (with a tongue depressor, for example), the *gag reflex* is produced. The fauces is surrounded by lymphoid tissue, which guards against the entrance of harmful microorganisms. The palatine tonsils are the largest of these lymphoid tissues.

In the mouth, food is *masticated* (chewed) by the ripping and grinding action of the teeth. As much pressure as 500 kg/sq cm (7000 lb/sq in.) can be exerted by the molars—considerably greater than the pressure under the tires of a moving car. At the same time, the food is moistened by saliva, which intensifies its taste and eases its passage down the delicate tissue of the esophagus. Saliva, secreted by the salivary glands, contains an enzyme that begins the breakdown of large molecules of starch into small molecules of sugar. This digestive enzyme is the first of many in the digestive tract.

Lips and Cheeks

The **lips** are the two fleshy, muscular folds that surround the opening of the mouth. They consist mainly of fibroelastic connective tissue and striated muscle fibers. The orbicularis oris muscle makes the lips capable of versatile movement. The lips are also extremely sensitive, and are abundantly supplied with blood vessels, lymphatic vessels, and sensory nerve endings from the trigeminal nerve.

Each lip is connected at its midline to the gum by a fold of mucous membrane called a *labial frenulum* (L. *frenum,* bridle) (see Figure 22.3). The superior and inferior labial frenula support the lips. The frenulum under the tongue is the *lingual frenulum.* It limits the backward movement of the tongue.

Lips, like the anus at the other end of the digestive system, form a transition between the external skin and the mucous membrane of the wet epithelial lining of the internal passageways. The skin of the outer lip surface (beneath the nose and above the chin) contains the usual sweat glands, sebaceous glands, and hair follicles. The *red free margin,* the portion we normally think of as "lips," is covered by a thin, translucent epidermis that allows the capillaries underneath to show through, giving the lips a reddish color. The skin of the red free margin does not contain sweat glands, sebaceous glands, or hair follicles.

The lips help to place food in the mouth and to keep it in the proper position for chewing. They also contain sensory receptors that make it relatively easy for us to identify specific textures of foods.

The wet mucous membrane of the inner surface of the lips leads directly to the mucous membrane of the inner surface of the cheeks. The **cheeks** are the fleshy part of either side of the face, below the eye and between the nose and ear. The mucous membrane lining the lips and cheeks is a thick stratified squamous epithelium (nonkeratinized). Such an epithelium is typical of wet epithelial surfaces that are subjected to a great deal of abrasive force. As soon as the surface cells are worn away, they are replaced by the rapidly dividing cells underneath.

The cheeks, like the lips, help to hold food in a position where the teeth can chew it conveniently. Also, the buccal muscles of the cheeks contribute to the chewing process.

Teeth and Gums

Six months or so after birth, the first **deciduous*** teeth (baby teeth, milk teeth) erupt through the gums (Figure 22.4). A normal child will eventually have 20 "baby" teeth, each jaw holding 10 teeth: 4 *incisors* (for cutting), 2 *canines* (for tearing), and 4 *premolars* (for grinding). The deciduous teeth are lost when the permanent teeth are ready to emerge. Both sets of teeth are usually present in the gums at birth, or shortly afterward, with the permanent teeth lying under the deciduous teeth (see Figure 22.4B). By the time a permanent tooth is ready to erupt, the root of the deciduous tooth above it has been completely resorbed by osteoclasts. The six permanent molars in each jaw have no deciduous predecessors. The shedding of deciduous teeth and the appearance of permanent teeth follow a fairly consistent pattern, as shown in Table 22.2.

The 32 permanent teeth (16 in each jaw) are arranged in two arches, one in the upper jaw (maxilla), and the other in the lower jaw (mandible). Each jaw holds 4 **incisors** (cutting teeth), 2 **canines** (cuspid, with one point or cusp in its crown), 4 **premolars** (bicuspids, each with two cusps), and 6 **molars** (millstone teeth). Because the upper incisors are wider than the lower ones, the lower grinding teeth are usually aligned slightly in front of the upper grinders. This arrangement enhances the grinding motion between the upper and lower teeth.

The teeth are held in their sockets by bundles of connective tissue called *periodontal ligaments* (see Figure 22.4C). The collagenous fibers of each ligament extend from the alveolar bone into the cement of the tooth, and allow for some normal movement of the teeth during chewing. Nerve endings in the ligaments monitor the pressures of chewing and relay the information to the brain centers involved with chewing movements.

Deciduous (dih-SIHDJ-oo-us) means "to fall off," and is typically used to describe trees that shed their leaves in the autumn.

FIGURE 22.3

The mouth, also known as the oral cavity or buccal cavity.

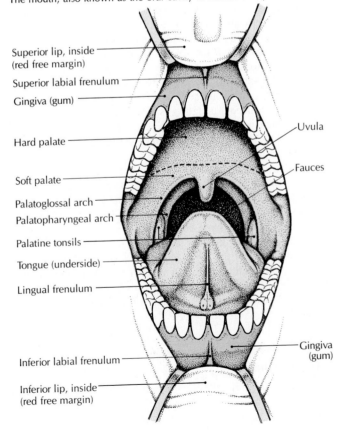

- Superior lip, inside (red free margin)
- Superior labial frenulum
- Gingiva (gum)
- Hard palate
- Soft palate
- Palatoglossal arch
- Palatopharyngeal arch
- Palatine tonsils
- Tongue (underside)
- Lingual frenulum
- Inferior labial frenulum
- Inferior lip, inside (red free margin)
- Uvula
- Fauces
- Gingiva (gum)

| Why are canine teeth called "eyeteeth"? | Early anatomical schemes often named body parts according to their relation to other structures or functions. Eyeteeth were so named probably because they lie directly under the eyes. (As another example, people in the Western world wear wedding rings on the fourth finger of the left hand, because it was believed that this finger was connected directly to the heart.) |

FIGURE 22.4

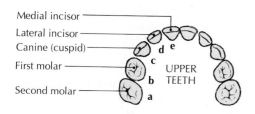

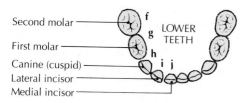

Medial incisor
Lateral incisor
Canine (cuspid)
First molar
Second molar

UPPER TEETH

Second molar
First molar
Canine (cuspid)
Lateral incisor
Medial incisor

LOWER TEETH

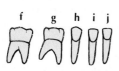

[A] DECIDUOUS "BABY" TEETH

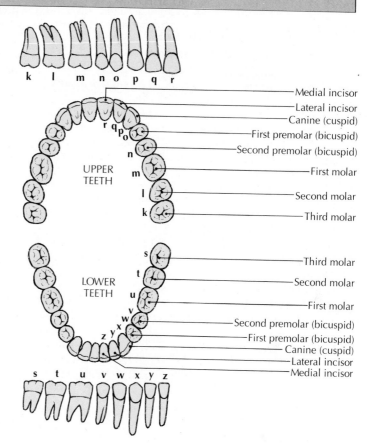

Medial incisor
Lateral incisor
Canine (cuspid)
First premolar (bicuspid)
Second premolar (bicuspid)
First molar
Second molar
Third molar

UPPER TEETH

Third molar
Second molar
First molar
Second premolar (bicuspid)
First premolar (bicuspid)
Canine (cuspid)
Lateral incisor
Medial incisor

LOWER TEETH

PERMANENT TEETH

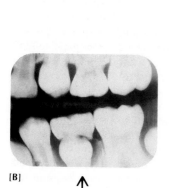

[B]

Teeth. (A) Deciduous and permanent teeth in the upper and lower jaws. (B) X ray showing the permanent second premolar (arrow) in place beneath the deciduous tooth of a 10-year-old child. (C) Vertical section of a tooth.

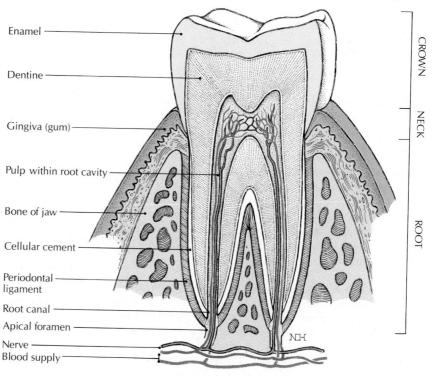

Enamel
Dentine
Gingiva (gum)
Pulp within root cavity
Bone of jaw
Cellular cement
Periodontal ligament
Root canal
Apical foramen
Nerve
Blood supply

CROWN
NECK
ROOT

[C]

TABLE 22.2   ERUPTION AND SHEDDING OF TEETH

| Type of tooth | Deciduous (baby) teeth | | Permanent teeth |
	Approximate time of eruption (months)*	Approximate time of shedding (years)*	Approximate time of eruption (years)*
Medial incisors†	6–12	6–7	6–8
Lateral incisors	9–16	7–8	7–9
Canines	16–23	9–12	9–12
First premolars	14–19	9–11	10–12
Second premolars	23–33	10–12	10–12
First molars	—	—	6–7
Second molars	—	—	11–13
Third molars (wisdom teeth)‡	—	—	17–21

*According to the American Dental Association.

†Deciduous medial incisors are sometimes present at birth.

‡Wisdom teeth often remain within the alveolar bone and do not erupt. In such cases they are said to be *impacted*, and are often removed surgically.

Chewing   Although chewing can certainly be voluntary, most of the chewing we do during a meal is an automatic rhythmic reflex that is triggered by the pressure of food against the teeth, gums, tongue, and hard palate. Such pressure causes the jaw muscles to relax and the jaw to drop slightly; then, as opposite muscles contract in an attempt to balance the relaxation, the jaw is pulled up again.

Parts of a tooth   All teeth, no matter what type, consist of the same three parts: (1) a *root*, embedded in a socket (alveolus) in the alveolar process of a jaw bone, (2) a *crown*, projecting upward from the gum, and (3) a narrowed *neck* (cervix) between the root and crown, which is surrounded by the gum (Figure 22.4C). The incisors, canines, and premolars have a single root, although the first upper premolar may initially have a double root. The lower molars have two flattened roots, and the upper molars have three conical roots. At the apex of each root is the *apical foramen* (root foramen), which leads successively into the *root canal* and the *root cavity* (pulp cavity).

Composition of teeth and gums   Each tooth is composed of dentine, enamel, cement, and pulp (see Figure 22.4C). The *dentine* is the extremely sensitive yellowish portion surrounding the pulp cavity. It forms the bulk of the tooth. The *enamel* is the insensitive white covering of the crown. It is the hardest substance in the body. (In order to cut through enamel, a dentist's drill spins at about half a million revolutions per minute.) The *cement* is the bonelike covering of

the neck and root. The *pulp* is the soft core of connective tissue that contains the nerves and blood vessels of the tooth.

Teeth are derived from the same tissue as the skin. The enamel is formed from the embryonic epidermis, and the dentine, pulp, and cement are formed from the embryonic dermis.

The *gum* (Old Eng. *goma*, palate, jaw), also called the *gingiva* (jihn-JYE-vuh), is the firm connective tissue covered with mucous membrane that surrounds the alveolar processes of the teeth (see Figure 22.4C). The stratified squamous epithelium of the gums is slightly keratinized to withstand friction during chewing. The gums are usually attached to the enamel of the tooth somewhere along the crown, but the gum line gradually recedes as we get older. In fact, the gum line may recede so far in elderly people that the gum is attached to the cement instead of the enamel (see Figure 22.4C).*

Tongue

The *tongue* functions in mechanical digestion, mainly in chewing and helping to move food from the mouth down into the throat. The front of the tongue is used to manipulate the food during chewing, and the base of the tongue aids in swallowing. It is also a sensitive tactile organ and plays an important role in speech. (In Chapter 15 we described the sensory functions of the tongue in tasting food.)

The tongue is composed mostly of striated muscle, and is covered by a smooth film of mucous membrane on the underside. The irregular dorsal (top) surface contains papillae, taste buds, and other structures associated with sensing different tastes, all of which were described in Chapter 15.

The mucous membrane covering the tongue is ordinarily divided into two sections, the *oral part* of the anterior two-thirds, and the *pharyngeal part* of the posterior third. The separate sections are delineated by the V-shaped sulcus terminalis (see Figure 15.2). The oral part, corresponding to the *body* of the tongue, contains the three types of papillae. The pharyngeal part, representing the *root* of the tongue, contains the lymphatic nodules of the lingual tonsil.

The interlacing muscles of the tongue are so arranged that it can be moved in any direction, and even slightly shortened or lengthened. The tongue contains three bilateral pairs of extrinsic muscles (muscles with attachments outside the tongue), and three pairs of intrinsic muscles (muscles wholly within the tongue). The extrinsic muscles move food within the mouth to form it into a round mass, or *bolus*, and the intrinsic muscles assist in swallowing. Both types of muscles are innervated by the hypoglossal cranial nerve (XII).

Palate

The *palate* (PAL-iht), or "roof of the mouth," is one of the many examples of structures that are perfectly suited to their functions. It has two sections (see Figure 22.3). (1) The an-

*The expression "long in the tooth," relating to elderly people or animals, refers to this phenomenon of the recession of the gums, which exposes more and more of the full length of the tooth.

terior *hard palate*, bordered by the upper teeth, is formed by a portion of the palatine bones and maxillae. Its upper surface forms the floor of the nasal cavity. (2) The posterior *soft palate* is continuous with the posterior border of the hard palate. It extends between the oral and nasal portions of the pharynx, with a small fleshy cone called the *uvula* (YOO-vyoo-luh; L. small grapes) hanging down from the center of its lowermost border. Extending laterally and downward from each side of the base of the soft palate are two curved folds of mucous membranes called the *palatoglossal* and *palatopharyngeal arches*. The palatine tonsils lie between the two arches.

When food is being chewed, and moistened with saliva, the tongue is constantly pushing it against the tough surface of the *hard palate*, crushing and softening the food before it is swallowed. The hard palate is covered with a firmly anchored mucous membrane and the same tough epithelium of keratinized stratified squamous epithelium as the cheeks (Figure 22.5).

The *soft palate* has a very different structure and function. It is composed of interlacing skeletal muscle that allows it to move back and forth over the nasal opening (nasopharynx). In this way it functions to close off the nasopharynx during swallowing, preventing food from being forced into the nasal cavity.

Salivary Glands

Many glands secrete saliva into the oral cavity, but the term *salivary glands* usually refers to the three largest pairs: the parotid, submandibular (formerly called submaxillary), and sublingual glands (Figure 22.6). *Saliva* is the watery, tasteless mixture of salivary and oral mucous-gland secretions. It lubricates chewed food, moistens the oral walls, contains salts to buffer chemicals in the mouth, and also contains amylase, the enzyme that begins the digestion of starches. The six salivary glands secrete more than a liter of saliva daily, with the parotid glands contributing 25 to 35 percent, the submandibular glands 60 to 70 percent, and the sublingual glands 3 to 5 percent.

Parotid glands The *parotid glands* (Gr. *parotis*, "near the ear") are the largest of the three main pairs of salivary glands.* They lie below the ears, covering the masseter muscle posteriorly (see Figure 22.6). The long ducts from the glands (called *parotid* or *Stensen's ducts*) pass forward over the masseter muscle (they can be felt as a ridge by moving the tip of a finger up and down over the muscle), and end in the vestibule alongside the second upper molar tooth. The parotid glands secrete water, salts, and starch-digesting amylase, but unlike the other salivary glands, they do not secrete *mucin* (MYOO-sihn), a protein that forms mucus when dissolved in water. As a result, the saliva is clear and watery.

*A specific viral infection of the parotid glands produces *mumps*. The incidence of this highly contagious disease has been considerably reduced since the advent of the combined measles, mumps, and German measles vaccine, which is routinely given to children when they are 15 months old.

Scanning electron micrograph of the epithelium of the hard palate. The epithelium is constantly being shed (arrows) to allow new cells to replace worn ones. [*Richard Kessel and Randy Kardon,* Tissues and Organs: A Text-Atlas of Scanning Electron Microscopy. *San Francisco, W. H. Freeman, 1979.*]

×995

The three major salivary glands: parotid, submandibular, and sublingual.

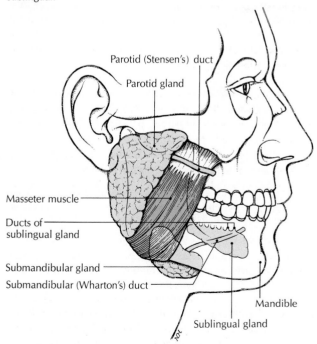

Parotid (Stensen's) duct

Parotid gland

Masseter muscle

Ducts of sublingual gland

Submandibular gland

Submandibular (Wharton's) duct

Mandible

Sublingual gland

Submandibular glands The *submandibular* ("under the mandible") *glands* are located on the medial side of the mandible (see Figure 22.6). They are about the size of a walnut, roughly half the size of the parotid glands. Their ducts, called *submandibular (Wharton's) ducts,* open into a papilla on the floor of the mouth beside the lingual frenulum behind the lower incisors. The submandibular glands secrete water, salts, amylase, and mucin. They secrete a thicker saliva than the parotid glands do, with less amylase.

Sublingual glands The *sublingual* ("under the tongue") *glands* are located in the floor of the mouth beneath the tongue* (see Figure 22.6). They are the smallest major salivary glands. Each gland drains through a dozen or so small ducts located on the summit of a fold on either side of the lingual frenulum, just behind the orifice of the submandibular duct. The sublingual glands secrete mostly water, salts, and mucin. Their secretion is the most viscous saliva of the three types of salivary glands. It is high in mucus and low in amylase.

Saliva and its functions The secretions of salivary glands contain about 99 percent water and 1 percent electrolytes and proteins. The proteins include *mucin,* which helps to form mucus, and at least one enzyme, salivary amylase or *ptyalin* (TIE-uh-lihn). The interaction of mucin with the water in saliva produces the highly viscous mucus that moistens and lubricates the food particles so they can slide down the pharynx and esophagus during swallowing. Ptyalin immediately acts on the polysaccharides of cooked starch, breaking them down into maltose and dextrin. However, it is inactivated by the stomach acids only minutes after it is secreted. Ptyalin is able to act on cooked starches because cooking disrupts their indigestible cellulose "coats."

Saliva also cleanses the mouth and teeth of cellular and food debris, helps keep the soft parts of the mouth supple, and buffers the acidity of the oral cavity (Table 22.3). The high bicarbonate concentration of saliva helps to reduce *dental caries* (cavities) by neutralizing the acidity of food in the mouth.[†] Remember also that taste buds cannot be stimulated until the food molecules are dissolved. Saliva provides the necessary solvent.

Control of salivary secretion Because the mouth and throat must be kept moist at all times, there is a continuous, spontaneous secretion of saliva, which is stimulated by mi-

---

*In human beings, part of the submandibular and sublingual glands merge to form a submandibular-sublingual complex.

[†]A substance in saliva called *sialin* seems to neutralize tooth-decaying acids formed by bacteria. People with an abundance of sialin appear to get fewer cavities than those with only a small amount. The amount of sialin in saliva is determined genetically.

| *What causes the dryness of "morning mouth"?* | *Ordinarily, a constant flow of saliva flushes out the mouth and keeps the papillae on the tongue short. But salivation is reduced during sleep, and the papillae grow longer, trapping food and bacteria, and producing "morning mouth."* |

nute amounts of acetylcholine released from the parasympathetic nerve endings that terminate in the salivary glands. In addition to this continuous low-level secretion, the presence of food in the mouth and stomach, and the act of chewing (which presses on chemoreceptors and pressoreceptors in the mouth), and even the smell, taste, sight, or thought of food usually stimulate the secretion of saliva (Figure 22.7).

Salivary secretion is entirely under the control of the autonomic nervous system, with no hormonal stimulation. (All other digestive secretions are regulated by both nervous and hormonal control.) The major stimulation of all salivary glands comes from the parasympathetic system. Unpleasant stimuli related to food (such as the smell of rotten fish) inhibit the parasympathetic system and cause the mouth and throat to become dry. In the same way, sympathetic stimulation reduces the amount of saliva. As a result, the mouth feels dry during stressful situations, when the sympathetic system is dominant. The production of saliva from the three major pairs of glands is intermittent and under nervous control, but the many small salivary glands appear to secrete continually.

Parasympathetic nerves controlling the salivary glands originate in the inferior and superior salivary nuclei in the medulla. Sympathetic nerves originate in the lateral horn of the first and second thoracic segments of the spinal cord.

| *Why do dental procedures stimulate salivation?* | *Manipulative activities in the mouth stimulate pressoreceptors that activate salivation by the unconditioned salivary reflex.* |

TABLE 22.3	COMPOSITION AND FUNCTIONS OF SALIVA
Major components	**Functions**
Water (about 99% of total composition of saliva)	Provides solvent for food in order for tasting and digestive reactions to occur. Moistens mouth. Aids speech.
Bicarbonates	Help maintain acidic pH of saliva at 6.35 to 6.85.
Chlorides	Activate salivary amylase, the starch-digesting enzyme.
Immunoglobulin A (IgA)	Part of salivary antibacterial system.
Lysozyme	Bacteria-destroying enzyme; prevents dental decay, infection of mucous membrane.
Mucin	A protein that helps form mucus.
Mucus	Lubricates food into bolus. Aids swallowing. Helps buffer acids and bases.
Phosphates	Help maintain pH of saliva.
Salivary amylase (ptyalin)	Catalyzes the breakdown of polysaccharides into disaccharides.
Urea, uric acid	No digestive function; waste products excreted via saliva.

Different types of stimuli produce different amounts or compositions of saliva. Sympathetic stimulation produces a thick, mucus-rich, viscous secretion, mostly from the submandibular glands, but also from the parotid glands. Parasympathetic stimulation acts on the submandibular glands to produce a profuse, watery secretion that is rich in enzymes.

Ask Yourself

1 What is the difference between digestion and absorption?

2 What are the parts of the alimentary canal?

3 What are the parts of a tooth, and what are they composed of?

4 How do the functions of the hard and soft palates differ?

5 What are the main functions of saliva?

PHARYNX AND ESOPHAGUS

The act of swallowing has a voluntary phase, which occurs in the mouth, and an involuntary phase, which involves the pharynx and esophagus.

Pharynx

Food moves from the mouth into the pharynx. The **pharynx** serves as both an air passage during breathing and a food passage during swallowing. As you saw in Chapter 21, it extends from the base of the skull to the larynx, where it becomes continuous inferiorly with the esophagus and anteriorly with the nasal cavity (Figure 22.8; see also Figure 22.1). Thus, the pharynx can be divided into three parts: (1) the *nasopharynx,* superior to the soft palate, (2) the *oropharynx,* from the soft palate to the epiglottis, and (3) the

---

FIGURE 22.7

Neural pathway for the salivary reflex.

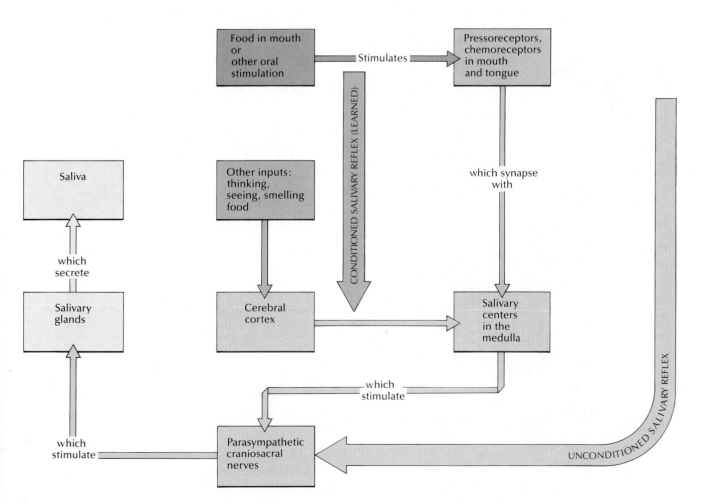

FIGURE 22.8

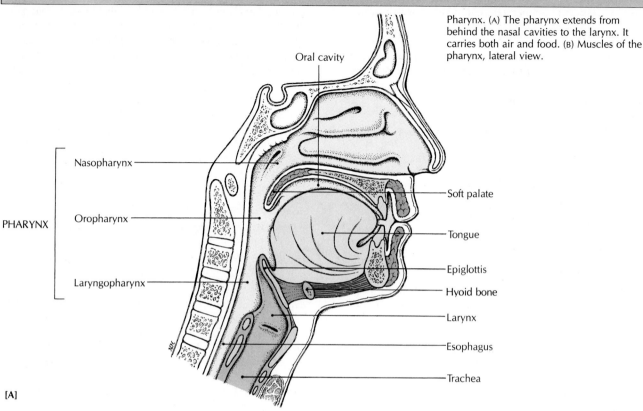

Pharynx. (A) The pharynx extends from behind the nasal cavities to the larynx. It carries both air and food. (B) Muscles of the pharynx, lateral view.

Oral cavity

Soft palate

Tongue

Epiglottis

Hyoid bone

Larynx

Esophagus

Trachea

PHARYNX

Nasopharynx

Oropharynx

Laryngopharynx

[A]

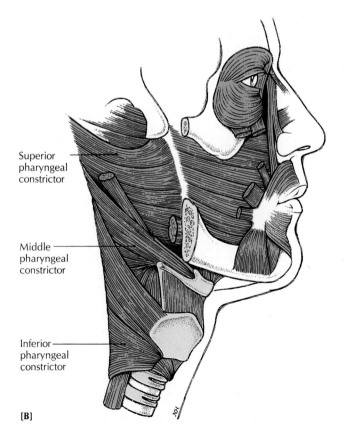

Superior pharyngeal constrictor

Middle pharyngeal constrictor

Inferior pharyngeal constrictor

[B]

laryngopharynx, posterior to the epiglottis, which joins the esophagus.

Since the nasopharynx conveys only air, which does not cause friction, it is lined with pseudostratified ciliated epithelium. But the oropharynx and laryngopharynx are subjected to friction during the passage of food, and they are lined with nonkeratinized stratified squamous epithelium. This tissue layer is moistened by mucus secreted by glands in the underlying connective tissue.

The muscles of the pharynx, called the *pharyngeal constrictors*, are striated (voluntary), and their contraction during swallowing is controlled by the somatic motor system. However, the musculature of the pharynx and the upper half of the esophagus, though striated in structure, is not under voluntary control. For this reason, swallowing is considered to be automatic, not autonomic.

Esophagus

The *esophagus* (ih-SOFF-uh-guss; Gr. gullet) is a muscular, membranous tube, about 25 cm (10 in.) long, through which food passes from the pharynx into the stomach (see Figure 22.1).

The inner *mucosa* of the esophagus is lined with nonkeratinized stratified squamous epithelium arranged in longitudinal folds. Several mucous glands in the mucosa and *submucosa* provide a film of lubricating mucus to ease the passage of food to the stomach (Figure 22.9). The submucosa also contains blood vessels. The middle *muscularis externa* consists wholly of striated voluntary muscle in the upper third of the esophagus, a combination of smooth and striated muscle in the middle third, and wholly smooth muscle in the lower third. The slow contractions of the smooth muscle in this area allow food to pass into the stomach without the force generated by skeletal muscle. The outer fibrous layer is called the *adventitia* (instead of serosa) because it lacks an epithelial layer (see Table 22.1).

The esophagus is located just in front of the vertebral column and behind the trachea. It passes through the lower neck and thorax before penetrating the diaphragm and joining the stomach.

Each end of the esophagus is closed by a sphincter muscle* when the tube is at rest and collapsed. The upper sphincter is the *superior esophageal sphincter*. Closing of this sphincter is caused not by active muscular contraction but rather by the passive elastic tension in the wall of the esophagus when the esophageal muscles are relaxed. The *lower esophageal sphincter* (also called the *gastroesophageal* or *cardiac sphincter*) is a band of smooth muscle that includes the last 4 cm of the esophagus just before it connects to the stomach. The lower sphincter relaxes only long enough to allow food and liquids to pass

---

*A *sphincter*, which comes from the Greek term for "to bind tight," is usually in a state of contraction, like the tightened drawstrings of a purse or a bag of marbles.

FIGURE 22.9

Scanning electron micrograph of a horizontal section of the esophagus, showing the central lumen, inner mucosa, submucosa, muscularis externa, and outer adventitia. [*Richard Kessel and Randy Kardon,* Tissues and Organs: A Text-Atlas of Scanning Electron Microscopy. *San Francisco, W. H. Freeman, 1979.*]

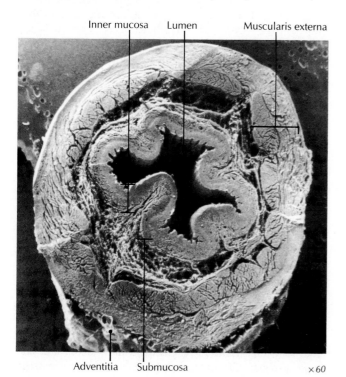

Inner mucosa   Lumen   Muscularis externa

Adventitia   Submucosa   ×60

---

into the stomach. The rest of the time it is contracted to prevent food and hydrochloric acid from being forced back into the esophagus when pressure increases in the abdomen. Such pressure typically increases when the abdominal muscles contract during the breathing cycle, during the late stages of pregnancy, and during the normal stomach contractions during digestion. If the lower esophageal sphincter does not close, the hydrochloric acid in the stomach can be forced up into the lower esophagus. The resultant irritation of the lining of the esophagus is known as *heartburn*, so called because the painful sensation is referred, and appears to be located near the heart.

Swallowing (Deglutition)

When a bit of food in the mouth is moistened and softened by saliva, it is known as a **bolus,** and is then ready to be swallowed. The first stage of swallowing, or **deglutition** (L. *deglutire,* to swallow down), is voluntary. In fact, it is the final voluntary digestive movement until waste products are expelled during defecation. The lips, cheeks, and tongue all help to form the food into a bolus, which is pushed by the tongue against the hard palate and into the pharynx. The voluntary phase of swallowing ends when the bolus touches the entrance to the pharynx. There it stimulates the glossopharyngeal (IX) nerve to trigger impulses in the "swallowing center" of the medulla. The medulla immediately responds by sending a regulated sequence of impulses to the muscles of the pharynx, esophagus, and stomach.

The involuntary stage of swallowing consists of the following three *simultaneous* movements (Figure 22.10):

1 *Voluntary oral phase.* Contractions of the mylohyoid and digastric muscles raise the hyoid bone and the tongue toward the roof (palate) of the mouth. In turn, the intrinsic tongue muscles elevate the tip of the tongue against the upper incisor teeth and maxilla, and then squeeze the bolus toward the pharynx (like toothpaste from a tube).

2 *Pharyngeal phase.* The pharyngeal phase is triggered in the region of the fauces by the stimulation of the glossopharyngeal nerve (gag reflex). The soft palate is raised to prevent the bolus from entering the nasopharynx and nasal cavity. (Once the pharyngeal stage of swallowing is initiated, it is impossible to breathe or speak.) The base of the tongue is then thrust backward (retracted), propelling the bolus into the oropharynx. The muscles at the back of the mouth contract, and together with the elevated and retracted base of the tongue, narrow the opening behind the bolus. The sequential contractions initiate a muscular wave called **peristalsis** (see Figure 22.10B) that propels the bolus along recesses on either side of the larynx on its way to the esophagus. The contractions of the stylohyoid and digastric muscles raise the hyoid bone and draw the larynx under the tongue so that the flap of cartilage called the *epiglottis* is pushed from a vertical to a horizontal position, and the larynx rises to close its opening against the epiglottis. The paired arytenoid cartilages "pivot" so that each vocal fold approaches its mate to close the larynx. This action can prevent the bolus from entering the trachea. Any food that slips into the trachea causes coughing and choking reflexes until the passage is cleared.

FIGURE 22.10

Swallowing consists of four simultaneous movements (A to D) that prevent food from entering the nasal passages, trachea, or larynx, while allowing its passage into the esophagus. (E) Food is conveyed through the esophagus by peristalsis.

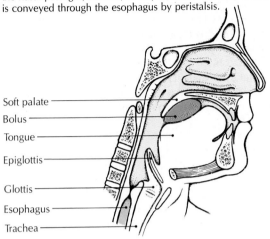

Soft palate
Bolus
Tongue
Epiglottis
Glottis
Esophagus
Trachea

[A]

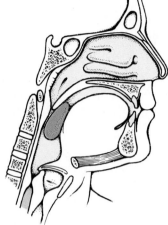

[B]

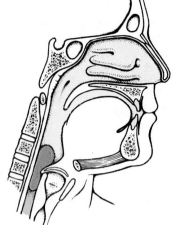

[C]

3 *Involuntary esophageal phase.* In this phase the inferior pharyngeal muscles contract, initiating a wave of peristalsis that propels the bolus through the esophagus and into the stomach in 4 to 10 seconds, or at the rate of 2.5 to 5 cm/sec (1 to 2 in./sec). (Peristalsis continues to assist the passage of food through the rest of the digestive tract.) The relaxation of the palatine, tongue, and pharyngeal muscles results in the opening of the passages between the nasal cavities and the pharynx, and between the oral and pharyngeal cavities. The larynx is drawn down by the contraction of the infrahyoid muscles, and the contraction of the hyoglossus and genioglossus muscles returns the tongue to the floor of the mouth. The epiglottis returns to its vertical position as the hyoid bone is lowered, and the laryngeal passage is opened as the vocal folds separate.

Two types of muscle contractions occur during peristalsis: (1) The diameter of the lumen is increased by the contractions of the longitudinal muscle layer. (2) The bolus is moved forward along the digestive tract by the contraction of the circular muscle layer in front of the widened area (see Figure 22.10c). The pressure of peristalsis is often higher than the typical arterial blood pressure. It is because of such a strong, one-way force that you can swallow while standing on your head, and an astronaut can swallow with little or no difficulty in a zero-gravity situation.*

As much as half a liter of air may be swallowed with a meal, and it is usually released by belching *(eructation)* before it travels farther than the stomach. Air that remains in the stomach or that reaches the small intestine may produce a gurgling sound called *borborygmus.* The sound is caused by the rapid movement of gas and liquid through the intestine. Excess intestinal gas is relieved by the passing of gas (flatus) from the anus long before the feces are ready to be expelled.

*The next time you see a bird taking a drink, notice how it throws its head back with each swallow. This is because it does not make use of peristalsis, and depends solely upon gravity for swallowing. Pigeons are an exception; they do use peristalsis.

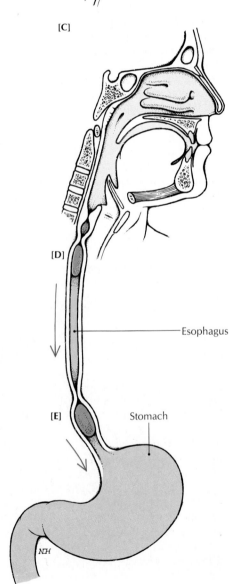

[D]

Esophagus

[E]

Stomach

NH

Ask Yourself

1 *What are the functions of the pharynx and esophagus?*

2 *What are the different regions of the pharynx?*

3 *What is heartburn?*

4 *What are the main stages of swallowing?*

5 *What types of muscular actions occur during peristalsis?*

ABDOMINAL CAVITY AND PERITONEUM

The **abdominal cavity** is the portion of the trunk cavities that lies inferior to the diaphragm (see Figures 22.1 and 1.23). If the pelvic cavity is included, it is referred to as the **abdominopelvic cavity,** with an imaginary plane separating the abdominal and pelvic cavities. The abdominal viscera (organs) include the liver, gallbladder, stomach, pancreas, spleen, kidneys, and small and large intestines. The pelvic viscera include the rectum, urinary bladder, and internal reproductive organs.

Serous membranes line the closed abdominal cavity and the viscera contained within the cavity. A serous membrane consists of a smooth sheet of simple squamous epithelium (called *mesothelium*) and an adhering layer of loose connective tissue containing capillaries. The serous membrane of the abdominal cavity is the **peritoneum** (per-uh-tuh-NEE-uhm). The *parietal peritoneum* lines the abdominal cavity, and the *visceral peritoneum* covers most of the organs in the cavity (Figure 22.11).

Between the parietal and visceral membranes is a space called the *peritoneal cavity* or *coelom.* It usually contains a small amount of *serous fluid,* secreted by the peritoneum and allowing for the nearly frictionless movement of the abdominal organs and their membranes.

Abdominal organs that lie posterior to the peritoneal cavity and are covered, but not surrounded, by peritoneum are called *retroperitoneal.* Retroperitoneal structures include the pancreas, most of the duodenum, the abdominal aorta, inferior vena cava, ascending and descending colons of the large intestine, and the kidneys. Retroperitoneal organs are more or less fixed in place.

Other organs in the abdominal cavity are suspended from the posterior abdominal wall by two fused layers of serous membrane called mesenteries or visceral ligaments (Figure 22.12). These are the *intraperitoneal organs,* and include the liver, stomach, spleen, and most of the small intestine. Intraperitoneal organs are not fixed firmly in position. In addition to providing a point of attachment for organs, the mesenteries carry the major arteries, veins, and nerves of the digestive tract, liver, pancreas, and spleen. The mesenteries may contain many lamellated (Pacinian) corpuscles, which function as mechanoreceptors that provide information about deform-

ing forces acting on the mesentery itself in addition to forces of gastrointestinal motility.

The *coelom* can be considered a *peritoneal sac,* subdivided into greater and lesser sacs by the stomach and two special mesenteries, the *greater* and *lesser omenta* (L. "fat skin"; singular, omentum). The greater and lesser sacs contain some fluid and a few cells within their potential spaces, but no organs. The **greater omentum** is an extensive folded membrane, extending from the greater curvature of the stomach to the back wall and down to the pelvic cavity. It contains large quantities of fat. Excess fat in the skin and greater omentum, plus a loss of muscle tone in the abdominal wall, produce the characteristic "pot belly."

The greater omentum hangs down like a "fatty" apron over the abdominal organs, protecting and insulating them

FIGURE 22.11

Female peritoneum (blue) and its relationship to some major structure in the abdominopelvic cavity, sagittal section.

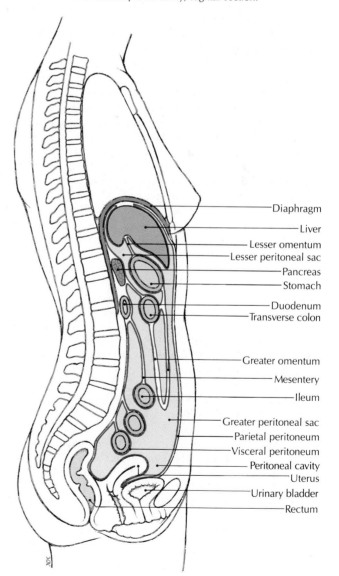

- Diaphragm
- Liver
- Lesser omentum
- Lesser peritoneal sac
- Pancreas
- Stomach
- Duodenum
- Transverse colon
- Greater omentum
- Mesentery
- Ileum
- Greater peritoneal sac
- Parietal peritoneum
- Visceral peritoneum
- Peritoneal cavity
- Uterus
- Urinary bladder
- Rectum

(see Figure 22.12). It also extends down to the pelvic cavity. The greater omentum is also known as the "guardian of the peritoneal cavity" because it is filled with plasma cells, eosinophils, macrophages, and monocytes. Numerous lymph nodes and lymphatic vessels in the peritoneum help to guard against infection, but if the peritoneum does become inflamed, the greater omentum wraps itself around the site of inflammation and walls off the infecting organisms. (See the discussion of *peritonitis* in Physiological and Anatomical Abnormalities at the end of this chapter.) The ***lesser omentum*** extends from the liver to the lesser curvature of the stomach; a small portion extends from the duodenum to the liver.

Ask Yourself

1 *What constitutes the abdominopelvic cavity?*

2 *What is the difference between the parietal and visceral peritoneums?*

3 *What is the function of the mesenteries?*

4 *What is the greater omentum? The lesser omentum?*

5 *What is the function of serous fluid?*

STOMACH

The bolus moves from the esophagus into the ***stomach,*** the saclike portion of the alimentary canal. The stomach is the most expandable part of the digestive tract, and functions to store, mix, and digest ingested nutrients.

Anatomy of the Stomach

The stomach is usually described as a J-shaped sac with a maximum length of 25 cm (10 in.), and a width of 15 cm (6 in.). It is sometimes said to resemble a boxing glove, with the wrist portion facing downward (Figure 22.13A). The average adult stomach has a capacity of about 1.5 L, but there are considerable variations in size and shape. As shown in Figure 22.13, the stomach has a *greater curvature* toward the left side of the body and a *lesser curvature* on the right.

The stomach is located much higher than some people think. Rather than lying behind the navel, it is directly under

**The word *stomach* comes from the Greek word for "throat." The Greek *gaster* for "stomach" gives us the stem *gastero-*, as in gastric.*

FIGURE 22.12

Mesenteries and greater and lesser omenta in the abdominal cavity. (A) Greater omentum, lifted with instruments to show the intestines underneath. (B) Mesentery, with greater omentum deleted for clarity. The intestines are lifted to reveal the mesentery.

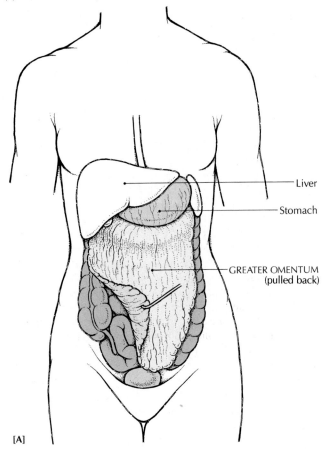

[A]

Liver
Stomach
GREATER OMENTUM (pulled back)

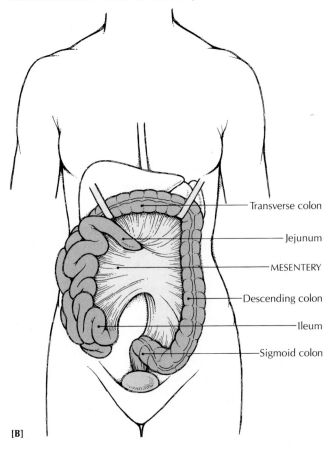

[B]

Transverse colon
Jejunum
MESENTERY
Descending colon
Ileum
Sigmoid colon

FIGURE 22.13

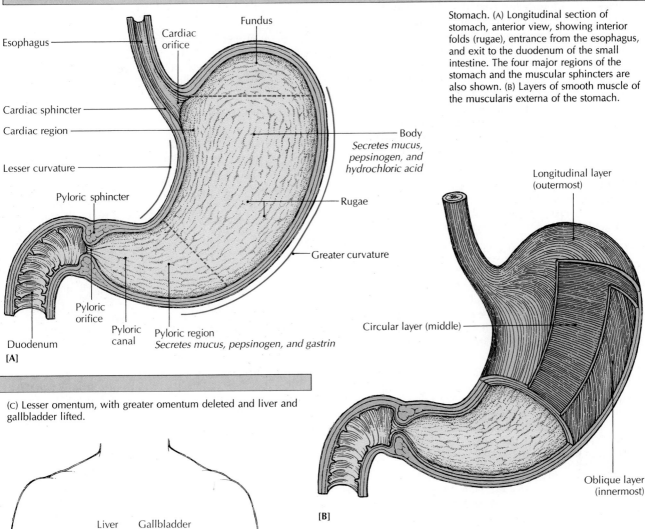

[A]

Esophagus

Cardiac orifice

Cardiac sphincter

Cardiac region

Lesser curvature

Pyloric sphincter

Pyloric orifice

Duodenum

Pyloric canal

Pyloric region
Secretes mucus, pepsinogen, and gastrin

Fundus

Body
Secretes mucus, pepsinogen, and hydrochloric acid

Rugae

Greater curvature

Stomach. (A) Longitudinal section of stomach, anterior view, showing interior folds (rugae), entrance from the esophagus, and exit to the duodenum of the small intestine. The four major regions of the stomach and the muscular sphincters are also shown. (B) Layers of smooth muscle of the muscularis externa of the stomach.

Longitudinal layer (outermost)

Circular layer (middle)

Oblique layer (innermost)

[B]

(C) Lesser omentum, with greater omentum deleted and liver and gallbladder lifted.

Liver Gallbladder

Duodenum

LESSER OMENTUM

Stomach

Transverse colon

Descending colon

Ascending colon

Sigmoid colon

[C]

the dome of the diaphragm, and is protected by the rib cage (see Figure 22.1).

The bolus of food from the esophagus enters the stomach through an opening called the ***cardiac*** (or ***esophageal***) ***orifice,*** so called because it is located near the heart. The cardiac orifice is usually 2 to 3 cm to the left of the median plane. Partially digested nutrients leave the stomach and enter the small intestine through an opening called the ***pyloric orifice*** (or *pylorus*),* usually located 2 to 3 cm to the right of the median plane. The cardiac and pyloric orifices both contain sphincters. The pyloric sphincter is more powerful than the cardiac sphincter, with a substantial amount of circular musculature. The pyloric orifice is usually opened slightly to permit fluids, but not solids, to pass into the duodenum.

The stomach is subdivided into four major regions (see Figure 22.13A): (1) The small ***cardiac region*** is near the cardiac orifice. (2) The ***fundus*** is a small, rounded area above

*The ancient anatomists were not totally without a sense of humor when it came to naming the parts of the body. *Pylorus* means "gatekeeper" in Greek.

the level of the cardiac orifice; it usually contains some swallowed air. (3) The *body* is the large central portion. (4) The *pyloric region* (or *antrum*), which includes the *pyloric canal*, is a narrow portion leading to the pyloric orifice.

The *muscularis externa* of the stomach consists of three layers of smooth-muscle fibers (see Figure 22.13B): (1) The outermost *longitudinal* layer is continuous with the muscles of the esophagus, and is most prominent along the curvatures of the stomach. (2) The middle *circular* layer is wrapped around the body of the stomach, and becomes thickened at the pylorus to form the pyloric sphincter. (3) The innermost *oblique* layer covers the fundus, and runs parallel with the lesser curvature of the stomach along the anterior and posterior walls.

When the stomach is empty, its inner mucous membrane contains branching wrinkles called *rugae* (ROO-jee; L. folds), which gradually flatten as the stomach becomes filled (see Figure 22.13A). The stomach is lined with a layer of simple columnar epithelium, which is indented by about 3.5 million *gastric pits*. Extending from each of these pits are three to eight tubular *gastric glands*.

There are three types of regional gastric glands: the *cardiac* glands near the cardiac orifice, the *pyloric* glands in the pyloric canal, and the *fundic* glands in both the fundus and body of the stomach. Of these three types, the fundic glands are the most numerous.

The six functionally active cell types lining the surface, pits, and glands of the stomach are shown in Figure 22.14 and described in Table 22.4.

Functions of the Stomach

The stomach has three main functions:

1 The stomach *stores ingested nutrients until they can be released into the small intestine* with steady, regulated spurts at a rate that is physiologically appropriate for the relatively slow processes of digestion and absorption to occur. The stomach is well suited for storage because its muscles have little tone, and it can expand up to 4 L if necessary.

2 The stomach *churns ingested nutrients, breaks them up into small particles, and mixes them with gastric juices* to form a soupy liquid mixture called *chyme* (KIME).

3 The stomach *secretes hydrochloric acid and enzymes that initiate the digestion of proteins*.

Digestive movements within the stomach Several types of muscular movements occur within the stomach during digestion, depending upon the location and volume of the chyme:

1 A few minutes after food enters the stomach through the cardiac orifice, slow peristaltic *mixing waves* start in smooth-muscle pacemaker cells in the fundus and body. The pacemaker cells generate action potentials at a rate of three to four per minute that sweep down the stomach toward the pyloric sphincter. This rhythmic rate is known as the *basic electrical rhythm* (BER) of the stomach. It is these rhythmic contractions that produce chyme and push it toward the pyloric region.

2 As the pyloric region begins to fill, *strong peristaltic waves* chop the chyme and propel it through the pyloric canal to-

TABLE 22.4		TYPES OF STOMACH CELLS
Type of cell	**Description**	**Function**
Surface mucous cells	Line the lumen surface of stomach and gastric pits of cardiac, fundic, and pyloric glands. Life span about three days; shed into lumen, replaced from below by undifferentiated cells in gastric pits.	Secrete alkaline mucus that protects stomach mucosa from pepsin and high acidity of gastric fluid; prevents ulceration of mucosa. "Parent" cells of all new cells of gastric mucosa.
Neck mucous cells	Line cardiac, pyloric, and fundic glands.	Secrete a more neutral mucus than that of surface mucous cells. Replace lost surface cells.
Parietal cells (oxyntic cells)	Pale, oval cells, found alongside chief cells in lining of fundic glands.	Secrete hydrochloric acid, which helps convert pepsinogen into pepsin. Also secrete intrinsic factor needed for absorption of vitamin B_{12}.
Chief cells (peptic cells, zymogenic cells)	Large pyramidal cells that line fundic glands.	Secrete the enzyme pepsinogen, a precursor of pepsin, the protein-digesting enzyme.
Enteroendocrine cells (argentaffin cells, enterochromaffin cells, argyrophylic cells)	Located near base of gastric glands. Secretory granules that release secretions into lumen rather than into laminia propria.	Secrete some substances (such as gastrin, secretin, cholecystokinin) that act as true peptide hormones. Secrete other peptides (including motilin) that are termed "candidate hormones," because their exact function is unclear.
Undifferentiated cells	Cells that are not specialized for any one particular function.	Replace other types of glandular cells when they die.

FIGURE 22.14

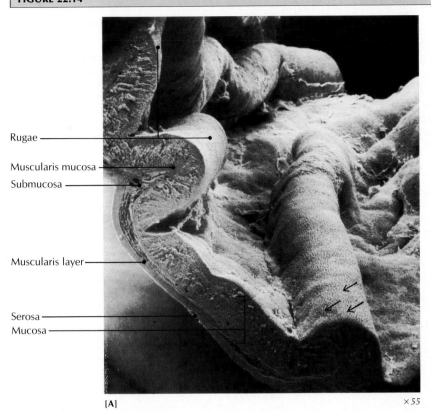

Rugae

Muscularis mucosa

Submucosa

Muscularis layer

Serosa

Mucosa

[A]   ×55

(A) Scanning electron micrograph of the stomach wall, showing the mucosa, which contains most of the gastric pits and glands, the muscularis mucosa, submucosa, muscularis layer, and outer serosa. Rugae can be seen, and the arrows point to the openings into gastric pits. (B) Transverse section through the stomach wall at the fundus. (C) Stomach cells near the body of the stomach at a higher magnification. [(A) *Gene Shih and Richard Kessel. Living Images: Biological Microstructures Revealed by SEM. Boston: Science Books International, 1982.*]

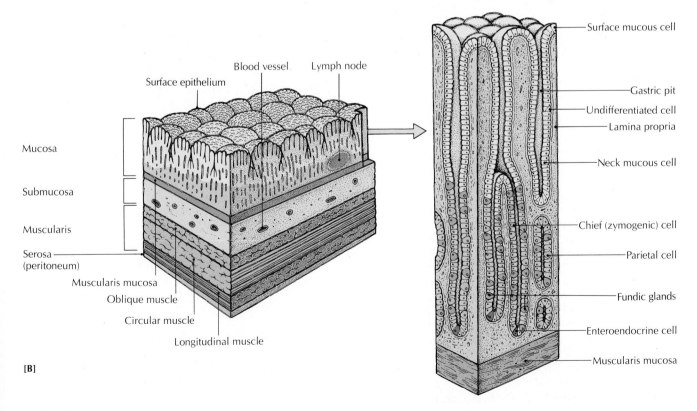

Surface epithelium   Blood vessel   Lymph node

Mucosa

Submucosa

Muscularis

Serosa
(peritoneum)

Muscularis mucosa

Oblique muscle

Circular muscle

Longitudinal muscle

[B]

Surface mucous cell

Gastric pit

Undifferentiated cell

Lamina propria

Neck mucous cell

Chief (zymogenic) cell

Parietal cell

Fundic glands

Enteroendocrine cell

Muscularis mucosa

[C]

FIGURE 22.15

Pathways regulating (A) the stimulation and (B) the inhibition of gastric emptying.

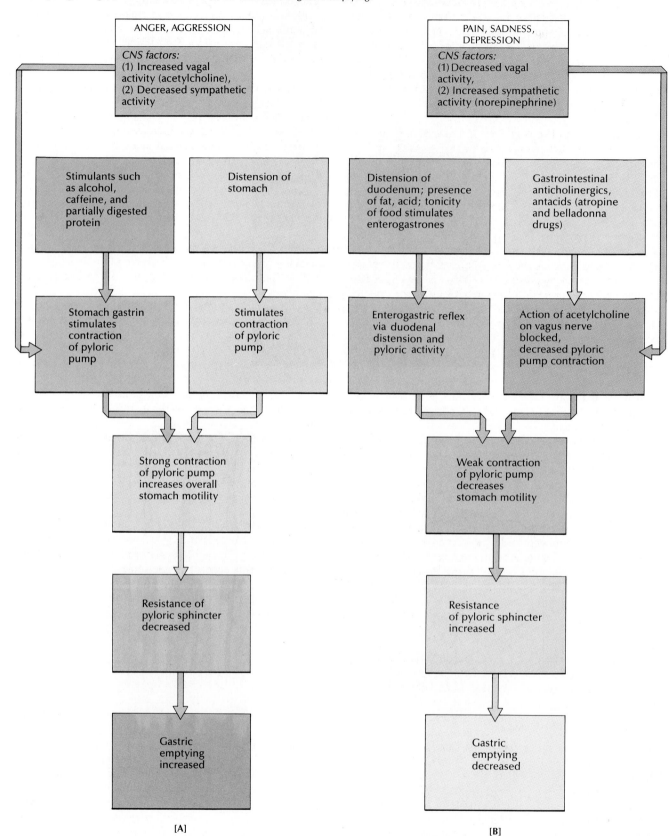

[A] [B]

ward the pyloric orifice. A few milliliters of chyme are pumped through the pyloric orifice with each wave. Because the orifice is very narrow and the pyloric sphincter around it is very strong, most of the chyme is sent back into the pyloric region for further chopping by peristaltic waves. The mechanism that forces chyme past the pyloric sphincter is often called the *pyloric pump.*

3 As the stomach empties, peristaltic waves move farther up the body of the stomach, ensuring that all the chyme is pushed into the pyloric region.

Regulation of gastric emptying The rate of *gastric emptying* (the movement of chyme from the stomach into the duodenum) is monitored by several factors in the stomach, duodenum, and central nervous system (Figure 22.15). Regulation ensures that the stomach does not become too full, and that it does not empty faster than the small intestine can process the incoming chyme.

The *stimulation* of gastric emptying occurs in the stomach. The stomach controls the pyloric pump through a relatively simple feedback system. The fuller the stomach, the stronger is the force of the pyloric pump. Gastric emptying is also stimulated by the presence of partially digested protein, and stimulants such as alcohol and caffeine. Distension of the stomach stimulates the vagus nerve, which strengthens the peristaltic waves and causes the secretion of the hormone *gastrin,* which stimulates stomach motility and further secretions of gastric juices. The pyloric sphincter relaxes, and chyme passes into the duodenum. As a result of this feedback system, we may have the urge to defecate immediately after a large meal. (Infants cannot control this urge, and often *do* defecate after a meal.)

Gastric emptying is *inhibited* primarily by the duodenum, which reacts to fat and acid in chyme with both neural and hormonal responses. The *neural response* is mediated by both the intrinsic nerve plexuses and autonomic nerves, producing reflexes collectively called the *enterogastric reflex,* which decreases stomach motility and gastric secretion. The *hormonal response* inhibits peristaltic contractions and gastric motility by releasing hormones collectively known as *enterogastrones.* The most important of these hormones are *secretin, cholecystokinin (CCK),* and *gastric inhibitory peptide (GIP).* (See Table 22.7 for a description of digestive hormones.)

Secretion of gastric juices *Gastric juice* is a clear, colorless fluid secreted by the stomach mucosa in response to food. Typically, more than 1.5 L of gastric juice are secreted daily. It is composed of *hydrochloric acid, mucus,* and several enzymes, especially *pepsinogen,* a precursor of the active enzyme *pepsin.* Small amounts of gastric *lipase* are secreted in the stomach, and the digestion of fats begins there, though only barely. Gastric juice also contains an *intrinsic factor* that combines with vitamin B_{12} (an *extrinsic factor*) from digested food to form an anti-anemic factor necessary for the formation of red blood cells. (See Table 22.6 for a more complete listing of digestive enzymes.)

The major function of gastric juice is to digest protein. Pepsinogen is synthesized by ribosomes attached to the endoplasmic reticulum of the chief cells, and is stored in secretory vesicles called *zymogen granules.* Pepsinogen is converted by the action of hydrochloric acid into the active protein-splitting enzyme *pepsin.* Pepsin, assisted by hydrochloric acid, breaks down large protein molecules into smaller molecules of peptones, proteoses, and amino acids. Hydrochloric acid also helps to kill bacteria that are swallowed with food. Recent evidence suggests that hydrochloric acid is actively secreted from intracellular canaliculi of stomach parietal cells into the lumen of the gastric pits (Figure 22.16).

Although the stomach wall is composed mainly of protein, it is normally not digested by gastric juices because it is covered with a protective, alkaline coat of mucus that is secreted by the surface epithelial cells and neck mucous cells. A break in the mucus lining often results in a stomach sore, or *gastric ulcer* (see Physiological and Anatomical Abnormalities at the end of the chapter). Also, because pepsin is secreted as the inactive pepsinogen, it cannot digest the cells that produce it. The lining of the stomach sheds about 500,000 cells per minute, and is completely renewed every three days. The stomach cells also appear to contain high levels of prostaglandins, which seem to be linked with the acid-neutralizing carbohydrates produced by the stomach.

The control of gastric juice secretion occurs in three overlapping phases:

1 *Cephalic ("head") phase.* When food is seen, smelled, tasted, chewed, or swallowed, the stomach is stimulated by activity of the vagus nerve, and a small amount of gastric juice is secreted. Vagal stimulation also causes the enteroendocrine cells of the pyloric gland area to release gastrin. Gastrin is carried by the blood back to the parietal chief cells to further stimulate the secretion of gastric juice. Gastrin is the strongest known stimulator of hydrochloric-acid secretion.

2 *Gastric ("belly") phase.* The stimuli involved in this phase are fragments of protein and peptide, distension of the stomach, and stimulants such as alcohol and caffeine, all of which increase gastric secretion by way of the intrinsic nerve plexuses in the stomach wall, the extrinsic vagal nerve, and the hormone gastrin.

3 *Intestinal phase.* The intestinal phase consists of an *excitatory component* followed by a dominant *inhibitory component.* During the excitatory component, the presence of protein in the duodenum releases intestinal gastrin, which stimulates further gastric secretion. The inhibitory component is activated by the accumulation of acidic chyme in the duodenum, which causes the production of CCK, secretin, and GIP, which are carried by the blood to the stomach, where they turn off the chief and parietal cells.

What are "hunger pangs"?	*When the stomach is empty, peristaltic waves cease, but after about 10 hours of fasting, new waves may occur in the pyloric region of the stomach. These waves can cause "hunger pangs" (or hunger pains) as sensory vagal fibers carry impulses to the brain.*

FIGURE 22.16

Secretion of hydrochloric acid (HCl) by a stomach parietal cell. The chloride ion (Cl$^-$) is actively transported from the plasma into the gastric pit lumen by the parietal cell. The hydrogen ion (H$^+$) that is secreted is not transported from the plasma but is derived from specific metabolic processes within the parietal cell. Since parietal cells contain a high concentration of the enzyme carbonic anhydrase, water (H$_2$O) readily combines with carbon dioxide (CO$_2$) to form carbonic acid (H$_2$CO$_3$), which rapidly dissociates into hydrogen (H$^+$) and bicarbonate (HCO$_3^-$) ions. Water also dissociates into hydrogen ions (H$^+$) and hydroxyl ions (OH$^-$). Bicarbonate passes into the plasma and the hydrogen ion into the gastric pit lumen by separate active-transport pumps. There the hydrogen and chloride ions combine to form hydrochloric acid.

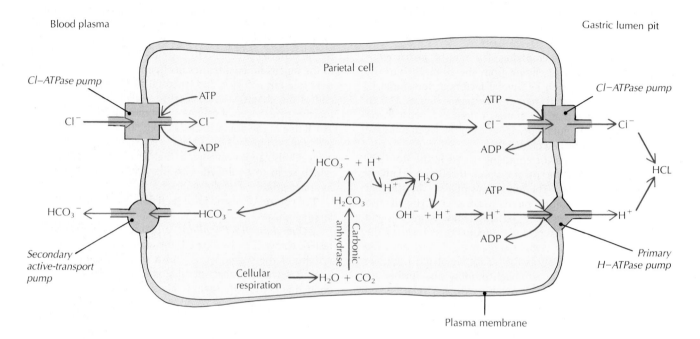

Ask Yourself

1 *What are the main parts of the stomach?*

2 *What are the six types of stomach cells?*

3 *What are the functions of the stomach?*

4 *How do the muscular layers of the stomach contribute to peristalsis?*

5 *What are the components of gastric juice?*

6 *What is the purpose of pepsin in the stomach?*

7 *What are the three stages in the control of the secretion of gastric juices?*

SMALL INTESTINE

After one to three hours in the stomach, the chyme moves into the *small intestine,* where further contractions continue to mix it. It takes 1 to 6 hours for chyme to move through the 6-m (20-ft) long small intestine (Figure 22.17). It is here that digestion of carbohydrates and proteins is completed, and where the digestion of most fats occurs. The small intes-

tine absorbs practically all of the digested molecules of food into the bloodstream and lymphatic system. The only exception is alcohol, which is absorbed by the stomach. Water passes through the stomach almost immediately, but remains longer in the small intestine, where much of it is absorbed.

Anatomy of the Small Intestine

The small intestine, like the large intestine, lies within the abdominopelvic cavity.* It is subdivided into three parts, the duodenum, jejunum, and ileum. Most of the remaining digestive processes take place in the duodenum, and most of the absorption of nutrients into the bloodstream and lymphatic system occur in the duodenum and jejunum.

Duodenum The *duodenum* (doo-oh-DEE-nuhm)† is the C-shaped initial segment of the small intestine (see Figure 22.17). It is about 25 cm (10 in.) long, and is the shortest of the three parts of the small intestine.

---

*The small intestine and large intestine are named for their relative diameters (about 4 cm and 6 cm, respectively), not their lengths. The small intestine is about 6 m (20 ft) long, and the large intestine is about 1.5 m (5 ft) long.

†*Duodenum* comes from the Greek word meaning "twelve fingers wide," referring to its length, and from the Latin *duodecim,* meaning "twelve."

The duodenum itself has four parts. The first, or *superior,* part is about 5 cm (2 in.) long. It is the most common site of duodenal (peptic) ulcers. The second, or *descending,* part is between 7 and 8 cm (3 in.) long. The third, or *horizontal,* part is about 10 cm (4 in.) long. It crosses the third lumbar vertebra, and connects to the *fourth* part, about 2.5 cm (1 in.) long, as it joins the jejunum.

Jejunum and ileum The *jejunum* (jeh-JOO-nuhm),* between the duodenum and ileum, is about 2.5 m (8 ft) long. The *ileum* (ILL-ee-uhm) extends from the jejunum to the cecum, the first part of the large intestine (see Figure 22.17). It is about 3.5 m (12 ft) long. Both the jejunum and ileum are suspended from the posterior abdominal wall by the mesentery.

The ileum joins the cecum at the *ileocecal valve,* a sphincter that ordinarily remains constricted, regulating the entrance of chyme into the large intestine and preventing the contents of the cecum from flowing back into the ileum. After a meal, the gastroileal reflex increases peristaltic activity in the ileum, and some of the chyme is propelled into the large intestine.

*Jejunum comes from a Latin word meaning "fasting intestine," so named because it was always found empty when a corpse was dissected.

Adaptations of the mucosa of the small intestine The wall of the small intestine is composed of the same four layers as the rest of the digestive tract (see Figure 22.2 and Table 22.1). However, the mucosa has three distinctive features that enhance the digestion and absorption processes that take place in the small intestine—plicae circulares, villi, and glands that secrete intestinal juice:

1 *Plicae circulares* (PLY-see sir-cue-LAR-eez) are circular folds that increase the surface area available for absorption. Unlike the rugae of the stomach, the plicae circulares are permanent, and do not disappear when the mucosa is distended. They are most abundant near the junction of the duodenum and jejunum.

2 The absorptive surface area of the mucosa is increased by millions of fingerlike protrusions called *villi* (VILL-eye; L. "shaggy hairs"), which look like the velvety pile of a rug (Figures 22.18A, B). The surface area is further increased by the infolding of the epithelium between the bases of the villi, forming tubular *intestinal glands (crypts of Lieberkühn)* (see Figures 22.18C, D). Each villus contains blood capillaries and a lymph vessel called a *lacteal* (see Figure 22.18D). Lacteals are important in the accumulation and transportation of lipids.

The villi are very effective structures. Only about 5 percent

FIGURE 22.17

Small and large intestines. The small intestine begins at the pyloric orifice, where it joins the stomach, and ends at the ileocecal valve, where it joins the large intestine.

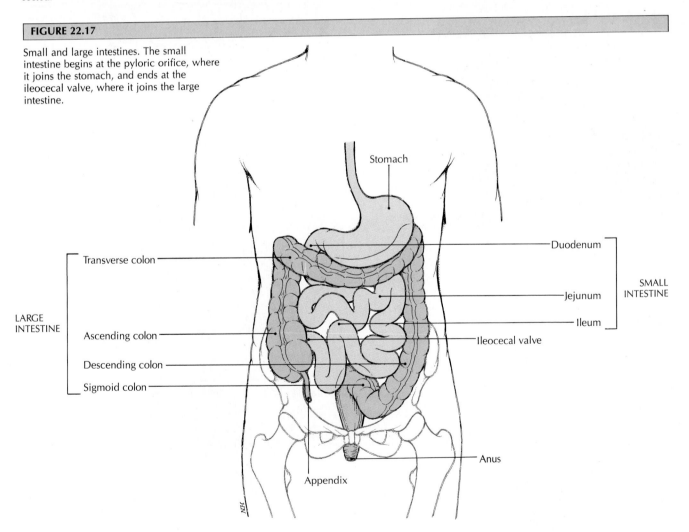

FIGURE 22.18

Wall and lining of the small intestine.
(A) Scanning electron micrograph of a
section of the small intestine, showing the
numerous villi lining the walls, lumen,
mucosa, and submucosa. (B) Lining of the
small intestine, showing plicae circulares
and villi. (C) Enlarged portion of transverse
section of the wall of the duodenum.
(D) Greatly enlarged drawing of the internal
structure of villus. Microvilli are shown
extending from the villus to form a brush
border. [(A) *Gene Shih and Richard Kessel.*
Living Images: Biological Microstructures
Revealed by SEM. *Boston: Science Books
International, 1982.*]

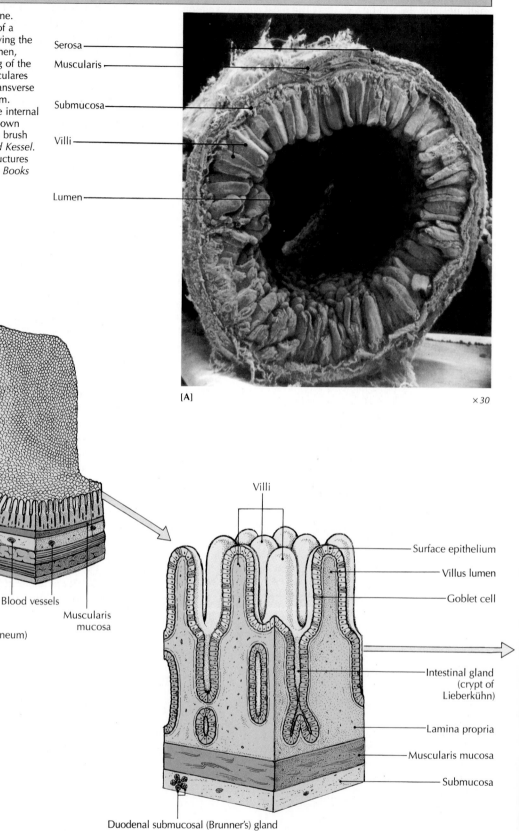

Serosa

Muscularis

Submucosa

Villi

Lumen

[A]

× 30

Plicae circulares

Villi

Submucosa

Circular muscle

Longitudinal muscle

Serosa (peritoneum)

Blood vessels

Muscularis mucosa

[B]

Villi

Surface epithelium

Villus lumen

Goblet cell

Intestinal gland (crypt of Lieberkühn)

Lamina propria

Muscularis mucosa

Submucosa

Duodenal submucosal (Brunner's) gland

[C]

of fats and 10 percent of proteins are not absorbed from the small intestine. In addition to increasing the surface area of the mucosa, the villi aid absorption by adding their constant waving motion to the peristaltic movement already occurring throughout the small intestine. The thousands of epithelial cells making up each villus are the units through which absorption from the small intestine takes place. The absorptive surface of each of these cells contains thousands of *microvilli*, which further increase the surface area of absorption by about 30-fold (see Figure 22.18D). The folded mucosa, villi, and microvilli of the small intestine increase the absorptive surface about 600 times more than a smooth-lined intestine would be.

3 Within the mucosa are simple, tubular glands that secrete several enzymes that aid the digestion of carbohydrates, proteins, and fats (see p. 704 and Table 22.9). The mucosal glands (intestinal glands or crypts of Lieberkühn) reach into the lamina propria (Figures 22.18C, D). Glands that reach into the submucosa, called *duodenal submucosal glands (Brunner's glands)*, are found only in the duodenum. Their viscous, alkaline mucus secretion presumably acts as a lubricating barrier to protect the mucosa from the acidic chyme from the stomach. They may also secrete a protein-digesting enzyme.

The lamina propria of the intestinal mucosa in the ileum has regions where lymphoid tissue is concentrated. Large lymphatic nodules form round or oval patches called *aggregated lymph nodules (Peyer's patches)*, which help to destroy microorganisms absorbed from the small intestine.

Cell types in the small intestine The epithelium of the intestinal mucosa is simple columnar epithelium. It contains several types of cells, including (1) columnar absorptive cells, (2) undifferentiated columnar cells, (3) mucous goblet cells, (4) Paneth cells, and (5) enteroendocrine cells.

Columnar absorptive cells are involved in the absorption of carbohydrates, proteins, and fats from the lumen of the small intestine. They also produce enzymes for the terminal digestion of carbohydrates and proteins (see p. 704).

Undifferentiated cells are found in the depths of the crypts. The entire epithelial surface of the small intestine is replaced about every five days. Undifferentiated cells at the base of the villi divide and become differentiated, and migrate upward to replace the other cell types as needed. As replaced cells disintegrate into the lumen of the intestine, they discharge digestive enzymes into the lumen.

The *mucous goblet cells* appear in the depths of the crypts and migrate upward, secreting and accumulating mucus until they swell into the bulbous shape of a goblet. After these cells release their mucus into the lumen, they die. Mucous goblet cells are especially abundant in the duodenum, where mucus protects the tissue from the high acidity of the chyme entering from the stomach. Because the mucous glands are inhibited by sympathetic stimulation, the duodenum, especially the superior portion, is particularly susceptible to peptic ulcers caused by nervous stress.

Paneth cells lie deep within the intestinal crypts. They contain zinc, and are thought to secrete enzymes (peptidases) for the digestion of proteins. They may also secrete lysozyme, an enzyme that destroys certain microorganisms.

Enteroendocrine cells (formerly called argentaffin, argyrophylic, or enterochromaffin cells) are found not only in the small intestine, but also in the stomach, large intestine, appendix, ducts of the pancreas and liver, and even respiratory passages. They are probably responsible for the synthesis of more than 20 gastrointestinal hormones, including gastric inhibitory peptide, cholecystokinin, and secretin.

Functions of the Small Intestine

The major functions of the small intestine, digestion and absorption, are made possible by movements of the intestinal muscles, and by the chemical action of enzymes. These enzymes are secreted not only by the small intestine itself, but also by the pancreas.

Digestive movements of the small intestine Rhythmic movements of circular and longitudinal muscles in the small intestine cause the chyme to be mixed, chopped, and ultimately pushed into the large intestine. The mixing and chopping movements reduce large particles of chyme to smaller ones, exposing as much of the chyme surface to digestive enzymes as possible. The two main types of muscular activity in the small intestine are *segmenting* and *peristaltic contractions*.

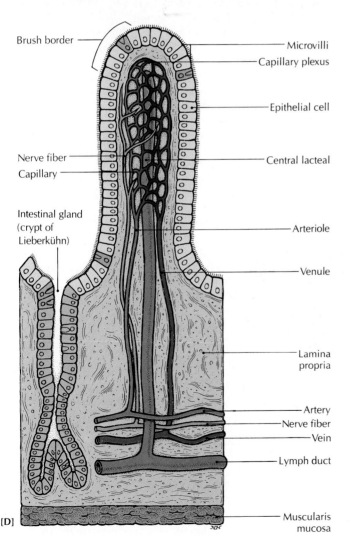

Brush border — Microvilli — Capillary plexus — Epithelial cell — Nerve fiber — Central lacteal — Capillary — Intestinal gland (crypt of Lieberkühn) — Arteriole — Venule — Lamina propria — Artery — Nerve fiber — Vein — Lymph duct — Muscularis mucosa

[D]

FIGURE 22.19

Segmenting contractions in the small intestine result from sharp contractions of the circular muscle.

Contracting circular smooth muscle

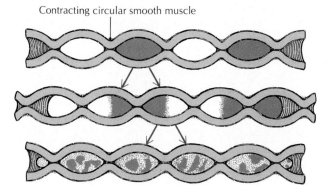

Direction of chyme bolus movement

Segmenting contractions divide the intestine into segments by sharp contractions of areas of the circular smooth muscle in the intestinal wall (Figure 22.19). The areas between the contracted segments contain the bolus of chyme, and are relaxed. Segmenting contractions are controlled by pacemaker cells in the longitudinal smooth muscle layer of the duodenum, which respond to the distension of the intestinal walls. The activated pacemaker cells send slow-wave electrochemical impulses through the longitudinal muscles from the duodenum at 11 cycles per minute, gradually decreasing to 8 cycles per minute at the ileum.

The duodenum and ileum start to segment simultaneously when chyme first enters the small intestine. Duodenal segmentation is caused by distension. Segmentation in the ileum is brought about by gastrin, which is secreted in response to chyme in the stomach. This latter control mechanism is known as the ***gastroileal reflex.*** Segmental activity is enhanced by parasympathetic stimulation and depressed by sympathetic stimulation.

Peristaltic contractions (propulsive contractions), or the *migrating motility complex,* propel chyme through the small intestine and into the large intestine with weak, repetitive waves that start at the stomach and move short distances down the intestine before dying out. Each wave takes 100 to 150 minutes to migrate from its initiation site to the end of the small intestine. After the peristaltic contractions reach the end of the small intestine, new waves replace them at the initiation site. It is believed that the hormone *motilin* regulates peristaltic contractions.

Digestive enzymes in the small intestine The exocrine glands in the mucosa of the small intestine secrete about 1.5 L of a dilute salt and mucus solution into the lumen daily. This solution *contains no enzymes,* and is called the *succus entericus,* or *intestinal juice.* it provides the necessary water in which most of the chemical reactions of the digestive process eventually takes place. Although no enzymes are secreted into the intestinal juice, the small intestine does synthesize digestive enzymes. These enzymes act within the cells or on

the borders of the epithelial cells lining the lumen, rather than actually being secreted into the lumen.

Digestion within the lumen of the small intestine is accomplished by enzymes secreted from the pancreas, with the digestion of lipids being enhanced by the release of bile (secreted by the liver) from the gallbladder. The pancreatic enzymes and bile both enter the small intestine through the common bile duct (see Figure 22.21B). The various salts in bile break down lipid droplets into particles small enough to be attacked by enzymes. The physical breakdown of lipid particles is called *emulsification* (the same process by which soap breaks down grease).

The bile and pancreatic juices help to change the chemical environment in the duodenum and entire small intestine from acidic to alkaline. The new alkaline environment stops the action of pepsin, and allows the intestinal enzymes to function properly. These enzymes initiate the breakdown of carbohydrates, proteins, and lipids. Carbohydrates are broken down into disaccharides and monosaccharides, proteins are broken down into small peptide fragments and some amino acids, and lipids are completely broken down into their absorbable units (glycerol and free fatty acids). The digestion of lipids is now completed, but carbohydrates and proteins need further digestive action.

The microvilli of the small intestinal epithelial cells (see Figure 22.18) have special actin-stiffened hairlike projections called the *brush border,* which contain three categories of enzymes:

1 *Enterokinase* converts the pancreatic enzyme trypsinogen into active trypsin, which, along with several other enzymes, completes the breakdown of peptides into their amino-acid components.

2 The ***disaccharidases***—sucrase, maltase, and lactase—complete carbohydrate digestion by converting the remaining disaccharides into simple monosaccharides, as follows: (a) sucrase converts sucrose into glucose and fructose, (b) maltase converts maltose into glucose, and (c) lactase converts lactose into glucose and galactose.

3 The ***aminopeptidases*** aid enterokinase by breaking down peptide fragments into amino acids.

The digestion of carbohydrates and proteins is now completed, and the products are ready to be absorbed from the small intestine into the bloodstream and lymphatic system. Figure 22.20 summarizes the digestion of lipids, proteins, and carbohydrates.

Absorption from the Small Intestine

All the products of carbohydrate, lipid, and protein digestion, as well as most of the ingested electrolytes, vitamins, and water, are normally absorbed by the small intestine, with most absorption occurring in the duodenum and jejunum. The ileum is the primary site for the absorption of bile salts and vitamin B_{12}.

Carbohydrates Monosaccharides (glucose, fructose, galactose) are readily absorbed through the microvilli on the border of columnar absorptive cells. From there they pass into the capillary network within the villi, and then into the blood

FIGURE 22.20

Compartmentalized digestion; portions of the digestive system are shown out of realistic proportion for the sake of clarity. The change from large to small letters P, C, and L indicates the breakdown of proteins, carbohydrates, and lipids into smaller particles. The letter L is still large in the stomach because the digestion of lipids scarcely begins there. Note also how long the food remains in each part of the tract. See Table 22.6 for a complete description of enzymatic action.

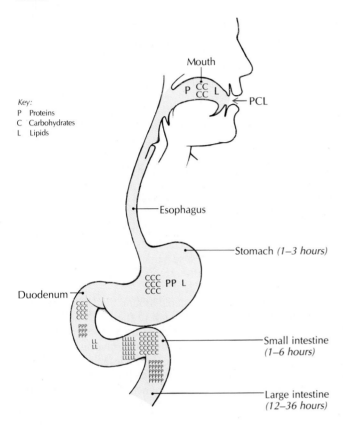

4 Bile salts also cause fatty acids, phospholipids, and glycerides to combine to form submicroscopic, *water-soluble* particles called **micelles** (my-SEHLZ; L. *mica*, grain).

5 Micelles, unlike lipids, can be absorbed into the villi of the absorptive cells.

6 Once the breakdown products of lipids pass through the membranes of the villi and are inside the endoplasmic reticulum of the cells, they are resynthesized into triglycerides.

7 The triglycerides, together with small amounts of phospholipids, cholesterol, and free fatty acids, are packaged into tiny, protein-coated droplets called **chylomicrons** (Gr. *chylos*, juice + *micros*, small).

8 The chylomicrons are released through the cell membranes and move into the central lacteal* of each villus (see Figure 22.17).

9 From the lacteal, the lipids move into larger lymphatic vessels, which eventually drain into the large main thoracic duct that empties into the left subclavian vein.

Thus, the actual transfer of monoglycerides, glycerol, and free fatty acids from the chyme, across the plasma membranes of the intestinal epithelial cells, is a passive process.

Water   Nearly all of the 5 to 10 L of water that enters the small intestine during a day is absorbed into the bloodstream, most of it from the duodenum. The remaining half liter or so enters the large intestine. Water absorption in the small intestine is largely dependent on the active transport of sodium ions, and to a lesser extent, on the osmotic gradient created by the end products of digestion. Sodium is actively transported into the lateral spaces between the epithelial cells of the villi, creating an area of high sodium concentration, and as a result, high osmotic pressure. This high-pressure area induces the movement of water from the intestinal lumen, through the plasma membrane of the epithelial cells, and through the cytoplasm of the cells (and also through the leaky tight junctions) into the lateral spaces. When water enters these spaces, it raises the fluid hydrostatic pressure, forcing water out of the spaces into the interior of the villi, where it moves by osmosis into the capillary network.

Other substances   Several other substances, including vitamins, nucleic acids, ions, and trace elements, are absorbed from the small intestine. The fat-soluble vitamins (A, D, E, K) are dissolved by the fat droplets that enter the small intestine from the stomach. They can be absorbed with micelles into the villi of the absorptive cells. The water-soluble vitamins are readily absorbed. Only vitamin B_{12} requires a specialized protein carrier, known as *gastric intrinsic factor*, for absorption in the terminal ileum by the process of endocytosis.

Nucleic acids in food are converted into smaller mononucleotides by the pancreatic enzyme nuclease. The nucleotides are then broken down further into free bases (adenine, guanine, cytosine, thymine, uracil) and monosaccharides, which are absorbed.

The absorption of ions and trace elements from the small intestine is outlined in Table 22.5.

*The *lacteals* received their name, which means "milky," because after a fatty meal they are filled with a milky suspension of chylomicrons.

vessels in the underlying lamina propria. The absorption of glucose and galactose involves an energy- and sodium-dependent active transport system. Fructose is absorbed by facilitated diffusion (passive carrier-mediated transport).

Proteins   After proteins have been broken down, their constituent amino acids are absorbed through columnar absorptive cells into capillary networks within the villi. An active transport mechanism involving sodium, like that used by glucose and galactose, is used for the absorption of amino acids.

Lipids   The absorption of lipids into the central lacteal of the intestinal villus involves a complex sequence of events:

1 Lipids enter the small intestine in the form of large *water-insoluble* triglyceride droplets.

2 Pancreatic lipase starts to break down the triglyceride droplets into free fatty acids, glycerol, and monoglycerides.

3 Bile salts speed the breakdown of triglycerides by emulsifying the lipid droplets into smaller ones, exposing additional surface area.

Ask Yourself

1 *What are the main parts of the small intestine?*

2 *In addition to digestion, what is the main purpose of the small intestine?*

3 *What are the main types of cells in the small intestine?*

4 *How do segmenting contractions differ from peristaltic contractions?*

5 *What are the main enzymes in the small intestine?*

6 *Why is the absorption of fats more complex than the absorption of proteins and carbohydrates?*

TABLE 22.5	ABSORPTION OF MAJOR IONS AND TRACE ELEMENTS BY THE SMALL INTESTINE
Substance	**Method of absorption**
Bicarbonate*	Absorbed by mucosal cells when level in lumen is high; secreted into lumen when blood level is high. Absorption rate also related to sodium absorption rate in order to maintain electrical neutrality.
Calcium	Active transport. Most rapid in duodenum. Rate of absorption varies with dietary intake and plasma levels; pregnancy increases rate of absorption, age decreases absorption. Absorption regulated in part by negative feedback with parathyroid hormone. Vitamin D increases calcium absorption.
Chloride*	Passive diffusion when following gradient path of sodium ions from lumen to blood. Active transport through cell coupled with sodium transport.
Copper	Active transport. Absorbed mostly by upper portion of small intestine. Excess amounts excreted in feces.
Iron	Active transport from lumen into epithelial cells. Absorbed in small intestine and stored briefly before being transferred to plasma in response to metabolic requirements. Most of transported iron is stored in the liver as ferritin, released when plasma levels of iron drop. Slow rate of absorption increased by ascorbic acid (vitamin C).
Phosphate	All portions of small intestine absorb phosphate, both actively and passively.
Potassium*	Passive diffusion from lumen to blood. Active transport across plasma membrane.
Sodium*	Passive diffusion when gradient is from lumen to blood through tight junctions. Active transport through cell, involving two different carrier systems.

*Readily absorbed from the small intestine.

PANCREAS AS A DIGESTIVE ORGAN

When the chyme is emptied into the small intestine, it is mixed with secretions of the pancreas and liver, in addition to the juices secreted by the intestinal cells. In the following sections we will discuss the important digestive roles of these accessory organs, along with the gallbladder, before continuing with a description of the large intestine.

The *pancreas* is a soft, pinkish-gray gland about 12 to 15 cm (5 to 6 in.) long. It lies transversely across the posterior abdominal wall, behind the stomach (Figure 22.21). It is completely retroperitoneal. The *endocrine* function of the pancreas is described in Chapter 16, but Figure 22.21 shows how little clumps of endocrine cells are clearly separated from the digestive exocrine cells.

The pancreas consists of three parts (see Figure 22.21A): (1) The broadest part, or *head*, lies on the right side of the abdominal cavity, and fits into the C-shaped concavity of the duodenum. (2) The main part, or *body*, lies between the head and tail, behind the stomach and in front of the second and third lumbar vertebrae. (3) The narrow *tail* extends to the left all the way to the spleen.

Pancreatic tissue is composed of both exocrine and endocrine cells. The *exocrine* cells form groups of cells called **acini** (ASS-ih-nye; L. grapes, so named because they resemble bunches of grapes), which secrete digestive juices into the small intestine (Figure 22.21C).*

The exocrine cells are clustered around tiny ducts that pass through the pancreas from the tail to the head. Each acinus has a central lumen that connects to the *main pancreatic duct (duct of Wirsung)*, which carries the digestive juices toward the descending part of the duodenum. The *accessory pancreatic duct (duct of Santorini)* empties a small amount of pancreatic juice into the duodenum, about an inch above the main duct. The main pancreatic duct does not empty its contents directly into the duodenum. Instead, it joins the common bile duct, and finally, a duct called the *hepatopancreatic ampulla (ampulla of Vater)*, just before it enters the wall of the duodenum (see Figure 22.21A).

When taste buds detect food in the mouth, they send impulses to the brain, which then alerts the pancreas via the vagus nerve that food is on the way. When the partially digested chyme passes from the stomach into the duodenum, the secretion of pancreatic juices is further stimulated by the intestinal secretion of two hormones, *secretin* and *cholecystokinin* (see Table 22.7).

The digestive pancreatic juice is alkaline. It contains approximately 15 enzymes or precursors of enzymes, which are synthesized by ribosomes attached to the endoplasmic reticulum and Golgi apparatus of the acinar cells, and stored as zymogen granules within the cytoplasm. The acinar cells secrete three types of enzymes capable of digesting all three food categories:

1 *Pancreatic lipase* is most effective when it is activated by bile from the liver. Lipase splits triglycerides (large-molecule fats) into smaller, absorbable glycerol and free fatty acids.

*The pancreas produces about 1.5 L of digestive juices daily, more than any other digestive organ.

2 *Pancreatic amylase* converts polysaccharides into monosaccharides and disaccharides, especially into maltose. Pancreatic amylase is more powerful than salivary amylase, and acts upon uncooked starches as well as cooked ones.

3 *Pancreatic proteolytic enzymes* consist of *trypsinogen*, *chymotrypsinogen*, and *procarboxypeptidase*. Each is secreted in an active form. After trypsinogen is secreted into the duodenal lumen, it is activated by the enzyme *enterokinase*, and

FIGURE 22.21

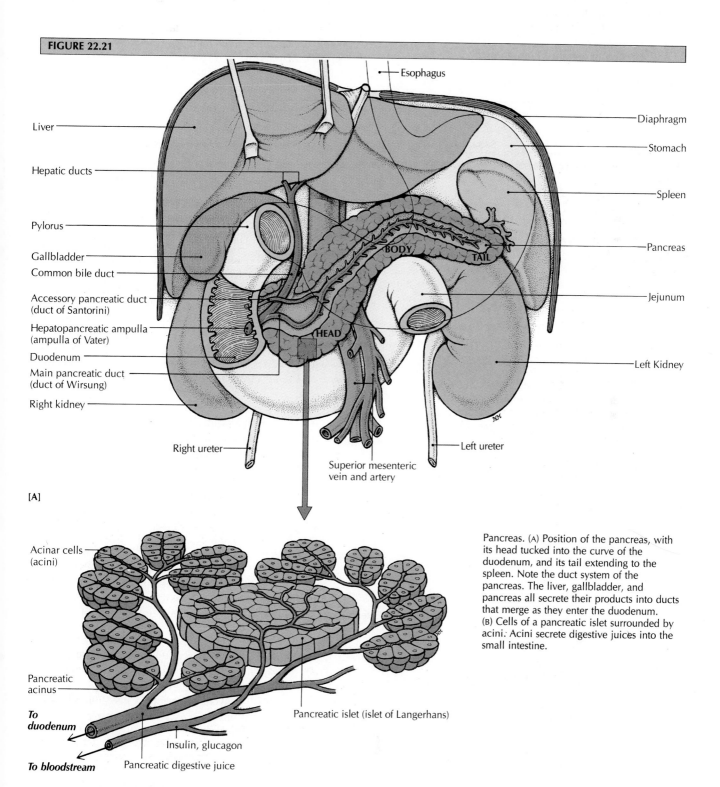

Liver
Hepatic ducts
Pylorus
Gallbladder
Common bile duct
Accessory pancreatic duct (duct of Santorini)
Hepatopancreatic ampulla (ampulla of Vater)
Duodenum
Main pancreatic duct (duct of Wirsung)
Right kidney
Right ureter

Esophagus
Diaphragm
Stomach
Spleen
Pancreas
Jejunum
Left Kidney
Left ureter
Superior mesenteric vein and artery

BODY
TAIL
HEAD

[A]

Acinar cells (acini)
Pancreatic acinus
To duodenum
Insulin, glucagon
To bloodstream
Pancreatic digestive juice
Pancreatic islet (islet of Langerhans)

[B]

Pancreas. (A) Position of the pancreas, with its head tucked into the curve of the duodenum, and its tail extending to the spleen. Note the duct system of the pancreas. The liver, gallbladder, and pancreas all secrete their products into ducts that merge as they enter the duodenum. (B) Cells of a pancreatic islet surrounded by acini. Acini secrete digestive juices into the small intestine.

becomes *trypsin*. Chymotrypsinogen and procarboxypeptidase are converted into their active forms, *chymotrypsin* and *carboxypeptidase*, in the intestinal lumen by the newly formed trypsin. Each of these proteolytic enzymes breaks different peptide linkages, resulting in a mixture of amino acids and small peptides.

See Table 22.6 for a summary of the major glands, secretions, and enzymes in the digestive system, and Table 22.7 for a description of digestive hormones.

Ask Yourself

1 *What is the exocrine function of the pancreas?*

2 *What two hormones stimulate the secretion of pancreatic juices?*

3 *What are the three main groups of pancreatic enzymes?*

TABLE 22.6   MAJOR DIGESTIVE GLANDS, SECRETIONS, AND ENZYMES

Place of digestion	Source	Secretion	Enzyme	pH	Digestive functions of secretion or enzyme
Mouth	Salivary glands.	Saliva.	Salivary amylase (ptyalin).	6–8	Begins carbohydrate digestion; breaks down cooked starch into maltose and dextrin. Inactivated by stomach hydrochloric acid.
	Mucous glands.	Mucus.		6–7	Lubricates.
Esophagus	Mucous glands.	Mucus.		6–7	Lubricates.
Stomach	Gastric glands.	Gastric juice.	Lipase.	7–9	Converts fats into fatty acids and glycerol.
			Pepsin.	1–3	Converts proteins into polypeptides (proteoses and peptones).
			Rennin (found only in children).*	5–6	Converts caseinogen into casein (milk protein), which is acted upon by pepsin.
	Gastric mucosa.	Hydrochloric acid.		0.1	Converts inactive pepsinogen into active pepsin. Dissolves minerals. Kills microorganisms.
	Mucous glands.	Mucus.		5–6	Lubricates.
Small intestine	Liver.	Bile.		7–9	Emulsifies fats. Activates lipase.
	Pancreas.	Pancreatic juice.	Amylase.	6–7	Converts starch into maltose and other disaccharides.
			Chymotrypsin.	7–9	Converts proteins into peptides and amino acids.
			Lipase	7–9	Converts fats into fatty acids and glycerol. Requires presence of bile salts.
			Nuclease.	7–9	Converts nucleic acids into mononucleotides.
			Trypsin.	7–9	Converts proteins into peptides and amino acids. Converts inactive chymotrypsinogen into active chymotrypsin.
	Intestinal glands (crypts of Lieberkühn).	Intestinal juice.	Enterokinase.	7–9	Converts inactive trypsinogen into active trypsin.
			Lactase.	7–9	Converts lactose into glucose and galactose.
			Maltase.	7–9	Converts maltose into glucose.
			Peptidase.	7–9	Converts polypeptides into amino acids.
			Sucrase.	7–9	Converts sucrose into glucose and fructose.
	Mucous glands.	Mucus.		7.5–8	Lubricates.
Large intestine	Mucous glands.	Mucus.		7.5–8	Lubricates.

*Some physiologists believe that rennin is found in the stomachs of young animals such as goats, but not in human beings of *any* age.

TABLE 22.7   MAJOR DIGESTIVE HORMONES

Hormone	Source	Primary stimulus for secretion	Major functions
Cholecystokinin (CCK)	Small intestine (duodenal mucosa).	Amino acids and fatty acids in duodenum.	Stimulates gallbladder to release bile and pancreas to secrete enzymes. Inhibits gastrin-stimulated secretion of acid. Promotes digestion of all foods.
Gastric inhibitory peptide (GIP)	Small intestine (duodenal mucosa).	Acid or peptides in duodenum.	Inhibits secretion of acid by stomach. Helps regulate movement of chyme into duodenum by decreasing gastric motility.
Gastrin	Stomach.	Distension of stomach, vagus nerve, protein in stomach.	Stimulates secretion of acid by oxyntic cells, which stimulates secretion of pepsinogen by chief cells. Increases motility of pyloric region.
Secretin	Small intestine (duodenal mucosa).	Acid or peptides in duodenum.	Stimulates pancreas and liver to secrete juices containing bicarbonate. Inhibits gastrin-stimulated secretion of acid. Inhibits motility of stomach and duodenum.

LIVER AS A DIGESTIVE ORGAN

The *liver** is a large compound, tubular gland, weighing about 1.5 kg (3 lb) in the average adult.[†] Though it lies outside the digestive tract, it has many functions that are relevant to digestion and absorption. It is involved with the "intermediate metabolism" of all types of food substances, modifying nutrients to make them usable by body tissues.

Anatomy of the Liver

The liver is reddish, wedge-shaped, and covered by a network of connective tissue called *Glisson's capsule.* It is located under the diaphragm in the upper region of the abdominal cavity, mostly on the right side. The undersurface faces the stomach, the first part of the duodenum, and the right side of the large intestine (Figure 22.22A; see also Figure 22.1). Although most of the liver is covered by peritoneum, it is primarily held in place, in part by its peritoneal attachments, but mostly by intra-abdominal pressure created by the tonus of the abdominal-wall musculature.

The liver is divided into two main lobes by the *falciform ligament,* a mesentery attached to the anterior midabdominal wall (see Figures 22.22B, C). The **right lobe,** which is about six times larger than the left lobe, is situated over the right kidney and the right colic (hepatic) flexure of the large intestine. The **left lobe** lies over the stomach. In the free border of the falciform ligament, extending from the liver to the umbilicus (navel), is the *ligamentum teres* (round ligament), a fibrous cord that is a remnant of the left fetal umbilical vein (see Figure 19.12).

*The Greek for "liver" is *hepar,* a word used in the combining form *hepato-* in such words as *hepatic* and *hepatitis.*

[†]An infant usually has a pudgy abdomen because of the disproportionately large size of its liver. In most children, the liver occupies about 40 percent of the abdominal cavity, and is responsible for approximately 4 percent of the total body weight. In an adult, the liver represents about 2.5 percent of the total body weight.

The right lobe is further subdivided into a small *quadrate lobe* and a small *caudate lobe* on its ventral (visceral) surface (see Figure 22.22C). The quadrate lobe is flanked by the gallbladder on the right and the ligamentum teres on the left. The quadrate lobe partially envelops and cushions the gallbladder. The caudate lobe is flanked by the inferior vena cava on the right and the *ligamentum venosus* on the left. The ligamentum venosus is a remnant of the fetal ductus venosus, a venous shunt from the umbilical vein to the inferior vena cava (see Figure 19.12).

Vessels of the liver The ***porta hepatis*** ("liver door") is the area through which the blood vessels, nerves, lymphatics, and ducts enter and leave the liver. It is located between the quadrate and caudate lobes, and contains the following vessels and ducts:

1 The *hepatic artery* is a branch of the celiac artery of the aorta. It supplies the liver with oxygenated arterial blood, which represents about 20 percent of the blood flow to the liver.

2 The *hepatic portal vein* drains venous blood into the liver from the entire gastrointestinal tract. It supplies the remaining 80 percent of the liver's blood. This blood contains the nutrients absorbed by the small intestine.

3 The *hepatic veins* receive venous blood from the liver, and drain it into the inferior vena cava.

4 Bile ducts, called **bile canaliculi,** are formed by the **bile capillaries** that unite after collecting bile from the liver cells (see Figure 22.24). Bile is secreted by the liver into the bile canaliculi, which drain into the **right** and **left hepatic ducts.** The ducts, in turn, converge with the **cystic duct** from the gallbladder to form the **common bile duct.** The common bile duct joins the main pancreatic duct, enlarges into the hepatopancreatic ampulla, and then joins the duodenal papilla, which opens into the second part of the duodenum (see Figure 22.25).

Figure 22.23 shows the four main vessels and ducts of the liver, with the hepatic artery and hepatic portal vein entering, and the hepatic vein and common bile duct leaving.

FIGURE 22.22

Liver. (A) Position of the liver. (B) Anterior view, showing right and left lobes. (C) Posteroinferior view, showing caudate and quadrate lobes.

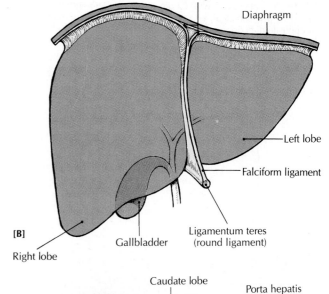

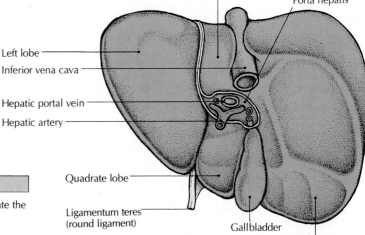

[B]

[C]

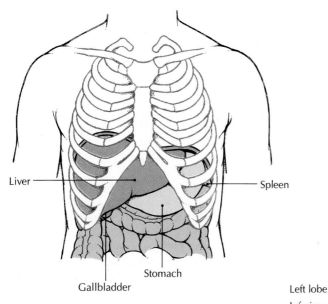

[A]

FIGURE 22.23

Vessels and ducts entering and leaving the liver. Arrows indicate the direction of flow.

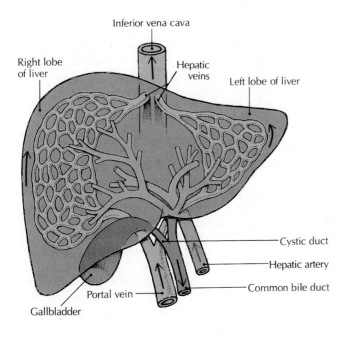

Microscopic anatomy of the liver The functional units of the liver are five- or six-sided *lobules* that contain a branch of the *hepatic vein* (the *central vein*) running longitudinally through it (Figure 22.24). Most lobules are about 1 mm (0.04 in.) in diameter. Liver cells, known as **hepatic cells** or hepatocytes, within the lobules are arranged in one-cell-thick platelike layers that radiate from the central vein to the edge of the lobule. Each corner of the lobule usually contains a *portal area*, a complex composed of branches of the portal vein, hepatic artery, bile duct, and nerve.

Between the radiating rows of cells are delicate blood channels called **sinusoids,** which transport blood from the portal vein and hepatic artery. Blood flows from the artery and vein in the portal areas into the sinusoids, and then to the central vein, which drains it from the lobule. The walls of the sinusoids are lined with endothelial cells. Attached to these lining cells are phagocytic stellate reticuloendothelial cells (Kupffer cells) that engulf and digest worn-out red and white blood cells, microorganisms, and other foreign particles passing through the liver.

FIGURE 22.24

Liver lobules. (A) Cross section, showing typical six-sided shape of lobules. (B) Cutaway view of lobule, showing channels of flow, including bile canaliculi and sinusoids, which convey their contents in opposite directions. (C) Scanning electron micrograph of a liver lobule, showing hepatocytes, or liver cells, radiating toward the central vein. Also visible is the conducting portal vein, which branches to form distributing portal veins. An arterial blood supply surrounds each lobule. [*Courtesy of Dr. M. Muto, Department of Anatomy, Niigata University Medical School.*]

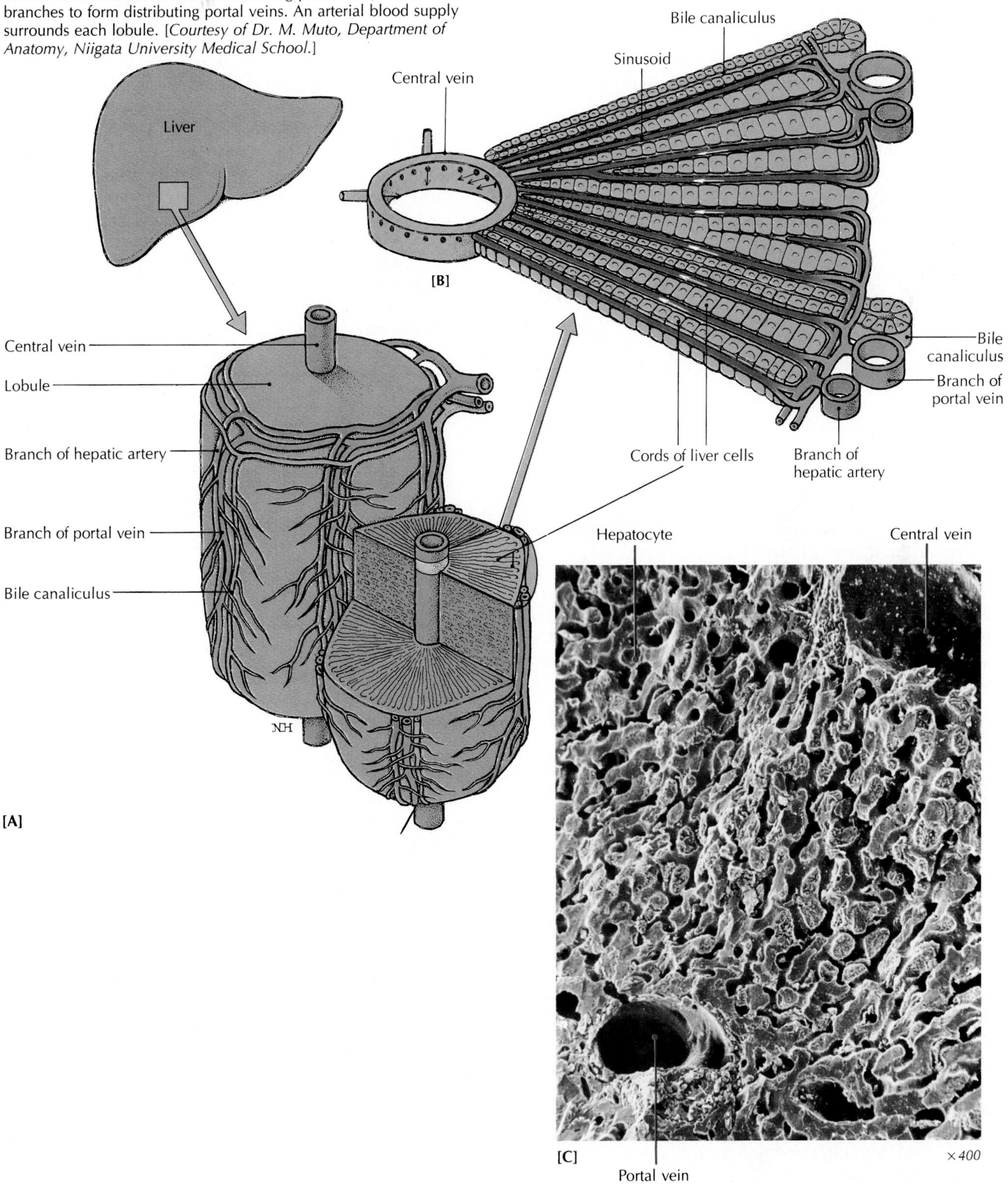

Functions of the Liver

The liver chemically converts the nutrients from food into usable substances and stores them until they are needed. In addition to the metabolic, storage, and secretory functions described here, the liver also is involved in excretion, the subject of Chapter 24.

Metabolic functions of the liver The liver performs so many metabolic functions that only the major ones can be presented here:

1 Removal of amino acids from organic compounds.

2 Formation of urea from worn-out tissue cells (proteins), and conversion of excess amino acids into urea (deamination) to decrease body levels of ammonia.

3 Homeostasis of blood by manufacturing most of the plasma proteins, forming fetal erythrocytes, destroying worn-out erythrocytes, storing the chemical hematin (needed for the maturation of red blood cells), removing bilirubin from the blood, manufacturing heparin, and helping to synthesize the blood-clotting agents prothrombin and fibrinogen from amino acids.

4 Synthesis of certain amino acids, including nonessential amino acids.

5 Conversion of galactose and fructose to glucose.

6 Oxidation of fatty acids.

7 Formation of lipoproteins, cholesterol, and phospholipids (essential constituents of plasma membranes).

8 Conversion of carbohydrates and proteins into fat.

9 Modification of waste products, toxic drugs, and poisons (detoxification). Some drugs, such as carbon tetrachloride, can damage the liver before they can be detoxified sufficiently.

10 Synthesis of vitamin A from carotene.

11 Maintenance of a stable body temperature by raising the temperature of the blood passing through it (its many metabolic activities make the liver the body's major heat producer).

Storage functions of the liver The liver is the body's main storage center. It stores glucose in the form of glycogen, and with the help of enzymes, it converts glycogen back into glucose as it is needed by the body. Because glucose is the body's main energy source, its storage is a particularly important function. The liver also stores the fat-soluble vitamins (A, D, E, and K), minerals such as iron from the diet, and antianemic factor. The liver can also store fats and amino acids, and convert them into usable glucose as required.

Secretion of bile One of the liver's main functions as a digestive organ is to secrete **bile,** an alkaline liquid containing water, sodium bicarbonate, bile salts, bile pigments, cholesterol, mucin, lecithin, and bilirubin. As much as 1 L of bile is secreted by the liver every day. It is stored in the gallbladder in a highly concentrated form until it is needed to break down fats or for any of its many other functions. Bile secretion may be increased by chemical, hormonal (secretin), or neural mechanisms.

Bile salts are derivatives of cholesterol. They are actively secreted into the bile and eventually pass into the duodenum along with other biliary secretions. After they participate in fat digestion and absorption, most of these salts are reabsorbed or taken up into the blood by a special active transport mechanism located in the terminal ileum. From there, bile salts are returned via the hepatic portal system to the liver, which can resecrete them into the bile. This recycling of bile salts—and other biliary secretions—between the small intestine and the liver is called the *enterohepatic circulation* of these substances.

Bile pigments give bile its color. They are derived from the hemoglobin of worn-out red blood cells from all over the body that are transported to the liver for excretion. The color of feces comes from one of the breakdown products of excreted bile pigments. The yellowish color of the skin called *jaundice* is caused by excessive amounts of the bile pigment *bilirubin* in extracellular fluids.

Ask Yourself

1 *Where is the liver located?*

2 *What are the vessels of the liver?*

3 *How does a liver lobule resemble a wheel?*

4 *What are the components of a portal area?*

5 *What are some of the major metabolic functions of the liver?*

6 *How does the liver make glucose available as it is required for energy?*

7 *What are some of the functions of bile?*

GALLBLADDER

The **gallbladder*** is a small, pear-shaped, saclike organ situated in a depression under the right lobe of the liver (Figure 22.25; see also Figure 22.21).

The gallbladder consists of an outer serous peritoneal coat, a middle muscular coat, and an inner mucous membrane that is continuous with the linings of the ducts. The mucous membrane secretes mucin, and absorbs water readily. It does not absorb bile salts or bile pigments, but does transport salt out actively, with water following osmotically. As a result, the bile is concentrated. In fact, the organic constituents of bile leaving the gallbladder may be as much as 10 times more concentrated than they were when bile entered from the liver.

---

*Gallbladder is derived from the Latin word *galbinus,* meaning greenish yellow, which is the usual color of *bile. Gall* is the archaic term for bile.

FIGURE 22.25

Gallbladder and its connecting ducts. Bile secreted from the liver reaches the gallbladder via the cystic duct.

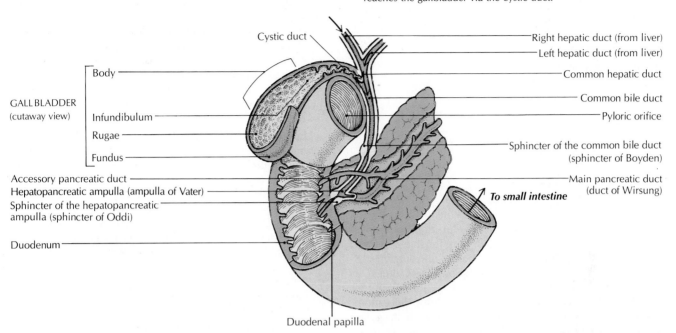

GALL BLADDER (cutaway view)

Body
Infundibulum
Rugae
Fundus

Cystic duct

Right hepatic duct (from liver)
Left hepatic duct (from liver)
Common hepatic duct
Common bile duct
Pyloric orifice
Sphincter of the common bile duct (sphincter of Boyden)
Main pancreatic duct (duct of Wirsung)

To small intestine

Accessory pancreatic duct
Hepatopancreatic ampulla (ampulla of Vater)
Sphincter of the hepatopancreatic ampulla (sphincter of Oddi)

Duodenum

Duodenal papilla

The gallbladder has an important storage function. Bile is secreted by the liver into the duodenum as it is needed for digestion. However, more bile is produced than is ordinarily required, and the excess is stored in the gallbladder until needed in the duodenum. The capacity of the gallbladder is between 30 and 60 mL. When the gallbladder is empty, its mucosa is thrown into folds (rugae), which permit the gallbladder to expand to hold the bile.

The so-called **biliary system** consists of (1) the *gallbladder*, (2) the *left* and *right hepatic ducts*, which come together as the *common hepatic duct*, (3) the *cystic duct*, which extends from the gallbladder, and (4) the *common bile duct*, which is formed by the union of the common hepatic duct and the cystic duct (see Figure 22.25).

The common bile duct and the main pancreatic duct join at an entrance to the duodenum about 10 cm (4 in.) from the pyloric orifice. They fuse to form the *hepatopancreatic ampulla*. The ampulla travels obliquely through the duodenal wall and opens into the duodenum through the *duodenal papilla*. A sphincter located at the outlet of the common bile duct is called the *sphincter of the common bile duct (sphincter of Boyden)*, and the muscle below it, near the duodenal papilla, is the *sphincter of the hepatopancreatic ampulla (sphincter of Oddi)* (see Figure 22.25). The sphincter of the common bile duct appears to be the stronger and more important of the two. About 30 minutes after a meal, or whenever chyme enters the duodenum, the sphincters relax, the gallbladder contracts, and bile stored in the gallbladder is squirted into the duodenum.

The contraction of the gallbladder, the relaxation of the sphincter of the hepatopancreatic ampulla, and the subsequent release of bile take place in the presence of cholecystokinin, a hormone released from the small intestine when fatty acids and amino acids reach the duodenum.

FIGURE 22.26

Section of gallbladder showing many gallstones. [*L. J. Schoenfield/ Visuals Unlimited.*]

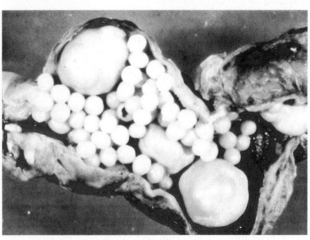

Bile is rich in cholesterol, a rather insoluble fatty substance, and concentrated bile in the gallbladder may form crystals of cholesterol that are commonly called *gallstones*. If these crystals grow large enough to block the cystic duct, they can block the flow of bile and produce a great deal of pain (Figure 22.26).

Is the gallbladder essential for the digestive process?

It is not, and a diseased gallbladder can be removed in a surgical procedure called a cholecystectomy. The hepatic ducts and common bile duct can dilate sufficiently to take over the storage function of the removed gallbladder.

Ask Yourself

1 What is the main function of the gall-bladder?

2 What components make up the biliary system?

3 How does cholecystokin contribute to the function of the gallbladder?

LARGE INTESTINE

The chyme remains in the small intestine for 1 to 6 hours. It then passes in liquid form through the ileocecal valve into the *cecum,* the first part of the large intestine. By now, digestion is complete, and the large intestine functions to remove more water and salts (Na$^+$, K$^+$, Cl$^-$) from the liquid matter. Removal of water, along with the action of microorganisms, converts liquid wastes into *feces* (FEE-seez; L. *faex,* dregs), a semisolid mixture that is stored in the large intestine until ready to be eliminated through the anus during defecation. The undigestible products of digestion remain in the large intestine from 12 to 36 hours.

Anatomy of the Large Intestine

The *large intestine* is the part of the digestive system between the ileocecal orifice and the anus. It consists of the cecum; vermiform appendix; ascending, transverse, descending, and sigmoid colons; and the rectum (Figure 22.27). The entire large intestine (sometimes called the *colon*) forms a rectangle that frames the tightly packed small intestine.

Just on the other side of the ileocecal orifice is the *cecum* (L. blind), a cul-de-sac pouch about 6 cm (2.5 in.) long. Opening into the cecum, about 2 cm below the *ileocecal valve,* is the *vermiform appendix* (L. *vermis,* worm, hence "worm-like"), popularly called the *appendix.* It is the narrowest part of the intestines, and can range in length from 5 to 15 cm (2 to 6 in.). The appendix of some animals is large enough to function as a special site for the prolonged digestion of cellulose, but the human appendix is too short, narrow, and twisted to serve a similar function. Because of these physical characteristics, bacteria and indigestible material may become trapped in the appendix, leading to inflammation (appendicitis) and one of the most common of all surgical procedures, an *appendectomy.* The appendix contains an abundance of lymphoid tissue, and may be involved with the immune system.

What is usually meant by "bowels"?

The bowels (L. botulus, sausage) are usually considered to be the large intestine and rectum, but they may also refer to the digestive tract below the stomach.

The *ascending colon* extends upward from the cecum. Under the liver it makes a right-angle bend known as the *right colic* (or *hepatic*) *flexure* (see Figure 22.27). It continues as the *transverse colon,* which extends across the abdominal cavity from right to left, where it makes a right-angle downward turn at the spleen, known as the *left colic* (or *splenic*) *flexure.*

The *descending colon* extends from the left colic flexure down to the rim of the pelvis, where it becomes the sigmoid colon. The *sigmoid colon* (Gr. *sigma,* the letter S) is usually S-shaped. It travels transversely across the pelvis to the right to the middle of the sacrum, where it continues to the rectum.

External anatomy of the large intestine Besides the differences in diameter and length between the large and small intestines, the large intestine has three distinctive structural differences (see Figure 22.27):

1 In the small intestine, an external longitudinal muscle layer completely surrounds the intestine; but in the large intestine, an incomplete layer of longitudinal muscle forms three separate bands of muscle called *taeniae coli* (TEE-nee-ee KOHL-eye; L. ribbons + Gr. intestine) along the full length of the intestine.

2 Because the taeniae coli are not as long as the large intestine itself, the wall of the intestine becomes puckered with bulges called *haustra.*

3 Fat-filled pouches called *epiploic appendages* are formed at the points where the visceral peritoneum is attached to the taeniae coli in the serous layer.

Microscopic anatomy of the large intestine The microscopic anatomy of the large intestine reflects its primary functions: the reabsorption of any remaining water and some salts, and the accumulation and movement (excretion) of undigested substances as feces. Elimination is aided by the secretion of mucus from the numerous goblet cells in the mucosal layer. Mucus acts as a lubricant and protects the mucosa from the semisolid dehydrated contents.

Because the mucosa of the large intestine does not contain villi or plicae circulares, its smooth absorptive surface is only about 3 percent of the absorptive surface of the small intestine. The lamina propria and submucosa contain lymphatic tissue in the form of many lymphoid nodules. As part of the body's immune system, this *gut-associated lymphoid tissue,* along with epithelial cells, helps defend against the ever-changing mixture of antigens, microorganisms, and potentially harmful substances that enter the digestive tract.

The simple columnar epithelium of the large intestine includes columnar absorptive cells, undifferentiated cells, mucous goblet cells, and enteroendocrine cells. The mucous cells are the most numerous.

Rectum, Anal Canal, and Anus

The terminal segments of the large intestine are the rectum, anal canal, and anus (Figure 22.28). The *rectum* (L. *rectus,* straight) extends about 15 cm (6 in.) from the sigmoid colon to the anus. Despite its name, the rectum is not straight. Three lateral curvatures occur because the longitudinal muscle coat of the rectum is shorter than the other layers of the wall. The

FIGURE 22.27

Large intestine. (A) Anterior view, with part of the cecum and ascending colon removed to show the ileocecal valve. (B) Transverse section of large intestine. (C) Scanning electron micrograph of rat intestine. [*Richard Kessel and Randy Kardon, Tissues and Organs: A Text-Atlas of Scanning Electron Microscopy. San Francisco, W. H. Freeman, 1979.*]

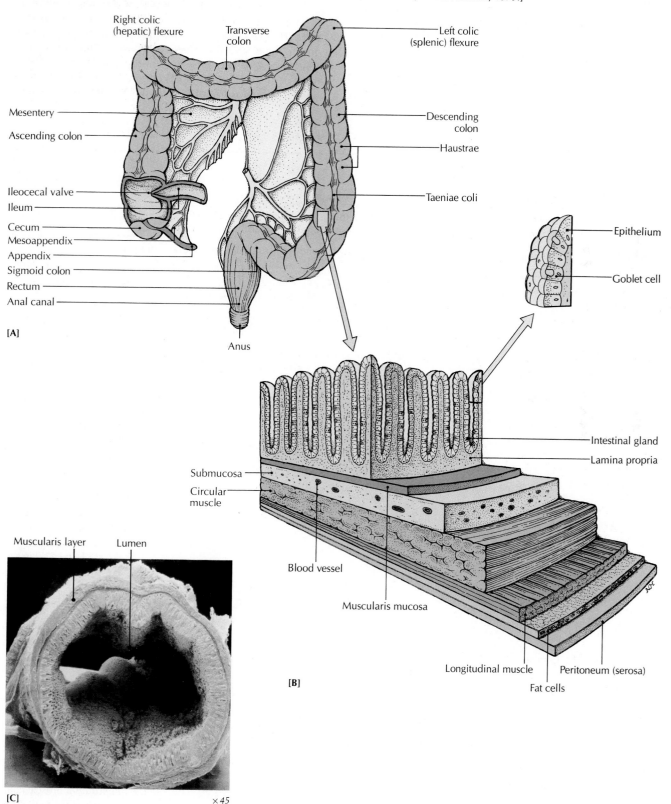

rectum is a retroperitoneal structure, and has no mesentery, appendixes, epiploic appendages, haustra, or taeniae coli. It begins at the third sacral vertebra, and continues along the downward and forward curve of the sacrum and coccyx. At that point, the rectum continues as the **anal canal**, pierces the muscular pelvic diaphragm, and turns sharply downward and backward. The anal canal is compressed by the anal muscles as it progresses for about 4 cm (1.5 in.) and finally opens to the outside as a slit, the **anus** (L. ring).

The anal canal and anus are open only during defecation. At all times they are held closed by an involuntary *internal anal sphincter* of circular smooth muscle and a complex *external anal sphincter* of *voluntary* striated muscle (see Figure 22.28). The external sphincter is under voluntary control.

The mucosa and circular muscle of the muscularis of the rectum form shelves within the tract called **plicae transversales**. The plicae must be avoided when diagnostic and therapeutic tubes and instruments are inserted into the rectum (Figure 22.29). The upper part of the anal canal contains 5 to 10 permanent longitudinal columns of mucous membrane known as **anal** (or **rectal**) **columns**. They are united by folds

TABLE 22.8

SUMMARY OF THE DIGESTIVE SYSTEM

Structure	Major digestive secretions	Major digestive functions
Lips and cheeks	None.	Hold food in position to be chewed. Help identify food textures. Buccal muscles contribute to chewing process.
Teeth	None.	Break food into small pieces. Expose extra surface area for digestive enzymes.
Tongue	None.	Assists chewing action of teeth. Helps shape food into bolus, pushes bolus toward pharynx to be swallowed. Contains taste buds.
Palate	None.	Hard palate helps crush and soften food. Soft palate closes off nasal opening during swallowing.
Salivary glands	Saliva—salivary amylase (ptyalin), mucus, water, various salts.	Salivary amylase (ptyalin) in saliva begins breakdown of cooked starches into soluble sugars maltose and dextrin. Helps form bolus and lubricates it prior to swallowing. Dissolves food for tasting. Moistens mouth. Helps prevent tooth decay.
Pharynx	None.	Continues swallowing activity when bolus enters from mouth, propels bolus from mouth into esophagus.
Esophagus	Mucus.	Propels food from pharynx into stomach.
Stomach	Hydrochloric acid, pepsin, mucus, some lipase, gastrin, intrinsic factor.	Stores, mixes, digests food, especially protein. Regulates even flow of food into small intestine. Helps kill bacteria with acid secretion. Lubricates and protects mucosal lining with mucus secretion.
Small intestine	Enzymes including enterokinase, peptidases, maltase, lactase, sucrase, intestinal amylase, intestinal lipase; salts, water, mucus, hormones—cholecystokinin, gastric inhibitory peptide, secretin.	Site of most chemical digestion and absorption of most water and nutrients into bloodstream.
Pancreas	Enzymes—trypsin, chymotrypsin, carboxypeptidase, pancreatic amylase, pancreatic lipase, nuclease; bicarbonate.	Secretes many digestive enzymes. Neutralizes stomach acid with alkaline bicarbonate secretion.
Liver	Bile, bicarbonate.	Secretes bile. Converts nutrients into usable substances and stores them for later use by body tissues. Converts fats into water-soluble substances. Detoxifies harmful substances. Neutralizes stomach acid with alkaline bicarbonate secretion.
Gallbladder	Mucin.	Stores and concentrates bile from liver and releases it when needed for digestion.
Large intestine	Mucus.	Removes salts and water from undigested food. Releases feces through anus. Aids synthesis of vitamins B_{12} and K through action of intestinal bacteria.
Rectum	None.	Removes solid wastes by process of defecation.

FIGURE 22.28

Longitudinal section of rectum, anal canal, and anus, anterior view.

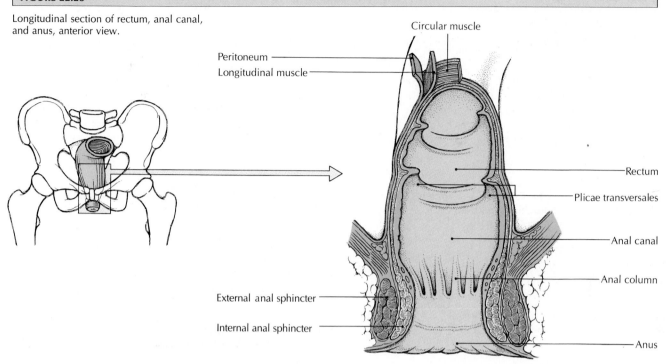

Circular muscle

Peritoneum

Longitudinal muscle

Rectum

Plicae transversales

Anal canal

Anal column

External anal sphincter

Internal anal sphincter

Anus

FIGURE 22.29

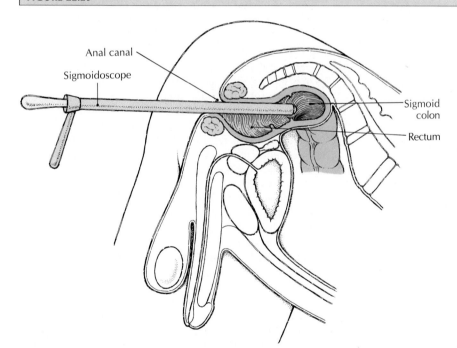

Anal canal

Sigmoidoscope

Sigmoid colon

Rectum

Sigmoidoscopy, a procedure for examining the anal canal, rectum, and sigmoid colon. Such an examination must be done carefully to avoid the delicate plicae transversales in the rectum. In this drawing the sigmoidoscope is shown as an inflexible instrument, but it may also be flexible.

called *anal valves*. The mucosa above the valves is lined by columnar epithelium, with a scattering of goblet cells and crypts. The mucosa and submucosa of the rectum contain a rich network of veins called the *hemorrhoidal plexus*. When these veins become enlarged, twisted, and blood-filled, the condition is called *hemorrhoids*.

Functions of the Large Intestine

By the time chyme reaches the large intestine, digestion is complete, and only some water and salts remain to be absorbed. The amount of chyme that enters the large intestine

TABLE 22.9   SUMMARY OF DIGESTION AND ABSORPTION

Region of digestive tract	Secretion	Food substance acted upon	Digestive activity	Absorptive activity
Mouth	*Saliva:* salivary amylase (ptyalin), mucus, lysozyme.	Proteins.	None.	None.*
		Carbohydrates.	*Salivary amylase* breaks down polysaccharides of cooked starches into disaccharide maltose and dextrin.	None.
		Fats.	None.	None.
Stomach	*Gastric juice:* hydrochloric acid, pepsinogen, lipase, mucus, intrinsic factor.	Proteins.	*Pepsin,* which results from activation of pepsinogen by *hydrochloric acid,* converts protein into polypeptides (proteoses, peptones, and amino acids). Hydrochloric acid acts to soften (denature) protein.	Little or none.†
		Carbohydrates.	Some digestion of carbohydrates continues, but salivary amylase is inactivated by stomach acid.	Little or none.
		Fats.	Gastric lipase hydrolyzes some emulsified fat into fatty acids and glycerol.	Little or none.

daily varies from less than 100 mL to 500 mL, and only about one-third of it is excreted as feces. The rest (mostly water) is absorbed back into the body in the ascending or transverse colons, and the unabsorbed water becomes part of the feces. Reabsorption of water in the large intestine helps to avoid dehydration.

Bacterial activity Intestinal bacteria are harmless as long as they remain in the large intestine; in fact, they are useful in synthesizing vitamins K and B_{12}. Bacteria, both living and dead, make up 25 to 50 percent of the dry weight of feces. Bacterial activity in the large intestine also contributes to the production of intestinal gas, or *flatus* (FLAY-tuss; L. a blowing), which causes *flatulence.* Flatus consists mostly of nitrogen, carbon dioxide, and hydrogen, with small amounts of oxygen, methane, and hydrogen sulfide. The cells in the intestinal glands (crypts of Lieberkühn) of the large intestine secrete large amounts of alkaline mucus, which helps to neutralize acids produced by intestinal bacteria, and also lubricates the lumen for the easy passage of feces.

Formation of feces Approximately 150 g of feces (about 100 g of water and 50 g of solids) are normally eliminated from the body daily. Besides containing water and bacteria, feces are composed of fat, nitrogen, bile pigments, undigested food such as cellulose, and other waste products from the blood or intestinal wall. The normal brown color of feces is caused by products of red blood cells in the form of bile pigments (bilirubin). Excessive fat causes feces to be a pale

color, and blood and other foods containing large amounts of iron will darken feces.

The characteristic odor of feces is caused by *indole* and *skatole,* two substances that result from the decomposition of undigested food residue, unabsorbed amino acids, dead bacteria, and cell debris. A high-protein diet increases the odor of feces because it results in the production of larger quantities of indole and skatole. Hydrogen sulfide and methane also contribute to the odor of feces.

Movements of the large intestine and defecation Most of the digestive movements of the large intestine are slow and nonpropulsive. The primary motility comes from *haustral contractions* that depend upon the autonomous rhythmicity of the smooth muscle cells. These movements are similar to the segmentations of the small intestine, but usually take 20 to 30 minutes between contractions. This slow movement allows for the final absorption of the remaining water and electrolytes.

Three to four times a day, generally following meals, motility increases markedly, as large segments of the ascending and transverse colons contract simultaneously, driving the feces one-half to three-quarters the length of the colon in a few seconds. These "sweeping" peristaltic waves are called *mass movements.* They drive the feces into the descending colon, where waste material is stored until defecation occurs.

The *duodenocolic reflex* moves the contents of the ileum into the large intestine. Mass movements in the colon are produced by the *gastrocolic reflex* when food enters the stomach.

Region of digestive tract	Secretion	Food substance acted upon	Digestive activity	Absorptive activity
Small intestine (with assistance of pancreatic and liver secretions)	*Succus entericus* (intestinal juice): Mucus, salt. Microvilli secretions: enterokinase, disaccharidases, aminopeptidases. Pancreatic juice: proteolytic enzymes (trypsinogen, chymotrypsinogen, procarboxypeptidase), pancreatic amylase, pancreatic lipase. Liver secretion: bile.	Proteins.	*Pancreatic proteolytic enzymes* break down different peptide linkages, producing amino acids and small peptides. (Trypsinogen is activated and converted into trypsin by enterokinase. Chymotrypsinogen and procarboxypeptidase activated and converted into chymotrypsin and carboxypeptidase by trypsin.) Amino peptidases convert peptides into amino acids.	Absorbed as amino acids into capillary network within villi by active transport.‡
		Carbohydrates.	*Pancreatic amylase* converts polysaccharides into monosaccharides and disaccharides, especially into maltose. *Dissacharide maltase* converts maltose into monosaccharide glucose. *Disaccharide lactase* converts lactose into monosaccharide glucose and galactose. *Disaccharide sucrase* converts sucrose into monosaccharide glucose and fructose. *Intestinal amylase* converts complex sugars into disaccharides.	Simple sugars are absorbed by active transport into capillary network within villi. Galactose is the most readily absorbed sugar. Fructose is absorbed by facilitated diffusion.
		Fats.	*Pancreatic lipase* converts fats into free fatty acids and glycerol. *Bile salts* convert fats into emulsified fats. *Intestinal lipase* converts fats into fatty acids and glycerol.	Most fats (60 to 70%) are emulsified by bile salts and absorbed as fatty acids and glycerol into lacteals of villi, from which they reach blood circulation through the lymphatic thoracic duct. Remaining fats, which have been broken down by lipases into fatty acids and glycerol, become water-soluble and enter villi on way to liver via hepatic portal system.
Large intestine	*Mucus.*	Proteins, carbohydrates, fats.	None.	Little or none.§

*Some drugs, such as nitroglycerine and cyanide, are absorbed through the oral mucosa.
†A few highly lipid-soluble substances, such as alcohol, and certain drugs such as aspirin, are absorbed through the stomach, but generally little absorption occurs there.
‡Electrolytes such as sodium, chloride, potassium, and calcium are absorbed through the wall of the small intestine by active transport.
§Most absorption is completed by the time chyme reaches the large intestine, but some water, electrolytes such as sodium and chloride, and some vitamins are absorbed from the initial segments of the large intestine. Glucose and some drugs can be absorbed when administered through the rectum.

Diet and Cancer

The federal government, the American Heart Association, the American Diabetes Association, and the National Cancer Institute all have issued dietary guidelines in recent years. The guidelines indicate a concern that certain foods may increase the incidence of cancer, especially in the breast, prostate gland, and large intestine and rectum. The dietary recommendations consistently include high fiber, high complex carbohydrates, low fat, low cholesterol, and low calories.

Studies have shown that the incidence of cancer varies according to dietary habits. For example, the incidence of stomach cancer is high in Japan, where salt-cured and smoked foods may produce elevated levels of such cancer-causing substances as hydrocarbons and nitrosamines. However, Japanese people who move to Hawaii often show a statistical increase in breast and colon cancer and a decrease in stomach cancer, apparently because the typical Hawaiian diet is high in fat and low in smoked foods. In Finland, where the diet contains twice as much fiber as the typical American diet, the incidence of colon cancer is one-third the United States rate.

Some dietary recommendations for decreasing the risk of cancer include the following. *Eat a low-fat diet,* which reduces the risk of cancer of the large intestine, breast, and prostate gland. *Eat a high-fiber diet,* which speeds up the passage of feces through the large intestine, and apparently reduces the risk of cancer of the colon. (Also, high-fiber fruits and vegetables contain large amounts of vitamin C and carotene, a precursor of vitamin A, which are believed to be natural inhibitors of the development of cancer.) *Eat fewer salt-cured and smoked foods,* including bacon, smoked fish, and ham. *Drink alcohol only in moderation,* since heavy drinking increases the risk of cancer of the mouth, pharynx, larynx, esophagus, liver, and urinary bladder.

The reflex is mediated from the stomach to the colon by the secretion of gastrin, and by the external autonomic nerves. The gastrocolic reflex pushes feces into the rectum, triggering the **defecation reflex,** the elimination of waste material from the anus.

The elimination of most of the body's waste products is conducted not by the digestive tract, but by the kidneys. Nevertheless, the removal of solid wastes in the form of feces is one of the most important functions of the large intestine. Defecation normally proceeds as follows:

1 When the rectum is distended by the accumulation of feces, the walls are stretched, pressure in the rectum rises, and receptors in the rectal walls are stimulated.

2 If a conscious decision is made to defecate, the defecation reflex is triggered. The internal anal sphincter relaxes at the same time that peristaltic waves in the descending colon are stimulated and the rectum and sigmoid colon contract vigorously.

3 Defecation is assisted by the contraction of the abdominal muscles and the diaphragm, and a forcible expiration against a closed glottis. These voluntary actions increase the abdominal pressure and create a pushing action that assists in the elimination of feces.

4 The defecation reflex, initiated in the medulla, passes impulses to the sacral portion of the spinal cord.

5 The external anal sphincter and other perineal muscles relax, and increased peristaltic waves in the sigmoid colon and rectum push the feces past the relaxed internal and external anal sphincters, and out of the body through the anus.

6 The initial propulsion of feces from the anus seems to initiate another defecation reflex, which stimulates the expulsion of feces from the sigmoid and descending colons as well as the rectum.

Normal "straining" to assist defecation causes a sharp increase in arterial pressure when the increased pressure in the thoracic cavity is transmitted to the heart. Consequently, the venous return of blood is stopped, and cardiac output and arterial pressure drop. Strokes and heart attacks have been known to occur when elderly people strain during defecation.

Defecation is a voluntary act, except in very young children and in adults whose spinal cord is severed at the sacral level or above, and can be inhibited by keeping the external anal sphincter contracted. *Voluntary inhibition* of defecation is accomplished in the following way. Once the decision is made *not* to defecate, neural impulses from the central nervous system cause the external anal sphincter to close tightly. Spinal cord impulses also cause the relaxation of the rectum and sigmoid colon, which decreases the tension in the rectal wall and prevents the stretch receptors from being activated. The defecation reflex is avoided, and defecation is delayed, either by voluntary control or until the rectum becomes distended to the point where defecation is absolutely necessary. If defecation is postponed voluntarily, water will continue to be absorbed from the feces, and constipation may result. Children younger than 1 or 2 years usually cannot inhibit defecation voluntarily because the appropriate motor neural pathways have not yet reached their proper stage of maturity.

Ask Yourself

1 What are the main parts of the large intestine?

2 What are taeniae coli?

3 What are the main functions of the large intestine?

4 How do bacteria in the large intestine serve a useful purpose?

5 What is the usual composition of feces?

6 What are the stages of defecation?

PHYSIOLOGICAL AND ANATOMICAL ABNORMALITIES

Anorexia nervosa and bulimia

Anorexia nervosa and bulimia are examples of eating disorders with a psychological basis. *Anorexia nervosa* (Gr. *an*, without + *orexis*, a longing) is characterized by self-imposed starvation and subsequent nutritional deficiency disorders. Victims of this disorder, most often adolescent females from upwardly mobile families, usually have a normal weight to begin with, but are convinced that they are grossly overweight. Excessive dieting is often accompanied by compulsive physical activity, resulting in a weight loss of 20 percent or more. The exact cause of anorexia nervosa is not known, but it is often brought on by anxiety, fear, or peer or family pressure that encourages thinness. Generally accompanying weight loss and chronic undernourishment are atrophy of skeletal muscle tissue, constipation, hypotension, dental caries, irregular or absent menstrual periods, and susceptibility to infection, among other disorders. If the victim does eat, it may be followed by self-induced vomiting or an excessive dose of laxatives or diuretics. Such tactics may lead to dehydration or metabolic alkalosis or acidosis. A drop of systolic pressure below 55 mm Hg may prove fatal, and electrolyte imbalance may lead to cardiac arrest. Anorexia nervosa is fatal in 5 to 15 percent of the cases, and feelings of despondency lead to a higher-than-normal suicide rate. Treatment aims to correct weight loss and malnutrition, correct the underlying psychological factors, and treat any organic disorders that may have developed.

Bulimia (Gr. *bous*, ox; *limos*, hunger) also occurs mainly in upwardly mobile young females of normal weight. It is also called *gorge-purge syndrome* and *bulimia nervosa*. It is characterized by an insatiable appetite and gorging on food four or five times a week (ingesting as many as 40,000 calories), followed by self-induced vomiting or an overuse of laxatives or diuretics that may lead to dehydration and metabolic imbalances as in anorexia nervosa. It may also impair kidney and liver function and cause dry skin, frequent infections, muscle spasms, and other disorders. Appar-

ently, bulimarectics choose this course to handle stress the way an alcoholic chooses alcohol. Studies have shown that 20 to 30 percent of college women are bulimarectics. Standard treatment includes group therapy, behavior modification, and individual psychotherapy, as well as nutritional guidance and medical treatment if necessary.

Cholelithiasis (gallstones)

When the proportions of cholesterol, lecithin, and bile salts in bile are altered, *gallstones* (biliary calculi) may form in the gallbladder, a condition known as *cholelithiasis* (bile, gall + stone). Gallstones are composed of cholesterol and bilirubin, and are usually formed when there are insufficient bile salts to dissolve the cholesterol in micelles. Cholesterol, being insoluble in water, crystallizes and becomes hardened into "gallstones." Gallstones often pass out of the gallbladder and lodge in the hepatic and common bile ducts, obstructing the flow of bile into the duodenum and interfering with fat absorption. Jaundice may occur if a stone blocks the common bile duct because bile pigments cannot be excreted, and surgery may be required if stones become impacted in the cystic duct, where pain is usually maximal because the contractions of the gallbladder press on the stones. Other common sites of gallstones are the pancreatic ducts and the hepatopancreatic ampulla.

A typical gallbladder attack begins with acute pain in the upper right quadrant, often in the middle of the night following a high-fat meal. The production of gallstones is sometimes initiated by pregnancy, diabetes mellitus, celiac disease, cirrhosis of the liver, pancreatitis, or the use of oral contraceptives. Occasionally, gallstones may be dissolved when the naturally occurring bile acid chenodeoxycholic acid (CDCA) is taken orally, but CDCA is expensive and has the undesirable side effect of increasing LDL cholesterol and decreasing HDL cholesterol, increasing the likelihood that the patient will develop atherosclerosis. In most cases, gallstones lodged in ducts are removed surgically, and sometimes the gallbladder itself is removed (cholecystectomy).

A recent nonsurgical technique for removing gallstones employs a device called a *lithotripter* (Gr. "stone crusher"), which uses high-pressure shock waves to pulverize the gallstones after they are located with ultrasound. The shock waves, repeated hundreds of times, are directed through soft tissue until they hit the gallstones, breaking them up into pieces small enough to pass through the digestive system. The procedure requires a hospital stay of 3 days or less, compared with about 8 days for surgery, plus a recovery period of about a month. Lithotripsy is also used to crush kidney stones (see Chapter 24).

Cirrhosis of the liver

Cirrhosis (sih-ROE-siss; L. "orange-colored disease," from the color of the diseased liver) is a chronic liver disease characterized by the destruction of hepatic cells and their replacement by abnormal fibrous connective tissues. These changes disrupt normal blood and lymph flow, and eventually result in blockage of portal circulation, portal hypertension, liver failure, and death. Cirrhosis is most prevalent among chronic alcoholics over 50, especially if their diets are poor. The mortality rate is high. Among the several types of cirrhosis, the most common is *Laennec's type* (also called portal, nutritional, or alcoholic cirrhosis), which accounts for up to half of all persons suffering from the disease. Of these people, about 90 percent are heavy consumers of alcohol.

Early symptoms include gastrointestinal problems such as loss of appetite, indigestion, and constipation or diarrhea. Later symptoms include respiratory difficulties, disorders of the central nervous system such as slurred speech and paranoia, bleeding, skin irregularities, edema of the legs, and jaundice. Many patients die within five years after the onset of the disease.

Colon-rectum cancer

More than half of all cancers of the large intestine (colon) are found in the rectum, and about 75 percent of intestinal cancers are located in the rectum and the intestinal region (sigmoid colon) just above it. Although the number of *colon-rectum cancer* cases is second only to lung cancer, early diagnosis and treatment make it possible to save 4 out of 10 patients. The exact cause of colon-rectum cancer is unknown, but increasing evidence points to a diet containing excessive animal fat (especially beef) and low in fiber (see the essay on p. 720). Early symptoms are usually vague and unnoticed, but an obstruction may soon develop, producing constipation or diarrhea, rectal bleeding, and dull pain. During later stages, the primary tumor may spread, or metastasize, to other organs in the area, such as the bladder, prostate, uterus, and vagina, or even to the liver or lungs.

Colon-rectum cancer can be diagnosed by several fairly simple methods, including a barium x ray, digital examination, and proctoscopy or sigmoidoscopy (examination with tubular instruments that enable the physician to see into the rectum and lower intestine).

Constipation and diarrhea

Constipation (L. "to press together") is a condition in which feces move through the large intestine too slowly. As a result, too much water is reabsorbed, the feces become hard, and defecation is difficult. Constipation may be caused by spasms that reduce intestinal motility, nervousness, a temporary low-fiber diet, and other factors. Constipation may produce abdominal discomfort and distension, loss of appetite, and headaches, but it does *not* increase the amount of toxic matter in the body.

Diarrhea (Gr. "a flowing through") is a condition in which watery feces move through the large intestine rapidly and uncontrollably in a reaction sometimes called a *peristaltic rush.* Among the causes are viral, bacterial, or protozoan infections, extreme nervousness, ulcerative colitis, and excessive use of cathartics, all of which increase fluid secretion and intestinal motility. Severe diarrhea is dangerous (especially in infants) because it increases the loss of body fluids and ions, especially sodium and potassium.

Crohn's disease

Crohn's disease (named after Dr. Burrill Crohn, who first diagnosed the disease in the early 1900s) is a chronic intestinal inflammation that most often affects the ileum. It extends through all layers of the intestinal wall, and is generally accompanied by abdominal cramping, fever, and diarrhea. The most common victims are 20- to 40-year-old adults, especially Jews, and is thought to be caused either by a virus or autoimmunity.

Diverticulosis and diverticulitis

Diverticula are bulging pouches in the gastrointestinal wall that push the mucosal lining through the muscle in the wall. Diverticula most commonly form in the sigmoid colon, but other likely sites are the duodenum near the hepatopancreatic ampulla and the jejunum. Diverticula appear most often in men over 40, especially if their diet is low in roughage.

In **diverticulosis,** pouches are present but do not present symptoms. In **diverticulitis,** the diverticula are inflamed and may become seriously infected if undigested food and bacteria become lodged in the pouches. The subsequent blockage of blood can lead to infection, peritonitis, and possible hemorrhage.

Food poisoning

The term *food poisoning* is commonly used to describe gastrointestinal diseases caused by eating food contaminated with either infectious microorga-

TABLE 22.10   FOOD POISONINGS ASSOCIATED WITH MICROORGANISMS AND TOXINS

Disease (in order of incidence)	Causative microorganism	Symptoms of disease	Foods implicated	Appearance of symptoms and duration of illness
Salmonellosis	*Salmonella.*	Fever, chills, cramps, nausea, vomiting.	Undercooked meat, raw milk.	Symptoms appear in 12–48 hours, duration 4–7 days.
Staph food poisoning	*Staphylococcus aureus.*	Sudden nausea, vomiting, diarrhea, cramps.	Mayonnaise, unrefrigerated creams, skim milk, fish, stuffing, pork products.	Symptoms in 1–6 hours, duration 12–24 hours.
Perfringens food poisoning	*Clostridium perfringens.*	Abdominal pain, diarrhea.	Unrefrigerated meat.	Symptoms 8–15 hours, duration 6–24 hours.
Botulism	*Clostridium botulinum* (exotoxins A–E).	Nerve paralysis, fatigue, double vision.	Improperly canned goods (low-acid foods), meat, fish, honey.	Symptoms 18–48 hours, duration 1–7 months.
Gastroenteritis	*Campylobacter jejuni.*	Fever, malaise, diarrhea, cramps, hemorrhaging.	Contaminated water or raw milk, raw poultry.	Symptoms 2–7 days, duration 1–2 weeks.
Shigellosis	*Shigella.*	Diarrhea, vomiting, abdominal cramps.	Shellfish, unpasteurized milk, high-carbohydrate salads, tuna.	Symptoms 36–72 hours, duration 1–3 days.
B. cereus food poisoning	*Bacillus cereus* toxins.	Nausea, vomiting, diarrhea.	Unrefrigerated starchy foods.	Symptoms 1–15 hours, duration 6–24 hours.

nisms or toxic substances produced by the microorganisms. Most food poisonings are associated with the production of toxins. These disorders are summarized in Table 22.10.

Hepatitis

Hepatitis, an infection of the liver, may have either a viral or a nonviral cause. Among the symptoms of *viral hepatitis* are loss of appetite, nausea, jaundice, and abdominal pain. Other symptoms such as headache, dizziness, fever, and rash may be chemically related. Viral hepatitis has three forms: type A, type B, and type non-A non-B. *Type A* is infectious, occurs most commonly during the fall and winter in children and young adults, and is transmitted through contaminated food (especially seafood), water, milk, semen, tears, and feces. Onset is sudden, and the incubation period is relatively short (15 to 45 days), but the overall prognosis for a complete cure is good. A person who contracts type A viral hepatitis does not become a carrier, but a patient afflicted with type B does carry the disease for an indefinite period. *Type B* viral hepatitis can affect people of any age, at any time of the year, and usually has a slow prolonged onset (40 to 100 days). It is transmitted through serum, blood and blood products, semen, and feces. Unlike type A, type B becomes worse with age, and permanent immunity does not result. *Type non-A non-B* hepatitis usually results from commercial blood donations. Because liver cells are capable of regeneration, liver cell de-

struction resulting from hepatitis is normally overcome.

Nonviral hepatitis is usually caused by exposure to chemicals or drugs such as carbon tetrachloride, poisonous mushrooms, and vinyl chloride. Symptoms usually resemble those of viral hepatitis.

Hernia

A **hiatal hernia** (L. gap + rupture), also called *hiatus hernia*, is a defect in the diaphragm that permits a portion of the stomach to protrude into the chest area. The common cause of a hernia is connected with a typical weakening of muscle tissue that often comes with aging. The symptoms of hiatal hernia resemble those of a peptic ulcer, with pain in the upper abdomen, and "heartburn," especially when the person is lying down. An **inguinal hernia** (L. *inguen,* groin) is the protrusion of the small or large intestine, omentum, or urinary bladder through the inguinal canal in the area of the groin.

Jaundice

Ordinarily, yellow bile pigment (bilirubin) is excreted by the liver in the bile, and not enough of it circulates in the blood to affect the color of the skin. Occasionally, however, pigment levels rise sufficiently to produce a yellowish tint in the skin, mucous membranes, and whites of the eyes. The condition is known as **jaundice** (Old Fr. *jaune,* yellow), or in common usage, *yellow jaundice.*

Jaundice may have one of three different causes: (1) When the liver cells

remove bilirubin from the blood by hemolysis (destruction) of red blood cells more slowly than it is produced, the pigment builds up in the blood and produces *hemolytic jaundice.* Although the feces usually darken, the urine remains a normal color. (2) In *hepatic jaundice,* the liver's ability to absorb bilirubin, process it metabolically, and secrete it becomes impaired. The urine darkens and the color of the feces gradually becomes lighter. (3) *Obstructive jaundice* is produced when obstructions in the liver's duct system cause bilirubin to flow backward from the liver cells into the sinusoids. The urine becomes very dark, while the feces are usually the color of light clay.

In each case, the underlying cause is treated rather than the jaundice itself.

Peptic ulcers

Peptic ulcers (Gr. *peptein,* to digest) are lesions in the mucosa of the digestive tract (see photos below). They are caused by the digestive action of gastric juice, especially hydrochloric acid. Most peptic ulcers occur in the duodenum, before gastric hydrochloric acid is neutralized by the alkaline secretions of the small intestine.

People who are most susceptible to *duodenal ulcers* secrete high amounts of gastric juices between meals, when there is little or no food in the stomach to buffer the acidity. Men between the ages of 20 and 50 are most vulnerable. *Gastric ulcers,* in contrast, seem to be most common in men over 50 and among people whose stomach mucosa has a reduced

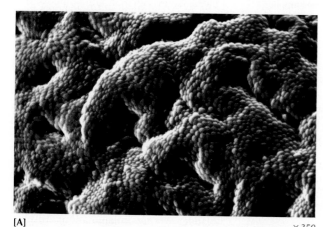

[A] ×350

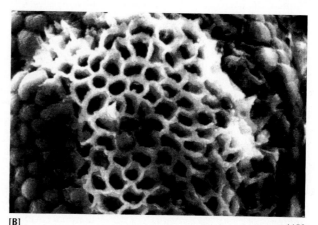

[B] ×1150

Scanning electron micrographs of (A) normal stomach lining, with a continuous coating of mucus, and (B) damaged stomach tissue that results in ulcers. [(A *and* B) *Dr. Gene M. Riddle, Director, Rheumatology Research Lab, Henry Ford Hospital, Detroit, Michigan.*]

resistance to digestion rather than an overabundance of gastric secretions. Large amounts of aspirin and alcohol decrease the resistance of the mucosa, and thus lead to gastric ulcers.

Symptoms of ulcers usually include heartburn and indigestion, pain after eating, weight loss, and gastrointestinal bleeding. Treatment may include rest, reduction of strain, antacids, and drug therapy. Hemorrhaging requires emergency treatment to reduce bleeding, and surgery is necessary in severe cases.

Peritonitis

Peritonitis is an inflammation of the peritoneum, usually resulting from a perforation in the gastrointestinal tract that allows bacteria to enter the normally sterile peritoneum. Such a perforation often results from a ruptured appendix, but can also be caused by diverticulitis, a peptic ulcer, or any other disease or physical trauma that breaks through the wall of the digestive tract. Peritonitis is often limited to a localized abscess rather than a general inflammation because the peritoneum is so resistant to infection. Severe pain, weakness, fever, and decreased intestinal motility are among the more obvious symptoms.

Tooth decay (dental caries)

Bacteria that cause **dental caries,** or tooth decay, produce a gluelike enzyme called *glucosyl transferase* (GTF) that converts ordinary sugar into *dextran,* a sticky substance that clings to the tooth's enamel. GTF also helps the bacteria adhere to the enamel. Dextran is involved in the formation of *plaque,* a destructive film that builds up on teeth. Once plaque is formed, the bacteria on the teeth pro-

duce another enzyme that promotes erosion of the enamel.* (Incidentally, apples may keep the doctor away, but not the dentist. They contain more than 10 percent fermentable carbohydrate, a potent producer of tooth decay.)

Vomiting

Vomiting (emesis) is the forceful expulsion of part or all of the contents of the stomach and duodenum through the mouth, usually in a series of involuntary spasms. The action is a coordinated reflex controlled by the vomiting center in the medulla. The vomiting center is activated either by receptors in the stomach or duodenum that respond to foreign constituents in the chyme, or by receptors near the brainstem that respond to chemicals carried by the blood, and to other external stimuli such as motion sickness or extreme pain.

The complex act of vomiting is preceded by increased salivation and sweating, accelerated or irregular heart rate and breathing, discomfort, and nausea, all of which are characteristic of a generalized discharge of the autonomic nervous system. During vomiting, the pyloric region of the stomach goes into spasm, and the usual downward motion of peristalsis is reversed. Meanwhile, the body of the stomach and cardiac sphinc-

*Our prehistoric ancestors had cavities too, but a study of skeletons that have been dated to before the Iron Age indicates that humans had about 2 to 4 percent decay then, compared to 40 to 70 percent today. However, public health officials predict the virtual end of tooth decay among children and young adults by the end of this century, due to improved dental technology and the widespread use of fluoridation in city water supplies.

ter are relaxed. The duodenum also goes into spasm, forcing its contents into the stomach.

The final thrust of vomit comes when the abdominal muscles contract, lifting the diaphragm and increasing the pressure in the abdomen. At the same time, slow, deep breathing with the glottis partially closed reduces the pressure in the thorax. As a result, the stomach is squeezed between the diaphragm and compressed abdominal cavity, forcing the contents of the stomach and duodenum past the cardiac sphincter into the esophagus, and out through the mouth. When this series of events is repeated several times, without vomiting, it is called *retching.*

During actual vomiting, the breathing rate decreases, the glottis is closed, and the soft palate is raised, all ensuring that the vomit does not enter the breathing passages and cause suffocation.

Although it seems clear that vomiting may be the body's way of ridding itself of a harmful food substance, it is not fully understood why motion sickness or pain should induce vomiting. Also, it is not certain why some women become nauseous and vomit during early pregnancy ("morning sickness"). It may be caused by hormonal changes, especially the increased secretion of human chorionic gonadotropin, and it may also be related to the changes in carbohydrate metabolism. Both factors are probably involved.

Excessive vomiting, like prolonged diarrhea, can lead to a dangerous depletion of fluids and salts. The loss of acid may lower the overall acidity of the body beyond the point where normal feedback systems can maintain homeostasis.

MEDICAL TERMINOLOGY

ACHOLIA (*a,* without + Gr. *khole,* bile) Absence of bile secretion.

ANTIEMETIC (Gr. *anti,* against + *emetos,* vomiting) A substance used to control vomiting.

ANTIFLATULENT (Gr. *anti,* against + L. *flatus,* a breaking wind) A drug that pre-

vents the retention of air in the digestive tract.

ASCITES (uh-SEE-teez; Gr. "bag") An accumulation of serous fluid in the abdominal cavity.

BARIUM ENEMA EXAMINATION Fluoroscopic and radiographic examination of

the large intestine after administering a barium sulfate mixture via the rectum. Commonly called *lower GI series.*

CHOLECYSTITIS (Gr. *khole,* bile + *cystis,* bladder + *-itis,* inflammation) Inflammation of the gallbladder.

CHOLESTEROSIS A condition in which an

abnormal amount of cholesterol is deposited in tissues.

COLECTOMY (L. *colon*, intestine + *-ectomy*, removal) Surgical removal of the colon. A *hemicolectomy* removes half the colon.

COLITIS Inflammation of the colon.

COLOSTOMY (L. *colon*, intestine + Gr. *stoma*, opening) Surgical creation of an opening in the large intestine, through which feces can be eliminated. Also, the opening thus created.

CREPITUS (L. *crepitare*, to crackle) Discharge of flatus from the intestine.

DYSPEPSIA (Gr. *dus*, faulty + *pepsia*, digestion) A condition in which digestion is difficult. Commonly called *indigestion*.

DYSPHAGIA (Gr. *dus*, faulty + *phagein*, to eat) Difficulty in swallowing.

EMETIC (Gr. *emetos*, vomiting) A substance used to induce vomiting.

ENTERITIS (Gr. *enteron*, intestine + inflammation) Inflammation of the intestines, particularly the small intestine.

GASTRECTOMY (Gr. *gaster*, belly + *-ectomy*, removal) Surgical removal of part or all of the stomach.

GASTROENTEROLOGIST A physician specializing in treating stomach and intestinal (gastrointestinal) disorders.

GASTROSCOPY (Gr. *gaster*, belly + *skopein*, to examine) A procedure in which a viewing device is inserted into the stomach for exploratory purposes.

GASTROSTOMY (Gr. *gaster*, belly + *stoma*, opening) Surgical creation of an opening in the stomach, usually for the purpose of inserting a feeding tube.

GINGIVITIS (L. *gingiva*, gum + inflammation) Inflammation of the gums.

HALITOSIS (L. *halitus*, breath) Bad breath.

HYPEROREXIA (Gr. *hyper*, over + *orexis*, a longing) Abnormal appetite.

HYPOCHLORHYDRIA A deficiency of hydrochloric acid in the stomach.

IRRITABLE BOWEL SYNDROME A generalized gastrointestinal disorder, characterized by a spastic colon, alternating diarrhea and constipation, and excessive mucus in the feces.

LAPAROTOMY (Gr. *lapara*, flank + *tome*, incision) Cutting into the abdomen in order to explore the abdominal cavity.

LAXATIVE (L. *laxus*, loose) A substance used to loosen the contents of the large intestine so that defecation can occur.

MALOCCLUSION (L. *mal*, badly + *occludere*, to close) A condition in which the upper and lower teeth do not fit properly when the jaws are closed.

ORTHODONTIA (Gr. *orthos*, straight + *odous*, tooth) The branch of dentistry that specializes in aligning teeth properly.

PORTAL HYPERTENSION A condition in which there is high blood pressure in the portal circulation of the liver.

PROCTOCELE (Gr. *proktos*, anus + *koilos*, hollow) Hernia of the rectum.

PROCTOSCOPY (Gr. *proktos*, anus + *skopein*, to examine) A procedure in which a viewing device is inserted into the anus and rectum for exploratory purposes.

PROLAPSE OF THE RECTUM A condition where the weakened walls of the rectum fall outward or downward.

PRURITIS ANI (L. *prurire*, to itch + anus) Chronic itching of the anus.

PYORRHEA (L. *pyo*, pus + Gr. *rrhoos*, flowing) A condition in which pus oozes from infected gums.

SIALITIS Inflammation of a salivary duct or gland.

SMALL INTESTINE SERIES Serial radiograms of the small intestine during the passage of ingested barium. Commonly called *upper GI series*.

STOMATITIS (Gr. *stoma*, mouth + inflammation) Inflammation of the mouth.

TRACHEOSTOMY (trachea + opening) Surgical creation of a permanent opening in the trachea.

TRENCH MOUTH (VINCENT'S INFECTION) A painful inflammation of the gums, usually including fever, ulcerations, and bleeding.

SUMMARY

Food is ingested into the digestive tract and undergoes **digestion,** the process of breaking down large food molecules into smaller ones. However, small nutritive molecules must leave the digestive system and enter the body proper before they can be used. This is accomplished by a second process, called **absorption,** when the small food molecules pass through the plasma membranes of the small intestine into the bloodstream.

Introduction to the digestive system

1 The digestive process takes place in the **digestive tract,** or **alimentary canal,** which extends from the lips to the anus.

2 The digestive system is compartmentalized, with each part adapted to a specific function. The alimentary canal consists of the mouth, pharynx, esophagus, stomach, small intestine, large intestine, rectum, anal canal, and anus. The associated structures include the teeth, lips, cheeks, salivary glands, pancreas, liver, and gallbladder.

3 Parts of the digestive tract have specialized functions, but all are composed of the same basic layers of tissue. The wall of the tube (from the inside out) is composed of the **mucosa, submucosa, muscularis externa,** and **serosa.**

Mouth

1 The **mouth (oral** or **buccal cavity)** consists of two parts, the small, outer **vestibule,** and the *oral cavity* proper.

2 The **lips** and **cheeks** aid the digestive process by holding the food in position to be chewed.

3 There are 20 **deciduous** (''baby'') teeth and 32 **permanent** teeth, with special- ized shapes for cutting, tearing, and grinding food. Each tooth consists of a **root, crown,** and **neck,** and is composed of **dentine, enamel, cement,** and **pulp.** The **gum** is the connective tissue that surrounds the aveolar processes of the teeth.

4 The **tongue** aids in digestion by helping to form the moistened, chewed food into a **bolus,** and pushing it toward the pharynx to be swallowed. It also contains taste buds.

5 The tongue also pushes the food against the **hard palate,** where it is crushed and softened. The movable **soft palate** prevents food from entering the nasal passages by closing over the nasopharynx.

6 The three largest pairs of **salivary glands** are the **parotid, submandibular,** and **sublingual** glands. They secrete saliva, which contains water, salts, proteins, and at least one enzyme, *salivary amylase (ptyalin)*, which begins the digestion of cooked starch. Saliva also moistens and lubricates food so it can be swallowed easily, and allows us to taste food by dissolving food molecules.

Pharynx and esophagus

1 The *pharynx*, or throat, leads from the mouth to the esophagus. It is the common pathway for the passage of food (swallowing) and air. The *esophagus* is the tube that carries food from the pharynx to the stomach.

2 Swallowing action, or *deglutition*, in the pharynx initiates a muscular wave called *peristalsis*, which pushes the food along the esophagus and into the stomach. Peristalsis continues to assist the passage of food through the rest of the digestive system.

Abdominal cavity and peritoneum

1 The *abdominal cavity* is the portion of the trunk cavities that lies inferior to the diaphragm. When considered as including the pelvic cavity, it is called the *abdominopelvic cavity*.

2 The serous membrane of the abdominal cavity is the *peritoneum*. The *pariteal peritoneum* lines the abdominal cavity, and the *visceral peritoneum* covers most of the organs within the cavity.

3 The space between the parietal and visceral membranes is the *peritoneal cavity*. It usually contains a small amount of *peritoneal fluid*, which reduces the friction as abdominal organs move.

4 Abdominal organs lying posterior to the peritoneal cavity and that are covered but not surrounded by the peritoneum are called *retroperitoneal*. Those surrounded by the peritoneum are called *intraperitoneal*.

5 Organs in the abdominal cavity are suspended from the cavity wall by the *mesenteries*, or *visceral ligaments*. The mesentery connected to the stomach is the *omentum*.

Stomach

1 The *stomach* is the most expandable portion of the alimentary tract. It stores, mixes, and digests food.

2 Food enters the stomach from the esophagus through the *cardiac orifice* and empties into the small intestine through the *pyloric orifice*.

3 The stomach is divided into the *cardiac region, fundus, body,* and *pyloric region.*

4 The stomach stores large quantities of food, uses a churning action to mix food into a soupy mixture called *chyme*, releases the chyme in regular spurts into the small intestine, and secretes gastric juices, which initiate the digestion of protein.

5 The *muscularis externa* of the stomach has three layers of smooth muscle: the outermost *longitudinal* layer, the middle *circular* layer, and the innermost *oblique* layer.

6 *Gastric juice* is composed of *hydrochloric acid, mucus,* and several *enzymes*, including *pepsinogen* (a precursor of the active enzyme *pepsin*). Small amounts of gastric *lipase* are also secreted, initiating the digestion of fat. The major function of gastric juice is to digest protein.

7 The control of gastric juice secretion occurs in three overlapping phases: *cephalic*, initiated by the stimulation of seeing or tasting food; *gastric*, in response to distension of the stomach; and *intestinal*, with an inhibitory component to prevent oversecretion of gastric juices.

Small intestine

1 The *small intestine* is subdivided into the *duodenum*, where most of the remaining digestion takes place, and the *jejunum* and *ileum*, where most of the absorption of nutrients and water into the bloodstream and lymphatic system occurs.

2 The absorptive surface of the small intestine is increased substantially by protrusions called *villi*, which contain additional protrusions called *microvilli*, and circular folds called *plicae circulares*.

3 *Mucosal* and *submucosal glands* (crypts of Lieberkühn) secrete *intestinal juice* and several enzymes that aid the digestion of carbohydrates, proteins, and fats. Large lymphatic nodules *(Peyer's patches)* combat microorganisms.

4 The two main types of muscular activity in the small intestine are *segmenting contractions* and *peristaltic contractions.*

5 The *intestinal juice* contains *water, salts,* and mucus. Enzymes produced in the small intestine include *enterokinase, lactase, lipase, maltase, peptidase,* and *sucrase.*

6 The products of carbohydrate, protein, and lipid digestion, as well as most ingested electrolytes, vitamins, and water, are absorbed by the small intestine. After carbohydrates and proteins have been broken down by digestion they are readily absorbed through columnar absorptive cells in the small intestine.

7 The absorption of lipids is more complex than that of carbohydrates and proteins. It involves the breakdown of water-insoluble triglyceride droplets into water-soluble particles called *micelles*, which are absorbed by cells. Once inside cells, the breakdown products of lipids are resynthesized into triglycerides, and are packaged into tiny droplets called *chylomicrons*, which move from the cells into lymphatic and blood vessels for distribution throughout the body.

Pancreas as a digestive organ

1 The *pancreas* consists of a head, body, and tail. It is composed of both exocrine and endocrine secretory cells. The exocrine cells form groups of cells called *acini*, which secrete digestive juices into the small intestine.

2 Secretion of pancreatic juices is stimulated by the detection of food by taste buds, and by the secretion of the digestive hormones *secretin* and *cholecystokinin* (CCK) after the chyme passes into the duodenum.

3 The three main types of pancreatic digestive enzymes are *pancreatic lipase*, which acts on fats, *pancreatic amylase*, which acts on carbohydrates, and *pancreatic proteolytic enzymes* (trypsinogen, chymotrypsinogen, procarboxypeptidase), which break down proteins into amino acids and small peptides.

Liver as a digestive organ

1 The liver, the largest glandular organ in the body, is divided into two main lobes by the *falciform ligament*, with the *right lobe* being six times larger than the *left lobe.* The right lobe is further subdivided into a *quadrate lobe* and a *caudate lobe.*

2 The *porta hepatis* is the door through which blood vessels, nerves, and ducts enter and leave the liver. The *hepatic artery* and *hepatic portal vein* enter, and the *hepatic vein* and *bile duct* exit.

3 The functional units of the liver are *lobules* that contain **hepatic cells** arranged in plates that radiate from a *central vein*. Between the rows of cells are blood channels called **sinusoids,** and in the corners of the five- or six-sided lobules are *portal areas* that contain branches of the portal vein, hepatic artery, bile duct, and nerve.

4 The many metabolic functions of the liver include removal of amino acids from organic compounds, conversion of excess amino acids into urea, homeostasis of blood, synthesis of certain amino acids, and conversion of carbohydrates and proteins into fat.

5 The liver is the body's main storage area. One of its major functions is to store glucose in the form of glycogen to be released for energy as needed.

6 Another major function is the secretion of **bile,** an alkaline liquid that breaks down fats.

Gallbladder

1 The **gallbladder** concentrates and stores bile from the liver until it is needed for digestion.

2 The **biliary system** consists of the gallbladder, hepatic ducts, cystic duct, and common bile duct.

3 The contraction of the gallbladder and the subsequent release of bile take place in the presence of cholecystokinin, which is secreted when fatty acids and amino acids reach the duodenum.

Large intestine

1 When chyme leaves the small intestine, digestion is complete, and the **large intestine** (also called the colon) functions to remove water and salts from the liquid chyme. Removal of water converts liquid wastes into **feces.**

2 The large intestine consists of the **cecum** (which contains the **vermiform appendix**), **ascending colon, transverse colon, descending colon, sigmoid colon,** and rectum.

3 The large intestine has an incomplete layer of longitudinal muscle that forms three separate bands of muscle called **taeniae coli.** The intestinal wall contains bulges called **haustra** and fat-filled pouches called **epiploic appendages.**

4 The terminal segments of the large intestine are the **rectum, anal canal,** and **anus.** The elimination of feces from the anus is called **defecation,** the only voluntary digestive act since the initial stage of swallowing.

Physiological and anatomical abnormalities

1 *Anorexia nervosa* is a lack of appetite and a self-imposed starvation, often accompanied by compulsive physical activity. *Bulimia* is characterized by food gorging, followed by self-induced vomiting or other measures intended to remove the gorged food from the body.

2 *Cholelithiasis* is a condition producing gallstones, which are composed of cholesterol and bilirubin. They can obstruct the flow of bile into the duodenum and interfere with fat absorption.

3 *Cirrhosis* is a serious liver disease in which hepatic cells are destroyed, and replaced by connective tissue. It is prevalent among chronic alcoholics.

4 *Colon-rectum cancer* may be related to diet. It often spreads to other organs in the area.

5 *Constipation* is a condition in which feces move through the large intestine too slowly. In contrast, *diarrhea* is a condition in which watery feces move through the large intestine rapidly and uncontrollably.

6 *Crohn's disease* is a chronic intestinal inflammation thought to be caused by a virus or autoimmunity.

7 *Diverticulosis* is the formation of bulging pouches in the intestinal wall. *Diverticulitis* is inflammation of the pouches.

8 *Food poisoning* describes several gastrointestinal diseases caused by eating food contaminated with infectious microorganisms or toxic substances produced by the microorganisms.

9 *Hepatitis* may be either viral or nonviral. Viral hepatitis takes three forms: type A, type B, and type non-A non-B. Nonviral hepatitis is caused by exposure to toxic chemicals or drugs.

10 A *hiatal hernia* is the protrusion of a portion of the stomach through a defective diaphragm into the chest area. An *inguinal hernia* is the protrusion of the intestine, omentum, or urinary bladder through the inguinal canal.

11 *Jaundice* is caused by excess bilirubin in the blood, producing a yellowish tint to the skin. Three types are hemolytic, hepatic, and obstructive jaundice.

12 *Peptic ulcers* are lesions of the digestive tract mucosa caused by the action of gastric juice.

13 *Peritonitis* is an inflammation of the peritoneum caused by the intrusion of bacteria.

14 Tooth decay (*dental caries*) is caused by bacteria that produce the enzyme *glucosyl transferase,* which converts sugar into *dextran.* Dextran helps bacteria adhere to tooth enamel, and is involved in the formation of *plaque,* a destructive film that promotes the production of an enamel-eroding enzyme.

15 *Vomiting* is the forceful expulsion of the contents of the stomach or duodenum through the mouth. It is a coordinated reflex controlled by the medulla.

UNDERSTANDING THE FACTS

1 Which organs form the digestive tract? What is the basic function of each?

2 What are the two main processes involved in converting food into raw materials for body cells?

3 List the major accessory organs and associated structures of the digestive system. How are they related to digestion?

4 Differentiate between chemical and mechanical digestion.

5 Define mucosa.

6 Which teeth have no deciduous predecessors? (Supply number and type.)

7 List the three largest pairs of glands that secrete saliva.

8 Define bolus. How is it formed?

9 Name the structure that regulates the flow of material from the stomach to the duodenum.

10 Which cells secrete hydrochloric acid? How do hydrochloric acid and pepsinogen interact?

11 What are the major functions of gastric juice?

12 What is the most common site of ulcers? Why?

13 What are the functions of the ileocecal valve?

14 What is meant by emulsification, and what substance is involved in this process?

15 In what form are lipids released across the plasma membranes of intestinal epithelial cells?

16 What is the origin of bile pigments?

17 Does the gallbladder produce bile? Explain.

18 Distinguish between diverticulosis and diverticulitis.

19 Give the site of action and the function of the following enzymes (be specific):
  a trypsin
  b maltase
  c lipase
  d ptyalin
  e pepsin

UNDERSTANDING THE CONCEPTS

1 If the small intestine were to be cut into, through what tissue layers would the scalpel pass?

2 What digestive activities occur in the mouth? What structures and secretions are involved?

3 Describe the typical series of events that lead to tooth decay.

4 How important is gravity to the process of swallowing? Explain.

5 Describe the movements of the stomach during digestion, and explain how they are regulated.

6 How does the arrangement of the muscle fibers in the walls of the stomach facilitate its functions?

7 Why do we normally not digest our stomach wall?

8 Other than its great length, what structural factors contribute to the great absorptive surface of the small intestine?

9 Sketch, label, and discuss the gross anatomy of the small intestine.

10 Discuss the sources and functions of the major digestive hormones.

11 How does the liver contribute to the homeostasis of the body?

12 Trace fat from the time it enters the mouth until it has been absorbed. Include the action of enzymes and other secretions.

13 Explain the main triggering mechanism that brings about the release of bile from the gallbladder.

14 Which phases of the digestive process are voluntary and which are involuntary? How is the nervous system involved?

15 Why may severe diarrhea be a serious problem in infants?

16 What may be the relationship of aspirin and alcohol to gastric ulcers?

SUGGESTED READING

BARON J. H., "Current Views on Pathogenesis of Peptic Ulcer." *Scandinavian Journal of Gastroenterology,* 17 (1982): 1–10.

BLISS, V. M., "Fat Absorption and Malabsorption." *Archives of Internal Medicine,* 141 (1981): 1213–1215.

COHEN, LEONARD A., "Diet and Cancer." *Scientific American,* November 1987.

DAVENPORT, HORACE W., *A Digest of Digestion,* 2nd ed. Chicago: Year Book Medical Publishers, 1978.

DAVENPORT, HORACE W., *Physiology of the Digestive Tract,* 5th ed. Chicago: Year Book Medical Publishers, 1982.

DAVENPORT, HORACE W., "Why the Stomach Does Not Digest Itself." *Scientific American,* January 1972.

FERNSTROM, JOHN D., AND RICHARD J. WURTMAN, "Nutrition and the Brain." *Scientific American,* February 1974.

JOHNSON, L. R., ed., *Gastrointestinal Physiology,* 3rd ed. St. Louis: Mosby, 1985.

KAPPAS, ATTALLAH, AND ALVITO P. ALVARES, "How the Liver Metabolizes Foreign Substances." *Scientific American,* June 1975.

KRETCHMER, NORMAN, "Lactose and Lactase." *Scientific American,* October 1972.

LIEBER, CHARLES S., "The Metabolism of Alcohol." *Scientific American,* March 1976.

MOOG, FLORENCE, "The Lining of the Small Intestine." *Scientific American,* November 1981.

PETERS, M. N., AND C. T. RICHARDSON, "Stressful Life Events, Acid Hypersecretion, and Ulcer Disease." *Gastroenterology,* 84 (1983): 114–119.

YOUNG, VERNON R., AND NEVIN S. SCRIMSHAW, "The Physiology of Starvation." *Scientific American,* October 1971.

23
Metabolism, Nutrition, and the Regulation of Body Heat

LEARNING OBJECTIVES

1 Describe the six classes of essential nutrients and their functions in the body.

2 Define metabolism, anabolism, catabolism, oxidation, and reduction, and give examples of each.

3 Outline the main events of cellular respiration, including glycolysis, the citric-acid cycle, and the electron transport system, and tell how they produce energy for the body.

4 Explain the storage and release of glucose in the body.

5 Describe the conversion of noncarbohydrates into glucose as a source of energy.

6 Discuss the anabolism and catabolism of proteins and lipids.

7 Explain what happens to carbohydrates, proteins, and lipids during absorptive and postabsorptive states.

8 List the federal dietary guidelines for good nutritional balance.

9 List the major vitamins and minerals needed by the body, their sources, and their functions in the body.

10 Describe the roles of hormones and enzymes in the control of metabolism.

11 Describe the factors that affect metabolism, and define basal metabolic rate.

12 Describe the mechanisms used by the body to regulate temperature.

13 Explain the homeostatic control of body temperature by the nervous and endocrine systems.

14 Discuss obesity, malnutrition, and metabolic and temperature-control disorders.

iving cells are continuously involved in the many chemical and energy transformations that take place during the moment-to-moment, day-to-day activities that keep our bodies alive and healthy. The general term that describes all these activities is **metabolism.** Metabolism maintains life and promotes homeostasis. It includes the conversion of nutrients into the usable energy contained in ATP, the production and replication of nucleic acids, the synthesis of proteins, the physical construction of cells and cell parts, the elimination of cellular wastes, and the production of heat, which helps regulate the temperature of the body.

Everything a human body has or does depends upon the **nutrients,** or chemical components, of food, which supply the energy and physical materials a body needs. The food we eat contains six kinds of *essential nutrients* that are vital to our well-being: carbohydrates, proteins, lipids, minerals, vitamins, and water. Nutrients must be supplied by the diet because (1) we break down food to get energy, and (2) many vital substances cannot be synthesized by the body. Each type of nutrient provides something different, as you will see in detail later. In general, proteins and some mineral salts are the body-building foods; carbohydrates and fats are fuels that provide energy; and vitamins and some minerals help to regulate cellular activity by enhancing the effects of the other essential nutrients.

The energy in food exists in *potential* form. It cannot be used unless it is released by the actions of enzymes in cells. **Cellular respiration*** is the process by which potential energy, in the form of chemical bonds in food molecules, is released. It may be used to make biologically useful molecules, especially ATP, or it may be released as heat. The ATP can be used by cells to do the work that keeps the whole body functioning. As shown in Figure 23.1, when energy in ATP is released from the cell, ADP is produced. When inorganic phosphate (P_i) and energy are added to ADP, ATP is formed.

---

*When we speak of respiration in the sense of cellular activity, we specifically mean *cellular respiration,* not "breathing."

REVIEW OF CHEMICAL REACTIONS

To give you a better understanding of metabolic processes, we begin with a review of several kinds of chemical reactions. Metabolism includes both anabolic and catabolic processes (Figure 23.2). **Anabolism** involves the chemical changes that allow cells to *build up* and repair themselves, or to synthesize macromolecules from smaller precursors, as when amino acids are arranged into complex three-dimensional proteins. (Anabolism is also called *biosynthesis.*) Because anabolic reactions synthesize molecules of increased size and complexity, more energy is required because more components are bonded in these macromolecules, and each bond formation *requires energy.* The required energy is provided by the breakdown of ATP.

Catabolism includes any chemical reactions in which macromolecules are broken down into smaller molecules, and energy is released. As you learned in the previous chapter, ingested carbohydrates, proteins, and fats must be mechanically and chemically broken down into smaller molecules before they can be absorbed by cells. Enzymes are required to complete the catabolism of nutrients during digestion. A more complex form of catabolism is *cellular respiration,* which occurs as a series of step-by-step reactions, each controlled by a separate enzyme. This process releases energy in small, usable amounts (see Figure 23.2).

In the process of catabolism, *heat is also released.* As you will see later in this chapter, some of this "waste" heat is used to keep the body warm. The energy and simple molecules produced during catabolism are both needed for anabolic processes.

It is important to understand that anabolic and catabolic reactions commonly take place at the same time in a cell, as structural components are constantly being broken down and replaced. Catabolism releases energy, which is used in anabolism to build up amino acids, sugars, fatty acids, and nitrogenous bases into large molecules such as proteins and nucleic acids.

In living cells, many energy transfers of cellular metabolism involve oxidation-reduction reactions. **Oxidation** occurs when atoms or molecules lose electrons or hydrogen ions. **Reduction** occurs when atoms or molecules gain electrons or hydrogen ions. Oxidation and reduction always occur together, because when one atom or molecule gains electrons or hydrogen ions, another must lose them. When electrons leave a molecule, that molecule is *oxidized.* There must be a receiver for the electrons, and when a receiver gains electrons, it is *reduced.* The oxidized molecule gives up energy, and the reduced molecule receives it. (See p. 39 for a detailed description of oxidation-reduction reactions.)

FIGURE 23.1

Overall mechanism of cellular respiration. The all-important production of ATP molecules provides energy for cellular activities.

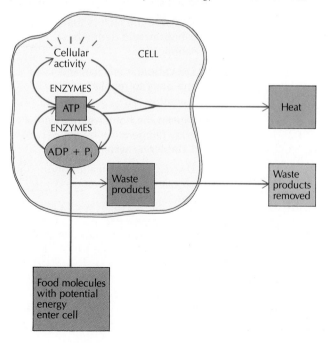

THE DYNAMIC BODY

The Industrious Cell

Cells are the chemical factories of our bodies, carrying on the many chemical activities that provide the energy for us to live. Within the industrious cell, specialized structures carry on an incredible number of diversified functions. In the *nucleus,* DNA is replicated, and some RNAs are synthesized. Ribosomal RNA is synthesized within the *nucleolus* of the nucleus. Through the *plasma membrane,* sodium, potassium, and amino acids are carried into and out of the cell, so that the supply of materials needed for all the cell's activities is renewed constantly.

In the *cytoplasm,* complex carbohydrates are broken down into simple glucose molecules, our most usable energy source; fatty acids are synthesized; and amino acids are activated for protein synthesis. In *lysosomes* within the cytoplasm, enzymes such as ribonuclease and phosphatase break down large molecules. In the *Golgi apparatus,* secretory vesicles are formed, and in the *endoplasmic reticulum,* certain lipids are synthesized, and materials needed for the synthesis of various substances are channeled into and out of the cell. *Ribosomes* are engaged in the synthesis of most proteins, and *mitochondria* are especially involved with the many chemical reactions that generate energy in the form of ATP, and build up smaller molecules into certain proteins.

Besides its many other functions, a cell is constantly repairing itself. Experiments with radioactive amino acids show that proteins in a cell are constantly being replaced, so that in time the cell is completely rebuilt with new molecules (see photo). If you could pull a few old threads out of a shirt and weave new ones in, and if you did that frequently enough, you would always have a "new" shirt. So it is with living cells. Some of the food taken in by the body is used for the endless process of replacing old molecules as they become worn out. The constant replacement and repair of cells is an essential part of the body's general metabolism.

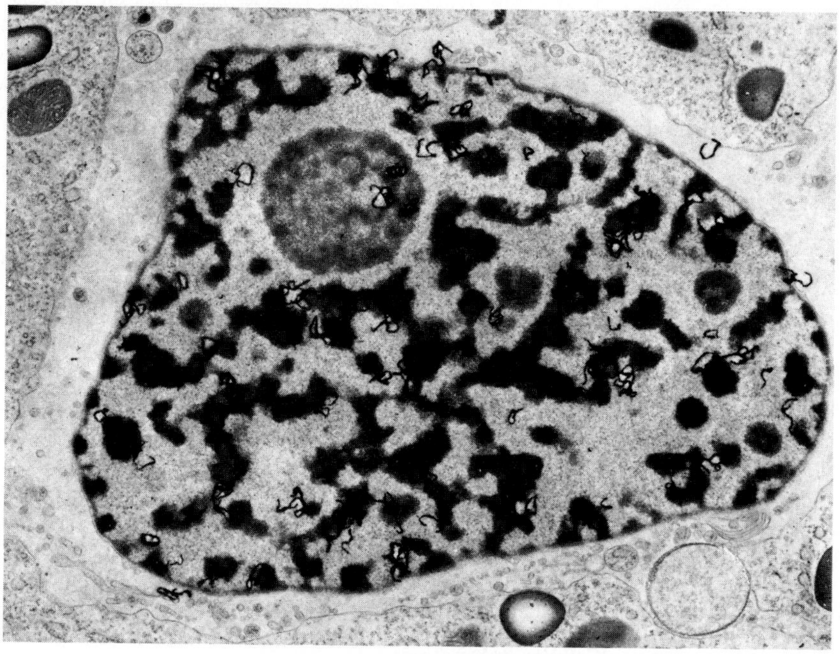

$\times 20,000$

This radiograph shows a cell that has been "fed" a radioactive amino acid. The spots indicate that the cell is synthesizing new protein. [K. G. Murti/Visuals Unlimited.]

Ask Yourself

1 *Give an example of what anabolic and catabolic reactions accomplish.*

2 *Is energy mainly required or released in catabolic reactions?*

3 *What happens during an oxidation-reduction reaction?*

INTRODUCTION TO CARBOHYDRATE METABOLISM

Carbohydrates have potential energy stored in chemical bonds. Within the bonds, it is *electrons* that actually carry the energy. During the metabolism of foods, chemical bonds are broken, and high-energy electrons are transferred to nucleotides such as NAD^+ (nicotinamide adenine dinucleotide) and FAD (flavin adenine dinucleotide), that is, these nucleotide molecules actually carry the high-energy electrons. In the process of receiving the electrons, NAD^+ and FAD also accept hydrogen ions (H^+). We say that the nucleotides are *reduced* (receive electrons), and the foods are *oxidized* (lose electrons). In this way, NAD^+ is converted into $NADH + H^+$, and FAD is converted into $FADH_2$. Each nucleotide receives two electrons and two hydrogen ions. The energy in the reduced nucleotides ($NADH + H^+$ and $FADH_2$) is most often used by mitochondria to make ATP, but may also be used in energy-requiring anabolic reactions.

Carbohydrate catabolism occurs in four stages:

1 Carbohydrate macromolecules in food are broken down

FIGURE 23.2

Energy relationships between catabolism and anabolism.

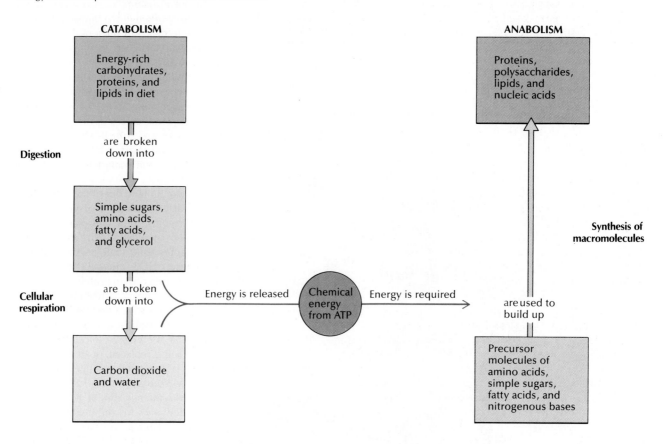

into their subunits, especially glucose. This is the process of *digestion,* which consumes energy.

2 *Glycolysis* occurs in the cytoplasm, and results in the breakdown of one molecule of glucose into two molecules of pyruvic acid. During this process, two molecules of NADH and two molecules of usable ATP are produced from every glucose molecule broken down.

3 The *citric-acid cycle* (Krebs cycle) generates large amounts of NADH + H$^+$, FADH$_2$, and some ATP.

4 The *electron transport system* converts the energy in NADH + H$^+$ and FADH$_2$ into ATP energy.

CARBOHYDRATE METABOLISM

As you learned in Chapter 2, carbohydrates are classified as monosaccharides (simple sugars), disaccharides, and polysaccharides. During digestion, the larger units are broken down into smaller, more usable units, such as glucose. Glucose, a monosaccharide, is an important energy source for the body. The breakdown of glucose in a complex series of metabolic reactions leads to the production of energy. (The breakdown of fats and proteins also contributes to the production of energy, which will be described later in the chapter.)

Cellular Respiration

In cellular respiration, cells use the potential energy stored in the chemical bonds in food molecules to create ATP from ADP and P$_i$. The process releases (generates) carbon dioxide and water. Although proteins and fats may enter this process at different points, we begin with the catabolism of glucose. The main steps in glucose catabolism are: (1) *glycolysis,* the initial breakdown of glucose and the release of ATP and NADH + H$^+$ and FADH$_2$; (2) the *citric-acid* or *Krebs cycle,* in which carbon dioxide, ATP, and hydrogen are released; and (3) the *electron transport system,* in which ATP and water are produced and oxygen is consumed (see Figure 23.12 for a summary of metabolic processes of all types of nutrients).

Glycolysis *Glycolysis* is the splitting *(lysis)* of a glucose molecule with six carbon atoms into two 3-carbon molecules of pyruvic acid. It occurs in the cytoplasm of a cell in a series of enzymatically controlled steps that do not require oxygen (Figure 23.3). Because glycolysis proceeds without oxygen, it is referred to as *anaerobic respiration.* (*Anaerobic* means "without air-life.") Most usable energy is obtained later, from the citric-acid cycle and the electron transport system.

Before a molecule of glucose is to be broken down into smaller units, the molecule is *activated,* and its energy level is raised. This activation is achieved by the transfer of a phosphate group from an ATP molecule to glucose, as ATP becomes ADP. (The transfer of a phosphate group from one compound to another is called *phosphorylation.*) The process of forming ATP during glycolysis and the citric-acid cycle is called *substrate-level phosphorylation.* The ATP-forming process that occurs during electron transport is called *oxidative phosphorylation* (see Figure 23.6).

Figure 23.3 shows the 10 steps of glycolysis. In steps 1 and 3, energy in the form of ATP is used to activate the breakdown of glucose and the rearrangement of the molecules. In step 4, the 6-carbon molecule of fructose 1,6-diphosphate is split into two 3-carbon molecules: dihydroxyacetone phosphate and glyceraldehyde 3-phosphate. Each of these 3-carbon molecules may be converted into the other (step 5), and some glyceraldehyde 3-phosphate will continue down the pathway. In step 6, the glyceraldehyde 3-phosphate picks up an inorganic phosphate (P_i) from the cytoplasm, and is also oxidized to form 1,3-diphosphoglyceric acid. The hydrogen released in the oxidation is picked up by the coenzyme nicotinamide adenine dinucleotide (NAD^+) to form $NADH + H^+$, the *reduced* form of NAD^+. During steps 7 through 10, four molecules of ATP are generated. Since two molecules of ATP were required to start the activation of the glucose molecule (in steps 1 and 3), there is a net gain of only two ATP molecules for each glucose molecule metabolized. In step 10, the formation of two molecules of **pyruvic acid** (pyruvate) marks the end of glycolysis.

Note that when a cell has broken down a glucose molecule as far as pyruvic acid, it has not only generated a net of two ATP molecules, but also two molecules of reduced, energy-carrying $NADH + H^+$. The cell has not released any carbon dioxide or used any oxygen. Most of the chemical-bond energy originally present in the glucose is still present in the two molecules of pyruvic acid.

Once a cell has made pyruvic acid, it can send it down various pathways. A *pathway* is a series of chemical reactions that generates a specific product; glycolysis is a pathway that generates pyruvic acid. Pyruvic acid can then be used in pathways that generate amino acids, that oxidize it to carbon dioxide and water, or that produce lactic acid.

Citric-acid (Krebs) cycle In one pathway of glucose catabolism, each of the two molecules of pyruvic acid combines with a molecule of coenzyme A (CoA) to form a 2-carbon compound called acetyl-coenzyme A (acetyl-CoA). This process releases a molecule of carbon dioxide and two hydrogen atoms. The acetyl-CoA begins the next pathway, a series of reactions called the *citric-acid cycle* or *Krebs cycle,** which occur in the mitochondria of the cell (Figure 23.4).

In step 1 of Figure 23.4A, the two carbons of acetyl-CoA combine with oxaloacetic acid, a 4-carbon compound already present in mitochondria. The reaction produces a 6-carbon molecule of citric acid. From here on around the citric-acid cycle, there is an ordered sequence of reactions that serve

several purposes. Some reactions are simple rearrangements that prepare molecules for later steps, some release carbon dioxide, some release hydrogen, and some release both. *The hydrogen-releasing reactions are the most important for energy transfer.* The other steps are means to that end.

In steps 2 and 3, the citric acid is rearranged to form isocitric acid, which *loses a pair of hydrogen atoms and a molecule of carbon dioxide,* which is immediately converted into the 5-carbon α-ketoglutaric acid. The α-ketoglutaric acid in step 5 *loses a pair of hydrogen atoms and a molecule of carbon dioxide,* producing 4-carbon succinyl-coenzyme A, which is broken down in step 6 to form succinic acid. This reaction ultimately generates a molecule of ATP. In step 7, the succinic acid *loses a pair of hydrogen atoms,* resulting in the formation of 4-carbon fumaric acid.

Step 8 rearranges the fumaric acid to yield the 4-carbon malic acid, which in step 9 *loses a final pair of hydrogen atoms,* producing oxaloacetic acid, the same kind of 4-carbon acid that accepted the 2-carbon acetyl group to begin the cycle. The 4-carbon oxaloacetic acid combines with the second 2-carbon acetyl group from the original glucose molecule to start another "turn" of the cycle.

Each of the reactions in the citric-acid cycle is assisted by one or more specific enzymes, which are bound to the mitochondrial membranes. Without these enzymes, the reactions would not take place fast enough to maintain homeostasis.

Electron transport system So far, the single glucose molecule has yielded only four new molecules of ATP: two from glycolysis and two from two turns of the citric-acid cycle. (Recall that glycolysis results in the formation of *two* molecules of pyruvic acid made available for the citric-acid cycle.) Where is the rest of the energy that is released from glucose? Also, what has happened to all the hydrogen atoms that have left the molecules during the citric-acid cycle?

Most metabolic energy is released by the **electron transport system,** when electrons are transferred from molecules of $NADH + H^+$ and $FADH_2$ to molecules of oxygen, (Figure 23.5). In the process, some of the chemical energy released is used to synthesize ATP, and the rest is given off as heat.

Most of the energy transfer from $NADH + H^+$ and $FADH_2$ actually involves a transfer of the *electrons* from hydrogen atoms to the accepting nitrogens in the NADH and FAD molecules. Electrons are transferred in a series of oxidation-reduction reactions in the inner membrane of mitochondria. At each step along the way, electrons fall to a lower energy state, until they are finally transferred to oxygen. Water forms when oxygen accepts hydrogen and electrons. Some of the energy released as the electrons move from high energy levels to lower ones is ultimately used to generate ATP inside the

---

*The Krebs cycle is named after the British biochemist Sir Hans Adolf Krebs (1900–1981), who during the 1930s first outlined the steps in the complete breakdown of pyruvic acid. Krebs won a Nobel Prize in physiology or medicine in 1953 for his work.

Why do I wake up cold when I fall asleep uncovered?

When you are asleep, your body is not as active as when you are awake. As a result, overall metabolism slows down, and less body heat is generated. There is also an increase in heat loss because of the dilation of blood vessels in the skin.

FIGURE 23.3

Glycolysis. (A) The 10 steps of glycolysis produce two molecules of 3-carbon pyruvic acid. The enzymes controlling the reactions are numbered to correspond to the 10 steps. Note that each of the intermediates in steps 6 through 10 (glyceraldehyde 3-phosphate to pyruvic acid) are made up of *two* molecules. (B) Simplified version of glycolysis, showing the major events.

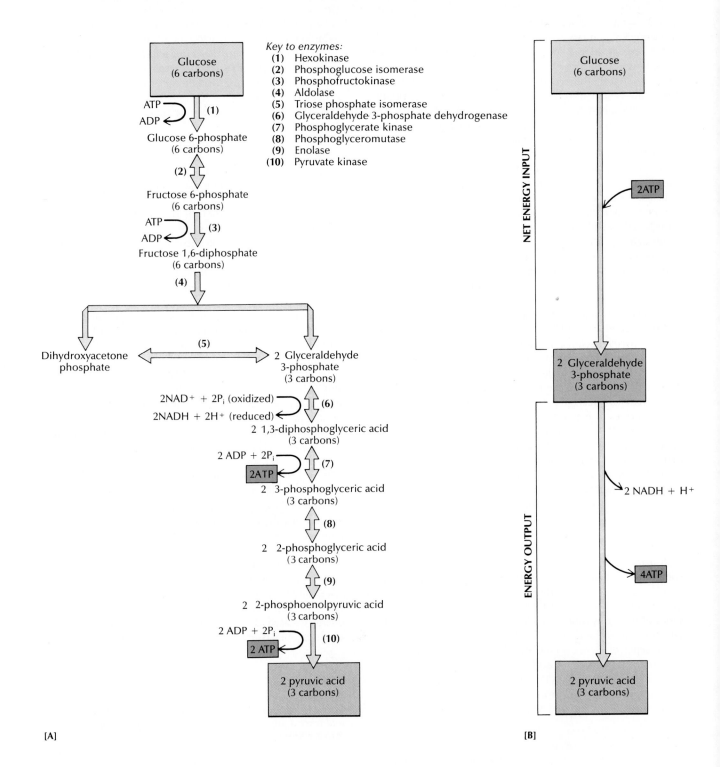

FIGURE 23.4A

Citric-acid (Krebs) cycle. (A) In this representation, the chemical structures and enzymes involved in the process are shown. Each time an acetyl group is oxidized (one turn of the cycle), 3 NADH + H^+, 1 $FADH_2$, and 1 ATP are produced. The GTP shown here is a high-energy compound that precedes the production of ATP. (A simplified version of the citric-acid cycle appears on the following page.)

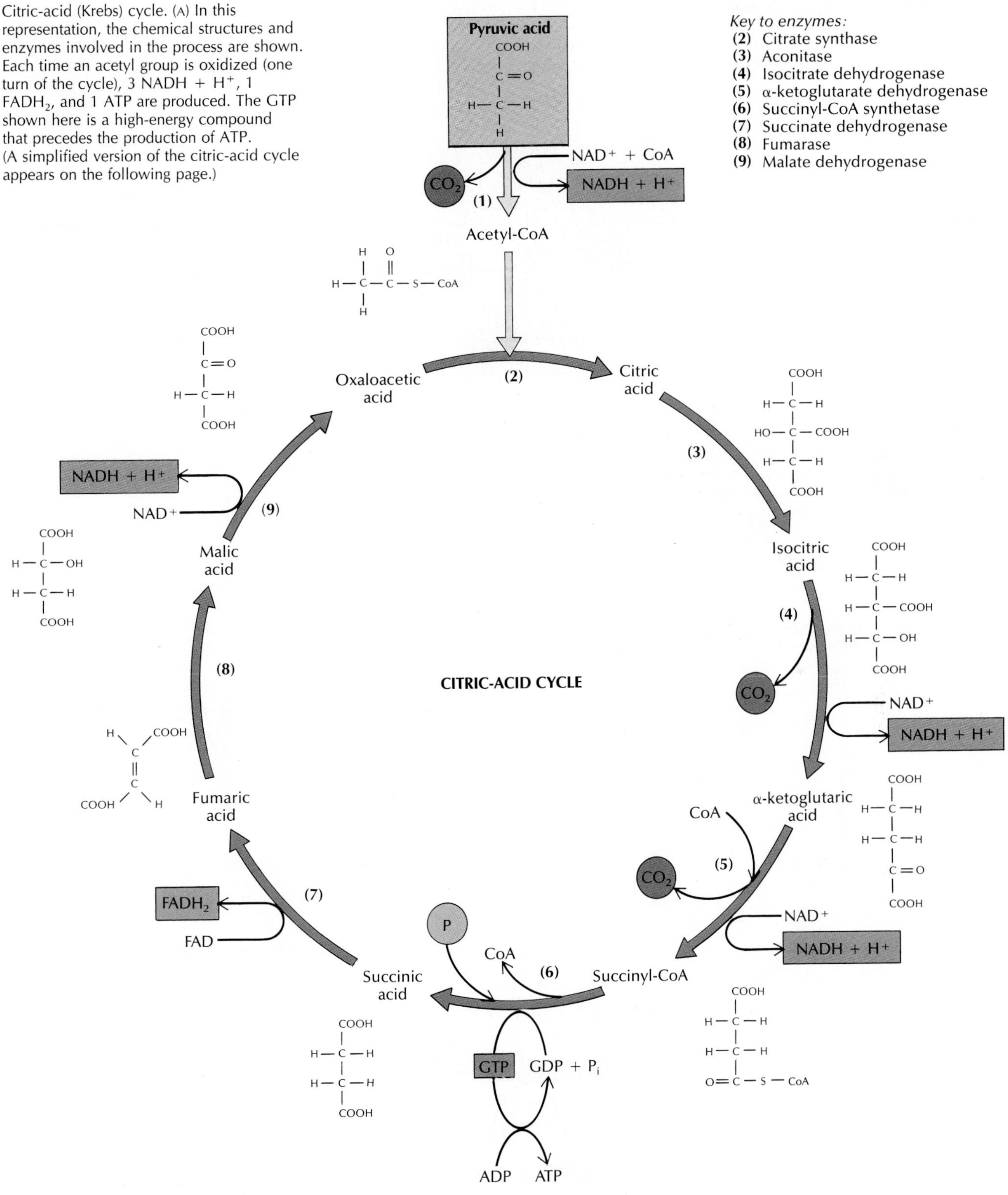
Pyruvic acid
NAD+ + CoA
NADH + H+
CO2
(1)
Acetyl-CoA
Oxaloacetic acid
(2)
Citric acid
(3)
Isocitric acid
(4)
CO2
NAD+
NADH + H+
α-ketoglutaric acid
CoA
(5)
CO2
NAD+
NADH + H+
Succinyl-CoA
(6)
CoA
P
Succinic acid
GTP
GDP + Pi
ADP
ATP
(7)
FADH2
FAD
Fumaric acid
(8)
Malic acid
(9)
NADH + H+
NAD+
CITRIC-ACID CYCLE
Key to enzymes:
(2) Citrate synthase
(3) Aconitase
(4) Isocitrate dehydrogenase
(5) α-ketoglutarate dehydrogenase
(6) Succinyl-CoA synthetase
(7) Succinate dehydrogenase
(8) Fumarase
(9) Malate dehydrogenase

[A]

FIGURE 23.4B

(B) Simplified version of the citric-acid cycle.

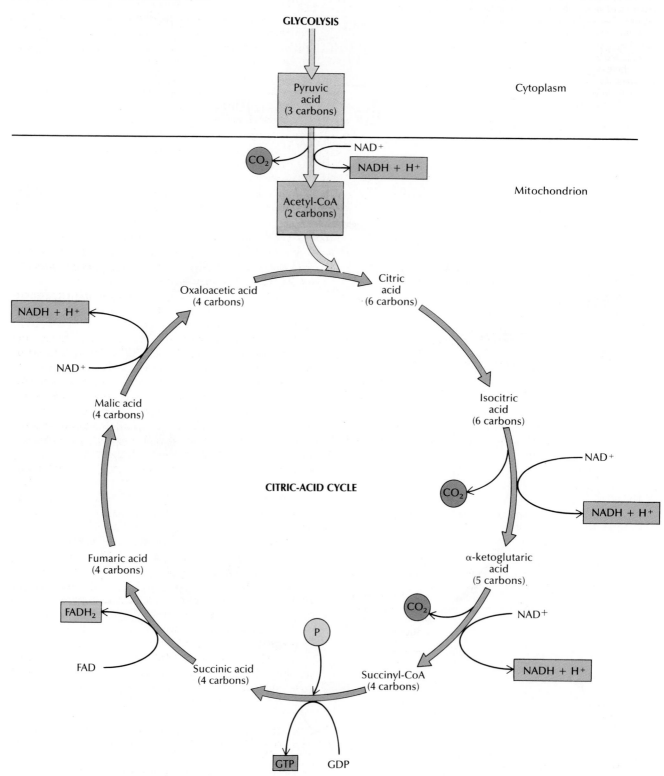

[B]

FIGURE 23.5

Electron transport system. The transfer of hydrogen atoms and electrons results in the formation of ATP at three places in the system.

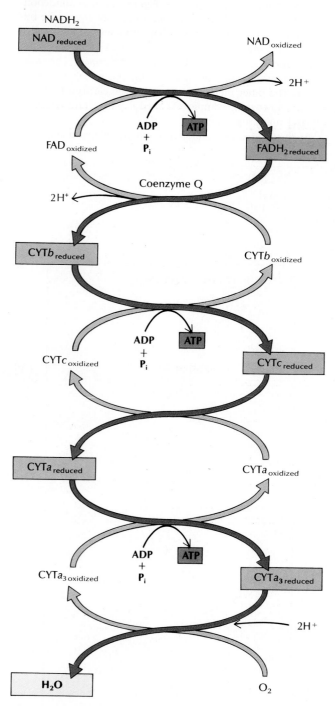

Key:
NAD = Nicotinamide Adenine Dinucleotide
FAD = Flavin Adenine Dinucleotide

mitochondrion. The ATP then moves throughout the cell, readily available to power the cell's many different metabolic reactions.

If we look at the electron transport system in greater detail (see Figure 23.5), we see that when electrons from the hydrogen atoms leave a molecule to be taken on by another molecule, energy goes with them. The electrons are transferred to an orderly arrangement of hydrogen acceptors called *cytochromes* (protein-plus-pigment molecules containing iron). Some of the cytochromes are accompanied by specific enzymes. Cytochromes can be alternatively reduced (by taking up electrons) and oxidized (by giving up electrons) on the iron group, with an energy loss or gain accompanying any electron transfer.

As electrons are passed from cytochrome to cytochrome, some energy is necessarily lost as heat, but some of it is caught up in coupled reactions that bring ADP and inorganic phosphate together to form ATP. The final hydrogen acceptor is oxygen, and when an oxygen atom accepts two hydrogen atoms, the result is water. *The final, biologically useful product of cellular respiration is ATP.* The carbon dioxide and water can be regarded as waste products.

If oxygen is present to act as the final hydrogen acceptor, each pair of hydrogen atoms released by cellular respiration contributes to the production of ATP. There are six reactions in the entire respiration process that liberate hydrogen: one during glycolysis, one during the formation of acetyl-CoA from pyruvic acid, and four during the citric-acid cycle.

Hydrogens from glyceraldehyde 3-phosphate and from pyruvic, isocitric, α-ketoglutaric, and malic acids reduce NAD^+ to $NADH + H^+$. Passage of each pair of hydrogen atoms from $NADH + H^+$ along the electron transport system generates three molecules of ATP. Hydrogen from succinic acid, however, reduces FAD to $FADH_2$. Therefore, since the first step is bypassed, only two molecules of ATP per passage are produced (see Figure 23.5).

Enough hydrogen comes from one molecule of glyceraldehyde 3-phosphate to reduce five molecules of NAD^+ to $NADH + H^+$, and to reduce one molecule of FAD to $FADH_2$. Five $NADH + H^+$ yield 14 molecules of ATP (not the expected 15, because one is used in glycolysis). One $FADH_2$ yields two molecules of ATP, for a total of 16. Since one molecule of glucose yields two molecules of glyceraldehyde 3-phosphate, the ATP figure is doubled, to provide 32 molecules* of ATP. Besides, each turn of the citric-acid cycle also yields one ATP molecule without the use of the electron transport system. Thus, the total number of ATP molecules produced from one molecule of glucose is 36,[†] with 2 from glycolysis, 2 from two turns of the citric-acid cycle, and 32 from the electron transport system (Figure 23.6).

*This number is a theoretical maximum, and is not realized in the cell because of the energy required to shuttle molecules into and out of the mitochondria.

[†]The number of ATP molecules produced from the oxidation of one glucose molecule may be 38 in some cases, depending upon cellular conditions. The additional two ATP molecules are produced in the electron transport system from the $NADH + H^+$ produced during glycolysis (see Figure 23.6).

Chemiosmotic theory of ATP production For many years, scientists tried to explain exactly how the energy released during the electron transport system was harnessed to produce ATP. In 1961, British biochemist Peter Mitchell proposed an explanation that is now commonly accepted. In fact, it was supported by such a growing mass of evidence that Mitchell received the 1978 Nobel Prize in chemistry for his work. Mitchell's hypothesis, now known as the *chemiosmotic theory* of ATP production, or *chemiosmosis,* is based on the fact that the production of ATP by mitochondria involves both chemical and transport processes ("chemi" + "osmosis") across a selectively permeable mitochondrial membrane (Figure 23.7). Chemiosmosis involves (1) the accumulation of electrochemical energy built up from an electrochemical proton gradient, and (2) the release of that energy to make ATP from ADP and P_i.

The chemiosmotic theory explains how mitochondria produce ATP, and the structure of mitochondria is important in understanding the process (see Figure 23.7). A mitochondrion has two membranes, an outer one that covers the mitochondrion and an inner one that forms several folds called *cristae*. The inner membrane is separated from the outer membrane by a space called the *outer compartment*. The space enclosed by the inner membrane is the *inner compartment,* or *matrix.* The inner membrane contains the electron transport system and enzyme complexes called F_1 *particles* (or ATP synthetase) that project into the matrix.

In the course of passing through the electron transport

FIGURE 23.6

Summary of the yield of ATP from the oxidation of 1 molecule of glucose. Although a total yield of 36 ATP molecules is shown here (2 from glycolysis, 2 from the citric-acid cycle, and 32 from the electron transport system; red circles), the actual yield may vary, depending on the cellular conditions. The dashed circle on the right indicates that the ATP yield from the NADH + H$^+$ produced during glycolysis may be 6 molecules instead of 4.

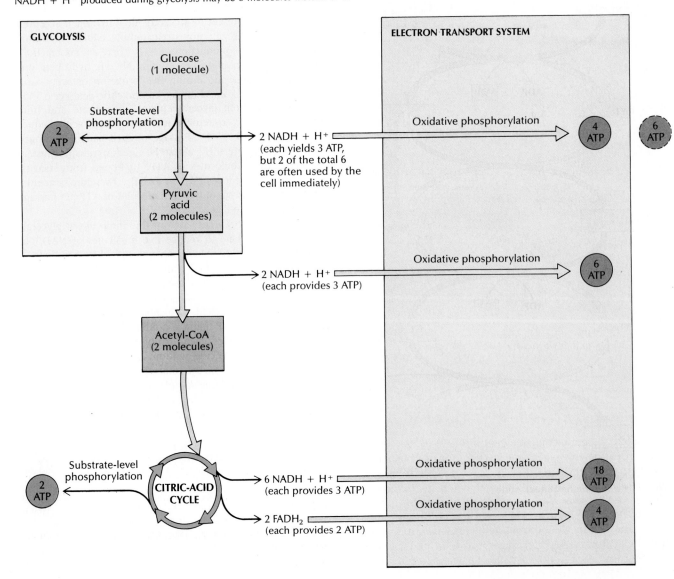

FIGURE 23.7

Chemiosmosis, the mechanism that produces ATP as a result of the electron transport system. (A) Schematic drawing of a mitochondrion, the site of ATP production. Note how the inner and outer membranes form the matrix and outer compartment. (B) Enlarged schematic drawing of mitochondrial membranes. Energy derived from an electrochemical proton (H^+) gradient is used to drive the enzyme ATP synthetase to catalyze the conversion of ADP and P_i into usable ATP.

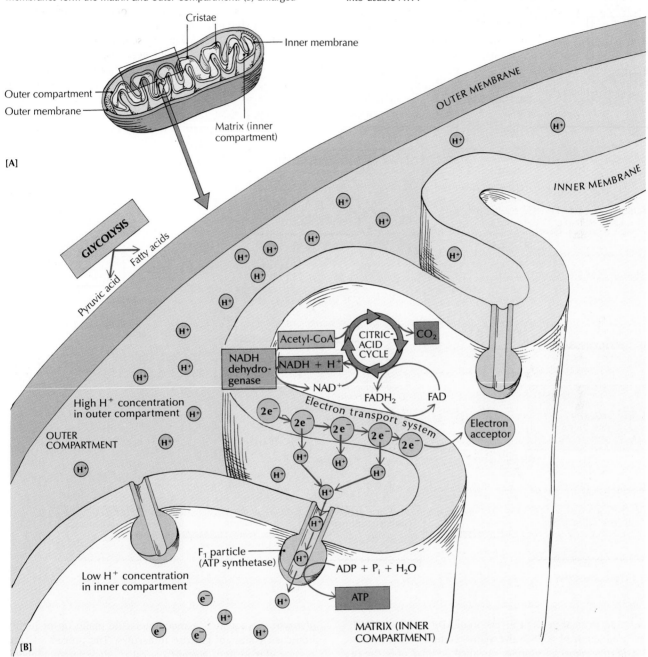

system in the inner membrane, electrons fall to lower energy levels, releasing energy. Hydrogen carriers in the membrane transport hydrogen ions (H^+) (protons) and electrons (e^-) from NADH in the matrix to the inner membrane (see Figure 23.7B). (NADH generated by glycolysis is also fed into the electron transport system.) Electron carriers in the inner membrane pick up the electrons and return them to the matrix; the hydrogen ions, however, are released into the outer compartment. As a result, there is a higher concentration of protons in the outer compartment than in the matrix. The energy released by the electrons in the electron transport system is harnessed to pump protons across the inner membrane into the outer compartment. The result is an *electrochemical proton gradient* between the inner and outer compartments. In order for the protons to move back into the inner compartment to re-establish equilibrium, they must pass through channels in the F_1 particles. Within the F_1 particles, most of the energy stored in the gradient powers an enzyme, *ATP*

FIGURE 23.8

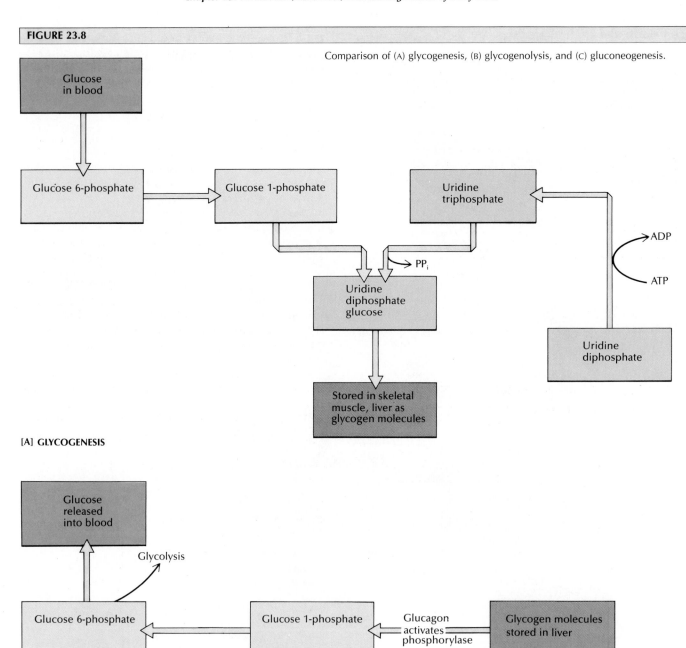

Comparison of (A) glycogenesis, (B) glycogenolysis, and (C) gluconeogenesis.

[A] GLYCOGENESIS

[B] GLYCOGENOLYSIS

synthetase, to catalyze the conversion of ADP and P$_i$ into ATP. The production of ATP in the matrix completes the process of *oxidative phosphorylation,* so called because only the reactions that take place within the inner membrane make direct use of oxygen.

Conversion of Glucose into Glycogen (Glycogenesis)

Glucose that is not needed immediately is removed from the blood and stored in the liver and skeletal muscle cells as *glycogen.* Glycogen is a macromolecule made up of highly branched chains of glucose molecules. The process of converting glucose into glycogen is called *glycogenesis* ("glycogen production") (Figure 23.8A). In glycogenesis, excess blood glucose is converted into glucose-6-phosphate, which in turn is converted into glucose-1-phosphate. Glucose-1-phosphate combines with uridine triphosphate (UTP) to form uridine diphosphate glucose (UDP glucose), which transfers glucose moieties to preformed glycogen chains. Glycogenesis is stimulated by insulin from the pancreas, and all of the reactions are assisted by specific enzymes. Excess glucose that cannot be stored in the liver or muscle cells is usually converted into fat and stored in adipose tissues.

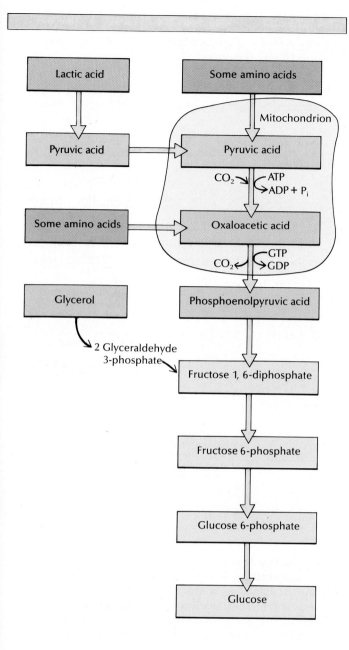

zyme into glucose-6-phosphate, which can be utilized directly for energy production (see Figure 23.3) or hydrolyzed into free glucose to permit its distribution via the bloodstream.

Production of Glucose from Noncarbohydrates (Gluconeogenesis)

Gluconeogenesis is the "production of new glucose" from noncarbohydrate sources such as lactic acid (lactate), glycerol, and some amino acids (Figure 23.8c). It occurs mainly in the liver, but also in bone cells and the cortex of the kidneys. When the level of blood glucose is low, and carbohydrates are not readily available, the glucocorticoid hormone cortisol from the adrenal cortex diverts some amino acids from body cells to the liver, where liver cells convert them into glucose. Each amino acid is converted by a slightly different chemical process. Thyroxine may also divert fats from adipose tissue to the liver, where the glycerol portion of the fat molecules is converted into glucose. Only about 60 percent of the amino acids in the body proteins can be converted into glucose.

Another very important starting molecule for gluconeogenesis is the lactic acid that originates in skeletal muscle cells as a normal waste product. Once lactic acid arrives in the liver via the bloodstream, it can be converted into glucose.

Ask Yourself

1 What is cellular respiration?

2 What does glycolysis produce?

3 Which phase of glucose catabolism produces the main supply of ATP?

4 What is the maximum number of ATP molecules that can be obtained from one molecule of glucose?

5 What is the main difference between glycogenesis and glycogenolysis?

6 Under what major condition is gluconeogenesis employed?

PROTEIN METABOLISM

The metabolic pathways for carbohydrates, proteins, and lipids are quite similar, differing only in the initial stages of cellular respiration. After proteins are broken down to amino acids, the amino acids undergo certain molecular changes, and then via acetyl-coenzyme A (acetyl-CoA) enter directly into the citric-acid cycle, skipping glycolysis.

After dietary proteins are converted into amino acids by enzymatic action in the digestive tract, the amino acids enter the bloodstream from the small intestine, and are carried to the liver via the portal vein. Amino acids are absorbed mainly by the liver, and may take several different metabolic routes (Figure 23.9):

Conversion of Glycogen into Glucose (Glycogenolysis)

When body tissues need extra glucose for energy, the glycogen stored in the liver is reconverted into glucose and released into the blood. This process is called *glycogenolysis* ("splitting of glycogen") (Figure 23.8B). The breakdown of glycogen occurs in liver cells under the endocrine control of glucagon from the pancreas and epinephrine from the adrenal medulla, which cause the activation of the enzyme phosphorylase to convert glycogen into glucose-1-phosphate. The glucose-1-phosphate is then reconverted by another en-

FIGURE 23.9

Metabolic pathways of amino acids in the liver.

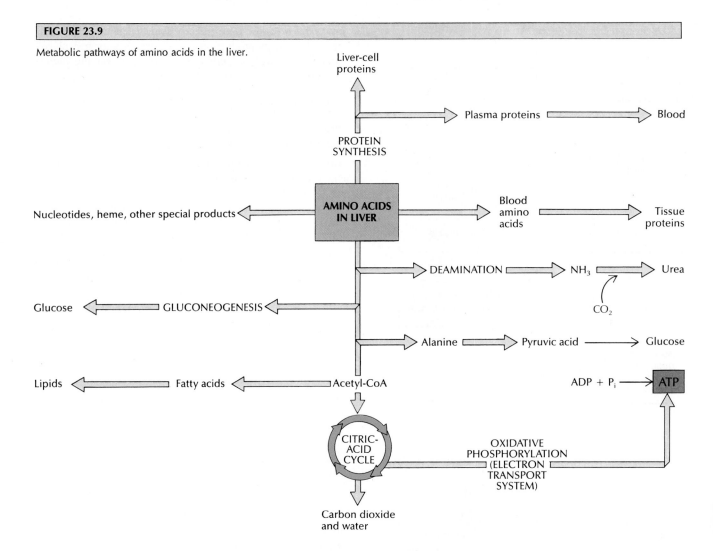

1 They can pass into the blood to be used by organs in the synthesis of tissue proteins.

2 They can be used to renew liver-cell proteins, or be used by the liver to synthesize plasma proteins.

3 They can be degraded and then converted into glucose or glycogen to produce ATP via the citric-acid cycle, or be converted into stored lipids.

4 They can be used in the *glucose-alanine cycle* in the liver to maintain the homeostatic balance of blood glucose between meals by converting them into pyruvic acid and then into glucose.

5 They can be used as building blocks to synthesize the nucleotides of nucleic acids, antibodies, hormones, or other nitrogen-containing compounds.

Some reactions of protein metabolism can be carried out by other cells, but not as efficiently or rapidly as by liver cells.

Protein Catabolism

After proteins have been broken down into amino acids, they must be further modified before they can be used for cellular respiration. One type of change in amino acids is called *oxidative deamination.* In the liver, amino acids can undergo oxidative deamination, in which the carbon chain is oxidized to a *keto acid* (α-ketoglutaric acid) and the amino group is released as free ammonia (NH_3). Then, the α-ketoglutaric acid can be converted into pyruvic acid and enter the citric-acid cycle (Figure 23.10). The leftover NH_3 group becomes incorporated into *urea,* and is excreted. For example, the amino acid alanine can be converted into pyruvic acid, the amino acid aspartic acid becomes oxaloacetic acid, and glutamine becomes α-ketoglutaric acid. Other more complex amino acids undergo their own special molecular rearrangements, and can also be used in cellular respiration to yield energy in the form of ATP.

Keto acids can be used to produce carbon dioxide and ATP, they can be intermediates in the formation of glucose, or they can be converted into acetyl-CoA to synthesize fatty acids. Because protein is not stored in cells, such alternate uses of protein are at the expense of its tissue-building function. When there is a protein deficiency, the prolonged use of protein as a source of glucose, for example, usually causes muscle cells to become smaller.

FIGURE 23.10

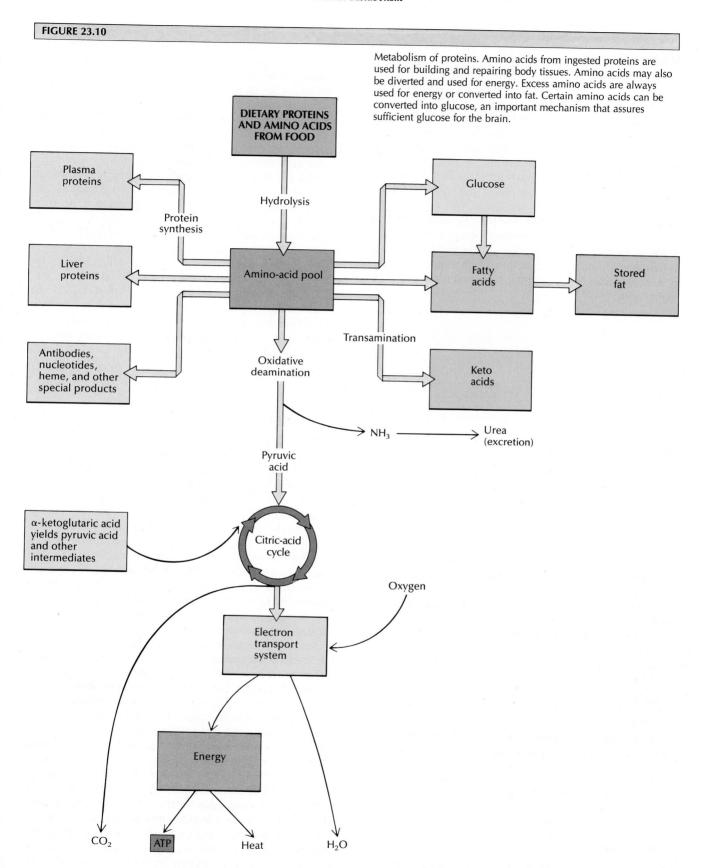

Metabolism of proteins. Amino acids from ingested proteins are used for building and repairing body tissues. Amino acids may also be diverted and used for energy. Excess amino acids are always used for energy or converted into fat. Certain amino acids can be converted into glucose, an important mechanism that assures sufficient glucose for the brain.

Protein Anabolism

Protein anabolism is carried on by most cells in the body as they synthesize proteins from amino acids. Half of the 20 usable amino acids can be synthesized by cells. These are called *nonessential amino acids.* The other half, which cannot be synthesized, and must be supplied by the diet, are called *essential amino acids.* Nonessential amino acids may also be supplied in part by proteins in the diet. The protein foods that contain all 10 essential amino acids are called *complete proteins.* These include eggs, milk, and meat, which are further classified as *first-class complete proteins* because they contain large quantities of the 10 essential amino acids.

A reaction used to synthesize nonessential amino acids is *transamination,* a process in which the amino group (NH_2) is transferred from an amino acid to a keto acid (a precursor of an amino acid), such as pyruvic acid, producing a different amino acid:

$$\begin{array}{cccc}
\underset{\text{amino acid}}{R_1\!-\!\overset{\displaystyle H}{\underset{\displaystyle NH_2}{\overset{|}{\underset{|}{C}}}}\!-\!COOH} + \underset{\text{keto acid}}{R_2\!-\!\overset{\displaystyle O}{\overset{\|}{C}}\!-\!COOH} & \longrightarrow & \underset{\substack{\text{new}\\\text{keto acid}}}{R_1\!-\!\overset{\displaystyle O}{\overset{\|}{C}}\!-\!COOH} + \underset{\substack{\text{new}\\\text{amino acid}}}{R_2\!-\!\overset{\displaystyle H}{\underset{\displaystyle NH_2}{\overset{|}{\underset{|}{C}}}}\!-\!COOH}
\end{array}$$

Amino groups are usually transferred from the nonessential amino acid glutamine, which is abundant in cells, but can also be obtained from three other nonessential amino acids: asparagine, glutamic acid, and aspartic acid. Transamination requires enzymes called *transaminases.*

Protein synthesis is regulated by several hormones: (1) *growth hormone* (GH) increases the rate of synthesis; (2) *insulin* speeds up the transport of amino acids into cells and also increases the available glucose; (3) *glucocorticoids* promote protein catabolism, and thus increase the amount of amino acids in body fluids, and also act on ribosomes to increase translation efficiency; and (4) *thyroxine* increases the rate of protein synthesis when adequate carbohydrates and lipids are available as energy sources, and also degrades proteins to be used for energy when the other nutrients are present in insufficient amounts.

Ask Yourself

1 *What is oxidative deamination?*

2 *What is transamination?*

3 *What is the difference between an essential and a nonessential amino acid?*

4 *What are some of the major hormones that regulate protein synthesis?*

LIPID METABOLISM

Although we usually use the term *fat* in everyday conversation, *lipid* (Gr. *lipos,* fat) is a more inclusive term for the numerous organic substances that share common solubility properties. The most abundant lipids are the *triacylglycerols* (called *neutral fats* because they are electrochemically uncharged), which are composed of three molecules of fatty acids linked to one molecule of glycerol. Triacylglycerols are synthesized and stored mainly in adipocytes, specialized cells in adipose tissue (see Chapter 4). Other types of lipids include fatty acids; some alcohols; sterols, including cholesterol and ergosterol; hydrocarbons, such as carotenoids; steroid hormones, such as cortisol and aldosterone; and the fat-soluble vitamins (A, D, E, and K). Lipids are generally insoluble in water, but they are soluble in organic solvents such as acetone and alcohol. (See Chapter 2 for a discussion of the structures and types of lipids.)

Digested fats are broken down into glycerol and fatty acids (Figure 23.11). The glycerol is first converted into an intermediate, dihydroxyacetone phosphate, and then into glyceraldehyde 3-phosphate. Because the latter is on both the glycolytic and the gluconeogenic pathways, glycerol can be converted into pyruvic acid or glucose, and then continue through the citric-acid cycle and the electron transport system in the same way as carbohydrates. But the fatty acids go directly into the citric-acid cycle via acetyl-CoA, bypassing glycolysis.

Lipid Catabolism (Lipolysis)

Adipocytes synthesize and store triacylglycerols. The stored triacylglycerols are hydrolyzed into fatty acids and glycerols needed for the synthesis of phospholipids or glycolipids (see Lipid Anabolism below), and as fuel molecules. During the catabolism of fatty acids (*beta oxidation*), fatty acids are activated to acetyl-CoA, transported across the inner mitochondrial membrane, and degraded in the mitochondrial matrix by a recurring sequence of four reactions: oxidation linked to FAD, hydration, oxidation linked to NAD^+, and conversion to acetyl-CoA. The $FADH_2$ and NADH formed in the oxidation steps transfer their electrons to oxygen by the electron transport chain (see Figure 23.5), and the acetyl-CoA formed normally enters the citric-acid cycle by condensing with oxaloacetic acid.

After beta oxidation, excess acetyl-CoA not required by the liver is condensed into *acetoacetic acid,* which is then converted into β-*hydroxybutyric acid* and *acetone.* Such substances derived from acetyl-CoA are collectively called *ketone bodies.* The formation of ketone bodies is called *ketogenesis.* Ketone bodies are circulated through the bloodstream to body cells, where they enter the citric-acid cycle. During periods when only fats are being metabolized, there may be an excessive accumulation of ketone bodies, a condition called *ketosis* (see p. 750).

The acetyl-CoA derived from fatty-acid catabolism enters the citric-acid cycle and is broken down in exactly the same way as the acetyl-CoA from pyruvic acid during the catabolism of glucose. (The catabolism of a typical fatty acid—one containing about 18 carbon atoms—provides the energy for the synthesis of approximately 146 molecules of ATP.) The metabolism of carbohydrates, protein, and lipids is summarized in Table 23.1 and Figure 23.12.

Lipid Anabolism (Lipogenesis)

The formation of new fatty acids and their incorporation into the fats of adipose tissue utilize enzymes and pathways similar to those found in the liver. Because adipose cells are not supplied as abundantly with cytoplasm or mitochondria as liver cells are, the process of *lipogenesis* (synthesis of fats) in adipose tissue relies primarily on precursor molecules, intermediates, and coenzymes. This is specifically the case in the associated reactions of carbohydrate metabolism leading to

the generation of acetyl-CoA, ATP, NADH, and α-glycerophosphate. This is the reason why it is not so much the amount of fat as the amount of carbohydrates in the diet that determines the amount of lipids deposited in the adipocytes.

Ask Yourself

1 *What are triacylglycerols?*

2 *What are the beginning and end products of beta oxidation?*

4 *What are the beginning and end products of fatty-acid synthesis?*

ABSORPTIVE AND POSTABSORPTIVE STATES

Although we may eat several times a day, with periods of fasting in between, our cells require a continuous supply of ATP energy. The mechanisms that cope with these alternating periods of feeding and fasting are controlled by hormones, especially insulin and glucagon from the pancreas. The feeding and fasting states are referred to as absorptive and postabsorptive states, as described below.

Absorptive State

The nutrients from food are broken down in the digestive tract. During the *absorptive state,* the nutrients are absorbed from the digestive tract into the cardiovascular and lymphatic systems (Figure 23.13). Following a meal, monosaccharides and amino acids enter the hepatic portal vein and are carried to the liver. But before fats can be absorbed into the general circulation, they must first enter the lymphatic system.

Monosaccharides After carbohydrates* are absorbed from the digestive tract, they enter liver cells, where they can be converted into and stored as glycogen, converted into fat, or used for other purposes. Some fat is stored in the liver, but most is transported by the blood in particles containing pro-

*Absorbed carbohydrates consist of monosaccharides such as galactose, fructose, and glucose. For the sake of simplicity, we will consider only glucose in this discussion.

FIGURE 23.11

Metabolism of fats. Fats are hydrolyzed into glycerol and fatty acids, which can be used immediately for energy or stored as fat to be used when energy is required. The red pathway indicates the most common one.

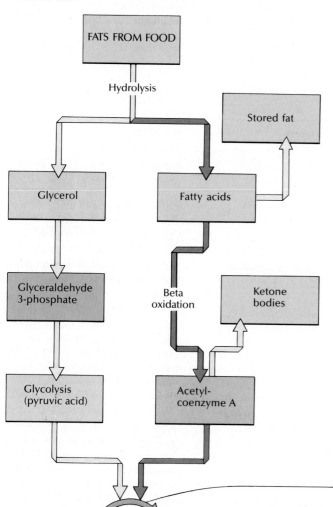

tein and other components, called *very low-density lipoproteins (VLDL).* Some of the fat is then transferred to adipose cells. Glucose that does not enter the liver is used almost immediately by most body cells as their main energy source during the absorptive state. Part of the remaining glucose is stored as muscle glycogen in skeletal muscle.

Triacylglycerols Before ingested lipids are absorbed into the bloodstream, they enter the lymphatic system as *chylomicrons,* which are small lipid droplets composed mainly of fatty acids, with small amounts of cholesterol and phospholipids, and a thin coating of protein. Chylomicrons are synthesized in intestinal mucosal cells from the products of lipid digestion, and are released into lymphatic vessels (see Chapter 22). When fatty acids are finally absorbed, they are probably taken up by all cells, since they are needed to make membrane phospholipids. During the absorptive state, cells may synthesize triacylglycerols and other fatty acids from glucose, or receive them directly from the VLDLs manufactured in the liver.

Amino acids Many of the absorbed amino acids enter the liver cells directly, where they are converted into keto acids by the removal of their NH_2 group. This process can also produce urea, which is excreted in urine. The keto acids enter the citric-acid cycle, and supply most of the liver's energy during the absorptive state. They may also be converted into triacylglycerols, which are stored mostly in adipose cells. The amino acids that are not taken up by the liver cells enter other cells to replace the protein that is being broken down continuously. Excess amino acids are stored as carbohydrate or fat.

Postabsorptive State

In the *postabsorptive* or *fasting state,* the energy requirements of body cells must be satisfied by the nutrients taken in during the absorptive state, and from the body's stored nutrients (Figure 23.14). Most importantly, the plasma glucose concentration must be maintained at a homeostatic level

TABLE 23.1   SUMMARY OF METABOLISM

Process	Description	Process	Description
CARBOHYDRATE METABOLISM		Glycogenolysis	Splitting of glycogen polymer chain to yield free glucose.
Cellular respiration (glucose catabolism)	Complete oxidation of glucose in cells to produce ATP as energy source; 1 molecule of glucose yields 36 molecules of ATP. Consists of glycolysis, citric-acid cycle, electron transport system.	Gluconeogenesis	Synthesis of glucose from lactic acid, pyruvic acid, or oxaloacetic acid, or from noncarbohydrates.
		PROTEIN METABOLISM	
Glycolysis (anaerobic respiration)	Initial breakdown of glucose and conversion into pyruvic acid to produce a net yield of 2 molecules of ATP. Takes place in cytosol. Does not require oxygen.	Protein catabolism	Oxidative deamination of amino acids to form keto acids, which can be used to produce glucose or fatty acids.
Citric-acid cycle (Krebs cycle)	Series of enzymatic reactions that release hydrogen from oxidized acids as a means of energy transfer. Produces CO_2, H_2O, and 2 ATP molecules from 2 "turns" of cycle. Takes place in mitochondria. Requires oxygen.	Protein anabolism	Directed by DNA in the nucleus and carried out by RNA in the cytoplasm. Synthesis by transamination (enzymatic transfer of an amino group from an amino acid to an α-keto acid).
Electron transport system	Series of oxidation-reduction reactions transferring electrons in hydrogen atoms to accepting cytochromes and oxygen. Produces most of the ATP of glucose catabolism (32 molecules). Occurs in mitochondria.	**LIPID METABOLISM**	
		Lipid catabolism (lipolysis)	Triacylglycerols hydrolyzed to glycerol, which enters glycolytic pathway, and fatty acids, which are catabolized by beta oxidation to acetyl-CoA, which enters the citric-acid cycle to produce ATP.
Glycogenesis	Conversion of glucose into glycogen by the polymerization of glucose molecules.	Lipid anabolism (lipogenesis)	Synthesis of fats by condensation of acetyl-CoA molecules and reduction to fatty acids and esterification of fatty acids to form triacylglycerols.

FIGURE 23.12

Summary of the metabolism of proteins, carbohydrates, and fats to produce ATP. All types of food can be digested to simpler compounds that can be converted into molecules that can be taken up in the citric-acid cycle and made to yield ATP. (Compare with Table 23.1.)

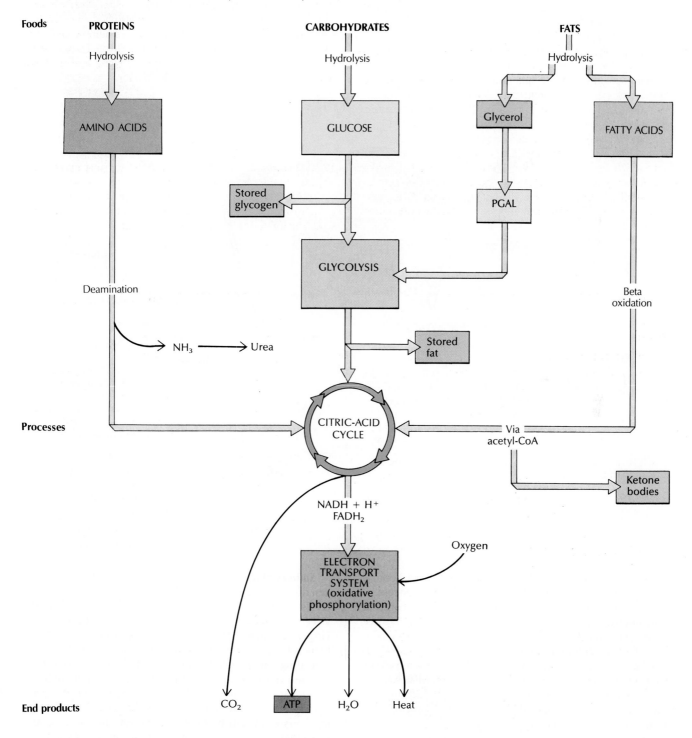

FIGURE 23.13

Metabolic pathways of the absorptive state.

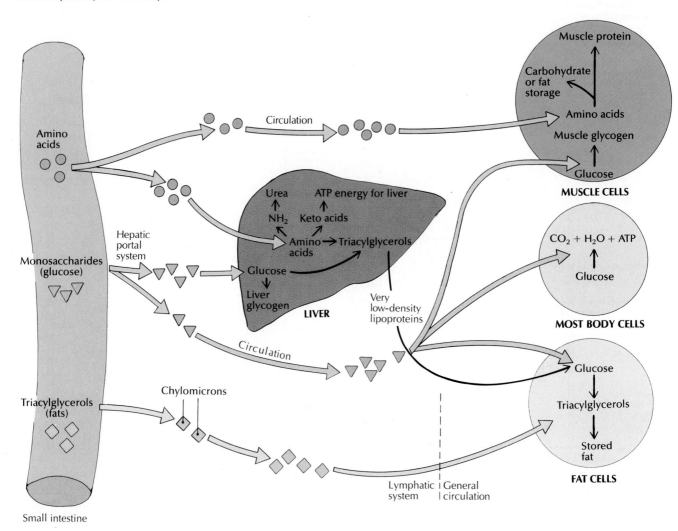

of 70 to 110 mg per 100 mL. This point is crucial because the brain, kidney medulla, and erythrocytes use glucose as their *only* energy source, but glucose is stored as glycogen in only limited amounts. The body solves this problem in three ways. (1) It breaks down glycogen stored in the liver (glycogenolysis). (2) It forms new glucose from protein and lipids (gluconeogenesis). (3) It shifts the metabolism of most cells so that they produce ATP from fatty-acid intermediates rather than from glucose, thus conserving glucose. In fact, most of the body's energy supply during the postabsorptive state comes from the oxidation of stored fat.

The postabsorptive state in human beings usually occurs in the late morning just before lunch, in the late afternoon just before dinner, and from the late evening until breakfast the next day. The complete absorption of a heavy meal takes about four hours, so the absorptive and postabsorptive states each occupy about 12 hours of the day.

Sources of blood glucose The glycogen in the liver is broken down directly to supply glucose to the blood. Triacylglycerols are broken down into glycerol and fatty acids. The glycerol can be converted into glucose by the liver, but the fatty acids cannot. The fatty acids are used directly by other body cells as an ATP energy source.

During long periods of fasting, the body uses protein from muscles, the liver, skin, and other parts of the body as an additional source of glucose. Because much of this protein is not essential for the function of muscles and other contributing tissues, it can be broken down into amino acids and converted into glucose. Ordinarily, however, very little structural protein is used for energy production.

Glucose sparing and fat usage With the exception of nerve cells, most body cells shift their metabolism during periods of fasting to use fat instead of glucose, sparing glucose for the

FIGURE 23.14

Metabolic pathways of the postabsorptive state.

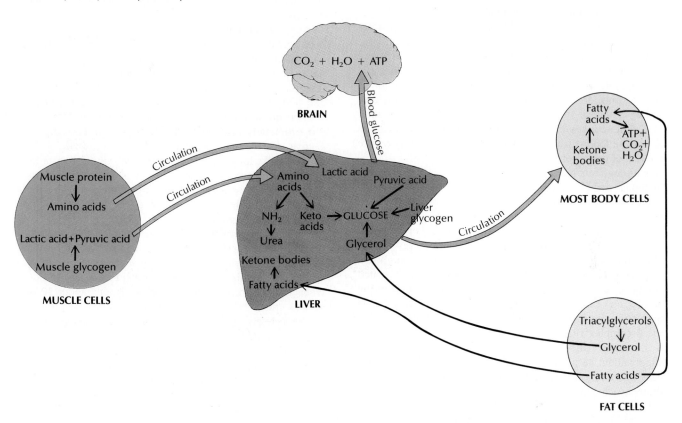

nervous system. The triacylglycerols of adipose tissue are hydrolyzed into glycerol and fatty acids, absorbed into the blood, and taken up by tissue cells to produce ATP during the citric-acid cycle. Fatty acids in the liver, however, are converted into ketone bodies such as acetone instead of entering the citric-acid cycle. The ketone bodies are then released into the blood, to be used by tissue cells during the citric-acid cycle to produce carbon dioxide, water, and ATP.

As a result of glucose sparing and fat usage, the concentration of blood glucose is reduced only a few percent, and homeostasis is maintained.

Ask Yourself

1 *What nutrients are the body's major energy sources during the absorptive state?*

2 *In what form do lipids enter the lymphatic system?*

3 *Why is it important for the body's glucose level to remain fairly constant?*

4 *What time of the day do the postabsorptive states usually occur?*

5 *How is the brain's glucose supply kept constant?*

NUTRITIONAL NEEDS AND BALANCE

The foods we eat contain six classes of *essential nutrients* that are vital to our well-being: carbohydrates, proteins, lipids (fats), vitamins, minerals, and water. Each type of food provides something different, and a well-balanced diet is necessary in order to ensure that we receive all types of nutrients in sufficient amounts to satisfy our various needs and maintain our health.

Calories and Energy

The energy value of food is measured in terms of calories or Calories. A *calorie* (L. *calor,* heat) is the amount of energy required to raise the temperature of 1 gram of water 1 degree Celsius. A calorie, with a small *c,* is also called a *gram calorie.* A **kilocalorie,** also known as a **Calorie** or *kilogram calorie (kcal),* is equal to 1000 calories (*kilo* = thousand). In popular usage, we talk about calories but actually mean Calories, because the larger unit is more useful in measuring the energy values of food. (If an advertisement says that a so-called light beer contains 95 calories per 12 oz., it really means 95,000 calories, 95 Calories, or 95 kcal.)

Dietary Carbohydrates

With the exception of lactose, which comes from milk, and glycogen, which is synthesized by the liver, all carbohydrates come from plant sources. Certain fruits supply glucose, fructose, sucrose, and starch.* Vegetables provide starch and some sugars. Syrups and honey supply glucose, fructose, and maltose, and ordinary table sugar is sucrose. Grain products and legumes are good sources of starch. Many plants also supply cellulose, an indigestible polysaccharide (see Chapter 2). Cellulose is sometimes called dietary fiber. It assists in the passage of foods through the large intestine, and may reduce the risk of cancer of the colon.

Although fats are the most concentrated source of energy in foods, carbohydrates provide the most readily available energy source. Special functions of stored liver glycogen include providing an important source of energy, maintaining the blood sugar level, and serving as a detoxifying and protective agent. Glucose is the major source of energy for the brain and nervous tissue. Galactosides, which are composed of galactose, a fatty acid, and a nitrogen base, are essential components of brain and nervous tissue. Lactose aids in the absorption of calcium, and promotes the growth of desirable bacteria in the small intestine that are involved with some of the B-complex vitamins.

Currently, carbohydrates in the average American diet supply about 46 percent of the body's energy. About 22 percent of these carbohydrates come from complex carbohydrates (such as starches), 6 percent from naturally occurring sugars, and 18 percent from refined and processed sugars. Sucrose alone may provide about 15 percent of the total of refined and processed sugars, a figure that is considered too high because sucrose has little nutritive value, and contributes to obesity, dental caries, and other homeostatic imbalances.

Dietary Proteins

The animal sources of protein include lean meat, poultry, fish and other seafoods, eggs, and milk and milk products, including yogurt and cheese. The plant sources include dried beans and peas (including tofu), nuts and peanut butter, cereals, bread, and pasta.

Protein is the most abundant of the body's organic compounds. Thousands of different types of proteins exist in cells, each cell carrying out a specific function regulated by its genes. (See Chapter 2 for a discussion of the structures and types of proteins.) Proteins form an important part of the blood. They are also important regulators of osmotic pressure and water balance within the body, and help to maintain the body's hydrogen-ion concentration.

Amino acids from dietary proteins are used for building and repairing body tissues. Some amino acids are diverted and used for energy in the process of gluconeogenesis. Pro-

*Starch and carbohydrates are often confused. Starch *is* one example of a carbohydrate, and is, in fact, the most common carbohydrate in your diet. Some high-starch foods are rice, bread, potatoes, and pasta.

teins and carbohydrates each produce about the same amount of energy, yielding about 4 Calories (kcal) per gram of food metabolized. Excess protein is always used for energy or converted into fat.

Although proteins are a vital part of our diet, we actually need to eat relatively little—about 0.8 gram per kilogram of weight for adults. This amount of protein supplies the essential amino acids, plus sufficient nitrogen, to make the nonessential amino acids. The need for protein varies with age and body weight. Ordinarily, men need more protein than women do because men are heavier. Young children need proportionately more protein than adults, because they use a great deal of their protein for the rapid growth of body tissue, while fully grown children and adults use protein mainly to replace the nitrogen lost as body protein is turned over. Pregnant women and nursing mothers require extra protein to provide for the growth of the fetus, and for the synthesis of milk protein.

Some dietary proteins provide more amino acids than others do because they have an amino-acid composition similar to that of body protein. For this reason, less meat, milk, and cheese are required than plant protein such as found in wheat. Although the protein in wheat is only about 30 percent as effective as the protein in milk, the plant protein is used effectively by the body because it contains some amino acids not found in other protein foods. Dietary protein is most beneficial when it is balanced with many types of food to provide all 20 necessary amino acids.

Dietary Lipids

Neutral fats or triacylglycerols are contained in cooking fats and oils, butter, margarine, salad dressings and oils, fat in meat and dairy products, nuts, chocolate, and avocados. Fat is visible in foods such as butter, cream, salad dressings, and meats. It is less obvious in foods such as egg yolk, cheese, nuts, and wheat germ.

Fats are the body's most concentrated source of food energy. They produce about 9 Calories (kcal) of usable energy per gram, more than twice as much energy available from an equal weight of carbohydrate or protein. However, lipids are utilized completely only if they are oxidized along with sugar. Without sufficient sugar, the body is forced to burn stored fat for energy. When only fats are being metabolized, however, there is an excessive accumulation of the breakdown products (ketone bodies)—acetoacetic acid, β-hydroxybutyric acid, and acetone—producing a condition called *ketosis.* The sweet smell of the acetone that is formed can frequently be detected on the breath when this condition exists. Ketosis occurs frequently if the diet is low in carbohydrates and high in fats, in starvation, and in uncontrolled diabetes mellitus, when the body switches to the metabolism of fats because glucose is not being metabolized properly (see Chapter 16).

Some ketone bodies are strong acids and so require large amounts of the body's alkaline supply to buffer them. If they continue to build up in the blood and interstitial fluids, they can easily use up available buffers and cause a drop in the blood pH. In this way, ketosis can lead to *acidosis,* one of the body's most serious acid-base imbalances, which through depression of the nervous system can lead to serious brain damage, coma, and death.

Adipose tissue is a compact storage site, consisting chiefly of triacylglycerol reserves. Large quantities of fat can be stored in the tissues, producing obesity. But not only lipids produce fatty tissue. Excess glucose can also be converted into fat. Obesity results whenever the number of calories in the diet exceeds the body's needs, regardless of whether the extra calories come from fats, carbohydrates, proteins, or alcohol.

Lipids have several functions. They supply a concentrated amount of usable energy in the form of calories, and they insulate the body and help to maintain a constant body temperature. Some lipids also supply or aid the absorption of the fat-soluble vitamins (A, D, E, K). Lipids add flavor to foods, cushion some internal organs, and help to promote and maintain healthy skin, normal growth, and reproductive ability. Cholesterol (a precursor of vitamin D) and phospholipids help to build cell membranes and intracellular structures. Essential fatty acids are precursors of prostaglandins.

Triacylglycerols contain three types of fatty acids: saturated (such as palmitic and stearic acids), unsaturated (such as oleic acid), and polyunsaturated (such as linoleic acid). Saturated fatty acids are solid at room temperature, while unsaturated fatty acids are liquid at room temperature and are generally termed "oils."

The American Heart Association suggests that 30 to 35 percent of our calories should come from fat. Most physicians and dietitians recommend that we consume more polyunsaturated fat than saturated fat, and that we restrict our cholesterol intake to 250 mg per day. A low-fat diet may reduce the risk of cancer and atherosclerosis, decrease fluid retention in the body, and promote weight loss.

Federal Dietary Guidelines

In 1980, the United States government issued the first federal dietary guidelines, which were published jointly by the Department of Agriculture and the Department of Health, Education, and Welfare. The guidelines specify:

1 Eat a variety of foods.

2 Maintain ideal weight.

3 Avoid too much fat, saturated fat, and cholesterol.

4 Eat foods with adequate starch and fiber.

5 Avoid too much sugar.

6 Avoid too much sodium (salt).

7 If you drink alcohol, do so in moderation.

Earlier, in 1977, the Senate Select Committee on Nutrition and Human Needs analyzed the typical American diet and set forth new dietary goals (Table 23.2). The Federal Dietary Guidelines recommend that we increase our consumption of carbohydrates to 58 percent of our total diet, with complex carbohydrates and naturally occurring sugars accounting for 48 percent, and refined and processed sugars dropping to only 10 percent.

TABLE 23.2   DIETARY GOALS FOR THE UNITED STATES

Nutrient	Current diet (%)	Dietary goals (%)
Carbohydrate	**46**	**58**
Complex carbohydrates and naturally occurring sugars	28	48
Refined and processed sugars	18	10
Protein	**12**	**12**
Fat	**42**	**30**
Saturated fat	16	10
Unsaturated fat	19	10
Polyunsaturated fat	7	10
	100	**100**

Source: The Senate Select Committee on Nutrition and Human Needs, 1977.

Vitamins

*Vitamins** are organic compounds required by the body in small amounts for the regulation of metabolism. As you saw in Chapter 2, many vitamins are converted into coenzymes after they enter the body. They play such an important role in supporting enzymes, especially those involved in providing energy for metabolism, that vitamin deficiencies can cause many different disorders (Table 23.3).

Because most vitamins cannot be synthesized by the body, they must be obtained from the diet. However, vitamin K is synthesized by bacteria in the large intestine, and some other vitamins can be synthesized in the body if certain building materials, called *provitamins*, are available in the diet. For example, carotene is a provitamin in the production of vitamin A. Its chemical structure is similar to that of vitamin A, making its conversion to vitamin A by the liver relatively easy.

Vitamins A, D, E, and K are usually classified as *fat-soluble*, and vitamin C and the B vitamins are *water-soluble*. A year's supply of vitamin A can be stored in the liver, and the other fat-soluble vitamins can be stored also,[†] but water-soluble vitamins must be replenished constantly. In fact, water-soluble vitamins are often lost during cooking because they dissolve in the cooking water or are degraded by heat. Water-soluble vitamins are easily absorbed, but the fat-soluble vitamins must be emulsified by bile salts in the gastrointestinal tract before they can be absorbed into the bloodstream. Once vitamins are absorbed, they are usually combined with more complex molecules, frequently in the liver.

*The term *vitamine* was suggested in 1912 to describe the life-promoting organic compounds we now call *vitamins*. The original term was a combination of *vita*, meaning life, and *amine*, because all vitamins were thought to be amines. When it was discovered that not all vitamins are amines, the final *e* was discarded.

[†]Fat-soluble vitamins may accumulate in the body in toxic amounts if they are taken in large, regular doses *(hypervitaminosis)*.

TABLE 23.3

Vitamin	Major sources	Major functions	Daily requirement	Effects of deficiency	Effects of excess
WATER-SOLUBLE VITAMINS					
B$_1$ (thiamine)*	Organ meats, whole grains, yeast, nuts, molasses, yogurt, oysters.	Formation of co-carboxylase needed to convert pyruvic acid into acetyl-CoA in citric-acid cycle. Helps regulate carbohydrate metabolism. Aids production of HCl.	1.5 mg	Impairs carbohydrate metabolism; beriberi, peripheral nerve changes, edema, heart failure, mental disturbance, paralysis, constipation, anorexia.	None reported.
B$_2$ (riboflavin)	Milk, eggs, beef and veal, liver, whole grains and cereals, spinach, nuts, yeast, molasses, beets.	Forms two flavin nucleotide enzymes (FAD, FMN) involved in oxidative phosphorylation. Aids cellular respiration. Releases energy to cells.	1.5–2.0 mg	Sensitivity to light, eye lesions, fissuring of skin, cataracts, vomiting, diarrhea, muscular spasticity.	None reported.
B$_6$ (pyridoxine)	Whole grains, liver, milk, molasses, leafy green vegetables, bananas, yeast, tomatoes, corn, yogurt.	Coenzyme (pyridoxal phosphate) for amino acid metabolism; transport of amino acids across plasma membranes.	1.0–2.0 mg	Dermatitis, nervous disorders, learning disabilities, anemia, kidney stones, fatty liver.	Sensory neuropathy.
B$_{12}$ (cyanocobalamin)	Organ meats, milk, cheese, eggs, yeast, fish, oysters. Synthesis by intestinal bacteria.	Coenzyme needed for RNA synthesis and for erythrocyte formation to prevent anemia.	2.0–5.0 mg	Pernicious anemia, nervous disorders (demyelination), malformed red blood cells, general weakness.	None reported.
Biotin	Liver, yeast, vegetables, eggs. Synthesis by intestinal bacteria.	Coenzyme concerned with nucleic acid synthesis, CO_2 fixation, transamination, nitrogen metabolism. Aids cell growth, fatty-acid production.	150–300 mg	Scaly dermatitis, muscle pains, weakness, insomnia, depression, fatigue.	None reported.
Folic acid (folacin, pteroylglutamic acid)	Organ meats, leafy green vegetables, eggs, milk, salmon. Synthesis by intestinal bacteria.	Needed for transfer of carbon units in nucleic acid metabolism and DNA formation. Red blood cell formation. Aids growth, reproduction, digestion.	0.5 mg or less	Failure of red blood cells to mature, anemia, gastrointestinal disturbances, diarrhea.	None reported.
Inositol	Fruits, nuts, vegetables, milk, yeast.	Aids in metabolism. Reduces cholesterol levels. Slows hardening of the arteries.	Not known.	Fatty liver, constipation, hair loss, eczema.	None reported.
Niacin (B$_3$, nicotinic acid)	Whole grains, liver, chicken, yeast, seafood, peanuts, dried beans and legumes.	Constituent of two coenzymes in hydrogen transport (NAD$^+$, NADP). Aids carbohydrate metabolism. Aids production of sex hormones. Reduces cholesterol level.	17.0–20.0 mg	Pellagra (skin and gastrointestinal lesions, mental disorders, digestive disturbances, mucous membrane inflammation, hemorrhage).	Flushing, burning, tingling around neck, face, hands.

Vitamin	Major sources	Major functions	Daily requirement	Effects of deficiency	Effects of excess
Pantothenic acid (B$_5$)	Food sources include yeast, liver, eggs, nuts, salmon, green vegetables, cereals. Synthesis by intestinal bacteria.	Forms part of CoA. Promotes proper growth, vitamin utilizaton, energy utilization.	8.5–10.0 mg	Fatigue, sleep disorders, neuromotor disorders, cardiovascular disorders, gastrointestinal distress, eczema.	None reported.
C (ascorbic acid)	Citrus fruits, butter, tomatoes, green peppers, broccoli, potatoes.	Necessary for oxidation reactions, synthesis and maintenance of collagen and other intercellular ground substance. Aids bone and tooth formation, healing.	75 mg	Scurvy, failure to form normal connective tissue fibers, anemia, low resistance to infection, bruises.	Kidney stones.
Choline	Leafy vegetables.	Part of phospholipids, precursor of acetylcholine.	Not known.	Not known; deficiency is unlikely.	None reported.

FAT-SOLUBLE VITAMINS

Vitamin	Major sources	Major functions	Daily requirement	Effects of deficiency	Effects of excess
A (carotene)	Not present in plant foods. Provitamins for the formation of vitamin A found in egg yolk, carrots, leafy vegetables, fruits, liver, oils, milk products.	Needed for synthesis of mucopolysaccharides, growth differential of epithelium, maintenance of normal epithelial structure. Formation of visual pigments (major constituent of rhodopsin).	5000 IU[†]	Night blindness, xerophthalmia (keratinization of corneal tissue), skin lesions, allergies, dry hair, fatigue, failure of skeletal growth, reproductive disorders.	Vomiting, headache, anorexia, irritability, skin peeling, swelling of long bones, enlarged spleen and liver, patchy hair loss, bone pain and thickening.
D[‡]	Fish oils, liver, milk, egg yolk. Synthesis of vitamin D$_3$ (cholecalciferol) results from ultraviolet radiation (sunlight) of 7-dehydrocholesterol in skin.	Increases calcium and phosphorus absorption from digestive tract. Helps control calcium deposition in bones and teeth. Promotes proper heart action.	400 IU	Rickets in children, osteomalacia in adults, diarrhea, insomnia, nervousness.	Vomiting, diarrhea, weight loss, kidney damage, calcification of soft tissues.
E (tocopherols)	Leafy green vegetables, wheat germ oil, liver, peanuts, eggs.	Helps red blood cells resist hemolysis. Aids in muscle and nerve maintenance. Acts as antioxidant to prevent cell membrane damage of unsaturated fats in organelles.	Not known.	Increased fragility of red blood cells, dry hair, hemolytic anemia in newborns.	None reported.
K (phylloquinone)	Yogurt, molasses, safflower oil, liver, leafy green vegetables. Synthesis by intestinal bacteria.	Aids in prothrombin synthesis (clotting factors VII, IX, X) in liver.	Not known.	Failure of blood to clot; severe bleeding, hemorrhages.	May cause jaundice, anemia, gastrointestinal problems.

*The B group of vitamins includes B$_1$ through pantothenic acid as listed here.
†An *IU,* or *International Unit,* is a quantity of a vitamin, hormone, antibiotic, or other biological substance that produces a specific internationally accepted biological effect.
‡The general term *vitamin D* actually refers to a group of steroid vitamins, including vitamin D$_3$ (cholecalciferol), vitamin D$_2$ (calciferol or ergocalciferol), and AT$_{10}$ (dihydrotachysterol). Vitamin D$_2$ and AT$_{10}$ are derived from ergosterol, a steroid found in plants. Although cholecalciferol (vitamin D$_3$) is the natural animal form of the vitamin, the *active* form of vitamin D is considered to be 1,25-dihydroxycholecalciferol, which is a modified form of cholecalciferol. The active form is 13 times more powerful than cholecalciferol.

Minerals

Minerals are naturally occurring, inorganic elements such as sodium, calcium, and potassium that are used for the regulation of metabolism (Table 23.4). Minerals make up about 4 to 5 percent of the adult body weight.

Some minerals, such as iodine, copper, zinc, cobalt, iron, and manganese, are needed in the body in such small amounts that they are called *microminerals* or *trace elements*. Although these microminerals are available in the body and in food in only minute amounts, their presence is essential nevertheless for proper metabolism. Microminerals function primarily as catalysts in enzyme systems. In contrast, calcium, phosphorus, sodium, chlorine, potassium, and sulfur are known as *macrominerals* because they are found in significant amounts in body tissue. (Calcium is the most abundant mineral in the body.) At least 100 milligrams of each macromineral are required in the daily diet. Minerals are frequently bonded chemically with an organic substance, such as iron in hemoglobin and iodine in thyroxine.

Water as a Nutrient

No nutrient is more important than water. About 60 percent of the body is water, so it is a major structural component. Water participates in many essential biochemical reactions, such as cellular respiration and the hydrolysis of carbohydrates, lipids, and proteins. It transports important chemicals throughout the body. It acts as a solvent and a lubricant. The important role of water in temperature regulation is discussed later in this chapter, and its regulatory role in excretion is covered in Chapters 24 and 25.

Ask Yourself

1 *What is the difference between a calorie and a Calorie?*

2 *What are the main sources of carbohydrates? Proteins? Lipids? What are some of their important functions?*

3 *Can you list the vitamins, giving some important sources and functions of each?*

4 *How do each of the macrominerals function in the body?*

5 *Why is water an essential nutrient?*

METABOLIC RATE AND TEMPERATURE CONTROL

As you have seen, the energy stored in the chemical bonds of food molecules such as sugar is eventually converted through a series of chemical reactions into energy-producing molecules such as ATP. The rest of the energy is dissipated as heat. Rather than being wasted, however, much of this by-product of heat helps to keep our bodies at a constant temperature. One way of measuring energy expenditure in the body as a whole is to determine the metabolic rate.

Factors Affecting the Metabolic Rate

The amount of energy used by the body during a specific period is the *metabolic rate*. It is measured in terms of the calories used up. Among the factors that affect the metabolic rate are type of food eaten, amount of exercise, hormones, and age.

Food The energy used in digesting a meal is known as the *specific dynamic action* of food. Eating a protein-rich meal generally raises the metabolic rate more than eating a meal high in lipids or carbohydrates. The greater increase in metabolic rate after eating protein is probably caused by the relatively rapid deamination of amino acids, among other factors.

Exercise Equally important is the amount of physical activity a person does, compared with the energy value of the food consumed, as measured in Calories. The number of Calories said to be contained in any food is actually the number of Calories that the body is able to metabolize and release. If the food you eat contains more Calories than you are able to use up by exercising, you will gain weight. If you use up more Calories than are provided in your diet, your body will begin to consume stored body fat, and you will probably lose weight. The metabolic rate decreases by about 10 percent when a person is asleep.

Hormones Several hormones are involved in a feedback system that helps maintain a steady metabolic rate. For example, when the metabolic rate is too low, the secretion of epinephrine and thyroxine increases. Thyroxine, secreted by the thyroid gland, affects metabolism more than any other hormone. When secretions of thyroxine increase, the metabolic rate may be raised as much as 100 percent above normal, and a decreased secretion may cause the metabolic rate to be lowered by as much as 40 percent below normal. Long-term thyroid disturbances can cause serious problems, as described in Chapter 16. The metabolic rate is raised by anxiety and lowered by depression. In situations of fear or anger, for example, the adrenal glands are stimulated to secrete extra epinephrine, which increases the metabolic rate. The metabolic rate is also affected by growth hormone and sex hormones.

Age and other factors One very important determinant of the metabolic rate is age. The metabolic rate of a child is generally higher than that of an adult, because the child requires more energy during its complex growth period. A pregnant or nursing woman also has an increased metabolic rate.

Other factors also affect the metabolic rate. Fever increases the rate. Muscular people tend to have a higher rate than people with a large amount of fatty tissue. Race and climate, which were once thought to affect the metabolic rate, are no longer considered important factors.

Basal Metabolic Rate

The *metabolic rate* is the total energy used by the body per unit of time to sustain the minimal normal cellular activity. Because the metabolic rate can be altered by many factors (including age, diet, exercise, and temperature), a person's metabolic rate is usually measured under standard, or basal, conditions. The metabolic rate computed under such conditions is called the **basal metabolic rate (BMR).** It is usually expressed as kilocalories of energy consumed per square meter of body surface area* per hour; BMR = $kcal/m^2/hr$.

There are several ways to calculate the BMR. The usual technique is to use an instrument called a *respirometer* or *metabolator* to determine the rate of oxygen consumed per unit of time. This method is based on the concept that because 95 to 100 percent of the energy available to the body is derived from cellular reactions involving oxygen, a person's BMR can be estimated accurately from oxygen consumption. It has been determined that for every liter of oxygen consumed, the amount of heat energy released is 4.825 kcal.

The person being tested should have no food for at least 12 hours after a light dinner the night before the test. A good night's sleep, complete rest on the morning of the test, and a moderate environmental temperature (20° to 26.7°C) are important. The person must be in a supine position for at least 20 minutes prior to the test, and should be relieved of all mental and physical stimuli that might cause excitement.

A sample calculation of the BMR for a 20-year-old male may proceed as follows. The respirometer shows that 10 liters of oxygen are consumed in one hour. When 10 is multiplied by the standard figure of 4.825 kcal liberated in one hour, the result is 48.25. This figure is then divided by the total body surface area, which has been calculated to be 1.5 m^2, producing a BMR of 32.2. According to standardized charts, the normal value for a 20-year-old male is 25.7, so the BMR of 32.2 is 25.3 percent above normal, and is therefore written as +25.3. If the BMR had been the same percentage below normal, it would have been written as −25.3. Since a BMR of plus or minus 15 percent is judged to be normal, the 20-year-old male being tested is considered to have an abnormally high BMR.

*The body surface area may be calculated based on the height and weight of the individual being tested, as follows:

Body surface area = $Weight^{0.425} \times Height^{0.725} \times 0.007184$

Ordinarily, published charts are available to estimate a person's surface body area.

Why do I feel warmer on hot, humid days than on hot, dry days?	*On hot, humid days the air is already so full of moisture that most of the sweat on your skin cannot evaporate. If the humidity is 90 percent, it means that the air contains 90 percent of the water vapor it can hold. As a result, your evaporation cooling system operates inefficiently. If the humidity reaches 100 percent, your body temperature will actually begin to rise when the outside temperature rises above 32°C (90°F).*

Temperature-Regulating Mechanisms

If the amount of heat usually generated by the resting body of a healthy individual were allowed to accumulate, the temperature of the body would rise about 30°C a day. Such a temperature increase would normally take place in a matter of hours, and during periods of vigorous exercise, it would occur in a matter of minutes. Vigorous exercise can increase heat production 20 times above normal in a very short time. Naturally, if we are to maintain homeostasis, such deviations in body temperature must be prevented. In fact, they do not occur because the body has certain built-in regulatory devices.

Several control mechanisms are available to regulate body heat. These include the three so-called *gradient mechanisms*—radiation, conduction, and convection—and the evaporation of sweat from the skin (perspiration) and water from surfaces in the upper respiratory tract. Metabolism always produces body heat, and evaporation always removes body heat, but gradient mechanisms can either warm the body or cool it. However, as you will see, the gradient mechanisms cannot cool the body when the temperature of the air is higher than the temperature of the skin.

Radiation When your body temperature rises, the temperature control center in your brain is alerted. (See the section on page 758 on the role of the hypothalamus.) Your brain causes the dilation of surface blood vessels, increasing the blood flow to the surface of the skin and carrying body heat along with it. Once the heat is concentrated near the body surface, it can be lost to the atmosphere through the process of **radiation,** the transfer of infrared heat rays from one object to another without physical contact between the objects.

In a normally heated room, 50 to 60 percent of the body's heat loss occurs through radiation (Table 23.5). Radiation can also work to warm the body when it is too cool. For example, although you are not in contact with the sun or the fire in your fireplace, you can still be warmed by their heat rays.

Conduction In contrast to radiation, **conduction** is the transfer of heat directly from one object to another. In order for conduction to occur, the two objects must be at different temperatures, and they must be in contact with each other. For example, the heat from your body warms the chair you sit in. Also, you warm your cold feet by putting them directly *on* a hot water bottle, not six inches away. Only about 3 percent of the body's heat transfer is accomplished through conduction, an almost negligible amount.

Convection *Convection* is the transfer of heat along a moving gas or liquid. For example, cool air next to the body is heated and then moved away on currents of air. The heated air removes heat from the body, and makes way for more cool air, which in turn is heated and moved away. (Because of the importance of convection in aiding heat loss from the body, the *wind-chill index* was developed. This quantitates the cooling effects of the combination of wind velocity and air temperature.)

TABLE 23.4   MAJOR MINERALS

Mineral	Major sources	Major functions	Daily requirement	Effects of deficiency	Effects of excess
MACROMINERALS					
Calcium	Dairy products, eggs, fish, soybeans.	Necessary for proper bone structure, normal heart action, blood clotting, muscle contraction, excitability, nerve synapses, mental activity, buffer systems, glycogen metabolism.	About 1 g	Tetany of muscles, loss of bone minerals, rickets in children, osteoporosis in adults.	Calcium deposits, calcification of soft tissues, heart failure.
Chloride	All foods, table salt.	Principal anion of extracellular fluid. Necessary for acid-base balance, osmotic equilibria.	2.0–3.0 g	Alkalosis, muscle cramps.	Acidosis, edema.
Magnesium	Green vegetables, milk, meats, nuts.	Necessary for proper bone structure, regulation of nerve and muscle action. Catalyst for intracellular enzymatic reactions, especially those related to carbohydrate metabolism.	About 1.2 g	Tetany.	None reported.
Phosphorus	Dairy products, meats, fish, poultry, beans, grains, eggs.	Combines with coenzymes in various metabolic processes. Necessary for proper bone structure, intermediary metabolism, buffers, membranes. Phosphate bonds essential for energy production (ATP), nucleic acids.	About 1.5 g	Extremely rare; related to rickets, loss of bone mineral.	None reported.
Potassium	All foods, especially meats, vegetables, milk.	Major component of intracellular fluid. Necessary for buffering, muscle contraction, nerve impulse transmission.	1.0–2.0 g	Changes in heart function, alteration in muscle contraction, muscle weakness, alkalosis.	Heart block.
Sodium	Most foods, table salt.	Major component of extracellular fluid. Necessary for ionic equilibrium, osmotic gradients, nerve impulse conduction, buffer systems.	About 1.0–3.0 g in adults	Dehydration, muscle cramps, kidney failure.	Edema, hypertension.
Sulfur	All protein-containing foods.	Structural, as amino acids are made into proteins.	1.2 g	None reported.	None reported.

Mineral	Major sources	Major functions	Daily requirement	Effects of deficiency	Effects of excess
MICROMINERALS (TRACE ELEMENTS)					
Chromium	Meats, vegetables, yeast, beer, unrefined wheat flour, corn oil, shellfish.	Necessary for glucose metabolism, formation of insulin for proper blood-sugar level.	0.05–0.2 mg	Impaired ability to metabolize glucose.	Industrial exposure may cause skin and kidney damage.
Cobalt	Meats.	Necessary for formation of red blood cells.	Not known.	May cause pernicious anemia.	Industrial exposure may cause dermatitis, diseases of erythrocytes.
Copper	Liver, meats, oysters, margarine, eggs, wheat products.	Necessary for hemoglobin formation, maintenance of certain copper-containing enzymes, proper intestinal absorption of iron.	2.0 mg in adults	Anemia, bone disease (rare), lack of white blood cells.	Wilson's disease (rare metabolic condition).
Fluorine	Fluoridated water, toothpastes, milk, tea.	Hardens bones and teeth. Suppresses bacterial action in mouth.	0.7 part/million in water (optimum)	Tendency toward dental caries, osteoporosis.	Mottling of teeth.
Iodine	Iodized table salt, fish, seaweed.	Necessary for synthesis of thyroxine, which is essential for maintenance of normal cellular respiration.	0.1–0.2 mg in adults	Goiter, cretinism in newborns.	Inhibits activity of thyroid.
Iron	Liver, eggs, red meat, shellfish, beans, nuts, raisins.	Component of hemoglobin, myoglobin. Necessary for transport of oxygen to tissues, cellular oxidation.	16 mg in adults	Iron-deficiency anemia, fatigue.	Cirrhosis of liver, hemosiderin deposits, blood diarrhea.
Manganese	Meats, bananas, bran, beans, leafy vegetables, whole grains, nuts.	Necessary for formation of hemoglobin, activation of enzymes.	0.4 g	Subnormal tissue respiration, growth retardation, bone and joint abnormalities, nervous system disturbances, reproductive abnormalities.	Muscular weakness, nervous system disturbances.
Molybdenum	Organ meats, legumes, green leafy vegetables.	Component of several enzymes.	0.15–0.5 mg	None reported.	Inhibited enzyme activity.
Selenium	Most foods, especially liver, other meats, seafoods.	Enzymes, lipid metabolism, antioxidant (protects plasma membranes from breaking down).	0.05–0.2 mg	Anemia (rare).	Gastrointestinal disorders, lung irritation.
Zinc	Meats, liver, seafood, eggs, legumes, milk, green vegetables.	Part of many enzymes.	15 mg	Impaired cell growth and repair, poor wound healing, impaired sense of taste, small reproductive glands.	Fever, nausea, vomiting, diarrhea.

TABLE 23.5   HOW THE BODY LOSES HEAT UNDER NORMAL CONDITIONS

Activity	Approximate percentage of total heat loss
Radiation	54
Conduction	3
Convection	15
Evaporaton from skin	15
Evaporation from respiratory tract (including lungs)	7
Warming air taken into lungs	3
Urination, defecation	3
	100

Evaporation   When the temperature of the air is higher than the temperature of the surface of the body, radiation, conduction, and convection cannot remove heat from the body. The only useful method under that condition is **evaporation,** the conversion of water from a liquid to a gas (such as water vapor). This process requires heat energy.

Evaporation takes place not only on the skin surface, but also in respiratory passages, including the nose, mouth, and lungs. For each kilogram of water evaporated from the body, 580 Calories are removed (0.58 Calorie for every gram of water). When you are physically active, you sweat, and are aware of being cooled as the sweat evaporates from your skin. Evaporation is effective because heat is required to convert sweat into water vapor. The necessary heat is given up by the body, producing a cooling effect. But you are actually losing body heat through evaporation all the time, either by evaporaton from the respiratory tract or through *insensible perspiration,* the undetected evaporation of sweat from your skin.*

---

*You lose about 7 Calories an hour through evaporation from the respiratory tract, and about 10 Calories an hour through insensible perspiration. During an entire day you lose about 0.7 L of water, enough to liberate approximately 400 Calories.

What are some ways in which the "normal" body temperature can vary?	*The body temperature of infants is usually higher than 37°C (98.6°F), that of elderly people lower. Adults wake up with a somewhat lower temperature, and go to sleep with a somewhat higher temperature. The normal temperature of a woman is generally 1°F higher about two days after ovulation. Rectal temperature usually is about 1°F higher than oral temperature, and the armpit is about 1°F cooler. In England, the "normal" temperature is 98.4°F.*

Homeostatic Control of Body Temperature

The temperature of the body is regulated in great part by negative feedback mechanisms within the nervous system, especially in the hypothalamus. The system works admirably, with the temperature of the body usually varying no more than about half a degree above or below the accepted "normal" temperature of 37°C (or about 1°F above 98.6°F). This negative feedback system has three major components: (1) *temperature receptors* that sense the existing body temperature, (2) *effector organ systems* that control heat production or heat loss, and (3) an *integrator* or *controller* that compares the sensed temperature with a "normal" or "reference" temperature. If the body temperature is too high or too low, the controller activates the appropriate effector system, returning the body temperature to normal.

It should be noted that normal temperature refers to the *core temperature*, or temperature of the interior of the body, rather than the *surface temperature*, or temperature of the arms, legs, and tissues immediately under the skin. Usually in referring to the core of the body we mean the head and trunk, as well as the vital organs housed within those areas. Even though the temperature of the skin and limbs fluctuates, it is important for the temperature of organs such as the heart, brain, and kidneys to remain constant. Interestingly, the temperature of the liver, where many chemical activities take place, is usually higher than that of other organs.

Role of the hypothalamus   The hypothalamus plays an important role in preventing the body from overheating. Heat-sensitive neurons in the preoptic area of the hypothalamus speed up their firing rate when the temperature of the body increases. Signals from the preoptic area are combined in the posterior hypothalamus with signals from the rest of the body to bring about a compensating heat loss that restores the normal temperature.

The feedback mechanisms in the hypothalamus cannot operate properly without complementary detectors in the body that are sensitive to temperature changes. The complete heat-controlling mechanism of the hypothalamus and its support systems is known as the **hypothalamic thermostat.**

Some cold-sensitive neurons are found in the hypothalamus, but they are thought to play a minor role in the regulation of body temperature. Instead, cold receptors in the skin, spinal cord, and other areas are sensitive to changes at the cold end of the scale, and help prevent the body from becoming too cool.

The hypothalamic thermostat *reduces* body heat in three ways:

1 Blood vessels in the skin are dilated sufficiently so that heat is transferred to the skin from the blood about eight times faster than normal. Such a massive dilation is caused by the inhibition of nerve centers in the posterior hypothalamus that usually cause the constriction of blood vessels.

2 Perspiration is increased, and the resulting increased evaporation of sweat cools the skin. If the body temperature

What Is Fever?

Fever is an elevation of body temperature above the normal 37°C (98.6°F). It may be thought of, at least initially, as a specific defense against infection, which is the usual cause of fever. Among the many other causes of fever are drugs, hormonal imbalances, congestive heart failure, anxiety, strenuous physical exertion, disturbances of the immune system or central nervous system, tumors, and injury to tissues. Whatever the cause, metabolic activity increases (about 7 percent per degree Celsius), body heat is retained, and core temperature rises.

Any substance capable of producing a fever is called a **pyrogen** (Gr. "fire-producing"). Pyrogens are classified as *exogenous* (*exo* = outside) if they function outside the hypothalamic thermo-regulatory control center, or *endogenous* (*endo* = inside) if they operate within the hypothalamus.

A fever often indicates that the body is fighting a viral or bacterial infection. When harmful bacteria or viruses enter the body, white blood cells (macrophages) release the hormone *interleukin-1* (an exogenous pyrogen), which travels to the temperature-regulating center in the hypothalamus. The hypothalamus then causes heat-sensitive preoptic neurons to release prostaglandins (endogenous pyrogens), which reset the body's "hypothalamic thermostat" at a temperature above normal, producing a fever. (Aspirin, which is often used to reduce fever, is a known inhibitor of prostaglandin synthesis.) Apparently, both the fever and the interleukin-1 increase the effectiveness of the immune system in fighting certain infections, sometimes increasing the production of T cells and antibodies as much as 20-fold.

A fever is usually preceded by a body chill and shivering, and cutaneous vasoconstriction causes the skin to remain cool even though the core temperature may be already rising. Shivering increases the production of heat, and at the same time, mechanisms for the reduction of heat are inhibited. As a result, the body temperature is raised and the skin becomes warm.

When the cause of the fever is removed, the original "set point" of the hypothalamus is restored, and normal thermoregulation resumes. As the body temperature begins to drop, the skin is flushed, and the cardiac output may be increased as excess heat is dissipated. Skin temperature begins to decrease, and the onset of perspiration indicates that the rate of heat loss is greater than the rate of heat production. Normal body temperature is soon restored (see graph.) The period of maximum elevation is called the *stadium* or *fastigium*. It may last from three days to three weeks. A sudden drop in temperature to normal is called the *crisis*, while a more gradual return to normal is known as *lysis*.

As long as the body temperature is abnormally high, the metabolic rate speeds up, and damaged tissues are repaired faster than at normal temperatures. The heat of a fever also helps to kill or incapacitate the foreign bacteria or viruses. But when a fever reaches approximately 40°C (104°F), it begins to harm the person as well.

Recent studies indicate that antibiotics are more effective at fever temperatures. High temperature may make the plasma membrane more permeable, so that it is easier for antibiotics to enter the cell. Or it may weaken the bacteria or virus, so that it is less able to resist the antibiotics.

Metabolic activity generally increases in proportion to the extent of the fever, and the body uses increased quantities of carbohydrates, fats, and proteins. A high-calorie, high-protein diet is desirable while the fever persists, especially since some body protein is destroyed during the early stages of a fever.

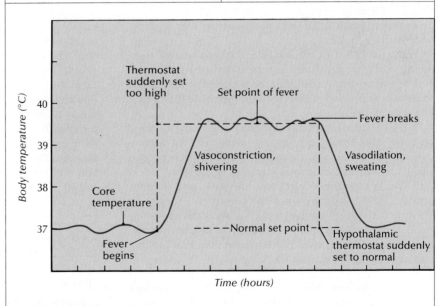

Major events of a fever.

TABLE 23.6   SUMMARY OF HORMONAL REGULATION OF METABOLISM

Hormone	Effect on blood glucose	Effect on carbohydrate metabolism	Effect on protein metabolism	Effect on lipid metabolism
Cortisol	Raises level.	Increases gluconeogenesis and glycogen formation.	Decreases protein synthesis.	Increases lipolysis. Decreases ketogenesis and lipogenesis.
Glucagon	Raises level.	Increases glycogenolysis and gluconeogenesis. Decreases glycogen formation.	None.	Increases lipolysis and ketogenesis.
Growth hormone	Raises level.	Increases gluconeogenesis and glycogen formation. Decreases glucose utilization.	Increases protein synthesis and amino-acid transport into muscle cells.	Increases lipolysis and ketogenesis. Decreases lipogenesis.
Epinephrine	Raises level.	Increases glycogenolysis and gluconeogenesis. Decreases glycogen formation.	None.	Increases lipolysis and ketogenesis.
Insulin	Lowers level.	Increases glycogen formation. Decreases glycogenolysis and gluconeogenesis.	Increases protein synthesis and amino-acid transport into muscle cells.	Increases lipogenesis. Decreases lipolysis and ketogenesis.
Testosterone	None.	None.	Increases protein deposition into muscle cells.	None.
Thyroxine	None.	Increases gluconeogenesis and glucose utilization.	Increases protein synthesis.	Increases lipolysis.

rises 1°C above normal, the rate of perspiration increases enough to remove 10 times the normal body heat.

3 Shivering and other muscular activities that increase body heat are inhibited.

When the body becomes too cold, the hypothalamic thermostat uses three regulatory devices to *increase* body heat:

1 The temperature centers in the posterior hypothalamus cause the constriction of blood vessels, thereby reducing heat loss through the skin.

2 The hypothalamus causes the hairs on the skin to "stand on end." This activity *(piloerection),* along with the accompanying "goose bumps," increases the production of heat slightly by increasing the muscle activity of the skin.

3 The hypothalamus stimulates shivering and thermogenesis. Shivering occurs when the *primary motor center for shivering* in the posterior hypothalamus stimulates the brainstem, spinal cord, and motor neurons that ultimately cause an increase in muscle tone. Shivering can raise heat production as much as five times above normal.

Roles of the endocrine system and sympathetic nervous system When the preoptic area of the hypothalamus is cooled, the secretion of thyrotropin-releasing hormone (TRH) from the hypothalamus is increased. TRH, in turn, stimulates the secretion of thyroid-stimulating hormone (TSH) from the pituitary gland. TSH then stimulates the thyroid gland to secrete thyroxine, which increases cellular metabolism and body heat (Table 23.6).

Increased secretions of epinephrine and norepinephrine from the adrenal glands help to raise the level of cellular metabolism also. This effect, known as *chemical thermogenesis,* is caused by the stimulation of the sympathetic nervous system. The effectiveness of chemical (nonshivering) thermogenesis seems to be directly related to the amount of brown fat in the body. Because brown fat contains many mitochondria and many small globules of fat instead of one large globule as in white fat, it is capable of a high level of cellular metabolism. An adult body contains little or no brown fat, so it can increase heat production only 10 to 15 percent through chemical thermogenesis. The infant body does contain some brown fat, so it can increase heat production 50 to 100 percent through chemical thermogenesis.

Ask Yourself

1 *What are some important factors that affect the metabolic rate?*

2 *What is the basal metabolic rate?*

3 *What are the so-called gradient mechanisms of body-heat regulation?*

4 *How does evaporation help cool the body?*

5 *What is the hypothalamic thermostat?*

PHYSIOLOGICAL AND ANATOMICAL ABNORMALITIES

Obesity

Approximately 50 to 70 million Americans suffer from **obesity** (L. *obesus,* "grown fat by eating"), the condition of being 20 to 25 percent over the recommended body weight. Also, approximately 65 percent of 50- to 60-year-old women and 38- to 48-year-old men are at least 10 percent over their ideal body weight. Obesity is usually caused by eating too much and exercising* too little (taking in more calories than are used up), rather than by metabolic disturbances. It is estimated that only about 5 percent of all cases of obesity are caused by hormonal imbalances. It should be noted, however, that the female hormone estrogen affects the deposition of fat to establish body contours, and may contribute to obesity in some women.

Why are so many people overweight? The typical American diet is high in animal fat, salt, and simple sugars. Also, modern labor-saving devices have made it unnecessary for us to expend energy. As a result, we burn off fewer calories. It has been estimated that if we didn't have machines working for us, we would need about 80 servants to be able to live the way we do.

Protein-calorie malnutrition

Protein-calorie malnutrition occurs when the body has been deprived of sufficient amino acids and calories for a long period. Two forms of malnutrition are marasmus (protein-calorie deficiency) and kwashiorkor (protein deficiency), both of which are common in underdeveloped countries and in areas where dietary protein is inadequate for proper growth and tissue building. *Marasmus* is caused by a diet that is very low in both calories and protein. It affects infants who are not receiving sufficient breast feeding, or who suffer from chronic diarrhea or other debilitating conditions. It is characterized by tissue wasting and impaired physical growth. *Kwashiorkor* occurs when the diet consists primarily of carbohydrates, and is very low in protein. It was originally

*"Exercise" is meant to include all physical activity—walking, gardening, housecleaning, and any other activity that uses up calories.

described in young children in Ghana whose diet consisted of a maize (corn) gruel. Such a diet lacks the essential amino acids lysine and tryptophan. Kwashiorkor does not affect body growth, but it does cause a decrease in adipose tissue, as fat is metabolized as the main energy source. Abdominal tissue edema gives the patient a puffed-up appearance, although muscle and adipose tissues are actually wasting away. Kwashiorkor is said to afflict as many as 25 percent of African children. Its name comes from a West African word meaning "sickness a child has when another child is born." The shrunken muscles, potbelly, skin sores, bleached and falling hair, and nausea and vomiting frequently begin when a child is weaned from the breast milk of its mother. It can be cured, sometimes with dramatic speed, by the addition of sufficient dietary protein containing the proper balance of amino acids.

Excessive vitamin intake

Most physicians generally agree that megadoses of fat-soluble vitamins over a prolonged period are toxic. Even water-soluble vitamins may be toxic in very large doses. The following conditions have been authenticated. Megadoses of vitamin B_6 can produce peripheral nerve damage. The symptoms of **hypervitaminosis A** include anorexia, headache, irritability, enlarged spleen and liver, scaly dermatitis, patchy hair loss, and bone pain and thickening. **Hypervitaminosis D** is characterized by weight loss, calcification of soft tissues, and kidney failure, and **hypervitaminosis K** is associated with anemia, jaundice, and gastrointestinal problems.

Some common metabolic disorders

Cystic fibrosis *Cystic fibrosis* is a metabolic disease typified by a deficiency of the pancreatic enzymes trypsin, amylase, and lipase. It is the most common fatal genetic disease of Caucasian children, and it is estimated that 1 out of every 25 Caucasians in Central Europe and North America is a carrier of the disease. The intestinal tract may be blocked at birth because the neonate

does not excrete a viscous intestinal substance called meconium, and older infants have poor intestinal digestion, weight loss, and enlarged abdomens. The young child has poor lipid absorption, and develops respiratory problems due to an accumulation of thick secretions in the bronchioles and alveoli. Chronic pneumonia, emphysema, or atelectasis (the incomplete expansion of lobules in the alveoli) may eventually cause death.

Phenylketonuria (PKU) *Phenylketonuria (PKU)* is a congenital disease caused by a DNA *mutation,* or genetic error. The mutation prevents the adequate formation of the enzyme phenylalanine hydroxylase, which normally converts the amino acid phenylalanine into the amino acid tyrosine. When phenylalanine is not converted at normal rates, its metabolites begin to accumulate in the blood, generally causing cerebral damage, mental retardation, decreased muscle control, and other problems. Exactly how this disorder leads to retardation is not known. PKU can sometimes be controlled if foods containing phenylalanine are kept out of the diet of a susceptible child.

Tay-Sachs disease When lysosomes are deficient in the enzyme hexosaminidase, they cannot break down fatty substances called gangliosides as they normally would. Gangliosides are products of normal cell growth. As these fatty deposits accumulate around neurons, the neurons become unable to transmit nerve impulses. Neurons that store too much ganglioside become swollen and eventually die. The disorder, which is hereditary, is called *Tay-Sachs disease* after the physicians who first reported it.

An afflicted baby appears normal at birth, but after six months or so, the large number of affected neurons begin to cause brain damage. A reddish spot appears on the retina, muscles weaken, vision and hearing are impaired, and mental ability is diminished. Because there is still no effective treatment for the disease, death usually comes before the child is five years old.

Tay-Sachs disease occurs most often among Ashkenazi Jews from eastern

Europe. About 1 out of every 30 Ashkenazi Jews is a carrier of the disease. A child is affected only if both parents carry the Tay-Sachs gene; carriers do not have the disease themselves. It has been calculated that 1 child out of every 3600 born to Ashkenazi parents will inherit Tay-Sachs disease.* Modern techniques make it possible to screen people to find out if they are carriers and if unborn babies have Tay-Sachs disease. Many carriers decide not to have children.

Disorders of temperature regulation

Fever Fever may not always be a "disorder." In fact, it is often beneficial, as you saw in the essay on page 759.

Frostbite What we call *frostbite* is a localized, direct, mechanical injury to plasma membranes. It occurs when water molecules in tissues are taken out of solution and form ice crystals. Blood vessels, especially delicate capillaries, are usually damaged, and the subsequent release of fibrinogen may cause red blood cells to become trapped in a blood clot. The skin in the affected area usually turns white until it thaws, and then it may be purplish-blue or black. The deep color usually indicates the death of tissue, and possibly gangrene. Untreated frostbite will almost certainly lead to gangrene, and amputation may be necessary. Treatment usually consists of gently warming the injured area, but never rubbing it, since rubbing can lead to further tissue damage. Frostbite occurs most often in the earlobes, fingers, and toes, where the temperature of the body is at its lowest.

Heat exhaustion/heat prostration *Heat exhaustion* or **heat prostration** is caused by a depletion of plasma volume

*One out of 3600 is a high ratio. Color blindness, for example, appears only once in every 17,850 births.

due to sweating, and hypotension in response to reflexively dilated blood vessels in the skin. Decreased blood pressure often causes fainting. Though the heart rate rises and blood pressure falls, the body temperature usually remains normal and the skin is cool. Heat exhaustion is accompanied by thirst, fatigue, giddiness, nausea, cramps, and possibly delirium. Death is rare; it is likely only if water depletion is greater than 20 percent of the total body water (20 percent is about 40 pt). Treatment consists of replacing the lost water and salt.

Heatstroke/sunstroke Unlike heat exhaustion, where the body temperature usually does not rise above normal, *heatstroke* or *sunstroke* causes core temperatures above 40.5°C (105°F). When the hypothalamus becomes overheated, its heat-regulating system breaks down, and sweating almost stops altogether. Excessive loss of water and electrolytes may already have occurred before sweating is halted, and dizziness, nausea, delirium, and coma are likely. If the rising temperature is not controlled quickly, brain damage can result, the cardiovascular system can collapse (circulatory shock), and the person can die. Heatstroke is most common in hot, humid weather, especially after a long period of strenuous physical activity.

Heatstroke should be treated as soon as the symptoms become apparent, usually by loosening or removing the person's clothing and applying cold compresses directly to the skin. Cooling is usually stopped when the core temperature drops to 39°C (102°F) to prevent shivering. A drink containing balanced electrolytes (such as Gatorade) may be given if the person is conscious, and cramped muscles may be massaged.

Hypothermia *Hypothermia* (*hypo*, under + *thermia*, heat) results when the

body mechanisms for producing and conserving heat are exceeded by exposure to severe cold. In contrast to frostbite, which is a localized injury, hypothermia affects the whole body. It is most common among the elderly, the young, heavy drinkers and drug users (because alcohol and barbiturates move blood into capillaries away from body organs), people with cardiovascular disease or an underactive thyroid gland, and very thin people who lack a sufficient layer of insulating fat. Body temperature may be reduced to 35 to 36.6°C (95 to 98°F) in mild cases and as low as 25 to 28°C (77 to 82.4°F) in severe cases.

Victims exhibit symptoms ranging from shivering in mild cases, to slurred speech in moderate cases, to muscle rigidity and shock in serious cases, to ventricular fibrillations (abnormal heart rhythms) and the appearance of death in the most severe cases. Hypothermia slows down the activity of most organ systems, and if pulse and respiration appear to cease, prolonged cardiopulmonary resuscitation become necessary. Gentle warming, starting with the core area to avoid fatal ventricular fibrillations, is recommended, and the condition requires the immediate attention of a physician.

The so-called *dive reflex* is a protective response that explains why children who fall into frozen lakes often survive even though breathing stops. The reflex is triggered by the splash of cold water on the face, causing the heart rate to drop and increasing the blood flow to the brain and heart. A child's small size also helps. When the icy water is inhaled it enters the lungs and spreads quickly through the relatively short bloodstream. The blood is chilled, cooling the brain and reducing its need for oxygen. As a result, a child may survive for as long as 40 minutes after breathing stops.

SUMMARY

Metabolism is the overall term for the many chemical and energy transformations that take place within cells to promote homeostasis. *Nutrients,* or chemical components of food, supply the body

with energy and materials. *Potential energy* stored in the chemical bonds of nutrient molecules is transformed into usable energy in the form of ATP. ATP is used by cells to carry on metabolic work.

Review of chemical reactions

1 *Anabolic reactions* are the chemical changes that allow cells to build and repair tissues through the synthesis of complex molecules from simpler building blocks. These reactions require energy.

2 *Catabolic reactions* break down complex molecules into simpler ones, and release energy and heat. Enzymes are usually involved.

3 Anabolic and catabolic reactions take place at the same time within a cell, and structural components are constantly being broken down and replaced.

4 *Oxidation-reduction reactions* are reactions in which an atom or molecule loses hydrogen atoms or electrons (oxidation) or gains hydrogen atoms or electrons (reduction). They always occur together, because when one atom or molecule gains electrons, another loses them.

Carbohydrate metabolism

1 Carbohydrates have potential energy stored in chemical bonds. Within the bonds, it is *electrons* that carry the energy. During the metabolism of foods, the electrons are transferred to nucleotides such as NAD^+ and FAD.

2 When the nucleotides accept electrons, they also receive hydrogen ions, converting NAD^+ into $NADH + H^+$ and FAD into $FADH_2$. The reduced nucleotides may be used as direct sources of energy, or they may be used to make ATP.

3 Glucose is usually the body's preferred nutrient energy source. Energy is released when glucose is catabolized in a series of complex reactions.

4 *Cellular respiration* is the process by which cells are able to release the chemical energy stored in food. The process starts with a food such as glucose; the end products are carbon dioxide, water, and ATP. The main steps in the process are *glycolysis,* the *citric-acid cycle,* and the *electron transport system.*

5 *Glycolysis* is the splitting of a 6-carbon molecule of glucose into two 3-carbon molecules of pyruvic acid in a series of anaerobic, enzymatic reactions in the cytosol of a cell. There is a net gain of two molecules of ATP.

6 The *citric-acid cycle* (*Krebs cycle*) is a series of enzymatic reactions that release hydrogen atoms from oxidized acids as a means of energy transfer. It takes place in the mitochondria, and its end products after two ''turns'' of the cycle are carbon dioxide, water, two molecules of ATP, six molecules of NADH, and two molecules of FADH.

7 The *electron transport system* is a series of oxidation-reduction reactions in which electrons are transferred to accepting cytochromes and oxygen. It occurs in mitochondria, and produces most of the ATP (32 molecules) from glucose catabolism.

8 The *chemiosmotic theory of ATP production* states that the production of ATP by mitochondria involves the accumulation of electrochemical energy in the form of an electrochemical proton gradient by selectively pumping protons out of the matrix into the inner compartment. The energy in this gradient is then used to drive the activity of the ATP synthetase to form ATP from ADP + P_i.

9 *Glycogenesis* is the conversion of blood glucose into glycogen, which is stored in the liver and skeletal muscle cells for later use. When the body needs extra glucose, the process of *glycogenolysis* converts glycogen stored in the liver into glucose and releases it into the bloodstream. *Gluconeogenesis* is the production of glucose from noncarbohydrate sources such as lactic acid, glycerol, and amino acids.

Protein metabolism

1 *Protein catabolism* is the breakdown of proteins into amino acids, which are modified further by *oxidative deamination* (the removal of the amino group and its replacement with oxygen to form a *keto acid*) before they enter the cellular respiratory system. Keto acids can be used to produce carbon dioxide and ATP on the way to forming glucose, or they can be converted into acetyl-CoA to synthesize fatty acids.

2 *Protein anabolism* is the synthesis of protein from amino acids, under the direction of nucleic acids. It takes place on ribosomes, and is regulated by several hormones.

3 *Nonessential amino acids* are synthesized by cells in the processes of *transamination. Essential amino acids* cannot be synthesized by cells, and therefore must be supplied by dietary protein.

Lipid metabolism

1 *Lipids* are the body's most concentrated source of food energy. The most abundant lipids are the *triacylglycerols.* Digested lipids are separated into glycerol and fatty acids.

2 Fatty acids undergo a series of reactions called **beta oxidation** that form

acetyl-coenzyme A, which enters the citric-acid cycle to produce ATP.

Absorptive and postabsorptive states

1 During the *absorptive state,* nutrients are absorbed from the gastrointestinal tract into the cardiovascular and lymphatic systems.

2 After absorption, carbohydrates enter liver cells, where they are either stored as glycogen or converted into fat. Glucose that does not enter the liver is used almost immediately by most body cells as an energy source.

3 Before absorption, some lipids enter the lymphatic system as *chylomicrons,* which contain triacylglycerols. Many of the absorbed amino acids enter the liver directly, where they are converted into carbohydrates and used for energy, or converted into fat, which is stored mainly in the liver.

4 In the *postabsorptive* or *fasting state,* the energy requirements of body cells (including the glucose supply of the brain) must be satisfied by nutrients taken in during the absorptive state, and from the body's stored nutrients.

5 Body cells shift their metabolism during periods of fasting to use more fat instead of glucose, sparing glucose for the nervous system.

Nutritional needs and balance

1 The food we eat contains six kinds of *essential nutrients:* carbohydrates, proteins, lipids, vitamins, minerals, and water.

2 A *calorie* (cal) is the amount of energy required to raise the temperature of 1 g of water 1°C. A *kilocalorie* (kcal) or *Calorie* is equal to 1000 calories.

3 Currently, carbohydrates in the average American diet supply about 46 percent of the body's energy, with 18 percent coming from refined and processed sugars.

4 The body needs relatively little dietary protein to supply the essential amino acids and nitrogen in a form that can be used to synthesize the nonessential amino acids. Dietary protein is most beneficial when it is balanced with all types of food, providing all 20 of the necessary amino acids. The need for protein varies with age and body weight.

5 *Triacylglycerols* may contain three types of fatty acids: saturated, unsaturated, and polyunsaturated. Most physicians and dietitians recommend that we consume more polyunsaturated fat than saturated fat.

6 *Vitamins* are organic compounds required by the body in small amounts for the regulation of metabolism. They are important in supporting the work of enzymes. Vitamins are usually classified as *fat-soluble* (A, D, E, K) or *water-soluble* (B group, C).

7 *Minerals* are naturally occurring, inorganic elements that are used for the regulation of metabolism. Minerals needed in minute amounts are called *microminerals* or *trace elements*. Those found in significant amounts in body tissue are *macrominerals*.

Metabolic rate and temperature control

1 The *metabolic rate* (number of calories used in a specific time period) is affected by the type of food eaten, amount of exercise, hormones, age, and other factors.

2 The amount of energy a person uses daily, under standardized conditions, to maintain only essential bodily functions is the *basal metabolic rate (BMR)*.

3 Control mechanisms that regulate body heat are the *gradient mechanisms* (radiation, conduction, and convection) and the evaporation of water from the skin and respiratory tract. *Radiation* is the transfer of infrared rays from one object to another without contact between the objects, *conduction* is the transfer of heat directly from one object to another, and *convection* is the transfer of heat along air currents. *Evaporation*

is the conversion of a liquid (such as sweat) to a gas (such as water vapor).

4 The complete heat-controlling mechanism of the hypothalamus and its support system is called the *hypothalamic thermostat*. It reduces heat by dilating skin blood vessels, increasing sweating, and inhibiting shivering and other muscular activities that increase body heat. It increases body heat by constricting blood vessels, causing piloerection, and stimulating shivering and thermogenesis.

5 The secretion of TRH from the hypothalamus starts a series of *hormonal events* that increases body metabolism and heat through the secretion of thyroxine. Metabolism is also increased by the secretion of epinephrine and norepinephrine from the adrenal glands.

6 *Fever* is the condition in which body temperature rises above normal due to pyrogenic substances acting on the hypothalamic thermostat. A moderate fever appears to help the body to fight certain viral and bacterial infections.

Physiological and anatomical abnormalities

1 *Obesity,* the condition of being 20 or more percent over the recommended body weight, is usually caused by eating too much and exercising too little, but there are occasional metabolic, hormonal, or genetic causes.

2 *Protein-calorie malnutrition* is the result of a diet deficient in amino acids and calories. It may occur as *marasmus* or *kwashiorkor.*

3 Excessive intake of fat-soluble vitamins over an extended period can lead to toxicity.

4 *Cystic fibrosis,* a common metabolic disease, involves a deficiency of the pancreatic enzymes trypsin, amylase, and lipase.

5 *Phenylketonuria (PKU)* is a congenital disease caused by a DNA mutation tha impairs the formation of the enzyme phenylalanine hydroxylase, which normally converts the amino acid phenylalanine into tyrosine.

6 *Tay-Sachs disease* is a hereditary disorder in which lysosomes are deficient in the enzyme hexosaminidase, which ordinarily breaks down fatty substances called gangliosides. The accumulation of gangliosides causes damage to the nervous system.

7 *Frostbite* is a localized injury caused when ice crystals form in the intracellular spaces of tissues, disrupting plasma membranes, the metabolic and enzymatic functions of cells, blood vessels, and the blood-clotting mechanism.

9 *Heat exhaustion* occurs when heavy sweating reduces plasma volume, and dilated blood vessels in the skin cause hypotension. *Heatstroke* occurs when the body temperature rises high above normal, sometimes breaking down the heat-regulating mechanism in the hypothalamus.

10 *Hypothermia* results when the body's mechanisms for producing and conserving heat are exceeded by exposure to severe cold.

UNDERSTANDING THE FACTS

1 What are the six kinds of essential nutrients?

2 Why does oxidation always occur together with reduction?

3 What are the most important sources and uses of carbohydrates?

4 What end product marks the completion of glycolysis?

5 In what parts of a cell do glycolysis, the citric-acid cycle, and the electron transport system proceed?

6 How many "turns" of the citric-acid cycle are involved in the catabolism of a single molecule of glucose?

7 What are cytochromes?

8 What stage of catabolism produces the most ATP?

9 Define glycogenesis, glycogenolysis, and gluconeogenesis.

10 Why are essential amino acids called "essential"?

11 What substance makes up the most abundant lipid group?

12 What is lipogenesis?

13 What are some differences between the absorptive state and the postabsorptive state?

14 How does the body maintain its supply of glucose when carbohydrates are scarce?

15 What are the federal dietary guidelines?

16 What is the difference between the metabolic rate and the basal metabolic rate?

UNDERSTANDING THE CONCEPTS

1 Why is anabolism also called biosynthesis?

2 In what ways can a cell use the pyruvic acid produced during glycolysis?

3 How is the energy released during each step of the electron transport system used?

4 What happens to hydrogen during catabolism?

5 Why is it efficient for acetyl-CoA to be produced in the mitochondria of a cell?

6 What does beta oxidation accomplish?

7 What are some important roles of the liver during the absorptive state?

8 What is the metabolic significance of glucose sparing?

9 Why do you feel warmer on a hot, humid day than on a hot, dry day?

10 Why is wiping sweat off your body not effective in cooling you?

11 How does the hypothalamus oversee the regulation of body temperature?

SUGGESTED READINGS

ANDERSON, K. E., "Nutrient Regulation of Chemical Metabolism in Humans." *Federal Proceedings,* 44 (1985):130.

BENZINGER, T. H., "Heat Regulation: Homeostasis of Central Temperature in Man." *Physiology Review,* 49 (1969):671.

BROWNELL, K. D., "The Psychology and Physiology of Obesity." *Journal of American Dietary Association,* 84 (1984):406.

CHRISTENSEN, H. N., "The Regulation of Amino Acid and Sugar Absorption by Diet." *Nutrition Review,* 42 (1984):237.

FRASER, D. R., "Regulation of Metabolism of Vitamin D." *Physiology Review,* 60 (1980):551.

FRISCH, ROSE E., "Fatness and Fertility." *Scientific American,* March 1988.

HELLER, H. CRAIG, LARRY I. CRAWSHAW, AND HAROLD T. HAMMEL, "The Thermostat of Vertebrate Animals." *Scientific American,* August 1978.

HINKLE, PETER C., AND RICHARD C. MCCARTY, "How Cells Make ATP." *Scientific American,* March 1978.

Human Nutrition: Readings from Scientific American. San Francisco: Freeman, 1978.

HUNT, S. M., *Nutrition: Principles and Clinical Practice.* New York: Wiley, 1980.

LECHTMAN, MAX D., BONITA ROOHK, AND ROBERT J. EGAN, *The Games Cells Play: Basic Concepts of Cellular Metabolism.* Reading, Mass.: Benjamin/Cummings, 1979.

LEHNINGER, ALBERT L., *Bioenergetics,* 2nd ed. New York: Benjamin-Cummings, 1971.

LEHNINGER, ALBERT L., *Principles of Biochemistry.* New York: Worth, 1982.

MAGNEN, J. L., "Body Energy Balance and Food Intake: A Neuroendocrine Regulatory Mechanism." *Physiology Review,* 63 (1983):314.

WILSON, EVA D., KATHERINE H. FISHER, AND PILAR A. GARCIA, *Principles of Nutrition,* 4th ed. New York: Wiley, 1979.

24
The Urinary System

LEARNING OBJECTIVES

1 Name the major components of the urinary system, and describe their functions.

2 Explain how the kidney (or nephron) functions to maintain homeostasis in the body.

3 Describe the location and external anatomy of the kidney.

4 Discuss the internal anatomy of the kidney.

5 Describe the blood, lymphatic, and nerve supply of the kidney.

6 Relate the components of the nephron to their specific functions.

7 Describe the components and function of the juxtaglomerular apparatus.

8 Discuss the process of glomerular filtration, effective filtration pressure, and the factors affecting the filtration rate.

9 Explain the active and passive processes involved in tubular reabsorption.

10 Discuss the process of tubular secretion and its effects on urine formation.

11 Explain the process of plasma clearance.

12 Describe the mechanism of the countercurrent multiplier system and its effects.

13 Discuss the role of the kidney in acid-base regulation.

14 Explain how kidney function changes from birth to old age.

15 Relate the structure of the ureters, urinary bladder, and urethra to their functions.

16 Discuss the constituents and physical properties of urine and the factors affecting them.

17 Describe the factors that affect the volume of urine, including the hormones.

18 Explain the neural control of micturition (or urination).

Human beings are somewhere in the middle of the broad spectrum of animals, from fish to desert dwellers, that have to deal with the problem of maintaining a specific amount of water in the body. Living totally on land as we do, we have the permanent task of obtaining and conserving water. We must also have a way of removing excess salts and metabolic wastes, especially the nitrogenous end products of protein metabolism, such as urea, uric acid, ammonia, and creatinine. Our urinary system accomplishes both of these crucial tasks: it removes waste products in urine, and it carefully regulates the liquid content of the body.

What we excrete has much more influence on homeostasis than what we eat. *Excretion* (L. *excernere*, to sift out) is the elimination of the waste products of metabolism, and the removal of surplus substances from body tissues. The urinary system eliminates the wastes of protein metabolism in the urine, but it does much more than that. It also regulates the amount of water in the body and, in turn, the amount of salts in the blood, including sodium, potassium, calcium, phosphate, and chloride. These regulatory functions are accomplished by the kidneys, which form and eliminate varying amounts of urine containing varying amounts of salts and water. In addition, many foreign substances, including drugs, are excreted by the kidneys. The kidneys function in acid-base balance, and so maintain the proper ionic composition, pH, and osmotic concentrations. The kidneys also have an endocrine function, releasing various hormones and altering others. Large amounts of prostaglandins are also released by the kidneys.

Excretion is performed in several ways by several organs. The lungs excrete carbon dioxide and water when we exhale. The skin releases some salts and organic substances such as urea and ammonia during periods of heavy sweating. The large intestine excretes a small amount of water, salts, and some microorganisms and undigested food. However, the prime regulator of the proper balance between water and other substances is the urinary system, with the kidneys as its main component.

THE URINARY SYSTEM: COMPONENTS AND FUNCTIONS

The **urinary system**, also called the *renal system* (L. *renes*, kidneys), consists of (1) two *kidneys*, which remove dissolved waste and excess substances from the blood and form urine, (2) two *ureters*, which transport urine from the kidneys to (3) the *urinary bladder*, which acts as a reservoir for urine, and (4) the *urethra*, the duct through which urine from the bladder flows to the outside of the body during urination (Figure 24.1).

Every day the kidneys filter about 1700 L of blood. Each kidney contains over a million *nephrons*, the functional units. Fortunately, we have many more nephrons than we actually need for these purposes, and if necessary, we could lead a normal life with only one kidney.

How do the nephrons accomplish such a controlled regulation? First, they *filter* water and soluble components from the blood. Second, they selectively *reabsorb* some of the components back into the blood to maintain a homeostatic blood concentration. Finally, they selectively *secrete* wastes into the urine that were not filtered from the blood efficiently. The water and other substances not reabsorbed are the true wastes, which constitute the urine.

Ask Yourself

1 *What are some of the major excretory organs?*

2 *What are the basic components of the urinary system, and what are their functions?*

3 *How do nephrons function to maintain homeostasis?*

ANATOMY OF THE KIDNEYS

The average adult takes in about 2.7 L of water every day, most of it by ingesting foods and liquids.* Under normal conditions, the same amount of water is given off daily, about a third of it evaporated from the skin and exhaled from the lungs, and more than half excreted in urine (Table 24.1). The remarkable organs that maintain this constant water balance are the *kidneys,* a pair of reddish-brown, bean-shaped organs, each one about the size of a large bar of soap. They are usually about 11.25 cm (4.4 in.) long, 5 cm (2 in.) wide, and 2.5 cm (1 in.) thick.

Location and External Anatomy

The kidneys are located in the posterior wall of the abdominal cavity, one on each side of the vertebral column. They usually

---

*Only about 47 percent of a person's daily water intake comes from drinking. As much as 39 percent is supplied by so-called solid food (meats are 50 to 70 percent water, most vegetables contain more than 90 percent water, and bread is about 35 percent water), and about 14 percent as a metabolic product. (Our driest food is the baked sunflower seed, which contains about 5 percent water. The wettest food is watermelon, with 97 percent water.)

TABLE 24.1   WATER INTAKE AND OUTPUT PER DAY*

Fluid intake (mL)		Water output (mL)	
Ingested liquid	1200–1500	Urine	1200–1700
Ingested food	700–1000	Feces	100–250
Metabolic oxidation	200–400	Sweat	100–150
		Insensible losses	
		Skin	350–400
		Lungs	350–400
Total	2100–2900	Total	2100–2900

*Water intake and output are approximately equal in a healthy person. The figures above are for an average 70-kg adult whose diet provides adequate calories.

span the distance from vertebra T12 to vertebra L3. The paired kidneys are protected, at least partially, by the last pair of ribs, and are capped by the *adrenal* ("upon the kidney") glands (see Figure 24.1).

Just as the stomach is higher than most people think, so are the kidneys. Rather than being located in the small of the back, the upper portion of each kidney is in contact with the diaphragm, and the left kidney also touches the spleen (see Figure 24.1). The right kidney is in extensive contact with the liver, and because of the liver's large size, the right kidney

---

FIGURE 24.1

The urinary system. (For differences in the lower portions of the male and female systems, see Figure 24.18.)

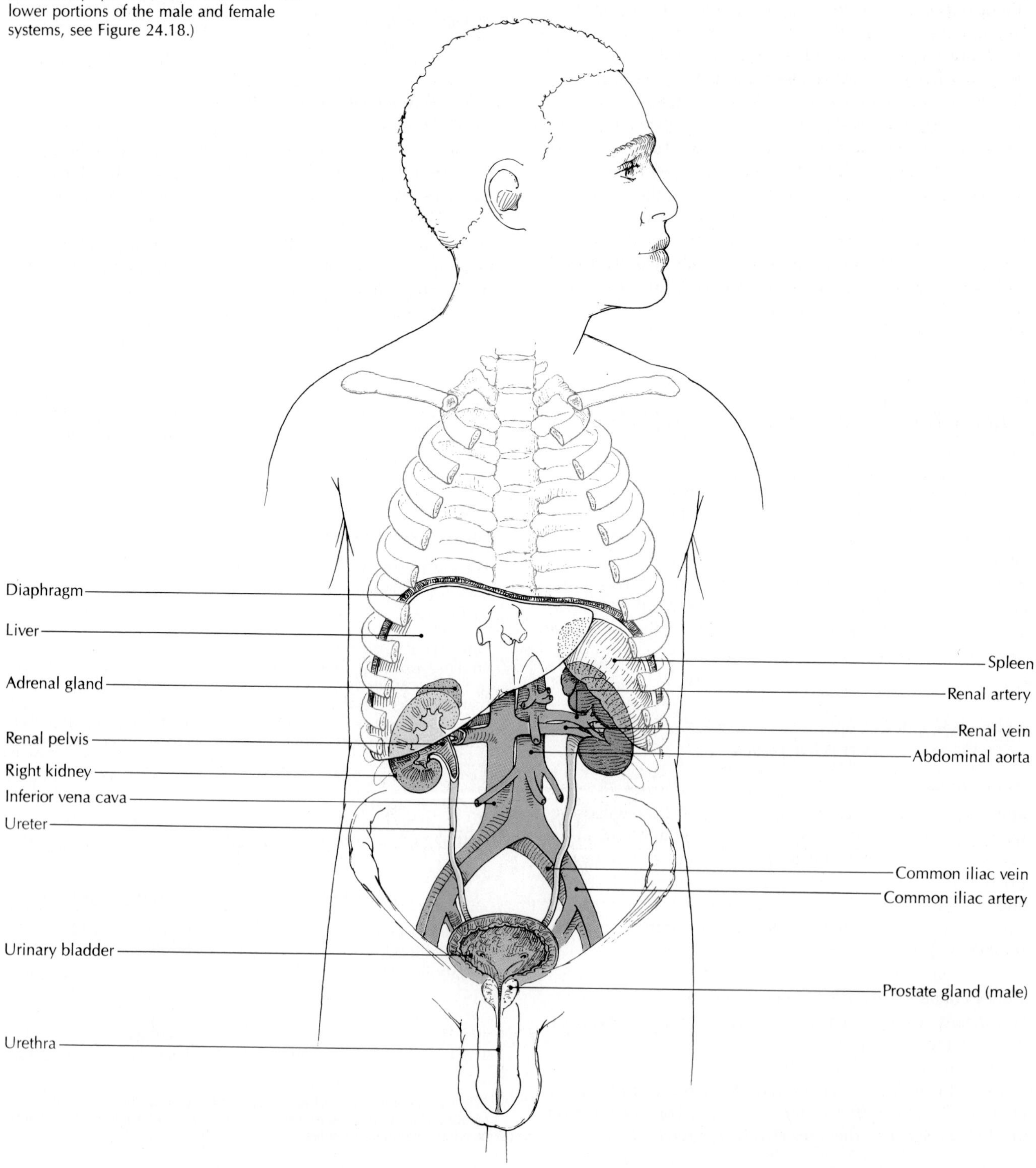

is slightly lower than the left. The kidneys are retroperitoneal (Figure 24.2), and are well supplied with blood vessels, lymphatics, and nerves.

The bean shape of the kidney is medially concave and laterally convex. On the medial concave border is the **hilus** (HYE-luhss), a small, indented opening where arteries, veins, nerves, and the ureter enter and leave the kidney (Figure 24.3).

Covering and supporting each kidney are three layers of tissue. (1) The innermost layer is a tough, fibrous material called the *renal capsule.* It is continuous with the surface layer of the ureters. (2) The middle layer is the *adipose capsule,* composed of pararenal ("beside the kidney") fat, which gives the kidney a protective cushion against impacts and jolts. (3) The outer layer is subserous fascia called the *renal fascia.* It is divided into two continuous parts, the *anterior* and the *posterior renal fasciae.* The renal fascia is composed of connective tissue that surrounds the kidney and attaches it firmly to the posterior abdominal wall. Renal fasciae have enough flexibility to permit the kidneys to shift slightly as the diaphragm moves during breathing. They also tend to prevent kidney infections from spreading to other structures. Although most pararenal fat is located in the middle layer, just beneath the outermost renal fasciae, some may also lie outside of the renal fasciae.

Internal Anatomy

A sagittal section of a kidney (see Figure 24.3) reveals three distinct regions called (from the inside out) the pelvis, medulla, and cortex. The **renal pelvis** is the large collecting space within the kidney, formed from the expanded upper portion of the ureter. It connects the structures of the medulla with the ureter. The pelvis branches into two smaller cavities, the **major calyces** and **minor calyces** (KAY-luh-seez; singular, *calyx,* KAY-licks; Gr. cup). There are usually 2 to 3 major calyces and 8 to 18 minor calyces in each kidney.

The **renal medulla** is the middle portion of the kidney. It consists of 8 to 18 **renal pyramids,** which are longitudinally striped, cone-shaped areas. The base of each pyramid is adjacent to the outer cortex. The apex of each renal pyramid ends in the *papilla,* which is pointed toward the hilus, where it opens into a minor calyx (see Figure 24.3). Renal pyramids consist of tubules and collecting ducts of the nephrons. The tubules of the pyramids are involved with the reabsorption of filtered materials. Urine passes from the medulla to the minor calyces, major calyces, and renal pelvis. From there, the urine drains into the ureter and is transported to the bladder (see Figures 24.3 and 24.1).

FIGURE 24.2

Transverse section through the upper abdomen, showing the retroperitoneal location of the kidneys. Note the layers of adipose tissue and fasciae surrounding the kidneys.

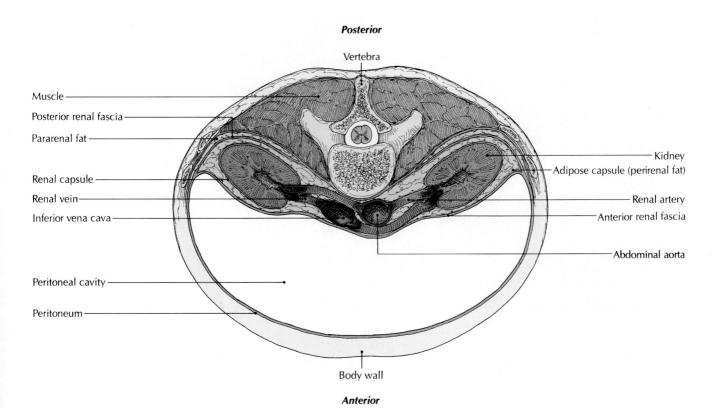

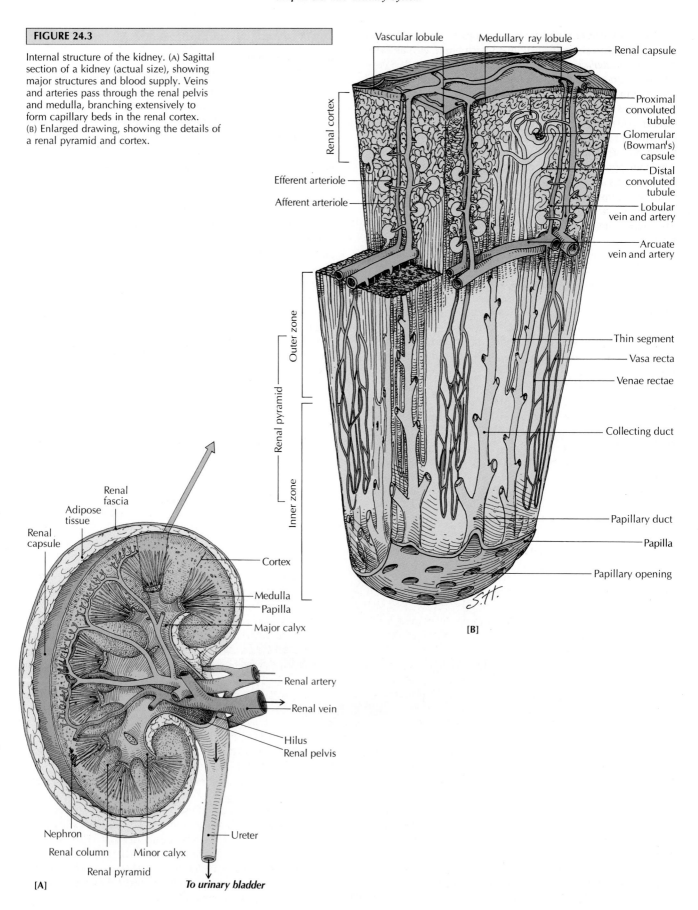

FIGURE 24.3

Internal structure of the kidney. (A) Sagittal section of a kidney (actual size), showing major structures and blood supply. Veins and arteries pass through the renal pelvis and medulla, branching extensively to form capillary beds in the renal cortex. (B) Enlarged drawing, showing the details of a renal pyramid and cortex.

The **renal cortex** is the outermost portion of the kidney. It is divided into two regions, the outer *cortical region* and the inner *juxtamedullary* ("next to the medulla") region. The cortex has a granular, textured appearance, and extends from the outermost *capsule* to the base of the renal pyramids (see Figure 24.3). Within the cortex, the granular appearance is caused by spherical bundles of capillaries and associated structures of the nephron that help to filter blood. The cortical tissue that penetrates the depth of the renal medulla between the renal pyramids forms **renal columns** (see Figure 24.3). The columns are composed mainly of **kidney tubules** that drain and empty the urine into the lumen of a minor calyx. Each minor calyx receives urine from a *lobe* of the kidney. Each lobe consists of a pyramid and its associated cortical tissue.

Blood Supply

The kidneys receive more blood, in proportion to their weight, than any other organ in the body. About 20 to 25 percent of the cardiac output goes to the kidneys. Approximately 1.2 L of blood pass through the kidneys every minute, and the body's entire blood supply is filtered through the kidneys 60 times a day. Little of this blood supplies the kidneys' nutritive needs. Instead, this large blood flow through the kidneys enables the kidneys to maintain homeostasis of the blood and, in turn, the body fluids.

Blood comes to the kidneys directly from the abdominal aorta through the **renal artery** (Figure 24.4; see also Figure 24.3). Once inside the kidney, the renal artery branches into the **interlobar arteries**, which pass through the renal columns. When the arteries reach the juncture of the cortex and medulla, they turn and run parallel to the bases of the renal pyramids. At the turning point, the arteries are the **arcuate arteries** (L. *arcuare*, to bend like a bow), which make small arcs around the boundary between the cortex and medulla (see Figure 24.4). The arcuate arteries branch further into the **interlobular arteries**, which ascend into the cortex to supply the renal corpuscle. Further branching produces numerous small **afferent arterioles**, which carry blood *to* the site of filtration (the glomerulus).

Each afferent arteriole branches extensively to form a tightly coiled ball (tuft) of capillaries called a **glomerulus** (glow-MARE-yoo-luhss; L. ball) (Figure 24.5). This capillary bed is where the blood is filtered. Glomerular capillary loops eventually join together to form a single **efferent arteriole**, which carries blood *away from* the glomerulus. The efferent arteriole is narrower than the afferent arteriole, and this difference tends to increase blood pressure in the glomerulus. The increased blood pressure produces an efficient filtration in the glomerulus (see Figure 24.11).

The efferent arteriole eventually branches to form a second capillary bed (the glomerulus was the first one), made up of the **peritubular** ("around the tubules") **capillaries** (Figure 24.6C). These capillaries reabsorb some of the water and solutes that were filtered from the blood in the glomerulus. The peritubular capillaries unite to form the **interlobular veins**, which carry blood out of the cortex to the **arcuate veins**. The

FIGURE 24.4

Schematic drawing of blood pathway through kidney. Veins and arteries beyond the incoming and outgoing renal artery and vein are actually inside the kidney; the actual locations of vessels are shown in Figure 24.3A.

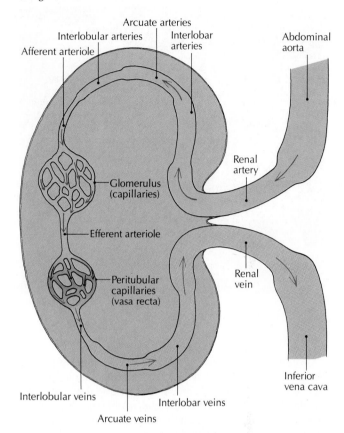

small arcuate veins join to form the larger **interlobar veins** in the renal columns, and the interlobar veins eventually come together to form the single **renal vein** that carries cleansed blood from each kidney to the inferior vena cava and into the circulatory system (see Figure 24.1).

The **arteriae rectae** ("straight arteries") are extensions of the efferent arterioles that surround a portion of the nephron called the loop of the nephron (loop of Henle) (see Figure 24.6C). They follow the loop of the nephron deep into the medulla, and drain into the **venae rectae**, which carry blood to the interlobular veins. Only about 1 or 2 percent of the total renal blood flows through these vessels, collectively called the **vasa recta**. Nevertheless, the vasa recta plays an important role in the concentration of urine, which takes place in the medullary structures of the nephron, as will be seen later.

Like all arterioles, the afferent and efferent arterioles that join the glomerular and peritubular capillary beds contain smooth muscle in their walls (tunica media). This muscle permits the arterioles to constrict or dilate in response to nervous or hormonal stimulation. Constriction or dilation not only allows for the control of blood pressure in the glomerulus, it also helps to maintain filtration.

merulus, as well as blood distribution throughout the kidney. (We know that the kidneys can function adequately without these nerves, however, because the nerve fibers to the kidney are severed permanently during a kidney transplant.)

In a male, the nerves from the kidneys also communicate with nerves from the testes. This explains why kidney disorders often produce pain in the testes.

The Nephron

Each **nephron** (Gr. *nephros,* kidney) is an independent urine-making unit, and each kidney contains approximately 1 million nephrons. As the functional unit of the kidney, the nephron accomplishes the initial filtration of blood, the selective reabsorption back into the blood of filtered substances that are useful to the body, and the secretion of unwanted substances.

A nephron consists of a tubular and a vascular component. The *tubular component* starts with the glomerular capsule and includes the excretory tubules, which are the proximal convoluted tubule, loop of the nephron, and distal convoluted tubule (see Figures 24.3 and 24.6). The excretory tubules of a nephron are coiled and winding. All the tubules from all the nephrons in the body have a combined length of about 80 km (50 mi). Each of the excretory tubules leads into a large collecting duct (which is not part of the nephron) that transports the resulting renal filtrate. The entire tubular portion of the nephron is composed of a single layer of epithelial cells. All along the tubule, cell structure changes with changes in the functions of the tubule (see Figure 24.8).

The *vascular component* of a nephron is made up of blood vessels. These include the glomerulus and the peritubular capillaries, which surround the excretory tubules. The reabsorption of substances from the excretory tubules into the blood takes place in the peritubular capillaries and vasa recta.

Two types of nephrons are recognized, *cortical* and *juxtamedullary* (see Figure 24.6B). The tubular structures of the cortical nephron extend only into the base of the renal pyramid of the medulla, while the longer loop of the nephron of the juxtamedullary nephron projects deeper into the renal pyramid. Cortical nephrons are about seven times more numerous than juxtamedullary nephrons.

Glomerular (Bowman's) capsule The **glomerular (Bowman's) capsule** (originally named for Sir William Bowman, the nineteenth-century English anatomist who first described this structure) is the portion of the nephron that encloses the glomerulus like a hand wrapped around a ball (Figure 24.7). Together, the capsule and glomerulus form the **renal corpuscle.** The glomerular capsule is always located in the cortex of the kidney, and is the first part of the tubular component of a nephron. The inner and outer walls form a cavity called the *capsular space* (see Figure 24.7A). The outer layer of the glomerular capsule is the *parietal layer,* and is composed of simple squamous epithelial cells, which have a basement membrane. The inner *visceral layer* is composed of specialized epithelial cells, called **podocytes** ("footlike cells"), which surround the glomerular capillaries (Figure 24.7B; see also Figure 24.5).

Scanning electron micrograph, showing podocyte cell process and capillary loops in a glomerulus. [*David M. Phillips/Visuals Unlimited.*]

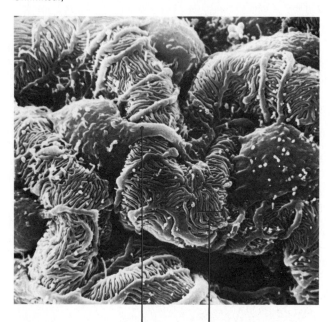

Podocyte cell process   Capillary loops

You would expect that if the systemic blood pressure falls, the pressure that drives glomerular filtration would also fall, but constriction of the efferent ateriotes slows down the flow of blood from the glomerulus, and maintains the pressure that drives glomerular filtration. The renal blood system, like the hepatic portal system, contains two separate capillary beds between artery and vein. This double capillary-bed system provides the kidneys with a built-in emergency mechanism for maintaining homeostasis, despite varying systemic conditions.

Lymphatics

Lymphatic vessels accompany the larger renal blood vessels, and are more prominent around the arteries than the veins. The lymphatic vessels converge in the renal sinus region into several large vessels that leave the kidney at the renal hilus.

Nerve Supply

Nerves from the **renal plexus** of the autonomic nervous system enter the kidney at the hilus, and follow the arteries to innervate the smooth muscle of the afferent and efferent arterioles. This vasomotor supply functions almost entirely as a vasoconstrictor. Generally, the afferent arteriole constricts more than the efferent arteriole. Changes in posture (gravity), physical activity, and stress also increase the activity of these sympathetic nerves. Overall, vasomotor nerves help to control kidney function by regulating blood pressure in the glo-

FIGURE 24.6

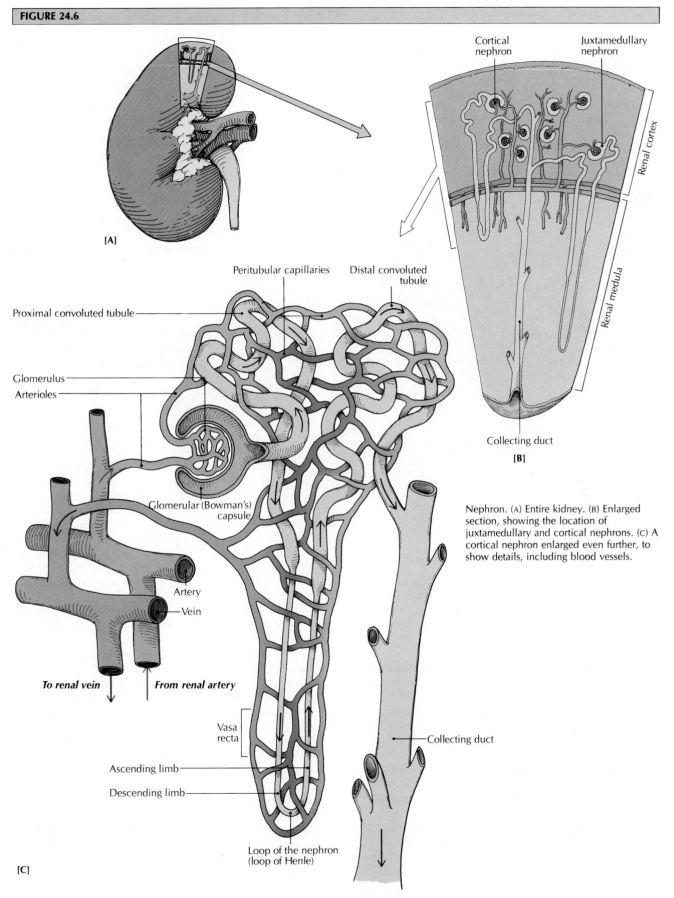

[A]

Cortical nephron

Juxtamedullary nephron

Renal cortex

Renal medulla

Collecting duct

[B]

Peritubular capillaries

Distal convoluted tubule

Proximal convoluted tubule

Glomerulus

Arterioles

Glomerular (Bowman's) capsule

Artery

Vein

To renal vein   *From renal artery*

Vasa recta

Ascending limb

Descending limb

Loop of the nephron (loop of Henle)

Collecting duct

[C]

To renal pelvis

Nephron. (A) Entire kidney. (B) Enlarged section, showing the location of juxtamedullary and cortical nephrons. (C) A cortical nephron enlarged even further, to show details, including blood vessels.

FIGURE 24.7

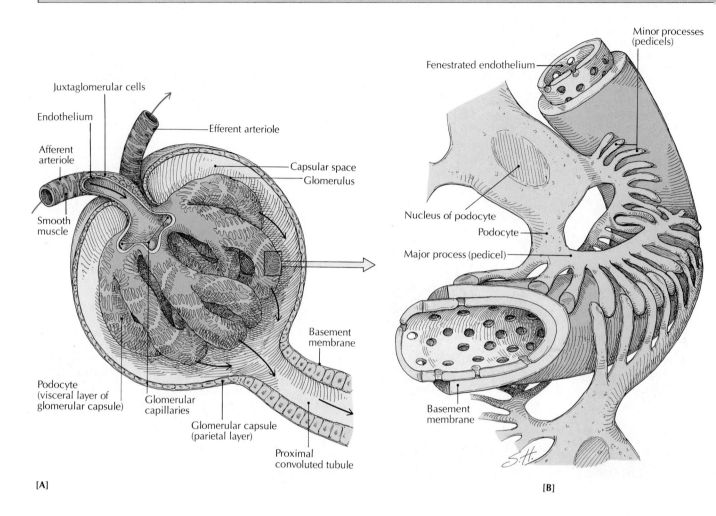

[A]

Juxtaglomerular cells

Endothelium

Afferent arteriole

Smooth muscle

Podocyte (visceral layer of glomerular capsule)

Glomerular capillaries

Glomerular capsule (parietal layer)

Proximal convoluted tubule

Efferent arteriole

Capsular space

Glomerulus

Basement membrane

[B]

Minor processes (pedicels)

Fenestrated endothelium

Nucleus of podocyte

Podocyte

Major process (pedicel)

Basement membrane

Filtration of the blood takes place in the renal corpuscle across three layers:

1 The first layer is the *endothelium* of the glomerulus, which contains tiny pores called *fenestrations* ("windows") (see Figure 24.7B).

2 The middle layer is the *basement membrane* (basal lamina) of the glomerulus. Like all capillary basement membranes, it is composed of fibrous proteins.

3 The third layer is the *visceral layer* of the glomerular capsule and the podocytes. The podocytes are relatively large, nucleated cells, with a nucleated cell body from which several cell processes spread. These processes eventually branch to form many smaller, fingerlike processes called *foot processes* or ***pedicels*** (L. "little feet") (see Figure 24.7C). The pedicels are the portions of the podocytes that are in contact with the glomerular capillaries. The small region between pedicels exposes the underlying basement membrane; it is called a ***filtration slit*** or *slit pore* (see Figure 24.7D). A thin *slit membrane* extends between the foot processes of adjacent cells, forming a barrier (the *filtration barrier*) that restricts the passage of

certain molecules through the filtration slit, but allows the passage of others. (The filtration barrier is actually made up of fenestrated endothelium, basement membrane, and filtration slits.)

Taken together, the three layers of the renal corpuscle constitute the ***endothelial capsular membrane.*** The filtration process involves the entire membrane. Although the cellular components of the blood and the large proteins do not normally pass through, water and dissolved solutes (electrolytes, sugars, urea, amino acids, and polypeptides) have no trouble passing from the blood into the capsular space of the glomerular capsule. The fluid filtered from the blood is called the ***glomerular filtrate.***

Proximal convoluted tubule From the glomerular capsule, the glomerular filtrate drains into the ***proximal convoluted tubule*** (see Figure 24.6). The name describes its location (it is the portion of the excretory tubule *proximal* to the glomerular capsule) and its appearance (it is coiled in an irregular, *convoluted* way).

Glomerular (Bowman's) capsule. (A) Drawing of the renal corpuscle, composed of a glomerulus and glomerular capsule. Arrows indicate direction of blood flow. (B) Enlarged drawing of part of a capillary loop of a glomerulus, showing its covering of podocytes and its underlying fenestrated endothelium, with pores that act as a sieve. A portion of a podocyte and capillary have been cut away to show the components of the filtration barrier. (C) Scanning electron micrograph, showing a transected glomerular capillary loop surrounded by podocytes of the visceral layer of the glomerular capsule. Extensive branching of tiny pedicels can be seen. Between the pedicels are filtration slits, the actual sites of filtration. (D) Electron micrograph of a portion of the wall of a glomerular capillary, showing pores (fenestrations) in the endothelium. The pedicels (foot processes of podocytes) rest on the outer surface of the basement membrane (basal lamina), with filtration slits between them. The filtration barrier is composed of the fenestrated endothelium, filtration slits, and basement membrane. [(C) *Reprinted with permission from Leon Weiss (ed.),* Histology: Cell and Tissue Biology, *5th ed. New York, Elsevier, 1973.* (D) *Daniel S. Friend, M.D.*]

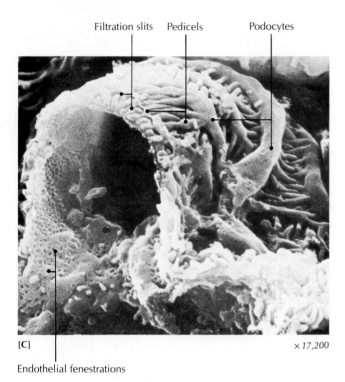

Filtration slits Pedicels Podocytes

[C]   × 17,200

Endothelial fenestrations

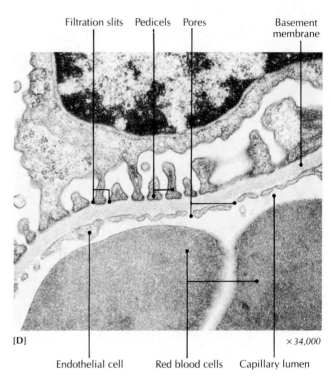

Filtration slits Pedicels Pores   Basement membrane

[D]   × 34,000

Endothelial cell   Red blood cells   Capillary lumen

The epithelial cells making up the tubule are cuboidal, and the surfaces that face the lumen of the tubule are lined with microvilli, forming a brush border (Figure 24.8). The microvilli enormously increase the epithelial surface area over which transport can take place. The proximal convoluted tubule is the site for the reabsorption of many substances filtered from the blood, such as water, sodium, potassium, chloride, glucose, some amino acids and polypeptides, and bicarbonate ions.

Loop of the nephron (loop of Henle) After passing through the proximal convoluted tubule, the glomerular filtrate enters a straightened portion of the excretory tubule called the *loop of the nephron* (*loop of Henle*), originally named for the nineteenth-century German anatomist Friedrich Henle (see Figure 24.6). The loop of the nephron is composed of descending and ascending limbs of cuboidal epithelium connected by a thin limb of simple squamous epithelium. The *descending* (thin) *limb* extends into the medulla of the kidney and becomes very narrow. At that point it makes an abrupt upward U-turn and widens again. This *ascending* (thick) *limb* passes through the medulla into the cortex. As you will see later (p. 784), the purpose of the loop of the nephron is to generate a concentration gradient.

Distal convoluted tubule After the glomerular filtrate passes through the entire length of the loop of the nephron, it moves into the *distal convoluted tubule* in the cortex (see Figure 24.6). As the name suggests, it is an irregularly shaped tubule located far from the origin of the excretory tubules at the glomerular capsule. The cuboidal epithelial cells of the distal tubule are similar in size to those of the proximal tubule, but they have very few microvilli. The cells are abundantly supplied with mitochondria near their basal surfaces, where potassium and hydrogen ions are actively transported into the glomerular filtrate.

Collecting duct Glomerular filtrate passes from the distal convoluted tubule into the *collecting duct* (see Figure 24.6). Actually, the distal tubules of many nephrons may empty into

FIGURE 24.8

Excretory tubule. (A) Electron micrograph of a portion of the tubular epithelium of the proximal convoluted tubule. Note the microvilli facing the tubular lumen, and the abundance of mitochondria. Both microvilli and mitochondria support the many active-transport functions of the proximal convoluted tubule. (B) Types of epithelial cells in different portions of the nephron. [(A) *C. Craig Tisher, M.D.*]

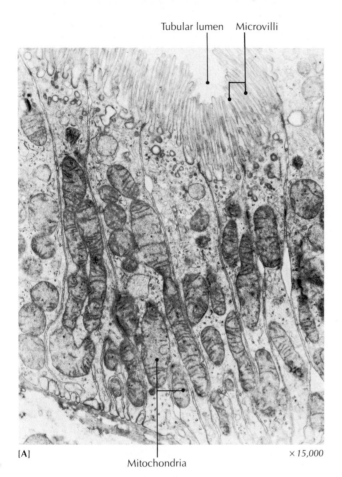

[A]   × 15,000

Tubular lumen   Microvilli

Mitochondria

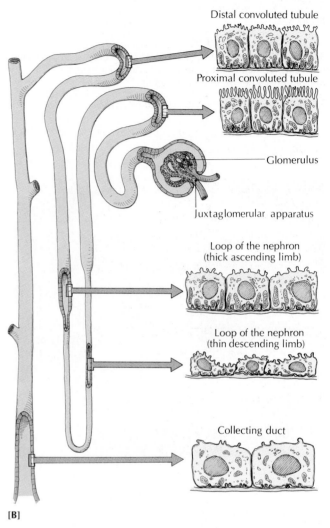

[B]

Distal convoluted tubule

Proximal convoluted tubule

Glomerulus

Juxtaglomerular apparatus

Loop of the nephron (thick ascending limb)

Loop of the nephron (thin descending limb)

Collecting duct

a single collecting duct, which travels through the medulla, roughly parallel to the limbs of the loop of the nephron. The collecting ducts join to form larger and larger tubes until they reach the minor calyx. From there, the final filtrate (now called urine) drains into the renal pelvis, which acts as a funnel that directs the urine into the ureter, and subsequently to the urinary bladder.

Juxtaglomerular apparatus In the cortex, the distal convoluted tubule makes intimate contact with the afferent and efferent arterioles of the glomerulus (Figure 24.9). Here, the smooth muscle cells of the tunica media of the arterioles have cytoplasm that contains more granules than myofilaments. These specialized smooth muscle cells, known as *juxtaglomerular cells,* are in contact with a group of epithelial cells of the distal tubule called the *macula densa* (L. "dense spot"). The cells of the macula densa are longer and narrower than

the epithelial cells of the distal tubule, and their nuclei are closer together.

Together, the juxtaglomerular cells, macula densa, afferent and efferent arterioles, and a pad of cells called the *extraglomerular mesangium* make up the **juxtaglomerular apparatus.** Also associated with these four structures are unmyelinated nerves.

Ask Yourself

1 *What is the relationship between the renal pelvis and the calyces?*

2 *Where are the renal pyramids located?*

3 *What are the two regions of the renal cortex?*

4 *What is the glomerulus? The vasa recta?*

5 *What are the main functions of a nephron?*

6 *What is the glomerular capsule, and how is it related to the proximal convoluted tubule, loop of the nephron, distal convoluted tubule, and collecting duct?*

7 *What is the juxtaglomerular apparatus?*

PHYSIOLOGY OF THE KIDNEYS

In the formation of urine, a series of events occurs. The end results of these events are the elimination of metabolic wastes from the body, regulation of total body water balance, cleansing and filtration of blood, control of the chemical composition of the blood and other body fluids (extracellular fluid volume), and control of acid-base balance. The urine that is excreted is much more concentrated than the glomerular filtrate that enters the nephron, because large amounts of water are reabsorbed into the circulatory system all along the permeable walls of the nephron tubules.

By three separate processes, the kidneys produce and modify the glomerular filtrate that is finally excreted from the body as urine (Figure 24.10):

1 The kidneys filter blood. When blood flows from the afferent arteriole into the glomerulus, it is under high pressure (about 75 mm Hg). This pressure forces *some* of the blood plasma into the glomerular capsule, but the blood cells and large proteins remain within the glomerulus, unable to pass through the endothelial capsular membrane. This process, known as ***glomerular filtration,*** forms the initial glomerular filtrate.

2 As the glomerular filtrate passes through the length of the nephron tubule, useful substances such as water, sodium ions, glucose, and amino acids that were initially lost from the blood during filtration are returned to the blood by active and passive transport. This process is called ***tubular reabsorption.***

FIGURE 24.9

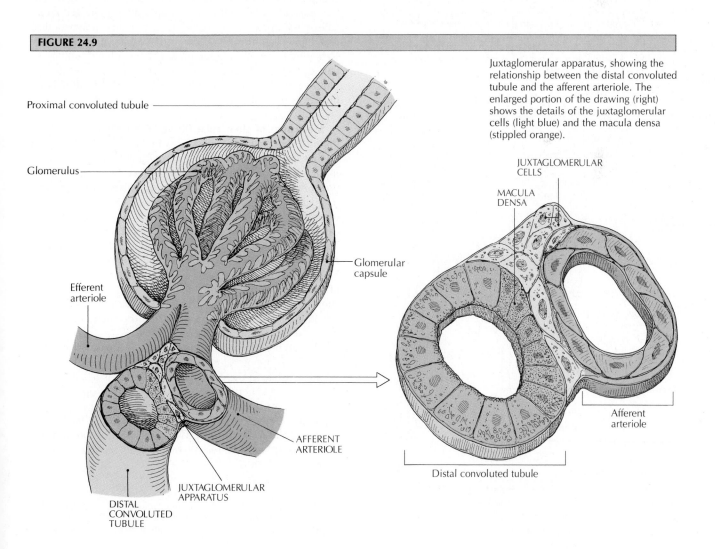

Proximal convoluted tubule

Glomerulus

Efferent arteriole

Glomerular capsule

AFFERENT ARTERIOLE

JUXTAGLOMERULAR APPARATUS

DISTAL CONVOLUTED TUBULE

Juxtaglomerular apparatus, showing the relationship between the distal convoluted tubule and the afferent arteriole. The enlarged portion of the drawing (right) shows the details of the juxtaglomerular cells (light blue) and the macula densa (stippled orange).

JUXTAGLOMERULAR CELLS

MACULA DENSA

Afferent arteriole

Distal convoluted tubule

3 Some unwanted ions and substances may be transported (secreted) from the blood in the peritubular capillaries *into* the glomerular filtrate as it passes through the excretory tubules. In this way, products such as potassium ions, hydrogen ions, certain drugs (penicillin, for example), and organic compounds may be excreted. This process is **tubular secretion.** It occurs in a direction opposite to tubular reabsorption.

In summary, glomerular filtrate flows through a nephron to the inner medulla of the kidney, and then back again to the cortex. In the process, essential substances such as water and glucose are reabsorbed into the blood. Finally, the glomerular filtrate moves to the medulla of the kidney again, where it is now called urine, and passes to the bladder through the ureter. The filtered blood is returned to the body through the renal vein.

FIGURE 24.10

Urine production in a nephron. In *glomerular filtration* (1) fluid passes from the blood into the glomerular capsule. In *tubular reabsorption* (2) essential substances such as glucose and water pass from the excretory tubule back into the blood of the peritubular capillaries. In *tubular secretion* (3) materials pass from the blood of the peritubular capillaries into the nephron tubule. The fluid that enters the collecting duct is urine. Note that tubular secretion and reabsorption actually take place simultaneously.

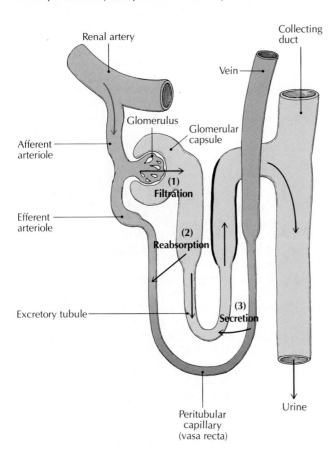

Renal artery

Collecting duct

Vein

Glomerulus

Glomerular capsule

Afferent arteriole

(1) Filtration

Efferent arteriole

(2) Reabsorption

Excretory tubule

(3) Secretion

Peritubular capillary (vasa recta)

Urine

Glomerular Filtration

The filtration that takes place in the renal corpuscle is a mechanical process driven by the high hydrostatic pressure within the glomerulus. Even though the endothelial capsular membrane of the glomerular capsule is composed of three layers (see Figure 24.7), it is 100 to 1000 times more permeable to water and compounds with low molecular weights than an ordinary capillary membrane. Small molecules pass through the pores (fenestrations) of the endothelium, the basement membrane, and the filtration slits of the visceral layer of the glomerular capsule. Normally, none of the blood cells can pass through this barrier.

Although the fenestrations are large enough to allow the passage of proteins, the fluid that enters the capsular space is free of large proteins. This exclusion occurs because negative charges on the large proteins hinder their passage through the negatively charged glycoproteins in the basement membrane of the capillaries. The large size and negative charges of the large plasma proteins may also restrict their movement through the filtration slits between the pedicels. Positively charged molecules pass through this barrier faster than neutral molecules of the same size, and negatively charged molecules pass more slowly. Some small proteins (polypeptides) are able to pass through, however. They are normally reabsorbed by pinocytosis in the proximal convoluted tubules, so that the filtrate is essentially protein-free. As a result, the filtrate and blood plasma are in equilibrium, and virtually no protein is present in the urine.

Mechanics of glomerular filtration *Filtration* is the forcing of a fluid, and the substances dissolved in it, through a membrane by pressure. The resulting fluid is called a *filtrate.* Filtration occurs in the kidneys because blood hydrostatic pressure forces water and dissolved solutes in the plasma through any openings large enough to allow passage. The resulting fluid, because it is formed in the glomerulus, is called **glomerular filtrate.** The pores, together with the high blood hydrostatic pressure, help the glomeruli to send huge amounts of glomerular filtrate into the nephron tubules. The blood pressure in the glomeruli is higher than in other parts of the body, partly because the renal afferent arterioles are short, straight branches of the interlobular arteries that can transmit a relatively high pressure, and the efferent arterioles have a relatively high resistance to flow.

The glomerular hydrostatic pressure is about 75 mm Hg. This force is opposed by the hydrostatic pressure of the glomerular filtrate in the capsular space, which is usually about 20 mm Hg. Glomerular hydrostatic pressure is opposed further by the oncotic pressure of the blood in the glomerular capillaries.* When blood enters the glomerulus, it has an oncotic pressure of 26 mm Hg. With the loss of water and solutes through filtration, the plasma proteins cause an increase

---

*Oncotic pressure is the portion of the colloidal osmotic pressure due specifically to plasma proteins. Osmotic pressure is the pressure that develops because of water movement into a contained solution.

in the oncotic pressure, so that when the blood leaves the glomerulus it has an oncotic pressure of about 34 mm Hg. This makes an average pressure of 30 mm Hg in the glomerulus.

If this value is applied to the filtration over the entire glomerulus, the net hydrostatic pressure for filtration is reduced from 75 to 25 mm Hg (Figure 24.11). This final pressure, called the *effective filtration pressure* (FP_{eff}), is determined by the Starling-Landis equation:

$$FP_{eff} = \begin{pmatrix} \text{Glomerular} \\ \text{blood} \\ \text{hydrostatic} \\ \text{pressure} \end{pmatrix} - \begin{pmatrix} \text{Capsular} & \text{Oncotic pressure} \\ \text{hydrostatic} + & \text{of the blood in} \\ \text{pressure} & \text{the glomerulus} \end{pmatrix}$$

By using the values above, a normal FP_{eff} can be calculated as follows:

FP_{eff} = 75 mm Hg − (20 mm Hg + 30 mm Hg)
  = 75 mm Hg − 50 mm Hg
  = 25 mm Hg

Thus, an effective filtration pressure of 25 mm Hg favors the movement of materials out of the glomerulus and into the glomerular capsule.

Glomerular filtration rate The amount of filtrate formed in the capsular space each minute is the *glomerular filtration rate* (GFR). In a normal 70-kg man, the renal blood flow is about 1200 mL/min, and the renal plasma flow is about 625 mL/min. The percentage of plasma entering the nephrons that actually becomes glomerular filtrate is called the *filtration fraction*. About 20 percent of the plasma is filtered at the glomeruli, giving a filtrate of about 125 mL/min, 7.5 L/hr, or

FIGURE 24.11

Effective filtration pressure. The effective filtration pressure is the net result of the glomerular hydrostatic pressure (orange arrow) favoring filtration, and the capsular hydrostatic pressure (interstitial pressure + tubular pressure) and glomerular blood oncotic pressure (dark green arrow) opposing filtration.

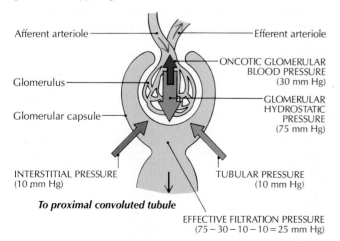

Afferent arteriole — — Efferent arteriole

Glomerulus —

Glomerular capsule —

ONCOTIC GLOMERULAR BLOOD PRESSURE (30 mm Hg)

GLOMERULAR HYDROSTATIC PRESSURE (75 mm Hg)

INTERSTITIAL PRESSURE (10 mm Hg)

TUBULAR PRESSURE (10 mm Hg)

To proximal convoluted tubule

EFFECTIVE FILTRATION PRESSURE (75 − 30 − 10 − 10 = 25 mm Hg)

180 L/day. However, the rate of glomerular filtration can vary considerably, depending upon nutrition and physical activity. It may be as low as 5 mL/min or as high as 200 mL/min. More than 99 percent of this filtrate is reabsorbed from the nephron tubule, producing a small volume of concentrated urine.

At the normal rate of 125 mL/min, the kidneys will produce 150,000 to 180,000 mL (150 to 180 L) of filtrate over a 24-hour period, but only 1000 to 2000 mL (1 to 2 L) of urine will be excreted (see Table 24.2). The total plasma volume is approximately 3 L, which means that the total plasma volume of a 70-kg man is filtered about 60 times a day, or once every half hour. Such a turnover allows for precise control of the body's internal environment.

Factors affecting the glomerular filtration rate The rate of glomerular filtration depends on several factors: (1) effective filtration pressure, (2) stress, (3) total surface area available for filtration, (4) capillary permeability, (5) intrinsic renal autoregulation, and (6) release of renin.

In general, the glomerular filtration rate increases when the effective filtration pressure in the glomerulus increases, and it decreases when the pressure decreases. The rate of blood flow through the nephrons affects the glomerular filtration rate by affecting the net colloidal oncotic pressure in the glomerulus. The glomerular filtration rate is also affected by the resistance offered by the afferent arterioles. For example, the constriction of the afferent arterioles decreases the glomerular filtration rate by decreasing both the plasma flow rate and pressure in the glomerulus. Conversely, when the afferent arteriole dilates, the flow of plasma and glomerular pressure both increase, raising the glomerular filtration rate (Figure 24.12).

Stress affects the glomerular filtration rate, since the input to the afferent arterioles is the sympathetic nervous system. In times of stress, sympathetic input decreases renal blood flow and glomerular filtration rate by constricting the afferent arterioles, which ensures an adequate blood flow to the brain and heart in times of stress.

The total number of nephrons and the surface area of the glomerulus available for filtration also set limits on the possible filtration rate in any given period. After birth, the number of nephrons and the glomerular surface area do not increase, but they may decrease as the result of kidney disease or damage such as acute glomerulonephritis and pyelonephritis (see Physiological and Anatomical Abnormalities at the end of the chapter). Certain kinds of surgery, such as a partial nephrectomy, also decrease the glomerular filtration rate by removing nephrons and surface area.

Capillary permeability does not change under normal circumstances, but is usually affected when the kidneys are diseased. For example, in certain kidney disorders, the capillary permeability to proteins increases, and proteins are excreted in the urine *(proteinuria)* along with large amounts of albumin *(albuminuria)*. In both situations, the glomerular filtration rate increases due to an increase of osmotic pressure in the glomerular capsule and a decrease in plasma oncotic pressure and, as a result, tends to pull (draw) more water from the blood.

FIGURE 24.12

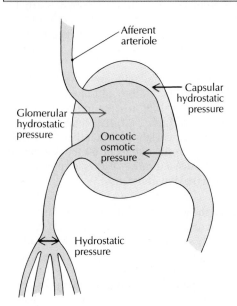

[A] EFFECTIVE FILTRATION PRESSURE NORMAL

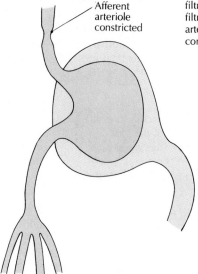

[B] EFFECTIVE FILTRATION PRESSURE DROPS

Effect of constriction of afferent arterioles on filtration pressure. (A) Normal effective filtration pressure with no constriction of arterioles. (B) Decrease in pressure with constriction.

Finally, the kidneys have the ability to maintain a relatively constant glomerular filtration rate in the face of fluctuating blood pressures via negative feedback mechanisms. The first negative feedback mechanism is called *intrinsic renal autoregulation.* It occurs primarily because of the effect of two negative feedback mechanisms on the afferent and efferent arterioles. The normal mean arterial pressure is approximately 100 mm Hg. When this pressure falls, there is a low glomerular filtration rate. This causes an excessive reabsorption of chloride ions (Cl^-) from the glomerular filtrate, lowering the concentration of chloride ions. In the presence of a low level of chloride ions, the cells of the macula densa of the juxtaglomerular apparatus automatically cause the afferent arterioles to dilate. This vasodilation increases the rate of blood flow into the glomerulus, and raises the glomerular filtration rate back to normal. When the pressure rises above 100 mm Hg, the negative feedback causes automatic afferent constriction. Both of these changes prevent either a major rise or fall, respectively, of blood pressure within the glomerular capillaries.

The second negative feedback system involves the release of the enzyme *renin,* also in response to a low level of chloride ions in the glomerular filtrate.

Within the arteriolar walls at their point of contact with the tubule (juxtaglomerular apparatus; see Figure 24.10), the smooth muscle cells are specialized to form *granular cells,* so called because they contain many granules. These granules are secretory vesicles containing the enzyme renin. When renin is released, it sets up a *renin-angiotensin pathway* (see p. 794), which causes the efferent arterioles to constrict. Constriction causes the glomerular pressure to rise, thereby increasing the glomerular filtration rate. Thus, even though systemic mean arterial pressure fluctuates daily, the kidneys can maintain a fairly stable glomerular filtration rate.

Tubular Reabsorption

About 99 percent of the glomerular filtrate formed in the glomerular capsule is eventually returned to the blood (Table 24.2). (The other 1 percent is excreted in the urine.) Water and selected solutes re-enter the blood in the peritubular capillaries and vasa recta. This **tubular reabsorption** from the tubules of the nephron is controlled by the epithelial cells of the nephron tubule. As shown in Figure 24.8, these cells have different shapes and functions, depending upon their location in the tubule. To some extent, reabsorption occurs throughout the entire length of the tubule, but there are important specializations.

Most of the nutritionally important substances, such as glucose, small proteins, amino acids, citric acid, and vitamins, are completely reabsorbed from the proximal convoluted tubule. As a result, these substances are virtually absent from

TABLE 24.2	COMPARISON OF FILTRATION, REABSORPTION, AND EXCRETION		
Substance	Amount filtered by glomeruli (daily)	Percentage reabsorbed by nephron tubules	Amount excreted in urine (daily)
Glucose	170.0 g	100.0	0.0 g
Water	150.0 L	99.0	1.5 L
Calcium	17.0 g	98.8	0.2 g
Salt	700.0 g	98.0	15.0 g
Phosphate	5.1 g	80.0	1.2 g
Urea	50.0 g	40.0	30.0 g
Sulfate	3.4 g	33.0	2.7 g

the loop of the nephron, distal convoluted tubule, and collecting duct.

Important ions, such as Na^+, Cl^-, K^+, and HCO_3^-, are reabsorbed from the proximal convoluted tubule and from the other segments of the excretory tubule. Table 24.2 shows the constituents of the filtrate and a comparison of filtration, reabsorption, and excretion.

Many by-products of metabolism are toxic to the body in high concentrations, and must be eliminated before they are allowed to build up. For example, in the process of metabolizing proteins, the body produces ammonia. Fish and other water animals can flush ammonia out to sea as soon as it is produced, but humans convert it to the less toxic urea instead. (Ammonia combines with carbon dioxide to form urea, ridding the body of two waste products at the same time.) Because our bodies can tolerate 100,000 times more urea than ammonia, we have time to incorporate the urea into urine and eliminate it when we urinate. Not only does the conversion from ammonia to urea protect us from a toxic waste product, it also allows us to eliminate urea with only 1/100,000 the amount of water we would have needed to eliminate the same amount of ammonia.

Besides urea, creatinine (a metabolic waste product of the creatine in muscle tissue), ammonia, sulfates, phosphates, and uric acid are reabsorbed back into the body in relatively small quantities. Because little water is reabsorbed at this point, however, these toxic substances become concentrated in the urine. If necessary, the nephron tubule can reabsorb some of these by-products in order to maintain optimum concentrations in the blood and extracellular fluids.

Transport mechanisms of tubular reabsorption Reabsorption of materials from the filtrate into the blood is achieved by both active and passive transport mechanisms. Recall that **active transport** requires metabolic energy to move a material against an electrochemical or concentration gradient. Also, active transport involves a *carrier molecule,* such as an enzyme, located in the plasma membrane. **Passive transport** is diffusion operating across one or more membrane barriers. As you saw in Chapter 3, passive transport *does not* require metabolic energy, because the material always moves in the direction of the concentration gradient, from a region of high concentration to a region of lower concentration. If a carrier molecule is involved, the passive transport is called *facilitated diffusion.* If no carrier is involved, the transport is by *simple diffusion.*

Both active and passive transport contribute to the reabsorption of specific substances in different segments of the nephron tubule. Figure 24.13 shows that the epithelial cells that make up the walls of the proximal convoluted tubule have two exposed sides. The *apical side* faces the lumen of the tubule; it contains many microvilli. The *basolateral side* (basal surface) faces the outside of the tubule, near the peritubular capillaries.

The concentration of sodium in the glomerular filtrate is the same as in the plasma. However, the tubular epithelial cells have a lower sodium concentration, which is due partially to the low plasma membrane permeability to sodium, and partially to the continual active transport of sodium out of the cell by the sodium-potassium pump. As a result, a

concentration gradient is established that causes sodium ions to diffuse into the tubular epithelial cells. The active transport of sodium takes place in the basolateral membrane of the proximal convoluted tubule cells. This raises the sodium concentration of the fluid in the interstitial spaces, and results in the diffusion of sodium ions into the adjacent peritubular capillaries.

As a result of the movement of sodium ions from the glomerular filtrate to the peritubular capillaries, an electrochemical gradient is created that allows chloride ions to follow the sodium ions by passive transport. The chloride and sodium ions raise the osmolarity and osmotic pressure of the surrounding tissue fluid above those of the tubular fluid, creating an osmotic gradient. Since the membranes of the

TABLE 24.3   MAJOR COMPONENTS AND FUNCTIONS OF THE NEPHRON

Structure	Major functions
Glomerulus	Vascular (capillary) component of renal corpuscle. Filters (by *hydrostatic pressure*) water, dissolved substances (minus most plasma proteins, blood cells) from blood plasma.
Glomerular (Bowman's) capsule	Initial tubular component of nephron. Transports glomerular filtrate to nephron tubules.
Proximal convoluted tubule	Reabsorbs (by *active transport*) Na^+, K^+, Ca^{2+}, amino acids, creatinine, uric acid, ascorbic acid, ketone bodies, glucose. Reabsorbs (by *electrochemical gradient*) PO_4^{3-}, Cl^-, SO_4^{2-}, HCO_3^-. Reabsorbs (by *osmosis*) water. Reabsorbs (by *diffusion*) urea. Actively secretes substances such as penicillin, histamine, organic acids, organic bases.
Descending loop of the nephron (loop of Henle)	Reabsorbs (by *active transport*) Na^+. Reabsorbs (by *electrochemical gradient*) PO_4^{3-}, Cl^-, SO_4^{2-}, HCO_3^-. Reabsorbs (by *osmosis*) water. Reabsorbs (by *diffusion*) urea.
Ascending loop of the nephron (loop of Henle)	Reabsorbs (by *active transport*) Na^+, Cl^-. Reabsorbs (by *electrochemical gradient*) PO_4^{3-}, SO_4^{2-}, HCO_3^-.
Distal convoluted tubule	Reabsorbs (by *active transport*) Na^+, K^+, amino acids, creatinine, uric acid. Reabsorbs (by *electrochemical gradient*) PO_4^{3-}, Cl^-, SO_4^{2-}, HCO_3^-. Reabsorbs (by *osmosis*) water. Reabsorbs (by *diffusion*) urea. Actively secretes H^+, NH_4^+.
Collecting duct	Reabsorbs (by *active transport*) NA^+. Reabsorbs (by *osmosis*, under control of ADH) water. Actively secretes H^+.

proximal convoluted tubule cells are very permeable to water, water now moves across the membrane. This passive transport (reabsorption) of water by simple diffusion is called *osmosis*. The passive transport of water (as from the proximal convoluted tubule) follows the active transport of sodium out of the filtrate into the peritubular capillaries. The osmotic movement of water is caused by the decreased solute concentration in the capillary blood. Since this reabsorption of water occurs because of osmosis, the process is called *obligatory reabsorption of water*.

When antidiuretic hormone (ADH) is present, the distal convoluted tubule and collecting duct show an increased permeability to water, and as a result, increased amounts of water are reabsorbed into the blood (Figure 24.14). Since this reabsorption of water occurs only when ADH is present, the process is called *facultative reabsorption of water*. This type of reabsorption accounts for approximately 20 percent of the water in the glomerular filtrate, and is a major mechanism for regulating the water content of the blood.

The movement of any substance—water, glucose, or an ion—from the nephron tubule into the blood of the peritubular capillaries requires the crossing of several membrane barriers. Substances pass from the nephron lumen through the tubular epithelial cells, into the interstitial fluid, and finally through the capillary endothelial cells before entering the capillary lumen. Not every crossing of a plasma membrane is an active process, but *if any step in the crossing sequence requires metabolic energy to move material against a concentration gradient, the entire process is classified as active transport*.

FIGURE 24.13

Active transport of sodium ions across the basal surface of epithelial cells of the proximal convoluted tubule. Subsequent movement of sodium ions may be by diffusion. Potassium is accumulated within epithelial cells of the distal convoluted tubule by an active transport system in the basal surface membrane. Potassium accumulates partly in exchange for sodium, which is pumped out actively. Potassium diffuses passively across the apical membrane of the tubule cell, entering the lumen of the tubule. Because one step in this process is an active transport, the entire phenomenon of potassium ion secretion is classified as active. Solid arrows indicate active transport; dashed arrows indicate passive transport.

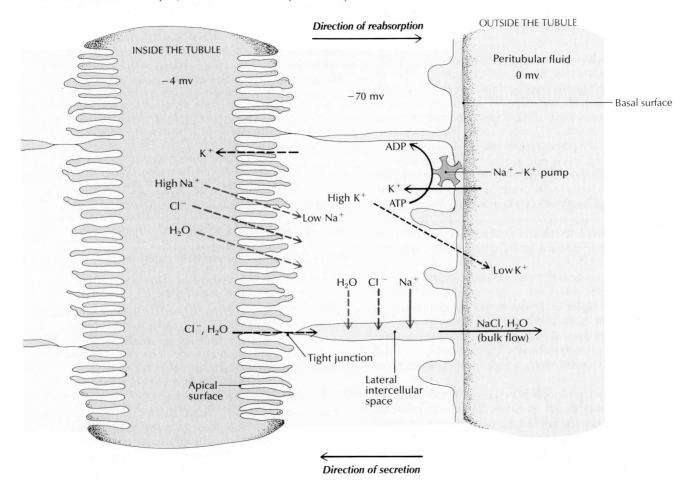

FIGURE 24.14

Secretion of ADH and the regulation of the osmolarity of blood.

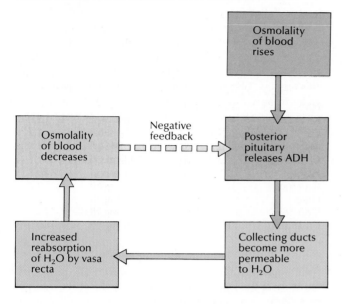

Tubular Secretion

The final composition of the urine excreted from the body depends not only upon filtration and reabsorption, but also upon the secretion of certain substances from the blood *into* the filtrate. This **tubular secretion,** which takes place in the distal convoluted tubule and collecting duct, allows the kidneys to increase their efficiency in clearing certain substances from blood plasma, and helps maintain the normal pH of blood between 7.35 and 7.45.

The collecting ducts concentrate the dilute fluid they receive by reabsorbing salts and water. Water is reabsorbed from the collecting duct in response to the high concentration of sodium ions in the interstitial fluid surrounding the duct. Water moves passively by osmosis, but the permeability of the epithelial cells to water is controlled by ADH. The secretion of ADH by the posterior lobe of the pituitary gland increases the permeability of the collecting duct to water. This results in the reabsorption of water, and the excretion of a concentrated urine. Eventually, the filtrate within the collecting ducts is in equilibrium with the blood. The composition of urine does not change after it leaves the collecting ducts. The renal pelvis, ureters, bladder, and urethra are merely storage and transportation units.

Potassium ion secretion Normal filtration processes rid the body of about 70 mEq (milliequivalents)* of potassium ions

---

*A milliequivalent (mEq) is a unit of measure that expresses the combining activity of an electrolyte. An equivalent weight is the amount of one electrolyte that will react with a given amount of hydrogen. One mEq of any cation will always react chemically with one mEq of an anion. The number of cations and anions must be equal for homeostasis to exist. Milliequivalents are always expressed as the number of ions in a liter of solution.

(K^+) daily. But because potassium is extremely abundant in most foods, it is not unusual to consume 100 to 200 mEq in a single day. Potassium is toxic in high concentrations *(hyperkalemia),* and can cause cardiac arrhythmias, fibrillation, muscle weakness, and cramps (see Chapter 18).

To assure the proper balance of potassium in the blood, the distal convoluted tubule secretes excess potassium, and operates as an exchange system in the following way: At the same time that sodium is actively transported out of the filtrate into the interstitial fluid, potassium is transported from the interstitial fluid into the epithelial cells of the tubule. Potassium then diffuses into the tubular lumen because of the high intracellular concentration of potassium ions in the epithelial cells. Apparently there is a one-for-one exchange system, with one sodium leaving for every potassium that enters.

Potassium secretion is under the control of the hormone **aldosterone,** secreted by the adrenal cortex in direct response to an increased intake of potassium. As aldosterone is increased, the cells of the distal convoluted tubule are stimulated to increase their secretion of potassium ions. In the absence of aldosterone, no potassium is excreted in the urine. Instead, all of the filtered potassium is reabsorbed from the distal tubule.

Hydrogen-ion secretion Not only do our diets include a great deal of acidic food, but our metabolic reactions also produce acids from carbohydrates and proteins. For these reasons, we must rid our bodies of excessive hydrogen ions to maintain our normal blood pH.

Increased acid in the blood directly stimulates epithelial cells of the proximal and distal convoluted tubules, as well as the collecting ducts, to secrete hydrogen ions. Apparently, there are no nervous or hormonal controls involved. (See the section on acid-base regulation, p. 786, for a complete discussion of H^+ secretion.)

The ability to secrete hydrogen ions depends upon the presence in the tubular filtrate of acceptors for these ions. The most important acceptors are phosphate (HPO_4^{2-}) and ammonia (NH_3). Phosphate comes to the filtrate directly from the plasma. Before reaching the distal convoluted tubule, some phosphate may be reabsorbed, but some remains. The ammonia, however, is formed by the tubular cells and diffuses into the lumen of the tubule. Both phosphate and ammonia are capable of accepting hydrogen ions, as shown in the following reactions:

$$HPO_4^{2-} + H^+ \longrightarrow H_2PO_4^-$$

$$NH_3 + H^+ \longrightarrow NH_4^+$$

Ammonia passes through plasma membranes easily, but the ammonium ion (NH_4^+) does not. Consequently, once ammonia has accepted a hydrogen ion, only excretion of the ammonium ion is possible. Both $H_2PO_4^-$ and NH_4^+ are excreted in the urine, and both are effective vehicles for ridding the body of excessive hydrogen ions. Because hydrogen ions (acids) are eliminated in this process, the urine is usually more acidic than blood. The pH of urine ranges from 5.0 to 8.0.

Plasma Clearance

As you have just seen, the processes of glomerular filtration, tubular reabsorption, and tubular secretion produce the excretory product urine. These processes cleanse the circulating blood of waste products that tend to upset homeostasis. The ability of the kidneys to clear wastes from blood plasma is measured by a process known as *plasma clearance.*

In order to quantitate the volume of plasma that can be cleared by the kidneys, two values must be determined. First, the *urinary excretion rate* must be known. This is the rate, measured in milligrams per minute, at which the kidneys can excrete the waste compound. Second, the *concentration of the compound* in the blood plasma must be known.

The excretion of creatinine serves as a good example of plasma clearance, since creatinine can be measured easily in the urine and blood, and is excreted from the body only by the kidneys. The plasma clearance of creatinine (or any other substance) can be calculated as follows: Simultaneous samples are taken of blood and urine excreted each minute, and the quantity of creatinine in each milliliter is analyzed chemically. Assuming that the kidneys excrete 2 mg of creatinine per minute, and the concentration of creatinine in the plasma is 0.01 mg/mL, we can calculate the plasma clearance:

$$\text{Plasma clearance} = \frac{\text{Urinary excretion rate (mg/min)}}{\text{Plasma concentration (mg/mL)}}$$

$$= \frac{2 \text{ mg/min}}{0.01 \text{ mg/mL}}$$

$$= 200 \text{ mL/min}$$

In other words, the kidneys can clear 0.01 mg/mL of creatinine from 200 mL of plasma each minute.

Because every compound excreted by the kidneys has a plasma-clearance value, one of the best ways to test overall kidney function is to measure the clearance of the excreted substances. Table 24.4 lists the plasma-clearance values for the typical constituents of urine.

Countercurrent Concentration of Urine

The filtrate in the proximal convoluted tubule is relatively dilute, but by the time it reaches the collecting duct it is usually highly concentrated. The final urine is said to be *hyperosmotic,* which means that it contains a larger amount of metabolic wastes and solutes than the same amount of blood does (Figure 24.15). When we drink large amounts of liquid, the kidneys keep a proper balance of water and solutes by producing a larger volume of dilute, hypo-osmotic urine, and excreting more urine. During periods of dehydration, the kidneys produce a more concentrated urine and excrete less water.

How is concentrated urine produced? The mechanism depends upon (1) the permeability of the different parts of the juxtamedullary nephron tubule (especially the descending and ascending limbs of the loop of the nephron, the vasa

TABLE 24.4 RELATIVE CONCENTRATIONS OF EXCRETED SUBSTANCES IN THE GLOMERULAR FILTRATE AND URINE, AND THEIR PLASMA-CLEARANCE VALUES

Substance	Concentration		Plasma clearance per minute (conc. urine/conc. plasma)
	Glomerular filtrate	Urine	
Na^+	142 mEq/L	128 mEq/L	0.9
K^+	5 mEq/L	60 mEq/L	12.0
Ca^{2+}	4 mEq/L	5 mEq/L	1.2
Mg^{2+}	3 mEq/L	15 mEq/L	5.0
Cl^-	103 mEq/L	134 mEq/L	1.3
HCO_3^-	28 mEq/L	14 mEq/L	0.5
Glucose	100 mg/dL	0 mg/dL	0.0
Urea	26 mg/dL	1820 mg/dL	70.0
Uric acid	3 mg/dL	42 mg/dL	14.0
Creatinine	1.1 mg/dL	196 mg/dL	140.0

recta capillaries, and the collecting duct), (2) the overall structure of the loop of the nephron, (3) the active transport of sodium ions, and (4) the concentration gradient in the renal medulla.

The function of the loop of the nephron is to generate a concentration gradient. The descending limb does not actively transport any substance, including sodium ions.* However, sodium ions may enter passively from the surrounding tissue (see Figure 24.16). The descending limb is also highly permeable to water, which moves out by osmosis. As the glomerular filtrate moves down the descending limb, it loses water passively, and becomes more and more concentrated (hyperosmotic). Note in Figure 24.16 that sodium ions have been entering the descending limb as water has been lost, which helps to make the filtrate more concentrated. By the time the glomerular filtrate reaches the deep medullary region, it is concentrated to 1200 to 1400 mOsm/L.

At this point, the structure of the loop of the nephron becomes important. At its most concentrated stage, the glomerular filtrate suddenly makes a sharp upward turn into the ascending limb, which runs parallel to the descending limb and close enough to it to allow an exchange of materials. The ascending limb actively transports sodium ions across the basolateral membrane via the sodium-potassium pump into the surrounding tissue fluid. The chloride ions (Cl^-) follow the sodium ions (Na^+) passively, due to an electrical attraction. As you just saw, these ions then diffuse into the adjacent descending limb. As the glomerular filtrate moves along the ascending limb, it continues to lose sodium ions, and it becomes less and less concentrated (hypo-osmotic) as it approaches the distal convoluted tubule. As the glomerular filtrate moves through the nephron, the vasa recta also removes the solutes and water that have been reabsorbed from the loops of the nephron and the collecting ducts.

---

*The epithelial cells of the descending limb are small, thin, and flattened (see Figure 24.8). They have none of the microvilli or numerous mitochondria present in epithelial cells that are engaged in active transport.

FIGURE 24.15

Changes in the osmolarity of glomerular filtrate in different portions of the nephron. The epithelium of the thickened wall of the ascending limb of the loop of the nephron is relatively impermeable to water. The presence of ADH makes the filtrate become hyperosmotic. Numerical values are in milliosmoles per liter.

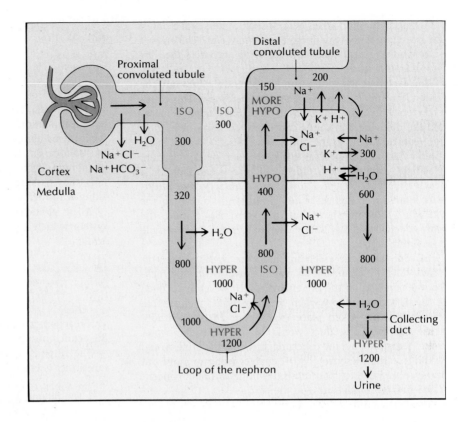

FIGURE 24.16

Operation of the countercurrent multiplier system to produce concentrated urine. Different segments of the loop of the nephron have different permeabilities and transport capacities that affect the countercurrent exchange. Numerical values are in milliosmoles per liter.

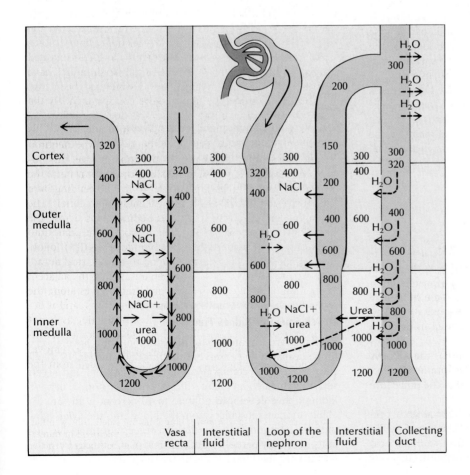

A crucial question arises here: If sodium ions move out of the ascending limb, why doesn't water move in? The answer is: It would, except that *the walls of the ascending limb are impermeable to water.*

The exchange of sodium from the ascending limb into the descending limb is called a *countercurrent exchange.* This countercurrent flow (flow in opposite directions) helps to maintain the differences in solute concentration from one end of the nephron tubule to the other. When such a system is assisted by active transport, as it is in the loop of the nephron, it actually multiplies the differences in concentration by a positive feedback system known as a ***countercurrent multiplier system*** (see Figure 24.16). That is, the more salt the ascending limb transports into the surrounding fluid, the more concentrated the fluid will be that returns to it from the descending limb.

In contrast to the ascending limb, the walls of the collecting duct *are* permeable to water. However, this permeability depends upon the presence of ADH, which is secreted by the posterior pituitary when neural messages indicate that dehydration has taken place (see Figure 24.14). Without ADH, the collecting duct would be impermeable, water would remain within the collecting duct, and the urine would be diluted. The presence of ADH enables water to move out of the collecting duct by osmosis, resulting in a final concentrated (hyperosmotic) urine in the bladder.

Finally, in order for this countercurrent multiplier system to be efficient, only a small number of sodium ions that are transported out of the ascending limb can be carried away by the surrounding peritubular capillaries. A hyperosmotic condition must be maintained in the renal medulla by the vasa recta, which forms long loops parallel to the loops of the nephron of the juxtaglomerular nephrons. The sodium (and, to a lesser extent, chlorine) that diffuses into the arteriae rectae subsequently diffuses passively out of the blood vessels in the ascending segments (venae rectae) of these vessels. Thus, the diffusion of sodium and water, first into the arteriae rectae and then out of the venae rectae, helps maintain the hyperosmotic condition of the interstitial fluid in the renal medulla.

Acid-Base Regulation

The kidneys help regulate the acid-base balance of the blood, primarily by simultaneously excreting hydrogen ions (H^+) and reabsorbing bicarbonate ions (HCO_3^-). These regulatory processes help maintain a homeostatic blood pH of 7.35 to 7.45. They are summarized here and discussed in detail in the next chapter.

When the blood becomes acidic (acidosis), the kidneys attempt to maintain homeostasis by eliminating excess hydrogen ions, and thus produce an acidic urine. The procedure occurs in three different ways:

1 The filtered hydrogen ions can be excreted as such, and the tubular cells can also generate additional hydrogen by the following reaction involving the formation of carbonic acid (H_2CO_3) from carbon dioxide (CO_2) and water (H_2O):

$$\overset{\text{Calcium}}{CO_2 + H_2O} \longrightarrow H_2CO_3 \longrightarrow H^+ + HCO_3^-$$

From cellular metabolism — Actively transported into tubular lumen — Reabsorbed into peritubular capillaries

2 Ammonia (NH_3) is formed in the tubular cells from protein deamination, and diffuses into the glomerular filtrate, where it combines with the filtered and secreted hydrogen ions to form ammonium ions (NH_4^+). Because the basement membrane of the tubular cells is permeable to ammonium ions, they pass into the urine, carrying the excess acid (H^+) along with them.

3 The glomerular filtrate contains the phosphate buffer system, which acts in the following way when the blood is acidic:

$$H^+ + HPO_4^{2-} \longrightarrow H_2PO_4^-$$

Acid — Base monohydrogen phosphate ions — Acid dihydrogen phosphate ions Salt

This reaction causes the urine to become more acidic, that is, the acidic phosphate ions increase.

The response of the kidneys to an increased blood pH (alkalosis) is essentially the opposite of the response to acidosis:

1 The filtered hydrogen ions are reabsorbed back into the blood from the filtrate, and the tubular cells secrete potassium ions in an equal exchange to preserve electrical neutrality. The reabsorbed hydrogen ions are neutralized within the tubular cells as follows:

$$\overset{\text{Calcium}}{H^+ + HCO_3^-} \longrightarrow H_2CO_3 \longrightarrow CO_2 + H_2O$$

Absorbed from the tubular lumen — From the blood — To the blood

2 Very little ammonia is formed by the tubular cells. This prevents hydrogen ions from being bound in the glomerular filtrate, and allows them to be reabsorbed.

3 The phosphate buffer system in the glomerular filtrate shifts as follows and allows more hydrogen ions to be reabsorbed to bring the blood pH back to normal:

$$H_2PO_4^- \longrightarrow HPO_4^{2-} + H^+$$

Development of Kidney Function

If infants are threatened by dehydration, they have a very limited capacity to conserve water through the kidneys. Diarrhea is the most common cause of dehydration, and it can seriously threaten an infant's life. At birth, the normal infant's kidneys have developed only 20 to 40 percent of the adult's ability to filter substances from the blood into the urine (glomerular filtration). During the first three months of life, infants have only partial ability to reabsorb substances from the filtrate into the blood (tubular reabsorption) and to excrete

Embryonic Development of the Kidneys

Each kidney develops from an embryonic **metanephros** ("hindkidney"), which is preceded by transitional structures called the *pronephros* ("forekidney") and *mesonephros* ("midkidney"). The nonfunctional pronephros forms early in the fourth week after fertilization, and degenerates late in the fourth week (drawing A). Most of the pronephric duct is used by the newly formed mesonephros, which appears caudally to the degenerated pronephros. The mesonephros possibly functions for about a month, although this is not certain.

Mesonephric ducts and tubules form late in the fourth week, and the metanephros begins to form early in the fifth week.* The metanephros develops from the *metanephric diverticulum* (ureteric bud) and the *metanephric mesodermal mass* (metanephrogenic blastema) (see drawing A). The metanephric diverticulum induces the formation of many important structures of the kidney. As the metanephric diverticulum extends dorsocranially, the metanephric mesodermal mass forms a cap over it (drawing B). This metanephric cap becomes larger, and internal differentiation occurs rapidly. At the same time, the stalk of the metanephric diverticulum and its extended cranial end form the *ureter* and *renal pelvis,* respectively (drawing C). The pelvis then divides into the *major* and *minor calyces* (drawing D), with *collecting tubules* soon emerging from the minor calyces (drawing E). The tubules branch repeatedly as they move closer to the outer edges of the metanephros.

At about 8 weeks, the metanephros begins to function, as mesodermal cells become arranged in small vesicular masses at the blind end of the collecting tubules (drawing F). These vesicular masses differentiate into a urine-carrying tubule that drains into its nearest collecting tubule. By week 13, the proximal portion of the S-shaped vesicular mass develops into the *distal* and *proximal convoluted tubules* and the *loop of the nephron* (loop of Henle), and the distal end of the vesicular mass becomes the *glomerulus* and *glomerular* (Bowman's) *capsule* (drawing G). The glomeruli are fully developed by week 36, about a month prior to birth.

*Although the mature kidneys are not yet developed, urine is formed throughout the fetal period. The urine mixes with the amniotic fluid surrounding the fetus in the uterus, which the fetus ingests. Then the urine is absorbed by the intestines and eventually deposited in the mother's bloodstream and eliminated in her urine.

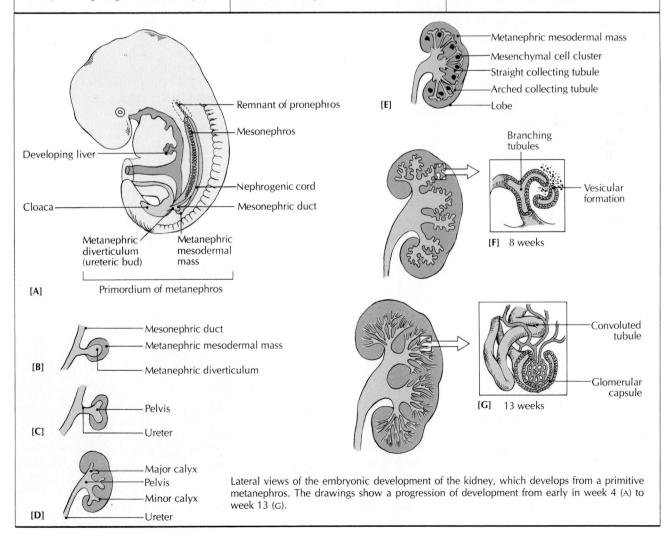

[A]
- Remnant of pronephros
- Mesonephros
- Developing liver
- Nephrogenic cord
- Mesonephric duct
- Cloaca
- Metanephric diverticulum (ureteric bud)
- Metanephric mesodermal mass
- Primordium of metanephros

[B]
- Mesonephric duct
- Metanephric mesodermal mass
- Metanephric diverticulum

[C]
- Pelvis
- Ureter

[D]
- Major calyx
- Pelvis
- Minor calyx
- Ureter

[E]
- Metanephric mesodermal mass
- Mesenchymal cell cluster
- Straight collecting tubule
- Arched collecting tubule
- Lobe

[F] 8 weeks
- Branching tubules
- Vesicular formation

[G] 13 weeks
- Convoluted tubule
- Glomerular capsule

Lateral views of the embryonic development of the kidney, which develops from a primitive metanephros. The drawings show a progression of development from early in week 4 (A) to week 13 (G).

a concentrated urine. At this early stage, infants also have a limited capacity to excrete any overload of water, salt, or acid that they consume. Consequently, careful selection and preparation of an infant's food are very important.

After one year, an infant's kidney function is equivalent to that of an adult (with allowances for differences in size and weight). This level of kidney function is maintained until about the age of 40, after which there is usually a progressive loss of function. Glomerular filtration, blood flow, number of nephrons, and concentrating ability all decrease with advancing age. By the age of 90, a reduction of 50 percent may occur. However, even in old age, the kidneys retain their capacity to excrete acid and help regulate acid-base balance.

Ask Yourself

1 What are the three processes used by the kidneys during the overall formation of urine?

2 What is meant by effective filtration pressure?

3 What is the glomerular filtration rate, and what factors affect it?

4 Why is tubular secretion important?

5 What is meant by plasma clearance? How is it calculated?

6 What is the countercurrent multiplier system?

7 How do the kidneys help regulate the acid-base balance of the blood?

8 At what point in the excretory tubule does the urine become concentrated?

ACCESSORY EXCRETORY STRUCTURES

Urine is produced in the kidneys, but accessory structures are required to store it and eventually eliminate it from the body. These structures are the two ureters, the urinary bladder, and the urethra (see Figure 24.1).

Ureters

Attached to each kidney is a tube about 25 to 30 cm (10 to 12 in.) long. This is the *ureter,* which transports urine from the renal pelvis to the urinary bladder. The ureters pass between the parietal peritoneum and the body wall to the pelvic cavity, where they enter the urinary bladder on the posterior lateral surfaces. Narrow at the kidneys, the ureters widen to about 1.7 cm (0.5 in.) near the bladder.

In cross section, the lumen of the ureter has a star shape, and three layers of tissue (Figure 24.17):

1 The innermost layer, the *tunica mucosa,* faces the lumen, and is made up of transitional epithelium and connective tissue.

2 The middle layer, the *tunica muscularis,* consists of two layers of smooth muscle: an inner longitudinal layer and an outer circular layer. In the lower third of the ureter, an additional outer longitudinal layer is present. Periodic peristaltic contractions of the smooth muscle, along with hydrostatic pressure and gravity, help to move urine into the bladder.

3 The outermost layer is the fibrous *tunica adventitia.* It consists of connective tissue that holds the ureters in place.

Before the ureters join with the cavity of the bladder, they run obliquely for a short distance within the bladder wall. This portion of the ureter is compressed by the wall of the bladder when the bladder is distended with accumulated urine. The compression of the ureter helps prevent the backflow of urine from the bladder into the ureters.

The arteries that supply the ureters are branches from the renal, testicular, internal iliac, and inferior vesical. There are no specific veins. The nerves are derived from the inferior mesenteric, testicular, and pelvic plexuses.

Urinary Bladder

The *urinary bladder* is the hollow, muscular sac that collects urine from the ureters and stores it until it is excreted from the body through the urethra. It usually accumulates 300 to 400 mL of urine before emptying, but it can expand enough to hold 600 to 800 mL. The bladder is located on the floor of the pelvic cavity, and like the kidneys and ureters, it is retroperitoneal. In males, it is anterior to the rectum and above the prostate gland (Figure 24.18B). In females, it is

---

FIGURE 24.17

Cross section through the ureter, showing the star-shaped lumen and the layers of mucosa, muscularis, and adventitia.
[*Stan Elems/Visuals Unlimited.*]

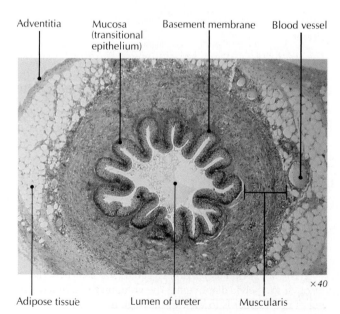

Adventitia   Mucosa (transitional epithelium)   Basement membrane   Blood vessel

Adipose tissue   Lumen of ureter   Muscularis

×40

located much lower, anterior to the uterus and upper vagina (Figure 24.18A).

The bladder is made up of four coats. The outermost is the **tunica serosa.** It is derived from the peritoneum and covers only the upper and lateral surfaces of the bladder. It is a moist tissue that lubricates the upper surface, eliminating friction when the distended bladder presses against other organs. The **tunica muscularis** consists of three layers of smooth muscle: an external layer of longitudinal fibers, a middle layer of circular fibers, and an inner layer of longitudinal fibers. Collectively, the three muscle layers are called the *detrusor muscle.* The third coat, the **tela** (L. web) **submucosa,** consists of connective tissue that holds the muscular coat to the fourth and innermost coat, the **tunica mucosa.** The mucous coat is composed of transitional epithelium that allows the bladder to stretch and contract. When the bladder is empty, the tunica mucosa is thrown into folds called *rugae.*

The openings of the ureters and urethra into the cavity of the bladder outline a triangular area, the **trigone** (Figure 24.19). At the site where the urethra leaves the bladder, involuntary smooth muscle in the bladder wall forms an *internal urethral sphincter* (involuntary muscle). The sphincter surrounds the opening, as sphincters do, and is usually contracted to prevent the bladder from emptying prematurely. It is now agreed that there is no involuntary sphincter at the exit of the urethra from the urinary bladder (below the internal sphincter). Instead, the voluntary striated muscle fibers of the bladder that surround the exit serve as a voluntary *external urethral sphincter* that holds back the urine until urination is convenient. (The involuntary muscle fibers are so oriented and arranged that, when relaxed, the opening of the bladder into the urethra is closed.)

The arteries supplying the bladder are the superior, middle, and inferior vesical. All are branches of the internal iliac artery. In the female, additional branches arise from the uterine and vaginal arteries. The veins form a plexus that drains into the iliac veins. The nerves arise from the third and fourth sacral nerves and the hypogastric plexus.

Urethra

The **urethra** is a tube of smooth muscle lined with mucosa. It joins the bladder at its lowest surface and transports urine outside the body during urination. In the female, the urethra is only about 4 cm (1.5 in.) long, and opens to the exterior at an orifice between the clitoris and the vaginal opening (see Figure 24.18A). In the male, the urethra is about 20 cm (8 in.) long, extending from the bladder to the external urethral orifice at the tip of the penis (see Figure 24.18B).

The male urethra passes through three different regions, and its portions have been named after those regions (see Figure 24.18B):

1 The **prostatic portion** passes through the prostate gland and is joined by the ejaculatory duct. Distal to this juncture, the male urethra serves both reproductive and excretory functions.

2 The **membranous portion** is a short segment that passes through the pelvic diaphragm. The voluntary external opening of the bladder is located in this region.

3 The longest portion is the **spongy portion,** which extends the length of the penis from the lower surface of the pelvic diaphragm to the external urethral orifice. The spongy portion is joined by ducts from the *bulbourethral glands* (Cowper's glands), which, together with the prostate gland, secrete fluids into the semen during ejaculation. (Mucus-secreting glands also empty into the urethra along its full length.)

Ask Yourself

1 *What is the purpose of the ureters? The bladder?*

2 *What are the functions of the internal urethral sphincter and the external muscular opening in the bladder?*

3 *What is the urethra? How does it differ in the male and female?*

URINE AND URINATION

The laboratory examination of urine (urinalysis) provides an important noninvasive "window" through which a physician can evaluate the condition of many body processes. The composition and physical characteristics of urine are influenced by the diet and nutritional status of the body, the condition of the body's metabolic processes, and the ability of the kidneys to process material brought to them by the blood.

Composition of Urine

The constituents of urine, and their concentrations, can vary greatly in a healthy person, depending upon diet, exercise, and water consumption, among other factors. In general, however, **urine** is composed mostly of water, urea, chloride, potassium, sodium, creatinine, phosphates, sulfates, and uric acid (Table 24.5).

Normal constituents Water is the major constituent of urine, averaging about 1100 mL per day. The next most abundant constituent is urea, a waste product of amino acid metabolism. Most of the ammonia from excess amino acids is converted by the liver into urea, which is excreted in the urine. However, a small amount of ammonia does appear in the urine.

Abnormal constituents Under normal conditions, there should be no protein in urine. Bleeding in the lower urinary tract, and other kidney disorders, however, usually produce large quantities of protein in the urine, a condition called **proteinuria,** or **albuminuria,** if the protein is primarily albumin. Kidney diseases often cause an increased permeability of the glomerular membrane. Because more plasma protein is filtered than can be reabsorbed in the proximal tubule, large quantities appear in the urine. In extreme cases, so much protein may be lost that the oncotic pressure of the plasma decreases, causing a net loss of water from the sys-

FIGURE 24.18

Bladder and urethra. (A) Sagittal section of the female pelvis, showing the bladder and urethra. Because of the location of these organs, an enlarged uterus during pregnancy pushes down on the bladder. (B) Sagittal section of the male pelvis. The longer male urethra also serves a reproductive function, carrying semen from the testes and accessory reproductive organs through the full length of the penis. (C) Section through the bladder wall, showing the three layers of tissue. (D) Scanning electron micrograph of the mucosa of the bladder. The folded tissue is capable of shrinking and expanding to accommodate the changing volume of the urine in the bladder. The ability to stretch is an important characteristic of transitional epithelium. The layer of smooth muscle is also capable of great ranges of stretching. [(C) *Michael H. Ross, Ph.D.* (D) *Fujita, Tanaka, and Tokunaga,* SEM Atlas of Cells and Tissues. *Tokyo, Igaku-Shoin, 1981.*]

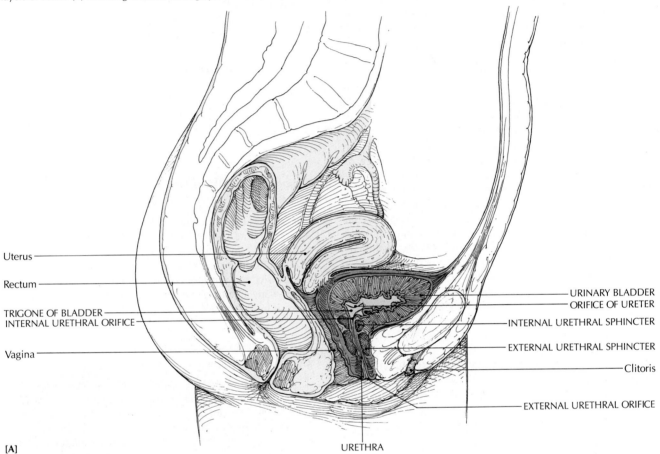

Uterus

Rectum

TRIGONE OF BLADDER
INTERNAL URETHRAL ORIFICE

Vagina

URINARY BLADDER
ORIFICE OF URETER
INTERNAL URETHRAL SPHINCTER
EXTERNAL URETHRAL SPHINCTER
Clitoris
EXTERNAL URETHRAL ORIFICE

URETHRA

[A]

temic capillaries. The water that is lost accumulates in the tissue spaces, producing a generalized swelling and puffiness called *edema* (see Chapter 25).

Another substance normally absent from urine is glucose, which is totally reabsorbed into the blood from the proximal convoluted tubules. Glucose appears in the urine of diabetics, who are unable to produce enough insulin to facilitate the uptake of glucose by the body cells. As a result, glomerular filtration produces a filtrate containing more glucose than can be reabsorbed by the proximal convoluted tubule. The presence of glucose in the urine is called **glycosuria.**

If vitamin C is taken in large doses, it appears in the urine, and high-protein meals produce large amounts of hydrogen ions in the urine. If no protein is consumed for an extended period, and the body metabolizes fat instead, the urine may contain large quantities of acetoacetic acid, β-hydroxybutyric acid, and acetone, collectively known as *ketone bodies.* This condition, known as **ketonuria,** may also be caused by diabetes mellitus, prolonged diarrhea, vomiting, or starvation.

Diseases of the kidney itself can alter its capacity to filter, reabsorb, or secrete properly. When the kidneys are diseased, they excrete abnormal amounts of protein, are unable to regulate electrolytes and acid, or excrete toxic end products that may cause a generalized deterioration of the function of other organs. If complete kidney failure results, death occurs unless the patient is maintained by an artificial kidney machine (see the essay on p. 793).

Other abnormal constituents of the urine are casts and calculi. **Casts** are decomposed blood cells, tubule cells, or fats that form tiny bits of hard material in the nephron tubules. **Calculi** (kidney stones) are formed when the salts in the kidney tubules, ureters, or bladder precipitate to form insoluble masses. They are most common when the diet is consistently high in minerals, is abnormally acidic or alkaline, or is low in water, and when the parathyroid glands are overactive. (See Physiological and Anatomical Abnormalities, p. 797.)

One final abnormal constituent of urine should be noted. About four to six weeks after conception, a woman's urine

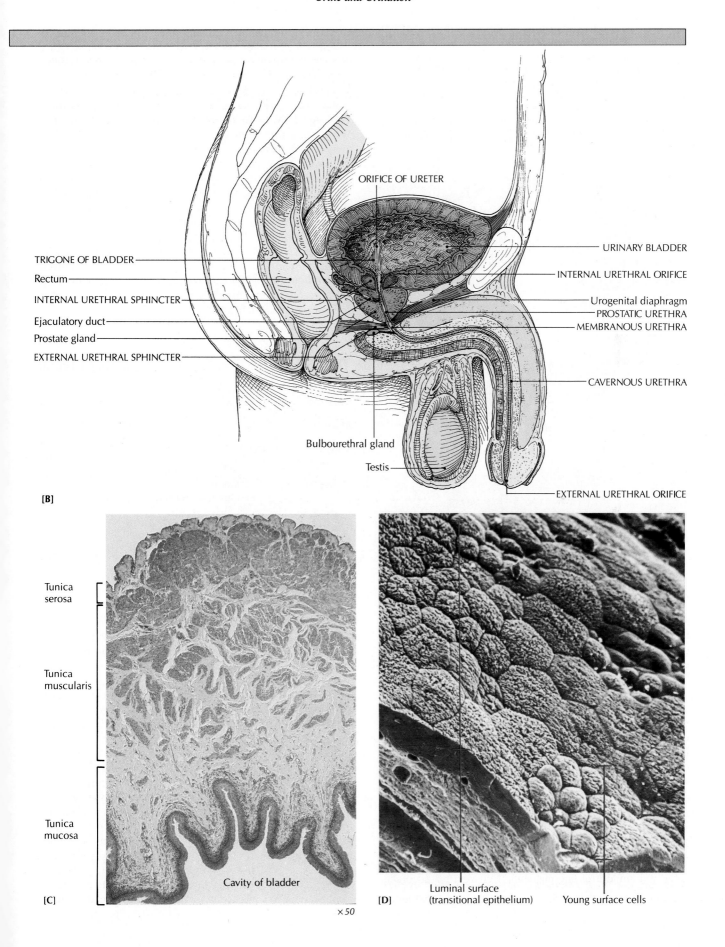

ORIFICE OF URETER

TRIGONE OF BLADDER

Rectum

INTERNAL URETHRAL SPHINCTER

Ejaculatory duct

Prostate gland

EXTERNAL URETHRAL SPHINCTER

URINARY BLADDER

INTERNAL URETHRAL ORIFICE

Urogenital diaphragm

PROSTATIC URETHRA

MEMBRANOUS URETHRA

CAVERNOUS URETHRA

Bulbourethral gland

Testis

EXTERNAL URETHRAL ORIFICE

[B]

Tunica serosa

Tunica muscularis

Tunica mucosa

Cavity of bladder

[C]

× 50

[D]

Luminal surface (transitional epithelium)

Young surface cells

FIGURE 24.19

FIGURE 24.19

Trigone and urethral sphincters. The trigone is a triangular area (dashed lines) inside the bladder formed by the two openings of the ureters and the urethral orifice.

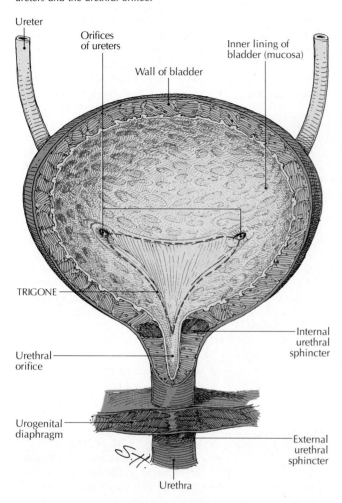

diet, fever, and therapeutic drugs (including some antibiotics) can alter the color to a dark yellow, brown, or even red. Other factors also influence the color of urine. For example, reddish urine may indicate the presence of hemoglobin from red blood cells (a condition called *hematuria*), and urine with a milky appearance may contain fat droplets or pus (*pyuria*). Any sign of blood or pus in the urine should be taken seriously, since it may indicate the early stages of kidney infection or malfunction.

Specific gravity is the ratio of the weight of a unit volume of a substance to the weight of an equal volume of distilled water. The specific gravity of distilled water is designated as 1.000. The value for urine ranges from 1.001 to 1.035, depending on the total amount of solute in the urine. The more solutes, the higher the specific gravity.

Urine has an aromatic but not unpleasant odor when it is fresh and free of microorganisms. Stale urine that is allowed to stand outside the body, however, is usually contaminated by microorganisms that convert urea into ammonia, and it acquires a harsh, stale smell.

TABLE 24.5 | MAJOR CONSTITUENTS OF URINE

Substance	Origin	Amount
WATER	Diet, metabolism.	1500.0 mL
ORGANIC SUBSTANCES		
Urea	Protein deamination.	30.0 g
Creatinine	Metabolism of creatinine in muscle.	1.6 g
Hippuric acid	Liver detoxification of benzoic acid.	0.7 g
Uric acid	Catabolism of nucleic acids.	0.7 g
Ketone bodies	Lipid metabolism.	0.04 g
INORGANIC SUBSTANCES		
Sodium	Diet.	1.5 g (NaCl, 15.0)
Chloride	Diet.	7.0 g
Potassium	Diet.	4.0 g
Phosphates	Diet, metabolism of phosphate-containing compounds (amino acids).	2.5 g
Sulfates	Diet, metabolism of sulfate-containing compounds.	2.5 g
Ammonia	Deamination of amino acids.	0.7 g
Calcium	Diet.	0.3 g
Magnesium	Diet.	0.1 g
Total solids		About 50 g
Total volume		About 1550 mL

will contain a hormone called *human chorionic gonadotropin* (hCG). It is produced by the placenta and is the basis for pregnancy tests (see Chapter 27).

Physical Properties of Urine

The pH, appearance, specific gravity, and volume of urine may have diagnostic importance when they are considered along with other symptoms. The pH of urine ranges from 5.0 to 8.0, but is usually slightly acidic. Daily variations are closely related to the food eaten. A person who eats meat regularly might have a pH of 5 or 6, and a vegetarian or someone who normally eats a high-fiber, low-protein diet might have a slightly alkaline pH of 7 or 8.

Urine is normally translucent (clear, not cloudy), and yellow or amber in color. The color is caused partly by the presence and amount of bile pigments. However, water intake,

Artificial Kidneys

If the kidneys fail, toxic wastes build up in the body until cells and organs begin to deteriorate and eventually die. Many tens of thousands of Americans suffer from kidney failure, but a large number of them are leading relatively normal lives because of the successful use of the "artificial kidney," or **hemodialysis therapy** (Gr. *dialuein*, to tear apart).

The principle of diffusion is basic to the working mechanism of the artificial kidney. Very simply, the artificial kidney is a machine that pumps 5 to 6 L of blood from the body through a hollow fiber dialyzer (see drawing). The blood is rinsed by a briny solution, and sodium, potassium, and waste products such as urea, uric acid, excess water, and creatinine diffuse through the dialyzer by osmotic pressure. The cleansed blood is then routed back into the body.

The full procedure is as follows: One tube is permanently linked to an artery (usually in the arm), and another is attached to a nearby vein. During dialysis, the tubes are hooked up to the machine. Blood is pumped from the artery through an oxygenated salt solution similar in ionic concentration to body plasma. Because the concentration of wastes is higher than the normal concentration of the plasmalike fluid, the wastes automatically diffuse through the semipermeable membrane of the tubes into this rinsing fluid. The membrane is porous to all blood substances except proteins and red blood cells. The wastes are eliminated from the body, and the purified blood is free to flow back into the body.

Dialysis is sometimes also used to add nutrients to the blood. For instance, large amounts of glucose may be added to the salt solution, so that the glucose may be diffused into the blood at the same time that wastes are being removed.

It takes about 6 hours and 20 passes through the bathing fluid to complete a full cycle of dialysis, and most patients receive treatment two or three times a week. Unfortunately, even this highly successful machine, which can remove wastes from the blood 30 times faster than a natural kidney, provides only partial relief for kidney-failure victims. All patients remain *uremic* (Gr. "urine in the blood") to some degree.

Another method is **peritoneal dialysis,** which uses the patient's peritoneal membrane instead of an artificial membrane for diffusion. The "clean" fluid (dialysis fluid) is injected into the peritoneal cavity each day for three or four days. Three daytime exchanges remain in the body for 5 hours each before they are removed, and the fourth overnight exchange remains in the body for 8 to 10 hours.

Some of the advantages of peritoneal dialysis are that it allows the patient to be less immobilized, it requires less equipment than hemodialysis, it is less expensive, and it puts less strain on the cardiovascular system. The main disadvantage of peritoneal dialysis is that there is a high risk of peritonitis. It also takes longer than hemodialysis and is generally less effective in removing urea.

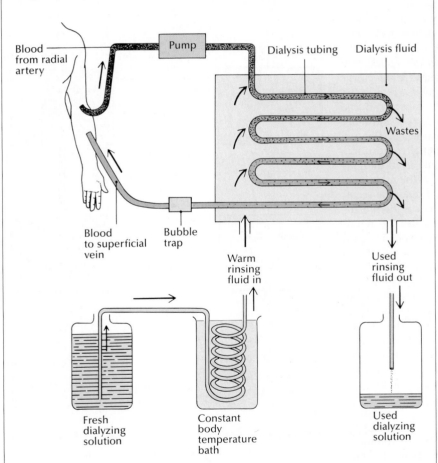

The mechanism of artificial kidneys.

Volume of Urine

To maintain the proper osmotic concentrations of the extracellular fluids (including blood), to excrete wastes, and to maintain proper kidney function, the body must excrete *at least* 450 mL of urine per day, or about 19 mL per hour. A healthy person with normal physical activity and water intake will eliminate from 1000 to 1800 mL of urine every day.

The volume of urine excreted can be influenced by water consumption, diet, external temperature, hormonal and enzymatic actions, blood pressure, diuretics, drugs, and emotional state. If water consumption is limited, the kidneys reabsorb water and excrete a low volume of concentrated urine. Excess water is excreted in diluted urine.

When the outside temperature is high, the body sweats, and loses a considerable amount of water (see Table 24.1). As a result, urine volume is reduced in hot weather. The opposite happens in cold temperatures. Peripheral and surface blood vessels constrict, and the flow of blood through the glomeruli is increased. Also increased are the hydrostatic pressure in the glomeruli, the rate of filtration, and the volume of urine. That is why you feel the urge to urinate more often than usual on a cold, wintery day.

The volume and concentration of urine are regulated by two hormones and an enzyme. The hormones are *antidiuretic hormone* (ADH) and *aldosterone;* the enzyme is *renin.* **ADH** controls the permeability of the walls of the collecting ducts. When dehydration occurs, more ADH is released, and more water is withdrawn from the urine. The opposite effect occurs during overhydration. **Aldosterone,** an adrenal cortical hormone, stimulates the reabsorption of sodium from the kidneys, thus reducing the amount of sodium in the urine.

In response to a decrease of extracellular fluid, the secretory granules of the juxtaglomerular cells release the proteolytic enzyme **renin** (REE-nihn), which activates the renin-angiotensin pathway.* Renin acts on a plasma globulin called *angiotensin,* releasing the inactive peptide *angiotensin I.* An enzyme in the lungs converts angiotensin I into the peptide *angiotensin II,* which causes the adrenal cortex to release aldosterone. Aldosterone then stimulates the distal tubule of the nephron to reabsorb sodium ions in exchange for either hydrogen or potassium ions, which increases water uptake, and consequently, the volume of extracellular fluid. Renin is important for the regulation of sodium balance, and apparently it also plays a role in water balance by stimulating the thirst reflex in the brain. However, thirst is also indicated by other signals, such as a dryness of the mouth.

Because the kidneys play a major role in the regulation of systemic blood pressure, which can alter the glomerular filtration rate and urine volume, they minimize the volume of urine excreted each day. The regulatory mechanism, which

*Do not confuse *renin* with the digestive enzyme *rennin.*

Why does beer stimulate the need to urinate?   *Alcohol in the beer inhibits ADH secretion, which then stimulates urine production. Also, another component of beer, lupulin from hops, is a diuretic.*

FIGURE 24.20

How the kidneys help to regulate blood pressure.

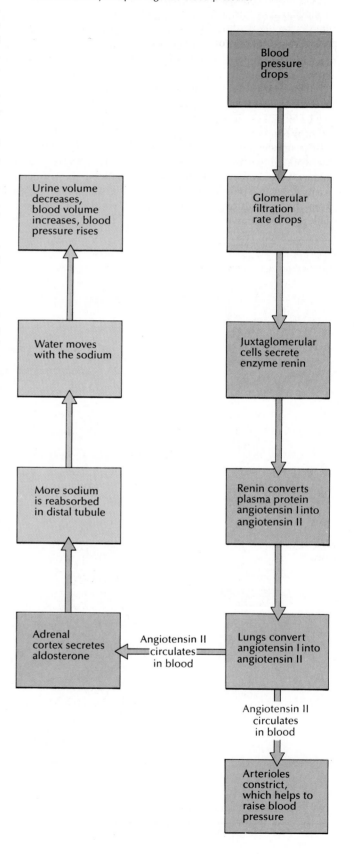

involves the renin-angiotensin pathway and aldosterone, is shown in Figure 24.20.

Certain foods, such as mustard, pepper, tea, coffee, and alcohol, increase urine production, and so does vitamin C taken in large quantities. In contrast, nicotine decreases the production of urine. Fortunately, urine production decreases while the body is asleep, so that we are usually not disturbed during the night.

Therapeutic drugs called **diuretics** (Gr. "to urinate through") can increase the volume of urine by decreasing the water reabsorbed from the collecting duct. (Remember that reabsorption is under the control of ADH.) Some diuretic drugs (*thiazide* and *thiazidelike diuretics*) increase the excretion of sodium and water by inhibiting sodium reabsorption in the ascending limb of the loop of the nephron. Other diuretics *(loop diuretics)* inhibit the reabsorption of sodium and chloride at the proximal portion of the ascending limb, increasing water excretion. Still other diuretics *(carbonic anhydrase inhibitors)* block enzymes to increase the kidney's excretion of sodium, potassium, bicarbonate, and water. Among several miscellaneous types of diuretics, mannitol increases the osmotic pressure of the glomerular filtrate, inhibiting the tubular reabsorption of water and electrolytes. Water, being a natural diuretic, decreases the reabsorption of water by inhibiting the secretion of ADH. Diuretics may act by altering the constriction of arterioles or the secretion of aldosterone or ADH, or by a combination of these factors.

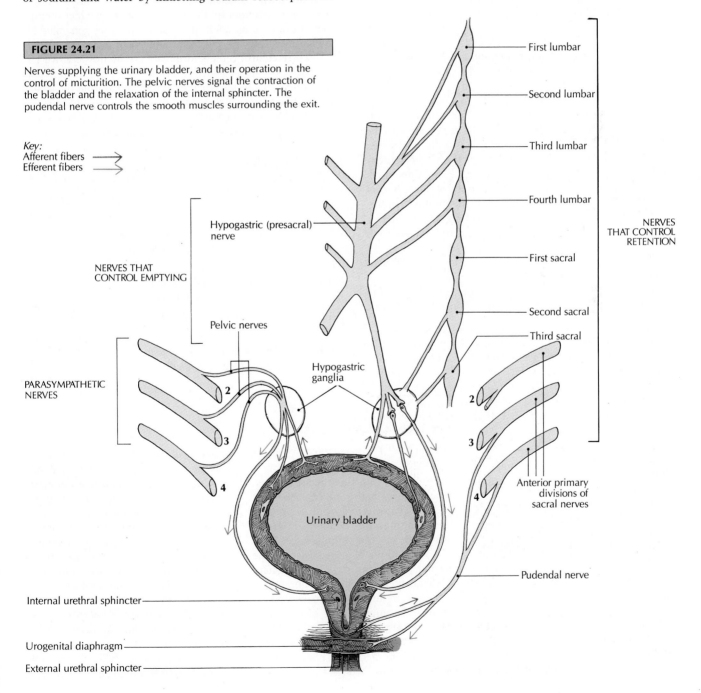

FIGURE 24.21

Nerves supplying the urinary bladder, and their operation in the control of micturition. The pelvic nerves signal the contraction of the bladder and the relaxation of the internal sphincter. The pudendal nerve controls the smooth muscles surrounding the exit.

Key:
Afferent fibers
Efferent fibers

Urine volume can also be altered by nervous and emotional states that stimulate the endocrine glands or autonomic nervous system. Increased stress or fear may cause increased urine production. The production of urine may not always increase during stress or excitement, but the urge to urinate may occur anyway.

Pregnant women also know that as the fetus develops, it puts more and more pressure on the woman's bladder, causing the need for frequent urination.

Urination (Micturition)

Micturition (L. *micturire,* to want to urinate) is the emptying of the urinary bladder. It is commonly called **urination.** As you saw earlier, about 150 to 180 L of water filter through the glomeruli each day, but only about 1.5 L are excreted in urine. The difference, about 99 percent, is reabsorbed.

Urine from the kidneys flows through the ureters to the urinary bladder, which has a capacity of about 600 to 800 mL. However, the bladder rarely fills to its total capacity. Every 10 or 20 seconds, peristaltic waves force small amounts of urine from the renal pelvis through the ureter and into the bladder. The bladder becomes distended as it fills, and when an adult accumulates about 300 mL (9 fl oz), the sensory endings of the pelvic nerve in the bladder wall are stimulated. These stretch receptors send impulses to the sacral region of the spinal cord, where parasympathetic neurons are stimulated, and a conscious urge to urinate results (Figure 24.21).

The mechanism of voluntary adult micturition proceeds as follows:

1 The musculature of the pelvic floor relaxes, and stretching of the detrusor muscle initiates a reflex action.

2 As the external sphincter relaxes, the bladder begins to empty, assisted by synchronous contractions of the thoracic and abdominal musculature, together with the closure of the glottis, which increases the thoracoabdominal pressure.

3 As urination ceases there are vigorous contractions of the external sphincter and muscles of the pelvic floor. In the female, gravity removes all the urine from the urethra, but in the male, voluntary contractions of the bulbocavernosus muscle expel the last drops of urine.

In an adult, the urge to urinate can be counteracted by consciously controlling the voluntary muscle fibers of the external voluntary sphincter, but an infant lacks the development of the higher brain centers, and cannot control the opening. Urination in an infant is a spinal reflex action initiated by the distension of the bladder; the parasympathetic neurons cause the bladder to contract, the striated muscle fibers at the external sphincter relax, and the bladder is emptied of urine. In an adult, those same impulses generated by stretch receptors are sent to the brainstem and cerebral cortex. These control centers formulate the impressions we experience as a "full bladder" and the "need to urinate," and we consciously decide whether to inhibit or permit the spinal reflex that controls urination. Conscious control of micturition involves learning how to counteract the spinal reflex and how to relax the external voluntary sphincter.

How often does the average adult urinate each day?

If it were convenient, we would urinate approximately 10 times a day.

Ask Yourself

1 *What are the main components of urine?*

2 *What is proteinuria?*

3 *What are some of the major physical properties of urine?*

4 *What enzyme and hormones regulate the volume and concentration of urine?*

5 *How do the kidneys play a part in the regulation of blood pressure?*

6 *What mechanism produces the urge to urinate before the bladder is filled?*

PHYSIOLOGICAL AND ANATOMICAL ABNORMALITIES

Acute and chronic renal failure

Acute renal failure is the total or near-total stoppage of kidney function. Little or no urine is produced, and substances that are normally eliminated from the body are retained. It is often caused by a diminished blood supply to the kidneys, which may be brought on by a serious blood loss due to an injury or hemorrhage, a heart attack, or a thrombosis. Another common cause of acute renal failure is a high level of toxic materials such as mercury, arsenic, carbon tetrachloride, and insecticides that build up in the kidneys. Other causes include obstruction (kidney stones, for example) and damage to the kidneys themselves. The course of the disease is often divided into three phases: onset, renal failure, and recovery. The first phase is characterized by an accumulation of nitrogenous waste in the blood *(azotemia),* electrolyte imbalance, pain, and a low output of urine *(oliguria).* The second phase may include pulmonary edema caused by sodium retention, increasing amounts of nitrogen, potassium, and sodium in the blood, water retention, and the presence of red blood cells or hemoglobin in any urine that is excreted. The recovery phase usually takes from 7 to 10 days and can be assisted by dialysis. Although complete recovery is not uncommon, residual kidney damage may lead to chronic renal failure.

In contrast to acute renal failure, **chronic renal failure** develops slowly and progresses over many years. Its most common causes are bacterial inflammation of the interstitial area and renal pelvis *(pyelonephritis),* kidney inflammation involving the structures around the renal pelvis or glomeruli *(glomerulonephritis),* and renal damage due to high blood pressure or obstructions in the lower urinary tract. The condition is characterized by progressive destruction of nephrons, which may lead to reduced amounts of urine, dilute urine, thirstiness, pigment retention, electrolyte imbalances, severe high blood pressure caused by the excessive retention of renin, poor appetite and vomiting, frequent urination, depletion of bone calcium, coma, and convulsions. A low-protein diet is usually prescribed, and dialysis may be necessary in some cases.

Glomerulonephritis

Acute glomerulonephritis (also known as AGN or Bright's disease) is an inflammation of the glomeruli that often follows a streptococcal infection. The body's normal response to this infection produces antibodies that cause an autoimmune inflammation in the glomerulus and damage the endothelial capsular membrane. Symptoms include presence of blood cells and plasma proteins in the urine (hematuria and proteinuria), reduced glomerular filtration rate, and retention of water, sodium, potassium, and hydrogen ions. Most patients recover fully (especially children), but chronic renal failure is not uncommon.

Chronic glomerulonephritis (CGN) is a progressive disease that usually leads to renal failure. It is usually irreversible by the time it produces symptoms, which include, first, hematuria, proteinuria, and hypertension, and later, nausea, vomiting, difficulty in breathing, and fatigue. The kidneys atrophy as the disease progresses, and recovery is unlikely.

Pyelonephritis

Acute *pyelonephritis* is one of the most common kidney diseases. It is caused by a bacterial infection, usually from the intestinal bacterium *Escherichia coli*, spreading from the bladder to the ureters, and then to the kidneys. The infection begins as an inflammation of the renal tissue between the nephrons, and progresses to the glomeruli, tubules, and blood vessels. The disease occurs most often in females, probably because their urethras are short and closer to the rectal and vaginal openings. Symptoms include a fever of 44°C (101°F) or higher, back pain, increased leukocytes in the blood, painful urination *(dysuria)*, cloudy urine with an ammonia smell, and presence of bacteria in the urine. The disease is further complicated for diabetics, whose glucose in the urine *(glycosuria)* provides a ready energy source for bacterial growth.

Renal calculi

Renal calculi are commonly called *kidney stones.* They may appear anywhere in the urinary tract, but are most common in the renal pelvis or calyx. The most usual type of calculus, which accounts for about 80 percent of all cases, forms from the precipitated salts of calcium (calcium oxalate and calcium phosphate). Kidney stones formed from calcium oxalate and trapped in the ureter are usually the ones that cause an intense, stabbing pain because of their jagged shapes. Calcium phosphate stones, in contrast, grow quickly and may occupy a large part of the renal pelvis. Kidney stones may also form from the salts of magnesium, uric acid, or cysteine.

The exact cause of the formation of stones is not known, but several conditions are implicated: dehydration, renal infection, obstruction in the urinary tract, hyperparathyroidism, renal tubular acidosis, high levels of uric acid, excessive intake of vitamin D or calcium, and ineffective metabolism of oxalate or cysteine.

Renal calculi are usually revealed in a *pyelogram,* an x ray of the kidney and ureters after an opaque dye has been introduced into the urinary system. Most renal calculi are small enough to pass out of the urinary system on their own, but in other cases the treatment includes a greatly increased water intake, antibiotics, analgesics, diuretics, a low-calcium diet, surgery, and *extracorporeal* ("outside the body") *shock-wave lithotripsy* that reduces kidney stones to passable particles without harming the body. Lithotripsy is usually effective in the removal of renal calculi less than 2 cm in diameter, which includes most stones. The characteristic extreme pain caused by the passage of stones is called *renal colic,* because it resembles intestinal (colic) pain.

Infection of the urinary tract

Two forms of a lower urinary tract infection are *cystitis,* inflammation of the bladder, and *urethritis,* inflammation of the urethra. Both are much more prevalent in women than in men; older men are usually affected as they begin to encounter prostate problems. Such inflammations may be trivial, but some urinary infections persist, causing permanent discomfort. The most common cause is an intestinal bacterium such as *Escherichia coli* or *Proteus mirabilis,* but some infections are caused by several different bacteria. Symptoms include frequent urges to urinate, spasms of the bladder, discharge from the penis in males, pain during urination, and excessive urination during the night *(nocturia).*

A form of upper urinary tract infection is *pyelitis,* an inflammation of the renal pelvis and calyces.

Urinary incontinence

The inability to retain urine in the urinary bladder and control urination is called **urinary incontinence.** Temporary incontinence may be caused by emotional stress. Permanent incontinence usually involves an injury to the nervous system, bladder infections, or tissue damage to either the bladder or urethra.

Kidney and bladder cancer

Kidney cancer occurs most often between the ages of 50 and 60 and is twice as prevalent in men as in women. Early symptoms are blood in the urine *(hematuria)*, pain in one side or the other, and a firm, painless growth of tissue. Kidney cancer often spreads, or *metastasizes,* to other parts of the body. Children under 6 or 7 may develop a variation of adult kidney cancer called *Wilms' tumor,* which causes the abdomen to swell noticeably. It is thought that Wilms' tumor originates in the embryo and then remains dormant for several years.

Bladder cancer appears most often in industrial cities, where such environmental carcinogens as benzidine, nitrates, tobacco smoke, and other chemical inhalants are common. Its symptoms resemble those of kidney cancer, and like kidney cancer, it develops most often in people over 50, especially in men.

Nephroptosis (floating kidney)

When a kidney is no longer held in place by the peritoneum, it usually begins to move to the abdominal area above. This condition is called **nephroptosis,** or floating kidney. The moving kidney sometimes twists its ureter, which may lead to incontinence.

Congenital abnormalities of the urinary system

Abnormalities of the urinary system occur in approximately 12 percent of all newborns. Some common abnormalities include the absence of a kidney (*renal agenesis*), location of a kidney in the abdominal region (*renal ectopia*), and fusion of the kidneys across the midline (*horseshoe kidney*). When a kidney contains many cysts it is called a *polycystic kidney*. The most common abnormality of the urethra occurs when the male urethra opens on the ventral surface of the penis instead of the glans (*hypospadias*). When the urethra fails to close on the dorsal surface of the penis, the condition is called *epispadias*.

Duplication of the ureters (two ureters from each kidney) occurs in about 1 in 200 births.

MEDICAL TERMINOLOGY

ANURIA The absence of urine, such as when the ureters are obstructed.

CYSTECTOMY Surgical removal of the urinary bladder.

CYSTEINURIA An inborn, genetic disorder of amino acid transport that causes excessive excretion of cysteine and other amino acids.

CYSTO- A prefix meaning "bladder."

CYSTOMETRY An examination of the bladder to evaluate the efficiency of the bladder.

CYSTOSCOPY An examination of the interior of the bladder with a fiberoptic scope.

CYSTOURETHROGRAPHY X-ray examination, after introducing a contrast dye, to determine the size and shape of the bladder and urethra.

GOUT A metabolic disease characterized by painful urate deposits, usually in the feet and legs.

NEPHRECTOMY Surgical removal of a kidney.

NOCTURNAL ENURESIS Involuntary urination during sleep. Commonly called "bed wetting."

POLYURIA Frequent urination.

RENAL ANGIOGRAPHY Examination of renal blood vessels, using contrast dye injected into a catheter in the femoral artery or vein.

RENAL HYPERTENSION High blood pressure of the kidney.

RENAL INFARCTION Formation of a clotted area of dead kidney tissue as a result of blockage of renal blood vessels.

RENAL VEIN THROMBOSIS Blood clotting in the renal vein.

RENOVASCULAR HYPERTENSION A rise in systemic blood pressure as a result of blockage of renal arteries.

UREMIA A condition in which waste products that are normally excreted in the urine are found in the blood.

VESICOURETERAL REFLUX A backflow of urine from the bladder into the ureters and renal pelvis.

SUMMARY

Excretion, the elimination of metabolic waste products, is accomplished in part by the lungs, skin, and large intestine, but the prime regulator of water balance and waste elimination is the urinary system.

The urinary system: components and functions

1 The *urinary system* consists of two kidneys, two ureters, the urinary bladder, and the urethra. Urine is formed in the kidneys, carried by the ureters, stored in the urinary bladder, and expelled through the urethra.

2 Each kidney contains over a million *nephrons,* the functional units that filter water and soluble components from the blood, selectively reabsorb some of them back into the blood to maintain a proper balance, and selectively secrete wastes into the urine.

Anatomy of the kidneys

1 The paired *kidneys* are retroperitoneal, located in the posterior part of the abdomen, lateral to the vertebral column.

2 The medial concave border of each kidney contains a *hilus,* an indented opening where blood vessels, nerves, and ureter join the kidney.

3 The innermost layer of the kidney is the *renal capsule;* the middle layer is the *adipose capsule;* the outer layer is the *renal fascia,* which attaches the kidney to the abdominal wall.

4 The kidney contains three regions: the innermost *renal pelvis,* which branches into the *major* and *minor calyces;* the middle *renal medulla,* consisting of several *renal pyramids,* which open into the calyces; and the outermost *renal cortex.*

5 Blood enters the kidney through the *renal artery,* which branches into *interlobar arteries, arcuate arteries, interlobular arteries,* and then *afferent arterioles,* which carry blood to the filtration site.

6 Each afferent arteriole branches extensively to form a ball of capillaries called a *glomerulus,* where filtration starts. Glomerular capillaries join to form an *efferent arteriole,* which carries blood away from the glomerulus.

7 The efferent arteriole branches to form the *peritubular capillaries,* which unite to form the *interlobular veins, arcuate veins, interlobar veins,* and the *renal vein,* which carries waste-free blood from the kidney to the inferior vena cava.

8 *Vasa recta* are extensions of the efferent arterioles that provide the kidney with an emergency system to maintain blood pressure and urine concentration.

9 Nerves from the *renal plexus* help regulate blood pressure in the glomerulus.

10 Each *nephron* is an independent urine-making unit. It consists of a *vascular component* (the glomerulus) and a *tubular component,* including a glomerular (Bowman's) capsule, proximal convoluted tubule, loop of the nephron (loop of Henle), distal convoluted tubule, and collecting duct. The *renal corpuscle* consists of the glomerulus and the *glomerular capsule.*

11 Filtration of blood takes place through the three layers of the renal corpuscle (constituting the *endothelial capsular membrane*) from the capillaries of the glomerulus into the glomerular capsule.

12 From the glomerular capsule, the fluid filtered from the blood (*glomerular filtrate*) moves into the *proximal convoluted tubule* where glucose, proteins, and certain other solutes filtered from the blood are absorbed.

13 The filtrate passes from the proximal convoluted tubule to the *loop of the nephron,* which is responsible for the reabsorption of water and the concentration of urine.

14 From the loop of the nephron, the filtrate moves into the *distal convoluted tubule,* where potassium and hydrogen ions are actively secreted into the filtrate.

15 The filtrate moves from the distal tubule into the *collecting duct,* where the dilute filtrate is concentrated and passed on to the minor calyx. The permeability of the walls of the collecting duct to water is controlled by *antidiuretic hormone* (ADH).

16 The *juxtaglomerular apparatus* is made up of *juxtaglomerular cells* and the *macula densa.* When the composition of the filtrate or the glomerular pressure changes, the juxtaglomerular apparatus secretes the enzyme *renin,* which alters the systemic blood pressure to reinstate normal conditions.

Physiology of the kidneys

1 The kidneys utilize three processes to produce and modify urine: glomerular filtration, tubular reabsorption, and tubular secretion.

2 *Glomerular filtration* is the process that forces plasma fluid from the glomerulus into the glomerular capsule. In the process, the filtration of blood is begun. The final hydrostatic pressure in the glomerulus is the *effective filtration pressure.* The amount of filtrate formed in the capsular space each minute is the *glomerular filtration rate.*

3 The rate of glomerular filtration depends on the effective filtration pressure, stress, total surface area available for filtration, capillary permeability, intrinsic

renal autoregulation, and release of renin.

4 *Tubular reabsorption* returns useful substances such as water, some salts, and glucose to the blood by active transport.

5 *Tubular secretion* is the process of collecting waste products such as potassium and hydrogen ions and certain drugs into the filtrate as it nears its final movement out of the kidney into the ureter.

6 The ability of the kidneys to clear wastes from blood plasma is measured by a process called *plasma clearance.*

7 The *countercurrent multiplier system* results from a countercurrent flow in the limbs of the loop of the nephron that helps to regulate the solute concentration. It assures that the urine will be more concentrated at the end of the nephron tubule than it was at the beginning.

8 The kidneys help regulate the acid-base balance of the blood, primarily by simultaneously excreting hydrogen ions and reabsorbing bicarbonate ions.

9 A child's renal function is not fully operative until about a year after birth. After 40 there is usually a progressive loss of kidney function.

Accessory excretory structures

1 The paired *ureters* carry urine from the renal pelvis of the kidney to the urinary bladder. Their tissue layers are the innermost *tunica mucosa,* the middle *tunica muscularis,* and the outermost *tunica adventitia.*

2 The muscular *urinary bladder* is an expandable sac that collects and stores urine until it is excreted. Its tissue layers resemble those of the ureter. The involuntary *internal urethral sphincter* and voluntary *external urethral sphincter* keep the urine from leaving the bladder until it is time to urinate.

3 The *urethra* is the tube that transports urine from the bladder to the outside during urination. It is much longer in males than in females. The male urethra contains three *portions,* designated as *prostatic, membranous,* and *spongy.*

Urine and urination

1 *Urine* is composed of water, urea, chloride, potassium, creatinine, phosphates, sulfates, and uric acid. Abnormal constituents include protein, glucose, ketone bodies, casts, and calculi.

2 Urine is usually slightly acidic, with the pH ranging from 5.0 to 8.0. It is normally clear and yellowish, but its color can vary greatly in a healthy person. Its specific gravity ranges from 1.008 to 1.030.

3 A healthy person excretes between 1.0 and 1.8 L of urine daily. In order to maintain homeostasis, an adult must excrete at least 0.45 L of urine daily. The volume and concentration of urine are influenced by diet, diuretics, and other factors and are regulated by ADH, aldosterone, and renin.

4 *Micturition,* or *urination,* is the emptying of the bladder. Urination in an infant is a spinal reflex action initiated by the distension of the bladder. In an adult, the impulses generated by stretch receptors in the bladder are sent to the brainstem and cerebral cortex. Conscious control of micturition must be learned.

Physiological and anatomical abnormalities

1 *Acute renal failure* is the stoppage of kidney function due to injury, obstruction, or toxic build-ups. *Chronic renal failure* is the progressive destruction of nephrons.

2 *Acute glomerulonephritis* is inflammation of the glomeruli, often following a streptococcal infection. *Chronic glomerulonephritis* is a progressive disease that may lead to renal failure.

3 *Pyelonephritis* is a bacterial infection, often spreading from the bladder to the ureters and kidneys.

4 *Renal calculi,* or kidney stones, usually are formed from precipitated salts of calcium.

5 Common infections of the urinary tract are *cystitis,* inflammation of the bladder; *urethritis,* inflammation of the urethra; and *pyelitis,* inflammation of the renal pelvis and calyces.

6 *Urinary incontinence* is the inability to control urination. Permanent incontinence is usually due to an injury to the nervous system or tissue damage.

7 *Kidney* and *bladder cancer* occur most often in people between the ages of 50 and 60, and are often caused by environmental influences.

8 *Nephroptosis,* or floating kidney, occurs when a kidney is no longer held in place by the peritoneum.

9 Congenital abnormalities of the urinary system occur in approximately 12 percent of all newborns.

UNDERSTANDING THE FACTS

1 What makes up the renal pyramids?

2 What are the renal columns, and what is their composition?

3 Within the kidney, where does the actual filtration of blood occur?

4 What makes up the renal corpuscle?

5 What structural modifications in the proximal convoluted tubule fit it for its function of reabsorption?

6 At what point in the urinary system has the final composition of urine been determined?

7 What is the source of the energy for filtration?

8 How does the composition of the glomerular filtrate differ from the composition of plasma?

9 What is the typical effective filtration pressure within the glomerulus?

10 Of the water filtered, what percentage is reabsorbed?

11 What portion of the nephron is the most important in the reabsorption of nutritionally important substances?

12 What is the major difference between simple diffusion and facilitated diffusion?

13 Where does tubular secretion take place?

14 Potassium secretion is under the control of which hormone?

15 What is meant by hypertonic urine?

16 List the accessory organs of the excretory system.

17 How are the ureters protected from the acids and concentrated solutes that they transport?

18 What prevents the backflow of urine from the bladder into the ureters?

19 Give the location of the following (be specific):

   a hilus   **e** ureters
   b renal plexus **f** trigone
   c vasa recta **g** juxtaglomerular
   d podocytes   apparatus

20 What are some common causes of acute renal failure?

21 What is acute glomerulonephritis?

22 What is a "floating kidney"?

UNDERSTANDING THE CONCEPTS

1 Describe the locaton of the kidneys.

2 The fact that the efferent artery leading from the glomerulus is smaller than the afferent artery has considerable physiological significance. Explain.

3 What is the significance of the fact that the efferent artery leading from the glomerulus contains smooth muscle in its walls?

4 What role does the autonomic nervous system play in regulating kidney function?

5 Discuss the structure of the nephron.

6 Explain how the structure of the endothelial capsular membrane is related to its main function.

7 Which factors are the most responsible for the glomerular filtration rate?

8 Show how the FP_{eff} of the glomerulus is calculated.

9 What would be the consequences of glomerular filtration without any reabsorption?

10 Why are the arterioles associated with the glomerulus so important in the regulation of glomerular hydrostatic pressure?

11 Discuss tubular secretion and its importance.

12 Distinguish between active and passive reabsorption.

13 Why may diarrhea threaten the life of an infant?

14 What is the role of ADH in urine production?

15 Discuss the abnormal constituents of urine and their significance.

16 How does the body handle the toxic ammonia produced in the metabolism of proteins?

17 Why may faulty kidney function lead to edema?

18 Why on a cold, wintery day does one feel the urge to urinate more often?

19 What is meant by a diuretic? Explain the action of diuretics.

20 Why is an infant unable to control urination?

SUGGESTED READING

ARENDSHORST, W. J., AND C. W. GOTTSCHALK, "Glomerular Ultrafiltration Dynamics: Historical Perspective." *American Journal of Physiology,* 248 (1985):F163.

ARONSON, P. S., "Mechanism of Active H^+ Secretion in the Proximal Tubule." *American Journal of Physiology,* 245 (1983):647.

BEEUWKES, R., "The Vascular Organization of the Kidney." *Annual Review of Physiology,* 42 (1980):531.

BRICKER, N. S., *The Kidney: Diagnosis and Management.* New York: Wiley, 1984.

BULGER, R. E., AND D. C. DOBYAN, "Recent Advances in Renal Morphology." *Annual Review of Physiology,* 45 (1983):533.

LASSITER, W. E., "Kidney." *Annual Review of Physiology,* 37 (1975):371.

PITTS, R. F., *Physiology of the Kidney and Body Fluids,* 3rd ed. Chicago: Year Book Medical Publishers, 1974.

SMITH, HOMER W., "The Kidney." *Scientific American,* January 1953.

SULLIVAN, L. P., AND J. J. GRANTHAM, *Physiology of the Kidney,* 2nd ed. Philadelphia: Lea & Febiger, 1978.

SUTTON, R. A., "Diuretics and Calcium Metabolism." *American Journal of Kidney Diseases,* 5 (1985):4.

TANAGHO, EMIL A., AND JACK W. MCANINCH, *Smith's General Urology,* 12th ed. Norwalk, CT: Appleton & Lange, 1988. (Paperback)

WALKER, L. A., AND H. VALTIN, "Biological Importance of Nephron Heterogeneity." *Annual Review of Physiology,* 44 (1982):203.

25

Regulation of Body Fluids, Electrolytes, and Acid-Base Balance

LEARNING OBJECTIVES

1 Differentiate between intracellular and extracellular fluids, and identify the major extracellular fluids.

2 Explain the forces that control the movement of body water from one compartment to another.

3 Discuss the causes and consequences of fluid imbalances in the body.

4 State the basic functions of water in the body.

5 Explain the usual means by which fluid intake and output are accomplished and how both are controlled by the body.

6 State three major functions of electrolytes in the body.

7 Explain how hormones control the balance of water and electrolytes in the body.

8 Discuss the balance of sodium, potassium, chloride, calcium, phosphate, and magnesium ions in the body fluids.

9 Explain the roles of weak acids and bases in acid-base buffer systems.

10 Describe three acid-base buffer systems and their functions in the body.

11 Discuss respiratory and renal regulation of pH.

12 Define acidosis and alkalosis, and describe the respiratory and metabolic causes of each.

13 Describe the causes of imbalances of electrolytes in the body fluids.

14 Explain the use of intravenous infusion.

The largest single constituent of the body is water. In fact, about 60 percent of your total body weight is fluid, mostly water. Your very life depends on maintaining the proper amount of body water, the correct proportion of water and electrolytes in the fluids, and the proper acid-base balance. Nowhere else in the body is homeostasis seen so vividly, and nowhere are imbalances more serious. For example, excessive vomiting or prolonged diarrhea could result in a dangerous condition if fluids are not replaced quickly. (For this reason, diarrhea in an infant is much more harmful than constipation.) A loss of less than 10 percent of the total body water usually produces lethargy, fever, and an overall feeling of dryness, especially in the mucous membranes. A loss of about 20 percent of the total body water is usually *fatal*. A person with third-degree burns may die, not necessarily from tissue damage, but from a loss of body fluid that seeps through burned areas.

Throughout the discussion, unless otherwise noted, the term *body fluid* refers to the body's water and its dissolved substances; *total body water* (TBW) refers to water minus the dissolved substances. Total body fluid is divided into an extracellular compartment (ECF or *extracellular fluid*) and an intracellular compartment (ICF or *intracellular fluid*). An approximate distribution of total body water is shown in Figure 25.1.

BODY FLUIDS: COMPARTMENTS AND COMPOSITION

All body fluids are either **intracellular** (inside the cell) or **extracellular** (outside the cell) **fluids** (see Figure 25.1). Cells contain intracellular fluid (ICF), which constitutes about two-thirds of all body fluid. All the rest of the fluids are lo-

cated outside cells. Several extracellular fluids are found in distinct areas: interstitial fluid (the immediate environment of body cells); blood plasma and lymph; cerebrospinal fluid; synovial fluid; fluids of the eyes and ears; pleural, pericardial, and peritoneal fluids; gastrointestinal fluids; and the glomerular filtrate of the kidneys. The most abundant and important of these extracellular fluids are the interstitial fluid and the blood plasma.

Water is the most abundant constituent in all body fluids, and is the vehicle for supporting and transporting the other constituents. The other constituents of body fluids are more variable. Figure 25.2 compares the electrolyte and protein concentrations of blood plasma, interstitial fluid, and intracellular fluid, using milliequivalents per liter (see Appendix B, Solutions, p. A.3).

The interstitial fluid occupies the spaces around body cells. It is derived directly from the blood plasma through capillaries and from the lymphatic system. For this reason, plasma and interstitial fluid are more similar to each other in their solute composition than to the intracellular fluid. The most important difference between the plasma and interstitial fluid is the presence of soluble proteins in the plasma and their near absence in the interstitial fluid. Under normal conditions, the walls of the capillaries are impermeable to these proteins, but will permit the other dissolved solutes to move more or less freely from the blood plasma into the interstitial fluid or in the opposite direction. Thus, proteins remain in the plasma.

Ask Yourself

1 *What is the definition of body fluid? Body water?*

2 *What are the two main categories of body fluids?*

3 *What are some examples of body fluids found in separate areas?*

4 *What solute is present in blood plasma but not in interstitial fluid?*

MOVEMENT OF BODY WATER

Movement of water from one body compartment to another is controlled by two forces: *hydrostatic pressure* and *osmotic pressure*. **Hydrostatic pressure** is the force exerted by a fluid against the surface of the compartment containing the fluid. **Osmotic pressure** is the potential pressure developed by a solution separated from pure water by a differentially permeable membrane. It is measured as the pressure required to stop the osmotic movement of water into a solution; it is therefore an index of the solute concentration of a solution. Osmotic pressure is sometimes called *osmotic potential*. The plasma membranes of cells provide for both the selective permeability and the constraint that permits a tension (hydrostatic pressure) to develop in cells, as a result of the pressure of fluids inside the cell. The walls of capillaries function in this capacity for the vascular system.

FIGURE 25.1

Proportion of fluids to body weight. Fluid normally makes up about 60 percent of the total body weight, with most of it located in the cells. Values are for a 70-kg adult male.

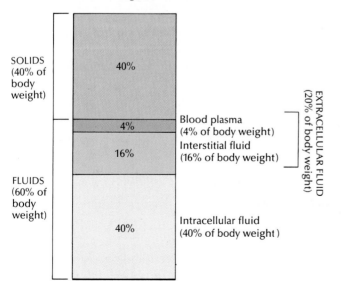

SOLIDS (40% of body weight)

40%

4% — Blood plasma (4% of body weight)

16% — Interstitial fluid (16% of body weight)

40% — Intracellular fluid (40% of body weight)

FLUIDS (60% of body weight)

EXTRACELLULAR FLUID (20% of body weight)

Because the movement of water is dependent on the total solute concentration, concentration changes in the major solutes of the plasma or interstitial fluid will have some effect on water distribution. Conversely, conditions that directly affect water balance (loss or gain in a compartment) will also alter the concentrations of the dissolved solutes. Water and solutes are closely interrelated. This important principle underlies both normal functioning and pathological imbalances.

Fluid Balance and Solute Concentration

The total concentration of all the dissolved solutes for interstitial fluid (281 mOsmole/L) is slightly less than that for plasma (282 mOsmole/L). In other words, the water in the interstitial fluid is maintained at a slightly higher concentration than the water in the plasma (Figure 25.3). The plasma proteins are chiefly responsible for this difference. As shown in Figure 25.4, fluid (water and nonprotein solutes) is filtered from the plasma into the interstitial fluid at the arterial end of each capillary. The force driving this filtration is hydrostatic pressure. The movement of fluid out of the capillary at its arterial end creates a more concentrated plasma (the plasma

becomes *hyperosmotic*) in the capillary and lowers the hydrostatic pressure of the blood vessel. The higher concentration of plasma proteins then causes the colloidal osmotic pressure to rise. At some point along the capillary, the change in hydrostatic pressure and the rising colloidal pressure cause water to move from the interstitial compartment back into the plasma (the fluid becomes *hypo-osmotic*) from a compartment of high concentration to one of lower concentration. As the return of water dilutes the capillary content, the electrolytes filtered from the plasma at the arterial end of the capillary move back into the plasma. The consequences of this movement are that the overall concentration of the electrolyte solutes in the plasma and the interstitial fluid is maintained at a homeostatic level, despite the exchanges that occur.

The movement of fluid between cells and the interstitial fluid results from similar forces. Since the intracellular and extracellular fluids have the same osmolarity (281 mOsmole/L), the hydrostatic and osmotic pressures within the cells and the surrounding intracellular fluid are also equal. As a result, the movement that does occur is the result of changes in osmotic pressure (see Figure 25.4).

Any change in the concentration of a major solute, such as sodium in the interstitial fluid, will have a profound effect

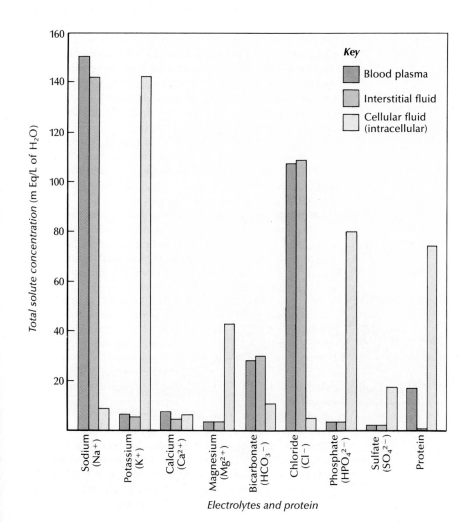

FIGURE 25.2

Comparison of the concentrations of soluble protein and electrolytes in the three major body fluids. The fluids with the relatively smaller total solute concentrations have higher water concentrations.

FIGURE 25.3

Movement of fluid as a result of changes in osmotic pressure. Fluids are absorbed from the small intestine into the interstitial compartment in tissue spaces. Ordinarily, there is a state of equilibrium between the extracellular and the intracellular fluid (bottom double arrows). If there is excessive sodium in the interstitial fluid, it is hyperosmotic to the cells (which are hypo-osmotic) and has a higher osmotic pressure. Water moves by osmosis from the cells, and they lose their shape. As water leaves cells, solutes within the cells become more concentrated (hyperosmotic), and essential electrolytes diffuse from the cells into the interstitial fluid.

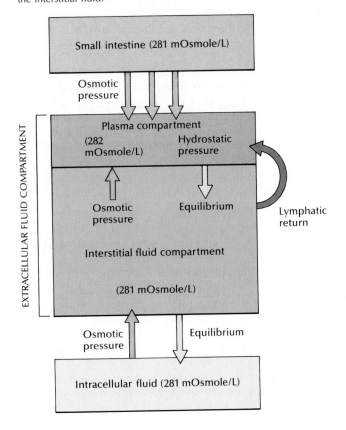

FIGURE 25.4

Movement of fluid between the plasma and interstitial fluid.

on the water balance between cells and the interstitial fluid. If sodium increases outside the cells (the fluid becomes hyperosmotic), the cells will lose water (they are hypo-osmotic) and become dehydrated. Consider the consequences of prolonged intake of a 3 percent salt solution, such as seawater. After ingestion, this concentrated salt solution would be absorbed from the small intestine, raising the plasma electrolyte concentration. This would result in an increase of the electrolyte concentration in the interstitial fluid (it becomes hyperosmotic), with a consequent reverse in the water balance between cells (which are now hypo-osmotic) and the interstitial fluid. The cells would lose water by osmosis, and all the cellular reactions that require homeostatic conditions would cease to function normally. Cell death would soon occur.

Edema: An Imbalance of Fluid Movement

Edema is the abnormal increase of water within the interstitial space. Such an increase in the volume of interstitial fluid produces distension of the tissue, which appears as puffiness on the surface of the body.

Edema may have one or several causes: (1) Plasma protein may leak across the capillary wall when the capillary lining is damaged. (2) In liver disease, protein synthesis is decreased, plasma protein concentration is reduced, plasma water increases, and net filtration into the interstitial space thus increases. (3) An increase in capillary or venous hydrostatic pressure may occur, which causes increased filtrations. (4) Lymphatic vessels may become obstructed. (5) In an inflammatory reaction, in response to infection or tissue damage, capillaries become more permeable.

A decrease in the concentration of plasma protein lowers the osmotic pressure of the plasma and reduces the return of water from the interstitial fluid to the plasma in the venule end of the capillary. The resulting fluid accumulation in the interstitial space produces the edema.

In varicose veins, the improper closing of venous valves produces edema. When blood accumulates within the veins, there is an increase in the hydrostatic pressure at the venule end of the capillary. This opposes the colloidal osmotic pressure that normally returns water from the interstitial fluid, and the volume of the interstitial fluid increases.

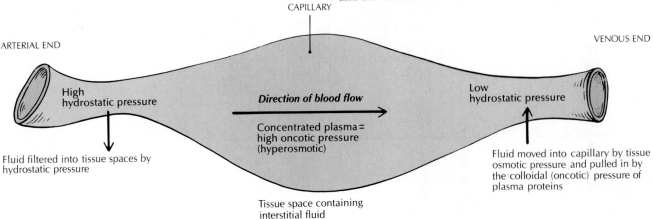

Obstruction of the lymph vessels that connect the interstitial compartment and drain part of its fluid back into the vascular system also produces edema. Such an obstruction is seen in the parasitic disease called *elephantiasis,* where the parasitic worms actually plug the vessels.

Localized tissue damage (burns) or infection that results in an inflammatory reaction and its attendant release of histamine may cause edema (see Chapter 20). Histamine makes capillaries more permeable, so that fluid moves from the capillaries into the interstitial area faster than it can be eliminated by the lymph vessels or returned to the venule capillaries.

Edema is usually a warning sign. The condition may have some beneficial side effects, however. The shift of fluid into the interstitial compartment in the vicinity of a bacterial infection may dilute bacterial toxins. Greater fluid in the vicinity of an infection may help the body's defense cells in fighting the infection.

Ask Yourself

1 What are the two main forces that move water between compartments?

2 How does water move between the plasma and the interstitial fluid?

3 If the concentration of sodium increases in the plasma, how will the water balance of cells be affected?

4 What is edema, and what conditions can contribute to it?

WATER

Total body water represents from 47 to 77 percent of the total body weight, depending on sex, age, and amount of body fat (Table 25.1). Adult females have less fluid per unit weight than adult males because females normally have proportionately more body fat than men. (This is a result of hormonal influences.) The percentage of water in the body decreases with age. A newborn infant is nearly 77 percent water. This fact is of great importance in caring for infants. Because so much of the infant's weight is water, any imbalance will have much more severe consequences than for the adult.

Functions of Water

Water has many important and interrelated functions in the body. Some of the major ones are:

1 It transports nutrients to cells and carries waste products from cells.

2 It provides a liquid medium for intracellular chemical reactions, including overall cellular metabolism.

3 It acts as a solvent for electrolytes and other solutes.

4 It helps to maintain body temperature, aids digestion, and promotes proper excretion.

5 It transports enzymes, hormones, blood cells, and many other substances.

Intake of Water

Most water is taken into the body by drinking water or other liquids that contain water, such as tea, coffee, or milk. A smaller amount enters as part of food (preformed water), and an even smaller amount is a by-product of the oxidation of food in cells (metabolic water). The important mineral solutes (electrolytes) of the body fluids enter the body through food or drink (see Table 24.1, p. 767).

Under normal conditions, water is taken into and excreted from the body, so that intake is matched by loss in order to maintain homeostasis. The main source of water needed by the body normally enters through the mouth (drinking), where intake is regulated by the nervous system. Dryness in the mucosa of the mouth and pharynx causes the flow of saliva to decrease. The feeling of dryness in the mucosa of the mouth and pharynx is conveyed to the brain, where it is interpreted as the sensation of thirst. Also, cells of the supraoptic nucleus of the hypothalamus function as *osmoreceptors,* sensing changes in the osmolarity of the extracellular fluid. As a result, the sensation of thirst is registered in the higher brain centers, which causes the person to want to drink. In addition, the posterior pituitary releases *antidiuretic hormone* (ADH), which helps to conserve water in the body (see Chapter 16).

TABLE 25.1   TOTAL BODY WATER AS
                                  A PERCENTAGE OF BODY WEIGHT

Age	Total body water (% body weight)
Newborn	77
6 months	72
2 years	60
16 years	60
20–39 years	
Male	60
Female	50
40–59 years	
Male	55
Female	47

What causes a blister?	A blister results when localized damage to capillary walls allows plasma proteins to leak into the interstitial space and into areas where epidermal layers separate because of thermal or mechanical damage. Such a leakage reduces the protein concentration difference between the plasma and interstitial fluid, and water moves into the interstitial space, forming a blister.

TABLE 25.2   MAJOR ELECTROLYTES IN THE BODY

Basic functions	Major homeostatic regulators
CALCIUM (CA^{2+}) 1 Required for building strong and durable bones and teeth. 2 Essential for blood coagulation. 3 Decreases neuromuscular irritability. 4 Promotes normal transmission of nerve impulses. 5 Establishes thickness and strength of plasma membranes. 6 Assists in absorption and utilization of vitamin B_{12}. 7 Activates enzymes that in turn activate chemical reactions within the body. 8 Needed for muscle contraction.	1 Parathyroid hormone raises serum level by increasing activity of osteoclasts. 2 Thyrocalcitonin hormone lowers serum level by inhibiting osteoclastic action. 3 Level of serum phosphate affects level of serum calcium. 4 Vitamin D is necessary for absorption and utilization of calcium.
CHLORIDE (CL^-) 1 Combines with hydrogen in gastric mucosal glands to form HCl. 2 Diffusion between ECF and ICF helps regulate osmotic pressure differences between compartments. 3 Assists in transmission of nerve impulses.	1 Aldosterone regulates Na^+ reabsorption, and chloride follows passively.
HYDROGEN (H^+) 1 Necessary for healthy cellular function. 2 Promotes efficient functioning of enzyme systems. 3 Necessary for the binding of oxygen by hemoglobin. 4 Determines relative acidity or alkalinity of body fluids.	1 Buffering, principally by the carbonic acid–bicarbonate buffer system, regulates concentration (normal ratio of carbonic acid to sodium bicarbonate is 1:20). 2 Lungs regulate carbonic acid side of the ratio. 3 Kidneys regulate sodium bicarbonate side of the ratio.
MAGNESIUM (MG^{2+}) 1 Activates many enzymes, in particular those associated with vitamin B metabolism and the utilization of potassium, calcium, and protein. 2 Promotes regulation of serum calcium, phosphorus, and potassium levels. 3 Essential for integrity of neuromuscular system and function of heart.	1 Parathyroid hormone increases absorption into blood from intestine.

Control of the amount of water consumed is exerted through the gastrointestinal tract, where distension of the stomach and intestine sends a nerve message to the hypothalamus, and thirst is inhibited. This mechanism appears to operate from the gastrointestinal tract well before all the water can be reabsorbed. It prevents ingestion of more water than the body needs.

Output of Water

Under normal conditions, the kidneys are the organs responsible for excreting most of the water from the body. Smaller but significant amounts of water are lost through the skin as sweat or insensible perspiration, through the gastrointestinal tract as feces, and through the lungs as water vapor in exhaled breath. In extreme heat and during prolonged exercise, loss of water through the skin is significant. Severe diarrhea or vomiting increases the amount of fluid (water and certain solutes) lost through the gastrointestinal tract and mouth. These conditions will alter the amount and composition of the fluid lost through the kidneys.

The kidneys have a great capacity for purifying blood, and in turn, other fluids. The kidney nephron is the site of control and regulation of water excretion. Fluid flows from the blood plasma in the glomerulus into the glomerular capsule to be filtered at a rate of 125 mL/min, or approximately 180 L/day (see Chapter 24). The amount of water actually excreted as urine depends on the amount of water reabsorbed back into the blood from the kidney collecting tubules. This reabsorption is under the direct control of ADH (see Chapter 24).

Ask Yourself

1 *What conditions influence the percentage of the water composition of the body?*

Basic functions	Major homeostatic regulators

POTASSIUM (K$^+$)

1 Regulates water and electrolyte content of ICF.

2 Helps promote transmission of nerve impulses, especially within the heart.

3 Helps promote skeletal muscle function.

4 Assists in transforming carbohydrates into energy, and restructuring amino acids into proteins.

5 Assists in regulation of acid-base balance by cellular exchange with H$^+$.

Major homeostatic regulators (Potassium)

1 Sodium pump conserves cellular K$^+$ by actively excluding Na$^+$.

2 Kidneys (which excrete 80 to 90 percent of K$^+$) conserve K$^+$ when cellular K$^+$ becomes depleted.

PROTEIN

1 Vital constituent of living cells.

2 Required for growth and development, and for maintenance and repair of tissue.

3 Forms the bulk of muscle, visceral, and epithelial tissue and is a constituent of plasma and hemoglobin.

4 Required for the manufacture of enzymes, hormones, many antibodies, and some vitamins.

5 Holds water within blood vessels and recovers water that leaks from blood vessels through colloid oncotic pressure.

SODIUM (NA$^+$)

1 Regulates fluid volume within ECF.

2 Increases plasma membrane permeability.

3 Maintains blood volume and controls size of vascular space.

4 Controls body water distribution between ECF and ICF.

5 Acts as a buffer base (sodium bicarbonate), thereby helping to regulate H$^+$ concentration.

6 Stimulates conduction of nerve impulses.

7 Helps maintain neuromuscular irritability.

8 Assists in controlling contractility of muscles, in particular, heart muscle.

Major homeostatic regulators (Sodium)

1 Aldosterone controls excretion and retention.

2 Atriopeptin stimulates excretion.

Source: Adapted from Karen Creason Sorensen and Joan Luckman, *Basic Nursing* (Philadelphia: Saunders, 1979), p. 483. Used with permission.

2 *How is water intake regulated by the nervous system?*

3 *What kidney structure controls water excretion?*

4 *What role does ADH play in maintaining water balance?*

ELECTROLYTES

Electrolytes are compounds that dissociate into ions when in solution, thus becoming capable of conducting an electrical current. The positive ions are called *cations,* and the negative ions are *anions.* A *nonelectrolyte* is a compound that does not form ions in solution. Most organic compounds are nonelectrolytes. Acids, bases, and salts* are electrolytes (Table 25.2). Although some of the body's electrolytes are found attached to proteins and some are deposited as solids to form bone and teeth, most electrolytes are dissolved in the body fluids. The most physiologically important electrolytes are the cat-

*When an acid is neutralized by a base, water and another product called a salt are produced:

$$HCl + KOH \longrightarrow H_2O + KCl$$

hydrochloric acid (acid)　potassium hydroxide (base)　water　potassium chloride (salt)

Some common salts are sodium chloride (found in intercellular and extracellular spaces and in table salt), potassium chloride (found in intracellular spaces), sodium bicarbonate (used as an antacid), magnesium sulfate (Epsom salt), and calcium sulfate (plaster of Paris).

Effects of Rapid Blood Loss on Fluid Balance

Hemorrhage is the rapid loss of blood from the blood vessels. It results in a rapid drop in blood pressure. Loss of as much as 1.5 L of blood can be compensated for by shifting the distribution of body fluids. Losses greater than 1.5 L produce shock, and if untreated, can become irreversible and fatal.

If a person has a limited loss of only about one liter of blood, the following redistribution of body fluids takes place.

As blood volume falls, blood pressure (hydrostatic pressure) drops, and interstitial fluid moves into the vascular system to compensate for the loss of plasma. Because interstitial fluid contains no proteins, it dilutes the plasma protein concentration (increases the plasma water concentration). This, in turn, limits further movement of water into the plasma. However, a rapid synthesis of new plasma proteins restores the os-

motic concentration of the plasma within a few hours and re-establishes the movement of water from the interstitial fluid into the plasma. Restoration of fluid to the plasma partially restores normal blood pressure and acts as a short-term aid to recovery. The full recovery and restoration of homeostasis requires the intake of water and electrolytes to replace those lost during hemorrhage.

TABLE 25.3   HORMONES CONTROLLING THE BALANCE OF WATER AND ELECTROLYTES

Hormone	Substance controlled	Mode and site of action	Source of secretion
Antidiuretic hormone (ADH)	Water.	Increases reabsorption in kidney tubules.	Hypothalamus; released by neurohypophysis.
Aldosterone	Sodium. Potassium.	Increases reabsorption in kidney tubules. Increases secretion in kidney tubules and into body cells.	Adrenal glands.
Atriopeptin	Renin. Aldosterone. Sodium.	Inhibits secretion in renin-angiotensin system. Inhibits secretion from adrenals. Stimulates glomerular filtration and excretion in urine.	Heart.
Parathyroid hormone (PTH)	Calcium.	Increases reabsorption from intestine and in kidney tubules. Increases release from bone.	Parathyroid glands.

ions sodium (Na^+), potassium (K^+), calcium (Ca^{2+}), magnesium (Mg^{2+}), and hydrogen (H^+), and the anions bicarbonate (HCO_3^-), chloride (Cl^-), phosphate (HPO_4^{2-}), and sulfate (SO_4^{2-}).

Electrolytes have three major functions in the body: (1) many are necessary for cell metabolism and contribute to body structures. (2) They facilitate the movement of water between the body compartments. (3) Together with the soluble proteins, they help to maintain the hydrogen-ion concentration (acid-base balance) of the body.

The major electrolytes (ions) have different concentrations in the various body fluids (see Figure 25.2). Sodium, potassium, and chloride ions are present in the highest concentrations, and these three electrolytes are particularly important in maintaining body function and normal water distribution among the fluid compartments. Sodium, potassium, and chloride have many essential roles, but their involvement in the transmission of nerve impulses is particularly significant. Calcium is essential to the development of hard tissue (bone

and teeth) and plays an important role in muscle and nerve action (see Chapters 9 and 11).

Each electrolyte indicated in Figure 24.2 is maintained at an optimal normal concentration in each fluid. The maintenance of this concentration directly or indirectly requires energy, and is an important example of homeostasis. The body can excrete an electrolyte that becomes too abundant, or it can restrict excretion and thus conserve an electrolyte that falls below optimal concentration. While restricting excretion will limit loss, return to the proper concentration requires ingestion.

Electrolytes are lost from the body through the skin in sweat, from the gastrointestinal tract in feces, and through the kidneys in urine. The kidneys are the organs most responsible for maintaining electrolyte homeostasis (see Chapter 24). The kidney nephron maintains the proper concentration of sodium, potassium, chloride, and calcium by regulating the excretion or reabsorption of these ions. All these electrolytes are readily filtered from the plasma in the glomerulus to enter the glomerular capsule and the nephron tubule. The machinery for controlling the actual excretion is located in the distal tubule of the nephron. For different electrolytes, active reabsorption, active secretion, or both processes may be involved. The nervous system and hormones play a key role in the control of the proper concentration of electrolytes in the plasma (Table 25.3).

Sodium

An adult body contains about 58 mEq of sodium per kilogram of body weight, or about 4000 mEq in a 70-kg (154-lb) body, or 142 mEq/L of interstitial fluid. The intake of sodium in a typical American diet greatly exceeds the need. Since sodium is actively removed from the cells of the body by the operation of the sodium-potassium pump (see Chapter 11), the intracellular fluid has a low sodium concentration (2 percent of the total). In contrast, sodium is highly concentrated (98 percent of the total) in the extracellular fluids and is their single most important ion. Because sodium is a very important ion in osmotic regulation, changes in the sodium concentration of the plasma and interstitial fluid result in dramatic changes in water distribution between cells and their fluid environment. A low sodium concentration is called *hyponatremia*, and a high concentration is *hypernatremia* (see Physiological and Anatomical Abnormalities at the end of the chapter).

Homeostasis of the total body sodium is regulated by mechanisms acting on its excretion from the kidneys. Sodium is readily filtered from the plasma in the glomerulus and constitutes part of the glomerular filtrate that passes through the nephron tubules. In the distal convoluted tubule, sodium is actively transported into the peritubular capillaries, where it again becomes part of the plasma. Reabsorption of sodium is an ongoing process necessary to return filtered sodium to the plasma, thereby maintaining its homeostatic concentration.

Sodium reabsorption can be increased by the steroid hormone aldosterone, which is secreted by the adrenal cortex. When sodium concentration falls below normal, when potassium becomes high (*hyperkalemia*), or when the renin-angiotensin system is activated, there is a need to conserve sodium, and aldosterone is secreted into the blood and transported to the kidneys (Figure 25.5). The renin-angiotensin system is the main regulator of aldosterone secretion when the sodium concentration becomes low. Angiotensin II acts in the adrenal cortex to cause the release of aldosterone. In the kidney, aldosterone acts on the cells of the distal convoluted tubule to increase reabsorption and thereby reduce the excretion of sodium in the urine. If the body's need for sodium still exists, it must be provided by ingestion. The return of sodium to normal levels inhibits aldosterone secretion, and sodium excretion balances sodium intake.

As you saw in Chapter 16, the peptide hormone **atriopeptin** is secreted by cardiac muscle cells in the atria of the heart when excess sodium accumulates in the body and cause the plasma volume to increase. An increase in plasma volume causes the atria to contain more plasma, stretching the atrial walls and stimulating stretch receptors there. In response,

FIGURE 25.5

Regulation of sodium reabsorption and potassium secretion by aldosterone.

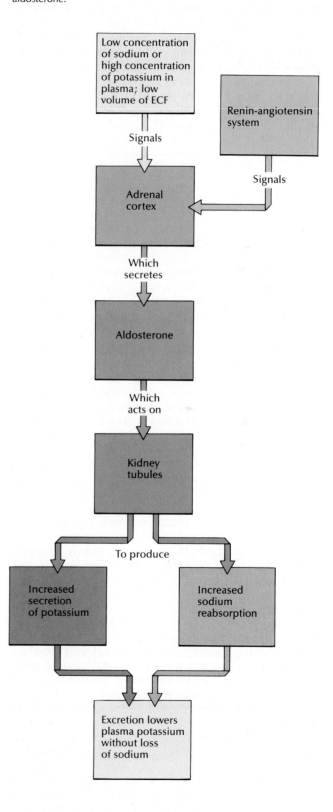

cardiac muscle cells secrete atriopeptin, which inhibits the secretion of renin in the renin-angiotensin system, and also inhibits the secretion of aldosterone from the adrenal glands. Both measures reduce the amount of sodium in the body. In addition, atriopeptin produces an increase in glomerular filtration in the kidneys, which causes the excretion of both sodium and urine. As a result of these combined factors, sodium is excreted and the plasma volume is reduced.

An increase in the sodium concentration of the plasma does not require hormonal regulation. Under conditions of excess, sufficient sodium is reabsorbed to keep the proper plasma concentration. Excess sodium is excreted in the urine.

Potassium

An adult body contains about 42 mEq of potassium per kilogram of body weight, or about 4.5 mEq/L. Potassium has a higher concentration within the cells of the body than in the extracellular fluids. In plasma, potassium has a relatively small concentration, and thus is not as important as sodium in osmotic regulation. However, the maintenance of potassium concentrations in the interstitial fluid and within cells is essential. Among other factors, proper nerve function and electrical impulse conduction in the heart depend directly on the concentration of this ion (see Chapter 11).

As with sodium, the potassium balance of the body is controlled by excreting an amount equal to the amount ingested. While some potassium may be lost in sweat or feces, most is excreted in the urine, and control of excretion is accomplished by the nephron of the kidney.

Like sodium, potassium is readily filtered from the plasma. Under conditions of normal or low plasma concentrations of potassium (*hypokalemia*), potassium is reabsorbed by active transport from the distal convoluted tubule into the plasma. If plasma potassium concentration rises above normal (*hyperkalemia*), excretion of the excess is accomplished by the passive diffusion of potassium into the tubular urine from the peritubular capillaries. Potassium secretion is under the control of aldosterone, the same hormone that controls sodium reabsorption (see Table 25.3). A rise in plasma potassium concentration signals the release of aldosterone and increased tubular secretion followed by increased excretion.

It is important to realize that aldosterone has opposite effects on the concentration of sodium and potassium. It acts to increase potassium excretion and to decrease sodium excretion. This common controlling hormone helps to ensure that the relative concentration of these two ions (high sodium/low potassium) in the extracellular fluid is maintained.

Chloride

An adult body contains about 33 mEq of chloride per kilogram of body weight, or about 103 mEq/L. Chloride is a

major anion in extracellular fluids. Red blood cells and the gastric mucosal cells that secrete hydrochloric acid (HCl) also contain large amounts of chloride. Chloride plays a major role in maintaining water balance. As a negatively charged ion, its movement follows the movement of positively charged sodium. Under most circumstances, where sodium goes, chloride follows. Since the total positive charges must equal the total negative charges, chloride preserves the ionic equivalence when cations are moved. When sodium is actively reabsorbed, chloride follows passively. When potassium is secreted, chloride is believed to accompany it. Control of chloride secretion is thus indirectly under the control of aldosterone. A low concentration of chloride in the blood is called *hypochloremia*; a high concentration is *hyperchloremia*.

Calcium

The calcium concentration in extracellular fluid is about 2.4 mEq/L. A 70-kg adult body contains about 65,000 mEq of calcium. Although calcium plays extremely important structural and functional roles in the body, such as being the structural component of bones and teeth, and being a crucial factor in blood clotting, muscle contraction, and nerve-impulse transmission, its concentration within most cells is low. Calcium concentration in extracellular fluids is higher than within cells, but nowhere near the magnitude of sodium and chloride. Compared to sodium and chloride, calcium has little effect on osmotic relationships. Chemically, calcium is closely associated with negatively charged phosphate. Where calcium goes, phosphate goes also.

Although the intake of calcium is necessary for proper body functioning, large stores of calcium (as solid calcium phosphate) in bones and teeth may provide a source for this electrolyte if reduced intake is prolonged.

Unlike the other electrolytes discussed, calcium is actively transported from the intestinal lining into the blood. Regulation of this absorption is one way in which calcium concentrations are regulated in extracellular fluids. Normally, only a small fraction of the calcium ingested is absorbed; the rest is lost with the feces. Vitamin D is necessary for the active absorption of calcium into the blood.

Calcium is excreted by the kidneys, where calcium associated with phosphate is filtered from the blood. Reabsorption takes place in the distal convoluted tubule.

Calcium concentration is under the control of *parathyroid hormone* (*PTH* or *parathormone*), which is produced by the parathyroid glands and secreted into the blood for transport (see Figure 16.10). Low calcium concentration (*hypocalcemia*) stimulates hormone production and release; high concentration (*hypercalcemia*) inhibits this function. Another hormone, calcitonin, also affects plasma concentrations of calcium, but its contribution to calcium homeostasis is small compared to parathyroid hormone.

Parathyroid hormone acts on living bone cells to move calcium from its solid reserve in bone into solution in the plasma. It also acts to increase the absorption of calcium from the gastrointestinal tract into the plasma, and to increase the reabsorption of calcium from the glomerular filtrate into the plasma.

Why do high-sodium diets tend to cause weight gain?

Because the body retains water in an effort to balance the excess sodium.

Phosphate

Phosphate is present in the body as inorganic and organic phosphate compounds. Most of it (about 85 percent) is found in calcium salts in bones and teeth.

The concentration of phosphate ions (HPO_4^{2-}) is regulated by PTH in two ways. First, PTH promotes bone resorption, and causes the release of large amounts of phosphate ions from bone salts into the extracellular fluid. Second, PTH decreases the transport of phosphate ions by the kidney tubules so that more phosphate ions are lost in the urine.

Magnesium

The average adult body contains about 2000 mEq of magnesium, about two-thirds of it in bone, and the normal plasma concentration is about 2 mEq/L. Magnesium ions (Mg^{2+}) are reabsorbed by all portions of the kidney tubules. However, magnesium ions also directly affect the tubular cells by decreasing reabsorption when the concentration of ions in the extracellular fluid is high. Thus, excess ions are excreted. When the magnesium concentration is low, magnesium ions are conserved.

Overall, magnesium is primarily an intracellular electrolyte that plays an important role in the sodium-potassium pump, as well as in the production of ATP energy in mitochondria. Most enzymes that use ATP and ADP in their reactions also need magnesium.

Ask Yourself

1 *What is an electrolyte?*

2 *What three important roles do electrolytes play in the body?*

3 *What is the major organ responsible for controlling electrolyte secretion?*

4 *How does aldosterone regulate the excretion of sodium and potassium?*

5 *At what three sites can the plasma concentration of calcium be controlled?*

ACID-BASE BALANCE

When referring to the regulation of the acid-base balance in the body, what is actually meant is the regulation of hydrogen ions in the body fluids, especially the extracellular fluids. In this section we will consider the main mechanisms involved in the regulatory processes that maintain acid-base homeostasis.

Any molecule that dissociates in solution to release a hydrogen ion (H^+) or proton is called an *acid* (see Chapter 2). Any molecule capable of accepting a hydrogen ion or proton is called a *base*. The hydrogen-ion concentration (measured by the pH scale) can greatly influence every chemical reaction and process in the human body (see Table 2.6). Enzymes, hormones, and the distribution of ions can all be affected by the concentration of hydrogen ions, so it is not surprising that the concentration is rigorously controlled by the body. The pH of the blood and interstitial fluid is maintained between 7.35 and 7.45. An increase or decrease in the pH value by only a few tenths of a pH unit can be disastrous. (A person can live for only a few hours at the lower or upper limits of 6.8 and 8.0.)

Homeostatic maintenance of an acceptable pH range in the extracellular fluid is accomplished by three mechanisms: (1) specific chemical buffer systems of the body fluids (react very rapidly, in less than a second); (2) respiratory regulation (reacts rapidly, in seconds); and (3) renal regulation (reacts slowly, in minutes to hours). Before discussing these pH-maintenance methods, we will review some fundamental properties of acids and bases, as well as equilibrium principles.

Strong Acids, Weak Acids, and Equilibrium

A *strong acid* is any molecule that completely dissociates in water to yield a hydrogen ion. Hydrochloric acid (HCl) is such an acid. It dissociates almost completely to yield hydrogen and chloride ions according to the equation*

$$HCl \rightleftharpoons H^+ + Cl^-$$

In water, only H^+ and Cl^- exist. There is virtually no intact HCl.

A *weak acid* is any molecule that only partially dissociates in water (Figure 25.6). Acetic acid is such an acid. In water, only a small fraction of the acetic acid molecules dissociates to yield hydrogen ions and the negatively charged acetate ions:

$$\underset{\text{acetic acid}}{CH_3COOH} \rightleftharpoons \underset{\text{hydrogen ion}}{H^+} + \underset{\text{acetate ion}}{CH_3COO^-}$$

In a water solution of acetic acid, all three components of the reaction above will be present. There will be a great deal of acetic acid and much smaller amounts of hydrogen ions and acetate ions.

In the dissociation of any weak acid, a stable but dynamic equilibrium is established. The equilibrium represented by the equation above is stable because the processes of dissociation and reassociation are constantly going on at equal rates. The equilibrium is dynamic because if one component is changed, the others will change also. For example, if the hydrogen ion (in the form of HCl) is added to the solution of acetic acid, the hydrogen ion will reassociate with the acetate ion, and the equilibrium is said to "shift to the left," forming acetic acid.

*The preferred direction of the reaction is indicated by the longer arrow.

FIGURE 25.6

Diagrammatic representation of (A) strong and (B) weak acids. Strong acids are almost completely dissociated, whereas weak acids are barely dissociated. Note that the same amount of solvent (water) is involved in each case.

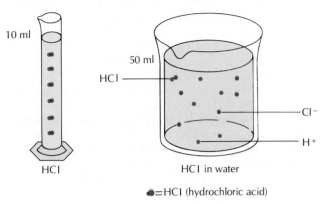

[A] STRONG ACID

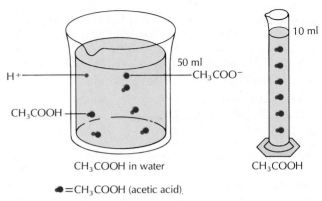

[B] WEAK ACID

In the reassociation of acetate ion and hydrogen ion, the acetate ion acts as a base, accepting a hydrogen ion. In any dissociation of a weak acid to form hydrogen ions, a weak base is also formed. This is called the **conjugate base** of the weak acid. In the acetic acid solution, if more base is added in the form of acetate (sodium acetate, $NaCH_3COO^-$), the increased concentration of the conjugate base acetate will associate with the hydrogen ions, forcing a shift in the equilibrium to the left, the formation of acetic acid molecules, a reduction of the hydrogen-ion concentration, and a neutralization of the acid (and an increase in pH).

Weak acids are important for two reasons. First, weak acids contribute hydrogen ions to the body and will alter the pH if not regulated. Many molecules are weak acids. For example, anaerobic respiration of glucose produces lactic acid, which is a weak acid. Aerobic respiration of glucose produces carbon dioxide, which reacts with water to produce the weak acid carbonic acid. The hydrolysis of fats generates fatty acids. Second, weak acids, together with their conjugate bases, form acid-base buffer systems that are the body's mechanism for coping with large or abrupt changes in the hydrogen-ion concentration (pH).

Acid-Base Buffer Systems

All of the **acid-base buffer systems** in the body work by the same principle to resist changes in pH. When a weak acid, such as carbonic acid (H_2CO_3), and its conjugate base, sodium bicarbonate ($NaHCO_3$), both exist in large concentrations in the body fluids, they constitute a buffer. Addition or depletion of hydrogen ions will result in equilibrium shifts, absorbing added hydrogen ions or releasing hydrogen ions so that the pH remains nearly unchanged. How does this work?

Bicarbonate buffer system The **bicarbonate buffer system** (also called the *carbonic acid-bicarbonate buffer*) helps maintain the pH of the blood. Both carbonic acid and bicarbonate are present in higher concentrations than the hydrogen ion because they are provided by the body. Carbonic acid dissociates to form hydrogen ion and bicarbonate:

$$\underset{\text{carbonic acid}}{H_2CO_3} \rightleftharpoons H^+ + \underset{\text{bicarbonate}}{HCO_3^-}$$

The normal ratio of carbonic acid to bicarbonate in the body is 1:20. The equation represents a stable, but dynamic equilibrium.

If hydrogen ions are generated by metabolism or by ingestion, they react with the bicarbonate ion (a base) to form more carbonic acid, and the equilibrium shifts toward the formation of the acid:

$$H^+ + HCO_3^- \longrightarrow H_2CO_3$$

The reaction produces slightly more carbonic acid than there was before, and slightly less bicarbonate base, but the concentration of hydrogen ions remains unchanged.

If hydrogen ions are withdrawn (by vomiting, for example), more carbonic acid dissociates,

$$H_2CO_3 \longrightarrow H^+ + HCO_3^-$$

yielding replacement hydrogen ions and bicarbonate ions. The equilibrium shifts, but the concentration of hydrogen ions remains unchanged.

When a strong acid, such as hydrochloric acid, is added to the buffer solution, the following reaction takes place:

$$\underset{\text{strong acid}}{HCl} + \underset{\text{weak base}}{NaHCO_3} \rightleftharpoons \underset{\text{weak acid}}{H_2CO_3} + \underset{\text{salt}}{NaCl}$$

From this equation it can be seen that the strong acid is converted into a weak acid, H_2CO_3. In contrast, if a strong base, such as NaOH, is added to the buffer solution, the following reaction takes place:

$$NaOH + H_2CO_3 \rightleftharpoons NaHCO_3 + H_2O$$

strong base   weak acid   weak base   water

The bicarbonate buffer system is important in the body because the concentration of each of its two components can be regulated. Carbon dioxide is regulated by the respiratory system, and the bicarbonate ion is regulated by the kidneys. As a result, the pH of the blood remains relatively constant.

Phosphate buffer system The *phosphate buffer system* regulates the pH within cells (red blood cells, for example), because as seen earlier, the concentration of phosphate in the intracellular fluid is many times that of the extracellular fluid, and within the kidney tubules, where phosphate usually becomes greatly concentrated. In the phosphate buffer, NaH_2PO_4 (sodium dihydrogen phosphate) plays the role of the weak acid, while $Na_2HPO_4^-$ (sodium monohydrogen phosphate) is the conjugate weak base:

$$H_2PO_4^- \rightleftharpoons H^+ + HPO_4^{2-}$$

When a strong acid, such as HCl, is added to the phosphate buffer system, the following reaction occurs:

$$HCl + Na_2HPO_4 \rightleftharpoons NaH_2PO_4 + NaCl$$

strong acid   weak base   weak acid   salt

The net result of this reaction is the exchange of a strong acid for a weak acid, with a relatively small change in pH. In contrast, if a strong base, such as NaOH, is added to the phosphate buffer system, the following reaction occurs:

$$NaOH + NaH_2PO_4 \rightleftharpoons Na_2HPO_4 + H_2O$$

strong base   weak acid   weak base   water

In this reaction, sodium hydroxide is buffered to form a weak base and water, allowing the pH to change only slightly.

Protein buffer system The *protein buffer system* is the most abundant and important in body cells and within the plasma because of the high concentration of proteins. As discussed in Chapter 2, proteins are composed of amino acids bound together by peptide linkages. Some of the amino acid side chains contain a carboxyl group (COOH) that can act as an acid by donating protons as follows:

The proton (H^+) can thus react with any excess hydroxyl ion (OH^-) to form water. (The dotted line signifies a weak bond, and R represents side chains of amino acids.) The amine

group (NH_2) tends to act as a base by accepting protons as follows:

Thus, a single amino acid can act as both an acidic and a basic buffer.

It usually takes several hours for the bicarbonate and phosphate buffer systems to diffuse through plasma membranes before they are effective. The high concentration and many side chains of amino acids within cells allow the protein buffer system to work almost instantaneously, making it the most powerful buffer system in the body.

Respiratory Regulation of Acid-Base Balance

Carbon dioxide is being formed continuously within the cells of the body by various metabolic processes. Any increase in the concentration of carbon dioxide in the body fluids as a result of cellular respiration lowers the pH (makes them more acidic), according to the following equation:

$$\text{Cellular respiration} \longrightarrow CO_2 + H_2O$$

                      carbon   water
                      dioxide

$$\rightleftharpoons H_2CO_3 \rightleftharpoons H^+ + HCO_3^-$$

           carbonic   hydrogen   bicarbonate
           acid

In contrast, a decrease in carbon dioxide concentration raises the pH toward the alkaline side by decreasing the amount of free hydrogen ions.

By excreting carbon-dioxide from the lungs, respiration plays a major role in maintaining hydrogen ion concentration. An increase in the rate of breathing increases the amount of carbon dioxide exhaled, thereby reducing the carbon dioxide level in the blood (Figure 25.7). This, in turn, reduces the amount of carbonic acid formed, and the concentration of hydrogen ions is also reduced.

Adjustment of the pH by breathing takes one to three minutes, much slower than the action of buffer systems. However, the breathing rate can be increased up to eight times the normal rate. This makes the respiratory control of hydrogen-ion concentration very important over brief periods.

Renal Regulation of pH

The body normally consumes more acid-producing foods than base-producing foods and not only must adjust the pH, but must also excrete hydrogen ions. This task is accomplished in the kidney tubules, where hydrogen and ammonium ions are secreted into the urine.

FIGURE 25.7

Respiratory control of pH.

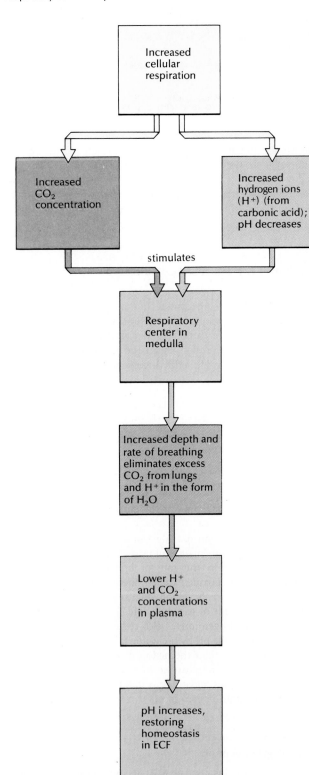

FIGURE 25.8

Kidney control of pH. (A) Regulation of acid-base balance by hydrogen-ion secretion in a kidney tubule cell, and sodium conservation, in exchange for hydrogen ions. Bicarbonate ions for

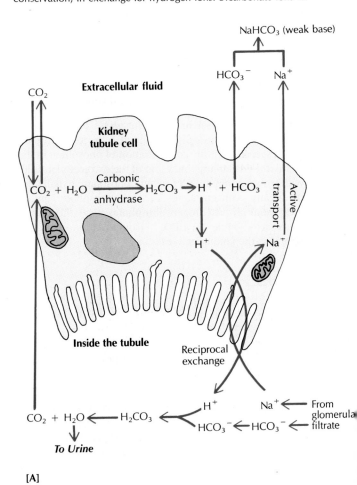

[A]

When a hydrogen ion is secreted into the tubular urine, a sodium ion is simultaneously exchanged. This exchange of H^+ for sodium is important because (1) it rids the body of excess H^+, and thus helps maintain pH levels in blood and extracellular fluid, (2) it conserves sodium, (3) it preserves ionic equivalence, and (4) it generates sodium bicarbonate for buffering (Figure 25.8A).

Another means for excreting hydrogen ions uses the base ammonia (NH_3) as a vehicle of hydrogen-ion acceptance, forming the ammonium ion (NH_4^+) (Figure 25.8B). As a result of H^+ and NH_4^+ secretion, the urine usually has a higher hydrogen-ion concentration (lower pH) than the blood. While blood is usually about 7.4, the pH of urine may be 6.0 or lower.

The third means by which the kidneys transport excess hydrogen ions and help to regulate pH is by using the phosphate buffer. Figure 25.8C illustrates how excess hydrogen ions are removed from the tubular fluid and how this system functions in the overall process of long-term renal acid-base balance.

buffering are generated in the body fluids. (B) Excretion of excess hydrogen ions using ammonia (NH_3). Ammonia secreted by the kidney tubule cell reacts with hydrogen ions to form ammonium ions and then ammonium chloride. This acidifies the urine and conserves $NaHCO_3$ for ECF buffering. (C) Transport of excess hydrogen ions into the urine by the phosphate buffer.

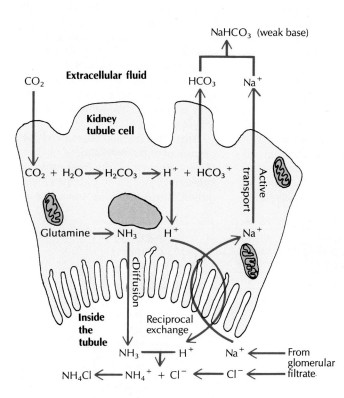

[B]

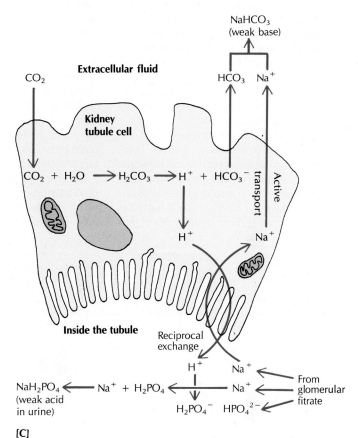

[C]

Ask Yourself

1 *What is the difference between a strong acid and a weak acid?*

2 *Why are weak acids important?*

3 *What is the buffer system and how does it operate?*

4 *What are the three most important buffer systems in the body?*

5 *What is the normal pH range for blood?*

PHYSIOLOGICAL AND ANATOMICAL ABNORMALITIES

Acid-base imbalances

If the pH of blood deviates from the range of 7.35 to 7.45, serious consequences for the body occur. When the blood falls below pH 7.35, the condition is known as *acidosis.* The pH may fall as low as 6.80 without irreversible damage, but acidosis below 6.80 is usually fatal. Acidosis can be due to respiratory or metabolic conditions.

Respiratory acidosis results from decreased carbon-dioxide removal from the lungs. Diseases such as emphysema and pulmonary edema can cause this condition. The retention of carbon dioxide in the lungs slows or "backs up" the removal of carbon dioxide from the

blood. Since carbon dioxide in the blood exists as carbonic acid, an increase in blood carbon dioxide results in an increase in hydrogen-ion concentration. As carbon dioxide accumulates, the breathing rate stimulates an increase in homeostasis-seeking activities. The nephron responds by increasing the excretion of hydrogen ions. If these efforts to restore homeostasis are insufficient, the pH of the blood continues to fall; a rapid pulse, mental disorientation, and loss of consciousness usually follow.

Metabolic acidosis occurs from the metabolic production of acids, or the loss of bases. The metabolic production of acids may result from the formation of

ketone bodies during increased fat metabolism. Metabolic acidosis may be observed in diabetes mellitus, malnutrition, or starvation. The loss of the bicarbonate base may cause acidosis or contribute to it, and can result from prolonged diarrhea or kidney disease. The lungs and kidneys attempt to rid the body of acid as described above. The symptoms and consequences of metabolic acidosis are the same as those for respiratory acidosis.

If the pH of the blood rises above 7.45 because of respiratory or metabolic conditions, *alkalosis* results. *Respiratory alkalosis* is caused by an increased loss of carbon dioxide from the lungs. The ab-

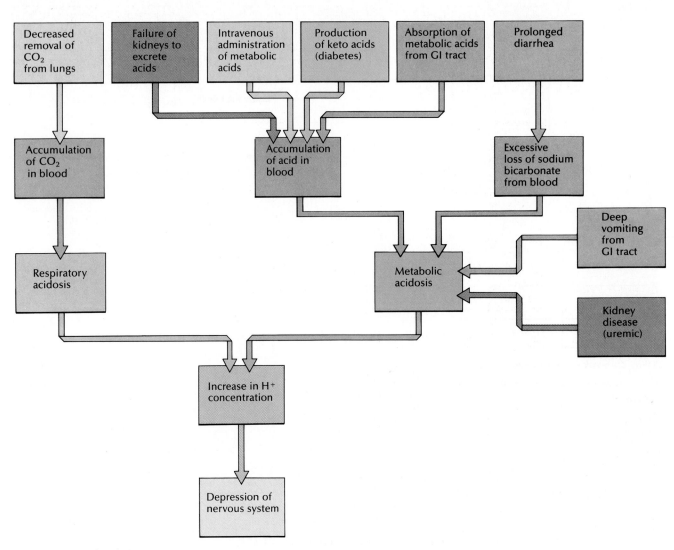

[A] A summary of acidosis.

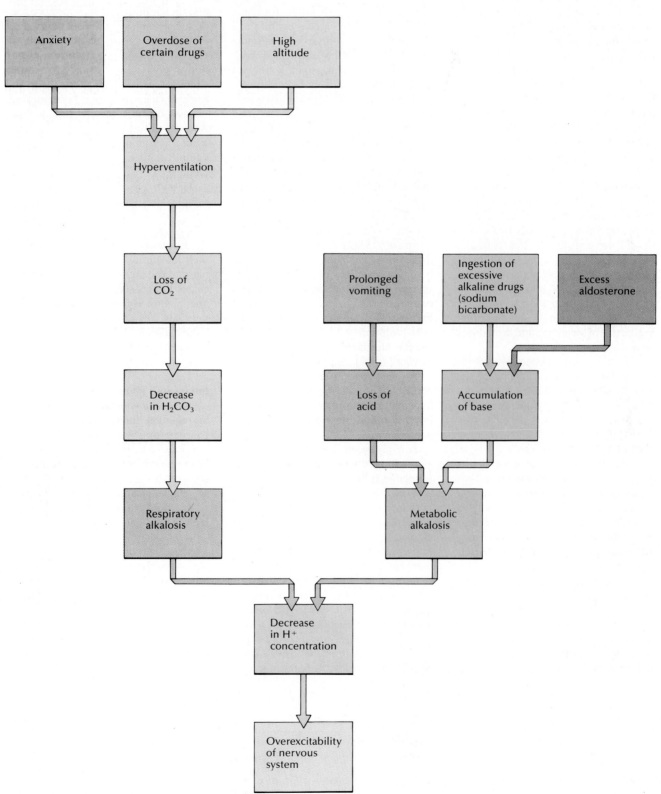

[B] A summary of alkalosis.

sence of a normal level of carbon dioxide in the lungs leads to excessive loss of carbon dioxide from the blood. This makes the blood alkaline (raises the pH). Emotional disturbances, an overdose of drugs such as aspirin, and high altitudes (low atmospheric pressure) may stimulate the rate of breathing and lead to respiratory alkalosis. The kidneys respond by retaining hydrogen ions, and by increasing bicarbonate excretion. Decreased carbon dioxide in the lungs will (or should) eventually slow the rate of breathing and permit a "normal" amount of carbon dioxide to be retained in the lungs.

Metabolic alkalosis results from an excessive loss of acid, or uptake of alkaline substances. Extensive vomiting of gastric secretions (especially hydrochloric acid) can contribute to this condition, as can excessive ingestion of sodium bicarbonate or other alkaline substances that can be absorbed into the blood. In an attempt to lower the pH, the body tries to retain carbon dioxide by reducing the rate of breathing, and the kidneys retain hydrogen ions.

The major clinical effect of alkalosis on the body is overexcitability of the central and peripheral nervous systems. Also, muscles can go into tetany, leading to respiratory paralysis.

Electrolyte imbalances

Imbalances of one or more of the electrolytes, particularly sodium and potassium, are fairly common under certain conditions. Severe imbalances, if left untreated, may cause death. Intravenous infusion of fluid may be necessary to restore depleted electrolytes and prevent serious consequences.

Low sodium concentration in the extracellular fluids is called *hyponatremia* (under + L. *natrium*, sodium). It may be due to excessive loss of sodium, reduced intake, or both. Prolonged sweating, vomiting, or diarrhea can deplete body sodium as well as water. Low sodium in

the interstitial fluid makes this fluid hypo-osmotic to the cells, and water moves into the cells, disrupting normal cell functions. Movement of water into cells can decrease the water in plasma and produce lower blood pressure and a weak pulse. Symptoms of hyponatremia include low blood pressure, rapid weak pulse, muscle cramps, small volume of urine, apprehension, and convulsions. Coma may occur if hyponatremia is prolonged.

The sudden addition of large amounts of sodium salts, such as by drinking seawater, causes *hypernatremia.* Because of the increase in the osmolarity of the plasma, water moves from the cells. The kidneys respond by increasing their sodium excretion in an attempt to maintain homeostasis. Unfortunately, water also leaves the body in the urine, and in the process, the body becomes seriously dehydrated. (This is why we can't drink salt water.) Hypernatremia also can occur by accidental intravenous infusion of hyperosmotic saline solution or the administration of sodium bicarbonate during cardiac resuscitation.

Plasma concentrations of sodium above 145 mEq/L are associated with hypernatremia, and concentrations below 135 mEq/L are associated with hyponatremia.

Low potassium concentration in the extracellular fluid is called *hypokalemia* (under + L. *kalium*, potassium). It can be caused by a diet poor in potassium, poor intestinal absorption, or increased loss of potassium. The first two problems can usually be treated with diet therapy. Excess loss is common and can result from kidney disease or excessive aldosterone production, as happens in Cushing's disease (see Chapter 16). Vomiting, diarrhea, or the use of diuretics can reduce plasma potassium. With the loss of potassium from the interstitial fluid, potassium diffuses from the cells to compensate, so that loss of potassium may

alter the acid-base balance in both intracellular and extracellular fluids. Symptoms of hypokalemia include a weakness in contractions of skeletal, smooth, and cardiac muscle, a decreased peristaltic activity, and a rapid weak pulse.

High potassium concentration is called *hyperkalemia.* Decreased excretion of potassium is usually the cause. Kidney disease, which affects the nephron's capacity to secrete potassium, or diseases of the adrenal cortex, such as Addison's disease, may cause hyperkalemia. Symptoms include small urine volume, nausea, weakness, and irregular pulse.

Plasma concentrations above 5.6 mEq/L are typical of hyperkalemia, while those below 3.5 mEq/L are typical of hypokalemia.

Hypermagnesemia is an abnormally high magnesium content in the blood plasma. The excessive concentration can cause disturbances of the central nervous system. *Hypomagnesemia,* an abnormally low magnesium level in the blood, affects both the muscular and nervous systems and can lead to irritability and convulsions.

Intravenous infusion

Intravenous infusion is widely used to maintain fluid balance, electrolyte concentration, and acid-base balance in hospitalized patients, and to correct imbalances that accompany disorders or damage. It is basically the introduction of relatively large amounts of a prescribed fluid at a controlled rate into the body of a patient.

A hypodermic needle is inserted into a vein (usually in the lower arm or dorsal surface of the hand), and attached by tubing to a bottle containing the fluid prescribed by a physician. The bottle of fluid is elevated above the patient and gravity ensures that the fluid flows. The rate of flow is controlled by a clamp on the tubing and can be measured by an attached drip-meter.

MEDICAL TERMINOLOGY

ANTACID An alkaline (basic) substance that neutralizes acids.

ANURIA (uh-NYOOR-ee-uh; L. *an,* not + urine) The inability to urinate.

DIURESIS (dye-yoo-REE-sis; Gr. *diourein,* to urinate) An increased excretion of urine.

KETONURIA Excessive ketone bodies

(from fat metabolism) in the urine.

KETOSIS Excessive ketone bodies in the body as a result of inadequate metabolism of carbohydrates.

SUMMARY

Body fluids: compartments and composition

1 All body fluids are either *intracellular* (inside the cell) or *extracellular* (outside). The intracellular fluid constitutes about two-thirds of all body fluid.

2 Distinct compartments contain several extracellular fluids, including interstitial fluid, plasma and lymph, and cerebrospinal fluid.

3 Plasma and interstitial fluid are more similar to each other in their solute composition than they are to intracellular fluid. However, plasma contains soluble proteins, which are nearly absent from interstitial fluid.

Movement of body water

1 Movement of water from one body compartment to another depends on *hydrostatic pressure* and *osmotic pressure.* Changes in the electrolyte (solute) concentration can alter the water distribution between compartments.

2 Because fluids are constantly moving into and out of body compartments, the overall concentration of solutes in the plasma and interstitial fluid is maintained.

3 *Edema* is the abnormal increase of fluid within the interstitial spaces. It may be caused by a decrease in plasma protein, increase in capillary or venous pressure, obstruction of lymphatic vessels, or inflammation and tissue damage.

Water

1 Water makes up between 47 and 77 percent of total body weight, the proportion varying with the sex, age, and amount of body fat.

2 Water transports nutrients to cells and removes wastes, provides a medium for intracellular chemical reactions, acts as a solvent, helps to maintain body temperature, aids digestion, promotes excretion, and transports enzymes and other substances.

3 Fluid intake and output are approximately equal in a healthy person. Most of the fluid is taken in as food or drink, and is excreted as urine.

4 Water intake is regulated by the hypothalamus.

5 The kidneys are the major organs for water excretion. *Antidiuretic hormone* (ADH) controls water excretion by regulating reabsorption of water from the kidney tubules into the plasma.

Electrolytes

1 *Electrolytes* are compounds that dissociate into ions when in solution. The most physiologically important electrolytes are sodium, potassium, calcium, magnesium, hydrogen, bicarbonate, chloride, phosphate, and sulfate. Each is maintained at an optimal concentration in each body fluid under normal conditions.

2 Electrolytes are required for cell metabolism, contribute to some body structures, facilitate movement of water between compartments, and help maintain acid-base balance.

3 Sodium, potassium, and chloride have the highest concentrations in different body fluids, and therefore have the greatest influence on osmotic relationships.

4 Electrolytes can be lost with water in sweat, feces, or urine. The kidney nephrons control electrolyte excretion through reabsorption and secretion.

5 The nervous and endocrine systems are essential for electrolyte regulation.

6 *Sodium* is low in concentration within cells, and high outside cells. Optimal plasma concentration is regulated by aldosterone and atriopeptin.

7 *Potassium* is high in concentration within cells, and low outside cells. Potassium secretion is regulated by aldosterone.

8 *Chloride* is the most abundant negative ion in the extracellular fluid, and is therefore important osmotically. It moves passively, following the movement of positive ions, especially sodium.

9 The concentration of *calcium* is maintained by absorption from the intestine, kidney excretion or reabsorption, and dissolution from bones, and is regulated by *parathyroid hormone* (PTH).

10 Fluid or water loss may produce imbalances in one or more electrolytes. Prolonged or severe imbalances of sodium or potassium result in an improper fluid balance.

11 The concentration of *phosphate* ions is regulated by PTH, which promotes bone resorption and decreases the transport of phosphate ions.

12 *Magnesium* is primarily an intracellular electrolyte that affects the sodium-potassium pump and the production of ATP energy in mitochondria.

Acid-base balance

1 *Acids* release hydrogen ions in water, and *bases* accept hydrogen ions. The hydrogen ion concentration (pH) of body fluids is maintained by buffer systems, respiratory regulation, and renal regulation.

2 *Weak acids* dissociate partially in water solutions, while *strong acids* dissociate completely. Weak acids and their conjugate bases form *acid-base buffer systems,* which resist changes in the pH when hydrogen ions are added or withdrawn.

3 The most important buffers are the *bicarbonate, phosphate,* and *protein systems.*

4 A change in the rate of breathing changes the carbon dioxide level in the blood and thus affects the hydrogen-ion concentration of body fluids.

5 The kidney nephrons can secrete hydrogen ions into the urine in exchange for reabsorbed sodium ions, can use ammonium ions for hydrogen-ion acceptance, and can use the phosphate buffer system.

Physiological and anatomical abnormalities

1 The normal pH range for blood is 7.35 to 7.45. A deviation below 7.35 is termed *acidosis;* a deviation above 7.45 is termed *alkalosis.*

2 Acidosis may be *respiratory,* caused by a decreased removal of carbon dioxide from the lungs, or it may be *metabolic,* caused by the production of acids, loss of a base, or failure to excrete acids.

3 Alkalosis may be *respiratory,* caused by an increased removal of carbon dioxide from the lungs, or *metabolic,* caused by an excessive loss of acids due to a body dysfunction or excessive uptake of alkaline substances.

4 Some electrolyte imbalances include *hyponatremia* and *hypernatremia* (sodium), *hypokalemia* and *hyperkalemia* (potassium), and *hypomagnesemia* and *hypermagnesemia* (magnesium).

5 *Intravenous infusion* is used to provide or restore proper electrolyte and fluid balance in hospitalized patients.

UNDERSTANDING THE FACTS

1 The loss of approximately what percentage of total body fluids is usually fatal?

2 Most solutes can pass through capillary walls. What is the major exception?

3 How does the percentage of water in an infant's body compare to that of an adult?

4 List the most physiologically important electrolytes.

5 What conditions may cause losses of extracellular fluid?

6 What is the relationship of sodium and potassium to intracellular and extracellular fluid?

7 The movement of calcium is closely associated with the movement of which other electrolyte?

8 What is a conjugate base?

9 Which buffer system contributes most directly to the maintenance of the pH of the blood, and which to the regulation of the pH within cells?

10 At what pH value is acidosis said to exist?

11 What is the purpose of intravenous infusion?

UNDERSTANDING THE CONCEPTS

1 Why is the cause of death in severe burn victims often not due to actual tissue destruction?

2 What are some of the functions of water in the body?

3 Distinguish between hydrostatic pressure and osmotic pressure.

4 What conditions can contribute to edema?

5 Give the major body functions of sodium, potassium, and calcium.

6 What is the relationship of parathyroid hormone to calcium and phosphate?

7 How does the respiratory system control the pH of body fluids?

8 What role do the kidneys play in the control of pH?

9 Explain the respiratory and metabolic causes of acidosis and alkalosis.

SUGGESTED READING

ANDERSON, B., "Regulation of Body Fluids." *American Review of Physiology,* 39(1977): 185.

AUKLAND, K., AND G. NICOLAYSEN, "Interstitial Fluid Volume: Local Regulatory Mechanisms." *Physiology Review,* 61(1981):556.

BURGESS, AUDREY, *The Nurse's Guide to Fluid and Electrolyte Balance,* 2nd ed. New York: McGraw-Hill, 1979.

GOLDBERGER, E., *A Primer of Water, Electrolyte, and Acid-Base Syndromes.* Philadelphia: Lea & Febiger, 1980.

GROLLMAN, A., "Body Fluids and Electrolytes." *Consultant,* May 1976.

GUYTON, A. C., AND T. G. COLEMAN, "Regulation of Interstitital Fluid Volume and Pressure." *Annals of the New York Academy of Science,* 150(1968):537.

KETTEL, L. J., "Acute Respiratory Acidosis." *Hospital Medicine,* February 1976.

METHENY, N. M., AND W. D. SNIVELY, *Nurse's Handbook of Fluid Balance,* 3rd ed. Philadelphia: Lippincott, 1979.

RECTOR, F. C., JR., AND M. G. COGAN, "The Renal Acidosis." *Hospital Practice,* April 1980.

REED, G. M., AND V. F. SHEPPARD, *Regulation of Fluid and Electrolyte Balance: A Programmed Instruction in Clinical Physiology.* Philadelphia: Saunders, 1977.

WELDY, NORMA JEAN, *Body Fluids and Electrolytes: A Programmed Presentation,* 3rd ed. St. Louis: Mosby, 1980.

V

REPRODUCTION AND DEVELOPMENT

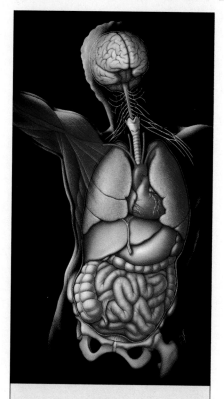

26
The Reproductive Systems

LEARNING OBJECTIVES

1 Name the principal male and female reproductive organs, and describe their structure and functions.

2 Describe the system of ducts and specialized cells within the testes, emphasizing their functional roles.

3 Describe the relationship of the male accessory ducts to one another, relating structure and function.

4 Explain the contribution of the three male accessory glands to the composition of semen.

5 Relate the structure of the penis to its functions.

6 Explain how specific hormones secreted by the anterior pituitary gland and by the testes affect the male reproductive system.

7 Explain how specific structures in the ovaries are related to the production of ova and the secretion of female hormones.

8 Describe the gross and microscopic structure of the uterus, and explain its functional role.

9 Describe the structure and functions of female accessory structures, including the vulva and mammary glands.

10 Explain how the specific hormones of the anterior pituitary gland and the ovaries control ovulation and menstruation.

11 Differentiate between mitosis and meiosis, between diploid and haploid cells, and between gametes and zygotes.

12 Outline the major steps in spermatogenesis and oogenesis, and compare them.

13 Describe the phases of sexual responses in males and females.

14 Explain the processes of fertilization and sex determination.

15 Compare the major methods of contraception.

16 Discuss several of the most common sexually transmitted diseases.

Reproduction, by means of sexual intercourse, produces new human beings, and allows hereditary traits to be passed from both parents to their children. Sexual reproduction always involves the union of two parental sex cells, an egg from the mother and a sperm from the father. This allows the hereditary material (DNA) from both parents to combine, forming a new individual with a unique combination of genes.

In this chapter we discuss the reproductive anatomy and physiology of males and females, with special attention to the female's rhythmic menstrual cycle. We also consider the hormonal control of reproductive activities, the formation of sperm and egg cells, the role of sexual intercourse, conception, and contraception.

MALE REPRODUCTIVE ANATOMY AND PHYSIOLOGY

The reproductive role of the male is to produce sperm cells and deliver them to the vagina of the female. These functions require four different types of structures (Figure 26.1):

1 The *testes* produce sperm cells and the primary male sex hormone, testosterone.

2 *Accessory glands* furnish a fluid for carrying the sperm cells to the penis. This fluid plus sperm cells is called *semen*.

3 *Accessory ducts* store and carry secretions from the testes and accessory glands to the penis.

4 The *penis* deposits semen into the vagina during sexual intercourse.

Testes

The paired **testes** (TESS-teez; singular, *testis*)* are the male reproductive organs (*gonads*), which produce *sperm,* or *spermatozoa*. Testes are often called *testicles* (from a Latin diminutive form of *testes*).

During fetal development, the testes are formed just below the kidneys inside the abdominopelvic cavity. By the third fetal month each testis has descended from its original site in the abdomen to the *inguinal* (groin) *canal*. During the seventh fetal month the testes pass through the inguinal canal (Figure 26.2). The inguinal canal is a passageway leading to the **scrotum** (SCROH-tuhm; L. *scrautum,* a leather pouch for arrows), an external sac of skin that hangs between the thighs (see Figure 26.1). The testes complete their descent into the scrotum shortly before or after birth. The inguinal canal is usually sealed off after the testes pass through. If the canal fails to close properly, or if the area is strained or torn, an *inguinal hernia* or *rupture* may result.

One testis (usually the left) hangs slightly lower than the other, so that the testes do not collide during normal activities. Because the testes hang outside the body, the temperature in the scrotum is about 3°F cooler than the body temperature. The lower temperature is necessary for active sperm production and survival. The sperm remain unviable if they

*In Latin, *testis* means "witness." The paired testes were believed to bear witness to a man's virility.

are retained inside the body cavity. For the same reason, a fever can kill hundreds of thousands of sperm. In warm temperatures, the skin of the scrotum hangs loosely, and the testes are held in a low position. In cold temperatures, the *dartos* muscles under the skin of the scrotum contract, and pull the testes closer to the warm body. In this way, the temperature of the testes remains somewhat constant. Sweat glands also help to cool the testes.

The interior of the scrotum is divided into two separate compartments by a fibrous *median septum.* One testis lies in each compartment. The line of the median septum is visible on the outside as a ridge of skin, the *perineal raphe* (RAY-fee; Gr. seam), which continues forward to the underside of the penis, and backward to the *perineum,* the diamond-shaped area between the legs and back to the anus.

Each testis is oval-shaped, weighs about 10 to 14 g, and measures about 4 to 5 cm (2 in.) long and 2.5 cm (1 in.) wide in an adult. It is enclosed in a fibrous sac called the **tunica albuginea** (al-byoo-JIHN-ee-uh; L. *albus,* white). This sac extends into the testis as *septae,* which divide it into compartments called *lobules* (Figure 26.3). The tunica albuginea is lined by the *tunica vasculosa* and covered by the *tunica vaginalis.* The tunica vasculosa contains a network of blood vessels, and the tunica vaginalis is a continuation of the membrane that lines the abdominopelvic cavity.

Each testis contains over 800 tightly coiled **seminiferous tubules** (see Figure 26.3), which produce thousands of sperm each second in healthy young men. The combined length of the seminiferous tubules in both testes is about 225 m (750 ft).

The walls of the seminiferous tubules are lined with *germinal tissue,* which contains two types of cells, spermatogenic cells and sustentacular (Sertoli) cells. The **spermatogenic cells,** including spermatogonia, spermatocytes, and spermatids, eventually develop into mature sperm (see Figure 26.3D). The development of sperm, or **spermatogenesis,** is discussed on page 844. The **sustentacular cells** provide nourishment for the germinal sperm as they mature. Sustentacular cells also secrete a fluid into the tubules that provides a liquid medium and an outward flow for the developing sperm. Recent evidence suggests that as the number of sustentacular cells decreases (usually after the age of about 60), the infertility rate increases proportionately.

Between the seminiferous tubules are clusters of endocrine cells called **interstitial endocrinocytes,** or *Leydig cells.* They secrete the male sex hormones, called **androgens,** of which **testosterone** is the most important.

Sperm (Spermatozoa)

Mature **sperm,** or **spermatozoa** (singular, spermatozoan; Gr. *sperma,* seed + *zoon,* animal), have a *head,* a *middle piece,* and a *tail.* At the tip of the head is an *acrosome* (formed from the Golgi body), containing several enzymes that help the sperm penetrate an egg. In the center of the head is a compact *nucleus,* containing the chromosomes (and therefore all the genetic material). The middle piece consists mainly of a coil of mitochondria, which supply ATP to provide the energy for movement. The beating movement of the undulating tail

FIGURE 26.1

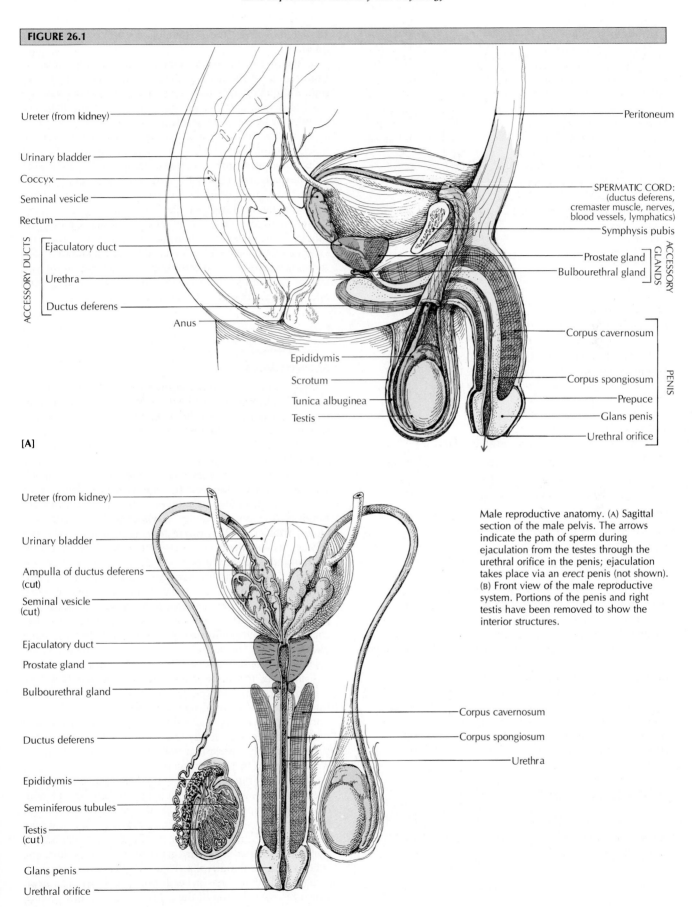

Ureter (from kidney)

Urinary bladder

Coccyx

Seminal vesicle

Rectum

ACCESSORY DUCTS

Ejaculatory duct

Urethra

Ductus deferens

Anus

Epididymis

Scrotum

Tunica albuginea

Testis

[A]

Peritoneum

SPERMATIC CORD:
(ductus deferens,
cremaster muscle, nerves,
blood vessels, lymphatics)

Symphysis pubis

ACCESSORY GLANDS

Prostate gland

Bulbourethral gland

Corpus cavernosum

Corpus spongiosum

Prepuce

Glans penis

Urethral orifice

PENIS

Ureter (from kidney)

Urinary bladder

Ampulla of ductus deferens
(cut)

Seminal vesicle
(cut)

Ejaculatory duct

Prostate gland

Bulbourethral gland

Ductus deferens

Epididymis

Seminiferous tubules

Testis
(cut)

Glans penis

Urethral orifice

[B]

Corpus cavernosum

Corpus spongiosum

Urethra

Male reproductive anatomy. (A) Sagittal section of the male pelvis. The arrows indicate the path of sperm during ejaculation from the testes through the urethral orifice in the penis; ejaculation takes place via an *erect* penis (not shown). (B) Front view of the male reproductive system. Portions of the penis and right testis have been removed to show the interior structures.

FIGURE 26.2

Descent of the testes from the abdominal cavity into the scrotum.

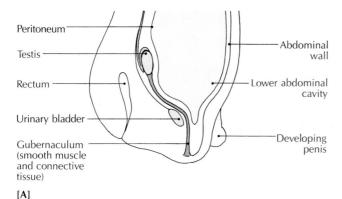

[A]

Peritoneum
Testis
Rectum
Urinary bladder
Gubernaculum (smooth muscle and connective tissue)
Abdominal wall
Lower abdominal cavity
Developing penis

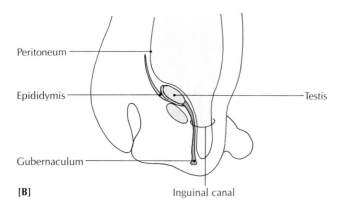

[B]

Peritoneum
Epididymis
Gubernaculum
Testis
Inguinal canal

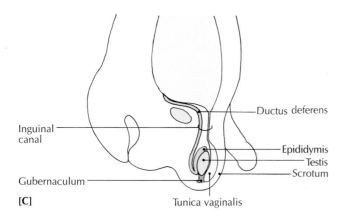

[C]

Inguinal canal
Gubernaculum
Ductus deferens
Epididymis
Testis
Scrotum
Tunica vaginalis

During a routine physical examination for a male, why does the physician insert a finger in the superficial inguinal ring under the scrotum and ask the patient to cough?

Coughing exerts pressure on the internal abdominal organs, forcing them against the inguinal canal. A loose inguinal ring, felt by the physician, may indicate a developing hernia.

(flagellum) drives the swimming sperm forward (Figure 26.4).

A sperm is one of the smallest cells in the body. Its length from the tip of the head to the tip of the tail is about 0.05 mm. Although it appears to be structurally simple—basically a nucleus with a tail—each sperm requires more than two months for its complete development.

Normally, 300 to 500 million sperm are released during an ejaculation. A male who releases less than about 20 to 30 million normal sperm is said to be **sterile.** Sterility may result from sexually transmitted diseases, which may interfere with sperm production, or by such diseases as mumps, which destroys the lining of the seminiferous tubules.

Sperm are constantly being produced in the seminiferous tubules, which always contain sperm in various stages of development (see Figures 26.3c–e). The final maturation of a sperm cell takes place in collection tubules lying on the surface of each testis. These tubules and other accessory ducts and glands are described in the following sections.

Accessory Ducts

The sperm produced in the testes are carried to the point of ejaculation from the penis by a system of ducts. These ducts lead from the testes into the abdominopelvic cavity, where they eventually join the urethra in the penis.

Epididymis The seminiferous tubules merge in the central posterior portion of the testis (see Figure 26.3a). This area is called the *mediastinum testis* (L. "being in the middle of the testis"). The seminiferous tubules straighten to become the *tubuli recti* ("straight tubules"), which open into a network of tiny tubules called the *rete testis* (L. *rete,* net). The rete testis, at its upper end, drains into 15 to 20 tubules called the *efferent ducts* (*ductuli efferentes*).

The efferent ducts penetrate the tunica albuginea at the upper posterior part of the testis, and extend upward into a convoluted mass of tubules that forms a crescent shape as it passes over the top of the testis and down along its side. This coiled tube is the *epididymis* (plural, *epididymides;* Gr. "upon the twin," the "twin" being the testis). The tightly coiled epididymis is bunched up along a length of only about 4 cm (1.5 in.), but if it were straightened out, it would extend about 6 m (20 ft). The epididymis has three main functions: (1) it stores sperm until they are mature and ready to be ejaculated. (2) It serves as a duct system for the passage of sperm from the testis to the ejaculatory duct. (3) It contains circular smooth muscle that helps propel mature sperm toward the penis by peristaltic contractions.

The efferent ducts are lined with alternating groups of ciliated high pseudostratified columnar epithelium and nonciliated low columnar epithelium. The movement of the cilia helps to propel sperm toward the epididymis. (Sperm do not become motile until they enter the vagina.) The epididymis itself has a thinner muscular coat than the efferent duct, and is lined with nonmotile cilia (Figure 26.5).

Each epididymis has a head, body, and tail. The *head,* which fits over the top of the testis, consists mostly of con-

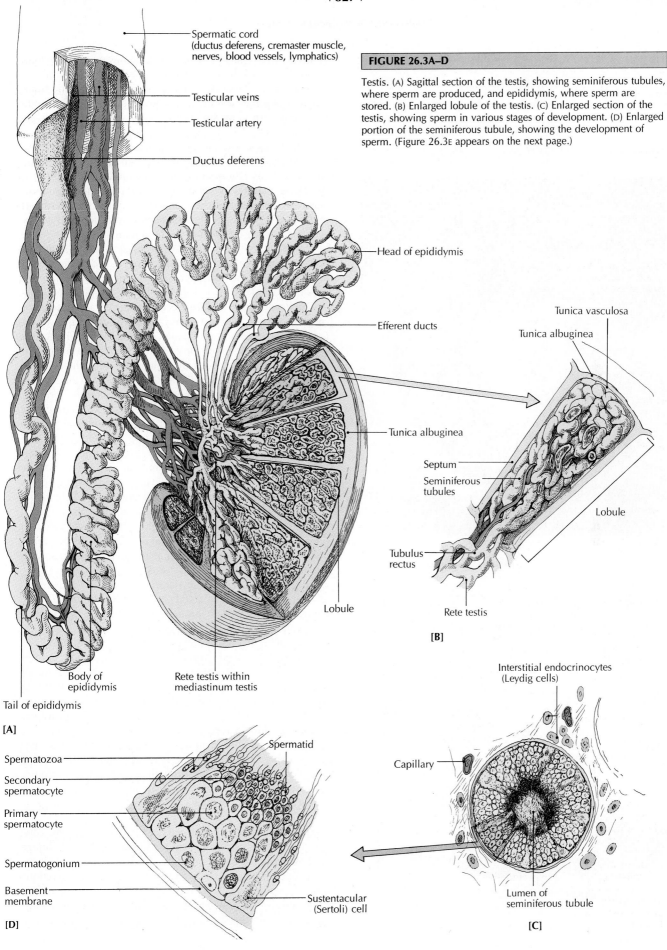

Spermatic cord
(ductus deferens, cremaster muscle,
nerves, blood vessels, lymphatics)

Testicular veins

Testicular artery

Ductus deferens

FIGURE 26.3A–D

Testis. (A) Sagittal section of the testis, showing seminiferous tubules, where sperm are produced, and epididymis, where sperm are stored. (B) Enlarged lobule of the testis. (C) Enlarged section of the testis, showing sperm in various stages of development. (D) Enlarged portion of the seminiferous tubule, showing the development of sperm. (Figure 26.3E appears on the next page.)

Head of epididymis

Efferent ducts

Tunica vasculosa

Tunica albuginea

Tunica albuginea

Septum

Seminiferous tubules

Lobule

Tubulus rectus

Rete testis

Lobule

[B]

Body of epididymis

Rete testis within mediastinum testis

Tail of epididymis

[A]

Interstitial endocrinocytes
(Leydig cells)

Spermatid

Spermatozoa

Capillary

Secondary spermatocyte

Primary spermatocyte

Spermatogonium

Basement membrane

Sustentacular
(Sertoli) cell

Lumen of
seminiferous tubule

[D]

[C]

FIGURE 26.3E

(E) Scanning electron micrograph of transverse section of a seminiferous tubule, showing spermatogonia, primary spermatocytes, spermatids, sustentacular (Sertoli) cells, and the long tails of sperm. [*Shih, Gene, and Richard Kessel. Living Images: Biological Microstructures Revealed by SEM. Boston: Science Books International, 1982.*]

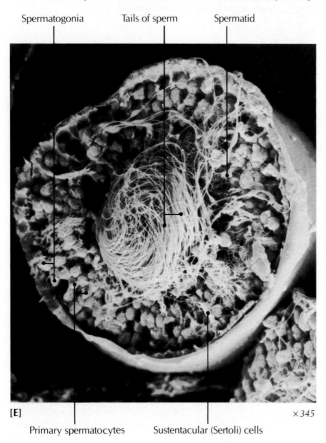

Spermatogonia   Tails of sperm   Spermatid

[E]   × 345

Primary spermatocytes   Sustentacular (Sertoli) cells

FIGURE 26.4

Sperm. (A) Schematic drawing of a sperm showing its internal structure, consisting of a head, middle piece, and tail. (B) Cross section of tail.

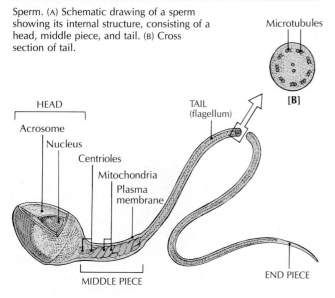

Microtubules

[B]

HEAD
Acrosome
Nucleus
Centrioles
Mitochondria
Plasma membrane

TAIL (flagellum)

MIDDLE PIECE

END PIECE

[A]

voluted efferent ducts. The *body,* extending down the posterolateral border of the testis, consists mainly of the *ductus epididymis.* The *tail* extends almost to the bottom of the testis, where it becomes less and less convoluted until it finally dilates and turns upward as the *ductus deferens* (see Figure 26.3A).

Maturing sperm leave the seminiferous tubules and move into the epididymis. The journey through the epididymis may take as little as 10 days or as long as 4 to 5 weeks, until the sperm are mature. While being stored, the sperm are nourished by the lining of the epididymis. When the sperm are mature, they enter the ductus deferens on their way to the seminal vesicles (see Figure 26.1). Sperm can remain fertile in the epididymis and ductus deferens for about a month. If they are not ejaculated during that time, they degenerate and are resorbed by the body.

Ductus deferens The ***ductus deferens*** (plural, *ductus deferentia;* L. *deferre,* to carry away), formerly called the *vas deferens* or *sperm duct,* is the dilated continuation of the ductus epididymis. It is about 45 cm (18 in.) long. The paired ductus deferentia extend between the epididymis of each testis and the ejaculatory duct (see Figure 26.1). As each one passes from the tail of the epididymis, it is covered by the *spermatic cord,* containing the testicular artery, veins, autonomic nerves, cremasteric muscle, lymphatics, and connective tissue from the anterior abdominal wall (see Figure 26.3A). Continuing upward after leaving the scrotum, the ductus deferens pierces the lower part of the abdominal wall by way of the inguinal canal (see Figure 26.7). There, the ductus deferens becomes free of the spermatic cord and passes behind the urinary bladder, where it travels alongside an accessory gland called the *seminal vesicle* and becomes the *ejaculatory duct.* If the anterior abdominal wall weakens at the site of the former inguinal canal, through which passes the spermatic cord, an inguinal hernia (*rupture*) may result.

Just before reaching the seminal vesicle, the ductus deferens widens into an enlarged portion called the *ampulla* (see Figure 26.1B). Sperm are probably stored in the ampulla prior to ejaculation.

The ductus deferens is the main carrier of sperm. It is lined with pseudostratified columnar epithelium, and contains three thick layers of smooth muscle (Figure 26.6). Some of the sympathetic nerves from the pelvic plexus terminate on this smooth muscle. Stimulation of these nerves produces peristaltic contractions that move sperm forward, toward the ejaculatory duct near the base of the penis.

The thick coating of smooth muscle gives the ductus deferens a characteristic cordlike structure, and the initial portion of the duct can be felt through the skin of the scrotum. Because the ductus deferens can be located easily, it is the preferred site to cut during a vasectomy (see Figure 26.25A).

Ejaculatory duct The ampulla of the ductus deferens connects with the duct of the seminal vesicle at the ***ejaculatory duct*** (see Figures 26.1 and 26.7). Each of the paired ejaculatory ducts is about 2 cm (1 in.) long. They receive secretions from the seminal vesicles, pass through the prostate gland on its posterior surface, where they receive additional secretions, and finally join the single urethra (see Figure 26.9).

FIGURE 26.5

Scanning electron micrograph of the epididymis, showing the vascular connective tissue between the many ducts, sperm within a lumen, and pseudostratified columnar epithelium lining. [*Fujita, Tanaka, and Tokunaga, SEM Atlas of Cells and Tissues. Tokyo, Igaku-Shoin 1981.*]

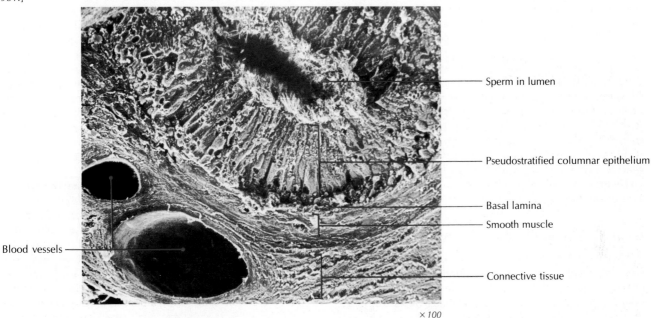

Sperm in lumen

Pseudostratified columnar epithelium

Basal lamina
Smooth muscle

Blood vessels

Connective tissue

× 100

FIGURE 26.6

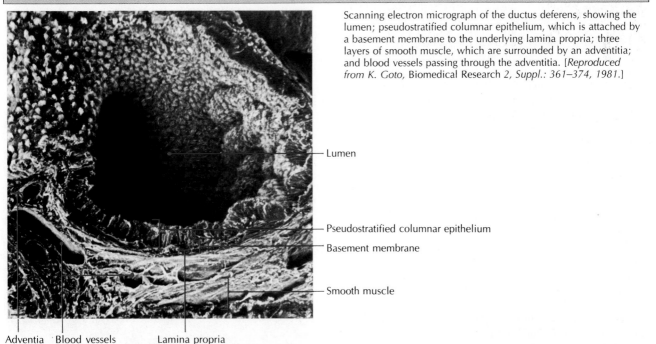

Scanning electron micrograph of the ductus deferens, showing the lumen; pseudostratified columnar epithelium, which is attached by a basement membrane to the underlying lamina propria; three layers of smooth muscle, which are surrounded by an adventitia; and blood vessels passing through the adventitia. [*Reproduced from K. Goto, Biomedical Research 2, Suppl.: 361–374, 1981.*]

Lumen

Pseudostratified columnar epithelium
Basement membrane

Smooth muscle

Adventia   Blood vessels   Lamina propria

Urethra The male *urethra* is the final section of the reproductive duct system. It leads from the urinary bladder, through the prostate gland, and into the penis (see Figures 26.1 and 26.9). Its reproductive function is to transport semen outside the body during ejaculation. As you may remember, the male urethra also carries urine from the urinary bladder during urination. However, it is physically impossible for a man to urinate and ejaculate at the same time because just prior to ejaculation the internal sphincter closes off the opening of the bladder. The sphincter does not relax until the ejaculation is completed. The closing of this internal sphincter prevents urine from entering the urethra, and also prevents the backflow of ejaculatory fluid into the urinary bladder.

The male urethra consists of prostatic, membranous, and spongy (cavernous) portions. The **prostatic portion** (about 2.5 cm) starts at the base of the urinary bladder and proceeds through the prostate gland. Here it receives secretions from the small ducts of the prostate gland and the two ejaculatory ducts. Upon its exit from the prostate gland, the urethra pierces the urogenital diaphragm, and is consequently called the **membranous portion** (about 0.5 cm). The external urethral sphincter muscle is located here. The **spongy,** or cavernous, **portion** (about 15 cm) of the urethra extends the full length of the spongy portion of the penis to the external urethral orifice on the glans penis, where either urine or semen leaves the penis. The wall of the urethra has a lining of mucous membrane and a thick, outer layer of smooth muscle. Within the wall are *urethral glands,* which secrete mucus into the urethral canal.

Accessory Glands

After the ductus deferens passes around the urinary bladder, several accessory glands add their secretions to the sperm as they are propelled through the remaining ducts. These accessory glands are the seminal vesicles, prostate gland, and bulbourethral glands. The fluid that results from the combination of sperm and glandular secretions is *semen,* or seminal fluid. Figure 26.1 shows the major complement of accessory ducts and glands.

Seminal vesicles The paired **seminal vesicles** are secretory sacs that lie next to the ampullae of the ductus deferentia (see Figure 26.1). Their alkaline secretions, which provide the bulk (about 60 percent) of the seminal fluid, contain mostly water, fructose, prostaglandins, and vitamin C. The secretions are produced by the mucous membrane lining of the glands. The seminal vesicles are innervated by sympathetic nerves from the pelvic plexus. Stimulation during sexual excitement and ejaculation causes the seminal fluid to be emptied into the ejaculatory ducts by muscular contractions of the smooth muscle layers. The seminal secretion provides an energy source for the motile sperm, and also helps to neutralize the natural acidity of the vagina.

The wall of each vesicle has an outer layer of connective tissue, a middle layer of smooth muscle, and a lining of mucosa. The mucosa contains foldings and outpockets that extend from a single folded and coiled tube. The folds provide a large area of cuboidal secretory epithelium (Figure 26.7).

Prostate gland The **prostate gland** lies inferior to the urinary bladder and surrounds the first 3 cm (1.2 in.) of the urethra (see Figure 26.9). It is a rounded mass about the size of a chestnut—about 4 cm (1.6 in.) across, 3 cm (1.2 in.) high, and 2 cm (0.8 in.) deep. The smooth muscles of the prostate can contract like a sponge to squeeze the prostatic secretions through tiny openings into the urethra. These secretions make sperm motile and help to neutralize vaginal acidity.

FIGURE 26.7

Seminal vesicle. (A) Drawing of seminal vesicle, ductus deferens, and ejaculatory duct. (B) Scanning electron micrograph of a seminal vesicle, showing large outpockets and numerous folds of the mucosa. [*Custom Medical Stock Photos.*]

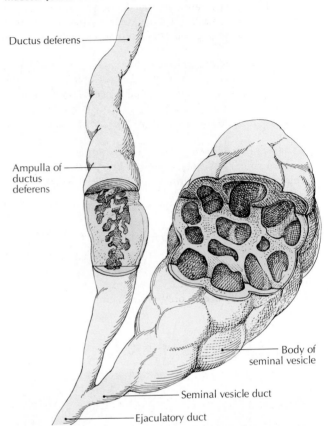

Ductus deferens

Ampulla of ductus deferens

Body of seminal vesicle

Seminal vesicle duct

Ejaculatory duct

[A]

Outpocket   Folds

[B]   ×50

Outpocket

As the urethra leaves the bladder, it travels through the prostate, where it receives prostatic secretions continually. Some of these secretions are passed off with the urine, but most of them are released with the semen during ejaculation. The secretions are released when smooth-muscle fibers in the wall of the prostate are stimulated by sympathetic nerves from the pelvic plexus. The ejaculatory ducts also pass through the prostate gland and receive its secretions during ejaculation.

The prostate is surrounded by a thin but firm capsule of fibrous connective tissue and smooth muscle. Inside, the prostate is made up of many individual glands, which release their secretions into the prostatic urethra through separate ducts (Figure 26.8). There are three types of glands inside the prostate:

1 The inner *mucosal glands* secrete mucus. These small glands sometimes become inflamed and enlarged (*prostatitis*) in older men and make urination difficult by pressing on the urethra.

2 The middle *submucosal glands*.

3 The *main* (external) *prostatic glands* supply the major portion of the prostatic secretions. The secretion contains mainly water, acid phosphatase (an enzyme of unknown function), cholesterol, buffering salts, phospholipids, and prostaglandins.* Cancer of the prostate, one of the more common types of cancer, usually occurs only in the main glands.

*Despite their name, prostaglandins are produced in the seminal vesicles and a wide variety of cells, but *not* in the prostate gland.

FIGURE 26.8

Cross section of the prostate gland, showing the three types of inner glands and their relationships to the urethra.

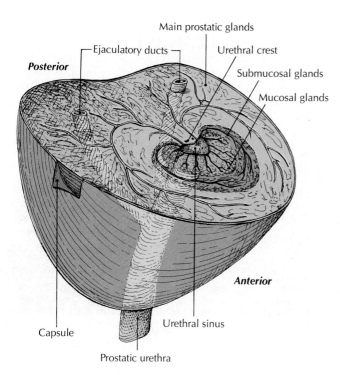

Bulbourethral glands The paired *bulbourethral glands* (also called *Cowper's glands*) are about the shape and size of a pea. They lie directly below the prostate gland, one on each side of the undersurface of the urethra (see Figures 26.1 and 26.9). Each gland has a duct that opens into the spongy part of the urethra.

With the onset of sexual excitement, bulbourethral glands secrete clear alkaline fluids into the urethra to neutralize the acidity of any remaining urine. These fluids also act as a lubricant within the urethra to facilitate the ejaculation of semen and to lubricate the tip of the penis prior to sexual intercourse.

Semen

Secretions from the epididymis, seminal vesicles, prostate gland, and bulbourethral glands, together with the sperm, make up the *semen* (SEE-muhn; L. seed), or *seminal fluid.* Sperm make up only about 1 percent of the semen. The rest is fluid from the accessory glands, which provides fructose to nourish the sperm, an alkaline medium to help neutralize urethral and vaginal acidity that could otherwise inactivate the sperm, and buffering salts and phospholipids that make the sperm motile.

Semen is over 90 percent water, but contains many substances, most notably energy-rich fructose. The known vitamins it contains are vitamin C and inositol, and trace elements include calcium, zinc, magnesium, copper, and sulfur. Semen also contains the highest concentration of prostaglandins in the body. The odor of semen is caused by amines (derivatives of ammonia), which are produced in the testes. The consistency of semen varies from thick and viscous to almost watery. The thinner consistency is usually the result of frequent ejaculations, but variations occur among men.

The average ejaculation produces about 3 or 4 mL of semen (about a teaspoonful) and contains 300 to 500 million sperm. Although it is not known why, semen with a high sperm count contains more Y ("male") sperm than X ("female") sperm, and a low sperm count has a greater proportion of X sperm.

Penis

The *penis* (PEE-nihss; L. tail) has two functions: (1) it carries urine through the urethra to the outside during urination, and (2) it transports semen through the urethra during ejaculation. In addition to the urethra, the penis contains three cylindrical strands of *erectile tissue:* two *corpora cavernosa,*

Is it possible for a woman to become pregnant if the man does not ejaculate during sexual intercourse?

Yes, it is, because secretions from the bulbourethral glands (released after sexual excitement, but before ejaculation) may contain sperm.

FIGURE 26.9

Penis. (A) Coronal section of penis. (B) Cross section of flaccid penis. (C) Cross section of erect penis.

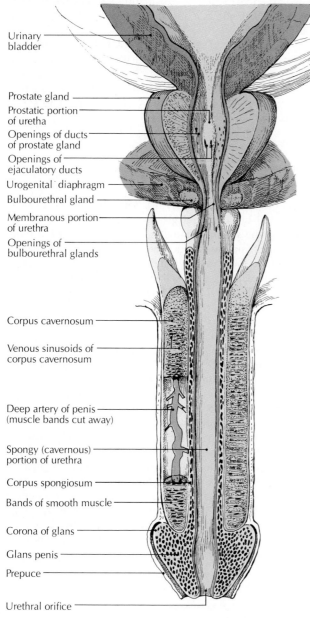

Urinary
bladder

Prostate gland
Prostatic portion
of uretha
Openings of ducts
of prostate gland
Openings of
ejaculatory ducts
Urogenital diaphragm
Bulbourethral gland
Membranous portion
of urethra
Openings of
bulbourethral glands

Corpus cavernosum

Venous sinusoids of
corpus cavernosum

Deep artery of penis
(muscle bands cut away)

Spongy (cavernous)
portion of urethra
Corpus spongiosum
Bands of smooth muscle

Corona of glans
Glans penis
Prepuce

Urethral orifice

[A]

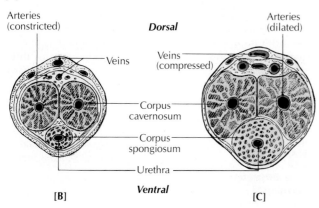

Arteries
(constricted)

Dorsal

Arteries
(dilated)

Veins

Veins
(compressed)

Corpus
cavernosum

Corpus
spongiosum

Urethra

[B]   **Ventral**   **[C]**

which run parallel on the dorsal part, and the ***corpus spongiosum*** (also called the *corpus cavernosum urethrae*), which contains the urethra. The corpora cavernosa contain numerous vascular cavities called *venous sinusoids* (Figure 26.9).

The corpus spongiosum extends distally beyond the corpus cavernosa, and becomes expanded into the tip of the penis, called the ***glans penis.*** Because the glans penis is a sensitive area containing many nerve endings, it is an important source of sexual arousal. The penile nerve endings are especially rich in the *corona,* the ridged proximal edge of the glans (see Figure 26.9).

The loosely fitting skin of the penis is folded forward over the glans to form the ***prepuce*** or ***foreskin. Circumcision*** is the removal, for religious or health reasons, of the prepuce. Today many circumcisions are performed in the belief that the operation may decrease the occurrence of cancer of the penis, although a controversy currently surrounds the practicality of circumcision.

Just below the corona, on each side of a ridge of tissue called the *frenum,* are the paired *Tyson's glands,* which are modified sebaceous glands. Their secretions, together with old cells shed from the glans and corona, form a cheeselike substance called *smegma.* An accumulation of smegma, which is more likely to occur in an uncircumcised male, may be a source of infection.

Ordinarily, the penis is soft and hangs limply. During tactile or mental stimulation, a parasympathetic reflex causes marked vasodilation within the arterioles, and the sinusoids of the corpora cavernosa become engorged with blood under high pressure. The distended sinusoids compress the veins that usually drain blood away, and the veins become constricted. This dual action prevents the blood from escaping, and the penis becomes enlarged and firm in an ***erection*** (see Figures 26.9B and C).

Two parts of the central nervous system control an erection, the hypothalamus in the brain and the sacral plexus of the spinal cord. Conscious thoughts within the cerebral cortex stimulate the erection center in the hypothalamus, which, in turn, causes parasympathetic vasodilation of the arterioles. Also, reflex responses in the sacral plexus can cause an erection in an infant or a sleeping adult.

When sexual stimulation ceases, the penile arteries constrict, pressure on the veins is relieved as blood is drained away, and the penis returns to its nonerect, *flaccid* (FLACK-sihd) state.* This return to the flaccid state is called *detumescence.*

Hormonal Regulation in the Male

Among the seminiferous tubules in the testes are small masses of ***interstitial endocrinocytes*** (*Leydig cells*). These cells secrete the male sex hormones testosterone, dihydrotestosterone, and androstenedione. The production and secretion of the main male hormone, ***testosterone,*** are controlled by

*The flaccid dimension of a penis bears little relation to the erect dimension, and the smaller flaccid penis erects to a proportionally larger size than a large flaccid penis. The greatest difference in penis size occurs in the *girth* of the erect penis, not the length.

FIGURE 26.10

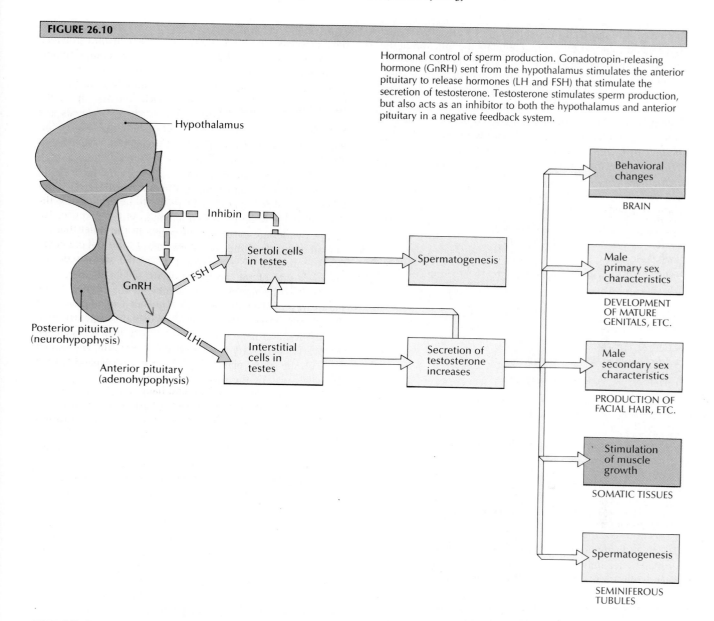

Hormonal control of sperm production. Gonadotropin-releasing hormone (GnRH) sent from the hypothalamus stimulates the anterior pituitary to release hormones (LH and FSH) that stimulate the secretion of testosterone. Testosterone stimulates sperm production, but also acts as an inhibitor to both the hypothalamus and anterior pituitary in a negative feedback system.

FSH (follicle-stimulating hormone) and *LH* (luteinizing hormone).

As you will see later in the chapter (see Table 26.1), FSH and LH are also major female sex hormones, and both hormones have exactly the same molecular structure in males and females. However, FSH and LH are named for their functions as female hormones; *follicle* and *luteinizing* are terms that refer to the female ovaries.

Both FSH and LH are produced by the anterior pituitary gland, which is controlled by the hypothalamus, and are chiefly responsible for stimulating spermatogenesis and the secretion of testosterone. While FSH and LH affect only the testes, testosterone affects not only spermatogenesis, but also the development of the sex organs and the appearance of such secondary male sex characteristics as a deep voice, facial and body hair, growth and skeletal proportions, and the distribution of body fat.

A deficiency of testosterone causes all the accessory male reproductive organs to decrease in size and activity. In most cases, both penile erection and volume of ejaculation diminish markedly. Testosterone deficiencies also produce an overall decrease in male sex drive and behavior, and affect secondary male sex characteristics.

Removal of the testes is called *castration*. Because a castrated male will not produce testosterone, he will eventually lose his sex drive. However, erection of the penis may still be possible after castration. A male who is castrated before puberty will gradually acquire such feminine characteristics as fatty deposits in the breasts and hips, lack of facial hair, and smooth skin texture. If an adult male is castrated, he will usually lose his facial hair, and his bones and muscles will probably diminish in thickness and size, but other feminine or childlike characteristics will be minimal.

Figure 26.10 shows the negative feedback system that regulates the production and secretion of testosterone. Some of the actions of testosterone are also shown. When the level of testosterone in the blood decreases, the hypothalamus is stimulated to secrete GnRH (gonadotropin-releasing hor-

834

Chapter 26: The Reproductive Systems

mone), which stimulates the secretion of FSH and LH, the two gonadotropic hormones that regulate the function of the gonads. FSH causes sustentacular cells in the seminiferous tubules of the testes to initiate spermatogenesis, and LH stimulates the interstitial endocrinocytes to secrete testosterone.

The cycle is completed when testosterone inhibits the secretion of LH, and another hormone, **inhibin,** is secreted from the sustentacular cells along with testosterone. Inhibin inhibits the secretion of FSH from the anterior pituitary. This feedback cycle maintains a constant rate (homeostasis) of spermatogenesis.

Ask Yourself

1 What are the basic reproductive functions of the male?

2 What are the main functions of the seminiferous tubules?

3 What are the three main functions of the epididymis?

4 How do the secretions of the seminal vesicles and prostate gland aid sperm?

5 What are the major constituents of semen?

6 What is the mechanism that produces an erect penis?

7 How do LH and FSH function in males?

FEMALE REPRODUCTIVE ANATOMY AND PHYSIOLOGY

The reproductive role of females is more complex than that of males. Not only do females produce egg cells (ova), but after fertilization they also nourish, carry, and protect the developing embryo. Then they nurse the infant for a time after it is born. Another difference between the sexes is the monthly rhythmicity of the female reproductive system.

The female reproductive system consists of a wide variety of structures with specialized functions (Figure 26.11):

1 Two *ovaries* produce ova* and the female sex hormones (a group of hormones called **estrogens**).

2 Two *uterine tubes,* also called *Fallopian tubes* or *oviducts,* one from each ovary, carry ova from the ovary to the uterus. Fertilization usually occurs in the uterine tubes.

3 The *uterus* houses and nourishes the developing embryo.

4 The *vagina* receives semen from the penis during sexual intercourse, is the exit point for menstrual flow, and is the canal through which the baby passes from the uterus during childbirth.

5 The *external genital organs,* collectively called the *vulva,* have protective functions and play a role in sexual arousal.

6 The *mammary glands,* contained in the paired *breasts,* produce milk for the newborn baby.

Ovaries

The female gonads are the paired **ovaries** (L. *ovum,* egg), which produce ova and female hormones (see Figure 26.11). These elongated, somewhat flattened bodies are about 2.5 to 5.0 cm (1.0 to 2.0 in.) long, 1.5 to 3.0 cm (0.6 to 1.2 in.) wide, and 0.6 to 1.5 cm (0.24 to 0.6 in.) thick—about the size and shape of an unshelled almond.

The ovaries are located in the pelvic part of the abdomen, one on each side of the uterus. Hanging almost free, the ovaries are attached by a mesentery called the **mesovarium** to the back side of the **broad ligament,** a fold of mesentery attached to the uterus. A thickening in the border of the mesovarium called the **ovarian ligament** extends from the ovary to the uterus. The mesovarium contains veins, arteries, lymphatics, and nerves to and from the **hilum** (opening) of the ovary. Each ovary is situated in a depression, the *ovarian fossa,* on the lateral pelvic wall. The ovary is suspended from the pelvic wall by the **suspensory ligament.**

The ovaries are covered by a layer of specialized epithelial cells, called the **germinal layer.** Beneath it is the **stroma,** a mass of connective tissue which contains ova in various stages of maturity.

Follicles and corpus luteum A cross section of an ovary reveals a *cortex* and a vascular *medulla* (Figure 26.12). The cortex contains round epithelial vesicles called **follicles,** the actual centers of ovum production, or **oogenesis.** Each follicle contains an immature ovum called a **primary oocyte,** and follicles are always present in several stages of development (see Figure 26.12). The outer part of the cortex, directly beneath the epithelial coating, forms a zone of thin connective tissue called the **tunica albuginea.**

Follicles are classified as either primordial or vesicular ovarian (Graafian). **Primordial follicles** are not yet growing, and **vesicular ovarian follicles** are almost ready to release a secondary oocyte (commonly called an ovum) in the process called *ovulation.* Follicles are usually located directly beneath the cortex of the ovary, but once they begin to mature they migrate toward the inner medulla. The medulla consists of layers of soft stromal tissue, which contains a rich supply of blood vessels, nerves, and lymph vessels. After a secondary oocyte is discharged, the lining of the follicle grows inward, forming the **corpus luteum** ("yellow body"), temporary endocrine tissue that secretes female sex hormones.

The corpus luteum secretes *estrogen* and *progesterone,* which stop ovulation and stimulate the thickening of the uterine wall and the development of the mammary glands in anticipation of pregnancy. A high concentration of progesterone also inhibits uterine contractions. If pregnancy does not occur within 14 days after the formation of the corpus luteum, it degenerates into the *corpus albicans,* a form of scar tissue. Menstruation follows almost immediately. If pregnancy occurs, the corpus luteum persists for about two to

---

*As you will see later in this chapter, a secondary oocyte, and not a mature egg, is released from an ovary during ovulation. An egg does not become mature until after it is penetrated by a sperm. For simplicity, however, we use the term *ova* (singular, *ovum*) to refer to the liberated eggs.

FIGURE 26.11

Female reproductive anatomy. (A) Median longitudinal section through the female pelvis. The arrows trace the path of an ovum as it leaves the ovary on its way to the uterus. (B) Front view of female reproductive system.

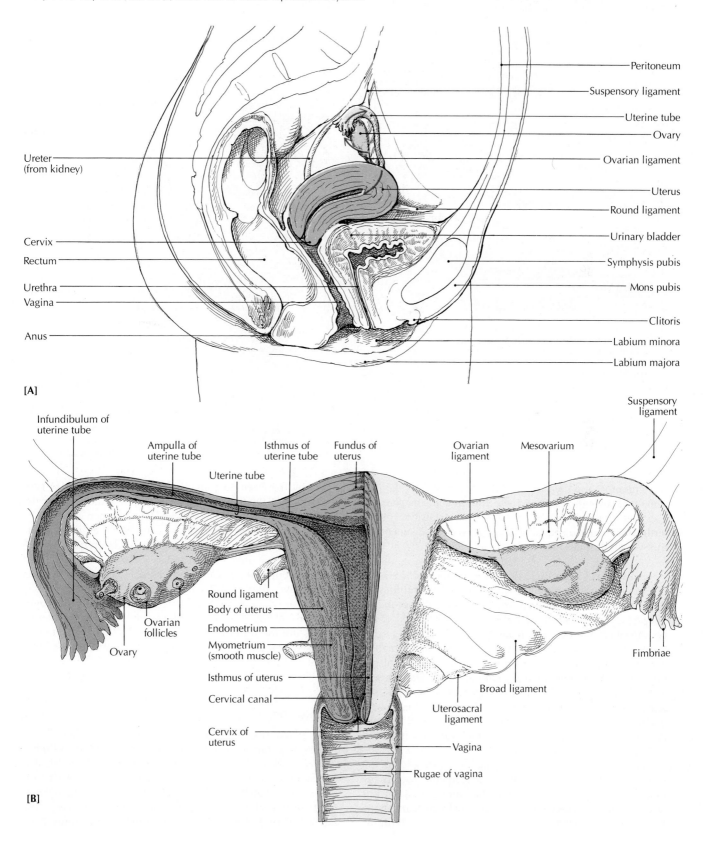

[A]

[B]

Peritoneum
Suspensory ligament
Uterine tube
Ovary
Ovarian ligament
Uterus
Round ligament
Urinary bladder
Symphysis pubis
Mons pubis
Clitoris
Labium minora
Labium majora

Ureter
(from kidney)

Cervix
Rectum

Urethra
Vagina

Anus

Infundibulum of
uterine tube
Ampulla of
uterine tube
Uterine tube
Isthmus of
uterine tube
Fundus of
uterus
Ovarian
ligament
Mesovarium
Suspensory
ligament

Ovarian
follicles
Ovary

Round ligament
Body of uterus
Endometrium
Myometrium
(smooth muscle)
Isthmus of uterus
Cervical canal
Cervix of
uterus

Fimbriae

Uterosacral
ligament
Broad ligament
Vagina
Rugae of vagina

FIGURE 26.12

Schematic diagram of an ovary, showing the development and eventual rupture of an ovarian follicle, and the subsequent formation and disintegration of a corpus luteum. The sequence of events begins with the primordial follicles and proceeds in the direction of the arrows.

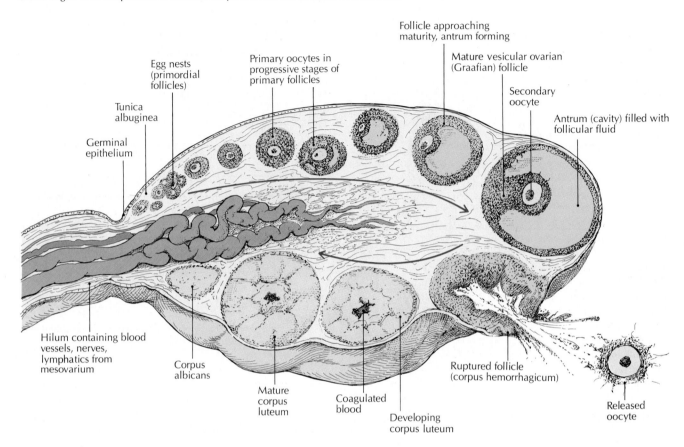

three months, and then it eventually degenerates as the placenta takes over its activities. The corpus luteum and placenta also secrete *relaxin.* Relaxin promotes the relaxation of the birth canal and the softening of the cervix and ligaments of the symphysis pubis in preparation for childbirth.

Ovulation In response to high concentrations of FSH and LH from the anterior pituitary, the mature follicle ruptures in the process called *ovulation.* During ovulation, the secondary oocyte is ejected from the ovary into the serous fluid of the peritoneal cavity near the uterine tube.

Beginning at puberty, about 20 ovarian follicles mature each month to the point of being ready for ovulation, but only one (rarely more) actually ruptures and releases a secondary oocyte. The other follicles degenerate, but only after releasing some progesterone and estrogen. After each secondary oocyte is released from the ovary, the site of the rupture heals. As a result, the surface of the ovary is scarred and pitted, especially in older women.

Although a woman will release no more than 400 to 500 secondary oocytes in her lifetime, she starts out with many more. A 7-month-old female fetus has about 1 million primordial follicles in each ovary, but most of them disintegrate, leaving about 200,000 to 400,000 in each ovary at birth. At

puberty (between 10 and 14 years of age), each ovary contains about 200,000 primary oocytes, with the number decreasing as a woman gets older. Most women cease to be fertile during their late forties or early fifties, although some women in their late fifties may still bear children.

Ordinarily, only one secondary oocyte is released each month, but several may be released, especially in women who have already had children, who have taken fertility drugs, or who are genetically predisposed toward the simultaneous release of more than one secondary oocyte. Fraternal twins are the result of the fertilization of two separate secondary oocytes. (See the discussion of multiple births in Chapter 27.)

Uterine Tubes

The paired tubes that receive the secondary oocyte from the ovary and convey them to the uterus are called *uterine tubes, Fallopian tubes,* or *oviducts* (see Figure 26.11B). In this discussion, we will refer to them as **uterine tubes.** The 10-cm-long (4-in.) tubes are not actually connected to the ovaries. The superior end opens into the abdominal cavity, very close to the ovary, and the inferior end opens into the uterus. Each uterine tube lies in the upper part of a transverse fold of peritoneum called the *broad ligament*. Three distinct portions of the tube are the funnel-shaped ***infundibulum*** (L. funnel) near the ovary, the thin-walled middle ***ampulla,*** and the ***isthmus,*** which opens into the uterus.

The wall of the uterine tube is made up of three layers:

1 The outer *serous membrane* is part of the visceral peritoneum.

2 The middle *muscularis* is composed of an inner layer of spirally oriented smooth muscle fibers and an outer longitudinal layer. Hormonal action produces peristaltic contractions in the muscularis close to the time when a secondary oocyte is released to help move it down the uterine tube to the uterus.

3 The inner *mucous membrane* has an epithelium made up of a single layer of columnar cells that alternate irregularly as ciliated and secretory cells. The secretory cells may provide nourishment for the ovum, and the cilia help propel the ovum toward the uterus.

The infundibulum is fringed with feathery ***fimbriae*** (FIHM-bree-ee; L. threads), which may actually overlap the ovary. Each month as a secondary oocyte is released, it is effectively swept across a tiny gap between the tube and the ovary into the infundibulum by the motion of cilia in the fimbriae. Few ova are ever lost in the abdominal cavity. Apparently there is no particular pattern that selects one ovary over the other each month.

Unlike sperm, the secondary oocyte is unable to move on its own. Instead, it is carried along the uterine tube toward the uterus by the peristaltic contractions of the tube and the waving movements of the cilia in the mucous membrane (Figure 26.13). Fertilization of the secondary oocyte usually occurs in the ampulla. An unfertilized oocyte will degenerate in the uterine tube. A fertilized oocyte (zygote) continues its journey toward the uterus, where it will become implanted. The journey takes four to seven days.

Occasionally, a fertilized ovum adheres to the wall of the uterine tube. This type of implantation is called an *ectopic pregnancy* (Gr. *ektopos,* out of place). Such a pregnancy is doomed because of a lack of nourishment and space to develop, and the embryo is discharged into the abdominal cavity through the opening of the ampulla or through a rupture in the wall of the uterine tube. This *miscarriage* is usually accompanied by hemorrhage. In some cases, surgery may be required to remove the implanted embryo, or to repair the ruptured uterine tube.

Uterus

The uterine tubes terminate in the ***uterus*** (L. womb), a hollow, muscular organ located in front of the rectum and behind the urinary bladder (see Figure 26.11). It is shaped like an inverted pear when viewed anteriorly, and it is pear-sized as well [about 7.5 cm (3 in.) long and 5 cm (2 in.) wide]. However, it increases three to six times in size during the nine months of pregnancy.

Just below the entrance of the uterine tubes are the ***round ligaments*** of the uterus. They help to keep the uterus tilted forward over the bladder. The uterus is attached to the lateral wall of the pelvis by two ***broad ligaments*** (double folds of peritoneum), which extend from the uterus to the floor and lateral walls of the pelvic cavity. Two ***uterosacral ligaments*** extend from the upper part of the cervix to the sacrum. The ***posterior ligament*** attaches the uterus to the rectum, and the ***anterior ligament*** attaches the uterus to the urinary bladder.

The wide upper portion of the uterus is called the ***fundus.*** The uterine tubes enter the uterus below the fundus (see

FIGURE 26.13

Uterine tube. (A) Cross section. (B) Scanning electron micrograph of the inner lining of a uterine tube showing the many protruding cilia. [(B) *E. F. Hafez.*]

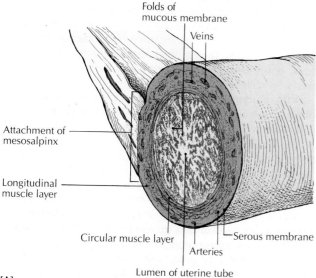

[A]

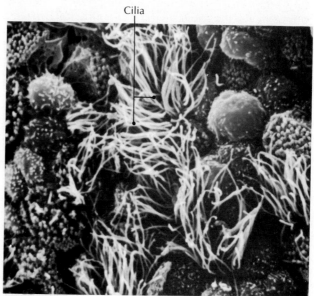

Cilia

[B]

FIGURE 26.14

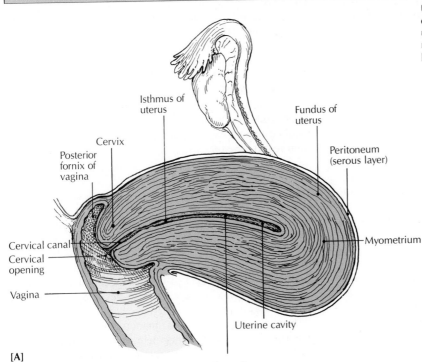

Uterus. (A) Sagittal view. (B) Scanning electron micrograph of cross section of uterus, showing lumen, endometrium, myometrium, serous layer, and blood vessels. [(B) *Shih, Gene, and Richard Kessel,* Living Images: Biological Microstructures Revealed by SEM. *Boston, Science Books International, 1982.*]

[A]

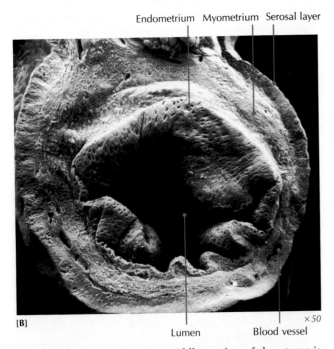

[B]   ×50

Figure 26.11B). The tapering middle portion of the uterus is the *body,* which terminates in the narrow *cervix* (L. neck), the juncture between the uterus and the vagina. The constricted region between the body and the cervix is the *isthmus.* The *cervical canal,* the interior of the cervix, opens into the vagina. The uterus of a woman who is not pregnant is somewhat flattened, so that the interior of the uterus, or *uterine cavity,* is just a slit (Figure 26.14). In its usual position, the flattened uterus leans forward over the urinary bladder,

at almost a right angle to the vagina. In this position, the uterus is said to be *anteflexed* (see Figure 26.11).

The uterus is made up of three layers of tissue (see Figure 26.14):

1 The outer *serous coat* of peritoneum extends to form the two broad ligaments that stretch from the uterus to the lateral walls of the pelvis.

2 The middle, muscular layer, called the ***myometrium*** (Gr. *myo,* muscle + *metra,* womb), makes up the bulk of the uterine wall. It is composed of three layers of smooth muscle fibers. From the outside in, they are arranged longitudinally, randomly in all directions, and both longitudinally and spirally. The interweaving muscles contract downward during labor with more force than any other muscle. These muscles are capable of stretching during pregnancy to accommodate one or more growing fetuses; they also contract during a woman's orgasm. The myometrium is almost a centimeter thick, but becomes even thicker during pregnancy in preparation for childbirth.

3 The innermost layer of the uterus is composed of a specialized mucous membrane called the ***endometrium,*** which is deep and velvety in texture. It contains an abundant supply of blood vessels, and is pitted with simple tubular glands. The endometrium is composed of two layers, the *stratum functionalis* and the *stratum basilis.* Every month, in response to estrogen secretion, the endometrium is built up in preparation for the possible implantation of a fertilized ovum (the beginning of *pregnancy*). Secretions of progesterone help the endometrium to develop into an active gland, rich in nutrients and ready to receive a fertilized ovum. If implantation does not occur, the *stratum functionalis* layer of the endometrium is shed together with blood and glandular secretions through the cervical canal and vagina. This breakdown of the endometrium makes up the **menstrual flow** (L. *mensis,* monthly), and the process is **menstruation.** The *stratum basilis* layer is permanent, and from it a new stratum functionalis regenerates after the three- to five-day menstrual period.

If fertilization and implantation do occur, the uterus houses, nourishes, and protects the developing fetus within its muscular walls. As the pregnancy continues, estrogen secretions develop the smooth muscle in the uterine walls in preparation for the expulsive action of childbirth.

Vagina

The uterus leads downward to the **vagina** (L. sheath), a muscle-lined tube about 8 to 10 cm (3 to 4 in.) long (see Figure 26.11). The vagina is the site where semen from the penis is deposited during sexual intercourse, the channel for the removal of menstrual flow, and the birth canal for the baby during childbirth. It lies behind the urinary bladder and urethra, and in front of the rectum and anus, and angles upward and backward.

The wall of the vagina is composed mainly of smooth muscle and fibroelastic connective tissue. It is lined with mucous membrane containing many folds called **rugae.** Stratified squamous nonkeratinizing epithelium covers the vagina and also the cervix of the uterus. Ordinarily the wall of the vagina is collapsed, but the vagina can enlarge to accommodate an erect penis during sexual intercourse or the passage of a baby during childbirth.

The mucous membrane secretes acids that help prevent infection, but the acids also create an environment hostile to sperm. (However, alkaline fluids from male accessory sex glands help to neutralize vaginal acidity.)

A fold of skin called the *hymen* (Gr. membrane*) partially blocks the vaginal entrance. The hymen is usually ruptured during the female's first sexual intercourse, but it may be broken earlier during other physical activities.

External Genital Organs

The **external genital organs,** or genitalia, include the mons pubis, labia majora, labia minora, vestibular glands, clitoris, and vestibule of the vagina (Figure 26.15). As a group, these organs are called the **vulva** (L. womb, covering).

The **mons pubis** (L. mountain + pubic), also called *mons veneris,* is a mound of fatty tissue that covers the symphysis pubis. At puberty, the mons pubis becomes covered with pubic hair. Unlike the pubic hair of a male, which may extend upward in a thin line as far as the navel, the upper limit of female pubic hair lies horizontally across the lower abdomen.

Just below the mons pubis are two longitudinal folds of skin, the **labia majora** ("major lips"), which form the outer borders of the vulva. They contain fat, smooth muscle, areolar tissue, sebaceous glands, and many sensory receptors. After puberty, their outer surface contains hairs.

The **labia minora** ("minor lips") are two smaller folds of skin that lie between the labia majora. Together with the labia majora, they surround and protect the vaginal and urethral openings. The labia minora contain sebaceous glands and many blood vessels, but no hair or fat. They also contain many nerve endings, and are sensitive to the touch. The labia merge at the top to form the **foreskin** or **prepuce** of the clitoris.

*Also, Hymen was the mythical Greek god of marriage.

FIGURE 26.15

Vulva, showing the borders of the diamond-shaped perineum.

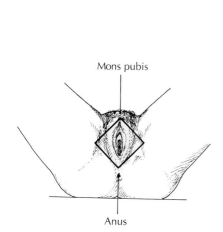

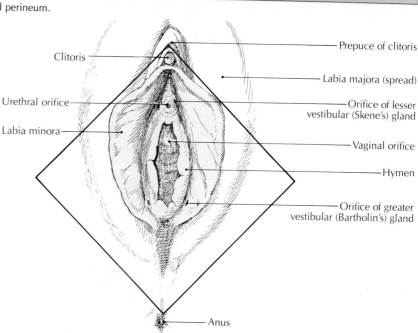

FORMATION OF SEX CELLS (GAMETOGENESIS)

The formation of *gametes,* or sex cells, is called *gametogenesis.* One phase of nuclear division involves mitosis, which was presented in Chapter 3. Another phase of nuclear division is called *meiosis,* which will be described here.

Meiosis

When we lose cells we replace them through mitosis, and when we grow from a single cell to an adult body with over 50 trillion cells, we also do it through mitosis. But when we reproduce, we use another process of cell division called *meiosis* (mye-OH-sihss; Gr. *meioun,* to diminish).

As you learned in Chapter 3, *mitosis* guarantees that new daughter cells receive exactly the same genetic information as the parent cell. If the parent cell has 23 *pairs* of chromosomes, or 46 chromosomes (as most body cells do), each daughter cell will also have 23 *pairs* of chromosomes (46 chromosomes). However, *gametes* cannot be formed by mitosis. If a sperm cell contained 46 chromosomes and an ovum contained 46 chromosomes, their combination during fertilization would produce a cell with 92 chromosomes. Such a cell would not be a normal human cell. The problem is solved when *meiosis reduces the number of chromosomes in gametes to half*—each sperm has 23 chromosomes (not 23 *pairs*), and each ovum has 23 chromosomes (not 23 *pairs*) (Figure 26.19)—so that when ovum and sperm unite, they produce a cell with 46 chromosomes. This cell is called a *zygote.*

Meiosis is divided into the same phases as mitosis: *prophase, metaphase, anaphase,* and *telophase.* However, the nucleus divides only once in mitosis, but *twice* in meiosis. In mitosis, the daughter cells have the same number of chromosomes as the parent cell, or the *diploid* (2*n*) number. The diploid number is 46 in human beings. Body cells, that is, all cells except sex cells, are diploid. In meiosis, the daughter cells have only half the parental number, or the *haploid* (1*n*) number. Gametes, with only 23 chromosomes, are haploid.

A critical part of meiosis occurs early in the sequence, when each double-stranded chromosome (each strand is called a *chromatid*) from the male parent lines up to a double-stranded chromosome from the female parent. Each male-female set of four chromatids is called a *homologous pair,* or a *tetrad* (Gr. *tetra,* four), because it consists of four chromatids. It is at this point that portions of a male chromatid may be exchanged with portions of its female homologue, resulting in a rearrangement of genes called *crossing-over.* The great variety among human beings is due to the unlimited possibilities of such genetic variation through meiosis and sexual reproduction.

Spermatogenesis

Haploid sperm cells are formed in the testes in a complex, precisely controlled series of events. This process, called *spermatogenesis* ("the birth of seeds"), continuously produces

mature sperm in the seminiferous tubules. It proceeds in the following stages (see Figure 26.20):

1 With the onset of puberty, when a boy is 11 to 14 years old, dormant, primitive, unspecialized germ cells called *spermatogonia* (singular, *spermatogonium*) become activated by secretions of testosterone.

2 Each spermatogonium divides through *mitosis* to produce two daughter cells, each containing the full complement of 46 chromosomes. (The term "daughter cells" has nothing to do with gender. These cells could just as easily be called "offspring cells.")

3 One of the two daughter cells is a spermatogonium, which continues to produce daughter cells. The other daughter cell is a *primary spermatocyte,* a large cell that moves toward the lumen of the seminiferous tubule.

4 The primary spermatocyte undergoes *meiosis* to produce two smaller *secondary spermatocytes,* each with 23 chro-

FIGURE 26.19

Comparison of mitosis and meiosis. In the simplified example shown here, there are four chromosomes: two long ones (one from the male parent and one from the female parent) and two short ones (one from the male parent and one from the female parent). Mitosis, with one division, results in two cells identical to the original one, with four chromosomes in each cell. Meiosis, with two divisions, results in four cells different from the original one, with two chromosomes in each cell.

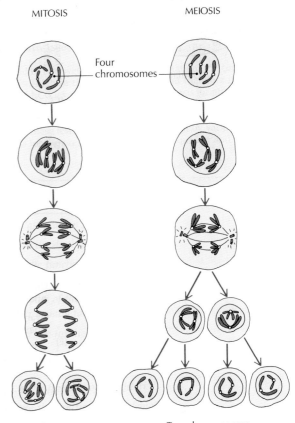

MITOSIS   MEIOSIS

Four chromosomes

Four chromosomes in each cell   Two chromosomes in each cell

mosomes: 22 body chromosomes and one X or Y sex chromosome.

5 Both secondary spermatocytes undergo a second mitotic division to form four final primitive germinal cells, the **spermatids,** which still have only 23 chromosomes.

6 The spermatids develop into mature **spermatozoa** (sperm cells) without undergoing any further cell division. Each spermatozoon has 23 chromosomes. The entire process of spermatogenesis takes about 64 days.

Oogenesis

Oogenesis, the maturation of ova in the ovary, differs from the maturation of sperm in several ways. Usually only one ovum at a time matures each month, but millions of sperm may mature at the same time. The process of cellular division

differs too. Whereas one primary spermatocyte yields four spermatozoa after meiosis, a primary oocyte yields only one ovum. Oogenesis proceeds through the following stages (Figure 26.21):

1 The **oogonium,** the diploid precursor cell of the ovum, is enclosed in a follicle within the ovary.

2 The oogonium develops into a **primary oocyte,** which contains 46 chromosomes. The primary oocyte undergoes meiosis, which produces two daughter cells of unequal size.

3 The larger of the daughter cells is the haploid **secondary oocyte.** It is perhaps a thousand times as large as the other cell, and contains most of the primary oocyte's cytoplasm, which provides nourishment for the developing ovum.

4 The smaller of the two daughter cells is the **first polar body.** It may divide again, but eventually it degenerates.

5 The large secondary oocyte leaves the ovarian follicle during ovulation, and enters the uterine tube. If the second-

---

FIGURE 26.20

Cellular events in spermatogenesis in the seminiferous tubules.

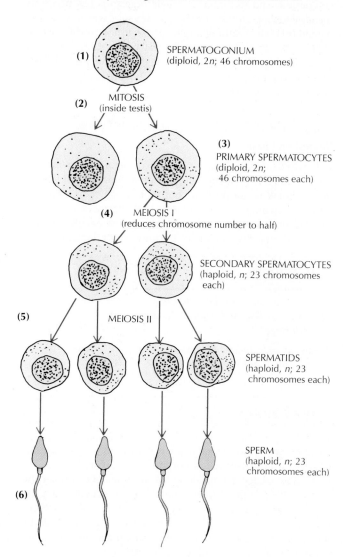

FIGURE 26.21

Cellular events in oogenesis in the ovary and uterine tube.

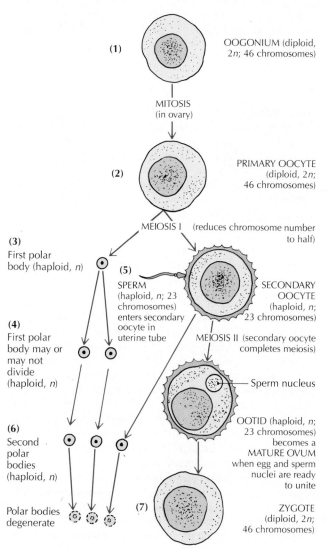

ary oocyte is fertilized, it begins to go through a second meiotic division, and a *second polar body* is "pinched off." It, too, is destined to die. If fertilization does not occur, menstruation follows shortly, and the cycle begins again.

6 During the second meiotic division, the secondary oocyte is completely reduced to the haploid number of 23 chromosomes, and is called an *ootid* (OH-oh-tihd). When the haploid sperm and ovum nuclei are finally ready to merge, the ootid is considered to have reached its final stage of nuclear maturity as a *mature ovum.*

7 The haploid nuclei of the ovum and sperm unite to form a diploid *zygote.*

Ask Yourself

1 *How does meiosis differ from mitosis?*

2 *What are the main steps in spermatogenesis? In oogenesis?*

3 *How does oogenesis differ from spermatogenesis?*

SEXUAL INTERCOURSE AND RESPONSES

The act of sexual intercourse, *coitus* (KOH-ih-tuhss; L. *coire,* to come together), is usually preceded by a period of preliminary sexual excitement, in which physiological and psychological changes occur in both sexes as a preparation for actual *copulation,* or coupling. In the past, it was thought that female sexual responses were different from and probably less intense than male sexual responses. However, an objective study of sexual response has shown that males and females experience quite similar feelings during sexual intercourse. Although coitus cannot really be divided neatly into discrete stages, we will describe it as if it could, in order to clarify the nature of the male and female responses.

Excitement and Plateau Phases

The initial foreplay of sexual intercourse may be referred to as the *excitement phase.* In the male, the penis becomes erect, fluids from the bulbourethral glands begin to be secreted, and the clear liquid oozes slowly from the penis. This viscous alkaline fluid neutralizes the acidity of any urine remaining in the urethra, and acts as a lubricant later when the penis is inserted into the vagina. In the female, the clitoris and the nipples become erect, and the glands of the cervix secrete mucuslike fluids that lubricate the vagina. In both sexes, heart rate, breathing rate, and muscular tension increase. A greater sensitivity to touch is apparent, along with reduced sensitivity to pain and outside stimuli.

The excitement phase lasts until the actual insertion of the penis into the vagina, which initiates the *plateau phase.* During this phase, the penis is thrust rhythmically within the vagina. The feeling of sexual pleasure usually increases for both partners, and both the glans of the penis and the clitoris are particularly stimulated. (The clitoris is massaged by the surrounding labia, not by the penis itself.) By now the semen has already been transported into the ductus deferens by peristalsis, and the scrotum has contracted and lifted the testes close to the body. Cervical secretions continue, and pelvic thrusting finally reaches a point of involuntary movement as sexual pleasure heightens and the next stage, orgasm, occurs.

Orgasm and Ejaculation

Orgasm, the climax of sexual excitement, is accompanied in males by *ejaculation* of the semen, which is expelled through the penis via the urethra by massive contractions of the ductus deferens, the seminal vesicles, and the other accessory glands, all of which contribute their secretions to the semen. Each normal ejaculation produces 3 to 5 mL (0.10 to 0.18 fl oz) of semen. The sperm cells begin to swim at speeds up to 5 cm (2 in.) an hour, wiggling their tails at 14 to 16 hertz (cycles per second). Some sperm cells may reach the uterus a few minutes after ejaculation, and may reach the uterine tubes as soon as 30 minutes after ejaculation.

Female orgasm does not necessarily accompany every copulation, and in humans, female orgasm is not necessary for fertilization. In the female, any sexual orgasm is accompanied by high breathing and heart rates, and flushing of the skin. Although multiple orgasms may occur, there is nothing in the female orgasm that represents an ejaculation. In both males and females, pelvic contractions during orgasm occur at intervals of 0.8 sec.

Resolution Phase

Following orgasm there is a latent or *resolution phase.* In men, a climax is usually followed by a period in which another climax is not possible, but there is no such refractory period in women. After orgasm, both partners undergo a rapid return to normal pulse, breathing, and circulation, and there is a subjective feeling of lassitude and relaxation.

Ask Yourself

1 *Into what stages can the sexual responses be divided, and what occurs during each stage?*

2 *How do male and female sexual responses differ, and how are they alike?*

CONCEPTION

Once every 28 days or so, a single secondary oocyte (ovum) emerges from a weakened portion of the ovary wall and is carried by ciliary action along the uterine tube toward the uterus. The ovum, moving much more slowly than the sperm, takes four to seven days to travel the 10 cm (4 in.) from the ovary to the uterus. During the first third of the journey, where fertilization is most advantageous, the ovum slows its pace, as if awaiting the sperm.

In Vitro *Fertilization and Surrogate Motherhood*

In vitro ("in a test tube") *fertilization* offers women who have difficulty conceiving a possible means of having a successful pregnancy. Essentially, an ovum is removed from the prospective mother's ovaries at the time of ovulation and fertilized in a glass dish or test tube with sperm from a male donor. Two or three days later, after cell division has begun, the embryo is implanted in the woman's uterus through her cervix. This technique has a 15 percent success rate. It was first done successfully in 1977, resulting in the birth of the first "test-tube baby" in 1978.

Two new techniques attempt to improve on the success rate of *in vitro* fertilization. In *gamete intrafallopian transfer (GIFT)*, the ovum and sperm are united in the uterine tube, where fertilization normally takes place. This procedure allows the fertilized ovum more time to develop before it reaches the uterus. Also, the removal of the ovum and its reimplantation take place on the same day, instead of several days apart as in *in vitro* fertilization. So far, the pregnancy rate is about 30 percent. Another technique, *transvaginal oocyte retrieval,* can be done in a physician's office. The physician obtains ova through the vaginal wall instead of through the abdominal wall. By this method, ova that would otherwise be hidden behind the uterus (a situation that occurs in about 5 to 10 percent of infertile women) can be retrieved.

Surrogate motherhood is an alternative for women who have difficulty either conceiving or carrying a child. For a woman who has a normal uterus but blocked uterine tubes that prevent the ova from being fertilized and traveling to the uterus, a female "surrogate" or donor can be artificially inseminated with sperm from the prospective father. If fertilization occurs, the embryo is transferred to the prospective mother. Also, a fertilized ovum from a woman who cannot produce the hormones necessary to sustain a pregnancy can be transferred to the uterus of another woman, the "surrogate mother," who will carry and give birth to the child.

Fertilization

The process of fertilization and the subsequent establishment of pregnancy are collectively referred to as *conception.* Fertilization must occur no more than 24 hours after ovulation, since the ovum will remain viable for only that period of time.* Sperm may remain viable in the female tract for up to 72 hours.†

During ejaculation, millions of flagellating sperm enter the female's vaginal canal. If coitus takes place at about the same time as ovulation, some of these sperm travel toward the opposite-moving ovum, but only one sperm may eventually enter and fertilize the ovum. If only one sperm may enter, why are millions discharged? Most likely, more than enough sperm are discharged to increase the chances that one will successfully complete its mission. The sperm must travel from the vagina all the way to the uterine tubes, a trip of 15 to 20 cm that may take as long as a few hours. They must resist the spermicidal acidity of the vagina and overcome opposing fluid currents in the uterus and uterine tubes. So it is not surprising that the mortality rate of the sperm is enormously high.

It is also assumed that the quantity of sperm discharged is so large so that several thousand will reach the ovum and act together to provide sufficient amounts of *hyaluronidase* (Figure 26.22). This enzyme, which is found in the head of the sperm, chemically dissolves just enough of the outer wall of the ovum, the *zona pellucida* (L. "transparent zone"), to allow a single sperm to enter.

The outer covering of the ovum is not its plasma membrane, but a special coating (over the zona pellucida), the *vitelline membrane,* on which only certain areas are receptive to sperm (Figure 26.23). Immediately after the entry of a sperm, the membrane thickens and is then called a *fertilization membrane.* It prevents other sperm from entering the

FIGURE 26.22

Scanning electron micrograph of the meeting of sperm and ovum. A sea urchin egg, the large sphere, is practically surrounded by hundreds of sperm. (The events are similar in human beings.) Although the ovum is enormous compared with the sperm, they both contribute exactly the same amount of genetic material to the zygote. [*M. J. Tegner and D. Epel, Science, 179: 687, February 16, 1973. © 1973 American Association for the Advancement of Science.*]

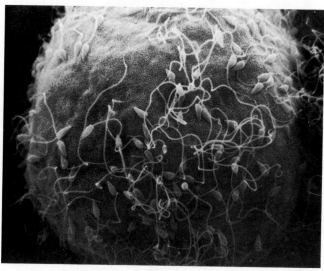

×1500

---

*It is ironic that after lying dormant for anywhere from 10 to 50 years, an unfertilized ovum released from an ovary has only about a day to live.

†If semen is stored in a laboratory or sperm bank at very cold temperatures, the sperm will remain viable for many months.

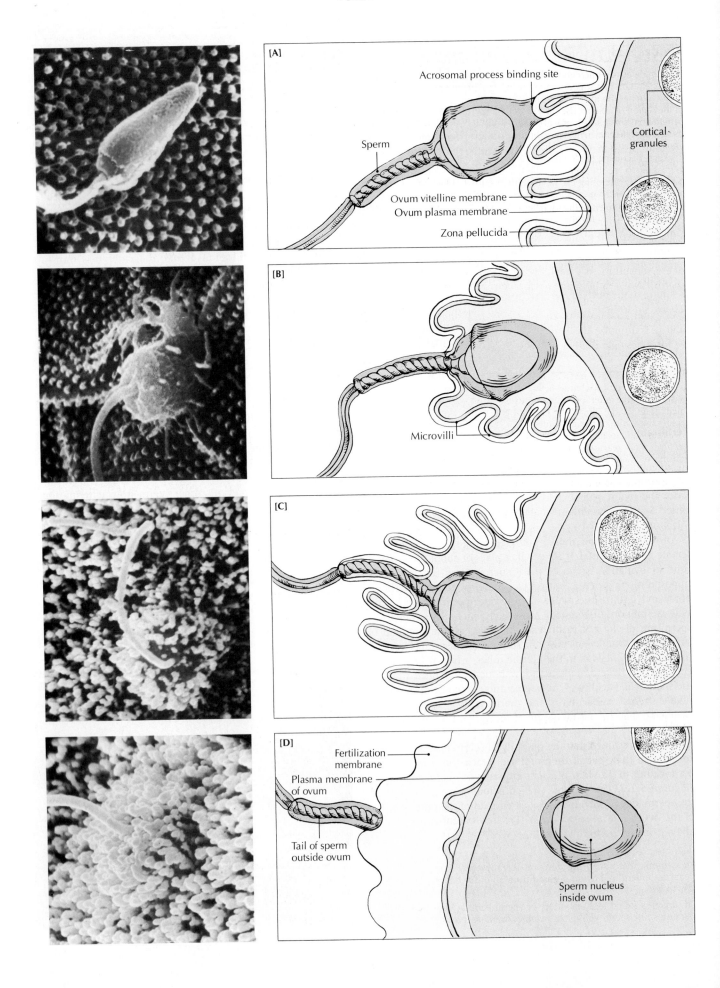

[A]

Acrosomal process binding site

Sperm

Cortical granules

Ovum vitelline membrane

Ovum plasma membrane

Zona pellucida

[B]

Microvilli

[C]

[D]

Fertilization membrane

Plasma membrane of ovum

Tail of sperm outside ovum

Sperm nucleus inside ovum

Scanning electron micrographs and comparative drawings showing the penetration of a sperm into an ovum. (A) The acrosome of the sperm touches the microvilli on the ovum surface. (B) The microvilli in the region of contact rise up to meet the sperm. Immediately after penetration, small sacs (cortical granules) just below the surface of the egg secrete enzymes that repel the extra sperm and create an impassable outer barrier. (C) The vitelline membrane rises and thickens, becoming the fertilization membrane, through which it is impossible to see. In the photo, the fertilization membrane has been peeled away, disclosing the sperm, almost buried in the ovum's cell membrane after three minutes. (D) The sperm sinks rapidly into the ovum cytoplasm, and a minute after the sperm body enters the plasma membrane only the tail remains outside. [(A, C, *and* D) *Mia Tegner, Scripps Institution of Oceanography.* (B) *Courtesy of Gerald Schatten.*]

ovum, ensuring that each ovum is fertilized by only one sperm.* Now embryonic development begins (Chapter 27).

Some aspects of fertilization are illustrated in Figures 26.22 and 26.23. About an hour after fertilization, the haploid (23 chromosomes) nuclei of the ovum and sperm fuse to form a single diploid (46 chromosomes) zygote. If more

*Recent studies have revealed that an ovum's outer membrane has a negative electrical charge before penetration by a sperm, and a positive charge after penetration. Apparently, the reversal of the membrane charge blocks the entry of any sperm after the first one enters the ovum. After a minute or so, the egg's outer membrane switches back to its original negative charge and initiates a *cortical reaction*. This reaction releases cortical granules from the ovum's plasma membrane. The granules release enzymes that destroy the glycoproteins serving as binding sites for sperm attachment, and at the same time cause the vitelline membrane to harden.

than one sperm somehow enters the ovum (*polyspermy*), the development of the zygote is quickly stopped. This cessation occurs because the extra mitotic spindles cause the abnormal segregation of chromosomes during cleavage.

Human Sex Determination

A new individual begins its development with a full complement of hereditary material. Each parent supplies 23 chromosomes, giving the zygote 23 pairs. In the male and female sets of chromosomes, 22 always match in size and shape, and determine the same traits. The 22 matching pairs are called **autosomes.** The other pair are the **sex chromosomes,** and they determine the sex of the new individual. Because the pair of sex chromosomes look alike in females, they are designated **XX.** Because the sex chromosomes do *not* look alike in males, they are designated **XY.**

After meiosis in males, half of the sperm contain 22 autosomes and an X chromosome, and the other half contain 22 autosomes and a Y chromosome (Figure 26.24). (After

| *Can women be allergic to sperm?* | *Yes, some women are allergic to their partner's sperm, and in fact, scientists are baffled as to why the immune systems of all women do not reject sperm as foreign bodies. Some men are infertile because they produce antibodies against their own sperm, especially following vasectomy.* |

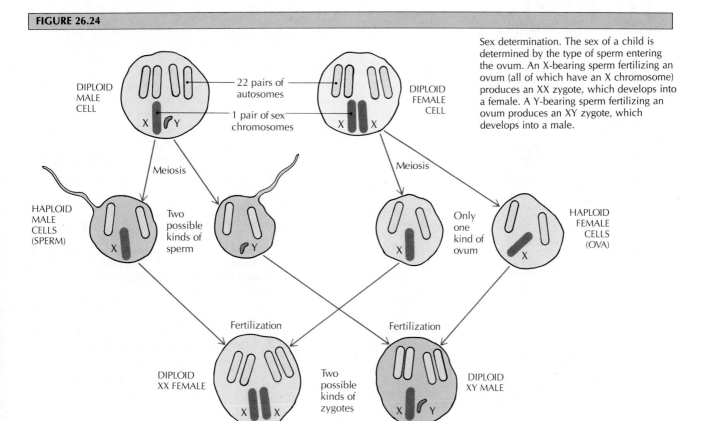

Sex determination. The sex of a child is determined by the type of sperm entering the ovum. An X-bearing sperm fertilizing an ovum (all of which have an X chromosome) produces an XX zygote, which develops into a female. A Y-bearing sperm fertilizing an ovum produces an XY zygote, which develops into a male.

TABLE 26.2   METHODS OF CONTRACEPTION CURRENTLY AVAILABLE IN THE UNITED STATES
(IN ORDER OF DECREASING EFFECTIVENESS)

Method	Mode of action	Effectiveness if used correctly	Action required at time of intercourse	Possible side effects or inconveniences
Sterilization				
Vasectomy	Prevents release of sperm.*	Very high.	None.	Irreversible.†
Tubal ligation	Prevents ovum from entering uterus.‡	Very high.	None.	Irreversible.†
Foam and condom	Foam kills sperm. Condom prevents sperm from entering vagina.	Very high.	Must be used before intercourse.	None usually. Foam may irritate. Condom may reduce sensation in male, may interrupt foreplay.
Oral contraceptive	Prevents follicle maturation and ovulation by altering hormones present; may also prevent implantation.	Very high.	None.	Early: some water retention, breast tenderness, nausea. Late: possible clots, hypertension, heart disease, breast cancer.
Intrauterine device (coil, loop)	Prevents implantation by stimulating inflammatory response.	High.	None.	Possible expulsion of device, menstrual discomfort, abnormal bleeding, infection, cramps.
Foam pad ("Today sponge")	Releases spermicide to kill sperm.	High.	Must be inserted before intercourse.	None as yet.
Diaphragm with spermicide	Diaphragm prevents sperm from entering uterus. Spermicide kills sperm.	High.	Must be inserted before intercourse.	None, but may cause overlubrication; cannot be fitted to all women.
Cervial cap with spermicide	Prevents sperm from entering uterus. Spermicide kills sperm.	High.	Must be inserted before intercourse.	May increase rate of abnormal PAP test. May be difficult for some to insert and remove.
Condom	Prevents sperm from entering vagina.	High.	Must be put on before intercourse.	Some reduction of sensation in male; may interrupt foreplay.
Temperature rhythm	Prevents sperm from being deposited when ovum is available for fertilization.	Medium.	None.	Requires abstinence during part of cycle; requires good record keeping to determine time of ovulation.
Calendar rhythm	Prevents sperm from being deposited when ovum is available for fertilization.	Medium to low.	None.	Requires abstinence during part of cycle; difficult to determine time of ovulation.
Spermicide (foam, jelly, cream)	Kills sperm.	Medium to low.	Must be inserted immediately before intercourse.	None usually; may irritate.
Withdrawal (coitus interruptus)	Prevents sperm from entering vagina.	Low.	Withdrawal.	Frustration in some.
Douche	Washes out sperm.	Lowest.	Must be done immediately after intercourse.	None, but requires immediate action.

*Vasectomy prevents the release of sperm, but does not alter the production of male hormones.

†Although new surgical techniques increase the possibility of reversing vasectomies and tubal ligations, both procedures should be considered irreversible at the time they are undertaken.

‡Tubal ligation prevents the passage of the mature egg cell to the uterus, but does not alter the production of female hormones.

Source: Adapted from Donald D. Ritchie and Robert Carola, *Biology*, 2nd ed. (Reading, Mass.: Addison-Wesley, 1983), p. 284. Used with permission.

meiosis in the female, *both* halves contain 22 autosomes and an X chromosome.) If an ovum is fertilized by an X-chromosome-bearing sperm, an XX zygote results, and the zygote will develop into a female. Fertilization of an ovum by a Y-bearing sperm produces an XY zygote, which will develop into a male. So the father actually determines the sex of the child, since only his sperm cells contain the variable, the Y chromosome. Also, it should be noted that the sex of the child is determined only at conception, never before or after.

Ask Yourself

1 *Why is the mortality rate of sperm high after they enter the female system?*

2 *What is hyaluronidase?*

3 *What is a fertilization membrane?*

4 *How does the genetic material of the mother and father combine to make a new individual?*

5 *How is the sex of a new individual determined?*

CONTRACEPTION

All **contraceptive** ("against conception") methods have one aim: the prevention of pregnancy. This aim can be achieved by preventing the production of ova or sperm, by keeping ova and sperm from meeting, or by preventing the implantation of an embryo in the uterus. All these methods are possible, but some are difficult to achieve, some are fallible, some are dangerous to the female, and some may be considered unacceptable because of psychological factors or religious beliefs. Table 26.2 describes the many methods of contraception now available in the United States. Two successful surgical methods of contraception, the male *vasectomy* and the female *tubal ligation,* are illustrated in Figure 26.25.

Ask Yourself

1 *Which contraceptive methods are most effective? Least effective?*

2 *What are some of the side effects of these methods?*

FIGURE 26.25

Surgical sterilization. (A) Male vasectomy removes a small portion of each ductus deferens to block the release of sperm from the testes. (B) Female tubal ligation removes a portion of each uterine tube to prevent the ovum from being fertilized and reaching the uterus.

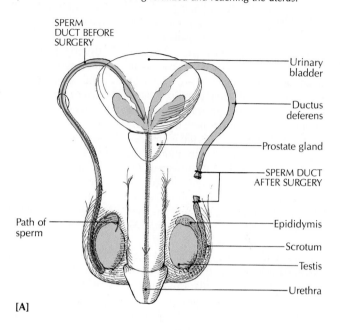

[A]

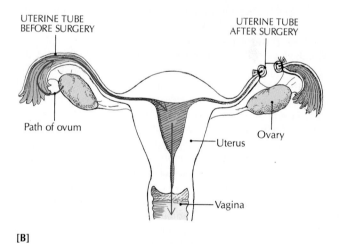

[B]

PHYSIOLOGICAL AND ANATOMICAL ABNORMALITIES

Sexually transmitted diseases

Some microorganisms are transferred from person to person mainly by sexual contacts. The diseases they cause are known as **sexually transmitted diseases** (STD) or **venereal** ("of Venus") **diseases** (VD). Contrary to some beliefs, these diseases are rarely contracted by casual, dry contact with persons or objects. The most common sexually transmitted diseases are *chlamydia,* caused by a bacterium; *type II herpes simplex,* caused by a virus; and *gonorrhea* and *syphilis,* both caused by bacteria.*

*As you saw in Chapter 20, AIDS can also be transmitted sexually, but it is not exclusively a sexually transmitted disease.

The most common venereal disease in the United States is the little-known *chlamydia* (klah-MIHD-ee-uh), named for the bacterium that causes it, *Chlamydia trachomatis*. Although it afflicts between 3 and 10 million Americans every year, chlamydia is not well known because the bacterium was isolated only recently, its symptoms are difficult to diagnose, and the disease is easily confused with gonorrhea.

Chlamydia in men produces infection in the urethra, prostate gland, and epididymis. If left untreated, it can cause sterility. Women are afflicted with a cervical infection and inflammation of the uterine tubes, which, if untreated, can result in scarring, sterility, or ectopic pregnancy. Women who acquire the disease during pregnancy may transmit it to their babies as they pass through the birth canal. Affected infants may have conjunctivitis or pneumonia, and there is some evidence that the risk of premature or stillborn infants is increased. Both men and women may be affected by conjunctivitis, pneumonia, and enlarged lymph nodes in the groin.

Penicillin is ineffective. The preferred treatment is an extended dose of erythromycin until the infection is gone, usually in about a week.

Type II herpes (Gr. "a creeping") *simplex* (also called herpes genitalis) ordinarily appears as blisterlike sores on or near the external genitalia about a week after intercourse with an infected partner. Fever, muscle aches, and swollen lymph nodes may also be apparent. When the blisters rupture, they produce painful ulcers and release millions of infectious viruses. The blisters usually heal after a week or two, but the viruses retreat to nerves near the lower spinal cord, where they remain dormant until the next attack. Infected people may harbor the infectious viruses for months, years, or a lifetime.

Perhaps the most serious complication of type II herpes simplex is the infection of a baby as it passes through the birth canal. For this reason, if a pregnant woman is known to have type II herpes simplex, the baby is usually delivered by cesarean section.

Gonorrhea (Gr. "flow of seed") is primarily an infection of the urinary and reproductive tracts (especially the urethra and cervix), but any moist part of the body, especially the eyes, may suffer. The causative bacterium, *Neisseria gonorrhoeae*, can be recognized by microscopic examination. Symptoms include urethritis (inflammation of the urethra) and urethral discharge in males, and a greenish-yellow cervical discharge in females. Itching, burning, redness, and dysuria (painful urination) are also common symptoms, but some infected people, especially women, are without symptoms. A complication of gonorrhea in men is epididymitis (inflammation of the epididymis), with an accompanying swelling of the testes. The leading cause of arthritis in young adults of both sexes is gonorrhea. If left untreated, gonorrhea becomes difficult to cure, but it can usually be cured by antibiotics, especially if it is diagnosed early.

Before routine treatment of the eyes with 1 percent silver nitrate in the newborn was established, thousands of babies were blinded by gonorrheal infection at the time of passage through the vaginal canal.

Syphilis* is a more dangerous disease than gonorrhea. It is caused by a motile, corkscrew-shaped bacterium, *Treponema pallidum*. It begins in the mucous membranes, and spreads quickly to lymph nodes and the bloodstream. The early symptom of syphilis is a sore, the hard chancre, at the place where infection occurred. Other symptoms may include fever, general body pain, and skin lesions. Sometimes these symptoms disappear, even without treatment, leaving the victim with the false impression that the disease is gone. But later, circulatory or nervous tissue may degenerate, so that paralysis, insanity, and death follow.

One of the most unfortunate aspects of sexually transmitted infection is the transfer of the microorganisms from a mother to her baby. The syphilis bacterium is able to cross the placenta during pregnancy, whereas the gonorrhea bacterium seems unable to do so. Thus, the developing fetus can contract syphilis early in its development and exhibit some of the serious symptoms of the disease at birth. If the baby contracts syphilis during the actual birth process, it is not likely to exhibit any symptoms at the time of birth. However, a baby infected with syphilis will grow poorly, be mentally retarded, and die early.

Like gonorrhea, syphilis can usually be cured by antibiotics, if treatment is started early. However, it is well to remember that the occurrence of sexually transmitted diseases in this country has reached epidemic proportions (more than 5 million cases of chlamydia and gonorrhea are reported annually, and the reported cases of syphilis in 1987 reversed a five-year trend of decreasing incidence), especially among teenagers and young adults. False feelings of security about not contracting the diseases, quick antibiotic cures, and lack of knowledge about the seriousness of the consequences of the diseases should not be encouraged. It is true that venereal diseases can be cured with greater ease than ever before, especially if they are reported and treated early, but the most effective treatment is still intelligent prevention.

Trichomoniasis is a sexually transmitted disease caused by the *Trichomonas vaginalis* protozoan. The infection affects the lower genitourinary tract, especially the vagina and urethra of females, and the lower urethra of males. Because the infecting organism is most productive in an alkaline environment, it is aided by excessive douching that disrupts the normal acidity of the vagina, pregnancy, and the use of oral contraceptives. Symptoms include itching, swelling, and frothy vaginal discharge in females and urethritis and dysuria in males.

Nongonococcal urethritis (NGU), also known as nonspecific urethritis (NSU), is a sexually transmitted disease characterized by an inflammation of the urethra accompanied by a discharge of pus. It is estimated that in the United States, 30 to 50 percent of the 3 million annual cases of acute urethritis (other than gonococcal) are caused by *Chlamydia trachomatis*. Other microorganisms, such as *Ureaplasma urealyticum*, can also cause NGU. The diagnosis of NGU is usually made when it is discovered that the infecting microorganism is not *N. gonorrhoeae*.

---

*Syphilis is the title character in a poem by Girolamo Fracastoro, a Veronese physician and poet. In the poem, published in 1530, Syphilis is stricken with a disease that was known at the time as the "great pox." Ever since 1530, however, the disease has been called *syphilis*.

Pelvic inflammatory disease

Pelvic inflammatory disease (PID) is a general term referring to inflammation of the uterus and uterine tubes (*salpangitis*) with or without inflammation of the ovary (*oophoritis*), localized pelvic peritonitis, and abscess formation (*tuboovarian abscess*). It is the most severe complication of the sexually transmitted diseases caused by *N. gonorrhoeae* and *C. trachomatis*. The spread of the infection can be controlled with antibiotics, if treatment is begun early. Severe cases of PID can lead to peritonitis.

Some common cancers of the reproductive systems

Breast cancer is a common malignant cancer in women, and is one of the leading causes of death by cancer among females. It is especially prevalent between the ages of 55 and 74, and usually does not occur before 35. Breast cancer kills three times as many women as uterine or ovarian cancer, spreading through lymphatics and blood vessels to other parts of the body, including the lungs, liver, bones, adrenal glands, and brain. The most common route is through the lymphatics that lead to the axillary lymph nodes.

The warning signs of breast cancer include a hard lump in the breast, a change in the shape or size of one breast, a change in skin texture, discharge from the nipple, itching or other changes in the nipple, a change in the skin temperature of the breast, and breast pain. Self-examination of the breasts and regular physical examinations are highly recommended by physicians as a way of early detection of these warning signs.

The type of surgical treatment for breast cancer depends upon the stage at which the cancer is diagnosed. If the diagnosis is early, the cancerous cells can often be removed successfully by a *lumpectomy*, in which the tumor and axillary lymph nodes are removed. The most drastic surgical treatment is a *radical mastectomy*, in which the entire breast and underlying fascia, the pectoral muscles, and all the axillary lymph nodes are removed. This is performed only as a last resort. Less drastic procedures include a *modified radical mastectomy*, in which the breast and axillary lymph nodes are removed, and a *simple mastectomy*, in which only the breast is removed.

Chemotherapy and radiation therapy are frequently used in conjunction with surgery. The current favored treatment for early breast cancer is a lumpectomy and postoperative radiation.

Cervical cancer (cancer of the cervix) is one of the most common cancers among females. When cancer cells invade the basement membrane and spread to adjacent pelvic areas or to distant sites through lymphatic channels, it is classified as *invasive*. When only the epithelium is affected, it is *preinvasive*. If detected early, preinvasive cancer is curable 75 to 90 percent of the time. While preinvasive cancer produces no apparent symptoms, invasive cancer is characterized by unusual vaginal bleeding or discharge, and postcoital pain or bleeding. The most effective method of detection is the Pap test (Papanicolau stain slide test), a microscopic examination of cells taken from the cervix. Advanced cases of invasive cervical cancer may call for a *hysterectomy*, the surgical removal of the uterus.

Prostatic cancer is one of the most common cancers among males. Most prostatic cancers originate in the posterior portion of the gland. When the cancer spreads beyond the prostate itself, it usually travels along the ejaculatory ducts in the spaces between the seminal vesicles. Because prostatic carcinomas rarely produce symptoms until they are well advanced, prostatic cancer is often fatal. Annual or semiannual rectal examinations may detect a small, hard nodule while it is still localized, and in such cases the recovery rate is high. Regular examinations are especially important in men over 40.

Prostate disorders

The prostate gland may be affected by inflammation (acute or chronic infections), enlargement, and benign growths. Benign enlargement of the prostate gland is known as *benign prostatic hypertrophy (BPH)*, or benign prostatic hyperplasia. It is the most common type of benign tumor, affecting about 75 percent of all men over 50. An enlarged prostate may be caused by a decrease in the secretion of male hormones, inflammation, decreased sexual activity, metabolic and nutritional imbalances, or other factors. When the prostate becomes enlarged, it compresses the urethra and obstructs normal urinary flow. Surgery that removes part or all of the prostate usually relieves the obstruction. A relatively new surgical technique does not sever the nerves near the prostate that influence erections. As a result, sexual potency is retained. Hormonal treatment is under investigation.

Ovarian cysts

Ovarian cysts generally occur either in the follicles or within the corpus luteum. Although most ovarian cysts are not dangerous, they must be examined thoroughly as possible sites of malignant cancer.

Follicular cysts are usually small, distended bubbles of tissue that are filled with fluid. Ordinarily, small follicular cysts do not produce symptoms unless they rupture, but large or multiple cysts may cause pelvic pain, abnormal uterine bleeding, and irregular ovulation. If follicular cysts are present at menopause, they secrete excessive amounts of estrogen in response to the increased menopausal secretions of LH and FSH.

Granulosa-lutein cysts are produced when an excessive amount of blood accumulates during menstruation. If they appear in early pregnancy, they may cause pain on one side of the pelvis, and if they rupture, there will be massive hemorrhaging within the peritoneum. Granulosa-lutein cysts in nonpregnant women may cause irregular menstrual periods and abnormal bleeding.

Because most ovarian cysts disappear of their own accord, typical treatment consists of observation to detect malignancies.

Endometriosis

Endometriosis is the abnormal location of endometrial tissue in sites such as the ovaries, pelvic peritoneum, and small intestine. Most cases probably develop as a result of retrograde passage of bits of menstrual endometrium through the opening of the uterine tube into the peritoneal cavity. This condition is usually associated with dysmenorrhea (painful menstruation), pelvic pain, infertility, and dyspareunia (painful coitus). Treatment ranges from symptomatic relief of pain to surgical removal of the endometrial implants, including the use of a laparascope and laser beam.

AMENORRHEA The absence of menstrual flow.

ANOVULATION A condition in which ovulation does not occur.

CRYPTORCHIDISM (krip-TOR-kye-dizm; Gr. *kryptos*, hidden + *orchis*, testis, because of the shape of the orchid root) A condition in which the testes do not descend from their fetal position inside the abdominal cavity into the external scrotum.

DILATION AND CURETTAGE A procedure that dilates the cervix in order to scrape the lining of the uterus. Also called *D and C.*

DYSMENORRHEA Painful or difficult menstruation.

FIBROADENOMA A fibroid breast tumor.

HERMAPHRODITISM A condition in which male and female sex organs are found in one person.

HYSTERECTOMY The surgical removal of the uterus. A *total* hysterectomy usually removes all the female reproductive organs (including the ovaries and uterine tubes) except the vagina.

IMPOTENCE The inability of a man to have an erection.

LEUKORRHEA A whitish, viscous discharge from the vagina or uterus.

OLIGOSPERMIA A condition in which only a small amount of sperm is produced.

OOPHORECTOMY The surgical removal of an ovary.

OOPHORITIS An inflammation of an ovary.

ORCHITIS An inflammation of the testes, produced by a bacillus or staphylococcus infection.

PREMENSTRUAL TENSION A condition characterized by nervousness, irritability, and abdominal bloating one to two weeks before menstruation.

PROSTATECTOMY The surgical removal of all or part of the prostate gland.

SALPINGECTOMY The surgical removal of a uterine tube.

SPONTANEOUS ABORTION The loss of a fetus through natural causes. Usually called a *miscarriage.*

STERILITY The inability to reproduce.

UTERINE LEIOMYOMA A smooth muscle uterine tumor, including myomas, fibromyomas, and fibroids.

VAGINITIS An inflammation of the vagina.

Male reproductive anatomy and physiology

1 The reproductive function of the male is to produce sperm and deliver them to the vagina of the female.

2 The paired *testes* are held outside the body in the *scrotum,* a sac of skin between the thighs. The *seminiferous tubules* in the testes produce sperm, and *interstitial endocrinocytes* (Leydig cells) secrete testosterone, the primary male sex hormone.

3 A mature *sperm* consists of a *head* that has an *acrosome,* containing enzymes to aid the sperm in penetrating an ovum, and a *nucleus,* containing DNA; a *middle piece* that contains the mitochondria that provide ATP energy; and a *tail* that drives the sperm forward.

4 The seminiferous tubules merge to form the *tubuli recti,* the *rete testis,* and then the *efferent ducts,* which pass into the tightly coiled epididymis in each testis. The *epididymis* stores sperm as they mature, serves as a duct for the passage of sperm from the testis to the ejaculatory duct, and propels sperm toward the penis with peristaltic contractions.

5 The paired *ductus deferentia* are the dilated continuations of the epididymides. They extend to the ejaculatory ducts and are the main carriers of sperm.

6 The paired *ejaculatory ducts* receive secretions from the seminal vesicles and prostate gland. They carry semen to the urethra during ejaculation.

7 The *urethra* is a single tube through the penis, consisting of *prostatic, membranous,* and *spongy* portions. It transports sperm outside the body during ejaculation.

8 Secretions of the paired *seminal vesicles* provide nourishment for the sperm and help to neutralize vaginal acidity. Secretions of the *prostate gland* make the sperm motile and help to neutralize vaginal acidity. Secretions of the paired *bulbourethral glands* neutralize any urine in the urethra prior to ejaculation and lubricate the urethra to facilitate ejaculation.

9 Secretions from the epididymis, seminal vesicles, prostate gland, and bulbourethral glands, together with the sperm, make up the *semen.* The average ejaculation produces about 3 to 4 mL of semen and contains 300 to 500 million sperm.

10 The *penis* carries urine to the outside during urination and transports semen during ejaculation. It contains the urethra and three strands of erectile tissue: two *corpora cavernosa* and the *corpus spongiosum.* The tip of the penis is the *glans penis.* During sexual stimulation, the penis becomes enlarged and firm in an *erection.*

11 The secretion of testosterone from the interstitial cells in the testes is controlled by *FSH* (follicle-stimulating hormone) and *LH* (luteinizing hormone) from the anterior pituitary gland, which is controlled by *GnRH* (gonadotropin-releasing hormone) from the hypothalamus.

Female reproductive anatomy and physiology

1 The reproductive functions of the female are to produce ova; to nourish, carry, and protect the developing fetus; and to nurse the newborn baby.

2 The female gonads are the paired *ovaries,* which produce ova and the female sex hormones. The centers of ovum production in the ovaries are the *follicles,* which are always present in various stages of development. *Primordial follicles* are not yet growing. *Vesicular ovarian* (Graafian) *follicles* are almost ready to release an ovum in the monthly process called *ovulation.* After a mature ovum is discharged from a ruptured follicle, the follicle becomes the *corpus luteum,* a temporary endocrine gland that secretes *estrogen, progesterone,* and *relaxin.*

3 The paired *uterine tubes* (*Fallopian tubes* or *oviducts*) receive mature ova from the ovary and convey them to the uterus. Ova are usually fertilized while they are still in the uterine tube, and the fertilized ovum (*zygote*) is then transported to the uterus.

4 The uterine tubes terminate in the *uterus,* a hollow, muscular organ that is the site of implantation of a fertilized egg. If pregnancy occurs, the uterus houses, nourishes, and protects the developing fetus. The inner lining of the uterus is the *endometrium,* which is built up every month in preparation for a possible pregnancy. If pregnancy does not occur, the endometrium is shed in the monthly process of *menstruation.*

5 The uterus leads downward to the *vagina,* a muscle-lined tube that is the site where sperm cells from the penis are deposited during sexual intercourse, the exit point for menstrual flow, and the birth canal for the baby during childbirth.

6 The *external genital organs,* collectively called the *vulva,* are the mons pubis, labia majora, labia minora, vestibular glands, clitoris, and vestibule of the vagina.

7 *Mammary glands* are modified sweat glands contained in the breasts. They produce and secrete milk for the newborn baby. The actual milk-producing hormone is *prolactin.*

8 The development of an oocyte in an ovary is influenced by *gonadotropin-releasing hormone (GnRH)* from the hypothalamus. *Follicle-stimulating hormone (FSH)* and *luteinizing hormone (LH)* are then released by the anterior pituitary to cause the oocyte's maturation and release from the ovary in the process of *ovulation.*

9 If pregnancy does not happen, the endometrium breaks down and menstruation occurs.

10 If pregnancy occurs, *human chorionic gonadotropin (hCG)* is released from the placenta and covering membranes of the embryo. The endometrium is maintained, and pregnancy continues.

11 The hormones *prolactin* and *oxytocin* induce the mammary glands to secrete and eject milk after childbirth. Oxytocin also stimulates uterine contractions during childbirth.

Formation of sex cells (gametogenesis)
1 *Meiosis* reduces the number of chromosomes in sex cells to half, with each sperm having 23 chromosomes and each ovum having 23. When ovum and sperm unite, they produce a cell with 46 chromosomes. A cell with the full number of chromosomes is *diploid,* and a cell with only half is *haploid.*

2 The process of forming haploid sperm cells in the testes is called *spermatogenesis.* A *spermatogonium* undergoes mitotic division to produce a primary spermatocyte and another spermatogonium. The *primary spermatocyte* undergoes meiosis to produce two *secondary spermatocytes,* each with 22 autosomes and one X or Y sex chromosome. A second mitotic division forms four haploid *spermatids,* the final primitive sex cells, which develop into mature *sperm.*

3 The maturation of ova in the ovaries is *oogenesis.* It begins with a basic cell, the *oogonium,* which develops into a diploid *primary oocyte* with 46 chromosomes. After meiosis, a haploid *secondary oocyte* and the first *polar body* are produced. The polar body disintegrates, and the secondary oocyte leaves the ovary and enters the uterine tube. If the secondary oocyte is fertilized, it undergoes a second meiotic division and produces a second polar body and a haploid *ootid.* When the haploid sperm and ovum nuclei merge, the ootid is considered to be a *mature ovum.*

Sexual intercourse and responses
1 Recent studies have shown that males and females experience similar responses during sexual intercourse, or *coitus.*

2 Coitus can be considered as being divided into the *excitement phase;* the *plateau phase,* when the penis is inserted into the vagina and thrust rhythmically; *orgasm,* the climax of sexual excitement; and the *resolution phase,* a period of relaxation.

Conception
1 Millions of flagellating sperm are ejaculated into the vagina during coitus. Although only one will penetrate the ovum, the *hyaluronidase* secretions of many are required to dissolve the outer shell of the ovum to allow penetration.

2 Once one sperm enters the ovum, an impenetrable membrane forms around the ovum, and no other sperm can enter.

3 About an hour after fertilization, the haploid (23 chromosomes) nuclei of the ovum and sperm fuse to form a single diploid (46 chromosomes) cell, the *zygote.*

4 From the parents, the child receives 22 pairs of matching chromosomes called *autosomes,* and another pair called the *sex chromosomes.* The pair of sex chromosomes look alike in females, and are designated *XX.* The sex chromosomes do not look alike in males, and are designated *XY.*

5 After meiosis in males, half of the sperm contain an X chromosome, and half contain a Y chromosome. An ovum fertilized by an X-bearing sperm produces an XX zygote, which develops into a female. Fertilization of an ovum by a Y-bearing sperm produces an XY zygote, which develops into a male.

Contraception
1 The contraceptive aim of preventing pregnancy can be achieved by preventing the production of ova and sperm (as by oral contraceptives), by keeping ova and sperm from meeting (as by vasectomy, tubal ligation, diaphragm, condom, spermicide, and rhythm), or by preventing the implantation of an embryo in the uterus (by IUD).

2 The most effective contraceptives are sterilization, vasectomy, tubal ligation, and oral contraceptives.

Physiological and anatomical abnormalities
1 The more common *sexually transmitted diseases* are *chlamydia, gonorrhea, type II herpes, syphilis, trichomoniasis,* and *nongonococcal urethritis.*

2 *Pelvic inflammatory disease* is a general term referring to inflammation of the uterus, uterine tubes, localized pelvic peritonitis, and abscess formation.

3 *Breast cancer* and *cervical cancer* are two common cancers in women, and *prostate cancer* occurs frequently in older men.

4 The prostrate gland may be affected by inflammation, enlargement, and benign tumors.

5 *Ovarian cysts,* which occur in the follicles or corpus luteum, are usually not dangerous, but may develop into malignancies.

6 *Endometriosis* is the abnormal location of endometrial tissue.

UNDERSTANDING THE FACTS

1 What are the two basic functions of the testes?

2 What is the function of the interstitial endocrinocytes? What cells of the testes provide nourishment for the sperm as they develop?

3 List the accessory reproductive organs of the male.

4 Give the location of the following (be specific):

    a corpora   **d** epididymis
    cavernosa   **e** prostaglandins
    b inguinal hernia   **f** bulbourethral
    c acrosome   glands

5 Distinguish between primordial and vesicular ovarian follicles.

6 What structure takes over the function of the corpus luteum after it degenerates?

7 Does more than one ovarian follicle ripen each month? Explain.

8 What factors may increase the probability that more than one ovum may be released at ovulation?

9 Describe the function of the fimbriae of the uterine tube.

10 Define vulva.

11 What is the site for the production of FSH, LH, estrogens, and progesterone?

12 Give the location of the following (be specific):

    a corpus luteum   **e** mons pubis
    b infundibulum   **f** areola
    c round ligaments   **g** hyaluronidase
    d cervix   **h** zygote

13 How many chromosomes does an ovum or sperm contain? Are the chromosomes paired?

14 What is meant by the diploid and the haploid number?

15 Trace the development of mature sperm from a spermatogonium.

16 What enhances sperm motility?

17 Define zona pellucida.

18 What is an autosome? A sex chromosome?

19 Why is a cesarean section often performed on a pregnant woman who has type II herpes simplex?

20 What is the most common site of cancer in the female? In the male?

UNDERSTANDING THE CONCEPTS

1 In what ways is the female reproductive system more complex than the male reproductive system?

2 Why can the scrotum be called a thermoregulatory organ?

3 Why is it important that the sperm contain mitochondria?

4 Trace the pathway of a sperm cell from the site of production until it leaves the body of the male.

5 Why is it important that some secretions of the male accessory glands be alkaline?

6 Summarize the functions of the various components of semen.

7 Describe the major effects of castration or a deficiency of testosterone in the male.

8 What is the significance of the fact that only one ovum is usually produced each month from a primary oocyte? If the process were similar to that in the development of a primary spermatocyte, how would the reproductive ability of a male be affected?

9 Why are the lymph nodes associated with the breast so important in breast cancer?

10 Summarize the role of specific hormones in the production of milk.

11 Explain why so large a number of sperm are contained in each ejaculation.

12 Discuss some of the current concepts that explain how entry into the ovum is blocked after penetration by one sperm.

13 List and compare the basic contraceptive methods.

14 Why do we say that the reproductive function in the female is more complex than in the male?

15 What is the significance of the fact that each ovum and each sperm contains the haploid number of chromosomes?

SUGGESTED READING

AMELAR, R. D., L. DUBIN, AND C. SCHOENFELD, "Sperm Motility." *Fertil Steril*, 34(1980):197–215.

EPEL, DAVID, "The Program of Fertilization." *Scientific American*, November 1977.

GORDON, J. W., AND F. H. RUDDLE, "Mammalian Gonadal Determination and Gametogenesis." *Science*, 211(1981):1265–1271.

GREEP, R. O., *Reproductive Physiology*. Baltimore: University Park Press, 1983.

GROBSTEIN, CLIFFORD, "External Human Fertilization." *Scientific American*, June 1979.

HATCHER, R. A., G. K. STEWART, F. GUEST, N. JOSEPHS, AND J. DALE, *Contraceptive Technology*. New York: Irvine, 1982.

JOHNSON, M., AND B. EVERTT, *Essential Reproduction*. Oxford: Blackwell, 1980.

JONES, R. E., *Human Reproduction and Sexual Behavior*. Englewood Cliffs, N.J.: Prentice-Hall, 1984.

MASTERS, W. H., AND V. E. JOHNSON, *Human Sexual Response*. Boston: Little, Brown, 1966.

MCCARY, JAMES LESLIE, *McCary's Human Sexuality*, 3rd ed. New York: Van Nostrand, 1978.

NAFTOLIN, F., "Understanding the Basis of Sex Differences." *Science*, 211(1981):1263–1264.

PARDRIDGE, W. M., "Androgens and Sexual Behavior." *Ann Int Med*, 96(1982):488–501.

ROSENFIELD, A., "The Pill: An Evaluation of Recent Studies." *Johns Hopkins Med J*, 150(1982):177–180.

SALISBURY, G. W., R. G. HART, AND J. R. LODGE, "The Spermatozoon." *Perspect Biol Med*, 20: 372–393.

WASSARMAN, PAUL M., "Fertilization in Mammals." *Scientific American*, December 1988.

WATERS, HARRY F., LUCY HOWARD, AND GEORGE HACKETT, "Herpes: The New Scarlet Letter." *Time*, August 2, 1982.

27
Human Growth and Development

LEARNING OBJECTIVES

1 Outline in sequence the events that take place from fertilization to birth.

2 Describe the stages of development during cleavage, with the related structural changes in the zygote.

3 Describe the process of implantation in the uterine wall.

4 Describe the three primary germ layers, and name the body structures derived from each.

5 Explain the structure and functions of the extraembryonic membranes.

6 Discuss the roles of the placenta and the umbilical cord in prenatal development.

7 Describe the major body structures that begin to form during the embryonic period.

8 Trace the development of major body structures during the fetal period.

9 Describe the important maternal changes that occur during pregnancy.

10 Explain the hormonal basis of pregnancy and of pregnancy tests.

11 Describe the process of childbirth, including the three stages of labor.

12 Explain some of the problems associated with premature birth and the size of the newborn.

13 Give some of the causes of multiple births, and contrast the development of fraternal and identical twins.

14 Explain two major adjustments that the newborn must make.

15 Define lactation, and explain the role of hormones in the process.

16 Discuss the changes that occur during each of the major stages of postnatal life from birth to old age.

17 Describe several important methods for examining and diagnosing problems in the developing embryo and fetus.

18 Discuss the uses of intrauterine transfusion and fetal surgery.

The adult human body consists of about 50 trillion cells, most of which are specialized in terms of structure and function. In fact, cells are so specialized structurally that the liver cells of a human and a horse are more alike than are human liver cells and other types of human cells. Cells and organs are also programmed to carry out specific *functions*. If they are healthy and normal, they will never do anything else. For instance, *all* vertebrate livers do basically the same things. Clearly, structure and function are closely related.

All the many specialized cells of an adult body are derived from a single fertilized ovum, which contains all the genetic information necessary to produce a complete, functional human being. As the fertilized ovum, now called a *zygote*, divides again and again, the resultant cells begin to differ more and more from their original forms and from one another. How does a cell "know" when and how to differentiate? And if mitosis creates daughter cells genetically identical to the parent cell, how is their identity altered to produce specialized, different cells? So far, no one has found all the answers.

The development of a human being may be divided into **prenatal** ("before birth") and **postnatal** ("after birth") periods. During the prenatal period, the developing individual is called a *zygote* for the first week of its existence, an *embryo* for the next seven weeks, and a *fetus* from nine weeks until birth. It becomes an *infant* at birth.

The time of prenatal activity is called the **gestation period** (L. *gestare*, to carry). For as long as the new individual lives, it will never again experience such a dramatic burst of growth and development. During the nine-month gestation period, the weight increases from the fertilized ovum to the newborn baby about six *billion* times. In contrast, the body weight increases only about 20 times from birth to age 20. During the first month of prenatal development, the embryo increases in weight about a million percent. From that time on, the rate of growth continues to decrease, though there is a growth spurt during the third month.

Prenatal development is usually divided into two distinct periods: the **embryonic period,** from fertilization to the end of the eighth week, and the **fetal period,** from the beginning of the ninth week until birth. Tables 27.2 and 27.3 present the highlights of embryonic and fetal development, but in this section we will take an even closer look at the stages of intrauterine development.

EMBRYONIC DEVELOPMENT

When a sperm penetrates an oocyte, it sheds its tail and works its way into the oocyte's cytoplasm. Meiosis in the oocyte, which had stopped at metaphase of the second division, is resumed after entry of the sperm. When meiosis is complete, the second polar body is pinched off, and the remaining haploid structure is the *ovum*. The ovum and sperm nuclei then fuse to form the diploid *zygote* nucleus. This nucleus contains all the genetic material, DNA, which subsequent mitotic divisions will distribute equally to all cells of the embryo.

Cleavage

As you saw in Chapter 26, after a sperm penetrates an ovum, the ovum immediately develops an impenetrable coat, called the *fertilization membrane*, which prevents any other sperm from entering the ovum. Fertilization is complete when the haploid nuclei of the sperm and ovum (called *pronuclei*) fuse to form a diploid zygote with a complete set of 23 pairs of chromosomes. Then a series of mitotic cell divisions called *cleavage* begins (Figure 27.1).

Cleavage involves dividing each existing cell in the zygote into two. The first cleavage is complete about 36 hours after fertilization, and subsequent cleavages take place about twice a day. The two-celled zygote that results from the first cleavage is approximately 0.1 mm in diameter, still only about the size of the period at the end of this sentence. The daughter cells are called **blastomeres** (Gr. *blastos*, bud + *meros*, a part). After about 50 hours there are four cells, and mitotic divisions continue. Because the cells do not receive additional nutrients during these early cleavages, they do not grow as they divide. As each blastomere divides, it forms two cells that are each half the size of the original cell. As a result, the overall size of the zygote stays approximately the same as the original single cell (see Figure 27.1).

The first six to eight cleavages occur while the zygote is still enclosed within the zona pellucida. About day 3 after fertilization, a solid ball of about 8 to 50 blastomeres is formed into a mulberry-shaped **morula** (MORE-uh-luh; L. *morum*, mulberry tree). By this time, the morula is completing its 10-cm (4-in.) journey along the uterine tube, and is approaching the entrance to the uterus.

About day 4 to 5 after fertilization, the morula develops into a fluid-filled hollow sphere called a **blastocyst** (Gr. *blasto*, germ + *kystis*, bladder), with an inner cavity called the **blastocoel** (BLASS-toh-seel; Gr. *koilos*, hollow) (see Figure 27.1). The distinguishing feature of the blastocyst is that it has differentiated at one pole into an **inner cell mass** from which the embryo will form, and a surrounding epithelial layer called the **trophectoderm,** which is composed of cells called

Why doesn't the mother's body reject the implanting blastocyst the way it would reject any other foreign tissue?

The National Institute of Child Health and Human Development has studied this phenomenon in rabbits. The findings indicate that two proteins may be responsible for preventing rejection. One protein, uteroglobin (which has not yet been found in human cells), folds over the foreign antigens on the surface of the embryo and effectively masks them from the mother's immune system. The other protein, transglutaminase (which is present in human cells), is a blood-clotting factor that encourages the masking action of uteroglobin. A similar mechanism may exist in human beings.

THE DYNAMIC BODY

The Role of the Fetus During Childbirth

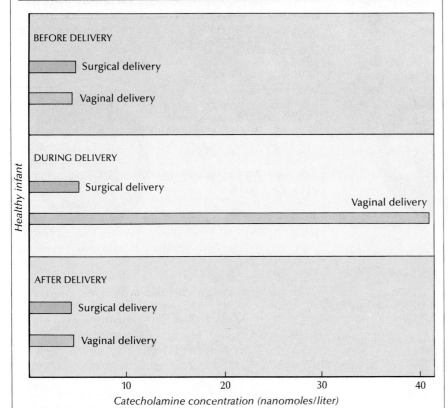

BEFORE DELIVERY
Surgical delivery
Vaginal delivery

DURING DELIVERY
Surgical delivery
Vaginal delivery

AFTER DELIVERY
Surgical delivery
Vaginal delivery

Healthy infant

10   20   30   40

Catecholamine concentration (nanomoles/liter)

The production of "stress hormones" associated with childbirth is highest during the actual delivery. Normal vaginal delivery elicits a far greater secretion than abnormal surgical (cesarean) delivery, which places relatively little stress on the fetus.

Until recently it was thought that the fetus played no active role in its own birth process. Now there is evidence that the fetus triggers the release of "stress" hormones during childbirth that help it survive the arduous process of childbirth and adjust to life outside its mother's uterus.

In addition to the pressure a fetus feels while passing through the birth canal, it is also deprived of oxygen (hypoxia) periodically when uterine contractions compress the umbilical cord and placenta. During such periods of stress, the fetus produces very high levels of epinephrine and norepinephrine, both of which are classified chemically as *catecholamines* (see graph). Catecholamines are generally produced to help an individual react favorably to incidents of extreme stress. For example, the secretion of epinephrine and norepinephrine allows the fetus to counteract hypoxia and other potentially harmful situations throughout most of the gestation period.

The stress situations are actually beneficial, since the presence of unusually high levels of catecholamines permits the newborn to adjust to new conditions outside its mother's uterus after delivery. Such postnatal adjustments include the ability to breathe, the breakdown of fat and glycogen into usable fuel for cells, the acceleration of heart rate and output, and the increase of blood flow to the heart, brain, and skeletal muscles.

The surge of catecholamines during childbirth also causes the newborn infant's pupils to become dilated, even when strong light is present. This alertness may help the infant to form an early bond with its mother.

The production of fetal catecholamines is a direct result of the adrenal glands' response to stress. In the adult, in contrast, the secretion of these hormones begins with the stimulation of the sympathetic nervous system. The adrenal glands of a fetus are proportionately larger than those in an adult, and fetuses and young children also have extra sources of norepinephrine in specialized tissues (known as paraganglia) near the aorta.

trophoblasts. The trophectoderm will later develop into the *chorion* (a fetal membrane that develops from trophoblasts) and become part of the membrane system that transports nutrients to the embryo and removes wastes from it (see Figure 27.6).

The blastocyst floats freely in the uterine cavity for one or two days while it continues to divide and grow, and is nourished by fluids secreted by the glands of the endometrium. Now the blastocyst is ready to shed its zona pellucida, allowing it to become attached to the maternal uterus. It will soon become embedded in the uterine wall in the process of **implantation.**

With the hormonal changes that accompany ovulation and the formation of the corpus luteum, the development of the endometrium continues. This building up of the uterine wall prepares the uterus to accept the blastocyst. An unfertilized ovum is not able to attach and implant itself into the endometrium.

[A] ZYGOTE BEFORE CLEAVAGE   ×200

Cleavage in an idealized embryo (not drawn to scale). (A) The single-celled zygote still has two polar bodies attached to it. Successive divisions result in two (B), four (C), eight (D) cells, and so on. Divisions continue to form a morula (E) and then a blastocyst (F, G). Cells resulting from these divisions are smaller than the zygote, and the whole blastocyst, consisting of many cells and a cavity (blastocoel), is scarcely larger than the one original zygote cell. [(A, G) *Biophoto Associates/Photo Researchers*. (B, C, E, *and* F) *Landrum B. Shettles, M.D.* (D) *Petit Format/Nestle/Photo Researchers*.]

Polar body — Nucleus

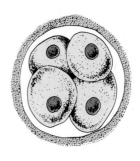

Zona pellucida — Blastomere   ×200

[B] 36 HOURS AFTER FERTILIZATION
    FIRST DIVISION, 2 CELLS

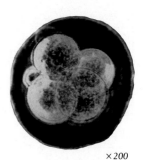

×200

[C] SECOND DIVISION, 4 CELLS

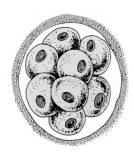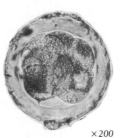

×200

[D] THIRD DIVISION, 8 CELLS

Zona pellucida

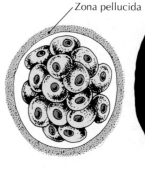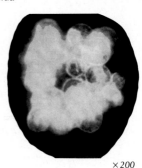

×200

[E] 3 TO 5 DAYS AFTER FERTILIZATION
    MORULA

Inner cell mass — Embryonic pole — Trophoblast — Degenerating zona pellucida

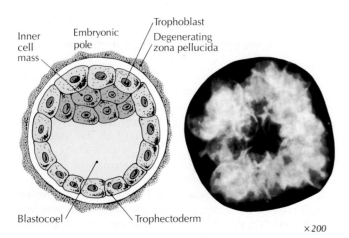

Blastocoel — Trophectoderm   ×200

[F] 5 TO 6 DAYS AFTER FERTILIZATION
    BLASTOCYST (EARLY)

Inner cell mass — Embryonic pole — Trophoblast

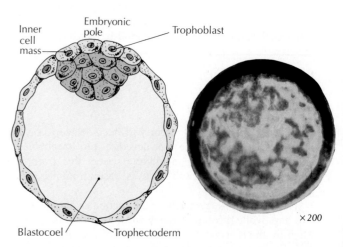

Blastocoel — Trophectoderm   ×200

[G] BLASTOCYST (LATE)

FIGURE 27.2

Implantation in the uterus. (A) About day 6 after fertilization, trophoblasts of the developing blastocyst attach to endometrial cells of the maternal uterine wall. (B) About day 7, implantation proper begins as the syncytiotrophoblast burrows into the endometrial epithelium as the blastocyst begins to move into the uterine wall. Hypoblasts of the inner cell mass will eventually differentiate into fetal tissues. (C) Implantation is almost complete about day 9 as the endometrial epithelium grows over the implanted blastocyst. The amnion, amniotic cavity, primitive yolk sac, bilaminar embryonic disk, and extraembryonic mesoderm are already present (see Figure 27.5).

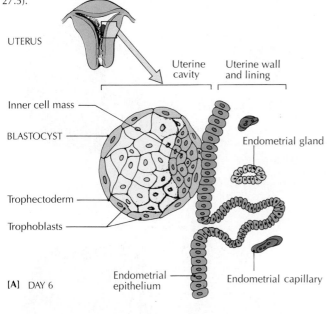

[A] DAY 6

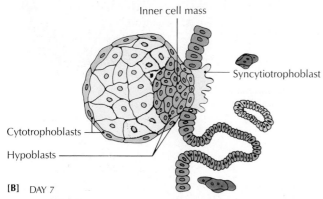

[B] DAY 7

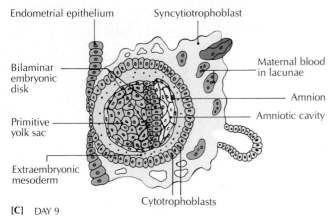

[C] DAY 9

Implantation in the Uterus

About day 6 to 7 after fertilization, implantation begins as the blastocyst attaches to the endometrium (Figure 27.2A). The inner cell mass faces toward the epithelium, while the trophectoderm actually attaches to the epithelium, usually to the upper, posterior wall of the uterus. At this point, the blastocyst is not yet completely implanted. Within 24 hours after the blastocyst becomes attached to the uterine wall, the trophoblasts differentiate into *cytotrophoblasts,* which surround the inner cell mass, and the *syncytiotrophoblast,* a large multinucleated structure. The syncytiotrophoblast implants the blastocyst in the uterine wall by invading the nutritious *inner* portion of the endometrium (Figure 27.2B). On about day 9, the blastocyst is completely enclosed by endometrial cells (Figure 27.2C), and it can continue to grow and develop. This complete implantation is the point at which *pregnancy* is considered to begin. Figure 27.3 shows an overall view of the processes of fertilization, cleavage, and implantation.

Maintenance of the endometrium Before implantation can begin, the trophoblasts must secrete the hormone **human chorionic gonadotropin (hCG),** which maintains the corpus luteum at a time when it would otherwise begin to degenerate. The corpus luteum continues its secretion of estrogen and progesterone, preventing menstruation and the breakdown of the endometrium, and allowing implantation to take place. Other physiological changes that facilitate implantation are induced largely by prostaglandins released by endometrial cells. The release of prostaglandins is triggered by some unknown ''messenger'' secreted by the blastocyst. Prostaglandins stimulate trophoblasts to secrete proteolytic enzymes that digest and liquefy endometrial cells, permitting the syncytiotrophoblast to invade the endometrium at the onset of implantation.

Eventually, the cytotrophoblasts and syncytiotrophoblast will form the *placenta* (see Figure 27.6), the organ that transfers nutrients from the pregnant woman to the fetus. In the meantime, the embryo receives nutrients from the many capillaries in the uterine wall.

As implantation continues during the second week, the inner cell mass changes shape, and the blastocyst assumes a flattened disk shape to form the **bilaminar** (two-layered) **embryonic disk.** These two layers are composed of endoderm and ectoderm. As the bilaminar disk is forming, several supporting membranes and other structures also develop. These are the amniotic cavity, yolk sac, body stalk, and chorion, which will be described in the next section. From the second week, when the bilaminar disk forms, until the end of the eighth week, the developing individual is called an **embryo.**

Development of Primary Germ Layers

Up to this point, repeated cell divisions have created hundreds of identical embryonic cells. However, once implantation begins, cell division stops temporarily. When cell division starts again, *all the cells produced are not alike.* **Differentiation,** the process by which cells develop into specialized tissues and organs, has begun. Soon the inner cell mass of

FIGURE 27.3

Overall view of fertilization, cleavage, and implantation. (A) Development from ovulation to implantation: (1) An ovum is released from the ovary when a mature follicle ruptures. The follicle becomes a corpus luteum. (2) The ovum is swept into the uterine tube. (3) Meiosis reduces the chromosome number to the haploid condition. A first polar body is formed. The corona radiata consists of follicle cells surrounding the mature ovum. (4) A sperm penetrates the ovum. A second meiosis occurs and a second polar body is formed. (5) The male and female nuclei fuse, forming the zygote. (6) The zygote undergoes its first cleavage division. (7, 8) Cells formed by cleavage divisions continue to divide until a morula is formed. (9) An early blastocyst is formed. (10) About four days after fertilization, a late blastocyst is formed. (11–13) About a week after fertilization, the blastocyst begins the process of implantation into the uterine wall. (14) Once implantation has been accomplished, the uterine wall starts contributing the outer portion of the placenta; the embryo itself contributes the inner part. (B) Photograph of the implanted blastocyst about six or eight days after fertilization. [(B) *Reprinted with permission from Weiss, Leon (ed.),* Histology: Cell and Tissue Biology, *fifth ed. New York: Elsevier Science Publishing Company, 1973. Photo by A. T. Hertig and J. Rock.*]

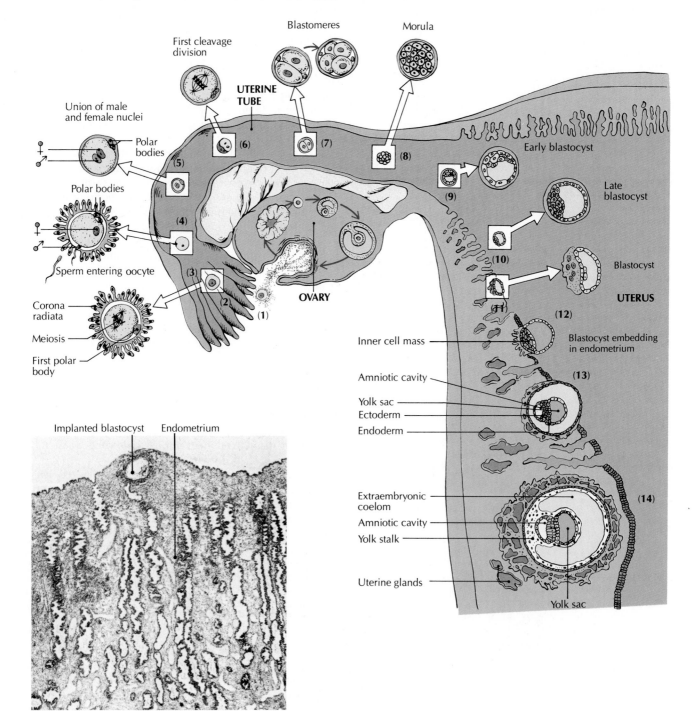

[B]   × 20

the blastocyst will be rearranged into three distinct *primary germ layers:* endoderm, ectoderm, and mesoderm (Figure 27.4 and Table 27.1).

The exact sequence of the formation of germ layers is not known, especially since the layers of ectoderm and endoderm form almost simultaneously. During the second week of embryonic development, the outer embryonic layer is formed. This is the *ectoderm* ("outside skin"), which will give rise to the outer layers of skin, hair, fingernails, tooth enamel, the nervous system, and other epithelial structures.

TABLE 27.1   MAJOR STRUCTURES DERIVED FROM PRIMARY GERM LAYERS

Endoderm	Mesoderm	Ectoderm

Epithelium of:
  Pharynx, larynx, trachea, lungs.
  Tonsils, adenoids.
  Thyroid, thymus, parathyroids.
  Esophagus, stomach, intestines, liver, pancreas, gallbladder; glands of alimentary canal (except salivary).
  Bladder (except trigone), urethra (except terminal male portion), prostate, bulbourethral glands.
  Vagina (partial), vestibule.
  Inner ear, auditory tubes.
	All muscle tissue (cardiac, smooth, skeletal), except in iris and sweat glands.
	
	All connective tissue (fibrous, adipose, cartilage, bone, bone marrow).
	
	Synovial and serous membranes.
	
	Lymphoid tissue:
	  Tonsils, lymph nodes.
	
	Spleen.
	
	Blood cells.
	
	Reticuloendothelial system.
	
	Dermis of skin.
	
	Teeth (except enamel).
	
	Endothelium of heart, blood vessels, lymphatics.
	
	Epithelium of:
	  Gonads, reproductive ducts.
	  Adrenal cortex.
	  Kidneys, ureters.
	  Coelom, joint cavities.
		All nervous tissue.
		
		Epidermis of skin; hair follicles, nails.
		
		Epithelium and myoepithelial cells of sweat glands, sebaceous glands, mammary glands.
		
		Lens of the eye.
		
		Receptor cells of sense organs.
		
		Enamel of teeth.
		
		Adrenal medulla.
		
		Anterior pituitary.
		
		Epithelium of:
		  Salivary glands, lips, cheeks, gums, hard palate.
		  Nasal cavity, sinuses.
		
		Lower third of anal canal.
		
		Terminal portion of male urethra.
		
		Vestibule of vagina and vestibular glands.
		
		Plain muscle of iris.

FIGURE 27.4

Formation of primary germ layers: ectoderm, endoderm, mesoderm.

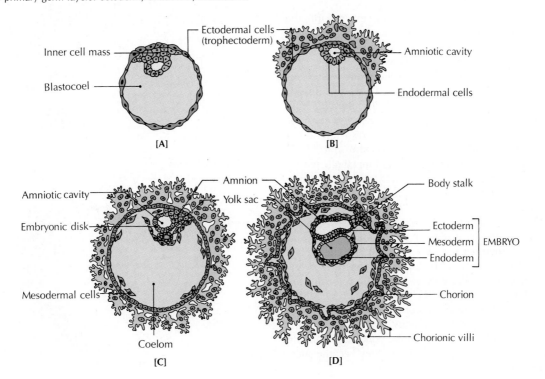

Also during the second week, a flattened layer of cells appears on the inner cell mass, facing the blastocoel. This layer of cells is the embryonic *endoderm* ("inside skin"). From this inner endodermal layer will develop the innermost organs, including the lining of the digestive tract and the glands and organs associated with it, the lining of the respiratory tract, and the lining of the bladder, urethra, and vagina.

The third week is a period of rapid development for the embryo, and it is also the time when the female will miss her menstrual period and suspect that she is pregnant. About the fifteenth day, the inner cell mass develops into the *trilaminar* (three-layered) *embryonic disk* as cells migrate from a linear band in the embryoblast called the *primitive streak.* These cells fill the area between the endoderm and ectoderm (the *coelom* or body cavity of the embryo) and eventually form the bulk of the embryo. This final layer is the *mesoderm* ("middle skin"), the forerunner of connective tissue in lower skin layers, bone, muscle, blood, and the epithelium of some internal organs (see Table 27.1).

Extraembryonic Membranes

During the first two weeks, the embryo does not have a functional circulatory system. During the implantation process, proteolytic enzymes released by the trophoblasts destroy some tiny maternal capillaries in the endometrium. Blood from these capillaries comes into direct contact with the trophoblasts, thus providing a temporary source of nutrition. In the third week, four membranes begin to form from tissues that were once the trophoblast. Because these membranes are not part of the embryo itself, they are called *extraembryonic* or *fetal membranes.* They include the yolk sac, amnion, allantois, and chorion.

As the trophoblast grows, it branches and extends into the tissue of the uterus. The two kinds of tissue, embryonic and maternal, grow until there is sufficient surface contact to ensure the adequate passage of nourishment and oxygen from the mother, as well as the removal of metabolic wastes, including carbon dioxide, from the embryo. After the embryo is implanted into the endometrium, small *chorionic villi* grow outward into the maternal tissue from the *chorion,* the protective sac around the embryo (Figure 27.5). The chorion makes up most of the placenta. It contains many villi, which allow the exchange of nutrients, gases, and metabolic wastes between mother and embryo. Eventually, the villi remain only where the embryo actually makes contact with the endometrium.

The *yolk sac* is a primitive respiratory and digestive system that is reduced by the sixth week of development to the thin *yolk stalk,* which then becomes incorporated into the umbilical cord via the body stalk. The yolk sac appears to be involved in transporting nutrients to the embryo during the second and third weeks, before the placental transfer is fully developed. It also is the initial site for the formation of blood cells until the liver assumes that responsibility during the fifth week. Finally, *primordial germ cells* in the yolk sac eventually migrate to the developing gonads, where they become spermatogonia or oogonia.

FIGURE 27.5

Development of extraembryonic membranes: chorion, yolk sac, amnion. (A) Week 3. (B) Week 4. (C) Week 10. (D) Week 20.

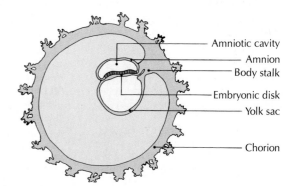

[A] WEEK 3

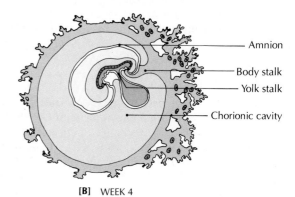

[B] WEEK 4

[C] WEEK 10

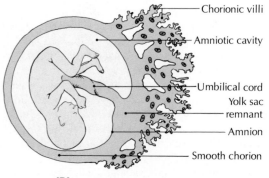

[D] WEEK 20

The **amnion** (Gr. sac) is a tough, thin, transparent membrane that envelops the embryo. Its interior space, the **amniotic cavity,** becomes filled with a watery **amniotic fluid,** which is made up of shed fetal epithelial cells, protein, carbohydrates, fats, enzymes, hormones, and pigments. Later, it also contains fetal excretions. The amniotic fluid suspends the embryo in a relatively shock-free environment, prevents the embryo from adhering to the amnion and producing malformations, permits the developing fetus to move freely (which aids the proper development of the muscular and skeletal systems), provides a relatively stable temperature for the embryo, and helps to dilate the mother's cervix at the start of labor.

The **allantois** (uh-LAN-toh-ihz; Gr. "sausage") usually appears early in the third week as a small fingerlike outpocket, or diverticulum, of the caudal wall of the yolk sac. It is involved with the formation of blood cells and the development of the urinary bladder, and its blood vessels become the vein and arteries of the umbilical cord. The allantois is eventually transformed into the median umbilical ligament, which extends from the urinary bladder to the navel.

Placenta and Umbilical Cord

The essential exchange mechanism between mother and embryo is in place by the beginning of the fourth week. At the site of implantation, the chorion joins intimately and intricately with the endometrium to develop into the **placenta** (L. flat cake, because of its disklike shape) (Figure 27.6). The placenta grows rapidly until the fifth month, when it is almost fully developed. A full-term placenta is about 2.5 cm (1 in.) thick and 20.5 cm (8 in.) in diameter. It weighs about 0.45 kg (1 lb).

The side of the placenta facing the fetus is relatively smooth, with the umbilical cord usually attached somewhere near the center. The side of the placenta that faces the mother has grooves and protuberances. The irregular surface increases the area for the interchange between fetal and maternal circulation. The surface area of the rough side is about 13 sq m (140 sq ft), more than three times greater than the smooth side. The connection between the fetal part of the placenta and the maternal part is close, with thousands of microvilli on the fetal part embedded in the maternal part, adding to the contact surface enormously. The fetal capillaries come close to the maternal capillaries to effect the exchange of substances. There is no actual blood flow between the fetal and maternal circulation, and there is no nervous connection.

The placenta has three main functions:

1 It transports materials between the mother and embryo, using simple diffusion, facilitated diffusion, active transport, and pinocytosis. The transported materials include *gases* (such as oxygen, carbon dioxide), *nutrients* (such as water, vitamins, glucose), *hormones* (especially steroids such as testosterone), *antibodies* (which bestow passive immunity), *wastes* (such as carbon dioxide, urea, uric acid, bilirubin), *drugs* (most drugs pass easily, especially alcohol), and *infectious agents* (such as rubella, measles, encephalitis, poliomyelitis, and AIDS viruses).

2 It synthesizes glycogen and fatty acids, and probably contributes nutrients and energy to the embryo, especially during the early stages of pregnancy.

3 It secretes hormones, especially the *protein hormones* human chorionic gonadotropin (hCG) and human chorionic somatomammotropin (hCS) and, with the cooperation of the fetus, the *steroid hormones* progesterone and estrogen.

The inner lining of the placenta is made up of the extensive blood vessels and connective tissue of the chorion. These blood vessels are formed from the embryo and are connected to the embryo by way of the **umbilical cord** (Figure 27.7). The umbilical cord (L. *umbilicus,* navel) is formed from the body stalk, yolk stalk, and other extraembryonic membranes during the fifth week. The cord contains two arteries, which carry carbon dioxide and nitrogen wastes from the embryo to the placenta, and a vein, which carries oxygen and nutrients from the placenta to the embryo. A gelatinous cushion of embryonic connective tissue surrounds the vessels of the umbilical cord. This resilient pad, together with the pressure of blood and other liquids gushing through the cord, prevents the cord from twisting shut when the fetus becomes active enough to turn around in the uterus. An umbilical cord is usually 1 to 2 cm (0.4 to 0.8 in.) in diameter and 50 to 55 cm (19 to 22 in.) in length.

There is normally no direct connection between the embryonic and the maternal tissue, at least no actual blood flow and no nerve connection. Sugars, water, oxygen, and hormones can cross the placental barrier, as can infectious agents, toxic substances (such as lead and insecticides), and drugs. Because these substances can pass into the fetal blood, the fetus can be infected, poisoned, or become addicted to drugs such as heroin. In fact, a newborn baby can show drug withdrawal symptoms if its mother used heroin during pregnancy. Nevertheless, the growing embryo is well insulated from most of the possibly harmful influences to which the mother is exposed. The placenta is eventually shed after the baby is born as part of the **afterbirth.***

Weeks 1 through 8: Formation of Body Systems

As already described, the first week of prenatal development is devoted to cleavage, while during the second week of prenatal life, the cells start to differentiate structurally and functionally, as the primary germ layers begin to take form. Within the developing chorionic villi, the first blood vessels are growing. Table 27.2 summarizes embryonic development.

Rapid development occurs during the third week. Differentiation of the endoderm, mesoderm, and ectoderm is complete, and the body, head, and tail can be distinguished. The primitive streak, a thickened dorsal longitudinal strip of ectoderm and mesoderm in the embryo, appears about day 15. The notochord also becomes apparent at this time. It is the dorsal, rodlike structure that runs the length of the embryo and serves as its internal skeleton. The neural tube will eventually form dorsal to the notochord and develop into the brain, spinal cord, and the rest of the nervous system (Figure

---

*The umbilical cord is usually disposed of after childbirth, but recently umbilical veins have been used successfully as grafts for bypass operations. The veins are removed from the cord, treated with a preservative, and reinforced with a polyester mesh before being used.

FIGURE 27.6

Development of the placenta. (A) The blastocyst is implanted in the wall of the uterus. (B) Enlarged view, showing the trophoblasts and the penetration of chorionic villi into the endometrium. (C) The trophectoderm differentiates into two layers, and chorionic villi reach the maternal blood supply. (D) The yolk stalk of the developing embryo will eventually become the umbilical cord. (E) At about 5 weeks, the umbilical blood vessels reach the embryo. (F) At about 8 weeks, the chorion and amnion form, and the placenta and umbilical cord are highly developed. The drawings are not meant to show actual sizes.

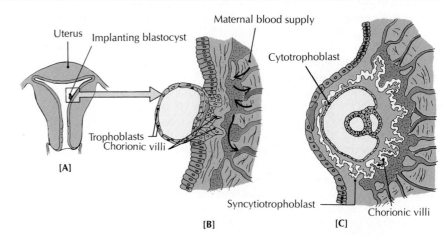

Uterus
Implanting blastocyst
Maternal blood supply
Cytotrophoblast
Trophoblasts
Chorionic villi
Syncytiotrophoblast
Chorionic villi
[A]
[B]
[C]

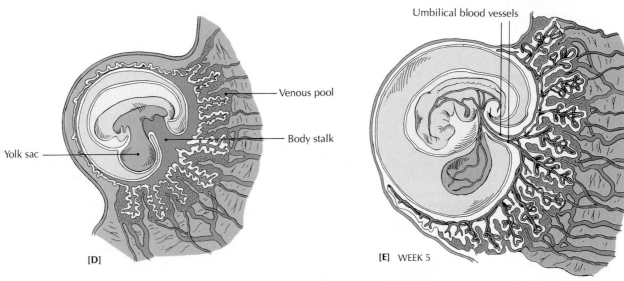

Venous pool
Body stalk
Yolk sac
[D]

Umbilical blood vessels
[E] WEEK 5

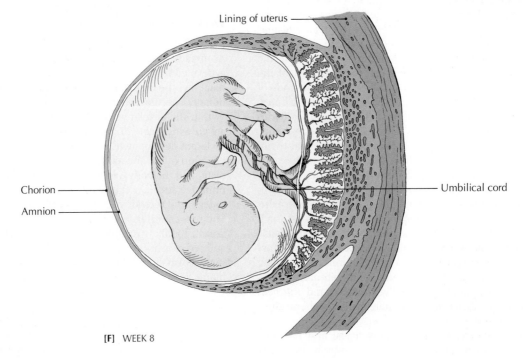

Lining of uterus
Chorion
Amnion
Umbilical cord
[F] WEEK 8

FIGURE 27.7

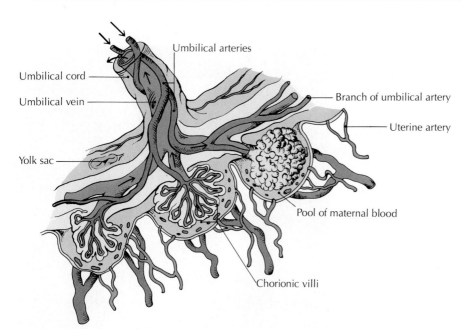

Umbilical arteries

Umbilical cord

Umbilical vein

Branch of umbilical artery

Uterine artery

Yolk sac

Pool of maternal blood

Chorionic villi

Umbilical cord and circulation in the placenta. A section of uterine wall is shown at the place where the umbilical cord is connected. The left and center microvilli are shown in a cutaway view to reveal the internal structure of the blood vessels. In addition to the exchange of respiratory gases, the placenta makes possible the elimination of metabolic wastes from the fetus through the umbilical arteries (downward arrows), and allows the entry into the fetus of foods, vitamins, and electrolytes through the umbilical vein (upward arrows).

TABLE 27.2   MAJOR DEVELOPMENTS OF THE EMBRYONIC PERIOD

Time period	Major developments
Week 1	Parental nuclei fuse during fertilization. Cleavage in uterine tube forms blastomeres. Morula is formed about day 3 and enters uterus. Blastocyst begins to implant on day 5 or 6.
Week 2	Ectoderm and endoderm begin to form. Bilaminar embryonic disk forms. Amniotic cavity, yolk sac, yolk stalk, chorion develop.
Week 3	Mesoderm begins to form. Primitive streak appears about day 15. Trilaminar embryonic disk forms. Notochord, neural tube, primitive body cavities, cardiovascular system form.
Week 4	Heart is beating. Upper limb buds, primitive ears are visible about day 26. Lower limb buds, primitive eye lenses appear about day 28. Three primary brain vesicles form.
Week 5	Head grows disproportionately as brain develops rapidly. Hand plates develop.
Week 6	Limb buds differentiate noticeably. Retinal pigment accentuates eyes.
Week 7	Yolk sac is reduced to yolk stalk. Limbs differentiate rapidly.
Week 8	Fingers, toes lengthen. Tail bud disappears. Embryo appears human. External ears are visible. External genitalia are visible, but are not distinctly male or female. The major organ systems are nearing completion.

FIGURE 27.8

Three-layered embryo after three weeks, posterior view. The drawing shows the actual size of the embryo. [*Landrum B. Shettles, M.D.*]

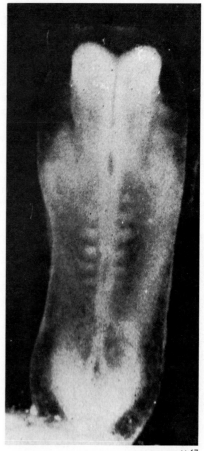

×47

27.8). The primitive body cavities and the cardiovascular system also take form. Villi continue to develop for an improved exchange between the embryo and mother. The embryo now measures about 2.5 mm (all measurements are from crown to rump).

By the fourth week, the embryo is C-shaped. The U-shaped heart has four chambers, and pumps blood through a simple system of vessels. Upper limb buds and primitive ears are visible about day 26. Lower limb buds and primitive eye lenses appear about day 28. The umbilical cord begins to develop. The intestine is a simple tube, and the liver, gallbladder, pancreas, and lungs begin to form. Also forming are the eyes, nose, and brain. The embryo is about 5 mm from crown to rump, about 10,000 times larger than the fertilized ovum. The relative size increase and the extent of physical change are greater in the first month than at any other time during gestation.

During the fifth week, the head grows disproportionately large as the brain develops rapidly. The forearm is shorter than the hand plate, and finger ridges begin to form about day 33. Primordial nostrils are present, and the tail is prominent (Figure 27.9). Primordial kidneys, the upper jaw, and the stomach begin to form. The nose continues to develop. The intestine elongates into a loop, and the genital ridge bulges. Primordial blood vessels extend into the head and limbs, and the spleen is visible. Spinal nerves are formed, and cranial nerves are developing. Premuscle masses appear in the head, trunk, and limbs. The epidermis is gaining a second layer. The cerebral hemispheres are bulging. At this stage, drugs taken by the mother, and such diseases as German measles may be transmitted to the embryo, affecting its development. The embryo measures about 8 mm.

By the sixth week, the components of the upper jaw are prominent but separate, and the lower jaw halves are fused. The limb buds differentiate noticeably, and the development of the upper limbs is more rapid than that of the lower limbs. The head is larger than the trunk, which begins to straighten

FIGURE 27.9

Embryo after five weeks. [*From Roberts Rugh, Landrum B. Shettles with Richard Einhorn,* From Conception to Birth: The Drama of Life's Beginnings. *New York, Harper & Row, 1971. Reprinted with permission from Donald D. Ritchie and Robert Carola,* Biology, *2nd ed. Reading, Mass., Addison-Wesley, 1983.*]

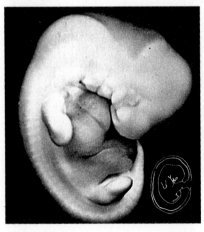

5–8mm

FIGURE 27.10

Embryo after six weeks. (The inset drawing shows the embryo at actual size.) [*Landrum B. Shettles, M.D.*]

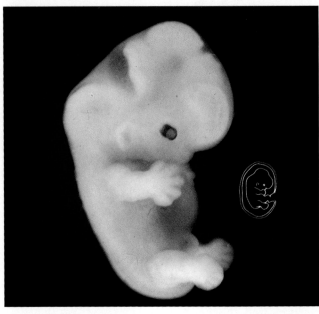

10–14mm

FIGURE 27.11

Embryo during the seventh week. [*From Roberts Rugh, Landrum B. Shettles with Richard Einhorn,* From Conception to Birth: The Drama of Life's Beginnings. *New York, Harper & Row, 1971. Reprinted with permission from Donald D. Ritchie and Robert Carola,* Biology, *2nd ed. Reading, Mass., Addison-Wesley, 1983.*]

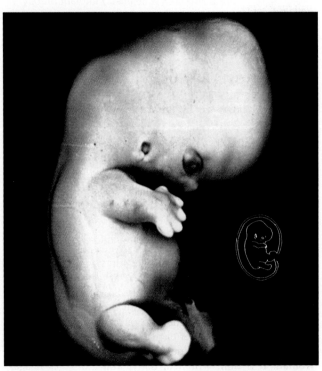

17–22mm

(Figure 27.10). The external ears appear, and the eyes continue to develop and become accentuated as the retinal pigment is added. Simple nerve reflexes are established. The heart and lungs acquire their definitive shapes. The embryo is about 12 mm (0.5 in.).

During the seventh and eighth weeks, the yolk sac is reduced to the yolk stalk. The face and neck begin to form, and the fingers and toes are differentiated. The back straightens, and the upper limbs extend over the chest (Figure 27.11). The tail is regressing, and the three segments of the upper limbs are evident. The jaws are formed and begin to ossify. The stomach is attaining its final shape, and the brain is becoming large. The muscles are differentiating rapidly throughout the body, assuming their final shapes and relations. The eyelids are beginning to form. At the end of the embryonic period, the embryo is about 17 mm (0.75 in.).

Ask Yourself

1 *What is cleavage, and when does it occur?*

2 *What is a morula? A blastocyst?*

3 *When does implantation take place?*

4 *What are the major differences among endoderm, mesoderm, and ectoderm?*

5 *What are extraembryonic membranes?*

6 *What is the difference between the chorion and the amnion?*

7 *What are the main functions of the placenta?*

8 *What is the vascular structure of the umbilical cord?*

9 *Describe the major developmental changes during the first eight weeks of prenatal life.*

FETAL DEVELOPMENT

Typically, the embryonic heart is already beating by the twenty-fifth day. By the end of the eighth week, all of the major external features (ears, eyes, mouth, upper and lower limbs, fingers, toes) are formed, and the major organ systems are nearing completion. Once this stage is reached, the embryo is referred to as a *fetus* (FEE-tuhss). The fetus, although only 3 cm (1.25 in.) long, looks recognizably human after two months (Figure 27.12). After three months in the uterus, the fetus is about 5 to 6 cm (2.0 to 2.4 in.) long, and contains all the organ systems characteristic of an adult.

The last six months of pregnancy are devoted to the increase in size and maturation of the organs formed during the first three months. (The rate of growth before birth is much faster than it is after birth.) By the time the fetus is 10 cm (4 in.) long, it can move and be felt by the pregnant woman. It is thin, wrinkled, hairy, and moist. As it ages,

FIGURE 27.12

Eight-week-old fetus in its amniotic sac after being removed from the chorionic sac. The fetus is actual size. [*Landrum B. Shettles, M.D.*]

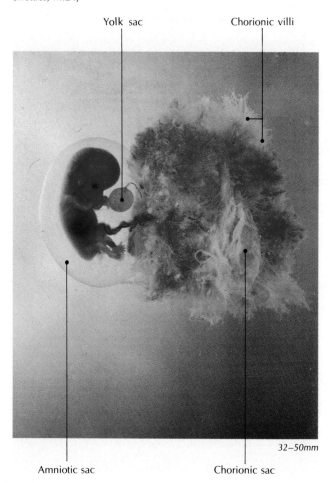

Yolk sac   Chorionic villi

32–50mm

Amniotic sac   Chorionic sac

the fetus loses most of its hair, its bones begin to ossify, it picks up fat, and it becomes mature enough to be born. It is said to have come *to term*. Table 27.3 (see p. 872) summarizes fetal development.

Third Lunar Month*

After 8 weeks the embryo is called a *fetus.* By the ninth week, its nose is flat, and the eyes are far apart. The tongue muscles are well differentiated, and the earliest taste buds form. The ear canals are distinguishable. The fingers and toes are well formed, and the head is elevating (Figure 27.13). The growth of the intestines makes the body evenly round. The small intestine is coiling within the umbilical cord, and intestinal villi are developing. The liver is relatively large, and the testes or ovaries are distinguishable as such (Figure 27.14). The

*Because the reproductive cycles of women are closer to the lunar (moon) month of 28 days than to the calendar month, fetal development is conventionally described in terms of lunar rather than calendar months.

FIGURE 27.13

Fetus after nine weeks. [*From Roberts Rugh, Landrum B. Shettles with Richard Einhorn,* From Conception to Birth: The Drama of Life's Beginnings. *New York, Harper & Row, 1971. Reprinted with permission from Donald D. Ritchie and Robert Carola,* Biology, *2nd ed. Reading, Mass., Addison-Wesley, 1983.*]

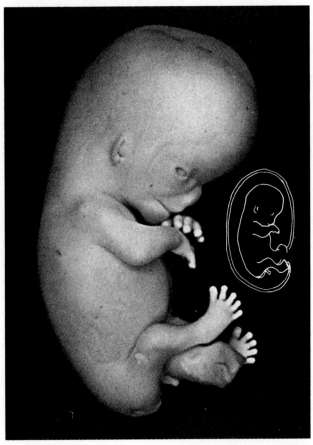

25–40mm

main blood vessels assume their final organization, and there is some bone formation. The muscles of the trunk, limbs, and head are well represented, and the fetus is capable of some movement. All five major subdivisions of the brain are formed, but the brain lacks the convolutions that are characteristic of later stages. The external, middle, and internal ears are assuming their final forms. The fetus is about 23 mm (1 in.).

All embryos appear female during early development because the genitals are not yet differentiated. Distinct male characteristics begin to differentiate through the action of male sex hormones (androgens) at about the sixth week of embryonic development. The differentiation of sexes is completed by the third month of fetal life (see Figure 27.14).

During the tenth to eleventh weeks, the head is held erect. The limbs are nicely molded (Figure 27.15), and the fingernail and toenail folds are indicated. The eyelids are fused, the lips have separated from the jaws, and the nasal passages are partitioned. The intestines withdraw from the umbilical cord, and the anal canal is formed. The kidneys begin to function, the urinary bladder expands as a sac, and urination is now

FIGURE 27.14

Development of the male and female reproductive organs.
(A) Development of the external genitals. The genitalia of the male fetus (left row) do not differ greatly from those of the female (right row) at 10 weeks. By 12 weeks in the female, the bud that derived from the genital tubercle is beginning to develop into the clitoris, the genital fold begins to develop into the labia, and the urogenital opening begins to divide into the separate openings for the urethra and vagina. By 12 weeks in the male, the bud is clearly developing into the penis, and the genital fold fuses over the urogenital opening to become the scrotum. The testes will begin to descend into the scrotum when the fetus is about 7 months old. At 34 weeks, male and female genitals look very much as they will at birth.
(B) Development of the internal reproductive systems. The male and female systems are still undifferentiated at 8 weeks (top), but have undergone distinctive changes by 10 weeks (middle row). The bottom row shows the internal male and female systems at birth.

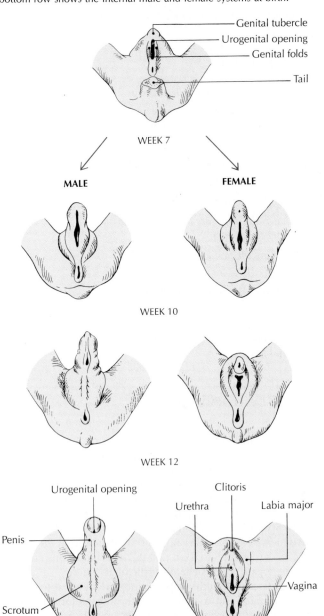

Genital tubercle
Urogenital opening
Genital folds
Tail

WEEK 7

MALE   FEMALE

WEEK 10

WEEK 12

Urogenital opening
Clitoris
Urethra
Labia major
Penis
Vagina
Scrotum
Anus

[A]   WEEK 34

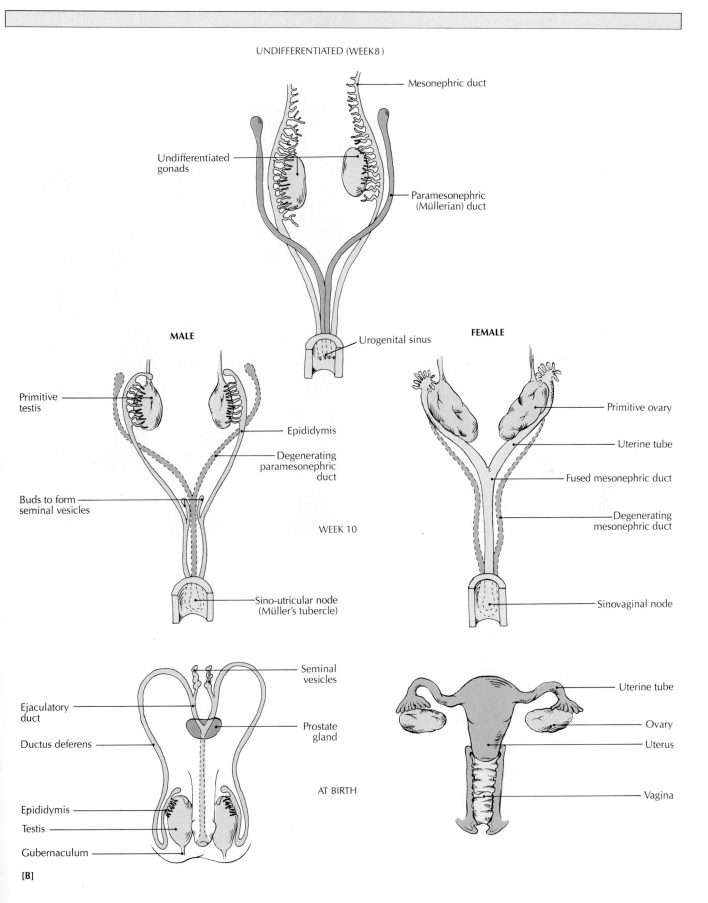

UNDIFFERENTIATED (WEEK 8)

Mesonephric duct

Undifferentiated gonads

Paramesonephric (Müllerian) duct

Urogenital sinus

MALE

FEMALE

Primitive testis

Epididymis

Degenerating paramesonephric duct

Buds to form seminal vesicles

WEEK 10

Sino-utricular node (Müller's tubercle)

Primitive ovary

Uterine tube

Fused mesonephric duct

Degenerating mesonephric duct

Sinovaginal node

Seminal vesicles

Ejaculatory duct

Prostate gland

Ductus deferens

AT BIRTH

Epididymis

Testis

Gubernaculum

Uterine tube

Ovary

Uterus

Vagina

[B]

possible. The vaginal sac and rudimentary sex ducts are forming. Early lymph glands appear, and enucleated red cells predominate in the blood. The earliest hair follicles begin to develop on the face, and tear glands are budding. During this period the spinal cord attains its definitive internal structure. A pulse is now detectable. The placenta is producing progesterone, which was formerly produced by the corpus luteum. The fetus is about 40 mm (1.5 in.) from crown to rump.

Fourth Lunar Month

The head is still dominant. The nose gains its bridge, tooth buds and bones form, the cheeks are represented, and the nasal glands form. The external genitalia attain their distinc-

TABLE 27.3 — MAJOR DEVELOPMENTS OF THE FETAL PERIOD

Time period	Major events
Weeks 9 to 12	Body growth accelerates as head growth slows down. Eyes close. Intestines develop in umbilical cord at week 9, in abdomen at week 10. External genitalia are distinguishable as male or female at week 12. Upper limbs are almost full length at end of week 12.
Weeks 13 to 16	Body growth is rapid. Legs lengthen. Skeleton ossifies rapidly. Scalp hair pattern is determined.
Weeks 17 to 20	Growth slows down. Lower limbs reach final proportions. Movement becomes active enough to be felt by mother. Downy body hair (lanugo) covers fetus at week 20. Eyebrows, head hair are visible. Brown fat forms.
Weeks 21 to 29	If born now, fetus might survive. Lungs, pulmonary blood vessels are highly developed. Central nervous system is mature enough to control breathing movements, body temperature. (Fetus is still not breathing on its own.) Eyes are open. Subcutaneous fat smooths out body wrinkles.
Weeks 30 to 34	Skin is pink and smooth. Upper and lower limbs begin to be chubby. White fat continues to increase. Testes are descending.
Weeks 35 to 40	Nails reach fingertips. Fetus still cannot hear. Some fetal circulatory passages close. Testes reach scrotum. Lanugo is shed. Fetal movements slow down, fetus turns head down. Maternal blood supplies antibodies. Placental blood vessels degenerate.
About 266 days from conception	Birth.

FIGURE 27.15

Fetus at about 10 weeks. [*From Roberts Rugh, Landrum B. Shettles with Richard Einhorn,* From Conception to Birth: The Drama of Life's Beginnings. *New York, Harper & Row, 1971. Reprinted with permission from Donald D. Ritchie and Robert Carola,* Biology, *2nd ed. Reading, Mass., Addison-Wesley, 1983.*]

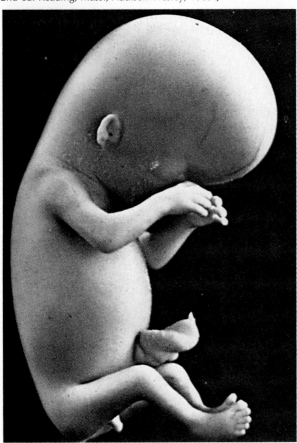

43–61mm

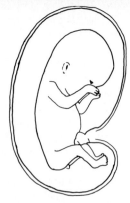

tive features. The lungs acquire their final shape, and in the male the prostate gland and seminal vesicles appear. Blood formation begins in the marrow, and some bones are well outlined. The epidermis is triple-layered, and the characteristic organization of the eyes is attained. The brain develops its general structural features (see the essay on p. 388 in Chapter 13). By the sixteenth week, all the vital organs are formed. The enlargement of the uterus can be felt by the mother, and the fetus measures about 56 mm (2.25 in.).

The Gender Gene

Until about the sixth week after fertilization, an embryo is not yet male or female. At that point, a genetic signal may occur that triggers the development of a male. Without this signal, the embryo develops into a female. It has been known for some time that a Y-bearing chromosome from the male parent produces a male child, and an X-bearing chromosome produces a female child, but until recently the exact mechanism of gender determination was unknown. Now David C. Page and his coworkers at the Whitehead Institute for Biomedical Research in Cambridge, Massachusetts, have proposed that sex determination depends on a single gene on the Y chromosome. The specific chromosomal segment is called *testis-determining factor (TDF)*.

Apparently, TDF regulates the activity of other genes (as yet unidentified) that set in motion a series of biochemical events that lead to the development of male primary sexual characteristics, and even differences in body systems. For example, it is well known that the nervous and immune systems are not identical in males and females.

TDF was discovered when Page and his team studied abnormal men with XX sex chromosomes and abnormal women with XY sex chromosomes. In these rare cases, where a person was genetically infertile but appeared normal, the XX men had a bit of a Y chromosome that contained TDF, and the Y chromosome of the XY women lacked TDF.

Fifth Lunar Month

At the beginning of the fifth lunar month the face looks "human," and hair appears on the head. Muscles become active spontaneously, and the body grows faster than the head. The hard and soft palates are differentiating, and the gastric and intestinal glands begin to develop. The kidneys attain their typical shape and plan. In the male, the testes are in position for their later descent into the scrotum. In the female, the uterus and vagina are recognizable as such. Blood formation is active in the spleen, and most bones are distinctly indicated throughout the body. Stretching movements by the fetus are now felt by the mother. The epidermis begins adding other layers to form the skin. Body hair starts developing, and sweat glands appear. The general sense organs begin to differentiate. The crown-to-rump measurement is about 112 mm (4.5 in.).

At the end of the fifth lunar month, the body is covered with downy hair called *lanugo*. The nasal bones begin to harden. In the female, the vaginal passageway begins to develop. Until birth, blood formation continues to increase in the bone marrow. Fetal heart sounds can be heard with a stethoscope, and the heart beats at twice the adult rate. The lungs are formed, but do not function. The gripping reflex of the hand begins to develop, and kicking movements and hiccupping may be felt by the mother. The fetus is about 160 mm (6.5 in.).

Sixth Lunar Month

The body is now lean and better proportioned than before. The internal organs occupy their normal positions, and the large intestine becomes recognizable. The nostrils open. The cerebral cortex now gains its typical layers. Thumb sucking may begin. Figure 27.16 shows the development of the hand from the fifth week to the eighth week.

FIGURE 27.16

Development of the hand. (A) At 5 weeks, the forearm is shorter than the hand, and finger ridges begin to form. (B) At 6 weeks, clearly delineated fingers have begun to develop. (C) At 7 to 8 weeks, the fingers, thumb, and fingerprints are well formed. [(A, B, and C) Courtesy of Carnegie Institute of Washington.]

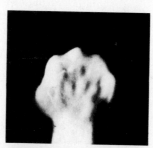

[A] 5 WEEKS

[B] 6 WEEKS

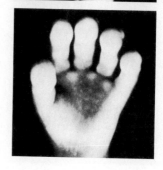

[C] 7–8 WEEKS

Seventh Lunar Month

The fetus is lean, wrinkled, and red (Figure 27.17). The eyelids open, and eyebrows and eyelashes form. Lanugo is prominent. The uterine glands appear. The scrotal sac develops and the testes begin to descend, concluding their descent in the ninth month. The nervous system is developed enough so that the fetus practices controlled breathing and swallowing movements. The brain enlarges, and cerebral fissures and convolutions are appearing rapidly. The retinal layers of the eyes are completed, so light can be perceived. If delivered at this stage, the baby would have at least a 10 percent chance of survival.

Eighth Lunar Month

The testes settle into the scrotum. Fat is collecting so that wrinkles are smoothing out, and the body is rounding. A sense of taste is present. The weight increase slows down. If delivered at this stage, the baby has about a 70 percent chance of survival.

FIGURE 27.17

Fetus at about 7 months. [*Landrum B. Shettles, M.D.*]

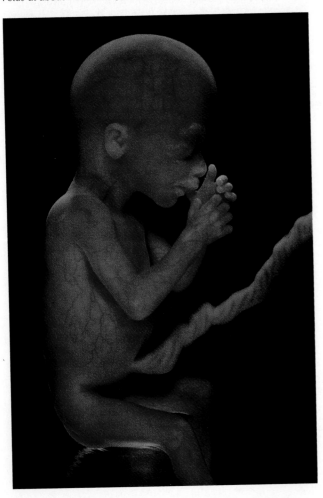

Ninth Lunar Month

The nails reach the tips of the fingers and toes. Additional fat accumulates.

Tenth Lunar Month

Pulmonary branching is only two-thirds complete, and the lungs do not function until birth. Some fetal blood passages are discontinued. The lanugo is shed. The digestive tract is still immature. The lack of space in the mother's uterus causes a decrease in fetal activity. Usually, the fetus turns to a head-down position (Figure 27.18). The maternal blood supplies antibodies. The placenta regresses, and placental blood vessels degenerate. The baby is now at *full term*, with a crown-to-rump measurement of about 350 mm (14 in.), and is ready to be born. (Measurements in the final months can vary greatly.) The newborn baby's eyes react to light, but will not assume their final coloration until about a month after birth. In developed countries about 99 percent of full-term infants survive.

Ask Yourself

1 *At what point is the developing individual considered to be a fetus?*

2 *When can the sex of the fetus first be distinguished?*

3 *At what point is the development of the hand nearly completed so that thumb sucking often begins?*

MATERNAL EVENTS OF PREGNANCY

The first sign of pregnancy is usually a missed menstrual period (*amenorrhea*) about three weeks after the coitus that resulted in conception. Because irregular menstruation may not be a drastic change for some women, amenorrhea is not a foolproof diagnostic tool. A test of the woman's urine may detect pregnancy about 12 days after the first missed period.

During the fifth or sixth week of pregnancy, some superficial veins may become prominent, the breasts may begin to enlarge and feel tender, and the areola may darken. The temporary nausea called "morning sickness" may begin about the seventh week, and "stretch marks" on the breasts may appear as the breasts continue to enlarge. At eight weeks, a physician can diagnose pregnancy through a physical examination by detecting an enlarged uterus and a softened cervix (Hagar's sign).

At the beginning of the ninth week, *colostrum* (a thin, milky fluid secreted by the mammary glands before milk is

FIGURE 27.18

Full-term fetus. Note that the drawing is actual size.

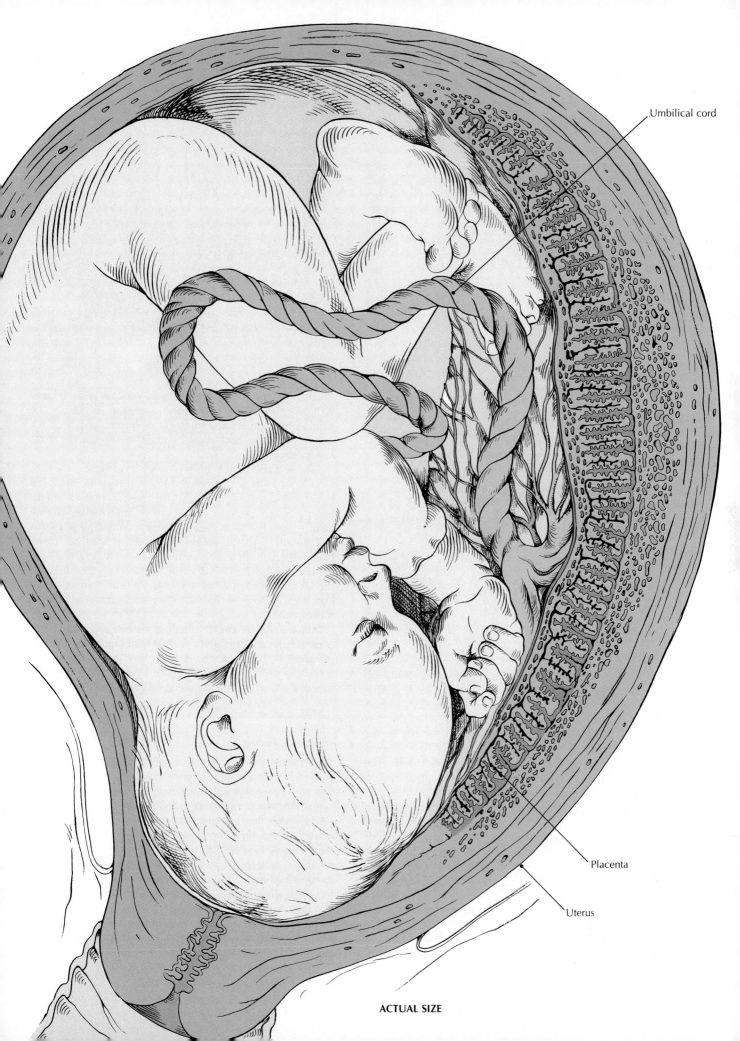

Umbilical cord

Placenta

Uterus

ACTUAL SIZE

FIGURE 27.19

(A) Maternal changes during pregnancy and shortly after childbirth. (B) The uterus usually returns to its nonpregnant size about six weeks after parturition.

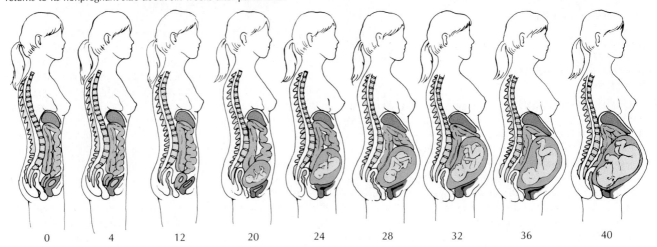

Weeks of pregnancy

[A]

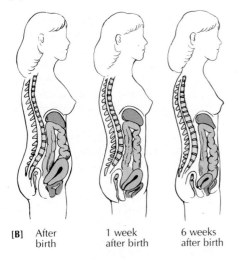

[B] After birth   1 week after birth   6 weeks after birth

released) may be squeezed from the breasts, and during the tenth week the physician may be able to feel weak uterine contractions. The nausea that started during the seventh week usually stops during the twelfth week.

During the sixteenth week, the abdomen usually begins to protrude as the uterus and fetus grow larger (Figure 27.19). The genitals may darken and turn bluish, and the forehead and cheeks may also darken in a pigmentation change called the "mask of pregnancy."

Some time between the seventeenth and twentieth weeks, the mother will probably feel the fetus moving in the uterus. By about the twenty-sixth week, the movements of the fetus may be noticeable from the outside. By the thirtieth week the abdomen protrudes considerably, the breasts are considerably enlarged, and an overall weight gain of almost 9 kg (20 lb) is not unusual. The cardiac output is now increased, and the heartbeat is accelerated. Many other changes are obvious, too. The pregnant woman tires more easily. Her hands feel clammy because of an increase of blood flow to them. A general feeling of hotness, increased thirst, and the need to

urinate more frequently are common. Varicose veins or hemorrhoids may develop as venous pressure increases. The woman may experience shortness of breath, and her back may ache as the back muscles strain to support the extra abdominal weight.

As the final month of pregnancy approaches, the ribs spread out to accommodate the lungs as they are displaced by the still-expanding uterus. As the room for internal organs becomes scarce, pressure may be put on some nerves, causing pain. At this point, or even earlier, the attending physician will usually urge the pregnant woman to limit her travel plans, especially air travel, because of reduced oxygen at high altitudes. In the thirty-sixth week, four weeks before delivery, the uterus has risen to the rib level, and the woman has to lean back to keep her balance. Uterine contractions become more frequent.

About 266 days after fertilization, uterine contractions become regular and strong, the uterine membranes rupture and the amniotic fluid is discharged, the cervix dilates, and after a variable period of labor, the baby is born. The placenta is expelled about 30 minutes later, and pregnancy is over.

Pregnancy Testing

The external signs of pregnancy are significant, but not always dependable, and a woman usually does not conclude that she is pregnant until laboratory tests confirm her suspicions. Pregnancy may be detected by some tests as early as 12 days after the first missed menstrual period, but tests are most accurate after the fortieth day of pregnancy. Most of these tests depend on the fact that *human chorionic gonadotropin* (hCG), a hormone produced by the placenta, is abundant in the blood and urine of a pregnant woman. The level of hCG is highest during the first three months of pregnancy and then decreases.

In testing, a drop of a chemical that neutralizes hCG (anti-hCG) is combined with a drop of the woman's urine on a glass slide, and a minute later, latex rubber particles are added. If the woman is not pregnant, the anti-hCG will attach itself to the latex particles, and curdlike lumps will form. If the woman *is* pregnant, the hCG will be neutralized by the anti-hCG before the latex particles are added, and no hCG will be left to form lumps.

Importance of the Maternal Environment

It has long been thought that oxygen passed through the placenta from the maternal blood to the fetus by passive diffusion. It now seems that molecules of oxygen may be shuttled across the placenta at an unusually fast rate by means of an unidentified carrier molecule. It also seems that certain chemicals, such as anesthetic gases, insecticides, tranquilizers, and the carbon monoxide in cigarette smoke, may interfere with the efficiency of the carrier. Fetuses that are deprived of this facilitated oxygen transport may develop into underweight children with a susceptibility to developmental defects. Severe oxygen deprivation may also cause miscarriages. It has been known for some time that female anesthetists and women who smoke heavily during pregnancy have more miscarriages and underweight children than do other pregnant women.

Probably the most dramatic evidence of how substances in the pregnant woman's body can affect her unborn child occurred in Europe during the early 1960s. A drug called *thalidomide* was put on the market in 1960 to relieve "morning sickness." Within two years, about 7000 children whose mothers had taken the drug during early pregnancy were born with serious birth defects. Like many other substances that get into a pregnant woman's circulatory system, thalidomide passed through the placenta into the body of the developing embryo. Depending on when the drug was taken, it interfered with the development of different parts of the embryonic body. Most of the damage was done between the thirty-fifth and fiftieth days of pregnancy, a critical developmental period. If taken early in this period, deformities of the ears resulted. If taken somewhat later, there was a severe shortening of the arms and then a deformation of the legs. By the end of the period, it affected only the thumbs.

Substances that chemically interfere with development and cause gross deformities without changing the embryonic DNA are called **teratogens** (Gr. *teras,* monster). Exposure to teratogens within the first two weeks of pregnancy usually causes the death of the embryo through a spontaneous abortion. Death may still result from the fifteenth to sixtieth days, but physical deformities are a more common effect. Later exposure, after the rudiments of all the essential organ systems have been formed, usually produces relatively minor abnormalities.

Two of the most common dangers for the unborn child are *smoking* and *alcoholic consumption* by the mother. Studies have shown that babies born to women who smoke during pregnancy weigh less than average. Smoking during pregnancy also appears to be related to a higher-than-normal incidence of miscarriage. Interestingly, the babies were lightest when *both* parents smoked, indicating that the mother was breathing in some of her husband's exhaled smoke and passing it on to the fetus. Smoking also reduces the oxygen supply of the fetus and may hinder mental development. Pregnant women who smoke a pack of cigarettes a day are twice as likely to give birth prematurely as nonsmoking pregnant women.

As for alcohol, the U.S. Surgeon General has warned that "alcohol consumption during pregnancy, especially during the early months, can harm the fetus." Women who drink heavily may produce children with a set of defects known collectively as **fetal alcohol syndrome,** which includes mental retardation, growth deficiencies, and abnormal facial features. The children may also have abnormal joints and heart disease. Presumably, the risks are less for women who consume small quantities of alcohol, but such problems as miscarriage and underweight babies still persist.

X rays are avoided during pregnancy since *radiation* can damage the fetal nervous system and the eyes, and may cause mental retardation. Transplacental infection of the fetus by the rubella (German measles) virus can cause cataracts and deafness and other ear defects. The antibiotic tetracycline may cause fetal tooth defects if taken by the mother during the final six months of pregnancy, and may cause cataracts if taken in large quantities.

The maternal environment can even affect the fetus as it is being born. For example, sexually transmitted diseases may be passed on to the baby as it moves through the birth canal. Also, if a woman has a very narrow pelvic bone structure the baby may be unable to fit through. In cases where venereal disease is suspected or the pelvis is too narrow, a *cesarean section* is performed, and the baby is lifted out through the abdominal incision.

Most physicians urge their pregnant patients to avoid *drugs* of all kinds during pregnancy. Drugs such as heroin, cocaine, and morphine may actually cause the fetus to be addicted to those drugs at birth, but even such supposedly safe drugs as aspirin, antibiotics, antihistamines, and tranquilizers may cause problems ranging from miscarriage to physical deformation.

Ask Yourself

1 *What is usually the first sign of pregnancy?*

2 *How is human chorionic gonadotropin related to pregnancy tests?*

3 *When is the fetus most susceptible to environmental factors?*

4 *How is a fetus affected when its mother smokes or drinks heavily during pregnancy?*

BIRTH AND LACTATION

Childbirth, or **parturition** (L. *parturire,* to be in labor), usually occurs about 266 days after fertilization, or about 280 days from the first day of the menstrual period preceding fertilization. During the last month of pregnancy, the myometrium of the uterus becomes increasingly sensitive to the hormone oxytocin, and a pregnant woman usually begins to experience irregular uterine contractions at that time.

Although the sensitivity to oxytocin may be caused by either an increased level of uterine prostaglandins or a decreased level of progesterone and estrogens, the exact cause is not known. The signal for the initiation of labor may come from the fetus itself, but no firm evidence supports this idea. One theory for the onset and maintenance of labor is that the fetus finally reaches a point in its development where its head begins to press down on the cervix. This pressure causes the cervix to dilate, which produces uterine contractions and the secretion of oxytocin. Oxytocin stimulates further uterine contractions until the baby is finally pushed down past the cervix and out through the vagina.

Process of Childbirth

The initial contractions stimulate the liberation of oxytocin from the pituitary gland (neurohypophysis), which further stimulates even more powerful uterine contractions. Waves of muscular contractions spread down the walls of the uterus, forcing the fetus toward the cervix (Figure 27.20). By now, the cervix is dilated close to its maximum diameter of about 10 cm (4 in.).

The amniotic sac may burst at any time during labor, or it may have to be ruptured by the attending physician. Either way, the ruptured sac releases the amniotic fluid. If this loss of fluid occurs very early, it may signal the onset of labor. Another possible indication that labor has begun is the release of the cervical plug of mucus from the vagina. Ordinarily, either the bursting of the amniotic sac or the release of the cervical plug (termed "show") will occur early enough to provide ample time to prepare for childbirth, but great variations exist among pregnant women.

During pregnancy, the muscle cells of the uterus grow to as much as 40 times their former size, transforming the uterus into an enormously powerful muscle. The sturdy walls of the uterus harbor and nourish the fetus during pregnancy, and finally, through the muscular contractions during labor, the uterus forcefully expels the fetus outward through the vaginal canal.

Babies are born head first, in the so-called *cephalic position,* about 95 percent of the time. Other possible birth positions almost always produce complications that require the intercession of the attending physician. Remember that the skull bones of a newborn are not yet fused, and the skull is still pliable. If it were not, the baby's head would not be small enough to pass through the vaginal canal.

After the baby is born, it is still attached to the placenta by the umbilical cord. Immediately after the baby is expelled from the uterus, the umbilical cord is clamped and cut below the clamp. Ordinarily, the umbilical cord is cut immediately after birth, but some physicians believe that the severing of the cord should take place only after the afterbirth is expelled and all of the placental blood drains into the baby's circulation. Such a procedure gives the baby an additional 80 to 90 mL of blood, about 25 to 30 percent more than it would receive otherwise.

FIGURE 27.20

Birth of a baby. The internal events are shown here in a series of six models. (Photographs of live births show only external events.) [The Birth Atlas, *Maternity Center Association.*]

[A]

[D]

Human fetuses are born long before they are able to care for themselves—in a sense, they are always born prematurely. But if gestation continued for longer than 9 or 10 months, the baby's relatively large head (and brain) could not pass through the vaginal canal.

Three stages of *labor* are usually described:

1 The *first stage of labor* starts with regular uterine contractions. Contractions occur every 20 or 30 minutes at first, and become more and more frequent until they occur every 2 or 3 minutes. The first stage usually lasts about 14 hours for the birth of a first child, and gets shorter for subsequent births. Its main function is to dilate the cervix to its maximum.

2 The *second stage of labor* is the actual birth of the child. It lasts anywhere from several hours to several minutes, with

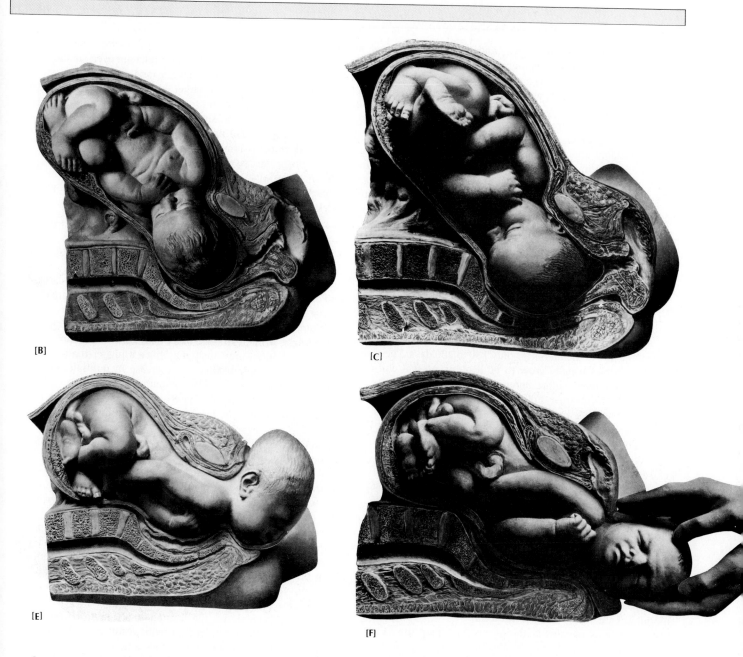

[B]

[C]

[E]

[F]

the average time being two hours. If the fetal membranes have not ruptured during the first stage, they do so now.

   3 The *third stage of labor,* which takes about 20 minutes, is the delivery of the placenta, fetal membranes, and any remaining uterine fluid, collectively called the *afterbirth.* It usually follows within 30 minutes of the expulsion of the baby, and is accomplished by further uterine contractions. These final contractions also help to close off the blood vessels that were ruptured when the placenta was torn away from the uterine wall. About half a liter (1 pt) of blood is usually lost at this time.

   The period after the placenta has been delivered is called the *puerperal period* (L. *puerperus,* bearing young; *puer,* child + *parere,* to give birth to). It is the time when the mother's body reverts back to its nonpregnant state, and it usually lasts

at least a month. For example, the uterus and vagina both revert back to their normal sizes about six weeks after childbirth (see Figure 27.19). If a woman is not breastfeeding her baby, it will take anywhere from 6 to 24 weeks for the first menstruation.

Why are some babies born prematurely?	*Babies can be born prematurely for many reasons, including fetal malformations and genetic defects, but the most common cause is the mother's poor health. Maternal factors include malnutrition, alcohol or drug abuse, smoking, and hypertension.*

Some Interesting Birth Facts

More babies are born in August, and fewer in April, than any other months.

The most common birth time is between 3 A.M. and 4 A.M., and the least common time is 3 P.M.

Babies born during the first six months of the year have a higher rate of infant mortality and congenital defects than those born in the latter half of the year.

For every 100 girls born, 106 boys are born. But childhood death is more common among boys, and by adolescence the numbers have evened out.

On the average, girl babies are carried one day longer than boys. The average length of pregnancy for white women is 265 days after fertilization for baby boys and 265.9 for girls. Black women carry their male offspring for 259.3 days on the average and their female offspring for 260.7 days. It may seem that the shorter gestation time for black children is reasonable because black newborns are usually smaller than white ones, but the newborns in India are even smaller, and they have a longer gestation time than the larger white children.

A normal woman, with a normal pregnancy, gains about 12.5 kg (27.5 lb) during pregnancy. Of those 12.5 kg, about 4 kg cannot be accounted for. The average weight of a newborn child is 3.3 kg (7 lb 4 oz).

On the average, twins are born 19 days earlier than a single child.

Premature and Late Birth

On the average, pregnancy lasts for about 266 days after fertilization, but many variations occur, and babies have survived after being in the uterus for only 180 to 200 days. These *premature* babies usually grow to be healthy, normal children, but they may require about three years to catch up developmentally to their full-term contemporaries. Babies born after 240 days are not considered to be premature as long as they weigh at least 2.4 kg (5.5 lb).

Premature babies are usually poorly proportioned compared with healthy full-term babies. They usually have a weak cry, poor temperature regulation, wrinkled and dull red skin, closed eyes, and short fingernails and toenails. Because their muscular system is not fully developed, their movements are labored, and breathing may be difficult. Yet, a baby born a month early has at least a 70 percent chance of survival. It is interesting that a 1.8-kg (4-lb) baby born after 34 or 35 weeks, and nursed by its mother, has a better chance of being normal than a full-term baby of the same weight. This is so because when a baby is born prematurely its mother's milk contains more protein, antibodies, sodium, and chloride than it would have at full term.

When a fetus remains in the uterus for longer than the normal full term, the attending physician may decide to induce labor with intrauterine injections of oxytocin or other methods, or to perform a cesarean section. Many physicians prefer to let the baby be born whenever it is ready, as long as there are no complications. Babies born late who weigh more than 4 or 4.5 kg (9 or 10 lb) usually are not as healthy as normal-sized full-term babies, probably because the mother's body cannot provide adequate nourishment in such circumstances.

Multiple Births

Multiple births are always an exception in human beings. Twins are born once in 86 births, triplets once in 7400 births, and quadruplets once in 635,000 births. The chances of quintuplets being born are about 1 in 55 million. In most cases, one or more of the children born as quintuplets or sextuplets do not survive beyond infancy.

Women who have taken "fertility drugs" (which stimulate the ovaries), women who are older than 35, and women who have had children previously may have a higher-than-average rate of multiple births. This increase in multiple births may be caused by irregular ovulation patterns in older women and the resultant release of more than one ovum at a time. Apparently, the tendency to release more than one ovum at a time is an inherited trait. However, the cell separation that produces identical twins is not due to a hereditary factor, and its cause is not known.

Twins may be identical or fraternal. *Fraternal (dizygotic) twins* are formed when more than one ovum is released from the ovary or ovaries at the same time and two ova are fertilized (Figure 27.21A). Each fraternal twin usually has its own placenta, umbilical cord, chorion, and amnion. Fraternal twins are the most common multiple births. About 70 percent of all twins are fraternal, and 30 percent are identical. Aside from being the same age, fraternal twins resemble one another no more than any other brothers or sisters, and may be the same sex or different sexes.

Figure 27.21B shows the formation of *identical (monozygotic) twins.* Because identical twins result from the same fertilized ovum, they are always the same sex, and are genetically identical. Apparently, identical twins are formed when a fertilized ovum divides into two identical cells before implantation (about day 6 or 7), but there is also some evidence that identical twins may also form after implantation. Because human identical twins develop from a single fertilized ovum, they are contained within a common chorionic sac and have a common placenta. However, the umbilical cords are separate and the amnions are also individual, except in rare instances.

Triplets may be either identical or fraternal. In the case shown in Figure 27.21c, more than one ovum was fertilized, creating fraternal male and female twin embryos, and then the male fertilized ovum split apart into two identical embryos, forming the identical boys. Triplets may also be formed in other ways.

FIGURE 27.21

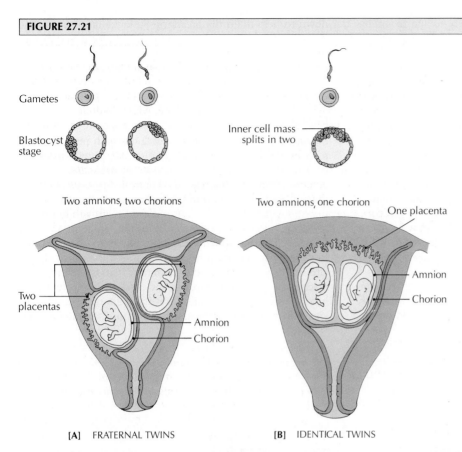

Multiple births. (A) Fraternal twins result when two ova are fertilized. Each usually has its own placenta and chorion. (B) Identical twins result when the inner cell mass of one fertilized ovum divides in half. They share the same placenta and chorion. (C) The triplets in the photo consist of identical twin boys and their fraternal "twin" sister. [(C) *Courtesy of Marc Anthony DeGruccio. Reprinted with permission from Donald D. Ritchie and Robert Carola,* Biology, *2nd ed. Reading, Mass., Addison-Wesley, 1983.*]

Two amnions, two chorions

Two amnions, one chorion

[A] FRATERNAL TWINS

[B] IDENTICAL TWINS

[C]

Conjoined twins, commonly called *Siamese twins,* * are identical twins whose embryonic disks do not separate completely. Sometimes the twins share only skin or tissue, but frequently they share an organ, such as a liver or an anus. Siamese twins usually die before they reach old age. They are

more often females than males, and are more common than quintuplets.

Adjustments to Life Outside the Uterus

Most of the body systems of the fetus are developed and ready to function, but they are not used until the baby is born. Although the sudden shift from total dependency in the uterus to near-independence upon birth may seem drastic, the newborn, or ***neonate,*** usually adapts to its new environment smoothly and without apparent trauma. The major changes affect breathing and circulation.

*In 1811, Eng and Chang were born as conjoined twins in Bangkok, the capital of Siam (now Thailand). They shared a liver, and were joined together at the lower end of the sternum. When they were 18, they were taken to the United States by P. T. Barnum, who made the twins famous in his circus. Eng and Chang (which mean "left" and "right" in Siamese) were not actually Siamese, since their parents were Chinese. The Siamese called them the "Chinese twins."

At birth, the baby must adapt quickly to life in air by making several rapid physiological and anatomical adjustments. It must shift from a fetal circulation, which depends on the placenta, to an essentially adult circulation, which depends on the lungs for gas exchange. Before birth, the fetal heart has an opening, the *foramen ovale,* between the left and right atria that allows blood from the venous and arterial systems to mix. Also, the *ductus arteriosus* carries blood from the pulmonary artery to the aorta (see Chapter 19), thus bypassing the lungs (see Figure 19.12).

Both the atrial opening and the pulmonary by-pass close at birth, allowing the blood to pass through the pulmonary artery to the newly functioning lungs. If either the foramen ovale or ductus arteriosus does not close off, the oxygen content of the blood will be low, and the baby's skin will appear slightly blue. A neonate with this condition is therefore called a "blue baby." Usually a defective ductus arteriosus must be corrected surgically a few days after birth.

The start of actual breathing can be helped if it does not begin immediately at birth. Fluid can be cleaned from the nose and mouth of the neonate, the baby can be held upside down to help drain the breathing passage, and if necessary, it can be given a smack on its bottom to shock it into crying, its first gasp of air. Today it is more common to stimulate breathing with a small whiff of carbon dioxide.

Lactation

Lactation (L. *lactare,* to suckle) includes both production of milk by the mammary glands and release of milk from the breasts. Milk production is controlled by *prolactin,* from the anterior pituitary. Milk release is controlled by *oxytocin,* from the posterior pituitary. During pregnancy, the breasts enlarge in response to increasing levels of estrogen, growth hormone, prolactin, adrenal glucocorticoids, insulin, progesterone, and human chorionic somatomammotropin from the placenta. But the most important hormone is prolactin, which stimulates milk production in the mammary glands. Before birth, placental secretions of estrogen and progesterone actually inhibit the production of milk from the mammary glands, but after the placenta has been expelled from the uterus, and

FIGURE 27.22

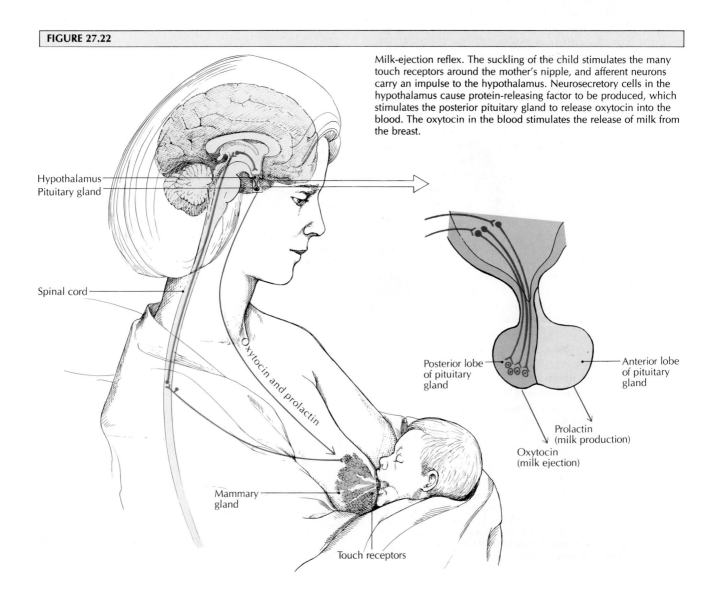

Milk-ejection reflex. The suckling of the child stimulates the many touch receptors around the mother's nipple, and afferent neurons carry an impulse to the hypothalamus. Neurosecretory cells in the hypothalamus cause protein-releasing factor to be produced, which stimulates the posterior pituitary gland to release oxytocin into the blood. The oxytocin in the blood stimulates the release of milk from the breast.

estrogen and progesterone levels drop, the breasts begin to produce copious amounts of milk. The continued production of milk requires most of the mother's other hormones, but most important are growth hormone, cortisol, and parathyroid hormone. These are necessary to provide the amino acids, fatty acids, and calcium required for milk production.

The actual release of milk from the mother's breasts does not occur until one to three days after the baby is born.* During those three days, the suckling baby receives *colostrum,* a high-protein fluid that is present in the breasts at birth. Colostrum may help strengthen the baby's immune system, and may also function as a laxative, removing fetal wastes called *meconium* (mucus, bile, and epithelial threads) from the intestines. Not only does the mother's milk introduce disease-fighting antibodies into the baby's body, but it also contains hormones that aid the proper development of the digestive system, stimulating the growth of cells that line the stomach and intestines.

The level of prolactin from the pituitary continues to rise in the mother's blood from the fifth week of pregnancy until birth. Following birth, the level returns to normal quickly. However, each time the mother nurses the infant nervous signals from the sucking action on the nipples pass to the mother's hypothalamus, and cause a surge of prolactin release that lasts about an hour (Figure 27.22). This prolactin acts on the mammary glands to provide milk for the next nursing period. The infant's sucking stimulates the hypothalamus, which in turn stimulates the posterior pituitary to secrete oxytocin. The oxytocin causes the smooth muscles in the breast to contract, which expels the milk

Breasts that are not emptied soon cease to produce milk. It also appears that the nursing of the baby helps the uterus return to its normal size and shape by the suppression of gonadotropic and ovarian hormones. Because of this stimulation by the suckling infant, lactation may be prolonged for two years or more if nursing continues, although the rate of milk production decreases considerably within seven to nine months.

On the average, nursing mothers produce about 1 L of milk each day. Multiple births result in larger secretions, usually about 1 L for each baby engaged in suckling. Ordinarily, milk production is greatest when the breasts enlarge during pregnancy. The initial enlargement of the mammary glands and tissues occurs only after the third month of pregnancy.

Although menstruation may not take place while the mother is nursing her baby, ovulation usually occurs about 50 percent of the time anyway. The baby's sucking action on the nipple stimulates the secretion of prolactin. Prolactin, in turn, inhibits the secretion of GnRH from the hypothalamus, which suppresses the release of pituitary gonadotropic hormones (FSH, LH). As a result, the levels of estrogen and progesterone drop, and ovulation is often inhibited. Despite tales to the contrary, a woman can indeed become pregnant during the lactation period. Nursing is certainly not an effective method of birth control. Women who do not nurse their babies usually menstruate about six weeks after childbirth.

*Abortions after the fourth month of pregnancy also stimulate a surge of milk production and secretion, indicating that the expulsion of the fetus stimulates the mammary glands.

Human milk is usually bluish-white (the blue comes from protein and the white from fat), sweet, and slightly heavier than water. It is composed mostly of water (88 percent) and also contains carbohydrate (6.5 to 8 percent), fat (3 to 5 percent), protein (1 to 2 percent), and salts (0.2 percent).

Ask Yourself

1 *What are the three stages of labor?*

2 *What are some of the problems that premature babies often face?*

3 *What are some of the physiological adjustments the neonate has to make?*

4 *How do fraternal and identical twins differ?*

5 *What hormones are most involved with lactation?*

POSTNATAL LIFE CYCLE

From the time of fertilization, human development is a highly organized combination of three processes: growth, morphogenesis, and cellular differentiation. *Growth* is an increase in overall size, and usually an increase in the number of cells as the result of mitosis. These cells must be arranged into specific structures that ultimately form the tissues and organs of the body. The cellular movements that bring about tissue and organ development are called *morphogenesis* (Gr. *morphe,* shape + *gennan,* to produce). After cells have become organized into specific forms, they must become specialized in order to perform the specific functions of tissues. This specialization process is called *cellular differentiation. Development* refers to the changes produced by these processes in successive phases of the life cycle, which begins with fertilization and ends with death. In the earlier sections of this chapter, we focused on prenatal development and birth. Here we will outline the stages of *postnatal* ("after birth") development, from the first weeks of life to old age.

Neonatal Stage

The newborn baby is called a *neonate* during its first four weeks of life. This period is one of many adjustments. In addition to the adjustments in the respiratory and circulatory systems already described, a neonate may face some special functional problems:

1 Because its bones are still being transformed from a skeleton of cartilage, a neonate requires an adequate supply of calcium, and also of vitamin D, which increases the absorption of calcium. It needs iron to ensure that its liver will continue to synthesize red blood cells, and it needs vitamin C (which is not stored) to aid the proper formation of cartilage, bone, and intercellular materials.

2 The neonatal rates of respiration, cardiac output, and body metabolism are about twice as high as adult rates, in proportion to body weight, and as a result, these body functions may be unstable.

3 Because the temperature-regulating system of a neonate is unreliable, body temperature can fluctuate greatly.

4 Fluid intake and excretion are about seven times greater in a neonate than an adult, making diarrhea and vomiting major problems. Also, twice as much acid is produced, creating a tendency toward acidosis.

5 The liver may not function adequately during the first postnatal week, causing a lowered concentration of plasma proteins, a subnormal excretion of bilirubin (producing jaundice), an insufficient synthesis of glucose, and poor blood coagulation.

6 Secretions of pancreatic amylase are too low for the proper digestion of starches, and fat absorption is poor.

7 Although the neonate has acquired antibodies from its mother during intrauterine development, its own production of antibodies is not fully operating for at least a year. When the infant does begin to produce antibodies, severe allergies may develop, resulting in serious eczema, gastrointestinal abnormalities, or even anaphylaxis.

Infancy

The neonatal period is followed by *infancy,* which continues until the infant can sit erectly and walk, usually between 10 and 14 months after birth. The body weight generally triples during the first year, from an average birth weight of 5.4 kg (7 lb 8 oz) for both boys and girls, to an average year-old weight of 9.5 kg (21 lb) for boys and 9.1 kg (20 lb) for girls. The average height for boys and girls is 45.7 cm (18 in.) at birth, and 73.7 cm (29 in.) at 1 year.

Gradual physiological changes occur during infancy, especially in the cardiovascular and respiratory systems. Myelinization of the nervous system begins during this period, and motor activities become more and more coordinated. Vision and other functions of the cerebral cortex are refined during the first few postnatal months. The size of the brain increases from one-quarter to one-half of adult size, and the incisor teeth erupt. The immaturity of the liver and kidneys may create an inability to excrete toxic substances, and in general, the infant is susceptible to viral and bacterial diseases before its immune system is functioning at its total capability.

Childhood

Childhood is the period from the end of infancy to the beginning of adolescence (12 to 13 years). In early childhood, ossification is rapid. Bone growth slows down in late childhood and then leaps forward again during the prepubertal period. Deciduous teeth erupt and are replaced by permanent teeth during late childhood, motor coordination becomes more fully developed, and language and other intellectual skills are refined.

Adolescence and Puberty

The hormone-secreting gonads of both sexes are inactive until the final maturation of the reproductive system. This period of maturation, development, and growth is called *adolescence* (L. *adolescere,* to grow up). It spans the years from the end of childhood to the beginning of adulthood. Adolescence and puberty are not the same. *Puberty* (L. *puber,* adult) is the period when the individual becomes physiologically capable of reproduction. It usually takes place during the early years of adolescence. The average age of puberty is 12 for girls and 14 for boys.

Before puberty, between the ages of 7 and 10, there is usually a slow increase in the secretion of estrogens and androgens. The concentrations of these sex hormones remain inadequate, however, to promote the development of *secondary sex characteristics* such as body hair or enlarged genital organs. At the start of puberty, the hypothalamus begins to release luteinizing hormone-releasing hormone (LHRH), which stimulates the secretion of FSH and LH from the pituitary. It is these pituitary gonadotropic hormones that stimulate the testes and ovaries to secrete androgens and estrogens at full capacity. The present belief is that during the maturation process, the amygdala portion of the brain stimulates the hypothalamus to begin secreting LHRH.

New evidence suggests that the pineal gland is an important factor in the hormonal changes that initiate puberty. The pineal gland secretes the hormone melatonin, which inhibits the production of sex hormones during early childhood. At the onset of puberty, however, there is an abrupt drop in the secretion of melatonin, allowing estrogens and androgens to promote sexual changes.

Sequence of body changes The events of puberty (Table 27.4) usually occur in a definite sequence. The typical pattern for a boy begins about the age of 11, when he may begin to get spontaneous erections, with no apparent cause. He may also accumulate deposits of fat prior to the actual changes of puberty. The events of puberty may take as long as four years. The average 18-year-old American boy is 1.76 m (5 ft 9.5 in.) tall, and weighs 68 kg (150 lb). He will probably stop growing when he is about 22, having added about 1.27 cm (0.5 in.) to his height.

The changes that accompany puberty in a girl start about two years earlier than those for boys, and they also happen more rapidly and closer together. The whole sequence of changes, starting with the breast buds increasing in size and the nipples beginning to protrude, usually takes two and half to three years. When an average American girl reaches physical maturity at about 18, she is 1.64 m (5 ft 4.5 in.) tall, and weighs 55.8 kg (123 lb).

Menarche The first menstrual period, the *menarche* (meh-NAR-kee; Gr. "beginning the monthly"), comes during the latter stages of puberty, though ovulation may not take place until a year or so later. The menarche seems to occur when a girl reaches a weight of about 48 kg (106 lb), rather than at a certain age or height. At the start of adolescence, a female has five times as much lean tissue as fat tissue, but the amount of fat more than doubles by the time of menarche. Should there be a shortage of food, the extra fat is enough to nourish a fetus through nine months of pregnancy, and to provide milk for the newborn baby for about a month.

The average age of menarche is 12.8 years, but it usually occurs later (13.1 years) in areas of high altitude. Contrary to some beliefs, the menarche is not hastened by a warm climate, but good nutrition seems to encourage an early menarche, and is probably the most important cause of the increasingly lower ages for menarche.

Adulthood

Adulthood spans the years between about 18 or 25 and old age. The peak of sexual capacity for males, and the maximum secretion of testosterone, are reached between the ages of 18 and 20, and muscular strength reaches its peak at about 25.

Typical age changes Although individuals vary, certain changes are common at various ages. Scalp hair usually remains thick throughout the 20s, but height may begin decreasing by 25. The level of thymic hormones decreases steadily.

Between the ages of 30 and 40, the functional capacity of the body declines by about 0.8 percent per year, although women usually reach their sexual peak during this age span. Heart muscles begin to thicken, and the skin loses some elasticity, producing wrinkles. Vertebral disks begin to deteriorate, causing the vertebrae to move closer together. Hearing, which was most efficient at 10, begins to decline.

Between 40 and 50, the back may begin to hunch over. A 40-year-old man is probably 4.5 to 9 kg (10 to 20 lb)

heavier and one-eighth of an inch shorter than at 20. Hair begins to turn gray and becomes thinner. Most people become farsighted in their late 40s. Lymphocytes are less effective against cancer cells, and the immune system declines.

Between 50 and 55, the skin continues to loosen and wrinkle, and the sense of taste declines. Nearsighted people may have temporarily normal eyesight as their eyes become less effective at close range. Diabetes becomes more prevalent with the decrease of trypsin and insulin production.

Rapid changes begin between the ages of 55 and 60. Muscles and other tissues begin to deteriorate, and body weight may decrease. As the metabolism slows, fat may accumulate to balance the weight loss. Although billions of neurons in the brain have deteriorated by now, memory loss is minimal in a healthy person. The volume of semen decreases, and male voices usually sound higher as the vocal cords stiffen and vibrate at a higher frequency.

Menopause Women between 45 and 55 usually stop producing and releasing ova, and the monthly menstrual cycle stops. This cessation of menstrual periods is called the *menopause* ("ceasing the monthly"), and it signals the end of reproductive ability. The average age for menopause is currently 52, and is increasing. Menopause is not an abrupt change; it usually takes about two years of irregular menstrual periods before menstruation and ovum production stop permanently.

The decrease of estrogen that accompanies menopause is thought to be the cause of several physical problems, including "hot flashes" of the skin, caused by changes in the vasomotor system that dilate blood vessels and increase blood flow. Estrogen decrease may also cause osteoporosis and other irregularities in bone metabolism, as well as dizziness,

TABLE 27.4	MAJOR BODY CHANGES AT PUBERTY: SECONDARY SEX CHARACTERISTICS	
Area of body	**Description of changes (male)**	**Description of changes (female)**
General bodily changes	Shoulders broaden, muscles thicken, height increases. Body odor from armpits, genitals. Skeletal growth ceases by about age 21.	Pelvis widens. Fat distribution increases in hips, buttocks, breasts. Skeletal growth ceases by about age 18.
External genital organs	Penis increases in size, becomes more pigmented. Scrotum becomes enlarged, more pigmented, more wrinkled.	Breasts enlarge. Vagina enlarges and vaginal walls thicken.
Internal genital organs	Testes enlarge. Sperm production increases in testes. Seminal vesicles, prostate gland, bulbourethral glands enlarge, begin to secrete.	Uterus enlarges. Ovaries secrete estrogens. Ova in ovaries begin to mature. Menstruation begins.
Skin	Secretions of sebaceous glands thicken and increase, often causing acne. Skin thickens.	Estrogen secretions keep sebaceous secretions fluid, inhibit development of acne and blackheads.
Hair growth	Hair appears on face, pubic area, armpits, chest, around anus. General body hair increases. Hairline recedes in the lateral frontal regions.	Hair appears on pubic area, armpits. Scalp hair increases with childhood hairline retained.
Voice	Voice becomes deeper as larynx enlarges and vocal cords become longer and thicker.	Voice remains relatively high pitched as larynx grows only slightly.

fatigue, headaches, chest and neck pains, and insomnia. Before menopause, diseases such as hardening of the arteries are quite rare in women, probably because estrogen lowers cholesterol in the blood. But as estrogen secretions continue to diminish after menopause, the incidence of cardiovascular disease becomes almost equal in men and women. Psychological disturbances, including depression, may accompany the physical problems. Some of the symptoms of menopause may be relieved by small doses of estrogen, administered under the strict supervision of a physician.

Although males usually experience a gradual decrease in testosterone secretion after they reach 40 or 50, they do not experience as drastic hormonal changes as women do at that age. The most likely cause of psychological problems during this period is not hormonal, but rather the fear of impotency and old age. Despite the decrease in testosterone, normal males may retain sexual potency in old age.

Senescence

The indeterminate period when an individual is said to grow old is called **senescence** (L. *senescere*, to grow old). By the age of 70, height is usually a full inch less than it was in the 20s or 30s. Between 70 and 80, body strength decreases to half of what it was at 25, lung capacity decreases to half, and about 65 percent of a person's taste buds become inactive. The nose, ears, and earlobes are up to a half inch longer. Life expectancy is currently 71.4 for males and 78.7 for females. (See Chapter 28 for a detailed discussion of the cellular changes that occur during old age.)

Ask Yourself

1 *What are the three main processes of human development?*

2 *Define neonate, infant, child, and adult.*

3 *What is the difference between adolescence and puberty?*

4 *How is the hypothalamus related to the initiation of puberty?*

5 *What changes occur during puberty in boys and girls?*

PHYSIOLOGICAL AND ANATOMICAL ABNORMALITIES

During the nine-month gestation period, *anything* can go wrong, but fortunately, everything of importance usually goes right. Several developmental problems have already been described in this chapter, so we will concentrate on how embryonic and fetal problems are detected and even corrected.

Amniocentesis

Amniocentesis (am-nee-oh-sehn-TEE-sihss; Gr. sac + puncture) is the technique of obtaining cells from a fetus. First, the fetus is located by bouncing high-frequency sound waves off it and recording the echoes. This is a relatively safe procedure, unlike the use of potentially dangerous x rays. Then, a hypodermic needle is inserted into the amnion, usually directly through the abdominal wall of the mother. The amnion surrounding the fetus is filled with fluid in which a number of loose cells float. Because all of these cells are derived from the original zygote, they are genetically alike. Some of the amniotic fluid may be grown as a tissue culture (see drawings). The cells in the fluid may be studied directly to save time, but cultured cells can be made to yield more information.

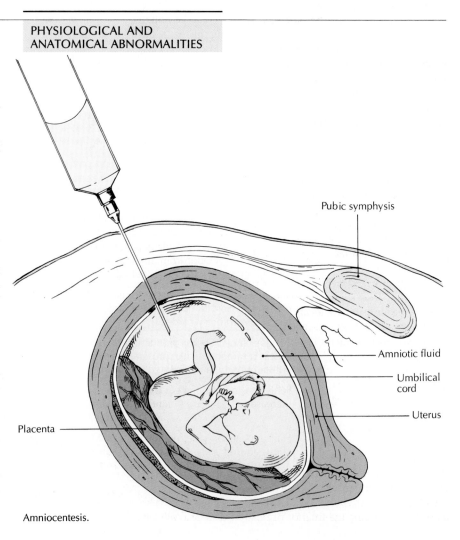

Pubic symphysis

Amniotic fluid

Umbilical cord

Uterus

Placenta

Amniocentesis.

By measuring the amount of protein that leaks from the neural tube into the amniotic fluid, such diseases as spina bifida and anencephaly (absence of the forebrain) may be detected. Hemolytic diseases (including hemolytic disease of the newborn) may be detected. Inborn errors of metabolism may be identified by studying cell cultures derived from the amniotic fluid.

Since chromosomes are examined, the sex of the fetus can be determined, but medically it is more important to discover any chromosomal abnormalities that might indicate possible disorders. The specific chromosome characteristics, or *karyotype,* can be read with fair accuracy (see Figure 29.9). For example, if three chromosomes of chromosome number 13 are present instead of the usual pair, the fetus, if allowed to come to term, will probably not survive infancy. The parents may decide to terminate the pregnancy by having an abortion. If there is a history of abnormality in the family, or the parents are known to be at risk for some disorder, amniocentesis may reassure the parents by showing that the chromosomes are normal.

Since the earlier an abortion is performed, the easier it is on the mother, amniocentesis is best done as soon as any question arises. (It is difficult to perform an amniocentesis before the fourteenth week, however.) The procedure of amniocentesis is reasonably safe for both fetus and mother. Amniocentesis is advisable if a mother is known to carry a chromosomal aberration or if she is over 40, since women over 40 have a disproportionately high number of babies with chromosomal defects.

A promising new technique provides more information than amniocentesis by allowing physicians to locate the fetus by ultrasound and place a needle into the blood vessels of the umbilical cord to take samples of fetal blood, to inject drugs, or to perform transfusions. This technique can be used as early as the eighteenth week of pregnancy.

Fetoscopy

Fetoscopy ("seeing the fetus") allows the physician to view the fetus directly. The fetus is located with ultrasound waves, and then an *endoscope* (an instrument for examining the interior of hol-

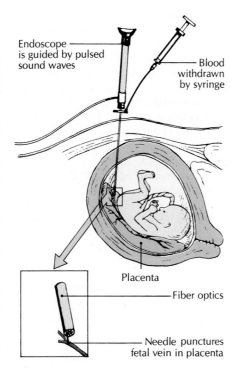

Fetoscopy.

low organs) is inserted into the uterus through a small abdominal incision (see drawing). Physicians can also insert tiny forceps to take a fetal skin sample from the fetal blood vessels that lie on the surface of the placenta. Like any other intrauterine examination, fetoscopy may damage fetal or maternal tissue if done improperly.

Chorionic villi sampling

Chorionic villi sampling is a technique that can be performed in a physician's office as early as the fifth week of pregnancy. A catheter is inserted through the mother's cervix into the uterus, where a sample of chorionic villi

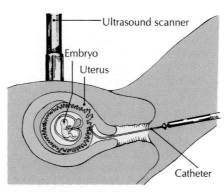

Chorionic villi sampling.

tissue is collected (see drawing). The tissue is identical to fetal tissue, and its DNA can be examined for such diseases as Huntington's chorea, Down's syndrome, muscular dystrophy, and sickle-cell anemia. The technique appears to have a low risk factor, and can be performed six weeks earlier than amniocentesis.

Alpha-fetoprotein test

Alpha-fetoprotein is a substance that is excreted by the fetus into the amniotic fluid and enters the maternal bloodstream. It can be measured by the *alpha-fetoprotein test* or a sample of blood from the pregnant woman. Abnormally high levels of alpha-fetoprotein in the maternal blood indicate that the neural tube of the fetus has not closed completely, allowing large quantities of alpha-fetoprotein to "leak." The test is administered when the fetus is 16 weeks old. It may detect such neurological disorders as *anencephaly,* a condition in which the newborn child has a primitive brain or no brain at all, and *spina bifida,* a condition that exposes a portion of the spine (see Chapter 7). Neural-tube defects occur in about 1 out of every 1000 births.

Intrauterine transfusion

An *intrauterine transfusion* is an injection of a concentrate of red blood cells into the peritoneal cavity of the fetus. The concentrate is prepared from Rh-negative whole blood to help the fetus combat hemolytic anemia (such as hemolytic disease of the newborn), a possibly fatal condition. The cells pass into the fetal circulation by way of the diaphragmatic lymphatics.

Ultrasonography

Ultrasonography is a technique that uses high-frequency sound waves to locate and examine the fetus. Ultrasound images are projected onto a viewing monitor. Ultrasonography allows the physician to see the embryonic sac during the embryonic period, the size of the fetus and its placenta, multiple fetuses, and any abnormal fetal positions. Serious abnormalities of the central nervous system, such as hydrocephaly, can also be detected.

Recent refinements in ultrasonography permit physicians to see *inside* the fetal heart. The new technique can detect severe congenital problems such as abnormal heart chamber and valve formation, and any other condition that would usually produce a ''blue baby.'' This technique allows the attending physician to prepare for treatment soon after birth.

Critics of electronic fetal monitoring, which includes ultrasonography, suggest that ultrasound waves may produce cell damage, and they recommend that sound waves be used only when complications are suspected.

Fetal surgery

Fetal surgery, surgery performed on the fetus *before* it is born, is still in its experimental stages, but the early results have been promising. In 1982 the first successful operation on a fetal kidney was performed at the University of Connecticut Health Center. The kidney was threatened by an accumulation of excess urine because of a blockage of the ureter. The kidney was drained through the pregnant woman's abdominal wall with the guidance of an ultrasound video scan.

About three weeks later, another

sonogram showed that fluid was accumulating again, and doctors decided to perform a cesarean section to deliver the 8-month-old fetus so that conventional corrective surgery could be performed. The surgery was successful, and the infant left the hospital a month later with two healthy kidneys.

Fetal surgery has corrected hydrocephalus and other diseases. In an extreme case, a 5-month fetus was removed from the uterus, operated on for a urinary blockage, and then returned to the uterus. It was born healthy four months later.

MEDICAL TERMINOLOGY

ABORTION (L. *aboriri,* to disappear) The fatally premature expulsion of an embryo or fetus from the uterus.

ABRUPTIO PLACENTAE The premature separation of the placenta from the uterus.

CESAREAN SECTION An incision through the abdominal wall and the uterine wall (*hysterotomy*) to remove a fetus.

CONGENITAL DEFECT A condition existing at birth, but not hereditary.

DYSTOCIA Abnormal or difficult childbirth.

EPISIOTOMY The cutting of the vulva at the lower end of the vaginal orifice at the time of delivery in order to enlarge the birth canal.

FETAL MONITORING The continuous recording of the fetal heart rate, using a transcervical catheter.

FETOLOGY The study of the fetus.

GALACTORRHEA A persistence of lactation.

GYNECOMASTIA Breast development in a male.

LAPAROSCOPE (Gr. *lapara,* flank) A long, thin, light-bearing instrument used during a *laparotomy,* which is the surgical incision into any part of the abdominal wall.

LOCHIA (LOE-kee-uh; Gr. *lokhios,* of childbirth) The normal discharge of blood, tissue, and mucus from the vagina after childbirth.

PERINATOLOGY (Gr. *peri* around + L. *natus,* born) The study and care of the fetus and newborn infant during the perinatal period, usually considered to be from about the fourth month of pregnancy to about a month after childbirth.

PLACENTA PREVIA A condition in which the placenta is closer to the cervix than the fetus is.

PROLAPSE OF THE UMBILICAL CORD Expulsion of the umbilical cord before the fetus is delivered, reducing or cutting off the fetal blood supply.

TOXEMIA An abnormal condition caused by toxic substances in the mother's blood that may be harmful to the fetus.

SUMMARY

Embryonic development

1 The fusion of ovum and sperm nuclei produces the diploid *zygote.* The daughter cells, produced by a process of cell division called *cleavage,* are *blastomeres.* During the third day, about the time the zygote is entering the uterus, a solid ball of about 16 blastomeres is formed, which is called the *morula.*

2 After two or three days, the morula develops into a fluid-filled hollow sphere called a *blastocyst.* The blastocyst is composed of about 100 cells. Its outer

covering of cells (*trophoblasts*) is the *trophectoderm;* the blastocyst cavity is the *blastocoel;* a grouping of cells at one pole is the *inner cell mass* (*embryoblast*), from which the embryo will grow.

3 About a week after fertilization, the blastocyst becomes implanted in the uterine lining. *Implantation* is the actual start of pregnancy.

4 During week 2, the blastocyst develops into the *bilaminar embryonic disk,* composed of the endoderm and ectoderm. The zygote is now an *embryo.*

5 The *differentiation* of cells into specialized tissues and organs begins shortly after the start of implantation. The inner cell mass of the blastocyst is rearranged into the *primary germ layers:* endoderm, ectoderm, and mesoderm.

6 *Endoderm* develops mainly into the innermost organs, *ectoderm* into the outer layers of skin and the nervous system, and *mesoderm* into connective tissue, bone, muscle, blood, and some inner epithelium.

7 The *extraembryonic membranes* include the yolk sac, amnion, allantois, and chorion.

8 The *placenta* synthesizes glycogen and other nutrients, transports materials between the mother and embryo, and secretes protein and steroid hormones. The blood vessels in the *umbilical cord* carry nutrients from the maternal placenta to the fetus, and wastes from the fetus to the placenta.

9 The embryonic heart is beating by the fourth week, and by the end of eight weeks, the major organ systems have become established. The embryo is now called a *fetus.*

Fetal development

The fetal period begins with the ninth week. Growth is rapid at first but slows down between the seventeenth and twentieth weeks, as development continues and systems become mature enough to sustain life outside the uterus.

Maternal events of pregnancy

1 The first maternal sign of pregnancy is usually a missed menstrual period. Pregnancy tests, which depend on the abundance of hCG in the blood and urine of pregnant women, confirm or deny pregnancy.

2 The pregnant woman's body goes through a series of physical and hormonal changes to accommodate the developing and enlarging fetus. Weak uterine contractions may begin about the tenth week, becoming stronger as the pregnancy progresses and frequent during the thirty-sixth week.

3 About 266 days after fertilization, uterine contractions become regular and strong, the uterine membranes rupture and the amniotic fluid is discharged, the cervix dilates, and after a variable period of labor, the baby is born. Pregnancy officially ends about 30 minutes later, with the expulsion of the placenta.

4 Substances that interfere chemically with normal embryonic development are called *teratogens.* They are usually introduced by the mother through the placenta. Drugs, smoking, and alcohol are common sources of developmental problems.

Birth and lactation

1 *Parturition* is the birth of the child. The initial uterine contractions stimulate the pituitary gland to secrete oxytocin, which increases the force of the contractions until the baby is born.

2 The three stages of *labor* are usually described as the start and continuation of regular uterine contractions, the birth of the child, and the delivery of the afterbirth. The period from the delivery of the afterbirth until the mother's body reverts to its nonpregnant state is the *puerperal period.*

3 A baby is considered *premature* if it is born before 240 days and weighs less than 2.4 kg.

4 *Multiple births* are always an exception in human beings. *Fraternal* (dizygotic) *twins* are formed when two ova are released and are fertilized. *Identical* (monozygotic) *twins* result from the same fertilized ovum and are genetically identical and the same sex. *Conjoined twins* are identical twins whose embryonic disks do not separate completely.

5 Among the adjustments the newborn makes to extrauterine life, the major ones are breathing on its own and the shift from a fetal circulation to an adult circulation, which depends on the lungs for gas exchange.

6 *Lactation* includes both milk secretion and milk release from the breasts. The most important hormones for lactation are *prolactin,* which stimulates milk production, and *oxytocin,* which stimulates the release of milk from the breasts.

Postnatal life cycle

1 Human development consists of *growth,* an increase in size and number of cells as a result of mitosis; *morphogenesis,* the cellular movements that bring about tissue and organ development; and *cellular differentiation,* the specialization of cells that allow them to perform the specific functions of tissues.

2 The first four weeks after birth are the *neonatal period.* It is followed by *infancy,* which continues until the infant can sit erectly and walk, usually at 10 to 14 months. *Childhood* is the period from the end of infancy to the beginning of adolescence at 12 or 13 years.

3 *Adolescence* is the period of matura-tion, development, and growth from the end of childhood to the beginning of adulthood. *Puberty* is the period during early adolescence when an individual becomes physiologically capable of reproduction. It occurs about two years earlier in females than in males.

4 At the start of puberty, the release of GnRH from the hypothalamus stimulates the secretion of FSH and LH from the pituitary. These pituitary hormones stimulate the testes and ovaries to secrete androgens and estrogens at full capacity.

5 The events of puberty usually occur in a definite sequence. The major body changes are called *secondary sex characteristics.* The first menstrual period, or *menarche,* usually occurs during the latter stages of puberty.

6 *Adulthood* spans the years between about 18 or 25 and old age. The cessation of menstrual periods, *menopause,* usually occurs between the ages of 45 and 55.

7 The indeterminate period when an individual is said to grow old is *senescence.*

Physiological and anatomical abnormalities

1 *Amniocentesis* is the technique of obtaining cells from an unborn fetus by inserting a hypodermic needle through the abdominal wall of the mother into the amniotic cavity.

2 *Fetoscopy* is the viewing of a fetus through an endoscope inserted into the uterus through the mother's abdomen.

3 *Chorionic villi sampling* is a technique in which chorionic villi tissue is obtained through a catheter inserted through the mother's cervix into the uterus.

4 The *alpha-fetoprotein test,* a measurement of the alpha-fetoprotein level in maternal blood, indicates the possibility of neural tube defects.

5 An *intrauterine transfusion* is an injection of a concentrate of red blood cells into the peritoneal cavity of the fetus in order to help the fetus combat blood diseases.

6 *Ultrasonography* uses high-frequency sound waves to locate and examine the fetus.

7 *Fetal surgery* is surgery performed on the fetus before it is born.

UNDERSTANDING THE FACTS

1 Into what two main periods may the development of a human be divided?

2 During what period in human development do we use the term zygote? Embryo? Fetus?

3 About how long after fertilization does the first cell division occur?

4 Define trophoblast, and tell what it will develop into.

5 When does pregnancy actually begin?

6 Into what structure does the blastocyst develop just after implantation?

7 From which primary germ layer does the bulk of the body develop?

8 What is the significance of the chorionic villi?

9 In which structure are blood cells initially developed?

10 Briefly describe the role of the placenta.

11 To what is the last six months of pregnancy devoted?

12 Give two or three indications that labor is imminent or has just begun.

13 Why are so many infants jaundiced during the first few days after birth?

14 What role do the placental hormones play in milk release?

15 What is colostrum?

16 List some of the factors that increase the probability of multiple births.

17 Is it true that all embryos are female during early development?

18 Why do we usually state development periods in lunar instead of calendar months?

19 Give the approximate survival percentages for a fetus born after the seventh lunar month, eighth lunar month, and tenth lunar month.

20 Most pregnancy tests are based on the presence of which hormone in the urine of the pregnant woman?

21 Give the location of the following (be specific):
  a morula   **d** ductus arteriosis
  b blastocoel   **e** lanugo
  c foramen ovale

UNDERSTANDING THE CONCEPTS

1 What is the significance of the fact that the overall size of the zygote when it consists of several blastomeres is approximately the same as that of the original fertilized egg?

2 The mother's body does not reject the blastocyst, which is foreign tissue. Why would the discovery of the reason for this fact be of possible medical value?

3 What is meant by differentiation?

4 What are several benefits of having the developing individual suspended in the amniotic fluid?

5 The vein within the umbilical cord transports oxygen and nutrients and the arteries transport carbon dioxide and nitrogen wastes. Can you reconcile this fact with the usual function of arteries and veins?

6 Explain how a baby can be born with an addiction to a drug, say, heroin.

7 What is the significance of the foramen ovale and the ductus arteriosus?

8 What are some of the benefits of breast feeding for the infant?

9 Why are identical twins often not as similar in weight as fraternal twins?

10 Why may hemorrhoids and varicose veins often develop during late pregnancy?

11 Why are teratogens more dangerous during the early weeks of pregnancy?

12 What are the hormonal changes that initiate puberty and what structures may be involved in triggering these changes?

13 Discuss the fetal alcohol syndrome.

14 List some of the information that can be obtained by amniocentesis.

15 What are some uses of intrauterine transfusions? Fetal surgery?

SUGGESTED READING

BALINSKY, B. I., *An Introduction to Embryology,* 4th ed. Philadelphia: Saunders, 1975.

BEACONSFIELD, PETER, GEORGE BIRDWOOD, AND REBECCA BEACONSFIELD, ''The Placenta.'' *Scientific American,* August 1980.

CARLSON, BRUCE M., *Patten's Foundations of Embryology,* 5th ed. New York: McGraw-Hill, 1988.

COWAN, W. MAXWELL, ''The Development of the Brain.'' *Scientific American,* September 1979.

EDELMAN, GERALD M., ''Cell-Adhesion Molecules: A Molecular Basis for Animal Form.'' *Scientific American,* April 1984.

FUCHS, FRITZ, ''Genetic Amniocentesis.'' *Scientific American,* June 1980.

GORDON, RICHARD, AND ANTONE G. JACOBSON, ''The Shaping of Tissues in Embryos.'' *Scientific American,* June 1978.

GROBSTEIN, CLIFFORD, ''External Human Fertilization.'' *Scientific American,* June 1979.

LAGERCRANTZ, HUGO, AND THEODORE A. SLOTKIN, ''The 'Stress' of Being Born.'' *Scientific American,* April 1986.

MILLER, JULIE ANN, ''The Littlest Babies.'' *Science News,* October 15, 1983.

MILLER, JULIE ANN, ''Small-Baby Biology.'' *Science News,* October 22, 1983.

MOORE, KEITH L., *The Developing Human,* 2nd ed. Philadelphia: Saunders, 1977.

NILSSON, LENNART, MIRJAM FURUHJELM, AXEL INGLEMAN-SUNDBERG, AND CLAES WIRSÉN, *A Child Is Born,* rev. ed. New York: Delacorte/Seymour Lawrence, 1977.

RUGH, ROBERTS, AND LANDRUM B. SHETTLES, *From Conception to Birth: The Drama of Life's Beginnings.* New York: Harper & Row, 1971.

TODD, JAMES T., LEONARD S. MARK, ROBERT E. SHAW, AND JOHN B. PITTENGER, ''The Perception of Human Growth.'' *Scientific American,* February 1980.

WESSELS, NORMAN K., AND WILLIAM J. RUTTER, ''Phases in Cell Differentiation.'' *Scientific American,* March 1969.

28
The Body in Transition: Aging and Cancer

LEARNING OBJECTIVES

1 Describe some of the changes that occur in cells as they die.

2 Distinguish between senescence and senility.

3 Discuss the major hypotheses about aging.

4 Explain some effects of aging on each of the body systems, and list the major causes of death in older persons.

5 Distinguish between benign and malignant neoplasms.

6 Name the three major types of malignant neoplasms, and the kind of tissue where each originates.

7 Explain how cancer spreads from its original site in the body.

8 Name five or more carcinogens, and indicate the parts of the body most often affected by each.

9 Briefly explain the main hypotheses about how cancer develops.

10 Name the most common sites in the body where cancer develops, and discuss the incidence and survival rates for these cancers.

11 Describe the major treatments for cancer.

12 Name the four categories of approved cancer-fighting drugs.

t is *normal* to grow old and die. Most of us will grow old, and certainly all of us will die. Much evidence points to the idea that aging, the accumulation of many little unnoticed, overlapping "losses," is a normal, genetically programmed process. And yet some people persist in thinking of aging as abnormal, something that could be prevented somehow. But there is no doubt that aging is normal. Death is normal. A body that performs less well at 60 than it did at 30 is normal.

What is *abnormal*? Leukemia that kills a 6-year-old is abnormal. Getting hit by a truck at any age is abnormal. Cancer is always abnormal. *Cancer,* by definition, means that cells are out of control, and in the human body anything that is out of control is abnormal. In this chapter, you will see how thin the line is between a normal, dying cell and an abnormal one that causes death in other cells. You will also confront another piece of irony: cancer cells have achieved the ultimate dream—they can live forever.

PHYSIOLOGY OF HUMAN AGING

Scientists have speculated that if the three main causes of death in old age—heart disease, cancer, and stroke—were eliminated, the life expectancy of human beings would be extended only 5 to 10 years beyond the present 72 (Figure

FIGURE 28.1

Average human life expectancy at birth. The life *expectancy* has improved steadily since the time of classical Greece, but scientists believe that the human life *span* has not changed appreciably since human beings lived in caves. It appears that a human being is genetically programmed to live a maximum of 120 years. We are coming closer and closer to the maximum life span of 120 years, but at present the average age we can expect to live is still below 80. Note that the life expectancy for females (F) is higher than for males (M) and that the gap between the sexes is widening; the difference between the life expectancies of females and males was 3 years in 1900, 6 years in 1980, and is expected to be 10 years in 2080. [*Sources: National Center for Health Statistics, Social Security Administration, and* Encyclopaedia Britannica.]

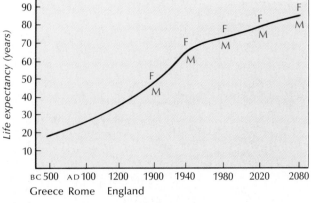

28.1). If this is true, there must be a cause of aging and death besides disease. As more and more people reach old age, and as the population of the world continues to increase, the study of aging (**gerontology**) becomes more and more important. Unfortunately, we know less about the processes of aging than we know about the surface of the moon.

Today, what aging *does* is much better known than what it *is* and when it occurs. For example, skin wrinkles with age because of changes in the skin—skin cells become dehydrated, the supportive cushion of fat cells beneath the skin becomes thinner, and elastic cells in the skin become less resilient. But the causes of these changes in cells—which scientists speculate are most likely changes in DNA—are still unknown.

Part of the aging process includes the death of cells, or **necrobiosis** (Gr. *nekros,* corpse, death + *biosis,* way of life). Necrobiosis is a natural death, as opposed to **necrosis,** which is the death of a cell or tissue resulting from irreversible damage caused by disease or accident. Changes in the nucleus are more significant than changes in organelles in indicating that a cell is dead. Three types of nuclear changes accompany cell death (Figure 28.2): (1) *pyknosis* is the condensation and shrinking of nuclear material (Figure 28.2A); (2) *karyorrhexis* is the disintegration of the nucleus into fragments of nuclear "dust"; and (3) *karyolysis* is the acceleration of cell death by the dissolution of the nucleus.

When the natural, hereditary changes of necrobiosis occur in the structure and composition of cells and, of course, in the body, the process is called *primary aging, biological aging,* or **senescence** (L. *senex,* old). Changes that take place because of disease or accidental injury (necrosis) are part of *secondary aging,* or **senility** (see Chapter 13).

An important question is: Do all elderly people become senile? In the technical sense, senility does not mean a loss of mental awareness and the typical forgetfulness of some elderly people. Aging is definitely *not* always accompanied by a loss of mental capabilities. In fact, there is usually little, if any, decline in intellectual abilities and skills unless the brain has been damaged by a stroke or other serious disease. People often use the term *senile* to describe an elderly person whose mental faculties have definitely deteriorated. However, what they are actually describing is a condition known as *senile dementia* (L. *dementia,* madness), a fairly common occurrence in senescence.

Aging is usually not a simple, single process, but a series of interrelated cellular events that accumulate until a change becomes noticeable and permanent. (It has been estimated that at least 7000 different genes are involved in the aging process.)

One way to describe aging is the inability of the body to adapt to the environment. Homeostatic mechanisms are no longer as efficient in old age. This is especially true when the body is under stress. Under normal conditions most body functions appear to be unaffected, but aging usually reduces our resistance to physical and mental stresses. For example, the level of sugar in the blood is usually about the same for young people as it is for old people, but once the level is raised in an elderly person, it takes longer to return to normal.

Hypotheses of Aging

The process of aging is still a mystery, but there are several hypotheses that attempt to explain the basic mechanism. At least some of these mechanisms are probably at work at the same time to produce what we call aging. In other words, there is probably no *single* hypothesis that can explain aging.

Free-radical damage Physiological activity takes place at the cellular level. Aging is no exception. One of the most important features of the aging process is the *free-radical damage* that occurs within cells (Figure 28.3). Catabolic activity in mitochondria produces superoxide radicals (O_2^-), negatively charged oxygen molecules that are normally converted into hydrogen peroxide (H_2O_2) by the enzyme superoxide dismutase. The potentially dangerous hydrogen peroxide is deactivated by peroxisomes in the cell by the enzyme catalase, and converted into harmless oxygen and water, which are excreted as wastes.

As cells age, however, superoxide dismutase and catalase are not synthesized fast enough, and superoxide radicals and hydrogen peroxide begin to accumulate. The excess radicals and hydrogen peroxide are highly reactive, and can cause harmful free-radical chain reactions to occur within a cell. For example, the unsaturated phospholipids in membranes may be attacked, causing the unsaturated phospholipid fatty acids to produce lipid peroxides and highly reactive malondialdehyde molecules. These molecules form cross-linked units that react with other phospholipids, proteins, nucleic acids, and enzymes, impairing their normal functions.

This cross-linking of the organic macromolecules produces fluorescent compounds called age pigments, which are primarily composed of lipofuscin. Lipofuscin accumulates mostly in the lysosomes of aging cells, especially very active cells such as cardiac muscle and nerve cells, where it impairs their functions. The accumulation of lipofuscin is used as an index of physiological aging at the cellular level.

Laboratory experiments have shown that antioxidants such as vitamins C and E halt the action of free radicals. Diets lacking in antioxidants may accelerate the aging process by providing inadequate protection against free radicals. However, researchers warn that excessive amounts of antioxidants can promote oxidation, resulting in more rapid aging. At the present time, the optimum amounts of dietary antioxidants are not known.

DNA mutations Occasional *mutations** occur as the genetic message in DNA is passed along to RNA. Many of these

---

*A mutation is a change in genetic material that alters the characteristics of a cell, making the cell abnormal.

FIGURE 28.2

Electron micrographs of cells of a cancerous neoplasm. (A) The cells at top left are still alive, and have healthy nuclei. In contrast, the nuclei undergoing pyknosis have shrunk into a condensed mass, and the nuclei undergoing karyorrhexis have already broken up into fragments. (B, C) Electron micrographs of a rat pancreas treated with a chemical that destroys cells in pancreatic islets (the light area in the center of the pictures). In (B), the nuclei are fairly normal, but some have already undergone pyknosis. In (C), most nuclei have dissolved, demonstrating karyolysis. [(A) *A. W. Ham and D. H. Cormack,* Histology, *9th ed. Philadelphia, J. B. Lippincott, 1987.* (B and C) *A. W. Ham and D. H. Cormack,* Histology, *8th ed. Philadelphia, J. B. Lippincott, 1979.*]

Cells undergoing karyorrhexis

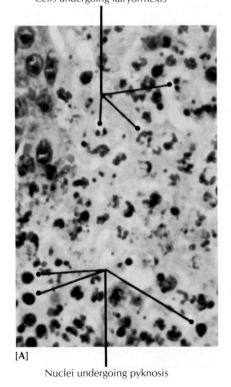

[A]

Nuclei undergoing pyknosis

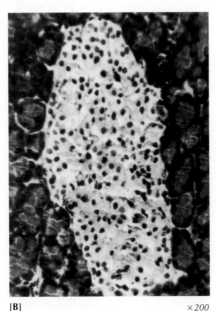

[B]   ×200

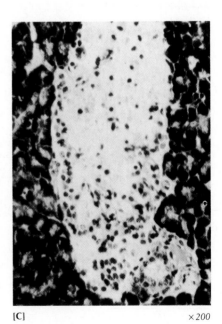

[C]   ×200

mutations are corrected during RNA processing, but some are expressed in the production of a defective enzyme or other protein. Defective enzymes in a cell cause the cell to operate inefficiently, and eventually to die. The more often an error is copied, the more severe it becomes. DNA has a built-in repair mechanism that corrects most of these defects before they become serious enough to be noticed, let alone deadly. However, the "repair DNA" itself does not remain totally effective for an indefinite time. Eventually, the repair mechanism breaks down, but the mutations persist. Cells die, and "aging" occurs.

Amount of "extra" DNA Redundant, or "extra," DNA exists in the nucleus of a cell. If a gene (a specific nucleotide sequence of a DNA molecule) is damaged, it can be replaced immediately by an identical sequence of DNA, and no per-

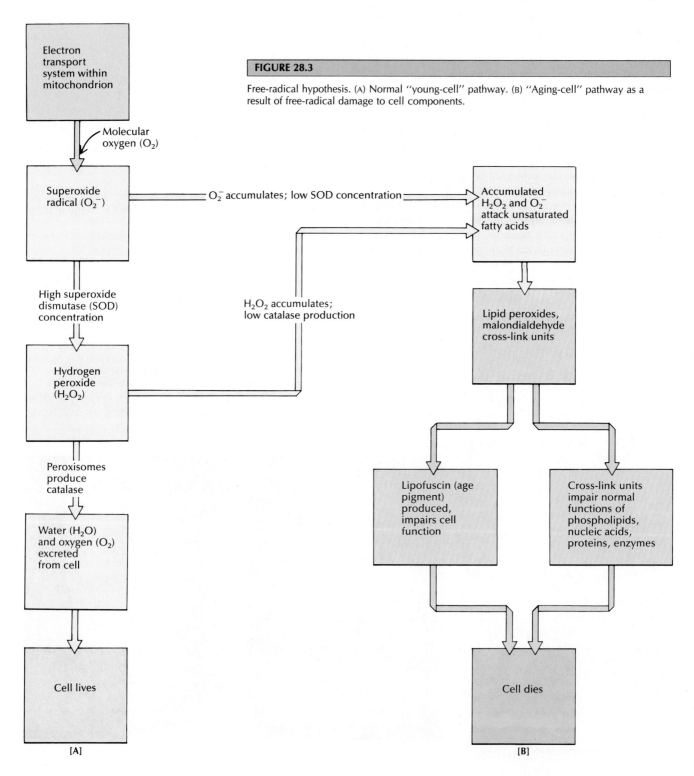

FIGURE 28.3

Free-radical hypothesis. (A) Normal "young-cell" pathway. (B) "Aging-cell" pathway as a result of free-radical damage to cell components.

manent damage is done. After a period of time, however, the redundant DNA is used up, and harmful or damaged genes are free to function. Eventually, changes associated with aging appear in the body. According to this hypothesis, the length of time a person lives might be dependent upon the amount of redundant DNA in the cells.

Genetic clock Aging is part of the normal genetic message. Normal events such as menopause and graying hair, harmless in themselves, indicate that cells are not operating as efficiently as they once were. With decreased cell efficiency comes increased susceptibility to certain diseases. Aging occurs as the "genetic clock" begins to slow down, and when the "genetic clock" finally stops, first the cells die, and then the body does.

Limits on cell division More than two decades ago, Paul Moorhead and Leonard Hayflick determined that certain cells are capable of a predetermined number of divisions (Figure 28.4). When fibroblasts (connective tissue cells) from human embryos are allowed to double 20 times and then are artificially stopped by freezing, they will double 30 more times after they are thawed. If cell division is stopped at 10 doublings, it will resume for 40 more after thawing. Obviously, these particular embryonic cells have the capacity to double 50 times. When similar cells from adults were used, they doubled anywhere from 14 to 29 times, significantly less than the younger cells. After some cells reach their genetically determined limit of cell division, they die. Other cells, such as neurons and muscle cells, persist for the life of the individual, even after they have lost their ability to divide. Overall, most dead cells are replaced by new ones, but as the body grows older, the replacement possibilities become fewer. No *normal* cells have been found that can divide without limit.*

Strain on cells As cells die and are not replaced (muscle cells, for example), other cells of that type must work harder to keep the organism functioning normally. This increased strain on the cells contributes to their aging.

Accumulation of wastes The older a cell is, the more metabolic wastes and harmful substances it accumulates. When such wastes cannot be disposed of in the usual manner, they interfere with the normal functioning of the cell, and can cause structural changes or gene mutations in chromosomes. Over time, these mutations hamper the cell's ability to produce enzymes and other proteins the cell needs to function properly. As a result, the cell deteriorates and dies.

Autoimmunity Under normal conditions, the body's immune system attacks and destroys foreign invaders such as viruses and bacteria. If gene mutations occur within the immune system, however, it may begin to produce autoantibodies, which destroy normal cells instead of foreign ones. By killing normal cells and ignoring harmful foreign cells, the deviant immune system greatly increases the aging and death

*Some cancer cells, given the opportunity for nourishment in a laboratory, can live forever, dividing indefinitely.

FIGURE 28.4

Limits on cell division. (A) Young connective tissue cells are well organized, with distinctive nuclei (dark ovals). (B) Aging connective tissue cells after 50 cell divisions show degenerative changes. [(A *and* B) *Dr. Leonard Hayflick, University of California, San Francisco, Medical School.*]

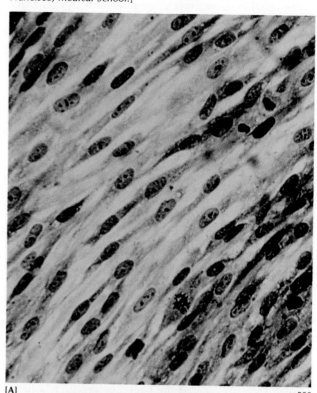

[A]   ×500

[B]   ×500

processes. Autoimmune reactions also occur in certain disorders, such as various cardiovascular diseases, diabetes, rheumatoid arthritis, some cancers, and myasthenia gravis.

Thickening of collagen   The most common protein in the body is the connective tissue *collagen*. As the body ages, elastin decreases and collagen thickens, resulting in hardening of the arteries, stiffening of the joints, sagging muscles, wrinkling of the skin, and other changes that eventually slow down and kill the body. The thickening of collagen may be caused by *cross-linking*, in which large molecules both inside and outside the cell begin to stick together instead of remaining separate. (Some cross-linking is caused by free radicals.) Tissues and cells become clogged eventually, and normal functions are impaired.

The effect of glucose on aging: glycosylation   Another process related to cross-linking is *glycosylation*, the chemical attachment of glucose to proteins without the aid of enzymes. It has been suggested that glycosylation adds glucose randomly to growing polypeptide chains, and results in the formation of cross-links between protein molecules. If this hypothesis is correct, it would help to explain the cross-linking of proteins that contributes to the stiffening and nonresilience of aging tissues. It has also been suggested that the nonenzymatic attachment of glucose to nucleic acids may cause a gradual, and increasing, damage to DNA as tissues age.

Hormone imbalance   As hormone levels become unbalanced, harmful changes in tissues occur. The decrease in sex hormones, for instance, probably contributes to tissue malfunction in many parts of the body not directly related to reproduction.

Cell starvation   Some cells starve to death when their nutrients are blocked by an increased growth of fibrous tissue, or when the blood supply is cut off by a blocked blood vessel.

Effects of Aging on the Body

Elderly people have a lowered resistance to disease, and about 75 percent of people over 65 have at least one chronic disease. The most common diseases of old age are cardiovascular diseases, cancer, arthritis and similar joint diseases, metabolic disorders, autoimmune diseases, diabetes, and diseases of the nervous system that affect the brain and spinal cord. Many diseases are directly related to arteriosclerosis (hardening of the arteries). Cancer among the elderly is not unusual, and people who live beyond 85 often have one or more cancers.

The most lethal diseases of old age are heart diseases, cancer, and cerebral hemorrhage (stroke). Elderly people die as a result of a disease or an accident. "Old age" by itself is never a cause of death. Table 28.1 illustrates how body functions are affected by age, and the following sections point out specific age-related diseases of the various systems of the body.

Integumentary system   With old age, the skin usually wrinkles as the elastic tissue becomes less resilient, and the fatty layer and supportive tissue beneath the skin decrease in thickness. As oil glands and sweat glands decrease their activity, the skin becomes dehydrated. Dry, brittle skin leads to more frequent bacterial skin infections. Also, sensitivity to changes in temperature increases as the skin becomes less able to maintain a constant body temperature. Blotching of skin color often occurs because of the irregular growth of pigment cells, probably due to exposure to sunlight and other ultraviolet radiation. Smooth, flat, brown areas called *lentigines* (commonly, and incorrectly, called "liver spots") may form on the face or back of the hands, and damaged capillaries may give rise to small, bright red bumps, known as cherry angiomas, and also to larger, purple blotches. Raised brown or black dime-sized growths called *seborrheic keratoses* may appear on the epidermal layer of the skin.

In addition to the typical inherited baldness in some men, there is a thinning of the hair in both sexes because of the atrophy of hair follicles. Hair usually becomes gray and brittle, and more hair appears in the nostrils, ears, and eyebrows. Postmenopausal women may grow longer hair on the upper lip and chin due to an increase in androgens and a decrease in estrogens.

Skeletal and muscular systems   As people age, bones may soften because of a decrease in calcium utilization. This condition makes bones brittle, and allows old bones to break

TABLE 28.1   HOW BODY FUNCTIONS AND CHARACTERISTICS DIMINISH WITH AGE

Body function or characteristic	Percentage of function remaining for average 75-year-old man*
Nerve conduction velocity	90
Body weight	88
Basal metabolic rate	84
Body water content	82
Blood flow to brain	80
Output of heart at rest	70
Filtration rate of kidneys	69
Number of nerve trunk fibers	63
Maximum usable lung capacity	56
Number of glomeruli in kidneys	56
Hand grip	55
Sperm count	50
Visual acuity	50
Vital capacity (total amount of air one can breathe in and out in a single breath)	50
Maximum breathing capacity (voluntary)	43
Blood flow to kidneys	42
Maximum oxygen uptake (during exercise)	40
Number of taste buds	36

*The body functions of an average 30-year-old man are set at 100 percent.

Source: From data in Nathan W. Shock, "The Physiology of Aging," *Scientific American,* January 1962, and other sources.

more easily than young ones. The most common fractures occur in the femur, wrist, and humerus. Fractures in the wrist and arm often occur when the person attempts to break a fall. Bone marrow decreases. *Osteoporosis* is a metabolic disorder, especially prevalent in postmenopausal women, that leads to a loss of skeletal mass and density (see Chapter 6).

With age, there is usually a decrease in the synovial fluid in the joints, and the cartilage in joints also becomes thinner. Arthritis, bursitis, and other diseases of the joints are prevalent. Ligaments become shorter and less flexible, bending the spinal column into a hunched-over position. Muscles begin to weaken and become smaller, flabby, and dehydrated. Fibrous tissue appears, and muscles become filled with fat (Figure 28.5). Muscle reflexes slow down. Nocturnal leg cramps are common, especially among women. Cramps may be caused by a deficiency of sodium, improper blood flow, or other factors.

For many reasons, elderly people tend to move about less than they used to, and less than they should. As a result of general reduced mobility, some of the following problems may arise: (1) blood clots (embolisms) within blood vessels; (2) loss of muscle tone, cramps, and osteoporosis; (3) digestive disturbances such as diarrhea, constipation, and general indigestion; (4) accumulations of secretions within the lungs and the rest of the respiratory tract, disrupting the balance between carbon dioxide and oxygen; (5) kidney stones and infections of the urinary tract; and (6) bedsores in bedridden patients whose movements are restrained.

Nervous system During old age, neurons continue to die, but there is usually no shortage. Some neurons lose their processes, and other neurons are replaced by fibrous cells,

but there is only a 10-percent decrease in the velocity of nerve impulses. However, reflexes do slow down. The many convolutions of the brain smooth out somewhat, and the blood supply to the brain is reduced. Although the number of brain cells is reduced, there is no appreciable loss of brain function unless the blood supply is cut off temporarily by a stroke or other problem. A stroke may lead to the progressive loss of intellectual abilities, personality, and memory.

Short-term memory may be impaired somewhat in old age, especially if the blood supply to the brain is decreased by diseased arteries. There is usually little or no change in learning ability, although many elderly people have been conditioned to think that their mental faculties are seriously diminished by the aging process.

Several age-related diseases affect brain functions. During *Alzheimer's disease*, which often appears between 40 and 60, some spaces between parts of the brain are enlarged, degenerative changes occur, and memory loss is common (see Chapter 13). *Parkinson's disease*, a motor disability, is commonly called shaking palsy because it is characterized by tremors of the head and hands, slow movements, rigid joints, and sagging facial muscles. It is a common ailment that usually appears between 55 and 70, and occurs more often in men than in women. The symptoms of Parkinson's disease are caused by the destruction of dopaminergic fibers that project from the substantia nigra and associated lesions of the globus pallidus (see Chapter 13). It is not yet known what causes this fiber destruction, but these neurons are the only ones in the brain that release dopamine, and the symptoms of Parkinson's disease appear to be the result of a decreased input of dopamine.

Eyesight is usually impaired because of the degeneration of fibers in the optic nerve and an accumulation of injuries

---

FIGURE 28.5

Effect of aging on muscle tissue. (A) Leg muscle from a normal adult rat, showing firm muscle fibers. (B) Leg muscle from an old rat, illustrating how connective tissue replaces dead muscle fibers. [(A *and* B) *Warren Andrew, Ph.D., M.D., The Anatomy of Aging in Man and Animals. New York, Grune and Stratton, 1971.*]

Connective tissue

[A]   ×570   [B]   ×570

to the eyes. Common ailments are *presbyopia* (loss of the ability to focus on close objects), *cataract* (opaque film on the lens of the eye), and *glaucoma* (excessive fluid pressure inside the eyeball). Color perception is reduced as cones in the retina degenerate. People over 60 usually need twice as much light as a 40-year-old because rods in the retina degenerate.

Hearing, smell, and taste are all reduced. Most elderly people have adequate hearing, but high-pitched sounds become increasingly difficult to hear as the eardrum loses its elasticity and fibers in the auditory nerve degenerate. Hearing may also be impaired by the accumulation of hardened wax in the outer ear. *Otosclerosis* is a disease of the small bones in the middle ear in which the bones fuse together, making it impossible for them to vibrate properly and amplify sound waves that must reach the auditory nerve. The abilities to smell and taste begin to deteriorate at about 60, as the lining of the mucous membrane becomes thinner and less sensitive, and as the number of active taste buds is reduced.

Elderly people often sleep up to two hours less than they used to, and wake up more often during the night. The more-frequent awakenings may be due to breathing problems and, in men, the need to urinate because of an enlarged prostate gland.

Endocrine system Although the endocrine system is relatively unaffected by aging, some changes do occur. For example, the thyroid becomes smaller, and the basal metabolism rate (the amount of energy used up at rest) is lowered. The metabolism of protein and sugar is diminished as the pancreas secretes less of the necessary enzymes. The body burns fuel less rapidly, but efficiency is not decreased since less fuel is needed.

Cardiovascular system Although the heart usually does not decrease in size, its pumping efficiency is reduced to about 70 percent because some muscle and valve tissues are replaced by fibrous tissue. Blood pressure is usually raised, and the heart rate does not compensate as well in response to stress as it used to. If hardening of the arteries occurs, the arterial walls become harder and less elastic, and the heart must work harder to pump the same amount of blood. As a result, the heart becomes enlarged, and if the heart reaches its maximum limits, heart failure occurs.

Cerebral arteriosclerosis is a disease in which the arteries that supply blood to the brain harden and narrow. Decreased blood flow to the brain may impair the memory of recent events or make speech difficult (aphasia). If blood vessels leading to the brain become blocked or leak, so that the brain is denied blood for even a short amount of time, serious brain damage and even death may occur. *Arteriosclerosis* and *hypertension* (high blood pressure) are the chief causes of *cerebral hemorrhage* (stroke). In the United States, about 98 percent of the 100,000 people who die of a stroke each year are over 50.

Respiratory system As people grow older, some air sacs in the lungs are replaced by fibrous tissue, and the exchange of oxygen and carbon dioxide is reduced. Breathing capacity and usable lung capacity (vital capacity) are also reduced when muscles in the rib cage are weakened and the lungs lose their full elasticity. Aging also causes a decrease in gaseous exchange across pulmonary membranes.

Emphysema is a fairly common disease, especially among heavy smokers who live in air-polluted cities. It is the result of reduced diffusion surface because of alveolar fusion (see Chapter 21). Exhalation becomes difficult when the lungs lose their elasticity, and air accumulates in the lungs. *Lung cancer, chronic bronchitis,* and *pneumonia* also become more frequent with age.

Digestive system With age, digestion may be impaired as the stomach produces less hydrochloric acid, and other factors contribute to a general breakdown of the normal digestive process. Peristalsis slows down, diseased gums (peridontal disease) and loss of teeth make chewing difficult, and even swallowing becomes more difficult. Constipation is a frequent result. The intestines produce fewer digestive enzymes, and the intestinal walls are less able to absorb nutrients. Defecation may be slower and less frequent as the muscles of the rectum weaken.

About 25 percent of all people over 65 have *diverticulosis,* the formation of small pockets (diverticuli) in the walls of the large intestine. When food becomes lodged in the pockets, and the pockets become inflamed, the disease is called *diverticulitis. Hemorrhoids* (varicose veins of the lower rectum and anus) frequently occur in people over 50, especially if constipation has been a chronic problem.

Urinary system About half of the kidneys' filtering ability is lost, but the kidneys still retain almost 70 percent of their total function. *Kidney stones* (renal calculi) will reduce both of those rates. The renal plasma flow to the kidneys is progressively reduced to about half by age 70. Although some kidney tissue may be lost (Figure 28.6), each kidney needs only about 25 percent of its original tissue to be able to function properly.

Frequent urination in elderly men may be due to enlargement of the prostate gland, which occurs in about 75 percent of men over 55. *Incontinence* (lack of control over urination) may result when muscles that control the release of urine from the bladder become weakened. Infections of the urinary tract may also become common.

Reproductive system As women grow older, the ovaries decrease in weight and begin to atrophy (Figure 28.7), and menopause usually occurs in the late 40s or early 50s. The vagina also decreases in length and width, its lining usually becomes less moist, and infections and vaginal discharges may result. Fibrous tissue becomes abundant, ovarian cysts are relatively common, and blood vessels harden. The uterus weighs about half as much at age 50 as it did at 30, and its elasticity is lost as elastic tissues are replaced by clumps of fibrous tissue. Ligaments that hold the uterus, bladder, and rectum in place may weaken in older women and allow these organs to drop down. A *cystocele* is a protrusion of the bladder into the vagina, *prolapse* of the uterus is its protrusion into the vagina, and a *rectocele* is a protrusion of the rectum into the vagina.

Ligaments supporting the breasts become lax, fibrous cells replace milk glands, fat tissue is lost, and the breasts sag and become flattened. The amount of pubic hair gradually decreases, as does the layer of fat under the skin in the pubic region.

Reproductive changes due to age are less evident in men than in women. Although the testes do not necessarily decrease in size and weight, the specialized cells that produce and nourish sperm cells gradually become fewer and less active. Sperm count gradually decreases to about 50 percent, but fertility is frequently retained past the age of 80. Accessory glands and organs such as the seminal vesicles and prostate gland begin to atrophy, and the secretion of testosterone decreases. Sexual desire in men and women is usually not affected by old age.

Certain cancers become more prevalent in elderly men and women. Women over 60 are more likely to get cancer of the breast or uterus than younger women, and men over 60 are particularly prone to cancer of the prostate gland.

A Final Word About Growing Old

Many people in our society have been conditioned to regard old age as a time when many functions such as sexual desire, memory, and learning ability are inevitably decreased or even lost altogether. However, such drastic changes in mental and physical prowess are usually the result of serious disease or psychological problems, and are not necessarily a part of the normal aging process. Proper diet, physical activity, and mental attitude will overcome many of the routine problems of old age. Sometimes we give up before our bodies do.

Ask Yourself

1 *Why do we say that there is no single cause of aging?*

2 *Is senility an inevitable part of growing old?*

3 *Why is movement sometimes difficult in old age?*

4 *How does old age affect the ability to learn?*

5 *Why is constipation a common problem with elderly people?*

6 *How does aging affect the respiratory, urinary, and reproductive systems?*

FIGURE 28.6

Effect of aging on kidney tissue. (A) Electron micrograph of normal adult rat kidney. (B) Kidney tissue from aged rat, with the glomerulus wasted away by the accumulation of fibrous tissue. [(A *and* B) *Warren Andrew, Ph.D., M.D.,* The Anatomy of Aging in Man and Animals. *New York, Grune and Stratton, 1971.*]

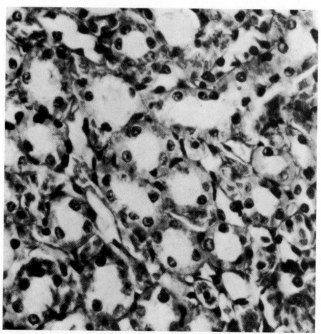

[A]   ×470

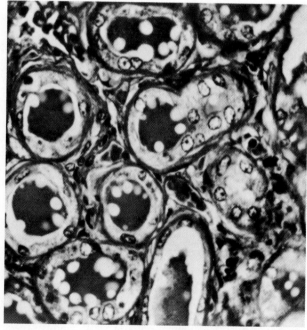

[B]   ×470

FIGURE 28.7

Scanning electron micrographs of the progressive atrophy of a human ovary. An ovary in a 3-year-old (A), a 14-year-old (B), a 27-year-old (C), and a 56-year-old, showing the scarring caused by the many ovulations since puberty (D). A ruptured ovarian follicle (arrow) or corpus hemorrhagium is filled with blood (E) and will subsequently turn yellow and be called a corpus luteum. [*Lennart Nilsson*, Behold Man. *Boston, Little, Brown, 1974.*]

[A]

[B]

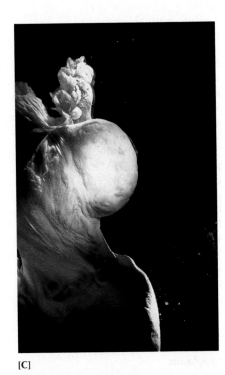

[C]

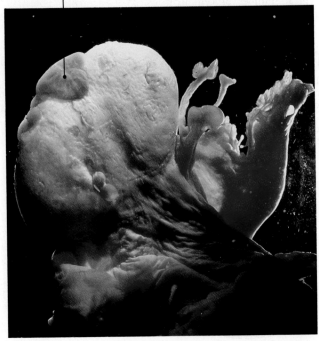

Ruptured ovarian follicle

[D]

[E]

CANCER: CELLS OUT OF CONTROL

Nearly 75 million Americans will get cancer in their lifetime, and about 500,000 will die of cancer this year (one death every minute). Next to cardiovascular diseases, cancer is the leading killer of adults, and only accidental death exceeds cancer as a cause of death in children. Although we have much to learn about cancer, we do know *something* about its prevention and treatment. What we know for certain is that *cancer* occurs when cells grow abnormally and then spread beyond the original site. As you learned earlier, normal human body cells usually divide about 50 times before they die and are replaced by new cells. The new cells develop at exactly the rate required for them to replace the dying ones. Cancer cells are different. They lack a controlling device that "tells" them when to stop reproducing.

Only a tiny fraction of cells that go out of control actually progress far enough to become harmful. Most abnormal cells die simply because they *are* abnormal, since abnormal cells are not equipped to survive. Other abnormal cells die early because the body's immune system recognizes them as "different" and destroys them. The cancer cells that survive, despite the body's built-in defenses against them, are hearty. They grow and spread so steadily that they eventually consume all the available nutrients, literally starving the normal cells to death.

Neoplasms

When cells continue to reproduce instead of differentiating, and when they do not die and stop reproducing after the typical number of cell divisions, they form an abnormal growth of new tissue called a **neoplasm** (Gr. *neos*, new +

L. *plasma*, form). Neoplasms are either benign or malignant. In contrast to a neoplasm, a *tumor* is any abnormal lump or swelling; it is not necessarily malignant.

The cells of a **benign neoplasm** (L. *bene*, well) do not differ greatly in structure from those in normal tissue. Although the cells grow abnormally, their nuclei divide almost like normal cells, and there are usually few chromosomal changes. A benign neoplasm remains safely enclosed within a capsule of thick connective tissue and does not spread beyond its original site (Figure 28.8A). A benign neoplasm can generally be located rather easily, and removed by surgery or other appropriate treatment without further problems. Benign neoplasms occasionally become malignant.

Of all the characteristics of a **malignant neoplasm** (L. *malus*, bad), the most important one is the ability of its cells to break out of their connective-tissue capsule and invade neighboring tissue (Figures 28.8B, C). Malignant neoplasms grow rapidly in an uncontrollable and disorderly pattern that can be recognized by the obvious changes in cell and tissue structure (Figure 28.9) and the many chromosomal changes.

Malignant neoplasms are usually classified according to the type of body tissue in which they originate:

1 A **carcinoma** (L. *carcinoma*, cancerous ulcer; *oma*, tumor) originates in *epithelial tissues*, the tissues that form the skin and the linings of inner organs. Carcinomas are the most common type of cancer, occurring most often on the skin and in the lungs, breasts, intestines, stomach, mouth, and uterus. They are usually spread by way of the lymphatic system.

2 A **sarcoma** (Gr. *sarkoun*, to make fleshy) originates in *connective tissue*, the tissue that forms the body's structural and support system, including bones and cartilage. Sarcomas are usually spread through the bloodstream, and may be found anywhere in the body.

3 A **mixed-tissue neoplasm** derives from tissue that is capable of differentiating into either epithelial or connective tissue and, as a result, is composed of several types of cells.

FIGURE 28.8

Difference between a benign neoplasm and a malignant neoplasm. (A) A benign neoplasm is made up of cells that grow abnormally, but remain enclosed within a sturdy capsule of connective tissue. (B) The cells of a malignant neoplasm burst out of the capsule and can spread to other areas of the body. (C) Cells that break away from a malignant neoplasm may enter the bloodstream through small openings in blood vessels. Once in the bloodstream, cancer cells may establish secondary neoplasms far from the primary neoplasm.

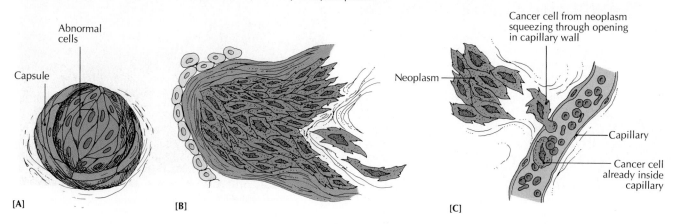

[A]   [B]   [C]

FIGURE 28.9

Changes in cell and tissue structure in malignant neoplasms. (A) Normal cells. (B) Cancerous cells. (C) Normal bronchial tissue. (D) Cancerous bronchial tissue. [(A *and* B) *Dr. Cecil Fox, NCI.* (C *and* D) *Boehringer Ingelheim GmbH. Photos by Lennart Nilsson.*]

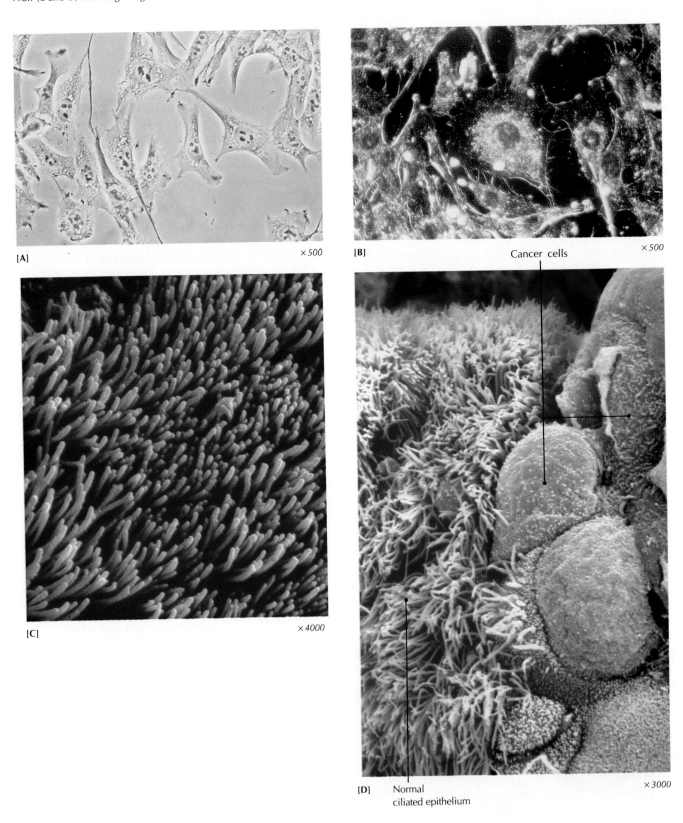

[A] ×500

[B] ×500

Cancer cells

[C] ×4000

[D] Normal ciliated epithelium ×3000

How Cancer Spreads

Because a benign neoplasm does not spread, it usually does not do any harm. In contrast, a malignant neoplasm is *cancer.* The difference lies in the ability of the malignant neoplasm to spread to other parts of the body. This spread is called **metastasis** (muh-TASS-tuh-siss; Gr. *meta,* involving change + *stasis,* state of standing).

Cancer starts to spread, or *metastasize,* when cancer cells break away from the original site, or *primary neoplasm,* and extend into normal tissues instead of merely pushing them aside the way a benign neoplasm does. Once cancer cells move from the original site, they may enter a body cavity such as the chest cavity or abdominal cavity. Or they may enter the bloodstream or the lymphatic system and be transported away from the primary neoplasm. They can then establish new malignant *secondary neoplasms* elsewhere in the body (Figure 28.10).

As cancer cells metastasize, they encounter defensive cells of the body's immune system and are sometimes killed (Figure 28.11). Cancer is lethal when the body can no longer fight off the ever-increasing number of invading malignant cells, and the malignant cells totally disrupt and starve healthy ones.

Although cancer cells are able to establish secondary neoplasms, they do not always do so. In fact, most cancer cells that reach the bloodstream die quickly in the apparently hostile environment. Cancer cells that do survive in the bloodstream are most effective in forming a new colony when they are able to group into large clumps of cells that adhere to the cells that line the blood vessels. Such clumps of cancerous cells are held together and protected by fibrin, the same fibrous protein that helps form blood clots. Apparently, the cancer cells actually stimulate the production of fibrin.

In order for transported cancer cells to form secondary neoplasms, they must first escape from the blood vessels. Cancer cells seem to have the ability to cause the cells that line blood vessels to separate slightly, producing spaces in the vessels that are large enough for the cancer cells to squeeze through. Cancer cells enter blood vessels in the same way (see Figure 28.8c). Table 28.2 shows some of the differences between benign and malignant neoplasms.

| *Why is cancer called by that name?* | *The word ''cancer'' is the Latin word for crab. The disease was originally named cancer because the swollen veins surrounding the affected part resemble the limbs of a crab. That is understandable, but what is noteworthy is that today we know that cancer looks (and acts) like a crab even on a cellular level. (The early Romans would have had no way of knowing that.) The scanning electron micrograph shows a malignant cancer cell, ''creeping'' along in the manner of a crab. [AP/Wide World Photos.]* |

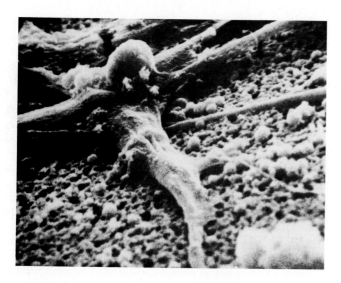

FIGURE 28.10

Metastasis of a melanoma, a cancer of a pigment-containing cell in the skin. Cancer spreads when malignant cells break away from the primary neoplasm and enter the circulatory system through lymphatic vessels or blood vessels. Malignant cells may be temporarily slowed down in lymph nodes before they move on to other tissues and organs. Melanoma frequently spreads to the brain, lungs, liver, and ovaries.

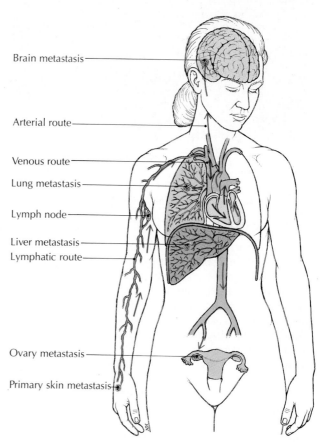

Brain metastasis

Arterial route

Venous route

Lung metastasis

Lymph node

Liver metastasis
Lymphatic route

Ovary metastasis

Primary skin metastasis

FIGURE 28.11

Death of a cancer cell, shown in a series of scanning electron micrographs. (A) A mouse cancer cell. (B) A macrophage anchored by its tendril-like filopodia. (C) The macrophage attacking the cancer cell. The typically wrinkled cancer cell begins to smooth out as it weakens. (D) The dying cancer cell. After about six hours of attack by the macrophage, holes develop in the surface of the cancer cell, indicating that it is dead or close to death. [(A–D) *Courtesy of The Upjohn Company.*]

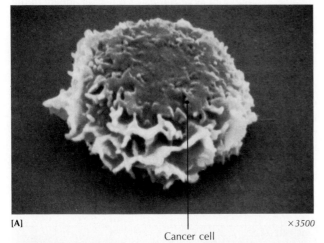

[A]   ×3500

Cancer cell

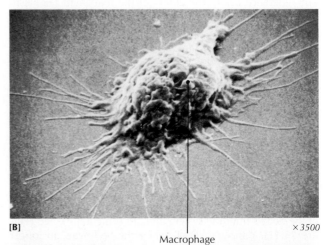

[B]   ×3500

Macrophage

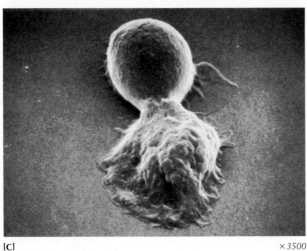

[C]   ×3500

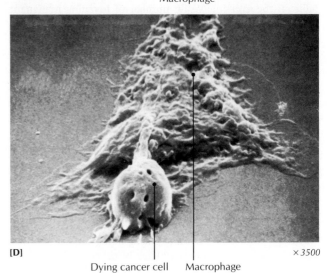

[D]   ×3500

Dying cancer cell   Macrophage

Causes of Cancer

In some cases, the tendency to develop cancer may be inherited, but cancer researchers believe that most cancers are caused by repeated exposures to cancer-causing agents, called **carcinogens.** The main categories of carcinogens are *chemicals, radiation,* and *viruses.* Carcinogens disrupt the homeostasis of normal cells and may eventually cause them to become cancerous. As the affected cells divide, they produce cells that are also cancerous, and as the condition of the cells becomes more and more abnormal, the cells produce their cancerous offspring faster and faster.

Most cancers are probably the result of prolonged exposure to a *combination* of several carcinogens (Table 28.3). For instance, both tobacco smoke and asbestos have been implicated as likely carcinogens. A heavy smoker who works as an installer of asbestos insulation has a far greater risk of developing lung cancer than an office worker who smokes heavily. The risk is increased even more if the heavy-smoking asbestos worker lives in a city with a high rate of air pollution.

Cancer of the large intestine may be related to diet. Richer countries, where fatty red meat is consumed regularly, generate more cancer deaths than countries where more cereal than meat is eaten. Studies done on women subjects indicate that deaths due to cancer of the large intestine are 40 times more prevalent among women in New Zealand (where red meat is plentiful) than in Nigeria, and 30 times more prevalent among American women than among Japanese women. However, Japanese women who move to the United States and adopt a typical American diet show an increased incidence of cancer of the large intestine. The same is true of Japanese men.

The World Health Organization has reported that approximately 85 percent of all cancer cases result from such environmental causes as smoking, overeating, overexposure to

TABLE 28.2 ESSENTIAL DIFFERENCES BETWEEN BENIGN AND MALIGNANT NEOPLASMS

Characteristic	Benign	Malignant
Growth	Slow expansion; push aside surrounding tissue, but do not infiltrate.	Usually infiltrate surrounding tissues rapidly, expanding in all directions.
Limitation	Frequently encapsulated.	Seldom encapsulated; often poorly delineated.
Recurrence	Rare after surgical removal.	Frequent after surgical removal due to infiltration into surrounding tissues.
Structure of cells	Closely resemble cells of tissue of origin.	May differ considerably from cells of tissue of origin.
Differentiation	Well differentiated.	Poor or no differentiation.
Cell division	Slight.	Extensive.
Metastasis	None.	Spread throughout body cavities and via blood and lymph systems to sites far away from primary neoplasm; establish secondary neoplasms.
Tissue destruction	Usually slight.	Extensive due to infiltration and metastatic lesion.
Effect on body	Usually not fatal, but may obstruct vital organs, exert pressure, produce excess hormones; can become malignant.	Cachexia typical;* fatal if untreated.

*Cachexia is a general wasting away of the body, involving anemia, emaciation, weakness, and similar signs.

Source: Adapted from *Diseases* (Horsham, Pa.: Nursing 81 Books, Intermed Communications, Inc., 1981), p. 183. Used with permission.

How soon after exposure to a carcinogen does cancer appear?	*If cancer does occur (and not all exposures to a carcinogen produce cancer), it usually develops slowly. Cancers of the liver, lungs, or bladder may not appear until 30 years after exposure to vinyl chloride, asbestos, or benzidine. Cigarette smokers usually develop lung cancer about 20 years after they start smoking.*

TABLE 28.3 SOME COMMON ENVIRONMENTAL CARCINOGENS

Substance	Source	Target tissues in humans
Arsenic	Mining and smelting industries.	Skin, lungs, liver.
Asbestos	Brake linings, construction sites, insulation, powerhouses.	Lungs, pleurae, peritoneum.
Benzene	Solvents, oil refineries, insecticides.	Bone marrow and other hemopoietic tissue.
Benzidine	Rubber making, dyes.	Urinary bladder.
Coal-combustion products (soot, tars, oils)	Steel mills, petrochemical industry, asphalt, coal tar.	Lungs, bladder, scrotum.
Nickel compounds	Metal industry, alloys.	Lungs, nasal sinuses.
Radiation	Ultraviolet rays from the sun, medical therapy.	Bone marrow, skin, thyroid.
Synthetic estrogens	Drugs.	Vagina, cervix, uterus.
Tobacco smoke	Cigarettes, pipes, cigars.	Lungs, urinary bladder, mouth, esophagus, pharynx, larynx, pancreas, kidneys.
Vinyl chloride	Plastics industry.	Liver, brain, lungs.

sunlight and other radiation, overdrinking, and exposure to dangerous chemicals such as asbestos and vinyl chloride.

In the United States, cigarette smoking is responsible for almost half of the cancers known to be caused by environmental carcinogens. It could very well be, then, that humans are actively elevating cancer to epidemic status. We know for certain that high doses of x rays and similar radiation can cause leukemia and other cancers. Clearly, human actions are responsible for the extremely high incidence of leukemia and cancer of the skin, breasts, intestines, and brain in Hiroshima and Nagasaki. At least one possibility for cancer prevention is clear: humans can prevent the cancers they cause.

Mutation hypothesis One hypothesis for the development of cancer suggests that carcinogens cause mutations in DNA (Figure 28.12). Current thinking is that certain genes help prevent cancer until they are made ineffective by a mutation. The chances of a successful mutation are increased after repeated exposure to carcinogens. Therefore, the probability of DNA mutations increases as we get older. If several mutations

Mutated cancer cells. (A) Normal human cells. (B) Irregular human cells mutated by radiation. [(A *and* B) *Dr. T. T. Puck, Eleanor Roosevelt Institute for Cancer Research.*]

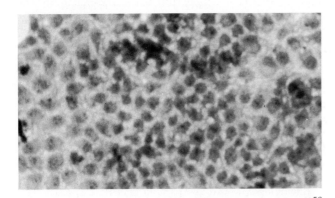

[A]   × 50

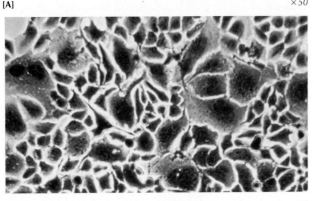

[B]   × 50

are required to finally cause cancer, then the cancer may have actually begun 10 or even 30 years before it becomes evident.

A variation of the mutation hypothesis suggests that an extensive rearrangement of genetic material, rather than a single gene, produces effective mutations. This hypothesis proposes that such massive changes are caused by chemicals and viruses.

"Cancer-gene" hypothesis Recently discovered genes, called ***oncogenes*** (Gr. *onkos,* tumor), are believed to be present in all normal cells. They usually function in keeping cell division turned on during the early stages of development, and are shut off permanently afterward. The "cancer-gene," or oncogene, hypothesis proposes that cancer occurs when oncogenes are reactivated by one or more processes, such as the following:

1 Regulatory genes that normally keep oncogenes inactive become ineffective through mutations.

2 Mutated oncogenes cannot be kept inactivated.

3 Viruses introduce uncontrollable oncogenes into normal cells.

4 Chromosome or gene breaks separate oncogenes from the genes that control them, allowing oncogenes to become reactivated.

Apparently, reactivated oncogenes trigger the growth of cancerous cells through several mechanisms. They may activate other genes that stimulate the excessive production and release of "growth factors," which cause cells to divide uncontrollably. Oncogenes may also function by causing changes in surface receptors on the plasma membranes of cells. The changes cause the cell to act as though it were activated by a growth factor. Growth-factor genes are sometimes altered to become oncogenes that cause the continuous production of growth factor.

Virus hypothesis Considerable evidence has been accumulated to support the idea that at least some types of cancer are caused by viruses. It is thought that viruses may be involved in causing Burkitt's lymphoma (Epstein-Barr virus), warts (benign papillomas of the skin), cervical cancer (herpes type II and certain papilloma viruses), and hepatocellular cancer (hepatitis B virus).

An important question about the viral origin of cancer is whether the cancer-causing virus is already present in the body, or whether it invades the body from outside. Either way, viruses have no ability to reproduce on their own. Instead, they insert their DNA into a host cell, where the DNA replicates, and new viruses eventually burst forth from the now-dead host cell on their way to injecting other host cells. An alternative besides death exists for a host cell that is invaded by a virus. The DNA of the virus may contain genetic information that causes the DNA of the host cell to reproduce uncontrollably, essentially transforming it into a cancer cell.

Some Common Cancer Sites

Cancer can strike anywhere in the body, but it usually occurs most often where cell division is rapid and frequent, such as in the lining of the digestive and respiratory tracts. The most common cancer sites in men are the lungs, prostate gland, and the colon and rectum, and the most frequently struck sites in women are the breasts, colon and rectum, uterus, and lungs (Tables 28.4 and 28.5). Some of the most common cancers have been discussed in earlier chapters.

Treatment of Cancer

Living in a polluted environment as we do, cancer cells form in our bodies frequently. Fortunately, the T cells of the immune system continuously screen and monitor the body for cancer cells, killing nearly all of them before they can multiply and spread. The current hypothesis about the effectiveness of T cells in monitoring the body for cancer cells proposes that the plasma membrane of cancer cells is altered enough so that wandering T cells can recognize the cells as "nonself" antigens and destroy them before they do harm. Without the constant surveillance of T cells, cancer would probably be much more prevalent than it is. However, the immune system may not always be at peak efficiency, and sometimes it does not recognize the cancer cells as "nonself." When the immune system fails, and the cancer cells begin to multiply and spread, we must turn to medical treatments.

The main treatments for cancer are surgery, radiation therapy, and drug therapy (chemotherapy). Although new forms of treatment are constantly being tested, these three remain the most effective, and survival rates continue to increase as the current methods become even more refined (Table 28.6).

Surgery Benign neoplasms can be removed easily by *surgery,* but the cells of malignant neoplasms may spread enough so that they cannot be completely removed with certainty. Surgery is useless when cancer cells have already metastasized to other parts of the body. Ordinarily, when a malignant neoplasm is removed, the tissue and lymph nodes adjacent to it are removed also, just in case some cancer cells have invaded them. Surgery is becoming more and more precise because of improved diagnostic techniques and surgical equipment. One highly effective tool is the *laser,* which produces an intense beam of light that can destroy or remove the neoplasm, usually without producing the extensive bleeding that accompanies major surgery. Surgery is still the main method of treating cancers of the breast (a *mastectomy* is the surgical removal of a breast), colon and rectum (*colec-*

tomy), lung (*lobectomy*), stomach (*gastrectomy*), and uterus (*hysterectomy*).

Radiation therapy *Radiation therapy* uses x rays or rays from radioactive substances such as cobalt or radium to kill cancer cells. If used properly, radiation therapy will not harm a significant number of normal cells. This form of therapy is used most often to treat cancer of the bladder, cervix, skin, and parts of the head and neck. Improvements in radiation equipment have made radiation therapy increasingly effective, and machines such as the cyclotron, supervoltage x-ray machine, and linear accelerator can penetrate deeper than earlier machines with less harm to healthy tissue. Side effects such as nausea, diarrhea, and local skin burns sometimes occur, and there may also be long-term effects such as genetic mutations and increased susceptibility to leukemia.

Chemotherapy Drug therapy, or **chemotherapy,** has become increasingly effective in treating cancer, especially leukemia and lymphoma. At least 50 drugs have been used to kill or retard the growth of cancer cells. Chemotherapy researchers continue to find new drugs, or combinations of drugs, that selectively kill cancer cells with less disruption of normal cells, and with fewer undesirable side effects such as loss of hair, anemia, vomiting, and gastrointestinal ulceration and bleeding. Side effects can be so serious that some patients are killed by the chemicals instead of the cancer.

Cancer-killing drugs approved by the Food and Drug Administration (FDA) fall into four categories: (1) *alkylating agents* prevent cell division by disrupting the pairing process of DNA; (2) *antimetabolites* resemble vitamins, but impair a cell's metabolism by substituting for substances the cell needs to survive; (3) *antibiotics* upset the synthesis of RNA; and (4) *steroids* are thought to prevent the production of key enzymes or other proteins.

Combination therapy *Combination therapy* uses more than one type of therapy to treat cancer. Surgery and radiation therapy have been used together for some time, and physicians are now including follow-up chemotherapy, especially for cancer of the breast, bones, colon and rectum, lung, and stomach.

Future cancer therapy Future cancer drug therapies may rely on the production of *monoclonal antibodies* (see Chapter 20) against specific cancer cells (Figure 28.13). A small amount of cancerous tissue is removed from the patient, and the cancerous cells are injected into a laboratory mouse. The immune system of the mouse recognizes the human cancer cells as "nonself" antigen, and the antibody-producing spleen lymphocytes (B cells) of the mouse begin to multiply. After several days, the mouse B cells are removed and fused (mixed) with myeloma cells (malignant white blood cells) of another laboratory mouse. The myeloma cells are easily grown in a tissue culture. When the different types of cells fuse, they form a *hybridoma,* a hybrid line of cells that has the desired properties of both original cell types: specific antibody production from the mouse B cells and rapid cell division from the myeloma. The hybridomas are then grown in large numbers to form *clones,* cells that are identical to the

TABLE 28.4 PERCENT OF CANCER INCIDENCE AND DEATHS BY SITE AND SEX*

Site	Incidence		Deaths	
	Male	Female	Male	Female
Skin	3	3	2	2
Mouth	4	2	3	1
Breast	—	27	—	18
Lung	20	11	36	20
Colon and rectum	14	16	11	14
Pancreas	3	3	5	5
Ovary	—	4	—	5
Uterus	—	10	—	4
Prostate	20	—	10	—
Urinary tract	10	4	5	3
Blood and lymphoid tissue (leukemia and lymphomas)	8	7	9	9
All other	18	13	19	19

*Based on 1987 estimates; includes nonmelanoma skin cancer and carcinoma in situ.
Source: American Cancer Society.

TABLE 28.5 THE SIX MOST COMMON CANCER SITES (AMERICAN MEN AND WOMEN)

Site	Number of new cases in 1987	Number of deaths in 1987
Lung	150,000	136,000 (91%)
Colon and rectum	145,000	60,000 (41%)
Breast	130,000	41,300 (32%)
Prostate gland	96,000	27,000 (28%)
Uterus	48,000	9,700 (20%)
Urinary bladder	45,000	10,600 (24%)

Source: American Cancer Society.

908

Chapter 28: The Body in Transition: Aging and Cancer

original cells. The cloned cells are tested for the desired antibody, and only the cells producing the antibody are cloned further.

Because all of the desired cloned cells produce the same antibody (hence the term *monoclonal antibody*), large amounts of antibodies specific to the patient's cancer can be produced. Also, potentially lethal cancer drugs or toxic molecules can be attached to the antibodies, which are then injected into the patient. The antibodies, with their "passenger" drugs, recognize only cancer cells, destroying them with little or no harm to normal body cells.

Ask Yourself

1 *What is the difference between a benign neoplasm and a malignant neoplasm?*

2 *What is a carcinoma?*

3 *What is metastasis?*

4 *What is a carcinogen? What are the main categories of carcinogens?*

5 *What is the "cancer-gene" hypothesis?*

6 *What are the three main types of cancer treatment?*

7 *How are monoclonal antibodies used to fight cancer?*

FIGURE 28.13

Production of monoclonal antibodies to treat cancer.

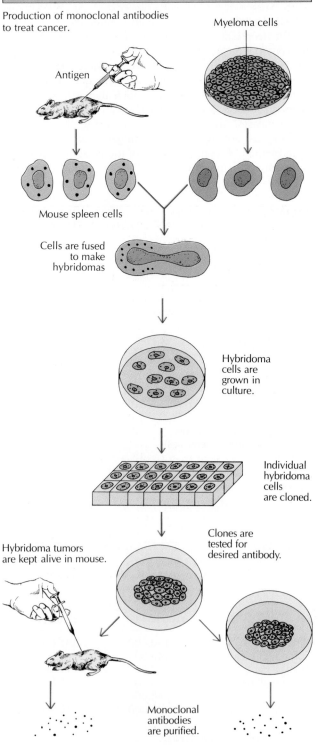

TABLE 28.6   FIVE-YEAR CANCER SURVIVAL RATES

Type of cancer	Survival rate (percent of cases diagnosed)*		
	1960–1963	1970–1973	1974–1982[†]
Urinary bladder			75
Males	53	61	
Females	53	60	
Breast (females)	63	68	74
Colon and rectum			52
Males	42	47	
Females	44	50	
Hodgkin's disease			73
Males	34	66	
Females	48	69	
Children	52	90	
Leukemia (acute)			33
Males	4	27	
Females	3	29	
Children	4	34	
Lung			13
Males	7	9	
Females	11	14	
Melanoma			80
Males	51	62	
Females	68	75	
Ovary	32	36	38
Testis	63	72	89

*Survival rates are based on the number of patients who are alive, without the cancer recurring, five years after treatment. Survival rates are expected to continue to rise as more effective cancer treatments are developed.
[†]Statistics for 1974 to 1982 are not available in separate categories for males, females, and children.
Source: National Cancer Institute, American Cancer Society.

MEDICAL TERMINOLOGY

ADENOMA (Gr. *aden,* gland + *-oma,* tumor) An epithelial tumor of glandular origin, usually benign or of low-grade malignancy.

ANAPLASIA (Gr. *ana,* reversion + *plasia,* growth) Reversion of cells to a less differentiated form.

BIOPSY (Gr. *bios,* life + *-opsy,* examination) The microscopic examination of living tissue removed from the body.

CARCINOMA IN SITU (*in situ,* in position) A malignant epithelial tumor in which no immediate invasion of adjacent tissue occurs.

GERIATRICS (Gr. *geras,* old age + *iatrikos,* physician, pertaining to a specific kind of medical treatment) The medical study of the physiology and ailments of old age.

HYPERPLASIA (Gr. *hyper,* above + *plasia,* growth) Enlargement of an organ or tissue due to an increase in the number of cells.

HYPERTROPHY (Gr. *hyper,* above + *trophe,* nourishment) Enlargement of an organ or tissue due to an increase in the size of the cells.

MELANOMA (Gr. *melas,* black + *-oma,* tumor) Neoplasma made up of cells containing a dark pigment, usually melanin.

METAPLASIA (Gr. *meta,* involving change + *plasia,* growth) The change of cells from a normal state to an abnormal one.

ONCOLOGY (Gr. *onkos,* mass, tumor + *logy,* the study of) The study of neoplasms; the study of cancer.

SUMMARY

Physiology of human aging

1 Part of the aging process includes the natural death of cells, or *necrobiosis.* The three types of nuclear changes that accompany cell death are *pyknosis, karyorrhexis,* and *karyolysis.*

2 *Senescence,* or *primary aging,* involves the natural, hereditary changes in the structure and composition of cells that occur with age. *Senility,* or *secondary aging,* involves changes that take place because of disease or accidental injury.

3 Aging is not a single process, but a series of interrelated events that accumulate until a change becomes noticeable and permanent.

4 No single theory can explain aging. Among the mechanisms that produce aging are DNA mutations, free-radical damage, loss of redundant DNA, genetically determined decreased cell efficiency, genetically determined limit on cell division, strain on cells, accumulation of wastes, autoimmunity, thickening of collagen, glyosylation, hormone imbalance, and cell starvation.

5 The most lethal diseases of old age are cardiovascular diseases, cancer, and stroke. Old age by itself is not a cause of death.

6 Every body system is affected adversely in some way by the aging process. Among the most common age-related ailments are cardiovascular diseases, cancer, arthritis and other joint diseases, metabolic disorders, and diseases of the central nervous system.

Cancer: cells out of control

1 When cells reproduce faster than the normal rate, and when they do not die after the typical number of cell divisions, they form an abnormal growth of tissue called a *neoplasm.*

2 The cells of a *benign neoplasm* grow abnormally, but do not differ significantly in structure from those in normal tissue and do not spread beyond their original site. The most important characteristic of a *malignant neoplasm,* or *cancer,* is the ability of its cells to invade surrounding tissues.

3 Malignant neoplasms are usually classified according to the type of body tissue in which they originate. *Carcinomas* originate in epithelial tissue, *sarcomas* in connective tissue, and *mixed-tissue neoplasms* in tissue that is capable of differentiating into either epithelial or connective tissue.

4 The spread of cancer cells beyond the primary neoplasm and the establishment of secondary neoplasms in other parts of the body is called *metastasis.*

5 The main categories of *carcinogens* (cancer-causing agents) are *chemicals, radiation,* and *viruses.* It is thought that most cancers are caused by repeated and prolonged exposures to a *combination* of several carcinogens.

6 Hypotheses to explain the basic cause of cancer involve DNA mutations, oncogenes (cancer genes), and viruses.

7 The most common cancer sites are the lungs, colon and rectum, breasts, prostate gland, uterus, and urinary bladder.

8 The main treatments of cancer are *surgery, radiation therapy,* and *chemotherapy. Combination therapy* uses more than one type of therapy to treat cancer.

9 The future treatment of cancer may involve the use of *monoclonal antibodies.*

UNDERSTANDING THE FACTS

1 What are the three main causes of death in old age?

2 Distinguish between necrobiosis and necrosis.

3 What is the difference between primary and secondary aging?

4 What kinds of changes occur within the nucleus that cause the death of a cell?

5 Which cells, if any, have the ability to be immortal?

6 Which body functions diminish the most with age? The least?

7 What is the free-radical theory of aging?

8 What is Parkinson's disease?

9 List some of the more common visual problems of older people.

10 What is probably the most fundamental difference between cancer cells and normal cells?

11 Malignant neoplasms are usually classified according to the tissues in which they originate. List these three categories.

12 Name some of the more important carcinogens.

13 What are the four approved types of cancer-fighting drugs?

14 What are the advantages of laser techniques over conventional surgery?

UNDERSTANDING THE CONCEPTS

1 What is gerontology, and why is more emphasis being placed on this area of medicine today than ever before?

2 What are some of the reasons the elderly are more susceptible to disease?

3 What age-related changes occur in the cardiovascular system?

4 Why do elderly people find it difficult to read a menu in a candlelit restaurant?

5 Discuss the following hypotheses about aging: (a) DNA mutations, (b) redundant DNA, (c) genetic clock, (d) limit on cell divisions, (e) collagen thickening.

6 Is "old age" a cause of death? Explain.

7 Why is exercise of great importance to the elderly? Consider both psychological and direct physical benefits.

8 Do you think the expression "mind over matter" can relate to aging in any way?

9 Some cancers take many years to develop after exposure to a carcinogen or group of carcinogens. How does this fact relate to the mutation hypothesis?

10 Why is cancer more prevalent in areas of the body where cell division is rapid and frequent?

11 Why has lung cancer increased among women during the last few years?

12 Explain how cancer spreads from the primary site to secondary sites in the body.

13 How can you explain the fact that the five-year survival rate for lung cancer (see Table 28.6) has not shown much improvement despite better detection and treatment?

14 What are some of the ways in which an individual can lower the risk of developing cancer?

15 Discuss the following possible causes of cancer: (a) mutation hypothesis, (b) "cancer-gene" hypothesis, (c) virus hypothesis.

SUGGESTED READING

CAIRNS, JOHN, "The Treatment of Diseases and the War against Cancer." *Scientific American,* November 1985.

Cancer Facts and Figures. New York: American Cancer Society, 1984.

COHN, JEFFREY P., "The Molecular Biology of Aging." *Bioscience,* February 1987.

CROCE, CARLO M., AND GEORGE KLEIN, "Chromosome Translocations and Human Cancer." *Scientific American,* March 1985.

CROCE, CARLO M., AND HILARY KOPROWSKI, "The Genetics of Human Cancer." *Scientific American,* February 1978.

FELDMAN, MICHAEL, AND LEA EISENBACH, "What Makes a Tumor Cell Metastatic?" *Scientific American,* November 1988.

FINCH, C. E., AND G. MOMENT, in J. A. Behnke, ed., *The Biology of Aging.* New York: Plenum, 1978.

FRIEDBERG, ERROL C., ed., *Cancer Biology: Readings from Scientific American.* New York: Freeman, 1986.

HAYFLICK, LEONARD, "The Cell Biology of Human Aging." *Scientific American,* January 1980.

HUNTER, TONY, "The Proteins of Oncogenes." *Scientific American,* August 1984.

KARTNER, NORBERT, AND VICTOR LING, "Multidrug Resistance in Cancer." *Scientific American,* March 1989.

KUPCHELLA, CHARLES E., *Dimensions of Cancer.* Belmont, Calif.: Wadsworth, 1987.

LEAF, ALEXANDER, "Getting Old." *Scientific American,* September 1973.

NICOLSON, GARTH L., "Cancer Metastasis." *Scientific American,* March 1979.

OLD, LLOYD J., "Tumor Necrosis Factor." *Scientific American,* May 1988.

SACHS, LEO, "Growth, Differentiation and the Reversal of Malignancy." *Scientific American,* January 1986.

WEINBERG, ROBERT A., "A Molecular Basis for Cancer." *Scientific American,* November 1983.

WEINBERG, ROBERT A., "Finding the Anti-Oncogene." *Scientific American,* September 1988.

29
Human Genetics

LEARNING OBJECTIVES

1 Define genetics, and describe the relationship between chromosomes and genes.

2 Discuss the following concepts: dominant versus recessive alleles, genotype versus phenotype, and homozygous versus heterozygous.

3 Explain the construction and use of the Punnett square.

4 Explain the concept of multiple alleles, and give an example of how they are expressed.

5 Describe the causes and effects of mutations, and compare germinal and somatic mutations.

6 Give examples of autosomal dominant and recessive inheritance.

7 Discuss polygenic inheritance, and give an example of it.

8 Define sex-linked inheritance, and cite an example to show how a particular trait may be transmitted from parents to children.

9 Describe the following alterations in the arrangement of chromosomes: deletion, duplication, inversion, and translocation.

10 Define nondisjunction, and give an example of it.

11 Give some examples of aberrations in sex chromosomes.

12 Discuss the functions and techniques of genetic counseling.

G enetics is the science of heredity. It examines the normal transmission and expression of characteristics from one generation to the next, and also the *variation* of that normal transmission. The science of genetics is a relatively new one. The very word *genetics* was first used in 1906 by the English biologist William Bateson, and not until three years later was the word *gene* introduced by Wilhelm Johannsen. It wasn't until 1956 that we knew for certain that most human cells have 46 chromosomes, arranged in 23 pairs. Until that time, all the textbooks said the number was 48.

Until fairly recently, the study of human genetics was limited to a few inherited features we could see easily: the color of our eyes, the shape of our nose, the color of our hair. Today, improved laboratory techniques that are acceptable for the study of human beings are yielding a growing knowledge of human genes and how they operate to influence the many traits that are passed on from one generation to the next.

Much of the study of human genetics has been based on diseases or chromosomal abnormalities that are inherited. This is not because of a morbid interest in diseases and abnormalities, but because by starting with the relatively few "abnormal" conditions, we can learn about the many more "normal" conditions by trying to determine what causes things to "go right" most of the time. If we know what causes things to go right, we may be able to correct the things that "go wrong."

INTRODUCTION TO MENDELIAN GENETICS

The science of genetics was started in the middle of the nineteenth century by an obscure Czechoslovakian monk named Gregor Mendel. It was not until 1900, however, that Mendel's contributions were recognized. We can present only an outline of his monumental findings here as a basis for understanding genetics as it applies to human beings.

Dominance and Recessiveness

Working with garden peas, Mendel realized that a characteristic for a given trait (tall, for stem length, for example) could appear in all individuals in the first generation, masking the other stem-length characteristic (dwarf). However, the masked characteristic (dwarf) was still present somewhere in some of the first-generation plants, because it appeared in the second generation. Mendel called the characteristic that showed (tall) **dominant** because it seemed to be more powerful than the temporarily hidden, or **recessive,** characteristic (dwarf). Examples of dominant and recessive characteristics in human beings are normal red blood cells and sickle-cell trait, and free earlobes and attached earlobes (Table 29.1).

Chromosomes, Genes, and Alleles

A **chromosome** (Gr. "colored body," because it colors readily when a cell's nucleus is stained) is a threadlike nucleoprotein

TABLE 29.1	SOME DOMINANT AND RECESSIVE HUMAN GENETIC TRAITS
Dominant trait	**Recessive trait**
Large eyes	Small eyes
Nearsightedness	Normal vision
Farsightedness	Normal vision
Normal color vision	Color blindness
Glaucoma	Normal eye pressure (no glaucoma)
Normal night vision	Night blindness
Astigmatism	Normal vision
Long eyelashes	Short eyelashes
Nonred hair	Red hair
Dark hair	Light hair
Premature baldness (male)	Normal hair
Normal pigmentation	Albinism
Freckles	No freckles
Dimples in cheeks	No dimples in cheeks
Free earlobes	Attached earlobes
Ability to curl tongue	Inability to curl tongue
Convex nose bridge	Concave or straight nose bridge
Achondroplasia (dwarfism)*	Normal height
Polydactylism (extra fingers or toes)	Normal fingers or toes
Syndactylism (webbed fingers or toes)	Normal fingers or toes
Brachydactylism (very short fingers)	Normal fingers
Normal arches in feet	Flat feet
Huntington's chorea	Normal nervous system
Normal sugar metabolism	Diabetes mellitus
Blood group A, B, AB	Blood group O
Normal red blood cells	Sickle-cell trait
Normal blood clotting	Hemophilia
Rh antigen	No Rh antigen
Migraine headaches	Normal

*Although such traits as achondroplasia and polydactylism are dominant, they are also rare. For this reason, most people are of normal height and have 10 fingers and 10 toes.

structure within the nucleus of a cell that contains hereditary information in the form of DNA (deoxyribonucleic acid). Most cells in the body contain DNA (mature red blood cells do not). Each strand of DNA contains segments called **genes** that control specific cellular functions, either by synthesizing enzymes or other proteins, or by regulating the action of other genes. Because enzymes regulate most of the chemical reactions in cells, we can say that genes affect the *behavior* of cells. The sequential action of DNA synthesis (see Chapter 3) keeps the molecular structure of each gene intact.

Chromosomes are arranged in pairs, as are genes. The chromosome pairs and the genes located on them separate during meiosis (see Chapter 26). Chromosome pairs form again when male and female nuclei fuse during fertilization. Gene pairs also form again because genes are physical units located on chromosomes. Each member of a pair of genes is called an **allele** (uh-LEEL), and each gene has its own specific place, or *locus*, on a chromosome. There are at least two alleles for each trait. For example, for purposes of this discussion, the pair of alleles for nose shape may be written as *Aa*, where *A* stands for a convex nose bridge and *a* stands for a concave or straight nose bridge. Each parent passes on one allele (*A* or *a*) to the child. The *A* allele is considered to be dominant, which means that it overrides the recessive *a* allele.

The alleles of a pair separate, or *segregate*, during the formation of gametes (sex cells) without blending or being altered in any way. This represents Mendel's *Law of Segregation*. Prior to Mendel's concept of a clean separation of alleles during sexual reproduction, it was believed that the hereditary units were like two colors of paint, and that once the colors were blended together, the original colors were lost forever.

Homozygosity and Heterozygosity

When a person has a pair of identical alleles, such as *AA* or *aa* for a given trait, that person is said to be **homozygous** for that trait. In contrast, a person is **heterozygous** for a given trait when the pair of alleles does not match, such as *Aa*.

Genotypes and Phenotypes

The genetic make-up, sometimes hidden, of a person is that person's **genotype** (JEEN-oh-tipe), which may be either homozygous (*AA* or *aa*) or heterozygous (*Aa*) for any one characteristic. The allele that is dominant will be *expressed*, that is, it will determine the appearance and physiology of the individual. The expression of a genetic trait (how a person looks or functions) is that person's **phenotype** (FEE-noh-tipe; Gr. "that which shows") (Figure 29.1). The phenotype is determined by the interaction of genes *and* the environment. For example, a person's growth is programmed in the chemical message of the genes, but the proper growth can take place only with the necessary environmental contributions of

water, food, minerals, and vitamins. In another example, a person may have relatively dark hair, but the hair may be lightened by constant exposure to the sun. In both examples, the change is only superficial, and does not affect the genes or genotype.

The Punnett Square

When geneticists know the genotypes of parents for any given trait, they can determine the possible genotypes of the children for that trait by constructing a **Punnett square,** named for the geneticist who first devised the scheme (Figure 29.2). A master square is drawn, and the squares inside it represent the various possible combinations of parental alleles that can occur when a zygote is formed during fertilization. The genotype of the female parent (showing the alleles) is usually placed at the top of the master square, with the male parental genotype at the left side. When only one trait is being tested, with only one pair of alleles for each parent, four squares are drawn inside the master square, since four possible combinations are possible (see Figure 29.2A). Such a combination of parental alleles indicates a *monohybrid cross*. When two different traits are being tested, with two pairs of alleles for each parent, there are 16 possible combinations and thus 16 interior squares (Figure 29.2B). This is a *dihybrid cross*.

MULTIPLE ALLELES

Many traits are affected by more than two alleles. Such alleles are called **multiple alleles.** (Note, however, that a person can have only two alleles, one on each chromosome, at any one specific locus.) A well-known example of multiple alleles in human beings is the ABO blood-grouping system. In this system, at least three alleles are available to determine the blood type of any person. However, it is only two of these alleles that combine to produce four different blood types: A, B, AB, or O. As you learned in Chapter 17, the different blood types are classified according to the presence or absence of two proteins (agglutinogens) on the surface of red blood cells. These proteins are labeled A or B. If only the A protein is present, the blood type is A. If only the B protein is present, the blood type is B. If both A and B are present, the blood type is AB. If neither protein is present, the blood type is O.

If we let *I* equal a dominant trait and *i* equal a recessive trait, we can say that blood type A may have a genotype of $I^A I^A$ or $I^A i$, blood type B may have a genotype of $I^B I^B$ or $I^B i$, blood type AB can only have a genotype of $I^A I^B$, and blood type O can only have a genotype of $i^O i^O$. Alleles I^A and I^B are *codominant* (characteristics of both phenotypes appear) when paired with each other (hence the AB type), but both I^A and I^B are dominant over the recessive allele i^O.

If we apply simple genetic rules to these genotypes and phenotypes, we can determine the inheritance of ABO blood types (Table 29.2). For example, a woman with type A blood may be $I^A I^A$ or $I^A i$. If she marries a man with type B blood ($I^B I^B$ or $I^B i$), she may have children with type A ($I^A i$), type B

FIGURE 29.1

Genotypes and phenotypes, using nose shape as an example. Note that a person who is heterozygous for the trait (*Aa*) has a convex nose bridge, but carries the recessive allele that can produce a child with a concave or straight nose bridge.

Phenotypes	Convex nose bridge	Convex nose bridge	Concave or straight nose bridge
Genotypes	*AA* (homozygous dominant)	*Aa* (heterozygous)	*aa* (homozygous recessive)
Possible gametes	All *A*	½ *A*, ½ *a*	All *a*

FIGURE 29.2

How geneticists use a Punnett square. (A) A monohybrid cross. Several combinations of homozygosity and heterozygosity are possible in mating, depending on the specific parental genotypes. In this example, one parent (father) is homozygous and the other parent (mother) is heterozygous for one pair of alleles. (B) A dihybrid cross, with both parents heterozygous for two pairs of alleles. Such a mating produces a distinctive 9:3:3:1 pattern of four phenotypes.

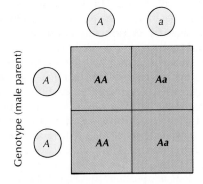

[A] MONOHYBRID CROSS

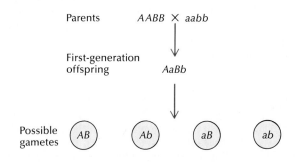

[B] DIHYBRID CROSS

TABLE 29.2	INHERITANCE OF ABO BLOOD TYPES	
Blood type of parents	**Children's blood type possible**	**Children's blood type not possible**
A + A	A, O	AB, B
A + B	A, B, AB, O	—
A + AB	A, B, AB	O
A + O	A, O	AB, B
B + B	B, O	A, AB
B + AB	A, B, AB	O
B + O	B, O	A, AB
AB + AB	A, B, AB	O
AB + O	A, B	AB, O
O + O	O	A, B, AB

Source: From Donald D. Ritchie and Robert Carola, *Biology,* 2nd ed. (Reading, Mass.: Addison-Wesley, 1983), p. 258. Used with permission.

($I^B i$), type AB ($I^A I^B$), or type O ($i^O i^O$). If she married a man with type B blood, and then had a child with type O blood, it would determine that both she and her husband were heterozygous for the ABO series.

The Rh factor is known to be a heritable characteristic of blood, but it follows a complex pattern that is not fully understood. One hypothesis proposes that the Rh factor is controlled by three closely linked loci (three distinct genes). Another hypothesis proposes that only one locus, with at least eight different alleles, is involved. For the sake of simplicity, we will assume the most common model of one locus with two alleles, with allele *R* determining the presence of the Rh antigen on the surface of erythrocytes and being dominant over the recessive allele *r*. Thus, *RR* and *Rr* individuals are Rh$^+$, since they contain the Rh antigen, and *rr* individuals are Rh$^-$, since they lack the antigen.

MUTATIONS

A change in the genetic material that brings about a change in the amino-acid sequence in a polypeptide chain is called a **mutation** (L. *mutare,* to change). The change may occur in the sequence of nucleotides in a DNA molecule (a *gene* mutation), or it can be a gross rearrangement of chromosomes (a *chromosome* mutation).

Experiments with bacteria and viruses have shown that genes mutate at different rates. A gene that is chemically stable will mutate less often than an unstable one. It is not unusual for a gene to mutate only once in 1 million DNA replications. However, an unstable gene may mutate once in 2000 replications, while an exceptionally stable gene may remain unaltered through billions of replications.

What is the difference between a congenital disease and a genetic disease?

Both are controlled by genes and are hereditary, but a congenital (L. congenitus, born together with) disease is one that is present at birth.

Causes and Effects of Mutations

Gene mutations happen when nucleotides in a DNA molecule are added, deleted, or rearranged. Just one substitution of one amino acid for another or one change in the sequence of amino acids in a protein (the gene product) can make the difference between normality and malfunction, or even between life and death. For example, if one of the hundreds of amino acids in hemoglobin is changed, the action of the whole hemoglobin molecule may be changed. A number of such changes are known, one of the most common being the substitution of valine for glutamate to produce sickle-cell anemia. That seemingly small change makes an enormous difference to the life of the person in whose blood it occurs. This one mutation puts a "sticky" patch on the surface of the hemoglobin molecule. When oxygen concentrations in the blood are low, the sticky patches interact, cause the hemoglobin molecules to aggregate into rods (distorted blood cells), and cause them to clump inside blood vessels.

Artificial (induced) mutations may be brought about by any radiation with enough energy to cause chemical changes in the molecules of cells. Among the most effective types of radiation are x rays, gamma rays, ultraviolet light, and radiation from radioactive elements, as in the dust of nuclear explosions.

Mutations in gametes may be induced by abnormally high temperatures or by chemical mutagens, such as those in LSD, automobile exhaust, mercury, many pesticides, food additives, certain dyes, and tobacco smoke. Viruses and chemical compounds foreign to a cell may also cause mutations.

Fate of Mutations

Mutations allow the genetic variability that makes adaptation and evolution possible as the environment changes. However, many mutations are harmful. If organisms were not already fairly well adapted to their environments, they probably would not have survived. Thus, changes in an organism could make it less well suited to its environment. However, over a long period of time, certain mutations have been favorable, and they are the basis for the genetic variability that makes evolution work.

In most cases, mutations are also recessive. For that reason, most new mutations are not expressed, unless, through chance mating, two identical mutant recessive alleles happen to form a pair. Because mutations are relatively rare, and because a homozygous pair of mutant alleles is not likely to be brought together, many mutations may exist in a population for many generations without ever being expressed. Also, if someone with a gene mutation dies without passing it along to the next generation, the mutation may also die out without ever having been expressed at all.

Mating between close relatives greatly increases the chance of both parents having identical alleles. As a result, there is an increased probability that their children will carry recessive homozygous alleles, which may express the recessive, and maybe harmful, traits. Such close mating is called *inbreeding*, and its harmful effects are called *inbreeding depres-*

sion. The effects of inbreeding depression increase proportionately to the genetic similarities of the mated couple. Harmful effects are most likely with mated siblings, less with first cousins, even less with second cousins, and so on.

Germinal and Somatic Mutations

If a mutation occurs in a cell and is not repaired or eliminated, it will be passed along to its daughter cells. But what will eventually happen to it depends on the kind of cell in which the mutation first appears. *Germinal mutations* occur in sex cells (or *germ cells*), and *somatic mutations* occur in body cells (or *somatic cells*). Only the germinal mutations have any real chance of being passed on to later generations, since a mutation in a somatic cell will be lost when the affected person dies. But a mutation in the germ cells of a man or woman may be included in an ovum or sperm cell, and later in the offspring.

Ask Yourself

1 *What is a mutation?*

2 *How can mutations be induced artificially?*

3 *Why are germinal mutations more important than somatic ones?*

AUTOSOMAL INHERITANCE

As we discussed in Chapter 26, human beings have two types of chromosomes: *sex chromosomes* (which differ in males and females and which carry the genetic instructions for determining the sex of offspring), and chromosomes other than sex chromosomes, called *autosomes* (which do not differ according to the sex of the individual). Human beings have 46 chromosomes, arranged in 23 pairs. Twenty-two of these pairs are autosomes, and the remaining pair are the sex chromosomes.

Chromosomal abnormalities can cause genetic disorders, which appear infrequently. For example, the chance of being born with galactosemia, a metabolic disorder, is about 1 in 40,000. As a rule, we are born without serious defects, genetic or otherwise, and most of us live a relatively healthy life. But because we learn more from the exceptional than the ordinary, we present a brief discussion of genetic disorders here, under the general headings of autosomal dominant inheritance and autosomal recessive inheritance.

Autosomal Dominant Inheritance

Autosomal dominant traits are inherited in equal numbers by both sexes, and are always inherited from at least one parent showing the trait. Consider foot shape. Normal arches are inherited as a dominant trait; flat feet, when they are

FIGURE 29.3

Some typical inheritance patterns. (A) Autosomal dominant. (B) Autosomal recessive. (C) Sex-linked recessive.

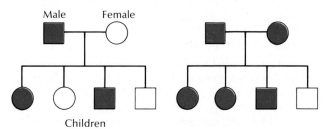

Parents

Male   Female

Children

One heterozygous parent shows trait; ½ children show trait, regardless of sex.

Both heterozygous parents show trait; ¾ children show trait, regardless of sex.

[A] AUTOSOMAL DOMINANT TRAITS

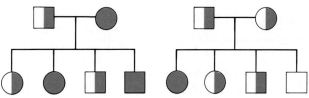

One parent shows trait, the other is carrier; ½ children show trait, ½ children carriers.

Both parents carriers; ¼ children show trait, ½ children carriers regardless of sex, ¼ children homozygous dominant.

[B] AUTOSOMAL RECESSIVE TRAITS

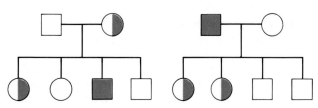

Mother carrier; ½ sons show trait, ½ daughters carriers.

Father shows trait, mother homozygous dominant; no son shows trait, all daughters carriers

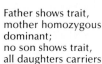
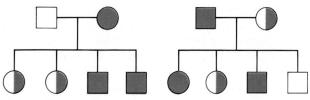

Mother shows trait, father does not; all sons show trait, all daughters carriers.

Father shows trait, mother carrier; ½ sons show trait, ½ sons do not; ½ daughters show trait, ½ daughters carriers.

[C] SEX-LINKED RECESSIVE TRAITS (ON X CHROMOSOME)

genetically determined, are the result of a recessive allele. Typically, two flat-footed parents, who must both be homozygous recessives, cannot have normal-arched children. (Check this conclusion by constructing a Punnett square.)

Many autosomal traits are known, some of them unimportant and some causing a physiological dysfunction. Those causing a dysfunction have received special attention because they have clinical or social consequences, and have been noted in the literature on human genetics. Abnormal dominant alleles are usually rare because they are likely to kill the individual possessing them, and thus they keep down the numbers of that allele in the population. Abnormal *recessive* alleles, if rare enough, may be present in a population but are hardly ever expressed, because there is too slim a chance of making a homozygous zygote (Figure 29.3).

An example of an autosomal dominant trait is *dentinogenesis imperfecta,* a dental defect that occurs about once in every 8000 births. The teeth have a brownish color, and their crowns wear down easily. Another example of an autosomal dominant trait is *Huntington's chorea* (Huntington's disease). This incurable nervous disorder generally strikes between the ages of 25 and 55. It produces uncoordinated movements and mental deterioration sometimes mistaken for drunkenness. Typically, it results in death about 10 to 15 years after its onset. The folk singer Woody Guthrie died of Huntington's chorea, and the disease is sometimes called Guthrie's disease. Because of the relatively late onset of the disease (and its frequent misdiagnosis), some victims of Huntington's chorea have children before realizing that they have the disease and could pass it on. (Woody Guthrie's son Arlo had a child of his own before it was known that Woody had the disease.)

A rare autosomal dominant trait, made famous because it apparently afflicted Abraham Lincoln, is *Marfan's syndrome,* or arachnodactyly (Gr. "spider fingers"). It is characterized by skeletal deformities that include eye problems (caused by an abnormal position of the lens), cardiovascular problems, weak joints, and abnormally long arms, legs, fingers, and toes. Lincoln's case (if he indeed had Marfan's syndrome)* seemed to be a mild one, and was an example of a trait that is not expressed fully.

Autosomal dominant traits tend to vary widely in their degree of *expressivity.* An example of a genetic abnormality that shows a wide range of expressivity is *brachydactyly* (Gr. "short fingers"), in which the second bone in the fingers is shorter than normal, either severely or only slightly. A final example of an autosomal dominant trait is *achondroplastic dwarfism* (Figure 29.4).

Autosomal Recessive Inheritance

People with **autosomal recessive** traits are *homozygous recessive* (*aa,* for example) for that trait. Heterozygous individuals do not show autosomal recessive traits because their dominant allele masks the recessive one. Such individuals are carriers, however (see Figure 29.3B). Homozygous recessives are more likely to occur in cases of inbreeding.

*Lincoln's reading glasses show no indication of severe nearsightedness, a typical symptom of Marfan's syndrome.

More autosomal recessive disorders are known than autosomal dominant ones, and some of them have relatively high allele frequencies. One of the most dramatic is *albinism,* the absence of pigmentation in the skin, hair, and iris (Figure 29.5). The frequency of albinism in blacks is about 1 in 22,000 births, but about 1 in 71 black people is a carrier.

FIGURE 29.4

Achondroplastic dwarfism, the result of an autosomal dominant gene, in brothers and sisters of the Owitch family. Achondroplastic dwarfism does not affect fertility. [*UPI/Bettmann Newsphotos.*]

FIGURE 29.5

Albinism, an autosomal recessive trait. It is sometimes not obvious among Caucasians, but it can be a stunning phenomenon when it occurs within a black population. [*AP/Wide World Photos.*]

Among Caucasians, the disorder affects about 1 in 38,000. Because albinism is recessive, the afflicted children must inherit a recessive allele (*a*) from each parent for the trait to be visible. (In this case, both parents must carry the recessive allele in their genotype, but need not be albinos themselves.)

Until recently it was thought that albinism was caused by a defective allele of the gene that produced tyrosinase, an enzyme that is responsible for melanin production. It is now believed that as many as six different genes may be involved, depending on what step in melanin synthesis is defective.

About 500 autosomal recessive diseases are known. *Cystic fibrosis* is a pancreatic enzyme deficiency that affects the function of mucous and sweat glands (see p. 922 and Chapter 23). *Tay-Sachs disease* occurs mostly in infants and results in the deterioration of the nervous system (see p. 922 and Chapter 23). *PKU (phenylketonuria)* is a liver enzyme deficiency that leads to a harmful build-up of the amino acid phenylalanine (see Chapter 23). *Cooley's anemia* (thalassemia) is a blood disorder that blocks the synthesis of hemoglobin. *Sickle-cell anemia* is a disease of red blood cells that prevents sufficient amounts of oxygen from being carried in the bloodstream. *Galactosemia* is a deficiency in the enzyme that is needed to metabolize lactose, or milk sugar. In these examples, parents who are normal can be carriers of the defective allele. If both parents are heterozygous and carry the defective allele, each child will have a 25-percent chance of being normal, a 50-percent chance of receiving a single defective allele and being a carrier, and a 25-percent chance of showing the trait (see Figure 29.3B).

Ask Yourself

1 *What is the main difference between sex chromosomes and autosomes?*

2 *Explain how autosomal dominant traits are inherited.*

3 *What is Huntington's chorea?*

4 *Are people with autosomal recessive traits homozygous for the trait? Dominant or recessive?*

5 *Is it possible for an albino child to have parents with normal skin and hair?*

POLYGENIC INHERITANCE

In most instances, a visible characteristic can be influenced by more than one gene pair affecting the same trait. When several pairs of genes act on a single characteristic, this is called *polygenic inheritance,* and the genes are called *polygenes.* Together, the small effect of each polygene produces a wide range in traits such as the amount of pigment synthesized or body size. Consider, for example, the overall growth of the body, where there is no specific allele for tallness or dwarfness as occurs in some plants. Most people are neither giants nor dwarfs. A few are very short, many are of medium height, and a few are very tall. Polygenic features are also affected by environmental factors. For example, body

FIGURE 29.6

Polygenic inheritance of skin color. If pigmentation is determined by four pairs of melanin-producing genes, then a couple with medium coloration might have four melanin-producing and four neutral genes. Their children would have nine possibilities, ranging from eight melanin-producing genes to none, as shown in the diagram. It is thus possible for moderately pigmented parents to produce children who are very dark or very light, or any of seven intermediate shades.

Parents

First-generation children

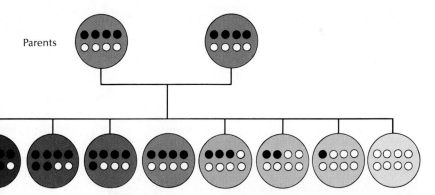

height can certainly be influenced by nutrition and growth hormone.

Just as people are generally neither giants nor dwarfs, so their skin color is neither pure black nor pure white. The degree of lightness or darkness depends on the thickness of the skin and the amount of melanin present in the outer layers. The more melanin, the darker the skin. Current thought is that there are at least four genes involved in the expression of skin color, with the resulting possibility of nine different skin shades (Figure 29.6).

SEX-LINKED INHERITANCE

The parental sex chromosomes transmit not only sexual characteristics, but also certain nonsexual traits in what is known as *sex-linked inheritance.* Most of the sex-linked genes appear on the X (maternal) chromosome because it is larger than the Y (paternal) chromosome, and thus has room for more genes. Y-linked traits are passed on from fathers to sons, never to daughters. In contrast, X-linked inheritance is more widespread, with more than 100 X-linked traits known.

Y-Linked Inheritance

Little is known about Y-linked traits. Fathers determine the sex of a child by contributing either an X (for female) or Y (for male) sex chromosome. Besides the sex-determining genes, only a few other trait-determining genes are present on male chromosomes. When Y-linked traits *do* occur, they are passed from fathers to sons, are inherited by all the sons of a father, and are not inherited by any daughters or children of daughters. Daughters cannot even be carriers.

X-Linked Inheritance

X-linked traits can be dominant or recessive. Dominant traits appear in the phenotypes of both men and women who carry the corresponding allele. Recessive traits always appear in the

phenotype of a man carrying the allele, but usually only in the genotype of a woman with the allele, since a woman must receive recessive alleles from *both* parents in order to show the trait phenotypically. A woman with only one recessive allele is a carrier, however, and can transmit the trait she receives from her father to her son, for example. This pattern is called *skip-generation inheritance,* and it is one of the strongest indicators of an X-linked gene. Having *only* sons with a trait from the mother but not from the father is another indicator of an X-linked gene.

Two of the most famous X-linked disorders are the result of recessive genes: hemophilia and color blindness.

Hemophilia *Hemophilia* is sometimes called the "bleeder's disease," because people afflicted with it lack one of the several protein factors needed to clot blood properly (see Chapter 17). Queen Victoria of England apparently carried the mutant hemophilia gene, which she transmitted to some of her children (Figure 29.7). (Since hemophilia was not present in the royal families of Europe before Queen Victoria, the allele was probably a mutation in a sex cell of one of Victoria's parents or in one of her own sex cells.) Several types of hemophilia are known. Victoria transmitted the one called *classic hemophilia,* or *hemophilia A,* an example of *X-linked recessive inheritance.* *

Queen Victoria, being heterozygous for the hemophilia gene, had a normal allele that was dominant over the abnormal one. As a result, she was a carrier, but was not herself a hemophiliac.

Color blindness Several kinds of **color blindness** are X-linked, and they all involve a deficiency in one or another of the eye pigments necessary to see certain colors. One example is *green-weakness color blindness,* which results from a reduced amount of the retinal pigment that is sensitive to light in the green wavelengths of the spectrum. About 5 per-

*All forms of hemophilia are sex-linked. In classic hemophilia the genetic defect is in clotting factor VIII, while other types of hemophilia involve other clotting factors.

FIGURE 29.7

Hemophilia in the descendants of Queen Victoria. The chart shows the relation of the queen to several royal houses into which her children married, and the results of those marriages. Note that 4 of Victoria's 9 children were either carriers or actually afflicted, and that of her other descendants (26 grandchildren and 34 great-grandchildren), 10 males were hemophiliacs and 4 females were carriers. Because it has no afflicted members, the British royal family of today is not represented on the chart. [*Culver Pictures.*]

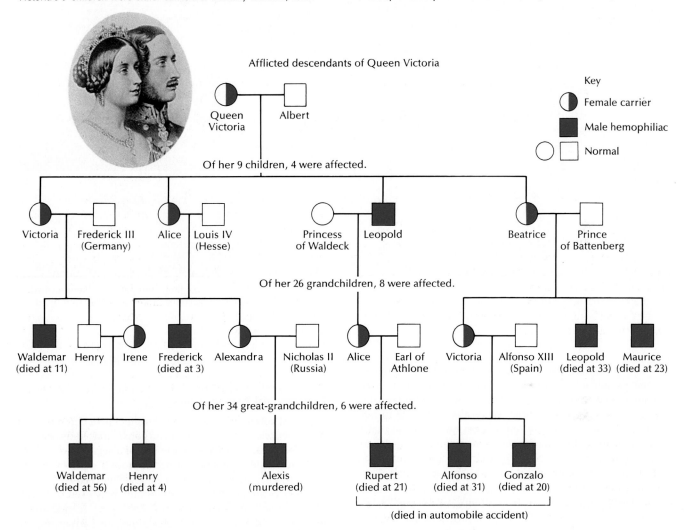

Afflicted descendants of Queen Victoria

cent of northern European males have green-weakness color blindness. To them, bright greens appear tan, olive greens appear brown, and reds appear reddish-brown.

The green-weakness allele (*g*) occurs on the X chromosome, and as a result, color blindness occurs in homozygous females (*gg*). In heterozygous females (*Gg*), the dominant allele codes for enough pigment to allow for near-normal color vision, but the female is a carrier of the color-blindness trait. All males who have the recessive allele are color-blind, because they have no dominant allele on their Y chromosome to counteract the recessive allele. When a normal male ($X^G Y$) and a homozygous recessive female (*gg*) have children, all daughters will have normal vision but will be carriers of the recessive allele, and all sons will be afflicted with green-weakness color blindness (Figure 29.8A). If the mother is heterozygous normal ($X^G X^g$) and the father is normal ($X^G Y$), the green-weakness trait will show up in half the sons, and half of the daughters will be carriers (Figure 29.8B).

Ask Yourself

1 *What are sex-linked genes?*

2 *How does Y-linked inheritance differ from X-linked inheritance?*

3 *Why are hemophilia and color blindness considered X-linked disorders?*

CHROMOSOMAL ABERRATIONS

Chromosomal aberrations (alterations) take place less often than gene mutations, but some mistakes do occur. Any deviation from the normal is likely to have a harmful effect on the development of the affected individual.

FIGURE 29.8

Inheritance patterns of green-weakness color blindness. (A) Normal father and color-blind, homozygous recessive mother. (B) Normal father and normal vision, heterozygous mother.

Key X^G = Normal green vision allele (normal color vision)
X^g = Green–weakness allele (green–weakness color-blind)
Y   = No allele for color blindness on Y chromosome

Normal father (X^GY)

	X^G	Y
X^g	X^GX^g (carrier)	X^gY (color-blind)
X^g	X^GX^g (carrier)	X^gY (color-blind)

Homozygous recessive color-blind mother (X^gX^g)

All daughters are carriers, but are not color-blind | All sons are color-blind

[A]

Normal father (X^GY)

	X^G	Y
X^G	X^GX^G (normal)	X^GY (normal)
X^g	X^GX^g (carrier)	X^gY (color-blind)

Heterozygous normal mother ($X^G X^g$)

½ daughters normal, ½ carriers | ½ sons normal ½ color-blind

[B]

Chromosomal aberrations can be recognized by the microscopic examination of a complete chromosome set of an individual, or *karyotype* (Figure 29.9). A karotype is prepared from a cultured tissue sample. When the nuclei are in metaphase of mitosis, the centromeres are spread out so that it is possible to see the individual chromosomes. The nuclei are stained and photographed, and the chromosomes are cut out of the photograph and put in an orderly arrangement. They are numbered from 1 to 22, from the longest to the shortest. The sex chromosome is labeled separately as either XX or XY.

FIGURE 29.9

Chromosomal analysis, or karyotype. The karyotype shown is of a normal male (XY). [Dr. J. H. Tjio.]

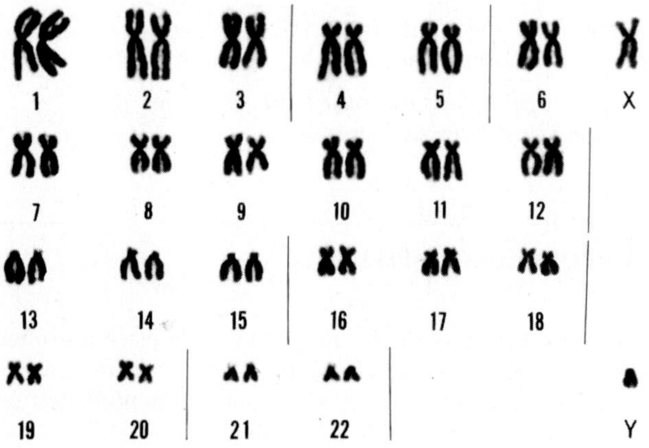

Chromosomal Rearrangements

The most common physical rearrangements in chromosomes are deletions, duplications, inversions, and translocations (Figure 29.10). A *deletion* is the loss of a segment of a chromosome. When the deletion occurs from a midsection of a chromosome, the broken ends heal together (Figure 29.10B). Such losses are likely to be harmful (and may even be lethal) because the individual becomes *hemizygous* (containing a gene or chromosome without a homologous partner) for the lost portion, which permits the expression of harmful alleles. The best-known condition caused by a deletion is *cri du chat syndrome*, which occurs on the short arm of chromosome number 5. Afflicted individuals have microcephaly, wide-set eyes, severe mental retardation, and in infants, a distinctive catlike cry.

A *duplication* is the repetition of a segment of a chromosome one or more times (Figure 29.10C). Duplication often happens in the long arm of chromosome number 14 and can be seen in various lymphomas.

An *inversion* occurs when a chromosome breaks in two places, and the broken-off segment is reinserted backwards (Figure 29.10D). Ordinarily, inversions do not produce gross changes in phenotypes. However, if crossing-over (reciprocal exchange of corresponding segments between homologous chromatids) occurs within the inverted segment, the gametes that contain the inverted chromosomal portion may produce harmful effects in offspring or may even be incapable of producing viable offspring.

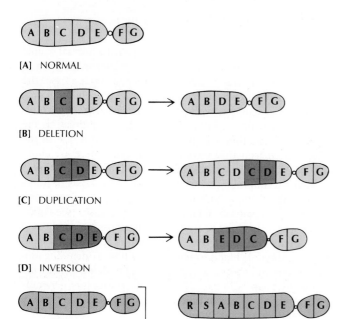

FIGURE 29.10

The most common chromosomal aberrations. (A) Chromosome before replication. (B) Deletion. (C) Duplication. (D) Inversion. (E) Translocation.

[A] NORMAL

[B] DELETION

[C] DUPLICATION

[D] INVERSION

[E] TRANSLOCATION

A *translocation* occurs when a segment breaks off one chromosome and is transferred to another, nonhomologous, chromosome (Figure 29.10E). Recently, it was discovered that patients suffering from *chronic myelogenous leukemia* had most of chromosome 9 fused to a piece of chromosome 22. The resultant aberrant chromosome has been named the *Philadelphia chromosome*, since it was first discovered by researchers in that city.

Nondisjunction in Chromosomes

As described earlier, chromosomes are arranged in pairs, and each member of the pair is a *homolog*. In cases of **nondisjunction**, however, a pair of homologs fails to separate at meiosis. The resulting zygote has either only one copy of the chromosome (a condition called *monosomy*) or, if the final gamete unites with a normal gamete (that is, one with only one homolog), the zygote will have three copies, a condition called *trisomy*.

The most common trisomic condition is **trisomy 21,** also known as **Down's syndrome** (and previously called Mongolism) (Figure 29.11). The condition is caused by the presence of an extra copy of chromosome 21. About 5 percent of people with Down's syndrome have a chromosomal translocation unrelated to meiotic nondisjunction. In these individuals, a small piece of chromosome 21 attaches itself to another

chromosome, causing the fragment to appear in daughter cells along with the normal chromosome 21. Upon fertilization, trisomy occurs.

People with Down's syndrome are mentally retarded (it is the most common cause of mental retardation in the United States), are short, have stubby hands and fingers, a rounded face, a lowered resistance to common diseases, some increased risk of vision impairment because of defective lenses, elevated levels of purines that can lead to nervous system disorders and a deficient immune system, an increased risk of developing leukemia, and usually suffer respiratory problems and heart defects. In the past, most people with Down's syndrome died early in childhood or rarely lived beyond their twenties. Antibiotics, improved surgical techniques, and special education have allowed many of the afflicted people to live relatively normal lives. Their average life span is now about 30 years, with about 25 percent reaching 50. However, about one-third still die before they are 10 years old. In addition, about 80 percent of fetuses with Down's syndrome are lost by spontaneous abortion.

The average incidence of Down's syndrome is about 1 in 650 to 700 live births, but the frequency increases dramatically in older mothers. A 20-year-old mother has about 1

FIGURE 29.11

A girl with trisomy 21 (Down's syndrome). [*Terry Gibbon.*]

chance in 2000 of having a child with Down's syndrome. The odds for a 40-year-old mother increase to about 1 in 80, and the odds for a 45-year-old mother go all the way up to about 1 in 45. Although women over 35 account for less than 14 percent of all pregnancies, they give birth to about 45 percent of all children born with Down's syndrome. For this reason, most physicians recommend amniocentesis for pregnant women over 35 (see p. 886). Many studies have been undertaken in an effort to find out if older fathers are also implicated in the transmission of Down's syndrome, and it appears that they are not.

Aberrations in Sex Chromosomes

Abnormalities in sex chromosomes can cause such conditions as Turner's syndrome and Klinefelter's syndrome. A person with **Turner's syndrome** has received one X chromosome from her mother or father and none from the other parent, producing an **XO** monosomic condition. (Turner's syndrome is the only viable monosomic condition.) The person is always a sterile female, with abnormal ovaries and permanently infantile sex organs. She fails to develop normal secondary sex characteristics such as pubic hair, is much shorter than average, and has a broad chest. One of the most noticeable characteristics is a thick fold of skin on each side of the neck. Turner's syndrome occurs in about 1 in 4000 live female births, but in addition, most Turner's syndrome fetuses are lost in spontaneous abortions (miscarriages). It is estimated that only about 5 percent of all XO embryos survive long enough to be born.

Some of the symptoms of Turner's syndrome can be alleviated somewhat by timely hormone therapy. If estrogen treatments are begun around the average time of puberty, secondary sex characteristics will appear, and growth will also occur, although not to the full extent of a normal young person. Body stature remains less than average, and sterility persists, but menstruation does take place if hormone therapy is continued throughout early adulthood.

An individual with **Klinefelter's syndrome** has an extra X chromosome, which produces an **XXY** genotype, and a total of 47 chromosomes. The condition occurs when an ovum with an extra X chromosome is fertilized by a Y-bearing sperm, or when an XY-bearing sperm fertilizes a normal ovum. In most cases, a person with Klinefelter's syndrome is phenotypically male, but has underdeveloped genitals, even after puberty. The patient has a short trunk and long legs, and the body generally looks rather feminine, with sparse body hair and broad hips. Enlarged breasts are evident in about half the cases. The affected person is sterile and may be mentally retarded. Klinefelter's syndrome occurs in about 1 in 600 male births.

Ask Yourself

1 *What is a karyotype?*

2 *What are some of the effects of the physical rearrangement of chromosomal segments?*

3 *What chromosomal abnormality causes Down's syndrome? Turner's syndrome? Klinefelter's syndrome?*

GENETIC COUNSELING

The most common reason for a phenotypically normal couple to seek out the services of a genetic counselor is the birth of a first child who has a genetic disorder. In such a case, the parents are justifiably concerned about the probability of producing other children with genetic defects. A genetic counseling service helps prospective parents by estimating the probability that they will produce children with one of the 3000 known genetically controlled defects. There are about 600 such counseling centers in the United States at this time.*

Genetic counseling relies mainly on three kinds of information. The first is a study of the family *pedigree*, a description of the known patterns of familial traits or diseases (see Figure 29.7). The second is a *karyotype*, or chromosomal analysis of the parents (see Figure 29.9). The third is a *biochemical analysis* of embryonic blood, urine, or body tissue (including amniocentesis and chorionic villi sampling) to detect genetic abnormalities (see Chapter 27). Genetic counselors generally look for several types of inheritance patterns in trying to detect abnormalities in a family.

In the relatively simple case of autosomal recessive inheritance, a phenotypically normal couple may produce a child displaying the recessive trait, for example, albinism, and want to know the probability of producing other albino children. A counselor will know that each parent must be carrying a hidden, recessive allele for albinism, and can estimate that the probability of producing another albino child is one out of four (see Figure 29.3B).

More complicated and serious defects may actually be predicted *before* a child is born with a genetic disorder. **Tay-Sachs disease** is an autosomal recessive disease that causes mental and physical deterioration. Usually, the afflicted child dies before the age of 5 years. Tay-Sachs disease is caused by an enzyme deficiency that impairs the storage of lipids, and is expressed when both parents are heterozygous carriers of the recessive defective allele. A biochemical test is available that uses cultured cells to detect altered enzyme levels that reveal if a person is a carrier of the Tay-Sachs allele. If both members of a couple are carriers, they can be advised of the possibility of producing a homozygous Tay-Sachs child.

Because Tay-Sachs disease occurs almost exclusively in Ashkenazic Jews from central or eastern Europe, a genetic counselor can offer advice to a couple about the disease after examining the family histories. Also, the disease can usually be detected by a "genetic marker" (see below) when a woman is pregnant, either through amniocentesis or chorionic villi sampling.

Cystic fibrosis is the most common genetically transmitted disease in the United States. About half of those afflicted die

*To obtain a list of genetic counseling centers, write to the Professional Education Department, The National Foundation of the March of Dimes, in your area.

before the age of 20. (It is the most common fatal genetic disease among Caucasian children.) Afflicted individuals produce thick, viscous mucus and other secretions that cause respiratory and digestive problems. It is believed that mucus accumulates because the passage of sodium and chlorine through plasma membranes is somehow affected.

The disease is almost always inherited with one particular DNA variation that can be detected when the affected chromosome is exposed to specific enzymes called *restriction enzymes*. These enzymes cleave the DNA at those points where the defective gene is located, providing a "genetic marker" for the gene. Prospective parents who already have a child with cystic fibrosis, or who suspect the possibility of producing a first child with the disease, can request a test of fetal cells that usually predicts with about 95-percent accuracy whether or not a defective gene is present in the fetus.

Some other genetic disorders that can be detected before birth are Duchenne's muscular dystrophy, fragile X syndrome, hemophilia, Huntington's chorea, polycystic kidney disease, sickle-cell anemia, and beta thalassemia. No cures are available for these diseases, but in the future, genetic engineering techniques may be able to correct genetic abnormalities right on the DNA strand itself.

Currently, the only "treatment" for most genetic disorders is abortion, and many couples feel that such a measure is not justified, even if an abnormal child is predicted with certainty. Actually, more than 95 percent of all fetuses tested by amniocentesis are normal, and abortions resulting from genetic testing account for less than 1 percent of all abortions in this country.

When genetic studies first showed how various genes exist in populations, and when genetic diseases were identified as such, some people began to think that the human species could be relieved of such diseases by the systematic elimination of certain alleles. *Eugenics* (Gr. well-born) is the advocacy of trying to improve humanity by applying certain genetic control measures. (The measures vary according to who wants to make the rules.) Fraught with legal and ethical ramifications, the practice of eugenics will probably remain only as a serious point of discussion, at least for many years to come, especially since its scientific effectiveness is questioned.

MEDICAL TERMINOLOGY

AGAMMAGLOBULINEMIA A sex-linked recessive condition in which affected individuals do not produce B cells and thus cannot produce antibodies.

CLEFT LIP, CLEFT PALATE A developmental abnormality, sometimes caused by chromosomal aberrations, in which the embryonic lip and palate are imperfectly fused.

CRISSCROSS INHERITANCE A pattern of transmission of X-linked alleles in which the allele is transmitted from male to female and then from female to male in succeeding generations.

EUPHENICS The manipulation of a phenotype in an attempt to nullify a specific undesirable trait.

EXPRESSIVITY The extent to which an individual demonstrates particular phenotypic effects of a gene.

F_1 The letter designation used in constructing pedigrees for progeny produced by the first mating, with F standing for filial generation. Subsequent generations are labeled F_2, F_3, etc.

MOSAICISM A condition in which some cells show the normal number of chromosomes and other cells show an abnormal number of one too few (such as Turner's syndrome mosaic) or one too many (such as trisomy 21 mosaic).

POLYPLOID A cell or individual possessing an entire extra set of chromosomes.

SUMMARY

Genetics is the science of heredity. It examines the normal transmission and expression of characteristics from one generation to the next, and also the *variation* of that normal transmission.

Introduction to Mendelian genetics
1 The science of genetics was started in the middle of the nineteenth century by Gregor Mendel.
2 *Dominant* genetic characteristics tend to mask *recessive* characteristics.
3 A *chromosome* is a threadlike nucleoprotein structure within the nucleus of a cell that contains hereditary information in the form of DNA. Each strand of DNA contains segments called *genes* that control specific cellular functions.
4 Genes are arranged in pairs. Each member of a pair of genes is an *allele*, an alternative form of a gene at a specific locus on a chromosome. There are at least two alleles for each trait. Dominant alleles mask recessive ones.
5 Mendel's Law of Segregation states that the alleles of a pair separate during gamete formation without blending or being altered in any way.
6 A person with a pair of identical alleles for a given trait (*AA, aa*) is *homozygous* for that trait. A person with a pair of nonidentical alleles for a given trait (*Aa*) is *heterozygous* for that trait.
7 The genetic make-up of a person is that person's *genotype*, which may be either homozygous or heterozygous for any one characteristic. The expression of a genetic trait (how a person looks or functions) is that person's *phenotype*.

Multiple alleles
Many traits are regulated by more than two alleles, or *multiple alleles*. ABO blood grouping is an example of a trait regulated by multiple-allele inheritance.

Mutations

1 A *mutation* is a change in the genetic material that changes the end product of protein synthesis. A *gene mutation* is a change in the sequence of nucleotides in a DNA molecule; a *chromosome mutation* is a gross rearrangement of chromosomes.

2 The cause of natural mutations is unknown, but artificial mutations can be brought about by radiation, drastic temperature changes, or chemical mutagens.

3 Most mutations are recessive and nonbeneficial.

4 *Germinal mutations* are mutations that occur in sex cells; *somatic mutations* occur in body cells. Only germinal mutations may be passed on to subsequent generations.

Autosomal inheritance

1 *Sex chromosomes,* designated XX in females and XY in males, carry the genetic instructions for determining the sex of offspring. Chromosomes other than sex chromosomes are *autosomes.*

2 *Autosomal dominant* traits are inherited in equal numbers by both sexes and are always inherited from at least one parent showing the trait. Examples are *dentinogenesis imperfecta, Marfan's syndrome,* and *Huntington's chorea.*

3 *Autosomal recessive* traits are expressed in homozygous individuals and are carried but masked in heterozygous individuals. Examples are *albinism* and *cystic fibrosis.*

Polygenic inheritance

1 *Polygenic inheritance* is the regulation of a single characteristic by several pairs of genes. The interacting genes are called *polygenes.*

2 Traits influenced by polygenic inheritance, such as skin color and body height, occur in wide ranges.

Sex-linked inheritance

1 *Sex-linked inheritance* is the transmission of nonsexual traits, in addition to the determination of sex, by the sex chromosomes. Most sex-linked genes appear on the maternal X chromosome, and X-linked inheritance is more complex and widespread than Y-linked inheritance.

2 Little is known about *Y-linked traits,* which are passed on from fathers to sons, never to daughters. *X-linked traits* can be dominant or recessive. Two well-known X-linked recessive disorders are *hemophilia* and *color blindness.*

3 *Skip-generation inheritance* is the transmission of a trait expressed in one generation through a carrier second generation to a third generation in which it is expressed again. The occurrence of the pattern indicates an X-linked gene, for example, passed from father to carrier daughter to the daughter's son.

Chromosomal aberrations

1 *Chromosomal aberrations* occur less often than gene mutations. The most common physical rearrangements in chromosomes are *deletion* (a segment of a chromosome is lost, and the broken ends heal together), *duplication* (a segment of a chromosome is repeated one or more times), *inversion* (a chromosome breaks and rejoins with part of the chromosome reinserted backward), and *translocation* (a segment of a chromosome is transferred to another, nonhomologous chromosome).

2 *Nondisjunction* is the failure of a pair of chromosomal homologs to separate during meiosis. A zygote resulting from the gametes produced may have either only one copy of the chromosome (*monosomy*) or three copies (*trisomy*). *Down's syndrome* is the most common trisomic condition.

3 *Turner's syndrome* is an aberration of sex chromosomes in which an individual receives a chromosome from one parent and none from the other parent, producing an *XO* monosomic condition. The victim is always a sterile female.

4 An individual with *Klinefelter's syndrome* has an extra X chromosome, producing an *XXY* genotype.

Genetic counseling

1 *Genetic counseling* provides prospective parents with estimates of the probability that they will produce children with a genetically controlled disease.

2 Genetic counseling relies mainly on information obtained from a family *pedigree* (description of known patterns of familial traits or diseases), a *karyotype* (number, form, and type of chromosomes) for both parents, and *biochemical analysis* of embryonic blood, urine, or body tissue (including amniocentesis and chorionic villi sampling) to detect genetic abnormalities.

3 *Eugenics* is the advocacy of applying genetic control measures to improve humanity.

Metric Conversion Factors

APPENDIX A

English to metric

WEIGHT (MASS)
Ounces × 28.3495 = grams
Pounds × .4536 = kilograms
Tons × 1.1023 = tonnes (1000 kg)

LENGTH
Inches × 25.4 = millimeters
Inches × 2.54 = centimeters
Feet × .3048 = meters
Yards × .9144 = meters
Miles × 1.6093 = kilometers

AREA
Square inches × 6.4515 = square centimeters
Square feet × .0929 = square meters
Square yards × .8361 = square meters
Square miles × 2.59 = square kilometers
Acres × .405 = hectares
Square miles × 259.07 = hectares

MEASURE
Fluid ounces × 29.47 = milliliters
Quarts × .9433 = liters
Gallons × 3.774 = liters
Cubic inches × 16.4 = cubic centimeters
Cubic feet × .02832 = cubic meters
Cubic yards × .7645 = cubic meters

PRESSURE
Pounds per square inch × .0703 = kilograms per square centimeter

TEMPERATURE
When you know the Fahrenheit temperature:

$$°C = \frac{(°F - 32)}{1.8}$$

°C		°F
100	Water boils	212
90		194
80		176
70		158
60		140
50		122
40		104
37	Normal body temperature	98.6
30		86
20		68
10		50
0	Water freezes	32

Metric to English

WEIGHT (MASS)
Grams × .03527 = ounces
Kilograms × 2.2046 = pounds
Tonnes × .9072 = tons

LENGTH
Millimeters × .0394 = inches
Centimeters × .3937 = inches
Meters × 3.2809 = feet
Meters × 1.0936 = yards
Kilometers × .621377 = miles

AREA
Square centimeters × .155 = square inches
Square meters × 10.7641 = square feet
Square meters × 1.196 = square yards
Square kilometers × .3861 = square miles
Hectares × 2.471 = acres
Hectares × .00386 = square miles

MEASURE
Milliliters × .0339 = fluid ounces
Liters × 1.06 = quarts
Liters × .265 = gallons
Cubic centimeters × 0.06 = cubic inches
Cubic meters × 35.3156 = cubic feet
Cubic meters × 1.308 = cubic yards

PRESSURE
Kilograms per square centimeter × 14.2231 = pounds per square inch

TEMPERATURE
When you know the Celsius temperature:

$$°F = (°C × 1.8) + 32$$

APPENDIXES / GLOSSARY / INDEX

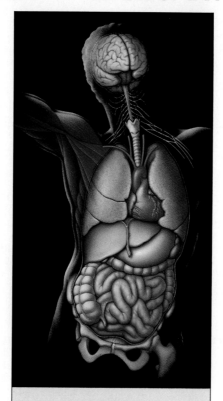

APPENDIXES

A

Metric Conversion Factors

B

Solutions

C

Prefixes, Combining Word Roots, and Suffixes in Medical Terminology

D

Reclassification of Eponymous Terms

E

Normal Laboratory Values for Blood and Urine

F

An Overview of Regional Human Anatomy

GLOSSARY

INDEX

UNDERSTANDING THE FACTS

1 Define genetics.

2 What is the relationship between chromosomes and genes?

3 What is an allele? Distinguish between dominant and recessive alleles.

4 What is the difference between homozygosity and heterozygosity?

5 What is a genotype? A phenotype?

6 How is a Punnett square used by geneticists?

7 Give an example of how multiple alleles are expressed in inheritance.

8 Do all genes have a similar mutation frequency? Explain.

9 Distinguish between germinal and somatic mutations.

10 Give examples of an autosomal dominant trait and an autosomal recessive trait.

11 Give an example of polygenic inheritance.

12 How are Y-linked traits passed from parents to children?

13 Give a short definition of the four main types of chromosomal rearrangements.

14 Define nondisjunction, and give an example of a condition resulting from it.

15 How are pedigrees, karyotypes, and biochemical analyses used in genetic counseling?

UNDERSTANDING THE CONCEPTS

1 Briefly explain how genes control cellular functions.

2 What are the possible ABO blood types of children if the mother is type AB and the father type O?

3 If the mother is type A and the child type B, what blood type(s) could the father be?

4 Distinguish between gene mutations and chromosomal mutations.

5 Can mutations in somatic cells be passed on to future generations?

6 Why must people with autosomal recessive traits be homozygous recessive for that trait?

7 Suppose that an albino man marries a woman who is normal (homozygous). Describe the possible genotypes and phenotypes of their children.

8 If one of the children from problem 7 marries an individual with a similar genotype, how might their children be affected?

9 If one of the male children from problem 8 marries an albino woman, what could be expected in their children?

10 If a color-blind man marries a normal woman (homozygous), how might their children be affected?

11 If one of the daughters of problem 10 marries a normal male, what could be expected of their offspring?

12 How can a female acquire color-vision defects?

SUGGESTED READING

ANDERSON, W. F., AND E. G. DI-ACUMAKOS, "Genetic Engineering in Mammalian Cells." Scientific American, July 1981.

AYALA, FRANCISCO J., AND JOHN A. KIGER, JR., *Modern Genetics*, 2nd ed. Menlo Park, Calif.: Benjamin, 1984.

CAVALLI-SFORZA, L. L., *Elements of Human Genetics*, 2nd ed. Menlo Park, Calif.: Benjamin, 1977. (Paperback)

COHEN, S. N., AND J. A. SHAPIRO, "Transposable Genetic Elements." Scientific American, February 1980.

FRIEDMANN, T., "Prenatal Diagnosis of Genetic Disease." Scientific American, November 1971.

FUCHS, FRITZ, "Genetic Amniocentesis." Scientific American, June 1980.

GOODENOUGH, U., *Genetics*, 3rd ed. Philadelphia: Saunders, 1984.

JENKINS, JOHN B., *Human Genetics*. Menlo Park, Calif.: Benjamin, 1983.

LURIA, S. E., *36 Lectures in Biology*. Cambridge, Mass.: MIT Press, 1975.

MANGE, A. P., AND E. J. MANGE, *Genetics: Human Aspects*. New York: Raven Press, 1977.

MOORE, JOHN A., *Heredity and Development*, 2nd ed. New York: Oxford, 1972. (Paperback)

MULLER, H. J., "Radiation and Human Mutation." Scientific American, November 1955.

PATTERSON, DAVID, "The Causes of Down Syndrome." Scientific American, August 1987.

SINGER, SAM, *Human Genetics*, 2nd ed. New York: Freeman, 1985. (Paperback)

THOMPSON, JAMES S., AND MARGARET W. THOMPSON, *Genetics in Medicine*, 4th ed. Philadelphia: Saunders, 1986.

WINCHESTER, A. M., *Human Genetics*, 2nd ed. Columbus, Ohio: Merrill, 1975. (Paperback)

Solutions

APPENDIX B

Some of the most interesting and important substances in nature are solutions. The oceans, which cover approximately three-quarters of the earth's surface and contain a multitude of fascinating life forms, are actually *liquid solutions*. However, solutions are not limited to liquids. The air we breathe is a *gaseous solution*. Our digestive processes are carried out in solutions, and our blood is a solution. Overall, the homeostasis of all living things is determined by the properties of solutions.

A *solution* is a mixture containing at least two substances, each of which is dispersed evenly throughout. It is a *homogeneous* mixture because the composition of the solution is the same in every part of the whole.

SOLVENT AND SOLUTE

In solutions, the substance responsible for dissolving the other is called the *solvent*. The *solute* is the substance being dissolved. For example, if water and sugar are mixed, the water is the solvent, and the sugar is the solute. At times it is difficult to determine which is the solute and which is the solvent, as, for example, in a mixture of alcohol and water. Usually, however, the solvent is present in larger amounts.

SOLUBILITY

The amount of substance dissolved in a given volume of solvent is known as the *concentration* of the solution. The ability of a substance to dissolve in a given volume of solvent is called its *solubility* (L. *solvere*, to loosen). Solubility is expressed in terms of grams of the solute that can be dissolved in 100 mL of solvent; for example, 100 grams when the solvent is water (100 g/100 mL or 1 g/mL). A *dilute solution* is one that contains a small amount of the dissolved substance. A *concentrated solution* is one that contains a large amount of the dissolved substance.

SATURATION

As a solute dissolves, the solution becomes increasingly populated with the molecules of the solute. When the solvent holds all the solute that it can take up directly from undissolved solute, the solution is said to be *saturated* (L. *saturare*, to fill) at that particular temperature and under other particular conditions. Any solution that contains a smaller amount of solute is said to be *unsaturated*.

EXPRESSING CONCENTRATION OF SOLUTIONS

To determine the amount of solute in solution, the following methods are used:

Ratio	*Example:* 1:1000 *or* 1 part per 1000
Weight per unit volume	*Example:* 10 g/L *or* 10 g in 1000 mL
Percentage	*Example:* 5 percent salt solution *or* 5 g of NaCl diluted in up to 100 mL of water
Volume-volume	*Example:* 5 mL of phenol + 95 mL of water *or* 1:20 dilution (5 + 95 = 100 ÷ 5 = 20) *or* 5 percent concentration (5 ÷ 100 = 0.05 g *or* 5 ÷ 100 = 5 percent)

MOLARITY OF SOLUTIONS

Strengths or concentrations of solutions used in physiology are commonly expressed in terms of *molarity*. A *molar solution* is one that contains 1 gram molecular weight (1 mole) of a solute dissolved in 1 liter of solvent. To determine the molecular weight of a compound, let us take the example of sodium chloride (NaCl). In the Periodic Table of the Elements, find the atomic weights of sodium (Na) and chlorine (Cl), round each to the nearest whole number, and add them:

Na = 23   Cl = 35
Molecular weight of NaCl = 58

The molarity of a solution is designated by the letter M. A 1 M NaCl solution contains 1 mole of NaCl dissolved in 1 liter (1000 mL) of water. So, 58 g of NaCl dissolved in up to 1000 mL of water is a 1-molar solution. A 0.1 M NaCl solution contains 1/10 of a mole (or 0.1 of the molecular weight) dissolved in 1 liter of water. A formula that is helpful in remembering this relationship of solutions is

$$\text{Molar solution} = \frac{\text{moles of solute}}{1000 \text{ mL of solvent}}$$

MOLAL SOLUTIONS

A *molal solution* contains 1 mole of a solute for every 1000 grams of solvent.* This volume, which is *not* a liter, varies with the volume of the formula weight of dissolved solute.

Generally, molal solutions are used whenever the *relative numbers* of solute and solvent molecules are determining factors. That is, molal solutions provide a definite ratio of solute to solvent molecules.

Molal solutions are used in studying various physical properties of solutions, such as vapor pressure, boiling and freezing points, and osmotic pressure.

NORMALITY

Normality (*N*) is the number of equivalents (eq) or gram equivalent masses of solute dissolved in 1 liter of solution. It is expressed by the formula

$$N = \frac{\text{equivalents of solute}}{\text{1 liter of solution}}$$

To prepare 1 liter of a normal (1.00 *N*) solution of a substance, weigh out 1 gram equivalent mass of solute, place the weighed solute in a 1-liter flask, and add enough solvent to make the *final volume* of the solution 1 liter. The procedure and concept are the same as for molarity, except that 1 gram equivalent mass (1.00 eq) of solute is weighed out instead of 1 gram molecular mass (1.00 mole).

MILLIEQUIVALENT

A *milliequivalent* (mEq) is the same as a *millimole* (0.001 mmol, or 1 formula weight in milligrams) when the *valence*, or the number of charges on the ion, is 1. The milliequivalent is 1/1000 of an equivalent. An *equivalent* (eq) or *equivalent weight* (eq wt) is the atomic weight divided by the valence. The number of milliequivalents of a specific ion in solution (1 L) can also be expressed by the formula

$$\text{mEq/L} = \frac{\dfrac{\text{milligrams of ion}}{\text{per liter of solution}}}{\text{atomic weight}} \times \frac{\text{number of charges}}{\text{on the ion (valence)}}$$

Using the formula above, let us calculate the milliequivalents of calcium in normal blood. In 1 liter of plasma there are 100 mg of calcium, and the Ca^{2+} has two charges (indicated by the $^{2+}$). Using these values in the formula, we have

$$\text{mEq/L} = \frac{100}{40} \times 2 = 5$$

*A mola*l* solution should not be confused with a mola*r* solution, which contains the gram molecular weight of solute in 1 liter of solution.

THE MOLE CONCEPT

Because atoms, molecules, and formula units are so small compared to everyday objects and units, chemists and physiologists prefer to think of them in very large quantities, and they use a special number. This number is called **Avogadro's number** (in honor of the Italian physicist and chemist Amedeo Avogadro, 1776–1856). It is equal to 6.02×10^{23} atoms, molecules, or formula units of a particular chemical substance.

A **mole** (mol) is the amount of substance represented by 6.02×10^{23} atoms, molecules, or formula units of that particular substance. The *gram atomic mass* of an element is the mass in grams of 1.000 mol or 6.02×10^{23} atoms of a naturally occurring mixture of its isotopes. It is numerically the same as the atomic mass in atomic mass units (amu). The *gram molecular mass* in grams of a mainly covalent compound is 1.000×10^{23} molecules. The *gram formula mass* of a mainly ionic compound is the mass in grams of 1.000 mol or 6.02×10^{23} formula units.

OSMOLARITY

The term **osmolarity** defines particle concentration. An *osmole* is 6.02×10^{23} molecules of a solution or mixture of solutes of any kind. Thus, an osmole of solute dissolved in 1000 mL (1 L) of solution constitutes a *1-osmolar solution*. When determining osmolarity, it is important to consider the effective concentration of all particles. For example, a 0.1-osmolar solution could be a mixture of salts, sugars, amino acids, and proteins. The osmolarity of a solution is always measured in an ideal situation, for example, with the solution separated from pure water by a perfectly semipermeable membrane. Solution A and solution B are *isomolar* if their effective particle concentrations are equal. On the other hand, if solution A has a higher concentration of osmotically active particles than solution B, A is *hyperosmolar* to B, and B is *hypoosmolar* to A.

The terms *osmolar* and *osmolarity* refer to osmoles per liter of solution. *Osmolal* and *osmolality* refer to osmoles per kilogram of water. (The distinction is similar to that between the terms *molar* and *molal*.)

Prefixes, Combining Word Roots, and Suffixes in Medical Terminology

APPENDIX C

PREFIXES AND COMBINING WORD ROOTS

a-, an- without, absent (*apnea:* temporary absence of respiration).

ab- away from (*abduct:* move away from).

abdomin(o)- abdomen (*abdominal:* pertaining to the abdomen).

acou- hearing (*acoustics:* study of sound).

acr(o)- extremity, tip (*acrocyanosis:* bluish color of hands and feet).

ad- to, toward, near (*adduct:* move toward).

aden- gland (*adenoma:* glandular tumor).

af- to, toward (*afferent:* toward a given point).

alba- white (*albino:* lacking color, appearing white).

alg- pain (*neuralgia:* nerve pain).

alve- channel (*alveolus:* air channel in lung).

ambi-, amphi- both, on both sides (*ambidextrous:* using both hands; *amphiarthrosis:* an articulation permitting only slight motion).

ambly- dull (*amblyopia:* diminished vision).

ana- up, back again (*anabolism:* building up).

andr(o)- male (*androgen:* male sex hormone).

angi(o)- vessel, duct, usually a blood vessel (*angiology:* study of blood and lymph vessels).

ankyl(o)- crooked, bent, fused, stiff (*ankylosed:* fused, as a joint).

ante- before, in front of (*antepartum:* before delivery).

anti- against, counteracting (*antidote:* treatment to counteract effects of a poison).

arthr(o)- joint (*arthritis:* inflammation of a joint).

atel- imperfect, incomplete (*atelectases:* imperfect expansion of lungs).

auto- self (*autoregulatory:* self-regulating).

bi- two, twice (*binocular:* pertaining to both eyes).

bio- life (*biochemistry:* chemistry of living matter).

blasto- growth (*blastocyst:* hollow ball of embryonic cells).

brachi(o)- arm (*brachialis:* muscle for flexing forearm).

brachy- short (*brachydactyly:* abnormal shortness of fingers or toes).

brady- slow (*bradycardia:* slow heart rate).

bronch- windpipe (*bronchitis:* inflammation of the bronchus).

cac(o)- bad, ill (*cachexia:* generally poor condition of the body).

capit- head (*capitulum:* small eminence, or little head, on a bone by which it articulates with another bone).

carcin(o)- cancer (*carcinogenic:* cancer causing).

cata-, kata- down, through (*catabolism:* breaking down).

caud- tail (*caudal:* toward the posterior part of the body).

centi- 1/100, 100 (*centimeter:* 1/100 meter).

cephal(o)- head (*encephalitis:* inflammation of the brain).

cerebr(o)- brain (*cerebral:* pertaining to the brain).

cervic- neck (*cervix:* neck-shaped outer end of the uterus).

chondr(o)-, chondri-, chondrio- cartilage (*perichondrium:* membrane surrounding cartilage).

chromo-, chromato- color (*chromocyte:* colored cell).

circum- around (*circumduction:* circular motion around a joint).

cirrh- yellow (*cirrhosis:* disease causing yellowing of the liver).

co-, com-, con- with, together (*congenital:* born with).

coel- cavity within a body or body organ (*coelom:* cavity formed by splitting the mesoderm into two layers).

contra- opposite, against (*contralateral:* opposite side).

crani(o)- skull (*craniotomy:* surgical opening in the skull).

cryo- cold (*cryosurgery:* surgery performed with application of extreme cold to tissues).

crypt(o)- hidden (*cryptorchidism:* undescended testicle).

cut- skin (*subcutaneous:* under the skin).

cyan(o)- blue (*cyanosis:* bluish discoloration of the skin).

cyst- sac, bladder (*cystitis:* inflammation of the bladder).

cyt(o)- cell (*cytology:* study of cells).

de- away, not, down (*dehydrate:* take water away from).

deca- 10 (*decaliter:* 10 liters).

deci- 1/10 (*deciliter:* 1/10 liter).

dent(i)- tooth (*dentiform:* tooth-shaped).

derm- skin (*dermatitis:* inflammation of the skin).

dextro- right (*dextrogastria:* displacement of the stomach to the right side of body).

di- two, twice, doubly (*diphasic:* occurring in two stages).

dia- apart, across, through, between, completely (*diastasis:* separation of normal joined parts).

diplo- double (*diplopia:* double vision).

dis- apart, away from (*disarticulate:* come apart at the joints).

dorso-, dorsi- back (*dorsiflexion:* bending to the back of hand, foot, or spine).

duct- lead, conduct (*ductus deferens:* tube that conducts sperm from testes to ejaculatory duct).

dys- diseased, difficult (*dysmenorrhea:* difficult menses).

ecto- outside (*ectoderm:* outer germ layer).

edem- swelling (*edema:* swelling).

em-, en- in (*encapsulated:* enclosed in a capsule).

endo- within (*endoderm:* inner germ layer).

enter(o)- intestine (*gastroenteritis:* inflammation of the digestive system).

epi- on, over, upon (*epidermis:* outer layer of skin).

erythr(o)- red (*erythropoiesis:* formation of red blood cells).

eu- good, normal (*eupnea:* easy, normal breathing).

ex- out of (*exhale:* breathe out).

exo- outside (*exocytosis:* process of expelling large particles from a cell).

extra- outside of, in addition to, beyond (*extracellular:* outside a cell).

fasci- band (*fascia:* sheet of fibrous tissue).

fore- before, in front (*forebrain:* front part of brain).

gastr(o)- stomach (*gastritis:* inflammation of the stomach).

glyco-, gluco- sugar, sweet (*glycolysis:* breakdown of glucose).

gyn(o), gyne(co)- female, woman (*gynecology:* study of the female reproductive system).

haplo- single, simple (*haploid:* having a single set of chromosomes).

hema-, hemato-, hemo- blood (*hematocyst:* a cyst containing blood).

hemi- one-half (*hemiplegia:* paralysis of one side of the body).

hepat(o)- liver (*hepatitis:* inflammation of the liver).

hetero- other, different (*heterosexual:* other sex).

hist(o)- tissue (*histology:* study of tissues).

homeo- unchanged (*homeostasis:* state of inner stability of the body).

homo- same (*homolateral:* on the same side).

hydr(o)- water (*hydrocele:* accumulation of serous fluid in a body cavity).

hyper- above, excessive (*hyperactivity:* excessive activity).

hypo- beneath, under, deficient (*hypogastric:* below the gastric region).

idio- peculiar to the individual, distinct (*idiopathic:* disease with no apparent cause).

in-, im- in, within, into; not (*indigestion:* difficulty in digesting food; *immerse:* dip into).

infra- below (*infracostal:* below the ribs).

inter- between, among (*intercellular:* between cells).

intra- within (*intracellular:* within a cell).

ir- against, into, toward (*irradiate:* emit rays into).

is(o)- equal, same (*isopia:* equal vision in both eyes).

labi- lip (*labia major:* external folds or lips of the vulva).

lacri- tears (*nasolacrimal apparatus:* tear-producing and -carrying apparatus).

lact(o)- milk (*lactose:* milk sugar).

leuk(o)- white (*leukocyte:* white blood cell).

lip(o)- fat, fatty (*lipoma:* benign tumor of fatty cells).

macr(o)- large (*macrophage:* large phagocytic cell).

mal- bad (*malabsorption:* impaired absorption).

medi- middle (*medial:* near the midline of the body).

mega- large (*megakaryocyte:* large cellular forerunner of platelet).

melan(o)- black (*melanin:* dark skin pigment).

meso- middle (*mesoderm:* middle layer of skin).

meta- beyond, next to (*metacarpal:* bone of the hand that is next to the wrist).

micr(o)- small (*photomicrograph:* photograph of a very small object as seen through a microscope).

milli- 1/1000 (*millimeter:* 1/1000 meter).

mon(o)- one (*monocular:* pertaining to one eye).

morph(o)- form, shape (*morphology:* study of the form and structure of organisms and parts of organisms).

multi- many (*multinuclear:* having many nuclei).

my(o)- muscle (*myocardium:* heart muscle).

myel(o)- marrow, spinal cord (*myeloma:* malignant tumor of bone marrow).

neo- new, strange (*neonatal:* newborn).

nephr(o)- kidney (*nephrectomy:* surgical removal of a kidney).

neur(o)- nerve (*neuron:* nerve cell).

non- not (*noninfectious:* not able to spread).

ob- against (*obstruction:* blocking of a structure).

ocul(o)- eye (*oculomotor:* pertaining to eye movement).

olig(o)- little, few, scanty (*oliguria:* secretion of a small amount of urine).

oo- egg (*oocyte:* egg cell).

ophthalmo- eye (*ophthalmology:* study of the eye).

orth(o)- straight, normal, correct (*orthodontic:* pertaining to the straightening of teeth).

oste(o)- bone (*osteitis:* inflammation of the bone).

ot(o)- ear (*otology:* study of the ear).

ovi-, ovo- egg (*ovum:* egg cell).

para- beside, beyond (*paraspinal:* beside the spine).

path(o)- disease, suffering (*pathogenic:* capable of producing disease).

ped(ia)- child (*pediatrician:* physician who specializes in childhood diseases).

pend- hang down (*appendicular skeleton:* portion of the skeleton that hangs from the shoulder and hip girdles).

per- through (*permeate:* to pass through).

peri- around (*pericardium:* membrane surrounding the heart).

phag(o)- eat, consume (*phagocyte:* cell that ingests other cells or particles).

phleb(o)- vein (*phlebitis:* inflammation of a vein).

pleuro-, pleura- rib, side (*pleurisy:* inflammation of pleura, the membrane covering the lungs and lining the thoracic sac).

plur(i)- many, more (*pluriglandular:* several glands).

pneumato-, pneuma- air (*pneumatometer:* instrument for measuring pressure of inspiration and expiration of lungs).

pneumo-, pneumono- lung, respiratory organs (*pneumonia:* chronic lung infection).

pod- foot (*podiatrist:* a specialist in care and treatment of the foot).

poly- many (*polycystic:* having many cysts).

post- after (*postpartum:* after delivery).

pre-, pro- before (*prenatal:* before birth).

pro- for, in front of, before (*prophylaxis:* preventive treatment, to keep guard before).

proct- rectum, anus (*proctoscope:* instrument for examining the rectum).

pseudo- false (*pseudocyst:* false cyst).
pulmo- lung (*pulmonary:* pertaining to the lungs).
py(o)- pus (*pyocyst:* pus-filled cyst).

quadr(i)- four (*quadriplegia:* paralysis of all four limbs).

re- back, again (*reflex:* bend back).
ren- kidney (*adrenal:* relating to the position of the adrenal gland atop the kidney).
retro- behind, backward (*retrograde:* going backward).
rhin(o)- nose (*rhinitis:* inflammation of nasal mucous membranes).

sclero- hard (*sclerosis:* hardening of tissue).
semi- one-half, partly (*semilunar valve:* half-moon–shaped valve).
steno- contracted, narrow (*stenosis:* condition of being narrowed).
sub- beneath (*subcutaneous:* beneath the skin).
super- above (*superior:* toward the head).
supra- above, over (*suprapubic:* above the pubic bone).
sym-, syn- together, union (*synapse:* meeting between two or more nerve cells).

tachy- fast (*tachycardia:* abnormally fast heart rate).
telo- end (*telophase:* final phase of mitosis).
tetr(a)- four (*tetrad:* group of four chromosomes formed during meiosis).
therap- treatment (*therapeutic:* having a healing effect).
therm(o)- heat, warmth (*thermometer:* device to measure heat).
thorac(o)- chest (*thoracic cavity:* chest cavity).
thromb(o)- lump, clot (*thrombosis:* formation of a blood clot in a vessel or heart cavity).
tox- poison (*toxemia:* blood poisoning).
trans- across, through (*transepidermal:* occurring through or across the skin).
tri- three, thrice (*triceps:* three-headed muscle of the posterior of arm).

ultra- excessive, extreme (*ultrasonic:* sound waves beyond the audio-frequency range).
un(i)- one (*unilateral:* one side).
uria-, uro- urine, urinary tract (*polyuria:* production of excess urine).

vas(o)- vessel (*cardiovascular:* vessel related to the heart).
viscera-, viscero- organ (*visceral:* pertaining to a body organ).

SUFFIXES

-ac pertaining to (*celiac:* pertaining to the abdominal region).
-ad toward, in the direction of (*cephalad:* toward the head).
-agra severe pain, seizure (*podagra:* severe pain in the foot).
-al pertaining to (*digital:* pertaining to the finger or toe).

-an, -ian pertaining to, characteristic of, belonging to (*ovarian:* pertaining to the ovary).
-ar of or relating to, being, resembling (*valvular:* relating to a valve).
-ary of or relating to, connected with (*biliary:* relating to bile).
-ase enzyme (*lactase:* enzyme that catalyzes conversion of lactose into glucose and galactose).
-ate that which is acted upon, marked by having, to act on (*substrate:* substance acted on by enzyme; *lobate:* having lobes; *separate:* to keep apart).
-atresia abnormal closure (*proctatresia:* closed anus).

-blast sprout, growth (*osteoblast:* cell from which bone develops).

-cele swelling, tumor (*cystocele:* hernia of urinary bladder).
-centesis puncture of a cavity (*paracentesis:* puncture of a space around an organ or within a cavity to remove fluid).
-cide kill (*germicide:* killer of germs).
-cis cut (*excise:* cut out).
-clasis, -clasia breaking up (*bacterioclasis:* breaking up of bacteria).
-cleisis closure (*colpocleisis:* operation for closure of vagina).

-desis binding, fusion (*tenodesis:* fixation of a loose tendon).
-dynia pain (*pleurodynia:* pain on the side of the chest).

-ectasia, -ectasis dilation, expansion (*telangiectasis:* dilation of capillaries).
-ectomy excision, cutting out (*laryngectomy:* removal of larynx).
-emia condition of the blood (*leukemia:* blood condition characterized by excess leukocytes).

-facient making or causing to become (*febrifacient:* causing a fever).
-ferent carrying (*efferent:* carrying away from).
-form structure, shape (*ossiform:* resembling the form of bone).
-fugal driving or traveling away from (*centrifugal:* moving away from the center).
-ful full of, characterized by (*stressful:* causing stress).

-gen, -gene producer (*mutagen:* substance that increases the frequency of mutation).
-genesis production of, origin of (*glycogenesis:* production of glycogen).
-gram record (*electrocardiogram:* record of electrical activity in the heart).
-graph instrument for making records (*electroencephalograph:* apparatus for recording brain waves).

-ia condition (*anuria:* condition of lack of urine).
-iasis a diseased condition (*cholelithiasis:* condition of bile stones).
-ic pertaining to (*colonic:* pertaining to the colon).
-ician, -ist specialist, practitioner (*pediatrician:* specialist in childhood diseases).
-ile having the qualities of or capability for (*febrile:* feverish).

-ion act of, state of, result of (*incision:* act of cutting or result of cutting).

-ism condition, act of, process of (*dwarfism:* condition of being a dwarf).

-itis inflammation (*phlebitis:* inflammation of a vein).

-ity quality of, state of (*obesity:* state of being obese).

-ive that which performs or tends toward (*tardive disease:* disease with late-appearing symptoms).

-logy study of (*cardiology:* study of the heart).

-lysis breaking down (*glycolysis:* breakdown of glucose).

-malacia softening (*osteomalacia:* softening of bone).

-meter instrument or means of measuring (*thermometer:* instrument to measure heat).

-ness quality of, state of (*illness:* state of being ill).

-oid having the appearance of (*cuboid:* resembling a cube).

-oma tumor (*carcinoma:* malignant tumor).

-opia eye disorder (*myopia:* nearsightedness).

-or that which, one who (*receptor:* that which receives).

-ory process of, pertaining to, function of (*sensory:* pertaining to the senses).

-ose full of, having the qualities of (*comatose:* having the qualities of a coma).

-osis action, state, process, condition (*halitosis:* condition of having bad breath).

-ostomy creation of an opening (*colostomy:* creation of a new opening between the bowel and abdominal wall).

-otomy cutting into (*tracheotomy:* cut into the windpipe).

-ous, -ious having the qualities of, capable of, full of (*infectious:* capable of being transmitted).

-pathy disease (*cardiopathy:* heart disease).

-penia deficiency, lack (*leukopenia:* lack of white blood cells).

-pexy fixation (*sigmoidopexy:* fixation of large intestine).

-philia love of, tendency toward (*hemophilia:* "love of blood," a blood disease).

-phobia abnormal fear (*claustrophobia:* fear of confinement).

-plasty molding or shaping (*osteoplasty:* plastic surgery on bone).

-plegia paralysis (*paraplegia:* paralysis of lower half of body).

-poiesis production, formation of (*hematopoiesis:* production of blood cells).

-ptosis dropping, downward displacement (*carpoptosis:* wrist drop).

-rrhagia bursting forth, excessive discharge (*hemorrhage:* escape of blood from vessels).

-rrhaphy closure of by suturing, repair (*cystorrhaphy:* suture of bladder).

-rrhea flow, discharge (*galactorrhea:* excessive or spontaneous flow of milk).

-scope looking at, examining (*bronchoscope:* instrument used to view inside breathing tubes).

-sect cut (*dissect:* cut into parts).

-sis process, action (*dialysis:* separation of particles through a semipermeable membrane).

-stasis state of being at a standstill (*hemostasis:* stopping of blood flow).

-sthenia strength (*asthenia:* loss of strength).

-stomy surgical opening (*colostomy:* surgical opening in the colon).

-tion act of, result of, process of (*elongation:* process of making longer).

-tomy to cut (*lobotomy:* incision into a lobe).

-tonia stretching, putting under tension (*hypertonia:* excessive tension).

-tripsy rubbing, crushing (*lithotripsy:* surgical crushing of kidney stones).

-trophic related to nutrition, growth, development (*dystrophic:* faulty nutrition).

-tropic turning toward, changing (*hydrotropic:* turning toward water).

-y having the nature or quality of (*gouty:* goutlike).

Reclassification of Eponymous Terms

APPENDIX D

An *eponym* is a person for whom something is named, and an *eponymous term* is the term that uses an eponym—*Achilles tendon,* for example. Because eponymous terms are not anatomically or physiologically descriptive, they have been replaced in this book with currently accepted terms that help the student understand the meaning of the term. Eponymous terms are shown in parentheses the first time the current term is used. In a few cases, because the eponym is used so extensively, it has been retained, with the new term in parentheses—*Broca's speech area (anterior speech area),* for example. Two eponymous terms have been retained because of the lack of an appropriate substitution: *Tyson's glands* and *corpuscles of Ruffini.* (For the record, the champion eponym is the French chemist Georges Deniges, with 78 eponymous terms named after him.)

Eponymous term	Preferred term
Achilles reflex	plantar reflex
Achilles tendon	calcaneal tendon
Adam's apple	thyroid cartilage
ampulla of Vater	hepatopancreatic ampulla
aqueduct of Sylvius	cerebral aqueduct
axis of Henke	subtalar axis
Bartholin's glands	greater vestibular glands
Bowman's capsule	glomerular capsule
Bowman's glands	olfactory glands
Broca's area	Broca's speech area (anterior speech area)
Brodmann area 4	primary motor cortex
Brunner's glands	duodenal submucosal glands
bundle of His	atrioventricular bundle
canal of Schlemm	scleral venous sinus
circle of Willis	cerebral arterial circle
Cooper's ligaments	suspensory ligaments of the breast
Cowper's glands	bulbourethral glands
crypts of Lieberkühn	intestinal glands
duct of Santorini	accessory pancreatic duct
duct of Wirsung	pancreatic duct
Eustachian tube	auditory tube
Fallopian tube	uterine tube
fissure of Rolando	central sulcus
fissure of Sylvius	lateral cerebral sulcus
foramen of Monro	interventricular foramen
Golgi tendon organ	tendon organ

Eponymous term	Preferred term
Graafian follicle	vesicular ovarian follicle
Graves' disease	hyperthyroidism
gyrus of Heschl	transverse temporal gyrus
Hassal's corpuscles	thymic corpuscles
Haversian canal	central canal
Haversian system	osteon
islet of Langerhans	pancreatic islet
Krause's end bulbs	bulbous corpuscles (of Krause)
Krebs cycle	citric-acid cycle
Kupffer cells	stellate reticuloendothelial cells
Langerhans' cell	nonpigmented granular dendrocyte
Langer's lines	cleavage lines
Leydig cells	interstitial endocrinocytes
loop of Henle	loop of the nephron
Malpighian corpuscles	splenic nodules
Meissner's corpuscles	tactile corpuscles (of Meissner)
Merkel's disks	tactile corpuscles (of Merkel)
Müllerian duct	paramesonephric duct
Müller's tubercle	sino-utricular node
Nissl bodies	chromatophilic substance
nodes of Ranvier	neurofibral nodes
organ of Corti	spiral organ
Pacinian corpuscle	lamellated (Pacinian) corpuscle
Peyer's patches	aggregated lymph nodules
Purkinje fibers	cardiac conducting myofibers (Purkinje fibers)
Ruffini's corpuscles	corpuscles of Ruffini
Schwann cell	neurolemmocyte
Schwann sheath	neurilemma cell
Sertoli cells	sustentacular cells
Sharpey's fibers	periosteal perforating fibers
Skene's glands	lesser vestibular glands
sphincter of Boyden	sphincter of the common bile duct
sphincter of Oddi	sphincter of the hepatopancreatic ampulla
Stensen's ducts	parotid ducts
Volkmann's canal	perforating canal, nutrient canal
Wernicke's area	Wernicke's area (posterior speech area)
Wharton's duct	submandibular duct
Wharton's jelly	mucous connective tissue
Wormian bone	sutural bone

Normal Laboratory Values for Blood and Urine
APPENDIX E

The following abbreviations are used for normal values:

↑	increased	L	liter	mμ	millimicron	μg	microgram
↓	decreased	M	molar	ng	nanogram	μL	microliter
>	greater than	mol	mole	nmol	nanomole	mm Hg	millimeter of mercury
<	less than	m^2	square meter	pg	picogram	mmol	millimole
cm^3	cubic centimeter	mCi	millicurie	S	second	mOsm	milliosmole
mm^3	cubic millimeter	mEq	milliequivalent	SI	international	mU	milliunit
dL	deciliter	mg	milligram		system of units*	WBCs	white blood cells
g	gram	mg/dL	milligram per deciliter	U	unit		
IU	international unit	mL	milliliter	μ	micron		
kg	kilogram	RBCs	red blood cells	$μ^3$	cubic micron		

*SI units are *Système International* units that are used uniformly in the European literature. The benefits consist of scientific standardization in reporting.

TABLE E.1 NORMAL LABORATORY ADULT BLOOD (CHEMISTRY) VALUES

Determination (test)	Normal values	Clinical significance
Acetone (ketone bodies)	Acetone: 0.3–2.0 mg/dL Ketones: 2–4 mg/dL	Values increase in diabetic ketoacidosis (dka), starvation, malnutrition, heat stroke, diarrhea.
Acid phosphatase (ACP)	0.1–2 U/dL (Gutman)	Values increase in prostate cancer, sickle-cell anemia, cirrhosis renal failure, myocardial infarction; values decrease in Down's syndrome.
Alanine amino-transferase (ALT, SGPT)	5–35 U/mL (Frankel)	Values increase in viral hepatitis, cirrhosis, congestive heart failure; values decrease in exercise.
Albumin	3.5–5 g/dL	Values increase in kidney disease, fever, trauma, and myeloma; values decrease in severe burns (proteinuria), starvation, leukemia, chronic liver disease.
Alkaline phosphatase (ALP)	159–400 mg/dL; 1.0–1.6 g/L	Values increase in biliary disease, cancer of liver, cirrhosis, hepatitis, arthritis, ulcers; values decrease in malnutrition, pernicious anemia.
Ammonia	3.2–4.5 g/dL; 32–45 g/L (SI units)	Values increase in hepatic failure, hemolytic disease of the newborn, congestive heart failure, emphysema; values decrease in renal failure, hypertension.
Amylase	60–160 U/dL (Somogyi); 111–296 U/L (SI units)	Values increase in pancreatitis, ulcers, cancer, diabetic acidosis, mumps, renal failure, burns, pregnancy; values decrease in necrosis of liver, chronic alcoholism, hepatitis, severe burns.
Aspartate amino-transferase (AST, SGOT)	5–40 U/mL; 4–36 IU/L	Values increase in acute myocardial infarction, hepatitis, liver necrosis, trauma, liver cancer, angina; values decrease in pregnancy, diabetic ketoacidosis, beriberi.
Bilirubin (total and direct)	Total: 0.1–1.2 mg/dL; 1.7–20.5 μmol/L (SI units) Direct: 0.1–0.3 mg/dL; 1.7–5.1 μmol/L (SI units)	Values increase in obstructive biliary disease, liver disease, liver cancer.
Blood urea nitrogen (BUN)	Male: 10–25 mg/dL Female: 8–20 mg/dL	Values increase in dehydration, high protein intake, GI bleeding, prerenal failure, diabetes mellitus, myocardial infarction, kidney disease; values decrease in severe liver damage, low-protein diet, malnutrition.

TABLE E.1 (Continued)

Determination (test)	Normal values	Clinical significance
Calcium (Ca)	4.5–5.5 mEq/L; 9–11 mg/dL; 2.3–2.8 mmol/L (SI units)	Values increase in hypervitaminosis D, bone cancer, lung cancer, multiple fractures, renal calculi, alcoholism; values decrease in diarrhea, laxative abuse, extensive infections, burns, hypoparathyroidism, alcoholism, pancreatitis.
Carbon dioxide combining power (CO_2)	22–30 mEq/L; 22–30 mmol/L (SI units)	Values increase in respiratory diseases, intestinal obstruction, vomiting; values decrease in acidosis, nephritis, diarrhea, starvation.
Chloride (Cl)	95–105 mEq/L; 95–105 mmol/L (SI units)	Values increase in nephritis, Cushing's syndrome, hyperventilation; values decrease in diabetic acidosis, burns.
Cholesterol (total) HDL (α) LDL (β) VLDL (pre-β)	150–250 mg/dL; 3.90–6.50 mmol/L (SI units) May increase with age. 29–77 mg/dL 62–185 mg/dL 0–40 mg/dL	Values increase in acute myocardial infarction, atherosclerosis, hyperthyroidism, biliary obstruction, diabetes mellitus, stress; values decrease in Cushing's disease, starvation, anemia, malabsorption.
Cholinesterase	0.5–1.0 units (RBC); 3–8 units/mL plasma; 6–8 IU/L (RBC); 8–18 IU/L (plasma)	Values increase in nephrotic syndrome; values decrease in insecticide poisoning, liver disorders, acute infection, anemia, carcinomatosis.
Cortisol (hydrocortisone, compound F)	8–10 A.M.: 5–23 µg/dL 4–6 P.M.: 3–13 µg/dL	Values increase in adrenal cancer, stress, pregnancy, obesity, myocardial infarction, diabetic acidosis, hyperthyroidism; values decrease in Addison's disease, respiratory distress syndrome, hypothyroidism.
Creatine phosphokinase (CPK)	Male: 5–35 µg/mL; 15–120 IU/L Female: 5–25 µg/mL; 10–80 IU/L	Values increase in muscular dystrophy, trauma, hypokalemia, myocardial infarction, hemophilia, tetanus, venom toxin; values decrease in pregnancy.
Creatinine (Cr)	0.6–1.2 mg/dL; 53–106 µmol/L (SI units)	Values increase in renal failure, shock, systemic lupus erythematosus, cancer, hypertension, myocardial infarction, diabetic nephropathy, high-protein diet; values decrease in pregnancy, eclampsia.
Gamma-glutamyl transpeptidase (GGTP)	Male: 10–38 IU/L Female: 5–25 IU/L	Values increase in cirrhosis of the liver, alcoholism, viral hepatitis, liver cancer, mononucleosis, diabetes mellitus, myocardial infarction, congestive heart failure, epilepsy.
Glucose (fasting)	70–110 mg/dL	Values increase in diabetes mellitus, liver disease, stress, nephritis, pregnancy, hyperthyroidism; values decrease in hypothyroidism, Addison's disease, pancreatic cancer.
Iron	50–150 µg/dL; 10–27 µmol/L (SI units)	Values increase in hemochromatosis, anemia, liver damage, lead poisoning; values decrease in iron-deficiency anemia, malignancies, arthritis, ulcers, renal failure.
Lactic acid	Arterial blood: 0.5–2.0 mEq/L Venous blood: 0.5–1.5 mEq/L Panic: > 5 mEq/L	Values increase in shock, dehydration, ketoacidosis, severe infections, neoplastic conditions, hepatic failure, renal disease, alcoholism.
Lactic dehydrogenase (LDH/LD)	150–450 U/mL (Wroblewski-LaDue method)	Values increase in myocardial infarction, ketoacidosis, severe infections, neoplastic conditions, hepatic failure, renal disease, alcoholism.
Lipids (fasting) Cholesterol Phospholipids Total fatty acids Total lipids Triglycerides	 120–220 mg/100 mL 9–16 mg/100 mL 190–420 mg/100 mL 450–1000 mg/100 mL 40–150 mg/100 mL	Values increase in hyperlipoproteinemia, myocardial infarction, hypothyroidism, diabetes mellitus, eclampsia; values decrease in chronic obstructive lung disease.
Osmolality	280–300 mOsm/kg H_2O	Values increase in diabetes insipidus, dehydration, hypernatremia, hyperglycemia, uremia; values decrease in excessive fluid intake, bronchus and lung cancer, adrenal cortical hypofunction.
Phosphorus (P_i)	1.7–2.6 mEq/L; 2.5–4.5 mg/dL	Values increase in renal insufficiency, renal failure, hypocalcemia, acromegaly; values decrease in starvation, malabsorption syndrome, hypercalcemia, hyperparathyroidism, diabetic acidosis, myxedema.

(Table E.1 continues on the following page.)

TABLE E.1 (Continued)

Determination (test)	Normal values	Clinical significance
Potassium (K)	3.5–5.0 mEq/L	Values increase in dehydration, starvation, renal failure, diabetic acidosis, Cushing's disease; values decrease in acute renal failure, diarrhea.
Protein		Total protein values increase in severe dehydration, shock; values decrease in hemorrhage and severe malnutrition.
Albumin	3.5–5.0 g/dL	
Globulin	1.5–3.5 g/dL	
AIG ratio	1.5:1 to 2.5:1	
Total	6.0–7.8 g/dL	
Sodium (Na)	135–145 mEq/L; 135–142 mmol/L (SI units)	Values increase in dehydration, severe vomiting and diarrhea, congestive heart failure, Cushing's disease, high-sodium diet; values decrease in vomiting, diarrhea, low-sodium diet, burns, renal disease.
Thyroxine (T_4)	4.5–11.5 μg/dL (by column)	Values increase in hyperthyroidism, viral hepatitis, myasthenia gravis, pregnancy; values decrease in hypothyroidism, protein malnutrition, anterior pituitary hypofunction, exercise.
Uric acid	Male: 3.5–7.8 mg/dL Female: 2.8–6.8 mg/dL	Values increase in gout, alcoholism, leukemia, cancer, diabetes mellitus; values decrease in anemia, burns, pregnancy.

TABLE E.2 LABORATORY ADULT BLOOD (SEROLOGY) VALUES

Determination (test)	Normal values	Clinical significance
Carcinoembryonic antigen (CEA)	< 2.5 mg/mL	Values increase in cancer of the GI system, heart disease, colitis, inflammatory disease.
Heterophile antibody	< 1:28 titers*	Values increase in mononucleosis, serum sickness, viral infections.
Immunoglobulins (Ig)		
Ig	900–2200 mg/dL	Values increase in malnutrition, liver disease, rheumatic fever; values decrease in leukemia, preclampsia, amyloidosis.
IgG	800–1800 mg/dL	Same as Ig.
IgA	100–400 mg/dL	Values increase in autoimmune disorders, rheumatic fever, chronic infections; values decrease in leukemia, malignancies.
IgM	50–150 mg/dL	Values increase in parasitic infections; values decrease in leukemia.
IgD	0.5–3 mg/dL	Values increase in chronic infections and myelomas.
IgE	0.01–0.05 mg/dL	Values increase in allergies.
Rheumatoid factor	< 1:20 titers	Values increase in rheumatoid arthritis, SLE, scleroderma, mononucleosis, tuberculosis, leukemia, hepatitis, syphilis, old age.

*A titer is the concentration of antibody in serum.

TABLE E.3 NORMAL ADULT BLOOD (GAS) VALUES

Arterial O_2 saturation	$\dfrac{O_2 \text{ content}}{O_2 \text{ capacity}} \times 100 = 95\%$	Mixed venous O_2 content	10–16 mL/100 mL of blood
		HCO_3^- content	24–28 mEq/L
Pulmonary arterial O_2 saturation	$\dfrac{O_2 \text{ content}}{O_2 \text{ capacity}} \times 100 = 75\text{–}80\%$	pH arterial plasma	7.35–7.45
		pH venous plasma	7.32–7.43
Whole blood O_2 capacity	17–21 mL of O_2/100 mL of blood	CO_2 combining power (venous plasma)	21–30 mEq/L
		Arterial CO_2 content (whole blood)	20–25 mEq/L
Arterial O_2 content	16.5–20.0 mL/100 mL of blood	Arterial and alveolar CO_2 tension (Pa_{CO_2})	37–41 mm Hg
P_{O_2} arterial content	80–95 mm Hg	P_{CO_2} arterial tension	35–45 mm Hg
P_{O_2} venous content	35–40 mm Hg	P_{CO_2} venous tension	38–50 mm Hg

Appendix E: Normal Laboratory Values for Blood and Urine

TABLE E.4

NORMAL LABORATORY ADULT BLOOD (HEMATOLOGY) VALUES

Determination (test)	Normal values	Clinical significance
Bleeding time	Duke method: 1–5 min Ivy method: 2–7 min	Time increases in liver disease, anemia, leukemia, thrombocytopenia; used to evaluate platelet function.
Capillary fragility	5 or less	Values increase in faulty capillary endothelial integrity, easy bruising, bleeding.
Coagulation time (Lee-White)	5–15 min; average 8 min	Time increases in afibrinogenemia and hyperheparanemia; used to evaluate blood-clotting system.
Erythrocyte sedimentation rate (ESR) (zeta sedimentation ratio; ZSR)	Under 50 years old (Westergren): Male: 0–15 mm/hr Female: 0–20 mm/hr Over 50 years old (Westergren): Male: 0–20 mm/hr Female: 0–30 mm/hr	Values increase in infection, inflammation, tissue necrosis, pregnancy, malignancy; values decrease in sickle-cell anemia, spherocytosis, congestive heart failure.
Hematocrit (HCT) or packed cell volume (PCV)	Male: 40–54%; 0.40–0.54 SI units Female: 36–46%; 0.36–0.46 SI units	Values increase in polycythemia, shock, severe dehydration; values decrease in anemia, blood loss, leukemia, liver disease, hyperthyroidism.
Hemoglobin (Hb or Hgb)	Male: 13.5–18 g/dL Female: 12–16 g/dL	Values increase in obstructive pulmonary disease, congestive heart failure, polycythemia, high altitudes; values decrease in anemia, blood loss, hemolysis, hyperthyroidism, liver disease, pregnancy, excessive fluid intake.
Platelet (thrombocyte) count	150,000–400,000/mm³ (mean 250,000/mm³) SI units: 0.15–0.4 × 10¹²L	Values increase in some anemias, polycythemia, cancer, liver disease, trauma; values decrease in leukemia, aplastic anemia, allergies, cancer chemotherapy.
Prothrombin time (PT)	11–15 S or 70–100%	Values increase in certain hemorrhagic diseases, liver disease, vitamin K deficiency, various drug use, congenital deficiencies, disseminated intravascular coagulation.
Red blood cell count (RBC)	Male: 4.6–6.0 million/mm³ Female: 4.0–5.0 million/mm³	Values increase in dehydration, posthemorrhaging, polycythemia, cor pulmonale, cardiovascular disease; values decrease in anemia, Addison's disease, hemorrhage, chronic infections, leukemia, chronic renal failure, pregnancy.
Reticulocyte count	0.5–1.5% of all RBCs; 25,000–75,000 mm³ (absolute count)	Values increase in hemolytic and sickle-cell anemia, thalassemia, hemorrhage, iron deficiency, leukemia, hemolytic disease of the newborn, pregnancy; values decrease in pernicious and aplastic anemias, radiation therapy, liver cirrhosis, anterior pituitary hypofunction.
White blood cells (WBCs) or leukocyte total	5000–10,000 mm³	Values increase in acute infections, myocardial infarction, cirrhosis, burns, cancer, arthritis, gout; values decrease in aplastic and pernicious anemias, viral infections, malaria, alcoholism, uncontrolled diabetes.

White blood cell differential
  Neutrophils
    Segments
    Bands
	
	50–70% of total WBCs
	50–65%
	0–5%
		Values increase in acute infections, inflammatory disease, lung disease, myocardial infarction.

Eosinophils	0–3%	Values increase in allergies, parasitic diseases, cancer, asthma, kidney disease; values decrease in stress, burns.
Basophils	1–3%	Values increase in inflammation, leukemia; values decrease in stress.
Lymphocytes	25–35%	Values increase in leukemia, viral infections; values decrease in cancer, neurological disorders, renal failure.
Monocytes	2–6%	Values increase in viral diseases, parasitic diseases, leukemia, cancer, arthritis, some anemias; values decrease in lymphocytic leukemia, aplastic anemia.

Appendix E: Normal Laboratory Values for Blood and Urine

TABLE E.5   NORMAL LABORATORY ADULT URINE VALUES

Determination (test)	Normal values	Clinical significance
Acetone plus acetoacetate (ketone bodies)	Negative	Values increase in ketonuria, diabetic acidosis, ketoacidosis of alcoholism and diabetes mellitus, fasting, starvation, high-protein diet.
Addis count (12 hr)	Casts: 0–5000 Erythrocytes: 0–500,000 Leukocytes: < 1,000,000	Evaluation of severity and course of glomerulonephritis, during which values increase.
Albumin (quantitative protein)	0–< 20 mg/24 hr	Values increase in renal disease, diabetes mellitus, glomerulonephritis, amyloidosis; values decrease following severe burns.
Ammonia (nitrogen)	20–70 mEq/L	Values increase in diabetes mellitus, liver disease.
Amylase	4–30 IU/2 hr; 24–76 U/mL	Values increase in carcinomas of pancreas, pancreatitis.
Bile (bilirubin)	Negative	Values increase in biliary and liver disease, obstructive disease of biliary tract, cancers.
Calcium	< 300 mg/24 hr	Values increase in hyperparathyroidism; values decrease in hypoparathyroidism.
Concentration test	Specific gravity > 1.026 or 850 mOsm/kg	Used to evaluate renal concentrating ability, a test of tubular function; values decrease in renal disease.
Creatinine	1–2 g/24 hr	Values increase in bacterial infections; values decrease in muscular dystrophy, leukemia, anemia.
Creatinine clearance	150–180 L/day; 100–140 mL/min	Values increase in renal disease.
Fat (lipid)	Negative	Values increase in nephrotic syndrome, renal tubular necrosis, poisoning.
Glucose (sugar)	Negative	Values increase in renal disease; helpful in evaluating glucosuria, renal tubular defects, and in managing diabetes mellitus.
Hemoglobin (free)	Negative	Presence indicates hemoglobinuria as a result of hemolysis.
Nitrite	Negative	Presence indicates bacterial infection, urinary tract infection.
Occult blood	Negative	Presence of blood in urine indicates hematuria, hemoglobinuria.
pH	4.8–7.8	Used to determine subtle presence of distal tubular disease or pyelonephritis, kidney stones. Values increase in vegetarian diet, urinary tract infection, diabetes mellitus, alkalosis; values decrease in acidosis, diabetes mellitus, starvation, emphysema, dehydration.
Phenylalanine screen (phenylpyruvic acid)	Negative	Values increase in phenylketonuria (PKU).
Specific gravity	Range: 1.001–1.035 Normal fluid intake: 1.016–1.022	Values increase (> 1.020) in dehydration, fever, diarrhea, diabetes mellitus, congestive heart failure; values decrease (< 1.009) in diuresis, hypothermia, diabetes insipidus, glomerulonephritis.
Urea	25–35 g/24 hr	Values increase in excessive protein breakdown; values decrease in impaired renal function.
Urea clearance	> 40 mL of blood cleared of urea/minute	Values increase in renal diseases.
Uric acid	0.6–1.0 g/24 hr as urate	Values increase in gout; values decrease in kidney disease.
Urobilinogen (2 hr urine)	Male: 0.3–2.1 mg Female: 0.1–1.1 mg	Values increase in hemolytic anemias, myocardial infarction, liver disease, hyperthyroidism; values decrease in bile duct obstruction, antibiotic overload, hepatitis, starvation.

An Overview of Regional Human Anatomy

APPENDIX F

The following photographs of dissected, preserved human cadavers convey a sense of the actual appearance and relations of grossly observable anatomical structures. The photographs are in contrast to the drawings in textbooks and atlases that, no matter how accurately illustrated, are merely representations of the actual body. The photographs in this appendix provide the student with an opportunity to view an actual dissection of the human body. The photographs begin with external views, proceed to views of the superficial and deep muscle layers, and then show the thoracic viscera, the posterior abdominal and thoracic wall, the abdominal cavity with viscera exposed, and finally, male and female pelvic cavities with reproductive organs exposed.

These photographs are reminders of the profound role played by the dissection of the human body in the advancement of science. Leonardo da Vinci (1452–1519) produced his legendary (but not always accurate) anatomical sketches during the Renaissance, and in 1542 a scientific breakthrough occurred with the publication of *De corporis humani fabrica* ("On the Structure of the Human Body") by Andreas Vesalius (1514–1564). The anatomically accurate drawings in this monumental work were based on dissections of the human body, and the illustrations of the muscles in particular were so accurate that they are still used by students of anatomy and medical illustration. Coincidentally, in the same year, the Polish astronomer Nicolas Copernicus (1473–1543) published his great book *De revolutionibus orbium coelestium libri VI* ("Six Books On the Revolutions of the Celestial Spheres") that proposed that the earth and other planets revolve around the sun in relatively fixed orbital paths. Together, the two books provided the foundation for the birth of the Scientific Revolution.

FIGURE F.1

External view of body, with skin (anterior view).

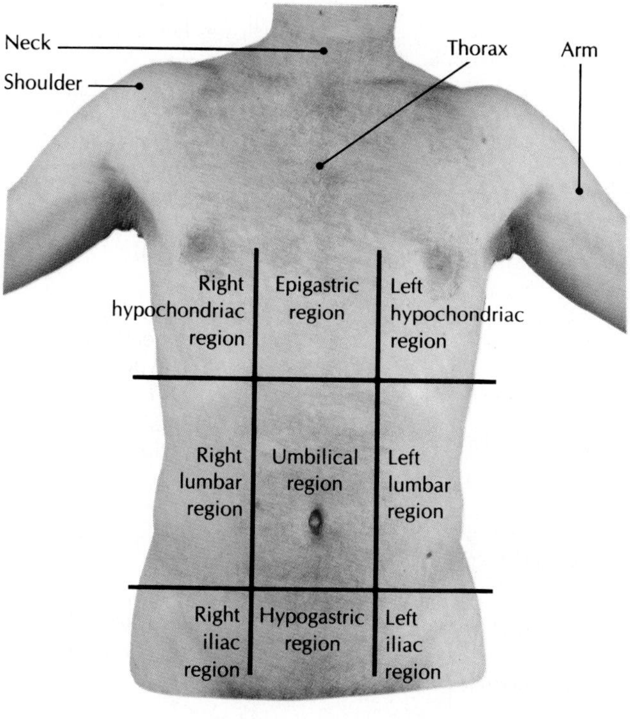

FIGURE F.2

External view of body, with skin (posterior view).

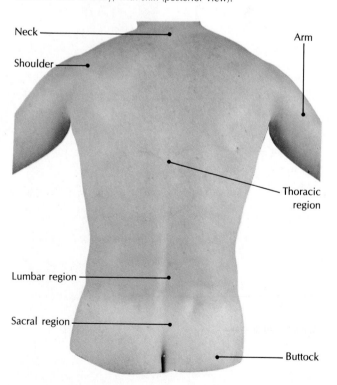

[Figures F.1 through F.11 McGraw-Hill photographs by Custom Medical Stock Photo.]

FIGURE F.3

Superficial muscle layer exposed
(anterior view).

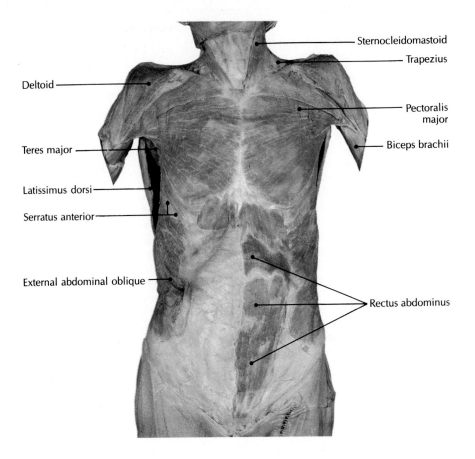

Deltoid

Teres major

Latissimus dorsi

Serratus anterior

External abdominal oblique

Sternocleidomastoid

Trapezius

Pectoralis
major

Biceps brachii

Rectus abdominus

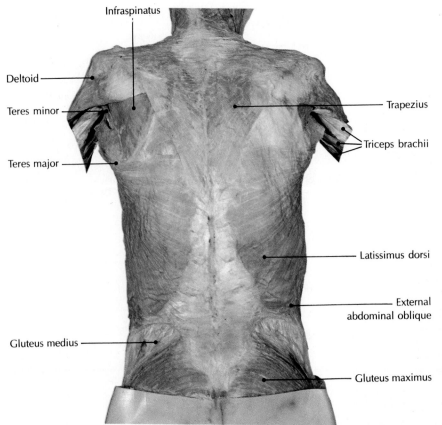

Infraspinatus

Deltoid

Teres minor

Teres major

Gluteus medius

Trapezius

Triceps brachii

Latissimus dorsi

External
abdominal oblique

Gluteus maximus

FIGURE F.4

Superficial muscle layer exposed
(posterior view).

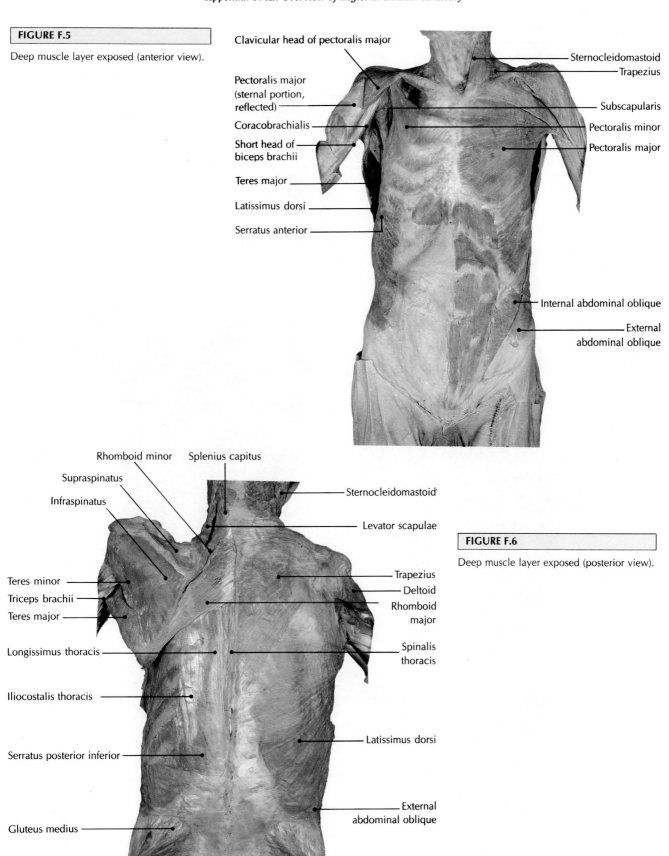

FIGURE F.5

Deep muscle layer exposed (anterior view).

Clavicular head of pectoralis major

Pectoralis major (sternal portion, reflected)

Coracobrachialis

Short head of biceps brachii

Teres major

Latissimus dorsi

Serratus anterior

Sternocleidomastoid

Trapezius

Subscapularis

Pectoralis minor

Pectoralis major

Internal abdominal oblique

External abdominal oblique

Rhomboid minor

Supraspinatus

Infraspinatus

Splenius capitus

Sternocleidomastoid

Levator scapulae

Teres minor

Triceps brachii

Teres major

Longissimus thoracis

Iliocostalis thoracis

Serratus posterior inferior

Gluteus medius

Gluteus maximus

Trapezius

Deltoid

Rhomboid major

Spinalis thoracis

Latissimus dorsi

External abdominal oblique

FIGURE F.6

Deep muscle layer exposed (posterior view).

FIGURE F.7

Thoracic viscera exposed (anterior view).

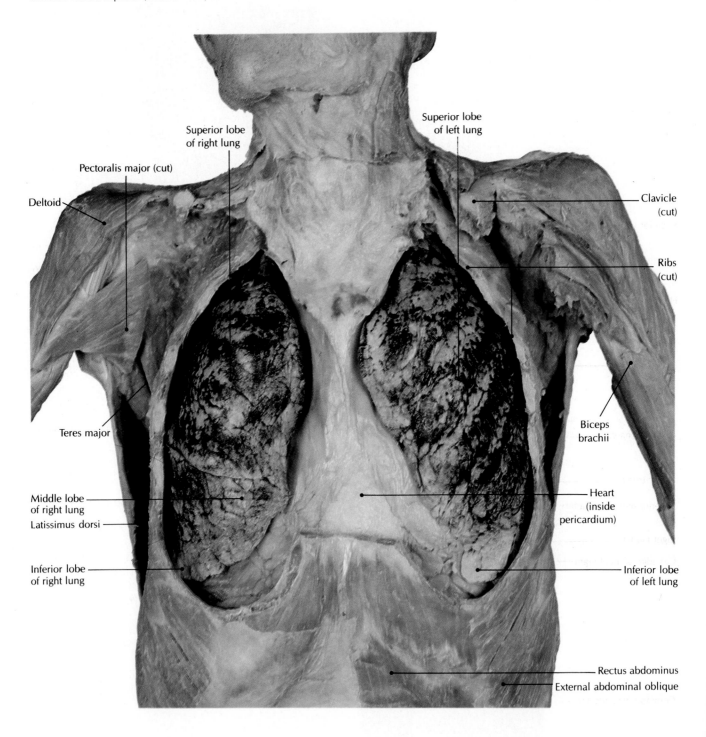

Superior lobe of right lung

Superior lobe of left lung

Pectoralis major (cut)

Deltoid

Clavicle (cut)

Ribs (cut)

Teres major

Biceps brachii

Middle lobe of right lung

Latissimus dorsi

Heart (inside pericardium)

Inferior lobe of right lung

Inferior lobe of left lung

Rectus abdominus

External abdominal oblique

FIGURE F.8

Posterior wall of thorax and abdomen (anterior view).

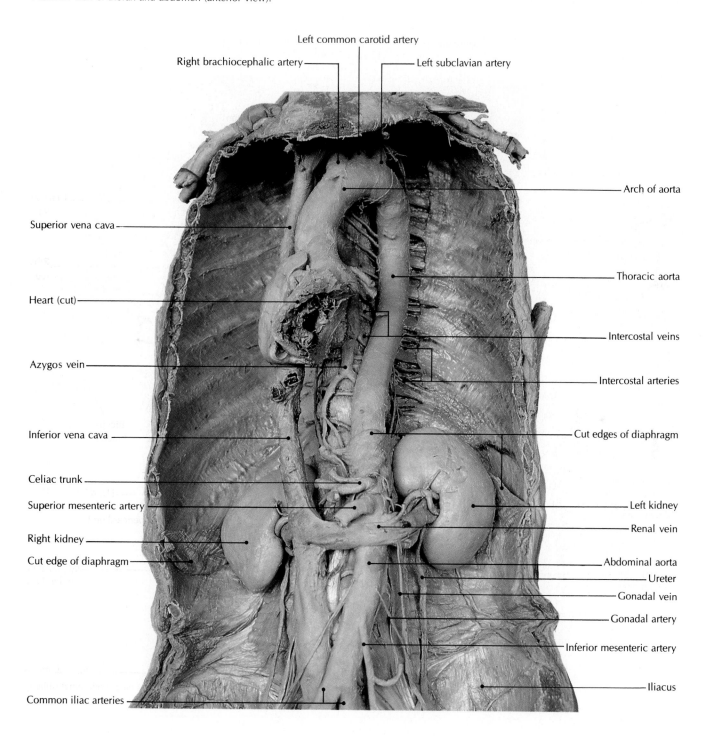

Left common carotid artery

Right brachiocephalic artery

Left subclavian artery

Arch of aorta

Superior vena cava

Thoracic aorta

Heart (cut)

Intercostal veins

Azygos vein

Intercostal arteries

Inferior vena cava

Cut edges of diaphragm

Celiac trunk

Superior mesenteric artery

Left kidney

Right kidney

Renal vein

Cut edge of diaphragm

Abdominal aorta

Ureter

Gonadal vein

Gonadal artery

Inferior mesenteric artery

Iliacus

Common iliac arteries

FIGURE F.9

Abdominal cavity, with viscera exposed (anterior view).

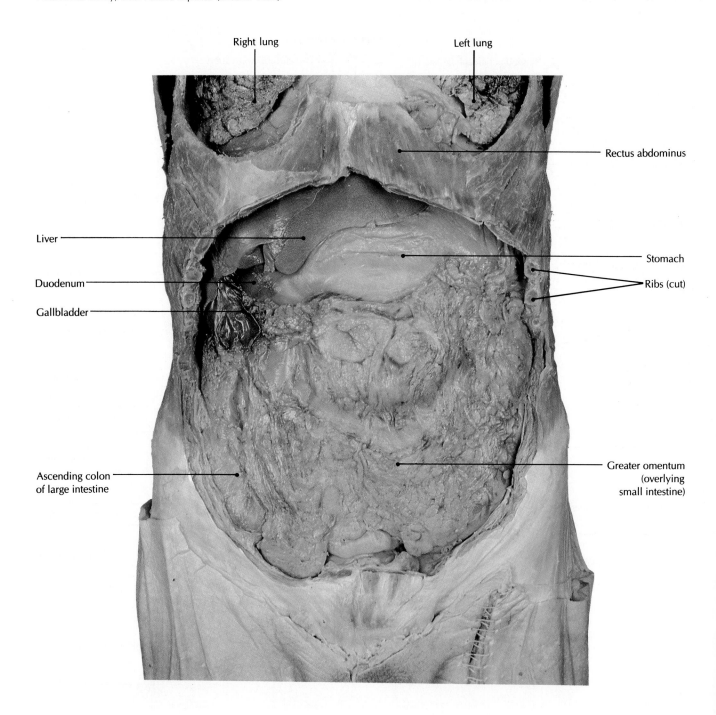

Right lung

Left lung

Rectus abdominus

Liver

Stomach

Duodenum

Ribs (cut)

Gallbladder

Ascending colon
of large intestine

Greater omentum
(overlying
small intestine)

FIGURE F.10

Male pelvic cavity, with reproductive organs exposed (left sagittal view).

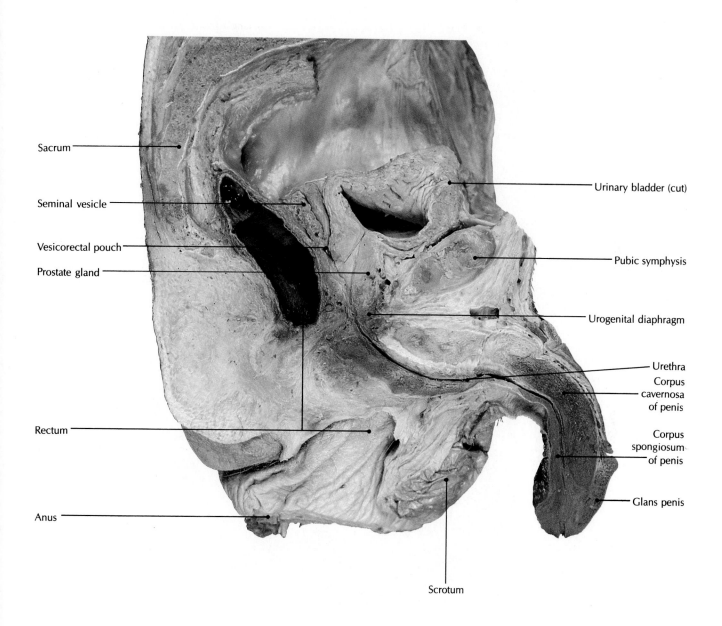

Sacrum

Seminal vesicle

Vesicorectal pouch

Prostate gland

Rectum

Anus

Urinary bladder (cut)

Pubic symphysis

Urogenital diaphragm

Urethra

Corpus cavernosa of penis

Corpus spongiosum of penis

Glans penis

Scrotum

FIGURE F.11

Female pelvic cavity, with reproductive organs exposed (left sagittal view).

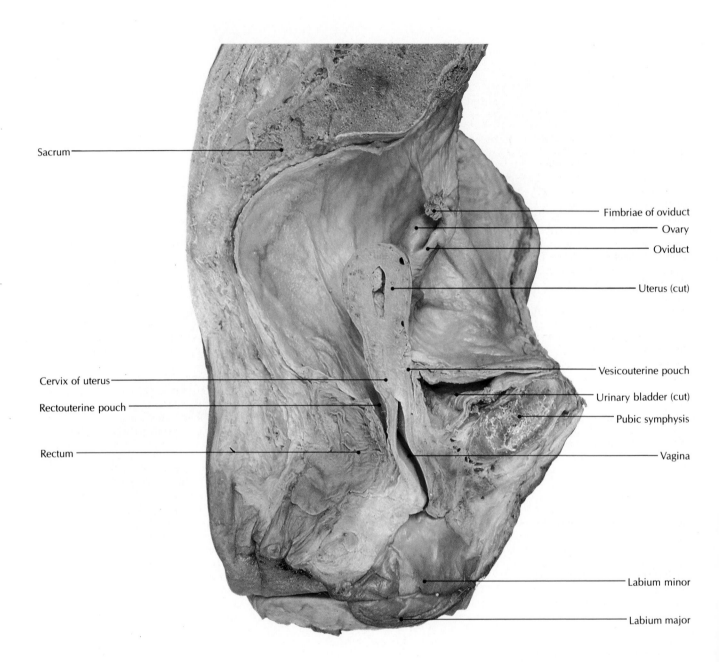

Sacrum

Fimbriae of oviduct

Ovary

Oviduct

Uterus (cut)

Vesicouterine pouch

Cervix of uterus

Urinary bladder (cut)

Rectouterine pouch

Pubic symphysis

Rectum

Vagina

Labium minor

Labium major

GLOSSARY

A

ABDOMINAL CAVITY The superior portion of the abdominopelvic cavity. It contains the liver, gallbladder, stomach, pancreas, spleen, and intestines.

ABDOMINOPELVIC CAVITY The portion of the ventral cavity below the diaphragm; divided by an imaginary line into *abdominal* and *pelvic cavities.*

ABDUCTION (L. *ab,* away from) Movement of a limb away from the midline of the body, or of fingers or toes from the medial longitudinal axis of the hand or foot.

ABORTION (L. *ab,* badly + *oriri,* be born) The fatally premature expulsion or removal of an embryo or fetus from the uterus.

ABSORPTION The passage of usable nutrient molecules from the small intestine into the bloodstream and lymphatic system.

ABSORPTIVE STATE The metabolic state during which nutrients are broken down in the digestive tract and absorbed into the cardiovascular and lymphatic systems; see *postabsorptive state.*

ACCOMMODATION A reflex that adjusts the lens of the eye to permit focusing of images from different distances.

ACETABULUM [ass-eh-TAB-yoo-luhm; L. vinegar cup] The socket of the ball-and-socket joint of the femur.

ACETYLCHOLINE [us-SEET-uhl-KOH-leen] A chemical neurotransmitter released by motor neurons at a neuromuscular junction; may be excitatory or inhibitory; plays a major role in muscle contraction.

ACETYLCHOLINESTERASE An enzyme that breaks down acetylcholine into acetic acid (acetate) and choline to halt continuous muscle contractions.

ACID (L. *acidus,* sour) A substance that releases hydrogen ions (H^+) when dissolved in water; any molecule that dissociates in solution to release a hydrogen ion or proton; see *base.*

ACID-BASE BUFFER SYSTEM A regulatory system in the body that helps resist changes in pH.

ACIDOSIS An abnormal metabolic condition in which acidic ketone bodies accumulate in blood and interstitial fluids, requiring a large amount of alkaline buffers, and thus lowering the blood pH below 7.35.

ACINAR GLAND (L. *acinus,* grape, berry) See *alveolar gland.*

ACINI [ASS-ih-nye; sing. **ACINUS;** L. grapes] Clustered exocrine cells in the pancreas that secrete digestive juices.

ACNE VULGARIS (common acne) The most common adolescent skin disorder; it occurs when increased hormonal activity causes sebaceous glands to overproduce sebum, and dead keratin cells become clogged in a follicle.

ACQUIRED IMMUNE DEFICIENCY SYNDROME (AIDS) A disease characterized by a deficiency of natural killer T cells that kill virus-infected and tumor cells, and virus-specific killer T cells that kill cells containing a particular virus; victims may die from slight microbial infections.

ACROMEGALY [ak-roh-MEG-uh-lee; Gr. *akros,* extremity + *megas,* big] A form of *giantism* (see) in which growth hormone is oversecreted after the skeletal system is fully developed.

ACROSOME (Gr. *akros,* topmost, extremity + *soma,* body) A region at the tip of the head of a sperm cell, containing several enzymes that help the sperm penetrate an ovum.

ACTIN A protein that makes up part of the thin myofilament of muscle fibers; forms light I bands of skeletal muscle.

ACTION POTENTIAL The spread of an impulse along the axon following *depolarization* of the plasma membrane; also called *nerve impulse.*

ACTIVE PROCESSES Methods of moving substances through a selectively permeable membrane from an area of low concentration to an area of higher concentration.

ACTIVE TRANSPORT The active process of moving molecules through a membrane against the *concentration gradient;* requires energy from the cell.

ACUTE RENAL FAILURE The total or near-total stoppage of kidney function.

ADDISON'S DISEASE A condition caused by an overactive adrenal cortex; characterized by anemia, weakness and fatigue, increased blood potassium levels, and decreased blood sodium levels.

ADDUCTION (L. *ad,* toward) Movement of a limb toward or beyond the midline of the body, or of fingers or toes toward the midline of a body part.

ADENOHYPOPHYSIS The anterior lobe of the pituitary (hypophysis), containing many secretory cells.

ADENOSINE TRIPHOSPHATE (ATP) [uh-DEN-uh-seen try-FOSS-fate] An organic compound containing adenine, ribose, and three phosphate groups; stores energy in chemical bonds, and thus serves as an energy source for chemical reactions in cells.

ADIPOSE CELL (L. *adeps,* fat) A fixed connective tissue cell that synthesizes and stores lipids; also called *fat cell.*

ADIPOSE TISSUE Tissue composed almost entirely of clustered adipose cells supported by strands of collagenous and reticular fibers; provides a reserve food supply, cushions organs, and helps prevent excessive loss of body heat.

ADOLESCENCE (L. *adolescere,* to grow up) The period from the end of childhood to the beginning of adulthood.

ADRENAL CORTEX The outer larger portion of the adrenal gland. It secretes *glucocorticoids,* a *mineralocorticoid,* and *gonadocorticoids.*

ADRENAL GLAND [uh-DREEN-uhl; L. "upon the kidneys"] A paired endocrine gland, one member of the pair resting upon each kidney.

ADRENAL MEDULLA The inner, smaller portion of the adrenal gland. Its main secretion is *epinephrine.*

ADRENALINE See *epinephrine.*

ADRENERGIC Pertaining to neurons and fibers of the sympathetic division of the autonomic nervous system; so called because the postganglionic neurons and fibers release norepinephrine (also called nor*adrenaline*); see *cholinergic.*

ADRENOCORTICOTROPIC HORMONE (ACTH) A hormone secreted by the adenohypophysis; it stimulates the production and secretion of adrenal cortex steroids; also called *corticotropin* or *adrenocorticotropin.*

ADRENOCORTICOTROPIN See *adrenocorticotropic hormone.*

ADRENOGENITAL SYNDROME An abnormal condition caused by an adrenocortical tumor that stimulates the excessive production of male sex hormones (androgens); characterized by male characteristics in a female, and accelerated sexual development in a male.

AFFERENT ARTERIOLES Branchings of the *interlobular arteries* in the kidney; they carry blood to the glomerulus.

AFFERENT NEURON (L. *ad*, toward, + *ferre*, to bring) A nerve cell that conveys nerve impulses from sensory receptors in the body to the central nervous system; also called *sensory neuron.*

AFTERBIRTH The collective name for the placenta, fetal membranes, and any remaining uterine fluid after their postnatal delivery.

AGGLUTININ A protective antibody in blood plasma; also called *isoantibody.*

AGGLUTINOGEN [uh-GLOOT-in-oh-jehn] An A or B *antigen* on the surface of an erythrocyte; it is the basis for the ABO grouping and Rh system of blood classification; also called *isoantigen.*

AGGREGATED LYMPH NODULES Clusters of unencapsulated lymphoid tissue found in the intestine and appendix; thought to generate plasma cells that secrete antibodies in response to antigens from the intestine; also called *Peyer's patches, gut-associated lymphoid tissue (GALT).*

AGGREGATION The clumping of *platelets.*

AGONIST (Gr. *agonia*, contest, struggle) A muscle that is primarily responsible for a movement; also called *prime mover.*

AIDS See *acquired immune deficiency syndrome.*

ALBINISM (L. *albus*, white) An inherited lack of normal skin pigmentation.

ALDOSTERONE The main mineralocorticoid hormone secreted by the adrenal cortex; it acts upon the kidneys to control sodium retention and potassium loss in urine.

ALIMENTARY CANAL The digestive tract, from mouth to anus.

ALKALOSIS A condition in which the blood pH rises above 7.45; see *acidosis.*

ALL-OR-NONE LAW The tendency of muscle fibers to contract fully or not at all, and of neurons to fire at full power or not at all.

ALLANTOIS [uh-LAN-toh-ihz; Gr. "sausage"] A small fingerlike outpocket of the caudal wall of the embryonic yolk sac; involved with blood-cell formation and the development of the urinary bladder.

ALLERGEN An antigen involved in an allergic response.

ALLERGY See *hypersensitivity.*

ALPHA-FETOPROTEIN TEST A measurement of alpha-fetoprotein excreted by the fetus into the amniotic fluid to detect possible neurological disorders.

ALVEOLAR [al-VEE-uh-lur; L. *alveolus*, hollow, cavity] Pertaining to a rounded portion.

ALVEOLAR GLAND An exocrine gland whose secretory portion is rounded; also called *acinar gland.*

ALVEOLI [al-VEE-oh-lie; sing. **ALVEOLUS**; L. a cavity] The functional units of the lungs; the sites where gas transfer takes place.

ALZHEIMER'S DISEASE A neurodegenerative disease characterized by a progressive loss of memory and intellectual function.

AMENORRHEA The absence of menstrual flow.

AMINO ACID [uh-MEE-noh] A chemical compound containing an amino group (—NH_2) and a carboxyl group (—COOH), plus a variable nonacid organic radical; the structural unit of proteins.

AMNIOCENTESIS [am-nee-oh-sehn-TEE-sihss; Gr. sac + puncture] The technique of obtaining cells of a fetus from its amniotic fluid.

AMNION (Gr. sac) A tough, thin, transparent membrane enveloping the embryo.

AMPHIARTHROSIS (Gr. *amphi*, on both sides, + *arthrosis*, articulation) A slightly movable joint.

AMPULLA A small dilation in a canal or duct.

AMYOTROPHIC LATERAL SCLEROSIS The most common motor neuron disease of muscular atrophy; also called *Lou Gehrig's disease.*

ANABOLISM [uh-NAB-uh-lihz-uhm; Gr. *ana*, upward progression] The metabolic process (synthesis) that combines two or more atoms, ions, or molecules into a more complex substance.

ANAL CANAL The continuation of the rectum, piercing the pelvic diaphragm and turning sharply downward and backward, opening into the anus.

ANAL COLUMNS About 5 to 10 permanent longitudinal columns of mucous membrane in the upper part of the anal canal; also called *rectal columns.*

ANALGESIC A pain-relieving drug, which may be either *nonopioid* or *opioid (opiate).*

ANAPHASE The third and shortest stage of mitosis; chromosome pairs separate and move toward opposite poles.

ANATOMICAL POSITION The universally accepted position from which locations of body parts can be described. The body is standing erect and facing forward, feet together, arms hanging at sides with palms facing forward.

ANATOMY (Gr. *ana-* + *temnein*, to cut) The study of the body and its parts.

ANEMIA An abnormal condition in which the number of red blood cells, the concentration of hemoglobin, or the *hematocrit* is below normal.

ANEURYSM (Gr. "to dilate") A balloon-like, blood-filled dilation of a blood vessel.

ANGINA PECTORIS (L. strangling + chest) An abnormal condition that occurs when insufficient blood reaches the heart because of a damaged or blocked artery, producing chest pain.

ANOREXIA NERVOSA (Gr. *an*, without + *orexis*, a longing) A psychological eating disorder characterized by self-imposed starvation and subsequent nutritional deficiencies.

ANTAGONIST ("against the agonist") A muscle that opposes the movement of a prime mover; see *agonist.*

ANTERIOR (L. *ante*, before) A relative directional term: toward the front of the body; also called *ventral.*

ANTIBODY A protein produced by lymphocytes in response to an antigen, which the antibody destroys through a complex defensive process.

ANTIDIURETIC HORMONE (ADH) A peptide hormone secreted by the neurohypophysis; it helps regulate the body's fluid balance; also called *vasopressin.*

ANTIGEN A substance against which an antibody is produced.

ANUS (L. ring) An opening from the rectum to the outside; the site of feces excretion.

AORTA The major arterial trunk emerging from the left ventricle of the heart; it supplies blood to the body.

AORTIC BODY A chemoreceptor in the sinuses of the aorta that responds to a lowered blood pH by increasing the respiratory rate.

APNEUSTIC AREA A breathing-control area in the pons which, when stimulated, causes strong inhalations and weak exhalations; see *respiratory center.*

APOCRINE GLAND [APP-uh-krihn; Gr. *apo*, away from, off; *krinein*, to separate] An exocrine gland that loses a small portion of its cytoplasm and plasma membrane along with its secretions, but can restore the damaged parts so that the cell is not destroyed; also called *odiferous glands.*

APONEUROSIS A broad, flat sheet of dense connective tissue that attaches to two or more muscles that work together, or to the covering of a bone.

APPENDICULAR (L. *appendere*, to hang from) Pertaining to the *appendicular part* of the body, which includes the *upper and lower extremities.*

APPENDICULAR SKELETON The bones of the upper and lower extremities; includes the shoulder and pelvic girdles.

APPENDIX See *vermiform appendix.*

AQUEOUS HUMOR A thin, watery fluid in the posterior and anterior chambers of the eye; helps to maintain a constant pressure within the eyeball, and provides nourishment for the avascular lens and cornea.

ARACHNOID (Gr., cobweblike) The delicate middle layer of the protective meninges covering the brain and spinal cord.

ARBOR VITAE [VYE-tee; L. tree of life] A branched arrangement of white matter within the vermis of the cerebellum.

AREOLA (L. *dim.* of *area*, open place) A small space in a tissue; a small, dark-colored area around a center portion; the pigmented area around a nipple.

AREOLAR TISSUE [uh-REE-uh-lur; L. *areola*, open place] Most common connective tissue, containing tiny extracellular spaces usually filled with ground substance and tissue fluids.

ARRECTOR PILI The muscle that contracts to pull a follicle and its hair to an erect position, elevating the skin above and producing a "goose bump."

ARRHYTHMIA See *cardiac arrhythmia.*

ARTERIOLE A small artery that branches before reaching capillary networks.

ARTERY A blood vessel that usually carries blood away from the heart to capillary beds throughout the body.

ARTHRITIS (Gr. *arthron*, joint + inflammation) Inflammation of a joint; a general term for many specific forms of arthritis.

ARTHROSCOPY A diagnostic and surgical technique in which a small fiberoptic scope is used to look into a joint.

ARTICULAR CAPSULE A fibrous capsule that lines the synovial cavity in the noncartilaginous parts of the joint, permitting considerable movement.

ARTICULAR CARTILAGE The smooth cartilage that caps the bones facing the synovial cavity; also called *hyaline cartilage.*

ARTICULAR DISK A fibrocartilage disk that (1) acts as a shock absorber for a joint, (2) adjusts the uneven articulating surfaces, and (3) allows two kinds of movements to occur simultaneously.

ARTICULATION (L. *articulus*, joint) A joint, the place where bones meet, or

where cartilages or bones and cartilages meet.

ASCENDING COLON The portion of the large intestine extending upward from the cecum.

ASSOCIATION FIBERS Nerve fibers that link one area of the cerebral cortex to another area of the cortex of the same cerebral hemisphere.

ASTHMA (Gr. no wind) A general term for difficulty in breathing, often relating to bronchial constriction.

ASTIGMATISM (Gr. "without focus") A condition in which the curvature of the lens or cornea is not uniform, producing an image that is partially unfocused.

ASTROCYTE The largest, most numerous glial cell; sustains neurons.

ATHEROSCLEROSIS (Gr. "porridgelike hardening") An abnormal condition characterized by deposits of fat, fibrin, cellular debris, and calcium on the inside of arterial walls.

ATLAS The first cervical vertebra (spinal bone), which supports the head.

ATOM (Gr. *atomos*, indivisible) The basic unit of all matter, consisting of protons, neutrons, and electrons; the smallest unit of an element that retains the chemical characteristics of that element.

ATOMIC NUMBER The number of protons in the nucleus of an atom; it can be used to identify an element.

ATOMIC WEIGHT The relative weight of an element compared with that of carbon.

ATP See *adenosine triphosphate.*

ATRIAL NATRIURETIC FACTOR (ANF) See *atriopeptin.*

ATRIOPEPTIN A peptide hormone secreted by secretory granules in cardiac muscle cells; it helps maintain the homeostatic balance of fluids and electrolytes, and lowers blood pressure and volume; formerly called *atrial natriuretic factor* or *ANF.*

ATRIOVENTRICULAR BUNDLE A group of conducting fibers in the interventricular septum that branch into each ventricle, finally branching into hundreds of tiny specialized muscle fibers called *cardiac conducting myofibers* (Purkinje fibers); formerly called *bundle of His.*

ATRIOVENTRICULAR (AV) NODE A mass of specialized heart tissue that delays the electrical activity of the heart a few hundredths of a second before allowing it to pass into the ventricles.

ATRIOVENTRICULAR (AV) VALVE Heart valves that allow blood to flow in one direction, from atria to ventricles.

ATRIUM [AY-tree-uhm; L. porch, antechamber] One of two upper heart chambers; see *ventricle.*

ATROPHY (Gr. "without nourishment") A condition in which the diameter of muscle fibers decreases as the result of a lack of physical activity.

AUDITORY CORTEX The portion of the temporal lobe of the cerebrum involved with basic sounds and the feeling of dizziness.

AUDITORY OSSICLES The three small bones (malleus, incus, stapes) of the middle ear, which help to transmit sound from the external ear to the inner ear.

AUDITORY TUBE A tube in the anterior wall of the *tympanic cavity* leading to the nasopharynx above the soft palate; permits air to pass from the nasal cavity into the middle ear, maintaining equal air pressure on both sides of the tympanic membrane; commonly called the *Eustachian tube.*

AURICLE (L. *auris*, ear) The external part of the ear, composed of a thin plate of fibrocartilage covered by a tight layer of skin; also called *pinna.*

AUTOIMMUNITY A condition in which the tolerance to "self" antigens breaks down and the body forms antibodies to its own antigens; results in *autoimmune diseases* such as myasthenia gravis.

AUTONOMIC GANGLIA Clusters of cell bodies and their dendrites, with synapses that occur outside the central nervous system (CNS); relay stations between the CNS and effectors.

AUTONOMIC NERVOUS SYSTEM (ANS) The efferent motor division of the visceral nervous system; composed of nerve fibers from the brain and spinal cord that inhibit or excite smooth muscle, cardiac muscle, or glands; consists of *sympathetic* and *parasympathetic nervous systems*; also called *visceral efferent motor system.*

AUTONOMIC PLEXUSES (L. braids) Branched, interlaced networks in the thoracic, abdominal, and pelvic cavities; includes cardiac, pulmonary, celiac, hypogastric, and enteric plexuses.

AUTOSOMES The 22 matching pairs of chromosomes that determine genetic traits, but not gender.

AXIAL (L. *axis*, hub, axis) Pertaining to the axis, or trunk, of the body. The *axial part* is composed of the head, neck, thorax, abdomen, and pelvis.

AXIAL SKELETON The portion of the skeleton forming the longitudinal axis of the body; includes the skull, vertebral column, sternum, and ribs.

AXIS The second cervical vertebra, on which the *atlas* rests, allowing the skull to move in a "no" motion.

AXON A long, specialized process of a neuron that carries nerve impulses away from the cell body to the next neuron, muscle cell, or gland; also called *axis cylinder*.

B

B CELL A type of lymphocyte that produces specific antibodies; also called *B lymphocyte*.

B LYMPHOCYTE See *B cell*.

BABINSKI'S REFLEX A diagnostic reflex in which the stroking of the lateral part of the sole results in the big toe pointing up and the toes to fan.

BALL-AND-SOCKET JOINT A multiaxial joint in which the globular head of one bone fits into a cuplike cavity of another bone, such as in the hip.

BARORECEPTORS (Gr. "pressure receivers") Clusters of cells in the walls of the aortic arch and carotid artery sinuses; they act as sensory receptors, and help maintain a homeostatic flow of blood to the brain by responding to changes in blood pressure; also called *pressoreceptors*.

BARTHOLIN'S GLANDS See *greater vestibular glands*.

BASAL Pertaining to, located at, or forming a base.

BASAL GANGLIA (sing. GANGLION) Deep, large cores of gray matter beneath the white matter of each cerebral hemisphere; cell bodies of neuron clusters that help coordinate muscle movements. Basal ganglia are actually *nuclei*.

BASAL LAMINA See *basement membrane*.

BASAL METABOLIC RATE (BMR) The total energy used by the body per unit of time to sustain the minimal normal cellular activity, measured under standardized (basal) conditions.

BASE A chemical substance that releases hydroxyl ions (OH^-) when dissolved in water; any molecule capable of accepting a hydrogen ion, or proton; see *acid*.

BASEMENT MEMBRANE A thin layer composed of tiny fibers and nonliving polysaccharide material produced by epithelial cells; anchors epithelial tissue to underlying connective tissue, provides elastic support, and acts as a partial barrier for diffusion and filtration; also called *basal lamina*.

BASOPHIL A type of white blood cell that is involved in allergic reactions and inflammation.

BELLY The bulging part of a muscle between its two ends.

BENIGN NEOPLASM (L. *bene*, well) An abnormal growth of tissue containing cells that appear and act almost normally, and are enclosed within a sta-

tionary capsule of thick connective tissue.

BICUSPID VALVE The left atrioventricular heart valve; also called the *mitral valve*.

BILAMINAR EMBRYONIC DISK A bilayered structure formed when the blastocyst assumes a flattened disk shape during the second embryonic week; composed of endoderm and ectoderm.

BILATERAL Pertaining to both sides of the body.

BILE An alkaline liquid secreted by the liver; aids in digestion of lipids.

BIOFEEDBACK The conscious self-regulation of some "involuntary" bodily responses typically under the control of the autonomic nervous system.

BIOPSY (Gr. *bios*, life + -*opsy*, examination) The microscopic examination of living tissue removed from the body.

BLASTOCOEL [BLASS-toh-seel; Gr. bud + *koilos*, hollow] The fluid-filled inner cavity of a *blastocyst*.

BLASTOCYST (Gr. *blasto*, germ + *kystis*, "bladder") A fluid-filled hollow sphere of cells formed about 4 to 5 days after fertilization.

BLASTOMERE (Gr. *blastos*, bud + *meros*, a part) One of two daughter cells resulting from the first cleavage of a zygote; a cell formed by the repeated division of a fertilized ovum.

BLOOD FLOW The quantity of blood flowing through a vessel during a specific period of time.

BLOOD PRESSURE The force (energy) with which blood is pushed against the walls of blood vessels and circulated throughout the body when the heart contracts.

BLOOD VELOCITY The distance blood moves along a vessel during a specific time period.

BLOOD-BRAIN BARRIER A system of tight junctions in the endothelial cells of brain capillaries that forms a semipermeable membrane, allowing only certain substances to enter the brain; also called *hematoencephalic barrier*.

BONE See *osseous tissue*.

BONE MARROW Tissue filling the porous *medullary cavity* of the *diaphysis* of bones; also called *myeloid tissue*.

BOWMAN'S CAPSULE See *glomerular capsule*.

BRACHIAL PLEXUS The ventral rami of the lower four cervical and first thoracic nerves in the lower neck and axilla.

BRAINSTEM The stalk of the brain, relaying messages between the spinal cord and brain; consists of the medulla oblongata, pons, and midbrain.

BREATHING The mechanical process

that moves air into and out of the lungs; includes *inspiration* and *expiration* (see); also called *ventilation* or *pulmonary ventilation*.

BROAD LIGAMENTS Paired double folds of peritoneum that attach the uterus to the lateral wall of the pelvis.

BROCA'S (MOTOR SPEECH) AREA (area 44) A motor area in the frontal lobe of the cerebrum; involved in formulating spoken words; also called *anterior speech area*.

BRONCHI [BRONG-kee; sing. **BRONCHUS**; Gr. throat, windpipe] Branches of the respiratory tree that emerge from the trachea into the lungs.

BRONCHIOLES [BRONG-kee-ohlz] Small tubes emerging from branching bronchi in the lungs; they continue to branch until they end as *terminal bronchioles*.

BUCCAL CAVITY See *oral cavity*.

BUFFER A substance that regulates acid-base balance; combinations of weak acids or bases and their respective salts in solution that help body fluids resist changes in pH when small amounts of strong acids or bases are added.

BULBOURETHRAL GLANDS Paired male reproductive glands that open into the urethra, secreting alkaline fluids that neutralize the acidity of urine, and act as a lubricant at the tip of the penis prior to sexual intercourse; also called *Cowper's glands*.

BULBOUS CORPUSCLES (OF KRAUSE) Sensory receptors in the skin believed to be sensors for touch-pressure, position sense of a body part, and movement; probably variants of *lamellated corpuscles (of Pacini)*; formerly called *Krause's end bulbs*.

BULIMIA (Gr. *bous*, ox + *limos*, hunger) A psychological eating disorder characterized by an insatiable appetite and gorging on food several times a week, followed by self-induced vomiting or an overuse of laxatives or diuretics that may lead to metabolic imbalances.

BUNDLE OF HIS See *atrioventricular bundle*.

BUNION (Old Fr. *buigne*, bump on the head) A lateral deviation of the great toe toward the second toe, accompanied by the formation of a bursa and callus on the bony prominence of the first metatarsal.

BURN Damaging of skin tissues by heat, electricity, radioactivity, or chemicals.

BURSAE [BURR-see; sing. **BURSA**, BURR-sah; Gr. purse] Flattened sacs filled with synovial fluid to help eliminate friction in areas where a muscle or tendon rubs against another muscle, tendon, or bone.

BURSITIS Inflammation of a bursa.

C

CALCITONIN A thyroid hormone that lowers calcium and phosphate levels in blood; also called *thyrocalcitonin*.

CALLUS (L. hard skin) An area of the skin hardened by repeated external pressure or friction.

CALORIE (L. *calor*, heat) The amount of energy required to raise the temperature of 1 g of water 1° Celsius; also called *gram calorie*; see *kilocalorie*.

CALVARIA (L. skull) The roof (vault) of the cranium, composed of the brow portion of the frontal bone, the parietal bones, and the occipital bone.

CALYCES [KAY-luh-seez; sing. **CALYX**; KAY-licks; Gr. cups] Two small cavities (major and minor calyces) in the kidney formed from the branching of the *renal pelvis*.

CANALICULI [KAN-uh-lick-yuh-lie; L. dim. *canalis*, channel] Small channels radiating from *lacunae* in bone; transport materials by diffusion.

CANCER (L. crab) Any of the various malignant neoplasms that spread to new sites.

CAPILLARY (L. *capillus*, hair) Tiny blood vessels that connect the arterial and venous systems.

CAPILLARY FLUID SHIFT A mechanism that regulates arterial pressure by altering blood volume.

CAPUT (L. head) See *head*.

CARBOHYDRATE A molecule composed of carbon, hydrogen, and oxygen in a ratio of 1:2:1; the main source of body energy.

CARCINOGEN [kar-SIHN-uh-jehn; Gr. *karkinos*, cancer, crab + *-gen*, producing] A cancer-causing agent.

CARCINOMA [kar-suh-NOH-muh; L. cancerous ulcer + *oma*, tumor] A malignant neoplasm that originates in epithelial tissues and spreads via the lymphatic system.

CARDIAC (Gr. *kardia*, heart) Pertaining to the heart.

CARDIAC ARRHYTHMIA A general condition referring to abnormal electrical conduction in heart tissue, or changes in heart rate and rhythm.

CARDIAC CHEMORECEPTORS Chemically sensitive receptors in the carotid sinuses and aortic arch that help regulate the heart rate by responding to changes in the blood levels of oxygen, carbon dioxide, or hydrogen ions.

CARDIAC CONDUCTING MYOFIBERS Modified nerve fibrils in the walls of the ventricles of the heart; help produce a coordinated pumping effort; formerly called *Purkinje fibers*.

CARDIAC CYCLE The carefully regulated sequence of steps commonly referred to as the beating of a heart; it includes contraction (systole) and relaxation (diastole).

CARDIAC MUSCLE TISSUE (Gr. *kardia*, heart) Specialized muscle tissue found only in the heart.

CARDIAC OUTPUT (CO) The quantity of blood pumped by either ventricle, but not both, in 1 min.

CARDIAC SKELETON Tough connective-tissue in the heart that provides attachment sites for the valves and corresponding muscle fibers.

CARDIOPULMONARY RESUSCITATION (CPR) A physical technique used to help re-establish respiration and circulation when a victim's breathing stops; may include mouth-to-mouth resuscitation or *external cardiac compression*.

CARDIOREGULATORY CENTER Sympathetic nerve centers in the medulla oblongata that regulate the heart rate.

CAROTID BODY A chemoreceptor in the sinuses of the carotid arteries that responds to a lowered blood pH by increasing the respiratory rate.

CARPUS (Gr. *karpos*, wrist) The eight short bones connected by ligaments in each wrist.

CARTILAGE A specialized type of connective tissue that provides support and aids movement at joints.

CARTILAGINOUS JOINT A joint in which bones are joined by hyaline cartilage or a fibrocartilaginous disk; it allows little or no movement. It includes *synchondroses* and *symphyses* (see).

CASTRATION The surgical removal of the testes.

CAT SCAN See *computer-assisted tomography*.

CATABOLISM [kuh-TAB-uh-lihz-uhm; Gr. *katabole*, a throwing down] The metabolic process (decomposition) that breaks down large molecules into two or more atoms, ions, or molecules.

CATARACT A condition in which the lens of the eye becomes opaque, possibly producing blindness.

CAUDA EQUINA [KAW-duh ee-KWY-nuh; L. horse's tail] The collection of spinal nerve roots passing caudally below the conus terminalis of the spinal cord.

CECUM (L. blind) A cul-de-sac pouch on the distal side of the ileocecal orifice.

CELL The smallest independent unit of life; the component of tissues.

CELL CYCLE The period from the beginning of one cell division to the beginning of the next; the life span of a cell.

CELL-MEDIATED IMMUNITY A protective method that involves direct contact between T cells and antigens.

CELLULAR RESPIRATION The sum of biochemical events by which the chemical energy of foods is released to provide energy for life's processes; see *external respiration, internal respiration*.

CEMENT The bonelike covering of the neck and root of a tooth.

CENTER OF OSSIFICATION The site at which a ring of cells forms around a blood vessel in the development of bone tissue; also called *ossification center*.

CENTRAL CANAL A longitudinal channel in the osteon (Haversian system); contains nerves and lymphatic and blood vessels; also called *Haversian canal*.

CENTRAL LOBE See *insula*.

CENTRAL NERVOUS SYSTEM (CNS) The brain and spinal cord.

CENTRIOLES Two small organelles in the *centrosome*, a specialized region of the cytoplasm of a cell; involved with the movement of chromosomes during cell division; form basal body of *cilia* and *flagella* (see).

CEPHALIC [suh-FAL-ihk; Gr. *kephale*, head] Pertaining to the head.

CEREBELLAR CORTEX A surface layer of gray matter covering the cerebellar lobes; composed of a network of billions of neurons.

CEREBELLAR PEDUNCLES [peh-DUNG-kuhlz; L. little feet] Three nerve bundles that attach the cerebellum to the brainstem; composed of the inferior, middle, and superior cerebellar peduncles.

CEREBELLUM (L. little brain) The second-largest part of the brain, composed of the vermis, two small flocculonodular lobes, and two large lateral lobes; located behind the pons in the posterior cranial fossa; refines and coordinates muscular movements.

CEREBRAL CORTEX A surface mantle of gray matter over the cerebrum; it is thin and convoluted, containing about 50 billion neurons and 250 billion glial cells.

CEREBRAL PALSY Impaired muscular power and coordination as a result of brain damage, usually occurring at, before, or shortly after birth.

CEREBRAL PEDUNCLES Two nerve fiber bundles composed of the pyramidal tract and corticobulbar and corticopontine fibers.

CEREBROSPINAL FLUID (CSF) A clear, watery ultrafiltrate solution formed from blood in the capillaries; bathes the ventricles of the brain and cavity of the spinal cord; cushions the brain and spinal cord, and is involved in maintaining homeostasis by helping to control the chemical environment of the central nervous system.

CEREBROVASCULAR ACCIDENT (CVA) A sudden withdrawal of sufficient blood supply to the brain; commonly called a *stroke*.

CEREBRUM [suh-REE-bruhm; L. brain] The largest and most complex structure of the nervous system; consists of two cerebral hemispheres and the diencephalon; each hemisphere is composed of a cortex (gray matter), white matter, and basal ganglia; the cortex is divided into five lobes. All conscious living depends on the cerebrum.

CERVICAL PLEXUS The ventral rami of C1 to C4 nerves in the neck region.

CERVICAL VERTEBRAE The seven small neck bones between the skull and thoracic vertebrae (spinal bones); they support the head and allow it to move.

CERVIX (L. neck) Any neck-shaped anatomical structure; the narrow juncture between the uterus and vagina.

CESAREAN SECTION An incision through the abdominal and uterine walls to remove a fetus.

CHEMICAL SYNAPSE A junction by which two cells communicate by way of a chemical neurotransmitter.

CHEMIOSMOSIS The accumulation of electrical energy built up from an electrochemical proton gradient, and the release of that energy to make ATP from ADP and P_i; occurs in mitochondria.

CHEMORECEPTOR (L. "chemical receiver") A sensory receptor that responds to chemical stimuli.

CHEMOTHERAPY Therapy that uses drugs in treating cancer.

CHILDHOOD The period from the end of infancy to the beginning of adolescence.

CHLAMYDIAL INFECTION [klah-MIHD-ee-uh] The most common sexually transmitted disease in the United States, caused by the bacterium *Chlamydia trachomatis*.

CHLORIDE SHIFT The condition that results when bicarbonate ions are removed from red blood cells, developing a positive charge inside the cell and allowing chloride ions from sodium chloride in the plasma to diffuse into the cell.

CHOLECYSTOKININ (CCK) [koh-lee-sis-TOE-kine-in] A digestive hormone secreted from the duodenal wall; it stimulates gallbladder contraction and the release of pancreatic enzymes.

CHOLELITHIASIS (bile, gall + stone) A condition in which hard deposits form in the gallbladder, usually caused by insufficient bile salts to dissolve the

cholesterol in *micelles* (see); the cholesterol crystallizes and hardens into "gallstones" (biliary calculi).

CHOLINERGIC Pertaining to preganglionic neurons, or their fibers, of the sympathetic nervous system that release *acetylcholine*.

CHOLINESTERASE An enzyme that causes the breakdown of acetylcholine into choline and acetic acid (acetate).

CHONDROCYTE (Gr. *khondros*, cartilage) A cartilage cell embedded in small cavities (lacunae) within the matrix of cartilage connective tissue.

CHORION The protective sac around the embryo.

CHORIONIC VILLI Fingerlike projections growing outward from the embryonic chorion into maternal tissue; allow for the exchange of nutrients, gases, and metabolic wastes between mother and embryo.

CHORIONIC VILLI SAMPLING A technique in which a sample of chorionic villi tissue is collected and examined for genetic diseases.

CHOROID The posterior two-thirds of the vascular layer of the eye, composed of blood vessels and connective tissue between the sclera and retina.

CHROMATOPHILIC SUBSTANCE Rough endoplasmic reticulum and free ribosomes in the cell bodies of neurons; involved in protein synthesis; formerly called *Nissl bodies*.

CHROMOSOME (Gr. "colored body") A threadlike nucleoprotein structure within the nucleus of a cell that contains the DNA.

CHRONIC RENAL FAILURE The progressive destruction of nephrons, possibly leading to the total, or near-total, stoppage of kidney function.

CHYLOMICRON A tiny, protein-coated droplet containing triglycerides, phospholipids, cholesterol, and free fatty acids.

CHYME [KIME] A soupy liquid mixture formed in the stomach from gastric juices and particles from broken-down nutrients.

CILIA [SIHL-ee-uh; sing. **CILIUM**; L. eyelid, eyelash] Short processes extending from the surface of some cells; often capable of producing a rhythmic paddling motion; see *flagella*.

CILIARY BODY The thickened part of the vascular layer of the eye, connecting the choroid with the iris.

CIRCULATORY SHOCK A condition in which there is an inadequacy of blood flow throughout the body.

CIRCUMCISION (L. *circumcidere*, to cut around) The surgical removal of the *prepuce* (see).

CIRCUMDUCTION A movement in

which the distal end of a bone moves in a circular path while the proximal end remains stable.

CIRRHOSIS OF THE LIVER [sih-ROE-siss; L. "orange-colored disease"] A chronic liver disease characterized by the destruction of liver cells and their replacement by abnormal fibrous connective tissue, eventually resulting in liver failure and death.

CITRIC-ACID CYCLE A series of intercellular metabolic reactions that begins with pyruvic acid and releases a molecule of carbon dioxide and two hydrogen atoms; also called *Krebs cycle*.

CLEAVAGE A series of mitotic cell divisions occurring in a zygote immediately after fertilization.

CLITORIS [KLIHT-uh-rihss; Gr. small hill] A small erectile organ at the upper end of the vulva, below the mons pubis, where the labia minora meet; a major source of sexual arousal.

CLONAL SELECTION THEORY A theory that states that an antigen influences the amount of antibody produced but does not affect the combining sites on the antibody's three-dimensional structure. It also proposes that each B cell is genetically programmed to react only to a particular antigen.

COAGULATION (L. *coagulare*, to curdle) The process of blood clotting.

COCCYGEAL PLEXUS The coccygeal nerve plus communications from S4 and S5.

COCCYX The three to five fused vertebrae (spinal bones) at the end of the vertebral column, below the sacrum; commonly called *tailbone*.

COCHLEA [KAHK-lee-uh; Gr. *kokhlos*, snail] A spiral bony chamber beyond the semicircular ducts of the inner ear; divided into the scala vestibuli, scala tympani, and scala media.

COENZYME An organic compound or metal required for enzyme function.

COITUS [KOH-uh-tuhss; L. *coire*, to come together] The act of sexual intercourse; also called *copulation*.

COLLAGEN [KAHL-uh-juhn; Gr. *kolla*, glue + *genes*, born, produced] The protein found in the fibers of bone, cartilage, and connective tissue proper.

COLLATERAL BRANCHES The side branches of an axon.

COLLECTING DUCT The portion of the renal excretory tubule that receives the glomerular filtrate after it has passed through the *distal convoluted tubule*.

COLOR BLINDNESS A deficiency in one or another eye pigment necessary to see certain colors.

COLOSTRUM A high-protein fluid present in the mother's breasts before milk production starts about 3 days later.

COMMISSURAL FIBERS The axons that project from a cortical area of one cerebral hemisphere to a corresponding cortical area of the opposite hemisphere.

COMPACT BONE TISSUE The very hard and dense portion of bone.

COMPLEMENT A group of 11 proteins and 8 additional factors found in blood serum; it enhances the body's defensive actions.

COMPOUND (L. *componere*, to put together) Molecules made up of atoms of two or more elements; usually possesses different chemical properties from the elements that compose it.

COMPUTER-ASSISTED TOMOGRAPHY (CAT) A scanning procedure that combines x rays with computer technology to show cross-sectional views of internal body structures; the device is called a *CAT scanner.*

CONCENTRATION GRADIENT The difference in solute concentration on either side of the plasma membrane of a cell.

CONCEPTION The process of fertilization and the subsequent establishment of pregnancy.

CONDYLE (L. knuckle) A rounded, knuckle-shaped projection on a bone; may be concave or convex.

CONE A color-sensitive photoreceptor cell concentrated in the retina.

CONGENITAL DEFECT A condition existing at birth, but not necessarily hereditary.

CONGESTIVE HEART FAILURE (CHF) A condition that occurs when either ventricle fails to pump blood out of the heart as quickly as it enters the atria, or when the ventricles do not pump equal amounts of blood.

CONJOINED TWINS Identical (monozygotic) twins whose embryonic disks do not separate completely; commonly called *Siamese twins.*

CONJUNCTIVA (pl. **CONJUNCTIVAE**; L. connective) The transparent mucous membrane lining the inner surface of the eyelid and exposed surface of the eyeball.

CONJUNCTIVITIS Inflammation of the conjunctiva.

CONNECTIVE TISSUE A supportive and protective tissue consisting of fibers, ground substance, cells, and some extracellular fluid; the most abundant type of tissue.

CONSTIPATION (L. "to press together") A condition in which feces move through the large intestine too slowly and too much water is reabsorbed, resulting in hard feces and difficult defecation.

CONTRACEPTION ("against conception") The prevention of conception.

CONTRACTILITY The ability of muscle tissue to contract; the basic physiological property of muscle tissue.

CONTRALATERAL Pertaining to opposite sides of the body.

CONUS TERMINALIS The cone-shaped lowermost end of the spinal cord; also called *conus medullaris.*

CONVERGENCE (L. *convergere*, to merge) 1. A condition in which the receptive segment of a postsynaptic neuron is excited or inhibited by the axon terminals of many presynaptic neurons. 2. The coordinated inward turning of the eyes to focus on a nearby object, so that both images fall on the corresponding points of both retinas.

CONVOLUTIONS See *gyri.*

COPULATION (L. *copulare*, to fasten together) See *coitus.*

CORNEA (L. *corneus*, horny tissue) The transparent anterior part of the outer layer of the eye, a uniformly thick, convex nonadjustable lens; light enters the eye through the cornea.

CORONAL See *frontal.*

CORONARY ARTERY DISEASE (CAD) An abnormal condition usually brought on by atherosclerosis, and characterized by a reduced supply of oxygen and nutrients to the *myocardium.*

CORONARY CIRCULATION The flow of blood that supplies the heart itself.

CORONARY SINUS A large vein that receives blood from the veins that drain the heart, emptying it into the right atrium.

CORPUS CALLOSUM (L. hard body) The larger of two cerebral commissures that connect the two cerebral hemispheres.

CORPUS LUTEUM (L. "yellow body") A temporary ovarian endocrine tissue that secretes female sex hormones; formed as the lining of a follicle grows inward after the discharge of an ovum.

CORPUS STRIATUM (L. furrowed body) The largest mass of gray matter in the cerebral basal ganglia; composed of the *caudate* and *lentiform nuclei.*

CORPUSCLES OF RUFFINI Sensory receptors in skin believed to be sensors for touch-pressure, position sense of a body part, and movement; probably variant of *lamellated corpuscles (of Pacini).*

CORTEX (L. bark, shell) The outer layer of an organ or part.

CORTICOSPINAL TRACTS See *pyramidal tracts.*

CORTICOTROPIN See *adrenocorticotropic hormone.*

COUNTERCURRENT MULTIPLIER SYSTEM A positive feedback system that results from a countercurrent flow in the loop of the nephron that helps to regulate the solute concentration.

COVALENT BOND A strong chemical bond formed when atoms gain stability by sharing one or more pairs of electrons.

CRANIAL NERVES Twelve pairs of peripheral nerves carrying sensory signals to the brain, and/or motor impulses to various places in the body.

CRANIAL PARASYMPATHETIC OUTFLOW Preganglionic fibers from the cranial portion of the parasympathetic nervous system that leave the brainstem via cranial nerves III, VII, IX, and X.

CRANIOSACRAL DIVISION Visceral efferent nerve fibers that leave the central nervous system via cranial nerves III, VII, IX, and X from the brainstem, or spinal nerves S3 and S4; part of *peripheral autonomic nervous system.*

CREATINE PHOSPHATE (CP) A high-energy phosphate similar to ATP. When ATP is needed, the high-energy phosphate bond of CP is broken, releasing energy that is used to synthesize ATP. CP is also called *phosphocreatine.*

CRETINISM [CREH-tin-ism] A condition caused by underactivity of the thyroid during fetal development after the twelfth week; characterized by mental retardation and irregular development of bones and muscles.

CRYPTORCHIDISM [krip-TOR-kye-dizm; Gr. *kryptos*, hidden + *orchis*, testis] A condition in which the testes do not descend from their fetal position into the scrotum after birth.

CRYPTS OF LIEBERKÜHN See *intestinal glands.*

CUSHING'S DISEASE A condition caused by an overactive adrenal cortex; characterized by fattening of the face, chest, and abdomen, and a tendency toward diabetes; also called *Cushing's syndrome.*

CUTANEOUS MEMBRANE The skin.

CYANOSIS (Gr. dark blue) The development of a bluish color of the skin, especially the lips, resulting from the build-up of deoxygenated hemoglobin.

CYCLIC AMP The end result of the *fixed-membrane-receptor mechanism* (see); it causes a target cell to perform its specific function.

CYCLIC GUANOSINE MONOPHOSPHATE (CGMP) An intracellular messenger hydrolyzed in rod cells in the retina after the activation of the protein transducin; prevents sodium ions from entering the rods, thus producing hyperpolarization and the transmission of neural signals and action potentials to the brain via the optic nerve, causing light to be perceived.

CYST A mass of fluid-filled tissue, extending to the dermis or hypodermis; over 1 cm in diameter.

CYSTIC FIBROSIS A genetic disease characterized by thick mucus secretions that impair breathing and digestion.

CYSTITIS Inflammation of the urinary bladder.

CYTOKINESIS (Gr. *cyto*, cell + movement) The separation of the cytoplasm into two parts during this final stage of cell division following mitosis; two genetically identical daughter cells are formed.

CYTOLOGY [sigh-TAHL-uh-jee; Gr. *kytos*, hollow vessel; *cyto*, cell] The microscopic study of cells.

CYTOPLASM The portion of a cell outside the nucleus, where metabolic reactions take place; the fluid portion is *cytosol*, which contains subcellular *organelles* (see).

CYTOPLASMIC INCLUSIONS Solid particles temporarily in cells; usually either basic food material or the stored products of the cell's metabolic activities.

CYTOSKELETON The flexible cellular framework, interconnecting *microfilaments*, intermediate filaments, and other organelles in a structural and functional unity; involved with support and cell movement; site for the binding of specific enzymes; also called *microtrabecular lattice*.

D

DECUSSATION (L. *decussis*, the number ten, X, indicating a crossing over) The crossing over of the axons of sensory and motor pathways from one side of the spinal cord or brainstem to the other.

DEEP A relative directional term of the body: farther from the surface of the body; opposite of *superficial*.

DEFECATION The discharge of solid wastes in the form of feces from the rectum through the anus.

DEGLUTITION (L. *deglutire*, to swallow down) The act of swallowing.

DELAYED HYPERSENSITIVITY An allergic reaction that takes several days or longer to develop and be expressed.

DENDRITE (Gr. *dendron*, tree) Short, threadlike extensions of the cell body of a neuron; they may be excited or inhibited, and conduct nerve impulses toward the cell body.

DENTAL CARIES Tooth decay ("cavities") initiated by bacteria that promote the production of an enamel-eroding enzyme.

DENTINE The extremely sensitive yellowish portion of a tooth surrounding the pulp cavity; it forms the bulk of a tooth.

DEOXYRIBONUCLEIC ACID (DNA) [dee-AHK-see rye-boh-noo-KLAY-ihk] A double-stranded nucleic acid that is a constituent of chromosomes; contains hereditary information coded in specific sequences of *nucleotides*.

DEPOLARIZATION A reversal of electrical charges on the plasma membrane of a neuron, giving the inner side of the membrane a positive charge (of about + 30 mV) relative to the outer side.

DEPRESSION A movement that lowers a body part, such as opening the mouth.

DERMATOLOGIST A physician specializing in the treatment of skin disorders.

DERMATOME (Gr. *derma*, skin + *tomos*, a cutting) 1. An instrument used in cutting thin slices of the skin, as in skin grafting. 2. A segment of skin with sensory fibers from a single spinal nerve; used in locating injuries to dorsal roots of the spinal cord.

DERMIS (Gr. *derma*, skin) The strong, flexible connective tissue meshwork of collagenous, reticular, and elastic fibers that makes up most of the skin.

DESCENDING COLON The portion of the large intestine extending from the left colic flexure of the transverse colon down to the rim of the pelvis, where it becomes the sigmoid colon.

DESMOSOME (Gr. *desmos*, binding) A junction with no direct contact between adjacent plasma membranes; common in skin; also called *spot desmosome*.

DIABETES INSIPIDUS [in-SIPP-ih-duhss] A condition in which ADH is undersecreted, and excessively large amounts of water are excreted in the urine.

DIABETES MELLITUS [MELL-ih-tuhss] A hereditary disease that results when insufficient insulin is produced, and glucose accumulates in the blood.

DIALYSIS A passive-transport process in which small particles are separated from large ones when the small particles diffuse through a selectively permeable membrane, leaving the larger particles behind; does not occur in the body naturally.

DIAPHYSIS [die-AHF-uh-siss; Gr. *dia*, between + growth] A tubular shaft of compact bone in most adult long bones; the center of the shaft is filled with marrow, and there is a spongy *epiphysis* at each end of the shaft.

DIARRHEA (Gr. "a flowing through") A condition in which watery feces move through the large intestine rapidly and uncontrollably.

DIARTHROSIS (Gr. *dia*, between + *arthrosis*, articulation) A freely movable joint.

DIASTOLE [die-ASS-toe-lee] The relaxation of the atria and ventricles during the *cardiac cycle* (see).

DIASTOLIC PRESSURE The portion of a blood-pressure measurement that represents the pressure during the interval between heartbeats; it is the second number shown in a blood-pressure reading; see *systolic pressure*.

DIENCEPHALON (L. "between brain") The deep part of the cerebrum, connecting the midbrain with the cerebral hemispheres; houses the third ventricle, and is composed of the thalamus, hypothalamus, epithalamus, and ventral thalamus (subthalamus).

DIFFERENTIATION The process by which cells develop into specialized tissues and organs.

DIFFUSION See *simple diffusion*.

DIGESTION The chemical and mechanical conversion of large nutrient particles into small, absorbable molecules.

DIGITAL SUBTRACTION ANGIOGRAPHY (DSA) (Gr. *angeion*, vessel) A noninvasive exploratory technique that uses a digital computer to produce three-dimensional pictures of blood vessels.

DISACCHARIDE [dye-SACK-uh-ride; Gr. *di*, two + *sakkharon*, sugar] A carbohydrate formed when two monosaccharides are united chemically. Disaccharides have the general formula $C_{12}H_{22}O_{11}$.

DISLOCATION (L. *luxare*, to put out of joint) A displacement of bones in a joint so that two articulating surfaces become separated; also called *luxation*.

DISSOCIATION The tendency for some molecules to break up into ions in water.

DISTAL (from *distant*) A relative directional term: away from the trunk of the body (away from the attached end of a limb).

DISTAL CONVOLUTED TUBULE The portion of the renal excretory tubule that receives the *glomerular filtrate* (see) after it has passed through the entire length of the *loop of the nephron* (see).

DIURETIC (Gr. "to urinate through") A therapeutic drug that increases the volume of urine by decreasing the water reabsorbed from the renal collecting duct.

DIVERGENCE (L. *divergere*, to bend) A condition in which the transmissive segment of a presynaptic neuron branches to have many synaptic connections with the receptive segments of many other neurons; see *convergence*.

DIVERTICULA (sing. **DIVERTICULUM**) Bulging pouches in the gastrointestinal wall that push the mucosal lining through the muscle in the wall; see *diverticulitis*.

DIVERTICULITIS An abnormal condition in which *diverticula* (see) in the large intestine become inflamed and may become seriously infected if food and bacteria become lodged in the diverticula; in *diverticulosis*, diverticula (pouches) are present but do not present symptoms.

DIVERTICULOSIS See *diverticulitis*.

DIZYGOTIC TWINS Twins formed when more than one ovum is released from the ovary or ovaries at the same time and two ova are fertilized; also called *fraternal twins*.

DNA See *deoxyribonucleic acid*.

DOMINANT A genetic characteristic that is expressed to the exclusion of a less-powerful *recessive* (see) characteristic.

DORSAL (L. *dorsalis*, back) 1. A relative directional term: toward the back of the body; also called *posterior;* opposite of *ventral*. 2. The smaller of two main body cavities; contains the cranial and spinal cavities. 3. The upper surface of the hand or foot.

DORSIFLEXION Flexion of the foot at the ankle joint.

DOWN'S SYNDROME See *trisomy 21*.

DUCTUS ARTERIOSUS A shunt in the aortic arch, where fetal blood enters and mixes with blood from the left ventricle; allows fetal blood to bypass the nonfunctional fetal lungs.

DUCTUS DEFERENS (pl. **DUCTUS DEFERENTIA**; L. *deferre*, to bring to) The dilated continuation of the ductus epididymis, extending between the epididymis of each testis and the ejaculatory duct; formerly called *vas deferens* or *sperm duct*.

DUCTUS VENOSUS A branch of the umbilical vein (which carries oxygenated blood), which bypasses the fetal liver.

DUODENUM [doo-oh-DEE-nuhm; Gr. "twelve fingers wide"] The C-shaped initial segment of the small intestine.

DURA MATER [DYOOR-uh MAY-ter; L. hard mother] The tough, fibrous outermost layer of the meninges.

DYNAMIC EQUILIBRIUM See *vestibular apparatus*.

DYNAMIC SPATIAL RECONSTRUCTOR (DSR) A scanning device that produces three-dimensional computer-generated pictures of the active brain. It can be used to view the flow of blood through the brain.

DYSLEXIA (L. *dys*, faulty + Gr. *lexis*, speech) An extreme difficulty in identifying printed words, usually manifested as a reading and writing disability.

DYSMENORRHEA Painful or difficult menstruation.

DYSURIA Painful urination.

E

ECTODERM (Gr. *ektos*, outside + *derma*, skin) The outermost of the three primary germ layers of an embryo, developing into mature epithelial and nervous tissue.

ECTOPIC PREGNANCY (Gr. *ektopos*, out of place) An abnormal implantation of a fertilized ovum into the wall of the uterine tube, for example, instead of the uterine wall.

EFFECTOR A muscle or gland that receives nerve impulses from the central nervous system that result in physical activity.

EFFERENT ARTERIOLE (L. *ex*, away from + *ferre*, to bring) A blood vessel in the kidney formed from the joining of glomerular capillary loops; carries blood away from the *glomerulus* (see).

EFFERENT NEURON A nerve cell that conveys nerve impulses away from the central nervous system to the effectors; also called *motor neuron*.

EJACULATION (L. *ejaculari*, to throw) The explusion of semen through the penis via the urethra; the expulsion accompanies orgasm in males.

EJACULATORY DUCT A duct continuing from the ductus deferens, transporting sperm and secretions from the seminal vesicles and prostate gland to the urethra.

ELECTRICAL SYNAPSE A gap junction by which two cells are electrically coupled by tiny intercellular channels.

ELECTROCARDIOGRAM (EKG OR ECG) The visual recording of the electrical waves of the heart, as registered on an *electrocardiograph*.

ELECTROENCEPHALOGRAM (EEG) ("electric writing in the head") The tracing, in the form of waves, produced by an electroencephalograph; shows the electrical activity in the brain.

ELECTROENCEPHALOGRAPH (EEG) An instrument used to record the electrical activity of the brain.

ELECTROLYTE Any substance whose solution conducts electricity.

ELECTRON A negatively charged subatomic particle that moves around the nucleus of an atom.

ELECTRON TRANSPORT SYSTEM A metabolic process occurring in mitochondria, where electrons are transferred from molecules of $NADH + H^+$ and $FADH_2$ to molecules of oxygen, producing some chemical energy to synthesize ATP.

ELEMENT A chemical substance that cannot be broken down into simpler substances by ordinary chemical means.

ELEVATION A movement that raises a body part.

EMBOLISM (Gr. *emballein*, to throw in) A traveling blood clot, air bubble, or other blockage that reduces the flow of blood through a blood vessel.

EMBRYONIC PERIOD The prenatal period from fertilization to the end of the eighth week; see *fetal period*.

EMPHYSEMA (Gr. blown up) A condition in which the alveoli in the lungs fail to contract properly, thus expelling insufficient air.

ENAMEL The insensitive white covering of the crown of a tooth.

ENCEPHALITIS A severe inflammation of the brain, usually involving a virus.

ENDOCARDIUM ("inside the heart") The fibrous layer covering the inside cavities of the heart and all associated valves and muscles.

ENDOCHONDRAL OSSIFICATION (Gr. *endon*, within + *khondros*, cartilage) The process by which bone tissue develops by replacing *hyaline cartilage* (see).

ENDOCRINE GLAND (Gr. *endon*, within + *krinein*, separate or secrete) An organ having specialized secretory cells but no ducts; the cells release their secretions directly into the bloodstream.

ENDOCYTOSIS (Gr. *endon*, within + *cyto*, cell) The active process that moves large molecules or particles through a plasma membrane when the membrane forms a pocket around the material, enclosing it and drawing it into the cytoplasm within the cell. Includes *pinocytosis, receptor-mediated endocytosis,* and *phagocytosis*.

ENDODERM (Gr. *endon*, within + Gr. *derma*, skin) The innermost of the three primary germ layers of an embryo, developing into mature epithelial tissue.

ENDOLYMPH The thin, watery fluid in the membranous labyrinth of the inner ear.

ENDOMETRIOSIS The abnormal location of endometrial tissue in sites such as the ovaries, pelvic peritoneum, and small intestine.

ENDOMETRIUM (inside + Gr. *metra*, womb) The innermost tissue layer of the uterus, composed of specialized mucous membrane.

ENDOMYSIUM (Gr. *endon*, inside, within + muscle) A connective-tissue sheath surrounding each muscle fiber.

ENDONEURIUM ("within the nerve") Interstitial connective tissue separating individual nerve fibers.

ENDOPLASMIC RETICULUM (ER) A complex labyrinth of flattened sheets, sacs, tubules, and double membranes that branch and spread throughout the cytoplasm, creating a series of channels for intracellular transport.

ENDORPHINS ("endogenous morphine-like substances") Naturally occurring peptides found in several regions of the brain, spinal cord, and pituitary gland; have the pain-killing effects of opiates by suppressing synaptic activity leading to pain sensation.

ENDOSTEUM [end-AHSS-tee-uhm; Gr. *endon,* inside + *osteon,* bone] The membrane lining the internal cavities of bones.

ENKEPHALINS [en-KEFF-uh-lihnz; "in-the-head substances"] Morphinelike breakdown products of *endorphins.*

ENTEROGASTRIC REFLEX A nervous reflex that decreases stomach motility and gastric secretion.

ENTEROKINASE An enzyme that converts the pancreatic enzyme trypsinogen into active trypsin, which helps complete the breakdown of peptides into amino acids.

ENZYME (Gr. *enzumos,* leavened) A protein catalyst that increases the rate of a chemical reaction.

EOSINOPHIL A type of white blood cell that helps modulate allergic inflammatory reactions and destroys antibody-antigen complexes.

EPENDYMAL CELL A glial cell (see *neuroglia*) that helps form part of the inner membranes of the neural tube during embryonic growth; secretes *cerebrospinal fluid* (see).

EPIDERMIS (Gr. *epi,* over + *derma,* skin) The outermost layer of the skin.

EPIDIDYMIS (Gr. "upon the twin") A coiled tube that stores immature sperm until they mature, and serves as a duct for the passage of sperm from the testis to the ejaculatory duct.

EPIDURAL SPACE The space between the dura mater of the spinal cord and the periosteum of the vertebrae; contains blood vessels and fat.

EPIGLOTTIS A flap of cartilage that folds down over the opening into the larynx during swallowing and swings back up when swallowing ceases.

EPILEPSY (Gr. *epilambanein,* seize upon) A nervous disorder characterized by recurrent attacks of motor, sensory, or psychological malfunction, with or without unconsciousness or convulsions.

EPIMYSIUM (*epi,* over, upon + muscle) The connective-tissue sheath below the deep fascia surrounding a muscle.

EPINEPHRINE [ep-ih-NEFF-rihn] The main secretion of the adrenal medulla; increases pulse rate, blood pressure, and heart rate; stimulates the contraction of smooth muscle; and increases blood sugar levels by stimulating the breakdown of glycogen; also called *adrenaline.*

EPINEURIUM ("upon the nerve") A sheath of connective tissue containing blood and lymphatic vessels that together surround a bundle of nerve fascicles.

EPIPHYSEAL PLATE A thick plate of hyaline cartilage that provides the framework for the construction of spongy bone tissue within the *metaphysis* (see); also called *growth plate.*

EPIPHYSIS [ih-PIHF-uh-siss; Gr. "to grow upon"] The roughly spherical end of a long bone; composed of spongy bone tissue.

EPIPLOIC APPENDAGES Fat-filled pouches in the large intestine formed where the visceral peritoneum is attached to the taeniae coli in the serous layer.

EPISIOTOMY The cutting of the mother's vulva at the lower end of the vaginal orifice at the time of delivery in order to enlarge the birth canal.

EPITHELIAL TISSUE Groups of cells that cover or line something; secretes substances that lubricate or take part in chemical reactions; also called *epithelium.*

EPITHELIUM See *epithelial tissue.*

EPONYCHIUM [epp-oh-NICK-ee-uhm; Gr. *epi,* upon + *onyx,* nail] The thin layer of epidermis covering the developing nail; commonly called the *cuticle* in the mature nail.

ERECTION The state of the clitoris or penis when its spongy tissue becomes engorged with blood, usually during erotic stimulation.

ERYTHEMA A diffuse redness of the skin.

ERYTHROCYTES [ih-RITH-roh-sites; Gr. *eruthros,* red + cell] Red blood cells, which constitute about half the total volume of blood; contain *hemoglobin.*

ERYTHROPOIESIS [ih-RITH-roh-poy-EE-sis; Gr. red + to make] The production of erythrocytes in bone marrow.

ERYTHROPOIETIN (Gr. *eruthros,* red + *poiesis,* making) A glycoprotein hormone produced mainly in the kidneys; it is the main controller of the rate of erythrocyte production.

ESOPHAGUS [ih-SOFF-uh-guss; Gr. gullet] A muscular, membranous tube through which food passes from the pharynx into the stomach.

ESSENTIAL AMINO ACIDS The amino acids that cannot be synthesized by the body, and therefore must be supplied in the diet.

ESTROGENS A class of ovarian hormones that help regulate the menstrual cycle and the development of the mammary glands and female secondary sex characteristics; include estrin, estrone, and estradiol.

EUGENICS (Gr. well-born) The advocacy of trying to improve humanity by applying certain genetic control measures.

EUSTACHIAN TUBE [yoo-STAY-shun] See *auditory tube.*

EVERSION A movement of the foot in which the great toe is turned downward and toward the midline of the body; opposite of *inversion* (see).

EXCITABILITY The capacity of a nerve or muscle cell to receive and respond to a stimulus.

EXCITATORY POSTSYNAPTIC POTENTIAL (EPSP) A partial depolarizing effect that lowers the membrane potential of a neuron, but not enough to generate an action potential.

EXCRETION (L. *excernere,* to sift out) The elimination of the waste products of metabolism, and the removal of surplus substances from body tissues.

EXHALATION See *expiration.*

EXOCRINE GLANDS Organs with specialized cells that produce secretions, and with ducts that carry the secretions to body surfaces.

EXOCYTOSIS (Gr. *exo,* outside + *cyto,* cell) The active process in which the endosome with its undigested particles fuses with the plasma membrane and expels unwanted particles from a cell; opposite of *endocytosis* (see).

EXPIRATION The process in which oxygen-poor venous blood is forced back to the lungs, with some waste products being exhaled back into the atmosphere; also called *exhalation.*

EXPIRATORY RESERVE VOLUME The quantity of air remaining in the lungs after an ordinary exhalation.

EXTENSION A straightening motion that increases the angle of a joint.

EXTERNAL A relative directional term of the body: outside; opposite of *internal.*

EXTERNAL AUDITORY CANAL A slightly curved canal separating the external ear from the middle ear; also called *external auditory meatus.*

EXTERNAL AUDITORY MEATUS See *external auditory canal.*

EXTERNAL EAR The visible part of the ear, composed of a thin plate of fibrocartilage covered with skin; also called *auricle.*

EXTERNAL NARES [NAIR-eez; L. nostrils] The two external openings of the nostrils.

EXTERNAL RESPIRATION A form of gas exchange in which oxygen from the lungs moves into the blood, and carbon dioxide and water move from the blood to the lungs.

EXTEROCEPTOR (L. "received from the

outside") A sensory receptor that responds to external stimuli that affect the skin directly.

EXTRACELLULAR FLUID Fluid that surrounds and bathes the body's cells.

EXTRAEMBRYONIC MEMBRANES Four membranes that form outside the embryo during the third week of embryonic development, including the yolk sac, amnion, allantois, and chorion; also called *fetal membranes*.

EXTREMITIES The extremities, or appendages, of the body are the *upper extremities* (shoulders, upper arms, forearms, wrists, hands) and *lower extremities* (thighs, legs, ankles, feet).

EXTRINSIC BLOOD-CLOTTING PATHWAY A rapid blood-clotting system activated when blood vessels are ruptured and tissues are damaged; see *intrinsic blood-clotting pathway*.

F

FACET (Fr. little face) A small flat surface on a bone.

FACILITATED DIFFUSION The passive-transport process in which large molecules need carrier proteins to pass through the protein channels of a plasma membrane.

FALLOPIAN TUBE See *uterine tube*.

FASCIA [FASH-ee-uh; pl. **FASCIAE**; L. band] A sheath of fibrous tissue enclosing skeletal muscles, holding them together; may be superficial, deep, or subserous (visceral).

FASCICLE A bundle of nerve or muscle fibers.

FASCICULI [fah-SICK-yoo-lie; L. little bundles] Bundles of fibers divided into tracts within each *funiculus* (see) in the white matter of the spinal cord.

FAT An energy-rich molecule that is a source of reserve food or long-term fuel; provides the body with insulation, protection, and cushioning.

FATTY ACID The part of a fat molecule that contains carboxyl groups (COOH); see *glycerol*.

FECES [FEE-seez; L. *faex*, dregs] An indigestible semisolid mixture stored in the large intestine until ready to be eliminated through the anus during defecation.

FEEDBACK CONTROL SYSTEM A regulatory system in which changes in the body or environment are fed back through a circular system into a central control unit such as the brain, where initial adjustments are made to maintain homeostasis.

FERTILIZATION MEMBRANE The thickened membrane that forms over an ovum immediately after the entry of a sperm, preventing other sperm from entering.

FETAL ALCOHOL SYNDROME A set of defects in the newborn caused by an excessive intake of alcohol by the mother during pregnancy.

FETAL MEMBRANES See *extraembryonic membranes*.

FETAL PERIOD The prenatal period from the beginning of the ninth week until birth; see *embryonic period*.

FETOSCOPY ("seeing the fetus") A procedure that allows the physician to view the fetus directly by inserting an endoscope into the mother's uterus.

FETUS [FEE-tuhss] The embryo after 8 weeks of development.

FIBRIN The insoluble, stringy plasma protein whose threads entangle blood cells to form a clot during the blood-clotting process.

FIBRINOGEN A soluble plasma protein converted into insoluble fibrin during blood-clotting.

FIBRINOLYSIS ("fibrin breaking") Blood-clot destruction.

FIBROBLASTS (L. *fibro*, fiber + Gr. *blastos*, growth) The most common connective tissue cells, and the only cells found in tendons; synthesize the matrix materials, and are considered to be secretory; assist in wound-healing.

FIBROUS JOINT A joint that lacks a joint cavity and has its bones united by fibrous connective tissue. Includes *sutures, syndesmoses*, and *gomphoses* (see).

FILTRATION A passive-transport process that forces small molecules through selectively permeable membranes with the aid of hydrostatic pressure or some other externally applied force.

FILUM TERMINALE A nonneural fibrous filament extending caudally from the conus terminalis of the spinal cord; attaches to the coccyx.

FIMBRIAE [FIHM-bree-ee; L. threads] Ciliated fringes that help sweep an ovum released by an ovary into the *infundibulum* (see).

FISSURE (L. *fissio*, split) A groove or cleft, as in bones and the brain.

FIXATOR MUSCLE A muscle that provides a stable base for the action of the prime mover; also called *postural muscle*.

FIXED-MEMBRANE-RECEPTOR MECHANISM A hormonal mechanism in which water-soluble hormones (amines and proteins) regulate cellular responses; formerly called the *second-messenger concept*.

FLACCID [FLACK-sihd; L. *flaccus*, hanging, flabby] The loose, nonerect state of the clitoris or penis; lacking muscle tone.

FLAGELLA (sing. **FLAGELLUM**; L. "whip") Threadlike appendages of certain cells, usually numbering no more than one or two per cell; used to propel the cell through a fluid environment; see *cilia*.

FLEXION A bending motion that decreases an angle at a joint.

FLEXOR REFLEX ARC A withdrawal reflex involving sensory receptors, afferent neurons, interneurons, alpha motor neurons, and voluntary muscles.

FOLLICLE (of hair) (L. *follicus*, little bag) The tubular structure enclosing the hair *root* and *bulb* (see).

FOLLICLE-STIMULATING HORMONE (FSH) A hormone secreted by the anterior pituitary; it stimulates spermatogenesis and the secretion of testosterone in males, and the secretion of estrogens in females.

FOLLICLES The actual centers of oogenesis in the cortex of the ovaries.

FONTANEL A large membranous area between incompletely ossified bones.

FORAMEN [fuh-RAY-muhn; pl. **FORAMINA**; L. opening] A natural opening into or through a bone.

FORAMEN OVALE ("oval window") An opening in the fetal interatrial septum, which closes at birth.

FORESKIN See *prepuce*.

FORMED ELEMENTS The elements that make up the solid part of blood; include erythrocytes, leukocytes, and thrombocytes.

FOSSA (L. trench) A shallow depressed area, as in bones.

FOVEA [FOE-vee-uh; L. small pit] A depressed area in the *macula lutea* (see) near the center of the retina; contains only cones; image formation and color vision are most acute here.

FRACTURE (L. *fractura*, broken) A broken bone.

FRATERNAL TWINS See *dizygotic twins*.

FREE-RADICAL DAMAGE A part of the aging process in which highly reactive superoxide radicals and hydrogen peroxide build up due to enzyme deficiencies, and harmful free-radical chain reactions occur within cells.

FRONTAL A plane dividing the body into anterior and posterior sections formed by making a lengthwise cut at right angles to the midsagittal plane. Also called *coronal*.

FRONTAL LOBE A cerebral lobe involved with the motor control of voluntary movements (including those associated with speech), and the control of a variety of emotional expressions and moral and ethical behavior; also called the *motor lobe*.

FROSTBITE A localized, direct, mechanical injury to plasma membranes that occurs when water molecules in tissues form ice crystals.

FSH See *follicle-stimulating hormone.*

FULCRUM (L. *fulcire,* to support) The point or support on which a lever turns.

FUNCTIONAL RESIDUAL AMOUNT The expiratory reserve volume plus the residual volume; about 1.8 L in women, 2.2 L in men.

FUNDUS (L. bottom) The inner basal surface of an organ farthest away from the opening; the wide upper portion of the uterus.

FUNICULI [fyoo-NICK-yoo-lie; sing. **FUNICULUS**; L. little ropes] Three pairs of columns of myelinated fibers that run the length of the white matter of the spinal cord.

G

GALLSTONES See *cholelithiasis.*

GAMETE (L. *gamos,* marriage) A sex cell; the female gamete is an ovum (egg cell), and the male gamete is a sperm cell.

GAMETE INTRAFALLOPIAN TRANSFER (GIFT) An *in vitro* procedure that unites an ovum and sperm in the uterine tube, where fertilization normally takes place.

GAMETOGENESIS The formation of gametes (sex cells).

GAMMA-AMINOBUTYRIC ACID (GABA) The major inhibitory neurotransmitter of the small local circuit neurons in such structures as the cerebral cortex, cerebellum, and upper brainstem.

GANGLIA (sing. **GANGLION**) Groups of cell bodies located outside the central nervous system.

GAP JUNCTION A junction formed from several links of channel protein connecting two plasma membranes; found in interstitial epithelia.

GASTRIC (Gr. *gaster,* belly, womb) Pertaining to the stomach.

GASTRIN A polypeptide hormone secreted by the stomach mucosa; it produces digestive enzymes and hydrochloric acid in the stomach.

GASTROINTESTINAL (GI) TRACT The part of the digestive tract below the diaphragm.

GENE [JEEN; Gr. *genes,* born, to produce] A segment of DNA that controls a specific cellular function, either by determining which proteins will be synthesized or by regulating the action of other genes; a hereditary unit that carries hereditary traits.

GENERAL SENSES The senses of touch-pressure, heat, cold, pain, and body position; also called *somatic senses.*

GENOTYPE [JEEN-oh-tipe] The genetic make-up, sometimes hidden, of a person.

GERIATRICS (Gr. *geras,* old age + *iatrikos,* physician) The medical study of the physiology and ailments of old age.

GERMINAL EPITHELIUM A layer of specialized epithelial cells covering the ovaries and lining the seminiferous tubules of the testes; also called *germinal layer.*

GESTATION (L. *gestare,* to carry) The period of carrying developing offspring in the uterus during pregnancy.

GIANTISM A hormonal disorder in which a person grows larger than normal because of an oversecretion of growth hormone during the period of skeletal development.

GINGIVA [jihn-JYE-vuh] See *gum.*

GLANS CLITORIS (L. acorn) The small mass of sensitive tissue at the tip of the clitoris.

GLANS PENIS (L. acorn) The sensitive tip of the penis.

GLAUCOMA [glaw-KOH-muh] A condition that occurs when the aqueous humor of the eye does not drain properly, producing excessive pressure within the eyeball; results in blindness when retinal neurons are destroyed by the increased pressure.

GLIAL CELLS See *neuroglia.*

GLIDING JOINT A small biaxial joint that usually has only one axis of rotation, permitting side-to-side and back-and-forth movements.

GLOMERULAR CAPSULE The portion of the nephron enclosing the glomerulus; also called *Bowman's capsule.*

GLOMERULAR FILTRATE The fluid filtered from the blood in the kidney.

GLOMERULAR FILTRATION The renal process that forces plasma fluid from the glomerulus into the glomerular capsule.

GLOMERULAR FILTRATION RATE The amount of glomerular filtrate formed in the capsular space each minute.

GLOMERULUS [glow-MARE-yoo-luhss; pl. **GLOMERULI**, glow-MARE-you-lie; L. ball] A coiled ball of capillaries in the kidney formed by the branching of an afferent arteriole; the site of blood filtration.

GLUCAGON [GLOO-kuh-gon] A peptide hormone secreted by alpha cells in the pancreas; it causes blood glucose levels to rise by stimulating *glycogenolysis* and *gluconeogenesis.*

GLUCOCORTICOIDS [gloo-koh-KORE-tih-koidz] A group of hormones secreted by the adrenal cortex; include cortisol, cortisone, corticosterone, and 11-deoxycorticosterone; the glucocorticoids affect metabolism, growth, and levels of blood sugar.

GLUCONEOGENESIS The process in

which glucagon stimulates the formation of glucose from noncarbohydrate sources.

GLYCEROL The part of a fat molecule containing three hydroxyl (OH^-) groups; also called *glycerin.*

GLYCOGENESIS ("glycogen production") The metabolic process of converting glucose into glycogen, which can be stored in the liver and skeletal muscle cells for later use.

GLYCOGENOLYSIS The process by which glucagon stimulates the liver to convert glycogen to glucose.

GLYCOLYSIS The anaerobic cellular metabolic process that splits a glucose molecule into two molecules of pyruvic acid.

GLYCOSURIA The presence of glucose in the urine.

GOITER See *hypothyroidism.*

GOLGI APPARATUS Flattened stacks of disklike membranes found in cytoplasm of most cells; packages glycoproteins for secretion; also called *Golgi complex, Golgi body.*

GOLGI TENDON ORGAN An encapsulated sensory receptor that monitors the tension and stretch in muscles and tendons; involved with voluntary muscle reflexes.

GOMPHOSIS (Gr. *gomphos,* bolt) A fibrous joint in which a peg fits into a socket, such as teeth in the maxilla or mandible.

GONADOCORTICOID A type of steroid hormone secreted by the adrenal cortex; consists mainly of weak male hormones (*androgens*) and small amounts of female hormones (*estrogens*), both of which have a slight effect on the gonads; also called *adrenal sex hormones.*

GONADOTROPIN-RELEASING HORMONE (GNRH) A hormone secreted by the hypothalamus to control pituitary secretion; in the female, it affects the timing of oocyte development by stimulating the secretion of FSH and LH.

GONADS (Gr. *gonos,* offspring) Sex organs: *ovaries* in females, and *testes* in males.

GONORRHEA (Gr. "flow of seed") A sexually transmitted disease caused by the bacterium *Neisseria gonorrhoeae;* it is primarily an infection of the urinary and reproductive tracts.

GRAAFIAN FOLLICLE See *vesicular ovarian follicle.*

GRADED LOCAL POTENTIAL An electrical potential that spreads its effects passively, fading out a short distance from the site of stimulation; includes *postsynaptic potentials* (see).

GRAY COMMISSURE (L. "joining together") The pair of anterior horns that forms the "cross bar" of the H-shaped gray matter in the spinal cord; functions in cross reflexes.

GRAY MATTER The central part of the spinal cord, consisting of nerve cell bodies and dendrites of association and efferent neurons, unmyelinated axons of spinal neurons, sensory and motor neurons, and axon terminals.

GRAY RAMI COMMUNICANTES [RAY-mee] Unmyelinated nerve fibers containing postganglionic sympathetic fibers.

GREATER OMENTUM (L. "fat skin") An extensive folded, fatty membrane extending from the greater curvature of the stomach to the back wall and down to the pelvic cavity; filled with plasma cells and other defensive cells, it protects and insulates abdominal organs; see *lesser omentum.*

GREATER VESTIBULAR GLANDS Paired glands located in the floor of the vaginal vestibule; during sexual arousal they secrete an alkaline mucus solution that provides some lubrication and offsets some vaginal acidity; also called *Bartholin's glands.*

GROSS ANATOMY Any branch of anatomy that can be studied without a microscope.

GROUND SUBSTANCE A homogeneous, extracellular material of tissues that provides a suitable medium for the passage of nutrients and wastes between cells and the bloodstream.

GROWTH HORMONE (GH) A hormone secreted by the adenohypophysis; it affects parts of the body associated with growth; also called *somatotropin* or *somatotropic hormone (STH).*

GROWTH PLATE See *epiphyseal plate.*

GUM (Old Eng. *goma,* palate, jaw) The firm connective tissue covered with mucous membrane that surrounds the alveolar processes of the teeth; also called *gingiva.*

GYNECOLOGY (Gr. *gune,* woman + study of) The medical science of female disease, reproductive physiology, and endocrinology.

GYRI [JYR-rye; sing. **GYRUS;** L. circles] Raised ridges of the cerebral cortex; also called *convolutions.*

H

HAIR A specialization of the skin that develops from the epidermis; composed of cornified threads of cells, covering almost the entire body.

HAIR CELLS Specialized proprioceptor cells of the vestibular apparatus in the inner ear; convert a mechanical force into an electrical signal that is relayed to the brain.

HAUSTRA Puckered bulges in the wall of the large intestine caused by the uneven pull of the *taeniae coli* (see).

HAVERSIAN CANAL See *central canal.*

HAVERSIAN SYSTEM See *osteon.*

HEAD The expanded, rounded surface at the proximal end of a bone; often joined to the shaft by a narrow neck; also called *caput.*

HEART (Gr. *kardia,* L. *cor*) The hollow muscular organ that pumps blood through the circulatory system.

HEAT EXHAUSTION A condition caused by a depletion of plasma volume due to sweating and hypotension due to reflexively dilated blood vessels in the skin; also called *heat prostration.*

HEAT PROSTRATION See *heat exhaustion.*

HEATSTROKE A condition in which the hypothalamus becomes overheated and its heat-regulating system breaks down, halting perspiration; also called *sunstroke.*

HEIMLICH MANEUVER A physical maneuver that dislodges a particle stuck in the glottis or larynx by compressing the remaining air in the lungs violently enough to blow the particle free.

HEMATOCRIT (Gr. "to separate blood") The volume percentage of red blood cells in whole blood.

HEMATOENCEPHALIC BARRIER See *blood-brain barrier.*

HEMATOMA [hee-muh-TOE-muh; L. *hemato,* blood + *oma,* tumor] A localized swelling filled with blood; a blood clot.

HEMATURIA The presence of hemoglobin in the urine.

HEMIPLEGIA Unilateral paralysis of upper and lower limbs; usually results from damage to only one side of the spinal cord, or serious brain damage on the opposite side.

HEMODIALYSIS THERAPY (Gr. *dialvein,* to tear apart) The use of an artificial device to filter blood or to add nutrients to the blood in the absence of total functioning kidneys.

HEMODYNAMICS The study of the principles governing blood flow.

HEMOGLOBIN [HEE-moh-gloh-bihn; Gr. *haima,* blood + L. *globulus,* little globe] A globular iron-containing protein found in erythrocytes; it transports oxygen from the lungs and helps remove carbon dioxide.

HEMOLYTIC DISEASE OF THE NEWBORN An abnormal condition in which the fetus's agglutinated erythrocytes break up and release hemoglobin into the blood; formerly called *erythroblastosis fetalis.*

HEMOPHILIA An affliction in which one of several protein clotting factors is absent; also called "bleeder's disease."

HEMOPOIETIC TISSUE [hee-muh-poy-ET-ihk; Gr. *haima,* blood + *poiein,* to make] A bone tissue that produces red blood cells, platelets, and certain white blood cells.

HEMORRHAGE (Gr. *hamia,* blood + *rhegnunai,* to burst forth) The rapid loss of blood from blood vessels.

HEMOSTASIS The stoppage of bleeding.

HEPARIN An acidic mucopolysaccharide that inhibits blood clotting.

HEPATIC (Gr. *hepatikos,* liver) Pertaining to the liver.

HEPATIC PORTAL SYSTEM (Gr. *hepatikos,* liver) A system of vessels that moves blood from capillary beds of the intestines to sinusoidal beds of the liver.

HEPATITIS A liver infection, with either a viral or a nonviral cause.

HERNIA (L. protruded organ) The protrusion of any organ or body part through the muscular wall that usually contains it; also called *rupture.*

HERNIATED DISK A condition in which the pulpy center of an intervertebral disk protrudes through a weakened or torn surrounding outer ring; the pulpy center pushes against a spinal nerve or even the spinal cord; also called *ruptured* or *slipped disk.*

HETEROZYGOUS Possessing two different alleles at a given locus on homologous chromosomes, for example, *Aa.*

HIATAL HERNIA See *hernia.*

HIGH BLOOD PRESSURE See *hypertension.*

HILUS [HYE-luhss; L. *hilum,* trifle] A small, indented opening on the medial concave border of the kidney, where arteries, veins, nerves, and the ureter enter and leave the kidney; also called the *hilum.*

HINGE JOINT A joint that resembles hinges on the lid of a box. The convex surface of one bone fits into the concave surface of another bone, permitting only a uniaxial movement around a single axis, such as at the knee joint; also called *ginglymus joint.*

HISTOLOGY [hiss-TAHL-uh-jee; Gr. *histos,* web] The microscopic study of tissues.

HODGKIN'S DISEASE A form of cancer characterized by the presence of large, multinucleate cells in the affected lymphoid tissue.

HOLOCRINE GLAND (Gr. *holos,* whole) An exocrine gland that releases its secretions by the detaching and dying of whole cells, which become the secretion; sebaceous glands are probably the only holocrine glands in the body.

HOMEOSTASIS [ho-mee-oh-STAY-siss; Gr. *homois,* same + *stasis,* standing still] A state of inner balance and stability in the body, which remains relatively constant despite external environmental changes.

HOMOZYGOUS Possessing two identical alleles at a given locus on homologous chromosomes; for example, *AA* or *aa*.

HORMONE A chemical "messenger" produced and secreted by endocrine cells or tissues; circulates via the bloodstream to "target" cells or tissues, affecting their metabolic activity in a specific way.

HUMAN CHORIONIC GONADOTROPIN (HCG) A hormone released by the placenta and covering membranes of the embryo; it prevents the corpus luteum from disintegrating during part of the pregnancy, stimulating it to secrete estrogen and progesterone.

HUMORAL IMMUNITY A type of immunity that involves B cells in the production of antibodies; most active against bacteria, viruses, toxins, and other soluble foreign proteins.

HYALINE CARTILAGE [HYE-uh-lihn; Gr. *hyalos*, glassy] The most prevalent type of cartilage, containing collagenous fibers scattered in a network filled in with ground substance.

HYALURONIDASE An enzyme found in the head of sperm; it chemically dissolves enough of the outer wall of an ovum to allow a single sperm to enter.

HYDROCEPHALUS (Gr. *hudor*, water + head) A condition in which there is an excess of cerebrospinal fluid within the ventricles, causing the skull to enlarge and put pressure on the brain; mental retardation is common; commonly called "water on the brain."

HYDROLYSIS (Gr. *hudor*, water + *lusis*, loosening) The chemical process by which a molecule of water interacts with a reactant, thereby breaking the bonds of the reactant and rearranging it into different molecules.

HYDROSTATIC PRESSURE The force exerted by a fluid against the surface of the compartment containing the fluid.

HYMEN (Gr. membrane; Hymen was the Greek god of marriage) A fold of skin partially blocking the vaginal entrance.

HYPEREXTENSION Excessive extension beyond the straight (anatomical) position.

HYPERKALEMIA (*hyper*, over + L. *kalium*, potassium) A condition in which there is a high potassium concentration in extracellular fluids.

HYPERMAGNESEMIA An abnormally high magnesium level in the blood.

HYPERMETROPIA (Gr. "beyond measure") Farsightedness, occurring when the focus occurs beyond the retina.

HYPERNATREMIA (*hyper*, over + L. *na-*

trium, sodium) A condition in which there is a high sodium concentration in the extracellular fluids.

HYPERSENSITIVITY An overreaction to an allergen (antigen); also called *allergy*.

HYPERTENSION A condition in which systolic or diastolic pressure is above normal all the time; commonly called *high blood pressure.*

HYPERTHYROIDISM A condition characterized by the overactivity of the thyroid gland.

HYPERTONIC (Gr. *hyper*, above + *tonos*, tension) A solution in which the solute concentration is higher outside a cell than inside.

HYPERTROPHY (Gr. "overnourished") A condition in which the diameter of muscle fibers is increased as the result of physical activity; opposite of *atrophy*.

HYPODERMIS (Gr. *hypo*, under + *derma*, skin) The layer of loose, fibrous connective tissue lying below the dermis; also called *subcutaneous layer.*

HYPOGLYCEMIA An abnormal condition caused by the excessive secretion of insulin; also called *low blood sugar.*

HYPOKALEMIA (*hypo*, under + L. *kalium*, potassium) A condition in which there is a low potassium concentration in extracellular fluids.

HYPOMAGNESEMIA An abnormally low magnesium level in the blood.

HYPONATREMIA (*hypo*, under + L. *natrium*, sodium) A condition in which there is a low sodium concentration in the extracellular fluids.

HYPOPHYSEAL PORTAL SYSTEM A system of vessels through which the blood flows from the capillary bed of the hypothalamus by way of veins to the capillary bed of the pituitary gland.

HYPOPHYSIS (Gr. undergrowth) See *pituitary gland.*

HYPOTHALAMIC-HYPOPHYSEAL PORTAL SYSTEM An extensive system of blood vessels connecting the hypothalamus and the adenohypophysis.

HYPOTHALAMUS ("under the thalamus") The part of the brain located under the thalamus, forming the floor of the third ventricle; regulates body temperature, some metabolic processes, and other autonomic activities.

HYPOTHERMIA ("under-heat") A nonlocal condition in which the heat-producing and conserving mechanisms are exceeded by severe cold.

HYPOTHYROIDISM A condition characterized by the underactivity of the thyroid gland; it is usually associated with a *goiter* (L. "throat"), a swelling in the neck caused by an insufficiency of iodine in the diet.

HYPOTONIC (Gr. *hypo*, under) A solu-

tion in which the solute concentration is lower outside a cell than inside.

HYSTERECTOMY The surgical removal of the uterus.

I

IDENTICAL TWINS See *monozygotic twins.*

ILEUM [ILL-ee-uhm] The portion of the small intestine extending from the *jejunum* to the cecum, the first part of the large intestine.

IMMEDIATE HYPERSENSITIVITY An allergic reaction that occurs within minutes after exposure to an allergen.

IMMUNE SYSTEM The overall defensive system of lymphocytes and lymphatic tissues and organs.

IMMUNITY (L. *immunis*, exempt) An overall protective mechanism that forms antibodies to help protect the body against foreign substances.

IMMUNOGLOBULIN An antibody in the globulin group of proteins; involved with the immune response; five classes are: IgA, IgD, IgE, IgG, and IgM.

IMMUNOSUPPRESSIVE DRUG A drug that helps to suppress an immune reaction.

IMPETIGO (*impetigo contagiosa*) (ihm-puh-TIE-go; L. an attack) A contagious infection of the skin, caused by staphylococcal or streptococcal bacteria, and characterized by small red macules that become pus-filled.

IMPLANTATION The process by which the blastocyst becomes embedded in the uterine wall.

IMPOTENCE The inability of a man to have an erection.

IN VITRO ("IN A TEST TUBE") FERTILIZATION A procedure in which an ovum is removed from a prospective mother's ovary at ovulation and fertilized in a glass dish or test tube with sperm from the male donor. The embryo is later implanted in the woman's uterus; a successful procedure results in the birth of a "test-tube baby."

INCONTINENCE See *urinary incontinence.*

INFANCY The period between the first 4 weeks of life and the time the child can sit erectly and walk.

INFECTIOUS MONONUCLEOSIS A viral disease caused by the Epstein-Barr virus, a member of the herpes group.

INFERIOR (L. low) A relative directional term: toward the feet; below.

INFERIOR VENA CAVA A large vein that drains blood from the abdomen, pelvis, and lower limbs, emptying it into the right atrium; see *superior vena cava.*

INFLAMMATION RESPONSE The sequence of events that follows tissue injuries such as punctures, abrasions, burns, and infections.

INFLAMMATORY RESPONSE Part of the

healing process, including redness, pain, swelling, scavenging by *neutrophils* and *monocytes*, and tissue repair by *fibroblasts*.

INFUNDIBULAR STALK (L. funnel) A stalk of nerve cells and blood vessels that connects the pituitary and the hypothalamus; also called *infundibulum*.

INFUNDIBULUM (L. funnel) The funnel-shaped portion of the uterine tube near the ovary; see *infundibular stalk*.

INGESTION The taking in of nutrients by eating or drinking.

INGUINAL HERNIA See *hernia*.

INHALATION See *inspiration*.

INHIBIN A male hormone secreted by the testes; it inhibits the secretion of FSH from the anterior pituitary, maintaining a constant rate of spermatogenesis.

INHIBITING FACTORS Substances secreted by the hypothalamus that inhibit secretions from the adenohypophysis; see *releasing factors*.

INHIBITORY POSTSYNAPTIC POTENTIAL (IPSP) An inhibitory condition that occurs when a neurotransmitter interacts with a postsynaptic receptor site, increasing the negative potential inside the plasma membrane of a neuron above the resting level.

INNER EAR (internal ear) The portion of the ear that includes the vestibule, the semicircular ducts and canals, and the spiral cochlea; also called *labyrinth*.

INORGANIC COMPOUND A chemical compound composed of relatively small molecules usually bonded ionically; see *organic compound*.

INSERTION The point of attachment of a muscle to the bone it moves.

INSPIRATION The process in which oxygen-rich, carbon dioxide-poor air travels from the atmosphere through the respiratory tract to the terminal portion of the lungs; also called *inhalation*.

INSPIRATORY CAPACITY The inspiratory reserve volume plus the tidal volume; about 2.4 L in women, 3.8 L in men.

INSPIRATORY RESERVE VOLUME The excess air taken into the lungs by the deepest inhalation beyond a normal inspiration.

INSULA (L. island) The cerebral lobe beneath the parietal, frontal, and temporal lobes; it appears to be associated with gastrointestinal and other visceral activities; also called *central lobe*.

INSULIN A peptide hormone secreted by beta cells in the pancreas; it helps to lower the blood glucose concentration by facilitating glucose transport across plasma membranes and stimulating glycogen breakdown.

INTERCALATED DISKS (L. *intercalatus*, to insert between) Thickenings of the sarcolemma that separate adjacent cardiac muscle fibers.

INTERFERON Small protein molecules produced by virally infected cells; protect other cells by preventing viruses from multiplying within them.

INTERNAL A relative directional term of the body: inside; opposite of *external*.

INTERNAL RESPIRATION The process in which body cells exchange carbon dioxide for oxygen in the blood.

INTEROCEPTOR (L. "received from inside") A specialized sensory receptor responding to stimuli originating in internal organs; also called *visceroceptor*.

INTERPHASE The period between cell divisions during which the activities of growth, cellular respiration, RNA and protein synthesis, and DNA replication take place.

INTERSTITIAL CELLS OF LEYDIG See *interstitial endocrinocytes*.

INTERSTITIAL ENDOCRINOCYTES Clusters of endocrine cells among the seminiferous tubules in the testes that secrete the male sex hormones, including *testosterone*; also called *Leydig cells*.

INTERSTITIAL FLUID The fluid between cells that bathes and nourishes surrounding body tissues.

INTESTINAL GLANDS Tubular glands in the small intestine, formed by the infolding of the epithelium between the bases of the villi; formerly called *crypts of Lieberkühn*.

INTRACELLULAR FLUIDS (ICF) Fluids inside a cell.

INTRAFUSAL MUSCLE FIBERS Tiny muscles within a neuromuscular spindle.

INTRAMEMBRANOUS OSSIFICATION The process by which bone tissue develops directly from embryonic connective tissue.

INTRAUTERINE DEVICE (IUD) A plastic or metal device placed inside the uterus, where it prevents implantation by stimulating an inflammatory response.

INTRAUTERINE TRANSFUSION An injection of a concentrate of red blood cells into the peritoneal cavity of the fetus to help the fetus combat hemolytic anemia.

INTRAVENOUS INFUSION The controlled introduction of relatively large amounts of prescribed fluid into the body of a patient in order to maintain fluid balance, electrolyte concentration, and acid-base balance.

INTRINSIC BLOOD-CLOTTING PATHWAY A blood-clotting mechanism activated when the inner walls of blood vessels become damaged or irregular.

INVERSION A movement of the foot in which the great toe is turned upward and away from the midline of the body; opposite of *eversion* (see).

INVOLUNTARY MUSCLE See *smooth muscle tissue*.

ION [EYE-ahn; Gr. "going particle"] An atom that has acquired an electrical charge by gaining or losing electrons. A positive ion is a *cation*, and a negative ion is an *anion*.

IONIC BOND A chemical bond formed when an atom or group of atoms develops an electrical charge and subsequently becomes attracted to an atom or group of atoms with an opposite charge.

IPSILATERAL Pertaining to the same side of the body.

IPSILATERAL REFLEX ("same side") A reflex occurring on the same side of the body and spinal cord as where the stimulus is received.

IRIS (Gr. rainbow) The colored portion of the eye.

ISLETS OF LANGERHANS See *pancreatic islets*.

ISOANTIBODY See *agglutinin*.

ISOANTIGEN See *agglutinogen*.

ISOMETRIC CONTRACTION (Gr. *isos*, equal + *metron*, length) A muscle contraction in which tension increases but the muscle remains the same length.

ISOTONIC (Gr. *isos*, equal) A solution in which the solute concentration is the same inside and outside a cell.

ISOTONIC CONTRACTION (Gr. *isos*, equal + *tonos*, tension) A muscle contraction in which the muscle becomes shorter and thicker, but the tension stays constant.

ISOTOPE (Gr. *isos*, equal + *topos*, place) One or more different atomic forms of an element. Isotopes of the same element have the same number of protons and electrons, but different numbers of neutrons.

ISTHMUS [ISS-muhss] A narrow passage connecting two larger cavities.

J

JAUNDICE (Fr. "yellow") A syndrome characterized by bile pigment in the skin and mucous membranes, resulting in a yellow appearance.

JEJUNUM [jeh-JOO-nuhm; L. "fasting intestine"] The 2.5-m-long portion of the digestive tract that extends between the duodenum and the ileum.

JUNCTIONAL COMPLEXES Specialized parts that hold cells together, enabling groups of cells to function as a unit; include *desmosomes, gap junctions* (see), and *tight junctions*.

JUXTAGLOMERULAR APPARATUS The portion of the kidney composed of juxtaglomerular cells (specialized smooth muscle cells), the macula densa (a group of epithelial cells of the distal tubule), afferent and efferent arterioles, and a pad of cells called the *extraglomerular mesangium.*

K

KERATIN (Gr. *keras,* horn) A tough protein forming the outer layer of hair and nails; it is soft in hair, and hard in nails.

KETOGENESIS The formation of *ketone bodies* (see).

KETONE BODIES Substances such as acetone, acetoacetic acid, and β-hydroxybutyric acid produced from excess acetyl-CoA during lipid catabolism.

KETOSIS An abnormal condition in which only fats are being metabolized, producing an excess of *ketone bodies.*

KIDNEY A paired organ located in the posterior wall of the abdominal cavity, on either side of the vertebral column; maintains a constant water and salt balance in the blood.

KILOCALORIE A unit of measurement equal to 1000 *calories* (see); also known as a *Calorie.*

KINESIOLOGY The study of motion.

KINESTHESIA (Gr. *kinema,* motion + sensory ability) The sense of body movement.

KLINEFELTER'S SYNDROME A chromosomal aberration in which an individual has an extra X chromosome, producing an XXY genotype.

KRAUSE'S END BULBS See *bulbous corpuscles (of Krause).*

KREBS CYCLE See *citric-acid cycle.*

KYPHOSIS (Gr. "hunchbacked") A condition in which the spine curves backward abnormally, usually at the thoracic level.

L

LABIA MAJORA ("major lips") Two longitudinal folds of skin just below the mons pubis, forming the outer borders of the *vulva.*

LABIA MINORA ("minor lips") Two relatively small folds of skin lying between the larger *labia majora.*

LABYRINTH (Gr. maze) The intricate interconnecting chambers and passages that makes up the inner ear.

LACRIMAL APPARATUS [LACK-ruh-mull; L. tear] An apparatus consisting of lacrimal gland and sac, and nasolacrimal duct.

LACRIMAL GLAND The tear gland of the eye, whose secretions keep the eye moist and clean, combat microorga-

nisms, and distribute water and nutrients to the cornea and lens.

LACTATION (L. *lactare,* to suckle) The process that includes both the production of milk by the mammary glands and the release of milk from the breasts.

LACTEALS (L. *lacteus,* of milk) Specialized lymphatic capillaries that extend into the intestinal villi.

LACTIC ACID A toxic waste substance that builds up in muscles as they become fatigued during the anaerobic respiration that accompanies vigorous physical activity.

LACTOGENIC HORMONE See *prolactin.*

LACUNAE [luh-KYOO-nee; sing. **LACUNA;** L. cavities, pods] Small cavities within the connective-tissue matrix containing *chondrocytes* (see) in cartilage, and *osteocytes* (see) in bone.

LAMELLAE (L. "thin plates") Concentric layers of bone that make up cylinders of calcified bone called *osteons* (see).

LAMELLATED CORPUSCLE (OF PACINI) A sensory receptor in skin, muscles, tendons, joints, and body organs; involved with vibratory sense and firm pressure (touch-pressure) on the skin; formerly called *Pacinian corpuscle.*

LANUGO (L. *lana,* fine wool) Fine, downy hair covering the fetus by the fifth month; shed before birth, except on the eyebrows and scalp, where it becomes thicker.

LARGE INTESTINE The part of the digestive system between the ileocecal orifice of the small intestine and the anus; removes salt and water from undigested food and releases feces through the anus.

LARYNGOPHARYNX The lowest part of the pharynx, extending downward into the larynx.

LARYNX An air passage at the beginning of the respiratory tract where the vocal cords (folds) are located; commonly called *voice box.*

LATERAL (L. *lateralis,* side) A relative directional term: away from the midline of the body (in other words, toward the *side* of the body); the eyes are lateral to the nose; opposite of *medial* (see).

LATERAL ROTATION A twisting movement, in which the anterior surface of a limb or bone moves away from the body's medial plane; opposite of *medial rotation* (see).

LENS An elastic, colorless, transparent body of epithelial cells behind the iris; its shape is adjustable to focus on objects at different distances.

LESSER OMENTUM (L. "fat skin") A fatty membrane extending from the liver to the lesser curvature of the stomach.

LESSER VESTIBULAR GLANDS Paired glands with ducts that open into the anterior part of the vaginal vestibule, between the urethral and vaginal orifices; during sexual arousal they secrete an alkaline mucus solution that provides some lubrication and offsets some vaginal acidity; also called *Skene's glands.*

LEUKEMIA (Gr. *leukos,* white and *-emia,* blood) A condition in which a malignant neoplasm originates in blood-forming cells of the bone marrow; characterized by uncontrolled reproduction of white blood cells.

LEUKOCYTE [LOO-koh-site; Gr. *leukos,* clear, white] A white blood cell, usually of the scavenger type, that ingests foreign material in the bloodstream and tissues.

LEYDIG CELLS See *interstitial endocrinocytes.*

LH See *luteinizing hormone.*

LIGAMENT A fairly inelastic fibrous thickening of an articular capsule that joins a bone to its articulating mate, allowing movement at the joint.

LIGHT TOUCH The sense that is perceived when the skin is touched, but not deformed.

LIMBIC SYSTEM An assemblage of structures in the cerebrum, diencephalon, and midbrain involved in memory and emotions, and the visceral and behavioral responses associated with them.

LIPID [LIHP-ihd; Gr. *lipos,* fat] Organic compound that is insoluble in water but can be dissolved in organic solvents; includes body fats.

LOOP OF HENLE See *loop of the nephron.*

LOOP OF THE NEPHRON A straightened portion of the kidney excretory tubule, which receives *glomerular filtrate* (see) from the proximal convoluted tubule; also called *loop of Henle.*

LORDOSIS (Gr. "bent backward") An exaggerated forward curve of the spine at the lumbar level; also called "swayback."

LOW BLOOD SUGAR See *hypoglycemia.*

LOWER EXTREMITIES See *extremities.*

LUMBAR PLEXUS The ventral rami of L1 to L4 nerves in the interior of the posterior abdominal wall.

LUNG One of the paired organs of respiration, on either side of the heart in the thoracic cavity.

LUTEINIZING HORMONE (LH) A hormone secreted by the anterior pituitary; it stimulates ovulation and progesterone secretion in females, and testosterone secretion in males.

LUTEOTROPIC HORMONE (LTH) See *prolactin.*

LYMPH (L. "clear water") *Interstitial fluid* inside *lymphatic capillaries.*

LYMPH NODES Bodies of lymphoid tissue situated along collecting lymphatic vessels; they filter foreign material from lymph on its way to the bloodstream.

LYMPHATIC Pertaining to lymph.

LYMPHATIC CAPILLARIES Capillaries within the lymphatic system that collect excess interstitial fluid in tissues.

LYMPHATIC SYSTEM The body system that collects and drains fluid that seeps from the bloodstream and accumulates in the spaces between cells.

LYMPHATICS Collecting vessels that carry lymph from lymphatic capillaries to veins in the neck, where it is returned to the bloodstream.

LYMPHOCYTE A wandering connective tissue cell found under moist epithelial linings of respiratory and intestinal tracts; main producers of antibodies; also called *plasma cell.*

LYMPHOMA [lihm-FOH-muh; L. *lympha,* water, pertaining to the watery appearance of lymph + *-oma,* tumor] A malignant neoplasm originating in lymph nodes.

LYSOSOME (Gr. "dissolving body") A small, membrane-bound organelle containing digestive enzymes; protects cell against harmful microorganisms, and clears away dead or damaged cell parts.

M

MACROPHAGE CELL [MACK-roh-fahj; Gr. *makros,* large; *phagein,* to eat] A connective tissue cell that is an active *phagocyte;* can be fixed or wandering; also called *macrophage.*

MACULA The receptor region of utricles and saccules in the ear, containing *hair cells* (see) embedded in the gelatinous otolithic membrane.

MACULA DENSA (L. "dense spot") See *juxtaglomerular apparatus.*

MACULA LUTEA (L. "yellow spot") An area in the center of the retina, containing only cones.

MAGNETIC RESONANCE IMAGING (MRI) A noninvasive exploratory diagnostic technique that uses a strong magnetic field to detect differences in healthy and unhealthy tissues; also called *nuclear magnetic resonance (NMR).*

MALIGNANT MELANOMA The most serious form of skin cancer, involving the pigment-producing melanocytes; usually starting as small, dark growths resembling moles.

MALIGNANT NEOPLASM (L. *malus,* bad) An abnormal growth whose uncontrolled cells break out of their connective-tissue capsule to invade neighbor-

ing tissue and continue to grow rapidly in an uncontrolled pattern.

MAMMARY GLANDS (L. *mammae,* breasts) Paired female modified sweat glands that produce and secrete milk for a newborn child.

MARROW See *bone marrow.*

MAST CELL A wandering connective tissue cell often found near blood vessels; contains secretory granules that produce *heparin* and *histamine.*

MASTECTOMY The surgical removal of a breast, usually to prevent a malignant neoplasm from spreading.

MATRIX [MAY-triks; L. womb, mother] 1. The extracellular fibers and ground substance in connective tissues. 2. The thick layer of skin beneath the root of a nail, where new cells are generated for nail growth and repair.

MEAN ARTERIAL PRESSURE (MAP) The average pressure that drives blood through systemic circulatory system.

MEATUS [mee-AY-tuhss; L. passage] A large, tubular opening, not necessarily through a bone.

MECHANORECEPTOR ("mechanical receiver") A sensory receptor responding to and monitoring such physical stimuli as touch-pressure, muscle tension, air vibrations, and head movements.

MEDIAL [MEE-dee-uhl; L. *medius,* middle] A relative directional term: toward the midline of the body.

MEDIAL ROTATION A twisting movement, in which the anterior surface of a limb or bone moves toward the medial plane of the body.

MEDIASTINUM [mee-dee-as-TIE-nuhm; L. *medius,* middle] The mass of tissues and organs between the lungs. It contains all the contents of the thoracic cavity except the lungs.

MEDULLA [meh-DULL-uh; L. marrow] The inner core of a structure; see *medulla oblongata.*

MEDULLA OBLONGATA (L. elongated marrow) The lowermost portion of the brainstem, continuous with the spinal cord; also called *medulla.*

MEDULLARY CAVITY [MED-uh-lehr-ee; L. *medulla,* marrow] The marrow cavity inside the shaft of a long bone.

MEDULLARY RHYTHMICITY AREA A portion of the respiratory center in the brain that regulates inspiration and expiration.

MEIOSIS [mye-OH-sihss; Gr. *meioun,* to diminish] A process that reduces the number of chromosomes in gametes to half.

MEISSNER'S CORPUSCLES See *tactile corpuscles (of Meissner).*

MELANIN [MEHL-un-nihn; Gr. *melas,* black] A dark pigment produced by

specialized cells called *melanocytes;* contributes to skin color.

MELANOCYTE-STIMULATING HORMONE (MSH) A chemical substance, probably a precursor to an active hormone, secreted by the fetal pituitary gland; it is involved in pigmentation.

MELANOMA (Gr. *melas,* black + *-oma,* tumor) A neoplasm composed of cells containing a dark pigment, usually melanin.

MEMBRANES Thin, pliable layers of epithelial and/or connective tissue that line body cavities and cover or separate regions, structures, and organs.

MENARCHE [meh-NAR-kee; Gr. "beginning the monthly"] The first menstrual period, usually occurring during the latter stages of puberty.

MENINGES [muh-NIHN-jeez; Gr. pl. of *meninx,* membrane] Three layers of protective membranes (dura mater, arachnoid, pia mater) surrounding the brain and spinal cord.

MENINGITIS Inflammation of the meninges of the brain or spinal cord.

MENOPAUSE (Gr. "ceasing the monthly") The cessation of menstrual periods.

MENSTRUAL CYCLE A monthly series of events that prepares the endometrium of the uterus for pregnancy and then discharges the sloughed-off endometrium, mucus, and blood in the menstrual flow if pregnancy does not occur.

MENSTRUATION (L. *mensis,* monthly) The monthly breakdown of the endometrium of a nonpregnant female.

MERKEL'S DISKS See *tactile corpuscles (of Merkel).*

MEROCRINE GLAND (Gr. *meros,* divide) An exocrine gland that releases its secretions via *exocytosis* (see), without breaking the plasma membrane.

MESENCHYME [MEHZ-uhn-kime; Gr. *mesos,* middle + L. *enchyma,* cellular tissue] Embryonic *mesoderm* (see) that develops into connective tissue.

MESENTERY [MEZZ-uhn-ter-ee; Gr. *mes,* middle + *enteron,* intestines] Fused layers of visceral peritoneum that attach abdominopelvic organs to the cavity wall; also called *visceral ligament.*

MESODERM (Gr. *mesos,* middle) The embryonic germ layer between the *ectoderm* and *endoderm* (see), developing into mature epithelial, connective, and muscle tissue.

MESOVARIUM A mesentery that attaches the ovaries to the broad ligament.

METABOLISM [muh-TAB-uh-lihz-uhm; Gr. *metabole,* change] The overall physical and chemical processes involved in the chemical activities of the body; may either build up or break down substances (see *anabolism, catabolism*).

METACARPAL BONES (L. behind the wrist) The five miniature long bones constituting the palm of each hand; also called *metacarpus*.

METACARPUS See *metacarpal bones*.

METAPHASE The second stage of mitosis; centromeres double, one going to each chromatid, each of which is now a single-stranded chromosome.

METAPHYSIS [muh-TAHF-uh-siss; Gr. "to grow beyond"] The area where longitudinal growth continues after birth, between the *epiphyseal (growth) plate* (see) and the *diaphysis* (see).

METASTASIS [muh-TASS-tuh-siss; Gr. *meta*, involving change + *stasis*, state of standing] The spread of malignant neoplasm cells from the primary site to other parts of the body.

METATARSAL BONES The five miniature long bones in each foot between the ankle, heel, and toes.

MICELLE [my-SELL; L. *mica*, grain] A submicroscopic water-soluble particle composed of fatty acids, phospholipids, and glycerides.

MICROFILAMENT A solid, rodlike structure containing the protein *actin*; provides cellular support and aids movement.

MICROGLIA The smallest glial cell (see *neuroglia*); a macrophage that removes disintegrating products of neurons.

MICROTUBULE A slender subcellular structure that helps support the cell; involved with organelle movement, cellular shape changes, and intracellular transport.

MICROVILLI [my-krow-VILL-eye; Gr. *mikros*, small; L. *villus*, shaggy hair] Microscopic fingerlike projections protruding from plasma membranes of some cells.

MICTURITION (L. *micturire*, to want to urinate) The process of emptying the urinary bladder; also called *urination*.

MIDBRAIN The portion of the brainstem located between the pons and diencephalon; connects the pons and cerebellum with the cerebrum; also called *mesencephalon*.

MIDDLE EAR A small chamber between the tympanic membrane and inner ear; composed of the tympanic cavity, and contains auditory ossicles.

MIDGET See *pituitary dwarf*.

MIDSAGITTAL The plane that divides the left and right sides of the body lengthwise along the midline into externally symmetrical sections.

MINERAL A naturally occurring, inorganic element used for the regulation

of metabolism; *microminerals (trace elements)* are required in only minute amounts.

MINERALOCORTICOID A steroid hormone secreted by the adrenal cortex.

MINUTE RESPIRATORY VOLUME The total volume of air taken into the lungs, measured in L/min.

MISCARRIAGE See *spontaneous abortion*.

MITOCHONDRION (pl. **MITOCHONDRIA**; Gr. *mitos*, a thread) A double-membraned, saclike organelle in the cytoplasm of a cell; produces most of the energy (in the form of ATP) for cellular metabolism.

MITOSIS (Gr. *mitos*, a thread) The process of nuclear division; it arranges cellular material for equal distribution to daughter cells and divides the nuclear DNA equally to each new cell.

MITRAL VALVE [MY-truhl] See *bicuspid valve*.

MIXED-TISSUE NEOPLASM A malignant neoplasm derived from tissue capable of differentiating into either epithelial or connective tissue.

MOBILE-RECEPTOR MECHANISM A hormonal mechanism in which lipid-soluble hormones (steroids) regulate cellular activity through the target cell's protein synthesis.

MOLECULE [MAHL-uh-kyool; L. *moles*, mass, bulk] The chemical combination of two or more atoms; the simplest unit that displays the physical and chemical properties of a compound.

MONOCLONAL ANTIBODY An antibody produced by laboratory-produced clones (identical copies) of B cells fused with cancer cells.

MONOCYTE A type of white blood cell that becomes a mobile phagocytic macrophage that ingests and destroys harmful particles.

MONOSACCHARIDE [mahn-oh-SACK-uh-ride; Gr. *monos*, single + *sakkharon*, sugar] A single-sugar carbohydrate that cannot be decomposed by uniting it with water.

MONOZYGOTIC TWINS Twins that result from the same fertilized ovum that divides into two identical cells before implantation; also called *identical twins*.

MONS PUBIS (L. mountain + pubic) A mound of fatty tissue covering the female symphysis pubis; also called *mons veneris*.

MONS VENERIS (L. mountain + Venus, the Roman goddess of love and beauty) See *mons pubis*.

MORPHOGENESIS (Gr. *morphe*, shape + *gennan*, to produce) The cellular development that brings about tissue and organ development.

MORULA [MORE-uh-luh; L. *morum*, mul-

berry tree] A solid ball of about 8 to 50 cells produced by cell divisions of a single fertilized ovum.

MOTOR END PLATE The point of contact (a synapse) between a muscle fiber and a motor neuron.

MOTOR NEURON See *efferent neuron*.

MOTOR UNIT A motor neuron and the muscle fibers it innervates.

MUCOUS MEMBRANE The membrane that lines body passageways that open to the outside of the body.

MUCUS [MYOO-kuhss] The thick, protective liquid secreted by glands in the mucous membranes.

MULTIPLE SCLEROSIS (MS) A progressive demyelination of neurons that interferes with the conduction of nerve impulses and results in impaired sensory perceptions and motor coordination.

MULTIPOLAR NEURON A nerve cell with many dendrites radiating from the cell body, but with only one axon.

MUSCLE (L. *musculus*, "little mouse") A collection of muscle fibers that can contract and relax to move body parts.

MUSCLE FIBER A collection of specialized, individual muscle cells that make up skeletal muscle tissue; muscle fibers have a long, cylindrical shape and several nuclei.

MUSCULAR DYSTROPHY The general name for a group of inherited diseases resulting in progressive weakness due to the degeneration of muscles.

MUTATION (L. *mutare*, to change) A change in genetic material that alters the characteristics of a cell, making the cell abnormal.

MYASTHENIA GRAVIS (Gr. *mus*, muscle + *astheneia*, without strength + L. *gravis*, weighty, serious) An autoimmune disease caused by antibodies directed against acetylcholine receptors, producing muscular weakness.

MYELIN [MY-ih-linn; Gr. *myelos*, marrow] A laminated lipid sheath covering an axon.

MYELIN SHEATH A thick pad of insulating myelin surrounding an axon; also called *medullary sheath*.

MYELINATED FIBER A nerve fiber covered with a myelin sheath.

MYOCARDIAL INFARCTION (L. *infercire*, to stuff) Commonly called a *heart attack*, it occurs when blood flow through a coronary artery is reduced and the *myocardium* (see) is deprived of oxygen leading to death of the heart tissue.

MYOCARDIUM ("heart muscle") The middle layer of muscle in the heart wall.

MYOFIBRIL Small units or fibers within individual threadlike muscle fibers; suspended in the sarcoplasm along with mitochondria and other multicellular material.

MYOFILAMENT A muscle filament composed of thick and thin threads that make up a *myofibril* (see).

MYOGLOBIN A form of hemoglobin found in muscle fibers.

MYOMETRIUM (muscle + Gr. *metra*, womb) The middle, muscular tissue layer of the uterus.

MYONEURAL JUNCTION See *neuromuscular junction*.

MYOPIA (Gr. "contracting the eyes") Nearsightedness, which occurs when light rays come to a focus before they reach the retina.

MYOSIN A fairly large protein that makes up the thick myofilaments of muscle fibers.

MYXEDEMA [mix-uh-DEE-muh] A condition caused by the underactivity of the thyroid during adulthood.

N

NAIL A modification of the epidermis, composed of hard keratin overlying the tips of fingers and toes.

NASAL CAVITY The cavity that fills the space between the base of the skull and the roof of the mouth.

NASAL SEPTUM A vertical wall dividing the nasal cavity into two bilateral halves.

NASOLACRIMAL DUCT (L. "nose" + "tear") A duct leading from each eye to the nasal cavity; it drains excessive secretions of tears into the nose.

NASOPHARYNX The part of the pharynx above the soft palate.

NECROBIOSIS (Gr. *nekros*, corpse, death + *biosis*, way of life) The natural degeneration and death of cells and tissues; see *necrosis*.

NECROSIS (Gr. *nekros*, corpse, death) The death of cells or tissues due to disease or injury.

NEGATIVE FEEDBACK SYSTEM A regulatory feedback system that produces a response that changes or reduces the initial stimulus.

NEONATE A newborn child during the first 4 weeks after birth.

NEOPLASM (Gr. *neos*, new + L. *plasma*, form) An abnormal growth of new tissue, which may be benign or malignant.

NEPHRECTOMY The surgical removal of a kidney.

NEPHRON (Gr. *nephros*, kidney) The functional unit of a kidney, each of the approximately 1 million in each kidney operating as an independent urine-making unit.

NEPHROPTOSIS A condition in which a kidney moves when it is no longer

held securely in place by the peritoneum; commonly called *floating kidney*.

NERVE A bundle of peripheral nerve fibers enclosed in a sheath.

NERVE IMPULSE See *action potential.*

NEURILEMMA The outer layer, or sheath, of a Schwann cell (see).

NEUROFIBRAL NODES Regular gaps in a myelin sheath around a nerve fiber; formerly called *nodes of Ranvier.*

NEUROFILAMENT A semirigid tubular structure in cell bodies and processes of neurons; also called *microfilament.*

NEUROGLIA (Gr. nerve + glue) Nonconducting cells of the central nervous system that protect, nurture, and support the nervous system; also called *glial cells.*

NEUROHYPOPHYSIS The posterior lobe of the pituitary.

NEUROLEMMOCYTE A type of peripheral nerve cell that forms the myelin sheath as rolled layers of the plasma membrane; formerly called *Schwann cell.*

NEUROMUSCULAR JUNCTION The site where a motor-nerve ending contacts a muscle fiber; also called *myoneural junction.*

NEURON (Gr. nerve) A cell specialized to transmit impulses; all neurons have properties of *excitability* and *conductivity.*

NEUROPEPTIDE A type of chemical neurotransmitter including somatostatin, endorphins, and encephalins.

NEUROTRANSMITTER A chemical substance that is synthesized by neurons, stored in secretory vesicles, and released into the *synaptic cleft* of a synapse; may produce an excitatory or inhibitory response in a receptor.

NEUROTUBULE A threadlike protein structure in cell bodies and processes of neurons; involved in intracellular transport of proteins and other substances.

NEUTROPHIL A type of phagocytic white blood cell.

NISSL BODIES See *chromatophilic substance.*

NOCICEPTOR [NO-see; L. "injury receiver"] A sensory receptor responding to potentially harmful stimuli that produce pain.

NODES OF RANVIER See *neurofibral nodes.*

NONESSENTIAL AMINO ACIDS Usable amino acids that can be synthesized by the body; see *essential amino acids.*

NONGONOCOCCAL URETHRITIS (NGU) A sexually transmitted disease characterized by inflammation of the urethra accompanied by a discharge of pus; also called *nonspecific urethritis (NSU).*

NONSPECIFIC URETHRITIS (NSU) See *nongonococcal urethritis.*

NORADRENERGIC Pertaining to neurons or fibers containing the neurotransmitter *norepinephrine* (see).

NOREPINEPHRINE [nor-ep-ih-NEFF-rihn] A hormone secreted by the adrenal medulla; it constricts arterioles and increases the metabolic rate; also called *NE, noradrenaline.*

NUCLEAR MAGNETIC RESONANCE (NMR) See *magnetic resonance imaging.*

NUCLEIC ACID [noo-KLAY-ihk] Any of two groups of complex compounds composed of bonded units called *nucleotides* (see); the carrier of hereditary material.

NUCLEOLUS [new-KLEE-oh-luhss; pl. **NUCLEOLI;** "little nucleus"] A somewhat spherical mass in the nucleus of a cell; contains genetic material in the form of DNA and RNA; a preassembly point for ribosomes.

NUCLEOPLASM The material within the nucleus of a cell.

NUCLEOTIDE [NOO-klay-uh-tide] A small structural unit of nucleic acids, composed of a phosphate group, a sugar, and a nitrogenous base.

NUCLEUS (L. nut, kernel) 1. The central portion of an atom, containing positively charged protons and uncharged neutrons. It is surrounded by negatively charged electrons. 2. The central portion of a cell, containing chromosomes. 3. A collection of nerve cells inside the central nervous system that processes afferent inputs.

NUTRIENT The chemical component of foods, supplying the energy and physical materials a body needs.

O

OBESITY (L. *obesus*, "grown fat by eating") The condition of being 20 to 25 percent over the recommended body weight.

OCCIPITAL LOBE The posterior cerebral lobe; composed of several areas organized for vision and its associated forms of expression.

OIL GLANDS See *sebaceous glands.*

OLFACTION (L. *olfacere*, to smell) The sense of smell.

OLFACTORY Pertaining to the sense of smell.

OLFACTORY BULB A stemlike extension of the olfactory region of the brain; receives impulses from nerve fibers that have been stimulated by an odiferous substance.

OLFACTORY TRACT Axons of mitral and tufted cells that carry impulses from the olfactory bulb posteriorly to the olfactory cortex in the brain.

OLIGODENDROCYTE A relatively small glial cell (see *neuroglia*) similar to a *neurolemmocyte* (see); produces and nurtures myelin sheath segments of many nerve fibers, provides a supportive framework, and supplies nutrition for neurons.

OLIGOSPERMIA (Gr. *oligos*, few + sperm) A condition in which only a small amount of sperm is produced.

ONCOGENE (Gr. *onkos*, tumor) A gene that apparently becomes reactivated and causes the growth of cancerous cells.

ONCOLOGY (Gr. *onkos*, tumor + *logy*, the study of) The study of neoplasms; the study of cancer.

OOGENESIS The monthly maturation of an ovum in the ovary.

OOPHORECTOMY The surgical removal of an ovary.

OPPOSITION The angular movement of the thumb pad touching and opposing a finger pad; occurs only at the carpometacarpal joint of the thumb; opposite of *reposition*.

OPTIC (Gr. *optikos*, visible) Pertaining to the eye, vision, and related topics.

OPTIC CHIASMA [kye-AZ-muh; after the X-shaped Greek letter *chi*, KYE] A point in the cranial cavity where half the fibers of each optic nerve of each eye cross over to the other side.

OPTIC TRACT Nerve fibers after they have passed through the *optic chiasma*.

ORAL CAVITY The mouth; also called *buccal cavity*.

ORAL CONTRACEPTIVE A hormonal pill taken daily by a woman that prevents follicle maturation and ovulation by altering hormones present; may also prevent implantation.

ORGAN An integrated collection of two or more kinds of tissue that combine to perform a specific function.

ORGAN OF CORTI See *spiral organ (of Corti)*.

ORGANELLE ("little organ") Various subcellular structures with specific structures and functions.

ORGANIC COMPOUND A chemical compound containing carbon and hydrogen, usually bonded covalently. See *inorganic compound*.

ORGANISM The product of all the body systems specialized within themselves and coordinated with each other; the body.

ORGASM (Gr. *orgasmos*, to swell with excitement) The climax of sexual excitement.

ORIGIN The end of a muscle attached to the bone that does not move.

OROPHARYNX The part of the pharynx at the back of the mouth, extending from the soft palate to the epiglottis, where the respiratory and alimentary tracts separate.

OSMOSIS (Gr. "pushing") The passive-transport process occurring when water (or another solvent) passes through a selectively permeable membrane from an area of high concentration to an area of lower concentration.

OSMOTIC PRESSURE The potential pressure developed by a solution separated from pure water by a differentially permeable membrane.

OSSEOUS TISSUE (L. *os*, bone) A tissue composed of cells embedded in a matrix of ground substance, inorganic salts, and collagenous fibers; constitutes the bony skeleton; also called *bone*.

OSSICLES The bones of the ear.

OSSIFICATION CENTER See *center of ossification*.

OSTEOBLAST (Gr. *osteon*, bone + *blastos*, bud, growth) A bone cell capable of synthesizing and secreting new bone matrix as needed; usually found on growing portions of bones.

OSTEOCLAST (Gr. *osteon*, bone + *klastes*, breaker) A multinuclear bone-destroying cell; usually found where bone is resorbed during normal growth.

OSTEOCYTE (Gr. *osteon*, bone + cell) A main cell of mature bone tissue; regulates the concentration of calcium in body fluids by helping to release calcium from bone tissue into the blood.

OSTEOGENIC CELL (Gr. *osteon*, bone + *genes*, born) A bone cell capable of being transformed into an *osteoblast* or *osteoclast*.

OSTEOMALACIA [ahss-teh-oh-muh-LAY-shee-uh; Gr. *osteon*, bone + *malakia*, soft] A skeletal defect caused by a deficiency of vitamin D.

OSTEOMYELITIS (Gr. *osteon*, bone + *myelos*, marrow) An inflammation of bone, and/or bone-marrow infection, frequently caused by bacteria.

OSTEON (Gr. bone) Concentric cylinders of calcified bone that make up compact bone; also called *Haversian system*.

OSTEOPOROSIS (Gr. *osteon*, bone + *poros*, passage) A bone disorder occurring most often in the elderly; the bones grow porous and crumble under ordinary stress.

OTIC (OH-tick; Gr. *otikos*, ear) Pertaining to the ear.

OTITIS MEDIA Inflammation of the middle ear.

OTOLITHS ("ear stones") Calcium carbonate crystals piled on top of the otolithic membrane; assist in maintaining *static equilibrium*.

OVARY (L. *ovum*, egg) A paired female gonad that produces ova and female hormones.

OVIDUCT See *uterine tube*.

OVULATION The monthly process in which a mature ovum is ejected from the ovary into the serous fluid of the peritoneal cavity near the uterine tube.

OXIDATION A chemical reaction in which atoms or molecules lose electrons or hydrogen ions.

OXIDATION-REDUCTION REACTION The simultaneous loss of electrons or hydrogen atoms *(oxidation)* by one substance and the gain of electrons or hydrogen atoms *(reduction)* by another substance during a chemical reaction.

OXIDATIVE PHOSPHORYLATION Chemical reactions in the inner compartments of mitochondria, using oxygen to produce ATP.

OXYHEMOGLOBIN The compound formed when oxygen is bound to hemoglobin.

OXYTOCIN A hormone secreted by the pituitary that stimulates uterine contractions during labor and milk ejection from the mammary glands after childbirth.

P

P WAVE The first activity in an *electrocardiogram*, caused by the depolarization and contraction of both atria.

P-R SEGMENT The recording in an *electrocardiogram* (see) of the passage of the electrical *P wave* (see) between the atria and ventricles.

PACINIAN CORPUSCLE [pah-SIHN-ee-an] See *lamellated corpuscle (of Pacini)*.

PAGET'S DISEASE A progressive bone disease in which a pattern of excessive bone destruction followed by bone formation contributes to bone thickening; also called *osteitis deformans*.

PALATE [PAL-iht] The roof of the mouth, divided into the anterior hard palate and the posterior soft palate.

PALMAR A relative directional term of the body: surface of the palm of the hand; also called *volar*. See *dorsal 3*.

PALPITATIONS Skipping, pounding, or racing heartbeats.

PANCREAS (Gr. "all flesh") An endocrine organ located posterior to the stomach; it secretes the hormones *insulin* and *glucagon*; it also functions as an exocrine gland.

PANCREATIC ISLETS Clusters of endocrine cells in the pancreas that produce insulin and glucagon; also called *islets of Langerhans*.

PAPANICOLAOU STAIN SLIDE TEST A diagnostic procedure that tests cells scraped from the female genital epithelium to detect malignant and premalignant conditions; also called *Pap test, Pap smear.*

PAPILLAE [puh-PILL-ee, sing. **PAPILLA;** L. nipple; dim. of *papula,* pimple] Projections on the tongue surface, palate, throat, and posterior surface of the epiglottis, which contain *taste buds.*

PARANASAL SINUS An air cavity within a bone, in direct communication with the nasal cavity.

PARAPLEGIA The motor or sensory loss of function in both lower extremities, resulting from damage to the spinal cord.

PARASYMPATHETIC NERVOUS SYSTEM The portion of the *autonomic nervous system* (see) that directs activities associated with the conservation and restoration of body resources.

PARATHORMONE (PTH) A parathyroid hormone that increases blood levels of calcium and decreases blood levels of phosphate; also called *parathyroid hormone.*

PARATHYROID GLANDS Small endocrine glands embedded in the posterior thyroid; their secretions affect blood levels of calcium and phosphate.

PARATHYROID HORMONE See *parathormone.*

PARIETAL [puh-RYE-uh-tuhl; L. *paries,* wall of a room] Pertaining to the outer wall of a body cavity.

PARIETAL LOBE The cerebral lobe that lies between the frontal and occipital lobes; concerned with the evaluation of the general senses (including an awareness of our body and its relation to the world around us), and of taste.

PARIETAL PLEURA See *pleural cavity.*

PARKINSON'S DISEASE A motor disability characterized by tremors and stiff posture; it results from a deficiency of dopamine; also called *Parkinsonism* and *shaking palsy.*

PAROTID GLANDS (Gr. *parotis,* "near the ear") The largest of the three main pairs of salivary glands; located below the ears.

PARTURITION (L. *parturire,* to be in labor) Childbirth, usually occurring about 266 days after fertilization.

PASSIVE TRANSPORT The movement of molecules across cell membranes from areas of high concentration to areas of lower concentration, without the use of cellular energy.

PATELLA REFLEX (KNEE JERK) A diagnostic reflex in which the tapping of the patellar tendon produces the contraction of the quadriceps femoris muscle, causing the lower leg to jerk upward.

PATENT DUCTUS ARTERIOSUS A condition that occurs when the *ductus arteriosus* (see) of the fetus, which carries blood from the pulmonary artery to

the descending aorta, does not shut down at birth.

PATHOLOGICAL ANATOMY (Gr. *pathos,* suffering + study) The study of changes in diseased cells and tissues; also called *pathology.*

PECTORAL GIRDLE The upper limb girdle, consisting of the clavicle and scapula; also called *shoulder girdle.*

PEDICEL (L. little foot) A footlike process in contact with glomerular capillaries.

PELVIC CAVITY The inferior portion of the abdominopelvic cavity. It contains the urinary bladder, rectum, anus, and internal reproductive organs.

PELVIC GIRDLE The paired hip bones (ossa coxae), formed by the fusion of the *ilium, ischium,* and *pubis;* also called *lower limb girdle.*

PELVIC INFLAMMATORY DISEASE (PID) A general term referring to inflammation of the uterus and uterine tubes, with or without inflammation of the ovary, localized pelvic peritonitis, and abscess formation.

PELVIS (L. basin) The bowl-shaped bony structure formed by the sacrum and coccyx posteriorly, and the two hip bones anteriorly and laterally.

PENIS [PEE-nihss; L. tail] The male copulatory organ, which transports semen through the urethra during ejaculation, and also carries urine through the urethra to the outside during urination.

PENNATE MUSCLE (L. *penna,* feather) A muscle with many short fascicles set at an angle to a long tendon.

PEPTIC ULCERS (Gr. *peptein,* to digest) Lesions in the mucosa of the digestive tract, caused by the digestive action of gastric juice, especially hydrochloric acid.

PEPTIDE BOND The strong covalent bond formed as a result of the union of two or more amino acids.

PERFORATING CANAL A branch running at a right angle to the *central canal* (see), extending the system of nerves and vessels outward to the *periosteum,* and inward to the *endosteum* (see) of the bony marrow cavity; also called *nutrient canal, Volkmann's canal.*

PERICARDIAL (Gr. *peri,* around + heart) Pertaining to the membranes enclosing the heart and lining the pericardial cavity.

PERICARDIUM ("around the heart") A protective sac around the heart; also called *pericardial sac.*

PERICHONDRIUM (Gr. *peri,* around + *khondros,* cartilage) A fibrous covering enclosing hyaline and elastic cartilage.

PERILYMPH The thin, watery fluid in the bony labyrinth of the inner ear.

PERIMYSIUM (Gr. *peri,* around + mus-

cle) A connective tissue layer extending inward from the *epimysium* (see); it encloses bundles of muscle fibers.

PERINEURIUM ("around the nerve") A thick connective-tissue sheath surrounding a primary bundle of nerve fibers; found where no *epineurium* (see) is present.

PERIOSTEUM [pehr-ee-AHSS-tee-uhm; Gr. *peri,* around + *osteon,* bone] A fibrous membrane covering the outer surfaces of bones (except joints); contains bone-forming cells, nerves, and vessels.

PERIPHERAL A relative directional term used to describe structures other than internal organs that are located or directed away from the central axis of the body.

PERIPHERAL AUTONOMIC NERVOUS SYSTEM A motor system of the nervous system, consisting of the *sympathetic* and *parasympathetic divisions* (see); each division sends efferent nerve fibers to the muscle, gland, or organ it innervates.

PERIPHERAL NERVOUS SYSTEM (PNS) The cranial nerves associated with the brain, and the spinal nerves associated with the spinal cord, may be subdivided on a functional basis into the *somatic* and *visceral nervous systems* (see).

PERIPHERAL RESISTANCE The impediment, by friction, to blood flow through a blood vessel.

PERISTALSIS The involuntary, sequential muscular contractions that move food along the digestive tract.

PERITONEUM [per-uh-tuh-NEE-uhm] The serous membrane that lines the abdominal cavity and covers most of the abdominal organs.

PERITONITIS An inflammation of the peritoneum.

PEROXISOME A membrane-bound organelle containing oxidative enzymes that carry out metabolic reactions and destroy toxic hydrogen peroxide.

PET SCAN See *positron-emission tomography.*

PHAGOCYTOSIS (Gr. *phagein,* to eat + *cyto,* cell) The active movement in which large molecules and particles are taken into the cytoplasm of a cell through the plasma membrane; a form of *endocytosis* (see).

PHALANGES [fuh-LAN-jeez; sing. **PHALANX;** FAY-langks, Gr. line of soldiers] The 14 finger bones in each hand; also the 14 toe bones in each foot.

PHARYNX A tube leading from the internal nares (nostrils) and mouth to the larynx and esophagus; serves as an air passage during breathing, and a food passage during swallowing; commonly called the *throat*.

PHENOTYPE [FEE-noh-tipe; Gr. "that which shows"] The expression of a genetic trait (how a person looks or functions).

PHENYLKETONURIA (PKU) A congenital metabolic disease caused by a mutation that prevents the adequate formation of phenylalanine hydroxylase, which normally converts the amino acid phenylalanine into tyrosine.

PHOTORECEPTOR (Gr. "light receiver") A light-sensitive sensory receptor in the retina.

PHYSIOLOGY The study of how the body and its parts function.

PIA MATER (L. tender-mother) The thin, highly vascular innermost layer of the meninges.

PINEAL GLAND [PIHN-ee-uhl; L. *pinea*, pine cone] A small gland in the midbrain that converts a signal received through the nervous system into an endocrine signal; it produces *melatonin*, but its exact function is uncertain; also called *pineal body, epiphysis cerebri*.

PINOCYTOSIS (Gr. *pinein*, to drink + *cyto*, cell) The nonspecific uptake by a cell of small droplets of extracellular fluid; a form of *endocytosis* (see).

PITUITARY DWARF A person whose skeletal development is halted at about the stage of a 6-year-old child because of an undersecretion of growth hormone; also called *midget*.

PITUITARY GLAND An endocrine gland consisting of two lobes, the anterior *adenohypophysis*, containing many secretory cells, and the posterior *neurohypophysis*, containing many nerve endings; also called *hypophysis*.

PIVOT JOINT A uniaxial joint that rotates only around a central axis.

PLACENTA A thick bed of tissues and blood vessels embedded in the wall of a pregnant woman's uterus; it contains arteries and veins of mother and fetus and provides an indirect connection between mother and fetus.

PLANTAR A relative directional term of the body: surface of the sole of the foot; see *dorsal (3)*.

PLANTAR FLEXION The downward bending of the foot at the ankle.

PLASMA The clear, yellowish liquid part of blood.

PLASMA CELL See *lymphocyte*.

PLASMA MEMBRANE The bilayered outermost boundary of a cell.

PLASMA PROTEIN One of the proteins found dissolved in blood plasma; include *albumins, fibrinogen,* and *globulins*.

PLATELET A type of blood cell that is

important in blood clotting; also called *thrombocyte*.

PLATELET AGGREGATION The process in which platelets aggregate, plugging a small hole in a damaged blood vessel; also called *platelet plug*.

PLATELET PLUG See *platelet aggregation*.

PLEURA [PLOOR-uh; Gr. side, rib] The serous membrane lining the *pleural cavity* and covering the lungs; also lines the thoracic wall.

PLEURAL CAVITY The moisture-filled potential space between the *visceral pleura* on the lung surface and the *parietal pleura* lining the chest cavity.

PLEURISY An infection of the pleurae covering the lungs and lining the chest cavity.

PLEXUS (L. braid) A complex network of interlaced nerves.

PLICAE CIRCULARES [PLY-see sir-cue-LAR-eez] Circular folds in the small intestine that increase the area available for absorption.

PNEUMONIA A general term for any condition that results in filling the alveoli with fluid.

PNEUMOTAXIC AREA A breathing-control area in the pons which, when stimulated, reduces inhalations and increases exhalations; see *respiratory center*.

PODOCYTE ("footlike cell") A specialized epithelial cell surrounding glomerular capillaries.

POLAR BODY The smaller of two cells resulting from the uneven distribution of cytoplasm during oogenesis.

POLAR MOLECULE A molecule whose two ends have different electrical charges; water is a polar molecule.

POLARITY (L. *polus*, pole) The tendency of a molecule to have different electrical charges at its opposite ends.

POLARIZATION A condition of the plasma membrane of a neuron, when the intracellular fluid is negatively charged relative to the positively charged extracellular fluid.

POLIOMYELITIS A contagious viral infection affecting both the brain and spinal cord, sometimes causing the destruction of neurons; damage to motor neurons in the spinal cord causes paralysis.

POLYPEPTIDE A protein formed by the chemical bonding of many amino acids.

POLYSACCHARIDE [pah-lee-SACK-uh-ride; Gr. *poly*, many + *sakkharon*, sugar] A carbohydrate made up of more than two simple sugars.

POLYURIA An excessive production of urine; frequent urination.

PORTA HEPATIS ("liver door") The area through which the blood vessels,

nerves, lymphatics, and ducts enter and leave the liver.

POSITIVE FEEDBACK SYSTEM A reaction that reinforces a stimulus rather than changing it. It can disrupt homeostasis, so it is rare.

POSITRON-EMISSION TOMOGRAPHY (PET) A scanning procedure that produces pictures that reveal the metabolic state of the organ being viewed; the device is called a *PET scanner*.

POSTABSORPTIVE STATE The metabolic state during which the body's energy requirements must be satisfied by the nutrients taken in during the *absorptive state* and from stored nutrients; also called *fasting state*.

POSTERIOR (L. *post*, behind, after) See *dorsal 1 and 2*; opposite of *anterior*.

POSTGANGLIONIC NEURON The second neuron in a two-neuron sequence in the autonomic nervous system, with its cell body in an autonomic ganglion, and an unmyelinated axon terminating in a motor ending associated with smooth or cardiac muscle, or glands.

POSTSYNAPTIC NEURON ("after the synapse") A nerve cell that carries impulses away from a synapse.

POSTSYNAPTIC POTENTIAL (PSP) An electrical potential on the postsynaptic (receptor) membrane of a neuron, muscle fiber, or gland cell.

POTENTIAL DIFFERENCE The difference in the electrical charge on either side of the plasma membrane of a neuron at any given point along the membrane.

PREGANGLIONIC NEURON ("before the ganglion") The first neuron in a two-neuron sequence in the autonomic nervous system, with its cell body in the brainstem or spinal cord, and a myelinated axon that terminates in an autonomic ganglion located outside the central nervous system.

PREGNANCY A series of events initiated by the fertilization of an ovum and its implantation in the uterine wall, continuing through fetal intrauterine development and ending with childbirth.

PREMENSTRUAL SYNDROME (PMS) A condition characterized by nervousness, irritability, and abdominal bloating 1 or 2 weeks before menstruation.

PREPUCE [PREE-pyoos] The loose-fitting skin folded over the glans of the clitoris and an uncircumcized penis; also called *foreskin*.

PRESBYOPIA (Gr. *presbus*, old man + eye) A condition in which the lens of the eye becomes less elastic with age, making it difficult to focus on near objects.

PRESSORECEPTORS See *baroreceptors*.

PRESYNAPTIC NEURON ("before the synapse'') A nerve cell carrying impulses toward a synapse; it initiates a response in the receptive segment of a *postsynaptic neuron* (see).

PREVERTEBRAL GANGLIA Autonomic ganglia lying in front of the vertebrae; also called *collateral ganglia*.

PRIMARY CENTER OF OSSIFICATION The site near the middle of what will become the diaphysis, where bone-cell development occurs by the second or third prenatal month.

PRIMARY GERM LAYERS Three layers of embryonic tissue called endoderm, mesoderm, and ectoderm, which form the organs and tissues of the body.

PRIMARY MOTOR CORTEX A motor area in the frontal lobe of the cerebrum controlling specific voluntary muscles or muscle groups; also called *Brodmann area 4*.

PRIMARY SOMESTHETIC AREA The portion of the parietal lobe of the cerebrum that receives information about the general senses from receptors in the skin, joints, muscles, and body organs; also called *general sensory area*.

PRIMARY VISUAL CORTEX (area 17) An area in the occipital lobe of the cerebrum that receives visual images from the retina; conveys visual information to cerebral areas 18 and 19 for further processing and evaluation.

PRIME MOVER See *agonist*.

PRIMORDIAL FOLLICLE An ovarian follicle that is not yet growing.

PROGESTERONE A female hormone secreted by the corpus luteum; it stimulates thickening of the uterine wall and the formation of mammary ducts.

PROGESTINS A class of female sex hormones that regulate the menstrual cycle and the development of the mammary glands, and aid in the formation of the *placenta* (see) during pregnancy.

PROLACTIN A hormone secreted by the adenohypophysis; stimulates the duct system of the mammary glands during pregnancy, and milk production after childbirth; also called lactogenic hormone and *luteotropic hormone (LTH)*.

PROLAPSE A condition in which a body organ, especially the uterus, falls or slips out of place.

PRONATION A pivoting movement of the forearm that turns the palm downward or backward, crossing the radius diagonally over the ulna; opposite of *supination* (see).

PROPERDIN [pro-PER-dihn; L. *pro*, acting as + *perdere*, to give away, hence ''to destroy''] A protein in blood serum; apparently, it functions to enhance the activation of *complement* (see).

PROPHASE The first stage of mitosis;

centriole pairs move to opposite poles of the nucleoplasm, microtubules form a spindle from pole to pole, and chromatid pairs move to the center of the spindle.

PROPRIOCEPTOR (L. ''received from one's self'') A sensory receptor responding to stimuli within the body, such as those from muscles and joints.

PROSTAGLANDINS Hormonelike substances made from fatty acids in plasma membranes; actions include increasing or decreasing effect of cyclic AMP on target cells, raising or lowering blood pressure, and regulating digestive secretions.

PROSTATE GLAND (Gr. *prostates*, standing in front of) A male secretory gland whose secretions pass into the semen to make sperm motile and help neutralize vaginal acidity.

PROSTATECTOMY The surgical removal of all or part of the prostate gland.

PROTEIN (Gr. *protos*, first) Any of a group of complex organic compounds that always contain carbon, hydrogen, oxygen, and nitrogen; their basic structural units are *amino acids* (see).

PROTEIN-CALORIE MALNUTRITION A condition that occurs when the body has been deprived of sufficient amino acids and calories for a long period; *marasmus* and *kwashiorkor* are two forms.

PROTHROMBIN An inactive plasma protein that is converted into the active enzyme thrombin during the blood-clotting process.

PROTRACTION A forward pushing movement; opposite of *retraction* (see).

PROXIMAL (L. *proximus*, nearest) A relative directional term: nearer the trunk of the body (toward the attached end of a limb); used with extremities; opposite of *distal* (see).

PROXIMAL CONVOLUTED TUBULE The portion of the coiled excretory tubule proximal to the glomerular capsule; it receives *glomerular filtrate* (see) from the glomerular capsule.

PSORIASIS [suh-RYE-uh-siss; Gr. *psorian*, to have the itch] A skin disease occurring when skin cells move from the basal layer to the stratum corneum in only 4 days instead of the usual 28, causing red, dry lesions covered with silvery, scaly patches.

PUBERTY (L. *puber*, adult) The developmental period when the person becomes physiologically capable of reproduction.

PULMONARY (L. *pulmo*, lung) Pertaining to the lungs.

PULMONARY ARTERIES Blood vessels carrying oxygen-poor blood from the heart to the lungs.

PULMONARY CIRCULATION The system

of blood vessels that carries deoxygenated blood from the heart to the lungs, where carbon dioxide is removed and oxygen is added; see *systemic circulation*.

PULMONARY TRUNK The major arterial trunk emerging from the right ventricle; it carries blood to the lungs.

PULMONARY VEINS Large veins that drain blood from the lungs into the left atrium.

PULP The soft core of connective tissue that contains the nerves and blood vessels of a tooth.

PULSE The alternating expansion and elastic recoil of the arterial wall that can be felt where an artery lies close to the skin.

PULSE PRESSURE The difference between *systolic* and *diastolic pressure* (see).

PUPIL (L. doll) The opening in the iris that opens and closes reflexively to adjust to the amount of available light.

PURKINJE FIBERS See *cardiac conducting myofibers*.

PYRAMID A bilateral elevated ridge in the ventral surface of the medulla; composed of fibers of motor tracts from the motor cerebral cortex to the spinal cord.

PYRAMIDAL DECUSSATION (L. *decussare*, from *dec*, ten; the Latin symbol for 10 is X, representing the crossing over of the pyramidal tracts) The crossing over of the *pyramidal tracts* (see) in the lower part of the medulla to the opposite side of the spinal cord, causing each side of the brain to control the opposite side of the body.

PYRAMIDAL SYSTEM Tracts from the cerebral motor cortex that terminate in the brainstem corticobulbar fibers and terminate in the spinal cord (pyramidal tracts).

PYRAMIDAL TRACTS Descending fibers of motor tracts from the motor cerebral cortex to the spinal cord; also called *corticospinal tracts*.

PYROGEN (Gr. ''fire-producing'') Any substance capable of producing a fever.

PYURIA The presence of fat droplets or pus in the urine.

Q

QRS COMPLEX A triple-wave activity recorded during an *electrocardiogram* as the ventricles are depolarized.

QUADRIPLEGIA Paralysis of all four extremities, as well as any part of the body below the level of injury to the spinal cord; usually results from injury at the C8 to T1 level.

R

RADIATION THERAPY Therapy that uses x rays or rays from radioactive substances to kill cancer cells.

RADIOGRAPHIC ANATOMY The study of the structures of the body using x rays.

RAMI COMMUNICANTES [RAY-mee; sing. **RAMUS COMMUNICANS** Myelinated or unmyelinated branches of a spinal nerve; composed of sensory and motor nerve fibers associated with the autonomic nervous system.

RAMUS [RAY-muhss; pl. **RAMI,** RAY-mye] A branch of a spinal nerve.

RECEPTOR (L. *recipere,* to receive) The peripheral end of the dendrites of afferent sensory neurons, specialized to receive stimuli and convert them into nerve impulses.

RECEPTOR-MEDIATED ENDOCYTOSIS An active process of movement across a plasma membrane that involves a specific receptor on the membrane that "recognizes" an extracellular macromolecule and binds to it; a form of *endocytosis* (see).

RECESSIVE A genetic characteristic that is temporarily hidden by the presence of a dominant characteristic.

RECTUM (L. *rectus,* straight) The 15-cm duct in the digestive tract between the sigmoid colon and the anus; it removes solid wastes by the process of defecation.

RED BLOOD CELLS See *erythrocytes.*

REDUCTION 1. The gaining of electrons or hydrogen atoms by an atom or molecule. 2. The act of restoring dislocated bones to their normal positions in a joint.

REFERRED PAIN A visceral pain felt subjectively in a somatic area away from the actual source of pain.

REFLEX (L. to bend back) A predictable involuntary response to a stimulus.

REFLEX ARC A sequence of events leading to a *reflex* (see).

REFRACTION The bending of light waves as they pass from one medium to another with a different density.

REFRACTORY PERIOD The brief period after the firing of a nerve impulse when the plasma membrane of a neuron cannot generate another impulse.

REGIONAL ANATOMY The anatomical study of specific regions of the body.

RELAXIN A female hormone secreted by the ovaries; it dilates the symphysis pubis and cervix during childbirth, and increases sperm motility.

RELEASING FACTORS Substances secreted by the hypothalamus that stimulate secretions from the adenohypophysis.

RENAL (L. *renes,* kidneys) Pertaining to the kidney.

RENAL CALCULI Kidney stones, usually formed from precipitated calcium salts.

RENAL COLUMN A column in the kidney formed by the tissue of the renal cortex that penetrates the depth of the *renal medulla* (see) between *renal pyramids* (see); composed mainly of *kidney tubules* that drain and empty urine.

RENAL CORPUSCLE The portion of a kidney consisting of the glomerulus and glomerular capsule.

RENAL CORTEX The outermost portion of the kidney, divided into the outer cortical region and the inner juxtamedullary region.

RENAL FASCIA The outer tissue layer of the kidney.

RENAL MEDULLA The middle portion of the kidney, consisting of 8 to 18 *renal pyramids* (see).

RENAL PELVIS The large collecting space within the kidney, formed from the expanded upper portion of the ureter; connects structures of the *renal medulla* (see) with the ureter.

RENAL PYRAMID A longitudinally striped, cone-shaped area within the *renal medulla* (see); consists of tubules and collecting ducts of the nephrons; involved with the reabsorption of filtered materials; 8 to 18 pyramids make up one renal medulla.

RENIN [REE-nihn] The renal enzyme, secreted by the juxtaglomerular apparatus, which alters the systemic blood pressure to maintain homeostasis.

REPOLARIZATION The restoration of a relatively positive charge outside the plasma membrane of a neuron.

RESIDUAL VOLUME The volume of air remaining in the lungs after a forceful expiration.

RESPIRATION The overall exchange of oxygen and carbon dioxide between the atmosphere, blood, lungs, and body cells.

RESPIRATORY CENTER The complete breathing-control center, including the inspiratory and expiratory circuits in the medulla and the apneustic and pneumotaxic areas in the pons.

RESTING MEMBRANE POTENTIAL The potential for electrical activity along the plasma membrane of a neuron that is in a state of *polarization* (see).

RETICULAR (L. *rete,* net) Resembling a network.

RETICULAR ACTIVATING SYSTEM (RAS) A network of branched nerve cells in the brainstem; involved with the adjustment of many behavioral activities, including the sleep-wake cycle, awareness, levels of sensory perception, emotions, and motivation; also called the *arousal system.*

RETICULAR CELL A flat, star-shaped cell that forms the cellular framework of bone marrow, lymph nodes, the spleen, and other lymphoid tissues involved in the immune response.

RETICULAR FIBERS Delicately branched networks that make up some connective tissues; similar to collagenous fibers, but not as elastic.

RETICULAR FORMATION A network of nerve cells and fibers throughout the brainstem; consists of ascending and descending pathways and cranial nerves; regulates respiratory and cardiovascular centers as well as the brain's awareness level.

RETINA [REH-tin-uh; L. *rete,* net] The innermost layer of the eye, containing a thick layer of specialized photoreceptor cells and other nervous tissue called the *neuroretina,* and a thin layer of pigmented epithelium that prevents reflection.

RETINAL A photosensitive derivative of vitamin A found in rods in the retina; 11-cis retinal initiates physiology of vision when it absorbs a photon of light and changes molecular configuration to all-*trans* retinal.

RETRACTION A backward movement, such as a backward pull of the shoulders; opposite of *protraction* (see).

RETROPERITONEAL (L. *retro,* behind + peritoneum) Located behind the abdominopelvic cavity and the peritoneum.

Rh FACTOR A blood factor characterized by an inherited *agglutinogen* (see) on the surface of erythrocytes.

RHEUMATISM (Gr. *rheumatismos,* to suffer from a flux or stream) Any of several diseases of muscles, tendons, joints, bones, or nerves.

RHODOPSIN A reddish, light-sensitive pigment in rod cells within the retina; contains the light-absorbing organic molecule 11-*cis* retinal and the protein *scotopsin;* also called *visual purple.*

RIBONUCLEIC ACID (RNA) [rye-boh-noo-KLAY-ihk] A single-stranded nucleic acid containing the sugar ribose; transcribed from DNA; found in both the nucleus and the cytoplasm of a cell.

RIBOSOME A subcellular structure containing RNA and protein; the site of protein synthesis.

RICKETS (variant of Gr. *rhakhitis,* disease of the spine) A childhood disease caused by a deficiency of vitamin D; progresses to skeletal deformity.

RIGHT LYMPHATIC DUCT A large duct that drains lymph from the right side of the head; right upper extremity, thorax, and lung; right side of the heart; and the upper portion of the liver into the right subclavian vein.

RNA See *ribonucleic acid.*

ROD A specialized photoreceptor cell in the retina; not sensitive to color, but very sensitive to light.

ROTATION A pivoting movement that twists a body part, arm, or leg on its long axis.

ROUND LIGAMENTS Paired bands of fibrous connective tissue just below the entrance of the uterine tubes into the uterus; help to keep the uterus tilted forward over the urinary bladder.

RUFFINI CORPUSCLES See *corpuscles of Ruffini.*

RUGAE [ROO-jee; L. folds] Folds or creases of tissue, as in the stomach and vagina.

S

SACRAL PLEXUS The ventral rami of L4, L5, and S1 to S3 nerves in the posterior pelvic wall.

SADDLE JOINT A multiaxial joint in which opposing articular surfaces of both bones are shaped like a saddle.

SAGITTAL (L. *sagitta,* arrow) An off-center longitudinal plane dividing the body into asymmetrical left and right sections.

SALIVA The secretion of salivary glands, composed of about 99 percent water and 1 percent electrolytes and proteins (including *mucin*), and the enzyme, salivary amylase, or *ptyalin.*

SALIVARY GLANDS The three largest pairs of glands that secrete saliva into the oral cavity: *parotid, submandibular,* and *sublingual glands.*

SALT The compound (other than water) formed during a neutralization reaction between acids and bases.

SALTATORY CONDUCTION (L. *saltare,* to jump) Conduction along a myelinated nerve fiber, where the *action potential* (see) appears to jump from one neurofibral node to the next.

SARCOLEMMA (Gr. *sarkos,* flesh + *lemma,* husk) A thin membrane enclosing each skeletal-muscle fiber.

SARCOMA (Gr. *sarkoun,* to make fleshy) A malignant neoplasm that originates in connective tissue and spreads through the bloodstream.

SARCOMERE (Gr. *sarkos,* flesh + *meros,* part) The fundamental unit of muscle contraction, composed of a section of muscle fiber extending from one Z line to the next.

SARCOPLASM (Gr. *sarkos,* flesh) A specialized form of cytoplasm found in skeletal-muscle fibers.

SARCOPLASMIC RETICULUM A specialized type of endoplasmic reticulum containing a network of tubes and sacs containing calcium ions.

SCHWANN CELL See *neurolemmocyte.*

SCIATICA [sye-AT-ih-kuh] Nerve inflammation characterized by sharp pains along the sciatic nerve and its branches.

SCLERA (Gr. *skleros,* hard) The posterior segment of the outer supporting layer of the eyeball; the "white" of the eye.

SCOLIOSIS (Gr. "crookedness") An abnormal lateral curvature of the spine in the thoracic, lumbar, or thoracolumbar region.

SCOTOPSIN The protein component of the photopigment *rhodopsin* (see); coupled with the light-sensitive organic molecule 11-*cis* retinal in rhodopsin.

SCROTUM [SCROH-tuhm; L. *scrautum,* a leather pouch for arrows] An external sac of skin that hangs between the male thighs; it contains the testes.

SEBACEOUS GLANDS [sih-BAY-shuhss; L. *sebum,* tallow, fat] Simple, branched alveolar glands in the dermis that secrete *sebum* (see); their main functions are lubrication and protection; also called *oil glands.*

SEBUM (L. tallow, fat) The oily secretion found at the base of a hair follicle.

SECOND-MESSENGER CONCEPT See *fixed-membrane-receptor mechanism.*

SECONDARY SEX CHARACTERISTICS Sexually distinct characteristics such as body hair and enlarged genitals that develop during puberty.

SECRETIN A polypeptide hormone secreted by the duodenal mucosa; it stimulates the release of pancreatic juice to neutralize stomach acid.

SELECTIVE PERMEABILITY (L. *permeare,* to pass through) The quality of cellular membranes that allows some substances into the cell while keeping others out.

SELLA TURCICA [SEH-luh TUR-sihk-uh; L. *sella,* saddle + Turkish] A deep depression within the body of the sphenoid bone, which houses and protects the pituitary gland.

SEMEN [SEE-muhn; L. seed] Male ejaculatory fluid containing sperm and secretions from the epididymis, seminal vesicles, prostate gland, and bulbourethral glands; also called *seminal fluid.*

SEMILUNAR VALVES Heart valves that prevent blood in the pulmonary artery and aorta from flowing back into the ventricles.

SEMINAL FLUID See *semen.*

SEMINAL VESICLES Paired secretory sacs whose secretions provide an energy source for motile sperm, and help to neutralize the acidity of the vagina.

SEMINIFEROUS TUBULES Tightly coiled tubules in the testes that produce sperm.

SENESCENCE (L. *senescere,* to grow old) The indeterminate period when an individual is said to grow old; also called *primary aging, biological aging.*

SENILE DEMENTIA (L. *senex,* old + madness) A progressive, abnormally accelerated deterioration of mental faculties in old age; commonly called *senility.*

SENILITY See *senile dementia.*

SENSORY NEURON See *afferent neuron.*

SEPTUM (L. *sepire,* to separate with a hedge) A wall between two cavities, such as in the heart or nose.

SEROUS (L. *serosus,* serum) Pertaining to the secretion of a serumlike fluid.

SEROUS MEMBRANE A double layer of loose connective tissue covered by a layer of simple squamous epithelium that lines some of the walls of the closed thoracic and abdominopelvic cavities, and covers organs lying within these cavities; includes *peritoneum, pericardium,* and *pleura.*

SERTOLI CELL See *sustentacular cell.*

SEX CHROMOSOMES A pair of chromosomes that determine the sex of the new individual; an XX pair produces a female, and an XY pair a male.

SEXUALLY TRANSMITTED DISEASE (STD) A disease that is caused by the transferral of infectious microorganisms from person to person, mainly by sexual contact; also called *venereal disease.*

SHINGLES Acute inflammation of the dorsal root ganglia along one side of the body; also called *herpes zoster.*

SIAMESE TWINS See *conjoined twins.*

SICKLE-CELL ANEMIA A hereditary form of anemia characterized by crescent-shaped red blood cells that do not carry or release sufficient oxygen.

SIGMOID COLON (Gr. *sigma,* the letter "S") The S-shaped portion of the large intestine immediately following the descending colon; it travels transversely across the pelvis to the right to the middle of the sacrum, where it continues to the rectum.

SIMPLE DIFFUSION (L. *diffundere,* to spread) A passive-transport process in which molecules move randomly from areas of high concentration to areas of lower concentration until they are evenly distributed.

SINOATRIAL (SA) NODE A mass of specialized heart muscle where the electrical stimulation starts and controls the heartbeat.

SKELETAL MUSCLE Muscle tissue that can be contracted voluntarily; it is attached to the skeleton, making the skeleton move; also known as *striated* and *voluntary muscle.*

SKENE'S GLANDS See *lesser vestibular glands.*

SKULL Bones of the head, with or without the mandible.

SLIDING-FILAMENT THEORY The theory that proposes that as muscle fibers are stimulated by nerve endings, the actin and myosin myofilaments slide past each other, with cross bridges pulling the muscle into a contracted state; the cross-bridge connections are broken when the muscle relaxes.

SMALL INTESTINE The 6-m-long portion of the digestive tract between the stomach and the large intestine; the site of most chemical digestion and absorption.

SMOOTH MUSCLE TISSUE Nonstriated muscle tissue, controlled by the autonomic nervous system; forms sheets in the walls of large, hollow organs; also called *involuntary muscle.*

SODIUM-POTASSIUM PUMP Part of a self-regulating transport system within the plasma membrane of a neuron; it helps regulate the concentration of sodium and potassium ions inside and outside the membrane.

SOLUTE (L. *solvere,* to loosen) A substance capable of being dissolved in another substance; see *solvent.*

SOLUTION A homogeneous mixture of a solvent and the dissolved solute.

SOLVENT A liquid or gas capable of dissolving another substance; see *solute.*

SOMATIC NERVOUS SYSTEM The portion of the *peripheral nervous system* (see) composed of a motor division that excites skeletal muscles, and a sensory division that receives and processes sensory input from the sense organs.

SOMATOTROPIC HORMONE (STH) See *growth hormone.*

SOMATOTROPIN See *growth hormone.*

SONOGRAPHY See *ultrasound.*

SPECIAL SENSES The senses of sight, hearing, equilibrium, smell, and taste.

SPERM (Gr. *sperma,* seed) Mature male sex cells; also called *spermatozoa* (Gr. *zoon,* animal).

SPERMATIC CORD A cord covering the ductus deferens as it passes from the tail of the epididymis; contains the testicular artery, veins, autonomic nerves, cremaster muscle, lymphatics, and connective tissue.

SPERMATOGENESIS ("the birth of seeds") The continuous process that forms haploid sperm cells in the testes.

SPERMICIDE A chemical agent that kills sperm.

SPHINCTER (Gr. "that which binds tight") A circular muscle that helps keep an opening or tubular structure closed.

SPHYGMOMANOMETER [sfig-moh-muh-NOM-ih-ter] The instrument used for measuring blood pressure.

SPINA BIFIDA [SPY-nuh BIFF-uh-duh; L. *bifidus,* split into two parts] A congeni-

tal disease in which the two sides of the neural arch of one or more vertebrae do not fuse during embryonic development; commonly called *cleft spine.*

SPINAL CORD The part of the central nervous system extending caudally from the foramen magnum; has 31 pairs of nerves, and is the connecting link between the brain and most of the body.

SPINOUS PROCESS A sharp, elongated process of a bone, such as the spine of a vertebra.

SPIRAL ORGAN (OF CORTI) The organ of hearing; formerly called *organ of Corti.*

SPLEEN The largest lymphoid organ, located below the diaphragm on the left side; it filters blood and produces phagocytic lymphocytes and monocytes.

SPONTANEOUS ABORTION The loss of a fetus through natural causes; also called *miscarriage.*

SPRAIN A tear of ligaments following the sudden wrenching of a joint.

STATIC EQUILIBRIUM See *vestibular apparatus.*

STEREOGNOSIS [STEHR-ee-oh-NO-siss; Gr. *stereos,* solid, three-dimensional + *gnosis,* knowledge] The ability to identify unseen objects by handling them.

STERILITY A condition in which a mature person is unable to conceive or produce offspring.

STIMULUS (L. a goad) A change in the external environment, or within the body itself, that is sensed by receptors and conveyed via nerves to the brain and spinal cord, where the input is integrated,

STOMACH The distensible sac that churns ingested nutrients into small particles, stores them until they can be released into the small intestine, and secretes hydrochloric acid and enzymes that initiate the digestion of proteins.

STRATUM BASALE (L. *basis,* base) The layer of the epidermis resting on the basement membrane next to the dermis.

STRESS (Mid. Eng. *stresse,* hardship) Any factor or factors that put pressure on the body to make an adaptive change in order to maintain homeostasis.

STROKE See *cerebrovascular accident.*

STROKE VOLUME The volume of blood ejected by either ventricle in one *systole* (see).

SUBARACHNOID SPACE The space between the arachnoid and pia mater layers of the meninges covering the brain and spinal cord; contains cerebrospinal fluid and blood vessels.

SUBCUTANEOUS LAYER (L. *sub,* under + *cutis,* skin) See *hypodermis.*

SUBDURAL SPACE The potential space between the dura mater and arachnoid meninges of the brain and spinal cord; contains no cerebrospinal fluid.

SUBLINGUAL GLANDS ("under the tongue") Paired salivary glands located in the floor of the mouth beneath the tongue.

SUBMANDIBULAR GLANDS ("under the mandible") Paired salivary glands located on medial side of the mandible.

SUDORIFEROUS GLANDS (L. *sudor,* sweat) Commonly called sweat glands, either *apocrine* (see) or *eccrine* types .

SULCUS [pl. **SULCI,** SUHL-kye; L. groove] A deep furrow on the surface of a bone or other surface, especially on the brain; also called *groove.*

SUMMATION OF TWITCHES The tension achieved when a muscle receives repeated stimuli at a rapid rate so that it cannot relax completely between contractions.

SUPERFICIAL A relative directional term of the body; nearer the surface of the body; opposite of *deep.*

SUPERIOR (L. *superus,* situated above) A relative directional term: toward the head; above; the head is superior to the neck; opposite of *inferior.*

SUPERIOR VENA CAVA The large vein that drains blood from the head, neck, upper limbs, and thorax, emptying it into the right atrium; see *inferior vena cava.*

SUPINATION A pivoting movement of the forearm that turns the palm forward or upward, making the radius parallel with the ulna; opposite of *pronation* (see).

SURFACTANT A phospholipid that reduces surface tension in the alveoli; produced by the lungs.

SUSTENTACULAR CELL A type of cell in the seminiferous tubules of testes that provides nourishment for the germinal sperm as they mature; also called *Sertoli cell.*

SUSTENTACULUM A process of a bone that supports.

SUTURAL BONES Separate small bones in the sutures of the calvaria of the skull; also called *Wormian bones.*

SUTURE (L. *sutura,* seam) A seamlike joint that connects skull bones, making them immovable; see *fontanel.*

SWALLOWING See *deglutition.*

SWEAT GLANDS See *sudoriferous glands.*

SYMPATHETIC NERVOUS SYSTEM The division of the autonomic nervous system that stimulates activities that are mobilized during emergency and stress situations; also called *thoracocolumbar division* or *adrenergic division.*

SYMPHYSIS (Gr. "growing together") A cartilaginous joint in which two bony surfaces are covered by thin layers of hyaline cartilage, and are cushioned by fibrocartilaginous disks; also called *secondary synchondrosis.*

SYNAPSE [SIN-apps; Gr. a connection] The electrochemical junction between neurons.

SYNAPTIC BOUTON A tiny swelling on the terminal ends of *telodendria* (see) at the distal end of an axon.

SYNAPTIC CLEFT The narrow space between the terminal ending of a neuron and the receptor site of the postsynaptic cell. In relation to a muscle, the space between the axon terminal and sarcolemma.

SYNAPTIC DELAY The period during which a neurotransmitter bridges a synaptic cleft.

SYNAPTIC GUTTER The invaginated area of the sarcolemma under and around the axon terminal; also called *synaptic trough.*

SYNARTHROSIS (Gr. *syn*, together + *arthrosis*, articulation) An immovable joint.

SYNCHONDROSIS (Gr. "together with cartilage") A cartilaginous joint whose main use is to allow growth, not movement; a temporary joint of cartilage that joins the epiphysis and diaphysis of a growing bone; it is eventually replaced by bone; also called *primary cartilaginous joint.*

SYNDESMOSIS (Gr. "to bond together"; *syn* + *desmos*, bond) A fibrous joint in which bones are held close together, but not touching, by collagenous fibers or interosseous ligaments.

SYNERGISTIC MUSCLE [SIHN-uhr-jist-ihk; Gr. *syn*, together + *ergon*, work] A muscle that complements the action of a prime mover (agonist).

SYNOVIAL [sin-OH-vee-uhl; Gr. *syn*, with + L. *ovum*, egg] Pertaining to the thick, lubricating fluid secreted by membranes in joint cavities.

SYNOVIAL CAVITY The space between two articulating bones, not including the articular cartilage; also called *joint cavity.*

SYNOVIAL FLUID A viscous fluid that lubricates synovial joints.

SYNOVIAL JOINT An articulation in which bones move easily on each other; most permanent joints of the body are synovial.

SYNOVIAL MEMBRANE A membrane lining the cavities of joints and similar areas where friction needs to be reduced; composed of loose connective and adipose tissues covered by fibrous connective tissue.

SYPHILIS A sexually transmitted disease caused by the bacterium *Treponema pallidum;* may produce degeneration of circulatory or nervous tissue, leading to paralysis, insanity, and death.

SYSTEM A group of organs that work together to perform a major body function.

SYSTEMIC ANATOMY The anatomical study of the systems of the body.

SYSTEMIC CIRCULATION The system of blood vessels that supplies the body with oxygen-rich blood, and also returns oxygen-poor blood from the body to the heart.

SYSTEMIC LUPUS ERYTHEMATOSUS (SLE) A disease that affects the lining of joints and other connective tissue; may be hereditary and caused by a breakdown in the immune system.

SYSTOLE [SISS-toe-lee] The contraction of the atria and ventricles during the *cardiac cycle.*

SYSTOLIC PRESSURE The portion of blood pressure measurement that represents the highest pressure reached during ventricular ejection; it is the first number shown in a blood pressure reading; see *diastolic pressure.*

T

T CELL A type of lymphocyte that attacks specific foreign cells; able to differentiate into helper T cell, suppressor T cell, or killer T cell; also called *T lymphocyte.*

T LYMPHOCYTE See *T cell.*

T WAVE A recorded representation during an *electrocardiogram* (see) indicating a recovery electrical wave from the ventricles to the atria as the ventricles are repolarized.

TACTILE CORPUSCLES (OF MEISSNER) Sensory receptors in the skin that detect light pressure; formerly called *Meissner's corpuscles.*

TACTILE CORPUSCLES (OF MERKEL) Sensory receptors of light touch, located in the deep epidermal layers of the palms and soles; formerly called *Merkel's disks.*

TAENIAE COLI [TEE-nee-ee KOHL-eye; L. ribbons + Gr. intestine] Three separate bands of longitudinal muscle along the full length of the large intestine.

TARGET CELL A cell with surface receptors that allow it to be affected by a specific hormone.

TARSUS The seven proximally located short bones of each foot.

TAY-SACHS DISEASE A hereditary metabolic disease in which lysosomes are deficient in the enzyme hexosaminidase, preventing the breakdown of fatty substances (gangliosides) that accumu-

late around neurons and make them unable to transmit nerve impulses.

TELERECEPTOR (Gr. "received from a distance") A sensory receptor located in the eyes, ears, and nose that detects relatively distant environmental stimuli.

TELOPHASE The final stage of mitosis; chromosomes arrive at the poles and are covered by a new nuclear envelope, and nucleoli are formed; mitosis is followed by *cytokinesis.*

TEMPORAL LOBE The cerebral lobe closest to the ears; has critical functional roles in hearing, equilibrium, and to a certain degree, emotion and memory.

TEMPORAL SUMMATION A condition in which one presynaptic neuron can increase its effect on one postsynaptic neuron by firing repeatedly.

TENDINITIS Inflammation of a tendon and tendon sheath.

TENDON A strong cord of fibrous connective tissue that attaches muscle to the periosteum of bone.

TERATOGEN (Gr. *teras*, monster) A substance that chemically interferes with embryonic development and causes gross deformities without changing the embryonic DNA.

TEST-TUBE BABY See *in vitro fertilization.*

TESTES [TESS-teez; sing. TESTIS, L. witness] The paired male reproductive organs, which produce sperm.

TESTIS-DETERMINING FACTOR (TDF) A specific chromosomal segment, or gene, on the Y chromosome that apparently determines the sex of offspring.

TESTOSTERONE The most important of the male sex hormones; it helps to stimulate sperm production, and is necessary for the development of male sex organs and behavior.

TETANUS A more or less continuous contraction of a muscle; also called *tetanic contraction.*

THALAMUS (Gr. inner chamber) Two masses of gray matter covered by a thin layer of white matter; located directly beneath the cerebrum and above the hypothalamus; forms the walls of the third ventricle; the intermediate relay point and processing center for all sensory impulses (except smell) ascending to the cerebral cortex from the spinal cord, brainstem, cerebellum, basal ganglia, and other sources.

THERMORECEPTOR (Gr. "heat receiver") A sensory receptor that responds to temperature changes.

THORACIC (Gr. *thorax*, breastplate) Pertaining to the chest.

THORACIC DUCT The largest of the lymphatic vessels; drains the lymph not drained by the *right lymphatic duct* (see) into the left subclavian vein; also called the *left lymphatic duct.*

THORACOLUMBAR DIVISION [thuh-RASS-oh-LUM-bar] The portion of the *peripheral autonomic nervous system* (see) with visceral efferent nerve fibers that leave the central nervous system through thoracic and lumbar spinal nerves.

THORACOLUMBAR OUTFLOW Myelinated nerve fibers emerging from the spinal cord in the ventral nerve roots of the 12 thoracic and first two or three lumbar spinal nerves.

THORAX (Gr. breastplate) The chest portion of the *axial skeleton*, including 12 thoracic vertebrae, 12 pairs of ribs, 12 costal (rib) cartilages, and the sternum.

THRESHOLD STIMULUS A stimulus strong enough to initiate an impulse in a neuron.

THROMBIN The active enzyme converted from prothrombin during the blood-clotting process; it acts as a catalyst to convert the soluble plasma protein fibrinogen into the insoluble plasma protein fibrin.

THROMBOCYTE See *platelet.*

THROMBOPHLEBITIS (Gr. *thrombos*, clot + *phleps*, blood vessel + *-itis*, inflammation) An acute condition characterized by clot formation and inflammation of deep or superficial veins.

THROMBOPLASTIN An enzyme involved in the blood-clotting process.

THROMBUS (Gr. a clotting) A blood clot obstructing a blood vessel or heart cavity; the condition is *thrombosis.*

THYMUS GLAND A ductless mass of flattened lymphoid tissue situated behind the top of the sternum; it forms antibodies in the newborn, and is involved in the development of the immune system.

THYROCALCITONIN See *calcitonin.*

THYROID GLAND An endocrine gland involved with the metabolic functions of the body.

THYROID-STIMULATING HORMONE (TSH) A hormone secreted by the adenohypophysis; stimulates the production and secretion of thyroid hormones; also called *thyrotropin, thyrotropic hormone.*

THYROTROPIC HORMONE See *thyroid-stimulating hormone.*

THYROTROPIN See *thyroid-stimulating hormone.*

THYROXINE A thyroid hormone that increases body metabolism and the sensitivity of the cardiovascular system to the nervous system, and affects the maturation and homeostasis of skeletal muscle; also called T_4.

TIDAL VOLUME The normal amount (usually about 0.5 L) inhaled and exhaled into and out of the lungs with each breath.

TISSUE An aggregation of many similar cells that perform a specific function; generally classified as epithelial, connective, muscle, or nervous.

TOMOGRAPHY (Gr. *tomos*, a cut or section + *graphein*, to write or draw) A technique for making pictures of a section of a body part, as in *computer-assisted tomography* (see). The picture produced is a *tomogram.*

TONSILS Aggregates of lymphatic nodules enclosed in a capsule of connective tissue; they help destroy foreign microorganisms that enter the digestive and respiratory systems; include *pharyngeal, palatine,* and *lingual* tonsils.

TOTAL LUNG CAPACITY The vital capacity plus residual volume; about 4.2 L in women, 6.0 L in men.

TRABECULAE [truh-BECK-yuh-lee; L. dim. *trabs*, beam] Tiny spikes of bone tissue surrounded by calcified bone matrix; prominent in the interior structure of spongy bone tissue.

TRACHEA [TRAY-kee-uh] An open tube extending from the base of the larynx to the top of the lungs, where it forks into two *bronchi;* commonly called *windpipe.*

TRACT A bundle of nerve fibers and their sheaths within the central nervous system.

TRANSDUCE (L. *transducere*, to lead across) To convert one form of energy into another.

TRANSDUCIN A protein activated by the conversion of 11-*cis* retinal into all-*trans* retinal during the absorption of light in the retina; activates the enzyme phosphodiesterase, which hydrolyzes the intracellular messenger *cyclic guanosine monophosphate (cGMP).*

TRANSVAGINAL OOCYTE RETRIEVAL An *in vitro* procedure in which the physician obtains through the vaginal wall ova that would otherwise be hidden behind the uterus, and unites them with sperm from the male donor; the fertilized ova are then implanted in the uterus.

TRANSVERSE A plane that divides the body horizontally into superior and inferior sections made at right angles to midsagittal, sagittal, and frontal planes. Also called *horizontal.*

TRANSVERSE COLON The portion of the large intestine immediately after the ascending colon, extending across the abdominal cavity from right to left, making a right angle downward turn at the spleen.

TRANSVERSE TUBULES A series of tubes crossing the sarcoplasmic reticulum at right angles within a muscle fiber; also called *T tubules.*

TREPPE (L. *trepidus*, alarmed) A type of muscle contraction in which the first few contractions increase in strength when a rested muscle receives repeated stimuli over a prolonged period.

TRIAD The combination of a transverse tubule and a terminal cisterna of a sarcoplasmic reticulum.

TRICHOMONIASIS A sexually transmitted disease caused by the *Trichomonas vaginalis* protozoan.

TRICUSPID VALVE The right *atrioventricular valve.*

TRIGONE A triangular area in the urinary bladder outlined by the openings of the ureters and urethra into the cavity of the bladder.

TRIIODOTHYRONINE A thyroid hormone that increases body metabolism and the sensitivity of the cardiovascular system and affects the maturation and homeostasis of skeletal muscle; also called T_3.

TRISOMY 21 A chromosomal aberration in which a pair of chromosomal homologs fails to separate at meiosis, producing an extra copy of chromosome 21 and causing mental retardation and physical abnormalities; also called *Down's syndrome,* previously called *mongolism.*

TROCHANTER Either of the two large, rounded processes found below the neck of the femur.

TROPHECTODERM The surrounding epithelial layer of a blastocyst, composed of cells called *trophoblasts;* later develops into a fetal membrane system for transporting nutrients and wastes to and from the fetus.

TROPHOBLAST See *trophectoderm.*

TUBAL LIGATION A surgical contraceptive procedure that removes a portion of each uterine tube to prevent the ovum from being fertilized and reaching the uterus.

TUBERCLE (L. small lump) A small, roughly rounded process of a bone.

TUBEROSITY (L. lump) A medium-sized, roughly rounded elevated process of a bone.

TUBULAR REABSORPTION The renal process that returns useful substances such as water, some salts, and glucose to the blood by active transport.

TUBULAR SECRETION The renal process that collects waste products such as potassium and hydrogen ions and certain drugs into the *glomerular filtrate* (see) before it leaves the kidney.

TUMOR (L. *tumere*, to swell) An abnormal growth of tissue, in which cells reproduce at a faster rate than normal; also known as *neoplasm*. (Note: *-oma* = tumor.)

TUNICA ADVENTITIA ("outermost covering") The outermost covering of an artery, composed mainly of collagen fibers and elastic tissue.

TUNICA ALBUGINEA [al-byoo-JIHN-ee-uh; L. *albus*, white] A fibrous sac enclosing a testis in the male, and forming a thin connective tissue over the outer portion (cortex) of an ovary in the female.

TUNICA INTIMA ("innermost covering") The innermost lining of the lumen of an artery.

TUNICA MEDIA ("middle covering") The middle, and thickest, layer of the arterial wall in large arteries.

TURNER'S SYNDROME A chromosomal aberration in which an individual receives an X sex chromosome from one parent and none from the other, producing an XO condition and a sterile female.

TWITCH A momentary spasmodic contraction of a muscle fiber in response to a single stimulus; the simplest type of recordable muscle contraction.

TYMPANIC CAVITY (Gr. *tumpanon*, drum) A narrow, irregular, air-filled space in the temporal bone; separated from the external auditory canal by the tympanic membrane, and from the inner ear by the posterior bony wall; also called *middle-ear cavity*.

TYMPANIC MEMBRANE A deflectable membrane between the external and middle ear; vibrates in response to sound waves entering the ear; popularly called the *eardrum*.

U

ULCERS See *peptic ulcers*.

ULTRASONOGRAPHY A technique using high-frequency sound waves to locate and examine a fetus or other internal structures.

ULTRASOUND A noninvasive exploratory technique that sends pulses of ultrahigh-frequency sound waves into designated body cavities; images are formed from echoes of the sound waves; also called *sonography*.

UMBILICAL CORD (L. *umbilicus*, navel) A connecting tube between the embryo and placenta, carrying carbon dioxide and nitrogen wastes from the embryo, and oxygen and nutrients to the embryo.

UNIPOLAR NEURON A nerve cell with one process dividing into two branches; one branch extends into the brain or spinal cord, while

the other extends to a peripheral sensory receptor in a distal part of the body.

UPPER EXTREMITIES See *extremities*.

UREMIA A condition in which waste products normally excreted in the urine are present in the blood.

URETER A paired tube that carries urine from the renal pelvis of each kidney to the urinary bladder.

URETHRA A tube of smooth muscle lined with mucosa that transports urine from the bladder to the outside during urination.

URETHRA The final section of the male reproductive duct system, leading from the urinary bladder, through the prostate gland, and into the penis, carrying semen outside the body during ejaculation, and urine during urination; in females it conveys urine from the urinary bladder.

URETHRITIS Inflammation of the urethra.

URINARY BLADDER A hollow, muscular sac that collects urine from the ureters and stores it until it is excreted from the body through the urethra.

URINARY INCONTINENCE The inability to retain urine in the urinary bladder and control urination.

URINARY SYSTEM The body system that eliminates the wastes of protein metabolism in the urine and regulates the amount of water and salts in the blood; composed of two kidneys and ureters, the urinary bladder, and the urethra; also called the *renal* system.

URINE (Gr. *ourein*, to urinate) The excretory fluid produced by the kidneys, composed mainly of water, urea, chloride, potassium, creatinine, phosphates, sulfates, and uric acid.

UTERINE TUBE A paired tube that receives the mature ovum from the ovary and conveys it to the uterus; also called *Fallopian tube, oviduct*.

UTEROSACRAL LIGAMENTS Two fibrous bands of tissue that attach the uterus to the rectum and urinary bladder.

UTERUS (L. *womb*) A hollow, muscular organ behind the urinary bladder that is the site of menstruation, and during pregnancy houses, nourishes, and protects the developing fetus; also called *womb*.

V

VAGINA (L. sheath) A muscle-lined tube from the uterus to the outside; the site where semen is deposited during sexual intercourse, the channel for menstrual flow, and the birth canal for the baby during childbirth.

VALVULAR HEART DISEASE A condition

in which one or more cardiac valves operate improperly.

VARICOSE VEINS (L. *varix*, swollen veins) Abnormally dilated and twisted veins, most often affecting the saphenous veins and their branches in the legs.

VAS DEFERENS See *ductus deferens*.

VASA VASORUM ("vessels of the vessels") Small blood vessels that supply nutrients to the walls of arteries and veins.

VASCULAR (L. *vasculum*, dim. of *vas*, vessel) Pertaining to blood vessels.

VASECTOMY A surgical contraceptive procedure that removes a small portion of each ductus deferens to block the release of sperm from the testes.

VASOCONSTRICTION The process of constricting blood vessels.

VASODILATION The process of dilating blood vessels.

VASOMOTION The collective term for *vasoconstriction* and *vasodilation* (see).

VASOMOTOR CENTER A regulatory center in the lower parts of the pons and medulla oblongata; one of its main functions is to regulate the diameter of blood vessels, especially arterioles.

VASOPRESSIN See *antidiuretic hormone*.

VEIN A blood vessel that usually carries blood from the body to the heart.

VENEREAL DISEASE (from *Venus*, Roman goddess of love) See *sexually transmitted disease*.

VENTILATION See *breathing*.

VENTRAL (L. *venter*, belly) 1. A relative directional term: toward the front of the body; the toes are ventral to the heel; also called *anterior*; opposite of *dorsal*. 2. The larger of two main body cavities; separated into the thoracic and abdominopelvic cavities by the diaphragm; also called *anterior*; see *dorsal (2)*.

VENTRICLE (L. "little belly") 1. A cavity in the brain filled with cerebrospinal fluid. 2. The left or right inferior heart chamber.

VENULE [VEHN-yool] A tiny vein into which blood drains from capillaries.

VERMIFORM APPENDIX (L. *vermis*, worm) The short, narrow, wormlike region of the digestive tract opening into the cecum; may be involved with the immune system; commonly called the *appendix*.

VERMIS (L. worm) The midline portion of the cerebellum, separating the lateral lobes (hemispheres); involved, together with the flocculonodular lobes, with maintaining muscle tone, equilibrium, and posture.

VERTEBRAE [VER-tuh-bree; L. "something to turn on"] The 26 individual bones in the spinal column. The bones and connecting intervertebral disks form a strong, flexible support for the neck and trunk, also protecting the spinal cord within.

VERTEBRAL COLUMN Commonly called the spinal column.

VESICULAR OVARIAN FOLLICLE An ovarian follicle that is almost ready to release a mature ovum in the process of *ovulation* (see).

VESTIBULAR APPARATUS (L. *vestibulum,* entrance) Specific parts of the inner ear that signal changes in the *motion* of the head *(dynamic equilibrium)* and the *position* of the head with respect to gravity *(static equilibrium,* or posture).

VESTIBULAR GLANDS See *greater* and *lesser vestibular glands.*

VESTIBULE 1. Any cavity, chamber, or channel serving as an approach or entrance to another cavity. 2. The space between the labia minora. 3. The central chamber of the labyrinth in the middle ear.

VILLI [VILL-eye; sing. **VILLUS;** L. "shaggy hairs"] Tiny fingerlike protrusions in the mucosa of the small intestine that increase the absorptive surface area.

VISCERA [VISS-ser-uh; L. body organ] The internal organs of the body.

VISCERAL Pertaining to an internal organ or a body cavity, or describing a membrane covering an internal organ.

VISCERAL NERVOUS SYSTEM The portion of the *peripheral nervous system* composed of a motor division (autonomic nervous system) that may inhibit or excite smooth muscle, cardiac muscle, or glands, and a sensory division that receives afferent input from internal organs.

VISCERAL PLEURA See *pleural cavity.*

VITAL CAPACITY The inspiratory reserve volume plus the tidal volume plus the expiratory reserve volume; about 3.1 L in women, 4.8 L in men.

VITAMIN An organic compound required by the body in small amounts for the regulation of metabolism; classified as *water-soluble* (C and B group) and *fat-soluble* (A, D, E, K).

VITELLINE MEMBRANE The special outer covering of an ovum, on which only certain areas are receptive to sperm.

VITREOUS HUMOR A gelatinous substance within the large vitreous chamber of the eyeball; keeps the eyeball from collapsing as a result of external pressure.

VOCAL CORDS Paired strips of stratified squamous epithelium at the base of the larynx that produce sound when vibrated.

VOICE BOX See *larynx.*

VOLAR See *palmar.*

VOLKMANN'S CANAL See *perforating canal.*

VOLUNTARY MUSCLE See *skeletal muscle.*

VOMITING The forceful expulsion of part or all of the contents of the stomach and duodenum through the mouth, usually in a series of involuntary spasms; also called *emesis.*

VULVA (L. womb) The collective name for the external female genital organs, including the mons pubis, labia majora and minora, vestibular glands, clitoris, and the vestibule of the vagina.

W

WANDERING CELLS Connective tissue cells usually involved with short-term activities such as protection and repair.

WART A benign epithelial tumor caused by various papilloma viruses; also called *verruca.*

WHITE BLOOD CELL See *leukocyte.*

WHITE MATTER The portion of the spinal cord consisting mainly of whitish myelinated nerve fibers.

WHITE RAMI COMMUNICANTES Myelinated preganglionic fibers of the *thoracolumbar outflow* (see) of the spinal cord that form small nerve bundles.

WINDPIPE See *trachea.*

WOMB See *uterus.*

WORMIAN BONES See *sutural bones.*

Y

YOLK SAC An extraembryonic membrane that is a primitive respiratory and digestive system before the development of the placenta; it becomes nonfunctional and incorporated into the umbilical cord by the sixth or seventh week.

Z

ZONA PELLUCIDA (L. *perlucere,* to shine through, thus "transparent zone") The outer wall of an ovum.

ZYGOTE (Gr. *zugotos,* joined, yolked) The cell formed by the union of male and female gametes.

INDEX

Page references in **boldface** introduce or define the term; page references in *italic* indicate illustrations; page references in ***boldface italic*** indicate an illustration and term.

G